3 tools designed to help you learn with
What Is Life? A Guide to Biology with Physiology

1	2	3
Book Companion Website	Student Success Guide	Prep-U

Free Book Companion Website
www.whfreeman.com/phelanphys1e

All students have FREE, unrestricted access to the Companion Website which offers a variety of study tools that reflect the textbook's coverage and goals. Free website Tools include:

- Q Animations
- Podcasts
- Lecture Companion Art
- Chapter Quizzes
- Key Terms Flashcards

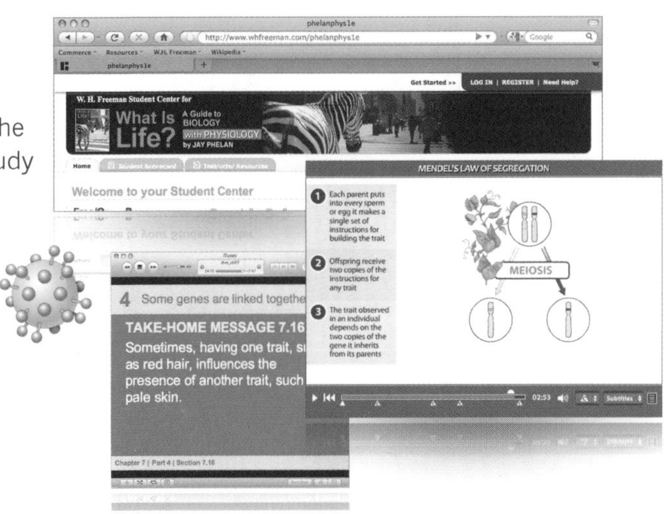

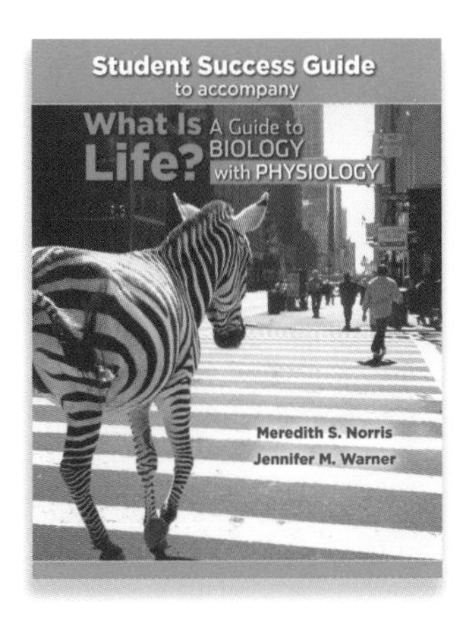

Printed Student Success Guide

The Student Success Guide contains over 200 pages of study tools to help you better understand key concepts and improve your exam scores. Including:

- Multiple-choice and short-answer practice questions
- Learning objectives
- Outlines of key chapter concepts fully illustrated with text figures
- Visual glossary of key terms

To order a copy of the Student Success Guide online, go to **www.whfreeman.com/phelanphys1e**

MORE! Learn about Prep-U, the online student quizzing tool that comes free with this textbook...

Prep-U is the online quizzing system that comes **free*** with this textbook created by Jay Phelan, author of **What Is Life? A Guide to Biology with Physiology.**

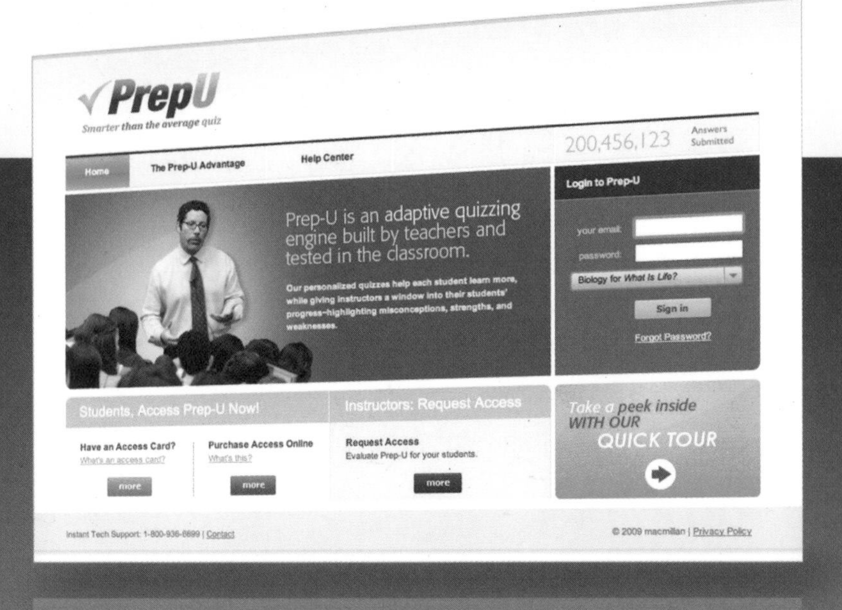

Take custom quizzes and find out what you don't know *before* the exam!

Take Practice Quizzes

Get ready for exams by quickly building quizzes over the chapter/s you choose. Prep-U contains over 5,000 high-quality questions prepared specifically for this textbook.

Get Personalized Results

Instant quiz results help you learn from your mistakes. Your results also determine the questions you will see on your next quizzes. Quiz questions are generated to fit your specific needs.

See How You Are Doing

Your "How Am I Doing?" page shows your overall Prep-U performance compared to your classmates, reveals your strengths and weaknesses, and tells you which chapters you need to study more.

Your Prep-U **activation code** is on the card facing this page.
Log-in at www.prep-u.com now!

* Prep-U comes free with new copies of this textbook. If this is a used book and the activation code has been revealed, you can purchase Prep-U access at www.prep-u.com.

What Is Life? A Guide to BIOLOGY with PHYSIOLOGY

What Is Life? A Guide to BIOLOGY with PHYSIOLOGY

Jay Phelan

University of California, Los Angeles

W. H. FREEMAN AND COMPANY
NEW YORK

Publisher: Peter Marshall
Senior Development Editor: Elizabeth Howe
Consulting Development Editor: Jane Tufts
Art Development Editor: Tommy Moorman
Illustrations: Tommy Moorman, Erin Daniel
Senior Media Editor: Patrick Shriner
Associate Director of Marketing: Debbie Clare
Managing Editor for First Editions: Elaine Palucki, PhD
Associate Editor for First Edition Supplements: Beth McHenry
Assistant Editor: Marni Rolfes
Managing Editor: Philip McCaffrey
Project Editor: Dusty Friedman
Art Director: Diana Blume
Text Design: Lissi Sigillo
Photo Editor: Christine Buese
Photo Researchers: Deborah Anderson, Elyse Rieder
Production Manager: Julia DeRosa
Composition and Layout: Sheridan Sellers
Printing and Binding: Worldcolor Dubuque

Library of Congress Control Number: 2009942967

Student Edition: ISBN-13: 978-1-4292-4666-8
ISBN-10: 1-4292-4666-9

Printed in the United States of America

First Printing

W. H. Freeman and Company
41 Madison Avenue, New York, NY 10010
Houndsmills, Basingstoke RG21 6XS, England

www.whfreeman.com

BRIEF CONTENTS

PART 5 Plant Life

PART 6 Health and Physiology

CONTENTS

PART 1 The Facts of Life

1 • Scientific Thinking

2 • Chemistry

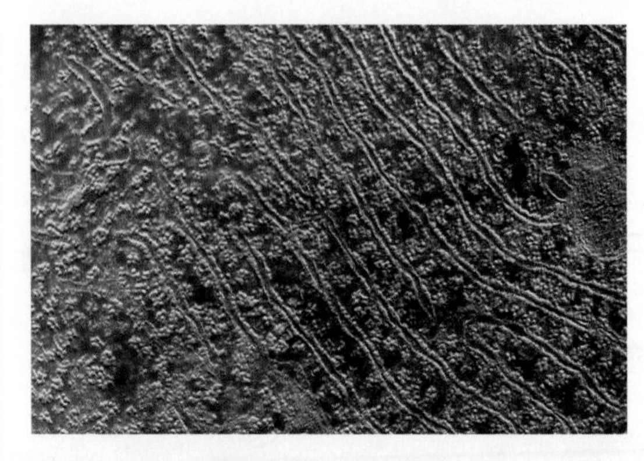

3 • Cells

PART 2 Genetics, Evolution, and Behavior

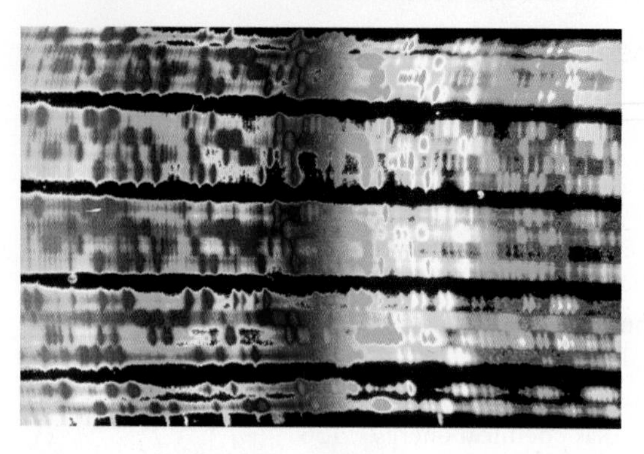

5 • DNA, Gene Expression, and Biotechnology

6 • Chromosomes and Cell Division

7 • Mendelian Inheritance

8 • Evolution and Natural Selection

9 • Evolution and Behavior

PART 3 Evolution and the Diversity of Life

10 • The Origin and Diversification of Life on Earth

11 • Animal Diversification

12 • Plant and Fungi Diversification

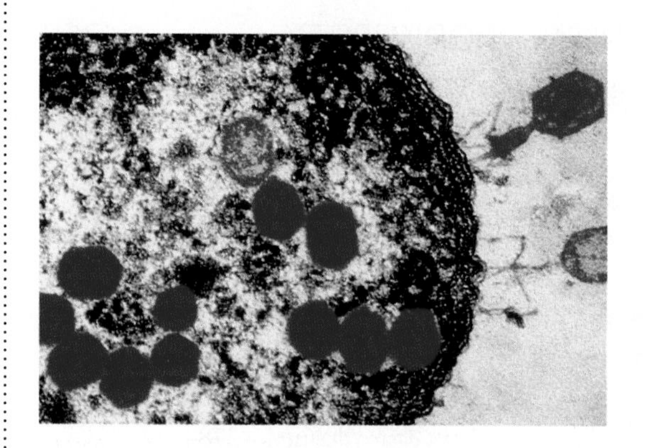

13 • Evolution and Diversity Among the Microbes

PART 4 Ecology and the Environment

14 • Population Ecology

15 • Ecosystems and Communities

16 • Conservation and Biodiversity

PART 5 Plant Life

17 • Plant Structure and Nutrient Transport

18 • Growth and Reproduction in Plants

19 • Plants Respond to Their Environments

REGULATING AND DEFENDING WHILE ROOTED IN THE GROUND 690

PART 6 Health and Physiology

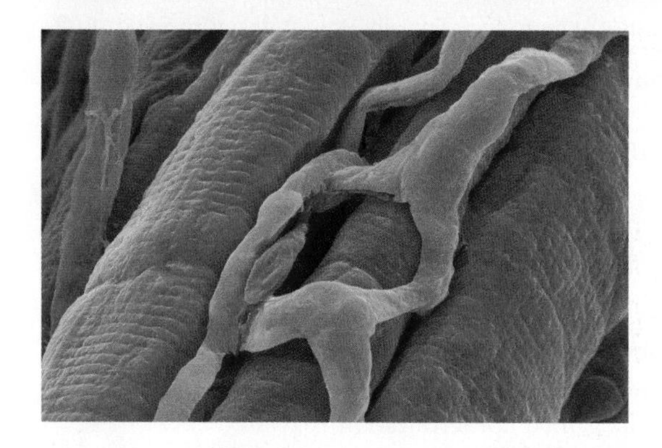

20 • Introduction to Animal Physiology

PRINCIPLES OF ANIMAL ORGANIZATION AND FUNCTION 714

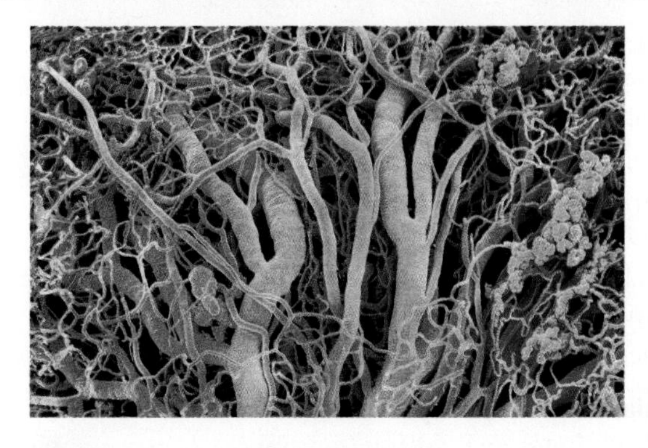

21 • Circulation and Respiration

TRANSPORTING FUEL, RAW MATERIALS, AND GASES INTO, OUT OF, AND AROUND THE BODY 744

22 • Nutrition and Digestion

AT REST AND AT PLAY: OPTIMIZING HUMAN PHYSIOLOGICAL FUNCTIONING 790

23 • Nervous and Motor Systems

24 • Hormones

25 • Reproduction and Development

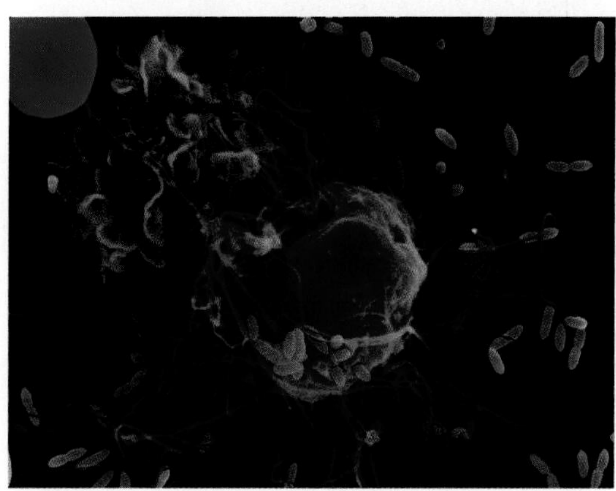

26 • Immunity and Health

To Julia

About the Author

Jay Phelan teaches biology at UCLA, where he has taught introductory biology to more than 8,000 majors and non-majors students over the past twelve years. He received his Ph.D. in evolutionary biology from Harvard in 1995, and his master's and bachelor's degrees from Yale and UCLA. His primary area of research is evolutionary genetics, and his original research has been published in *Evolution*, *Experimental Gerontology*, and the *Journal of Integrative and Comparative Biology*, among other journals. His research has been featured on *Nightline*, CNN, the BBC, and National Public Radio; in *Science Times* and *Elle*; and in more than a hundred newspapers. He is the recipient of more than a dozen teaching awards.

With economist Terry Burnham, Jay is the co-author of the international best-seller *Mean Genes: From Sex to Money to Food—Taming Our Primal Instincts*. Written for the general reader, *Mean Genes* explains in simple terms how knowledge of the genetic basis of human nature can empower individuals to lead more satisfying lives.

Dear Readers,

Biology is about you and it touches on your life every day, in dozens of ways. It's creative. It's fun. And understanding it is increasingly essential to your life. I wrote this book because I want to help you become biologically literate: to talk confidently and knowledgeably about science with your friends, and to understand and appreciate both the beauty and the utility of this subject.

There are two versions of *What Is Life? A Guide to Biology.* One of them, *What Is Life? A Guide to Biology with Physiology,* includes all of the chapters of the other version, with an additional ten chapters on plant and animal physiology. It's not always possible to include these chapters in a one-term course, but they include a rich introduction to the importance of biology to understanding the processes central to how many organisms work, with particular significance for human health.

You will find here an overview of the key themes in biology as well as the specifics about topics meaningful to your life. You'll find answers to the questions that you're curious about, whether you are a future lawyer, teacher, entrepreneur, parent, consumer, citizen, or all of the above:

- Why are humans one of the only species to have friendships?
- Is DNA fingerprinting foolproof?
- Why do dieters lose so much "water weight" during the beginning of a diet?
- Why isn't it always wise to take aspirin or other medicine when you have a fever?
- How does caffeine combat fatigue?

In this book, as in life, there are **interesting questions everywhere.** As you read, red Qs pose questions with real relevance to you, and point toward specific passages where you can uncover the answers. If you look at those bits of the text and don't think the answer is there, look again and think some more. You may be surprised at what you know. The **StreetBio Knowledge You Can Use** section at the end of each chapter unpacks biology-related questions with a practical twist, such as *How clean is that food you just dropped?* (See Chapter 13.)

There's much more to this book than just words. Flip through its pages and look at the **photographs.** For twenty years, I have searched for the best photos to illustrate biological concepts. I hand-picked every photo here with a special purpose in mind—it should provoke, engage, and inspire you, while helping you to make connections between complex ideas.

Because the subject is so varied, I don't want you to lose sight of "the big picture." And so, in organizing each chapter I have broken down the topics into **discrete sections.** At the end of each, I highlight the **Take-Home Message** in clear and concise terms. While there may be a lot of material, in these manageable chunks you can master it.
The impact of science on your life is increasing every day. I hope that you embrace this exciting world of knowledge and that you gain as much satisfaction and enjoyment reading this book as I have received in putting it together for you.

Sincerely,

Jay Phelan

Acknowledgments

As a new graduate student at Harvard, I heard from experienced teaching fellows that if you were interested in learning how to be an effective teacher, it was essential to seek out extraordinary mentors. Based on word of mouth, I became involved with E. O. Wilson's course in Evolutionary Biology and Irven DeVore's course in Human Behavioral Biology. Both were known to be unusually provocative, challenging, and entertaining classes for non-science majors.

I aggressively pursued teaching positions in both classes—which I held onto tightly for 12 semesters. Working under these legendary instructors, I was set on a course that inspired and prepared me to write this book. The two courses were quite different from each other, but at their core both were built on two beliefs that are central to this book and to my thinking about education: (1) Biology is creative, interesting, and fun. (2) Biology is relevant to the daily life of every person. There was a palpable sense that, in teaching non-science majors especially, we had a responsibility to provide our students with the tools to thrive in a society increasingly permeated by scientific ideas and issues, and that one of our most effective strategies would be to convey the excitement we felt for biology and the enormous practical value it has to help us understand the world.

I thank Professors Wilson and DeVore for all that they have shared with me.

My development as a scientist and, particularly, my appreciation for rigorous and methodical thinking have been shaped by the kind support and wise guidance of Richard Lewontin. I have also been fortunate to have as a long-time mentor and collaborator Michael Rose, who has instilled in me a healthy skepticism about any observation in life that is not fivefold replicated. And for almost daily insightful input on matters relating to scientific content, teaching, writing, and more, I thank Terry Burnham.

There are many other friends and colleagues I wish to thank for helping me with *What Is Life? A Guide to Biology with Physiology.* In researching and writing the book and in developing the numerous courses I teach, I have benefited from more than a decade of perceptive and valuable contributions too numerous to list from Glenn Adelson, Alon Ziv, Michael Cooperson, and Alicia Moretti. I am tremendously appreciative of all they have done for me.

For a project covering so many topics and years, it is essential to have a close group of trusted, tolerant, and knowledgeable colleagues, I am grateful to Harold Owens, Greg Graffin, Jeff Egger, Andy Tobias, Elisabeth Tobias, Joshua Malina, Melissa Merwin-Malina, Bill U'ren, Chris Bruno, Michelle Richmond, and Meredith Dutton, who have offered advice, guidance, and support, far beyond the call of duty. Numerous colleagues at UCLA provided assistance and support, including Steve Strand, Cliff Brunk, Fred Eiserling, Emil Reisler, Deb Pires, Lianna Johnson, Gaston Pfluegl, Bob Simons, Frank Laski, Jeff Thomas, and Tracy Newman.

I owe a tremendous debt to Sara Tenney, without whose vision and knowledge this project could never have been begun. Thank you, Sara. W. H. Freeman is an extremely author-centric publisher. Throughout the process of creating this book, from the first inception of the idea through the production of all the supplementary materials, Liz Widdicombe, Brian Napack, and John Sargent have been tremendously supportive. I am grateful for their welcoming me into their publishing family. Publisher Peter Marshall has been a tenacious, versatile, and skillful manager of the entire team. I am very fortunate to have such a wise leader overseeing all aspects of this project.

The team of editors that worked with me on this book—and two people in particular—improved it immeasurably. I cannot adequately convey my gratitude to development editor Jane Tufts, whose meticulous attention to detail, commitment to accuracy, and almost obsessive drive to create a thorough and readable book are apparent on every page. And development editor Beth Howe, who oversaw every aspect of the writing and production of the book, attending to issues of content and design while making insightful contributions throughout, expertly managed the thousand details necessary to put everything together.

It is impossible to teach biology without illustrations. My deepest gratitude goes to Tommy Moorman for creating such innovative and effective figures for the book. Tommy's vision for an elegant and beautiful art program completely integrated with the text is apparent on every page. Working with him to develop each illustration in this book has been one of my most enjoyable and satisfying professional collaborations. Thanks also go to Erin Daniel and Alison Kendall for assisting with the creation of the illustrations. For excellent assistance with photo research, thanks to Deborah Anderson, Christine Buese, and Elyse Rieder.

I wish to thank Harvey Pough for providing excellent drafts of Chapters 11–13 and contributing to Chapter 16. Thanks are due to Kelly Hogan for writing a fine draft of chapter 26.

For creating the innovative media and print materials that accompany the book, I am thankful for the vision of my excellent media editor, Patrick Shriner, and for the work of supplements editor Beth McHenry. I thank all of the authors who have created instructor and student resources for this book. I also appreciate the contributions of Renee Altier, Kimberly White, and Larry Jankovic to Prep-U. Sheri Snavely provided significant guidance in developing pedagogical strategies throughout the book; I also appreciate her thoughtful and smart advice at nearly every step in the publishing process. Copyeditor Linda Strange helped to ensure consistency and readability throughout the text. I thank Chris Hunt for compiling the thorough index, and Elaine Palucki for coordinating reviews. The rest of the life sciences editorial team at W. H. Freeman, too, have been knowledgeable and supportive, particularly Kate Parker, Marc Mazzoni, Jerry Correa, Susan Winslow, Susan Moran, Lisa Samols, and Marni Rolfes.

For their efficiency and commitment to producing a beautiful book, I am most grateful to the W. H. Freeman production team: Mary Louise Byrd, Diana Blume, Julia DeRosa, Philip McCaffrey, and Ellen Cash. Sheridan Sellers did an artistic job of page composition. Thank you! Thanks are also due to Dusty Friedman for keeping the book on track.

The people on the marketing team at W. H. Freeman have contributed enormously in helping with the challenging task of introducing a new book to students and instructors across the country. Debbie Clare, Steve Rigolosi, Brenda Bravener, and John Britch have been enthusiastic and dedicated in creating materials and strategies to assist instructors in evaluating the ways that *What Is Life? A Guide to Biology with Physiology* can aid them as they develop their own courses and strategies for success.

Finally, I thank my family—Kevin Phelan, Patrick Phelan, Erin Enderlin, and my parents—for their unwavering support and interest as I wrote this book. Reading draft after draft and following each revision, they made valuable contributions at every stage. I thank Jack, Charlie, and Sam, too. Most of all, for her generous and passionate support of this project from day one, her substantive contributions to both the content and presentation of ideas, and so much more, I thank Julia.

Contact the author with your feedback.

The content of this book has been greatly improved through the counsel of reviewers and students. Your comments, suggestions, and criticism are also welcome; they are essential in guiding its ongoing evolution. Please contact the author at **jay@jayphelan.com.** (He's serious about this.)

We thank the many reviewers who aided in the development of this text.

Text Development

Stephanie J. Aamodt, University of Louisiana–Shreveport
M. Stephen Ailstock, Anne Arundel Community College
Sylvester Allred, Northern Arizona University
Jessica K. Baack, Montgomery Community College–Rockville
Marilyn Banta, Texas State University–San Marcos
Sarah Barlow, Middle Tennessee State University
Rebecca A. Bartow, Western Kentucky University
Brian J. Baumgartner, Trinity Valley Community College
José Bava, El Camino Community College
Cynthia Bida, Henry Ford Community College
William Blanton, St. Phillip's College
Lisa L. Boggs, Southwestern Oklahoma State University
Charlotte Borgeson, University of Nevada–Reno
Susan L. Bower, Pasadena City College
Peggy Brickman, University of Georgia
Jason Brown, Young Harris College
Carole L. Browne, Wake Forest University
Neil J. Buckley, State University of New York–Plattsburgh
Anne E. Bunnell, East Carolina University
Joseph Burdo, Bridgewater State University
David Byres, Florida Community College–Jacksonville
William Caire, University of Central Oklahoma
Michael S. Carr, Oakton Community College
Kelly Carrillo Burke, College of the Canyons
Jeannie Chapman, University of South Carolina–Spartansburg
Thomas Chen, Santa Monica College
Roger Choate, Oklahoma City Community College
Thomas F. Chubb, Villanova University
Nira Clark, Southwestern College
Erica Cline, University of Washington
George Clokey, University of Wisconsin–Whitewater
Sandi Connelley, Rochester Institute of Technology
Erica Corbett, Southeastern Oklahoma State University
Anthony D. Cornett, Valencia Community College–Osceola
Becky Croteau, Lincoln Land Community College
Karen Curto, University of Pittsburgh
Don C. Dailey, Austin Peay State University
Michael S. Dann, Pennsylvania State University
Douglas W. Darnowski, Indiana University–Southeast
Garry Davies, University of Alaska
Renee Dawson, University of Utah
Lewis E. Deaton, University of Louisiana–Lafayette
Tom Deaton, Auburn University
Elizabeth Anne Desy, Southwestern Minnesota State University
Sandra G. Devenny, Delaware County Community College
Doreen R. Dewell, Whatcom Community College
Mary E. Dominiecki, Slippery Rock University
Danielle DuCharme, Waubonsee Community College
Charles Dunn, Austin Community College
Frank DuRoy, Essex County College
Kari Eamma, Tarrant County Community College–Northeast
Heidi Engelhardt, Brandon University
Marirose T. Ethington, Genesee Community College
Rebecca A. Fetherson, University of Dayton
Anita Flick, North Carolina State University
Robert C. Frankis, University of New Mexico
Monica C. Frazier, Columbus State University
Anne Galbraith, University of Wisconsin–La Crosse
Elizabeth Gerbec, University of Wisconsin–River Falls
Sandra Gibbons, Moraine Valley College
Phil Gibson, University of Oklahoma
Philip G. Gibson, Gwinnett Technical College
Andrew Goliszek, North Carolina A & T State University
Stephen M. Gómez, Central New Mexico Community College
Becky C. Graham, University of West Alabama
Cara Gubbins, Butte College
Charles (Billy) Gunnels IV, Florida Gulf Coast University
Janelle M. Hare, Morehead State University
Carla Hass, Pennsylvania State University
Colleen Hatfield, California State University–Chico
J. L. Henriksen, Bellevue Community College
Deena Hergert, Rock Valley College
John A. Hnida, Peru State College
Berta Hopkins, Spartanburg Community College
Thomas Horvath, State University of New York–Oneonta
Laurie Host, Harford Community College
Adam W. Hrincevich, Louisiana State University
Catherine Hurlbut, Florida Community College at Jacksonville
Dianne Jennings, Virginia Commonwealth University
Mitrick A. Johns, Northern Illinois University
Robert D. Johnson Jr., Pierce College
Victoria Johnson, San Jose State University

Martin A. Kapper, Central Connecticut State University
Arnold J. Karpoff, University of Louisville
Todd Kelson, Brigham Young University–Idaho
Stephen T. Kenny, Yakima Valley Community College
Jennifer M. Kilbourne, Community College of Baltimore County
Joanne Kilpatrick, Auburn University–Montgomery
Brenda Knotts, Eastern Illinois University
Michael Koban, Morgan State University
Catherine Koo, Caldwell College
Olga R. Kopp, Utah Valley University
Anna Koshy, Houston Community College
Jerome A. Krueger, South Dakota State University
Jim Krupa, University of Kentucky
Josephine Kurdziel, University of Michigan
Archana Lal, Independence Community College
Ellen Shepard Lamb, University of North Carolina–Greensboro
Thomas Landefeld, California State University–Dominquez Hills
Leah LaPerle Larkin, University of New Mexico
Lynn Larsen, Portland Community College
Kathleen H. Lavoie, State University of New York–Plattsburgh
David E. Lemke, Texas State University
Gregory P. Lewis, Furman University
Harvey Liftin, Broward Community College
Tammy J. Liles, Lexington Community College
Cynthia W. Littlejohn, University of Southern Mississippi
Jonathan Lochamy, Georgia Perimeter College
Melanie Loo, California State University–Sacramento
Juan M. López-Bautista, University of Alabama
Blasé Maffia, University of Miami
Jose Maldonado, El Paso Community College–Valle Verde
Jeffrey M. Marcus, Western Kentucky University
Floyd Douglas Martin, St. Edward's University
Betsy Maxim, Austin Community College
M. Victoria McDonald, University of Central Arkansas
Karen McFarland, Austin Peay State University
Mark A. McGinley, Texas Technical University
Malinda McMurry, Morehead State University
Cassandra Meeks, Oklahoma City College
Karen Meisch, Austin Peay State University
John Mersfelder, Sinclair Community College
Shahroukh Mistry, Westminster College
Jeanne Mitchell, Truman State University
Beth A. Montelone, Kansas State University
Cynthia L. Morin, Lincoln University of Missouri
Royden Nakamura, California Polytechnic–San Luis Obispo
Rocky Nation, Southern Wesleyan University
Allan D. Nelson, Tarleton State University
Kim Nelson, Pennsylvania State University
Tania Nezrick, Kishwaukee College
David Niebuhr, College of William and Mary
Meredith Sommerville Norris, University of North Carolina–Charlotte
Alexander E. Olvido, Virginia State University
Betsy Ott, Tyler Junior College
Joanna M. Padolina, Virginia Commonwealth University
Randi Papke, Southwestern Illinois College
Mary Paulson, Central Washington University
Krista Peppers, Central Arkansas University
John M. Pleasants, Iowa State University
Michael Plotkin, Mt. San Jacinto College
Gregory J. Podgorski, Utah State University
Ronald Porter, Pennsylvania State University
Melissa Presch, California State University–Fullerton
Eric Rabitoy, Citrus College
Karen Raines, Colorado State University
Erin Rempala, San Diego Mesa College
Dawn Ranish, Nova Southeastern University
Angela M. Reevely, Lonestar College
Melody Ricci, Victor Valley College
Kathleen Richardson, Portland Community College
Dave Rintoul, Kansas State University
Steven Rissing, The Ohio State University
Robert R. Robbins, Utah Valley State College
Bill Rogers, Ball State University
Troy T. Rohn, Boise State University
Thomas Rooney, Wright State University
Heather Rushforth, University of North Carolina–Greensboro
Susan T. Rouse, Southern Wesleyan University
Lynette Rushton, South Puget Sound Community College
Michael L. Rutledge, Middle Tennessee State University
Shamili Ajgaonkar Sandiford, College of DuPage
Amanda Schaetzel, Front Range Community College
John Schampel, Phoenix College (AZ)
Fayla Schwartz, Everett Community College
Jennifer Scoby, Illinois Central College
Erik P. Scully, Towson University
Juanita C. Sharpe, Chicago State University
Marilyn Shopper, Johnson County Community College
Michele Shuster, New Mexico State University
Neil E. Simister, Brandeis University
Colleen Sinclair, Towson University
Anu Singh-Cundy, Western Washington University
Jennifer Skillen, College of Southern Nevada
Kerri M. Skinner, University of Nebraska–Kearny
Pamela Skoubis, DePaul University
Marc Allen Smith, Sinclair Community College
Nancy G. Solomon, Miami University of Ohio
Roberta Lynn Soltz, Central Washington University
Bryan G. Spohn, Florida Community College at
 Jacksonville–Kent Campus
Amanda Starnes, Emory University
Leo Sternberg, University of Miami
Joan C. Stover, South Seattle Community College
Christine Stracey, University of Florida
Sukanya V. Subramanian, Collin County Community College
Franklyn Tan Te, Miami Dade Community College
William J. Thieman, Ventura College
Pamela Thinesen, Century College
Elizabeth M. Thomas, West Virginia University
Gene Thomas, Solano Community College
Michael W. Thompson, Middle Tennessee State University

Jeffry Thornsberry, Northwest Missouri State University
William Unsell, The University of Central Oklahoma
Paul Ustach, San Joaquin Delta College
Martin A. Vaughan, Indiana University/Purdue University at Indianapolis
Kim Vietti, Illinois Central College
Hung Vu, Tarrant County College
Jack Waber, West Chester University of Pennsylvania
Jennifer M. Warner, University of North Carolina–Charlotte
John E. Whitlock, Hillsborough Community College
Elizabeth Willott, University of Arizona
Clifford Wilson, Chicago State University
Mark Woelfle, Vanderbilt University
Carol Wymer, Morehead State University
Calvin Young, Fullerton College
Kerry Yurewicz, Plymouth State University
Michelle Zimmerman, Indiana University Southeast
Brenda Zink, Northeastern Junior College

Art Development

Ann Aguanno, Marymount Manhattan College
Neil Baker, Ohio State University
Marilyn Banta, Texas State University–San Marcos
Christine Barrow, Prince George's Community College
Mark Belk, Brigham Young University
Kristen Byrd, Tarrant Community College (Northeast)
Kelly Carrillo Burke, College of the Canyons
William Caire, University of Central Oklahoma
Claudia Cash, Tarrant Community College (Northeast)
Debra Chapman, Wilkes University
Genevieve C. Chung, Broward Community College
Jeffrey Scott Coker, Elon University
Sandi Connelley, Rochester Institute of Technology
Michael S. Dann, Pennsylvania State University
Renee Dawson, University of Utah
Francisco Delgado, Pima Community College
Sandra G. Devenny, Delaware County Community College
Doreen R. Dewell, Whatcom Community College
Hartmut G. Doebel, George Washington University
Diane Lynn Doidge, Grand View College
James Doyle, College of the Holy Cross
Kari Eamma, Tarrant Community College (Northeast)
Dave Eakin, Eastern Kentucky University
Gerald Farr, Texas State University–San Marcos
Paul Farnsworth, University of New Mexico
Brandon L. Foster, Wake Technical and Community College
Patrick Galliart, North Iowa Community College
David Gordon, Pittsburg State University
Charles (Billy) Gunnels IV, Florida Gulf Coast University
Joe Harsh, Butler University
Bernard Hauser, University of Florida
Jane J. Henry, Baton Rouge Community College
Mark A. Holland, Salisbury University
Laurie Host, Harford Community College
Virginia E. Irintcheva, Black Hawk College–Quad City Campus
Philip Jardim, City College of San Francisco
Denim M. Jochimsen, University of Idaho
Judy Kaufman, Monroe Community College
Ariel R. Krakowski, Laney College
Carissa Krane, University of Dayton
Dan E. Krane, Wright State University
Maria Kretzman, Glendale College
Kathleen H. Lavoie, State University of New York–Plattsburgh
Mary LeFever, Columbus State Community College
Jeff Kiggans, Monroe Community College
Jose Maldonado, El Paso Community College–Valle Verde
Stephen Matheson, Calvin College
James McCaughern-Carucci, Quinnipiac University
Karen McFarland Meisch, Austin Peay State University
Dorian R. McMillan, College of Charleston
Caroline McNutt, Schoolcraft College
Lori A. Pitkofsky, Rampano College
Joel Piperburg, Millersville University
Michael P. Robinson, Miami University
Wendy F. Rothwell, University of California–Santa Cruz
Matt Schmidt, SUNY–Stonybrook
Jason F. Schreer, State University of New York–Potsdam
Jennifer J. Scoby, Illinois Central College
Dawn Sherry, Macon State University
Cara Shillington, Eastern Michigan University
Patricia L. Smith, Valencia Community College
Jill M. Tall, Youngstown State University
Jeffrey Thomas, California State University–Northridge
Bruce Tomlinson, State University of New York–Fredonia
Joy B. Trauth, Arkansas State University
Alan R. Wasmoen, Metropolitan Community College
Brad Wetherbee, University of Rhode Island
Michael Windelspecht, Appalachian State University
David E. Wolfe, American River College
Kerry Yurewicz, Plymouth State University

Accuracy Reviewers

Adrienne Alaie-Petrillo, Hunter College
Sylvester Allred, Northern Arizona University
Brian Bagatto, University of Akron
Christine Bezotte, Elmira College
Kelly Carrillo Burke, College of the Canyons
Michael Carr, Oakton Community College
Michelle Cawthorn, Georgia Southern University
Erica Champion, Yale University
William Crampton, University of Central Florida
Michael S. Dann, Pennsylvania State University
Renee Dawson, University of Utah
Jean DeSaix, University North Carolina at Chapel Hill
D. Michael Denbow, Virginia Polytechnic Institute and State University
Sherri Graves, Sacramento City College

Janelle M. Hare, Morehead State University
Bernard Hauser, University of Florida
Albert Herrera, University of Southern California
John A. Hnida, Peru State College
Kelly Hogan, University of North Carolina–Chapel Hill
Jane Horlings, Saddleback College
Anne Houtman, California State University–Fullerton
Victoria Johnson, San Jose State University
Nancy Kinkaid, Troy University
Ariel R. Krakowski, Laney College
Paul Marshall, Northern Essex Community College
Jennifer Metzler, Ball State University
Michael Silva, El Paso Community College
Ignatius Tan, New York University
Pamela Thinesen, Century College
Jeffrey Thomas, California State University–Northridge
Christopher Thompson, Loyola University Maryland
Martin A. Vaughan, Indiana University/Purdue University at Indianapolis
William A. Velhagen, New York University
Michael Wenzel, California State University–Sacramento
John E. Whitlock, Hillsborough Community College
David Wolfe, American River College

Focus Group Attendees

Text and Content Development

(2004, Los Angeles)

Garen Baghdasarian, Santa Monica College
Thomas Chen, Santa Monica College
Elizabeth Ciletti, Pasadena City College
Mary Colavito, Santa Monica College
Ana Ester Escandon, Los Angeles Harbor College
Phyllis Hirsh, East Los Angeles College
Melody Ricci, Victor Valley College

(2004, Memphis)

T. Wayne Barger, Tennessee Technical University
Don Baud, University of Memphis
Jay Blundon, Rhodes College
Martha P. Brown, University of Memphis
Dave Eakin, Eastern Kentucky University
Stan Eisen, Christian Brothers University
Jack Grubaugh, University of Memphis
Min-Ken Lao, Furman University

(2004, Chicago)

Sandra Bobick, Community College of Alleghany
Christopher Dobson, Grand Valley State University
Sondra Dubowsky, Allen County Community College
Eileen Gregory, Rollins College
Tracy Harris, Florida Keys Community College
Robert Krasner, Providence College
Bernard A. Marcus, Genesee Community College
Laurel B. Roberts, University of Pittsburgh
Lyndell P. Robinson, Lincoln Land Community College
Brian Schmaefsky, Kingwood College
Janet Vigna, Grand Valley State University
Daniel W. Ward, Waubonsee Community College

(2006, Florida)

Peggy Brickman, University of Georgia
Anne E. Bunnell, East Carolina University
Cara Gubbins, Butte College
Janelle M. Hare, Morehead State University
John A. Hnida, Peru State College
Mitrick A. Johns, Northern Illinois University
Kathleen H. Lavoie, State University of New York–Plattsburgh
Blasé Maffia, University of Miami
Gregory J. Podgorski, Utah State University
Susan T. Rouse, Southern Wesleyan University
Jennifer M. Warner, University of North Carolina–Charlotte

Physiology

(2009, San Francisco)

Dan Alex, Chabot College
Dennis Anderson, Oklahoma City Community College
Sean Brumbaugh, Santa Rose Junior College
Tom Chen, Santa Monica College
Genevieve Chung, Broward Community College (Central)
Melinda Downing, Merritt College
Danielle DuCharme, Waubonsee Community College
Gerald Farr, Texas State University–San Marcos
Ari Krakowski, Laney College
Jennifer Lange, Chabot College
Pramila Sen, Houston Community College
Michael Windelspecht, Appalachian State University

Art Development

(2007, Atlanta)

Renee Dawson, University of Utah
Charles (Billy) Gunnels IV, Florida Gulf Coast University
Bernard Hauser, University of Florida
Arnold J. Karpoff, University of Louisville
Stephen R. Kelso, University of Illinois–Chicago
Meredith Sommerville Norris, University of North Carolina–Charlotte
Therese M. Poole, Georgia State University
Mary Celeste Reese, Mississippi State University
Bill Rogers, Ball State University
Kim Cleary Sadler, Middle Tennessee State University
Amy S. Wernette, Hazard Community College

Media and Supplements

(2007, Santa Monica)

Michael Bucher, College of San Mateo
Kelly Carrillo Burke, College of the Canyons
John A. Hnida, Peru State College

Anne Houtman, California State University–Fullerton
Hinrich Kaiser, Victor Valley College
Melanie Loo, California State University–Sacramento
Gregory J. Podgorski, Utah State University
Melissa Presch, California State University–Fullerton
Calvin Young, Fullerton College

(2008, Miami)

Barbara Blonder, Flagler College
Charlotte Borgeson, University of Nevada–Reno
Kris Curran, University of Wisconsin–Whitewater
Kari Eamma, Tarrant Community College
Evelyn Frazier, Florida Atlantic University
John Janovy, University of Nebraska–Lincoln
Jennifer Schramm, Chemeketa Community College
Greg Sievert, Emporia University
Heather Vance Chalcraft, East Carolina University
Martin A. Vaughan, Indiana University/Purdue University at Indianapolis

Comparative Reviewers

Zulfiqar Ahmad, East Tennessee State University
Julie H. Aires, Florida Community College at Jacksonville
William Anyonge, Xavier University
Tami Asplin, North Dakota State University
Randy Brewton, University of Tennessee–Knoxville
Dustin Brisson, University of Pennsylvania
Jeannie Chapman, University of South Carolina–Upstate
Thomas Chen, Santa Monica College
Craig Clifford, Northeastern State University
William Gareth Richard Crampton, University of Central Florida
Chris Davison, Long Beach City College
Tiffany Doan, Central Connecticut State University
Ernest F. DuBrul, University of Toledo
Eugene Fenster, Metropolitan Community College–Longview
Rebecca Fetherson, University of Dayton
Teresa G. Fischer, Indian River State College
Anita Pardue Flick, North Carolina State University
Paul Florence, Jefferson Community and Technical College
Evelyn Frazier, Florida Atlantic University
Diane Wilkening Fritz, Northern Kentucky University
Dennis W. Fulbright, Michigan State University
Douglas C. Gayou, University of Missouri–Columbia
Betsy Gerbec, University of Wisconsin–River Falls
Oliver Ghobrial, Santa Fe College
Julie Gibbs, College of DuPage
Florence K. Gleason, University of Minnesota
Bruce Griffis, Kentucky State University
Tim Grogan, Valencia Community College
Charles J. Grossman, Xavier University
Luis S. Guerra, South Texas College
Pieter de Haan, Berkeley City College
Mary F. Haskins, Rockhurst University
Keith R. Hench, Kirkwood Community College
Sherry Hickman, Hillsborough Community College
Mark Hollier, Georgia Perimeter College
Robert A. Holmes, Hutchinson Community College
Laurie Host, Harford Community College
Robert Iwan, Inver Hills Community College
Philip Jardim, City College of San Francisco
Dianne Jennings, Virginia Commonwealth University
David Knowles, East Carolina University
Kim Lackey, University of Alabama
Ellen Lamb, University of North Carolina at Greensboro
David Loring, Johnson County Community College
Bill Mackay, Edinboro University of Pennsylvania
Barbara Mania-Farnell, Purdue University–Calumet
Tom J. McConnell, Ball State University
Wallace M. Meyer, College of the Redwoods
Thelma Miller-Anderson, Hillsborough Community College
Michael R. Millward, University of Cincinnati
Brenda Moore, Truman State University
Michael Muller, University of Illinois–Chicago
Zia Nisani, Antelope Valley College
Tanya Noel, York University
James Nolan, Georgia Gwinnett College
Laura Palmer, Pennsylvania State University–Altoona
Louis Pech, University of Wisconsin Centers–Marathon County
Joel B. Piperberg, Millersville University
Karen Plucinski, Missouri Southern State University
Aggie Posthumus, Olivet Nazarene University
Bonnie Ripley, Grossmont College
Troy Rohn, Boise State University
Matthew Rowe, Sam Houston State University
Yelena Rudayeva, Palm Beach Community College
Arthur Sandquist, Washburn University
Donald Slish, Plattsburgh State University
Christine Stracey, University of Florida
Anthony J. Stancampiano, Oklahoma City Community College
Tim Strakosh, State University of New York–Fredonia
Richard P. Stringer, Harrisburg Area Community College
Mark Sturtevant, Oakland University
Sukanya Subramanian, Collin County Community College
Rob Swatski, Harrisburg Area Community College
Kimberly Taugher, Diablo Valley College
Jeffrey Thornsberry, Northwest Missouri State
Nina Thumser, Indiana University of Pennsylvania
Jonathan Titus, State University of New York–Fredonia
Mike Tveten, Pima Community College Northwest
Katherine M. Van de Wal, Community College of Baltimore County–Essex Campus
Leslie VanderMolen, Humboldt State University
Fred Vogt, Elgin Community College
William A. Wehbi, University of Pennsylvania

John E. Whitlock, Hillsborough Community College
Jennifer Wiatrowski, Pasco-Hernando Community College
Christina Willis, Rockhurst University
Donald S. Wood, Odessa College
Aimee Wurst, Lincoln University of Missouri
Lan Xu, South Dakota State University

We thank the hundreds of students and faculty at the schools listed below for class-testing elements of the text.

Q Questions

Borough of Manhattan Community College
Broward Community College
Central New Mexico Community College
Colorado Mountain College
Florida Community College–Jacksonville
Heartland Community College
Johnson County Community College
Kutztown University of Pennsylvania
Lincoln Land Community College
Metropolitan Community College
Mt. San Jacinto College
Portland State University
Rowan University
Seattle Central Community College
Suffolk County Community College–Ammerman
Tarrant County Community College (Northeast)
University of Northern Colorado
Wake Technical Community College

Chapters

(2007)

College of San Mateo
Georgia Perimeter College
Georgia State University
Hazard Community College
Southwestern College
University of Arizona
University of Louisville
University of North Carolina–Charlotte
University of Utah
Utah State University
Virginia Commonwealth University

(2008)

Bluegrass Community College
California State University–Northridge
Chaffey College
Clark Atlanta University
Columbus State University
Daytona Community College
Edinboro University
Genesee Community College
George Washington University
Glendale College
Lake Land College
Long Beach Community College
Los Angeles Harbor College
Macon State University
Marshall University
McPherson College
Millersville University of Pennsylvania
Milwaukee Area Technical College
Owensboro Community College
Oxnard College
Philadelphia University
Prince George's Community College
Quinnipiac University
Rock Valley College
Sacramento City College
Salem State College
San Jose State University
Sinclair Community College
Southwestern College
St. Clair Community College
University of New Mexico
University of Rhode Island
Valencia Community College

(2009)

Central Carolina Community College
Florida Community College at Jacksonville
Maple Woods Community College
Morton College
Virginia Commonwealth University
Virginia Polytechnic Institute and State University

The media and supplements package was created with the input of many dedicated and experienced non-majors biology instructors.

Reviewers

Ann Aguanno, Marymount Manhattan College
Charlotte E. Borgeson, University of Nevada–Reno
Kelly Carillo Burke, College of the Canyons
Michelle Cawthorn, Georgia Southern University
G. Chung, Broward Community College
Michael S. Dann, Pennsylvania State University
Gerald Farr, Texas State University–San Marcos
Debra B. Folsom, Pasadena City College
Julie V. Gibbs, College of DuPage
Stephen M. Gómez, Central New Mexico Community College
Jane J. Henry, Baton Rouge Community College
Jane Horlings, Saddleback College
Carina Endres Howell, Lock Haven University
John Janovy Jr., University of Nebraska–Lincoln
Hinrich Kaiser, Victor Valley College
Judy Kaufman, Monroe Community College
Charlease Kelly-Jackson, Claflin University
Patrick J. Lewis, Sam Houston State University
Cindy S. Malone, California State University–Northridge
Jon Milhon, Azusa Pacific University
Meredith Sommerville Norris, University of North Carolina–Charlotte
Greg Podgorski, Utah State University
Lawrence F. Roberge, Florida Community College–Jacksonville
William D. Rogers, Ball State University
Georgianna Saunders, Missouri State University
Jason Schreer, SUNY–Potsdam
Marilyn Shopper, Johnson County Community College
Carol St. Angelo, Hofstra University
Eric Stavney, North Seattle Community College
Alicia Steinhardt, Hartnell College
Jamey Thompson, Hudson Valley Community College
Helen Walter, Diablo Valley College
Jennifer M. Warner, University of North Carolina–Charlotte
Michael Windelspecht, Appalachian State University

The ***What Is Life?*** media and supplements package was created with the input of hundreds of non-majors biology instructors and through the hard work of our experienced and committed media and supplements contributor team.

Instructor's Manual—Classroom Catalysts

Verona A. Barr, Heartland Community College
Stephen M. Gómez, Central New Mexico Community College
Paul H. Marshall, Northern Essex Community College
Lawrence Roberge, Laboure College
David E. Wolfe, American River College

Connecting Biology to Life: Application Books

Mike Tveten, Pima Community College, Tucson

eBook and Non-Majors Biology Study Tools

Robert Iwan, Inver Hills Community College
Eric Stavney, North Seattle Community College

Figure Conversion Engine Tool

Ann Aguanno, Marymount Manhattan College

Image Bank Advisor

Jane J. Henry, Baton Rouge Community College

Keynote Lecture Presentation

Michael C. Bucher, College of San Mateo

PowerPoint Lecture Outlines

Kristen L. Curran, University of Wisconsin–Whitewater
Danielle DuCharme, Waubonsee Community College
Jennifer Lange, Chabot College
Mark S. Manteuffel, St. Louis Community College

Prep-U Instructor's Test Bank and Prep-U for Students

Glenn Adelson, Lake Forrest College
Jay Phelan, University of California, Los Angeles
Alon Ziv, Prep-U

Q Animations

Anne Bunnell, East Carolina University
G. Chung, Broward College
Johnny El-Rady, University of South Florida
Eric Stavney, North Seattle Community College
Julie V. Gibbs, College of DuPage

Student Success Guide for Phelan's *What Is Life?*

Meredith S. Norris, University of North Carolina–Charlotte
Jennifer M. Warner, University of North Carolina–Charlotte

Questions About Life Reader

Heather Vance-Chalcraft, East Carolina University

Student Worksheets

Michael C. Bucher, College of San Mateo

Hands-On Biology:Laboratories for Distance Learning

Mimi Bres, Arnold Weisshaar, and Cassandra Moore-Crawford, Prince George's Community College (all)

Exploring Biology: Case Studies

Michelle Cawthorn, Georgia Southern University
Jennifer L. Holzman, Emory University

What Is Life? A Guide to BIOLOGY with PHYSIOLOGY

1

Scientific Thinking

Your best pathway to understanding the world

❶ Science is a collection of facts and a process for understanding the world.

Already a scientist? It starts with curiosity.

1•1

What is science? What is biology?

You are already a scientist. You may not have realized this yet, but it's true. Because humans are curious, you have no doubt asked yourself or others questions about how the world works and wondered how you might find the answers.

- Does the radiation released by cell phones cause brain tumors?
- Are antibacterial hand soaps better than regular soap?
- Do large doses of vitamin C reduce the likelihood of getting a cold?

These are all important and serious questions. But you've probably also pondered some less weighty issues, too.

- Why is morning breath so stinky? And can you do anything to prevent it?
- Why is it always windy on streets with tall buildings?
- Does taking aspirin before drinking alcohol lead to faster intoxication?

And if you really put your mind to the task, you will start to find questions all around you whose answers you might like to know (and, like those above, whose answers you will learn as you read this book).

- Why is it that seaside towns are not as hot in the summer or as cold in the winter as inland communities?
- Why do so few women get into barroom brawls?
- What is "blood doping" and does it really improve athletic performance?
- Why is it so much easier for an infant to learn a complex language than it is for a college student to learn biology?

Still not convinced you're a scientist? Then here's some good news: science doesn't require advanced degrees or secret knowledge dispensed over years of technical training. It does, however, require an important feature of our species—a big brain—as well as curiosity and a desire to learn about the world around us and inside us. But curiosity, casual observations, and desire can only take you so far. In today's world there are important and pressing issues that require some understanding of science, what it can and can't do, and what it can and can't explain. Nutritional claims on foods and dietary supplements, human behavior, health and disease, how drugs can cure or poison you, the interactions of plants, animals, and their environments, global warming, the continuity and diversity of life in all its forms—by learning more about science, you can think about, learn about, and understand issues like these and things about yourself, your life, and the world around you.

To satisfy your curiosity and to explain seemingly unexplainable, mysterious, or magical things, curiosity and the desire to learn are not enough. Explaining how something works or why something happens requires methodical, objective, and rational observation and analysis that are not clouded with emotions or preconceptions about what is being studied and observed.

Science is not simply a body of knowledge or a list of facts to be remembered. It is an intellectual activity, encompassing observation, description, experimentation, and explanation of natural phenomena. Put another way, science is a pathway by which we can come to discover and better understand the world around us. Later in this chapter, we explore specific ways in which we can most effectively use scientific thinking in our lives. But first let's look at how our understanding of the world can be enhanced by asking the single question that underlies scientific thinking:

How do you know that is true?

Once you begin asking this question—of others and of yourself—you are on the road to a better understanding of the world.

The following story about a popular and successful over-the-counter medicine shows the importance of questioning the truth of many "scientific" claims you see on merchandise packages or read in newspaper or on the internet. For more than 10 years, a product called "Airborne" was marketed and sold to millions of customers. On the packaging and in advertisements, the makers asserted that Airborne tablets could ward off colds and boost your immune system (**FIGURE 1-1**). Not surprisingly, Airborne was a great success and generated more than $100 million in revenue, and became one of the fastest selling health products ever. Then some consumers posed a reasonable question to the makers of Airborne: how do you know that it wards off colds?

Q Can we trust the packaging claims that companies make?

To prove their claims, the makers of Airborne pointed to the results of a "double-blind, placebo-controlled study" conducted by a company specializing in clinical drug trials. We'll discuss exactly what those terms mean later in the chapter; for now we just need to note that as a result of a class-action lawsuit, it became clear that no such study had been conducted and that there was *no* evidence to back up Airborne's claims. The Airborne company removed the claims from the packaging and agreed to refund the purchase price to anyone who had bought Airborne. It also removed any reference to its "clinical trials," with the company's CEO saying that people "are really not scientifically minded enough to be able to understand a clinical study."

Are you insulted by the CEO's assumption about your intelligence? You should be. Did you or your parents (along with millions of other people) fall for Airborne's false claims? Possibly. But here's some more good news: you don't have to be at the mercy of slick packaging, false advertising, cranks, or charlatans that claim a product can ward off colds, make you lose 10 pounds in 7 days, make you healthier by cleansing your intestines, or increase or decrease the size of various body parts. You can learn to be skeptical and suspicious (in a good way) of such claims. You can learn exactly what it means to have scientific proof or evidence that something is absolutely true. And you can learn this by learning what it means to think scientifically.

Scientific thinking is an important and productive element in the study of a wide variety of topics: it can help you understand economics, psychology, history, and many other subjects. Our focus in this book is **biology,** the study of living things. Taking a scientific approach, we will investigate the facts and ideas in biology that are already known and will study the process by which we come to learn new things. As we move through the four parts of the book, we explore the most important questions in biology.

- What is the chemical and physical basis for life and its maintenance?
- How do organisms use genetic information to build themselves and to reproduce?
- What are the diverse forms that life on earth takes and how has that diversity arisen?
- How do organisms interact with each other and with their environment?

FIGURE 1-1 Some products claim to improve our health, but how do we know if they work?

In this chapter, we explore how to think scientifically and how to use the knowledge we gain to make wise decisions. Although we generally restrict our focus to biology, scientific thinking can be applied to nearly every endeavor, and so in this chapter we use a wide range of examples—including some examples from beyond biology—as we learn how to think scientifically. Although the examples will vary greatly, they all convey a message that is key to scientific thinking: it's okay to be skeptical and you don't need to take things on faith.

Fortunately, learning to think scientifically is not difficult—and it can be fun, particularly because it is so empowering. **Scientific literacy,** a general, fact-based understanding of the basics of biology and other sciences, is increasingly important in our lives, and literacy in matters of biology is especially essential. As we see in the next section, issues hinging on biological ideas and processes are turning up in nearly every facet of our lives, from medicine to nutrition to personal relationships to politics, and even to crime and justice.

TAKE-HOME MESSAGE 1•1

Through its emphasis on objective observation, description, and experimentation, science is a pathway by which we can discover and better understand the world around us.

1•2

Biological literacy is essential in the modern world.

In the past, it was possible to get through life with little or no knowledge of science, but now scientific literacy has become a necessity. A brief glance at any newspaper will reveal just how many important health, social, medical, political, economic, and legal issues pivot on complex scientific data and theories.

- Why are unsaturated fats healthier for you than saturated fats?
- What are allergies? Why do they strike children from clean homes more than children from dirty homes?
- Why is friendship so common among humans yet extremely rare elsewhere in the animal kingdom?
- Why do new agricultural pests appear faster than new pesticides?
- What does scientific testing reveal about the accuracy of eyewitness testimony in courtrooms?

Biological literacy is the ability to (1) use the process of scientific inquiry to think creatively about real-world issues that have a biological component, (2) communicate these thoughts to others, and (3) integrate these ideas into your

January 24, 2007

With DNA From Exhumed Body, Man Finally Wins Freedom

By FERNANDA SANTOS

AUBURN, N.Y., Jan. 23 — Roy Brown, who spent 15 years in prison on a murder conviction and uncovered evidence while there that linked another man to the crime, was released from prison on Tuesday after DNA tests on the other man's exhumed body matched saliva on a nightshirt at the crime scene.

FIGURE 1-2 In the news. Every day, news sources report on social, political, medical, and legal issues related to science.

decision making. Biological literacy doesn't involve just the big issues facing society or just abstract ideas (**FIGURE 1-2**). It also matters to you personally. Should you take aspirin when you have a fever? Are you using the wrong approach if you try to lose weight and, after some initial success, you find your rate of weight loss diminishing? Is it a good idea to consume moderate amounts of alcohol?

> "Scientific issues permeate the law. I believe [that] in this age of science we must build legal foundations that are sound in science as well as in law. The result, in my view, will further not only the interests of truth but also those of justice.
>
> — U.S. Supreme Court Justice Stephen Breyer, at the annual meeting of the American Association for the Advancement of Science, February 1998"

Biological issues—including global warming, fossil fuel use, stem cell research, and the proliferation of genetically modified foods—have also become important in political campaigns. And all around you, you will encounter products bearing too-good-to-be-true claims and technical-sounding language designed to lure you into purchasing them. Lack of biological literacy will put you at the mercy of "experts" who may try to confuse you or convince you of things in the interest of (their) personal gain. Scientific thinking will help you make wise decisions for yourself and for society.

TAKE-HOME MESSAGE 1•2

Biological issues permeate all aspects of our lives. To make wise decisions, it is essential for individuals and societies to attain biological literacy.

1•3

The scientific method is a powerful approach to understanding the world.

> "If science proves some belief of Buddhism wrong, then Buddhism will have to change.
>
> — The 14th Dalai Lama, *New York Times,* December 2005"

It's a brand new age, and science, particularly biology, is everywhere. We are called upon more and more frequently to make decisions that hinge on our abilities to grasp biological information and to think scientifically. To illustrate the value of scientific thinking in understanding the world, let's look at what happens in its absence, by considering some unusual behaviors in the common laboratory rat.

Rats can be trained without much difficulty to push a lever to receive a food pellet from a feeding mechanism (**FIGURE 1-3**). When the mechanism is altered so that there is a 10-second delay between the lever being pushed and the food pellet being dispensed, however, strange things start to happen. In one cage, the rat will push the lever and then, very methodically, run and push its nose into one corner of the cage. Then it moves to another corner and again pushes its nose against the cage. It repeats this behavior at the third and fourth corners of the cage, after which the rat stands in front of the feeder and the pellet is dispensed. Each time the rat pushes the lever it repeats the nose-in-the-corner sequence before moving to the food tray.

In another cage, with the same 10-second delay before the food pellet is dispensed, a rat pushes the lever and then

FIGURE 1-3 **"In the absence of the scientific method . . ."** Rats develop strange superstition-like behaviors if there is a 10-second delay between when they push a lever and when food is delivered.

proceeds to do three quick back-flips in succession. It then moves to the food tray for the food pellet when the 10 seconds have elapsed. Like the nose-in-the-corner rat, the back-flip rat will repeat this exact behavior each time it pushes the lever.

Q Why and when do people develop superstitions? Can animals be superstitious?

In cage after cage of rats with these 10-second-delay food levers, each rat eventually develops its own peculiar series of behaviors before moving to the food dish to receive the pellet. Why do they do this? Because it seems to work! They have discovered a method by which they can get a food pellet. To some extent, the rats' behaviors are reasonable. They associate two events—pushing the lever and engaging in some sequence of behaviors—with another: receiving food. In a sense, they have taken a step toward understanding their world, even though the events are not actually related to each other.

Humans can also mistakenly associate actions with outcomes in an attempt to understand and control their world. The irrational belief that actions that are not logically related to a course of events can influence its outcome is called **superstition.** In the absence of scientific thinking, individuals can develop incorrect ideas, such as superstitions about how the world works (**FIGURE 1-4**). For example, Nomar Garciaparra, a

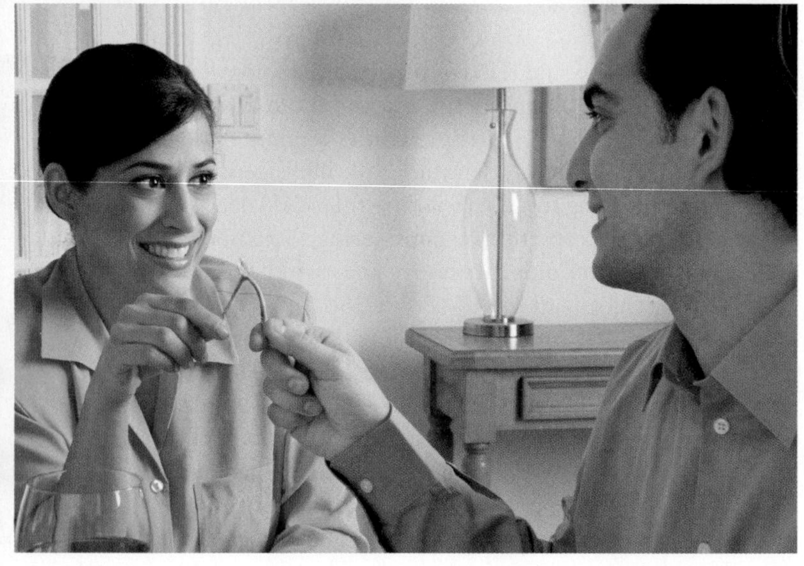

FIGURE 1-4 Superstitions abound. As comforting as myths and superstitions may be, they are no substitute for really understanding how the world works.

baseball player for the Oakland Athletics, engages in a precise series of toe taps and adjustments to his batting gloves before he bats.

Beyond superstitions, humans have developed a variety of ways of understanding their world. Ancient Greeks, for example, understood that the evils of the world flew out of Pandora's box when, despite Zeus's strict orders not to open the box, her curiosity got the best of her and she lifted the lid. In the Kalahari desert, the African Bushmen tell the tale that their creator, Kaang, ordered them not to build fires. Because their ancestors forgot these orders and built a fire, Kaang punished them by destroying their friendship with animals. Although Bushmen and animals could communicate with each other in ancient times, now the animals simply fear the Bushmen.

Thousands of different narratives from all over the world exist to help people understand the world around them. These stories explain everything from birth and death to disease and healing. On the creation of the earth and universe alone, more than 4,000 narratives have been catalogued.

As helpful and comforting as stories and superstitions may be (or as helpful as people *think* they are!), they are no substitute for really understanding how the world works: for really understanding, for example, that you are sick not because the gods are displeased with you but because the water you are drinking is contaminated, and that, if the water can be purified, then you won't get sick. This kind of understanding does not come all at once by some magical power; instead, it begins when someone wonders about why something is the way it is and then decides to try to find out the answer. This process of examination and discovery is called the **scientific method.**

The scientific method usually begins with someone observing a phenomenon and proposing an explanation for it. Next, the proposed explanation is tested through a series of experiments. If the experiments reveal that the explanation is accurate, and if the experiments can be done by others with the same result, then the explanation is considered to be valid. If the experiments do not support the proposed explanation, then the explanation must be revised or alternative explanations that more closely reflect experimental results must be proposed and tested. This process continues as better, more accurate explanations are found.

While the scientific method reveals much about the world around us, it doesn't explain everything. There are many other methods through which we can gain an understanding of the world. For example, much of our knowledge about plants and animals does not come from the use of the scientific method, but rather comes from systematic, orderly observation, without the testing of any explicit hypotheses. Other disciplines also involve understandings of the world based on non-scientific processes. Knowledge about history, for example, comes from the systematic examination of past events as they relate to humans, while the "truths" in other fields, such as religion, ethics, and even politics, often are based on personal faith, traditions, and mythology.

Scientific thinking can be distinguished from these alternative ways of acquiring knowledge about the world in that it is **empirical.** Empirical knowledge is based on experience and observations that are rational, testable, and repeatable. The empirical nature of the scientific approach makes it self-correcting: in the process of analyzing a topic, event, or phenomenon with the scientific method, incorrect ideas are discarded in favor of more accurate explanations. In the next sections, we look at how to put the scientific method into practice.

TAKE-HOME MESSAGE 1·3

There are numerous ways of gaining an understanding of the world. Because it is empirical, rational, testable, repeatable, and self-correcting, the scientific method is a particularly effective approach.

❷ A beginner's guide: what are the steps of the scientific method?

Scientific thinking relies on rational, testable, and repeatable observations.

1.4

Thinking like a scientist: how do you use the scientific method?

"Scientific method"—this term sounds like a rigid process to follow, much like following a recipe. In practice, however, the scientific method is not a single pathway that is always rigidly followed from start to finish. Rather, it is an adaptable process that includes many different methods. This flexibility makes the scientific method a more powerful process that can be used to explore a wide variety of thoughts, events, or phenomena, not only in science, but in other areas as well.

The basic steps in the scientific method are:

Step 1. Make observations.

Step 2. Formulate a hypothesis.

Step 3. Devise a testable prediction.

Step 4. Conduct a critical experiment.

Step 5. Draw conclusions and make revisions.

Once begun, though, the process doesn't necessarily continue linearly through the five steps until it is concluded (**FIGURE 1-5**). Sometimes, observations made in the first step can lead to more than one hypothesis and several testable predictions and experiments. And the conclusions drawn from experiments often suggest new observations, refinements to hypotheses, and, ultimately, increasingly precise conclusions.

THE SCIENTIFIC METHOD

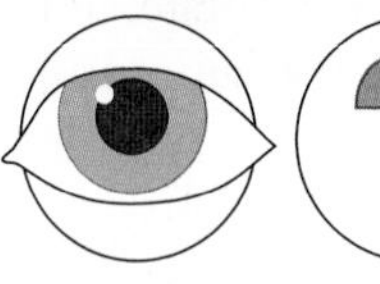

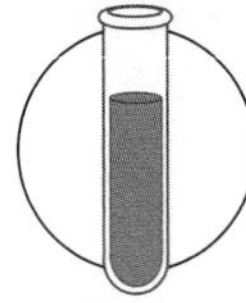

The scientific method rarely proceeds in a straight line. Conclusions, for example, often lead to new observations and refined hypotheses.

FIGURE 1-5 The scientific method: five basic steps and one flexible process.

Q What should you do when something you believe in turns out to be wrong?

An especially important feature of the scientific method is that its steps are self-correcting. As we continue to make new observations, a hypothesis about how the world works might change. If our observations do not support our current hypothesis, that hypothesis must be given up in favor of one that is not contradicted by any observations. This may be the most important feature of the scientific method: ***it tells us when we should change our minds.***

Because the scientific method is a general strategy for learning, it needn't be used solely to learn about nature or scientific things. In fact, we will investigate how it works by looking at a specific question that has recently begun to benefit from analysis using the scientific method:

- How reliable is eyewitness testimony in criminal courts?

For more than 200 years, courts in the United States have viewed eyewitness testimony as unassailable. Few things are seen as more convincing to a jury than an individual testifying that she can identify the person she saw commit a crime (**FIGURE 1-6**). Common sense has told us that events we see with our own two eyes are as they appear. But is eyewitness identification always right? Can the scientific method tell us whether this belief—or some other belief—is true? As we describe how to use the scientific method to answer questions about the world, it will become clear that the answer is a resounding *yes.* To show how the scientific method can be used to examine a wide variety of issues, we will also look at how the scientific method can be used to answer three additional questions:

- Does echinacea reduce the intensity or duration of the common cold?
- Does chemical runoff give rise to hermaphrodite fish?
- Does shaving hair from your face, legs, or anywhere else cause it to grow back coarser or darker?

FIGURE 1-6 "With your own two eyes . . ." How reliable is eyewitness testimony in criminal courts?

Hold the fries. *We apply an understanding of science when we choose foods from the menu that have fewer calories and less saturated fat.*

TAKE-HOME MESSAGE 1•4

The scientific method (observation, hypothesis, prediction, test, and conclusion) is a flexible, adaptable, and efficient pathway to understanding the world, because it tells us when we must change our beliefs.

1·5

Step 1: Make observations.

Scientific study always begins with observations. At the first stage in the scientific method, we simply look for interesting patterns or cause-and-effect relationships. This is where a great deal of the creativity of science comes from. In the case of eyewitness testimony, we know now that new technologies have made it possible to assess whether tissue such as hair or blood from a crime scene came from a particular suspect. Armed with these tools, the Justice Department recently reviewed 28 criminal convictions that had been overturned by DNA evidence. They found that in most of the cases, the strongest evidence against the defendant during the trial had been eyewitness identification. The observation here is that many defendants who are later found to be innocent were initially convicted based on eyewitness testimony.

Opportunities for other interesting observations are unlimited. Using the scientific method, we can (and will) answer our three other questions.

Many people have claimed that consuming extracts of the herb echinacea can reduce the intensity or duration of symptoms of the common cold. For this reason, echinacea is a widely used herbal treatment (**FIGURE 1-7**). Returning to our fundamental question underlying the scientific method, we can ask: how do you know that is true?

- Does taking echinacea reduce the intensity or duration of the common cold?

Some people have noted that chemicals in sewage runoff—particularly those related to the hormone estrogen—seem to cause male fish to turn into hermaphrodites, organisms that have the reproductive organs of both sexes. Is this true?

- Does chemical runoff give rise to hermaphrodite fish?

And finally, some people have suggested that shaving hair from your face, legs, or anywhere else causes the hair to grow back coarser and darker. Is this true?

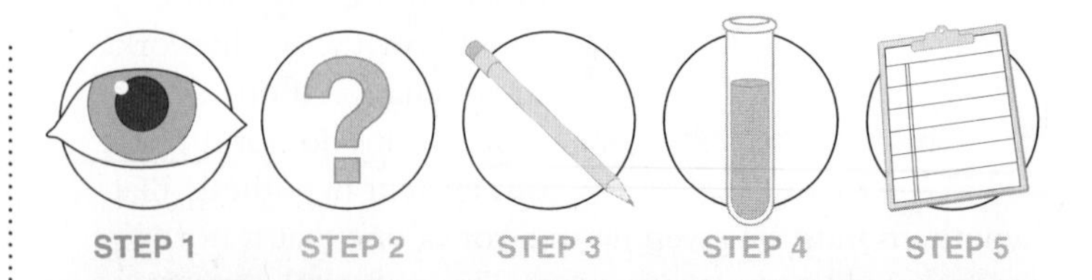

STEP 1: MAKE OBSERVATIONS

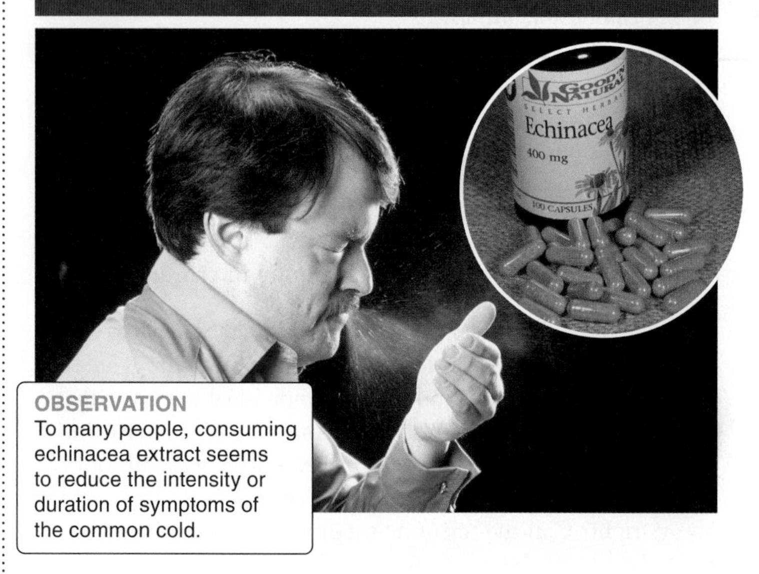

OBSERVATION
To many people, consuming echinacea extract seems to reduce the intensity or duration of symptoms of the common cold.

FIGURE 1-7 The first step of science: making observations about the world.

- Does hair that is shaved grow back coarser or darker?

Using the scientific method, we can answer all of these questions.

TAKE-HOME MESSAGE 1·5

The scientific method begins by making observations about the world, noting apparent patterns or cause-and-effect relationships.

1·6

Step 2: Formulate a hypothesis.

Based on observations, we can develop a **hypothesis** (*pl.* **hypotheses**), a proposed explanation for observed phenomena. What hypotheses could we make about the eyewitness testimony observations described in the previous section? We could start with the hypothesis: "Eyewitness testimony is always accurate." We may need to modify our hypothesis later, but this is a good start. At this point, we can't draw any conclusions. All we have done is summarize some preliminary observations into a possible explanation for what we have observed.

To be most useful, a hypothesis must accomplish two things.

1. It must clearly establish mutually exclusive alternative explanations for a phenomenon. That is, it must be clear that if the proposed explanation is not supported by evidence or further observations, a different hypothesis is a more likely explanation.

2. It must generate testable predictions (**FIGURE 1-8**). This characteristic is important because we can only evaluate the validity of a hypothesis by putting it to the test. A hypothesis that can never be shown, by some form of observation, to be untrue is not a useful hypothesis. Consider, for example, the hypothesis that your dog loves you. If, for any observation someone makes, you can explain how that observation is consistent with your hypothesis, then the hypothesis is not useful (even if your dog really does love you).

For example we could disprove the "Eyewitness testimony is always accurate" hypothesis by demonstrating that, in certain circumstances, individuals who have witnessed a crime might misidentify the criminal when asked to select the suspect from a lineup.

Often researchers will pose a hypothesis as a negative statement proposing that there is no relationship between two factors, such as "Echinacea has no effect on the duration and severity of cold symptoms." Or "There is no difference in the coarseness or darkness of hair that has been shaved." A hypothesis that states a *lack* of relationship between two factors is called a **null hypothesis.** These hypotheses are equally valid but are easier to disprove. This is because a single piece of evidence or a single new observation that contradicts a null hypothesis is sufficient to reject it and conclude that an alternative hypothesis is true, or that it is highly probable that an alternative is true. So, once you have one piece of solid evidence that your null hypothesis is not true, you gain little by collecting further data. Conversely, it is impossible to prove a hypothesis is absolutely and permanently true: all evidence or further observations that support a hypothesis are valuable, but they do not rule out the possibility that some future evidence or observation might show that the hypothesis is not true.

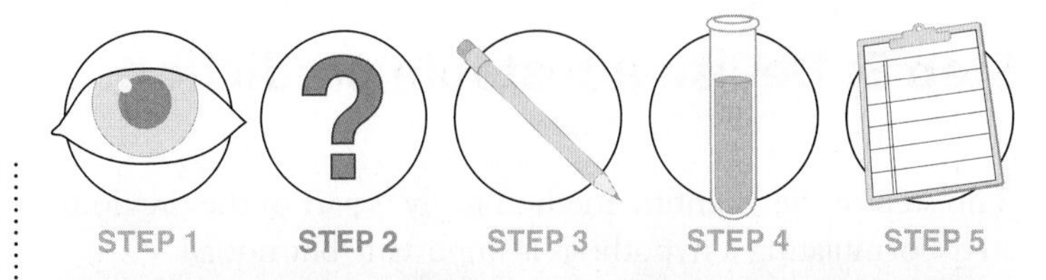

STEP 2: FORMULATE A HYPOTHESIS

FIGURE 1-8 Hypothesis: the proposed explanation for a phenomenon.

For our other observations, we could state our hypotheses in two different ways:

Hypothesis: Echinacea reduces the duration and severity of the symptoms of the common cold.

Null hypothesis: Echinacea has no effect on the duration or severity of the symptoms of the common cold.

Hypothesis: Estrogens in sewage runoff turn fish into hermaphrodites.

Null hypothesis: Estrogens in sewage runoff have no effect in turning fish into hermaphrodites.

Hypothesis: Hair that is shaved grows back coarser and darker.

Null hypothesis: There is no difference in the coarseness or color of hair that is shaved relative to hair that is not shaved.

TAKE-HOME MESSAGE 1·6

A hypothesis is a proposed explanation for an observed phenomenon.

1•7

Step 3: Devise a testable prediction.

This step of the scientific method really is part of the previous step. Formulating a hypothesis is important, but not all hypotheses are created equal. Some, in fact (such as the hypothesis about your dog loving you), are of no use at all when it comes to helping us better understand the world. For a hypothesis to be useful, it must generate a prediction. That is, it must suggest that under certain conditions we will be able to observe certain outcomes. Put another way, a good hypothesis helps us make predictions about novel situations. This is a powerful feature of a good hypothesis: it guides us to knowledge about new situations.

All of this is rather abstract. Let's get more concrete with the four hypotheses we are considering. Keep in mind that when you do not understand some aspect of the world, any one of several possible explanations could be true. In devising a testable prediction from a hypothesis, the goal is to propose a situation that will give a particular outcome if your hypothesis is true, but will give a different outcome if your hypothesis is not true.

Hypothesis: Eyewitness testimony is always accurate.

Prediction: Individuals who have witnessed a crime will correctly identify the criminal regardless of whether multiple suspects are presented one at a time or all at the same time in a lineup.

This is a good, testable prediction because if our hypothesis is true, then our prediction will always be true. On the other hand, if one method of presenting suspects consistently causes incorrect identification of the criminal, our hypothesis cannot be true and must be revised or discarded.

Hypothesis: Echinacea reduces the duration and severity of the symptoms of the common cold (**FIGURE 1-9**).

Prediction: If echinacea reduces the duration and severity of the symptoms of the common cold, then individuals taking echinacea should get sick less frequently than those not taking it, and when they do get sick, their illness should not last as long.

Hypothesis: Estrogens in sewage runoff turn fish into hermaphrodites.

Prediction: If estrogens in sewage runoff turn fish into hermaphrodites, then exposing fish to estrogens should cause more fish to turn into hermaphrodites than when fish are not exposed to estrogens.

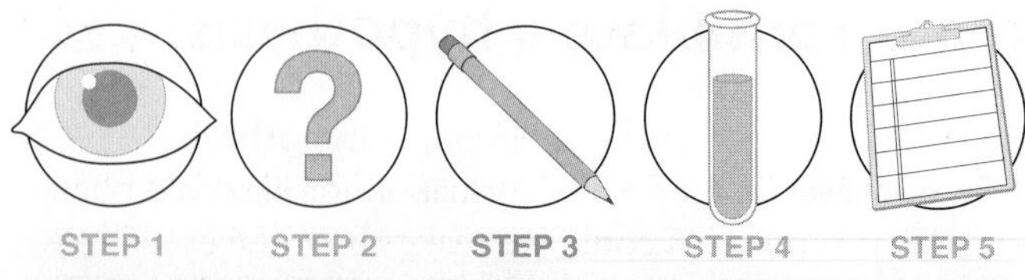

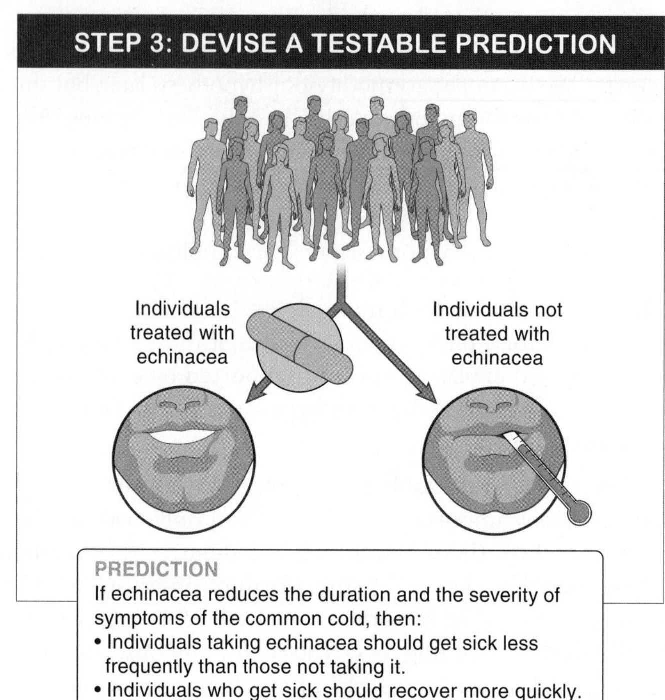

FIGURE 1-9 Devising a testable prediction. For a hypothesis to be useful, it must generate a testable prediction.

Hypothesis: Hair that is shaved grows back coarser and darker.

Prediction: If shaving leads to coarser, darker hair growing back, then if individuals shaved one leg only, the hair that re-grows on that leg should become darker and coarser than the hair growing on the other leg.

As you begin to think scientifically, you will find yourself making a lot of "if . . . then" types of statements: "*If* that is true," referring to some hypothesis or assertion someone makes, or perhaps to a claim made by the manufacturer of a new health product, "*then* I would expect that . . ." proposing your own prediction about a related situation. Once you've made a testable prediction, the next step is to go ahead and test it.

TAKE-HOME MESSAGE 1•7

For a hypothesis to be useful, it must generate a testable prediction.

Step 4: Conduct a critical experiment.

Once we have formulated a hypothesis that generates a testable prediction, we conduct a **critical experiment,** an experiment that makes it possible to decisively determine whether a particular hypothesis is correct. There are many crucial elements in designing a critical experiment, and Section 1-11 covers the details of this process. For now, it is important just to understand that, with a critical experiment, if the hypothesis being tested is not true, we will make observations that compel us to reject that hypothesis.

In this step of the scientific method, a bit of cleverness can come in handy. Suppose we were to devise a critical experiment to test our hypothesis "Eyewitness testimony is always accurate." First, we stage a mock crime such as a purse snatching in front of a group of observers who do not know the crime is staged. Next, we ask observers to identify the criminal. To one group of observers we might present six "suspects" all at once in a lineup. To another group of observers, we might present the six suspects one at a time. The beauty of this experiment is that we actually know who the "criminal" is. With this knowledge, we can evaluate with certainty whether an eyewitness's identification is correct or not.

This exact experiment was done with an additional, devious little twist: the researcher did not include the actual "criminal" in any of the lineups of six "suspects." If eyewitness testimony is always accurate, however, this slight variation should not matter. We would predict that the observers in both groups (the lineup group and the one-at-a-time group) would indicate that the criminal was not present. In the next section, we will see what happened. Right now, let's devise critical experiments for our other hypotheses.

Hypothesis: Echinacea reduces the duration and severity of the symptoms of the common cold. The critical experiment for the echinacea hypothesis has been performed, and it is about as close to a perfect experiment as possible. Researchers began with 437 people who volunteered to be exposed to viruses that cause the common cold. Exposure to the cold-causing viruses was a bit unpleasant: all of the volunteers had cold viruses (in a watery solution) dripped into their noses. Research subjects were then secluded in hotel rooms for five days, and doctors examined them for the presence of the cold virus in their nasal cavity and for any cold symptoms (**FIGURE 1-10**).

The subjects were randomly divided into four groups. In two of the groups, each individual began taking a pill each day for

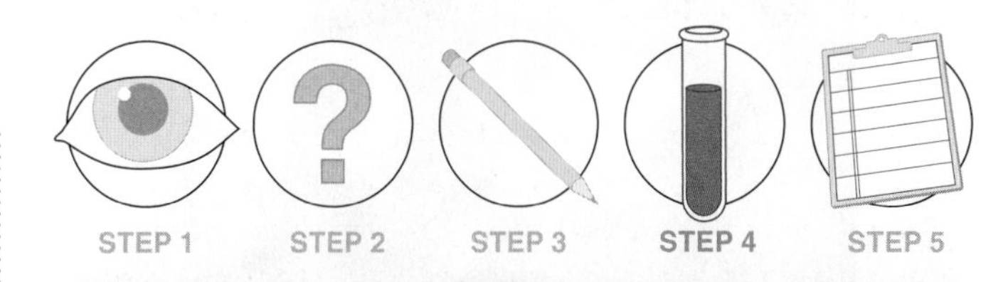

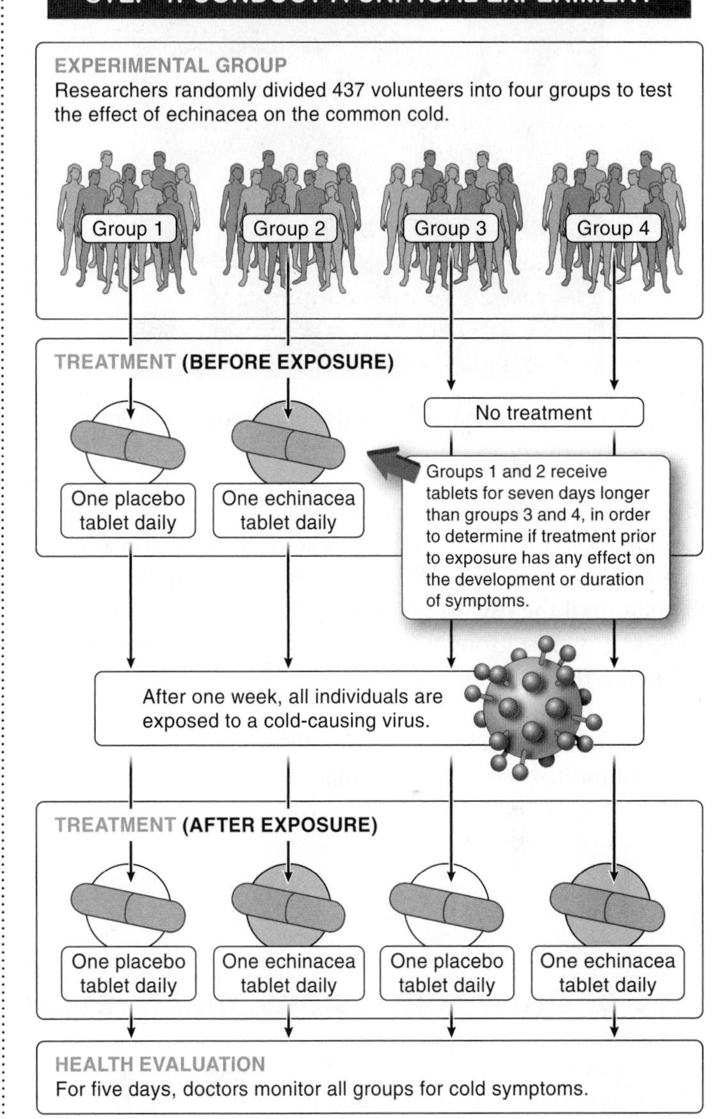

FIGURE 1-10 "When you need to know . . ." A critical experiment makes it possible to conclusively determine whether a hypothesis is correct.

a week prior to exposure to the cold virus. Those in one group received echinacea tablets while those in the other took a **placebo,** a pill that looked identical to the echinacea pill but contained no echinacea or other active ingredient. Neither

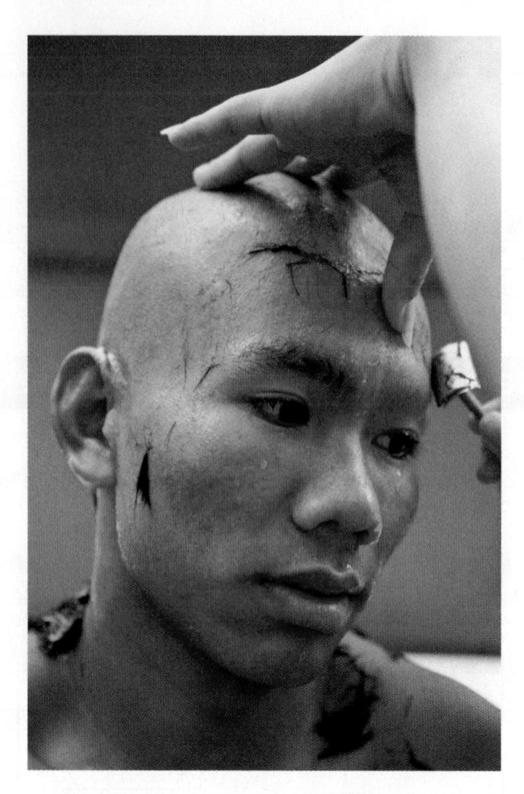

FIGURE 1-11 Does shaving hair cause it to grow back coarser or darker?

the subject nor the doctor administering the pills (and later checking for cold symptoms) knew what the pills contained. In the other two groups, the individuals did not begin taking the pills until the day on which they were exposed to the cold virus. Again, one group got the echinacea pill and the other group got the placebo.

Hypothesis: Estrogens in sewage runoff turn fish into hermaphrodites. To try to determine the validity of this hypothesis, researchers placed cages containing male trout downstream from sewage treatment plants on several rivers in England. They then examined the fish for the presence of eggs in their testes and tested their blood for the presence of an egg-production hormone that is not normally found in males.

Q Does shaving or cutting hair make it grow back more thickly?

Hypothesis: Hair that is shaved grows back coarser and darker. A critical experiment does not have to be complex or high-tech. All that matters is that it can decisively determine whether or not a hypothesis is correct. This point is illustrated by an experiment published in the scholarly journal *Archives of Facial Plastic Surgery.* Researchers decided to test the hypothesis that shaving hair causes it to grow back coarser or darker. A group of volunteers had one of their eyebrows selected at random and completely shaved off. The subjects were then evaluated for eyebrow re-growth over the course of the next six months, with observers also analyzing photographs of the individuals (**FIGURE 1-11**).

In the next section we'll see how our hypotheses survive being confronted with the results of the critical experiments described above and how we can move toward drawing conclusions.

TAKE-HOME MESSAGE 1•8

A critical experiment is one that makes it possible to decisively determine whether a particular hypothesis is correct.

1•9

Step 5: Draw conclusions, make revisions.

Q Is eyewitness testimony in courts always right?

Once the results of the critical experiment are in, they are pulled apart, examined, and analyzed. Researchers look for patterns and relationships in the evidence they've gathered from their experiments; they draw conclusions and see whether their findings and conclusions support their hypotheses. If an experimental result is not what you expected, that does not make it a "wrong answer." Science includes a great deal of trial and error, and if the conclusions do not support the hypothesis, then you must revise your hypothesis, which often spurs you to conduct more experiments. This step is a cornerstone of the scientific method because it demands that you must be open-minded and ready to change what you think.

The results of the purse-snatching experiment were surprising. When the suspects (which, as you'll recall, did not include the actual "criminal") were viewed together in a lineup, the observers/witnesses erroneously identified someone

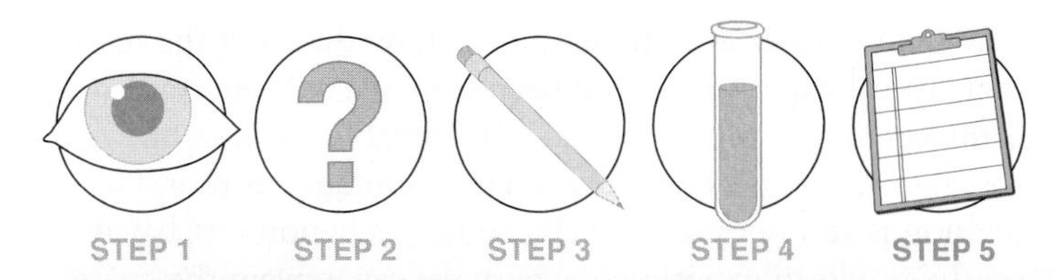

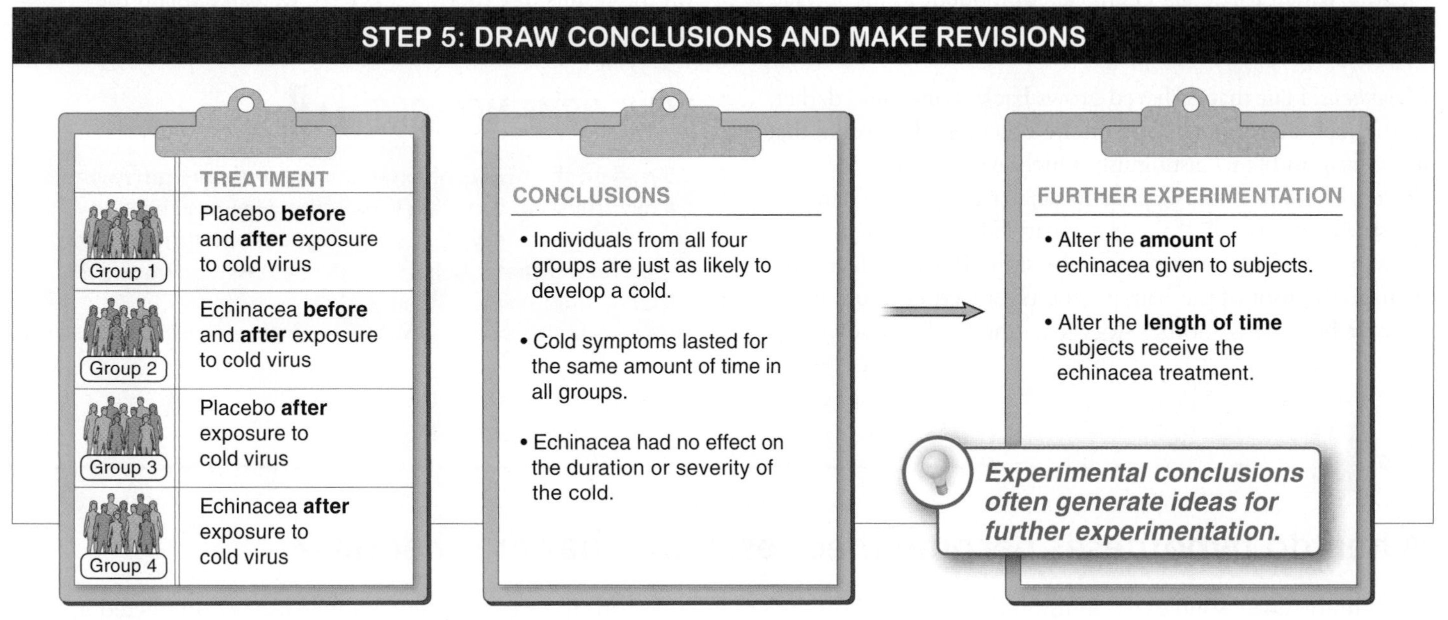

FIGURE 1-12 Drawing conclusions and making revisions.

as the purse snatcher about a third of the time. When the suspects were viewed one at a time, the observers made a mistaken identification less than 10% of the time.

In this or any other experiment, it does not matter whether we can imagine a reason for the discrepancy between our hypothesis and our results. What is important is that we have demonstrated that our initial hypothesis—"Eyewitness testimony is always accurate"—is not supported by the data. Our observations suggest that, at the very least, the accuracy of an eyewitness's testimony depends on the method used to present the suspects. Based on this result, we might then adjust our hypothesis to: "Eyewitness testimony is more accurate when suspects are presented to witnesses one at a time."

We can then devise new and more specific testable predictions to try to further refine a hypothesis. In the case of eyewitness testimony, further investigation suggests that when suspects or pictures of suspects are placed side by side, witnesses compare them and tend to choose the suspect that *most resembles* the person they remember committing the crime. When viewed one at a time, suspects can't be compared in this way and witnesses are less likely to make misidentifications.

Q Does echinacea help prevent the common cold?

Hypothesis: Echinacea reduces the duration and severity of the symptoms of the common cold (**FIGURE 1-12**). In the echinacea study, the results were definitive. Those who took the echinacea were just as likely to catch a cold, and, once they caught the cold, the symptoms lasted for the same amount of time. In short, echinacea had no effect at all. Several similar studies have been conducted, all of which have shown that echinacea does not have any beneficial effect. As one of the researchers commented afterward, "We've got to stop attributing any efficacy to echinacea."

Although it seems clear that our initial hypothesis that echinacea prevents people from catching colds and reduces the severity and duration of cold symptoms is not correct, further experimentation might involve altering the amount of echinacea given to the research subjects or the length of time they take echinacea before exposure to the cold-causing viruses.

Hypothesis: Estrogens in sewage runoff turn fish into hermaphrodites. The results of the sewage experiment showed that, in every single river tested, male fish downstream from the runoff developed eggs in their testes and had significant levels of the egg-production hormone in their blood. Further research stimulated by these results has investigated a variety of common industrial chemicals to see whether they can stimulate production of the egg-development hormone in male fish under laboratory conditions, in an attempt to identify which particular chemical component in the sewage causes the male fish to change.

Hypothesis: Hair that is shaved grows back coarser and darker. In the hair-shaving experiment, the observers discovered that it was impossible to distinguish which eyebrow had been shaved. This was not a surprise to dermatologists evaluating the study because all of the living parts involved in hair growth are below the surface of the skin. (Plucking hairs can damage the root of the hair, potentially affecting future hair growth, but that is another story, awaiting further study.)

The outcomes in all of these studies show that after the results of a critical experiment have been gathered and interpreted, it is important not just to evaluate the initial hypothesis but also to consider any necessary revisions or refinements to it. This revision is an important step; by revising a hypothesis, based on the results of experimental tests, we can explain the observable world with greater and greater accuracy.

TAKE-HOME MESSAGE 1•9

Based on the results of experimental tests, we can revise a hypothesis and explain the observable world with increasing accuracy. A great strength of scientific thinking, therefore, is that it helps us understand when we should change our minds.

1•10

When do hypotheses become theories, and what are theories?

It's an unfortunate source of confusion that, among the general public, the word "theory" is often used to refer to a hunch or a guess or speculation—that is, something we are not certain about—while to scientists, the word means nearly the opposite: a hypothesis of which they are most certain. To reduce misunderstandings, we examine two distinct levels of understanding that scientists use in describing our knowledge about natural phenomena.

Hypothesis As we have seen, hypotheses are at the very heart of scientific thinking. A hypothesis is a proposed explanation for a phenomenon. A good hypothesis leads to testable predictions. Commonly, when non-scientists use the word "theory"—as in, "I've got a theory about why there's less traffic on Friday mornings than on Thursday mornings"—they actually mean that they have a hypothesis.

Theory A **theory** is an explanatory hypothesis for natural phenomena that is exceptionally well supported by the empirical data. A theory can be thought of as a hypothesis that has withstood the test of time and is unlikely to be altered by any new evidence. Like a hypothesis, a theory is testable; but because it has already been repeatedly tested and no observations or experimental results have contradicted it, a theory is viewed by the scientific community with nearly the same confidence as a fact. For this reason, it is inappropriate to describe something as "just a theory" as a way of asserting that it is not likely to be true.

Theories in science also tend to be broader in scope than hypotheses. In biology, two of the most important theories (which we explore in more detail in Chapters 3 and 8) are *cell theory,* that all organisms are composed of cells and all cells come from pre-existing cells, and the *theory of evolution by natural selection,* that species can change over time and all species are related to each other through common ancestry.

TAKE-HOME MESSAGE 1•10

Scientific theories do not represent speculation or guesses about the natural world. Rather, they are hypotheses—proposed explanations for natural phenomena—that have been so strongly and persuasively supported by empirical observation that the scientific community views them as very unlikely to be altered by new evidence.

③ Well-designed experiments are essential to testing hypotheses.

Controlled experiments increase the power of our observations. Here, flasks containing Panax (Asian ginseng) are prepared.

1•11

Controlling variables makes experiments more powerful.

From our earlier discussion of critical experiments, you have a sense of how important it is to have a well-planned, well-designed experiment. Some experiments are just better than others when it comes to figuring out how the world works. At their most basic level, experiments help us figure out the cause-and-effect relationship between two things. In the previous sections, for example, we hypothesized about the relationship between witnessing a crime and being able to identify the person who committed it, between taking echinacea and catching a cold and having it for a period of time, between water pollution and hermaphrodite fish, and between shaving and the subsequent coarseness and color of hair. In performing experiments, our goal is to figure out whether one thing influences another thing; if an experiment enables us to draw a correct conclusion about that cause-and-effect relationship, it is a good experiment.

In our initial discussion of experiments, we just described what the experiment was without examining why the researchers chose to perform the experiment the way they did. In this section, we explore some of the ways to maximize an experiment's power, and we'll find that, with careful planning, it is possible to increase an experiment's ability to discern causes and effects.

First, let's consider some elements common to most experiments. These include:

1. Treatment. This is any experimental condition applied to the research subjects. It might be the shaving of one of an individual's legs, or the pattern used to show "suspects" (all at once or one at a time) to the witness of a staged crime, or a dosage of echinacea given to an individual.

2. Experimental group. This is a group of subjects who are exposed to a particular treatment—for example, the individuals given echinacea rather than placebo in the experiment described above. It is sometimes referred to as the "treatment group."

3. Control group. This is a group of subjects who are treated identically to the experimental group, with one exception: they are not exposed to the treatment. An example would be the individuals given placebo rather than echinacea.

4. Variables. These are the characteristics of an experimental system that are subject to change. They might be, for example, the amount of echinacea a person is given, or a measure of the coarseness of an individual's hair. When we speak of "controlling" variables—the most important feature of a good experiment—we are describing the attempt to minimize any differences (which are also called "variables") between a control group and an experimental group other than the treatment itself. That way, any differences in the outcomes we observe between the groups are most likely due to the treatment.

Let's look at an example that illustrates the importance of considering all these elements when designing an experiment.

Stomach ulcers are erosions of the stomach lining that, due to the highly acidic condition of that part of the digestive tract, can be very painful. In the late 1950s, a doctor reported in the *Journal of the American Medical Association* that stomach ulcers could be effectively treated by having a patient swallow a balloon connected to some tubes that circulated a refrigerated fluid. He argued that by super-cooling the stomach, acid production was reduced and the ulcer symptoms relieved. He had convincing data to back up his claim: in all 24 of his patients who received this "gastric freezing" treatment, their condition improved. As a result, the treatment became widespread for many years (**FIGURE 1-13**).

Although there was a clear hypothesis ("Gastric cooling reduces the severity of ulcers") and some compelling observations (all 24 patients experienced relief), this experiment falls short of qualifying as a good example of the scientific method. Why? Because it was not designed well. In particular, there was no clear group with whom to compare the patients who received the treatment. In other words, who is to say that just going to the doctor or having a balloon put into your stomach doesn't improve ulcers? The results of this doctor's experiment do not rule out these interpretations.

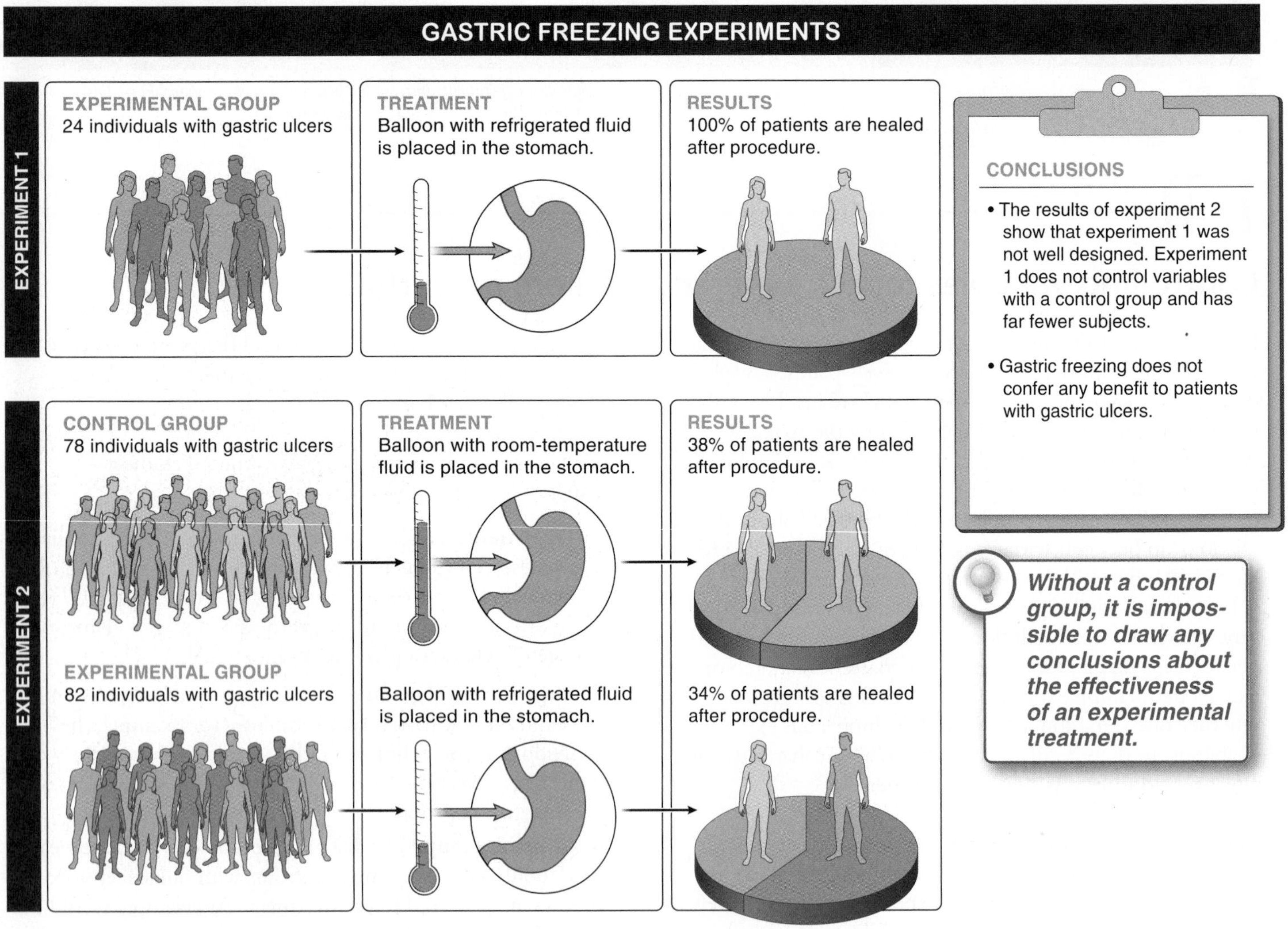

GASTRIC FREEZING EXPERIMENTS

EXPERIMENT 1

EXPERIMENTAL GROUP
24 individuals with gastric ulcers

TREATMENT
Balloon with refrigerated fluid is placed in the stomach.

RESULTS
100% of patients are healed after procedure.

EXPERIMENT 2

CONTROL GROUP
78 individuals with gastric ulcers

TREATMENT
Balloon with room-temperature fluid is placed in the stomach.

RESULTS
38% of patients are healed after procedure.

EXPERIMENTAL GROUP
82 individuals with gastric ulcers

Balloon with refrigerated fluid is placed in the stomach.

34% of patients are healed after procedure.

CONCLUSIONS

- The results of experiment 2 show that experiment 1 was not well designed. Experiment 1 does not control variables with a control group and has far fewer subjects.
- Gastric freezing does not confer any benefit to patients with gastric ulcers.

Without a control group, it is impossible to draw any conclusions about the effectiveness of an experimental treatment.

FIGURE 1-13 No controls. A poorly designed experiment: gastric freezing and stomach ulcers.

A few years later, another researcher decided to do a more carefully controlled study. He recruited 160 ulcer patients and gave 82 of them the gastric freezing treatment. The other 78 received a similar treatment in which they swallowed the balloon but had room-temperature water pumped in. The latter was an appropriate control group because the subjects were treated exactly like the experimental group, with the exception of only a single difference between the groups—whether they experienced gastric freezing or not. The new experiment could test for an effect of the gastric freezing, while controlling for the effects of other, lurking variables that might affect the outcome.

Surprisingly, although the researcher found that 34% of those in the gastric freezing group improved, he also found that 38% of those in the control group improved. These results indicated that gastric freezing didn't actually confer any benefit when compared with a treatment that did not involve gastric freezing. Not surprisingly, the practice was abandoned. (We will discuss below why the control group might have had such a high rate of improvement.)

Q Is arthroscopic surgery for arthritis beneficial for the 300,000 people in the United States who have it each year? How do we know?

More recently, a well-controlled experiment demonstrated that arthroscopic knee surgery, a treatment for osteoarthritis performed on more than 300,000 people in the United States each year, produced moderate benefits to the patient. But the study demonstrated even greater benefits to patients in a control group who were subjected not to the arthroscopic procedure but rather to "sham" surgery, in which the surgeon made three superficial incisions in the knee, manipulated the knee a bit, and clanged the instruments around in the operating room. As a consequence of this study, many doctors have begun abandoning the use of arthroscopic surgery for the treatment of osteoarthritis—which once accounted for more than a billion dollars in surgical procedures each year. (This type of surgery is still useful for the treatment of some other knee problems, including the repair of ligament and cartilage damage.)

A surprising result from both the gastric freezing studies and the arthroscopic surgery study was that they demonstrated the **placebo effect,** the frequently observed, poorly understood, phenomenon in which people respond favorably to *any* treatment, regardless of whether it is as simple as swallowing a sugar pill or as unpleasant as having to swallow a balloon and have it inflated in their stomach. The placebo effect highlights the need for comparison of treatment effects with an appropriate control group. We want to know whether the treatment is actually responsible for any effect seen; if the control group receiving the placebo or sham treatment has an outcome like that of the experimental group, we can conclude that the treatment itself does not have an effect.

Another pitfall to be aware of in designing an experiment is to ensure that the person(s) conducting the experiment don't influence the experiment's outcome. An experimenter can often unwittingly influence the results of an experiment. For example, eyewitnesses to crimes more frequently identify someone that the police suspect rather than other individuals in the lineup when the police officer conducting the lineup knows which individual is the suspect.

This phenomenon is seen in the story of a horse named Clever Hans. Hans was considered clever because his owner claimed that Hans could perform remarkable intellectual feats, including multiplication and division. When given a numerical problem, the horse would tap out the answer number with his foot. Controlled experiments, however, demonstrated that Hans was only able to solve problems when he could see the person asking the question and when that person knew the answer (**FIGURE 1-14**). It turned out that the questioners revealed, unintentionally and through very subtle body language, the answers.

FIGURE 1-14 Math whiz or ordinary horse? The horse Clever Hans was said to be capable of mathematical calculations, until a controlled experiment demonstrated otherwise.

The Clever Hans phenomenon highlights the benefits of instituting even greater controls when designing an experiment. In particular, it highlights the value of **blind experimental design,** in which the experimental subjects do not know which treatment (if any) they are receiving, and **double-blind experimental design,** in which neither the experimental subjects nor the experimenter knows which treatment a subject is receiving.

Another hallmark of an extremely well designed experiment is that it combines the blind/double-blind strategies we've just described in a **randomized,** controlled, double-blind study. In this context, "randomized" refers to the fact that, as in the echinacea study described above, the subjects are randomly assigned into experimental and control groups. In this way, researchers and subjects have no influence on the composition of the control and treatment groups.

The use of randomized, controlled, double-blind experimental design can be thought of as an attempt to imagine all the possible ways that someone might criticize an experiment and to design the experiment so that the results cannot be explained by anything but the effect of the treatment. In this way, the experimenter's results either support the hypothesis or invalidate it—in which case, the hypothesis must be rejected. If multiple explanations can be offered for the observations and evidence from an experiment, then it has not succeeded as a critical experiment.

Suppose you want to know whether a new drug is effective in fighting the human immunodeficiency virus (HIV), the virus that leads to AIDS. Which experiment would be better, one in which the drug is added to HIV-infected cells in a test tube under carefully controlled laboratory conditions, or one in which the drug is given to a large number of HIV-infected individuals? There is no definitive answer. In laboratory studies, it is possible to control nearly every environmental variable. In their simplicity, however, lab studies may introduce difficulties. Complex and naturally varying factors in human subjects, such as nutrition and stress, may have important interactions with the experimental drug that influence its effectiveness. These interactions will not be present and taken into account in the controlled lab study. Good experimental design is more complex than simply following a single recipe. The only way to determine the quality of an experiment is to assess how well the variables that were not of interest were controlled, and how well the experimental treatment tested the relationship of interest.

TAKE-HOME MESSAGE 1•11

To draw clear conclusions from experiments, it is essential to hold constant all those variables we are not interested in. Control and experimental groups should vary only with respect to the treatment of interest. Differences in outcomes between the groups can then be attributed to the treatment.

1•12

Repeatable experiments increase our confidence.

Q Can science be misleading? How can we know?

In 2005, a study showed that some patients being treated for HIV infection who were also taking an epilepsy drug called valproic acid had significantly reduced numbers of HIV particles in their blood. Newspaper headlines announced this finding—"AIDS Cure Possible, Study Suggests"—and raised many people's hopes. Two years later, however, a study of people who were already taking valproic acid and anti-HIV drugs concluded that those on valproic acid had not benefited at all from the drug (**FIGURE 1-15**). It's not certain why the later study didn't produce the same results as the first study, but the second study did address two serious shortcomings of the first study—namely, that the first study was much smaller, involving only four patients, and that no control group was used. This pair of studies reveals the importance of repeatability in science.

Q Do megadoses of vitamin C reduce cancer risk?

A powerful way to demonstrate that observed differences between a treatment group and a control group truly reflect the effect of the treatment is for the researchers to conduct the experiment over and over again. Even better is to have other researcher groups repeat the experiment and get the same results. Researchers describe this desired characteristic of experiments by saying that an experiment must be "reproducible" and "repeatable."

Email Print A A A Text Size

8/7/08

Texas Tech Researchers May Have Found AIDS Cure

Posted: Aug 7, 2008 09:59 PM

Texas Tech University in partnership with Harvard University is on the research forefront of a possible vaccine for HIV, the virus that causes AIDS.

To give you some idea of how big this story is, on Thursday it was the number one story on Newsweek's website and it will be the lead in their next magazine.

NewsChannel 11 met with the president of the Texas Tech Health Sciences Center, Doctor John Baldwin and Texas Tech Chancellor Kent Hance. Both men are optimistic about this research and its possible impact on the world and Texas Tech.

Repeatability is essential! Scientific conclusions are more reliable when experiments have been repeated (and modified, if necessary).

Valproic Acid Does Not Reduce Latent HIV Reservoir in Resting CD4 Cells

By Liz Highleyman

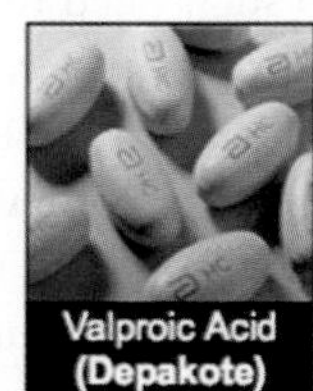

Some prior studies have suggested that valproic acid (Depakote), a drug used to treat epilepsy and bipolar disorder, might help eradicate HIV from the body. Two studies published in the June 19, 2008 issue of *AIDS*, however, indicate that it has little long-term effect.

Study 1

David Margolis and colleagues from the University of North Carolina previously reported that valproic acid plus intensified antiretroviral therapy could "flush" HIV out of long-lived latently infected CD4 T-cells and potentially reduce the resting CD4 cell reservoir. They later conducted a study to test the ability of valproic acid to deplete resting CD4 cell infection in patients receiving standard HAART.

Infection of resting CD4 cells was measured in 11 patients with stable undetectable HIV viral load using a standard HIV RNA assay twice rior to, and twice after, adding 1000 mg valproic acid to their standard antiretroviral regimen. Resting CD4 T-cell infection frequency was measured by outgrowth assay, and low-level viremia was quantified in 9 patients using a highly sensitive single-copy plasma HIV RNA assay.

FIGURE 1-15 Once is not enough. Experiments and their outcomes must be repeatable for their conclusions to be valid and widely accepted.

An experiment that can be done over and over again by a variety of researchers with the same results is an effective defense against biases (which we discuss in the next section) and reflects a well-designed experiment. Experiments whose results cannot be confirmed by repeated experiments or by experiments performed by other researchers are the downfall of many dramatic claims. Even though chemist Linus Pauling had not one, but two Nobel prizes, he was never able to convince the scientific community that megadoses of vitamin C are an effective treatment for cancer. Every time other individuals tried to repeat Pauling's studies, properly matching the control and treatment groups, they found no difference. Vitamin C just isn't effective against cancer—a conclusion that the vast majority of the medical community now believes. The scientific method is profoundly egalitarian: more important than a scientist's credentials are sound and reproducible results.

When a study is repeated (also referred to as "replicating" a study), sometimes a tiny variation in the experimental design can lead to a different outcome; this can help us isolate the variable that is primarily responsible for the outcome of the experiment. Alternatively, when experiments are repeated and the same results are obtained, our confidence in them is increased.

TAKE-HOME MESSAGE 1•12

Experiments and their outcomes must be repeatable for their conclusions to be considered valid and widely accepted.

1•13

We've got to watch out for our biases.

Q Can scientists be sexist? How would we know?

In 2001, the journal *Behavioral Ecology* changed its policy for reviewing manuscripts that were submitted for publication. Its new policy instituted a double-blind process, whereby neither the reviewers' nor the authors' identities were revealed. Previously, the policy had been a single-blind process in which reviewers' identities were kept secret but the authors' identities were known to the reviewers. In an analysis of papers published between 1997 and 2005, it turned out that after 2001, when the double-blind policy took effect, there was a significant increase in the number of published papers in which the first author was female (**FIGURE 1-16**). Analysis of papers published in a similar journal that maintained the single-blind process over that period revealed no such increase.

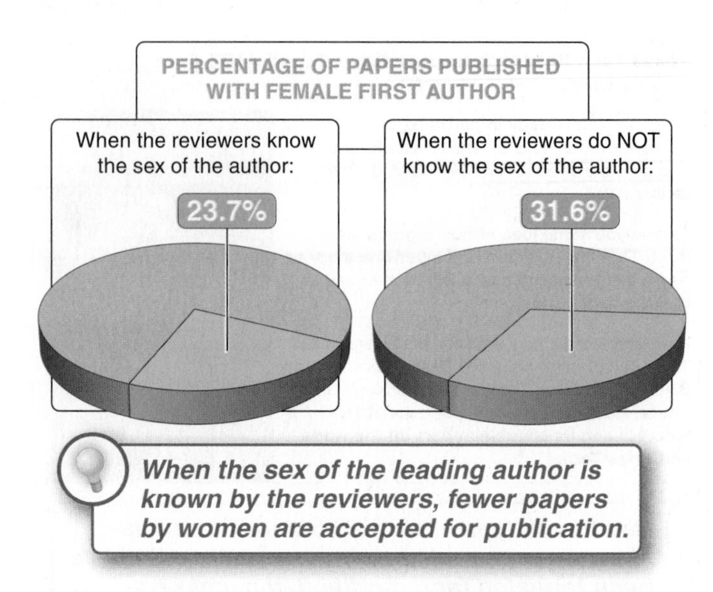

FIGURE 1-16 Bias against female scientists? The journal *Behavioral Ecology* accepted more papers from female authors when the reviewers were not aware of the author's sex.

This study reveals that people, including scientists, may have biases—sometimes subconscious—that influence their behavior. It also serves as a reminder of the importance of proper controls in experiments. If knowing the sex of the author influences a reviewer's decision on whether a paper should be published, it is possible that researchers' biases can creep in and influence their collection of data and analysis of results. (Even the decision of what—and what not—to study can be influenced by our biases.)

It can be hard to avoid biases. Consider a study that required precise measuring of the fingers of the left and right hands—comparing the extent to which people in different groups were physically symmetrical. The researcher noted that when she measured individuals from the group she predicted would be more symmetrical, she felt a need to re-measure if the reading on her digital ruler indicated a big asymmetry. Because she was thinking that the person's fingers *should* be symmetrical, she assumed that she had made an error. She felt no such need to re-measure for subjects in the other group, when asymmetries confirmed her hypothesis. To control for this bias, the researcher connected her digital ruler to a computer and, without ever having the number displayed, transmitted the measurement directly to the computer when she pushed a button. In this way, she was able to make each measurement without introducing any regular bias.

TAKE-HOME MESSAGE 1•13

Biases can influence our behavior, including our collection and interpretation of data. With careful controls, it is possible to minimize such biases.

④ The scientific method can help us make wise decisions.

What can you believe? Reading labels is essential to evaluating products and the claims about them.

1•14

Statistics can help us in making decisions.

In Section 1-12 we saw that researchers repeatedly found, through experimentation, that megadoses of vitamin C do not reduce cancer risk. If you put yourself in the researchers' shoes, you might wonder how you'd figure out that the vitamin C did not reduce cancer risk. Perhaps there were 100 individuals in the group receiving megadoses of vitamin C: some of them developed cancer and some of them did not. And among the 100 subjects in the group not receiving the megadoses, some of the individuals developed cancer and some did not. How do you decide whether the vitamin C actually had an effect? This knowledge comes from a branch of mathematics called **statistics,** a set of analytical and mathematical tools designed to help researchers gain understanding from the data they gather. To understand statistics, let's start with a simple situation.

Suppose you measure the height of two people. One is a woman who is 5 feet 10 inches tall. The other is a man who is 5 feet 6 inches tall. If these were your only two observations of human height, you might conclude that female humans are taller than males. But suppose you measure the height of 100 women and 100 men chosen randomly from a population. Then you can say, "of the 100 men, the average man is 5 feet 9.5 inches, and of the 100 women, the average woman is 5 feet 4 inches." Better still, the data can illuminate for you not only the average, but also some measure of how much variation there is from one individual to another. Statistical analysis can tell you not only that the average man in this study is 5 feet 9.5 inches tall, but that two-thirds of the men are between 5 feet 6.5 inches tall (3 inches less than the average) and 6 feet 0.5 inches tall (3 inches more than the average). You will often see this type of range stated as "5 feet 9.5 inches ± 3 inches" ("plus or minus 3 inches"). Similarly, the data might show that the female subjects in the study are 5 feet 6 inches ± 3 inches," indicating that two-thirds of the women are between 5 feet 3 inches and 5 feet 9 inches tall. As we discussed in Section 1-11, on experimental design, and as this example shows, larger numbers of participants are

HUMAN HEIGHT AND STATISTICS

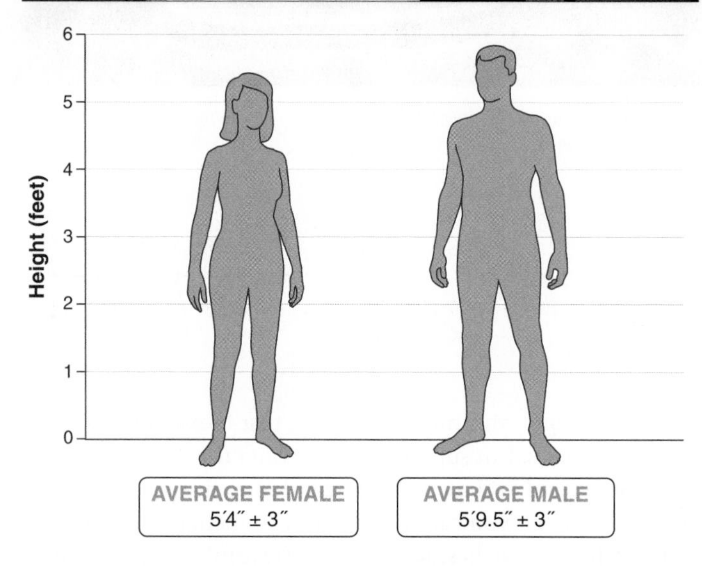

FIGURE 1-17 Drawing conclusions based on limited observations is risky. Measuring a greater number of people will generally help us draw more accurate conclusions about human height.

better than smaller numbers if you want to draw general conclusions about natural phenomena such as the height of men and women (**FIGURE 1-17**).

Using data to describe the characteristics of individuals participating in a study is useful, but often we want to know whether data support (or do not support) a hypothesis. If the scientific method is to be effective in helping us understand the world, it must help us make wise decisions about concrete things. For example, suppose we want to know whether having access to a textbook helps a student perform better in a biology class. Statistics can help us answer this question. After conducting a study, let's say we find that students who had access to a textbook scored an average of 81% ± 8% on their exams, while those who did not scored an average of 76% ± 7%.

In this example, it is difficult to distinguish between the two possible conclusions.

Possibility 1: Students having access to a textbook DO perform better in biology classes. In other words, our sampling of this class revealed a true relationship between the two variables, textbook access and class performance.

Possibility 2: Students having access to a textbook DO NOT perform better in biology classes. The variation in the scores in the two groups may be too large to allow us to notice any effect of having access to a textbook. Instead, the difference in average scores for the two groups may mean that more of the high-performing students just happened, through random chance, to be in the group given access to a textbook.

But what if the students with access to a textbook scored, on average, 95% ± 5%, while those without access scored only 60% ± 5%. In this case, we would be much more confident that there is a significant effect of having access to a textbook (**FIGURE 1-18**), because even with the large variation in the scores seen in each group, the averages are still very different from each other.

Statistical methods help us to decide between these two possibilities and, importantly, to state how confident we are that one or the other is true. The greater the difference between two groups (95% versus 60% is a greater difference than 81% versus 76%), and the smaller the variation in each group (± 5% in the 95% and 60% groups, versus ± 8% and ± 7% in the 81% and 76% groups), the more confident we are of the conclusion that there is a significant effect of the treatment (having access to a textbook, in this case). In other words, in the case where the groups of students scored 95% or 60%, depending on whether or not they had access to a textbook, it is possible that having a textbook did not actually improve performance and that this observed difference was just the result of chance. But this conclusion is very, very unlikely.

Statistics can also help us identify relationships (or the lack of relationships) between variables. For example, we might note that when there are more firefighters at a fire, the fire is larger and causes more damage. This is a **positive correlation,** meaning that when one variable (the number of firefighters) increases, so does the other (the severity of the fire). Should we conclude that firefighters make fires worse? No. While correlations can reveal relationships between variables, they

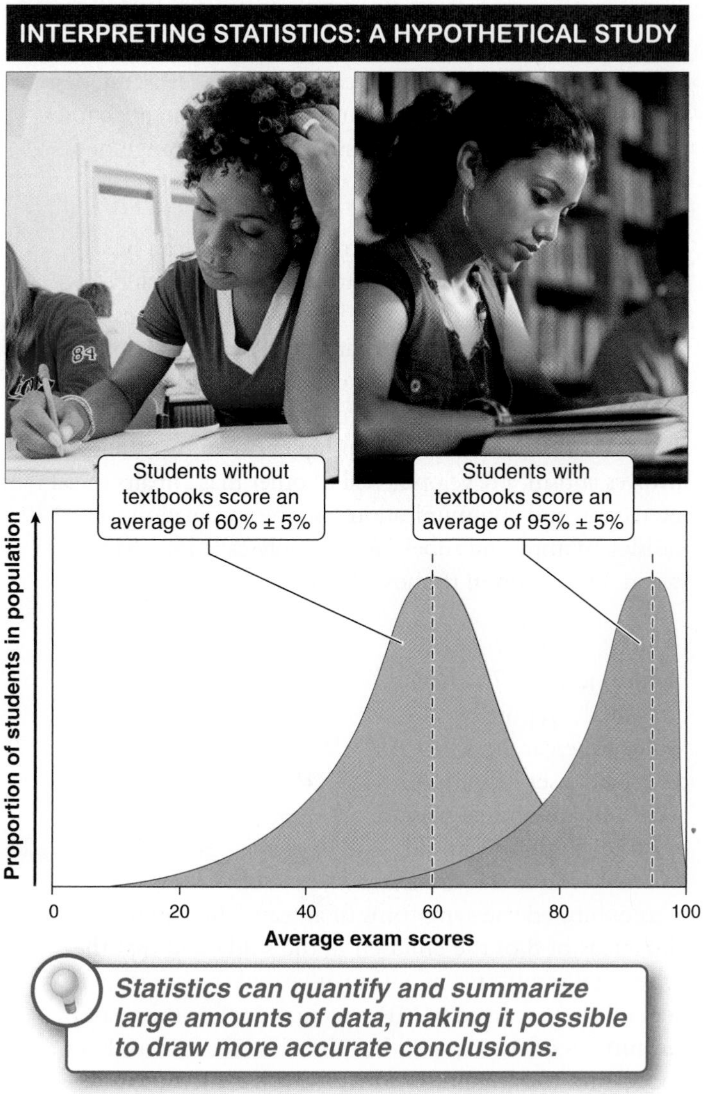

INTERPRETING STATISTICS: A HYPOTHETICAL STUDY

Students without textbooks score an average of 60% ± 5%

Students with textbooks score an average of 95% ± 5%

Proportion of students in population

0 20 40 60 80 100

Average exam scores

Statistics can quantify and summarize large amounts of data, making it possible to draw more accurate conclusions.

FIGURE 1-18 Drawing conclusions based on statistics. Statistical analyses can help us organize and summarize the observations that we make and the evidence we gather in an experiment.

don't tell us *how* the variables are related or whether change in one variable causes change in another. (You may have heard or read the phrase "correlation is not causation," which refers to this sort of situation.) Before drawing any conclusions about more firefighters causing larger fires, we need to know about the type of fire and its size when the firefighters arrived, because those factors will significantly influence the ultimate amount of damage. To estimate the effect of the number of firefighters on the amount of damage, we would need to compare the amount of damage from fires of similar size that are fought by different numbers of firefighters.

Ultimately, statistical analyses can help us organize and summarize the observations that we make and the evidence we gather in an experiment. These analyses can then help us decide whether any differences we measure between experimental and control groups are likely to be the result of the treatment, and how confident we can be in that conclusion.

TAKE-HOME MESSAGE 1•14

Because much variation exists in the world, statistics can help us evaluate whether any differences between a treatment group and control group can be attributed to the treatment rather than random chance.

1•15

Pseudoscience and misleading anecdotal evidence can obscure the truth.

One of the major benefits of being familiar with the scientific way of thinking is that it can prevent you from being taken in or fooled by false claims. There are two types of "scientific evidence" that frequently are cited in the popular media and are responsible for people erroneously believing that links between two things exist, when in fact they do not. These are:

1. **Pseudoscience,** in which individuals make scientific-sounding claims that are not supported by trustworthy, methodical scientific studies.
2. **Anecdotal observations,** in which, based on only one or a few observations, people conclude that there is or is not a link between two things.

Pseudoscience is all around us, particularly in the claims made on the packaging of consumer products and food (**FIGURE 1-19**). Beginning in the 1960s, for example, consumers encountered the assertion by the makers of a sugarless gum that "4 out of 5 dentists surveyed recommend sugarless gum for their patients who chew gum." If you ask yourself the question

FIGURE 1-19 Pseudoscientific claims are often found on food products.

"How do they know what they know?" and can't answer it, you may be looking at pseudoscience. Maybe the statement is factually true, but the general relationship it implies may not be. How many dentists were surveyed? If the gum makers surveyed only five dentists, then the statement may not represent the proportion of *all* dentists who would make such a recommendation. And how were the dentists sampled? Were they at a shareholders meeting for the sugarless gum company? What alternatives were given—perhaps gargling with a tooth-destroying acid? You just don't know. That's what makes it pseudoscience.

Pseudoscience capitalizes on a belief shared by most people: that scientific thinking is a powerful method for learning about the world. The problem with pseudoscience is that the scientific bases for a scientific-sounding claim are not clear. The claims generally sound reasonable, and they are persuasive in convincing people to purchase one product over another. But one of the beauties of real science is that you never have to just take someone's word about something. Rather, you are free to evaluate people's research methods and results and decide for yourself whether their conclusions and claims are appropriate.

We are all familiar with anecdotal evidence. Striking stories that we read or hear, or our own experiences, can shape our views of cause and effect. We may find compelling parallels between suggestions made in horoscopes and events in our lives, or we may think that we have a lucky shirt, or we may be moved by a child whose cancer went into remission following treatment involving eating apricot seeds. Yet, despite lacking a human face, data are more reliable than anecdotes, primarily because they can illustrate a broader range of observations, capturing the big picture.

Anecdotal observations can seem harmless and can be emotionally powerful. But because they do not include a sufficiently large and representative set of observations of the world, they can lead people to draw erroneous conclusions, often with disastrous consequences. One important case of anecdotal evidence being used to draw general conclusions about a cause-and-effect relationship between two things involves autism, a developmental disorder that impairs social interaction and communication, and the vaccination for measles, mumps, and rubella (commonly called the MMR vaccine) that is given to most children.

Q Does the measles, mumps, and rubella vaccine cause autism?

In 1998, the prestigious medical journal *The Lancet* published a report by a group of researchers that described a set of symptoms (diarrhea, abdominal pain, bloating) related to bowel inflammation in 12 children who exhibited the symptoms of autism. The parents or physicians of 8 of the children in the study said that the behavioral symptoms of autism appeared shortly after the children received MMR vaccination. For this reason, the authors of the report recommended further study of a possible link between the MMR vaccine, the bowel problems, and autism.

In a press conference, one of the paper's authors suggested a link between autism and the MMR vaccine, recommending single vaccines rather than the MMR triple vaccine until it could be proved that the MMR vaccine did not trigger autism. This statement caused a major health scare that only recently has begun to subside. Noting that there had been a significant increase in the incidence of autism in the 1990s and early 2000s, the press and many people took the claims made by the researcher at the press conference as evidence that the MMR vaccine causes autism. Over the course of the next few years, the number of children getting the MMR vaccine dropped significantly, as parents sought to reduce the risk of autism in their children (**FIGURE 1-20**).

Unfortunately, this is a notable case of poor implementation of the scientific method, with numerous flaws leading to an incorrect conclusion. Most important among these is that

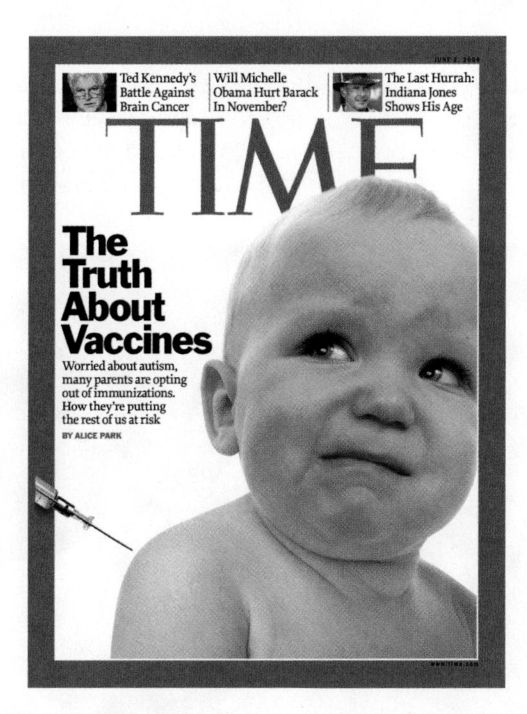

FIGURE 1-20 Headline news. A misguided fear of "catching" autism has caused some parents to decline immunizations for their children.

the study was small (only 12 children), the sample was carefully selected rather than randomized (that is, the researchers selected the study participants based on the symptoms they showed), and no control group was included for comparison (for example, children who had been vaccinated but did not exhibit autism symptoms). On top of this, the study, even flawed as it was, did not actually find or even report a link between autism and the MMR vaccine. That purported link only came from the unsupported, suggestive statements made by one of the study's authors at the press conference.

On their own, the design flaws in the study would be sufficient to invalidate the one author's claims about the link between autism and the MMR vaccine, but there are more weaknesses in the research. It later was discovered that before the paper's publication, the study's lead author (the one who spoke at the press conference) had received large sums of money from lawyers seeking evidence to use in lawsuits against the MMR vaccine manufacturers. It also came to light that he had applied for a patent for a vaccine that was a rival to the most commonly used MMR vaccine. In the light of these undisclosed biases, 10 of the paper's 12 authors published a retraction of their original interpretation of their results.

In the years since this paper was published, many well-controlled large-scale studies have conducted critical experiments on this subject, and all have been definitive in their conclusion that there is no link between the MMR vaccine and autism. Here are a few of these studies:

1. A study of all children born in Denmark between 1991 and 1998 found no difference in the incidence of autism among the 440,655 children who were vaccinated with the MMR vaccine and the 96,648 children who were not vaccinated.
2. A 2005 study in Japan showed that after use of the MMR vaccine was stopped in 1993, the incidence of autism continued to increase.
3. A study of 1.8 million children in Finland who were followed up for 14 years after getting the MMR vaccine found no link at all between the occurrence of autism and the vaccine.

At this point, the consensus of the international scientific community is that there is no scientific evidence for a link between the MMR vaccine and autism.

> "Science is a way to call the bluff of those who only pretend to knowledge. It is a bulwark against mysticism, against superstition, against religion misapplied to where it has no business being. If we're true to its values, it can tell us when we're being lied to."
>
> — Carl Sagan, in *The Demon-Haunted World: Science as a Candle in the Dark,* 1997

So what explains the observation that there are more autism cases now than in the past? It seems that the increased number of autism cases is a function of better identification of the condition by doctors and of changes in the process by which autism is diagnosed. Another reason for the perceived link between autism and the MMR vaccine was simply the coincidence that most children receive the vaccine at around 18–19 months of age, which happens to be the age at which the first symptoms of autism are usually noticed. In the end, what we learn from this is that we must be wary that we do not generalize from anecdotal observations or let poorly designed studies obscure the truth.

TAKE-HOME MESSAGE 1•15

Pseudoscience and anecdotal observations often lead people to believe that links between two phenomena exist, when in fact there are no such links.

1•16

There are limits to what science can do.

The scientific method is a framework that helps us make sense of what we see, hear, and read in our lives. Science and scientific thinking guide us in recognizing facts and help us in the interpretations of data, analyses of hypotheses, and drawing of conclusions. In doing so, scientific thinking reveals and illuminates explanations about how to think about various events and phenomena, and it can help us make decisions in diverse areas of our lives, not just "scientific" areas. There are, however, limits to what science can do.

The scientific method will never prove or disprove the existence of God. Nor is it likely to help us understand the mathematical elegance of Fermat's last theorem or the beauty of Shakespeare's sonnets. As one of several approaches to the acquisition of knowledge, the scientific method is, above all, empirical. It differs from non-scientific approaches such as mathematics and logic, history, music, and the study of artistic expression in that it relies on *measuring* phenomena in some way. The generation of value judgments and other types of non-quantifiable, subjective information—such as religious assertions of faith—falls outside the realm of science. Despite all of the intellectual analyses the scientific method gives rise to and the objective conclusions it makes possible, it does not, for example, generate moral statements and it cannot give us insight into ethical problems. What "is" (i.e., what we observe in the natural world) is not necessarily what "ought" to be (i.e., what is morally right). It may or may not be.

Further, much of what is commonly considered to be science, such as the construction of new engineering marvels or the heroic surgical separation of conjoined twins, is not scientific at all. Rather, these are technical innovations and developments. While they frequently rely on sophisticated scientific research, they represent the application of research findings to varied fields such as manufacturing and medicine to solve problems.

The application of science to solve a problem can result in sophisticated technical innovations, like this thought-controlled prosthetic arm.

As we begin approaching the world from a more scientific perspective, we can gain important insights into the facts of life, yet must remain mindful of the limits to science.

TAKE-HOME MESSAGE 1•16

Although the scientific method may be the most effective path toward understanding the observable world, it cannot give us insights into the generation of value judgments and other types of non-quantifiable, subjective information.

⑤ On the road to biological literacy: what are the major themes in biology?

Finding unity in all the diversity: the Hall of Biodiversity, American Museum of Natural History, New York.

1•17

A few important themes tie together the diverse topics in biology.

Driven forward by the power of the scientific method, modern biology is in the midst of an explosion of new and exciting developments. Now more than ever, biology is influencing and even transforming our lives. Its increasing relevance outside the world of scientists is felt in a multitude of areas, including breakthroughs in agriculture, innovative developments in criminology, and increased understanding and awareness of environmental issues and the need for widespread changes in behavior.

In this guide to biology, as we explore the many facets of biology and its relevance to life in the modern world, you will see two unifying themes recurring throughout.

- **Hierarchical organization.** Life is organized on many levels within individual organisms including atoms, cells, tissues, and organs. And in the larger world, organisms themselves are organized into many levels: populations, communities, and ecosystems within the biosphere.
- **The power of evolution.** Evolution, the change in genetic characteristics of individuals within populations over time, accounts for the diversity of organisms, but also explains the unity among them.

These two major themes connect the diverse topics, as we make our way through four chief areas of focus:

1. The chemical, cellular, and energetic foundations of life
2. The genetics, evolution, and behavior of individuals
3. The staggering diversity of life and the unity underlying it
4. Ecology, the environment, and the subtle and important links between organisms and the world they inhabit

Armed with an appreciation for the scientific method and the broader themes you'll encounter in your study of biology, let's continue our exploration of life!

TAKE-HOME MESSAGE 1•17

Although the diversity of life on earth is tremendous, the study of life is unified by the themes of hierarchical organization and the power of evolution.

streetBio

Knowledge You Can Use

When should you change what you think? Risky behaviors while driving.

Q: Should drunk driving be illegal? Why? Almost everyone believes that people should not drive when they've been drinking. Even after just one or two drinks, a driver is more than twice as likely to cause a fatal traffic accident as a sober driver. With three to five drinks, the risk is almost 10 times greater than when sober. People see the data and understand; the conclusion is not difficult: you put yourself and others at risk when driving drunk.

Q: What about people who claim that they can drive safely even when they are drunk? People believe that their own risks are much lower than they really are. Based on an analysis of U.S. National Highway Traffic Safety data, traffic risks among 16- to 21-year-olds after consuming alcohol are 2–10 times greater than their perceptions of the risks. For instance, people often feel that their reaction times are just as quick whether they are sober or drunk.

Q: If observations and analyses revealed an equally risky and preventable behavior on the road, would it be advisable to prohibit that behavior, too? They have! Such observations and analyses have been made and have resulted in many states banning the use of handheld cell phones and text-messaging while driving. Let's consider two studies.

In a study using a driving simulator, individuals were evaluated when talking on a cell phone or after they had consumed alcohol to the point of intoxication (they used vodka and orange juice in this University of Utah study). The results? The driving of individuals using cell phones was as much impaired as the driving of those who were drunk. One very interesting and perhaps surprising result: handheld and hands-free phones impaired driving equally; it seems that it is inattention, rather than reduced dexterity, that is responsible for the impairment.

Another study, analyzing the cell phone records of individuals who were in real traffic collisions, came to the same conclusion, finding that cell phone users were 4–5 times more likely to get into an accident than non-distracted drivers. When you use a cell phone while driving, you put yourself and others at the same level of risk as if you were driving drunk.

Q: What observations could be made that would cause *you* not to use your cell phone while driving?

What can you conclude? Scientific observations can reveal unexpected truths about the world and, in doing so, can cause us to change what we think and how we behave. If we consider one risk (such as driving drunk) unacceptable, should we consider an equal risk (such as driving while using a cell phone) also unacceptable? Science can't answer that, but it can help you to understand the world better and to think more consciously about decisions you may be making without even realizing it.

1 Science is a collection of facts and a process for understanding the world.

Through its emphasis on objective observation, description, and experimentation, science is a pathway by which we can discover and better understand the world around us. Biological issues permeate all aspects of our lives. To make wise decisions, it is essential for individuals and societies to attain biological literacy. There are numerous ways of gaining an understanding of the world. Because it is empirical, rational, testable, repeatable, and self-correcting, the scientific method is a particularly effective approach.

2 A beginner's guide: what are the steps of the scientific method?

The scientific method (observation, hypothesis, prediction, test, and conclusion) is a flexible, adaptable, and efficient pathway to understanding the world because it tells us when we must change our beliefs. The scientific method begins by making observations about the world, noting apparent patterns or cause-and-effect relationships. A hypothesis is a proposed explanation for a phenomenon. For a hypothesis to be useful, it must generate a testable prediction. A critical experiment is one that makes it possible to decisively determine whether a particular hypothesis is correct. Based on the results of experimental tests, we can revise our hypothesis and explain the observable world with increasing accuracy. A great strength of scientific thinking, therefore, is that it helps us understand when we should change our minds. Scientific theories do not represent speculation or guesses about the natural world. Rather, they are hypotheses—proposed explanations for natural phenomena—that have been so strongly and persuasively supported by empirical observation that the scientific community views them as very unlikely to be altered by new evidence.

3 Well-designed experiments are essential to testing hypotheses.

To draw clear conclusions from experiments, it is essential to hold constant all those variables we are not interested in. Control and experimental groups should vary only with respect to the treatment of interest. Differences in outcomes between the groups can then be attributed to the treatment. Experiments and their outcomes must be repeatable for their conclusions to be valid and widely accepted. Biases can influence our behavior, including our collection and interpretation of data. With careful controls, we can minimize such biases.

4 The scientific method can help us make wise decisions.

Because much variation exists in the world, statistics can help us evaluate whether differences between a treatment group and control group can be attributed to the treatment rather than random chance. Pseudoscience and anecdotal observations often lead people to believe that links between two phenomena exist, when in fact there are no such links. Although the scientific method may be the most effective path toward understanding the observable world, it cannot give us insights into the generation of value judgments and other types of non-quantifiable, subjective information.

5 On the road to biological literacy: what are the major themes in biology?

Although the diversity of life on earth is tremendous, the study of life is unified by the themes of hierarchical organization and the power of evolution.

KEY TERMS

anecdotal observation, p. 25
biological literacy, p. 4
biology, p. 3
blind experimental design, p. 20
control group, p. 17
critical experiment, p. 13
double-blind experimental design, p. 20
empirical, p. 7
experimental group, p. 17
hypothesis (*pl.* hypotheses), p. 11
null hypothesis, p. 11
placebo, p. 13
placebo effect, p. 19
positive correlation, p. 24
pseudoscience, p. 25
randomized, p. 20
science, p. 3
scientific literacy, p. 4
scientific method, p. 7
statistics, p. 23
superstition, p. 6
theory, p. 16
treatment, p. 17
variable, p. 18

CHECK YOUR KNOWLEDGE

1. Science is:

a) a field of study that requires certain "laws of nature" to be taken on faith.
b) both a body of knowledge and an intellectual activity encompassing observation, description, experimentation, and explanation of natural phenomena.
c) a process that can be applied only within the scientific disciplines, such as biology, chemistry, and physics.
d) the only way to understand the natural world.
e) None of the above.

2. All of the following are elements of biological literacy except:

a) the ability to use the process of scientific inquiry to think creatively about real-world issues having a biological component.
b) reading the most important books in biology.
c) the ability to integrate into your decision making a consideration of issues having a biological component.
d) the ability to communicate with others about issues having a biological component.
e) All of the above are elements of biological literacy.

3. Superstitions are:

a) held by many humans, but not by any non-human species.
b) just one of many possible forms of scientific thinking.
c) true beliefs that have yet to be fully understood.
d) irrational beliefs that actions not logically related to a course of events influence its outcome.
e) proof that the scientific method is not perfect.

4. In a recent study, patients treated with a genetically engineered heart drug were able to walk on a treadmill 26 seconds longer than those not receiving the drug, and they showed no side effects from taking the drug. Can we conclude that this drug is an effective treatment for heart disease?

a) No. Genetically engineered drugs cannot be tested by the scientific method; they require comparative observations.
b) No. It is not clear how many subjects were in the study.
c) No. It is not clear that the drug is not a placebo.
d) Yes.
e) No. It is not clear that the proper controls were included.

5. Empirical results:

a) rely on intuition.
b) are generated by theories.
c) are based on observation.
d) cannot be replicated.
e) must support a tested hypothesis.

6. In a well-designed experiment:

a) the prediction will be highly probable if the experiment shows the explanation is correct.
b) the prediction will be highly improbable if the experiment shows the explanation is incorrect.
c) the null hypothesis will not be tested.
d) the prediction will most likely be correct.
e) Both a) and b) are correct.

7. Which of the following statements is correct?

a) A hypothesis that does not generate a testable prediction is not useful.
b) Common sense is usually a good substitute for the scientific method when trying to understand the world.
c) The scientific method can be used only to understand scientific phenomena.
d) It is not necessary to make observations as part of the scientific method.
e) All of the above are correct.

8. The placebo effect:

a) is the frequently observed, poorly understood phenomenon that people tend to respond favorably to any treatment.
b) reveals that sugar pills are generally as effective as actual medications in fighting illness.
c) reveals that experimental treatments cannot be proven effective.
d) demonstrates that most scientific studies cannot be replicated.
e) is an urban legend.

9. Before experimental drugs can be brought to market, they must undergo many rigorous trials to ensure they deliver their medical benefits effectively and safely. One method commonly used in this process is to compare the effect of a drug with that of a placebo in double-blind tests. Which of the following correctly describes a double-blind test?

a) The researchers apply two-layered blindfolds to the study participants so they cannot see whether they are receiving the drug or a placebo.
b) Neither the researchers nor the study participants know who is receiving the drug and who is receiving the placebo.
c) The researchers know who is receiving the drug and who is receiving the placebo, but do not know what the supposed effects of the drug should be.
d) The researchers do not know who receives the drug or the placebo, but the participants know and tell them later.
e) None of the above.

10. In controlled experiments:

a) one variable is manipulated while others are held constant.
b) all variables are dependent on each other.
c) all variables are held constant.
d) all variables are independent of each other.
e) all critical variables are manipulated.

11. If a researcher collects data by using the same experimental setup as in another study, but using different research subjects, the process is considered:

a) an uncontrolled experiment.
b) intuitive reasoning.
c) extrapolation.
d) replication.
e) exploration.

12. You hear the males of a particular bird species calling in their natural environment and question why they call. You design an experiment to try to answer your question. You cage one male bird and record his calling rate in response to four treatments, varying the number of conspecific (of the same species) males and the number of conspecific females that the male subject can see. The treatments are: no conspecific birds; 10 conspecific males; 10 conspecific females; 5 conspecific males and 5 conspecific females. Your results are: the rate at which the male subject calls is the same across all four treatments. Which of the following is a null hypothesis that your experimental design could reject?
    a) Males of the species do not call.
    b) Males of the species call more often than do females.
    c) Males of the species call at a rate that is independent of the sex composition of the audience.
    d) Males of the species call less often than do females.
    e) Males of the species call when called to.

13. Statistical methods make it possible to:
    a) prove any hypothesis is true.
    b) determine how likely it is that certain results have occurred by chance.
    c) unambiguously learn the truth.
    d) reject any hypothesis.
    e) test non-falsifiable hypotheses.

14. Anecdotal evidence:
    a) is a more efficient method for understanding the world than the scientific method.
    b) tends to be more reliable than data based on observations of large numbers of diverse individuals.
    c) is a necessary part of the scientific method.
    d) is often the only way to prove important causal links between two phenomena.
    e) can seem to reveal links between two phenomena, but the links do not actually exist.

15. A relationship between phenomena that has been established based on large amounts of observational and experimental data is referred to as:
    a) a theory.
    b) a fact.
    c) an assumption.
    d) a conjecture.
    e) a hypothesis.

16. Which of the following issues would be least helped by application of the scientific method?
    a) developing more effective high school curricula
    b) evaluating the relationship between violence in videogames and criminal behavior in teens
    c) determining the most effective safety products for automobiles
    d) formulating public policy on euthanasia
    e) comparing the effectiveness of two potential antibiotics

17. What is the meaning of the statement "correlation does not imply causation"?
    a) Just because two variables vary in a similar pattern does not mean that changing one variable causes a change in the other.
    b) It is not possible to demonstrate a correlation between two variables.
    c) When a change in one variable causes a change in another variable, the two variables are not necessarily related to each other in any way.
    d) It is not possible to prove the cause of any naturally occurring phenomenon.
    e) Just because two variables vary in a similar pattern does not mean that they have any relationship to each other.

Short-Answer Questions

1. A pharmaceutical company plans to test a potential anti-cancer drug on human subjects. The drug will be administered in pill form. How should this study be designed so that appropriate controls are in place?

2. Lately there have been many claims concerning the health benefits of green tea. Suppose you read a claim that alleges drinking green tea causes weight loss. You are provided with the following information about the studies that led to this claim.

   People were weighed at the beginning of the study.

   People were asked to drink two cups of green tea every day for 6 weeks.

   People were weighed at the end of the study.

   People who drank green tea for 6 weeks lost some weight by the end of the study.

   It was concluded that green tea is helpful for weight loss.

   This study obviously had some holes in its design. Assuming no information other than that provided above, indicate at least four things that could be done to improve the experimental design.

3. Use of the word "theory" can imply different things to different people. Statements referring to evolution as "just a theory" imply that a scientific theory is much like a hunch, something that can easily be disproved. How is a scientific theory different from any other type of "theory"? What requirements must be met before a hypothesis is considered to be a scientific theory?

See Appendix for answers. For additional study questions, go to www.prep-u.com.

1 Atoms form molecules through bonding.

- **2.1** Everything is made of atoms.
- **2.2** An atom's electrons determine how (and whether) the atom will bond with other atoms.
- **2.3** Atoms can bond together to form molecules or compounds.
- **2.4** A molecule's shape gives it unique characteristics.

2 Water has features that enable it to support all life.

- **2.5** Hydrogen bonds make water cohesive.
- **2.6** Water has unusual properties that make it critical to life.
- **2.7** Living systems are highly sensitive to acidic and basic conditions.

3 Carbohydrates are fuel for living machines.

- **2.8** Carbohydrates include macromolecules that function as fuel.
- **2.9** Simple sugars are an effective, accessible source of energy.
- **2.10** Complex carbohydrates are time-released packets of energy.
- **2.11** Not all carbohydrates are digestible.

4 Lipids store energy for a rainy day.

- **2.12** Lipids are macromolecules with several functions, including energy storage.
- **2.13** Fats are tasty molecules too plentiful in our diets.
- **2.14** Cholesterol and phospholipids are used to build sex hormones and membranes.

5 Proteins are body-building molecules.

- **2.15** Proteins are versatile macromolecules that serve as building blocks.
- **2.16** Proteins are an essential dietary component.
- **2.17** Proteins' functions are influenced by their three-dimensional shape.
- **2.18** Enzymes are proteins that initiate and speed up chemical reactions.

6 Nucleic acids store the information on how to build and run a body.

- **2.19** Nucleic acids are macromolecules that store information.
- **2.20** DNA holds the genetic information to build an organism.
- **2.21** RNA is a universal translator, reading DNA and directing protein production.

2

Chemistry

Raw materials and fuel for our bodies

❶ Atoms form molecules through bonding.

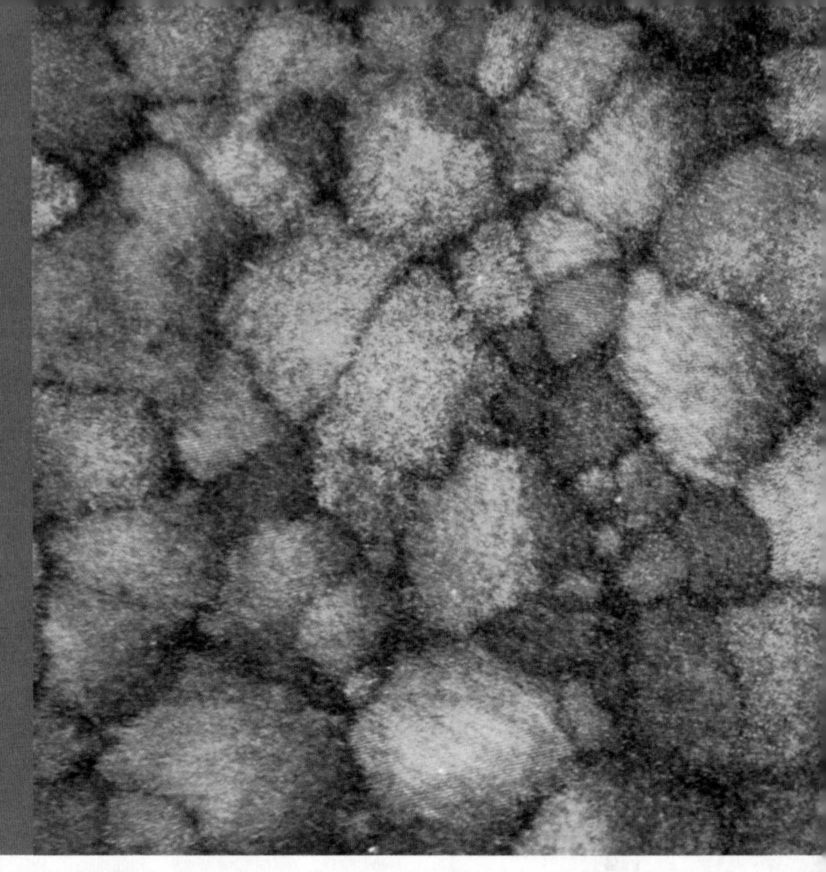
A collection of salt crystals, made from cesium chloride.

2.1

Everything is made of atoms.

A little bit of chemistry goes a long way in the study of biology and in understanding a great deal about your everyday life. Will eating those beans in your soup keep you up all night in gastrointestinal distress? Just knowing whether the beans are lentils or lima beans—each contains slightly different types of sugar molecules—will give you an answer. Will the butter you spread on your toast sabotage your efforts to lose weight? Understanding a little about the carbon-hydrogen connections in the fat molecules can help you decide.

The chemistry that is most important in biology revolves around a few important elements. An **element** is a substance that cannot be broken down chemically into any other substances. Gold and carbon and aluminum are elements you might be familiar with. Whatever the element, if you keep cutting it into ever smaller pieces, each of the pieces behaves exactly the same as any other piece. The smallest piece of pure gold will still have the softness, reflectivity, and malleability characteristic of that element (**FIGURE 2-1**).

If you could continue cutting, you would eventually separate the gold into tiny pieces that could no longer be divided

FIGURE 2-1 Familiar elements. Elements are substances that cannot be broken down chemically into any other substances.

without losing their gold-like properties. These are the individual component pieces of an element, called **atoms.** An atom is a bit of matter that cannot be subdivided any further without losing its essential properties. The word "atom" is from the Greek for "indivisible."

Everything around us, living or not, can be reduced to atoms. All atoms—whether of the element gold or some other element such as oxygen or aluminum or calcium—have the same basic structure. At the center of an atom is a **nucleus,** which is usually made up of two types of particles, called protons and neutrons. **Protons** are particles that have a positive electrical charge and **neutrons** are particles that have no electrical charge. The amount of matter in a proton or neutron, its **mass,** is about the same (**FIGURE 2-2**).

Whirling in a cloud around the nucleus of every atom are negatively charged particles called **electrons.** An electron weighs almost nothing—less than one-twentieth of one percent of the weight of a proton. (The mass of an atom—its **atomic mass**—is made up of the combined mass of all of its protons and neutrons; for our purposes here, electrons are so light that their mass can be ignored.) Particles that have the same charge repel each other; those with opposite charges are attracted to each other. Because all electrons have the same charge, the electrons in an atom repel each other. But because they are negatively charged, they are attracted to the positively charged protons in the nucleus. This attraction holds electrons close enough to the nucleus to keep them from flying away, while the energy of their fast movement keeps them from collapsing into the nucleus.

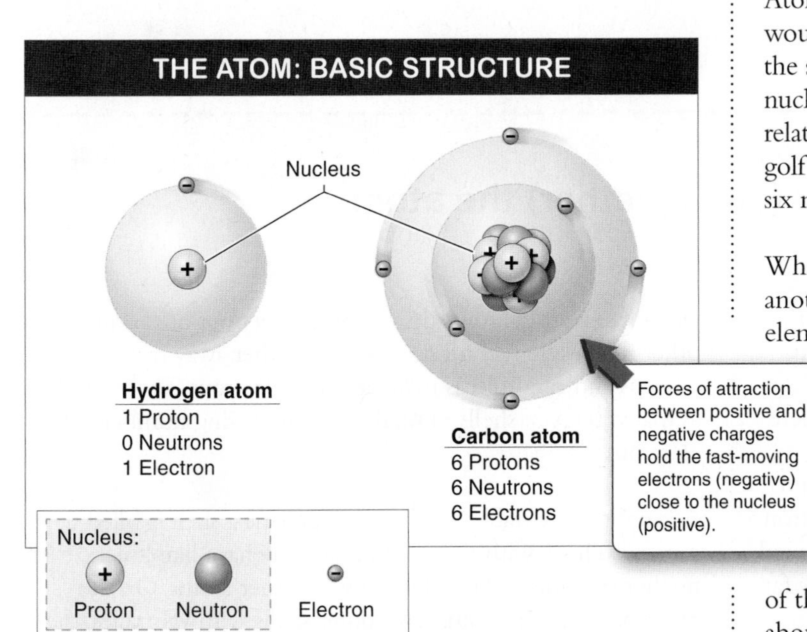

FIGURE 2-2 The atom. At the center of an atom is a nucleus containing protons and (in all elements except hydrogen) neutrons. The nucleus is surrounded by electrons whirling about in a cloud.

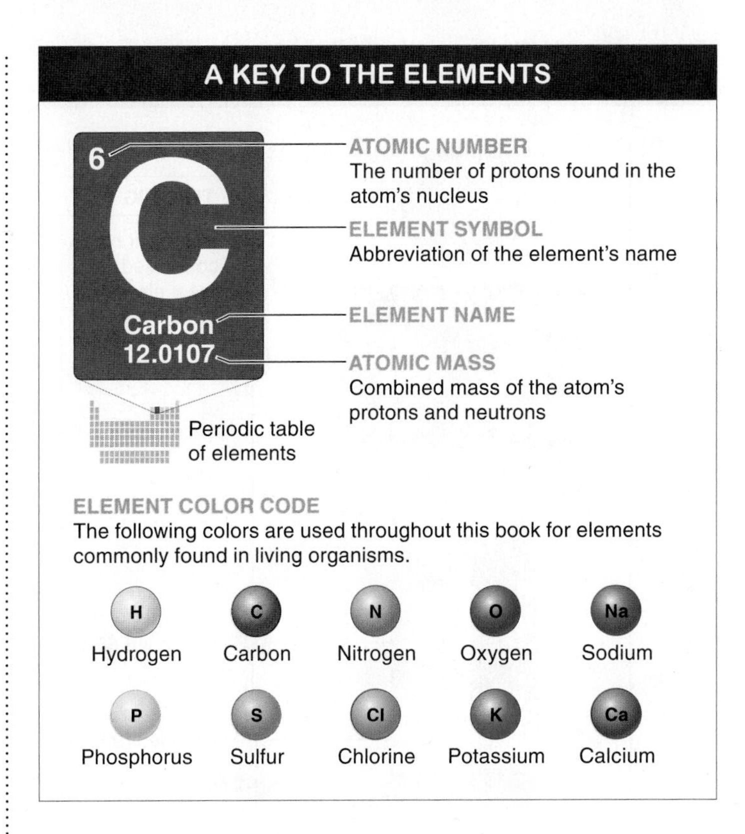

FIGURE 2-3 The vital statistics of atoms. A guide to reading the periodic table.

Atoms are tiny. Enlarge an atom by a billion times and it would only be the size of a grapefruit. Paradoxically, most of the space taken up by an atom is empty. That is, because the nucleus is very small and compact, the electrons zip about relatively far from the nucleus. If the nucleus were the size of a golf ball, the electrons would be anywhere from half a mile to six miles away.

What distinguishes one element, such as chlorine, from another, such as neon or oxygen? Atoms of different elements have a different number of protons in the nucleus. A chlorine atom has 17 protons, a neon atom has 10 protons, and an oxygen atom has 8 protons. Each element is given a name (and an abbreviation, such as O for oxygen and C for carbon) and an **atomic number** that corresponds to how many protons it has (**FIGURE 2-3**). All the known elements can be arranged in a scheme, in the order of their atomic number, called the "periodic table." So far, about 90 elements have been discovered that are present in nature and about 25 others can be made in the laboratory. Everything you see around you is made up of some combination of those naturally occurring elements.

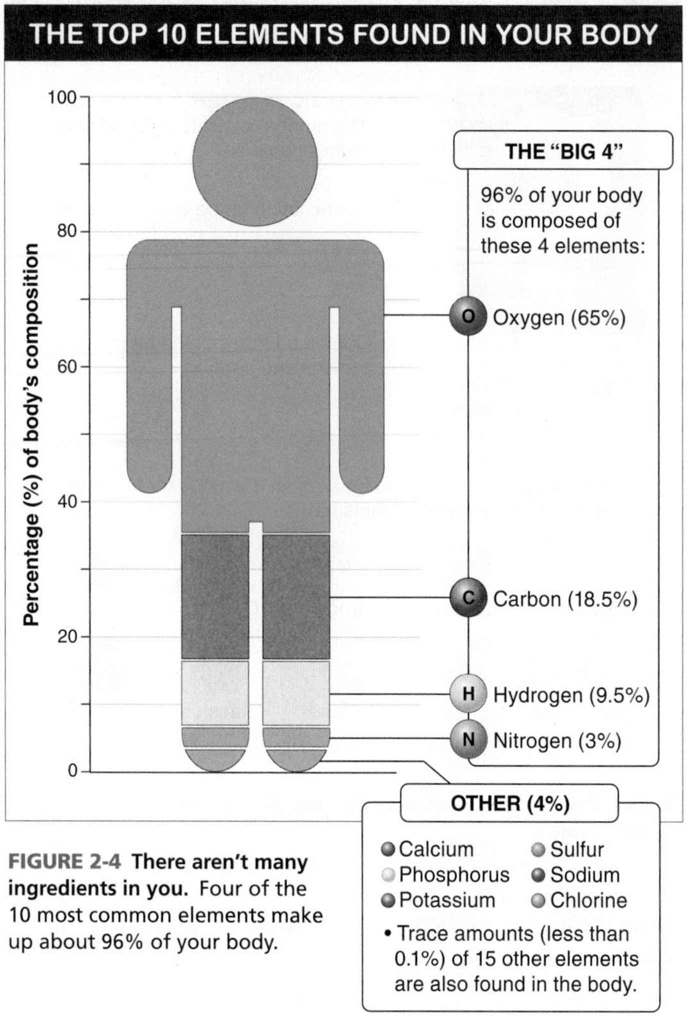

THE TOP 10 ELEMENTS FOUND IN YOUR BODY

Percentage (%) of body's composition

THE "BIG 4"

96% of your body is composed of these 4 elements:

Oxygen (65%)

Carbon (18.5%)

Hydrogen (9.5%)

Nitrogen (3%)

OTHER (4%)

Calcium, Phosphorus, Potassium, Sulfur, Sodium, Chlorine

• Trace amounts (less than 0.1%) of 15 other elements are also found in the body.

FIGURE 2-4 There aren't many ingredients in you. Four of the 10 most common elements make up about 96% of your body.

Of all the elements found on earth, only 25 are found in your body. The "Top 10" most common of these make up 99.9% of your body (**FIGURE 2-4**). And just four elements make up more than 96% of your body. These "Big 4" are oxygen, carbon, hydrogen, and nitrogen. With knowledge about the Big 4, you can understand a huge amount about nutrition and physiology (how your body works), so we'll focus on the properties of these four elements later in this chapter.

The number of protons in an atom's nucleus determines what element it is. The number of neutrons in the nucleus is usually equal to the number of protons. Since protons and neutrons have approximately the same mass, the mass of the atom is about double its atomic number. The element oxygen, for example, has the atomic number 8 because it has 8 protons, so we expect it to have an atomic mass of 16, simply the mass of the 8 protons and 8 neutrons added together—which it generally does. (As noted above, we can ignore the mass of the electrons.)

TAKE-HOME MESSAGE 2•1

Everything around us, living or not, is made up of atoms, the smallest unit into which material can be divided without losing its essential properties. All atoms have the same general structure. They are made up of protons and neutrons in the nucleus, and electrons, which circle far around the nucleus.

2•2

An atom's electrons determine how (and whether) the atom will bond with other atoms.

While the number of protons identifies an element, it is an atom's electrons that determine how (and whether) it bonds with other atoms. Electrons move so quickly that it is impossible to determine, at any given moment, exactly where an electron is. Electrons are not just moving about haphazardly, though. Speeding around the nucleus, they tend to stay within a prescribed area. This area is called an "electron shell," and an atom may have several shells, each shell occupied by its own set of electrons. Within a shell, the electrons stay far apart because their negative charges repel one another.

The first electron shell is closest to the nucleus and can hold two electrons (**FIGURE 2-5**). If an atom has more than two electrons, as most atoms do, the other electrons are arranged in other shells. The second shell is a bit farther away from the nucleus and can hold as many as eight electrons. There can be as many as seven shells in total, holding varying numbers of electrons.

Atoms become stable when their outermost shell is filled to capacity. Those with this configuration behave like loners, neither reacting nor combining with other atoms. On the other hand, when atoms have outer shells with vacancies, they are likely to interact with other atoms, giving, taking, or sharing electrons to achieve that desirable state: a full outer shell of electrons. In fact, based on the number of electron

ELECTRON SHELLS AND ATOM STABILITY

ELECTRON SHELLS

Electrons move around the nucleus in designated areas called electron shells. An atom can have as many as seven electron shells in total.

O

First electron shell (capacity: 2 electrons)

Second electron shell (capacity: 8 electrons)

Vacancy

Oxygen atom

The chemical characteristics of an atom depend upon the number of electrons in its outermost shell.

ATOM STABILITY

Atoms become stable when their outermost shell is filled to capacity. Stable atoms tend not to react or combine with other atoms.

Unstable atoms

Stable atoms

H

Hydrogen atom

He

Helium atom

N

Nitrogen atom

Ne

Neon atom

Only when atoms have electron vacancies in their outermost shell are they likely to interact with other atoms.

FIGURE 2-5 Electrons and the shells they inhabit.

vacancies in the outermost shell of an atom, it's possible to predict how amenable to bonding that atom will be, and even which other atoms its likely bonding partners will be.

IONS ARE CHARGED ATOMS

An atom that loses one or more electrons becomes positively charged, while an atom that acquires electrons becomes negatively charged. This transfer of electrons is driven by the fact that atoms with full outer electron shells are more stable.

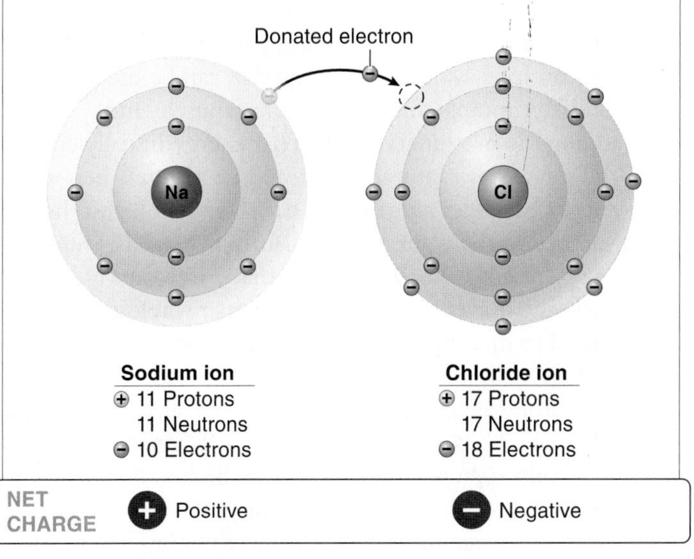

FIGURE 2-6 Ions are electrically charged atoms.

Normally, an atom has the same number of electrons as protons. Sometimes an atom may have one or more extra electrons or may lack one or more electrons relative to the number of protons. An atom with extra electrons becomes negatively charged, and an atom lacking one or more electrons is positively charged. Such a charged atom is called an **ion** (**FIGURE 2-6**). Due to their electrical charge, ions behave very differently from the atoms that give rise to them. As we'll see later in the chapter, ions are more likely to interact with other, oppositely charged ions.

TAKE-HOME MESSAGE 2•2

The chemical characteristics of an atom depend on the number of electrons in its outermost shell. Atoms are most stable and least likely to bond with other atoms when their outermost electron shell is filled to capacity.

Atoms can bond together to form molecules or compounds.

When you eat a meal, it is like filling your car's tank with gasoline; you have a source of energy that can be used to fuel activities like running, thinking, building muscle, and maintaining the machinery of life. That energy initially comes from the sun and is captured and stored by plants. When we eat plants or eat other animals that eat plants, we ingest the energy stored in the plant material. But how exactly is energy stored in plants? It is stored in bonds that join atoms together. Eventually the bonds are broken, the energy is released, and it can be used to fuel the body's activities.

Groups of atoms held together by bonds are called **molecules.** It usually requires energy to build bonds and "glue" atoms together. Conversely, when molecules are broken down—such as when animals (including humans) consume molecules in their diet—bonds are broken, releasing the energy that was used to create them. In a sense, molecules are created as a short-term storage of energy that can be harnessed later.

There are three principal types of bonds that hold multiple atoms together. The type of bonding that any atom is likely to take part in depends almost entirely on the number of electrons in its outermost shell.

Covalent Bonds **Covalent bonds** are strong bonds formed when two atoms share electrons. The simplest example of a covalent bond is the bonding of two hydrogen atoms to form H_2 (**FIGURE 2-7**), the simplest of all molecules. A hydrogen atom has an atomic number of 1: it has a single proton in its nucleus and a single electron circling around the nucleus in the first shell. Because the atom is most stable when the first shell has two electrons, two hydrogen atoms can each achieve a complete outermost shell by sharing electrons. The nuclei come close together (but not too close, since they are positively charged) and the two electrons circle around both of the nuclei, almost in a figure 8. The new H_2 molecule is very stable because, now that both atoms have two electrons in their outermost shell, they are no longer likely to bond with other atoms.

Oxygen, with an atomic number of 8, has two electrons in its innermost shell and six in its outermost shell. Consequently it needs to form two covalent bonds to fill its outermost shell. Sometimes two oxygen atoms join together to form O_2. Each oxygen atom shares two electrons with the other, filling the outermost shell of both. The sharing of two electrons between two atoms is called a **double bond** (see Figure 2-7). O_2 is the most common form in which we find oxygen in the world.

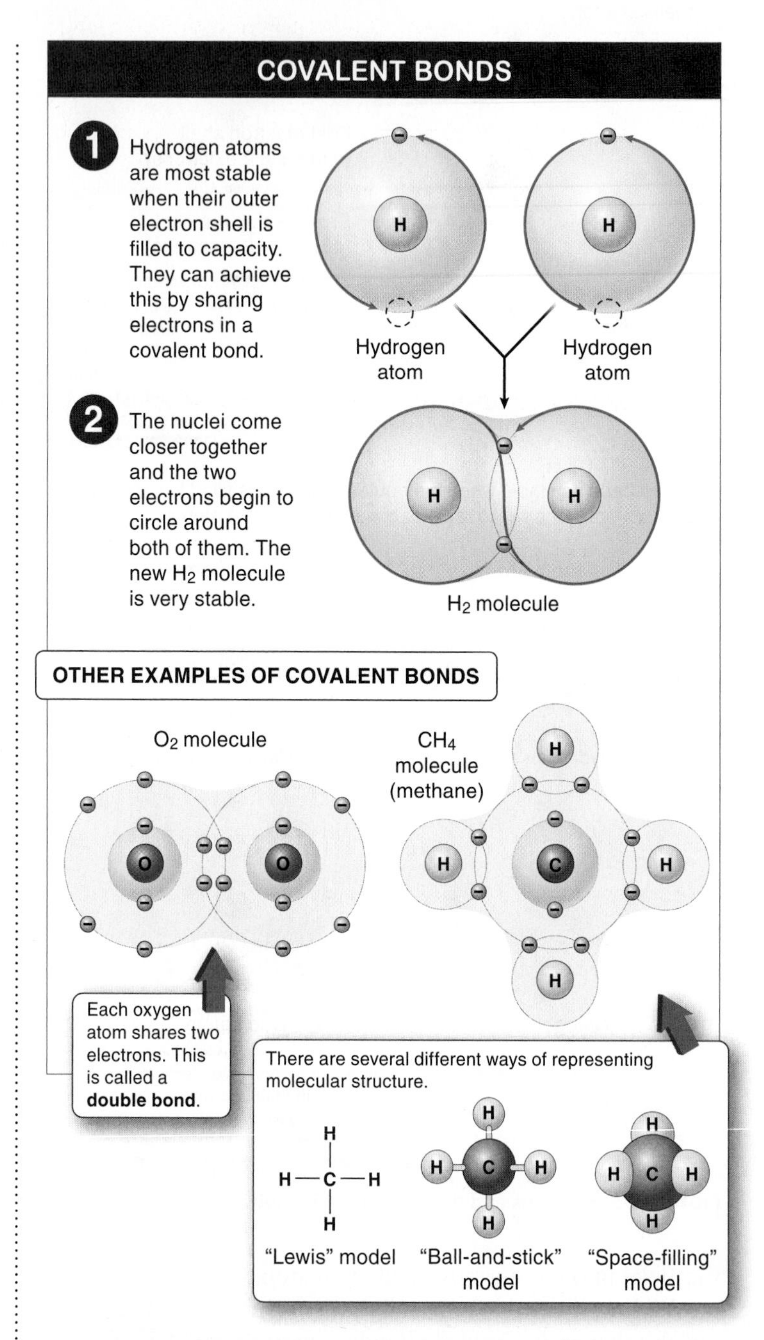

FIGURE 2-7 Covalent bonds: strength through electron sharing. The molecular structures formed by covalent bonding can be represented on paper in several different ways.

Carbon is a particularly extroverted molecule. It has an atomic number of 6, meaning that two electrons fill its first shell and four remain to occupy the second shell. Because four electron vacancies are left, carbon can, and frequently does, form four covalent bonds, joining up with other atoms in a wide variety of molecules. Methane, the chief component of natural gas, is formed when one atom of carbon covalently bonds with four atoms of hydrogen (see Figure 2-7). It has the chemical formula CH_4.

Ionic Bonds Atoms can also bond together without sharing electrons. In **ionic bonds,** one atom transfers one or more of its electrons completely to another. As a result of this transfer, each atom becomes an ion, since each has an unequal number of protons and electrons. The atom gaining electrons becomes negatively charged, while the atom losing electrons becomes positively charged. Rather than forming molecules, ionic bonds form **compounds,** in which the two oppositely charged ions attract each other. Unlike the much stronger covalent bonds, in ionic bonds each electron circles around a single nucleus. Ions of equal and opposite charges are attracted to each other and the compound is neutral—that is, it has no charge (**FIGURE 2-8**).

Hydrogen Bonds Ionic and covalent bonds link two or more atoms together. **Hydrogen bonds,** on the other hand, are important in bonding multi-atom molecules together. A hydrogen bond is formed between a hydrogen atom in one molecule and another atom, often an oxygen or nitrogen atom, in another molecule (or, in a very large molecule, in another part of the same molecule). This bond is based on the attraction between positive and negative charges (**FIGURE 2-9**).

The atoms taking part in hydrogen bonds are not ions, so where do the electrical charges come from? The hydrogen atom is already covalently bonded to another atom in the

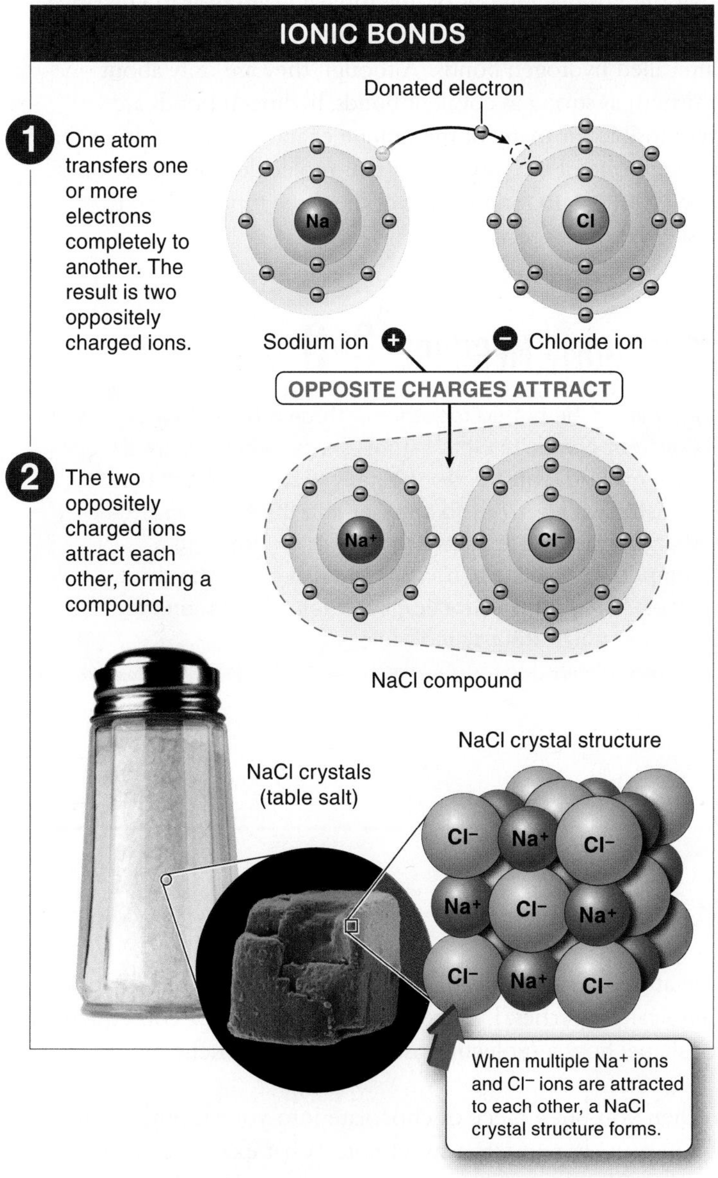

FIGURE 2-8 Ionic bonds: transfer of electrons from one atom to another. The resulting charged atoms (ions) attract each other.

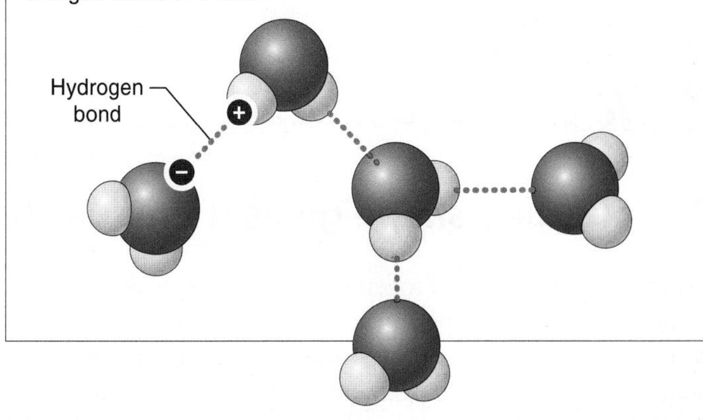

FIGURE 2-9 Hydrogen bonds: attraction between a polar atom or molecule and a hydrogen atom. As we'll see later in the chapter, the hydrogen bonding between H_2O molecules gives water some of its characteristic properties.

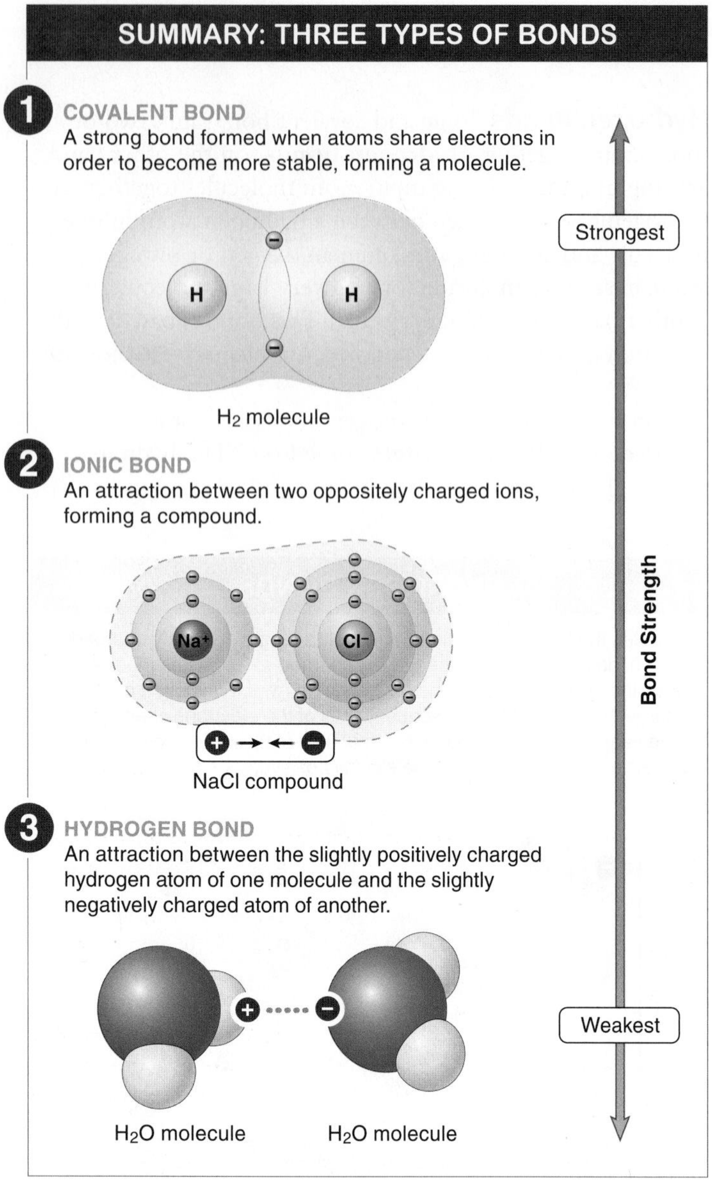

SUMMARY: THREE TYPES OF BONDS

1 COVALENT BOND
A strong bond formed when atoms share electrons in order to become more stable, forming a molecule.

H H

H_2 molecule

2 IONIC BOND
An attraction between two oppositely charged ions, forming a compound.

Na^+ Cl^-

NaCl compound

3 HYDROGEN BOND
An attraction between the slightly positively charged hydrogen atom of one molecule and the slightly negatively charged atom of another.

H_2O molecule H_2O molecule

Strongest

Bond Strength

Weakest

FIGURE 2-10 Three ways in which atoms and molecules are bound.

same molecule and shares its electron. That electron circles both the hydrogen nucleus and the nucleus of the other atom, but the electron is not shared equally. Given that the other atom always has more than the one proton found in the hydrogen nucleus, the electron contributed by hydrogen spends more of its time near the other, more positively charged nucleus than near the hydrogen nucleus. Having an extra electron nearby causes the larger atom to be slightly negatively charged, while the hydrogen atom becomes slightly positively charged.

In a sense, the covalently bonded molecules become like a magnet, with distinct positive and negative sides (see Figure 2-9). Magnet-like molecules with distinct positive and negative regions like this are said to be "polar." They are attracted to other polar molecules, lining up in particular orientations such that the positive regions of one molecule are near the negative regions of another. These attractions between polar molecules are called hydrogen bonds. Although they are only about one-thirtieth as strong as covalent bonds, hydrogen bonds are responsible for many of the unique characteristics of water that make it one of the most important molecules for life on earth. See **FIGURE 2-10** for a review of the three types of bonds discussed in this section.

TAKE-HOME MESSAGE 2•3

Atoms can be bound together in three different ways. Covalent bonds, in which atoms share electrons, are the strongest. In ionic bonds, the next strongest, one atom transfers its electrons to another and the two oppositely charged ions are attracted to each other, forming a compound. Hydrogen bonds, the weakest of the three types, involve the attraction between a hydrogen atom and another polar atom or molecule.

2•4

A molecule's shape gives it unique characteristics.

When many atoms bond together, the resulting molecule or compound acquires a "personality" that results from the accumulated properties of the atoms and their interactions with each other. The arrangement of atoms within a molecule also determines the molecule's shape: the atoms may be lined up in a row or may stack in a pyramid or may pack in a more globular formation, like a piece of popcorn. Whatever form the atoms take, a molecule's shape will determine many of its physical properties. Let's consider two important consequences of shape: how a molecule tastes and how it smells.

When you put a piece of chocolate into your mouth, your brain quickly registers a sweet taste. What exactly does that mean? Why does chocolate taste sweet, while a piece of fish may taste salty? The human tongue has various types of taste receptors; some detect bitter tastes, some sweet, some sour, some salty, and others savory. A molecule of the sugar glucose in that piece of chocolate triggers a sensation of sweetness when it binds to one of our sweet-taste receptors.

How sweet something tastes depends on the shape of its molecules—how well the molecules fit into the sweet-sensing receptors. Glucose molecules fit very well—slightly more snugly than the sucrose molecules in table sugar (**FIGURE 2-11**). Molecules of fructose, a sugar found in fruits and some vegetables, fit even better.

Food chemists have been able to create artificial sweeteners that have a better and better fit with receptors. Aspartame and saccharin, for example, are 150 times and 450 times sweeter than table sugar. Splenda is 600 times sweeter. This means that with very tiny amounts of artificial sweetener, we can sense a sweetness equivalent to a much greater amount of table sugar.

Insect mating behavior is also heavily influenced by molecular interactions that depend on molecular shape. Male gypsy moths, for example, will fly as far as two miles just because a molecule they encounter fits into one of their receptors and tells them that a sexually receptive female is nearby (**FIGURE 2-12**). At the heart of this behavior is something chemically similar to how taste receptors sense the sweetness of a piece of chocolate. Rather than being in a piece of food, though, the molecules the insects "taste" are airborne.

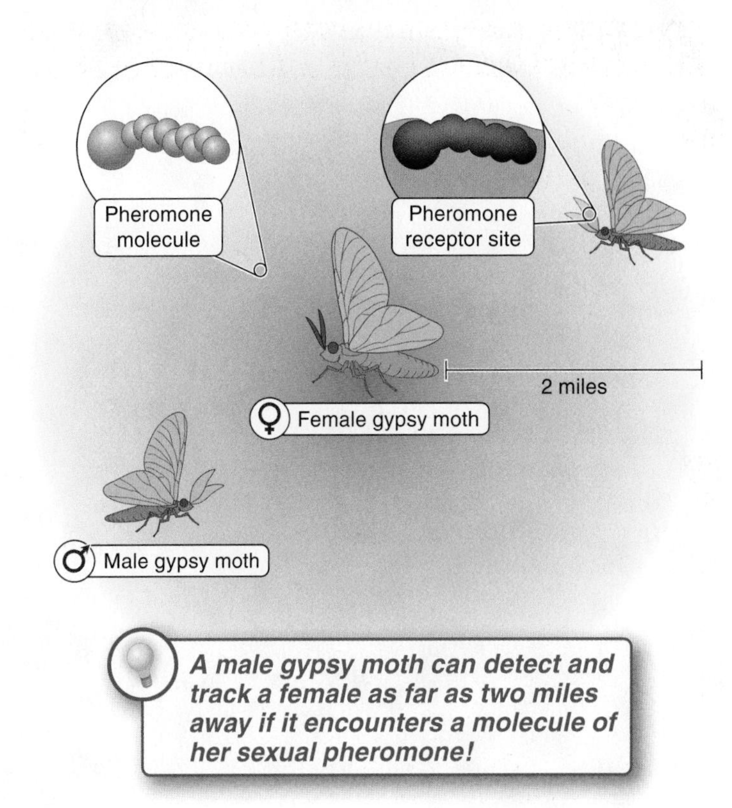

FIGURE 2-12 Molecular shape and function. A molecule's shape can give it a particular smell—and even be essential to its function as a sexual attractant.

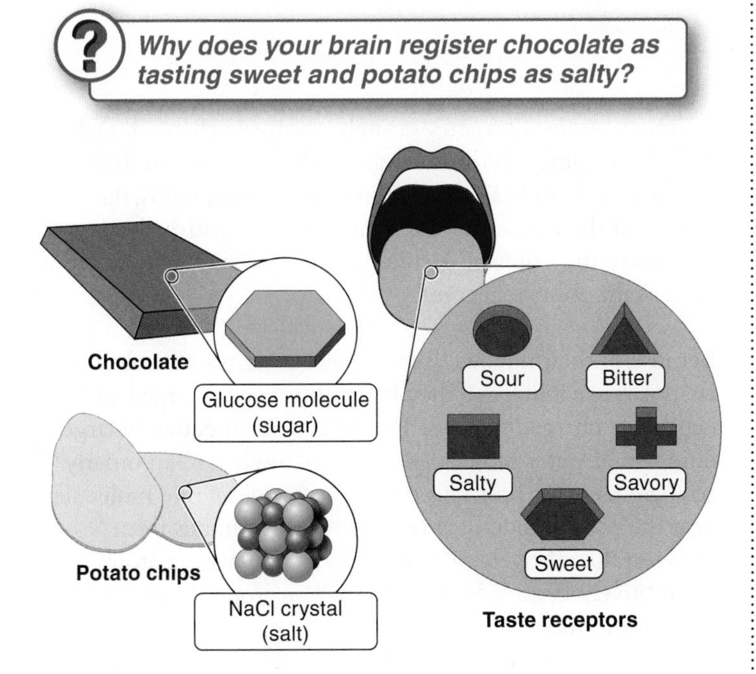

FIGURE 2-11 The shape of a molecule matches the shape of its taste receptors.

Taste is just one special case illustrating the chemical importance of a molecule's shape. More generally, every interaction between molecules is influenced by the fit between the molecules. Next we will look at the shape and bonding characteristics of one of the most important biological molecules, water, and how they give that molecule superhero-like properties.

TAKE-HOME MESSAGE 2•4

The chemical and physical properties of a molecule—such as how the molecule tastes or smells, or its likelihood of bonding with other molecules—depend on the shape in which the atoms are linked together and on the electrons exposed in their outermost shells.

❷ Water has features that enable it to support all life.

Water flowing down a Japanese waterfall.

2•5

Hydrogen bonds make water cohesive.

Predators chasing after green basilisk lizards in Central America sometimes get the shock of a lifetime (and lose their lunch in the process). When chased, the lizards run quickly across rocks and dirt—nothing unusual for a lizard. Then, when they come to the edge of a stream or river, these fleeing lizards show their uniqueness and earn their nickname, the "Jesus lizard." Without stopping, a Jesus lizard runs right across the surface of the water, leaving stunned predators behind at the water's edge. How do these animals do it?

The Jesus lizard, like numerous insects such as the water strider, makes use of the fact that water molecules have tremendous cohesion. That is, they stick together with unusual strength. This molecular cohesiveness is due to hydrogen bonds between the water molecules.

Each water molecule is V-shaped (**FIGURE 2-13**; see also Figure 2-9). The hydrogen atoms are at the ends of the two arms and the oxygen is at the bottom end of the V, between the two hydrogen atoms. Oxygen's strongly positively charged nucleus pulls the circling electrons toward itself and holds on to them for more than its fair share of the time. Consequently, the oxygen at the bottom of the V has a slight negative charge and the end of the water molecule containing the hydrogen atoms has a slight positive charge.

Because of their unequally shared electrons, water molecules are polar; like a magnet, they have a positively charged and a negatively charged side (see Figure 2-9). Consequently, large numbers of water molecules orient themselves in an orderly arrangement that positions the negative side of one molecule near the positive side of another. Hydrogen bonds form between the relatively positively charged hydrogen atoms and the relatively negatively charged oxygen atoms of *adjacent* water molecules.

Hydrogen bonds are much weaker than covalent bonds and they don't last very long. Nonetheless, to the Jesus lizard, the

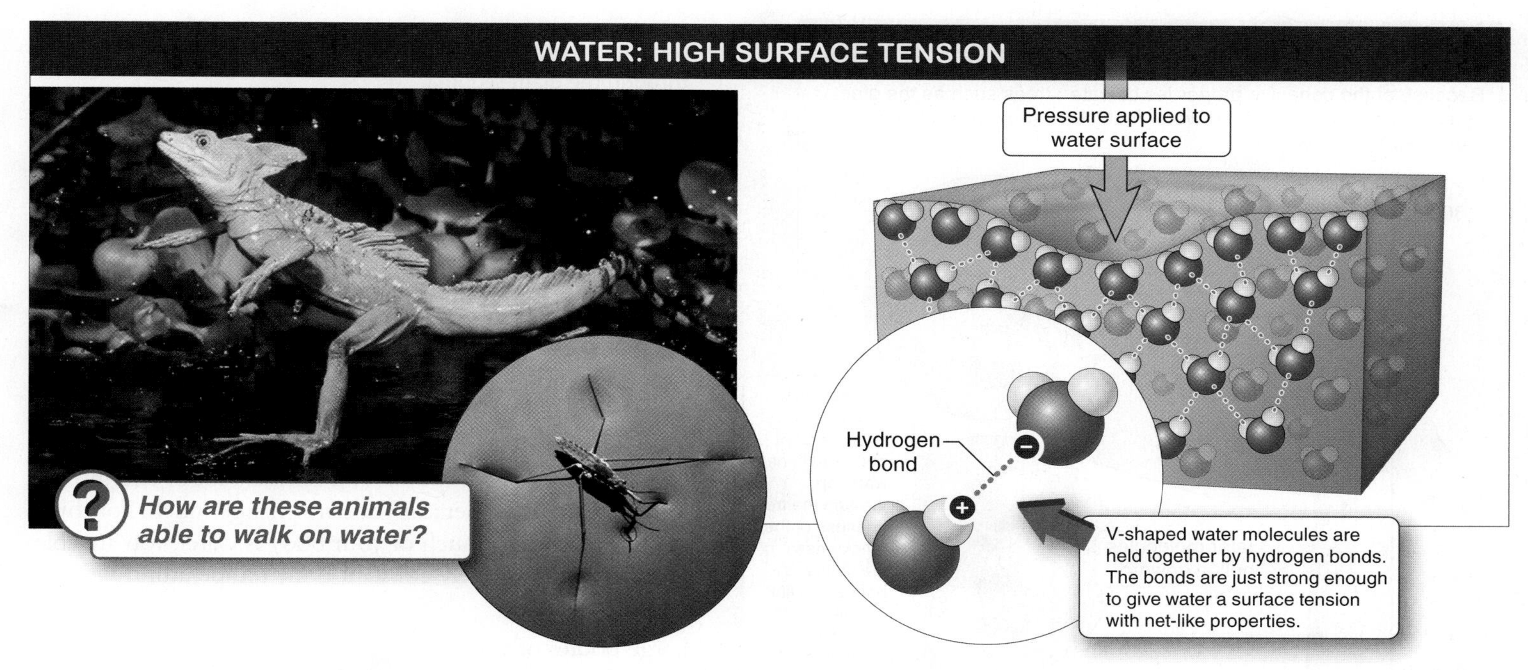

FIGURE 2-13 Walking on water! Hydrogen bonds make this possible for some animals.

cumulative effect of all the hydrogen bonds in water is to link together all the water molecules in the stream just enough to give the water a surface tension with some net-like properties. If the lizard were too fat or too slow, its weight would overwhelm the hydrogen bonds and the lizard would push the molecules apart and sink like a rock.

TAKE-HOME MESSAGE 2•5

Water molecules easily form hydrogen bonds, giving water great cohesiveness.

2•6

Water has unusual properties that make it critical to life.

All life on earth depends on water; organisms are made up mostly of water and require it more than any other molecule. Hydrogen bonding among water molecules gives water several important properties that contribute to its crucial role in the biology of all organisms.

1. Cohesion. We saw in the previous section how the connection of water molecules through hydrogen bonds makes water cohesive, resulting in, for example, high surface tension. The cohesiveness of water molecules also makes it possible for tall trees to exist (**FIGURE 2-14**). Leaves need water. Molecules of water in the leaves are continually lost to the atmosphere through evaporation or are used up in the process of photosynthesis. In order to get more water, plants must pull it up from the soil. The problem for many plants, such as the giant sequoia trees, is that the soil may be 300 feet below the leaf. Hydrogen bonds, however, allow water molecules to pull up adjacent water molecules to which they have hydrogen-bonded. The chain of linked molecules extends all the way down to the soil, where another water molecule is pulled in via the roots each time a water molecule evaporates from a leaf far above.

2. Large heat capacity. Walking across a sandy beach on a hot day, you can feel how easily sand heats up. By comparison, stepping from the beach into the cooler ocean reveals that water resists warming. It takes a lot of energy to change the temperature of water even a small amount. Why? Again, we must look to hydrogen bonding for our answer.

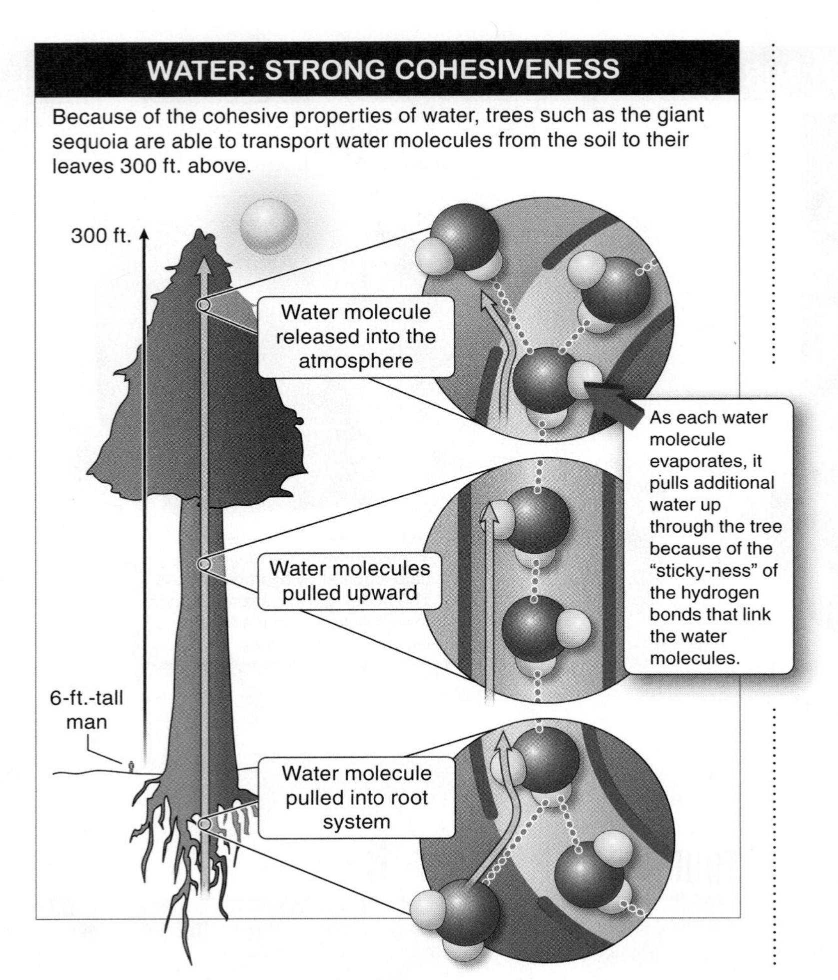

FIGURE 2-14 Like a giant straw. Hydrogen bonds cause water molecules to "stick" together, so that they can be pulled up through the giant sequoia.

The temperature of a substance is a measure of how quickly all of the molecules are moving. The molecules move more quickly when energy is added in the form of heat. When we heat water, the added energy doesn't immediately increase the movement of the individual water molecules. Rather, it disrupts some of the hydrogen bonds between the molecules (**FIGURE 2-15**). As quickly as they can be disrupted, though, hydrogen bonds form again somewhere else. And since the water molecules themselves don't increase their movement, the temperature doesn't increase. The net effect is that even if you release a lot of energy into water, the temperature doesn't change much. For this reason, because so much of your body is water, you are able to maintain a relatively constant body temperature.

Q Why do coastal areas have milder, less variable climates than inland areas?

Large bodies of water, especially oceans, can absorb huge amounts of heat from the sun during warm times of the year, reducing temperature increases on the land. Similarly, during cold times of the year the ocean slowly cools, giving off heat that reduces the temperature drop on shore.

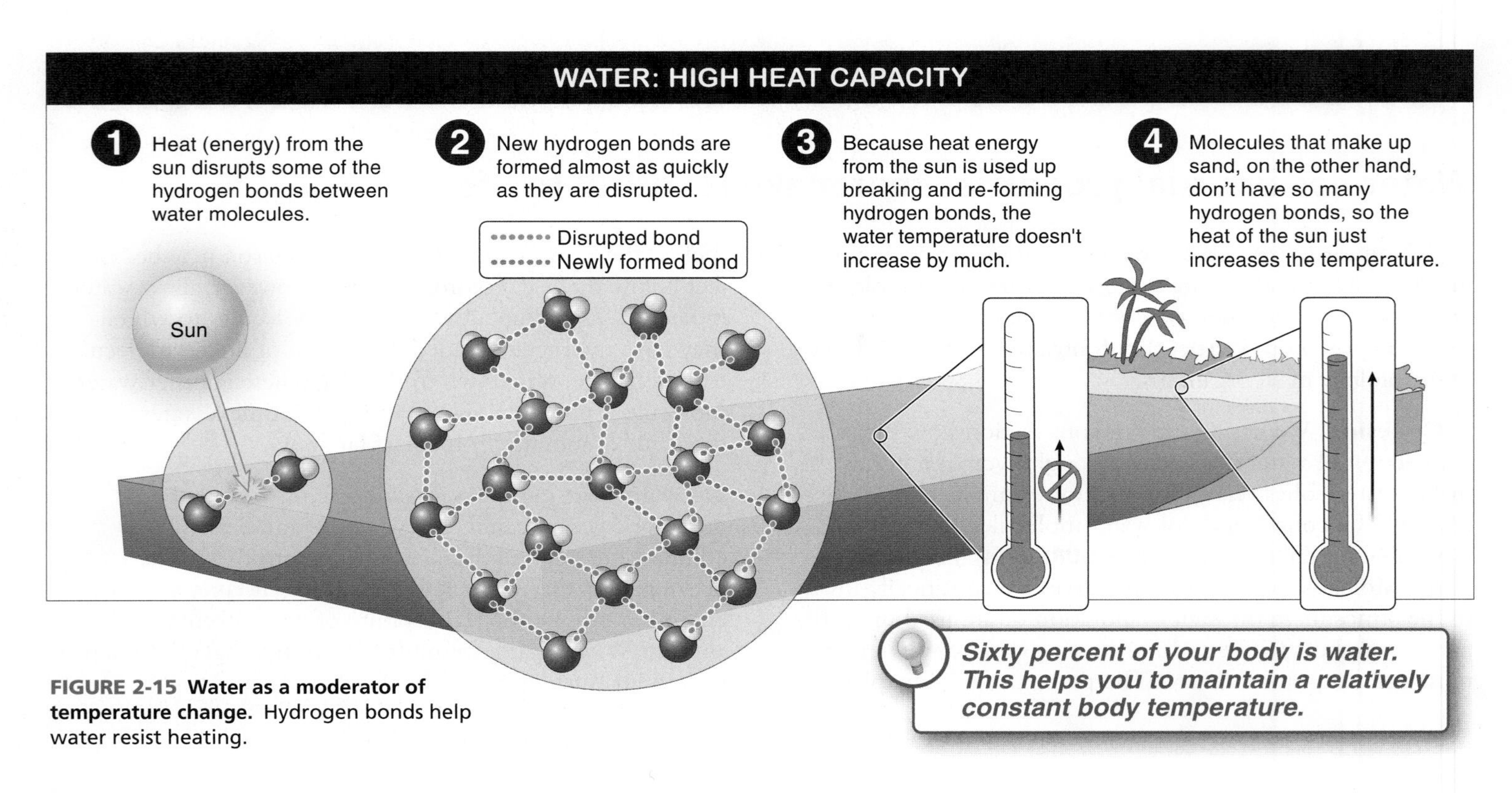

FIGURE 2-15 Water as a moderator of temperature change. Hydrogen bonds help water resist heating.

Sixty percent of your body is water. This helps you to maintain a relatively constant body temperature.

3. Low density as a solid. Ice floats. This is unusual, since most substances *increase* in density when frozen; as the molecules slow down, they pack together more and more efficiently—and densely. Consequently, the solid sinks. Water, however, becomes less dense and, as you might expect by now, this is due to hydrogen bonding. As the temperature drops and water molecules slow down, rather than becoming more and more tightly packed, they become less so. Each V-shaped water molecule bonds with four partners, via hydrogen bonds, forming a crystalline lattice in which the molecules are held slightly farther apart than in the liquid, causing ice to be less dense than water (**FIGURE 2-16**).

4. Good solvent. If you put a pinch of table salt into a glass of water, it will quickly dissolve. This means that all the charged sodium (Na^+) and chloride (Cl^-) ions that were ionically bonded together become separated from one another. The sodium and chloride ions were initially attracted to each other because they are polar molecules (ions), each carrying a slight charge. Water is able to pry them apart because, as a polar molecule, it, too, carries charges. The positively charged sodium ions are attracted to the negatively charged side of the water molecule, and the negatively charged chloride ions are attracted to the positively charged side (**FIGURE 2-17**). Many substances are, like water, polar. That is why, like salt, they easily dissolve in water.

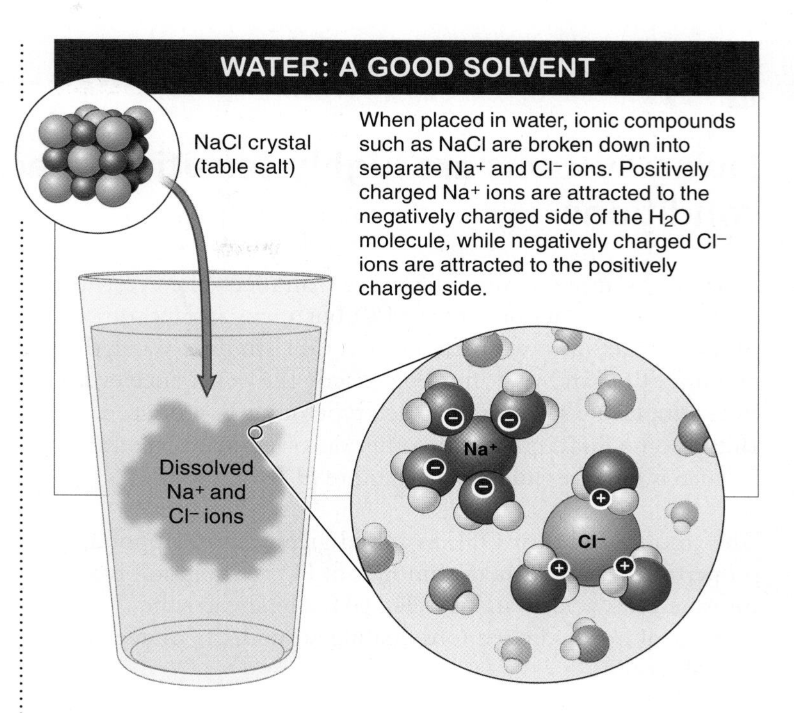

FIGURE 2-17 Solutions. Water pries apart ionic bonds, dissolving ionic compounds.

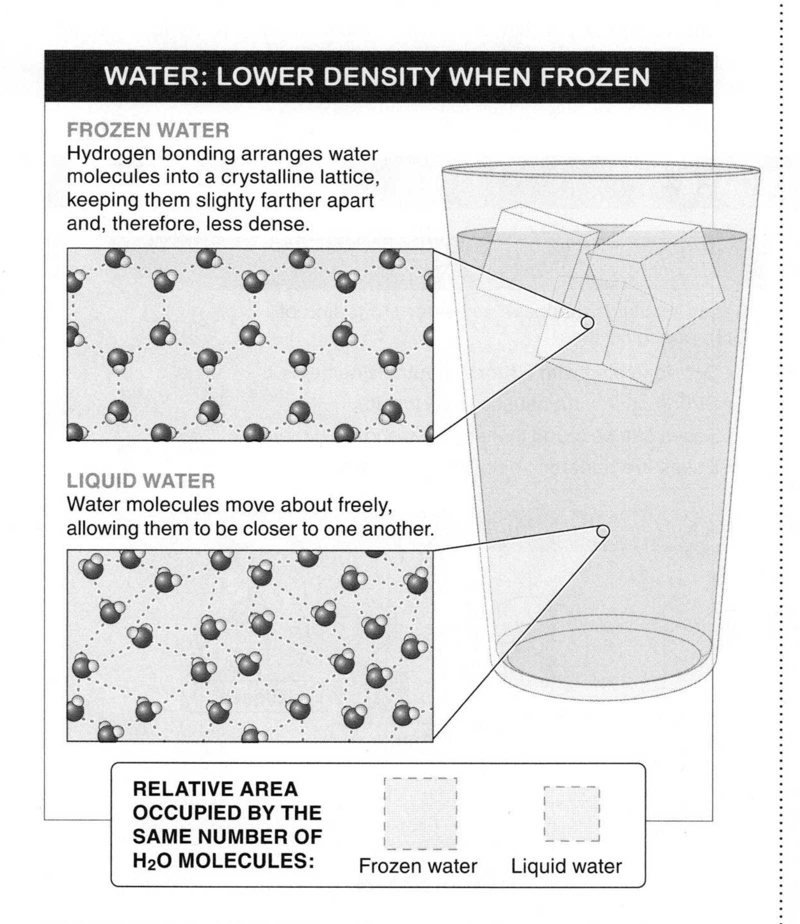

FIGURE 2-16 Ice floats. When frozen, water becomes less dense.

Non-polar molecules (such as oil) have neither positively charged regions nor negatively charged regions. Consequently, the polar water molecules are not attracted to them. Instead, when oil is poured into a container of water, the water molecules distance themselves from the oil, leaving the oil molecules in isolated aggregations that never dissolve.

Q Why don't oceans freeze as easily as freshwater lakes?

Because so much salt is dissolved in the oceans, many of the water molecules have their positively charged sides facing Cl^- ions. Simultaneously, many molecules of water are turned the other way, with their negatively charged sides facing Na^+ ions. Consequently, the orderly lattices of hydrogen bonds found in ice cannot form in salt water, and it does not freeze well.

TAKE-HOME MESSAGE 2·6

The hydrogen bonds between water molecules give water several of its most important characteristics, including cohesiveness, a reduced density as a solid, the ability to resist temperature changes, and broad effectiveness as a solvent.

2•7

Living systems are highly sensitive to acidic and basic conditions.

There's a lot more going on in water than meets the eye. Most of the molecules are present as H_2O, but at any instant some of them break into two parts: H^+ and OH^-. In pure water, the amount of H^+ and OH^- must be exactly the same, since every time a molecule splits, one of each type of ion is produced. But in some fluids containing other dissolved materials, this balance is lost: the fluid can have more H^+ or more OH^-.

The amount of H^+ or OH^- in a fluid gives it some important properties. In particular, the amount of H^+ in a solution is a measure of its acidity and is called **pH.** The greater the number of free hydrogen ions floating around, the more acidic the solution is.

Pure water is in the middle of the pH scale, with a pH of 7.0. Any fluid with a pH below 7.0 has more H^+ ions (and fewer OH^- ions) and is considered an **acid.** Any fluid with a pH above 7.0 has fewer H^+ ions (and more OH^- ions) and is considered a **base.** The pH scale, like the Richter scale for earthquakes, is logarithmic, although in the case of pH, the *lower* the number the *greater* the acidity: a decrease of 1 on the scale represents a 10-fold increase in the hydrogen ion concentration (**FIGURE 2-18**). A decrease of 2 represents a 100-fold increase. This means that a cola, with a pH of about 3.0, is 10,000 times (!) more acidic than a glass of water, with a pH of 7.0.

H^+ ions are essentially free-floating protons. Acids can donate their H^+ ions to other chemicals. In fact, H^+ is a very reactive little ion. Its presence gives acids some unique properties. For instance, the hydrogen ions in acids can bind with atoms in metals, causing them to corrode. That's why you can dissolve nails by dumping them in a bucket of acid (or cola).

Your stomach produces large amounts of hydrochloric acid (HCl) and has a pH between 1 and 3. (HCl is acidic because most of the hydrogen ions split off from the chlorine, raising the H^+ concentration of the fluid.) The acid in your stomach helps to kill most bacteria that you ingest. It also greatly enhances the breakdown of the chemicals in the food you eat and the efficiency of digestion and absorption. You may have

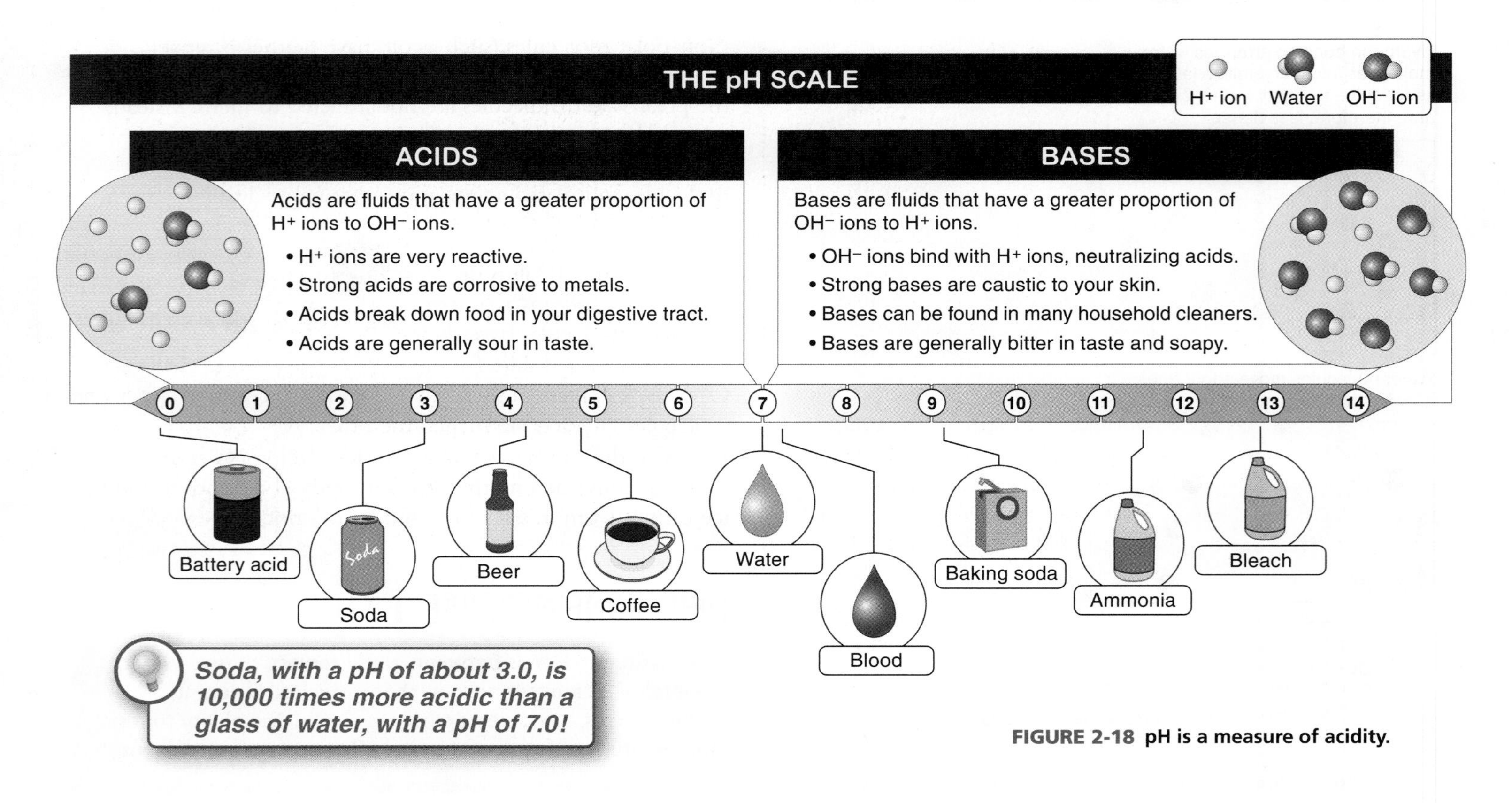

FIGURE 2-18 **pH is a measure of acidity.**

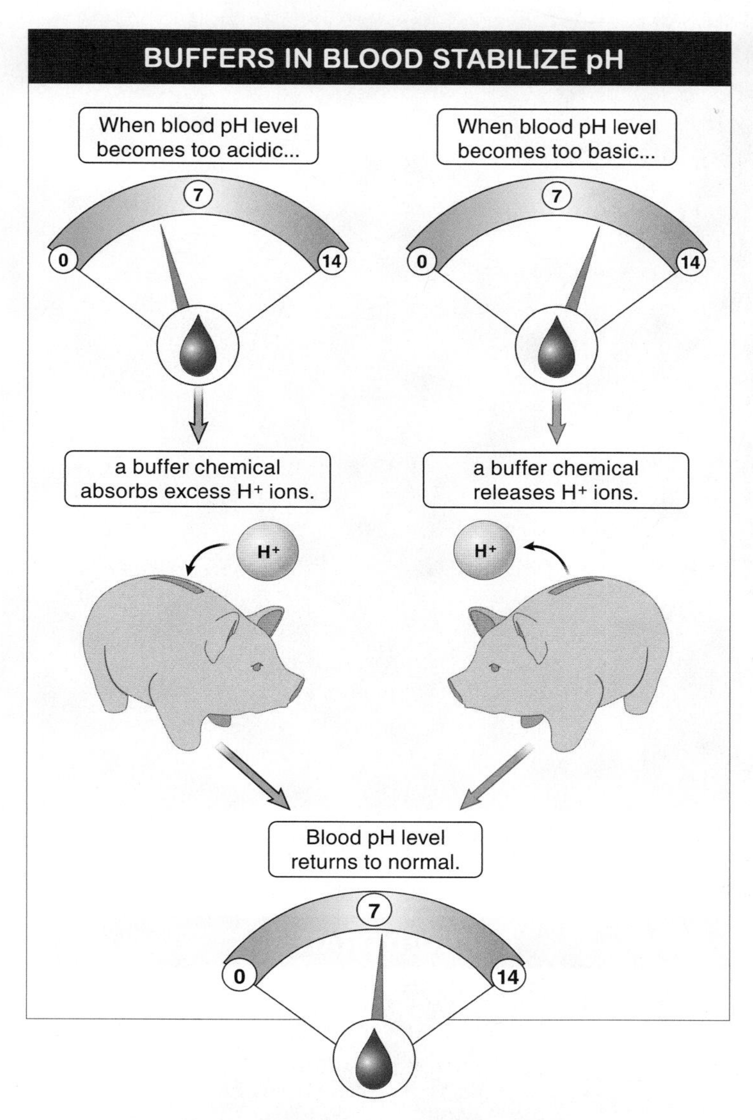

FIGURE 2-19 Maintaining a constant internal environment. Buffers in blood prevent potentially damaging pH swings.

learned firsthand of the high acidity of your stomach fluids if you have experienced heartburn or the sour taste of vomit.

Bases have very low concentrations of H^+ and relatively high concentrations of OH^-. Baking soda is a common basic substance. Some bases are called "antacids" because the OH^- ions in bases can bind with excess H^+ ions in acidic solutions, neutralizing the acid. Base-containing products such as Alka-Seltzer and Milk of Magnesia reduce the unpleasant feeling of

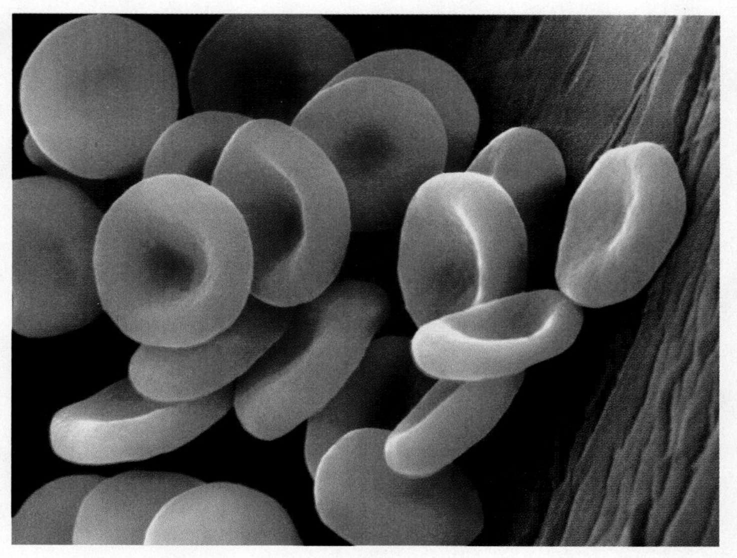

Buffers keep the pH of blood close to 7.4.

heartburn and acid indigestion that sometimes arise from the overproduction of acids by the stomach. Bases are commonly used in household cleaning products and generally have a bitter taste and a soapy, slippery feel.

The pH of blood is usually 7.4. Given that most cellular reactions produce or consume H^+ ions, there ought to be great swings in the pH of our blood. Unfortunately, our bodies can't tolerate such swings. Most of the chemicals that aid in the chemical reactions in our blood or cells stop functioning properly if the pH swings up or down by less than half a point. Fortunately, our bodies contain some chemicals that act like bank accounts for H^+ ions (**FIGURE 2-19**). Called **buffers,** these chemicals can quickly absorb excess H^+ ions to keep a solution from becoming too acidic, and they can quickly release H^+ ions to counteract any increases in OH^- concentration. Therefore, buffers are chemicals that act to resist changes in pH.

TAKE-HOME MESSAGE 2•7

The pH of a fluid is a measure of how acidic or basic the solution is and depends on the concentration of dissolved H^+ ions present. Acids, such as vinegar, can donate protons to other chemicals; bases, including baking soda, bind with free protons.

③ Carbohydrates are fuel for living machines.

Worldwide, more corn is produced (by weight) than any other grain.

2•8

Carbohydrates include macromolecules that function as fuel.

Hardly a day goes by without an item in a magazine or newspaper or on TV talking about whether carbohydrates are good or bad for us. Sports drinks such as Gatorade are filled with carbs, and in the days before a big game or race, athletes often "carbo-load" by eating bowls of pasta. Yet other dietary supplements and "power" bars tout that they are low-carb and effective in weight loss by causing the body to resort to using fat for energy. Meanwhile, nutritionists, doctors, and many diet programs exhort people to increase the amount of fiber—another type of carbohydrate—in their diet. What exactly are they all talking about?

There are four types of **macromolecules**—large molecules made up from smaller building blocks or subunits—in living organisms: carbohydrates, lipids, proteins, and nucleic acids. **Carbohydrates** are molecules that contain mostly carbon, hydrogen, and oxygen: they are the primary fuel for running all of the cellular machinery and also form much of

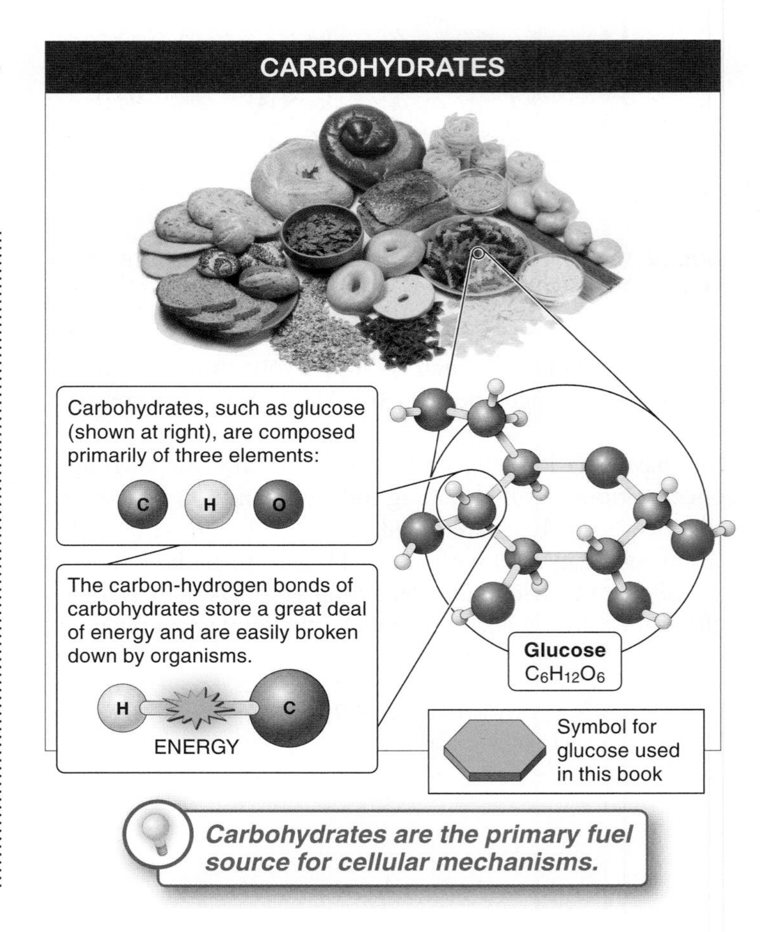

FIGURE 2-20 **All carbohydrates have a similar structure and function.**

the structure of cells in all life forms. Sometimes they contain atoms of other elements, but they must have carbon, hydrogen, and oxygen to be considered a carbohydrate (**FIGURE 2-20**). Further, a carbohydrate generally has approximately the same number of carbon atoms as it does H_2O units. For instance, the best-known carbohydrate, glucose, has the composition $C_6H_{12}O_6$ (6 carbons and, as a little math will show us, 6 H_2O units; notice that $6 \times H_2 = H_{12}$ and $6 \times O = O_6$). A carbohydrate called maltose has the composition $C_{12}H_{22}O_{11}$.

Carbohydrates function well as fuels because their many carbon-hydrogen bonds store a great deal of energy. These C–H bonds are easily broken, and organisms can capture the energy released and put it to use.

TAKE-HOME MESSAGE 2•8

Carbohydrates are the primary fuel for running all cellular machinery and also form much of the structure of cells in all life forms. Carbohydrates contain carbon, hydrogen, and oxygen, and generally have the same number of carbon atoms as they do H_2O units. The C–H bonds of carbohydrates store a great deal of energy and are easily broken by organisms.

2•9

Simple sugars are an effective, accessible source of energy.

Carbohydrates are classified into several categories, based on their size and their composition. The simplest carbohydrates are the **monosaccharides** or **simple sugars.** These simple sugars contain anywhere from three to seven carbon atoms and, when they are broken down, the products usually are not carbohydrates. Two common monosaccharides are glucose, found in the sap and fruit of many plants, and fructose, found primarily in fruits and vegetables, as well as in honey. Fructose is the sweetest of all naturally occurring sugars. The suffix *-ose* tells us that a substance is a carbohydrate.

SOME COMMON MONOSACCHARIDES

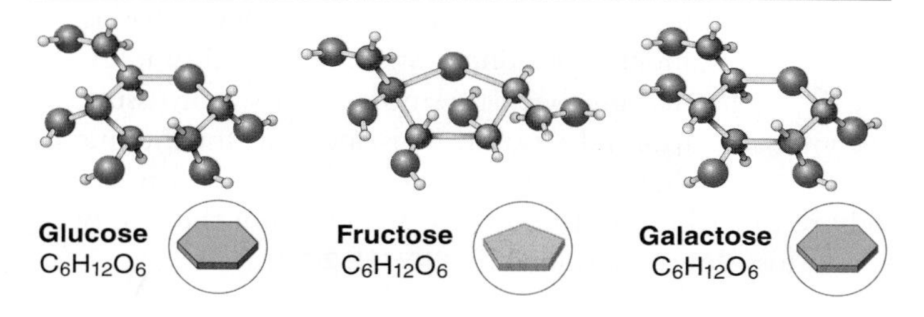

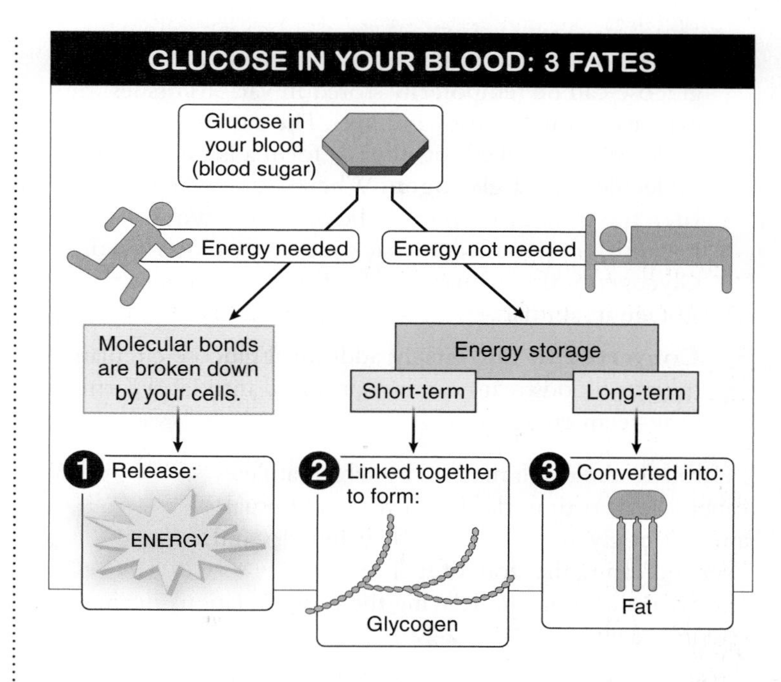

FIGURE 2-21 What happens to sugar in your blood?

The carbohydrate of most importance to living organisms is glucose. This simple sugar is found naturally in most fruits, but most of the carbohydrates that you eat, including table sugar (called sucrose) and the starchy carbohydrates found in bread and potatoes, are converted into glucose in your digestive system. The glucose then circulates in your blood at a concentration of about 0.1%. Circulating glucose, also called "blood sugar," has one of three fates (**FIGURE 2-21**):

1. **Fuel for cellular activity.** Once it arrives at and enters a cell, glucose can be used as an energy source. Through a series of chemical reactions, the cell breaks the bonds between the atoms of the glucose molecule (a process explained in detail in Chapter 4) and then uses the released energy to fuel cellular activity, including the muscle contractions that enable you to move and the nerve activities that enable you to think.
2. **Stored temporarily as glycogen.** If there is more glucose circulating in your bloodstream than is necessary

FIGURE 2-22 Some competitive athletes carbo-load to maintain high energy levels.

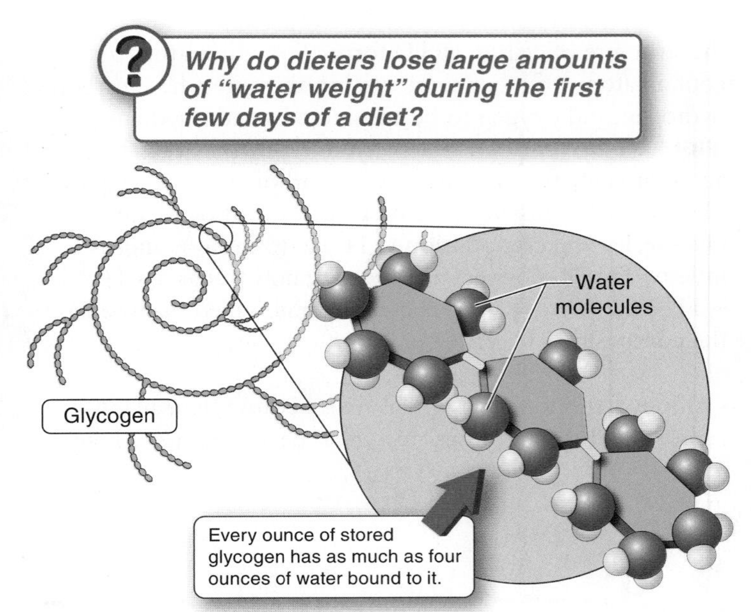

FIGURE 2-23 Water weight. Water molecules bound to glycogen account for much of the weight lost early in a diet.

to meet your body's current energy needs, the excess glucose can be temporarily stored in various tissues, primarily your muscles and liver. The stored glucose molecules are linked together to form a large web of molecules called **glycogen.** When you need energy later, the glycogen can easily be broken down to release glucose molecules back into your bloodstream. Glycogen is the primary form of short-term energy storage in animals.

3. **Converted to fat.** Finally, additional glucose circulating in your bloodstream can be converted into fat, a form of long-term energy storage.

What is "carbo-loading"?

"Carbo-loading" is a method by which athletes can, for a short time, double or triple the usual amount of glycogen stored in their muscles and liver, increasing the store of fuel available for extended exertion and delaying the onset of fatigue during an endurance event (**FIGURE 2-22**).

Carbo-loading is usually done in two phases, a depletion phase and a loading phase. The depletion phase begins six or seven days before a competition. In this phase, the combination of a super-low carbohydrate intake and exhaustive exercise depletes glycogen in the muscles. The loading phase takes place during the two days before the competition. During this phase, a super-high carbohydrate diet is combined with reduced exercise to achieve a higher blood glucose than is necessary, so that much of the excess glucose is stored as glycogen.

Glycogen also plays a role in the initial rapid weight loss people experience when dieting. If you reduce your caloric intake—perhaps as part of a low-carbohydrate diet—such that your body is burning more calories than you are consuming, your body must use stored forms of energy. The first, most accessible molecules that can be broken down for energy in the absence of sufficient sugar in your bloodstream are glycogen molecules in your muscles and liver. Large amounts of water are bound to glycogen. In fact, every ounce of stored glycogen has as much as 4 ounces of water bound to it. As that glycogen is removed from your tissue, so, too, is the water. This accounts for the initial dramatic weight loss that occurs before your body resorts to using stored fat, at which point the rate of weight loss slows considerably (**FIGURE 2-23**).

TAKE-HOME MESSAGE 2•9

The simplest carbohydrates, including glucose—the most important carbohydrate to living organisms—are monosaccharides or simple sugars. They contain from three to seven carbon atoms.

2•10

Complex carbohydrates are time-released packets of energy.

In contrast to the simple sugars, **polysaccharides** contain more than one sugar unit or building block. For example, two simple sugars can be joined together into a **disaccharide,** such as sucrose (table sugar) and lactose (the sugar found in milk). When more than two simple sugars are joined together—sometimes as many as 10,000—the resulting polysaccharide is called a **complex carbohydrate** (**FIGURE 2-24**). Depending on how the simple sugars are bonded together, polysaccharides may function as "time-release" stores of energy or as structural materials that may be completely indigestible by most animals. An example of such a structural material is the polysaccharide cellulose—the primary component of plant cell walls.

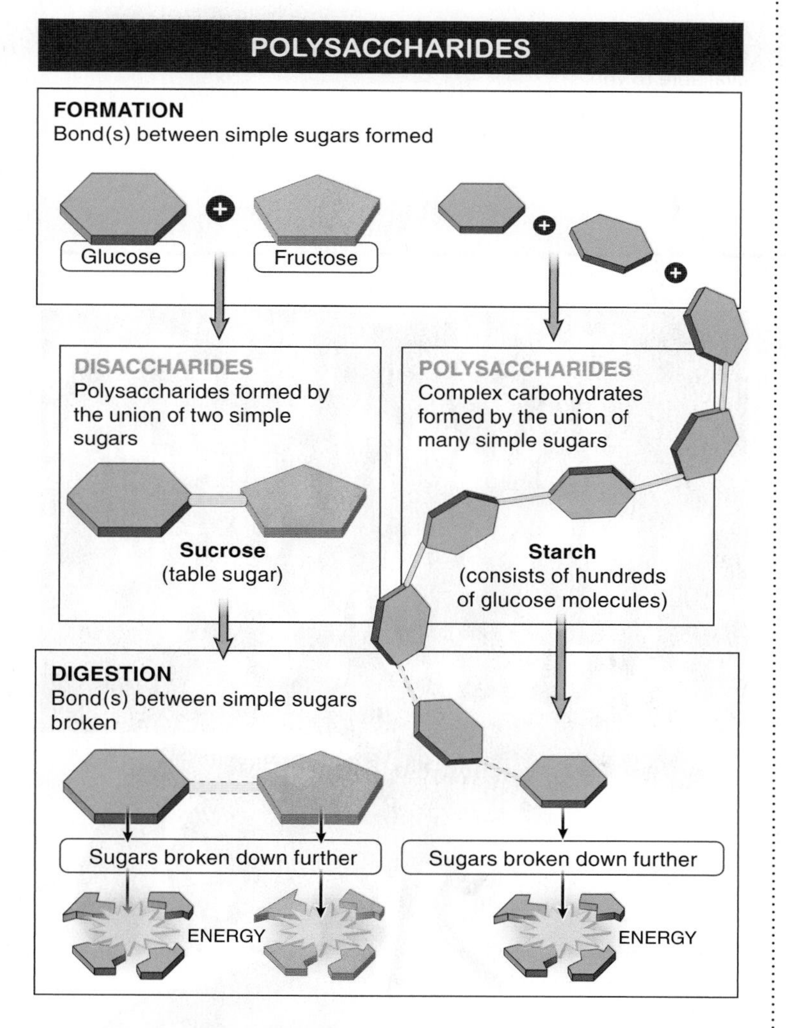

FIGURE 2-24 Chains of sugars. Polysaccharides are made from simple sugars bound together.

Like simple sugars, many disaccharides and polysaccharides are important sources of fuel. Unlike simple sugars, however, disaccharides and polysaccharides must undergo some preliminary processing before the energy can be released from their bonds. Let's look at what happens when we eat some sucrose, common table sugar. Sucrose is the primary carbohydrate in plant sap. It is a disaccharide composed of two simple sugars, glucose and fructose, linked together. Because humans can't directly use sucrose, the body must first break the bond linking the glucose and fructose. Only then can the individual monosaccharides be broken down into their component atoms and the energy from the broken bonds be harvested and used. Similarly, lactose is a disaccharide made up of a molecule of glucose and a molecule of galactose bound together. As with sucrose, we must break the bond before we can extract any usable energy from a molecule of lactose.

Energy can also be stored in a complex carbohydrate called **starch,** which consists of a hundred or more glucose molecules joined together in a line. In plants, starch is the primary form of energy storage, found in roots and other tissues (see Figure 2-24). Commonly cultivated grains such as barley, wheat, and rye are high in starch content, and corn and rice are more than 70% starch. Although it is composed exclusively of glucose molecules linked together, starch does not taste sweet. Because of its molecular shape it does not stimulate the sweetness receptors on the tongue. The glycogen that stores energy in your muscles and liver is also a complex carbohydrate, so it is sometimes referred to as "animal starch" (although it has a more branched structure than starch and carries more glucose units linked together).

Complex carbohydrates are like "time-release" fuel pellets. The glucose molecules become available slowly, one by one, as the bonds between glucose units are broken. As an interesting demonstration of the gradual release of glucose from complex carbohydrates, put a piece of potato or bread into your mouth and let it rest on your tongue. Initially, it does not have a sweet taste. After a few minutes, though, you will begin to taste some sweetness as the chemicals in your saliva break the starch down into glucose.

The relative amounts of complex carbohydrates and simple sugars in foods cause them to have very different effects when you eat them. Oatmeal (along with rice and pasta), for example, is rich in complex carbohydrates. Fresh fruits, on the other hand, are rich in simple sugars, such as fructose.

Before heading to the library for a long study session, students would be wise to consume oatmeal rather than fresh fruits. Why?

Consequently, although the fruit will give a quick burst of energy as the sugars are almost immediately available, the fuel will soon be gone from the bloodstream. The simple sugars in the oatmeal will become available only gradually, as the complex carbohydrates of the oats are slowly broken down into their simple sugar components (**FIGURE 2-25**).

TAKE-HOME MESSAGE 2•10

Multiple simple carbohydrates are sometimes linked together into more complex carbohydrates. Types of complex carbohydrates include starch, which is the primary form of energy storage in plants, and glycogen, which is a primary form of energy storage in animals.

Depending on their structure, dietary carbohydrates can lead to quick-but-brief or slow-but-persistent increases in blood sugar.

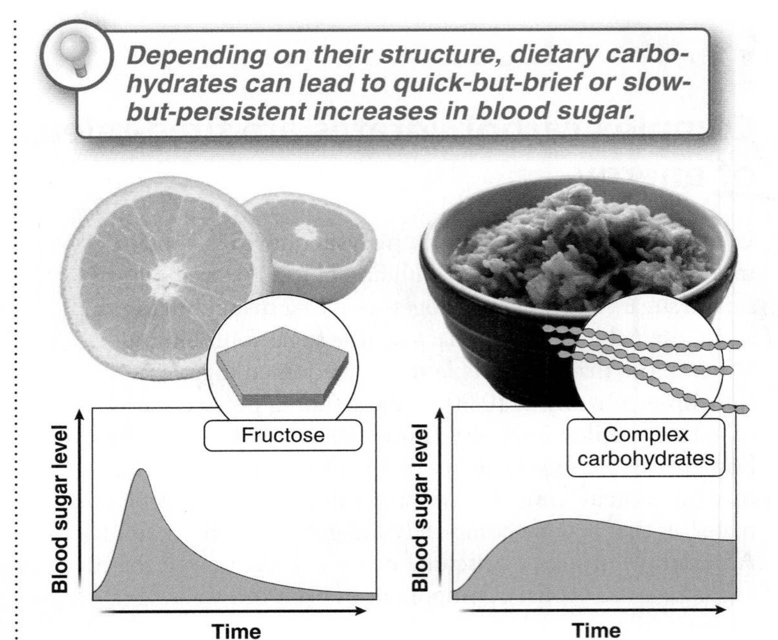

FIGURE 2-25 Short-term versus long-term energy? Complex carbohydrates and simple sugars differ in the way they make energy available to you.

2•11

Not all carbohydrates are digestible.

Despite their general importance as a fuel source for humans, not all carbohydrates can be broken down by our digestive system. Two different complex carbohydrates—both indigestible by humans—serve as structural materials for invertebrate animals and plants: **chitin** (pronounced KITE-in) and **cellulose** (**FIGURE 2-26**). Chitin forms the rigid outer skeleton of most insects and crustaceans (such as lobsters and crabs). Cellulose forms a huge variety of plant structures that are visible all around us. We find cellulose in trees and the wooden structures we build from them, in cotton and the clothes we make from it, in leaves and in grasses. In fact, it is the single most prevalent compound on earth.

Surprisingly, cellulose is almost identical in composition to starch. Nonetheless, because of one small difference in the chemical bond between the simple-sugar units, cellulose has a slightly different three-dimensional structure. And, as we noted earlier, even tiny differences in the shape of a molecule can have a huge effect on its behavior. In this case, the

FIGURE 2-26 Carbohydrates can serve as structural materials.

> In the last third of his life, there came over Laszlo Jamf . . . a hostility, a strangely personal hatred, for the covalent bond. . . . That something so mutable, so soft, as a sharing of electrons by atoms of carbon should be at the core of life, *his* life, struck Jamf as a cosmic humiliation. *Sharing?* How much stronger, how everlasting was the ionic bond—where electrons are not shared, but *captured. Seized!* and held! polarized plus and minus, these atoms, no ambiguities . . . how he came to love that clarity: how stable it was, in much mineral stubbornness!
>
> — Thomas Pynchon, *Gravity's Rainbow*
>
> [Was Laszlo Jamf mistaken in his understanding of the relative strengths of ionic and covalent bonds?]

FIGURE 2-27 Fiber. It's not digestible but it's still important for our diet.

difference in shape makes it impossible for humans to digest cellulose as they can starch. Consequently, the cellulose we eat passes right through our digestive system unused.

Although it is not digestible, cellulose is still important to human diets. The cellulose in our diet is known as "fiber" (**FIGURE 2-27**). It is also appropriately called "roughage" because, as the cellulose of celery stalks and lettuce leaves passes through our digestive system, it scrapes the wall of the digestive tract. Its bulk and the scraping stimulate the more rapid passage of food and the unwanted, possibly harmful products of digestion through our intestines. That is why fiber reduces the risk of colon cancer and other diseases (but it is also why too much fiber can lead to diarrhea.)

Unlike humans, termites have some microorganisms living in their gut that are able to break down cellulose. That's why they can chew on wood and, with the help of the cellulose-digesting boarders in their gut, can break down the cellulose and extract usable energy from the freed glucose molecules.

TAKE-HOME MESSAGE 2•11

Some complex carbohydrates, including chitin and cellulose, cannot be digested by most animals. Such indigestible carbohydrates in the diet, called fiber, aid in digestion and have many health benefits.

④ Lipids store energy for a rainy day.

A well-insulated harbor seal, in Alaska.

2•12

Lipids are macromolecules with several functions, including energy storage.

Lipids are a second group of macromolecules important to all living organisms. Lipids, just like carbohydrates, are made primarily from atoms of carbon, hydrogen, and oxygen, but the atoms are in different proportions. In particular, while carbohydrates usually have two hydrogen atoms for every oxygen, lipids have more hydrogen atoms for each carbon. As a result, lipids have significantly more C–H bonds and contain significantly more stored energy.

> **Q** Why does a salad dressing made with vinegar and oil separate into two layers shortly after you shake it?

What exactly is a lipid? That's not as easy to answer as you might expect. Lipids come in a wide variety of structures. They don't have any unique subunits (such as the simple sugars that make up disaccharides and polysaccharides) or particular ratios of atoms that serve as defining features. Consequently, lipids are defined based on their physical characteristics. Most notably, lipids do not dissolve in water and are greasy to the touch—think of salad dressings.

Lipids are insoluble in water because, in sharp contrast to water, they consist mostly of hydrocarbons, which are non-polar. Non-polar molecules (or parts of molecules) tend to minimize their contact with water and are considered **hydrophobic** ("water-fearing"). Lipids cluster together when mixed with water, never fully dissolving. Molecules that readily form hydrogen bonds with water, on the other hand, are considered **hydrophilic** ("water-loving").

One familiar type of lipid is *fat,* the type most important in long-term energy storage and insulation (**FIGURE 2-28**). (Penguins and walruses can maintain relatively high body temperatures, despite living in very cold habitats, due to

INTRODUCTION TO LIPIDS

FIGURE 2-28 The many purposes of lipids.

TYPICAL FEATURES OF LIPIDS
- Non-polar molecules that do not dissolve in water
- Greasy to the touch
- Significant source of energy storage

THREE TYPES OF LIPIDS

FATS

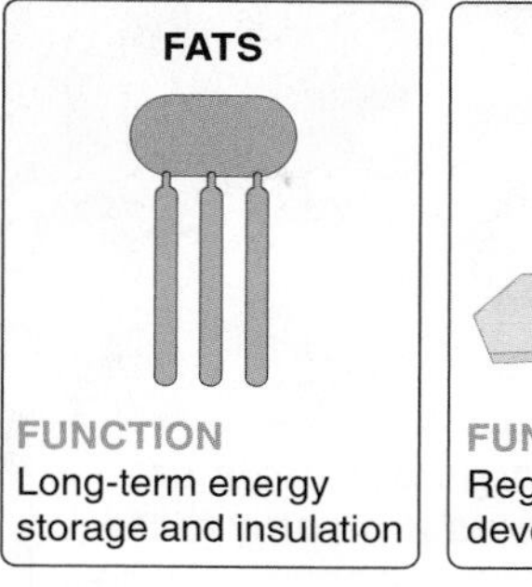

FUNCTION
Long-term energy storage and insulation

STEROLS

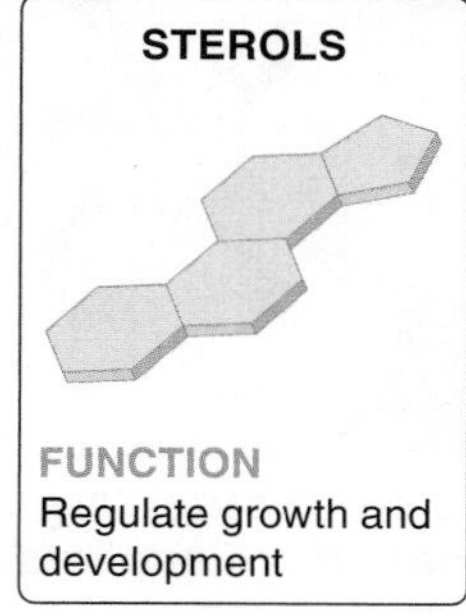

FUNCTION
Regulate growth and development

PHOSPHOLIPIDS

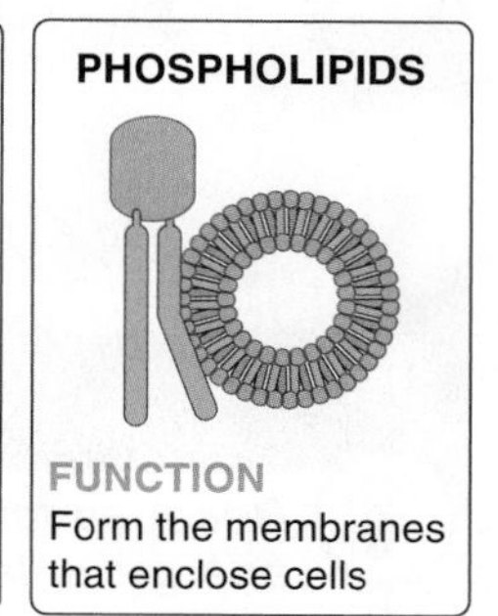

FUNCTION
Form the membranes that enclose cells

their thick layer of insulating fat.) Lipids also include *sterols,* which include *cholesterol* and many of the sex hormones that play regulatory roles in animals, and *phospholipids,* which form the membranes that enclose cells. We examine each type of lipid below.

TAKE-HOME MESSAGE 2•12

Lipids are insoluble in water and greasy to the touch. They are valuable to organisms for long-term energy storage and insulation, in membrane formation, and as hormones.

2•13

Fats are tasty molecules too plentiful in our diets.

All fats have two distinct components: they have a "head" region and two or three long "tails" (**FIGURE 2-29**). The head region is a small molecule called **glycerol.** It is linked to "tail" molecules known as **fatty acids.** A fatty acid is simply a long hydrocarbon—that is, a chain of carbon molecules, often a dozen or more, linked together with one or two hydrogen atoms attached to each carbon.

The fats in most foods we eat are **triglycerides,** which are fats having three fatty acids linked to the glycerol molecule. For this reason, the terms "fats" and "triglycerides" are often used interchangeably. Triglycerides that are solid at room temperature are generally called "fats," while those that are liquid at room temperature are called "oils."

Fat molecules contain much more stored energy than carbohydrate molecules. That is, the chemical breakdown of fat

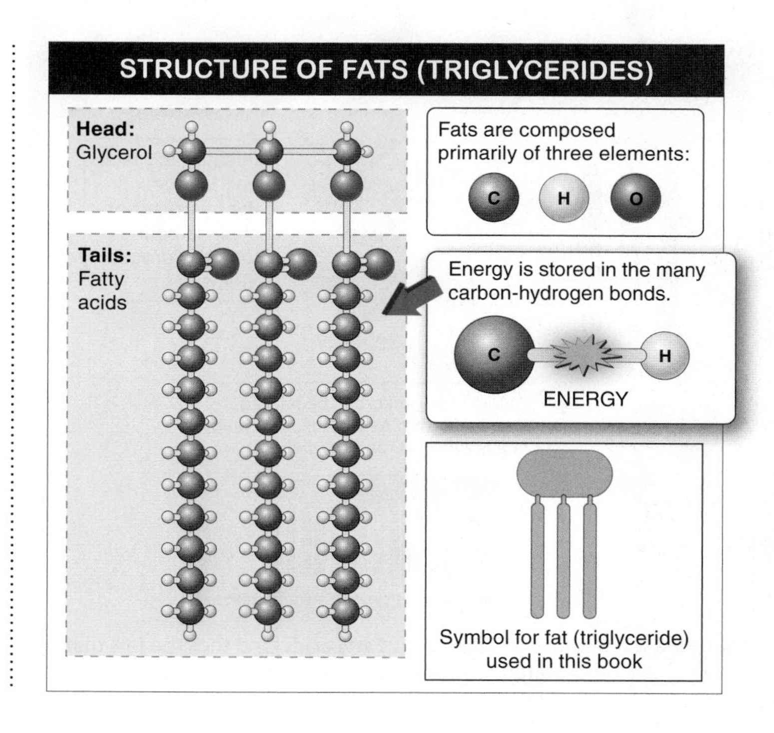

FIGURE 2-29 Triglycerides have glycerol heads and fatty acid tails.

FIGURE 2-30 **Animals (including humans!) prefer the taste of fats.**

molecules releases significantly more energy. A single gram of carbohydrate stores about 4 calories of energy, while the same amount of fat stores about 9 calories—not unlike the difference between a $5 bill and a $10 bill. Because fats store such a large amount of energy, animals have evolved a strong taste preference for fats over other energy sources (**FIGURE 2-30**). Organisms evolving in an environment of uncertain food supply will build the largest surplus by consuming molecules that hold the most amount of energy in the smallest mass. This feature helped the earliest humans to survive, millions of years ago, but today puts us in danger from the health risks of obesity now that fats are all too readily available.

An important distinction is made between "saturated" and "unsaturated" fats (**FIGURE 2-31**). These terms refer to the hydrocarbon chain in the fatty acids. If each carbon atom in the hydrocarbon chain of a fatty acid is bonded to two hydrogen atoms, the fat molecule carries the maximum

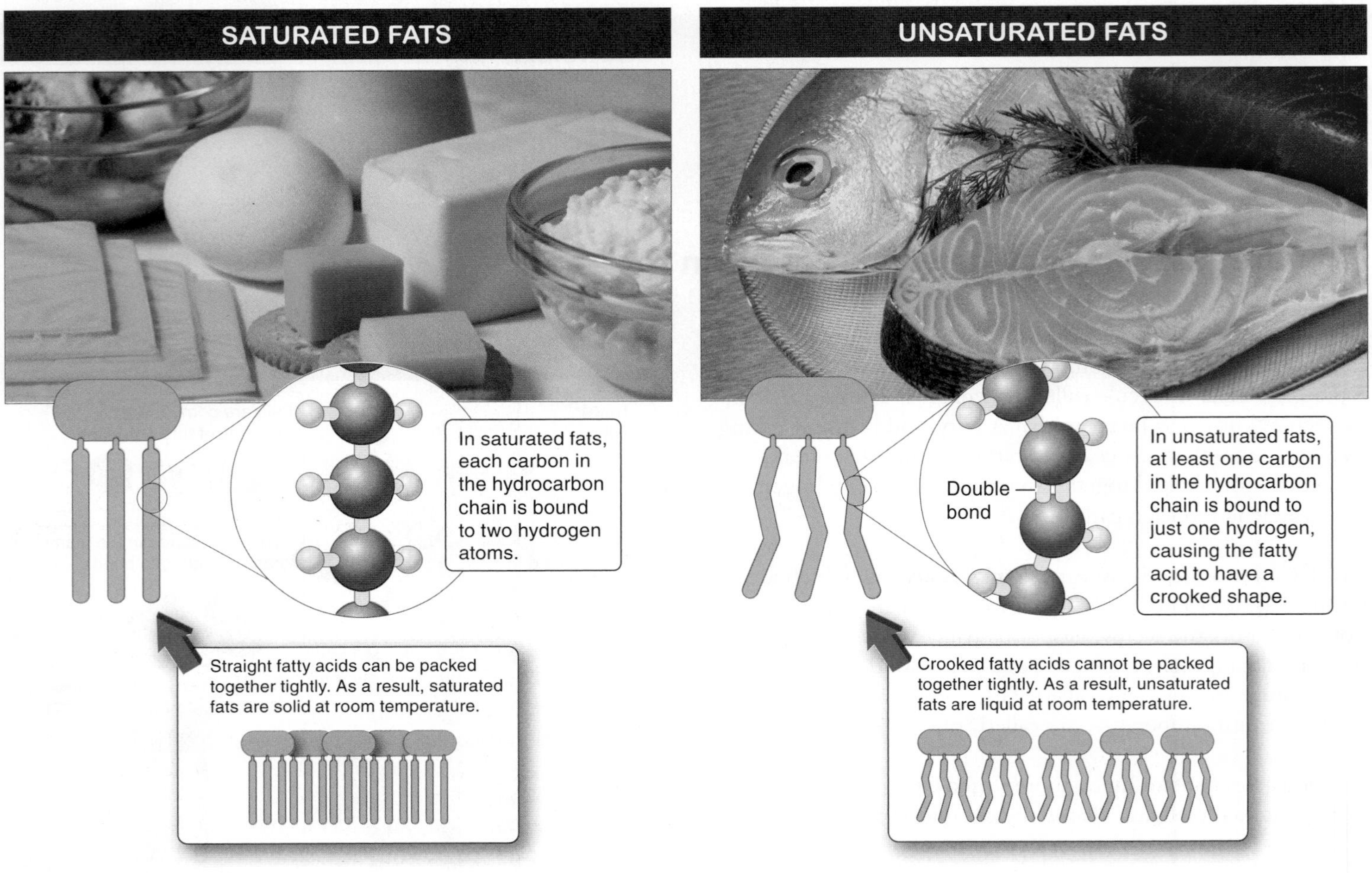

SATURATED FATS

In saturated fats, each carbon in the hydrocarbon chain is bound to two hydrogen atoms.

Straight fatty acids can be packed together tightly. As a result, saturated fats are solid at room temperature.

UNSATURATED FATS

Double bond

In unsaturated fats, at least one carbon in the hydrocarbon chain is bound to just one hydrogen, causing the fatty acid to have a crooked shape.

Crooked fatty acids cannot be packed together tightly. As a result, unsaturated fats are liquid at room temperature.

FIGURE 2-31 **Degrees of saturation.** Fatty acids (and thus the fats that contain them) can be unsaturated or saturated; unsaturated fats can be saturated artificially, making them tastier but less healthful.

Q Chocolate chip cookie recipes call for some lipids. How will the "chewy-ness" of the cookies differ depending on whether you use butter or vegetable oil as the lipid? Which cookies will be better for your health?

number of hydrogen atoms and is said to be a **saturated fat.** Most animal fats, including those found in meat and eggs, are saturated. They are not essential to your health and, because they accumulate in your bloodstream and can narrow the vessel walls, they can cause heart disease and strokes.

An **unsaturated fat** is one in which some of the carbon atoms are bound to only a single hydrogen. Most plant fats are unsaturated. Unsaturated fats may be *mono-unsaturated* (if a fatty acid hydrocarbon chain has only one pair of neighboring carbon atoms in an unsaturated state—that is, has only one double bond) or *polyunsaturated* (if more than one pair of carbons is unsaturated—there's more than one double bond). Unsaturated fats are still high in calories, but because they can lower cholesterol, they are generally preferable to saturated fats. Foods high in unsaturated fats include avocados, peanuts, and olive oil. Relative to other animals, fish tend to have less saturated fat.

The shapes of unsaturated fat molecules and saturated fat molecules are different. When saturated, the hydrocarbon tails of the fatty acids all line up very straight and the fat molecules can be packed together tightly. The tight packing causes the fats, such as butter, to be solid at room temperature. When unsaturated, the fatty acids have kinks in the hydrocarbon tails and the fat molecules cannot be packed together as tightly (see Figure 2-31). Consequently, unsaturated fats, such as canola oil and vegetable oil, do not solidify as easily and are liquid at room temperature.

The ingredient list for many snack foods includes "partially hydrogenated" vegetable oils. The hydrogenation of an oil means that a liquid, unsaturated fat has had hydrogen atoms added to it so that it becomes more saturated. This can be useful in creating a food with a more desirable texture, since increasing a fat's degree of saturation changes its consistency and makes it more solid at room temperatures. By attaining just the right degree of saturation, it is possible to create foods, such as chocolate, that are near the border of solid and liquid and "melt in your mouth" (**FIGURE 2-32**). Unfortunately, hydrogenation also makes the food less

Q Many snack foods contain "partially hydrogenated" vegetable oils. Why might it be desirable to add hydrogen atoms to a vegetable oil?

HYDROGENATION

Hydrogenation is the artificial addition of hydrogen atoms to an unsaturated fat. This can improve a food's taste, texture, and shelf-life.

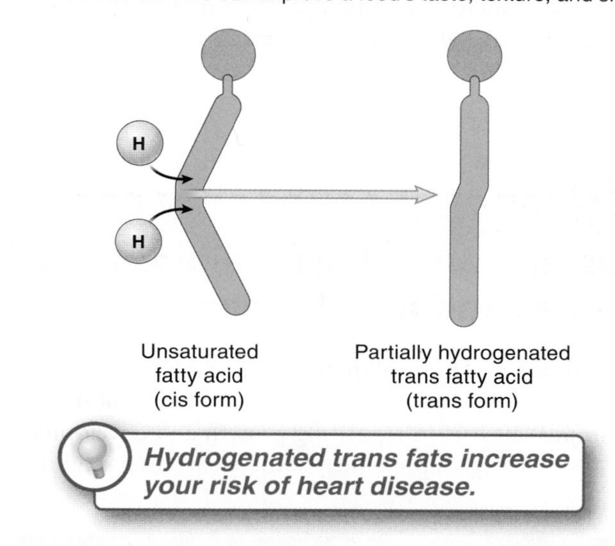

FIGURE 2-32 Hydrogenation improves a food's taste, texture, and shelf-life (but at a cost).

healthful because saturated fats are less reactive—your body is less likely to break them down—and so are more likely to accumulate in your blood vessels, increasing the risk of heart disease.

Hydrogenation of unsaturated fats is problematic from a health perspective because it creates **trans fats,** the "trans" referring to the unusual orientation of the added hydrogen atoms, which differs from that in other dietary fats—which have their hydrogens in an orientation called "cis." Trans fats cause your body to produce more cholesterol, further raising the risk of heart disease, and they also reduce your body's production of a type of cholesterol that protects against heart disease.

Olestra is a recently developed "fake fat" chemical that gives foods the taste of fat, without adding the calories of fats. What chemical structure might make this possible?

Because of the well-documented links between dietary fats and heart disease, many people are trying to reduce their fat intake. "Fake fats" make this possible. They are designed to be similar to fats in taste and texture, but have one big difference: they cannot be digested by humans. One such "fake fat" is olestra. Olestra, instead of being a triglyceride fat with three fatty acid tails linked to a glycerol molecule, has eight separate hydrocarbon fatty acids attached to a molecule of sucrose. This octopus-like molecule allows the fatty acids to stimulate their usual taste buds on your tongue, telling your brain that you are eating a fat. The complex shape of the molecule, however, prevents your body's digestive chemicals from grabbing onto it and breaking it down. As a consequence, it passes through your digestive system without being digested. It's not a perfect solution, however. Olestra reduces absorption of some vitamins, and in some people causes abdominal cramping.

TAKE-HOME MESSAGE 2•13

Fats, including the triglycerides common in the food we eat, are one type of lipid. Characterized by long hydrocarbon tails, fats effectively store energy in the many carbon-hydrogen and carbon-carbon bonds. Their caloric density is responsible for humans' preferring fats to other macromolecules in the diet, and is responsible for their association with obesity and illness in the modern world.

2•14

Cholesterol and phospholipids are used to build sex hormones and membranes.

Not all lipids are fats, nor do lipids necessarily function in energy storage. A second group of lipids, called the **sterols,** plays an important role in regulating growth and development (**FIGURE 2-33**). This group includes some very familiar lipids: cholesterol and the steroid hormones such as testosterone and estrogen. These molecules are all modifications on one basic structure formed from four interlinked rings of carbon atoms.

STEROLS

CHOLESTEROL
- Important component of cell membranes in animals
- Can attach to vessel walls and cause them to thicken, which may lead to high blood pressure, stroke, and heart attack

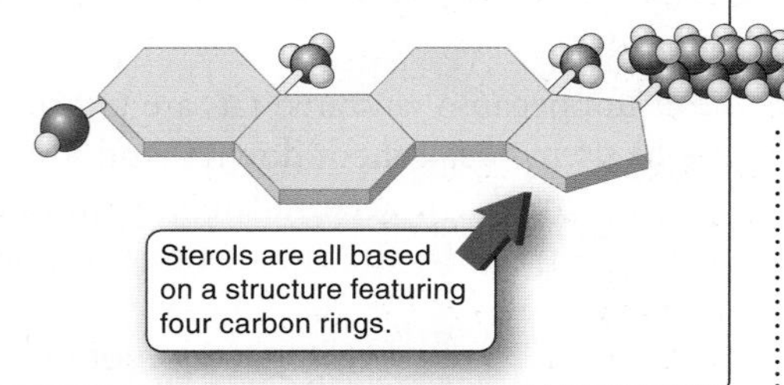

STEROID HORMONES
- Regulate sexual development, maturation, and sex cell production
- Estrogen influences memory and mood
- Testosterone stimulates muscle growth

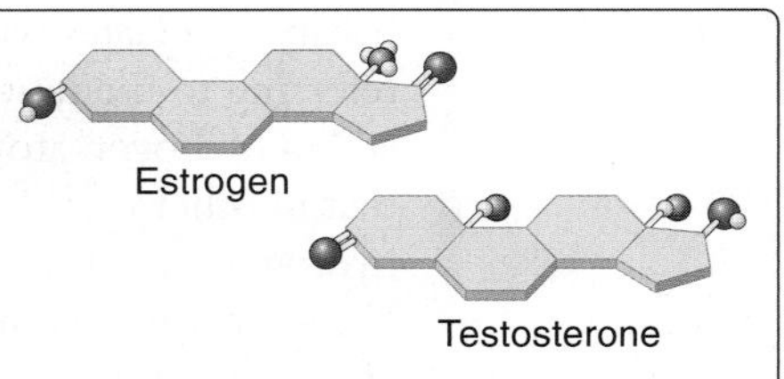

FIGURE 2-33 Not all lipids are for energy storage. Cholesterol, estrogen, and testosterone are all lipids.

Cholesterol is an important component of most cell membranes. For this reason, it is an essential molecule for living organisms. It has a bad reputation in most Western cultures, though, that is mostly well deserved. When we ingest too much cholesterol (present in animal-based foods such as egg yolks, red meat, and cream) and high levels of cholesterol circulate in our bloodstream, the cholesterol can attach to blood vessel walls and cause them to thicken. In turn, this thickening can lead to high blood pressure, a major contributor to strokes and heart attacks. For these reasons, nutritionists advise limiting the consumption of foods high in cholesterol.

Surprisingly, however, most of the cholesterol in our blood doesn't come directly from the food we consume. Instead, cells in our liver produce almost 90% of the circulating cholesterol. How do they do it? By transforming the saturated fats in our diet. The best way to reduce cholesterol levels, then, may be to consume fewer saturated fats.

The steroid hormones estrogen and testosterone are built through slight chemical modifications to cholesterol. These hormones are among the primary molecules that direct and regulate sexual development, maturation, and sperm and egg production. In both males and females, estrogen influences memory and mood, among other traits. Testosterone has numerous effects, one of which is to stimulate muscle growth. As a consequence, athletes often take synthetic variants of testosterone to increase their muscularity. But the use of these supplements is often accompanied by dangerous side effects, including extreme aggressiveness (" 'roid rage"), high cholesterol, and, following long-term use, cancer. As a consequence, nearly all athletic organizations have banned their use (**FIGURE 2-34**).

Phospholipids and waxes are also lipids. **Phospholipids** (**FIGURE 2-35**) are the major component of the membrane that surrounds the contents of a cell and controls the flow of chemicals into and out of the cell. They have a structure similar to fats, but with two differences: they contain a phosphorus atom (hence *phospho*lipids) and they have two fatty acid chains rather than three. We will explore the significant role of phospholipids in cell membranes in the next chapter.

FIGURE 2-34 Dangerous bulk. Steroids can increase muscularity, but with serious health consequences.

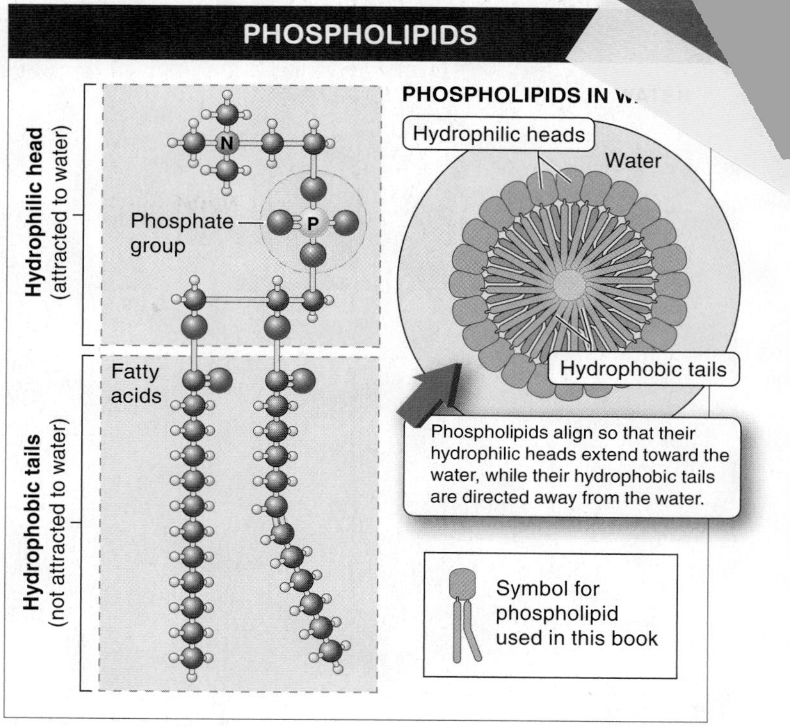

PHOSPHOLIPIDS

Hydrophilic head (attracted to water)

Hydrophobic tails (not attracted to water)

N

Phosphate group

P

Fatty acids

PHOSPHOLIPIDS IN W...

Hydrophilic heads

Water

Hydrophobic tails

Phospholipids align so that their hydrophilic heads extend toward the water, while their hydrophobic tails are directed away from the water.

Symbol for phospholipid used in this book

FIGURE 2-35 Lipid versatility. Phospholipids have important roles in many organisms.

Waxes resemble fats but have only one long-chain fatty acid linked to the glycerol head of the molecule. Because the fatty acid chain is highly non-polar, waxes are strongly hydrophobic; that is, these molecules do not mix with water but repel it. Their water resistance accounts for their presence as a natural coating on the surface of many plants and in the outer coverings of many insects. In both cases, the waxes prevent the plants and animals from losing the water essential to their life processes. Many birds, too, have a waxy coating on their wings, keeping them from becoming water-logged when they get wet.

TAKE-HOME MESSAGE 2•14

Cholesterol and phospholipids are lipids that are not fats. Both are important components in cell membranes. Cholesterol also serves as a precursor to steroid hormones, important regulators of growth and development.

⑤ Proteins are body-building molecules.

Hair and feathers are built from proteins.

2•15

Proteins are versatile macromolecules that serve as building blocks.

You can't look at a living organism and not see proteins (**FIGURE 2-36**). Inside and out, **proteins** are the chief building blocks of all life. They make up skin and feathers and horns. They make up bones and muscles. In your bloodstream, proteins fight invading microorganisms and stop you from bleeding to death from a shaving cut. Proteins control the levels of sugar and other chemicals in your bloodstream and carry oxygen from one place in your body to another. And in just about every cell in every living organism, proteins called **enzymes** initiate and assist every chemical reaction that occurs.

Although proteins perform several very different types of functions, all are built in the same way and from the same raw materials in all organisms. In the English language, every sentence is made up of words and every word is formed from one or more of the 26 letters of the alphabet. With 26 letters we can write anything, from sonnets to cookbooks to biology textbooks. Proteins, too, are constructed from a sort of alphabet. Instead of 26 letters there are 20 molecules, known as **amino acids.** Unique combinations of these 20 amino acids are strung together, like beads on a string, and the resulting protein has a unique structure and chemical behavior.

Let's look more closely at the structure of the amino acids in the protein alphabet. They all have the same basic two-part structure: one part is the same in all 20 amino acids, and the other part is unique, differing in each of the 20 amino acids.

Proteins contain the same familiar atoms as carbohydrates and lipids—carbon, hydrogen, and oxygen—but differ in an important way: they also contain nitrogen. At the center of every amino acid is a carbon atom, with its four covalent bonds (**FIGURE 2-37**). One bond attaches the carbon to something called a **carboxyl group,** which is a carbon bonded to two oxygen atoms. The second bond attaches the central carbon to

PROTEIN DIVERSITY

Proteins perform a variety of different functions. They all, however, are built the same way and from the same raw materials in organisms.

STRUCTURAL
Hair, fingernails, **feathers**, horns, cartilage, tendons

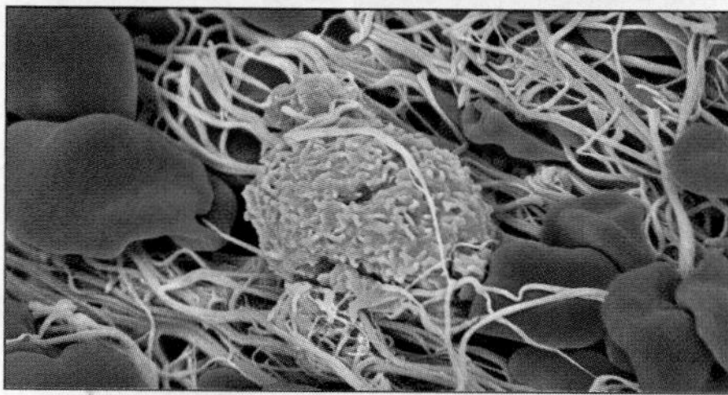

PROTECTIVE
Help fight invading microorganisms, coagulate **blood**

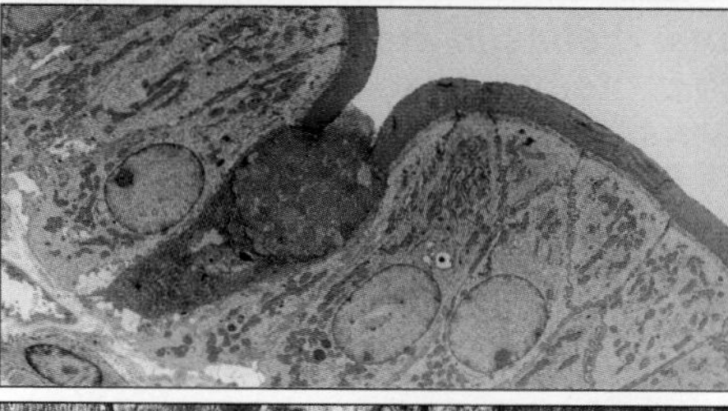

REGULATORY
Control cell activity, constitute some **hormones**

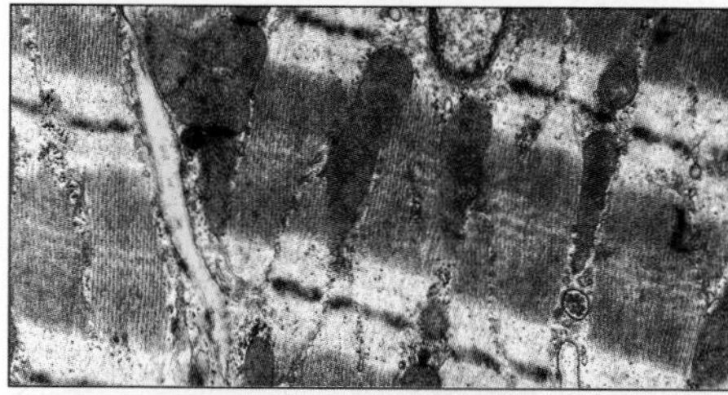

CONTRACTILE
Allow **muscles** to contract, heart to pump, sperm to swim

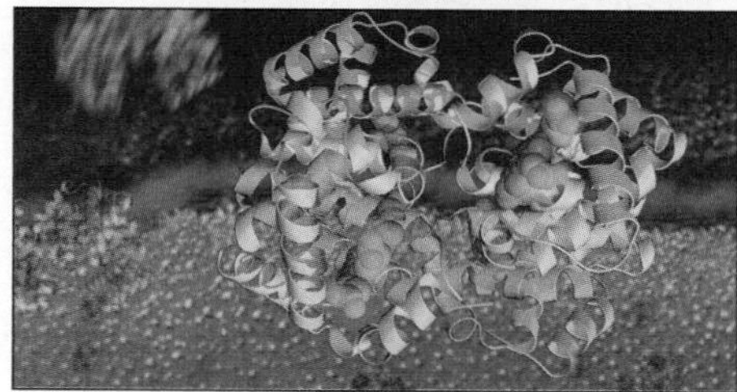

TRANSPORT
Carry molecules such as **oxygen** around your body

FIGURE 2-36 Proteins everywhere! Proteins are the chief building blocks of all organisms.

a single hydrogen atom. The third bond attaches the central carbon to an **amino group,** which is a nitrogen atom bonded to three hydrogen atoms. These components—the central carbon with its attached hydrogen atom, carboxyl group, and amino group—are the foundation that identifies a molecule as an amino acid and, as multiple amino acids are joined together, forms the "backbone" of the protein.

AMINO ACIDS

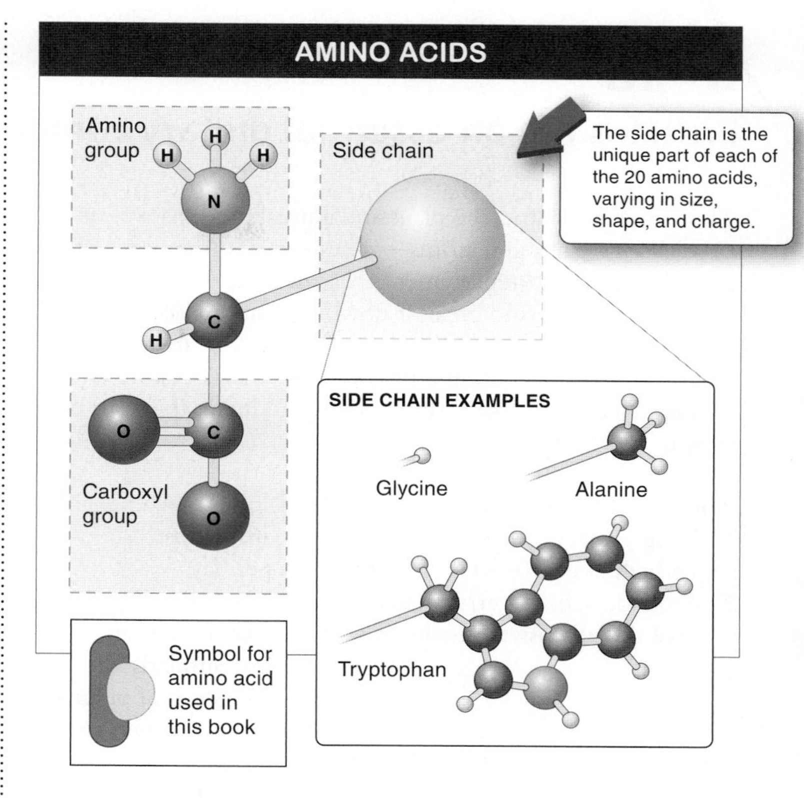

FIGURE 2-37 Amino acid structure. Amino acids are made up of an amino group, carboxyl group, and a side chain.

The fourth bond of the central carbon attaches to a functional group or side chain. This side chain is the unique part of each of the 20 amino acids. In the simplest amino acid, glycine, for example, the side chain is simply a hydrogen atom. In other amino acids, the side chain is a single CH_3 group or three or four such groups. Most of the side chains include both hydrogen and carbon, and a few include nitrogen or sulfur atoms. The side chain determines an amino acid's chemical properties, such as whether the amino acid molecule is polar or non-polar.

TAKE-HOME MESSAGE 2•15

Unique combinations of 20 amino acids give rise to proteins, the chief building blocks of the physical structures that make up all organisms. Proteins perform myriad functions, from assisting chemical reactions to causing blood clotting to building bones to fighting microorganisms.

2•16

Proteins are an essential dietary component.

The atoms present in the plant and animal proteins we eat—especially the nitrogen atoms—are essential to the constant growth, repair, and replacement that take place in our bodies. As we eat protein and break it down into its amino acids through digestion, our bodies are collecting the amino acids needed for various building projects. Proteins also store energy in their bonds and, like carbohydrates and lipids, they can be used to fuel living processes.

The amount of protein we need depends on the extent of the building projects underway at any given time. Most individuals need 40–80 grams of protein per day. Bodybuilders, however, may need 150 grams a day or more to achieve the extensive muscle growth stimulated by their training; similarly, the protein needs of pregnant or nursing women are very high.

Q Food labels indicate an item's protein content. Why is this insufficient for determining whether you are protein deficient, even if your protein intake exceeds your recommended daily amount?

Contrary to the impression you might get from the labels you see on food packaging, all proteins are not created equal (**FIGURE 2-38**). Every different protein has a different composition of amino acids. And while our bodies can manufacture certain amino acids as they are needed, many other amino acids must come from our diet. Those that we must get from our diet—about half of the 20 amino acids—are called "essential amino acids." For this reason, we shouldn't just speak of needing "*x* grams of protein per day." We need to consume all of the essential amino acids every day.

Many foods, containing "complete proteins," have all of the essential amino acids. Animal products such as milk, eggs, fish, chicken, and beef tend to provide complete proteins. Most vegetables, fruits, and grains, on the other hand, more often contain "incomplete proteins," which do not have all the essential amino acids. If you consume only one type of incomplete protein in your diet, you may be deficient in one or more of the essential amino acids. But two incomplete proteins that are "complementary proteins," when eaten together, can provide all the essential amino acids. Traditional dishes in many cultures often include such pairings (see Figure 2-38). Examples are corn and beans in Mexico, rice and lentils in India, and rice and black-eyed peas in the southern United States.

Traditional dishes in many cultures combine proteins, bringing together all essential amino acids.

FIGURE 2-38 All proteins are not created equal. Some foods have "complete proteins" with all the essential amino acids. Other foods have "incomplete proteins" and we must consume proteins from multiple sources to get all the essential amino acids.

TAKE-HOME MESSAGE 2•16

Twenty amino acids make up all the proteins necessary for growth, repair, and replacement of tissue in living organisms. Of these amino acids, eight are essential for humans: they cannot be synthesized by the body so must be consumed in the diet. Complete proteins contain all eight essential amino acids, while incomplete proteins do not.

2•17

Proteins' functions are influenced by their three-dimensional shape.

Proteins are formed by linking individual amino acids together with a **peptide bond,** in which the amino group of one amino acid is bonded to the carboxyl group of another. Two amino acids joined together form a *dipeptide* and several amino acids joined together form a *polypeptide.* The sequence of amino acids in the polypeptide chain is called the **primary structure** of the protein and can be compared to the sequence of letters that spells a specific word (**FIGURE 2-39**).

Amino acids in a polypeptide chain don't remain in a simple straight line like beads on a string. The chain begins to fold as side chains come together and hydrogen bonds form between various atoms in the chain. The two most common patterns of hydrogen bonding between amino acids cause the chain to either twist in a corkscrew-like shape or form a zigzag folding pattern. This hydrogen bonding between amino acids gives a protein its **secondary structure.**

The protein eventually folds and bends upon itself, and additional bonds continue to form between atoms in the side chains of amino acids that are near each other. Eventually, the protein folds into a unique and complex three-dimensional shape called its **tertiary structure.** The exact form comes about as the secondary structure folds and bends, bringing together amino acids that then form bonds such as hydrogen bonds or covalent sulfur-sulfur bonds (see Figure 2-39).

Some protein molecules have a **quaternary structure** in which two or more polypeptide chains are held together by bonds between amino acids in the different chains. Hemoglobin, the protein molecule that carries oxygen from the lungs to the cells where it is needed, is made from four polypeptide chains, two "alpha" chains and two "beta" chains.

Some proteins are attached to other types of macromolecules. *Lipoproteins,* for example, circulate in the bloodstream carrying fats. They are formed when cholesterol and a triglyceride (both lipids) combine with a protein. *Glycoproteins* are combinations of carbohydrates and proteins. These are found on the surfaces of nearly all animal cells and play a role in helping the immune system to distinguish between your own cells and foreign cells. (We learn more about glycoproteins in the next chapter, which discusses cells.)

STRUCTURE OF PROTEINS

PRIMARY STRUCTURE
The sequence of amino acids in a polypeptide chain, similar to the sequence of letters that spell out a specific word

SECONDARY STRUCTURE
The corkscrew-like twists or pleated folds formed by hydrogen bonds between amino acids in the polypeptide chain

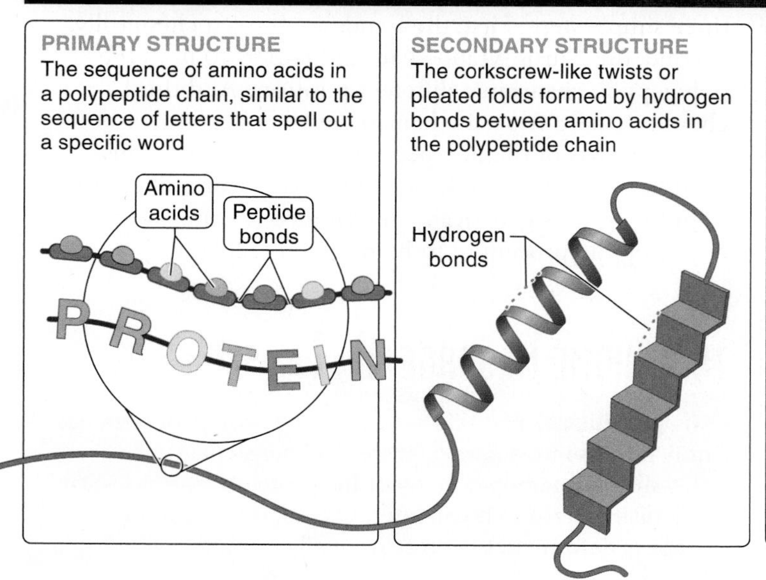

TERTIARY STRUCTURE
The complex three-dimensional shape formed by multiple twists and bends in the polypeptide chain based on interactions between the side chains

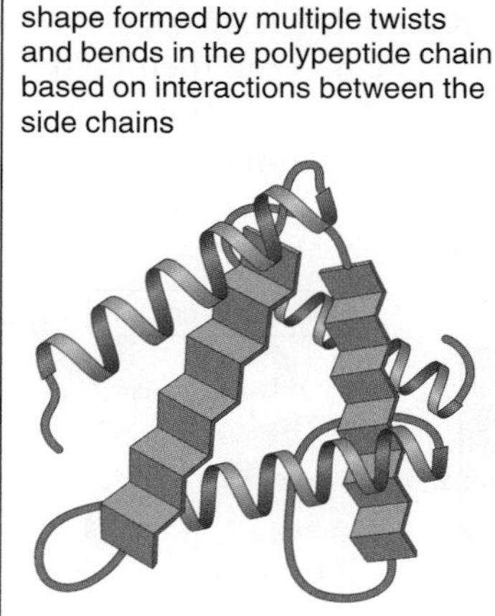

QUATERNARY STRUCTURE
Two or more polypeptide chains bonded together

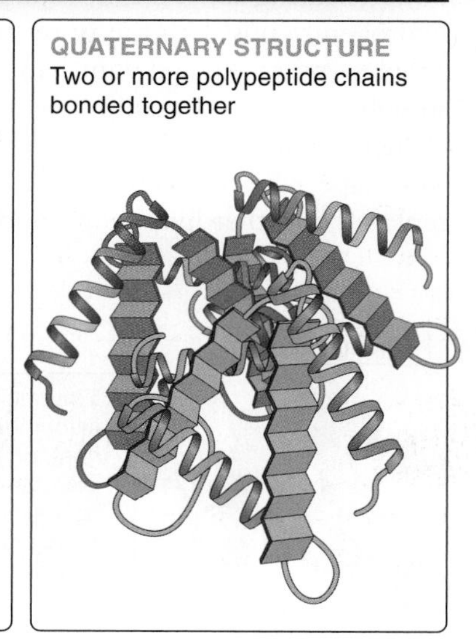

FIGURE 2-39 **Protein structure.** The functions of proteins are influenced by their three-dimensional shape.

Egg whites contain a lot of protein. Why does beating them change their texture, making them stiff?

The overall shape of a protein molecule determines its function—how it behaves and the other molecules it interacts with. For proteins to function properly, they must retain their three-dimensional shape. When their shapes are deformed, they usually lose their ability to function. We can see proteins deformed when we fry an egg. The heat breaks the hydrogen bonds that give the proteins their shape. The proteins in the clear egg white unfold, losing their secondary and tertiary structure. This disruption of protein folding is called **denaturation** (**FIGURE 2-40**).

Almost any extreme environment will denature a protein. Take a raw egg, for instance, and crack it into a dish containing baking soda or rubbing alcohol. Both chemicals are sufficiently extreme to turn the clear protein opaque white, as in fried egg whites.

Why is wet hair easier to style than dry hair?

Hair is a protein whose shape most of us have modified at one time or another. Styling hair—whether curling or straightening it—involves altering some of the hydrogen bonds between the amino acids that make up the hair protein, changing its tertiary structure. When your hair gets wet, the water is able to disrupt some of the hydrogen bonds, causing some amino acids in the protein to form hydrogen bonds with the water molecules instead. This enables you to change your hair's shape—making it straighter or, if you manipulate it around curlers, making it curlier—if you style it while it's wet. The hair can then hold this shape when it dries as the hydrogen bonds to water are replaced by other hydrogen bonds between amino acids of the hair protein as the water evaporates. Once your hair gets wet again, however, unless it is combed, brushed, or wrapped in a different style, it will return to its natural shape.

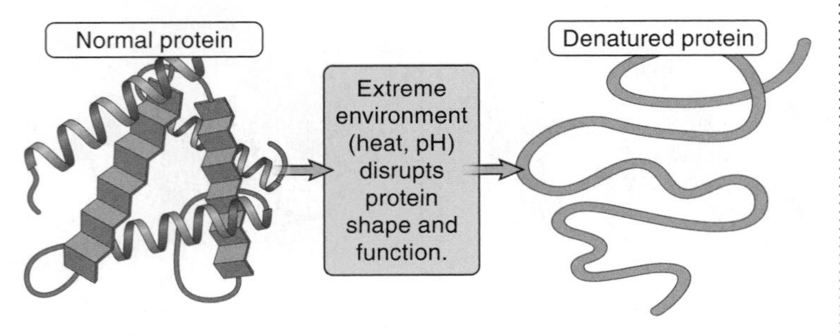

FIGURE 2-40 Denaturation. When proteins are unfolded, they lose their function.

FIGURE 2-41 Curly or straight? Proteins determine it!

Why do some people have curly hair and others have straight hair?

Whether your hair is straight or curly or somewhere in between also depends on your hair protein's amino acid sequence and the three-dimensional shape it confers (**FIGURE 2-41**). This amino acid sequence is something you're born with (that is, it's genetically determined). The chains are more or less coiled, depending on the extent of covalent and hydrogen bonding between different parts of the coil. Many hair salons make use of the ability to alter covalent bonds to change hair texture semi-permanently. They are able to do this in three simple steps. First, the bonds are broken chemically. Second, the hair is wrapped around curlers to hold the polypeptide chains in a different position. And third, chemicals are put on the hair to create new covalent bonds between parts of the polypeptide chains. The hair thus becomes locked in a new position. (New hair will continue to grow with its genetically determined texture, of course, requiring the procedure to be repeated regularly.)

TAKE-HOME MESSAGE 2·17

The particular amino acid sequence of a protein determines how it folds into a particular three-dimensional shape. This shape determines many of the protein's features, such as which molecules it will interact with. When a protein's shape is deformed, the protein usually loses its ability to function.

2•18

Enzymes are proteins that initiate and speed up chemical reactions.

Protein shape is particularly critical in **enzymes,** molecules that help initiate and accelerate the chemical reactions in our bodies. Enzymes emerge unchanged—in their original form—when the reaction is complete and thus can be used again and again. Here's how they work.

Think of an enzyme as a big piece of popcorn. Its tertiary or quaternary structure gives it a complex shape with lots of nooks and crannies. Within one of those nooks is a small area called the "active site" (**FIGURE 2-42**). Based on the chemical properties of the atoms lining this pocket, the active site provides a place for the participants in a chemical reaction, the reactants or **substrate** molecules, to nestle briefly.

Enzymes are very choosy: they bind only with their appropriate substrate molecules, much like a lock that can be opened with only one key (see Figure 2-42). The exposed atoms in the active site have electrical charges that attract rather than repel the substrate molecules, and only the substrate molecules can fit into the active-site groove.

Once the substrate is bound to the active site, a reaction can take place—and usually does so very quickly. An enzyme can help to bring about the reaction in a variety of ways. These include:

1. Stressing, bending, or stretching critical chemical bonds, increasing the likelihood of their breaking.
2. Directly participating in the reaction, perhaps temporarily sharing one or more electrons with the substrate molecule, thereby giving it chemical features that increase its ability to make or break other bonds.
3. Creating a "micro-habitat" that favors the reaction. For instance, the active site might be a water-free, non-polar environment, or it might have a slightly higher or lower pH than the surrounding fluid. Both of these slight alterations might increase the likelihood that a particular reaction occurs.
4. Simply orienting or holding substrate molecules in place so that they can be modified.

Sometimes a protein "word" is misspelled—that is, the sequence of amino acids is incorrect. If an enzyme is altered even slightly, the active site may change and the enzyme no longer functions. Slightly modified, non-functioning enzymes

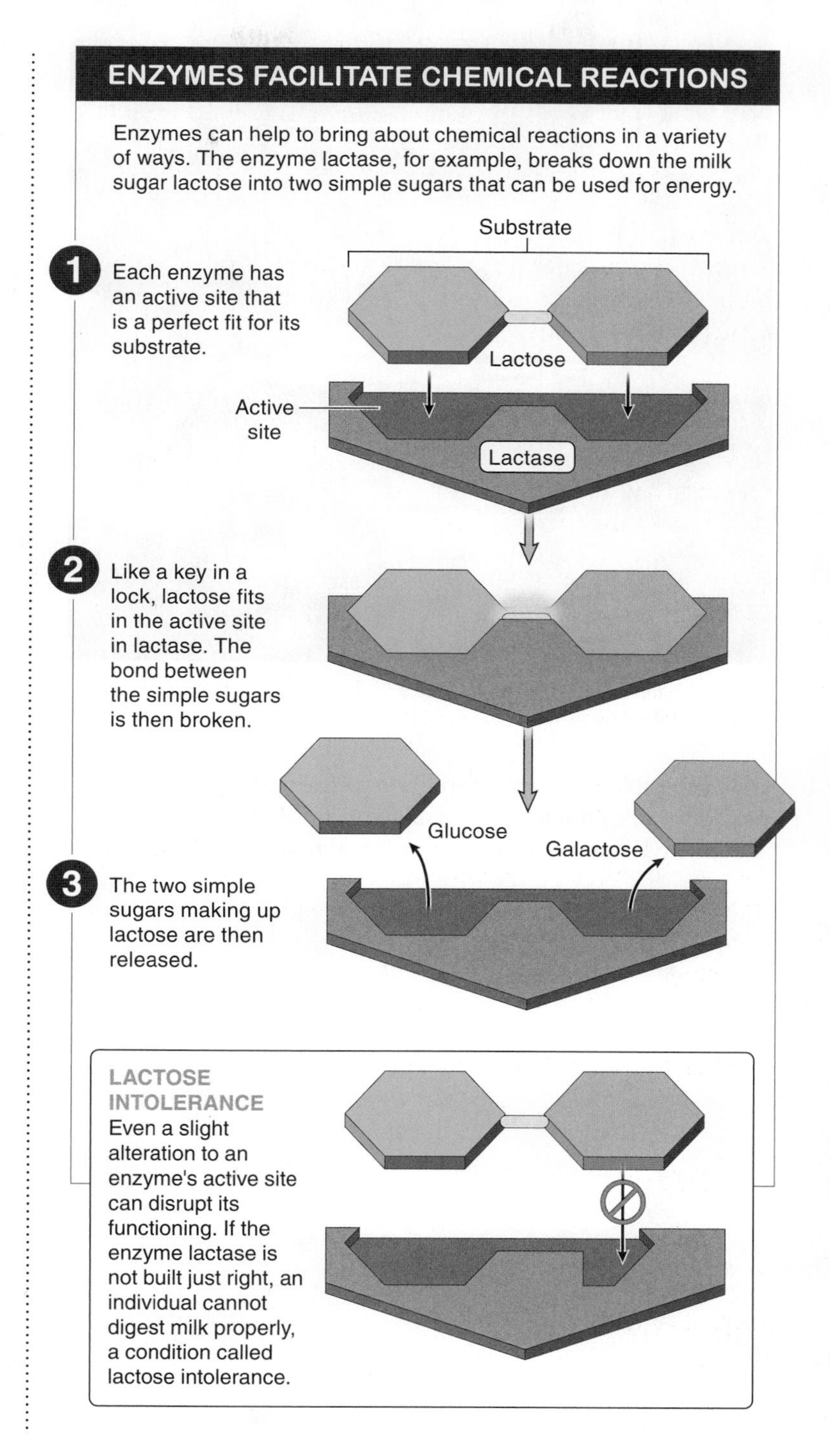

FIGURE 2.42 Lock and key. Enzymes are *very* specific about which molecules and reactions they will catalyze.

Without a functioning version of the enzyme lactase, some people are unable to break down the disaccharide lactose in milk.

are responsible for a large number of diseases and physiological problems; an example is the inability to break down the amino acid phenylalanine (in a condition known as phenylketonuria).

Q Why do some adults get sick when they drink milk?

One protein "misspelling" is responsible for the condition called lactose intolerance. Normally, during digestion, the lactose in milk is broken down into its component parts, glucose and galactose (see Figure 2-42). These simple sugars are then used for energy. But some people are unable to break the bond linking the two simple sugars because they lack a functioning version of the enzyme lactase that assists in this process. Consequently, the lactose passes through their stomach and small intestine undigested. Then, when it reaches the large intestine, bacteria living there consume the lactose. The problem is that, as they break down the lactose, they produce some carbon dioxide and other gases. These gases are trapped in the intestine and lead to severe discomfort.

These unpleasant symptoms can be avoided by not consuming milk, cheese, yogurt, ice cream, or any other dairy products, but they can also be avoided by taking a pill containing the enzyme lactase. It doesn't matter how the enzyme gets into your digestive system; as long as it's there the lactose in the milk can be broken down.

TAKE-HOME MESSAGE 2•18

Enzymes are proteins that help initiate and speed up chemical reactions. They aren't permanently altered in the process, but rather can be used again and again.

⑥ Nucleic acids store the information on how to build and run a body.

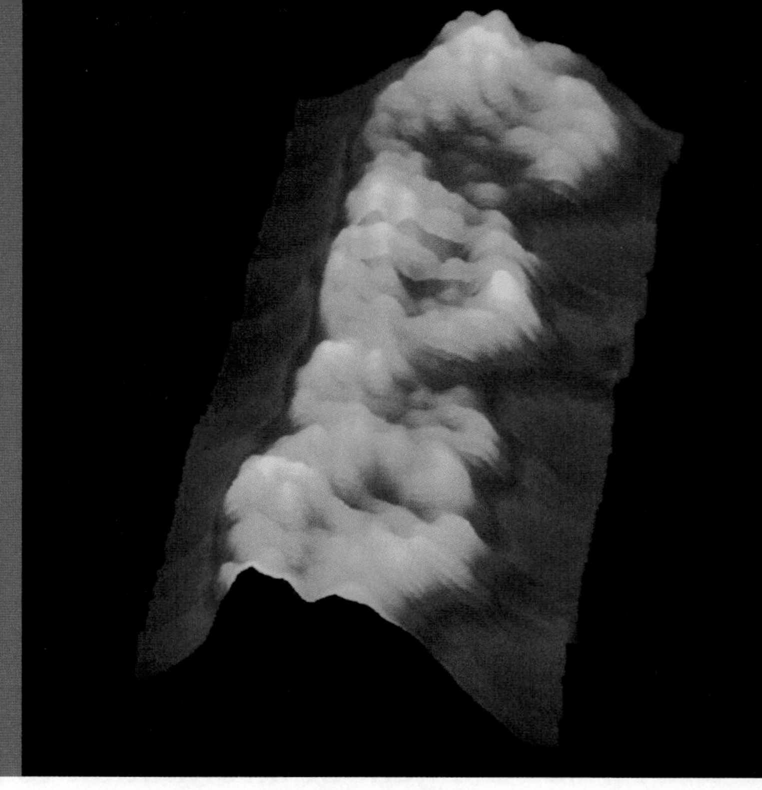

Three turns of the DNA double helix (from a scanning tunneling micrograph).

2•19

Nucleic acids are macromolecules that store information.

We have examined three of life's macromolecules: carbohydrates, lipids, and proteins. We turn our attention now to the fourth: **nucleic acids,** macromolecules that store information and are made up of individual units called **nucleotides.** All nucleotides have three components: a molecule of sugar, a phosphate group (containing a phosphorus atom bound to four oxygen atoms), and a nitrogen-containing molecule (**FIGURE 2-43**).

There are two types of nucleic acids: **deoxyribonucleic acid** (**DNA**) and **ribonucleic acid** (**RNA**). Both play central roles in directing the production of proteins in living organisms, and by doing so play a central role in determining all of the inherited characteristics of an individual. In both types of nucleic acids, the molecule has the same type of backbone: a sugar molecule attached to a phosphate group attached to another sugar, then another phosphate, and so on. Attached to each sugar is one of the nitrogen-containing molecules called DNA **bases** (so named because of their chemical structure).

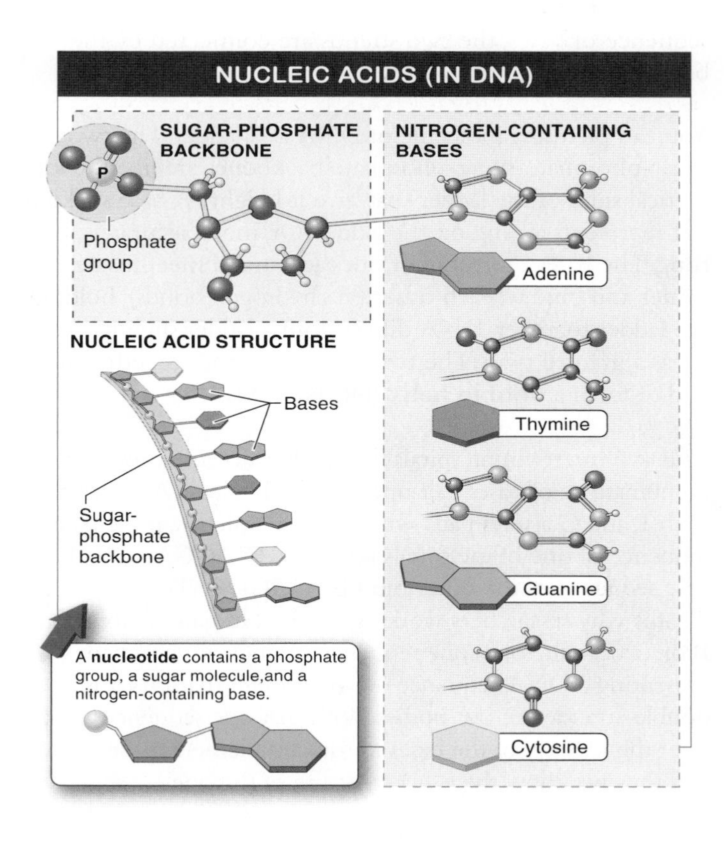

FIGURE 2-43 The molecules that carry genetic information. The structure of nucleic acids.

A 10-unit nucleic acid therefore would have 10 bases, one attached to each sugar within the sugar-phosphate-sugar-phosphate backbone. But the base attached to each sugar is not always the same. It can be one of several different bases. For this reason, a nucleic acid is often described by the sequence of bases attached to the sugar-phosphate-sugar-phosphate backbone.

Nucleic acids are able to store information by varying which base is attached at each position in the molecule. At each position in a molecule of DNA, for example, the base can be any one of four possible bases: adenine (A), thymine (T), guanine (G), or cytosine (C). Just as the meaning of a sentence is determined by which letters are strung together, the information in a molecule of DNA is determined by its sequence of bases. One molecule may have the sequence adenine, adenine, adenine, guanine, cytosine, thymine, guanine—abbreviated as AAAGCTG. Another molecule may have the sequence CGATTACCCGAT. Because the information differs in each case, so, too, does the protein for which the sequence codes, as we'll see.

TAKE-HOME MESSAGE 2•19

The nucleic acids DNA and RNA are macromolecules that store information in their unique sequences of nucleotides, their building-block molecules. Both nucleic acids play central roles in directing protein production in organisms.

2•20

DNA holds the genetic information to build an organism.

A molecule of DNA has two strands, each a sugar-phosphate-sugar-phosphate backbone with a base sticking out from each sugar molecule. The two strands wrap around each other, each turning in a spiral. Although each strand has its own sugar-phosphate-sugar-phosphate backbone and sequence of bases, the two strands are connected by the bases protruding from them.

You can picture a molecule of DNA as a ladder. The two sugar-phosphate-sugar-phosphate backbones are like the long vertical sides of the ladder that give it height. A base sticking out represents a rung on the ladder. Or, more accurately, half a rung. The bases protruding from each strand meet in the center and bind to each other (via hydrogen bonds), holding the ladder together. DNA differs from a ladder slightly, in that it has a gradual twist. The two spiraling strands together are said to form a **double helix** (**FIGURE 2-44**).

The two intertwining spirals fit together because only two combinations of bases pair up together. The base A always pairs with T, and C always pairs with G. Consequently, if the base sequence of one of the spirals is CCCCTTAGGAACC, the base sequence of the other must be GGGGAATCCTTGG. That is why researchers working on the Human Genome Project describe only one sequence of nucleotides when presenting a DNA sequence—even though that DNA is double-stranded in our bodies. With that one sequence, we can infer the identity of the bases in the complementary sequence and thus we know the exact structure of the nucleic acid.

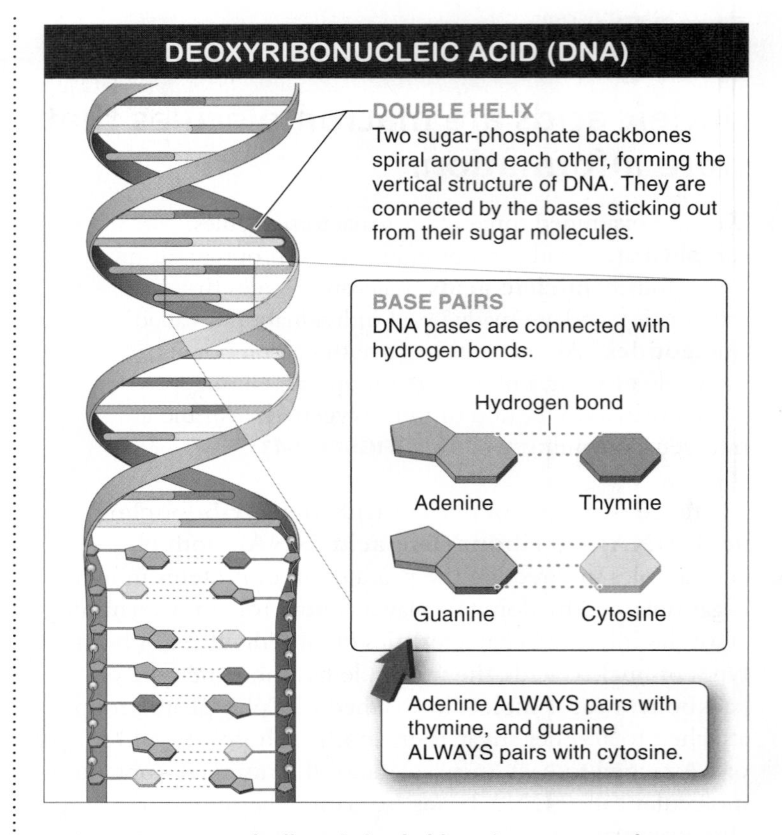

FIGURE 2-44 A gradually twisting ladder. The structure of DNA.

The sequences of bases containing the information about how to produce a particular protein have anywhere from a hundred to several thousand bases. In a human, all of the DNA in a cell, containing all of the instructions for every protein that a human must produce, contains about 3 billion base pairs. This DNA is generally in the nucleus of a cell.

TAKE-HOME MESSAGE 2•20

DNA is like a ladder in which the long vertical sides of the ladder are made from a sequence of sugar-phosphate-sugar-phosphate molecules and the rungs are pairs of nucleotide bases. The sequence of nucleotide bases contains the information about how to produce a particular protein.

2•21

RNA is a universal translator, reading DNA and directing protein production.

The process of building a protein from a DNA sequence is not a direct one. Rather, it incorporates a middleman, RNA, that is also a nucleic acid (**FIGURE 2-45**). Segments of the DNA are read off, directing the production of short strips of RNA that contain the information, taken from the DNA, about the amino acid sequence in a protein. The RNA moves to another part of the cell and then directs the piecing together of amino acids into a three-dimensional protein. We explore this in greater detail in Chapter 5.

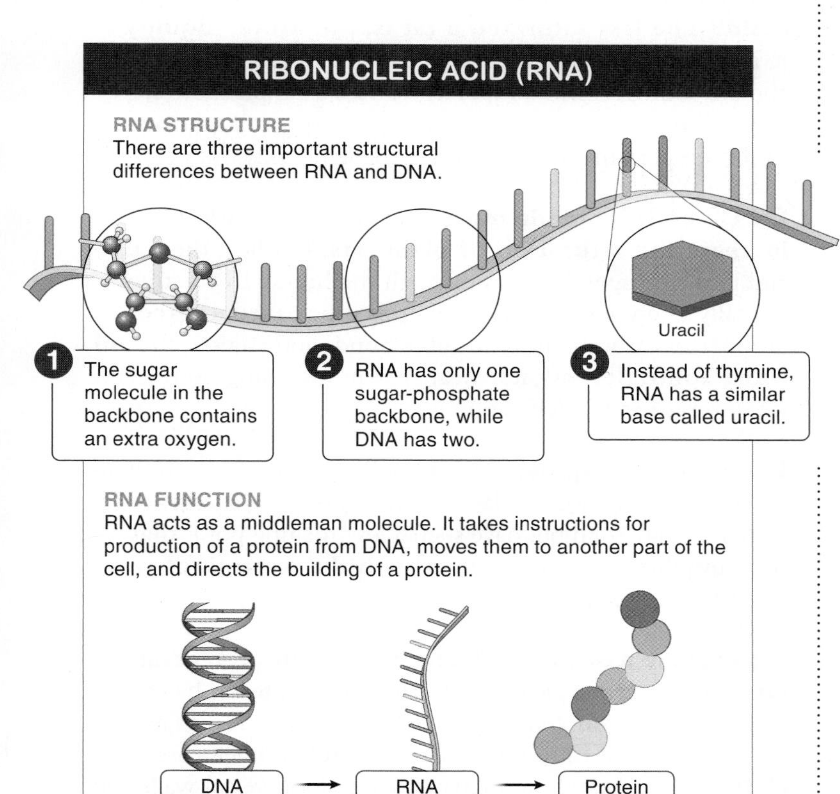

FIGURE 2-45 The middleman between DNA and protein. The structure of RNA.

RNA differs from DNA in three important ways. First, the sugar molecule of the sugar-phosphate-sugar-phosphate backbone differs slightly, containing an extra atom of oxygen. Second, RNA is single stranded. The sugar-phosphate-sugar-phosphate backbone is still there, as are the bases that protrude from each sugar. The bases, however, do not bind with anything else—do not form base pairs with another RNA strand. And third, while RNA has the bases A, G, and C, it replaces the thymine with a similar base called uracil (U).

Whether we're looking at the nucleotides that make up RNA and DNA or the lipids used to build sex hormones and cell membranes, we see a recurring theme in the construction of biological macromolecules: from relatively simple sets of building blocks linked together, infinitely complex molecules can be formed. Complex webs of one simple sugar, bonded together as glycogen, for instance, provide fuel for organisms. Similarly, sequences of amino acids of 20 different types, joined together, specify the structure of all of the proteins found in every species on earth.

TAKE-HOME MESSAGE 2•21

RNA acts as a middleman molecule—taking the instructions for protein production from DNA to another part of the cell where, in accordance with the RNA instructions, amino acids are pieced together into proteins.

Knowledge You Can Use

Did you know? Melt-in-your-mouth chocolate may not be such a sweet idea.

Food chemists have figured out how to make chocolate that melts in your mouth. Is that a good thing?

Q: Why are some fats "liquidy," like oil, and others solid? The less saturated a fat is, the more "liquidy" it is at room temperature. Most animal fats are saturated and are solid at room temperature. Best example: butter. Most plant fats are polyunsaturated fats and are liquid at room temperature. Best example: vegetable oils.

Q: Can oils be made more solid? It's possible to increase the saturation of plant fats. Just heat them up and pass hydrogen bubbles through the liquid. In creating partially hydrogenated plant oils, this process reduces the number of carbon–carbon double bonds and **makes the oil more solid.** (It's easy, it's cheaper than just using butter, and it increases foods' shelf-life.)

Q: Does that improve their taste? By precisely controlling the level of saturation in plant fats, **it is possible to create foods** that are solid but have such a low melting point **that they quickly melt on contact with the warmth of your mouth.** This seems great, but . . .

Q: Is there a downside? The saturation of vegetable fats creates trans fats, due to the position taken by the newly added hydrogen atoms in the molecule. Trans fats increase levels of LDL ("bad") cholesterol and decrease HDL ("good") cholesterol, narrowing blood vessel walls and increasing the risk of heart disease and strokes.

Conclusion: Partially hydrogenated vegetable oils can give food a perfect texture and a pleasing feel in your mouth. But the **creaminess comes with a high cost** when it comes to your health.

1 Atoms form molecules through bonding.

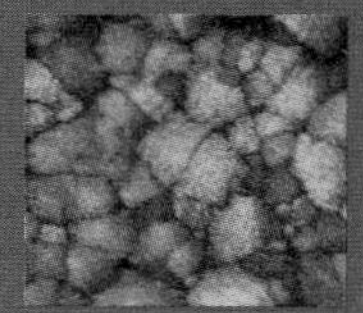

An atom is the smallest unit into which material can be divided without losing its essential properties. Atoms are made up of protons and neutrons in the nucleus and electrons that circle far around that nucleus. Molecules are atoms linked together.

2 Water has features that enable it to support all life.

Water molecules easily form hydrogen bonds, giving water great cohesiveness and the ability to resist temperature changes, and making it a versatile solvent. The pH of a fluid is a measure of how acidic or basic a solution is.

3 Carbohydrates are fuel for living machines

Carbohydrates, made up of carbon, oxygen, and hydrogen, are the primary fuel for all cellular machinery and also form much of the structure of cells. The simplest carbohydrates, including glucose, are called monosaccharides or simple sugars. Multiple simple carbohydrates sometimes link together into more complex carbohydrates, including starch, which is the primary form of energy storage in plants, and glycogen, which is a primary form of energy storage in animals. Some carbohydrates are not digestible by humans.

4 Lipids store energy for a rainy day.

Lipids are macromolecules—made up primarily of carbon, hydrogen, and oxygen—that are insoluble in water. Lipids are important for energy storage, hormones, and membrane structure. Their breakdown of dietary fats releases more energy per gram than other macromolecules.

5 Proteins are body-building molecules.

Cells and tissues are primarily built from proteins, sequences of amino acids that fold into complex, three-dimensional shapes. The atoms, especially nitrogen, present in the plant and animal proteins that an organism eats are essential to the organism's growth and repair. The amino acid sequence of a protein determines how it folds into a three-dimensional shape. This shape determines many of the protein's features, such as which molecules it will interact with. Enzymes are proteins that help initiate and speed up chemical reactions. They aren't permanently altered in the process.

6 Nucleic acids store the information on how to build and run a body.

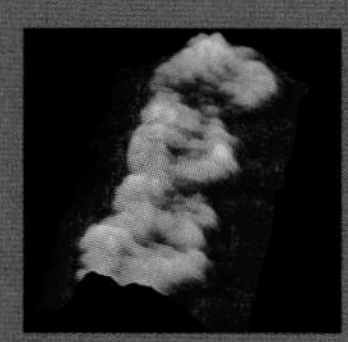

The nucleic acids DNA and RNA are macromolecules that store information by having unique sequences of molecules. Both play central roles in directing protein production in organisms. The sequence of nucleotide bases contains the information about how to produce a particular protein. RNA acts like a universal translator of the genetic code into proteins. It reads DNA sequences and directs the production of a sequence of amino acids.

KEY TERMS

CHECK YOUR KNOWLEDGE

1. The atomic number of carbon is 6. Its nucleus must contain:
   a) 6 neutrons and 6 protons.
   b) 3 protons and 3 neutrons.
   c) 6 neutrons and no electrons.
   d) 6 protons and no electrons.
   e) 6 protons and 6 electrons.

2. The second orbital shell of an atom can hold ___ electrons.
   a) 2
   b) 3
   c) 4
   d) 6
   e) 8

3. A covalent bond is formed when:
   a) two non-polar molecules associate with each other in a polar environment.
   b) a positively charged particle is attracted to a negatively charged particle.
   c) one atom gives up electrons to another atom.
   d) two atoms share electrons.
   e) two polar molecules associate with each other in a non-polar environment.

4. When you put a piece of chocolate on your tongue, your brain registers a sensation of sweetness. What aspect of molecules is responsible for their having a particular taste?
   a) the total number of protons in the molecule
   b) the number of hydrogen bonds in the molecule
   c) the total number of electron shells in the atoms of the molecule
   d) the ratio of covalent bonds to ionic bonds joining the atoms of the molecule
   e) the molecule's shape

5. Which of the following phenomena is most likely due to the high cohesiveness of water?
   a) Lakes and rivers freeze from the top down, not the bottom up.
   b) The "Jesus lizard" can run across the surface of liquid water for short distances.
   c) Adding salt to snow makes it melt.
   d) The temperature of Santa Monica Bay, off the coast of Los Angeles, fluctuates less than the air temperature throughout the year.
   e) All of the above are due to the cohesiveness of water.

6. Water can absorb and store a large amount of heat while increasing only a few degrees in temperature. Why?
   a) The heat must first be used to break the hydrogen bonds rather than raise the temperature.
   b) The heat must first be used to break the ionic bonds rather than raise the temperature.
   c) The heat must first be used to break the covalent bonds rather than raise the temperature.
   d) An increase in temperature causes an increase in adhesion of the water.
   e) An increase in temperature causes an increase in cohesion of the water.

7. A chemical compound that releases H^+ into a solution is called:
   a) a proton.
   b) a base.
   c) an acid.
   d) a hydroxide ion.
   e) a hydrogen ion.

8. Which of the following foods is not a significant source of complex carbohydrates?
   a) fresh fruit
   b) rice
   c) pasta
   d) oatmeal
   e) All of the above are significant sources of complex carbohydrates.

9. Sucrose (table sugar) and lactose (the sugar found in milk) are examples of:
   a) naturally occurring enzymes.
   b) simple sugars.
   c) monosaccharides.
   d) disaccharides.
   e) polyunsaccharide sugars.

10. Which of the following statements about starch is incorrect?
    a) Starch is the primary form of energy storage in plants.
    b) Starch consists of a hundred or more glucose molecules joined together in a line.
    c) Starch tastes sweet because it is made from glucose.
    d) Starch is a polysaccharide.
    e) All of the above statements about starch are correct.

11. Which of the following statements about fiber is incorrect?
    a) Dietary fiber reduces the risk of colon cancer.
    b) Fiber in the diet slows the passage of food through the intestines.
    c) Humans are unable to extract any caloric value from fiber.
    d) The cellulose of celery stalks and lettuce leaves is fiber.
    e) Fiber passing through the digestive system scrapes the wall of the digestive tract, stimulating mucus secretion and aiding in the digestion of other molecules.

12. In an unusually detailed dissection of your dinner, you isolate an unknown fatty acid. It is a liquid at room temperature (that is, has a low melting point) and contains carbon-carbon double bonds. What are you most likely to be eating?
    a) a plant
    b) a cow
    c) a pig
    d) a chicken
    e) a lamb

13. An unsaturated fatty acid is one in which:
    a) carbon-carbon double bonds are present in the hydrocarbon chain.
    b) the hydrocarbon chain has an odd number of carbons.
    c) the hydrocarbon chain has an even number of carbons.
    d) no carbon-carbon double bonds are present in the hydrocarbon chain.
    e) not all of the carbons in the hydrocarbon chain are bonded to hydrogen atoms.

14. Which statement about phospholipids is incorrect?
a) Because their phosphate groups repel each other, they are used as organisms' chief form of short-term energy.
b) They are hydrophobic at one end.
c) They are hydrophilic at one end.
d) They are a major constituent of cell membranes.
e) They contain glycerol linked to fatty acids.

15. Proteins are an essential component of a healthy diet for humans (and other animals). Their most common purpose is to serve as:
a) raw material for growth.
b) fuel for running the body.
c) organic precursors for enzyme construction.
d) long-term energy storage.
e) inorganic precursors for enzyme construction.

16. Dietary proteins:
a) are considered "complete" only if they contain the eight essential amino acids required by humans.
b) consist of all 20 amino acids required in the human body.
c) are considered "complete" only if they contain the 12 non-essential amino acids required by humans.
d) are nutritionally identical, since all are broken down into their constituent amino acids in the digestive system.
e) can be obtained from animal sources but not plant sources.

17. The primary structure of proteins is often described as amino acids connected like beads on a string. In this same vein, which of the following images best describes a protein's quaternary structure?
a) threads in a cloth
b) needles in a haystack
c) rungs on a ladder
d) links on a chain
e) coils in a spring

18. Which of the following statements about enzymes is incorrect?
a) Enzymes can initiate chemical reactions.
b) Enzymes speed up chemical reactions.
c) Enzymes often induce conformational changes in the substrates that they bind.
d) Enzymes contain an active site for binding of particular substrates.
e) Enzymes undergo a permanent change during the reactions they promote.

19. Which of the following nucleotide bases are present in equal amounts in DNA?
a) adenine and cytosine
b) thymine and guanine
c) adenine and guanine
d) thymine and cytosine
e) adenine and thymine

20. Which type of macromolecule contains an organism's genetic information?
a) polysaccharide
b) monosaccharide
c) fatty acid
d) DNA
e) phospholipid

21. All of the following are features of RNA except:
a) a sugar-phosphate-sugar-phosphate backbone.
b) a coiled double-stranded structure.
c) a different type of sugar than is found in DNA.
d) an ability to direct protein production.
e) a different type of base than is found in DNA.

Short-Answer Questions

1. An atom of sodium has an atomic number of 11 and an atom of chlorine has an atomic number of 17.
a) Indicate the number of electrons that are present in the outermost electron shell of each atom.
b) Predict what type of bond would form between these atoms, and explain why.

2. Explain how hydrogen bonding plays a part in each of the unique properties of water and, in essence, allows for life on earth.

3. There are many cleaning products on the market that advertise their ability to get rid of stubborn, oily dirt. Why is water alone not sufficient for removing oily dirt?

4. Some who are watching what they eat may opt for a salad when they are dining out. A dinner salad may include lettuce, celery, strawberries, walnuts, chicken, and a salad dressing of oil and vinegar. Explain which category of major macromolecules each of these individual salad components falls into. Then explain the general purpose or function of each major macromolecule.

See Appendix for answers. For additional study questions, go to www.prep-u.com.

1 What is a cell?

3.1 All organisms are made of cells.

3.2 Prokaryotic cells are structurally simple, but there are many types of them.

3.3 Eukaryotic cells have compartments with specialized functions.

2 Cell membranes are gatekeepers.

3.4 Every cell is bordered by a plasma membrane.

3.5 Molecules embedded in the plasma membrane help it perform its functions.

3.6 Faulty membranes can cause disease.

3.7 Membrane surfaces have a "fingerprint" that identifies the cell.

3 Molecules move across membranes in several ways.

3.8 Passive transport is the spontaneous diffusion of molecules across a membrane.

3.9 Osmosis is the passive diffusion of water across a membrane.

3.10 In active transport, cells use energy to move small molecules into and out of the cell.

3.11 Endocytosis and exocytosis are used for bulk transport of particles.

4 Cells are connected and communicate with each other.

3.12 Connections between cells hold them in place and enable them to communicate with each other.

5 Nine important landmarks distinguish eukaryotic cells.

3.13 The nucleus is the cell's genetic control center.

3.14 Cytoplasm and the cytoskeleton form the cell's internal environment, provide its physical support, and can generate movement.

3.15 Mitochondria are the cell's energy converters.

3.16 Lysosomes are the cell's garbage disposals.

3.17 The endoplasmic reticulum is the site where cells build proteins and disarm toxins.

3.18 The Golgi apparatus is the site where the cell processes products for delivery throughout the body.

3.19 The cell wall provides additional protection and support for plant cells.

3.20 Vacuoles are multipurpose storage sacs for cells.

3.21 Chloroplasts are the plant cell's power plant.

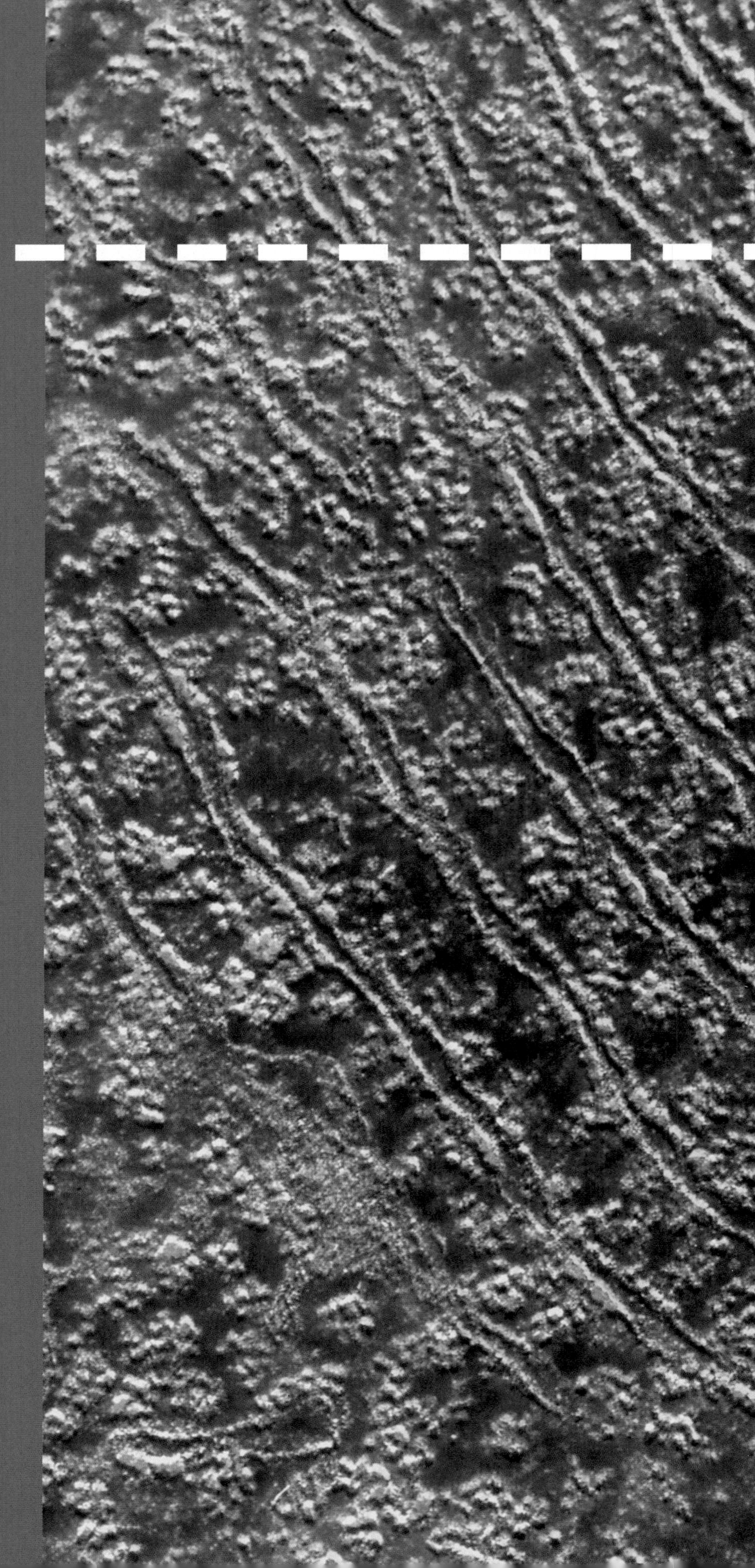

3

Cells

The smallest part of you

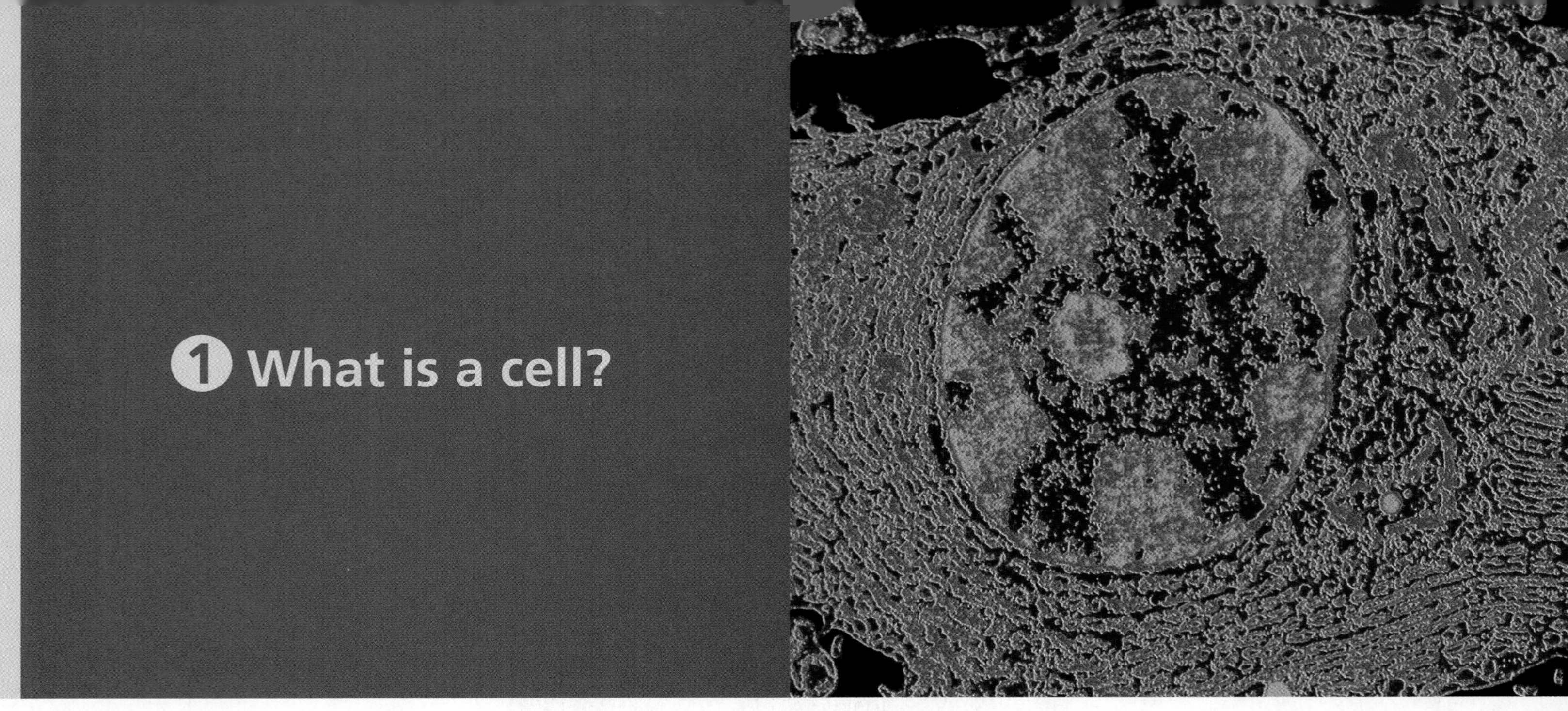

Human cell, packed with organelles.

3•1

All organisms are made of cells.

Where do we begin if we want to understand how organisms work? Given their complexity, this task can be daunting. Fortunately, as with most complex things, a strategy of divide and conquer comes in handy. In the case of organisms, whether we are studying a creature as small as a flea or as large as an elephant or giant sequoia, they can be broken down into smaller units that are more easily studied and understood (**FIGURE 3-1**). The most basic unit of any organism is the **cell,** the smallest unit of life that can function independently and perform all the necessary functions of life, including reproducing itself. Understanding cell structure and function is the basis for our understanding of how complex organisms are organized.

The term "cell" was first used in the mid-1600s by Robert Hooke, a British scientist also known for his contributions to philosophy, physics, and architecture. When he was made Curator of Experiments for the Royal Society of London, Hooke suddenly had access to many of the first microscopes available, and he began to examine everything he could get his hands on. Because Hooke thought the close-up views of a very thin piece of cork resembled a mass of small, empty rooms, he named these compartments *cellulae,* Latin for "small rooms."

Today, we know that a cell is a three-dimensional structure, like a fluid-filled balloon, in which many of the essential chemical reactions of life take place (such as the breakdown of carbohydrates for energy, the modification of cholesterol to create testosterone and estrogen, and the translation of the genetic code into protein production). Generally, these reactions involve transporting raw materials and fuel into the cell and exporting finished materials and waste products out of the cell. In addition, nearly all cells contain DNA (deoxyribonucleic acid), a molecule that contains the information that directs the chemical reactions in a cell, the formation of various cellular products within the cell, and the cell's ability to reproduce itself. (A few types of cells, including mammalian red blood cells and cells called "sieve tube elements" that form part of the plant circulatory system, lose their nuclei after they are created and are unable to divide.) We explore all of these features of cell functioning in this and the next three chapters.

To see a cell, you don't have to work in a lab or use a microscope. Just open your refrigerator. Chances are you've got a dozen or so visible cells in there: eggs. Although most cells are too small to see with the naked eye, there are a few exceptions, including hens' eggs from the supermarket. As long as they are unfertilized, which most store-bought hens' eggs are, each egg tends to contain just one cell. The ostrich egg, weighing about three pounds, contains the largest of all animal cells. (It should be noted, however, that by the time the ostrich lays a fertilized egg, the embryo inside has already gone through multiple divisions.) In addition to being among the largest cells around, eggs are also the most valuable. Almas caviar, eggs from the beluga sturgeon, sells for nearly $700 per ounce. This value is exceeded only by that

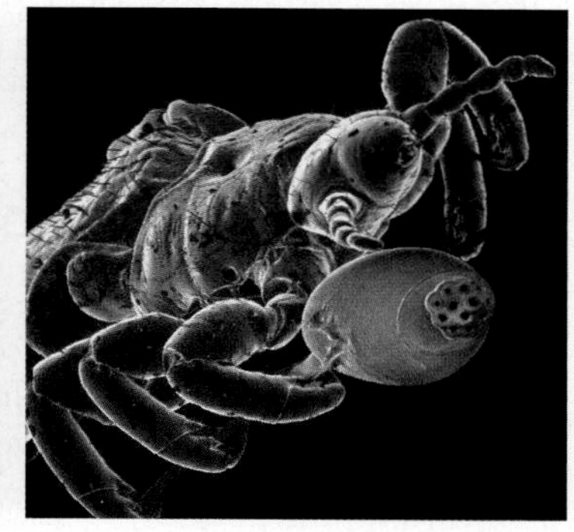

FIGURE 3-1 What do these diverse organisms have in common? Cells.

of human eggs, which currently fetch as much as $25,000 for a dozen or so eggs on the open market (**FIGURE 3-2**). (Human sperm cells command only about a penny per 20,000 cells!)

Most cells are much smaller than hens' eggs and ostrich eggs. Consider that, at this very moment, there are probably more than seven billion bacteria in your mouth—even if you just brushed your teeth! This is more than the number of people on earth. It is possible to squeeze so many bacteria in there because most cells (not just bacteria) are really, really tiny—so tiny that 2,000 red blood cells, lined up end to end, would just extend across a dime.

After sufficient improvements were made to early microscopes in the 19th century, the central role of the cell in biology could be understood. As scientists began putting everything they studied under their microscopes, the importance and universality of the cell finally dawned on them. By the 1830s, scientists realized that all plants and animals were made entirely from cells. Subsequent studies revealed that every cell seemed to arise from the division of another cell. You, for example, are made up of at least 60 trillion cells, all of which came from just one cell: the single fertilized egg produced when an egg cell from your mother was fertilized by a sperm cell from your father.

The facts that (1) all living organisms are made up of one or more cells and (2) all cells arise from other, pre-existing cells are the foundations of **cell theory,** one of the unifying theories in biology, and one that is universally accepted by all biologists. As we will see in Chapter 10, the origin of life on earth was a one-time deviation from cell theory: the first cells on earth probably originated from free-floating molecules in the oceans early in the earth's history (about 3.5 billion years ago). Since that time, however, all cells and thus all life have been produced as a continuous line of cells, originating from these initial cells.

In this chapter, we investigate the two different kinds of cells that make up all of the organisms on earth, the processes by which cells control how materials move into and out of the cell, and how cells communicate with each other. We also explore some of the important landmarks found in many cells. We look, too, at the specialized roles these structures play in a variety of cellular functions and learn about some of the health consequences that occur when they malfunction.

TAKE-HOME MESSAGE 3•1

The most basic unit of any organism is the cell, the smallest unit of life that can function independently and perform all of the necessary functions of life, including reproducing itself. All living organisms are made up of one or more cells, and all cells arise from other, pre-existing cells.

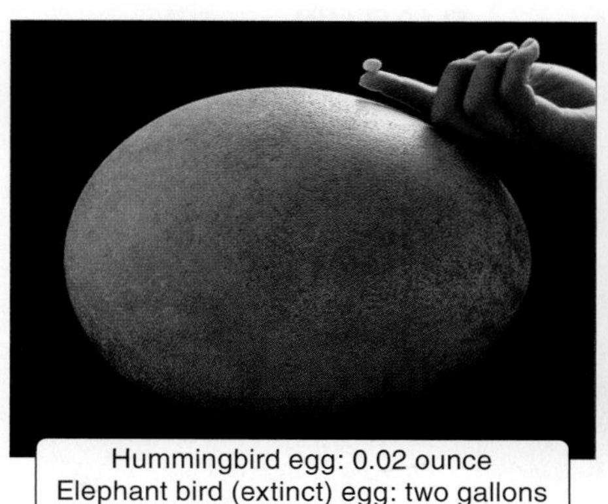
Hummingbird egg: 0.02 ounce
Elephant bird (extinct) egg: two gallons

Beluga sturgeon eggs: $700 per ounce

Human eggs: Thousands of dollars per egg

FIGURE 3-2 Not all cells are tiny. And some cells are extremely valuable!

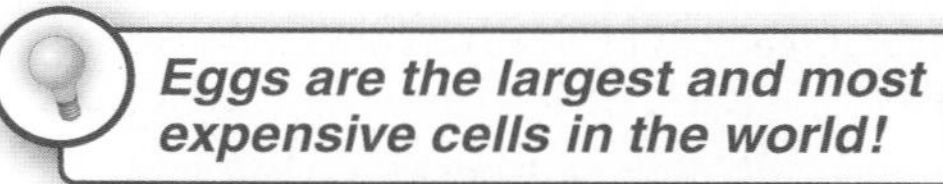

3•2

Prokaryotic cells are structurally simple, but there are many types of them.

Although there are millions of diverse species on earth, billions of unique organisms alive at any time, and trillions of different cells in many of those organisms, every cell on earth falls into one of two basic categories.

A **eukaryotic cell** (from the Greek for "good" and "kernel") has a central control structure called a nucleus, which contains the cell's DNA. Organisms composed of eukaryotic cells are called **eukaryotes.**

A **prokaryotic cell** (from the Greek for "before" and "kernel") does not have a nucleus; its DNA simply resides in the middle of the cell. An organism consisting of a prokaryotic cell is called a **prokaryote.**

We begin our study of the cell by exploring the prokaryotes. Prokaryotes were the first cells on earth, making their appearance about 3.5 billion years ago, and, for a long time (1.5 billion years), they had the planet to themselves.

All prokaryotes are one-celled organisms and are thus invisible to the naked eye. Prokaryotes have four basic structural features (**FIGURE 3-3**).

1. A **plasma membrane** encompasses the cell (and is sometimes simply called the "cell membrane"). For this reason, anything inside the plasma membrane is referred to as "intracellular," and everything outside the plasma membrane is "extracellular."
2. The **cytoplasm** is the jelly-like fluid that fills the inside of the cell.
3. **Ribosomes** are little granular bodies where proteins are made; thousands of them are scattered throughout the cytoplasm.
4. Each prokaryote has one or more circular loops or linear strands of DNA.

In addition to the characteristics common to all prokaryotes, some prokaryotes have additional, unique features. Many have a rigid **cell wall,** for example, that protects and gives shape to the cell. Some have a slimy, sugary capsule as their outermost layer. This sticky outer coat provides protection and enhances the prokaryotes' ability to anchor themselves in place when necessary.

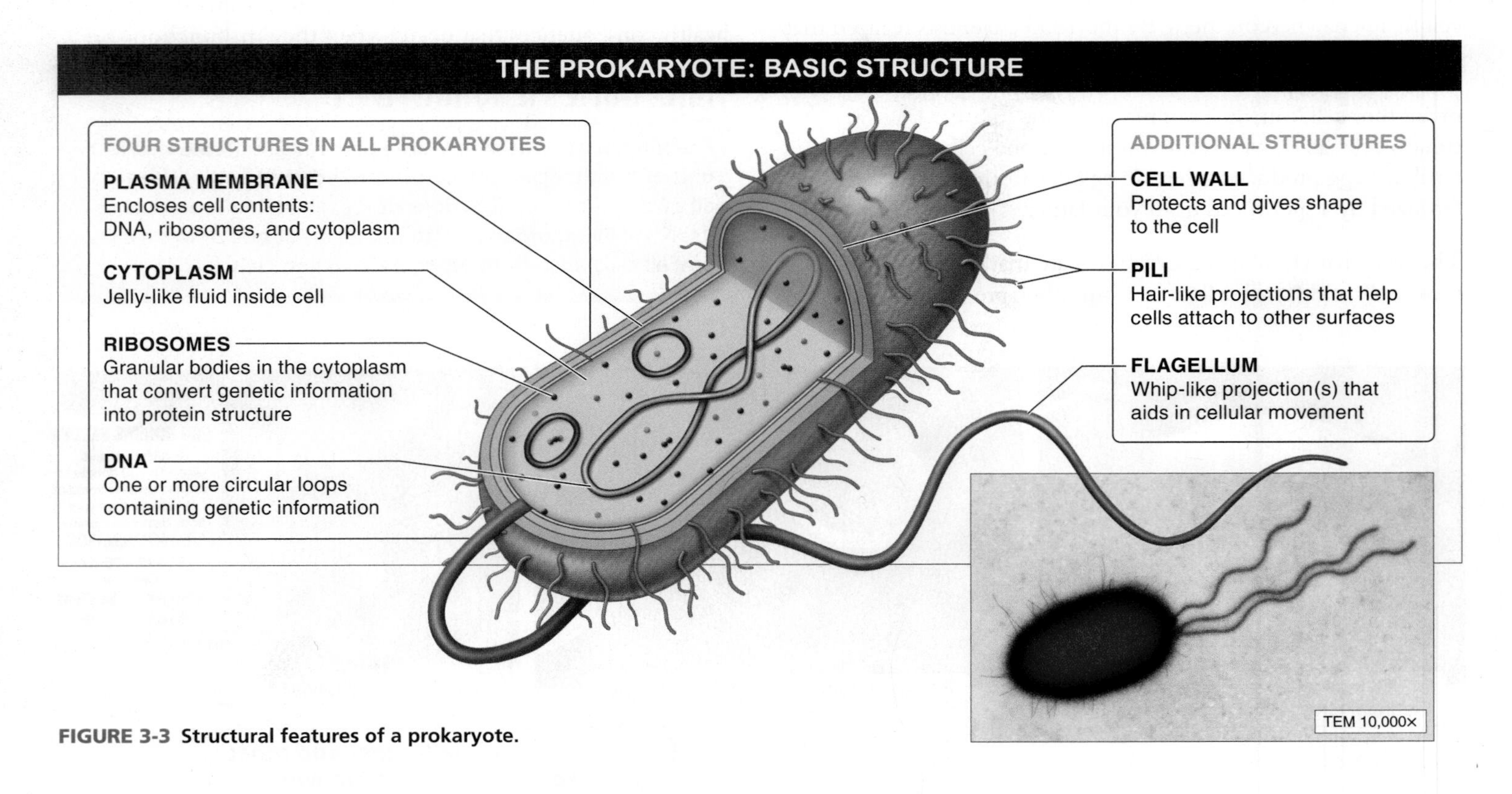

FIGURE 3-3 **Structural features of a prokaryote.**

Many prokaryotes have a **flagellum** (*pl.* **flagella**), a long, thin, whip-like projection that rotates like a propeller and moves the cell through the medium in which it lives. Other appendages include **pili** (*sing.* **pilus**), much thinner, hair-like projections that help prokaryotes attach to surfaces.

From our human perspective, it is easy to underestimate the prokaryotes. Although they are smaller, evolutionarily older, and structurally more simple than eukaryotes, prokaryotes are fantastically diverse metabolically (i.e., in the way they break down and build up molecules). Among many other innovations, bacteria can fuel their activities in the presence or absence of oxygen, using almost any energy source on earth, depending on the type of bacteria—from the sulfur in deep-sea hydrothermal vents to hydrogen to the sun.

You may think you've never encountered a prokaryote, but you are already familiar with the largest group of prokaryotes: the bacteria. All bacteria are prokaryotes, from those such as *Escherichia coli* (*E. coli*), which live in your intestine and help your body make some essential vitamins, to those responsible for illness, such as *Streptococcus pyogenes,* which causes strep throat. Another recently discovered group of organisms, the archaea, are also prokaryotes.

TAKE-HOME MESSAGE 3•2

Every cell on earth is either a eukaryotic or a prokaryotic cell. Prokaryotes, which have no nucleus, were the first cells on earth. They are all single-celled organisms. Prokaryotes include the bacteria and archaea and, as a group, are characterized by tremendous metabolic diversity.

3•3

Eukaryotic cells have compartments with specialized functions.

Eukaryotes showed up about 1.5 billion years after prokaryotes, and in the 2 billion years that they have been on earth, they have evolved into some of the most dramatic and interesting creatures, such as platypuses, dolphins, giant sequoias, and the Venus fly trap. Because all prokaryotes are single-celled and thus invisible to the naked eye, every organism that we see around us is a eukaryotic organism. (**FIGURE 3-4**). All fungi, plants, and animals are eukaryotes, for

FIGURE 3-4 Diversity of the eukaryotes. Every organism that we can see without magnification is a eukaryotic organism.

EUKARYOTES vs. PROKARYOTES

TYPICAL EUKARYOTIC CELL FEATURES

- DNA contained in nucleus
- Internal structures organized into compartments
- Larger than prokaryotes—usually 10 times bigger
- Cytoplasm contains specialized structures called organelles

Compartments
Nucleus
Organelles
TEM 6,000×

TYPICAL PROKARYOTIC CELL FEATURES

- No nucleus—DNA is in the cytoplasm
- Internal structures not organized into compartments
- Much smaller than eukaryotes

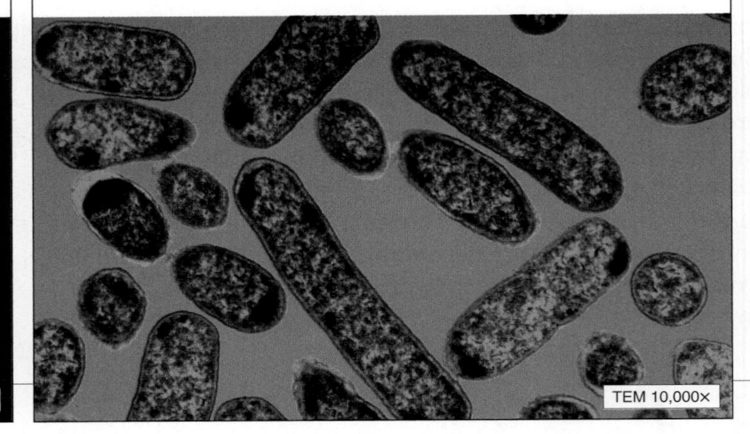

FIGURE 3-5 Comparison of eukaryotic and prokaryotic cells.

instance. Not all eukaryotes are multicellular, however. There is a huge group of eukaryotes, called the Protista (or protists), nearly all of which are single-celled organisms visible only with a microscope.

The chief distinguishing feature of eukaryotic cells is the presence of a **nucleus,** a membrane-enclosed structure that contains linear strands of DNA. In addition to a nucleus, eukaryotic cells usually contain in their cytoplasm several other specialized structures, called **organelles,** many of which are enclosed separately within their own lipid membranes. Eukaryotic cells are also about 10 times larger than prokaryotes. All of these physical differences make it easy to distinguish eukaryotes from prokaryotes under a microscope (**FIGURE 3-5**).

Organelles enable eukaryotic cells to do many things that a prokaryotic cell cannot do. Most important, the creation of physically separate compartments means that there can be distinct areas within the same cell in which different chemical reactions can occur simultaneously. In the non-compartmentalized interior of a prokaryotic cell, random molecular movements quickly blend the chemicals throughout the cell, reducing the ease with which different reactions can occur simultaneously.

FIGURE 3-6 illustrates a generalized animal cell and a generalized plant cell. Because they share a common, eukaryotic ancestor, they have much in common. Both can have a plasma membrane, nucleus, cytoskeleton, and ribosomes, and a host of organelles, including rough and smooth endoplasmic membranes, Golgi apparatus, and mitochondria. Animal cells have centrioles, which are not present in most plant cells. Plant cells have a rigid cell wall (as do fungi and many protists) and chloroplasts (also found in some protists). Plants also have a vacuole, a large central chamber (only occasionally found in animal cells). We explore each of these plant and animal organelles in detail later in this chapter.

When you compare a complex eukaryotic cell with the structurally simple prokaryotic cell, it's hard not to wonder how eukaryotic cells came about. How did they get filled up with so many organelles? We can't go back 2 billion years to watch the initial evolution of eukaryotic cells, but we can speculate about how it might have occurred. One very appealing idea, called the **endosymbiosis theory,** has been developed to explain the presence of two organelles in eukaryotes: chloroplasts in plants and algae, and mitochondria in plants and animals. Chloroplasts help plants and algae convert sunlight into a more usable form of energy. Mitochondria help plants and animals harness the energy stored in food molecules. (Chapter 4, on energy, explains the details of both of these processes.)

According to the theory of endosymbiosis, two different types of prokaryotes may have set up close partnerships with each

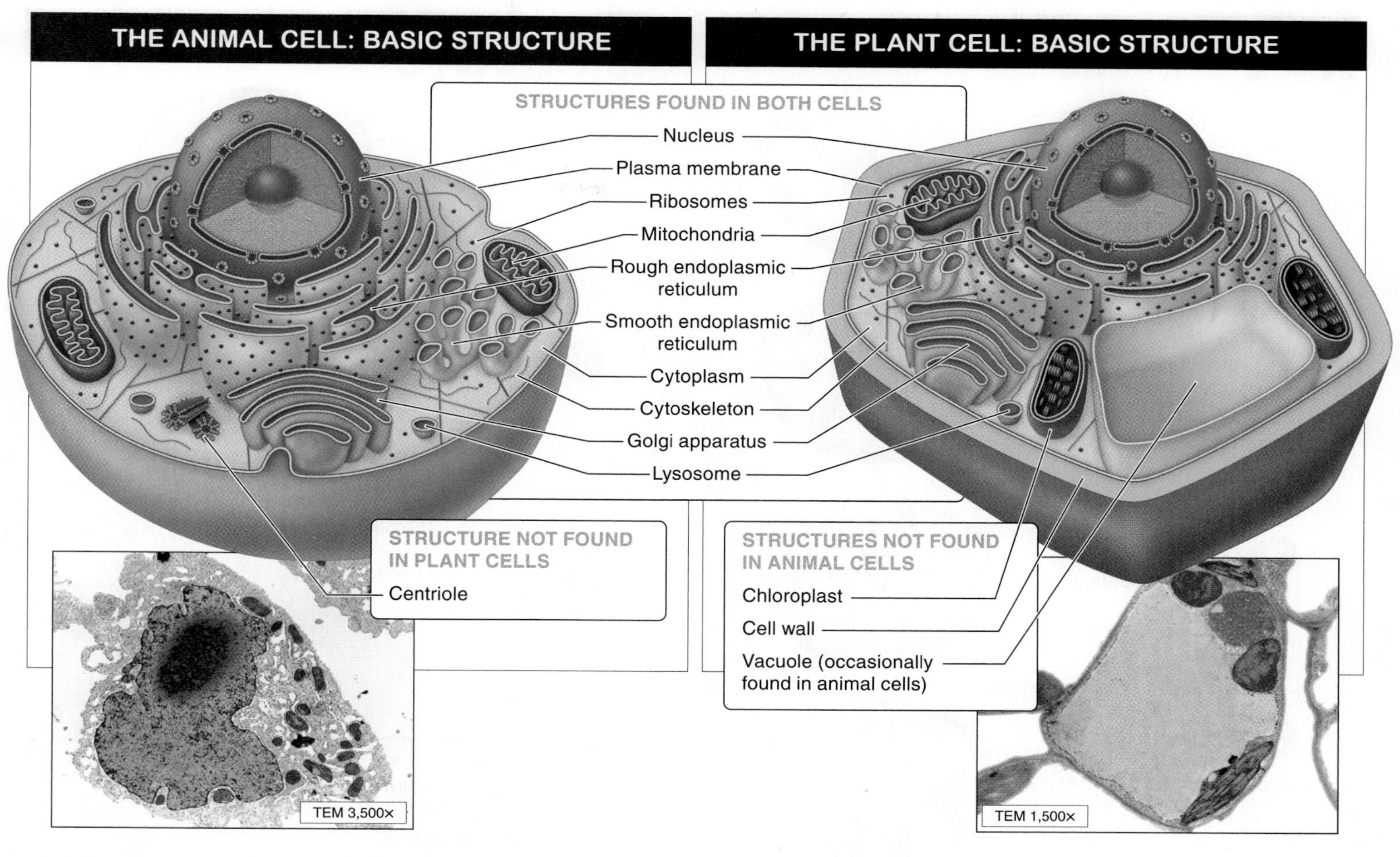

FIGURE 3-6 Structures found in animal and plant cells.

other. For example, some small prokaryotes capable of performing photosynthesis (the process by which plant cells capture light energy from the sun and transform it into the chemical energy stored in food molecules) may have come to live inside a larger "host" prokaryote. The photosynthetic "boarder" may have made some of the energy from its photosynthesis available to the host.

Q Humans, deep down, may be part bacteria. How can that be?

After a long while, the two cells may have become more and more dependent on each other until neither cell could live without the other (they became "symbiotic"), and they became a single, more complex organism. Eventually, the photosynthetic prokaryote evolved into a **chloroplast,** the organelle in plant cells in which photosynthesis occurs. A similar scenario might explain how a prokaryote unusually efficient at converting food and oxygen into easily usable energy took up residence inside another, host prokaryote and evolved into a **mitochondrion,** the organelle in plant and animal cells that converts the energy stored in food into a form usable by the cell (**FIGURE 3-7**).

The idea of endosymbiosis is supported by several observations.

1. Chloroplasts and mitochondria are similar in size to prokaryotic cells.
2. Chloroplasts and mitochondria have small amounts of circular DNA, similar to the circular DNA in prokaryotes and in contrast to the linear DNA strands found in a eukaryote's nucleus.
3. Chloroplasts and mitochondria divide by splitting (fission), just like prokaryotes.
4. Chloroplasts and mitochondria have internal structures called ribosomes that are similar to those found in bacteria.

In addition to the theory of endosymbiosis as an explanation for the existence of chloroplasts and mitochondria, another

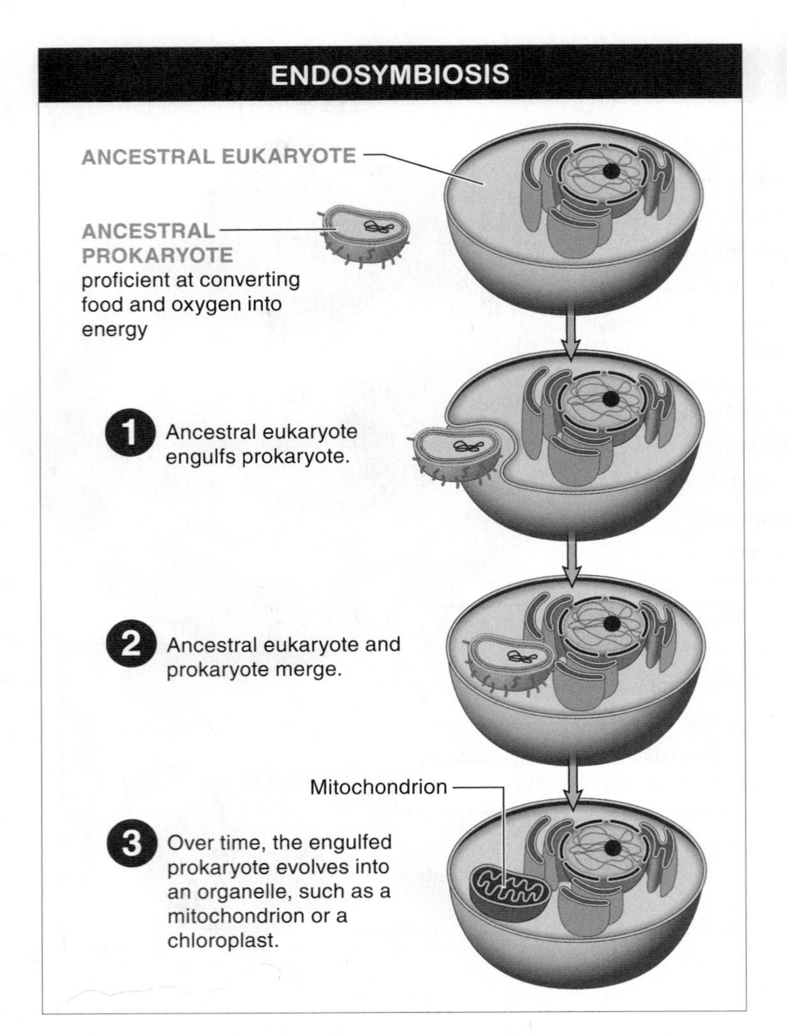

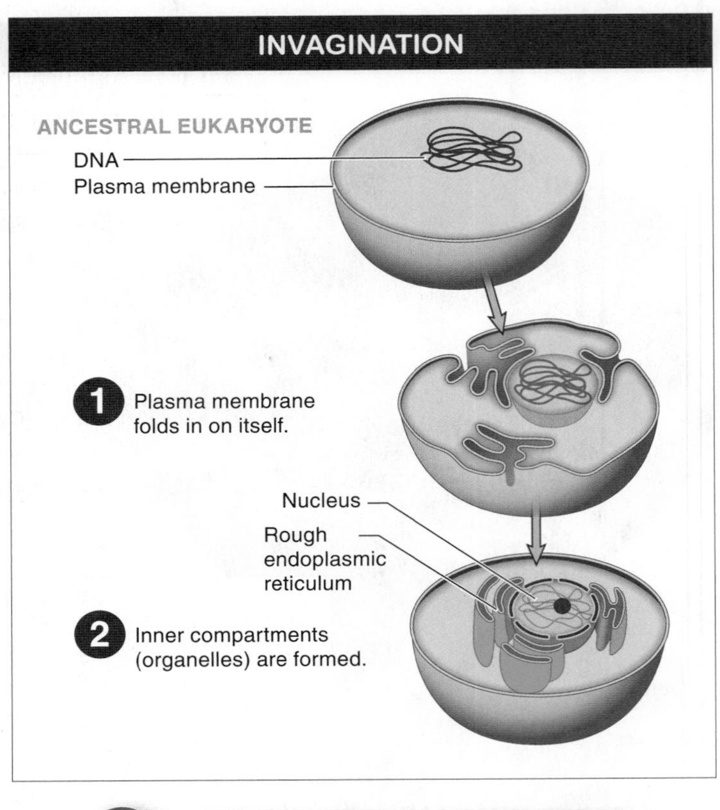

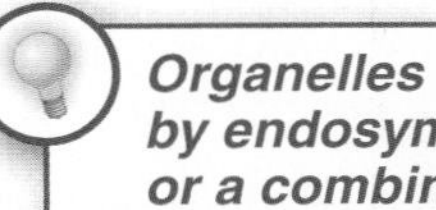

Organelles may have developed by endosymbiosis or invagination or a combination of the two.

FIGURE 3-7 How did eukaryotic cells become so structurally complex? Two theories.

theory about the origin of organelles in eukaryotes is the idea that the plasma membrane around the cell may have folded in on itself (a process called **invagination**) to create the inner compartments, which subsequently became modified and specialized (see Figure 3-7). It may turn out that organelles arose from both processes. Perhaps mitochondria and chloroplasts originated from endosymbiosis, and all of the other organelles arose from plasma membrane invagination. We just don't know.

TAKE-HOME MESSAGE 3·3

Eukaryotes are single-celled or multicellular organisms consisting of cells with a nucleus that contains linear strands of genetic material. The cells also commonly have organelles throughout their cytoplasm; these organelles may have originated evolutionarily through endosymbiosis or invagination, or both.

Cell membranes control the movement of material into and out of the cell.

3•4

Every cell is bordered by a plasma membrane.

Just as skin covers our bodies, every cell of every living thing on earth, whether a prokaryote or a eukaryote, is enclosed by a plasma membrane, a two-layered membrane that holds the contents of a cell in place and regulates what enters and leaves the cell. Plasma membranes are thin (a stack of a thousand would be only as thick as a single hair) and flexible, and in photos or diagrams the membranes often resemble simple plastic bags, holding the cell contents in place. This image is a gross oversimplification, however. Membranes are indeed thin and flexible, but they are far from simple: a close look at a plasma membrane will reveal that its surface is filled with pores, outcroppings, channels, and complex molecules floating around within the two layers of the membrane itself (**FIGURE 3-8**).

Why are plasma membranes such complex structures? It's because they perform several critical functions beyond simply acting as the boundary for a cell's interior contents. Cells, after all, are perpetually interacting with their external environment. They take in food and nutrients and dispose of waste products. They take in water. They build and export molecules needed elsewhere in the body. Sometimes they absorb heat from the outside environment, whereas at other times they dissipate excess heat generated by cell activities. Like the booths or checkpoints at a country's borders that control the flow of people into and out of the country, plasma membranes serve as gatekeepers that control the flow of molecules into and out of the cell.

The foundation of all plasma membranes is a layer of lipid molecules all packed together. These are a special type of lipid, called **phospholipids,** which, as you may recall from Chapter 2, have what appear to be a head and two long tails. The head consists of a molecule of **glycerol** linked to a molecule containing phosphorus (**FIGURE 3-9**). This head region is said to be **polar,** because it has an electrical charge. As you may also recall from Chapter 2, water is also a polar molecule and, for this reason, other polar molecules mix easily with water.

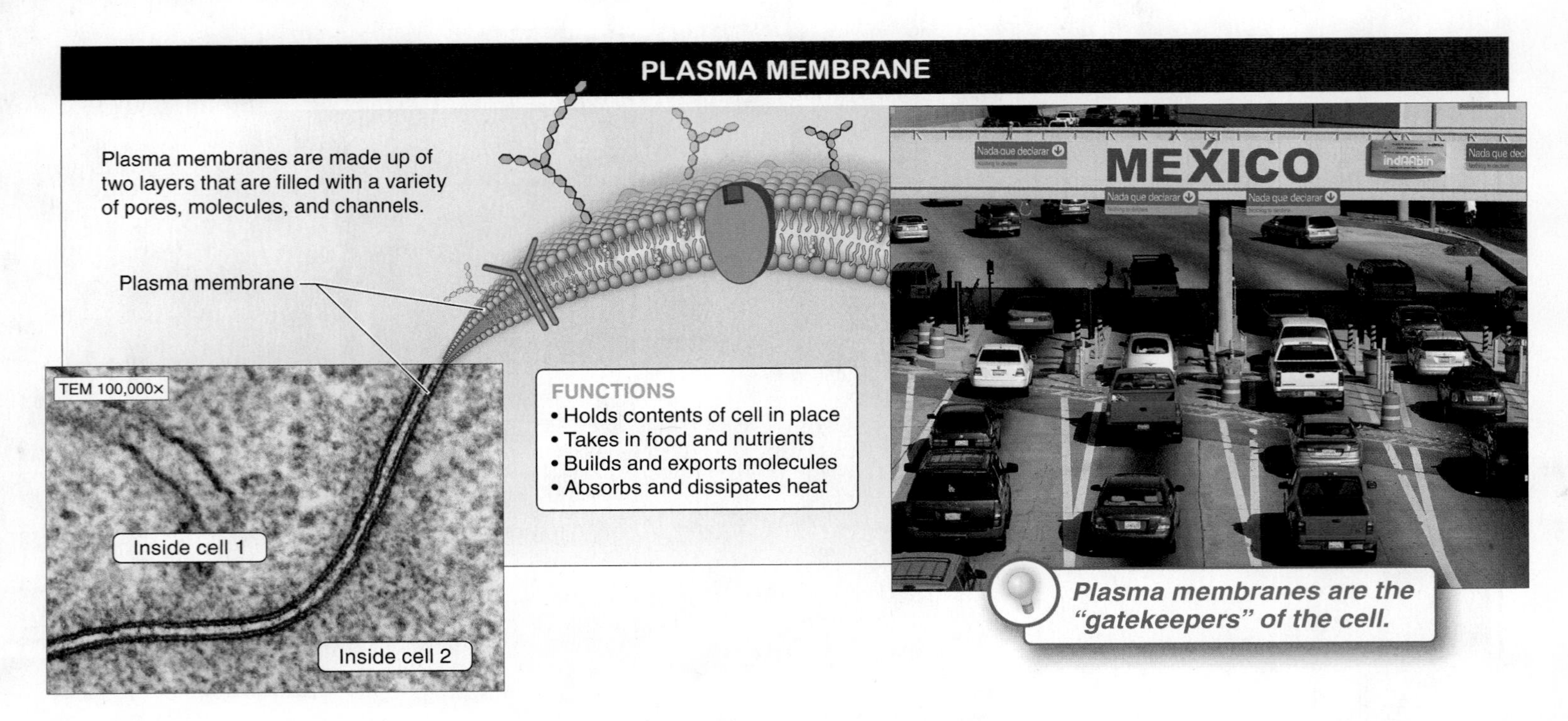

FIGURE 3-8 More than just an outer layer. The plasma membrane performs several critical functions beyond simply enclosing a cell's interior contents.

Molecules that can mix with water are described as **hydrophilic** ("water-loving") molecules. The two tails of the phospholipid are long chains of carbon and hydrogen atoms. Because they have no electrical charge, the carbon-hydrogen chains are **non-polar.** And because they are non-polar, these tails do not mix with water and are said to be **hydrophobic** ("water-fearing"). The chemical structure of phospholipids gives them a sort of split personality: their hydrophilic head

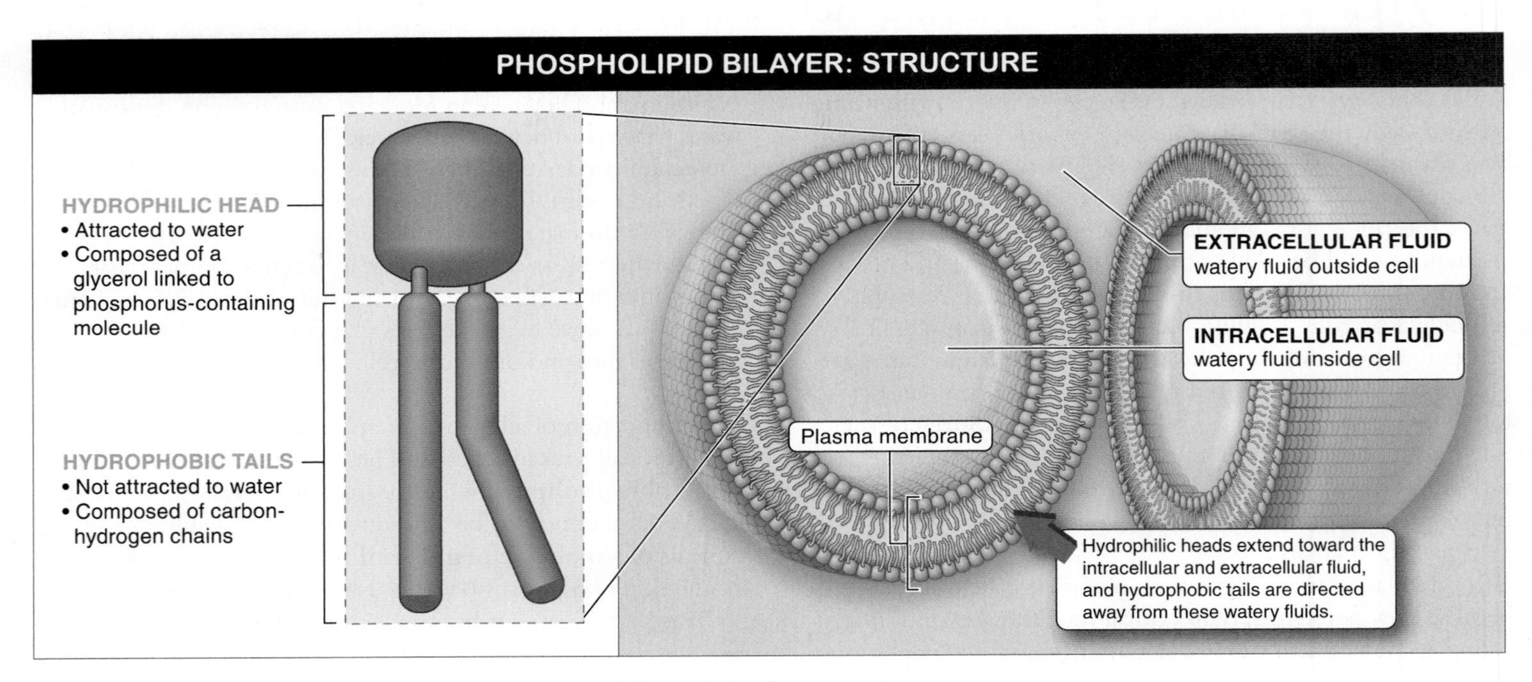

FIGURE 3-9 Good membrane material. The phospholipid bilayer of the plasma membrane prevents fluid from leaking out of the cell.

region mixes easily with water, while their hydrophobic tail region does not mix with water.

The split personality of phospholipids makes them good membrane material. Once a large number of phospholipids are packed together with all of their heads facing one way and their tails the other, we have a sheet with one side that is hydrophilic and one that is hydrophobic. In a cell's plasma membrane, two of these sheets of phospholipids are arranged so that the hydrophobic tails are all in contact with one another and the hydrophilic heads are in contact with the watery solution outside and inside the cell (see Figure 3-9). This arrangement gives us another way to describe the structure of the plasma membrane: as a **phospholipid bilayer.**

The phospholipids are not locked in place in the plasma membrane; they just float around in their side of the bilayer. They cannot pop out of the membrane, or flop from one side to the other, because their hydrophobic tails always line up away from any watery solution. Just as similarly charged sides of two magnets push away from each other, so do the hydrophobic tails in the center of the membrane push away from and avoid coming into contact with water molecules. Because the center part of the bilayer membrane is made up of hydrophobic lipids, the solution on one side of the membrane cannot leak across into the solution on the other side. In this way, the plasma membrane forms a boundary around the cell's contents.

TAKE-HOME MESSAGE 3•4

Every cell of every living organism is enclosed by a plasma membrane, a two-layered membrane that holds the contents of a cell in place and regulates what enters and leaves the cell.

3•5

Molecules embedded in the plasma membrane help it perform its functions.

While the plasma membrane's phospholipid bilayer construction forms the cell's basic boundary with its environment, there is much more to a plasma membrane. Embedded within or attached to the phospholipid bilayer are different types of protein, carbohydrate, and lipid molecules (**FIGURE 3-10**).

The proteins found in the plasma membrane enable it to carry out most of its gatekeeping functions. For every 50 to 100 phospholipids in the membrane, there is one protein molecule. Some of these proteins, called **transmembrane proteins,** penetrate right through the lipid bilayer, from one side to the other. Others, called **surface proteins,** reside primarily on the inner or outer surface of the membrane.

What determines whether a protein resides on the surface or extends through the bilayer? Its tertiary structure. Remember from Chapter 2 that all the amino acids that make up each protein have side chains that differ from one another chemically. Some of these side chains are hydrophobic, others are hydrophilic, and as a protein is assembled into its final shape, these side chains can cause parts of the protein to be attracted to hydrophobic or hydrophilic regions. Because a

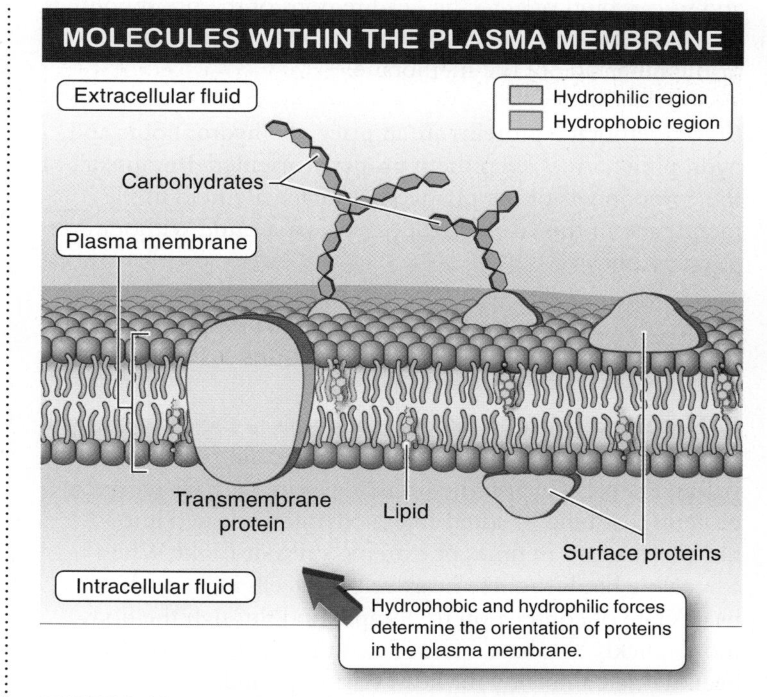

FIGURE 3-10 Protein and carbohydrate molecules are embedded in the plasma membrane.

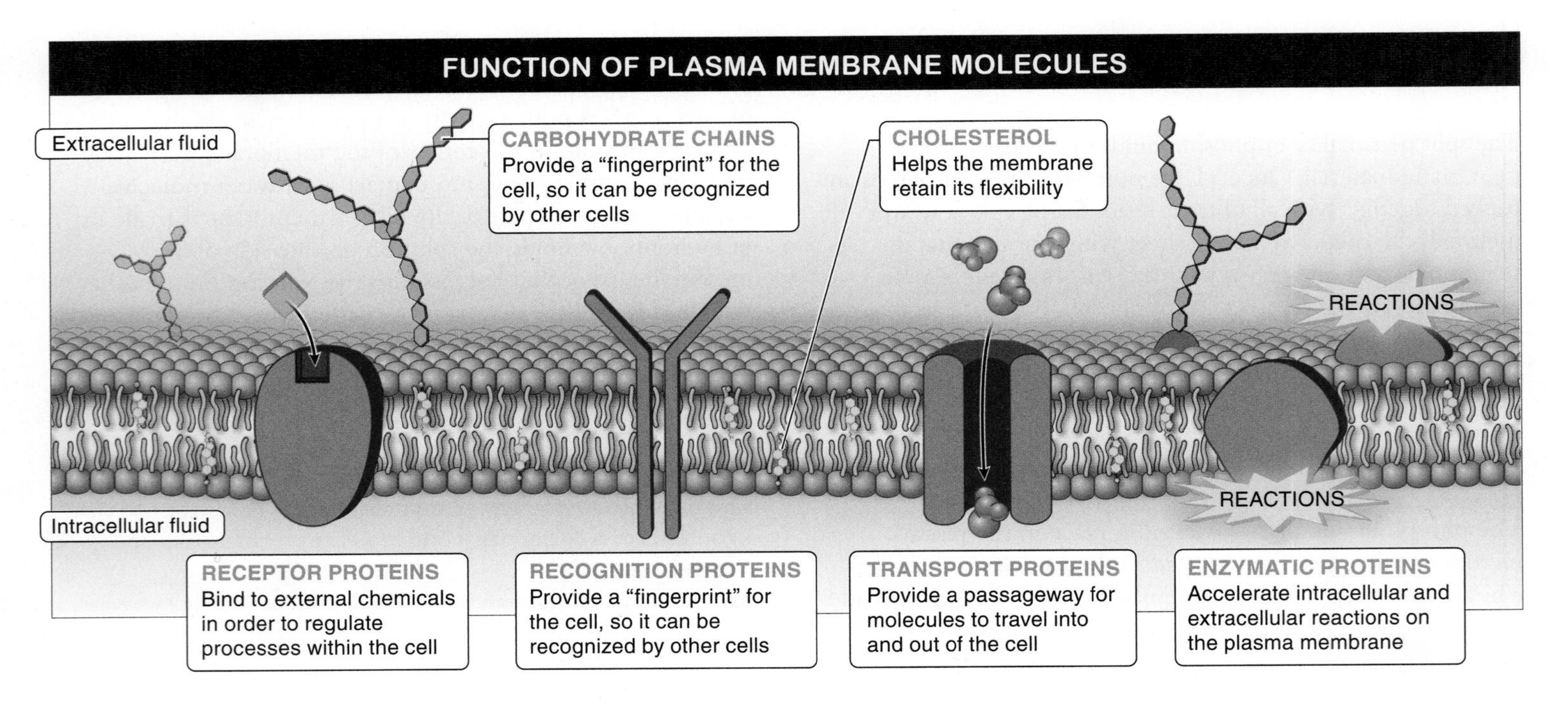

FIGURE 3-11 Plasma membrane molecules serve diverse roles.

transmembrane protein has both hydrophobic and hydrophilic regions, part of the protein can be positioned in the hydrophobic region in the center of the membrane while the other part resides in the hydrophilic regions. Peripheral membrane proteins, on the other hand, reside on the membrane surface and have an entirely hydrophilic structure and so can bind only to the head regions of the phospholipids. As a consequence, they can be positioned on either the outer or the inner side of the membrane.

Once membrane proteins are in place, the hydrophobic and hydrophilic forces keep them properly oriented. Because all of the components of the plasma membrane are held in the membrane in this manner, they can float around without ever popping out.

There are four primary types of membrane proteins, each of which performs a different function (**FIGURE 3-11**).

1. Receptor proteins bind to chemicals in the cell's external environment and, by doing so, regulate certain processes within the cell. Cells in the heart, for example, have receptor proteins that bind to adrenaline, a chemical released into the bloodstream in times of extreme stress or fright. When adrenaline binds to these heart cells, the cells increase the heart's rate of contraction to pump blood through the body more quickly. You have experienced this reaction if you've ever been startled and felt your heart start to pound.

2. Recognition proteins give each cell a "fingerprint" that makes it possible for the body's immune system (which fights off infections) to distinguish the cells that belong inside your body from those that are invaders and need to be attacked. (Note that carbohydrates also play a role in recognition.) Recognition proteins also can help cells bind to or adhere to other cells or molecules.

3. Transport proteins are transmembrane proteins that help large and/or strongly charged molecules pass through the plasma membrane. Transport proteins come in a variety of shapes and sizes, making it possible for a wide variety of molecules to be transported.

4. Enzymatic proteins (enzymes) accelerate chemical reactions on the plasma membrane's surface (a variety of enzymatic proteins exist, with some accelerating reactions on the inside of the plasma membrane and others accelerating reactions on the outside of the plasma membrane).

In addition to the various kinds of membrane proteins, two other types of molecules can be incorporated in a cell's plasma membrane.

1. Short, branched carbohydrate chains that are attached to proteins or to phospholipid heads on the outside of the cell membrane serve as part of a membrane's fingerprint, along with recognition proteins. This fingerprint allows the cell to be recognized by other cells, such as those of the immune system.

2. The plasma membrane also can contain **cholesterol,** a lipid that helps the membrane maintain its flexibility. It prevents the membrane from becoming too fluid or floppy at moderate temperatures and acts as a sort of antifreeze, preventing the membrane from becoming too rigid at freezing temperatures. The membranes of some cells are about 25% cholesterol; other plasma membranes, such as those of most bacteria and plants, have no cholesterol at all.

As we've seen, the plasma membrane is made up from several different types of molecules, like a mosaic, and many of those molecules float around, held in a proper orientation by hydrophobic and hydrophilic forces, but not always anchored in place. For these reasons, the plasma membrane is often described as a **fluid mosaic.**

TAKE-HOME MESSAGE 3•5

The plasma membrane is a fluid mosaic of proteins, lipids, and carbohydrates. Proteins found in the plasma membrane enable it to carry out most of its gatekeeping functions. The proteins act as receptors, help molecules gain entry into and out of the cell, and catalyze reactions on the inner and outer cell surfaces. In conjunction with carbohydrates, some plasma membrane proteins identify the cell to other cells. And, in addition to the phospholipids that make up most of the plasma membrane, cholesterol is an important lipid in some membranes, influencing fluidity.

3•6

Faulty membranes can cause disease.

With all the complexity of plasma membranes, it's not surprising that there are many ways in which they can malfunction. One disease that results from an improperly functioning membrane is cystic fibrosis, the single most common fatal inherited disease in the United States. At any given time, about 30,000 people in the United States have cystic fibrosis.

Cystic fibrosis occurs when an individual inherits from both parents incorrect genetic instructions for producing one type of transmembrane protein. This protein occurs primarily in the membranes of cells in the lungs and digestive tract. When functioning normally, the protein serves as a passageway that allows one type of molecule—chloride ions—to get into and out of cells.

There are more than a thousand different ways in which these genetic instructions can be defective, but in every case the result is the same: the lack of properly working chloride passageways in a cell's membrane. This defect leads to gradual accumulation of chloride ions within cells. Although it is not clear exactly how the chloride ion accumulation leads to the symptoms of cystic fibrosis, in nearly all cases two primary effects occur: an improper salt balance in the cells and a buildup of thick, sticky mucus—particularly in the lungs. Normal mucus helps to protect the lungs by trapping dust and bacteria. This mucus is then moved out of the lungs (helped along by coughing). The mucus produced by someone with cystic fibrosis, however, is too thick and sticky to be moved out of the lungs, so it collects there, where it impairs lung function and increases the risk of bacterial infection. Because of the improper cellular salt balance, one way to test for cystic fibrosis is to measure the concentration of salt in the sweat—abnormally high concentrations indicate that the person has the disease.

Although many high-tech treatments have been promised for the sufferers of cystic fibrosis and a great deal of research is being done on this disease, one of the most common treatments is decidedly low-tech. Parents help their children with cystic fibrosis clear the mucus out of their lungs by holding them on a steep slant, almost upside down, and vigorously patting or thumping their chest and back to shake loose the mucus in their lungs and move it to a place where they can cough it up. With careful treatment, the life expectancy of someone with cystic fibrosis can be 35—40 years or longer (**FIGURE 3-12**).

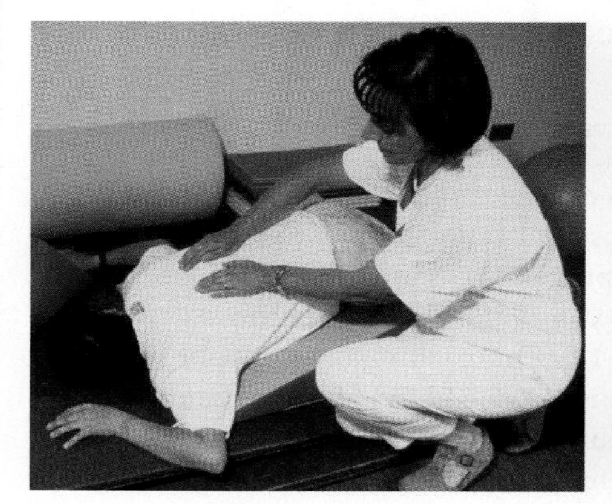

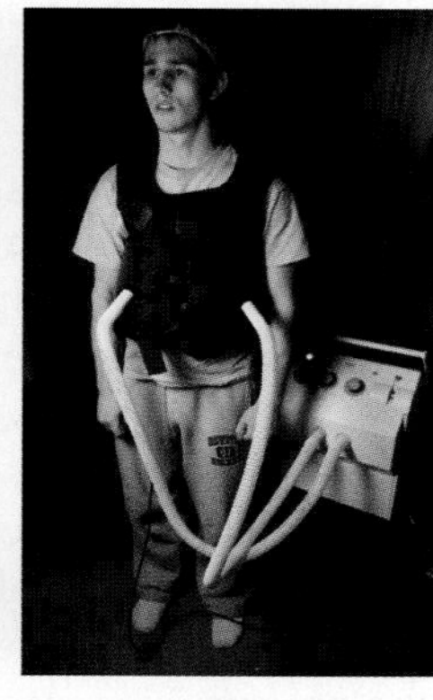

FIGURE 3-12 Moving mucus manually, or through an inhalation vest. The thick and sticky mucus produced by someone with cystic fibrosis collects in the lungs, impairing lung function and increasing the risk of bacterial infection. Thumping on the chest and back can loosen the mucus. The vest, by inflating and deflating rapidly, can have a similar effect in the course of a 20-minute session.

BETA-BLOCKERS

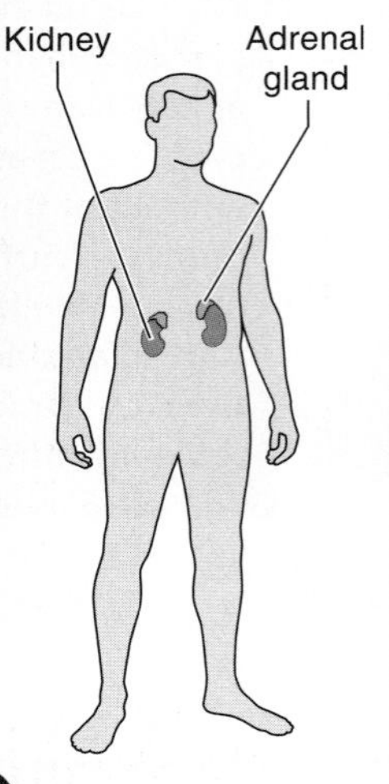

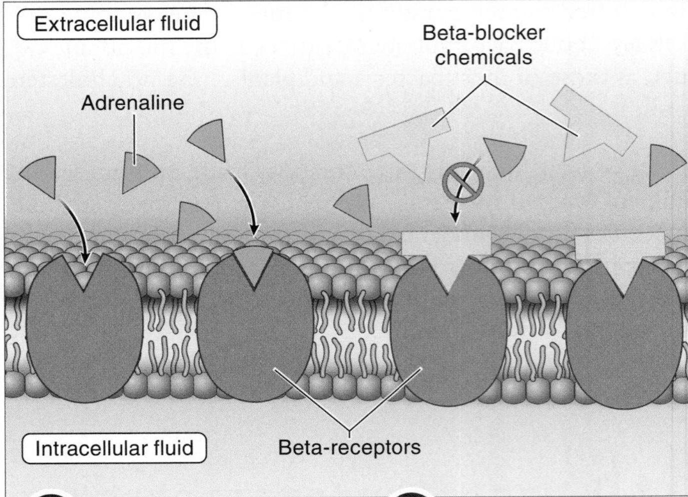

1 In stressful situations, the adrenal glands pump out adrenaline.

2 Adrenaline binds with beta-receptors on cells, causing a faster heartbeat and increased blood pressure.

3 Beta-blocker chemicals bind to receptors and prevent adrenaline from binding to the cell.

By binding to adrenaline receptors, beta-blockers reduce anxiety symptoms.

FIGURE 3-13 Reducing anxiety through beta-blockers.

While faulty membranes can cause a disorder as serious as cystic fibrosis, pharmaceutical tinkering with cell membranes can alter cellular function in ways that have beneficial effects. For example, a group of drugs called "beta-blockers" is extremely effective at reducing anxiety. This effect was discovered almost accidentally—the drugs were actually developed as a treatment for high blood pressure.

Q Why do "beta-blockers" reduce anxiety?

Here's how beta-blockers work. Many cells in your body, particularly the cells of the heart, have receptor proteins on their plasma membranes that can bind to *adrenaline,* a chemical manufactured in the adrenal glands (located atop the kidneys) that helps your body cope with stressful situations. These receptor proteins are called "beta-receptors." In stressful situations, your adrenal glands pump out adrenaline (**FIGURE 3-13**). On reaching cells in your heart (among other locations in your body) and binding to beta-receptors, the adrenaline promptly causes your heart to beat faster and more forcefully, increasing your blood pressure in the process. This reaction is fine in a short-term fight-or-flight situation, but it is not healthy over the long run because the increased pressure can damage blood vessels. Depending on its severity, this reaction can also be problematic if you are giving a presentation or taking a test, or if you're in any other anxiety-producing situation.

When you take a beta-blocker pill, the pill dissolves and the chemicals travel throughout your body until they encounter the beta-receptors. They bind to the receptors, hold on, and block the adrenaline from doing its job. This outcome slows your heart rate, causes a reduction in blood pressure, and can bring great relief to those suffering from the sweating and trembling associated with anxiety.

TAKE-HOME MESSAGE 3•6

Normal cell functioning can be disrupted when cell membranes—particularly the proteins embedded in them—do not function properly. Such malfunctions can cause health problems, such as cystic fibrosis. But disruption of normal cell membrane function can also have beneficial, therapeutic effects, such as in the treatment of high blood pressure and anxiety.

3•7

Membrane surfaces have a "fingerprint" that identifies the cell.

As mentioned earlier, every cell in your body has a "fingerprint" made from a variety of molecules on the outside-facing surface of the cell membrane. Some of these membrane molecules vary, depending on the specific function of the cell. Others are common to all of your cells and convey to your immune system: "I belong here." Cells with an improper fingerprint are recognized as foreign and are attacked by your body's defenses.

Throughout our evolutionary history, this system has been tremendously valuable in helping our bodies fight infection. In some cases, however, this vigilance is a problem, like a car alarm that goes off even when you don't want it to. Suppose you receive a liver (or any other organ) transplant. Even if the donor is a close relative, the molecular fingerprint on the cells of the donated liver is not identical to your own. Consequently, your body sees the new organ as a foreign object and puts up a fight against it (**FIGURE 3-14**). Because your body will naturally try to reject the new organ, doctors must administer drugs that suppress your immune system. Immune suppression helps you tolerate the new liver, but, as you can imagine, it leaves you without some of the defenses to fight off other foreign invaders, such as bacteria that may cause infection. The presence or absence of certain molecular markers on plasma membranes is also responsible for a person's blood type and can lead to problems with simple blood transfusions.

Q Why is it extremely unlikely that a person will catch HIV from casual contact—such as shaking hands—with an infected individual?

The AIDS-causing virus, HIV, uses the molecular markers on plasma membranes to infect an individual's cells. These same molecular markers are also the reason that it is extremely unlikely that you can catch an HIV infection from casual contact with an infected individual, such as shaking his or her hand. The specific molecular markers involved in infection by HIV belong to a group of identifying markers called "clusters of differentiation." Abbreviated as "CD markers" and having names such as CD1, CD2, and CD3, these marker molecules are proteins embedded in the plasma membrane that enable a cell to bind to outside molecules and, sometimes, transport them into the cell.

One CD marker, called the CD4 marker, is found only on cells deep within the body and in the bloodstream, such as immune system cells and some nerve cells. It is the CD4 marker, in conjunction with another receptor, that is targeted by HIV. If the virus can find a cell with a CD4 marker, it can infect you, and because the CD4 markers *never* occur on the surface of your skin cells, casual contact such as touching is very unlikely

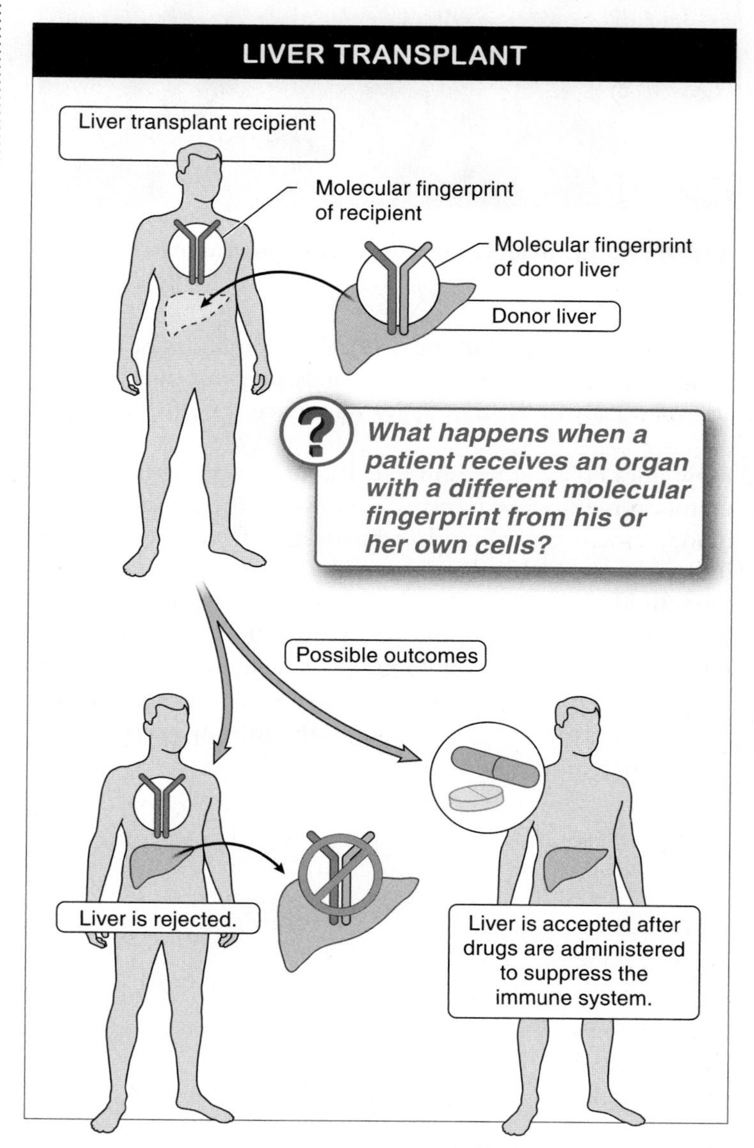

FIGURE 3-14 Mismatched molecular fingerprints can cause difficulty in organ transplantation.

HIV TRANSMISSION

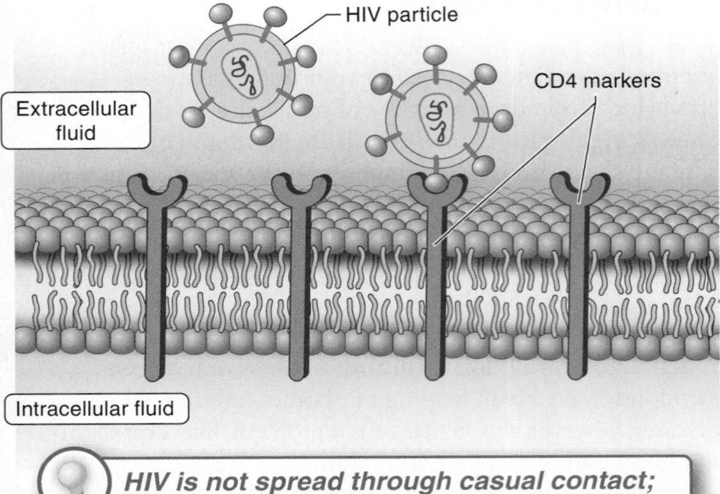

FIGURE 3-15 HIV is not transmitted through casual contact.

to transmit the virus (**FIGURE 3-15**). Even if millions of HIV particles are present on a person's hands, they just can't gain access into any of the other person's surface cells.

How does HIV get transmitted? Far and away, the most common methods of transmission involve the transfer of blood, semen, vaginal fluid, or breast milk from one individual to another. Particles of the virus are present in these fluids, as are cells already infected by the virus. As a consequence, the chief routes of transmission are from an infected mother to her child in breast milk, from an infected mother to her baby at birth, the use of contaminated needles, and unprotected sexual intercourse. Because an open sore or cut might expose some of the cells in your bloodstream to the outside world, however, it is not impossible for casual transmission of HIV to occur.

TAKE-HOME MESSAGE 3•7

Every cell in your body has a "fingerprint" made from a variety of molecules on the outside-facing surface of the cell membrane. This molecular fingerprint is key to the function of your immune system.

③ Molecules move across membranes in several ways.

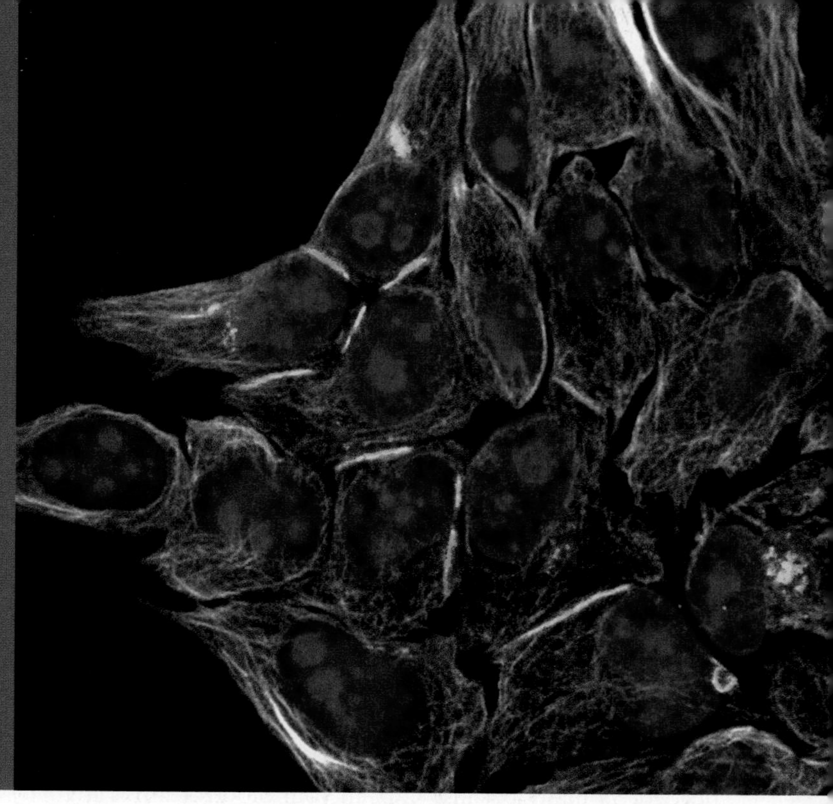

Passage of proteins between cells can be tracked (junctions between these human cancer cells are stained red and green).

3•8

Passive transport is the spontaneous diffusion of molecules across a membrane.

To function properly, cells must take in food and/or other necessary materials and must move out metabolic waste and molecules produced for use elsewhere in the body. In some cases, this movement of molecules requires energy and is called active transport. (We cover active transport later in this chapter.) In other cases, the molecular movement occurs spontaneously, without the input of energy, and is called **passive transport.** There are two types of passive transport: diffusion and osmosis. (Osmosis is the topic of the next section.)

Diffusion is passive transport in which a particle, called a **solute,** is dissolved in a gas or liquid (a **solvent**) and moves from an area of high solute concentration to an area of lower concentration (**FIGURE 3-16**). (When we are talking about cells, "diffusion" usually refers to the situation in which the solute moves across a membrane.) In the absence of other forces, molecules of a substance will always tend to move from

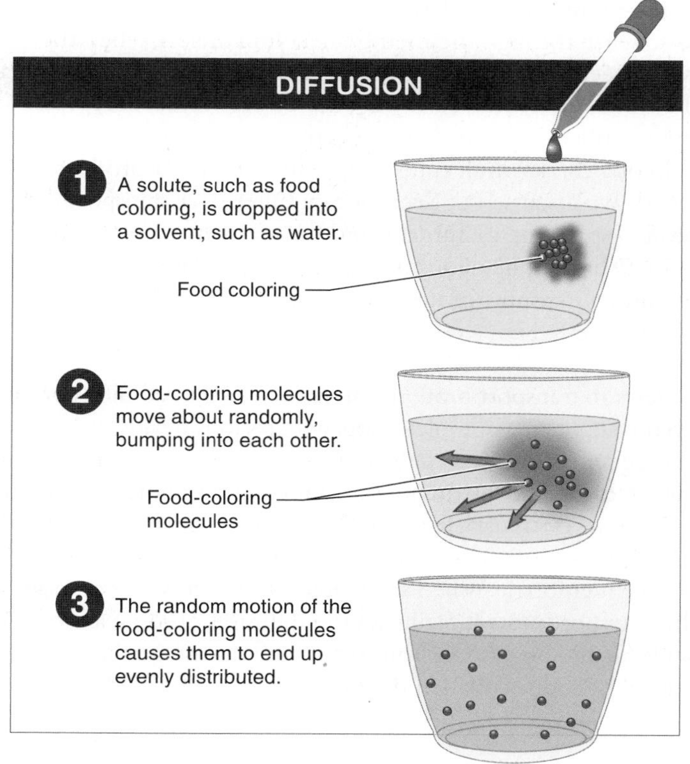

FIGURE 3-16 Diffusion: a form of passive transport that results in an even distribution of molecules.

where they are more concentrated to where they are less concentrated. This movement occurs because molecules move randomly and are equally likely to move in any direction. And when certain molecules are highly concentrated, they keep bumping into each other and eventually end up evenly distributed. For this reason, we say that molecules tend to *move down their concentration gradient*. For a simple illustration, drop a tiny bit of food coloring into a bowl of water and wait for a few minutes. The molecules of dye are initially clustered together in a very high concentration. Gradually, they disperse down their concentration gradient until the color is equally spread throughout the bowl.

Molecules such as oxygen (O_2) and carbon dioxide (CO_2) that are small and carry no charge can pass directly through the phospholipid bilayer of the membrane without the assistance of any other molecules, in a process called **simple diffusion.** Each time you take a breath, for example, there is a high concentration of O_2 molecules in the air you pull into your lungs. That oxygen diffuses across the plasma membranes of the lung cells and into your bloodstream, where red blood cells pick it up and deliver it to parts of your body where it is needed. Similarly, because CO_2 in your bloodstream is at a higher concentration than in the air in your lungs, it diffuses from your blood into your lungs and is released to the atmosphere when you exhale (**FIGURE 3-17**).

Most molecules, however, can't get through plasma membranes on their own. They may be electrically charged (polar) and, hence, repelled by the hydrophobic middle region of the phospholipid bilayer. Or the molecules may be too big to squeeze through the membrane. Nonetheless, when there is a concentration gradient from one side of the membrane to the other, these molecules may still be able to diffuse across the membrane, down their concentration gradient, with the help of a carrier molecule—one of the transport proteins, which we discussed in Section 3-5. Often, this transport protein spans the membrane and functions like a revolving door. When spontaneous diffusion across a plasma membrane requires a transport protein, it is called **facilitated diffusion** (see Figure 3-17).

Defects in transport proteins can reduce facilitated diffusion or even bring it to a complete stop, with serious health consequences. Many genetic diseases are the result of inheriting incorrect genetic instructions for building transport proteins. In the disease cystinuria, for example, incorrect genetic instructions result in a malformed transport protein in the plasma membrane. When structured and functioning properly, this transport protein facilitates the diffusion of some amino acids (including cystine, from which the disease gets its name) out of the kidneys. When the proteins are malformed, they cannot facilitate this diffusion and these amino acids build up in the kidneys, forming painful and dangerous kidney stones.

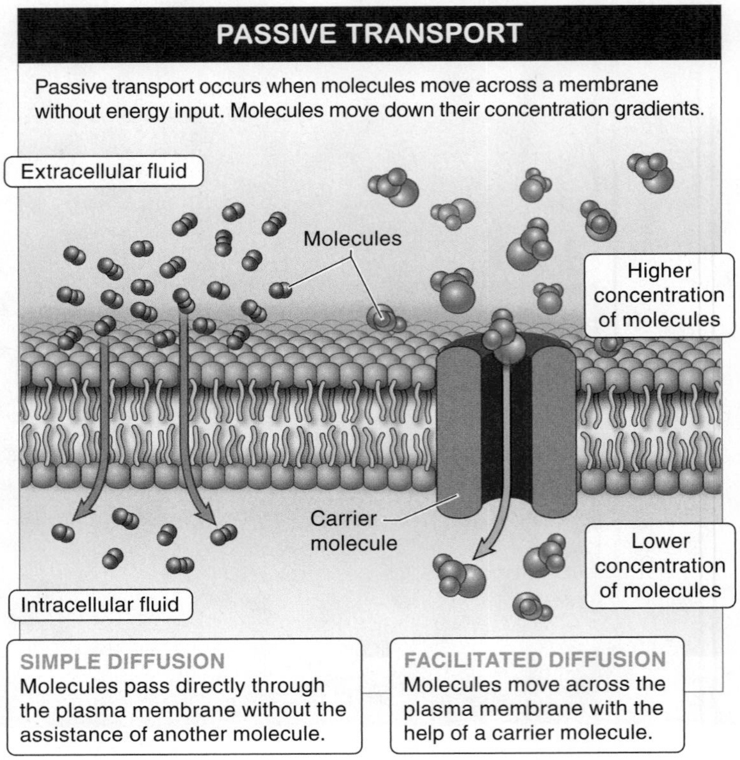

FIGURE 3-17 Simple and facilitated diffusion.

Diffusion across membranes doesn't just occur in animals, we see it in all organisms on earth. As we'll see in Chapter 4, one of the most important biological processes on earth is the diffusion of CO_2 from the atmosphere (an area of relatively high concentration) into the leaf cells of plants (areas of relatively low concentration), where it can be attached to other molecules, forming sugars through photosynthesis. At the same time, O_2 diffuses out of the leaves and into the atmosphere.

TAKE-HOME MESSAGE 3•8

For proper functioning, cells must acquire food molecules and/or other necessary materials from outside the cell. Similarly, metabolic waste molecules and molecules produced for use elsewhere in the body must move out of the cell. In passive transport—which includes simple and facilitated diffusion and osmosis—the molecular movement occurs spontaneously, without the input of energy. This generally occurs as molecules move down their concentration gradient.

3•9

Osmosis is the passive diffusion of water across a membrane.

Just as solute molecules will passively diffuse down their concentration gradients, water molecules will also move from areas of high concentration to areas of low concentration to equalize the concentration of water inside and outside the cell. The diffusion of water across a membrane is a special type of passive transport called **osmosis** (**FIGURE 3-18**). As solute molecules diffuse across a plasma membrane, molecules of water also move across the membrane, equalizing the water concentration inside and outside the cell.

Osmosis can have some dramatic effects on cells. It turns out that many molecules just can't move across a cell membrane. They may be too big or too charged, or there may not be any carrier molecules in the membrane that can move them. Let's imagine, for instance, that there is a high concentration of a large molecule inside a cell and that it can't pass through the cell membrane. Although the molecule can't move out of the cell and down its concentration gradient, water can move into the cell down *its* concentration gradient. As water diffuses into the cell, the cell will get larger.

When a cell is in a solution, we describe the concentration of solutes outside the cell relative to that inside the cell as the **tonicity** of the solution (see Figure 3-18). A **hypertonic** solution has a greater concentration of solutes than the solution inside the cell. As a consequence, if the cell membrane is not permeable to the solutes, water will move out of the cell into the surrounding solution, equalizing the solute concentration inside and out, and the cell shrivels. A **hypotonic** solution has a lower solute concentration than the solution inside the cell. If the cell membrane is not permeable to the solutes, water will move into the cell and it will swell. In an **isotonic** solution, the concentration of solutes is the same inside and outside the cell.

You can see osmosis in action in your own kitchen (**FIGURE 3-19**). Take a stalk of celery and leave it on the counter for a couple of hours. As water evaporates from the cells of the celery stalk, it will shrink and become limp. You can make it crisp again by placing it in a solution of distilled water. Why distilled water? Because it contains very few dissolved molecules—it is a hypotonic solution. The cells in the celery stalk, on the other hand, contain many dissolved molecules, such as salt. Because those dissolved molecules can't easily pass across the plasma membrane of the celery stalk cells, the water molecules move into the cells (moving from the pool of distilled water where they are at super-high concentration to the interior of the celery cells where water is scarcer). Would the celery regain its crispness if you placed it in concentrated salt water? No, because in the hypertonic salt solution there are more dissolved molecules (and fewer water molecules) outside the celery cells, so what little water remains in the celery cells actually moves, via osmosis, out of the celery, causing it to shrivel even more.

OSMOSIS

Osmosis is a type of passive transport by which water diffuses across a membrane, in order to equalize the concentration of water inside and outside the cell. The direction of osmosis is determined by the total amount of solutes on either side of the membrane. Water will always move toward the side that has a greater concentration of solutes.

HYPERTONIC SOLUTION
- Solute concentrations are higher in the extracellular fluid.
- Water diffuses out of cells.

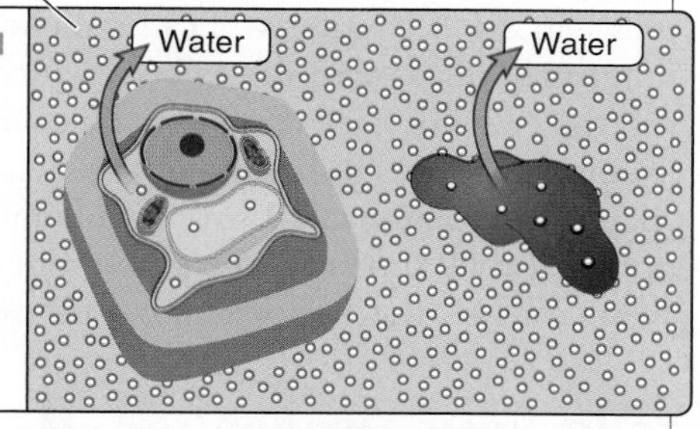

HYPOTONIC SOLUTION
- Solute concentrations are lower in the extracellular fluid.
- Water diffuses into cells.

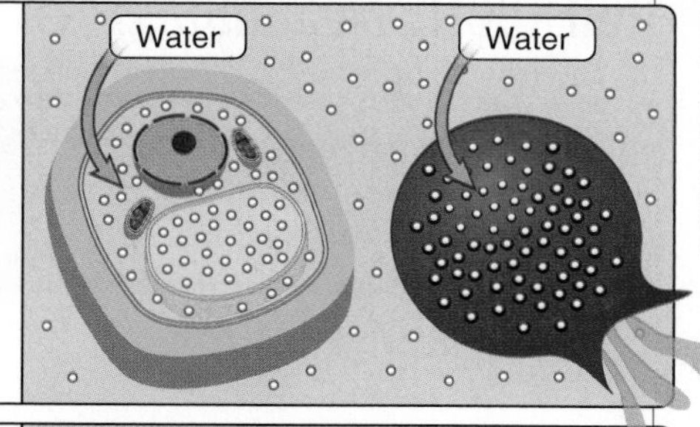

ISOTONIC SOLUTION
- Solute concentrations are balanced.
- Water movement is balanced.

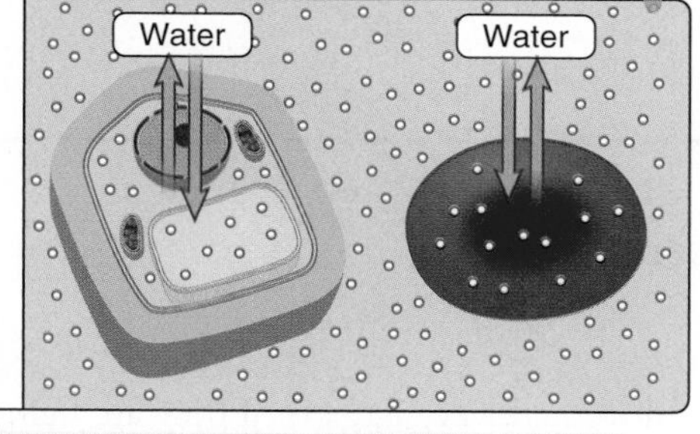

Unlike animal cells, plant cells generally do not explode in hypotonic solutions, because their rigid cell walls limit cellular expansion.

FIGURE 3-18 Osmosis overview.

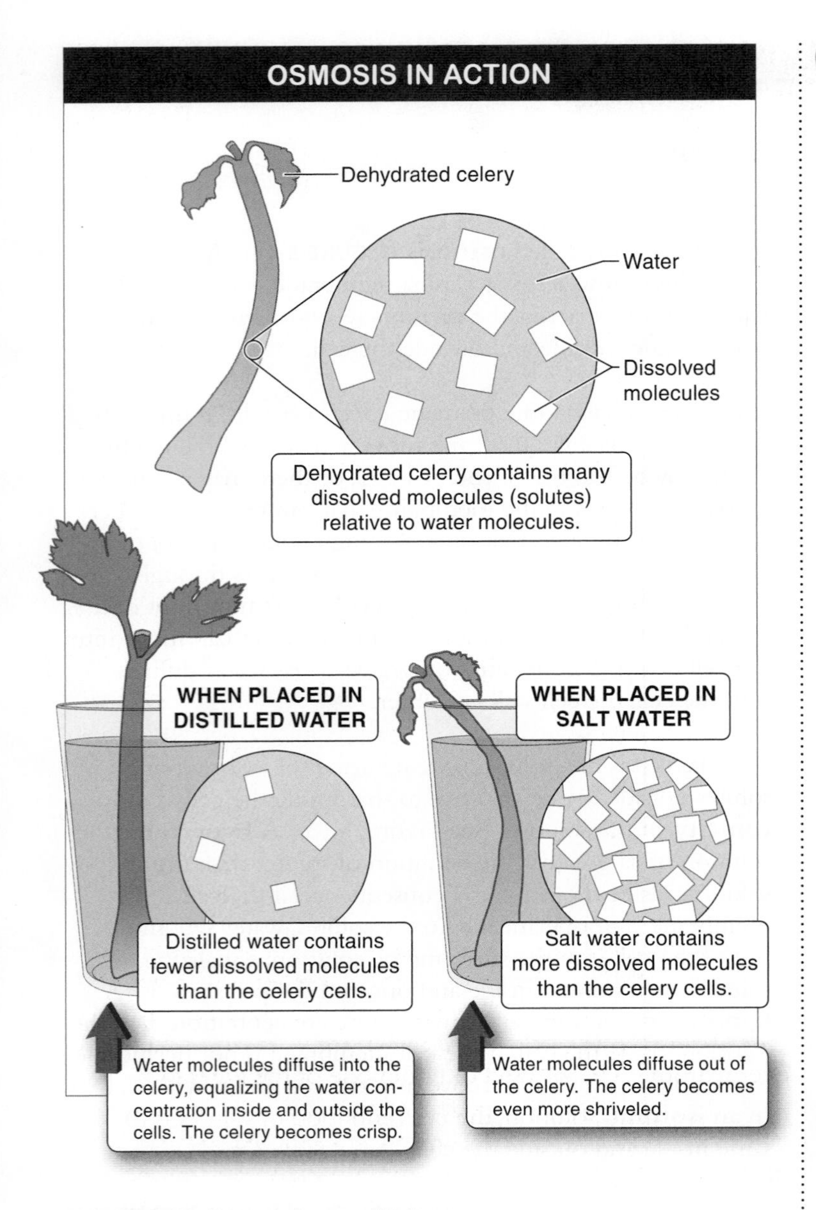

FIGURE 3-19 **Osmosis in your kitchen.**

Q How do laxatives relieve constipation?

A more practical use of osmosis is seen with the laxative Milk of Magnesia. This product contains magnesium salts, which are poorly absorbed from the digestive tract. As a consequence, after a dose of the laxative, water moves via osmosis from the surrounding cells into the intestines. The water softens the feces in the intestines and increases the fecal volume, thereby relieving constipation.

It's important to note that the direction of osmosis is determined only by a difference in the *total concentration* of all the molecules dissolved in the water: it does not matter what the solutes are, only how many molecules of solutes there are. To determine which way the water molecules will move, you need to determine the total amount of "dissolved stuff" on either side of the membrane. The water will move toward the side with more stuff dissolved in it. For this reason, even small increases in the salinity of lakes can have disastrous consequences for the organisms living there, from fish to bacteria. Conversely, putting animal cells—such as red blood cells—in distilled water will cause them to explode, because water will diffuse into the cell (which contains more solutes) and the cell will swell and burst. This will not generally happen to plant cells because (as we will see later in this chapter) the plant cell plasma membrane is surrounded by a rigid cell wall that limits the amount of cell expansion possible when water moves in by osmosis.

TAKE-HOME MESSAGE 3•9

The diffusion of water across a membrane is a special type of passive transport called osmosis. Water molecules move across the membrane until the concentration of water inside and outside the cell is equalized.

3•10

In active transport, cells use energy to move small molecules into and out of the cell.

Molecules can't always move spontaneously and effortlessly in and out of cells. Sometimes their transport needs energy, in which case the process is called **active transport.** Such energy expenditures may be necessary if the molecules to be moved are very large or if they are being moved *against* their concentration gradient. In all active transport, proteins embedded in the membrane act like motorized revolving doors, pushing molecules into the cell regardless of the

ACTIVE TRANSPORT

Active transport occurs when the movement of molecules into and out of a cell requires the input of energy. For example, in response to eating, the cells lining your stomach use ATP to pump large numbers of H+ ions into the stomach.

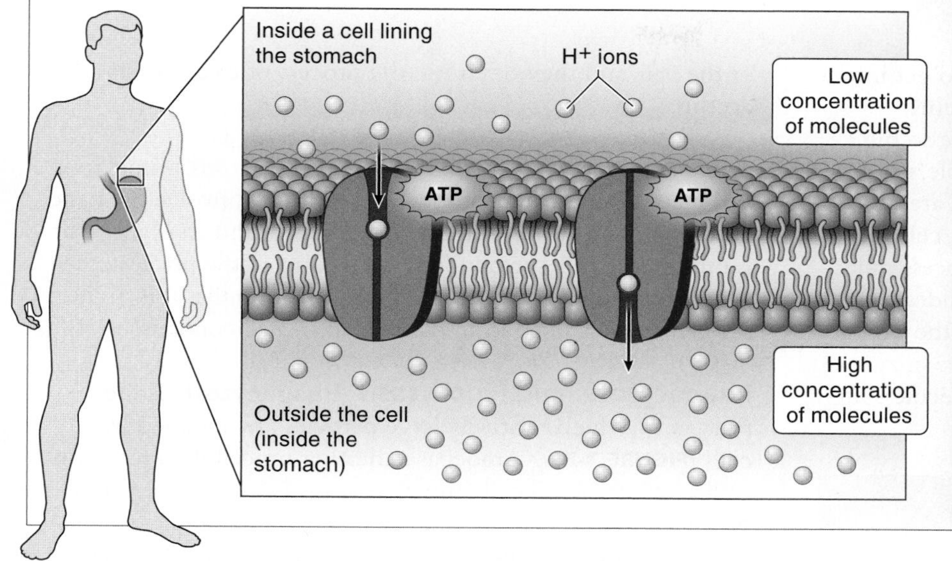

FIGURE 3-20 Active transport. Some energy expenditure is required to move molecules into and out of cells when the molecules are very large or are being moved against their concentration gradient.

concentration of those molecules on either side of the membrane. There are two distinct types of active transport, primary and secondary, which differ only in the source of the fuel that keeps the revolving doors spinning.

Every time you eat a meal, you start the motors that spin the revolving-door transport proteins in your stomach (**FIGURE 3-20**). To help break down the food into more digestible bits, the cells lining your stomach create an acidic environment by pumping large numbers of H^+ ions (also called protons) into the stomach contents, against their concentration gradient. (That is, there are more H^+ ions in the stomach contents than there are inside the cells lining the stomach.) All of this H^+ pumping increases your ability to digest the food but comes at a great energetic cost—in the form of usage of the high-energy molecule ATP—because the protons would not normally flow into a region against their concentration gradient. When active transport uses energy directly from ATP to fuel the revolving door, the process is called **primary active transport.** (We explore ATP in more detail in Chapter 4.)

Many transporter proteins use an indirect method of fueling their activities rather than using energy released directly from ATP. In the process of **secondary active transport,** the transport protein simultaneously moves one molecule against its concentration gradient while letting another flow down its concentration gradient. Although no ATP is used directly in this process, at some other time and in some other location, energy from ATP was used to pump one of the types of molecule involved in the secondary transport process against its concentration gradient. The process is a bit like using energy to pump water up to the top of a high water tower. Later, the water can be allowed to run out of the tower over a water wheel, which can, in turn, power a process such as grinding wheat into flour. Our bodies frequently use the energy from one reaction that occurs spontaneously to fuel another reaction that requires energy.

TAKE-HOME MESSAGE 3•10

In active transport, moving molecules across a membrane requires energy. Active transport is necessary if the molecules to be moved are very large or if they are being moved against their concentration gradient. Proteins embedded in the plasma membrane act like motorized revolving doors to actively transport (pump) the molecules.

3•11

Endocytosis and exocytosis are used for bulk transport of particles.

Many molecules needed by cells are just too big to get into a cell by passive or active transport. After all, a protein embedded in a thin plasma membrane can only be so big, and some cells in your immune system, for example, must even ingest (and destroy) entire bacterial cells that are invading your body. To absorb such large particles, cells engulf them with their plasma membrane in a process called **endocytosis.** Cells also have another method besides passive and active transport for moving molecules out of the cell. Cells that manufacture molecules (such as digestive enzymes) for use elsewhere in the body must get those molecules out of the cell, and they often use the process of **exocytosis** to do this.

There are three types of endocytosis: phagocytosis, pinocytosis, and receptor-mediated endocytosis. All three involve the basic process of the plasma membrane oozing around an object that is outside the cell, surrounding it, forming a little pocket called a **vesicle,** and then pinching off the vesicle so that it is inside the cell but separated from the rest of the cell contents.

Phagocytosis and Pinocytosis **Phagocytosis** is the process by which relatively large particles are engulfed by cells (**FIGURE 3-21**). Amoebas and other unicellular protists, as well as white blood cells, use phagocytosis to consume entire organisms either as food or as their way of defending against pathogens (disease-causing organisms or substances). Whereas "phagocytosis" comes from the Greek for "eat" and "container," the term **pinocytosis** comes from the Greek for "drink" and "container" and describes the process of cells taking in dissolved particles and liquids. The two processes are largely the same, except that the vesicles formed during pinocytosis are generally much smaller than those formed during phagocytosis.

Receptor-Mediated Endocytosis The third type of endocytosis, **receptor-mediated endocytosis,** is much more specific than either phagocytosis or pinocytosis. In this process, receptor molecules on the surface of a cell sit waiting until the one type of molecule they recognize bumps into them. For one receptor it might be insulin, for another it might be cholesterol. Many receptors of the same type are often clustered together in a cell's plasma membrane. When the appropriate molecule binds to each of the receptor proteins, the membrane begins to fold inward, first forming a little pit and then completely engulfing the molecules, which are still attached to their receptors.

One of the most important examples of receptor-mediated endocytosis involves cholesterol (**FIGURE 3-22**). Most cholesterol that circulates in the bloodstream is in the form of particles called low-density lipoproteins, or LDL. Each molecule of LDL is a cholesterol globule coated by phospholipids. Proteins embedded within the LDL's phospholipid coat are recognized by receptor proteins built into the plasma membranes of liver cells. Once bound to the receptors, the LDL molecule is consumed by the cell via endocytosis. Inside the cell, the

PHAGOCYTOSIS

TEM 4,700×

Phagocytosis is a type of endocytosis by which cells engulf large particles.

Plasma membrane

Large particle

Extracellular fluid

Intracellular fluid

1 The plasma membrane forms a pocket-like vesicle around a large particle.

2 The particle is transported into the cell in a vesicle.

Vesicle

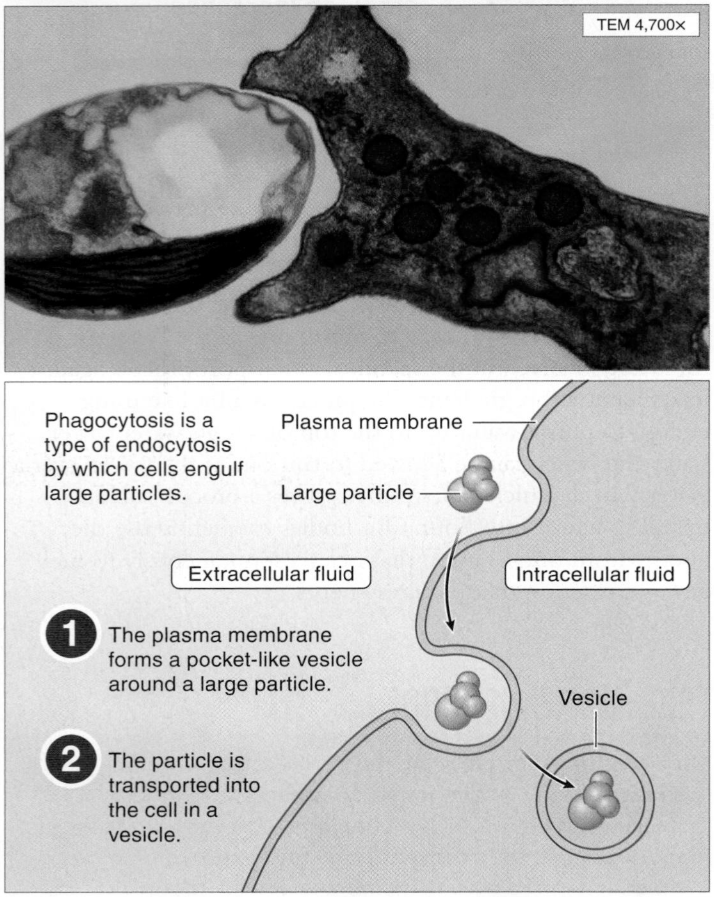

FIGURE 3-21 Through phagocytosis, amoebas and other unicellular protists, as well as white blood cells, consume other organisms for food or for defense.

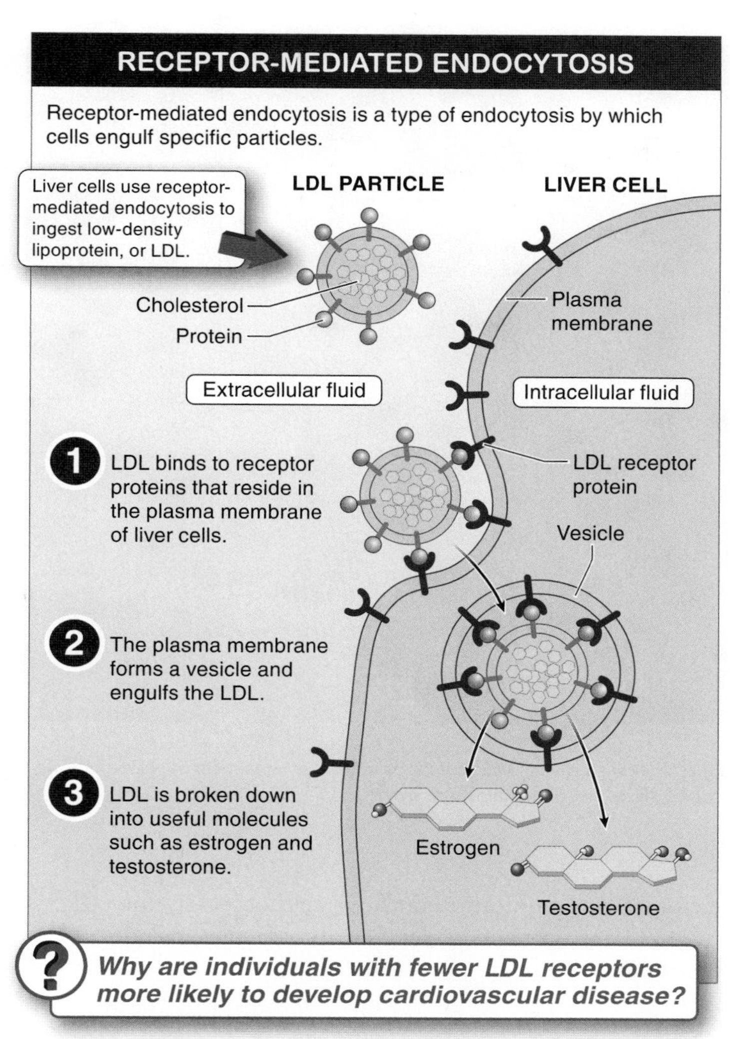

FIGURE 3-22 Receptor proteins aid in endocytosis. This process is important in the ability of your liver to remove cholesterol from your bloodstream.

cholesterol is broken down and used to make a variety of other useful molecules, such as the hormones estrogen and testosterone.

> **Q** Faulty cell membranes are a primary cause of cardiovascular disease. What modification to these membranes might be an effective treatment?

Circulating cholesterol often builds up on the walls of arteries, reducing the area (that is, the diameter of the blood vessel) available for blood flow and causing the artery to harden. Because too much circulating cholesterol in LDL molecules can lead to cardiovascular disease and death (**FIGURE 3-23**), it's good to remove as much cholesterol from your bloodstream as quickly as possible. Individuals lucky enough to have large numbers of LDL receptors on the plasma membranes of their liver cells have a significantly lower risk of cardiovascular disease.

Conversely, individuals who consume food laden with too much cholesterol (such as egg yolks, cheese, and sausages), or those who have the misfortune of inheriting genes that code for faulty liver cell membranes that have only a few LDL receptors, can have a heavy load of circulating blood cholesterol. This outcome can result in the early onset of cardiovascular disease. In the extreme case where an individual is born with no LDL receptors at all (a disorder called familial hypercholesterolemia), circulating cholesterol accumulates in the arteries so rapidly that cardiovascular disease begins to develop even before puberty, and death from a heart attack can occur before the age of 30.

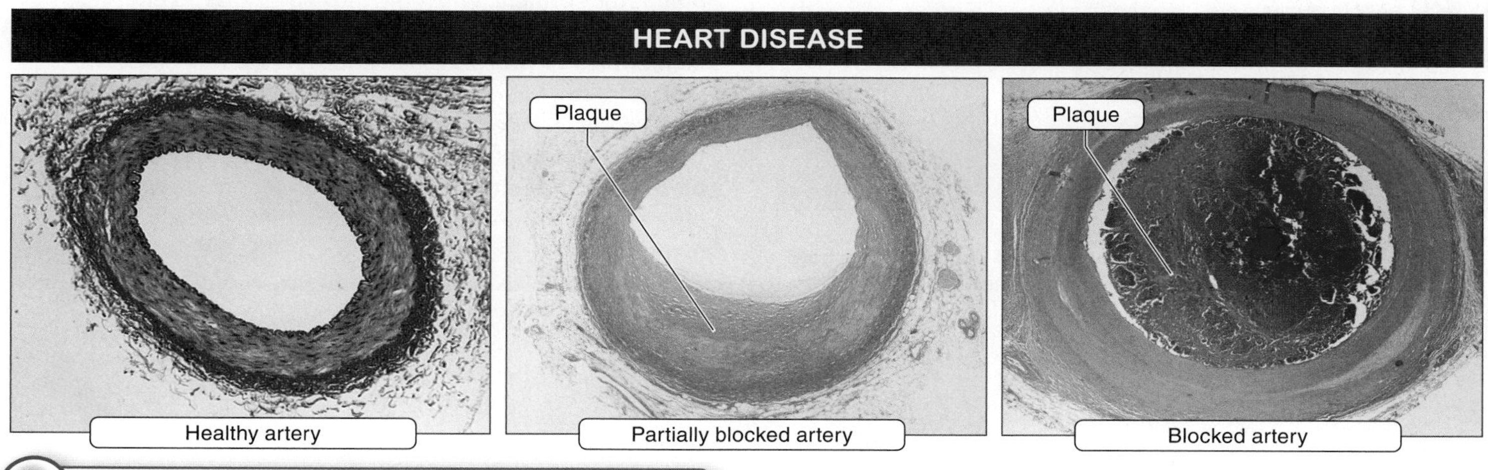

LDL receptors bind to LDL molecules, removing cholesterol from the bloodstream and reducing the risk of cardiovascular disease.

FIGURE 3-23 Comparison of a healthy artery and arteries choked by the buildup of cholesterol.

Exocytosis Cells in the pancreas produce a chemical called insulin that moves throughout the circulatory system, informing body cells that there is glucose in the bloodstream that ought to be taken in and utilized for energy. The insulin molecule is much too large to pass out through the plasma membranes of the cells in which it is manufactured. As a result, after molecules of insulin are produced, they are coated with a phospholipid membrane to form a vesicle. The insulin-carrying vesicle then moves through the cytoplasm to the inner surface of the cell's plasma membrane. Once there, the phospholipid membrane surrounding the insulin and the phospholipid membrane of the cell fuse together, dumping the contents of the vesicle out of the cell and into the bloodstream (**FIGURE 3-24**).

Exocytosis, the movement of molecules out of a cell, occurs throughout the body and is not restricted to large molecules. In the brain and other parts of the nervous system, for example, communication between cells occurs as one cell releases large numbers of very small molecules, called neurotransmitters, via exocytosis.

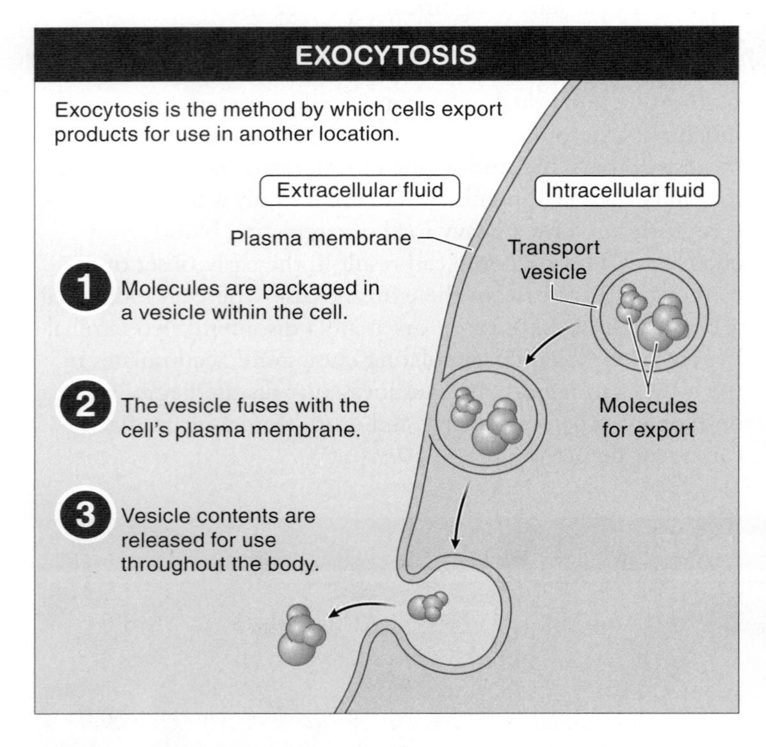

FIGURE 3-24 Exocytosis moves molecules out of the cell. This is how insulin is exported from the cells where it is synthesized.

After drinking a glass of orange juice, your blood concentrations of glucose increase, causing the cells of your pancreas to secrete insulin through exocytosis.

TAKE-HOME MESSAGE 3•11

When materials cannot get into a cell via diffusion or a pump (for example, when the molecules are too big), cells can engulf the molecules or particles with their plasma membrane in a process called endocytosis. Similarly, molecules can be moved out of a cell via exocytosis. In both processes, the plasma membrane moves to surround the molecules or particles and forms a little vesicle that is pinched off inside the cell (endocytosis) or fuses with the plasma membrane and dumps its contents outside the cell (exocytosis).

4 Cells are connected and communicate with each other.

Cells reach out and connect; no telephone lines required.

3•12

Connections between cells hold them in place and enable them to communicate with each other.

So far in this chapter we have examined the cell as a free-living and independent entity. The majority of cells in any multicellular organism, however, are connected to other cells. Let's consider how they are connected and how they communicate with each other. There is a variety of methods by which cells adhere to other cells, involving numerous types of protein and glycoprotein adhesion molecules. We examine three primary types of connections between animal cells: (1) tight junctions, (2) desmosomes, and (3) gap junctions (**FIGURE 3-25**); and one type of connection between plant cells: plasmodesmata.

Tight junctions form continuous, water-tight seals around cells and also anchor cells in place. Much like the caulking around a tub or sink that keeps water from leaking into the surrounding walls, tight junctions prevent fluid flow between cells. Tight junctions are particularly important in the small intestine, where digestion occurs. Cells lining the small intestine absorb nutrients from the watery fluid moving through your gut. If the fluid—and the resident bacteria—inside the intestine were to leak between the cells and into your body cavity, you would not be able to extract sufficient energy and nutrients from your food, and the bacteria would make you sick. The tight junctions instead force fluid to pass *into* the cells that line the intestine, where the nutrients can be utilized.

Desmosomes are like spot welds or rivets that fasten cells together into strong sheets. They occur at irregular intervals and function like fastened Velcro: they hold cells together but are not water-tight, allowing fluid to pass around them. Desmosomes are found in much of the tissue that lines the cavities of animal bodies. They also are found in muscle tissue, holding fibers together. Genetic disorders that reduce cells' ability to form desmosome proteins or lead to destruction of desmosomes by the immune system result in the formation of blisters, where layers of skin separate from each other.

Finally, **gap junctions** are pores surrounded by special proteins that form open channels between two cells (see Figure 3-25). Functioning like secret passageways, these junctions are large enough for salts, sugars, amino acids, and the chemicals that carry electrical signals to pass through, but are too small for the passage of organelles or very large molecules such as proteins and nucleic acids. Gap junctions

THREE PRIMARY CONNECTIONS BETWEEN ANIMAL CELLS

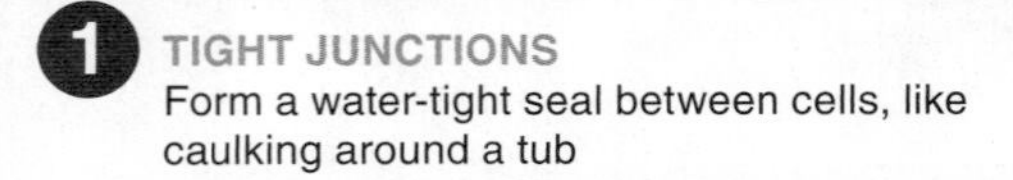

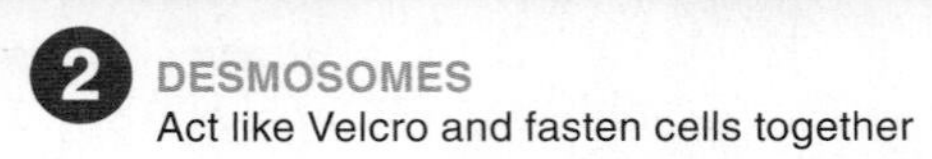

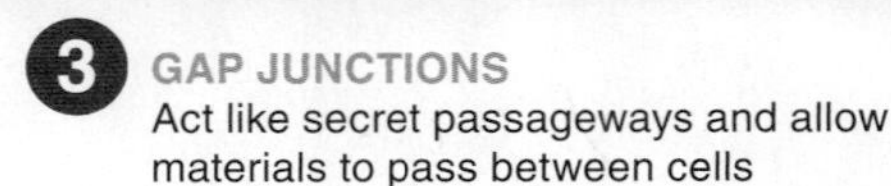

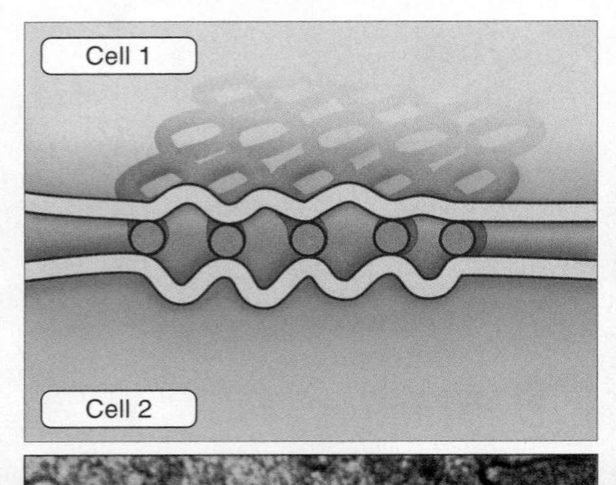

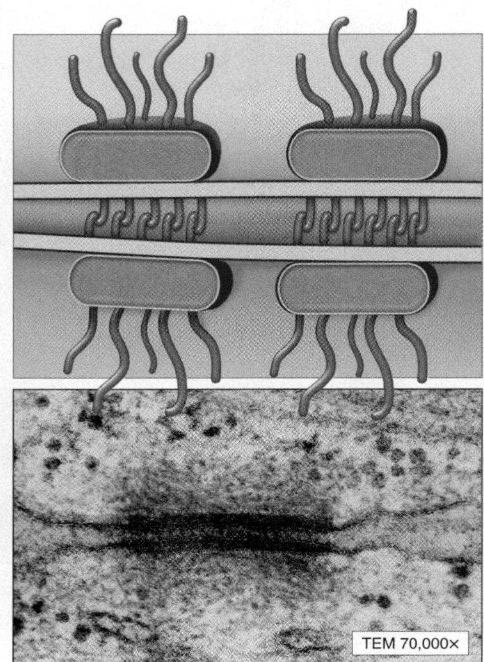

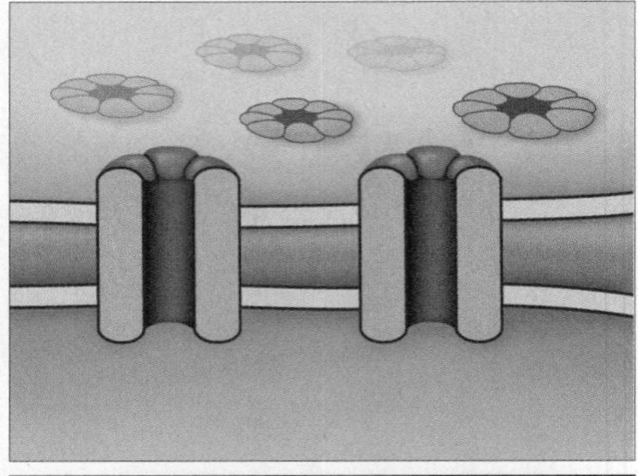

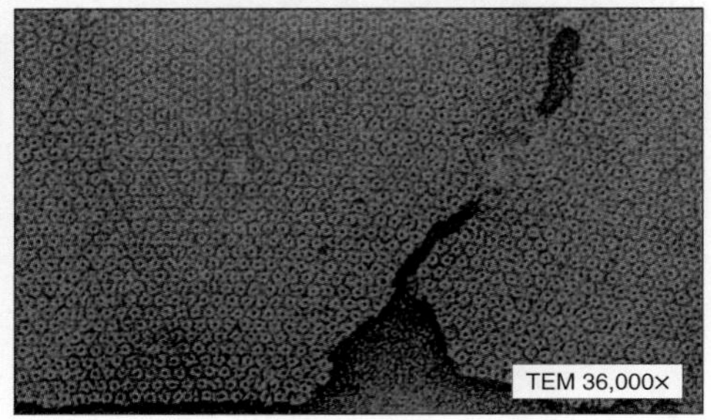

FIGURE 3-25 Cell connections: tight junctions, desmosomes, and gap junctions.

are an important mechanism for cell-to-cell communication. In the heart, for example, the electrical signal telling muscle cells to contract is passed from cell to cell through gap junctions. Gap junctions are also important in allowing a cell to recognize that it has bumped up against another cell; chemicals flowing from one cell to the next can signal the body to stop producing cells of a particular type.

> **Q** How can a lack of communication between cells lead to cancer?

Normal cells generally stop dividing when they bump up against other cells, a phenomenon called **contact inhibition.** Cancer cells, however, do not have this feature. Instead, they divide continuously, eventually forming a mass of cells, called a tumor, which can cause serious health problems. Cancer cells differ from other cells in several important ways; one of these ways is that they usually have reduced numbers of gap junctions. With fewer of these "secret passageways" of communication, it may be harder for a cell to detect that it has bumped up against other cells, in which case the cell is more likely to continue dividing, possibly contributing to its eventually becoming a tumorous mass of dividing cells.

In most plants, the cells have anywhere from 1,000 to 100,000 microscopic tube-like channels, called **plasmodesmata** (*sing.* **plasmodesma**) (see Figure 3-36), connecting the cells to each other and enabling communication and transport between them. In fact, because so many of the cells are connected to one another, sharing cytoplasm and other molecules, some biologists have wondered whether we should consider a plant as just one big cell. How would you respond to that idea?

TAKE-HOME MESSAGE 3•12

In multicellular organisms, most cells are connected to other cells. The connections can form a water-tight seal between the cells (tight junctions), can hold sheets of cells together while allowing fluid to pass between the cells (desmosomes), or can function like secret passageways, allowing the movement of cytoplasm, molecules, and other signals between cells (gap junctions). In plants, plasmodesmata connect cells and enable communication and transport between them.

❺ Nine important landmarks distinguish eukaryotic cells.

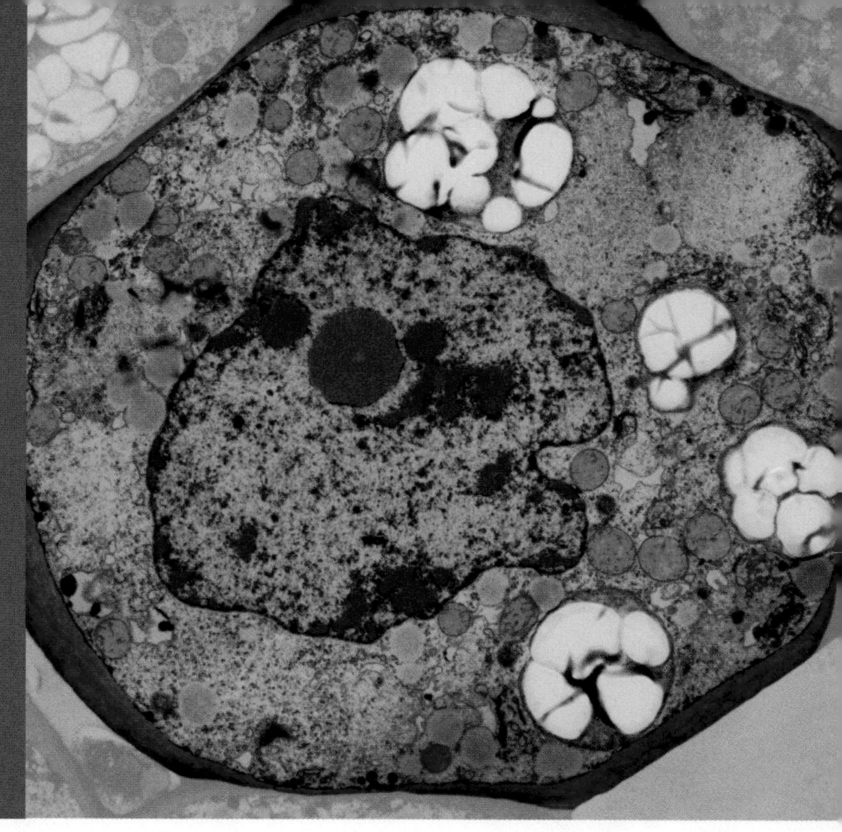

Cell of the voodoo lily. How many different organelles can you spot in this plant cell?

3•13

The nucleus is the cell's genetic control center.

As we have seen, all cells are surrounded by a complex plasma membrane that actively regulates what materials get into and out of a cell. Now it's time to look at the cellular contents surrounded by the plasma membrane in eukaryotic cells.

The nucleus is the largest and most prominent organelle in most eukaryotic cells. In fact, the nucleus is generally larger than any prokaryotic cell. If a cell were the size of a large lecture hall or movie theater, the nucleus would be the size of a big-rig 18-wheeler truck parked in the front rows. The nucleus has two primary functions, both related to the fact that most of the cell's DNA resides in the nucleus. First, the nucleus is the genetic control center, directing most cellular activities by controlling which molecules are produced, and in what quantity. Second, the nucleus is the storehouse for hereditary information.

Three important structural components stand out in the nucleus (**FIGURE 3-26**). First is the **nuclear membrane,** sometimes called the nuclear envelope, which surrounds the nucleus and separates it from the cytoplasm. Unlike most plasma membranes, however, the nuclear membrane consists of *two* bilayers, one on top of the other, much like the double-bagging of groceries at the market. The nuclear membrane is not a sack, though. It is perforated, covered with tiny pores made from multiple proteins embedded in the phospholipid membranes and spanning both bilayers. These pores enable large molecules to pass from the nucleus to the cytoplasm and from the cytoplasm to the nucleus.

The second prominent structure in the nucleus is the **chromatin,** a mass of long, thin fibers consisting of DNA with some proteins attached to it that keep it from getting impossibly tangled. Most of the time, as the DNA directs cellular activities, the chromatin resembles a plate of spaghetti. When it's time for cell division (a process described in detail in Chapter 6), the chromatin coils up and the threads become shorter and thicker until they become visible as chromosomes, the compacted, linear DNA molecules on which all of the organism's hereditary information is carried.

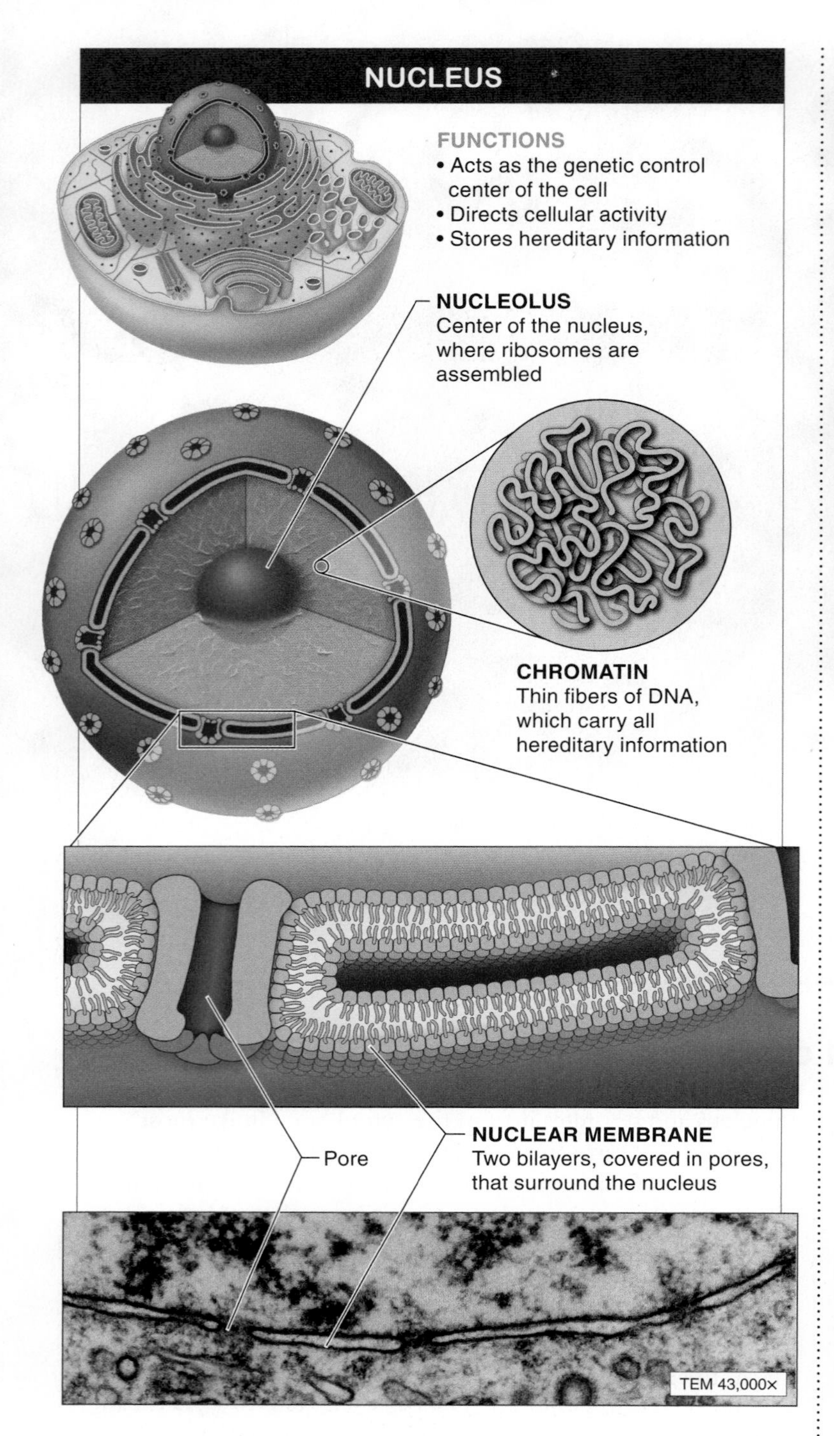

FIGURE 3-26 The nucleus: the cell's genetic control center.

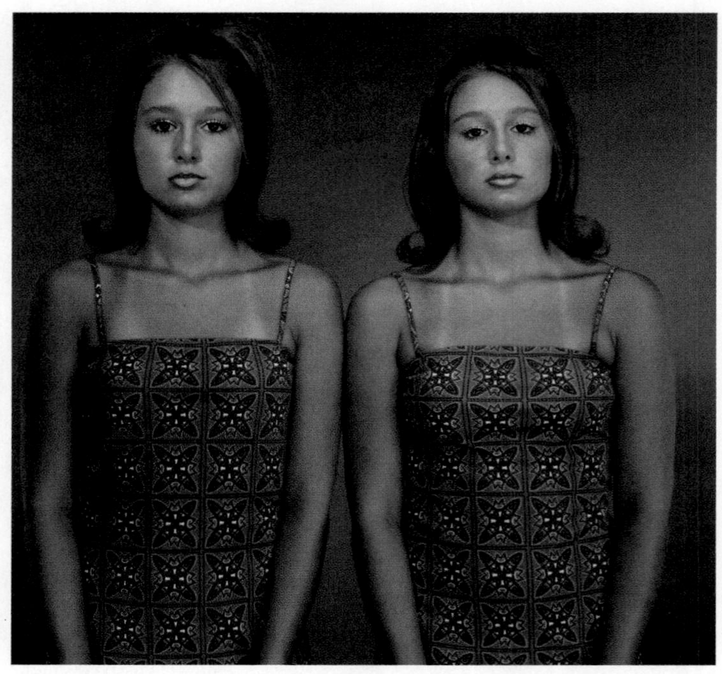

The nucleus holds the genetic information that makes it possible to build organisms. (That's why identical twins, carrying the same genetic information, look so similar.)

A third structure in the nucleus is the **nucleolus,** an area near the center of the nucleus where subunits of the ribosomes, a critical part of the cellular machinery, are assembled. Ribosomes are like little factories that copy bits of the information stored in the DNA and use them to construct proteins, including enzymes and the proteins that make up tissues such as the bark of trees or the bone of vertebrates. The ribosomes are built in the nucleolus but pass through the nuclear pores and into the cytoplasm before starting their protein-production work.

TAKE-HOME MESSAGE 3·13

The nucleus is usually the largest and most prominent organelle in the eukaryotic cell. It directs most cellular activities by controlling which molecules are produced and in what quantity. The nucleus is also the storehouse for all hereditary information.

3•14

Cytoplasm and the cytoskeleton form the cell's internal environment, provide its physical support, and can generate movement.

If you imagine yourself inside a cell the size of a big lecture hall, it might come as a surprise that you can barely see that big rig of a nucleus parked in the front rows. Visibility is almost zero, not only because the room is filled with jelly-like cytoplasm, but also because there is a dense web of thick and thin, straight and branched, ropes, strings, and scaffolding, running every which way throughout the room.

This inner scaffolding of the cell, which is made from proteins, is the **cytoskeleton** (**FIGURE 3-27**). It has three chief purposes. First, it gives animal cells shape and support—making red blood cells look like little round doughnuts (without the hole in the middle) and giving neurons their very long, thread-like appearance. Plant cells are shaped primarily by their cell wall (a structure we discuss later in the chapter), but they also have a cytoskeleton. Second, the cytoskeleton controls the intracellular traffic flow, serving as a series of tracks on which a variety of organelles and molecules are guided across and around the inside of the cell. And third, because the elaborate scaffolding of the cytoskeleton is dynamic and can generate force, it gives all cells some ability to control their movement.

The cytoskeleton also plays a more direct and obvious role in helping some cells to move—whether moving the cell through its environment, or moving stuff in the environment over the cell surface. **Cilia** (*sing.* **cilium**) are short projections that often occur in large numbers on a single cell (**FIGURE 3-28**). Cilia beat

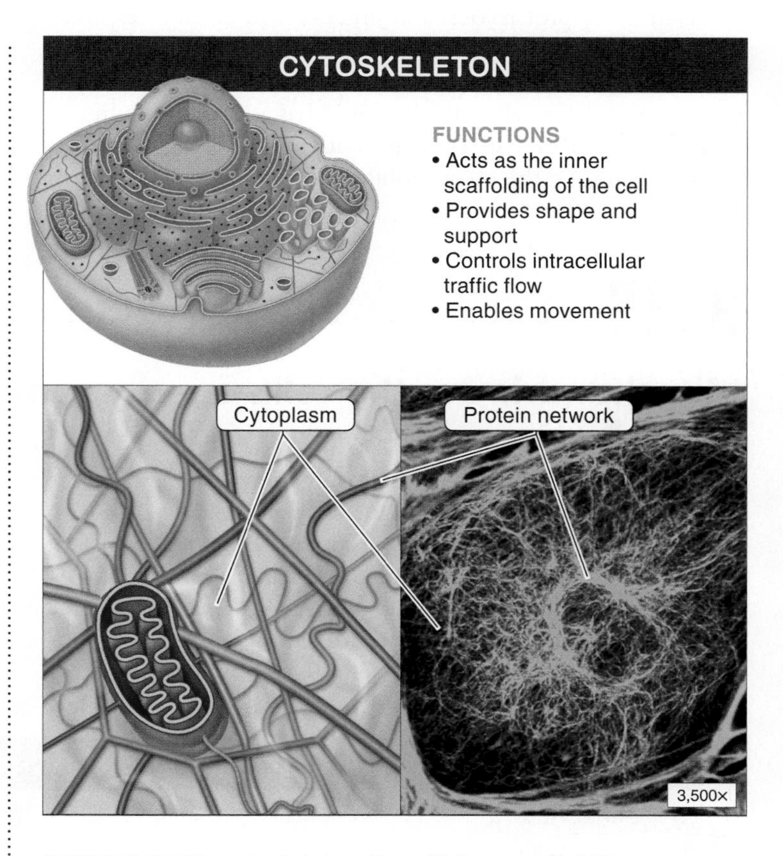

FIGURE 3-27 The cytoskeleton: the cell's inner scaffolding.

CELLULAR MOVEMENT

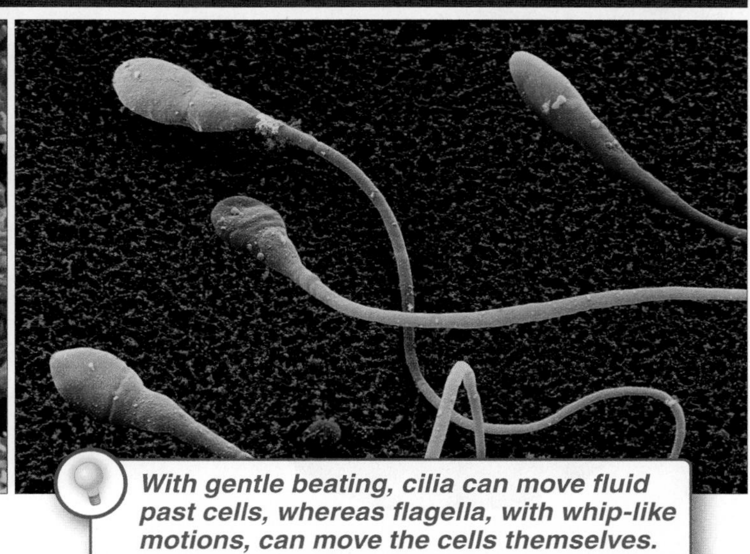

FIGURE 3-28 Cilia and flagella assist the cell with movement.

swiftly, often in unison and in ways that resemble blades of grass in a field, blowing in the wind. Cilia can move fluid along and past a cell. This movement can accomplish many important tasks, including sweeping the airways to our lungs to clear them of debris (such as dust) in the air we breathe.

Flagella (*sing.* **flagellum**) are much longer than cilia. Flagella occur in many prokaryotes and single-celled eukaryotes, and many algae and plants have cells with one or more flagella. But in animals, cell types with one or more flagella are very rare. One of these cell types, however, has a critical role in every animal species: sperm cells. With a flagellum for a tail, sperm are among the most mobile of all animal cells. Some spermicidal birth control methods prevent conception by disabling the flagellum and immobilizing the sperm cells.

TAKE-HOME MESSAGE 3•14

The inner scaffolding of the cell, which is made from proteins, is the cytoskeleton. It gives animal cells their shape and support, gives cells some ability to control their movement, and serves as a series of tracks on which organelles and molecules are guided across and around the inside of the cell.

3•15

Mitochondria are the cell's energy converters.

Cars have it easy: we generally put only a single type of fuel in them, and it is exactly the same fuel every time. With human bodies it's a different story. During some meals we put in meat and potatoes. Other times, it's juice and fruits. Still others, bread or vegetables, popcorn and gummy bears, pizza and ice cream. Yet no matter what we eat, we expect our body to utilize the energy in these various foods to power all the reactions in our bodies that make it possible for us to breathe, move, and think. The mitochondria are the organelles that make this possible (**FIGURE 3-29**).

Mitochondria (*sing.* **mitochondrion**) are the cell's all-purpose energy converters, and they are present in virtually all plant cells, animal cells, and every other eukaryotic cell.

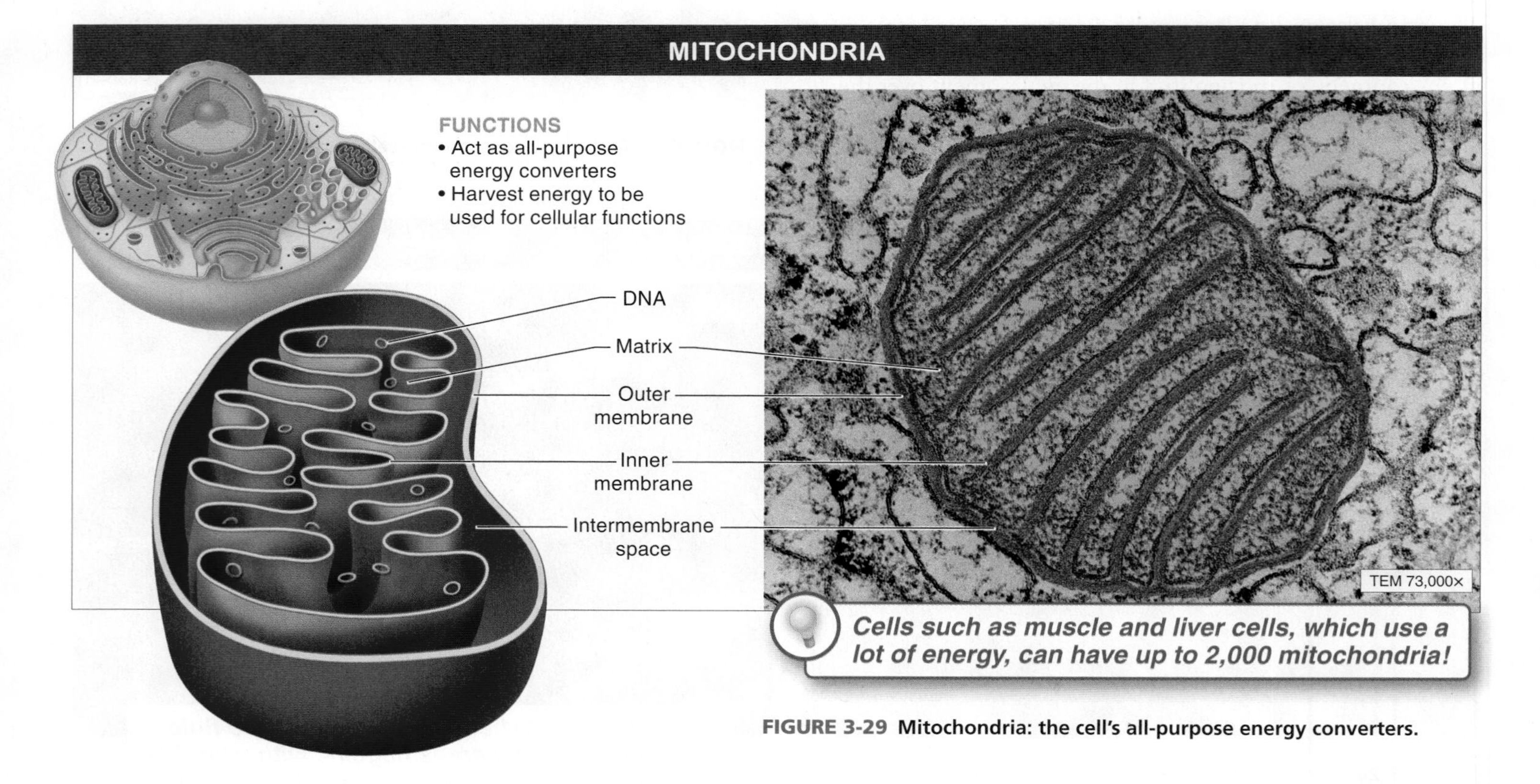

FIGURE 3-29 Mitochondria: the cell's all-purpose energy converters.

Although we consume a variety of foods, our mitochondria allow us to convert the energy contained in the chemical bonds of the carbohydrates, fats, and proteins in the food into carbon dioxide, water, and ATP—the molecule that is the energy source all cells use to fuel all their functions and activities. (ATP and how it works are described in detail in Chapter 4.) Because this energy conversion requires a significant amount of oxygen, organisms' mitochondria consume most of the oxygen used by each cell. In humans, for example, our mitochondria consume as much as 80% of the oxygen we breathe. They give a significant return on this investment by producing about 90% of the energy our cells need to function.

Q We all have more DNA from one parent than the other. Who is the bigger contributor: mom or dad? Why?

As with so many aspects of our bodies, form follows function. Cells that are not very metabolically active, such as fat cells in humans, have very few mitochondria. Other cells, such as muscle, liver, and sperm cells in animals and fast-growing root cells in plants—all of which have large energy requirements—are packed densely with mitochondria. Some of these cells have as many at 2,000 mitochondria!

To visualize the structure of a mitochondrion, imagine a plastic sandwich bag. Now take another, bigger plastic bag and stuff it inside the sandwich bag. That's the structure of mitochondria: there is a smooth outer membrane and a scrunched up inner membrane. This construction creates two separate compartments within the mitochondrion: a region outside the inner plastic bag (called the **intermembrane space**) and another region, called the **matrix,** inside the inner plastic bag. This bag-within-a-bag structure has important implications for energy conversion. And having a heavily folded inner membrane that is much larger than the outer membrane provides a huge amount of surface area on which to conduct chemical reactions. (We discuss the details of mitochondrial energy conversion and the role of mitochondrial structure in Chapter 4.)

As we learned in our earlier discussion of endosymbiosis, mitochondria may very well have existed, billions of years ago, as separate single-celled bacteria-like organisms. They are similar to bacteria in size and shape, and may have originated when symbiotic bacteria took up permanent residence within other cells. Perhaps the strongest evidence for this is that mitochondria have their own DNA. Mixed in among the approximately 3,000 proteins in each mitochondrion, there are anywhere from 2 to 10 copies of its own little ring-shaped DNA. This DNA carries the instructions for making 13 important mitochondrial proteins necessary for metabolism and energy production.

We're always taught that our mothers and fathers contribute equally to our genetic composition, but this isn't quite true: the mitochondria in every one of your cells (and the DNA that comes with them) come from the mitochondria that were initially present in your mother's egg, which, when fertilized by your father's sperm, developed into you. In other words: all of your mitochondria are descended from your mother's mitochondria. The tiny sperm contributes DNA, but no cytoplasm and, hence, no mitochondria. Consequently, mitochondrial DNA is something that we inherit exclusively from our mothers. This is true not only in humans but in most multicellular eukaryotes. As the fertilized egg develops into a two-celled, then four-celled, embryo, the mitochondria split themselves by a process called fission, the same process of division and DNA duplication used by bacteria—so there are always a sufficient number of mitochondria for the newly produced cells. This similarity between mitochondria and bacteria is another characteristic that supports the theory that mitochondria were originally symbiotic bacteria.

Given the central role of mitochondria in converting the energy in food molecules into a form that is usable by cells, you won't be surprised to learn that mitochondrial malfunctions can have serious consequences. Recent research has focused on possible links between defective mitochondria and diseases characterized by fatigue and muscle pain. It seems, for instance, that many cases of "exercise intolerance"—extreme fatigue or cramps after only slight exertion—may be related to defective mitochondrial DNA.

TAKE-HOME MESSAGE 3•15

In mitochondria, which are found in virtually all eukaryotic cells, the energy contained in the chemical bonds of carbohydrate, fat, and protein molecules is converted into carbon dioxide, water, and ATP, the energy source for all cellular functions and activities. Mitochondria may have their evolutionary origins as symbiotic bacteria living inside other cells.

3·16

Lysosomes are the cell's garbage disposals.

Garbage. What does a cell do with all the garbage it generates? Mitochondria wear out after about 10 days of intensive activity, for starters. And white blood cells constantly track down and consume bacterial invaders, which they then have to dispose of. Similarly, the thousands of ongoing reactions of cellular metabolism produce many waste macromolecules that cells must digest and recycle. Many eukaryotic cells deal with this garbage by maintaining hundreds of versatile floating "garbage disposals" called lysosomes (**FIGURE 3-30**).

> **Q** Why is Tay-Sachs disease like a strike by trash collectors?

Lysosomes are round, membrane-enclosed, acid-filled vesicles that function as garbage disposals. Lysosomes are filled with about 50 different digestive enzymes and a super-acidic fluid, a corrosive broth so powerful that if a lysosome were to burst, it would almost immediately kill the cell by rapidly digesting all of its component parts. The selection of enzymes in the lysosome represents a broad spectrum of chemicals designed for dismantling macromolecules that are no longer needed by cells or are generated as by-products of cellular metabolism. Some of the enzymes break down lipids, others carbohydrates, others proteins, and still others nucleic acids. Consequently, lysosomes are frequently also a first step when, via phagocytosis, a cell consumes and begins digesting a particle of food or even an invading bacterium. And, ever the efficient system, the cell releases most of the component parts of molecules that are digested, such as amino acids, back into its cytoplasm, where they can be reused by the cell as raw materials.

With 50 different enzymes necessary for lysosomes to carry out their metabolic salvaging act, malfunctions sometimes occur. A common genetic disorder called Tay-Sachs disease is the result of just such a genetic mishap. In Tay-Sachs disease, an individual inherits an inability to produce a critical lipid-digesting enzyme. Even though the lysosomes cannot digest certain lipids, the cells continue to send lipids to the lysosomes, where they accumulate, undigested. This leads to a sort of lysosome constipation. The lysosome swells until it bursts and digests the whole cell or until it chokes the cell to death. Within the first few years of life, this process occurs in large numbers of cells and eventually leads to the child's death.

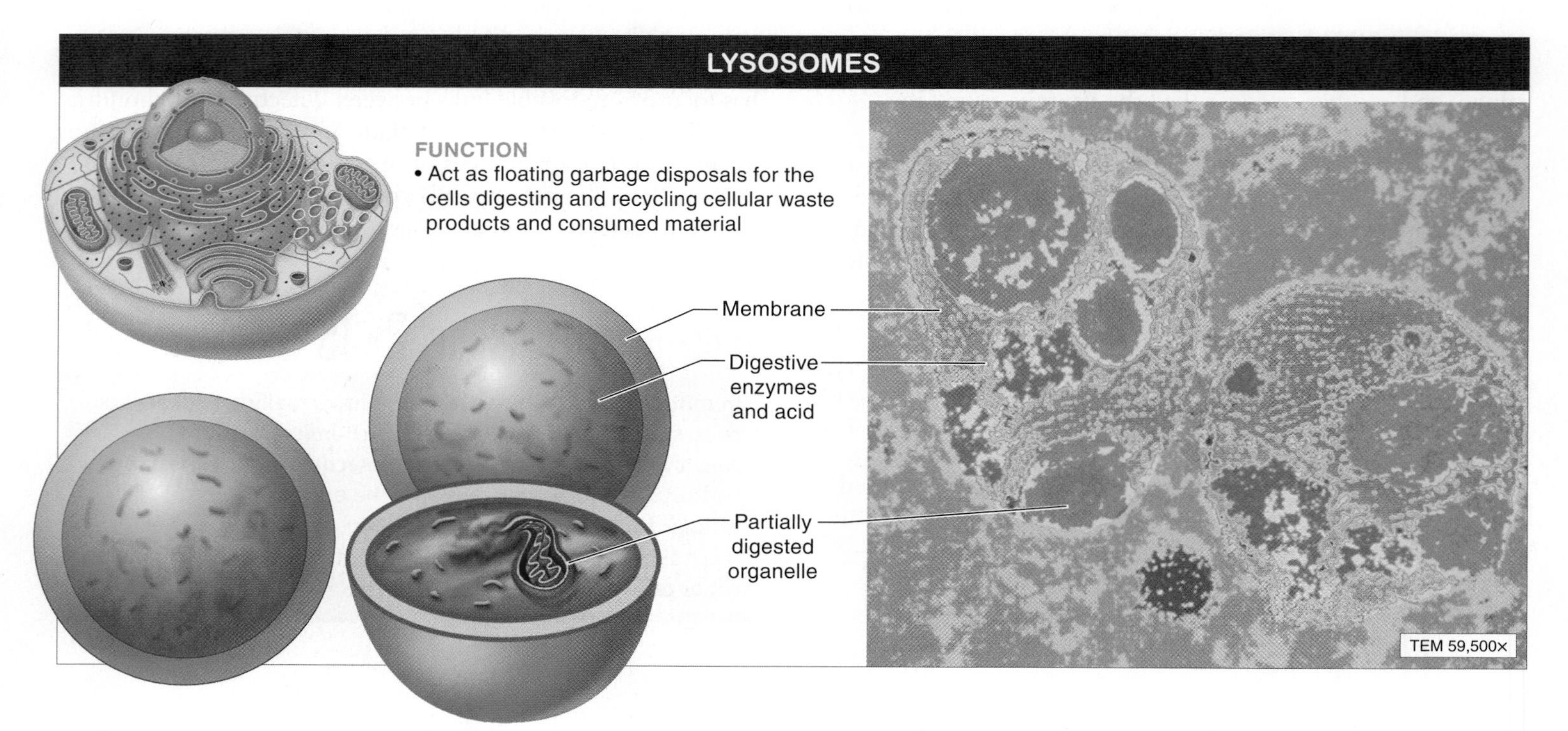

FIGURE 3-30 Lysosomes: digestion and recycling of the cell's waste products.

Although it has long been debated among biologists, it does seem that plant cells contain lysosomes, compartments with similar digestive broths that carry out the same digestive processes. Among animals, some immune system cells tend to have particularly large numbers of lysosomes, most likely because of their great need for disposing of the by-products of disease-causing bacteria.

TAKE-HOME MESSAGE 3•16

Lysosomes are round, membrane-enclosed, acid-filled organelles that function as a cell's garbage disposals. They are filled with about 50 different digestive enzymes and enable a cell to dismantle macromolecules, including disease-causing bacteria.

3•17

The endoplasmic reticulum is the site where cells build proteins and disarm toxins.

The information for how to construct the molecules essential to a cell's smooth functioning and survival is stored in the DNA found in the cell's nucleus. The energy used to construct these molecules and to run cellular functions comes primarily from the mitochondria. The actual production and modification of biological molecules, however, occurs in a system of organelles called the **endomembrane system** (**FIGURE 3-31**). This mass of interrelated membranes spreads out from and surrounds the nucleus, forming chambers within the cell that contain their own mixture of chemicals. The endomembrane system takes up as much as one-fifth of the cell's volume and is responsible for many of the fundamental functions of the cell.

Lipids are produced within these membrane-enclosed compartments, products for export to other parts of the body are modified and packaged here, polypeptide chains are assembled into functional proteins here, and many of the toxic chemicals that find their way into our bodies—from recreational drugs to antibiotics—are broken down and neutralized in the endomembrane system.

Because the process of protein production follows a somewhat linear path, we explore the endomembrane system sequentially, beginning just outside the nucleus with the endoplasmic reticulum.

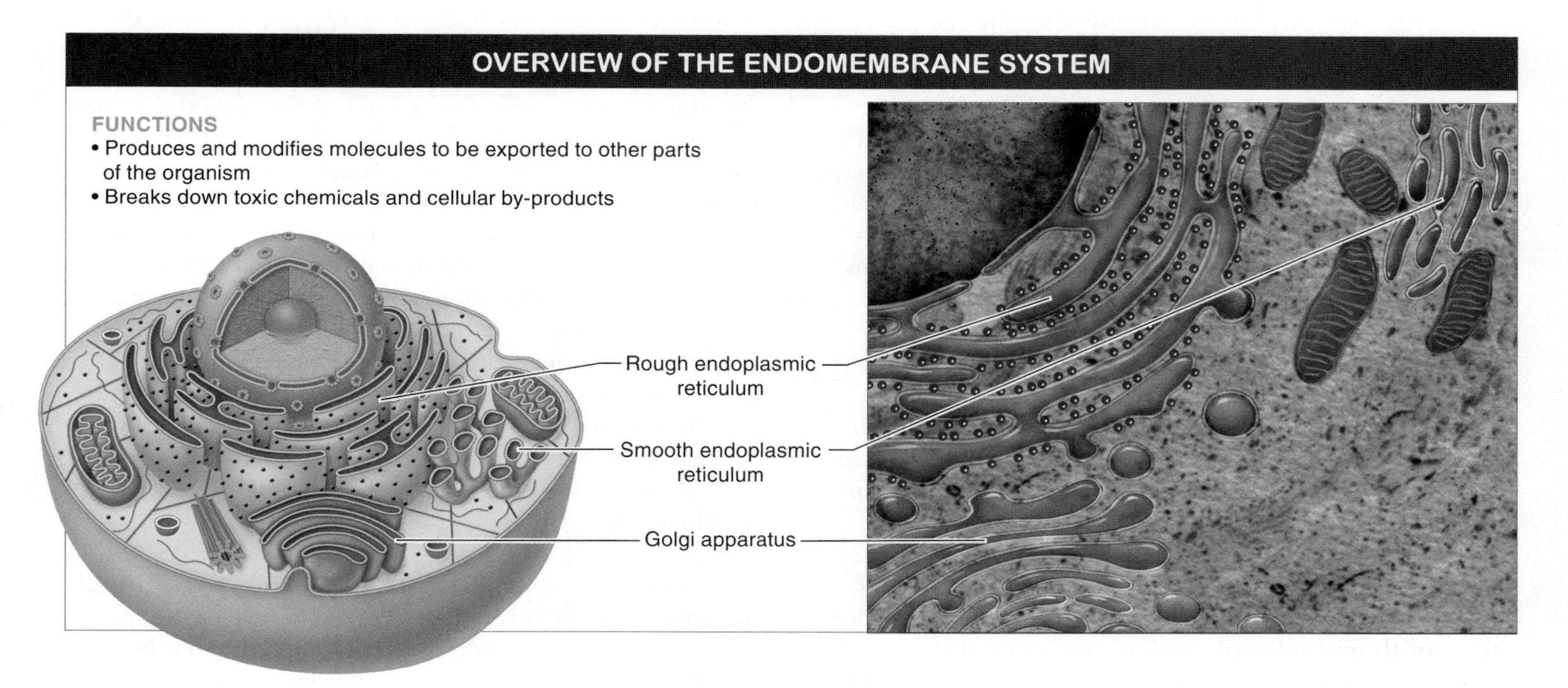

FIGURE 3-31 The endomembrane system: the smooth endoplasmic reticulum, rough endoplasmic reticulum, and Golgi apparatus.

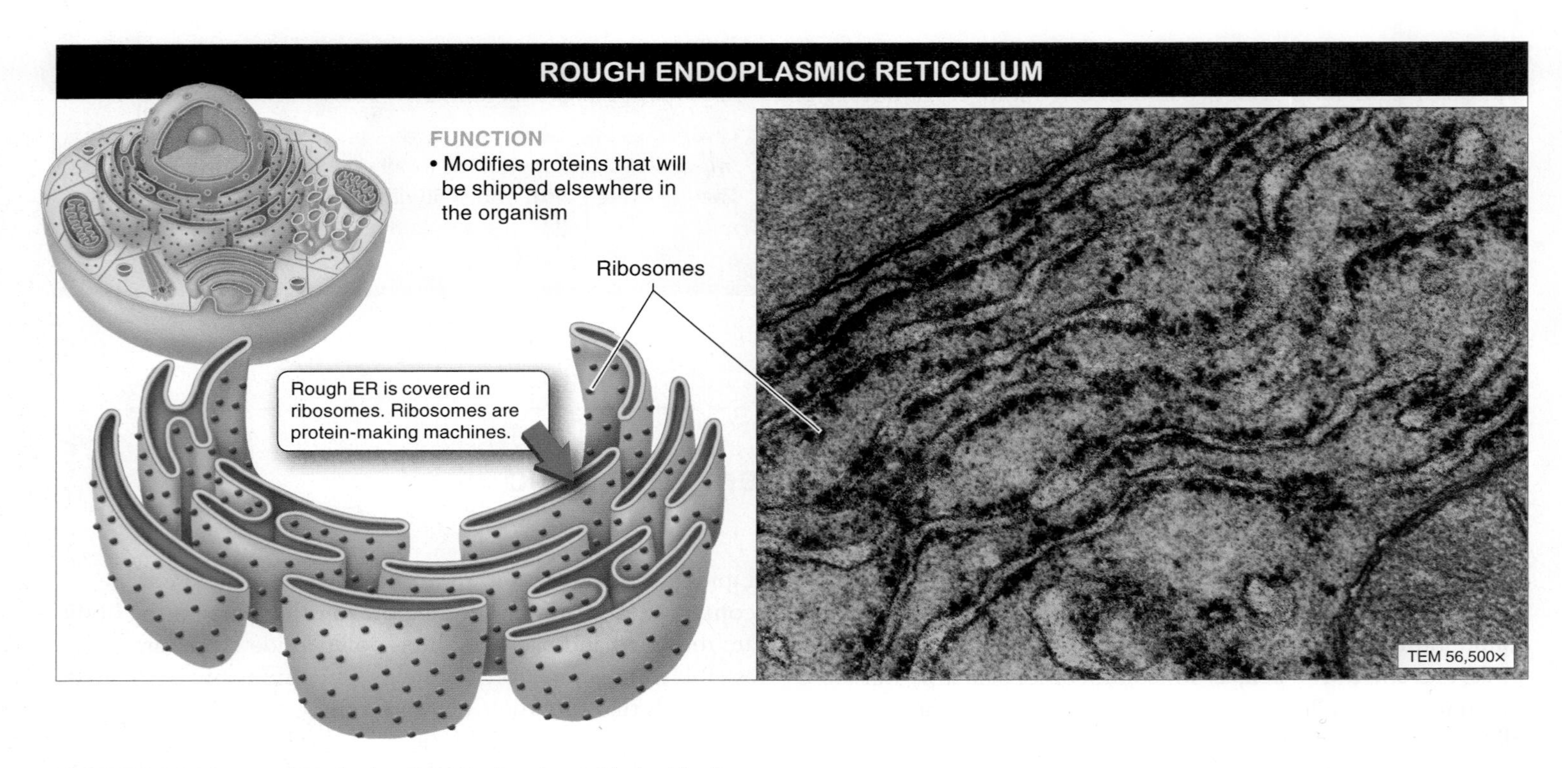

FIGURE 3-32 The rough endoplasmic reticulum is studded with ribosomes.

Rough Endoplasmic Reticulum Perhaps the organelle with the most cumbersome name, the **rough endoplasmic reticulum,** or **rough ER** ("endoplasmic reticulum" is derived from the Greek for "within" and "anything molded" and the Latin for "small net"), is a large series of interconnected, flattened sacs that look like a stack of pancakes. These sacs are connected directly to the nuclear envelope. In most eukaryotic cells, the rough ER almost completely surrounds the nucleus (**FIGURE 3-32**). It is called "rough" because its surface is studded with little bumps. These bumps are ribosomes, the cell's protein-making machines, and generally cells with high rates of protein production have large numbers of ribosomes We cover the details of ribosome structure and protein production in Chapter 5.

The primary function of the rough ER is to fold and package proteins that will be shipped elsewhere in the organism. Poisonous frogs, for example, package their poison in rough ER of the cells where it is produced before transporting it to the poison glands on their skin. Proteins that are used within the cell itself are generally produced on free-floating ribosomes in the cytoplasm.

Smooth Endoplasmic Reticulum As its name advertises, the **smooth endoplasmic reticulum** is part of the endomembrane network that is smooth because there are no ribosomes bound to it (**FIGURE 3-33**). Although it is connected to the rough ER, it is farther from the nucleus and, besides lacking ribosomes, differs slightly in appearance. Whereas the rough ER looks like stacks of pancakes, the smooth ER sometimes looks like a collection of branched tubes.

The smooth surface gives us the first hint that smooth ER has a different job than rough ER. Because ribosomes are absolutely essential for protein production, we can be certain that, without them, the smooth ER is not involved in folding or packaging proteins. It isn't. Instead, it synthesizes lipids such as fatty acids, phospholipids, and steroids. Exactly which lipids are produced varies throughout the organism and across plant and animal species. Inside the smooth ER of mammalian ovaries and testes, for example, the hormones estrogen and testosterone are produced. Inside the smooth ER of liver and fat cells, lipids are produced. As in the rough ER, following their production these lipids are packaged in transport vesicles and are then sent to other parts of the cell or to the plasma membrane for export.

Another critical responsibility of the smooth ER—particularly the smooth ER in human liver cells—is to help protect us from the many dangerous molecules that get into our bodies. Alcohol, antibiotics, barbiturates, amphetamines or other

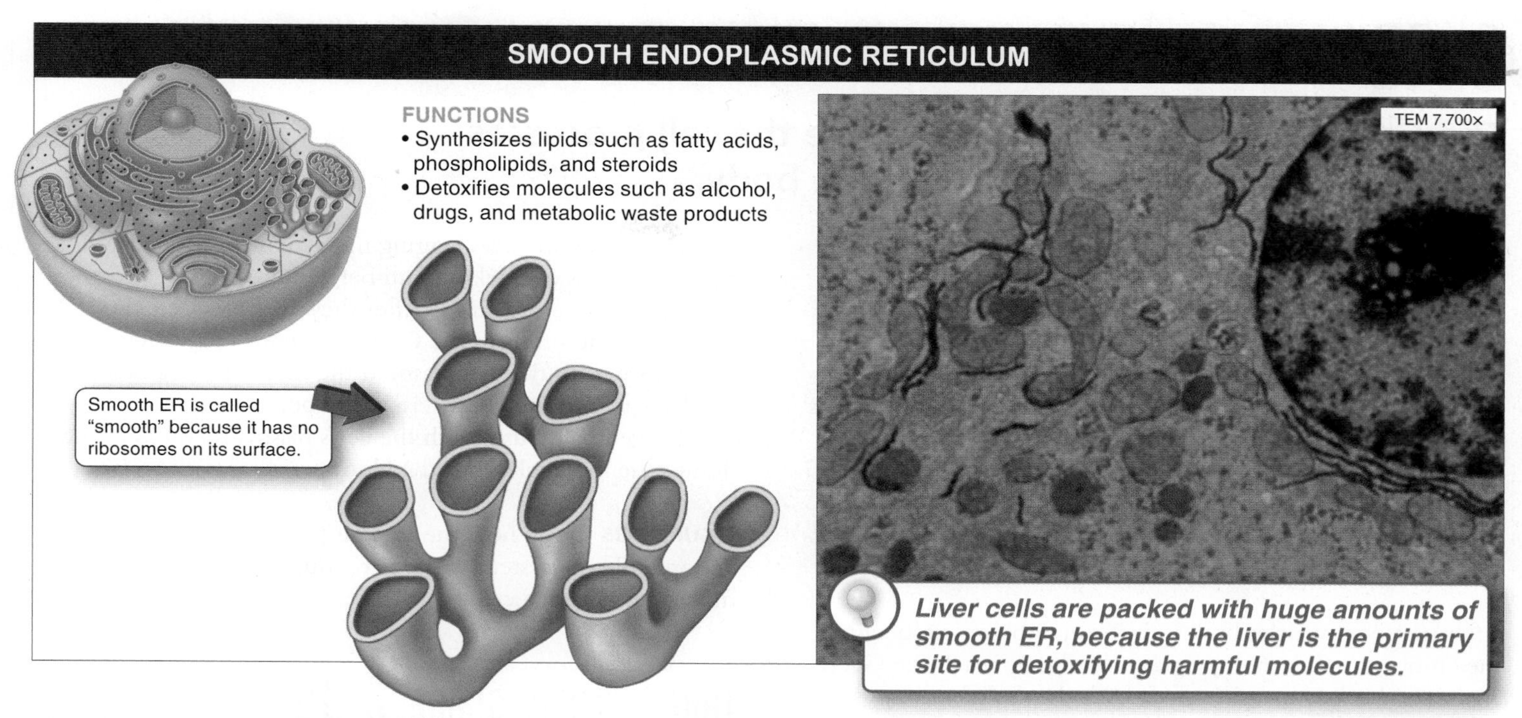

FIGURE 3-33 In the smooth endoplasmic reticulum, lipids are synthesized and alcohol, antibiotics, and other drugs are detoxified.

stimulants that we may consume, along with many toxic metabolic waste products formed in our bodies, are made less harmful by the detoxifying enzymes in the smooth ER. And just as a body builder's muscles get bigger and bigger in response to weightlifting, our smooth ER proliferates in cells that are exposed to large amounts of particular drugs. This proliferation can be beneficial because it increases the capacity for detoxification. But as the cells become more and more efficient at detoxification, our tolerances to the very drugs the smooth ER is trying to destroy can increase.

Q How can long-term use of one drug increase your resistance to another, different drug that you have never encountered?

Chronic exposure to many drugs (from antibiotics to heroin) can induce a proliferation of smooth ER, particularly in the liver, and the smooth ER's associated detoxification enzymes. This proliferation in turn increases tolerance to the drugs, necessitating higher doses to achieve the same effect. Moreover, the increased detoxification capacity of the cells often enables them to better detoxify other compounds, even if the person has never been exposed to these chemicals. This increased detoxification capacity can lead to problems. An individual who has been exposed to large amounts of certain drugs, for example, may end up responding less well to antibiotics.

As we see over and over in living organisms, form follows function. Not surprisingly, then, just as cells with high rates of protein production have large numbers of ribosomes, we find huge amounts of smooth ER in liver cells because they are the primary sites of molecular detoxification. Other cells that are packed with ER (both rough and smooth) include plasma cells in the blood that produce immune system proteins for export, and pancreas cells that secrete large amounts of digestive enzymes. On the other hand, there is relatively little smooth ER in cells such as fat-storage cells, which produce little or no protein for export.

TAKE-HOME MESSAGE 3•17

The production and modification of biological molecules in eukaryotic cells occurs in a system of organelles called the endomembrane system, which includes, among other organelles, the rough and smooth endoplasmic reticulum. In rough ER, proteins that will be shipped elsewhere in the body are folded and packaged. In the smooth ER, lipids are synthesized and alcohol, antibiotics, and other drugs are detoxified.

3•18

The Golgi apparatus is the site where the cell processes products for delivery throughout the body.

Moving farther outward from the nucleus, we encounter another organelle within the endomembrane system: the Golgi (GOHL-jee) apparatus (**FIGURE 3-34**). The **Golgi apparatus** processes molecules synthesized in the cell—primarily proteins and lipids—and packages those that are destined for use elsewhere in the body. The Golgi apparatus is also a site of carbohydrate synthesis, including the complex polysaccharides found in many plasma membranes. The Golgi apparatus, which is not connected to the endoplasmic reticulum, is a flattened stack of membranes (each of which is called a Golgi body) that are *not* interconnected.

After transport vesicles bud from the endoplasmic reticulum, they move through the cytoplasm until they reach the Golgi apparatus. Here, the vesicles fuse with the Golgi apparatus membrane and dump their contents into a Golgi body. There are about four successive Golgi body chambers to be visited, and the molecules get passed from one to the next. In each Golgi body, enzymes make slight modifications to the molecule (such as the addition or removal of phosphate groups or sugars). The processing that occurs in the Golgi apparatus often involves tagging molecules (much like adding a postal address or tracking number) to direct them to some other part of the organism. After they are processed, the molecules bud off from the Golgi apparatus in a vesicle, which then moves into the cytoplasm. If the molecules are destined for delivery and use elsewhere in the body, the transport vesicle eventually fuses with the cell's plasma membrane and dumps the molecules into the bloodstream via exocytosis.

FIGURE 3-35 shows how the various parts of the endomembrane system work to produce, modify, and package molecules within a cell.

TAKE-HOME MESSAGE 3•18

The Golgi apparatus—another organelle within the endomembrane system—processes molecules synthesized in a cell and packages those that are destined for use elsewhere in the body.

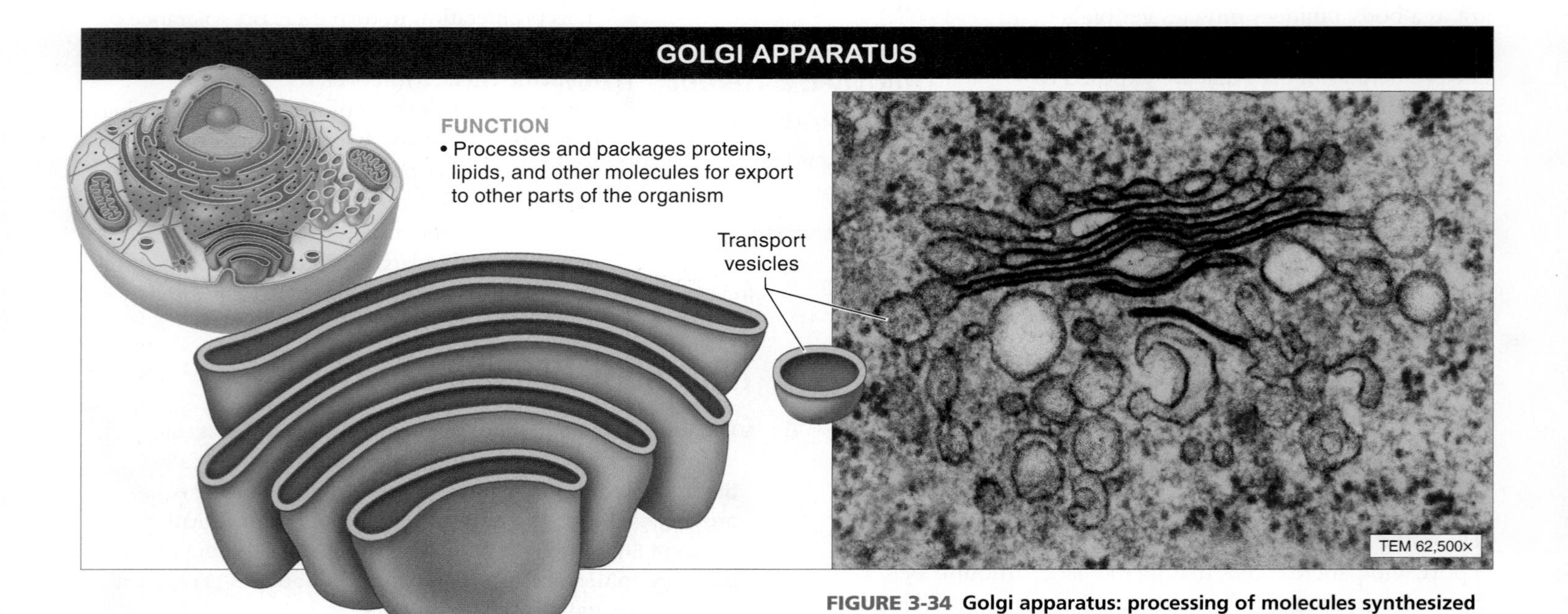

FIGURE 3-34 Golgi apparatus: processing of molecules synthesized in the cell and packaging of those molecules destined for use elsewhere in the body.

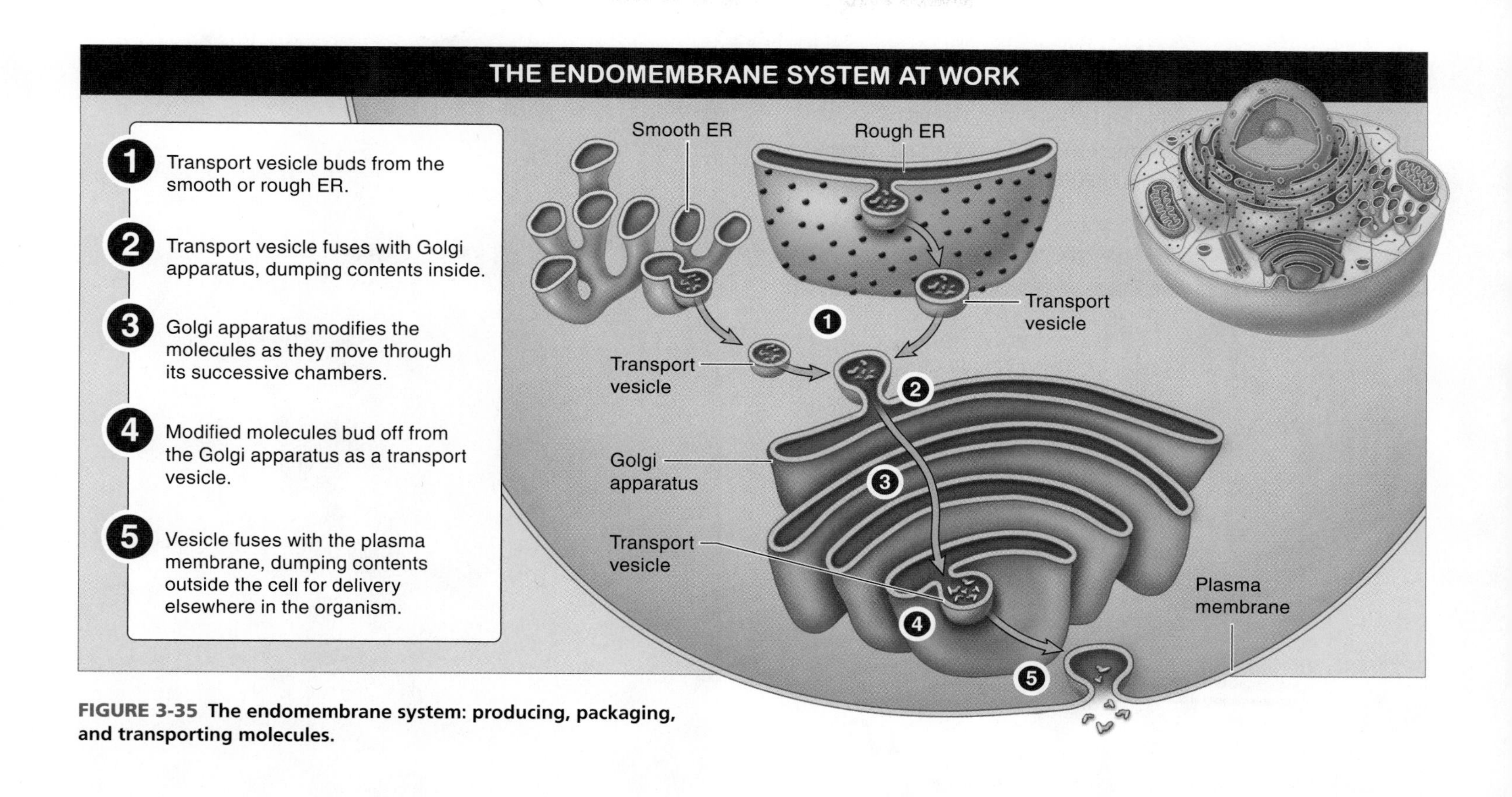

FIGURE 3-35 The endomembrane system: producing, packaging, and transporting molecules.

3•19

The cell wall provides additional protection and support for plant cells.

Up until now, we've discussed organelles and structures that are common to both plants and animals. Now we're going to look at several structures that are not found in all eukaryotic cells. Because plants are stuck in the ground, unable to move, they have several special needs beyond those of animals. In particular, they can't outrun or outmaneuver their competitors to get more food. Nor can they outrun organisms that want to eat them. If they want more sunshine, they have few options beyond simply growing larger or taller. And if they want to resist plant-eating animals, they must simply grow tougher outer layers.

One organelle that helps plants achieve both of these goals is the cell wall, a structure that surrounds the plasma membrane. The cell wall is made from polysaccharides in which another carbohydrate, called cellulose, is embedded. Note that although animal cells do not have cell walls, some non-plants such as archaea, bacteria, protists, and fungi also have cell walls, but the chemical composition of their cell walls differs from that found in plants.

In plants, cell wall production can be a multistage process. Initially, when the cell is still growing, a primary cell wall is laid down, along with some glue that helps adjacent cells adhere to each other. Sometime later in life, a secondary cell wall is laid down. This secondary cell wall usually contains a complex molecule called lignin, which gives the wall much of its strength and rigidity. Oddly, this strength remains even after the plant cell dies—hence wood's great value to humans as a construction material, among other things.

The cell wall is nearly 100 times thicker than the plasma membrane, and the tremendous structural strength it confers on plant cells enables some plants to grow several hundred feet tall. Cell walls also help to make plants more water-tight—an important feature in organisms that cannot reduce water loss through evaporation by moving out of the hot sun—and provide some protection from insects and other animals that might eat them (**FIGURE 3-36**).

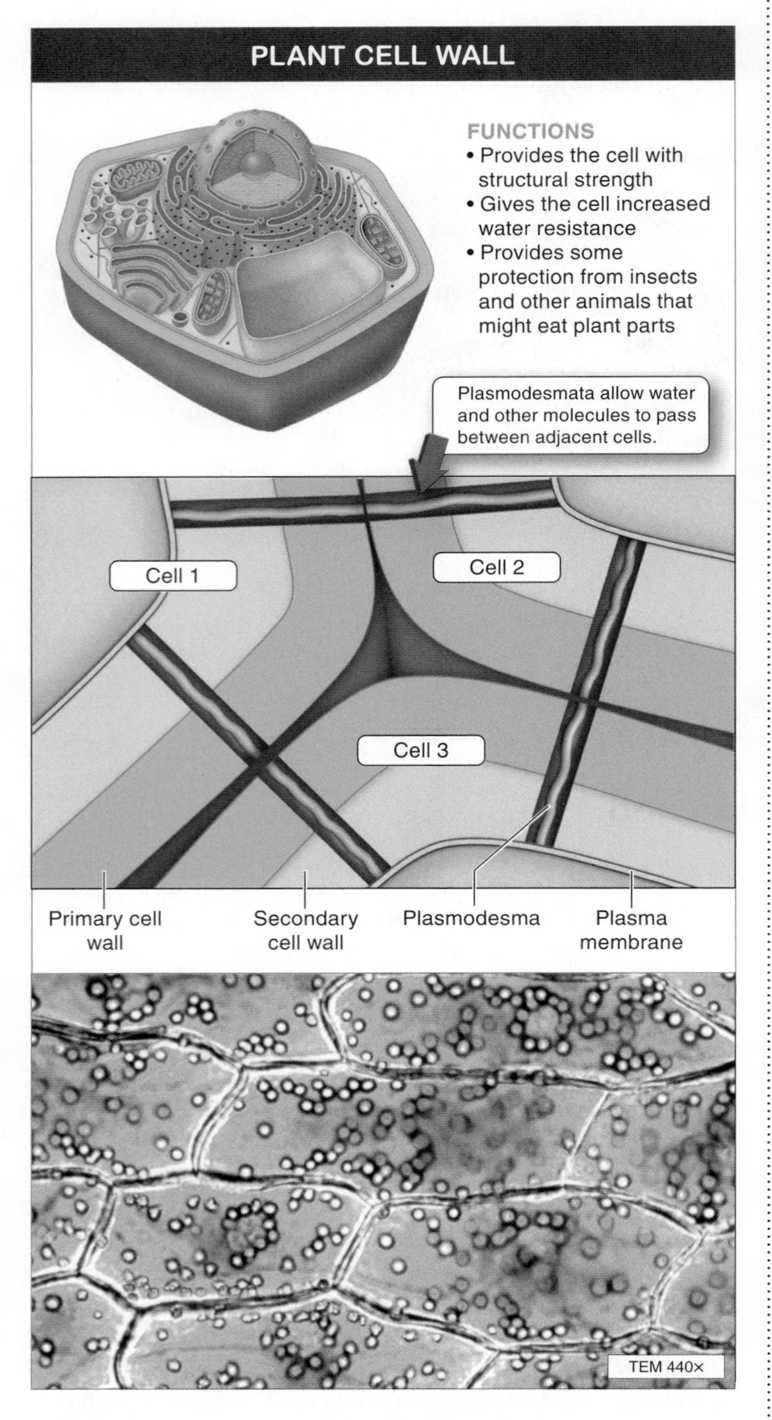

FIGURE 3-36 The plant cell wall: providing strength.

The rigid walls of the plant cell allow plants to grow tall.

Surprisingly, despite its great strength, the cell wall does not completely seal off plant cells from one another. Rather, it is porous, allowing water and solutes to reach the plasma membrane. It also has plasmodesmata, the channels connecting adjacent cells (as described in Section 3-12) and allowing the passage of some molecules between them.

TAKE-HOME MESSAGE 3•19

The cell wall is an organelle found in plants (and in some other non-animal organisms). It is made primarily from the carbohydrate cellulose and it surrounds the plasma membrane of the cell. The cell wall confers tremendous structural strength on plant cells, gives plants increased resistance to water loss, and provides some protection from insects and other animals that might eat them.

3•20

Vacuoles are multipurpose storage sacs for cells.

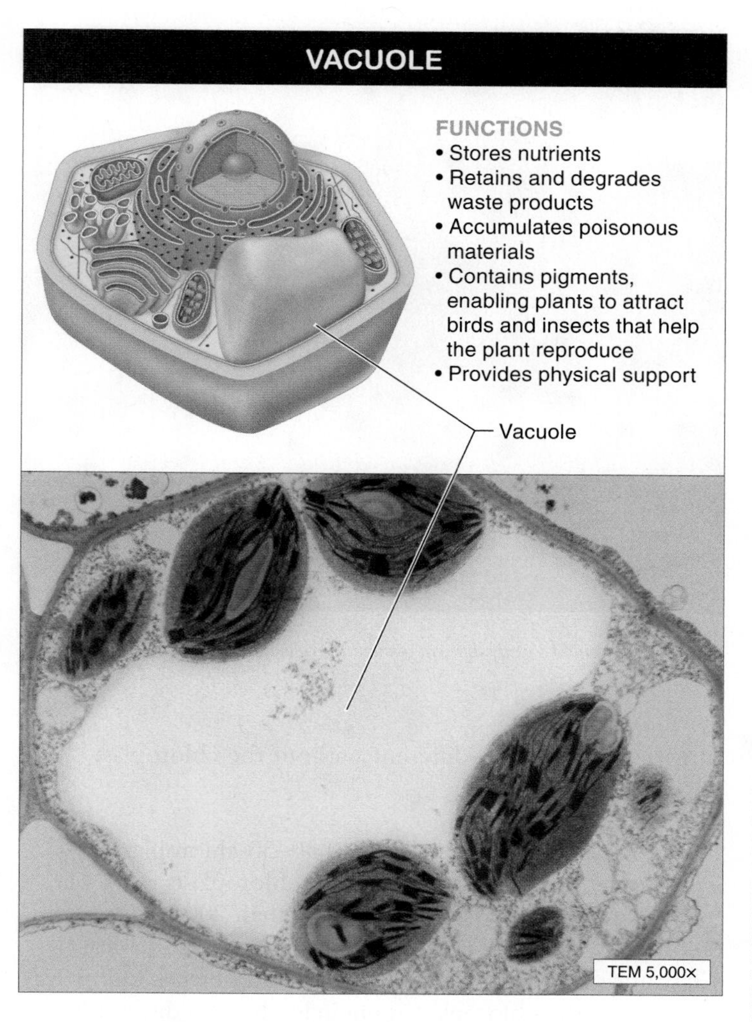

FIGURE 3-37 The vacuole: multipurpose storage.

If you look at a mature plant cell through a microscope, one organelle, the **central vacuole,** usually stands out more than all the others because it is so huge and appears empty (**FIGURE 3-37**). Although it may look like an empty sac, the central vacuole is anything but. Surrounded by a membrane, filled with fluid, and occupying from 50% to 90% of a plant cell's interior space, the central vacuole can play an important role in five different areas of plant life. (Vacuoles are also found in some other eukaryotes, including some protists, fungi, and animals, but they tend to be particularly prominent in plant cells.)

1. **Nutrient storage**—the vacuole stores hundreds of dissolved substances, including amino acids, sugars, and ions.
2. **Waste management**—the vacuole retains waste products and degrades them with digestive enzymes, much like the lysosome in animal cells.
3. **Predator deterrence**—the poisonous, nasty-tasting materials that accumulate inside the vacuoles of some plants make a powerful deterrent to animals that might try to eat parts of the plant.
4. **Sexual reproduction**—the vacuole may contain pigments that give some flowers their red, blue, purple, or other colors, enabling them to attract birds and insects that help the plant reproduce by transferring pollen.
5. **Physical support**—high concentrations of dissolved substances in the vacuole can cause water to rush into the cells through the process of osmosis. The increased fluid pressure inside the vacuole can cause the cell to enlarge a bit and push out the cell wall. This process is responsible for the pressure (called **turgor pressure**) that allows stems, flowers, and other plant parts to stand upright. The ability of non-woody plants to stand upright is due primarily to turgor pressure. Wilting is the result of a loss of turgor pressure.

TAKE-HOME MESSAGE 3•20

In plants, vacuoles can occupy most of the interior space of the cell. Vacuoles are also present in some other eukaryotic species. They function as storage spaces and play a role in nutrition, waste management, predator deterrence, reproduction, and physical support.

3•21

Chloroplasts are the plant cell's power plant.

It would be hard to choose the "most important organelle" in a cell, but if we had to, the chloroplast would be a top contender. The chloroplast, an organelle found in all plants and eukaryotic algae, is the site of photosynthesis—the conversion of light energy into the chemical energy of food molecules, with oxygen as a by-product. Because all photosynthesis in plants and algae takes place in chloroplasts, these organelles are directly or indirectly responsible for everything we eat and for the oxygen we breathe. Life on earth would be vastly different without the chloroplast (**FIGURE 3-38**).

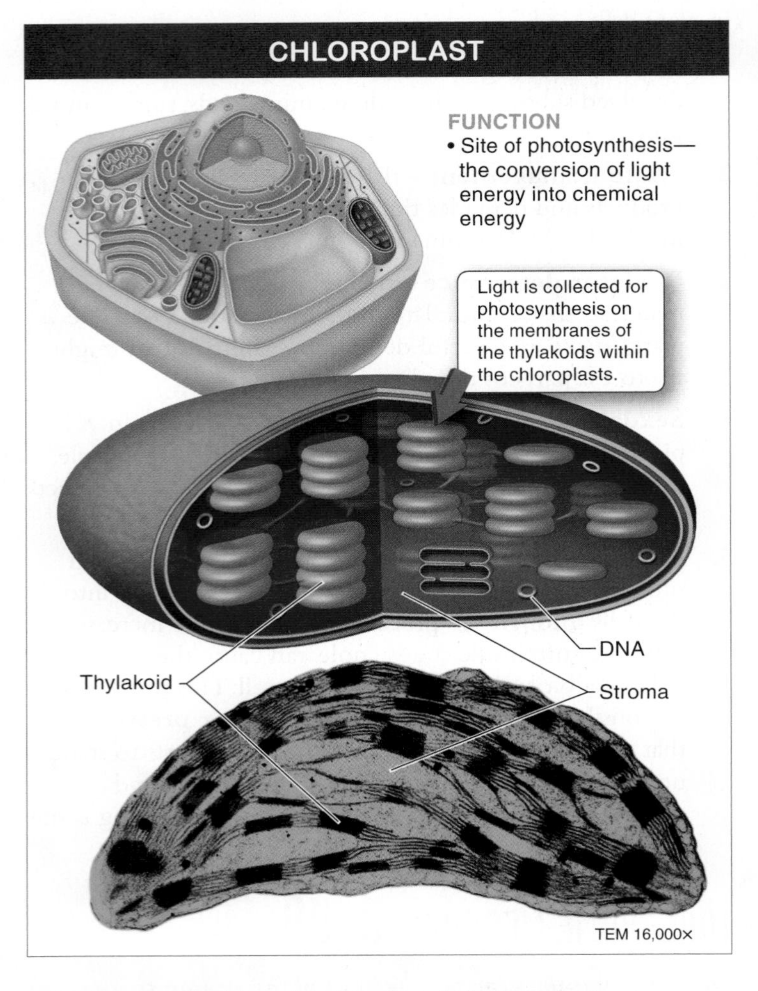

FIGURE 3-38 The chloroplast: location of photosynthesis.

Life on earth would be vastly different without the chloroplasts found in photosynthetic plants.

In a green leaf, each cell has about 40–50 chloroplasts. This means there are about 500,000 chloroplasts per square millimeter of leaf surface (that's more than 200 million chloroplasts in an area the size of a small postage stamp). Chloroplasts are oval, somewhat flattened objects. On the outside, the chloroplast is encircled by two distinct layers of membranes (similar to the structure of mitochondria). The fluid in the inside compartment, called the **stroma,** contains some DNA and much protein-making machinery.

With a simple light microscope, it is possible to see little spots of green within the chloroplasts. Up close, these spots look like stacks of pancakes. Each stack consists of numerous, interconnected little flattened sacs called **thylakoids,** and it is on the membranes of the thylakoids that the light-collecting for photosynthesis occurs. (In Chapter 4, we discuss the details

ORGANELLE REVIEW			
STRUCTURE	ANIMALS	PLANTS	FUNCTION
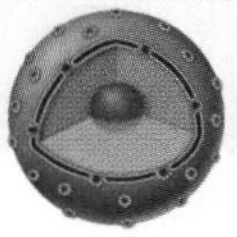 Nucleus	Yes	Yes	Directs cellular activity and stores hereditary information
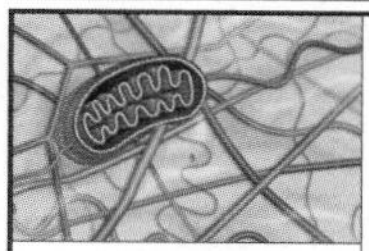 Cytoskeleton	Yes	Yes	Provides structural shape and support and enables cellular movement
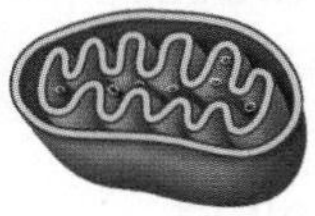 Mitochondrion	Yes	Yes	Harvests energy for cellular functions
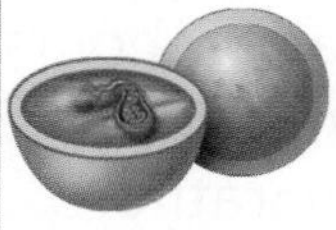 Lysosomes	Yes	Yes	Digests and recycles cellular waste products and consumed material
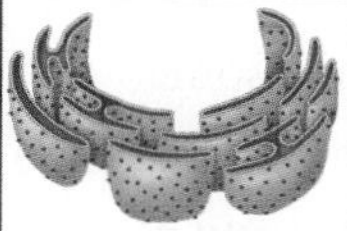 Rough ER	Yes	Yes	Modifies proteins that will be shipped elsewhere in the organism
 Smooth ER	Yes	Yes	Synthesizes lipids and detoxifies molecules
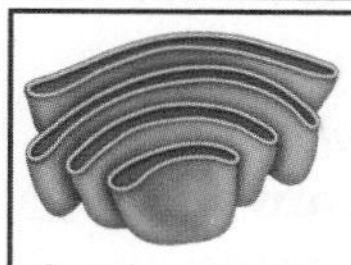 Golgi apparatus	Yes	Yes	Processes and packages proteins, lipids, and other molecules
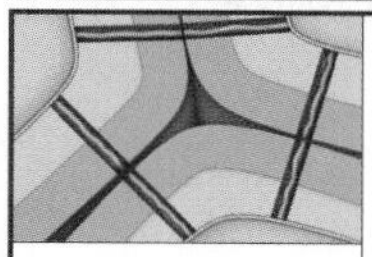 Cell wall	No	Yes	Provides structural strength, protection, and increased resist-ance to water loss
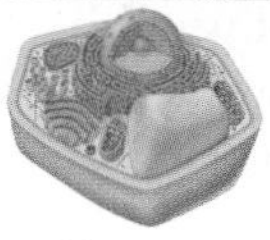 Vacuole	Sometimes	Yes	Stores nutrients, degrades waste products, provides pigments and structural support
 Chloroplast	No	Yes	Performs photosynthesis

of how chloroplasts convert the energy in sunlight into the chemical energy stored in sugar molecules.)

A peculiar feature of chloroplasts, mentioned in Section 3-3, is that they resemble photosynthetic bacteria, particularly with their circular DNA (which contains many of the genes essential for photosynthesis). Also, the dual outer membrane of the chloroplast is consistent with the idea that, long ago, a cell engulfed a photosynthetic bacterium, enveloping it with its plasma membrane in endocytosis. These features have given rise to the belief that chloroplasts might have originally been bacteria that were engulfed by a predatory cell. According to the endosymbiosis theory (see Section 3-3), the bacteria remained alive and, rather than becoming a meal, became the cell's meal ticket, providing food for the cell in exchange for protection.

FIGURE 3-39 summarizes all of the parts of a cell, in animals and plants, and their functions.

TAKE-HOME MESSAGE 3•21

The chloroplast is the organelle in plants and algae that is the site of photosynthesis—the conversion of light energy into chemical energy, with oxygen as a by-product. Chloroplasts may have originally been bacteria that were engulfed by a predatory cell by endosymbiosis.

FIGURE 3-39 Review of the structures and functions of cellular organelles.

Knowledge You Can Use

Did you know? Drinking too much water can be dangerous!

People have long understood the risks of becoming dehydrated, particularly during activities such as marathon running in which a great deal of fluid can be lost through sweating. Only recently, however, has it been recognized that too much water—called "water intoxication"—can also be dangerous. In 2007, a 28-year-old woman competing in a water-drinking contest died just a few hours after consuming about two gallons of water without urinating. Similarly, the death of a runner during the 2002 Boston marathon was attributed to water intoxication.

Q: When people sweat, what do they lose? Sweat consists of water and some dissolved solids, primarily salt. As the sweat evaporates, the body is cooled.

Q: How do people respond to nausea and dizziness, the warning signs of dehydration? By drinking water. Because the water they drink lacks salt and other solutes, this can lead to sodium imbalances in the body, particularly in the bloodstream and the fluid around cells (extracellular fluid).

Q: What is osmosis? Osmosis is the diffusion of water across a membrane toward areas where there is a greater concentration of solutes and, consequently, a lower concentration of water molecules.

Q: If you drink lots of water (two gallons or more)—which has few or no solutes—where is it going to move once it gets into your bloodstream and extracellular fluid? It moves by osmosis into your cells, where there is a higher concentration of solutes (primarily sodium) and a lower concentration of water molecules.

Q: What happens to your cells if they absorb too much water? They swell, sometimes to the point of rupturing. Such swelling and rupturing in the enclosed brain cavity can cause vomiting and confusion. Disastrously, these symptoms are often mistaken for the symptoms of dehydration and so the consumption of water is recommended, making the situation worse. Further swelling can lead to seizures, coma, and even death. (The reverse situation occurs when people drink too much seawater. Their bloodstream and extracellular fluid gain in solutes, particularly salt, so water flows out of cells by osmosis. Cell dehydration and death can occur—usually from a shrinkage of brain cells.)

What should you do? Many marathon organizers have been reducing the number of water stops they offer runners during the race and testing symptomatic runners' sodium levels in medical tents along the course. Consuming salt tablets and salty snacks helps keep blood sodium levels in a healthy range. Also, when possible, runners should weigh themselves during a race to make sure they are only replacing the fluid they have lost through sweating and are not gaining weight.

1 What is a cell?

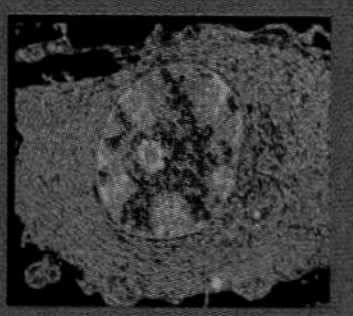

The most basic unit of any organism is the cell, the smallest unit of life that can function independently and perform all of the necessary activities of life, including reproducing itself. All living organisms are made up of one or more cells, and all cells arise from other, pre-existing cells. Every cell on earth is either a eukaryote or a prokaryote. Prokaryotes, which have no nucleus and are single-celled, were the first cells on earth. Eukaryotes are single-celled or multicellular organisms whose cells have a nucleus that contains linear strands of genetic material.

2 Cell membranes are gatekeepers.

Every cell of every living organism is enclosed by a plasma membrane, a two-layered membrane that holds the contents of a cell in place and regulates what enters and leaves the cell. The plasma membrane is a fluid mosaic of proteins, lipids, and carbohydrates. Every cell in your body has a "fingerprint" made from a variety of molecules on the outside-facing surface of the cell membrane that is key to the functioning of your immune system.

3 Molecules move across membranes in several ways.

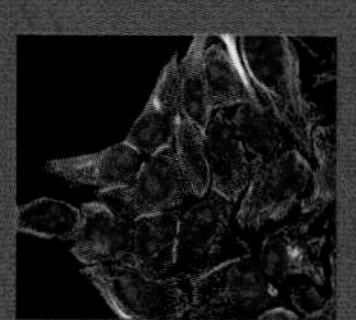

Cells must acquire food molecules and other necessary materials from outside the cell. Similarly, metabolic waste molecules and molecules produced for use elsewhere must move out of the cell. In passive transport the molecular movement occurs spontaneously, without the input of energy. The diffusion of water across a membrane is a special type of passive transport called osmosis. Active transport is necessary if the molecules to be moved are very large or if they are being moved against their concentration gradient. It relies on proteins embedded in the plasma membrane which act like motorized revolving doors. Cells can also engulf particles with their plasma membrane in a process called endocytosis. Similarly, molecules can be moved out of a cell via exocytosis.

4 Cells are connected and communicate with each other.

In multicellular organisms, most cells are connected to other cells. The connections can form a water-tight seal between the cells (tight junctions), can hold sheets of cells together while allowing fluid to pass between the cells (desmosomes), or can function like secret passageways between cells (gap junctions). In plants, plasmodesmata connect cells and enable communication and transport between them.

5 Nine important landmarks distinguish eukaryotic cells.

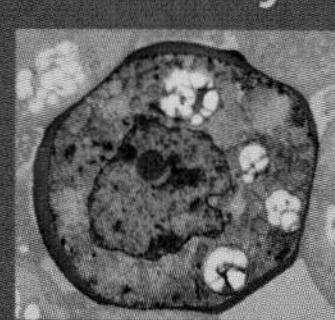

Eukaryotic cells contain many distinct structures, each of which performs a specific life-sustaining function. The nucleus directs all protein production and is also the storehouse for all hereditary information. The cytoskeleton gives animal cells shape and support. In mitochondria, energy from macromolecules is converted into usable energy in the form of ATP. Lysosomes are like garbage disposals for dismantling waste and bacterial pathogens. The production and modification of molecules, including drug detoxification, occurs in the rough and smooth endoplasmic reticulum, and the Golgi apparatus processes molecules synthesized in the cell and packages those destined for use elsewhere in the body. In plants, the cell wall confers structural strength, herbivore protection, and resistance to water loss. Vacuoles function as storage spaces, and chloroplasts are the site of photosynthesis, the conversion of light energy into food molecules.

KEY TERMS

Check Your Knowledge

1. Which of the following statements about the cell theory is correct?
   a) All living organisms are made up of one or more cells.
   b) All cells arise from other, pre-existing cells
   c) All eukaryotic cells contain symbiotic prokaryotes.
   d) All prokaryotic cells contain symbiotic eukaryotes.
   e) Both a) and b) are correct.

2. Which of the following statements about prokaryotes is incorrect?
   a) Prokaryotes appeared on earth before eukaryotes.
   b) Prokaryotes have circular pieces of DNA within their nuclei.
   c) Prokaryotes contain cytoplasm.
   d) Prokaryotes contain ribosomes.
   e) Some prokaryotes can conduct photosynthesis.

3. Which of the following facts supports the claim that mitochondria developed from bacteria that, long ago, were incorporated into eukaryotic cells by the process of phagocytosis?
   a) Mitochondria have flagella for motion.
   b) Mitochondria have proteins for the synthesis of ATP.
   c) Mitochondria are the "powerhouses" of the cell.
   d) Mitochondria are small and easily transported across cell membranes.
   e) Mitochondria have their own DNA.

4. Hydrophobic molecules can pass freely through the plasma membrane, but molecules with electrical charges (such as ions and polar molecules) are impeded by the hydrophobic core. For this reason, plasma membranes can be considered:
   a) partially permeable.
   b) impermeable.
   c) hydrophobic.
   d) hydrophilic.
   e) None of these terms properly describe plasma membranes.

5. Drugs called beta-blockers do all of the following except:
   a) reduce high blood pressure.
   b) block signaling through adrenaline receptors.
   c) reduce outward symptoms of anxiety.
   d) bind to the cytoplasmic side of a receptor protein.
   e) reduce the effects of adrenaline on the heart.

6. Cellular "fingerprints":
   a) are exposed on the cytoplasmic side of the membrane.
   b) are made from cholesterol.
   c) are "erased" by the HIV virus.
   d) can help the immune system distinguish "self" from "non-self."
   e) All of the above are correct.

7. The movement of molecules across a membrane from an area of high concentration to one of low concentration is best described as:
   a) active transport.
   b) inactivated transport.
   c) passive transport.
   d) channel-mediated diffusion.
   e) electron transport.

8. The transport of water across a membrane from a solution of lower solute concentration to a solution of higher solute concentration is best describe as:
   a) osmosis.
   b) facilitated diffusion.
   c) receptor-mediated transport.
   d) active transport.
   e) general diffusion.

9. In an experiment, you measure the concentration of a polar molecule inside and outside a cell. You find that the concentration is high and gradually increasing inside the cell. You also measure the ATP concentration inside the cell and find that it is dropping. What is your best hypothesis for the process you are observing?
   a) facilitated diffusion
   b) passive transport
   c) simple diffusion
   d) active transport
   e) endocytosis

10. Which of the following allows the passage of small molecules between animal cells?
    a) nucleoli
    b) tight junctions
    c) desmosomes
    d) gap junctions
    e) black holes

11. The largest structure in a eukaryotic cell is the ______ and it is surrounded by ______ membrane(s).
    a) nucleus; one
    b) nucleus; two
    c) Golgi apparatus; one
    d) mitochondrion; two
    e) mitochondrion; one

12. The cytoskeleton:
    a) is a viscous fluid found in all cells.
    b) fills a cell's nucleus but not the other organelles.
    c) gives an animal cell shape and support, but cannot control movement.
    d) helps to coordinate intracellular movement of organelles and molecules.
    e) All of the above are correct.

13. Which of the following statements about mitochondria is correct?
    a) Mitochondria are found in both eukaryotes and prokaryotes.
    b) There tend to be more mitochondria in fat cells than in liver cells.
    c) Most plant cells contain mitochondria.
    d) Mitochondria may have originated evolutionarily as photosynthetic bacteria.
    e) All of the above are correct.

14. Which of the following organelles is not present in animal cells?
    a) lysosome
    b) Golgi apparatus
    c) rough endoplasmic reticulum
    d) mitochondrion
    e) chloroplast

15. Given that a cell's structure reflects its function, what function would you predict for a cell with a large Golgi apparatus?
    a) movement
    b) secretion of digestive enzymes
    c) transport of chemical signals
    d) rapid replication of genetic material and coordination of cell division
    e) attachment to bone tissue.

16. Cell walls:
    a) only occur in plant cells.
    b) are not completely solid, having many small pores.
    c) confer less structural support than the plasma membrane.
    d) dissolve when a plant dies.
    e) are made primarily from phospholipids.

17. Which one of the following organelles is not found in both plant and animal cells?
    a) nucleus
    b) rough endoplasmic reticulum
    c) mitochondrion
    d) smooth endoplasmic reticulum
    e) central vacuole

18. In plant cells, chloroplasts:
    a) serve the same purpose that mitochondria serve in animal cells.
    b) are the site of conversion of light energy into chemical energy.
    c) play an important role in the breakdown of plant toxins.
    d) have their own linear strands of DNA.
    e) Both a) and b) are correct..

Short-Answer Question

1. Name and note the function of the organelles shown in the diagram.

See Appendix for answers. For additional study questions, go to www.prep-u.com.

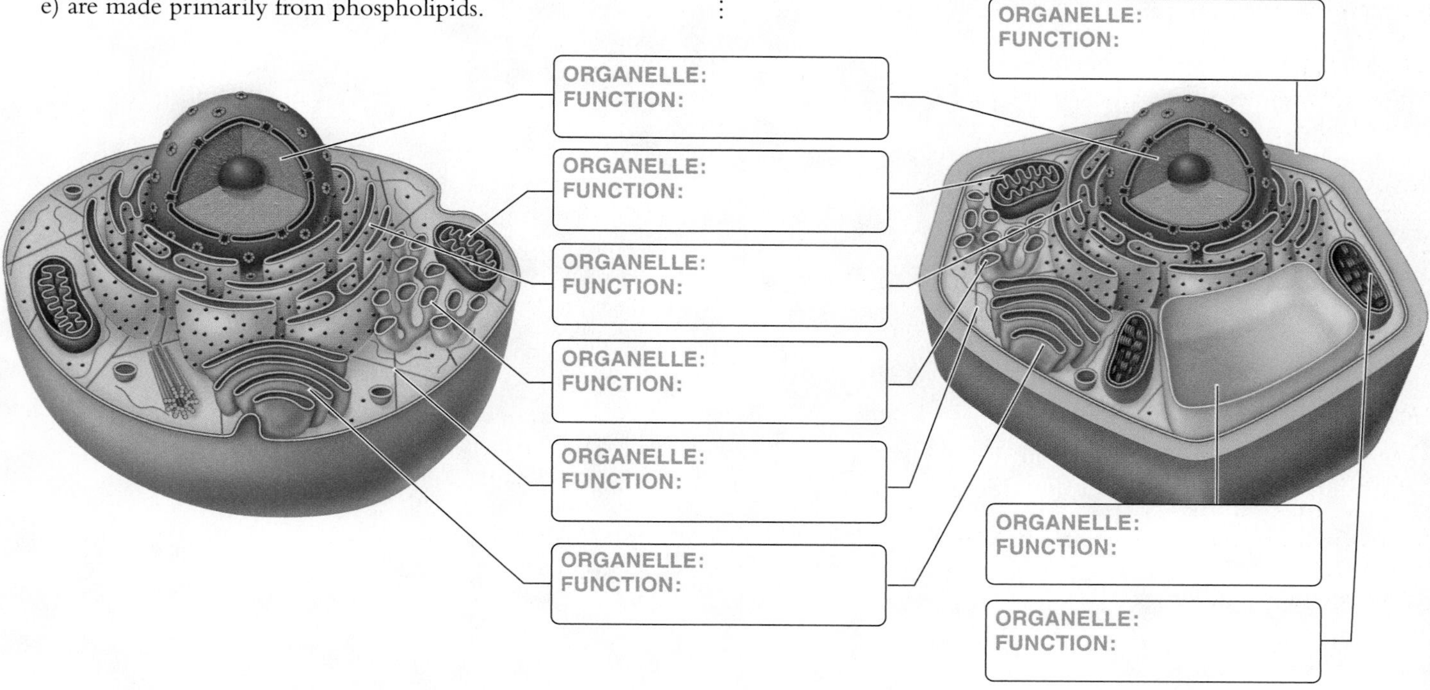

1 Energy flows from the sun and through all life on earth.

- **4.1** Cars that run on french fry oil? Organisms and machines need energy to work.
- **4.2** Energy has two forms: kinetic and potential.
- **4.3** As energy is captured and converted, the amount of energy available to do work decreases.
- **4.4** ATP molecules are like free-floating rechargeable batteries in all living cells.

2 Photosynthesis uses energy from sunlight to make food.

- **4.5** Where does plant matter come from? Photosynthesis: the big picture.
- **4.6** Photosynthesis takes place in the chloroplasts.
- **4.7** Light energy travels in waves: plant pigments absorb specific wavelengths.
- **4.8** Photons cause electrons in chlorophyll to enter an excited state.
- **4.9** Photosynthesis in detail: the energy of sunlight is captured as chemical energy.
- **4.10** Photosynthesis in detail: the captured energy of sunlight is used to make food.
- **4.11** The battle against world hunger can use plants adapted to water scarcity.

3 Cellular respiration converts food molecules into ATP, a universal source of energy for living organisms.

- **4.12** How do living organisms fuel their actions? Cellular respiration: the big picture.
- **4.13** The first step of cellular respiration: glycolysis is the universal energy-releasing pathway.
- **4.14** The second step of cellular respiration: the Krebs cycle extracts energy from sugar.
- **4.15** The third step of cellular respiration: ATP is built in the electron transport chain.

4 There are alternative pathways to energy acquisition.

- **4.16** Beer, wine, and spirits are by-products of cellular metabolism in the absence of oxygen.
- **4.17** Eating a complete diet: cells can run on protein and fat as well as on glucose.

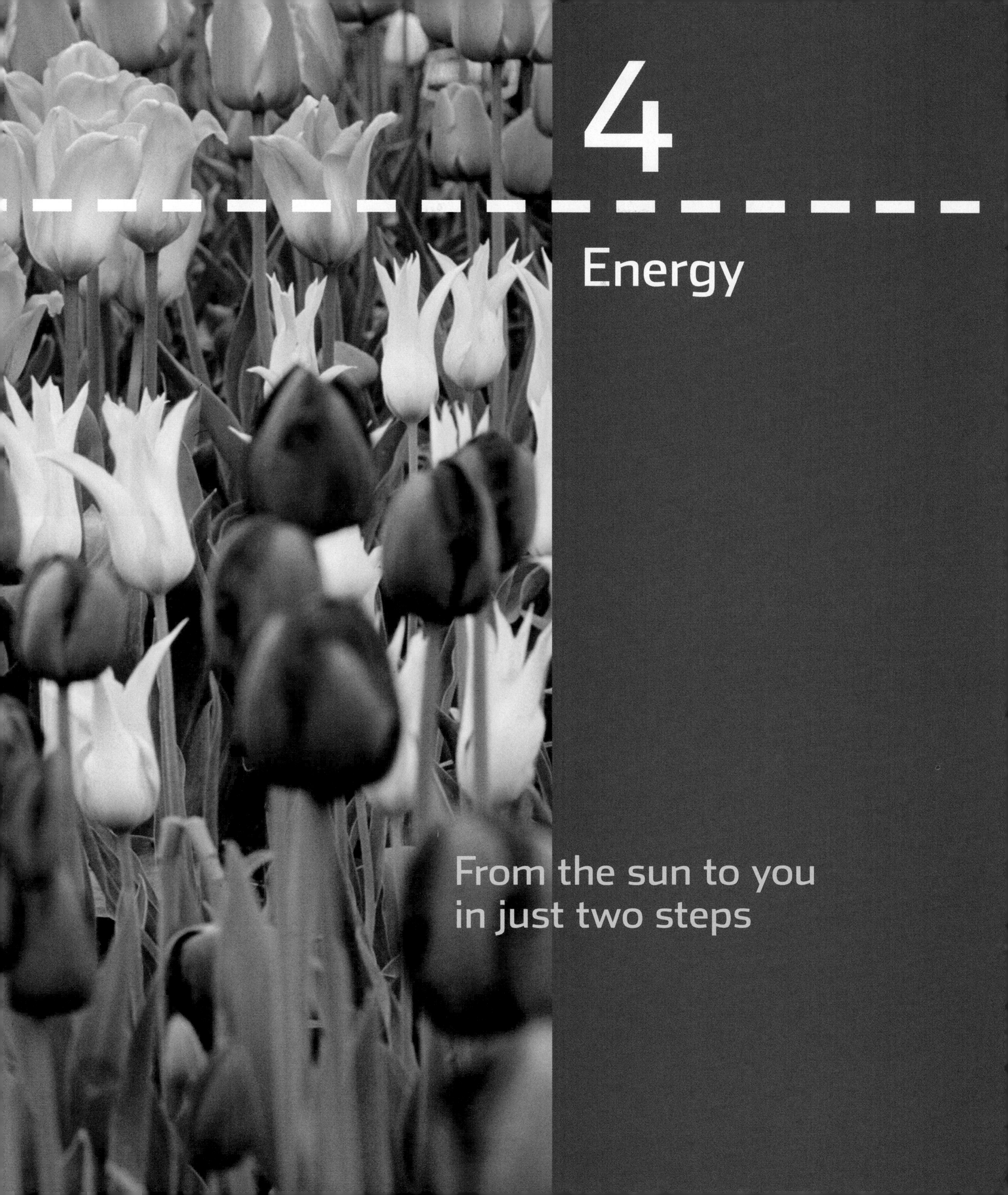

4

Energy

From the sun to you in just two steps

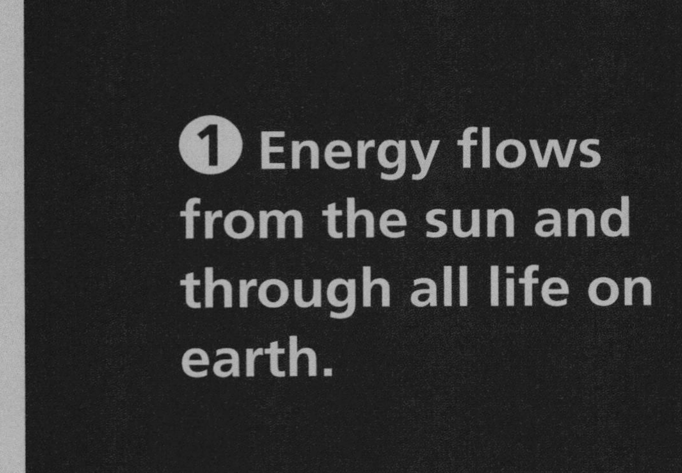

Sunlight filters through the northern California redwoods.

4•1

Cars that run on french fry oil? Organisms and machines need energy to work.

Imagine that you are on a long road trip. The fuel gauge of your car is nearing empty, so you pull off the highway. Instead of driving into a gas station, however, you head to a fast-food restaurant. You drive around to the back where the restaurant stores used cooking materials and fill your car's fuel tank with recycled cooking grease. You head back to the highway, ready to drive several hundred miles before needing another pit stop.

Q Humans can get energy from food. Can machines?

Fast food for your car? Yes! The idea isn't as far-fetched as it sounds. In fact, on the roads of America today many vehicles run on **biofuels,** fuels produced from plant and animal products (**FIGURE 4-1**). Most vehicles, however, run on **fossil fuels** such as gasoline. These fuels (which also include oil, natural gas, and coal) are produced from the decayed remains of ancient plants and animals modified over the course of millions of years by heat, pressure, and bacterial processes.

It turns out that biofuels, fossil fuels, and the food fuels that supply energy to most living organisms are chemically similar. This fact is not surprising because energy from the sun is the source of the energy stored in the chemical bonds between the atoms in all these fuels. Let's investigate how fuels provide energy.

When we burn gasoline, long chains of carbon and hydrogen atoms separate, releasing carbon dioxide (CO_2), water, and a lot of energy that was stored in the chemical bonds holding each gasoline molecule together. (An interesting fact: the energy released in burning one gallon of gas is equivalent to the caloric content of 15 large cheese pizzas.) In an automobile engine, some of this released energy is harnessed to push pistons, spin a crankshaft, turn wheels, and move the car.

Animal fats and the oils in many plants—such as those used to cook french fries—share an important chemical feature with

FIGURE 4-1 Biofuel technology. Biofuels are chemically similar to fossil fuels.

gasoline. Like gasoline, these fats and oils contain chains of carbon and hydrogen atoms bound together, and just as with gasoline, breaking these bonds releases large amounts of energy (and water and CO_2). If this released energy can be captured efficiently, it, too, can be used to push pistons and turn car wheels.

Cars that run on biofuels are more than just a technological trick. The production of biofuels requires only the plant or animal source, sunlight, air, water, and a relatively short amount of time—a few months or years, depending on the source. On the other hand, the production of fossil fuels such as coal or crude oil requires plant and animal remains and millions of years. This difference gives biofuels an important advantage over fossil fuels: they are a renewable resource. For this reason, they point the way toward a future of reduced dependence on fossil fuels, the supplies of which are dwindling and the combustion of which has many harmful consequences, such as increasing global warming and the release of cancer-causing particles into the atmosphere. Are we at the point yet where all our cars can run on biofuels? Not quite: there are several significant drawbacks to the increased use of biofuels, chief among them the destruction of forests, wetlands, and other ecologically important habitats from the increased use of land to grow these fuels. This is why the search for better fuels continues.

In this chapter we explore how plant, animal, and other living "machines" run on energy stored in chemical bonds. Just as the forward motion of a car is fueled by breaking the chemical bonds in gasoline and harnessing the released energy, the activities of living organisms are fueled by breaking the chemical bonds in food molecules and harnessing the released energy.

All life depends on capturing energy from the sun and converting it into a form that living organisms can use. This

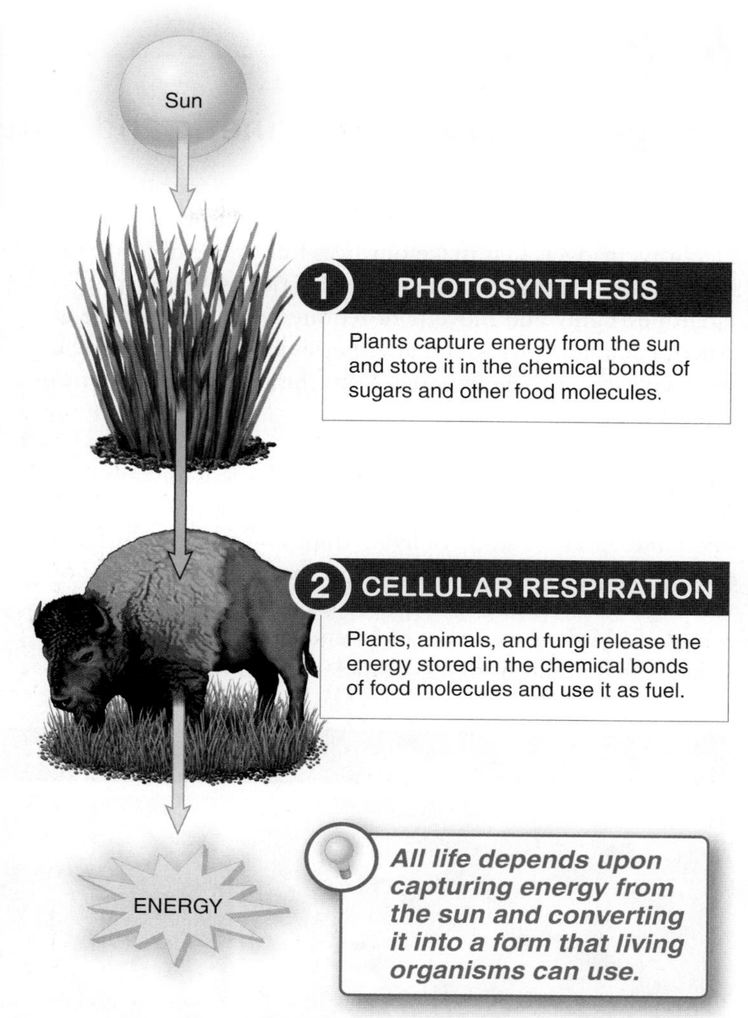

FIGURE 4-2 Photosynthesis and cellular respiration.

energy capture and conversion occurs in two important processes that mirror each other: (1) **photosynthesis,** the process by which plants capture energy from the sun and store it in the chemical bonds of sugars and other food molecules they make, and (2) **cellular respiration,** the process by which all living organisms release the energy stored in the chemical bonds of food molecules and use it to fuel their lives (**FIGURE 4-2**). The sun to you in just two steps!

TAKE-HOME MESSAGE 4•1

The sun is the source of the energy that powers all living organisms and other "machines." The energy from sunlight is stored in the chemical bonds of molecules. When these bonds are broken, energy is released regardless of whether the bond is in a molecule of food, a fossil fuel, or a biofuel like the oil in which french fries are cooked.

4•2

Energy has two forms: kinetic and potential.

"Batteries not included." For a child, those are pretty depressing words. We know that many of the toys and electronic gadgets that make our lives fun or useful (or both) need energy—usually in the form of batteries. Generating ringtones, lights, and movement requires energy. The same is true for humans, plants, and all other living organisms: they need energy for their activities, from thinking to moving to reproducing.

Energy is the capacity to do work. And work is anything that involves moving matter against an opposing force. The energy encountered in the study of living things is of two types: kinetic energy and potential energy. **Kinetic energy** is the energy of moving objects. Legs pushing bike pedals and the flapping wings of a bird are examples of kinetic energy (**FIGURE 4-3**). Heat, which results from lots of molecules moving rapidly, is another form of kinetic energy. Because it comes from the movement of high-energy particles, light is also a form of kinetic energy—probably the most important form of kinetic energy on earth. (We explore the nature of sunlight and how it is harnessed for the work of producing food molecules when we look at photosynthesis later in this chapter.)

An object does not have to be moving to have the capacity to do work; it may have **potential energy,** which is stored energy or the capacity to do work that results from an object's location or position. Water behind a dam, for example, has potential energy. If a hole in the dam is opened, the water can flow through, and perhaps spin a water wheel or turbine. A concentration gradient, which we discussed in Chapter 3, also has potential energy: if the molecules in an area of high concentration move toward an area of lower concentration,

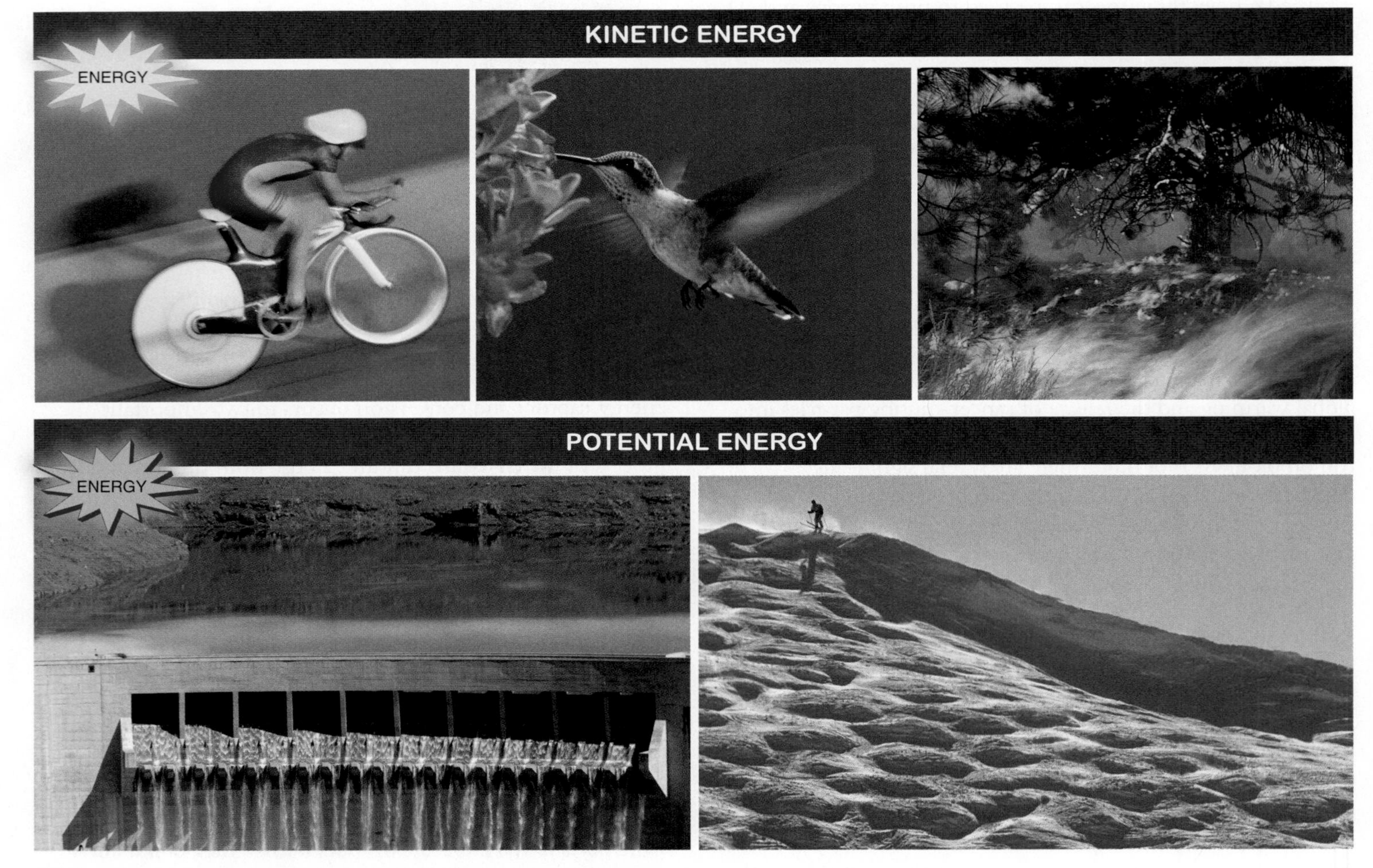

FIGURE 4-3 Kinetic and potential energy. Kinetic energy is the energy of motion; potential energy is energy stored in an object, such as water trapped behind a dam.

the potential energy of the gradient is converted to the kinetic energy of molecular movement, and this kinetic energy can do work. And **chemical energy,** the storage of energy in chemical bonds, is also a type of potential energy.

Because potential energy doesn't involve movement, it is a less obvious form of energy than kinetic energy. An apple has potential energy, as does any other type of food (**FIGURE 4-4**). Why? Because the chemical energy stored in the chemical bonds making up the food can be broken and the energy released during cellular respiration, enabling you to run, play, and work. We explore cellular respiration, the energy-releasing breakdown of molecules, later in this chapter. But first we need to know more about the nature of energy.

TAKE-HOME MESSAGE 4•2

Energy, the capacity to do work, comes in two forms. Kinetic energy is the energy of moving objects, while potential energy, such as chemical energy, is stored energy or the capacity to do work that results from the position or location of an object.

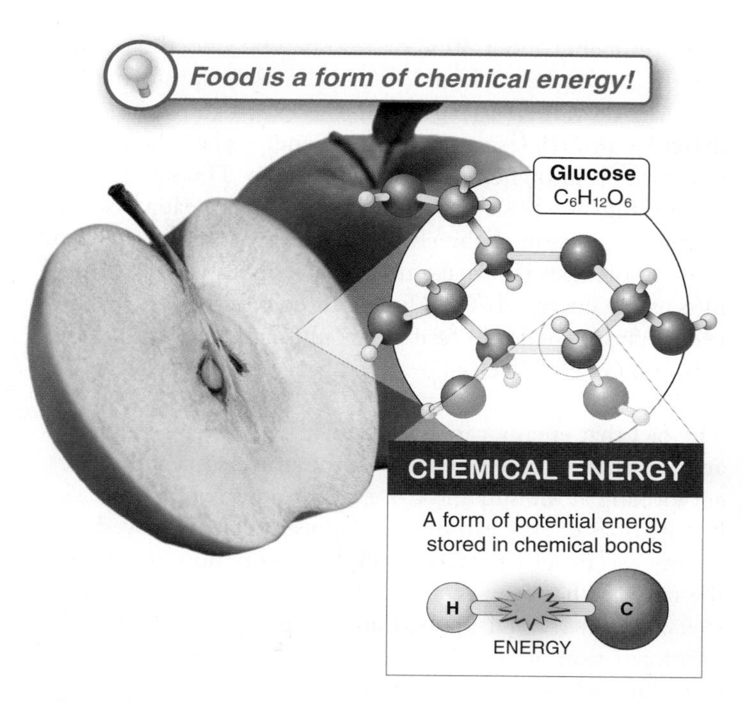

FIGURE 4-4 Chemical energy. The chemical bonds in food molecules are a form of potential energy.

4•3

As energy is captured and converted, the amount of energy available to do work decreases.

Every minute of every day—even on cloudy ones—the sun shines brightly, releasing tremendous amounts of energy. Organisms on earth cannot capture every single bit of energy released by the sun; indeed, most plants capture only about 1% of the available energy. What happens to the other 99%? This unused energy does not simply disappear. Accountants would love to monitor the flow of energy because, as in a good accounting ledger, all of the energy numbers add up perfectly. All energy from the sun can be accounted for: some (probably less than 1%) is captured and transformed into usable chemical energy by organisms through photosynthesis (**FIGURE 4-5**). The rest of the energy from the sun is reflected back into

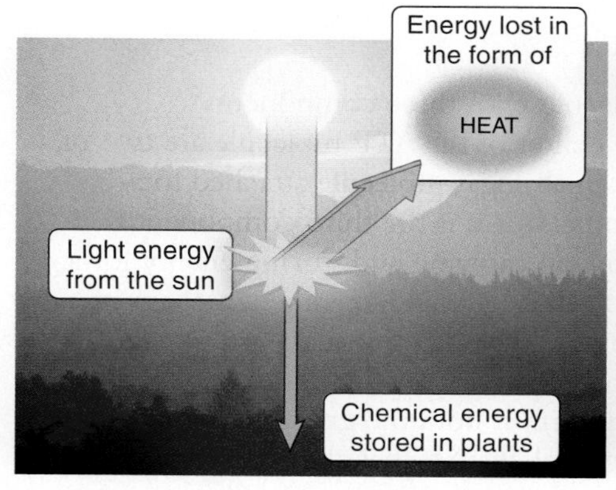

FIGURE 4-5 Energy efficiency and heat loss. As energy is converted to do work, some energy is released as heat.

space (probably about 30%) or absorbed by land, the oceans, and the atmosphere (about 70%), and mostly transformed into heat. Heat is not easily harnessed to do work, however, and is therefore a much less useful form of energy than the energy transformed into chemical energy in plants. The same accounting also exists on a smaller level. If you eat a bowl of rice, some portion of the chemical energy stored in the bonds of the molecules that make up the rice grains is transformed into usable energy that can fuel your cells' activities. All the rest is transformed into heat and is ultimately lost into the atmosphere.

The fact that energy can change form but never disappear is an important feature of energy in the universe, whether we are looking at the sun and the earth or a human and her rice bowl. Just as energy can never disappear or be destroyed, energy can never be created. All the energy now present in the universe has been there since the universe began, and everything that has happened since then has occurred by the transformation of one form of energy into another. In all our eating and growing, driving and sleeping, we are simply transforming energy. The study of the transformation of energy from one type to another, such as from potential energy to kinetic energy, is called **thermodynamics,** and the **first law of thermodynamics** states that energy can never be created or destroyed. It can only change from one form to another.

Because plants capture less than 1% of the sun's energy, it might seem like they are particularly inefficient. But we humans are also rather inefficient at extracting the chemical energy of plants when we eat them. These inefficiencies occur because every time energy is converted from one form to another some of the energy is always converted to heat. When a human converts the chemical energy in a plate of spaghetti into the kinetic energy of running a marathon or when a car transforms the chemical energy of gasoline into the kinetic energy of forward motion, in both cases some energy is converted to heat, the least usable form of kinetic energy. In automobiles, for example, about three-quarters of the energy in gasoline is lost as heat. Thus, for every $10 spent on gas, only about $2.50 goes toward moving the car!

The **second law of thermodynamics** states that every conversion of energy is not perfectly efficient and invariably includes the transformation of some energy into heat. Although heat is certainly a form of energy, because it is not easily harnessed to do work it is almost completely useless to living organisms for fueling their cellular activity. Put another way, the second law of thermodynamics tells us that although the quantity of energy in the universe is not changing, its quality is. Little by little, the amount of energy that is available to do work decreases. Now that we understand that organisms on earth cannot capture every single bit of energy released by the sun—and that energy conversions are inefficient—we can look at the chief energy currency of the cell: ATP.

TAKE-HOME MESSAGE 4•3

Energy is neither created nor destroyed but can change forms. Each conversion of energy is inefficient and some of the usable energy is converted to less useful heat energy.

4•4

ATP molecules are like free-floating rechargeable batteries in all living cells.

Much of the work that cells do requires energy. But even though light energy from the sun carries energy, as do molecules of sugar, fat, and protein, none of this energy can be used directly to fuel chemical reactions in organisms' cells. First it must be captured in the bonds of a molecule called **adenosine triphosphate (ATP),** a free-floating molecule found in cells that acts like a rechargeable battery that temporarily stores energy that can then be used for cellular work in plants, animals, bacteria, and all the other organisms on earth. The use of ATP solves an important timing and coordination problem for living cells: a supply of ATP guarantees that the energy required for energy-consuming reactions will be available when it's needed.

ATP is a simple molecule with three components (**FIGURE 4-6**). At the center of the ATP molecule are two of these components: a small sugar molecule attached to a molecule called adenine. But it is the third component that makes ATP so effective in carrying and storing energy for a short time: attached to the sugar and adenine is a chain of three negatively charged phosphate molecule groups (hence the "tri" in "triphosphate"). Because the bonds between these three phosphate groups must hold the groups together in the face of the three electrical charges that all repel one another, each of these bonds contains a large amount of energy and is stressed and unstable. The instability of these high-energy bonds makes the three phosphate groups like a tightly coiled

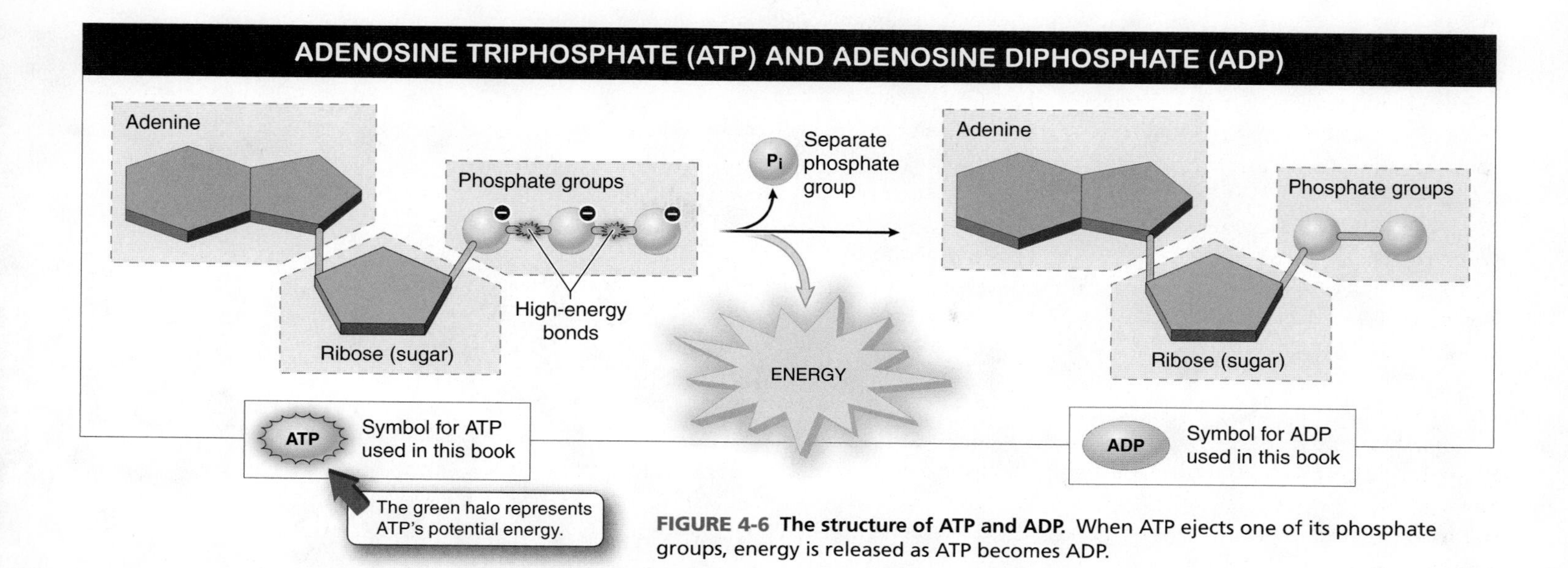

FIGURE 4-6 The structure of ATP and ADP. When ATP ejects one of its phosphate groups, energy is released as ATP becomes ADP.

spring or a twig that is bent almost to the point of breaking. With the slightest push, one of the phosphate groups will pop off, releasing a little burst of energy that the cell can use.

It is precisely because each molecule of ATP is always on the brink of ejecting one of its phosphate groups that ATP is such an effective energy source inside a cell. As long as plenty of ATP molecules are around, they can energize the chemical reactions that make it possible for the cell to carry out reactions that require work, such as building muscle tissue, repairing a wound, or growing roots. Each time a cell expends one of its ATP molecules to pay for an energetically expensive reaction, a phosphate is broken off and energy is released. What is left is a molecule with two phosphates, called ADP (adenosine *di*phosphate), and a separate phosphate group (labeled P_i).

An organism can then use ADP, a free-floating phosphate, and an input of kinetic energy to rebuild its ATP stocks

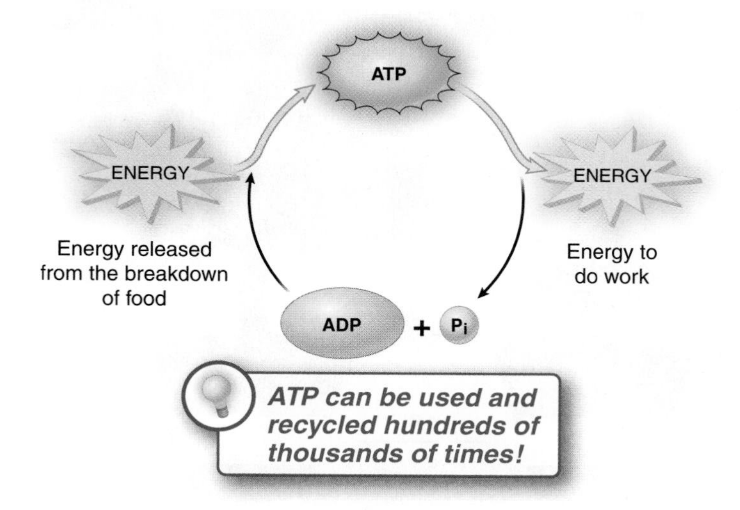

FIGURE 4-7 ATP is like a rechargeable battery.

(**FIGURE 4-7**). The kinetic energy is converted to potential energy when the free phosphate group attaches to the ADP molecule and makes ATP. In this manner, ATP functions like a rechargeable battery. Where does the input of energy for recharging ATP come from? When we discuss photosynthesis, we'll see that plants, algae, and some bacteria directly use light energy from the sun to make ATP from ADP and free-floating phosphate groups. Animals use the energy contained in the bonds of their food molecules. In either case, the energy is used to re-create the unstable bond in the triphosphate chain. When energy is needed, the organism can again release it by breaking the bond holding the phosphate group to the rest of the molecule. Our bodies recycle ATP molecules in this way tens of thousands of times a day.

Here's the ATP story in a nutshell. Breaking down a molecule of sugar—in a glass of orange juice, for example—leads to a miniature burst of energy in your body. The energy from the mini-explosion is put to work building the unstable high-energy bonds that attach phosphate groups to ADP molecules, creating new molecules of ATP. Later—perhaps only a fraction of a second later—when an energy-consuming reaction is needed, your cells can release the energy stored in the new ATP molecules.

TAKE-HOME MESSAGE 4•4

Cells temporarily store energy in the bonds of ATP molecules. This potential energy can be converted to kinetic energy and used to fuel life-sustaining chemical reactions. At other times, inputs of kinetic energy are converted to the potential energy of the energy-rich but unstable bonds in the ATP molecule.

② Photosynthesis uses energy from sunlight to make food.

Sunflowers capture light energy and convert it to organic matter (including tasty seeds).

4•5

Where does plant matter come from? Photosynthesis: the big picture.

Watching a plant grow over the course of a few years can seem like watching a miracle, or at least a very subtle magic trick. Of course it's neither, but the process is nonetheless amazing. Consider that in five years a tree in a big planter can increase its weight by 150 pounds (68 kg) as it grows (**FIGURE 4-8**). Where does that 68 kg of new tree come from?

Our first guess might be the soil. Could that be it? It's easy enough to weigh the soil in the pot when first planting the tree and again 5 years later. After 5 years, though, we find that the soil in our planter has lost less than a pound, nowhere near enough to explain the massive increase in the amount of plant material. Perhaps the new growth comes from the water? Wrong again. Although the older and much larger tree holds more water in its many cells, the water provided to the plant does not come close to accounting for the increase in the dry weight of the plant.

FIGURE 4-8 Where does a plant come from?

Q When humans grow, the new tissue comes from food we eat. When plants grow, where does the new tissue come from?

The amazing truth is that most of the new material comes from an invisible gas in the air. In the process of photosynthesis, plants capture carbon dioxide gas (CO_2) from the air and, using energy they get from sunlight, along with water and small amounts of chemicals usually found in soil, they produce solid, visible (and often tasty) sugars and other organic matter that are used to make plant structures such as leaves, roots, stems, flowers, fruits, and seeds. In the process, the plants give off oxygen (O_2), a by-product that happens to make all animal life possible.

Although plants are the most visible organisms that have evolved the ability to capture light energy and convert it to organic matter, they are not the only organisms capable of photosynthesis. Some bacteria and many other unicellular organisms are also capable of using the energy in sunlight to produce organic materials (**FIGURE 4-9**).

There are three inputs to the process of photosynthesis (**FIGURE 4-10**): light energy (from the sun), carbon dioxide (from the atmosphere), and water (from the ground). From these three inputs, the plant produces sugar and oxygen. As we will see, photosynthesis is best understood as two separate events: a "photo" segment, during which light is captured, and a "synthesis" segment, during which sugar is synthesized.

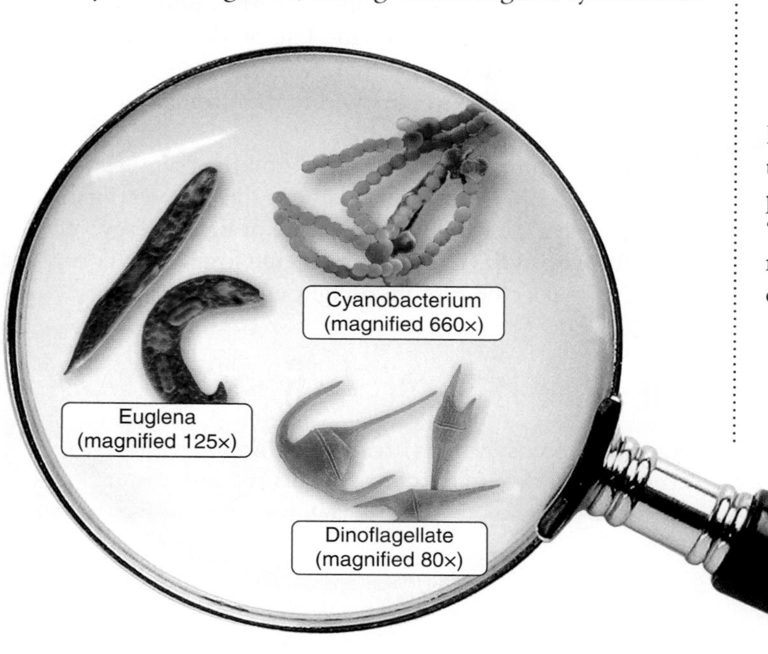

FIGURE 4-9 Plants aren't the only photosynthesizers. Some bacteria and single-celled organisms are capable of photosynthesis.

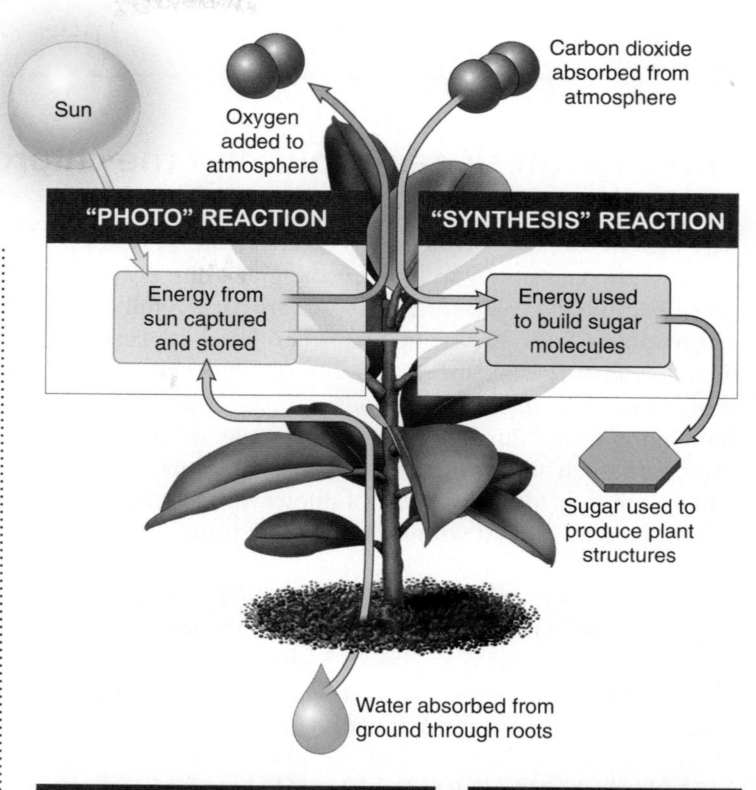

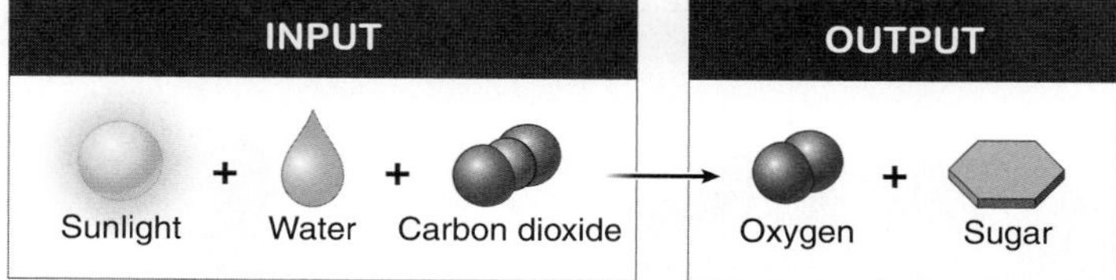

FIGURE 4-10 Photosynthesis: the big picture.

In the "photo" reactions, light energy is captured and temporarily stored in energy-storage molecules. During this process, water molecules split and produce oxygen. In the "synthesis" reactions, the energy in the energy-storage molecules is used to build sugar molecules from carbon dioxide in the air.

TAKE-HOME MESSAGE 4•5

Through photosynthesis, plants use water, the energy of sunlight, and carbon dioxide gas from the air to produce sugars and other organic materials. In the process, photosynthesizing organisms also produce oxygen, which makes all animal life possible.

4•6

Photosynthesis takes place in the chloroplasts.

The best way to know which parts of a plant are photosynthetic is simply to look for the green bits. Leaves are green because the cells near the surface are packed full of **chloroplasts,** light-harvesting organelles found in plant cells, which make it possible for the plant to use the energy from sunlight to make sugars (their food) and other plant tissue (much of which animals use for food) (**FIGURE 4-11**). Other plant parts, such as stems, may also contain chloroplasts (in which case they, too, are capable of photosynthesis), but most chloroplasts are located within the cells in a plant's leaves.

Let's take a closer look at chloroplasts (**FIGURE 4-12**). The sac-shaped organelle is filled with a fluid called the **stroma.** Floating in the stroma is an elaborate system of interconnected membranous structures called **thylakoids,** which often look like stacks of pancakes. Once inside the chloroplast, you can be in one of two places: in the stroma or inside the thylakoids. The conversion of light energy to chemical energy—the "photo" part of photosynthesis—occurs inside the thylakoids. The production of sugars—which are made in the "synthesis" part of photosynthesis—occurs within the stroma.

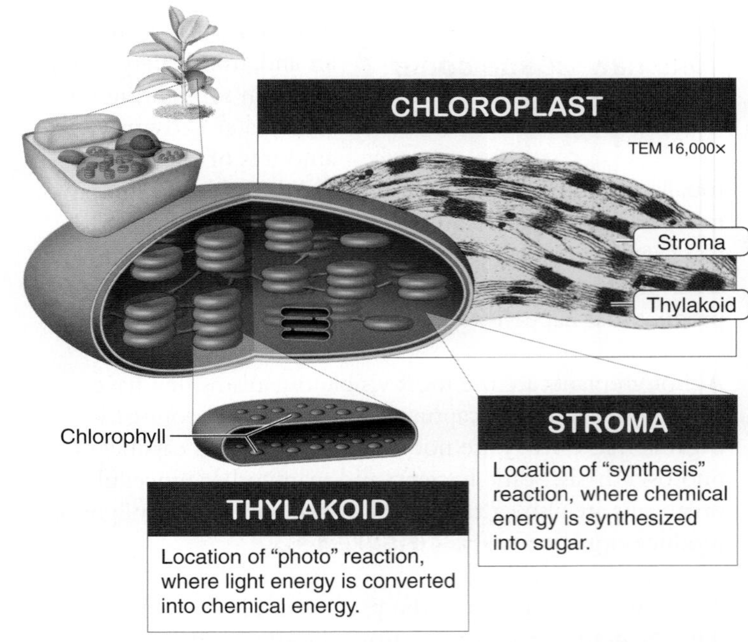

FIGURE 4-12 The chloroplast. The chloroplast is where photosynthesis takes place in a plant.

FIGURE 4-11 Photosynthesis factories. Cells near the leaf's surface are packed with photosynthetic chloroplasts.

We examine both the "photo" and the "synthesis" processes in greater detail later in this chapter. First, however, we examine the nature of light energy and of **chlorophyll,** the special molecule found in chloroplasts that makes the capture of light energy possible.

TAKE-HOME MESSAGE 4•6

In plants, photosynthesis occurs in chloroplasts, green organelles packed in cells near the plants' surfaces.

4•7

Light energy travels in waves: plant pigments absorb specific wavelengths.

You can't eat sunlight. That's because sunlight is *light* energy rather than the *chemical* energy found in the bonds of food molecules. Photosynthesis is powered by **light energy,** a type of kinetic energy made up of little energy packets called **photons,** which are organized into waves. Photons can do work as they bombard surfaces such as your face (heating it) or a leaf (enabling it to build sugar from carbon dioxide and water).

Photons have various amounts of energy, and the length of the wave in which they travel corresponds to the amount of energy carried by the photon. The shorter the wavelength, the more energy the light carries. Within a ray of light, there are super-high-energy packets of photons (those with short wavelengths), relatively low-energy packets (those with longer wavelengths), and everything in between. This range, which is called the **electromagnetic spectrum,** extends from extremely short high-energy gamma rays and X rays, with wavelengths as short as 1 nanometer (a human hair is about 50,000 nanometers in diameter), to very long low-energy radio waves, with wavelengths as long as a mile (**FIGURE 4-13**).

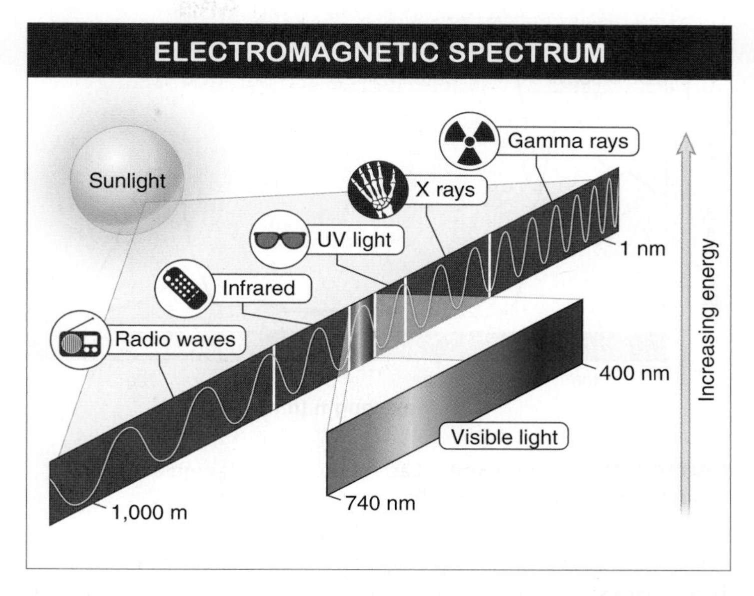

FIGURE 4-13 A spectrum of energy. A ray of light emits high-energy photons, low-energy photons, and everything in between. Plants use only a fraction of the light's available energy.

Just as we can't hear some super-high-pitched frequencies of sound (even though many dogs can), there are some wavelengths of light that are too short or too long for us to see. The light that we can see, visible light, spans all the colors of the rainbow. Humans (and some other animals) can see colors because our eyes contain light-absorbing molecules called **pigments.** These pigments absorb wavelengths of light within the visible range. The energy in these light waves excites electrons in the pigments, which in turn stimulate nerves in our eyes. These nerves then transmit electrical signals to our brains. We perceive different wavelengths within the visible spectrum as different colors. The pigments in the human eye absorb many different wavelengths pretty well: that's why we can see so many colors. When plants use sunlight's energy to make sugar during photosynthesis, they also use the visible portion of the electromagnetic spectrum. Unlike our eyes, however, plant pigments (the energy-capturing parts of a plant) absorb and use only a portion of visible light wavelengths.

Chlorophyll is the pigment molecule in plants that absorbs light energy from the sun. Chlorophyll molecules are embedded in the thylakoid membrane of chloroplasts, which are found primarily in plants' leaves. Just as light energy excites electrons in the pigments responsible for color vision in humans, electrons in a plant's chlorophyll can become excited by certain wavelengths of light and can capture a bit of this light energy.

Plants produce several different light-absorbing pigments (**FIGURE 4-14**). The primary photosynthetic pigment, called **chlorophyll *a*,** efficiently absorbs blue-violet and red wavelengths of light. Every other wavelength generally travels through or bounces off this pigment. Because chlorophyll *a* cannot efficiently absorb green light and instead reflects those wavelengths, our eyes and brain perceive the reflected light waves as green, and so the pigment (and the leaves in which it is found) appears green. Another pigment, **chlorophyll *b*,** is similar in structure but absorbs blue and red-orange wavelengths. Chlorophyll *b* reflects yellow-green wavelengths. A related group of pigments called **carotenoids** absorbs blue-violet and blue-green wavelengths and reflects yellow, orange, and red wavelengths.

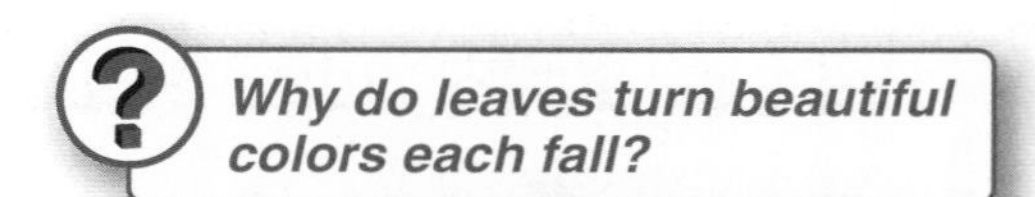

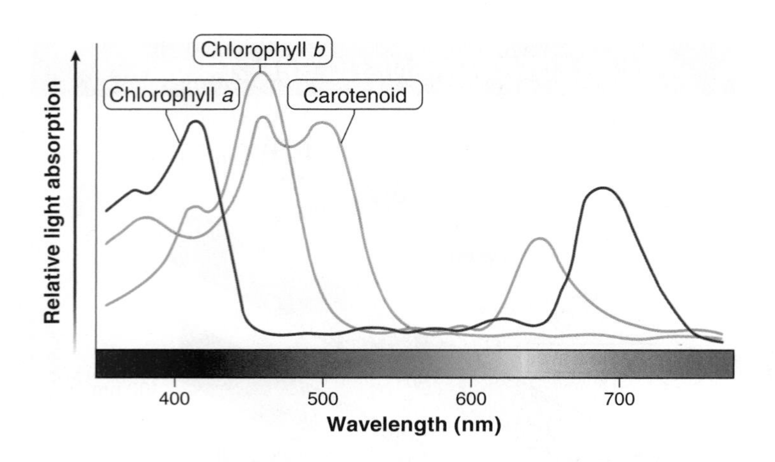

FIGURE 4-14 Plant pigments. Each photosynthetic pigment absorbs and reflects specific wavelengths.

> **Q** Why do the leaves of some trees turn beautiful colors each fall?

In the late summer, cooler temperatures cause some trees to prepare for the winter by shutting down chlorophyll production and reducing their photosynthesis rates, and they go into a state that resembles an animal's hibernation. Gradually, the chlorophyll *a* and *b* molecules present in the leaves are broken down and their chemical components are stored in the branches. As the amounts of chlorophyll *a* and *b* in the leaves decrease relative to the remaining carotenoids, the striking colors of the fall foliage are revealed (see Figure 4-14). During the rest of the year, chlorophyll *a* and *b* are so abundant in leaves that green masks the colors of the other pigments.

TAKE-HOME MESSAGE 4•7

Photosynthesis is powered by light energy, a type of kinetic energy made of energy packets called photons. Photons hit chlorophyll and other light-absorbing molecules near the green surfaces of plants. These molecules capture the light energy and harness it to build sugar from carbon dioxide and water.

4•8

Photons cause electrons in chlorophyll to enter an excited state.

An organism can use energy from the sun only if it can convert the light energy of the sun into the chemical energy in the bonds between atoms. The most important molecule in this conversion is the pigment chlorophyll (**FIGURE 4-15**). When chlorophyll is hit by photons of certain wavelengths, the light energy bumps an electron (e^-) in the chlorophyll molecule to a higher energy level, an *excited* state. Upon absorbing the photon, the electron briefly gains energy, and the potential energy in the chlorophyll molecule increases.

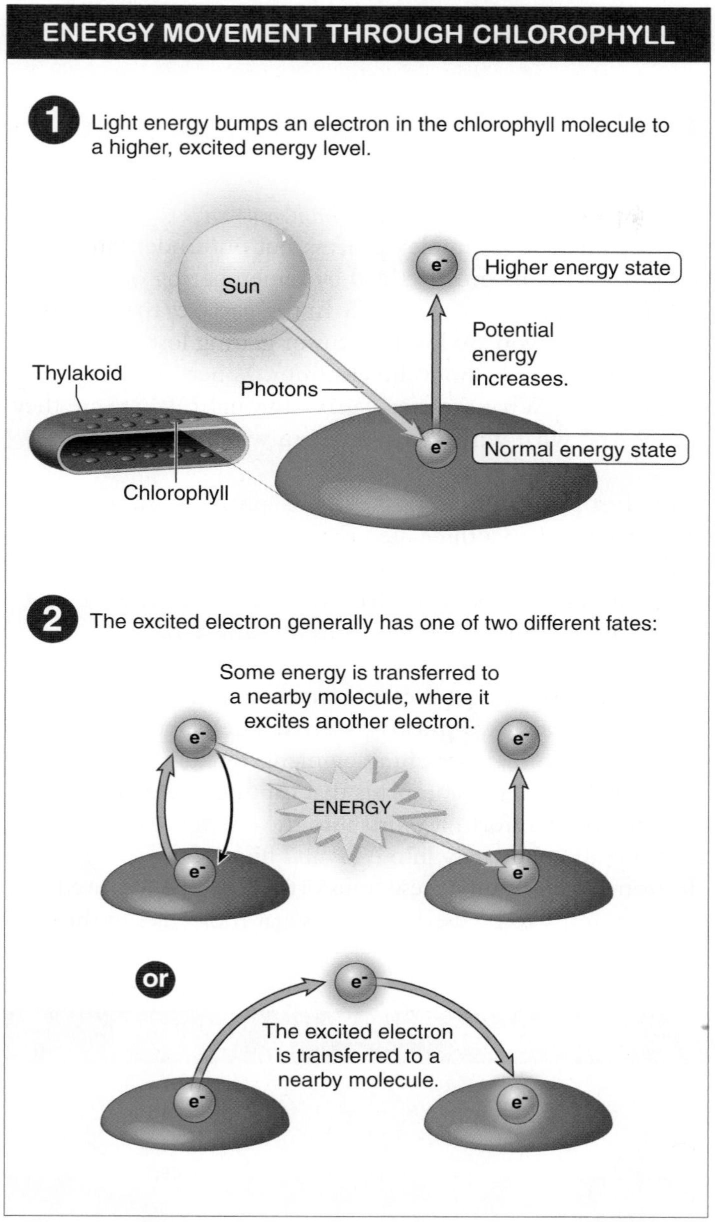

FIGURE 4-15 Capturing light energy with excited electrons. Chlorophyll electrons are excited to a higher energy state by light energy.

An electron in a photosynthetic pigment that is excited to a higher energy state generally has one of two fates: (1) the electron returns to its resting, unexcited state and, in the process, energy is released, some of which may be transferred to a nearby molecule, bumping electrons in that molecule to a higher energy state (and the rest of the energy is dissipated as heat), or (2) the excited electron itself is passed to another molecule.

The passing of electrons from molecule to molecule is one of the chief ways that energy moves through cells. Many molecules carry or accept electrons during cellular activities. All that is required is that the acceptor have a greater attraction for electrons than the molecule from which it accepts them. This receiver molecule, in turn, hands off electrons to another acceptor with an even greater attraction for them. A molecule that gains electrons always carries greater energy than it did before receiving the electron(s). For this reason, the passing of electrons from one molecule to another can be viewed as a passing of potential energy. In this way, energy moves through cells.

This transfer of electrons is one of the first steps of photosynthesis, the process that enables a plant to harness light energy from the sun efficiently and convert it to the more readily usable chemical energy. As we see later in this chapter, the dismantling of food molecules such as glucose to generate energy is also a story of breaking and rearranging chemical bonds as electrons pass from one atom or molecule to another.

Q Suppose a large meteor hit the earth. How could smoke and soot in the atmosphere wipe out life far beyond the area of direct impact?

Because any particles in the atmosphere can block the light from the sun and reduce the excitation of electrons in chlorophyll molecules, photosynthesis depends on a relatively clean atmosphere. Any reduction in the available sunlight can have serious effects on plants. Scientists believe that if a large meteor hit the earth—as one did when the dinosaurs were wiped out 65 million years ago—smoke, soot, and dust in the atmosphere could block sunlight to such an extent that plants in the region, or even possibly all of the plants on earth, could not conduct photosynthesis at high enough levels to survive. And when plants die off, all of the animals and other species that rely on them for energy die as well. All life on earth is completely dependent on the continued excitation of electrons by sunlight.

TAKE-HOME MESSAGE 4•8

When chlorophyll is hit by photons, the light energy excites an electron in the chlorophyll molecule, increasing the chlorophyll's potential energy. The excited electrons can be passed to other molecules, moving the potential energy through the cell.

4•9

Photosynthesis in detail: the energy of sunlight is captured as chemical energy.

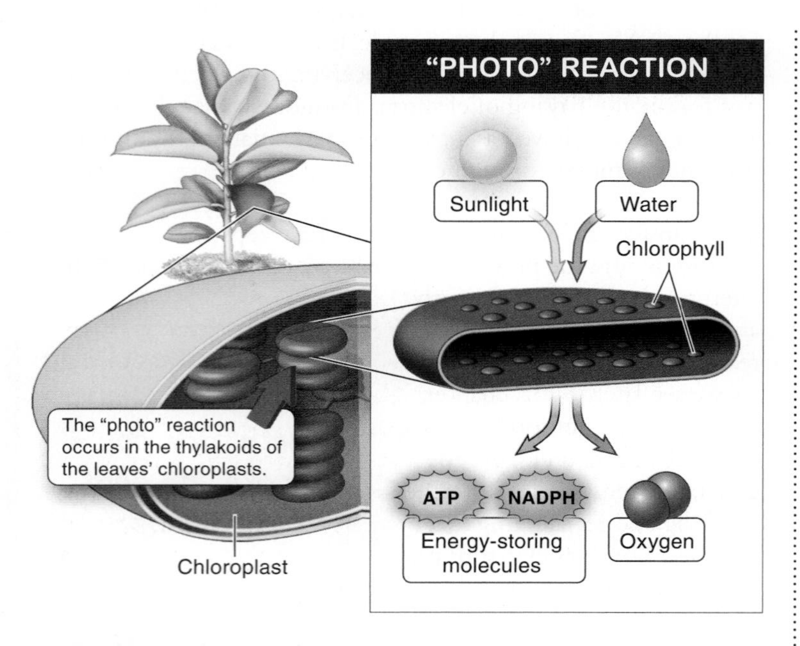

FIGURE 4-16 Overview of the "photo" portion of photosynthesis. Light energy is captured in the "photo" portion of photosynthesis. It is later used to power the building of sugar molecules.

Photosynthesis is a complex process, but our understanding of this process can be greatly aided by remembering one phrase: FOLLOW THE ELECTRONS. In the passages that follow, if you feel that you are losing focus or getting lost, just remember to think about the electrons: Where are they coming from? What are they passing through? Where are they going? And what will happen to them when they get there?

In the first part of photosynthesis, the "photo" part, sunlight hits a plant and, in a three-step process, the energy in this sunlight is ultimately captured and stored in an ATP molecule and in another molecule (called NADPH) that stores energy by accepting high-energy electrons (**FIGURE 4-16**).

The energy-capturing process occurs in a series of steps that are carried out in two **photosystems.** These photosystems are arrangements of light-catching pigments (including chlorophyll) within chloroplasts that capture energy from sunlight and transform it—first into the energy of excited electrons and ultimately into ATP and high-energy electron carriers. After these transformations, the captured energy is ready to be used to make sugar molecules in the

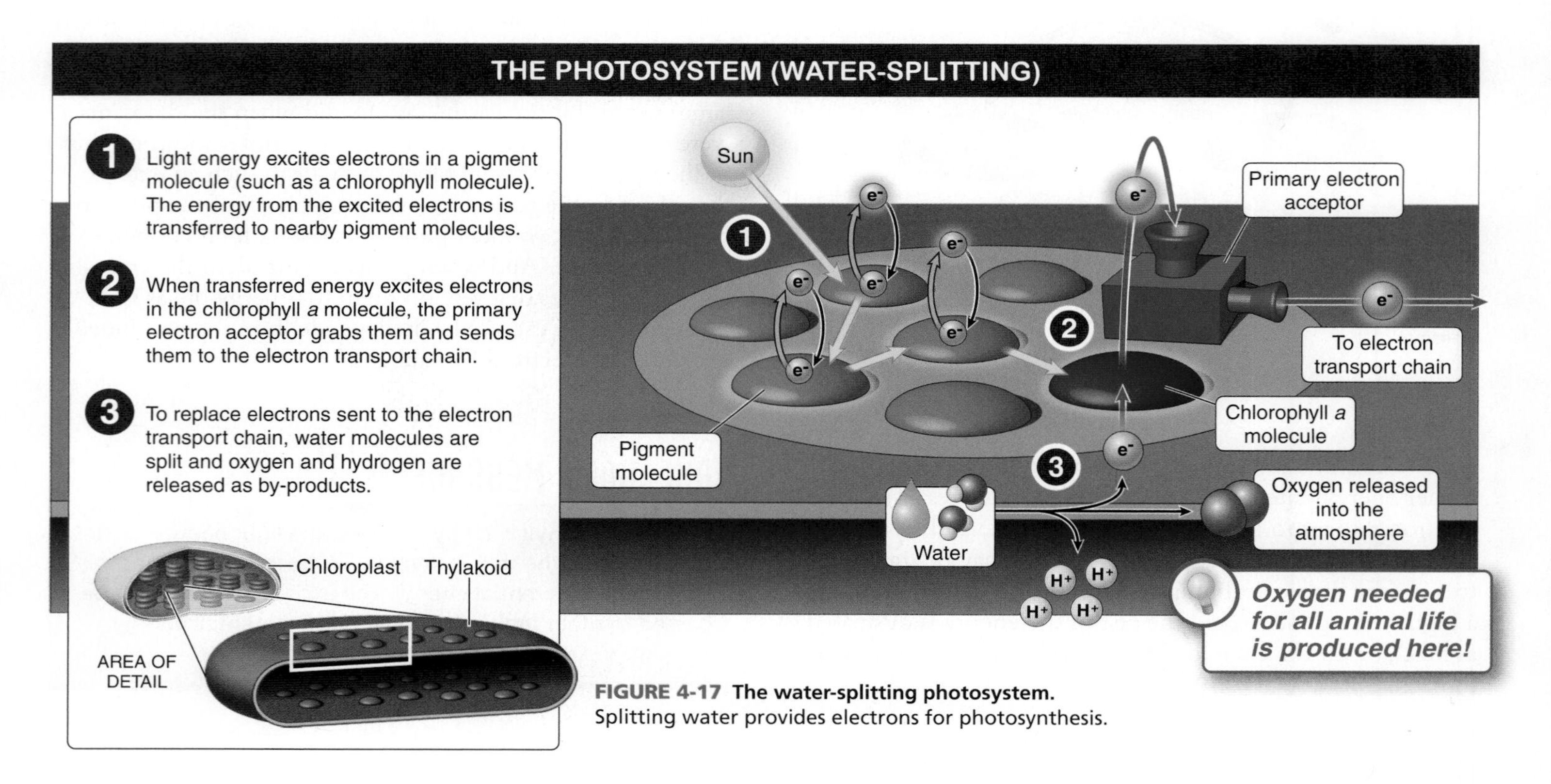

FIGURE 4-17 The water-splitting photosystem. Splitting water provides electrons for photosynthesis.

"synthesis" part of photosynthesis. We begin following electrons as sunlight hits a leaf and the chloroplasts in it (**FIGURE 4-17**).

Q Why must plants get water for photosynthesis to occur?

The chloroplasts in a leaf convert light energy from the sun into chemical energy. This conversion begins in a collection of chlorophyll molecules and other light-absorbing pigments arranged within a chloroplast's thylakoid membrane. As these pigments absorb photons from the sunlight that hits the leaves, electrons in the pigments become excited and then return to their resting state. As the electrons return to their resting state, energy (but not the electrons) is transferred to neighboring pigment molecules. This process continues until the transferred energy from many pigment molecules excites the electrons in a chlorophyll *a* molecule at the center of the photosystem (see Figure 4-17). This is where the electron journey begins.

Each energy burst boosts chlorophyll's electrons to an excited state, but the chlorophyll *a* molecule at the center of the photosystem is special, differing from the other pigment molecules in one key feature. When its electrons are boosted to an excited state they do not return to their resting, unexcited state. Instead, the special chlorophyll *a* continually loses its excited electrons to a nearby molecule, called the **primary electron acceptor,** which acts like an electron vacuum.

As electrons keep getting taken away from the special chlorophyll *a* molecule, the electrons must be replaced. The replacement electrons come from water. As long as photosynthesis is occurring, a constant supply of replacement electrons is required. Molecules of water inside the thylakoid are continuously split in the chloroplast, near the special chlorophyll *a* molecule in the thylakoid membrane. This split causes electrons from the water to fly off the water molecule and replenish chlorophyll *a's* electron supply. A convenient and life-sustaining by-product of the splitting of water in photosynthesis is the oxygen that is released from the cell. Only a by-product—but a by-product essential for all animal life.

Once the primary electron acceptor gets hold of the high-energy electrons from chlorophyll *a,* it passes them like hot potatoes to another molecule, which passes them to another, which in turn passes them to yet another in what is called an **electron transport chain** (**FIGURE 4-18**). At each step in the electron transport chain's sequence of electron handoffs, the electrons fall to a lower energy state and a little bit of energy is released. These bits of energy are harnessed to power pumps in the thylakoid membrane that move protons (H^+ ions) from the stroma to the inside of the thylakoid. These pumps pack the protons inside the thylakoid sac at higher and higher concentrations. Think of a pump pushing water into an elevated tank, creating a store of potential energy that can gush out of the tank with great force and kinetic energy. Similarly, the protons eventually rush out of

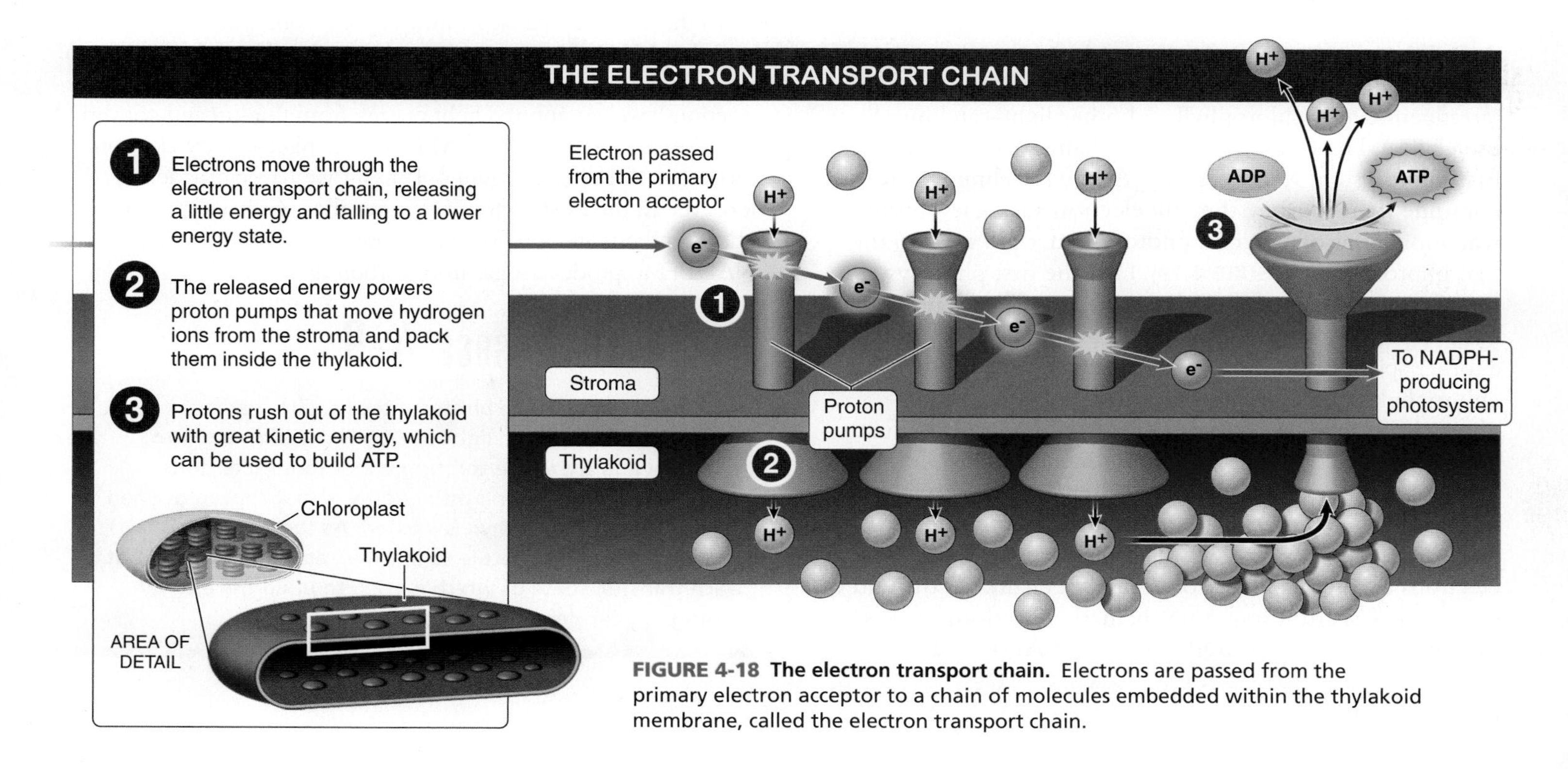

FIGURE 4-18 The electron transport chain. Electrons are passed from the primary electron acceptor to a chain of molecules embedded within the thylakoid membrane, called the electron transport chain.

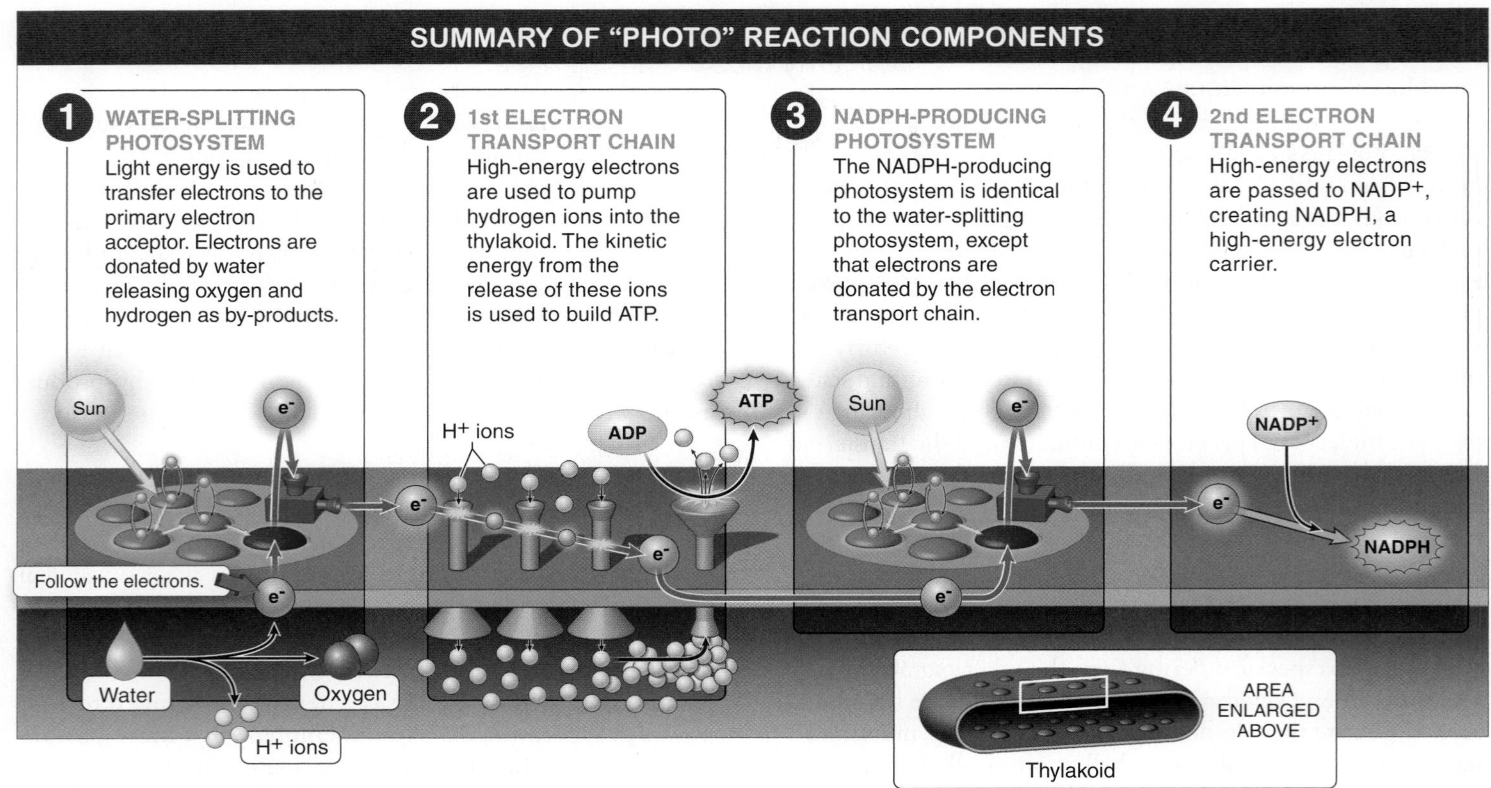

FIGURE 4-19 Summary of the "photo" reactions.

the thylakoid sacs with great force—and that force is harnessed to build energy-storing ATP molecules, one of the two products of the "photo" portion of photosynthesis.

Recall that the energy-capturing and transforming process of the "photo" reactions occurs in two photosystems (arrangements of chlorophyll and other light-catching molecules). The electron transport chain physically links the first photosystem to the second. As the traveling electrons continue their journey, they fill electron vacancies in the reaction center of a second photosystem, right next to the first photosystem (**FIGURE 4-19**). Like the first photosystem, the second photosystem also has numerous pigments that harness photons from the sun and pass the light energy to another special chlorophyll *a* molecule. The special chlorophyll *a* molecule at the center of this second photosystem has electron vacancies because, as in the first photosystem, when electrons in the special chlorophyll *a* molecule are boosted to an excited state, they are whisked away from the chlorophyll molecule by another primary electron acceptor. This electron acceptor then passes the electrons to a second electron transport chain. At the end of this second electron transport chain, the electrons are passed to a molecule called $NADP^+$, creating **NADPH,** a high-energy electron carrier. NADPH is the second important product of the "photo" portion of photosynthesis.

With the electrons' passage through the second photosystem and arrival in NADPH, we now have the final products of the "photo" part of photosynthesis (which are also called the light-dependent reactions): we've captured light energy from the sun and converted it to the chemical energy of ATP and the high-energy electron carrier, NADPH (see Figure 4-19). But we haven't made any food yet. In the next section, we cover the "synthesis" part of photosynthesis and see how plants use the energy in ATP and NADPH to produce sugar from carbon dioxide.

TAKE-HOME MESSAGE 4•9

There are two parts to photosynthesis. The first is the "photo" part, in which light energy is transformed into chemical energy, while splitting water molecules and producing oxygen. Sunlight's energy is first captured when an electron in chlorophyll is excited. As this electron is passed from one molecule to another, energy is released at each transfer, some of which is used to build the energy-storage molecules ATP and NADPH.

4•10

Photosynthesis in detail: the captured energy of sunlight is used to make food.

The "synthesis" part of photosynthesis takes place in a series of chemical reactions called the **Calvin cycle.** All the Calvin cycle reactions occur in the stroma of the leaves' chloroplasts, outside the thylakoids. Plants carry out these reactions using the energy stored in the ATP and NADPH molecules that are built in the "photo" portion of photosynthesis. This dependency links the light-gathering ("photo") reactions with the sugar-building ("synthesis") reactions (**FIGURE 4-20**).

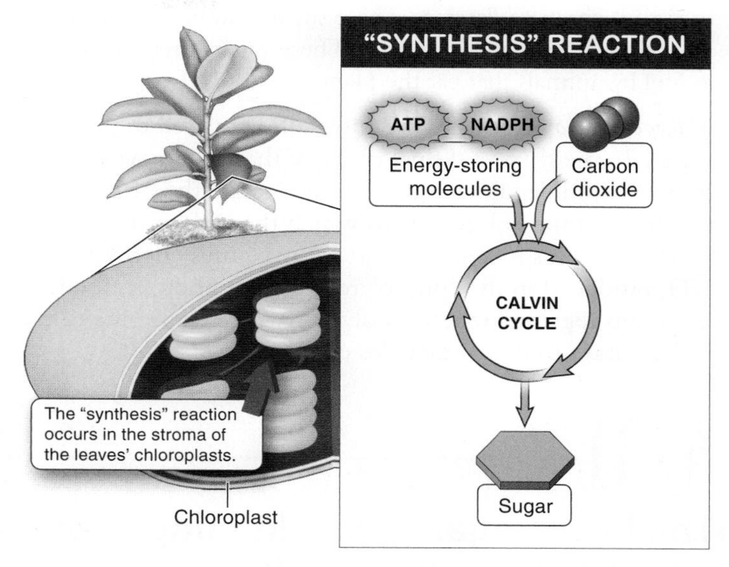

FIGURE 4-20 Overview of the "synthesis" reactions of photosynthesis.

If there is any part of photosynthesis that appears magical, it is the Calvin cycle. Just as a magician seems to make a rabbit appear from thin air, the Calvin cycle takes invisible molecules of CO_2 from the air and uses them to assemble visible—even edible—molecules of sugar. The processes in the Calvin cycle occur in three steps (**FIGURE 4-21**).

1. ***Fixation.*** First, using an enzyme called **rubisco,** plants pluck molecules of carbon from the air where they occur in the form of carbon dioxide and then attach, or "fix," them to a visible organic (carbon-containing) molecule within the chloroplast. Not surprisingly, given its role as the critical chemical that enables plants to build food molecules, rubisco is the most abundant protein on earth.

2. ***Sugar creation.*** Next, the newly built molecule is chemically modified: a phosphate from ATP is added to it and it also receives some high-energy electrons from

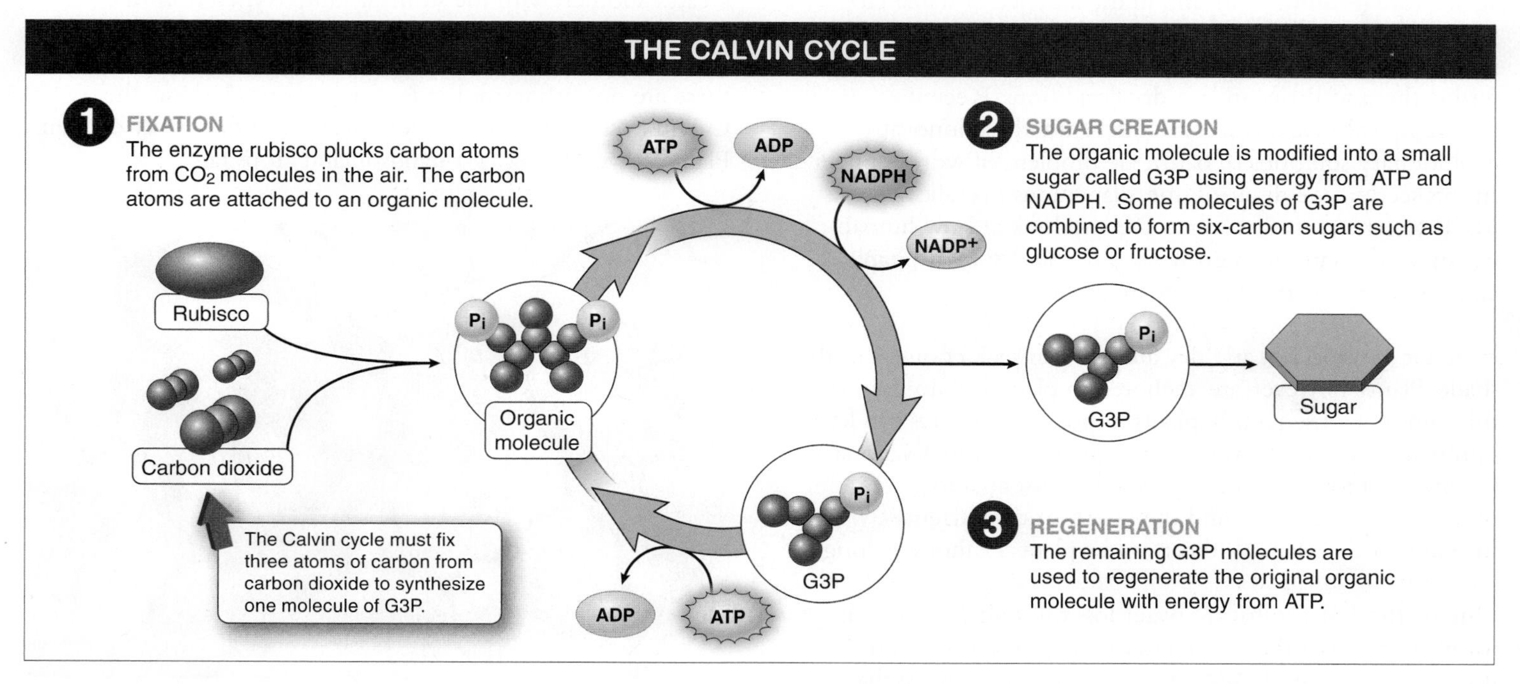

FIGURE 4-21 The Calvin cycle. In the Calvin cycle—the "synthesis" phase of photosynthesis—sugars are produced.

NADPH. This product of the Calvin cycle is a small sugar called glyceraldehyde 3-phosphate (G3P). To synthesize one molecule of the three-carbon sugar G3P, the Calvin cycle must fix three molecules of carbon from carbon dioxide to the initial organic molecule.

Some G3P molecules are combined to make the six-carbon sugars glucose and fructose. These sugars can be used as fuel by the plant, enabling it to grow. These sugars can also be used as fuel by animals that eat the plant.

3. ***Regeneration.*** Not all G3P molecules are used to produce sugars. In the third and final phase of the Calvin cycle, some G3P molecules are used to regenerate the original molecule in the chloroplast to which the carbon from CO_2 is attached. This regeneration process requires energy from ATP produced in the "photo" reactions of photosynthesis. With this regeneration, the Calvin cycle can continue to fix carbon and produce molecules of G3P.

Ultimately, to synthesize one molecule of G3P, the Calvin cycle must fix three molecules of carbon from carbon dioxide; this process consumes nine molecules of ATP and six molecules of NADPH generated in the "photo" reactions of photosynthesis.

TAKE-HOME MESSAGE 4•10

The second part, or "synthesis" part, of photosynthesis is the Calvin cycle, which occurs in the stroma of the chloroplast. During this phase, carbon from CO_2 in the atmosphere is attached (fixed) to molecules in chloroplasts, sugars are built, and molecules are regenerated to be used again in the Calvin cycle. The fixation, building, and regeneration processes consume energy from ATP and NADPH (the products of the "photo" part of photosynthesis).

4•11

The battle against world hunger can use plants adapted to water scarcity.

The Sudan. Ethiopia. India. Somalia. Many of the world's regions with the highest rates of starvation also are places with the hottest, driest climates. This is not a coincidence. These climate conditions present difficult challenges for sustaining agriculture (**FIGURE 4-22**), and in the absence of stable crop yields, food production is unpredictable and the risk of starvation high. But evolutionary adaptations in some plants enable them to thrive in hot, dry conditions. Recent technological advances in agriculture use these innovative evolutionary solutions to battle the problem of world hunger. In this section, we discuss some adaptations that allow plants to thrive when water is scarce. We also look at how humans use these adaptations to grow food in the dry, inhospitable climates where starvation rates are highest.

When it gets too hot and dry, animals can seek coolness in the shade. Plants, however, are anchored in place and do not have this option. Consequently, plants in hot, dry climates can lose significant amounts of water through evaporation. Evaporation is a problem for plants because water is essential to photosynthesis, growth, and the transport of nutrients throughout the plant. Without water, plants cannot live long.

One method of combating water loss through evaporation is for plants to close their **stomata,** small pores usually on the underside of leaves (**FIGURE 4-23**). These openings are the primary sites for gas exchange in plants: carbon dioxide for photosynthesis enters through these openings and oxygen generated as a by-product in photosynthesis exits through them. When open, the stomata also allow water to evaporate from the plant. Closing their stomata, however, solves one problem for plants (too much water evaporation) while it creates another: with the stomata shut, oxygen from the "photo" reactions of photosynthesis cannot be released from the chloroplasts, and carbon dioxide cannot enter them. If there are no carbon molecules for sugar production, the Calvin cycle tries to fix carbon but instead finds only oxygen. Plant growth comes to a standstill and crops fail.

FIGURE 4-22 Nowhere to hide. Plants that lose too much water can't always survive in extremely hot, dry weather.

FIGURE 4-23 Plant stomata. Carbon dioxide enters a plant through stomata (*sing.* stoma), but water can be lost through the same openings.

Some plants, including corn and sugarcane, have evolved a process that minimizes water loss but still enables them to make sugar when the weather is hot and dry. In the process called **C4 photosynthesis,** these plants add an extra set of steps to the usual process of photosynthesis (**FIGURE 4-24**). In these steps, the plants produce an enzyme that functions like the ultimate "CO_2-sticky tape." This enzyme has a tremendously strong attraction for carbon dioxide; it can find

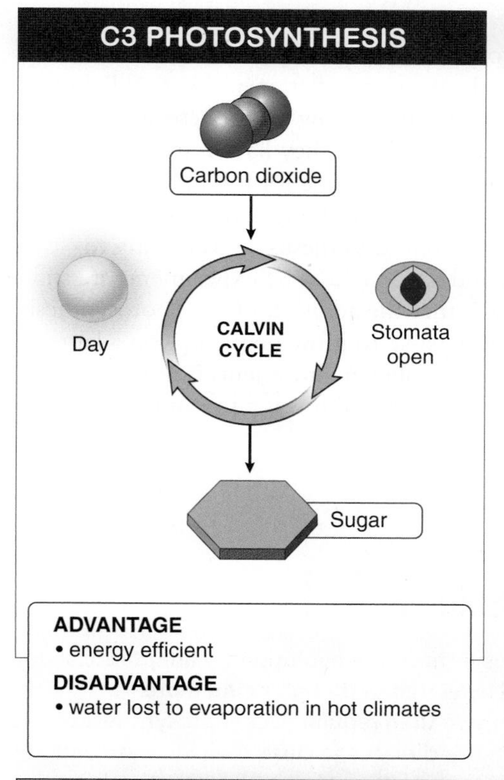

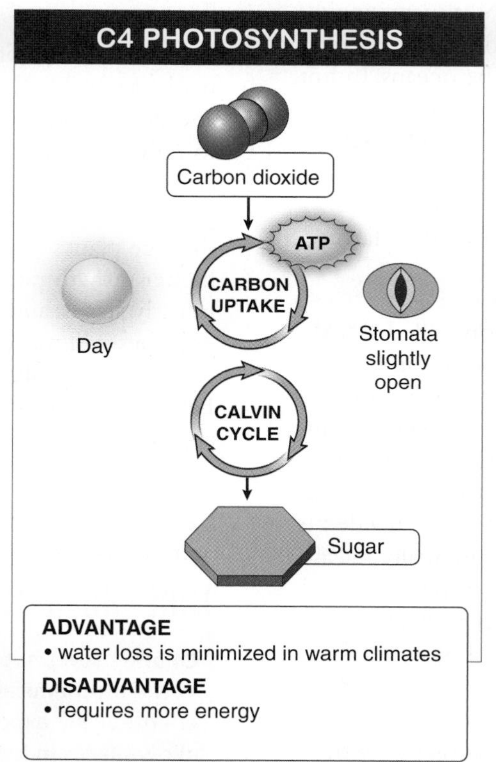

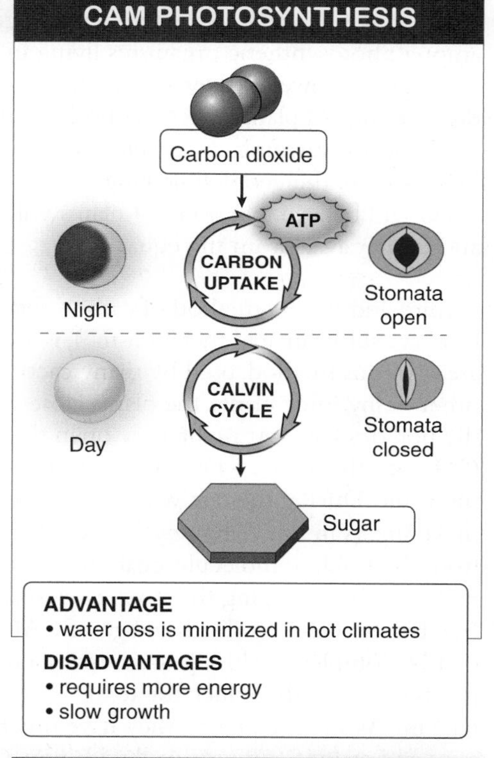

FIGURE 4-24 C3, C4, and CAM photosynthesis.

and bind carbon even when CO_2 concentration is very low. (In contrast, rubisco, the usual enzyme that plants use to pluck carbon from the atmosphere, functions poorly when CO_2 is scarce.) As a consequence, stomata can be opened just a tiny bit, and let in just a little CO_2. Having the stomata closed also reduces evaporation and conserves water for the plant.

This seems like such a good solution that we would expect all plants to use it. There is a catch, though. The extra steps in C4 photosynthesis require the plant to expend additional energy. Specifically, every time the plant generates a molecule of the "CO_2-sticky tape" enzyme, it uses one molecule of ATP. It is acceptable to pay this energy cost only when the climate is so hot and dry that the plant would otherwise have to close its stomata and completely shut down all sugar production. If the climate is mild, however, plants conducting the more energetically expensive C4 photosynthesis will be out-competed by the more efficient plants conducting standard photosynthesis (called C3 photosynthesis). Not surprisingly, we see few C4 plants in the temperate regions of the world or among photosynthetic organisms living in the oceans. In hot, dry regions, however, they are the dominant plants and displace the C3 plants wherever both occur (**FIGURE 4-25**). With global warming, many scientists expect to see a gradual expansion of the geographic ranges over which C4 plants occur and believe that non-C4 plants will be pushed farther and farther away from the equator.

A third and similar method of carbon fixation, called **CAM** (for "crassulacean acid metabolism"), is also found in hot, dry areas. In this method, used by many cacti, pineapples, and other fleshy, juicy plants, the plants close their stomata during the hot, dry days. At night, they open the stomata and let CO_2 into the leaves, where it binds temporarily to a holding molecule. During the day, when a carbon source is needed to make sugars in the Calvin cycle, the CO_2 is gradually released from the holding molecule, enabling photosynthesis to proceed while keeping the stomata closed to reduce water loss (see Figure 4-24). A disadvantage of CAM photosynthesis is that by completely closing their stomata during the day, CAM plants significantly reduce the total amount of CO_2 they can take in. As a consequence, they have much slower growth rates and cannot compete well with non-CAM plants under any conditions other than extreme dryness.

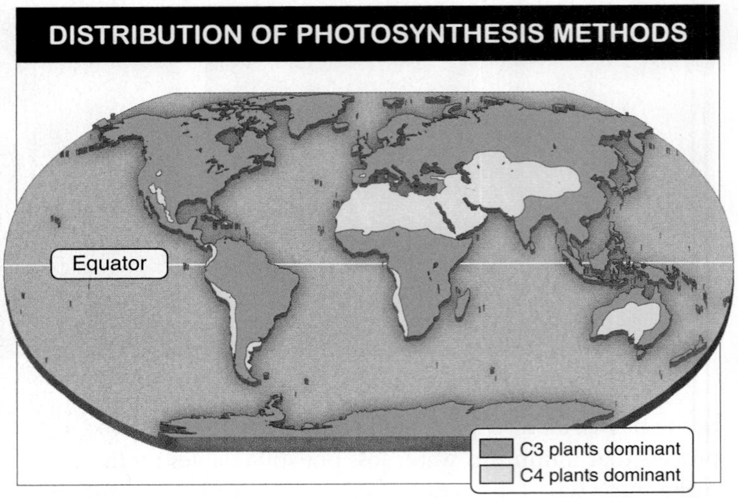

FIGURE 4-25 Map of C3 and C4 photosynthesis.

C4 and CAM photosynthesis originally evolved because they made it possible for plants to grow better in the world's hot and dry regions. Researchers are now using these adaptations to fight world hunger. Specifically, they have introduced from corn into rice several genes that code for the C4 photosynthesis enzymes. Once in the rice, these genes increase the rice plant's ability to photosynthesize, thus leading to higher growth rates and food yields. The experiments are still in early stages, and whether the addition of C4 photosynthesis enzymes will make it possible to grow new crops on a large scale in previously inhospitable environments is not certain. Early results suggest, however, that this is a promising approach.

TAKE-HOME MESSAGE 4•11

C4 and CAM photosynthesis are evolutionary adaptations at the biochemical level that, although being more energetically expensive than regular (C3) photosynthesis, allow plants in hot, dry climates to close their stomata and conserve water without shutting down photosynthesis.

③ Cellular respiration converts food molecules into ATP, a universal source of energy for living organisms.

Hamburger powered! We extract energy from all the foods we eat.

4•12

How do living organisms fuel their actions? Cellular respiration: the big picture.

Food is fuel. And all of the activities of life—growing, moving, reproducing—require fuel. Plants, most algae, and some bacteria obtain their fuel directly from the energy of sunlight, which they harness through photosynthesis. Less self-sufficient organisms, such as humans, alligators, and insects, must extract the energy they need from the food they eat. This energy comes from photosynthetic organisms either directly (from eating plants) or indirectly (from eating animals that eat plants) (**FIGURE 4-26**).

All living organisms extract energy from the chemical bonds of molecules (which can be considered "food") through a

FIGURE 4-26 Living organisms require fuel.

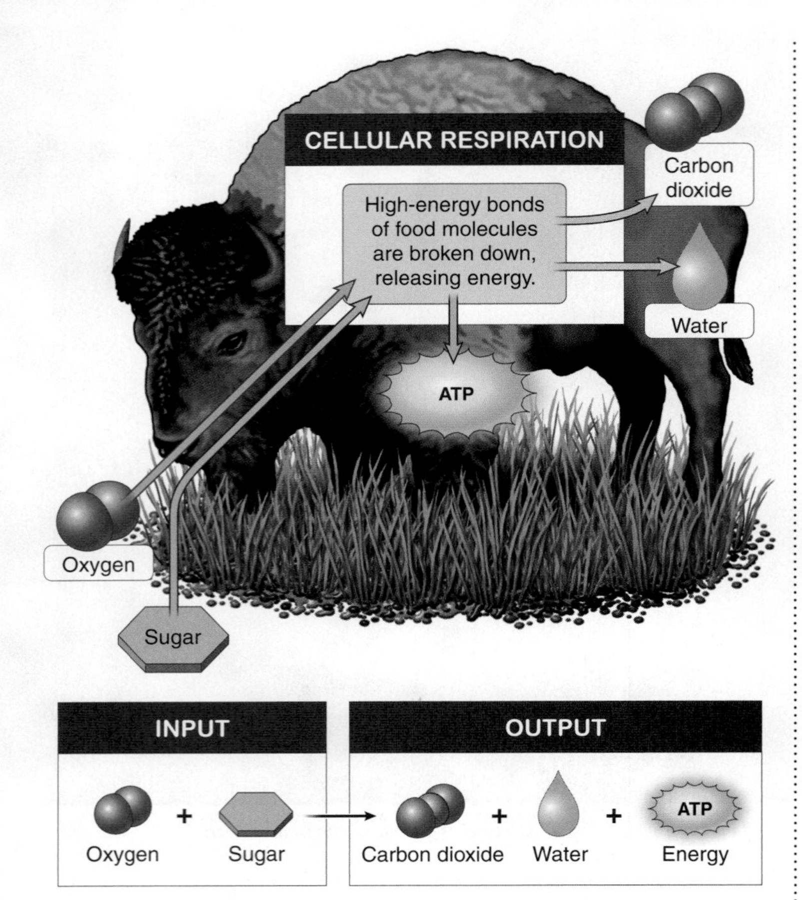

FIGURE 4-27 Cellular respiration: the big picture.

process called cellular respiration (**FIGURE 4-27**). This process is a bit like photosynthesis in reverse. In photosynthesis, the energy of the sun is captured and used to build molecules of sugars, such as glucose. In cellular respiration, plants and animals break down the chemical bonds of sugar and other energy-rich food molecules (such as fats and proteins) to release the energy that went into creating them. (Don't confuse cellular respiration with the act of breathing, which is also called *respiration*.) As energy is released by these reactions, cells capture it and store it in the bonds of ATP molecules. This plentiful, readily available stored energy can then be tapped as needed to fuel the work of the life-sustaining activities and processes of all living organisms.

In humans and other animals, cellular respiration starts after we eat food, digest it, absorb the nutrient molecules into the bloodstream, and deliver them to the cells of our bodies. At this point, our cells begin to extract some of the energy stored in the bonds of the food molecules. We focus here on the breakdown of glucose, but later in this chapter we'll see that the process is similar for the breakdown of fats or lipids. Ultimately, when a food molecule has been completely processed, the cell has used the food molecule's stored energy (along with oxygen) to create a large number of high-energy-storing ATP molecules (which supply energy to power the cell's activities), water, and carbon dioxide (which is exhaled into the atmosphere).

TAKE-HOME MESSAGE 4•12

Living organisms extract energy through a process called cellular respiration, in which the high-energy bonds of sugar and other energy-rich molecules are broken, releasing the energy that went into creating them. The cell captures the food molecules' stored energy in the bonds of ATP molecules. This process requires fuel molecules and oxygen and it yields ATP molecules, water, and carbon dioxide.

4•13

The first step of cellular respiration: glycolysis is the universal energy-releasing pathway.

To generate energy, fuels such as glucose and other carbohydrates and proteins and fats are broken down in three steps: (1) glycolysis, (2) the Krebs cycle, and (3) the electron transport chain. **Glycolysis** means the splitting (*lysis*) of sugar (*glyco-*) and it is the first step all organisms on the planet take in breaking down food molecules; for many single-celled organisms, this one step is sufficient to provide all of the energy they need (**FIGURE 4-28**).

As **FIGURE 4-29** illustrates, glycolysis is a sequence of chemical reactions (there are 10 in all) through which glucose is broken down to yield two molecules of a substance called **pyruvate.** Glycolysis has two distinct phases: an "uphill" preparatory phase and a "downhill" payoff phase.

Just as you sometimes have to spend money to make money, before any energy can be extracted from glucose, some

All organisms extract energy through glycolysis!

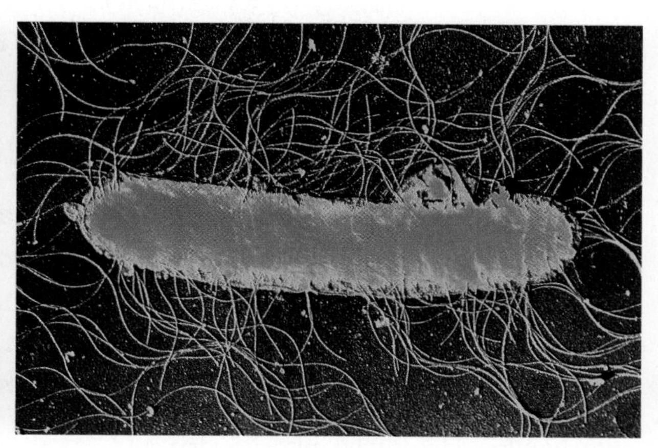

FIGURE 4-28 Plants, animals, bacteria, and all other organisms undergo glycolysis.

energy must be added to the molecule. This addition occurs during the "uphill" phase. The additional energy (which comes from ATP) makes the glucose molecule less stable and ripe for chemical breakdown. Once the glucose has been destabilized, it can be broken down chemically, and the energy stored in its bonds can be harnessed as the bonds are broken.

GLYCOLYSIS

1 PREPARATORY PHASE

2 PAYOFF PHASE

ATP (2)

P_i

P_i

ATP (4)

Unstable molecule prepared to be broken down

NADH (2)

Glucose

Water

−2 ATP

Pyruvate (2)

Gycolysis takes place in the cell's cytoplasm.

+4 ATP +2 NADH

FIGURE 4-29 Glycolysis up close.

Three of the 10 steps in glycolysis yield energy. In two of these three, as bonds from the sugar are broken, the energy released is quickly harnessed by the attachment of phosphate groups to molecules of ADP to create energy-rich ATP molecules. In the third energy-yielding step of glycolysis, electrons originally from the glucose are transferred to NAD^+ to become the high-energy electron carrier NADH. Later (in the electron transport chain) this energy will be converted to ATP. The net result of glycolysis is that each glucose molecule is broken down into two molecules of pyruvate. During this breakdown, some of the released energy is captured in the production of energy-rich ATP molecules and molecules of the high-energy electron carrier NADH. Two molecules of water are also produced during glycolysis.

In the absence of oxygen and in many yeasts and bacteria, glycolysis is the only game in town for fueling activity. Because single-celled organisms have much lower energy needs, they can function solely on the yields of glycolysis. For many organisms (including humans), however, glycolysis is a springboard to further energy extraction. The additional energy payoffs come from the Krebs cycle and the electron transport chain.

TAKE-HOME MESSAGE 4•13

Glycolysis is the initial phase in the process by which all living organisms harness energy from food molecules. Glycolysis occurs in a cell's cytoplasm and uses the energy released from breaking chemical bonds in food molecules to produce high-energy molecules, ATP and NADH.

4•14

The second step of cellular respiration: the Krebs cycle extracts energy from sugar.

Cells could stop extracting energy when glycolysis ends, but they rarely do because that would be like leaving most of your meal on your plate. This is because in glycolysis only a small fraction of the energy stored in sugar molecules is recovered and converted to ATP and NADH. Cells get much more of an "energy bang" for their "food buck" in the steps following glycolysis, which occur in the mitochondria. This is why mitochondria are considered ATP "factories." In the mitochondria, the molecules produced from the breakdown of glucose during glycolysis are broken down further, during two steps that are dramatically more efficient at capturing the energy in food molecules: the Krebs cycle (the subject of this section) and the electron transport chain (the subject of the next section). In breaking down the products of glycolysis, the **Krebs cycle** produces some additional molecules of ATP and, more important, captures a huge amount of chemical energy by producing high-energy electron carriers.

Q Aerobic training can cause our bodies to produce more mitochondria in cells. Why is this beneficial?

Before the Krebs cycle can begin, however, the end products of glycolysis—two molecules of pyruvate for every molecule of glucose used—must be modified. To start this modification process, the pyruvate molecules move from the cytoplasm into the mitochondria. Once in the mitochondria, the pyruvate molecules undergo three quick modifications that prepare them to be broken down in the Krebs cycle (**FIGURE 4-30**):

Modification 1. Each pyruvate molecule passes some of its high-energy electrons to the energy-accepting molecule NAD^+, building two molecules of NADH.

Modification 2. Next, a carbon atom and two oxygen atoms are removed from each pyruvate molecule and released as carbon dioxide. The CO_2 molecules diffuse out of the cell and, eventually, out of the organism. In humans, for example, these CO_2 molecules pass into the bloodstream and are transported to the lungs, from which they are eventually exhaled.

Modification 3. In the final step in the preparation for the Krebs cycle, a giant compound known as coenzyme A attaches itself to the remains of each pyruvate molecule, producing two molecules called acetyl-CoA. Each acetyl-CoA molecule is now ready to enter the Krebs cycle.

There are eight separate steps in the Krebs cycle, but our emphasis here is on its three general outcomes (**FIGURE 4-31**):

Outcome 1. *A new molecule is formed.* Acetyl-CoA adds its two-carbon acetyl group to a molecule of the starting material of the Krebs cycle, a four-carbon molecule called oxaloacetate. This process creates a six-carbon molecule.

Outcome 2. *High-energy electron carriers (NADH) are made and carbon dioxide is exhaled.* The six-carbon molecule then gives electrons to NAD^+ to make the high-energy electron carrier NADH. The six-carbon molecule releases two carbon atoms along with four oxygen atoms to form two carbon dioxide molecules. This CO_2 is carried by the bloodstream to the lungs from which it is exhaled into the atmosphere.

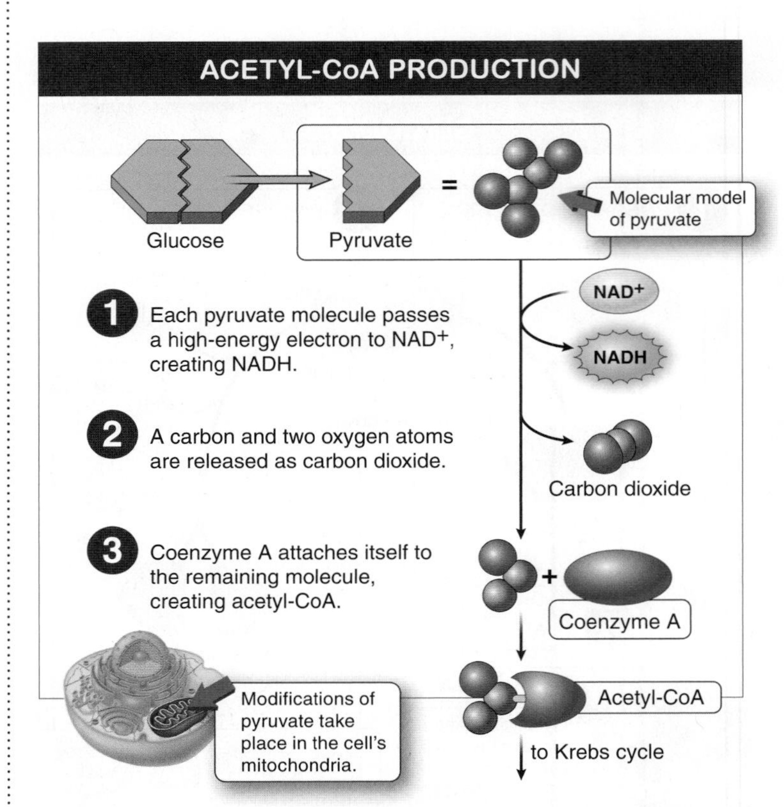

FIGURE 4-30 Preparation of pyruvate. In the mitochondria, pyruvate must be modified before it can be broken down in the Krebs cycle.

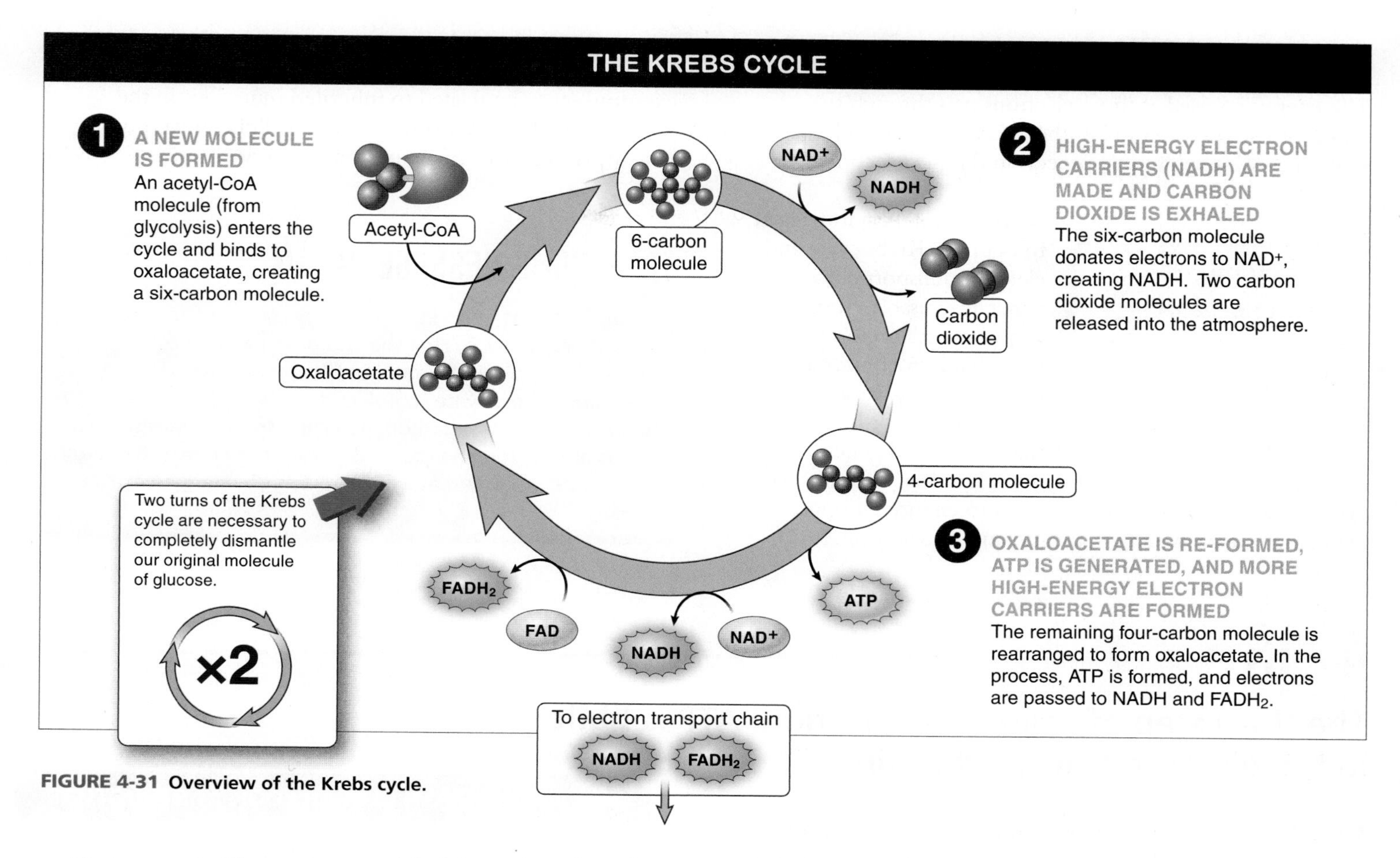

FIGURE 4-31 **Overview of the Krebs cycle.**

Outcome 3. *The starting material of the Krebs cycle is re-formed, ATP is generated, and more high-energy electron carriers are formed.* After the CO_2 is released, the four-carbon molecule that remains from the original pyruvate-oxaloacetate molecule formed in Outcome 1 is modified and rearranged to once again form oxaloacetate, the starting material of the Krebs cycle. In the process of this reorganization, one ATP molecule is generated and more electrons are passed to one familiar high-energy electron carrier, NADH, and a new one, $FADH_2$. The formation of these high-energy electron carriers increases the energy yield of the Krebs cycle. One oxaloacetate is re-formed, and the cycle is ready to break down the second molecule of acetyl-CoA. Two turns of the cycle are necessary to completely dismantle our original molecule of glucose.

We started this process of cellular respiration with a molecule of glucose, and now that we have seen the Krebs cycle in its entirety, let's trace the path of the original six carbons in that glucose molecule.

1. Glycolysis: the six-carbon starting point. Glucose is broken down into two molecules of pyruvate. No carbons are removed.
2. Preparation for the Krebs cycle: two carbons are released. Two pyruvate molecules are modified to enter the Krebs cycle and they each lose a carbon atom, in the form of two molecules of carbon dioxide.
3. Krebs cycle: the last four carbons are released. A total of four carbon molecules enter the Krebs cycle in the form of acetyl Co-A. For each turn of the Krebs cycle, two molecules of carbon dioxide are released. So the two final carbons are released into the atmosphere during the second turn of the wheel. Poof! The six carbon atoms that were originally present in our single molecule of glucose are no longer present.

In a sense, the carbon atoms that were first plucked from the atmosphere to make sugar during photosynthesis have been exhaled back into the atmosphere as six molecules of carbon dioxide. At this point, we've come full circle. In photosynthesis, carbon atoms from the air were used to build sugar molecules, which had energy stored in the bonds between their carbon, hydrogen and oxygen atoms. In cellular respiration, the energy previously stored in the bonds of the sugar is converted to molecules of ATP, NADH, and $FADH_2$. Carbon atoms from sugar are exhaled back into the air as CO_2 and water is produced. What happens to the high-energy electron carriers, NADH and $FADH_2$? They eventually give up their high-energy electrons to the electron transport chain. The energy released as those electrons pass through the transport chain is captured in the bonds of more ATP molecules. We explore that process in the next section.

Why might mitochondria malfunctions play a role in lethargy or fatigue?

As we explore the important roles of the Krebs cycle and electron transport chain in generating usable energy for organisms, it's important to note that mitochondria malfunctions have serious consequences for health. More than a hundred genetic mitochondrial disorders have been identified, all of which can lead to energy shortage, including muscle weakness, fatigue, and muscle pain. It appears, for instance, that many cases of "exercise intolerance"—extreme fatigue or cramps after only slight exertion—may be related to inherited mutations in the mitochondrial DNA. Damage to our cellular powerhouses, not surprisingly, can create personal energy crises.

TAKE-HOME MESSAGE 4•14

A huge amount of additional energy can be harvested by cells after glycolysis. First the end product of glycolysis, pyruvate, is chemically modified. Then, in the Krebs cycle, the modified pyruvate is broken down step by step. This breakdown releases carbon molecules to the atmosphere as bonds are broken and captures some of the released energy in two ATP molecules and numerous high-energy electron carriers.

4•15

The third step of cellular respiration: ATP is built in the electron transport chain.

How do we finally get a big payoff of usable energy from our glucose molecule? Glycolysis and the Krebs cycle produce a few molecules of ATP for each molecule of glucose broken down, but it is the energy held in the high-energy electron carriers NADH and $FADH_2$ generated in these processes that ultimately generates the largest amount of usable energy in the form of ATP. In fact, almost 90% of the energy payoff from a molecule of glucose is harvested in the final step of cellular respiration when the electrons from NADH and $FADH_2$ move along the electron transport chain. This process, like the Krebs cycle, takes place in the mitochondria.

In a manner that is similar to that seen in the chloroplast during photosynthesis, mitochondria convert kinetic energy (from electrons) into potential energy (a concentration gradient of protons). Two structural features of mitochondria are essential to their impressive ability to harness energy from molecules.

Feature 1. Mitochondria have a "bag-within-a-bag" structure that makes it possible to harness the potential energy in the bonds of NADH and $FADH_2$ molecules to produce ATP (**FIGURE 4-32**). Material inside the mitochondrion can lie in one of two spaces: (1) the

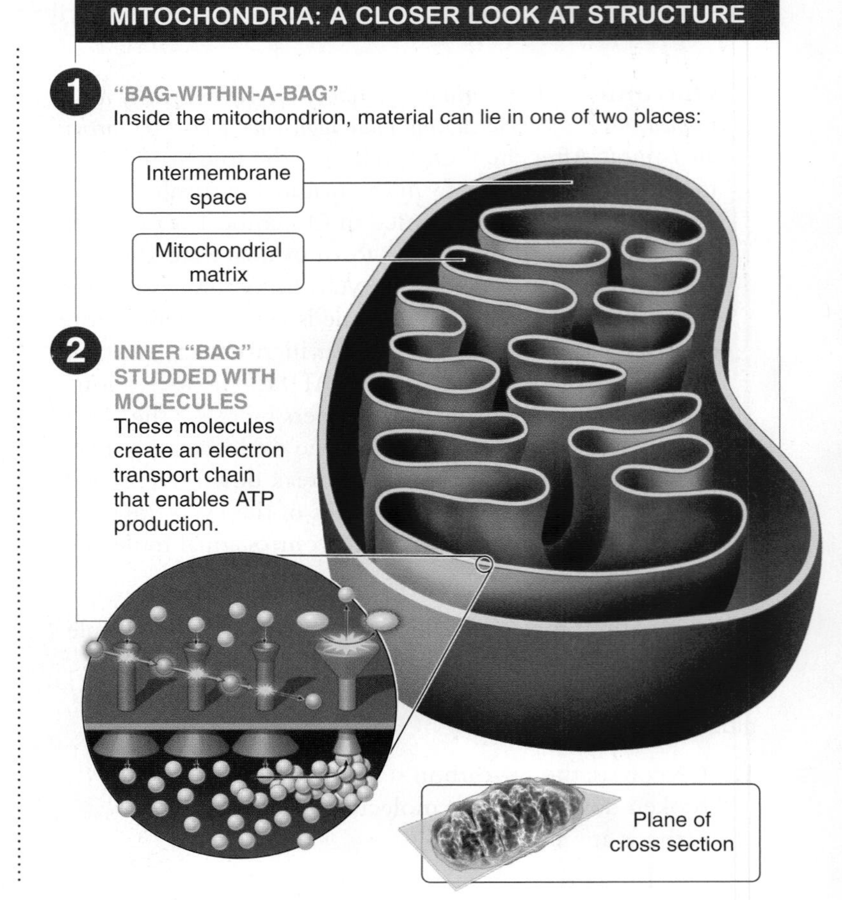

FIGURE 4-32 "A bag within a bag." The structure of mitochondria makes possible their impressive ability to harness energy from food molecules.

intermembrane space, which is outside the inner bag, or (2) the **mitochondrial matrix,** which is inside the inner bag.

With two distinct regions separated by a membrane, the mitochondrion can create higher concentrations of molecules in one area or the other. And because a concentration gradient is a form of potential energy—molecules move from the high-concentration area to the low-concentration area the way water rushes down a hill—once a gradient is created, the energy released as the gradient equalizes itself can be used to do work. In the electron transport chain, this energy is used to build the energy-rich molecule ATP.

Feature 2. The inner bag of the mitochondria is studded with molecules. These molecules, mostly electron carriers, are sequentially arranged within the inner membrane of the mitochondrion, hence their description as a "chain." This arrangement makes it possible for the molecules to hand off electrons in an orderly sequence.

Q Over-the-counter NADH pills provide energy to sufferers of chronic fatigue syndrome. Why might this be?

Now let's explore how these features of mitochondria make it possible to harness energy from high-energy electron carriers (**FIGURE 4-33**).

Step 1 of the electron transport chain begins with NADH and $FADH_2$ in the mitochondrial matrix (inside the inner bag) moving to the inner mitochondrial membrane. There, the high-energy electrons they carry are transferred to molecules embedded within the membrane. After they donate their electrons, the molecules that remain, NAD^+ and FAD^+, are recycled back to the Krebs cycle.

The embedded molecules pass the electrons to the next carrier, which passes the electrons to the next, and so on. At each handoff, a bit of energy is released. Thus, as electrons move from one carrier to another through the electron transport chain, they lose energy at each handoff.

At the end of the chain (step 2), the lower-energy electrons are handed off to oxygen, which then combines with free H^+ ions in the mitochondrial fluid to form water.

As shown in step 3 of Figure 4-33, most of the energy released at each handoff from one electron carrier to another in the electron transport chain is used to pump protons (H^+ ions) from the mitochondrial matrix across the membrane and into the intermembrane space. As more and more protons are pumped across the membrane and packed in the

THE MITOCHONDRIAL ELECTRON TRANSPORT CHAIN

High-energy electrons are passed from the carriers NADH and $FADH_2$ to a series of molecules embedded in the inner mitochondrial membrane called the electron transport chain.

1. At each step in the electron transport chain's sequence of handoffs, the electrons fall to a lower energy state, releasing a little bit of energy.
2. At the end of the chain, the lower-energy electrons are handed off to oxygen, which then combines with free H^+ ions to form water.
3. The energy is used to power proton pumps, which pack hydrogen ions from the mitochondrial matrix into the intermembrane space.
4. The protons rush back into the mitochondrial matrix with great kinetic energy, which can be used to build ATP.

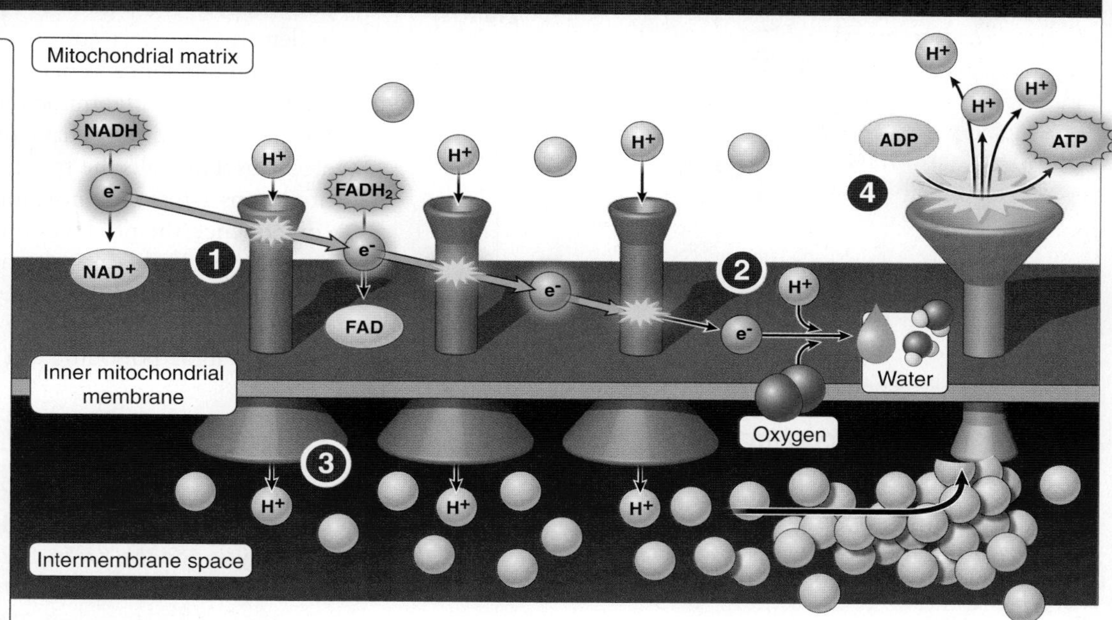

FIGURE 4-33 The big energy payoff. Most of the energy harvested during cellular respiration is generated by the electron transport chain in the mitochondria.

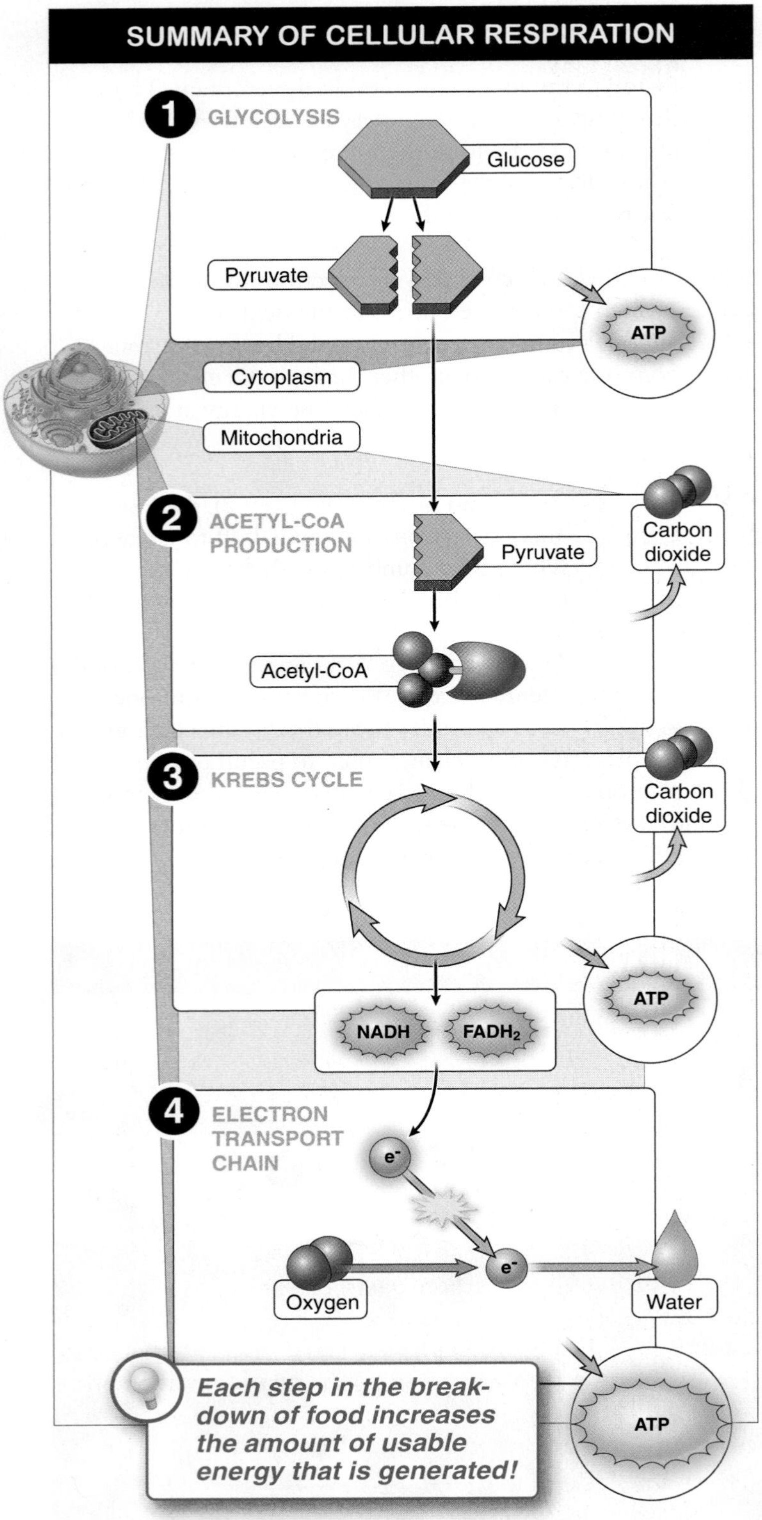

FIGURE 4-34 **The steps of cellular respiration: from glucose to usable energy.**

intermembrane space, a concentration gradient is created. This gradient represents a significant source of potential energy.

If this description seems familiar, it is. In chloroplasts, during photosynthesis, great numbers of protons are pumped from the stroma outside the thylakoid sacs to the inside of the thylakoids. We likened this potential energy to the potential energy of water in an elevated tower, which can be released with great force. Similarly, in step 4 of the mitochondrial electron transport chain, the protons pumped into the intermembrane space rush back to the inner mitochondrial matrix through channels in the inner mitochondrial membrane. And as the protons pass through, the force of their flow fuels the attachment of free-floating phosphate groups to ADP to produce ATP.

In the end, the number of ATP molecules generated from the complete dismantling of one molecule of glucose is about 36, most of which are produced with the energy harnessed from high-energy electron carriers as they pass their electrons down the electron transport chain (**FIGURE 4-34**).

> **Q** *Cyanide blocks the passage of electrons to oxygen in the electron transport chain. Why does this make it a toxic poison?*

Given the central role of the electron transport chain in the generation of usable energy from the breakdown of food molecules, any interference in its functioning has dire consequences. And in fact, murder by cyanide poisoning, an old tradition in detective stories, is just such an interference. When cyanide gets into the mitochondria, it binds to a molecule in the electron transport chain, preventing it from accepting electrons. This halts the transfer of electrons and the pumping of protons across the mitochondrial membrane. As a consequence, the production of ATP that would occur when protons rush back across the membrane down their concentration gradient ceases. Halting the production of ATP removes a cell's energy source, starving it very quickly. For this reason, cyanide poisoning can cause death within minutes.

TAKE-HOME MESSAGE 4•15

The largest energy payoff of cellular respiration comes as electrons from NADH and $FADH_2$ produced during glycolysis and the Krebs cycle move along the electron transport chain. The electrons are passed from one carrier to another and energy is released, pumping protons into the intermembrane space. As the protons rush back to the inner mitochondrial matrix, the force of their flow fuels the production of large amounts of ATP.

④ There are alternative pathways to energy acquisition.

Unlike cars, living organisms (such as this cheetah) can use multiple different types of fuel.

4•16

Beer, wine, and spirits are by-products of cellular metabolism in the absence of oxygen.

Every beer brewery, the entire wine industry, and all distilleries of whiskey, vodka, tequila, and other alcoholic beverages owe their existence to microscopic yeast cells scrambling to break down their food for energy under stressful conditions. To better understand how yeast metabolism produces the alcohol we drink, it helps to investigate first what happens when humans and other animals try to metabolize energy from sugar molecules under some stressful conditions.

FIGURE 4-35 Energy production without oxygen. Organisms have a backup method for breaking down sugar when oxygen is not present.

If you run or swim as fast as you can, you soon feel a burning sensation in your muscles. What is happening? Your muscle cells are suddenly becoming very acidic. This acid buildup occurs when we demand of our bodies bursts of energy beyond that which they can sustain (**FIGURE 4-35**). (The next-day muscle soreness is not due to acid buildup, which goes away in a matter of minutes or hours; it is caused by damage to the muscle fibers. They break down a bit before growing stronger.)

With rapid, strenuous exertion, our bodies soon fall behind in delivering oxygen from the lungs to the bloodstream to the cells and finally to the mitochondria. Oxygen deficiency then limits the rate at which the mitochondria can break down fuel and produce ATP (**FIGURE 4-36**). This slowdown in ATP production occurs because the electron transport chain requires oxygen as the final acceptor of all the electrons that are generated during glycolysis and the Krebs cycle. If oxygen is in short supply, the electrons from NADH (and $FADH_2$) have nowhere to go. Consequently, the regeneration of NAD^+ (and FAD^+) in the electron transport chain is halted, leaving no recipient for the high-energy electrons harvested from the breakdown of glucose and pyruvate, and the whole process of cellular respiration can grind to a stop. Organisms don't let this interruption last long, though; most have a backup method for breaking down sugar.

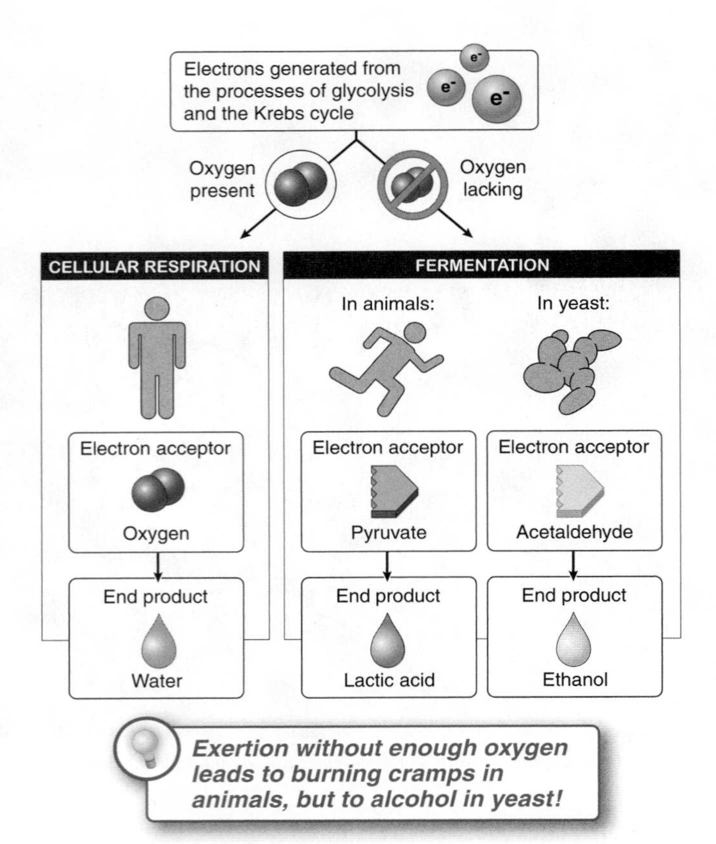

FIGURE 4-36 Energy production compared: with and without oxygen. Energy production without oxygen (anaerobic) is also called fermentation.

Among animals, there is one willing acceptor for the NADH electrons in the absence of oxygen: pyruvate, the end product of glycolysis. When pyruvate accepts the electrons, it forms lactic acid (see Figure 4-36). Once the NADH gives up its electrons, NAD^+ is generated, and glycolysis can continue. But as lactic acid builds up, it causes a burning feeling in our muscles. It's not an ideal situation, but if you are trying to escape a predator or are otherwise engaged in strenuous activity, the two ATP molecules generated from each glucose molecule during glycolysis are better than nothing.

Like humans, yeast normally use oxygen during their breakdown of food. And like humans, they have a backup method when oxygen is not available. But yeast make use of a different electron acceptor than do humans when oxygen isn't available, and the resulting reaction leads to the production of all drinking alcohol.

In these single-celled organisms, after glycolysis pyruvate is usually converted to a molecule called *acetaldehyde,* generating bubbles of CO_2 (which, when yeast is used in baking, allows bread to rise). In the absence of oxygen, this molecule accepts the electrons released from NADH, allowing glycolysis to resume. Acetaldehyde's acceptance of NADH's electrons results in the production of **ethanol,** the molecule that gives beer, wine, and spirits their kick. Ethanol can also be used as a fuel source because it can be combusted in much the same way as the biofuels generated from animal fats and plant oils described earlier in this chapter.

The process by which glycolysis occurs in the absence of oxygen, in which alternative molecules are used as electron acceptors, is called **fermentation.** Interestingly, although ethanol is always the alcohol produced by fermentation, the flavor of the output of fermentation depends on the type of sugar metabolized by the yeast. Fruits, vegetables, and grains all give different results. If the sugar comes from grapes, wine is produced. If the sugar comes from a germinating barley plant, beer is produced. Potatoes, on the other hand, are the sugar source usually used to produce vodka.

Because yeast prefer the more efficient process of aerobic respiration, they produce alcohol only in the absence of oxygen. That is why the fermentation tanks used in producing wine, beer, and other spirits are built specifically to keep oxygen out (**FIGURE 4-37**). Only in this circumstance will yeast be forced to use their backup respiratory pathway of fermentation.

TAKE-HOME MESSAGE 4•16

Oxygen deficiency limits the breakdown of fuel because the electron transport chain requires oxygen as the final acceptor of the electrons generated during glycolysis and the Krebs cycle. When oxygen is unavailable, yeast resort to fermentation, in which they use a different electron acceptor, pyruvate, generating in the process ethanol, the alcohol in beer, wine, and spirits.

FIGURE 4-37 Just a by-product of metabolism under stressful conditions. Beer, wine, and spirits are by-products of cellular metabolism in the absence of oxygen.

4•17

Eating a complete diet: cells can run on protein and fat as well as on glucose.

Most automobiles can run on only one type of fuel, gasoline. If you run out of fuel and cannot get to a gas station, you are out of luck. In this chapter, we have examined the steps by which plants, animals, and other organisms use glucose as fuel. But living organisms are more flexible than autos when it comes to fuel sources. They also have more complex needs than an automobile because their fuel must also provide raw materials for growth.

Evolution has built humans and other organisms with the metabolic machinery that allows them to extract energy and other valuable chemicals from proteins, fats, and a variety of carbohydrates (**FIGURE 4-38**). For that reason, we are able to consume and efficiently utilize meals comprising various combinations of molecules.

Sugars. In the case of dietary sugars, many are polysaccharides—multiple simple sugars linked together—rather than solely the simple sugar glucose. Before they can be broken down by cellular respiration, the polysaccharides must first be separated by enzymes into glucose or related simple sugars that can be broken down by cellular respiration.

Lipids. Dietary lipids are broken down into their two constituent parts: a glycerol molecule and fatty acids. The glycerol is chemically modified into one of the molecules produced during one of the 10 steps of glycolysis. It then enters glycolysis at that step in the process and is broken down to yield energy. The fatty acids, meanwhile, are chemically modified into acetyl-CoA, at which point they enter the Krebs cycle.

Proteins. Proteins are chains of amino acids. Upon consumption they are broken down chemically into their constituent amino acids. Once that is done, each amino acid is broken down into (1) an amino group that may be used in the production of tissue or excreted in the urine and (2) a carbon compound that is converted to one of the intermediate compounds in glycolysis or the Krebs cycle, allowing the energy stored in its chemical bonds to be harnessed.

In the end, humans (indeed, all animals) are able to harvest energy from a variety of food sources beyond simple glucose. Whether a meal contains carbohydrates, lipids, proteins, or some combination thereof, the nutrients are chemically modified in some preliminary steps and are then fed into one of the intermediate steps in glycolysis or the Krebs cycle to furnish usable energy for the organism (see Figure 4-38).

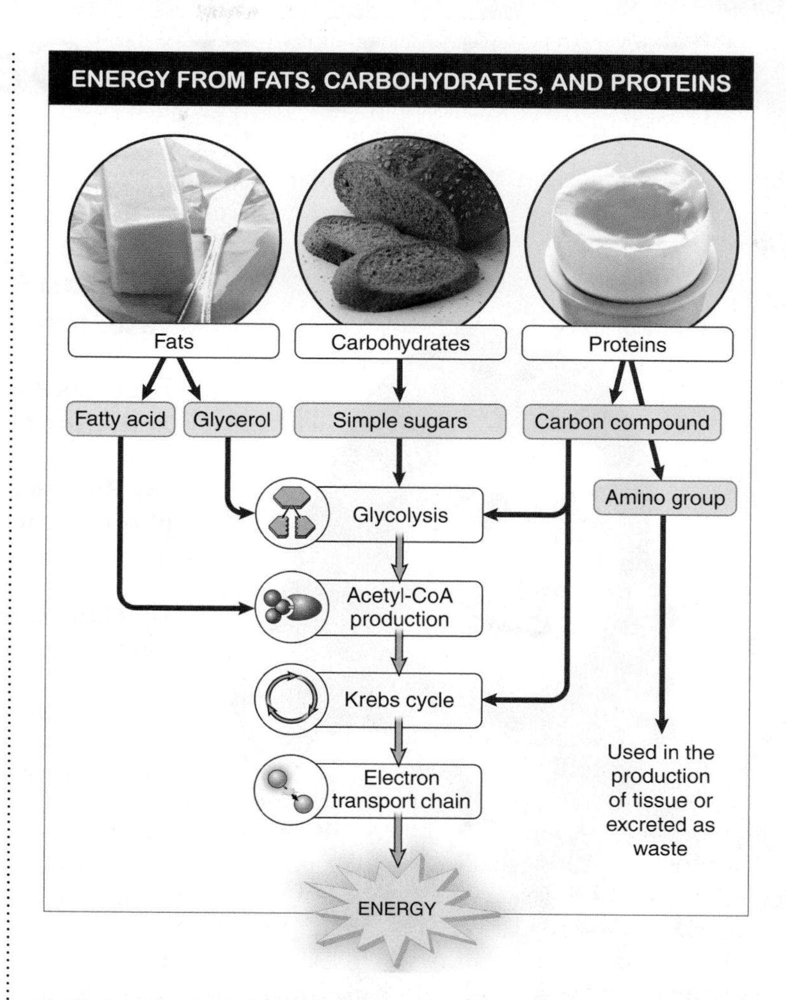

FIGURE 4-38 "More flexible than autos." Animals are able to harvest energy from proteins, carbohydrates, and lipids.

TAKE-HOME MESSAGE 4•17

Humans and other organisms have metabolic machinery that allows them to extract energy and other valuable chemicals from proteins, fats, and carbohydrates in addition to the simple sugar, glucose.

streetBio

Knowledge You Can Use

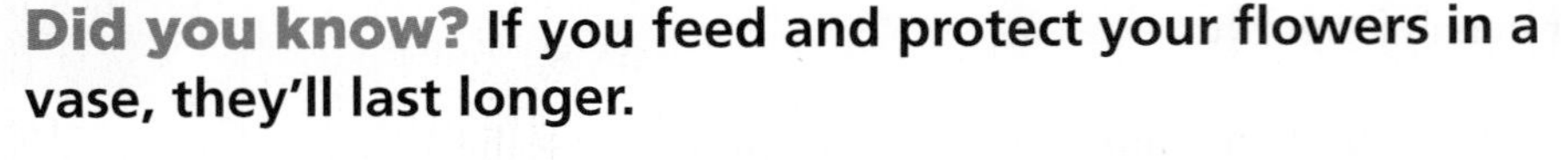

Did you know? If you feed and protect your flowers in a vase, they'll last longer.

Fact: Like animals, plants require fuel for the cellular activities that keep them alive.

Q: How do plants get the energy they need to stay alive? Plants use photosynthesis to harness light energy, converting it to sugar molecules that serve as their food.

Q: Can plants still photosynthesize once they're in a vase in your house? Yes, but humans don't make it easy. Once cut, plants generally cannot produce sufficient sugar through photosynthesis. **Light levels in houses tend to be a bit too low, and the loss of many or most plant leaves reduces the number of chloroplasts in which photosynthesis can occur.** So they're starving to death.

Q: Can you slow their demise? Yes. **Plants are able to take up sugar in the vase water and use it as an energy source** for cell activities. So adding a bit of sugar is like putting fuel in their tank.

Q: Is it that easy? No. Putting sugar—a molecule with lots of energy stored within its chemical bonds—in the vase is like offering a free lunch. And many, many organisms are looking for a free lunch. Unfortunately, **when you add sugar, bacteria on the stems can grow rapidly, blocking the water-conducting tubes in flower stems.** This slows the flow not just of sugar, but of water as well.

Q: What should you do? With the addition to vase water of both sugar and an antibacterial chemical such as chlorine bleach, you can feed and protect your flowers, significantly increasing their longevity in vases.

Conclusion: Here's a useful summary from Martha Stewart (www.marthastewart.com): "Flowers last longer if their stems are cut under water. Use a sharp knife, and cut the stem on a slant so it can absorb as much water as possible. Next, place [the flowers] in a flask filled with 2 inches of warm water (no hotter than 110 degrees); this expedites the flow of water into the flower. Add either cut-flower food or a teaspoon of sugar and a couple drops of liquid bleach to a vase filled with cool water up to the point where the flower foliage will hit. After 5 minutes, transfer [the flowers] to the vase."

BIG IDEAS IN ENERGY

1 Energy flows from the sun and through all life on earth.

Energy is neither created nor destroyed but can change forms. Energy from the sun fuels all life on earth as it is converted to different forms. Each conversion is inefficient and some of the usable energy is converted to less useful heat energy. Cells temporarily store energy in the bonds of ATP molecules. This potential energy can be converted to kinetic energy and used to fuel life-sustaining chemical reactions.

2 Photosynthesis uses energy from sunlight to make food.

In the "photo" part of photosynthesis, chloroplasts transform light energy into the chemical energy of ATP and NADPH, while splitting water molecules and producing oxygen. In the Calvin cycle, the "synthesis" part of photosynthesis, carbon from CO_2 in the atmosphere is attached to molecules in chloroplasts to build sugars. This production of sugars consumes ATP and NADPH generated in the "photo" part of photosynthesis. C4 and CAM photosynthesis are evolutionary adaptations at the biochemical level that, although having competitive drawbacks, allow plants in hot, dry climates to close their stomata and conserve water without shutting down photosynthesis.

3 Cellular respiration converts food molecules into ATP, a universal source of energy for living organisms.

In cellular respiration, glucose and oxygen are converted to carbon dioxide, water, and energy that is stored in the high-energy bonds of ATP molecules. In the electron transport chain, high-energy electrons from NADH and $FADH_2$ pass from carrier to carrier, releasing energy with each step. This energy fuels the creation of a proton concentration gradient. The energy released as the protons move back down their concentration gradient is used to produce ATP. Overall, the cellular respiration of one glucose molecule yields about 36 molecules of ATP.

4 There are alternative pathways to energy acquisition.

Organisms are built with multiple systems for acquiring energy. They can generate ATP even when oxygen is lacking or when organic molecules other than glucose, such as lipids, proteins, and polysaccharides, are consumed.

KEY TERMS

CHECK YOUR KNOWLEDGE

1. Animal fats and plant oils are sometimes used as sources of fuel for automobile engines. How is energy harvested from these molecules?
   a) They contain long chains of hydrogen and carbon that, when broken, release the energy stored in the bonds linking the atoms together.
   b) They contain hydrogen and carbon tails linked by disulfide bridges that, when broken, release chemical energy.
   c) They contain multiple phosphate groups that each release energy when "liberated" from the molecule chemically.
   d) They contain long hydrophobic regions that, when mixed with water, generate explosive resistances.
   e) They contain long carbon tails and each atom has unpaired electrons that are released on exposure to extreme heat and pressure.

2. A cyclist rides her bike up a very steep hill. Pick from the following choices the statement that properly describes this example in energetic terms.
   a) Potential energy in food is converted to kinetic energy as the cyclist's muscles push her up the hill.
   b) Kinetic energy is highest when the cyclist is at the crest of the hill.
   c) The cyclist produces the most potential energy as she cruises down the hill's steep slope.
   d) Potential energy is greatest when the cyclist is at the top of the hill.
   e) Both a) and d) are correct.

3. Every time a source of energy is converted from one form to another:
   a) the potential energy of the system increases.
   b) heat is required.
   c) the second law of thermodynamics is violated.
   d) the total amount of energy in the universe is reduced by a tiny amount.
   e) some of the energy is converted to heat, which is not a very usable form of kinetic energy.

4. In your body, when energy is released from the breakdown of a molecule such as glucose:
   a) adenosine monophosphate is created.
   b) adenosine diphosphate is created.
   c) some of the energy may be harnessed by building the unstable high-energy bonds that attach phosphate groups to ADP.
   d) molecules of ATP are required to capture and absorb the heat generated by the reaction.
   e) all of the energy is lost as heat.

5. A green plant can carry out photosynthesis if given nothing more than:
   a) water, light, and carbon dioxide.
   b) water, light, and oxygen.
   c) carbon dioxide.
   d) oxygen.
   e) oxygen and carbon dioxide.

6. The actual production of sugars during photosynthesis takes place:
   a) within the outer membrane of the chloroplast.
   b) within the stroma, inside the thylakoids of the chloroplast.
   c) within the stroma, outside the thylakoid, but still inside the chloroplast.
   d) just outside the chloroplast, within the mitochondria.
   e) within the thylakoid membrane of the chloroplast.

7. The leaves of plants can be thought of as "eating" sunlight. From an energetic perspective this makes sense because:
   a) light energy, like chemical energy released when the bonds of food molecules are broken, is a type of kinetic energy.
   b) both light energy and food energy can be converted to kinetic energy without the loss of heat.
   c) the carbon-oxygen bonds within a photon of light release energy when broken by the enzymes in chloroplasts.
   d) the carbon-hydrogen bonds within a photon of light release energy when broken by the enzymes in chloroplasts.
   e) photons are linked together by hydrogen bonds, which release energy when striking the surface of a leaf.

8. A molecule of chlorophyll increases in potential energy:
   a) when it binds to a photon.
   b) when one of its electrons is boosted to a higher-energy excited state on being struck by a photon of light.
   c) when it loses an electron.
   d) only in the presence of oxygen.
   e) None of the above. The potential energy of a molecule cannot change.

9. Photosynthesizing plants rely on water:
   a) to provide the protons necessary to produce chlorophyll.
   b) to concentrate the beams of light hitting a leaf, focusing them on the reaction center.
   c) to replenish oxygen molecules that are lost during photosynthesis.
   d) to replace electrons that are excited by light energy and passed from molecule to molecule down an electron transport chain.
   e) to serve as a high-energy electron carrier.

10. During photosynthesis, which step is most responsible for a plant's acquisition of new organic material?
    a) the "building" of NADPH during the Calvin cycle
    b) the excitation of chlorophyll molecules by photons of light
    c) the "plucking" of carbon molecules from the air and fixing of them to organic molecules within the chloroplast
    d) the loss of water through evaporation
    e) the production of ATP during the light reactions

11. During C4 photosynthesis:
    a) plants use less ATP when producing sugar.
    b) plants are able to continue producing sugars even when they must close their stomata to reduce water loss on hot days.
    c) plants are able to generate water molecules to cool their leaves.
    d) plants are able to reduce water loss by producing more rubisco.
    e) plants are able to produce sugars without any input of carbon dioxide.

12. Cellular respiration is the process by which:
a) oxygen is used to transport chemical energy throughout the body.
b) oxygen is produced during metabolic activity.
c) light energy is converted to kinetic energy.
d) ATP molecules are converted to water and sugar.
e) energy from the chemical bonds of food molecules is captured by an organism.

13. During cellular respiration, most of the energy contained within the bonds of food molecules is captured in:
a) the conversion of the kinetic energy of food to the potential energy of ATP.
b) the Krebs cycle and electron transport chain.
c) digestion.
d) glycolysis.
e) None of the above. Energy is lost, not gained, during cellular respiration.

14. Which of the following energy-generating processes is the only one that occurs in all living organisms?
a) the Krebs cycle
b) glycolysis
c) combustion
d) photosynthesis
e) None of the above. There are no energy-generating processes that occur in all living organisms.

15. During the Krebs cycle:
a) the products of glycolysis are further broken down, generating additional ATP and the high-energy electron carrier NADH.
b) high-energy electron carriers pass their energy to molecules of sugar, which store them as potential energy.
c) the products of glycolysis are further broken down, generating additional ATP and the high-energy electron carrier NADPH.
d) cellular respiration can continue even in the absence of oxygen.
e) the products of glycolysis are converted to acetyl-CoA.

16. Mitochondria have a "bag within a bag" structure. This structure is necessary in order to:
a) create two distinct regions with a concentration gradient, a form of potential energy.
b) keep molecules of ADP in close proximity to molecules of the ATP-synthesizing enzyme.
c) allow light-reactive accessory pigments to be embedded within the membranes.
d) segregate the most toxic digestive enzymes from molecules of ATP and $NADP^+$.
e) None of the above.

17. All alcoholic beverages are produced as the result of:
a) cellular respiration by bacteria that occurs in the absence of oxygen.
b) cellular respiration by bacteria that occurs in the absence of free electrons.
c) cellular respiration by yeast that occurs in the absence of free electrons.
d) cellular respiration that occurs in the absence of sugar.
e) cellular respiration by yeast that occurs in the absence of oxygen.

18. In harvesting the chemical energy of the molecules in food:
a) all macromolecules must first be converted to glucose.
b) all macromolecules must first be converted to their hydrocarbon chains.
c) all macromolecules must first be converted to glucose or another simple sugar.
d) organisms can use sugars, lipids, and proteins.
e) all macromolecules must first be converted to proteins or free-form amino acids.

Short-Answer Questions

1. A plant and a mouse are placed in a sealed container with a watering system. For a period of time, both will survive. Explain what the two organisms can provide for each other in order to allow them to survive temporarily.

2. It is possible to measure the rate of the "photo" reactions in plants by measuring O_2 levels produced by chloroplasts. If bicarbonate solution, a source of CO_2, is provided to the plants, the rate of the "photo" reactions increases.

a) Explain why measuring O_2 levels provides an indication of the rate of the "photo" reactions.

b) Being as specific as possible, describe how the plants will use the CO_2 provided by the bicarbonate solution.

c) Propose an explanation of how increasing the CO_2 source for the plants will affect the rate of the "photo" reactions.

3. Some plants grow much more quickly than others. For a plant that is growing at a rapid rate, propose a hypothesis as to whether cellular respiration or photosynthesis is occurring at a faster rate. Justify your hypothesis.

See Appendix for answers. For additional study questions, go to www.prep-u.com.

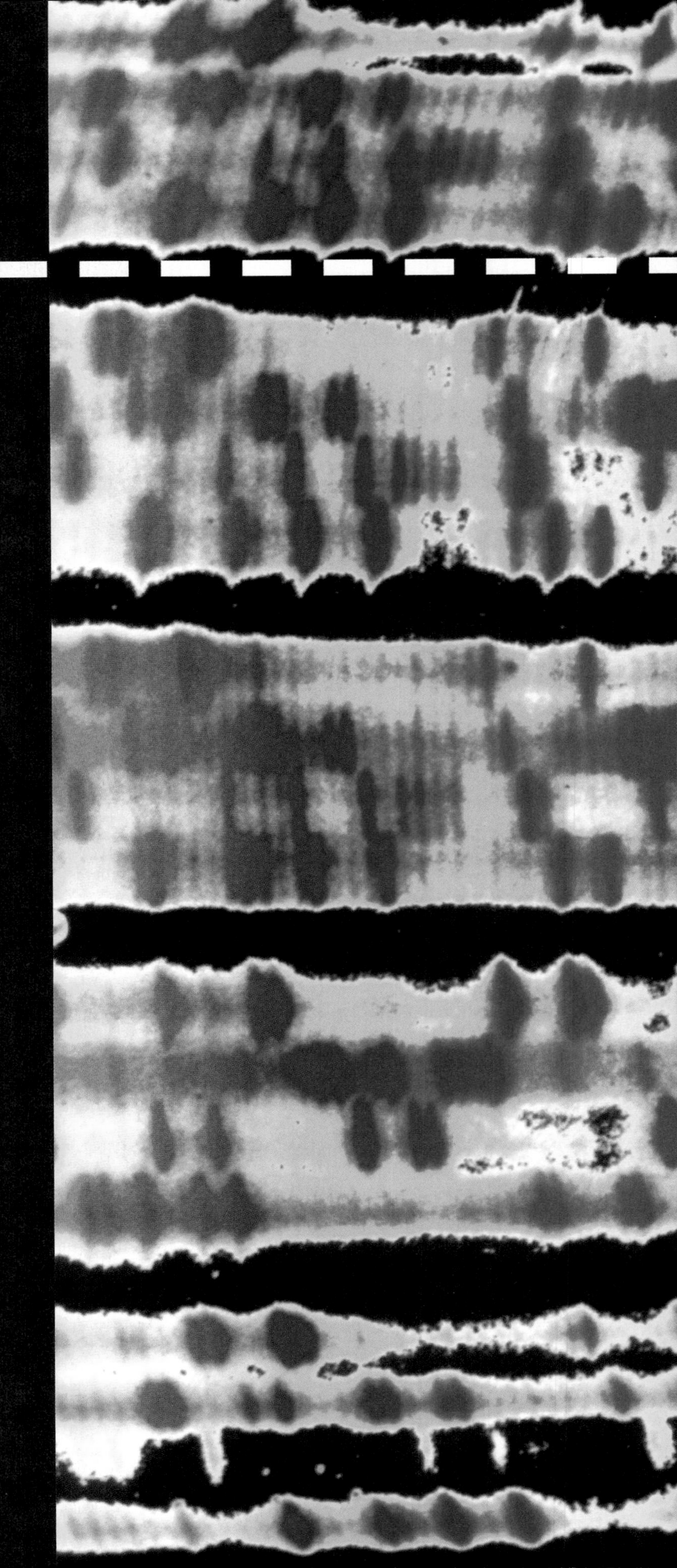

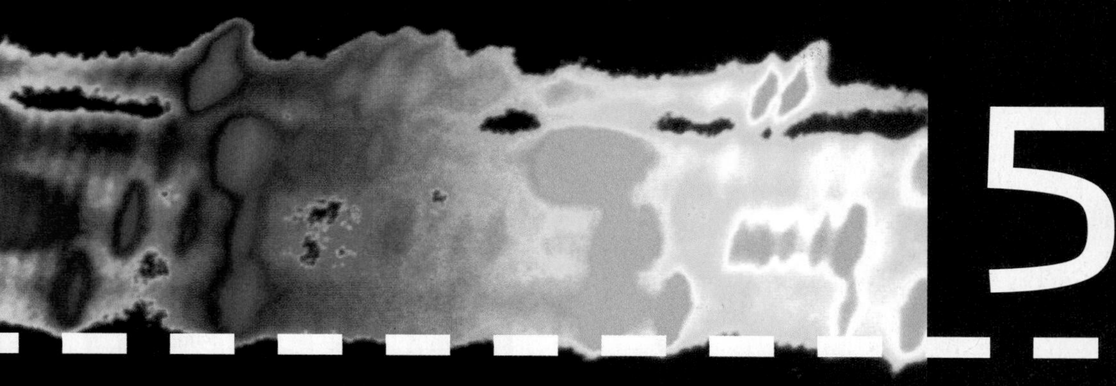

5

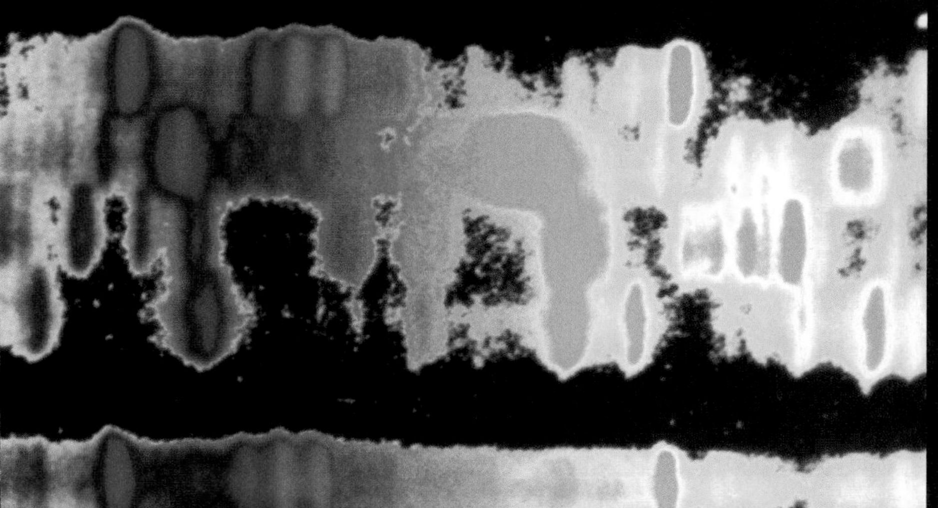

DNA, Gene Expression, and Biotechnology

What is the code and how is it harnessed?

❶ DNA: What is it and what does it do?

DNA microarrays measure the expression levels of many genes simultaneously.

5•1

"The DNA 200"—knowledge about DNA is increasing justice in the world.

In 1981, Julius Ruffin, a 27-year-old maintenance worker at Eastern Virginia Medical School, got onto an elevator and lost more than 20 years of his life. Several weeks earlier, a nursing student at the school had been raped by an attacker who broke into her apartment. When Ruffin got on the elevator, the student thought she recognized him as her attacker and called the police, who immediately arrested him.

Ruffin's girlfriend testified that he was with her at the time of the attack, but on the basis of the victim's eyewitness testimony, a jury found Ruffin guilty and sentenced him to life in prison. Ruffin maintained his innocence, to no avail, until 2003. At that time, the state's Division of Forensic Science performed a DNA analysis on a swab of evidence that remained from the investigation. The analysis revealed that Ruffin was not the attacker. Rather, the analysis showed that the DNA perfectly matched that of a man who was already in a Virginia prison, serving time for rape. Ruffin was freed, but only after having served more than two decades in prison (**FIGURE 5-1**).

Q What is the most common reason why DNA analyses overturn incorrect criminal convictions?

Julius Ruffin's case is tragic, but it is not unusual. He is one of a group of 200 people in the United States who have been freed from prison as a result of DNA analyses. Called "the DNA 200," these unjustly imprisoned people spent an average of 12 years behind bars. Eighty percent had been convicted of sexual assault, 28% had been convicted of murder. In more than three-quarters of the cases, inaccurate eyewitness testimony played an important role in the guilty verdict. (Recall, from Chapter 1, the experiments that revealed the unreliability of eyewitness identification.)

FIGURE 5-1 Vindicated by DNA evidence. Julius Ruffin was released from prison after two decades of wrongful incarceration.

In this chapter, we take a close look at DNA, the molecule responsible for Julius Ruffin's exoneration and the deferred justice served to the DNA 200. It is a molecule that all living organisms—people, plants, animals, bacteria, and otherwise—carry in almost every cell in their body (with just a few exceptions, such as red blood cells). Like a social security number, DNA is unique in virtually every person. In addition to being contained in our living cells, our DNA exists in what we leave behind. It is in our saliva, hair, blood, and even the dead skin cells that fall from our bodies. This is why DNA can serve as an individual identifier, a trail we all leave that is increasingly being used to ensure greater justice in our society.

The importance of DNA goes far beyond its function as an individual identifier, however. The information carried within this molecule, which is organized into individual units called genes, is among the most important of all biological knowledge. And, as witnessed by the following excerpts, it is often in the news.

> "Selfish dictators may owe their behaviour partly to their genes, according to a study that claims to have found a genetic link to ruthlessness."
>
> —*Nature,* April 2008

> "Whether a man has one type of gene versus another could help decide whether he's good 'husband material,' a new study suggests."
>
> —*Washington Post,* September 2008

By all accounts, we live in the "Age of the Gene." It's nearly impossible to open a newspaper or watch a news report these

FIGURE 5-2 Genetics issues are in the news.

days without being informed that yet another complex human characteristic or trait has been linked to a newly discovered gene (**FIGURE 5-2**). But what does that actually mean? What is this entity that apparently exerts influence over everything from our marriage readiness to our ruthlessness? We explore these issues and more in this chapter, beginning with a look at the structure of DNA itself and how it contains the information for producing organisms. In later sections of the chapter, we learn how modern manipulations of DNA are having far-reaching implications not just for human health but for agriculture as well.

TAKE-HOME MESSAGE 5·1

DNA is a molecule that all living organisms carry in every cell in their body. Because every person's DNA is unique and because we leave a trail of DNA behind us as we go about our lives, DNA can serve as an individual identifier. This fact is the basis of the DNA analysis that is being used increasingly to ensure greater justice in our society, such as by establishing the innocence of individuals wrongly convicted of crimes.

5•2

The DNA molecule contains instructions for the development and functioning of all living organisms.

Beginning in the 1900s and continuing through the early 1950s, a series of experiments revealed two important features of DNA. First, molecules of DNA are passed down from parent to offspring. Second, the instructions on how to create a body and control its growth and development are encoded in the DNA molecule.

Once scientists knew that DNA encoded instructions and that these instructions were passed on from one generation to the next, DNA became something of a celebrity molecule. Given that it must be able to hold the instructions for how to produce every possible type of structure in every living organism, scientists were crazed to learn all they could about it. There was a spirited race to determine the chemical structure of DNA and to understand how the molecule was assembled and shaped so that it could hold and transmit so much information.

Many formidable scientists rose to the challenge of determining the structure of DNA. American Linus Pauling had already won a Nobel prize in chemistry for his work on elucidating the structure of molecules when he began investigating the structure of DNA. Simultaneously, Maurice Wilkins and Rosalind Franklin in England devoted their research to this task as well, and produced X-ray pictures of DNA that were critical to decoding its shape. But it was Englishman Francis Crick and American James Watson, working in Cambridge, England, who happened to put all the pieces together and deduce the exact structure of DNA (**FIGURE 5-3**). As we will see later in this chapter, their discovery was more than just a description of a molecule. As soon as they figured out DNA's structure, the answers to several other thorny problems in biology—such as how DNA might be able to duplicate itself—became apparent immediately. We explore that process later in this chapter, but first we need to examine the structure of DNA in more detail.

FIGURE 5-3 Watson and Crick. James Watson and Francis Crick figured out the exact structure of DNA. The first sketch of the double helix (lower right) was rendered by Crick's wife, Odile.

DNA (**deoxyribonucleic acid**) is a **nucleic acid,** a macromolecule that stores information. It consists of individual units called **nucleotides,** which have three components: a molecule of sugar, a phosphate group (containing four oxygen atoms bound to a phosphorus atom), and a nitrogen-containing molecule called a **base.** The physical structure of DNA is frequently described as a "double helix." What exactly is a double helix? Picture a long ladder twisted around like a spiral staircase and you'll have a good idea of what a DNA molecule looks like (**FIGURE 5-4**). The molecule has two distinct strands, like the vertical sides of a ladder. These are the "backbones" of the DNA molecule, and each is made from two alternating molecules: a sugar, then a phosphate, then a sugar, then a phosphate, and so on. The sugar is always deoxyribose and the phosphate molecule is always the same, too. It is the shapes of the backbones that cause the DNA "ladder" to twist.

The alternating sugars and phosphates hold everything in place, but they play only a supporting role. The rungs of the ladder are where things get interesting. Attached to each sugar, and protruding like half of a rung on the ladder, is one of the nitrogen-containing bases. Each sugar always has a base attached to it, and there are four different kinds of bases—adenine, thymine, cytosine, and guanine. When discussing DNA, these bases are usually referred to by their first letter: A, T, C, and G.

Both backbones of the ladder have a base protruding from each sugar. The base on one side of the ladder binds, via hydrogen bonds, to a base on the other side, and together these **base pairs** form the rungs of the ladder. They don't just pair up at random, though. Every time a C protrudes from one side, it forms hydrogen bonds with a G on the other side (and

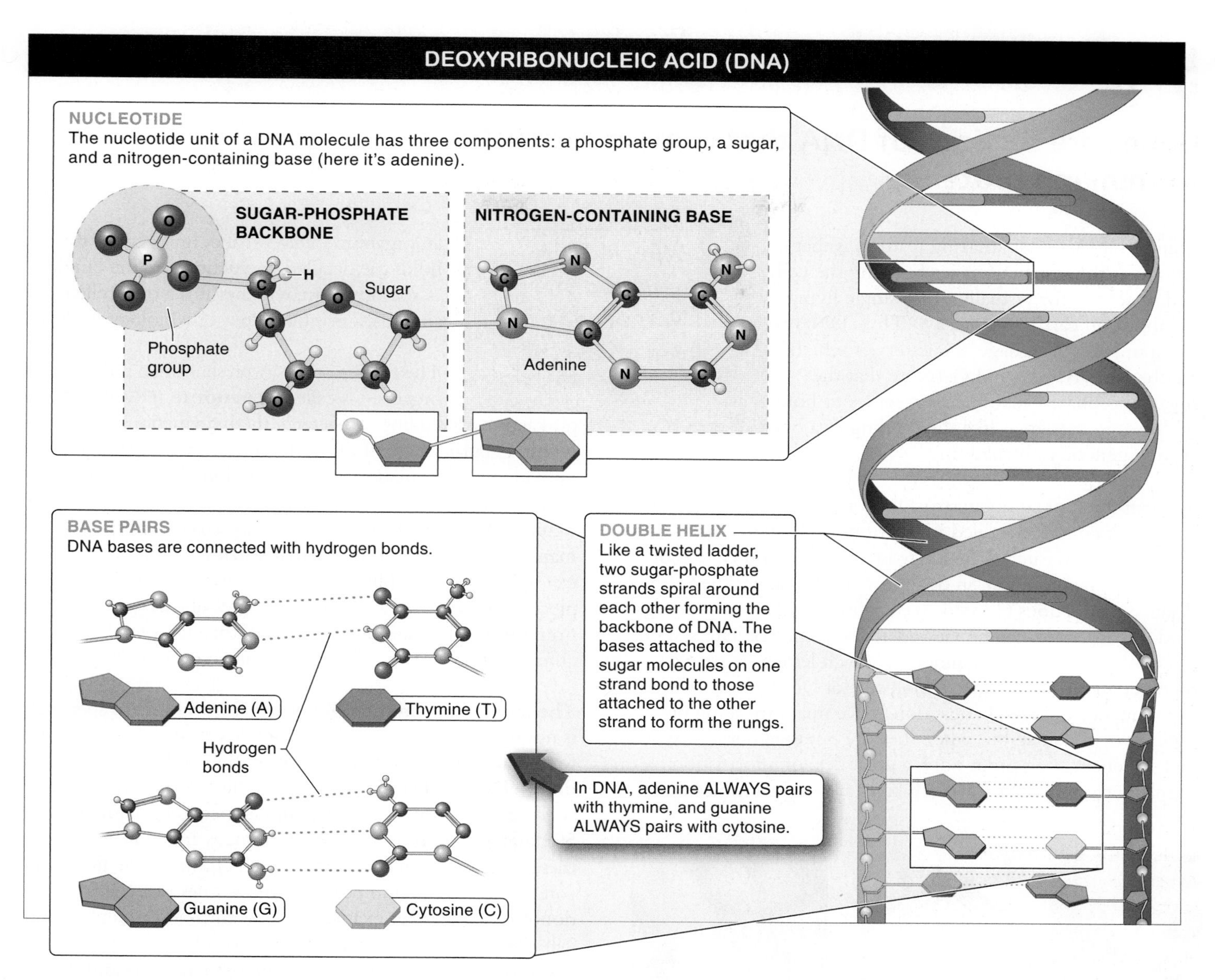

FIGURE 5-4 Overview of the structure of DNA.

vice versa: a G always bonds to a C). Similarly, every time a T protrudes from one side, it forms hydrogen bonds with an A on the other side (and vice versa). For this reason, each DNA molecule always has the same number of Gs and Cs, and the same numbers of As and Ts. Because of these base-pairing rules, it also is true that if we know the base sequence for one of the strands in a DNA molecule, we know the sequence in the other. For this reason, a DNA sequence is described by writing the sequence of bases on only one of the strands.

If our DNA molecule were really a spiraled ladder, it would be a very, very long one. One molecule of DNA can have as many as 200 million base pairs, or rungs. This length would make it difficult to squeeze the molecule into the cell. But the rungs are small and the twisting of the molecule—twists upon twists—shortens it considerably: think about how, if you repeatedly twist your shoe lace around and around, it becomes shorter and shorter. That's what happens with DNA. Let's now investigate how DNA's structure—the rungs of the ladder, in particular—enables it to carry information.

TAKE-HOME MESSAGE 5•2

DNA is a nucleic acid, a macromolecule that stores information. It consists of individual units called nucleotides: a sugar, a phosphate group, and a nitrogen-containing base. DNA's structure resembles a twisted ladder, with the sugar and phosphate groups serving as the backbones of the molecule and base pairs serving as the rungs.

5•3

Genes are sections of DNA that contain instructions for making proteins.

Q Why is DNA considered the universal code for all life on earth?

One of DNA's most amazing features is that it embodies the instructions for building the cells and structures for virtually every single living organism on earth (**FIGURE 5-5**). Thus, DNA is like a universal language, the letters of which are the bases A, T, C, and G. (Note that the sugar-phosphate backbone only serves to hold the bases in sequence, like the binding of a book. It does not convey genetic information.)

You can think of an organism's DNA as a cookbook. Just as a cookbook contains detailed instructions on how to make a variety of foods (such as french toast, macaroni and cheese, or chocolate chip cookies), an organism's DNA carries the detailed instructions to build an organism and keep it running. And just as a book can be viewed as a sequence of letters, with the book's meaning determined by which letters are strung together and in what order, a molecule of DNA can be viewed as a sequence of bases. Letters don't have much meaning on their own, of course, but when they are put together into words and sentences, their order holds a great deal of information. Similarly, the sequence in which bases appear in a molecule of an organism's DNA makes up a "code" that holds the detailed instructions for the building of the organism, whether it is a one-celled amoeba, a giant oak tree, or a biology student.

DNA provides the instructions for building virtually every organism on earth!

FIGURE 5-5 DNA is the universal code for all life on earth.

The full set of DNA present in an individual organism is called its **genome** (**FIGURE 5-6**). In prokaryotes, including all bacteria, the information is contained within circular pieces of DNA. In eukaryotes, including humans, this information is laid out in long linear strands of DNA in the nucleus. Rather than being one super-long DNA strand, eukaryotic DNA exists as many smaller, more manageable pieces, called **chromosomes.** Humans, for example, have three billion base pairs, divided into 23 unique pieces of DNA (and we have two copies of each piece, one from our mother and one from our father, for a total of 46 chromosomes and six billion base pairs in every cell).

The number of chromosomes varies from species to species, but is not related to how simple or complex an organism is. Corn, for example, has 10 unique chromosomes and fruit flies have only 4. Dogs and chickens have 39 different chromosomes, while goldfish put most other animal species to shame with 50 different chromosomes. As in humans, individuals in each of these species inherit one copy of each chromosome from each parent, so the total number of chromosomes in a cell is doubled— a dog, for example, has 78 chromosomes in each cell. Chromosomes vary in length, too. In humans, the longest chromosome has more than 200 million base pairs and the shortest has fewer than 50 million.

Within the long sequences of bases are relatively short sequences, on average about 3,000 bases long, that are the actual genes. "Gene" may seem like an impossibly nebulous concept because the word is often used casually in the media as if it were some magical, irresistible, and mysterious force that controls our bodies and behavior. Beyond these vague descriptions, though, the word has a literal meaning. A **gene** is a sequence of bases (strictly, as we've noted, base pairs) in a DNA molecule that carries the information necessary for producing a functional product, usually a protein or RNA molecule. Nothing more, nothing less.

Remember that a DNA molecule is like a ladder in which half of each rung is any one of the four bases— A, T, G, or C. Strung together, a segment might be read as "AAAGGCTAGGC . . ." continuing on for another

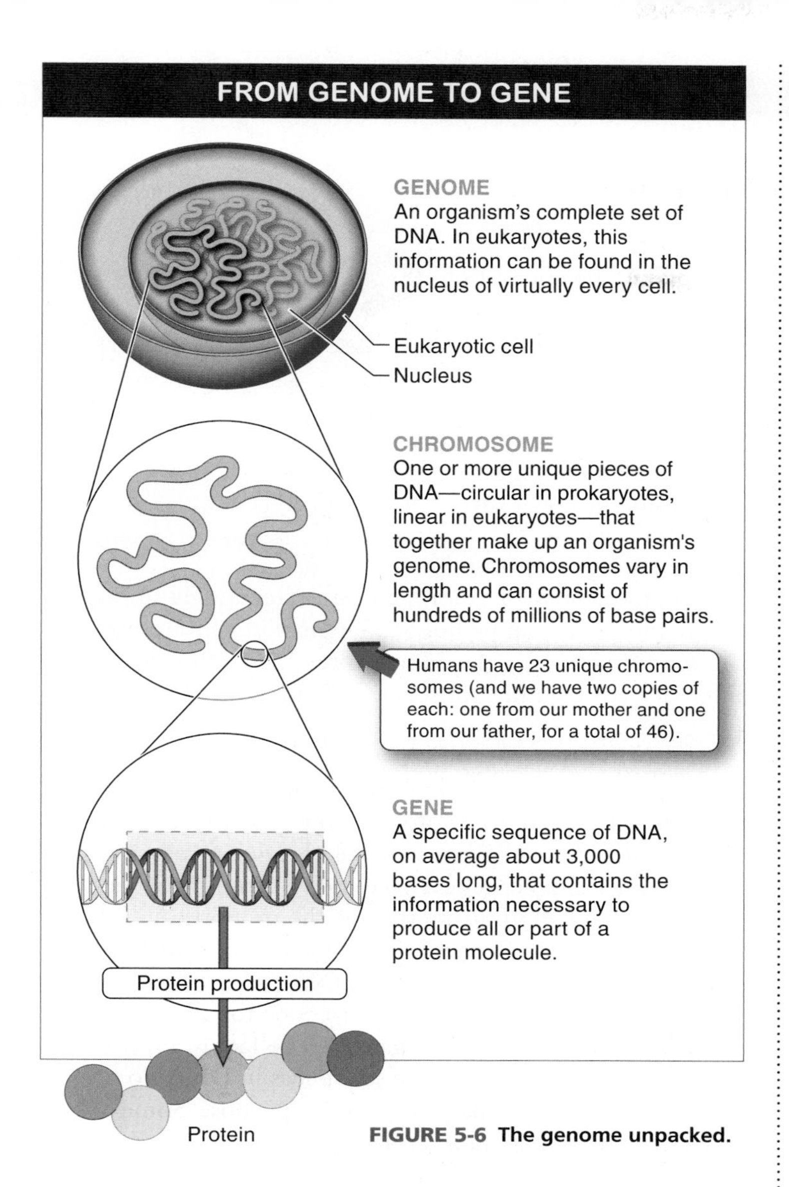

FIGURE 5-6 The genome unpacked.

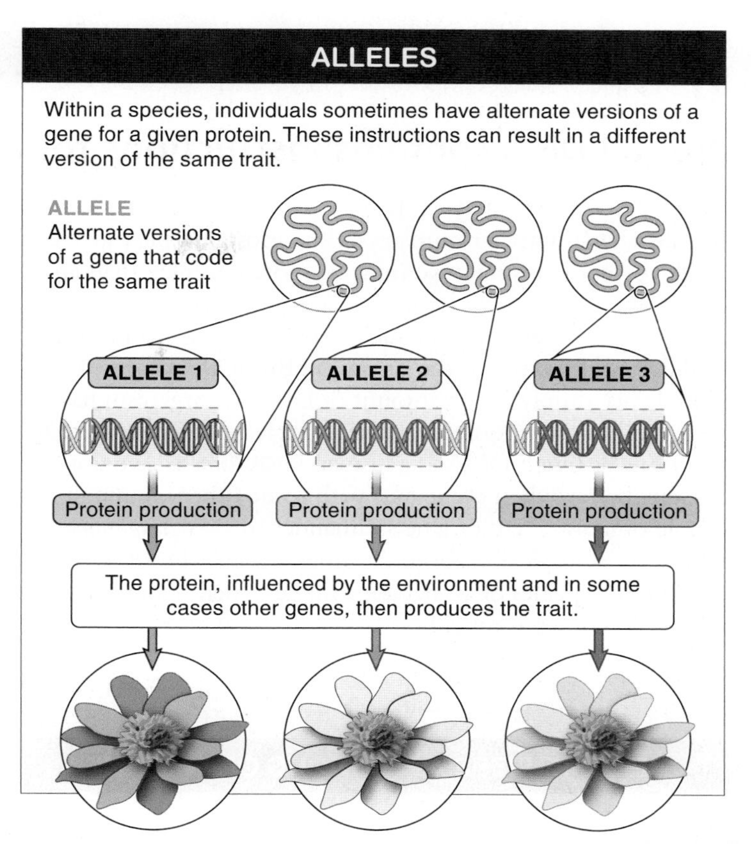

FIGURE 5-7 "Different versions of the same thing." Alleles are alternative versions of a gene.

3,000 or so bases. Returning to our cookbook analogy, this segment can be thought of as a recipe, which is also a series of letters. And just as a particular sequence of letters in a cookbook may be read and understood as the directions for baking chocolate chip cookies, the sequence of bases in DNA also carries information. Perhaps the sequence spells out part of the instructions for producing a red blood cell, for constructing the keratin that will form part of a curly strand of hair, or for assembling a chemical that alters your brain chemistry so that you exhibit a mood disorder or suicidal behavior.

Each gene is the instruction set for producing one particular molecule, usually a protein. For example, there is a gene in silk moths that codes for fibroin, the chief component of silk. And, there is a gene in humans that codes for triglyceride lipase, an enzyme that breaks down dietary fat. Within a species, individuals sometimes have slightly different instruction sets for a given protein, and these instructions can result in a different version of the same characteristic. These alternative versions of a gene that code for the same feature are called **alleles** (**FIGURE 5-7**). And any single characteristic or feature of an organism is referred to as a **trait.** A simple hypothetical example will clarify the meaning of these terms. The color of a daisy's petals is a trait. The instructions for producing this trait are found in a gene that controls petal color. However, this gene may have many different alleles; one allele may specify the trait of red petals, another may specify white petals, and yet another may specify yellow petals. Similarly, one allele for eye color in fruit flies may carry the instructions for producing a red eye, while another, slightly different allele may have instructions for brown eyes. (Ultimately, the trait may be influenced not just by the genes an individual carries but the way those genes interact with the environment, too.)

TAKE-HOME MESSAGE 5•3

DNA is a universal language that provides the instructions for building all the structures in all living organisms. The full set of DNA an organism carries is called its genome. In prokaryotes, the DNA occurs in circular pieces. In eukaryotes, the genome is divided among smaller, linear strands of DNA called chromosomes. A gene is a sequence of bases in a DNA molecule that carries the information necessary for producing a functional product, usually a protein or RNA molecule.

5•4

Not all DNA contains instructions for making proteins.

It is debatable whether humans are the most complex species on the planet, but surely we must be more complex than an onion. "Complexity" is somewhat subjective and can be assessed in a variety of ways, such as by counting the number of different cell types in the organism. But if we measured complexity simply as the amount of DNA an organism has, we'd have to say an onion is more complex—it has more than five times as much DNA as a human (**FIGURE 5-8**)! We don't fare any better when compared with some other seemingly simple organisms, either. The salamander species *Amphiuma means,* for example, has about 25 times as much DNA as we do, and one species of amoeba—a single-celled organism—has almost 200 times as much!

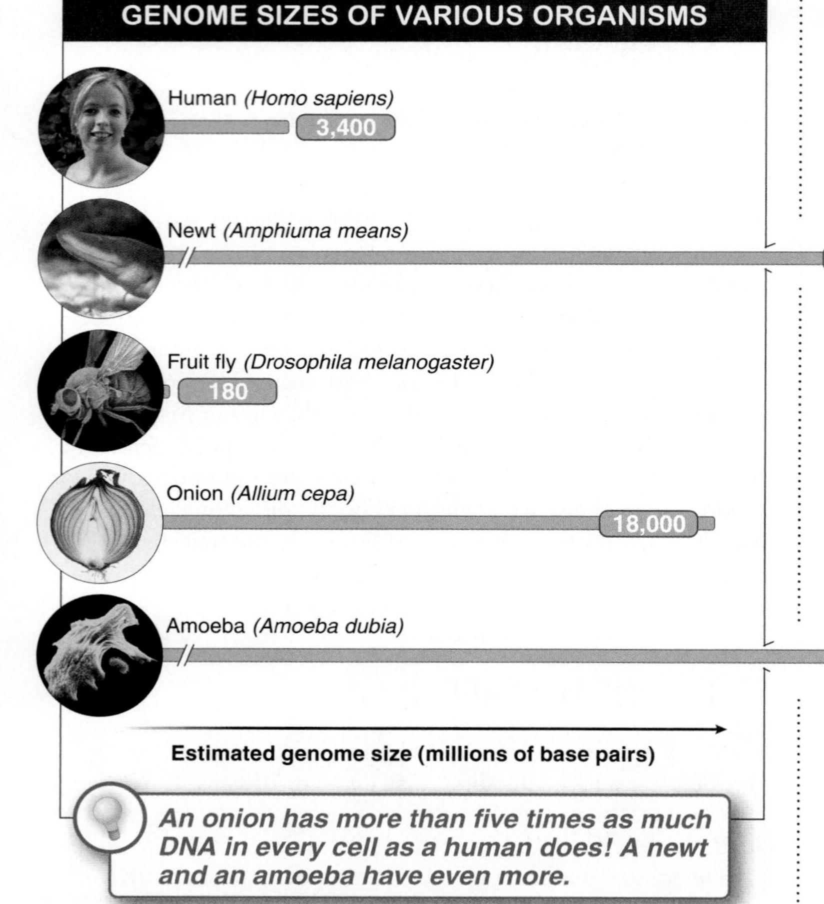

FIGURE 5-8 **Is the size of an organism's genome related to its complexity?**

Q An onion has five times as much DNA as a human. Why doesn't that make onions more complex than us?

Comparing the amount of DNA present in various species, in terms of both numbers of chromosomes and numbers of base pairs, reveals a paradox: there does not seem to be any relationship between the size of an organism's genome and the organism's complexity. Moreover, among species of seemingly similar complexity, there can be huge differences in genome size (among fish, for example, there is more than a threefold difference among different species of sturgeon). What is going on?

The description earlier in this chapter about what DNA is and how genes code for proteins is logical and tidy, but it doesn't completely explain what we observe in cells. In humans, for example, genes make up less than 5% of the DNA (**FIGURE 5-9**). In many species, the proportion of the DNA that consists of genes is even smaller. In virtually all eukaryotic species, the amount of DNA present far exceeds the amount necessary to code for all of the proteins in the organism. The fact is, a huge proportion of base sequences in DNA do not code for anything and have no obvious purpose. Some biologists even call it "junk DNA."

In what types of organisms do we find the most "junk DNA"? Bacteria and viruses tend to have very little non-coding DNA; genes make up 90% or more of their DNA. It is in the eukaryotes (with the exception of yeasts) that we see an explosion in the amount of non-coding DNA, about 25% of which occurs within genes and 75% between genes (**FIGURE 5-10**).

Non-coding regions of DNA sometimes take the form of short sequences that are repeated thousands (or even hundreds of thousands) of times. Other times, the non-coding regions are slightly longer repeated sequences. Occasionally, the non-coding DNA consists of gene fragments, duplicate versions of genes, and pseudogenes (sequences very similar to actual genes but with a few slight alterations that make them lose their protein-coding ability). Non-coding regions occur both within genes—in which case they are called **introns**—and between genes.

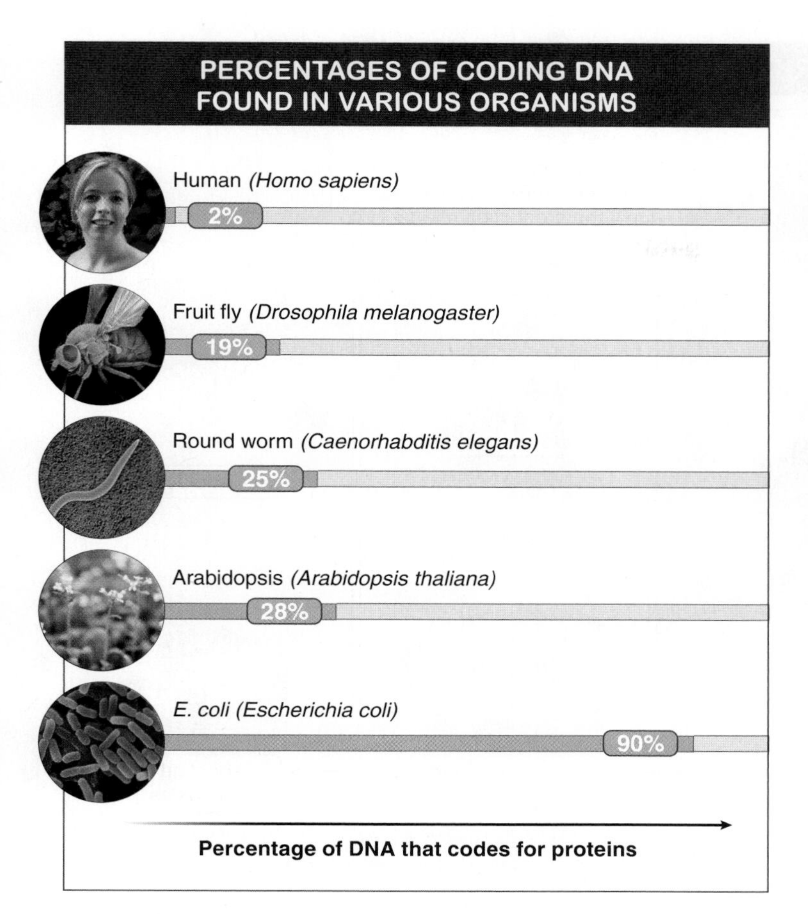

FIGURE 5-9 "Junk DNA"? The proportion of DNA that codes for proteins or RNA varies greatly among species. Which species seems to have the least "junk DNA"?

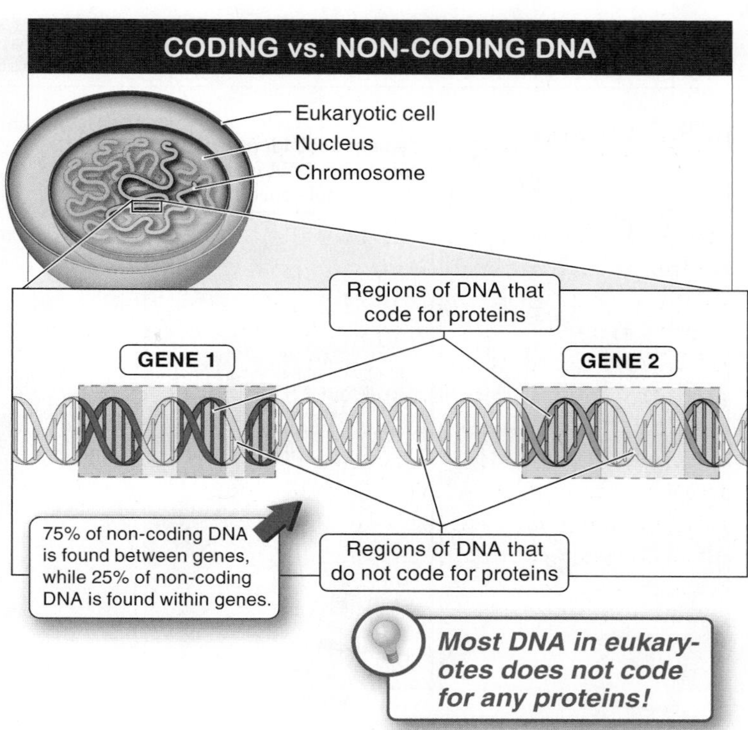

FIGURE 5-10 Non-coding regions of DNA. These regions are found both between and within genes.

In the end, the presence of all this non-coding DNA remains a big mystery. All the extra DNA may serve some purpose. Perhaps it is a reservoir of potentially useful sequences. Or it may have some function in regulating when genes are turned on or off. Because we are not certain what non-coding DNA does, the label "junk DNA" is probably a bad description for it. The jury is still out and it is too early to assume that it has no function.

TAKE-HOME MESSAGE 5•4

Only a small fraction of the DNA in eukaryotic species is in genes, coding for proteins or RNA; the function of the rest is still a mystery.

5•5

How do genes work? An overview.

Just as having a recipe for chocolate chip cookies is not the same thing as having the actual cookies, having a wealth of hereditary information about how to build muscle cells or root cells or leaf cells is not sufficient to produce an organism. Think about it: every cell in an organism contains all of the information needed to manufacture every protein in its body. This means that the skin cells on your arm contain the genes for producing proteins only found in liver cells or red blood cells or muscle tissue—but they don't produce them. Having the instructions is not the same as having the products.

The genes in strands of DNA are a storehouse of information, an instruction book, but they are only one part of the process by which an organism is built. If the genes that an organism

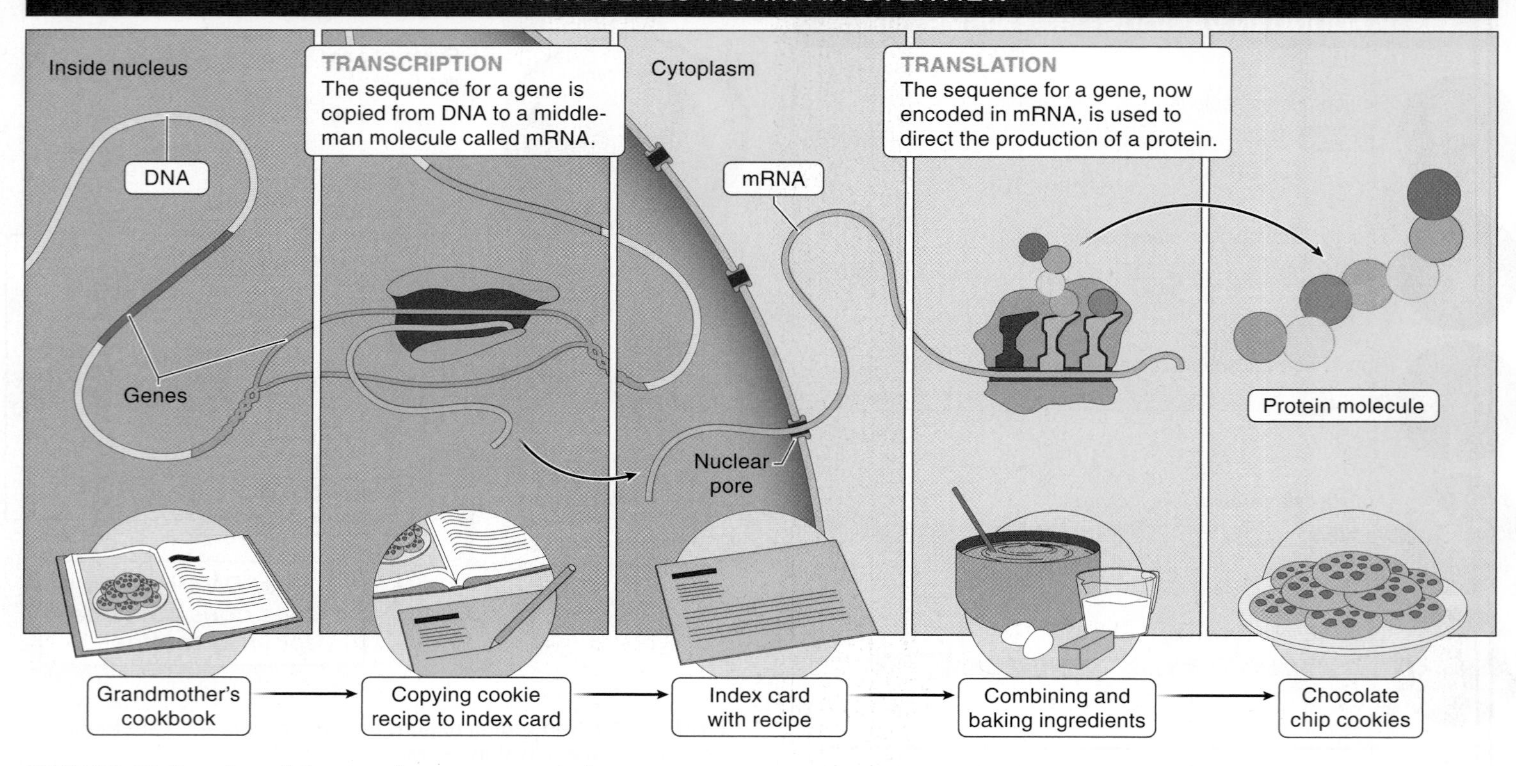

FIGURE 5-11 Overview of the steps from gene to protein.

carries for a particular trait—its **genotype**—are like a recipe in a cookbook, the physical manifestation of the instructions—the organism's **phenotype**—is the cookie, the macaroni and cheese, the french toast, or any of the other foods described by the recipes. And, just as you have to assemble ingredients, mix them, then bake the dough to get a cookie, there are several steps in the production of the molecules, tissues, and even behaviors that make up a phenotype.

How does a gene (a sequence of bases in a section of DNA) affect a flower's color or the shape of a nose or the texture of a dog's fur (the phenotype)? The process occurs in two main steps: **transcription,** in which a copy of a gene's base sequence is made, and **translation,** in which that copy is used to direct the production of a protein.

FIGURE 5-11 presents an overview of the processes of transcription and translation. In transcription, which in eukaryotes occurs in the nucleus, the gene's base sequence or **code** is copied into a middleman molecule called **messenger RNA** (**mRNA**). This is like copying the information for the chocolate chip cookie recipe out of the cookbook and onto an index card. In translation, the mRNA moves out of the nucleus and into the cytoplasm of the cell, where the messages encoded in the mRNA molecules are used to build proteins.

TAKE-HOME MESSAGE 5•5

The genes in strands of DNA are a storehouse of information, an instruction book. The process by which this information is used to build an organism occurs in two main steps: transcription, in which a copy of a gene's base sequence is made, and translation, in which that copy is used to direct the production of a protein.

❷ Building organisms: through gene expression, information in the DNA directs the production of the molecules that make up an organism.

Transcription and translation are a bit like copying a cookie recipe, then baking the cookies.

5•6

In transcription, the information coded in DNA is copied into mRNA.

If DNA is like a cookbook filled with recipes, transcription and translation are like cooking. In cooking, you use information about how to make chocolate chip cookies to actually produce the cookies. In an organism, the information about putting together proteins is used to build the proteins the organism needs in order to function. In this section, we examine transcription, the first step in the two-step process by which DNA regulates a cell's activity and its synthesis of proteins (see Figure 5-11). In transcription, a single copy of one specific gene within the DNA is made. Continuing our cookbook analogy, transcription is like copying a single recipe from the cookbook onto an index card. It happens in four steps (**FIGURE 5-12**).

Step 1. Recognize and bind. To start the transcription process, a large molecule, the enzyme RNA polymerase, recognizes a **promoter site,** a part of the DNA molecule that indicates the start of a gene, and, in effect, tells the RNA polymerase, "Start here." RNA polymerase binds to the DNA molecule at the promoter site and unwinds it just a bit, so that only one strand of the DNA can be read. Then, like a court reporter transcribing everything that is said in a courtroom, RNA polymerase begins to read the gene's message.

Step 2. Transcribe. As the DNA strand is processed through the RNA polymerase, the RNA polymerase builds a copy—called a "transcript"—of the gene from the DNA molecule. This copy is called messenger RNA (mRNA) because, once this copy of the gene is created, it can move elsewhere in the cell and its message can be translated into a protein. Throughout transcription, DNA is unwound ahead of the RNA polymerase so that a single strand of the DNA can be read, and is rewound after the polymerase passes.

The mRNA transcript is constructed from four different molecules called ribonucleotides (which are almost identical to DNA nucleotides, consisting of a sugar-phosphate complex with a nitrogen-containing base attached). Each

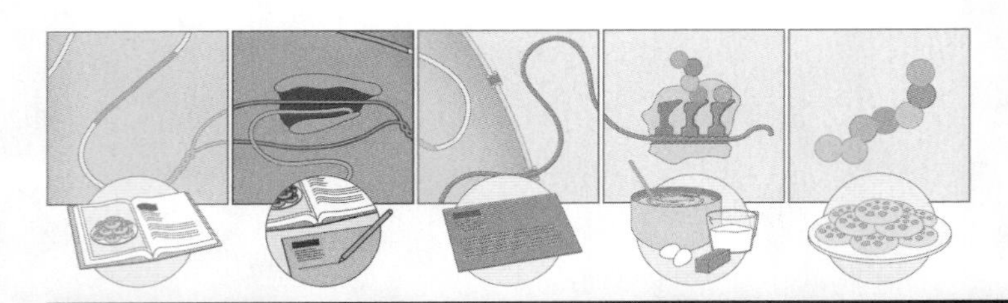

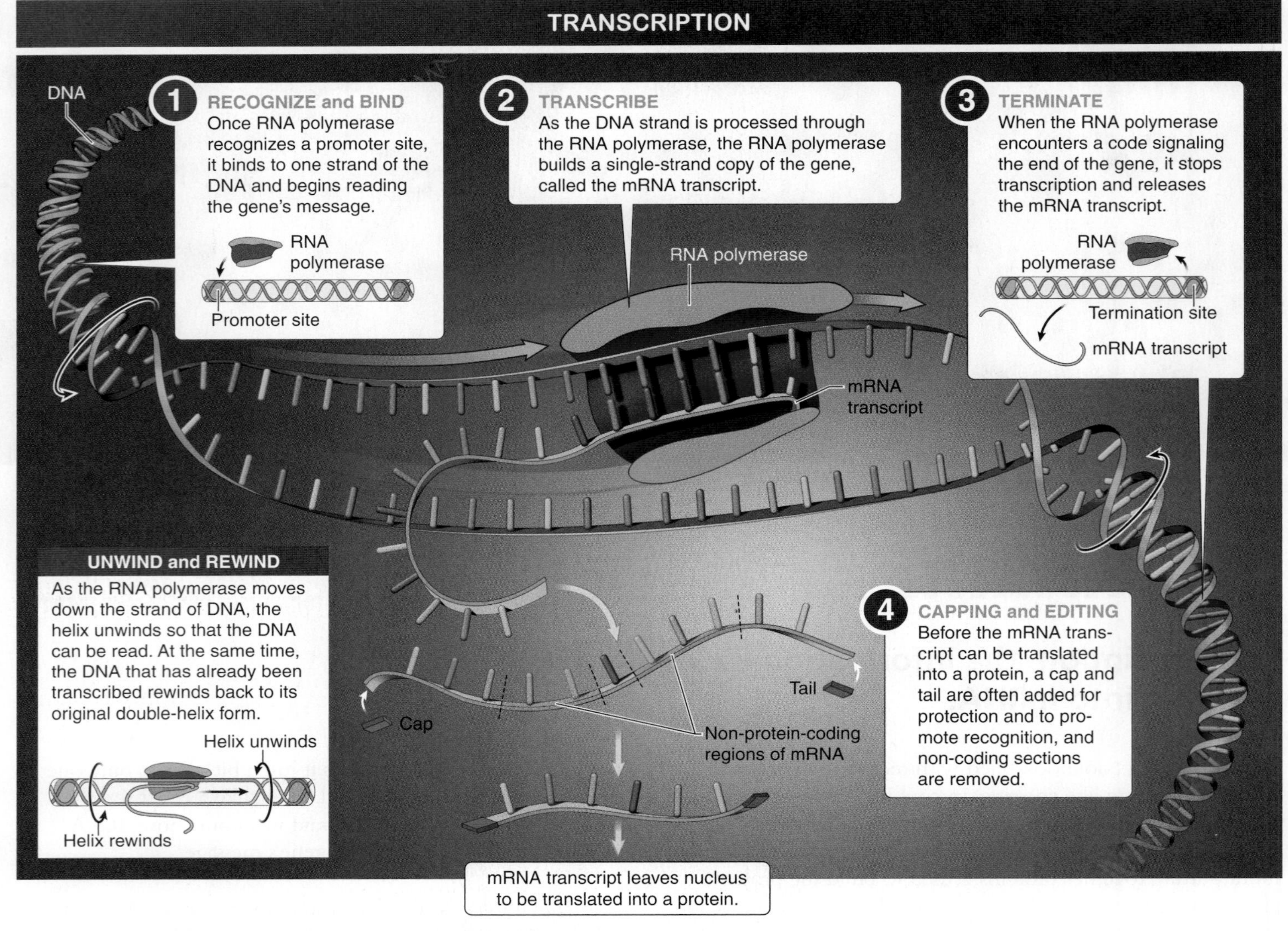

FIGURE 5-12 Transcription: copying the base sequence of a gene. The first step in a two-step process by which DNA regulates a cell's activity and synthesis of proteins.

ribonucleotide pairs up with an exposed base on the now unwound and separated DNA, following these rules:

If the DNA strand has a thymine (T), an adenine (A) is added to the mRNA.

If the DNA strand has an adenine (A), a uracil (U) is added to the mRNA.

If the DNA strand has a guanine (G), a cytosine (C) is added to the mRNA.

If the DNA strand has a cytosine (C), a guanine (G) is added to the mRNA.

Because our court reporter (RNA polymerase) transcribes a specific sequence of DNA letters (the gene), the mRNA transcript carries this specific bit of the DNA's information. And because it is separate from the DNA, the mRNA transcript can move throughout the cell, to the places where the information is needed, while leaving the original information within the DNA.

Step 3. Terminate. When the RNA polymerase encounters a sequence of bases on the DNA at the end of the gene (called a termination sequence), it stops creating the transcript and detaches from the DNA molecule. After termination, the mRNA molecule is released as a free-floating, single-strand copy of the gene.

Step 4. Capping and editing. In prokaryotic cells, once the mRNA transcript separates from the DNA it is ready to be translated into a protein (it doesn't have a nuclear membrane to cross). In most eukaryotes, however, the transcript must first be edited in several ways. First, a cap and a tail may be added at the beginning and end of the transcript. Like a front and back cover to a book, these serve to protect the mRNA from damage and help the protein-making machinery recognize the mRNA. Second, because (as we saw earlier in the chapter) there may have been non-coding bits of DNA transcribed into mRNA, those sections—the introns—are snipped out. Once the mRNA transcript has been edited, it is ready to leave the nucleus for the cytoplasm, where it will be translated into a protein.

TAKE-HOME MESSAGE 5•6

Transcription is the first step in the two-step process by which DNA directs the synthesis of proteins. In transcription (occurring in the nucleus in eukaryotic cells), a single copy of one specific gene in the DNA is made, in the form of a molecule of mRNA, and this moves to the cytoplasm where it can be translated into a protein.

5•7

In translation, information from DNA is used to build functional molecules.

Once the mRNA molecule moves out of the cell's nucleus and into the cytoplasm, the translation process begins. In translation, the information carried by the mRNA is read and ingredients present in the cytoplasm are used to produce a protein (see Figure 5-11). The process of translation is like the combining and baking of the ingredients listed in our chocolate chip cookie recipe to produce a cookie.

Several ingredients must be present in the cytoplasm for translation to occur. First, there must be large numbers of free amino acids floating around. Recall from Chapter 2 that amino acids are the raw materials for building proteins and an essential component of our diet. Second, there must be molecules called **ribosomal subunits,** the protein-production factories where amino acids are linked together in the proper order to produce the protein. Finally, there must also be molecules that can read the mRNA code and translate it from a sequence of bases into a protein. These molecules, called **transfer RNA (tRNA),** interpret the mRNA code and link specific base sequences on the mRNA with specific amino acids (**FIGURE 5-13**). Because tRNAs play such a central role in translation, we examine them more closely.

Human translators can hear one a word in one language and say the same word in a different language. Transfer RNA molecules function in a similar manner. Picture a two-sided

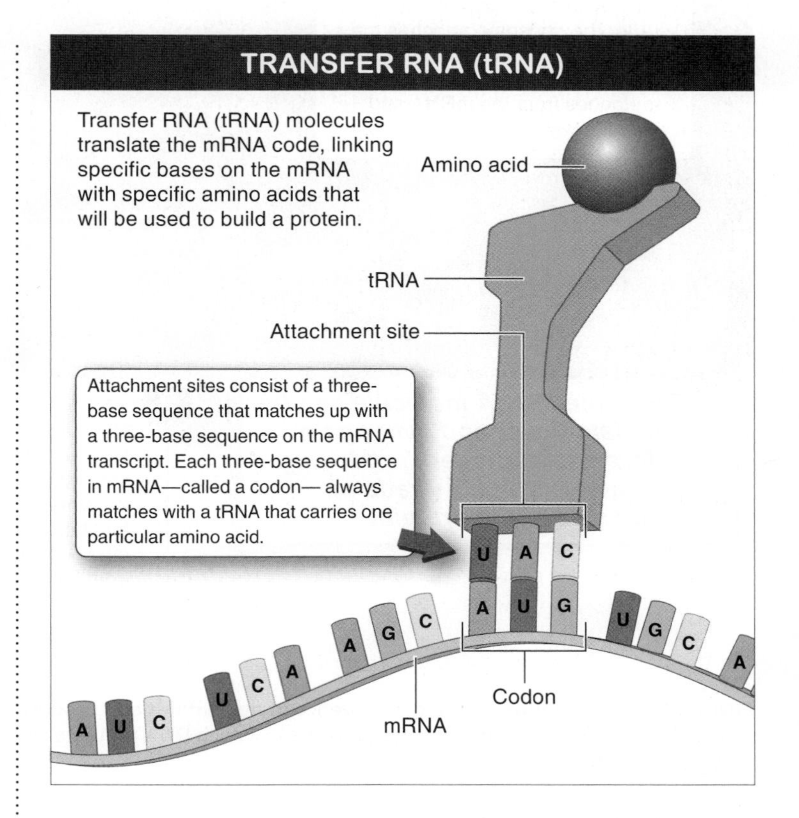

FIGURE 5-13 Each transfer RNA attaches a particular amino acid to the mRNA.

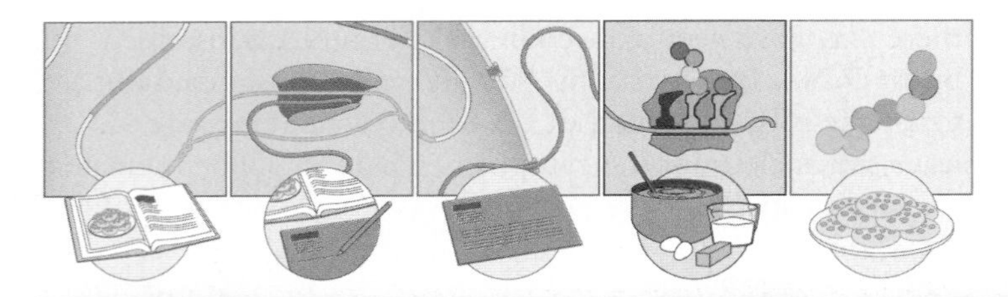

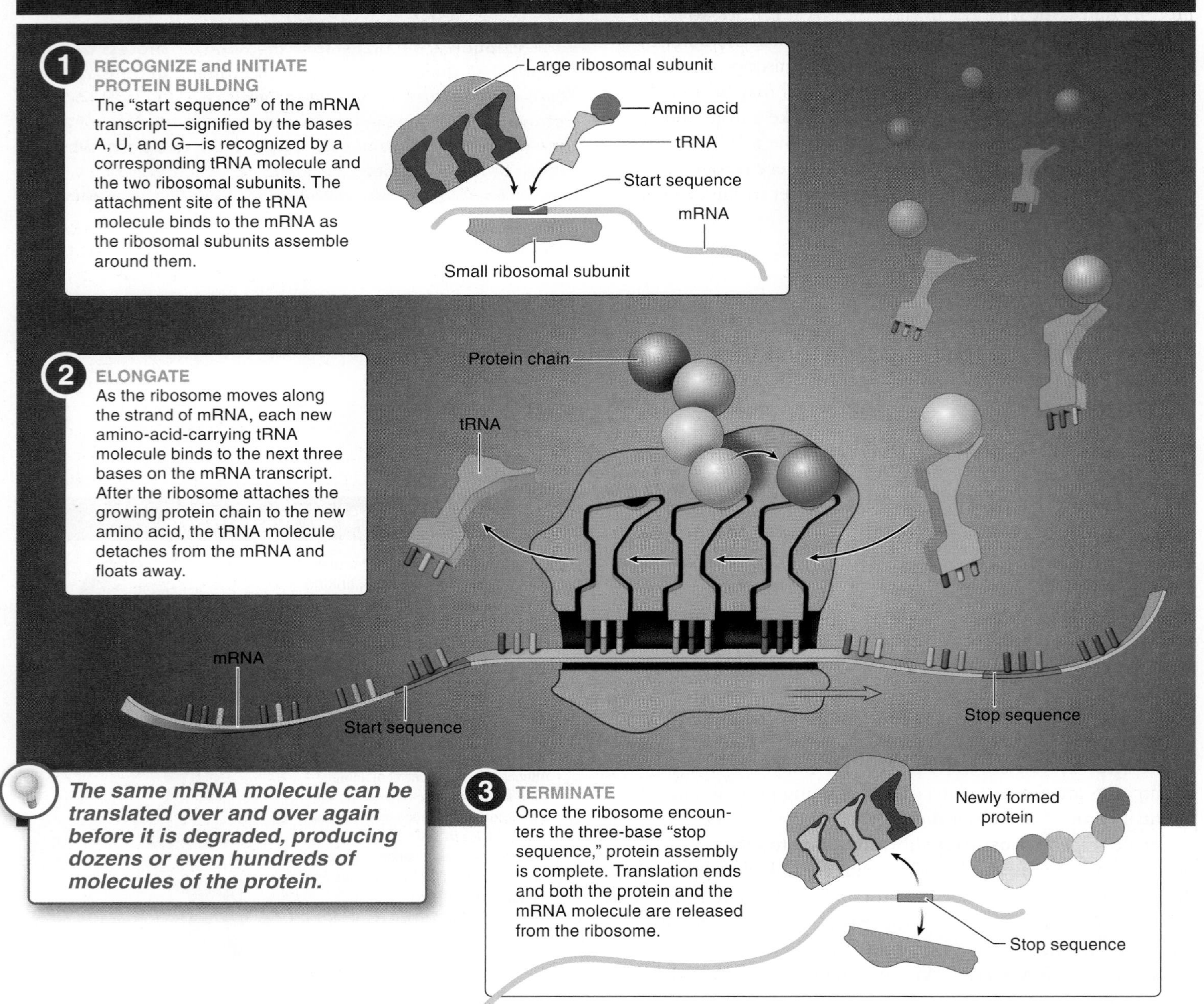

FIGURE 5-14 Translation: reading a sequence of nucleotides and producing protein. The second step in a two-step process by which DNA regulates a cell's activity and synthesis of proteins.

molecule. On one side a particular amino acid is attached. This amino acid can be any one of the 20 different amino acids that make up proteins. In other words, there are different tRNA molecules for each of the 20 different amino acid molecules, which are floating around in the cytoplasm in great numbers. On the other side of the two-sided tRNA molecule is an attachment site, consisting of a three-base sequence that matches up with a three-base sequence on the mRNA transcript. (Each three-base sequence in mRNA—called a **codon**—always matches with a tRNA that carries one particular amino acid.) This matchup enables the tRNA molecule to attach to the mRNA. Thus, the tRNA translates the code on the mRNA into an amino acid that will be used to build a protein. Translation occurs in three steps (**FIGURE 5-14**).

Step 1. Recognize and initiate protein building. Translation begins in the cell's cytoplasm when the subunits of a ribosome, essentially a two-piece protein-building factory, recognizes and assembles around a "start sequence" on the mRNA transcript. This start sequence is always the sequence A, U, and G (AUG). As the ribosomal subunits assemble themselves into a ribosome, one side of a tRNA molecule also recognizes the start sequence on the mRNA and binds to it. This initiator tRNA has the amino acid methionine bound to its other side. This will be the first amino acid in the protein that is to be produced (although occasionally, in eukaryotes, it is eventually "edited out").

Step 2. Elongate. After the mRNA start sequence (AUG), the next three bases on the mRNA specify which amino acid–carrying tRNA molecule should bind to the mRNA. If the next three bases on the mRNA transcript are GUU, for example, a tRNA molecule that recognizes that sequence, and carrying its particular amino acid, will attach to the mRNA at that point. The ribosome then facilitates the connection of this second amino acid to the first. After the amino acid carried by a tRNA molecule is attached to the preceding amino acid, the tRNA molecule detaches from the mRNA and floats away.

As this process continues, the protein grows. The next three bases on the mRNA specify the next amino acid to be added to the first two. And the three bases after that specify the fourth amino acid, and so on. This process is called **protein synthesis** because all proteins are chains of amino acids, like beads on a string.

The mRNA continues to be "threaded" through the ribosome, with the ribosome moving down the mRNA strand reading and translating its message in little three-base chunks. Each three-base sequence specifies the next amino acid, lengthening the growing amino acid strand.

Step 3. Terminate. Eventually, the ribosome arrives at the three-base sequence on the mRNA that signals the end of translation. Once the ribosome encounters this stop sequence, the assembly of the protein is complete. Translation ends, and the amino acid strand and mRNA molecule are released from the ribosome. As it is produced, the amino acid sequence folds and bends, based on the chemical features in the amino acid side chains (as we saw in Chapter 2), acquiring a three-dimensional structure. The completed protein—such as a membrane protein or insulin or a digestive enzyme—may be used within the cell or packaged for delivery via the bloodstream to somewhere else in the body where it is needed.

Following the completion of translation, the mRNA strand may remain in the cytoplasm to serve as the template for producing another molecule of the same protein. In bacteria, an mRNA strand may last from a few seconds to more than an hour; in mammals, mRNA may last several days. Depending on how long it lasts, the same mRNA strand may be translated hundreds of times. Eventually, it is broken down by enzymes in the cytoplasm.

TAKE-HOME MESSAGE 5•7

Translation is the second step in the two-step process by which DNA directs the synthesis of proteins. In translation, the information from a gene that has been encoded in the nucleotide sequence of an mRNA is read and ingredients present in the cell's cytoplasm are used to produce a protein.

③ Damage to the genetic code has a variety of causes and effects.

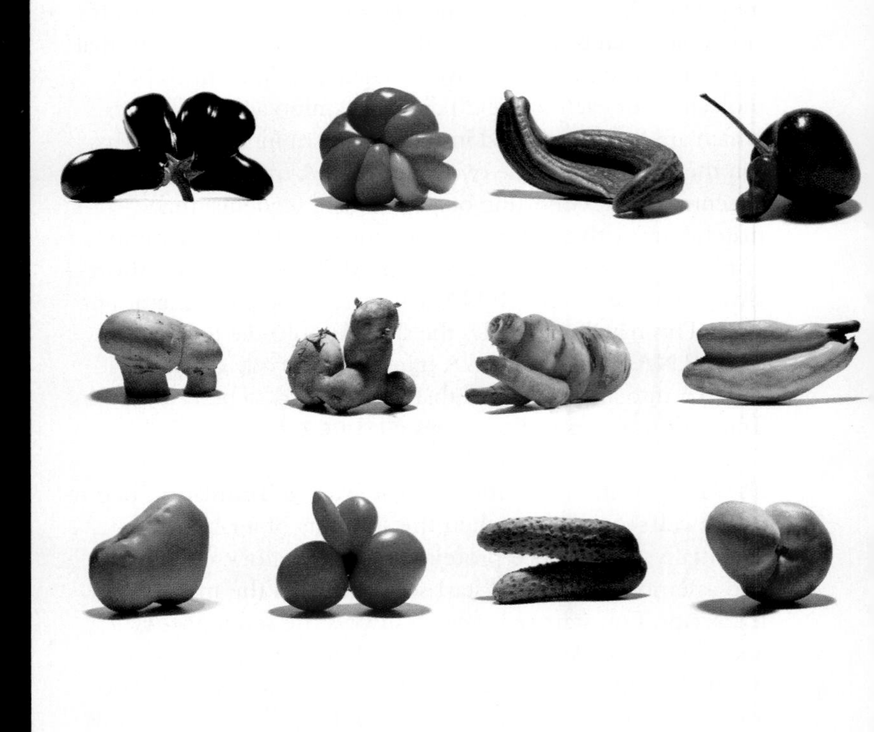

Damage to the genetic code can interfere with normal development.

5•8

What causes a mutation and what are its effects?

Through the two-step process of transcription and translation, an organism converts the information held in its genes into the proteins necessary for life. But the process is only as good as the organism's underlying genetic information. Sometimes, however, something occurs to alter the sequence of bases in an organism's DNA. Such an alteration is called a **mutation,** and it can lead to changes in the structure and function of the proteins produced. Mutations can have a range of effects. Sometimes they result in a serious, even deadly, problem for an organism. Sometimes they have little or no detrimental effect. And, occasionally—but very rarely—they may even turn out to be beneficial to the organism.

As an example of how mutations can affect organisms, consider the case of breast cancer in humans. When two human genes, called BRCA1 and BRCA2, are functioning properly, they help to prevent breast cancer by deterring cells from dividing uncontrollably. If the DNA sequence of either of these genes is altered through mutation and the gene's normal function is lost, the person carrying the gene has a significantly increased risk of developing breast cancer. (Because a variety of other factors, including environmental variables, are involved in development of cancer, it's impossible to know for certain whether these individuals will definitely develop breast cancer.) Currently, more than 200 different changes in the DNA sequences of these genes have been detected, each of which results in an increased risk of developing breast cancer.

Given the havoc they can cause for an organism, it's not surprising that mutations have a bad reputation. After all, because they can change the protein produced, they tend to disrupt normal processes and usually harm the individual (**FIGURE 5-15**). Fortunately, mutations are very, very rare. More surprising, though, is the paradoxical fact that mutations are essential to evolution. Those mutations that don't kill an organism, or reduce its ability to survive and reproduce, can be beneficial. Every genetic feature in every organism was, initially, the result of a mutation. In Chapter 8, we will explore the relationship between mutation and evolution.

FIGURE 5-15 Wreaking havoc. Mutations can change the protein produced by the altered gene, with disastrous consequences.

It's important to note that mutations can occur in an organism's gamete-producing cells (that is, cells that produce sperm or eggs) as well as in its non-sex cells (such as skin cells or cells in the lungs). Mutations in the non-sex cells generally have bad health consequences for the person carrying them. Many forms of cancer, such as lung cancer or skin cancer, result from such mutations. On the upside, though, non-sex-cell mutations are not passed on to your children. Mutations in the sex cells (gametes), on the other hand, do not have any adverse health effects on the person carrying them, but they can be passed on to offspring, with terrible effects such as the induction of a miscarriage or the occurrence of birth defects. Individuals inheriting certain mutations from a parent—because the mutation occurred in the parent's sex cells or the parent inherited the mutation from his or her parent—can be at increased risk for certain diseases such as breast cancer or cardiovascular disease.

Mutations generally fall into two types: point mutations and chromosomal aberrations (**FIGURE 5-16**). In point mutations, one base pair is changed, whereas in chromosomal aberrations, entire sections of a chromosome are altered.

Point mutations are mutations in which one nucleotide base pair in the DNA is replaced with another, or a base pair is inserted or deleted. Insertions and deletions can be much more harmful than substitutions because the amino acid sequence of a protein is determined by reading the three-base sequence (codon) on an mRNA molecule and attaching the particular amino acid specified by that sequence. If a single base is added or removed, the three-base groupings get thrown off and the sequence of amino acids stipulated "downstream" from that point will be all wrong—the reading frame is shifted. It's almost like putting your hands on a computer keyboard, but offset by one key to the left or right, and then typing what should be a normal sentence. It comes out as gibberish.

Chromosomal aberrations are changes to the overall organization of the genes on a chromosome. Chromosomal aberrations are like the manipulation of large chunks of text when you are working on a paper. They can involve the complete deletion of an entire section of DNA, the moving of a gene from one part of a chromosome to elsewhere on the same chromosome or to a different chromosome, or the duplication of a gene, with the new copy inserted elsewhere on the chromosome or on a different chromosome. Whatever the type of aberration, a gene's expression—the production of the protein that the gene's sequence codes for—can be altered when it is moved, as can the expression of the genes around it.

Given the potentially hazardous health consequences of mutations, it is advisable to minimize their occurrence. Can this be done? Yes and no. There are three chief causes of mutation and, although one of them is beyond our control, the other two can be significantly reduced (**FIGURE 5-17**).

1. *Spontaneous mutations.* Some mutations arise by accident as long strands of DNA are duplicating themselves—at the rate of more than a thousand bases a minute in humans—when cells are dividing (you'll read more details on this process

TYPES OF MUTATIONS

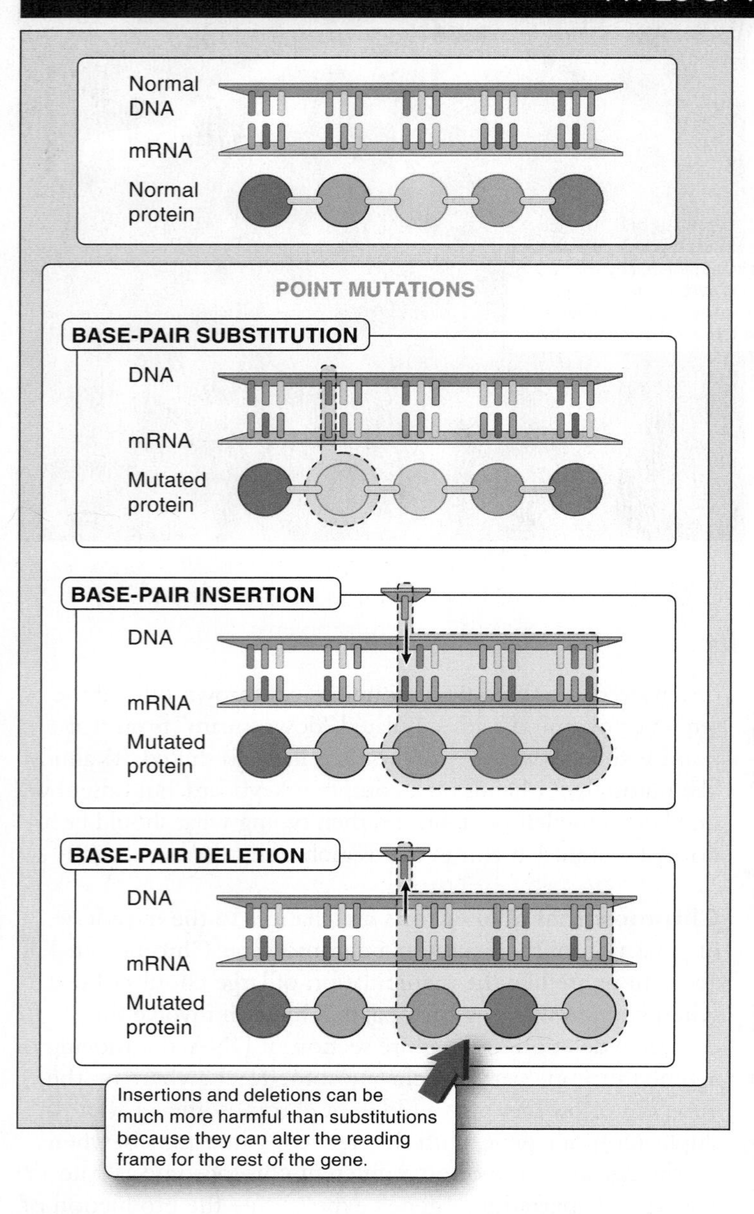

FIGURE 5-16 Point mutations and chromosomal aberrations.

in Chapter 6). Most errors are repaired by DNA repair enzymes, but some still slip by and there's not much we can do about them.

2. *Radiation-induced mutations.* Ionizing radiation is radiation with enough energy to disrupt atomic structure—even breaking apart chromosomes—by removing tightly bound electrons. Sources of ionizing radiation include X rays and the ultraviolet (UV) rays of the sun. When you lie in the sun, for example, its ultraviolet rays can induce a mutation in one of your skin cells that can transform the cell into a cancer cell. This is why long-term sun exposure can contribute to the development of skin cancer. Because ionizing rays cannot pass through lead, the lead apron a doctor or dentist puts over you when you get an X ray protects your body from the ionizing radiation.

Q Why do dentists put a heavy apron over you when they X-ray your teeth?

Another source of dangerous radiation is found in the core of nuclear power plants, where radioactive atoms are used and

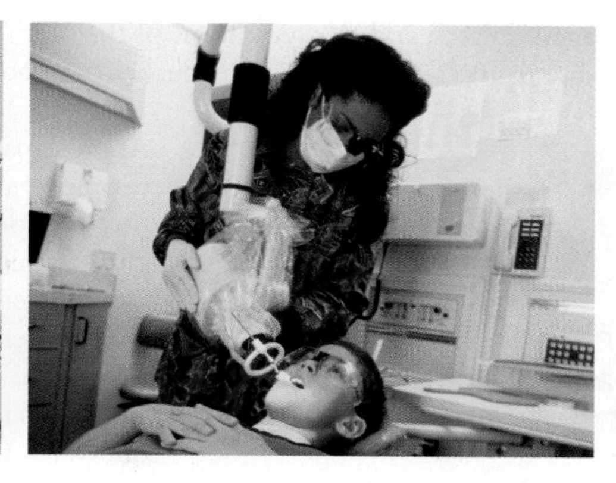

FIGURE 5-17 Gambling with mutation-inducing activities? You can increase or decrease your risk of mutations with your behavior: smoking, sun worshipping, and radiation (but notice the protective lead apron).

Q Why is it dangerous to be near the core of a nuclear power plant?

produced in energy-generating reactions. While it is the high energy of the radioactivity that fuels the production of usable energy, this radioactivity is potentially harmful because it can pass through your body and disrupt your DNA, causing point mutations. Radioactivity can also cause larger-scale chromosomal aberrations. With the proper safety precautions, however, nuclear power plant workers can minimize their exposure to harmful radiation.

3. *Chemical-induced mutations.* Many chemicals, such as those found in cigarette smoke and in exhaust from internal combustion engines, can also react with the atoms in DNA molecules and induce mutations.

In the next section, we examine how even tiny changes in the sequence of bases in DNA can lead to errors in protein production and profound health problems.

TAKE-HOME MESSAGE 5•8

Mutations are alterations in a single base or changes in large segments of DNA that include several genes. They are rare, but when they do occur in a gene this usually disrupts normal physiological functioning. Mutations play an important role in evolution.

5•9

Faulty genes, coding for faulty enzymes, can lead to sickness.

Isabella joins her friends in sipping wine during a dinner party. As the meal progresses, her companions become tipsy. Their conversations turn racy, their moods relaxed. They refill their glasses, reveling in a little buzz. Not so for Isabella. Before her first glass is empty, she experiences a "fast-flush" response: her face turns crimson, her heart begins to race, and her head starts to pound. Worse still, she soon feels the need to vomit.

How can people respond so differently to alcohol? It comes down to a difference in a single base pair in their DNA, a single difference that can influence dramatically a person's behavior, digestion, respiration, and general ability to function. The base-pair change leads to the production of a non-functional enzyme, and the lack of a functional version of this enzyme leads to physical illness. Let's look at the details.

When we consume alcohol, our bodies start a two-step process to convert the alcohol molecules from their intoxicating form into innocuous molecules. Each of the two steps is made possible by a specific enzyme, whose assembly instructions are coded in the DNA of most people.

Fast-flushers like Isabella complete the first step of breaking down alcohol, but cannot complete the breakdown because they carry defective genetic instructions for making aldehyde dehydrogenase, the enzyme that makes possible the second step of the process. A poisonous substance subsequently accumulates, and the symptoms of the "fast-flush" reaction are due to this substance's toxic effects in the body.

Q Why do many Asians have unpleasant experiences associated with alcohol consumption?

Approximately half of the people living in Asia carry a non-functional form of the gene for aldehyde dehydrogenase, a mutation that may confer a greater benefit than harm. In a study of 1,300 alcoholics in Japan, not a single one was a fast-flusher, even though half of all Japanese people are fast-flushers. The minor change in the genetic code that makes alcohol consumption an unpleasant experience may be responsible for the lower incidence of alcoholism among Japanese and other Asian people.

In many other cases—perhaps in the majority of genetic diseases—the link between a particular defective DNA sequence and physical illness is equally direct. Recall from Chapter 3 the case of Tay-Sachs disease. In Tay-Sachs disease, an individual inherits genes with a mutation that causes an inability to produce a critical lipid-digesting enzyme in their lysosomes, the cellular garbage disposals. Because these organelles cannot digest certain lipids, the lipids accumulate, undigested. The lysosomes swell until they eventually choke the cell to death. This occurs in numerous cells in the first few years of life, and ultimately leads to the child's death.

Although the details differ from case to case, the overall picture is the same for many, if not most, inherited diseases. The pathway from mutation to illness includes just four short steps (**FIGURE 5-18**).

1. A mutated gene codes for a non-functioning protein, commonly an enzyme.
2. The non-functioning enzyme can't catalyze the reaction as it normally would, bringing the reaction to a halt.
3. The molecule with which the enzyme would have reacted accumulates, just as half-made products would pile up on a blocked assembly line.
4. The accumulating chemical causes sickness and/or death.

The fact that most genetic diseases involve illnesses brought about by faulty enzymes suggests some strategies for treatment. These include administering medications that contain the normal-functioning version of the enzyme. For instance, lactose-intolerant individuals can consume the enzyme lactase, which for a short while gives them the ability to digest lactose. Alternatively, lactose-intolerant individuals can reduce their consumption of lactose-containing foods to keep the chemical from accumulating, thus reducing the problems that come from its overabundance.

FROM MUTATION TO ILLNESS

1. A mutated gene codes for a non-functioning protein, commonly an enzyme.

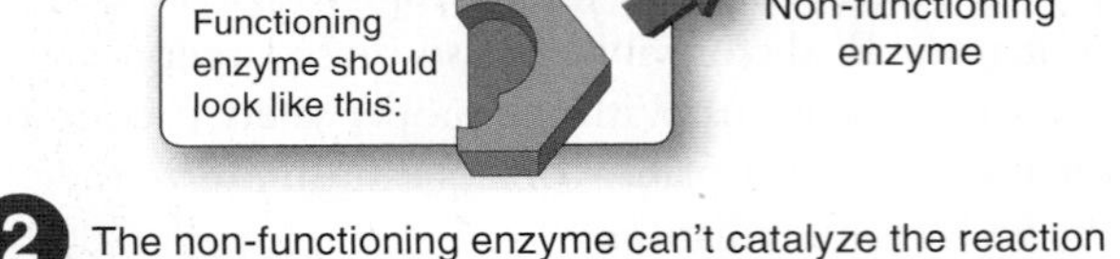

2. The non-functioning enzyme can't catalyze the reaction as it normally would.

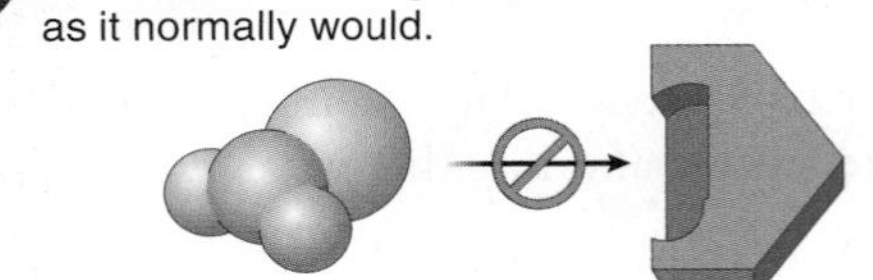

3. The molecule it would have reacted with accumulates.

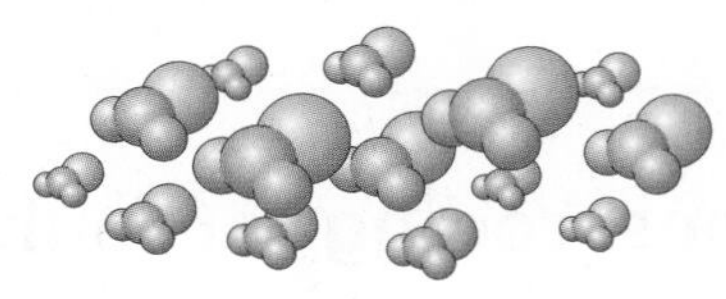

4. The accumulating chemical causes sickness (fast-flushing, for example) or death (Tay-Sachs disease, for example).

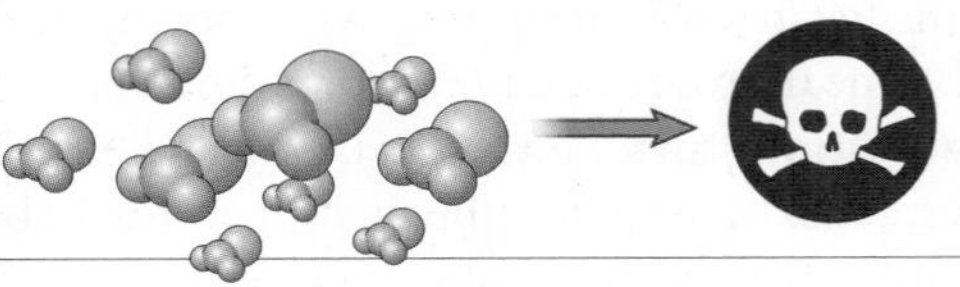

FIGURE 5-18 Faulty enzymes can interfere with metabolism.

TAKE-HOME MESSAGE 5•9

Most genetic diseases result from mutations that cause a gene to produce a non-functioning enzyme, which in turn blocks the functioning of a metabolic pathway.

④ Biotechnology has the potential for improving human health.

A visual representation of the human genome. (Each color represents a different nucleotide.)

5•10

What is biotechnology?

Understanding a phenomenon is nice, but controlling it takes the satisfaction to an entirely new level. This may explain why there is so much excitement surrounding the field of **biotechnology,** in which organisms, cells, and their molecules are modified to achieve practical benefits. As we have seen in this chapter, the structure and function of DNA are pretty well understood. Biotechnology uses this knowledge to develop medicines for fighting viral infections, produce more nutritious food while reducing reliance on pesticides, and analyze forensic evidence as part of criminal investigations.

The modern emphasis in biotechnology is on **genetic engineering,** the manipulation of organisms' genetic material by adding, deleting, or transplanting genes from one organism to another. Such manipulation may create a better-tasting tomato or improve the treatment of a human disease.

This section introduces the most important tools and techniques used in biotechnology. We then explore the two primary areas in which biotechnology is applied: human health and agriculture. Biotech advances in human health fall into three categories: (1) producing medicines to treat diseases, (2) curing diseases, and (3) preventing diseases from occurring in the first place. In agriculture, biotechnology is used to create more nutritious and healthful food as well as to increase the efficiency and reduce the ecological impact of farming (**FIGURE 5-19**).

FIGURE 5-19 Biotechnology in agriculture and human health.

How would you create a plant resistant to being eaten by insects? Or a colony of bacteria that can produce human insulin? Although there are many different uses of biotechnology, it employs a surprisingly small number of processes and tools. Each of these applications, for example, would use a similar sequence of steps and applications (**FIGURE 5-20**).

1. **Chop** up the DNA from a donor organism that exhibits the trait of interest.
2. **Amplify** the small amount of DNA into more useful quantities.
3. **Insert** the different DNA pieces into bacterial cells or viruses.
4. **Grow** separate colonies of the bacteria or viruses, each of which contains a different inserted piece of donor DNA.
5. **Identify** the colonies that have received the DNA containing the trait of interest.

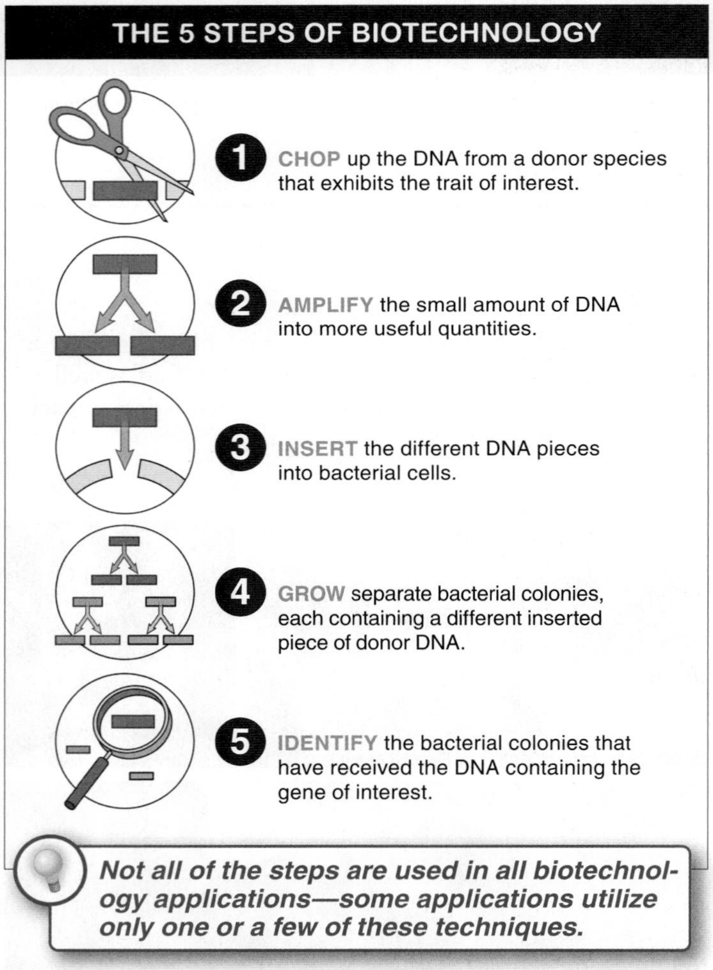

THE 5 STEPS OF BIOTECHNOLOGY

1. CHOP up the DNA from a donor species that exhibits the trait of interest.
2. AMPLIFY the small amount of DNA into more useful quantities.
3. INSERT the different DNA pieces into bacterial cells.
4. GROW separate bacterial colonies, each containing a different inserted piece of donor DNA.
5. IDENTIFY the bacterial colonies that have received the DNA containing the gene of interest.

Not all of the steps are used in all biotechnology applications—some applications utilize only one or a few of these techniques.

FIGURE 5-20 Five important tools and techniques of most biotechnology procedures.

In other words, we can reduce decades of complex research down to five steps: Chop, Amplify, Insert, Grow, and Identify. Then we're ready to utilize the products. This may be as simple as harvesting a product of interest or may require inserting the gene into another species, possibly a crop plant or maybe a sick human. To be sure, it's easier said than done, but these steps capture the essence of most modern biotechnology. Not all of the steps are used in all biotech applications—some use only one or a few of these techniques—nonetheless, they all are mainstays of today's biotechnology world. Now let's explore each step in more detail.

Step 1. Chopping up DNA from a donor organism. To begin, researchers select an organism that has a trait they want to make use of. For example, they might want to produce human growth hormone in large amounts. Their first step would be to obtain a sample of human DNA and cut it into smaller pieces. Cutting DNA into small pieces requires the use of **restriction enzymes** (**FIGURE 5-21**).

Restriction enzymes have a single function: when they encounter DNA, they cut it up into small pieces. These enzymes evolved to protect bacteria from attack by viruses. Immediately upon encountering DNA from an invading virus, a restriction enzyme recognizes and binds to a particular sequence of four to eight bases and cuts the DNA there, thus making it impossible for the virus to reproduce within the bacterial cell. Dozens of different restriction enzymes exist, each of which recognizes a different sequence of bases and cuts the DNA molecule at a different location.

Restriction enzymes are one of the key tools in biotechnology because the first step of nearly all biotech applications involves cutting the DNA from a donor species that has a gene or genes of interest into shorter lengths (without actually cutting anywhere *within* the gene of interest).

Step 2. Amplifying DNA pieces into more useful quantities. In many situations, only a small amount of DNA will be available for analysis or for some other biotechnology use. The **polymerase chain reaction (PCR)** is a laboratory technique that allows a tiny piece of DNA (perhaps one that has been cut from a larger piece by restriction enzymes, or one that has been recovered from a crime scene) to be duplicated repeatedly, producing virtually unlimited amounts of that piece of DNA. PCR is especially useful in forensics, because investigators often find only small amounts of DNA at crime scenes, which must then be amplified before they can be analyzed by using the technique of DNA fingerprinting (which we discuss below).

In the first step of PCR, the DNA of interest is heated, which separates the double-stranded DNA into separate, single strands (**FIGURE 5-22**). Next, the DNA is cooled and an

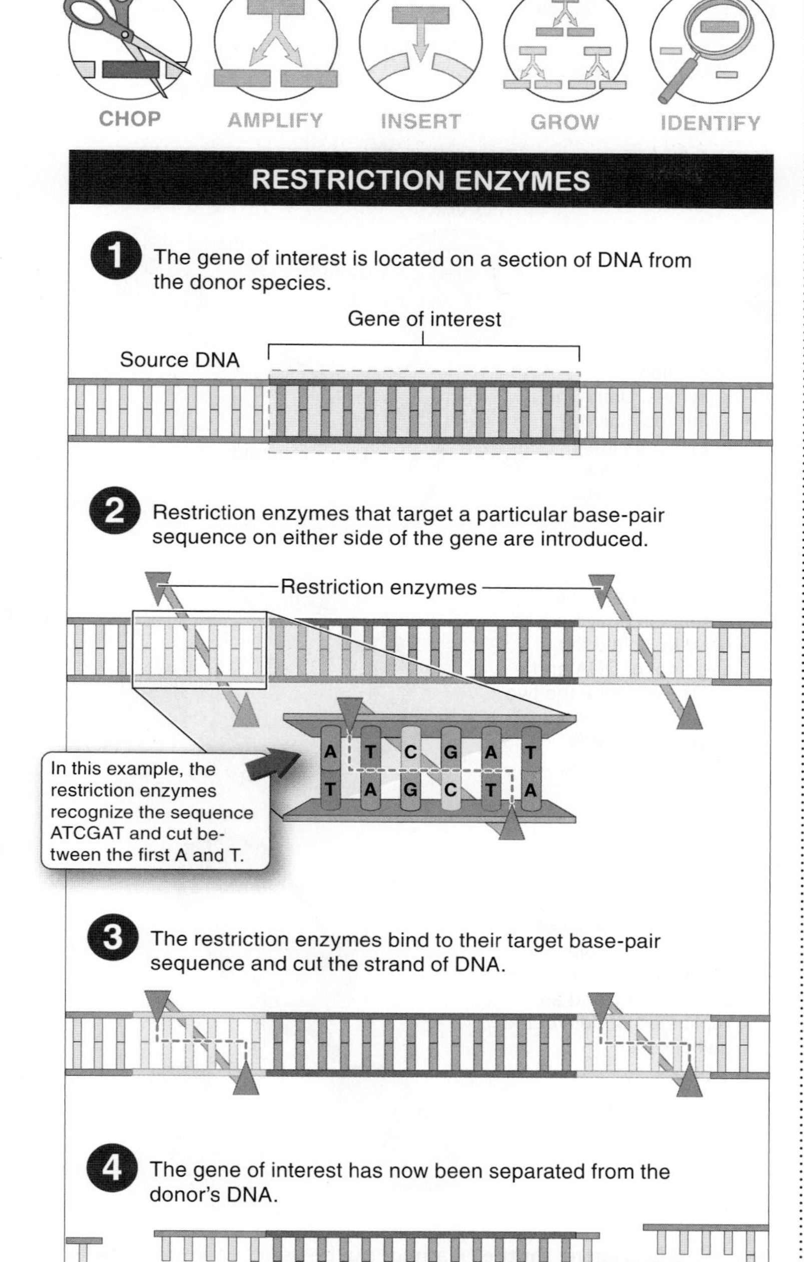

FIGURE 5-21 Restriction enzymes are used to isolate a gene of interest.

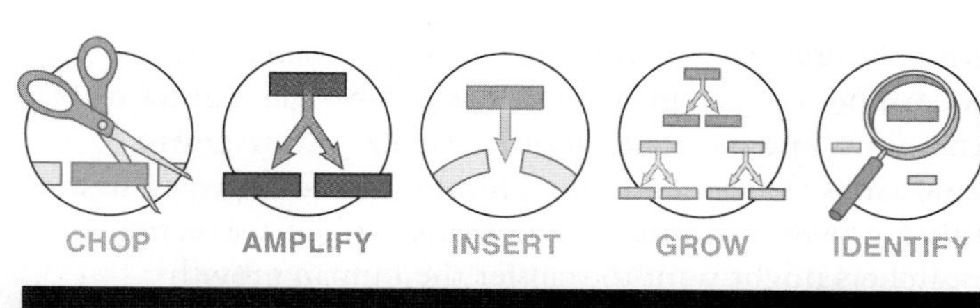

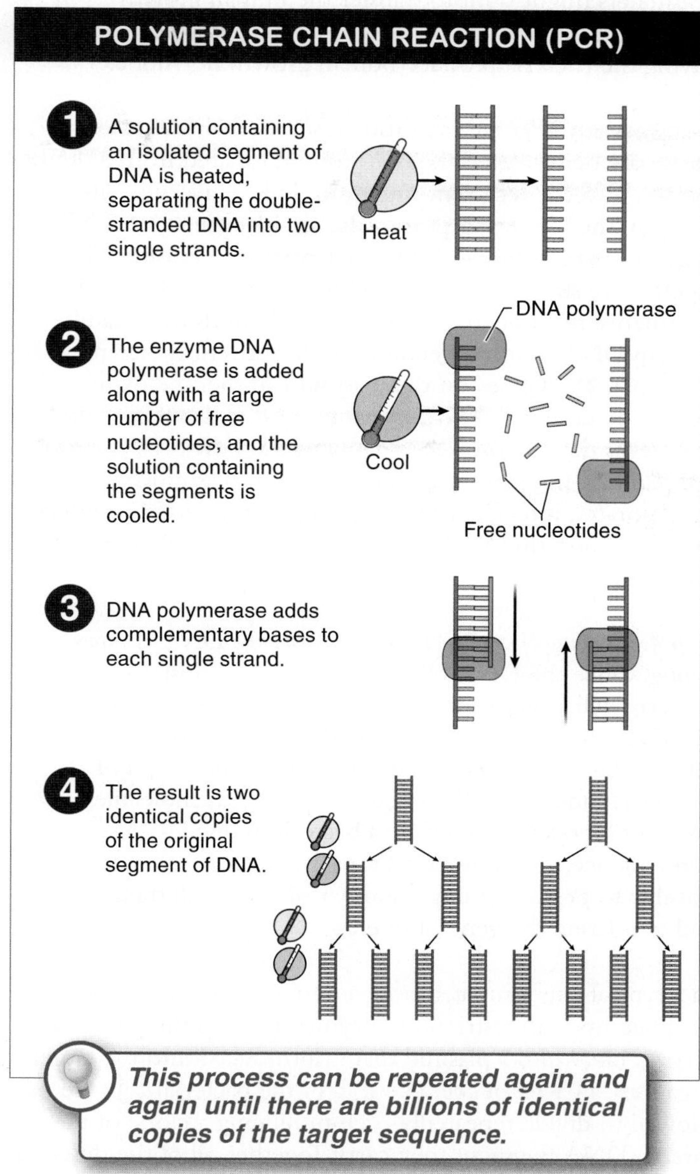

FIGURE 5-22 The polymerase chain reaction can duplicate a small strand of DNA into billions of copies.

enzyme called DNA polymerase and a large number of free nucleotides are added to the DNA mixture. DNA polymerase uses one of the two strands of DNA as a template to build a complementary strand, using the free nucleotides as building blocks. This makes each of the single strands double-stranded again, resulting in two complete copies of the DNA of interest. This process of heating and cooling can be repeated again and again until there are billions of identical copies of the target sequence. Depending on the experiment, PCR can be done so that it will only amplify a DNA fragment with a specific sequence, for example the gene for human growth hormone, which can later be separated from the original DNA. In other experiments, scientists want to amplify all the available DNA fragments so they can be surveyed later on.

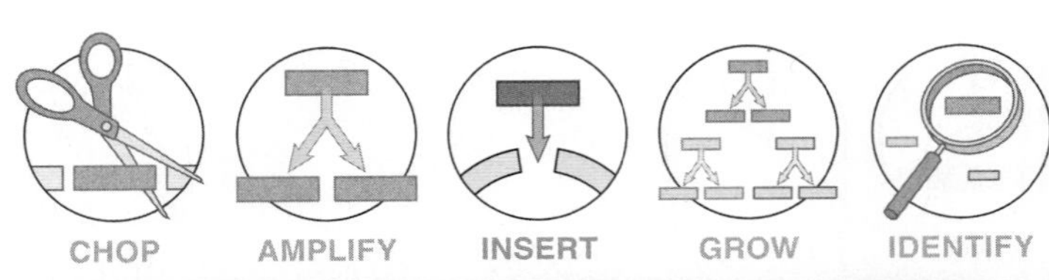

Step 3. Inserting foreign DNA into the target organism. Most modern biotechnology applications involve the transfer of genes from one species to another, creating **transgenic organisms** that contain DNA from the donor species. In the human growth hormone example mentioned above, the researchers might want to transfer the human growth hormone gene into the bacterium *E. coli,* with the goal of having the bacteria produce human growth hormone.

To create a transgenic organism, researchers have to physically deliver the DNA of interest, isolated from a donor species, into the recipient organism. This delivery is often accomplished by using **plasmids,** circular pieces of DNA that can be incorporated into a bacterium's genome. This is an effective strategy because, following certain chemical treatments, researchers can construct plasmids that readily take up DNA that has been cut with restriction enzymes (**FIGURE 5-23**). Genes on the plasmid can then be expressed in the bacterial cell. The gene of interest is also replicated whenever the cell divides, so both of the new cells contain the plasmid. In some cases, rather than in plasmids, genes are incorporated into viruses, which can then be used to infect organisms and thus transfer the genes of interest to those species.

INSERTING DNA BY USING PLASMIDS

1. A target segment of source DNA is isolated using restriction enzymes. Using the same restriction enzyme, a single cut is made in a bacterial plasmid.

Source cell

Bacterial cell

Gene of interest

Plasmid

2. Because the same restriction enzyme was used to isolate the segment of DNA and cut the plasmid, the two share complementary base pairs and fit perfectly together.

3. The plasmid—now including the gene of interest—is inserted back into the bacterial cell. Genes on the plasmid can be expressed in the bacterial cell and are replicated whenever the cell divides.

FIGURE 5-23 A gene shuttle. Plasmids can be used to transfer DNA from one species to another.

Step 4. Growing bacterial colonies that carry the DNA of interest: cloning. Once the DNA of interest has been transferred to the bacterial cell, every time the bacterium divides, it creates a **clone,** a genetically identical cell that contains the DNA of interest. The term **cloning** describes the production of genetically identical cells, organisms, or DNA molecules, a process that occurs each time a bacterium divides. As a consequence, with numerous rounds of cell division, it is possible to produce a huge number of clones, all transcribing and translating the gene of interest.

In a typical experiment, a large amount of DNA may be chopped up with restriction enzymes. Each of the pieces is incorporated into a plasmid that is introduced into the target organism, a bacterial cell. Then, all of the bacterial cells are allowed to divide repeatedly, each producing a clone of the foreign DNA fragment they carry. Together, all of the different bacterial cells containing all of the different fragments of the original DNA are called a **clone library** or a **gene library** (**FIGURE 5-24**).

Step 5. Identifying bacterial colonies that have received the gene of interest. Although a clone library contains a huge amount of potentially useful genetic information, the information is in no particular order, much like a library of books, uncatalogued and in complete disarray. If a researcher is interested in working with bacterial cells that contain the one human gene that produces human growth hormone, how can the bacteria containing that one particular piece of DNA be identified and separated from the millions of other clones in the library?

The gene of interest is usually found by using a method called **hybridization** (**FIGURE 5-25**). Hybridization requires the use of a **DNA probe,** a short sequence of single-stranded DNA that (1) contains a small part of the sequence of the gene of interest and (2) has had some of its nucleotides modified so that they carry radioactive elements. Bacteria that may carry the gene of interest are washed with a chemical that separates

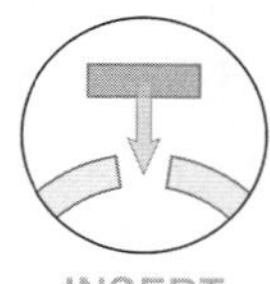

CREATING A GENE LIBRARY

❶ To create a gene library, a large amount of DNA is chopped up using restriction enzymes. ❷ Each piece is inserted into a plasmid that is then introduced into a bacterial cell. ❸ The bacteria are allowed to divide repeatedly, each producing a clone of the foreign DNA fragment it carries. ❹ Together, all of the different bacterial cells contain all of the different fragments of the original DNA.

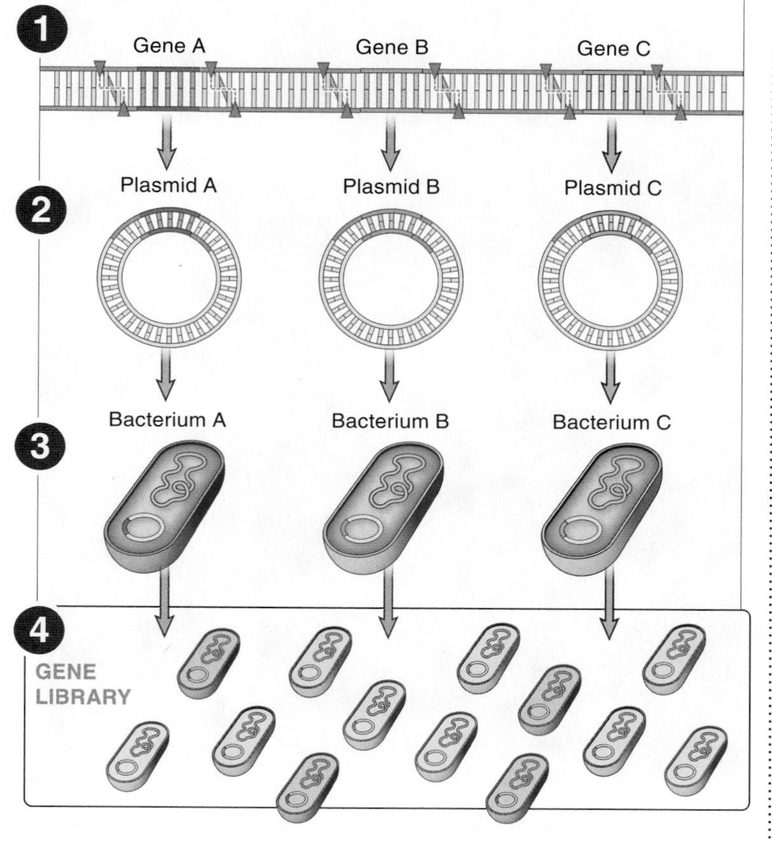

FIGURE 5-24 DNA collection. A gene library (or clone library) is a collection of cloned DNA fragments.

the DNA strands, making single-stranded DNA. Then the radioactive probe is washed over the single-stranded DNA. Bacteria with DNA containing the sequence complementary to the probe will bind to the probe and glow with radioactivity. Once the desired clone in a library is identified, these cells can be grown in large numbers—for example, vats of *E. coli* that produce human growth hormone.

In the remainder of this chapter, we learn how these tools and techniques are being used to develop products. Keep in mind,

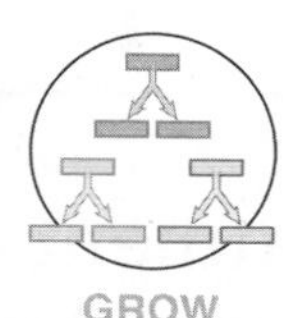

IDENTIFYING A GENE BY USING A DNA PROBE

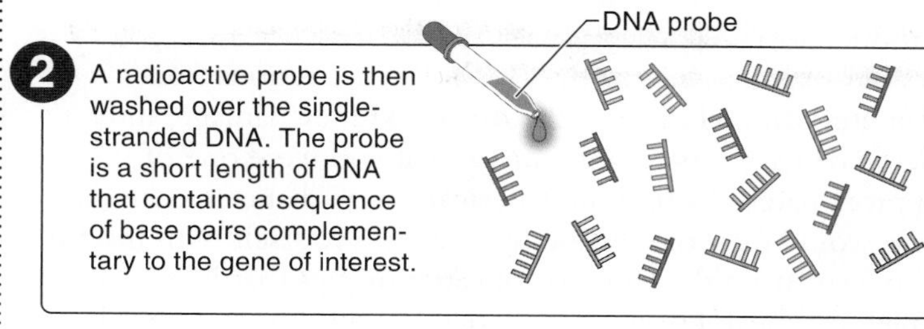

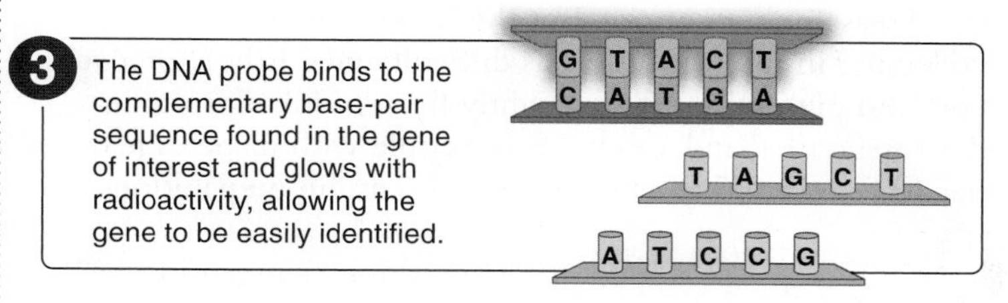

3 The DNA probe binds to the complementary base-pair sequence found in the gene of interest and glows with radioactivity, allowing the gene to be easily identified.

G T A C T
C A T G A
T A G C T
A T C C G

FIGURE 5-25 Finding the needle in a haystack. A DNA probe is used to locate the desired clone in a gene library.

however, that even though biotechnology is frequently in the news and has had many important successes, the field is still in its infancy.

TAKE-HOME MESSAGE 5•10

Biotechnology is the use of technology to modify organisms, cells, and their molecules to achieve practical benefits. Modern molecular methods make it possible to cut and copy DNA from one organism and deliver it to another, not necessarily of the same species. With these methods, biotechnology has achieved some important successes in medicine, agriculture, and forensics, but the field is still in its infancy.

5•11

The treatment of diseases and production of medicines are improved with biotechnology.

You can't always get what you want. In the best of all worlds, biotechnology would *prevent* humans from ever getting debilitating diseases. Next best would be to *cure* diseases once and for all. But these noble goals are not always possible, so biotechnology often is directed at the more practical goal of *treating* diseases, usually by producing medicines more efficiently and more effectively than they can be produced with traditional methods. Biotechnology has had some notable successes in achieving this goal. The treatment of diabetes is one such success story.

Diabetes is a chronic disease in which the body cannot produce insulin, a chemical that allows cells to take up and break down sugar from the blood. Complications from diabetes can include vascular disease, kidney damage, and nerve damage. As recently as 1980, if you were one of the approximately 15 million Americans with diabetes, each day you would treat the disease by injecting yourself with insulin extracted from the pancreas of cattle or pigs that had been killed for meat. For most diabetics, the insulin injections kept the disease under control. But the traditional process of collecting insulin this way was difficult and costly. Moreover, cow and pig insulins differ slightly from human insulin in their structures, and about 5% of people with diabetes had negative reactions to the cow and pig insulin preparations.

Q Why do some bacteria produce human insulin?

Everything changed in 1982 when a 29-year-old entrepreneur, Bob Swanson, joined scientist Herbert Boyer to transform the potential of **recombinant DNA technology,** the combination of DNA from two or more sources, into a product. In doing so, they started the biotech revolution. Working with the scientist Stanley Cohen, Swanson and Boyer used restriction enzymes to snip out the human DNA sequence that codes for the production of insulin. They then inserted this sequence into the bacterium *E. coli.* After cloning the new, transgenic bacteria, the team was able to grow vats and vats of the bacteria, all of which churned out human insulin (**FIGURE 5-26**). The drug could be produced efficiently in huge quantities and made available for patients with diabetes. This was the first genetically engineered drug approved by the Food and Drug Administration and it continues to help millions of people every day.

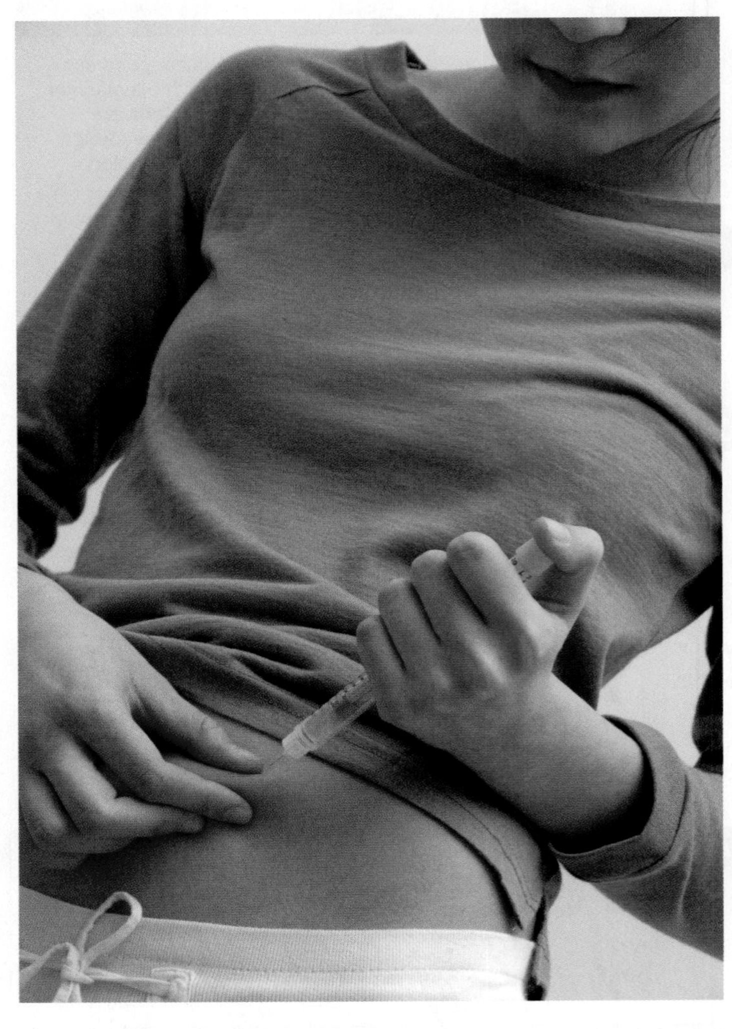

FIGURE 5-26 Life-saving insulin. Human insulin is engineered through recombinant DNA technology.

Perhaps even more significant than providing a better source of insulin, Swanson, Boyer, and Cohen's application of biotechnology revealed a generalized process for genetic engineering. It instantly opened the door to a more effective method of producing many different medicines and treating diseases. Today, more than 1,500 companies work in the recombinant DNA technology industry, and their products generate more than $40 billion in revenues each year.

Several important achievements followed the development of insulin-producing bacteria. Here are just two examples.

1. *Human growth hormone (HGH).* Produced by the pituitary gland, human growth hormone has dramatic effects throughout the body. It stimulates protein synthesis, increases utilization of body fat for energy to fuel metabolism, and stimulates the growth of virtually every part of the body (**FIGURE 5-27**). Insufficient growth hormone production, usually due to pituitary malfunctioning, leads to dwarfism.

When treated with supplemental HGH, individuals with dwarfism experience additional growth. HGH is also used to combat weight loss in people with AIDS. Until 1994, HGH treatment was prohibitively expensive because the growth hormone could only be produced by extracting and purifying it from the pituitary glands of human cadavers. Through the creation of transgenic bacteria, using a technique similar to that used in the creation of insulin-producing bacteria, HGH can now be created in virtually unlimited supplies and made available to more people who need it.

The availability of HGH, which can increase strength and endurance, may be irresistibly tempting to some athletes—even at $7,500 for a month's supply. Recent sporting scandals suggest that the illegal use of HGH occurs frequently among elite swimmers, cyclists, and other athletes.

FIGURE 5-27 Bulking up with a little (illegal) help. Sylvester Stallone—age 61 in this photo—used human growth hormone (along with strength training) to develop larger muscles for a film role.

2. *Erythropoietin.* Produced primarily by the kidneys, erythropoietin (also known as EPO) is a hormone that regulates the production of red blood cells. Numerous clinical conditions (nutritional deficiencies and lung disease, among others) and treatments (such as chemotherapy) can lead to anemia, a lower than normal number of red blood cells, which reduces an individual's ability to transport oxygen to tissues and cells.

Cloned in 1985, recombinant human erythropoietin (rhu-EPO) is now produced in large amounts in hamster ovaries. It is used to treat many forms of anemia. Worldwide sales of EPO exceeded $10 billion in 2004.

Q What is "blood doping"? How does it improve some athletes' performance?

EPO has been at the center of several "blood doping" scandals in professional cycling. Why might a cyclist be interested in having elevated levels of this hormone? EPO increases the oxygen-carrying capacity of the blood, so some otherwise healthy athletes have used EPO to improve their athletic performance. It can be very dangerous, though. By increasing the number of red blood cells, the blood can become much thicker and this can strain the heart and increase the risk of heart attack.

Beyond these and other medicines currently produced via transgenic organisms, plans are under way to create a variety of other useful products—including potatoes that produce antibodies enabling more effective response to illnesses, and other food plants that fight tooth decay. Remember, however, that putting human genes in bacteria or other organisms makes it possible to harness the gene products to treat symptoms of diseases, but it doesn't cure or prevent the diseases. In the next section we examine the as yet less successful attempts to *cure* diseases through biotechnology.

TAKE-HOME MESSAGE 5•11

Biotechnology has led to some notable successes in treating diseases, usually by producing medicines more efficiently and more effectively than they can be produced with traditional methods.

5•12

Gene therapy: the use of biotechnology to cure diseases is having limited success.

When it comes to curing a disease by using biotechnology, there is good news and bad news. The good news is that, in the 1990s, a handful of humans with a normally fatal genetic disease called severe combined immunodeficiency disease (SCID) were completely cured through the application of biotechnology. The bad news is that these promising techniques could not be applied to other diseases.

It's not for a lack of trying: there have been more than 500 other clinical trials for **gene therapies** designed to treat or cure a variety of diseases by inserting a functional gene into an individual's cells to replace a defective version of the gene. These trials have included attempts to cure cystic fibrosis, hemophilia, and an inherited form of heart disease, among many others. But there have been no clear successes. Not one. Moreover, recent evidence has even diminished the good news about the SCID treatment, which has been suspended following the deaths of two patients from illness related to their treatment.

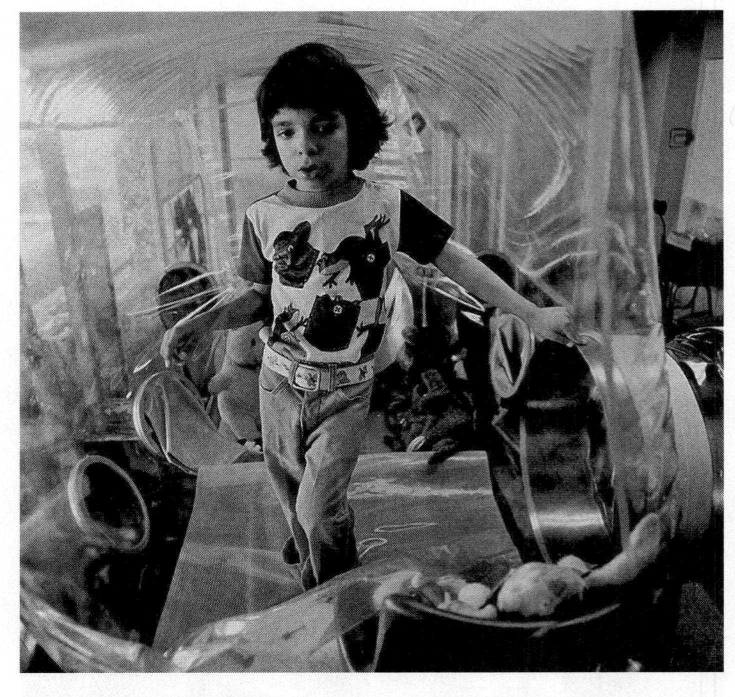

FIGURE 5-28 Protected by a bubble. This child with severe combined immunodeficiency lived 12 years in a protective bubble at Texas Children's Hospital.

The case of SCID is worth examining in some detail, however, because it proves that biotechnology can indeed be used to exchange working genes for malfunctioning disease-causing genes in a way that permanently cures a disease in a human individual.

SCID is a genetic disorder in which a baby is born with an immune system unable to properly produce a type of white blood cell, leaving the infant vulnerable to most infections. The disease usually causes death within the first year of life. It is sometimes called the "bubble boy" disease because in the 1970s, a boy suffering from SCID survived for 12 years by living in a sterile, germ-free bubble (**FIGURE 5-28**).

Q Why has gene therapy had such a poor record of success in curing diseases?

Researchers have been able to cure the disease in some individuals by using gene therapy that is simple in concept but difficult in practice. First, the researchers removed some cells from an affected baby's bone marrow. These were **stem cells,** cells that have the ability to develop into any type of cell in the body—a feature which makes them very useful in biotechnology. In bone marrow, they normally produce white blood cells, but in individuals with SCID, a malfunctioning gene disrupts normal white blood cell production. Next, in a test tube, the bone marrow stem cells were infected with a transgenic virus carrying the functioning gene. If all went well, the virus inserted the good gene into the DNA of the stem cells, which were then injected back into the baby's bone marrow. There, the cells could produce normal white blood cells, thereby permanently curing the disease.

Difficulties with gene therapy have been encountered in several different areas, usually related to the organism used to transfer the normal-functioning gene into the cells of a person with a genetic disease.

1. *Difficulty getting the working gene into the specific cells where it is needed.* For example, gene therapy for cystic fibrosis—a disease in which thick mucus accumulates in the lungs—attempts to use viruses like those responsible for the common cold to insert a normal gene into malfunctioning lung cells. Unfortunately, because the lungs have evolved to keep foreign objects out, the body is very effective at mounting immune responses against the viruses used to deliver the working gene to the lung cells.

2. *Difficulty getting the working gene into enough cells and at the right rate to have a physiological effect.* In the two deaths related to gene therapy for SCID, for example, the treatment apparently caused leukemia due to problems with the insertion of the gene disrupting normal functioning in white blood cells, triggering them to become cancerous.

3. *Problems with the transfer organism getting into unintended cells.* Viruses can cause serious side effects if they infect cells other than the target cells. Their delivery to target cells must therefore be precise.

Beyond these common problems, gene therapy faces additional difficulties because, for many diseases, the malfunctioning gene has not been identified or the disease is caused by more than one malfunctioning gene. And finally, it is important to keep in mind that gene therapy targets cells in the body other than sperm and eggs. Consequently, while the disease can, in theory, be cured in an individual, the techniques do not prevent those individuals from passing on the disease-causing gene(s) to their offspring. It's not clear what the future holds for gene therapy, but many scientists and patients remain hopeful and a great deal of research is still in progress.

TAKE-HOME MESSAGE 5•12

Gene therapy has had a poor record of success in curing human diseases. This stems primarily from technical difficulties in transferring normal-functioning genes into the cells of a person with a genetic disease.

5•13

Biotechnology can help prevent disease.

Would you want to know? Once, this was just a hypothetical question: if you carried a gene that meant you were likely to develop a particular disease later in your life, would you want to know about it? Or another question: would you want to know if there's a good chance that your future children will be born with a genetic disease? Now, for better or for worse, these are becoming real questions that we all must address. And there is more at stake than simply peace of mind. As biotechnology develops the tools to identify some of the genetic time-bombs that many of us carry, it also carries the danger that such information may become the basis for greater discrimination than we have ever known.

Intervening to prevent diseases by using biotechnology focuses on answering questions such as those posed above at three different points in time.

1. *Is a given set of parents likely to produce a baby with a genetic disease?* Many genetic diseases occur only if an individual inherits two copies of the disease-causing gene, one from each parent. This is true for Tay-Sachs disease, cystic fibrosis, and sickle-cell anemia, among others. Individuals with only a single copy of the disease-causing gene never fully show the disease, but they may pass on the disease gene to their children. Consequently, two healthy parents (that is, having no disease symptoms) may produce a child with the disease. In these cases, it can be beneficial for the parents to be screened to determine whether they carry a disease-causing copy of the gene. Such screening, combined with genetic counseling and testing of embryos following fertilization, can reduce the incidence of the disease dramatically. This has been the case with Tay-Sachs disease, for example. Since screening began in 1969, the incidence of Tay-Sachs disease has been reduced by more than 75% (**FIGURE 5-29**).

FIGURE 5-29 **Genetic screening can determine the presence of the Tay-Sachs gene.**

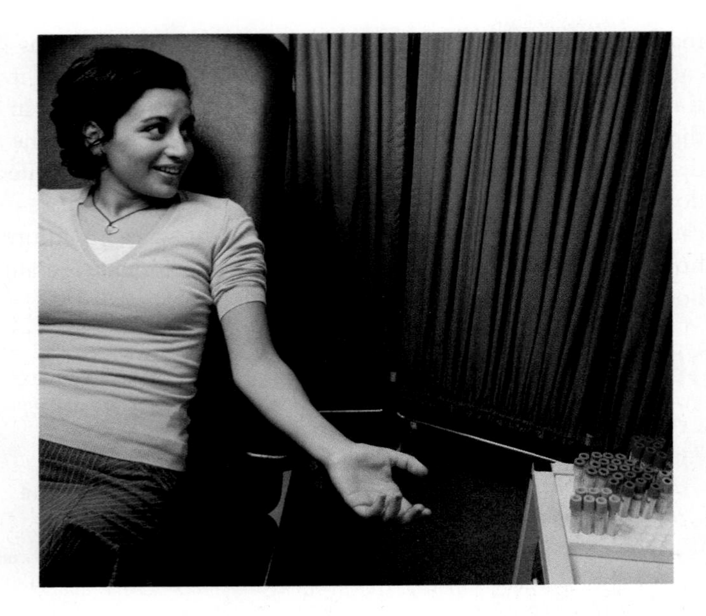

Possessing mutations in the genes called BRCA1 and BRCA2 (short for breast cancer 1 and breast cancer 2) can make women more susceptible to breast cancer. A genetic screening for the gene is done with a simple blood test.

2. *Will a baby be born with a genetic disease?* Once fertilization has occurred, it is possible to test an embryo or developing fetus for numerous genetic problems. Prenatal genetic screening can detect cystic fibrosis, sickle-cell anemia, Down syndrome, and other disorders. The list of conditions that can be detected is growing quickly.

To screen the fetus, doctors must sample some of the fetal cells and/or the amniotic fluid, which carries many chemicals produced by the developing embryo. This is usually done via amniocentesis or chorionic villus sampling (CVS), techniques that we explore in detail in Chapter 6. Once collected, the cells can be analyzed by a variety of means.

3. *Is an individual likely to develop a genetic disease later in life?* DNA technology can also be used to detect disease-causing genes in individuals who are currently healthy but are at increased risk of developing an illness later. Early detection of many diseases, such as breast cancer, prostate cancer, and skin cancer, can greatly enhance the ability to treat the disease and reduce the risk of more severe illness or death.

While biotechnology may reduce suffering and the incidence of diseases, these potential benefits come with significant potential costs, including ethical dilemmas. People who have a gene that puts them at increased risk of developing a particular disease, for example, might be discriminated against, even though they are not currently sick and may never suffer from the particular disease for which they are at heightened risk. Although a federal Genetic Information Nondiscrimination Act was signed into law in 2008, the law does not cover all forms of genetic discrimination. It does not, for example, cover life insurance, disability insurance, and long-term care insurance. Insurance companies have already cancelled or denied such insurance coverage on discovering that an individual carries a gene that puts them at increased risk of disease. Additionally, parents who discover that their developing fetus will develop a painful, debilitating, or fatal disease soon after birth are confronted with the difficult question of how to proceed.

TAKE-HOME MESSAGE 5•13

Using biotechnology, many tools have been developed that enable us to identify whether a given set of parents is likely to produce a baby with a genetic disease, whether a baby is likely to be born with a genetic disease, and whether an individual carries certain disease-causing genes that may have their effect later in life. These tools can help us to reduce suffering and the incidence of diseases, but they also come with significant potential costs, particularly the risk of discrimination.

⑤ Biotechnology is producing improvements in agriculture.

Genetically modified corn growing in a lab.

5•14

More nutritious and better foods are being produced with biotechnology.

Your breakfast cereal is probably fortified with vitamins and minerals. And for snacking you may have switched from candy bars to protein bars that have as much protein as a full chicken breast. It shouldn't come as a surprise, then, that farmers have begun using biotechnology to improve on the natural levels of vitamins, minerals, and other nutrients in the fruits, vegetables, and livestock they produce.

Using our knowledge of biology for agricultural improvements is nothing new. For thousands of years humans have been practicing a relatively crude and slow—but still very effective—form of genetic engineering by carefully selecting which plants or animals they use as the breeders for maintaining their crop or animal populations. Through this process of selecting plants or animals with certain desirable traits and breeding the organisms that have them, farmers and ranchers have improved virtually every product of agriculture, producing meatier turkeys, seedless watermelons, and big, juicy corn kernels (**FIGURE 5-30**). But what used to take many generations of breeding can now be accomplished in a fraction of the time, using recombinant DNA technology. Agriculture has entered into exciting but unknown territory.

FIGURE 5-30 Ancestral corn and modern corn—selected for larger, juicier kernels—compared.

The process begins with the identification of a new characteristic, such as larger size or faster ripening time, that farmers would like in a particular crop. Traditionally, plant breeders would then search for an organism within the same species that had the desirable trait, breed it with their crop organisms, and hope that the offspring would express the trait in the desired way. With recombinant DNA technology, the desired trait can come from *any* species so the pool of organisms from which the trait can be taken becomes much larger. Organisms produced with recombinant DNA technology are referred to in the media as genetically modified organisms, or GMOs.

Although the rewards are potentially huge, in practice the process of creating transgenic species that are more nutritious or have other desirable traits turns out to be difficult. Nonetheless, the results so far hint at a rich and fruitful marriage of agriculture and technology.

Q How might a genetically modified plant help 500 million malnourished people?

Nutrient-Rich "Golden Rice." Almost 10% of the world's population suffers from vitamin A deficiency, which causes blindness in a quarter-million children each year and a host of other illnesses in people of all ages. These nutritional problems are especially severe in places such as southern Asia and sub-Saharan Africa, where rice is a staple of most diets. Addressing this global health issue, researchers have developed what may be the model for solving problems with biotechnology. It involves the creation of a new crop called "golden rice."

Mammals generally make vitamin A from beta-carotene, a substance found in abundance in most plants (it's what makes carrots orange). Beta-carotene is also found in rice plants, but not in the edible part of the rice grains. Researchers set out to change this by inserting into the rice genome three genes that code for the enzymes used in the production of beta-carotene: two genes from the daffodil plant and one bacterial gene (**FIGURE 5-31**). It's clear that the transplanted genes are working because the normally white rice grain takes on a golden color from the accumulated beta-carotene. The rice doesn't yet supply a full-day's requirement of vitamin A in one serving, but it does provide a significant amount. Since golden rice was first developed in 1999, new lines have been produced, now having only two transplanted genes, yet producing almost 25 times the vitamin A found in the original strains. Field tests of golden rice are still under way, so it is not yet being used widely. But this is viewed as one of the most promising applications yet of biotechnology.

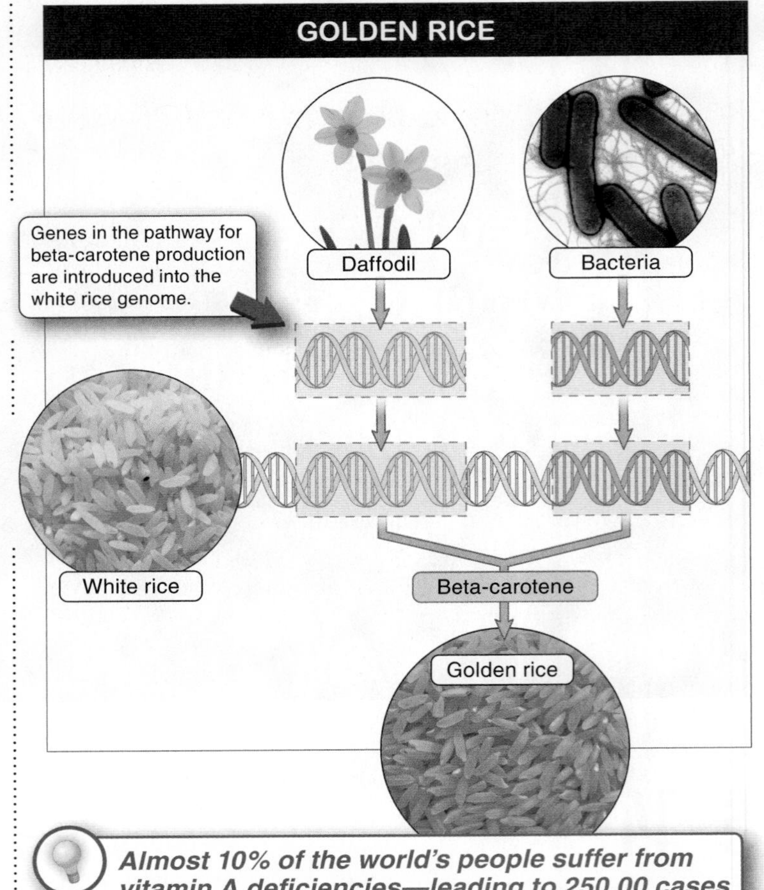

FIGURE 5-31 The potential to prevent blindness in 250,000 people each year. Engineering rice to prevent blindness by increasing its vitamin A content.

TAKE-HOME MESSAGE 5•14

Biotechnology has led to important improvements in agriculture by using transgenic plants and animals to produce more nutritious food.

5•15

Farming can be made more efficient and eco-friendly through biotechnology.

> Italians come to ruin most generally in three ways, women, gambling, and farming. My family chose the slowest one.
>
> —Pope John XXIII

We are in the midst of a revolution—a green revolution—and few people are aware of it. The possibilities of biotechnology playing a role in treating, curing, and preventing human diseases may garner all the headlines, but the actual results pale in comparison with the current impact of biotechnology on agriculture. And although biotechnology can improve agriculture by making more nutritious food (as in the case of golden rice), the real revolution is the extent to which biotechnology has reduced the costs, both environmental and financial, of producing the plants and animals we eat.

The extent of biotechnology's success in the area of producing plants with built-in insecticides and genetically engineered resistance to herbicides is staggering. Currently, more than 170 million acres worldwide are planted with genetically modified crops, two-thirds of which are in the United States, representing more than a fortyfold increase in the past 10 years. The financial benefits to farmers—at least in the short run—are so great that more and more of them are embracing the genetically modified crops.

Q What genetically modified foods do most people in the U.S. consume (usually without knowing it)?

The numbers are surprising: 45% of all corn grown in the United States—including corn grown in every single state in the continental United States—is genetically modified. Seventy-six percent of all cotton grown is genetically modified. And 85% of all soybeans grown is genetically modified (**FIGURE 5-32**). Two factors explain much of the rapid expansion and extensive adoption of genetically modified plants in U.S. agriculture. (1) Many plants have had insecticides and herbicide resistance engineered into them, which can reduce the amounts of insecticides and herbicides used in agriculture. (2) Because herbicide- and insect-resistant plants can reduce the amount of plowing required around crops to remove weeds,

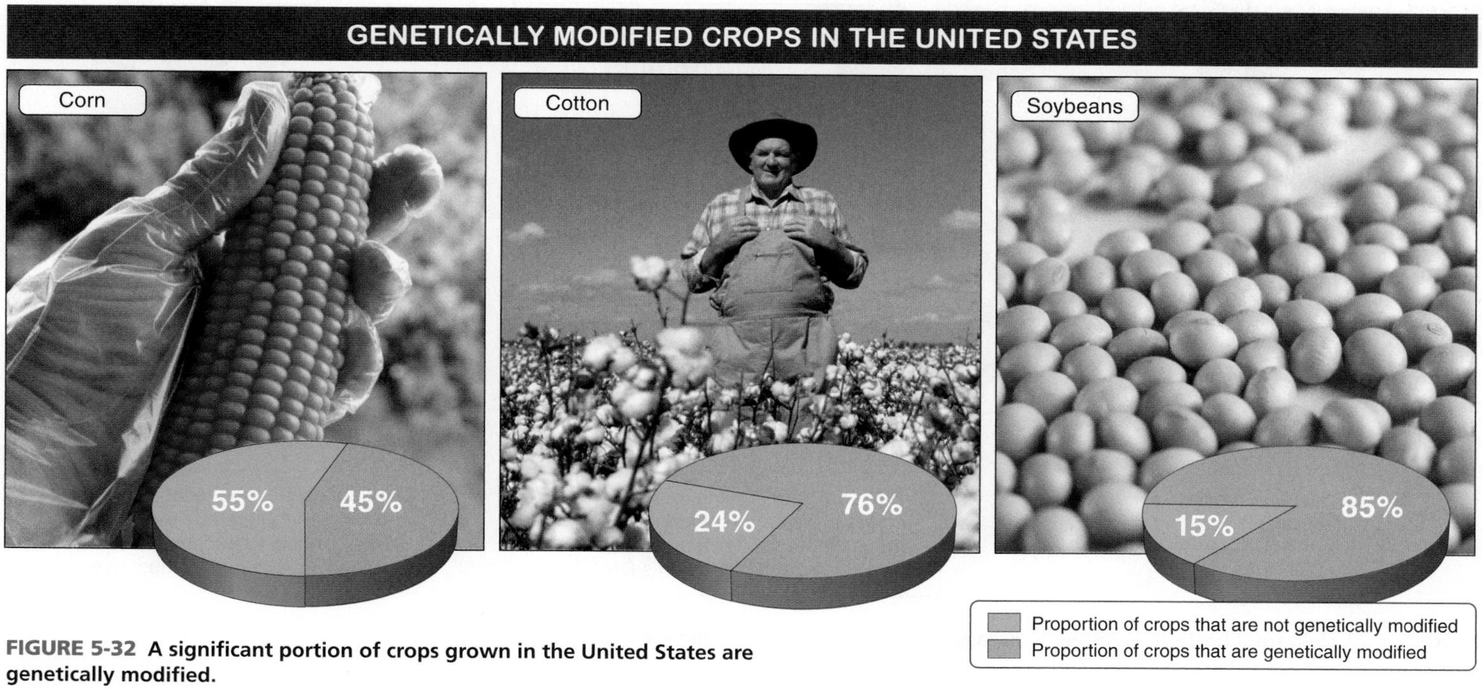

FIGURE 5-32 A significant portion of crops grown in the United States are genetically modified.

FIGURE 5-33 Competing with us for food. Insects can ravage crops.

genetically modified plants can reduce the costs of producing food and the loss of topsoil to erosion.

Several of the biggest successes of the application of recombinant DNA technology to agriculture are described below.

Insect Resistance Insect pests have a field day on agricultural crops (**FIGURE 5-33**). Crops that are planted at high densities and nurtured with ample water and fertilizer represent a huge potential food resource for insects. Every year, about 40 million tons of corn are unmarketable as a consequence of insect damage. Increasingly, however, farmers have been enjoying greater success in their battles against insect pests, primarily through the use of genetically modified crops.

Farmers owe much of this success to soil-dwelling bacteria of the species *Bacillus thuringiensis.* These bacteria produce spores containing crystals that are highly poisonous to insects but harmless to the crop plants and to people. Within an hour of ingesting "Bt" crystals, the insects' feeding is disrupted. The toxic crystals cause pores to develop throughout the insects' digestive systems, paralyzing their gut and making them unable to feed. Within a few days the insects die from a combination of tissue damage and starvation.

> **Q** How can genetically modified plants lead to reduced pesticide use by farmers?

Beginning in 1961, the toxic Bt crystals were included in the pesticides sprayed on crop plants. In 1995, however, recombinant DNA technology led to a huge reduction in the application of this toxic spray. The gene coding for the production of the Bt crystals was inserted directly into the DNA of many different crop plants, including corn, cotton, and potatoes, so that the plants themselves produced the crystals. As a consequence, it is no

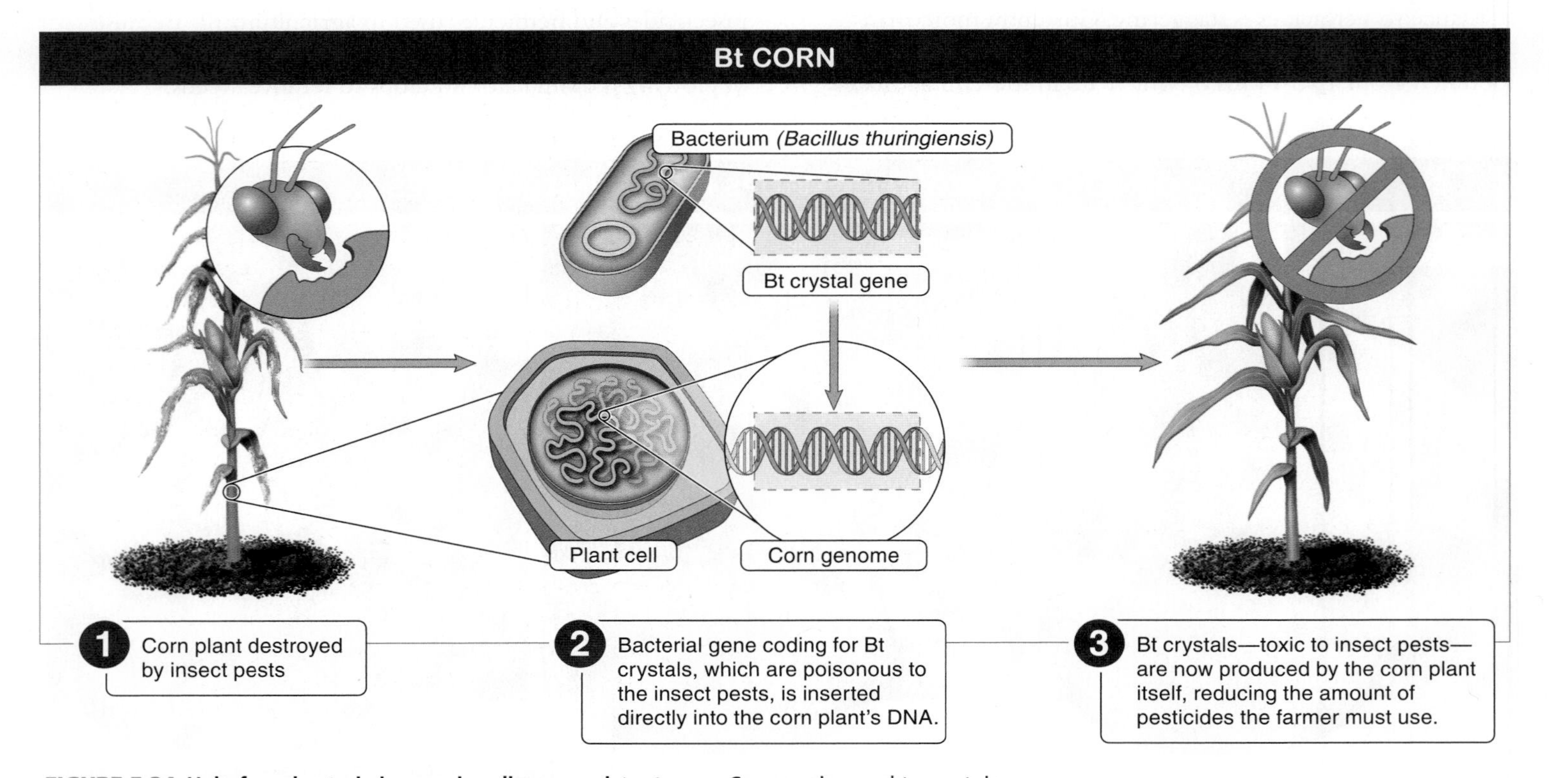

FIGURE 5-34 Help from bacteria in growing disease-resistant corn. Corn engineered to contain spores of the bacterium *Bacillus thuringiensis* (Bt) kills insect pests but does not harm humans.

longer necessary for farmers to apply huge amounts of Bt-containing pesticides (**FIGURE 5-34**). Instead, the plants do that work themselves. The insects that try to eat the genetically modified plants ingest the toxin and soon die. There is no evidence that Bt crystals have any harmful effects on humans at all, even when they are exposed to very high levels. Because pesticides can be damaging to the environment, particularly when they flow from the farm and into the water supply as runoff, reduction in pesticide use can be beneficial for the environment.

Herbicide Resistance Bacteria have come to the aid of farmers in their fight against pests in another way, too. Consider a seemingly impossible challenge that farmers face when they attempt to kill weeds that harm crop plants by competing for light, water, and soil nutrients. Herbicides, chemicals that kill plants, can be applied to kill the weeds (**FIGURE 5-35**). These chemicals usually work by blocking the action of an enzyme that enables plants (as well as fungi and bacteria) to build three critical amino acids. Without these amino acids, the organisms soon die. The problem is that because herbicides affect *all* plants similarly, they are generally toxic to the crop plant as well as the weeds. This is where bacteria come in.

In the 1990s, researchers discovered bacteria that can resist the effect of herbicides. The gene that gives the bacteria resistance to herbicides was identified and introduced into crop plants. Integration of this gene into the plants' DNA gives them resistance to the herbicides and allows farmers to kill weeds with herbicides without harming the crop plants, greatly increasing yields.

FIGURE 5-35 Crop duster. Herbicides like the one applied by this crop duster must kill weeds while leaving the crop unharmed.

FIGURE 5-36 Bigger salmon. The larger salmon carries a growth hormone gene that keeps it growing year-round rather than in the summer only.

Faster Growth and Bigger Bodies Agriculture includes the cultivation not just of plants but also of animals. Currently, researchers are developing transgenic salmon that grow significantly faster and reach a much larger size than normal salmon (**FIGURE 5-36**). These salmon carry a version of the growth hormone gene that functions year round, rather than primarily in the summer. It was isolated from a cold-water fish species, called arctic pout, and injected into the egg of a salmon with genetic engineering techniques. The super-fish can be raised much more quickly and using significantly less feed than normal salmon, reaching market size within 18 months rather than the usual 24–30 months. In taste tests, consumers cannot tell the difference between the transgenic and non-transgenic salmon.

Quite troubling, though, is the fact that researchers expected growth rate increases of about 25% but found that the genetically modified fish grew about 500% faster. Additionally, if the larger, faster-growing fish were to escape from their breeding nets back into their natural habitat—something that experts agree is inevitable—the fish might harm populations of other species because they can consume more resources and may grow too large to be consumed by their natural predators. It is unclear what the outcome would be.

TAKE-HOME MESSAGE 5•15

There has been a revolution in the extent to which biotechnology has reduced the environmental and financial costs of producing food, through the creation of herbicide-resistant and insect-resistant crops.

5•16

Fears and risks: are genetically modified foods safe?

Chickens without feathers look ridiculous (**FIGURE 5-37**). But such a genetically modified breed was developed with a valuable purpose in mind: "naked" birds are easier and less expensive to prepare for market, benefiting farmers by lowering their costs and benefiting consumers by lowering prices. Such chickens, however, turned out to be unusually vulnerable to mosquito attacks, parasites, and disease, and ultra-sensitive to sunlight. They also have difficulty mating, because the males are unable to flap their wings. Researchers currently are working to address these problems.

These chickens teach us an important lesson about genetically modified plants and animals. Although the new breed of featherless chickens was produced by traditional animal husbandry methods—the cross-breeding of two different types of chickens—rather than by recombinant DNA technology, the new breed ended up having not just the desired trait of no feathers but also some unintended and undesirable traits. Now, as more genetically modified foods are created using modern methods of recombinant DNA technology, the same risks of unintended and potentially harmful traits occurring must be weighed. For these and other reasons—some legitimate and rational, others irrational—many people have concerns about the production and consumption of genetically modified foods (**FIGURE 5-38**).

Featherless birds are cheaper for farmers and consumers. But there are unintended consequences, including vulnerability to mosquitoes and other parasites.

FIGURE 5-37 "Naked" birds. The breeding of featherless chickens has benefited farmers but led to some unanticipated consequences.

FIGURE 5-38 Consumer fears. Protesters voice opposition to the use of genetically modified organisms (GMOs).

Fear 1. Organisms that we want to kill may become invincible. Weed-resistant canola plants were cultivated in Canada, making it possible for farmers to apply herbicides freely to kill the weeds but not the canola plants. But the weed-resistant canola plants accidentally spread to neighboring farms and grew out of control, because traditional herbicides could not kill them. Similarly, there is the possibility that insect pests will develop resistance to the Bt that genetically modified crops produce, which will also make these pests resistant to Bt pesticides applied to crops that are not genetically modified.

Fear 2. Organisms that we don't want to kill may be killed inadvertently. Monarch butterflies feed on milkweed plants. Recent research has demonstrated that if pollen from plants genetically modified to contain the insect-killing Bt genes accidentally lands on milkweed plants and is consumed by monarch butterflies, the butterflies can be killed, which may significantly reduce their populations. Although such an incident has not occurred outside experimental fields, it illustrates a risk that may be hard to control.

Fear 3. Genetically modified crops are not tested or regulated adequately. It is impossible to ever really know whether a new technology has been tested adequately. Still, scientists and lawmakers have been working toward an organized and responsible set of policies designed to ensure sufficient safety testing is done. For example, laboratory procedures for working with recombinant DNA have been established, and researchers have developed techniques that make it impossible for most genetically engineered organisms to survive outside the specific conditions for which they are developed.

As an example of the degree of testing used in the field of genetically engineered foods, the Monsanto Company has had its strain of herbicide-resistant soybeans evaluated and approved by 31 different regulatory agencies in 17 different countries, including, in the United States, the Department of Agriculture, the Food and Drug Administration, and the Environmental Protection Agency. In a recent report on genetically modified animals, however, an expert committee from the U.S. National Academy of Sciences warned that genetically modified organisms still pose risks that the government is unable to evaluate. Technology is moving so fast that it is difficult to even know what the new risks might be.

> "And he gave it for his opinion, 'that whoever could make two ears of corn, or two blades of grass, to grow upon a spot of ground where only one grew before, would deserve better of mankind, and do more essential service to his country, than the whole race of politicians put together.'"
>
> —Jonathan Swift, *Gulliver's Travels*, 1726

Fear 4. Eating genetically modified foods is dangerous. In the 1990s, a gene from Brazil nuts was used to improve the nutritional content of soybeans. The genetically modified soybeans had better nutritional content, but they also acquired some allergy-causing chemicals previously present in the Brazil nuts but not in soybeans. This outcome illustrates the risk that some unwanted features might be passed from species to species in the creation of transgenic organisms. In this case, all of the genetically modified soybeans were destroyed and this research program was suspended. To date, no evidence has appeared to suggest that consumption of any genetically modified foods is dangerous.

Fear 5. Loss of genetic diversity among crop plants is risky. As increasing numbers of farmers stop using non-genetically modified crops in favor of one or a few genetically modified strains of crops, there is a reduction in the genetic diversity of the crops. This can make them more vulnerable to environmental changes or pests. The Irish Potato Famine is an example of the value of genetic diversity in crops. In the mid-1800s, much of the population of Ireland depended on potatoes. Because most of the potato crops had been propagated from cuttings from the same plant, they were all genetically the same. When the crops were infected by a rot-causing mold, all of the potato plants were susceptible and most were wiped out, causing a famine responsible for the deaths of more than a million people.

Fear 6. Hidden costs may reduce the financial advantages of genetically modified crops. When seed companies create genetically modified seeds with crop traits desirable to farmers, the companies also engineer sterility into the seeds. As a consequence, the farmers must purchase new seeds for each generation of their crops. Such increases in the long-term costs and dependency on seed companies must be factored in by the farmers.

Among the less rational fears is that genetically modified foods are not "natural" and therefore must be harmful. This should not be a cause for concern. Smallpox, HIV, poison ivy, and cyanide, after all, are natural. The smallpox vaccine, on the other hand, is unnatural. Innumerable other valuable technological developments are equally unnatural. There simply is no value in knowing whether something is natural or unnatural when evaluating whether it is good and desirable or not.

In the end, we must compare the risks of producing genetically modified foods with the benefits. For example, farm workers will greatly benefit from spending less time applying pesticides to genetically modified, pest-resistant crops. The benefits of the workers' reduced pesticide exposure will be significantly greater in the less-developed countries, where safety regulations for pesticide use are more frequently ignored. The cost-benefit analyses will have to include the potential to reduce food costs and the ability to reduce environmental degradation from agriculture.

TAKE-HOME MESSAGE 5•16

More and more genetically modified foods are being created using modern methods of recombinant DNA technology. Some legitimate fears among the public remain, however, about the safety of these foods, given that their development relies on such new technology, and about the long-term financial advantages they offer.

⑥ Biotechnology today and tomorrow: DNA sequences are rich in information.

With the cloning of a sheep, a new era in biotechnology began.

5•17

DNA is an individual identifier: the uses and abuses of DNA fingerprinting.

In another time, Colin Pitchfork, a murderer and rapist, would have walked free. But in 1987, he was captured and convicted, betrayed by his DNA, and is now serving two life sentences in prison. Pitchfork's downfall began when he raped and murdered two 15-year-old high-school girls in a small village in England in the 1980s. The police thought they had their perpetrator when a man confessed, but only to the second murder. He denied any involvement with the first murder, though, which perplexed the police because the details of the two crimes strongly suggested that the same person committed both.

At the time, British biologist Alec Jeffreys made the important discovery that there were small pieces of DNA in every person's chromosomes that were tremendously variable in their base sequences. In much the same way each person has a driver's license number or social security number that differs from everyone else's, these DNA fragments are so variable that it is extremely unlikely that two people would have identical sequences at these locations. Thus, a comparison between these regions in a person's DNA sample and in DNA-containing evidence left at a crime scene would enable police to determine whether the evidence came from that person.

Jeffreys analyzed DNA left by the murderer/rapist on the two victims and found that it did indeed come from a single person, and that that person was *not* the man who had confessed to one of the crimes. That original suspect was released and has the distinction of being the first person cleared of a crime through DNA fingerprinting. To track down the criminal, police then requested blood samples from all men in the area who were between 18 and 35 years old—a practice many viewed as an invasion of privacy—collecting and analyzing more than 5,000 blood samples. This led them to Colin Pitchfork, whose DNA matched perfectly the DNA left on both of the victims, and ultimately was the evidence responsible for his conviction. (He almost slipped through, having persuaded a friend to give a blood sample in his name.

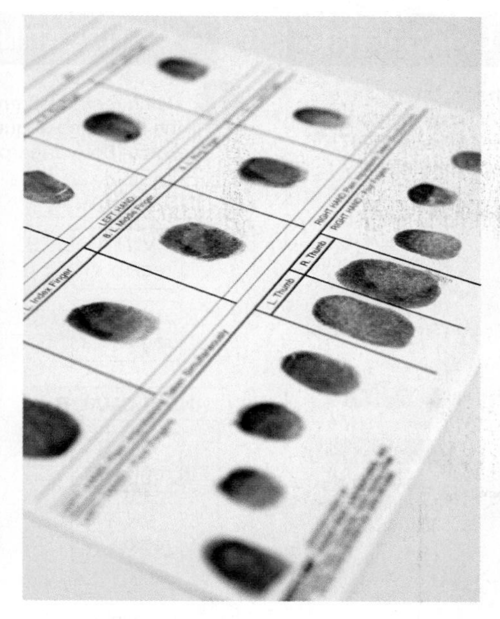

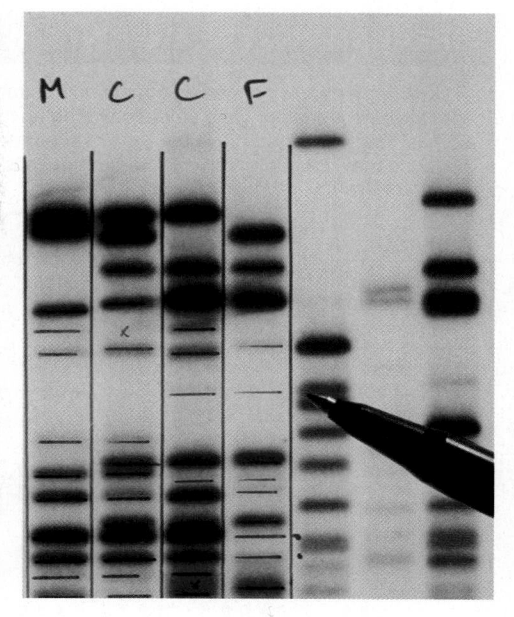

FIGURE 5-39 Betrayed by his DNA. Colin Pitchfork, the first criminal brought to justice with DNA fingerprinting.

But when the friend was overheard telling the story in a pub, police tracked down Pitchfork to get a blood sample.)

DNA fingerprinting is now used extensively in forensic investigations, in much the same way that regular fingerprints have been used for the past 100 years (**FIGURE 5-39**). But traditional fingerprinting is limited in its usefulness for many crimes because, often, no fingerprints are left behind. DNA fingerprinting, on the other hand, is not so limited because DNA samples more frequently are left behind, usually in the form of semen, blood, hair, skin, or other tissue. As a consequence, this technology has been directly responsible for bringing thousands of criminals to justice and, as we saw at the beginning of the chapter, for establishing the innocence of more than 200 people wrongly convicted of murder and other capital crimes. Let's examine how DNA fingerprinting is done, why it is such a powerful forensic tool, and why it is not foolproof.

The DNA from different humans is almost completely identical. More than 99.9% of the sequences of two individuals are the same because we are all from the same species and thus share a common evolutionary history. Even so, in a genome of three billion base pairs, even a tenth of a percent difference translates to about three million base-pair differences. These differences are responsible for the fact that all individuals have their own unique genome. (The lone exception? Identical twins, whose DNA is exactly the same.) Thus, when we are trying to evaluate whether the DNA from a crime scene matches that from a suspect, the analysis focuses on the parts of our DNA that differ. There are several dozen of these highly variable regions in the human genome.

Some of the most highly variable parts of the DNA are regions called **VNTR**s, because they contain a **v**ariable **n**umber of **t**andem **r**epeats. A VNTR region, we'll call it VNTR 1, might have a short sequence of 15 to 100 bases that repeats over and over again. Imagine that a person's genome is a book. The VNTR would be like the insertion of the word "green" over and over again on one of the pages. An individual—we'll call her Individual A—has two copies of each chromosome, one from her mother and one from her father, and the two copies of the VNTR 1 site that she carries are likely to differ from each other—maybe there are 14 repeats in one case and 3 repeats in the other. Individual B also has two copies of the VNTR 1 region, and the number of times the sequence repeats in those regions of Individual B's DNA might be 5 and 11 (**FIGURE 5-40**). Elsewhere in a person's genome are more than a dozen other VNTR regions that can be compared similarly.

Why are VNTRs so useful for forensic scientists? Because the existence of these regions means that forensic scientists don't need to look at all three billion bases in a person's entire DNA sequence. Instead, they only have to look at how many times a sequence is repeated in a suspect's two copies of VNTR 1, let's say 14 and 3 times. Then they look at how many times the VNTR 2 sequence is repeated in that person, let's say 7 and 34, and VNTR 3 might have 42 and 16 repeats. So the individual in question could be thought of as having a DNA fingerprint

CREATING DNA FINGERPRINTS BY USING VNTRs

A DNA fingerprint is created by counting the number of times that a repetitive sequence of base pairs occurs on an individual's chromosomes. Regions of repetitive base-pair sequences are called VNTRs because they contain a **V**ariable **N**umber of **T**andem **R**epeats.

INDIVIDUAL A

VNTR

Number of times the repeating sequence occurs

Chromosome from mother — 14

Chromosome from father — 3

14 / 3

Individual A's DNA fingerprint for this region

INDIVIDUAL B

Chromosome from mother — 5

Chromosome from father — 11

5 / 11

Individual B's DNA fingerprint for this region

For a given region, it's the same sequence that repeats, but the number of times it repeats differs from individual to individual (and differs on the maternal and paternal copies of a chromosomes that an individual carries).

FIGURE 5-40 VNTRs in forensics. Forensic scientists use VNTRs to genetically link a person of interest with DNA-containing evidence left at a crime scene.

CREATING A DNA FINGERPRINT GEL

1 Using restriction enzymes, DNA samples are cut on either side of the repeating sequences. Only two VNTR regions are shown here. Actual DNA fingerprints are created by sampling up to eight regions.

INDIVIDUAL A

VNTR region

VNTR region #1

VNTR region #2

Restriction enzymes cut here

INDIVIDUAL B

VNTR region #1

VNTR region #2

2 The isolated DNA sequences are then poured into a gel and an electrical charge is applied. Because DNA is a negatively charged molecule, the pieces of DNA move toward the bottom of the gel. This process causes the pieces of DNA to be separated by size. Smaller pieces—those with a small number of repeats—move more quickly than larger pieces and are found closer to the bottom.

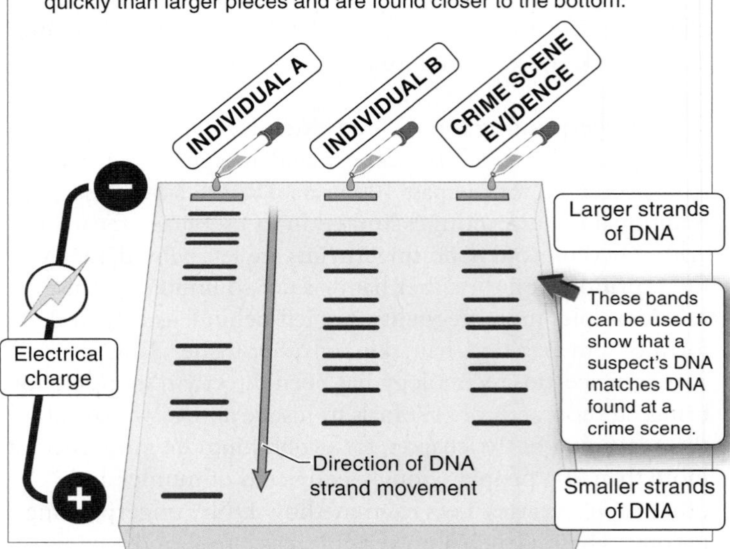

FIGURE 5-41 Your unique identifier: a DNA fingerprint. DNA fingerprints are being used to match a suspect's DNA to DNA found at the scene of a crime.

of 14/3, 7/34, 42/16. It's like checking whether two books are identical by looking at the first word on pages 16, 48, 123, and 200. It's a lot easier than comparing every word, but still can produce the correct answer. In the U.S., an individual's complete DNA fingerprint is generally built by sampling their DNA at more than ten different VNTR regions.

In court, DNA fingerprints are usually shown as photos of dark bands in a gel to demonstrate that the banding pattern of a suspect's DNA either does or doesn't match the banding pattern of DNA found at the crime scene (**FIGURE 5-41**).

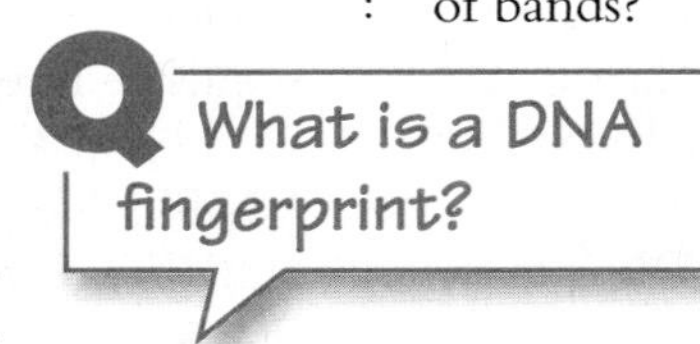

How do we get from an entire genome to a simple photo of bands?

First, the DNA in an evidence sample is chopped with a restriction enzyme that makes cuts on either side of the area where the repeating sequence occurs. The DNA is then poured into a gel and an electrical charge is applied to induce the DNA to migrate toward the bottom of the gel. Because smaller pieces move more quickly than larger pieces, this process causes different-size pieces of DNA to separate by size. The smaller pieces—those with a small number of repeats—are found closer to the bottom; the larger pieces—those with a large number of repeats—are found toward the top. At this point, because the DNA still is not visible, the repeating DNA sequence is stained with a radioactively labeled chemical and the gel is placed in contact with X-ray film. The film records dark bands at the point in the gel to which the DNA fragments have migrated.

DNA samples from different people will form different bands, whereas different samples of DNA from one person will have exactly the same pattern of dark bands.

In the end, despite universally accepted methods, DNA fingerprinting is not foolproof, because the possibility for human error remains. Even though we should not blindly draw conclusions solely from this one type of evidence, however, DNA fingerprinting is an increasingly valuable tool for law enforcement. The FBI, for example, has reported that nearly a third of their suspects are cleared immediately by DNA testing, and that many more criminals, because of DNA fingerprinting, now plead guilty to the crimes they have committed.

TAKE-HOME MESSAGE 5•17

Comparisons of highly variable DNA regions can be used to identify tissue specimens and determine the individual from whom they came.

5•18

DNA sequences reveal evolutionary relatedness.

It's hard to get somewhere new without a map. Just as explorers of the New World created maps to get the lay of the land, genetics researchers have been working since 1990 to map the genomes of humans and other species (**FIGURE 5-42**). Today, useful drafts of these sequences have been completed for more than 180 species, including humans and various species of bacteria, yeast, plants, worms, and fruit flies. The list is expected to top 1,000 species in a few years. Now that we have all of these data and are acquiring more, the big question is, what do we do with all of this information? Progress is being made in numerous different directions. We'll touch on just two of them.

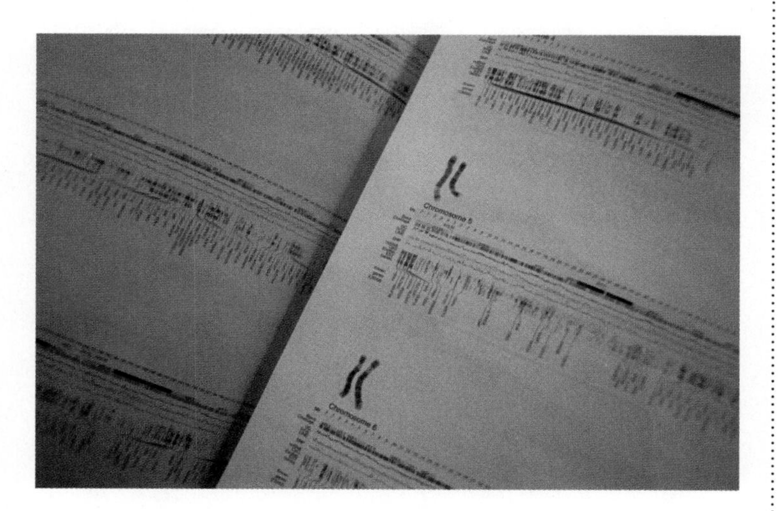

FIGURE 5-42 The Human Genome Project: mapping the genetic landscape.

1. *Mapping genetic landscapes.* The **Human Genome Project** is a project to decode the three billion base pairs in the human genome and to identify all the genes present in it. The project neared completion in 2000, and subsequent efforts have focused on determining some basic information, including how many genes there are, where they are, and how they work. In fact, the first big surprise of the Human Genome Project was the discovery that the number of genes in human genetic material is much smaller than expected, falling somewhere between 20,000 and 25,000. This is not much different from the number of genes in the genomes of fruit flies (13,000) or nematode worms (19,000) or the small plant *Arabidopsis* (26,000). We saw earlier that the total amount of DNA an organism carries is not related to the organism's complexity. It seems that there also is no relationship between the number of genes an organism has and its complexity, either.

2. *Building earth's family tree.* Humans crave order. To organize our cities we use address systems, to manage our digital information we use folders, subfolders, and files on our computers, and to keep track of our books we use the Library of Congress cataloging system. Not surprisingly, humans desire order when it comes to the billions of living organisms on earth. For this, we create **phylogenetic trees,** a way of grouping organisms in a hierarchical system that reflects the evolutionary history and relatedness of all organisms. Phylogenetic trees are a useful organizational system that helps us understand the diversity of life on earth and the nature of how species are related to one another.

Traditionally, phylogenetic trees were built on the basis of similarities among organisms' physical structures and embryological development. We'll examine the details of how this was done in Chapter 10, but the net result was a system of six kingdoms—bacteria, archaea, protists, plants, animals, and fungi—that reflected reasonably well the consensus hypothesis about the evolutionary history of the different groups of organisms. Today, however, DNA sequences are being analyzed as a feature that can be traced over time, and this analysis is revolutionizing the construction of phylogenetic trees. The new DNA-sequence-based phylogenetic trees reflect much more accurately the branching patterns of different groups of organisms as they evolved (**FIGURE 5-43**).

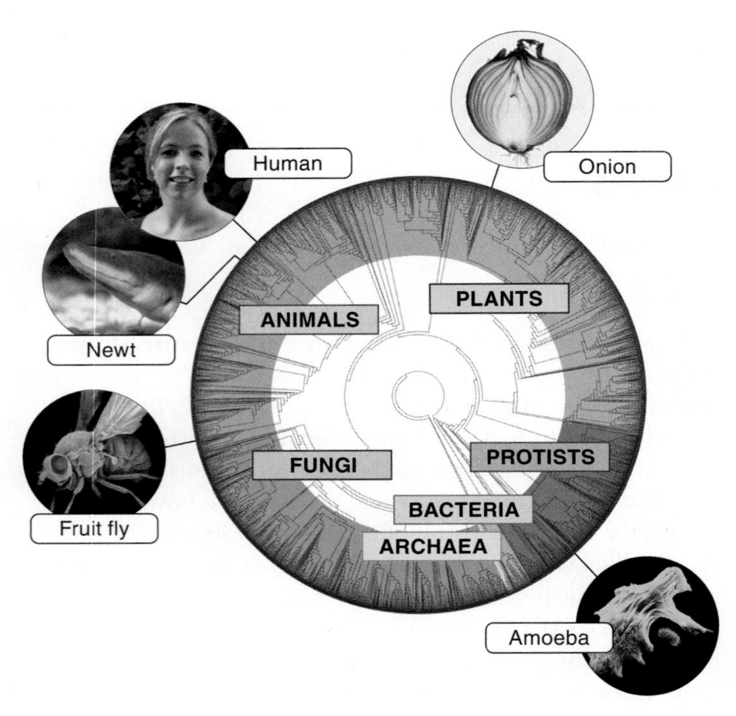

FIGURE 5-43 The tree of life. Our understanding of the evolutionary relationships among species is made more accurate through analysis and comparison of DNA sequences.

The rationale behind constructing phylogenies based on DNA sequences is straightforward. Organisms inherit DNA from their ancestors. As species evolve in their own unique ways, their DNA sequences diverge, becoming increasingly different. The more time that passes following the splitting off of one species from another, the greater the difference between their DNA sequences. Phylogenies are then constructed that place species in positions that reflect their relative genetic similarity.

Q When we say humans and chimps are genetically 96% identical, what do we mean?

Measures of overall DNA similarity between chimps and humans reveal that 96% of our base pairs are the same. (This is actually a comparison with only one chimp, a male named Clint, from the Yerkes Primate Research Center in Atlanta.) This is a greater similarity than humans share with any other species and indicates that chimps are our closest living relatives. Although most DNA doesn't code for proteins, when comparing only the sequences that code for proteins, we find that 29% of the genes code for the same exact proteins in chimps and humans.

For the purpose of comparison:

1. The difference between human DNA and mouse DNA is 60 times greater than the human-chimp difference.
2. The difference between mouse and rat DNA is 10 times greater than the human-chimp difference.
3. The difference between human and chimp DNA is 10 times greater than the average difference in DNA between two humans.

TAKE-HOME MESSAGE 5•18

Comparisons of DNA sequence similarities across species reveal their evolutionary relatedness and make it possible to construct detailed evolutionary trees.

5•19

Cloning—from organs to individuals—offers both promise and perils.

Cloning. Perhaps no scientific word more readily conjures horrifying images of the intersection of curiosity and scientific achievement. But is fear the appropriate emotion to feel about this burgeoning technology? Perhaps not.

For starters, let's clarify what the word means. "Cloning" actually refers to a variety of different techniques. To be sure, cloning can refer to the creation of new individuals that have exactly the same genome as the donor individual—a process called "whole organism cloning." That is, a clone is like an identical twin, except that it may differ in age by years or even decades. It is also possible to clone tissues (such as skin) and entire organs from an individual's cells. And, as we have already seen in this chapter, it is possible to clone genes.

Cloning took center stage in the public imagination in 1997 when Ian Wilmut, a British scientist, and his colleagues first reported that they had cloned a sheep, named Dolly. Their research was based on ideas that went back to 1938, when Hans Spemann first proposed the experiment of removing the nucleus from an unfertilized egg and replacing it with the nucleus from the cell of a different individual. Although the process used by Wilmut and his research group was difficult and inefficient, it was surprisingly simple in concept (**FIGURE 5-44**). They removed a cell from the mammary gland of a grown sheep, put its nucleus into another sheep's egg from which the nucleus had been removed, induced the egg to divide, and transplanted it into the womb of a surrogate mother sheep. Out of 272 tries, they achieved one success. But that was enough to show that the cloning of an adult was possible.

Shortly after news of Dolly's birth, teams set about cloning a variety of other species, including mice, cows, pigs, and cats (**FIGURE 5-45**). Not all of this work was driven by simple curiosity. For farmers, cloning could have real value. It can take a long time to produce animals with desirable traits from an agricultural perspective—such as increased milk production in cows. And with each successive generation of breeding, it can be difficult to maintain these traits in the population. But through the process of transgenic techniques and whole-animal cloning, large numbers of valuable animals with such traits can be produced and maintained.

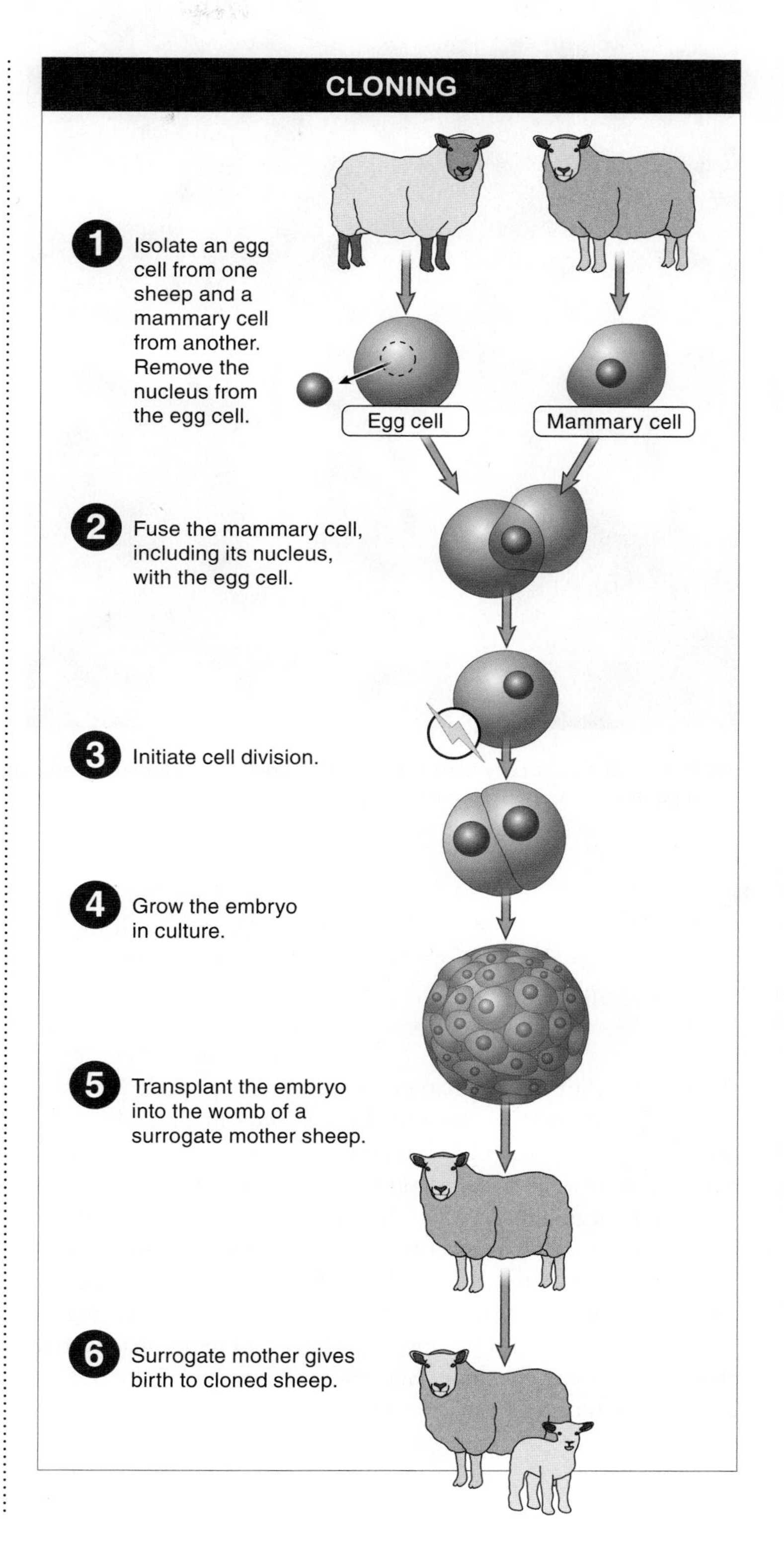

FIGURE 5-44 No longer science fiction. The steps used in the cloning of Dolly the sheep.

FIGURE 5-45 Genetically identical cloned animals. The cloning of animals can maintain desirable traits from generation to generation.

Are there any medical justifications for cloning?

Medical researchers, too, see much to gain from cloning. In particular, transgenic animals containing human genes—such as those discussed earlier in this chapter—can be very valuable. For instance, sheep that produce the human protein alpha-1 antitrypsin that is used to treat cystic fibrosis, or hamsters that produce human erythropoietin to treat kidney disease, are needed in large numbers. By cloning the transgenic animals, large colonies of pharmaceutical-drug-producing animals can be created. Pig clones, too, are being created with the hope of using their organs for transplantation into humans. Such pigs are unusual and useful because they have had genes inactivated that would normally cause their tissue to be rejected by the immune systems of human recipients.

Is it possible to clone a dinosaur? How could it be done?

Movies have raised the question of whether it is possible to clone a dinosaur (**FIGURE 5-46**). While the short answer is *yes,* that is in theory only. Several practical steps would be necessary, each of which might prove impossible. The first step would be retrieving the DNA of the dinosaur to be cloned. The most likely route would be to salvage fossilized DNA from the stomach of a mosquito, preserved since the time when dinosaurs lived. Mosquitoes from those days have actually been found, preserved in amber, a hardened, fossilized tree sap, but no ingested blood has been recovered from them. Then the dinosaur DNA would have to be extracted from the blood—assuming the mosquito had bitten a dinosaur. The biggest problem at this step is that it would be necessary to extract

FIGURE 5-46 Can a dinosaur be cloned?

and isolate all of the dinosaur's DNA, not just a few fragments. In the unlikely event that this could be done, this DNA would need to be inserted into the egg from a living organism, perhaps a crocodile, to create an embryo. At this point the embryo would have to be put in some sort of artificial egg where it could be incubated and fed as it grows, until it could finally hatch.

And if a dinosaur, why not a human? At this point, it is almost certain that the cloning of a human will be possible. Many people wonder, though, whether such an endeavor should be pursued. There is near unanimity among scientists that human cloning should not be attempted, and governments are struggling to develop wise regulations for this new world.

TAKE-HOME MESSAGE 5•19

Cloning of individuals has potential benefits in agriculture and medicine, but ethical questions linger.

Knowledge You Can Use

Did you know? Mixing aspirin and alcohol can lead to metabolic interference and unexpected inebriation.

Ethanol is the form of alcohol found in cocktails, beer, and wine. When you consume alcohol, the first step in your body's metabolic breakdown of the ethanol is carried out by the enzyme alcohol dehydrogenase.

Q: What happens to ethanol molecules *not* broken down? Any ethanol molecules in the bloodstream not broken down by alcohol dehydrogenase make their way to the brain. Once there, they cause you to feel a bit inebriated—or drunk, if the amount of ethanol is sufficiently large. Aspirin has the unintended side effect of disabling alcohol dehydrogenase, thus interfering with its normal functioning.

Q: What do you think happens to someone who takes two aspirin and then drinks several alcoholic beverages? By blocking the activity of alcohol dehydrogenase, aspirin interferes with a person's ability to break down ethanol. As a consequence, the ethanol remains in the body longer and has a more pronounced effect on the brain, producing greater inebriation (26% greater, according to one study).

What can you conclude? Medications, even over-the-counter products, can have unexpected (and unintended) physiological consequences, particularly when mixed with alcohol consumption. Exercise great caution when taking them.

1 DNA: What is it and what does it do?

DNA, the genetic material for all living organisms, is a double-helix molecule consisting of two "backbone" chains of sugar and phosphate molecules linked by pairs of nucleotide bases. Genes are sequences of bases in DNA that contain the code for the construction of protein molecules. Only a small fraction of the DNA in eukaryotes codes for genes.

2 Building organisms: through gene expression, information in the DNA directs the production of the molecules that make up an organism.

Information in the genes is transcribed into the sequences of mRNA molecules. These mRNA sequences then direct the construction of proteins.

3 Damage to the genetic code has a variety of causes and effects.

Mutations are rare alterations in a single base or changes in large segments of DNA that include several genes. When they occur in a gene, normal physiological functioning is usually disrupted. Mutations play an important role in evolution. Most genetic diseases result from a mutation that causes a gene to produce a non-functioning enzyme, resulting in a blocked metabolic pathway.

4 Biotechnology has the potential for improving human health.

Biotechnology is the use of technology to modify organisms, cells, and their molecules to achieve practical benefits. With modern molecular methods, researchers can cut and copy DNA and deliver it to new organisms, not necessarily of the same species. Biotechnology has had some notable successes in treating diseases, usually by producing medicines more efficiently and more effectively than with traditional methods. Gene therapy, however, has so far had a poor record of success in curing human diseases. With biotechnology, many tools have been developed that enable us to identify individuals carrying disease-causing genes. These tools can help reduce the symptoms and incidence of diseases, but come with significant potential costs, particularly the risk of discrimination.

5 Biotechnology is producing improvements in agriculture.

Biotechnology has led to important advances in agriculture by using transgenic plants and animals to produce more nutritious food. Even more dramatic is the extent to which biotechnology has reduced the environmental and financial costs of producing food through the creation of herbicide- and insect-resistant crops. Some legitimate fears among the public remain, however, as to the safety of genetically modified foods

6 Biotechnology today and tomorrow: DNA sequences are rich in information.

Advances in biotechnology have led to automated methods for analyzing DNA sequences. Comparisons of highly variable DNA regions have forensic value in identifying tissue specimens and determining the individual from whom they came. Comparisons of sequence similarities across species reveal evolutionary relatedness and allow construction of detailed evolutionary trees. Cloning of individuals has potential benefits in agriculture and medicine, but ethical questions linger.

KEY TERMS

allele, p. 165
base, p. 162
base pair, p. 162
biotechnology, p. 179
chromosomal aberration, p. 175
chromosome, p. 164
clone, p. 182
clone library, p. 182
cloning, p. 182
code, p. 168
codon, p. 173
deoxyribonucleic acid (DNA), p. 162
DNA probe, p. 182
gene, p. 164
gene library, p. 182
gene therapy, p. 186
genetic engineering, p. 179
genome, p. 164
genotype, p. 168
Human Genome Project, p. 199
hybridization, p. 182
intron, p. 166
messenger RNA (mRNA), p. 168
mutation, p. 174
nucleic acid, p. 162
nucleotide, p. 162
phenotype, p. 168
phylogenetic tree, p. 199
plasmid, p. 182
point mutation, p. 175
polymerase chain reaction (PCR), p. 180
promoter site, p. 169
protein synthesis, p. 173
recombinant DNA technology, p. 184
restriction enzymes, p. 180
ribosomal subunits, p. 171
stem cell, p. 186
trait, p. 165
transcription, p. 168
transfer RNA (tRNA), p. 171
transgenic organism, p. 182
translation, p. 168
VNTRs (variable number of tandem repeats), p. 197

CHECK YOUR KNOWLEDGE

1. A person's unique DNA is carried in:
a) muscle cells.
b) hair.
c) saliva.
d) skin cells.
e) All of the above contain a person's DNA.

2. Which of the following are always the same in every DNA molecule?
a) the sugar
b) the base
c) the phosphate group
d) Only a) and b) are always the same.
e) Only a) and c) are always the same.

3. The full set of an individual organism's DNA is called its:
a) complement.
b) genome.
c) nucleosome.
d) nucleotide.
e) chromosome.

4. In humans, genes make up ______ of the DNA.
a) about 75%
b) 100%
c) less than 5%
d) about 10%
e) about 50%

5. Genotype is to phenotype as:
a) cookie is to oven.
b) cookie is to recipe.
c) cookbook is to cookie.
d) recipe is to cookie.
e) oven is to cookie.

6. To start the transcription process, a large molecule, ______, recognizes a ______.
a) RNA polymerase; messenger RNA
b) DNA polymerase; termination site
c) DNA polymerase; promoter site
d) RNA polymerase; promoter site
e) DNA polymerase; messenger RNA

7. During transcription, at the point where the DNA strand being copied has an adenine, a(n) ______ is added to the ______.
a) thymine; tRNA
b) cytosine; DNA
c) uracil; tRNA
d) adenine; mRNA
e) uracil; mRNA

8. There are different ______ molecules for each of the 20 different amino acids that are used in building proteins.
a) ribosomal subunit
b) tRNA
c) mRNA
d) DNA
e) elongation

9. Deletions and substitutions are two types of point mutations. Which type is more likely to cause mistranslations of proteins?
a) Substitutions, because they shift the reading frame and cause downstream amino acids to be changed.
b) Substitutions, because one protein is substituted for another protein.
c) Deletions, because they shift the reading frame and cause downstream amino acids to be changed.
d) Deletions, because one protein is deleted.
e) None of the above is correct.

10. Which of the following statements about the metabolism of ethanol (which is present in alcoholic beverages) is incorrect?
a) Individuals who produce non-functioning aldehyde dehydrogenase exhibit "fast-flushing."
b) The process requires two enzymes: alcohol dehydrogenase and isopropyl dehydrogenase.
c) Individuals who are "fast-flushers" are less likely to become alcoholics.
d Aspirin interferes with the action of alcohol dehydrogenase.
e) All of the above are correct.

11. The polymerase chain reaction (PCR):
a) makes it possible to create huge numbers of copies of tiny pieces of DNA.
b) enables researchers to determine the sequence of a complementary strand of DNA when they have only single-stranded DNA.
c) utilizes RNA polymerase to build strands of DNA.
d) can create messenger RNA molecules from small pieces of DNA.
e) All of the above are correct.

12. Which of the following is not a difficulty that medicine has encountered in its attempts to cure human diseases through gene therapy?
a) The transfer organism—usually a virus—may get into unintended cells and cause disease.
b) It is difficult to get the working gene into the specific cells where it is needed.
c) It is difficult to get the working gene into enough cells at the right rate to have a physiological effect.
d) For many diseases, a malfunctioning gene has not been identified.
e) All of the above are difficulties encountered in attempts to cure human diseases through gene therapy.

13. Golden rice:
  a) grows without a husk, thereby reducing the processing required before it can be consumed.
  b) can make vitamin A without beta-carotene.
  c) could help prevent blindness in the 250,000 children who lose their sight each year because of vitamin A deficiency.
  d) supplies more vitamin A in one serving than an individual needs in a full week.
  e) is one of the most recent developments in organic farming.

14. Which of the following statements about Bt crystals is correct?
  a) They are produced by soil-dwelling bacteria of the species *Bacillus thuringiensis.*
  b) The gene coding for the production of Bt crystals has been genetically engineered into the genome of dairy cows, increasing their milk production sixfold.
  c) They are produced by the polymerase chain reaction (PCR).
  d) They are produced by most weedy species of plants.
  e) All of the above are correct.

15. A VNTR is:
  a) a gene that codes for a biochemical trait, rather than a structural trait.
  b) used in biotechnology when creating a clone.
  c) produced when a mutation occurs in a non-sex cell.
  d) a highly variable section of an individual's DNA.
  e) produced when a mutation occurs in a sperm-producing or egg-producing cell.

16. Ninety-six percent of the base-pair sequences in chimps and humans are the same. This finding indicates that:
  a) humans are more closely related to chimps than any two mice are related to each other.
  b) 96% of the genes in humans have identical counterparts in chimps.
  c) humans are more closely related to chimps than to any other species on earth.
  d) 4% of the DNA we carry contains mutations.
  e) 96% of the proteins produced by chimps is identical to proteins produced by humans.

Short-Answer Questions

1. After sequencing a molecule of DNA, you discover that 20% of the bases are cytosine. What percentage of the bases would you expect to be thymine? Explain why.

2. A cell is in the process of transcription. The single strand of DNA that is being copied has the sequence ATC GAC GGA TCC. Indicate the mRNA strand that will be synthesized from this strand, listing the correct sequence of bases. Explain why this step is important in the process of producing protein.

3. What is being "translated" in translation? What is the role of transfer RNA molecules in this step?

4. The genetic disorder known as sickle-cell disease results in an altered form of the protein hemoglobin, which ultimately can affect the structure of red blood cells. A portion of the sequence for the normal hemoglobin allele is:

   CTG ACT CCT GAG GAG AAG TCT

   Individuals with the sickle-cell disorder have the following sequence:

   CTG ACT CCT GTG GAG AAG TCT

   Identify the type of mutation and explain how it would result in an altered (in this case non-functional) protein.

5. Describe three major applications of biotechnology and their impact on human health.

See Appendix for answers. For additional study questions, go to www.prep-u.com.

6

Chromosomes and Cell Division

Continuity and variety

❶ There are different types of cell division.

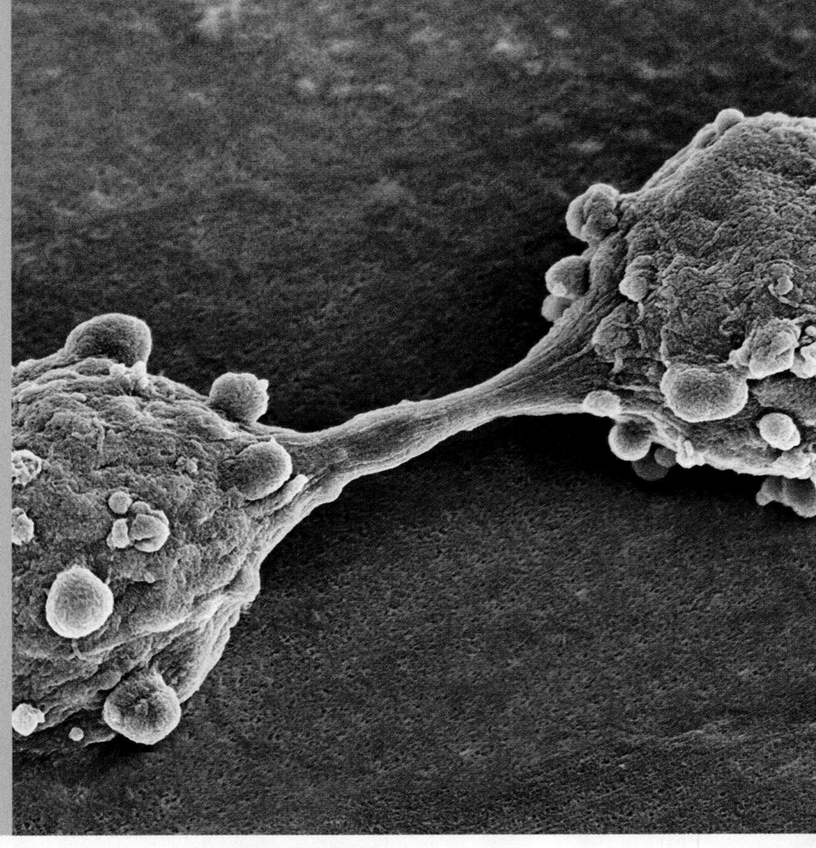

Bladder cancer cells in the final stage of cell division.

6•1

Immortal cells can spell trouble: cell division in sickness and in health.

Once you are fully grown, do you have just one set of cells that live as long as you do? The answer is *no:* your cells are continually dying off and the ones that remain divide and replace the cells you've lost, in an ongoing process. But how long can this last? Can it go on forever? And does a cell even know how old it is?

Actually, a cell does have a measure of how old it is. Just as a car comes with an odometer, which keeps track of how far the car has been driven, animal cells have a sort of counter that keeps track of how many times a cell has divided. This "counter" is a section of DNA called the **telomere,** which is located at each tip of every chromosome, right next to the genes that direct the processes that keep an organism alive. The telomere functions like a protective cap at the end of the DNA. Every time a cell divides, making an exact copy of itself, its DNA divides as well. However, each time the DNA divides, the telomere gets a bit shorter. Eventually, after some critical number of cell divisions, the telomere can become so short that additional cell divisions cause the loss of functional, essential DNA, and that means almost certain death for the cell (**FIGURE 6-1**).

At birth, the telomeres in most human cells are long enough to support about 50 cell divisions. Occasionally, however, individuals are born with telomeres that are much shorter than normal. In these people, the normal functioning of many genes is disrupted, and their cells and tissues begin to appear aged very soon after birth (**FIGURE 6-2**). As a consequence, those born with this disorder rarely live beyond the age of 13.

Just as someone might think it's a good idea to disconnect his or her car's odometer, you might think that disconnecting the cellular odometer by continually rebuilding the telomere would be an effective strategy for allowing a cell and its descendants to function for a longer time than normal. Such

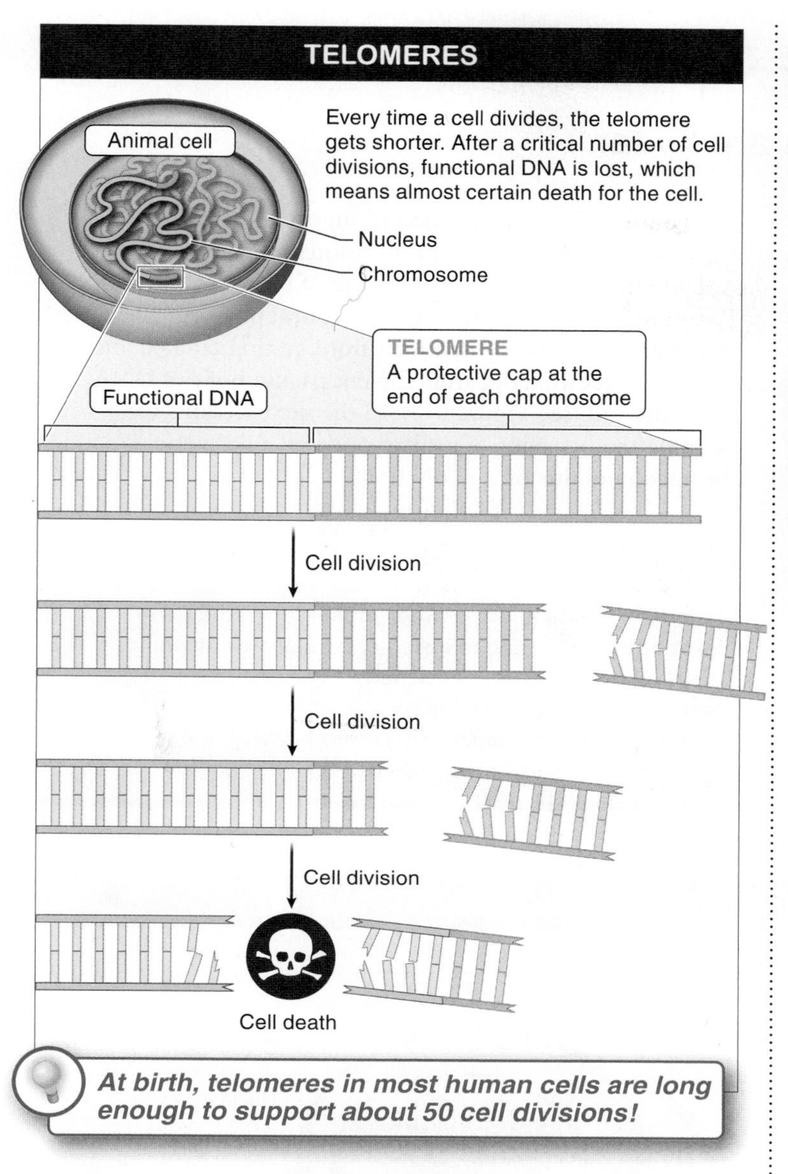

FIGURE 6-1 A cellular "odometer." Telomeres limit cells to a fixed number of divisions.

a line of cells would never die—so you'd think that the continually rebuilding telomere would be like a fountain of youth. It's not. We know this because there *are* some cells with a sort of disconnected odometer. These cells rebuild their telomeres after each cell division, restoring the protective cap. Unfortunately, most cells that rebuild their telomeres with each cell division present a big problem: they don't know when to stop dividing. There is a word for such cells: cancer.

Because cancer is runaway cell division, discovering a cure for cancer will necessarily involve a deep understanding of cell division. In this chapter, we explore the processes that have evolved that enable cells to divide and create new cells.

FIGURE 6-2 Just a child. Individuals with the genetic condition Hutchinson-Gilford progeria syndrome are born with shorter-than-normal telomeres and appear to age at a rapid rate. The inset shows chromosomes (in blue), with the telomeres highlighted in yellow.

> "Cancer cells are those which have forgotten how to die.
>
> —Harold Pinter, playwright, from the poem "Cancer Cells," 2002"

Prokaryotes have one method. It is called binary fission and it serves all of their needs. Eukaryotes, on the other hand, have two methods, mitosis and meiosis, each of which has a specific purpose. In addition to discussing what happens when cell division goes unchecked, we also will explore some of the difficulties that arise when cells wind up with too many or too few chromosomes after cell division.

TAKE-HOME MESSAGE 6•1

Within cells there is a protective section of DNA called the telomere that gets a bit shorter every time the cell divides. If a telomere becomes too short, additional cell divisions cause the loss of essential DNA and cell death. Cells that rebuild the telomere with each division can become cancerous.

6•2

Some chromosomes are circular, others are linear.

As a method for storing genetic information, DNA has complete market saturation. All life on earth uses it. This is pretty remarkable considering the tremendous diversity of life that exists on earth—from bacteria to plants and animals. One way in which different organisms' DNA varies is in how it is organized into chromosomes.

In most bacteria and the other prokaryotes, the genetic information is carried in a single, circular chromosome, a strand of DNA that is attached at one site to the cell membrane (**FIGURE 6-3**). Eukaryotes, on the other hand, have much more DNA than do bacteria and organize it into free-floating linear chromosomes within the nucleus. As we saw in Chapter 5, a eukaryotic organism may have as few as 2 chromosomes or as many as 1,000. For plants and animals, the number is usually between 10 and 50.

The most important part of a eukaryotic chromosome is the DNA molecule, which carries information about how to carry out the processes needed to support the life of the organism. But eukaryotic chromosomes (and some prokaryotic chromosomes) are made of more than just DNA. The long, linear strand is wrapped around proteins called **histones,** which keep the DNA from getting tangled and enable an orderly, tight, and efficient packing of the DNA inside the cell (see Figure 6-3). In the next sections we examine the behavior of prokaryotic and eukaryotic chromosomes during cell division.

TAKE-HOME MESSAGE 6•2

In most bacteria and the other prokaryotes, the genetic information is carried in a single, circular chromosome, a strand of DNA that is attached at one site to the cell membrane. Eukaryotes have much more DNA than do bacteria and organize it into free-floating linear chromosomes within the nucleus.

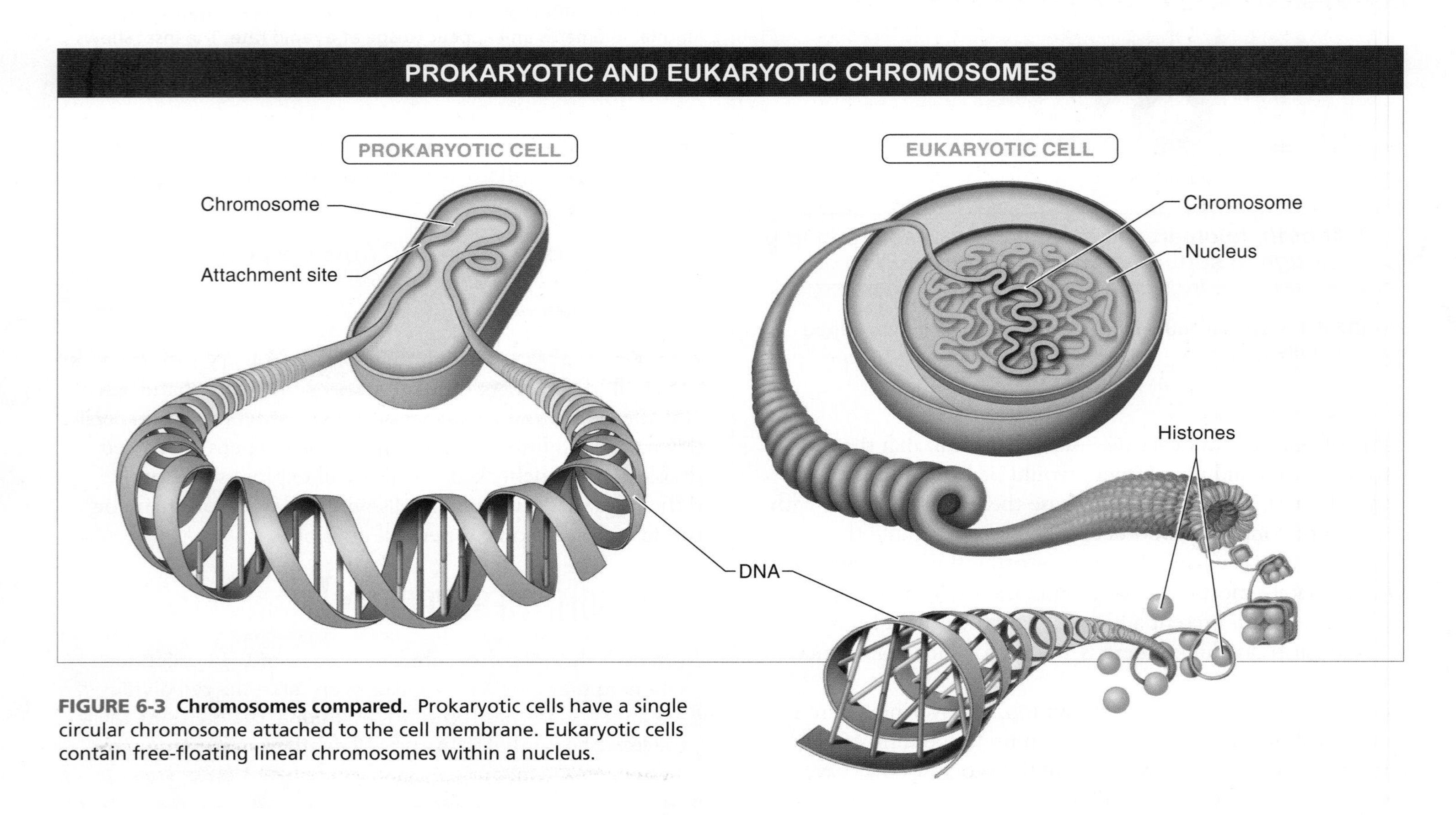

FIGURE 6-3 Chromosomes compared. Prokaryotic cells have a single circular chromosome attached to the cell membrane. Eukaryotic cells contain free-floating linear chromosomes within a nucleus.

6•3

Prokaryotes divide by binary fission.

When it is time for bacteria and the other prokaryotes to reproduce, they use a method called **binary fission,** which means "division in two" (**FIGURE 6-4**). This process begins with **replication,** the process by which a cell's DNA duplicates itself. Replication begins as the double-stranded DNA molecule unwinds from its coiled-up configuration. Once the strands are uncoiled, they split apart like a zipper. As the double-stranded molecule unzips, free-floating nucleotide bases attach to the bases exposed on each of the two separated, single-stranded circular molecules of DNA, matching A to T and G to C to create two identical double-stranded DNA molecules. The newly created circular chromosomes of DNA attach to the inside of the plasma membrane at a different spot than the original. The original cell, called the **parent cell,** then pinches into two new cells, called **daughter cells,** so each of the daughter cells has an exact and precise two-stranded copy of the original two-stranded circular chromosome.

In some prokaryotes, such as the *E. coli* that live in our digestive system, the complete process of binary fission can occur very quickly—often in as little as 20 minutes. Binary fission is considered **asexual reproduction** because the daughter cells inherit their DNA from a single parental cell. In **sexual reproduction,** a combination of DNA from two separate individuals is passed on to offspring, resulting in offspring that are genetically different both from one another and from their parents. We will discuss sexual reproduction in more detail later in this chapter.

TAKE-HOME MESSAGE 6•3

Bacteria divide by a type of asexual reproduction called binary fission: first the circular chromosome duplicates itself and then the cell splits into two identical new cells.

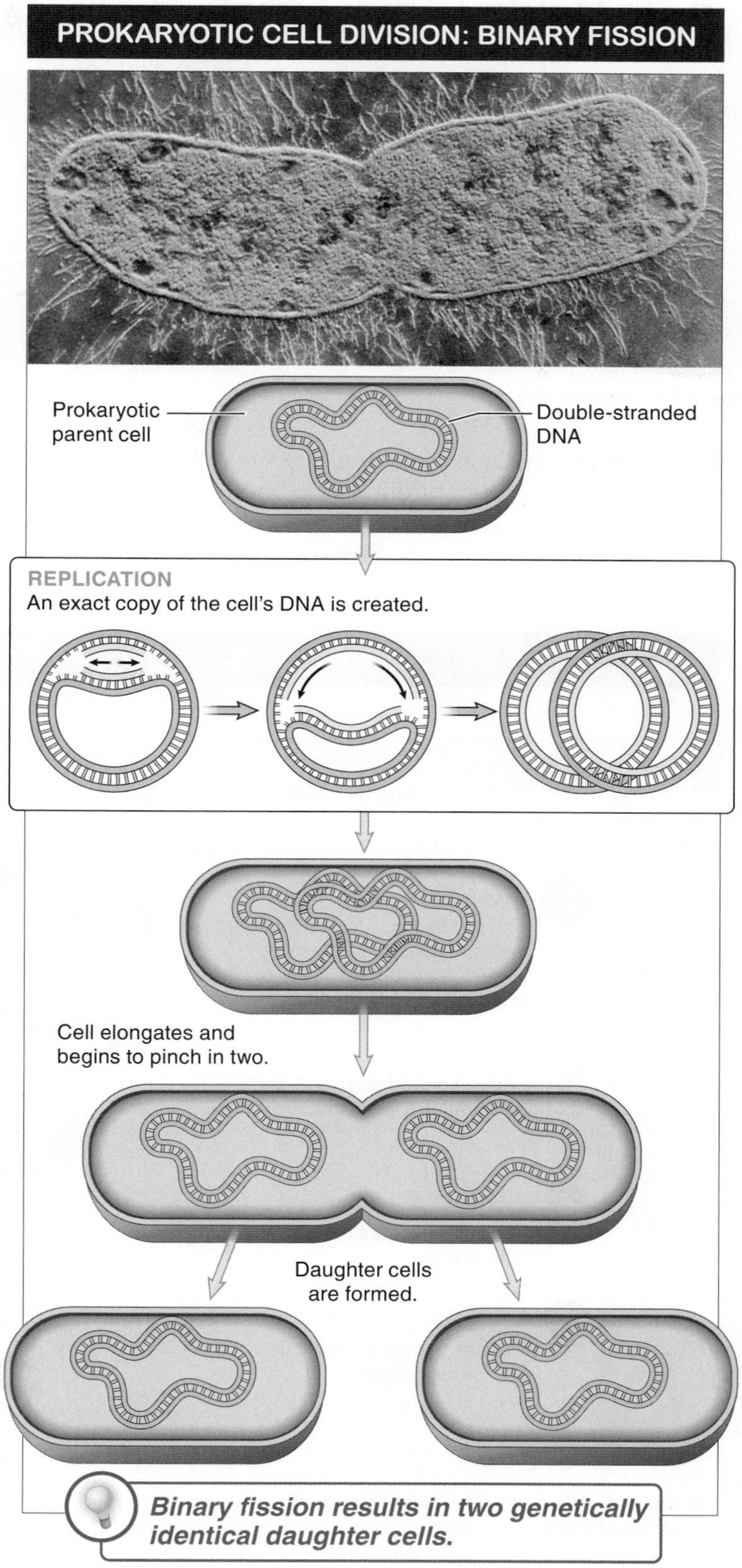

PROKARYOTIC CELL DIVISION: BINARY FISSION

Prokaryotic parent cell
Double-stranded DNA
REPLICATION
An exact copy of the cell's DNA is created.
Cell elongates and begins to pinch in two.
Daughter cells are formed.
Binary fission results in two genetically identical daughter cells.

FIGURE 6-4 "Dividing in two." Asexually reproducing prokaryotes reproduce rapidly through binary fission.

6•4

There is a time for everything in the cell cycle.

In life, we often go through phases that are defined by the focus of our interests and activities. For many years we go to school. For a while we may tend to our career. Then, for a spell, our personal life may take center stage. Most eukaryotic cells do this, too. They may spend a long period of time occupied with activities relating solely to growth of the cell and then suspend those activities as they segue into a period devoted exclusively to reproducing themselves. This alternation of activities between processes related to cell division and processes related to growth is called the **cell cycle** (FIGURE 6-5).

Before we go any further in discussing the cell cycle, we need to note an important distinction among the cells of the body. All of the cells of a multicellular eukaryotic organism can be divided into two types: **somatic cells** are the cells forming the body of the organism; **reproductive cells** are the sex cells, or **gametes.** The processes of cell division by which somatic cells and reproductive cells are produced differ from each other. In this section, we focus on somatic cells. Later in the chapter, we examine cell division leading to the production of sex cells.

The cell cycle describes the series of phases that leads to somatic cell division. There are two main phases in the cell cycle: **interphase,** during which the cell grows and prepares to divide, and the **mitotic phase** (or **M phase**), during which first the nucleus and genetic material within the cell divides, and then the rest of the cellular contents divides. Interphase is further divided into three distinct subphases, described below. A eukaryotic somatic cell moves through the phases in precise order and, at any given moment, a somatic

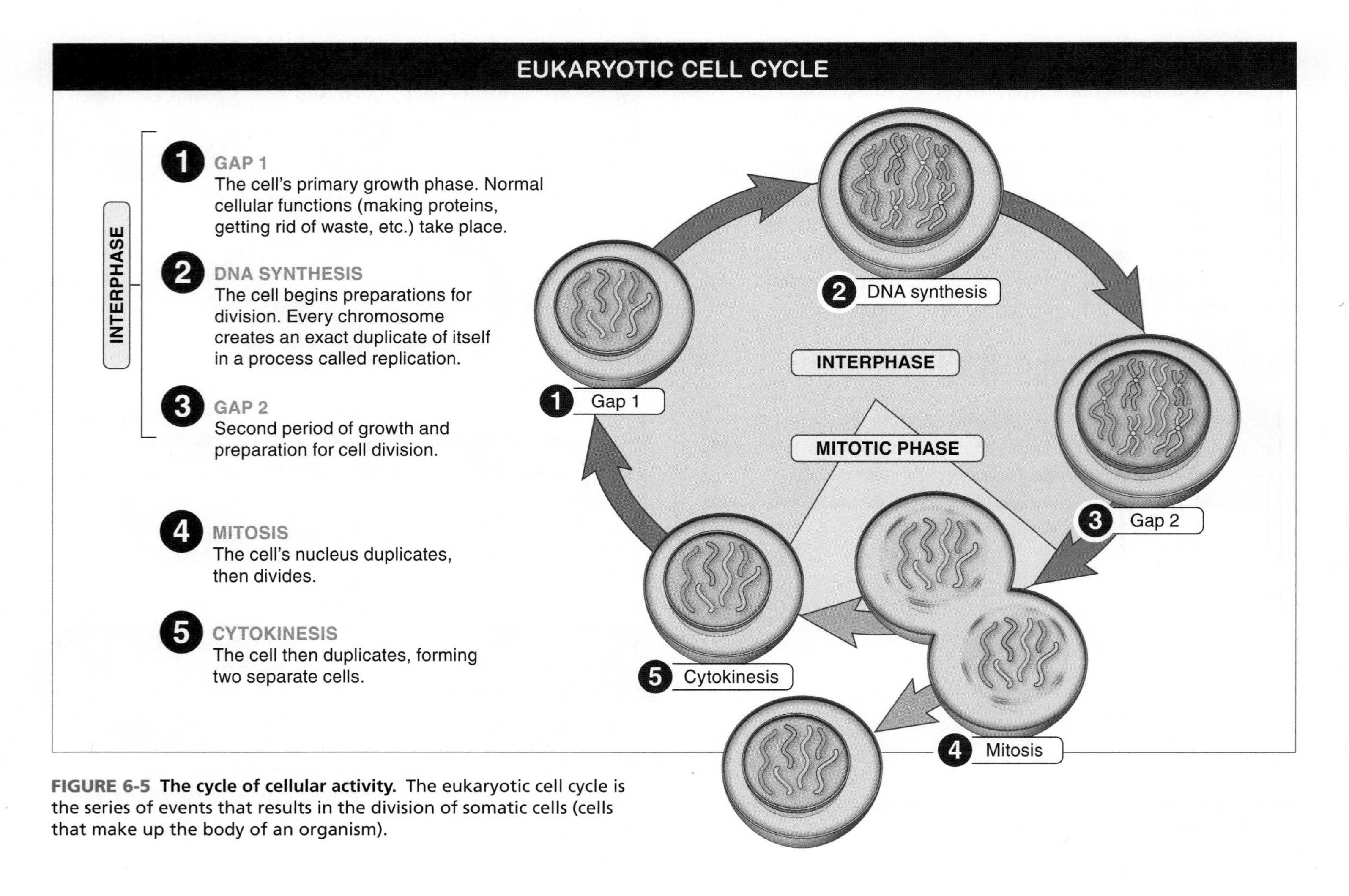

FIGURE 6-5 The cycle of cellular activity. The eukaryotic cell cycle is the series of events that results in the division of somatic cells (cells that make up the body of an organism).

cell is at some point in this cycle. The phases and subphases can be summarized as follows.

Interphase

Gap 1. During this period, a cell grows and performs all cellular functions (making proteins, getting rid of waste, and so on). Cells that divide infrequently, such as most neurons and heart muscle cells, spend most of their time in the Gap 1 phase.

DNA synthesis. During this phase, the cell begins to prepare for cell division. Every chromosome creates an exact duplicate of itself by replication. Before replication, each chromosome is a long linear strand of genetic material. After replication, each chromosome has become a pair of identical long linear strands, held together near the center; this region, where the chromosomes are in contact, is called the **centromere.**

Gap 2. In this phase, the cell continues to grow and prepare for cellular division. This phase differs from Gap 1 because the genetic material has now been duplicated. Gap 2 is usually much shorter than Gap 1.

Mitotic phase

This phase begins with **mitosis,** a process in which the nucleus of the parent cell duplicates. Mitosis is generally followed by **cytokinesis,** during which the cytoplasm, organelles, and the rest of the materials in the cell duplicate and the cell separates into two daughter cells, each of which has a complete set of the parent cell's DNA and other cellular structures. (Note that each daughter cell becomes a parent cell in the next cell cycle.) The mitotic phase is usually the shortest period in the eukaryotic cell cycle.

The transition from one phase to the next is triggered by specific events. For instance, a cell moves from the Gap 1 phase (denoted G_1) to DNA synthesis only if environmental conditions outside the cell are favorable for cell division. Until conditions are favorable, the cell remains in G_1. Likewise, a cell begins mitosis only when the genetic material has been successfully duplicated. Later in this chapter we'll see that errors in this orderly, step-by-step process can lead to out-of-control cell division and cancer.

TAKE-HOME MESSAGE 6•4

Eukaryotic somatic cells alternate in a cycle between cell division and other cell activities. The cell division portion of the cycle is called the mitotic phase. The remainder of the cell cycle, called interphase, consists of two gap phases separated by a DNA synthesis phase during which the genetic material is replicated.

6•5

Cell division is preceded by replication.

In one of the great understatements in the scientific literature, James Watson and Francis Crick wrote the following sentence in their paper describing the structure of DNA: "It has not escaped our notice that the specific pairing we have postulated immediately suggests a possible copying mechanism for the genetic material." The researchers' ability to explain how DNA copied itself was a critical feature of their description of DNA's structure. After all, every single time any cell in any organism's body divides, that cell's DNA must first duplicate itself so that each of the two new cells has all of the genetic material of the original parent cell. The process of DNA duplication, as we've seen, is called replication.

What is it about DNA's structure that makes duplication so straightforward? Watson and Crick were referring to the feature of DNA called **complementarity,** the characteristic that in the double-stranded DNA molecule, the base on one strand always has the same pairing partner (called the **complementary base**) on the other strand: A pairs with T (and vice versa), and G pairs with C (and vice versa).

With this consistent pattern of pairing, when the DNA molecule separates into two strands, it is possible to perfectly reconstruct all the information on the missing half because one strand carries all the information needed to construct its complementary strand. Just before cells divide, the DNA molecule unwinds and "unzips" and each half of the unzipped molecule serves as a template on which the missing half is reconstructed. At the end of the process of reconstructing the missing halves, there are two DNA molecules, each identical to the original DNA molecule, one for each of the two new cells.

The process of DNA replication occurs in two steps: unwinding and rebuilding (**FIGURE 6-6**).

1. Unwinding. Replication begins when the coiled, double-stranded DNA molecule unwinds and separates into two strands, like a zipper unzipping.

2. Rebuilding. In the rebuilding process, each of the single strands becomes a double strand as an enzyme connects the appropriate complementary base to the exposed base. If an A is exposed, a T binds to it; if a C is exposed, a G binds to it.

The end result of replication is two double-stranded DNA molecules, each identical to the cell's original double-stranded DNA molecule. DNA replication occurs in all types of cells, somatic and reproductive, prior to cell division.

A variety of mutations (such as mismatched bases and added or deleted segments) can occur during replication, but many

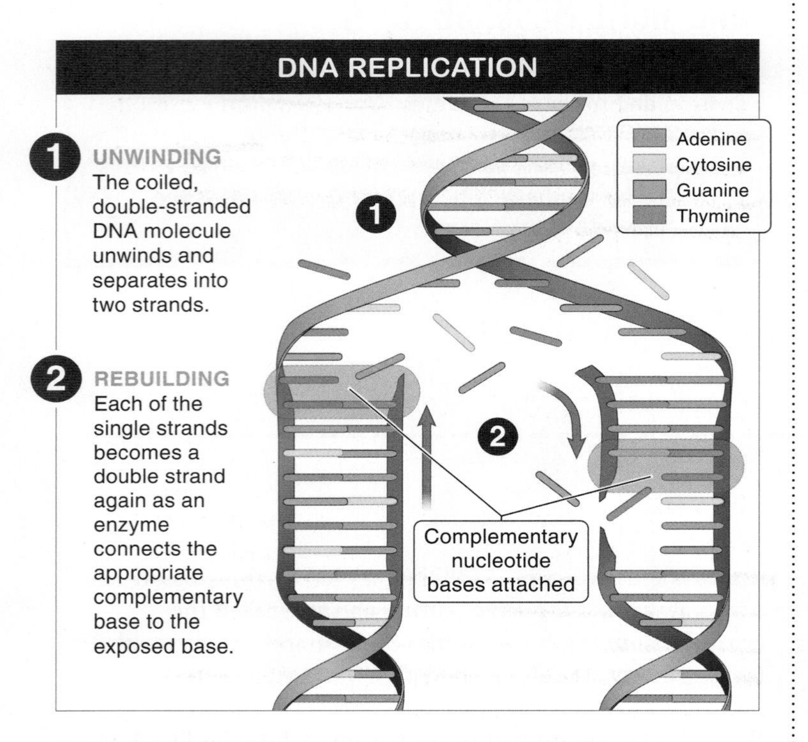

FIGURE 6-6 How DNA replicates: unwinding and rebuilding.

This artist is carefully duplicating a painting. Similarly, DNA must be duplicated prior to cell division.

of these errors are caught and repaired after replication. If an error remains, however, the sequences in a replicated DNA molecule (including the genes) can be different from those in the parent molecule. A changed sequence may then produce a different mRNA sequence, which may in turn produce a different amino acid sequence, and, hence, a different protein. In this way, new genes can enter a population and be acted on by evolution. We explore the relationship between mutation and evolution in more detail in Chapter 8.

Q Errors sometimes occur when DNA duplicates itself. Why might that be a good thing?

TAKE-HOME MESSAGE 6•5

Every time a cell divides, that cell's DNA must first duplicate itself so that each of the two new cells has all the genetic material of the original parent cell. The process of DNA duplication is called replication. Errors in replication can lead to mutations, which are changes in the DNA sequence.

❷ Mitosis replaces worn-out old cells with fresh new duplicates.

For most of the cell cycle, chromosomes resemble a plate of spaghetti more than the familiar condensed, replicated, X shape seen in photos.

6•6

Most cells are not immortal: mitosis generates replacements.

Q What is dust? Why is it your fault?

Look around your dorm room. Dust is everywhere. What is it? It is primarily dead skin cells. In fact, you (and your roommates) slough off millions of dead skin cells each day—yet your skin doesn't disappear. How can that be? Obviously, your body is replacing the sloughed-off cells. The cells have simply worn out, so your body creates replacements. It does this through the process of mitosis.

Mitosis has just one purpose: to enable existing cells to generate new, genetically identical cells. There are two different reasons for this need (**FIGURE 6-7**).

1. Growth. During development, organisms get bigger. Growth happens in part through the creation of new cells. In fact, if you want to see cell division in action, one surefire place to look is at the tip of a plant root: at a growth rate of about half an inch per day, the root is one of the fastest-growing parts of a plant.

2. Replacement. Cells must be replaced when they die. The wear and tear that come from living can physically damage cells. The daily act of shaving, for example, damages thousands of cells on a man's face. It's nothing to worry about, though. Microscopic views of human skin reveal several distinct layers, with the outermost layers—the layers under assault during shaving—made up primarily of dead cells. These cells help protect us from infection and also reduce the rate at which the underlying living cells dry out. The living cells that exist just below the layers of dead cells are being produced at a high rate by mitosis; they can also be harmed if you're not careful.

FIGURE 6-7 Reasons for mitosis. Mitosis is important in organism growth and the replacement of cells.

Some other cells that must be replaced actually die on purpose, in a planned process of cell suicide called **apoptosis.** This seemingly counterproductive strategy is employed in parts of the body where the cells are likely to accumulate significant genetic damage over time and are therefore at high risk of becoming cancer cells (a process described later in this chapter). Cells targeted for apoptosis include many of the cells lining the digestive tract as well as those in the liver, two locations where cells are almost constantly in contact with harmful substances.

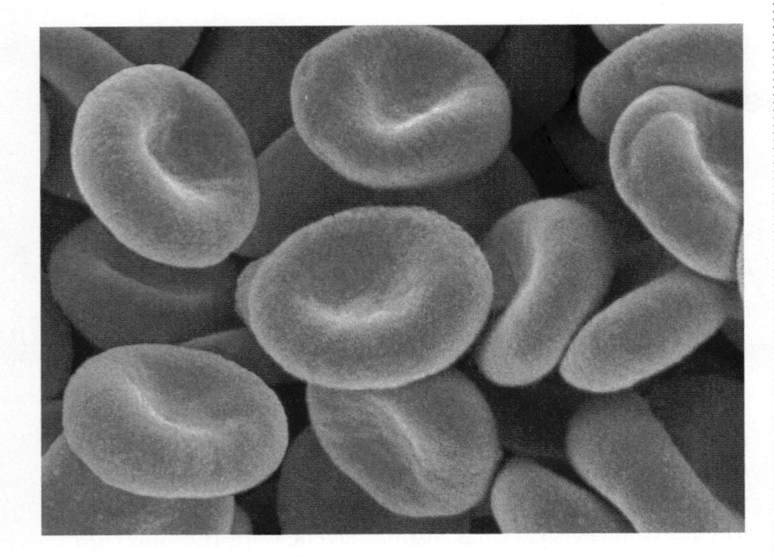

The red blood cells in your body carry oxygen to the body's organs and tissues and are replaced about every 120 days.

Every day, a huge number of cells in an individual must be replaced by mitosis. In humans this number is in the billions. Nearly all the somatic cells of the body—that is, everything other than sperm- and egg-producing cells—undergo mitosis. There are a few notable exceptions. Brain cells and heart muscle cells, in particular, do not seem to divide, or, if they do divide, they do so at very, very slow rates. (It is not known why this is so.)

The rate at which mitosis occurs varies dramatically for different types of cells. The most rapid cell division occurs in the blood and in the cells lining various tissues and organs. The average red blood cell, for example, is in circulation only for about two to four months and then must be replaced. The cells lining the intestines are replaced about every three weeks. Hair follicles, too, are among the most rapidly dividing cells.

TAKE-HOME MESSAGE 6·6

Mitosis enables existing cells to generate new, genetically identical cells. This makes it possible for organisms to grow and to replace cells that die.

6•7

Overview: mitosis leads to duplicate cells.

For mitosis to begin, the parent cell duplicates its DNA, creating a duplicate copy of each chromosome. Once this task is completed, the other materials in the cell duplicate and the cell divides into two new duplicate cells, the daughter cells.

Mitosis occurs in just four steps. It begins with the condensing of the chromosomes, which are all tangled during interphase. Next, all of the duplicated and condensed chromosomes move to the center of the cell. Each chromosome is then pulled apart from its duplicate. In the final step, new membranes form around each complete set of chromosomes to create the new nuclei, and the cytoplasm and rest of the cell duplicate as well. Where once there was one parent cell, now there are two daughter cells (**FIGURE 6-8**).

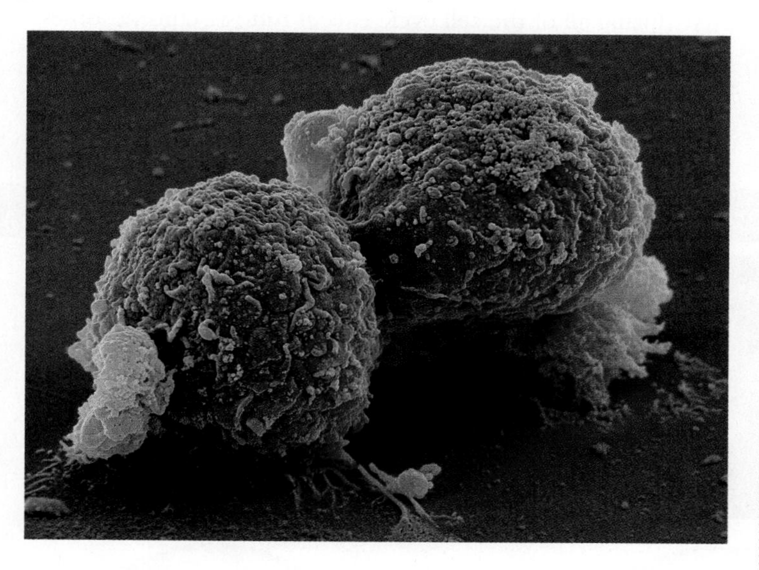

These two human embryonic stem cells have just completed mitosis.

TAKE-HOME MESSAGE 6•7

Mitosis is the process by which cells duplicate themselves. It occurs in just four steps, following the replication of chromosomes. It leads to the production of two daughter cells from one parent cell.

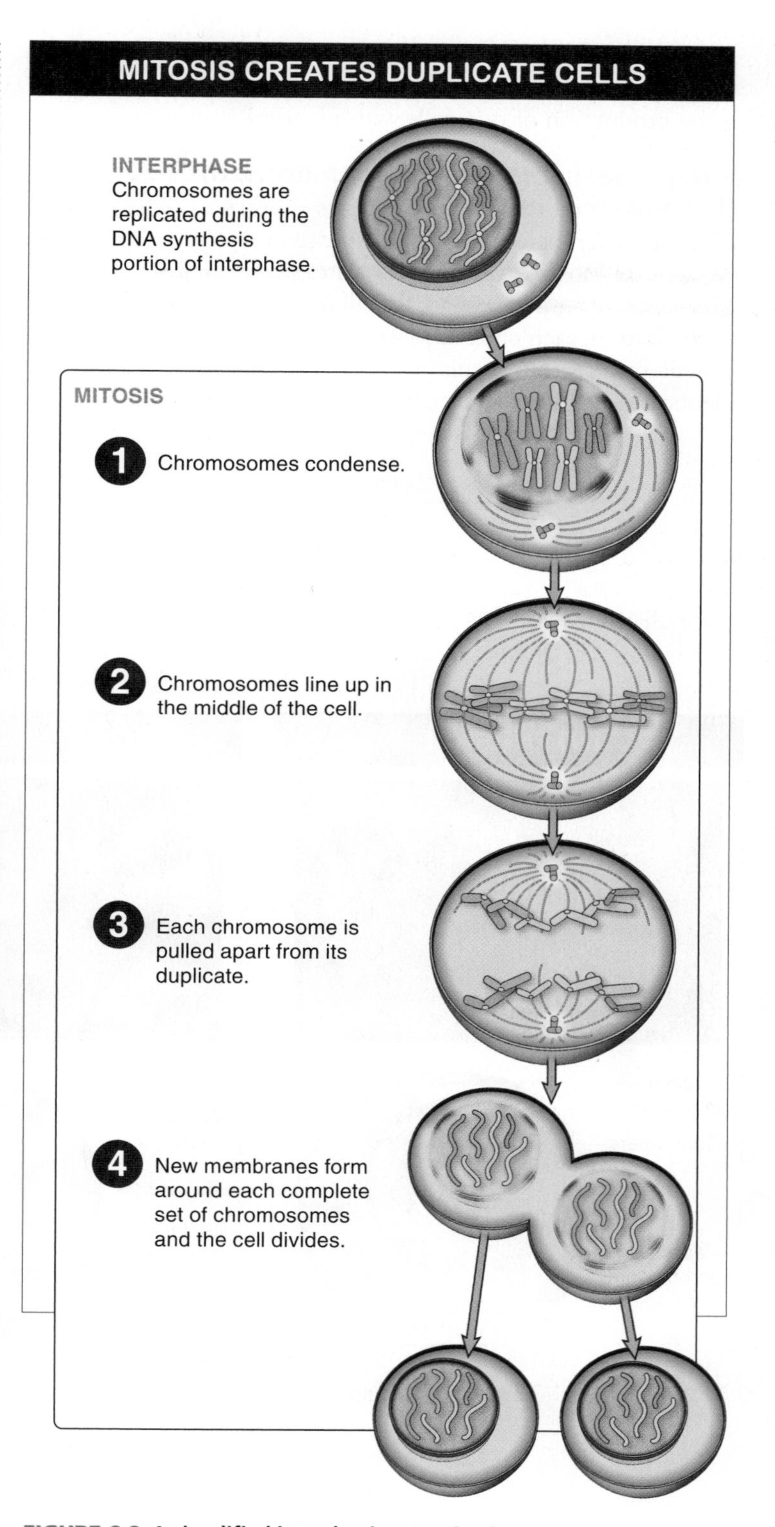

FIGURE 6-8 **A simplified introduction to mitosis.**

6•8

The details: mitosis is a four-step process.

Let's look at the process of mitosis in a bit more detail, keeping in mind that the ultimate consequence of the process is the production of two cells with identical chromosomes.

Interphase: in preparation for mitosis, the chromosomes replicate Processes essential to cell division take place even before the mitotic phase of the cell cycle begins. During the DNA synthesis part of interphase, every chromosome replicates itself. Recall from Section 6-5 that prior to replication, each chromosome is just a long, linear strand of genetic material; after replication, each chromosome is a *pair* of identical long, linear strands, held together at the center.

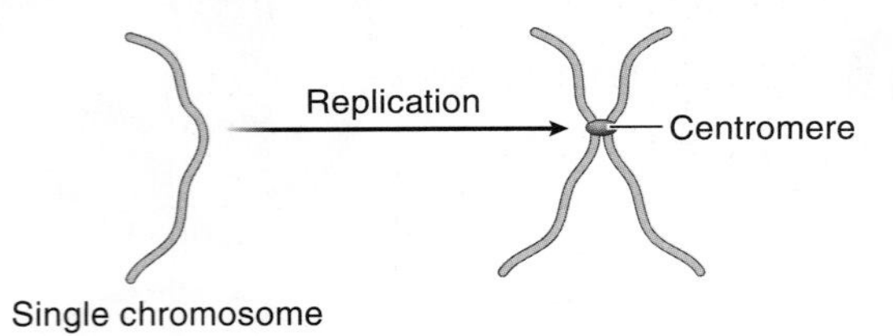

Mitosis The actual process of cell division occurs in four steps (**FIGURE 6-9**).

1. Prophase: the long, linear chromosomes that have replicated condense. When you look at a cell through a microscope, you generally won't see any chromosomes because the cells you're looking at are in some part of their interphase. At this point, the chromosomes are strung out so thin that they are not visible. Mitosis begins with prophase, when the chromosomes in the cell's nucleus become more and more tightly coiled. As they condense, they become thick enough to be seen through a light microscope.

At this point, in prophase, each chromosome looks like the letter X. This appearance is misleading for a couple of reasons. First, during all of the cell cycle except mitosis, chromosomes are uncoiled and spread out in a diffuse way like a mass of spaghetti. Because they are so stretched out, they are not dense

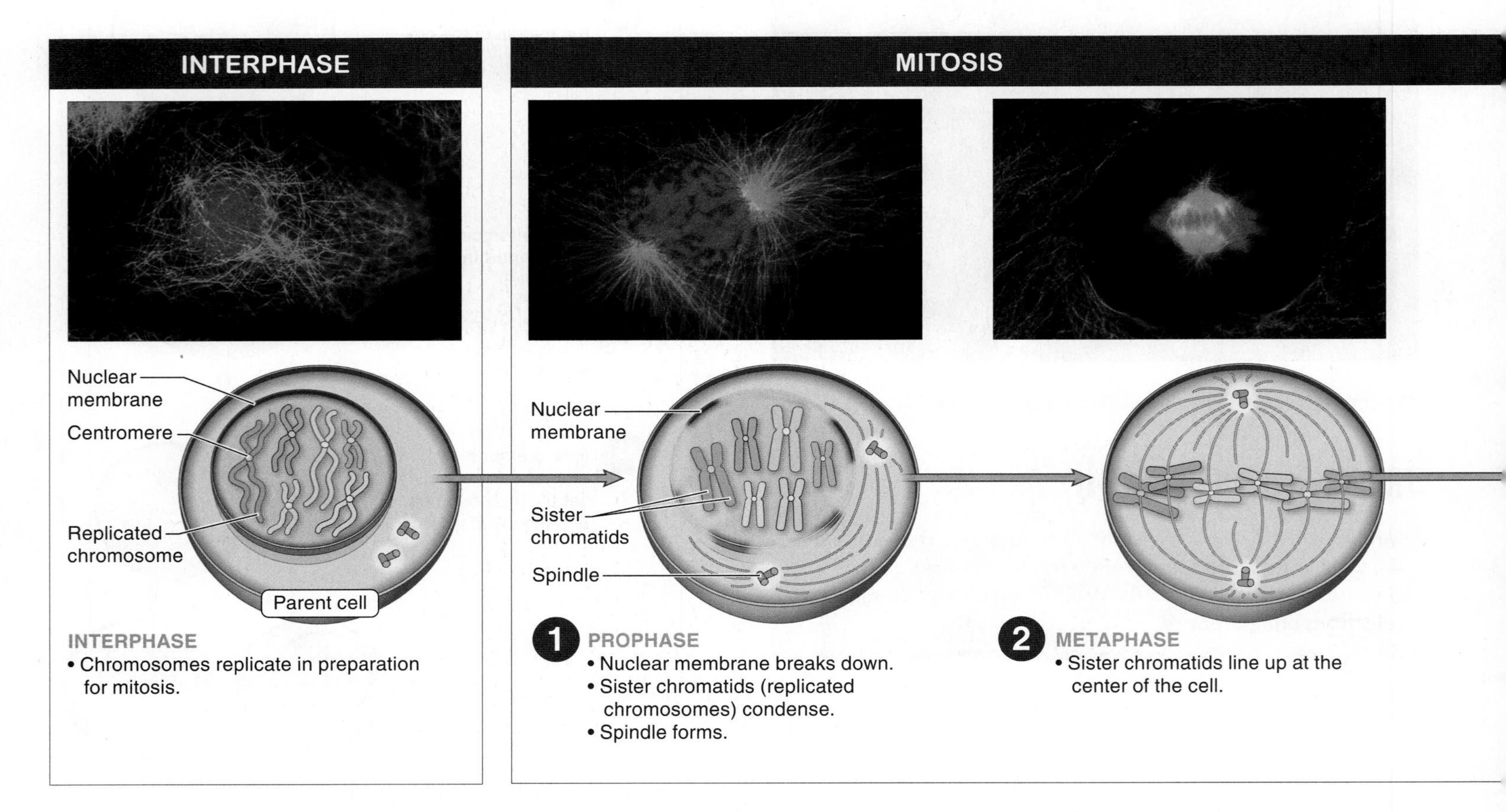

FIGURE 6-9 Mitosis: cell duplication, step by step.

enough to be visible. Second, chromosomes are not actually X-shaped. They are *linear.* Each X is really two identical linear DNA molecules: a chromosome and its identical replicated copy, joined at the centromere. (Each of the identical DNA molecules is called a **chromatid;** together, the pair are called **sister chromatids.**) The reason for the X shape in most photos of chromosomes is that the only time the chromosome is coiled tightly and thus thick enough to be seen is during cell division.

Q Animal chromosomes are linear. So, in pictures, why do they look like the letter X?

Centromere

Sister chromatids

Because a lot of room is required for separation of the duplicated chromosomes, the membrane around the nucleus is dismantled and disappears near the end of this first stage of mitosis. At the same time, a structure called the **spindle** is assembled. The spindle, which is part of the cell's cytoskeleton, can be thought of as a group of parallel threads stretching from one pole of the cell to the other. These threads (known as **spindle fibers**) pull the sister chromatids to the middle of the cells and will eventually be used to pull the chromatids apart as cell division proceeds.

2. Metaphase: the chromosomes congregate at the cell center. After condensing, the chromatids seem to move aimlessly around the cell, but eventually they line up at the cell's center, pulled by spindle fibers attached to the centromere. Metaphase can be considered the half-time of mitosis. At the end of metaphase, all the chromatids are lined up in an orderly fashion, straddling the center in a "single-file" congregation that is called the metaphase plate. The chromatids are at their most condensed during this part of mitosis.

3. Anaphase: the chromatids separate and move in opposite directions. In anaphase, the fibers attached to the centromeres begin pulling each chromatid in the sister chromatid pairs toward opposite poles of the cell. From each pair of sister chromatids, one strand of DNA is pulled

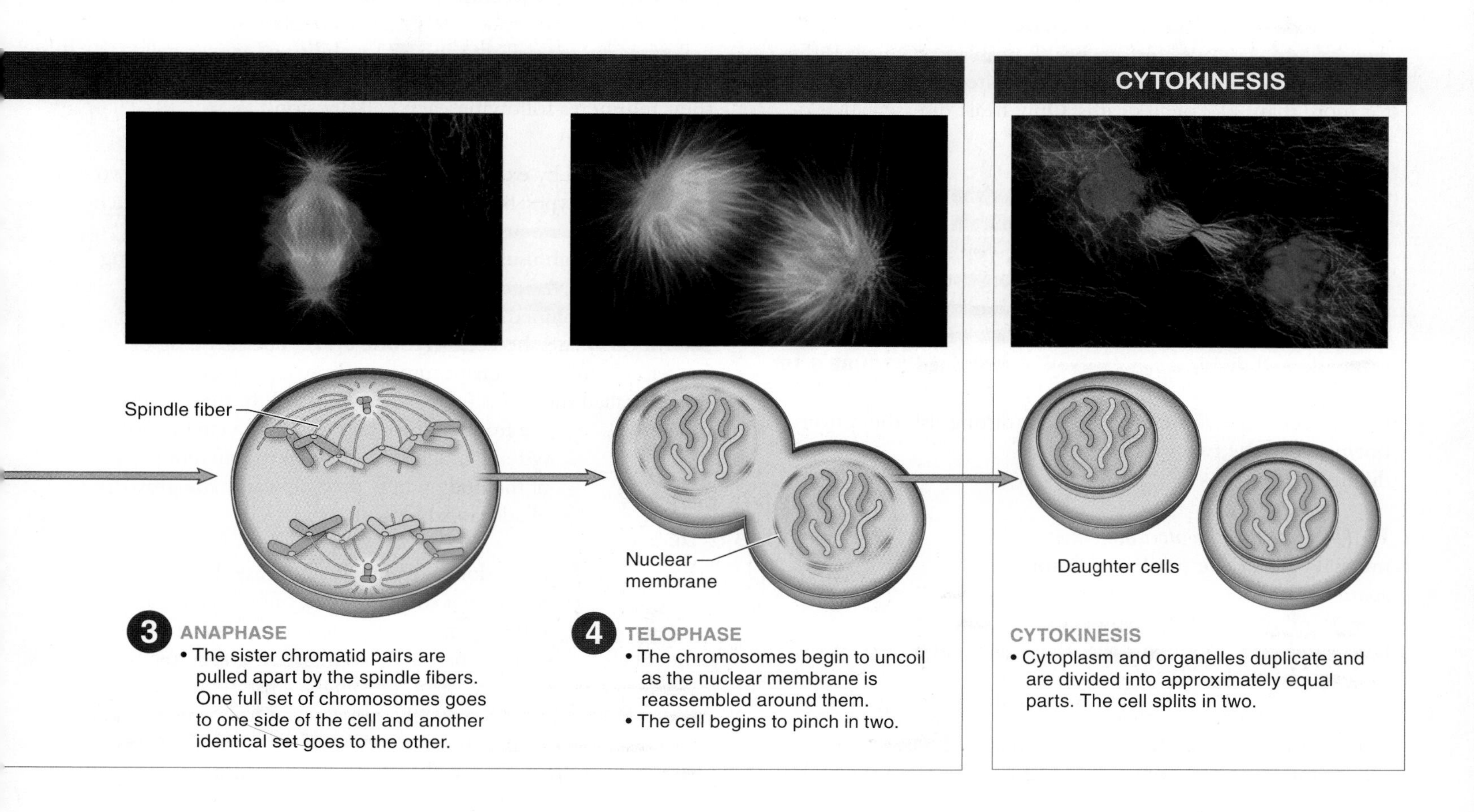

in one direction and the other, identical strand is pulled in the opposite direction. At the end of anaphase one full set of chromosomes is at one end of the cell and another identical full set is at the other end. These chromosome sets will eventually reside in the nucleus of each of the two new daughter cells that result from the cell division.

4. Telophase: new nuclear membranes form around the two complete chromosome sets. With two full, identical sets of chromosomes collected at either end of the cell, the cell is prepared to divide into two genetically identical cells. In this last step, called telophase, the chromosomes begin to uncoil and fade from view, nuclear membranes reassemble, and the cell begins to pinch into two.

The process of mitosis, the division of a cell's nucleus into two identical sets, is generally followed by cytokinesis, during which the cell's cytoplasm is also divided into approximately equal parts, with some of the organelles going to each of the two new cells. Only when cytokinesis is complete do the two new daughter cells, each with an identical nucleus containing identical genetic material, enter interphase and begin the business of being cells.

TAKE-HOME MESSAGE 6•8

The ultimate consequence of mitosis and cytokinesis is the production of two genetically identical cells.

6•9

Cell division out of control means cancer.

Too much of a good thing can be bad. This is especially true when cell division runs amok, a situation that can lead to cancer. **Cancer** is defined as unrestrained cell growth and division and can cause serious health problems. Cancer is the second leading cause of death in the United States, responsible for more than 20% of all deaths. Only heart disease causes more deaths.

Cancer occurs when some disruption of the DNA in a normal cell interferes with the cell's ability to regulate cell division. DNA disruption can be caused by chemicals that mutate DNA or by sources of high energy such as X rays, the sun, or nuclear radiation. Cancer can even be caused by some viruses. However it begins, once a cell loses control over its cell cycle, cell division can proceed unrestrained (**FIGURE 6-10**).

Cancer cells have several features that distinguish them from normal cells; the two most significant differences are:

Q What is cancer? How does it usually cause death?

1. *They lose their "contact inhibition."* Most normal cells divide until they bump up against other cells or collections of cells. At that point, they stop dividing. Cancer cells, however, ignore the signal that they are at high density and continue to divide.

2. *Cancer cells can divide indefinitely.* As we saw in Section 6-1, normal cells can divide approximately 50 times. After that point, they may continue living but they lose the ability to divide. Cancer cells, on the other hand, never lose their ability to divide and continue to do so indefinitely, even in the presence of conditions that normally would halt the cell cycle before cell division. (The ability of cancer cells to divide indefinitely is made possible because they rebuild their telomeres following each cell division.)

Tumors caused by excessive cell growth and division are of two very different types: benign and malignant. Benign tumors, such as many moles, are just masses of normal cells that do not spread. They can usually be removed safely without any lasting consequences. Malignant tumors, on the other hand, are the result of unrestrained growth of cancerous cells. They grow continuously and shed cells (**FIGURE 6-11**). The shedding of cancer cells from malignant tumors is how cancer spreads, a process called metastasis (meh-TASS-tuh-siss). In this process, cancer cells separate from a tumor and invade the circulatory system, which spreads them to different parts of the body where they can cause the growth of additional tumors.

How does cancer actually cause death? Somewhat surprisingly, it's not because of some chemical or genetic property of the cancer cells. It's simpler than that. As a tumor gets larger, it takes up more and more space, pressing against neighboring cells and tissue (**FIGURE 6-12**). Eventually, the tumor may block other cells and tissue from carrying out their normal functions and even kill them. This cell dysfunction or cell death can

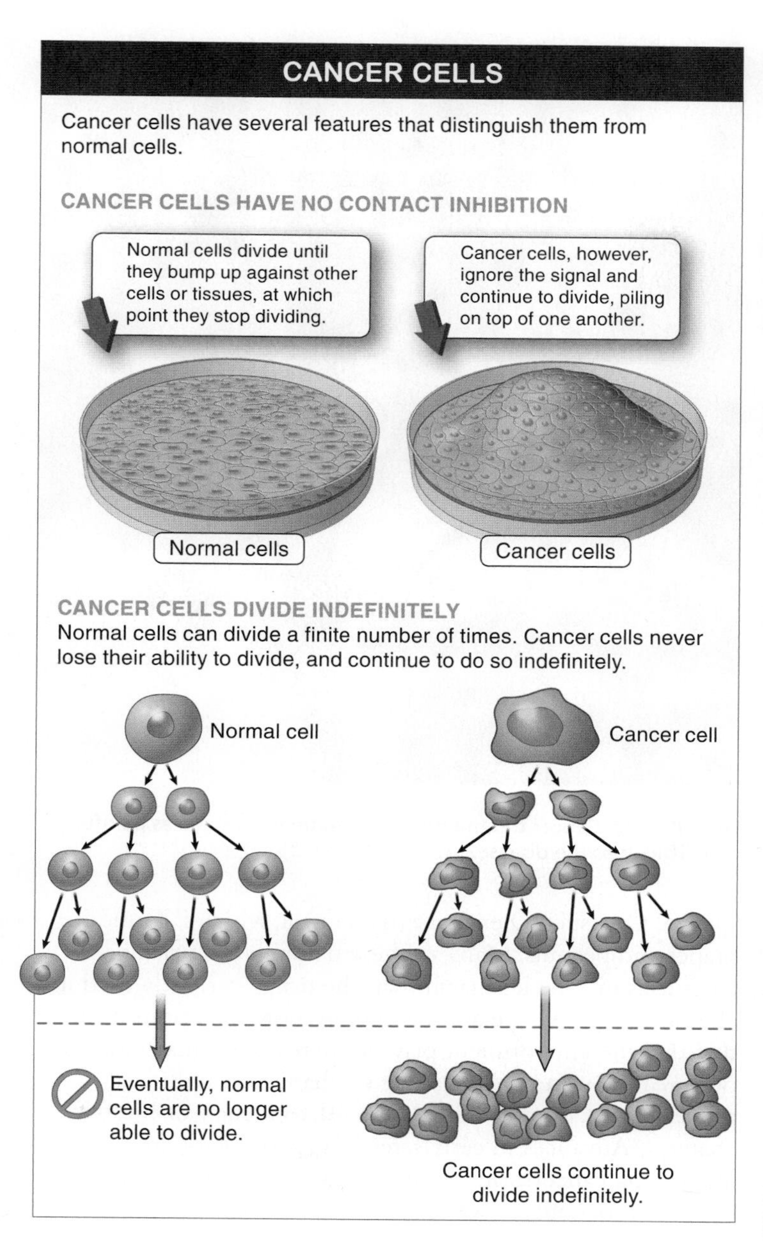

FIGURE 6-10 Uncontrolled cell division. Cancer occurs when there is a disruption in cells' ability to regulate cell division.

have disastrous consequences when the affected normal tissue controls processes critical to life, such as breathing, heart function, or the detoxification processes in the liver.

To treat cancer, the rapidly dividing cells must be removed surgically, killed, or their division at least slowed down. Currently, the killing and slowing down are done in two ways: by chemotherapy and by radiation. In chemotherapy, drugs that interfere with cell division are administered, slowing down the growth of tumors. Because these drugs interfere with rapidly dividing cells throughout the body (not just the rapidly dividing

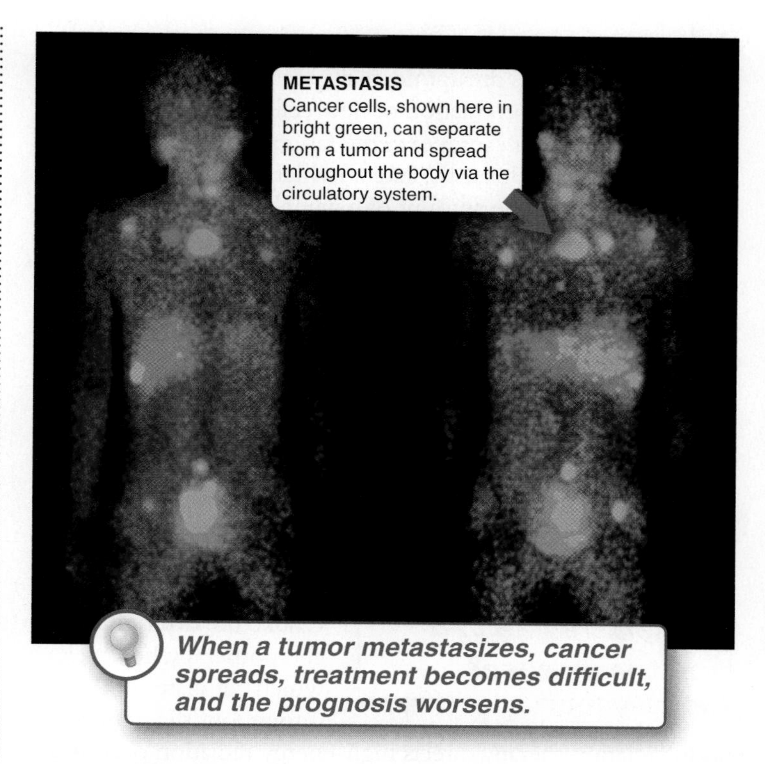

FIGURE 6-11 Multiple tumors. Metastasis spreads cancer cells throughout the body.

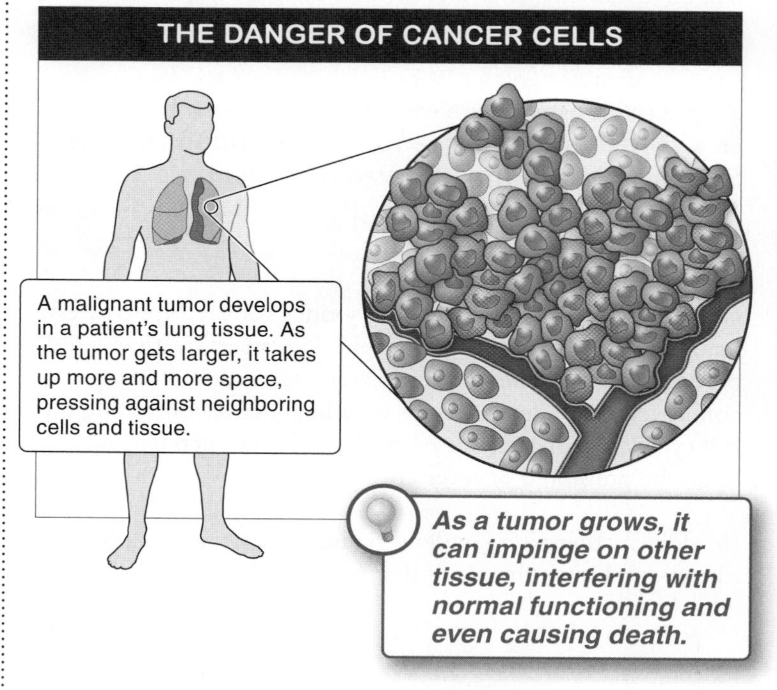

FIGURE 6-12 Cancer's effects. Cancer cells harm the body by crowding and disrupting normal cells.

cancer cells), they can have terrible side effects. In particular, chemotherapy drugs disrupt normal systems that rely on the rapid and constant production of new cells. For instance, chemotherapy often causes extreme fatigue and shortness of breath because it reduces the rate at which red blood cells are produced, limiting the amount of oxygen that can be transported throughout the body. By interfering with the division of bone marrow stem cells, chemotherapy also reduces the production of platelets and white blood cells and thus increases bruising and bleeding, and increases susceptibility to infection. Another location of rapidly dividing cells commonly affected by chemotherapy and radiation is the hair follicle. Many people lose their hair when undergoing chemotherapy (it usually grows back when treatments stop).

Q Why is the treatment for cancer often considered as bad as the disease?

Like chemotherapy, radiation works by disrupting cell division. Unlike the drugs used in chemotherapy, however, which circulate throughout the entire body, radiation therapy directs high-energy radiation only at the part of the body where a tumor is located. Like chemotherapy, the radiation process is not perfect and nearby tissue is often harmed as well. The significant negative effects of chemotherapy and radiation treatment on normal tissue and the suffering this leads to has caused some patients to comment that the treatment for cancer is worse than the disease itself—especially because patients may be asymptomatic at the time of treatment (**FIGURE 6-13**).

> "Bitterness is like cancer. It eats upon the host. But anger is like fire. It burns it all clean."
>
> —Maya Angelou, U.S. poet and writer

Is there hope for a complete cure soon? Not yet. Nonetheless, many potentially successful therapies for cancer treatment and prevention are on the horizon. Extensive research is being conducted on the mechanisms by which genes controlling the cell cycle are damaged, for example, and how such damage might be prevented or reversed.

In addition, there is a rapidly growing field of research into cancer-inhibiting chemicals that have evolved in many plant species. One such chemical is resveratrol, which is found in grapes and peanuts. Many of these naturally occurring chemicals may be less harsh than the drugs currently used in chemotherapy. We are also becoming more conscious of the role that the environment plays in increasing cancer risk; our greater awareness is leading to changes in industrial processes and, on the individual level, to dietary or lifestyle changes. Advances in early screening, particularly in identifying cancer prior to metastasis, continue to have significant benefits as well.

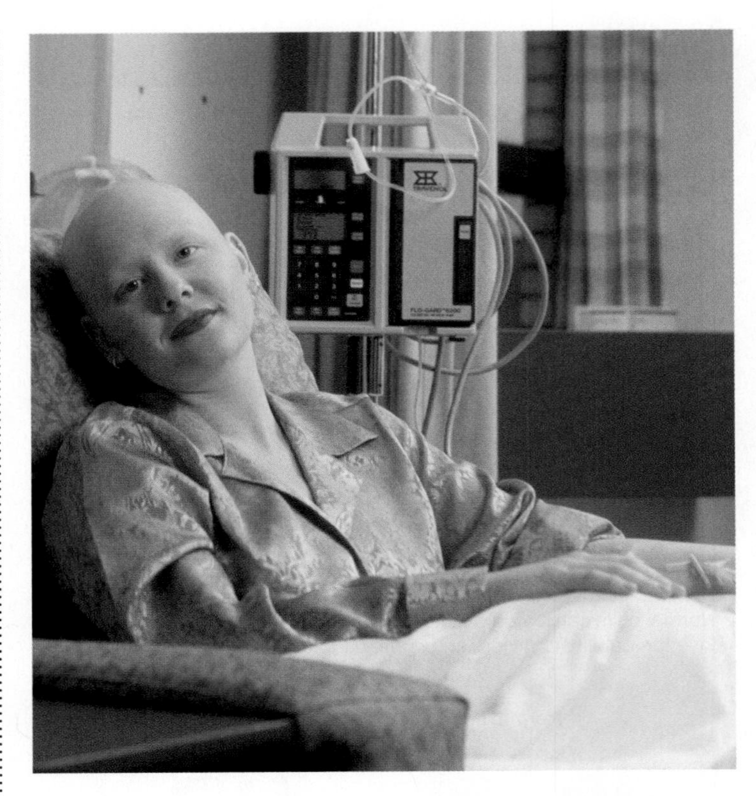

FIGURE 6-13 Cancer treatments. The treatments can be as painful and difficult as the disease.

TAKE-HOME MESSAGE 6•9

Cancer is unrestrained cell growth and division, leading to large masses of cells called malignant tumors that can cause serious health problems. Treatment focuses on killing or slowing down the fast-growing and dividing cells, usually using chemotherapy and/or radiation.

③ Meiosis generates sperm and eggs and a great deal of variation.

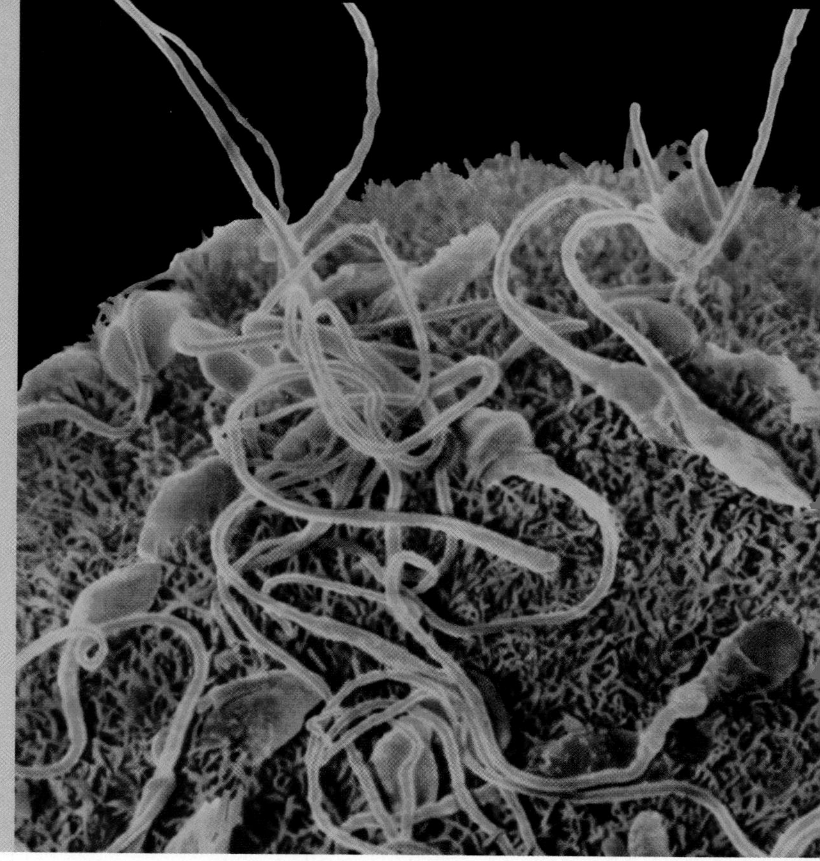

Human gametes: sperm attempting to fertilize an egg.

6•10

Overview: sexual reproduction requires special cells made by meiosis.

There are two ways in which an organism can reproduce. Many organisms, including bacteria, fungi, and even some plants and animals, undergo asexual reproduction, in which a single parent produces identical offspring. Other organisms, including most animals and plants, undergo sexual reproduction, in which offspring are produced by the fusion of two reproductive cells in the process of **fertilization.**

Sexually reproducing organisms, humans included, have a problem to solve when it comes to cell division. When offspring are created, they carry the genetic material from two individuals. But think about the difficulties this presents. If reproductive cells were produced through mitosis, both parents would contribute a full set of genes—that is, 23 pairs of chromosomes in humans—to create a new individual, and the new offspring would inherit 46 pairs of chromosomes in all. And when that individual reproduced, if he or she contributed 46 pairs of chromosomes and his or her mate also contributed 46 pairs, their offspring would have 92 pairs of chromosomes. Where would it end? The genome would double in size every generation. That wouldn't work at all. At the very least, within a few generations cells would be so overloaded with chromosomes that they would explode.

Sexually reproducing organisms have evolved a way to avoid chromosome overload. The solution is **meiosis,** a process that enables organisms, prior to fertilization, to make special reproductive cells, the gametes, which have only half as many chromosomes as the rest of the cells in the organism's body. In other words, in anticipation of combining one individual's genome with another's, meiosis reduces each individual's genome by half in the gametes. In humans, for example, gamete cells have only one set of 23 chromosomes, rather than two sets. In genetics, the term **diploid** refers to cells that have two copies of each chromosome, and the term **haploid** refers to cells that have one copy of each chromosome. Thus, somatic cells are diploid and gametes, the cells produced in meiosis, are haploid. At fertilization, two haploid

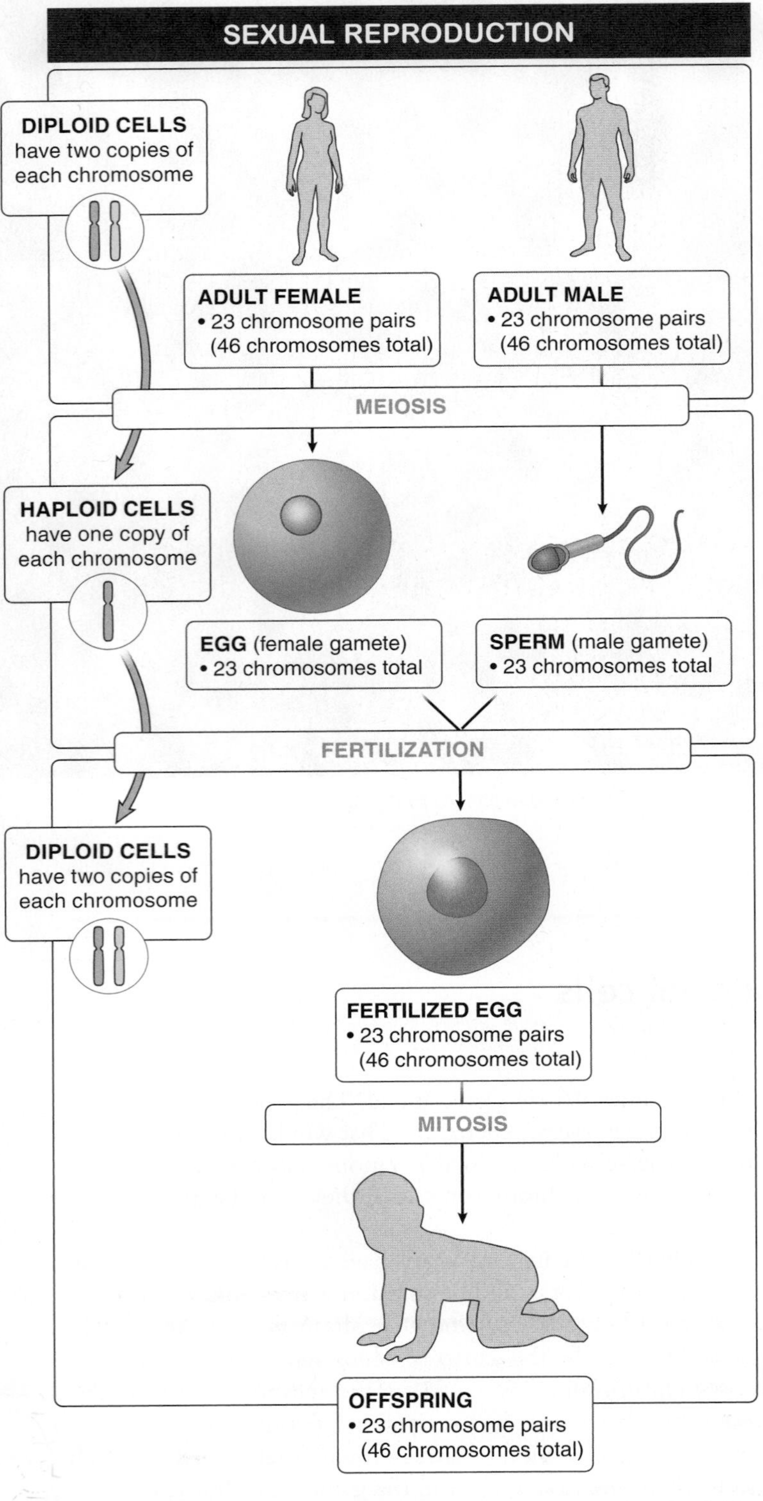

FIGURE 6-14 Sexual reproduction. Diploid organisms produce haploid gametes that fuse at fertilization and return to the diploid state.

cells, each with one set of 23 chromosomes, merge and create a new individual with the proper diploid human genome of 46 chromosomes. And when the time comes to reproduce, this new individual also will produce haploid gametes through meiosis that have only a single set of 23 chromosomes. With sexual reproduction, then, diploid organisms produce haploid gametes that fuse at fertilization to restore the diploid state (**FIGURE 6-14**). This process repeats perpetually and maintains a stable genome size in a species.

Although there are some variations on this pattern of alternation between the haploid and diploid states, most multicellular animals produce simple haploid cells for reproduction. And after two of those gametes come together to form a diploid fertilized egg, multiple cell divisions via mitosis again produce a diploid multicellular animal.

Meiosis achieves more than just a reduction in the amount of genetic material in gametes. As a diploid individual, you have two copies of every gene in every somatic cell: one from your mother and one from your father. When making haploid cells from cells that are diploid, an individual creates cells that have one allele for each trait rather than two. Which of these two alleles is included in each gamete cannot be predicted.

Each egg or sperm is produced with a varied combination of maternal and paternal alleles so that all of the gametes an individual makes carry a unique set of alleles. As a consequence, each offspring inherits a slightly different assemblage of that parent's alleles. All offspring resemble their parents, yet none resembles them in exactly the same way. This variation among offspring can be seen among three of the Kennedy brothers in the photo that opens this chapter (pages 208–209). Edward Kennedy is in the middle, with Robert on the left and John on the right.

In all, meiosis has two important outcomes:

1. It reduces the amount of genetic material in gametes.
2. It produces gametes that differ from one another with respect to the combinations of alleles they carry.

In the next sections, we examine the exact steps by which these outcomes occur and investigate the consequences of all the variation that meiosis and sexual reproduction generate.

TAKE-HOME MESSAGE 6•10

Meiosis is the process by which gametes are produced in sexually reproducing organisms. It results in gametes that have only half as much genetic material as the parent cell and that differ from one another with respect to the combinations of alleles they carry.

6•11

Sperm and egg are produced by meiosis: the details, step by step.

Mitosis is an all-purpose process for cell division. It occurs all over the body, all the time. Meiosis, on the other hand, is a special-purpose process. It occurs in just a single place: the **gonads** (the ovaries and testes in sexually reproducing animals). And it occurs for just a single reason: the production of gametes (reproductive cells). To better understand meiosis, let's examine in detail the steps by which a diploid cell creates haploid cells, each of which is different from the other.

Meiosis starts with a diploid cell—not just any diploid cell, but one of the specialized diploid cells, found in the gonads, that is capable of undergoing meiosis. Thus, for humans, meiosis starts with a cell that has 46 chromosomes: a maternal copy and a paternal copy of each of the 23 human chromosomes. Together, the maternal and paternal copies of a chromosome are called a **homologous pair,** or **homologues.** This means that there 23 homologous pairs in a diploid human cell (**FIGURE 6-15**).

Before meiosis can occur (and just as we saw with mitosis), all of these 46 chromosomes are duplicated. So now, instead of having just a maternal and paternal copy of chromosome 1, we have two copies of the maternal copy of chromosome 1, bound together as sister chromatids, and a pair of sister chromatids for the paternal copy of chromosome 1. This means that when meiosis begins, we have 92 (2×46) strands of DNA after replication. Remember that this duplication of the chromosomes occurs during the interphase portion of the cell cycle, before actual cell division has begun.

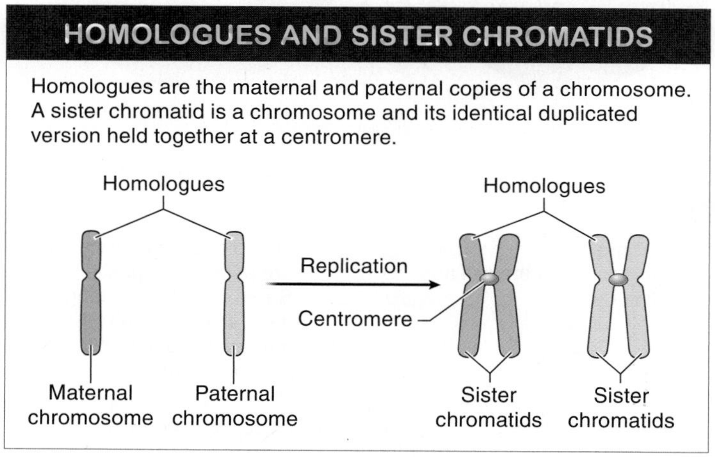

HOMOLOGUES AND SISTER CHROMATIDS

Homologues are the maternal and paternal copies of a chromosome. A sister chromatid is a chromosome and its identical duplicated version held together at a centromere.

FIGURE 6-15 Chromosome vocabulary: homologues and sister chromatids.

Unlike mitosis, which has only one cell division, cells undergoing meiosis divide twice. In the first division, the homologues separate. In other words, for each of the 23 chromosome pairs, the maternal sister chromatids and the paternal sister chromatids separate into two new cells. In the second division, each of the two new cells divides again, and the sister chromatids separate from each other into two newer cells. At the end of meiosis, there are four new cells, each of which has 23 strands of DNA—that is, 23 chromosomes (**FIGURE 6-16**).

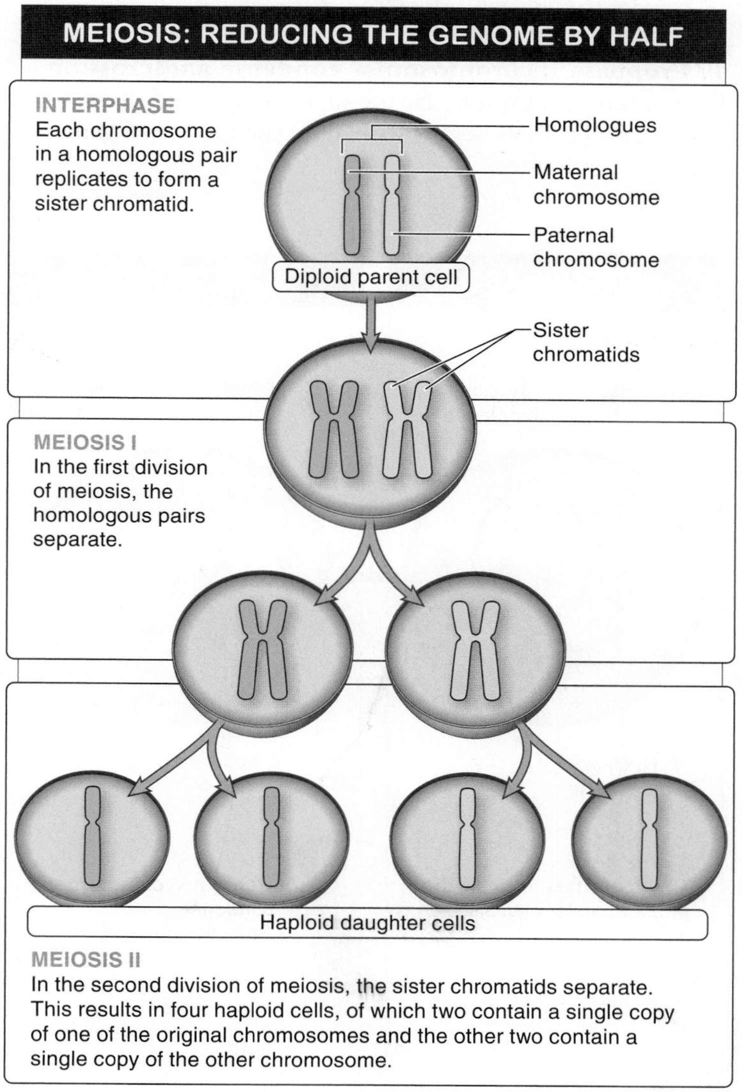

MEIOSIS: REDUCING THE GENOME BY HALF

INTERPHASE
Each chromosome in a homologous pair replicates to form a sister chromatid.

MEIOSIS I
In the first division of meiosis, the homologous pairs separate.

MEIOSIS II
In the second division of meiosis, the sister chromatids separate. This results in four haploid cells, of which two contain a single copy of one of the original chromosomes and the other two contain a single copy of the other chromosome.

FIGURE 6-16 Meiosis reduces the genome by half in anticipation of combining it with another genome.

Interphase: in preparation for meiosis, the chromosomes replicate Meiosis is a one-way process: cells undergoing meiosis produce gametes that do not themselves undergo meiosis, so the cells that undergo meiosis do not have a cell cycle. Before meiosis occurs, however, preparations for cell division occur that are similar to those occurring in interphase of the mitotic cell cycle. In particular, there is a DNA synthesis phase during which every chromosome creates an exact duplicate of itself by replicating. As in mitosis, before replication each chromosome is a long, linear strand of genetic material, while after replication each chromosome is a *pair* of identical long, linear strands, held together at the centromere.

Meiosis begins following replication and occurs in eight steps (**FIGURE 6-17**).

Meiosis division I: the homologues separate

1. Prophase I: chromosomes condense and crossing over occurs. This is by far the most complex of all the phases of meiosis. It begins with all of the replicated genetic material condensing.

As the sister chromatids become shorter and thicker, the homologous chromosomes come together. That is, the maternal and paternal sets of sister chromatids for chromosome 1 come together so there are four versions of this chromosome lined up together. The maternal and paternal sets of chromosome 2, 3, 4, and so on, also line up. Under a microscope, the two pairs of sister chromatids appear as pairs of X's lying on top of each other.

At this point, the sister chromatids that are next to each other do something that makes every sperm or egg cell genetically unique: they swap little segments. In other words, when this is happening inside of you, some of the genes that you inherited from your mother get swapped onto the strand of DNA you inherited from your father, and the corresponding bit from your father is inserted into the DNA strand from your mother. This event is called **recombination** or **crossing over,** and it can take place at several (up to dozens) of spots. As a result of crossing over, every sister chromatid ends up having a unique mixture of the genetic material that you received from your two parents. Note that crossing over occurs only during the production of gametes, in meiosis. It does not occur during

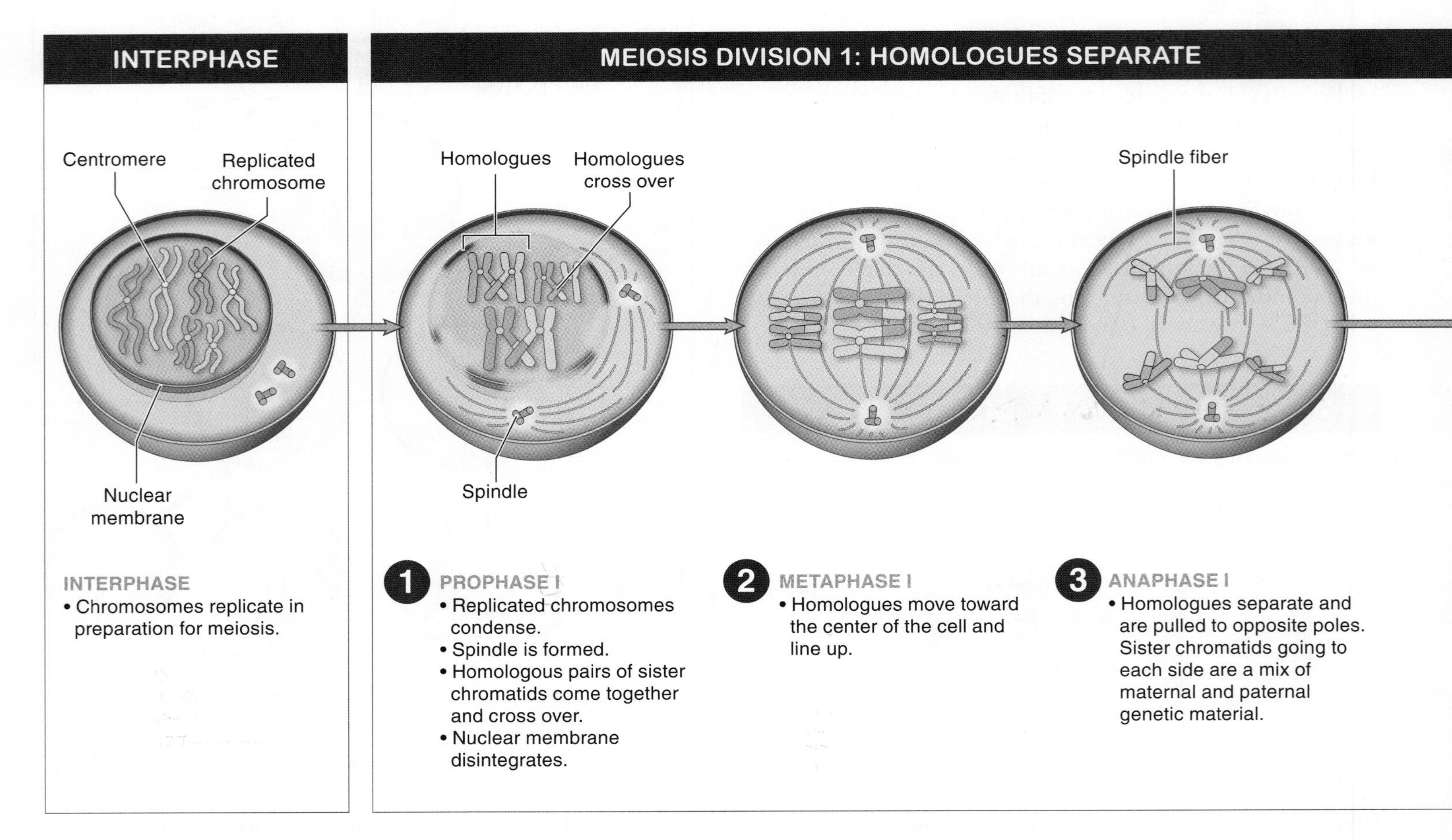

FIGURE 6-17 Meiosis: generating reproductive cells, step by step.

mitosis. Following crossing over, the nuclear membrane disintegrates.

We explore crossing over in more detail in Section 6-13. The remaining steps of meiosis are relatively straightforward.

2. Metaphase I: chromosomes all line up at the center of the cell. After crossing over has occurred, each pair of homologous chromosomes (that is, the pairs of X's lying on top of each other) moves to the center of the cell, pulled by the spindle fibers to form the arrangement called the metaphase plate. Remember that each pair of homologous chromosomes includes the maternal and paternal version of the chromosome *and* the replicated copy of each—four strands in all.

3. Anaphase I: homologues are pulled to either side of the cell. This is the beginning of the first cell division that occurs during meiosis. In this phase, the homologues are pulled apart toward opposite sides of the cell. One of the homologues (consisting of two sister chromatids) goes to the top pole, the other to the bottom. At this point, something else occurs that contributes, along with crossing over, to making all the products of meiosis genetically unique. The maternal and paternal sister chromatids are pulled to the ends of the cell in a random fashion. As a result, the sister chromatids gathered at each pole are a mix of maternal and paternal sister chromatids. Imagine all the different combinations that can occur in a species with 23 pairs of chromosomes. For chromosome pair 1, perhaps the maternal homologue goes to the "top" pole and the paternal to the "bottom" pole. And for pair 2, perhaps the maternal homologue also goes to the top and the paternal to the bottom. But maybe for pair 3 it is the paternal homologue that goes to the top and the maternal homologue to the bottom. For each of the 23 pairs, you cannot predict which will go to the top and which will go to the bottom. There is a *huge* number of possible combinations. (Can you figure out how many? Here's a hint: if there were 2 chromosome pairs, the answer would be 4; if there were 3 chromosome pairs, the answer would be 8...)

4. Telophase I and cytokinesis: nuclear membrane reassembles around sister chromatids and two daughter cells form. After the chromatids arrive at the two poles of the cell, the nuclear membrane re-forms, then cytokinesis occurs and the cytoplasm divides, and the cell membrane pinches the cell into two daughter cells, each with its own nucleus,

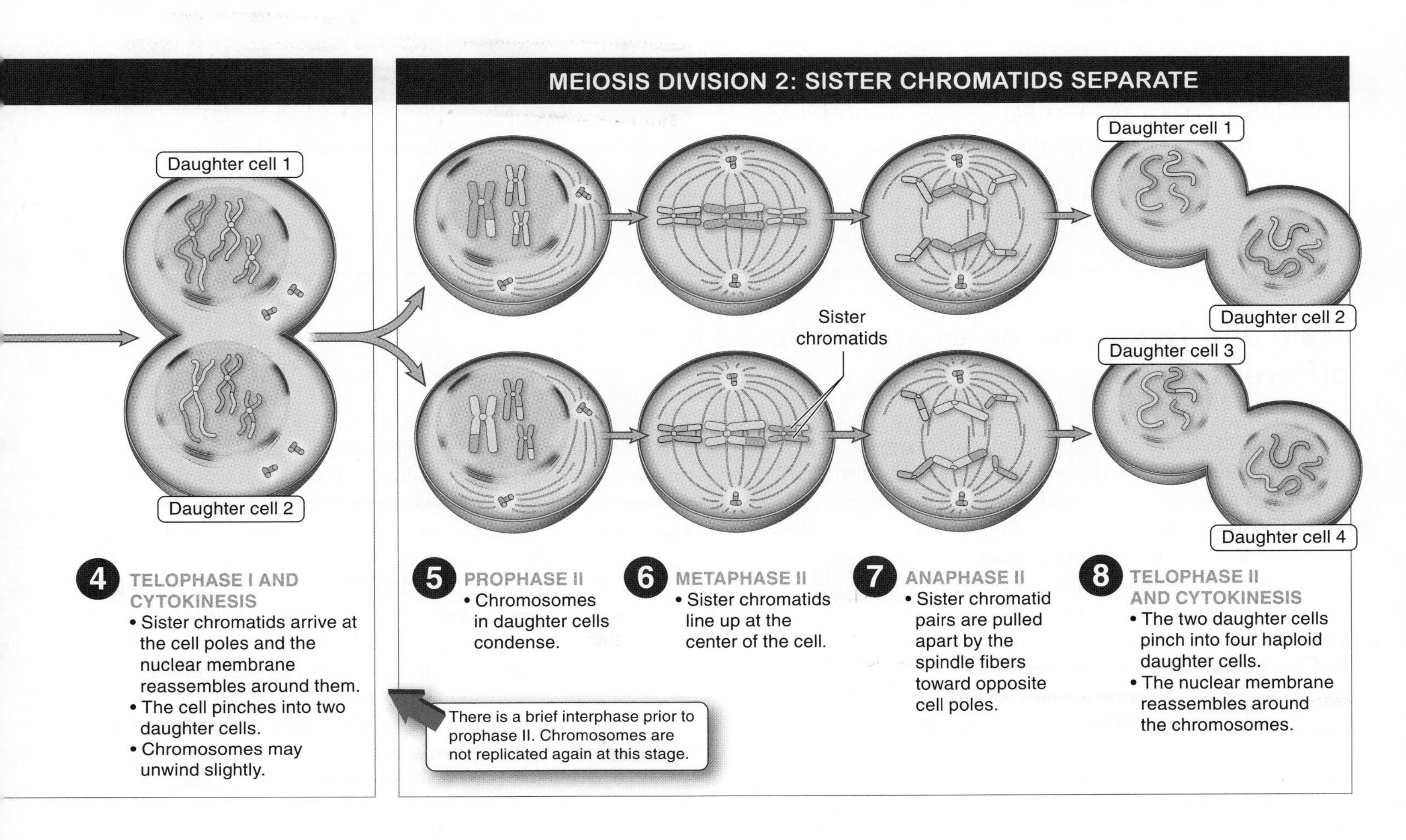

containing the genetic material—two sister chromatids for each of the 23 chromosomes in humans.

Meiosis division II: separating the sister chromatids There is a brief interphase after the first division of meiosis. In some organisms, the DNA molecules (now in the form of chromatids) briefly uncoil and fade from view. In others, the second part of meiosis begins immediately. *It is important to note that in the brief interphase prior to prophase II, there is no replication of any of the chromosomes.*

5. Prophase II: chromosomes re-condense. The second part of meiosis, like the first division of meiosis, is a four-phase process. It begins with prophase II. In this phase, the genetic material in each of the two daughter cells once again coils tightly, making the chromatids visible under the microscope.

6. Metaphase II: sister chromatids line up at the center of the cell. In each of the two daughter cells, the sister chromatids (each appearing as an X) move to the center of the cell, pulled by spindle fibers attached to the centromere, where the sister chromatids are held together. The congregation of all the genetic material in the center of each daughter cell is visible as a flat metaphase plate.

7. Anaphase II: sister chromatids are pulled to opposite sides of the cell. This phase starts with 46 pieces of DNA in each cell created by meiosis I: 23 sister chromatid pairs. During this phase, the fibers attached to the centromere begin pulling each chromatid in the 23 sister chromatid pairs toward opposite poles of each daughter cell. When anaphase II is finished, each of what will become the four daughter cells (you could think of them as granddaughter cells) has one single copy of each of the 23 chromosomes.

8. Telophase II and cytokinesis: nuclear membranes reassemble and the two daughter cells pinch into four haploid gametes. Finally, the sister chromatids for all 23 chromosomes have been pulled to opposite poles. The cytoplasm then divides, the cell membrane pinches the cell into two new daughter cells, nuclear membranes begin to re-form, and the process comes to a close.

In humans, the outcome of one diploid cell undergoing meiosis is the creation of four haploid daughter cells, each with just one set of 23 chromosomes. These chromosomes contain a combination of traits from the individual's diploid set of chromosomes.

TAKE-HOME MESSAGE 6•11

Meiosis occurs only in gamete-producing cells. It is preceded by DNA replication and consists of two rounds of cellular division, one in which homologous pairs of sister chromatids separate and a second in which the sister chromatids separate. The final product of meiosis in a diploid organism is four haploid gametes.

6•12

Male and female gametes are produced in slightly different ways.

No matter how small or large they are, no matter whether they're plants or animals, there is just one way to distinguish males from females in all sexually reproducing organisms. Regardless of the species, the defining feature is always the same (and if you're thinking of anything visible to the naked eye, you're wrong). When there are two sexes—as there are in nearly every sexually reproducing plant and animal species—the females are the sex that produces the larger gamete, and the males produce the smaller, more motile, gamete (**FIGURE 6-18**). Because gametes are all produced through meiosis, it is a slight variation in how meiosis works in males and females that leads to this difference.

Q How do you distinguish a male from a female?

The size difference between the male gamete and the female gamete (sperm and eggs in animals) all comes down to the fact that sperm cells have very little cytoplasm, while eggs have a huge amount. During the production of sperm, the two divisions occur just as described in Section 6-11, resulting in four evenly sized cells that become sperm. During the production of eggs, things are a bit different. The first division occurs just as described in Section 6-11, except that in telophase I, as the cell divides, instead of forming two cells of the same size, the division is lopsided. The genetic material is evenly divided, but nearly all of the cytoplasm goes to one of the cells and almost none goes to the other. Then, in the second meiotic division of the

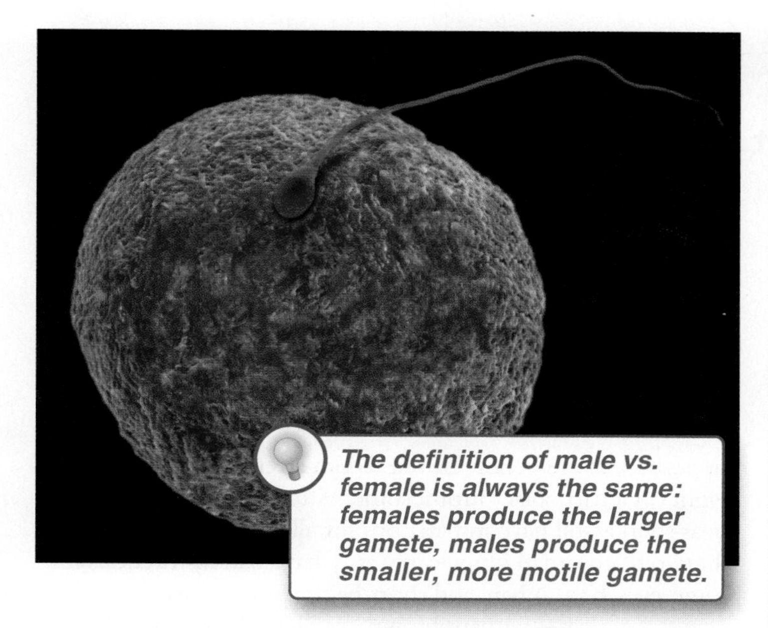

FIGURE 6-18 Egg and sperm. In sexually reproducing organisms, the female produces the larger gamete.

larger cell, there is again an unequal division of cytoplasm. As in the first division, one of the cells gets nearly all of the cytoplasm and the other gets almost none. The net result of meiosis in the production of eggs is one large egg with lots of cytoplasm and three small cells with very little cytoplasm. These small cells degrade almost immediately and never function as gametes (**FIGURE 6-19**).

Ultimately, whether it is eggs or sperm that are being produced through meiosis, each gamete ends up with just one copy of each chromosome. That way, the fertilized egg that results from the fusion of sperm and egg carries two complete sets of chromosomes, so the developing individual will be diploid. The extra cytoplasm carried by the egg contains a large supply of nutrients and other chemical resources to help with the initial development of the organism following fertilization.

TAKE-HOME MESSAGE 6•12

In species with two sexes—including nearly every sexually reproducing plant and animal species—females are the sex that produces the larger gamete, and males produce the smaller gamete. Whether it is male or female gametes that are being produced, each gamete ends up with just one copy of each chromosome.

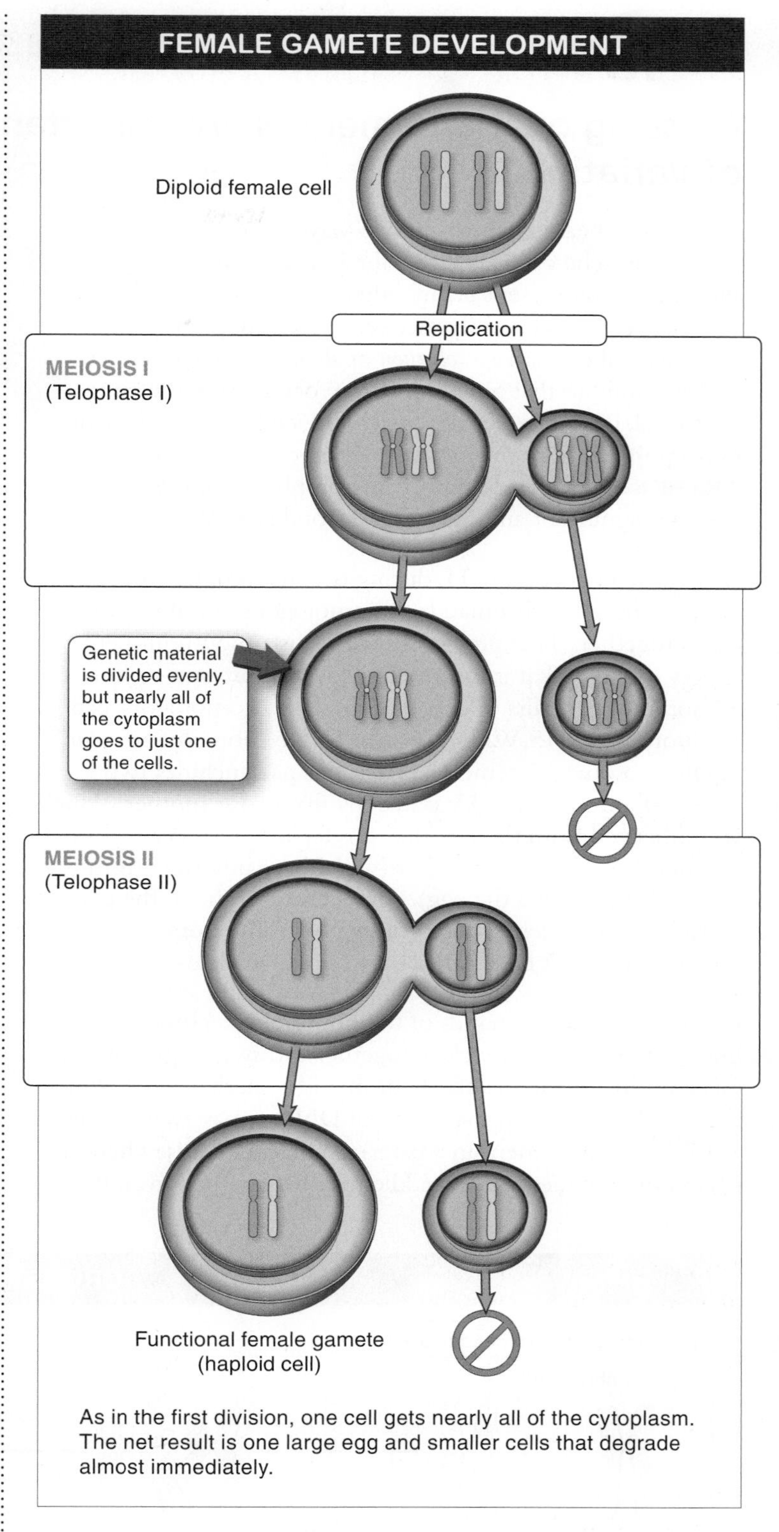

FIGURE 6-19 Unequal distribution of cytoplasm results in one large egg.

6·13

Crossing over and meiosis are important sources of variation.

Genetically speaking, there are two ways to create unique individuals. The obvious way is for an organism to carry an allele that is not present in any other individuals. Alternatively—and equally successful in creating uniqueness—an individual can carry a *collection* of alleles, no single one of which is unique, that has never before occurred in the same individual. Both types of novelty introduce important variation into a population of organisms. The process of crossing over (**FIGURE 6-20**), which occurs during prophase I in meiosis, creates a significant amount of the second type of variation.

As we saw in Section 6-11, during the first prophase of meiosis, the sister chromatids of homologous chromosomes all come together. That previous sentence is a mouthful! Let's review exactly what it means. Let's look at just one of your 23 homologous pairs of chromosomes, the homologous pair of chromosome 15. When we refer to the "homologous pair" of chromosome 15, remember that this pair includes two copies of chromosome 15: one copy from your mother (which you inherited from the egg that was fertilized to create you) and one copy from your father (which you inherited from the sperm that fertilized the egg). Each chromosome in the pair carries the same genes, but because they came from different people, they don't necessarily have the same alleles.

Once the sister chromatids of the homologous chromosomes line up, regions that are close together can swap segments. A piece of one of the maternal chromatids—perhaps including the first 100 genes on the strand of DNA—may swap places with the same segment in a paternal chromatid. Elsewhere, a stretch of 20 genes in the middle may be swapped from the other maternal chromatid with one of the paternal chromatids. Every time a swap of DNA segments occurs, an identical amount of genetic material is exchanged, so all four chromatids still contain the complete set of genes that make up the chromosome. The *combination* of alleles on each chromatid, though, is now different.

Suppose there are genes relating to eye color, hair color, and height on a particular chromosome. After crossing over, the linear strands still have instructions for all of those traits. But where one chromatid previously may have had instructions for brown eyes, brown hair, and short height, it may now have some differences—perhaps brown eyes, blond hair, and tall height. All of the alleles from your parents are still carried on one DNA molecule or another. But the *combination of traits* that are linked together on a single chromatid is new. And when a gamete, let's say it's an egg, carrying a new combination of alleles is fertilized by a sperm, the developing individual will carry a completely novel set of alleles. And so without creating any new traits (perhaps yellow eyes or purple hair), crossing over still creates gametes with collections of alleles that may never have existed together. In Chapter 8, we will see that this variation is tremendously important for evolution.

TAKE-HOME MESSAGE 6·13

Although it doesn't create any new traits, crossing over during the first prophase of meiosis creates gametes with collections of traits that may never have existed before; this variation is important for evolution.

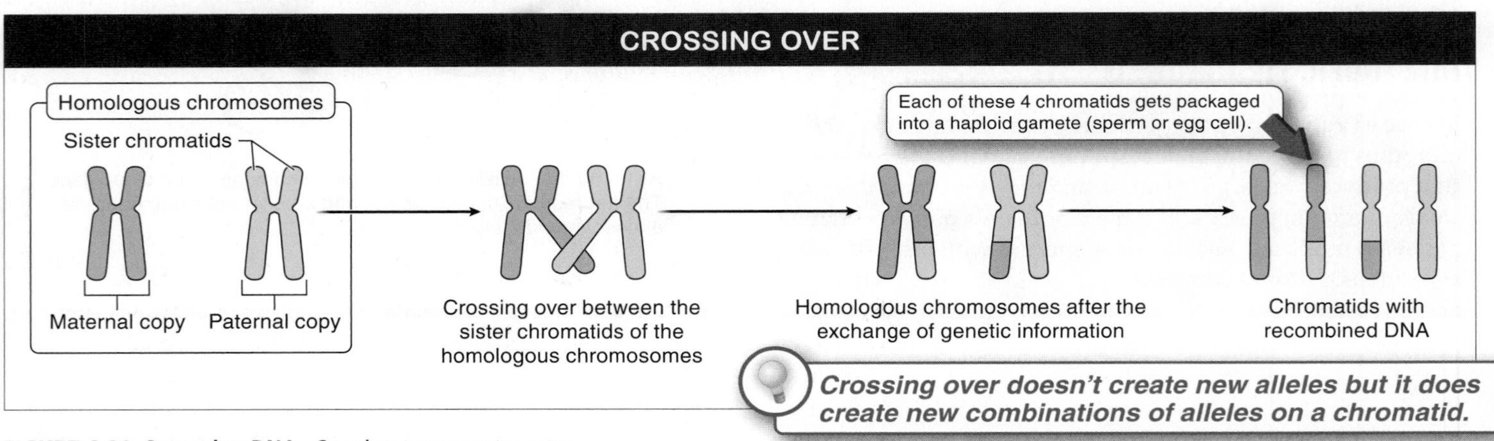

FIGURE 6-20 Swapping DNA. Crossing over creates new combinations of alleles on each chromatid.

6•14

What are the costs and benefits of sexual reproduction?

There are two fundamentally different ways by which cells and organisms can reproduce. On the one hand there is mitosis and asexual reproduction via binary fission. On the other hand there is meiosis and sexual reproduction. Is one method better than the other? It depends. In fact, the more appropriate question is, what are the advantages and disadvantages of each method and under what conditions do the benefits outweigh the costs?

What are the advantages of sexual reproduction?

Sexual reproduction leads to offspring that are all genetically different from one another and from either parent, through three different routes (**FIGURE 6-21**).

1. *Crossing over during the production of gametes.* As we saw, crossing over during prophase I of meiosis causes every chromosome in a gamete to carry a mixture of that individual's maternal and paternal genetic material.

2. *Shuffling and reassortment of homologues during meiosis.* Recall that as the homologues for each chromosome are pulled to opposite poles of the cell during the first division of meiosis, maternal and paternal homologues are randomly pulled to each pole. This means that there is an extremely large number of different combinations of maternal and paternal homologues that might end up in each gamete.

3. *Combining alleles from two parents at fertilization.* First and foremost, with sexual reproduction a new individual comes from the fusion of gametes from two different people. Each of these parents comes with his or her own unique set of genetic material.

Q Bacteria reproduce asexually, whereas most plants and animals reproduce sexually. Which is the better method?

The variability among the offspring produced by sexual reproduction enables populations of organisms to cope better with changes in their environment. After all, if the environment is gradually changing from one generation to the next, individuals producing many offspring increase the likelihood that one of their offspring will carry a set of genes particularly suited to the new environment. Over time, populations of sexually reproducing organisms can quickly adapt to changing environments. It's like buying lottery tickets—the more different tickets you buy, the more likely it is that one of them will be a winner.

SOURCES OF GENETIC VARIATION

There are multiple reasons why offspring are genetically different from their parents and one another.

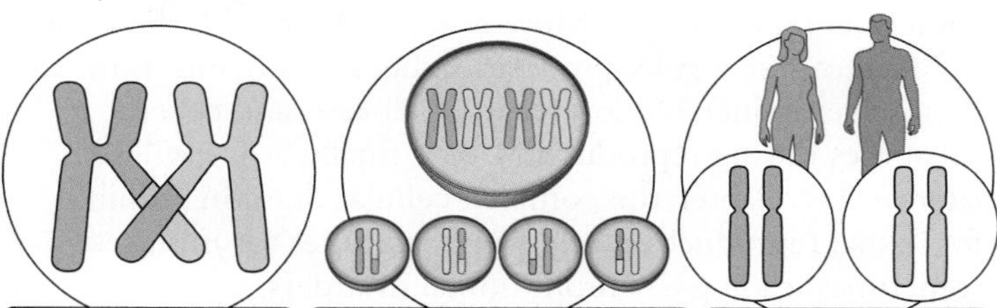

CROSSING OVER	REASSORTMENT OF HOMOLOGUES	ALLELES COME FROM TWO PARENTS
Crossing over during meiosis produces a mixture of maternal and paternal genetic material on each chromatid.	The homologues and sister chromatids distributed to each daughter cell during meiosis are a random mix of maternal and paternal genetic material.	Each parent donates his or her own set of genetic material.

FIGURE 6-21 Crossing over and meiosis: creating many different combinations of alleles. As a result of crossing over and meiosis, each individual born from a separate fertilized egg is genetically unique.

What are the disadvantages of sexual reproduction? In the yellow dung fly (*Scathophaga stercoraria*), males sometimes wrestle each other for mating access to a female (**FIGURE 6-22**). The female awaits the outcome of the battle in a pile of dung. Occasionally, females drown in the dung pile as they wait. Dangers associated with mating are just one of the downsides to sexual reproduction. There are several others. First, when one parent reproduces, only half of its offsprings' alleles will come from that organism.

FIGURE 6-22 Drowning in dung. A downside to sexual reproduction.

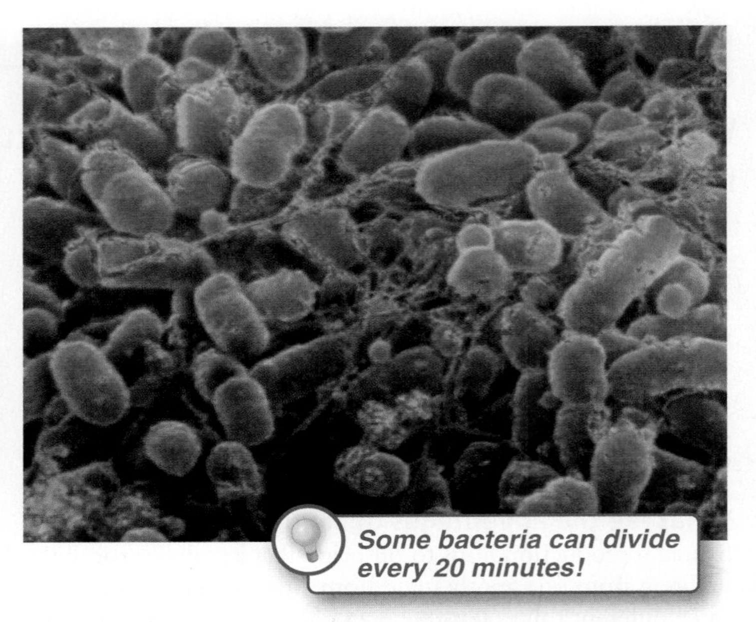

Some bacteria can divide every 20 minutes!

FIGURE 6-23 Pluses and minuses. This asexually reproducing bacterial colony can reproduce quickly and efficiently, but because each new cell is an exact duplicate of the parent genome, there is no genetic variety to enable adaptation should the environment change.

The other half will come from the other parent. With asexual reproduction, parents produce nearly identical offspring, so there is a very efficient transfer of genetic information from one generation to the next. Also, with sexual reproduction it takes time and energy to find a partner. This is energy that asexual organisms can devote to additional reproduction. Moreover, as we see with the dung flies, sex can be a risky proposition because organisms make themselves vulnerable to predation, disease, and other calamities during reproduction. And finally, as we will see later in this chapter, the complex cellular division required for sexual reproduction offers opportunities for mistakes, sometimes leading to chromosomal disorders.

What are the advantages of asexual reproduction? With asexual reproduction, the advantages and disadvantages are more or less reversed. Because asexual reproduction involves only a single individual, it can be fast and easy. Some bacteria can divide, forming a new generation, every 20 minutes (**FIGURE 6-23**). And for organisms in isolated habitats or when establishing new populations, asexual reproduction can be advantageous as well. Asexual reproduction is efficient, too. Offspring carry all of the genes that their parent carried—they are genetically identical. If the environment is stable, it is beneficial for organisms to produce offspring as similar to themselves as possible.

What are the disadvantages of asexual reproduction? The downside to asexual reproduction is that the more closely an offspring's genome resembles its parent's, the less likely it is that the offspring will be suited to the environment when it changes.

In the end, we still see large numbers of species using asexual reproduction and large numbers, including the vast majority of animals and plants, that reproduce sexually. It seems that conditions favoring each occur in the great diversity of habitats of the world. That we see both sexual and asexual reproduction also highlights the recurring theme in biology that there often is more than one way to solve a problem.

TAKE-HOME MESSAGE 6•14

There are two fundamentally different ways that cells and organisms can reproduce: (1) mitosis and asexual reproduction via binary fission, and (2) meiosis and sexual reproduction. Asexual reproduction can be fast and efficient, but leads to genetically identical offspring that carry all of the genes that their parent carried. Sexual reproduction leads to offspring that are all genetically different from one another and from either parent, but takes more time and energy and can be risky.

❹ There are sex differences in the chromosomes.

XX and XY: Only one of the 23 chromosome pairs determines sex in humans.

6•15

How is sex determined in humans?

> **Q** Which parent determines a baby's sex? Why?

In humans the sex of a baby is determined by its father. The complex sequence of events involved in sex determination is instigated by one special pair of chromosomes called the sex chromosomes. The sex chromosomes carry information that directs a growing fetus to develop as a male or as a female. An individual's sex depends on the sex chromosome inherited from his or her father.

Let's take a closer look at the sex chromosomes. In humans, we noted that there are 23 pairs of chromosomes in every somatic cell. These can be divided into two different types: there is one pair of sex chromosomes and there are 22 pairs of non-sex chromosomes. The human sex chromosomes are called the **X and Y chromosomes** (**FIGURE 6-24**).

How do the X and Y chromosomes differ from the other chromosomes? As we described earlier, all of the genetic information is stored on the chromosomes in all cells of an organism's body. But most of this information is not sex specific—that is, if you are building an eye or a neuron or a skin cell or a digestive enzyme, it doesn't matter whether it is for a male or for a female; the instructions are the same for both sexes. Some genetic information, however, instructs the body to develop into one sex or the other. That information—which *does* differ dramatically depending on whether it is the genetic information for building a female or a male—is found on the sex chromosomes.

An individual has two copies of all of the non-sex chromosomes. One copy is inherited from the mother, one from the father. Individuals also have two copies of the sex chromosomes, but not always two copies of the same kind. Males have one copy of the X chromosome and one copy of the Y chromosome. Females, on the other hand, don't have a Y chromosome but instead have two copies of the X chromosome.

So how does the father determine the sex of the baby? During meiosis in females, the gametes that are produced carry only one copy of each chromosome. This is true for

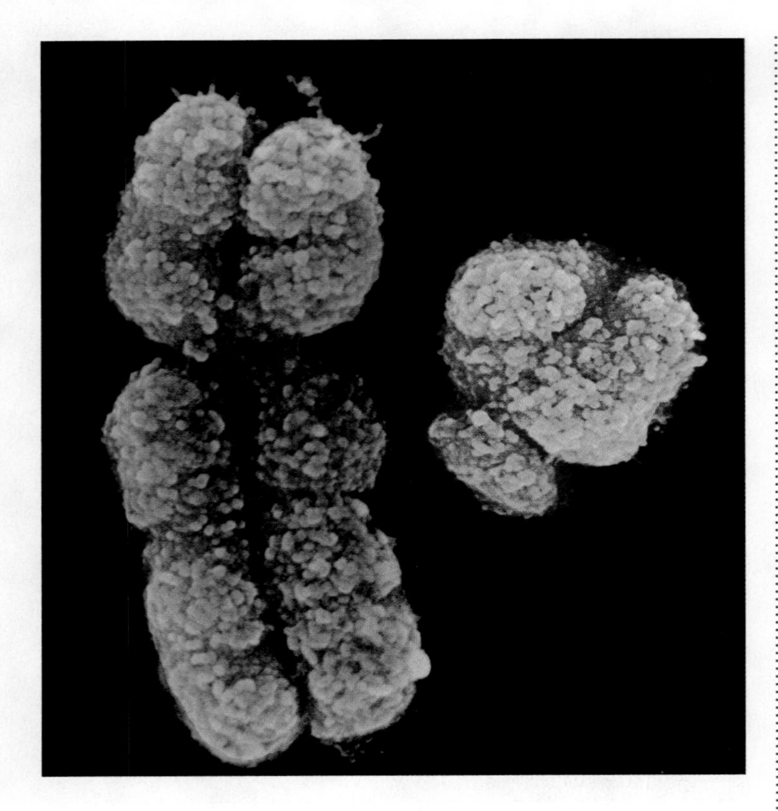

FIGURE 6-24 X and Y: the human sex chromosomes.

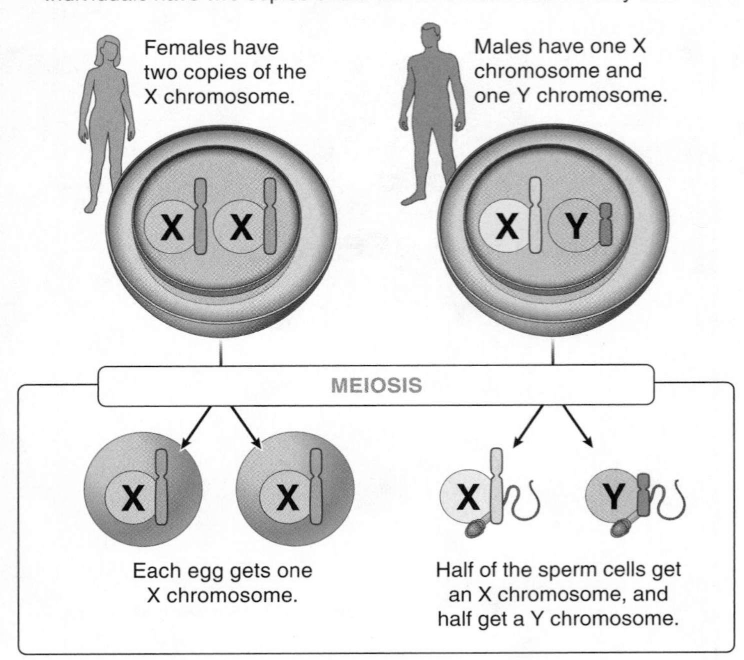

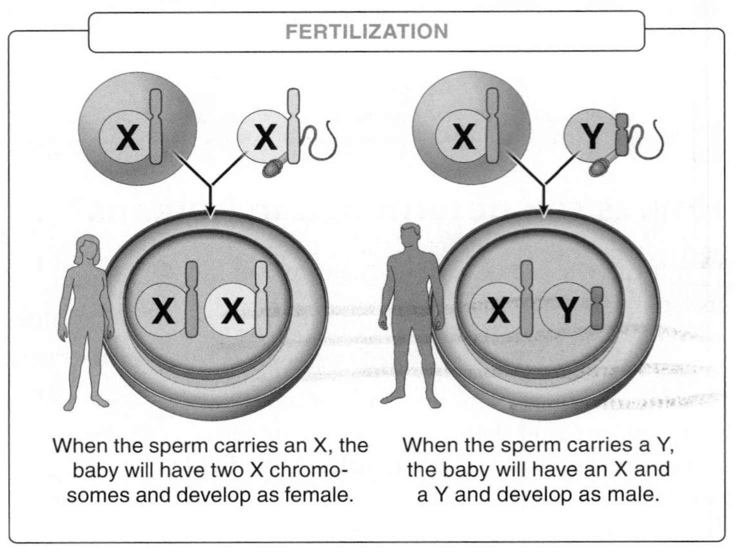

FIGURE 6-25 Sex determination in humans. The sex of offspring is determined by the father.

the sex chromosomes, too. So from the two copies of the X chromosome carried by females, half of the gametes end up with a copy of one of those X chromosomes while the rest of the gametes inherit the other X chromosome. Thus, every egg has an X for its one sex chromosome. During meiosis in males, the sperm that are produced also carry one copy of each chromosome, including the sex chromosomes, but in this case, half of the sperm produced inherit the X chromosome and the other half inherit the Y chromosome. At fertilization, an egg bearing a single X chromosome (and one copy of all the non-sex chromosomes) is fertilized by a sperm bearing one copy of all the non-sex chromosomes and *either* an X chromosome or a Y chromosome. When the sperm carries an X, the baby will have two X chromosomes and will develop as a female. When the sperm carries a Y, the baby will have an X and a Y and so will develop as a male (**FIGURE 6-25**).

Q Why can we be certain that there is no essential genetic information carried on the Y chromosome?

Given the distribution of X and Y chromosomes into males and females, it must be true that there is no essential genetic information carried on the Y chromosome. Why?

Because females don't have a Y chromosome in any of their cells, yet they are able to develop and live normal, healthy lives. For this reason, we know that nothing on the Y chromosome is absolutely necessary.

Physically, the X and Y chromosome look very different from each other. The X chromosome is relatively large and carries a great deal of genetic information relating to a large number of non-sex-related traits. The Y chromosome, on the other hand, is tiny and carries genetic information only about a very small number of traits. The small set of genetic instructions on the Y chromosome instructs the fetal gonads to develop as testes rather than ovaries. Once this is done, very little additional genetic input from the Y chromosome is necessary. Instead, the hormones produced by the testes or ovaries generally direct the rest of the body to develop as a male or female. In the next sections we investigate some of the ramifications of males having two different sex chromosomes and look at some of the systems for sex determination that have evolved in other groups of organisms.

TAKE-HOME MESSAGE 6•15

In humans, the sex chromosomes carry information that directs a growing fetus to develop as either a male (if a Y chromosome is present) or a female (if no Y chromosome is present). The determination depends on the sex chromosomes inherited from the parents.

6•16

The sex of offspring is determined in a variety of ways in non-human species.

For something as fundamental to a species as sex determination, you might imagine that one method evolved and all species use it. The world is more diverse than that. In most plants there aren't even distinct males and females. In corn, for example, every individual produces both male and female gametes. All earthworms and garden snails are also capable of producing both male and female gametes. Such organisms are called **hermaphrodites** (from the names of the Greek god Hermes and goddess Aphrodite) because both male and female gametes are produced by an individual. But even among the species with separate males and females there are several different methods of sex determination. The human method, in which having both an X and a Y chromosome makes you male and having two X chromosomes makes you female, is common among eukaryotes and is seen in all mammals. But it is only one of at least four different methods used by organisms for determining sex. We'll examine the three others, used by different groups, here (**FIGURE 6-26**).

SEX DETERMINATION IN VARIOUS OTHER SPECIES

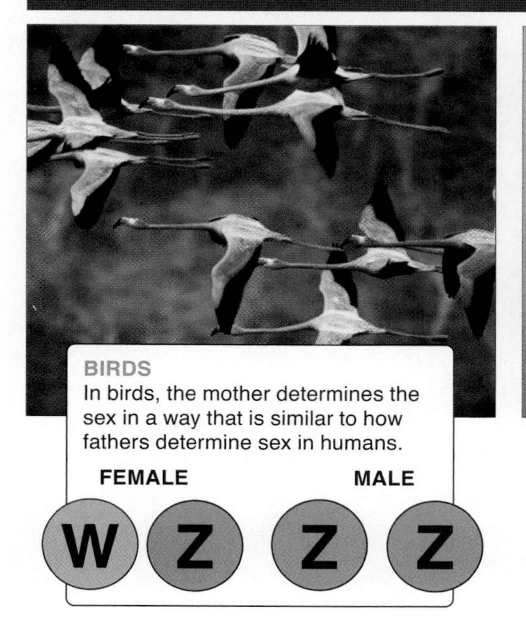

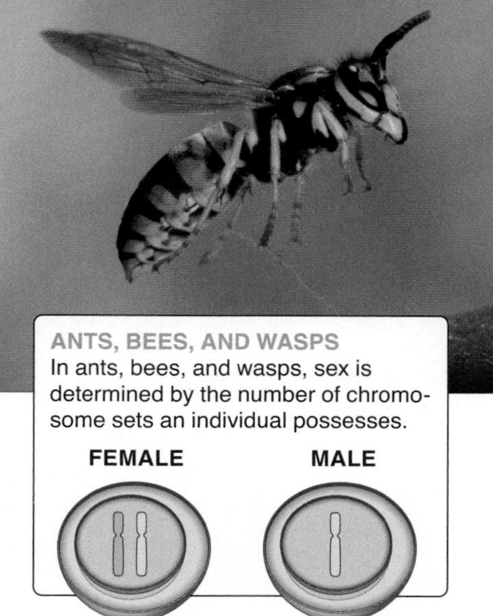

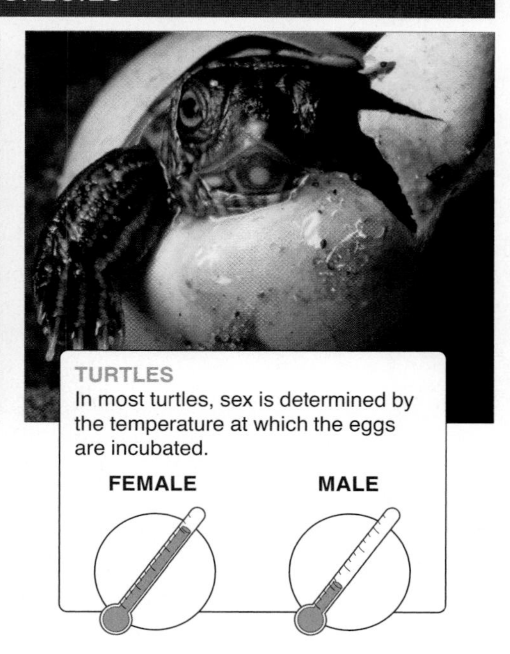

FIGURE 6-26 Other methods of sex determination. In some species, sex is determined by the type of chromosomes, the number of chromosomes, or incubation temperature of the eggs.

Some biologists are concerned that global warming will harm turtles, like this loggerhead sea turtle. Since the sex of the animal is determined by the temperature of the incubating eggs, a warming of the nesting beaches could result in all female offspring.

Birds In birds, the mother determines the sex in a way that is similar to the way fathers determine sex in humans. Females have one copy of two different sex chromosomes, called the Z and the W chromosomes. Males, on the other hand, have only one type, carrying two copies of the Z chromosome. Consequently, the sex of bird offspring is determined by the mother rather than the father. This method is also found in some fish and butterfly species.

Ants, bees, and wasps In these insects, sex is determined by the number of chromosome sets an individual possesses. Males are haploid, having only a single set of chromosomes, and females are diploid, carrying two full sets of chromosomes. In this unusual method, females produce haploid eggs by meiosis. They then mate with males and store the sperm in a sac. As each egg is produced, the female can allow it to be fertilized by some of the sperm she has stored, in which case the offspring will be diploid and female. Alternatively, she can lay the unfertilized egg, which can develop into a haploid, male individual. Just think, in these species, males don't have a father, yet they do have a grandfather!

Turtles In some species, sex determination is controlled by the environment rather than by the number or types of chromosomes an organism has. In most turtles, for example, offspring sex is determined by the temperature at which the eggs are kept. Eggs that are kept relatively hot during incubation become females, while the eggs incubated at slightly cooler temperatures become males. The sex of some lizards and crocodiles is also determined by the temperature at which the eggs are incubated.

TAKE-HOME MESSAGE 6•16

A variety of methods are used for sex determination across the world of plant and animal species. These include the presence or absence of sex chromosomes, the number of chromosome sets, and environmental factors such as incubation temperature.

⑤ Deviations from the normal chromosome number lead to problems.

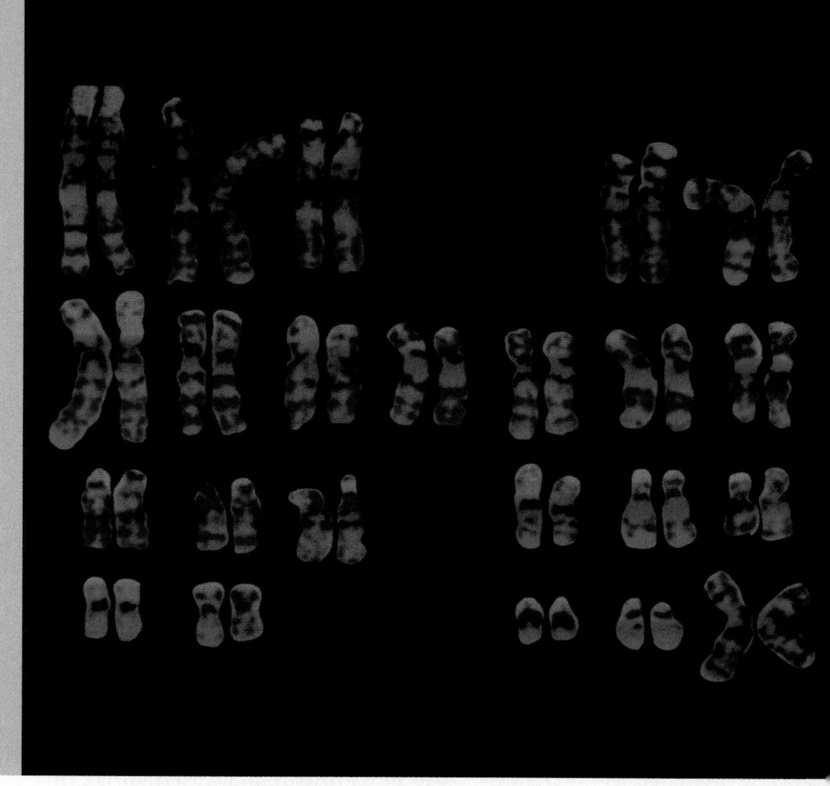

Karyotype displaying the chromosome pairs of a normal human female.

6•17

Down syndrome can be detected before birth: karyotypes reveal an individual's entire chromosome set.

It becomes riskier for women to have children as they become older. Increasingly, their gametes contain incorrect numbers of chromosomes or chromosomes that have been damaged. These problems can lead to effects that range from minor to fatal.

To find out if their offspring may have some disorders associated with incorrect numbers of chromosomes, parents can request a quick test for some common genetic problems even before their baby is born. This information is available from an analysis of an individual's **karyotype,** a visual display of a complete set of chromosomes. A karyotype can be made for adults or children, but it is most commonly done for a fetus. A karyotype is a useful diagnostic tool because it can be prepared very early in the fetus's development to assess whether there is an abnormality in the number of chromosomes or in their structure. And because it shows all of the chromosomes, even the sex chromosomes, it also reveals the sex of the individual.

Preparing a karyotype takes five steps. (1) First it is necessary to obtain some cells from the individual. (2) The cells are then encouraged to divide by culturing them in a test tube with nutrients. (3) After a few days to two weeks of cell division, the cells are treated with a chemical that stops them exactly midway through cell division—a time when the chromosomes are coiled thickly and more visible than usual. (4) Then the cells are placed on a microscope slide and a stain is added that binds to the chromosomes, making them visible. (5) And finally, the chromosomes are arranged by size and shape and displayed on a monitor or in a photograph (**FIGURE 6-27**).

The first step—obtaining cells to prepare a karyotype from—is relatively easy in an adult or child. Usually cells are collected from a small blood sample. Collecting cells from a fetus, on the other hand, poses a special problem. Two different methods for collecting cells have been developed.

1. *Amniocentesis.* This procedure can be done approximately three to four months into a pregnancy (**FIGURE 6-28**). In one quick motion (and without using anesthetic), a 3- to 4-inch needle is pushed through the abdomen, through the amniotic sac, and into the amniotic fluid that surrounds and protects the fetus. Using ultrasound for guidance, the doctor

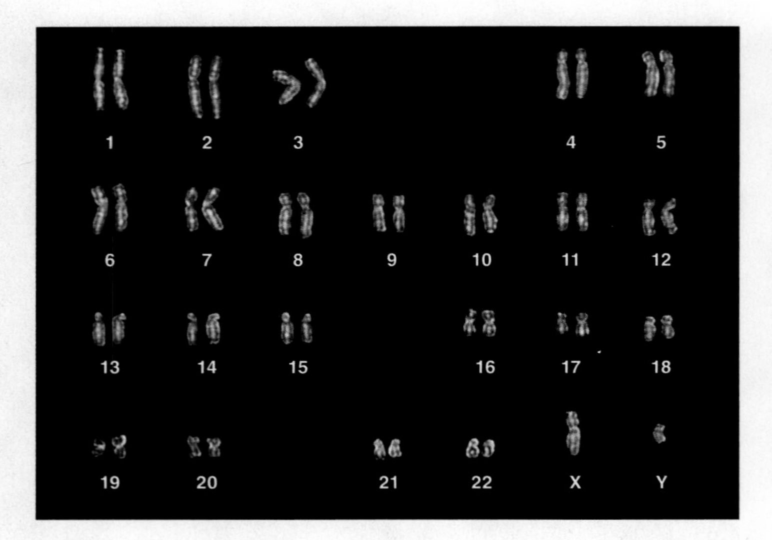

FIGURE 6-27 A human karyotype. A visual display of a complete set of chromosomes.

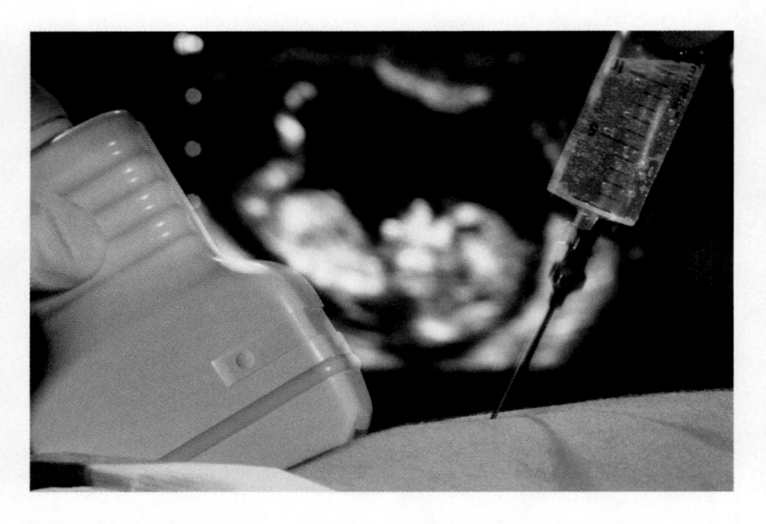

FIGURE 6-28 Amniocentesis. Fluid surrounding a fetus (containing some of its cells) is extracted and analyzed to determine the genetic composition of the developing fetus.

aims for a small pocket of fluid as far as possible from the fetus and withdraws about 2 tablespoons of fluid. This fluid contains many cells from the fetus, which can then be used for karyotype analysis. Any chromosomal abnormalities in the fetus will be present in these cells.

2. *Chorionic villus sampling (CVS).* In this procedure, rather than sampling cells from the amniotic fluid, a small bit of tissue is removed from the **placenta,** the temporary organ that allows the transfer of gases, nutrients, and waste products between a mother and fetus. A needle is inserted either through the abdomen or through the vagina and cervix, again using ultrasound for guidance. Then a small piece of the finger-like projections from the placenta are removed via the syringe. Because the fetus and placenta both develop from the same fertilized egg, their cells contain the same genetic composition. The chief advantage of CVS over amniocentesis is that it can be done several weeks earlier in the pregnancy, usually between the 10th and 12th weeks.

The resulting karyotype, whether from amniocentesis or CVS, will reveal whether the fetus carries an extra copy of any of the chromosomes or lacks a copy of one or more chromosomes. Of all the chromosomal disorders detected by karyotyping, Down syndrome is the most commonly observed. Named after John Langdon Down, the doctor who first described it in 1866, the syndrome is revealed by the presence of an extra copy of chromosome 21 (**FIGURE 6-29**). (For this reason, the condition that causes Down syndrome is called "trisomy 21.") Striking about 1 of every 1,000 children born, Down syndrome is

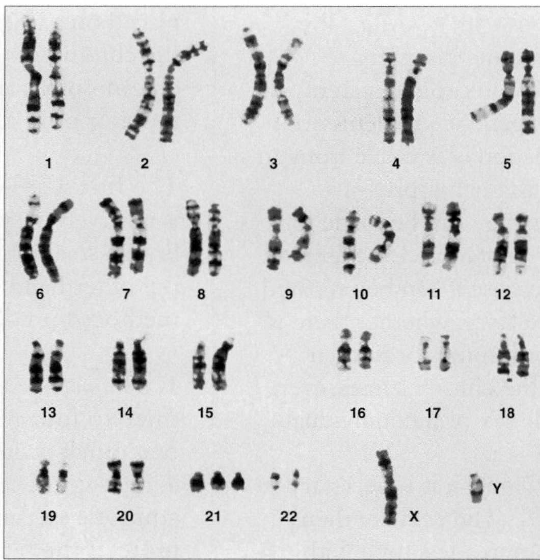

FIGURE 6-29 Trisomy 21. An extra copy of chromosome 21 causes Down syndrome.

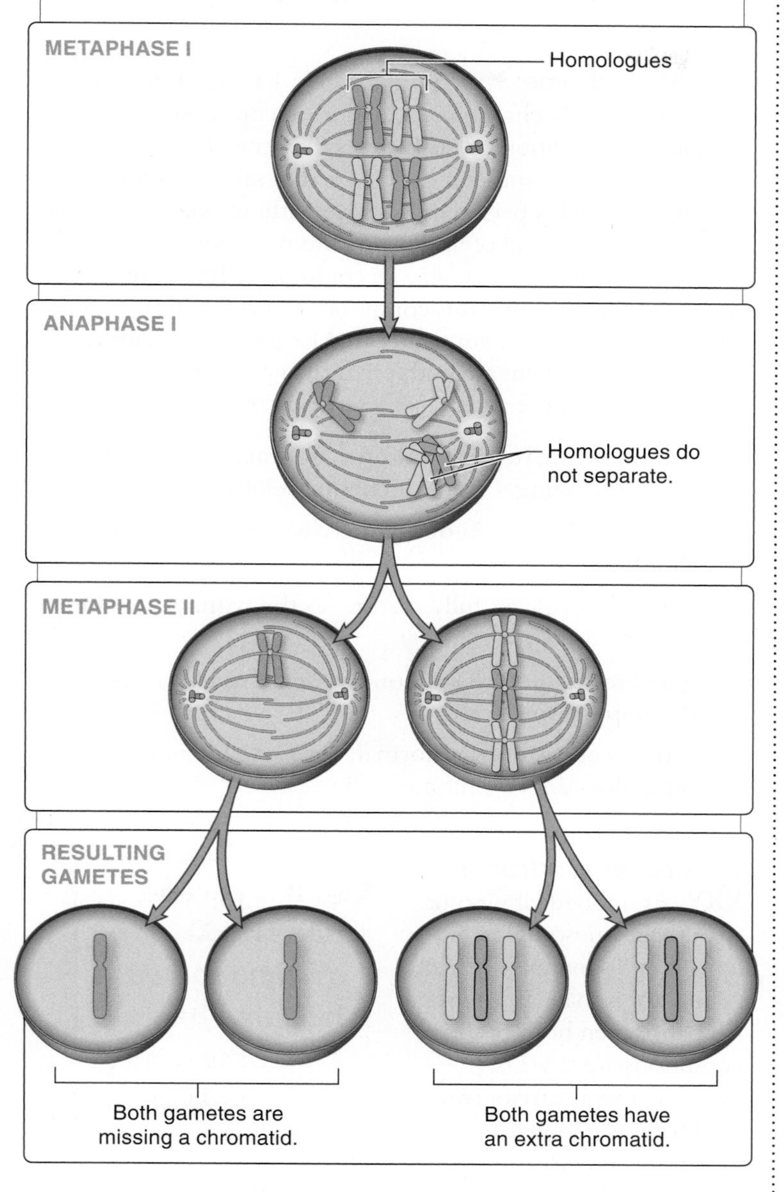

FIGURE 6-30 Nondisjunction: forming gametes with too many or too few chromosomes.

characterized by a suite of physical and mental characteristics that includes learning disabilities, a flat facial profile, heart defects, and increased susceptibility to respiratory difficulties.

Down syndrome and other disorders caused by a missing chromosome or an extra copy of a chromosome are a consequence of **nondisjunction,** the unequal distribution of chromosomes during meiosis. Nondisjunction can occur at two different points in meiosis: homologues can fail to separate during meiosis I or sister chromatids can fail to separate during meiosis II (**FIGURE 6-30**). In both cases, nondisjunction results in an egg or sperm with zero or two copies of a chromosome rather than a single copy. Each of the chromosomes is approximately equally likely to fail to separate during cell division, but the ramifications of trisomy (having an extra copy of one chromosome in every cell) are greater for chromosomes with larger numbers of genes. When trisomy occurs for chromosomes with greater numbers of genes, the likelihood that the developing embryo survives to birth is reduced. Consequently, we tend to see cases of trisomy that involve only the chromosomes with the fewest genes, such as chromosomes 13, 15, 18, 21, and 22. In fact, actual observations show that trisomy 1 is never seen (all such fertilized eggs die before implantation in the uterus), trisomy 13 occurs in 1 in 20,000 newborns (and most die soon after birth), while trisomy 21 occurs in 1 in 1,000 newborns (many of whom live long lives).

> **Q** Why do older women have more babies with Down syndrome?

As we mentioned at the beginning of this section, it becomes riskier for women to have children as they become older. Older women, for example, have more babies with Down syndrome. Why does the risk of having a baby with Down syndrome or some other disorder that results from trisomy increase with increasing age for women but not for men? As women age, their gametes tend to have more errors. The problem is that those eggs *began* meiosis near the time the woman was born and they may not complete it until she is 40 or more years old. During those decades, the cells may accumulate impairments that interfere with normal cell division. In men, the cells that are undergoing meiosis are relatively young because new sperm-producing cells are produced every couple of weeks after puberty.

Whereas lacking a non-sex chromosome or having an extra chromosome usually has serious consequences for health, we'll see in the next section that lacking or having an extra X or Y chromosome has consequences that are much less severe.

TAKE-HOME MESSAGE 6•17

A karyotype is a visual display of a complete set of chromosomes. A karyotype is a useful diagnostic tool because it can be prepared very early in the fetus's development to assess whether there is an abnormality in the number of chromosomes or in their structure, such as in the case of Down syndrome. Down syndrome is caused by having an extra copy of chromosome 21.

6•18

Life is possible with too many or too few sex chromosomes.

It is usually fatal to have one too many or one too few of the non-sex chromosomes. When it comes to the sex chromosomes, however, the situation is less extreme. Many individuals are born lacking one of the sex chromosomes or with an additional X or Y chromosome, and they usually survive. We can get a glimpse into the role of the sex chromosomes by looking at the physical and mental consequences of having an abnormal number of sex chromosomes. In each of the cases below, the condition is caused by nondisjunction during the production of sperm or eggs (**FIGURE 6-31**).

TOO MANY OR TOO FEW SEX CHROMOSOMES

GAMETE GENOTYPE	OFFSPRING GENOTYPE	SYNDROME CHARACTERISTICS
Egg X + sperm _ ; Egg _ + sperm X	X _	TURNER SYNDROME • Short height • Web of skin between neck and shoulders • Underdeveloped ovaries; often sterile • Some learning difficulties
Egg X X + sperm Y ; Egg X + sperm X Y	X X Y	KLINEFELTER SYNDROME • Underdeveloped testes • Lower testosterone levels; usually infertile • Development of some female features • Long limbs and slightly taller than average
Egg X + sperm Y Y	X Y Y	"SUPER MALES" • Taller than average • Moderate to severe acne • Intelligence may be slightly lower than average
Egg X X + sperm X	X X X	"METAFEMALES" • May be sterile • No obvious physical or mental problems

FIGURE 6-31 Characteristics of individuals with too many or too few sex chromosomes.

Turner syndrome: X_ Approximately 1 in 5,000 females carry only one X chromosome and no Y chromosome, exhibiting a condition called Turner syndrome. This condition, denoted as X_ (or sometimes XO), is the only condition in humans in which a person can survive without one of a pair of chromosomes. Instead of having 46 chromosomes in every cell, these individuals have only 45. As common as Turner syndrome is, 98% of the time that this condition occurs in a fertilized egg, the egg is spontaneously aborted long before a fetus can come to term. There are both physical and mental consequences of the absence of a second sex chromosome.

- Women with Turner syndrome are usually relatively short, averaging 4 feet 8 inches in height.
- They develop a web of skin between the neck and shoulders.
- The ovaries never fully mature, so the women are almost always sterile.
- The breasts and other secondary sex characteristics develop incompletely.
- Intelligence is usually normal, but some learning difficulties are common.

Q If a person has two X chromosomes but also has a Y chromosome, is the individual male or female?

Klinefelter syndrome: XXY An individual who has two X chromosomes is female; an individual with an X and a Y is male. But what happens when both of those conditions are true? An individual who carries two X chromosomes *and* a Y chromosome, a condition known as Klinefelter syndrome, develops as male. This is because, as we saw earlier, the Y chromosome carries genetic instructions that, if present, cause fetal gonads to develop as testes; if absent, the fetal gonads develop as ovaries. The extra X chromosome, however, does cause Klinefelter males to be somewhat feminized, although this effect can be reduced through treatments such as testosterone supplementation. Approximately 1 in 1,000 males have the genotype XXY, making this one of the most common genetic abnormalities in humans. Klinefelter syndrome has physical and mental consequences:

- Affected individuals have testes, but they are smaller than average. Because the testes are small, levels of testosterone are low and the individuals are almost always infertile.
- They develop some female features, including reduced facial and chest hair, and some breast development.
- They have long limbs and are slightly taller than average (about 6 feet on average).
- They learn to speak at a later age than average and tend to have language impairments.
- Sometimes individuals have further additional X chromosomes, exhibiting the karyotypes XXXY or even XXXXY. These individuals also exhibit Klinefelter syndrome but more frequently have mental retardation.

Q Do human hermaphrodites exist?

A hermaphrodite is an individual with functioning male and female reproductive organs capable of producing both male and female gametes. Often it is mistakenly assumed that a person with Klinefelter syndrome must be a hermaphrodite because he has both two X chromosomes (which would usually make an individual a female) and an X and a Y chromosome (usually making the individual a male). But although hermaphroditism is common among invertebrates and occurs in some fish and other vertebrates, contrary to urban legends, human hermaphrodites do not exist. Although some men with Klinefelter syndrome may have some features of the opposite sex, they do not produce female gametes and so are not hermaphrodites.

XYY males There is no official name for the condition in which an individual has one X chromosome and two Y chromosomes, although they are sometimes referred to as "super males." This chromosomal abnormality occurs in approximately 1 in 1,000 males (and in about 1 in 325 males who are 6 feet or taller). There are no distinguishing features at birth to indicate that an individual carries an extra Y chromosome, and the vast majority live their lives normally without knowing they have an extra chromosome in every cell. Several consequences of the XYY condition have been well documented.

- XYY males are relatively tall (about 6 feet 2 inches on average).
- They tend to have moderate to severe acne.
- Their intelligence usually falls within the same range as that of XY males, although the average may be slightly lower.

Because in the United States about 20 times as many XYY males are found in prisons as in the population as a whole, there has been tremendous controversy over whether the extra Y chromosome predisposes individuals to criminal behavior. Because these observations were not part of randomized, controlled, double-blind experiments, however, they are open to numerous alternative interpretations. One large study, for example, concluded that the increased representation of XYY males in prison was more likely due to the fact that taller males and males of lower intelligence are more likely to be apprehended, regardless of which chromosomes they carry. The controversy remains unresolved, and the majority of XYY males are *not* in prison.

Although an error during meiosis in either the father *or* the mother causes Klinefelter or Turner syndrome, it is only an error in meiosis in a male that gives rise to an XYY child. Why? Because only males carry Y chromosomes. Consequently, an XYY child must receive both of his Y chromosomes from his father. This means that the father's sperm cell must have contained two copies of the Y, a result of nondisjunction during meiosis in the father.

XXX females Sometimes called "metafemales," individuals with three X chromosomes occur at a frequency of about 1 in 1,000 women. Very few studies of this condition have been completed, although initial observations suggest that some XXX females are sterile but otherwise have no obvious physical or mental problems.

TAKE-HOME MESSAGE 6•18

Although it is usually fatal to have one too many or one too few of the non-sex chromosomes, individuals born lacking one of the sex chromosomes or with an additional X or Y chromosome usually survive, though usually with physical and/or physiological problems.

streetBio

Knowledge You Can Use

Can you select the sex of your baby? (Would you want to?)

Some fertility clinics now promise that for about $200 they can allow a couple to choose the sex of their baby.

Q: Given our knowledge of the behavior of the X and Y chromosomes, is it more likely that these sex-selection techniques involve manipulations of sperm or of eggs? Why? Because all eggs have X chromosomes, the X chromosome doesn't play a role in determining the sex of a baby. Instead, the fertility clinics must somehow sort and separate sperm cells based on whether or not they carry an X or a Y chromosome (men produce approximately equal numbers of each).

Q: What feature of sperm must be used in separating them? This process must be based on distinguishing between sperm carrying an X chromosome and sperm carrying a Y chromosome. After the two types of sperm are distinguished, they must then be separated.

Q: What techniques might make this possible? One method involves determining, by weighing sperm, which cells have more DNA. The heavier sperm must be carrying the X rather than the Y chromosome because the X is so much larger. Another method is to add a fluorescent dye to sperm that temporarily attaches to DNA. Because sperm with an X chromosome have more DNA, more of the dye attaches to them and they are more fluorescent. A machine then sorts the sperm one by one, and at ovulation, insemination is performed using the sperm with the desired sex chromosome.

Q: Does it work? The technique is still fairly new, but initial results suggest that the procedures have a success rate of 70% to 90%.

Concerns for you to ponder. What are the ethical concerns raised by selecting the sex of one's baby? Are there some circumstances in which such a selection would be more acceptable than in others?

BIG IDEAS IN CHROMOSOMES AND CELL DIVISION

1 There are different types of cell division.

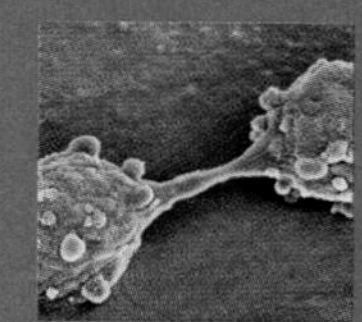

Telomeres are protective sections of DNA that get shorter every time a cell divides. If a telomere gets too short, additional cell divisions cause loss of essential DNA and cell death. Cells that rebuild telomeres with each division can become cancerous. In most prokaryotes, genetic information is carried in a single, circular DNA strand (chromosome). Eukaryotes have much more DNA, organized into linear chromosomes in the nucleus. Bacteria divide by a type of asexual reproduction, binary fission. In the cell cycle, eukaryotic somatic cells alternate between division, called the mitotic phase, and other cell activities, called interphase, consisting of two gap phases separated by a DNA synthesis phase. Every time a cell divides, its DNA must first replicate so that each of the two new cells has all the genetic material of the parent cell.

2 Mitosis replaces worn-out old cells with fresh new duplicates.

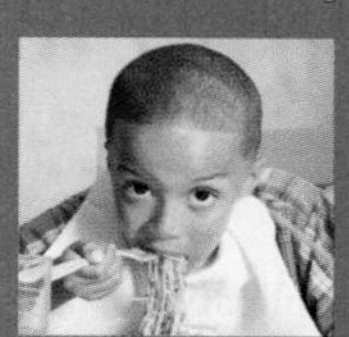

In mitosis, cells generate new, genetically identical cells, enabling organisms to grow and to replace cells that die. Mitosis occurs in four steps, following replication of the chromosomes, to produce two genetically identical daughter cells from one parent cell. Cancer is unrestrained cell growth and division, leading to large masses of cells (tumors). Treatment focuses on killing or slowing down cell growth and division, by chemotherapy and/or radiation.

3 Meiosis generates sperm and eggs and a great deal of variation.

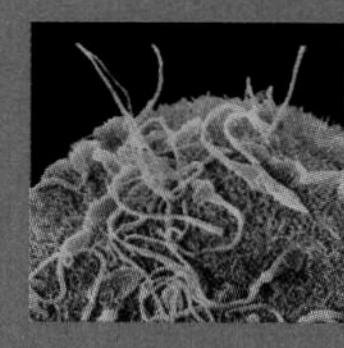

Meiosis occurs only in gamete-producing cells of sexually reproducing organisms. Gametes have only half as much genetic material as the parent cell and they differ from one another in the combinations of alleles they carry. Meiosis is preceded by DNA replication and consists of two rounds of cell division: separation of homologous pairs of sister chromatids, then separation of the sister chromatids. The final product in a diploid organism is four haploid gametes. In species with two sexes, the female is the sex producing the larger gamete, while males produce the smaller gamete. Every gamete has just one copy of each chromosome. Crossing over during the first prophase of meiosis creates gametes with collections of alleles that may never have existed together before; this variation is important for evolution. Cells and organisms can reproduce by mitosis and asexual reproduction via binary fission or by meiosis and sexual reproduction.

4 There are sex differences in the chromosomes.

In humans, the sex chromosomes carry information that directs a fetus to develop as male (if a Y chromosome is present) or female (no Y present). The sex is determined by the sex chromosome inherited from the parents. Methods for sex determination in plant and animal species include the presence or absence of sex chromosomes, the number of chromosome sets, and environmental factors such as incubation temperature.

5 Deviations from the normal chromosome number lead to problems.

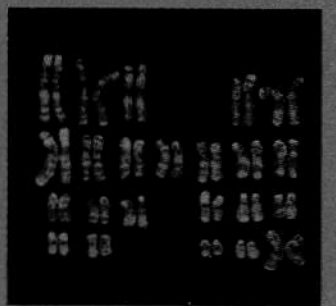

A karyotype, a visual display of a complete set of chromosomes, is a useful diagnostic tool. It can be prepared very early in fetal development to detect abnormalities in the number or structure of chromosomes. Having one too many or one too few non-sex chromosomes is usually fatal, but individuals born lacking one of the sex chromosomes or with an additional X or Y usually survive, but often with physical and/or physiological problems.

KEY TERMS

anaphase, p. 221
apoptosis, p. 218
asexual reproduction, p. 213
binary fission, p. 213
cancer, p. 222
cell cycle, p. 214
centromere, p. 215
chromatid, p. 221
complementarity, p. 215
complementary base, p. 215
crossing over, p. 228
cytokinesis, p. 215
daughter cell, p. 213
diploid, p. 225
fertilization, p. 225
gamete, p. 214
gonads, p. 227
haploid, p. 225
hermaphrodite, p. 237
histone, p. 212
homologous pair (homologues), p. 227
interphase, p. 214
karyotype, p. 239
meiosis, p. 225
metaphase, p. 221
mitosis, p. 215
mitotic phase (M phase), p. 214
nondisjunction, p. 241
parent cell, p. 213
placenta, p. 240
prophase, p. 220
replication, p. 213
reproductive cell, p. 214
sexual reproduction, p. 213
sister chromatids, p. 221
somatic cell, p. 214
spindle, p. 221
spindle fiber, p. 221
telomere, p. 210
telophase, p. 222
X and Y chromosomes, p. 235

Check Your Knowledge

1. Which of the following statements about telomeres is incorrect?
   a) They function like a counter, keeping track of how many times a cell has divided.
   b) At birth, they are long enough to permit approximately 50 cell divisions in most cells.
   c) They are slightly shorter in prokaryotic cells than in eukaryotic cells.
   d) They function like a protective cap on chromosomes.
   e) They contain no critical genes.

2. Prokaryotic cells can divide via:
   a) mitosis.
   b) binary fission.
   c) meiosis.
   d) both mitosis and binary fission.
   e) None of the above.

3. In multicellular organisms, cells that undergo mitotic division but not meiotic division are called ______ cells.
   a) somosis
   b) skin
   c) interphase
   d) somatic
   e) germ

4. The ______ marks the break in the cell cycle between the end of mitosis and beginning of the DNA synthesis stage.
   a) mitotic phase
   b) synthesis phase
   c) Gap 2 phase
   d) Gap 1 phase
   e) None of the above.

5. Mitosis results in:
   a) daughter cells with twice as much genetic material as the parent cell and a unique collection of alleles.
   b) eight daughter cells.
   c) daughter cells with the same number and composition of chromosomes.
   d) gametes.
   e) daughter cells with the same number of chromosomes but different combinations of alleles.

6. Using a light microscope, it is easiest to see chromosomes:
   a) during mitosis and meiosis, because the condensed chromosomes are thicker and therefore more prominent.
   b) during interphase, when they are concentrated in the nucleus.
   c) in the mitochondria, because the chromosomes are circular.
   d) during asexual reproduction.
   e) during interphase, because they are uncoiled and have a more linear structure.

7. The division of the cytoplasm during cell division is referred to as:
   a) cytoplasm splicing.
   b) cytokinesis.
   c) vegetative growth.
   d) cytodivision.
   e) hybridization.

8. Which of the following statements about tumors is incorrect?
   a) Benign tumors pose less of a health risk than do malignant tumors.
   b) Malignant tumors shed cancer cells that can spread to other parts of the body.
   c) Tumors are caused by excessive cell growth and division.
   d) Cancer cells from malignant tumors can travel to other parts of the body in a process called metastasis.
   e) They contain cells with abnormally high contact inhibition.

9. During meiosis but not during mitosis:
   a) haploid gametes are produced that are all identical in their allelic composition.
   b) division of the cytoplasm occurs.
   c) chromosomes line up in the center of the cell during metaphase.
   d) genetic variation among the daughter cells is increased.
   e) two identical daughter cells are produced.

10. Sister chromatids are:
    a) the result of crossing over.
    b) identical molecules of DNA resulting from replication.
    c) homologous chromosomes.
    d) produced in meiosis but not in mitosis.
    e) single-stranded.

11. Which of the following events does not occur during prophase I of meiosis?
    a) Spindle fibers form.
    b) Homologous pairs of chromosomes align at the metaphase plate.
    c) Homologues come very close together.
    d) Crossing over occurs between sister chromatids.
    e) Chromosomes condense.

12. Which of the following is the best way to distinguish male from female?
    a) Males are larger.
    b) Males are more brightly colored.
    c) Males produce motile gametes.
    d) Males are more aggressive.
    e) All of the above statements are correct.

13. A potential disadvantage of asexual reproduction is that:
    a) it increases the time required to find a mate.
    b) it allows perpetuation of a population even when the members are isolated.
    c) it allows population size to increase rapidly.
    d) it produces genetically uniform populations.
    e) it requires additional rounds of meiosis, which is energetically much more costly than mitosis.

14. Which of the following is not a method of chromosomal sex determination that occurs in nature?

a) In birds, the mother's sex chromosomes determine the sex of the offspring.

b) In sea turtles, eggs laid in hot sand become females and eggs laid in cooler sand become males.

c) In humans, the presence of a Y chromosome makes an individual male, even if he also possesses two X chromosomes.

d) In bees, the eggs that the queen allows to be fertilized become females and the eggs she doesn't fertilize become males.

e) All of the above are naturally occurring examples of sex determination.

15. A karyotype reveals:

a) the shape of the spindle.

b) 23 pairs of chromosomes.

c) the number, shapes, and sizes of chromosomes in an individual cell.

d) the sex chromosomes but not the non-sex chromosomes.

e) the non-sex chromosomes but not the sex chromosomes.

16. What type of chromosomal abnormality leads to the phenotype known as Down syndrome?

a) trisomy 21

b) monosomy 21

c) monosomy 22

d) trisomy 22

e) disomy 21

17. Nondisjunction:

a) is the unequal division of the genetic material during cell division.

b) occurs during meiosis but not mitosis.

c) occurs in males but not females among mammals, and in females but not males among birds.

d) leads to a missing chromosome or extra chromosome.

e) Both a) and d) are correct.

18. Of all the varying types of sex chromosome abnormalities, one possibility—that of an individual having one Y but no X—has never been reported. Which of the following statements is the most likely reason that medical researchers have yet to report a person having a karyotype with one Y and no X?

a) There is no meiotic event that could produce an egg that does not have an X chromosome.

b) Individuals are probably born with this karyotype; however, because the phenotypic abnormalities caused are so slight, they live out their lives normally and are unaware of this genetic anomaly.

c) The Y chromosome needs the X chromosome to be present in the cell in order for its genes to be expressed.

d) The X chromosome contains genes that are unrelated to sex determination and are essential for life.

e) Our current technology for karyotyping would not reveal such an abnormality.

Short-Answer Questions

1. Seventy to ninety percent of the genetic material in a sperm or egg cell made in your body could be inherited from your mother. Why isn't there a 50/50 split (half from your mother and half from your father)? Is the reverse situation (more genetic material from your father than from your mother) possible?

2. Healthy individuals may have just one sex chromosome, as long as it is an X chromosome. Why can't a person survive with a Y chromosome and no X chromosome?

3. Some organisms can switch between sexual and asexual reproduction. Under what conditions would you expect them to reproduce sexually? What conditions would be more conducive to asexual reproduction? How would you test your hypotheses under laboratory conditions?

See Appendix for answers. For additional study questions, go to www.prep-u.com.

1 Why do offspring resemble their parents?

- **7.1** Family resemblance: your mother and father each contribute to your genetic makeup.
- **7.2** Some traits are controlled by a single gene.
- **7.3** Mendel learned about heredity by conducting experiments.
- **7.4** Segregation: you've got two copies of each gene but put only one copy in each sperm or egg.
- **7.5** Observing an individual's phenotype is not sufficient for determining its genotype.

2 Probability and chance play central roles in genetics.

- **7.6** Chance is important in genetics.
- **7.7** A test-cross enables us to figure out which alleles an individual carries.
- **7.8** We use pedigrees to decipher and predict the inheritance patterns of genes.

3 The translation of genotypes into phenotypes is not a black box.

- **7.9** Incomplete dominance and codominance: the effects of both alleles in a genotype can show up in the phenotype.
- **7.10** What's your blood type? Some genes have more than two alleles.
- **7.11** Multigene traits: how are continuously varying traits such as height influenced by genes?
- **7.12** Sometimes one gene influences multiple traits.
- **7.13** Why are more men than women color-blind? Sex-linked traits differ in their patterns of expression in males and females.
- **7.14** Environmental effects: identical twins are not identical.

4 Some genes are linked together.

- **7.15** Most traits are passed on as independent features: Mendel's law of independent assortment.
- **7.16** Red hair and freckles: genes on the same chromosome are sometimes inherited together.

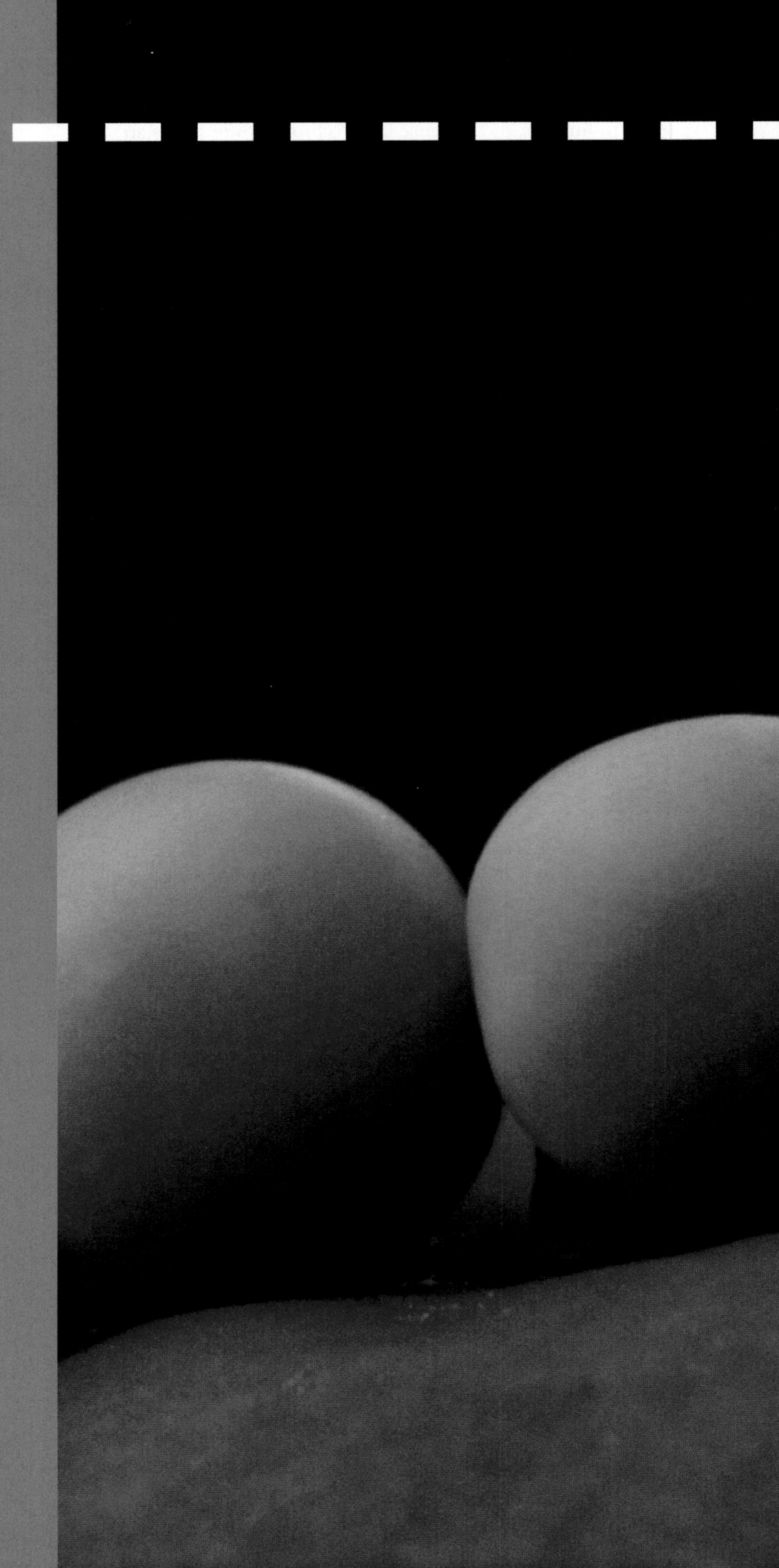

7

Mendelian Inheritance

Family resemblance: how traits are inherited

❶ Why do offspring resemble their parents?

Identical twins inherit the same genetic material. (These twins were not told what to do with their hands when the photographer took the photo. Note the similarities in their hand positions.)

7•1

Family resemblance: your mother and father each contribute to your genetic makeup.

Q How can a single bad gene make you smell like a rotten fish?

Can a gene be cruel? Of course not. But if one could, consider this candidate: in humans, there is a gene for an enzyme called FMO3 (flavin-containing mono-oxygenase-3), which breaks down a chemical in our bodies that smells like rotting fish. Some unfortunate individuals inherit a defective FMO3 gene and can't break down the noxious chemical. Instead, their urine, sweat, and breath excrete it, causing them to smell like rotting fish. Worst of all, because the odor comes from within their bodies, they cannot wash it away no matter how hard they try. Called "fish odor syndrome," this disorder often causes those afflicted by it to suffer ridicule, social isolation, and depression.

For individuals born with this malady, beyond their own suffering there looms a scary question: "Will I pass this condition on to my children?" They might also wonder how they came to have the disorder, particularly if neither of their parents suffered from it.

Fortunately, most of us do not have to worry about fish odor syndrome, but we do wonder about many other inherited traits that we may or may not pass down to our children. Who do I blame for my straight hair (or curly hair or big feet)? Hearing a new baby described as having "his father's nose" and "his grandmother's lips" could make someone think that we're all cobbled together from a bunch of used parts. Where do we begin and our parents end (**FIGURE 7-1**)?

In many cases, the answers to questions about heredity are simple. Fish odor syndrome, for example, is just such a case. Recall from Chapter 6 that sexually reproducing organisms generally carry two copies of every chromosome in every cell (except their sex cells), because they inherit one copy of each chromosome from each parent. Humans, for example, have 23 pairs of chromosomes (46 individual chromosomes). At each location on the two chromosomes of a pair is the same gene: one copy from the mother and one from the father (**FIGURE 7-2**). As we noted in Chapter 5, each of the two copies of the gene is call an **allele.** The gene for FMO3 is on chromosome 1. There is a normal version—or allele—of the gene for FMO3, which most people carry, and there is a rare, defective version that is responsible for fish odor syndrome. As long as a person has at least one normal version of the FMO3 gene, he or she will produce enough of the enzyme to break down the fishy chemical. But if a person inherits two copies of the defective version of the fish odor gene, one from each parent, that person will inherit the disorder (**FIGURE 7-3**).

When it comes to having children, there is a bright side, at least. Although individuals with fish odor syndrome carry two defective copies of the allele and will pass on one set of the bad instructions in every sperm or egg cell that they make, their children won't necessarily inherit the disease. As long as the other parent supplies a normal version of the FMO3 gene, the child will not have fish odor syndrome. The unaffected children, though, will carry a silent copy of the fish odor allele,

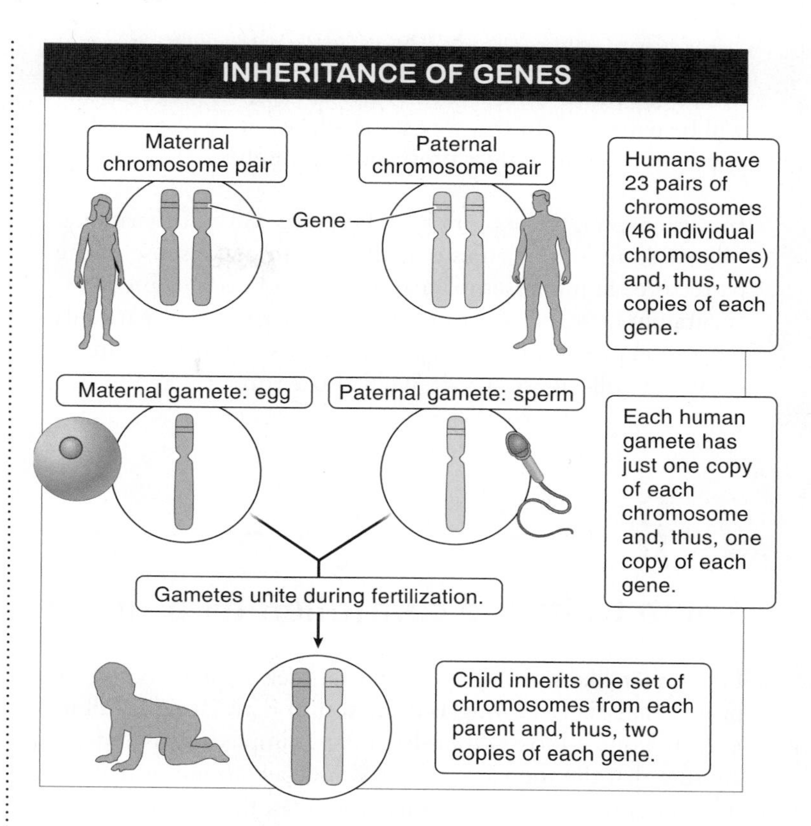

FIGURE 7-2 How we inherit our genes. Each human offspring inherits one maternal set of 23 chromosomes and one paternal set.

FIGURE 7-1 Where do we begin and our parents end? Family resemblances reveal that many traits are inherited.

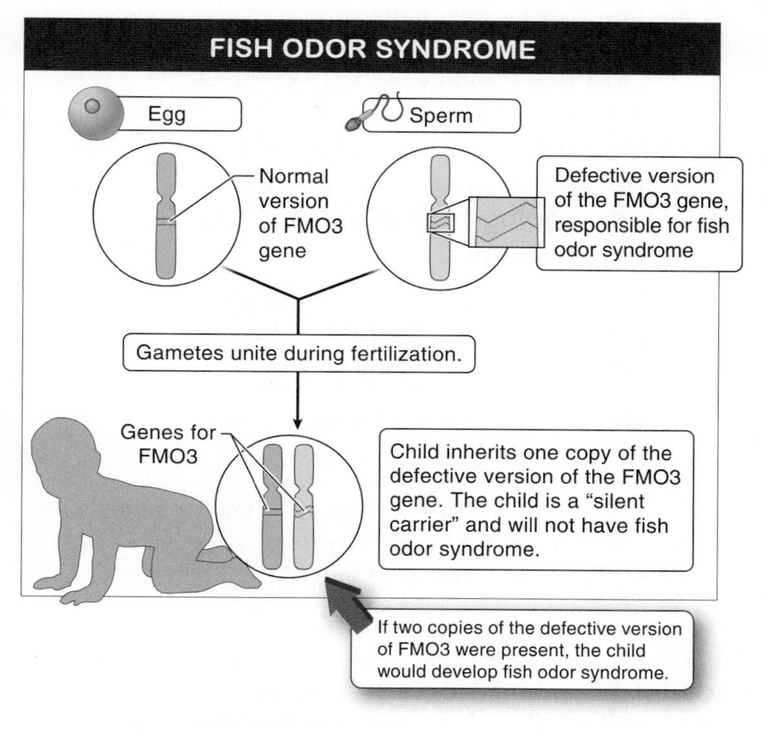

FIGURE 7-3 Unlucky catch. A baby inheriting two copies of the defective FMO3 gene will develop fish odor syndrome.

and if, when they have children, it comes together with a defective FMO3 allele from another person, the fish odor trait will be expressed. In this way, some alleles can exist in a population without always revealing themselves.

In this chapter, we explore how heredity works. Inheritance follows some simple rules that allow us to make sense of the patterns of family resemblance we see, and even to predict accurately how many offspring are likely to inherit particular traits—or, put another way, predict the likelihood that an offspring will inherit a trait. We'll also examine why the behavior of some traits is easy to predict while many other traits have less straightforward patterns of inheritance, yet still can be studied experimentally and yield predictable patterns.

TAKE-HOME MESSAGE 7•1

Offspring resemble their parents because they inherit genes—instruction sets for biochemical, physical, and behavioral traits, some of which are responsible for diseases—from their parents.

7•2

Some traits are controlled by a single gene.

Like father, like son. No one will ever seek a paternity test to prove that Michael Douglas is the son of Kirk Douglas. All it takes is a quick look at their legendary dimpled chins and we are sure that the men are related. We see all around us that offspring resemble their parents more than they resemble other random individuals in the population—a consequence of the passing of characteristics from parents to offspring through their genes. This is **heredity.**

> HEREDITY
>
> I AM the family face;
> Flesh perishes, I live on,
> Projecting trait and trace
> Through time to times anon,
> And leaping from place to place
> Over oblivion.
>
> The ears-heired feature that can
> In curve and voice and eye
> Despise the human span
> Of durance—that is I;
> The eternal think in man,
> That heeds no call to die.
>
> —Thomas Hardy, *Moments of Vision and Miscellaneous Verses*, 1917

Observing heredity is easy. Elucidating how it works is not. For thousands of years before the mechanisms of heredity were discovered and understood, plant and animal breeders recognized that there is a connection from parents to offspring across generations. In ancient Greece, for example, the poet Homer extolled the tremendous benefits to society that came from the skillful breeding of horses. The awareness of beneficial breeding practices enabled farmers to systematically create strains of crops, livestock, and even pets with desirable traits (**FIGURE 7-4**). From flocks of sheep and herds of reindeer to litters of puppies and fields of wheat, the similarity among relatives is clear. Once breeders recognized the existence of heredity, they began selecting individual plants or animals with the desired traits to breed with each other, in the hope that their offspring would also have the desirable traits. And since then, it has been possible to create a rich world of sweeter corn, loyal dogs, docile livestock, beautiful flowers, and more.

Once breeders recognized the existence of heredity, selective breeding—such as for increased body size in cattle—became possible.

FIGURE 7-4 Have a cow! Selective breeding was used to produce this steer.

EXAMPLES OF SINGLE-GENE TRAITS

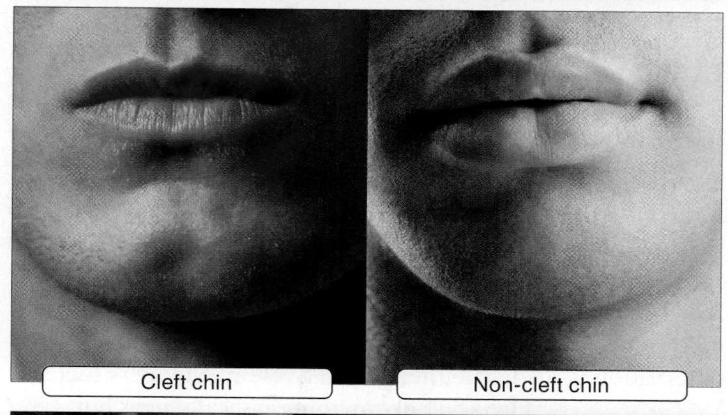

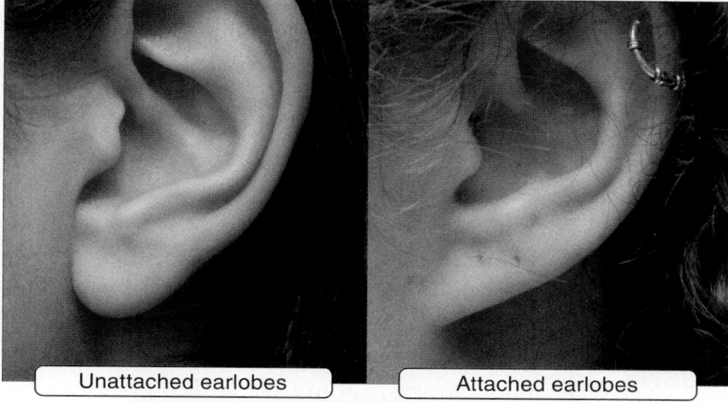

FIGURE 7-5 Chins, ears, and hairlines: single-gene traits. Some traits are determined by the instructions a person carries on one gene.

And these breeders did it all without ever really understanding how heredity works.

Practical successes were many, but people didn't understand exactly how the outcomes were achieved. They knew the result of selective breeding, but why and how it worked, or didn't always work, was a mystery. Patterns of similarity among related individuals are impossible to make sense of without an underlying understanding of *how* heredity works. This is the province of the field of genetics.

Q Are any human traits determined by a single gene?

Mechanisms of heredity are like a puzzle, and they become a bit easier to grasp when we focus on aspects of family resemblances that are commonplace. Virtually everyone with a cleft chin, for instance, has a similarly dimpled parent. Everyone who has earlobes that hang from their ears rather than being attached directly to their head has at least one parent with the same feature. Everyone who has a widow's peak has a parent with the same hairline. Everyone who is farsighted has a similarly farsighted parent. The list goes on and on. In fact, there are more than 9,000 human traits that exhibit these straightforward patterns of inheritance. Traits that are determined by the instructions a person carries on one gene are called **single-gene traits** (**FIGURE 7-5**).

How single-gene traits such as the cleft chin pass from parent to child is the easiest pattern of inheritance to decipher. For this reason, we first explore the mechanisms by which single-gene traits are inherited. Then we investigate how to expand our model of heritability to account for the passing on of more complex traits.

TAKE-HOME MESSAGE 7•2

More than 9,000 human traits are determined by the instructions a person carries on a single gene, and the traits exhibit straightforward patterns of inheritance.

7•3

Mendel learned about heredity by conducting experiments.

What do parents "give" their offspring that confers similarity? In other words, how is it that a physical entity is inherited—that is, how is it passed from parent to offspring? It wasn't until the mid-1800s that any real headway was made on these questions. At that time, Gregor Mendel, a monk living in what is now the Czech Republic, began some studies that not only shed light on these questions but practically answered them completely. Mendel understood the essence of the genetics puzzle and set out to piece it all together.

Q Individuals who share common ancestry are called "blood relatives," yet they don't actually share any blood. How might the phrase "blood relatives" be a reflection of early conceptions of inheritance?

When Mendel turned to these questions about heredity, there were no obvious answers and the prevailing state was confusion. There were a couple of existing ideas, but they had obvious problems. One postulated that perhaps an entire, pre-made human—albeit a very, very tiny one—was contained in every sperm cell (**FIGURE 7-6**). That idea had been popular since the late 1600s. It doesn't take much imagination to see why it ran into difficulty, though, because children resemble both their mothers and their fathers, not just their fathers. Another popular idea was that offspring reflect a simple blending of their two parents' traits via the blood. But this couldn't explain how brown-eyed parents could give birth to blue-eyed children. Or why one tall and one short parent sometimes produced a tall child, rather than always producing children of intermediate height.

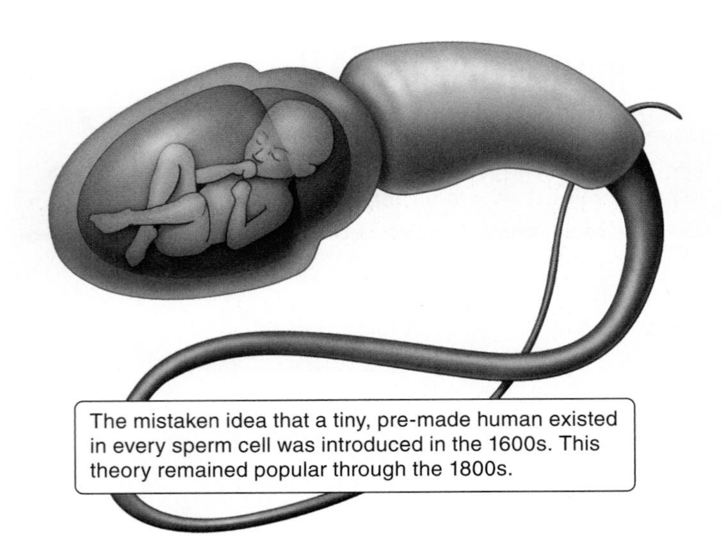

FIGURE 7-6 The "pre-made human." Tiny humans in every sperm cell? Why would children have any resemblance to their mothers?

Sometimes scientific breakthroughs are made because an innovative new technique is invented or a lucky, chance observation is made. Neither played a role in Mendel's success. He didn't do anything radically new, but simply applied the tried and true methodical experimentation and scientific thinking (as described in Chapter 1) that had been a boon to chemists and physicists. Three features of Mendel's research, in particular, were critical to its success (**FIGURE 7-7**).

1. He chose a good organism to study: the garden pea. It's not that Mendel had a particular fondness for vegetables. In fact, his goals were to understand inheritance in all organisms, not just plants. Pea plants were simply the right tool. Cats and dogs or even mice wouldn't have served his purposes very well, because they would have been too hard to take care of in the large numbers—thousands and thousands of individuals—that Mendel required. Humans, too, would have made a terrible study organism. We take too long to breed (and won't produce offspring on command). Pea plants, on the other hand, have a lot going for them as experimental organisms. They are relatively easy to fertilize manually by "pollen dusting." It is easy to collect dozens or even hundreds of offspring from a single **cross,** in which male pollen is used to fertilize female eggs. And, on top of all this, pea plants are fast enough breeders that Mendel could conduct experiments that lasted for multiple generations.

2. Mendel chose to focus on easily categorized traits. For instance, shape: all peas of the variety that Mendel studied are either round or wrinkled in shape, with nothing in between. In addition, all peas are either yellow or green in color, never any intermediate shade. In all, Mendel looked at seven different traits but, for each trait, only two variants ever appeared, so he and his research assistants could easily observe and unambiguously identify them.

3. Mendel began his studies by first repeatedly breeding together similar plants until he had many distinct populations, each of which was unvarying for a particular

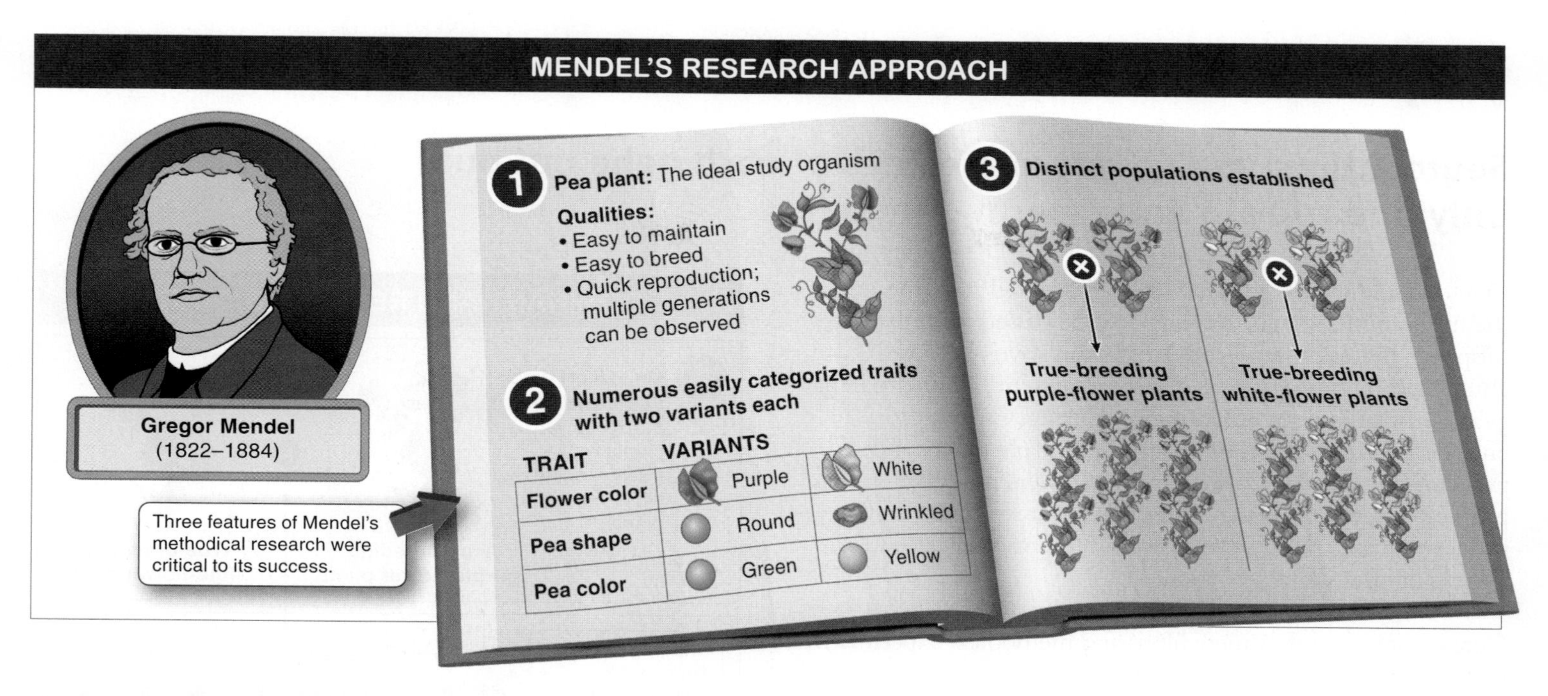

FIGURE 7-7 Mendel's notebook. Careful planning and a well-selected study organism contributed to the success of Mendel's experiments.

trait. He described these plants as **true-breeding** for that trait because they always produced offspring with the same variant of the trait as the parents. True-breeding round-pea plants always produced plants with round peas when they were crossed together. True-breeding purple-flowered pea plants always produced purple-flowered offspring, while true-breeding white-flowered pea plants always produced white-flowered offspring. It took a lot of prep work to establish these populations, but once Mendel had them, he was in a position to set up all of the different crosses that enabled him to piece together the genetics puzzle.

Once he had obtained his groups of true-breeding plants, Mendel began a straightforward process of experimentation. He crossed plants with different traits—for example, he crossed plants with green peas and plants with yellow peas—and observed large numbers of their offspring, counting the numbers of green peas and yellow peas produced. (This cross, written as "green × yellow cross," always resulted in plants with nothing but yellow peas.) Mendel devised a hypothesis that would explain his observations and generate predictions about the outcome of further crosses. Then he conducted those crosses to see whether the predictions generated by his hypothesis were borne out.

There was a simple elegance to Mendel's work: his predictions were always in the form of simple questions that could be answered by his experiments. His crosses would come out either one way or the other. Everything was black and white. This process of performing well thought out, rigorous experiments was a radical innovation for biology. In the end, Mendel's patience, perseverance, and cleverness paid off. As we explore his results in the next section, we see that they allowed a definitive interpretation.

TAKE-HOME MESSAGE 7•3

In the mid-1800s, Gregor Mendel conducted studies that helped us understand how traits are inherited. He applied methodical experimentation and rigorous hypothesis testing, focusing on easily observed and categorized traits in garden peas.

7.4

Segregation: you've got two copies of each gene but put only one copy in each sperm or egg.

One odd and recurring result spurred Mendel to figure out the mechanism by which traits could be passed from parent to offspring. Just as brown-eyed parents can have blue-eyed children, sometimes traits that weren't present in either parent pea plant would show up in their offspring. Crosses that entailed the fertilization of plants with purple flowers by pollen from other plants with purple flowers produced *mostly* purple-flowered offspring, for instance, but *sometimes* they produced plants with white flowers. How was that possible? Where did the whiteness come from?

Here's where Mendel's meticulous and methodical experiments paid off. First, he started with some true-breeding white-flowered plants. Then he got some true-breeding purple-flowered plants. He wondered: which color wins out when a white-flowered plant is crossed with a purple-flowered plant? The answer was definitive: purple wins (**FIGURE 7-8**). All of the offspring from these crosses were purple, every time. For this reason, Mendel called the purple-flower trait **dominant,** and he considered the white-flower trait to be the **recessive** trait. In general, a dominant trait masks the effect of a recessive trait when an individual carries both the dominant and the recessive versions of the instructions for the trait.

Things got a bit more interesting when Mendel took the purple-flowered plants that came from the cross between true-breeding purple- and white-flowered plants and bred these purple offspring with each other. He found that these mixed-parentage plants were no longer true-breeding. Occasionally, they would produce white-flowered offspring. (To be exact, of the 929 plants Mendel examined, 705 had purple flowers, and 224 had white flowers.) Apparently, the directions for building white flowers—last seen in one of their grandparents—were still lurking inside the purple-flowered parent plants. The existence of traits that could disappear for a generation and then show up again was quite perplexing. Mendel devised a simple and perfect hypothesis to explain these observations (**FIGURE 7-9**).

Mendel's hypothesis for explaining this pattern of inheritance incorporated three ideas that then helped him (and now help us) make predictions about crosses he hadn't yet done.

1. Rather than passing on the trait itself, **each parent puts into every sperm or egg it makes a single set of instructions for building the trait.** Today, we call that instruction set a *gene.*

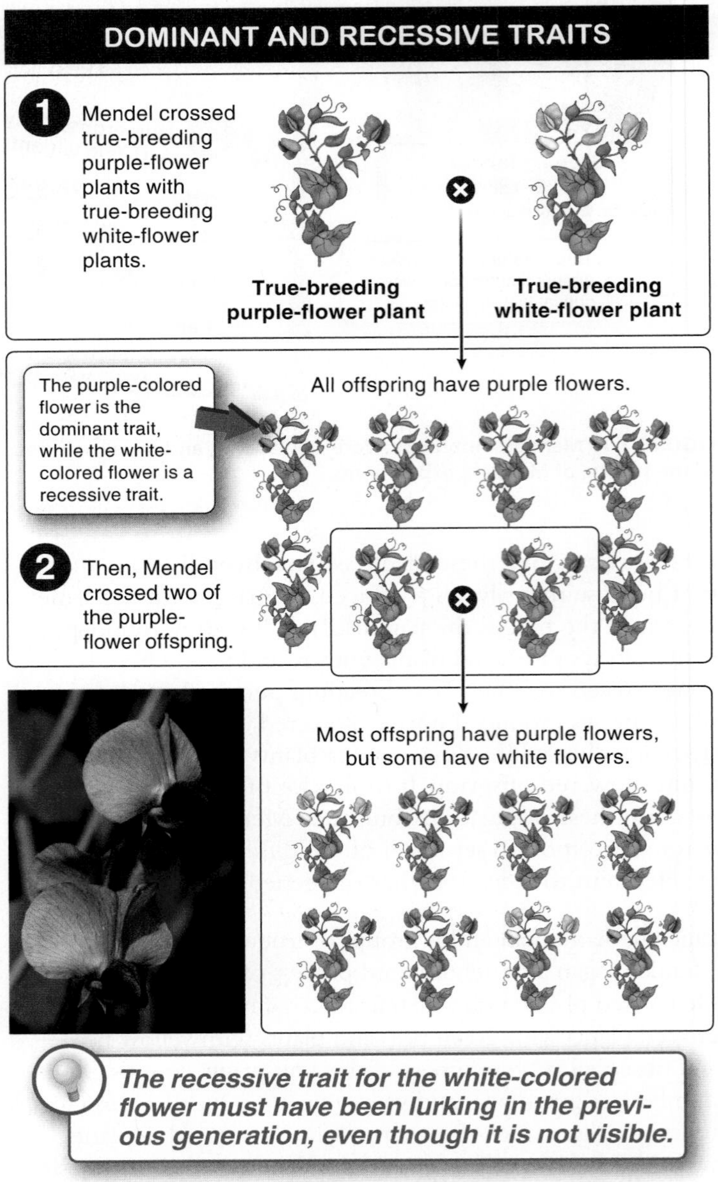

DOMINANT AND RECESSIVE TRAITS

1 Mendel crossed true-breeding purple-flower plants with true-breeding white-flower plants.

True-breeding purple-flower plant

True-breeding white-flower plant

All offspring have purple flowers.

The purple-colored flower is the dominant trait, while the white-colored flower is a recessive trait.

2 Then, Mendel crossed two of the purple-flower offspring.

Most offspring have purple flowers, but some have white flowers.

The recessive trait for the white-colored flower must have been lurking in the previous generation, even though it is not visible.

FIGURE 7-8 White or purple? By careful and repeated crosses among pea plants, Mendel determined that there were "dominant" and "recessive" traits.

2. Offspring receive two copies of the instructions for any trait. Often, both sets of instructions are identical and the offspring produce the traits according to those instructions. Other times, though, each parent contributes a slightly different set of instructions—that is, a different

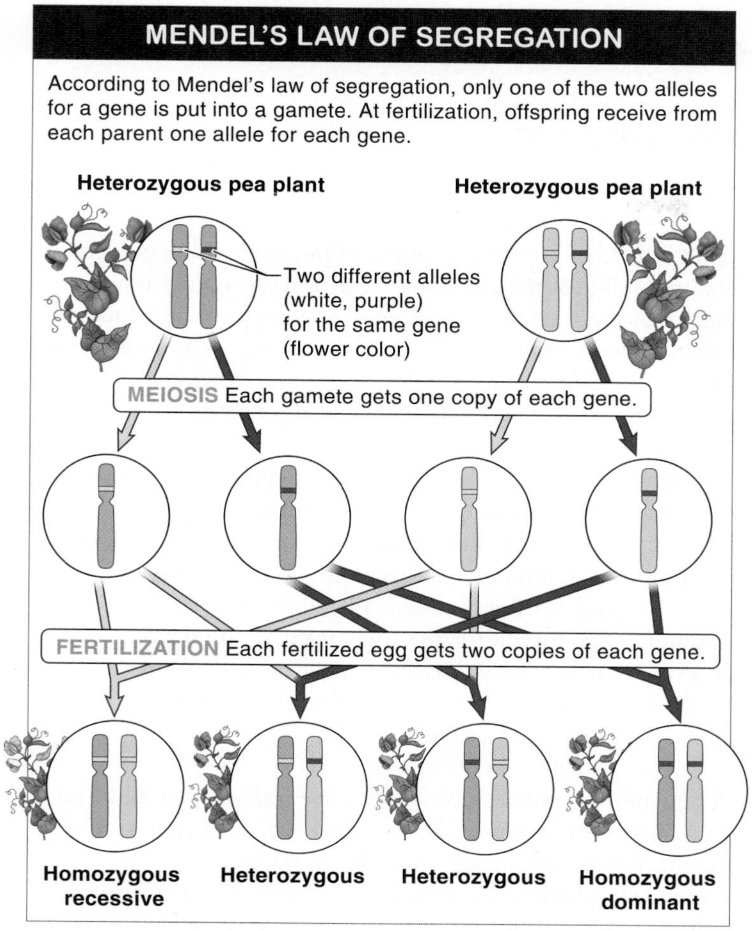

MENDEL'S LAW OF SEGREGATION

According to Mendel's law of segregation, only one of the two alleles for a gene is put into a gamete. At fertilization, offspring receive from each parent one allele for each gene.

Heterozygous pea plant
Heterozygous pea plant
Two different alleles (white, purple) for the same gene (flower color)
MEIOSIS Each gamete gets one copy of each gene.
FERTILIZATION Each fertilized egg gets two copies of each gene.
Homozygous recessive
Heterozygous
Heterozygous
Homozygous dominant

FIGURE 7-9 Mendel's law of segregation. Organisms have two copies of each gene but place only one copy into each gamete during the process of meiosis (as described in Chapter 6).

allele—for the trait. So, for example, in pea plants there are two alleles for flower color: a purple-flower allele and a white-flower allele. Each is a different allele, but because both specify instructions for producing flowers, they are both genes for flower color. Same gene (flower-color gene), different alleles (purple and white).

3. The trait observed in an individual depends on the *two* copies of the gene it inherits from its parents. When an individual inherits the same two alleles for this gene, the individual's genotype for that gene is said to be **homozygous** and the individual shows the trait specified by the instructions embodied in those alleles. When an individual inherits a different allele from each parent, the individual's genotype for that gene is said to be **heterozygous.** Dominant and recessive alleles are defined by their action when they are in the heterozygous state: if one of the alleles is fully apparent in the individual while the other is not, the first is called the dominant allele and is said to "mask" the effect of the other allele, which is called the recessive allele.

> "'There were no questions.'
>
> —Entry in the meeting notes following Mendel's first public presentation of his ideas on how heredity works. (*No one in the audience of scientists had any idea what he was talking about. It was about 40 years before the world understood his discoveries.*)"

Later, when the offspring grows up and gets ready to have its own offspring, it only gets to contribute one of its two copies of the gene to its offspring. The other parent will contribute the other allele. So, when sperm and eggs are made, each sex cell gets only one copy of the gene—as opposed to the two copies present in every other cell in the body. For a male who is heterozygous, for example, it means that half of the sperm he produces will have one of the alleles and half will have the other. The idea that, of the two copies of each gene everyone carries, only one of the two alleles gets put into each gamete is significant and important enough that it is called **Mendel's law of segregation.**

TAKE-HOME MESSAGE 7•4

Each parent puts into every sperm or egg it makes a single set of instructions for building a particular trait. This instruction set is what today we call a gene. The trait observed in an individual depends on the two copies (alleles) of the gene it inherits from its parents.

7•5

Observing an individual's phenotype is not sufficient for determining its genotype.

Things are not always as they appear. Take skin coloration, for example. Humans and many other mammals have a gene that contains the information for producing melanin, one of the chemicals responsible for giving our skin its coloring (**FIGURE 7-10**). Unfortunately, there is also a defective, non-functioning version of the melanin gene that is passed along through some families. An individual who inherits two copies of the defective version of the gene cannot produce pigment and has a condition known as albinism, a disorder characterized by little or no pigment in the eyes, hair, and skin. But it is impossible to tell whether a normally pigmented individual carries one of these defective alleles just by looking at his or her appearance—we would need to get a genetic analysis done to discover this information.

The inability to deduce an individual's genetic makeup through simple observation is a general problem in genetics: physical appearances don't always exactly reflect the underlying genes. A normally pigmented individual may have two copies of the pigment-producing allele or may have only one copy. In either case, the individual will look the same. The outward appearance of an individual is called its **phenotype.** A phenotype includes features visible to the naked eye such as flashy coloration, height, or the presence of antlers. A phenotype also includes less easily visible characteristics such as the chemicals an individual produces to clot blood or digest lactose. An individual's phenotype even includes the behaviors it exhibits.

PHENOTYPE: Little or no pigment in the eyes, hair, and skin
GENOTYPE: Homozygous for the recessive allele for albinism

FIGURE 7-10 An albino deer stands out from its peers. Carrying two non-functioning versions of the gene that produces the pigment melanin causes albinism.

Underlying the phenotype is the **genotype.** This is an organism's genetic composition. We usually speak of an individual's genotype in reference to a particular trait. For example, an individual's genotype might be described as "homozygous for the recessive allele for albinism." Another individual's genotype for the melanin gene might be described as "heterozygous." Occasionally, the word "genotype" is also used as a way of referring to *all* of the genes an individual carries.

When an organism exhibits a recessive trait, such as albinism, we know with certainty what its genotype is. When it shows the dominant trait, on the other hand, it's impossible to distinguish whether the organism carries two copies of the dominant allele (homozygous dominant) or carries one copy of the dominant allele and one of the recessive (heterozygous). Because it's not possible to discern the genotypes of two individuals with the same phenotype just by looking at them, much of genetic analysis must make use of clever experiments and careful record-keeping to determine organisms' genotypes. Let's explore a useful method of analyzing the relationships between genotypes and phenotypes.

How do we analyze and predict the outcome of crosses? First, we must assign symbols to represent the different variants of a gene. Generally we use an upper-case letter for the dominant allele and lower-case letter for the recessive allele. In the case of the albino giraffe, we represent the albino individual as *aa,* because it must carry two copies of the recessive allele, *a*. A giraffe that is pigmented must have the genotype *AA* or *Aa*. If we don't know which of the two possible genotypes the pigmented individual is, we can write *A_*, where _ is a placeholder for the unknown second allele, whose identity we aren't certain of.

We can trace the possible outcomes of a cross between two individuals using a handy tool called the **Punnett square.** In **FIGURE 7-11** we illustrate the cross between a true-breeding

PUNNETT SQUARE: ALBINISM

A Punnett square is a useful tool for determining the possible outcomes of a cross between two individuals.

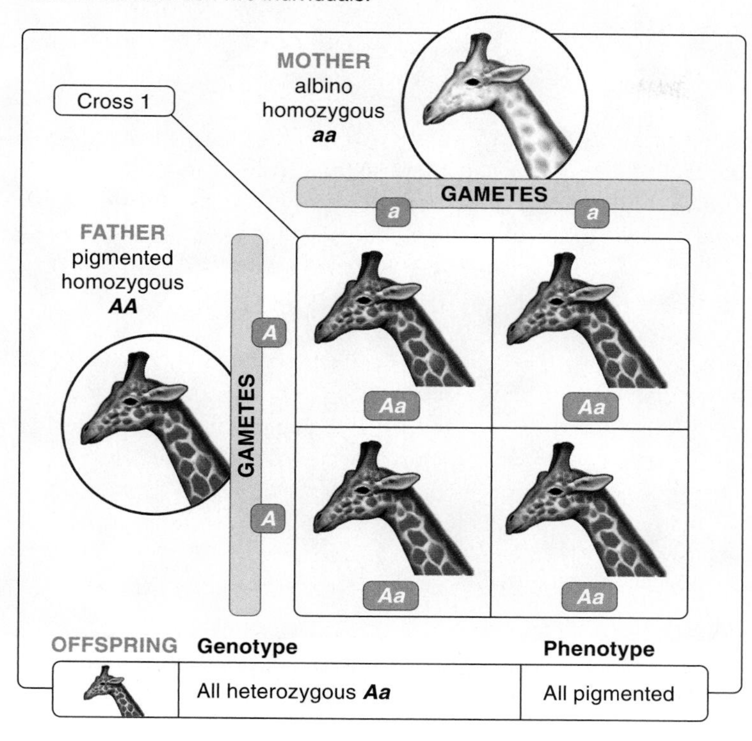

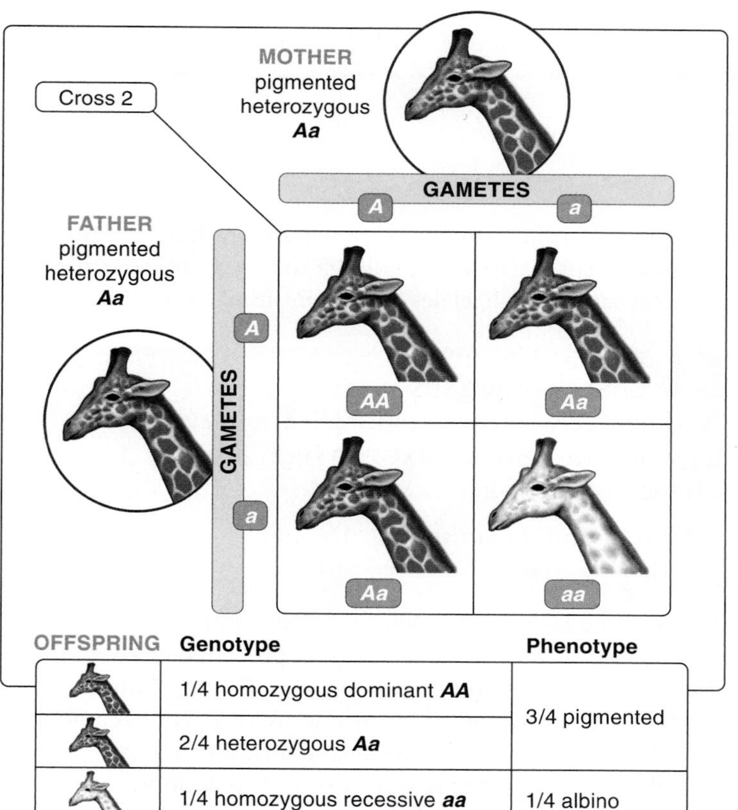

FIGURE 7-11 Predicting the outcome of crosses. A Punnett square helps in determining the likelihood of inheriting albinism.

pigmented individual, *AA*, and an albino, *aa*. Along the top of the square we list, individually, the two alleles that one of the parents produces, and along the left side of the square we list the two alleles that the other parent produces. We split up an individual's two alleles in this way because, although the individual carries two alleles, only one of the alleles is contained in each sperm or egg cell that it produces.

In the four cells of the Punnett square, we enter the genotypes of all the possible offspring resulting from our cross—a cell contains the allele given at the head of the column, which is one parent's contribution, and the allele at the left of the row, which is the other parent's contribution. The two gametes that come together at fertilization produce the genotype of the offspring. In Cross 1 illustrated in Figure 7-11, every possible offspring would be heterozygous and would be normally pigmented, because it receives a dominant allele from the pigmented parent and a recessive allele from the albino parent.

In the bottom half of Figure 7-11 (Cross 2), we trace the cross between two heterozygous individuals. Note that each parent produces two kinds of gametes, one with the dominant allele and one with the recessive allele. This cross has four possible outcomes: one-quarter of the time the offspring will be homozygous dominant (*AA*), one-quarter of the time the offspring will be homozygous recessive (*aa*), and the remaining half of the time the offspring will be heterozygous (*Aa*). Phenotypically, three-quarters of the offspring will be normally pigmented (*AA* or *Aa*) and one-quarter will be albino (*aa*).

TAKE-HOME MESSAGE 7•5

It is not always possible to determine an individual's genetic makeup, known as its genotype, by observation of the organism's outward appearance, known as its phenotype. For a particular trait, an individual may carry a recessive allele whose phenotypic effect is masked by the presence of a dominant allele. Much genetic analysis therefore makes use of clever experiments and careful record-keeping, using Punnett squares, to determine organisms' genotypes.

❷ Probability and chance play central roles in genetics.

Because of the role of chance in genetics, we cannot always predict the exact outcome of a genetic cross.

7•6

Chance is important in genetics.

Sometimes genetics is a bit like gambling. Even with perfect information, it can still be impossible to know the genetic outcome with certainty. It's like flipping a coin: you can know every last detail about the coin, but you still can't know whether the coin will land on heads or tails. The best you can do is define the probability of each possible outcome.

The rules of probability (the same ones that govern coin tosses and the rolling of dice) have a central role in genetics, for two reasons. The first is a consequence of segregation, the process Mendel described, in which each gamete an individual produces receives only one of the two copies of each gene that the individual carries in most of its other cells. As a result, it is equally likely that the haploid gamete—the sperm or egg—will include one or the other of the two alleles that the individual carries. It is impossible to know which allele it will be. The second reason probability plays a central role in genetics is that fertilization, too, is a chance event. All of the sperm or eggs produced by an individual are different from one another, and any one of those gametes may be the gamete involved in fertilization. Thus, knowing everything about the alleles a parent carries is not enough to be able to determine with certainty which alleles his or her offspring will carry.

> Chaos umpire sits,
> And by decision more embroils the fray
> By which he reigns: next him high arbiter
> Chance governs all.
>
> —John Milton, *Paradise Lost,* 1667

Let's explore how we can make predictions in games of chance, as well as in matings, based on probabilities. As with the probability of getting heads in a coin flip, any gamete produced by an individual heterozygous for a trait has a 50% probability of carrying the dominant allele and a 50% probability of carrying the recessive allele. If an individual is homozygous for a trait, 100% of his or her gametes will carry that allele. Let's use these probabilities in an example.

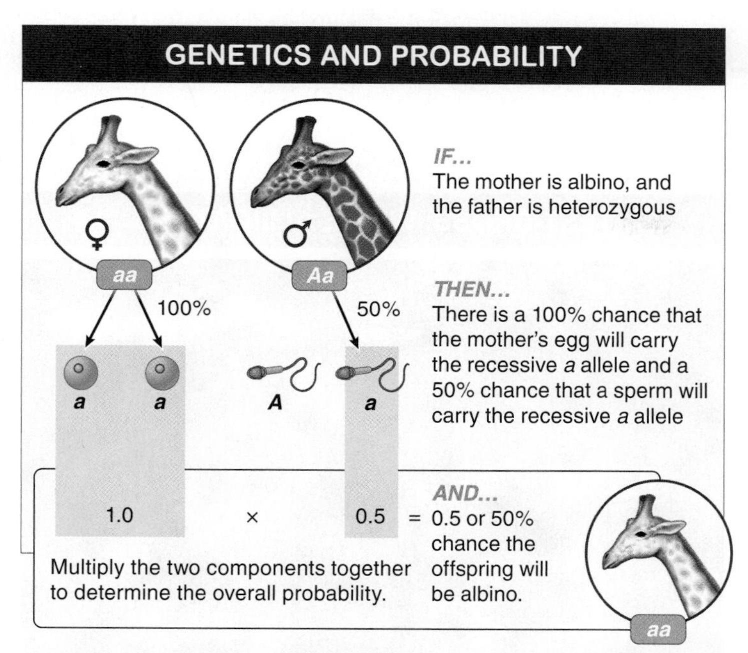

FIGURE 7-12 Using probability to determine the chance of inheriting albinism.

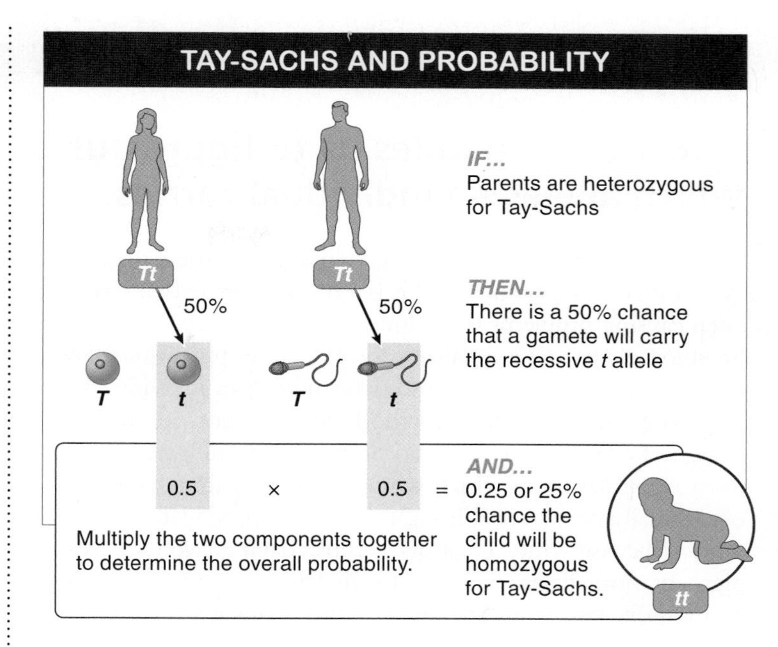

FIGURE 7-13 The chance of inheriting Tay-Sachs disease. The probability that a child inherits Tay-Sachs disease from two heterozygous parents is 25%.

If a male is heterozygous for albinism (*Aa*) and a female is albino (*aa*), what is the probability that their child will be homozygous for albinism (*aa*)? In order to get the *aa* outcome, two events must occur. First, the father's gamete must carry the recessive allele (*a*) and, second, the mother's gamete must carry the recessive allele (*a*). In this case, the probability of a homozygous recessive offspring is 0.5 (the probability that the father's gamete carries *a*) times 1.0 (the probability that the mother's gamete carries *a*), for a probability of 0.5. This is a general rule when determining the likelihood of a complex event occurring: if you know the probability of each component that must occur, you simply multiply all the probabilities together to get the overall probability of that complex event occurring (**FIGURE 7-12**).

Consider another example, involving Tay-Sachs disease. As described in Chapter 3, Tay-Sachs is a disease caused by malfunctioning lysosomes that do not digest cellular waste properly. It leads to death in early childhood. Tay-Sachs occurs if a child inherits two recessive alleles for the Tay-Sachs gene. If each parent is heterozygous for the Tay-Sachs gene, what is the probability that their child will have Tay-Sachs disease?

To solve this, just break down the event of the child having Tay-Sachs into two separate events: first, it must inherit a recessive allele from its heterozygous father. The probability of that is 0.5. Second, it must also inherit a recessive allele from its heterozygous mother. The probability of that is also 0.5. So the overall probability of the child having Tay-Sachs disease is $0.5 \times 0.5 = 0.25$, or 1 in 4 (**FIGURE 7-13**). Of course, if the couple has only one child, we can't predict with certainty whether this child will have Tay-Sachs. Mathematicians, on the other hand, might be satisfied just to be able to say that if the couple had an infinite number of children, we could be certain that one-fourth of them would have Tay-Sachs. Still, this is not much help for heterozygous couples trying to decide whether to have a child or not.

TAKE-HOME MESSAGE 7•6

Probability plays a central role in genetics. In segregation, each gamete that an individual produces receives only one of the two copies of each gene the individual carries in its other cells, but it is impossible to know which allele goes into the gamete. Chance plays a role in fertilization, too: all of the sperm or eggs produced by an individual are different from one another, and any one of those gametes may be the gamete involved in fertilization.

7•7

A test-cross enables us to figure out which alleles an individual carries.

How can you see something that is invisible to the naked eye? When it comes to genetics, this is a practical issue that has been causing problems for as long as humans have been breeding plants and animals for food or other purposes. Genes are too small to be seen, and so determining an individual's genotype requires indirect methods. Suppose you are in charge of the alligators at a zoo. Some of your alligators come from a population in which white, albino alligators have occasionally occurred, although none of your alligators are white. Because white alligators —those having two recessive pigmentation alleles, *mm* ("m" for melanin)—are popular with zoo visitors, you would like to produce some at your zoo through a mating program.

The problem is that you cannot be certain of the genotype of your alligators. They might be homozygous dominant, *MM,* or they might be heterozygous, *Mm*. In either case, their phenotype is normal coloration. How can you figure out which of these two possibilities is the genotype of a particular alligator? This is a challenge to animal breeders, but not an insurmountable one. Genes may be invisible, but their identity can be revealed by a simple tool called the **test-cross.**

In the test-cross, you take an individual exhibiting a dominant trait but whose genotype is unknown. You cross (i.e., mate) that individual with an individual that is homozygous recessive and look at the phenotypes of their offspring. In the case of your breeding program for albino alligators, you could borrow an albino alligator from another zoo and breed your unknown-genotype alligator (genotype: *M_*) with that albino alligator (genotype: *mm*). There are two possible outcomes, and they will reveal the genotype of your unknown-genotype alligator (**FIGURE 7-14**). If your alligator is homozygous dominant (*MM*), it will contribute a dominant allele, *M,* to every offspring. Even though the albino alligator will contribute the recessive allele, *m,* to all its offspring, all the offspring will be heterozygous, *Mm,* and none of them will be albino.

If, on the other hand, your unknown-genotype alligator is heterozygous, *Mm,* half of the time it will contribute a recessive allele, *m,* to the offspring. In every one of those cases, the

TEST-CROSS: WHITE ALLIGATORS

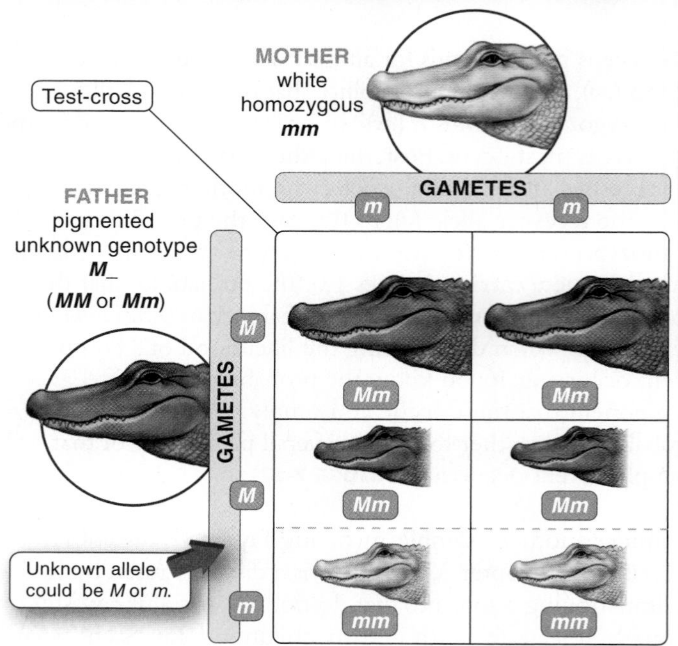

OFFSPRING (if unknown genotype is ***MM***)

	Genotype	Phenotype
	All heterozygous ***Mm***	All pigmented

OFFSPRING (if unknown genotype is ***Mm***)

	Genotype	Phenotype
	2/4 heterozygous ***Mm***	2/4 pigmented
	2/4 homozygous recessive ***mm***	2/4 white

FIGURE 7-14 A test-cross can reveal an unknown genotype. In this test-cross, a homozygous white female alligator is bred with a normally colored male of unknown genotype. The color of their offspring will help identify whether the male is homozygous dominant or heterozygous.

Study Guide for Dr. Ripley's Material
BIOL 101

FINAL EXAM REVIEW

EXAM: **WEDNESDAY, MAY 4 8-10AM**
THERE WILL BE NO MAKE-UP EXAMS!

COVERS: All lecture notes, assigned readings, and homework for the ENTIRE semester. The final is cumulative with 75 questions on material covered over the entire semester. Half of the questions will cover material presented by Dr. Ripley and the other half will be material covered by Dr. Salihu.

This list is only a study guide, **NOT** a complete list of all the material on the test. You may also use the study guides from Exams 1-4 as a source of review questions.

You may review your other exams in my office—come on in!

Scientific Method & Characteristics of Living Things

1. What are the steps of the scientific method? Describe these steps and be able to identify them from a description of an experiment.
2. Define theory and hypothesis. How are they related? How can a hypothesis become a theory?
3. Can a theory or hypothesis be proven? Explain.
4. What characteristics of a good experiment have we emphasized in class? Know about controls and variables (independent & dependent). You should be able to identify missing components from a description of an experiment.
5. What characteristics define "living?" That is, what are the characteristics of living things? If something moves, is it alive? If it breathes? If it reproduces? Responds to the environment?

Population and Community Ecology

6. Describe the terms population, community, and ecosystem as they are related to one another.
7. How does the environment influence population size? You should be able to define carrying capacity, density-dependent factors and density-independent factors.
8. What's the difference between exponential and logistic growth? What type of growth do most species display?
9. A biological community is shaped by interactions between species. Review the types of interactions between species: competition, predator-prey, parasite-host, commensalism, and mutualism. How do these interactions affect both of the species involved—do they benefit or harm each participant? Can plants participate in all these interactions?
10. Discuss some of the adaptations utilized in predator-prey interactions including camouflage, warning coloration, chemical defenses, mechanical defenses, and behavioral defenses.
11. What is competitive exclusion? What is resource partitioning? What kind of interaction between species causes them?
12. Explain how the shape of the earth, and air and oceanic circulation determine the weather and hence influence the geographical distribution of biomes on the Earth.
13. What are the natural reservoirs (storage sites) for carbon, phosphorus, and nitrogen? How are each recycled?
14. What is the ultimate source of energy for our planet? How does the sun influence the size of food webs?
15. Discuss the 10% rule of energy transfer through food webs.
16. What is biodiversity? What are biodiversity hotspots? What 2 criteria define biodiversity hotspots?

17. You have a pet and don't want it any more. Why not just release it into the wild? Use what you know about ecology and introduced species to justify your answer.
18. List as many examples from class as you can: human impact on the biosphere.
19. How does succession influence biodiversity?
20. What are the byproducts of the burning of fossil fuels and what are the impacts of these byproducts on the ecosystem?
21. Describe global warming. What causes it? How does deforestation contribute to global warming?

Genetic Inheritance

22. Understand the terms dominant, recessive, homozygous, and heterozygous.
23. What gametes can be produced if an organism's genotype is XxYY? That is, you need to be able to FOIL correctly.
24. Determine the inheritance of autosomal recessive traits utilizing monohybrid and dihybrid crosses. For instance, you predict the probability that fry (i.e. baby fish) will have shiny scales if shiny scales are dominant to dull scales and both parents are heterozygous. You should also be able to determine the phenotypic ratio of scale shininess and color (pink scales are dominant to white scales), if one fish is homozygous dominant for both traits and the other fish is heterozygous for scale shininess but homozygous recessive for scale color.
25. What is the phenotypic ratio of a dihybrid cross in which both parents are heterozygous for both traits?
26. Distinguish incomplete dominance, codominance, polygenic and pleiotrophic traits. Identify examples of each type of inheritance pattern.
27. Use a Punnett square to predict the cross of a sex-linked trait. For example, color blindness is a sex-linked recessive condition. A color blind woman marries a man who is not color blind. What is the probability that their child will be color blind? Is this child a boy or girl?
28. What is the genotype of a carrier of a autosomal recessive condition? Autosomal dominant condition? Sex-linked recessive condition? Sex-linked dominant condition?

Evolution

29. Define evolution. Do individuals or populations evolve? Is evolution's goal to make a perfect organism? Is natural selection another term for evolution?
30. Be able to calculate allele frequencies and determine if the population has evolved. For instance, let's say that claw length in crabs is determined by a single gene. Three individuals in the population are homozygous dominant for large claw length, 12 individuals are heterozygous and 4 individuals are homozygous recessive for small claw length. In the next generation, the number of individuals is 5, 10, and 2 respectively. Has the population evolved?
31. Be able to describe how each of the following lines of evidence support evolution. Be able to list specifically which each shows us.
    - the fossil record
    - biogeography
    - comparative anatomy and embryology
    - molecular biology
    - laboratory and field experiments
32. What is the difference between homologous and analogous structures?
33. What are vestigial structures?
34. Distinguish mutations, genetic drift (including the founder effect and bottleneck), gene flow and natural selection.
35. What 3 criteria are required for evolution to occur by natural selection?

offspring will be homozygous recessive and thus albino. So the cleverness of the test-cross is that when you cross your unknown-genotype organism with an individual showing the recessive trait, if it sometimes produces offspring with the recessive trait (and if it does, it will do so half the time, on average), its genotype must be heterozygous. If it never produces offspring with the recessive trait, it must be homozygous for the dominant allele. In order to be confident in concluding that the unknown-genotype alligator has the *MM* genotype, though, you'd have to observe as many offspring as possible. After all, even if its genotype is *Mm,* quite a few offspring in a row might be normally pigmented, with the genotype *Mm*. Eventually, however, a heterozygous individual is likely to produce an offspring with the homozygous recessive genotype, *mm,* and the albino phenotype.

TAKE-HOME MESSAGE 7•7

In a test-cross, an individual that exhibits a dominant trait but has an unknown genotype is mated with an individual that is homozygous recessive. The phenotypes of the offspring reveal whether the unknown-genotype individual is homozygous dominant (all of the offspring exhibit the dominant trait) or heterozygous (half of the offspring show the dominant trait and half show the recessive trait).

7•8

We use pedigrees to decipher and predict the inheritance patterns of genes.

People want to know things about the future, such as: what is the likelihood that I will have a child with a particular genetic disease, say hemophilia? Or what is my own risk of developing a genetic disease, such as Huntington's disease, later in my life? Geneticists who study these and other diseases want to know other things: how is a particular disease inherited? Is it recessive or dominant? Is it carried on the sex chromosomes or on one of the other chromosomes? A **pedigree** is a type of family tree that can help answer these questions.

In a pedigree, information is gathered from as many related individuals as possible across multiple generations (**FIGURE 7-15**). Starting from the bottom, each row in the chart represents a generation, listing all of the children in their order of birth and whether or not they express a particular trait. Working up the pedigree, their parents are indicated and, above them, their parents' parents, for as far back as data are available. Squares represent males and circles represent females, and these shapes are shaded to indicate that an individual exhibits the trait of interest. Sometimes the genotype (as much of it as is known) is also listed for each individual.

By analyzing which individuals manifest the trait and which do not, it may be possible to deduce the pattern of inheritance for the trait—or, at least, rule out certain patterns. For example, if an individual exhibits a trait that neither of his or her parents exhibits, the trait is recessive. For dominant traits, all affected individuals must have at least one parent who exhibits the trait. In contrast, an individual can exhibit a recessive trait even if both parents are unaffected (i.e., do not exhibit the trait). In this case, the individual's parents must be heterozygous for that trait, each carrying one dominant and one recessive allele. Similarly, it is sometimes possible to determine whether a trait is carried on the sex chromosomes (X or Y, which we discuss in Section 7-13) or on one of the non-sex chromosomes (also called "autosomes"). Traits that are controlled by genes on the sex chromosomes are called **sex-linked traits.** Recessive sex-linked traits, for example, appear more frequently in males than females, whereas dominant sex-linked traits appear more frequently in females. These patterns may become obvious only on inspection of a large pedigree.

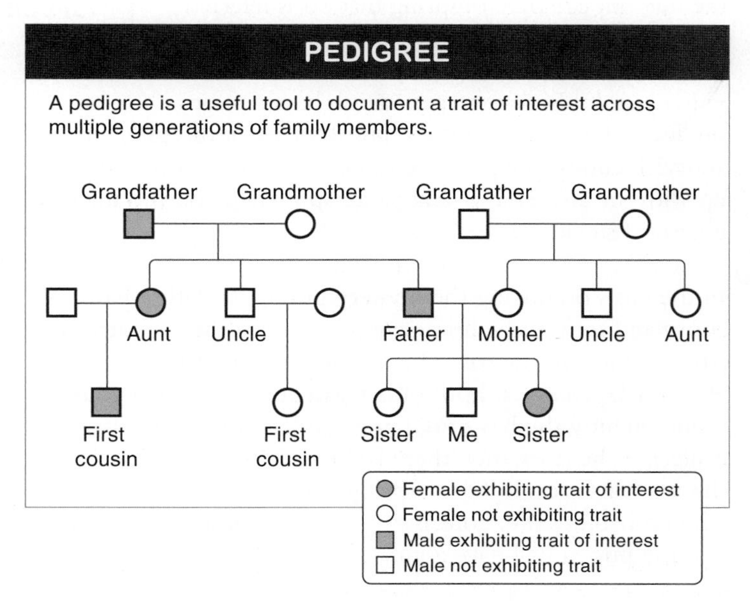

FIGURE 7-15 Family tree. A pedigree maps the occurrence of a trait in a family.

ANURY

Anury is a condition in dogs and other animals in which they are born without a tail. The condition is inherited as a recessive trait.

Can you figure out the genotype of the individual labeled 1?

IF...
A dog has no tail, but his father was normal

THEN...
The normal-looking mother (1) of the tailless dog must be heterozygous.

PEDIGREE

1

?

Female with anury
Female not exhibiting anury
Male with anury
Male not exhibiting anury

Can you now determine the probability of this puppy having anury?

FIGURE 7-16 Pedigree puzzle: tracing the occurrence of anury in a family of dogs.

An example of how pedigrees can help determine how traits are inherited is given in **FIGURE 7-16**. Anury is a condition seen in dogs and some other animals in which the animal has no tail. The pedigree reveals that anury is inherited as a recessive trait, because unaffected parents can have offspring with the disorder.

Can you figure out the genotype of the individual labeled "1"? That female must be heterozygous (*Aa*) for anury. (An individual who carries one allele for a recessive trait, and so does not exhibit the trait but can have offspring that do, is referred to as **carrier** of the trait.) The same is true for her mate. On the same pedigree, note that two individuals in the second generation have a puppy (indicated by "?"). What is the probability that this puppy has anury? Examine the pedigree carefully and see if you can come up with the answer; the next paragraph will guide you through if you get stuck.

Q Why do breeders value "pedigreed" horses and dogs so much?

In the cross producing the mystery puppy, the father has anury and so must be homozygous recessive; he will definitely pass on one *a* to the son. The mother's genotype can be *AA*, *Aa*, or *aA*, given that both of her parents are heterozygous. Consequently, she has a $\frac{2}{3}$ (2 in 3) probability of carrying the *a* allele. If she does, then there is a 1 in 2 (50%) chance that she will pass it on to her offspring and he will have anury. The probability, therefore, is $\frac{2}{3} \times \frac{1}{2} = \frac{1}{3}$—or a 1 in 3 chance that the puppy will have anury.

As we will discuss in later sections of this chapter, some traits may not show complete dominance and many traits are also influenced by the environment, so it is not always completely obvious what a trait's mode of inheritance is. In such cases, the more individuals we can include in the pedigree, the more accurate the analysis. Some human pedigrees contain thousands of individuals and stretch back six or more generations. With some other species it is possible to analyze tens of thousands of individuals per generation for a dozen or more generations. This is why plants and small insects (among other organisms) are excellent for studying inheritance patterns.

Dog and horse breeders often speak of "pedigreed" animals, a feature that adds tremendous value to the animals. This is because, with knowledge of an animal's family tree or complete lineage, it is much less likely that any genetic surprises will occur as the animal develops and reproduces.

Once you have an idea about how a trait is inherited, it is possible to make informed predictions about a couple's risk of having a child with a particular disorder. This makes earlier prenatal testing and treatment possible. It also enables the identification of individuals who are "carriers" of a recessive trait.

TAKE-HOME MESSAGE 7•8

Pedigrees help scientists, doctors, animal and plant breeders, and prospective parents determine the genes that individuals carry and the likelihood that the offspring of two individuals will exhibit a given trait.

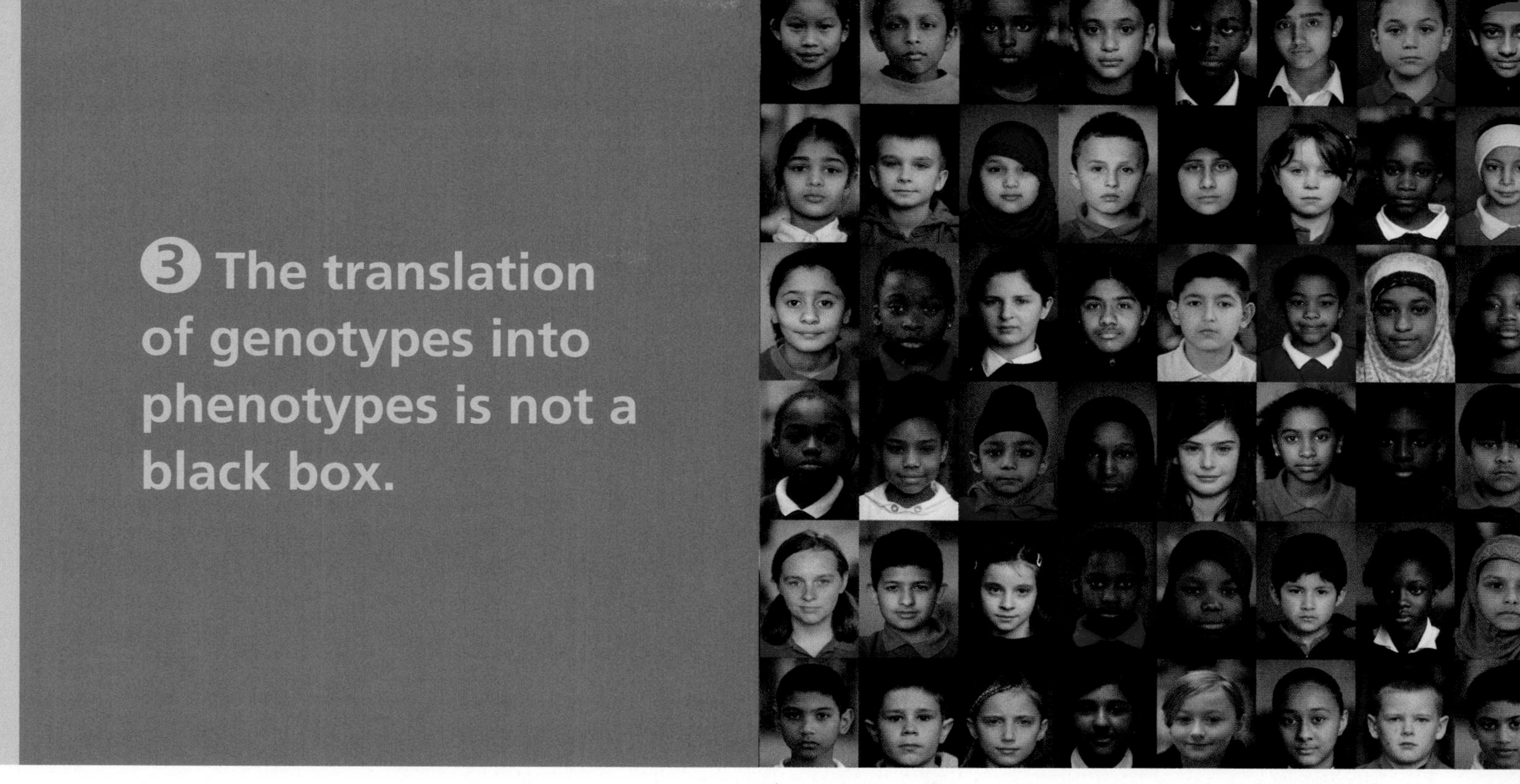

❸ The translation of genotypes into phenotypes is not a black box.

Phenotypic diversity has multiple sources and is all around us. (The photo shows schoolchildren from London.)

7•9

Incomplete dominance and codominance: the effects of both alleles in a genotype can show up in the phenotype.

As Mendel saw it, the world of genetics was straightforward and simple. We should be so lucky. Unfortunately, the world in which each trait is coded for by a single gene with two alleles—one completely dominant and one recessive—and with no environmental effects at all doesn't quite capture the complexity of the world beyond Mendel's pea plants. So, here and in the next sections, we'll build up a more complex model of how genes influence the building of bodies. We begin with the observation that, sometimes, the phenotype of heterozygous individuals differs from that of either of the homozygotes, and instead reflects the influence of both alleles rather than a clearly dominant allele.

One situation in which complete dominance is not observed is called **incomplete dominance,** in which the heterozygote appears to be intermediate between the two homozygotes. The quintessential example of incomplete dominance is the flower color of snapdragons (**FIGURE 7-17**). We can obtain true-breeding (homozygous) lines of snapdragons with red flowers, and true-breeding (homozygous) lines that produce only white flowers. When plants from these two populations are crossed, we would expect—if one allele were dominant over the other—either all red or all white flowers. Instead, such crosses always produce plants with pink flowers. Interestingly, when we cross two plants with pink flowers, we get ¼ red-flowered plants, ½ pink-flowered plants, and ¼ white-flowered plants.

How can we interpret this cross? It seems that the plants with white flowers have the genotype C^WC^W and produce no pigment. At the other extreme, the plants with red flowers have the genotype C^RC^R and produce a great deal of pigment. The letter "C" refers to the fact that the gene codes for **c**olor, and the superscript "W" or "R" refers to an allele producing no pigment (**w**hite) or **r**ed pigment. We use these designations for the genotypes because it isn't clear that either white or red is dominant over the other, and so neither should be upper-case or lower-case. The pink flowers receive one of the pigment-producing C^R alleles and one of the no-pigment-producing C^W alleles, and so produce an intermediate amount

INCOMPLETE DOMINANCE: SNAPDRAGONS

Incomplete dominance occurs when a heterozygote exhibits an intermediate phenotype between the two homozygotes.

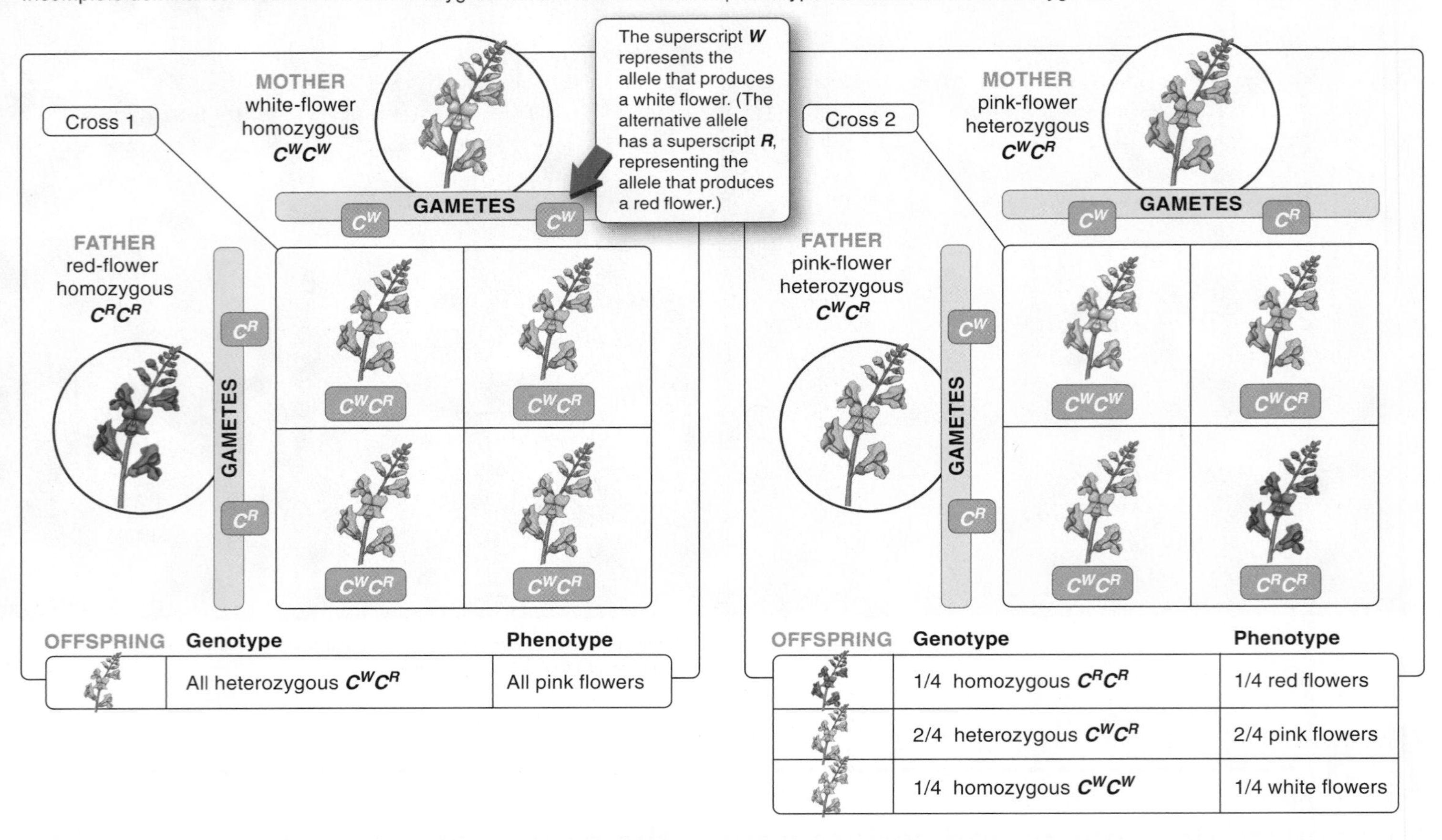

FIGURE 7-17 Pink snapdragons demonstrate incomplete dominance. When true-breeding white and red snapdragons are crossed, offspring have pink flowers.

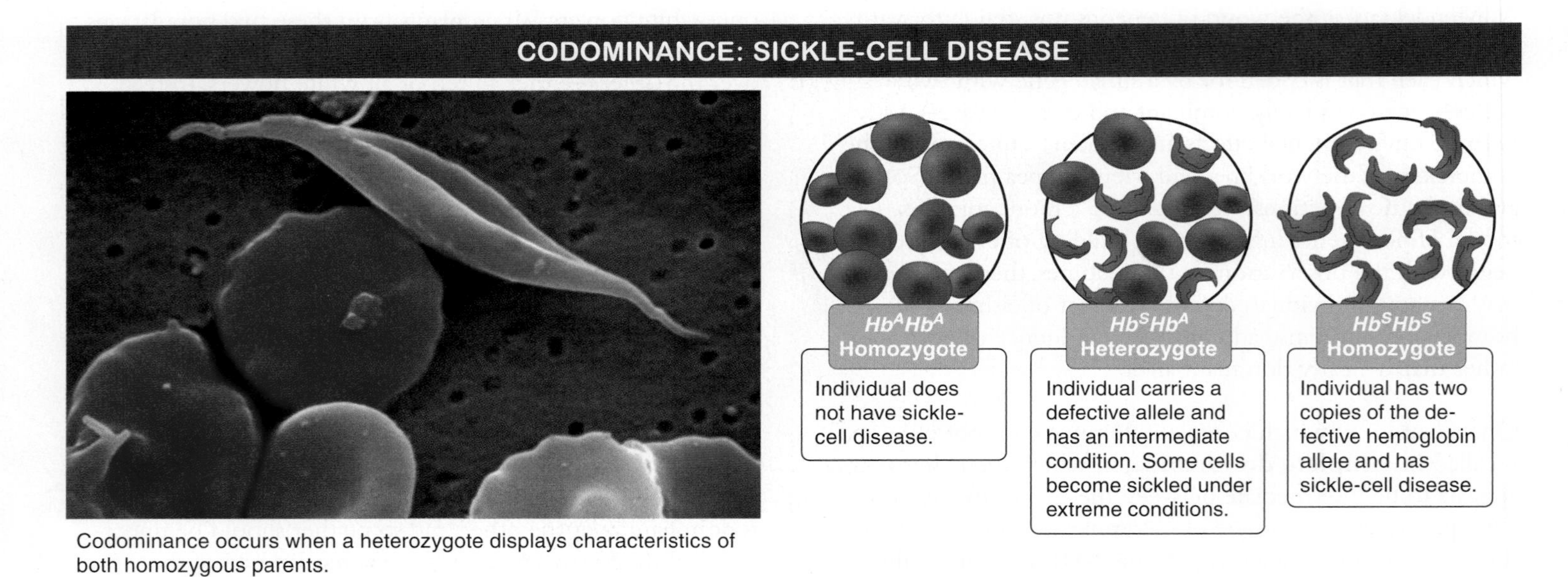

Codominance occurs when a heterozygote displays characteristics of both homozygous parents.

FIGURE 7-18 With codominance, a heterozygous individual shows features of both alleles.

of pigment. Ultimately, the intensity of pigmentation just depends on the amount of pigment chemical made by the flower-color gene.

A second situation in which complete dominance is not observed is called **codominance,** in which the heterozygote displays characteristics of both homozygotes. In this situation, the alleles can be thought of as codominant because neither masks the effect of the other. An example of codominance occurs with sickle-cell disease (**FIGURE 7-18**), a condition in which individuals produce defective red blood cells that change their shape, becoming sickle-shaped, when they lose the oxygen they carry. The defective blood cells cannot effectively transport oxygen to tissues, and they accumulate in blood vessels, causing extreme pain. Individuals with sickle-cell disease suffer shortness of breath and numerous other problems that lead to a significantly reduced lifespan.

Sickle-cell disease seems, at first inspection, to be a condition seen only in individuals with two copies of the sickle-cell allele for hemoglobin, Hb^S; recall that hemoglobin is the oxygen-carrying molecule in red blood cells. Individuals with two copies of the normal hemoglobin allele, Hb^AHb^A, and those who are heterozygous, Hb^SHb^A, seem to be normal. However, closer observation reveals that heterozygotes—who are described as having "sickle-cell trait"—carry many red blood cells that become sickled under extreme conditions of physical exertion or low atmospheric oxygen. In other words, heterozygous individuals show an intermediate phenotype between the two homozygotes.

TAKE-HOME MESSAGE 7•9

Sometimes the effects of both alleles in a heterozygous genotype are evident in the phenotype. With incomplete dominance, a heterozygote appears to be intermediate between the two homozygotes. With codominance, a heterozygote displays characteristics of both homozygotes.

7•10

What's your blood type? Some genes have more than two alleles.

What is your blood type? It can be O, A, B, or AB. Each of these different blood types indicates something about the physical characteristics of your red blood cells and has implications for blood transfusions—both giving and receiving blood. The blood groups are also interesting from a genetic perspective, because they illustrate a case of **multiple allelism,** in which a single gene has more than two alleles. Each individual still carries only two alleles—one from the mother and one from the father. But if you survey all of the alleles present in the population, you will find more than just two alleles.

Inheritance of the ABO blood groups provides the simplest example of multiple allelism, because there are only three alleles. We can call these alleles *A, B,* and *O.* The *A* and *B* alleles are both completely dominant to *O,* so individuals are considered to have blood type A whether they have the genotype *AA* or *AO* (**FIGURE 7-19**). Similarly, an individual with the genotype *BB* or *BO* is considered to have blood type B. If you carry two copies of the *O* allele, you are blood type

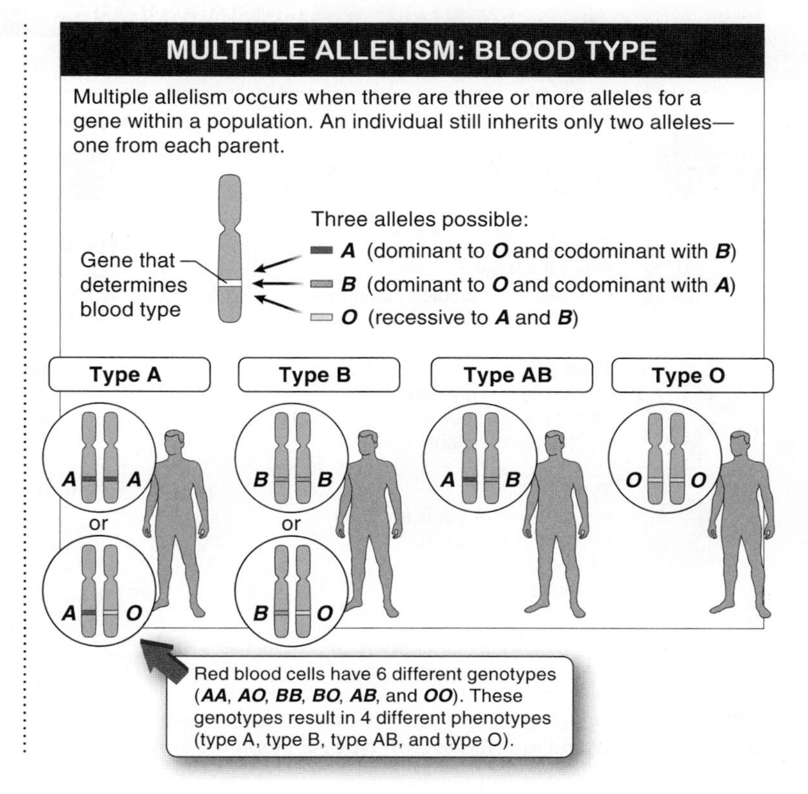

FIGURE 7-19 Multiple allelism. There are three different alleles—*A, B,* and *O*—for blood type.

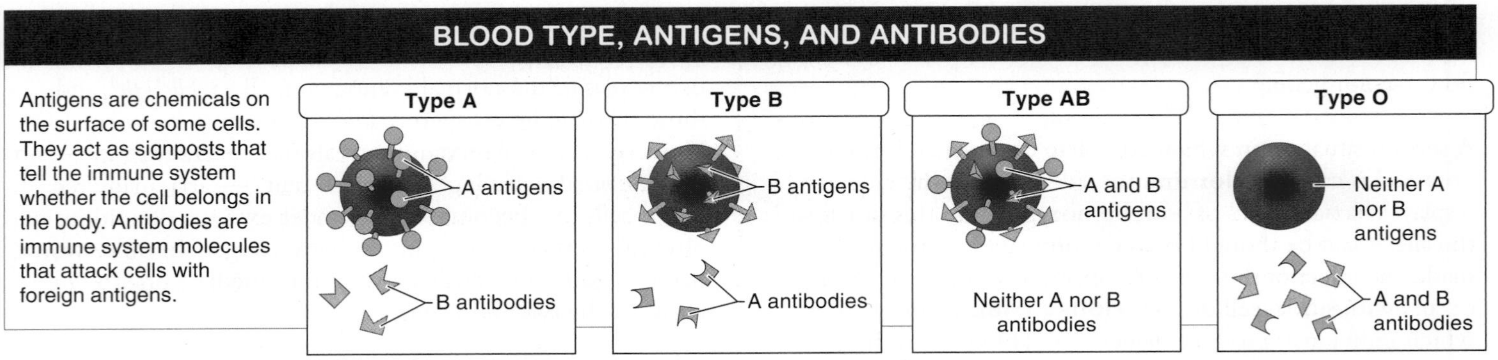

BLOOD TYPE, ANTIGENS, AND ANTIBODIES

Antigens are chemicals on the surface of some cells. They act as signposts that tell the immune system whether the cell belongs in the body. Antibodies are immune system molecules that attack cells with foreign antigens.

Type A	Type B	Type AB	Type O
A antigens	B antigens	A and B antigens	Neither A nor B antigens
B antibodies	A antibodies	Neither A nor B antibodies	A and B antibodies

FIGURE 7-20 Friend or foe? Antigens are signposts that tell the immune system whether or not a cell belongs in the body.

O. The *A* and *B* alleles are codominant with each other, so the genotype *AB* gives rise to blood type AB. Consequently, with these three alleles in the population, individuals can be one of four different blood types: A, B, AB, and O.

What are the phenotypes of these alleles? An individual's blood type alleles carry instructions that direct construction of a specific set of chemicals, called *antigens,* that protrude from every red blood cell. Antigens are proteins that jut from the surface of a cell and can "turn on" its defenses against foreign invaders. The *A* allele directs the production of A antigens all over the surface of red blood cells. Similarly, the *B* allele directs the production of B antigens on all red blood cells. The *O* allele does not code for either type of antigen. This means that individuals with blood type AB have red blood cells with both A and B antigens, while individuals with blood type O have red blood cells with a "smoother" surface because they produce no A or B antigens (**FIGURE 7-20**).

Antigens on red blood cells play a role in the body's disease-fighting immune system. Antigens are like signposts, telling the immune system whether a cell belongs in the body or not. If a red blood cell with the wrong antigens were to enter your bloodstream, your immune system would recognize it as a

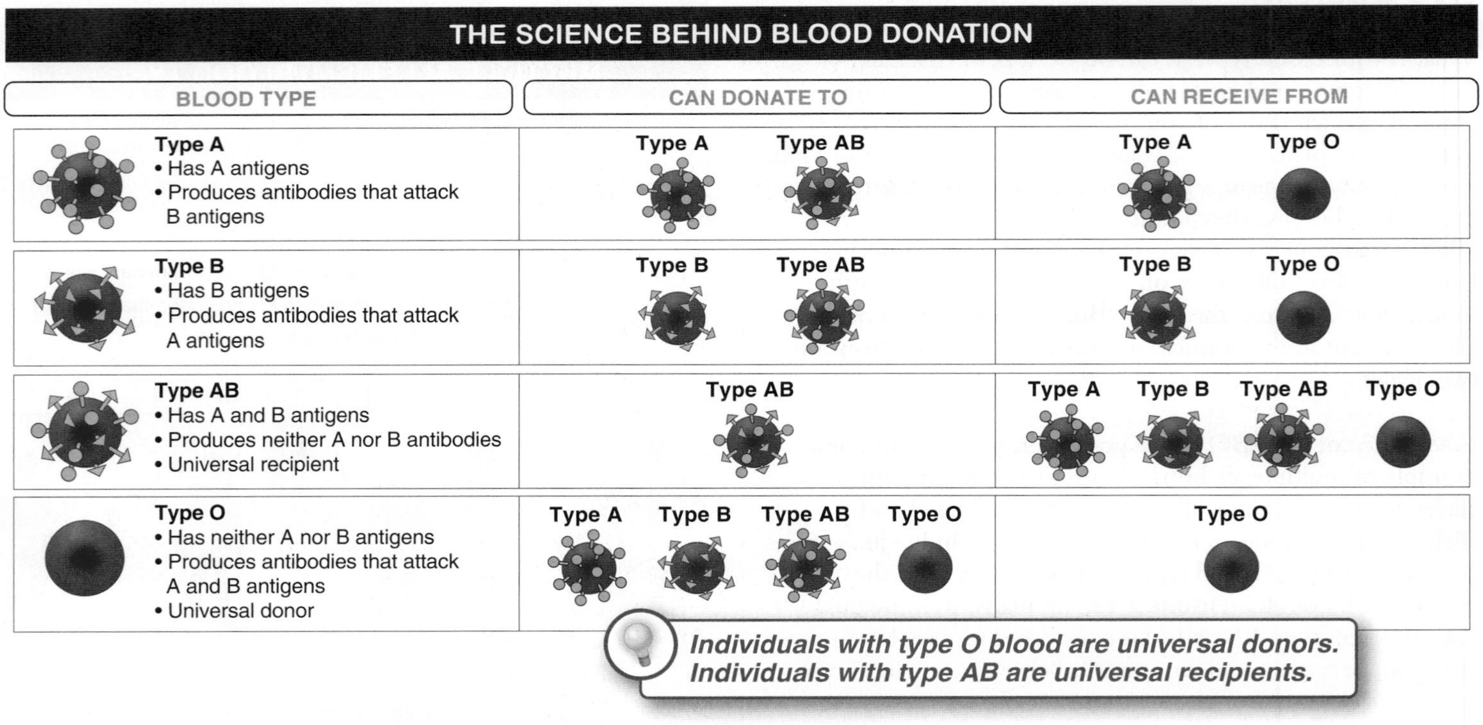

THE SCIENCE BEHIND BLOOD DONATION

BLOOD TYPE	CAN DONATE TO	CAN RECEIVE FROM
Type A • Has A antigens • Produces antibodies that attack B antigens	Type A, Type AB	Type A, Type O
Type B • Has B antigens • Produces antibodies that attack A antigens	Type B, Type AB	Type B, Type O
Type AB • Has A and B antigens • Produces neither A nor B antibodies • Universal recipient	Type AB	Type A, Type B, Type AB, Type O
Type O • Has neither A nor B antigens • Produces antibodies that attack A and B antigens • Universal donor	Type A, Type B, Type AB, Type O	Type O

Individuals with type O blood are universal donors. Individuals with type AB are universal recipients.

FIGURE 7-21 Mapping blood compatibility. Individuals with type O blood are said to be universal donors, and individuals with type AB blood are universal recipients.

foreign invader and destroy it. Such an attack is initiated by molecules in the bloodstream called *antibodies,* which attack only foreign antigens. Individuals with only A antigens on their red blood cells produce antibodies that attack B antigens. If they encounter a red blood cell with B antigens, they attack it. Such an immune response can lead to destruction of red blood cells, low blood pressure, and even death. Under normal circumstances, antibodies would not encounter a red blood cell with foreign antigens—unless, that is, red blood cells with foreign antigens are accidentally injected into the person's bloodstream in a transfusion.

Q Why are people with type O blood considered "universal donors"? Why are those with type AB considered "universal acceptors"?

Individuals with only B antigens on their red blood cells produce antibodies that attack A antigens. Individuals with blood type O, who have neither A nor B antigens on their red blood cells, produce antibodies that attack both A and B antigens. Individuals with blood type AB don't produce either type of antibody (or else they would have antibodies that attacked their own blood cells). From this information, we can deduce which blood types can be used in transfusions. This is shown in **FIGURE 7-21**.

Another marker on the surface of red blood cells is the Rh blood group marker. This marker results from the effects of a single gene with two alleles. Individuals' red blood cells carry the Rh cell surface marker if they have one or two copies of the dominant Rh marker allele. This is then noted in their blood type as a "+," or "positive," in addition to their ABO blood type, as in "O-positive" or "B-negative." Individuals who have two copies of the recessive allele for this gene do not have any Rh markers, and they are described as "negative," as in "O-negative" or "A-negative." If individuals who are Rh-negative are exposed to Rh-positive blood, their immune system attacks the Rh antigens as foreign invaders—an immune response that can vary from mild, and passing unnoticed, to severe, which can lead to death. Thus, the Rh marker further restricts whose blood a person can receive in a transfusion.

Beyond the ABO and the Rh marker groups, there are many, many genes with multiple alleles—a dozen or even more alleles in some cases. In fact, one gene for eye color in fruit flies has more than 1,000 different alleles!

TAKE-HOME MESSAGE 7•10

In multiple allelism, a single gene has more than two alleles. Each individual still carries only two alleles, but in the population, more than just two alleles occur. This is the case for the ABO blood groups in humans.

7•11

Multigene traits: how are continuously varying traits such as height influenced by genes?

When babies are born, the parents are often curious about how tall their child will grow to be. Old wives' tales suggest a couple of ways for predicting height. If the baby is a boy, they say to add 5 inches to the mother's height and average that with the father's height. If it is a girl, subtract 5 inches from the father's height and average that with the mother's height. Alternatively, the lore says, just take the child's height at two years of age and double it.

While these methods can be surprisingly accurate at making predictions, they don't really help us understand the underlying reason *why* height can be predicted so well. The reason these methods work is that genes play a strong role in influencing height, and so offspring do resemble their parents in measurable ways. But unlike the simple case of Mendel's peas, where a single gene with two alleles determines the height of the plant, for humans and most animals, adult height is influenced by many different genes. Such a trait is said to be **polygenic.**

Recent research has identified two genes that help determine height in humans. Individuals with the "tall" allele for both genes are taller than those with the tall allele for only one. Individuals with one tall allele, in turn, are taller than people who do not have a tall allele for either gene. The term **additive effects** describes what happens when the effects of alleles from multiple genes all contribute to the ultimate phenotype. Traits such as skin color in humans also work this way. In the case of height genes, one of the genes is on chromosome 15 and codes for an enzyme that converts testosterone to estrogen. This enzyme influences height because estrogen helps bones to fuse at their ends and thus stop growing. In all, the two height genes identified in humans account for only about 2 inches of a person's height, but they exemplify how multiple genes can influence a single characteristic. The variety of heights from very short to very tall, with every height in between, reflects that height is a trait influenced by contributions from multiple genes as well as the environment (**FIGURE 7-22**).

Q Why might computer nerds be more likely to have autistic children?

Many other physical traits are influenced by multiple genes, including eye color and skin color in humans. Eye color, long believed to be controlled by a single gene with a dominant brown-eye allele and a recessive blue-eye allele, now seems to be the result of the interactions of at least two genes and possibly more—a situation that makes more sense, given the significant continuous variation in eye color seen among adults.

POLYGENIC TRAITS

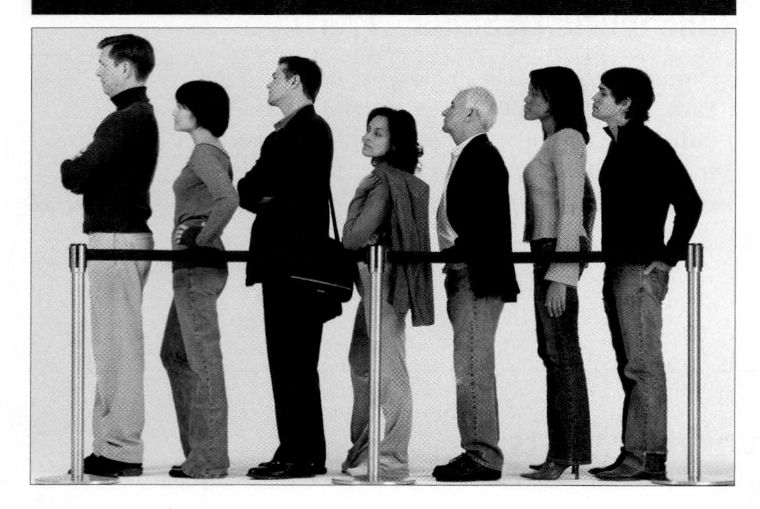

FIGURE 7-22 From many genes, one trait. Height and skin color are multigene traits.

Many behavioral traits are influenced by multiple genes. The developmental disorder known as autism, for example, seems to be the result of alterations in numerous—perhaps as many as 10 or even 20—different genes. Individuals with autism have difficulty interacting with others, particularly in making emotional connections. Autistic individuals also tend to have narrowly focused and repetitive interests. Autism is notoriously difficult to study, though, because its symptoms are varied, as is their intensity. It turns out that this may be because different combinations of the many genes involved in autism cause different variations of the disorder, much as different alleles for several height genes may work together to produce a variety of heights.

Interestingly, the genes responsible for autism may also influence some desirable characteristics. Behaviors that cause problems when manifest in the extreme may turn out to be beneficial when less extreme, such as unusual abilities of perception, analytical skills, and focus. This idea—called the "geek theory of autism"—was first suspected when an upsurge in cases of autism in children was noticed in areas with large numbers of high-tech workers, including Silicon Valley in California and near Cambridge, England, in the heart of the United Kingdom's high-technology industry. Perhaps alleles for genes that contribute to making individuals good at computer programming and solving complex technology problems may produce autism when an individual carries too many of them.

TAKE-HOME MESSAGE 7•11

Many traits, including continuously varying traits such as height, eye color, and skin color, are influenced by multiple genes.

7·12

Sometimes one gene influences multiple traits.

Q What is the benefit of "almost" having sickle-cell disease?

Just as multiple genes can influence one trait, some individual genes can influence multiple unrelated traits, a phenomenon called **pleiotropy.** In fact, this may be true of nearly all genes. For example, earlier we described the sickle-cell allele for hemoglobin production. This allele causes sickle-cell disease when in the homozygous state, but in the heterozygous state it has the effect of conferring resistance to the parasite that causes malaria. The resistance is due to the fact that the malarial parasite—which lives in red blood cells—cannot survive well in cells that carry the defective version of the hemoglobin gene. Because individuals who are heterozygous for sickle-cell anemia have a significant number of such red blood cells, their bloodstream is just not a hospitable environment for the malarial parasite (**FIGURE 7-23**).

Another example of pleiotropy is the SRY gene. Named for "**s**ex-determining **r**egion on the **Y** chromosome," this gene causes fetal gonads to develop as testes shortly after fertilization. Following the gonads' secretion of testosterone, a cascade of other developmental changes then occurs, such as development of the internal and external male reproductive structures. Ultimately, the SRY gene is responsible for numerous behavioral characteristics as well. These are described further in Chapter 9.

TAKE-HOME MESSAGE 7·12

In pleiotropy, one gene influences multiple unrelated traits. Most, if not all, genes may be pleiotropic.

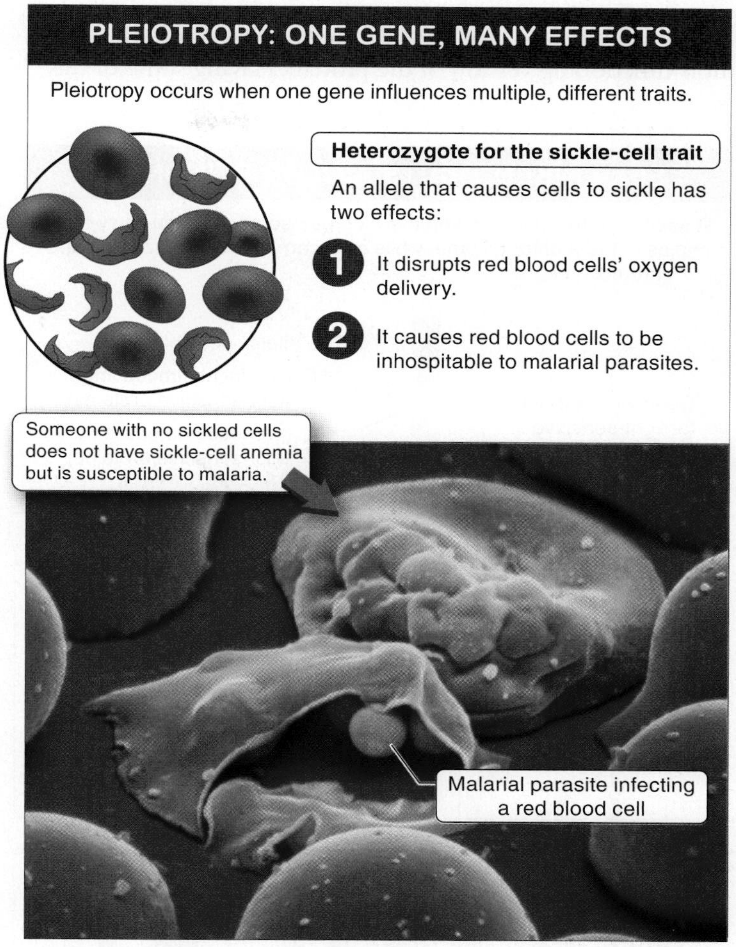

PLEIOTROPY: ONE GENE, MANY EFFECTS

Pleiotropy occurs when one gene influences multiple, different traits.

Heterozygote for the sickle-cell trait

An allele that causes cells to sickle has two effects:

1 It disrupts red blood cells' oxygen delivery.

2 It causes red blood cells to be inhospitable to malarial parasites.

Someone with no sickled cells does not have sickle-cell anemia but is susceptible to malaria.

Malarial parasite infecting a red blood cell

FIGURE 7-23 From one gene, multiple traits. The allele for sickle-cell disease is pleiotropic: it causes red blood cells to form an unusual, sickled shape, and it also provides resistance to malaria.

7·13

Why are more men than women color-blind? Sex-linked traits differ in their patterns of expression in males and females.

The patterns of inheritance of most traits do not differ between males and females. When a gene is on an autosome (one of the non-sex chromosomes), both males and females inherit two copies of the gene, one from their mother and one from their father. The likelihood that an individual inherits one genotype versus another does not differ between males and females.

Traits coded for by the sex chromosomes, on the other hand, have different patterns of expression in males and females. One of the most easily observed examples of this phenomenon is red-green color-blindness. On the X chromosome in humans, there is a gene that carries the instructions for producing light-sensitive proteins in the eye that, when produced normally, respond differently to red wavelengths and green wavelengths of light. These proteins make it possible to distinguish between the colors red and green. As long as an individual has at least one functioning copy of the gene carrying the instructions for production of these light-sensitive proteins, he or she produces sufficient amounts of the protein to have normal color vision.

There is a rare allele for this gene, however, that produces a non-functioning version of the protein. Having some of this non-functioning protein is not a problem as long as the person also carries another, normal version of the gene and produces some of the functioning protein. But this poses a problem for men, because they have only one X chromosome. In contrast, females inherit two X chromosomes.

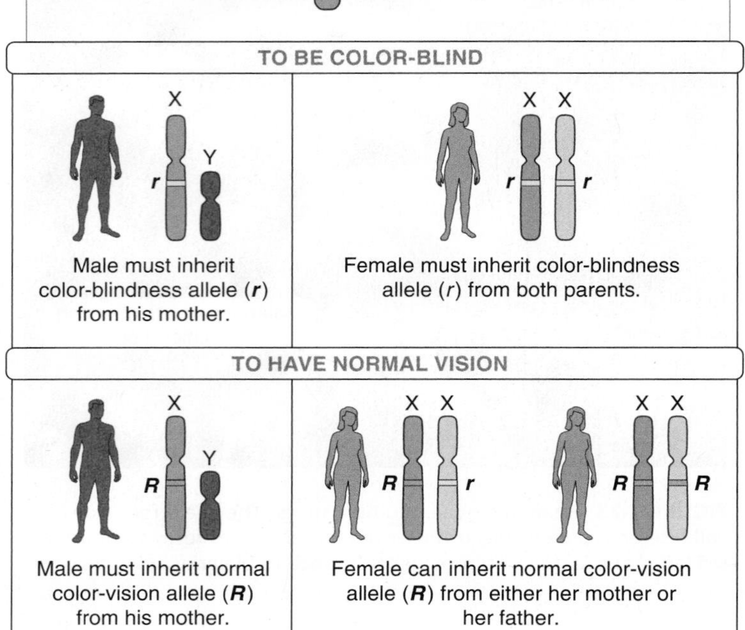

FIGURE 7-24 Sex-linked traits such as color-blindness are not expressed equally among males and females.

Here's the problem: men only get one chance to inherit the normal version of the gene that codes for red-green color vision. It must be on the X chromosome they inherit from their mother. Women get two chances. Although a woman may inherit the defective allele from one parent, she can still inherit the normal allele from the other parent. As long as she inherits the normal gene from one parent, she will have normal color vision. As we would predict, then, the frequency of red-green color-blindness is significantly greater in males than in females (**FIGURE 7-24**). Approximately 7% to 10% of men exhibit red-green color-blindness, while fewer than 1% of women are red-green color-blind.

> **Q** If a man is color-blind, did he inherit this condition from his mother, his father, or both parents?

While males exhibit sex-linked recessive traits more frequently than do women, the situation is reversed for sex-linked dominant traits. In these cases, because females have two chances to inherit the allele that causes the trait, they are more likely to have the allele and thus exhibit the trait than are males, who have only one chance to inherit the allele.

TAKE-HOME MESSAGE 7•13

The patterns of inheritance of most traits do not differ between males and females. However, when a trait is coded for by a gene on a sex chromosome, such as color vision on the X chromosome, the pattern of expression differs for males and females.

7•14

Environmental effects: identical twins are not identical.

It is a very serious warning, in boldfaced capital letters: **"PHENYLKETONURICS: CONTAINS PHENYLALANINE."** But it's not next to a skull and crossbones on a glass bottle in a chemistry lab, it's on cans of diet soda intended for human consumption (**FIGURE 7-25**). Most of us either don't notice the warning or ignore it. Still, it's always there, and to some people it's a matter of life and death. What does it mean?

Q Drinking diet soda can be deadly if you carry a single bad gene. What gene is it and why is it so deadly?

At the most fundamental level, this warning is on products with phenylalanine because an organism's phenotype is a product of its genes in combination with the environment. Put another way, nature and nurture interact to produce bodies. In this case, the interaction is related to how humans metabolize a particular chemical called phenylalanine, one of the amino acids. Sometimes our bodies use phenylalanine directly to build proteins, adding it to a growing amino acid chain. At other times, phenylalanine is chemically converted into another amino acid, called tyrosine. Tyrosine is then used to build proteins and in a variety of other ways by the body.

The problem is this: at birth, some people carry two copies of a mutant version of the gene that is supposed to produce the enzyme that converts phenylalanine into tyrosine. The mutant gene produces a malfunctioning enzyme, and none of the body's phenylalanine is converted into tyrosine. Perhaps this doesn't seem as though it should be a big problem, but it is. Little by little, as these individuals consume phenylalanine, it builds up in their bodies because none of it is converted into tyrosine. Eventually, so much accumulates that it reaches toxic levels and poisons the brain, leading to mental retardation and death. The disease is called phenylketonuria, or PKU.

FIGURE 7-25 Change the environment, "cure" the disease. The ingestion of phenylalanine in diet soda can be fatal to individuals with the inherited condition phenylketonuria (PKU).

Here's where the warning label comes in. By limiting the amount of phenylalanine in their diet (from diet soda and other sources), individuals with the two mutant alleles for processing phenylalanine can avoid the toxic buildup of the amino acid in their brain. In essence, they modify their environment so that it contains little or no phenylalanine. And in that environment, the PKU mutant alleles are harmless. This example highlights the fact that genes by themselves do not "code" for physical characteristics. Rather, genes interact with the environment to produce physical characteristics. Unless you have information about both the genes *and* the environment, it is not usually possible to know what the phenotype will be.

Q Could you create a temporarily spotted Siamese cat with an ice pack? Why?

A less life-threatening illustration of the interaction of genes with the environment can be seen at a pet shop. Siamese cats (as well as the Himalayan rabbit) carry genes that produce dark pigmentation. These genes interact strongly with the environment and are heat-sensitive. Dark pigment is produced only in relatively cold areas of the animal's body, while warm areas remain very light in color. This is why the fur on the coldest parts of the body—the ears, paws, tail, and tip of the face—becomes the darkest, while the fur on the rest of the body remains cream-colored or white (**FIGURE 7-26**). For Siamese cats living in cold climates that spend a lot of time outside, it's interesting to notice that they become significantly darker in color during the winter months. Those that lounge indoors all winter remain lighter in color.

There are thousands of other cases in which genes' interactions with the environment influence their ultimate effects in the body. The very fact that identical twins—who inherit exactly the same set of alleles at the time of

FIGURE 7-26 Heat-sensitive fur color. Some pigment-producing genes produce dark fur only under cold conditions. That's why these animals have darker patches of fur on their extremities.

fertilization—don't die on the same day at the same moment and from the same cause, for example, reveals that environmental variation influences the expression of genes. The scope of environmental influences ranges from traits with large and obvious environmental effects, such as body weight and its relationship with caloric intake, to traits such as eye color that are barely influenced by environmental effects, to traits with complex and subtle interactions with the environment, such as intelligence or personality.

Because of the role of environmental factors in influencing phenotypes, DNA is not like a blueprint for a house. This metaphor, though frequently used, is inappropriate, because there is nearly always a significant interaction between the genotype and the environment that influences the exact phenotype produced. The use of this metaphor might even be considered irresponsible, to the extent that it suggests that the phenotype is determined solely by the genotype. If that were the case, there would be no reason to invest in better schools, physical fitness regimens, nutritional monitoring, self-help efforts, or any other process by which individuals or societies try to improve people's lives (i.e., alter phenotypes) by enriching their environment.

Q James Watson, the co-discoverer of DNA, once wrote that when we completed the Human Genome Project, we would have "the complete genetic blueprint of man." Why was he wrong?

TAKE-HOME MESSAGE 7•14

Genotypes are not like blueprints that specify phenotypes. Phenotypes are generally a product of the genotype in combination with the environment.

④ Some genes are linked together.

Different genes influence red hair and freckles, so why are they often inherited together?

7•15

Most traits are passed on as independent features: Mendel's law of independent assortment.

Sometimes you can be right about something for the wrong reason. This happened to Gregor Mendel. He didn't know that genes were carried on chromosomes. He believed that the units of heredity were just free-floating entities within cells. Given this perspective, it made sense to him that the inheritance pattern of one trait wouldn't influence the inheritance of any other trait. He believed that all genes behaved independently.

It helps to consider an example. Earlier in this chapter we saw that the characteristic of having a dimpled chin is caused by a single dominant gene. Imagine that you had a true-breeding population of people with dimpled chins (remember, "true-breeding" for a trait means that all offspring always manifest the trait; in the case of dimples, all have the genotype *DD*). Now suppose that an individual in this population had children with someone from a population of true-breeding, non-dimple-chinned people (everyone has the genotype *dd*). In this case, all of their children would have dimpled chins, but they would be heterozygous (*Dd*), getting a dominant allele from their dimple-chinned parent and a recessive allele from the other parent. If two of those heterozygotes then had children together, they would produce three-quarters dimpled and one-quarter non-dimpled children, with genotypes in the ratio of ¼ *DD,* ½ *Dd,* and ¼ *dd*. That is just what Mendel observed for traits in pea plants.

But what if we concurrently observed another characteristic of these people? Suppose the original true-breeding population of dimpled-chin people (*DD*) was also true-breeding for albinism (all *aa*), a disorder caused by carrying two recessive alleles for a single gene. And suppose that individuals from the normal-chinned population (*dd*) were also true-breeding for normal pigmentation (all *AA*). The question is, do the alleles an individual inherits for the dimpled-chin trait influence which alleles that individual inherits for pigmentation? And the answer is that, in most cases, they do not (**FIGURE 7-27**). Rather, the first cross of a dimple-chinned albino with a normal-chinned, normally pigmented individual would result in a child heterozygous for both traits, expressing each of the dominant traits. Phenotypically, this and all subsequent offspring would be dimple-chinned and normally pigmented. Then, in a mating between two of these doubly heterozygous individuals, three-quarters of their children would have the dominant trait and one-quarter would have the recessive trait,

MENDEL'S LAW OF INDEPENDENT ASSORTMENT

Mendel's law of independent assortment states that one trait does not influence the inheritance of another trait.

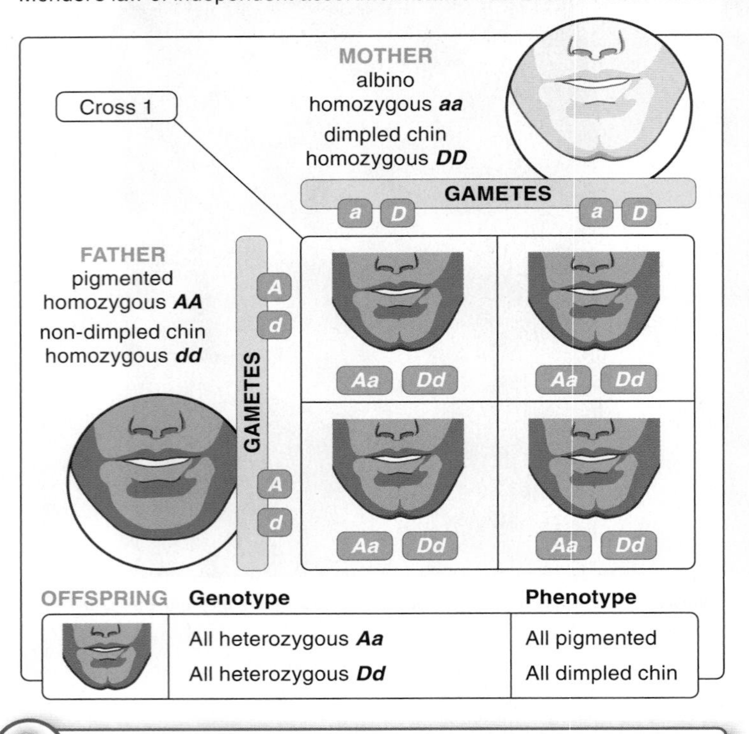

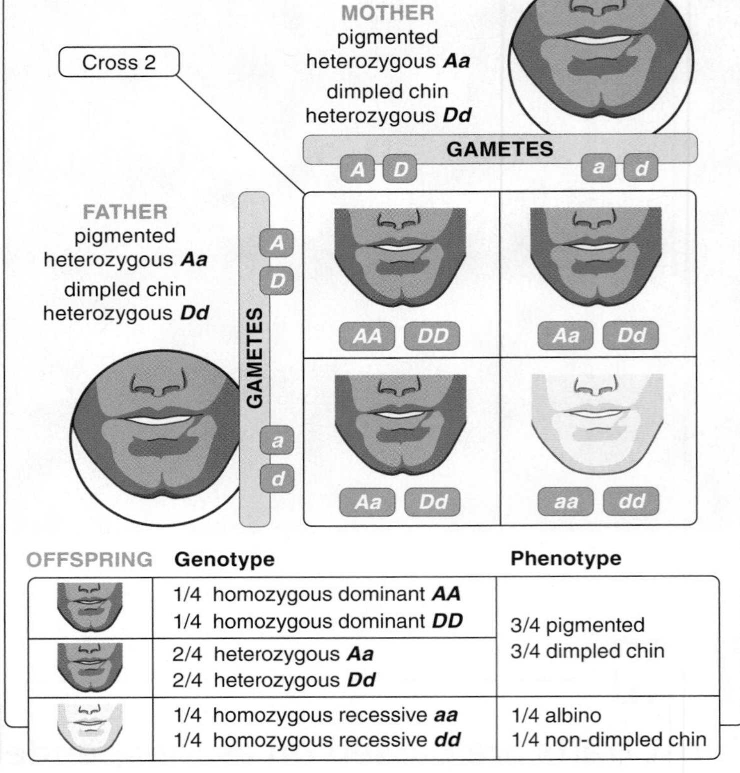

FIGURE 7-27 Independent assortment of genes.

regardless of which trait you are tallying. In other words, neither trait influences the inheritance pattern for the other trait; all traits are inherited independent of each other. This is known as **Mendel's law of independent assortment.**

In the next section we'll see that, despite Mendel's correct understanding that separate traits are inherited independently, his belief that this happened because all genes just float freely around in the cell was not correct. The genes, as we now know, are carried on chromosomes. And this fact leads occasionally to some situations in which independent assortment does *not* occur.

TAKE-HOME MESSAGE 7•15

Genes tend to behave independently, such that the inheritance pattern of one trait doesn't usually influence the inheritance of any other trait.

7•16

Red hair and freckles: genes on the same chromosome are sometimes inherited together.

Most redheads have pale skin and freckles. This simple observation is problematic for the law of independent assortment as Mendel imagined it. After all, in his law, he asserted that the inheritance of one trait does not influence the inheritance of another. But clearly that doesn't happen in this case. Having one of the traits, such as red hair, seems to influence the presence of

FIGURE 7-28 Violating the law of independent assortment. Red hair and freckles are often inherited as a package deal.

another trait, pale skin (**FIGURE 7-28**). Strictly speaking, Mendel's second law is not true for *every* pair of traits. Sometimes the alleles for two genes are inherited together and expressed almost as a package deal. Let's investigate why that is so.

Q Why do most redheads have pale skin?

For starters, how many traits make up an individual? Of course, this depends on the species. Nonetheless, the number will almost always extend into the thousands, if not tens of thousands. In the case of humans, for example, there are about 25,000 genes in our genome. Yet we have only 23 unique chromosomes (two copies of each). Thus, genes influencing some different traits must be on the same chromosome, maybe even right next to each other. When they are close together, we say that they are **linked genes.** One 2008 study, for example, demonstrated a link between human genes that influence hair color and skin pigmentation.

Why are linked genes inherited together? To answer this, we must revisit the behavior of chromosomes during the production of gametes, discussed in Chapter 6. Imagine that you have a child. Let's focus on your gamete that took part in the fertilization to produce that child. When you made that sperm or egg by meiosis, only one of the two copies of each of your chromosomes ended up in the gamete. It may have been the one from your mother or it may have been the one from your father. In either case, all of the alleles that were on the chromosome from that one parent were passed on as a group to the child that resulted from the fertilization involving that gamete. This process continues generation after generation. The linked alleles never get split up unless, during meiosis, recombination occurs between them, moving one or more to the other chromosome in the pair so that they now become linked with the alleles on that chromosome (**FIGURE 7-29**).

> Four be the things I'd been better without: Love, curiosity, freckles, and doubt.
>
> —Dorothy Parker, *Inventory,* 1926

When alleles are linked closely on the same chromosome, Mendel's second law doesn't hold true. It is very surprising—and was fortunate for Mendel—that of the seven pea traits he studied, none of them were close together on the same chromosome. For this reason they all behaved as if they weren't linked, and in his crosses of different pea plants Mendel never noticed any linked genes.

TAKE-HOME MESSAGE 7•16

Sometimes, having one trait, such as red hair, influences the presence of another trait, such as pale skin. This is because the alleles for two genes are inherited and expressed almost as a package deal when the genes are located close together on the same chromosome.

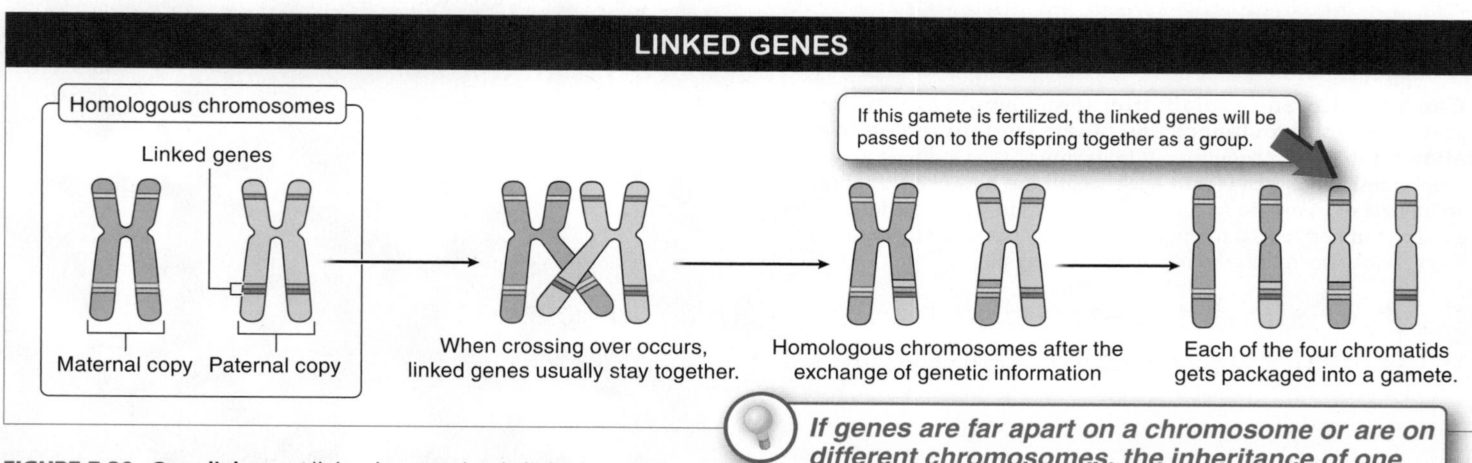

FIGURE 7-29 Gene linkage. Alleles that are closely linked on the same chromosome will be passed on to offspring in one bundle.

streetBio

Knowledge You Can Use

Where did you come from? Where are you going (to dinner)? World migrations, novelty-seeking, and a single gene that influences both.

On human chromosome 11 there is a gene called D4DR. This gene carries the instructions for building a protein embedded in the membranes of many of your brain cells. The protein is a receptor for the brain chemical dopamine, which is active in several areas of the brain called "pleasure centers" and influences initiative and motivation.

Although all humans have some form of the D4DR gene, there are several different alleles that a person might carry (though only two per person, of course). Each type of allele can be described by the length of the sequence of bases that makes up its DNA: some sequences are short, others are long.

Q: Can a single gene influence your personality? Several studies have reported that individuals carrying one or two long alleles of the D4DR gene, as opposed to two short alleles, are more likely to exhibit certain personality traits, including novelty-seeking and engaging in impulsive behaviors. On personality questionnaires, several differences emerge:

People carrying two short alleles are slightly more likely to be:	People carrying one or more long alleles are slightly more likely to be:
Reflective	Impulsive
Rigid	Exploratory
Stoic	Excitable
Slow-tempered	Quick-tempered
Frugal	Extravagant

Individuals carrying the long alleles even show a preference for spicy foods!

Q: Can a single gene explain why some people are willing to migrate over long distances and others remain rooted in one location? A study of 39 populations from around the world reported a strong relationship between D4DR allele length and how likely a population was to migrate or to remain stationary. Populations in which the long alleles were more common tended to migrate long distances, while those in which the short alleles were more common tended to stay put.

Population	Frequency of the long allele
Chinese and Japanese	5%
African	16%
European	20%
North American	32%
Central American	42%
South American	69%

Q: Do you want to know which alleles you carry for the D4DR gene? As more and more companies offer personalized genetic testing, it may soon be possible for you to learn which alleles you carry for the D4DR gene. Would you want to know? What would you do with this information?

Q: Do you want your employers or insurance companies to know which alleles you carry for the D4DR gene? Carrying one or more long alleles has also been linked with increased use and abuse of alcohol and other drugs. Information about your genotype could put you at risk for genetic discrimination. Moreover, the personality differences influenced by D4DR are relatively slight, they seem to be influenced by numerous other genes, and there is a tremendous variation in personality traits that is unrelated to the D4DR gene. So it's just not clear what it means for you to have short or long D4DR allele.

What can you conclude? Possessing the long D4DR alleles does not necessarily mean you will want to go sky-diving or feel an urge to move out of the country. Conversely, possessing the short D4DR alleles may not discourage you from climbing K2. **But knowledge about the makeup of your D4DR gene might place you at risk for employment or insurance discrimination.**

1 Why do offspring resemble their parents?

Offspring resemble their parents because they inherit genes—instruction sets for biochemical, physical, and behavioral traits, some of which are responsible for diseases—from their parents. Gregor Mendel conducted studies in garden peas that helped us understand how traits are inherited, by applying methodical experimentation and rigorous hypothesis testing and focusing on easily observed and categorized traits in the pea plants. Each parent puts into every sperm or egg it makes a single set of instructions for building a particular trait. The trait observed in an individual depends on the two copies of the gene that it inherits from its parents. It is not always possible to determine an individual's genotype through observation of its phenotype. Much genetic analysis therefore makes use of clever experiments and careful record-keeping to determine organisms' genotypes.

2 Probability and chance play central roles in genetics.

Probability plays a central role in genetics. In segregation, each gamete an individual produces receives only one of the two copies of each gene the individual carries in its other cells; it is impossible to know which allele it will be. Chance plays a role in fertilization, too: all of the sperm or eggs produced by an individual are different from one another, and any one of those gametes may be the gamete involved in fertilization. Test-crosses and pedigrees help scientists, doctors, plant and animal breeders, and prospective parents determine which genes individuals carry and the likelihood that the offspring of two individuals will exhibit a given trait.

3 The translation of genotypes into phenotypes is not a black box.

Sometimes the effects of both alleles in a heterozygous genotype are seen in the phenotype. Many traits are influenced by multiple genes, and many genes may influence multiple unrelated traits. When a trait is coded for by a gene on the sex chromosomes, the pattern of expression differs for males and females. Genotypes are not like blueprints that specify phenotypes. Rather, phenotypes are a product of the genotype in combination with the environment.

4 Some genes are linked together.

Many genes tend to behave independently, such that the inheritance pattern of one trait doesn't usually influence the inheritance of any other trait. When genes are located close together on the same chromosome, however, the alleles for two genes are inherited and expressed almost as a package deal.

KEY TERMS

additive effects, p. 269
allele, p. 251
carrier, p. 264
cross, p. 254
codominance, p. 267
dominant, p. 256
genotype, p. 258
heredity, p. 252
heterozygous, p. 257
homozygous, p. 257
incomplete dominance, p. 265
linked gene, p. 277
Mendel's law of independent assortment, p. 276
Mendel's law of segregation, p. 257
multiple allelism, p. 267
pedigree, p. 263
phenotype, p. 258
pleiotropy, p. 271
polygenic, p. 269
Punnett square, p. 258
recessive, p. 256
sex-linked trait, p. 263
single-gene trait, p. 253
test-cross, p. 262
true-breeding, p. 255

CHECK YOUR KNOWLEDGE

1. Most genes come in alternative forms called:
a) alleles.
b) heterozygotes.
c) gametes.
d) chromosomes.
e) homozygotes.

2. Traits that are determined by a single gene:
a) occur in single-celled organisms, but not in humans.
b) are common in humans.
c) must occur on the X chromosome, because males have only a single X and so the gene must be able to function in the absence of its homologous allele.
d) include eye color and skin color.
e) can have only two alleles.

3. Pea plants were well suited for Mendel's breeding experiments for all of the following reasons except:
a) Peas exhibit variations in a number of observable characteristics, such as flower color and seed shape.
b) Mendel and his staff could control the pollination between different pea plants.
c) It is easy to obtain large numbers of offspring from any given cross.
d) Many of the characteristics that vary in pea plants are not linked closely on the same chromosome.
e) Peas have a particularly long generation time.

4. The law of segregation states that:
a) the transmission of genetic diseases within families is always recessive.
b) an allele on one chromosome will always segregate from an allele on a different chromosome.
c) gametes cannot be separate and equal.
d) the number of chromosomes in a cell is always divisible by 2.
e) each of two alleles for a given trait segregate into different gametes.

5. In pea plants, purple flower color is dominant to white flower color. If two pea plants that are true-breeding for purple flowers are crossed, in the offspring:
a) all of the flowers will be purple.
b) three-quarters of the flowers will be purple and one-quarter will be white.
c) half of the flowers will be purple and one-quarter will be white.
d) one-quarter of the flowers will be purple and three-quarters will be white.
e) all of the flowers will be white.

6. Pea flowers may be purple (*P*) or white (*p*). Pea seeds may be round (*R*) or wrinkled (*r*). What proportion of the offspring from a cross between purple-flowered, round-seeded individuals will have both white flowers and round seeds?
a) 1/16
b) 1/2
c) 3/16
d) 3/4
e) 9/16

7. The test-cross:
a) makes it possible to determine the genotype of an individual of unknown genotype that exhibits the dominant version of a trait.
b) is a cross between an individual whose genotype for a trait is not known and an individual homozygous recessive for the trait.
c) sometimes requires the production of multiple offspring to reveal the genotype of an individual whose genotype is unknown (but who exhibits the dominant phenotype).
d) Only a) and b) are correct.
e) Choices a), b), and c) are correct.

8. Which of the following statements is correct regarding pedigree analysis?
a) Darkened squares or circles always represent individuals with the trait being traced.
b) White squares or circles always represent heterozygous individuals.
c) Horizontal lines connect siblings.
d) The length of the vertical lines is dependent on the relatedness between two individuals.
e) Squares represent females, and circles represent males.

9. All of the offspring of a black hen and a white rooster are gray. The simplest explanation for this pattern of inheritance is:
a) multiple alleles.
b) codominance.
c) incomplete dominance.
d) incomplete heterozygosity.
e) sex linkage.

10. A woman with type B blood and a man with type A blood could have children with which of the following phenotypes?
a) AB only
b) AB or O only
c) A, B, or O only
d) A or B only
e) A, B, AB, or O

11. Which of the following traits shows a polygenic method of inheritance?
a) flower color in snapdragons
b) blood type in humans
c) seed color in peas
d) sickle-cell disease in humans
e) skin color in humans

12. The impact of a single gene on more than one characteristic is called:
a) incomplete dominance.
b) environmentalism.
c) balanced polymorphism.
d) pleiotropy.
e) codominance.

13. A rare, X-linked dominant condition in humans, congenital generalized hypertrichosis, is marked by excessive hair growth all over a person's body. Which of the following statements about this condition is incorrect?
   a) The son of a woman with this disease has just slightly more than a 50% chance of having this condition.
   b) The daughter of a man with this condition has a 100% chance of having this condition.
   c) The daughter of a woman with this disease has just slightly more than a 50% chance of having this condition.
   d) The son of a woman with this disorder has a 100% chance of having this condition.
   e) Males with this condition almost always have a more severe phenotype, because women usually have at least one normal allele that dilutes the effect of the mutated allele.

14. Individuals carrying two non-functioning alleles for the gene producing the enzyme that converts phenylalanine into tyrosine:
   a) will develop phenylketonuria (PKU).
   b) may or may not develop phenylketonuria, because the environment (diet, in this case) also plays a role in determining their phenotype.
   c) will develop both PKU and Tay-Sachs disease, because the genes causing the two traits are on the same chromosome.
   d) will develop both PKU and Tay-Sachs disease, because one pleiotropic gene influences both traits.
   e) do not need to worry about their phenylalanine consumption, because PKU depends on both genes and the environment.

15. Because of Mendel's law of independent assortment:
   a) individuals with red hair are more likely to have freckles.
   b) skin color and hair texture tend to be inherited together.
   c) we can deduce that genes cannot exist as free-floating entities within a cell but must be carried on chromosomes.
   d) the alleles coding for one trait do not usually influence the inheritance pattern for another trait.
   e) Both a) and b) are correct.

16. Thousands (or even tens of thousands) of different traits make up an individual. For this reason:
   a) in a species with 23 different chromosomes, some traits must be coded for by genes on the same chromosome.
   b) the environment must influence more than half of our traits.
   c) all (or nearly all) genes must be pleiotropic.
   d) knowing an individual's phenotype is not sufficient for determining the person's genotype.
   e) All of the above are correct.

Short-Answer Questions

1. Two healthy parents have two children, both of whom have a degenerative muscular disorder called SMA (spinal muscular atrophy).
   a) Explain how SMA must be inherited and what the genotype of the parents must be.
   b) The parents wish to have a third child and are contemplating adopting versus having a biological child. They seek the guidance of a genetic counselor and ask what the probability of having another child with SMA would be. What does the genetic counselor tell them? Explain.

2. Mackenzie's favorite flower is the hydrangea. She plants a large hydrangea bush in her yard and enjoys the big purple blossoms. Her sister, Anna, would like some hydrangeas for her yard as well, so they transplant a small portion of the large bush to Anna's yard. Anna is surprised when, the first time the plant in her yard blooms, the blossoms are pink instead of purple. If the genotype of this plant has not changed, what might have influenced a change in the phenotype?

3. Your friend explains to you that her father has hemophilia, a genetic disorder that affects the process of blood clotting, but her mother does not.
   a) Knowing that the inheritance pattern for this disorder is sex-linked, explain to your friend why it is more common for males to inherit a sex-linked trait.
   b) Using a Punnett square, indicate the probability that a brother of your friend would have hemophilia, assuming that her mother carries two functional copies of the gene involved in this sex-linked recessive disorder. Explain your answer.

See Appendix for answers. For additional study questions, go to www.prep-u.com.

1 Evolution is an ongoing process.

8.1 We can see evolution occurring right before our eyes.

2 Darwin journeyed to a new idea.

8.2 Before Darwin, most people believed that all species had been created separately and were unchanging.

8.3 A job on a 'round-the-world survey ship allowed Darwin to indulge and advance his love of nature.

8.4 Observing geographic similarities and differences among fossils and living plants and animals, Darwin developed a theory of evolution.

8.5 In 1858, after decades of mulling and procrastinating, Darwin published his thoughts on natural selection.

3 Four mechanisms can give rise to evolution.

8.6 Evolution occurs when the allele frequencies in a population change.

8.7 Mutation—a direct change in the DNA of an individual—is the ultimate source of all genetic variation.

8.8 Genetic drift is a random change in allele frequencies in a population.

8.9 Migration into or out of a population may change allele frequencies.

8.10 When three simple conditions are satisfied, evolution by natural selection is occurring.

4 Through natural selection, populations of organisms can become adapted to their environment.

8.11 Traits causing some individuals to have more offspring than others become more prevalent in the population.

8.12 Organisms in a population can become better matched to their environment through natural selection.

8.13 Natural selection does not lead to perfect organisms.

8.14 Artificial selection is a special case of natural selection.

8.15 Natural selection can change the traits in a population in several ways.

8.16 Natural selection can cause the evolution of complex traits and behaviors.

5 The evidence for the occurrence of evolution is overwhelming.

8.17 The fossil record documents the process of natural selection.

8.18 Geographic patterns of species distributions reflect their evolutionary histories.

8.19 Comparative anatomy and embryology reveal common evolutionary origins.

8.20 Molecular biology reveals that common genetic sequences link all life forms.

8.21 Laboratory and field experiments enable us to watch evolution in progress.

8

Evolution and Natural Selection

Darwin's dangerous idea

What big eyes you have! The glass wing butterfly, Cithaerias aurorina, *from Peru.*

8•1

We can see evolution occurring right before our eyes.

What's the longest that you've ever gone without food? Twenty-four hours? Thirty-six hours? If you've ever gone hungry that long, you probably felt as if you were going to die of starvation, but humans can survive days, even weeks, without food. In 1981, for example, 27-year-old Bobby Sands, a political prisoner in Northern Ireland, went on a hunger strike. Forsaking all food and consuming only water, he gradually deteriorated and ultimately died—after 66 days without food.

Each hour without food is even more dangerous when you're tiny. How long do you think a fruit fly can last without food? The answer is just under a day—20 hours, give or take a few (**FIGURE 8-1**). Fruit flies can't live long without food because their tiny bodies don't hold very large caloric reserves. But could you breed fruit flies that could live longer than 20 hours on average? Yes.

First, start with a population of 5,000 fruit flies. In biology, the term "population" has a specific meaning: a **population** is a group of organisms of the same species living in a particular geographic region. The region can be a small area like a test tube or a large area like a lake. In your experiment, your population occupies a cage in your laboratory.

From these 5,000 fruit flies, you choose only the "best" fruit flies to start the next generation, with "best" meaning the 20% of flies that can survive the longest without food. Choosing these long-lived, hungry flies can be accomplished in three simple steps:

1. Remove the food from the cage with the population of 5,000 flies.
2. Wait until 80% of the flies starve to death and then put a container of food into the cage.

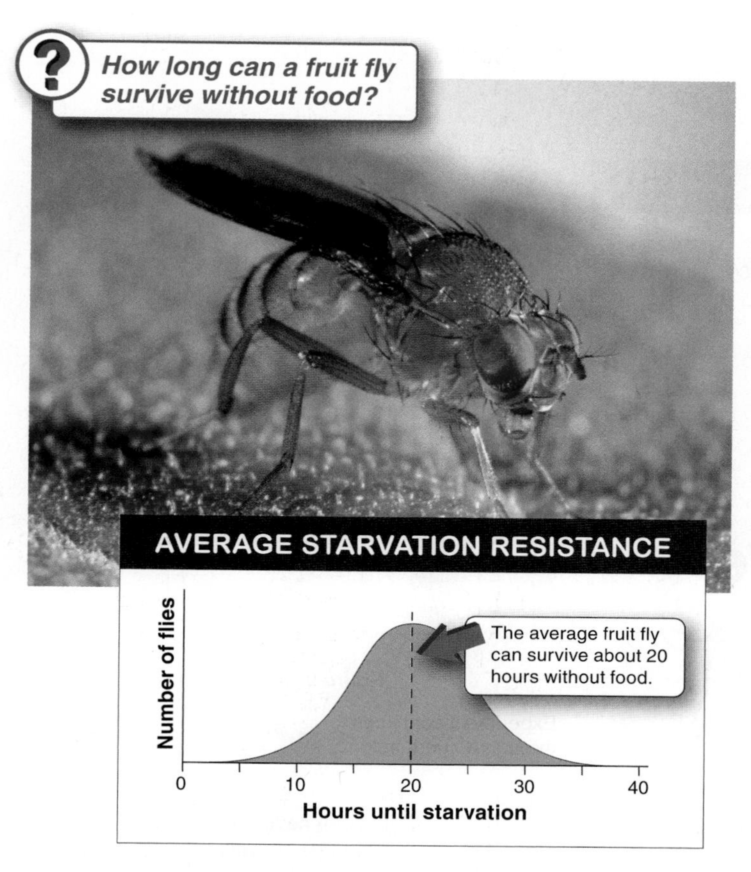

FIGURE 8-1 Starvation resistance in fruit flies.

3. After the surviving flies eat, they'll have the energy to reproduce. When they do, collect the eggs they lay and transfer those eggs to a new cage.

You can now start a new generation populated only by the offspring of those fruit flies able to survive without food for the longest amount of time. When these eggs hatch, do you think the flies in this new generation will live longer than 20 hours without food? When this experiment is done, the flies do indeed show increased resistance to starvation. The average fly in the new generation can live for about 23 hours without food. And again, some of the new flies survive for more than 23 hours, some for less. The surviving flies tend to be those that pack on the fat when the food is available. The new starvation-resistance curve for this second generation is similar to that of the original flies, but shifted to the right a bit so the highest part of the curve occurs at 23 hours, indicating that most flies of the new generation survive for an average of 23 hours without food.

What if you kept repeating these three steps? Start each new generation using eggs only from the fruit flies in the top 20% of starvation resistance. After five generations, would the average starvation-resistance time for a fly from your population be even higher? In fact, it would. With each new generation, you would see a slight increase, and with the fifth generation, the population's average survival would be about 32 hours.

One benefit to conducting research on fruit flies is that they have very short lives—they reach maturity at about two weeks of age and live for one month, on average. An experiment such as this one can be continued for many generations. After 60 generations of allowing the flies that are "best" at surviving without food to reproduce, how has the population changed? Amazingly, the *average* fly in the resulting population can survive for more than 160 hours without food (**FIGURE 8-2**). In 60 generations, the flies have gone from a starvation resistance of less than a day to one of nearly a week! At the point where the food is removed, the flies in generation 60 are noticeably fatter than the flies in generation 0 were.

What happened? In a word: **evolution.** That is, there was a genetic change in the population of fruit flies living in the cage. Every fly in the generation 60 population, even the fly with the worst starvation resistance, is still more than seven times better at resisting starvation than the best fly in the original population. This evolution is the result of **natural selection.** We'll discuss natural selection in more detail later in this chapter, but, in short, it is the consequence of certain individual organisms in a population being born with characteristics that enable them to survive better and reproduce more than the offspring of other individuals in the population. In this experiment, the 20% of fruit flies that were the most starvation resistant had a huge reproductive advantage over less-resistant flies because they were the *only* flies in the population that survived to reproduce.

This experiment answers a question that is sometimes perceived as complex or controversial: does evolution occur? The answer is an unambiguous *yes.* We can watch it happen in the lab whenever we want. Recall from our discussion of the scientific method (see Chapter 1) that for an experiment's results to be valid, they must be reproducible. To be certain of these results, researchers carried out the starvation-resistance experiment described above five separate times. The results were the same every time.

In this experiment we changed starvation resistance, but what if we used dogs instead of flies and, instead of allowing the most starvation-resistant individuals to reproduce, we allowed only the smallest dogs to reproduce? Would the average body size of individual dogs decrease over time? Yes. What if the experiment were done on rabbits and only the fastest rabbits escaped death from predation by foxes? Would the average running speed in rabbits increase over time? Yes. We know

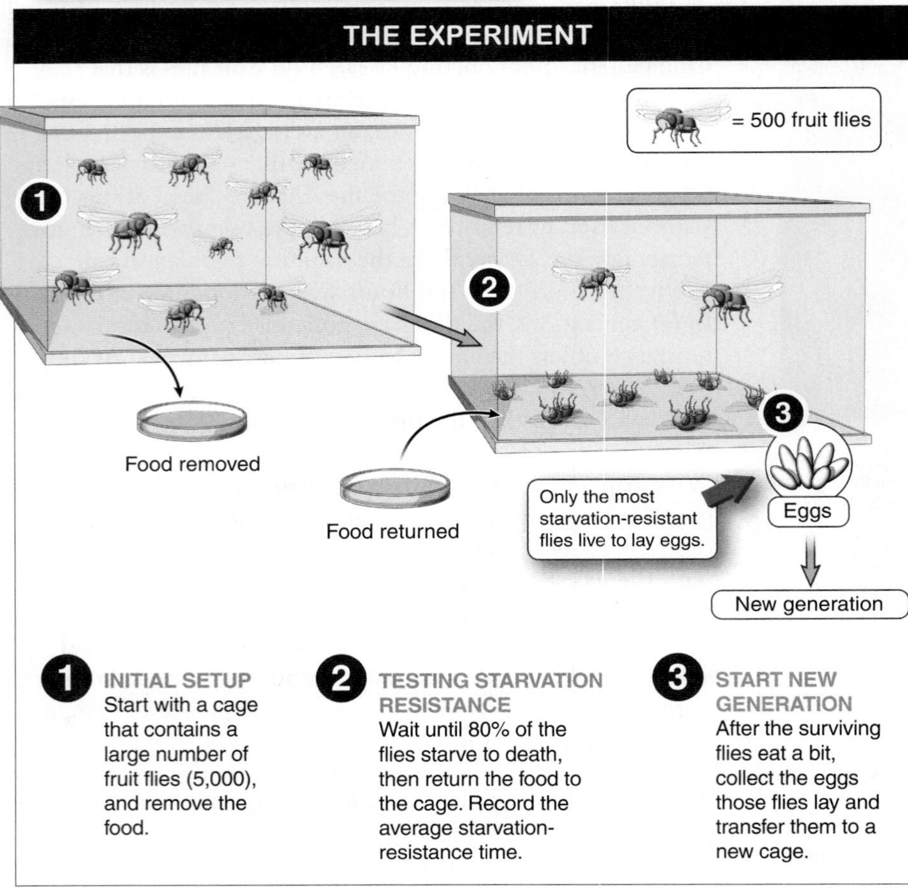

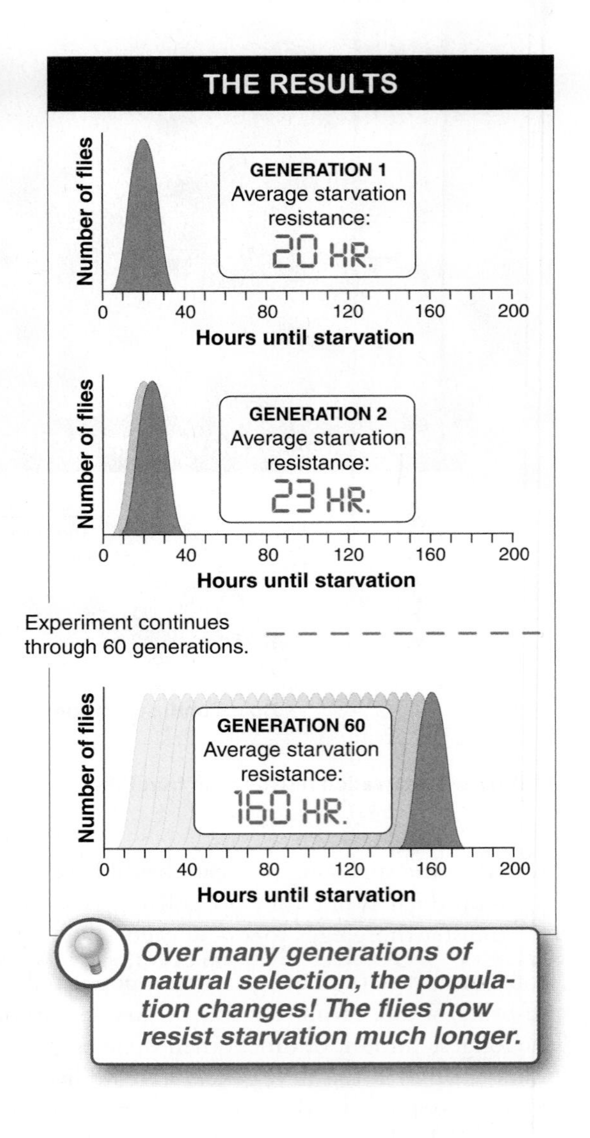

FIGURE 8-2 Evolution in action: increasing fruit fly resistance to starvation.

these answers because each of these experiments has already been done.

In this chapter we examine how evolution can occur and the type of changes it can cause in a population. We also review the five primary lines of evidence that indicate that evolution and natural selection are processes that help us to clarify all other ideas and facts in biology. Let's begin our investigation with a look at how the idea of evolution by natural selection was developed. This knowledge will help us better understand why this idea is regarded as one of the most important ideas in human history, why it is sometimes considered a dangerous idea, and why it generates emotional debate.

TAKE-HOME MESSAGE 8•1

The characteristics of individuals in a population can change over time. We can observe such change in nature and can even cause such change to occur.

❷ Darwin journeyed to a new idea.

The Galápagos tortoise, Geochelone elephantopus.

8•2

Before Darwin, most people believed that all species had been created separately and were unchanging.

Charles Darwin grew up in an orderly world. When it came to humans and our place in the world, in the mid-19th century the beliefs of nearly everyone were virtually unchanged from the beliefs of people who had lived more than 2,000 years earlier. Biblical explanations were sufficient for most natural phenomena: the earth was thought to be about 6,000 years old. With the occasional exception of a flood or earthquake or volcanic eruption, the earth was believed to be mostly unchanging. People recognized that organisms existed in groups called species. (In Chapter 10 we discuss in more detail what a species is; for now, we'll just say that individual organisms in a given species can interbreed with each other but not with members of another species.) People also believed that all species, including humans, had been created at the same time and that, once created, they never changed and never died out. This was pretty much what Aristotle had believed more than two millennia earlier.

Before he left this world, Darwin had thrown into question long and dearly held beliefs about the natural world and forever changed our perspective on the origins of humans and our relationship to all other species. He didn't smash the worldview to pieces all at once, though, and he didn't do it by himself (**FIGURE 8-3**).

In the 1700s and 1800s, scientific thought was advancing at a rapid pace. In 1778, the respected French naturalist Georges Buffon began to shake things up by suggesting that the earth must be about 75,000 years old. He arrived at this age by estimating that 75,000 years was the minimum time required for the planet to cool from a molten state. In the 1790s, Georges Cuvier began to explore the bottoms of coal and slate mines and found fossil remains that had no obvious similarity to any living species. The implications of Cuvier's discoveries were unthinkable, since biblical accounts did not allow for species to be wiped out. Cuvier's publications documented giant fossils (including the Irish elk, the mastodon, and the giant ground sloth) that bore no resemblance to any currently living animals (**FIGURE 8-4**). These fossils allowed only one explanation: extinction was a fact. Troubling as this observation was for the prevailing worldview, it was only the beginning.

DARWIN'S INFLUENCES

Georges Buffon
(1707–1788)
Suggested that the earth was much older than previously believed.

Georges Cuvier
(1769–1832)
By documenting fossil discoveries, showed that extinction had occurred.

Jean-Baptiste Lamarck
(1744–1829)
Suggested that living species might change over time.

Charles Lyell
(1797–1875)
Argued that geological forces had gradually shaped the earth and continue to do so.

FIGURE 8-3 Scientists who shaped Darwin's thinking.

Not only was it starting to seem that species could disappear from the face of the earth, but the biologist Jean-Baptiste Lamarck suggested in the early 1800s that living species might change over time. Although his ideas about the mechanisms by which this change might occur were wrong—he thought that change came about through the use or disuse of features—his willingness to question previously sacred "truths" contributed to an atmosphere of unfettered scientific thought in which it was possible to challenge convention.

Perhaps the heretical ideas that most inspired Darwin were those of the geologist Charles Lyell. In his 1830 book, *Principles of Geology,* Lyell argued that geological forces had shaped the earth and were continuing to do so, producing mountains and valleys, cliffs and canyons, through gradual but relentless change. This idea that the physical features of the earth were constantly changing would most closely parallel Darwin's idea that the living species of the earth, too, were gradually—but constantly—changing.

TAKE-HOME MESSAGE 8•2

In the 18th and 19th centuries, scientists began to overturn the commonly held beliefs that the earth was only about 6,000 years old and that all species had been created separately and were unchanging. These gradual changes in scientists' beliefs helped shape Charles Darwin's thinking.

Fossils of organisms no longer found on earth means that extinction occurs!

FIGURE 8-4 Extinction occurs. Deep in coal mines, Cuvier discovered the fossilized remains of very large animals no longer found on earth.

8.3

A job on a 'round-the-world survey ship allowed Darwin to indulge and advance his love of nature.

Charles Darwin was born into a wealthy family in England in 1809 and his early life was unremarkable. Charles was never at the top of his class and professed to hate schoolwork. Nonetheless, at 16, he went to the University of Edinburgh to study medicine, following his father's steps. He was bored, though, and finally left at the end of his second year, when the prospect of watching gruesome surgeries (in the days before anesthesia) was more than he could bear.

Upon the abrupt ending of Charles's medical career, his father urged him to pursue the ministry and sent him to study theology at Cambridge University. Although he never felt great inspiration in his theology studies, Darwin was in heaven at Cambridge, where he could pursue his real love, the study of nature. While at Cambridge, he spent most of his time reading nature books, collecting specimens, and developing close relationships with many of his professors.

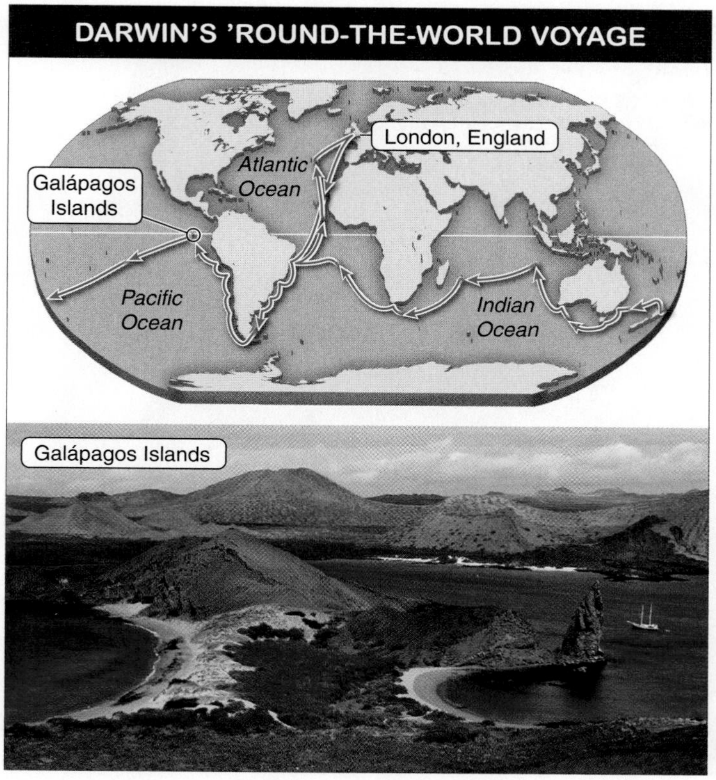

FIGURE 8-5 "Like a schoolboy on permanent summer vacation." In 1831, Darwin set out on the HMS *Beagle* on a five-year surveying expedition that took him around the globe.

FIGURE 8-6 Charles Darwin. Darwin found his life's calling in exploration and study of the natural world.

Shortly after graduation in 1831, Darwin—with significant help from his botany professor—landed his dream job, a position as a "gentleman companion" for the captain of the HMS *Beagle.* The *Beagle,* a 90-foot sailing vessel, was on a five-year, 'round-the-world surveying expedition (**FIGURE 8-5**). This job came as a huge relief to the young Darwin, who was not really interested in becoming a minister.

Once on the *Beagle,* Darwin found that he didn't actually like sea travel. He was seasick for much of his time on board, where he shared a room that was much tinier than a dorm room. He wrote to his cousin: "I hate every wave of the ocean with a fervor which you . . . can never understand." To avoid nausea, Charles spent as much time as possible on shore. It was there that he found his true calling (**FIGURE 8-6**).

Darwin loved fieldwork. At each stop that the *Beagle* made, he would eagerly investigate the new worlds he found. In Brazil, he was enthralled by tropical forests. In Patagonia, near the southernmost part of South America, he explored beaches and cliffs, finding spectacular fossils from huge extinct mammals. Elsewhere he explored coral reefs and barnacles, always packing up specimens to bring back to museums in England and recording his observations for later use. He was like a schoolboy on permanent summer vacation.

TAKE-HOME MESSAGE 8.3

After initially training in medicine and theology, Charles Darwin was able to focus on studying the natural world when, in 1831, he got a job on a ship conducting a five-year, 'round-the-world survey.

8•4

Observing geographic similarities and differences among fossils and living plants and animals, Darwin developed a theory of evolution.

The only book Darwin brought with him on the *Beagle* was Lyell's *Principles of Geology,* which he read again and again and discussed with anyone who would listen. Intrigued by the book's premise that the earth has been, and continues to be, constantly changing, Darwin's mind was ripe for fresh ideas—particularly as he began observing seemingly inexplicable things. How could he explain finding marine fossils high in the Andes Mountains, hundreds of miles from the nearest ocean? The idea that the earth was an ever-changing planet was seeping into Darwin's mind. That idea would serve him well when, a few years into its journey, the *Beagle* stopped at the Galápagos Islands, off the northwest coast of South America.

This group of volcanic islands was home to many unusual species, from giant tortoises to extremely docile lizards that made so little effort to run away that Darwin had to avoid stepping on them. Darwin was particularly intrigued by the wide variety of birds, especially the finches, which seemed dramatically more variable than those he had seen in other locations.

Darwin noticed two important and unexpected patterns on his voyage that would be central to his discovery of a mechanism for evolution. The first involved the finches he collected and donated to the Zoological Society of London. Darwin had assumed that they were the equivalent of tall and short, curly-haired and straight-haired people. That is, he thought that all the finches were of the same species but with different physical characteristics or **traits,** such as body size, beak shape, or feather color. The staff of the Zoological Society, however, could see from the birds' physical differences that the birds were not reproductively compatible and that there were 13 unique species—a different species for every one of the Galápagos Islands that Darwin had visited. Moreover, although the birds

FIGURE 8-7 **Darwin observed unexpected patterns.**

were different species, they all resembled very closely the single species of finch living on the closest mainland, in Ecuador (**FIGURE 8-7**).

This resemblance seemed a suspicious coincidence to Darwin. Perhaps the island finches resembled the mainland species because they used to be part of the same mainland population. Over time they may have separated and diverged from the original population and gradually formed new—but similar—species. Darwin's logic was reasonable, but his idea flew in the face of all of the scientific thinking of the day.

The second important but unexpected pattern Darwin noted was that, throughout his voyage, at every location there was a striking similarity between the fossils of extinct species and the living species in that same area. In Argentina, for instance, he found some giant fossils from a group of organisms called "glyptodonts." The extinct glyptodonts looked just like armadillos—a species that still flourished in the same area. Or rather, they looked like armadillos on steroids: the average armadillo today is about the size of a house cat and weighs 10 pounds, while the glyptodonts were giants. At 10 feet long and more than 4,000 pounds, they were larger than an automobile (see Figure 8-7).

If glyptodonts had lived in South America in the past, and armadillos were currently living there, why was only one of the species still alive? And why were the glyptodont fossils found only in the same places that modern armadillos lived? Darwin deduced that glyptodonts resembled armadillos because they were their ancestors. Again, it was a logical deduction, but it contradicted the scientific dogma of the day that species were unchanging and extinction did not occur.

The wheels were turning in Darwin's head, but he didn't have a great "Eureka!" moment until several years after his voyage, when he was back in England. He was reading "for amusement" a book called *Essay on the Principle of Population* by the economist Thomas Malthus. Malthus prophesied doom and gloom for populations—including humans—based on his belief that populations had the potential to grow much faster than food supplies could. Darwin speculated that, rather than the future holding certain catastrophe for all, maybe the best individuals would "win" in the ensuing struggle for existence and the worst would "lose." If so, he suddenly saw that "favourable variations would tend to be preserved, and unfavourable ones to be destroyed." Everything was falling into place for Darwin, who wrote: "Here then I had at last got a theory by which to work."

In 1842, Darwin finally hammered out a first draft of his ideas in a 35-page paper, written in pencil, and fleshed it out over the next couple of years. He knew that his idea was important—so important that he wrote a letter to his wife instructing her, in the case of his sudden death (an odd request, given that he was only 35 years old), to give his "sketch" to a competent person, with about $1,000 and all of his books, so that this person could complete it for him.

He also had an inkling that his ideas would rock the world. In a letter to a close friend, he wrote: "At last gleams of light have come, and I am almost convinced (quite contrary to the opinion I started with) that species are not (it is like confessing to murder) immutable." But "confessing to murder" was apparently more than Darwin was ready for. Inexplicably, he put his sketch into a drawer, where it remained for 14 years.

TAKE-HOME MESSAGE 8•4

Darwin noted unexpected patterns among fossils he discovered and living organisms he observed while on the voyage of the *Beagle*. Fossils resembled but were not identical to the living organisms in the same area in which they were found. And finch species on each of the Galápagos Islands differed from each other in small but significant ways. These observations helped Darwin to develop his theory of how species might change over time.

8•5

In 1858, after decades of mulling and procrastinating, Darwin published his thoughts on natural selection.

It came in the form of a letter from a young British biologist in the throes of malaria-induced mania in Malaysia. In the letter, Alfred Russel Wallace, who had also read Malthus's book on population growth, laid out a clear description of the process of evolution by natural selection (**FIGURE 8-8**). He asked Darwin to "publish it if you think it is worthy." Crushed at having been scooped, Darwin wrote that "all my originality will be smashed." He had numerous friends among the most prominent scientists of the day, however, and they arranged for a joint presentation of Wallace's and Darwin's work to the Linnaean Society of London. As a result, both Darwin and Wallace are credited for the first description of evolution by natural selection.

Darwin then sprang into action, rapidly putting together his thoughts and observations and completing, 16 months later, a full book. In 1859, he published *Origin of Species* (its full title is *On the Origin of Species by Means of Natural Selection, or the Preservation of Favoured Races in the Struggle for Life*) (**FIGURE 8-9**).

The book was an instant hit, selling out on its first day, provoking public discussion and debate and ultimately causing a wholesale change in the scientific understanding of evolution. Where once the worldview was of a young earth, populated by unchanging species all created at one time, with no additions or extinctions, now there was a new dynamic view of life on earth: species could and did change over time, and as some species split into new species, others became extinct.

FIGURE 8-8 Alfred Russel Wallace. Darwin and Wallace independently identified the process of evolution by natural selection.

WORLDVIEW BEFORE AND AFTER DARWIN

BEFORE:
- Organisms were all put on earth by a creator at the same time.
- Organisms are fixed: no additions, no subtractions.
- Earth is about 6,000 years old.
- Earth is mostly unchanging.

First published in 1859

AFTER:
- Organisms change over time.
- Some organisms have gone extinct.
- Earth is more than 6,000 years old.
- The geology of earth is not constant, but always changing.

FIGURE 8-9 Reconsidering the world. Commonly held ideas before and after Darwin.

Darwin's theory has proved to be among the most important and enduring contributions in all of science. It has stimulated an unprecedented diversity of theoretical and applied research programs and it has withstood repeated experimental and observational testing. With the background of Darwin's elegant idea in hand, we can now examine its details.

TAKE-HOME MESSAGE 8•5

After putting off publishing his thoughts on natural selection for more than 15 years, Darwin did so only after Alfred Russel Wallace independently came up with the same idea. The two men published a joint presentation on their ideas in 1858, and Darwin published a much more detailed treatment in *Origin of Species* in 1859, sparking wide debate and discussion of natural selection.

③ Four mechanisms can give rise to evolution.

Even with its snowy coloring, this snowshoe hare has been spotted by a cougar.

8•6

Evolution occurs when the allele frequencies in a population change.

Suppose you were put in charge of a large population of tigers in a zoo. Virtually all of them are orange and brown with black stripes. Occasionally, though, unusual all-white tigers are born. This white phenotype is the result of the possession of a rare pair of alleles that suppresses the production of most fur pigment (**FIGURE 8-10**). Because visitors flock to zoos to see the white tigers, you want to increase the proportion of your tiger population that is white. How would you go about doing that?

One possibility is that you could alter the population by trying to produce more animals with the desired white phenotype. In this case, your best strategy would be to try to breed the white tigers with each other. Over time, this would lead to the production of more and more of the white tigers you desire. And as the generations go by (this will take a while, since the generation time for tigers is about eight years), your population will include a higher proportion of white tigers. When this happens, you will have witnessed evolution, a change in the allele frequencies of the population.

Another way that you might increase the proportion of your white tiger population would be to acquire some white tigers from another zoo, if you could find any. By directly adding white tigers to your population, your population would include a higher proportion of white tigers (you might even trade some of your orange tigers, which would further increase the proportion of your population that is white). By adding white tigers or removing orange tigers, you will again have witnessed evolution in your population.

This example illustrates that evolution doesn't involve changing the genetics or physical features of *individuals*. Individuals do not evolve. Rather, you change the proportion of white fur-pigment alleles in the *population*. An individual tiger has two alleles for any gene, and within a population, the

THE EVOLUTION OF POPULATIONS

TIGER POPULATION

Allele frequencies:

- Proportion of orange fur-pigment alleles in the population
- Proportion of white fur-pigment alleles in the population

Evolution is a change in the allele frequencies of a population over time. For example, a change in the proportion of pigment alleles in the population of tigers means that evolution has occurred.

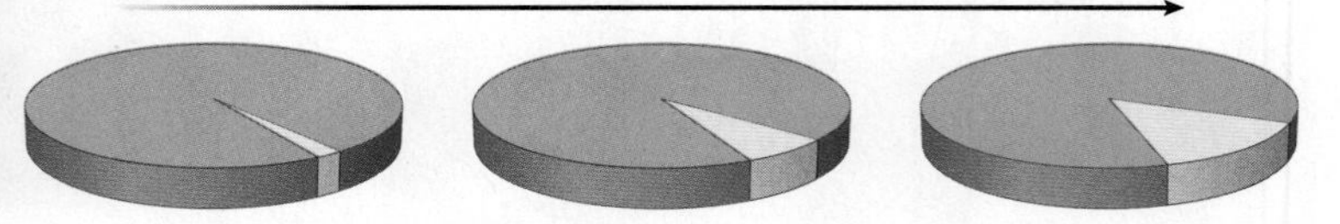

FIGURE 8-10 Evolution defined. Evolution is a change in allele frequencies in a population.

proportion of all the alleles coming from any one allele is that allele's frequency. It is helpful to think of each allele as having some "market share" of all of the alleles. In the tiger example, the white alleles initially had little market share, perhaps as small as 1%. Over time, though, they came to make up a larger and larger proportion of the total pigment alleles. And as this happened, evolution occurred.

Any time an allele's market share changes, evolution is occurring. Perhaps the most dramatic way we see this is in acts of predation. When one organism kills another, the dead animal's alleles will no longer be passed on in the population. And if certain alleles make it more likely that an animal will be killed by another—maybe they alter its coloration, or speed—or some other defense, those alleles are likely to lose market share. The converse is equally true, evolution occurs when alleles increase in frequency.

Darwin demonstrated that natural selection could be an efficient mechanism of evolution and a powerful force in adapting populations to their environment. Evolution and natural selection, however, are not the same thing. Natural selection is one way that evolution can occur, but it is not the only agent of evolutionary change. It is one of four. They are:

1. Mutation
2. Genetic drift
3. Migration
4. Natural selection

Keeping in mind that evolution is genetic change in a population, we'll now explore each of these four forces that are capable of causing such genetic changes.

TAKE-HOME MESSAGE 8•6

Evolution is a change in allele frequencies within a population. It can occur by four different mechanisms: mutation, genetic drift, migration, and natural selection.

8•7

Mutation—a direct change in the DNA of an individual—is the ultimate source of all genetic variation.

In describing the first of the four mechanisms of evolution, it is helpful to keep in mind our precise definition of evolution: a change in the allele frequencies found in a population. **Mutation** is an alteration of the base-pair sequence of an individual's DNA, and when this alteration occurs in a section of the DNA that codes for a particular gene, the change in the DNA sequence may change the allele. Say, for example, that a mutation changes one of the two blue-eye alleles that an individual possesses into a brown-eye allele. If this mutation occurs in the sperm- or egg-producing cells it can be passed on to the next generation; the offspring may be born carrying the brown-eye allele. When this happens, the proportion of blue-eye alleles in the population, their market share, is slightly reduced and the market share of brown-eye alleles is slightly increased. Evolution has occurred.

Q Tanning beds bombard the body with ultraviolet radiation. Can they cause mutations?

What causes mutations? For one thing, the incredibly complex process of cell division can go awry. Duplication of the 3 *billion* base pairs of a cell's DNA and separation of the new and original versions of the DNA into two new cells, after all, are not simple processes. Mutations also can be induced by environmental phenomena. Most environmentally induced mutations occur when the DNA of cells is exposed to high-energy sources or mutation-inducing chemicals (called **mutagens**). Radioactive isotopes, for example, emit high-energy particles or photons that can disrupt DNA. This is the reason that radiation therapy is both dangerous and effective as a treatment for cancer: the high-energy source kills most of the cells in its vicinity but can also mutate the DNA in other cells. Ultraviolet radiation, even from the sun, can cause mutations as well; sunscreen serves as a barrier, blocking the harmful radiation from reaching cells and the DNA they carry.

Q Mobile phones release radiation. Can they cause brain tumors?

Although environmental factors influence the rate at which mutation occurs, they do not generally influence exactly *which* mutations occur. Thus, mutations are random, and whether a mutation happens or not is unrelated to how useful or harmful that mutation might be. Treatment of some agricultural pest with harmful chemicals, for example, may increase the number of mutations occurring but does not increase the likelihood that a particular mutation is beneficial or detrimental.

There are many potential mutation-causing threats in our world today, including tanning beds, for example, which cause individuals to become tan by bombarding the skin with ultraviolet rays similar to those produced by the sun. These rays induce tanning just as exposure to the sun does: by stimulating pigment-producing cells to produce more melanin, a pigment that darkens skin color. But because the ultraviolet rays are high-energy waves, they can cause mutation just as the sun's rays can. Not surprisingly, researchers have found that the greater an individual's exposure to ultraviolet rays, the greater the incidence of mutations and, as a consequence, of skin cancer.

Another potential source of mutation-causing energy is of concern to many more people: mobile phones. The scientific community is still divided on this question, however. Although mobile phones do emit electromagnetic radiation, that radiation is sufficiently low level that it is generally unable to break chemical bonds and therefore cause mutations. The majority of studies have not found any consistent causal relationship between mobile phone usage and any short- or medium-term health hazards.

Since 2004, however, several studies have suggested that further investigation is warranted. A study from 12 laboratories found evidence of DNA damage in cell cultures exposed to levels of radiation similar to those produced by mobile phones. Another study found that individuals who had used mobile phones for 10 years or more had an increased risk of brain tumors. A third study discovered an increased risk of brain tumors and noted that the increased risk was only for the side of the head on which the phone was used. Stay tuned for more information on this important question.

In addition to being an agent of evolutionary change, mutation has another important role that is relevant to all agents of evolution, including natural selection: mutation is

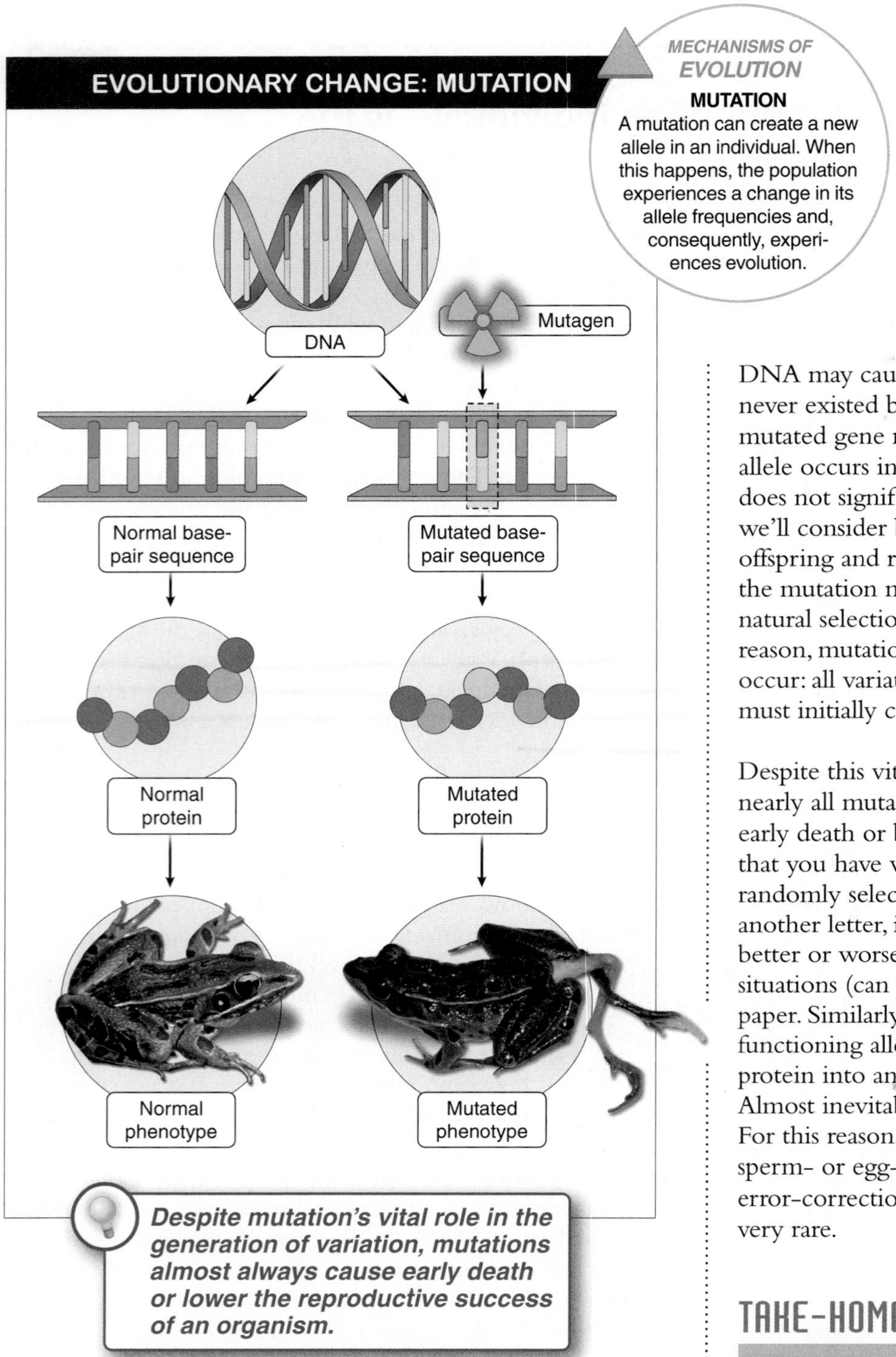

EVOLUTIONARY CHANGE: MUTATION

MECHANISMS OF EVOLUTION

MUTATION

A mutation can create a new allele in an individual. When this happens, the population experiences a change in its allele frequencies and, consequently, experiences evolution.

DNA

Mutagen

Normal base-pair sequence

Mutated base-pair sequence

Normal protein

Mutated protein

Normal phenotype

Mutated phenotype

Despite mutation's vital role in the generation of variation, mutations almost always cause early death or lower the reproductive success of an organism.

FIGURE 8-11 Agents of evolutionary change: mutation.

the ultimate source of genetic variation in a population. We saw that a mutation may lead to the conversion of one allele to another that is already found within the population, as in our blue-eye and brown-eye example. More important, though, a mutation may instead create a completely novel allele that codes for the production of a new protein (**FIGURE 8-11**). That is, a change in the base-pair sequence of a person's DNA may cause the production of a gene product that has never existed before: instead of blue or brown eyes, the mutated gene might code for yellow or red. If such a new allele occurs in the sperm- and egg-producing cells and if it does not significantly reduce an individual's "fitness" (which we'll consider below), the new allele can be passed on to offspring and remain in the population. At some future time, the mutation might even confer higher fitness, in which case natural selection may cause it to increase in frequency. For this reason, mutation is critically necessary if natural selection is to occur: all variation—the raw material for natural selection—must initially come from mutation.

Despite this vital role in the generation of variation, however, nearly all mutations reduce an organism's fitness by causing its early death or by reducing its reproductive success. Suppose that you have written a 10-page paper. If you were to randomly select one letter in the paper and change it to another letter, is the change more likely to make your paper better or worse? The answer is obvious: except in *very* rare situations (can you imagine one?) the change will hurt your paper. Similarly, mutations usually change a normally functioning allele that codes for a normally functioning protein into an allele that codes for a non-functioning protein. Almost inevitably, that outcome reduces an organism's fitness. For this reason, our bodies are built in ways that protect our sperm- or egg-producing DNA with a variety of built-in error-correction mechanisms. As a result, mutations are very, very rare.

TAKE-HOME MESSAGE 8•7

Mutation is an alteration of the base-pair sequence in an individual's DNA. Such an alteration constitutes evolution if it changes an allele that the individual carries. Mutations can be caused by high-energy sources or chemicals in the environment and also can appear spontaneously. Mutation is the only way that new alleles can be created within a population, and so generates the variation on which natural selection can act.

8•8

Genetic drift is a random change in allele frequencies in a population.

Along with mutation, another evolutionary agent is **genetic drift,** a random change in allele frequencies. Genetic drift is best illustrated with an example. Imagine that in a population there are two alleles for a particular trait, let's say a cleft chin. This is a dominant trait, so individuals with either one or two copies of the dominant allele (CC or Cc) exhibit the cleft chin (**FIGURE 8-12**). Now suppose that two heterozygous (Cc) people have one child. Which combination of alleles will that child receive? This is impossible to predict because it depends completely upon which sperm fertilizes which egg—the luck of the draw. If the couple's sole child inherited a recessive allele from each parent, would the *population's* allele frequencies change? Yes. After all, there is now another individual in the population, and that individual has two recessive alleles (cc). There are slightly more recessive alleles in the population. And because a change in allele frequencies has occurred, evolution has happened. It is equally likely that this couple's only child would have received two of the dominant alleles (CC), rather than the recessive alleles. In either case, because a change in allele frequencies has occurred, evolution has happened.

The important factor that distinguishes genetic drift from natural selection is that the change in allele frequencies is not related to the alleles' influence on reproductive success. A cleft chin does not affect an individual's ability to reproduce.

This impact of genetic drift is much greater in small populations than in large populations. In a large population, it is more likely that any one couple having a child with two recessive alleles is likely to be offset by another couple having one child with two dominant alleles. When that occurs, the

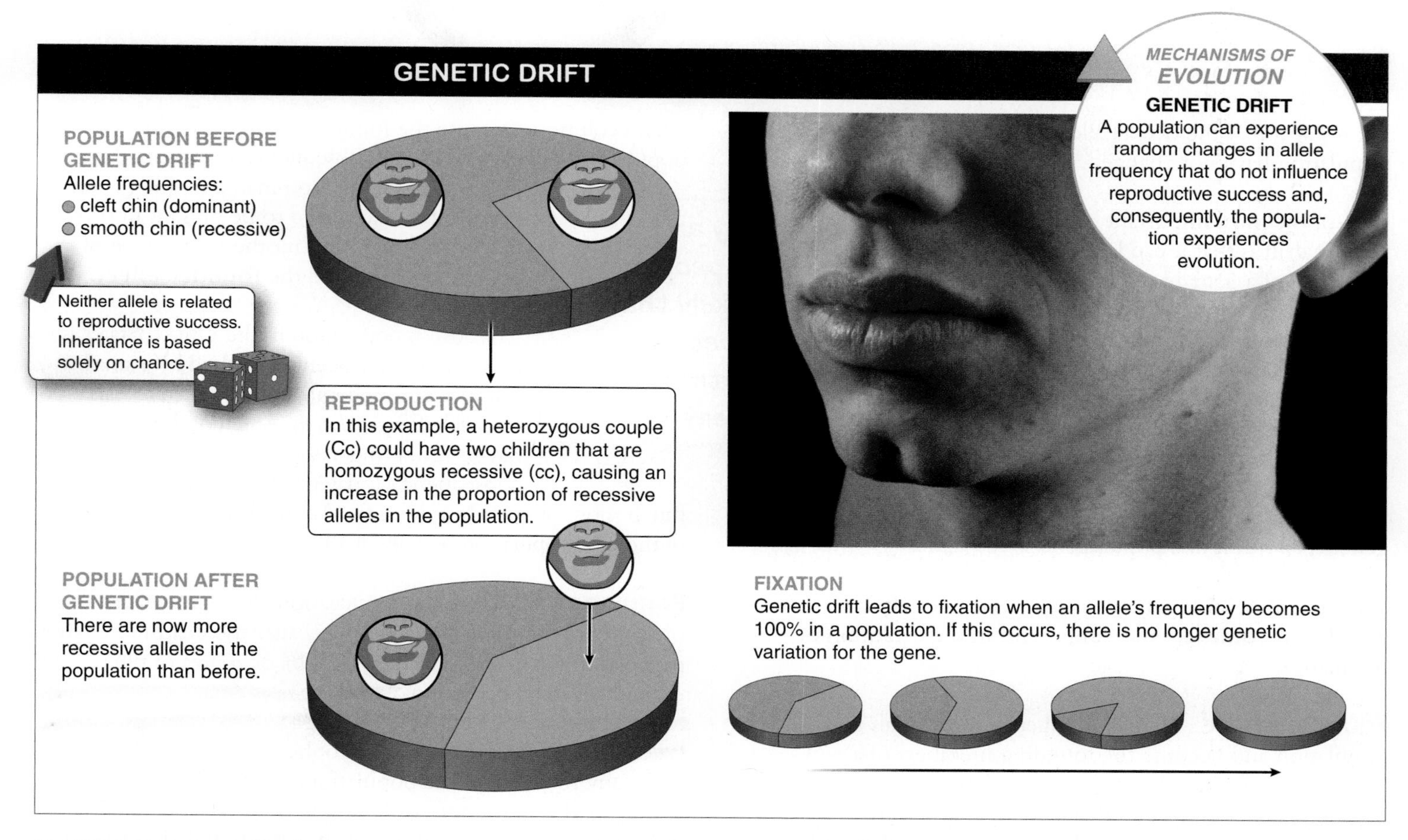

FIGURE 8-12 Agents of evolutionary change: genetic drift. Genetic drift has the greatest impact in small populations.

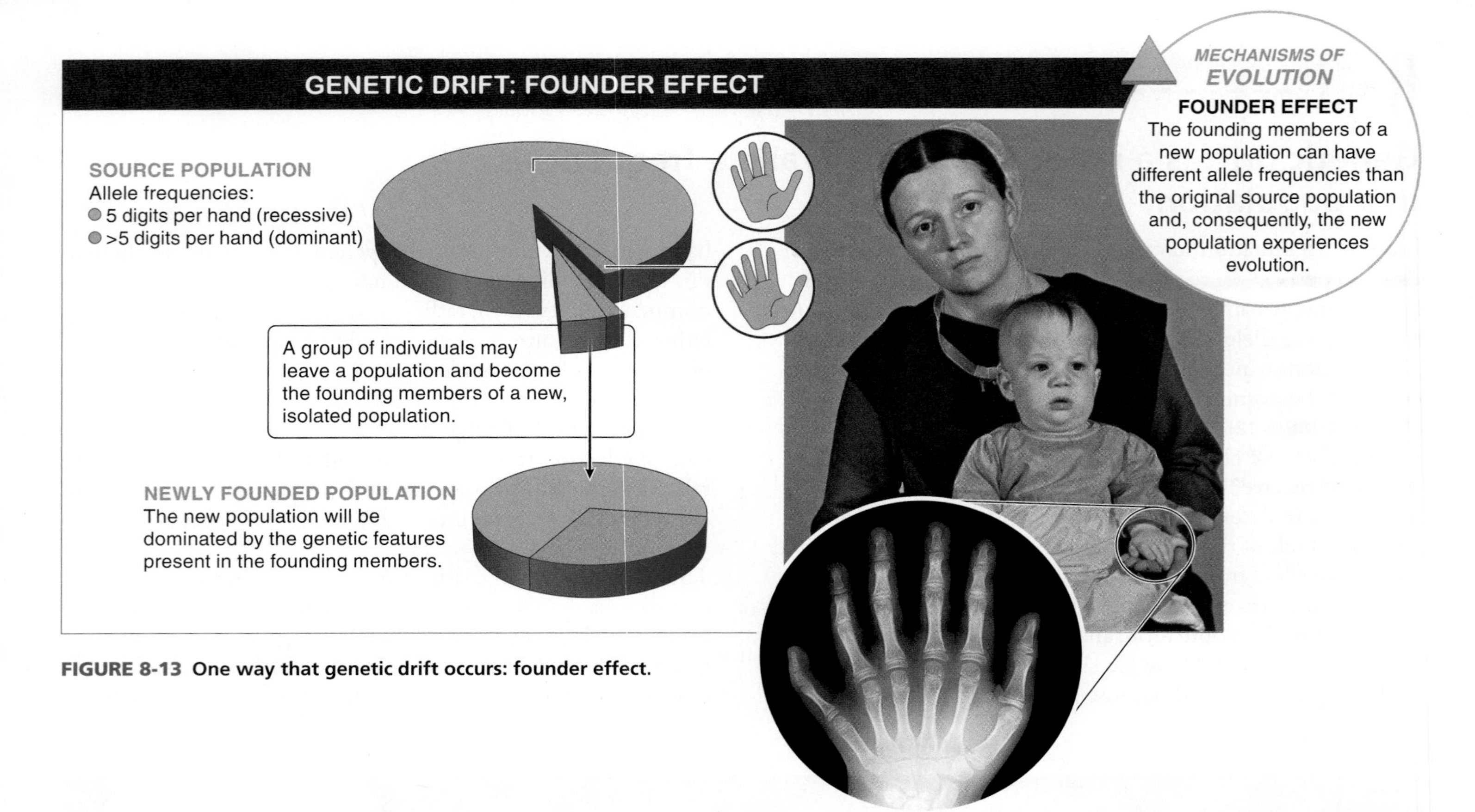

FIGURE 8-13 **One way that genetic drift occurs: founder effect.**

net result is that the overall allele frequencies in the population have not changed, and evolution has not occurred.

One of the most important consequences of genetic drift is that it can lead to **fixation** for one allele of a gene in a population (see Figure 8-12). Fixation is said to occur when an allele's frequency in a population reaches 100% (and the frequency of all other alleles of that gene becomes 0%). If this happens, there is no more variability in the population for this gene; all individuals will always produce offspring carrying only that allele (until new alleles arise through mutation). For this reason, genetic drift reduces the genetic variation in a population.

Q Why are Amish people more likely than other people to have extra fingers and toes?

Two special cases of genetic drift, the founder effect and population bottlenecks, are important in the evolution of populations.

Founder Effect A small number of individuals may leave a population and become the founding members of a new, isolated population. The founder population may have different allele frequencies than the original "source" population, particularly if the founders are a small sample. If this new population has different allele frequencies, there has been evolution. Because the founding members of the new population will give rise to all subsequent individuals, the new population will be dominated by the genetic features that happened to be present in those founding fathers and mothers. This type of genetic drift is called the **founder effect.**

The Amish population in the United States is believed to have been established by a small number of founders who happened to carry the allele for *polydactyly*—the condition of having extra fingers and toes. As a consequence, today this trait, while rare, occurs much more frequently among the Amish than among the rest of the U.S. population (**FIGURE 8-13**).

Population Bottlenecks Occasionally, a famine, disease, or rapid environmental change may cause the deaths of a large proportion (sometimes as many as 90% or more) of the individuals in a population. Because the population is quickly reduced to a small fraction of its original size, this reduction is called a **bottleneck.** If the catastrophe is equally likely to strike any member of the population, the remaining members are essentially a random, small sample of the original population. For this reason, the remaining population may not possess the same allele frequencies as the original population.

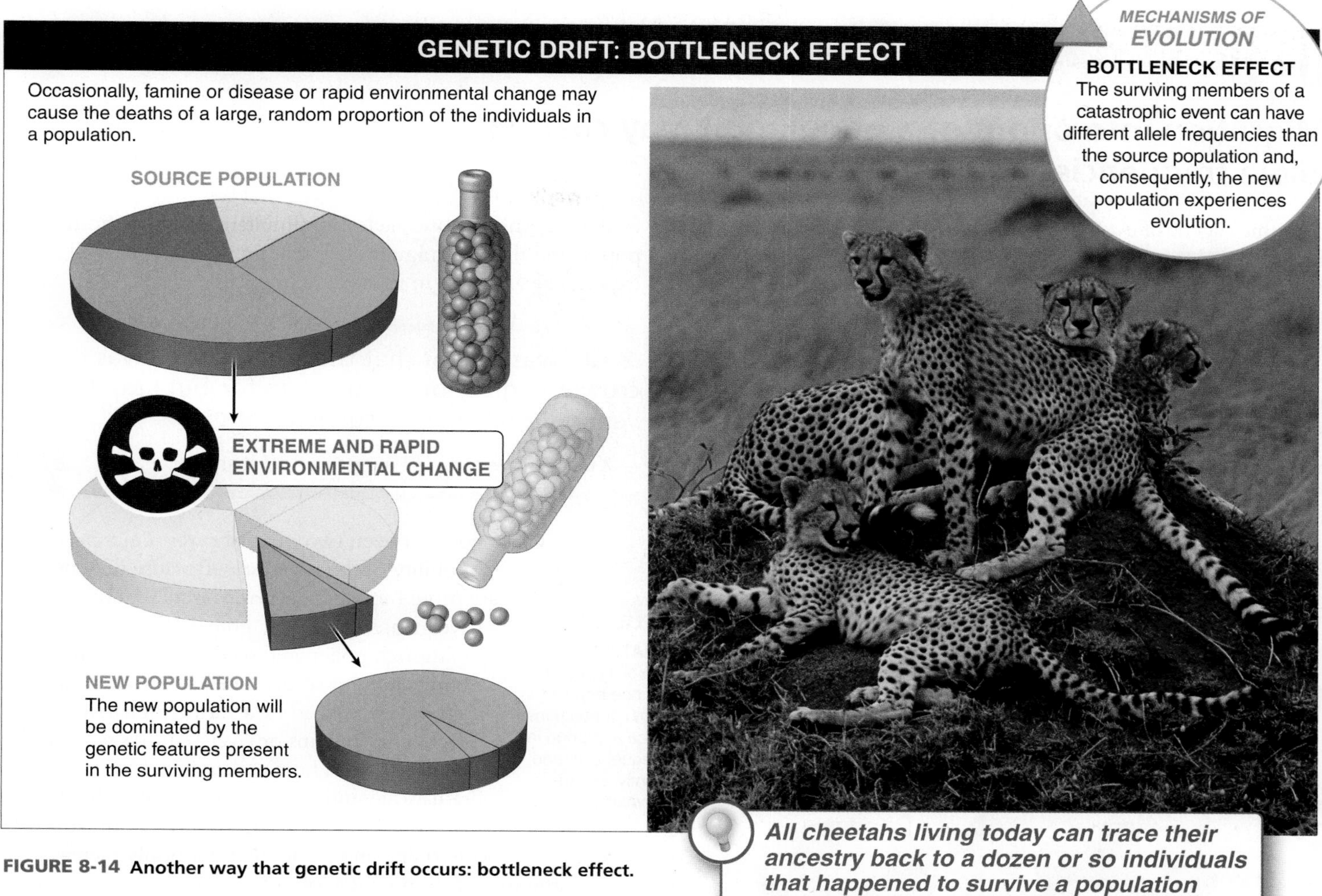

FIGURE 8-14 Another way that genetic drift occurs: bottleneck effect.

Thus, the consequence of such a population bottleneck would be evolution through genetic drift (**FIGURE 8-14**).

Just such a population bottleneck occurred with cheetahs near the end of the last ice age, about 10,000 years ago. Although the cause is unknown—possibly environmental cataclysm or human hunting pressures—it appears that nearly all cheetahs died. And although the population has rebounded, all cheetahs living today can trace their ancestry back to a dozen or so lucky individuals that survived the bottleneck. As a result of this past instance of evolution by genetic drift, there is almost no genetic variation left in the current population of cheetahs. (And, in fact, a cheetah will accept a skin graft from any other cheetah, much as identical twins will from each other.)

TAKE-HOME MESSAGE 8·8

Genetic drift is a random change in allele frequencies within a population, unrelated to the alleles' influence on reproductive success. Genetic drift is a significant agent of evolutionary change primarily in small populations.

8•9

Migration into or out of a population may change allele frequencies.

The third agent of evolutionary change is **migration.** Migration, also called **gene flow,** is the movement of some individuals of a species from one population to another (**FIGURE 8-15**). This movement from population to population within a species distinguishes migration from the founder effect, in which individuals migrate to a new habitat, previously unpopulated by that species. If migrating individuals can survive and reproduce in the new population, and if they also carry a different proportion of alleles than the individuals in their new home, then the recipient population experiences a change in allele frequencies and, consequently, experiences evolution. And because alleles are simultaneously lost from the population that the migrants left, that population, too, will experience a change in its allele frequencies.

> I was taught that the human brain was the crowning glory of evolution so far, but I think it's a very poor scheme for survival.
>
> —Kurt Vonnegut, Jr., American writer

MECHANISMS OF EVOLUTION

MIGRATION

After a group of individuals migrates from one population to another, both populations can experience a change in their allele frequencies and, consequently, experience evolution.

Gene flow between two populations is influenced by the mobility of the organisms. Because they usually are more mobile than plants, animals tend to have greater migratory potential than plants (although pollen and seeds can sometimes be transported great distances). Gene flow can also be influenced by a variety of barriers, such as mountains or rivers. And. as we'll see in Chapter 16, human activities, too, can dramatically influence gene flow, whether it is a new highway restricting movement of organisms or boats, trains, and planes that drastically increase the migratory potential of organisms.

MIGRATION (GENE FLOW)

1 BEFORE MIGRATION
Two populations of the same species exist in separate locations. In this example, they are separated by a mountain range.

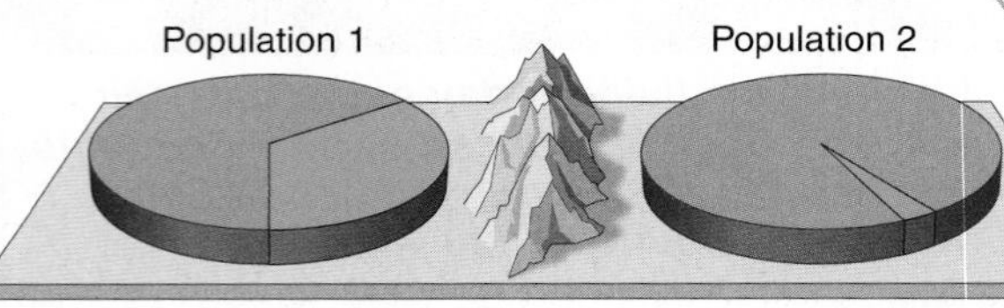

2 MIGRATION
A group of individuals from Population 1 migrates over the mountain range.

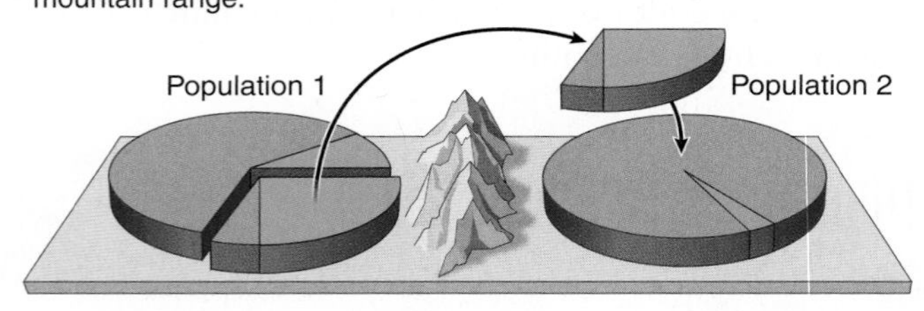

3 AFTER MIGRATION
The migrating individuals are able to survive and reproduce in the new population.

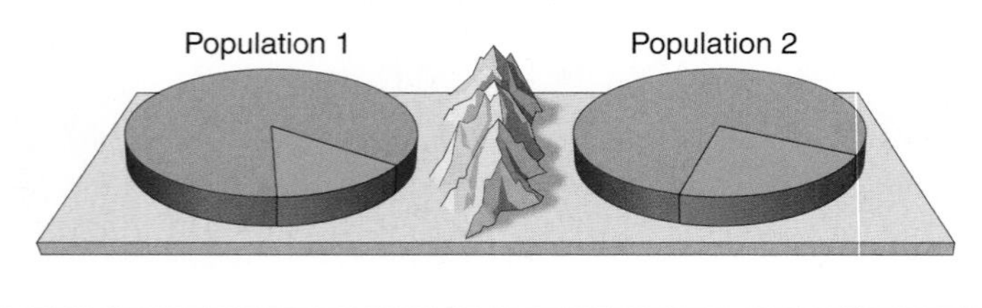

FIGURE 8-15 Agents of evolutionary change: migration (gene flow).

TAKE-HOME MESSAGE 8•9

Migration, or gene flow, leads to a change in allele frequencies in a population as individuals move into or out of the population.

8•10

When three simple conditions are satisfied, evolution by natural selection is occurring.

The fourth agent of evolutionary change is natural selection. This is the mechanism that Darwin identified in *The Origin of Species,* in which he noted that three conditions are necessary for natural selection to occur:

1. There must be variation for the particular trait within a population.
2. That variation must be inheritable (that is, it must be capable of being passed on from parents to their offspring).
3. Individuals with one version of the trait must produce more offspring than those with a different version of the trait.

Let's examine these conditions more closely.

Condition 1: Variation for a Trait Close your eyes and imagine a dog. What does the dog look like? Chances are, if 50 people were asked this question, we would get descriptions of 50 different dogs (**FIGURE 8-16**): Chihuahuas, Great Danes, sheepdogs, greyhounds, spaniels, and more. Some are big, some are small. Some have short hair, some long. They vary in just about every way that you can imagine. Similarly, if 50 people were to imagine a human face, a similarly broad range of images would pop into their heads. Variation is all around us. Beyond making the world an interesting place to live in, variation serves another purpose: it is the raw material on which evolution feeds.

Variation is not limited to physical features such as fur color and texture or face shape and eye color. Organisms vary in physiological and biochemical ways, too. Some people can quickly and efficiently metabolize alcohol, for example. Others find themselves violently ill soon after sipping a glass of wine. The same goes for digesting milk. Similarly, we vary in our susceptibility to poison ivy or diseases such as malaria. Behavioral variation—from temperament to learning abilities to interpersonal skills—is dramatic and widespread, too. So impressed was Darwin with the variation he observed throughout the world, he devoted the first two chapters of *The Origin of Species* to a discussion of variation in nature and among domesticated animals. Darwin knew that the variation he saw all around him was an essential component of evolution by natural selection. He considered it the first of three conditions necessary for natural selection.

Condition 2: Heritability The second condition that Darwin identified as necessary for natural selection was a no more complex discovery than the first: for natural selection to act on a trait within a population, offspring must inherit the trait from their parents. Although inheritance was poorly understood in Darwin's time, it was not hard to see that, for many traits, offspring look more like their parents than like some other random individual in the population (**FIGURE 8-17**). Animal breeders had long known that the fastest horses generally gave birth to the fastest horses. Farmers, too, understood that the plants with the highest productivity generally produced seeds from which highly productive plants grew. And everyone knew that children resembled their parents, from their appearance to their behavior to their temperament. It was enough to know that

FIGURE 8-16 Necessary conditions for natural selection: 1. Variation for a trait.

FIGURE 8-17 Necessary conditions for natural selection: 2. Heritability. Goldie Hawn and daughter Kate Hudson resemble each other.

this similarity between offspring and parents existed—it was not necessary to understand how it occurred or to be able to quantify just how great the similarity was. We call the transmission of traits from parents to their children through genetic information **inheritance** or **heritability.**

Q Most agricultural pests evolve resistance to pesticides. How does this happen?

Condition 3: Differential Reproductive Success

It would be nice to say that Darwin made a stunning and insightful discovery for the third of the three conditions necessary for natural selection to occur, but he didn't. Rather, he derived the third condition for natural selection from three fairly simple observations that he made:

1. More organisms are born than can survive.
2. Organisms are continually struggling for existence.
3. Some organisms are more likely to win this struggle and survive and reproduce. The struggle does not always involve direct physical contact, but in a world of scarce, limited resources, finding food or shelter is a zero-sum game: if one organism is feasting, another is more likely to be starving.

This three-part observation (which goes by the complex name **differential reproductive success**) led Darwin to his third condition for natural selection: from all the variation existing in a population, individuals with traits most suited to reproduction in their environment generally leave more offspring than do individuals with other traits (**FIGURE 8-18**). For example, in the experiments selecting for fruit flies that can survive for a long time without food, we saw that the fruit flies that inherit the ability to pack on the fat when food is available end up leaving more offspring than those inheriting a poor ability to pad their little fruit fly frames with fat deposits. The portly fruit flies have greater reproductive success than other individuals.

That's it. Natural selection—certainly one of the most influential and far-reaching ideas in the history of science—occurs when three basic conditions are met (**FIGURE 8-19**):

1. Variation for a trait
2. Heritability of that trait
3. Differential reproductive success based on that trait

When these three conditions are satisfied, evolution by natural selection is occurring. It's nothing more and nothing less; no mysterious black box is required. Over time, the traits that lead some organisms to have greater reproductive success than others will increase in frequency in a population while traits that reduce reproductive success will fade away.

When it comes to some traits, the reason they confer greater reproductive success is that they make the individual exhibiting them more attractive to the opposite sex. Such traits—including the brightly colored tail feathers seen in male peacocks, the large antlers seen in male red deer, and a variety of other "ornaments" that increase an individual's status or appeal—because they satisfy the three conditions for natural selection, increase in frequency just the same, in a phenomenon referred to as **sexual selection.**

Another way of looking at natural selection is to focus not on the winners (the individuals who are producing more offspring), but on the

The tiniest dog in a litter has reduced differential reproductive success. Its more robust siblings prevent access to the food it needs to grow and thrive.

FIGURE 8-18 Necessary conditions for natural selection: 3. Differential reproductive success.

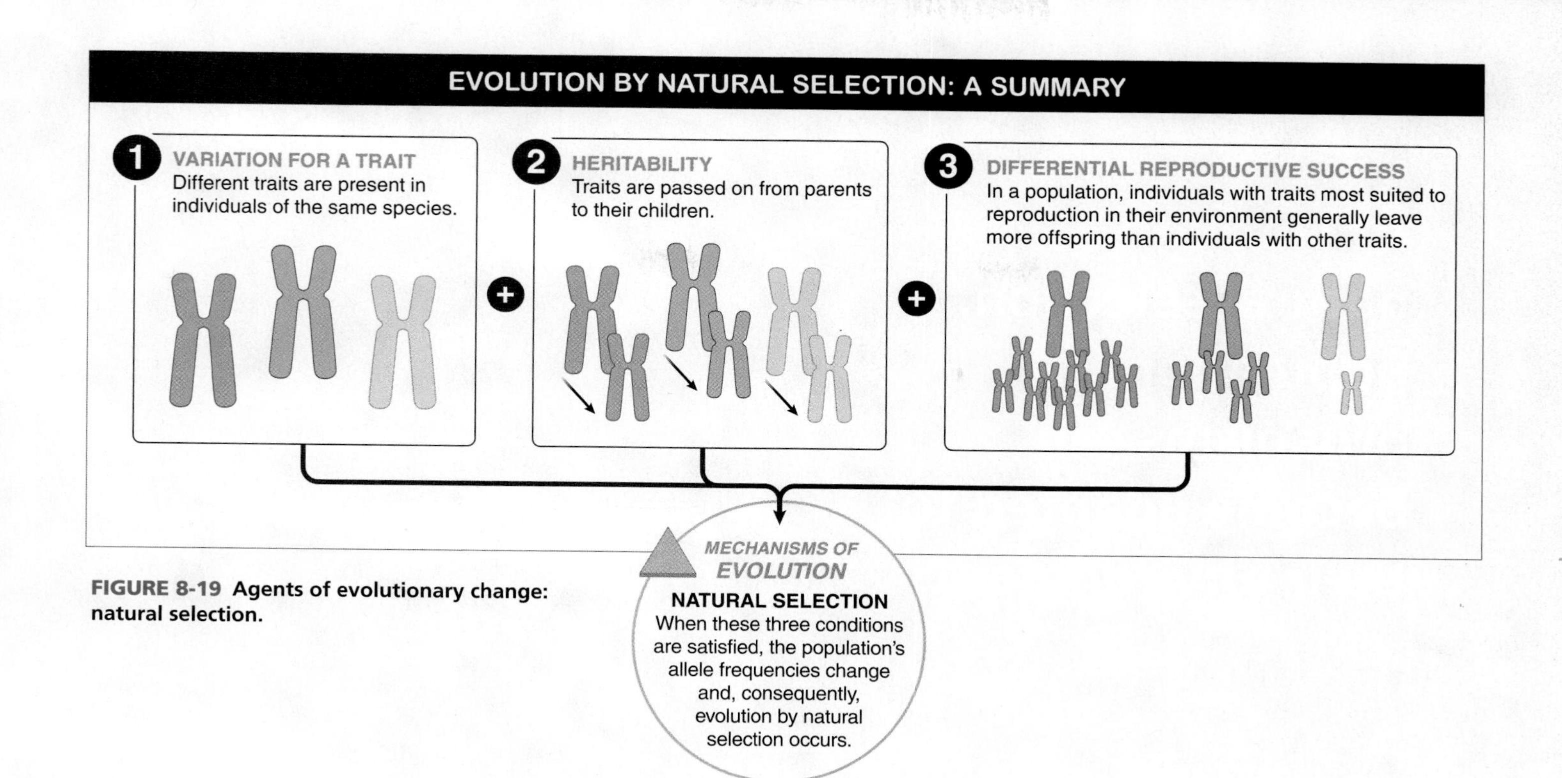

FIGURE 8-19 Agents of evolutionary change: natural selection.

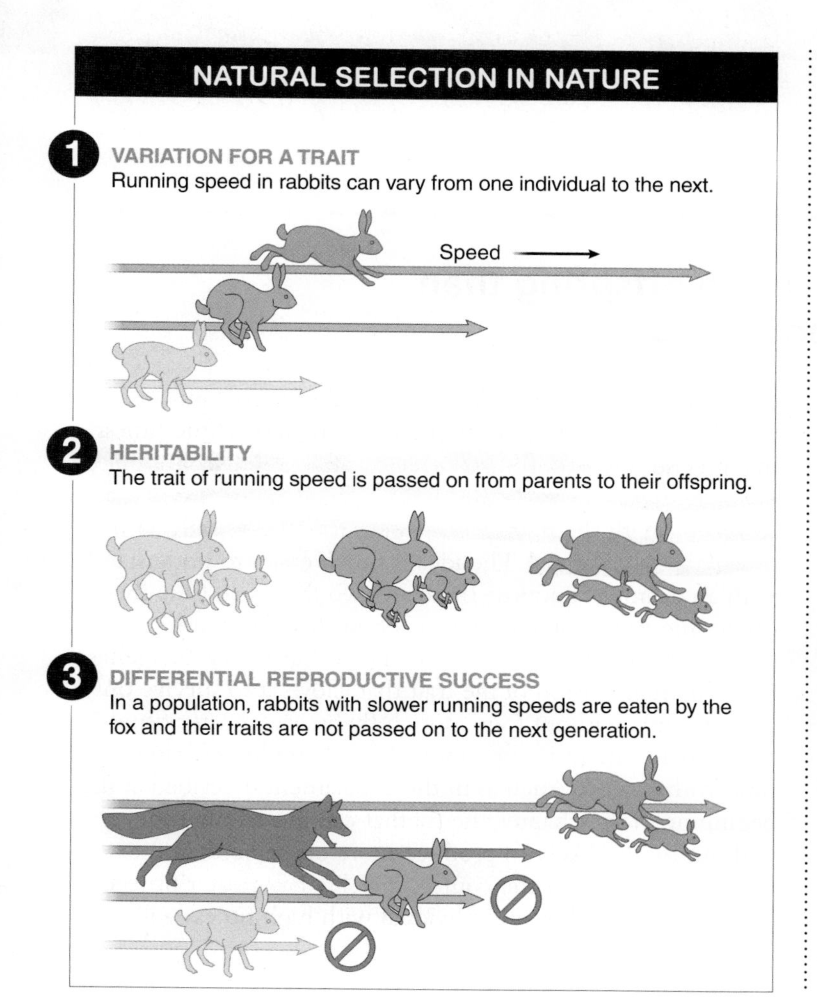

losers. Natural selection can be turned on its head and viewed as the elimination of some heritable traits from a population. If you carry a trait that makes you a slower-running rabbit, for example, you are more likely to be eaten by the fox (**FIGURE 8-20**). If running speed is a heritable trait (and it is), the next generation in a population contains fewer slow rabbits. Over time, the population is changed by natural selection. It evolves.

One of Darwin's contemporaries, Thomas Huxley, supposedly cursed himself when he first read *The Origin of Species,* saying that he couldn't believe that he didn't figure it out on his own. Each of the three basic conditions is indeed simple and obvious. The brilliant deduction, though, was to put the three together and appreciate the consequences.

TAKE-HOME MESSAGE 8•10

Natural selection is a mechanism of evolution that occurs when there is heritable variation for a trait and individuals with one version of the trait have greater reproductive success than do individuals with a different version of the trait. It can also be thought of as the elimination from a population of alleles that reduce the reproductive rate of individuals carrying those alleles relative to the reproductive rate of individuals who do not carry them.

FIGURE 8-20 Removing the losers. Natural selection can be thought of as the elimination from a population of traits that confer poor reproductive success.

④ Through natural selection, populations of organisms can become adapted to their environment.

Adapted to the hot, dry desert: A camel caravan in the Sahara.

8•11

Traits causing some individuals to have more offspring than others become more prevalent in the population.

"Survival of the fittest." This is perhaps the most famous phrase from Darwin's most famous book. Unfortunately, it can be a confusing and even misleading phrase. Taken literally, it seems to be a circular phrase: those organisms that survive *must* be the fittest if being fit is defined as the ability to survive. But we will see that when using the word "fitness" in discussing evolution, it has little to do with an organism's ability to survive or to its physical strength or health. Rather, fitness has everything to do with an organism's reproductive success.

Here's an interesting side note: the phrase "survival of the fittest" was coined not by Darwin but by Herbert Spencer, an influential sociologist and philosopher. Moreover, the phrase did not appear in *Origin of Species* when Darwin first published it. It wasn't until the fifth edition, 10 years later, that he used the phrase in describing natural selection.

Before we see how fitness affects natural selection and a population's adaptation to its environment, let's define fitness in a precise way. **Fitness** is a measure of the relative amount of reproduction of an individual with a particular phenotype, as compared with the reproductive output of individuals with alternative phenotypes. The idea is much easier to understand with an example. Suppose there are two fruit flies. One fly carries the genes for a version of a trait that allows it to survive a long time without food. The other fly has the genes for a different version of the trait that allows it to survive only a short while without food. Which fly has the greater fitness? If the environment is one in which there are long periods of time without food, such as in the experiment described at the beginning of the chapter, the fly that can live a long time without food is likely to produce more offspring than the other fly and so over the course of its life has greater fitness. The alleles carried by an individual with high fitness will increase their market share in a population over time and the population will evolve.

There are three important elements to an organism's fitness.

1. An individual's fitness is measured relative to other genotypes or phenotypes in the population. Those traits that confer on an individual the highest fitness will generally increase their market share in a population, and their increase will always come at the expense of the market share of alternative traits that confer lower fitness.

2. Fitness depends on the specific environment in which the organism lives. The fitness value of having one trait versus another depends on the environment an organism finds itself in. A sand-colored mouse living in a beach habitat will be more fit than a chocolate-colored mouse. But that same sand-colored mouse will practically call out to potential predators when living in the darker brush away from the beach. An organism's fitness, although genetically based, is not fixed in stone and unchanging—it can change over time and across habitats (**FIGURE 8-21**).

Q "Survival of the fittest" is a misnomer. Why?

3. Fitness depends on an organism's reproductive success compared with other organisms in the population. If you carry an allele that gives you the trait of surviving for 200 years, but that allele also causes you to be sterile and incapable of producing offspring, your fitness is zero; that allele will never be passed down to future generations and its market share will soon be zero. On the other hand, if you inherit an allele that gives you a trait that causes you to die at half the age of everyone else, but also causes you to have twice as many offspring as the average while you are alive, your fitness is increased. It is reproductive success that is most important in determining whether particular traits increase their market share.

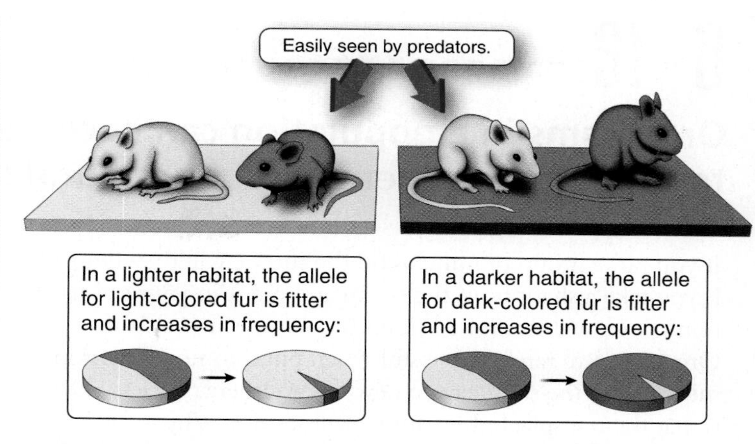

FIGURE 8-21 **An organism's fitness depends on the environment in which it lives.**

"Survival of the fittest" is a misleading phrase since it is the individuals that have the greatest *reproductive output* that are the most fit in any population. It becomes a more meaningful phrase if we consider it as a description of the fact that the alleles that increase an individual's fitness will "survive" in a population more than the alleles that decrease an individual's fitness.

TAKE-HOME MESSAGE 8•11

Fitness is a measure of the relative amount of reproduction of an individual with a particular phenotype, as compared with the reproductive output of individuals with alternative phenotypes. An individual's fitness can vary, depending on the environment in which the individual lives.

8•12

Organisms in a population can become better matched to their environment through natural selection.

If you put a group of humans on the moon, will they flourish? If you take a shark from the ocean and put it in your swimming pool, will it survive? In both cases, the answer is *no.* Organisms are rarely successful when put into novel environments. And the stranger the new environment, the less likely it is that the transplanted organism will survive. Why is that?

As Darwin noted over and over again during his travels, the organisms that possess traits that allow them to better exploit the environment in which they live will tend to produce more offspring than the organisms with alternative traits. With passing generations, a population will be made up of more and more of these fitter organisms. And, as a consequence, organisms will tend to be increasingly well matched or adapted to their environment.

Adaptation refers both to the process by which organisms become better matched to their environment *and* to the specific features that make an organism more fit. Examples of adaptations abound. Bats have an extremely accurate type of hearing (called echolocation) for navigating and finding food, even in complete darkness. Porcupine quills make porcupines almost impervious to predation (**FIGURE 8-22**). Mosquitoes produce strong chemicals that prevent blood from clotting, so that they can extract blood from other organisms.

FIGURE 8-22 Adaptations increase fitness. Quills are an adaptation that almost completely eliminates the risk of predation for porcupines.

TAKE-HOME MESSAGE 8•12

Adaptation—the process by which organisms become better matched to their environment and the specific features that make an organism more fit—occurs as a result of natural selection.

8•13

Natural selection does not lead to perfect organisms.

In Lewis Carroll's *Through the Looking-Glass,* the Red Queen tells Alice that "in this place it takes all the running you can do, to keep in the same place." She might have been speaking about the process of evolution by natural selection. After all, if the least fit individuals are continuously weeded out of the population, we might logically conclude that, eventually, fitness will reach a maximum and all organisms in all populations will be perfectly adapted to their environment. But this never happens. That is where the Red Queen's wisdom comes in.

Consider one of the many clearly documented cases of evolution: that of the beak size of Galápagos finches. Over the course of a multidecade study, the biologists Rosemary and Peter Grant closely monitored the average size of the finches' beaks. They found that the average beak size within a population fluctuated according to the food supply. During dry years—when the finches had to eat large, hard seeds—bigger, stronger beaks became the norm. During wet years, smaller-beaked birds were more successful since there was a surplus of small, softer seeds.

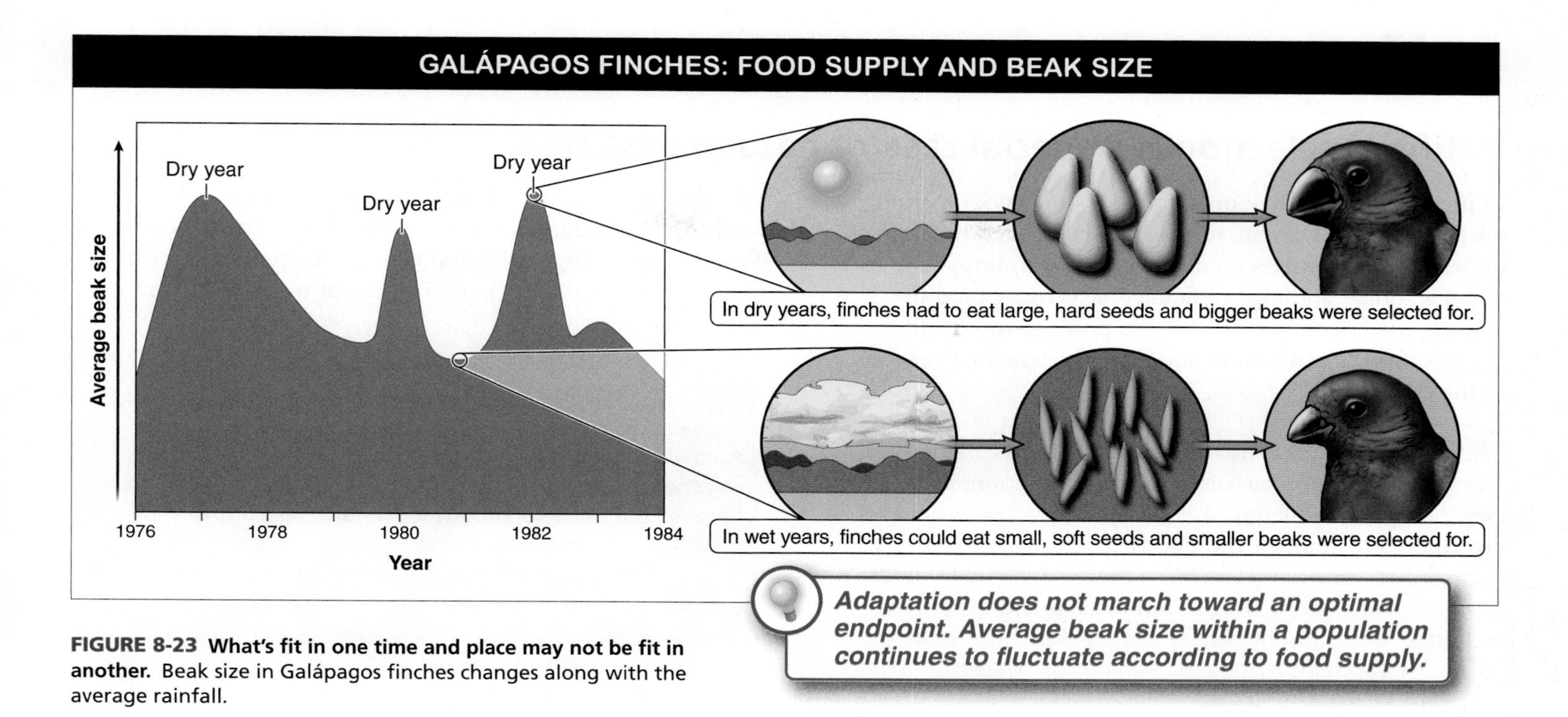

FIGURE 8-23 What's fit in one time and place may not be fit in another. Beak size in Galápagos finches changes along with the average rainfall.

The ever-changing "average" finch beak illustrates that adaptation does not simply march toward some optimal endpoint (**FIGURE 8-23**). Evolution in general, and natural selection specifically, do not guide organisms toward "better-ness" or perfection. Natural selection is simply a process by which, in each generation, the alleles that cause organisms to have the traits that make them most fit in that environment tend to increase in frequency. If the environment changes, the alleles causing the traits favored by natural selection may change, too.

Q Why doesn't natural selection lead to the production of perfect organisms?

In the next to last paragraph of *Origin of Species,* Darwin wrote: "We may look with some confidence to a secure future of great length. And as natural selection works solely by and for the good of each being, all corporeal [physical] and mental endowments will tend to progress towards perfection." In this passage, he overlooks several different factors that prevent populations from progressing inevitably toward perfection:

1. Environments change quickly. Natural selection may be too slow to adapt the organisms in a population to the constantly moving target that is the environment.

2. Variation is needed as the raw material of selection—remember, it is the first necessary condition for natural selection to occur. If a mutation creating a new, "perfect" version of a gene never occurs, the individuals within a population will never be perfectly adapted. Why, for instance, are there no mammals with wheeled appendages? That might be a great trait, but until the genes for it exist, natural selection cannot increase their market share in the population.

3. There may be multiple different alleles for a trait, each causing an individual to have the same fitness. In this case, each allele represents an equally fit "solution" to the environmental challenges.

TAKE-HOME MESSAGE 8•13

Natural selection does not lead to organisms perfectly adapted to their environment because (1) environments can change more quickly than natural selection can adapt organisms to them; (2) all possible alleles are not produced by mutation; (3) there is not always a single optimum adaptation for a given environment.

8•14

Artificial selection is a special case of natural selection.

In practice, plant and animal breeders understood natural selection before Darwin, they just didn't know that they understood it. Farmers bred crops for maximum yield, and dog, horse, and pigeon fanciers selectively bred the animals with their favorite traits to produce more and more of the offspring with more and more exaggerated versions of the trait.

The artificial selection used by animal breeders and farmers is also natural selection because the three conditions are satisfied, even though the differential reproductive success is being determined by humans and not by nature. Apple growers, for example, use artificial selection to produce the wide variety of apples available: green, yellow, and red, tart and sweet, large and small. What is important is that it is still differential reproductive success, and the results are no different. It was a stroke of genius for Darwin to recognize that the same process farmers were using to develop new and better crop varieties was occurring naturally in every population on earth and always had been.

TAKE-HOME MESSAGE 8•14

Animal breeders and farmers are making use of natural selection when they modify their animals and crops, because the three conditions for natural selection are satisfied. Since the differential reproductive success is determined by humans and not by nature, this type of natural selection is also called artificial selection.

8•15

Natural selection can change the traits in a population in several ways.

Certain traits are easily categorized—some people have blue eyes and others have brown eyes, just as some tigers have white fur and some have orange fur. Other traits, such as height in humans, are influenced by many genes and environmental factors so that a continuous range of phenotypes occurs (**FIGURE 8-24**). Whether a given trait is influenced by one gene or by a complex interaction of many genes and the environment, it can be subject to natural selection and be changed in any of several ways.

Directional Selection In **directional selection,** individuals with one extreme from the range of variation in the population have higher fitness. Milk production in cows is an example. There is a lot of variation in milk production from cow to cow. As you might expect, farmers select for breeding those cows with the highest milk production and have done so for many decades. The result of such selection is not surprising: between 1920 and 1945, average milk production increased by about 50% in the United States (**FIGURE 8-25**).

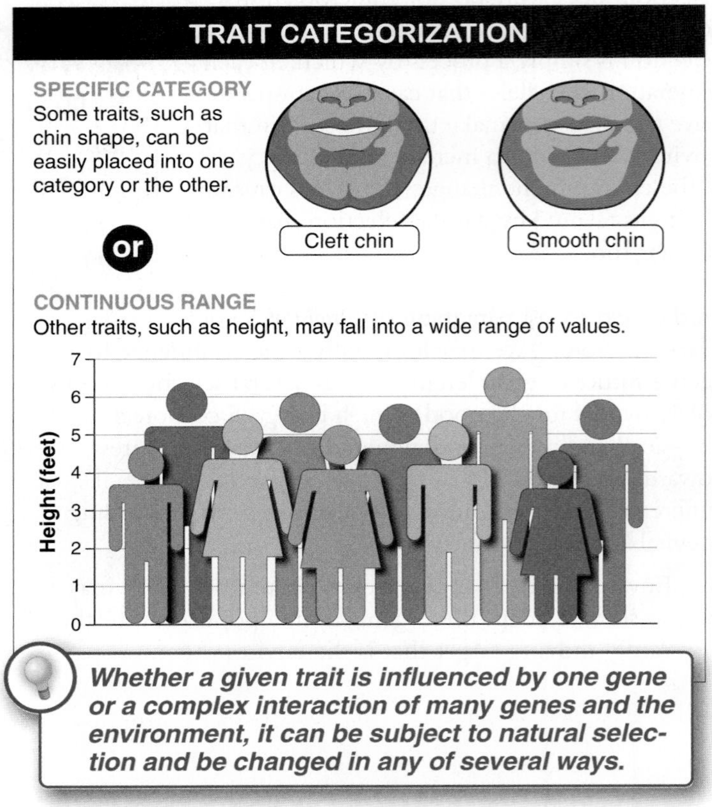

FIGURE 8-24 Some traits fall into clear categories, others range continuously.

FIGURE 8-25 More milk? Cows have been selected for their ability to produce more and more milk.

Those at the other end of the range, the cows that produce the smallest amounts of milk, have reduced fitness, since the farmers do not allow them to reproduce and therefore contribute their "less milk" alleles to the subsequent generation. (This is true whether the selection occurs naturally or through artificial selection by breeders.) In subsequent generations, the average value for the milk-production trait increases. Note, however, that the variation for the trait decreases a bit, too, because the alleles contributing to one of the phenotype extremes are eventually removed from the population. Although the experiment has never been done, we can confidently predict that directional selection for *reduced* milk production on these dairy farms would have worked equally well.

Q Turkeys on poultry farms have such large breast muscles that they can't get close enough to each other to mate. How can such a trait evolve?

Hundreds of experiments in nature and laboratories demonstrate the power of directional selection. In fact, it is one of the marvels of laboratory selection and animal breeding that nearly any trait they choose to exaggerate via directional selection responds dramatically—sometimes with absurd results (**FIGURE 8-26**). Turkeys, for instance, have been selected for increased size of their breast muscles, which makes them more valuable to farmers. Directional selection has been so successful that the birds' breast muscles are now so large that it has become impossible for them to mate. All turkeys on large-scale, industrial poultry farms must now reproduce through artificial insemination!

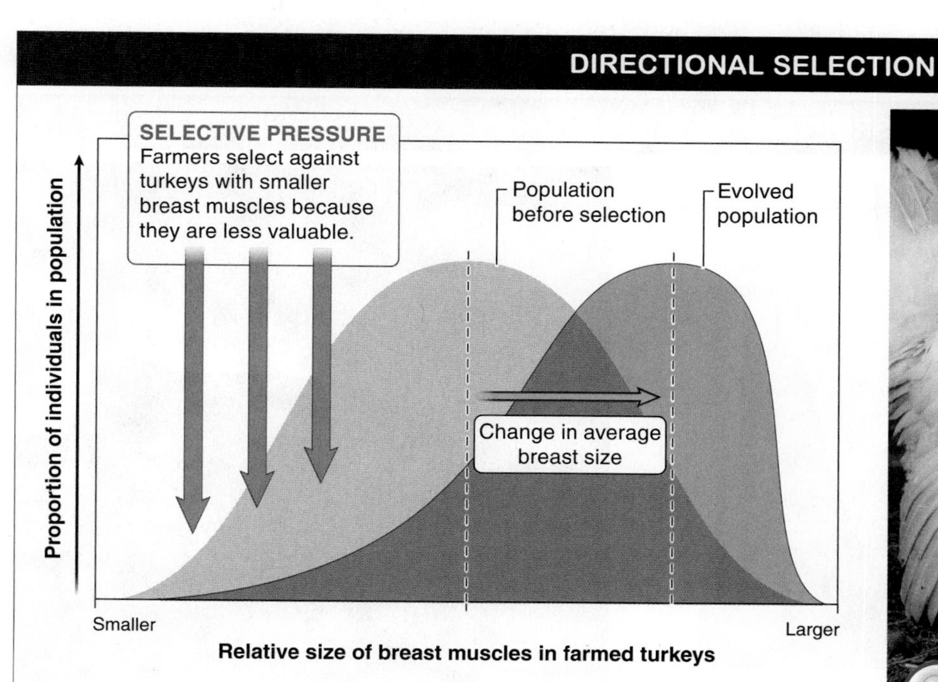

FIGURE 8-26 Patterns of natural selection: directional selection.

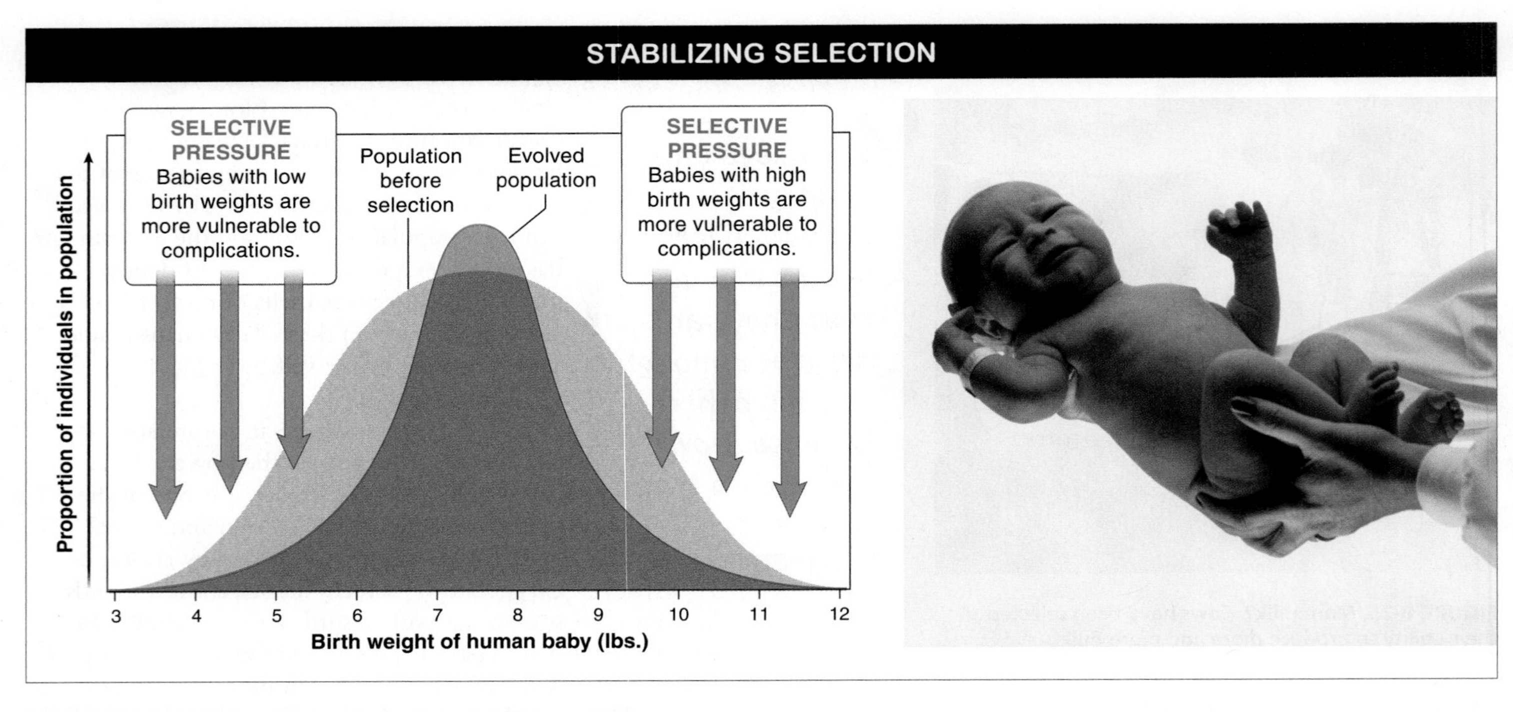

FIGURE 8-27 **Patterns of natural selection: stabilizing selection.**

Stabilizing Selection How much did you weigh at birth? Was it more than 10 pounds? More than 11? Was it close to the 22½ pounds that Christina Samane's child weighed when born in South Africa in 1982? Or was it 5 pounds or less?

Unlike directional selection, for which there is increased fitness at one extreme and reduced fitness at the other, **stabilizing selection** is said to occur when individuals with intermediate phenotypes are the most fit. The death rate

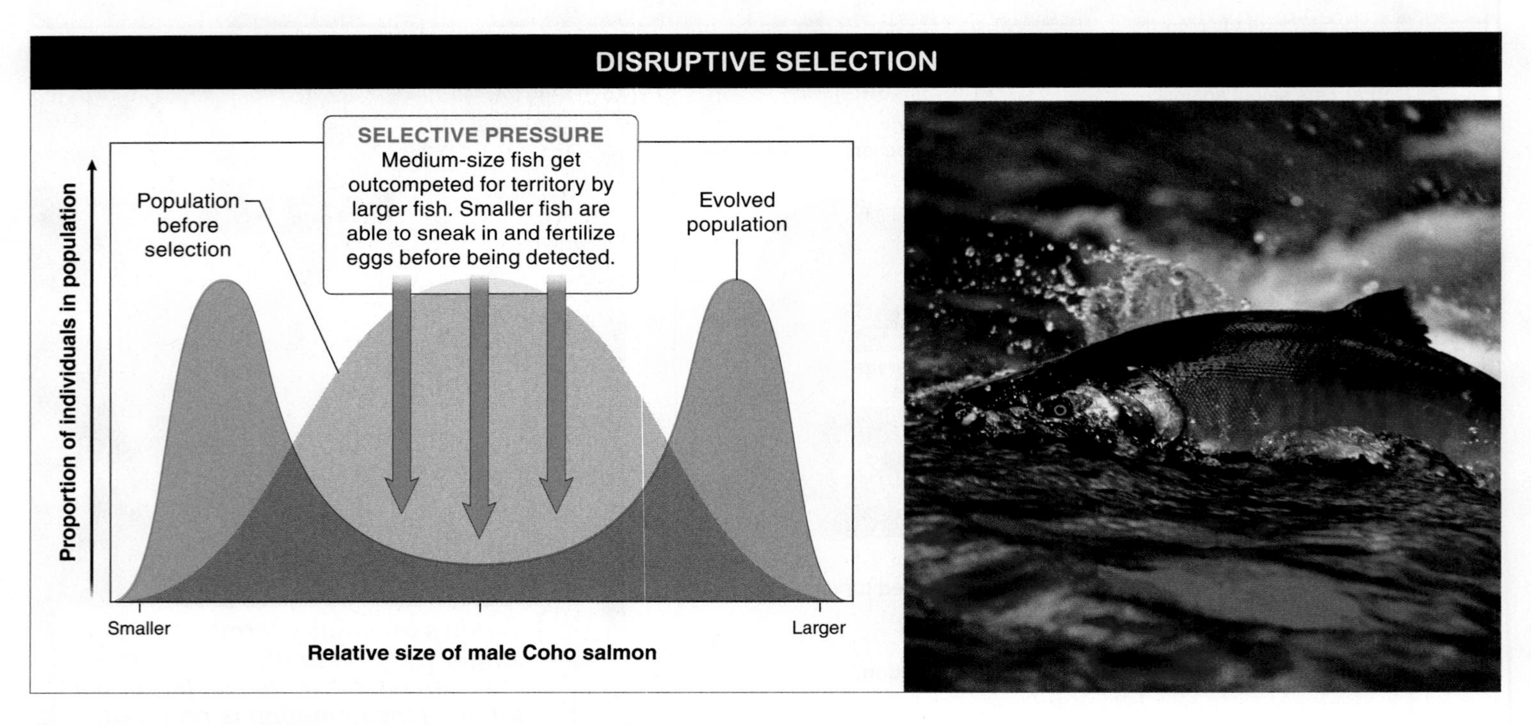

FIGURE 8-28 **Patterns of natural selection: disruptive selection.**

among babies, for example, is lowest between 7 and 8 pounds, rising among both lighter and heavier babies (**FIGURE 8-27**). This outcome has kept the average weight of a baby constant over many generations, but at the same time the variation in birth weight has decreased as stabilizing selection has reduced the market share of those genes associated with high and low birth weights.

Q How is medical technology undoing the work of natural selection in optimizing the number of babies with normal birth weights?

Modern technologies, including Caesarean deliveries and premature-birth wards, allow many babies to live who would not have survived without such technology. This has the unintended consequence of reducing the selection against any alleles causing those traits, reducing the rate of their removal from the population.

Disruptive Selection There is a third kind of natural selection, in which individuals with extreme phenotypes experience the highest fitness and those with intermediate phenotypes have the lowest. Although examples of this type of selection, called **disruptive selection,** are rare in nature, the results are not surprising. Among some species of fish—the Coho salmon, for instance—only the largest males acquire good territories. They generally enjoy relatively high reproductive success. While the intermediate-size fish regularly get run out of the good territories, some tiny males are able to sneak in and fertilize eggs before the territory owner detects their presence. Consequently, we see an increase in the frequency of small and large fish with a reduction in the frequency of medium-size fish (**FIGURE 8-28**).

TAKE-HOME MESSAGE 8•15

Acting on multigene traits for which populations show a large range of phenotypes, natural selection can change populations in several ways, including directional selection, in which the average value for the trait increases or decreases; stabilizing selection, in which the average value of a trait remains the same while extreme versions of the trait are selected against; and disruptive selection, in which individuals with extreme phenotypes have the highest fitness.

8•16

Natural selection can cause the evolution of complex traits and behaviors.

We have seen that natural selection can change allele frequencies and modify the frequency with which simple traits like fur color or turkey-breast size appear in a population. But what about complex traits, such as behaviors, that involve numerous physiological and neurological systems? For instance, can natural selection improve maze-running ability in rats?

The short answer is *yes*. Remember, evolution by natural selection is occurring, changing the allele frequencies for traits whenever (1) there is variation for the trait, (2) that variation is heritable, and (3) there is differential reproductive success based on that trait. These conditions can easily be satisfied for complex traits, including behaviors.

Q How can a wing evolve if 1% of a wing doesn't help an organism fly or glide at all?

In 1954, to address this question, William Thompson trained a group of rats to run through a maze for a food reward (**FIGURE 8-29**). There was a huge amount of variation in the rats' abilities: some rats learned much more quickly than others how to run the maze. Thompson then selectively bred the fast learners (called "maze-bright" rats) with each other and the slow learners (the "maze-dull" rats) with each other. Over several generations, he developed two separate populations: rats descended from a line of fast maze learners and rats descended from a line of slow maze learners. After only six generations, the maze-dull rats made twice as many errors as the maze-bright rats before mastering the maze, while the bright rats were adept at solving complex mazes that would give many humans some difficulty. Fifty years later, it still is unclear which actual genes are responsible for maze-running behavior, yet the selection experiment still demonstrates a strong genetic component to the behavior.

Natural selection can also produce complex traits in unexpected, roundabout ways. One vexing case involves the question of how natural selection could ever produce a

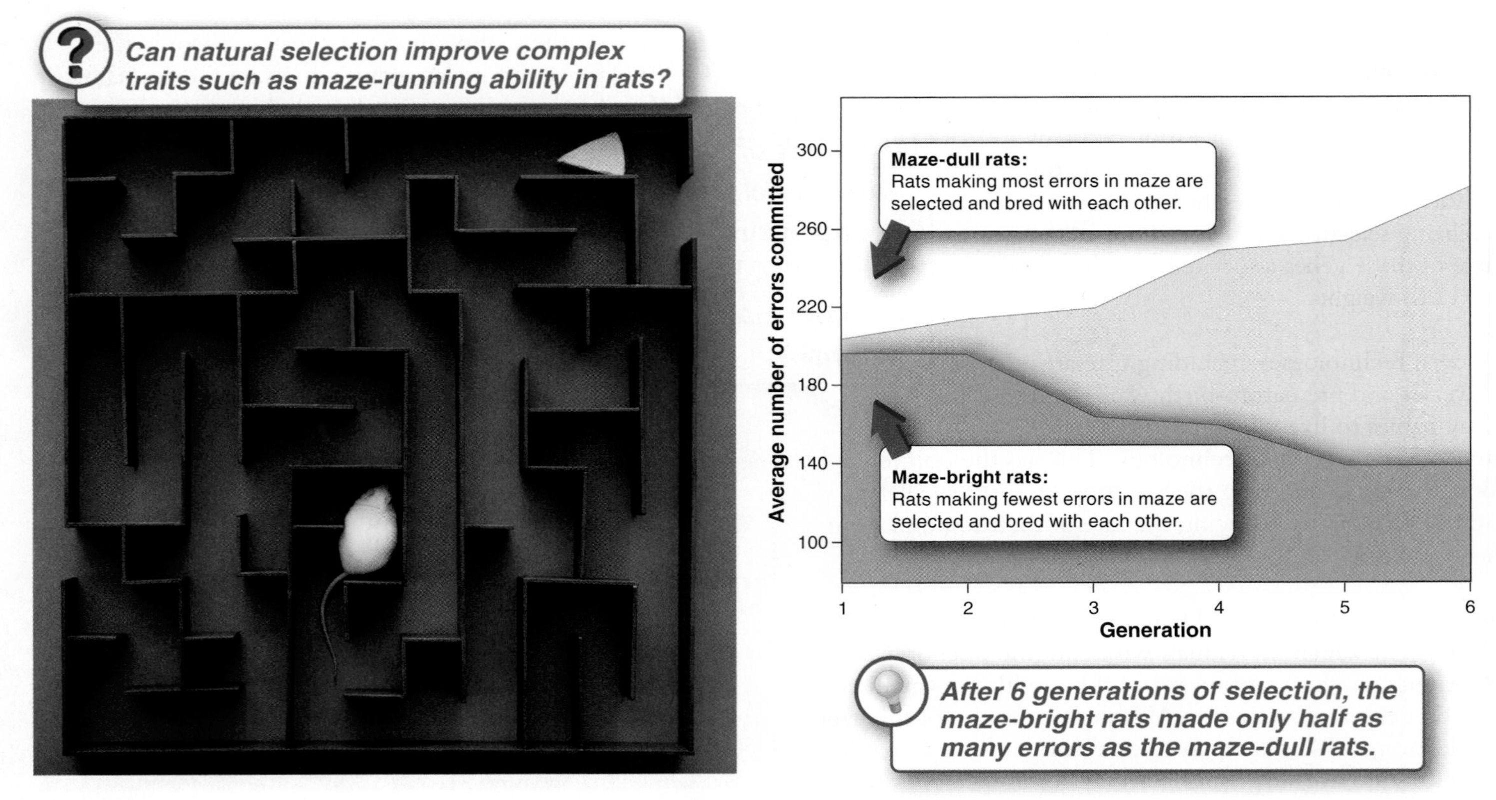

FIGURE 8-29 "Not too complex . . ." Natural selection can produce and alter complex traits, including behaviors.

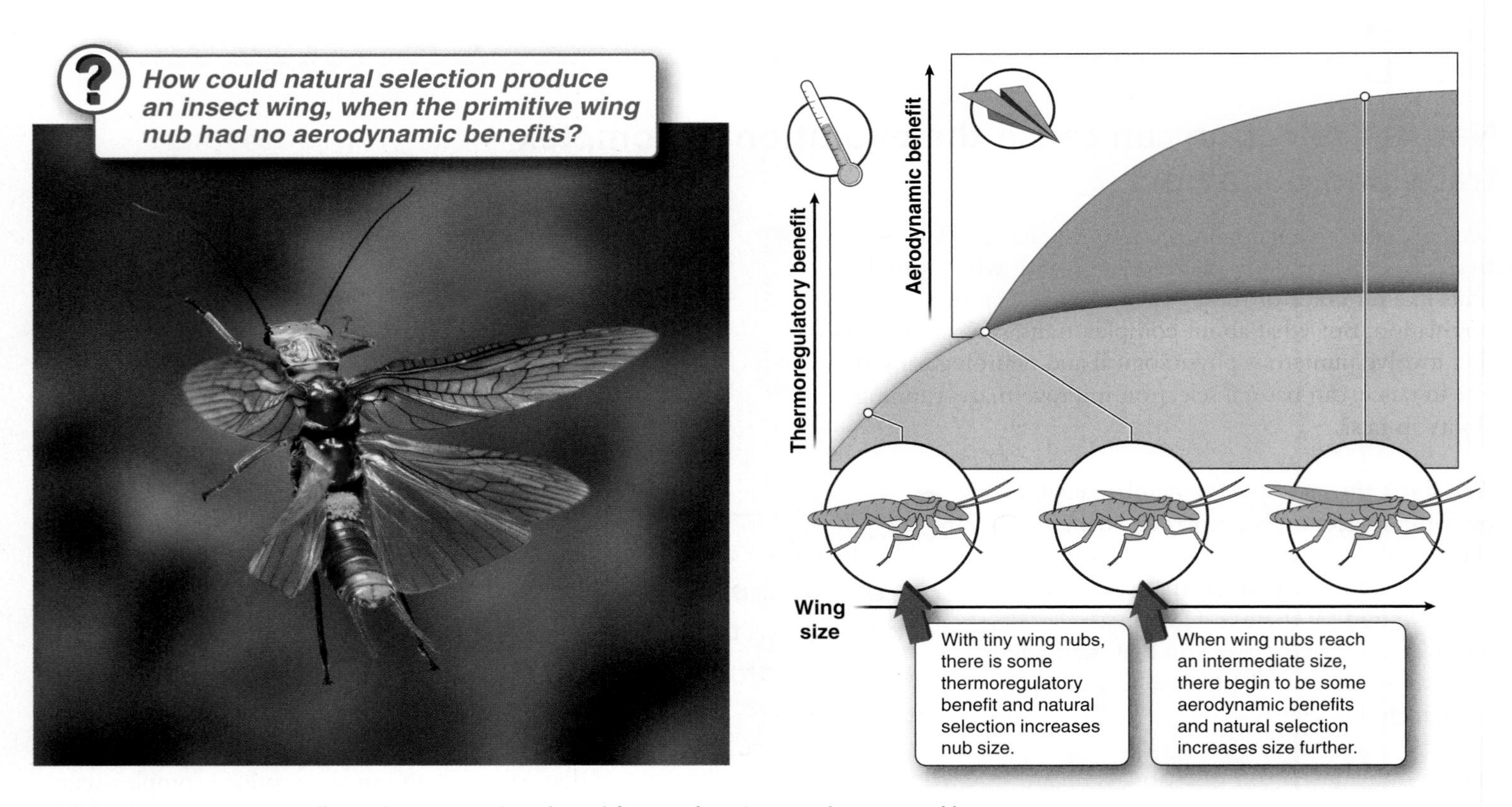

FIGURE 8-30 "Not necessarily a wing . . ." Traits selected for one function may be co-opted by natural selection for a new function.

complex organ such as a fly wing, when 1% or 2% of a wing—that is, an incomplete structure—doesn't help an insect to fly. In other words, while it is clear how natural selection can preserve and increase traits' frequencies in populations, how do these soon-to-be-useful traits increase during the early stages if they don't increase the organism fitness at all (**FIGURE 8-30**)?

> "All things must change
> To something new, to something strange.
> —Henry Wadsworth Longfellow"

The key to answering this question is that 1% of a wing doesn't actually need to function as a wing at all to increase an individual's fitness. Often, structures are enhanced or elaborated on by natural selection because they enhance fitness by serving some other purpose. In the case of little nubs or "almost-wings" on an insect, experiments using models of insects demonstrated that, as expected, the nubs confer no benefit at all when it comes to flying. (They don't even help flies keep their orientation during a "controlled fall.") The incipient wings do help the insects address a completely different problem, though. They allow much more efficient temperature control, allowing an insect to gain heat from the environment when the insect is cold and to dissipate heat when the insect is hot. Experiments on heat-control efficiency, in fact, show that as small nubs become more and more pronounced, they are more and more effective, probably conferring increased fitness on the individual in which they occur—but only up to a point.

Eventually, the thermoregulatory benefit stops increasing, even if the nub length continues to increase. But it is right around this point that the proto-wing starts to confer some aerodynamic benefits (see Figure 8-30). Consequently, natural selection may continue to increase the length of this "almost-wing," but now the fitness increase is due to a wholly different effect. Such functional shifts explain the evolution of numerous complex structures that we see today and may be common in the evolutionary process.

Wings may have originally evolved because they enabled an animal to absorb or release heat, thereby regulating body temperature more effectively.

TAKE-HOME MESSAGE 8•16

Natural selection can change allele frequencies for genes involving complex physiological processes and behaviors. Sometimes a trait that has been selected for one function is later modified to serve a completely different function.

⑤ The evidence for the occurrence of evolution is overwhelming.

The well-camouflaged veiled chameleon sits and waits for insect prey to come near.

8•17

The fossil record documents the process of natural selection.

> It is indeed remarkable that this theory [evolution] has been progressively accepted by researchers, following a series of discoveries in various fields of knowledge. The convergence, neither sought nor fabricated, of the results of work that was conducted independently is in itself a significant argument in favor of this theory.
>
> —Pope John Paul II, 1996

In the nearly 150 years since Darwin first published *The Origin of Species,* thousands of experiments involving his theory of evolution by natural selection have been conducted, both in the laboratory and in natural habitats. A wide range of modern methodologies has also been developed, all contributing to a much deeper understanding of the process of evolution. This ongoing accumulation of evidence overwhelmingly supports the basic premise that Darwin put forward, while filling in many of the gaps that frustrated him.

Here we review the five primary lines of evidence demonstrating the occurrence of evolution:

1. **The fossil record**—physical evidence of organisms that lived in the past
2. **Biogeography**—patterns in the geographic distribution of living organisms
3. **Comparative anatomy and embryology**—growth, development, and body structures of major groups of organisms
4. **Molecular biology**—the examination of life at the level of individual molecules
5. **Laboratory and field experiments**—implementation of the scientific method to observe and study evolutionary mechanisms

FIGURE 8-31 Evidence for evolution: the fossil record. Fossils can be used to reconstruct the appearance of organisms that lived long ago.

The first of the five lines of evidence is the fossil record. Although it has been central to much documentation of the occurrence of evolution, it is a very incomplete record. After all, the soft parts of an organism almost always decay rapidly and completely after death. And there are only a few unique environments (such as tree resin, tar pits, and the bottom of deep lakes) in which the forces of decomposition are reduced so that an organism's hard parts, including bones, teeth, and shells, can be preserved for thousands or even millions of years. These remains, called **fossils,** can be used to reconstruct what organisms must have looked like long ago. Such reconstructions often provide a clear record of evolutionary change (**FIGURE 8-31**).

The use of **radiometric dating** helps in further painting a picture of organisms' evolutionary history by telling us the age of the rock in which a fossil has been found (**FIGURE 8-32**). In Darwin's time, it was assumed that the deeper down in the earth a fossil was found, the older it was. Radiometric dating goes a step further, making it possible to determine not just the *relative* age of fossils, but also their *absolute* age. This is accomplished by evaluating the amounts of certain radioactive isotopes present in fossils. Radioactive isotopes in a rock begin breaking down into more stable compounds as soon as the rock is formed, and they do so at a constant rate. Nothing can alter this. By measuring the relative amounts of the radioactive isotope and the leftover decay product in the rock where a fossil is found, the age of the rock and thus of the fossil can be calculated.

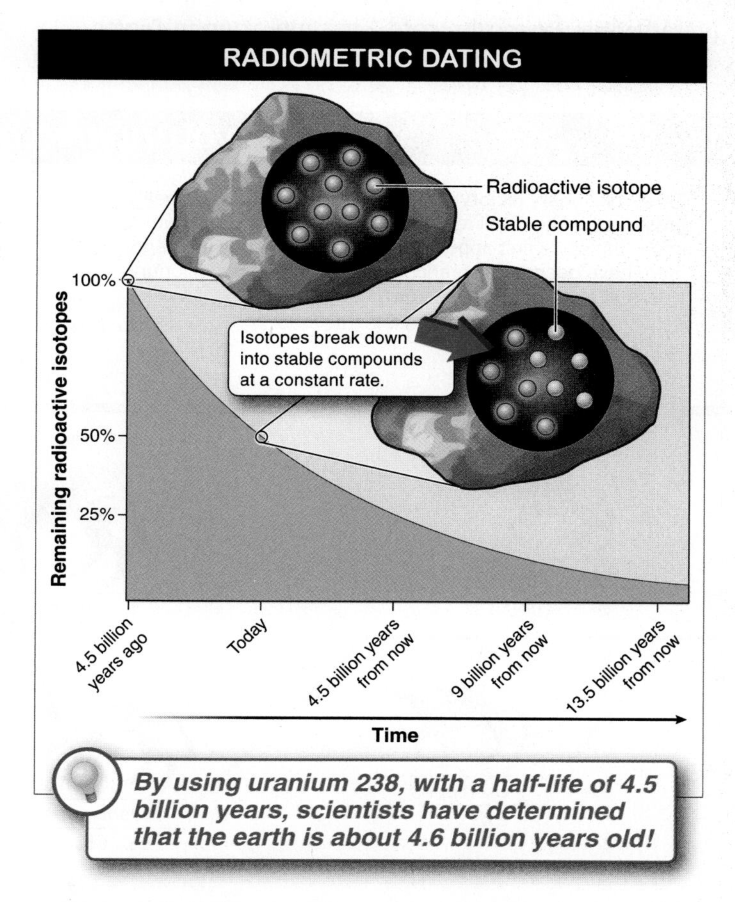

FIGURE 8-32 How old is that fossil? Radiometric dating of uranium 238 helps to determine the age of rocks and the fossils in them.

Radiometric dating confirms that the earth is very old. Rocks more than 3.8 billion years old have been found on all of the earth's continents, with the oldest so far found in northwestern Canada. By using the radioactive isotope uranium 238, with a half-life of 4.5 billion years, researchers have determined that the earth is about 4.6 billion years old and that the earliest organisms appeared at least 3.5 billion years ago. Radiometric dating also makes it possible to put the fossil record in order. By dating all the fossils discovered in one locale, paleontologists can learn if and how the organisms were related to each other and how groups of organisms changed over time.

Paleontologists must deal with the fact, however, that fossil formation is an exceedingly unlikely event, and when it does occur it represents only those organisms that happened to live in that particular area and that also had physical structures that could leave fossils. For this reason, the fossil record is annoyingly incomplete. Entire groups of organisms have left no fossil record at all and some others have numerous gaps. Still, fossils have been found linking all of the major groups of vertebrates.

The evolutionary history of horses is among the most well preserved in the fossil record. First appearing in North America about 55 million years ago, horses then radiated around the world, with more recent fossils appearing in Eurasia and Africa. These fossils exhibit distinct adaptations to those environments. Later, about 1.5 million years ago, much of the horse diversity—including all North American horse species—disappeared, leaving only a single remaining genus, or group of species, called *Equus.* Because there is now only one horse genus on earth, it is tempting to imagine a simple linear path from modern horses straight back through their 55-million-year evolutionary history. But that's just not how evolution works. In reality, there have been numerous branches of horses that have split off over evolutionary time, flourished for millions of years, and only recently gone extinct. What we see living today is only a single branch of a greatly branched evolutionary tree (**FIGURE 8-33**).

The fossil record provides another valuable piece of evidence for the process of natural selection in the form of "missing links." These are fossils that demonstrate a link between groups of species believed to have shared a common ancestor. One such "missing link" is *Tiktaalik* (**FIGURE 8-34**). First found in northern Canada and estimated to be 375 million years old, *Tiktaalik* fossils seem to represent a transitional phase between fish and land animals. These creatures, like fish, had gills, scales,

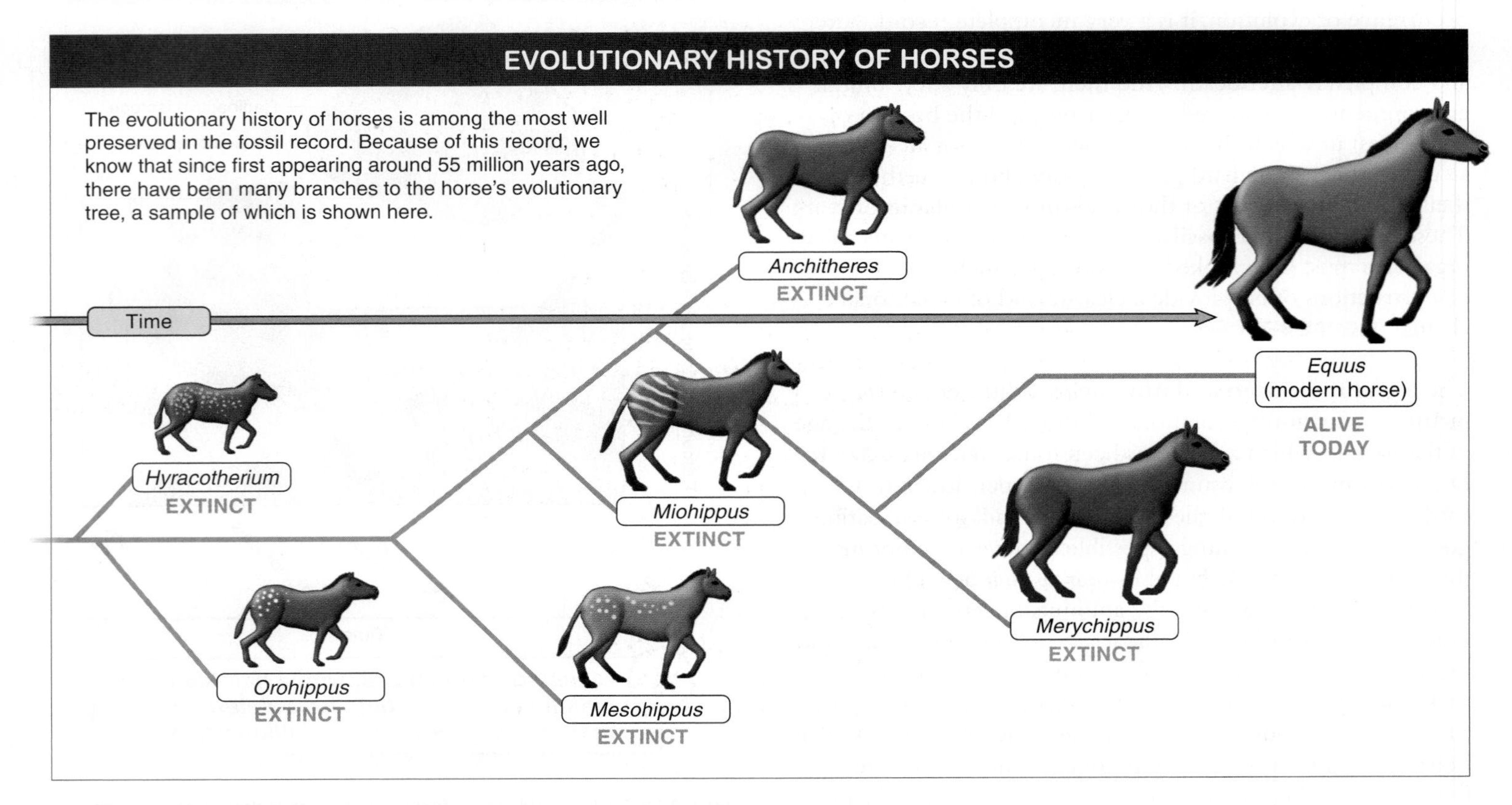

FIGURE 8-33 An evolutionary family tree. The branching evolutionary tree of the horse.

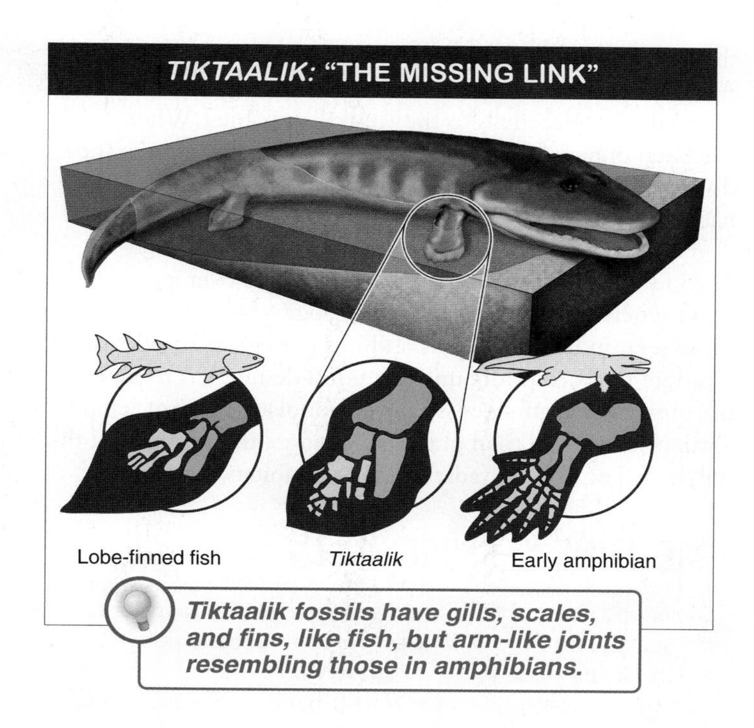

FIGURE 8-34 A missing link? *Tiktaalik* seems to be a transitional phase between fish and land animals.

and fins, but they also had arm-like joints in their fins and could drag their bodies across land in much the same way as a marine mammal such as a seal.

TAKE-HOME MESSAGE 8•17

Radiometric dating confirms that the earth is very old and makes it possible to determine the age of fossils. Analysis of fossil remains enables biologists to reconstruct what organisms looked like long ago, learn how organisms were related to each other, and understand how groups of organisms evolved over time.

8•18

Geographic patterns of species distributions reflect their evolutionary histories.

The study of the distribution patterns of living organisms around the world is called **biogeography.** This is the second line of evidence that helps us to see that evolution occurs and to better understand the process. The patterns of biogeography that Darwin and many subsequent researchers have noticed provide strong evidence that evolutionary forces are responsible for these patterns. Species often more closely resemble other species that live less than a hundred miles away but in radically different habitats than they do species that live thousands of miles away in nearly identical habitats. On Hawaii, for example, nearly every bird is some sort of modified honeycreeper (a small, finch-like bird). There are seed-eating honeycreepers, woodpecker honeycreepers, and curved-bill nectar-feeding honeycreepers. There are even parrot honeycreepers (**FIGURE 8-35**).

FIGURE 8-35 Evidence for evolution: biogeography. The Hawaiian honeycreepers resemble a common ancestor from mainland North America, but all have unique features.

When it comes to species distributions, history matters. Species were not designed from scratch to fill a particular niche. Rather, whatever arrived first—usually a nearby species—took up numerous different lifestyles in numerous different habitats, and the populations ultimately adapted to and evolved in each environment. In Hawaii, it seems that a finch-like descendant of the honeycreepers arrived 4–5 million years ago and rapidly evolved into a large number of diverse species. The same process has occurred and continues to occur in all locales, not just on islands.

Large isolated habitats also have interesting biogeographic patterns. Australia and Madagascar are filled with unique organisms that are clearly not closely related to organisms elsewhere. In Australia, for example, marsupial species, rather than placental mammals, fill all of the usual roles. There are marsupial "wolves," marsupial "mice," marsupial "squirrels," and marsupial "anteaters" (**FIGURE 8-36**). They physically resemble their placental counterparts for most traits, but molecular analysis shows that they are actually more closely related to each other, sharing a common marsupial ancestor. Their relatedness to each other is also revealed by similarities in their reproduction: females give birth to offspring at a relatively early state of development and the offspring finish their development in a pouch. The presence of marsupials in Australia does not simply mean that marsupials are better adapted than placentals to the Australian habitat. When placental organisms have been transplanted to Australia they do just fine, often thriving to the point of endangering the native species.

A good designer would use the best designs over and over again wherever they might fit. Biogeographic patterns such as those seen in honeycreepers and in the marsupials of Australia illustrate that evolution, unlike a good designer, is more of a tinkerer. Evolution takes whatever populations are at hand in a particular location then gradually changes their traits, and the species become better adapted to the habitats they occupy.

TAKE-HOME MESSAGE 8•18

Observing geographic patterns of species distributions, particularly noting similarities and differences among species living close together but in very different habitats and among species living in similar habitats but located far from one another, helps us to understand the evolutionary histories of populations.

FIGURE 8-36 Evidence for evolution: biogeography. Many Australian marsupials resemble placental counterparts, though they are not closely related.

8•19

Comparative anatomy and embryology reveal common evolutionary origins.

If you observe any vertebrate embryo, you will see that it passes through a stage in which it has little gill pouches on the sides of the neck. It will also pass through a stage in which it possesses a long bony tail. This is true whether it is a human embryo or that of a turtle or a chicken or a shark. These gill pouches disappear before birth in all but the fishes. Similarly, we don't actually find humans with tails. Why do they exist during an embryo's development? Such common embryological stages indicate that the organisms share a common ancestor, from which all have been modified (**FIGURE 8-37**). Study of these developmental stages and the adult body forms of organisms provides our third line of evidence, helping us to see that evolution occurs and to better understand the process.

Among adults, several features of anatomy reveal the ghost of evolution in action. We find, for example, that many related organisms show unusual similarities that can be explained only through evolutionary relatedness. The forelimbs of mammals are used for a variety of functions in bats, porpoises, horses, and humans (**FIGURE 8-38**). If each had been designed specifically for the uses necessary to that species, we would expect dramatically different designs. And yet in each of these species, we see the same bones—modified extensively—that betray the fact that they share a common ancestor. These features are called **homologous structures.**

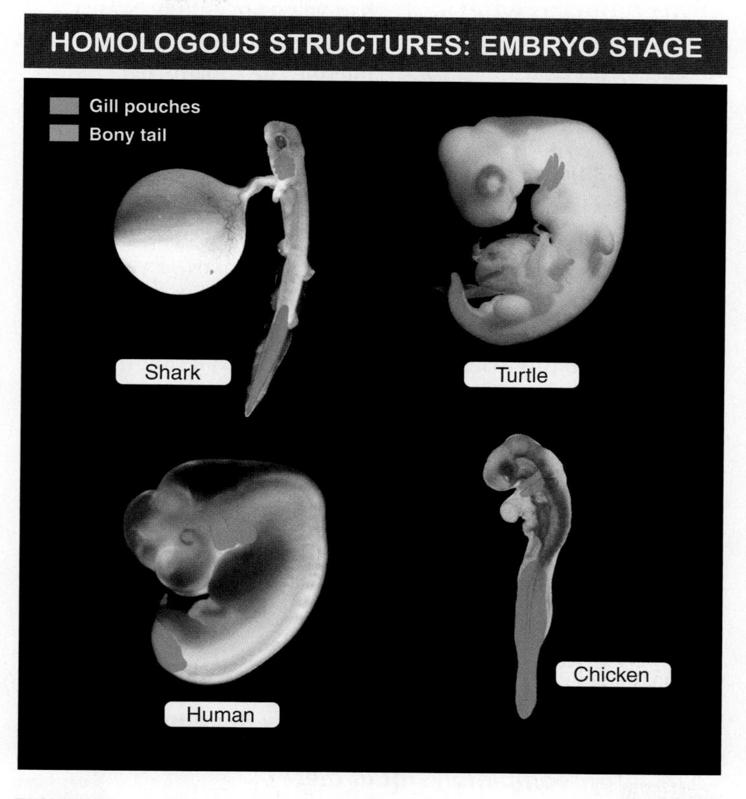

FIGURE 8-37 Evidence for evolution: embryology. Structures derived from common ancestry can be seen in embryos.

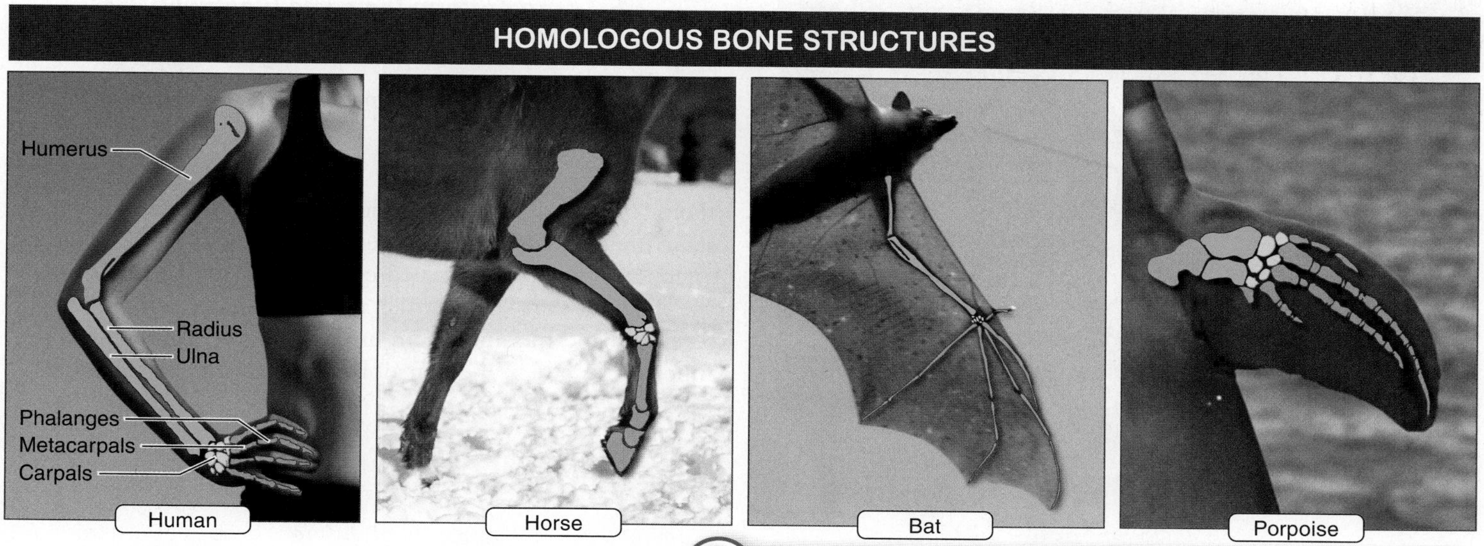

HOMOLOGOUS BONE STRUCTURES

FIGURE 8-38 Evidence for evolution: comparative anatomy. Homologous bone structures among some mammals.

The similarities in the bone structure of the forelimbs of mammals demonstrate common ancestry.

> **Q** The human appendix appears to serve no function. Why are we all born with one?

At the extreme, homologous structures sometimes come to have little or no function at all. Such evolutionary leftovers, called **vestigial structures,** exist because they had value ancestrally. Some vestigial structures in mammals include the molars that continue to grow in vampire bats, even though these bats consume a completely liquid diet (**FIGURE 8-39**); eye sockets (with no eyes) in some populations of cave-dwelling fish; and pelvic bones in whales that are attached to nothing (but serve as an important attachment point for leg bones in nearly all other mammals). Even humans may have a vestigial organ—the appendix. It is greatly enlarged in our relatives the great apes, in whom it hosts cellulose-digesting bacteria that aid in breaking down the plants in their diet, but in humans it appears to serve no purpose. (Recently, however, it has been suggested that the appendix may play a role in the immune system.)

Not all organisms with adaptations that appear similar actually share ancestors. We see flying mammals (bats) and flying insects (locusts) (**FIGURE 8-40**). Similarly, dolphins and penguins live in similar habitats and have flippers that help them to swim. In both examples, however, all of the analogous structures developed from different original structures. Natural selection—in a process called **convergent evolution**—uses the different starting materials available (such as a flipper or a forelimb) and modifies them until they serve similar purposes, much as we saw in the marsupial and placental mammals in Figure 8-36.

FIGURE 8-39 Evidence for evolution: vestigial structures.

FIGURE 8-40 Evidence for evolution: convergent evolution and analogous structures.

TAKE-HOME MESSAGE 8•19

Similarities in the anatomy of different groups of organisms and in their physical appearance as they proceed through their development can reveal common evolutionary origins.

8•20

Molecular biology reveals that common genetic sequences link all life forms.

The development of new technologies for deciphering and comparing the genetic code provides our fourth line of evidence that evolution occurs. First and foremost is the fact that all living organisms share the same genetic code. From microscopic bacteria to flowering plants to insects and primates, the molecular instructions for building organisms and transmitting hereditary information are the same: four simple bases, arranged in an almost unlimited variety of sequences. This discovery strongly suggests that all living organisms are related.

When we examine the similarity of DNA among related individuals within a species, we find that they share a greater proportion of their DNA than do unrelated individuals. This is not unexpected; you and your siblings got all of your DNA from the same two parents, while you and your cousins each got half of your DNA from the same two grandparents. The more distantly you and another individual are related, the more your DNA differs.

We can measure the DNA similarity between two species by comparing their DNA sequences for individual genes. For example, let's look at the gene that codes for the amino acids used to build the hemoglobin molecule. In vertebrate animals, hemoglobin is found inside red blood cells, where its function is to carry oxygen throughout the body. It is made of two chains of amino acids, the alpha chain and the beta chain. In humans, the beta chain has 146 amino acids. In rhesus monkeys, this beta chain is nearly identical: of the 146 amino acids, 138 are the same as those found in human hemoglobin, and only 8 are different. In dogs the sequence is still similar, but a bit less so, with 32 different amino acids. When we look at non-mammals such as birds, we see about 45 amino acid differences, and between humans and lamprey eels (still vertebrates, but it has been more than 500 million years since we shared a common ancestor), there are 125 amino acid differences.

The differences in the amino acid structure of the beta hemoglobin chain (and remember that this structure is governed by an allele or alleles of a particular gene) seem to indicate that humans have more recently shared a common ancestor with rhesus monkeys than with dogs. And that we have more recently shared an ancestor with dogs than with birds or lampreys. These findings are just as we would expect, based on estimates of evolutionary relatedness made from comparative anatomy and embryology as well as those based on the fossil record. It is as if there is a molecular clock that is ticking. The longer two species have been evolving on their own, the greater the number of changes in amino acid sequences—or "ticks of the clock" (**FIGURE 8-41**).

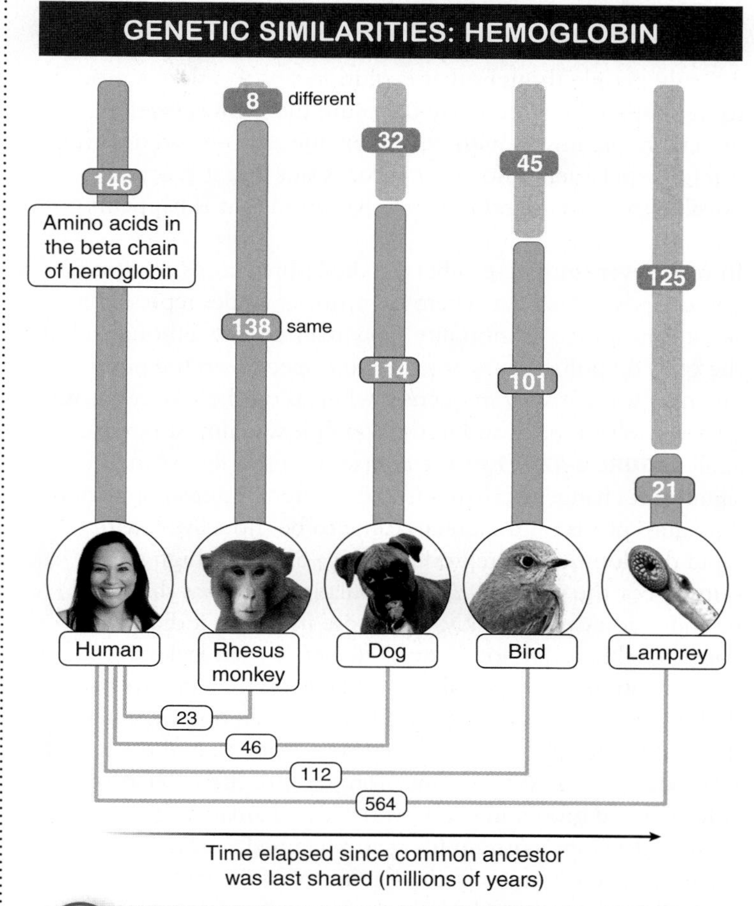

FIGURE 8-41 An evolutionary clock? Genetic similarities (and differences) demonstrate species relatedness.

TAKE-HOME MESSAGE 8•20

All living organisms share the same genetic code. The degree of similarity in the DNA of different species can reveal how closely related they are and the amount of time that has passed since they last shared a common ancestor.

8•21

Laboratory and field experiments enable us to watch evolution in progress.

A fifth line of evidence for the occurrence of evolution comes from multigeneration experiments and observations. Until recently, people thought that evolution was too slow a process to be observed in action. By choosing the right species—preferably organisms with very short life spans—and designing careful experiments, however, it turns out that it is actually possible to observe and measure evolution as it is happening.

In one clever study, researchers studied populations of grass on golf courses—a habitat where lawnmower blades represent a significant source of mortality. They realized that although all of the grass on golf courses was the same species, on the putting greens it was cut very frequently, whereas on the fairways it was cut only occasionally, and in the rough it was almost never cut at all (**FIGURE 8-42**). Over the course of only a few years, significant changes occurred in these different grass populations. The grass plants on the greens came to be short lived, with rapid development to reproductive age and very high seed output. For the plants in these populations, life was short and reproduction came quickly. (For those in which it did not come quickly, it did not come at all; hence their lack of representation in the population.) Plants on the fairways had slightly slower development and reduced seed output, while those in the rough had the slowest development and the lowest seed output of all. When plants from each of the habitats were collected and grown in greenhouses, the dramatic differences in growth, development, and life span remained, confirming that there had been changes in the frequency of the various alleles controlling the traits of life span and reproductive output in the three populations; in other words, evolution had occurred.

A more disturbing line of evidence for the occurrence of evolution in nature comes from the evolution of antibiotic-resistant strains of bacteria that cause illness in humans. For example, in the 1940s when penicillin was first used as a treatment for bacterial infections, it was uniformly effective in killing *Staphylococcus aureus.* Today, more than 90% of isolated *S. aureus* strains are resistant to penicillin (**FIGURE 8-43**). Because penicillin has become such a pervasive toxin in the environment of *Staphylococcus,* natural selection has led to an increase in the frequency of the alleles that make these strains resistant. As a consequence, humans are increasingly at risk for becoming infected with *Staphylococcus* and getting diseases such as pneumonia and meningitis. In the 1960s, the antibiotics methicillin and oxacillin were developed. Like penicillin, they initially had nearly complete effectiveness against *Staphylococcus.* Today, though, nearly a third of staph infections are resistant to these antibiotics. The meaning of such unintentional natural selection "experiments" is clear and consistent: evolution is occurring all around us.

And finally, let's return to the spectacularly starvation-resistant flies introduced at the beginning of this chapter. In them we saw an unambiguous demonstration that natural selection can

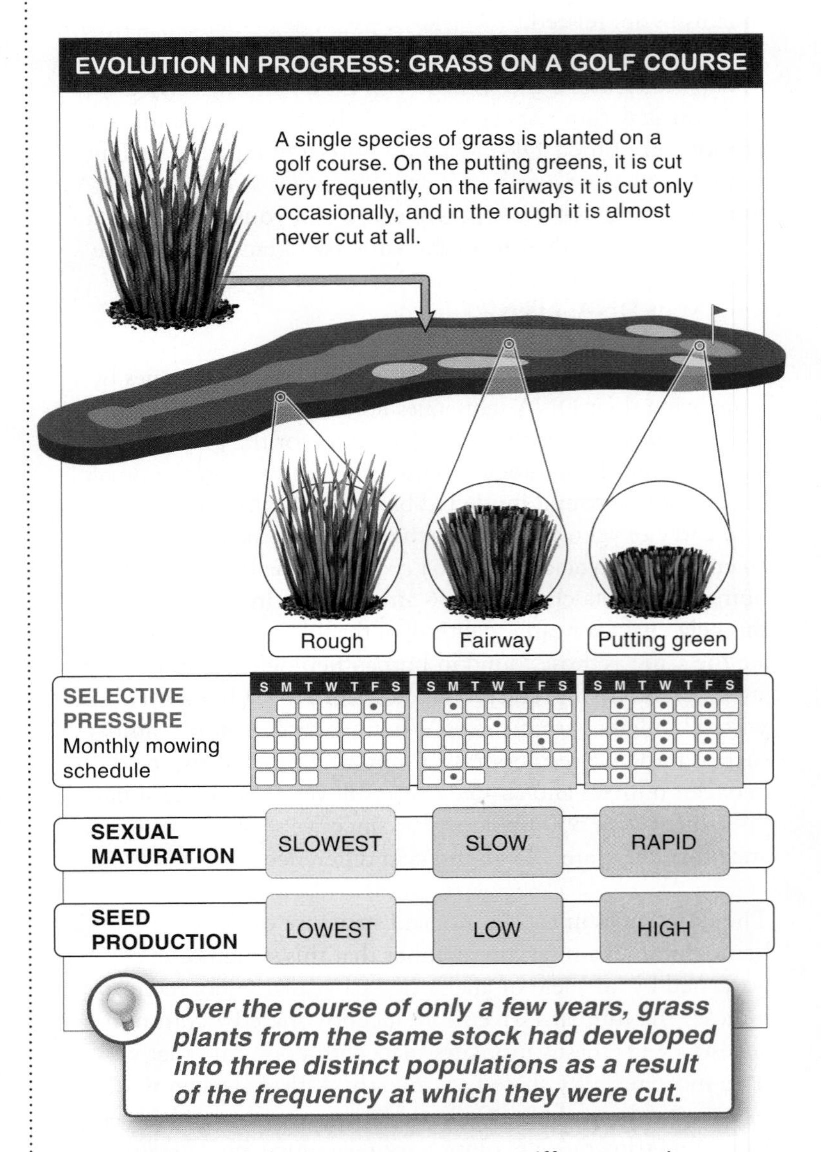

FIGURE 8-42 Evolution in progress: grasses. Different mowing patterns can cause evolution in the grass on a golf course.

EVOLUTION OF ANTIBIOTIC RESISTANCE

When first used as medicine in the 1940s, penicillin was uniformly effective in killing the bacterium *Staphylococcus aureus*. Today, natural selection has led to an increase in antibiotic-resistant alleles, and humans are increasingly at risk from untreatable *Staphylococcus* infections.

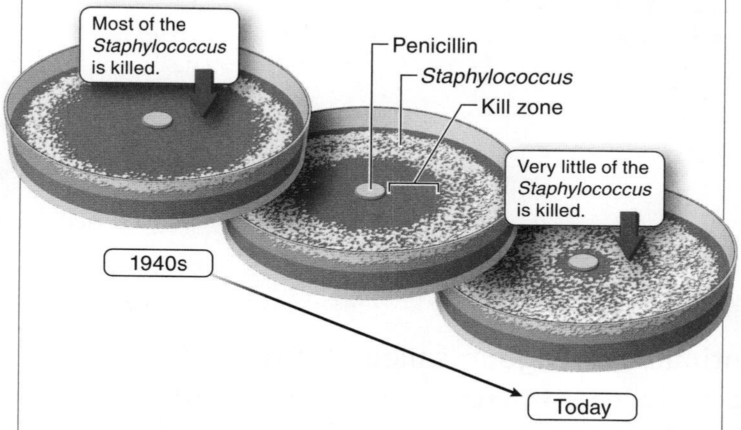

FIGURE 8-43 Evolution in progress: disease-causing bacteria. Antibiotic resistance has evolved in *Staphylococcus.*

occur, that it has the potential to produce dramatic changes in physical features and physiological processes, and that it can bring about these changes very quickly. And, perhaps most important, we saw that replicating the same evolutionary process over and over repeatedly produced the same predictable results. In the laboratory, the farm, the doctor's office, the bathroom sink, the deserts, streams, and forests, evolution is occurring.

Reflecting on both the process and products of evolution and natural selection, in the final paragraph of *The Origin of Species* Darwin eloquently wrote: "There is grandeur in this view of life . . . and that . . . from so simple a beginning endless forms most beautiful and most wonderful have been, and are being, evolved."

TAKE-HOME MESSAGE 8•21

Multiply replicated, controlled laboratory selection experiments and long-term field studies of natural populations—including observations on antibiotic-resistant strains of disease-causing bacteria—enable us to watch and measure evolution as it occurs.

Knowledge You Can Use

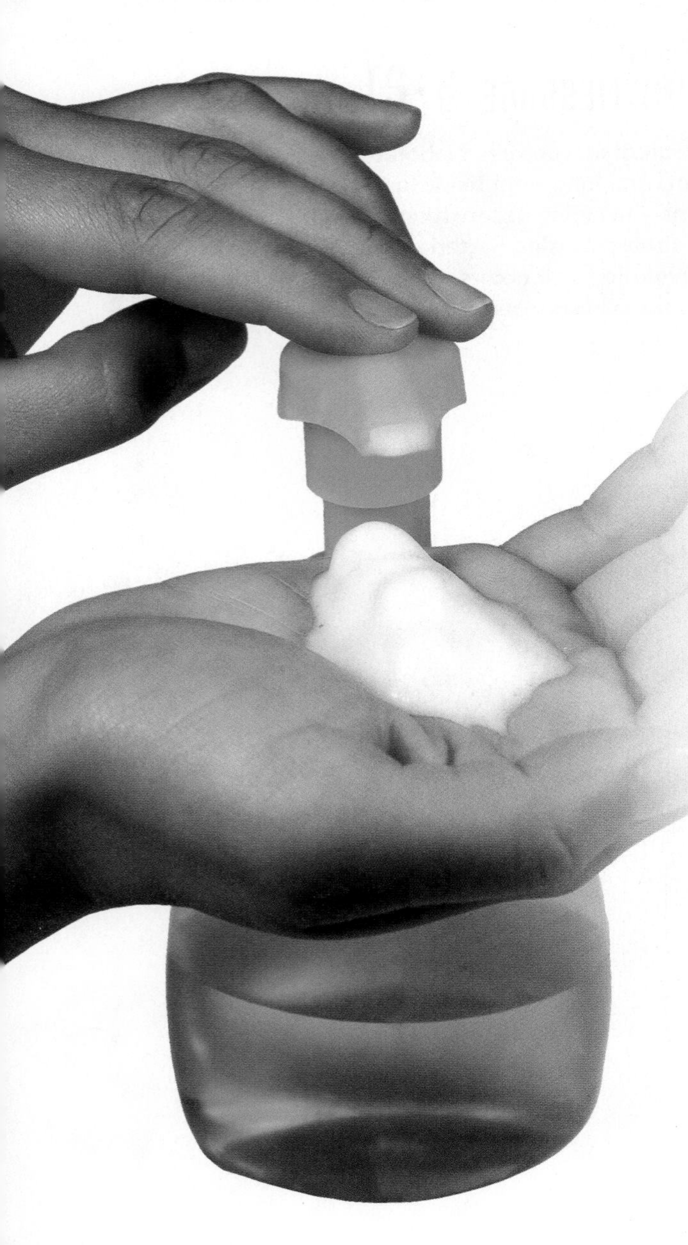

Did you know? Antibacterial soaps may be dangerous. And it may be your fault.

Fact: The antimicrobial agent, called triclosan, in most antibacterial soaps takes several minutes to kill bacteria.

Q: How long do you wash your hands? On average, **people spend less than a minute washing their hands.**

Q: If you're not killing all of the bacteria, which ones *are* you killing? Some bacteria are easily killed by the antimicrobial chemical in soaps. Others take longer. In the first minute you're killing only the weak bacteria (and not all of the 800,000 to 1,000,000 bacteria residing on your hands).

Q: After one minute, what is left? After a minute of hand-washing, only the strong bacteria—those that happen to carry triclosan-resistant genes—remain and continue reproducing. (And the next time you wash your hands, you may pick up those super-strength drug-resistant bacteria from the sink area.) **You're causing evolution to happen.** (And you may not want to.)

Q: What do you make of this? What can you do? The two simplest alternatives are to wash your hands for much longer or to stop using soaps with antibacterial agents in them.

BIG IDEAS IN EVOLUTION AND NATURAL SELECTION

1 Evolution is an ongoing process.

The characteristics of individuals in a population can change over time. We can observe such change in nature and can even cause such change to occur.

2 Darwin journeyed to a new idea.

Following years of observations on worldwide patterns of the living organisms and fossils of plant and animal species, Charles Darwin developed a theory of evolution by natural selection that explained how populations of species can change over time.

3 Four mechanisms can give rise to evolution.

Evolution—a change in the allele frequencies in a population—can occur via genetic drift, migration, mutation, or natural selection.

4 Through natural selection, populations of organisms can become adapted to their environment.

When there is variation for a trait, and the variation is heritable, and there is differential reproductive success based on that trait, evolution by natural selection is occurring. Fitness is an individual's relative reproductive output. The alleles that yield traits that confer the highest fitness on the individuals carrying them will increase their market share in a population over time, but this process does not necessarily lead to "perfect" organisms in an environment. Artificial selection, in which humans determine which individuals will have the highest reproductive success, is a type of natural selection. Directional, stabilizing, and disruptive selection can alter the average value of a trait and/or the variation for a trait in a population. Natural selection also can produce and modify complex traits, including behaviors. This modification may occur as traits selected for one purpose later confer increased fitness for a different purpose.

5 The evidence for the occurrence of evolution is overwhelming.

Many overwhelming lines of evidence document the occurrence of evolution and point to the central and unifying role of evolution by natural selection in helping us to better understand all other ideas and facts in biology.

KEY TERMS

adaptation, p. 306
biogeography, p. 317
bottleneck effect, p. 298
convergent evolution, p. 320
differential reproductive success, p. 302
directional selection, p. 308
disruptive selection, p. 311
evolution, p. 285
fitness, p. 304
fixation, p. 298
fossil, p. 315
founder effect, p. 298
gene flow, p. 300
genetic drift, p. 297
heritability, p. 302
homologous structure, p. 319
inheritance, p. 302
migration, p. 300
mutagen, p. 295
mutation, p. 295
natural selection, p. 285
population, p. 284
radiometric dating, p. 315
sexual selection, p. 302
stabilizing selection, p. 310
trait, p. 290
vestigial structure, p. 320

Check Your Knowledge

1. The average time to death from starvation in a fruit fly is about 20 hours. Selecting for increased starvation resistance in fruit flies:

a) has no effect because starvation resistance is a not a trait that influences fruit fly fitness.
b) has little effect because ongoing mutation continuously reduces starvation resistance, counteracting any benefits from selection.
c) cannot increase their survival time because there is no genetic variation for this trait.
d) has no effect because starvation resistance is too complex a trait, dependent on the effects of too many genes.
e) can produce populations in which the average time to death from starvation is 160 hours.

2. Georges Cuvier's discovery of fossils of Irish elk and giant ground sloths:

a) demonstrated that extinction must occur.
b) was possible because of Buffon's determination that the earth was more than 6,000 years old.
c) was possible only following Darwin's publication of *The Origin of Species.*
d) was made in deep ocean trenches.
e) suggested that species were immutable.

3. While on the voyage of the HMS *Beagle,* Darwin:

a) wrote *The Origin of Species.*
b) nurtured his love of studying nature, exploring plant and animal diversity, and collecting fossils.
c) discovered a love for sea travel.
d) studied Malthus's book *Essay on the Principle of Population.*
e) corresponded extensively with Alfred Russel Wallace about their ideas on evolution.

4. Which one of the following statements best describes the difference between artificial and natural selection?

a) Natural selection is limited to physical traits and artificial selection is not.
b) Artificial selection has produced many of the most delicious food items for humans; natural selection has not.
c) Natural selection acts without the input of humans; artificial selection requires human input.
d) Charles Darwin understood natural selection but was unaware of artificial selection in his time.
e) Natural selection works on all species; artificial selection works only on laboratory-raised species.

5. Which of the following statements about Charles Darwin is incorrect?

a) He spent five years traveling the world observing living organisms and collecting fossils.
b) He was under constant pressure from his father to make something of himself.
c) He dropped out of medical school.
d) He was enthusiastic about unleashing his theory of natural selection on the world as soon as he thought of it.
e) He and Alfred Russel Wallace independently came up with the theory of evolution by natural selection.

6. Evolution occurs:

a) only when the environment is changing.
b) only through natural selection.
c) almost entirely because of directional selection.
d) only via natural selection, genetic drift, migration, or mutation.
e) by altering physical traits but not behavioral traits.

7. Which of the following statements about mutations is incorrect?

a) Mutations are almost always random with respect to the needs of the organism.
b) A mutation is any change in an organism's DNA.
c) Most mutations are harmful or neutral to the organism in which they occur.
d) The origin of genetic variation is mutation.
e) All of these statements are correct.

8. The chief concern among conservation biologists trying to protect small populations is:

a) to preserve genetic diversity.
b) that mutation rates are much higher in small populations.
c) that natural selection can only operate in large populations.
d) to maximize rates of allele fixation.
e) to reduce the duration of genetic bottlenecks.

9. When a group of individuals colonizes a new habitat, the event is likely to be an evolutionary event because:

a) members of a small population have reduced rates of mating.
b) gene flow increases.
c) mutations are more common in novel environments.
d) new environments tend to be inhospitable, reducing survival there.
e) small founding populations are rarely genetically representative of the initial population.

10. To establish that evolution by natural selection is operating in a population, one must demonstrate variability for a trait, heritability of that trait, differential reproductive success based on that trait, and:

a) increased complexity of the organism.
b) random mating.
c) progress.
d) continuous change in the environment.
e) nothing else.

11. "Survival of the fittest" may be a misleading phrase to describe the process of evolution by natural selection because:

a) it is impossible to determine the fittest individuals in nature.
b) survival matters less to natural selection than reproductive success does.
c) natural variation in a population is generally too great to be influenced by differential survival.
d) fitness has little to do with natural selection.
e) reproductive success on its own does not necessarily guarantee evolution.

12. Adaptation:
  a) refers both to the process by which organisms become better matched to their environment and to the features of an organism that make it more fit than other individuals.
  b) cannot occur in environments influenced by humans.
  c) is possible only when there is no mutation.
  d) is responsible for the fact that porcupines are at unusually high risk of predation.
  e) occurs for physical traits but not behaviors.

13. Adaptations shaped by natural selection:
  a) are magnified and enhanced through genetic drift.
  b) are unlikely to be present in humans living in industrial societies.
  c) may be out of date, having been shaped in the past under conditions that differed from those in the present.
  d) represent perfect solutions to the problems posed by nature.
  e) are continuously modified so that they are always fitted to the environment in which an organism lives.

14. Artificial selection was used on corn to produce a single strain of corn with increased growth rates and greater resistance to a fungus. Although farmers have continued to select for these traits, the productivity of this strain is no longer increasing. This suggests that:
  a) the population size has been decreasing.
  b) all or most of the natural variation for these traits has been eliminated.
  c) gene migration is a major evolutionary agent in corn.
  d) long-term disruptive selection may lead to speciation.
  e) artificial selection is not as strong as natural selection.

15. In a population in which a trait is exposed to stabilizing selection over time:
  a) neither the average nor the variation for the trait changes.
  b) both the average value and variation for the trait increase.
  c) the average value increases or decreases and the variation for the trait decreases.
  d) the average value for the trait stays approximately the same and the variation for the trait decreases.
  e) the average value for the trait stays approximately the same and the variation for the trait increases.

16. Maze-running behavior in rats:
  a) is too complex a trait to be influenced by natural selection.
  b) is a heritable trait.
  c) is not influenced by natural selection because it does not occur in rats' natural environment.
  d) shows no variation.
  e) is influenced primarily by mutation in the laboratory.

17. A fossil is defined most broadly as:
  a) the preserved pieces of hard parts (e.g., shell or bone) of extinct animals.
  b) any preserved remnant or trace of an organism from the past.
  c) the preserved bones of vertebrates.
  d) a piece of an organism that has turned into rock.
  e) the process of preservation of intact animal bodies.

18. Which of the following are correct about marsupial mammals?
  a) fill many niches in Australia that are occupied by placental mammals in other parts of the world.
  b) are less fit than placental mammals.
  c) have gone extinct as a result of the greater fitness of placental mammals.
  d) are more closely related to each other than they are to placental mammals.
  e) Both a) and d) are correct.

19. Convergent evolution can occur only when two species:
  a) have a recent common ancestor.
  b) live in the same geographic area.
  c) are separated by a barrier such as a new river.
  d) evolve under similar selective forces.
  e) are both unpalatable to predators.

20. What can be concluded about comparing differences in molecular biology between different species?
  a) Extremely different species are fundamentally unrelated in any way.
  b) Only DNA sequences can be used to compare relatedness between species.
  c) Birds are more closely related to humans than dogs are.
  d) Genetic similarities and differences demonstrate species relatedness.
  e) The longer two species have been evolving on their own, the fewer the genetic differences between them.

21. Evolution:
  a) occurs too slowly to be observed in nature.
  b) can occur in the wild but not in the laboratory.
  c) is responsible for the increased occurrence of antibiotic-resistant bacteria.
  d) does not occur in human-occupied habitats.
  e) None of these statements is correct.

Short-Answer Questions

1. Distinguish between evolution and natural selection. (Restrict your answer to 30 words or fewer.)

2. What is genetic drift? Why is it a more potent agent of evolution in small populations rather than large populations?

3. In *The Origin of Species,* Charles Darwin wrote: "We may look with some confidence to a secure future of great length. And as natural selection works solely by and for the good of each being, all corporeal and mental endowments will tend to progress towards perfection." Describe three reasons why he is wrong.

4. How does the increasing frequency of antibiotic-resistant strains of bacteria represent an example of the occurrence evolution?

See Appendix for answers. For additional study questions, go to www.prep-u.com.

1 Behaviors are traits that can evolve.

- **9.1** Behavior has adaptive value, just like other traits.
- **9.2** Some behaviors are innate.
- **9.3** Some behaviors must be learned (and some are learned more easily than others).
- **9.4** Complex-appearing behaviors don't require complex thought in order to evolve.

2 Cooperation, selfishness, and altruism can be better understood with an evolutionary approach.

- **9.5** A general theory of "kindness" helps explain the evolution of apparent acts of altruism.
- **9.6** Apparent altruism toward relatives can evolve through kin selection.
- **9.7** Apparent altruism toward unrelated individuals can evolve through reciprocal altruism.
- **9.8** In an "alien" environment, behaviors produced by natural selection may no longer be adaptive.
- **9.9** Selfish genes win out over group selection.

3 Sexual conflict can result from disparities in reproductive investment by males and females.

- **9.10** There are big differences in how much males and females must invest in reproduction.
- **9.11** Males and females are vulnerable at different stages of the reproductive exchange.
- **9.12** Tactics for getting a mate: competition and courtship can help males and females secure reproductive success.
- **9.13** Tactics for keeping a mate: mate guarding can protect a male's reproductive investment.
- **9.14** Monogamy versus polygamy: mating behaviors can vary across human and animal cultures.
- **9.15** Sexual dimorphism is an indicator of a population's mating behavior.

4 Communication and the design of signals evolve.

- **9.16** Animal communication and language abilities evolve.
- **9.17** Honest signals reduce deception.

9

Evolution and Behavior

Communication, cooperation, and conflict in the animal world

❶ Behaviors are traits that can evolve.

This chimpanzee has learned to use a twig as a tool to capture termites.

9•1

Behavior has adaptive value, just like other traits.

Make a mental list of your favorite things to eat. Perhaps donuts or hot fudge sundaes top your list. Or maybe pizza, or cheeseburgers and french fries. One thing is almost certain about the food preferences of any human: the list is filled with calorie-rich substances. Put another way, when it comes to eating, humans show an almost complete aversion to dirt or pebbles or other substances from which no energy can be extracted (**FIGURE 9-1**).

FIGURE 9-1 Mud pie for dinner? Humans show an aversion to eating dirt (and other substances from which we cannot extract nutrition).

In controlled studies, humans demonstrate a clear and consistent preference for sweet and fatty foods. Specifically, when people are presented with food consisting of different mixtures, their preferences for the food increase as the sugar concentration in the food increases—up to a point, after which adding more sugar makes the food less preferable. But the preference becomes stronger and stronger as the food's fat content increases. In short, we prefer sweet (but not too sweet) foods that are packed with as much fat as possible (**FIGURE 9-2**).

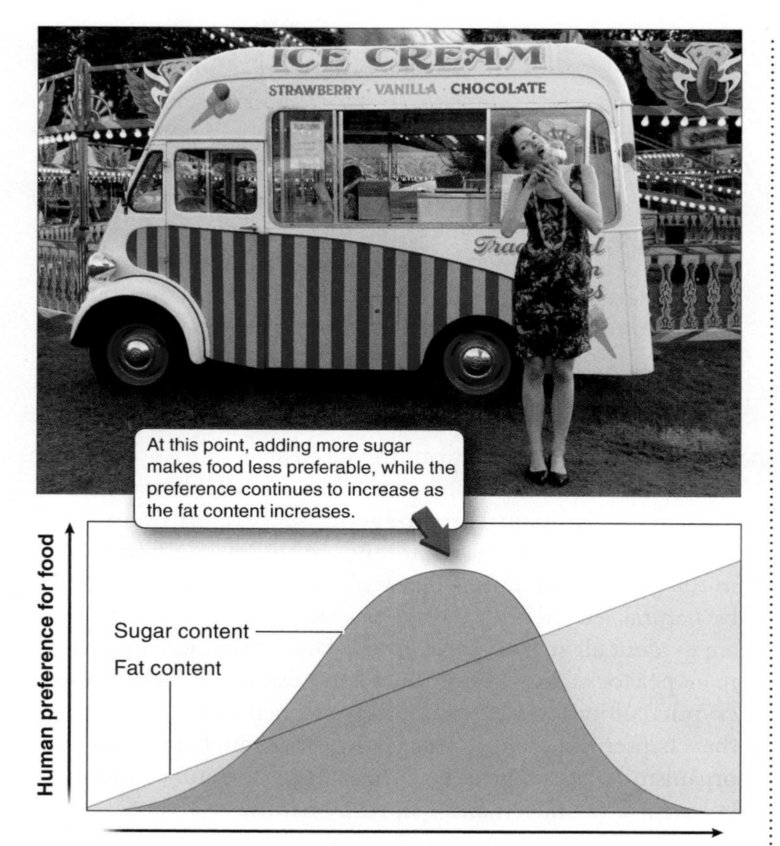

In controlled experiments, humans exhibit preferences for sugary foods, high in fat—and the more fat, the stronger the preference.

FIGURE 9-2 Fatty and sweet. Across cultures, humans prefer sweet, but not too sweet, foods that are packed with fat.

Watch birds in a field and you'll see that they, too, have preferences. A starling, for example, does not eat everything in its path. Rather, it walks through the grass, probing the ground and only occasionally picking up something to eat, such as a beetle larva. Shore crabs, too, have definite preferences. When presented with mussels of different sizes, shore crabs prefer the mussels that provide the greatest amount of energy relative to the amount of energy a crab must expend to open the mussel. The largest mussels take too long to open and so are less preferred; the smallest mussels are easily opened but don't provide enough caloric content to make the effort worthwhile (**FIGURE 9-3**). Thus, the research shows, shore crabs prefer medium-sized mussels.

Humans, starlings, shore crabs, and all other animals have taste preferences for the same reason: animals (including humans), unlike plants, cannot create their own food, so they must consume materials from which they can extract the most energy and acquire essential nutrients. If they don't, they die. The feeding choices of animals affect the number of offspring they can produce, and so taste preferences directly influence the evolutionary fitness of organisms. As a consequence, natural selection can shape feeding behaviors just as it can bring about changes in physical structures. We may believe that some things simply taste good while others do not, but these taste sensitivities and preferences are a direct result of the actions of molecules within our taste buds. And those molecules are produced by a large number of genes that have been shaped by natural selection.

Before we continue, let's define exactly what we mean by "behavior." **Behavior** encompasses any and all of the actions performed by an organism, often in response to its environment or to the actions of another organism. Feeding behavior is only one of many behaviors influenced by natural selection. Maternal care is another. In mice, an allele occasionally arises through mutation that causes a mother to show reduced maternal care. She leaves her pups unattended and neglects feeding them. Not surprisingly, this allele never increases in frequency in the mouse population, because most of the pups that inherit it fail to survive long enough to pass it

ENERGY EFFICIENT FEEDING BEHAVIOR HAS EVOLVED

The feeding behavior of shore crabs has evolved so that they preferentially choose mussels that provide the most energy relative to the effort it takes to open the shell.

FIGURE 9-3 Efficient eater. A shore crab feeds on a mussel.

on to their own offspring. The only reason this behavior ever occurs is that the mutation causing it occasionally occurs.

Another behavior influenced by natural selection is the singing behavior of songbirds. Among warblers, males sing complex and loud songs. They aren't singing for our enjoyment, though. They are singing to impress female warblers. Females prefer males with more complicated songs, apparently because only the healthiest, most well-nourished males are able to produce the most complex songs. Thus, in warblers, males' singing behavior has a strong effect on their evolutionary fitness.

The scope of animal behavior is vast and varied. Even a brief listing of well-studied topics in animal behavior is impressively long:

- Conflict, aggression, and territoriality
- Cooperation, alliance building, and sociality
- Competing for food and avoiding predation
- Migration and navigation
- Behavioral control of body temperature
- Courtship and mate choice
- Pair bonding and fidelity
- Breeding and parental behavior
- Communication
- Learning and tool use

In Chapter 8, we learned about the evolutionary origins of many *physical traits.* We did not investigate each and every physical trait that occurs in organisms, however. It was sufficient to look at a few examples, such as birth weight in babies, milk production in cows, and fur color in tigers, to understand the general process by which natural selection acts and how it can lead to the evolution of organisms adapted to their environments. Similarly, in this chapter, although we explore the evolution of *behavioral traits,* we will not investigate each and every behavioral trait in organisms. Rather, we will explore three broad areas of animal behavior—cooperation and conflict, communication, and mating and parenting—and focus on why those behaviors evolved. That is, we'll be answering this question: how do these behaviors contribute to an organism's fitness, its survival and reproduction?

Our approach to the study of evolution and behavior parallels our approach to the study of the evolution of non-behavioral traits. Recall from the discussion of natural selection in the last chapter that when a heritable trait increases an individual's reproductive success relative to that of other individuals, that trait tends to increase in frequency in the population. This is Darwin's mechanism for evolution by natural selection and, as we saw in Chapter 8, its effects are evident all around us, from the fancy ornamentation of male peacocks' feathers produced by sexual selection to the cryptic coloration that camouflages so many organisms. In this chapter, we see that *behavior* is just as much a part of an organism's phenotype as is its anatomical structures, and that behavior is produced and shaped by natural selection.

TAKE-HOME MESSAGE 9•1

Behavior encompasses any and all of the actions performed by an organism. When a heritable trait increases an individual's reproductive success relative to that of other individuals, that trait tends to increase in frequency in the population. Behavior is as much a part of an organism's phenotype as is an anatomical structure, and as such can be produced and shaped by natural selection.

9•2

Some behaviors are innate.

In his study of pea plants, Gregor Mendel described a single gene for plant height that caused a plant to be either tall or short. The production of a trait like plant height, however, is not completely genetically determined. Certain environmental conditions, such as the type of soil and the availability of water, nutrients, and sunlight, also have a role to play. Nearly all physical traits of all organisms are the products not only of genes, but also of environmental conditions. When it comes to the production of behaviors, the environment also plays an important role.

The degree to which a behavior depends on the environment for expression, however, varies a great deal. At one extreme are behaviors—called **instincts** or **innate behaviors**—that don't require any environmental input to develop. Innate behaviors are present in all individuals in a population and do not vary

FIGURE 9-4 No learning required. Egg-retrieval behavior in geese and the aggressive behavior of male sticklebacks are fixed action patterns.

much from one individual to another or over an individual's life span. An example of innate behavior is a **fixed action pattern.** Triggered in response to a specific signal called a **sign stimulus,** a fixed action pattern is a sequence of behaviors that requires no learning, does not vary, and, once started, runs to completion. Examples of fixed action patterns include:

Egg-retrieval in geese. When a goose spots an egg outside its nest, a fixed action pattern is triggered. The goose gets out of the nest and rolls the egg back, using a side-to-side motion, keeping the egg tucked underneath its bill. Once started, a goose continues the side-to-side egg-retrieval movement all the way back to the nest, even if the egg is taken away during the process (**FIGURE 9-4**).

Aggressive displays and attacks by stickleback fish. The bellies of male stickleback fish turn bright red when the breeding season arrives. During this time, the sight of a red belly on any other stickleback triggers aggressive behavior in a male stickleback. In fact, the males become aggressive at the sight of anything remotely resembling a red belly. One researcher even noticed that his sticklebacks performed their aggressive displays every day when a red mail truck drove by his window.

TAKE-HOME MESSAGE 9·2

The degree to which a behavior depends on the environment for expression varies a great deal. At one extreme are instincts, innate behaviors present in all individuals in a population that do not vary much from one individual to another or over an individual's life span. Innate behaviors don't require any environmental input to develop.

9•3

Some behaviors must be learned (and some are learned more easily than others).

In contrast to innate behaviors are those behaviors that are influenced to a much greater degree by the individual's environment. Requiring some degree of **learning,** these behaviors are altered and modified over time in response to past experiences. There is tremendous variation among behaviors that require learning: some come relatively easily and are learned by most individuals in a population, while other behaviors are less easily learned.

Consider a trait common to most primates, including humans: fear of snakes. In the wild, rhesus monkeys are afraid of snakes. The sight of a snake causes the monkeys to engage in fear-related responses, including making alarm calls and rapidly moving away from the snake.

Observations of monkeys in captivity, however, reveal that they aren't born with a fear of snakes. Captive monkeys will reach over a plastic model of a snake to get a peanut. Experiments also show that monkeys learn to fear snakes if they see another monkey terrified at the sight of a snake (**FIGURE 9-5**). Even if they see only a videotape of another monkey expressing fear at the sight of a snake, monkeys immediately and permanently become afraid of snakes. They will no longer reach over a snake model to get a peanut, even when they are very hungry. Instead, they scream and move as far away from the artificial snake as possible. Studies on humans show an equally easily learned fear of snakes. Rather than shrieking and cowering in the corner of a cage, however, we respond with sweaty palms and an increased heart rate.

Q Why is it so much easier for an infant to learn a complex language than for a college student to learn biology?

Behaviors that are learned easily and by all (or nearly all) individuals are called **prepared learning.** In addition to the snake-fearing behavior of monkeys, examples of prepared learning abound. The acquisition of language in humans is a dramatic example. Most children don't talk until they are a year old, but by the age of three they understand most rules of sentence construction, and the average six-year-old who is a native English-speaker already has a vocabulary of about 13,000 words. These skills are impressive, particularly given that, at this age, children are generally not competent in reading, writing, and fine motor coordination.

PREPARED LEARNING

FIGURE 9-5 Learning to be afraid. The monkey can easily learn to fear snakes by observing fear in other monkeys. This is a type of prepared learning.

Examples of prepared learning also underscore the fact that organisms don't learn everything with equal ease. For example, researchers demonstrated how easily monkeys learned to fear snakes, but in a related experiment, they demonstrated that the monkeys were not prepared to learn everything with equal ease. They altered the videotape that previously showed a monkey expressing fear on encountering a snake. On the modified tape, a monkey was shown having the same fear reaction in response to a flower or a toy rabbit. The captive monkeys—which had never seen flowers or rabbits or snakes—did not learn to fear flowers or rabbits. Only snakes.

Q Human babies quickly and easily develop a fear of snakes. Yet they don't easily develop a fear of guns. Why?

These observations point to an evolutionary basis for the acquisition of certain behaviors. It seems that organisms are well-prepared to learn behaviors that have been important to their ancestors' survival and reproductive success over the course of their evolutionary history, and are less prepared to learn behaviors irrelevant to their evolutionary success. Consider, for example, that although human babies quickly and easily develop a fear of snakes, they don't easily develop a fear of guns. Given that more than 30,000 people are killed by guns in the United States each year, while fewer than three dozen people are killed by snakes, it would seem that we ought to be very afraid of guns and relatively unconcerned about snakes. But we are built in just the opposite way (**FIGURE 9-6**). The likely explanation is that evolution can be slow in producing populations that are adapted to their environments—that is, evolutionary change cannot always keep up with a rapidly (in evolutionary terms) changing environment.

Like all genes, any genes involved in behaviors have been handed down to us from our ancestors. Snakes caused many human deaths over the course of our evolutionary history. In contrast, guns didn't kill a single person until very recently on the evolutionary time scale. Accordingly, we still fear our ancient enemy, the snake, and have no instinctual response to novel threats, regardless of how deadly.

FIGURE 9-6 Instincts out of date. A human infant is born with a fear of snakes.

TAKE-HOME MESSAGE 9•3

In contrast to innate behaviors are those behaviors that are influenced more by the individual's environment, requiring some learning, and are often altered and modified over time in response to past experiences. Organisms are well-prepared to learn behaviors that have been important to the reproductive success of their ancestors, and less prepared to learn behaviors irrelevant to their evolutionary success.

9•4

Complex-appearing behaviors don't require complex thought in order to evolve.

Why do humans and other animals have sex? Is it because they are thinking, robot-like: "Must maximize reproductive success. Must maximize reproductive success"? Of course not. Yet they do go about their lives behaving as if they were consciously trying to maximize their reproductive success.

Q Animals don't consciously try to maximize their reproductive success, yet they behave as if they do. How does this happen?

In actuality, natural selection produces organisms that exhibit relatively simple behaviors in response to certain situations or environmental conditions—behaviors that to us may seem complex and sophisticated. Let's look at an example. An animal that experiences pleasure when it has sex has the incentive to seek out additional opportunities to experience that pleasure. As long as reproductive success is enhanced in the process, it is not necessary (from the reproductive-success standpoint) for the animal to deliberately seek that outcome. In other words, natural selection doesn't have to produce animals consciously trying to maximize reproductive success. It only needs to produce animals that behave in a way that actually results in reproductive success, and this outcome will occur if organisms experience a pleasurable sensation by having sex. It's like an evolutionary short cut. Behaviors that lead to a specific outcome that increases the animal's relative reproductive success will be favored by natural selection.

We can investigate experimentally whether natural selection works in this way by tricking animals. Consider the example of egg-retrieval in geese described above. When an egg falls out of the nest, a goose retrieves it in exactly the same way every time. Is a goose able to do this because it vigilantly keeps tabs on all of its eggs? This seemed to be the case, until researchers decided to trick some geese. Researchers learned that putting *any* object that remotely resembles an egg near a goose's nest triggers the retrieval process. Geese will retrieve beer cans, door knobs, and a variety of other objects that will not increase the animal's reproductive success. Moreover, if multiple egg-like items are just outside the nest, the goose will retrieve the largest item first.

To test the limits of the retrieval behavior, researchers began putting larger and larger models of eggs outside the nests. When given a choice between an actual goose egg and an artificial egg the size of a basketball, the goose always tried to retrieve the giant egg. It seems that geese aren't actually keeping tabs on their eggs. Rather, they are following a rule of thumb dictating that they retrieve any nearby egg-like objects, and preferentially retrieve the larger objects first (**FIGURE 9-7**). Thus, natural selection can produce organisms that exhibit silly and clearly

RETRIEVING BEHAVIOR IN GEESE

NATURALLY OCCURRING FIXED ACTION PATTERN

When an egg rolls out of a nest, a goose retrieves it in exactly the same way every time.

EXPERIMENTAL FIXED ACTION PATTERN

A goose retrieves any object that remotely resembles its eggs.

A goose retrieves the largest egg-like object first instead of the actual goose eggs.

By tricking an animal, setting up a situation that would never occur in nature, we can reveal that it is following simple behavioral rules.

FIGURE 9-7 Programmed to retrieve. The goose will retrieve any egg-like object outside its nest.

maladaptive (i.e., fitness-decreasing) behaviors in experiments—the goose trying to retrieve a basketball-sized egg—if, in nature, individuals' performance of those simple actions nearly always leads to fitness-increasing behaviors.

Humans, too, may behave in ways that bolster reproductive success without their awareness. One way involves the avoidance of incest. Because sexual activity among close relatives such as siblings leads to a higher proportion of offspring with genetic defects—the closely related parents may have inherited a disease-causing allele from the same family member and then pass it to their offspring—humans and many other species have evolved an aversion to incest. The kibbutzim in Israel are large, communal systems of child-rearing and education. All of the children in a kibbutz are reared together, interacting constantly and sharing most of their daily experiences from infancy through maturity. Somewhat unexpectedly, there are no records of any children raised in the same kibbutz from infancy who marry each other as adults—even in the face of social encouragement. It seems that communal living during infancy and childhood triggers an incest aversion, the same aversion that is generally triggered in biological siblings living together in typical families.

Q Unrelated people who grow up together from infancy on the same kibbutz never marry each other. Why?

Four children change for bed after baths at the Kibbutz Baram's children's house in Israel. Children live in the children's house and are cared for by all community members.

Taking an evolutionary approach to the study of the behavior of animals, including humans, is not new. In 1859, Charles Darwin wrote in *On the Origin of Species:* "In the distant future I see open fields for far more important researches. Psychology will be based on a new foundation." And in the past several decades, Darwin's prophecy has increasingly proved true. In a synthesis of evolutionary biology and psychology, researchers taking an approach called "evolutionary psychology" have begun to view the human brain and human behaviors, including emotions, as traits produced by natural selection, selected as a result of their positive effects on survival and reproduction. So, too, has the evolutionary approach to studying behavior begun to influence the field of economics. The 2002 Nobel prize for economics was awarded to researchers who used insights from evolutionary biology and psychology in their analyses of human decision making.

In the rest of this chapter, working from the understanding that organisms' traits, including their behaviors, have evolved by natural selection, we examine the question of why organisms behave as they do. We begin with an exploration of selfishness and cooperation in animals, investigating in particular how natural selection has produced organisms that engage in apparent acts of altruism.

TAKE-HOME MESSAGE 9•4

If an animal's behavior in natural situations usually increases its relative reproductive success, the behavior will be favored by natural selection. The natural selection of such behaviors does not require the organism to consciously try to maximize its reproductive success.

② Cooperation, selfishness, and altruism can be better understood with an evolutionary approach.

Placing herself at risk, a female ground squirrel alerts the colony of approaching danger.

9•5

A general theory of "kindness" helps explain the evolution of apparent acts of altruism.

If we look closely, we see many behaviors in the animal world that *appear* to be **altruistic behaviors**—that is, they seem to be behaviors that come at a cost to the individual performing them while benefiting a recipient. When discussing altruism, we define costs and benefits in terms of their contribution to an individual's fitness.

Take the case of the Australian social spider (*Diaea ergandros*). After giving birth to about 50 hungry spiderlings, the mother's body slowly liquefies into a nutritious fluid that the newborn spiders consume. Over the course of about 40 days, her offspring literally eat her alive; they ultimately kill their mother, but start their lives well-nourished. The cost to the mother is huge—she is unable to produce any more offspring—but there is a clear benefit to the many recipients (**FIGURE 9-8**).

Such altruistic-appearing behavior is common enough in the natural world that it puzzled Darwin. Natural selection, he believed, generally works to produce selfish behavior. The alleles that cause the individual carrying them to have the greatest reproductive success should increase their market share—that is, should increase as a proportion of all the alleles for that gene within the population—in any given generation. At the same time, alleles that cause the individual carrying them to help increase other individuals' reproductive success should decrease their market share. Put another way, natural selection should never produce altruistic behavior. Darwin worried that if the apparent instances of altruism were indeed truly altruistic, they would prove fatal to his theory.

As it turns out, Darwin's theory is safe. Virtually all of the apparent acts of altruism in the animal kingdom prove, on closer inspection, to be not truly altruistic; instead, they have evolved as a consequence of either **kin selection** or **reciprocal altruism.**

Kindness toward close relatives: kin selection. Kin selection can lead to the evolution of apparently altruistic behavior toward close relatives. Suppose that, for one gene, you carry allele *K,* which causes you to behave in a way that

FIGURE 9-8 Parental care to the extreme. The female *Diaea ergandros* spider feeds her offspring with her own body.

increases the fitness of a close relative of yours, while decreasing your own fitness. Allele *K* might actually increase its market share in the population, because you and your relatives tend to share many of the same alleles—including allele *K*. Thus the increased fitness of the relatives you help might compensate for your own reduced fitness, because allele *K* will, overall, increase its frequency in the population when you help your relatives increase their reproductive output.

Kindness toward unrelated individuals: reciprocal altruism. Reciprocal altruism can lead to the evolution of apparently altruistic behavior toward unrelated individuals. Suppose that, for another gene, you carry allele *R*, which causes you to help individuals to whom you are not related. This help increases their fitness and decreases yours in the process. Allele *R* might still increase its market share in the population, if the individuals whom you help become more likely to return the favor and help you at some point in the future, increasing your fitness.

Seen in this light, both kin selection and reciprocal altruism can lead to the evolution of behaviors that are *apparently* altruistic but, in actuality, are beneficial—from an evolutionary perspective—to the individuals engaging in the behaviors.

In the next two sections, we explore kin selection and reciprocal altruism in more detail. We also investigate some of the many testable predictions about when acts of apparent kindness should occur, whom they should occur between, and how we could increase or decrease the frequency of their occurrence through a variety of modifications to the environment.

> "It is not from the benevolence of the butcher, the brewer, or the baker that we expect our dinner, but from their regard to their own interest. We address ourselves, not to their humanity, but to their self-love, and never talk to them of our own necessities, but of their advantages.
>
> —Adam Smith, *An Inquiry into the Nature and Causes of the Wealth of Nations,* 1776"

Occasionally, we see individuals engaging in behavior that *is* genuinely altruistic. We will examine how and why certain environmental situations cause individuals to behave in a way that decreases their fitness and, as such, is evolutionarily maladaptive. It is also important to note that even as genes play a central role in both fostering cooperation and spurring conflict, one of the hallmarks of humans is our ability to override the impulses that often push us toward selfishness. This capacity for self-control sets us apart from the rest of the animal kingdom and is responsible for much of the rich diversity in human behavior that we see around us.

TAKE-HOME MESSAGE 9•5

Many behaviors in the animal world appear altruistic. In almost all cases, the apparent acts of altruism are not truly altruistic and have evolved as a consequence of either kin selection or reciprocal altruism and, from an evolutionary perspective, are beneficial to the individual engaging in the behavior.

9•6

Apparent altruism toward relatives can evolve through kin selection.

The grasslands of the western United States are home to the Belding's ground squirrels. The squirrels form large colonies of hundreds or even thousands of individuals. They feed mostly on plants, and they live in underground burrows. Because the colonies are so large, they attract many birds of prey. When a bird comes to the colony, it succeeds in killing a squirrel about 10% of the time. Squirrels have a system for reducing predation risk that resembles a neighborhood watch program. When an aerial predator approaches, it is common for a squirrel standing on top of a burrow to produce a loud whistle-like call that serves as an alarm, warning squirrels of the impending danger and reducing their risk of death. On hearing an alarm call, squirrels quickly take cover in their burrows. Making an alarm call is a very dangerous activity: about half the time that an alarm call is made, the squirrel making the alarm call is killed by the predator (**FIGURE 9-9**).

Some squirrels see predators and make alarm calls, while other squirrels see predators and keep their mouths shut. This raises the question: why would any squirrel make an alarm call? And given that some do, which squirrels are most likely to engage in this altruistic-appearing behavior? It certainly isn't random: about 80% of the squirrels making alarm calls are female. Moreover, older females are five times more likely than young females to make alarm calls.

These differences between males and females, and between young and old, reveal that alarm calling is about protecting relatives. The more kin an individual is likely to have, the more likely that individual is to sound the alarm. Because males travel long distances to live in new colonies shortly after reaching maturity, most adult males don't live near their parents, siblings, or any other relatives besides their own offspring. Females, on the other hand, remain near the area where they were born and are likely to have many close relatives nearby. Older females, then, are likely to have the largest number of relatives.

The biologist W. D. Hamilton expressed the idea of "kin selection," that one individual assisting another could compensate for its own decrease in fitness if it helped a close

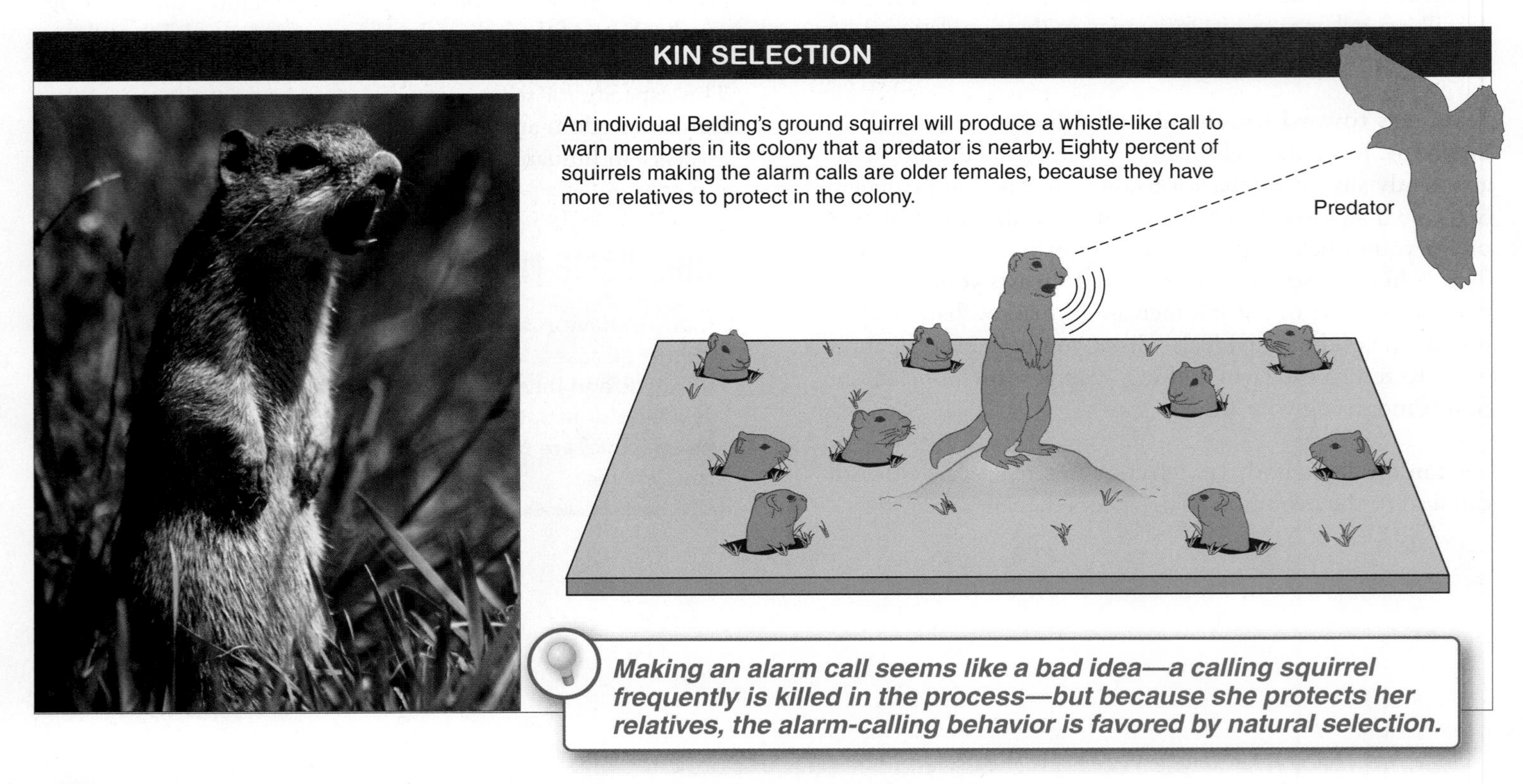

FIGURE 9-9 Protecting relatives by making an alarm call. At great risk to herself, a female Belding's ground squirrel will make an alarm call likely to save her family members from a predator.

INAPPROPRIATE ALARM CALLING IN UNNATURAL SITUATIONS

Researchers transplanted an older female Belding's ground squirrel from her original colony to a new colony, where she had no relatives. Even though this older female is not related to any individuals in the colony, she behaves as if she were and is likely to make an alarm call.

Predator

As with the egg-retrieval experiments in geese, by tricking an animal, setting up a situation that would never occur in nature, we can reveal that the animal is following simple behavioral rules.

FIGURE 9-10 Following simple behavioral rules. A transplanted female Belding's ground squirrel that is not related to the squirrel colony behaves as though she is.

relative in a way that increased the relative's fitness. After all, he realized, the recipient of the aid was likely to have at least some genes in common with the altruistic individual. And the more genes they shared in common (i.e., the more closely the individuals were related), the more likely it was that the alleles being propagated by the recipient of the altruistic-like behavior were the same alleles found in the altruistic individual.

Thus, altruistic-appearing behavior will most likely occur when the benefits to close relatives are greater than the cost to the individual performing the behavior. That is, the more closely related two individuals are, the more likely they are to act altruistically toward each other. It seems that individuals are acting selflessly/altruistically, but they are really acting in their genes' best interests. (The converse is also true: the less closely related two individuals are, the more likely they are to experience conflict.)

Belding's ground squirrels probably don't know exactly how closely they are related to the squirrels around them, so it is likely that natural selection has led to an evolutionary short-cut that allows them to behave as if they did know. In a clever experiment that tested this idea, researchers trapped adult female squirrels and relocated individuals to distant ground squirrel colonies in which they had no close relatives. If the transplanted females were able to determine exactly how closely they were related to the squirrels around them, they would not make alarm calls. To do so would put the caller at risk, but would not benefit her genes. What the researchers found, however, was that transplanted females were just as likely to sound the alarm for the colony of strangers as they would be if they had been left on their home territories, surrounded by close relatives (**FIGURE 9-10**). It seems that female ground squirrels have evolved to follow a simple rule that says "if I am an older female, behave as if I have many close relatives around me."

Q Who are people most likely to bequeath money to in their will?

Another example of apparent altruistic behavior toward kin is seen in how people give their money and property away when they die. Kin selection theory gives rise to the prediction that people will bequeath their assets based on their degree of relatedness to others. In an analysis of 1,000 randomly chosen wills, researchers tested two hypotheses: (1) that individuals would leave more of their estates to genetic kin, including their parents, children, and siblings (or spouses, who, the researchers assumed, would distribute the assets to the dead person's genetic kin), and (2) that closer relatives would be left greater amounts than more distant relatives. Both predictions were supported by the analysis: 92% of assets divvied up in the wills were given to genetic kin. And of this 92%, 46% was given to those individuals most closely related to the dead person (siblings and children), while only 8% was given to half-siblings, grandchildren, and nieces and nephews, and about 1% was given to cousins or more distant relations (**FIGURE 9-11**).

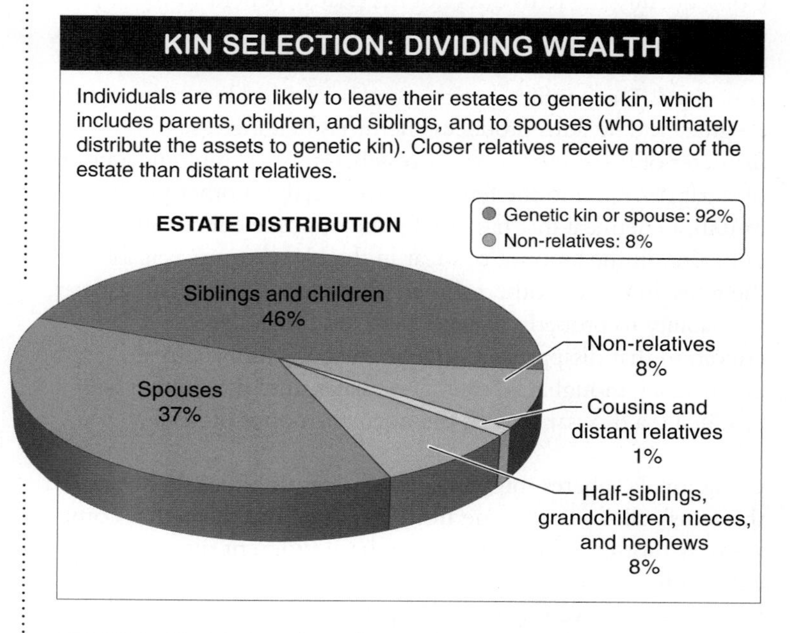

FIGURE 9-11 Reading the will. How likely are you to inherit money from a family member?

Based on the idea of kin selection, it is necessary to redefine an individual's fitness. An individual's fitness is not just measured by his or her total reproductive output (this is called **direct fitness**). Fitness also includes the reproductive output that individuals bring about through their seemingly altruistic behaviors toward their close kin. This is called an individual's **indirect fitness.** The sum of an individual's indirect and direct fitness is called **inclusive fitness.**

Q James Joyce wrote: "Whatever else is unsure in this stinking dunghill of a world, a mother's love is not." According to the concept of kin selection, he may not be completely right. Why?

No two individuals—with the exception of identical twins—are genetically identical. And because different individuals do not share all of the same alleles, we expect that they should experience some conflict. One disturbing example of such conflict takes place between a pregnant woman and her developing fetus.

A mother and fetus differ on the question of how much food—doled out as nutrients in the blood flow across the placenta—the fetus ought to get. There is a point at which it is in the mother's best interests to reduce the amount of glucose and other nutrients given to the fetus. Her genes gain if she saves some of her resources for future fetuses, to whom she will be related equally as to the present fetus.

Now consider the fetus's "point of view." Future siblings will carry some but not all of the same genes as the fetus. Consequently, the fetus does not necessarily benefit from sacrificing nutritional intake for the benefit of future siblings. In a sense, it is a sibling rivalry that starts before the sibling is even conceived! The conflict results in a physiological battle throughout pregnancy. The fetus produces chemicals that increase the diameter of the mother's blood vessels, thus increasing the amount of sugar delivered to the fetus. In response, the mother produces insulin, a chemical that has exactly the opposite effect, causing a reduction in the amount of sugar in the bloodstream available to the fetus. In some mothers this conflict causes gestational diabetes, an inability to properly regulate her blood sugar levels—a condition that disappears as soon as the baby is born. In all pregnancies, though, the conflict escalates until the mother is producing a thousand times the normal amount of insulin.

Gestational diabetes might also be the result of conflict between the fetus's alleles that come from its father and those that come from its mother. The parents may have different interests concerning how much to invest in the current offspring versus how much to save for future offspring.

Q Is a child living with one or more step-parents at greater risk of abuse than a child of the same age living with his or her biological parents?

The idea of kin selection gives rise to another prediction about conflict among relatives and non-relatives. Specifically, it generates the prediction that children in step-families will be injured or killed more frequently than those living with their biological families. And in fact, evidence from numerous studies across many different cultures, including evaluation of 20,000 reports from the American Humane Association in the United States, supports what has been called "the Cinderella syndrome." These findings included, for example, an estimate that the risk of abuse for a pre-schooler is about 1 in 3,000 for a child living with two biological parents (with whom he or she shares considerable genetic relatedness) and 40 in 3,000 for a child living with a step-parent (with whom he or she has no genetic relatedness). This difference in risk for children living in a home with a step-parent versus those living with their biological parents remains even when socioeconomic factors are taken into account.

Although the effect has been noted in multiple other cultures, the finding that living with a step-parent represents a significant risk factor for an individual has met with shock and has been criticized. Nonetheless, it is essential to note that while these data are unambiguous, abuse is still exceedingly rare—with homicide rates of approximately 0.06% per year among children 0–2 years of age living with a step-parent and 0.001% for those living with biological parents. Put simply, in the overwhelming majority of step-families no abuse occurs—a testament that "counter-Darwinian" behavior is also an important part of human nature. Ongoing research continues to explore a range of explanations for the effect. As would be expected, infanticide by unrelated "step-parent" males is also common in other species, such as lions. Males, upon taking over a pride, will usually kill the cubs that are under about nine months of age (although females defend their young aggressively), causing the females to become reproductively ready sooner than they would have if they had continued nursing their cubs.

TAKE-HOME MESSAGE 9•6

Kin selection describes apparently altruistic behavior in which an individual that assists its genetic relative compensates for its own decrease in direct fitness by helping increase the relative's fitness and, consequently, its own inclusive fitness.

9•7

Apparent altruism toward unrelated individuals can evolve through reciprocal altruism.

It is ironic that studies of altruism in the world reveal that natural selection has primarily produced selfish behavior among animals. And, as we saw in the previous section, when behavior appears altruistic, this is frequently because individuals are helping kin and, by doing so, are promoting the reproduction of the copies of the genes (i.e., the alleles) they carry that are also carried by their close relatives. Does any apparently altruistic behavior toward non-relatives (what humans might call "friendship") occur in the animal world? Yes, but there's not much of it. In this section we explore the conditions that give rise to reciprocal altruism, how it may have arisen, and why it is so common among humans but so rare among most other animal species.

We start by examining one species with well-documented altruistic-appearing behavior: the vampire bat. Vampire bats live in social groups of 8–12 adults, roosting primarily in caves and hollow trees. They feed by landing on large mammals such as cattle, horses, and pigs, piercing the skin with their razor-sharp teeth, and drinking the blood that flows from the wound. A chemical in their saliva keeps the blood from clotting, and thus allows the bats to feed for a longer period of time. Because of their small body size (about the size of your thumb) and their very high metabolic rate, vampire bats must consume almost their entire body weight in blood each night. If they go for more than about 60 hours without finding a meal, they are likely to die from starvation. Here's where the apparent altruism comes in: a bat that has not found food and is close to death will beg food from a bat that has recently eaten. In many cases, the bat that has just eaten will regurgitate some of the blood it has consumed into the mouth of the hungry bat, saving it from starvation. This act obviously has very high benefit for the recipient of the blood, but it comes at a cost to the sharing bat, which loses some of the caloric content of a meal it has just obtained (**FIGURE 9-12**).

RECIPROCAL ALTRUISM IN VAMPIRE BATS

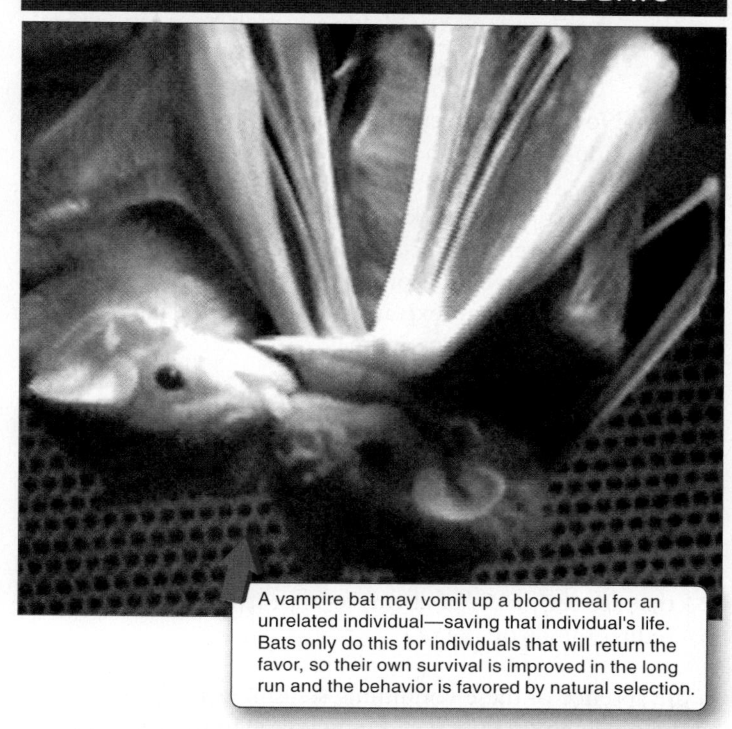

FIGURE 9-12 The gift of life. A starving vampire bat receives blood from another, well-fed bat. The recipient of the blood will return the favor one day.

Kin selection is responsible for some of the blood sharing (females often regurgitate blood for their own offspring), but, in many cases, individuals give blood to unrelated individuals. How might this behavior have arisen? To answer the question, it is important to note three other features of vampire bats. First, they are able to recognize more than a hundred distinct individuals. Second, bats that receive blood donations from non-relatives reciprocate significantly more than average with the bats that have shared blood with them. And third, bats that are not familiar with each other (and do not have a history of helping each other) generally do not regurgitate for each other.

One method proposed to explain the evolution of this apparent altruism is that the bats giving blood to other bats in need are repaid the favor when they are in need of blood. In other words, the act only seems to be selfless, when in actuality it is selfish. With such reciprocal altruism, both individuals (at different encounters) give up something of *relatively* low value in exchange for getting something of great value at a later time when they need it most. In other words, they are storing goodwill in another individual, in much the same way that a person might put money in a bank for a rainy day. In both cases, individuals are protected from some of the world's uncertainties.

We're all born with a spare kidney, yet virtually no one donates theirs to one of the thousands of non-relatives in need. Why?

Imagine that you are a parent and your child is in need of a kidney transplant to survive. Would you consider donating a kidney to save your son or daughter? Most people do agree to give a kidney in this situation, and this type of donation is an illustration of kin selection. Your action may indirectly improve your fitness (and also help someone you cherish.) Now, imagine a circumstance where you are asked to donate a kidney to someone you have never met. Would you?

You might be surprised to know that kidney donation to a stranger is an extremely uncommon occurrence; a consideration of the conditions under which reciprocal altruism evolves helps to explain why. Reciprocal altruism can evolve if certain conditions are met:

1. Repeated interactions among individuals, with opportunities to be both the donor and the recipient of altruistic-appearing acts
2. Benefits to the recipient that are significantly greater than the costs to the donor
3. The ability to recognize and punish cheaters, individuals that are recipients of altruistic-appearing acts but do not return the favor

In the absence of these conditions, selfishness is expected to be the norm among unrelated individuals. In the presence of all three conditions, on the other hand, reciprocity masquerading as altruism is likely to occur, as it does with blood sharing among the vampire bats.

The three conditions required for the evolution of reciprocal altruism are not satisfied in many animal species, which may be why altruistic-appearing behavior among unrelated individuals is rare. But there are some exceptions. The cleaner fish, for example, is a small fish that is allowed to swim right into the mouth of a grouper, a much larger fish, where it cleans the grouper of parasites by eating them. Stomach analyses reveal that groupers almost never eat the cleaner fish, opting to pass up this opportunity for a meal. And so, while the grouper is cleaned of parasites, the cleaner fish gets to eat a meal of parasites. In this case, it is easy to rule out kin selection as the reason for the altruistic-appearing behavior; the fish are from different species and can't possibly be close relatives. It is difficult to determine with certainty how this relationship got started, but many mutually beneficial relationships such as this seem to develop from unavoidable antagonistic encounters, particularly when the environment is stressful. In other words, when the conditions are right, behaviors can evolve that help organisms live more harmoniously with each other—when doing so improves their reproductive success.

Why are humans among the few species to have friendships?

Reciprocal altruism has also been documented in vervet monkeys. Monkeys that would otherwise be searching for food or, perhaps, a mate spend time grooming other monkeys (mostly by picking parasites from their fur). In return for their grooming behavior, the monkeys are more likely to receive assistance in response to their solicitations for aid (**FIGURE 9-13**).

As rare as it is in other species, reciprocal altruism is very common among humans. Friendship, among other human relationships, is built on reciprocity and is almost universal. The opportunity for friendship may be enhanced by our long life span and our ability to recognize thousands of faces and keep track of cheaters. These features are essential in individuals engaging in reciprocal altruism, because an individual becomes very vulnerable when he or she acts in an altruistic manner toward an unrelated individual. The risk is that the altruism will not be repaid, in which case the cheater enjoys greater fitness than the altruist.

RECIPROCAL ALTRUISM IN GROOMING BEHAVIOR

Time that could be spent finding food? Individuals that groom others are more likely to receive assistance if they ever are in danger—and so the grooming behavior is favored by natural selection.

FIGURE 9-13 You scratch my back, I'll scratch yours. One vervet monkey will groom another, but expects assistance in return.

Q Why is it easier to remember gossip than physics equations?

While cooperators, those who repay altruistic-appearing behavior, have evolutionary advantages over loners, this advantage disappears if the cooperators are the givers all or most of the time, never, or rarely, getting anything in return. Thus, it pays to avoid cheaters, individuals who accept altruistic behavior without repaying it. The importance of keeping track of cheaters and of identifying good potential reciprocity partners (i.e., other cooperators) may be why humans are so interested in social information and seemingly trivial gossip. As with many of the complex behaviors we have explored, humans probably use some rules of thumb when making decisions about reciprocal altruism. This might include keeping track of *all* available social information to best determine any promising individuals with whom to engage in reciprocal altruism. This includes information about health, generosity, social status, and reproduction. Researchers have even hypothesized that the evolution of maintaining an interest in social information, while valuable in most contexts, may now have maladaptive manifestations. We may find ourselves interested in and distracted by the social life of Julia Roberts, Tom Cruise, or other individuals whom we will most likely never even meet.

TAKE-HOME MESSAGE 9•7

In reciprocal altruism, an individual engages in an altruistic-appearing act toward another individual. Although giving up something of value, the actor does so only when likely to get something of value at a later time. Reciprocal altruism occurs only if individuals have repeated interactions and can recognize and punish cheaters, conditions satisfied in humans but in few other species.

9•8

In an "alien" environment, behaviors produced by natural selection may no longer be adaptive.

A transplanted female squirrel makes an alarm call, putting herself in grave danger to help out dozens of squirrels that share none of her genes. A person in the United States donates money to a fundraising drive to benefit refugees in Africa. Or someone finds it difficult to maintain her body weight, craving high-fat food even as she knows it may increase her risk of heart disease and shorten her life.

These situations have something in common. In each, genetically influenced traits, behaviors favored by natural selection over evolutionary time, cause individuals to behave in a way that *reduces* their fitness. But how can that be? How can natural selection lead to fitness-reducing behaviors?

To better understand these situations, it is helpful to return to our discussion of fitness, in Chapter 8. Recall that fitness not only is an individual's reproductive success, relative to that of other individuals in the population, but also depends on the specific environment in which the organism lives. An adult female Belding's ground squirrel that makes alarm calls generally increases her inclusive fitness by doing so, because all around her are individuals who share her genes. For this reason, such behavior has been favored by natural selection and has increased in frequency in the population.

The world in which alarm calling evolved did not include biologists with pickup trucks who trapped female squirrels, drove them long distances, and released them into colonies of unrelated individuals. (In a world where long-distance transplantation was common, alarm calling would not have evolved.) So, when a female squirrel suddenly finds herself in that world, her evolved behaviors can no longer be expected to be adaptive—no more than a human could be expected to survive on the moon. But an individual's genes and all the behaviors they influence cannot change overnight. Adaptation to a new environment takes time, and the more quickly an environment changes, the more likely it is that the evolved behaviors of a population will no longer be appropriate.

Let's return to the donations to refugees on another continent. To understand how natural selection could lead to this behavior, it is necessary to understand a bit about human evolutionary history. We know from archaeological deposits that for more than 2 million years, our ancestors lived as hunter-gatherers in small groups of a few hundred people, at most. Their success depended on joint efforts against predators and prey; being "nice" paid off when the prospect of hunting alone or sleeping outside the camp meant almost certain death. The loners died, so we are descended from those who could work well with others.

It is only recently, the blink of an eye in evolutionary time, that humans invented agriculture, industrialization, and the means of food production and distribution. As a result of these changes, population group size has increased dramatically; on a given day, you may see 10 or even 100 *times* as many people as a hunter-gatherer ancestor might have seen. Additionally, if you were a hunter-gatherer, every one of the people in your group would be a regular part of your life. You would have many opportunities to help them and, in turn, to be helped by them. Reciprocity paid. And we evolved so that "altruistic" acts gave us pleasure, stimulating parts of our brain in ways that made us want to repeat such actions.

It is with this brain that you approach the issue of refugees halfway across the world, or a homeless person somewhere in the United States. It is almost certain that you will not have repeated interactions with any of those people. Nor will you ever be in a position to be helped by them. From an evolutionary perspective, your action almost certainly will not increase your fitness. But in the world humans evolved in, such altruistic-appearing behavior would most likely have been reciprocated at some future time, and thus your instincts guide you to, and reward you for, your kindness (**FIGURE 9-14**). It feels right.

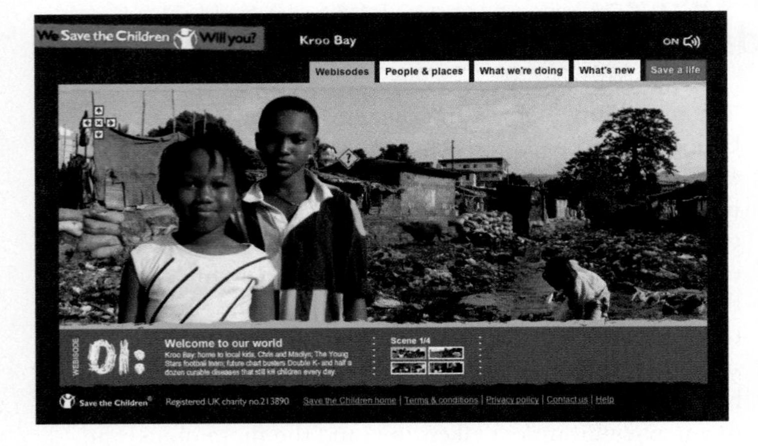

FIGURE 9-14 Charitable acts can give us pleasure. In the small hunter-gatherer groups in which humans evolved, altruistic behaviors would be reciprocated. Today, the pleasure we feel in response to such behaviors remains—even though the favor may never be returned.

In addition to small population sizes, our ancestors' hunter-gatherer world was characterized by unpredictable food sources that could not be stored for long periods of time. In response to this environment, humans evolved to have great appetites and taste preferences for calorie-laden foods. Fitness increased with food consumption and the storage of fat, helping to protect individuals during the times of starvation that occurred with regularity. From an evolutionary perspective, today's world is truly an alien environment: most of us have easy access to unlimited amounts of food, including fat-laden fast foods. Because humans have not yet adapted to this environment of plentiful food, our instincts to consume large amounts are strong, making it difficult for most people to control their body weight.

From battles against obesity to charitable contributions to alarm calling in transplanted Belding's ground squirrels, when organisms of any species find themselves in an environment that differs from the environment to which they are evolutionarily adapted, we expect (and see) behaviors that appear to be (and are) not evolutionarily adaptive. Still, understanding the process of natural selection can help us make sense of these seemingly nonsensical behaviors. It is important not to lose sight of the fact that just because a behavior such as making a donation to help people on a faraway continent is not adaptive from an evolutionary perspective, this aspect of the behavior is completely irrelevant when you're deciding whether or not it is something you want to do. When it comes to ethical decision making, there is nothing inherently good or important about the behaviors that natural selection has favored.

Can you think of other ways in which the modern industrial environment differs from the environment to which humans are adapted? How do these differences lead to situations in which our behavior, like the behavior of the transplanted Belding's ground squirrel, may not be adaptive from an evolutionary perspective?

TAKE-HOME MESSAGE 9•8

When organisms are in environments that differ from the environment to which they are adapted, the behaviors they exhibit are not necessarily evolutionarily adaptive.

9•9

Selfish genes win out over group selection.

Kin selection and reciprocal altruism can evolve in a population of animals, as we have seen in the previous three sections, and examples of both abound in nature. As a consequence, casual observers of nature frequently see individuals acting in ways that appear altruistic, even though, as we have seen, these individuals are truly acting—from the perspective of evolutionary fitness—in their own selfish interests. This nearly universal selfishness raises the question of whether evolution ever leads to behaviors that are good for the species or population but detrimental to the individual exhibiting the behavior, a process called **group selection.**

It might seem that evolution would favor individuals that behave in a manner that benefits the group, even if it comes at a cost to the individual's reproductive success. But this does not happen. Behaviors that reduce an individual's reproductive output (relative to that of other individuals in the population) are not likely to evolve. Natural selection favors individuals producing more offspring relative to those producing fewer offspring.

Let's look at an example. Imagine that a new allele appears in a population (perhaps through mutation) that causes the individual carrying the allele to double its reproductive output, even though this might spell doom for the species as individuals over-use their resources. An individual carrying this "selfish" allele will pass it to more offspring than an individual carrying the alternative allele, coding for the production of fewer offspring, will pass its allele—and this will lead to excessive, "selfish" consumption of the species' resources. And the selfish offspring also will pass on the selfish allele at a higher rate than the alternative allele is passed on. This scenario may lead to extinction of the species, yet it still occurs. With its market share perpetually increasing, the selfish allele eventually will be favored by evolution and will predominate (**FIGURE 9-15**).

Just because a behavior (determined by an "unselfish" allele) leads to a better outcome for the group doesn't mean that natural selection will favor that behavior. In the end, natural selection generally causes increases in the alleles that benefit the *individual* carrying them, even when this comes at the expense of the group. In some special situations, it is possible for natural selection to lead to group selection. But the stringent conditions necessary for this to occur are so rarely found in nature that we almost never see it.

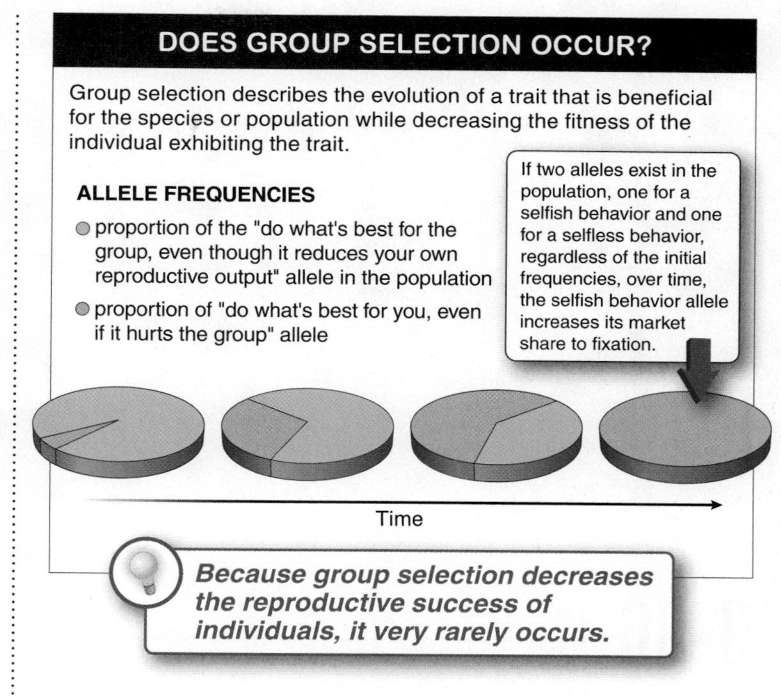

FIGURE 9-15 **Can a "selfless" gene increase in frequency in a population?**

TAKE-HOME MESSAGE 9•9

Behaviors that are good for the species or population but detrimental to the individual exhibiting such behavior are not generally produced in a population under natural conditions.

③ Sexual conflict can result from disparities in reproductive investment by males and females.

The wandering albatross extends its wings during a courtship display as a potential mate looks on.

9•10

There are big differences in how much males and females must invest in reproduction.

As we've seen in this chapter so far, many behaviors have evolved that influence the ways in which animals interact with each other, reducing conflict and sometimes leading to cooperation. One aspect of animal life that necessarily involves interaction among individuals—for sexually reproducing species—is reproduction, from courtship and mating to parental investment and the forming and breaking of ties between two animals. In this and the next few sections, we will explore these behaviors and the factors that influence them.

How many babies can a woman produce over her lifetime? Is the number larger or smaller than the maximum number of babies a man can produce? The answer to the first question is probably higher than you would guess: a Russian woman had 69 children (in 27 pregnancies). But even this high number is greatly exceeded by the 888 offspring produced by one man (the Emperor of Morocco from 1672 to 1727). This large difference between the number of offspring that can be produced by females and males is less surprising. Among other species of mammals, the pattern is consistent: a male elephant seal can produce 100 offspring, while the maximum produced by a female over her lifetime is 8; a male red deer can produce 24 offspring, while the maximum produced by a female is 14. Here we examine the physical differences between males and females and how they lead to the differences in male and female sexual behavior that have evolved.

The very definition of "male" and "female" hinges on a physical difference between the sexes. Recall from Chapter 6 that in species with two distinct sexes, a **female** is defined as the sex that produces the larger gamete, while a **male** produces the smaller gamete (**FIGURE 9-16**). At conception, the mother's material and energetic contribution to the offspring, the energy she will expend in the growth, feeding, and care of offspring—her **reproductive investment**—exceeds the father's. This is true for all animals, whether mammals, birds, insects, or sharks. (It is also true for plants.) Not only are female gametes larger, they tend to be relatively immobile and are produced in smaller numbers. Male gametes, on the other hand, while smaller, are more plentiful and very motile.

This discrepancy in size and quantity between the sperm produced by a male and the eggs produced by a female may seem trivial, but it sets the stage for evolutionary developments that magnify this initial difference in reproductive investment. For starters, the difference in the number of gametes that males

FIGURE 9-16 Many sperm, just one egg.

and females can produce means that males have the potential to produce many, many more offspring than females. Put another way, a male's **total reproductive output,** the lifetime number of offspring he can produce, tends to increase as the number of females he is able to fertilize increases. A female, on the other hand, does not generally increase her reproductive success by mating with additional males beyond the first (**FIGURE 9-17**).

Q Why do males usually compete for females rather than the opposite?

Because additional matings usually lead to greater increases in reproduction (and fitness) for males relative to females, selective pressure has resulted in the evolution of some differences in male and female reproductive behavior. For males, the most effective way to maximize reproductive success is often to find and gain access to mating opportunities with additional females. For females, an effective way to maximize reproductive success is often to put more effort into parenting their offspring and less effort into seeking mates.

Two physical differences that can exist between males and females are particularly important when it comes to reproduction. First, in species with internal fertilization, including most mammals, fertilization takes place in the female. The offspring also grow and develop within the female's body. Because of internal fertilization and development, the amount of energy invested in reproduction by females—their reproductive investment—is much greater than the investment by males. It also limits a female's reproductive output; she can only be pregnant once at a time. A second important physical difference between females and males occurs among the mammals: lactation occurs in females and not in males (with a

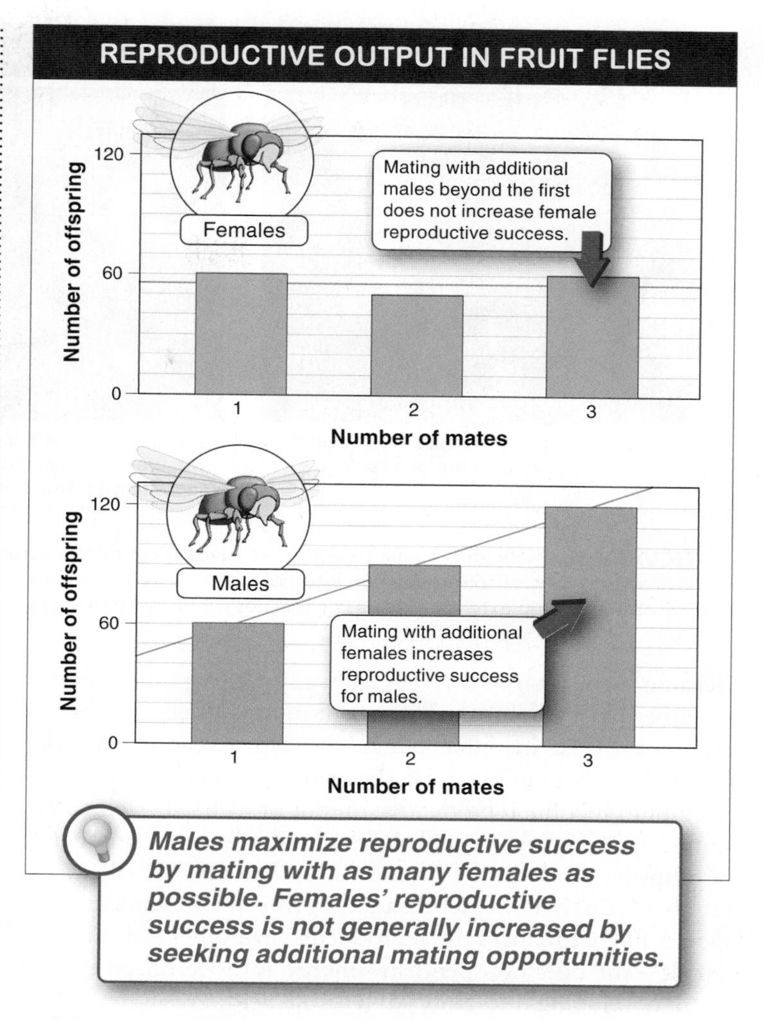

FIGURE 9-17 Maximizing reproductive success in fruit flies.

very small number of exceptions). In these species then, nurturing during both pregnancy and lactation can only be accomplished by the female. This difference in reproductive investment has led to the evolution of some very different reproductive behaviors between males and females.

Although the gamete size difference is consistent across all animals (i.e., the egg is always bigger and energetically more costly to make than the sperm), the physical differences between the sexes in the early nurturing (i.e., gestation and feeding) of offspring can vary considerably across animal species. In some cases, the early investment in reproduction is much greater for females than for males, but male and female investments become more nearly equal after fertilization. In birds, for example, after fertilization but prior to emergence of the chicks, much of the development of the fertilized egg is external: the female lays an egg, but either the male or the female can protect and incubate the developing embryo. Further, birds do not lactate. Once hatched, the chicks must be fed—a task that can be done by both parents.

REPRODUCTIVE INVESTMENT

The female mammal invests much energy in the production and care of offspring.

Amphibian eggs are left to develop on their own; there is very little maternal or paternal reproductive investment after fertilization.

Among most bird species, males and females have an equal reproductive investment.

FIGURE 9-18 Male and female reproductive investment differs across species. Female investment is greatest when offspring develop internally and are fed by lactation after birth. In animals whose offspring develop externally, males or females can provide the care.

Because incubation of the eggs and early feeding can be done by either parent, the early investment in reproduction in birds does not necessarily differ so dramatically between males and females as it does in mammals. In many bird species, the maximum lifetime reproductive output of males and females is similar. In the kittiwake gull, for example, the largest number of offspring produced by a male is 26, while for a female the number is 28. External fertilization in fish and amphibian species further reduces the reproductive investment of the female—she does not spend any energy as the fertilized eggs begin developing into embryos (**FIGURE 9-18**).

We have seen, then, that in species where fertilization occurs inside the female body, there is an unequal energetic investment in reproduction, and as a result the female expends more energy in growth and care of offspring (much more care for mammals; less so for birds). Another profound consequence of internal fertilization is that a male cannot be 100% certain that any offspring a female produces are his progeny. Because it is possible for a female to mate with multiple males, any of whom could be the father, male mammals and birds will always have some degree of **paternity uncertainty.**

In the next few sections, we'll continue to explore how the physical differences between males and females in reproductive investment, along with paternity uncertainty, have led to the evolution of differences in male and female reproductive behavior.

TAKE-HOME MESSAGE 9•10

In mammals and many other types of animals, there are important physical differences between males and females relating to reproduction. Fertilization usually takes place in the female. Lactation occurs only in the female. And in species where fertilization occurs inside the female, males cannot be certain that offspring are their progeny. These physical differences have led to the evolution of differences in male and female reproductive behavior.

9•11

Males and females are vulnerable at different stages of the reproductive exchange.

Suppose you are on your college campus and a person of the opposite sex comes up to you and says, "Hi. I have been noticing you around campus. I find you very attractive. Would you go out with me tonight?" What percentage of men would answer *yes*? What percentage of women would answer *yes*? In a study conducted at Florida State University in 1978 and 1982, this encounter was set up for both males and females, and the percentage answering yes was 50% for both males and females.

Now imagine a different version of the encounter. In this case, the question was: "Would you have sex with me tonight?" Among the men asked, 75% said *yes*; among the women, not a

FIGURE 9-19 The choosier sex. Males and females differ in their attitude toward mating opportunities. For the female more than the male, the choice of the wrong mate could have expensive consequences.

single one said *yes*. It is tempting to interpret such results as a consequence solely of the culture of Western society, characterized by significant differences in the expectations and tolerances of male and female sexual conduct. However, cultural expectations don't adequately explain the consistency of the pattern found across Western and non-Western societies: men and women differ in their approach to sexuality.

Gender roles with regard to sexual willingness are greatly influenced by socialization, and they can vary cross-culturally. Studies of male-female differences in selectiveness about sexual partners, however, drawing on a wide range of industrial and preliterate human cultures—including many as far removed from Western influence as possible, such as the Trobriand Islanders of Melanesia in the South Pacific—most commonly reveal a difference in men's and women's willingness to have sex. Why might this difference exist?

Humans, like nearly all mammals, are characterized by greater initial reproductive investment by females. As we saw in the college sex study, women are more hesitant to take a mating opportunity, while men are much more willing. For females, at the point of mating, the cost of a poor choice can have significant consequences, from an evolutionary perspective—pregnancy and lactation, with offspring from a low-quality male or a male who deserts her and does not provide any parental investment or access to valuable resources (**FIGURE 9-19**). For a male, the consequences of a poor choice are less dire—little beyond the time and energy involved in mating.

Two differences in the sexual behavior of males and females have evolved:

1. The sex with the greater energetic investment in reproduction will be more discriminating when it comes to mating.
2. Members of the sex with less energetic investment in reproduction will compete among themselves for access to the higher-investing sex.

A dramatic illustration of how a high reproductive investment leads to the evolution of choosiness in mating behavior comes from the insect world. When bush crickets mate, the male loses about a quarter of its body weight contributing a massive ejaculate, which the female then uses for energy (**FIGURE 9-20**). (It can represent up to one-tenth of her lifetime caloric intake.)

Not surprisingly, male crickets are very choosy when selecting a mate. Their contribution during sex would be the equivalent of nearly 50 pounds of semen in humans, a much greater up-front investment than usual. As a consequence, male bush crickets reject small females that would produce relatively few offspring. (Females, you won't be surprised to learn, spend a great deal of effort courting males.)

So far, we have been talking about the consequences of reproductive investment. We see how the sex with the greater reproductive investment must be choosy. This isn't to say that paternal investment will not or should not occur, however. In many species, particularly those in which the offspring are not

When bush crickets mate, the male loses about a quarter of his body weight contributing a massive ejaculate that the female uses for energy. Why has this led to the evolution of choosiness among male bush crickets?

FIGURE 9-20 A costly decision. Because of the investment the male bush cricket is required to make when fertilizing a female, he chooses a mate very carefully.

well-developed at birth (including humans), offspring survival is significantly increased with paternal investment. As a consequence, this benefit frequently outweighs the evolutionary risks associated with paternity uncertainty. The female's reproductive success is vulnerable because a poor choice of mate could be disastrous. Males are not made vulnerable through their mating choices, because the matings are of little energetic consequence. However, there is a point at which males *are* vulnerable in the reproductive exchange.

The point of greatest vulnerability for males comes when they provide parental care to offspring. Due to paternity uncertainty, there is some chance that the male may be investing in offspring that are not his own. This is an action that has significant evolutionary costs: rather than increasing his own fitness, he is increasing the fitness of another male. This isn't to say that paternal investment will not or should not occur, however. In many species, particularly those in which the offspring are not well-developed at birth (including humans), offspring survival is significantly increased with paternal investment. As a consequence, this benefit frequently outweighs the evolutionary risks associated with paternity uncertainty. Females, conversely, are not at all vulnerable at the point of providing parental care to offspring, because a female can be completely certain that the offspring she gives birth to are her own.

In the next two sections, we explore the evolution of reproductive behaviors that help males and females reduce their vulnerability and maximize their reproductive success. First we explore competition and courtship tactics used in selecting and securing a mate. And then we investigate the various strategies of "mate guarding" employed by males to increase their paternity certainty.

As we explore some general patterns of reproductive behavior among males and females, it is important to keep two critical points in mind:

1. There is tremendous variability among different species in male and female behaviors. For example, the use of DNA fingerprinting to identify the parents of bird offspring has lead to some surprising revelations. In significantly more cases than researchers originally predicted, offspring have different fathers than observers believed. It seems that things are not always as they appear to observing biologists. As a consequence, much of the new research calls into question some long-held assumptions about the behavioral consequences of physical differences in reproductive investment among birds versus mammals. There are general behavioral patterns among birds and among animals, but it is increasingly clear that these are not universal features of their biology, and a reasonable skepticism is important when identi-fying and interpreting broad trends across large groups of species.

2. Also, throughout history there have been many cases of people using observations and scientific findings to justify a wide variety of discriminatory thoughts and behaviors—for example, that if among mammals males "naturally" tend to be less faithful to their mates, such observations reduce individual responsibility for behaviors among humans. Such thinking ignores the tremendous variation in behavior among species and the power of cultural norms and socialization to influence and shape human behavior, while encouraging the inappropriate assumption that biology can supply meaningful insights into morality or ethical decision making.

TAKE-HOME MESSAGE 9•11

Differing patterns of investment in reproduction make males and females vulnerable to exploitation at different stages of the reproductive process. This has contributed to the evolution of differences in the sexual behavior of males and females. The sex with greater energetic investment in reproduction is more discriminating about mates, and members of the sex with less energetic investment in reproduction compete among themselves for access to the higher-investing sex.

9•12

Tactics for getting a mate: competition and courtship can help males and females secure reproductive success.

Throughout the animal kingdom, there is tremendous variation in the relative investment in reproduction among males versus females. A common pattern, however, is that females make greater investment in producing and raising offspring, while the male contribution is less. Accordingly, in these cases, males generally increase their reproductive success by mating with many females and have evolved to compete among themselves to get the opportunity to mate. Females, conversely, do not increase their reproductive success through extra matings. Instead, they increase their reproductive success by caring for their offspring and being choosy about selecting a mate.

Female choosiness (and the male-male competition that it leads to) tends to increase the likelihood that a female will select only those males that have plentiful resources or relatively high-quality genes, either of which is beneficial to the female, causing her to produce more offspring or better offspring—where "better" may mean increased disease resistance or simply the possession of physical traits that will be found attractive by future mates. Female choosiness is manifested by four general rules.

1. Mate only after subjecting a male to courtship rituals. In many bird species, including the western grebe, females require the male to perform an elaborate and time-consuming courtship dance before she will mate with him. For the grebe, this courtship dance involves fancy dives into water, graceful hovering, various head movements, and flamboyant twists and turns. The courtship process can go on for several days. But if the male passes the time-consuming audition, he can generally be counted on to stick around to see a brood through hatching and early care (**FIGURE 9-21**).

Among bower birds, a courting male builds a small thatched structure, at which he makes displays that can involve complex dance steps and dramatic poses, while courting a female. The female will walk around the structure and inspect it. Only if she deems it satisfactory will she mate with the male. Males that are rejected by multiple females will usually tear down their bower and rebuild it from scratch to try again to attract a female.

2. Mate only with a male who controls valuable resources. Territorial defense is a common means by which males compete for access to females. Among arctic ground squirrels, for example, a female chooses a mate based, in part, on the territory he defends, which is where she will reside after mating. With greater quality and quantity of resources in

FACTORS IN MATE SELECTION

COURTSHIP RITUALS
A female grebe requires the male to perform a courtship dance before she will mate with him.

GIFTS UP FRONT
A female hanging fly will not mate with a male unless he brings her a large offering of food.

CONTROL OF VALUABLE RESOURCES
A female arctic squirrel is attracted to a male squirrel that controls the best territory.

GOOD LOOKS
A female peacock is attracted to a male with the most beautiful tail feathers.

FIGURE 9-21 **Four factors that influence a female's choice of mate.**

his territory, a male is better able to attract females, whose reproductive success can be increased if the territory is rich in resources.

> It is a truth universally acknowledged, that a single man in possession of a good fortune must be in want of a wife.
>
> —Jane Austen, *Pride and Prejudice,* 1813

3. Mate only with a male who contributes a large parental investment up front. Better than a believable pledge to commit resources to future offspring is an actual, on-the-spot exchange in which a female requires a male to give her his parental investment up front, in the form of resources that will help her maximize her reproductive success. In the hanging fly, for example, a female will not mate with a male unless he brings her a big piece of food—usually a dead insect. If a male offers an acceptable food item or other token, called a **nuptial gift,** as part of his courtship behavior, she will mate with him (see Figure 9-21). The larger the food item, the longer she will mate; and the more she eats, the larger the number of eggs that she will lay. After about 20 minutes of mating, though, when a male has transferred all of the sperm that he can, he is likely to break off the mating and take back whatever remains of the "gift," which he may use to try to attract another mate. Nuptial feeding is common among birds and insects.

4. Mate only with a male that has a valuable physical attribute. Male-male competition for the chance to mate with females can also take a more literal form: actual physical contests. Across the animal kingdom, from dung beetles to hippopotamuses weighing more than 5,000 pounds, male-male contests determine the dominance rankings of males. Females then mate primarily with the highest-ranking males.

In a similar process, rather than choosing the best-fighting or largest males, females sometimes base their choice on some physical attribute, such as antler size in red deer, the bright red chest feathers of frigate birds, or the elaborate tail feathers of the male peacock (see Figure 9-21). In each case, the physical feature serves as an indicator to females of the relative quality of the male, possibly because the feature is correlated with the male's health.

Q Why do so few women get into bar-room brawls?

With so many examples of male-male fighting as part of the courtship rituals that have evolved for attracting a mate, it is reasonable to ask: why is it so rare for females to fight? And why do females generally not have to advertise their health with flashy feathers or other ornamentation? The answer is that as long as females are making the greater investment in reproduction, nearly any male will mate with them. Consequently, there is nothing more to be gained by trying to outcompete other females or otherwise attract the attention of males.

Moreover, among humans, social and cultural values also have powerful influences over mating behavior, complicating interpretation. We are not lumbering robots, destined to follow some genetic program. It's important to take into consideration that researchers have noted some subtle manifestations of female health and fertility, including waist-to-hip ratios and patterns of facial and body symmetry, in many different species. There is a rich and complex world of mating tactics, many of which we do not fully understand.

TAKE-HOME MESSAGE 9•12

As a consequence of male-female differences in initial reproductive investment, males tend to increase their reproductive success by mating with many females and have evolved to compete among themselves to get the opportunity to mate. Females increase their reproductive success not through extra matings but rather by caring for their offspring and being choosy about selecting a mate.

9•13

Tactics for keeping a mate: mate guarding can protect a male's reproductive investment.

If a male simply abandons a female after mating and searches for other mating opportunities, rather than making any investment in the potential offspring, he has no risk of investing further energy in offspring that are not his. This is one way to minimize the potential costs associated with paternity uncertainty. If you don't play, you can't lose.

This strategy, however, is not necessarily the most effective way for a male to maximize his reproductive success. If a male has a hundred or even a thousand matings but no offspring survive, that behavior is not evolutionarily successful. Consequently, in species for which offspring's survival can be enhanced with greater parental investment, there is an incentive for a male to provide some parental care, even though such behavior makes him vulnerable to paternity uncertainty.

Q *Why do so few females guard their mates as aggressively as males do?*

In situations in which males provide parental care, it is common for the male to reduce his vulnerability through some form of **mate guarding.** Mate guarding is a consequence of (and an attempt to reduce) paternity uncertainty. As long as the offspring emerge from the female's body, she can be certain that they contain her genes. In contrast, a male inhabits a "danger zone" that lasts as long as the female is fertile. If she mates with any other males during this time, the offspring she produces may not be his. If he is going to help raise the offspring, he benefits by minimizing his risk in the danger zone. It is during this period that mate guarding is particularly common.

Mate guarding methods range from the simple to the macabre. If a male wants to ensure that a female does not mate with another male, why bother to stop mating at all? In many species, males take this approach to reduce their risk in the danger zone. Among house flies, even though the male has completed the transfer of sperm to the female in 10 minutes of copulation, he does not separate from her for a full hour. Moths go even further and continue to mate for a full 24 hours. And in the extreme case of this strategy, certain frog species continue individual bouts of mating for several months (**FIGURE 9-22**). If humans mated for a similar percentage of their lives, they'd spend almost 10 years on a single round of intercourse.

Some frog species continue individual bouts of mating for several months continuously. If humans mated for a similar percentage of our lives, a single round of intercourse would last almost ten years!

FIGURE 9-22 Preventing paternity uncertainty. A prolonged period of mating prevents the female from accessing other males, assuring the male of reproductive success.

EXTREME MATE GUARDING

FIGURE 9-23 A reasonable trade-off? The male black widow spider ensures his paternity by an extreme form of mate guarding.

In a slightly subtler form of mate guarding that occurs in reptiles, insects, and many mammalian species, males block the passage of sperm into the female by producing a copulatory plug. Formed in the female reproductive tract from coagulated sperm and mucus, copulatory plugs can be very effective. Male garter snakes that encounter a female snake with a copulatory plug, for example, do not attempt to court or mate with her, treating her instead as if she were not available.

A much more extreme form of mate guarding occurs in the black widow spider: the male breaks off his sexual organ inside the female, preventing her from ever mating again. Interestingly, when the act is completed, the female usually kills and eats the male (**FIGURE 9-23**). In sealing his mate's reproductive tract, the male assures himself of fathering the offspring, and in consuming her mate's nutrient-filled body, the female gets resources that help her produce the offspring.

TAKE-HOME MESSAGE 9•13

Because paternity certainty is low in animals with internal fertilization, males that provide parental care are vulnerable to loss of their investment. As a consequence, mate guarding and other tactics have evolved to increase paternity certainty for species in which males contribute to parental care.

9•14

Monogamy versus polygamy: mating behaviors can vary across human and animal cultures.

As we continue our tour of animal mating behavior, we turn again to the elephant seal. During the breeding season, these animals appear on islands off the coast of northern California. In December of each year, the males begin to compete with each other for possession of the beach. Through bloody fights, they establish dominance hierarchies, with the biggest males—which are 13 feet long and weigh more than 2 tons—generally winning control of the beach.

In mid-January, the females arrive and are ready to mate. They congregate in large groups on just a few prime beaches. Because the females stick close together, the biggest males, who control the prime beaches, can dominate the other males in the competition for sexual access to females (**FIGURE 9-24**). In one study that observed 115 males, the 5 highest-ranking males fathered 85% of the offspring. While nearly every female will mate and produce offspring, the majority of males never get the chance to mate during the 10–20 years of their life.

The elephant seals' mating pattern exemplifies **polygamy,** a system in which some individuals attract multiple mates while other individuals attract none. Polygamous mating systems can be subdivided into **polygyny** and **polyandry.** In polygyny, individual males mate with multiple females; in polyandry, individual females mate with multiple males. Polygamy can be contrasted with **monogamy,** in which most individuals mate and remain with just one other individual. Polygamy and monogamy are two types of **mating systems,** which describe the patterns of mating behavior in a species. In this section, we explore the features of environments and species that influence mating systems and survey the range of mating systems observed in nature.

As we have seen, throughout the animal world, males tend to make less parental investment than females. As a consequence, females are choosy about whom they mate with and males compete for access to mating opportunities.

POLYGAMY IN ELEPHANT SEALS

FIGURE 9-24 King of the beach. The biggest, best-fighting male elephant seals control the beach and mate with as many females as possible.

In a polygamous mating system, some individuals have multiple mates, while others have few or none.

DOMINANT MALE ELEPHANT SEAL

LOW-RANKING MALE ELEPHANT SEALS

OFFSPRING

In an extremely polygynous mating system, such as in elephant seals, the vast majority of males have no reproductive success at all.

Not surprisingly, multiple females often end up selecting the same male—usually a male on a territory rich in resources or a male with unusually pronounced physical features, such as antler size. Such female selection is the reason for the fancy ornamentation we see in male peacocks. Although, the male peacock's tail is like a giant bull's-eye to predators, the number of eyespots on the tail is directly related to how well he can attract a mate: below 140 eyespots, he gets no mates; at 150, he gets two to three mates on average; and with 160 eyespots he gets six or more mates (see Figure 9-21). Given their greater parental investment, virtually *all* females are able to attract a mate, regardless of whether they are the most fit, high-status female or the least.

Mating systems are not as easy to define as the elephant seal example might lead us to believe. Three issues, in particular, complicate the task. First, there are often differences between animals' mating behavior and their bonding behavior. That is, it may seem that a male and female have formed a **pair bond**—in which they spend a high proportion of their time together, often over many years, sharing a nest or other "home" and contributing equally to parental care of offspring in what appears to be a monogamous relationship. Closer inspection (often including DNA analysis of the offspring) sometimes reveals, however, that the male and/or the female may be mating with other individuals in the population, and perhaps the mating system is better described as a variation on polygamy. A second difficulty in defining a species' mating system arises because the mating system may vary within the species. That is, some individuals may be monogamous, while others are polygamous. And the mating system may even change over the course of an individual's life. A third difficulty is that males and females often differ in their mating behavior. In the elephant seals described above, it could be said that the females are all mating monogamously, while the males are polygamous.

Within almost any large group of related animal species, there will be a variety of species with each type of mating system, so it is not usually possible to make generalizations. And in a recent surprising discovery, researchers were able to alter the behavior of males of a polygynous species of meadow voles by inserting into their brain a single gene from a related but monogamous species of prairie voles. With the new gene, the formerly polygynous males showed a preference for their current partner over other females, and their ability to form a long-term bond with their partner increased.

Examination of birds and mammals in general, however, reveals one sharp split. As we discussed in Section 9-10, the vast majority of female mammals have greater parental investment than male mammals. In birds, females and males have a more equal parental investment.

Does the difference in parental investment patterns in birds and mammals lead to different mating systems? Yes. In mammals, polygyny is the most common mating system across all large groups, from rodents to primates. Polygyny is a consequence of the significant female investment and lesser male investment. Males can generally benefit more by seeking additional mating opportunities, leading to male-male competition. In birds, the relatively equal parental investment by males and females has led to much less polygyny. Among the approximately 10,000 species of birds we know about, more than 90% appear to be monogamous (**FIGURE 9-25**).

Q Are humans monogamous or polygamous?

And what of humans? Across a variety of cultures, males consistently have greater variance in reproductive success than females—that is, some males have very high reproductive success while many others have little or none. **FIGURE 9-26**, for example, presents data from a study of the Xavante Indians of Brazil. The *average* number of children does not differ for men and women, but some men have very large numbers (as many as 23!) and many have none. There is significantly less variability in

MONOGAMOUS BIRDS

Of the approximately 10,000 species of birds, more than 90% appear to be monogamous.

FIGURE 9-25 Let's stay together. Parental investment that is roughly equal often leads to monogamous mating behavior in birds.

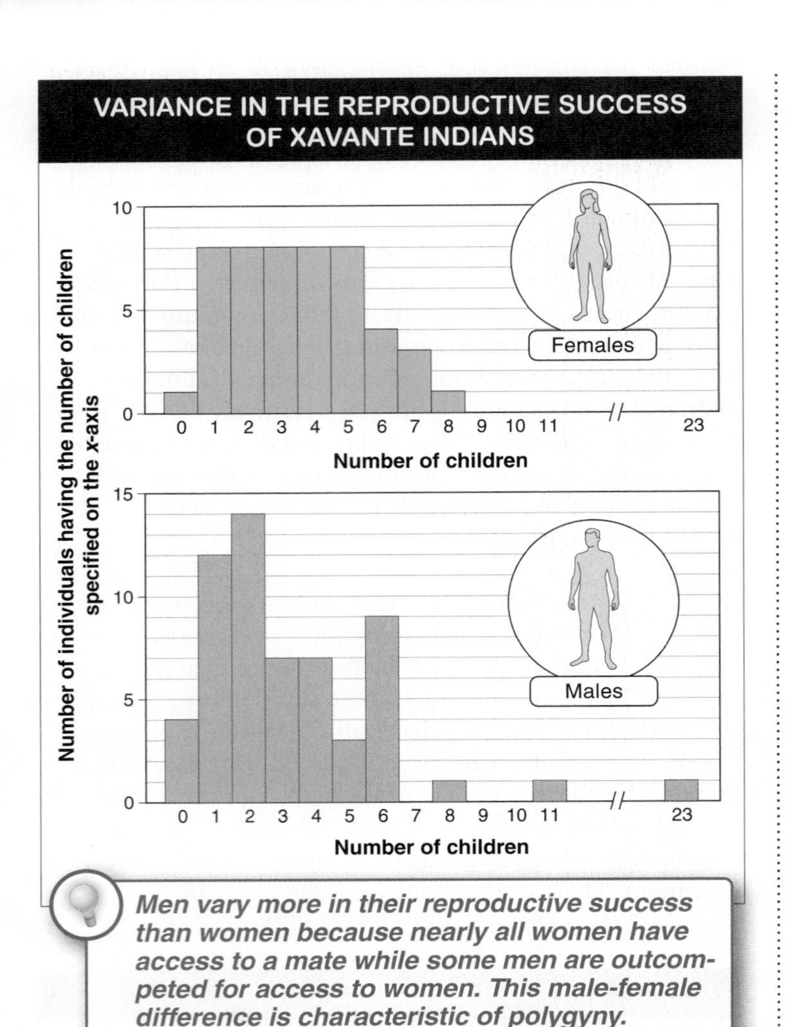

FIGURE 9-26 Larger male variance in reproductive success. Studies of the Xavante Indians of Brazil show that the range in number of offspring over their lifetime is much greater for men than for women.

reproductive success from one woman to another. Such differences in the variance in reproductive success among males and females, which can be significantly influenced by socialization and other cultural forces, are generally associated with polygynous mating systems, but the difference between the sexes found in the Brazilian study is quite small when compared with many other mammalian species, such as elephant seals, and is close to that seen in populations with a monogamous mating system. Humans, consequently, seem to have a mating system close to, but not completely, monogamous. Consistent with the Brazilian study is the fact that the majority of societies in the world permit a man to have more than one wife.

TAKE-HOME MESSAGE 9•14

Mating systems—monogamy, polygyny, and polyandry—describe the variation in number of mates and the reproductive success of males and females. They are influenced by the relative amounts of parental investment by males and females.

9•15

Sexual dimorphism is an indicator of a population's mating behavior.

Male elephant seals are three to four times the size of female elephant seals. In contrast, the males and females of most bird species are the same size. In these species, even expert bird-watchers often cannot distinguish between the sexes, except when the female bird is carrying eggs (**FIGURE 9-27**). When the sexes of a species differ in size or appearance, this is called **sexual dimorphism.** Why do species have such dramatic differences in the degree to which males and females resemble each other in size and appearance?

Q *It's almost impossible to distinguish males from females in most bird species. Why does this simple observation tell us a huge amount about their sexual behavior, including that they are monogamous?*

Body size is an important clue to behavior. We have seen how male elephant seals have a winner-take-all tournament for control of the beach, and as the tournament winner, a male has access to mating opportunities. Because the largest individual will have the most offspring, there is selection for larger and larger body size. The smaller males will have relatively low reproductive success. There is no selection for larger body size among female elephant seals, because females of any size can mate. Hence the dramatic size difference between the sexes.

SEXUAL DIMORPHISM

BEHAVIORS ASSOCIATED WITH SEXUAL DIMORPHISM

- One parent invests more in caring for the offspring.
- Mating system tends toward polygamy.
- One sex (usually females) is choosier when selecting a mate.
- One sex (usually males) competes for access to mating opportunities with the other sex.

SEXUAL MONOMORPHISM

BEHAVIORS ASSOCIATED WITH SEXUAL MONOMORPHISM

- Both parents invest (approximately) equally in caring for the offspring.
- Mating system tends toward monogamy.
- Both sexes are equally choosy when selecting a mate.

FIGURE 9-27 Size says a lot. Species with large males and small females are very likely to be polygamous. Species with similarly sized males and females are likely to be monogamous.

In addition to body size, coloration is also a clue to behavior. Because females of some polygynous species choose the males with the brightest or flashiest coloration, rather than the largest males, male-male competition sometimes results in differences in physical appearance between the sexes instead of (or in addition to) differences in size.

Q Men are bigger than women. What does that tell us about our evolutionary history of monogamy versus polygamy?

What happens when there is little male-male competition for mates? Among bird species, males can provide significant investment in offspring. Eggs have to be incubated, and many chicks emerge in a very poorly developed condition and it takes two parents to raise them. So a pair tends to stay together for a season or longer. If each female can only pair up with one male, there really isn't much competition for mates. And without male-male competition, there is little selection for increased size. As a consequence, there is little sexual dimorphism among most bird species.

In turn, we can predict a great deal about the parental practices of a species just by looking at a picture of a male and a female and examining the ratio of body sizes. If the two are dramatically different, as in elephant seals, you can be confident that the smaller sex is doing most of the care of the offspring and that the species is more likely to be polygamous than monogamous.

TAKE-HOME MESSAGE 9•15

Differences in the level of competition among individuals of each sex for access to mating opportunities can lead to the evolution of male-female differences in body size and other aspects of appearance. In polygynous species, this results in larger males that are easily distinguished from females visually. In monogamous species, there are few such differences between males and females.

COMPLEX FORMS OF ANIMAL COMMUNICATION

FIGURE 9-29 Dancing, signing, and speaking. Animals have evolved a variety of ways to convey complex information to others.

The evolution of language influences many of the other behaviors discussed throughout this chapter. The evolution of reciprocal altruism, for example, may be influenced as language makes it easier for individuals to convey both their needs and their resources to other individuals. Similarly, in courtship and the maintenance of reproductive relationships, language gives individuals a tool for conveying complex information relating to the resources and value they may bring to the interaction.

TAKE-HOME MESSAGE 9•16

Methods of communication—chemical, acoustical, and visual—have evolved among animal species, enabling them to convey information about a variety of features of their condition and situation. These abilities influence fitness and the evolution of virtually all other behaviors.

9•17

Honest signals reduce deception.

Among Natterjack toads, found in northern Europe, males produce booming calls to attract females. Because females desire large males and size determines the volume of the calls, females will push their way through murky swamps until they reach the loudest croaker. They are drawn by calls that can be heard a mile away and are often louder than the legal noise limit for a car engine (**FIGURE 9-30**).

It is not surprising that females desiring the largest possible male should select their mates based on the volume of the call. The Natterjack's call is an **honest signal,** a signal that cannot be faked and that is given when both the individual making the signal and the individual responding to it have the same interests. An honest signal is one that carries the most accurate information about an individual or situation, and animals that respond to signals of any sort have evolved to value most highly those signals that cannot be faked. The Natterjack's call, for example, cannot be faked by a small Natterjack toad. And so there is evolutionary pressure for animals to both produce and respond to honest signals, because these behaviors enable them to maximize their fitness by selecting the best possible mate.

When one animal can increase its fitness by deceiving another, deception can also be expected to evolve. An example might be baby birds begging for food from their parent. If an allele that causes a chick to exaggerate its need for food leads to faster growth and better health for that individual, the allele is likely to increase in frequency in the population.

Communication and signaling, therefore, are features of populations that are continually evolving. We would expect this evolutionary "arms race" between honest signals and deception

HONEST SIGNAL

Female Natterjack toads make mating decisions based in part on how loud a male's call is. Loud calls can only be made by large males—smaller males can't fake it—so it is an honest signal, carrying accurate information.

FIGURE 9-30 **Loud and true.** The female Natterjack toad is attracted to the biggest male with the loudest call—an "honest signal," because there is no way for a small toad to fake the call volume of a larger toad.

to lead, over time, to both increasingly unambiguous signals and ever more sophisticated patterns of deception.

We have only touched the surface of the rich field of animal behavior here. Still, we've seen how our understanding of the behavior of animals, including humans, has been greatly expanded by applying scientific thinking, an experimental approach, and careful consideration of the process by which the evolutionary mechanism of natural selection shapes populations. All of this has been made possible through the subtle but important shift in thinking that we highlighted at the beginning of this chapter: an organism's phenotype is not limited to its physical traits but includes the organism's behaviors, as well. And consequently, behaviors respond to selective pressures and evolve.

TAKE-HOME MESSAGE 9•17

Animals have evolved to rely primarily on signals that cannot easily be faked, in order to gain the maximum amount of information from them.

streetBio

Knowledge You Can Use

How to win friends and influence people.

As we have seen, in the animal world it is very, very rare to find individuals engaging in costly behaviors that benefit unrelated individuals; with few exceptions, we see little that resembles animal "friendship."

Humans are a rare exception—every day, in myriad ways, we see friendly behaviors. Just as we put money in the bank for a rainy day, we buffer ourselves from the world's uncertainties by storing goodwill in our neighbors. Cooperators have evolutionary advantages over loners.

Q: Why is "kindness" a risky behavior evolutionarily? From an evolutionary perspective, we remain vulnerable. With each costly act we perform for unrelated individuals, strangers and friends alike, there is the risk that our efforts will be in vain, our energies lost. For that reason, in our evolution as reciprocal altruists, we acquired a cautiousness, a hesitancy to stick our necks out.

Q: How can we help others feel less vulnerable and more willing to cooperate? Our ability to override the impulses that often push us toward selfishness is one of the hallmarks of being human. Still, there is no reason to put people in situations where they may be better off rejecting friendship. In each interaction there is an individual on the other side who is also feeling vulnerable. By taking steps, often absurdly simple, to address the unconscious vulnerabilities that others feel, we can increase the likelihood that they choose cooperation, reciprocity, and friendship.

- **Learn and use other people's names.** It tells them you recognize them specifically and, consequently, that you understand exactly who it is that you "owe." Smile and make eye contact—even when attempting to change lanes in traffic. These gestures feel like the beginning of a relationship, thus stimulating favor-granting instincts.
- **Embrace etiquette:** acknowledge your debts to others. Be effusive and public in your thanks for kindnesses done to you. Send thank-you cards.
- **Take the first step in reciprocity.** Human cooperation is so tied to reciprocal exchange that even tiny gestures of good faith can play an important role in building relationships. Whenever possible, give gifts. Even small ones. Even when they are not required. Especially when they are not required.
- **Develop a good reputation:** for generosity, for loyalty, for remembering and acknowledging kindnesses done to you.

Can you think of other steps?

BIG IDEAS IN EVOLUTION AND BEHAVIOR

1 Behaviors are traits that can evolve.

Behavior encompasses any and all of the actions performed by an organism. When a heritable trait increases an individual's reproductive success relative to that of other individuals, that trait tends to increase in frequency in the population. Behavior is as much a part of an organism's phenotype as is its anatomical structures, and as such can be produced and shaped by natural selection. Innate behaviors are present in all individuals in a population and do not vary much from one individual to another or over the life span of an individual. Organisms are well-prepared to learn behaviors that have been important to their reproductive success over the course of their evolutionary history and less prepared to learn behaviors irrelevant to their evolutionary success. It is not necessary that an organism consciously tries to maximize its reproductive success in order for a behavior to evolve.

2 Cooperation, selfishness, and altruism can be better understood with an evolutionary approach.

Many behaviors in the animal world appear altruistic. In almost all cases, the apparent acts of altruism are not truly altruistic and have evolved as a consequence of either kin selection or reciprocal altruism and, from an evolutionary perspective, are beneficial to the individual engaging in the behavior. Kin selection describes apparently altruistic behavior in which an individual assisting a genetic relative compensates for its own decrease in direct fitness by helping increase the relative's fitness and, consequently, its own inclusive fitness. Reciprocal altruism describes the situation in which an individual engages in an altruistic-appearing act toward another individual but is likely to get something of value at a later time. Behaviors that are good for the species or population but detrimental to the individual exhibiting the behavior are not generally produced under natural conditions.

3 Sexual conflict can result from disparities in reproductive investment by males and females.

In mammals and many other animal species, there are important physical differences between males and females relating to reproduction. These physical differences have led to the evolution of differences in male and female reproductive behaviors. Males tend to increase their reproductive success by mating with many females and have evolved to compete among themselves to get the opportunity to mate. Females do not increase their reproductive success through extra matings, but rather by caring for their offspring and being choosy when selecting a mate. Mate guarding and other tactics to increase paternity certainty evolve when males provide parental care. Mating systems describe the variation in mate number and reproductive success of males and females. Differences in the level of competition among the individuals of each sex for access to mating opportunities can lead to the evolution of male-female differences in body size and other aspects of appearance.

4 Communication and the design of signals evolve.

Methods of communication—chemical, acoustical, and visual—have evolved in animal species, enabling individuals to convey information about a variety of features of their condition and situation. These abilities influence fitness and the evolution of virtually all other behaviors. Animals have evolved to rely primarily on signals that cannot easily be faked, in order to gain the maximum amount of information from them.

KEY TERMS

altruistic behaviors, p. 338
behavior, p. 331
communication, p. 360
direct fitness, p. 342
female, p. 348
fixed action pattern, p. 333
group selection, p. 347
honest signal, p. 362
inclusive fitness, p. 342
indirect fitness, p. 342
innate behavior, p. 332
instinct, p. 332
kin selection, p. 338
language, p. 361
learning, p. 334
male, p. 348
mate guarding, p. 355
mating system, p. 356
monogamy, p. 356
nuptial gift, p. 354
pair bond, p. 357
paternity uncertainty, p. 350
pheromone, p. 360
polyandry, p. 356
polygamy, p. 356
polygyny, p. 356
prepared learning p. 334
reciprocal altruism, p. 338
reproductive investment, p. 348
sexual dimorphism, p. 358
total reproductive output, p. 349
waggle dance, p. 361

CHECK YOUR KNOWLEDGE

1. Why do human taste preferences exist?

a) They are culturally generated; consumers are influenced by suggestive advertising.
b) Human feeding choices influence energy intake and, consequently, fitness.
c) Humans can extract energy from a variety of non-food sources, and so taste preferences cause us to focus on foods that are more plentiful in our environment.
d) Fats actually do taste better than sugars.
e) Vitamins and minerals give food unusual tastes, which causes us to seek them out.

2. From an evolutionary perspective, behavior can best be viewed as:

a) a trait that arises by learning and not by natural selection.
b) non-heritable.
c) a trait subject to drift and mutation, but not natural selection.
d) part of the phenotype.
e) All of the above are correct.

3. Why is it so much easier for an infant to learn a complex language than for a college student to learn biology?

a) Language involves memorization only, without the need for understanding rules, while biology involves both.
b) Language is a feature with great evolutionary relevance for humans.
c) Learning biology was not a behavior with evolutionary relevance for humans.
d) Biology involves much more vocabulary than learning a complex language.
e) Both b) and c) are correct.

4. Babies in the United States quickly and easily develop a fear of snakes. Yet they don't easily develop a fear of guns. Why?

a) Humans cannot develop fears of inanimate objects.
b) Evolution can be slow in producing populations that are adapted to their environments.
c) Babies are more likely to encounter snakes than guns as they develop in the United States.
d) Fewer individuals are killed by guns than by snakes in the United States each year.
e) All of the above are correct.

5. In Belding's ground squirrels, why are females much more likely than males to engage in altruistic behavior by sounding alarm calls?

a) Belding's ground squirrels have a sex ratio that is biased toward females.
b) Females invest more in foraging and food storage, so they are more likely to lose their lives or their food if a predator attacks.
c) Belding's ground squirrels have a sex ratio that is biased toward males.
d) Females tend to remain in the area where they were born, so the females that call are warning their own kin.
e) Males forage alone, so their alarm calls are useless.

6. Gestational diabetes is thought to be the consequence of:

a) pregnant women decreasing their average daily amount of activity.
b) a mother withholding investment in future offspring in order to invest more in the current pregnancy.
c) a mother consuming too much sugar during gestation.
d) conflict between the mother and fetus with respect to how much food the fetus should be given; the mother is equally related to the fetus and to any future offspring, while the fetus is not equally related to itself and any subsequent siblings.
e) physiological constraints on the amount of insulin a mother can provide for the fetus.

7. Vampire bats:

a) sometimes regurgitate blood into the mouth of another bat that is close to starving, but the likelihood is a function of whether the individuals are genetically related.
b) are unusual in that they are one of the few animal species that exhibit kin selection.
c) sometimes regurgitate blood into the mouth of an unrelated bat that is close to starving.
d) exhibit reciprocal altruism but not kin selection.
e) There are no such things as vampire bats; they're found only in a Dracula novel.

8. Altruistic behavior in animals may be a result of kin selection, a theory maintaining that:

a) genes promote the survival of copies of themselves when behaviors by animals that possess those genes assist other animals that share those genes.
b) aggression within sexes increases the survival and reproduction of the fittest individuals.
c) companionship is advantageous to animals because in the future they can recognize those that have helped them and provide help to those individuals.
d) aggression between sexes increases the survival and reproduction of the fittest individuals.
e) companionship is advantageous to animals because in the future they can recognize those that have helped them and request help once again.

9. All of the following are necessary conditions for reciprocal altruism to evolve in a species except:

a) the ability to recognize different individuals.
b) the ability to punish cheaters who do not reciprocate.
c) repeated interactions with the same individuals.
d) at least one of the sexes must not disperse, so that some individuals always live near their kin.
e) None of the above is necessary for the evolution of reciprocal altruism.

10. In a situation in which males guard eggs and care for the young without help from the female, which of the following statements would most likely be correct?

a) Males are large and more brightly colored in order to attract the very best females.
b) Males and females are equally brightly colored, but males court females aggressively.
c) The population is monogamous with no sexual dimorphism.
d) A single male controls a harem of females to which he has exclusive reproductive access.
e) Females are more brightly colored than males and court males aggressively.

11. In mammals, as well as many other species, males generally compete for females. The best explanation for this phenomenon is:

a) males are more aggressive.
b) males, on average, have higher fitness.
c) females have a higher parental investment.
d) males are choosy.
e) females are better looking.

12. Mate guarding is a reproductive tactic that functions to:

a) reduce paternity uncertainty.
b) increase the female's investment in offspring.
c) reduce the male's reproductive investment.
d) reduce the female's fitness.
e) increase the number of mates to which a male has access.

13. Relative to birds, more mammalian species are:

a) polygynous.
b) monogamous.
c) polyandrous.
d) hermaphroditic.
e) sexually monomorphic.

14. In a species such as pigeons, in which males are almost indistinguishable in appearance from females, the most likely mating system is:

a) monomorphism.
b monogamy.
c) polygyny.
d) polyandry.
e) It is impossible to predict the mating system with only this information.

15. If you find a species of fish in which males are much more brightly colored and larger than females, what might you infer about their mating system?

a) The degree of sexual dimorphism does not give any information about the mating system.
b) They are simultaneous hermaphrodites.
c) They exhibit parallel monogamy.
d) They are serially monogamous.
e) They are polygynous.

16. Polygynous species:

a) usually employ external fertilization.
b) are usually sexually dimorphic, with males larger and more highly ornamented.
c) are usually sexually dimorphic, with females larger and more highly ornamented.
d) usually have males and females that are physically indistinguishable.
e) are more commonly found among birds than among mammals.

Short-Answer Question

1. Female moorhens are larger and more aggressive than males. They also compete among themselves for access to the smaller, fatter males. Which sex do you think provides more parental care? Explain your answer.

See Appendix for answers. For additional study questions, go to www.prep-u.com.

1 Life on earth most likely originated from non-living materials.

10.1 Complex organic molecules arise in non-living environments.

10.2 Cells and self-replicating systems evolved together to create the first life.

2 Species are the basic units of biodiversity.

10.3 What is a species?

10.4 How do we name species?

10.5 Species are not always easily defined.

10.6 How do new species arise?

3 Evolutionary trees help us conceptualize and categorize biodiversity.

10.7 The history of life can be imagined as a tree.

10.8 Evolutionary trees show ancestor-descendant relationships.

10.9 Similar structures don't always reveal common ancestry.

4 Macroevolution gives rise to great diversity.

10.10 Macroevolution is evolution above the species level.

10.11 The pace of evolution is not constant.

10.12 Adaptive radiations are times of extreme diversification.

10.13 There have been several mass extinctions on earth.

5 An overview of the diversity of life on earth: organisms are divided into three domains.

10.14 All living organisms are classified into one of three groups.

10.15 The bacteria domain has tremendous biological diversity.

10.16 The archaea domain includes many species living in extreme environments.

10.17 The eukarya domain consists of four kingdoms: plants, animals, fungi, and protists.

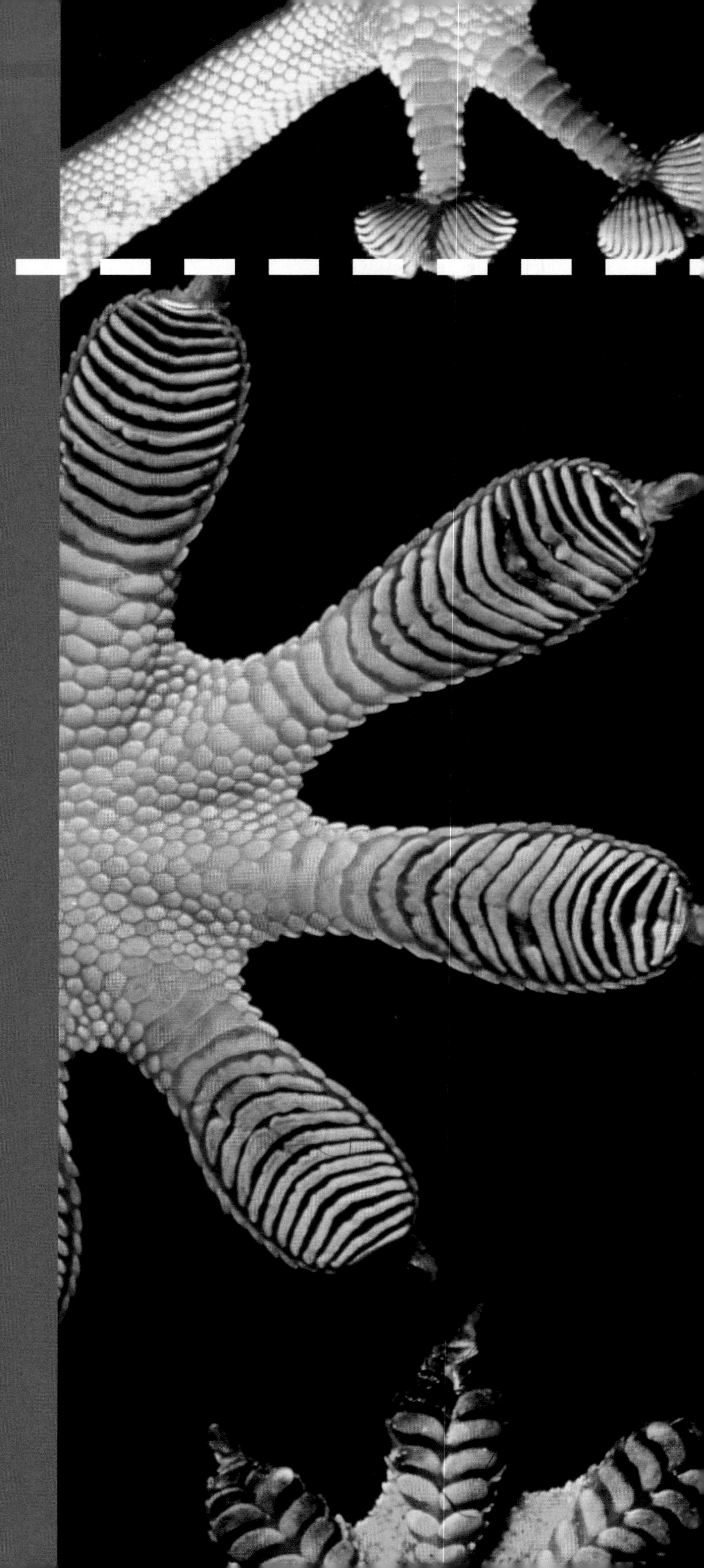

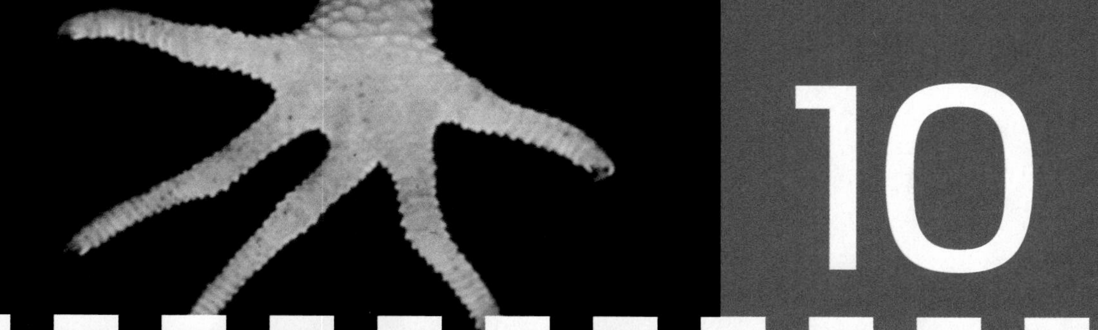

10

The Origin and Diversification of Life on Earth

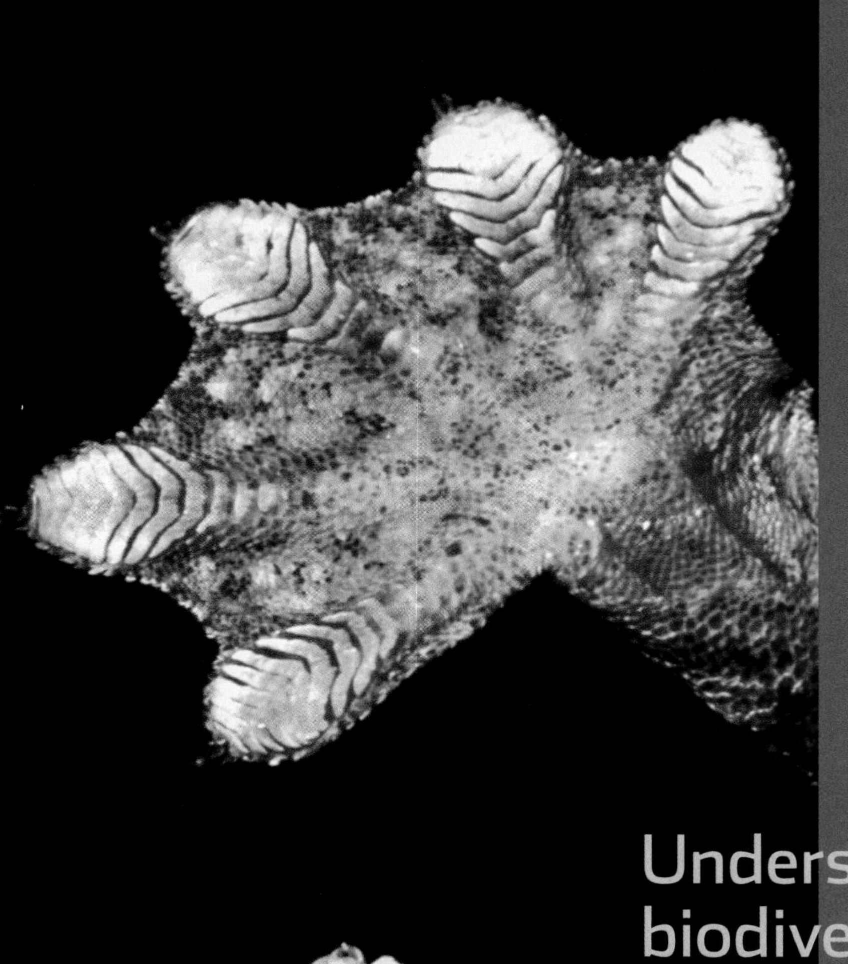

Understanding biodiversity

① Life on earth most likely originated from non-living materials.

Life on earth arose in an inhospitable environment. (Shown here: the formation of volcanic rocks.)

10•1

Complex organic molecules arise in non-living environments.

In the beginning, there was nothing. Now there is something. That, in a nutshell, describes one of the most important, yet difficult to resolve, questions in science: how did life on earth begin? We open our investigation of this question by describing a few things that we do know and exploring how they help us speculate about the things we do not yet know. But before we start, let's define what we mean by "life." Here we give just a basic definition: **life** is defined by the ability to replicate and by the presence of some sort of metabolic activity. In the next section we expand on this definition and explore the transition from non-living to living.

Earth formed about 4.5 billion years ago from clouds of dust and gases left over after the formation of the sun. The planet was initially super-hot, and as it very gradually cooled, a crust formed at the surface and condensing water formed the oceans. This probably took several hundred million years. The oldest rocks, found in Greenland, are about 3.8 billion years old. And the earliest life forms appeared not long after these rocks were formed: fossilized bacteria-like cells have been found in rocks

CONDITIONS ON EARTH AT THE TIME LIFE BEGAN

FIGURE 10-1 Darwin's "warm little pond." The first life on earth tolerated an atmosphere without oxygen.

that are 3.4 billion years old. (Note that 0.4 billion years—or 400 million years—is "not long" in geological terms.)

From that initial point at earth's formation, there is now tremendous **biodiversity**—variety and variability among all genes, species, and ecosystems—on the planet. In this chapter we explore in more detail how this biodiversity might have come to be, and how we name groups of organisms and determine the relatedness of these groups to each another. We begin by returning to the question of the origin of life on earth.

How did these first organisms arise? Some have suggested that life may have originated elsewhere in the universe and traveled to earth, possibly carried on a meteor. There is little evidence for this idea, however, and it is hotly debated whether microbes could even survive the multi-million-year trip to earth in the cold vacuum of space with no protection from ultraviolet and other forms of radiation. Experimental data have been unable to answer this question definitively. Most scientists believe, instead, that life originated on earth, probably in several distinct phases.

Phase 1: The formation of small molecules containing carbon and hydrogen

The conditions on earth around the time of the origin of life were very different from those of today. In particular, chemical analyses of old rocks reveal that no oxygen gas was present. The atmosphere included large amounts of carbon dioxide, nitrogen, methane, ammonia, hydrogen, and hydrogen sulfide. Most of these molecules were produced by volcanic eruptions. It was in this environment that some locations on earth probably served as the cradle of life, or what Darwin called the "warm little pond" (**FIGURE 10-1**).

Critical to the origin of life was the formation of small molecules containing carbon and hydrogen. Because of their chemical structures, these molecules bond very easily and in many ways, causing them to have a huge variety of forms and functions that make them indispensable for the chemical processes of life. There are several plausible scenarios for how the creation of small organic molecules may have occurred at this time. The most likely comes from some simple but revealing four-step experiments done in 1953 by a 23-year-old student named Stanley Miller and his advisor, Harold Urey (**FIGURE 10-2**).

1. They created a model of the "warm little pond" and their best estimate of earth's early atmosphere: a flask of water with H_2, CH_4 (methane), and NH_3 (ammonia).
2. They subjected their mini-world to sparks, to simulate lighting.
3. And they cooled the atmosphere so that any compounds formed in it would rain back down into the water.
4. Then they waited, and examined the contents of the water to see what happened.

THE UREY-MILLER EXPERIMENT

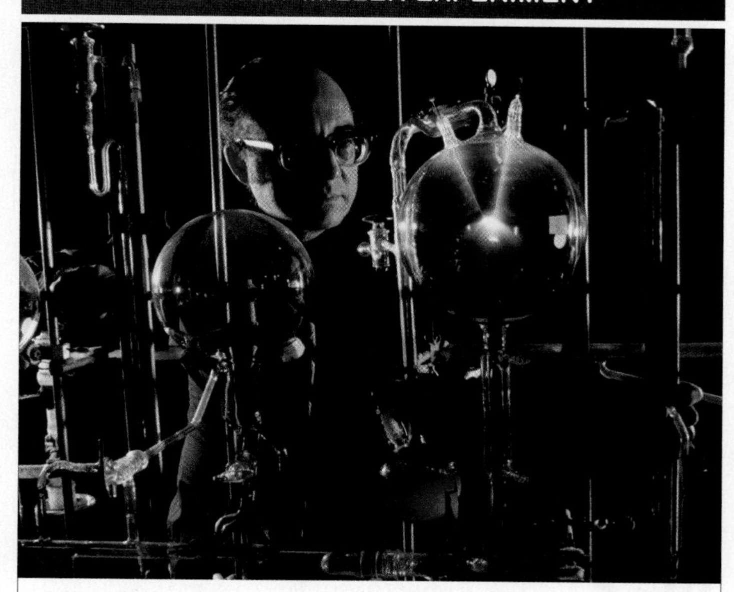

In 1953, Stanley Miller and Harold Urey developed a simple four-step experiment that demonstrated how complex organic molecules could have arisen in earth's early environment.

1. They created a model of the chemicals present in the "warm little pond" and atmosphere early in earth's history: H_2, CH_4 (methane), and NH_3 (ammonia).
2. The atmosphere was subjected to sparks, to simulate lightning.
3. The atmosphere was cooled so that any compounds in it would rain back down into the water.
4. They examined the water, looking for organic molecules.

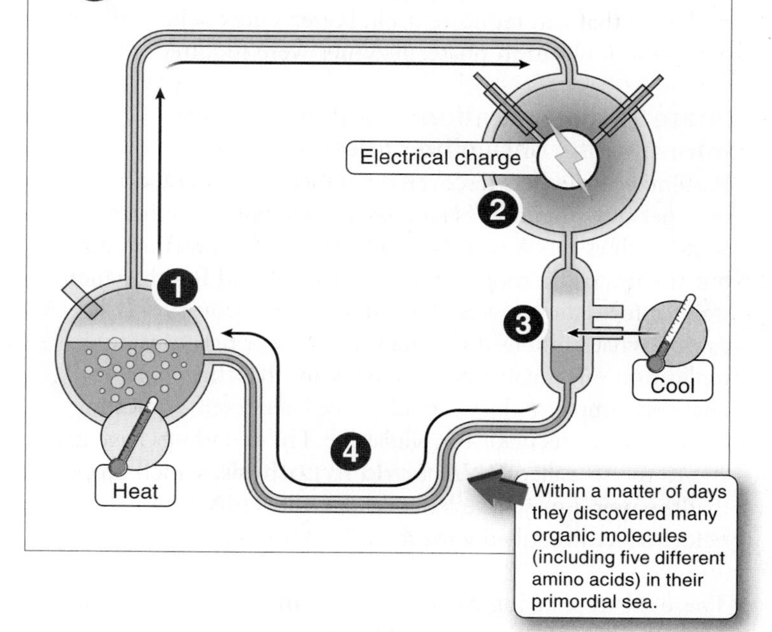

FIGURE 10-2 A promising first step. The Urey-Miller experiment generated organic molecules from hydrogen, methane, and ammonia.

They didn't have to wait long to get exciting results. Within a matter of days—not millions of years or even a few months—they discovered many organic molecules, including five different amino acids, in their primordial sea. (Using more sensitive equipment, re-analyses in 2008 of the residues from their original experiments revealed that all 20 amino acids present in living organisms were present in the residues.) This was the first demonstration that complex organic molecules could have arisen in earth's early environment. The Urey-Miller experiments are still far from proof that life arose from non-life, and questions remain, such as whether the environment they assumed to exist on the early earth was reasonable. Without oxygen gas, for example, earth's atmosphere would not have had an ozone layer protecting the early environment from the harmful ultraviolet radiation that breaks down methane and ammonia. Still, the experiments are a promising first step, suggesting that complex organic compounds—including amino acids, the primary constituents of proteins and an essential component of living systems—could have been produced from inorganic chemicals and lightning energy in the primitive environment of earth.

In the next section, we consider the final two phases by which these small organic molecules may have given rise to living organisms.

TAKE-HOME MESSAGE 10•1

Under conditions similar to those on early earth, small organic molecules form, and these molecules have some chemical properties of life.

10•2

Cells and self-replicating systems evolved together to create the first life.

After the generation of numerous organic molecules such as amino acids, the second phase in the generation of life from non-life was probably the assembly of these building block molecules into self-replicating, information-containing molecules. This is where things get a bit more speculative. It's complicated enough to generate a complex organic molecule, but it's a whole lot more complicated to generate an organic molecule that can replicate itself. Researchers believe that to get to the replication phase, enzymes were required.

Phase 2: The formation of self-replicating, information-containing molecules

Recently, researchers discovered a molecule that could unexpectedly function as an enzyme that links together nucleic acids (such as DNA or RNA). The molecule wasn't a protein but was, instead, a molecule of the nucleic acid RNA, which stores information just as DNA does. The discovery that RNA can do what proteins do—catalyze reactions necessary for replication—is notable because it means that this single, relatively simple molecule could have been a self-replicating system and a precursor to cellular life. These findings have given rise to the so-called **RNA world hypothesis,** which proposes that the world may have been filled with RNA-based life before it became filled with the DNA-based life we see today.

These self-replicating molecules raise an important question: when exactly was the threshold between living and non-living crossed? In the RNA world, self-replicating nucleic acid molecules carried the information of how to replicate *and* served as the machinery to actually carry out the replication. Is that enough? At this point, it is reasonable to ask, again: what exactly is "life"? How do we define it?

Fossils of 3.4-billion-year-old cells have been found in rocks from South Africa. These cells appear to be prokaryotic cells,

THE EARLIEST LIFE ON EARTH?

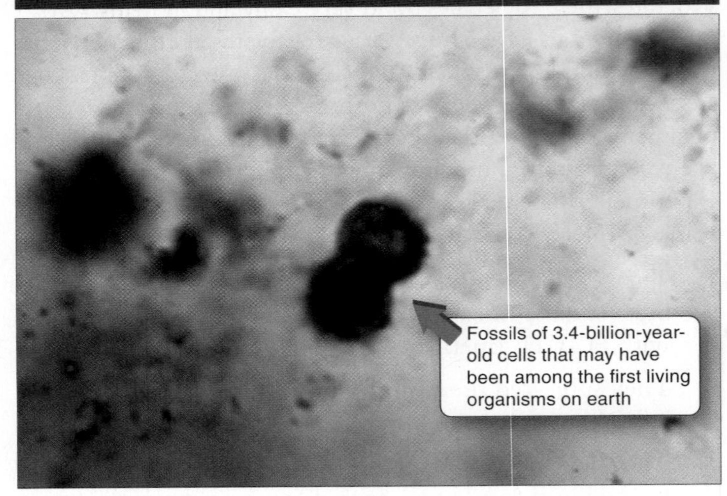

FIGURE 10-3 Ancient prokaryotic cells. These fossilized cells may have been among the first living organisms on earth.

similar to living bacterial cells, with no nucleus, no organelles, and a circular strand of genetic information (**FIGURE 10-3**). They even look as if they are in the process of dividing. Were these the first living organisms on earth? And were they descendants of earlier, self-replicating molecules of RNA? Because no earlier fossils have been found, it is difficult to answer these questions.

Among humans, it's obvious whether someone is living or non-living. But this distinction becomes difficult to make when discussing simpler collections of molecules, and the border between living and non-living can be surprisingly fuzzy. Although no perfectly acceptable, precise definition exists, most scientists agree with the general definition given at the start of this chapter—that life is defined by two characteristics: the ability to replicate and the ability to carry out some sort of metabolism. An organism has metabolic activity when it can acquire certain molecules and use them in controlled reactions that maintain the chemical and energetic conditions necessary for its life. By this definition, following the first two phases described above, the self-replicating RNA molecules were right on the border, satisfying the first condition for life (replicating) but not the second (carrying out metabolism). We now explore the critical third phase in the generation of life from non-life: the development of a membrane that separated these self-replicating small molecules from their surroundings, thus compartmentalizing them into cells and making metabolic activity possible.

Phase 3: The development of a membrane, enabling metabolism and creating the first cells

One way that self-replicating molecules could acquire chemicals and use them in the controlled reactions of metabolism was by packaging the molecules within membranes. As we saw in Chapter 3, cell membranes are semi-permeable barriers that separate the inside of the cell from its external environment. Membranes make numerous aspects of metabolism possible. In particular, they make it possible for chemicals inside the cell to be at higher concentrations than they are outside the cell. Differences in chemical concentrations inside and outside a cell are essential to most life-supporting reactions.

So, if we could combine a self-replicating molecule and some random metabolic chemicals into a unit, surrounded by a membrane, life would be possible. But how could this have happened initially? Some evidence suggests that the first cells may simply have come together spontaneously. Specifically, researchers have found that mixtures of phospholipids placed in water or salt solutions tend to spontaneously form small spherical units that resemble living cells. These units may even "sprout" new buds at their surface, appearing to divide. Because these cell-like units don't pass on any genetic material, however, they cannot be considered to be alive. But if at some point in the past, units like these incorporated some self-replicating molecules inside, maybe by forming around them, such **microspheres** may have been important in the third phase in the generation of life from non-life: the compartmentalization of self-replicating, information-containing molecules into cells. If this did occur, the final step in the creation of something from nothing would be complete (**FIGURE 10-4**).

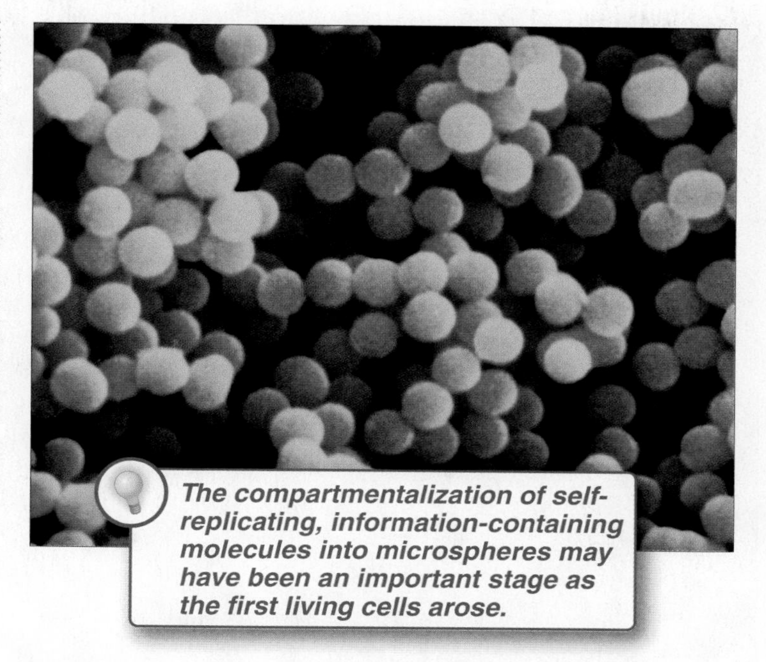

FIGURE 10-4 **Are microspheres a key stage in the origin of life?**

The exact process by which life on earth originated is still uncertain, and research continues. What is clear, however, is that life now abounds in great diversity. In the next sections, we begin to explore that diversity and how it is generated. We start with an investigation of what a species is and how individual species split and create additional species.

TAKE-HOME MESSAGE 10•2

The earliest life on earth, which resembled bacteria, appeared about 3.5 billion years ago, not long after the earth was formed. No one is certain how life arose, but evidence supports the idea that self-replicating molecules—possibly RNAs—may have formed in earth's early environment and later acquired or developed membranes, enabling them to replicate and making metabolism possible, the two conditions that define life.

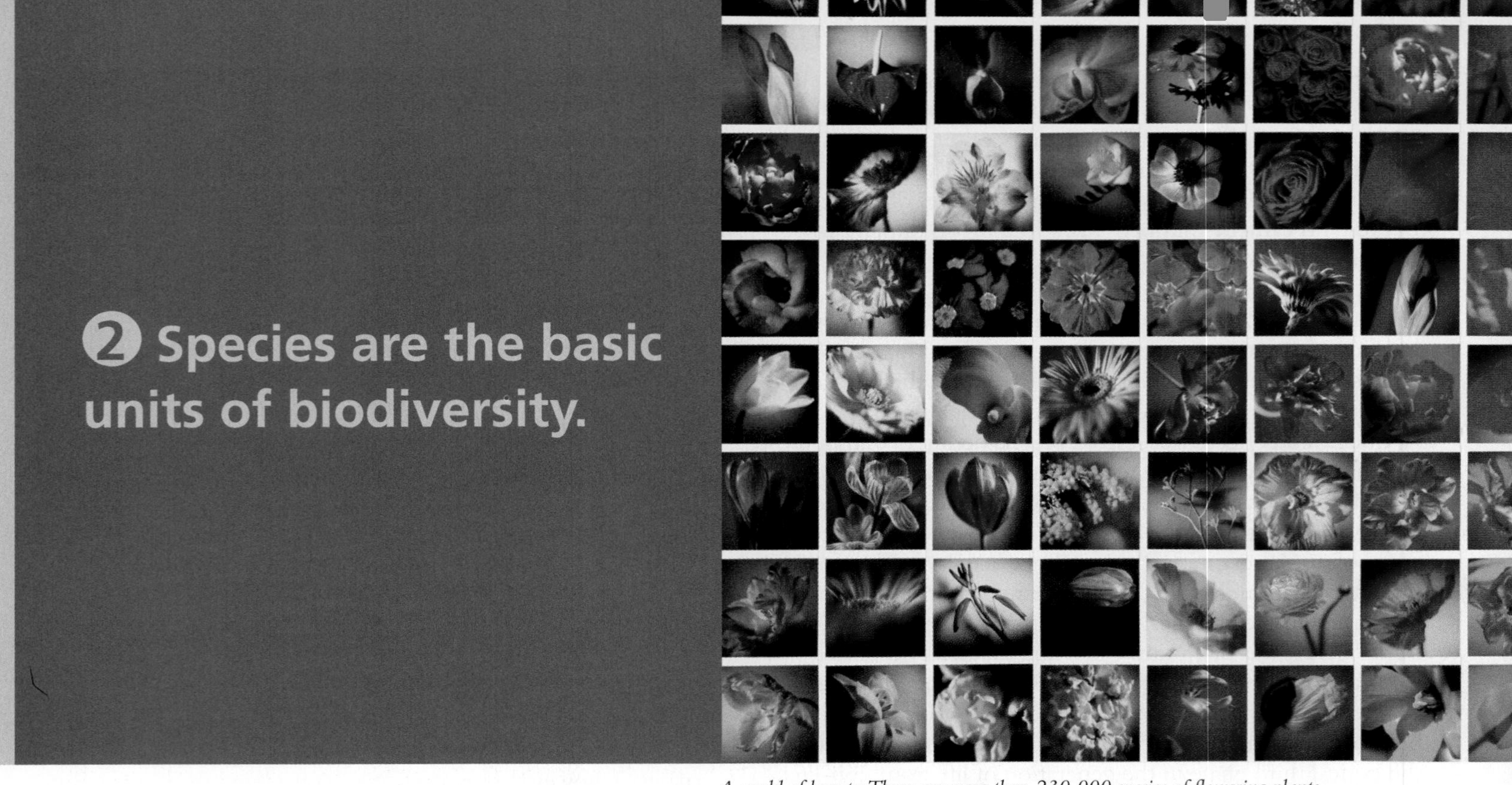

A world of beauty. There are more than 230,000 species of flowering plants.

10•3

What is a species?

A cat and a mouse are different things. Oprah Winfrey and Bill Gates, on the other hand, are different *versions* of the same thing. That is, although they are different individuals, both are humans. We generally distinguish between different kinds of living organisms: a rose versus a daisy, a wasp versus a fly, a snake versus a frog. Similarly, we lump some organisms into the same group and recognize them as different versions of the same thing: red roses and yellow roses, Chihuahuas and Dalmatians. How do we know when to classify two individuals as members of different groups and when to classify them as two individuals that are members of the same group (**FIGURE 10-5**)? And how do we classify two individuals that are somewhat similar but

Do similar-looking organisms belong to the same or different species? How do we know?

FIGURE 10-5 **The same but not the same.** We recognize Bill Gates and Oprah Winfrey as members of the same species, but can you tell whether the rose and the tulip or the wood mouse and the common brown rat are from the same species? (Hint: they're not.)

seem too different to classify as the same thing, such as a mouse and a rat, or a goat and a sheep?

Biologists use the word **species** to label different kinds of organisms. According to the **biological species concept,** species are populations of organisms that interbreed, or could possibly interbreed, with each other under natural conditions, and that cannot interbreed with organisms outside their own group (**FIGURE 10-6**).

Notice that the biological species concept completely ignores physical appearance when defining a species and instead emphasizes **reproductive isolation,** the inability of individuals from two populations to produce fertile offspring with each other, thereby making it impossible for gene exchange between the populations to occur.

Let's clarify two important features of the biological species concept. First, it says that members of a species are either actually interbreeding or *could possibly* interbreed. This emphasis means that just because two individuals are physically separated, they aren't necessarily in different species. A person living in the United States and a person living in Iceland, for example, may not be able to mate because of the distance between them. But if they were brought to the same location, they could mate if they wanted to, so we do not consider them to be reproductively isolated. Second, our definition refers to "natural" conditions. This distinction is important because occasionally, in captivity, individuals may interbreed that would not interbreed in the wild, such as lions and tigers (**FIGURE 10-7**).

FIGURE 10-7 Interbreeding is not enough. This "liger" is the offspring of a lion and a tiger, two species that will interbreed, but only in captivity.

WHAT MAKES A SPECIES?

SPECIES ARE

- populations of organisms that interbreed with each other,
- or could possibly breed, under natural conditions,
- and are reproductively isolated from other such groups.

FIGURE 10-6 An interbreeding population. This herd of Dall sheep lives in the remote regions of Denali National Park, Alaska.

There are two types of barriers that prevent individuals of different species from reproducing: prezygotic and postzygotic barriers (**FIGURE 10-8**). (Remember, an egg that has been fertilized by a sperm cell is a *zygote.*)

Prezygotic barriers make it impossible for individuals to mate with each other or, if they can mate, make it impossible for the male's reproductive cell to fertilize the female's reproductive cell. These barriers include situations in which the members of the two species have different courtship rituals or have sufficient physical differences that they are unable to mate. In some cases, individuals from two different species can and do mate, but physical or biochemical factors prevent the male gamete from fertilizing the female gamete.

Postzygotic barriers occur after fertilization and generally prevent the production of fertile offspring from individuals of two different species—such offspring are called **hybrids.** These barriers are responsible for the production of hybrid individuals that either do not survive long after fertilization or, if they do survive, are infertile or have reduced fertility. Mules, for example, are the hybrid offspring of horses and donkeys, and although they can survive, they cannot breed with each other or produce offspring.

At first glance, the biological species concept seems to make it possible to determine unambiguously whether individuals belong

REPRODUCTIVE ISOLATION: KEEPING SPECIES SEPARATE

PREZYGOTIC BARRIERS
- Individuals are physically unable to mate with each other.
  OR
- If individuals are able to mate, the male's reproductive cell is unable to fertilize the female's reproductive cell.

POSTZYGOTIC BARRIERS
- Matings produce hybrid individuals that do not survive long after fertilization.
  OR
- If hybrid offspring survive, they are infertile or have reduced fertility.

FIGURE 10-8 Barriers to reproduction. With postzygotic barriers to reproduction, even if fertilization does occur, the animal (such as the mule, on the right) is usually sterile.

to the same or different species. For plants and animals this is usually true. For many other organisms, however, it is impossible to apply the biological species concept. We examine those species in Section 10-5, and explore some alternative methods of identifying a species. Even though it is not perfect, however, the biological species concept gives us an important tool for conceptualizing and categorizing the tremendous diversity of organisms found on earth, and it is used by most biologists.

TAKE-HOME MESSAGE 10•3

Species are generally defined as populations of individuals that interbreed with each other, or could possibly interbreed, and that cannot interbreed with organisms outside their own group. This concept can be applied easily to most plants and animals, but not to many other types of organisms.

10•4

How do we name species?

Keeping track of a large group of anything requires an organizational system. Many libraries, for example, catalog their books by using the Dewey Decimal System, which organizes books into 10 different classes, further subdivides each class into 10 divisions, and each division into 10 sections. On your computer, you store your electronic files within virtual folders and subfolders, which can be organized by name, size, date, or type.

With the huge number of species on earth, such a classification system is particularly important. Biologists use the system developed by the Swedish biologist Carolus Linnaeus in the mid-1700s and published in his book *Systema Naturae* ("System of Nature").

Here's how it works (**FIGURE 10-9**). Every species is given a scientific name that consists of two parts, a **genus** (*pl.* genera) and a **specific epithet.** Linnaeus gave humans the name *Homo sapiens,* meaning "wise man." *Homo* is the genus and *sapiens* is the specific epithet. (The genus is capitalized, and the genus and specific epithet are both italicized.) The redwood tree has the name *Sequoia sempervirens.* The strength of Linnaeus's system is that it is a hierarchical system—that is, each element of the system falls under a single element in the level just above it.

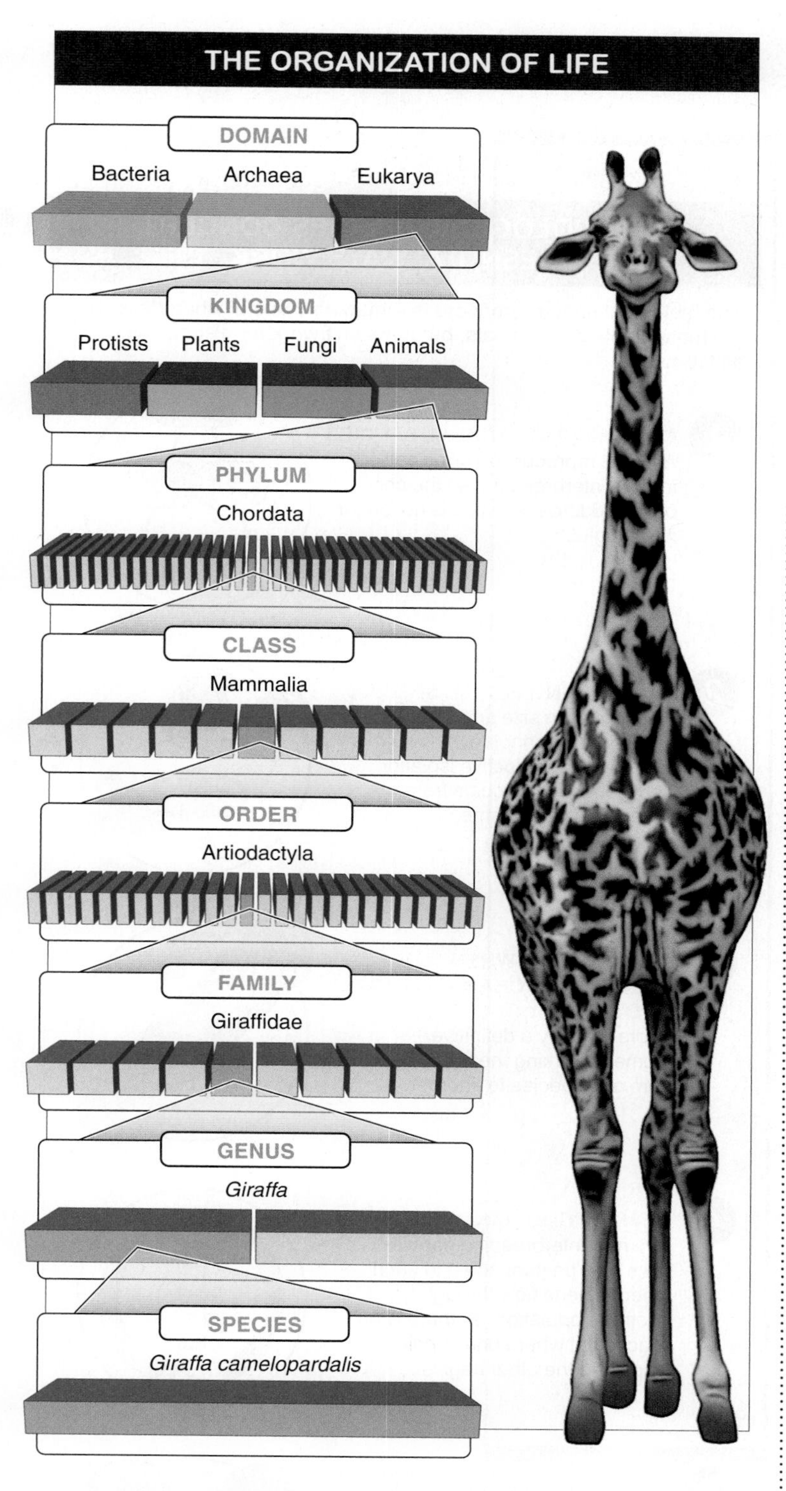

FIGURE 10-9 Name that giraffe. Each species on earth, like *Giraffa camelopardalis,* is given a scientific name and is categorized according to hierarchical groups.

In Linnaeus's system, the species is the narrowest classification for an organism. The specific epithet for every species within a genus is unique, but many different species may be placed within the same genus. Similarly, many genera are grouped within a broader group, called a **family.** And many families are grouped within an **order.** Orders are grouped within a **class.** Classes are grouped within a **phylum.** And at the highest level—though this has changed today, as noted below—all phyla were classified under one of three **kingdoms:** the animal kingdom, the plant kingdom, or the "mineral kingdom."

Today, many of the species classifications that Linnaeus described have been revised, and some of his designations, such as the "mineral kingdom," have been left out. Also, all of the kingdoms are now classified under an even higher order of classification, the **domain.** But Linnaeus's basic hierarchical structure remains, and all life on earth is still named using this system, with all organisms belonging to one of the three domains.

When new species are discovered, they are given names based on Linnaeus's system. The higher-level groups into which a new species should be classified are generally clear—that is, it's usually obvious whether something is in the animal or plant kingdom, or whether it is in the mammalian or amphibian class. But when scientists are assigning a specific epithet, they frequently have a little fun. In recent years, for example, an amphibian fossil was named *Eucritta melanolimnetes,* which translates roughly into "creature from the black lagoon." And in honor of Elvis Presley, a wasp species was named *Preseucoila imallshookupis.* A variety of other celebrities also have had species named after them, including the Beatles (several trilobite species), Mick Jagger (a mollusk species and a trilobite species), Kate Winslett (a beetle species), Steven Spielberg (a dinosaur species), and Greta Garbo (a wasp).

Plants and animals often are referred to by common names, which are different from their "official" names in the Linnaean system and are based on similarities in appearance. These common names, however, can cause confusion. "Fish" for example is used as part of the names of jellyfish, crayfish, and silverfish, none of which are closely related to the vertebrate group of fishes that includes salmon and tuna. Later in this chapter we'll see how an important goal of modern classification and the naming of organisms is to link an organism's classification more closely with its evolutionary history than with its physical characteristics.

TAKE-HOME MESSAGE 10•4

Each species on earth is given a unique name, using a hierarchical system of classification. Every species falls into one of three domains.

10•5

Species are not always easily defined.

Biologists, like all humans, can be biased. When investigating the natural world, for example, they often focus on plants and animals, to the exclusion of the rest of the earth's rich biodiversity. This gets them into trouble when it comes to an idea such as the biological species concept. While the biological species concept is remarkably useful when describing most plants and animals, it falls short of representing a universal and definitive way of distinguishing many life forms (**FIGURE 10-10**).

Difficulties in classifying asexual species The biological species concept defines species as groups of interbreeding individuals. But this is a useless distinction for the asexual species of the world. Recall from Chapter 6 that asexual reproduction, common among single-celled organisms (including all bacteria), many plants, and some animals, is a form of reproduction that doesn't involve fertilization or even two individuals. Rather, the cell (or cells) of an individual simply divides, creating new individuals. Asexually reproducing organisms don't have a partner or need a partner, they just divide. Because asexual reproduction does not involve any interbreeding, the concept of reproductive isolation is no longer meaningful and it might seem that every individual should be considered a separate species. Clearly, that's not a helpful rule to follow.

Difficulties in classifying fossil species When classifying fossil species, differences in the size and shape of fossil bones from different individuals can never reveal definitively whether there was reproductive isolation between those individuals. This makes it impossible to apply the biological species concept.

Difficulties in determining when one species has changed into another Based on fossils, it seems that modern-day humans, *Homo sapiens,* evolved from a related species called *Homo erectus* about 250,000–400,000 years ago. This seems reasonable, until you consider that your parents—who are in the species *Homo sapiens*—were born to your *Homo sapiens* grandparents, who were born to your *Homo sapiens* great-grandparents, and so on. If humans evolved from *Homo erectus,* at what *exact* point did *H. erectus* turn into *H. sapiens*? It may not be possible to identify the exact point at which this change occurred.

Difficulties in classifying ring species Living in central Asia are some small, insect-eating songbirds called greenish warblers, which are unable to live at the higher elevations of the Tibetan mountain range. Because of this limitation, the warblers

THE BIOLOGICAL SPECIES CONCEPT DOESN'T ALWAYS WORK

The biological species concept is remarkably useful when describing most plants and animals, but it doesn't work for distinguishing all life forms.

1 CLASSIFYING ASEXUAL SPECIES
Asexual reproduction does not involve interbreeding, so the concept of reproductive isolation is no longer meaningful.

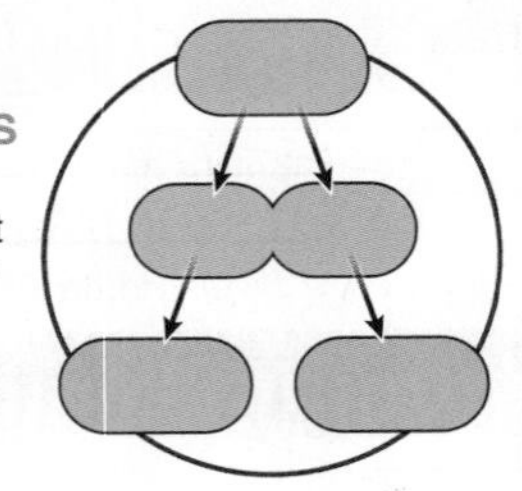

2 CLASSIFYING FOSSIL SPECIES
Differences in size and shape of fossil bones cannot reveal whether there was reproductive isolation between the individuals from whom the bones came.

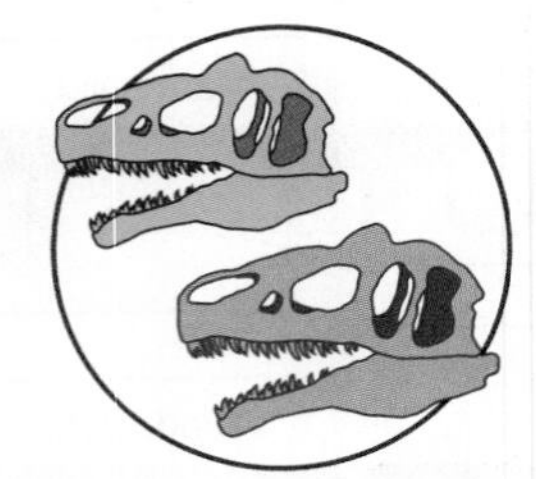

3 DETERMINING WHEN ONE SPECIES HAS CHANGED INTO ANOTHER
There is rarely a definitive moment marking the transition from one species to another.

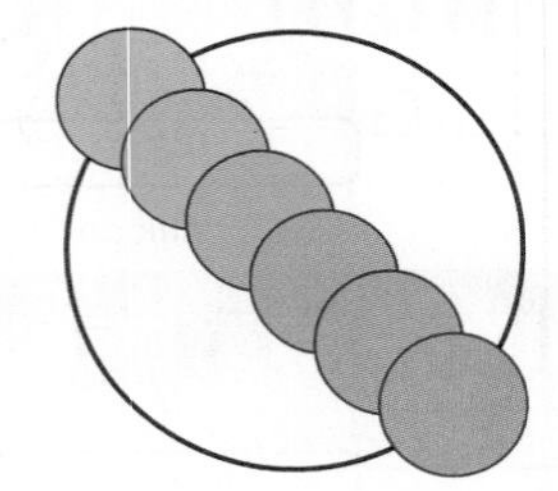

4 CLASSIFYING RING SPECIES
Two non-interbreeding populations may be connected to each other by gene flow through another population, so there is no exact point where one species stops and the other begins.

5 CLASSIFYING HYBRIDIZING SPECIES
Hybridization—the interbreeding of closely related species—sometimes occurs and produces fertile offspring, suggesting that the borders between the species are not clear cut.

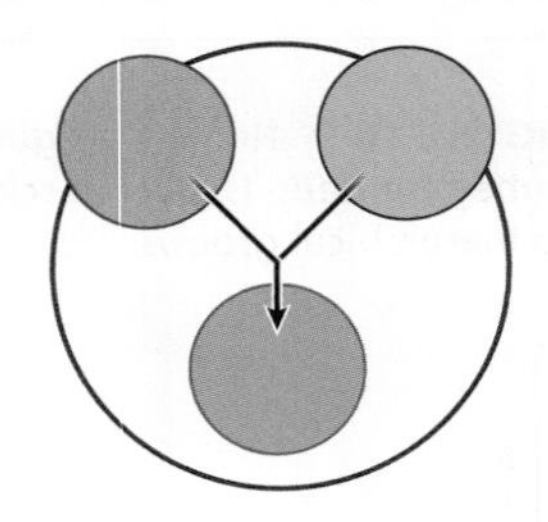

FIGURE 10-10 A useful concept that can't always be easily applied.

live in a ring around the mountain range. At the southern end of the ring, in northwest India, the warblers interbreed with each other. In more northern locations, along either side of the mountain range, the warbler population is split. On each side, the warblers interbreed, but warblers from one side do not interbreed with those from the other, because the mountain range separates them. Where the two "side" populations of warblers meet up again at the northernmost end of the ring, in the forests of Siberia, they can no longer interbreed.

Q Chihuahuas and Great Danes generally can't mate. Does that mean they are different species?

What has happened? Gradual changes in the warblers on each side of the mountain range have accumulated so that the two populations that meet up in Siberia are sufficiently different, physically and behaviorally, that they have become reproductively incompatible. But because the two non-interbreeding populations in the north are connected by gene flow through other populations farther south, there is no exact point at which one species stops and the other begins. So where do you draw the line? The greenish warblers are just one example of more than 20 such **ring species** that have been described.

Difficulties in classifying hybridizing species

Increasingly, **hybridization**—the interbreeding of closely related species—has been observed among plant species and among animal species. This phenomenon fits with the biological species concept as long as postzygotic barriers have evolved, so that the hybrids are weak and unable to reproduce. But in some cases, such as among butterflies in the genus *Heliconius,* the hybrids have high survival rates *and* are fertile, whether interbreeding with other hybrids or with individuals from either of the parental species. This suggests that the borders between the species are not clear cut.

All of these shortcomings have prompted the development of several alternative approaches to defining what a species is. These alternatives tend to focus on aspects of organisms other than reproductive isolation as defining features. The most commonly used alternative is the **morphological species concept,** which characterizes species based on physical features such as body size and shape. Although the choice of which features to use is subjective, an important feature of the morphological species concept is that it can be used effectively to classify asexual species. And because it doesn't require knowledge of whether individuals can actually interbreed, the morphological species concept is a bit easier to use than the biological species concept when observing organisms in the wild.

When it comes to species definitions, although the biological species concept is the most widely used and can be applied without difficulty to most plants and animals, we should not expect one size to fit all, and there will probably never be a universally applicable definition of what a species is. From asexual species to ring species to hybridizing species, the diversity of the natural world is simply too great to fit into neat, completely defined and distinct little boxes. Nonetheless, scientists can generally use a species definition that is satisfactory for a particular situation.

TAKE-HOME MESSAGE 10•5

The biological species concept is useful when describing most plants and animals, but it falls short of representing a universal and definitive way of distinguishing many life forms. Difficulties arise when trying to classify asexual species, fossil species, species arising over long periods of time, ring species, and hybridizing species. In these cases, alternative approaches to defining species can be used.

10•6

How do new species arise?

Biologists don't really have a clue about how many species there are on earth. Estimates of the number vary tremendously, from a low of 5 million to a high of 100 million. Biologists do know, however, the process by which all species arose.

The process of **speciation,** in which one species splits into two distinct species, occurs in two phases and requires more than just evolutionary change in a population. The first phase of speciation is *reproductive isolation* (which we've already discussed), through which two populations come to have independent evolutionary fates. The second phase of speciation is *genetic divergence,* in which two populations evolving as separate entities accumulate physical and behavioral differences over time as they become adapted differently to features of their separate environments, including to different predators and types and abundances of food available.

The initial reproductive isolation necessary for speciation often comes about, as we've seen, when two populations are geographically separated. Although this is an effective and common way for speciation to occur, speciation can occur with or without it.

Speciation with geographic isolation: allopatric speciation Suppose one population of squirrels is split into two separate populations because the climate in its habitat grows wetter and a river forms and splits the habitat in two. Because the squirrels cannot cross the river, the population on one side is reproductively isolated from the population on the other side. Over time, the two populations have different evolutionary paths as they adapt to particular features of their habitats, which may differ on either side of the river. Eventually, the two populations may genetically diverge enough that if the river separating them disappeared and the populations came back into contact, squirrels from the two groups might no longer be able to interbreed. Two species of antelope ground squirrels formed on the north and south rims of the Grand Canyon as a result of this type of speciation, known as **allopatric speciation** (**FIGURE 10-11**).

An example of allopatric speciation can also be seen in the various species of finches found in the Galápagos Islands. Individual finches from the nearest mainland, now Ecuador, in South America, originally colonized one or more of the islands. (And later, additional islands may have been colonized by birds from the mainland or from previously colonized islands.) But because the islands are far apart from one another, the finches tended not to travel between islands and these colonization events are rare, so the populations remained reproductively isolated from one another. Consequently, 14 different species of finches have evolved in the Galápagos Islands, each of which specializes in eating certain of the wide range of insects, buds, and seeds that occur on the islands (**FIGURE 10-12**). Only one species of finch is found in Ecuador.

The barrier doesn't have to be an expanse of water. A forming glacier could split a population into two or more isolated populations. Or a drop in the water level of a lake might expose strips of land that divide the lake into separate, smaller bodies of water, separating one large population of fish into two distinct populations. In each case, the result is the same: geographic isolation that enforces reproductive isolation.

Q Could you create a new species in the laboratory? How would you do it?

Researchers can easily create new species in the lab, using an analogous strategy. In one experiment, a single population of fruit flies (*Drosophila pseudoobscura*) was divided into two. The two populations were then maintained on different diets—one population received food made from the sugar maltose, while the other population was fed a starch-based food. After only eight generations of reproductive isolation and adaptation to their differing

ALLOPATRIC SPECIATION

Allopatric speciation occurs when a geographic barrier causes one group of individuals in a population to be reproductively isolated from another group.

1 INITIAL POPULATION

2 REPRODUCTIVE ISOLATION
Suppose a river forms through the squirrels' habitat, separating the population. Because they cannot cross the river, they are reproductively isolated.

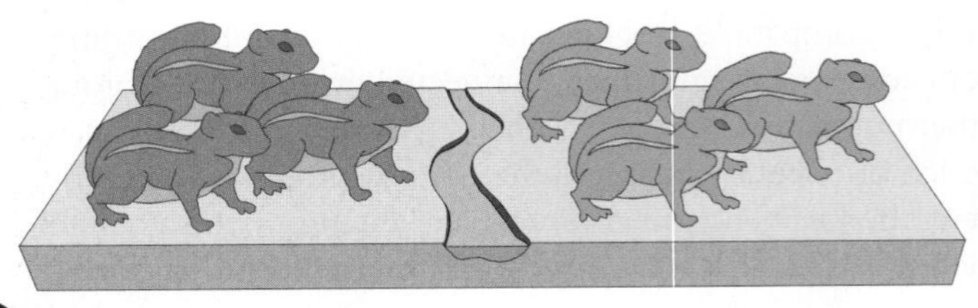

3 GENETIC DIVERGENCE
Over time, the populations on either side diverge enough genetically that they are no longer able to interbreed.

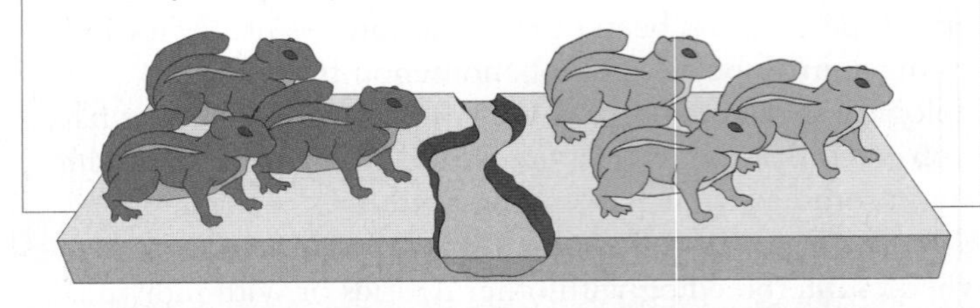

FIGURE 10-11 Geographic isolation can result in genetic divergence. Two species of antelope ground squirrel formed on the north and south rims of the Grand Canyon.

ALLOPATRIC SPECIATION: GALÁPAGOS ISLAND FINCHES

FIGURE 10-12 Five of the 14 Galápagos Island finch species. Each species of finch specializes in eating the insects, buds, and/or seeds found in its island habitat.

Due to allopatric speciation, fourteen different species of finches have evolved in the Galápagos Islands. Only one species of finch is found on the nearest mainland.

nutritional environments, the populations had diverged sufficiently to form separate species; when the populations were mixed, the fruit flies from one population would no longer interbreed with flies from the other population.

Speciation without geographic isolation: sympatric speciation Speciation can also occur among populations that overlap geographically. This type of speciation is called **sympatric speciation.** Among animals, it is rare for populations of the same animal to become reproductively isolated when they coexist in the same area, so this method of speciation is relatively uncommon. But it *is* common among plants, and it occurs in one of two ways (**FIGURE 10-13**).

During cell division in plants (both in reproductive cells and in other cells of the plant body), an error sometimes occurs in which the chromosomes are duplicated but a cell does not divide. This creates a new cell that may then grow into an individual with twice as many sets of chromosomes as the parent from which it came. The new individual may have four sets of chromosomes, for example, while the original individual had two sets. This doubling of the number of sets of chromosomes is called **polyploidy.** The individual with four sets can no longer interbreed with individuals having only two sets of chromosomes, because their offspring would have three sets (two sets from the parent that had four sets, and one set from the parent that had two sets), which could not divide evenly during cell division. The individual with four sets can, however, propagate through self-fertilization or by mating with other individuals that have four sets. As a consequence, the individuals with four sets of chromosomes have achieved instant reproductive isolation from the original population and are therefore considered a new species. Although rare in animals, speciation by polyploidy has occurred several times among some species of tree frogs.

A much more common method of speciation through polyploidy occurs when plants from different but closely related species interbreed, forming a hybrid. The hybrid may not be able to interbreed with either of the parental species but may be able to grow and propagate itself asexually—as many plants can. Subsequent errors in cell division and repeated hybridization, however, can ultimately produce fertile individuals with chromosomes from multiple different species (see Figure 10-13). This method of speciation, called **allopolyploidy,** has led to the production of a large number of important crop plants, including wheat, bananas, potatoes, and coffee.

When reproductive separation occurs, for whatever reason, the new populations are isolated from each other. But because the biological species concept considers individuals from populations that could *potentially* interbreed as still belonging

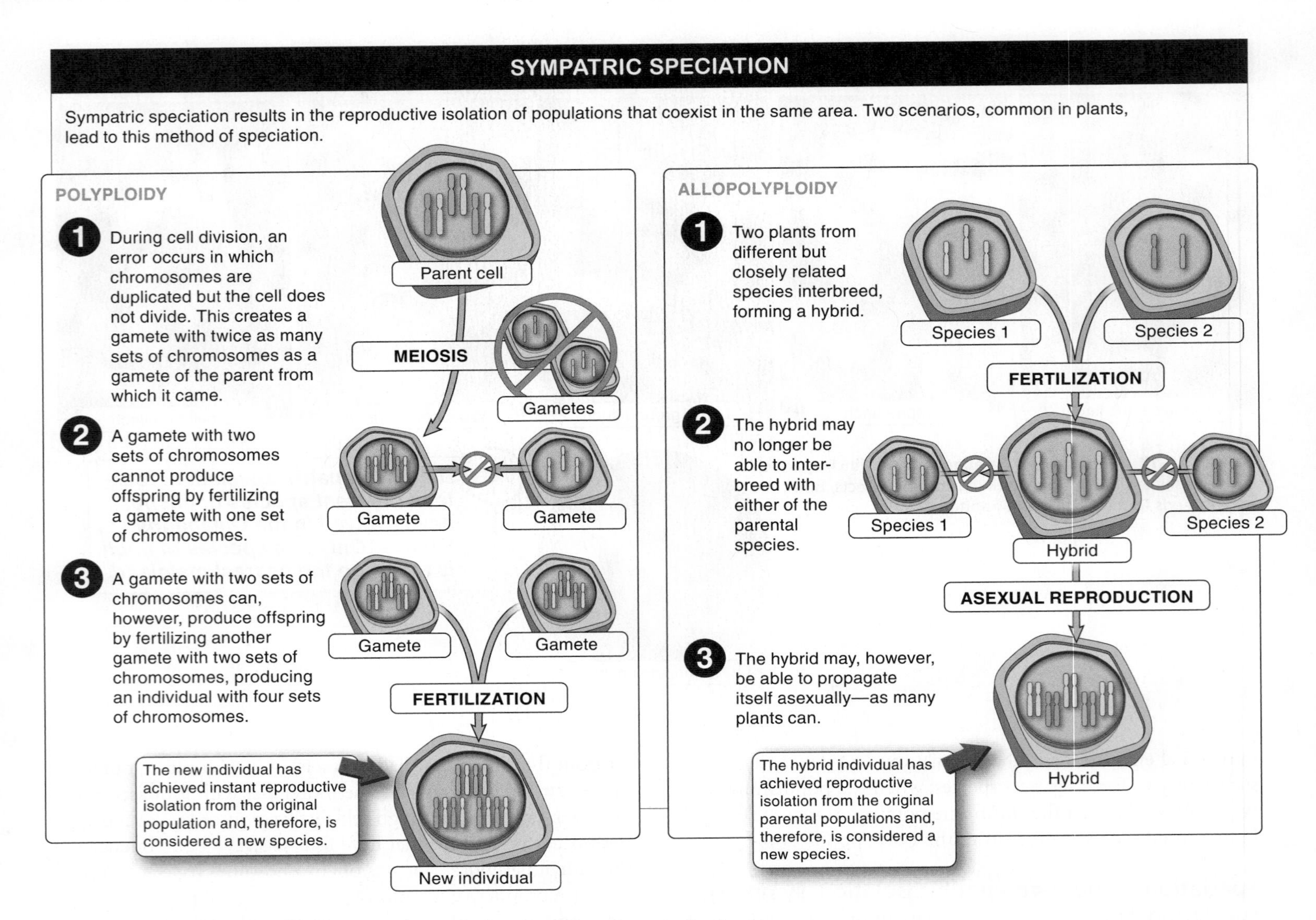

FIGURE 10-13 Speciation without geographic isolation. Plants may genetically diverge through sympatric speciation.

to the same species, speciation is not complete until sufficient differences have evolved in the two populations that they can no longer interbreed even if they do come in contact. It is in isolation from each other, however, that two populations generally diverge into separate species—that is, the prezygotic and/or postzygotic reproductive isolating mechanisms evolve that form the biological barriers to genetic exchange between the two new species.

Reproductive separation can occur relatively quickly—as long as it takes for a new river (or freeway) to divide one large population into two. But the genetic divergence that causes true reproductive isolation—no successful interbreeding even when individuals come into contact with each other—can take a very long time, sometimes thousands of years. For this reason, speciation can be difficult to study and observe.

TAKE-HOME MESSAGE 10•6

Speciation is the process by which one species splits into two distinct species that are reproductively isolated. It can occur by polyploidy or by a combination of reproductive isolation and genetic divergence.

③ Evolutionary trees help us conceptualize and categorize biodiversity.

The diversity of life on earth can be thought of as branching like a tree.

10•7

The history of life can be imagined as a tree.

Although Carolus Linnaeus's method for naming species was an important step in categorizing and cataloguing earth's biodiversity, his underlying assumption about where species came from was wrong. He believed, as did all biologists of his day, that all species had been created at the same time and were unchanging. For this reason, his classification of organisms into hierarchical groups was based on nothing more than his own evaluation of how physically similar various organisms appeared.

A hundred years later, however, Charles Darwin proposed and documented that species could, in fact, change and give rise to new species. With Darwin, the classification of species acquired a new goal and a more important function. In *The Origin of Species* Darwin wrote: "Our classifications will come to be, as far as they can be so made, genealogies." That is, Darwin proposed that the classifications of organisms would resemble family trees that link generations of parents and offspring over long periods of time. With these words, Darwin was the first to link classification with evolution.

Classification such as that used by Linnaeus involved placing organisms within groups as a function of their apparent similarity with each other. The modern incarnation of Darwin's vision of classification is called **systematics** and has the broader goal of reconstructing the **phylogeny,** or evolutionary history, of organisms. That is, through systematics, all species—even extinct species—are named and arranged in a manner that indicates the common ancestors they share and the points at which they diverged from each other. The common-ancestor points at which species diverge are called **nodes.** A complete phylogeny of all organisms is like a family tree for all species, past and present.

A phylogenetic tree not only shows the relationships among organisms but also presents a hypothesis about the evolutionary history of species. The hypothesis represented by a phylogenetic tree is an unfolding of a story (**FIGURE 10-14**). At the beginning of life on earth there was the first living organism, one that could replicate itself and that had a metabolism. Then a **speciation event** occurred. After this event, the population of the first living organism split into independent evolutionary lineages. The phylogenetic tree had its first branch, and there was now biodiversity. Over hundreds of millions of years, speciation events continued to occur, and today the tree has branches with millions (or possibly tens of millions) of tips that represent all the species on earth. Moving up a branch of any evolutionary tree from its trunk toward its

FIGURE 10-14 **Mapping common ancestors.** Evolutionary trees reveal the evolutionary history of every species and the sequence of speciation events that gave rise to them.

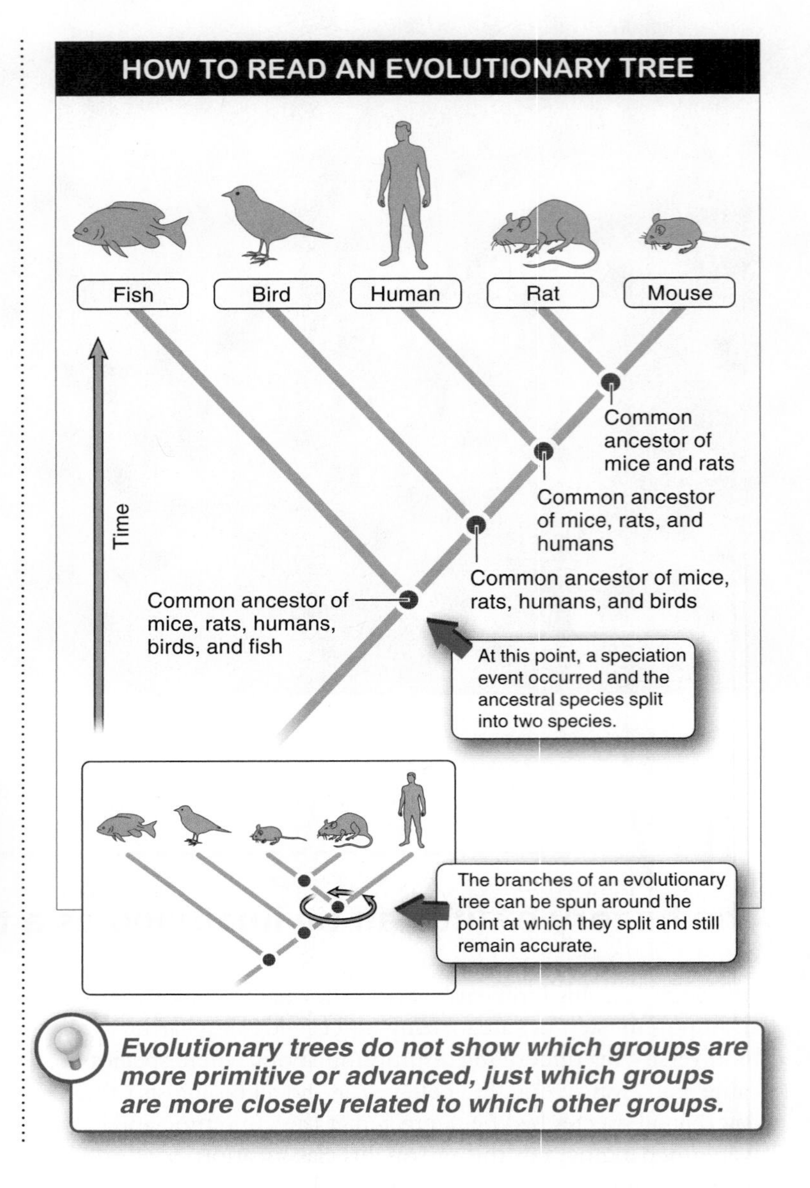

tips, we can see when groups split, with each branching point representing a speciation event. And it doesn't stop here—even today, speciation continues to add new branches all the time.

It is important to remember that phylogenies are hypotheses and, like any hypotheses, they are subject to revision and modification. In the next section we explore how phylogenetic trees are constructed and look at the rich information they convey about the history of life on earth. We also investigate the modern methods of molecular systematics and see how data from DNA sequences are illuminating phylogenies with greater accuracy than ever before. Based on this new molecular and DNA evidence, biologists are revising many earlier phylogenies that were based on observations of organisms' physical features.

TAKE-HOME MESSAGE 10•7

The history of life can be visualized as a tree; tracing from the branches back toward the trunk follows the pathway of descendants back to their ancestors. The tree reveals the evolutionary history of all species and the sequence of speciation events that gave rise to them.

10•8

Evolutionary trees show ancestor-descendant relationships.

Most human cultures and religions have a sacred "Tree of Life." The bodhi tree that the Buddha sat beneath for seven years to achieve the ultimate truth is the Tree of Enlightenment; the apple tree in the Garden of Eden is the Tree of Knowledge. The history of the relationship between all organisms that constitute life on earth is another tree of life. Like most trees, evolutionary trees have a trunk and branches. But these parts represent ancestor-descendant relationships that link living organisms with all life that has ever existed on earth. Although the evolutionary tree of life can be thought of as one giant tree, as a practical matter biologists often study only particular branches of the tree. These branches can be illustrated as big trees or little trees and can be expanded to include whatever organisms the biologist is studying: the tree might, for example, include all animals or just the rodents.

When drawing an evolutionary tree, it does not matter on which side you put a particular group (see Figure 10-14). Any branches can be spun around the point at which they split (the node) as if they were a pinwheel. This pinwheel effect means that you cannot assume that rats are more evolutionarily advanced than mice, or vice versa. They are equally advanced in the sense that both groups derived from the same speciation event.

> **Q** Are humans more advanced, evolutionarily, than cockroaches? Can bacteria be considered "lower" organisms?

What are the implications of this? Evolutionary trees tell us many things, but one thing they do not tell us is which groups are most primitive and which are most advanced. This property of phylogenetic trees can serve to undermine some of our most sacred beliefs, such as the one that humans are the pinnacle of evolution. Many trees can be drawn to support this idea, including Figure 10-14. But notice that if you rotate this tree around any one of its nodes, you can get a number of different trees.

What trees do tell us is which groups are most closely related to which other groups. One of the most interesting revelations of "tree thinking" is that—despite appearances— fungi, such as mushrooms, yeasts, and molds, are more closely related to animals than to plants. By analyzing the genetic composition of plants, fungi, and animals, biologists have been able to determine that the relationship between the three groups is that shown in **FIGURE 10-15**. We investigate this surprising fact in Chapter 11.

> "We all should know that diversity makes for a rich tapestry, and we must understand that all the threads of the tapestry are equal in value no matter what their color.
>
> —Maya Angelou, in *Wouldn't Take Nothing for My Journey Now,* 1993"

Biologists use the term **monophyletic** to describe a group in which all of the individuals are more closely related to each other than to any individuals outside that group. Monophyletic groups are determined by looking at the nodes of the trees. Animals and fungi, taken together, compose a monophyletic group because they share a more recent common ancestor (designated by node A in

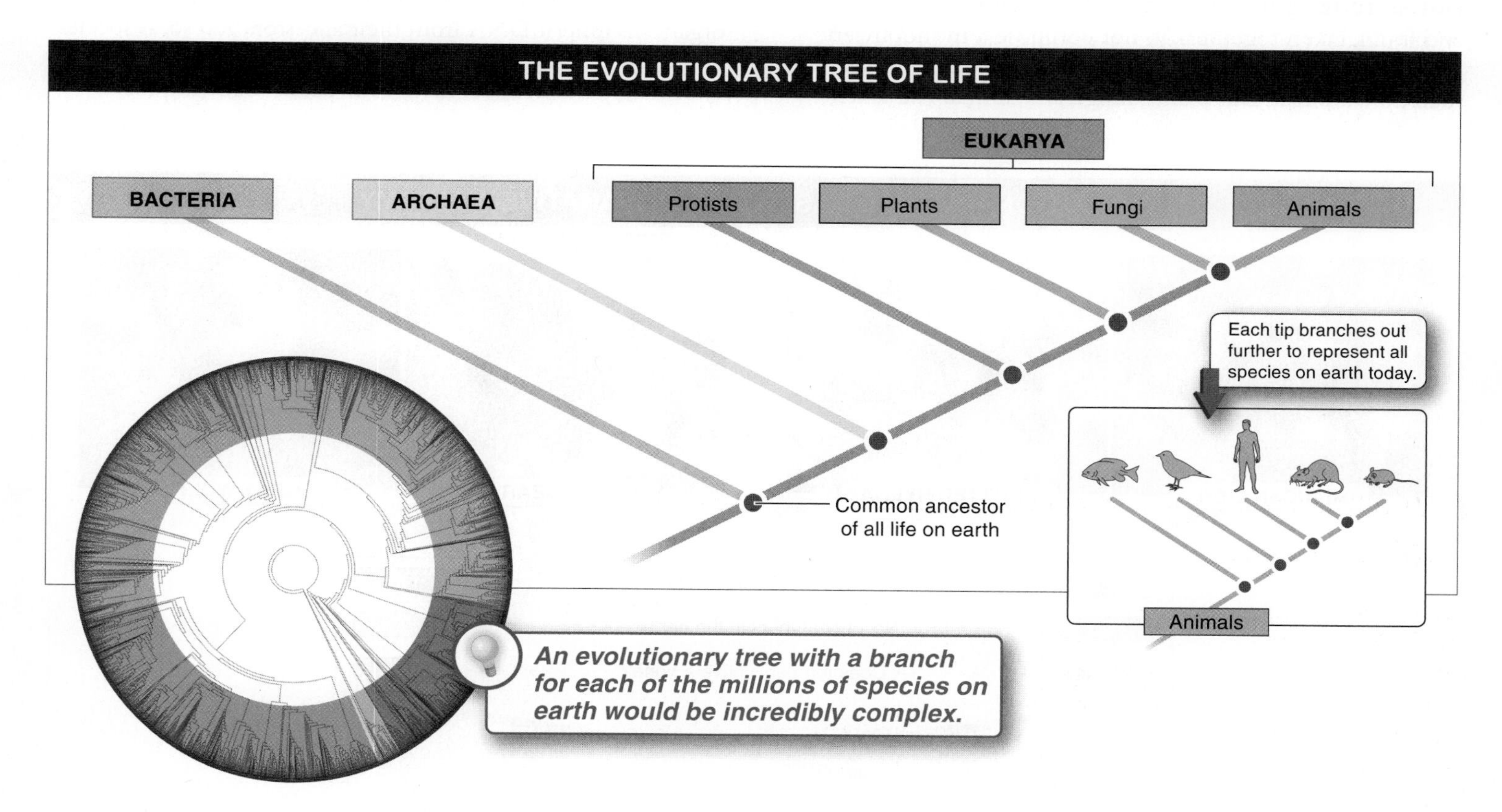

FIGURE 10-15 A growing tree. The evolutionary tree of life has branches with millions of tips representing all species on earth. As speciation events occur, new branches are added to the tree.

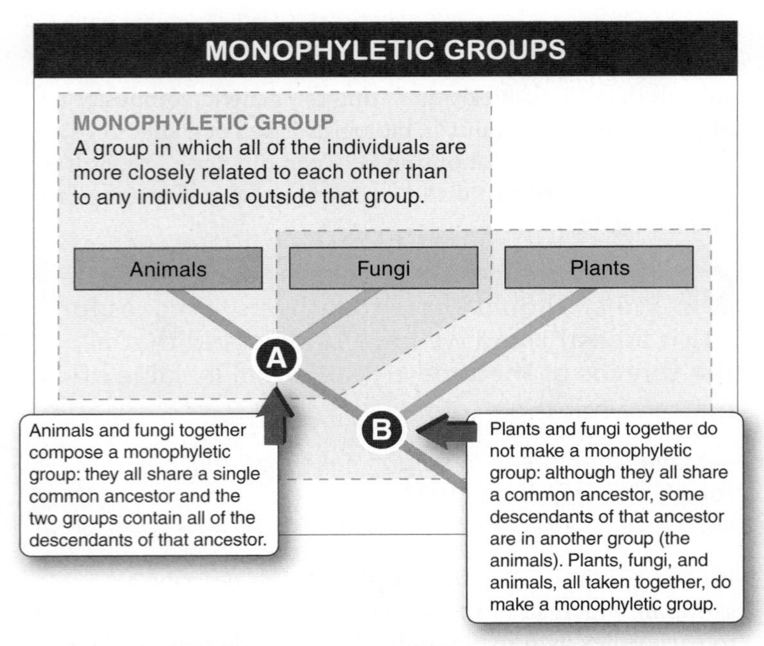

FIGURE 10-16 Members of a monophyletic group share a common ancestor and the group contains all of the descendants of that ancestor.

FIGURE 10-16) than either group shares with plants. Plants and fungi, taken together, do not compose a monophyletic group, because their common ancestor (at node B) is also shared by animals. But plants, fungi, and animals, all taken together, *do* compose a monophyletic group, by virtue of all three sharing the common ancestor at node B.

Constructing evolutionary trees by comparing similarities and differences between organisms

Reading an evolutionary tree reveals which groups are most closely related and approximately how long ago they shared a common ancestor. But how are evolutionary trees—which might hypothesize historical events, such as the cat-dog split, that happened 60 million years ago—constructed in the first place?

Until recently, these trees were assembled by looking carefully at numerous physical features of species and generating tables that compared these features across the species. **FIGURE 10-17** is a simple example of such a table. With only three traits, this table shows a clear split between the characteristics of the lion and the hyena, on the one hand, and those of the wolf and the bear, on the other. For most of the 20th century, biologists classifying organisms would often use 50 or more traits to generate a tree.

Then, beginning in the 1980s, biologists began using molecular sequences rather than physical traits to generate evolutionary trees. The rationale for this approach is that organisms inherit DNA from their ancestors and so, as species diverge, their DNA sequences also diverge, becoming increasingly different. As more time passes following the

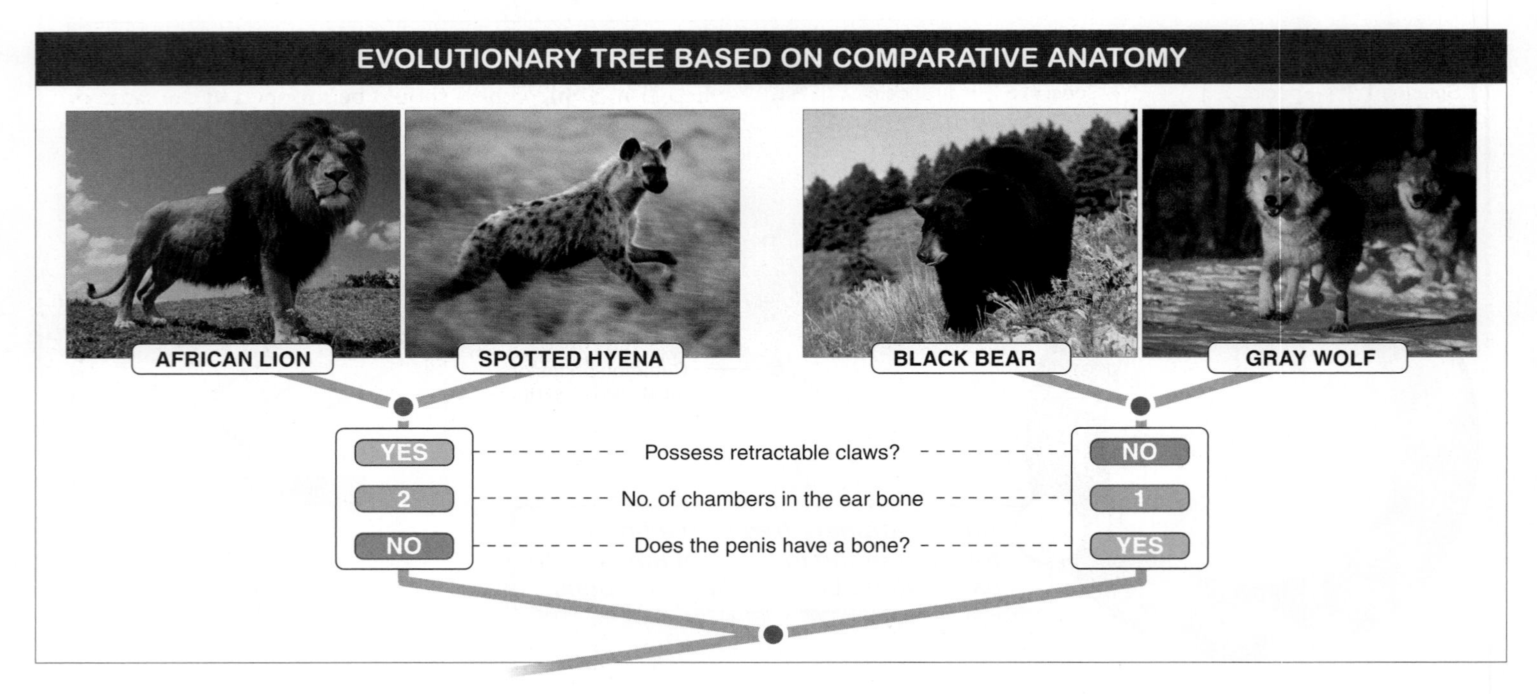

FIGURE 10-17 Looking for clues in body structures. Before DNA sequencing, physical features of species were used to determine evolutionary relatedness.

splitting of one species into two, the differences in their DNA sequences become greater. By comparing how similar the DNA sequences are between two groups, it is possible to estimate how long ago they shared a common ancestor (**FIGURE 10-18**).

In theory, the construction of evolutionary trees using DNA sequences is not really different from using physical traits. It is just a way to use many, many more traits in the comparison—each DNA base pair can be thought of as a "trait" being compared.

DNA-BASED EVOLUTIONARY TREES

By comparing how similar the DNA sequences are between two groups, it is possible to estimate how long it has been since they shared a common ancestor.

Difference in DNA base pairs

Species 1 and 2 are more similar to each other than they are to either species 3 or 4. This is because they probably share a more recent common ancestor.

Species 1

Species 2

Species 3

Species 4

A

T

C

G

The more similar their DNA sequences, the more closely two species are related.

FIGURE 10-18 DNA sequences reveal evolutionary relatedness.

The genome of Tasha, a boxer from upstate New York, was sequenced in 2004. Knowing the makeup of a dog's DNA helps researchers understand where genes responsible for specific traits or defects are located. Knowing the sequence of an animal's DNA allows comparisons with human DNA, improving understanding of both humans and animals.

Although most evolutionary trees are now produced by using molecular sequence comparisons, when it comes to actually naming species we still use Linnaeus's system, assigning a species name and fitting it within the other categories such as kingdoms, phyla, classes, orders, and families. This can be difficult to do because there is no objective way to decide how to assign groups above the level of species—for example, whether multiple genera should be grouped in the same or different families, or whether multiple families should be grouped in the same or different orders.

TAKE-HOME MESSAGE 10•8

Evolutionary trees constructed by biologists are hypotheses about the ancestor-descendant relationships among species. The trees represent an attempt to describe which groups are most closely related to which other groups based on physical features, usually DNA sequences.

10•9

Similar structures don't always reveal common ancestry.

Bats and many insects have wings; whales and sharks have fins. One of the goals of evolutionary biology is to figure out how the vast diversity of species and their characteristics evolved. Do bats have wings because they are most closely related to insects and inherited their wings from the common, winged ancestor they shared with insects, or did insects' wings and bats' wings evolve independently? As the evolutionary tree in **FIGURE 10-19** makes clear, insects' wings and bats' wings evolved independently. Bats did not inherit their wings from insects. Wings are an adaptation that arose separately on more than one occasion.

Did bats inherit their wings from insects or did they evolve independently?

REPEATED EVOLUTION OF THE WING

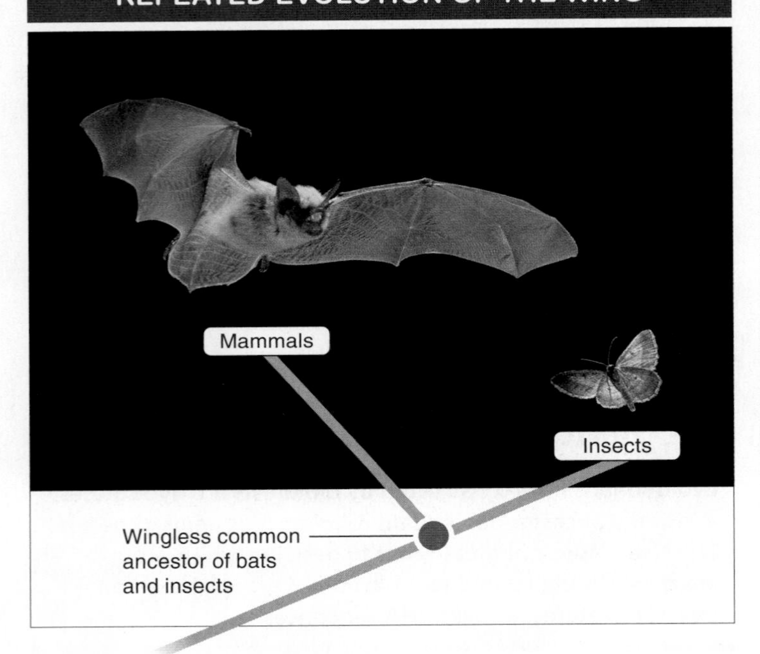

Bats and insects share a wingless common ancestor, proving that they evolved wings independently.

FIGURE 10-19 What do wings tell us? Because wings are an adaptation that has evolved independently many times in populations living in similar environments under similar selective pressures, the presence of wings does not necessarily indicate evolutionary relatedness.

The African golden mole resembles a shrew, but DNA sequence comparisons reveal that it is more closely related to an elephant.

FIGURE 10-20 Looks can be deceiving. Which two of these three animals are more closely related? Can you tell just by looking?

The mapping of species' characteristics onto phylogenetic trees provides us with the story of evolution. Before the 1980s, biologists had always tried to decode the process of evolution by comparing physical features of organisms, such as the presence or absence of wings. Then, with the advent of methods for comparing DNA sequences in the 1980s, tracing the history of life on earth became a more rigorous science. Let's look at a case that illuminates why the original methods were weaker (**FIGURE 10-20**).

Initially, biologists thought that African golden moles belonged in the order known as insectivores, which includes shrews, hedgehogs, and other moles. This belief seemed reasonable because the moles have many characteristics in common with these animals: they are small, they have long, narrow snouts, their eyes are tiny, and they live in underground burrows. Biologists thought that this group of characteristics evolved just once, and that every species in the insectivore order possessed these characteristics because they inherited them from a common ancestor.

Recent evidence from DNA sequencing, however, surprised everyone who studies these animals. The DNA evidence reveals that the African golden moles are actually more closely related to elephants than to the insectivores, including all of the other mole species! Why do they look so similar to the insectivores? Because of a phenomenon called **convergent evolution,** which occurs when populations of organisms live in similar environments and so experience similar selective forces. **Analogous traits** are characteristics (such as bat wings and insect wings) that are the same because they were produced by convergent evolution, not because they descended from a common structure in a shared ancestor.

Q: Humans have several fused vertebrae at the bottom of the spinal cord that look like a tiny, internal "tail." Are we evolving a tail or losing a tail?

Features that are inherited from a common ancestor are called **homologous features.** All mammals have hair because they inherited that trait from a common ancestor. Similarly, among all reptiles, having a mouth is a homologous trait because all reptiles share a common ancestor who passed on the trait to them.

These peppered moths are the same species, Biston betularia. *The dark coloration of the moth on the left acts as camouflage when it sits on the bark of trees blackened by industrial soot. The moth on the right is adapted to an environment free from soot, and its salt and pepper coloration perfectly blends into the bark of a tree.*

Analogous features are problematic when constructing evolutionary trees, because they are the result of natural selection rather than common ancestry. As a consequence, these traits should not be considered a sign of relatedness between the organisms that have them, and they should not be reflected as a sign of common ancestry when constructing an evolutionary tree. But, how do we know whether traits are homologous or analogous? Through DNA analysis. DNA sequences do not become similar during convergent evolution, so molecular phylogenies cannot be fooled. This is an important reason why molecular-based evolutionary trees are preferable to trees based on physical features.

Although analogous traits and the process of convergent evolution are not helpful in constructing evolutionary trees, that doesn't mean they are useless to biologists. Far from it. Convergence provides some of the best evidence for the power of natural selection. For example, from the 19th to the 20th century in industrialized parts of Europe, a large number of distinct butterfly and moth species became more and more darkly colored. This change was not the result of their sharing a common ancestor, but instead came about because industrialization had caused an accumulation of dark soot on the bark of many trees. This soot accumulation caused light-colored moths and butterflies to stand out against the background when they landed on trees, making them an easier target for predators than dark-colored individuals. As a consequence, natural selection led to a rapid evolution of body color, with many species converging on a darker coloration at the same time.

TAKE-HOME MESSAGE 10•9

Evolutionary trees are best constructed by comparing DNA sequences among organisms, because convergent evolution can cause distantly related organisms to appear much more closely related but it doesn't increase their DNA sequence similarity.

④ Macroevolution gives rise to great diversity.

More than 400 Anolis *lizard species exist, including this colorful lizard* (Anolis grahami)*, which lives in trees.*

10•10

Macroevolution is evolution above the species level.

When water runs over rocks, it wears them away. The process is simple and slow, yet it is powerful enough to have created the Grand Canyon. To be sure, water running over rocks does not always make a Grand Canyon. Nonetheless, no additional physical processes are necessary. The process of evolution has a lot in common with a stream of water running over rocks: in the short term, it produces small changes in a population, yet the accumulation of these changes can be "canyon-esque" (**FIGURE 10-21**). Let's consider some examples.

The production of 200-ton dinosaurs from rabbit-size reptile ancestors. The diversification of flowering plants from a single species into more than 230,000 species. The emergence of animals that live on land. These large-scale examples, which are the products of evolutionary change involving the origins of entirely new groups of organisms, are referred to as **macroevolution.** These examples can be contrasted with the increase in milk production in cows during the first half of the 20th century, or the gradual change in the average beak size of birds with changing patterns of rainfall, phenomena involving changes in allele frequencies in a population that are referred to as **microevolution** (**FIGURE 10-22**).

No additional processes necessary! Just as a stream of water running over rocks can create the Grand Canyon, accumulated microevolutionary changes can lead to dramatic macroevolutionary changes.

FIGURE 10-21 From a trickle of water to the Grand Canyon. The Grand Canyon is a product of small changes to the landscape over millions of years.

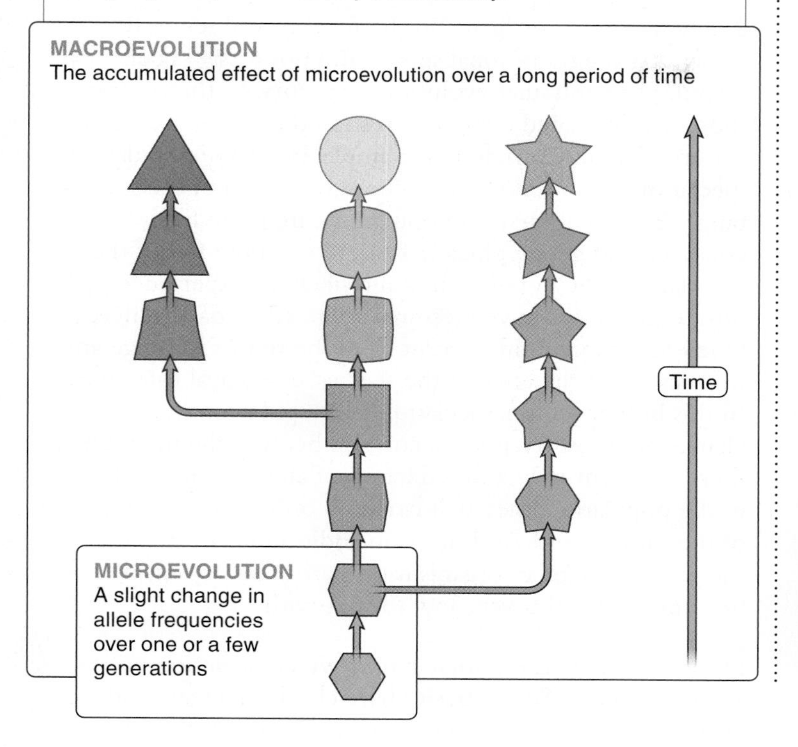

FIGURE 10-22 The scale of evolution. Microevolutionary changes in allele frequencies in a population over time can lead to macroevolution, changes on a grand scale, including vast diversification of species.

These micro and macro events might seem like two very different processes, but they are not. Evolution, whether at the micro or macro level, is one thing only: a change in allele frequencies over time. In the short term, over one or a few generations, evolution can appear as a slight and gradual change within a species. But over a longer period of time, the accumulated effects of this process, acting continuously and combined with reproductive isolation of populations, can lead to the dramatic phenomena described as macroevolution. Just as a trickle of water, given enough time, produced the Grand Canyon, so evolution has created the endless forms of diversity on earth. In a sense, microevolution is the process and macroevolution is the result.

TAKE-HOME MESSAGE 10•10

The process of evolution—changes in allele frequencies within a population—in conjunction with reproductive isolation is sufficient to produce speciation, diversification, and the rich diversity of life on earth.

10•11

The pace of evolution is not constant.

If you listened in on some scientists debating their research, you might hear one side describing "evolution by jerks" and the other side making light of "evolution by creeps." They're not gearing up for a fight out in the school yard, though. They're just describing two examples of the pace of evolution.

The traditional model of how evolutionary change occurred was that used in the previous section to describe microevolution. Populations changed slowly but surely, gradually accumulating sufficient genetic differences for speciation—hence the phrase "evolution by creeps." Spurred on by findings from the fossil record that do not always support this view, however, researchers have come to believe that evolution may often occur in a different way: brief periods of rapid evolutionary change immediately after speciation, followed by long periods with relatively little change—hence "evolution by jerks." This newer view of the pace of evolution, in which rapid periods of evolutionary change are punctuated by longer periods with little change, is called **punctuated equilibrium** (**FIGURE 10-23**).

In nature, we can find examples of both gradual change and the irregular pattern of punctuated equilibrium. It's not a situation in which one version is right and one is wrong. For many groups of organisms, including many mollusks and mammals, many fossils from a particular species exist and reveal that it seems to have persisted with little change for very long periods of time. Then, a large number of mollusk or mammal fossils appear in the fossil record that are similar enough to the older fossils that they appear to be in the same evolutionary lineage, yet there are no fossils representing the gradual transition between the newer and older fossils. For other groups, such transitional fossils do exist. Very complete sets of fossils reveal the transitional sequence from fox-like

THE TEMPO OF EVOLUTION

The pace of evolution varies for different species. Some species have evolved gradually over time, while others spend vast amounts of time with little change.

GRADUAL CHANGE

Evolution by creeps: The pace of evolution occurs gradually in incremental steps.

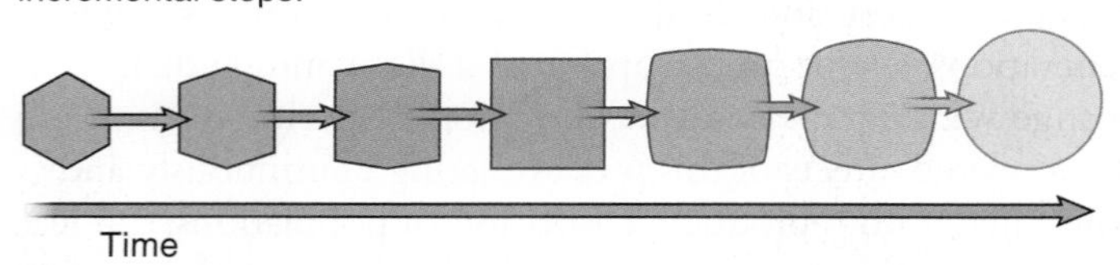

PUNCTUATED EQUILIBRIUM

Evolution by jerks: Rapid periods of evolutionary change are punctuated by longer periods with little change.

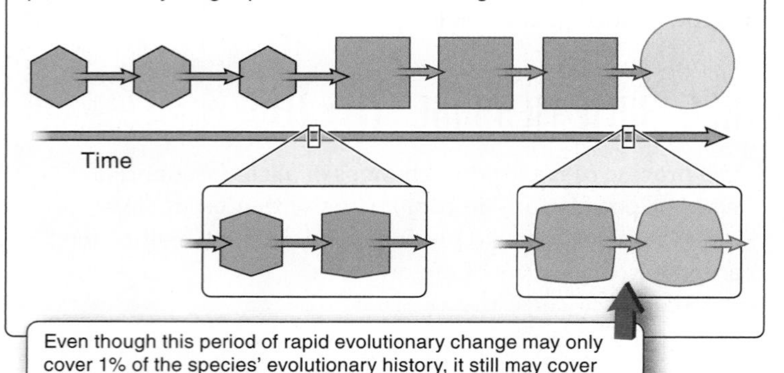

Even though this period of rapid evolutionary change may only cover 1% of the species' evolutionary history, it still may cover hundreds or thousands of generations. This could take tens of thousands of years in a primate or a matter of months in bacteria.

FIGURE 10-23 The pace of evolution can be rapid or slow.

ancient terrestrial mammals to modern whales, for example. And fairly complete sequences of fossils allow the tracing of modern horses back to similar small, fox-like mammals.

Punctuated equilibrium has erroneously been portrayed as a "problem" for evolutionary biology, for two primary reasons.

1. It is mistakenly believed that the rapid evolutionary changes of punctuated equilibrium are the result of some new mechanism of evolution—thus seeming to undermine evolutionary theory. In actuality, although the punctuated change appears lightning-fast on a geological time scale, the per-generation rates of change necessary to produce it are relatively slow and not at all beyond the rates of change that have been measured in evolving populations. As an example, a fossil species may appear unchanged for most of its 3- to 5-million-year history, save for a "rapid" burst during the first 50,000–100,000 years after the species appears (following a speciation event). Even though the period of change represents only about 1% of that species' evolutionary history, it still covers tens of thousands of generations—a more than ample period for even the most dramatic of morphological changes.

2. The existence of "gaps" in the fossil record seems to imply that the transitional species did not exist—again seeming to contradict evolutionary theory. To the contrary, though, punctuated equilibrium should generate exactly the patterns found. Consider an example. In a large population, speciation is most likely to occur at the edges of the species' range. At the edges, small populations are more likely to be separated and geographically isolated from the rest of the population. The population is also likely to experience the most extreme selective pressures at the edges of its range, where the habitat differs most from the rest of its range and may represent the edge of the species' ecological tolerance. In this border region, speciation and rapid evolutionary change are most likely to occur. But because the individuals there represent only a tiny fraction of all the individuals in the population, they will make up only a tiny fraction of the fossil record. And this is in addition to the fact that the probability of an organism (or part of an organism) becoming a fossil is very low to start with.

In the end, there is no single rate at which evolution occurs across all species. Some species have clearly changed gradually, over time, for their entire evolutionary history. Others have spent vast periods of time with little change at all. The coelacanth is a species of fish, for example, that seems to have undergone almost no physical changes over more than 300 million years. For any organisms, the rate of evolutionary change depends on the selective forces acting on the population. Strong and directional forces may act over long periods of time and lead to rapid change, while stabilizing selection may lead to very little change.

Darwin foreshadowed the discussion about the pace of evolution in *The Origin of Species,* writing, "the periods during which species have undergone modification, though long as measured in years, have probably been short in comparison with the periods during which they retain the same form."

TAKE-HOME MESSAGE 10•11

The pace at which evolution occurs can be rapid or very slow. In some cases, the fossil record reveals rapid periods of evolutionary change punctuated by longer periods with little change. In other cases, species may change at a more gradual but consistent pace.

10•12

Adaptive radiations are times of extreme diversification.

Even the greatest of success stories often owe something to a bit of luck. As mammals, we owe more than a little of our success to luck. Flash back to 65 million years ago. Our mammalian ancestors had their place among the organisms on earth, but it didn't remotely resemble our dominant position today. Our ancestors were rodent-size, insect-eating creatures, relegated to a nocturnal existence. It was the dinosaurs' time, and the giant reptiles dominated the earth.

All that changed in an instant when the earth was struck near what is now the eastern part of Mexico (the Yucatán Peninsula) by an asteroid about 6 miles (10 km) in diameter. This caused the environmental conditions on earth to change quickly and dramatically. We'll explore the details of this catastrophic event in the next section, but one outcome was definitive: all of the dinosaur species were wiped out. Our mammalian ancestors, on the other hand, survived and found themselves living on a planet where most of their competitors had suddenly disappeared. We were in the right place at the right time.

What followed was an explosive expansion of mammalian species. In a brief period of time, a small number of species diversified into a much larger number of species, able to live in a wide diversity of habitats. Called an **adaptive radiation,** such a large and rapid diversification has occurred many times throughout history (**FIGURE 10-24**).

Three different phenomena tend to trigger adaptive radiations. After one of these events, surviving species find themselves in locations where they suddenly have access to plentiful new resources.

1. *Mass extinction events.* With the disappearance of the dinosaurs, a world of "opportunities" opened up for the mammals. Where previously the dinosaurs had prevented mammals from utilizing resources, mammals suddenly had few competitors. Not surprisingly, the number of mammalian species increased from a very small number—perhaps a few hundred—to about 130 genera, with more than 4,000 species overall, including primates, bats, horses, rodents, and the first completely aquatic mammals. This happened in about 10 million years, barely the blink of an eye by geological standards. Following other large-scale extinctions, numerous other groups that suddenly lost most of their competitors experienced similar adaptive radiations.

ADAPTIVE RADIATION

Adaptive radiation occurs when a small number of species diversify into a larger number of species. Three phenomena tend to trigger adaptive radiation.

1 MASS EXTINCTION EVENTS
With their competition suddenly eliminated, remaining species can rapidly diversify.

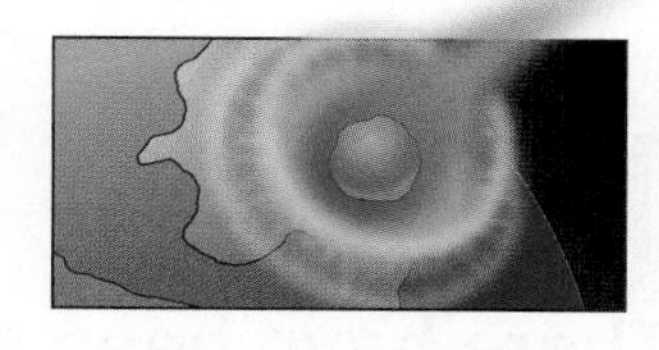

2 COLONIZATION EVENTS
Moving to a new location with new resources (and possibly fewer competitors), colonizers can rapidly diversify.

3 EVOLUTIONARY INNOVATIONS
With the evolution of an innovative feature that increases fitness, a species can rapidly diversify.

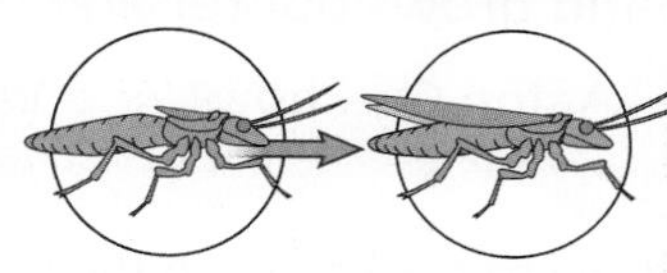

FIGURE 10-24 **A rapid diversification of species.**

2. *Colonization events.* In a rare event, one or a few birds or small insects will fly off from a mainland and end up on a distant island group, such as Hawaii or the Galápagos Islands. Once they make it there, they tend to find a large number of opportunities for adaptation and diversification. In the Galápagos, as we learned in Section 10-6, 14 finch species evolved from the single species of finch found on the nearest mainland, Ecuador, 600 miles away. In Hawaii, there are several hundred species of fruit flies, all believed to have evolved from one species that colonized the islands and experienced an adaptive radiation. Perhaps a few flies were blown there by a storm, or maybe they were carried there, stuck in the feathers of a bird.

3. *Evolutionary innovations.* In the world of computers, software developers are always looking for the "killer app"—the killer application, a new type of software so useful that it immediately leads to huge success, opening up a large new niche in the software market or greatly expanding an already-existing niche. The first spreadsheet, email program, and web browser were all killer apps. In nature, evolution sometimes produces killer apps, too. These are innovations such as the wings and rigid outer skeleton that appeared in insects and helped them diversify into the most successful group of animals, with more than 800,000 species today (more than a hundred times the number of mammalian species). The flower is an innovation that propelled the flowering plants to an explosion of diversity and evolutionary success relative to the non-flowering plants, such as ferns and pine trees. Today, about 9 out of 10 plant species are flowering plants.

TAKE-HOME MESSAGE 10•12

Adaptive radiations—brief periods of time during which a small number of species diversify into a much larger number of species—tend to be triggered by mass extinctions of potentially competing species, colonizations of new habitats, or the appearance of evolutionary innovations.

10•13

There have been several mass extinctions on earth.

> “Forests keep disappearing, rivers dry up, wild life's become extinct, the climate's ruined and the land grows poorer and uglier every day.
>
> —Anton Chekhov, *Uncle Vanya,* 1899

If the past is a guide to the future, we know this: no species lasts forever. Speciation is always producing new species, but **extinction,** the complete loss of all individuals in a species population, takes them away. Extinction, which is always occurring, is a risk faced by all species.

For any given time in earth's history, it is possible to estimate the rate of extinctions at that time, and the evidence reveals that these rates are far from constant (**FIGURE 10-25**). Although the particular details differ in most cases, extinctions generally fall into one of two categories: "background" extinctions or mass extinctions.

Background extinction is the term used to describe the extinctions that occur at lower rates during periods other than periods of mass extinction. Background extinctions occur mostly as the result of natural selection. Competition with other species, for example, may reduce a species' size or the range over which it can roam or grow. Or a species might be too slow to adapt to gradually changing environmental conditions, and becomes extinct as its individuals die off.

Mass extinctions are periods during which a large number of species on earth become extinct over a relatively short period of time. There have been at least five mass extinctions on earth and, during each of these extinctions, 50% or more of the animal species living at the time became extinct (see the red parts of Figure 10-25).

TWO CATEGORIES OF EXTINCTIONS

Extinctions generally fall into two categories:

- **MASS EXTINCTIONS**
  A large number of species become extinct over a short period of time due to extraordinary and sudden environmental change.
- **BACKGROUND EXTINCTIONS**
  These extinctions occur at lower rates during times other than mass extinctions.

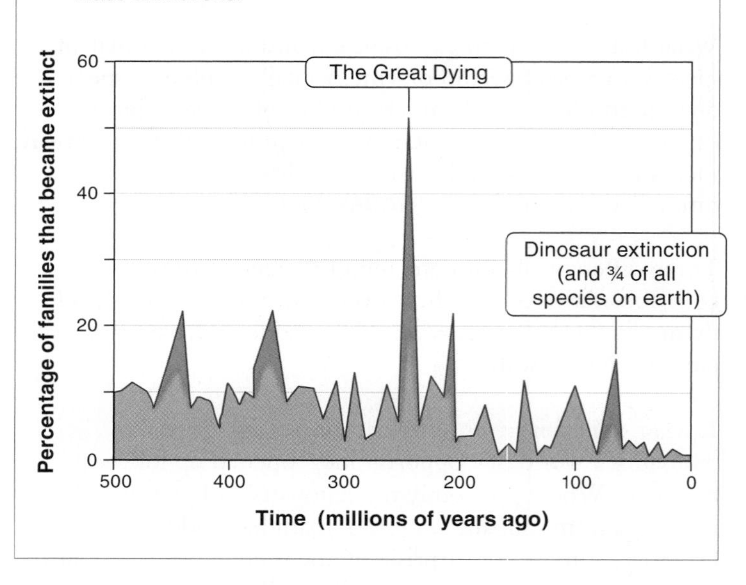

FIGURE 10-25 Extinction never sleeps.

There is a fundamental difference between background and mass extinctions that goes beyond differences in rates. They have different causes. Mass extinctions are due to extraordinary and sudden changes to the environment (such as the asteroid that brought about the dinosaurs' extinction). As a consequence, nothing more than bad luck is responsible for the extinction of species during mass extinctions; fit and unfit individuals alike perish.

Of the five mass extinctions during the past 500 million years, the most recent is also the best understood. Sixty-five million years ago, an almost unimaginable catastrophe occurred. As discussed in the previous section, a massive asteroid smashed into the Caribbean near the Yucatán Peninsula of Mexico (**FIGURE 10-26**). The impact left a crater more than 100 miles wide, and probably created an enormous fireball that caused fires worldwide, followed by a cloud of dust and debris that blocked all sunlight from the earth and disturbed the climate worldwide for months. There may also have been other asteroid impacts around this same time. In the aftermath of this catastrophe, about 75% of all species on earth were wiped out, including the dinosaurs, a tremendously successful group that had been thriving for 150 million years.

Global climate change! Worldwide mass extinction! These catastrophic events set the stage for explosive speciation of the remaining groups, including the mammals.

FIGURE 10-26 Catastrophic collision. Sixty-five million years ago, an asteroid 6 miles in diameter smashed into earth near the Yucatán Peninsula of Mexico and wiped out three-fourths of all species on earth, including the dinosaurs.

As bad as the dinosaur-exterminating asteroid event was, from a biodiversity perspective it is not the worst catastrophe in earth's history. That distinction falls to a mass extinction that took place 250 million years ago, called the Great Dying (see Figure 10-25). Although the cause is not clear—hypotheses include an asteroid impact, continental drift, a supernova (a star explosion), or extreme volcanic eruptions—more than 95% of all marine life became extinct, along with almost 75% of all terrestrial vertebrates. The causes of the three other mass extinctions also are poorly understood, but the evidence in the fossil record of the tremendous upheavals is clear.

It is interesting to note that although, from time to time, sudden and extreme events lead to mass extinctions that wipe out a large proportion of the species on earth, in every case these mass extinctions are followed by a time of explosive speciation in the groups that survive. An important question now being debated is whether we are currently in the midst of a human-caused mass extinction event. We will explore whether this is true—and why—in Chapter 16.

TAKE-HOME MESSAGE 10•13

As new species are being created, others are lost through extinction, which may be a consequence of natural selection or large, sudden changes in the environment. Mass extinctions are periods during which a large number of species on earth become extinct over a short period of time. These events are usually followed by periods of unusually rapid adaptive radiation and diversification of the remaining species.

❺ An overview of the diversity of life on earth: organisms are divided into three domains.

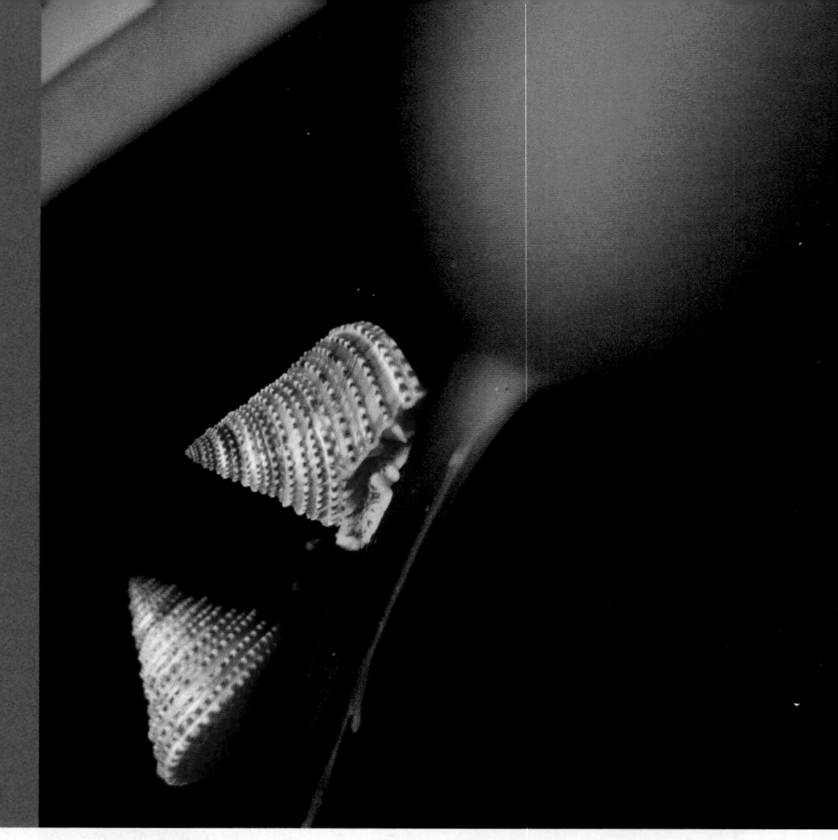

Purple-ring top snails (Calliostoma annulatum) *on a kelp bulb.*

10•14

All living organisms are classified into one of three groups.

Learning about biological diversity can be humbling. At the very least, it makes it harder to believe that humans are particularly special. We are not at the center of earth's "family tree." Nor are we at its peak. We are simply another branch. A newer branch, and definitely connected to the same trunk as all the other branches, but a bit small and off to one side, just the same.

When Linnaeus first put together his system of classification, he saw a clear and obvious split: all living organisms were either plant or animal. Plants could not move and they made their own food. Animals could move but could not make their own food. So in Linnaeus's original classification, all organisms were put in either the animal kingdom or the plant kingdom (his third, "mineral" kingdom, now abandoned, only included non-living matter).

With the refinement of microscopes and subsequent discovery of the rich world of **microbes**—microscopic organisms—the two-kingdom system was inadequate. Where did the microbes belong? Some could move, but many of those could also make their own food, seeming to put them somewhere between plants and animals. And the problems didn't stop with the microbes: mushrooms and molds, among other organisms originally categorized as plants, didn't move but they didn't make their own food either—they digested the decaying plant and animal material around them.

The two-kingdom system gave way in the 1960s to a five-kingdom system. At its core, the new system was a division based on the distinction between *prokaryotic* cells (those without nuclei) and *eukaryotic* cells (those with nuclei). The prokaryotes were put in one kingdom, the monera, and the only residents in this kingdom were the bacteria: single-celled organisms with no nucleus, no organelles, and genetic material in the form of a circular strand of DNA. The eukaryotes—having a nucleus, compartmentalized organelles, and individual, linear pieces of DNA—were divided into four separate kingdoms: plants, animals, fungi, and **protists,** a group that includes most of the single-celled eukaryotes, such as algae.

The classification of organisms took a huge leap forward in the 1970s and 1980s, and the five-kingdom system had to be discarded. Until that point, organisms had been classified primarily based on their appearance. But because the ultimate goal had changed to reconstructing phylogenetic trees that reflected the evolutionary history of earth's diversity, Carl Woese, an American biologist, and his colleagues began classifying organisms by examining the nucleotide sequences they carried.

Q All living organisms have probably evolved from a single common ancestor. How can the existence of one particular molecule support this claim?

Woese assumed that the more similar the genetic sequences were between two species, the more closely related they were, and he built phylogenetic trees accordingly. The only way Woese could compare the evolutionary relatedness of all the organisms present on earth today was by examining one molecule that was found in *all* living organisms and looking at the degree to which it differed from species to species. He discovered a perfect candidate for this role: a molecule called ribosomal RNA, which helps translate genes into proteins (see Chapter 5). Ribosomal RNA has the same function in all organisms on earth, almost certainly because it comes from a common ancestor. Over time, however, its genetic sequence (i.e., the DNA that codes for the ribosomal RNA) has changed a bit. Tracking these changes makes it possible to reconstruct the process of diversification and change that has taken place.

The trees that Woese's genetic sequence data generated had some big surprises. First and foremost, the sequences revealed that the biggest division in the diversity of life on earth was not between plants and animals. It wasn't even between prokaryotes and eukaryotes. The new trees revealed instead that the diversity among microbes was dramatically greater than ever imagined—particularly because of the discovery of a completely new group of prokaryotes called **archaea,** which thrive in some of the most extreme environments on earth and differ greatly from bacteria. The tree of life was revised to show three primary branches, called domains: the bacteria, the archaea, and the eukarya (**FIGURE 10-27**).

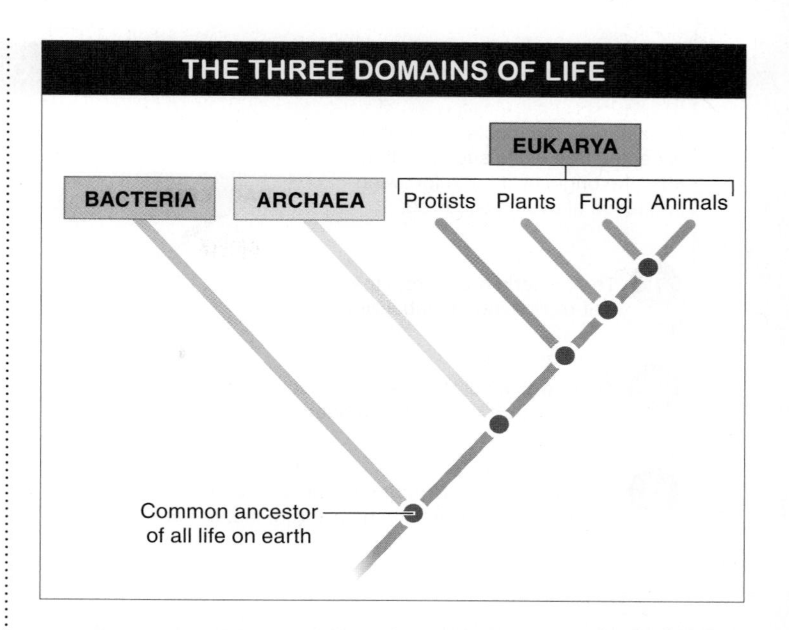

FIGURE 10-27 All living organisms are classified into one of three groups.

Woese put the domains above the kingdom level in Linnaeus's system. In Woese's new system, which is the most widely accepted classification scheme today, the bacteria and archaea domains each have one kingdom and the eukarya domain has four. Because both bacteria and archaea are microscopic, it can be hard to believe that the two domains are as different from each other as either domain is from the eukarya. Each of the three domains is monophyletic, however, meaning that each contains species that share a common ancestor and includes all descendants of that ancestor. Close inspection even reveals that the archaea are more closely related to the eukarya than they are to the bacteria.

The three-domain, six-kingdom approach is not perfect and is still subject to revision. Within the eukarya, for example, it seems that single-celled protists are so diverse that they should be split into multiple distinct kingdoms. Also problematic is the fact that bacteria sometimes engage in **horizontal gene transfer,** which means that, rather than passing genes simply from "parent" to "offspring," they transfer genetic material directly into another species. This process complicates the attempt to determine phylogenies based on sequence data, because it creates situations in which two organisms might have a genetic sequence that is similar not because they share a common ancestor, but as a result of a direct transfer of the sequence from one species to another.

Q Are viruses alive?

Additionally, a fourth group of incredibly diverse and important biological entities, the **viruses,** is not even included in the tree of life, because they are not considered to be living organisms. Viruses can replicate, but can only have metabolic activity by taking over the metabolic processes of another organism. Their lack of metabolic activity puts viruses just outside the definition of life we use in this book, but some scientists view viruses as living.

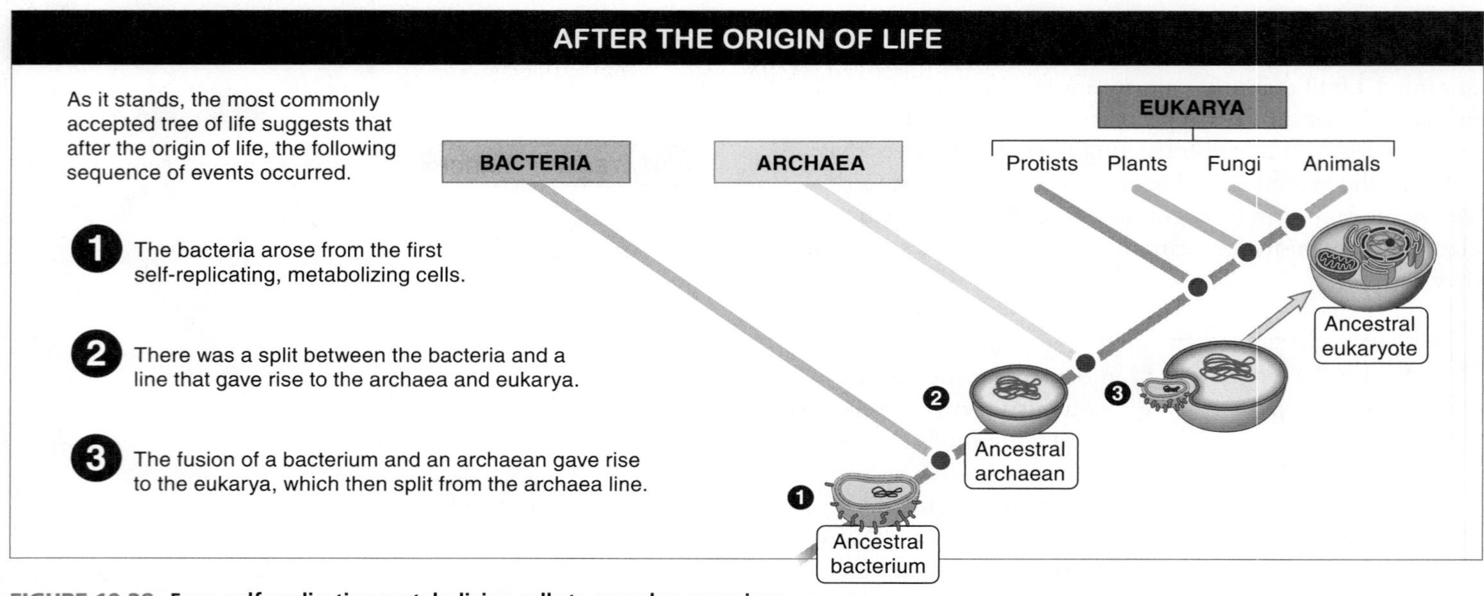

FIGURE 10-28 From self-replicating metabolizing cells to complex organisms. The biggest branches on the tree of life separate the three domains.

As it stands currently, the most commonly accepted tree of life suggests that, after the origin of life, the following sequence of events occurred (**FIGURE 10-28**):

1. The bacteria arose from the first self-replicating, metabolizing cells.
2. There was a split between the bacteria and a line that gave rise to the archaea and eukarya.
3. The fusion of a bacterium and an archaean gave rise to the eukarya, which then split from the archaea line.

The next three sections introduce each of the three domains, surveying the broad diversity that has evolved within each domain and the common features that link all the members of each. Chapters 11, 12, and 13 cover the domains in greater detail.

TAKE-HOME MESSAGE 10•14

All life on earth can be divided into three domains—bacteria, archaea, and eukarya—which reflect species' evolutionary relatedness to each other. Plants and animals are just two of the four kingdoms in the eukarya domain, encompassing only a small fraction of the domain's diversity.

10•15

The bacteria domain has tremendous biological diversity.

Q Why is morning breath so stinky?

Morning breath is stinky. And learning the cause of the offensive smell may make the situation even worse. You see, when you wake up, your mouth contains huge amounts of bacterial waste products. Perhaps the only consolation for this situation is that it gives us a glimpse into just how diverse and resourceful bacteria are.

At any given time, there are several hundred species of bacteria in your mouth—mostly residing on your tongue—all competing for the resources you put there (**FIGURE 10-29**). Some of the bacteria are aerobic, requiring oxygen for their metabolism, and others are anaerobic. At night, because the flow of saliva slows down and the oxygen content of your mouth decreases, the anaerobic bacteria start to get the upper hand in terms of growth and reproduction. These bacteria metabolize food bits in your mouth, plaque on your teeth and gums, and dead cells from the lining of your mouth, breaking down proteins in these materials to use as their energy source. As they do so, waste products accumulate. Because proteins are made from amino acids, some of which contain the smelly chemical sulfur, their breakdown leads to the bad smell.

In the morning, once you wake up, you breathe more and produce more saliva, both of which increase the oxygen level in

FIGURE 10-29 **Bacteria thrive on your tongue and in your mouth.**

your mouth. This tips the battle for space and food back in favor of the aerobic bacteria. Because aerobic bacteria prefer carbohydrates as their energy source and because carbohydrates don't contain sulfur, as the aerobic bacteria start to outcompete the anaerobic bacteria, the sulfur smell goes away. The aerobic bacteria are, of course, also filling your mouth with waste products, it's just that their waste products don't smell as bad.

On a small scale, your mouth reveals some of the tremendous biological versatility of the bacteria: hundreds of species can live in a tiny area—a teaspoon of soil, for example, is home to more than a billion bacteria—they can thrive in a variety of unexpected habitats, they can utilize a variety of food sources, and they can survive and thrive with or without oxygen.

FIGURE 10-30 **We're outnumbered.** Bacteria are everywhere and greatly outnumber eukarya.

Looking around the world, we find that the diversity in the bacteria domain is even greater. A survey of the numbers shows the clear dominance of bacteria on our planet. By any measure, it is their planet: the biomass of bacteria (if they were all collected, dried out, and weighed) exceeds that of all the plants and animals on earth. Bacteria live in soil, air, water, arctic ice, and volcanic vents (**FIGURE 10-30**). Many can even make their own food, utilizing light from the sun or harnessing energy from chemicals such as ammonia.

While they differ in many ways, the bacteria are a monophyletic group, sharing a common ancestor. For this reason, they all have a few features in common. All bacteria are single-celled organisms with no nucleus or organelles, with one or more circular molecules of DNA as their genetic material, and using several methods of exchanging genetic information. Because they are asexual—they reproduce without a partner, just by dividing—the biological species concept cannot be applied to bacteria when classifying them into narrower categories. Thus bacteria are classified on the basis of physical appearance or, preferably, genetic sequences.

It is rather unfair to introduce the bacteria by describing how they can cause bad breath, given that many people already think of bacteria as illness-causing organisms. While bacteria are responsible for many diseases, including strep throat, cholera, syphilis, pneumonia, botulism, anthrax, leprosy, and tuberculosis, disease-causing bacteria are only a small fraction of the domain, and bacteria seem to get less credit for their many positive effects on our lives. Consider that bacteria (*E. coli*) living in your gut help your body digest the food you eat and, in the process, make certain vitamins your body needs. Other bacteria (actinomycetes) produce antibiotics such as streptomycin. Still other bacteria live symbiotically with plants as small fertilizer factories, converting nitrogen into a form that is usable by the plant. Bacteria also give taste to many foods, from sour cream to cheese, yogurt, and sourdough bread. Increasingly, bacteria are used in biotechnology —from those that can metabolize crude oil and help in the cleanup of spills to transgenic bacteria used in the production of insulin and other medical products, as described in Chapter 5.

We explore the great diversity of the bacteria in more detail in Chapter 13.

TAKE-HOME MESSAGE 10•15

All bacteria share a common ancestor and have a few features in common. They are prokaryotic, asexual, single-celled organisms with no nucleus or organelles, with one or more circular molecules of DNA as their genetic material, and using several methods of exchanging genetic information. Bacteria have evolved a broad diversity of metabolic and reproductive abilities relative to the eukarya.

10•16

The archaea domain includes many species living in extreme environments.

In a bubbling hot spring in Yellowstone National Park, the temperatures range from the 212° F (100° C) of boiling water down to a relatively cool 165° F (74° C) at the surface. It would seem a most inhospitable place for life. Yet researchers have found 38 different species of archaea thriving there. Perhaps even more surprising, the genetic differences among these species are more than double the genetic differences between plants and animals.

In the freezing waters of Antarctica, too, archaea abound. More than a third of the organisms in the Antarctic surface waters are archaea. In swamps, completely devoid of oxygen, and in the extremely salty water of the Dead Sea, the story is the same: where once it was assumed that no life could survive, the archaea not only exist but thrive and diversify (**FIGURE 10-31**).

Q Will life be found elsewhere in the universe? How does the discovery of archaea alter the likelihood of this?

If biologists ever find themselves thinking that we have a handle on the breadth of life that is possible on earth, they would be wise to remember the archaea. Until relatively recently, our perception of life on earth was that there were bacteria and there were eukaryotes, such as the plants and animals. But in the past several decades, as researchers have explored some of the most unlikely of habitats, they have found archaea thriving. There are entire worlds of life on earth that we never even imagined.

FIGURE 10-31 Archaea can thrive in the most inhospitable-seeming places. Archaea live in a diversity of environments; they resemble bacteria but are actually more closely related to eukarya.

Analyses of genetic sequences indicate that the archaea and the bacteria diverged about 3 billion years ago. Although it is likely that some genetic exchanges continued to occur between them, they have evolved along largely independent paths ever since. Approximately 2.5 billion years ago, the eukarya split off from the archaea. The archaea are still grouped in one kingdom within the domain archaea, but we have no idea how many species exist. Given that archaea are the dominant microbes in the deep seas, it may very well be that these organisms—which we were completely ignorant of until recent decades—are the most common organisms on earth. It is still too early to tell.

We do know that, like the bacteria, all archaea are single-celled prokaryotes. For that reason, under a microscope they look very similar to bacteria. Several physical features distinguish them from the bacteria, however. Archaea have cell walls made from materials that are different from those found in the cell walls of bacteria and eukaryotes. Specifically, the archaean cell walls contain polysaccharides not found in either bacteria or eukaryotes. The archaea also have cell membranes, ribosomes, and some enzymes similar to those found in the eukarya.

The archaea exhibit tremendous diversity and are often divided into five groups based on their physiological features:

1. Thermophiles ("heat lovers"), which live in very hot places
2. Halophiles ("salt lovers"), which live in very salty places
3. High- and low-pH-tolerant archaea
4. High-pressure-tolerant archaea, found as deep as 4,000 meters (about 2.5 miles) below the ocean surface, where the pressure is almost 6,000 pounds

per square inch (compared with an air pressure of less than 15 pounds per square inch at sea level)

5. Methanogens, which are anaerobic, methane-producing archaea

There also seem to be large numbers of archaea living in relatively moderate environments that are also commonly home to bacteria. We explore the great diversity of the domain archaea in more detail in Chapter 13.

TAKE-HOME MESSAGE 10•16

Archaea, many of which are adapted to life in extreme environments, physically resemble bacteria but are more closely related to eukarya. Because they thrive in many habitats that humans have not yet studied well, including the deepest seas and oceans, they may turn out to be much more common than currently believed.

10•17

The eukarya domain consists of four kingdoms: plants, animals, fungi, and protists.

After exploring the archaea and bacteria domains, turning to the eukarya feels like coming home. We are most familiar with this domain: all of the living organisms that we can see are eukarya. Plants, mushrooms, slime molds, insects, fish, birds, and, of course, mammals, including humans. There are three kingdoms of eukarya that can be seen with the naked eye: plants, animals, and fungi. All are made up from eukaryotic cells—they have a membrane-enclosed nucleus—and each kingdom is almost entirely multicellular.

A fourth kingdom, called protista (protists), which is most often invisible to the naked eye, is a sort of grab bag that includes a wide range of mostly single-celled eukaryotic organisms, including amoebas, paramecia, and algae. Discovery of new species of protists, as with the other microscopic organisms on earth (the bacteria and archaea), continues at very high rate. It is likely that the protista will turn out to be the most diverse of all the eukarya kingdoms, or it may even be split into multiple kingdoms.

The eukarya split from the archaea about 2.5 billion years ago (**FIGURE 10-32**). At the time, the eukarya were single-celled and probably resembled modern protists more than any other modern eukarya. The split may have occurred when symbiotic bacteria became incorporated within an ancestor of the eukaryotes, resulting in what would become the mitochondria. Later, approximately 1.5 billion years ago, a second important symbiosis between bacteria and eukarya resulted in what would become chloroplasts (see Chapter 3).

Because they are so much easier to see than bacteria and archaea, a disproportionate number of the named species on earth are in the domain eukarya. In fact, of the 1.5 million named species, the majority are eukarya, with about half being insects. This is more a result of the interests and biases of biologists than a reflection of the relative numbers of actual species in the world. We explore the great diversity of the eukarya in Chapters 11 (on animals) and 12 (on plants and fungi).

TAKE-HOME MESSAGE 10•17

All living organisms that we can see with the naked eye (and many that are too small to be seen) are eukarya, including all plants, animals, and fungi. The eukarya are unique among the three domains in having cells with organelles.

FIGURE 10-32 All shapes and sizes. The slime mold, passion flower, fungi, and an orangutan reflect the diversity of eukaryotic organisms.

Knowledge You Can Use

Do racial differences exist on a genetic level?

In a fraction of a second we are aware of someone's gender, size, and age—and race. But what is race? Does race have biological meaning? And can humans be categorized into groups that can scientifically be called races?

Q: What is a race, biologically speaking? The biological species concept hinges on a straightforward question: can two individuals interbreed under natural conditions? The concept of a race, conversely, is not straightforward. There is no universally agreed upon definition of race. Because systematics has the goal of reconstructing the evolutionary relationships of organisms, however, biologists have proposed that races are groups of interbreeding populations phylogenetically distinguishable, but reproductively compatible with each other

It is important to note that religious, cultural, ethnic, and geographic groups (among others) are commonly referred to as races. These racial distinctions, though often poorly defined, are a deeply entrenched aspect of many legal, social, and political issues today. Moreover, regardless of whether these socially constructed groupings reflect underlying genetic differences, they continue to be a real part of our lives. But the question we raise here is whether any of these uses of "race" reflect meaningful genetic divisions.

Q: What do racial groupings reflect? Are blacks and whites genetically different? The answer is obviously yes; black-skin alleles differ from white-skin alleles. Moreover, the prevalence of some genetic diseases varies by race and ethnicity. Sickle cell anemia, for example, is relatively common among Africans and Southeast Asians because the genes that cause it also improve a person's resistance to malaria, a disease prevalent in Africa and Southeast Asia.

These two observations—racial differences in skin color and patterns of disease prevalence—reveal one important problem with racial groupings. The genetic resistance to malaria, and the consequent higher risk of sickle cell anemia, is shared by Africans to the south of the Mediterranean and Europeans to the north. Africans from the southernmost part of the continent, on the other hand, have no greater risk of sickle cell anemia than do the Japanese, because malaria is similarly rare in both of their homelands. Should we use this trait rather than skin color when determining a person's race? If we do, the southern Africans more closely resemble the Japanese than they do the North Africans, and so must be grouped differently.

Even if we decide to base race definitions on the traits we can see, how do we even decide who is white and who is black (or who is Basque and who is Native American)? Walking down Main Street, USA, this may seem like a simple observation—"they" look alike. But if you instead travel from tropical Africa up through Egypt into the Middle East and further north, it is impossible to discern where one race ends and the next begins. In a world in which people fall at every point along the continuum, grouping becomes arbitrary. Thus it may be more useful to ask a slightly different question.

Q: How much genetic difference exists between major geographic groups, besides the genes that are relevant to skin color? DNA-sequence comparisons among races reveal an important feature of the genetic variation in the human species. There is little rhyme or reason to how this variation is parceled out from one person to the next, and *similarity in skin color between two people does not tell you about overall genetic similarity.* There is, for example, huge variation in blood type among Africans: some are type O, some AB, others A or B. But the same goes for Asians and Turks, Russians and Spaniards. Yet we do not group people according to blood type. The same goes for variation in many other physiological and biochemical traits. But again, we do not group people according to these variations.

What is the future of race? Racial groupings such as "Hispanic" or "black or African American" or "Asian" do not reflect consistent genetic differences among the groups. Because no human race is completely isolated from all the others, human racial groups have not developed consistent genetic differences across a large number of genes. Still, even as the U.S. Census Bureau has defined race as not primarily biological or genetic, race remains an important part of individual social, cultural, and ethnic identity for many, and its use is mandated in the official collecting and presenting of data for federal agencies. For these reasons, it will remain a powerful social construct, even as scientific consensus emerges that racial groupings do not reflect meaningful genetic distinctions. Interestingly, with our extensive movements around the globe and frequent intermarriage between groups, it is likely that the existing genetic differences between races will shrink in the future.

1 Life on earth most likely originated from non-living materials.

Under conditions similar to those on early earth, small organic molecules form that have some chemical properties of life. Eventually, self-replicating molecules—possibly RNA—may have acquired or developed membranes, enabling them to replicate and making metabolism possible, the two conditions that define life. However it came about, the earliest life, which resembled bacteria, appeared about 3.5 billion years ago, not long after the earth was formed.

2 Species are the basic units of biodiversity.

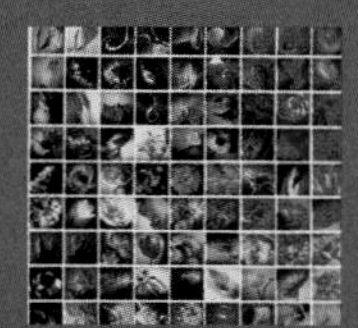

Species are distinct biological entities, named using a hierarchical system of classification. Reproductive isolation and genetic divergence together lead to speciation. The distinction between two species is not easily defined for all types of organisms.

3 Evolutionary trees help us conceptualize and categorize biodiversity.

The history of life can be visualized as a tree; tracing from the branches back toward the trunk follows the pathway of descendants back to ancestors. The tree reveals the evolutionary history of every species and the sequence of speciation events that gave rise to them. Evolutionary trees are best constructed by comparing genetic similarity among organisms, because convergent evolution can cause distantly related organisms to appear much more closely related.

4 Macroevolution gives rise to great diversity.

The process of evolution—changes in allele frequencies within a population—in conjunction with reproductive isolation is sufficient to produce speciation, diversification, and the rich diversity of life on earth. The pace at which this occurs can be rapid or very slow. As new species are being created, others are lost through extinction, which may be a consequence of natural selection or large, sudden changes in the environment. Mass extinctions are usually followed by periods of unusually rapid adaptive radiation and diversification of the remaining groups or organisms.

5 An overview of the diversity of life on earth: organisms are divided into three domains.

All life on earth can be divided into three domains, which reflect species' phylogeny, their evolutionary relatedness to each other. Bacteria have evolved a broad diversity of metabolic and reproductive abilities relative to eukarya. The eukarya are unique in having cells with organelles. Archaea, many of which are adapted to life in extreme environments, physically resemble bacteria but are more closely related to eukarya.

Key Terms

adaptive radiation, p. 393
allopatric speciation, p. 380
allopolyploidy, p. 381
analogous trait, p. 389
archaea, p. 397
background extinction, p. 394
biodiversity, p. 371
biological species concept, p. 375
class, p. 377
convergent evolution, p. 389
domain, p. 377
extinction, p. 394
family, p. 377
genus, p. 376
homologous feature, p. 389
horizontal gene transfer, p. 397
hybrid, p. 375
hybridization, p. 379
kingdom, p. 377
life, p. 370
macroevolution, p. 390
mass extinction, p. 394
microbe, p. 396
microevolution, p. 390
microsphere, p. 373
monophyletic, p. 385
morphological species concept, p. 379
node, p. 383
order, p. 377
phylogeny, p. 383
phylum, p. 377
polyploidy, p. 381
postzygotic barrier, p. 375
prezygotic barrier, p. 375
protist, p. 396
punctuated equilibrium, p. 391
reproductive isolation, p. 375
ring species, p. 379
RNA world hypothesis, p. 372
speciation, p. 379
speciation event, p. 383
species, p. 375
specific epithet, p. 376
sympatric speciation, p. 381
systematics, p. 383
virus, p. 397

CHECK YOUR KNOWLEDGE

1. In a set of classic experiments performed in the early 1950s, Urey and Miller subjected an experimental system composed of H_2, CH_4 (methane), and NH_3 (ammonia) to electrical sparks. A few days later, they found ______ in their system.
   a) amino acids
   b) DNA
   c) microspheres
   d) cells
   e) RNA

2. What kind of molecule is thought to have been the first genetic material?
   a) protein
   b) DNA
   c) carbohydrate
   d) RNA
   e) microsphere

3. According to the biological species concept, species are natural populations of organisms that have the potential to interbreed and are ______ isolated from other such populations.
   a) behaviorally
   b) prezygotically
   c) postzygotically
   d) geographically
   e) reproductively

4. The classification rank that includes genera but not orders is:
   a) species.
   b) domain.
   c) class.
   d) kingdom.
   e) family.

5. Populations of *Larus* gulls around the North Pole show an unusual pattern of reproductive isolation: each population is able to interbreed with its neighboring populations, but populations separated by larger geographic distances are not able to interbreed. *Larus* gulls are an example of a(n) ______ species.
   a) ring
   b) polyploid
   c) circular
   d) Escher
   e) linked

6. In animals, it is believed that the most common mode of speciation is:
   a) autopolyploidy.
   b) chromosomal.
   c) directional.
   d) allopatric.
   e) sympatric.

7. Polyploidy:
   a) arises only when there is an error in meiosis resulting in diploid gametes instead of haploid gametes.
   b) is a common method of sympatric speciation for plants and animals.
   c) arises when sympatric speciation causes plants to have more sets of chromosomes than their parent plants.
   d) is an increased number of sets of chromosomes.
   e) always results in sympatric speciation.

8. In the plant kingdom, all of the species are descended from a single common ancestor. In terms of phylogeny, what type of tree of life is this?
   a) monophyletic
   b) uniphyletic
   c) punctuated
   d) sympatric
   e) allopatric

9. Phylogenetic trees should be viewed as:
   a) true genealogical relationships of species.
   b) the result of vertical, but never horizontal, gene transfer.
   c) intellectual exercises, not to be interpreted literally.
   d) representations of allopatric speciation events.
   e) hypotheses regarding evolutionary relationships of groups of organisms.

10. The difference between microevolution and macroevolution is that:
    a) they take place on different time scales.
    b) macroevolution occurs with physical structures, whereas microevolution occurs with physiological traits.
    c) microevolution involves changes to individuals within a population, whereas macroevolution involves changes to all of the individuals within a population.
    d) microevolution has been proven, whereas macroevolution is very speculative.
    e) microevolution occurs in prokaryotes, whereas macroevolution takes place among eukaryotes.

11. The idea of punctuated equilibrium challenges which component of Darwin's theory of evolution?
    a) steady change
    b) gradualism
    c) species stasis
    d) Both a) and b) are correct.
    e) None of the above is correct.

12. Which of the following scenarios would best facilitate adaptive radiation?
    a) A population of birds native to an island archipelago is forced to relocate to the mainland by a storm.
    b) A population of cheetahs goes through an event in which all genetic diversity in the population is wiped out.
    c) Darker-colored moths have a selective advantage over lighter-colored moths due to industrial soot on trees.
    d) A population of birds becomes stranded on an island archipelago.
    e) All of the above would facilitate adaptive radiation equally.

13. The mass extinction that occurred on earth 65 million years ago was immediately followed by:
    a) the rise of archaea.
    b) the emergence of the first non-photosynthetic organisms.
    c) the rise of the reptiles, including the dinosaurs.
    d) an increase in atmospheric oxygen levels.
    e) the rapid divergence and radiation of modern mammals.

14. Prokaryotes are classified into ______ domain(s).
a) 1
b) 2
c) 3
d) 4
e) 5

15. Protists are all alike in that all are:
a) bacterial.
b) photosynthetic.
c) eukaryotic.
d) prokaryotic.
e) archaean.

16. Which of the following domains share the most recent common ancestor?
a) archaea and eukarya
b) bacteria and eukarya
c) archaea and bacteria
d) None of the above; all three domains evolved from different ancestors.
e) None of the above; all three domains are equally related to each other.

17. Which of the following groups would be placed nearest the fungi in an evolutionary tree based on DNA sequences?
a) plants
b) bacteria
c) animals
d) archaea
e) protists

Short-Answer Questions

1. Explain the difference between defining a species according to the biological species concept and according to the morphological species concept. Include an example of why one method might be more appropriate than the other for certain situations.

2. Evolutionary trees are important in helping to explain the history of life. Outline the difference between how evolutionary trees were constructed in the 20th century and how they are constructed today. Explain what types of information these trees can and cannot convey.

3. The shape of the skull, as well as the shape and size of teeth, often reflects the method of feeding in mammals. A skull with a more rounded jaw and shorter, narrower teeth indicates a mammal that is well adapted to pulling bunches of leaves off trees and grinding them up. A skull with a more square jaw indicates a mammal adapted to grazing and cropping vegetation at the ground level. Explain how the adaptations of long, flat teeth and a more square jaw or muzzle could have triggered adaptive radiation in horses.

See Appendix for answers. For additional study questions, go to www.prep-u.com.

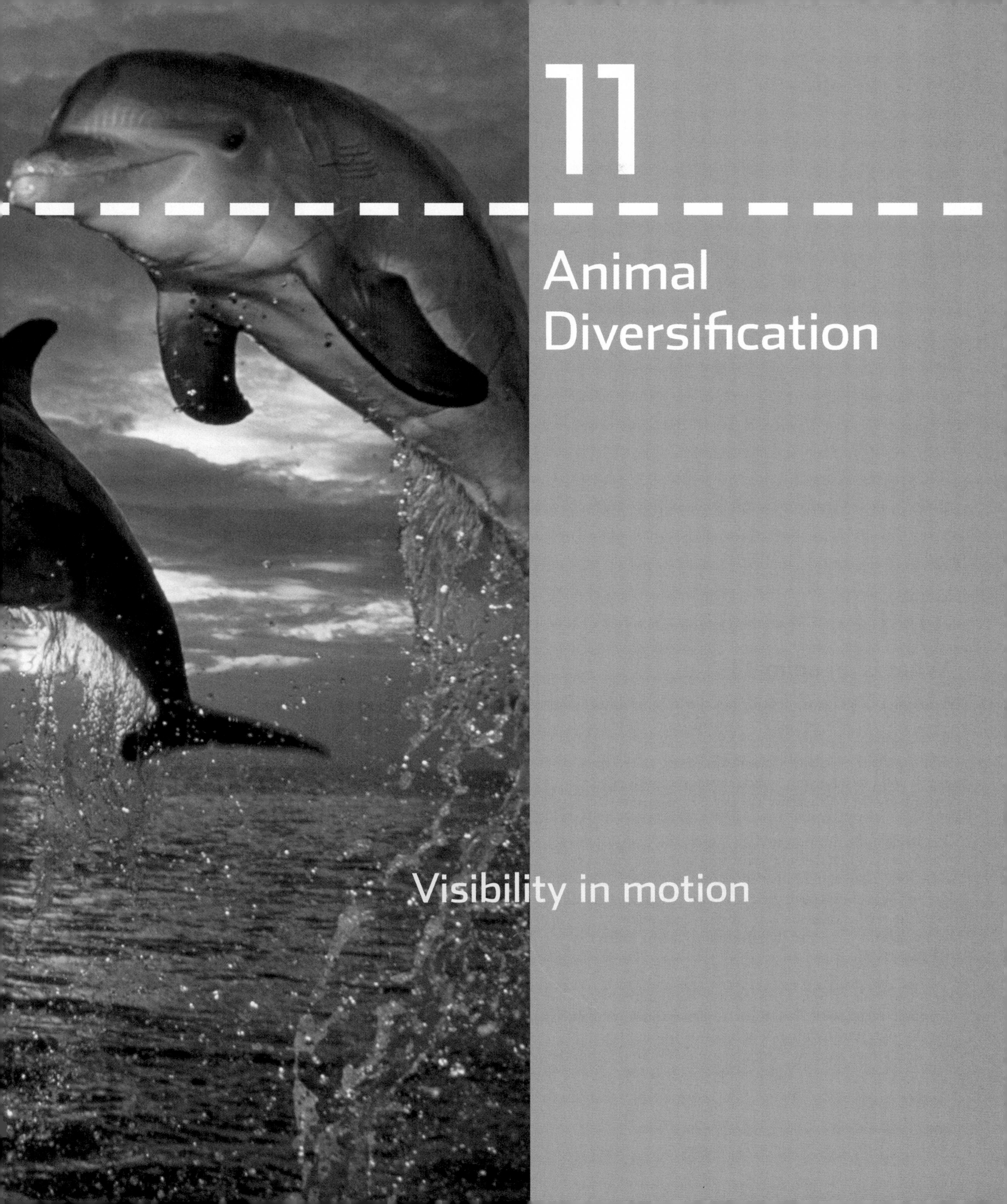

11

Animal Diversification

Visibility in motion

❶ Animals are just one branch of the eukarya.

A kangaroo rests in the sun at Koala Sanctuary, Brisbane, Australia.

11•1

What is an animal?

Looking at a kangaroo, it's obvious that it is an animal. The same goes for a mosquito, earthworm, or jellyfish. It gets a bit more difficult, though, if you look at a sponge. Is it an animal? Or is it a plant? Or is it something else? This is when it is helpful to have some guidelines for identifying and organizing the biodiversity we can see on earth.

For the sponge, the answer is *yes,* it is classified as an animal. Here's why. When defining what is or is not an animal, we look for certain features that are common to all animals. Surprisingly, considering the enormous differences among even the few animals mentioned above, just three characteristics are generally sufficient to define **animals** (**FIGURE 11-1**).

1. All animals eat other organisms. That is, rather than manufacturing their own food, as plants and some bacteria do through photosynthesis, animals consume plants, bacteria, other animals, fungi, or some combination of these. And because the best way to catch other organisms is to go to them or chase after them, that brings us to the second characteristic that sets animals apart from plants.

2. All animals move—*at least at some stage of their life cycle.* That qualifying phrase is important because some animals move only during a period called the "larval stage" that comes early in their lives, and when they turn into adults they fasten themselves to a surface. If you visit a rocky seashore, you can find two of these animals—mussels and barnacles—living on rocks in the tidal zone. Organisms that are fastened in place, such as adult mussels and barnacles, are said to be **sessile,** but even animals that are sessile as adults moved when they were larvae.

3. All animals are multicellular, and they generally have body parts that are specialized for different activities. For example, most sense organs (the eyes, ears, nose, tongue, antennae, whiskers, and so on) are at an animal's front end, where

WHAT IS AN ANIMAL?

1 ANIMALS EAT OTHER ORGANISMS All animals acquire energy by consuming other organisms.

2 ANIMALS MOVE All animals have the ability to move—at least at some stage of their life cycle.

3 ANIMALS ARE MULTICELLULAR Animals consist of multiple cells and have body parts that are specialized for different activities.

FIGURE 11-1 Characteristics of an animal. The cup coral (*left*) is eating a young octopus; a herd of spotted deer move through a forest; the animal with multiple appendages is a mantis.

it detects prey. Once an animal detects prey, the animal attempts to capture the prey, and the animal consumes it.

There are a few other general features of animals—most reproduce sexually, for example—but these three are the easiest criteria to use in determining whether or not you're looking at an animal.

TAKE-HOME MESSAGE 11•1

Animals are organisms that share three characteristics: all of them eat other organisms, all of them can move during at least one stage of their development, and all of them are multicellular.

11•2

Four key distinctions divide the animals.

More than two-thirds of the almost two million species on earth that have been identified are animals. In general, we see that closely related animals have more similar body forms than distantly related animals (although convergent evolution can obscure this; see Section 10-9). But with such a large number of animals to organize and identify, it is useful to divide them into a few large categories, based on their evolutionary relatedness.

Analyses of DNA and RNA sequences have helped biologists identify four key distinctions, into which we can group all of the animals. Each of these distinctions hinges on a particular adaptation and on the question of whether or not an animal descends from an ancestor with that adaptation (**FIGURE 11-2**).

1. Does the animal have defined tissues, with specialized cells? In the sponges, there is no distinctive tissue. Rather, the organisms are just aggregations of similar cells, with little coordination of activities between the cells. This is in contrast to the animals with defined tissues, such as humans, which have highly specialized cells such as skin cells, muscle cells, and sensory cells.

2. Does the animal develop with radial symmetry or bilateral symmetry? Among the animals with defined tissues, there is an important distinction between those with **radial symmetry** and those with **bilateral symmetry.** Radial symmetry describes organisms with a body structured like a pie. It is possible to make multiple slices, all going through

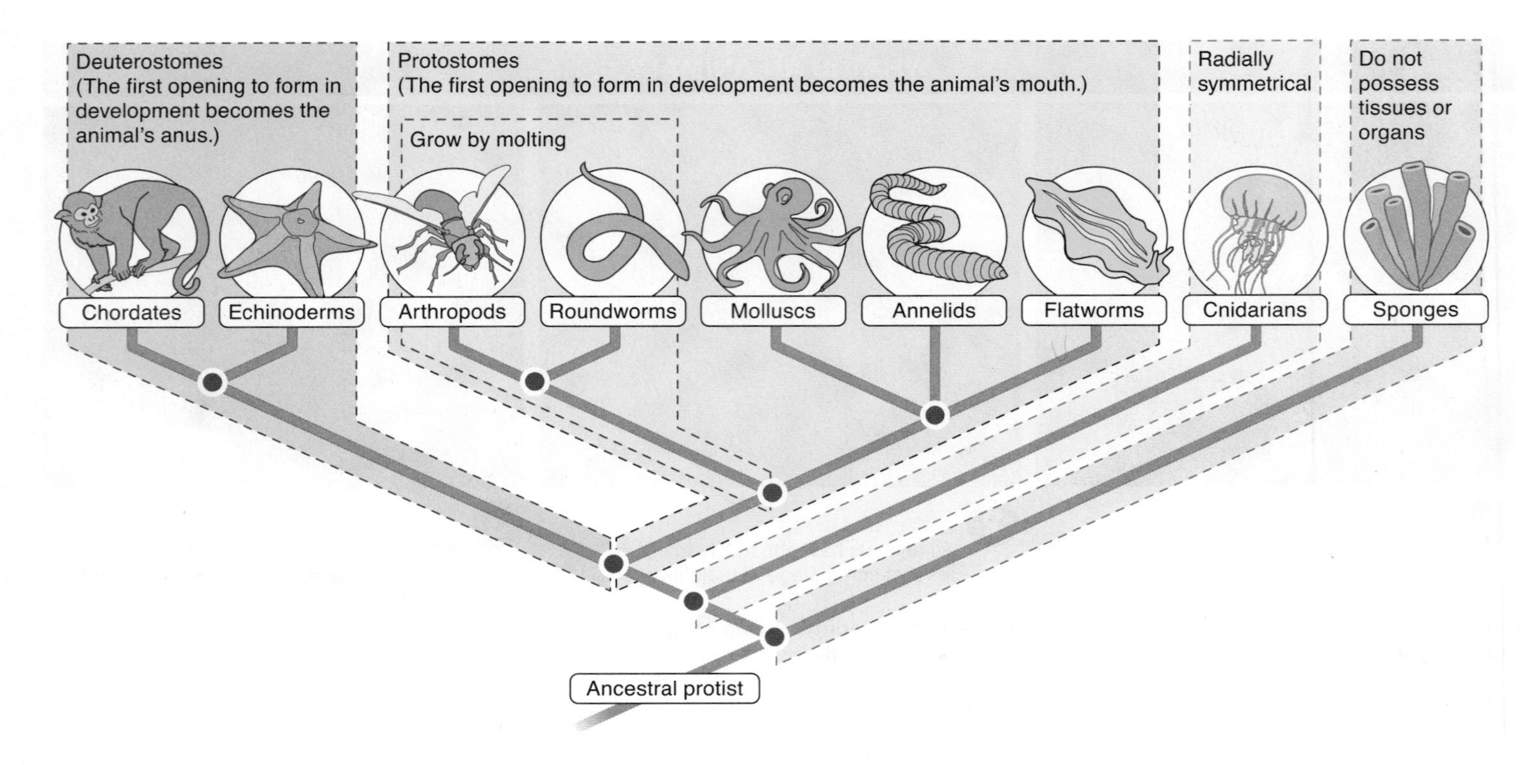

FIGURE 11-2 Innovations in body plan. The nine groups of animals in this figure represent just 25% of all animal phyla, but contain 96% of all animal species. Each of these groups will be explored in this chapter

the center, that divide the organism into identical pieces. This differs from bilateral symmetry, in which organisms have a left and right side, which are mirror images. Radially symmetrical organisms tend to be slow-moving or free-floating, while bilaterally symmetrical organisms tend to be faster-moving and more adept at searching for food and avoiding predators.

3. During development, does the animal's gut develop from front to back or back to front? Among the bilaterally symmetrical animals with defined tissues, there is an important distinction—at least with respect to determining the evolutionary lineage to which they belong—in how their early development occurs. Early in the evolution of animals—about 630 million years ago—a major split occurred that allows us to separate the bilaterally symmetrical animals with defined tissues into two distinct lineages: the **protostomes** (which can be translated as "mouth first") and the **deuterostomes** ("mouth second"). These names refer to the way the gut develops—from front to back or from back to front. In protostomes the gut develops from front to back, so the first opening that forms is the mouth of the adult animal, whereas in deuterostomes the gut develops from back to front and the second opening becomes the mouth. Although it's something you can see only when closely observing animals' embryonic development, the protostome/deuterostome split is the most basic division of animals, in terms of revealing their evolutionary relatedness to one other.

4. Does growth occur by molting or by adding continuously to the skeletal elements? Among the bilaterally symmetrical animals that are protostomes, there is an important distinction between animals that molt—shedding an exoskeleton (a hard outer layer) and replacing it with a larger one at regular intervals during their life—and animals that grow by adding to the size of their skeletal elements, in a more continuous manner than occurs in the molting animals.

While all of the animals share a common ancestor, an ancestral protist, the four distinctions above help us identify and organize all of the approximately 36 living phyla of animals.

TAKE-HOME MESSAGE 11•2

The animals probably originated from an ancestral protist. Four key distinctions divide the extant animals: (1) tissues or not, (2) radial or bilateral symmetry, (3) protostome or deuterostome development, and (4) growth through molting or through continuous skeletal enlargement.

11•3

Everything that is not extinct is evolutionarily successful.

"Never use the word higher or lower." This is a note that Charles Darwin wrote to himself in the margin as he was reading a book that proposed a theory of cosmic and biological evolution. It was an important insight, but one that can be difficult to adhere to—particularly as we turn our attention to the animals.

What is "success," evolutionarily speaking? There are two possible states for a species. It can be extant, meaning that it currently exists. Or it can be extinct. From an evolutionary perspective, every living species is successful. The fact that a species exists means that it is able to do the three things that all organisms must do: find food, escape predators, and reproduce. Earthworms and tigers live very different lives, and the characteristics that work well for one species would not work at all for the other, but the earthworm is as successful at being a worm as the tiger is at being a very large cat (**FIGURE 11-3**).

Throughout Darwin's writings, rather than viewing one species as higher or lower, as more or less advanced or primitive, he emphasized adaptation among the species he described and their relationship to their environment. One species was never better or worse than another. Rather, each species was differentiated from the others, showing specializations that adapted individuals of that species to the particular niche in which they occurred. Different environments posed different challenges. All extant species, as evidenced by their existence and persistence, are able to overcome those challenges.

From an evolutionary perspective, every living species is successful.

FIGURE 11-3 Two equally successful organisms: the earthworm and the tiger.

In most biology textbooks, the tour through the animal kingdom begins with animals having a simple body plan, such as sponges, jellyfish, and worms, continues through the more complex invertebrates, the molluscs and arthropods, and culminates in the complexity of the vertebrates (mammals, amphibians, reptiles [including birds], and fishes). But because phylogenetic trees do not provide any information about the relative success of any species, there is no reason to begin our tour with the simplest animals, building until we reach humans at the "crown of creation."

So, our tour begins with the vertebrates. We chose this group because humans are vertebrates and vertebrates are the animals we are most familiar with. If we followed that logic to its conclusion, though, we would start by discussing humans. But we decided to begin with jawless fishes and move across to humans, because that sequence makes the evolution of certain organs and body parts, such as lungs, legs, and the amniotic egg, easier to understand.

As we survey the many different kinds of animal, we will see that even as each extant species represents an evolutionary success, certain adaptations have led to the rapid and extensive diversification of particular groups of species. Paying attention to these adaptations, we will investigate nine of the most successful phyla (shown in Figure 11.2) and the evolutionary innovations that contributed to their success. Although the nine groups represent only a quarter of all animal phyla, they actually account for 96% of the animal species extant today.

TAKE-HOME MESSAGE 11•3

From an evolutionary perspective, all extant species are successful. Among the higher levels of taxonomic grouping, however, some groups are represented by more species than others. Of the 36 animal phyla, 9 phyla account for more than 96% of all described animal species.

❷ Vertebrates are animals with a backbone.

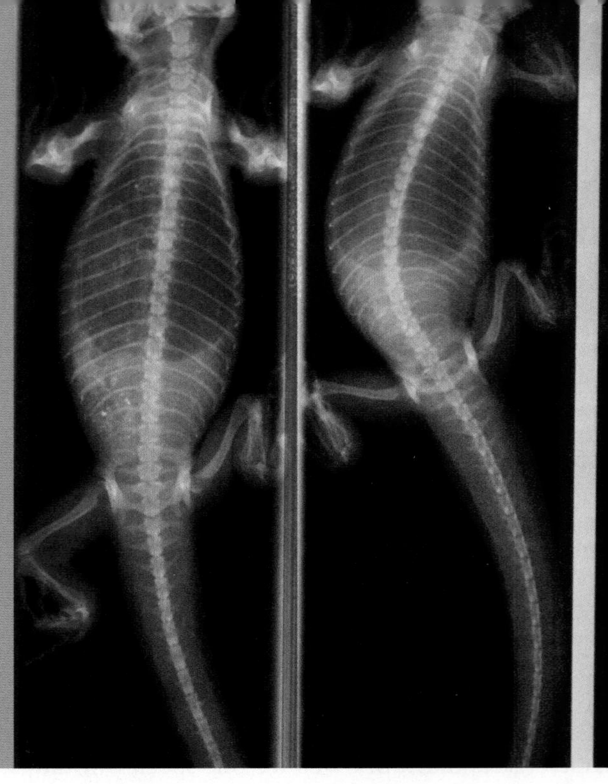

Animals in motion: these X-ray images show tuatara walking.

11•4

All vertebrates are members of the phylum Chordata.

All vertebrates are deuterostome animals, so they have defined tissues and are bilaterally symmetrical. Further, they are all part of a phylum called Chordata (the chordates), whose members are defined by four distinct body structures (**FIGURE 11-4**).

1. The **notochord,** a rod of tissue extending from the head to the tail, is the structure that gives chordates their name. The notochord stiffens the body when muscles contract during locomotion (movement). Primitive chordates retain the notochord throughout life, but in advanced chordates, such as vertebrates, the notochord is present only in early embryos and is replaced by the vertebral column (backbone).

2. A **dorsal hollow nerve cord** extends from head to tail. In vertebrates, this nerve cord forms the central nervous system, which consists of the spinal cord and the brain. Other kinds of animals (worms, insects, and so on) also have a nerve cord, but it lies in the lower portion of the front (ventral) part of the body and is solid instead of hollow.

3. The earliest chordates were aquatic and passed water through slits in the pharyngeal region (the area between the back of the mouth and the top of the throat) for breathing and feeding. **Pharyngeal slits** are present in the embryos of all chordates, even in the embryos of humans, which lost their gills far back in evolutionary time.

4. A **post-anal tail** is another characteristic of chordates. "Post-anal" means that the tail extends back beyond the end of the trunk. (The posterior end of the digestive system is the anus, and this structure marks the end of the trunk.) The anus of worms, insects, and other non-chordate animals lies at the extreme hind end of the body. Some vertebrates, including humans, have a tail, but only for a brief period of time during embryonic development.

The phylum Chordata contains three subphyla: the tunicates (Urochordata; with about 2,000 species), lancelets (Cephalochordata; with about 20 species), and vertebrates (Vertebrata; with about 56,000 species). All three of the subphyla have the four characteristic chordate structures for at

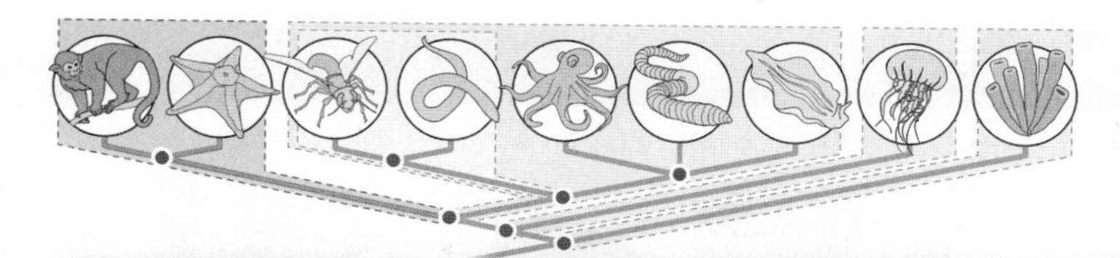

THE CHORDATES

COMMON CHARACTERISTICS
All chordates possess four common body structures, although in many chordates, these structures are only present during specific life stages.
- Notochord
- Dorsal hollow nerve cord
- Pharyngeal slits
- Post-anal tail

MEMBERS INCLUDE
- Tunicates (about 2,000 species)
- Lancelets (about 20 species)
- Vertebrates (about 56,000 species)

FIGURE 11-4 Understanding the chordates. Shown here (*from left*) are Pacific double saddle butterfly fish, a golden lion tamarin, a thorny devil lizard, and compound ascidians (tunicates).

least a portion of their lives, but they look very different from one another (**FIGURE 11-5**).

Tunicates are marine animals, and the adults, which are about the size of your thumb, look like balls of brownish green jelly. You can find them attached to docks and mooring lines of boats. It is the free-swimming larvae of tunicates that reveal their chordate characteristics.

Lancelets are slender, eel-like animals, about the length of your little finger, that live in coastal waters. Unlike tunicates, adult lancelets have all of the chordate characteristics. Lancelets (and

GROUPS OF CHORDATES

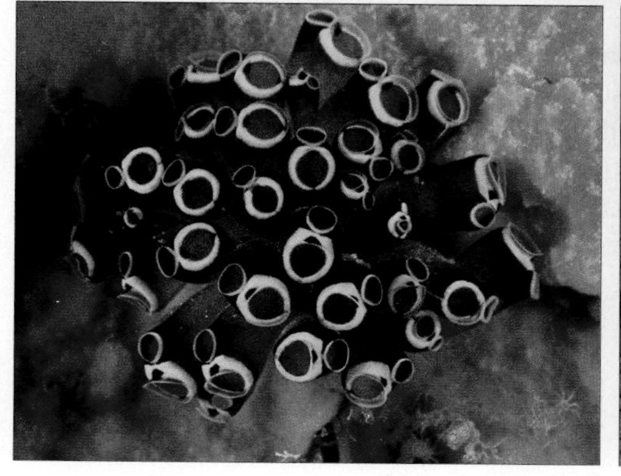

TUNICATES
- Filter-feeding marine animals
- Can often be found attached to docks
- Larvae are free-swimming

LANCELETS
- Filter-feeding marine animals
- Live in coastal waters

VERTEBRATES
- The most diverse subphylum of chordates
- Range in size from the tiny hummingbird to the enormous blue whale
- Can be found in most of the habitats on earth

FIGURE 11-5 Representing the subphyla of chordates: (*left to right*) singing sea squirt, lancelet, and hummingbird.

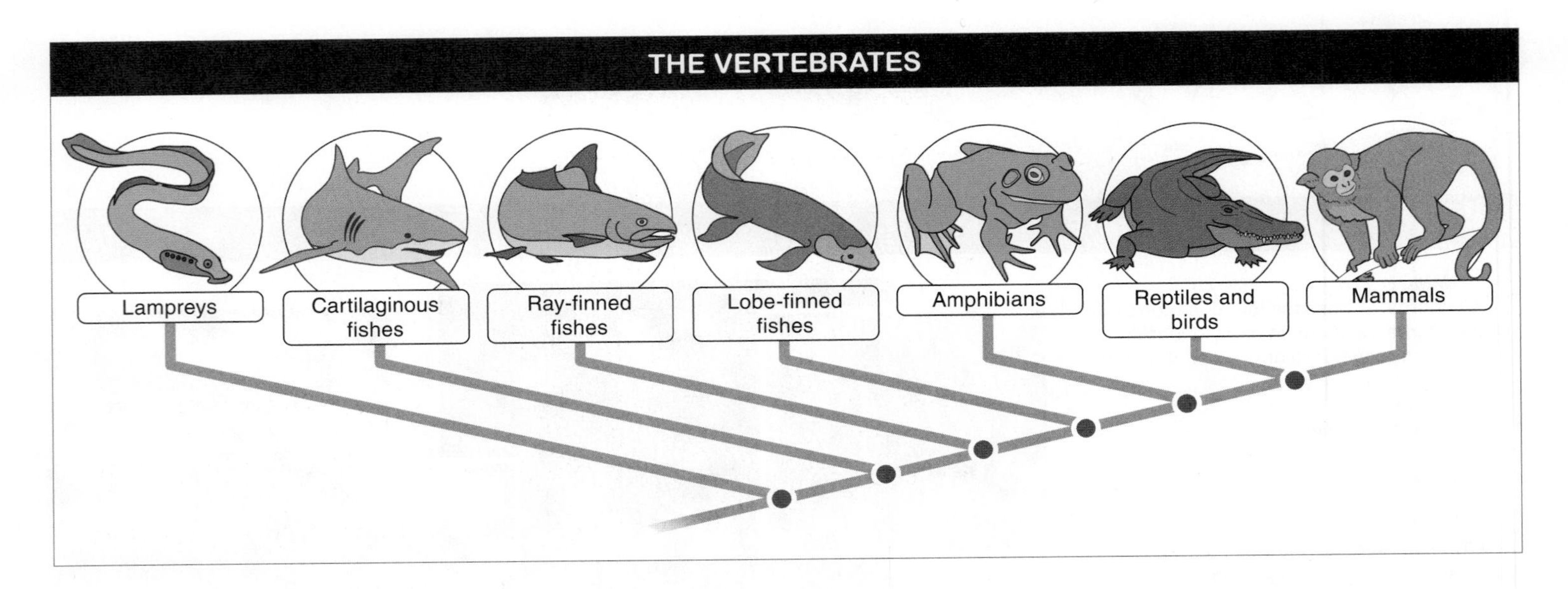

FIGURE 11-6 Vertebrate phylogeny.

tunicates) are filter-feeders: cilia draw a current of water through the mouth into the pharynx and out through the pharyngeal slits, and microscopic food items are trapped by the bars between the slits.

The capybara is a semi-aquatic vertebrate found in South America in forested areas near bodies of water, such as lakes, rivers, swamps, ponds, and marshes. It is the largest of all living species of rodents, and can weigh as much as 150 pounds.

Vertebrates are the most diverse subphylum of chordates (**FIGURE 11-6**). The size range of vertebrates extends from tiny hummingbirds and shrews that weigh only a tenth of an ounce (about 3 grams) to the blue whale, which weighs in at more than 130 tons (120,000 kilograms). Vertebrates move by swimming, burrowing, crawling, walking, running, and flying, and they have invaded all of the habitats on earth, from the depths of the sea to the tops of mountains, and from rain forests to deserts and the ice at the Poles. But despite all these differences, vertebrates still have the four chordate characteristics at some stage of their lives.

TAKE-HOME MESSAGE 11•4

All chordates have four characteristic structures: a notochord, a dorsal hollow nerve cord, pharyngeal gill slits, and a post-anal tail. The three subphyla of chordates are superficially very different, but all are united by possessing these four characteristics—although some of the subphyla reveal these characteristics only during specific life stages.

11.5

The evolution of jaws and fins gave rise to the vast diversity of vertebrate species.

Vertebrates eat other organisms, and they need a mouth to do so. But a mouth is just an opening through which food passes—jaws help an animal get the food into its mouth and chew it. The earliest vertebrates were fish-like animals that lived more than 500 million years ago and did not have jaws. Two kinds of jawless vertebrates still exist: the lampreys (with 41 species) and the hagfishes (with 43 species). Lampreys are parasitic; they attach to other fishes by creating suction with a circular oral disk that is studded with sharp spines. First, the spines scrape an opening through the scales and skin. Then, a protein produced by the lamprey keeps the wound from closing, and the lamprey feeds on the blood and body fluids that seep out. Hagfishes feed on dead animals, using two spiny dental plates that they embed in their prey.

Q Why are jaws needed?

Lampreys and hagfishes do the two things that animals do—feed on other organisms and use locomotion to reach their food—but they are about as basic as it is possible to be. And because they do not have jaws, they cannot seize prey or chew food before swallowing. It was the evolution of fins and jaws that set the stage for the evolutionary explosion of diversity of vertebrates.

Fish swim the same way that lampreys and hagfishes swim—that is, by bending the body from side to side to create an S-shaped wave that moves from the head of the fish toward its tail. A typical fish has a total of seven fins serving different purposes: to drive the fish forward, to minimize rolling from side to side, and for steering and stopping. The combined effects of the seven fins allow the fish to swim rapidly in a straight line when it is in open water or to weave its way through a dense stand of plants or corals, with great maneuverability.

The evolution of fins was paralleled by the evolution of jaws, because the two kinds of structure work together. Fins get you to the organism you are going to eat, and jaws capture and kill it (**FIGURE 11-7**).

THE EVOLUTION OF JAWS AND FINS

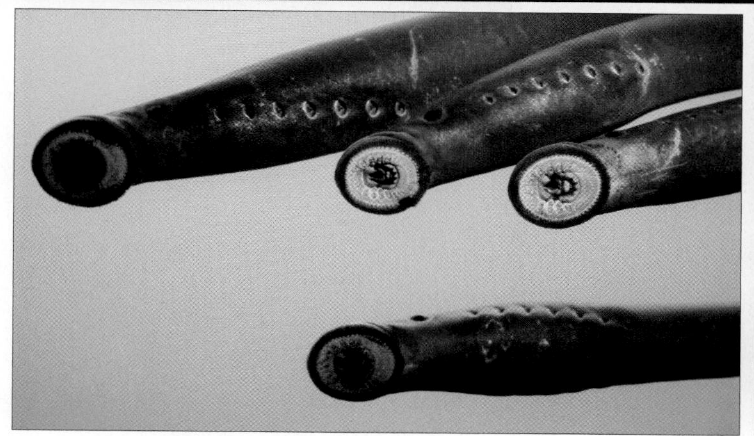

FISHES WITHOUT JAWS OR PAIRED FINS
- Tail fin propels organism through water
- Feed by attaching mouthparts to prey

FISHES WITH JAWS AND PAIRED FINS
- Paired fins provide controlled movement through water
- Jaws allow for seizing and chewing prey

FIGURE 11-7 The lamprey and the shark. The lamprey (*left*) has an oral disk that it uses for sucking in a meal. The shark has a powerful jaw lined with sharp teeth, and pairs of fins that allow it to move rapidly through the water and attack prey.

The development of fins and jaws set the stage for the evolutionary explosion of vertebrate diversity.

JAWED FISHES

CARTILAGINOUS FISHES
- Characterized by a skeleton made completely from cartilage
- Contain about 880 species, including sharks and rays

RAY-FINNED FISHES
- Characterized by having rigid bones and fins lined with hardened rays
- Possess a swim bladder, which aids in flotation
- Contain about 27,000 species, including almost everything you think of as "fish," from salmon to goldfish and guppies

LOBE-FINNED FISHES
- Characterized by having two pairs of sturdy fins on the underside of their body
- Contain six species of lungfish and two species of coelacanths

FIGURE 11-8 Fishes with jaws: cartilaginous fishes, ray-finned fishes, and lobe-finned fishes. Shown are stingray, rainbow trout, and Australian lungfish.

The development of jaws and fins allowed great diversification in vertebrate body shapes and composition. The **cartilaginous fishes,** with about 880 species, include sharks and rays (**FIGURE 11-8**). The cartilaginous fishes are characterized by a skeleton made completely from cartilage, the same tissue that gives your nose its shape. The largest group of jawed fish is also the most diverse group of vertebrates, the **ray-finned fishes,** with about 27,000 species, which includes almost everything you think of as "fish," from salmon to goldfish. Ray-finned fishes have rigid bones and a mouth at the narrow tip, or apex, of the body. They get their name from the fact that their fins are lined with hardened rays.

An important evolutionary development in the ray-finned fishes was the swim bladder, a gas-filled organ that keeps the fish from sinking. This bladder later evolved into the lung. The cartilaginous fishes, because they have no swim bladder, must constantly move around in the water or they will sink. A third group of jawed fishes, the **lobe-finned fishes,** are represented by just eight species. These fishes have sturdy fins on the underside of their body (see Figure 11-8), and as we will see in the next section, their features were useful in initiating the move onto land.

TAKE-HOME MESSAGE 11•5

The development of two structures—fins and jaws—set the stage for the enormous diversity of modern vertebrates.

11•6

The movement onto land required lungs, a rigid backbone, four legs, and eggs that resist drying.

Movement from water onto land required four major evolutionary innovations: lungs, a backbone, four legs, and eggs that won't dry out. All four of these characteristics are necessary for a land animal, and all four evolved in the immediate ancestors of terrestrial vertebrates, which were predatory fish that lived in shallow water.

Q Where did legs and lungs come from?

As we've noted, lungs developed from the swim bladder found in ray-finned fishes. All of the terrestrial (land-dwelling) vertebrates—amphibians, reptiles (including birds), and mammals—are descendants of the lobe-finned fishes that lived during the Devonian period, some 400 million years ago. At this time, the lobe-finned fishes had lungs and the rudiments of limbs, and they were still fully aquatic and living near the shore. The four sturdy fins on the underside of their body helped them move through shallow water, and lungs allowed them to breathe air when the oxygen concentration in the warm, stagnant water was low. Thus, these fishes already had some of the basic characteristics of terrestrial animals—they could walk using their fins, and they could get oxygen from air via their lungs instead of from water via their gills. The jointed bones in the fins of lobe-finned fishes don't look much like your arms and legs, but they are evolutionarily homologous structures.

To move onto land, a vertebrate needs more than just legs and lungs; it needs structural support to resist the pull of gravity. Each vertebra of a terrestrial vertebrate has projections that interlock with projections from the vertebra ahead of it and the vertebra behind it. These interlocking projections prevent the backbone from sagging under the pull of gravity, and the body weight is transmitted through the limbs to the ground.

The last innovation necessary to move onto land is an egg that resists drying out (**FIGURE 11-9**). When eggs are deposited on land, they are exposed to air and lose water by evaporation. The eggs of terrestrial animals need a waterproof covering—a membrane and eggshell—to prevent them from drying out before they hatch. The appearance of shelled eggs about 380 million years ago marks the final step in the evolution of entirely terrestrial vertebrates, the groups that evolved into mammals and reptiles (including birds). These eggs have a water-tight membrane that keeps the embryo surrounded by a bath of fluid.

FROM WATER TO LAND

The transition of vertebrates from life in water to life on land required overcoming three main obstacles. Four major evolutionary innovations allowed for this transition.

1 PROBLEM: RESPIRATION
Aquatic animals use gills to acquire dissolved oxygen from water. The transition onto land required the ability to breathe air.

SOLUTION: LUNGS
Gas exchange was transferred from gills to lungs, which evolved from the swim bladder found in ray-finned fishes.

Ray-finned fish

Early terrestrial vertebrate

2 PROBLEM: GRAVITY
The transition onto land required structural support to resist the pull of gravity.

SOLUTIONS: LIMBS and MODIFIED VERTEBRAE
Limbs evolved from the jointed fins found on the underside of lobe-finned fishes.

Vertebrae were modified to transmit the body weight through the limbs to the ground.

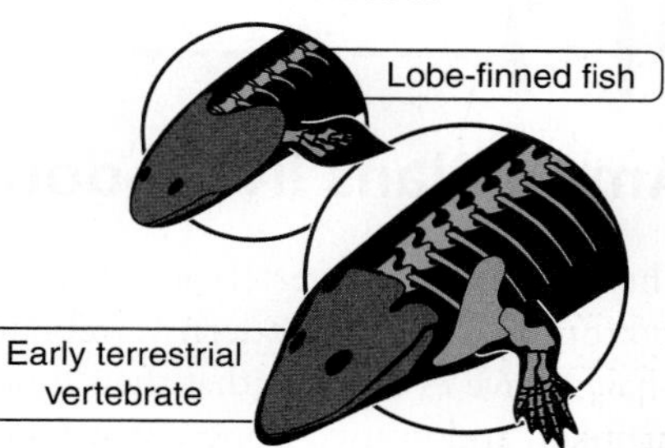

3 PROBLEM: EGG DESICCATION
The transition onto land required an egg that resisted drying out when exposed to air.

SOLUTION: AMNIOTIC EGG
Terrestrial animals developed a waterproof eggshell, which prevents eggs from drying out before they hatch.

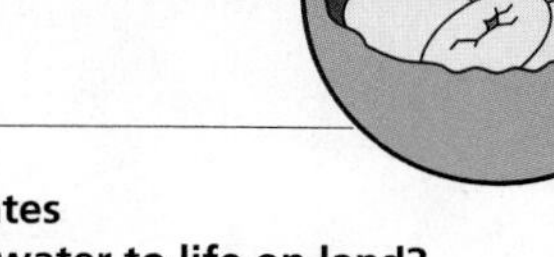

FIGURE 11-9 How did vertebrates make the transition from life in water to life on land?

TAKE-HOME MESSAGE 11•6

In the transition of vertebrates from life in water to life on land, fins were modified into limbs, vertebrae were modified to transmit the body weight through the limbs to the ground, and the site of gas exchange was transferred from gills to lungs. The only entirely new feature to appear in the early development of terrestrial vertebrates was an egg that resisted drying out.

❸ All terrestrial vertebrates are tetrapods.

A flap-necked chameleon crosses a salt flat in Botswana.

11•7

Amphibians live a double life.

The ancestors of all vertebrates that live on land had both lungs and four legs. Some modern vertebrates (such as snakes and whales) have evolved so that their limbs are reduced to a few shrunken and unused bones. Whales (and some snakes) have returned to living in water. And some salamanders have lost their lungs and breathe through the skin, but these are recent changes. Taxonomically, all terrestrial vertebrates are **tetrapods** (*tetra* = four; *poda* = feet). And it is the presence or absence of the amniotic egg that allows us to distinguish two large groups of terrestrial vertebrates: the non-amniotes (amphibians, or

THE AMPHIBIAN LIFE CYCLE

Amphibians, such as frogs, toads, and salamanders, are terrestrial vertebrates (tetrapods) with non-amniotic eggs. Most species live on land as adults, but develop in water.

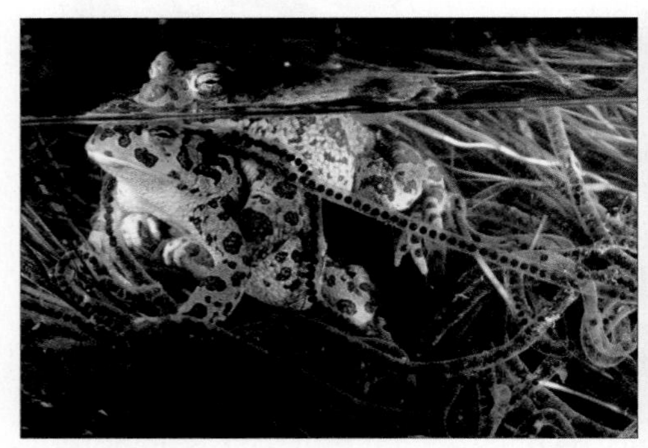

EGGS
Amphibians have non-amniotic eggs, which must be laid in water to prevent desiccation.

LARVAE
Amphibians spend their larval stage underwater and undergo metamorphosis to develop legs and lungs.

ADULTS
Only the adults are true land animals; however, most of the species in this group stay close to water to lay their eggs.

FIGURE 11-10 From egg, to larva, to adult: the amphibian life cycle.

tetrapods without desiccation-proof amniotic eggs) and the **amniotes** (reptiles and mammals, tetrapods with amniotic eggs) that are the subjects of this and the next three sections.

The first terrestrial vertebrates were **amphibians,** from the Greek word *amphibios,* meaning "living a double life." Only the adults are true land animals—the young of most amphibian species remain aquatic (**FIGURE 11-10**). Most of the 6,000 species in this group are still significantly tied to life in the water, because they must always be near water to lay their eggs, which are simple structures, not unlike fish eggs. Frogs and toads make up the vast majority of amphibians (about 5,400 species, with salamanders comprising another 550 species, and a group of legless burrowing amphibians, called caecilians, comprising 170).

The larval stage of the frog, the tadpole, lives in the water, lacks legs, and eats algae. During metamorphosis, the tadpole develops legs, lungs, and a digestive system fit for its adult life as a carnivore. Adult frogs have thin moist skin through which much of their gas (oxygen and carbon dioxide) exchange with the air takes place.

The last two decades have seen a stunning decline in amphibian populations throughout the world, attributable to a combination of causes: climate change, habitat degradation, fungal diseases, and increased pollution.

TAKE-HOME MESSAGE 11•7

Amphibians are terrestrial vertebrates (tetrapods), but most species still lay eggs in water and the eggs hatch into aquatic larvae (in frogs: tadpoles).

11•8

Birds are reptiles in which feathers evolved.

Soon after amniotic vertebrates appeared, two different evolutionary lineages began to diverge (**FIGURE 11-11**). One of these lineages is the mammals, the group to which humans belong. The other lineage is the one we have been calling "reptiles (including birds)." That is an awkward term, but the evolutionary tree (see Figure 11-11) shows why it is necessary. Birds (about 9,700 species) are merely one branch of the reptile lineage that also includes snakes and lizards (about 8,000 species), turtles (about 300 species), crocodiles and alligators (23 species), the New Zealand tuatara (2 species), as well as the extinct dinosaurs.

The characteristics that hold the bird-crocodile-dinosaur group together are mostly similarities in bones (especially the bones of the skull and legs) and their DNA sequences.

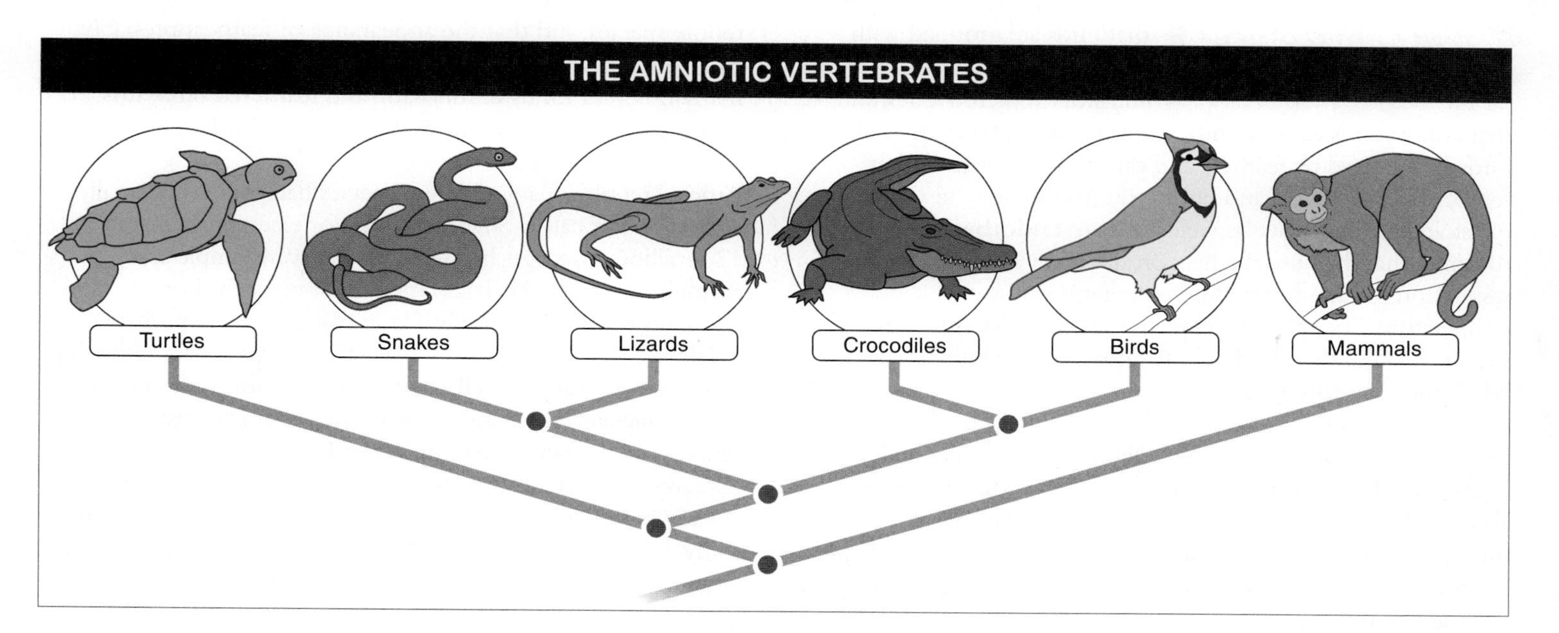

FIGURE 11-11 Amniotic vertebrate phylogeny.

REPTILES vs. BIRDS

REPTILES
- Skin is covered in scales.
- Body temperature is controlled by external conditions, such as air temperature (exothermic).
- Include snakes and lizards (about 8,000 species), turtles (about 300 species), crocodiles and alligators (23 species), and tuatara (2 species)

BIRDS
- Skin is covered in feathers, enabling flight and providing insulation.
- Body temperature is maintained by heat generated from cellular respiration (endothermic).
- Include about 9,700 species

FIGURE 11-12 Comparison of reptiles and birds. A Jackson's chameleon (*left*) and a scarlet macaw.

Reptiles are amniotes, and thus have amniotic eggs. You are familiar with the hard-shelled eggs of birds, and some lizard, snake, and turtle eggs also have hard shells. Other species in these groups have eggs with a parchment-like shell.

Most people find it confusing that birds are grouped with turtles, lizards and snakes, and alligators and crocodiles, and that confusion is easy to understand (**FIGURE 11-12**). After all, birds are covered by feathers and can fly, whereas the other reptiles have bare skin and do not fly. And there is an important physiological difference as well: birds are **endotherms**—meaning that they use the heat produced by their cellular respiration to raise their body temperature above air temperature. The other reptiles are **ectotherms**—they bask in the sun to raise their body temperature, and seek the shade when the air is too warm.

How can it be that animals as different as crocodiles and hawks are closely related? During the Mesozoic era, from about 250 million years ago to about 65 million years ago, dinosaurs were the dominant terrestrial vertebrates. The fossil record shows a remarkably clear series of forms that bridge the transition between bare-skinned dinosaurs and feathered birds—birds are, in fact, just one group of dinosaurs.

Q What were the first feathers used for?

The most apparent difference between reptiles (including most dinosaurs) and birds is the presence of feathers in birds. The fossil record reveals that feathers actually evolved before birds, appearing in many reptile species, and that the appearance of feathers probably had nothing to do with flight. In fact, feathers evolved in many different kinds of dinosaur, and feathered dinosaurs continued to live for millions of years after birds appeared.

Among reptiles, many different species had feathers. A small, agile dinosaur called *Sinosauropteryx,* for example, living about 120 million years ago, had feathers that were simple spiky filaments on the neck, back, and tail, along with shorter filaments covering the body. These filaments were probably brightly colored and used by male *Sinosauropteryx* for courtship displays to females, as well as for aggressive displays to other males. *Sinosauropteryx* and many of the other feathered dinosaurs were latecomers, however. The pigeon-sized *Archaeopteryx,* which was among the first bird species, had already, by 147 million years ago, developed feathers that it used to fly; by 140 million years ago, the skies were filled with birds.

If *Archaeopteryx* and a host of other early birds already had the specialized feathers needed for flight, why did later dinosaurs

FIGURE 11-13 ***Sinosauropteryx:* an early feathered, flightless dinosaur.**

still have primitive feathers? The answer is that feathers have many functions, of which flight is only one. *Sinosauropteryx,* as we saw above, probably used their feathers for courtship or territorial displays, and they may also have used them as insulation (**FIGURE 11-13**).

Another big difference between birds and reptiles is that, as we noted earlier, birds use internally generated heat to maintain a high body temperature, while the reptiles rely on the sun to heat their bodies. This evolutionary change may have been related to feathers. As feathers evolved for display and flight, they may also have provided some insulation. This insulation may then have enabled the evolution of high rates of cellular respiration and the maintenance of a high and constant body temperature.

TAKE-HOME MESSAGE 11•8

Birds are just one branch of the reptile lineage but possess feathers and can generate body heat. The complex anatomical and physiological systems that we see in living animals, such as feathers and endothermy in birds, are the products of hundreds of millions of years of step-by-step changes that began with simple structures. Feathers were originally colorful structures used for behavioral displays, and their current uses as insulation and in flight developed later.

11•9

Mammals are animals that have hair and produce milk.

Long before the first birds appeared, the mammals evolved from the reptiles, originally as small, nocturnal insect-eaters. These early mammals remained small and, judging from the scarce fossil record, they were never in great abundance while the dinosaurs were around. The early mammals had several features in common, however, that are still present in all mammals today. Two important features are **hair**—dead cells filled with the protein keratin—which serves as an insulator, and **mammary glands** in female mammals, which enable the nursing of young with nutritious, calorie-rich milk.

As the earliest mammals evolved, there was a gradual transition from bodies with short legs that projected out horizontally from the trunk, like the legs of an alligator, to bodies with long legs held vertically beneath the trunk, like those of a dog (**FIGURE 11-14**). These anatomical changes would have allowed the earliest mammals to run faster and farther to capture prey.

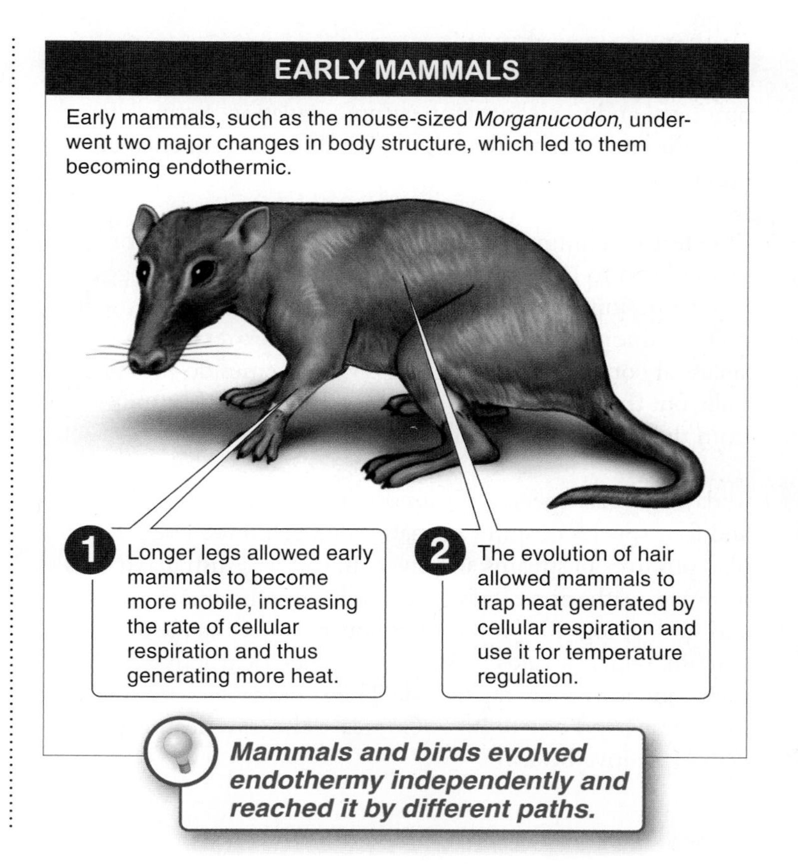

FIGURE 11-14 How did endothermy evolve in early mammals?

GROUPS OF MAMMALS

Mammals are endothermic vertebrates that have hair and produce milk for their young.

MONOTREMES

- Females lay eggs.
- Females produce milk, but do not have nipples—babies suck milk from the hairs on their mother's chest.
- Only five species survive—the platypus and four species of spiny animals called echidnas.

MARSUPIALS

- Females give birth after a short period of development.
- Females of most species have a pouch where the young complete their development.
- About 300 species survive, including kangaroos, koalas, wallabies, and possums.

PLACENTAL MAMMALS

- Females have a placenta that provides oxygen and nutrients to embryos in the uterus.
- About 4,500 species survive.

FIGURE 11-15 Three groups of mammals: monotremes, marsupials, and placentals. Shown here are a platypus, a kangaroo, and a bonobo.

But an increase in muscle activity for running had to be accompanied by an increase in the rate of cellular respiration to provide the energy to keep the leg muscles working. This increase in metabolism, along with the evolution of hair, would have allowed the early mammals to trap the heat produced and use it for temperature regulation, setting the stage for endothermy (but through a different sequence of events than in the birds).

One feature common to most mammals—**viviparity,** or giving birth to babies ("live birth") rather than laying eggs—is not a defining mammalian characteristic. This is because it isn't common to all mammals. **Monotremes** retain the ancestral condition of laying eggs. Monotremes do produce milk, but they do not have nipples—the babies suck milk from the skin and hairs on their mother's chest.

Today, only five species of monotremes survive—the platypus and four species of spiny animals called echidnas. The platypus lives in streams and rivers in eastern Australia. It uses its broad, leathery, electrosensitive bill to probe under rocks and sunken logs for prey such as aquatic insect larvae and crayfish, which it finds by sensing the electrical activity of their muscles as they try to hide. The echidnas—three species in Australia and one in New Guinea—also have leathery electrosensitive beaks.

The remaining two lineages of mammals, the **marsupials** and the **placentals,** are both viviparous, but the newborn young of marsupials and placentals are quite different (**FIGURE 11-15**). Marsupials are called the "pouched mammals" because the females of most marsupial species have a pouch where the young complete their development, following a short period of embryonic life in the uterus.

Placental mammals (including humans) take their name from the *placenta,* which is the structure responsible for the transfer of nutrients, respiratory gases, and metabolic waste products between the mother and the developing fetus. Dense capillary beds in the placenta send finger-like projections into the wall of the uterus, where they are surrounded by blood from the mother's circulation.

TAKE-HOME MESSAGE 11•9

The presence of hair and lactation are defining characteristics of mammals. Monotremes are egg-laying mammals. Marsupial mammals give birth after a short period of intrauterine development and have a pouch in which the baby completes its development. Placental mammals have a placenta that provides oxygen and nutrients to embryos that undergo a longer development in the uterus.

11.10

Humans tried out different life styles.

The evolutionary lineage to which humans belong, the primates, originated about 55 million years ago. We can get an idea of what the ancestral primate looked like from the modern species of tree-living mammals called prosimians—a group that includes the lemurs. Many of the anatomical characteristics of humans and the other primates can be traced to our arboreal origin. Our forward-directed eyes and binocular vision that allow us to judge distances accurately, our shoulder and elbow joints that allow our arms to rotate, and the retention of 10 fingers and 10 toes that allow us to grasp objects—all these are traits we inherited from our arboreal ancestors (**FIGURE 11-16**).

Humans are part of the anthropoid lineage of primates, which includes the New World and Old World monkeys (both of which have tails) and the apes (which lack tails). Among the apes, gibbons and orangutans live in pairs or alone, while gorillas and chimpanzees live in social groups that consist of one or more adult males and several females, the ancestral social structure for human societies. Within the apes, genetic and anatomical characteristics show that chimpanzees are our closest living relatives. Human and chimpanzee genes are very similar—their base sequences differ by about 1%, and one-third of human and chimpanzee genes are identical (**FIGURE 11-17**).

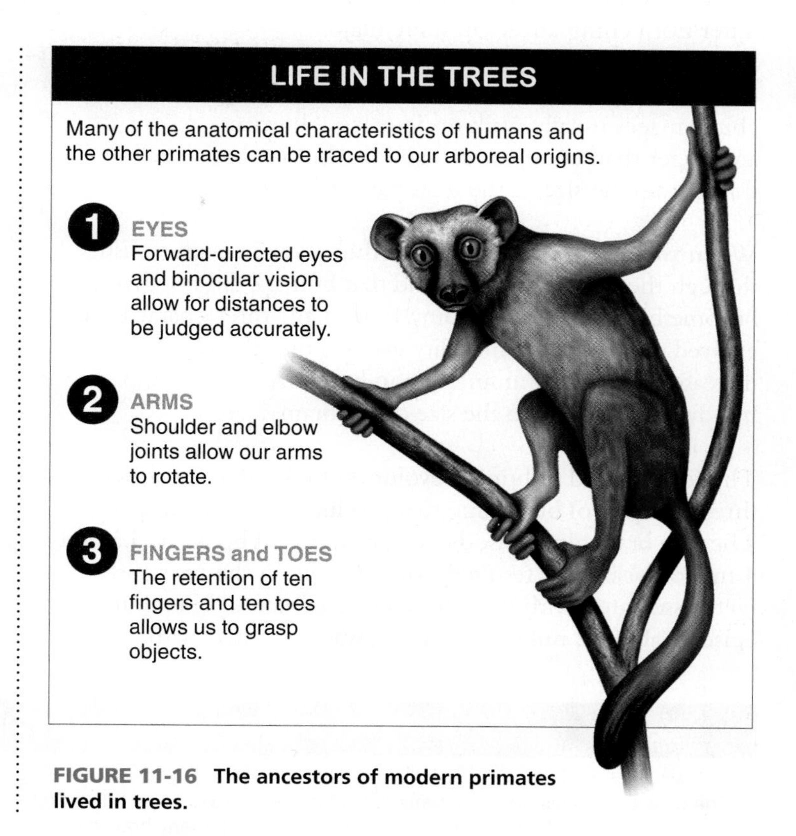

FIGURE 11-16 The ancestors of modern primates lived in trees.

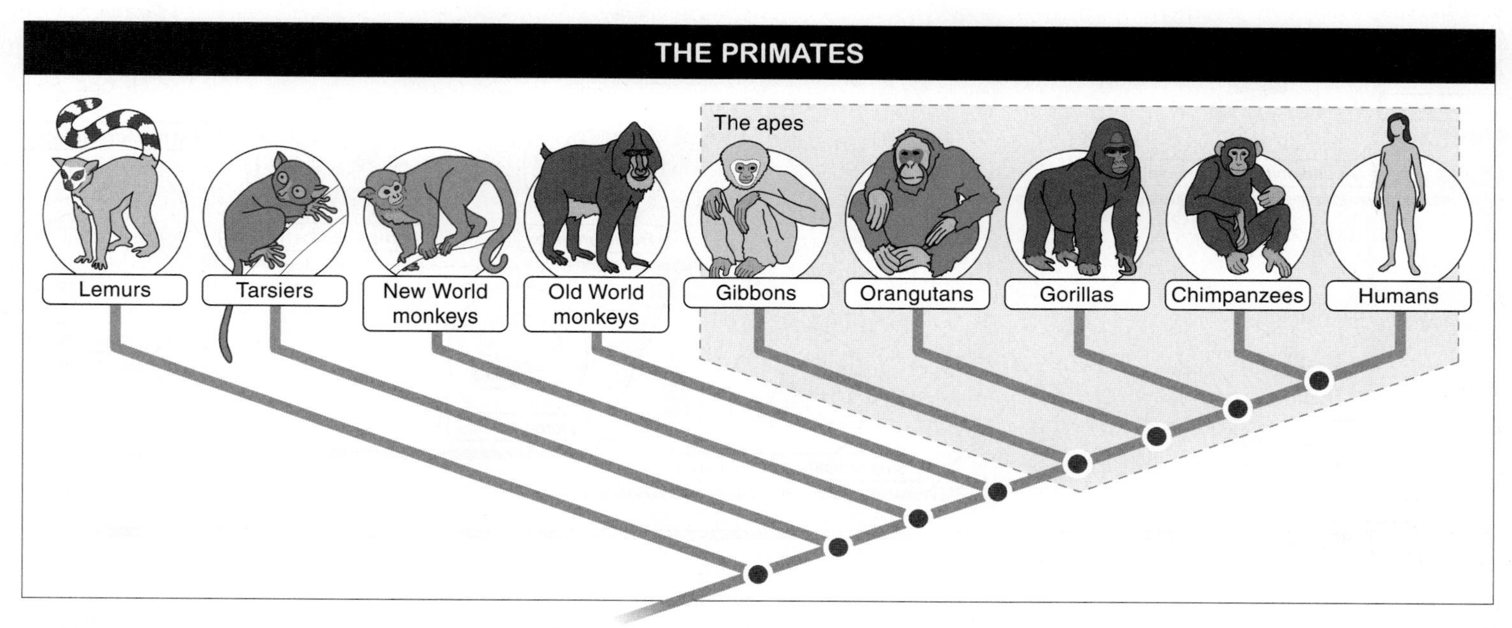

FIGURE 11-17 Phylogeny of primates.

The amount of genetic difference between humans and chimpanzees indicates that the chimpanzee and modern human lineages separated only 5 or 6 million years ago. Humans differ from chimpanzees in three major anatomical characteristics: humans are bipedal (we normally walk on two legs, whereas chimpanzees usually walk on four legs), humans are bigger than chimpanzees, and the human brain is about three times the size of the chimpanzee brain.

Q What are the advantages of walking on two feet rather than four?

When we trace the appearance of these human characteristics through the fossil record, we find that humans did not gradually become bipedal, big, and brainy. Instead, the three characteristics evolved one-by-one. Bipedality evolved first, then brain volume increased, and finally an increase in body size was accompanied by a further increase in the size of the brain.

The fossil record of human evolution looks like a bush with three episodes of branching that produced clusters of species. The first branching episode, which occurred between 3.5 and 4 million years ago, resulted from changes in the jaws and teeth associated with different diets. The second branching episode, about 2 million years ago, was associated with increased brain size and probably with the first use of tools. And a final branching event, half a million years ago, followed an increase in body size.

Shifting from walking on four legs to walking on two legs requires changes in several parts of the skeleton. The evolution of bipedality was not a simple process, but it was the first change in the human lineage and it seems to have set the stage for all the changes that followed. The primary advantage of bipedal locomotion for early humans was probably energetic efficiency. Bipedal locomotion at walking speed uses less energy than quadrupedal locomotion. Efficient walking locomotion thus could have been a valuable adaptation as the human lineage was moving out of the forest onto the savannas, while the chimpanzee lineage remained in the forest.

About 3.5 or 4 million years ago, several bipedal groups, the australopithecines, appeared (**FIGURE 11-18**). They were no larger than chimpanzees and they had the same brain volume, just a bit larger than a pint jar (350–400 cubic centimeters). The fossil known as Lucy was the first of many australopithecines to be discovered. An adult female, Lucy was only 3 feet (about

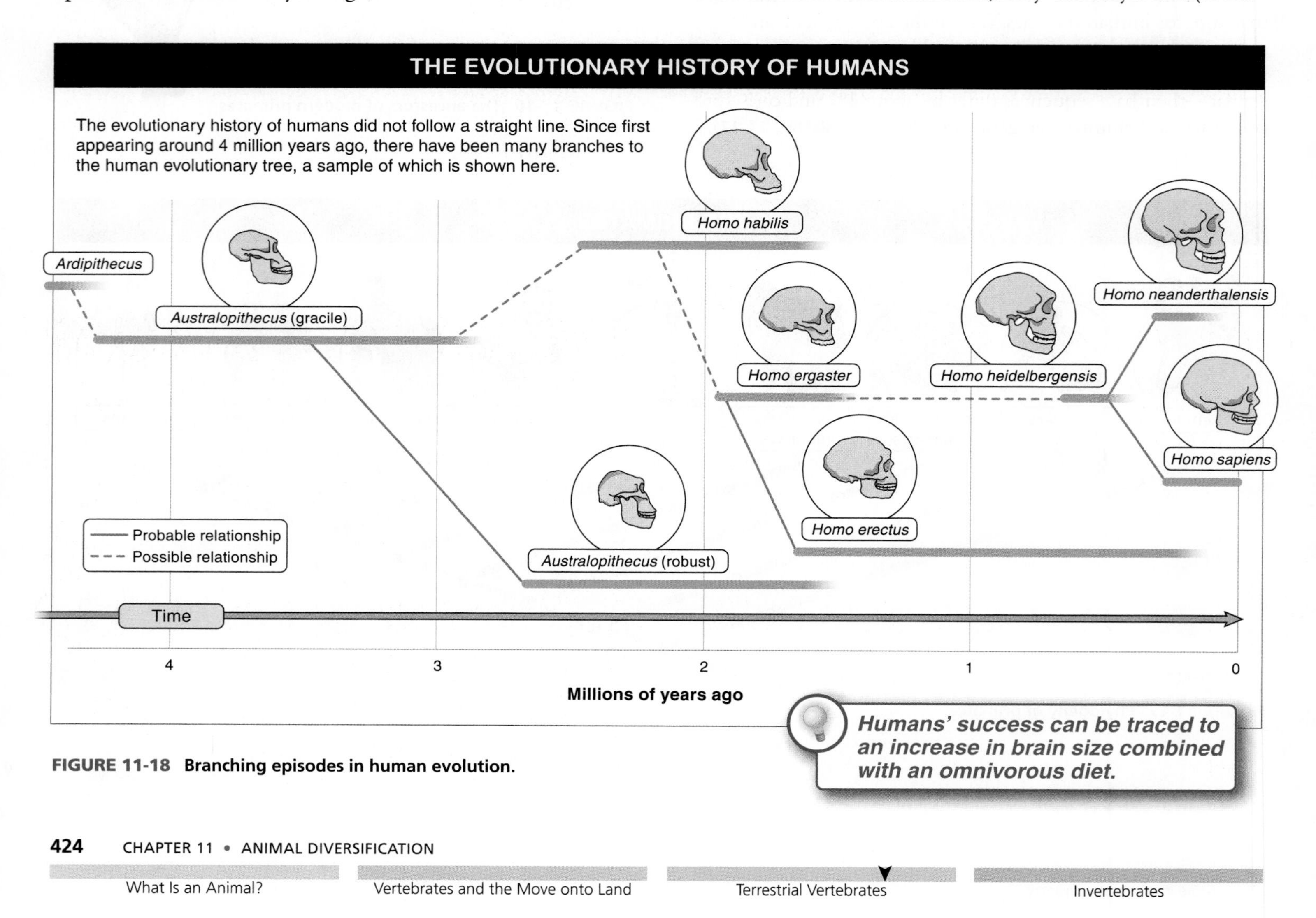

FIGURE 11-18 Branching episodes in human evolution.

1 meter) tall and probably weighed no more than 60 pounds (a bit less than 30 kilograms). The australopithecine group to which Lucy belonged lived in grassy habitats with scattered trees; their hands and feet retained the curved finger and toe bones that are characteristic of arboreal animals.

Probably, Lucy and the other closely related species of australopithecines foraged on the ground for food and climbed into trees to escape predators and perhaps at night to sleep. With jaws and teeth much like the jaws and teeth of chimpanzees, their diet was probably also like that of chimpanzees, a mixture of leaves, soft fruits, and nuts.

Following the appearance of bipedal primates, a second branching episode in human evolution originated among the australopithecines and produced the earliest species in our own genus, *Homo.* These earliest humans had brain volumes about twice those of chimpanzees, but little increase in body size. The species in this radiation had markedly smaller teeth than their australopithecine ancestors, a change that might indicate they had started to use tools instead of their teeth for the initial preparation of food. Stone tools are found in the same deposits as the fossils of *Homo habilis,* and this species may have been the first to use tools. Up to this time, all of human evolution had taken place in the southern and eastern parts of Africa, but about 2 million years ago, one branch, represented by *Homo erectus,* migrated out of Africa and spread through eastern Europe and Asia, while a second branch, represented by *Homo ergaster,* remained in Africa.

A third branching episode in human evolution (see Figure 11-18) gave rise to several species of advanced humans, including Neandertals (*Homo neanderthalensis*) and our own species (*Homo sapiens*). This radiation coincided with an increase in body size to approximately the height and weight of modern humans and an increase in brain volume to nearly twice that of earlier ancestors. In addition, the body form of the species that evolved during this branching episode looked like that of modern humans, with shorter arms and longer legs.

The evolutionary lineage to which *H. sapiens* belongs became bipedal, started using tools, and developed larger brains and larger bodies. These features, along with an omnivorous diet, turned out to be a winning strategy—other lineages closely related to that of *H. sapiens* became extinct while our lineage survived.

TAKE-HOME MESSAGE 11•10

Humans' forward-looking eyes, hands and feet with 10 fingers and 10 toes, and shoulder and elbow joints that allow the arms to rotate are characteristics retained from our arboreal ancestors. The early ancestors of humans left the trees and took up life on the ground, where they walked on two legs. Our success can be traced to an increase in brain size combined with an omnivorous diet.

11•11

How did we get here? The last 100,000 years of human evolution.

A bit more than 100,000 years ago, a new species of human evolved in Africa—the first modern *Homo sapiens.* One hundred thousand years is a ballpark figure based on a combination of molecular and fossil evidence. At that time, a small group—probably no more than 100—of these modern humans left Africa and ultimately spread across the earth.

Mitochondrial DNA shows that the initial human migration out of Africa followed three major pathways (**FIGURE 11-19**). One lineage turned west after leaving Africa and spread into Europe. A second group turned southeast and spread into southern Asia and through the Indo-Australia Archipelago to

HUMAN MIGRATION OUT OF AFRICA

About 100,000 years ago a small group of modern humans left Africa, and ultimately spread across the earth. Mitochondrial DNA shows that this migration followed three major pathways.

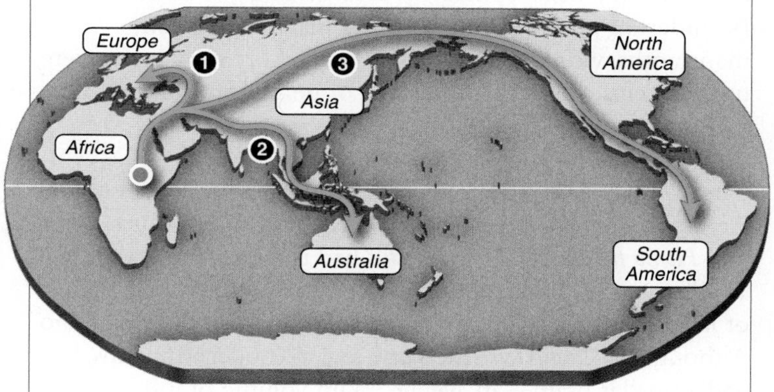

FIGURE 11-19 ***Homo sapiens*** **on the move: three migratory paths out of Africa.**

Australia. The third group moved northeast, populating northern Asia. This is the group that crossed the Bering Straits bridge about 15,000 years ago and spread southward through the Americas.

Were two species of humans ever alive at the same time? If so, what happened?

There was something about the world those *Homo sapiens* moved into as they left Africa that we would find bizarre—*they were not alone.* Populations of at least three other species of humans (i.e., species in the genus *Homo*) were already present in some of the areas that modern humans were moving into. That situation is totally different from the world we know today, in which we are the only species of human. What would it be like to live with species that were as similar to us as dogs are to wolves, coyotes, and jackals? How would you interact with an individual who was clearly human, but not human in exactly the same way that you are?

This is the situation that the first modern *Homo sapiens* encountered when they left Africa. Neandertals (*Homo neanderthalensis*) had already spread across Europe and the Middle East. Neandertals were about the same size as modern humans, but they were more robust and muscular. Fossils of Neandertals often include bones that had been broken and healed. These injuries provide two types of information about Neandertals.

1. Neandertals must have lived in organized groups that included a social support system, because these fossils reveal serious injuries. The injured individuals would have been incapacitated for days or even weeks until they healed, and family or clan members must have cared for them during this period. Ritual burials of Neandertals have also been found, and these provide additional support for the hypothesis that they lived in organized social groups.

2. The pattern of injuries found in Neandertal skeletons is distinctive—the only modern counterpart is found in professional rodeo bull and bronco riders, people who come in very close contact with large, angry animals. Neandertals probably hunted large mammals (bison, mammoths, wooly rhinoceroses) with short spears that were used for close-up jabbing instead of being thrown from a safe distance.

The *Homo sapiens* who migrated through the Indo-Australian Archipelago encountered two species of humans: *Homo erectus* and *Homo floresiensis. Homo erectus* had migrated to Asia at least a million years earlier, and they were still present on the island of Java when modern humans arrived about 50,000 years ago. Although *H. erectus* was the same size as modern humans, it had a smaller brain—an average brain volume of just over a quart (about 1,000 cubic centimeters) for *H. erectus* compared with a quart and a half (about 1,400 cubic centimeters) for *H. sapiens.* Nonetheless, *H. erectus* had technological skills. They may have been the first humans to use fire (the evidence on this is not definitive) and they almost certainly built boats that allowed them to move along coasts and from island to island.

In 2003, stone tools and fossils of a species of human only 3 feet (about 1 meter) tall and with a brain volume of just over a pint (about 350 cubic centimeters) were discovered on Flores Island, which lies east of Java. The paleontologists who discovered this species gave it the scientific name *Homo floresiensis* ("Flores man"), but the world press promptly called it the Hobbit because of its tiny size. At the time that the dwarf human *H. floresiensis* lived, Flores Island was also home to a dwarf elephant, which *H. floresiensis* probably ate, and a giant monitor lizard that probably ate *H. floresiensis.*

All three of the other humans disappeared after modern *Homo sapiens* spread into the areas where they occurred. Neandertals persisted until about 30,000 years ago, *Homo erectus* became extinct about 27,000 years ago, and *Homo floresiensis* disappeared about 12,000 years ago. Why did these three species vanish? Broadly speaking, there are two possible explanations. They might have interbred with *H. sapiens,* in which case modern humans would carry genes from the three recently extinct species. No evidence supports that possibility, however, and it is more likely that *H. sapiens* exterminated the other species, either by monopolizing access to food and living space or by killing them in battles for food and space.

Now we are the only extant species of human, a situation that is unique in the evolutionary history of humans. Ever since the first branching event about 4 million years ago, multiple species of closely related species have coexisted. We are the first species of human to be alone.

TAKE-HOME MESSAGE 11•11

Modern humans (*Homo sapiens*) evolved in Africa between 200,000 and 100,000 years ago, and all living humans are descended from that evolutionary radiation. About 100,000 years ago, a small group of modern humans moved out of Africa, and the descendants of this group ultimately populated Europe, Asia, and the Americas. Three other species of humans were living at this time. They all became extinct between 30,000 and 12,000 years ago, after modern humans had spread into the areas where they were living.

④ Invertebrates are animals without a backbone.

A brightly colored shield bug sits on a leaf.

11•12

Invertebrates are the largest and most diverse group of animals.

Up to this point, we have been discussing the evolution of deuterostomes, primarily one particular group within the phylum Chordata, the vertebrates. Although the vertebrates may be the most well understood animals, they are nowhere near the most diverse or abundant. The animal group that takes the prize for diversity and abundance is the invertebrates—defined simply as animal species without a backbone.

Although the invertebrate grouping has a convenient and easy-to-apply definition, the invertebrates are not a monophyletic group. There is one large group of invertebrates that, like the vertebrates, is made up of deuterostomes. Called echinoderms—and including sea stars (starfish), sea urchins, and sand dollars—this phylum of animals shows the same back-to-front gut development as the vertebrates, and these animals are the chordates' closest relatives. All of the other invertebrates are protostomes. Taken together, the invertebrates include more than 96% of the animal kingdom (**FIGURE 11-20**).

The invertebrate protostomes include eight separate evolutionary lineages with an enormous diversity of body forms, body sizes, habitats, and behaviors. Worms are

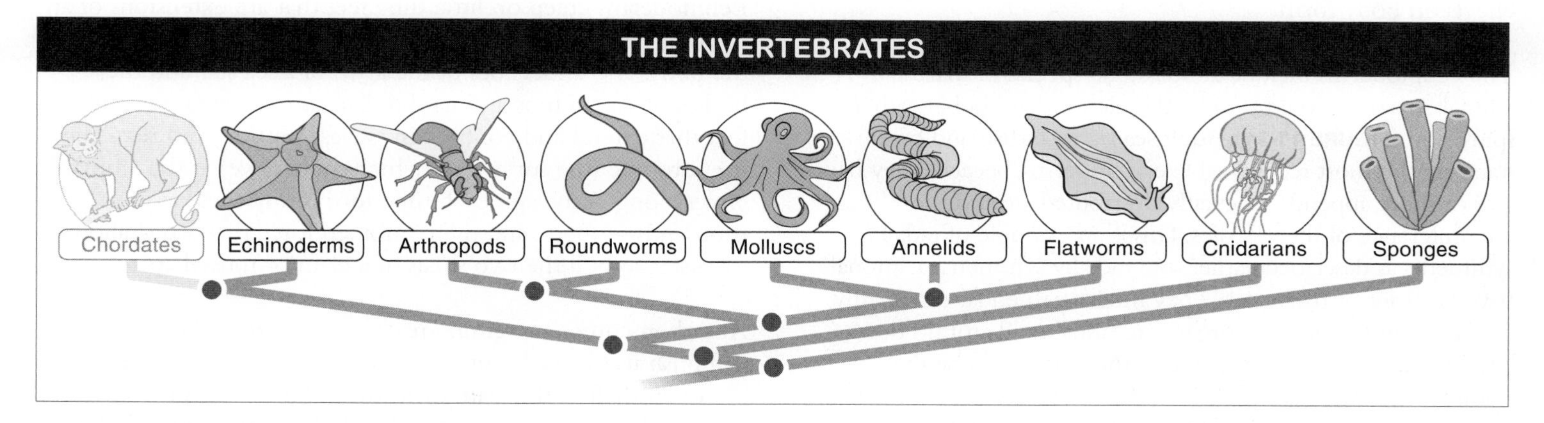

FIGURE 11-20 Phylogeny of invertebrates.

invertebrates, and they are long and flexible with a soft body covering, but molluscs (clams, oysters, and their relatives) are also invertebrates and they are round or oval and some are encased in rigid shells. Sponges are the most sedentary and plant-like of the invertebrates, and octopuses and squids are as active and exploratory as any vertebrate.

Some invertebrates (such as worms) have soft bodies, others (snails, for example) have a shell, and still others (including insects and crustaceans such as lobsters) have a rigid external covering called an **exoskeleton.**

No invertebrate can be considered to represent the group as a whole, but the phylum Arthropoda (the **arthropods**) has the largest number of species—about 75% of the animals on earth are arthropods. This phylum includes the most familiar invertebrates: insects, spiders, scorpions, lobsters, crabs, shrimp, and their relatives. We begin by examining the vertebrates' closest relatives, the deuterostome echinoderms, followed by a tour of the protostome invertebrates.

TAKE-HOME MESSAGE 11•12

Invertebrates, defined as animals without a backbone, are the largest and most diverse group of animals, comprising 96% of all living animal species. The invertebrates are not a monophyletic group, however, and include protostomes and deuterostomes.

11•13

Echinoderms are vertebrates' invertebrate relatives: sea stars, sea urchins, and sand dollars.

Appearances can be deceiving—take the sea star, for example. Considering the great diversity among the invertebrates—including insects, crabs, worms, and other bilaterally symmetrical organisms that live on land, many with an obvious head region, eyes, and mouth—it is a bit surprising that the sea star and other members of the phylum Echinodermata are, as deuterostomes, more closely related to humans than they are to any other invertebrate groups. They stand as an example of how classifying organisms based on evolutionary relatedness rather than physical similarity can reveal a great deal about how the force of natural selection, adapting organisms to their environment, can have dramatic effects on body form.

The echinoderms include about 6,000 species of marine animals, most enclosed by a hard skeleton that lies beneath a spiny skin (**FIGURE 11-21**). Adult sea stars (starfish and brittle stars) are the most recognizable echinoderms, because they have five or more appendages evenly distributed around the circumference of their bodies. This is an example of radial symmetry, as described earlier, and radially symmetrical animals have no front or back ends. They move with equal ease in any direction, and their sense organs are distributed around their circumference, because any side of the animal can be the leading edge. Adult sea urchins and sand dollars don't have projecting arms, but they are also radially symmetrical.

Although radial symmetry is characteristic of adult echinoderms, their larvae are bilaterally symmetrical, having a left and right side of the body. Bilaterally symmetrical animals generally move in one direction, with their sense organs grouped at the front end of the body. Echinoderms evolved from bilaterally symmetrical ancestors, and radial symmetry is an evolutionary specialization associated with their mode of locomotion and feeding specializations. The larvae of echinoderms have anatomical characteristics in common with the larvae of primitive chordates, and as we've noted, echinoderms and chordates are each other's closest relatives.

Echinoderms creep on little tube feet that are extensions of an internal system of water-filled tubes that radiate throughout the body. The undersides of the arms of a sea star and the bodies of sea urchins and sand dollars are carpeted with tube feet that extend and contract in waves, grasping and releasing the substrate (the surface on which they move) as the animal glides along. Although each tube foot is tiny, there are thousands of them, and the combined force they exert allows a sea star to pull the two shells of a clam or mussel apart.

And then comes something really unexpected. The sea star can push its stomach out through its mouth and insert it into the open shell of the clam or mussel. The stomach then secretes digestive enzymes that break down the tissue of the

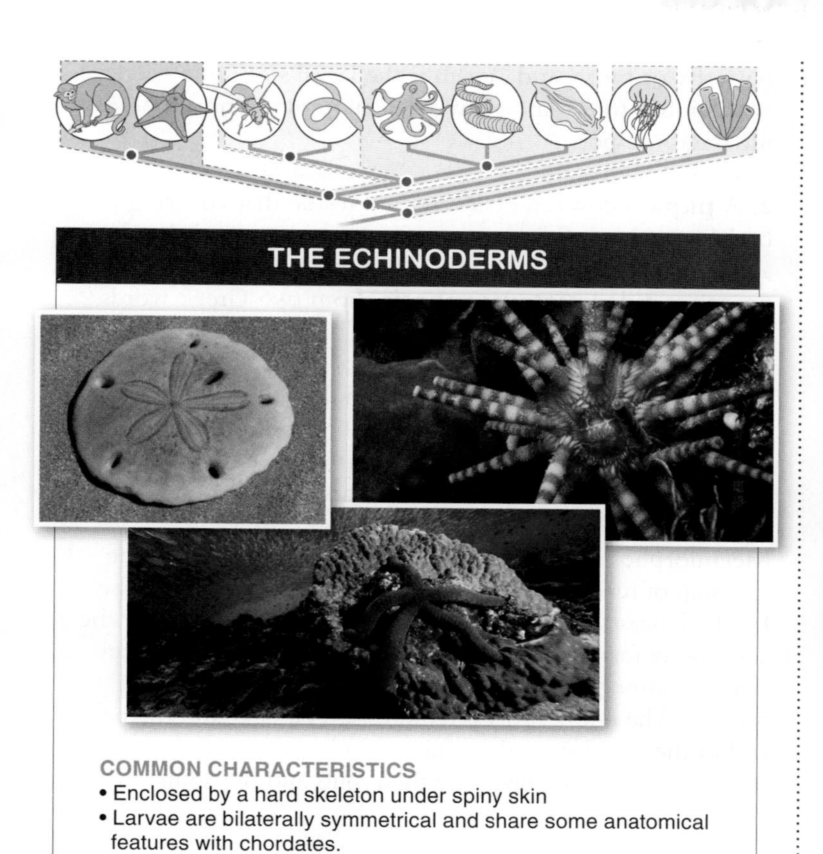

FIGURE 11-21 What is an echinoderm? Examples of echinoderms (*from top left, clockwise*) are a sand dollar, a club sea urchin, and a blue sea star.

prey, and the soup that results is absorbed by the sea star. Sea urchins feed on algae, scraping them loose from rocks with sharp tooth-like surfaces made of calcium, and sand dollars capture floating particles of algae and other organic matter by trapping them in streams of mucus that are moved toward their mouth by beating cilia.

TAKE-HOME MESSAGE 11•13

Echinoderms are, like chordates, deuterostomes, but they lack a backbone. Their aquatic larvae are bilaterally symmetrical and share some anatomical features with chordates, but adult echinoderms are radially symmetrical.

11•14

An external skeleton and metamorphosis produced the greatest adaptive radiation ever.

When it comes to biodiversity, the insects take the cake. They outnumber all other forms of life in species diversity. The flies, alone, comprise 90,000 species; the bees, wasps, and ants number 150,000 species; and the butterflies and moths have 120,000 species. But even those numbers are dwarfed by the beetles, with 320,000 species! To put it in perspective, there are more species of beetles than there are species of plants, and nearly six times as many species of beetles as there are species of vertebrates. In this section, we'll investigate why insects have become so diverse.

The protostome phylum of arthropods (Arthropoda) contains invertebrates that have segmented bodies (i.e., bodies with a distinct head, thorax, and abdomen), an exoskeleton made of a stiff carbohydrate called chitin, and legs with joints (think of all the joints in a lobster's claw). In all, the phylum contains almost 1 million recognized species, or approximately 75% of all animal species, and as many as 60% of *all* the species on earth (**FIGURE 11-22**). If diversification is a measure of evolutionary success, the arthropods are unmatched. One characteristic of insects has been central to their success: the way they cope with the change in body size as they grow.

Baby animals are smaller than adults—that's true of any kind of animal, and it can be a problem because a tiny baby cannot do the things that an adult does. In particular, babies cannot feed

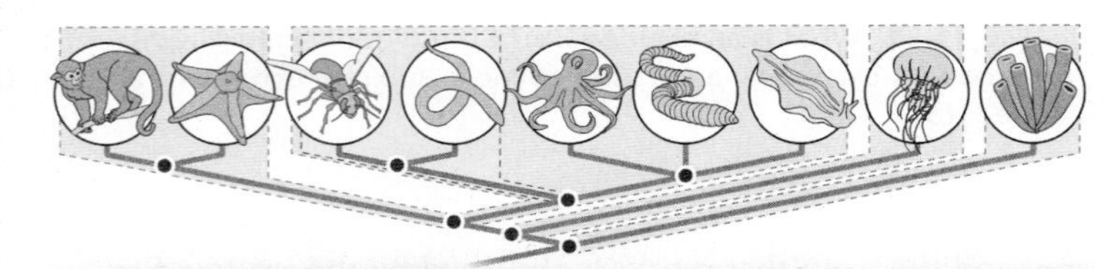

THE ARTHROPODS

COMMON CHARACTERISTICS
- Segmented body with a distinct head, thorax, and abdomen
- Exoskeleton made of chitin
- Jointed appendages

MEMBERS INCLUDE
- Insects (over 750,000 species)
- Arachnids (about 60,000 species)
- Crustaceans (about 52,000 species)
- Millipedes and centipedes (about 10,000 species)

Arthropods make up about 75% of the animal species on earth.

FIGURE 11-22 Exploring the arthropods. Shown here (*from top left, clockwise*) are a blue millipede, an elephant hawk moth, a scorpion sitting in the dust, and a hermit crab peeking out from its home in hard coral.

Mammals get bigger and bigger the more they eat. Why don't insects?

themselves. Mammals solve that problem by supplying milk to infants and providing parental care until the young animal has grown large enough to feed and take care of itself. Many insects have a different way of solving the problem—their life history is divided into three completely different stages (**FIGURE 11-23**).

1. An egg hatches into a **larva,** which looks completely different from an adult. A caterpillar is the larval stage of a butterfly or moth, and a grub is the larval stage of a beetle. A larva's job is to eat and grow large enough to enter the next life stage, the pupa.

2. A **pupa** is covered by a case, and inside that covering the body structures of the larva are broken down to molecules and then reassembled into the adult form. This rebuilding process is called **metamorphosis,** from two Greek words meaning "change of form."

3. When the **adult** form hatches from the pupa, it is at its full size. Insects do not grow after they emerge from their pupae—for this reason, some adults do not even eat. Rather than eating and growing, the job of an adult insect is reproduction.

Metamorphosis is the key to this life history pattern. In the first step of metamorphosis, the larva encloses itself in a case. Inside the case, the insect activates the genes that code for the adult body form. Proteins that made up the larva are broken down to amino acids, which are used to synthesize adult proteins. The entire process requires several days, and by the end of the pupal stage the adult insect is curled tightly inside the pupal case. The pupal case then splits and the adult

COMPLETE METAMORPHOSIS

Complete metamorphosis is the division of an organism's life history into three completely different stages.

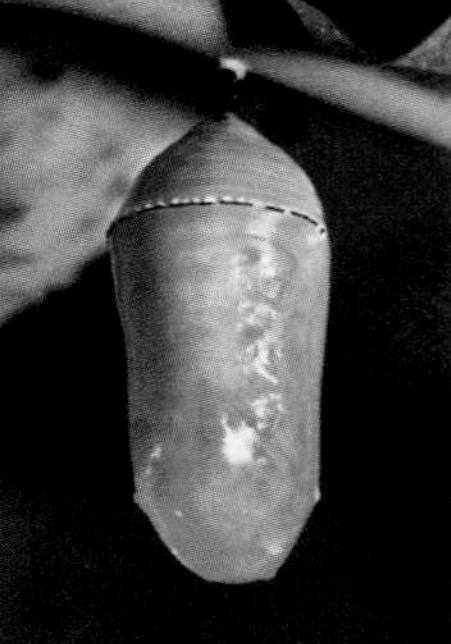

LARVA
An egg hatches into a larva, which eats to grow large enough to enter into the pupal stage.

PUPA
The larva encloses itself in a case, where its body structures are broken down into molecules that are reassembled into the adult form.

ADULT
The adult emerges from the pupa, and its primary job is to reproduce.

The separation of life stages has contributed to the enormous ecological diversity of insects.

FIGURE 11-23 From larva to adult monarch butterfly. Some insects have a life history divided into unique stages, each with a distinct purpose.

squirms out. The wings of butterflies and moths are tightly folded inside the pupa, and they require time to expand and stiffen before the adult is ready to fly.

"Complete metamorphosis," involving the three stages, allows the larva and adult to act as if they were different species, optimized for very specialized tasks. Caterpillars eat leaves, for example, and spend their entire larval period on a single plant. In contrast, butterflies feed on nectar that they collect by flying from flower to flower. As a larva, the organism must feed and grow. As an adult, it reproduces. The genes that control the larval body form are different from the genes that determine the adult form, so natural selection can act on the larval and adult stages independently.

Not all insects undergo the dramatic change (complete metamorphosis) from larva to adult that we see in beetles and butterflies. Juveniles of some primitive insects, such as grasshoppers and cockroaches, look and behave very much like the adults. These juveniles, which are called nymphs, do not have wings or reproductive organs, but they live in the same habitats as adults and eat the same foods.

About 83% of insect species have complete metamorphosis, and this separation of the life stages has helped insects diversify into nearly 20 times as many named species as vertebrates, and at least that many more species of insects have yet to be named.

TAKE-HOME MESSAGE 11•14

The arthropods are protostome invertebrates, including the insects. The life cycle of most insects includes a larval stage that is devoted to feeding and growth and an adult stage in which the insect reproduces. This separation of life stages has contributed to the enormous ecological diversity of insects and has produced remarkable specializations among the nearly one million species of insects that have been named so far.

11•15

Other arthropods include arachnids, crustaceans, millipedes, and centipedes.

The arthropod phylum is so large that even if we excluded all the insects, more species still would remain than are present in any other phylum. In this section we survey the other major groups of arthropods: the arachnids, crustaceans, and millipedes and centipedes (**FIGURE 11-24**).

Arachnids Arachnids are land-dwelling arthropods that include spiders, scorpions, mites, and ticks. Arachnids usually have four pairs of walking legs and a specialized feeding apparatus, such as the fangs of spiders and the grinding mouthparts of scorpions. Arachnids (like insects) are distinguished from other arthropods by the positioning of their legs: they have legs only on the middle section of their body (the thorax), whereas other arthropods have legs on the posterior section of their body (the abdomen) as well as on the thorax.

Of the 60,000 species of arachnids, spiders and scorpions are the ones you are most likely to see—and to fear. But their reputation as fear-worthy animals may not be fully deserved. Most species are small, harmless, and often beneficial, because they eat many insects that are pests in agriculture or in homes.

Arachnids are predators. Many spiders construct webs by spinning fibers of silk (which consists of interconnected protein molecules). Orb-weaving spiders construct intricate webs that trap (and sometimes even lure) flying insects, and they place their web across the routes followed by flying insects. A spider rests inconspicuously until an insect gets trapped in the web, then runs out to wrap the prey in silk threads.

When a spider captures prey, it uses its fangs to inject venom that contains two types of enzymes. The first set of enzymes disrupts the prey's nervous system and paralyzes it, and then other enzymes dissolve its internal organs. When the prey's innards have liquefied, the spider sucks out its contents. (Maybe their reputation *is* deserved.)

The venom of most spiders, however, usually is harmless to animals as large as humans, but two North American spiders *are* dangerous. The black widow spider is the better known, and is easily identified by a red hourglass marking on the bottom surface of the female. The brown recluse spider lacks distinctive markings, so it is harder to identify, but it is more dangerous. Both species are likely to move indoors, constructing webs in little-used spaces, such as in closets and cabinets and behind furniture that stands against a wall. Brown recluse bites are notorious for creating open wounds that can last for months or years and sometimes require repeated skin grafts to close.

GROUPS OF ARTHROPODS

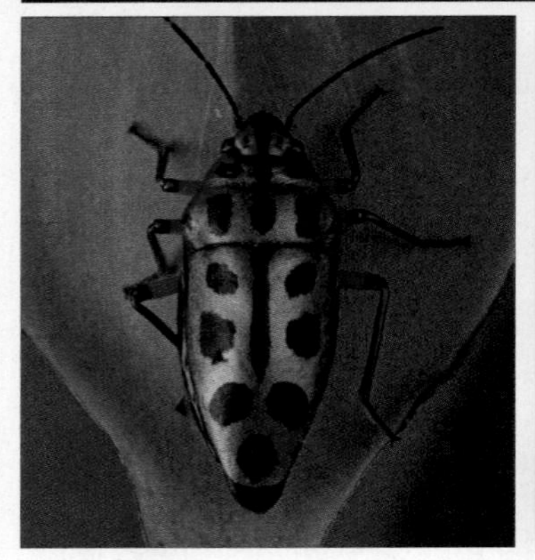

INSECTS
- Three pairs of walking legs
- Legs are located on the thorax
- Life cycle consists of separate life stages

ARACHNIDS
- Usually four pairs of walking legs
- Legs are located on the thorax
- Specialized mouthparts
- Predators

CRUSTACEANS
- Many pairs of legs
- Usually five pairs of appendages that extend from head
- Mostly aquatic

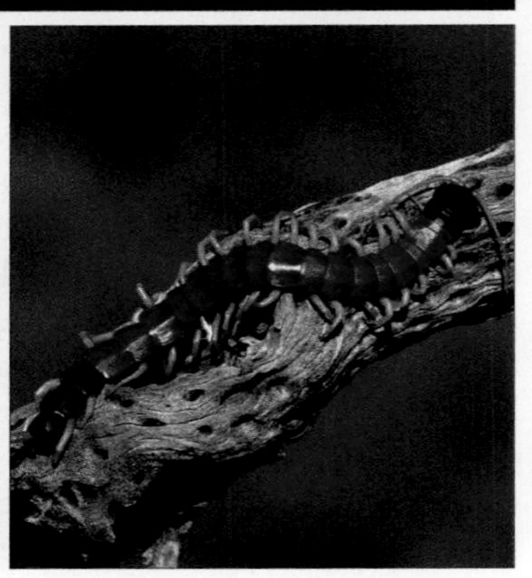

MILLIPEDES AND CENTIPEDES
- Many pairs of legs (one pair per segment)
- Long, segmented body

FIGURE 11-24 Arthropod diversity: insects, arachnids, crustaceans, and millipedes and centipedes. Shown here (*from left to right*) are a shield bug, a banded garden spider, a red swamp crayfish, and a giant desert centipede.

Scorpions look dangerous. With their large claws (which are modified legs) and curved tails, with the stinger prominently displayed, they clearly say, "Don't mess with me!" But being stung by most scorpions is no worse than being stung by a bee. Scorpions are hunters, searching at night for insects. Scorpions seize prey with their claws; the sting is not used unless the prey struggles. Some of the scorpion's mouth structures are covered with spines that grind the prey, while glands secrete enzymes that liquefy it. The scorpion sucks up the soup and discards the hard parts of the prey.

Crustaceans If arachnids are the most feared arthropods, the aquatic crustaceans—lobsters, crayfish, crabs, and shrimp—may be regarded as the most desirable arthropods, at least at meal time. In the United States, the top three fishery crops, by their annual market value, are shrimp, crabs, and lobsters. There are about 52,000 species of crustaceans, and although crustaceans can look very different—from sessile barnacles to mobile, tiny krill—they all have five pairs of appendages extending from their heads. Three of these appendages are used for feeding and two are antennae that sense the environment.

Crustaceans have many pairs of legs, and their legs are modified for many purposes. Shrimp and barnacles (which are essentially shrimp that spend their adult lives fastened to a rock and enclosed in a shell) have legs with comb-like projections that they use to capture tiny plankton, and free-living crustaceans (shrimp, crabs, lobsters) have modified limbs in the abdominal region that are used to hold eggs or newly hatched young.

Although most crustaceans are aquatic, the inconspicuous wood lice (also known as pill bugs, sow bugs, or roly-polies) are terrestrial forms that you can find when you turn over a rock in your garden. Wood lice are herbivores and play an important role in recycling dead plant material.

Millipedes and centipedes The names *millipede* and *centipede* translate to "a thousand feet" and "a hundred feet" and refer to the enormous numbers of legs on these long, skinny animals. They don't really have a thousand—or even a hundred—feet, but millipedes do have more legs than centipedes, and that is the easiest way to tell them apart. Millipedes and centipedes have long, segmented bodies that seem almost worm-like, but their jointed legs and hard exoskeletons are characteristics of arthropods. Both live among fallen leaves, especially in forests where the leaf litter layer is usually cool and moist. Millipedes feed on decaying plant material, while centipedes are predators that use fangs equipped with venom to kill insects and even small mammals. Unlike spiders and scorpions, centipedes use their jaws to tear prey into pieces that are small enough to swallow.

TAKE-HOME MESSAGE 11•15

Spiders and scorpions are predatory arthropods that eat insects and occasionally small vertebrates. Lobsters, crabs, shrimps, and barnacles are predatory marine crustaceans. Centipedes are predators with fangs that inject venom, while millipedes are herbivores that feed on dead plant material.

11•16

Most molluscs live in shells.

Molluscs (phylum Mollusca) are a large group of animals—nearly 100,000 named species and many more that have not yet been named—that live in the ocean, in fresh water, and on land. There are many familiar molluscs, including clams, snails, scallops, oysters, mussels, squids, and octopuses. They are so diverse that it is difficult to describe any single defining characteristic. The position of molluscs within the animal phylogeny reflects the fact that they have true tissue, are bilaterally symmetrical, and are protostomes that do not molt. Three features are common to many molluscs, however: a shell that protects the soft body, a mantle (the tissue that secretes calcium carbonate to form the shell), and a sandpaper-like tongue structure, called the radula, that is used during feeding (**FIGURE 11-25**).

There are three major groups of molluscs, sharing most of the common features but with varied body plans: **gastropods, bivalve molluscs,** and **cephalopods.**

Gastropods Snails and slugs are gastropod molluscs. They get their name because *gastropod* means "belly foot" and they have an expanded foot on the bottom of their body that allows them to climb a vertical surface as easily as they glide across a horizontal one. Snails have a one-piece, curled shell, and slugs are snails that have reduced the shell to a tiny remnant that is not even visible because it is covered by the mantle. Found in both aquatic and terrestrial environments, snails and slugs account for three-quarters of all molluscs.

The snail's shell is its primary protection against predators, but because both terrestrial slugs and sea slugs have very little shell material, other defense mechanisms are necessary. Slime is what a terrestrial slug relies on—when a slug is attacked, it secretes slime that sticks to the predator. Worse still, anything that touches the slime on the predator sticks to it, so a bird that attacks a slug quickly finds that its beak and face are covered by pieces of dead leaves, clods of soil, and twigs. Sea slugs have their own methods of defense. Some species synthesize toxic chemicals, and others, which feed on sponges, recycle sponge toxins into their own slime. One of the most remarkable uses of another animal's defense equipment is found among sea slugs that eat sea anemones. Anemones are protected by stinging cells, and the sea slugs that eat anemones transfer some of those cells from the anemones to their own skin, enabling them to sting other creatures.

Bivalve molluscs Clams, scallops, oysters, and mussels have a pair of shells that clamp together, and these animals are the bivalve molluscs. There are about 8,000 species of bivalves and most of them live in the ocean, although there are some freshwater species. Clams spend their lives buried in mud or sand, scallops live on the sea floor, and oysters and mussels fasten themselves to underwater objects, such as rocks, the pilings that support piers, and the hulls of boats. All bivalves are filter-feeders that draw a current of water through a tube called the "incurrent siphon," across their gills, where tiny food particles are captured, and out through the "excurrent siphon."

FIGURE 11-25 Molluscs: an overview. Some examples of molluscs (*from top left, clockwise*) are a giant clam, a banana slug, a Caribbean reef octopus, and a nudibranch.

When a grain of sand is trapped in the shell of a bivalve, the mantle may secrete layer after layer of a shell-like protein material that covers the sand grain and forms the iridescent gem called a pearl. Oysters are the best known source of pearls, but clams and other bivalve molluscs also form pearls.

Cephalopods The third major group of molluscs is the cephalopods (**FIGURE 11-26**). It includes 6 species of nautilus and about 600 species of squids and octopuses. The nautilus has an external shell that resembles a snail shell, except that it is not twisted. Squids have very small shells that are covered by the mantle, and octopuses have lost the shell entirely. Their tentacles appear to grow directly from the head, which explains the name "cephalopod" (translated as "head-footed").

The most obvious feature of cephalopods may be their tentacles, which they use to walk and to capture prey. Squids have eight short tentacles called arms and two long sucker-bearing tentacles that are used to capture prey. After a slow approach to its unsuspecting prey, a squid propels its tentacles forward with astonishing speed—accelerations of more than 800 feet per second per second (about 250 meters per second per second) have been measured, the equivalent of your car accelerating from 0 to 6,000 mph in 10 seconds! The suckers on the tentacles adhere to the prey and draw it back toward the squid, where the arms take over, turning and manipulating the prey as it is consumed by the squid's sharp beak.

Many species of squid are more than 3 feet (about 1 meter) long, and the giant squids reach lengths of 40 feet (about 13 meters). When a squid is in a hurry, it swims tail-end first, using jet propulsion. Water is expelled through the siphon at the tentacle end of the body, shooting the animal backward. Octopuses, which can have tentacles that spread 12 feet (about 4 meters), are bottom dwellers, living in coral reefs and rocky coasts. Because they have completely lost the shell, octopuses can squeeze through astonishingly small openings, and their ability to escape from even carefully covered tanks is a perpetual challenge for the keepers of zoos and aquaria. Keepers are accustomed to arriving in the morning and finding an octopus wandering around the room.

In the next section we further explore the cephalopods—do some of their skills of manual dexterity make them the smartest of the invertebrates?

TAKE-HOME MESSAGE 11·16

Molluscs are protostome invertebrates that do not molt. Including the snails and slugs, clams and oysters, and squids and octopuses, the molluscs are the second most diverse phylum of animals. Most feature a shell for protection, a mantle of tissue that wraps around their body, and a specialized tongue called a radula.

GROUPS OF CEPHALOPODS

SQUIDS
- 8 short tentacles and 2 long sucker-bearing tentacles
- Free-swimming
- Contains about 300 species

OCTOPUSES
- 8 short tentacles called arms
- Bottom dwellers, living in coral reefs and rocky coasts
- Contains about 300 species

NAUTILUSES
- Chambered shell used for protection
- Free-swimming
- Contains 6 species

FIGURE 11-26 Cephalopod diversity. Shown here (*from left*) an oval squid, big red octopus, and chambered nautilus.

11·17

Are some animals smarter than others?

Q Are octopuses smart?

Squids and octopuses are wide-ranging predators that must locate, capture, and subdue other animals. This is a far more active life style than that of bivalves or snails, and it requires quick movements and rapid responses to stimuli. All of these features are conspicuous elements of the behavior of an octopus as it moves through a coral reef, inserting a tentacle into every opening. In fact, as an expert multi-tasker, an octopus can capture a prey item with one of its tentacles, use a second tentacle to hold something else, and still keep the remaining six tentacles in motion. In addition, octopuses are very good at manipulating things with their tentacles. It doesn't take a captive octopus long to learn how to screw the top off a glass jar to reach a fish that's swimming inside—after all, that's just a minor variation on twisting a clam to open it (**FIGURE 11-27**).

FIGURE 11-27 An octopus attempts to open a jar.

Because of their manipulative skills and easily observed exploratory behavior, octopuses are sometimes featured on TV shows and in magazine articles, where they are often described as the smartest invertebrates, and their intelligence compared to that of mammals. But this is the sort of claim that must be viewed skeptically.

What is intelligence, after all? The concept of intelligence loses its meaning when animals that live in completely different worlds and respond to entirely different stimuli are compared. We can say that octopuses are very good at doing the things that octopuses need to do—searching for prey, capturing it, and manipulating it, all within their marine environment, for example. But is following a complex sequence of actions the same thing as intelligence? Spiders, for example, construct elaborate webs that are strategically placed in the flight paths of insects; they detect the impact of an insect with the web and rush out to wrap the insect in silk before it can escape. Does that require more or less intelligence than opening a glass jar to eat a fish? And what about the red squirrels of northern Europe, which hide more than 3,000 acorns, cones, and nuts each fall, ultimately recovering more than 80% of their hidden nuggets of food over the course of the winter—can a human manage that? An octopus can't.

These questions are analogous to the question asked earlier in the chapter about what makes a species evolutionarily successful. Recall that, from an evolutionary perspective, any non-extinct species must be considered a success. Similarly, the very concept of "intelligence" may actually be relevant only to humans, to the extent that intelligence is defined as the ability to do some problem-solving task valued by humans. Applying the question of intelligence to non-human species may just be an assessment of how well they can accomplish tasks. For this reason, questions of comparative animal intelligence are not generally useful. Rather, animals' abilities should be considered as evolutionary responses to particular selective pressures imposed by their environment, and the fact that a species currently survives should be taken as an indication of some mastery of its particular niche.

TAKE-HOME MESSAGE 11·17

The predatory behavior of octopuses involves exploration and manipulation behaviors often considered as intelligent by humans, but the concept of intelligence cannot be applied objectively to other species, which have evolved in response to the selective forces relevant only to their own particular niche.

⑤ Worms, jellies, and sponges are examples of invertebrate diversity.

A Lion's Mane jellyfish swims in the Atlantic waters off the coast of Nova Scotia, Canada.

11•18

Segmented worms, flatworms, and roundworms come in all shapes and sizes.

It's a good thing that biologists don't run bait shops—if they did, a simple trip to buy worms for fishing would be way too complicated. The name "worm" is applied to a long, skinny animal without a backbone, but you can find animals fitting that description in eight different phyla. You will never encounter most of them, and that's a good thing, because many are parasites that cause really unpleasant diseases. We consider here just three phyla—**segmented worms** (phylum Annelida), **flatworms** (phylum Platyhelminthes) and **roundworms** (phylum Nematoda)—that illustrate the diversity of the animals we call worms.

Segmented worms Segmented worms, also called annelids, number about 13,000 species and are easy to recognize by the grooves running around their bodies that mark the divisions between segments—if you've seen an earthworm, you are familiar with those grooves. The segmented worms are protostomes with defined tissues and do not molt. They are organized into three different groups: marine polychaetes (pronounced POL-ly-keets), terrestrial oligochaetes, and leeches (**FIGURE 11-28**).

Polychaetes are marine worms, living on the sea floor. *Polychaete* means "many bristles," and the combination of segments and bristles makes polychaetes easy to recognize. Some species burrow through the mud and extract organic material from it. Tube worms use sand grains or limestone to make a tube in which they live, with just their waving tentacles exposed. Small particles of food are trapped in mucus on the tentacles and transported to the worm's mouth.

Earthworms, which belong to the group called *oligochaetes* ("few bristles"), are the annelid worms you are most likely to have seen. The night crawler is a typical earthworm. It gets its name from its habit of emerging from its burrow on rainy nights to crawl across the surface of the ground. More than 4,000 species of earthworms have been named, ranging in size from less than half an inch (about a centimeter) long to the giant Gippsland earthworm, 7–10 feet (2–3 meters) long!

Earthworms are bulk-feeders—that is, as an earthworm burrows, it consumes particles of soil and organic material. The organic material is digested as it passes through the worm's gut, and the fecal material plus the inorganic part of the soil are excreted as feces, called castings. Earthworm castings are valued by gardeners as soil supplements, and an earthworm can produce its own weight in castings every day. Because they mix the soil components, creating a more even mixture of nutrients,

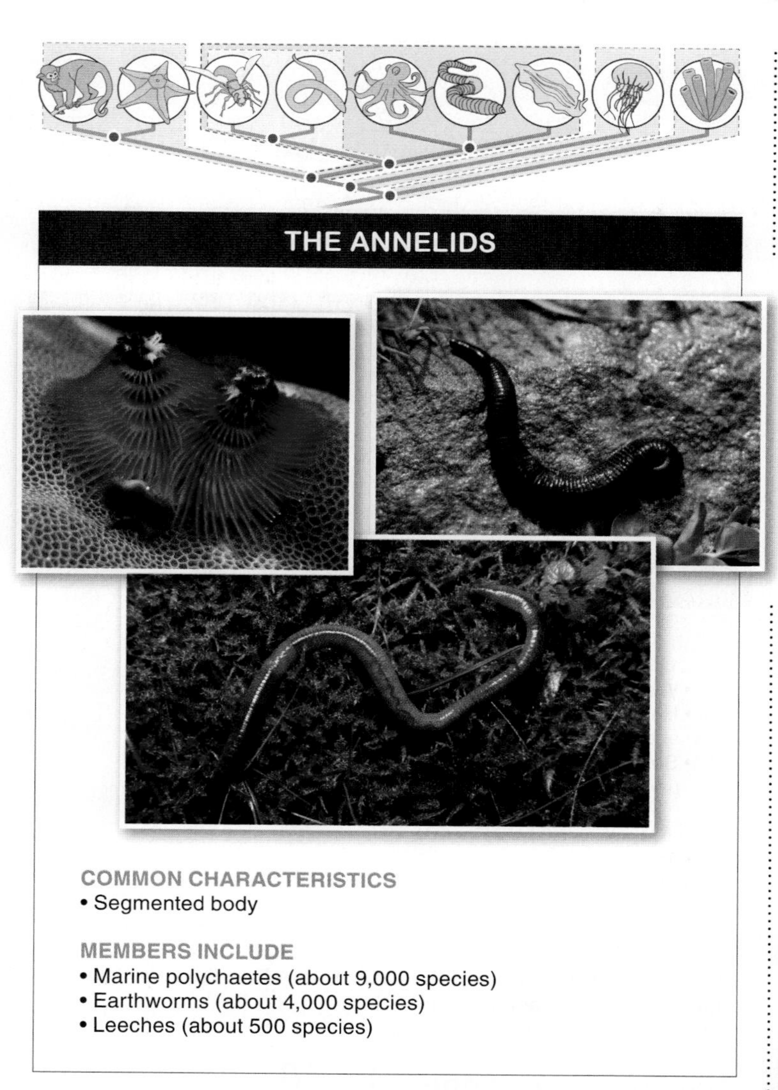

FIGURE 11-28 Annelid diversity. Shown here (*from top left, clockwise*) are a polychaete worm, a leech, and an earthworm.

and expedite the breakdown of organic materials in soil, making the nutrients available for plants, the economic value of earthworms to agriculture is enormous. A government study published in 2008 estimated the economic value of earthworms in Ireland alone to be more than $1 billion per year.

Have you ever waded in a pond or swamp and found a long, dark brown worm clinging to your leg when you emerged? If so, you have met the third major group of annelids, the *leeches.* Probably you pulled the leech off (not an easy thing to do, because leeches are both slippery and stretchy) and threw it as far as you could. If you had looked at it carefully, though, you would have noticed that, like earthworms, leeches have segmented bodies.

If you have had this close encounter with a leech, you'll also have noticed that the wound that it made on your leg continued to ooze blood for several hours. That's because the saliva of blood-sucking leeches contains an anticoagulant substance that prevents blood from clotting. If the leech is undisturbed while it is feeding, blood keeps flowing and the leech inflates like a small balloon.

Not all leeches are blood-suckers—in fact, more than half the species of leeches are predators. The horse leech, for example, which reaches a length of 8 inches (about 20 centimeters), feeds on smaller annelid worms, snails, and aquatic insect larvae.

Roundworms The roundworms are protostomes with defined tissues and, unlike the segmented worms and flatworms, they grow by molting. They are probably the most abundant animals on earth—a spoonful of garden soil contains several thousand individuals, and some species produce more than 200,000 eggs every day. More than 90,000 species of roundworms have been named, and there may be five times as many species that have not yet been identified (**FIGURE 11-29**)

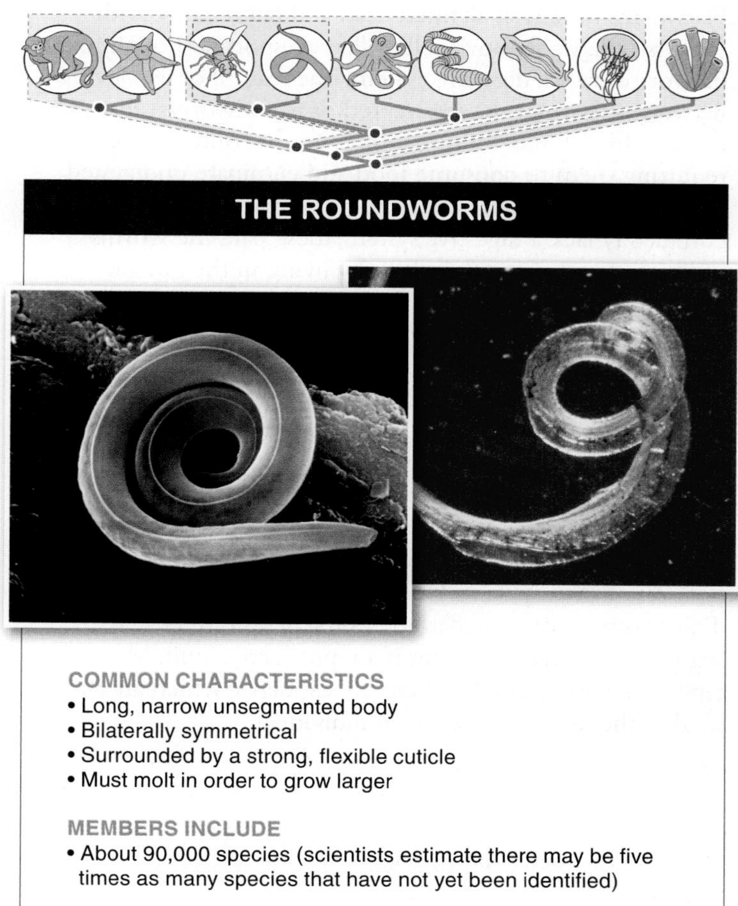

FIGURE 11-29 Roundworm diversity. A nematode worm (*left*) and a whipworm.

Nematodes are soil-dwelling roundworms that live in the roots of plants, causing damage or even death of the plant. About 15,000 species of roundworms are parasites of vertebrate animals, and roundworms are responsible for a large number of human diseases. Most roundworms are transmitted by fecal contamination of soil or food.

Tiny parasitic roundworms called *filariae* are responsible for several tropical diseases, including some that have an enormous economic and social impact in certain parts of the world. Elephantiasis, for example, is a disease in which filariae transmitted by the bite of a mosquito block the lymph ducts so that fluid accumulates in the limbs or scrotum, causing grotesque swelling. These filariae occur in India, Africa, South Asia, Pacifica, and tropical America.

Flatworms Like the segmented worms, but unlike the roundworms, the flatworms are protostomes that do not molt (**FIGURE 11-30**). Flatworms include more than 20,000 species and can be parasites (the flukes and the tapeworms) or brilliantly colored, free-living aquatic creatures.

Flatworms have well-defined head and tail regions, with clusters of light-sensitive cells called eyespots. Although many flatworms have a digestive system, one feature of free-living flatworms (but not the segmented or roundworms) that many humans find distasteful is that their gut has only one opening, requiring them to consume food and eliminate undigested food through the same opening. Tapeworms, by contrast, completely lack a digestive system; these parasitic worms utilize nutrients from their host by living in the gut and absorbing nutrients directly through their body wall. Most flatworms are hermaphroditic, each individual worm producing both male and female gametes, and they engage in both sexual and asexual reproduction.

All 5,000 species of tapeworms are parasites, and most have a two-stage life cycle that is split between two different host species. For example, the common tapeworm that infects dogs spends the other half of its life cycle in fleas, while the human pork tapeworm splits its life cycle between pigs and humans. Tapeworms have long, flat bodies made up of repeated segments, and each segment is a reproductive unit. Mature tapeworms spread by breaking off segments, which are then shed in the feces of an infected individual.

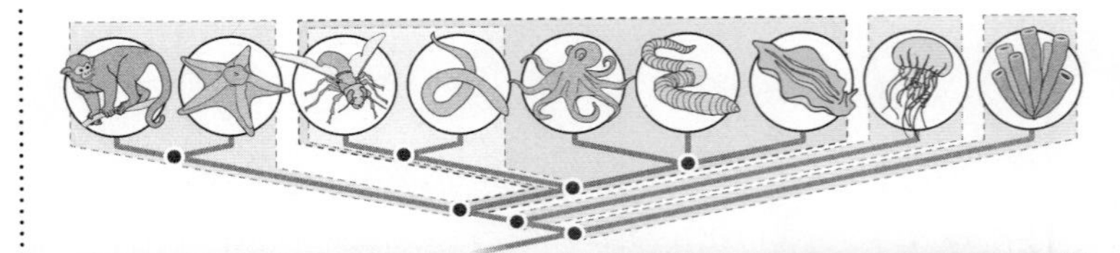

THE FLATWORMS

COMMON CHARACTERISTICS
- Well-defined head and tail regions
- Hermaphroditic and can engage in both sexual and asexual reproduction
- Some have a single opening in the body, which serves as a mouth and an anus

MEMBERS INCLUDE
- Tapeworms
- Flukes

FIGURE 11-30 Overview of the flatworms. A liver fluke (*left*) and a tapeworm of the mammalian intestine.

TAKE-HOME MESSAGE 11·18

Worms are found in several different phyla and are not a monophyletic group. All are bilaterally symmetrical protostomes with defined tissues. The segmented worms (annelids) and flatworms do not molt, while the roundworms do. Earthworms are annelids that play an important role in recycling dead plant material. Many roundworms are parasites of plants or animals and are responsible for several widespread human diseases. Flatworms include parasitic flukes and tapeworms, many of which infect humans.

11•19

Jellyfish and other cnidarians are among the most poisonous animals in the world.

If you feel a sting when you are swimming in the ocean, you have met a jellyfish. If you are lucky, it will be one of the thousands of species that feed on tiny floating organisms called plankton, because these species of jellyfish have only mild stings. If it is a species that feeds on fish or shrimp, however, you are definitely unlucky, because these jellyfish have powerful stings that can cause extreme pain or even death.

The jellyfish, along with sea anemones and corals, belong to the phylum Cnidaria (pronounced nigh-DARE-ee-ah), the cnidarians. This phylum includes about 11,000 species, all of which have defined tissues and radial symmetry and are generally simpler than the bilaterally symmetrical invertebrates we have discussed so far (**FIGURE 11-31**).

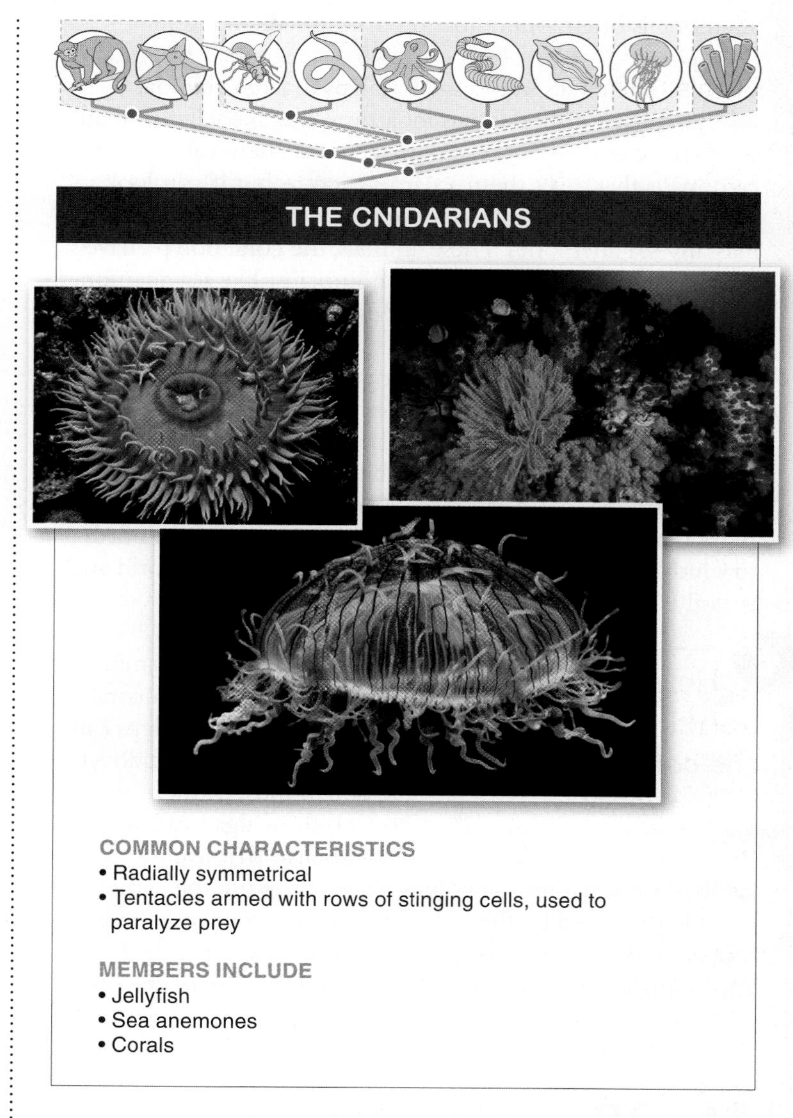

FIGURE 11-31 Understanding cnidarians. Shown here (*from top left, clockwise*) are a giant green sea anemone, a portion of coral reef, and an aptly named flower hat jellyfish.

There are two types of cnidarian bodies: a sessile "polyp" and a free-floating "medusa." In some species, individuals spend part of their life cycle as a polyp and part as a medusa. Other species occur only as medusas or, as in corals and sea anemones, only as polyps. All cnidarians can reproduce both sexually and asexually.

Cnidarians are carnivores, using their tentacles to capture and feed on a wide variety of marine organisms, from protists all the way to fish and shellfish. Their method of capturing prey relies on a structure unique to the cnidarians—a stinging cell called a cnidocyst. All cnidarians have tentacles that are armed with rows of stinging cnidocysts. Each cnidocyst has a coiled thread with barbs inside and a "trigger" on the outside. When something comes in contact with the trigger, perhaps a prey item (or perhaps your leg as you swim in the ocean), the coiled thread is ejected and, like a harpoon, can penetrate the prey, often injecting toxin. The Portuguese man-o'-war is one of the dangerously poisonous species of cnidarians. Being stung by a Portuguese man-o'-war is painful (extremely painful if the unlucky swimmer becomes entangled in the tentacles and receives hundreds of stings), but these encounters are rarely fatal. Although all cnidarians possess cnidocysts, there is great diversity among the three major groups of cnidarians.

Jellyfish Jellyfish range tremendously in size. The Asian giant jellyfish is more than 6 feet (about 2 meters) across and weighs nearly 500 pounds (more than 200 kilograms). When swarms of these giant jellyfish appear, they clog the nets of fishing boats with a glutinous mass of toxic stingers. At the opposite end of the size scale, the irukandji jellyfish is about the size of a hen's egg, but it is so deadly that when just five irukandji were sighted in Hervey Bay, Queensland, Australia, this completely halted the filming of a major Hollywood movie, which had to be completed in the safety of a studio.

Sea anemones As their name suggests, sea anemones look rather like flowers, and some of the colorful species are popular additions to saltwater aquaria. The polyp body form of a sea anemone resembles an upside-down medusa form of a jellyfish—the tentacles with their stinging cells are at the top of the body, surrounding the mouth. Sea anemones have a larval stage that swims freely and then settles on a rock and metamorphoses into the adult form. Even as adults, many sea anemones can crawl a few inches a day, and the wandering sea anemone is often found floating in seaweed.

Corals The corals live as small polyps in large colonial groups. At first glance, corals look even less like jellyfish than sea anemones do—because they secrete a chemical, calcium carbonate, that gives them a hard structure, but if you look inside the hard external structure you'll find animals that look like tiny sea anemones. These animals, the coral polyps, have stinging tentacles surrounding a mouth, just like sea anemones; they can use the tentacles to catch small fish and plankton. Corals grow in a wide variety of shapes—from free-standing spherical or branched forms to crusts that grow on rocks or on other corals. A coral reef creates a complex three-dimensional environment that is home to a greater diversity of species than any other marine habitat. The Great Barrier Reef, located in the Coral Sea off the coast of Queensland, extends more than 1,500 miles (about 2,600 kilometers) from north to south. It is the largest biological structure in the world and is easily visible from the International Space Station.

Q How is global warming affecting the coral reefs of the world?

The effects of global warming are already being seen in coral reefs. Although coral polyps can catch prey using their cnidocytes, they obtain most of their nutrition from algae, called zooxanthellae, that live symbiotically within the polyps. These algae use the carbon dioxide produced by the polyps to conduct photosynthesis. Strangely, when coral polyps are too hot they expel their zooxanthellae. The zooxanthellae are responsible for the colors of many corals, so when coral polyps expel their zooxanthellae, the coral turns white, a phenomenon called coral bleaching. The zooxanthellae return to the polyps when the water cools, but repeated bleaching events lead to the death of the coral polyps. The world's oceans are already warming, and coral bleaching events are becoming more and more frequent (**FIGURE 11-32**).

FIGURE 11-32 Coral bleaching.

TAKE-HOME MESSAGE 11•19

Jellyfish, sea anemones, and corals are radially symmetrical animals with defined tissues, in the phylum Cnidaria. All cnidarians are carnivores and use specialized stinging cells located in their tentacles to capture prey.

11•20

Sponges are animals that lack tissues and organs.

Sponges (phylum Porifera), though they scarcely resemble most of the organisms we think of as animals, are indeed animals, too (**FIGURE 11-33**). They have no specialized tissues or organs. Most lack any symmetry at all. Despite their simplicity, however, sponges are remarkably efficient at gathering food. You can think of the anatomy of a sponge as a cylinder with holes in it. Epidermal cells cover the outside of the sponge, and collar cells cover the inside. The middle layer is merely a gel with fibers that stiffen the sponge's body.

Each collar cell has a long, whip-like flagellum, and the beating of the flagella creates a current that carries water upward and

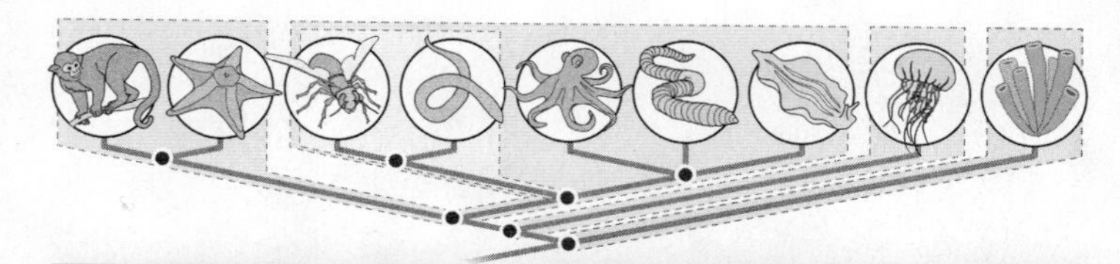

THE SPONGES

COMMON CHARACTERISTICS
- No tissues or organs
- Body consists of a hollow tube with pores in its wall
- Feed by pumping water, along with bacteria, algae, and small particles of organic material, through their pore cells
- Free-swimming larvae
- Sessile as adults

MEMBERS INCLUDE
- About 5,000 species

FIGURE 11-33 An overview of the sponges. A brown tube sponge (*left*) and a purple sponge.

out through the opening at the top of the sponge. That water is replaced by an inward flow of water through pore cells on the sides of the sponge, and the incoming water carries bacteria and algal cells, other microscopic organisms, and small particles of organic material into the sponge. The volume of water moving through a sponge daily can be as much as 20,000 times the volume of the sponge.

The collars of the collar cells are covered with sticky mucus that traps food particles. Cilia move the trapped food particles to the cell body, where they are absorbed by endocytosis. Sponges have no circulatory systems to move nutrients from one place to another. Instead, a special kind of cell, an amoebocyte cell, picks up food particles from the collar cells and moves around through the sponge, transferring food particles to the epidermal cells. Amoebocytes can move freely because sponges lack tight junctions (see Section 3-12) between their cells.

Sponges are hermaphrodites, each individual containing both male and female reproductive organs, but they produce only one kind of gamete (eggs or sperm) at a time. Sponges that are acting as males release a cloud of sperm that swim to other sponges and fertilize their eggs. The eggs develop into larvae, which are released into the water and then drift for a few days before settling on a rock or coral outcrop and developing into a new sponge.

Sponges also reproduce asexually—small sponges bud from the outside of the sponge and eventually break off, settle to the bottom, and grow into new sponges. Even a piece accidentally broken off a sponge by wave action or a collision with a passing fish will grow into a new sponge. Most remarkable is the ability of sponge cells to reassemble. Picture this experiment. You put a living sponge into a food blender and puree it, then strain the liquid through a fine sieve to remove all the chunks, leaving a suspension of individual cells, which you then dump into an aquarium. Within a day you will see small clumps of sponge forming as the individual cells move around and attach to each other when they meet, and within a week a new sponge will have formed. For an even more dramatic demonstration, you can repeat the experiment with two species of sponge of different colors. Not only will the sponges reassemble, but the cells from each species will assemble only with other cells of the same species, so you will have two sponges, each with only its correct cells.

Because sponges are superficially the most "plant-like" of the animals, they make a fitting transition to the subject of our next chapter: the diversity of plants. Although less numerous than animal species, plants are complex organisms that live on land and in water and are found on nearly every surface of the earth.

TAKE-HOME MESSAGE 11·20

Sponges are among the simplest of the lineages of animals. A sponge consists of a hollow tube with pores in its wall; it has no tissues or organs. Sponges reproduce asexually (by budding) and sexually by producing eggs and sperm. The fertilized eggs grow into free-swimming larvae that settle and metamorphose into sessile adult sponges.

Knowledge You Can Use

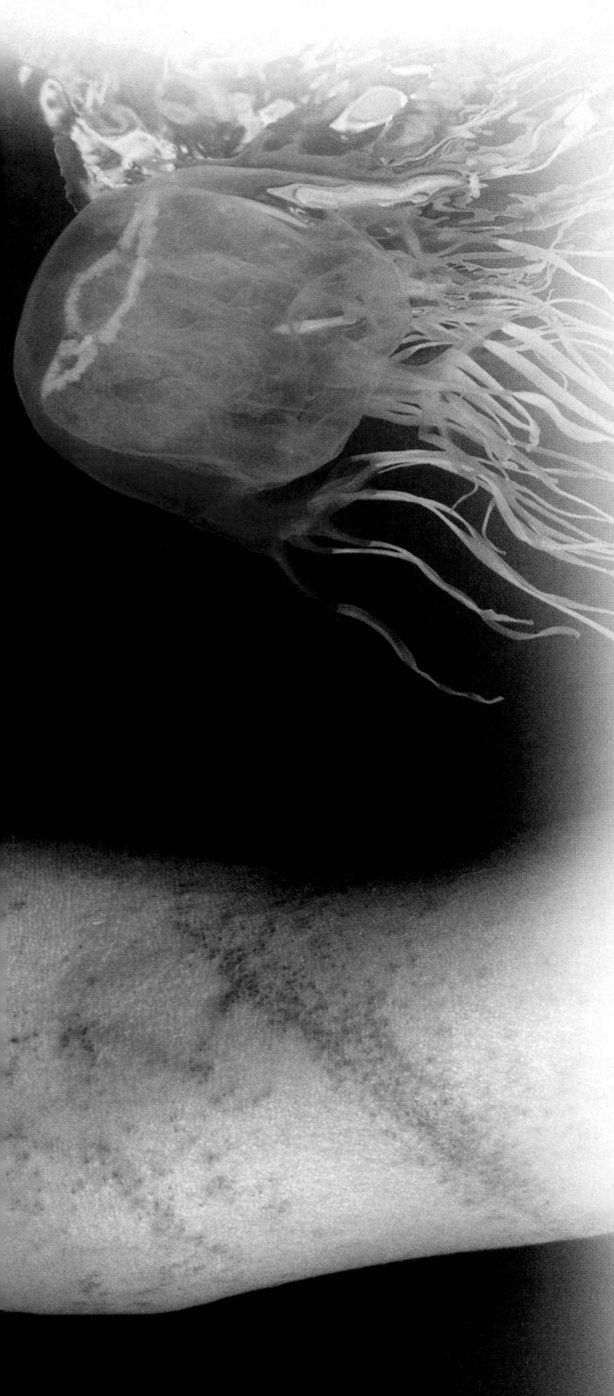

The box jellyfish: nature at its nastiest? When it comes to acquiring food, we expect strong selection for effective strategies.

Natural selection has produced some pretty lethal structures, from the razor-sharp teeth of the piranha, to the hypodermic, poison-injecting fangs of a viper, to the toxic sting of a hornet. But these animals all have an aura of danger and we are not surprised by their lethality. Everyone's got to eat, and their methods for catching prey are impressively effective.

Q: Can a gentle-looking animal with simple body plan be a lethal predator? Quietly swimming in the ocean are some animals that reach a whole new level of lethality. One of nature's most serene-appearing animals, the box jellyfish, has no teeth, no muscles, and no warning hiss. Throughout the waters north of Australia and extending east to the beaches of Hawaii, these animals simply contract and relax their bodies, gently propelling themselves forward.

Q: What is the fastest cellular process in nature? If you find yourself sharing the water with a box jellyfish, you may get a deadly demonstration of nature's fastest cellular process. Along the lengths of their tentacles, these jellyfish have high densities of the stinging cells known as cnidocytes. But these aren't just any stingers. Each cell has a coiled, barbed hair within it. The slightest contact with a trigger on the cell releases the stinger, like a dart. Because the stinger is held at such high pressure, when released, it is ejected 10–20 micrometers in less than a millionth of a second. In a study using a high-speed camera, the acceleration of the dart was measured at over one million times the force of gravity (for comparison, during a rocket launch, astronauts experience about five times the force of gravity). The cnidocyte dart can strike prey with the force of a bullet, easily penetrating the shell of a crustacean (or the skin of a human).

Q: What does the sting do to a human? Feeling the bullet-like strikes of the hundreds of cnidocytes on a single tentacle isn't even the worst of it. The toxin injected when the cnidocyte fires is among the top 10 most toxic animal products on earth. The pain of a box jellyfish sting is so excruciating that swimmers may go into shock or experience heart failure and drown before they can reach shore. More than a hundred people have been killed by box jellyfish stings. Even a non-fatal sting can produce severe scarring and pain that persists for weeks.

The best treatment is prevention? In the disastrous event of an encounter with a box jellyfish, what can you do? Immediately pouring vinegar on the sting and any still-attached tentacles is extremely helpful, disabling any cnidocytes that have not yet fired (though doing nothing for the pain). Many lifeguard stations on Australian beaches keep vinegar on hand for such occasions. But because little else can be done, the best strategy is to avoid swimming at beaches where box jellyfish have been sighted. Stand on the shore instead and safely appreciate one of natural selection's most deviously dangerous creations.

BIG IDEAS IN ANIMAL DIVERSIFICATION

1 Animals are just one branch of the eukarya.

Animals are multicellular organisms that feed on other organisms and can move during at least one stage of their life. Animals probably originated from an ancestral protist. Four key distinctions divide the extant animals: tissues or not, radial or bilateral symmetry, protostome or deuterostome development, and growth through molting or without molting.

2 Vertebrates are animals with a backbone.

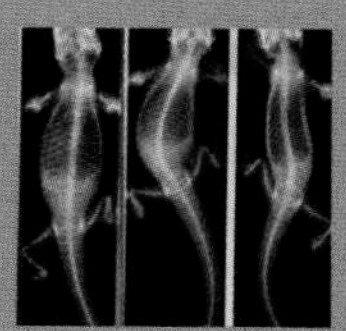

All chordates have four characteristic structures: a notochord, a dorsal hollow nerve cord, pharyngeal gill slits, and a post-anal tail. The development of fins and jaws set the stage for the enormous diversity of modern vertebrates. The transition of vertebrates from life in water to life on land was enabled by four developments: fins modified into limbs, vertebrae modified to transmit body weight through the limbs to the ground, transfer of the site of gas exchange from gills to lungs, and an egg resistant to drying out.

3 All terrestrial vertebrates are tetrapods.

Amphibians are terrestrial vertebrates (tetrapods), but most species still lay eggs in water and the eggs hatch into aquatic larvae. Feathers were originally colorful structures used for behavioral displays in reptiles. Their current uses as insulation and in flight developed later. Hair and lactation are defining characteristics of mammals. The evolutionary success of humans can be traced to an increase in brain size combined with an omnivorous diet. Modern humans evolved in Africa 200,000–100,000 years ago, and all living humans are descended from that evolutionary radiation.

4 Invertebrates are animals without a backbone.

Invertebrates, defined as animals without a backbone, are the largest and most diverse group of animals, comprising 96% of all living species of animals. The invertebrates are not a monophyletic group, however, and include protostomes and deuterostomes. Echinoderms, like chordates, are deuterostomes. The arthropods, protostome invertebrates, are the most diverse phylum on earth, with nearly 1 million species of named insects alone. Molluscs are protostome invertebrates that do not molt and include the snails and slugs, clams and oysters, and squids and octopuses. The concept of intelligence cannot be applied objectively to compare species, which have evolved in response to the selective forces relevant only to their own particular environments.

5 Worms, jellies, and sponges are examples of invertebrate diversity.

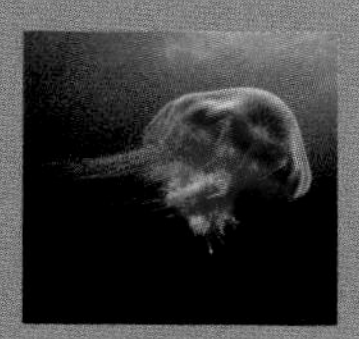

Worms fall into several different phyla. All are bilaterally symmetrical protostomes with defined tissues. The segmented worms (annelids) and flatworms do not molt; the roundworms do. Jellyfish, sea anemones, and corals are radially symmetrical animals with defined tissues, in the cnidarian phylum. All cnidarians are carnivores and use specialized stinging cells located in tentacles to capture prey. Sponges, lacking specialized tissues, organs, and symmetry, are the simplest lineage of animals.

KEY TERMS

adult, p. 430
amniote, p. 419
amphibian, p. 419
animal, p. 408
arthropod, p. 428
bilateral symmetry, p. 409
bivalve mollusc, p. 433
cartilaginous fishes, p. 416
cephalopod, p. 433
deuterostome, p. 410
dorsal hollow nerve cord, p. 412
ectotherm, p. 420
endotherm, p. 420
exoskeleton, p. 423
flatworm, p. 436
gastropod, p. 433
hair, p. 421
larva, p. 430
lobe-finned fishes, p. 416
mammary glands, p. 421
marsupial, p. 422
metamorphosis, p. 430
monotreme, p. 422
notochord, p. 412
pharyngeal slits, p. 412
placental, p. 422
post-anal tail, p. 412
protostome, p. 410
pupa, p. 430
radial symmetry, p. 409
ray-finned fish, p. 416
roundworm, p. 436
segmented worm, p. 436
sessile, p. 408
tetrapod, p. 418
viviparity, p. 422

Check Your Knowledge

1. Which of the following is not a characteristic of all animals?
   a) They are able to move at some point in their life.
   b) They get their energy by eating other organisms.
   c) They are multicellular.
   d) They are sexually reproducing at some point in their life cycle.
   e) All of the above are characteristics of all animals.

2. The lineage that first separated from the common ancestor of all animals, and retains many of those primitive features to this day, includes which of the following modern organisms?
   a) the grasshopper
   b) the jellyfish
   c) the sponge
   d) the earthworm
   e) the sea turtle

3. Which of the following extant species is the most evolutionarily successful?
   a) the house mouse (*Mus domesticus*)
   b) the housefly (*Musca domestica*)
   c) the Western gorilla (*Gorilla gorilla*)
   d) irukandji, the most toxic box jellyfish (*Carukia barnesi*)
   e) All of the above species are equally successful because they are not extinct.

4. Which of the following are chordates?
   a) fish
   b) humans
   c) frogs
   d) All of the above are chordates.
   e) Only a) and c) are chordates.

5. The two most important evolutionary innovations in vertebrates, which resulted in their eventual domination among the large animals, were:
   a) air-breathing lungs and a more efficient heart.
   b) opposable thumbs.
   c) jaws and amniotic eggs.
   d) the ability to walk or fly.
   e) air-breathing lungs and endothermy.

6. The molluscan mantle is used primarily for:
   a) feeding.
   b) gas exchange.
   c) producing the shell.
   d) excretion.
   e) reproduction.

7. Which of these animals is a tetrapod that does not produce amniotic eggs?
   a) salamander
   b) human
   c) monkey
   d) elephant
   e) python

8. Why is the amniotic egg considered a key evolutionary innovation?
   a) It prohibits external fertilization, thereby facilitating the evolutionary innovation of internal fertilization.
   b) It has an unbreakable shell.
   c) It greatly increases the likelihood of survival of the eggs in a terrestrial environment.
   d) It enables eggs to float in an aquatic medium.
   e) It extends the time of embryonic development.

9. Which came first, the chicken or the egg?
   a) The chicken, because the amniotic egg did not evolve until after the first chicken appeared.
   b) The egg, because the amniotic egg evolved well before the first birds.
   c) The chicken, because during speciation the adult stage always precedes the juvenile stage.
   d) The egg, because the chicken is not a real species.
   e) It is impossible to determine, because eggs leave no fossils.

10. Which of the following is not a reptile?
    a) sparrow
    b) snake
    c) frog
    d) turtle
    e) dinosaur

11. Marsupials and which of the following groups combine to make a monophyletic assemblage?
    a) birds
    b) carnivores
    c) primates
    d) monotremes
    e) placental mammals

12. According to the fossil record, the first humans appeared approximately ______ years ago.
    a) 6 million
    b) 6,000
    c) 190 million
    d) 100,000
    e) 1 million

13. Which of the following insect groups has the most species currently named by scientists?
    a) bees, wasps, and ants
    b) beetles
    c) grasshoppers and crickets
    d) flies
    e) butterflies and moths

14. Which is the only animal phylum to have over one million described species?
    a) chordates
    b) arthropods
    c) flatworms
    d) nematodes
    e) molluscs

15. The phylum Arthropoda includes all of the following kinds of animals except:
   a) snails.
   b) crabs.
   c) crayfish.
   d) butterflies.
   e) scorpions.

16. Which of the following traits is unique to arthropods?
   a) a ventral nerve cord
   b) an exoskeleton
   c) segmentation
   d) a large ratio of brain size to body size
   e) wheeled appendages

17. In cnidarians, cnidocytes are primarily used for:
   a) creation of water flow across the body wall.
   b) formation of free-living medusas.
   c) secretion of digestive enzymes.
   d) prey capture and defense.
   e) muscular contraction during movement.

18. Sponges are sessile, meaning that they:
   a) reproduce asexually.
   b) are parasitic and depend on their host for a constant supply of nutrients.
   c) have exoskeletons that they must shed as they grow.
   d) live within shells they find on the ocean floor.
   e) live attached to a solid structure and do not move around.

Short-Answer Questions

1. The cheetah is an amazingly fast cat that hunts its prey on the African savannas, sprinting for short distances at speeds approaching 70 miles per hour. The sloth in the Amazon rain forests of South America is a very slow creature that hangs from a tree by all four limbs and eats vegetation. How can it be explained, in evolutionary terms, that the cheetah can sprint at high speeds, while the sloth moves almost in slow motion? What modifications might occur to the cheetah's ability to run if it were to occupy the same niche as the sloth, assuming the cheetah has no natural enemies there?

2. Explain why an ectotherm, such as a snake, requires only an occasional meal, while an endotherm, such as a mouse, requires frequent meals to survive?

3. A caring pet owner brings her boa constrictor to the veterinarian because it has been more sluggish than normal and appears to be ill. The veterinarian is concerned when she sees that the owner has placed a thick sweater, specially knitted, on the snake. Why might the sweater worry the veterinarian?

4. You have a friend who confuses a terrestrial slug with a segmented worm. Describe some of the characteristics of the slug that you could point out to him that are not found in any segmented worms.

See Appendix for answers. For additional study questions, go to www.prep-u.com.

1 Plants are just one branch of the eukarya.

- **12.1** What makes you a plant?

2 The first plants had neither roots nor seeds.

- **12.2** Colonizing land brings new opportunities and new challenges for plants.
- **12.3** Mosses and other non-vascular plants lack vessels for transporting nutrients and water.
- **12.4** The evolution of vascular tissue made large plants possible.

3 The advent of the seed opened new worlds to plants.

- **12.5** What is a seed?
- **12.6** With the evolution of the seed, gymnosperms became the dominant plants on earth.
- **12.7** Conifers include the tallest and longest-living trees.

4 Flowering plants are the most diverse and successful plants.

- **12.8** Angiosperms are the dominant plants today.
- **12.9** A flower is nothing without a pollinator.
- **12.10** Angiosperms improve seeds with double fertilization.

5 Plants and animals have a love-hate relationship.

- **12.11** Fleshy fruits are bribes that flowering plants pay animals to disperse their seeds.
- **12.12** Unable to escape, plants must resist predation in other ways.

6 Fungi and plants are partners but not relatives.

- **12.13** Fungi are closer to animals than they are to plants.
- **12.14** Fungi have common structures, but exploit an enormous diversity of habitats.
- **12.15** Most plants have fungal symbionts.

12

Plant and Fungi Diversification

Where did all the plants and fungi come from?

1 Plants are just one branch of the eukarya.

The primitive land plant Psilotum nudum, *growing in Hawaii Volcanoes National Park, Hawaii.*

12•1

What makes you a plant?

An animal can move from place to place to get food or water, to avoid being eaten, or to find a mate. It's harder for a plant. It must do all the things an animal does—obtain food and water, protect itself from predators, and reproduce—*but* it can't move a centimeter in the process. Wherever a seed puts down roots, that's where the plant stays anchored for its entire life.

So plants are organisms that are fixed in place, but that's only one aspect of being a plant. A **plant** is a multicellular eukaryote that produces its own food by photosynthesis and has an embryo that develops within the protected environment of the female parent (**FIGURE 12-1**). Plants vary in size from less than 0.04 inch (1 millimeter) to 380 feet (117 meters) tall, and most plants live on land.

WHAT IS A PLANT?

PLANTS CREATE THEIR OWN FOOD
Plants carry out photosynthesis, using energy from sunlight to convert carbon dioxide and water into sugar.

PLANTS ARE SESSILE AND TERRESTRIAL
Plants are anchored in place at their bases.

PLANTS ARE MULTICELLULAR
Plants consist of multiple cells and have structures that are specialized for different functions.

FIGURE 12-1 Defining characteristics of a plant.

12

Plant and Fungi Diversification

Where did all the plants and fungi come from?

➊ Plants are just one branch of the eukarya.

The primitive land plant Psilotum nudum, *growing in Hawaii Volcanoes National Park, Hawaii.*

12•1

What makes you a plant?

An animal can move from place to place to get food or water, to avoid being eaten, or to find a mate. It's harder for a plant. It must do all the things an animal does—obtain food and water, protect itself from predators, and reproduce—*but* it can't move a centimeter in the process. Wherever a seed puts down roots, that's where the plant stays anchored for its entire life.

So plants are organisms that are fixed in place, but that's only one aspect of being a plant. A **plant** is a multicellular eukaryote that produces its own food by photosynthesis and has an embryo that develops within the protected environment of the female parent (**FIGURE 12-1**). Plants vary in size from less than 0.04 inch (1 millimeter) to 380 feet (117 meters) tall, and most plants live on land.

WHAT IS A PLANT?

PLANTS CREATE THEIR OWN FOOD
Plants carry out photosynthesis, using energy from sunlight to convert carbon dioxide and water into sugar.

PLANTS ARE SESSILE AND TERRESTRIAL
Plants are anchored in place at their bases.

PLANTS ARE MULTICELLULAR
Plants consist of multiple cells and have structures that are specialized for different functions.

FIGURE 12-1 Defining characteristics of a plant.

There are other organisms on earth that are multicellular and photosynthetic eukaryotes. These include many species of algae, for example the brown algae commonly known as seaweed, and include the closest relatives of land plants. They differ from plants, however, in that they live only in water or on very moist land surfaces. This is in sharp contrast to most plants, which can live even in deserts.

FIGURE 12-2 **Dodder: a parasitic, non-photosynthetic plant.**

As we have said, most plants make their own food—that is, they carry out photosynthesis, using energy from sunlight to convert carbon dioxide and water to sugar, and then convert some of the sugar to starch. However, a plant can't live on carbohydrate alone. A plant needs nitrogen to build proteins, phosphorus to make ATP, and salts to create concentration gradients between the inside and outside of cells. Plants use **roots,** the part of a plant below ground, to obtain these needed substances from the soil. Above ground, plants have a **shoot** that consists of a stem and leaves. The stem is a structure that supports the main photosynthetic organ of a plant: its leaves.

Q If a plant growing in the shade can't move to a new, sunnier location, how can it reach the available sunlight?

When we think of plants, we often think of them as "chlorophyll-containing," but some plants have no chlorophyll—their ancestors had chlorophyll, but they have lost almost all of it over evolutionary time—and can't carry out photosynthesis. Instead, they live as parasites that steal nutrients from other plants. Dodder is an example of a parasitic plant that almost completely lacks chlorophyll and gets its sugar from the host plant it grows on (**FIGURE 12-2**). You can probably find dodder growing on plants in a vacant lot or at a roadside near where you live.

Plants that carry out photosynthesis need sunlight, and that creates a challenge for some plants. Think about plants growing on the floor of a forest. It's mostly shady under the tree canopy, with just a few sunny spots where light penetrates the tree branches overhead. If a seed sprouts (germinates) in the shade, the young plant needs to somehow get to the nearest sunny spot. Plants can do that, of course, but not by moving. Instead, they grow toward the light (**FIGURE 12-3**). You may have seen a plant called "lucky bamboo" (it's not really a bamboo) for sale in a grocery or drug store. It catches your eye because it grows in spirals or loops. The growers have produced these shapes by moving the light source every few days, forcing the plant to bend to follow the light. If you buy a lucky bamboo, you must continue rotating the light source when you get it home or it will start to grow straight, like any other plant.

STATIONARY LIVING CREATES CHALLENGES

OBTAINING FOOD
Because plants can't move to reach sunlight, they bend in place and grow toward light.

FINDING A MATE
Male and female plants can't meet to reproduce, so they have developed ways of getting the male gamete to the female gamete, including alternating haploid and diploid life stages, and using other organisms to transport the gametes.

RESISTING PREDATION
Plants can't run from predators, so they have developed adaptations such as thorns to defend themselves.

FIGURE 12-3 **Plants must overcome the constraint of being immobile.**

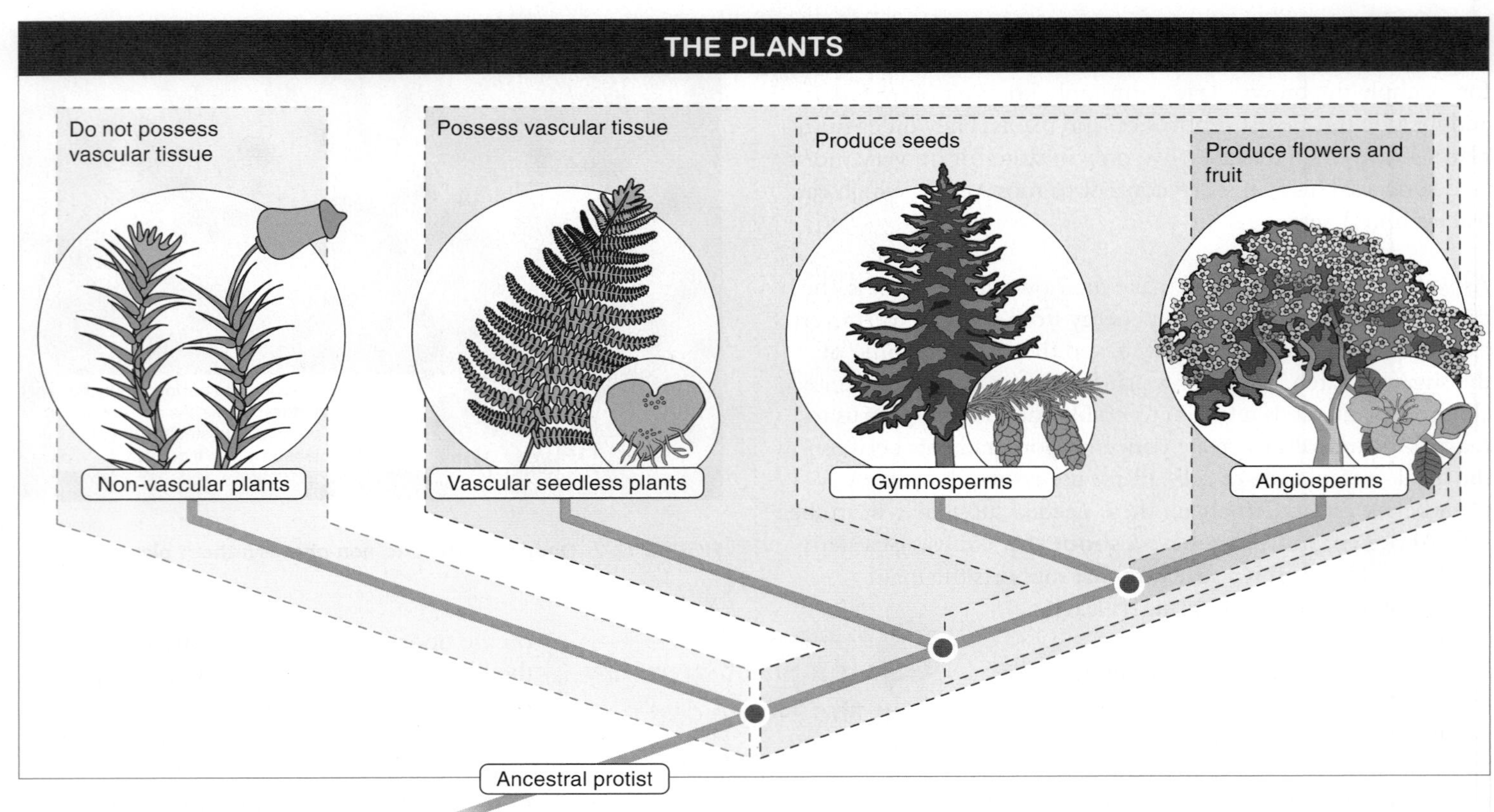

FIGURE 12-4 Phylogeny of the plants.

Sex can also be a challenge for an organism that is anchored in place by its roots. Not surprisingly, plants have a sex life that is very different from that of animals. Consider humans as a typical animal. From the moment an egg is fertilized by a sperm, humans have the diploid number of chromosomes—that is, two sets of chromosomes, one set from each parent. Only the egg and sperm are haploid—an egg or a sperm has only one set of chromosomes. Humans and other animals have no haploid stage in their lives (other than their gametes), but many plants do. The haploid stage is the one that allows the male and female plants to reproduce, even though they will never meet because they can't move. Some plants enlist the help of animals to carry the male gamete to the female gamete.

Resisting predators is another challenge for organisms that cannot move. Running away is how most animals react to a predator, but plants have found other ways to defend themselves. Everyone is familiar with thorns, which are an anatomical defense plants use to avoid predation (see Figure 12-3). Plants also use chemicals to deter predators, as many people learn when they develop an itchy rash from poison ivy.

The earliest land plants were the first multicellular organisms to live on land; these simple, non-vascular plants had no vessels to transport water and nutrients (**FIGURE 12-4**). The subsequent evolution of land plants was a series of radiations of forms with characteristics that made them increasingly independent of water. The evolutionary tree of plants shows these stages clearly: first the development of vessels to conduct water from the soil through the plant (the vascular plants), then seeds that provide nutrients to get the next generation off to a good start (the gymnosperms), and finally flowers that allow plants to entice or trick insects and birds to spread the plant's male gametes (the angiosperms). We'll look at each of these plant adaptations in this chapter. We also explore the ecologically important kingdom of fungi. Although fungi are not plants, they are very closely associated with plants.

TAKE-HOME MESSAGE 12•1

Plants are multicellular organisms that spend most of their lives anchored in one place by their roots. The inability of plants to move shapes the way they obtain food, reproduce, and protect themselves from predation. The evolutionary history of plants reveals the progressive appearance of characteristics that allow plants to succeed on land despite their inability to move.

❷ The first plants had neither roots nor seeds.

Close view of a filmy fern in Westland National Park, New Zealand.

12•2

Colonizing land brings new opportunities and new challenges for plants.

The aquatic ancestors of land plants were green algae. Like plants, green algae are multicellular, photosynthetic eukaryotes, but green algae live only in water or on very moist land surfaces. As water dwellers, green algae do not require specialized structures to obtain water and nutrients; water simply enters their cells by osmosis, and the nutrients they require are in solution in the water that surrounds them. Some green algae, such as sea lettuce and stoneworts, even look like plants—and there is enough uncertainty among taxonomists that some classify the green algae within the plant kingdom. Others, however, classify green algae within the protist kingdom and consider some green algae that look like slime on rocks—organisms called coleochaetes (pronounced KOH-lee-oh-keets)—to be the closest relatives of plants (**FIGURE 12-5**). An individual coleochaete is about the size of a pinhead and is only one cell thick. Coleochaetes can withstand exposure to air, so they survive when the water level in a lake falls and leaves them high and dry. This resistance to drying was the first evolutionary step that plants took as they moved from water to land.

FIGURE 12-5 Plants' closest relatives: a green alga called coleochaete.

MOVING ONTO LAND PRESENTS CHALLENGES

When plants emerged onto land, they faced the same two challenges that terrestrial animals faced 25 million years later.

1 **PROBLEM:** GRAVITY
SOLUTION: The earliest plants grew very close to the ground, as mosses do today, in order to resist the pull of gravity.

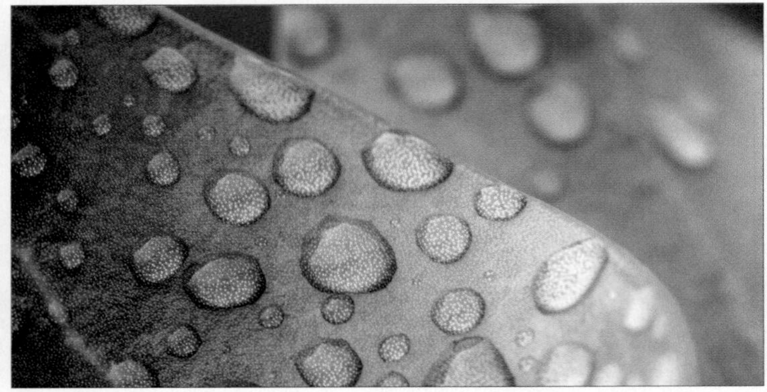

2 **PROBLEM:** DESICCATION
SOLUTION: Plants developed an outer waxy layer called a cuticle that covers their entire surface.

FIGURE 12-6 Leaving the water. With the transition to land, plants had to develop ways to resist the pull of gravity and—a more immediate need—to resist drying out.

The first land plants appeared about 475 million years ago. They weren't impressive-looking, just some patches of low-growing green stems at the water's edge. They did not have any of the structures we associate with plants today—no roots, leaves, or flowers. But from an evolutionary perspective, those early plants were enormously important because, until terrestrial plants evolved, there was nothing on land for other land organisms to eat. Thus, the first land plants not only set the stage for the tremendous diversity of plant life that we know today, but also paved the way for the evolution and diversification of land animals.

When plants emerged onto land, they faced the same two challenges that were to confront the first terrestrial animals, some 25 million years later: supporting themselves against the pull of gravity and reducing evaporation so they didn't dry out. The second of these problems was the more urgent. The earliest plants did not have to grow upward—they could creep along the ground—but they *did* have to avoid drying out (**FIGURE 12-6**). The material that protects land plants from drying is a waxy layer called the **cuticle** that covers the leaves and stem of the plant. The cuticle is what makes leaves shiny, and all terrestrial plants have a cuticle.

TAKE-HOME MESSAGE 12•2

The first land plants were small, had no leaves, roots, or flowers, and could grow only at the water's edge. Nonetheless, these inconspicuous little plants set the stage for the enormous diversity of terrestrial plants and animals on earth today.

12•3

Mosses and other non-vascular plants lack vessels for transporting nutrients and water.

The earliest land plants were low-growing for a reason—they had no structures that could transport water and nutrients from the soil upward into the plant. The only way that these substances could move was by diffusing from one cell into an adjacent cell, and then into the next cell, and so on. Diffusion is a slow process, and plants that rely on diffusion can grow only a few centimeters tall.

Despite the limitations of diffusion, three groups of plants, all known as **bryophytes,** still use diffusion to move substances, rather than having any sort of "circulatory system." They are the liverworts, hornworts, and mosses. Liverworts and hornworts are small (less than an inch in height), simple plants that grow in moist and shady places and resemble flattened moss. Mosses are the plants you are most likely to be familiar with, and more than 12,000 species of mosses grow in habitats extending from arctic and alpine regions to the tropics. These bryophyte plants are referred to as **non-vascular** because they do not have vessels to transport water and food. Water and nutrients are absorbed by projections from the outermost layer of cells that penetrate a few micrometers into the soil. Because these projections are so short, non-vascular plants must either live in places where the soil is always moist or become dormant when the soil surface dries out (**FIGURE 12-7**).

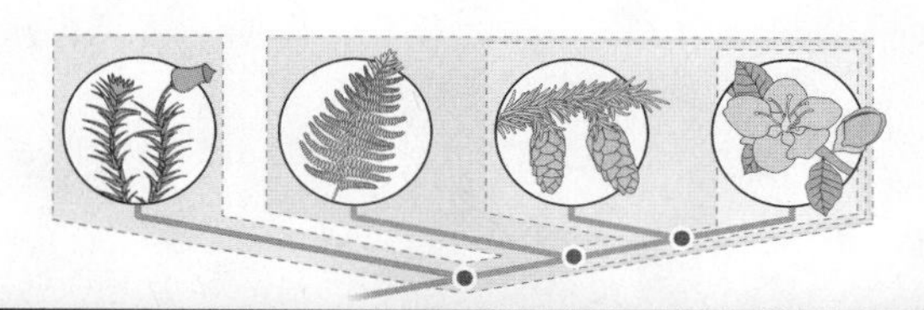

THE NON-VASCULAR PLANTS

COMMON CHARACTERISTICS
- Distribute water and nutrients throughout plant by diffusion
- Release haploid spores, which grow and produce gametes
- Life cycle with multicellular haploid and diploid phases

MEMBERS INCLUDE
- Mosses (about 12,000 species)
- Liverworts (about 8,000 species)
- Hornworts (about 100 species)

FIGURE 12-7 Overview of the non-vascular plants. The plants shown here (*from left, clockwise*) are moss, liverwort, and hornwort.

In order to adapt to land, the non-vascular plants had to develop a method of reproduction that protected the plant embryo from drying out and provided it with a source of nutrients. The innovation that made this possible was a life cycle of alternating haploid and diploid generations, in which the diploid embryo is protected by the haploid female parent. This "alternation of generations" in plants is radically different from the life cycle of human and most other animals. In animals, a diploid organism (the organism you see) produces haploid gametes (which are single-celled and remain that way, never visible to the naked eye). The haploid gametes, at fertilization, produce a new, diploid cell that becomes a multicellular embryo, eventually growing until we can see it. And then, once we have the adult animal, the process starts all over again.

The life cycle of a moss provides a useful example of how the alternation of generations works. When you look at a spongy mass of moss (or any other non-vascular plant), what you see is the *haploid* part of the life cycle (**FIGURE 12-8**). The adult moss plants are haploid plants—that is, all the cells have only one set of chromosomes. There are male and female moss plants that have male and female reproductive structures located among the feathery leaves at the tips of the stems. Water collects here during rainstorms, allowing the sperm to "swim" (or be splashed) from the male structures to fertilize eggs in the female structures. Once the egg is fertilized in the female structure, a diploid zygote is formed, and it divides to become an embryo. The embryo is sheltered within the female reproductive structure while it continues to divide by mitosis and to mature. The female reproductive structure provides water and nutrients for the growing embryo, and both the embryo and the reproductive structure enlarge. Eventually, the female structure elongates to such a degree that it breaks in two and forms a capsule extending over the top of the plant like a raised fist. Inside the capsule, haploid spores are formed by meiosis. A **spore** is a single cell, containing DNA, RNA, and a few proteins. When a capsule ruptures, it releases hundreds of spores. The spores that land in moist, sheltered spots grow into new (haploid) male or female moss plants.

Some non-vascular plants are economically or ecologically important. "Peat moss" is the name given to *Sphagnum* moss when it is harvested and sold as a soil enhancer for gardening. Peat bogs are found in many parts of the world and consist of partly decayed moss. Peat harvested from bogs is dried and burned as fuel in areas where trees are scarce, and peat bogs—such as those on the coasts of Malaysia and Indonesia—are important in flood control, because they can absorb enormous amounts of water during the monsoon season and release it over a period of months. Another commercial application: mounds of burning peat are used to dry the barley used to

MOSS LIFE CYCLE

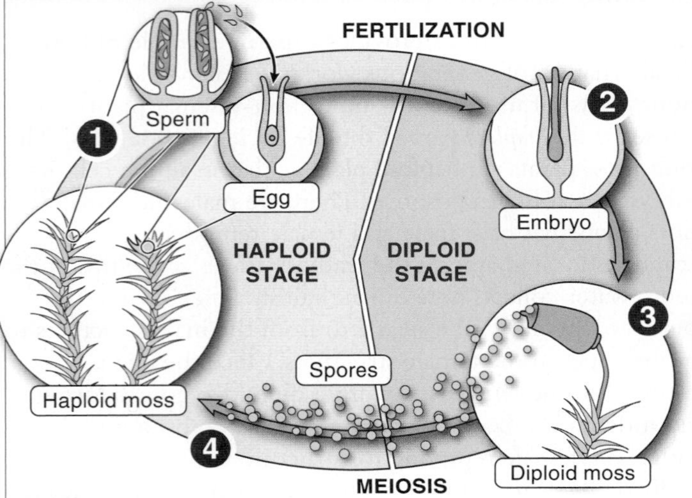

1. During rainstorms, sperm swim from male reproductive structures to female reproductive structures, where they fertilize the egg.
2. A diploid embryo forms and develops into an adult diploid moss.
3. The diploid moss develops a capsule, which bursts and releases haploid spores.
4. A spore lands on moist soil and grows into an adult haploid moss.

FIGURE 12-8 Alternation of generations in mosses. The spongy part of the moss, at ground level in the photograph, is the haploid form. Hovering above, on stalks, are the spore capsules, the diploid form.

produce Scotch whisky. This gives Scotch its distinctive smoky taste (**FIGURE 12-9**).

Unlikely as it seems, non-vascular plants can grow in deserts, and mosses are important as components of the biological crust that holds desert soils in place. This crust is composed of mosses, lichens, and cyanobacteria. It cements the soil particles together and allows the soil to resist wind erosion, but the crust is

FIGURE 12-9 Uses of peat. Peat can be dried and then burned as fuel to heat homes. In traditional Scotch whisky production, the burning of peat during the distillation process provides the smoky flavor.

extremely fragile and very slow to regenerate. When people or cattle walk over the crust, they break it into small pieces that cannot resist wind, and wind erosion can then strip a meter or more of soil from the unprotected ground surface in a few decades, leaving tree trunks supported in mid-air by their roots.

TAKE-HOME MESSAGE 12•3

Non-vascular plants—mosses, liverworts, and hornworts—have scarcely evolved beyond the stage of the earliest land plants. They lack roots and vessels to move water and nutrients from the soil into the plant, and they reproduce with spores that form when a sperm from a male reproductive structure "swims" through a drop of rainwater to fertilize the egg in a female reproductive structure.

12•4

The evolution of vascular tissue made large plants possible.

What a difference some tubes make. That's all vascular tissue is—a sort of infrastructure of tubes that begins in the roots and extends up the stem of a plant and out to the farthest tips of its leaves. The evolution of vascular tissue allowed early land plants to transport water and nutrients faster and more effectively than the cell-to-cell diffusion that non-vascular plants must rely on, and their roots penetrate far enough into the soil to reach moisture even when the soil surface is dry. Also, roots that reach deep into the soil provide the support that a plant needs to grow upward without falling over. As a consequence, the **vascular plants** can grow taller than non-vascular plants and are more successful in areas where the surface of the ground dries out between rainstorms (**FIGURE 12-10**).

Ferns are the most familiar of the primitive vascular plants. During the Carboniferous period, which extended from 360 to 300 million years ago, ferns were a major component of the huge swamp forests that eventually formed the coal deposits we are mining now. A related group of plants, the horsetails, like their Carboniferous ancestors, grow in wet habitats where the oxygen concentration around the roots is low because the spaces between soil particles are filled with water (rather than air). To adapt to this wet environment, horsetails developed hollow stems that allow oxygen from the air to diffuse down to the roots. The thin leaves of horsetails have a single nutrient- and water-carrying vessel extending from the base to the tip.

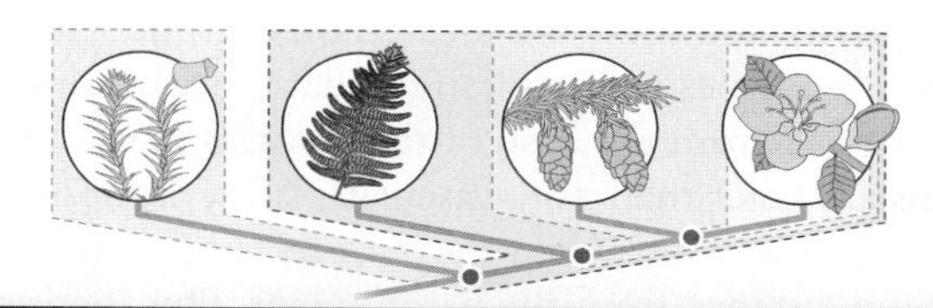

THE VASCULAR SEEDLESS PLANTS

COMMON CHARACTERISTICS
- Distribute water and nutrients throughout plant with a "circulatory system" of vascular tissue
- Release haploid spores, dispersed by the wind, which grow and produce gametes
- Life cycle (unlike in animals) with free-living, multicellular haploid and diploid phases

MEMBERS INCLUDE
- Ferns (about 12,000 species)
- Horsetails (about 15 species)

FIGURE 12-10 Snapshot of the vascular plants: ferns and horsetails. These are the groups of vascular plants that don't produce seeds. The plants shown here (*from left, clockwise*) are Christmas fern, horsetails, and ferns in a forest.

A fern is unfurling its leaves in the springtime. The young coiled leaves are called fiddleheads.

FERN LIFE CYCLE

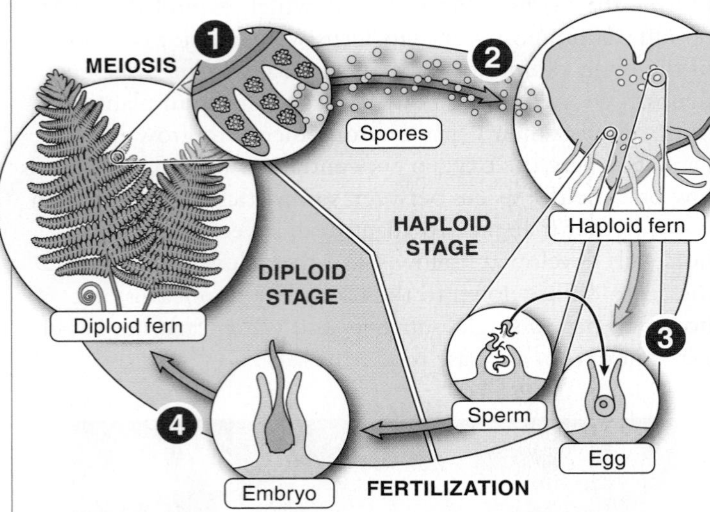

1. Adult ferns release haploid spores, which are carried by the wind to a new location.
2. A spore lands on moist soil and grows into a haploid fern called a prothallus.
3. During rainstorms, sperm swim from male reproductive structures to female reproductive structures, where they fertilize the egg.
4. A diploid embryo forms and continues to grow into an adult diploid fern.

FIGURE 12-11 Haploid and diploid life stages in ferns. The photograph shows the sporangia, the spore-producing bodies, located on the underside of most fern leaves—on the diploid plant. The diploid and haploid plants in the fern life cycle are very different in appearance.

Simply having vessels, as horsetails do, is an important evolutionary innovation, but ferns took an additional step. In addition to the central vessel in each leaflet, the leaves of ferns have vessels branching from the central vessel to the edges of the leaflet. This arrangement places a channel for the movement of water and nutrients close to each cell in the leaf.

Like non-vascular plants, horsetails and most ferns reproduce with spores. At the top of the central hollow stem of the horsetail is a conical structure where the haploid spores are produced and released. In many ferns, the structures in which the spores are produced are called **sporangia** (*sing.* **sporangium**), and these are located on the underside of the leaves. Some species have sporangia that branch from the fronds or grow from the base of the plant. Because horsetails and ferns are taller than non-vascular plants, the spores can be blown by the wind when they are released, and they may settle some distance from the parent plant. This increased dispersal ability was an important adaptation—the non-vascular plants are so low-growing that wind can't play much of a role in moving their spores.

A spore that lands on moist soil grows into a tiny heart-shaped structure called a **prothallus,** which is the free-living haploid life stage of a fern (**FIGURE 12-11**). The prothallus produces the haploid gametes: some cells produce eggs (with one set of chromosomes) and others produce sperm (also with one set of chromosomes). Sperm "swim" through drops of rainwater to fertilize an egg, and the fertilized egg (a diploid zygote with two sets of chromosomes) grows into an adult fern.

TAKE-HOME MESSAGE 12•4

Like a circulatory system, vessels are an effective way to carry water and nutrients up from the soil to the leaves, and the first vascular plants—including the earliest ferns and horsetails—were able to grow much taller than their non-vascular predecessors.

❸ The advent of the seed opened new worlds to plants.

Forests surrounding Mistaya Lake in Banff National Park, Alberta, Canada.

12•5

What is a seed?

As we've seen, the evolutionary development of vascular tissue allowed plants to grow large. Plants could now colonize areas where the soil at the surface of the ground was not always wet, because their roots could penetrate the soil to reach water and nutrients. The next big innovation in plant evolution was the **seed,** an embryonic plant with its own supply of water and nutrients encased within a protective coating (**FIGURE 12-12**).

Seeds are goodbye gifts that adult plants give to their offspring—they contain the materials needed to begin a new life. Unlike spores, which are single cells that contain only DNA, RNA, and

SEEDS: STRUCTURE AND GROWTH

STRUCTURE
Fertilization produces a diploid seed, which contains a multicellular embryo and a store of carbohydrate (endosperm) to fuel its initial growth.

Embryo
Protective coating
Endosperm

GROWTH
A seedling draws energy from the endosperm while it extends its leaves upward to begin photosynthesis and its roots downward into the soil to reach water and nutrients.

FIGURE 12-12 What is a seed? A seed is a package that contains a multicellular embryo and a store of carbohydrate, the endosperm, to fuel its initial growth. (This makes many seeds desirable as food sources—shown here are seeds for sale at a market in Jordan.)

a few proteins, seeds contain both a multicellular embryo and a store of nutrients, mostly starch. Called **endosperm** in angiosperms, this nutritive tissue can fuel the seed's initial growth. A seedling draws energy from the endosperm while it extends its leaves upward to begin photosynthesis and its roots downward into the soil to reach water and nutrients. There are two modern groups of seed-producing plants: **gymnosperms** (including pines, firs, and redwoods) and **angiosperms** (all of the flowering plants and trees).

Q How are seeds formed?

Seed plants have a very small life stage that produces haploid gametes (sperm and egg). This is analogous to an individual human, which produces haploid gametes. Botanists call this haploid plant form the **gametophyte. Pollen grains** and **ovules** are the male and female gametophytes, respectively, of seed plants. In brief (we expand on this later in the chapter), a haploid female gamete (egg) forms inside the ovule. When a pollen grain lands near the ovule, it produces a pollen tube that grows into the ovule. Sperm from the pollen grain move through the pollen tube into the ovule and fertilize the egg. The external layer of the ovule then forms the seed coat.

Another of the challenges that plants face is dispersing their seeds. The seed stage is the only opportunity most plants have to send their offspring away from home, and seeds and seed pods have many ways to do this. These range from the forceful send-off of exploding seed pods, to seeds that hitch a ride in or on passing animals, to those that are so small and light that they can float in water or almost fly (think of dandelions)—not to mention the use of fruits in seed dispersal. As we will see later in the chapter, many methods for dispersing seeds have evolved.

TAKE-HOME MESSAGE 12•5

Seeds are the way that plants give their offspring a good start in life and get them to leave home. A seed contains a multicellular embryo plus a store of carbohydrate and other nutrients. Seeds are distributed by wind, by attaching themselves to animals, or by floating through air or water.

12•6

With the evolution of the seed, gymnosperms became the dominant plants on earth.

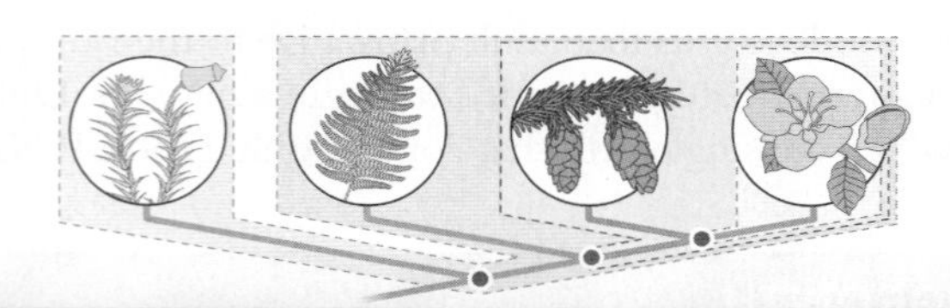

Gymnosperms include four major groups: the conifers, cycads, gnetophytes, and ginkgo (just one species) (**FIGURE 12-13**). All of the 900 or more species of gymnosperms are seed-bearing plants that produce ovules on the edge of a

THE GYMNOSPERMS

COMMON CHARACTERISTICS
- Distribute water and nutrients throughout plant with a "circulatory system" of vascular tissue
- Commonly have needle-like leaves
- Reproductive structures called cones produce the gametes.
- Fertilization produces seeds.

MEMBERS INCLUDE
- Conifers (about 600 species)
- Cycads (about 300 species)
- Gnetophytes (about 65 species)
- Ginkgo (1 species)

FIGURE 12-13 Overview of the gymnosperms: non-flowering plants with seeds. Shown (*from left*) are ginkgo, Douglas fir, welwitschia, and cycad.

GROUPS OF GYMNOSPERMS

CONIFERS
- Most commonly found in colder temperate and sometimes drier regions of the world
- Important source of timber
- Include pines, spruces, firs, cedars, hemlocks, yews, larches, cypresses

CYCADS
- Slow-growing gymnosperms of tropical and subtropical regions
- Most resemble palm trees
- Several species are facing extinction in the wild.

GNETOPHYTES
- Composed of 3 groups: *Gnetum, Ephedra,* and *Welwitschia*
- Most gnetophyte species are *Ephedra*, a shrub-like plant sometimes used as a herbal remedy for respiratory ailments.

GINKGO
- *Ginkgo biloba* is the only remaining species.
- Distinctive fan-shaped leaves
- The outer covering of the seeds emits a foul odor.

FIGURE 12-14 Major groups of gymnosperms: conifers, cycads, gnetophytes, and ginkgos.

cone-like structure. If you were a dinosaur living in the middle of the Mesozoic era, 160 million years ago, the forests you walked through would consist entirely of gymnosperms—pine trees, redwoods, cycads, and their relatives. Flowering plants (angiosperms) would not appear for another 35 million years. Indeed, when it comes to population size and the range of habitats in which they can thrive, pine trees and their relatives are among the most evolutionarily successful groups of plants on earth, growing on every ice-free continent and extending from sea level to the tree line on mountains (**FIGURE 12-14**).

Pines, spruces, firs, redwoods and their relatives—collectively, the conifers—are the most familiar gymnosperms, at least to residents of the temperate regions, and many have needle-like leaves—but that is not true of all gymnosperms. The ginkgo has distinctive fan-shaped leaves that are nearly identical in size and shape to fossils of ginkgo leaves (see Figure 12-14). Cycads have palm-like fronds with many small leaflets, and these, too, are nearly identical to fossils from the Mesozoic.

Q Why does tree pollen coat the windshield of cars parked outside in the spring?

The reproductive structures of gymnosperms—the cones—are male or female. The pine cones you are familiar with are the female cones, and these produce the ovules and, eventually, the seeds (**FIGURE 12-15**). The male cones are smaller and release pollen that is blown

CONES

MALE CONES
The male cone releases pollen grains that require wind to reach a female cone.

FEMALE CONE
The female cone has ovules on the protruding scales. They produce seeds when fertilized by pollen.

Wind dispersal of pollen is highly inefficient. For each grain that fertilizes an egg, billions of pollen grains are wasted. Nonetheless, pines have been using wind pollination successfully for more than 200 million years.

FIGURE 12-15 Cones are the reproductive structures of gymnosperms.

GYMNOSPERM LIFE CYCLE

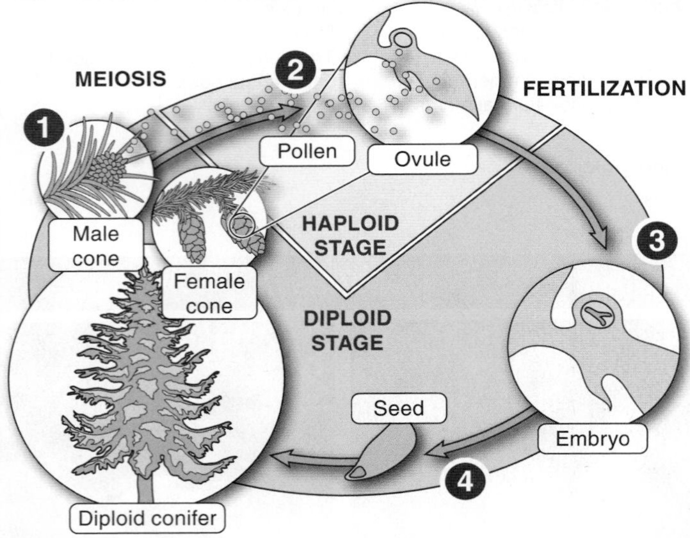

1. Male cones release pollen grains that are dispersed by the wind to ovules found beneath the scales of female cones.
2. Pollen grains release sperm that fertilize an egg within the ovule.
3. Fertilization creates a diploid embryo that matures into a seed.
4. Eventually, the seed is released from the female cone, and grows into an adult tree.

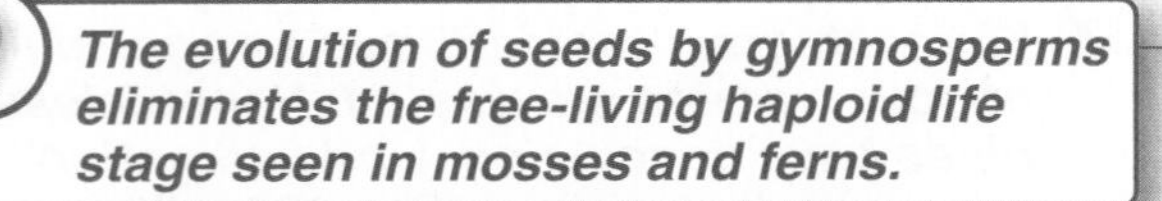

FIGURE 12-16 Assisted by the wind: the life cycle of the gymnosperms.

by the wind, and some of it reaches the ovules, which lie beneath the protruding scales of the female cones. The quantity of pollen released by conifers is beyond imagination—the air in a pine forest becomes hazy and every surface becomes coated with the yellow pollen. Wind dispersal is clearly an inefficient method of **pollination** (getting the pollen to the vicinity of the ovule): billions of pollen grains are wasted for each grain that lands on a female cone and produces sperm to fertilize an egg in an ovule. Nonetheless, this "brute force" method of ensuring fertilization works: pines have been using wind pollination successfully for more than 200 million years.

When the pollen arrives at the female cone, a pollen tube forms and transports the haploid sperm to the ovule, where fertilization occurs and a diploid embryo begins to grow. The embryo develops slowly within the female cone over the course of many months, until the scales of the cone open and release the seed, ready to sprout into a new plant (**FIGURE 12-16**). The evolution of seeds in gymnosperms almost completely eliminated the haploid life stage that we see in mosses and ferns. Non-vascular plants spend most of their lives in the haploid stage of their cycle, and ferns produce spores that grow into free-living haploid plants. With the appearance of gymnosperms, plants developed a life history with no free-living haploid stage.

This evolutionary change in life history probably reflects the different degrees of evolutionary fitness of haploid versus diploid organisms. A haploid organism has just one copy of each chromosome, which means it has just one copy of each allele. If the haploid organism carries a defective allele for a critically important gene, the organism is doomed because it has only that one, defective version of the gene. In contrast, diploid organisms have two sets of chromosomes and thus two copies of each allele. If one copy is defective, the second copy can function as a backup, enabling the plant to produce the gene product. As a result, mutations are much less likely to be lethal for diploid organisms than for haploid organisms.

TAKE-HOME MESSAGE 12•6

Gymnosperms (pine trees and their relatives) were the earliest plants to produce seeds, and this mode of reproduction offers advantages over the spores of earlier plants. Seeds gave gymnosperms the boost they needed to become the dominant plants of the early and middle Mesozoic era. Gymnosperms depend on wind to carry their pollen—not an efficient method of dispersal. Conifers protect the developing seeds in the female cone.

12•7

Conifers include the tallest and longest-living trees.

FIGURE 12-17 Towering and ancient gymnosperms.

In addition to being abundant, diverse, and widespread, the cone-bearing trees—specifically, the conifers—include both the tallest and the oldest living organisms on earth. The four tallest trees in the world are conifers: a coast redwood that is 380 feet (113 meters) tall, a Douglas fir and a Sitka spruce, each 318 feet (97 meters), and a Sierra redwood at 311 feet (95 meters). The oldest tree trunk belongs to a Great Basin bristlecone pine (**FIGURE 12-17**). The current record holder, the Methuselah tree, has lived for more than 4,800 years, and a slightly older tree was cut down in 1964. There's more to a tree than its trunk, though, and trunks can be replaced. That is what a Norway spruce in Sweden has been doing: the current trunk is about 600 years old, but the roots are 9,550 years old. This tree has persisted by sending up a new trunk each time the old one dies. But not all conifers are big; there also are miniature species of conifers, such as the shore pine, which can be just 20 centimeters tall.

Trees can grow large and live to great ages because woody plants can be exceptionally strong and resistant to attack by herbivores. A cross section of a tree trunk shows the structural characteristics that allow a tree to stand erect and transport water to leaves more than 100 meters above the ground. Rigidity is provided primarily by the heartwood, which is a core of dead tissue that contains complexly cross-linked molecules.

Bark covers the outside of the tree trunk and branches, and the outer layers of bark are dead tissue that can be shed without damage to the tree, protecting the living tissue from attack by plant-eating insects. Part of the success of conifers can also be traced to their ability to defend themselves by exuding a sticky pine pitch that can engulf and smother insects.

TAKE-HOME MESSAGE 12•7

Conifers are the success stories among gymnosperms, with more species and a larger geographic range than all of their relatives combined. Rigidity, an exterior layer of bark, and the ability to exude sticky pitch protect conifers, allowing them to grow taller and reach older ages than any other plants.

④ Flowering plants are the most diverse and successful plants.

Bleeding heart flowers, surrounded by neat, heart-shaped leaves.

12•8

Angiosperms are the dominant plants today.

The appearance of flowering plants (angiosperms) about 135 million years ago set the stage for the botanical world we know today, with flowering trees, flowering bushes, and all the grasses and herbaceous (non-woody) plants we see around us. The vast majority of plants on earth are flowering plants in the angiosperm group. Many of the early flowering plants would look familiar to us, and angiosperms dominate the plant world now, with some 250,000 species, compared with approximately 1,000 species of gymnosperms (**FIGURE 12-18**).

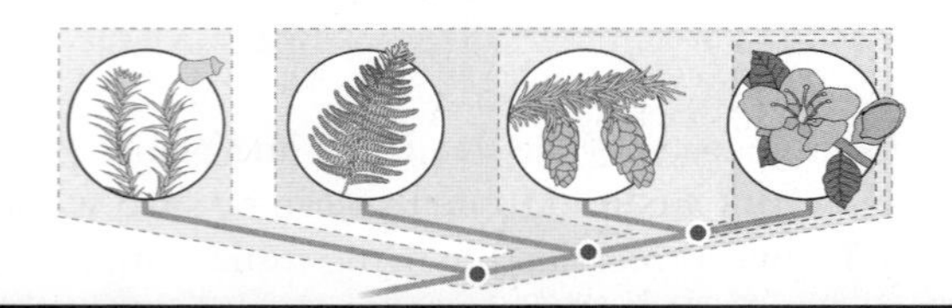

THE ANGIOSPERMS

COMMON CHARACTERISTICS
- Distribute water and nutrients throughout plant with a "circulatory system" of vascular tissue
- Produce flowers, which produce gametes
- Seeds are enclosed within an ovule

MEMBERS INCLUDE
- Flowering trees, bushes, herbs, and grasses (about 250,000 species)

FIGURE 12-18 Snapshot of the angiosperms. The angiosperms shown here (*from left*) are water lily, apple tree, Gewürztraminer grapes, and rhododendron in a forest.

Flowers come in a bewildering variety of sizes, shapes, and colors, but they all have similar structures: a supporting stem with modified leaves—the flashy petals and the sepals, which are a (usually) green wrapping that encloses the flower while it is in bud. Most plants combine the male and female reproductive structures in the same flower. The male structure is the **stamen** and includes the **anther,** which produces the pollen, and its supporting stalk, the **filament.** The female reproductive structure is the **carpel.** It has an enclosed chamber, called the **ovary,** at its base, which contains one or more ovules in which eggs develop; a stalk (the style), extending from the ovary; and a sticky tip (the stigma) (**FIGURE 12-19**).

Pollination, in angiosperms, is the transfer of pollen from the male reproductive structures to the female reproductive structures of a flower, sometimes the same flower, and sometimes a flower on a different individual. Pollination is a multi-step process in which a pollen grain sticks to the stigma, forms a tube that grows until it reaches the ovule, and thus provides a route for sperm to travel down the tube to fertilize the egg. In the next section, we explore the tremendously rich and varied methods by which plants get the male gametes to the female gametes.

Although most flowers contain both male and female structures, several thousand species of plants produce flowers that are only male or only female. Sometimes, male and female flowers are borne on different individual plants, but in quite a few plants, the same plant has some male flowers and some female flowers, often side by side. Maize (corn) is a familiar plant with male and female flowers in different places on the same plant, although you might not recognize the reproductive structures as flowers. The tassel at the top of the plant is the male flower, and the ear is formed by female flowers.

The lilies in this photo clearly show the carpel, the female reproductive structure in the center of the flower, surrounded by the male stamens, each topped with a brown anther.

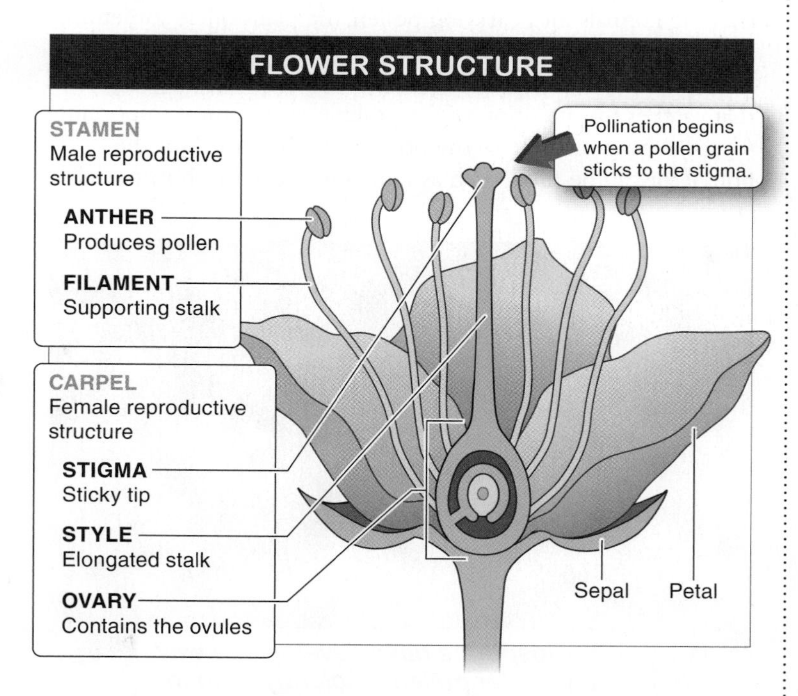

FIGURE 12-19 **A flower houses a plant's reproductive structures.**

TAKE-HOME MESSAGE 12•8

Flowering plants appeared in the Cretaceous period, about 100 million years ago, and diversified rapidly to become the dominant plants in the modern world. A flower houses a plant's reproductive structures. Most flowers have both male and female structures, but some species have flowers with only male or only female reproductive structures.

12•9

A flower is nothing without a pollinator.

Flowers may play only a tiny role in the sex lives of humans—did you give or receive any last Valentine's Day?—but they play the lead part in the sex lives of most plants. Here we examine how it is possible for individual plants, rooted in the ground in different locations, to bring their gametes together in sexual reproduction.

Picture a flower. What does it look like? Not everyone forms the same image in his or her mind, because flowers differ hugely from one to another, and we all have our favorites. There are giant sunflowers and colorful roses and some rotten-smelling orchids and lilies, just to name a few. The label "flower" would seem to indicate little beyond the fact that flowers are usually brightly colored, frequently have a prominent odor, and generally sit at the end of a plant's branches. Closer inspection, however, reveals that, with very few exceptions, all flowers have the same four fundamental structures introduced in Section 12-8—sepals, petals, stamens, and carpels.

Fertilization occurs when the male gamete merges with the female gamete. When it comes to plants, that's a lot easier said than done. Before fertilization can occur, the male gamete has to get to where the female gamete is. The first step toward this is for a pollen grain to make the journey to the stigma. This is the step referred to as pollination.

A small number of angiosperm species achieve pollination by simply releasing their pollen into the wind, as do the gymnosperms, or into water—on the slim chance that through luck (random movement in air or water), some of the pollen will land on the female reproductive organs of another plant. Given the really minuscule probability of any one pollen grain actually doing that, wind- and water-pollinated plants respond in the only reasonable way: they produce tremendous amounts of pollen. As we saw earlier, although this method is not particularly efficient, it works. Most angiosperms, though, have a different way of moving pollen from the anthers of one flower to the stigma of another: they use animals to carry it (**FIGURE 12-20**).

Q Why are flowers so flashy?

To ensure that animals will be willing to visit a flower, picking up and delivering their pollen cargo, two different and clever strategies for achieving pollination have evolved among the flowering plants.

1. *Trickery:* The plant deceives some animals into carrying its pollen from one plant to another. Among the tricksters are some orchid species that achieve pollination by producing flowers that resemble female wasps. The mimicry is so good that male wasps mount the flower and attempt to fly off and mate with it. Because the flower is attached to the plant, however, the male wasp resembles a cowboy riding a bucking bronco, twirling wildly on the flower and repeatedly whacking his head against the strategically located anthers. In the process, he doesn't end up having any reproductive success, but he does succeed in getting pollen stuck all over his head and body. That is not enough for the plant to achieve pollination—but it's a start. If that male wasp gets fooled again by a flower on another orchid, when he mounts that flower and tries to mate, he will inadvertently deposit some of the pollen from his body onto the also strategically placed stigma of that flower. In the end, the wasp looks a bit foolish and does not gain from his actions, but the orchids have used an effective system of pollination.

2. *Bribery:* The plant bribes some animals to carry its pollen from one plant to another. This more common strategy for achieving pollination is less selfish than trickery. Rather than tricking animals into carrying pollen, the plant offers something of value to the animal. For this bribery method of pollination to work, the plant must produce (a) a sticky pollen, (b) a flower that catches the attention of the pollinator, and, most important, (c) something of value to the pollinator. The payoff for the pollinator can be food, such as nutritious nectar rich in sugars

FIGURE 12-20 Delivering precious cargo (inadvertently). The bee in this photo is completely covered in yellow pollen grains.

COEVOLUTION: FLOWERS AND THEIR POLLINATORS

COLORS AND PATTERNS
WHITE: Nocturnal pollinators, such as moths and bats
BRIGHT: Visually oriented, diurnal pollinators, such as birds, butterflies, and bees

FLOWER STRUCTURE
TUBE: Pollinators with long tongues, such as moths
INTRICATE/CLOSED: Pollinators such as bees

ODORS
SWEET: Pollinators with a good sense of smell, such as moths, butterflies, and bees
STINKY: Pollinators, such as flies, looking for rotten meat on which to lay eggs
NO ODOR: Pollinators with a poor sense of smell, such as birds

NECTAR
ABUNDANT: Pollinators with high energy needs, such as bees, birds, and butterflies
NO NECTAR: Pollinators, such as flies, looking for a place to lay eggs, or such as beetles, looking for petals, pollen, and other parts to eat

FIGURE 12-21 Evolving together: plants and pollinators. You can often determine the type of animal that pollinates a flower just by examining the features of the flower.

and amino acids, or perhaps a safe, hospitable location for an insect to lay its eggs.

The variety of flower structures is tremendous. They differ in shape, color, smell, time of day they are open, whether or not they produce nectar, and whether or not their pollen is edible. And just as the variety of flower types is wide, so, too, is the variety of pollinators: birds (mostly hummingbirds), bees, flies, beetles, butterflies, moths, and even some mammals (mostly bats) (**FIGURE 12-21**). In each case, there has been strong coevolution between the plants and their pollinators: the plants get more and more effective at attracting the pollinators and deterring other species from visiting the flower, while the pollinators get more and more effective at exploiting the resources offered by the plants. Because of this strong coevolution between the plant and animal species, for most flowers we can determine, just by examining the features of the flower, the type of animal that will pollinate it.

Pollination is just one step toward fertilization. Fertilization itself doesn't happen until the male and female sex cells meet and fuse. This can only occur after the male gamete is taken to the female gamete by a pollen tube, which grows from the stigma, where the pollen grain lands, down into the ovary. We explore this process in the next section.

TAKE-HOME MESSAGE 12•9

Angiosperms have found a way to transfer pollen efficiently from the anthers of one flower to the stigma of another—let an animal carry it. Flowers are conspicuous structures that advertise their presence with colors, shapes, patterns, movements, and odors. Using these devices, plants are able to trick or bribe animals into transporting male gametes to female gametes, where fertilization can occur.

12•10

Angiosperms improve seeds with double fertilization.

The process of fertilization—the fusion of two gametes to form a zygote—*begins* when a pollen grain lands on a stigma of a flower, but that is far from the end. In angiosperms, the process is called **double fertilization** because there are two separate fusions of male nuclei (each carrying a complete set of the organism's genetic material) from the pollen grain with female nuclei in an ovule. As we will see, double fertilization is a more efficient system than fertilization in gymnosperms because whenever an embryo is produced at fertilization (and only then), so, too, is a more substantial, ready-made food source.

When we left off in the previous section, a pollen grain—which will deliver the haploid male gamete—had just arrived at the stigma, in the female reproductive structure. The pollen grain forms a pollen tube that extends downward through the stigma and style to the ovary, and ultimately enters an ovule. The pollen grain forms two haploid sperm by mitosis, both of which move down the pollen tube (**FIGURE 12-22**).

While the pollen tube is growing, a haploid spore in the ovule divides mitotically to form a total of seven haploid cells, but only four of these participate in fertilization and formation of a seed. The remaining three are degenerated. One of the five cells is the female gamete (the egg), and this remains near the place where the pollen tube will be guided into the ovule by two of the other haploid cells, which aid in fertilization. The remaining haploid cells of the four participating in fertilization move into the middle of the ovule. So, when the pollen tube reaches the ovule, one of the two sperm fuses with the egg to form a zygote. The other sperm fuses with the two nuclei in the middle of the ovule to form the endosperm, which has *three sets* of chromosomes (i.e., is triploid). The process is called double fertilization because two sperm enter the ovule and combine with haploid female cells in two separate events, forming (1) a zygote (with two sets of chromosomes) and (2) an endosperm (with three sets of chromosomes).

The final steps in producing a seed occur as the diploid zygote cell undergoes multiple mitotic divisions to form an embryo, while the triploid cells multiply mitotically to produce the endosperm that will provide nutritional support for the seedling through its initial growth stages. The outer layers of the ovule form the seed coat that will protect the seed until it sprouts. The enclosure of the seed within the ovary is a distinction between angiosperms and gymnosperms. The seeds of a gymnosperm are unenclosed and are sometimes referred to as "naked."

ANGIOSPERM LIFE CYCLE

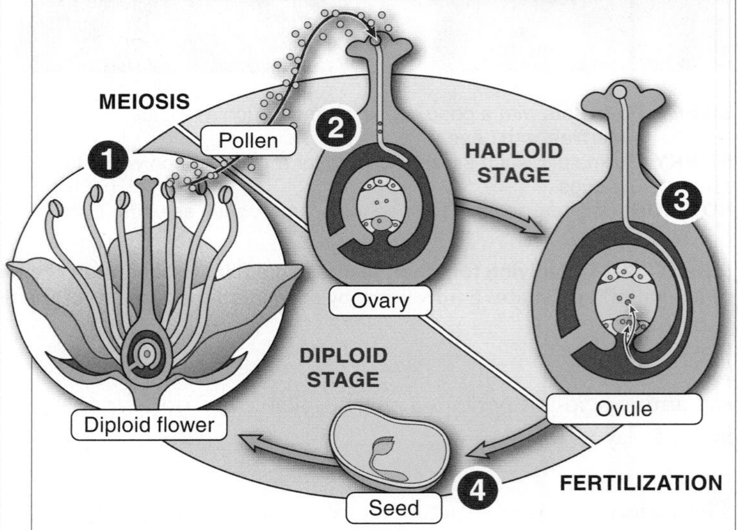

1 Male anthers release haploid pollen grains that are delivered to the stigma of another flower.

2 A pollen grain produces a tube that extends through the stigma to the ovary. Within the ovule of the ovary, a spore divides into seven haploid cells. One cell becomes the egg, another cell—with two nuclei—will form the endosperm following fertilization by a sperm cell.

3 Two sperm are released by the pollen grain. In a process called double fertilization, one sperm fuses with the egg to form a zygote, while the other sperm fuses with the two nuclei of the endosperm-forming cell.

4 The zygote and the endosperm continue to develop within the ovule, forming a seed that will eventually be released and grow into a mature plant.

FIGURE 12-22 Double fertilization fortifies the seeds of angiosperms.

What is the advantage of such a complex method of fertilization? Gymnosperms do not do this—a single sperm fuses with an egg to form a zygote in an ovule that contains hundreds of other female cells that are used as a source of nutrients for the resulting embryo.

Q What advantage does double fertilization give to angiosperms?

There are two important advantages to the angiosperm method.

1. Double fertilization initiates formation of endosperm only when an egg is fertilized. Waiting to be sure that an egg is fertilized is a good strategy, because making endosperm is a large energy investment for a plant. Gymnosperms invest that energy up front, and nutrients in ovules that are not fertilized are wasted because those "seeds" do not contain an embryo. In contrast, angiosperms do not waste energy forming endosperm in ovules that will not contain embryos.

2. Angiosperms can produce smaller gametes than gymnosperms, because the large energetic reserves will be produced only *after* fertilization occurs. The small size of the male and female gametes of angiosperms ensures that seeds are produced quickly. **FIGURE 12-23** shows that as different reproductive strategies have evolved in plants, their gametes have become progressively smaller. Rapid production of seeds allows angiosperms to grow as annual plants (i.e., plants that complete their life cycle from sprouting to seed production in one growing season), which is something gymnosperms cannot do.

HAPLOID AND DIPLOID LIFE STAGES

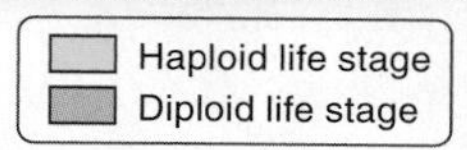

NON-VASCULAR PLANTS
The majority of the life cycle is spent in the haploid stage.

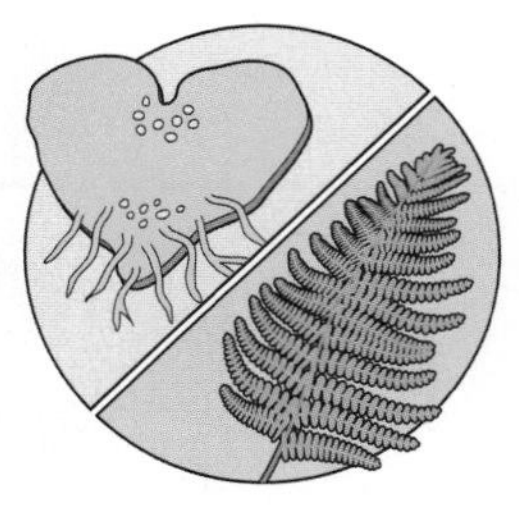

VASCULAR SEEDLESS PLANTS
The haploid and diploid stages are both multicellular and physically independent from one another.

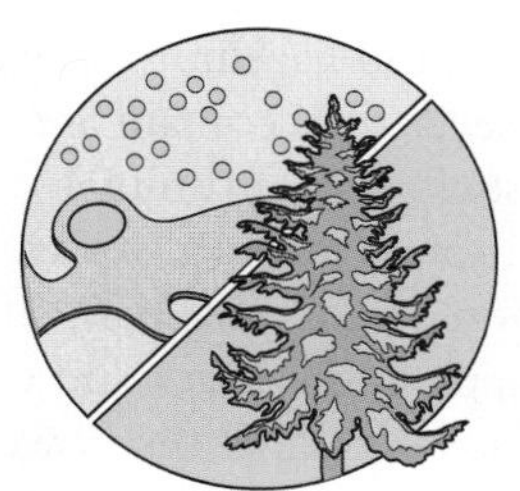

GYMNOSPERMS
The evolution of seeds in gymnosperms almost completely eliminates the prominent haploid stage seen in mosses and ferns.

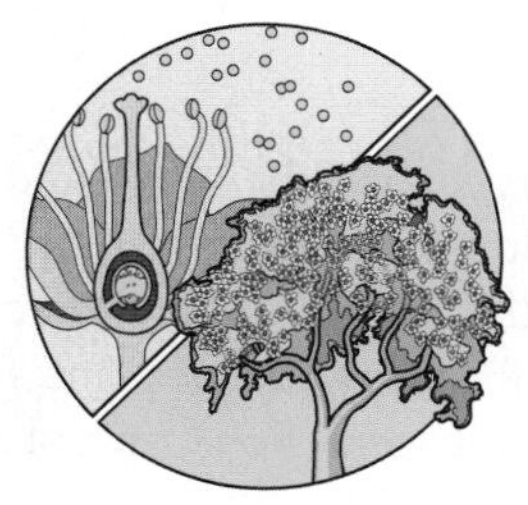

ANGIOSPERMS
Haploid gametes are further reduced in size, enabling more rapid seed production.

As plants have developed different reproductive strategies, they have progressed from having a prominent haploid stage of life to simply having haploid gametes.

FIGURE 12-23 Overview of the haploid and diploid stages of plant life cycles.

Outbreeding, the combination of haploid cells from two different individuals, produces offspring with greater genetic diversity (i.e., carrying a greater diversity of alleles) than offspring that result from inbreeding, the combination of a male and female gamete from the same individual. Plants have a variety of ways to ensure that only sperm from another individual will fertilize the female gamete. Plants that have male and female reproductive structures in different flowers or even on different individuals increase the chance that a pollen grain landing on a stigma comes from a different plant. But many flowers with both male and female parts also have mechanisms to prevent self-fertilization. For example, the anthers may mature before the stigma, so the stigma is not ready to receive pollen when the anthers are active. And when the stigma becomes functional, the anthers are no longer producing pollen. Other plants use a molecular recognition system to prevent inbreeding: proteins on the surface of the stigma will not allow pollen from the same individual to form a pollen tube.

Q Do flowers with both male and female structures fertilize themselves?

TAKE-HOME MESSAGE 12·10

Angiosperms undergo a process of double fertilization, which ensures that a plant does not invest energy in forming endosperm for an ovule that has not been fertilized. Angiosperms have also developed methods to reduce the occurrence of self-fertilization, thereby ensuring greater genetic variation among offspring than through inbreeding.

5 Plants and animals have a love-hate relationship.

Fruits like these peaches are structures that contain fertilized seeds.

12•11

Fleshy fruits are bribes that flowering plants pay animals to disperse their seeds.

Leaving home is an inevitable part of growing up. But as difficult as it may be, imagine you had to move away but didn't have a car or moving van—or even legs. How would you do it? This is yet another issue facing plants. Ingenious methods have evolved for transporting the male gametes to the female sex organs for fertilization, but that still results in all of the fertilized eggs, and all the new offspring they develop into, living right on the female plant—or in nearly the exact same physical location, if they manage to fall off the parent plant. This is a problem, because the new offspring must leave home. Otherwise, parent and offspring will end up competing for the same light, space, soil nutrients, and other resources.

The answer to this problem is the fruit, a structure that aids in dispersing seeds—the reproductive packets made up of the embryo, some food reserves, and a hard coat. The fruit usually develops from the ovary (and sometimes from nearby tissue), right around the seeds, which develop from the ovules. This should cause you to look at fields of flowers differently: given that the ovary is part of every flower, all of the flowers you see, after they are pollinated and fertilized, will turn into fruits. Or conversely: every fruit you eat used to be a flower.

Pea pods, sunflower seeds, corn kernels, and hazelnuts are fruits. These fruits are "dry fruits," and the seeds they contain are transported by wind or water or by sticking to an animal (**FIGURE 12-24**). Many angiosperms, though, make a **fleshy fruit** that consists of the ovary plus some additional parts of the flower. For example, the core of an apple is the ovary and the flesh of the apple is derived from adjacent parts of the flower. Blueberries, watermelons, oranges, tomatoes, and peaches are other examples of fleshy fruits.

The fleshy part of a fruit is often larger than the ovary, and a plant invests a lot of energy in producing fleshy fruit. The payoff comes when an animal eats the fruit and then defecates the seeds at a location far removed from the parent plant. In other words, fleshy fruit is the bait that some flowering plants use to get animals to disperse their seeds.

When you look at fruits, you can see several characteristics that help this system to work:

- Fruits are colorful—red is the most common color for fruits, followed by yellow and orange. All of those colors contrast with green foliage and make fruits conspicuous.

SEED DISPERSAL

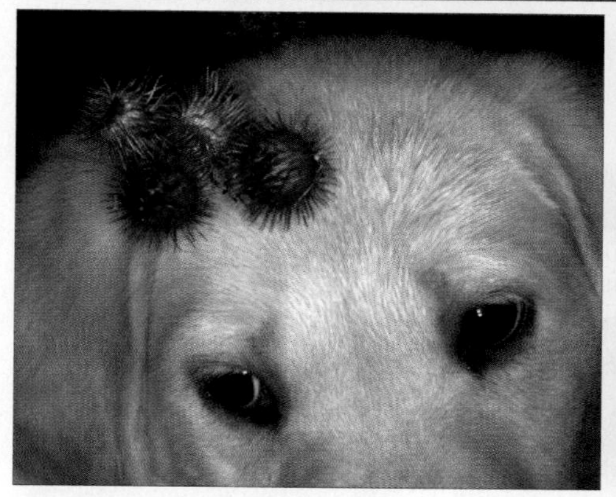

HITCHING A RIDE
Seed pods have spines or projections that attach them to passing animals.

FLYING AND FLOATING
The structure of the seed allows it to be carried away from the parent plant by wind or water.

PROVIDING A FOOD SOURCE
Fleshy fruit is a form of bait that lures an animal to eat the seed and carry it far from the parent plant before eliminating it.

FIGURE 12-24 Methods of seed dispersal.

- Fruits taste good—plants dump sugars into fruits, and their sweetness appeals to many animals.
- Fruit is good for animals—many birds incorporate the red and yellow colors of fruits into their own color patterns, and colorful male birds are appealing to female birds. Thus, male birds that eat fruit are likely to produce a lot of offspring who also eat fruit—a good deal for the plant.

Q Can seeds still sprout after being eaten by an animal?

Of course, there's one more requirement to make fruit a successful way for a plant to disperse its seeds: the seeds must survive passage through the animal's digestive tract and be able to germinate when they come out. That's not a problem—seeds are fully viable when they emerge. It's easy enough to test this yourself. The next time you eat fresh tomatoes, check the toilet about 24 hours later. The seeds that have passed through your digestive system are still intact. If you want to test their viability, that's easy—recover a few, rinse them off, and put them on a wet paper towel in a sealed plastic bag. If you want to do a controlled experiment, you can take some seeds directly from a tomato and put them in another bag. Put both bags in a warm, dark place, and check them in a couple of days. You'll find that both sets of seeds have sprouted. (Or, you can just take our word for it.)

Seeds not only survive being eaten by animals—some seeds will not germinate *unless* they are eaten. For some plants, the pulp that surrounds the seeds in the fruit inhibits germination and it must be digested away before the seed will sprout. The seeds of other plants have chemicals in the seed coats that must be removed by the acidity of an animal's stomach before the seeds will germinate. And there's another plus to this system: the seeds are deposited with a bit of manure that provides nutrients for the seedlings.

TAKE-HOME MESSAGE 12•11

Following pollination and fertilization, plants often use the assistance of animals to disperse their fruits, which contain the fertilized seeds, depositing them at a new location where the seedlings can grow. Fruits are made up from the ovary and, occasionally, some surrounding tissue.

12 • 12

Unable to escape, plants must resist predation in other ways.

The interactions of plants and animals are not simple. On the one hand, plants depend on animals to pollinate their flowers and disperse their seeds. On the other hand, many kinds of animals eat plants, and plants are vulnerable because they can't run away. If you have ever been poised above a dunking tank as your friends throw softballs at a target, hoping to dump you into the water, you know how vulnerable a plant feels.

Because plants can't run away from plant-eating animals, a host of defensive devices have evolved in plants, which fall into two categories: anatomical structures, such as thorns, and chemical compounds, including hallucinogens (**FIGURE 12-25**).

Spines, prickles, and thorns are a common way to discourage herbivores, and some of them are impressively large. The acacias, a group of plants that grow as large bushes or small trees, are notoriously spiny. In these plants, the deterrent effects of spines are sometimes enhanced by the presence of ants that live in the swollen bases of the spines and fight off other animals, from insects to mammals, including humans, that would feed on their host acacia. Spines on palm fronds carry pathogenic bacteria that infect the wounds that the spines create, providing a long-lasting reminder that it's not a good idea to try to eat a palm frond. In most plants, the surfaces of leaves and stems have microscopic hairs called trichomes. Most trichomes protect plants against insect herbivores by exuding a sticky fluid that traps the insect, but stinging plants such as nettles use larger trichomes to inject a potent fluid into any animal that brushes against a leaf.

Many chemical defenses are substances that make plants bitter or otherwise unpalatable, but some plants play hardball by producing chemicals that affect the nervous or reproductive systems of animals that eat them. Locoweeds, which occur all over western North America, contain substances called alkaloids that can cause permanent damage to the nervous system. When cattle and horses eat locoweed, they become lethargic and stop feeding. Other defensive chemicals are hallucinogens. Tetrahydrocannabinol, or THC, is found in the leaves and buds of marijuana plants, and the disorientation it produces causes mammals to stop eating.

PLANT DEFENSES

Because plants can't run from plant-eating animals, they have developed many defensive features.

ANATOMICAL STRUCTURES
Some plants have spines, spikes, and thorns that deter predators. Some trees have thick layers of bark that are shed in order to get rid of attacking insects.

STICKY TRAPS
Conifers exude pitch, a sticky substance that can engulf and smother attacking insects.

CHEMICAL COMPOUNDS
Some plants synthesize chemicals that induce physiological and behavioral changes in the animals eating them.

FIGURE 12-25 Spines, sticky traps, and toxic compounds. Plants have developed a range of defenses against predation.

Many of the chemicals that plants produce to protect themselves have physiological effects, and for thousands of years humans have used plants for medicinal purposes. The aboriginal peoples of North America and Eurasia, for example, knew that extracts of the bark from willows and aspens could relieve pain. Salicin is the compound in the bark that relieves pain, and we still use acetylsalicylic acid (aspirin) for pain relief. Opium poppies produce opium, a more powerful pain reliever; foxglove flowers contain digitalin, a compound that slows the heart rate; and the ipecacuanha tree produces ipecac, a compound that causes vomiting.

The term "bioprospecting" is applied to the search for plants with medicinal uses, and representatives of pharmaceutical companies trudge through rain forests and deserts to find compounds with medicinal applications. The rewards from such a discovery can be great. For example, artemisin from Chinese sweet wormwood is effective in treating malarial infections that are resistant to conventional pharmaceutical compounds.

Although plants can't move, they do send messages. They release volatile chemicals when they are attacked by insects—a volatile chemical is one that floats in the air—and the chemicals can warn nearby plants of the impending threat or can call protective insects to their aid.

When plants are nibbled by insects, they release the chemical methyl jasmonate (MJ), which is carried with the breeze to nearby plants. When these plants detect the presence of MJ, they ramp up production of defensive chemicals so that they are prepared if the insects move onto them. Insects can also detect MJ, and insects that prey on other insects are attracted to plants emitting the chemical, because it signals the presence of prey. Release of MJ by lima beans and apple trees that are being attacked by spider mites summons predatory mites. When corn and cotton plants are attacked by caterpillars, release of MJ summons parasitic wasps that deposit their eggs inside the caterpillars. When the eggs hatch, the wasp larvae eat the caterpillars from the inside out.

About 900 species of plants turn the tables and eat insects. They use the protein in their insect meals to supplement the nitrogen compounds they get from the soil. Insectivorous plants are most common in boggy areas, because boggy soil often has low nitrogen concentrations. (**FIGURE 12-26**). Pitcher plants are an example—their name comes from conspicuous pitcher-shaped structures that capture insects. It's easy for an insect to enter the open top of a pitcher, but hard to get out, because the inner walls are lined with hair-like structures that point downward, forcing the insect into a pool of liquid. Some pitcher plants merely absorb the nitrogen that is released when the trapped insects decay, but other species secrete digestive enzymes into the pool. These enzymes, which are basically the same as the enzymes in your stomach, digest the insects and make the nitrogen available more rapidly.

FIGURE 12-26 **Carnivorous plants feed on insects.**

TAKE-HOME MESSAGE 12·12

Plants have a wide range of defenses against herbivorous animals: from physical defenses such as thorns to chemicals that have complex effects on animals' physiology. Plants respond to insect attack by synthesizing chemicals that make the plant less palatable. Some plants living in soil that is deficient in nitrogen have switched roles, preying on insects.

⑥ Fungi and plants are partners but not relatives.

*An elegant purple fan: a view of the underside of the fungus amethyst deceiver (*Laccaria amethystea*).*

12•13

Fungi are closer to animals than they are to plants.

Think about your toes for a minute. How do they feel? Do they itch? Maybe there's a burning sensation between the toes? Hopefully, they feel fine, because itching and burning are the symptoms of athlete's foot. One study found that 15% of all people in the United Kingdom had athlete's foot, and almost everyone has it at some time during his or her life. The reason we bring up this unpleasant topic is that athlete's foot is a fungal infection of the skin, and it is the encounter with a fungus that many people remember most clearly.

Fungi make up their own monophyletic kingdom within the eukarya domain. Most fungi are multicellular, sessile decomposers (**FIGURE 12-27**). Although they were originally thought to be plants lacking chlorophyll, it turns out that they

WHAT IS A FUNGUS?

FUNGI ARE DECOMPOSERS OR SYMBIONTS
Fungi acquire energy by breaking down the tissues of dead organisms or by absorbing nutrients from living organisms.

FUNGI ARE SESSILE
Fungi are anchored to the organic material on which they feed.

FUNGI HAVE CELL WALLS MADE OF CHITIN
Fungi have cell walls made of chitin, a chemical that is also important in producing the exoskeleton of insects.

FIGURE 12-27 **Defining characteristics of fungi.**

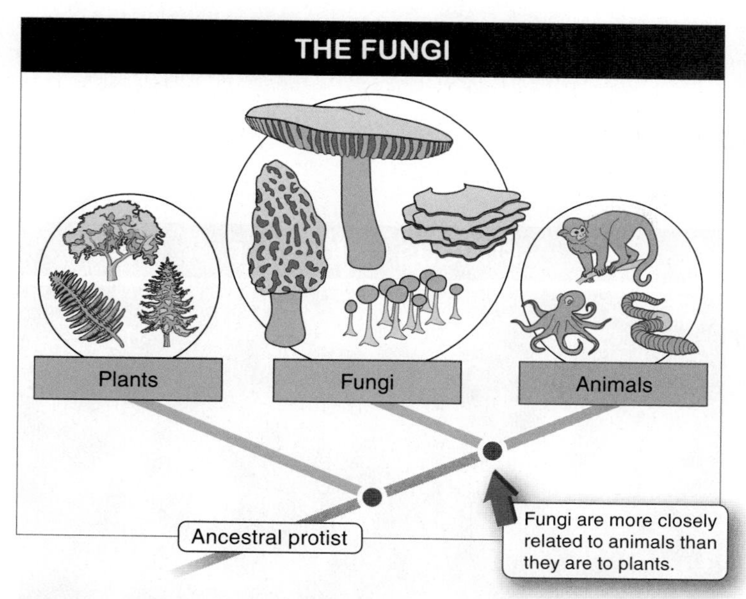

FIGURE 12-28 Phylogeny of the fungi.

have little in common with plants. In fact, DNA sequence comparisons reveal that fungi are more closely related to animals than they are to plants (**FIGURE 12-28**). As eukaryotes, fungi have all the basic cellular components you would expect to find: nuclei, mitochondria, an endomembrane system, and a cytoskeleton. They also have cell walls, but the cell walls, instead of including cellulose, as in plants, are made of the carbohydrate chitin, a chemical also important in producing the exoskeleton of insects.

The most commonly encountered fungi are mushrooms, molds, and yeasts (the only single-celled fungi) (**FIGURE 12-29**). The fungus that causes athlete's foot is multicellular and consists of thread-like structures made up of long strings of cells called **hyphae** (pronounced HIGH-fee). Because of the thinness and length of the hyphae (*sing.* hypha), the fungal cells have an enormous surface area. This means they are very good at taking up nutrients from your skin, but it also makes them very vulnerable to drying out. The fungus responsible for athlete's foot grows best in moist places—like on the skin between your toes.

Unless you actually have an active athlete's foot infection right now, your most recent encounter with a fungus (or, at least, with evidence that a fungus was once present) was most likely when you last ate bread. Bread rises because yeast cells are mixed into the dough, which is then kept in a warm place while the yeast consumes sugar and produces carbon dioxide through fermentation. The carbon dioxide released by the yeast makes the dough swell into the shape we associate with a loaf of bread.

TAKE-HOME MESSAGE 12·13

Fungi are eukaryotes with the same internal cellular elements other eukaryotes have—and one distinctive feature: a cell wall formed from the carbohydrate chitin. Some fungi, the yeasts, live as individual cells; most of the other fungi are multicellular.

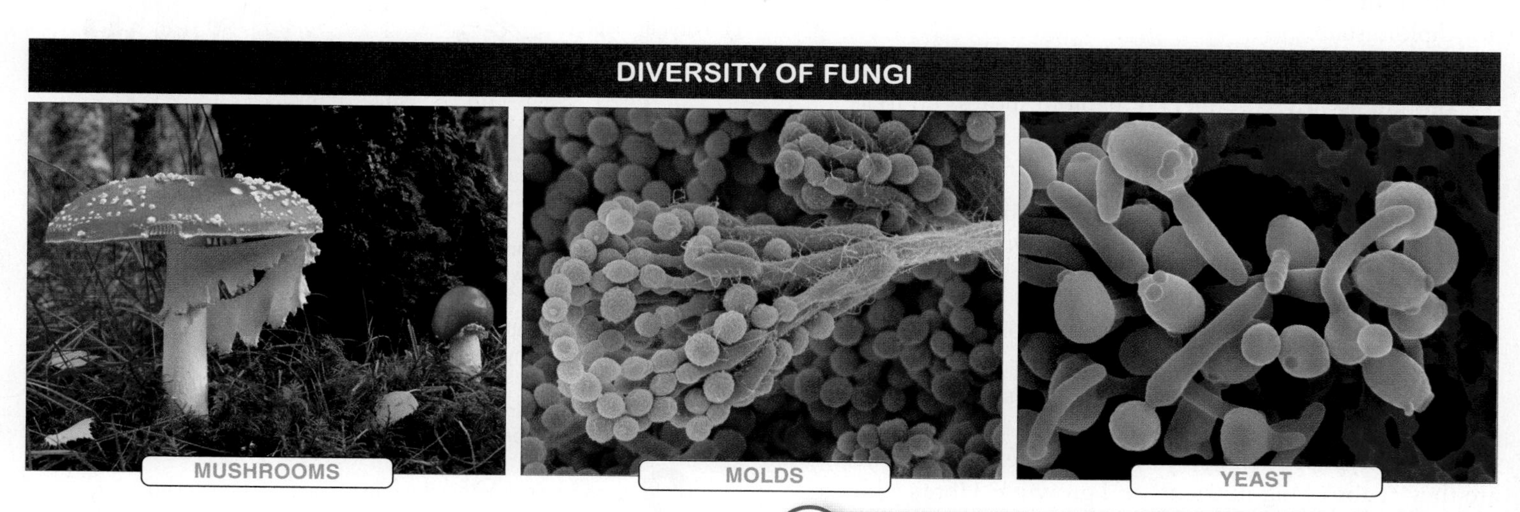

FIGURE 12-29 The fungi that humans know best. The mushroom shown here is a fly agaric; the mold is *Penicillium chrysogenum*; and the yeast is *Candida albicans.*

There are more than 1.5 million fungal species and they are found everywhere on earth: from the tropics to the arctic, on land and in the water.

12 • 14

Fungi have common structures, but exploit an enormous diversity of habitats.

As we've seen, most fungi are multicellular and are composed of long strings of barely visible, thread-like hyphal cells. The hyphae interconnect to form a mass of tissue called a **mycelium,** and this is the form in which a fungus spends most of its time, usually underground or in a decaying tree or log. Unless you dug down to reach the mycelium, you would not know a fungus was there.

The structure that most people associate with fungi is the mushroom. But a mushroom is just a temporary reproductive structure (or "fruiting body"), part of a complex reproductive cycle in some fungi (**FIGURE 12-30**):

1. Underground, genetically distinct haploid hyphae fuse. But the *nuclei* from these cells *don't* fuse, so instead of being diploid, they are "dikaryotic," meaning that each cell in the hyphae has two nuclei.
2. The dikaryotic hyphal state can last for years, with the dikaryotic mycelium growing and spreading.
3. At some point, a mushroom may form. It is produced from tightly packed dikaryotic hyphae.
4. In some cells in the mushroom, the haploid nuclei in dikaryotic hyphae fuse, putting the mushroom into a diploid state.
5. The diploid cells in the mushroom undergo meiosis, producing huge numbers of haploid spores (up to a billion in a single mushroom!), which are dispersed by wind or water or on the bodies of animals.
6. When the spores land in a hospitable place, they start growing as haploid hyphae and the cycle begins anew.

Fungi have an unusual and effective method of getting nutrition. Unlike humans, they digest their food outside their "body." While growing underground, hyphae secrete strong enzymes that break down the organic molecules around them. The hyphae then absorb those nutrients. How effective are they at absorbing nutrients and growing? Here's a better question: what is the largest living organism in the world? Not surprisingly (considering where we are asking the question), it is a fungus—specifically, a yellow honey mushroom fungus that covers an area of about 4 square miles

FUNGUS LIFE CYCLE

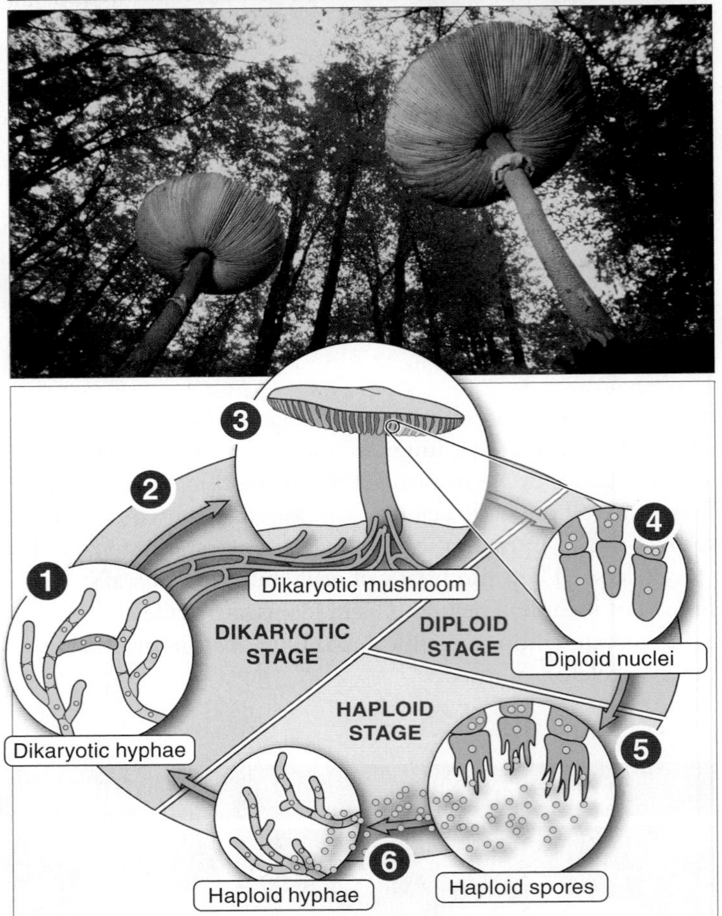

1. Underground, distinct haploid hyphae fuse to form dikaryotic hyphae, in which single cells have two distinct nuclei.
2. The dikaryotic hypha state can last for years, as the dikaryotic mycelium spreads throughout an area.
3. At some point, a reproductive structure—the mushroom—forms. It is produced from tightly packed dikaryotic hyphae.
4. In some cells within the mushroom, the haploid nuclei in dikaryotic hyphae fuse, becoming diploid.
5. The diploid cells within the mushroom then undergo meiosis, producing haploid spores.
6. When the spores land in a hospitable place, they start growing as haploid hyphae and the cycle is complete.

FIGURE 12-30 **The three-stage reproductive cycle of a fungus.**

(nearly 10 square kilometers) in eastern Oregon (**FIGURE 12-31**). This fungus is estimated to be at least 2,400 years old, and it may be more than 8,000 years old, which would place it in the category of the oldest living organisms. Huge as it is, though, the fungus is hard to detect because the mycelium is underground. The fungus appears on the surface only as mushrooms (edible, but not particularly tasty) and as hyphae in dying and dead trees. Indeed, no one even knew that this enormous fungus existed until 2000, when Oregon foresters sought an explanation for an epidemic of dead and dying trees in the Malheur National Forest.

Many mushrooms are gastronomic delicacies, and commercially grown portobello and shiitake mushrooms command high prices. The white button mushroom that you buy in plastic containers when you are making pizza is a species of *Agaricus*. Truffles are the underground reproductive structures of a fungus called *Tuber* that grows in forests. Truffles are not much to look at—they resemble small, dirt-covered potatoes—but the most prized variety, the white truffle, sells for $1,750 to $3,500 per pound. Because truffles grow underground, there is nothing on the surface to indicate their presence, and truffle hunters use trained pigs or dogs to locate them by scent.

Fungi can grow in many different habitats because, as **decomposers,** all they need for their nutrient supply is some sort of organic material that they can break down. Fungi play an enormously important ecological role in speeding the decay of organic material in forests. They don't need light, so they can grow underground or inside dead trees and logs. There, the hyphae release enzymes to digest organic material and then absorb the resulting nutrient-rich fluid.

FIGURE 12-31 The largest organism in the world: a yellow honey mushroom fungus. Shown here is just a small cluster of the fungus's temporary reproductive structures—what we call "mushrooms."

FIGURE 12-32 Cleaning out toxic mold. A building overrun with mold spores may cause its occupants to become ill.

Fungi can also thrive in poorly ventilated spaces in buildings (**FIGURE 12-32**). Molds are multicellular fungi that are responsible for many cases of "sick-building syndrome"—a situation in which people living or working in a building experience a variety of unpleasant symptoms. Burning or watering eyes, a runny nose, and itchy skin are allergic reactions to the proteins in fungal spores, whereas more severe effects, such as cancers and miscarriages, are probably produced by toxins released when the fruiting bodies of the fungi disintegrate. Curing a sick building can be so expensive that in some cases, the entire building is destroyed and rebuilt.

If you break down the tissues of dead organisms you are a decomposer, and if you break down the tissues of living organisms you are a parasite. Fungi are equally happy in either role. Fungi are happy to grow in and on people, and "mycosis" is a general term for a disease that is caused by a fungus. The fungi that cause athlete's foot, and the related fungi that infect other parts of the body (causing problems with names like jock itch, beard itch, scalp itch, ringworm, and toenail fungus), get their nutrition by digesting some of the organic molecules of your body! And all of these fungal diseases are quite contagious because they are spread by spores that can linger on moist surfaces and in clothing, so they tend to be a problem in places like dormitories, gymnasiums, and fitness centers.

Fungi also form mutually beneficial relationships with chlorophyll-containing bacteria and/or algae. These two-way or three-way partnerships are called **lichens,** and they grow on surfaces such as tree trunks and rocks. The fungus is fed by its photosynthetic partner and helps it absorb water and nutrients. In addition, the fungus provides nutrients for its partners by excreting enzymes that dissolve organic material and acid that dissolves rock.

TAKE-HOME MESSAGE 12•14

Fungi are decomposers, and all they need to thrive is organic material to consume and a moist environment so their hyphae don't dry out. Fungi can grow almost anywhere that is moist, and they can attain enormous sizes. Fungi have complex life cycles, with both sexual and asexual phases, and the parts of a fungus that are most often visible are its temporary spore-producing bodies.

12•15

Most plants have fungal symbionts.

If you examine the roots of a plant with a microscope, you will find round structures and fibers closely associated with the fine rootlets and root hairs. These are root fungi, or **mycorrhizae** (pronounced my-ko-RYE-zay) (**FIGURE 12-33**). Associations between roots and fungi are ancient—they have been found in 400-million-year-old fossils of early land plants—and nearly all species of modern plants have them. Some mycorrhizae, called ectomycorrhizae, have hyphae that press closely against the outer side of the walls of root hair cells. The endomycorrhizae, by contrast, actually send hyphae through the root cell walls and into the space between the cell wall and the plasma membrane. The association between mycorrhizal fungi and plants is beneficial to both partners. The fungus extracts phosphorus and nitrogen from the soil and releases them inside the roots of the plant, and the additional nutrients that the plant receives from the fungus allow it to grow faster and larger. The mycorrhizal fungus benefits by drawing sugar from the plant, which it uses to support its own cellular respiration.

Not all plants play fair, however. Some take nutrients from the fungus and give nothing in return. One of the best known of these mycorrhizal parasites is the ghost pipe, which grows in forests in the northern hemisphere (**FIGURE 12-34**). The ghost pipe is an angiosperm related to heaths and heathers. The plant's name comes from its ghostly white appearance, which makes it visible even at night when only moonlight shines on the forest floor. The roots of the ghost pipe have mycorrhizal fungi associated with them, and those same mycorrhizae are also connected to the roots of trees. The ghost pipe obtains all its nutrients from or through the

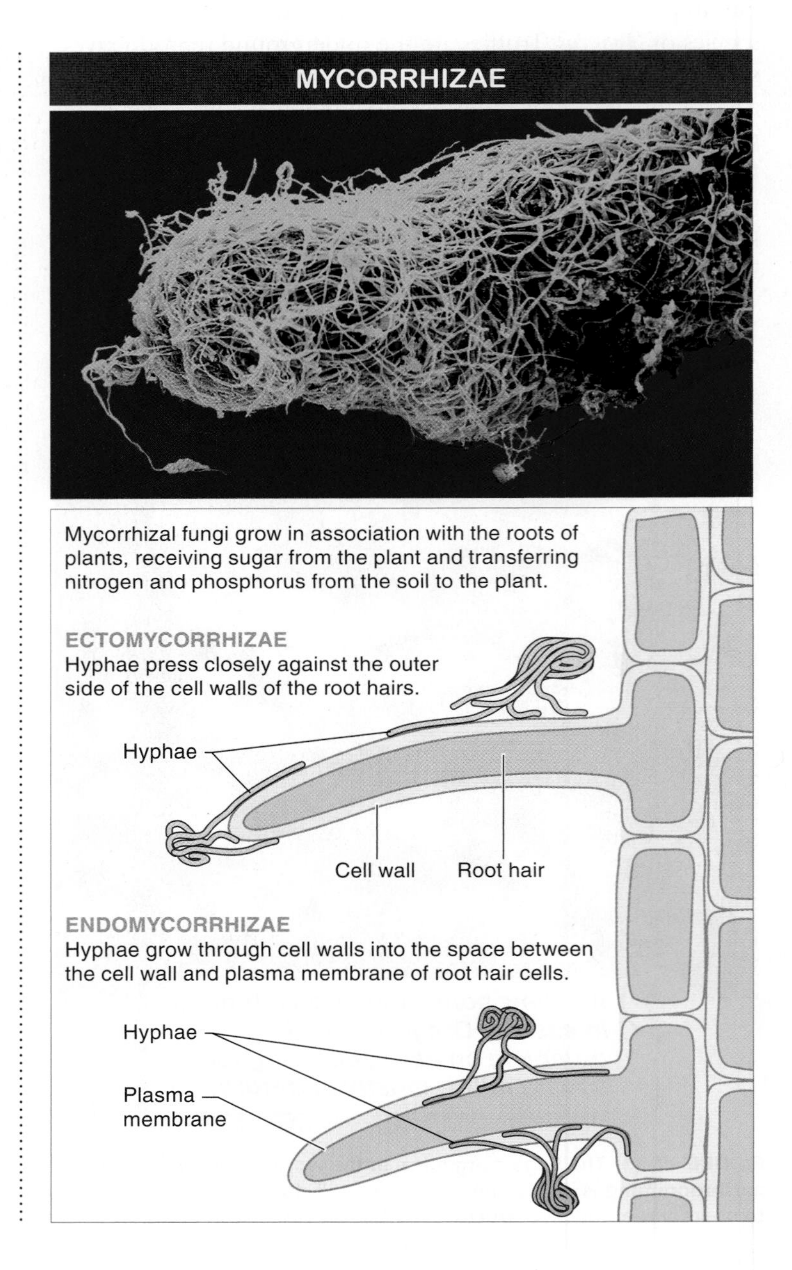

FIGURE 12-33 The association between mycorrhizal fungi and plants is beneficial to both.

FIGURE 12-34 Ghost pipe: a plant that parasitizes mycorrhizae for nutrients.

mycorrhizae—nitrogen and phosphorus are provided by the mycorrhizal fungus, and sugar travels from tree roots to the ghost pipes via the mycorrhizae.

The significance of mycorrhizae is appreciated in the horticultural industry. Lumber companies have found that they have greater success when replanting areas they have clear-cut if the conifer seedlings are first grown in soil that has been inoculated with mycorrhizal fungi. Home gardeners can buy potting soil with mycorrhizal spores or buy spores compressed into tablets that can be placed in the soil when seedlings are transplanted.

Some plants manipulate mycorrhizae to gain a competitive advantage over other plants—garlic mustard is an example. Garlic mustard was brought to North America by European colonists, because they liked using it to add a tangy flavor to salads. Initially it was planted in kitchen gardens, but, inevitably, it escaped into the wild where it has proved itself to be an extremely successful—and unusually diabolical—invasive species. Like many weedy plants, garlic mustard can thrive without mycorrhizal fungi, but garlic mustard goes further than other weeds. It excretes compounds from its roots that interfere with the partnership between mycorrhizal fungi and local native plants. Because the native plants *do* need mycorrhizal fungi to prosper, the destruction of the fungi by garlic mustard weakens the plants. This chemical warfare is so successful that garlic mustard has wiped out entire populations of native woodland plants and is threatening to damage the trees that make up the forest itself.

TAKE-HOME MESSAGE 12•15

Plants and fungi have a close and mutually beneficial association in mycorrhizae. Mycorrhizal fungi grow in intimate association with the roots of most plants, receiving sugar from the plant and transferring nitrogen and phosphorus to the plant.

streetBio

Knowledge You Can Use

Yams: nature's fertility food? Do you know any twins? Have you ever wondered how common twins are?

In the United States and Europe, about 12 in every 1,000 births are twin births. In Asia the number is slightly lower, at 8 sets of twins per 1,000 births. Elsewhere, the rates are similar—except for southwest Africa, particularly in Nigeria, a place that has been called "The Land of Twins." There, the rate of twin births among the Yoruba people is more than four times that in the United States, with about 50 pairs of twins per 1,000 births. (Triplet births are unusually common, too, occurring 16 times more frequently in Nigeria than the United States.)

Q: Why are so many twins born in Nigeria? Is it something in the water? Nope. But that's not too far off. It's the diet of the Yoruba people. A staple in their diet is the white yam, a starchy vegetable that looks a bit like a potato; many people eat yams several times each day.

Q: What's in the yams? Some preliminary studies suggest that an estrogen-related compound in yams is responsible for stimulating the ovaries, increasing the likelihood, in any given month, that more than one egg will be released.

Q: How can we be sure it's the yams? At this point, no one is certain, and the question awaits a good application of the scientific method. In one intriguing laboratory study, rats fed a diet rich in yams *doubled* their litter size. As yet, however, a definitive link between yams and increased rates of twin births in humans has not been demonstrated.

So, will it work for you? It's unclear. Anecdotes abound about women having twins after purposely increasing their yam consumption. But a randomized, controlled, double-blind study has yet to be conducted. How would you set up and analyze a study like that?

BIG IDEAS IN PLANT AND FUNGI DIVERSIFICATION

1 Plants are just one branch of the eukarya.

Plants are multicellular organisms that spend most of their lives anchored to one place by their roots. Plants' inability to move shapes the way they obtain food, reproduce, and protect themselves from predation. The evolutionary history of plants reveals the progressive appearance of characteristics that allow plants to succeed on land, despite their inability to move.

2 The first plants had neither roots nor seeds.

The first land plants were small, had no leaves, roots, or flowers, and could grow only near water. Non-vascular plants—mosses, liverworts, and hornworts—have scarcely evolved beyond the earliest land plants. They lack roots and vessels to move water and nutrients from the soil into the plant. Vessels are an effective way to carry water and nutrients from the soil to the leaves, and the first vascular plants—the earliest ferns, horsetails, and some related forms—were able to grow much taller than their non-vascular predecessors.

3 The advent of the seed opened new worlds to plants.

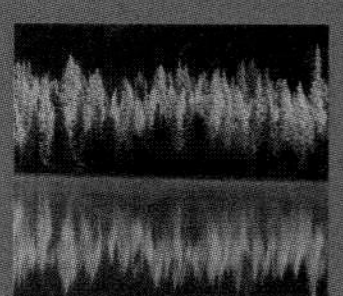

Seeds are the way plants give their offspring a good start in life and get them to leave home. A seed contains a multicellular embryo and a store of nutrients. Seeds are distributed by air, water, or animals. Gymnosperms were the earliest plants to produce seeds, which offer advantages over the spores of earlier plants, but they depend on wind to carry their pollen. Conifers are the most successful gymnosperms, with more species and a larger geographic range than all of their relatives combined.

4 Flowering plants are the most diverse and successful plants.

Flowering plants appeared in the Cretaceous period, about 100 million years ago, and diversified rapidly to become the dominant plants in the modern world. A flower houses a plant's reproductive structures and advertises their presence with colors, patterns, and odors, which it uses to trick or bribe animals into transporting male gametes to female gametes, where fertilization can occur. Double fertilization in angiosperms ensures that a plant does not invest energy in forming endosperm for an ovule that has not been fertilized.

5 Plants and animals have a love-hate relationship.

Overcoming one big challenge of being rooted in the ground, plants often use the assistance of animals to disperse their fruits, which contain fertilized seeds, depositing them at a new location where the seedlings can grow. Fruits are made up from the ovary and, occasionally, some surrounding tissue. Overcoming a second big challenge posed by immobility, plants have a wide range of defenses against herbivorous animals, from physical defenses such as thorns to chemicals with complex effects on animals' physiology. Some plants living in nitrogen-deficient soil have switched roles, preying on insects.

6 Fungi and plants are partners but not relatives.

Fungi are eukaryotes with a cell wall made of chitin. Some fungi (yeasts) live as individual cells, but most are multicellular. Fungi are decomposers, and all they need to thrive is organic material to consume and a moist environment. They can attain enormous sizes. Fungi have complex life cycles, with sexual and asexual phases, and the parts of fungi that are most often visible are their temporary spore-producing bodies. Plants and fungi have a close and mutually beneficial association. Mycorrhizal fungi grow in intimate association with the roots of most plants, receiving sugar from the plant and transferring nitrogen and phosphorus to the plant.

KEY TERMS

angiosperm, p. 458
anther, p. 463
bryophyte, p. 452
carpel, p. 463
cuticle, p. 452
decomposer, p. 475
double fertilization, p. 466
endosperm, p. 458
filament, p. 463
fleshy fruit, p. 468
flower, p. 463
gametophyte, p. 458
gymnosperm, p. 458
hyphae, p. 473
lichen, p. 476
mycelium, p. 474
mycorrhizae, p. 476
non-vascular plant, p. 452
ovary, p. 463
ovule, p. 458
plant, p. 448
pollen grain, p. 458
pollination, p. 460
prothallus, p. 456
root, p. 449
seed, p. 457
shoot, p. 449
sporangia, p. 455
spore, p. 454
stamen, p. 463
vascular plant, p. 455

CHECK YOUR KNOWLEDGE

1. In the plant kingdom, all the different species are descended from a single common ancestor. In terms of phylogeny, what type of tree is this?
   a) monophyletic
   b) unbranched
   c) biphyletic
   d) uniphyletic
   e) branched

2. Which of the following is a significant new challenge that plants faced when they moved from their aquatic environment onto the land?
   a) light availability
   b) desiccation
   c) nutrient availability
   d) predation
   e) the need for osmotic regulation

3. Which is the best brief description of the vascular system of the very first terrestrial plants?
   a) The first plants developed specialized vessel cells that conducted water.
   b) The first plants did not possess a vascular system.
   c) The first plants had a very basic vascular system with a simple method of internal transport.
   d) The first plants had only long, needle-like leaves, from which water could evaporate easily.
   e) None of the above answers is correct.

4. Which of the following statements about ferns is incorrect?
   a) Their sporophyte is dominant.
   b) They require liquid water for fertilization.
   c) Their seeds are dispersed via the wind.
   d) They have vascular tissue for distributing water and nutrients throughout the plant.
   e) Their spores are contained in sporangia.

5. Mosses and ferns differ in their reproductive strategies from gymnosperms and angiosperms in which of the following ways?
   a) Mosses and ferns rely on liquid water for fertilization, whereas angiosperms and gymnosperms do not need liquid water for fertilization.
   b) Mosses and ferns have much larger seeds than do angiosperms and gymnosperms.
   c) Mosses and ferns use wind pollination, whereas angiosperms and gymnosperms use insects for pollination.
   d) Mosses and ferns are primarily diploid in their adult (reproductive) form, whereas gymnosperms and angiosperms are primarily haploid.
   e) Mosses and ferns are primarily haploid in their adult form, whereas gymnosperms and angiosperms are primarily diploid.

6. Which of the following is characteristic of gymnosperms?
   a) They are more diverse than the angiosperms.
   b) They are wind-pollinated.
   c) The gametophyte generation is dominant.
   d) They are water-pollinated.
   e) All of the above are correct.

7. In terms of their adaptation to living on land, how are reptiles similar to the seed plants?
   a) Both reptiles and seed plants became completely independent of water.
   b) Reptiles eat plants.
   c) Seed plants and reptiles have developed structures such as cuticles and impermeable skin to minimize desiccation.
   d) Reptiles and seed plants have developed structures that house their gametes and protect them from the surrounding environment.
   e) Both c) and d) are correct.

8. Angiosperms and gymnosperms differ from each other in that:
   a) angiosperms have vascular systems, but gymnosperms do not.
   b) angiosperms produce seeds, but gymnosperms do not.
   c) angiosperms tend to rely on animal pollinators, whereas gymnosperms tend to rely on wind pollination.
   d) angiosperms use pollen, whereas gymnosperms use cones.
   e) angiosperms have a single fertilization process, whereas gymnosperms use double fertilization.

9. Anthers and stigmas can be found on:
   a) bryophytes.
   b) fungi.
   c) angiosperms.
   d) gymnosperms.
   e) all of the above.

10. Which of the following is not an example of a group of angiosperms?
    a) cacti
    b) cherry trees
    c) pine trees
    d) grasses
    e) orchids

11. Over the evolutionary history of plants:
    a) the sporophyte has become smaller, though more independent.
    b) the gametophyte and sporophyte have grown increasingly independent of each other.
    c) there has been a trend toward gametophyte dominance.
    d) there has been a trend toward gametophyte dependence.
    e) the gametophyte has become larger, though more dependent.

12. Which of the following comparisons and contrasts between fungi and plants is incorrect?
    a) Fungi cannot photosynthesize, but plants can.
    b) Both fungi and plants use chitin as a structural stabilizer.
    c) Fungi are heterotrophs, but plants are not.
    d) Both fungi and plants have cell walls.
    e) Both fungi and plants have a sexual stage in their reproductive cycle.

13. Dispersal of fungal spores is typically done by:
    a) movement of cilia.
    b) insects.
    c) hummingbirds.
    d) wind.
    e) movement of flagella.

14. You are taking a hike down a forest trail and see the familiar sight of a mushroom on the ground. This visible portion of a fungal body is the structure also referred to as the:
    a) hypha.
    b) fruiting body.
    c) thallus.
    d) spore sac.
    e) mycelium.

15. In most cases, the relationship between roots and fungi in mycorrhizae can best be described as:
    a) mycelium.
    b) competition.
    c) trickery.
    d) symbiosis.
    e) parasitism.

Short-Answer Questions

1. What distinguishes a vascular plant from a non-vascular plant? How do these differences affect their geographic distribution?

2. How do fungi benefit plants? What benefit do the fungi derive from this relationship?

3. How does development of a fleshy fruit benefit plants?

See Appendix for answers. For additional study questions, go to www.prep-u.com.

1. **There are microbes in all three domains.**
   - **13.1** Microbes are the simplest, but most successful organisms on earth.
   - **13.2** Not all microbes are evolutionarily related.
2. **Bacteria may be the most diverse of all organisms.**
   - **13.3** What are bacteria?
   - **13.4** Bacterial growth and reproduction is fast and efficient.
   - **13.5** Many bacteria are beneficial.
   - **13.6** Metabolic diversity among the bacteria is extreme.
   - **13.7** Bacteria cause many human diseases.
   - **13.8** Bacteria evolve drug resistance quickly.
3. **Archaea exploit some of the most extreme habitats.**
   - **13.9** Archaea are profoundly different from bacteria.
   - **13.10** Archaea thrive in habitats too extreme for most other organisms.
   - **13.11** Much archaean diversity has yet to be discovered.
4. **Most protists are single-celled eukaryotes.**
   - **13.12** The first eukaryotes were protists.
   - **13.13** There are animal-like protists, fungus-like protists, and plant-like protists.
   - **13.14** Some protists can make you very sick.
5. **Viruses are at the border between living and non-living.**
   - **13.15** Viruses are not exactly living organisms.
   - **13.16** Viruses are responsible for many health problems.
   - **13.17** Viruses infect a wide range of organisms.
   - **13.18** HIV illustrates the difficulty of controlling infectious viruses.

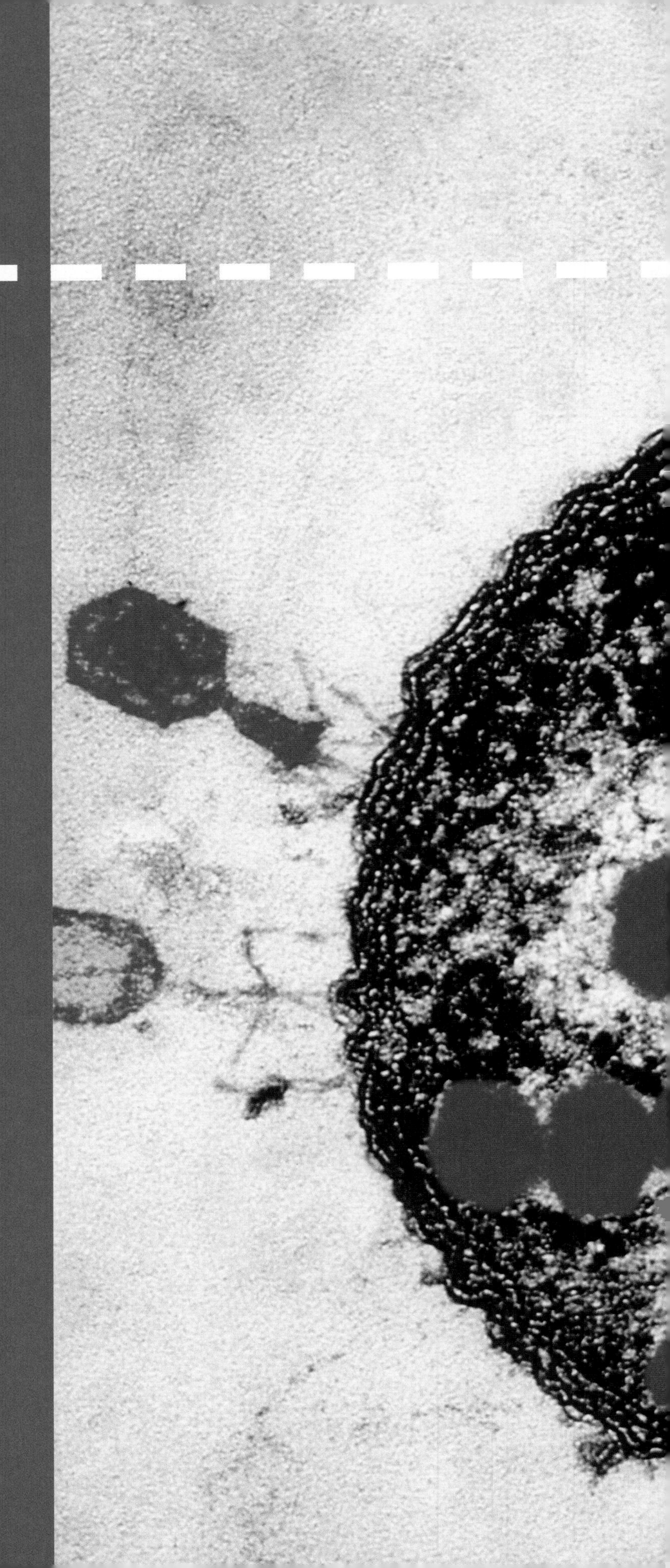

13

Evolution and Diversity Among the Microbes

Bacteria, archaea, protists, and viruses: the unseen world

❶ There are microbes in all three domains.

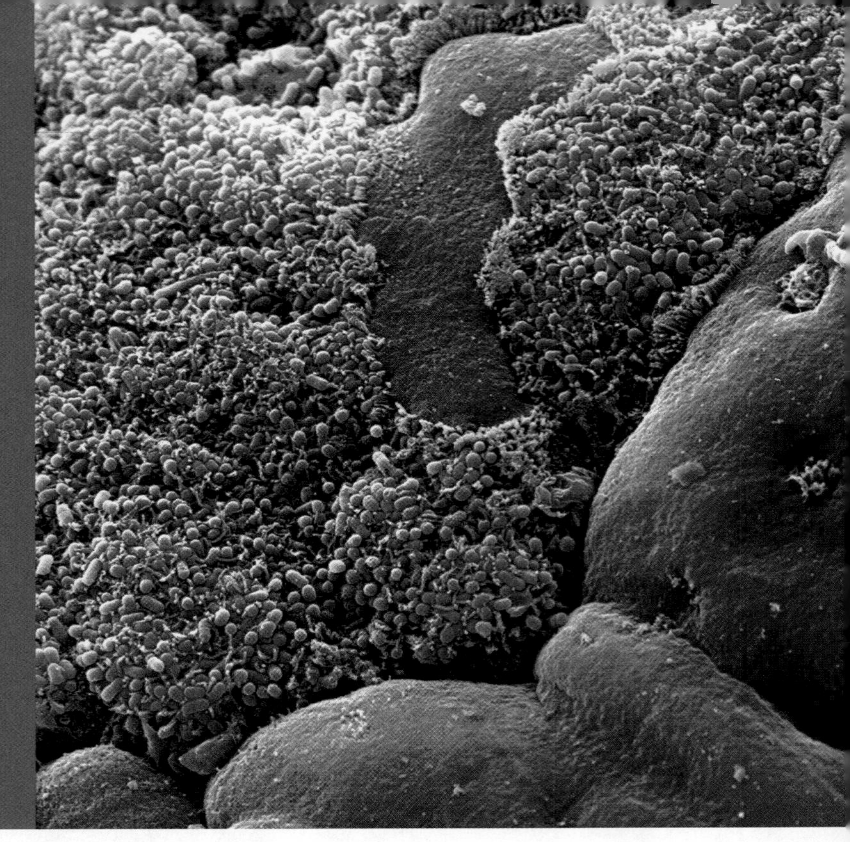

This strain of E. coli *produces a toxin that causes extreme gastrointestinal discomfort.*

13 • 1

Microbes are the simplest, but most successful organisms on earth.

Humans are large organisms, and being large comes with a lot of baggage that we don't usually think about. We need a skeletal system to support our weight against the pull of gravity, and a respiratory system to take in oxygen and get rid of carbon dioxide. We need a circulatory system to move oxygen and carbon dioxide and other molecules around our bodies, and a digestive system that includes a mouth to take in food and teeth to grind the food into small pieces that can be broken down by enzymes in our stomach and intestines. We even need a nervous system so our brain knows what distant parts of our body are doing.

Q How can a microbe function when its body is just a single cell?

Microbes—the most abundant organisms on earth—don't have skeletal, respiratory, circulatory, digestive, or nervous systems because they are too small to need them; in fact, "microbe"

RELATIVE SIZES OF MICROBES

Although almost all microbes are invisible to the naked eye, they vary tremendously in size. Here, the relative sizes of several microbes are shown proportionally, using everyday objects.

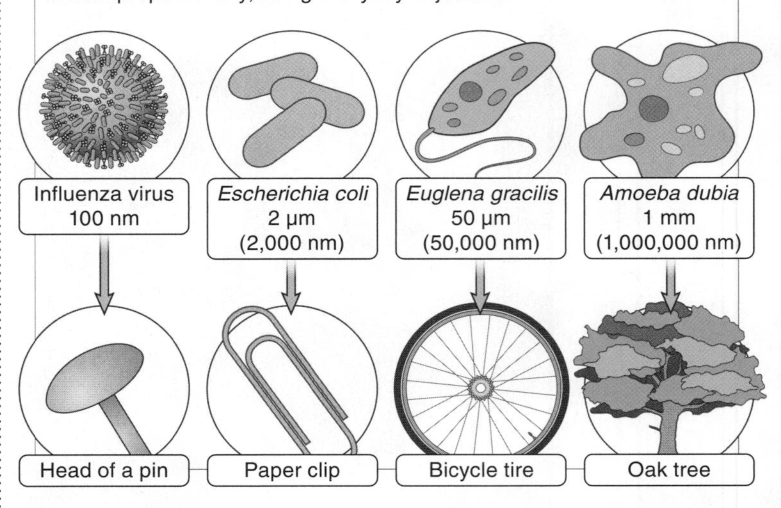

FIGURE 13-1 The most abundant organisms on earth are too small to see. If an influenza virus were the size of a pinhead, an amoeba would be the size of an oak tree.

is a combination of Greek words that means "small life," and the "small" part of the name fits them well. Take an amoeba, for example—it has a volume about a million billion (10^{15}) times smaller than a human. When you are that tiny, the force of gravity becomes trivial, so the amoeba needs no skeleton to support it. Plenty of oxygen diffuses inward across its cell membrane, and carbon dioxide diffuses outward, so an amoeba does not need a respiratory system; and because every part of its interior is close to the body surface, it doesn't need a circulatory system to transport gases. An amoeba eats by enclosing food items in a piece of its cell membrane and digests the food with the same enzymes it uses to recycle its own proteins, lipids, and carbohydrates, so it does not need a digestive system. And no part of an amoeba is far enough from another part to require a specialized nervous system for communication.

Most microbes are actually even smaller than an amoeba: a typical bacterium or archaean is about one thousand million billion (10^{18}) times smaller than a human, and an influenza virus is about one thousand billion trillion (10^{24}) times smaller than you are (**FIGURE 13-1**).

Q How would you decide whether a group of organisms can be considered successful?

They may be invisible, but microbes could never be considered unsuccessful just because of their (lack of) size. To the contrary, they are more successful than humans in almost every imaginable way.

Microbes are genetically diverse. More than 500,000 kinds of microbes have been identified by their unique nucleotide sequences, and further studies will almost certainly distinguish millions of additional species of microbes.

Microbes can live almost anywhere and eat almost anything. As you read these lines, more than 400 species of microbes are thriving in your intestinal tract, 500 more species thrive in your mouth, and nearly 200 species call your skin home (**FIGURE 13-2**). The microbes that live in you and on you eat mostly what you eat—some of the bacteria in your mouth and intestine compete with you, trying to digest your food before you can, and the others use the waste products you release after you have broken down the food. Others feed on the leftovers released by the breakdown of your cells during the normal process of cell renewal.

Living conditions in the human body are relatively moderate. Other microbes inhabit some of the toughest environments on earth—in the almost boiling water of hot springs, at depths a mile below the earth's surface, and more than a mile deep in the oceans where hydrothermal vents emit water at 400° C (750° F).

Microbes are abundant. Surface seawater contains more than 100,000 bacterial cells per milliliter, and diatoms (protists in the

MICROBES CAN LIVE ALMOST ANYWHERE

Human intestine — *Escherichia coli*

Hot spring — *Thermus aquaticus*

Hydrothermal vent — *Methanopyrus kandleri*

Microbes live in nearly every kind of environment, including water at temperatures of up to 750° F!

FIGURE 13-2 Microbes are everywhere on earth.

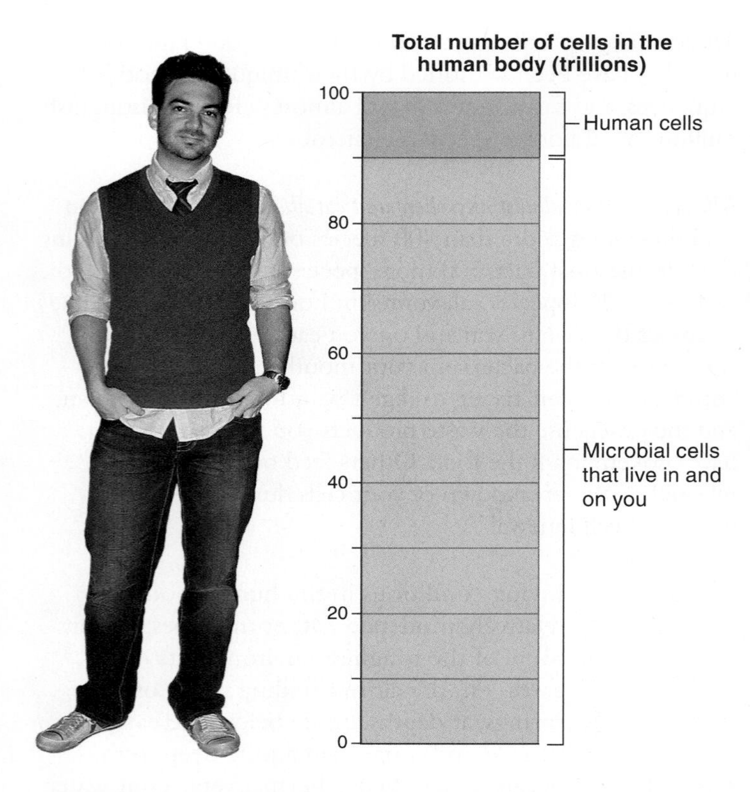

FIGURE 13-3 Microbe majority. Your body has more microbial cells than human cells.

Everywhere yet unseen. *Microbes outnumber all other species and make up most living matter, including this Malaysian rain forest.*

eukarya domain) are as abundant as bacteria. These densities translate to about 8,000 million billion trillion (8×10^{30}) individuals of just these two kinds of microbes in the world's oceans. Your own body is a testament to the abundance of microbes: it contains about 100 trillion cells, but only one-tenth of those cells are actually human cells—the remaining 90 trillion cells are the microbes that live in and on you (**FIGURE 13-3**). You're a minority in your own body.

TAKE-HOME MESSAGE 13•1

Microbes are very small, simple organisms, but they do everything that larger, multicellular organisms do. They can live anywhere, from moderate to extreme environments. There are millions of different kinds of microbes on earth, in enormous numbers.

13•2

Not all microbes are evolutionarily related.

"Microbes" is an appropriately descriptive name, but it's sloppy. We could use it to point to any one of many different types of organisms too small to see without magnification. In Chapters 3 and 10 we introduced two very different kinds of microbes: bacteria and viruses. In this chapter we explore bacteria and viruses in more detail, and also include two more kinds of microbes—the protists and the archaea. The amoeba, a kind of protist, may be familiar to you, but the archaea are probably very unfamiliar to you. That's not surprising, because archaea were only recently recognized as a distinct group of microbes. The group doesn't even have a common name (unless you like "the group formerly known as archebacteria"). We'll just call them archaea.

The sloppiness of the name "microbes" stems from the fact that not all microbes are evolutionarily related. As a consequence, it is hard to make generalizations about them as we can for other organisms. We can say of all animals, for example, that they have the ability to move, if only during one portion of their life cycle. Microbes, on the other hand, are grouped together simply because they are small, not because they all share a recent common ancestor. In fact, microbes occur in all three domains of life—bacteria, archaea, and eukarya—and so the various types of microbes could not be more widely separated (**FIGURE 13-4**). In this chapter we focus on the tiny organisms from each of these domains. The microbes in the domains bacteria and archaea are prokaryotic, although archaea have some characteristics that are like prokaryotes and others that are like eukaryotes. Protists are the microbe members of domain eukarya. And viruses, another type of microbe, are not classified into any domain at all, because they are only at the borderline of life.

TAKE-HOME MESSAGE 13•2

Microbes are grouped together only because they are small, not because of evolutionary relatedness. They occur in all three domains of life, and include the viruses, which are not included in any of the domains.

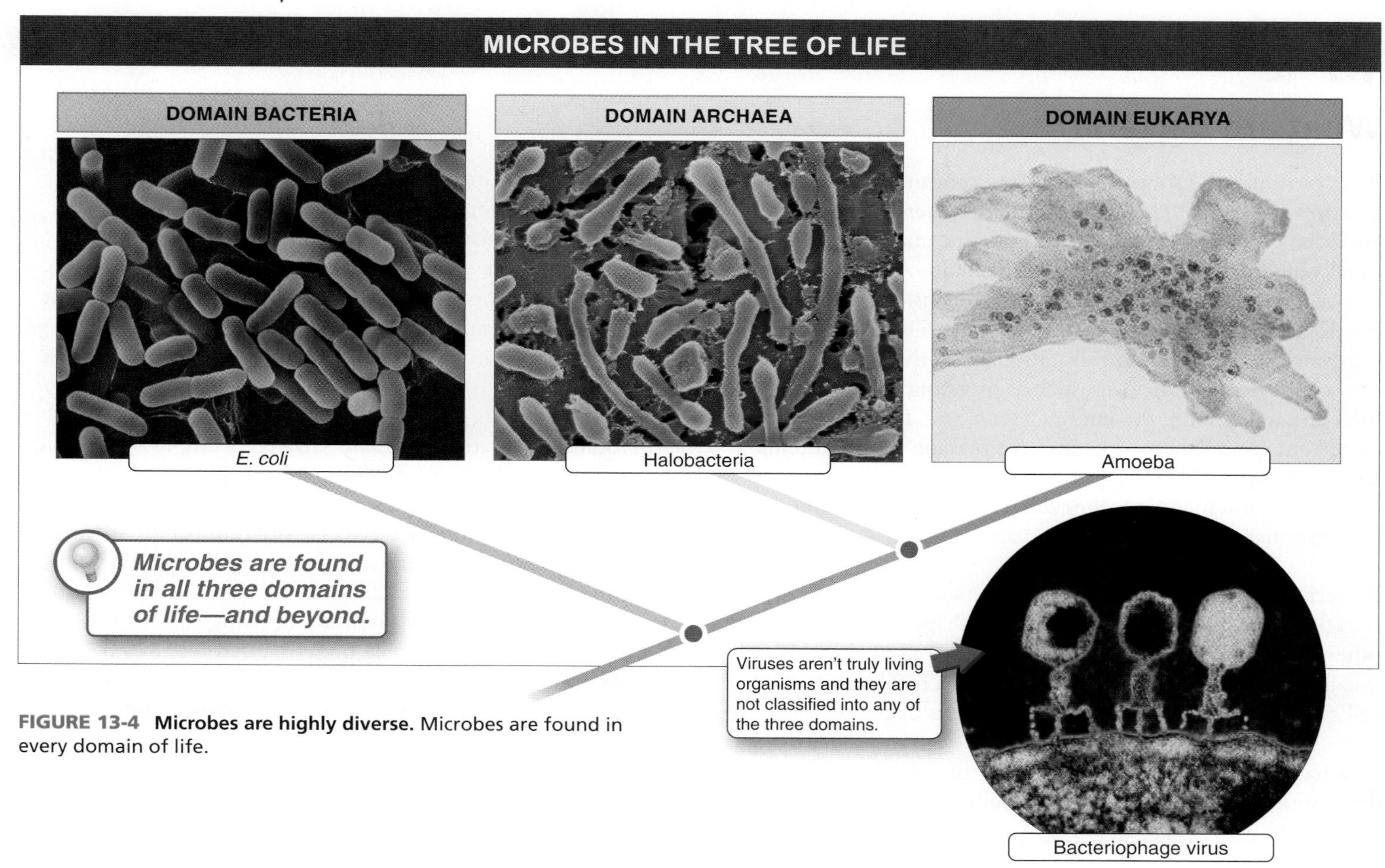

FIGURE 13-4 Microbes are highly diverse. Microbes are found in every domain of life.

❷ Bacteria may be the most diverse of all organisms.

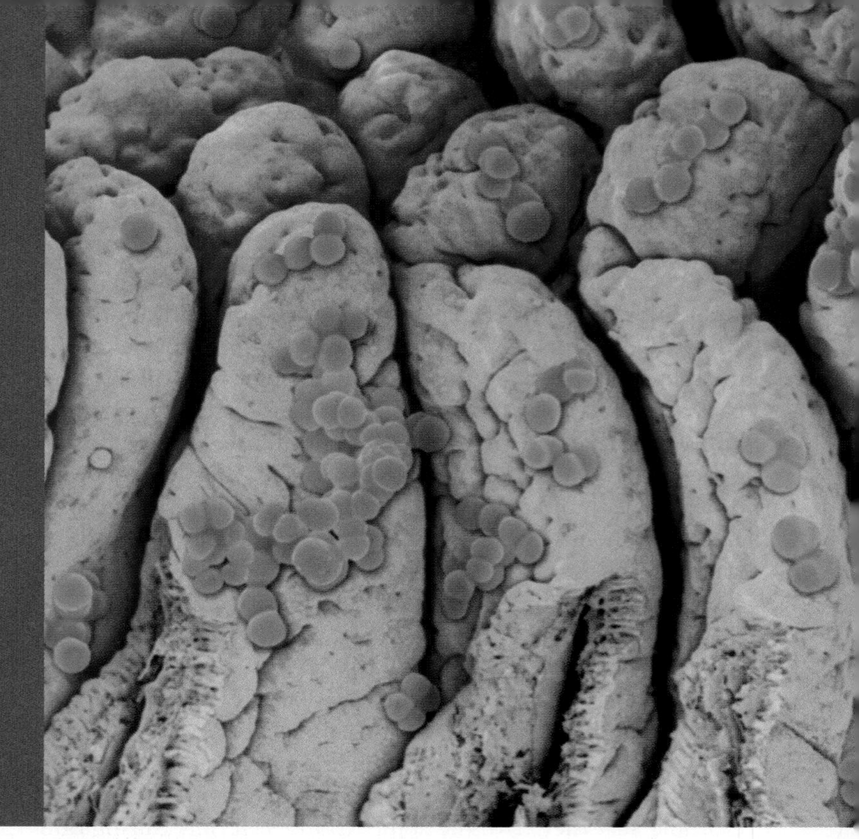

Staphylococcus aureus (green) on the surface of the small intestine.

13•3

What are bacteria?

A bacterium is the simplest organism you can imagine—and in many ways, it is the most efficient. A bacterium has a cell envelope consisting of a plasma membrane and, in most cases, a cell wall—that's what it needs to maintain conditions inside that are different from conditions outside. Inside the cell envelope it has cytoplasm—the liquid that fills all kinds of cells (including your own cells). Proteins in the cytoplasm carry out essential functions, such as digesting molecules of food and transferring the energy gained to ATP. DNA in the cytoplasm carries the instructions for making those proteins, and messenger RNAs carry this information to ribosomes, where the proteins are synthesized. That's it—everything an organism needs with no extra baggage.

Bacteria may be classified by their shape: some are spherical cells (known as the cocci), some are rod-shaped (the bacilli), and others are spiral-shaped (the spirilli) (**FIGURE 13-5**). Bacteria usually reproduce by binary fission, and in a few hours, a single cell can form a culture containing thousands of cells.

As a bacterial cell divides, the number of cells doubles every generation, producing a colony of cells, each of which is a clone of the original cell. Colonies of different species of bacteria look different. The familiar human intestinal bacterium *Escherichia coli,* for example, forms beige or gray colonies that have smooth margins and a shiny mucus-like covering. Species of *Proteus,* which are often responsible for spoiling food because they can grow at refrigerator temperatures, form colonies with a surface that looks like a contour map.

Microbiologists can often identify bacteria simply by looking at the colors and shapes of their colonies. They can get additional information by examining a single cell under a microscope, but living bacterial cells are transparent, so you can't see them with an ordinary microscope unless they have been dyed. In 1884, Hans Christian Gram, a Danish microbiologist, described a method of dying the cell walls of bacteria to make them visible under a microscope, and a **Gram stain** is still the first test microbiologists use when they are identifying an unknown bacterium (**FIGURE 13-6**).

Gram-positive bacteria are colored purple by the stain, because they have a thick layer of a glycoprotein called

BACTERIA: CLASSIFICATION AND STRUCTURE

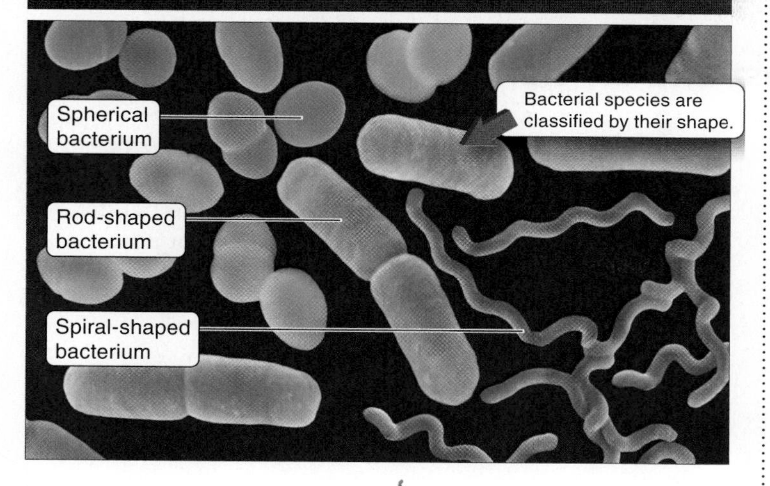

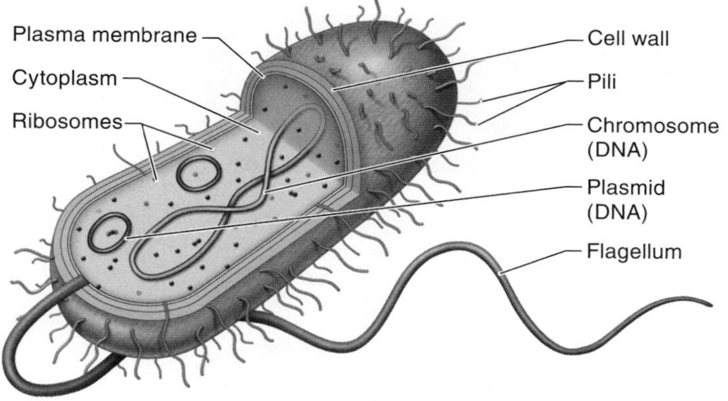

FIGURE 13-5 Bacteria basics. Bacteria are single-celled organisms that lack a nucleus. They are usually classified by shape.

peptidoglycan on the outside of the cell wall. In *Gram-negative bacteria,* the layer of peptidoglycan lies beneath a membrane and is not stained by the dye. Penicillin is effective in treating infections by Gram-positive bacteria because it interferes with the formation of peptidoglycan cross-links. Penicillin is less effective on Gram-negative bacteria because it does not pass through the outer membrane that covers the peptidoglycan layer.

Being able to distinguish these two groups of bacteria is a big help to microbiologists trying to identify a bacterium, but the peptidoglycan is there because it is important for the bacterial cells—not to make life easier for microbiologists. The extensive interlocking bonds of the long peptidoglycan molecules provide strength to the cell envelope. Many bacteria also have a **capsule** that lies outside the cell wall. This capsule may restrict the movement of water out of the cell and allow bacteria to live in dry places, such as on the surface of your skin. In other cases, the capsule may be important in allowing the bacteria to bind to solid surfaces such as rocks or to attach to human cells.

TAKE-HOME MESSAGE 13•3

Bacteria are efficient single-celled organisms, with an envelope surrounding the cytoplasm, which contains the DNA (they have no nuclei and no intracellular organelles). Bacterial cells undergo binary fission, and a single cell can grow into a colony of cells.

METHODS OF IDENTIFYING BACTERIA

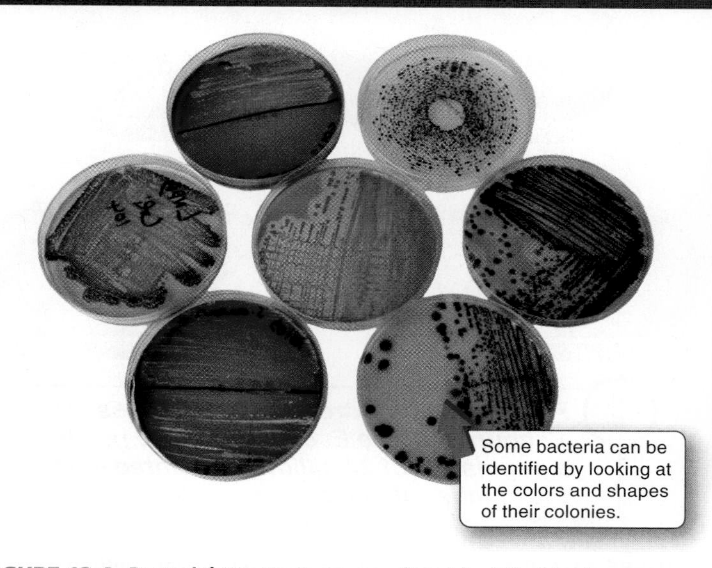

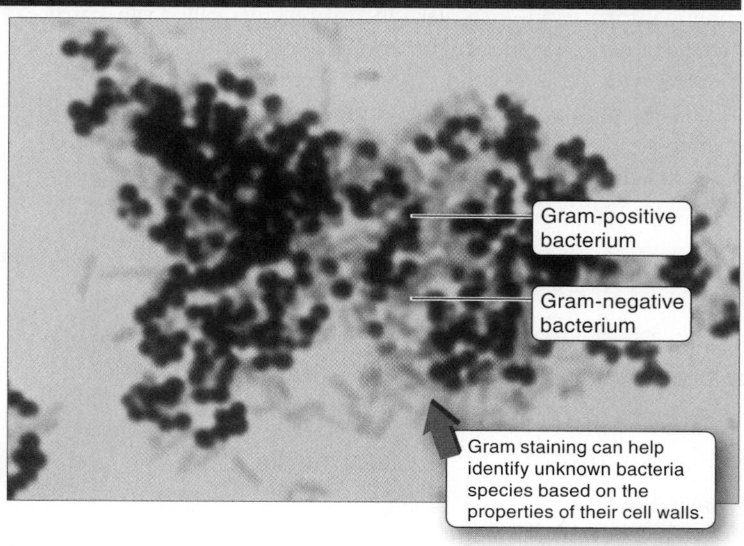

FIGURE 13-6 Bacterial IDs. Bacteria are often identified by the appearance of an entire colony or of the cells' response to Gram staining.

13·4

Bacterial growth and reproduction is fast and efficient.

The time it takes for a bacterium to reproduce can be very short; most bacteria have generation times between 1 and 3 hours, and some are even shorter. *Escherichia coli,* for example, has a generation time of 20 minutes in optimal conditions, and in under 12 hours a single *E. coli* could give rise to a population of 20 billion cells.

Bacteria carry genetic information in two structures: the genes that provide instructions for all of the basic life processes of a bacterium are usually located in a circular DNA molecule, the bacterial chromosome. Most bacteria have just one chromosome, but some have more than one. A bacterial chromosome is organized more efficiently than a eukaryotic chromosome, in two ways. First, in bacteria, the genes that code for proteins with related functions—enzymes that play a role in a pathway that breaks down food for energy, for example—are often arranged right next to each other on the chromosome. This makes it possible to efficiently control the transcription of all the genes together.

Additionally, unlike in eukarya, almost all the DNA in a bacterial chromosome codes for proteins, so bacteria do not use time and energy transcribing mRNA that will not be translated. As we saw in Chapter 5, as much as 90% of the DNA in the chromosomes of eukaryotes does not code for genes and is edited from mRNA after it has been transcribed and before it is translated into protein.

Besides their main chromosome, many bacteria have additional genetic information—circular DNA molecules called **plasmids** that carry genes for specific functions. These include *metabolic plasmids,* with genes enabling bacteria to break down specific substances, such as toxic chemicals; *resistance plasmids,* with genes enabling bacteria to resist the effects of antibiotics; and *virulence plasmids,* with genes that control how sick an infectious bacterium makes its victim. The strain of *E. coli* that occasionally sickens patrons of fast-food restaurants carries a virulence plasmid that magnifies the effects of a gene for a sometimes-lethal toxin. (*E. coli* without that virulence plasmid are normal components of the bacterial community of the human intestine.)

Q Would it be useful to be able to transfer genetic information from one adult human to another?

When a bacteria cell divides, it creates two new daughter cells, each carrying the genetic information that was present in the chromosome of the mother cell. Thus, binary fission transmits genetic information from one bacterial generation to the next (**FIGURE 13-7**). However, bacteria can also transfer genetic information laterally—in other words, not just to their "offspring," but to other individuals *within* the same generation—through any of three different processes: conjugation, transduction, and transformation.

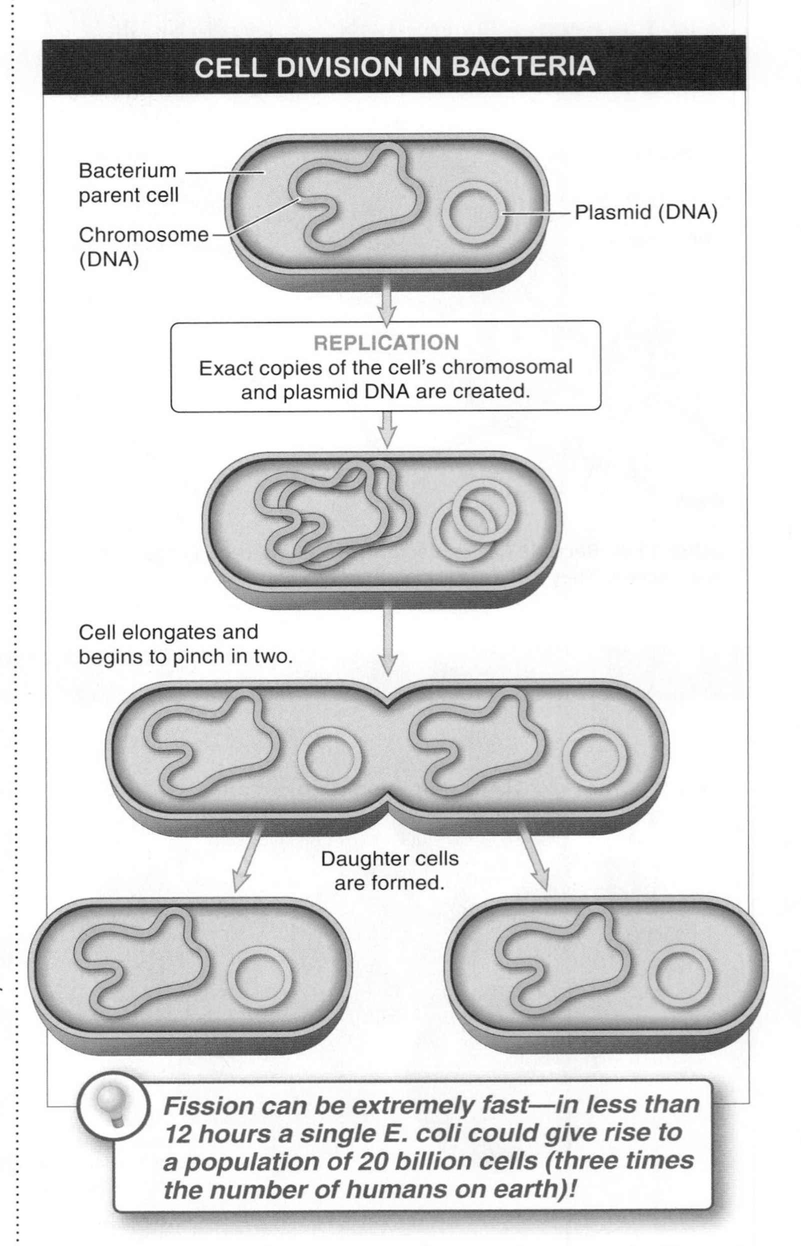

FIGURE 13-7 Binary fission. This asexual cell division method is used by prokaryotes.

METHODS OF GENETIC EXCHANGE IN BACTERIA WITHIN THE SAME GENERATION

CONJUGATION
A bacterium transfers a copy of some or all of its DNA (from the main chromosome or a plasmid) to another bacterium, giving the second bacterium genetic information it did not have before.

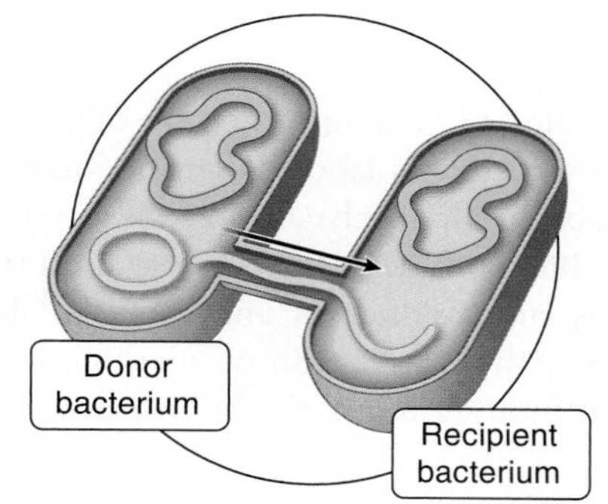

TRANSDUCTION
A virus containing pieces of bacterial DNA that it inadvertently picked up from its previous host infects a bacterial cell, and passes along new bacterial genes to the bacterium.

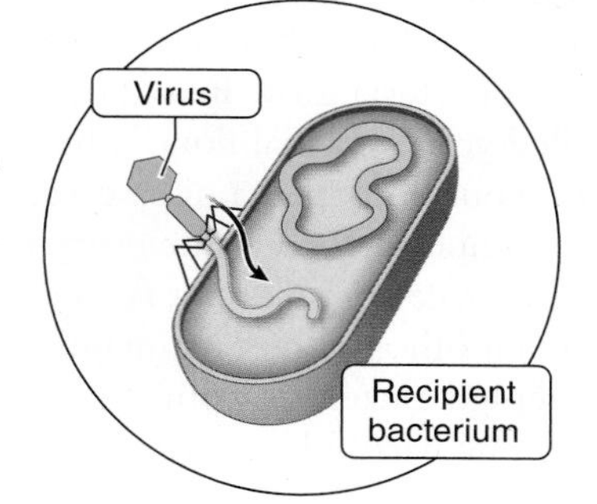

TRANSFORMATION
After a bacterial cell bursts open, short lengths of DNA can be taken up by a living bacterial cell and inserted into its own chromosome, potentially adding genes that it did not have before.

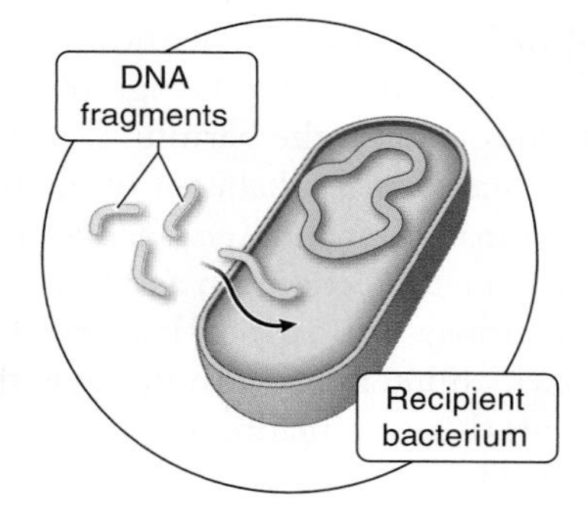

FIGURE 13-8 Lateral transfer of genetic information: conjugation, transduction, and transformation.

Conjugation is the process by which one bacterium transfers a copy of some of its genetic information to another bacterium—even when the two bacteria are different species. It's very much like plugging your iPod into a computer to transfer songs from the computer to the iPod. This is not reproduction, because you start with one iPod and one computer and that's what you have when you finish. But the iPod now contains songs that it did not have before. Plasmid transfer does exactly that for a bacterium—you still have just two bacterial cells, but the newly acquired plasmid has given the recipient bacterium genetic information it did not have before. These genes could enable the bacterium to make an enzyme that allows it to metabolize a new chemical or to defend itself against a new antibiotic (**FIGURE 13-8**).

Transduction occurs when a kind of virus called a bacteriophage (see bacteriophage photo, Figure 13-4) infects a bacterial cell. The virus reproduces inside the bacterial cell, and sometimes, inadvertently, the new virus particles are filled with pieces of bacterial DNA in addition to or instead of the viral DNA. When these viruses are released and infect new bacterial cells, the bacterial DNA can be inserted into the host bacterium's chromosome, passing new bacterial genes to that bacterium.

Transformation is the process by which bacterial cells scavenge DNA from their environment. This DNA comes from other bacterial cells that have burst open, releasing their cellular contents. The circular chromosomes break into short lengths of DNA, which can then be taken up by living bacterial cells and inserted into their own chromosomes, potentially adding genes they did not originally have.

TAKE-HOME MESSAGE 13•4

Bacteria grow rapidly. They have efficiently organized chromosomes—genes are organized in groups with related functions and virtually all the DNA codes for proteins. Bacteria sometimes carry genes for specialized traits on small DNA molecules called plasmids, which can be transferred from one bacterial cell to another by conjugation. DNA can also be transferred laterally between bacterial cells by transduction or transformation.

13•5

Many bacteria are beneficial.

Do you like the taste of yogurt? You can thank bacteria for that—*Lactobacillus acidophilus* and several other species of bacteria are added to milk to create yogurt. As the bacterial cells use the milk sugar for energy, the by-product is lactic acid, which reacts with the milk proteins to produce the characteristic taste and texture of yogurt. If you buy a brand of yogurt that is labeled as containing "live cultures," you are consuming living bacterial cells as you eat the yogurt (**FIGURE 13-9**). Bacteria are also used to produce many other foods, such as cheeses, and bacteria (along with yeasts) are used in the production of beer, wine, and vinegar. Industrial microbiology is a multi-billion-dollar industry.

FIGURE 13-9 Yogurt contains beneficial bacteria.

But you can thank bacteria for more than just tasty snacks—you owe your life to bacteria. As we've seen, hundreds of species of bacteria grow in and on your body; these microbes are called your "normal flora." The normal flora take up every spot on your body that a disease-causing bacterium could adhere to, and they consume every potential source of nutrition, making it difficult for a disease-causing bacterium to gain a foothold. Thus, maintaining a robust population of these benign bacteria is your first line of defense against infection by harmful bacteria.

Probiotic therapy is a method of treating infections by deliberately introducing benign bacteria in numbers large enough to swamp the harmful forms. *Lactobacillus acidophilus,* which is a normal inhabitant of the human body, is used to treat gastrointestinal upsets, such as traveler's diarrhea, and urinary tract infections. In addition to growing so vigorously that it crowds out harmful bacteria, *L. acidophilus* releases lactic acid, which interferes with the growth of other bacteria and prevents them from adhering to the walls of the urinary tract and bladder.

TAKE-HOME MESSAGE 13•5

Your body fights bacteria with bacteria. A disease-causing bacterium must colonize your body before it can make you sick, and your body is already covered with harmless bacteria. If the population of harmless bacteria is dense enough, it will preclude invading bacteria from gaining a foothold.

13•6

Metabolic diversity among the bacteria is extreme.

One important feature making bacterial diversity possible is that bacteria can metabolize almost anything. Some of them can even use energy from light to make their own food. Microbiologists place bacteria into "trophic" (feeding) categories that reflect their metabolic specialization.

Chemical organic feeders (**chemoorganotrophs**) are bacteria that consume organic molecules, such as carbohydrates. You probably see the products of organic feeders every time you take a shower—they are responsible for the pink deposits on the shower curtain and the floor of the shower (**FIGURE 13-10**). Most of the bacteria that live in and on your body are also organic feeders. Some compete with you to metabolize the food you eat. Others digest things you can't digest or can't eat.

Chemical inorganic feeders (**chemolithotrophs,** meaning "rock feeders") are able to use inorganic molecules such as ammonia, hydrogen sulfide, hydrogen, and iron as sources of energy. The most common inorganic feeders are the iron bacteria responsible for the brown stains that form on plumbing fixtures in regions where tap water contains high levels of iron. Sulfur bacteria are associated with iron bacteria, and are responsible for the slimy black deposits that you will probably find if you lift the stopper out of the drain in your bathroom sink. It's hard to understand how bacteria can live in that environment, because it contains little that we would consider food; but in iron compounds and other inorganic molecules, these bacteria utilize a completely different type of "food."

On a larger scale, inorganic feeders are responsible for acid mine drainage. The ore that is being mined usually makes up only a small part of the total amount of rock that is removed from a mine. The portion that does not contain ore is discarded on the ground surface in piles called "tailings." This material is often rich in minerals such as pyrites (iron sulfides). It's not of any nutritional value to us, but inorganic feeders can gain energy by oxidizing these minerals and, in the process, they release compounds that combine with rainwater to produce strong acids, such as sulfuric acid. When this acidic water drains into streams, it can kill fish and aquatic plants and insects.

"Light self-feeders" (**photoautotrophs**)—bacteria that use the energy from sunlight—contain chlorophyll and use the light energy to convert carbon dioxide to glucose by photosynthesis. The floating mats of gooey green material that you see in roadside ditches are a type of photoautotroph

METABOLIC DIVERSITY AMONG BACTERIA

CHEMOORGANOTROPHS
• Feed on organic molecules

CHEMOLITHOTROPHS
• Feed on inorganic molecules

PHOTOAUTOTROPHS
• Use energy from sunlight to produce glucose via photosynthesis

You may view them as "gross." But bacteria thrive in so many places because they are resourceful, able to extract "food" from a very broad range of sources, including organic molecules on your shower curtain, inorganic molecules in your pipes, and even using energy from the sun.

FIGURE 13-10 Resourceful feeders. Bacteria can metabolize almost anything.

called cyanobacteria. The cyanobacteria living today closely resemble the first photosynthetic organisms that appeared on earth about 2.6 billion years ago. That date was the start of a major global change, because until then the earth's atmosphere contained no free oxygen—instead, air consisted almost entirely of nitrogen and carbon dioxide. Cyanobacteria were the first organisms that could use solar energy to build organic compounds from carbon dioxide, and in the process break down water molecules to release free oxygen. The accumulation of oxygen released by cyanobacteria is called the **Oxygen Revolution.** Oxygen—which humans depend on—now makes up about 21% of the volume of air, and cyanobacteria still release important quantities of oxygen into the atmosphere.

TAKE-HOME MESSAGE 13•6

Some bacteria eat organic molecules, some eat minerals, and still others carry out photosynthesis. About 2.6 billion years ago, the photosynthesizing bacteria were responsible for the first appearance of free oxygen in the earth's atmosphere.

13•7

Bacteria cause many human diseases.

The number of **pathogenic** (disease-causing) bacteria is tiny compared with the total number of species of bacteria, but some pathogens kill millions of people annually, despite advances in medicine and sanitation. Some bacteria are always pathogenic (such as those that produce cholera, plague, and tuberculosis), but others (the ones responsible for acne, strep throat, scarlet fever, and "flesh-eating" necrotizing fasciitis) are normal parts of the communities of bacteria that live in or on humans. These bacteria in our "normal flora" become pathogenic only under special circumstances.

The great cholera epidemic that devastated London in 1854 became a milestone in epidemiology (the study of the occurrence of outbreaks of disease), when Dr. John Snow identified the Broad Street pump as the source of infection by mapping the pattern of deaths from cholera (**FIGURE 13-11**). When Dr. Snow persuaded the authorities to shut down the pump by removing the handle, new cases of cholera in the area dropped sharply.

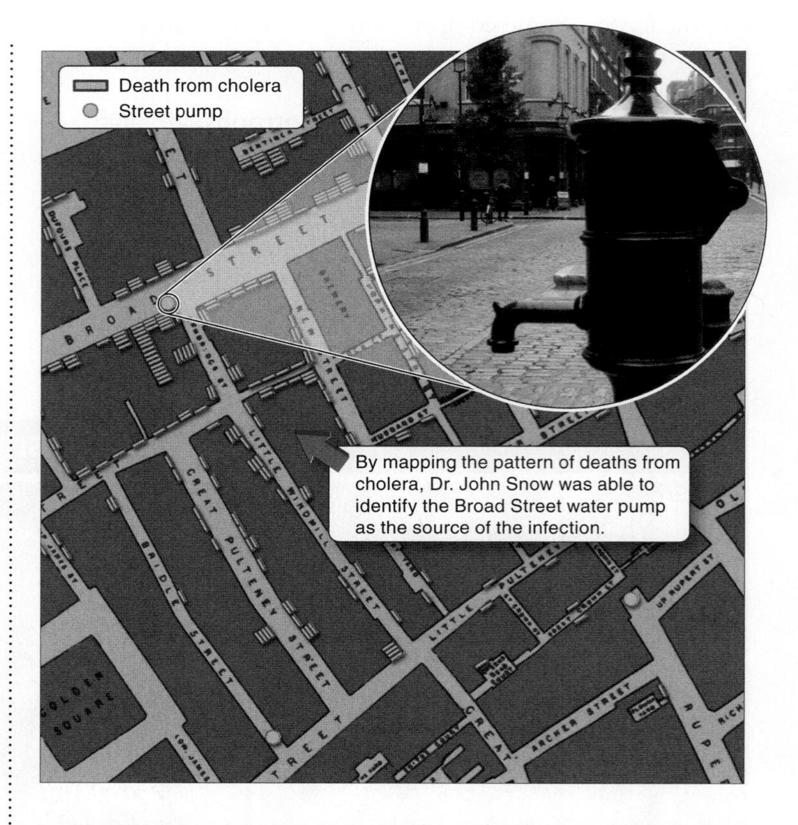

FIGURE 13-11 Contaminated! Water from the Broad Street pump in London was the source of a cholera outbreak that killed more than 500 people.

Cholera epidemics are a continuing threat, especially in areas without effective sanitation. Epidemics of cholera are occurring now in parts of Iraq where the sewage systems have been destroyed. The strains of cholera found in these areas are especially potent; as the bacteria multiply inside their hosts, they cause severe diarrhea with a massive loss of water. Victims of these severe strains of cholera are incapacitated and soon die, but before they die they release billions of cholera bacteria in diarrhea—an effective strategy for the cholera bacteria, because they contaminate water used for drinking or bathing and rapidly infect new victims.

Streptococcus pyogenes is a normal part of the bacterial community of your nose and mouth. Normally it is harmless, and many people pass their entire lives without experiencing any of the diseases it can cause. When the population of *S. pyogenes* is not held in check by competition with other members of the bacterial community, however, it can become a pathogen. Strep throat (more formally known as "streptococcal pharyngitis") is the most common disease caused by *S. pyogenes;* it produces a severe sore throat and a

Streptococcus pyogenes is usually harmless, but some strains are responsible for strep throat, scarlet fever, and necrotizing fasciitis (flesh-eating bacteria).

FIGURE 13-12 Strep throat is the most common disease caused by *S. pyogenes.*

distinctive rash of red abscesses with white pus at the top of the throat (**FIGURE 13-12**). Some strains of *S. pyogenes* produce a toxin that is released into the bloodstream and produces a red rash that spreads across the skin—the disease called scarlet fever. Most threatening is an infection caused by strains of *S. pyogenes* that have a toxin that allows them to enter body tissues, where they produce necrotizing fasciitis. These strains of *S. pyogenes* are known by their well-earned name, "the flesh-eating bacteria."

TAKE-HOME MESSAGE 13•7

Some bacteria always cause disease and others do no harm except under certain conditions. For example, *Streptococcus pyogenes* can be harmless, but under some conditions releases toxins that are responsible for strep throat, scarlet fever, and necrotizing fasciitis (caused by the flesh-eating strains).

13•8

Bacteria evolve drug resistance quickly.

Q: Where do antibiotics come from, and why do they so quickly lose their effectiveness?

Penicillin was the first antibiotic to be manufactured and used widely in the battle against illness-inducing bacteria. It came into use during the Second World War and caused a revolution in the care of the wounded. But antibiotic-resistant infections soon appeared, and the number of resistant bacteria has increased rapidly ever since (**FIGURE 13-13**). Now, 60 years after the first use of antibiotics, many bacteria are resistant to many, or even to most, antibiotics. In the United States, more people now die of *Staphylococcus aureus* infections that are resistant to many different antibiotics (MRSA infections) than die of HIV/AIDS. A few years ago, antibiotic-resistant staph infections were acquired only in hospitals, but now these infections have spread from hospitals to the community at large. Instead of contracting antibiotic-resistant staph infections in hospitals, patients (or their visitors) are now bringing the resistant bacteria into the hospital with them!

Microbes live everywhere they can, and compete for the best places to attach themselves and the richest sources of food to eat. This competition takes many forms: rapid growth to crowd competitors out of a living space, superior ability to take in nutrients to starve competitors, and—this is the key to both the benefits of antibiotics and the problem of antibiotic resistance for humans—production of chemicals that kill other microbes, or at least stop them from growing. Antibiotics are produced by microbes to help them compete with other microbes, and most of the antibiotics we use today are derived from microbes.

Bacteria and other microbes have developed a variety of ways to resist antibiotics. For example, some bacteria pump an antibiotic out of their cells as fast as it enters, so it never reaches a lethal concentration inside the bacterial cell. Other bacteria have proteins that bind to the antibiotic molecule and block its lethal effect. Still others carry that approach a step further—they have enzymes that break down the antibiotic molecules, which are then used as fuel to help the bacteria grow faster! Many of the genes that code for resistance are on plasmids. This means that a bacterium carrying a resistance gene can transmit the gene to other bacteria by conjugation; there's no need to wait for multiple generations of natural selection.

Q: Why is it essential to take every dose of an antibiotic prescribed by a doctor?

When an antibiotic is taken as prescribed—that is, at the times specified on the label and until all the pills have been consumed—the population of target bacteria is greatly reduced. All of the target bacteria that

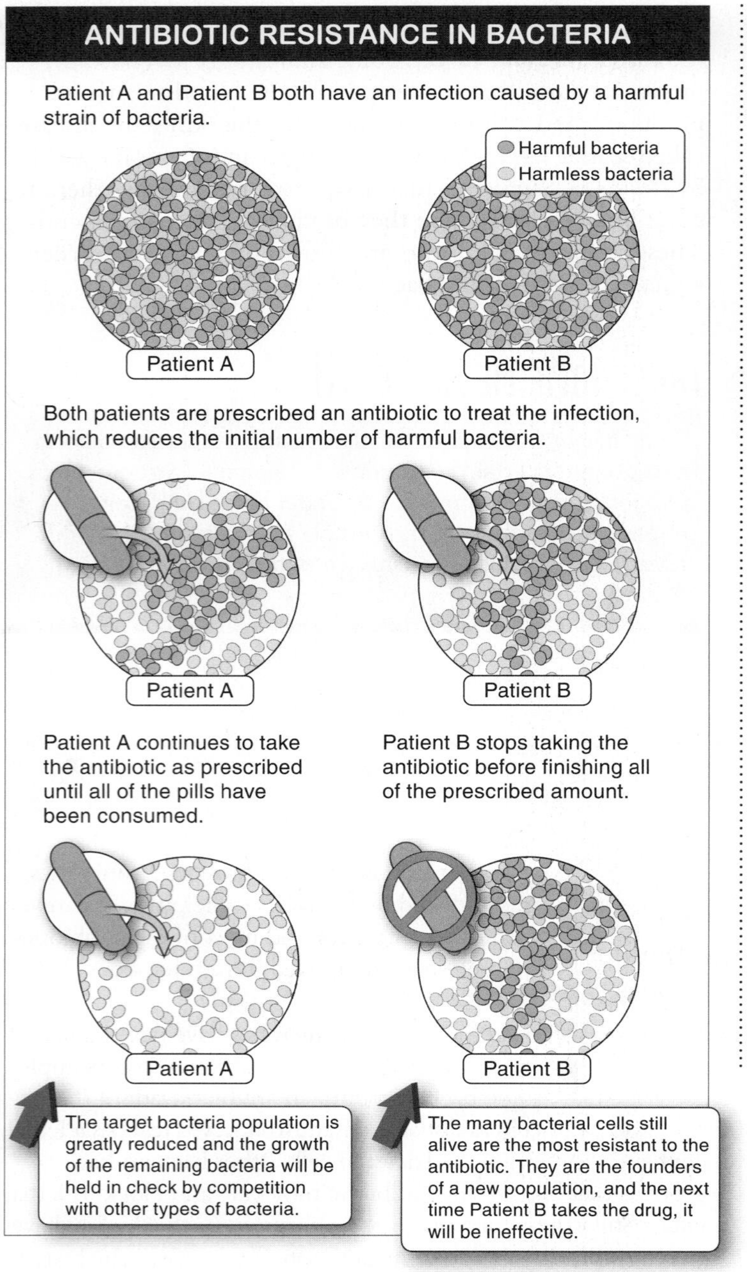

ANTIBIOTIC RESISTANCE IN BACTERIA

Patient A and Patient B both have an infection caused by a harmful strain of bacteria.

Harmful bacteria
Harmless bacteria

Patient A

Patient B

Both patients are prescribed an antibiotic to treat the infection, which reduces the initial number of harmful bacteria.

Patient A

Patient B

Patient A continues to take the antibiotic as prescribed until all of the pills have been consumed.

Patient B stops taking the antibiotic before finishing all of the prescribed amount.

Patient A

Patient B

The target bacteria population is greatly reduced and the growth of the remaining bacteria will be held in check by competition with other types of bacteria.

The many bacterial cells still alive are the most resistant to the antibiotic. They are the founders of a new population, and the next time Patient B takes the drug, it will be ineffective.

FIGURE 13-13 Evolving resistance. Sixty years after the appearance of antibiotics, many bacteria are resistant to many, or even most, antibiotics.

remain are resistant to the antibiotic, but there are not very many of them. The growth of these resistant bacteria will be held in check by competition with other types of bacteria. But, if you stop taking the antibiotic before you have finished all of the prescription, many of the target bacterial cells will remain alive, including the ones that are most resistant to the antibiotic. These resistant cells will be the founders of a new population of bacteria in your body, so the next time you take that drug it will be ineffective. Even worse, taking antibiotics when they are not needed—to treat a viral infection, for example—selects for resistant bacteria without providing any benefit, because antibiotics have no effect on viruses.

FIGURE 13-14 Antibiotics are used in agriculture. Livestock are given antibiotics to prevent diseases that are easily spread in crowded living conditions. The practice can cause the evolution of bacteria that are resistant to antibiotics.

The use of antibiotics in agriculture is another reason for the spread of antibiotic resistance. Low concentrations of antibiotics are routinely added to the feed for cattle, hogs, chickens, and turkeys. This can be beneficial in the short-term, promoting growth and minimizing disease in the crowded conditions of commercial meat and milk production. But in the long run it can have disastrous consequences, as the practice can lead to selection for bacteria resistant to the antibiotics. The antibiotics can also pass through the food chain to humans (**FIGURE 13-14**). Data gathered by the Union of Concerned Scientists indicate that agriculture in the United States uses about 25 million pounds of antibiotics each year—about eight times more than is used for all human medicine!

TAKE-HOME MESSAGE 13•8

Microbes routinely evolve resistance to antibiotics. The genes that allow bacteria to combat antibiotics are located on plasmids, and plasmid transfer allows an antibiotic-resistant bacterium to pass that resistance to other bacteria. Excessive use of antibiotics in medicine and agriculture has made several of the most important pathogenic bacteria resistant to every known antibiotic, and infections caused by these bacteria are nearly impossible to treat.

3 Archaea exploit some of the most extreme habitats.

A steamy pool in Rotorua, North Island, New Zealand, heated by volcanic vents and home to heat-tolerant archaea.

13•9

Archaea are profoundly different from bacteria.

When you look at the photographs of archaea in **FIGURE 13-15**, you may find it hard to believe that they are even a little bit different from bacteria, let alone *profoundly* different. And you would be right to wonder, because the differences between bacteria and archaea are not externally visible. Both live as single cells or colonies of cells, both are surrounded by a plasma membrane, and both have species with flagella that twirl like propellers. In fact, biologists considered the archaea to be bacteria until the 1970s. Only then did comparisons of DNA reveal that archaea are as different from bacteria as humans are.

Archaea look very much like bacteria. But closer inspection—of their physiology, biochemistry, and DNA—reveals them to be profoundly different from all bacteria.

FIGURE 13-15 Appearances are deceiving.

Those studies of archaean nucleotide sequences stimulated other comparisons, which identified additional differences among bacteria, archaea, and eukarya. For example, the chemical compositions of the plasma membranes, cell walls, and flagella of archaea are qualitatively different from those of bacteria. And beyond the large DNA sequence difference and differences in the plasma membrane, cell walls, and flagella, a third difference helps explain why archaea are placed between bacteria and eukarya on the tree of life (see Figure 13-4): eukarya have a distinct cell nucleus that is separated from the cytoplasm by a nuclear membrane, whereas bacteria and archaea do not have either a nucleus or a nuclear membrane. Having a nucleus provides protection for the chromosomes and allows the cell to control what molecules interact with its DNA.

TAKE-HOME MESSAGE 13•9

Archaea show a set of characteristics that places them between bacteria and eukaryotes on the tree of life. Archaea and bacteria may look similar, but they possess large and significant differences in their DNA sequences, as well as differences in their plasma membranes, cell walls, and flagella. Furthermore, neither archaea nor bacteria resemble eukarya in one key way. Eukarya alone possess a distinct cell nucleus and nuclear membrane.

13 • 10

Archaea thrive in habitats too extreme for most other organisms.

Archaea are famous for their ability to live in places where life would seem to be impossible, such as in water around hydrothermal vents that emerge from the seafloor that would be boiling if it were at sea level. At 2,000 meters below the surface, the water pressure reaches 200 atmospheres, or almost 3,000 pounds per square inch, and the 400° C water cannot boil. Yet the archaea thrive there. Most organisms die at temperatures between 40° and 50° C, because their protein molecules are denatured, so the ability of archaea to survive at temperatures above 100° C is truly remarkable.

Equally impressive is the ability of archaea to live in water as acidic as pH 0 or as salty as a saturated NaCl solution, and to metabolize an extraordinary range of compounds—including sulfur, iron, and hydrogen gas—to obtain energy (**FIGURE 13-16**). Some bacteria also can tolerate extreme physical and chemical conditions, and the organisms that can live in these conditions, both bacteria and archaea, are called **extremophiles** ("lovers of extreme conditions").

FIGURE 13-16 Living in the harshest environments. Archaea thrive in many extreme environments, such as this salt pond.

Archaea in your intestine break down a chemical bond found in beans (a bond humans cannot break)...but the process generates gas, which can cause discomfort as it tries to escape.

FIGURE 13-17 Helping you digest tough bonds. Intestinal archaea help you digest some chemical bonds in beans.

But not all archaea are extremophiles. Archaea live everywhere that bacteria do, and many of them live in places you would find perfectly comfortable yourself. In fact, you are home to many archaea and, on occasion, some of them make their presence all too obvious. Beans are notorious for their tendency to produce gas in the intestine, but the beans themselves are not really to blame—archaea are the culprits. Beans contain a couple of carbohydrates with chemical bonds that are not broken down very well by any human enzymes (**FIGURE 13-17**). Methane-producing archaea, on the other hand, have no such difficulties, producing an enzyme that targets those bonds. As a result, it is the archaea in your intestine that digest most of these carbohydrates. But in the process of breaking these bonds, they produce gases that, as they escape the digestive system, can cause considerable distress.

TAKE-HOME MESSAGE 13 • 10

Archaea can tolerate extreme physical and chemical conditions that are impossible for most other living organisms, but some live in moderate conditions and even in the human intestine.

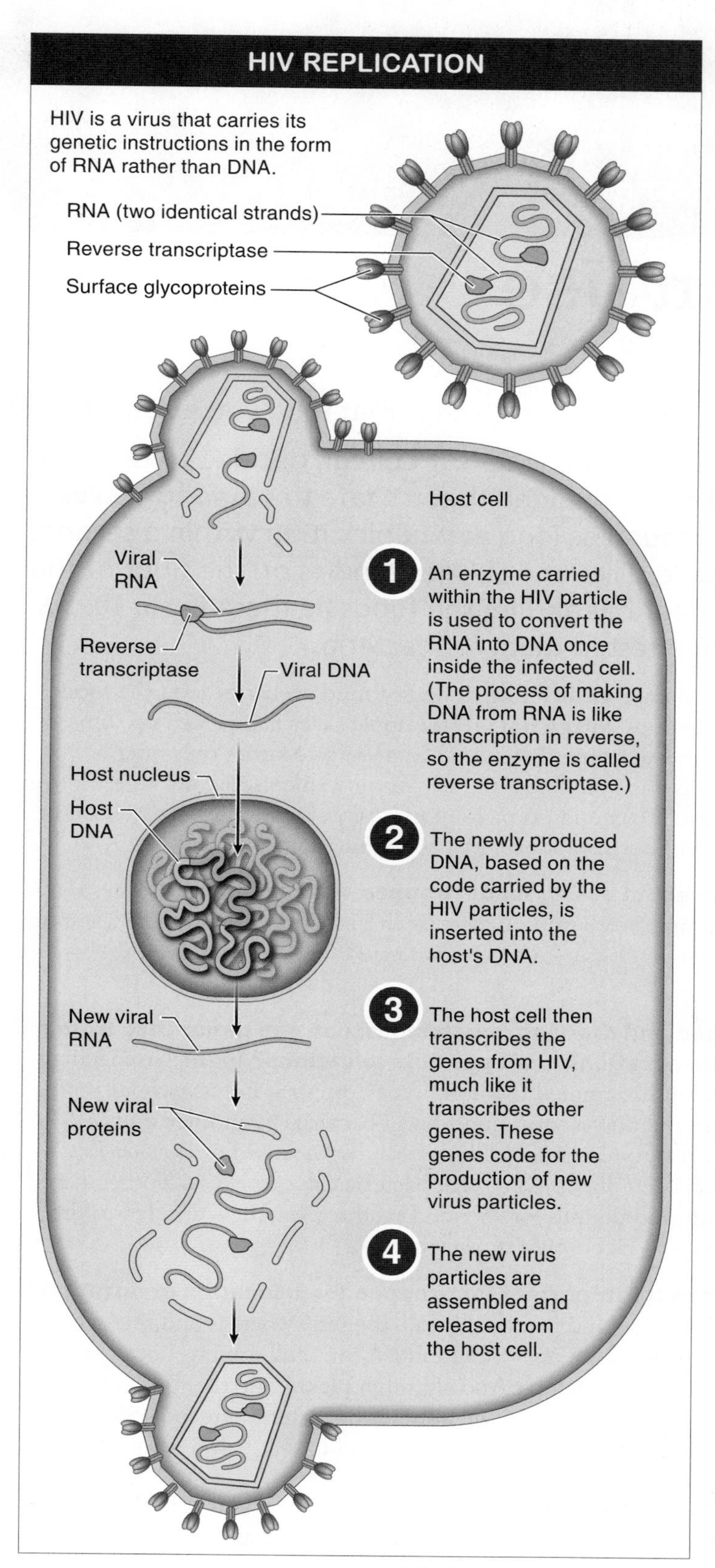

FIGURE 13-28 HIV infection. This RNA virus takes over the replicating machinery of a white blood cell to produce a new generation of HIV particles. HIV mutates rapidly during this process, because the reverse transcriptase enzyme is highly error-prone.

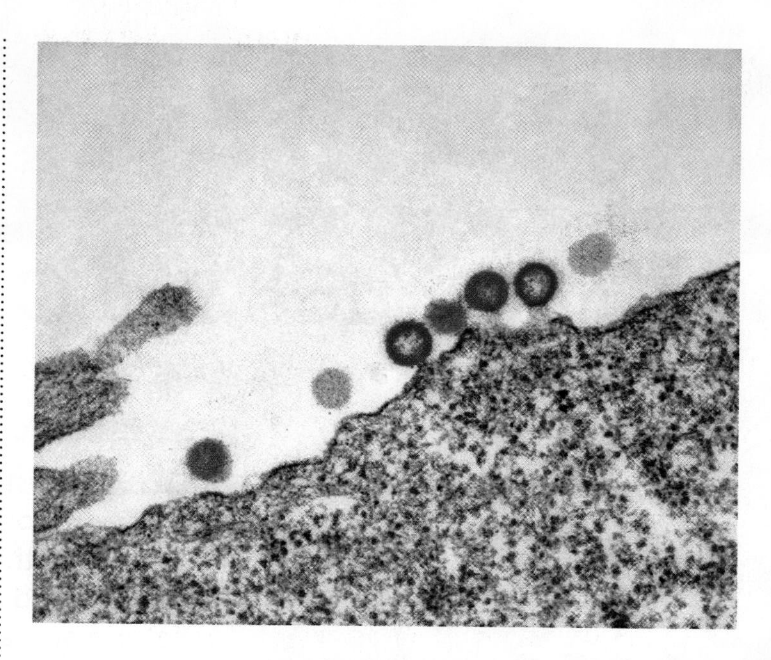

FIGURE 13-29 HIV attacks white blood cells essential for identifying foreign invaders.

Each time HIV infects another white blood cell, the reverse transcriptase makes errors in transcribing the RNA to DNA, and eventually one of the mutations allows the virus to bind to the glycoprotein on the surface of a specialized type of white blood cell—a bacteria- and virus-hunting white blood cell that is critically important in identifying disease-causing threats. A suitable mutation may occur in a couple of years, but more often it takes about 10 years or longer. But when it does happen, it signals a new stage in the HIV infection, the development of AIDS (**FIGURE 13-29**).

The immune system collapses. Normally, white blood cells all work together to identify and destroy cells that have been infected by a virus. When HIV begins to kill the cells that hunt for viruses and bacteria, the immune system begins to fail—it can no longer respond to HIV, or to any other infectious agent. Patients with AIDS develop multiple infections, bacterial and viral, as well as cancers, because they have lost the immune system cells that would normally have marked infected and cancerous cells for destruction.

TAKE-HOME MESSAGE 13•18

HIV infection is especially difficult to control. Mutations change the properties of the virus so that it is hard for the immune system to recognize it, and they produce variants that are resistant to the drugs used to treat the HIV infection.

Knowledge You Can Use

The five-second rule. How clean is that food you just dropped?

In 2007, two students at Connecticut College decided to test the "five-second rule": the idea that it's safe to eat food you've dropped on the ground as long as you pick it up within 5 seconds. They dropped apple slices and Skittles candies on the floor of the cafeteria and a snack bar and let the foods lie there for 5, 10, 30, or 60 seconds, then tested them for bacteria.

Q: Is the five-second rule valid? The students found no bacteria on the food picked up within 30 seconds. After a minute, the apple slices had picked up some bacteria, but the Skittles had none (they found bacteria on Skittles only after 5 minutes). The students concluded that the five-second rule should get an extension, proposing that you have 30 seconds to pick up moist foods and more than a minute to pick up dry foods without risk of bacterial contamination.

Q: Should we relax about eating foods dropped on the floor? This question is a serious one and warrants more study. Each year in the United States there are more than 76 million cases of illness caused by contaminated food, including more than 5,000 that are fatal.

Q: Scientific method and drawing conclusions: can you generalize from a small number of observations to all possible situations? In the Connecticut College study, the students examined two food types, dropped in just two locations, and onto surfaces with unknown concentrations of bacteria. In another study, published in the *Journal of Applied Microbiology,* researchers focused on *Salmonella* bacteria. Because it has been documented that bacteria can survive on clothes, hands, sponges, cutting boards, and utensils for several days, the researchers decided to drop food on surfaces known to be covered in bacteria.

Q: Which factor is more important for dropped food: location or duration? Armed with slices of bologna and pieces of bread, the researchers found that when dropped on surfaces covered with *Salmonella* and left for a full minute, both types of food took up 1,500 to 80,000 bacteria. And although picking up the food quickly—within just 5 seconds—reduced by 90% the number of bacteria present, 150 to 8,000 bacteria still had time to hitch a ride on the food in that first 5 seconds.

A grubby conclusion. It's probably safe to say that if you're in the Connecticut College cafeteria, you can eat your dropped food as long as you pick it up quickly. But for all other situations, a bit more caution is advised—particularly, taking notice of the "drop zone." Unsanitary surfaces likely to have microbes will contaminate your food almost immediately. So hold on tight.

1 There are microbes in all three domains.

Microbes are grouped together because they are small, not because they are evolutionarily related. In fact, there are microbes in all three domains of life: bacteria, archaea, and eukarya. And viruses, which are on the border between living and non-living, are not in any domain. Because they are so small, microbes can also be simple—they do not need skeletons to support them, circulatory systems to transport materials, or nervous systems to coordinate stimulus with response.

2 Bacteria may be the most diverse of all organisms.

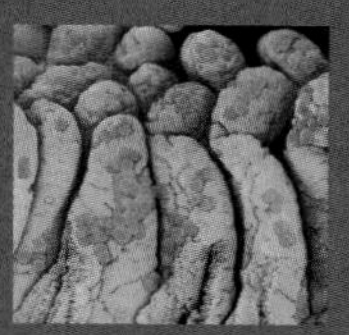

Bacteria are single-celled organisms with a very simple structure—they have no nucleus and no intracellular organelles. Despite their simplicity, bacteria are found virtually everywhere and can eat virtually anything. The entire bacterial chromosome (unlike eukaryotic chromosomes) consists of functional genes, and bacteria reproduce rapidly and efficiently by binary fission. Genes for specialized functions, such as antibiotic resistance, are carried on small DNA molecules called plasmids that can be transferred from one bacterium to anther. Bacterial generation times are measured in minutes or hours, and the combination of short generation times and plasmid transfer allows antibiotic resistance to spread rapidly.

3 Archaea exploit some of the most extreme habitats.

Archaea look very much like bacteria, but differ in several characteristics. Archaea thrive in extreme conditions—at pressures exceeding 200 atmospheres, at temperatures above 100° C, and in solutions as acidic as pH 0—and they can eat toxic substances. These characteristics make archaea potentially valuable for industrial applications, such as clearing deposits from pipes in oil refineries and power plants.

4 Most protists are single-celled eukaryotes.

Most protists are single-celled eukaryotes. But the only characteristic they all share is a nucleus. Amoebas and *Paramecium* are animal-like protists that eat bacterial cells. The so-called brown algae are multicellular protists. Some other protists are colonial—collections of cells that form large structures in which each cell carries out all the activities needed for life. Plasmodial slime molds are large single cells with multiple nuclei. *Plasmodium* is a parasitic protist that causes malaria. It constantly changes the way it appears and thus avoids detection by the host's immune system.

5 Viruses are at the border between living and non-living.

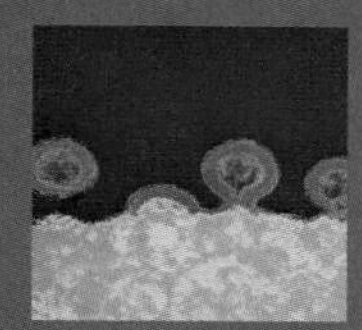

Viruses consist of genetic material (RNA or DNA) packed inside a protein coat. A virus must infect a living cell before it can carry out any of the activities characteristic of living organisms; the virus hijacks the machinery of the cell and uses it to make more viral genetic material and viral protein. RNA viruses (such as those that cause influenza) mutate rapidly. Human immunodeficiency virus (HIV) is an RNA retrovirus. The complex method by which retroviruses replicate their RNA is very error-prone, so HIV mutates especially fast, complicating the use of drug treatment for HIV infections. In addition, HIV infects two kinds of cells that are part of the immune system, eventually causing the system to fail. This condition, acquired immunodeficiency syndrome (AIDS), is characterized by the occurrence of multiple bacterial and viral infections and, often, cancer.

KEY TERMS

capsid, p. 504
capsule, p. 489
chemolithotroph, p. 493
chemoorganotroph, p. 493
conjugation, p. 491
extremophile, p. 498
Gram stain, p. 488
host, p. 503
microbe, p. 484
Oxygen Revolution, p. 494
parasite, p. 503
pathogenic, p. 494
peptidoglycan, p. 589
phagocytosis, p. 501
photoautotroph, p. 493
plasmid, p. 490
probiotic therapy, p. 492
retrovirus, p. 508
transduction, p. 491
transformation, p. 491

CHECK YOUR KNOWLEDGE

1. In the human body:
   a) there are more microbial cells than human cells.
   b) archaea can survive, but bacteria are killed by the immune system.
   c) bacterial cells can survive only inside the eukaryotic cells.
   d) the diversity of bacterial cells living in the mouth is exceeded by the diversity living on the skin.
   e) the bacterial cells are larger than the eukaryotic cells.

2. From an evolutionary perspective, the most basic division among all organisms on earth is among:
   a) plants, animals, and bacteria.
   b) bacteria, archaea, and eukarya.
   c) plants, animals, and eukarya.
   d) bacteria, animals, and eukarya.
   e) bacteria, archaea, and protists.

3. Which of the following statements about prokaryotes is incorrect?
   a) Prokaryotes contain ribosomes.
   b) Some prokaryotes can conduct photosynthesis.
   c) Prokaryotes have circular pieces of DNA within their nuclei.
   d) Prokaryotes appeared on earth prior to eukaryotes.
   e) Prokaryotes contain cytoplasm.

4. "Establishing a phylogeny for bacteria is more difficult than establishing a phylogeny for plants or animals." This statement is:
   a) correct, because of the quick generation time of bacteria.
   b) incorrect, because lateral gene transfer makes creating a phylogeny simpler.
   c) incorrect, because the metabolic diversity of bacteria makes classification impossible.
   d) correct, because it is difficult to obtain enough morphological data from bacteria.
   e) correct, because bacteria can engage in lateral gene transfer.

5. A population of *Escherichia coli* can double every ______ in an ideal laboratory culture.
   a) 20 seconds
   b) 2 minutes
   c) 2 days
   d) 20 minutes
   e) 2 hours

6. Plasmids containing genes for antibiotic resistance can be exchanged between bacterial cells in a culture by:
   a) conjugation.
   b) artificial exchange.
   c) cloning.
   d) transduction.
   e) conduction.

7. Which group of organisms utilizes the largest variety of energy sources?
   a) prokaryotes
   b) fungi
   c) protists
   d) invertebrate animals
   e) vertebrate animals

8. Which of the following statements about antibiotics is incorrect?
   a) Penicillin was the first antibiotic isolated from bacteria and widely used to fight bacterial infections.
   b) Antibiotics help microbes compete with other microbes.
   c) Antibiotic-resistant microbes are selected for in humans who are taking antibiotics.
   d) Antibiotics, though effective against viruses, are not effective against bacteria.
   e) Antibiotics are used not just in human health care but also in agriculture.

9. Which of the following domains are the most closely related, in that they share a unique common ancestor?
   a) archaea and bacteria
   b) archaea and eukarya
   c) bacteria and eukarya
   d) None of the above; all three domains evolved from different ancestors.
   e) None of the above; all three domains are equally related to each other.

10. Unlike other microbes, archaea are able to thrive:
    a) in very extreme (with respect to temperature, salinity, pressure, etc.) environments.
    b) in all the same places that bacteria are found.
    c) in the human body.
    d) in lakes and rivers (i.e., fresh water).
    e) in well-aerated soils.

11. *Taq* polymerase:
    a) is produced in viruses and bacteria.
    b) is denatured at temperatures above the normal human body temperature.
    c) is useful in catalyzing the polymerase chain reaction.
    d) can break down the sludge in oil refinery tanks.
    e) can reduce the mineral deposits from rusting pipes.

12. All protists are alike in that they all are:
    a) photosynthetic.
    b) eukaryotic.
    c) parasitic.
    d) colonial.
    e) multicellular.

13. Which of the following is not a protist?
    a) *Paramecium*
    b) slime mold
    c) brown alga
    d) diatom
    e) All of the above are protists.

14. Which of the following statements about *Plasmodium* is incorrect?
    a) It causes sickle-cell anemia.
    b) Because it changes its surface proteins frequently, it is largely invisible to the immune system.
    c) It is transmitted by *Paramecium*.
    d) It confers resistance to malaria.
    e) It is most common in North America.

15. Many biologists do not consider viruses to be alive. Which of the following characteristics of viruses leads to this conclusion?
a) Viruses lack a metabolic system.
b) Viruses do not respond to external stimuli.
c) Viruses are unable to reproduce on their own.
d) All of the above are correct.
e) Only a) and c) are correct.

16. The genetic information in all viruses is:
a) DNA.
b) RNA.
c) protein.
d) polymerase enzymes.
e) DNA or RNA.

17. Viruses are able to infect:
a) humans.
b) plants.
c) bacteria.
d) birds.
e) All of the above.

18. Which of the following statements about HIV is incorrect?
a) It is a virus.
b) It attacks red blood cells.
c) It mutates frequently.
d) It is derived from a simian immunodeficiency virus.
e) It contains RNA but not DNA.

Short-Answer Question

1. Complete the chart on the characteristics of microbes.

	Prokaryotes		Eukaryotes	Viruses
	Bacteria	Archaea	Protists	
Unicellular, colonial, or multicellular?				
Plasma membrane?				
Peptidoglycan in cell wall?				
Flagella? What type?				
Presence of nucleus?				
Intracellular organelles?				
Type of genetic material (DNA, RNA, or both possible)?				
Chromosome structure (linear or circular)?				

See Appendix for answers. For additional study questions, go to www.prep-u.com.

1 Population ecology is the study of how populations interact with their environments.

- **14.1** What is ecology?
- **14.2** A population perspective is necessary in ecology.
- **14.3** Populations can grow quickly for a while, but not forever.
- **14.4** A population's growth is limited by its environment.
- **14.5** Some populations cycle between large and small.
- **14.6** "Maximum sustainable yield" is a useful but nearly-impossible-to-implement concept.

2 A life history is like an executive summary of a species.

- **14.7** Life histories are shaped by natural selection.
- **14.8** Populations can be described quantitatively in life tables and survivorship curves.
- **14.9** There are tradeoffs between reproduction and longevity.

3 Ecology influences the evolution of aging in a population.

- **14.10** Things fall apart: what is aging and why does it occur?
- **14.11** What determines the average longevity in different species?
- **14.12** Can we slow down the process of aging?

4 The human population is growing rapidly.

- **14.13** Age pyramids reveal much about a population.
- **14.14** Moving from third world to first world a demographic transition occurs.
- **14.15** Human population growth: how high can it go?

14

Population Ecology

Planet at capacity: patterns of population growth

① Population ecology is the study of how populations interact with their environments.

Giraffe and zebra forage for food on the Masai Mara National Reserve, Kenya.

14•1

What is ecology?

Lobster tastes delicious. Many people consider it one of the finest delicacies. It's not surprising, then, that catching and selling lobsters is big business. Generating almost $200 million a year for the State of Maine alone, it is among the most economically important businesses in New England. Now imagine for a minute that you were placed in charge of the lobster industry. The 6,000 lobster fishermen in Maine depend on your managing the lobster fishery so that not only can they catch and sell enough lobsters to survive this year, but sufficient numbers of lobsters are left behind to ensure that there will be lobsters to catch in all the years to come. How would you do it? You would be faced with some tough questions:

- How many lobsters should you allow each fisherman to take each day? Each year?
- Should you require that certain lobsters be thrown back? If so, should they be the biggest or the smallest? Does it matter whether they are males or females?
- Is it best to increase the lobster population size or should it be maintained at current levels?

Ecologists have a challenging task managing natural populations when it's difficult to count the number of individuals present or determine how many the environment can support.

FIGURE 14-1 Managing valuable resources. Population growth models can help.

- But wait a minute. Forget about these obviously complex questions and start with a seemingly simpler question: how many lobsters currently live in the waters off the coast of Maine? Is it even possible to have a clue as to what this number is (**FIGURE 14-1**)?

Welcome to ecology.

These difficult questions, and many others like them, are all part of **ecology,** a subdiscipline of biology defined as the study of the interactions between organisms and their environments. But don't be fooled by the simple definition. Ecology encompasses a very large range of interactions and units of observation and is studied at different levels (**FIGURE 14-2**). These include:

- *Individuals:* How do individual organisms respond biochemically, physiologically, and behaviorally to their environment?
- *Populations:* How do groups of interbreeding individuals change, in terms of their growth rates, distributions, and genetic makeup, over time?
- *Communities:* How do the populations of different species in a locale interact with each other?
- *Ecosystems:* At the highest level of organization in ecology, how do the living and non-living elements interact in a particular area, such as a forest, desert, or wetland?

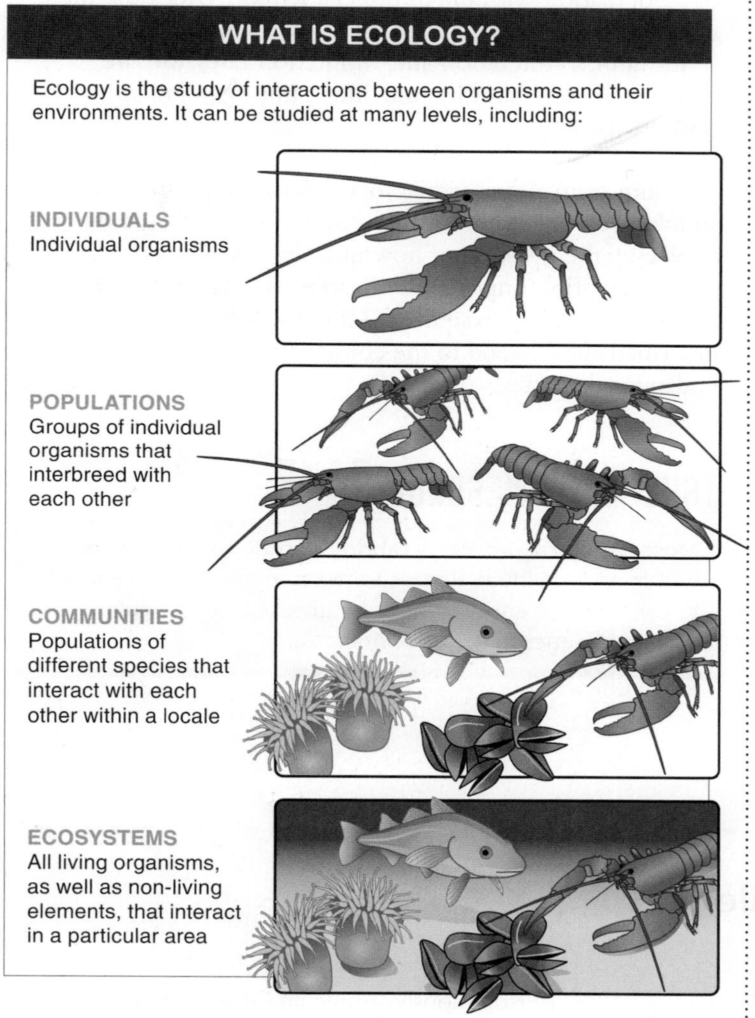

WHAT IS ECOLOGY?

Ecology is the study of interactions between organisms and their environments. It can be studied at many levels, including:

INDIVIDUALS
Individual organisms

POPULATIONS
Groups of individual organisms that interbreed with each other

COMMUNITIES
Populations of different species that interact with each other within a locale

ECOSYSTEMS
All living organisms, as well as non-living elements, that interact in a particular area

FIGURE 14-2 From individuals to ecosystems. Ecologists study living organisms and their relationship to the environment.

> I wandered lonely as a cloud
> That floats on high o'er vales and hills,
> When all at once I saw a crowd,
> A host, of golden daffodils.
>
> — William Wordsworth, *Daffodils,* 1804

We explore ecology at all of these levels throughout this and the next two chapters. We focus in this chapter on **population ecology,** a subfield of ecology that focuses on populations of organisms of a species and how they interact with the environment. We also examine the special case of human population growth and its impacts on the environment, and the ways in which ecological knowledge can give insight and answers that make possible effective conservation biology policies. In the next chapter, we move from populations to communities and ecosystems, where our focus will be on how species interact with each other and with their environments, and how this influences their own evolution and the environment itself.

TAKE-HOME MESSAGE 14•1

Population ecology is the study of the interactions between populations of organisms and their environments, particularly their patterns of growth and how they are influenced by other species and by environmental factors.

14•2

A population perspective is necessary in ecology.

As we begin our study of population ecology, a subtle but critical shift in perspective is required. It is a shift from viewing the individual as the primary focus of our attention to viewing the population—a group of organisms of the same species living in a particular geographic region. Most ecological processes cannot be observed or studied within an individual. Rather, they emerge when considering the entire group of individuals that regularly exchange genes in a particular locale.

Population ecology examines features that cannot be studied on an individual organism, such as population size.

FIGURE 14-3 A change in perspective. Ecology requires a focus on populations, rather than isolated individuals.

As an example of how a population rather than individual perspective is needed, consider the case of adaptations produced by natural selection. Genetic changes don't occur within an individual. That is, a longer neck did not evolve in one individual giraffe. Instead, the genetic changes occurred within a population of giraffes over time. As a consequence of differential reproductive success among the individuals of a population with different neck lengths, over time there came to be more individuals with longer necks. Birth rates, death rates, immigration and emigration rates, too, are features possessed not by individuals but by populations (**FIGURE 14-3**).

The shift from an individual to a population perspective should become clearer in the next sections. We begin by exploring the question of how quickly or slowly populations grow, a question important both to the management of consumable natural resources—as in the lobster fisheries described above—and to the conservation of rare and endangered species.

TAKE-HOME MESSAGE 14•2

Most ecological processes cannot be observed or studied within an individual. Rather, when studying them we need to consider the entire group of individuals that regularly exchange genes in a particular locale.

14•3

Populations can grow quickly for a while, but not forever.

Q Who leaves more surviving offspring, a pair of elephants or a pair of rabbits?

In stable populations, how many of the five million eggs that a female cod might lay over the course of her life will, on average, survive and grow to adulthood? How many babies that an elephant gives birth to will survive to adulthood and reproduce? These may seem like difficult questions to answer, requiring complex mathematical models of population growth. But actually, the most fundamental fact about population growth is ridiculously simple and doesn't require special knowledge about fish or elephants. Reasoning and logic will give us the answer. Think about those two questions: how many cod eggs and baby elephants produced by one female will make it to adulthood?

Just two. If a population is not growing, each individual is ultimately replacing itself with a new individual. This is true for all species. Two of those cod eggs will make it. And two of those baby elephants will make it. (For a stable population, two survive, rather than one, because both a male and a female contributed to each offspring. If each pair produced only one, the population would get smaller and smaller.)

> There is no exception to the rule that every organic being naturally increases at so high a rate that, if not destroyed, the Earth would soon be covered by the progeny of a single pair.
>
> —Charles Darwin, *The Origin of Species,* 1859

What would happen if more than two eggs or two babies survived? If there were more than one offspring per individual, the population would grow and grow and grow until the earth was covered with cod, elephants, and every other species. But that hasn't happened, and it can't. Let's investigate why.

We start by figuring out how a population grows if each individual does more than just "replace" itself. Two pieces of information will tell us how much a population is growing (or shrinking). The first is the per capita growth rate, abbreviated as *r*. This is simply the rate of births minus the rate of deaths.

If the rate of births is greater than the rate of deaths, the population gets bigger. If the death rate is bigger than the birth rate, *r* becomes a negative number and the population gets smaller. For example, if there are 500 individuals in a population and, over the course of a year, 125 offspring are born, the birth rate is 125/500, or 0.25 births per individual. And if 25 of the 500 individuals die during the course of the same year, the death rate is 25/500, or 0.05 deaths per individual. In this population, then, the growth rate is 0.25 – 0.05, or 0.20 individuals per individual. But how many individuals is that?

To calculate the **growth rate** of a population—that is, the change in the number of individuals in the population in some unit of time, such as a year—we need to know the number of individuals in the population now. The growth of the population in a year is the growth rate times the number of individuals present, written as $r \times N$.

In the population just described, we said there were 500 individuals. This population would increase by 0.20×500, or 100 people, during a year (the time period observed). After a second year, if it grew at the same rate, the population growth would be 0.20×600, or 120 new individuals. This would give the population 720 individuals.

With the same calculations, we could determine the population size for the next 10 or even 50 years—these numbers are plotted in **FIGURE 14-4.** When a population grows at a rate that is proportional to its current size—in other words, the bigger the population, the faster it grows—the growth is called **exponential growth.** As the graph reveals,

EXPONENTIAL GROWTH

Exponential growth occurs when each individual produces more than the single offspring necessary to replace itself. According to realistic (and moderate) estimates of birth and death rates, a population of just 500 elk would grow to more than a billion individuals within 80 years and eventually would cover the earth.

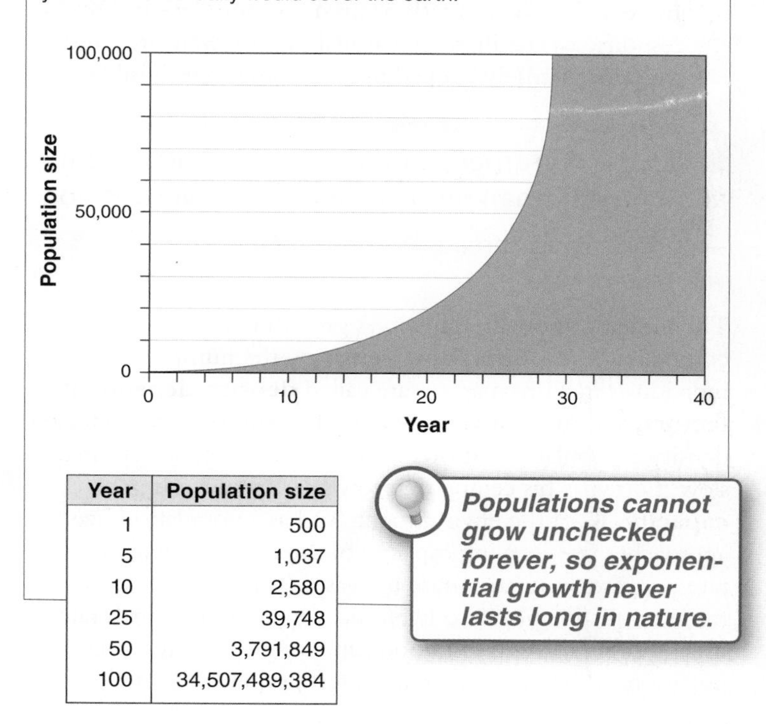

Year	Population size
1	500
5	1,037
10	2,580
25	39,748
50	3,791,849
100	34,507,489,384

FIGURE 14-4 Unchecked growth. Any animal or plant species that grew exponentially would eventually cover the planet.

the size of a population growing exponentially very quickly becomes astronomical. In 10 years, our population would have grown from 500 individuals to 2,580. In 50 years, it would reach almost four million. In fact, after 80 years, the population would pass a billion individuals. It's clear that exponential population growth ends badly.

But the world isn't overrun with people or cod or elephants, so, clearly, exponential growth doesn't occur for long. In the next section, we explore why, in the real world, populations cannot simply grow unchecked forever, and we add some realism to the exponential growth description.

TAKE-HOME MESSAGE 14•3

Populations tend to grow exponentially, but this growth is eventually limited.

14•4

A population's growth is limited by its environment.

Life gets harder for individuals when it gets crowded. Whether you are an insect, a plant, a small mammal, or a human, difficulties arise in conjunction with increasing competition for limited resources. In particular, as population size increases, organisms experience:

- Reduced food supplies, due to competition
- Diminished accessibility to places to live and breed, also due to competition
- Increased incidence of parasites and disease, which can spread more easily when their hosts live at higher density
- Increased predation risk, as predator populations grow in response to the increased availability of their prey and as the more densely packed prey become more visible

> "I would rather sit on a pumpkin and have it all to myself, than be crowded on a velvet cushion."
> —Henry David Thoreau, *Walden,* 1854

The limitations on a population's growth that are a consequence of **population density**—the number of individuals in a given area—are called **density-dependent factors,** and they cause more than discomfort. With increased density, a population's growth is reduced as limited resources slow it down. This ceiling on growth is the **carrying capacity, *K*,** of the environment. And as a population size approaches the carrying capacity of the environment, death rate increases, migration rate increases (as individuals seek more hospitable places to live), and a reduction in birth rate usually occurs, too, as low food supplies give rise to poor nutrition, which, in turn, reduces fertility (**FIGURE 14-5**).

DENSITY-DEPENDENT FACTORS LIMIT POPULATION SIZE

DENSITY-DEPENDENT FACTORS
- Food supply
- Habitat for living and breeding
- Parasite and disease risk
- Predation risk

FIGURE 14-5 Fighting over scarce resources. These wolves cannot miss an opportunity for a meal.

Here's how the carrying capacity of an environment influences a population's growth. We start with our exponential growth equation, which calculated a population's growth as $r \times N$. Then we multiply this growth rate by a term, $[(K - N)/ K]$, that varies between 0 and 1 and can slow down exponential growth.

If the new term, $[(K - N)/ K]$, is close to 1, we are multiplying normal exponential growth by 1, which doesn't change it. This happens when N, the population size, is small relative to the carrying capacity, K (meaning that there are plenty of resources to go around). But if the term $[(K - N)/ K]$ is less than 1 and close to 0, we are multiplying the exponential growth by that small number, so total growth is now only a small fraction of what it would have been. This occurs when N, the population size, is very close to the carrying capacity, K (in which case, the environment is full to capacity). When a

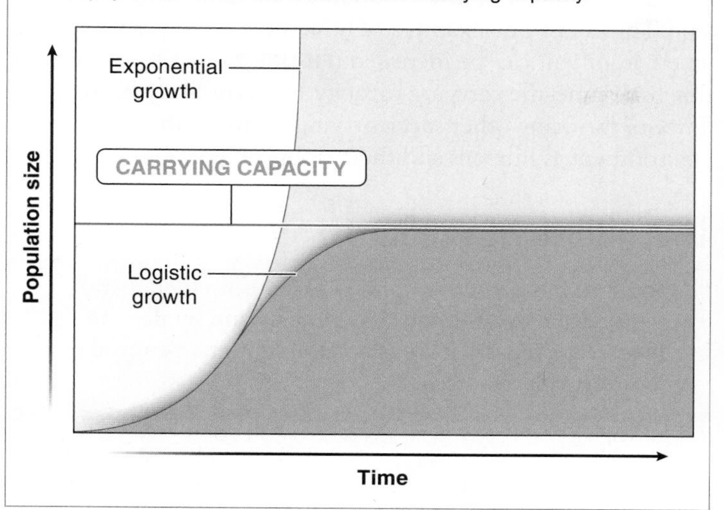

FIGURE 14-6 Lack of resources limits growth. Logistic growth is population growth that has stabilized because of limited resources.

population grows exponentially at first, but then its growth slows as the population size approaches the environment's carrying capacity—an S-shaped growth curve—the population growth is called **logistic growth** (**FIGURE 14-6**). Logistic growth is a much better approximation than exponential growth of how populations grow in the real world.

Density-independent factors can also knock a population down. These are forces that strike populations without regard for population size, increasing the death rate or decreasing the number and rate of offspring produced. Density-independent forces are like "bad luck" limits to population growth. They are mostly weather- or geology-based, including calamities such as floods, earthquakes, fires, and lightning. They also include habitat destruction by other species, such as humans. The population hit may be at its carrying capacity or it may be in the initial stages of exponential growth in which the population grows at an exponential growth rate, prior to having growth slowed by the carrying capacity of the environment. In either case, the density-independent force simply knocks down the population size. The population then resumes logistic growth.

In an environment in which these "bad luck" events repeatedly occur, a population might never have time to grow as high as the carrying capacity. Instead, the population might perpetually be in the exponential growth portion of the logistic growth curve, with periodic massive mortality events. The growth for this population would appear as a series of jagged curves, as seen in **FIGURE 14-7**.

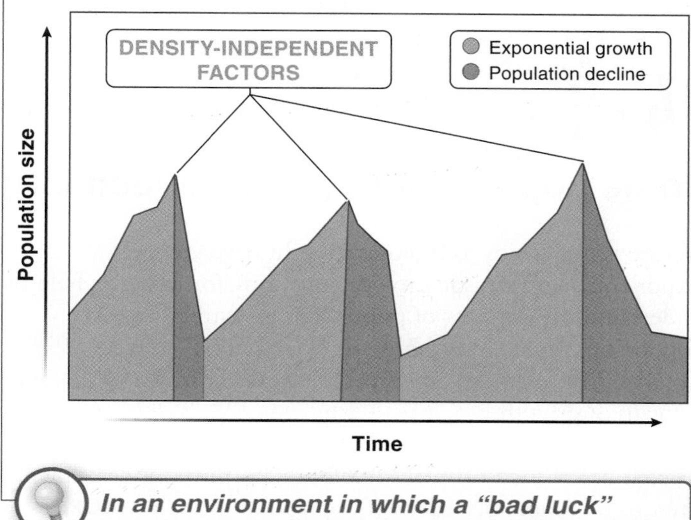

FIGURE 14-7 Events can limit population growth. Fires, floods, and earthquakes are density-independent forces that can dramatically reduce population size.

Q How many people can the earth support? Why does the number keep increasing?

The growth of populations doesn't always appear as a smooth S-shaped logistic growth curve. For some populations, particularly humans, the carrying capacity of an environment is not set in stone. Consider that, in 1883, an acre of farmland in the United States produced an average of 11.5 bushels of corn. By 1933, this was up to 19.5 bushels per acre. And by 1992, it had increased to 95 bushels per acre! How did this happen? The

FIGURE 14-8 Efficient crop production. Advances in agriculture make it possible to feed the world's growing population.

development of several agricultural technologies—including the use of vigorous hybrid varieties of corn, rich fertilizers, crop rotation, and effective pest management—has made it possible to produce more and more food from the same amount of land. This is just one example of how the carrying capacity of an environment can be increased (**FIGURE 14-8**). Of course, over this same time, the carrying capacity has most likely been *decreased* for many other species trying to live in the same environment as humans and their crops.

TAKE-HOME MESSAGE 14•4

A population's growth can be reduced both by density-dependent factors related to crowding and by density-independent factors such as natural or human-caused environmental calamities.

14•5

Some populations cycle between large and small.

Nature is not as tidy as biologists might have you believe. Exponential and logistic growth equations, for instance, help us understand the concept of population growth but real populations don't always show such "textbook" growth patterns. Instead, they sometimes vary greatly in their rates and patterns of growth.

Recent desert locust swarms of biblical proportions certainly attest to the unpredictability of population growth (**FIGURE 14-9**). In northwest Africa, the desert locusts (migrating grasshoppers) normally live as solitary individuals in relatively small, scattered populations. In 2004, though, the population size increased rapidly, probably due to unusually good rains and mild temperature. As the rainy season ended and green areas gradually shrank, however, the locusts became progressively concentrated in smaller and smaller areas. At this point, for reasons that are related to their overcrowding but are not completely understood, the locusts began behaving like a mob, rather than solitary individuals. Giant swarms—some including tens of millions of insects—began flying across huge expanses of land in search of food. They completely consumed huge swaths of farmland, causing more than $100 million in damage.

There is more than one way for populations to deviate from the standard population growth pattern. The explosive locust population growth, for example, occurs at unpredictable intervals. Another unusual pattern is the oscillations of the lynx and snowshoe hare populations of Canada. As seen in **FIGURE 14-10**, rather than smooth logistic growth, the populations of

FIGURE 14-9 Population explosion! Sometimes, populations don't grow logistically, as occasionally happens for desert locusts.

both the snowshoe hare and their predators, the lynx, have regular cycles between very large numbers and crashes to much smaller numbers. Thanks to the Hudson's Bay Company, which kept detailed records for decades on the number of pelts it purchased from fur trappers, this population cycling is well documented. Although its cause is not fully certain, to some extent, the predator and the prey cause their own cycling:

1. The hare population size grows,
2. providing more food for the lynx,
3. which then reproduce at a higher rate,

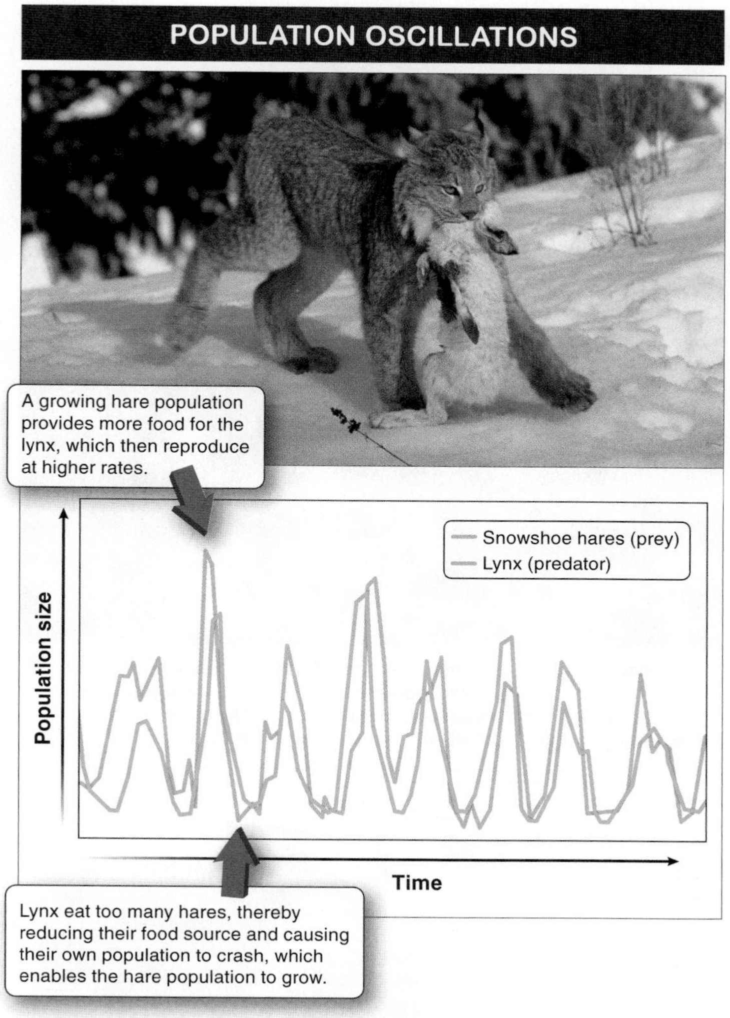

POPULATION OSCILLATIONS

A growing hare population provides more food for the lynx, which then reproduce at higher rates.

Snowshoe hares (prey)
Lynx (predator)

Population size

Time

Lynx eat too many hares, thereby reducing their food source and causing their own population to crash, which enables the hare population to grow.

FIGURE 14-10 Predator and prey. The snowshoe hare and lynx populations rise and crash on a 10-year cycle. The population numbers are based on the number of pelts sold to the Hudson's Bay Company by trappers.

4. causing them to eat too many of the hares,
5. thereby reducing their food source,
6. causing the lynx population to crash,
7. thus enabling the hare population to grow, and the cycle to begin anew.

In discussing these examples of unusual population growth, we must not lose sight of the fact that, for the most part, regularly cycling populations and populations with periodic huge outbreaks are more the exception than the rule. In general, the logistic growth pattern describes the general growth of populations better than any other model.

One population-growth myth that demands debunking is the story of lemmings and their supposed suicidal response to overcrowding. Do lemmings jump off cliffs, committing suicide, when their population size becomes too large? In a word: *no.*

FIGURE 14-11 An urban myth debunked. Lemmings do not jump off cliffs.

This is an urban myth. Lemming populations do occasionally experience large increases in size. Then, just as locust behavior changes when population density gets too great, lemming behavior changes a bit, and many individuals migrate in search of less crowded habitats and more food. As they enter unfamiliar territory, some lemmings may suffer increased rates of death. But these deaths are not suicidal, and they do not occur in large groups (**FIGURE 14-11**).

Q Do lemmings commit suicide by jumping off cliffs when their populations get too big?

How did the myth of lemming mass suicides become so prevalent? This can be attributed to a 1958 Disney documentary film, *White Wilderness.* In one sequence in this film, many lemmings were placed on a giant snow-covered turntable so that, when filmed, it would appear that they were migrating. In a later scene, many lemmings were filmed on a cliff overlooking a river, and then were chased into the water as if in a mad rush. None of these scenes were of actual lemming migrations, and in nature, lemmings never behave like this or do anything remotely resembling jumping off a cliff or committing suicide in any other way.

There are practical reasons for predicting population sizes and growth rates, as we'll see. But it can be difficult to translate simple growth models into feasible management practices.

TAKE-HOME MESSAGE 14•5

Although the logistic growth pattern is better than any other model for describing the general growth pattern of populations, some populations cycle between periods of rapid growth and rapid shrinkage.

14•6

"Maximum sustainable yield" is a useful but nearly-impossible-to-implement concept.

Suppose you are in charge of an eastern hardwood forest, working for a logging company, and responsible for selecting which trees to cut down. Or suppose you are in charge of the lobster fishery discussed at the beginning of this chapter, responsible for deciding how many lobsters should be harvested. In either case, your responsibilities would be very similar. In the case of the hardwood forest, how many trees would you cut down for maximum wood yield?

This is a trick question. The maximum yield on any given day would be obtained by cutting *everything* down (also known as clear-cutting). No other strategy could yield more right now. But such a harvest can be done only once, so it isn't really a management strategy. A more sensible strategy would include harvesting some individuals and leaving others for the future. With that type of strategy, which trees (or lobsters) would you harvest so as to minimally affect the population's growth? Ideally, you would harvest those that are post-reproductive. After all, they are no longer contributing to the population growth. But there may not be many post-reproductive individuals in the population, or it may be too hard to identify them.

What's the better solution? For long-term management, it is best to harvest some but leave others still growing and reproducing for harvest at a later time. With such a strategy, the population can persist indefinitely. The special case in which as many individuals as possible are removed from the population without impairing its growth rate is called the **maximum sustainable yield** (**FIGURE 14-12**). The value in such a harvest comes from the fact that it can be carried out forever, clearly yielding more than the shortsighted strategy of a complete, one-time harvest.

MANAGING NATURAL RESOURCES

Effective and sustainable management of natural resources requires the determination of a population's maximum sustainable yield, the point at which the maximum number of individuals are being added to the population (and so can be harvested or utilized).

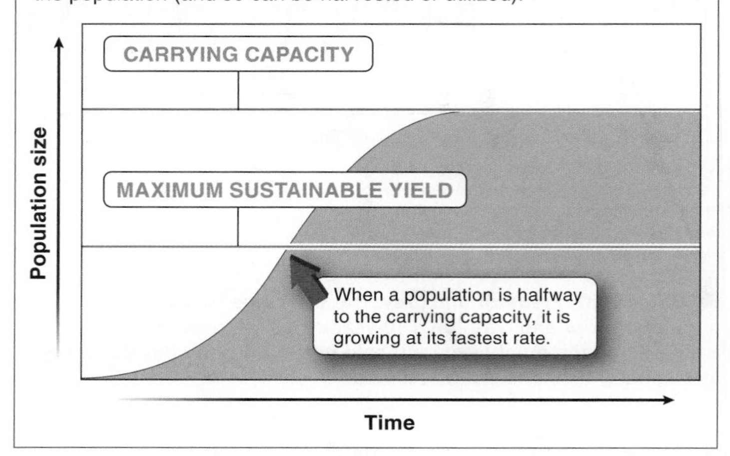

FIGURE 14-12 Tree harvest. Wood is a renewable resource, but how many trees can we cut down (and which ones?) without irreversibly depleting the supply?

Your first step as the manager of a hardwood forest or lobster fishery is to determine the maximum sustainable yield for the resource. This is actuallly straightforward. Maximum sustainable yield is calculated as that point at which the population is growing at its fastest rate. If we examine the logistic growth curve in Figure 14-6, we can see that the population is getting larger at the fastest rate when its size is equal to half the carrying capacity. At this midpoint, scarcity of mates is not a problem, as it can be at low population levels, and nor is competition, as it can be near the carrying capacity—one of the reasons that the population doesn't grow at all when it reaches the carrying capacity.

Q Virtually all natural resource managers working for the U.S. government fail to do their job exactly as mandated. Why?

Maximum sustainable yield is a useful concept, applicable not just to tree or lobster harvesting but also to livestock, agriculture, and nearly every other useful natural resource. And, in fact, there are 31 official U.S. government agencies that, in their charters, are mandated to utilize the maximum sustainable yield concept in determining harvest levels. But here's the rub: this is nearly always an impossible task. Why?

There are numerous reasons, most of which become apparent only when you leave the simplicity of the theoretician's desk and head into the messy real world. For starters, if maximum sustainable yield occurs when a population is at half its carrying capacity, do we first have to wait till the population stabilizes at its carrying capacity to determine what half of the carrying capacity would be? But isn't it inefficient to sit around waiting for the population to reach its carrying capacity, when we want to maintain it at half that size? Or can we just estimate carrying capacity and then calculate half of it? But won't that be difficult because lobsters live underwater (**FIGURE 14-13**)? Put simply, we rarely know the value of K.

And the problem gets worse. For starters, with many species, not only do we not know the carrying capacity, we don't even know the number of individuals alive. It is difficult to accurately count humans, so imagine how hard it is to count individuals of species that live underwater or fly or are microscopic.

And if we were to solve the mysteries of counting individuals and knowing carrying capacity, we would still have to figure out whether carrying capacity is stable from year to year. If carrying capacity depends on levels of resources, it may be cyclic, for instance. And even if we can figure out the carrying capacity and population size, we would not be certain which individuals ought to be harvested. Often, the individuals in a population are not contributing equally to population growth. The post-reproductive individuals, as mentioned above, do not contribute to this growth.

Maximum sustainable yield is a useful concept, but it is difficult to put into practice because it is hard to accurately determine population size and carrying capacity in nature.

FIGURE 14-13 How many lobsters can be harvested sustainably?

The concept of maximum sustainable yield generates insights into fighting biological pests such as cockroaches.

FIGURE 14-14 Taking over the kitchen. Population growth models don't just guide us to maximum production of valuable resources. They can help us efficiently reduce pest populations, too.

As a natural resource manager, you do your best, knowing that the theory behind the concept of maximum sustainable yield can almost never be put into practice perfectly.

Harvesting natural resources for maximum efficiency turns out to generate insights into fighting biological pests, such as cockroaches or termites (**FIGURE 14-14**). The problem is just turned on its head: which pest animals would you concentrate on killing so as to most effectively slow population growth? The pests to kill are those at the age of maturity, with the highest reproductive value, because they contribute most to the growth of the population. Similarly, because a population can still grow very rapidly with only a few males—given that any one male can fertilize a large number of females—it is most effective to focus pest control on females.

Patterns of population growth and the environmental features that influence them can lead to evolutionary changes in populations, shaping features such as life span, the age of reproduction, the number of offspring produced, and the amount of care given to them—the topics we turn to next.

TAKE-HOME MESSAGE 14•6

Based on models of population growth, it might seem easy to efficiently and sustainably utilize natural resources. In practice, however, difficulties such as estimating population size and carrying capacity complicate the implementation of such strategies.

② A life history is like an executive summary of a species.

A large population of flamingoes rests in Lake Nakuru National Park, Kenya.

14•7

Life histories are shaped by natural selection.

> **Q** Do any animals mate themselves to death? Why?

Some animals reproduce with a "big bang." *Antechinus* is a mouse-sized marsupial that lives in Australia, and male *Antechinus* are classic big-bang reproducers (**FIGURE 14-15**). At one year of age, they enter a two- to three-week period of intense mating activity, copulating for as long as 12 hours at a time. Shortly after this mating period, the males undergo rapid physical deterioration—they lose weight, much of their fur falls out, their resistance to parasites falls—and then they die.

Other animals are a bit—but only a bit—more restrained in their reproduction. The house mouse reaches maturity in about one month and produces litters of 6–10 offspring nearly every month, sometimes generating more than a hundred offspring in its first year of life.

And some animals could not be further from big-bang reproducers. The little brown bat is also mouse-sized, yet does not reach maturity until one year of age and typically produces only a single offspring per year. It can, however, live more than 33 years in its natural habitat.

Why all the variation? And is one strategy better than the others, evolutionarily? One of the most important recurring themes in biology is that evolution nearly always finds more than one way to solve a problem. As the marsupial mouse, house mouse, and little brown bat illustrate, there are many possible responses to the challenge of when to reproduce, how often to reproduce, and how much to reproduce in any given episode. The answers to these questions comprise an organism's **life history**—a sort of "executive summary" of the vital statistics of the species, including age at first reproduction, probabilities of survival and reproduction at each age, litter size and frequency, and longevity. Life histories are a little short on warm fuzzy stories about how plants and animals live, but they do tell us as much about a species as possible in a small amount of information.

VARIATION IN LIFE HISTORIES

BIG-BANG REPRODUCTION
- Reaches sexual maturity at one year
- Mates intensely over a three-week period
- Males die shortly after mating period

FAST, INTENSIVE REPRODUCTIVE INVESTMENT
- Reaches sexual maturity at one month
- Produces litters of six to ten offspring every month

SLOW, GRADUAL REPRODUCTIVE INVESTMENT
- Reaches sexual maturity at one year
- Produces about one offspring per year

FIGURE 14-15 Reproductive strategies. There are many strategies for reproducing, from the big bang of *Antechinus,* to the early, fast, abundant reproduction of the house mouse, to the slow but sure methods of the little brown bat.

Life histories vary from the big-bang reproductive strategy in *Antechinus,* in which all of the material and energetic contribution that an individual will make to its offspring, its **reproductive investment,** is made in a single episode, to strategies in which organisms have repeated episodes of reproduction, such as humans. Among plants, life histories have a similar range, from annuals such as peas, marigolds, and cauliflower, to perennials such as bananas and mint, which reproduce repeatedly.

Returning to the question posed above—which life history strategy is best?—we need to consider two questions.

1. *What is the cost of reproductive investment during any reproductive episode?* Producing offspring is risky. From the cost of producing gametes, to the risk of dying during the act of reproducing, to the wear and tear of the reproduction itself on an individual's body, reproduction takes its toll. The greater the reproductive investment in any given episode, the greater an individual's risk of dying. Thus, the number of offspring an organism produces can only go so high before it begins to be detrimental. An organism might produce four offspring in an episode, for example, without significantly increasing its risk of dying from the effort. But if, by chance, it produces five or six offspring, this may take such a toll that the individual is unlikely to live to have another litter. Thus, for many organisms, a smaller litter is favored by natural selection because it enables the individual to have additional litters in the future, maximizing its *lifetime* reproductive success.

2. *What is an individual's likelihood of surviving to have future reproductive episodes?* If predation rates or other sources of mortality are very high, an individual might not be alive in one month. If that is the case, it makes less sense for an individual to defer current reproduction. Why save for a future unlikely to come?

> **Q** Why do humans put off mating so much longer than do cats or mice?

Taken together, the answers to these questions help us understand the variety of life history strategies we see in nature. At the slow extreme, humans have such a low probability of dying in any given year that they can defer reproduction, or at least reproduce in small amounts—once every few years—without much risk, enabling them to have just one offspring at a time but to invest significant parental effort in maximizing its likelihood of surviving. Organisms with this type of life history generally have evolved in such a way that their competitive ability—that is, their likelihood of

FIGURE 14-16 **With such a low probability of dying in any given year, humans can defer reproduction.**

surviving—is high, although their rate of population growth is low. That is, not many offspring are produced, but those that are have a high likelihood of surviving (**FIGURE 14-16**).

Toward the fast extreme are rodents. They experience very high mortality rates in the wild, and natural selection has consequently favored early and heavy reproductive investment. Rodents provide relatively little parental care but produce numerous litters, each with many offspring. Organisms with this type of life history tend to have relatively poor competitive ability but a high rate of population growth; each individual's likelihood of surviving is low, but so many are produced that the odds of one surviving are good.

TAKE-HOME MESSAGE 14•7

An organism's investment pattern in growth, reproduction, and survival is its life history.

14•8

Populations can be described quantitatively in life tables and survivorship curves.

Biologists have learned some important lessons from insurance agents. Because insurance companies need to estimate how long an individual is likely to live, about a hundred years ago they invented something called the "life table." It ought to be called a death table, though, because in the table, insurers tally the number of people in a population within a certain age group, say 0 to 10 years old or 10 to 20 years old, and the number of individuals dying within that age range. From these numbers, they can predict an individual's likelihood of either dying within or surviving a particular age interval. For the insurance companies, their ability to make money hinges on making accurate life span predictions. For biologists, a life table is like a quick window into the lives of the individuals of a population, showing how long they're likely to live, when they'll reproduce, and how many offspring they'll produce.

From life tables, biologists can create **survivorship curves,** graphs showing the proportion of individuals of a particular age that are now alive in a population. Survivorship curves indicate an individual's likelihood of surviving through an age interval. And they reveal, at a quick glance, a huge amount of information about a population, such as whether most offspring produced die shortly after birth—think back to the five million eggs produced by some fish; or whether most survive and are likely to live long lives—think about humans in the United States.

Three distinct types of survivorship curves are given in **FIGURE 14-17**. The curves plot the proportion of individuals surviving at each age, across the entire range of ages seen for that species.

Type I: At the top of Figure 14-17, in purple, is the survivorship curve seen in most human populations, and shared by the giant tortoise. We have a very high probability of surviving every age interval until old age, then our risk of dying increases

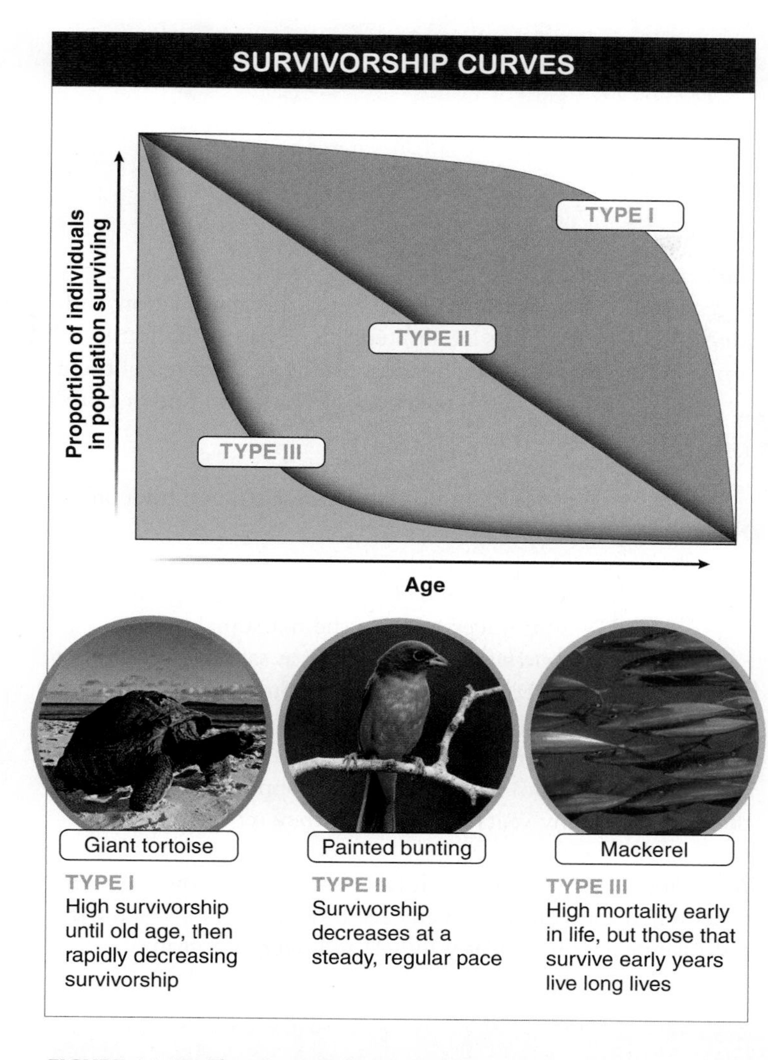

FIGURE 14-17 Three types of survivorship. Survivorship curves show the proportion of individuals of a particular age that are alive in a population.

dramatically. Species with this shape of survivorship curve, including most large mammals, usually have a few features in common. They have few natural predators, so they are likely to live long lives. They tend to produce only a few offspring—after all, most of these offspring will survive—and they invest significant time and effort in each of their offspring.

Type II: In the middle, in blue, is a survivorship curve seen in many bird species such as the painted bunting, and in small mammals such as squirrels. The straight line indicates that the proportion alive in each age interval drops at a steady, regular pace. In other words, the likelihood of dying in any age interval is the same, whether the bird is between 1 and 2 years old, or between 10 and 11.

Type III: At the bottom, in green, is a survivorship curve in which most of the deaths occur in the youngest age groups. Common in most plant and insect species, as well as in many marine species such as oysters and fish, this survivorship curve describes populations in which the few individuals lucky enough to survive past the first few age intervals then become likely to live a much longer time. Species with this type of survivorship curve tend to produce very large numbers of offspring, because most will not survive. They tend, however, not to provide much parental care, if any. A classic example might be a fish such as the mackerel. One female might produce a million eggs! Obviously, most of these (on average, 999,998) must not survive to adulthood or else the planet would be overrun with mackerel.

In reality, most species don't have survivorship curves that are definitively type I, II, or III. They may be anywhere in between. These three types, though, represent the extremes and help us to easily make predictions about reproductive rates and parental investment without extensive observations on individual behavior. Survivorship curves can also change over time. Humans in developing countries, for example, tend to have higher mortality rates in all age intervals—particularly in the earliest intervals—relative to individuals in industrialized countries.

TAKE-HOME MESSAGE 14•8

Life tables and survivorship curves summarize the survival and reproduction patterns of the individuals in a population. Species vary greatly in these patterns: the highest risk of mortality may occur among the oldest individuals or among juveniles, or mortality may strike evenly at all ages.

14•9

There are tradeoffs between reproduction and longevity.

If you were designing an organism, how would you structure its life history for maximum fitness? (Recall from Chapter 8 that fitness is defined as an organism's reproductive output relative to that of other individuals in the population.) Ideally, you would create an organism that could:

- produce many offspring,
- beginning just after birth,
- continuing every year,
- while growing tremendously large, to reduce the predation risk, and
- living forever.

Unfortunately, evolution operates with some constraints. Some of these traits are not possible because selection that changes one feature tends to adversely affect others.

Q With one simple surgical procedure, most men could add more than 10 years to their life span. Why doesn't anyone opt for it?

The impossibility of maximizing all of these features is a consequence of evolutionary tradeoffs. When evolution increases one life history characteristic, frequently another characteristic may decrease. Life history tradeoffs can be better understood in the context of the three areas to which an organism can allocate its resources—growth, reproduction, and survival. When resources are limited, life history evolution is similar to a zero-sum game: increased allocation to one of these areas tends to reduce allocation to one or both of the others.

Some examples of the best-studied tradeoffs include:

Reproduction and survival. Among big-bang reproducers such as the marsupial mouse, *Antechinus,* and fish such as salmon, investment in reproduction is exceedingly high, followed shortly by death (**FIGURE 14-18**). If individuals are prevented from reproducing by physically separating them from individuals of the opposite sex, however, they can live many years more than if they reproduced.

Reproduction and growth. Beech trees grow more slowly during years when they produce many seeds than in years when they produce no seeds. Similarly, female red deer kept from

LIFE HISTORY TRADEOFFS

REPRODUCTION AND SURVIVAL
Big-bang reproducers such as salmon make a single exceptionally high investment in reproduction, then die shortly afterward.

REPRODUCTION AND GROWTH
Beech trees grow much more slowly in the years when they produce many seeds than they do in years when they produce few seeds.

NUMBER AND SIZE OF OFFSPRING
Female lizards of the species *Uta stansburiana* produce medium-sized eggs as a compromise between a large number of small-sized eggs (with poor survival of offspring) and few large-sized eggs (with better survival of offspring).

FIGURE 14-18 Life history tradeoffs. When resources are limited, an organism's increased allocation of resources in one area tends to reduce allocation in another.

The Arctic hare lives in the tundra of Canada, on Arctic islands and Greenland. The female has one (and sometimes) two litters per year, with an average of slightly more than 5 babies per litter. Hares found in more moderate climates have several litters per year with smaller average litter sizes.

reproducing in their first year of life grow significantly larger during that year.

Number and size of offspring. The females of one lizard species, *Uta stansburiana,* can lay more eggs if the eggs are smaller, but a higher proportion of the offspring survive if the female lays larger eggs. This species has evolved to produce a happy medium, with an egg size that tends to maximize the number of offspring produced.

There are many other tradeoffs beyond these, such as number of offspring produced and the amount of parental care given. Additionally, among mammals, litter size increases with latitude (distance from the equator). This trend may be due to the tradeoff between litter size and the number of litters per year. Closer to the equator, there is less seasonal variation in weather and animals can breed for a longer portion of the year (year-round near the equator). Farther from the equator, the winters become more severe and the breeding season is shorter. To compensate for having fewer litters at higher latitudes, mammals can produce more offspring per litter.

Can you think of other tradeoffs? How would you test whether or not they exist?

TAKE-HOME MESSAGE 14•9

Because constraints limit evolution, life histories are characterized by tradeoffs between investments in growth, reproduction, and survival.

③ Ecology influences the evolution of aging in a population.

Elephants live many years but have few offspring.

14 • 10

Things fall apart: what is aging and why does it occur?

Attempting to define what was obscene, an exasperated justice of the United States Supreme Court once concluded that he might never be able to define obscenity, "but I know it when I see it." Aging is like that, as well. Picture your grandparents or someone in their seventies or eighties. Clearly they have aged. But what exactly does that mean? And does it differ from the way that people in their thirties or forties have aged? Sometimes it involves sagging skin. Sometimes it involves weakened muscles and bones. Sometimes it involves senility.

The difficulty comes when we realize that no two people age in exactly the same way, yet without fail, everyone does age (**FIGURE 14-19**). So we must retreat to a somewhat vague definition. For an individual, aging is the gradual breakdown of the body's machinery. With this definition, each individual can experience aging differently, but because this physiological deterioration makes a person more likely to die, aging can be seen much more clearly by examining a population as a whole.

From a population perspective, **aging** emerges as a definitive and measurable feature: it is simply an increased risk of dying with increasing age. For example, among humans, a person is more likely to die between the ages of 70 and 71 than between 60 and 61. Similarly, she is more likely to die between the ages of 44 and 45 than between 34 and 35. The causes of death at these ages often differ, but the shuffle toward death becomes more and more dire, with no respite. This is the most useful definition of aging for ecologists, and it doesn't apply only to humans. Aging can be measured and assessed for any species in this same way.

Jeanne Louise Calment lived for more than 122 years, the longest life span ever documented for a human. While this is a spectacularly long life relative to most organisms on earth (and is significantly longer than the average life span in the United States, 78 years), it pales in comparison with some of the longest-lived organisms. Bristlecone pine trees, for example, are the longest-lived, surviving for thousands of years. A counting of the growth rings of one bristlecone pine indicated that it was more than seven thousand years old!

Among the animals, lobsters and quahogs (a type of clam) are the longest-lived invertebrates (up to 100 and 200 years, respectively). Long-lived vertebrates include striped bass,

Disease. The list is long. And for different species, these risks are more or less serious. Rodents, for example, are at *very* high risk of predation nearly every minute of the day. Without perpetual vigilance, they become the meal of a bird of prey or some other animal. Even *with* perpetual vigilance, their risk of death is high every day. Consequently, they are likely to be dead within a few years (**FIGURE 14-21**).

Because death from some external source is likely to come so early, these organisms living in high-risk worlds must reproduce even earlier. Natural selection favors this strongly. (If they didn't reproduce earlier, they would not leave any descendants.) Now, when a harmful mutation arises that causes its damage at an early age—say, 1 or 2 years of age—the rodent is likely to have already reproduced and passed the mutation on. And so that harmful allele increases in frequency in the population, causing individuals to age earlier.

Humans and tortoises live in relatively low-risk worlds. Because we have few predators, and because tortoises have armored-car-like shells that protect them from danger, death from external sources is low. Early, intensive investment in reproduction is not necessary and so has not been favored by natural selection. In these species, when a harmful mutation arises that has its effect at 5 or 10 years of age, the individual carrying it is likely to die before it has reproduced. Consequently, the mutation isn't passed on; natural selection weeds it out of the population. Only if the harmful effects occur much later in life will the mutation be passed on to future generations.

The age at time of reproduction, then, is a key factor determining longevity. Factors that favor early reproduction will also favor early aging; factors that don't favor early reproduction—or actively favor later reproduction—favor later aging. But then, what determines when an organism reproduces?

We can think of each population as evolving in a world with a specific **hazard factor** (see Figure 14-21). These factors include the risks of death for individuals in that population from all types of external forces. When the hazard factor is low, individuals of a species tend to reproduce later, and so natural selection can weed out all harmful mutations except those that have their effects very late in life. Individuals in these species will live a long time before they succumb to aging. A high hazard factor, on the other hand, will lead to earlier reproduction, and therefore early aging and shorter life spans.

Can you predict which species should age more slowly in captivity, a porcupine or a guinea pig? The difference is large, revealing unambiguously that it's good to be a porcupine. Nobody wants to eat you. Or rather, nobody can eat you. Because of their sharp quills, porcupines live in a world with little risk of predation. This is another way of saying that as sharp quills evolved, the world in which porcupines have been evolving has for them a lower and lower hazard factor.

Guinea pigs, by contrast, live with a high risk of predation. Because they've been evolving in a world characterized by a high hazard factor for them, they reproduce much sooner, age more quickly, and die younger. A porcupine in captivity can live 21 years, and a captive guinea pig lives less than 10 years. In the wild the difference is even greater: 15 years for the porcupine and only 3 or 4 years for the guinea pig. Using the same reasoning, what difference would you expect between bats and mice, or between poisonous and non-poisonous snakes?

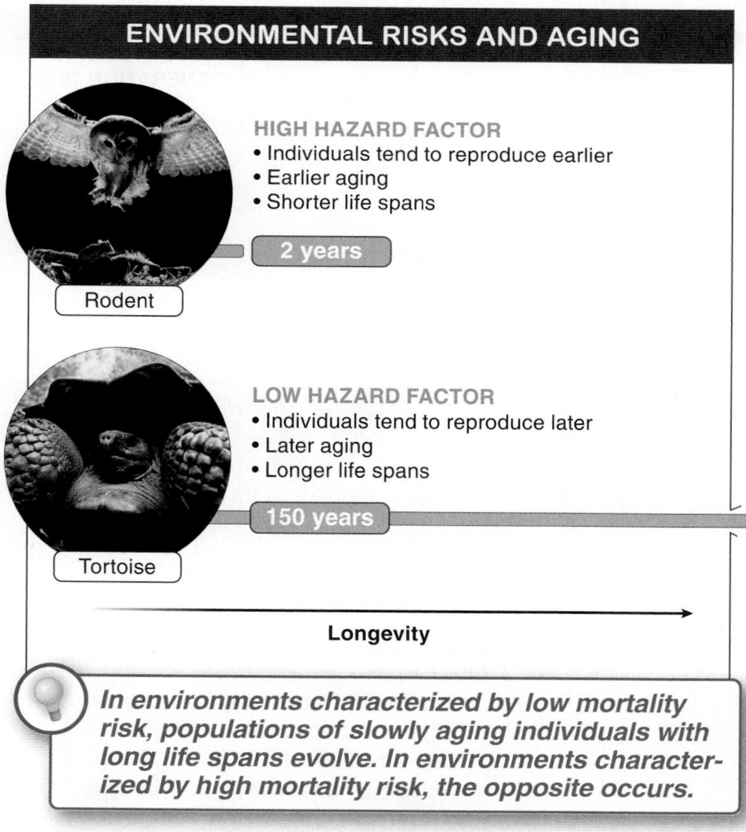

FIGURE 14-21 Hazard factor. The mouse and the tortoise face different risks of dying.

TAKE-HOME MESSAGE 14•11

The rate of aging and pattern of mortality for any species are determined by the hazard factor of that organism's environment. In environments characterized by low mortality risk, populations of slowly aging individuals with long life spans evolve. In environments characterized by high mortality risk, populations of early-aging, short-lived individuals evolve.

14•12

Can we slow down the process of aging?

Life extension is possible. Real, honest-to-goodness doubling of the normal life span. Not only is it possible, researchers have demonstrated it repeatedly. The only problem is that such life extension has been achieved only in fruit flies.

Although the results of these life-extension studies are incredible, the methods by which they were achieved are simple. Researchers began with several large populations of fruit flies maintained and propagated by a standard method in the laboratory. Each population lived in a small Plexiglas cage the size of a shoe box. There they would feed (and lay eggs) on a small plate of food consisting of bananas, corn syrup, and molasses. Every few days, a fresh plate of food would be put in the cage and the old one discarded. At 2 weeks, however, eggs were collected from the removed food plate and used to start the next generation of flies. The adult flies in the cage were then discarded. Propagated this way, the fly populations had been maintained for hundreds of generations. When put under ideal conditions outside the cage, the flies started to experience the physiological breakdowns associated with aging after a few weeks and, after about a month, they died.

One researcher had the idea that changing the force of natural selection on the flies would produce an evolution in the aging pattern. He did this in a clever way. Instead of collecting eggs at 2 weeks to start the new generation, he waited for 3 weeks (**FIGURE 14-22**). This meant that every fly in the new generation had a parent that must have lived for 3 weeks. If

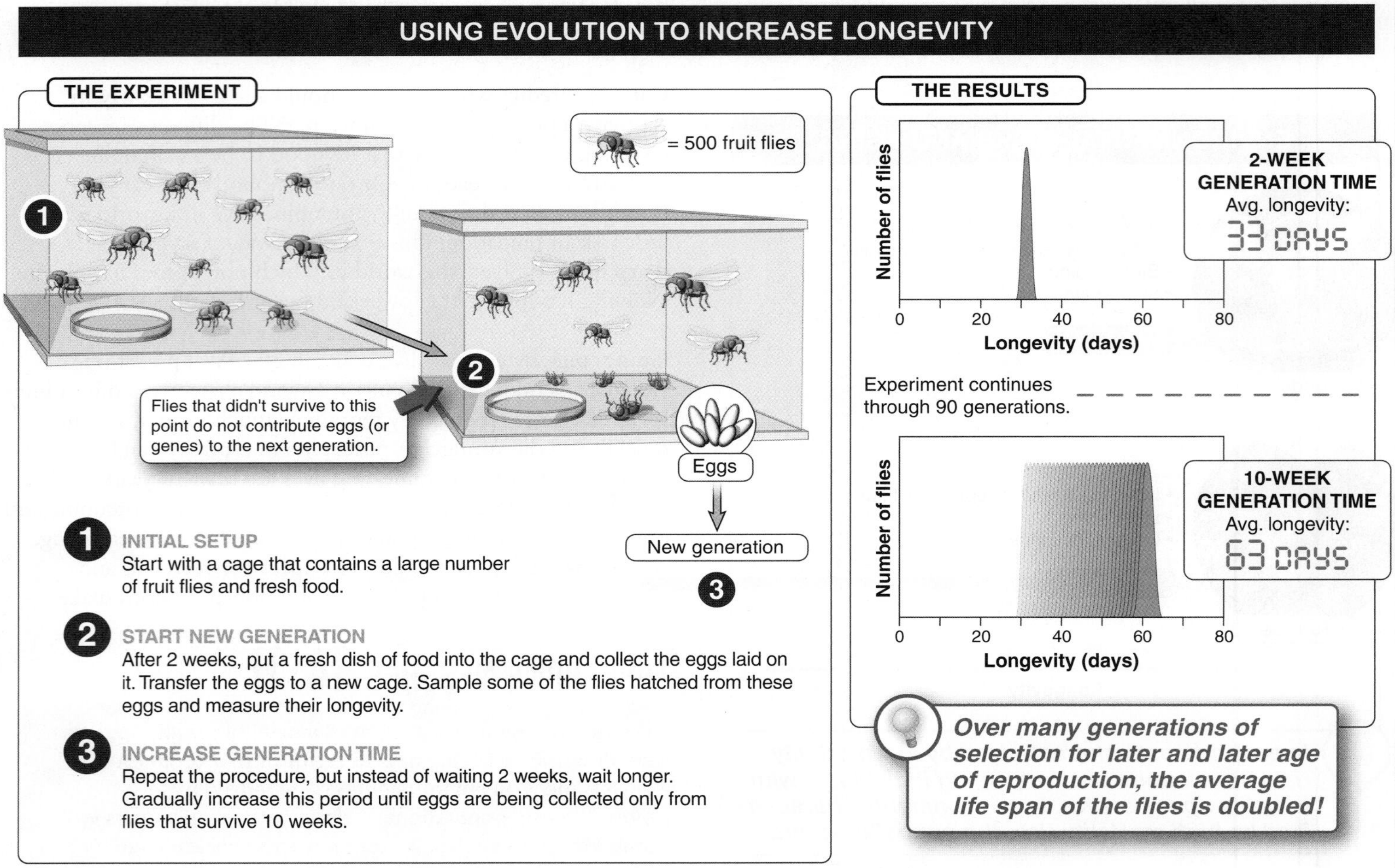

USING EVOLUTION TO INCREASE LONGEVITY

THE EXPERIMENT

= 500 fruit flies

Flies that didn't survive to this point do not contribute eggs (or genes) to the next generation.

Eggs

New generation

1 INITIAL SETUP
Start with a cage that contains a large number of fruit flies and fresh food.

2 START NEW GENERATION
After 2 weeks, put a fresh dish of food into the cage and collect the eggs laid on it. Transfer the eggs to a new cage. Sample some of the flies hatched from these eggs and measure their longevity.

3 INCREASE GENERATION TIME
Repeat the procedure, but instead of waiting 2 weeks, wait longer. Gradually increase this period until eggs are being collected only from flies that survive 10 weeks.

THE RESULTS

Number of flies
Longevity (days)
0 20 40 60 80
2-WEEK GENERATION TIME
Avg. longevity:
33 DAYS

Experiment continues through 90 generations.

Number of flies
Longevity (days)
0 20 40 60 80
10-WEEK GENERATION TIME
Avg. longevity:
63 DAYS

Over many generations of selection for later and later age of reproduction, the average life span of the flies is doubled!

FIGURE 14-22 Creating longer-lived organisms through evolution in the laboratory.

Can the life span of humans be extended? The longest-living human, Jeanne Louise Calment, lived more than 122 years.

any fly carried a mutation that caused it to die between 2 and 3 weeks, that fly did not contribute to the next generation and that mutation was not passed on.

After doing this for several generations, the researcher started collecting the eggs for the next generation at 4 weeks instead of 3 weeks. At this point, the flies making up the new generation all had parents that had survived for 4 weeks. Any flies that died prior to 4 weeks didn't contribute to the new generation. Natural selection was now able to reduce the prevalence of mutant genes responsible for early aging.

Q Is it possible, with our current knowledge, to double longevity in humans?

This experimental approach was continued, progressively increasing the generation time to 5 weeks, then 6, then 7, and ultimately to 10 weeks! At this point, only those flies with a genome free of mutations that might cause death in the first 10 weeks of life could contribute to the new generation. The net result of these experiments was the creation of flies with much "cleaner" genomes. When put under ideal conditions outside the cages, these "super" flies did not experience aging until long after the original flies would have been dead. In fact, the average life span of the new flies was more than 60 days—double that of the original flies! These experiments are the equivalent of decreasing the hazard factor of the fly populations.

TAKE-HOME MESSAGE 14•12

By increasing the strength of natural selection later in life, it is possible to increase the mean and maximum longevity of the individuals in a population. This occurs in nature and has also been achieved under controlled laboratory conditions.

④ The human population is growing rapidly.

At capacity? How many people can the earth support, living at a reasonable level of dignity and in harmony with other species?

14•13

Age pyramids reveal much about a population.

The "baby boom." It happened half a century ago, yet young people today may end up paying a price for it. What exactly was it? And why does it still matter so much? For insight into this question, consider another question. Imagine that you run a grade school. How many teachers will you need for the first grade in the new school year? That all depends on how many students will be entering. If it was 50 students last year, will it be 50 students again this year? Probably. The number of students is likely to be similar from year to year. But it can change. If we set aside changes due to migration, it depends on how many babies were born six years ago. If that number was higher than in the previous year—because more people had babies—you may need more teachers at your school. Or you may need fewer teachers if fewer babies were born.

Q What is the baby boom? Why is it bad news for young people today?

Increasing birth rates caused the baby boom. Beginning in the late 1940s, just after World War II, and continuing through the early 1960s, families in the United States had many more babies—about 30% more per month—than they would have had if they had followed the same pattern as their parents. Then the birth rates began to return to their earlier levels, where they have remained ever since. As the babies from the boom years grew to school age, schools had to increase in size to accommodate all the children. Then the schools had to decrease in size once the baby boomers had graduated. Now, as these individuals approach retirement age, the question of how their retirement and health care needs will be met is one of the biggest issues facing society.

Populations often vary across space—dense in the cities and more sparse in rural areas. The baby boom illustrates that populations aren't just distributed in space. They also have an "age distribution" as well. Just as there may be more individuals in some areas and fewer in others, there may be more individuals in certain age groups and fewer in others.

Describing populations in terms of the proportion of individuals in each age group reveals interesting population features. A population can be divided into the percentage of individuals that are in specific age groupings, called *cohorts:* for example, 0–5 years, 6–10 years, 11–15 years, and so on, in an "age pyramid." If two populations are the same size but have different age distributions, they will have some very different features. People have radically different likelihoods of dying or reproducing, for instance, based on their age. A 10-year-old isn't going to produce any offspring, but a 30-year-old has a relatively high likelihood of reproducing. And a 10-year-old is unlikely to die in the next year, whereas that is much more likely for an 80-year-old.

It can be useful, therefore, for a society to know the relative numbers of 10-, 30-, and 80-year-olds in its population. The age data can be used to determine whether a society would be better off investing in new schools or fertility wards or convalescent homes, among other social issues. Companies, too, rely on such demographic data to best plan their strategies for producing the goods that people will want and need.

Around the world, countries vary tremendously in the age pyramids describing their populations. Most industrialized countries are growing slowly or not at all. In these countries, the age pyramid is more rectangular than pyramid-shaped: because birth rates are low, the bottom of the pyramid is not very wide (**FIGURE 14-23**). And because death rates are low, too, the higher age classes in the pyramid don't get significantly narrower. Instead, they remain about the same size all the way into the late sixties and seventies, at which point high mortality rates finally cause them to become narrower. This gives the pyramid a rectangular shape. In these countries, most of the population is middle-aged or old.

A shape more like a pyramid is seen in the age pyramids of developing countries. Mexico and Kenya, for example, have very high birth rates, reflected as a large base of their age pyramid (see Figure 14-23). But high mortality rates, usually due to poor health care, cause a large and continuous reduction in the proportion of individuals in older age groups. In these countries, most of the population is in the younger age groups.

The shape of the age pyramid in the United States has economists worried that the social security system (including Social Security and Medicare) will not be able to offer older citizens sufficient benefits in the next 10 to 30 years. There is a "bulge" in the U.S. age pyramid (**FIGURE 14-24**). Because of the unusually large number of babies born about 50 years ago, there is now an unusually large number of people reaching retirement age. Since the baby boomers were born, there

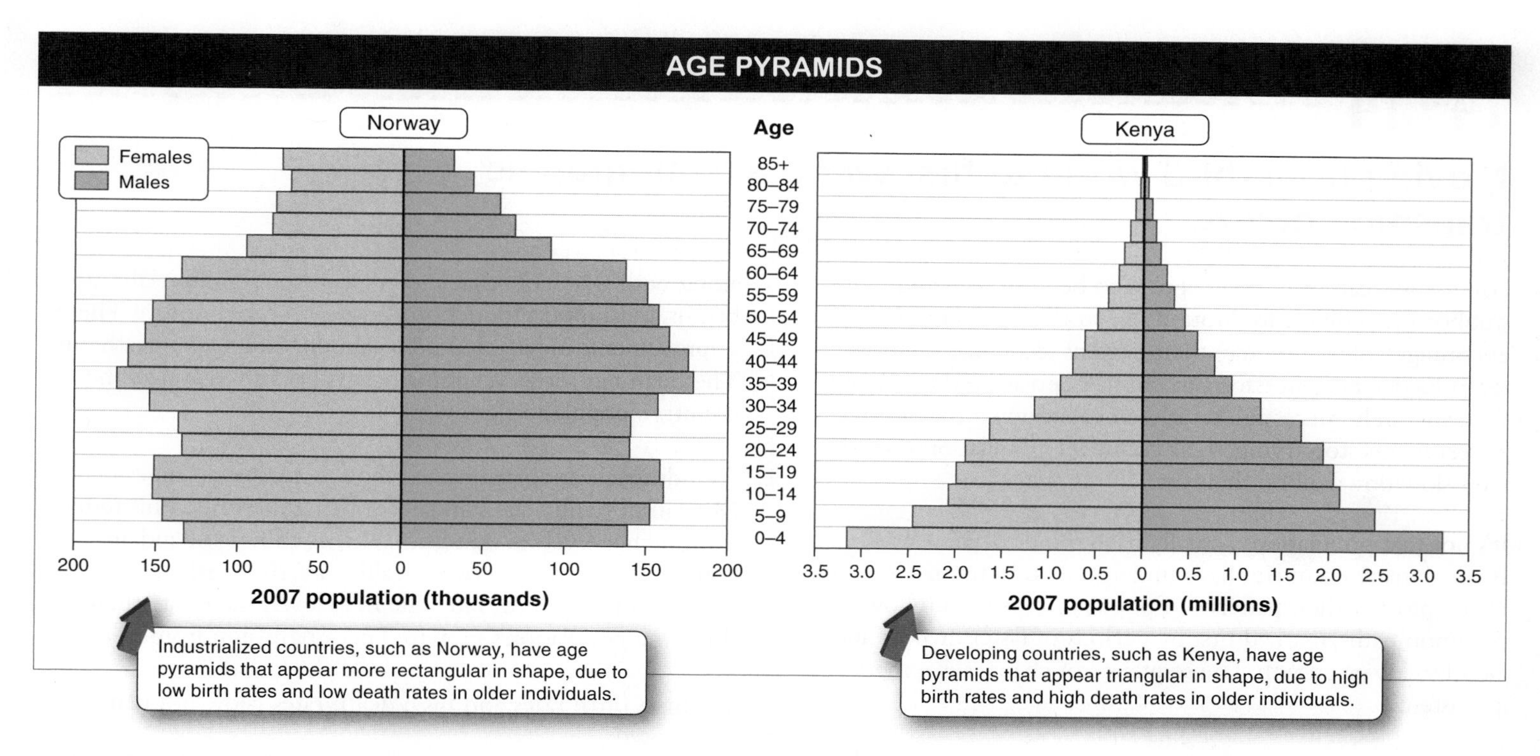

FIGURE 14-23 A visual representation of population growth. Age pyramids go from triangular to rectangular when population growth stabilizes.

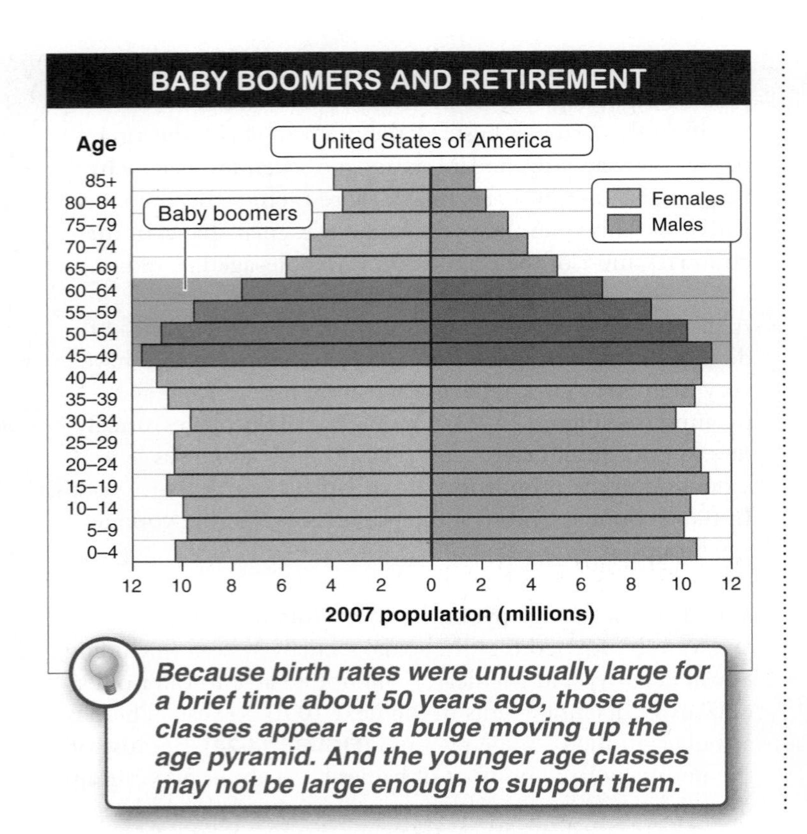

FIGURE 14-24 The retirement boom. As baby boomers reach retirement, the population bulge may strain the social security system in the United States.

hasn't been such a large cohort produced each year. This means that the current numbers of working individuals who contribute to the social security system are not sufficient to cover the payouts promised to the large number of retirees, and the baby boomers will be expensive to support as they reach retirement.

Possible solutions include investing the money held by the system in higher-risk investments that offer the possibility of higher rates of return, raising the amount put into the system by current workers, reducing the amount of the payouts to retirees, or the baby boomers postponing retirement.

TAKE-HOME MESSAGE 14•13

Age pyramids show the number of individuals in a population within any age group. They allow us to estimate birth and death rates over multi-year periods.

14•14

Moving from third world to first world a demographic transition occurs.

Populations can sometimes appear to be as strange and stubborn as individuals. Around the world, governments encourage (almost coerce) their citizens to reduce their reproductive rates in efforts to check population growth, but they are rarely successful. Yet it seems that when the governments stop trying, their countries' population growth rates slow down all on their own.

It's not that populations actually have minds of their own. Rather, many countries have undergone industrialization and, in the process, their patterns of population growth follow common paths, marked first by periods of faster growth and later by slower growth. The sequence of changes is remarkably consistent.

Start with a preindustrial country. Such countries usually have high birth rates and high death rates. These result from poor and inefficient systems of food production and distribution, along with a lack of reliable medical care. As industrialization begins, both food production and health care improve. These improvements inevitably lead to a reduction in the death rate. The birth rate, however, remains relatively high and so the country's population grows rapidly.

As industrialization continues, though, further changes occur. Most importantly, the standard of living increases. This results from higher levels of education and employment and, in conjunction with improved health care, this finally causes a reduction in the birth rate. The new, lower birth rate then slows the population's growth. The progression from:

1. high birth rates and high death rates (slow population growth) *to*
2. high birth rates and low death rates (fast population growth) *to*

3. low birth rates and low death rates (slow population growth)

is called the **demographic transition** (**FIGURE 14-25**).

The demographic transition can take decades to complete, so it is not always easy to identify it as it occurs. A survey of countries around the world, though, reveals many at different points on the transition. Sweden, for example, has a low fertility rate (1.7 children born per woman) and a low death rate (10.4 deaths per 1,000 people). At the other extreme, Nigeria and Ethiopia have very high fertility rates (5.5 and 5.3 children born per woman) and high death rates (17.2 and 15.1 deaths per 1,000). Mexico, in contrast, is in the midst of a clear demographic transition, with a moderately high fertility rate (2.45 children born per woman) yet a very low death rate (4.7 deaths per 1,000 people).

Q Population growth is alarmingly slow in Sweden and alarmingly fast in Mexico. Why is there a difference?

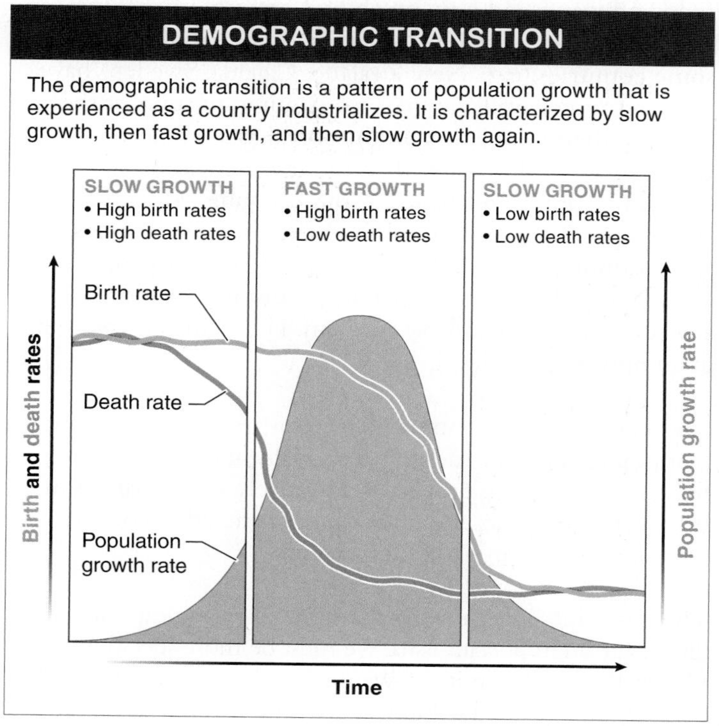

DEMOGRAPHIC TRANSITION

The demographic transition is a pattern of population growth that is experienced as a country industrializes. It is characterized by slow growth, then fast growth, and then slow growth again.

SLOW GROWTH
- High birth rates
- High death rates

FAST GROWTH
- High birth rates
- Low death rates

SLOW GROWTH
- Low birth rates
- Low death rates

Birth rate

Death rate

Population growth rate

Birth and death rates

Population growth rate

Time

FIGURE 14-25 With industrialization, death rates drop and, later, birth rates drop, too.

In the past few decades, the demographic transition has been completed in Japan, Australia, the United States, Canada, and most of Europe, leading to a slowing of population growth. In Mexico, Brazil, Southeast Asia, and most of Africa, on the other hand, the transition has not been completed and population growth is still dangerously fast.

The demographic transition illustrates how health, wealth, and education can lead to a reduction in the birth rate without direct government interventions. But because more than three-quarters of the world's population lives in developing countries and less than a quarter in the developed countries, the slowed population growth that generally occurs along with the demographic transition is unlikely to be sufficient to keep the world population at a manageable level. Instead, population growth will continue to rise quickly. What are the potential consequences of such explosive growth? We explore this next.

TAKE-HOME MESSAGE 14•14

The demographic transition tends to occur with the industrialization of countries. It is characterized by an initial reduction in the death rate, followed later by a reduction in the birth rate.

14•15

Human population growth: how high can it go?

Humans are a phenomenally successful species. The numbers don't lie: there are more than 6 billion people alive today (**FIGURE 14-26**). And current birth rates exceed death rates by so much that we add 80 million more people to the total each year. But for all of our success, the laws of physics and chemistry still apply. We all need food for energy and we need space to live. We need a variety of other resources, too, and we also need the capability of processing and storing all of the waste our societies generate. Because of these limits to perpetual population growth, we may become victims of our own success.

This we know for certain: human population growth cannot continue forever at the current rate. Like every other species, our environment has a carrying capacity beyond which the population cannot be maintained. The question is, how high

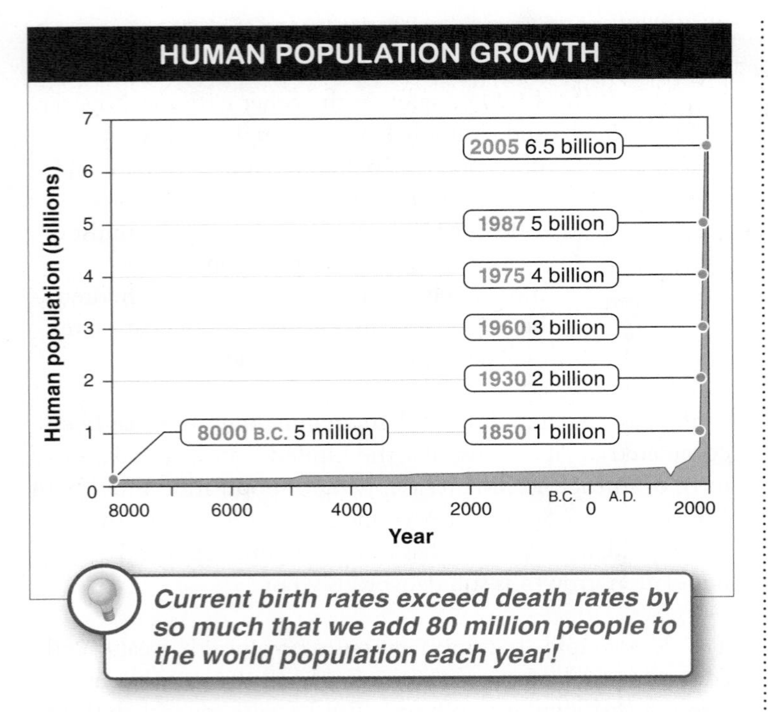

FIGURE 14-26 World population: a slow start, but rapidly picking up steam.

can it go? What is our carrying capacity? This, unfortunately, is a difficult question to answer—although not for lack of trying. For more than 300 years, biologists have been making estimates, starting when Anton van Leeuwenhoek (the inventor of the microscope) made an estimate of just over 13 billion people. The median of all the estimates, the old as well as the new, is just over 10 billion, and the United Nations conservatively suggests that the carrying capacity is somewhere between 7 and 11 billion.

There is huge variation from one biologist's estimate to the next. Why is it so hard to figure this out? After all, it's important that we know so that we can work to avoid a global catastrophe. The problem, it seems, goes back to the reasons behind the tremendous human success in the first place. We are so clever that we seem to keep increasing the carrying capacity before we ever bump into it. We do this in a variety of ways, all made possible by various technologies that we invent. In particular, we make advances on three fronts:

1. *Expand into new habitats.* With fire, tools, shelter, and efficient food distribution, we can survive almost anywhere on earth.

2. *Increase the agricultural productivity of the land.* With fertilizers and modern, mechanized agricultural methods, and selection for plants and animals with higher yields, fewer people can now produce much more food than was previously thought possible.

3. *Circumvent the problems that usually accompany life at higher densities.* Public health and civil engineering advances make it possible for higher and higher densities of people to live together with minimal problems from infectious diseases and waste management (**FIGURE 14-27**).

But the question remains: how high can the population go? Determining the earth's carrying capacity for humans is very difficult, for many reasons. The most difficult problem may be assessing just how many resources each person needs. It is possible to estimate the **ecological footprint** of the people in each country by evaluating how much land, how much food and water, and how much fuel, among other things, are necessary. The problem is, this method reveals that although some countries (e.g., New Zealand, Canada, Sweden) have more resources available than are required to support the needs of their population, in the world as a whole, resource use is significantly greater than the resources available, implying that we are already at our carrying capacity. From the perspective of ecological footprints, the populations of many countries—including the United States, Japan, Germany, and England—currently consume far more resources than are available to them. Their citizens are living at an unsustainable level.

Another difficulty in estimating the earth's carrying capacity is that populations can and do occasionally alter their fundamental growth properties. Relatively small changes in the worldwide birth rate can have dramatic effects on the ultimate population size of the planet.

The difficulties in estimating the earth's carrying capacity do illuminate an important issue. We must be more specific when we ask: how high can it go? We must add: and at what level of comfort, security, and stability, and with what impact on the other species on earth? There are, after all, tremendous tradeoffs.

> "There is in every American, I think, something of the old Daniel Boone—who, when he could see the smoke from another chimney, felt himself too crowded and moved further out into the wilderness.
>
> —Hubert Humphrey, U.S. 38th Vice President, in a speech at the University of Chicago, January 14, 1966"

The ecological footprint of the 1 billion people currently living in India, for example, is much smaller than that of Japan or Norway or Australia. It is possible, we know, to live with much less impact on the environment per person. But how much do we want to sacrifice in order to enable a larger number to live stably on the planet? Is our goal to maximize

LIMITLESS CARRYING CAPACITY?

One reason that the carrying capacity for the human population is difficult to estimate is that we can increase it in a variety of ways.

1 EXPANDING INTO NEW HABITATS

2 INCREASING THE AGRICULTURAL PRODUCTIVITY OF THE LAND

3 FINDING WAYS TO LIVE AT HIGHER DENSITIES

FIGURE 14-27 Three ways of increasing the carrying capacity for the human population.

FIGURE 14-28 How high can the population go? Humans are resourceful, but a rapidly growing population may strain the earth's ability to support us.

the number? Or to reduce the number and increase the resources available to each person? Or should our goal be something else entirely? The answers to these questions will influence whether the ultimate carrying capacity is even higher than 11 billion or perhaps lower than the current 6.8 billion people now alive (**FIGURE 14-28**).

The problem is indeed a difficult one. And it gets even more difficult because the world population currently has significant momentum. That is, even if today we immediately and permanently reduced our fertility rate to the replacement rate of just two children per couple, the world population would continue growing for more than 40 years, putting us up to at least 7 or 8 billion. This is because there are so many young people alive that each year more and more of them will enter the reproductive population. As we get closer to the carrying capacity of earth for humans (or exceed it), we should be better able to recognize it, but by then it may be too late to take preemptive measures—to avoid the resource depletion that may doom us to the much more unpleasant experience of natural population controls.

TAKE-HOME MESSAGE 14•15

The world population is currently growing at a very high rate, but limited resources will eventually limit this growth, most likely at a population size between 7 and 11 billion.

Knowledge You Can Use

Life history tradeoffs and a mini-fountain of youth.

What is the relationship between reproduction and longevity?

It's well documented that there is a tradeoff between reproduction and longevity. In a wide variety of animals, decreasing reproductive effort has been shown to increase longevity.

Q: Does this have any practical applications? Yes: spay or neuter your pet to increase its life span!

Q: Does this work? Yes! Data from more than 1,000 cat autopsies showed that non-sterilized females lived a mean of 3.0 years, while sterilized females lived significantly longer, with a mean life span of 8.2 years. The results for males were similar. (Perhaps the most dramatic example of the reproduction-longevity tradeoff is that of the marsupial mouse, *Antechinus stuartii,* in which castration leads to a doubling or even tripling of the usual life span. Similarly, following castration, salmon can live up to 18 years longer than usual.)

Not that you asked . . . In a study in the early 1900s of men institutionalized for mental retardation, those who were castrated lived 13 years longer (69.3 years vs. 55.7 years) than non-castrated men at the same institution, matched for age and intelligence!

Q: What is responsible for the tradeoff between reproduction and longevity? If it were a question of sterilized individuals living longer simply because they don't have to expend energy on reproduction, perhaps similar life span increases could be achieved by simply giving animals access to more energy. In practice, however, this doesn't increase life span (and usually *decreases* it). Rather, it seems that a significant part of the "cost" of reproduction, at least in mammals, is an increased incidence of cancer, caused by the higher levels of circulating hormones in "reproductively ready" (that is, fertile and receptive) animals.

Possibilities to think about.

The link between maintaining reproductive readiness and reduced longevity has led some researchers to contemplate the design of birth control pills that might have the additional effect of increasing longevity. Stay tuned.

1 Population ecology is the study of how populations interact with their environments.

Population ecology is the study of the interaction between populations of organisms and their environment, particularly their patterns of growth and how they are influenced by other species and by environmental factors. Populations tend to grow exponentially until limited resources cause the growth to slow to logistic rates. A population's growth can be reduced both by density-dependent factors related to crowding and by density-independent factors, such as floods, fires, and earthquakes. Difficulties such as estimating population size and carrying capacity complicate the management of populations for maximum sustainable yield.

2 A life history is like an executive summary of a species.

An organism's pattern of investment in growth, reproduction, and survival is its life history. Life tables and survivorship curves summarize the survival and reproduction patterns of the individuals of a population, varying greatly across species. Because constraints limit evolution, life histories are characterized by tradeoffs between investments in growth, reproduction, and survival.

3 Ecology influences the evolution of aging in a population.

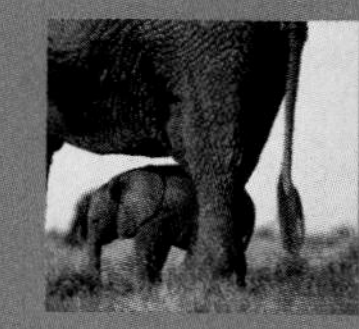

Natural selection cannot weed out bad alleles that do not diminish an individual's relative reproductive success. Consequently, these alleles accumulate in the genomes of organisms of nearly all species. This leads to the multiple physiological breakdowns that we see as aging. The rate of aging and pattern of mortality for a species are determined by the hazard factor of its environment. By increasing the strength of natural selection later in life, it is possible to increase the mean and maximum longevity of the individuals in a population.

4 The human population is growing rapidly.

Age pyramids show the number of individuals within any age group in a population. They allow us to estimate birth and death rates over multi-year periods. The demographic transition tends to occur as countries industrialize. It is characterized by an initial reduction in the death rate, followed later by a reduction in the birth rate. The world population is currently growing at a very high rate, but limited resources will eventually limit this growth, most likely at a population size between 7 and 11 billion.

Key Terms

aging, p. 532
carrying capacity, p. 520
demographic transition, p. 541
density-dependent factors, p. 520
density-independent factors, p. 521
ecological footprint, p. 542
ecology, p. 517
exponential growth, p. 519
growth rate, p. 519
hazard factor, p. 535
life history, p. 526
logistic growth, p. 521
maximum sustainable yield, p. 524
population density, p. 520
population ecology, p. 517
reproductive investment, p. 527
reproductive output, p. 533
survivorship curve, p. 528

Check Your Knowledge

1. Ecology is best defined as the study of:
   a) the relationships between all living organisms and their environments.
   b) the relationships between parasites and their hosts.
   c) aquatic organisms.
   d) interactions between predator and prey populations.
   e) the preservation of habitats.

2. On average, which leaves more offspring that survive to become adults and reproduce: a pair of elephants or a pair of rabbits?
   a) The pair of elephants, because elephants live much longer and have more breeding seasons.
   b) The pair of rabbits, because they have so many more offspring per breeding season than do elephants.
   c) The pair of rabbits, because they reach sexual maturity more rapidly.
   d) The pair of elephants, because any individual elephant born is more likely to survive to become an adult and reproduce.
   e) If both populations are stable, the pair of elephants and the pair of rabbits will leave the same number of offspring that survive to become adults and reproduce.

3. In a population exhibiting logistic growth, the rate of population growth is greatest when N is:
   a) $0.5K$.
   b) 0.
   c) above the carrying capacity.
   d) K.
   e) All of the above are correct; the rate of population growth is constant in logistic growth.

4. In a population, as N approaches K, the logistic growth equation predicts that:
   a) the carrying capacity of the environment will increase.
   b) the growth rate will approach zero.
   c) the population will become monophyletic.
   d) the population size will increase exponentially.
   e) the growth rate will not change.

5. The number of individuals that can be supported in a given habitat is the:
   a) innate capacity for increase.
   b) biotic potential.
   c) density-dependent effect.
   d) density-independent effect.
   e) None of the above are correct.

6. Which of the following statements about maximum sustainable yield is incorrect?
   a) The maximum sustainable yield for a population is the population growth rate at $K/2$.
   b) The maximum sustainable yield for a population can be difficult to determine because it is not always possible to accurately measure N.
   c) The maximum sustainable yield for a population can be difficult to determine because it is not always possible to accurately measure K.
   d) The maximum sustainable yield for a population is a useful management guideline for harvesting plant products such as timber, but is not helpful for managing animal populations.
   e) The concept of maximum sustainable yield can generate useful information for fighting the growth of pest species.

7. Dr. David Reznick has studied life history evolution in guppies that live in streams in Trinidad. Guppies are found in two different types of habitat: sites where predation is high, and sites where predation is low. Which of the following life history characteristics would you expect to evolve in a guppy population living in a high-predation site?
   a) bright colors and courtship displays
   b) increased egg number
   c) a female-biased sex ratio
   d) increased egg size
   e) delayed sexual maturation

8. The death rate of organisms in a population exhibiting a type III survivorship curve is:
   a) unrelated to age.
   b) usually correlated with density-independent causes.
   c) higher in post-reproductive than in pre-reproductive years.
   d) lower after the organisms survive beyond the earliest age groups.
   e) more or less constant throughout their lives.

9. Which of the following is a major tradeoff in life histories?
   a) size of offspring for amount of parental investment
   b) size of offspring for number of reproductive events
   c) growth for reproduction
   d) size for life span
   e) number of reproductive events for number of offspring per reproductive event

10. Natural selection:
   a) does not influence aging because aging is determined by an individual's environment.
   b) cannot reduce the frequency of alleles that cause mortality among individuals who have not yet reached the age of maturity.
   c) cannot weed out from a population any alleles that do not reduce an individual's relative reproductive success, even if these alleles increase an individual's risk of dying.
   d) can influence aging but not longevity.
   e) leads to an increase in the frequency of any illness-inducing alleles that have their effect when an organism can reproduce.

11. Which of the following statements about the hazard factor of a population is incorrect?
   a) It is a measure of organisms' risk of death from external sources.
   b) It is lower for a population of porcupines than for a population of guinea pigs.
   c) It is a measure of the ratio of mortality risk from external (environmental) causes relative to internal (genetic) causes.
   d) It is responsible for the rate of aging among the individuals of the population.
   e) It is a measure of how quickly the individuals in a population age.

12. Life extension:
   a) is not possible, because natural selection cannot weed out disease-causing alleles that have an effect only after the age when reproduction is no longer possible.
   b) can be achieved in species by selectively breeding individuals with the earliest age of maturity.
   c) works in insects but could not work in humans.
   d) has been achieved using laboratory selection for delayed reproduction.
   e) None of the above are true.

13. A population pyramid:
   a) represents the number of individuals in various age groups in a population.
   b) can be constructed from the data in a life table.
   c) directly predicts future age distributions of the population.
   d) shows the current birth and death rates of a population.
   e) predicts survival and mortality rates for an individual at a given age.

14. A primary difference between the age pyramids of industrialized and developing countries is that:
   a) mean longevity is significantly greater in developing countries.
   b) in developing countries, much larger proportions of the population are in the youngest age groups.
   c) developing countries show a characteristic "bulge" that indicates a baby boom.
   d) in developing countries, females live significantly longer than males, whereas in industrialized countries the reverse is true.
   e) developing countries have significantly more individuals than industrialized countries.

15. Which statement best describes expectations for the world's human population by the year 2010?
   a) It will exceed 10 billion if the current rate of increase continues.
   b) Negative growth in the United States and Europe will counterbalance positive growth in the developing countries.
   c) It will exceed 7 billion if the current rate of increase continues.
   d) It will drop below 6 billion if the current rate of decrease continues.
   e) The problem is too complex to make any predictions.

Short-Answer Questions

1. Explain the differences between density-dependent and density-independent factors that can limit a population's growth. Give an example of each type of factor.

2. Describe the characteristics of an organism with a type I survivorship curve. How would the life history change if this organism had a type III survivorship curve?

3. Explain why some species of organisms live very long lives and others do not. How could the life span be extended?

See Appendix for answers. For additional study questions, go to www.prep-u.com.

1 Ecosystems have living and non-living components.

15.1 What are ecosystems?

15.2 A variety of biomes occur around the world, each determined by temperature and rainfall.

2 Physical forces interacting create weather.

15.3 Global air circulation patterns create deserts and rain forests.

15.4 Local topography influences the weather.

15.5 Ocean currents affect the weather.

3 Energy and chemicals flow within ecosystems.

15.6 Energy flows from producers to consumers.

15.7 Energy pyramids reveal the inefficiency of food chains.

15.8 Essential chemicals cycle through ecosystems.

4 Species interactions influence the structure of communities.

15.9 Interacting species evolve together.

15.10 Each species' role in a community is defined by its niche.

15.11 Competition can be hard to see, but it still influences community structure.

15.12 Predation produces adaptation in both predators and their prey.

15.13 Parasitism is a form of predation.

15.14 Not all species interactions are negative: mutualism and commensalism.

5 Communities can change or remain stable over time.

15.15 Many communities change over time.

15.16 Some species are more important than others within a community.

15

Ecosystems and Communities

Organisms and their environments

① Ecosystems have living and non-living components.

A red kangaroo (Macropus rufus) *in its natural habitat, Finke Gorge National Park, Australia.*

15•1

What are ecosystems?

Picture a lush garden: some greenery, a bit of debris from rotting wood, and abundant wildlife. Grazing animals abound, while predators feed on other animals and their eggs. Parasites are poised, looking for hosts, and, just below the surface, scavengers find meals among the organic detritus. It would seem to be the quintessential **ecosystem,** a community of biological organisms plus the non-living components with which the organisms interact. But now imagine that the entire scene gets up and walks away! The "camera" in your mind pulls back to reveal that the entire scene is playing out on the back of a beetle no more than two inches long (**FIGURE 15-1**).

Ecosystems are found not just in obvious places like ponds and tropical rain forests, but also in some unexpected places, like on the back of a beetle!

FIGURE 15-1 A small-scale ecosystem. Not all ecosystems are as big as a desert or a forest. Tiny, unseen ecosystems can exist on the back of a beetle or in the gut of an individual human.

The host of this mini-ecosystem is a beetle from New Guinea called the large weevil. The weevil is camouflaged from its predators by lichens—fungi and photosynthetic algae living together—while the lichens are given a safe surface on which to live. And the garden of lichens supports a wide range of other organisms, from tiny mites to a variety of other microscopic invertebrates, some free-living and some parasitic.

Not all ecosystems are the obvious assemblages of plants and animals that we usually picture—ponds, deserts, or tropical forests. A similar scene can just as easily be found in your large intestine, where several hundred bacterial species flourish. These ecosystems are contained within ecosystems that are contained within ecosystems. The scale can vary tremendously. The closer you look, the more you find.

What is important is that the two essential elements of an ecosystem are present: the biotic environment and the physical (abiotic) environment (**FIGURE 15-2**).

1. The **biotic** environment consists of all the living organisms within an area and is often referred to as a **community.**
2. The physical environment, often referred to as the organisms' **habitat,** consists of:
    - the *chemical resources* of the soil, water, and air, such as carbon, nitrogen, and phosphorus, and
    - the *physical conditions,* such as the temperature, salinity (salt level), moisture, humidity, and energy sources.

Biologists view communities of organisms and their habitats as "systems" in much the same way engineers might, hence the term eco*system*. Biologists monitor the inputs and outputs of the system, tracing the flow of energy and various molecules as they are captured, transformed, and utilized by organisms and later exit the system or are recycled. They also study how the activities of one species affect the other species in the community—whether the species have a conflicting relationship, such as predator and prey, or a complementary relationship, such as flowering plants and their pollinators. On a small scale, such as the back of a beetle, making these measurements is easy. But there are also some well-studied giant ecosystems, such as the Hubbard Brook Experimental Forest in New Hampshire, which covers 7,600 acres. Regardless of size, the same principles of ecosystem study apply: observe and analyze organisms and their environments, while monitoring everything that goes into and comes out of the system.

Why should researchers bother with such methodical—and tedious—analysis of ecosystems? This straightforward approach has led to numerous valuable discoveries, from understanding how clear-cutting forests dramatically reduces soil quality, to understanding the link between the use of fossil fuels and the creation of destructive acid rain. As we see in this chapter and the next, humans, perhaps more than any other species in history, are significantly affecting most of the ecosystems on earth. And, in addition to improving our understanding of environmental issues, ecosystem studies can also tell us about the microbial ecosystems living inside humans and thus lead to advances in public health.

FIGURE 15-2 What makes up an ecosystem? Living organisms are not self-sufficient. They need energy and raw materials. Ecosystems are the living and non-living elements in a given area.

TAKE-HOME MESSAGE 15•1

An ecosystem is all of the living organisms in a habitat as well as the physical environment. Ecosystems are found not just in obvious places such as ponds, deserts, and tropical rainforests but also in some unexpected places, like the digestive tracts of organisms or the shell of a beetle.

15•2

A variety of biomes occur around the world, each determined by temperature and rainfall.

Dense vegetation surrounds you. Above you is a canopy of evergreen trees, 30–40 meters tall. Climbing vines hang from virtually all the trees. And dozens or even hundreds of species of insects are flourishing around you. Even if you've never been there, the description of a tropical rain forest is easy to recognize. But where exactly would you be if you were in this scenario? It could be South America or Africa or Southeast Asia. The species are different, but the general pattern of life forms is the same. The same holds for arctic tundra: whether you were in northern Asia or North America, the view would be similar. These are examples of the largest of the earth's ecosystems, the **biomes.**

Biomes cover huge geographic areas of land or water—the deserts that stretch almost all the way across the northern part of Africa, for example. Terrestrial (land) biomes are defined and usually described by the predominant types of plant life in the area. But looking at a map of the world's terrestrial biomes, it is clear that they are mostly determined by the

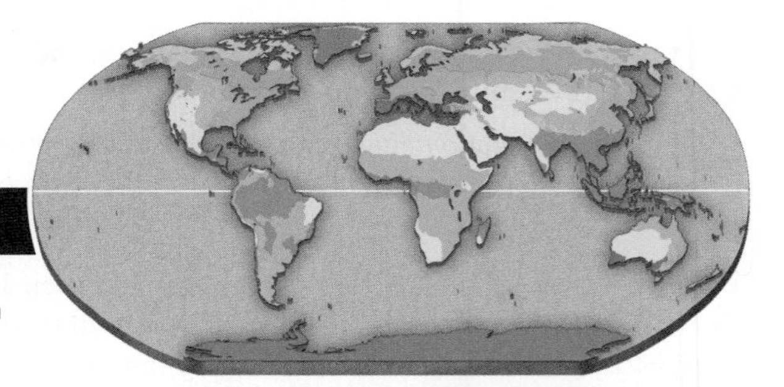

TERRESTRIAL BIOMES

Biomes are large ecosystems that cover huge geographic areas. The nine chief terrestrial biomes are determined by the temperature and amounts of precipitation in conjunction with the magnitude of seasonal variation in these factors.

FIGURE 15-3 Terrestrial ecosystem diversity. The terrestrial biomes depend on the amount of rainfall and average temperature.

FIGURE 15-4 Aquatic ecosystem diversity. Aquatic biomes are defined by variables such as salinity, water movement, and depth.

weather. Specifically, when defining terrestrial biomes, we ask four questions about the weather:

1. What is the average temperature?
2. What is the average rainfall (or other precipitation)?
3. Is the temperature relatively constant or does it vary seasonally?
4. Is the rainfall relatively constant or does it vary seasonally?

By knowing the answers to these questions, it is possible to predict, with great accuracy, the type of biome. This is because temperature and precipitation dictate the **primary productivity** levels, or the amount of organic matter produced, primarily through photosynthesis. And the numbers and types of **primary producers**—the organisms responsible for the primary productivity, such as grasses, trees, and agricultural crops—are, in turn, the chief determinants of the amount and breadth of other life in the region.

For example, where it is always moist and the temperature does not vary across the seasons, **tropical rain forests** develop. And where it is hot but with strong seasonality that brings a "wet" season and a "dry" season, **savannas** or **tropical seasonal forests** tend to develop. At the other end of the spectrum, in dry areas with a hot season and a cold season, **temperate grasslands** or **deserts** develop. **FIGURE 15-3** shows examples of the nine chief terrestrial biomes; all are determined, in large part, by the precipitation and temperature levels.

Aquatic biomes are defined a bit differently, usually based on physical features such as salinity, water movement, and depth. Chief among these environments are (1) lakes and ponds, with non-flowing fresh water; (2) rivers and streams, with flowing fresh water; (3) **estuaries** and wetlands, where salt water and fresh water mix in a shallow region characterized by exceptionally high productivity; (4) open oceans, with deep salt water; and (5) coral reefs, highly diverse and productive regions in shallow oceans (**FIGURE 15-4**).

If the terrestrial biomes of the world are determined by the temperature and rainfall amounts and seasonality, what determines those features? In other words, what makes the weather? We investigate next how the geography and landscape of the planet—from the shape of the earth and its orientation to the sun, to patterns of ocean circulation—cause the specific patterns of weather that create the different climate zones and the biomes characteristic of each. Then we'll see how energy and chemicals are made available for life to flourish in these biomes.

TAKE-HOME MESSAGE 15•2

Biomes are the major ecological communities of earth, characterized mostly by the vegetation present. Different biomes result from differences in temperature and precipitation, and the extent to which they vary from season to season.

❷ Physical forces interacting create weather.

Monsoon clouds forming over the Indian Ocean and Madagascar.

15•3

Global air circulation patterns create deserts and rain forests.

Temperature and rainfall, that's all. The type of terrestrial biome depends on little else besides these two aspects of the weather. But what determines the temperature and rainfall in a particular part of the world? As we'll see, differences in both ultimately result from one simple fact: the earth is round. Let's first explore how the earth's curvature influences temperature. Then we will examine how rainfall patterns are an inevitable consequence of the variation in temperature across the globe.

Wherever you are, begin walking toward the equator. As you get closer, does it get hotter or colder? Hotter, of course. Nearly everyone is aware of this universal trend (and the reverse as well: it gets colder as you move away from the equator and approach the North or South Pole). What is responsible for the increased warmth at the equator? This is most easily explained with a drawing (**FIGURE 15-5**).

? ***Why is it warm at the equator and cold at the Poles?***

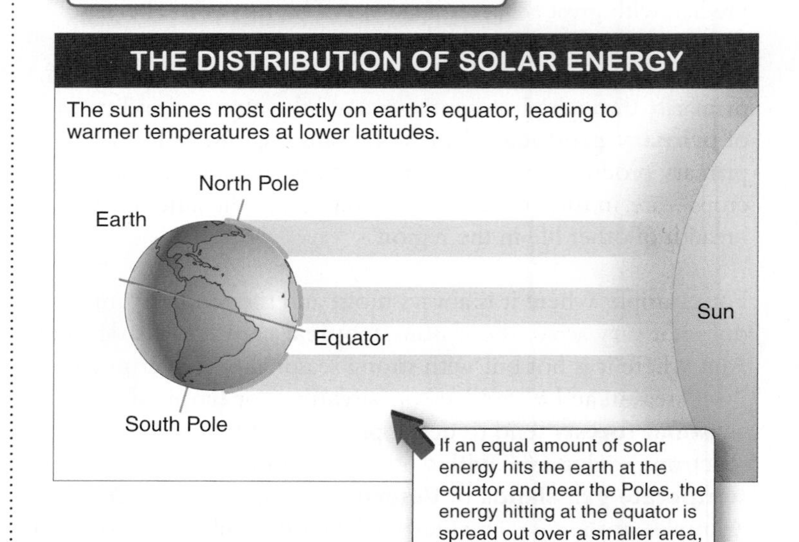

FIGURE 15-5 The curvature of the earth. The sun shines more directly on the equator. Closer to the Poles, the curvature of the earth causes sunlight to be spread out over a larger area, reducing the heat in any one spot.

> There was a hot desert wind blowing that night. It was one of those hot dry Santa Anas that come down through the mountain passes and curl your hair and make your nerves jump and your skin itch. On a night like that every booze party ends in a fight. Meek little wives feel the edge of a carving knife and study their husbands' necks. Anything can happen.
>
> —Raymond Chandler, "Red Wind," 1938

The sun shines most directly on the equator. That means that a given amount of solar energy hitting the earth at the equator is spread out over a relatively small area. Away from the equator, the earth curves toward the North and South poles. Because of this curvature, the same amount of solar energy hitting the earth at the Poles is spread out over a much larger area and also travels a greater distance through the atmosphere, which absorbs or reflects much of the heat. Because the energy is spread out over a larger area, there is less warmth at any one point on the earth's surface. It's similar to the fact that at noon, the sun's rays hit the earth at a more direct angle than they do later in the day. That is why the sun provides less warmth late or early in the day. It is also why the risk of sunburn and skin cancer is greatest around noon and the nearer you are to the equator.

Global patterns of rainfall can be predicted just as easily, by taking into account that warm air holds more moisture than cold air. We'll start at the equator again. This is where the greatest warming power of the sun hits the earth, and some of that energy radiates back, warming the air. This starts a three-step process that ends with rain. (1) Hot air rises. (2) As it rises high into the atmosphere, it cools. And (3) because cool air holds less moisture, as the air rises, clouds form and the moisture that can no longer be held in the air falls as rain (**FIGURE 15-6**). The equator is hot, but it is also very wet.

FORMATION OF RAIN

1 AIR IS HEATED AND RISES
When solar heat hits the earth, it warms the air at that point. The heated air rises.

2 RISING AIR COOLS
As hot air rises, getting farther from the warm earth, it cools.

3 COOLING AIR LOSES MOISTURE
Because cold air holds less moisture than warm air, clouds form and rain falls.

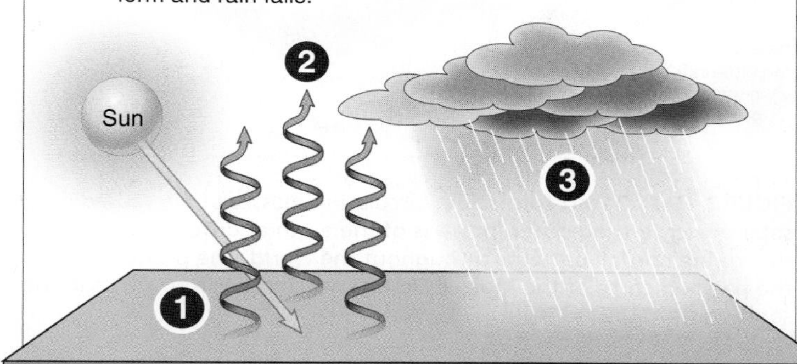

FIGURE 15-6 Rainmaking.

The cold air that just rained on the equator at 0° latitude is high in the atmosphere and expands outward, away from the equator, to about 30° north and south. Here—about a third of the way from the equator toward each Pole—the cold air, which is heavier than warm air, begins to fall downward toward the earth. The falling air becomes warmer as it gets closer to the earth's surface, which radiates back some of the heat from the sun. And as the air gets warmer, it can hold more and more moisture, making it less likely to rain. That is, moisture can be held in the air rather than being released as rain. In fact, the falling and rapidly heating air can hold so much moisture that it sometimes sucks up moisture from the land itself. For this reason, at around 30° north and south, around the world, there is very little rainfall, the ground is very dry, and deserts form (**FIGURE 15-7**). The Atacama Desert of South America, which lies at approximately 30° south of the equator, is an extreme example. In some parts of this large desert, *no rainfall* has ever been recorded. Other great deserts of the world are also at 30° latitude, including the Sahara, Kalahari, Mojave, and Australian deserts (although some deserts do occur at other latitudes). These areas stand in stark contrast to the equator, where it is not uncommon for an area to receive more than three meters (!) of rain in a year.

> **Q** Nearly all of the world's deserts occur a third of the way from the equator toward the Poles, at 30° latitude. Is this just a coincidence?

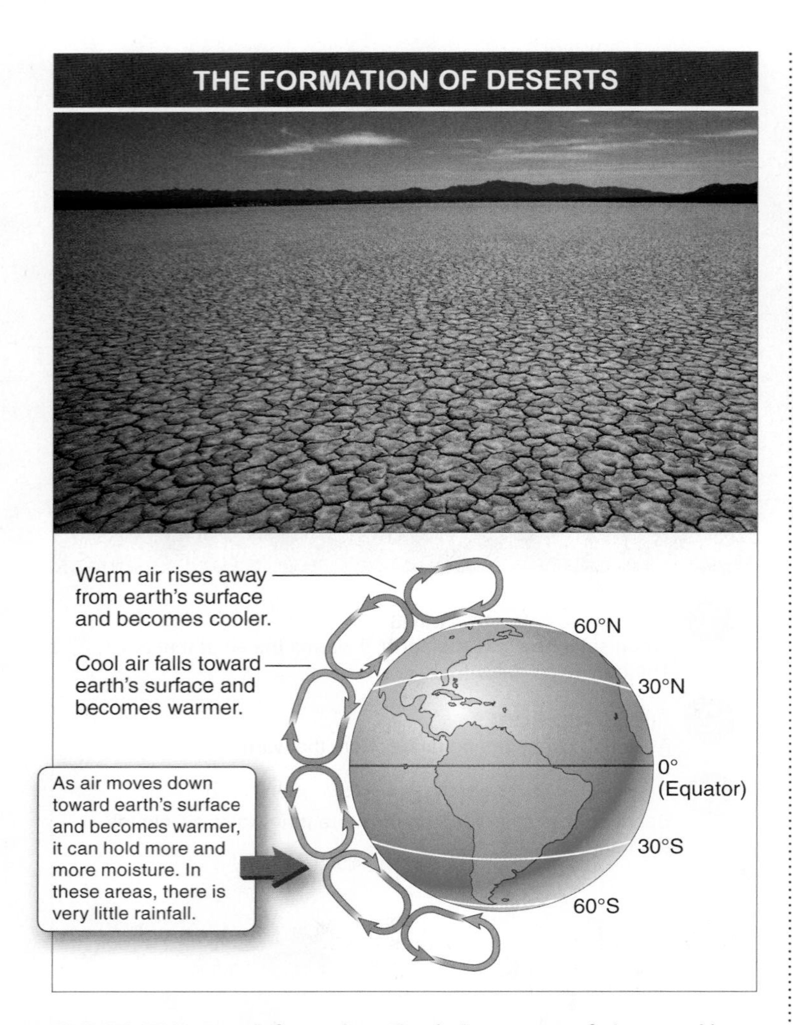

FIGURE 15-7 Desert formation. Circulating masses of air, caused by solar energy hitting different parts of the earth at different angles, determine rainfall patterns throughout the world. The photo shows the parched earth of the Mojave Desert, which spans four states in the southwestern United States.

As the air falls near 30° north and south and hits the earth, it spreads equally toward the equator and toward the Poles. Moving toward the Poles, a similar pattern is repeated. The air moving over the surface of the earth absorbs radiated heat and moves along the surface toward the Poles, gradually getting warmer, until, at around 60° latitude, it begins to rise because of its accumulated heat. Again, the rising air loses its moisture as rainfall. And so at 60° latitude, two-thirds of the way toward the Poles from the equator, it's not particularly warm but there is a great deal of rain. Not surprisingly, at these latitudes lie huge temperate forests with extensive plant growth. And finally, around the Poles, air masses again descend. And as they do, because they can hold more moisture, very little rain falls. The Poles are cold, but with little precipitation—they resemble a frozen desert.

Antarctica receives little annual rainfall (or other precipitation) and could be classified as a desert by this definition. The interior of the continent is particularly dry, although the coasts can receive heavy snowfall when storms pick up moisture from the surrounding sea and deposit it as snow along the coasts.

TAKE-HOME MESSAGE 15•3

Global patterns of weather are largely determined by the earth's round shape. Solar energy hits the equator at a more direct angle than at the Poles, leading to warmer temperatures at lower latitudes. This temperature gradient generates atmospheric circulation patterns that result in heavy rain at the equator and many deserts at 30° latitude.

15•4

Local topography influences the weather.

Why is it so windy on the sidewalk around tall buildings? And why does it rain all the time on one side of some mountain ranges, while deserts form on the other side? And is it actually warmer in the city than in the country? The answers to these questions hinge on the fact that physical features of the land, its **topography**—including features created by humans—can have dramatic but predictable effects on the weather. Let's explore some of these features, beginning with natural landscapes.

High altitudes have lower temperatures. With increasing elevation, the air pressure drops—this is because the weight of the atmosphere gets less and less as altitude increases. And when pressure is lower, the temperature drops. For each 1,000 meters above sea level, the temperature drops by about 6° C. This is why the changes in weather and vegetation that you see while climbing a mountain are similar to those you would see as you moved farther and farther away from the equator—it gets colder in both cases.

Rain shadows create deserts. When wind blows against a mountain, the air rises to get over the top. But because the air cools as it rises, it can't hold as much moisture. So clouds form and rain falls. After the air eventually passes over the top of the mountain, it can fall back down toward lower elevations. As it falls, it becomes warmer, and because warm air can hold more moisture than cold air, there is rarely any rain on the backside of the mountain. In fact, often the air will pull moisture from the ground, intensifying the already dry conditions, creating **rain shadow** deserts (**FIGURE 15-8**). Along the west coast of the United States, the Sierra Nevada and the Cascade mountain range are responsible for the Mojave Desert in California and the Great Sandy Desert in Oregon.

FORMATION OF RAIN SHADOWS

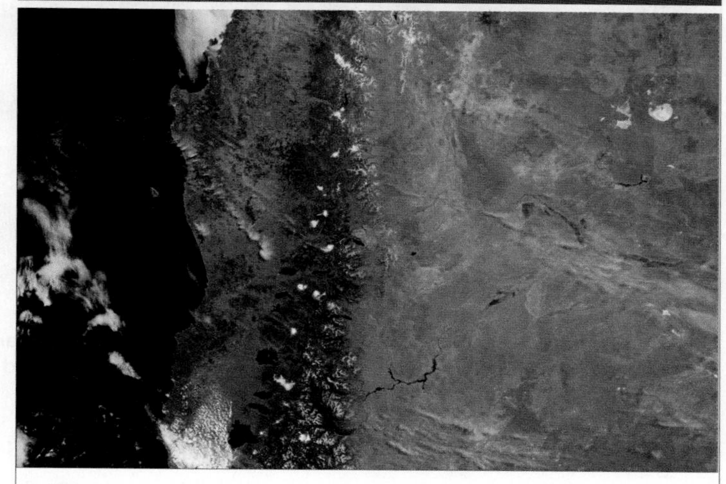

1. Wind blows from oceans toward land, rising when it hits mountains.
2. Rising air cools and holds less moisture, leading to cloud formation and rain.
3. Air passes over mountain top and falls, becoming warmer and increasing the moisture it can hold. This reduces rainfall and results in dry areas.

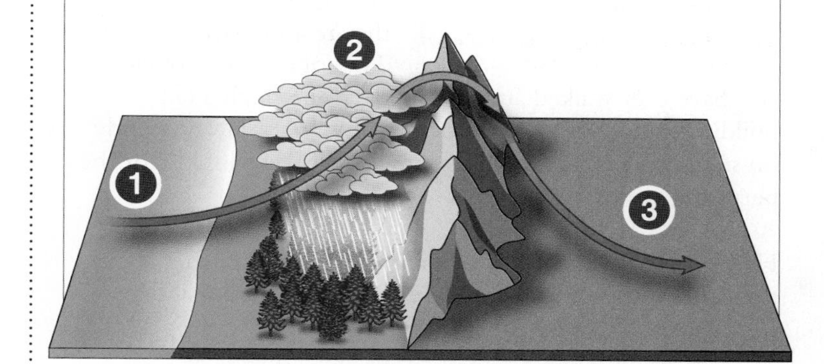

FIGURE 15-8 The rain shadow effect.

Q Is it warmer or cooler in urban areas relative to nearby rural areas?

Asphalt, cement, and tops of buildings absorb heat, raising the temperature. Modern landscapes also influence the weather, creating "urban heat islands," metropolitan areas where the asphalt and concrete lead to greater absorption, rather than reflection, of solar energy. When energy from the sun hits concrete or pavement or the dark roof of a building, some of it is reflected, heating the air around it, and most of the rest is absorbed by these man-made surfaces that function as a reservoir of heat energy. At night, the surfaces lose heat to the sky, which is colder by comparison, through radiation. This is very different from what happens when sunlight hits trees or other plant life. The solar energy evaporates water in the leaves. The ground surface doesn't get much hotter, and neither does the air. It's not surprising, then, that cities tend to be 1° to 6° C warmer than surrounding rural areas. And not only is it hotter in cities, but the rising warm air also alters rainfall patterns both in and around cities.

Because people who live in urban heat islands tend to use much more energy for air conditioning, a variety of steps are

MODERN LANDSCAPES INFLUENCE WEATHER

FIGURE 15-9 Human engineering. When humans alter the land, this can have unintended consequences, such as alterations in temperature and in wind speed and direction.

being taken to create "greener" cities that absorb and release less heat from the sun. These methods include the creation of buildings with lighter-colored rooftops, the planting of trees around buildings and along roads, and the development of rooftop gardens rich with vegetation.

Q Why is it so windy on streets with tall buildings?

Tall buildings channel wind downward. You may not have noticed that cities are getting warmer, because they're also getting windier. This, of course, is obvious if you have ever walked around a corner on which a tall building stands. It's not a coincidence that it is always windy on such corners. Tall buildings are actually responsible for the perpetual winds.

Here's why. For starters, winds are stronger the farther above the earth you go. This is because the earth's frictional drag slows winds as they get closer to the surface. That is, the plants, dirt, rocks, and water on the earth's surface slow the winds. Up higher, the winds blow more strongly. When these strong winds suddenly encounter tall buildings, they are deflected. Some of the wind goes up and over the building, but much of it is pushed downward, reaching double or even triple its initial speed by the time it reaches street level (**FIGURE 15-9**).

TAKE-HOME MESSAGE 15•4

Local features of topography influence the weather. With higher altitude, the temperature drops. On the windward side of mountains, rainfall is high; on the backside, descending air reduces rainfall, causing rain shadow deserts. Urban development increases the absorption of solar energy, leading to higher temperatures, and creates wind near the bottom of tall buildings.

15•5

Ocean currents affect the weather.

Weather is affected not just by circulating air masses. It is also affected by the oceans. The ocean is vast and deep, warmed almost exclusively by the sun beating down on its surface. But every molecule of water does not just sit in one place. Far from it. Just as there are global patterns of air circulation that produce broad climate patterns across the globe, there are also global patterns of circulation in the oceans. The water is continuously moving and mixing, due to a combination of forces including wind, the earth's rotation, the gravitational pull of the moon, temperature, and salt concentration. There are several large, circular patterns of flow in the world's oceans, as illustrated in **FIGURE 15-10**.

Q Why do beach communities have milder weather than more inland communities (not as hot in the summer, not as cold in the winter)?

Much of water's effect on weather stems from its great capacity to absorb and hold heat—a 10,000 times greater capacity than air. This means that temperatures fluctuate much more in air than in water. During a day at the beach, the air temperature can go from mild to very hot and back to mild over the course of a day, while the heat of the sun will have a tiny, almost negligible effect on the water temperature. Thus, during summers, much of the heat of the sun is absorbed by the ocean water in coastal towns, rather than causing hot air temperatures. Conversely, during cold winter months, heat from the water can reduce the coldness of the air.

One of the strongest ocean currents is the Gulf Stream. It travels north through the Caribbean, bringing warm water up the east coast of the United States and then across the Atlantic Ocean toward Europe. Because the current begins in a warm part of the globe, close to the equator, it is still warm when it reaches the east coast of the United States and then Europe. The warm water also warms the climate in these areas. In fact, if it weren't for the Gulf Stream, much of Europe—given its high latitudes—would have a climate more like Canada's. Because all ocean currents in the northern hemisphere rotate in a clockwise direction, water reaching the beaches of California, unlike water reaching east coast beaches, has just come from the north, near Alaska, where it gets very cold.

Beaches on the east coast of the United States have warmer water than west coast beaches at the same latitudes. Why?

OCEANIC CIRCULATION

There are several large, circular patterns of flowing water in the oceans due to a combination of forces including wind, the earth's rotation, the gravitational pull of the moon, temperature, and salt concentration.

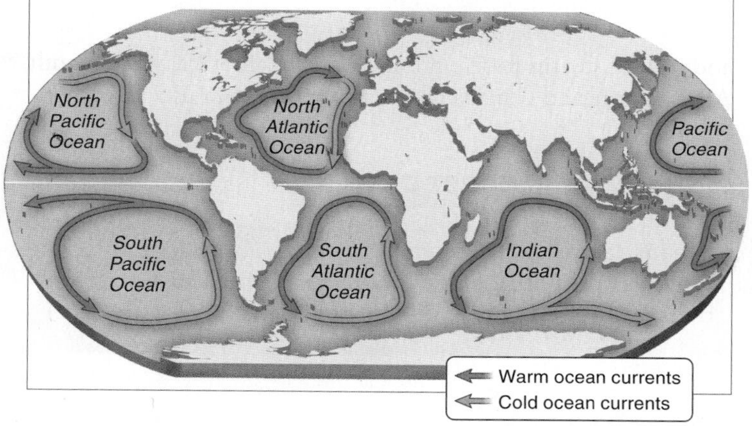

North Pacific Ocean
North Atlantic Ocean
Pacific Ocean
South Pacific Ocean
South Atlantic Ocean
Indian Ocean
Warm ocean currents
Cold ocean currents

FIGURE 15-10 Ocean currents have a powerful influence over the weather.

The most dramatic climate change driven by ocean currents is a phenomenon called **El Niño,** a sustained surface temperature change in the central Pacific Ocean (**FIGURE 15-11**). It occurs every two to seven years and is blamed for flooding, droughts, famine, and a variety of other extreme climate disruptions. We can more easily understand it if we contrast it with the more common pattern.

In a typical year:

1. A steady wind blows westward across the Pacific Ocean from South America toward Southeast Asia and Australia.
2. These winds push the warm surface water away from the coast of South America and raise the sea level in Southeast Asia and Australia by about half a meter. The warmth of the water heats the air above it, causing the air to rise and produce great tropical rainstorms. (To experience this effect, fill your bathtub with hot water and feel the room get warmer and moister.)
3. Off the coast of South America, an upwelling of colder water from deep in the ocean replaces the water that is pushed west. The cold upwelling cools the air above it, causing extremely dry weather.
4. The deep water brings up rich nutrients from below, enabling plankton to flourish and nourishing huge populations of fish.

In an El Niño year:

1. The normal South America–to–Southeast Asia winds ease just a bit. It is not known what causes this, but the effect is severe.
2. Without wind pushing it, the warm water piling up in the west flows back toward South America. This water warms the air, which rises and produces huge amounts of rainfall on the coast of South America where it is normally dry, often causing flooding.
3. Without the warm surface water being blown west, the normal upwelling of cool, deep water with high nutrient content cannot occur. The plankton levels drop to a fraction of their normal levels, and the typically rich fish stocks are wiped out.

EL NIÑO

El Niño occurs every two to seven years and is blamed for flooding, droughts, famine, and a variety of other extreme climate disruptions.

NORMAL CONDITIONS

A steady wind blows westward across the Pacific Ocean from South America toward Southeast Asia. These winds push warm surface water away from the coast of South America, heating the air above it, which causes air to rise and produce tropical rainstorms.

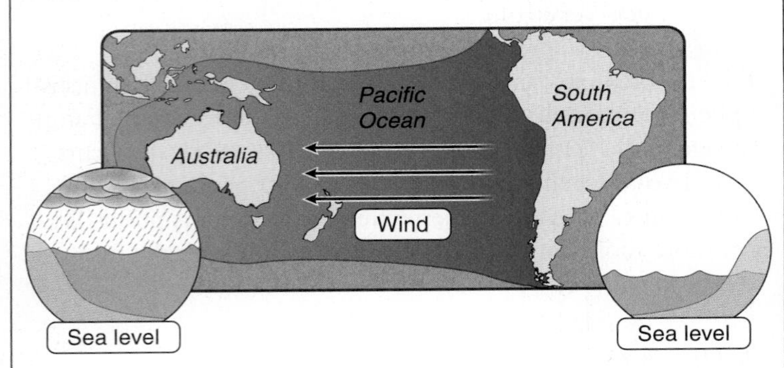

Off the coast of South America, colder water upwells from the ocean depths. The cold water cools the air above, causing extremely dry weather.

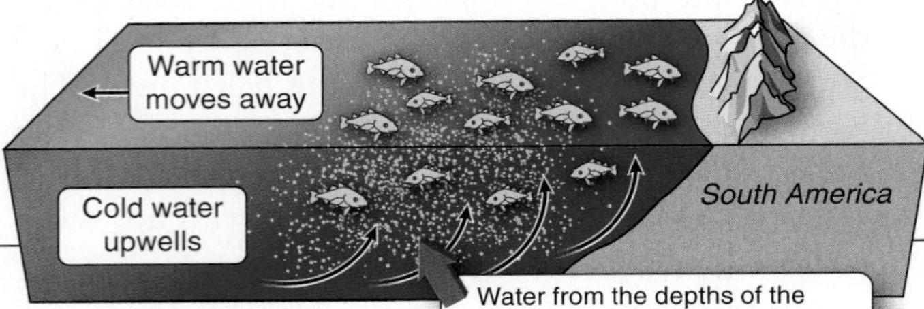

EL NIÑO CONDITIONS

The normal South America–to–Southeast Asia winds ease just a bit. Without the push of the wind, warm water flows back toward South America. The cooled water around Australia and the Philippines cools the air and causes dry weather that can lead to droughts.

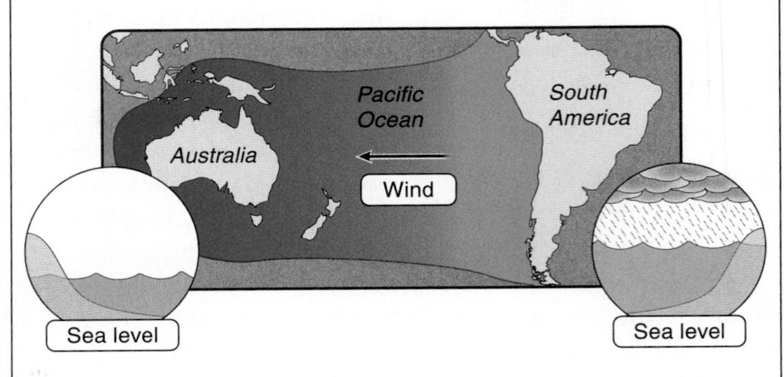

Without the warm surface water being blown west, colder water from the ocean depths cannot upwell. The warm surface water warms the air, resulting in rainfall.

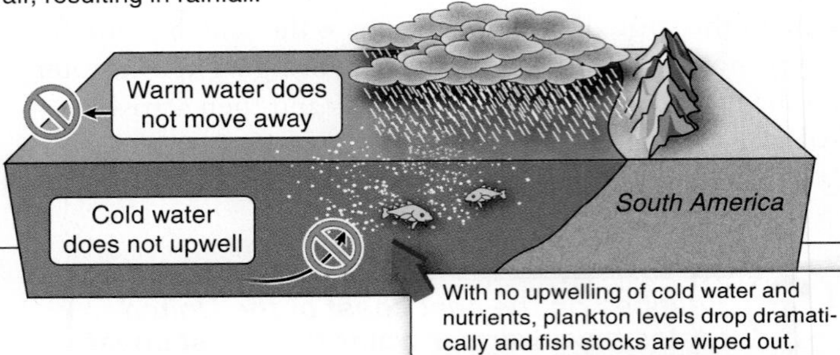

FIGURE 15-11 Domino effect. A simple reduction in the usual east-to-west ocean breeze can cause a cascade of disastrous weather. El Niño years are characterized by flooding in South America and droughts in Southeast Asia.

Changes during an El Niño event can start a chain reaction of unusual weather around the globe, including increased rainfall in the midwestern United States, serious storms off the coast of California, and warmer than usual weather in South Africa and Japan.

4. Meanwhile, in the absence of warm water from the east, Indonesia, the Philippines, and Australia experience drastic climate changes. The cooler water cools the air above and causes unusually dry weather. This frequently leads to droughts and, in extreme years, causes famines.

These predictable changes during an El Niño event can start a chain reaction of additional unusual weather throughout the globe. Although the mechanisms aren't exactly clear, in El Niño years more rain falls in the midwestern United States and serious storms form off the coast of California, while South Africa, Japan, and Canada all enjoy warmer than usual weather.

TAKE-HOME MESSAGE 15•5

Oceans have global circulation patterns. Disruptions in these patterns occur every few years and can cause extreme climate disruptions around the world.

❸ Energy and chemicals flow within ecosystems.

An African fish eagle (Haliaeetus vocifer) *has just caught a fish in the Okavango Delta, Botswana.*

15•6

Energy flows from producers to consumers.

Things are not always as complicated as they appear. Looking at an ecosystem such as a desert savanna, for example, trying to understand how all of the species, from grasses and trees to birds and mammals and worms, interact with each other, and what role each plays within the system, can seem overwhelming. Where is the order? But if we focus on just one aspect of the ecosystem—the pathways energy takes as it flows through the system—a simple and logical order becomes clear.

Everything, all life on that savanna and everywhere else on earth, is made possible because energy flows perpetually from the sun to the earth. This is where our pathway of energy flow begins. Most of the energy is absorbed or reflected by the earth's atmosphere or surface, but about 1% of it is intercepted and converted to chemical energy through photosynthesis. That intercepted energy is then transformed again and again by living organisms, making about four stops as it passes through an ecosystem. Let's examine what happens at each stop, known as **trophic levels** (**FIGURE 15-12**).

First stop: primary producers. When it comes to energy flow, all the species in an ecosystem can be placed in one of two groups: producers and consumers. Plants (along with some algae and bacteria) are, as we noted earlier, the *primary producers.* They convert light energy from the sun into chemical energy through photosynthesis, as discussed in Chapter 4. We use another word to describe that chemical energy: food.

Second stop: primary consumers—the herbivores. Cattle grazing in a field, gazelle browsing on herbs, insects devouring the leaves of a crop plant—these are the **primary consumers** in an ecosystem, the animals that eat plants. Plant material such as cellulose can be difficult to digest. Consequently, most **herbivores**—animals that eat plants—need a little help in digesting the plants they eat. Primary consumers, from termites to cattle, often have symbiotic bacteria living in their digestive system. These microorganisms break down the cellulose, enabling the herbivore to harness the energy held in the chemical bonds of the plants' cells.

Third stop: secondary consumers—the carnivores. Energy originating from the sun is converted into chemical energy within a plant's tissue. The herbivore that eats the plant breaks down the chemical bonds, releasing the energy. This energy fuels the

ENERGY FLOW WITHIN AN ECOSYSTEM

Energy from the sun is intercepted and converted into chemical energy, which passes through an ecosystem in about four stops.

Sun

1 PRIMARY PRODUCERS
Plants convert light energy from the sun into food through photosynthesis.

2 PRIMARY CONSUMERS
Herbivores are animals that eat plants.

3 SECONDARY CONSUMERS
Carnivores are animals that eat herbivores.

4 TERTIARY CONSUMERS
Top carnivores are animals that eat other carnivores.

FIGURE 15-12 Follow the fuel. Plants produce the chemical energy. Herbivores eat plants. Carnivores eat herbivores. And the top carnivores eat the animals that eat the animals that eat the plants that capture energy from the sun.

herbivore's growth, reproduction, and movement, but the energy doesn't remain in the herbivore forever. **Carnivores,** such as cats, spiders, and frogs, are animals that feed on herbivores. They are also known as **secondary consumers.** As they eat their prey, some of the energy stored in the chemical bonds of the carbohydrate, protein, and lipid molecules is again captured and harnessed for their own movement, reproduction, and growth.

Fourth stop: tertiary consumers—the "top" carnivores. In some ecosystems, energy makes yet another stop: the **tertiary consumers,** or "top carnivores." These are the "animals that eat the animals that eat the animals that eat the plants." They are several steps removed from the initial capture of solar energy by a plant, but the general process is the same. A top carnivore, such as a tiger, eagle, or great white shark, consumes other carnivores, breaking down their tissues and releasing energy stored in the chemical bonds of the cells. As in each of the previous steps, the top carnivores harness this energy for their own physiological needs.

This path from primary producers to tertiary consumers is called a **food chain.** We see later in this chapter why a food chain almost never extends to a fifth stop.

The food chain pathway from photosynthetic producers through the various levels of animals, though, is a slight oversimplification. In actuality, food chains are better thought of as **food webs,** because many organisms are **omnivores** and can occupy more than one position in the chain. When you eat a simple meal of rice and chicken, after all, you're simultaneously a carnivore and a herbivore. On average, about 30% to 35% of the human diet comes from animal products and the remaining 65% to 70% comes from plant products. Many other animals, from bears to cockroaches, also have diets that involve harvesting energy from multiple stops in the food chain.

All the while, as energy is transformed through the steps of a food chain, organic wastes are produced. Whether in the form of animals' waste products or the dead bodies of plants and animals, organic material accumulates in every ecosystem. But

Decomposers and detritivores break down organic wastes, releasing chemical components that can then be reused by plants and other primary producers.

FIGURE 15-13 Dust to dust. Eventually, organisms from every level in the food chain die, but even then they provide sustenance for other organisms. Their dead bodies are broken down by decomposers and detritivores, and important chemical components are recycled through the food chain.

this material does not go to waste. Another component of every food web harnesses the last bits of energy that remain. **Decomposers,** usually bacteria or fungi, and **detritivores,** including scavengers such as vultures, worms, and a variety of arthropods, break down the organic material, harvesting energy still stored in the chemical bonds (**FIGURE 15-13**). Because the decomposers are able to break down a much larger range of organic molecules, they are distinguished from the detritivores. Both groups, nonetheless, release many important chemical components from the organic material than can eventually be recycled and utilized by plants and other primary producers.

Energy flows from one stop to the next in a food chain, but not in the way that runners pass a baton in a relay race. The difference is that at every step in the food chain, much of the usable energy is lost as heat. An animal that eats five pounds of plant material doesn't convert that into five new pounds of body weight. Not by a long shot. In the next section, we'll see how this inefficiency of energy transfers ensures that most food chains are very short.

TAKE-HOME MESSAGE 15•6

Energy from the sun passes through an ecosystem in several steps. First, it is converted to chemical energy in photosynthesis. Herbivores then consume the primary producers, the herbivores are consumed by carnivores, and the carnivores, in turn, may be consumed by top carnivores. Detritivores and decomposers extract energy from organic waste and the remains of organisms that have died. At each step in a food chain, some usable energy is lost as heat.

15•7

Energy pyramids reveal the inefficiency of food chains.

Look out of the nearest window. What organisms can you see? Almost without fail, you will see green plant life. Maybe some trees, possibly bushes and grasses as well. You'll have to look a bit longer and harder to see any animals, but you'll probably see a few, most likely small animals and various insects that eat plants. On the other hand, you might stare out of the window all day and not see any animals (other than some fellow humans) that eat other animals. Why is that? And why are big, fierce animals so rare? And why are there so many more plants than animals?

The answers to these questions are closely related to our earlier observation that an animal consuming five pounds of plant material does not gain five pounds in body weight from such a meal. The actual amount of growth such a meal can support is far, far less. And when that herbivore is consumed by a carnivore, the carnivore, too, can convert only a small fraction of the energy it consumes into its own tissue. The fraction turns out to be about 10%, and it is fairly consistent across all levels of the food chain.

> **Q** In humans, why is vegetarianism more energetically efficient than meat-eating?

This means that only about 10% of the **biomass**—the total weight of all the living organisms in a given area—of plants in an ecosystem is converted into herbivore biomass. So the herbivore consuming five pounds of plant material is likely to gain only about half a pound in new growth, while the remaining 90% of the meal is either expended in cellular respiration or lost as feces. Similarly, a carnivore eating the herbivore converts only about 10% of the mass it consumes into its own body mass. Again, 90% is lost to metabolism and feces. And the same inefficiency holds for a top carnivore as well. Let's explore how this 10% rule limits the length of food chains and is responsible for the rarity of big, fierce animals outside your window and across the world.

> **Q** Why are big, fierce animal species so rare in the world?

Given the 10% efficiency with which herbivores convert plant biomass into their own biomass, how much plant biomass is necessary to produce a single 1,200-pound (500-kg) cow? On average, that cow would need to eat about 12,000 pounds (5,000 kg) of grain in order to grow to weigh 1,200 pounds. But that 1,200 pound cow, when eaten by a carnivore, could only add about 120 pounds of biomass to the carnivore, and only 12 pounds to a top carnivore. That's a huge amount of plant biomass required to generate only a very tiny amount of our top carnivore, which explains why big, fierce animals are so rare (and why vegetarianism is more energetically efficient than meat-eating). After all, multiply that 5,000 kg of grain by several hundred (or thousands, more appropriately) to get an idea of how much plant matter would be necessary to support even a small population of top carnivores: millions of kilograms of grain can support only a few top carnivores.

How much would be required to support an even higher link on the food chain? Ten times as much, so much that there may not be enough land in the ecosystem to produce enough plant material. And even if there were, the area required would be so large that the "top, top carnivores" might be so spread out and so busy trying to eat enough that they were unlikely to encounter each other in order to mate. Hence, the 10% rule limits the length of food chains.

We can illustrate the path of energy through the organisms of an ecosystem with an **energy pyramid,** in which each layer of the pyramid represents the biomass of a trophic level. In **FIGURE 15-14**, we can see that for terrestrial ecosystems, the biomass (in kilograms per square meter) found in the photosynthetic organisms, at the base of the pyramid, is reduced significantly at each step, given the incomplete utilization by organisms higher up the food chain. **FIGURE 15-15** illustrates the huge variation in primary productivity across a variety of ecosystems. It is highest in tropical rain forests, marshes, and algal beds, and lowest in deserts, tundra, and the open ocean. In each case, the shapes of the energy and biomass pyramids are similar. With a smaller base, though, the ability of an ecosystem to support higher levels of the food chain is reduced. One dramatic exception is seen in some aquatic ecosystems where the primary producers are plankton. Because plankton have such short life spans and rapid reproduction rates, a relatively small biomass can support a large biomass of consumers, giving rise to an inverted pyramid (see the bottom pyramid in Figure 15-15). If you quantify the amount of energy available to consumers (rather than biomass), though, the pyramid resembles those seen in terrestrial ecosystems.

ENERGY PYRAMID: INEFFICIENT ENERGY FLOW

Only about 10% of the biomass from each trophic level is converted into biomass in the next trophic level. This explains why there are so many more plants than animals, and why top carnivores are relatively rare.

TERTIARY CONSUMERS

10% converted to biomass

SECONDARY CONSUMERS

90% is used in cell respiration and lost as heat.

10% converted to biomass

PRIMARY CONSUMERS

10% converted to biomass

PRIMARY PRODUCERS

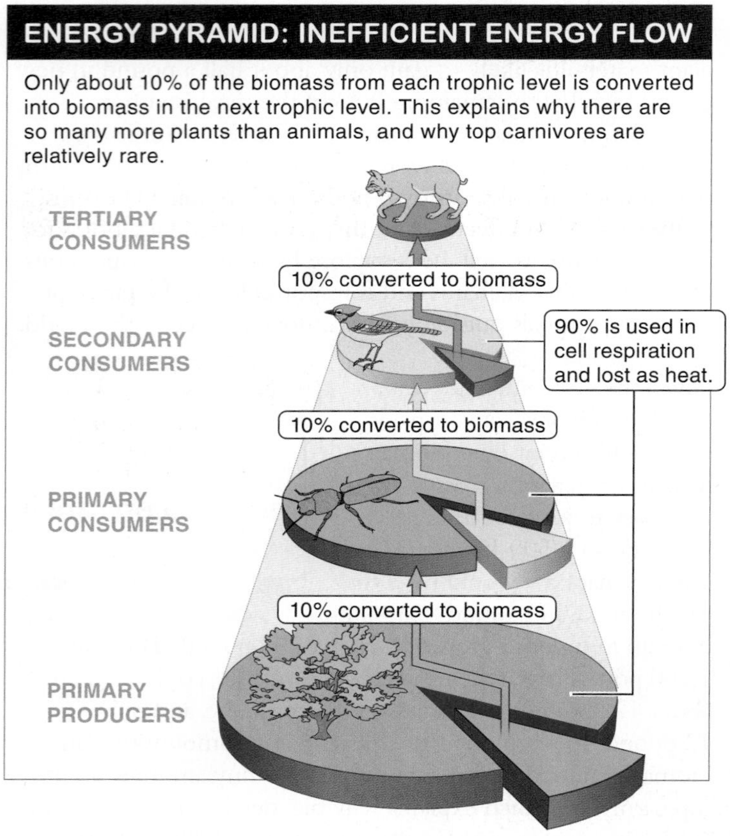

FIGURE 15-14 Inefficiencies in the transfer of biomass. The biomass of photosynthetic organisms at the base of the pyramid is significantly reduced at each step because of incomplete utilization by organisms at higher trophic levels.

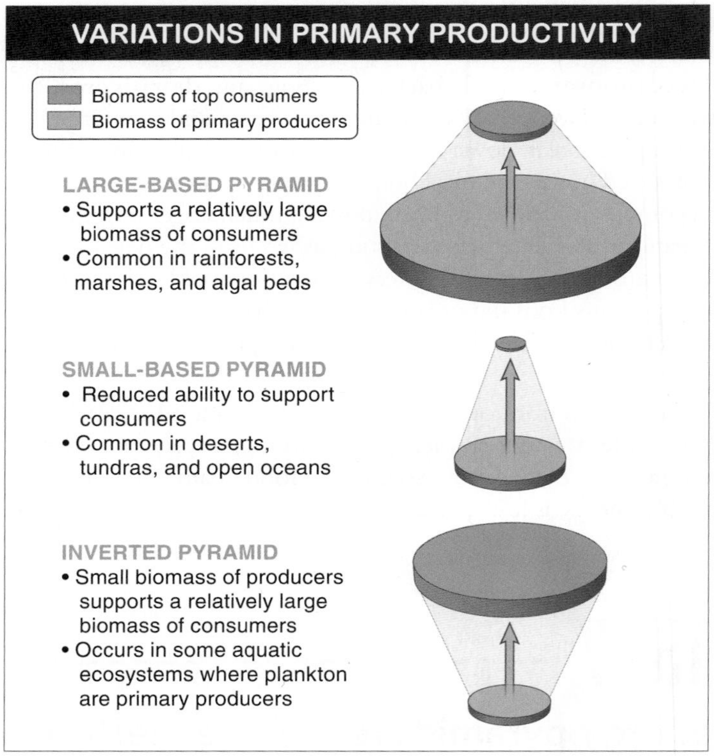

VARIATIONS IN PRIMARY PRODUCTIVITY

Biomass of top consumers
Biomass of primary producers

LARGE-BASED PYRAMID
- Supports a relatively large biomass of consumers
- Common in rainforests, marshes, and algal beds

SMALL-BASED PYRAMID
- Reduced ability to support consumers
- Common in deserts, tundras, and open oceans

INVERTED PYRAMID
- Small biomass of producers supports a relatively large biomass of consumers
- Occurs in some aquatic ecosystems where plankton are primary producers

FIGURE 15-15 Relative biomass of producers and consumers. Across ecosystems, there is huge variation in primary productivity.

TAKE-HOME MESSAGE 15•7

Energy from the sun passes through an ecosystem in several steps known as trophic levels. Energy pyramids reveal that the biomass of primary producers in an ecosystem tends to be far greater than the biomass of herbivores. Similarly, the biomass transferred at each successive step along the food chain tends to be only about 10% of the biomass of the organisms being consumed, due to energy lost in cellular respiration. As a consequence of this inefficiency, food chains rarely exceed four levels.

15•8

Essential chemicals cycle through ecosystems.

What is necessary for life? Energy and some essential chemicals top the list. New energy continually comes to earth from the sun, fulfilling the first need. And everything else is already here. The chemicals just cycle around and around, using the same pathway taken by energy—the food chain. Plants and other producers generally take up the molecules from the atmosphere or the soil. Then the chemicals move into animals as they consume plants or other animals, and thus move up the food chain. And as the plants and animals die, detritivores and decomposers return the chemicals to the abiotic environment. From a chemical perspective, life is just a continuous recycling of molecules.

Chemicals cycle through the living and non-living components of an ecosystem. Each chemical is stored in a non-living part of the environment called a "reservoir." Organisms acquire the chemical from the reservoir, the chemical cycles through the food chain, and eventually the chemical is returned to the reservoir.

This is a useful overview, but we can get a deeper appreciation of the functioning of ecosystems and the ecological problems that can occur when they are disturbed—particularly by humans—by investigating a few of these cycles in more detail. Here we'll explore three of the most important chemical cycles: carbon, nitrogen, and phosphorus.

Carbon For carbon, the most important reservoir is the atmosphere, where carbon is in the form of carbon dioxide (CO_2). As we saw in Chapter 4, plants and some microorganisms utilize carbon dioxide in photosynthesis, separating the carbon molecules from CO_2 and using them to build sugars. Carbon then moves through the food chain as organisms eat plants and are themselves eaten (**FIGURE 15-16**). A secondary reservoir of carbon is in the oceans. Here, many organisms utilize dissolved carbon to build shells (which later dissolve back into the water after the organism dies).

Most carbon returns to its reservoir as a consequence of organisms' metabolic processes. Organisms extract energy from food by breaking carbon-carbon bonds, releasing the energy stored in the bonds and combining the released carbon atoms with oxygen. They then exhale the end product as CO_2.

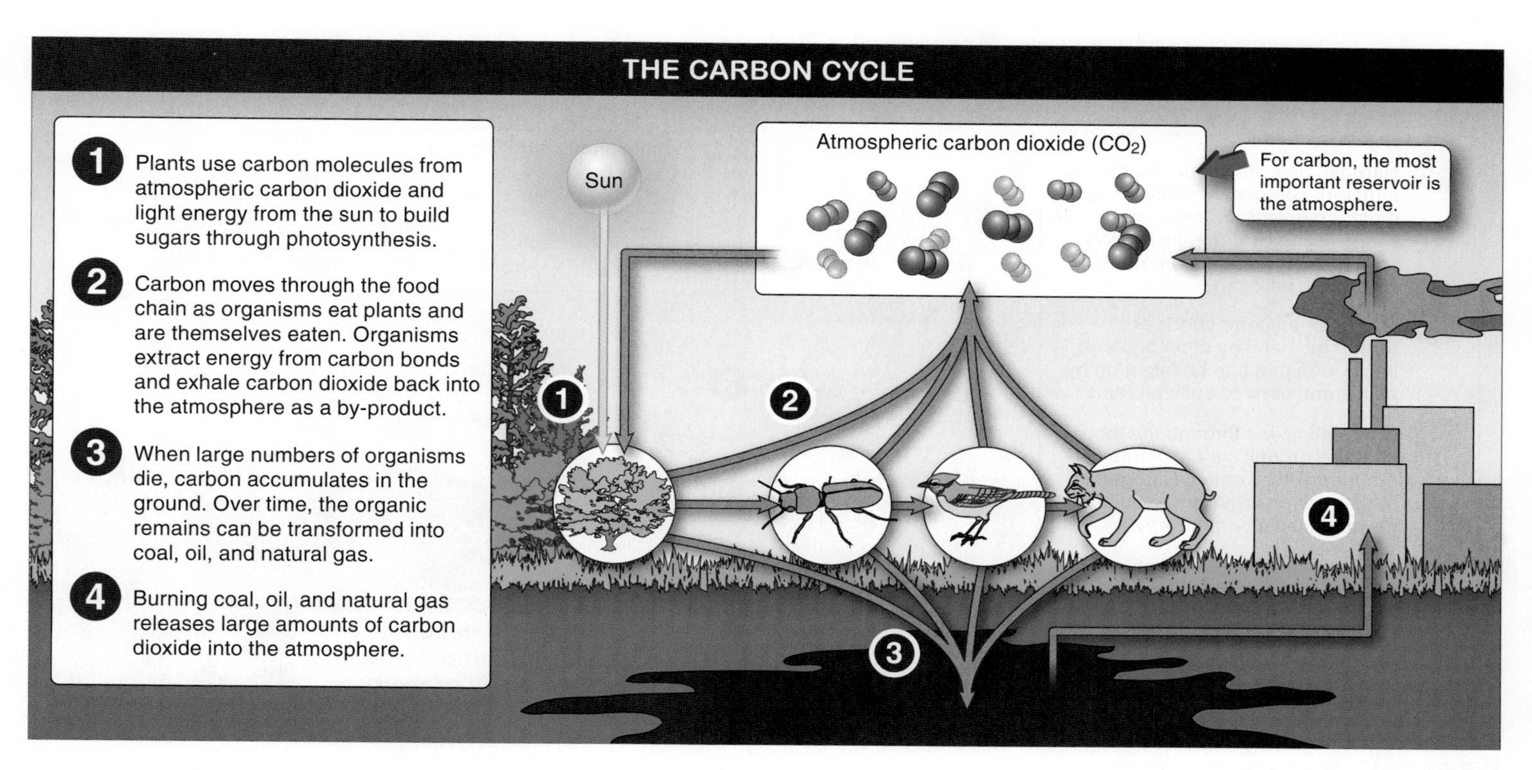

FIGURE 15-16 Element cycling: carbon.

> **Q** Why are global CO_2 levels rising?

Increasingly, another source of carbon is adding significantly to the atmospheric reservoir. Fossil fuels are created when large numbers of organisms die and are buried in sediment lacking oxygen. In the absence of oxygen, at high pressures, and after very long periods of time, the organic remains are ultimately transformed into coal, oil, and natural gas. Trapped underground or in rock, these sources of carbon played little role in the global carbon cycle until humans in industrialized countries began using fossil fuels to power various technologies. Burning coal, oil, and natural gas releases large amounts of carbon dioxide, thus increasing the average CO_2 concentration in the atmosphere—the current level of CO_2 in the atmosphere is the highest it has been in almost half a million years. This has potentially disastrous implications, as we will see in Chapter 16.

> **Q** Global CO_2 levels are rising in general, but they also exhibit a sharp rise and fall within each year. Why?

Although, on average, the level of CO_2 in the environment is increasing, there is a yearly cycle of ups and downs in the CO_2 levels in the northern hemisphere. This cycling is due to fluctuations in the ability of plants to absorb CO_2. Many trees lose their leaves each fall and, during the winter months, relatively low rates of photosynthesis lead to low rates of CO_2 consumption, causing an annual peak in atmospheric CO_2 levels. During the summer, leaves are present, sunlight is strong, and photosynthesis (consuming CO_2) occurs at much higher levels, causing a drop in the atmospheric CO_2 level.

Nitrogen Nitrogen is necessary to build a variety of molecules critical to life, including all amino acids, the components of every protein molecule. Like carbon, the chief reservoir of nitrogen is the atmosphere. But even though more than 78% of the atmosphere is nitrogen gas (N_2), for most organisms this nitrogen is completely unusable. The problem is that atmospheric nitrogen consists of two nitrogen atoms bonded tightly together, and these bonds need to be broken to make the nitrogen usable for living organisms. Only through the metabolic magic (chemistry, actually) of some soil-dwelling bacteria, the nitrogen-fixers, can most nitrogen enter the food chain (**FIGURE 15-17**).

These bacteria chemically convert or "fix" nitrogen by attaching it to other atoms, including hydrogen, producing ammonia and related compounds. These compounds are then further modified by other bacteria into a form that can be taken up by plants and used to build proteins. And once nitrogen is in plant tissues, animals acquire it in the same way they acquire carbon: by eating the plants. Nitrogen

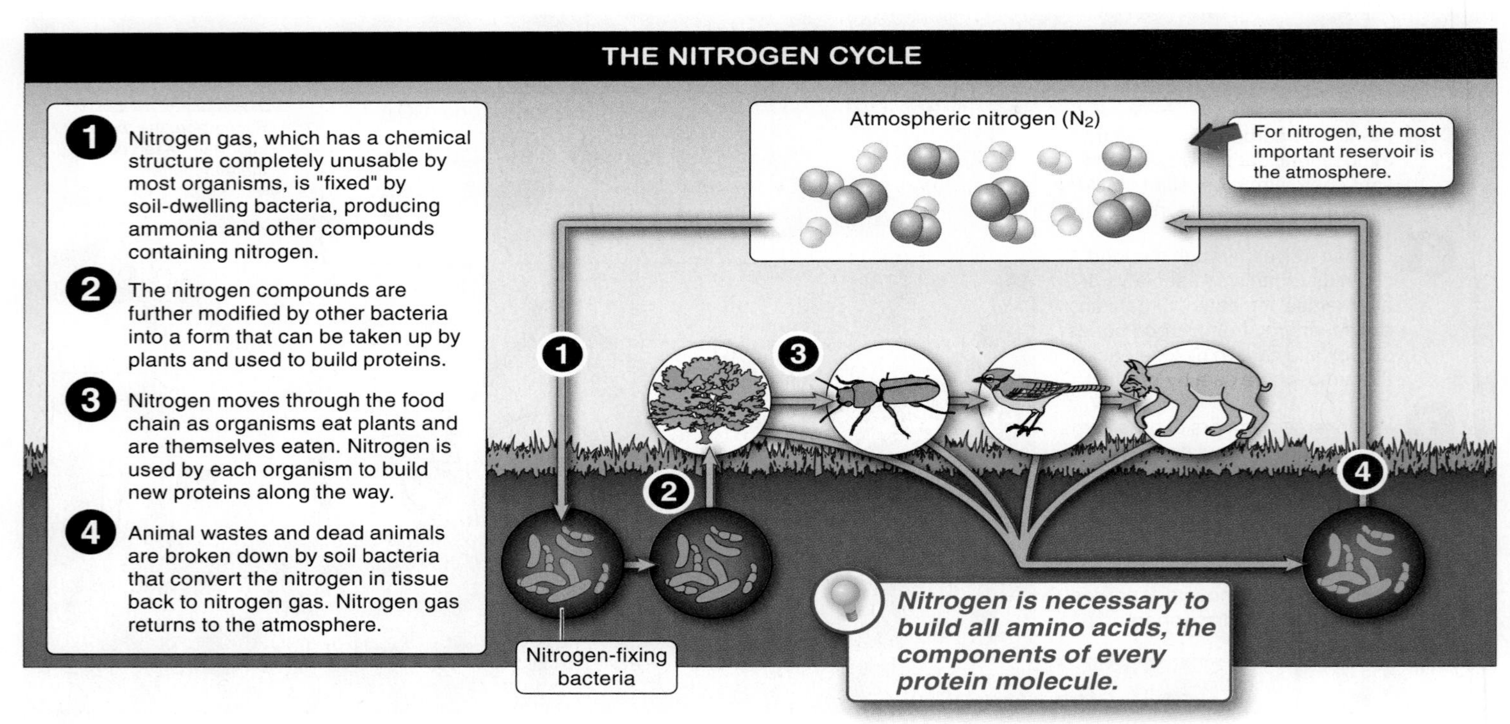

FIGURE 15-17 Element cycling: nitrogen.

returns to the atmosphere when animal wastes and dead animals are broken down by soil bacteria that convert the nitrogen compounds in tissues back to nitrogen gas.

Nitrogen is like a bottleneck limiting plant growth. Because it is necessary for the production of every plant protein, and because all nitrogen must first be made usable by bacteria, plant growth is often limited by nitrogen levels in the soil. For this reason, most fertilizers contain nitrogen in a form usable by plants.

Phosphorus Every molecule of ATP and DNA requires phosphorus. Virtually no phosphorus is available in the atmosphere, though. Instead, soil serves as the chief reservoir. Like nitrogen, phosphorus is chemically converted—"fixed"—into a form usable by plants (phosphate) and then absorbed by their roots. It cycles through the food chain as herbivores consume plants and carnivores consume herbivores. As organisms die, their remains are broken down by bacteria and other organisms, returning the phosphorus to the soil. The pool of phosphorus in the soil is also influenced by the much slower process of rock formation on the sea floor, its uplifting into mountains, and the eventual weathering of the rock, releasing its phosphorus (**FIGURE 15-18**).

Like nitrogen, phosphorus is often a limiting resource in soils, constraining plant growth. Consequently, fertilizers usually contain large amounts of phosphorus. This is beneficial in the short run, but it can have some disastrous unintended consequences. As more and more phosphorus (and nitrogen) is added to soil, some of it runs off with water and ends up in lakes, ponds, and rivers. In these habitats, also, it acts as a fertilizer, making spectacular growth possible for algae. But eventually the algae die and sink, creating a huge source of food for bacteria. The bacteria population increases and can use up too much of the dissolved oxygen, causing fish, insects, and many other organisms to suffocate and die.

> "An atom of phosphorus 'X' had marked time in the limestone ledge since the Paleozoic seas covered the land. *Time, to an atom locked in a rock, does not pass.* The break came when a bur-oak root nosed down a crack and began prying and sucking. In the flash of a century the rock decayed, and X was pulled out and up into the world of living things. He helped build a flower, which became an acorn, which fattened a deer, which fed an Indian all in a single year."
>
> —Aldo Leopold, *A Sand County Almanac*, 1949

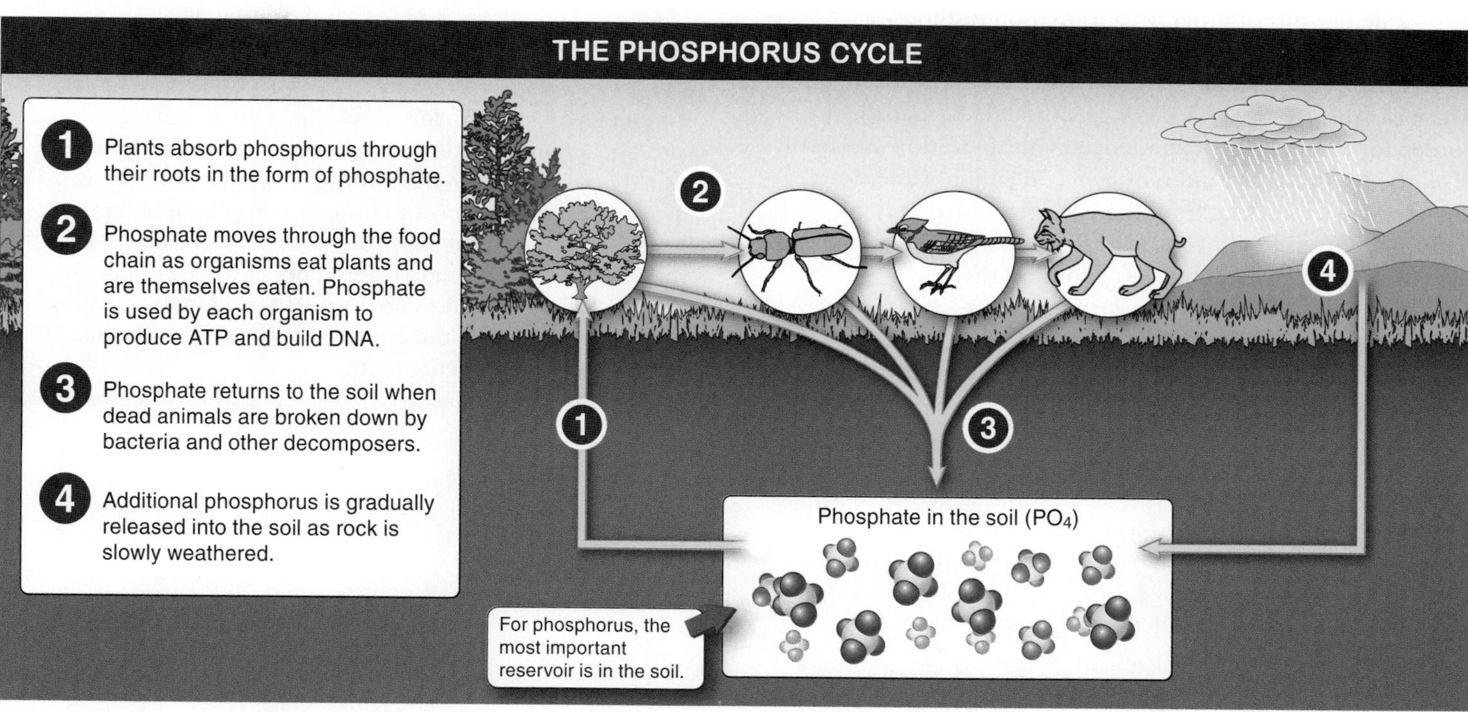

FIGURE 15-18 Element cycling: phosphorus.

EUTROPHICATION

Eutrophication—the increase in nutrients in an ecosystem, particularly nitrogen and phosphorus—often leads to the rapid growth of algae and bacteria in lakes and other aquatic ecosystems. These organisms then consume much of the oxygen, leading to large die-offs of animal life. This process often occurs as a result of runoff from fertilizer added to farmlands.

Due to the extensive use of fertilizers in agriculture, eutrophication now affects more than half of the lakes in Asia, Europe, and North America.

FIGURE 15-19 Too much of a good thing?

Water also cycles through ecosystems. It continually moves from the ocean to the air and land, then returns to the ocean in a cyclic pattern.

The increase in nutrients, particularly nitrogen and phosphorus, in an ecosystem, is called **eutrophication.** The consequent rapid growth of algae and bacteria, followed by large-scale die-offs of other organisms, is increasingly a problem in both small and large bodies of water, including more than half of the lakes of Asia, Europe, and North America (**FIGURE 15-19**). Lake Erie, on the U.S.-Canadian border, for example, has experienced eutrophication as a result of all the phosphorus- and nitrogen-containing waste water that drains into the lake from the extensive surrounding farmlands. Given the lower use of fertilizers in South America and Africa, eutrophication is less common there.

Water, too, cycles through ecosystems. Moving from the oceans to land via evaporation and rain, it then returns to the oceans through runoff.

TAKE-HOME MESSAGE 15•8

Chemicals essential to life—including carbon, nitrogen, and phosphorus—cycle through ecosystems. They are usually captured from the atmosphere, soil, or water by growing organisms, passed from one trophic level to the next as organisms eat other organisms, and returned to the environment through respiration, decomposition, and erosion. These cycles can be disrupted as human activities significantly increase the amounts of the chemicals utilized or released to the environment.

④ Species interactions influence the structure of communities.

Cleaning behavior: a blue-striped grunt (Haemulon sciurus) *presents its mouth for cleaning by a foureye butterflyfish* (Chaetodon capistratus).

15•9

Interacting species evolve together.

There is a moth in Madagascar with a tongue that is 11 inches long! This might seem absurdly long—until the moth approaches a similarly odd-looking orchid. The orchid's flower has a very long tube, also about 11 inches long, that has a bit of nectar at the very bottom. The moth's tongue, although usually rolled up, straightens out as fluid is pumped into it and can be inserted into the long nectar tube. As its tongue reaches the bottom, the moth slurps up the nectar, gaining nourishment and energy. The moth also gets a bit of pollen stuck to it in the process, pollen that gets brushed onto the reproductive parts of the next orchid flower it visits (**FIGURE 15-20**).

FIGURE 15-20 A perfect match? Long-tubed orchids and long-tongued moths in Madagascar each influence the evolution of the other.

It's clear that having an 11-inch tongue is useful, even necessary, to extract nectar from an 11-inch nectar tube. And it's clear that putting nectar at the bottom of an 11-inch tube is a strategy to restrict access to only those pollinators that will reliably pass pollen from plant to plant of the same species. But how did such a system ever originate? Which came first, the long-tongued moths or the long-tubed flowers? Each trait only seems to make sense in a world in which the other already exists. The answer is that neither came first. They both evolved—**coevolved**—together.

Natural selection causes organisms to become better adapted to their environment. As long as there is variation for a trait, and the trait is heritable, differential reproductive success will lead to a change in the population. It is easy to imagine populations becoming more and more efficient at utilizing non-living resources. But natural selection does not distinguish between biotic and abiotic resources as selective forces. Either can cause individuals with certain traits to reproduce at a higher rate than others, and so either can cause evolution. So small changes in the moth lead to small changes in the orchid, which select for further small changes in the moth . . . and the process continues on. In the end, species become adapted not just to their physical environment but also to the other species around them.

TAKE-HOME MESSAGE 15•9

In producing organisms better adapted to their environment, natural selection does not distinguish between biotic and abiotic resources as selective forces.

15•10

Each species' role in a community is defined by its niche.

Within a society, most humans seem to find their niche. Each person plays a particular role, defined by the nature of his or her work, activities, and interactions with others. Other species do the same thing. Within their *communities*—geographic areas defined as loose assemblages of species, sometimes interdependent, with overlapping ranges—each species has its own niche.

We can define an organism's **niche** in terms of the ways in which the organism utilizes the resources of the environment. More than just a *place* for living, a niche is a complete *way* of living. In other words, an organism's niche encompasses (1) the space it requires, (2) the type and amount of food it consumes, (3) the timing of its reproduction (its life history), (4) its temperature and moisture requirements, and virtually every other aspect that describes the way the organism uses its environment (**FIGURE 15-21**).

> "And NUH is the letter I use to spell Nutches,
> Who live in small caves, known as Niches, for hutches.
> These Nutches have troubles, the biggest of which is
> The fact there are many more Nutches than Niches.
> Each Nutch in a Nich knows that some other Nutch
> Would like to move into his Nich very much.
> So each Nutch in a Nich has to watch that small Nich
> Or Nutches who haven't got Niches will snitch.
>
> —Dr. Seuss, *On Beyond Zebra,* 1955"

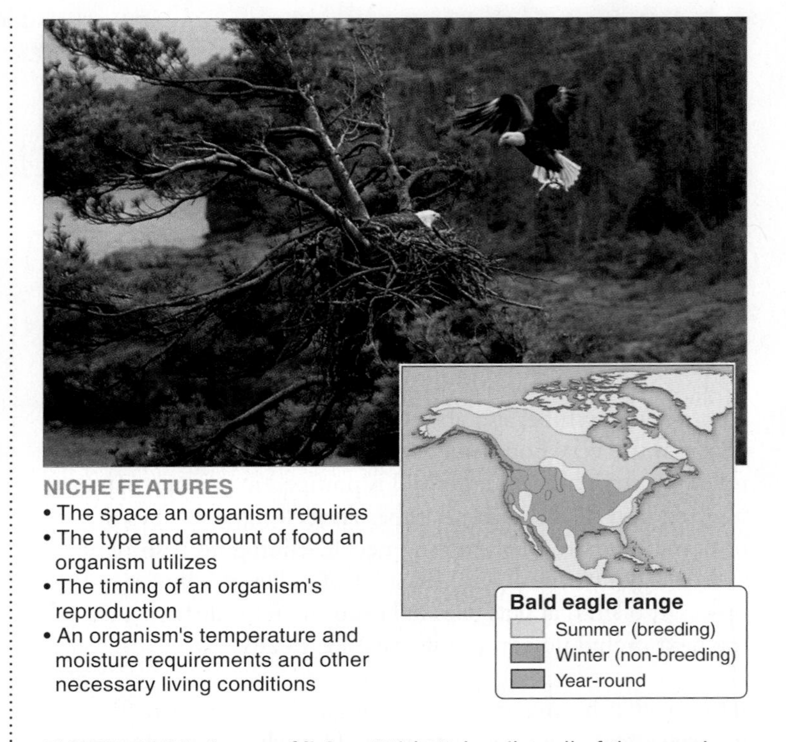

FIGURE 15-21 A way of living. "Niche" describes all of the ways in which an organism utilizes the resources of the environment.

Although a niche describes the role a species *can* play within a community, the species doesn't always get to have that exact role. Consider the rats of Boston. Until the 1990s, they lived in relative peace in the sewers beneath the city's streets. But when the city embarked on the largest underground highway construction project in U.S. history, engineers displaced and forcibly drove out thousands of rats from much of their habitat. In essence, there was suddenly an overlap between the rat niche and the human niche, and the rats were now restricted to just some portions of the sewer. As a result, the rats' **realized niche,** where and how they are actually living, is now just a subset of their **fundamental niche,** the full range of environmental conditions under which they can live.

It is common for species to find themselves competing with other species for parts of a niche. When this occurs, generally one of the species is restricted from its full niche. We'll examine next the various possible outcomes when such niche overlap occurs.

TAKE-HOME MESSAGE 15•10

A population of organisms in a community fills a unique niche, defined by the manner in which it utilizes the resources in its environment. Organisms do not always completely fill their niche; competition with species that have overlapping niches can reduce their range.

15•11

Competition can be hard to see, but it still influences community structure.

Some species, particularly closely related species, have similar niches. This can lead to conflict as they try to exploit the same resources. Almost invariably, when the fundamental niches of two species overlap, competition occurs. This competition doesn't last forever, though. Inevitably, one of two outcomes occurs: competitive exclusion or resource partitioning (**FIGURE 15-22**).

1. In **competitive exclusion,** two species battle for resources in the same niche until the more efficient of the two wins and the other species is driven to extinction in that location ("local extinction"). In the 1930s, this was demonstrated in simple laboratory experiments using *Paramecium,* a single-celled organism. Populations of two similar *Paramecium* species were grown either separately or together in test tubes containing water and their bacterial food source. When grown separately, each species thrived. When grown together, though, one species always drove the other to extinction.

2. Resource partitioning is an alternative outcome of niche overlap. Individual organisms and species can adapt to changing environmental conditions, and resource partitioning can result from an organism's behavioral change or a change in its structure. When this occurs, one or both species become restricted in some aspect of their niche, dividing the resource. In other experiments with *Paramecium,* for example, one of the two species was replaced with a different species. As in the initial experiment, either species thrived when grown alone. But when the two species were grown together in the same test tube, they ended up dividing the test tube "habitat." One species fed exclusively

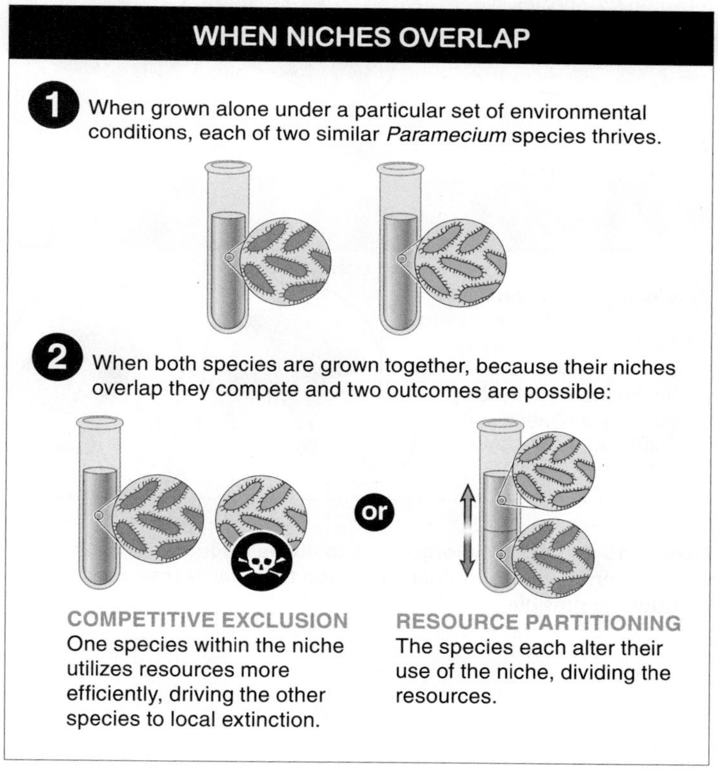

FIGURE 15-22 Overlapping niches: competitive exclusion and resource partitioning.

at the bottom of the test tube, and the other fed only at the top. Simple behavioral change made coexistence possible.

In many situations, resource partitioning is accompanied by **character displacement,** an evolutionary divergence in one or both of the species that leads to a partitioning of the niche. A clear example occurs among two species of seed-eating finches on the Galápagos Islands. On islands where both species live, their beak sizes differ significantly. One species has a deeper beak, better for large seeds, while the other has a shallower beak, better for smaller seeds, and they do not compete. On islands where either species occurs alone, beak size is intermediate between the two sizes (**FIGURE 15-23**).

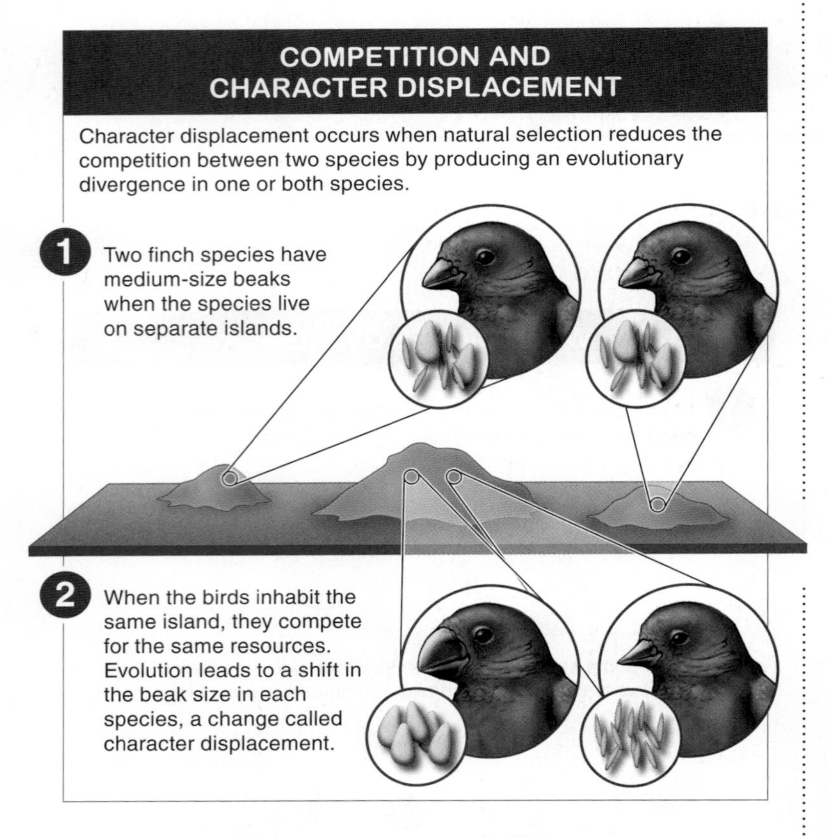

FIGURE 15-23 Allowing organisms to divide resources. Character displacement is an evolutionary adaption that makes resource partitioning possible.

Q Why is competition hard to see in nature?

Competition between species has one very odd feature: it is very hard to actually see it occurring because it causes itself to disappear. That is, after only a short period of competition, either one of the species becomes locally extinct or leaves the area where the niches overlap, or character displacement occurs, largely reducing the competition. In either case, the net result is that the level of competition is significantly reduced or wiped out altogether. For this reason, biologists often have to look for character displacement—the "ghost of competition past"—to identify areas where competition has occurred in the past. Moreover, even while it is occurring, competition tends to be indirect, rather than head-to-head battles. Like a game of musical chairs, it's more a question of both species trying to utilize a particular resource, with one being a bit better at it.

Biologists do have the opportunity to see competition, however, in the occasional natural experiments in which a new species is introduced to a community. This happened, for example, when the American mink, a weasel-like mammal, was introduced into Europe, where previously only the European mink had lived. Within 10 years, in the area where their ranges overlapped, the European mink increased in size while the American mink decreased in size, as biologists documented the evolution. In the next chapter, we'll investigate in greater depth why newly introduced species frequently win the competition with native species.

TAKE-HOME MESSAGE 15•11

Populations with completely overlapping niches cannot coexist forever. Competition for resources occurs until one or both species evolve in ways that reduce the competition, through character displacement, or until one becomes extinct in that location.

15•12

Predation produces adaptation in both predators and their prey.

Some words of advice in case you ever think about quickly approaching a horned lizard: be afraid. Be very afraid. Here's why: as you get close to the lizard, it may zap you with streams of blood squirted from its eyes. In all likelihood, you will flinch. And as you flinch, the horned lizard will scurry away. The display is shocking, but the fact that evolution has produced extreme and effective anti-predator adaptations is not.

Predation—an interaction between two species in which one species eats the other—is one of the most important forces shaping the composition and abundance of species in a community. Predation, though, isn't restricted to the obvious interactions involving one animal chasing down and killing another. Herbivores eating leaves, fruits, or seeds is a form of predation, even though it doesn't necessarily kill the plant. And predators are not necessarily physically imposing. Each year more than a million humans die as a result of disease from mosquito bites, compared with fewer than a dozen from shark attacks.

Q Why do exotic species often flourish when released into novel habitats, even though natural selection has not adapted them to this new environment?

Predators are a potent selective force: organisms eaten by predators tend to have reduced reproductive success. Consequently, in prey species, a variety of features have evolved (and continue to evolve)—including the blood-squirting-eyeball effect—that reduce their predation risk. But as prey evolve, so do predators. This coevolution is a sort of arms race with ever-changing and escalating predation-effectiveness adaptations causing more effective predator-avoidance adaptations, and vice versa. (In this light, it may seem unexpected that exotic species often flourish when released into novel habitats, even though natural selection has not adapted them to their new environment. As it turns out, just as these species are not fully adapted to their new environment, they also have few predators there. And with low predation risk, they often are able to flourish.) We'll examine some common adaptations of both predators and prey.

Prey adaptations for reducing predation There are two broad categories of defenses against predators: physical and behavioral.

Physical defenses include mechanical, chemical, warning coloration, and camouflage mechanisms.

1. *Mechanical defenses.* Predation plays a large role in producing adaptations such as the sharp quills of a porcupine, the prickly spines of a cactus, the wings of a bird or bat, or the tough armor protecting an armadillo or sow bug. These, as well as claws, fangs, stingers, and other physical structures, can reduce predation risk (**FIGURE 15-24**).

2. *Chemical defenses.* Further prey defenses can include chemical toxins that make the prey poisonous or unpalatable. Plants can't run from their predators, so chemical defenses are especially important to them. Virtually all plants produce some chemicals to deter organisms that might eat them. The toxins can be severe, such as strychnine from plants in the genus *Strychnos,* which kills most vertebrates, including humans, by stimulating non-stop convulsions and other extreme and painful symptoms leading to death. At the other end of the spectrum are chemicals toxic to some insects but relatively mild to humans, such as those found in cinnamon, peppermint, and jalapeño peppers. Ironically, many plants that evolved to produce such chemicals to deter predators are now cultivated and eaten specifically *for* the chemicals they produce. One organism's toxic poison is another's spicy flavor.

Some animals can also synthesize toxic compounds. The poison dart frog has poison glands all over its body, making it toxic to the touch. The fire salamander, too, is toxic, with the capacity to squirt a strong nerve toxin from poison glands on its back. Some animals, including milkweed bugs and monarch butterflies, are able to safely consume toxic chemicals from plants and sequester them in their tissues, becoming toxic to predators who try to eat them.

3. *Warning coloration.* Species protected from predation by toxic chemicals frequently have evolved bright color patterns to warn potential predators. They are essentially carrying a sign that says: "Warning, I'm poisonous, so keep away." To make it as easy as possible for predators to learn, different poisonous species often have the same color patterns. It's as if devising a single common sign saying "Danger: poison" is more effective than multiple species coming up with their own specific signs (see Figure 15-24).

In a clever twist on this, some species that are perfectly edible to their predators also have evolved the same bright colors, in

FIGURE 15-24 Prey defenses: some physical means for avoiding predation. Shown (*from left, clockwise*) are Cape porcupine, azure poison dart frog, praying mantis, and monarch butterfly.

a phenomenon known as **mimicry.** Their coloration mimics the same warning sign but without the toxins. As long as they are relatively rare compared with the toxic individuals they mimic—reducing the chance that predators might catch on to their trickery—the evolutionary ruse is quite successful.

4. *Camouflage.* An alternative to warning coloration, and one of the most effective ways to avoid being eaten, is simply to avoid being seen. An adaptation in many organisms is patterns of coloration that enable them to blend into their surroundings. These include insects that look like leaves or twigs and hares that are brown for most of the year but turn white when there is snow on the ground.

Behavioral defenses include both seemingly passive and active behaviors: hiding or escaping, or alarm calling or fighting back.

1. *Hiding or escaping.* Anti-predator adaptations need not involve toxic chemicals, physical structures, or special coloration. Many species excel at hiding and/or running. With vigilance, it is possible to get advance warning of impending predator attacks, and then quickly and effectively avoid predators. A variation of this strategy comes from safety in numbers: many species, including schooling fish and emperor penguins, travel in large groups to reduce their predation risk (**FIGURE 15-25**).

> **Q** On islands, animals frequently have no fear of humans (in contrast to having more skittish behavior on the mainland). Why?

On many islands, animals show no fear of humans. Rather than encountering

BEHAVIORAL DEFENSES THAT REDUCE PREDATION

FIGURE 15-25 Prey defenses: behavioral means for avoiding predation. Fish may travel in huge schools, finding safety in numbers. And fish may mob a predator, here a shark, to make it harder for the predator to attack any one member of the school. The bird can project vomit at a predator, frightening it away.

skittish lizards, for example, a human visitor to the Galápagos Islands must be careful not to step on them. The animals are not afraid because, given the small size of most islands, the number of predators is restricted. As a consequence, there has been no selection for skittishness and hyper-vigilance, traits that normally evolve in response to predation risk.

2. *Alarm calling and fighting back.* In many species, especially birds and mammals, individuals warn others with an alarm call. Although risky for the caller, such alarm calling can give other individuals—often close kin that are nearby—enough advance warning to escape (recall, from Chapter 9, the warning calls of Belding's ground squirrels). Some prey species also turn the tables, mobbing predators to keep them from successfully completing their task. This category might also include the blood-squirting lizard described above, or the fulmar, a seabird that defends its nest from attacks with projectile vomiting aimed at the intruder.

Predator adaptations for enhancing predation Just as prey use physical and behavioral features to reduce their risk of predation, predators evolve in parallel ways to increase their efficiency. As milkweed plants have evolved to sequester toxic chemicals within the structures of the plant that kill their predators, in milkweed bugs toxic-avoidance methods, have evolved that allow the bugs to eat the toxic plants without suffering harm. And as prey have become better at hiding and escaping, predators have developed better sensory perception to help them detect hiding prey, and faster running ability to catch them. Predators, too, make use of mimicry. The angler fish, the tasseled frogfish, and the snapping turtle all have physical structures that mimic something—usually a food item—that is of interest to potential prey. As the prey come closer to inspect, the predator snaps them up (**FIGURE 15-26**).

Q Why don't predators become so efficient at capturing prey that they drive the prey to extinction?

Although natural selection leads to predators with effective adaptations for capturing prey, the adaptations are rarely so efficient that the prey are driven to extinction. The explanation for this is referred to as

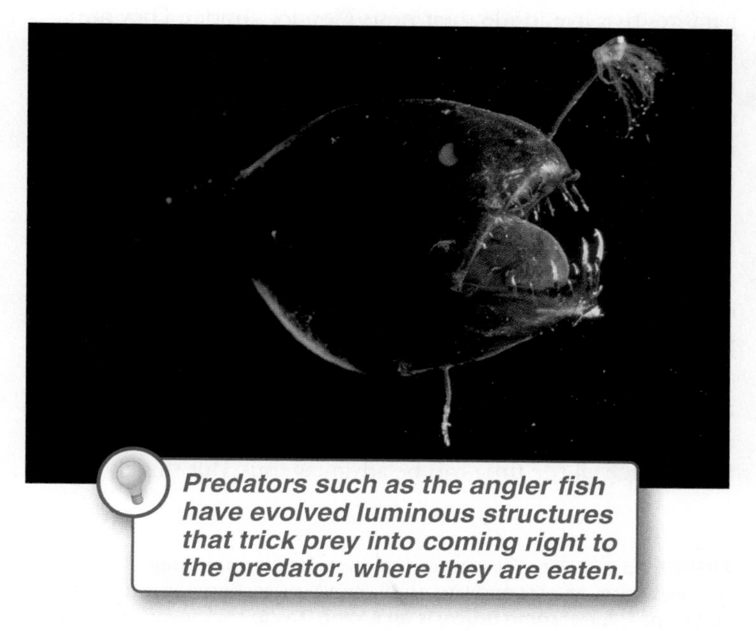

FIGURE 15-26 A better predator. Some adaptations enhance predation.

the "life-dinner hypothesis." It generally holds true because selection for "escape ability" in the prey is stronger than selection for "capture ability" in the predator. When a prey, such as a rabbit, for example, can't escape from a fox, the cost is its life—and it will never reproduce again. On the other hand, when a fox can't keep up with a rabbit, all it loses is a meal; it can still go on to capture prey and reproduce in the future. In other words, the cost of losing in the interaction is much higher for the rabbit.

TAKE-HOME MESSAGE 15•12

Predators and their prey are in an evolutionary arms race; as physical and behavioral features evolve in prey species to reduce their predation risk, predators develop more effective and efficient methods of predation. The coevolutionary process can result in brightly colored organisms, alarm calling, and many types of mimicry.

15•13

Parasitism is a form of predation.

For most people, the word "predation" conjures images of a large animal such as a cheetah, chasing and killing another large animal such as a gazelle. But there is another world of predation, largely unseen but equally deadly: **parasitism,** defined as a symbiotic relationship in which one organism (the **parasite**) benefits while the other (the **host**) is harmed. In fact, this hidden world of parasites is thriving with activity: parasites are probably the most numerous species on earth, with three or four times as many species as non-parasites!

Two general types of parasites make life difficult for most organisms. *Ectoparasites* ("ecto" meaning outside) include organisms such as lice, leeches, and ticks. One species of Mexican parrot is all too aware of ectoparasites: it has 30 different species of mites living on its feathers alone. And many of the parasites even have parasites themselves! *Endoparasites* are parasites that live inside their hosts ("endo," inside). They are equally pervasive. Endoparasites infecting vertebrates include many different phyla of both animals and protists, the single-celled eukaryotes (**FIGURE 15-27**). In all of these parasite-host interactions, as with all predator-prey interactions, the predator or parasite benefits and the prey or host is harmed.

Even though they are considered predators, parasites have some unique features and face some unusual challenges relative to other predators. The most obvious of these features is that in parasitism, the parasite generally is much smaller than its host and stays in contact with the host for extended periods of time, normally not killing the host but weakening it as the parasite uses some of the host's resources. Being located right on your food source all the time can be advantageous. But this also leads

PARASITES

Parasites are predators that benefit from a symbiotic relationship with their hosts.

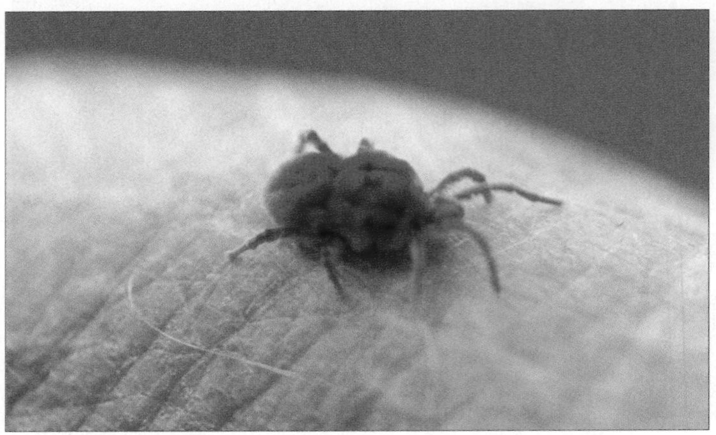

ECTOPARASITES
Parasites that live on their host

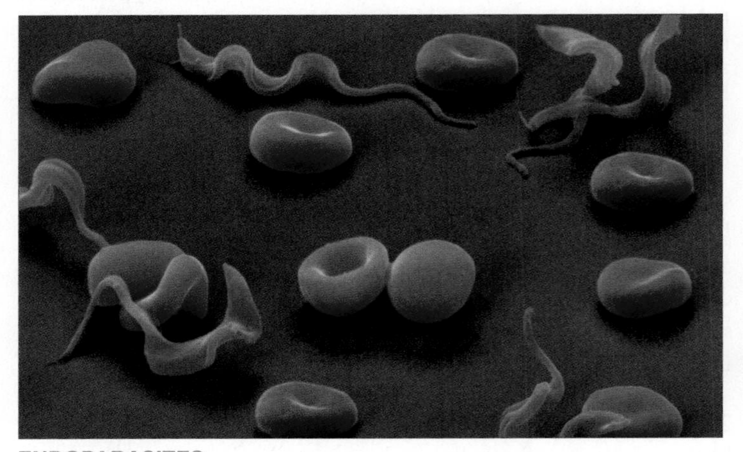

ENDOPARASITES
Parasites that live inside their host

FIGURE 15-27 Parasites: predators dwarfed by their prey. A velvet mite, an ectoparasite on human skin, is shown in extreme close-up. *Trypanosome brucei* is an endoparasite that invades a human host and is the cause of African sleeping sickness.

There may be three to four times as many parasitic species as non-parasitic species on earth!

to what is perhaps the greatest challenge that parasites face: how to get from one individual host to another. A parasite can't survive long once its host dies, after all. The methods by which parasites accomplish such dispersal are unexpected and surprising. Many of their complicated life cycles involve passing through two (or more) different species (and could have come about only through coevolution with each of the host species). These life cycles are likely to give us a new appreciation for the ingenuity of these microorganisms—or rather, for the evolutionary process that has produced them. Let's look at a few representative examples.

Case 1: Parasites can induce foolish, fearless behavior in their hosts. Rats fear cats. During their evolution, rats have developed a protective wariness of their feline predators and areas in which cats have been roaming. *Toxoplasma* is an organism that changes this. This single-celled parasite of rats must also spend part of its life cycle in cats. It does this by altering the brain of its rat host so that the rat no longer fears cats. In fact, *Toxoplasma*-infected rats not only lose their fear of cats, they become attracted to them. Is this an accident? No. This behavioral change, while quite dangerous for the rat, is exactly the change that increases the likelihood that the rat will be attacked by a cat, spreading the parasite in the process.

Case 2: Parasites can induce inappropriate aggression in their hosts. Rabid animals don't behave normally. They froth at the mouth and become unusually aggressive (**FIGURE 15-28**). Is this an accident? No. Rabies is caused by a virus that infects warm-blooded animals, mostly raccoons, skunks, foxes, and coyotes. It is passed from one host to another via saliva. Inducing these "rabid" behaviors, of course, is exactly the change in behavior that will most help the virus to spread.

Case 3: Parasites can induce bizarre and risky behavior in their hosts. The lancet fluke is a parasitic flatworm. It has also been described as a "zombie-maker." This fluke spends most of its life in sheep and goats, but the fluke's eggs pass into snails that graze on vegetation contaminated by sheep and goat feces. Once inside the snail, the fluke eggs grow and develop, eventually forming cysts that the snail surrounds with mucus and then excretes. Continuing on their complex life cycle, the fluke cysts find their way into ants that eat the snail mucus. The flukes' journey back to a sheep or goat is now expedited by the so-called zombie-making. In an infected ant, the flukes grow into the ant's brain, altering its behavior. Whereas ants normally remain low to the ground, when infected by the lancet fluke they climb to the tops of grass blades or plant stems and clench their mandibles on leaves. Is this an accident? No. This behavioral change puts the ants much more at risk of being eaten by a grazing mammal—a bad outcome for the ant, but just what the parasite needs to complete its life cycle.

These parasite-induced behavior modifications are among the most dramatic, but there are many others that are more subtle but equally effective at allowing a parasite to thrive attached to or within the host's body. This subset of predation is an active and exciting area of ecological and physiological investigation.

PARASITE INGENUITY

Parasites have evolved surprising methods to solve a challenging problem: how to get from one individual host to another.

FIGURE 15-28 Rabies. The rabies parasite (a virus) is more effectively passed from one animal to another by causing its host to foam at the mouth and aggressively bite.

TAKE-HOME MESSAGE 15•13

Parasitism is a symbiotic relationship in which one organism benefits while the other is harmed. Parasites face some unusual challenges relative to other predators, particularly how to get from one individual host to another, and some complex parasite life cycles have evolved.

15•14

Not all species interactions are negative: mutualism and commensalism.

It's easy to get the idea that all species interactions in nature are harsh and confrontational, marked by a clear winner and a clear loser. That is largely the case when it comes to competition and predation. Nonetheless, not all species interactions are combative. Every flower you see should be a reminder that evolution produces beneficial species interactions as well. These types of interactions fall into two categories: mutualism and commensalism.

Mutualism: everybody wins

- Coral that gain energy from photosynthetic algae living inside their tissues
- Termites capable of subsisting on wood, only with the assistance of cellulose-digesting microbes living in their digestive system
- Flowers pollinated by animals nourished by nectar

Each of these relationships is an example of **mutualism,** an interaction in which both species gain and neither is harmed (**FIGURE 15-29**). Mutualism is common in virtually all communities. Plants, particularly, have numerous such interactions: with nutrient-absorbing fungi, with nitrogen-fixing bacteria, and with animals that pollinate them and disperse their fruits.

Commensalism: an interaction with a winner but no loser Some species interactions are one-sided. The cases in which one species benefits and the other neither benefits nor is harmed are called commensal relationships, or **commensalism.** Cattle egrets have just such a relationship with grazing mammals such as buffalo and elephants (see Figure 15-29). As the large mammals graze through grasses, they stir up insects. The birds, which feed near the mammals—particularly near the forager's head—are able to catch more insects with less effort this way. The grazers are neither helped nor harmed by the presence of the birds.

TAKE-HOME MESSAGE 15•14

Not all species interactions are combative. Evolution produces beneficial species interactions as well, including mutualism, in which both species benefit from the interaction, and commensalism, in which one species benefits and the other is neither harmed nor helped.

NON-NEGATIVE SPECIES INTERACTION

MUTUALISM
Interaction in which both species benefit

COMMENSALISM
Interaction in which one species benefits and the other neither benefits nor is harmed

FIGURE 15-29 Not always "red in tooth and claw." Mutualisms and commensalisms abound in nature.

5 Communities can change or remain stable over time.

Prairie dogs are a keystone species in the prairie ecosystem.

15•15

Many communities change over time.

Human "progress" and development can completely transform an environment—turning a patch of desert into Las Vegas, for example. Urban landscapes, too, can obliterate any signs of the nature that was once there. This is why it can be surprising (and heartening) to observe what happens when humans abandon an area. Little by little, nature reclaims it. The area doesn't necessarily recover completely, and change is slow. Still, this process is almost universal and virtually unstoppable. Nature responds similarly to other disturbances, too, from a single tree falling in a forest, to a massive flood or fire, to massive volcanic eruptions.

The process of nature reclaiming an area and of communities gradually changing over time is called **succession.** It is defined specifically as a change in the species composition over time, following a disturbance. There are two types of succession. *Primary succession* is when the process starts with no life and no soil. *Secondary succession* is when an already established habitat is disturbed but some life and some soil remain.

Primary succession Primary succession can take thousands or even tens of thousands of years, but it generally occurs in a consistent sequence. It always begins with a disturbance that leaves an area barren of soil and with no life. Frequently the disturbance is catastrophic. The huge volcanic eruption on Krakatoa, Indonesia, in 1883, for example, completely destroyed several islands and wiped out all life and soil on others. Primary succession has also begun, in a less dramatic fashion, in regions where glaciers have retreated, such as Glacier Bay, Alaska. Although succession does not occur in a single, definitive order, several steps are relatively common (**FIGURE 15-30**).

- The first arrivals, or **colonizers,** to a lifeless, soil-less area are usually bacteria or fungi or other photosynthetic microorganisms, floating in the air.
- Lichens, a symbiotic association between a fungus and a photosynthetic alga or bacterium, are also common first colonizers. They can grow on rock and are able to generate energy through photosynthesis. As they grow, lichens change their environment, too, secreting acids that break down the rocks and release useful minerals, making the terrain more hospitable for other organisms.
- Mosses also tend to arrive in the early stages, utilizing many of the nutrients freed up by lichens. They trap moisture and can provide a hospitable site for germination of the seeds of other plants.
- Following mosses, small herbs often arrive. Later, some shrubs may begin to thrive, shading out the mosses and herbs.
- Shrubs in turn are eventually outcompeted and shaded out by small trees.
- And the first trees generally are outcompeted by taller, faster-growing trees.

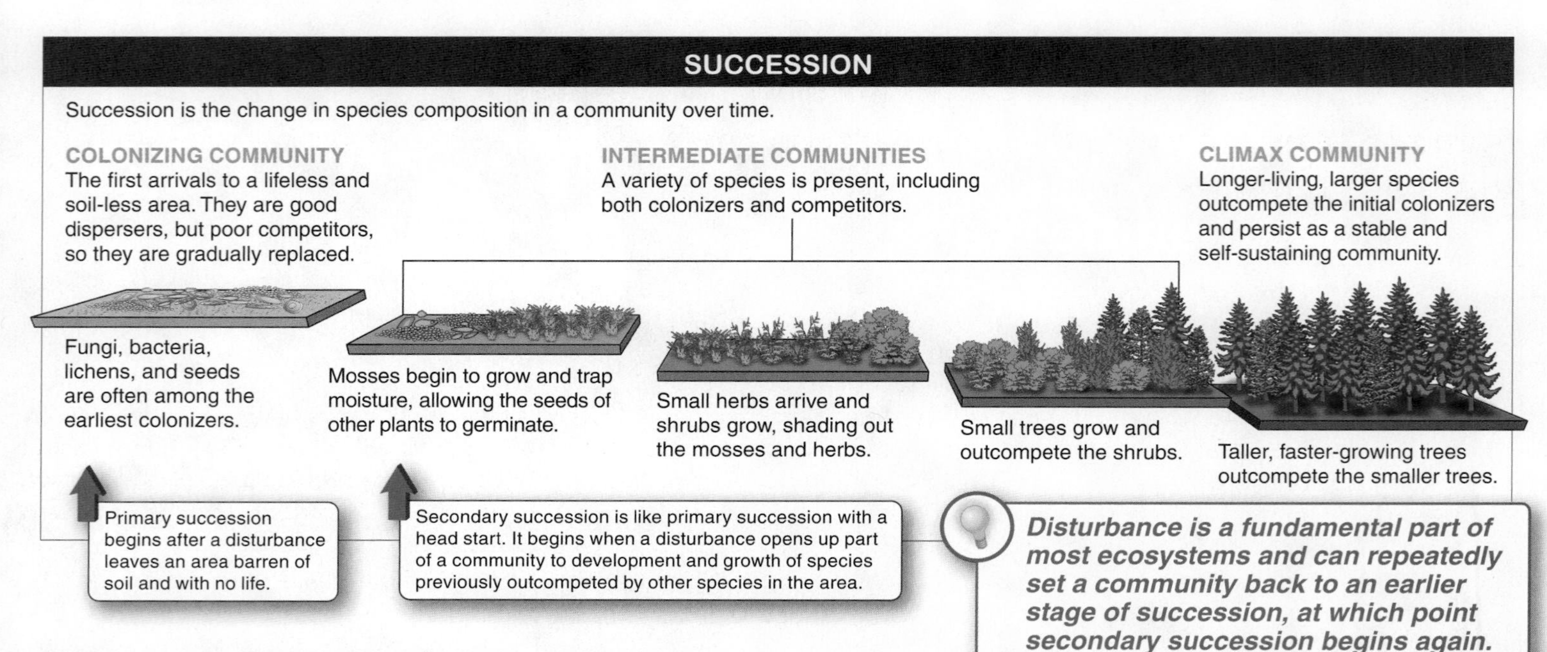

FIGURE 15-30 Species composition of a community changes over time.

An important feature of the colonizers seen in the earliest stages of primary succession is that while they are all good **dispersers,** able to move away from their original home (hence their early arrival to a newly available locale), they are not particularly good *competitors.* That's why they are gradually replaced.

Ultimately, it is the longer-living, larger species that tend to outcompete the initial colonizers and persist as a stable and self-sustaining community, called a **climax community.** The specific species present in the climax community depend on physical factors such as temperature and rainfall.

Secondary succession Secondary succession is much faster than primary succession because life and soil are already present. Rather than the thousands or even tens of thousands of years that primary succession may take, secondary succession can happen in a matter of centuries, decades, or even years. It frequently begins with organisms colonizing the decaying remains of dead organisms. It may involve fungi establishing themselves in the decaying trunk of a tree that has fallen, and these being replaced over time by different species of fungi.

Or secondary succession may begin with weeds springing up in formerly cultivated land that is left untended. If the weeds are allowed to grow, they eventually are outcompeted and replaced by perennial species, and then shrubs, and eventually larger trees. The process is similar to primary succession, but with a head start. Ultimately, if undisturbed, secondary succession also leads to establishment of a stable, self-sustaining climax community.

Q Why doesn't succession occur on front lawns?

Given that primary and secondary succession both progress toward climax communities, it might seem surprising that, at any given time, so many communities are in states far from climax. This apparent paradox reflects the fact that succession leads to climax communities in the *absence of disturbance.* But disturbance is a fundamental part of most ecosystems. Massive disturbances, such as the fires that destroyed a third of Yellowstone Park in 1988 or the volcanic eruptions of Mount St. Helens, can send a community back to square one. Minor disturbances, on the other hand, may undo just a small degree of the succession that has occurred. Interestingly, it is at intermediate levels of disturbance that communities tend to support the greatest number of species. With very high or very low disturbance levels, the community is restricted to the top colonizers—those with the best dispersal or fastest reproductive rates—or the top competitors, respectively. At the intermediate level, a larger variety of species with intermediate combinations of features can persist.

TAKE-HOME MESSAGE 15•15

Succession is the change in the species composition of a community over time following a disturbance. In primary succession, the process begins in an area with no life present; in secondary succession, the process occurs in an area where life is already present. But in both types, the process usually takes place in a predictable sequence.

15•16

Some species are more important than others within a community.

Not all species are created equal. In this chapter, we have examined the ways in which species interact with one other and with their habitats, noting the dependence of species on each other. But not all species have equal dependence on and influence over others. Within a community, the presence of some species, called **keystone species,** greatly influences which other species are present and which are not (**FIGURE 15-31**). That is, if a keystone species is removed, the species mix in the community changes dramatically. The removal of other species, conversely, causes relatively little change.

Bison are a keystone species. Experiments were set up in which a herd of bison was allowed to graze in certain parts of a prairie in Kansas, while being excluded from other areas. The areas without the bison were soon dominated by a single species of tall grass. The grass was like a bully to other plant species. The areas where the bison grazed, on the other hand, had significantly greater species diversity. When kept in check by the grazing bison, the bully grass was unable to outcompete the other plant species, which were able to thrive.

Sea stars, too, are a keystone species (see Figure 15-31). On the rocky seashores of the Pacific Northwest, they consume mussels, which tend to crowd out other species. In a five-year study, sea stars were removed and kept out of certain areas of the intertidal zone but allowed in other areas. As with the bison study, species diversity dropped dramatically where sea stars were excluded. In fact, ultimately, just the mussels remained. In the areas with sea stars, mussels also remained but in reduced numbers. More important, 28 other species of animals and algae were also able to live there.

Suppose you are in charge of conserving biodiversity in a large region of a country. With limited funds, you cannot simply purchase and restrict access to the entire region. It is not possible to preserve everything. Instead, you must make difficult decisions and prioritize species for preservation. Knowing that keystone species exist can make your job considerably easier. There is real practical value in identifying keystone species and focusing conservation efforts on these. After all, in the sea star experiment, preservation of the sea stars alone would have the "side effect" of preserving 28 other species.

Keystone species make it possible to get more "bang for your buck" when your aim is to conserve biodiversity. Consequently, identifying keystone species is an important part of conservation biology. Some other keystone species include dam-building beavers, elephants of the African savanna, and lichens in the desert. We'll explore these questions of conservation in greater detail in the next chapter.

Q When it comes to conservation, are some species more valuable than others? Why?

KEYSTONE SPECIES

A keystone species, such as the sea star, has an unusually large influence on the presence or absence of numerous other species in the community. When sea stars were removed from an area within an intertidal zone, species diversity decreased drastically—only mussels remained.

COMMUNITY WITH KEYSTONE SPECIES

COMMUNITY WITH KEYSTONE SPECIES REMOVED

Identifying keystone species is an important part of conservation biology–preserving just one keystone species has the effect of preserving many additional species at the same time.

FIGURE 15-31 Preserving biodiversity. Keystone species can keep aggressive species in check, allowing more species to coexist in a community.

TAKE-HOME MESSAGE 15•16

Keystone species have a relatively large influence on which other species are present in a community and which are not. Unlike other species, when a keystone species is removed from the community, the species mix changes dramatically. For this reason, preserving keystone species is an important strategy in maximizing the preservation of biodiversity.

Knowledge You Can Use

Did you know? In boosting plant productivity on farms, we've created a "dead zone" bigger than the size of Connecticut in the Gulf of Mexico.

In the water at the mouth of the Mississippi River, each spring and summer, there is so little oxygen that virtually no life can survive.

Q: Which chemical almost always limits plant growth? Plant growth requires the production of proteins, all of which contain nitrogen. Access to nitrogen nearly always determines how much a plant grows.

Q: If you're a farmer, how can you increase the productivity of your crops? Add nitrogen. This, in fact, is the chief component of all plant fertilizers.

Q: Where does the unused nitrogen end up? Spring and summer rains wash nutrients, dissolved organic materials, and other runoff (much of which comes from human sources) from the middle of the United States into nearby rivers and, from those rivers, into the Mississippi River. From there they flow into the Gulf of Mexico.

Q: What happens when large amounts of nitrogen are dumped into a body of water? Organisms such as plankton and algae living near the mouth of the Mississippi River grow like crazy with all the extra nitrogen and nutrients that are washed into the Gulf of Mexico. But the rapid increase in their productivity disrupts the food chain as they decay, increasing the organic matter that sinks to the bottom and feeds the bacteria there. And as the bacteria thrive, they consume ever increasing amounts of oxygen.

Q: What happens when bacteria use up too much oxygen? Excessive bacterial growth depletes the supply of dissolved oxygen in the water, starving all other life of oxygen—including fish, crabs, oysters, and shrimp, all of which die or move out of the area. This creates a "dead zone" and ruins local fisheries, with significant impact on the economies they support.

What can you conclude? Disrupting food chains, even increasing the productivity of certain trophic levels, can have unintended effects throughout an ecosystem, and even thousands of miles away.

What can be done? Reducing and reversing the Gulf of Mexico dead zone (and other dead zones throughout the world) can be accomplished by (1) reducing fertilizer use, (2) preventing animal wastes from getting into waterways, and (3) controlling the release of other sources of nutrients (phosphorus as well as nitrogen) from industrial facilities into rivers and streams.

BIG IDEAS IN ECOSYSTEMS AND COMMUNITIES

1 Ecosystems have living and non-living components.

Ecosystems are the living organisms and the physical environment within an area. The largest ecosystems in the world are called biomes. The characteristic organisms and physical features present in each terrestrial biome depend largely on the temperature and rainfall patterns.

2 Physical forces interacting create weather.

The greater amount of solar energy hitting earth at the equator causes global air circulation patterns responsible for heavy rainfall near the equator and deserts at 30° latitude. Global patterns of ocean circulation also influence patterns of temperature and rainfall. Local topography, from mountain ranges to human-made structures, can also influence rainfall, temperature, and wind patterns.

3 Energy and chemicals flow within ecosystems.

Energy pyramids describe the energy flow through ecosystems, from primary producers to herbivores to carnivores. At each step of the food chain, the rate of energy capture and conversion is inefficient, with about 90% lost to heat, respiration, and feces. While all energy comes from the sun and flows in a pathway through food chains, chemicals repeatedly cycle from the physical environment through living organisms and back into the environment.

4 Species interactions influence the structure of communities.

Interacting species in a community coevolve in a variety of ways, some antagonistic and others mutually beneficial. The organisms of a population fill a unique niche, defined by how they utilize the resources in their environment. Organisms with overlapping niches compete until character displacement or local extinction occurs.

5 Communities can change or remain stable over time.

Most communities change over time. The patterns of succession, in which species composition changes within an ecosystem, depend on the rate and magnitude of disturbances. In a community, species vary in the degree to which they influence diversity. Keystone species are the most influential.

KEY TERMS

biomass, p. 563
biome, p. 552
biotic, p. 551
carnivore, p. 562
character displacement, p. 572
climax community, p. 580
coevolution, p. 570
colonizer, p. 579
commensalism, p. 578
community, p. 551
competitive exclusion, p. 571
decomposer, p. 563
desert, p. 553
detritivore, p. 563
disperser, p. 580
ecosystem, p. 550
El Niño, p. 559
energy pyramid, p. 564
estuary, p. 553
eutrophication, p. 568
food chain, p. 562
food web, p. 562
fundamental niche, p. 571
habitat, p. 551
herbivore, p. 561
host, p. 576
keystone species, p. 581
mimicry, p. 574
mutualism, p. 578
niche, p. 570
omnivore, p. 562
parasite, p. 576
parasitism, p. 576
predation, p. 573
primary consumer, p. 561
primary producer, p. 553
primary productivity, p. 553
rain shadow, p. 557
realized niche, p. 571
resource partitioning, p. 571
savanna, p. 553
secondary consumer, p. 562
succession, p. 579
temperate grassland, p. 553
tertiary consumer, p. 562
topography, p. 557
trophic level, p. 561
tropical rain forest, p. 553
tropical seasonal forest, p. 553

CHECK YOUR KNOWLEDGE

1. An ecosystem consists of ______ in a given area.
   a) all the photosynthetic organisms
   b) all the living organisms
   c) all the abiotic factors that influence living organisms
   d) all the living organisms and non-living materials
   e) the plant life and climate

2. Earth's largest terrestrial ecosystems, the biomes, are defined primarily by:
   a) the average rainfall.
   b) the average temperature.
   c) the seasonal variability in temperature.
   d) the seasonal variability in rainfall.
   e) All of the above.

3. Why is it hotter at the equator than at the Poles?
   a) The angle at which sunlight hits the earth leads to a given amount of solar energy being spread over a smaller area at the equator than at the Poles.
   b) The increased rotational speed of the earth at the Poles creates strong winds that increase radiant cooling.
   c) Greater cloud cover at the equator leads to a greater "greenhouse" effect than at the Poles.
   d) The water in the oceans has a high heat capacity, and there is significantly more water near the equator than at the Poles.
   e) All of the above are responsible for the higher average temperature at the equator than at the Poles.

4. Why do beaches on the west coast of the United States have colder water than beaches on the east coast?
   a) The ocean currents in the northern hemisphere rotate in a clockwise direction, so water off the coast of California has just come from the cold north, while water off the coast of New York has just come from the warm south.
   b) West coast beaches are not colder than east coast beaches. Ocean temperatures depend only on latitude.
   c) The Pacific Ocean is deeper than the Atlantic Ocean, so more of the sun's heat is absorbed by the colder, deep water, reducing the water temperature at the west coast beaches.
   d) Trade winds blowing north from the Tropic of Cancer in the Pacific Ocean sweep the warm air away from the west coast, taking with it much of the warmth of the water.
   e) Deep water upwellings bring cool water to the west coast beaches, while the surface currents on the east coast cause downwellings, sending warm surface water to depths where it is less likely to be lost through evaporation.

5. When moist wind from an ocean blows onshore toward a mountain range:
   a) as the air rises, it pulls moisture from the ground, causing the higher elevations to be drier.
   b) as the air goes over the top of the mountain and falls back down toward lower elevations, it holds less moisture, creating a "rain shadow" zone of unusually high precipitation.
   c) as the air rises, it holds less moisture, causing the dissipation of all clouds.
   d) as the moist air goes over the top of the mountain and falls back down toward lower elevations, it holds even more moisture, creating a "rain shadow" desert with very little precipitation.
   e) None of the above are correct.

6. "Top" carnivores:
   a) are more common than secondary consumers.
   b) rely directly on primary producers for energy.
   c) consume primarily herbivores.
   d) consume primarily carnivores.
   e) rely on symbiotic bacteria living in their digestive systems to help digest cellulose.

7. The 10% rule of energy-conversion efficiency:
   a) explains why big, fierce animals are so rare.
   b) explains why the biomass of herbivores must exceed that of carnivores.
   c) limits the length of food chains.
   d) suggests that 90% of what an organisms eats is used in cellular respiration or is lost as feces.
   e) All of the above are correct.

8. Nitrogen enters the food chain:
   a) primarily through soil-dwelling bacteria that "fix" nitrogen by attaching it to other atoms.
   b) from the atmosphere when "fixed" by the photosynthetic machinery of plants.
   c) when rocks dissolved by rainwater become soil and are utilized by plants as they grown.
   d) through soil erosion followed by runoff into streams and ponds.
   e) through methane, produced by herbivores as a by-product of the breakdown of plant material.

9. Which of the following statements about an organism's niche is incorrect?
   a) It encompasses the space the organism requires.
   b) It includes the type and amount of food the organism consumes.
   c) It is not always fully exploited.
   d) It may be occupied by two species, as long as they are not competitors.
   e) It reflects the ways in which the organism utilizes the resources of its environment.

10. The "ghost of competition past" refers to the fact that:

a) competition often leads to character displacement, which remains even after direct competition is reduced.
b) competition cannot be seen in nature.
c) competition inevitably leads to the extinction of one of the competitors.
d) competition inevitably leads to the extinction of both competitors.
e) the fossil record is a record of competitive interactions.

11. Chemical defenses are more common among plants than animals because:

a) plants cannot move to escape predators and so must develop other deterrents.
b) the cell wall can contain the chemicals more effectively than a simple plasma membrane.
c) mechanical defenses against predators can evolve only in animals.
d) parasite loads in plants are significantly higher than in animals.
e) All of the above are correct.

12. Rabid animals:

a) froth at the mouth so as to increase the likelihood that the rabies-causing virus is passed on to another host.
b) have been infected by parasitic bacteria.
c) must die before their host can be passed on.
d) are infected by a viral ectoparasite.
e) often show reduced aggression, which increases their ability to get closer to other animals, thereby spreading the infection.

13. In a commensal relationship:

a) one species pollinates the other.
b) neither species benefits nor is harmed, but the community itself benefits.
c) one species provides nutrients, usually cellulose, for the other.
d) one species benefits while the other neither benefits nor is harmed.
e) the "loser" has reduced reproductive output.

14. Coevolution:

a) is responsible for all the beautiful flowers in the world.
b) is responsible for nectar production by plants.
c) reveals that both biotic and abiotic resources can serve as selective forces.
d) can produce an insect with a tongue as long as its body.
e) All of the above are correct.

15. The chief difference between primary and secondary succession is that:

a) primary succession occurs among the plants in a habitat, while secondary succession occurs among the animals.
b) primary succession begins with no life or soil, while secondary succession begins with both.
c) secondary succession alters the biotic environment, and primary succession alters the abiotic environment.
d) primary succession occurs more quickly than secondary succession.
e) primary succession can occur in terrestrial and aquatic habitats, while secondary succession can occur only in terrestrial habitats.

16. Keystone species:

a) occur only in intertidal zones.
b) play an unusually important role in determining the species composition in a habitat.
c) can be removed from a habitat without any impact on the remaining species in the community.
d) are primary producers and therefore usually are plants.
e) are more expendable than commensal species, from a conservation perspective.

Short-Answer Questions

1. What changes in climate occur on moving from the equator toward the Poles? Which two aspects of weather determine the type of biome found at any location?

2. Describe the energy flow that occurs through an ecosystem. How does this energy flow influence the number of top carnivores?

3. What is a niche? What is the potential fate of two species that occupy the same niche?

See Appendix for answers. For additional study questions, go to www.prep-u.com.

16

Conservation and Biodiversity

Human influences on the environment

1 Measuring and defining biodiversity is complex.

The Cape Floral Region of South Africa is a biodiversity hotspot known for its wide array of plants.

16•1

Biodiversity benefits humans in many ways.

In the Pacific Northwest of the United States, a medium-size tree grows. This conifer, the Pacific yew tree (*Taxus brevifolia*), isn't the biggest tree in the forest, nor is it the most common. It's not the sort of tree that usually warrants attention in a textbook. But within the bark of the Pacific yew there is a chemical, called taxol, that has some important properties—it acts as an anti-cancer agent, interfering with the division of cells that come into contact with it (**FIGURE 16-1**). The role taxol plays for the Pacific yew is not clear; it may reduce the rate at which other organisms feed on the plant. But in humans, taxol is effective in the treatment of ovarian cancer, breast cancer, and lung cancer (generating more than $1 billion a year in pharmaceutical sales).

Consider also the following:

- The chemicals vinblastine and vincristine, isolated from the Madagascar periwinkle (*Catharanthus roseus*), are so effective in treating leukemia and Hodgkin's lymphoma that both diseases, formerly incurable, now are curable in the vast majority of people.
- The chemical ancrod, found only in the Malayan pit viper (*Agkistrodon rhodostoma*), dissolves blood clots and is effective in treating some heart attack and stroke patients.

FIGURE 16-1 Taxol is cancer medicine derived from a tree.

- Epibatidine, a poisonous compound in the saliva of a small frog (*Epipedobates tricolor*) that lives in Ecuador, is 200 times more effective than morphine in relieving pain and is non-addictive.

These are just a few examples—there are many, many more—illustrating that medically important compounds often come from some unlikely organisms living in locations around the world. Usually, toxin production evolved in these organisms as a method of protecting them from other organisms. Co-opting the chemicals for human use reveals that one important value of biodiversity is as a sort of universal medicine cabinet.

Brocadia anammoxidans is another species that may turn out to have great value to humans. This bacterial species is able to break down an important component of human waste and, more important, it can do so without oxygen. For this reason, when treating human waste with *B. anammoxidans,* it is unnecessary to aerate the sewage sludge. Normal waste treatment utilizes bacteria that need oxygen and, consequently, requires costly, power-hungry equipment. Still other bacteria, with sulfur-metabolizing properties, are now being used to remove the stench of sewage. In these cases, as with the medicines described above, the value of biodiversity to humans is utilitarian: from other species we can obtain products or make use of processes that make our lives better.

But viewing the value of biodiversity only in the context of our abilities to extract medicines or to better process our waste products—the utilitarian view—is to take much too narrow a view. Biodiversity provides significant, perhaps less tangible, benefits by helping regulate the climate, in the form of trees that both create shade and take up carbon dioxide from the atmosphere. Biodiversity also benefits humans in numerous other ways that are difficult to quantify, but are no less valuable than medicinal plants or sludge-eating bacteria.

Have you ever gazed up at a 300-foot-tall redwood tree? Would you like to see a 100-foot-long, 180-ton blue whale? Do you like knowing that bald eagles live throughout North America? If you answered *yes* to any of these questions, you identified additional ways in which biodiversity can have value to humans. Although many of the ways that people value biodiversity are not easily quantified (in dollars, for example), they represent value just the same (**FIGURE 16-2**).

Some types of non-utilitarian values include:

- *Aesthetic value.* We think of a field of flowers or a deer in the woods as beautiful to look at and pleasant to experience.
- *Symbolic value.* Humans are so connected to the natural world that images of animals and plants can convey meaningful abstract ideas: the dove as a symbol of peace; the butterfly as a symbol of transformation; the owl as a symbol of wisdom; the rose as a symbol of love.
- *Naturalistic value.* This can be thought of as the satisfaction people can get from direct contact with nature, including the sense of awe and fascination that can come from gaining an understanding of nature's complexity and diversity.

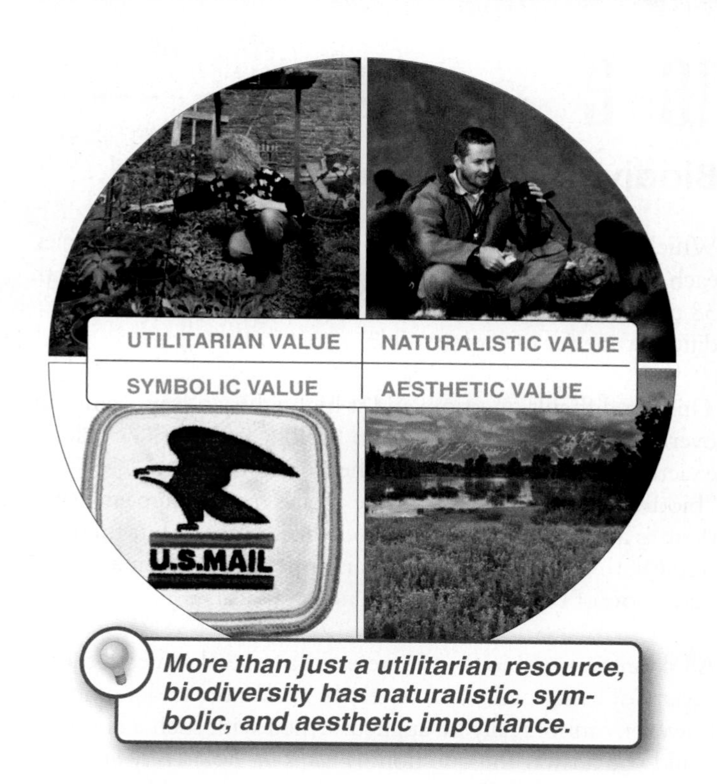

FIGURE 16-2 Biodiversity has value to humans in many different ways.

In recognizing that people can view the same object (or organism) and value it in different ways, we are on our way to making wiser conservation decisions. This knowledge, for example, can help us anticipate potential conflict when it comes time to invest limited conservation resources in the conservation of biodiversity, and can help us seek out methods for balancing conflicting desires. We explore some strategies for making these difficult decisions later in this chapter. First, we investigate the question of where the greatest amounts of biodiversity are to be found.

TAKE-HOME MESSAGE 16•1

Biodiversity provides significant benefits to humans in numerous utilitarian ways, such as the production of medicines and other products and processes. Biodiversity also provides important value to humans in less tangible ways, including aesthetic, symbolic, and naturalistic values.

16•2

Biodiversity is not easily defined.

Which habitat has greater biodiversity, one with 3 or 4 species each of birds, reptiles, mammals, plants, and insects or one with 38 unique species of fruit flies? Or one with 100 radically different species of prokaryotes?

This is a difficult question, and it highlights an important but overlooked issue in conservation biology today: it's not clear exactly what a person means when using the term "biodiversity." Discussions in the media make it appear that there is a consensus on what biodiversity means, but there is not. Or, to the extent that there is consensus, it is on a definition of biodiversity that is not practical.

All definitions of biodiversity refer, in a general sense, to the variety of living organisms on earth. Such a broad definition, however, cannot really be applied: armed with such a definition, could you answer the question: *What is the biodiversity in a particular patch of land?* Think about it—what would you count—the number of trees, or the number of tree species, or the number of insects that live in those trees? Could you compare this patch of land with another in order to determine which has the greater biodiversity?

By taking this approach—considering how to *apply* the concept of biodiversity—we can start to move toward a more concrete definition. Expanding on our general definition above, biodiversity can be considered at multiple levels, from entire ecosystems to species to genes and alleles (**FIGURE 16-3**). This makes it a bit more manageable as a concept.

Defining biodiversity as the variety of genes, species, and ecosystems on earth, however, is still not sufficiently practical. The construction of a meaningful inventory of earth's biodiversity is fraught with difficulty. As we saw in Chapter 5, the Human Genome Project, which has catalogued all of the genes in humans, was a multi-year, multi-billion-dollar undertaking. A similar undertaking for all species is clearly not practical. And as we saw in Chapter 10, the very concept of "species" is difficult (and sometimes downright impossible) to apply to many, if not most, of the organisms on earth. And as we saw in Chapter 15, it is even difficult to determine where one ecosystem ends and another begins.

Believe it or not, the difficulty of defining biodiversity, and the difficulty in applying the concept as part of any efforts to protect biodiversity, is not an insurmountable problem. To the contrary, defiining and applying the concept is a positive step, likely to lead to the effective preservation of earth's biodiversity.

WHAT IS BIODIVERSITY?

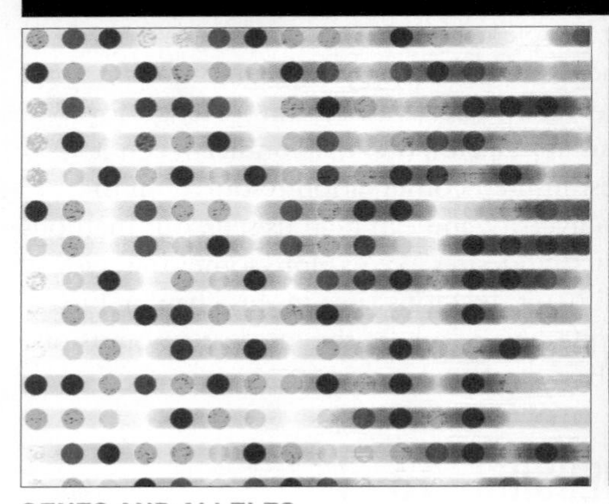

GENES AND ALLELES
The number of alleles in a species

SPECIES
The number of species in an ecosystem

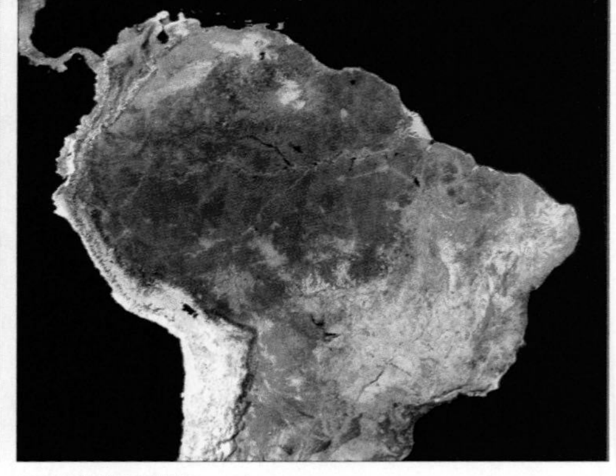

ECOSYSTEMS
The number of ecosystems in a region

Biodiversity is more than just a counting of species. It encompasses the genetic variability among organisms within a species, the variety of different species, and the variety of ecosystems on earth.

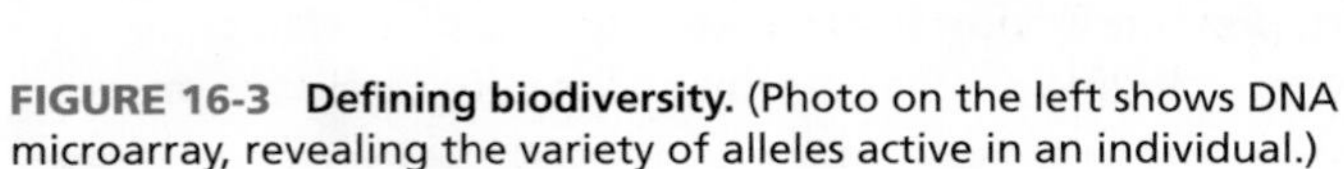

FIGURE 16-3 Defining biodiversity. (Photo on the left shows DNA microarray, revealing the variety of alleles active in an individual.)

CONSERVATION BIOLOGY

Conservation biology is a multidisciplinary science that addresses how to preserve the natural resources of earth and protect biodiversity.

FIGURE 16-4 Preserving the earth's natural resources.

Let's return to our original question. If you were given the task of overseeing the conservation of biodiversity in a region where these three habitats occur, which habitat would you protect first: the habitat with a few species of birds, reptiles, mammals, plants, and insects, or the habitat with many species of fruit flies, or the habitat with a huge number of prokaryotes?

In making your decision, you might first define biodiversity—maybe as the number of different classes or families of organisms in a habitat, or maybe as the number of distinct species in the habitat, which is, in fact, the most commonly used definition of biodiversity—to support your decision. In doing so, however, you are acknowledging that biodiversity can be defined in many ways and that the definition you are using (or that any decision maker uses) is not completely objective. Rather, it is a function of your values, biases, and interests. With this recognition, it becomes possible to discuss, manage, use, and conserve biodiversity in a practical and more effective manner.

Conservation biology is the interdisciplinary field that addresses how to preserve the natural resources of earth. Drawing on knowledge, data, and insights from biology, economics, psychology, sociology, and political science, one of the goals of conservation biology is the preservation and protection of biodiversity (**FIGURE 16-4**). For this reason, the complexities of defining biodiversity directly affect the decisions that conservation biologists make. In this chapter, we explore how the actions of humans can have significant, even disastrous, consequences for biodiversity and other natural resources, and we also explore the strategies that have been developed for effective conservation.

Underpinning many of the difficulties in conservation biology is the fact that biodiversity, as we've noted, can have value for humans in so many different ways. Throughout the chapter, we will see how that complexity makes it difficult to balance competing interests when they are in conflict.

TAKE-HOME MESSAGE 16•2

Defining biodiversity is difficult because it can be considered at multiple levels, from entire ecosystems to species to genes and alleles. In practice, it is most often defined as the number of distinct species in a habitat, which has important implications for conservation biology—the field that addresses questions of how to preserve the natural resources of earth.

16•3

Where is most biodiversity?

If you were to stand at the equator in South America and identify the number of land mammals, you would count about 400 different species. If you then started walking south, away from the equator, and assessed the diversity of land mammals when you got to 25° latitude, you would find about 300 different species. Continuing your long walk south, when you got to 30° latitude you would find only 200 different species. And the trend would continue for as far south as you could go, with 100 species at 40° latitude and fewer than 50 species at 50° latitude. On a similar walk north from the equator, through North America, you would observe the same trend, with more than 300 species of land mammals at 15° latitude

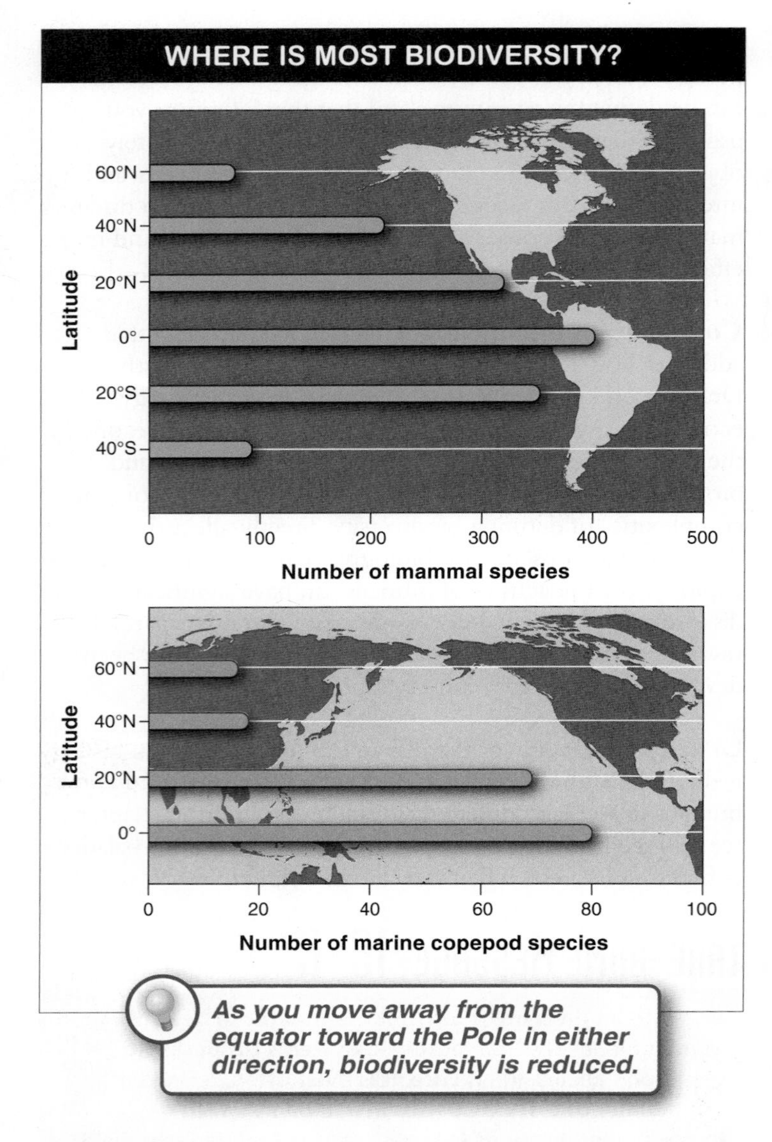

FIGURE 16-5 Biodiversity is unevenly distributed throughout the world.

reduced to only 15 species in northern Canada (**FIGURE 16-5**). Biodiversity is not evenly distributed throughout the world. And perhaps the strongest trend in biodiversity distribution is that as you move away from the equator in either direction, diversity is reduced with increasing latitude.

The strong biodiversity gradient from the tropics to the Poles occurs not just in land mammals but in nearly any group of plants or animals observed. Interestingly, even the diversity of organisms in the oceans follows this trend. One survey of marine copepods—tiny crustacean animals, including plankton—found the exact same trend moving north from the tropical region of the Pacific Ocean (80 species present) to the Arctic Ocean (fewer than 10 species).

Q Why are there more species in an acre of tropical rain forest than in an acre farther from the equator, such as in a temperate forest or prairie?

Numerous factors that influence species richness—that is, the number of species in an area—are responsible for producing the tropical-to-temperate-to-polar gradient in biodiversity. Three, in particular, play strong roles (**FIGURE 16-6**).

1. *Solar energy available.* Perhaps the simplest predictor of species diversity is climatic favorability, the amount of solar energy available in an area. Solar energy is, after all, the fuel for all life. As we saw in Figure 15-5 in Chapter 15, the sun shines most directly on the equator, and, as we move away from the equator, the earth's curvature causes the solar energy hitting the earth to be spread out over increasingly large areas. This leads to less solar energy at any one point as latitude increases. In a variety of species of plants and animals, researchers have documented strong relationships between energy availability and species richness.

2. *Evolutionary history of an area.* Communities diversify over time. Consequently, the more time that passes, generally speaking, the greater is the diversity in an area. A high level of

FACTORS THAT INFLUENCE BIODIVERSITY

1 SOLAR ENERGY AVAILABLE
Greater access to solar energy, the fuel for all life, provides increased species richness.

2 EVOLUTIONARY HISTORY OF AN AREA
Communities diversify over time. The more time that passes without a climatic event, such as an ice age, the greater the diversity in an area.

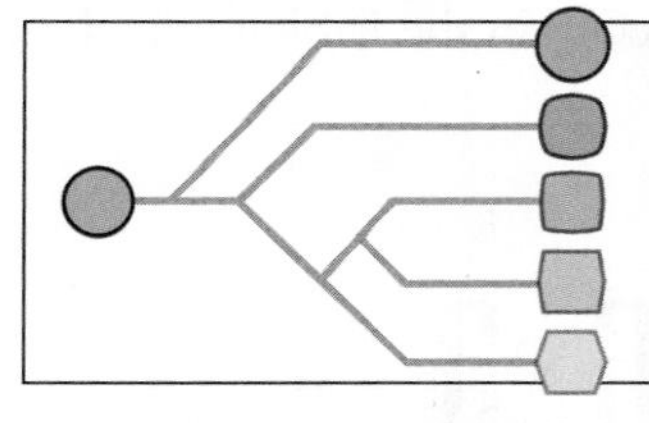

3 RATE OF DISTURBANCE
A habitat with an intermediate amount of disturbance tends to have the greatest species richness.

FIGURE 16-6 What determines species richness in an area?

biodiversity in an area, however, can be knocked back down by climatic disasters such as glaciations. This biodiversity decline may occur in temperate and polar regions.

3. *Rate of disturbance.* Over time, the best competitor for a resource is expected to outcompete other species, excluding them from the community and thus reducing species richness. Communities may be kept from reaching this state, though—in a number of ways. Factors such as predation, or environmental perturbations such as fires, can prevent any one species from excluding others. For this reason, a habitat with an *intermediate* amount of disturbance is expected to have the greatest species richness.

Conservation biologists are increasingly interested in **biodiversity hotspots,** those regions of the world having significant reservoirs of biodiversity that are under threat of destruction. Twenty-five biodiversity hotspots have been identified around the globe that, while covering less than 1% of the world's area, have 20% or more of the world's species (**FIGURE 16-7**). Habitats included among the biodiversity hotspots are tropical rain forests, coral reefs, and islands.

Tropical rain forests. These are common near the equator in Asia, Australia, Africa, South America, and Central America, as well as on many of the Pacific Islands. A particularly important, but still poorly studied, region of tropical rain forests is the canopy. This is the uppermost region of a forest, where nearly all of the sunlight is intercepted. As much as 30 feet in thickness, the canopy is home to most of the diversity in tropical rain forests, and estimates suggest that as many as half of the species on earth reside in the canopy of tropical rain forests. Due to its inaccessibility, the canopy is not easily studied and the extent of biodiversity there is still not fully known.

Coral reefs. Built by corals, tiny members of the animal kingdom, coral reefs occur in ocean areas with low levels of nutrients. The physical structures created by the corals provide a home and habitat for huge numbers of other

FIGURE 16-7 Biodiversity hotspots are significant reservoirs of biological diversity.

There are 25 biodiversity hotspots around the world that, while covering less than 1% of the world's area, have 20% or more of the world's species.

species. The Great Barrier Reef of Australia (which is not listed among the 25 biodiversity hotspots, because it is relatively well protected) is an example of just how much diversity can be supported by coral reefs. It is the largest coral reef system in the world and covers such a large area off the northeast coast of Australia that it can be seen from space. It is home to numerous species of whales, dolphins, sea turtles, sharks, and stingrays, as well as 1,500 species of fish and 5,000 species of molluscs. Five hundred species of algae are found there, too. Above the Great Barrier Reef, more than 200 species of birds are supported by the rich diversity of life below. Other coral reefs are not as well protected, and the world's second largest coral reef, in New Caledonia in the southwestern Pacific, is one of the 25 biodiversity hotspots.

Islands. Because of their generally smaller size, which limits the population size of organisms living on them, islands can be regions where biodiversity is at risk. Madagascar, located in the Indian Ocean off the southeastern coast of Africa, is the fourth largest island in the world. It is home to 5% of the world's species, 80% of which are **endemic** (not naturally found elsewhere)—a higher percentage of endemic plants and animals than in any comparably sized area on earth. Madagascar is an area unusually rich in species diversity.

The deep ocean regions are much less studied than the terrestrial parts of the world. Although they cover two-thirds of the earth's surface, the oceans have proved difficult to explore. Once thought to be completely devoid of life, the deepest regions of the ocean have recently been found to be teeming with biodiversity. One survey of the deepest parts of the Antarctic Ocean, sampling at a depth of more than 18,000 feet (6,000 meters), discovered 585 new species of crustaceans alone.

TAKE-HOME MESSAGE 16•3

Biodiversity is not evenly distributed. There is a strong latitudinal gradient of biodiversity from the tropics (highest biodiversity) to the poles for nearly all groups of plants and animals, both marine and terrestrial. Factors influencing the species richness in an area include the amount of solar energy available, the area's evolutionary history, and its rate of disturbance. Biodiversity hotspots are regions of significant biodiversity under threat of destruction.

16•4

Island biogeography helps us understand the maintenance and loss of biodiversity.

"Can you help me exterminate all the insects on a few islands?" When E. O. Wilson started calling exterminators in Florida, asking them to help him put tents over several islands and then pump ethyl bromide into them to kill all of the insects, he had trouble being taken seriously. Most of the exterminators opted to pass on the job, but eventually he found one (**FIGURE 16-8**).

Why would a biologist (of all people) want to wipe out all insect life on six islands? Robert MacArthur and Wilson had developed a theory, six years earlier, and now Wilson and one of his students, Daniel Simberloff, had decided it was time to put the theory to a test.

MacArthur and Wilson's theory was called the "theory of island biogeography." Designed to explain and predict the patterns of species diversity on islands, it had two fundamental components (**FIGURE 16-9**).

1. *The area effect.* Observations of islands in the West Indies revealed that there was a close relationship between the

FIGURE 16-8 Experiment in insect eradication and recolonization: explaining patterns of species diversity on islands.

THEORY OF ISLAND BIOGEOGRAPHY

The theory of island biogeography is designed to explain and predict patterns of species diversity on islands. It has two fundamental components:

THE AREA EFFECT

There is a close relationship between the number of species inhabiting an island and the island's area—the larger an island is, the more species it holds.

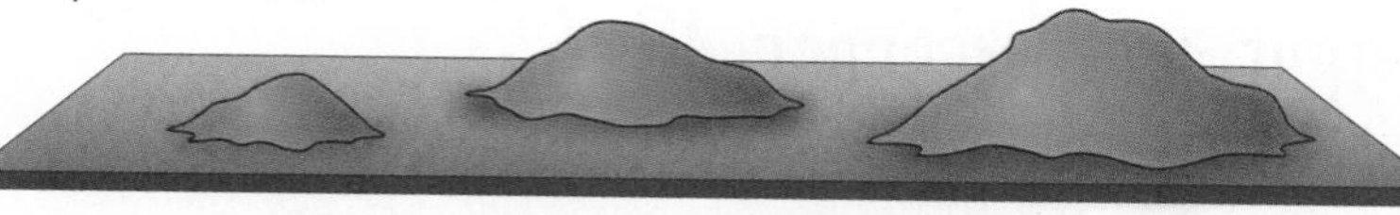

THE DISTANCE EFFECT

There is a close relationship between the number of species inhabiting an island and how isolated an island is—the farther an island is from the mainland, the fewer species it holds.

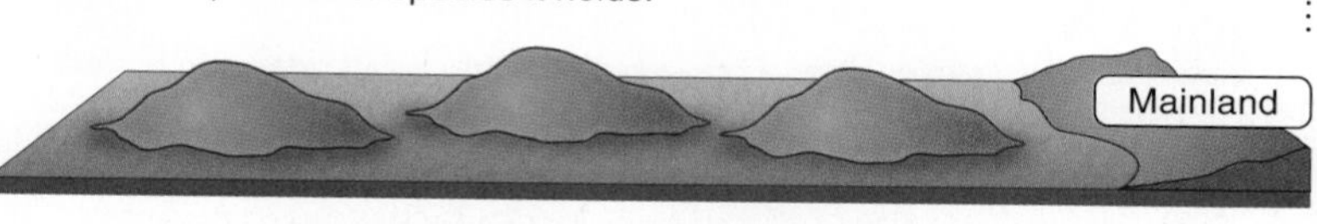

FIGURE 16-9 The lonely island. The theory of island biogeography generates predictions about the patterns of species diversity in isolated habitats.

number of species inhabiting the island and the island's area: the larger the island, the more species it held. This relationship held true for bird species as well as plants, beetles, and snails.

2. *The distance effect.* Observations of islands in the Pacific Ocean revealed a close relationship between how isolated an island was, measured as its distance from the nearest mainland, and the number of species that inhabited the island: the farther from the mainland the island was, the fewer species it held.

From these observations, MacArthur and Wilson concluded that the number of species on an island represents a balance between the rate of immigration of new species to the island and the rate of extinction of species on the island. They proposed that the rate of immigration to an island would primarily be determined by how close the island was to the mainland, with the mainland serving as a pool of new species. And they proposed that the rate of extinction on an island would primarily be determined by the island's size: smaller islands generally support a smaller population size, making the population more vulnerable to extinction.

Beyond enabling them to predict the number of species on an island, MacArthur and Wilson's theory also suggested to them that the composition of species on the island would not be fixed; rather, it might change over time. In other words, once it was "full," an island might have 10 species of birds, but exactly *which* 10 species occurred on the island could change over time. They called this a **dynamic equilibrium** (**FIGURE 16-10**).

DYNAMIC EQUILIBRIUM

EQUILIBRIUM

The number of species on an island represents a balance between the rate of immigration to the island by new species and the rate of extinction of species on the island.

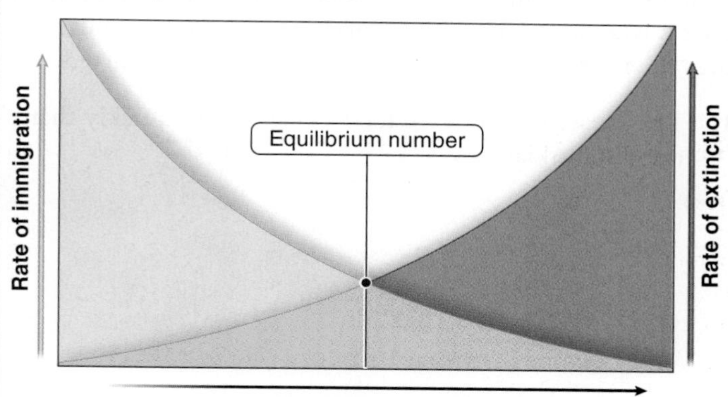

DYNAMIC EQUILIBRIUM

The composition of species on an island is not fixed; rather, it might change over time. An island might have five species of birds, but exactly which five species occurred on the island could change over time.

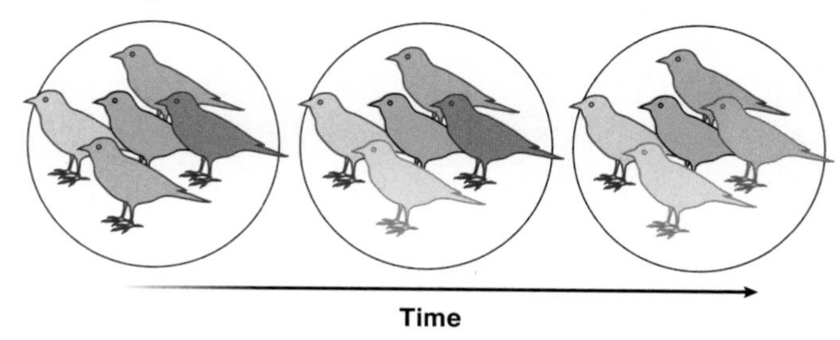

FIGURE 16-10 Offering insight into population dynamics: dynamic equilibrium.

Back to the island exterminations. Wilson's reason for exterminating all of the insects on multiple islands was so that he could monitor the islands as they were recolonized following extermination. He chose the mangrove islands in the Florida Keys because there are thousands of small islands there, some only a few meters across. He and Simberloff selected eight islands, 33–54 feet (11–18 meters) in diameter. The islands varied in how far they were from the nearest mainland, with the closest just 6 feet (2 meters) and the farthest more than half a mile (1,188 meters) from the mainland.

> "The moth settled onto the curtain and sat still. It was an astonishing creature, with black and white wings patterned in geometric shapes, scarlet underwings, and a fat white body with black spots running down it like a snowman's coal buttons. No human eye had looked at this moth before; no one would see its friends. So much detail goes unnoticed in the world.
>
> —Barbara Kingsolver, *Prodigal Summer,* 2000

After selecting two islands as control islands, the biologists and their assistants fumigated the others, verifying that all of the insects were killed. They then surveyed the islands regularly over the next year. Here's what they found—as predicted:

1. The closer islands were recolonized more quickly than the distant islands.
2. On all of the islands, the number of insect species after fumigation returned to about the same number of insect species prior to fumigation.
3. The composition of insects on each island changed over the course of the study, even on the two control islands.

The theory of island biogeography is not just about real islands, as it turns out. Lakes can be thought of as islands, too, from an immigration and extinction perspective. The theory makes testable predictions about the biodiversity levels that should occur in them. Mountain tops can be viewed similarly. And perhaps most important, conservation biologists can gain insights from island biogeography into the design of nature preserves. The importance of developing nature preserves as close as possible to sources of new species, for example, is one such insight. We explore others later in this chapter, when discussing the strategies for preservation of biodiversity.

TAKE-HOME MESSAGE 16•4

The theory of island biogeography generates predictions about the patterns of species diversity in isolated habitats. Both the size of the habitat and its distance from a pool of new species influence the equilibrium number of species present. Ideas gained from island biogeography have important implications for the development of nature preserves.

❷ Extinction reduces biodiversity.

The Bengal tiger has been hunted, captured, and poisoned to near extinction.

16•5

There are multiple causes of extinction.

Imagine that you are on a deserted island. Alone. It would be hard and it would be lonely. The hope of rescue someday, however, would probably make things bearable. "Martha," a passenger pigeon living in the Cincinnati zoo in the early 1900s, found herself in a situation of this sort—minus the hope. Passenger pigeons in the wild had been completely wiped out by a combination of factors, including hunting of the birds for meat, loss of their habitat due to deforestation in North America, and possibly disease. And so, except for the small flock that Martha was part of, the species was gone. One by one, those birds died as well, and when the second-to-last passenger pigeon died in 1912, Martha was all alone. She lived for two more years, with no possibility for ever reproducing; she died on September 1, 1914 (**FIGURE 16-11**).

On average, species persist for about 10 million years, although some last for hundreds of millions of years. In either case, when extinction occurs—the complete loss of all individuals in a species—it is the end. And almost any way you look at biodiversity, from a utilitarian or aesthetic or symbolic perspective, extinction is a tragic loss.

Once numbering in the millions, passenger pigeons were hunted to extinction. The last passenger pigeon on earth, named Martha, died at the Cincinnati zoo in 1914.

FIGURE 16-11 The passenger pigeon: a species completely lost to the earth.

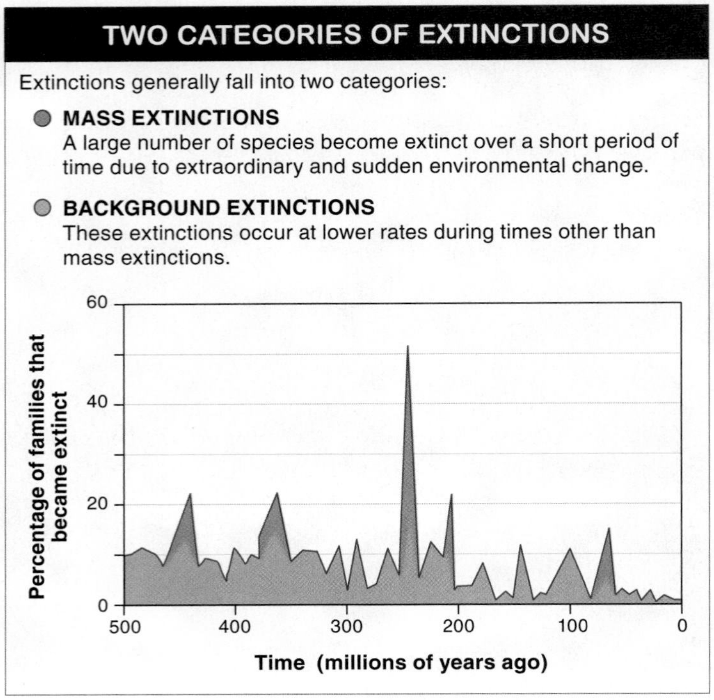

TWO CATEGORIES OF EXTINCTIONS

Extinctions generally fall into two categories:

- **MASS EXTINCTIONS**
  A large number of species become extinct over a short period of time due to extraordinary and sudden environmental change.
- **BACKGROUND EXTINCTIONS**
  These extinctions occur at lower rates during times other than mass extinctions.

Percentage of families that became extinct

60
40
20
0

500 400 300 200 100 0

Time (millions of years ago)

FIGURE 16-12 Background and mass extinctions.

As we discussed in Chapter 10, extinctions can be divided into two general categories: mass extinctions and background extinctions (**FIGURE 16-12**). They're distinguished by the numbers of species affected. When a large proportion of species on earth are lost in a short period of time—such as when an asteroid struck earth 65 million years ago and about 75% of all species on earth were wiped out—it's considered a mass extinction. These extinctions are above and beyond the normal rate of extinctions that occur in any given period of time, referred to as background extinctions.

Q Does extinction only happen to weak, unfit species?

In both background and mass extinctions, it's the same outcome for the species involved, but the causes tend to differ. In mass extinctions, the particular features of a species' biology don't really play a role in the extinction; rather, it's more like really bad luck. Background extinctions, by contrast, tend to be a conse-quence of one or more features of that particular species' biology.

For any species, there is always a risk of extinction. This risk can be larger or smaller, depending on certain features. Here we look at three important aspects of species' biology that can influence their extinction risk (**FIGURE 16-13**).

1. *Geographic range: extensive versus restricted.* Species restricted in their range—including those limited to small bodies of water and those confined to islands—are more vulnerable than those with extensive ranges. The Tasmanian devil is a marsupial carnivore about the size of a dog. Although these animals once thrived in Australia, they now are confined to the island of Tasmania, smaller than the state of Maine. Unable to expand their range, Tasmanian devils are more vulnerable to extinction than if their range was not limited.

2. *Local population size: large versus small.* Tigers and peregrine falcons, along with *Welwitschia,* a slow-growing, long-lived plant in southwest Africa, are examples of species that—as a consequence of their small population sizes—are at increased risk of extinction. With a small population size, a species is more susceptible to extinction due to a variety of factors that can kill individuals, including fire, diseases, habitat destruction, and predation. Put simply, the more individuals that are alive, the more likely it is that some of them will survive these events. With small population size, inbreeding is also increased, which generally reduces the fertility and longevity of individuals.

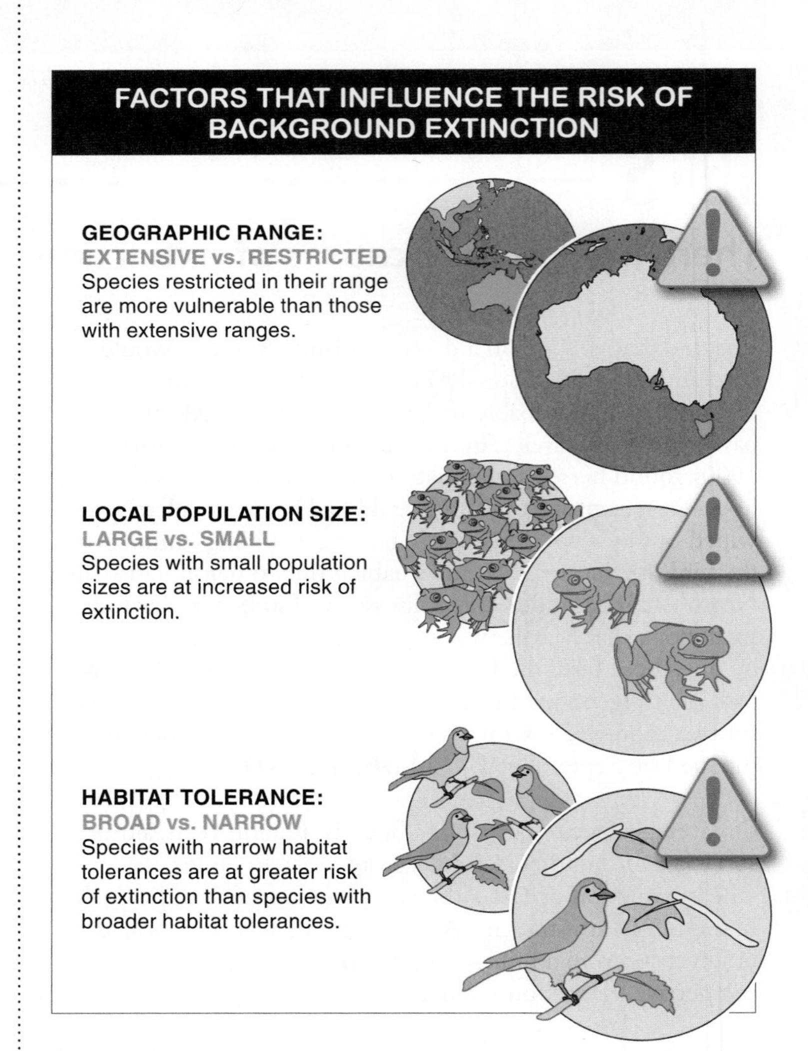

FIGURE 16-13 Range, population size, and tolerance influence risk of extinction.

3. *Habitat tolerance: broad versus narrow.* "Habitat tolerance" describes the breadth of habitats in which a species can survive. Some plant species, for example, can tolerate large swings in water availability or soil pH or temperature. Others are limited to very narrow habitat ranges. The now-extinct passenger pigeons could only build nests in a specific type of forest and needed large numbers of individuals, breeding communally, in order to breed successfully. As forests were cut down and as the birds were hunted, the size of their flocks diminished. Their narrow habitat tolerance made them extremely vulnerable to extinction. In general, because species with narrow habitat tolerance cannot adapt in the face of habitat degradation and loss, they are at greater risk than species with broader habitat tolerance.

One species that is enormously successful, thanks to its extensive geographic range, large local population sizes, and broad habitat tolerance, is our own. Humans are growing and expanding our range at a staggering rate, consuming resources in an unprecedentedly voracious fashion. By any measure, humans are one of the most successful species in earth's history. This success, unfortunately, is having an increasingly negative effect on the species with which we share the planet, causing extinctions at a significantly higher rate than ever before. We next explore in more detail the conflict between humans' success as a species and the survival of other species.

TAKE-HOME MESSAGE 16•5

Extinctions occur for fundamentally different reasons. Mass extinctions are more the result of bad luck than the particulars of a species' biology. Background extinctions tend to be a consequence of one or more features of the species' biology. Small population size, limited habitat range, and narrow zones of habitat tolerance contribute to background extinctions.

16•6

We are in the midst of a mass extinction.

Sometimes you can be so close to something that it's difficult to really see it. That may be the situation that humans are in right now when it comes to the global loss of biodiversity. But humans are becoming increasingly aware that we are in the midst of a mass extinction. In a recent survey by the American Museum of Natural History, 70% of biologists indicated that they believed this is true and that steps must be taken by governments and individuals to stop this massive loss of species.

Data on current rates of extinction in every well-studied group of plants and animals support the hypothesis that a mass extinction is under way. Among the mammals, 11% of all species are currently *endangered,* under imminent threat of extinction, and 14% are *threatened,* characterized as vulnerable to extinction. Among birds, 4% of all species are endangered and 50% are in decline. Almost a third of all amphibian species are threatened. Among fish, molluscs, insects, fungi, and plants, the story is the same.

Historically, background extinctions rates are about one extinction per million species per year. The numbers above suggest that current extinction rates are 1,000 times (or more) greater than usual. It is difficult to know the exact magnitude of the problem because, as we saw in Chapter 10, biologists don't have much of an idea about how many species there are on earth—the estimates range from 5 million to 100 million.

Unlike the last mass extinction event, in which the dinosaurs and most other species were wiped out in the wake of an asteroid's hitting earth, the cause of this current mass extinction seems to be the result of the activity of one species—humans. Ironically, it is the unprecedented success of humans that is responsible for this situation. The resource needs of so many people have inevitably interfered with the ability of other species to coexist with us.

The chief reason for the loss and impending loss of so many species is habitat loss and habitat degradation. Particularly harmful is habitat loss in earth's tropical rain forests, where biodiversity is greatest. In the past 25 years, half of the world's tropical rain forests have been destroyed, usually by burning to make way for agricultural use of the land or by logging (**FIGURE 16-14**). Urban development, too, is responsible for destruction of rain forests, as the growing human populations continue to expand. Intensive agriculture, livestock grazing, and the development of urban centers has led to the destruction and fragmentation of habitats worldwide.

The loss of biodiversity-supporting habitat is not restricted to the tropics. In the Pacific Northwest, as well, logging of forests

FIGURE 16-14 The devastation of deforestation.

FIGURE 16-15 Overexploitation of species: the bonobo.

has occurred at an unsustainable rate, leading to significant reductions in a diverse range of populations.

Beyond habitat loss, other factors reducing biodiversity include overexploitation of species, including those killed for food, furs, and medicinal products. Bonobos (*Pan paniscus*), for example, a type of chimpanzee and our closest living relative, are almost never seen in the wild, having been driven to extinction, hunted for food (**FIGURE 16-15**). And, as we'll see later in this chapter, the introduction of exotic species, too, is having a significantly adverse effect on biodiversity. Consequences of the loss of biodiversity, beyond the values of biodiversity discussed earlier, are largely unknown, but most likely include a serious reduction in the capacity of the environment to provide clean air and water and to recover from environmental and human-induced disasters.

Later in this chapter, we discuss some strategies by which the current high rate of biodiversity loss can be reduced.

TAKE-HOME MESSAGE 16•6

Most biologists believe that we are currently in the midst of a mass extinction, that it is the result of human activities, and that it poses a serious threat to the future survival of humans.

❸ Human interference generally reduces biodiversity.

An uneasy coexistence: Rio de Janeiro, Brazil, presses up against the Atlantic rain forest.

16•7

Some ecosystem disturbances are reversible, others are not.

When you fly over southern New England today, you look down on an undulating carpet of forest that covers hills and valleys. Roads traverse the forest, and here and there you see towns and cities or cultivated land and pastures, but most of what you see is the tops of trees.

> **Q** Once land is cultivated or developed, can it ever return to its natural state?

If you could have flown over the same route 200 years ago, things would have looked very different. In the early 1800s, most of the forest in southern New England had been cleared and the land was in cultivation. Homesteads consisting of farm houses, barns, stables, and other outbuildings were scattered across the countryside. Stone walls extended across hills and valleys, separating cropland from pastures.

Going back farther, the same flight 400 years ago would have taken you over mostly forested land. Here and there, Indian villages would have stood among cleared fields, but most of what you saw then would have looked much like what you see today.

The change from forest 400 years ago to cleared land 200 years ago was a major ecosystem disturbance, but now the area has returned to forest. Today, the stone fences are the only obvious trace of the agricultural history of the area. The fences still snake their way up and down hills, but now they run through solid forest instead of between fields and pastures.

By the middle 1800s, New England settlers were abandoning farms on marginal land and moving west. The fields they had cultivated began a process of ecological succession that returned them to forests (see section 15.15).

In the first year after a field is abandoned, its bare soil provides a harsh environment for plants. Sunlight blazes down, heating the ground surface and baking moisture from the soil. It takes a tough plant to sprout and grow under those conditions, but there are plants that can do it. We call those pioneering species of plants "weeds," and we see them growing in disturbed habitats—along roadsides, in vacant lots, and even in cracks in pavement.

As weeds grow and die, year after year, their organic matter enriches the soil and helps it retain moisture. And every year, the plants grow more densely so that they shade the ground surface, which no longer gets quite so hot and dry. Additional species of plants can grow in these milder conditions. These

SOME DISTURBANCES ARE REVERSIBLE

FIGURE 16-16 The forest returning. Just two decades after an explosive eruption, shrubs and small trees have returned to the plain at the foot of the active volcano, Mount St. Helens, in Washington State.

plants are taller, so they provide still more shade and the ground becomes still cooler and more moist. At this stage, the seeds of bushes and small trees can sprout. As these woody plants grow, they make the ground cool and shady enough to allow the seeds of forest trees (oak, maple, or beech) to sprout. More saplings of forest trees sprout and grow every year, and eventually they are so closely packed that little sunshine penetrates their leafy canopies to reach ground level. At this point, the forest stops changing, because the mature trees have created conditions in which only their own seeds can sprout and grow.

An ecosystem disturbance is reversible when the disturbance, although it alters the biotic and abiotic nature of the habitat, does not include the complete extinction of any species, so species can re-establish their populations. The example of New England forests is based on a disturbance that was caused by humans—clearing land for agriculture in the 1700s—but wind storms, forest fires, floods, and volcanic eruptions have been clearing forests throughout the history of life (**FIGURE 16-16**).

> "Any fool kid can step on a beetle. But all the professors in the world can't make one.
>
> —Arthur Schopenhauer, German philosopher (1788–1860)"

If, conversely, an ecosystem disturbance involves the complete loss of a species to extinction, the disturbance is irreversible. Each species is the result of a long and uninterrupted evolutionary history, involving an interplay of random and selective forces, and producing a unique genome. Once lost, a species can never be created again (**FIGURE 16-17**). For this reason, ecosystem disturbances involving the loss of species can have more devastating consequences than disturbances not leading to extinctions, no matter how great the changes to the abiotic environment might be.

BIOLOGICAL DISTURBANCES ARE IRREVERSIBLE

The blue whale is the largest mammal in the world and the largest animal ever known to have lived on earth. Once lost, a species can never be re-created.

FIGURE 16-17 The blue whale. Disturbances involving the loss of species are devastating.

TAKE-HOME MESSAGE 16•7

An ecosystem disturbance is reversible when the disturbance, although it alters the physical and biological nature of the habitat, does not include the complete extinction of any species, so they can re-establish their populations. An ecosystem disturbance that involves the complete loss of a species to extinction is irreversible, because a species, once lost, can never be created again.

16•8

Disruptions of ecosystems can be disastrous: 1. Introductions of exotic species.

Although the phrase suggests colorful birds and butterflies winging through a forest, **exotic species** refers to species introduced, by human activities, to areas other than the species' native range. Nonetheless, in the rain and sleet of a Chicago winter, the idea of colorful birds flying about is very appealing, and you might see exactly that, in the form of the monk parakeet. These small green members of the parrot family are about a foot (30 centimeters) from head to tail tip. Native to southern South America, monk parakeets were imported to the United States as pets in the 1960s and 1970s. Some of those pet birds escaped, and free-living populations of monk parakeets now occur in 11 states, from Oregon to Rhode Island and from Florida to New York (**FIGURE 16-18**).

The appearance of these free-living parakeets was greeted with alarm, because they are regarded as agricultural pests in South America. Records from the Inca civilization that flourished before Pizarro's invasion of Peru in 1532 attest to crop damage by monk parakeets, and when Charles Darwin visited Uruguay in 1833, he was told that monk parakeets attack fruit orchards and grain fields. Fortunately, the dire predictions that these small parakeets would devastate crops in the United States have not been fulfilled as yet.

FIGURE 16-18 A parakeet in the Midwest. The monk parakeet is an exotic species from South America, introduced into Chicago, Illinois.

But monk parakeets do make a nuisance of themselves in cities such as Chicago, where they build nests on utility poles, transformers, and floodlights. There is nothing delicate about a monk parakeet nest—it is a large mass of sticks and twigs, and a nest that has been soaked with rain is heavy enough to bring down wires, interrupting electrical service to entire neighborhoods.

Another exotic species is also notorious for short-circuiting power lines—the brown tree snake causes thousands of power outages on the island of Guam every year. This snake, which is native to northern Australia, Papua New Guinea, and the Solomon Islands, reaches a length of 10 feet (3 meters). It eats birds, which it captures as they roost in trees at night, killing them with constriction and by injecting venom with long, grooved teeth in the rear of its upper jaw.

Q Why should we worry about exotic species?

Introductions of exotic species can be intentional, as with the monk parakeet, or unintentional, as was the case with the brown tree snake, which reached Guam in the 1940s by stowing away in shipments of military equipment during the Second World War. In either case, two characteristics, in particular, make invasive species harmful (**FIGURE 16-19**):

1. Exotic species have no predators in their new habitat, so their populations grow unchecked.
2. Native plants and animals have no mechanisms to compete with or defend themselves against invading exotic species.

Guam, which lies in the middle of the South Pacific Ocean, has no native snakes and no predators that are specialized to eat snakes. But Guam did have a magnificent fauna of native birds that had evolved in isolation on the island for thousands of years. Because the species of birds on Guam had never experienced predation by snakes, they had no defensive mechanisms against them—and brown tree snakes have eradicated most of the native species of birds. Extraordinary efforts are being made to prevent brown tree snakes from

DISASTROUS DISRUPTIONS: EXOTIC SPECIES

PROBLEM
Exotic species can threaten native populations.

CAUSE
Often introduced, both accidentally and intentionally, by humans. Once introduced, exotic species can multiply, unchecked by predation, overwhelming competitors and irreversibly altering ecosystems.

STRATEGIES FOR SOLUTION
Better regulation and restriction of intentional introductions; better vigilance against accidental introductions

FIGURE 16-19 Exotic species: the problem and its cause.

spreading to other Pacific islands, such as Hawaii. Like Guam, these islands are home to species of birds that occur nowhere else in the world, and the islands have no native species of snakes and no native predators of snakes.

We don't have to look beyond the borders of the continental United States, though, to find examples of exotic species that are responsible for massive ecological shifts and economic costs. Purple loosestrife is an attractive flowering plant that is native to Eurasia. It was imported to the United States in the 1800s as a garden plant, and rapidly escaped from cultivation and invaded wetlands in every state except Florida. It produces thousands of seeds a year, and also spreads by sending out underground stems. This aggressive growth overwhelms native grasses, sedges, and flowering plants, replacing diverse wetland communities with monocultures of loosestrife that provide poor-quality habitats for bog-dwelling insects, birds, reptiles, amphibians, and mammals (**FIGURE 16-20**).

The title of Most Destructive Invaders in North America may belong to the zebra mussels and quagga mussels. These thumbnail-size freshwater mussels are native to eastern Europe and western Asia. In the early 1800s, the mussels spread to western Europe, and they came to North America in the 1950s, after completion of the Saint Lawrence Seaway allowed oceangoing ships to enter the Great Lakes. Female mussels produce up to five million eggs per year, and the larvae settle on any solid surface, even clogging the intake pipes of water-treatment plants and factories and the cooling systems of power plants.

The ecological threat that zebra mussels represent is even more serious than the damage they cause to industrial facilities. The Great Lakes fisheries are based on game species (lake trout, salmon, walleye) and commercial species (whitefish, perch, herring), and they produce revenues of more than $1.5 billion annually. Most of the commercially valuable fish feed on small species of fish, such as smelt, and the small fish feed on microscopic plants and animals (phytoplankton and zooplankton). Zebra and quagga mussels are exceptionally efficient at filtering phytoplankton and zooplankton from the water—so good that they are depleting

EXOTIC SPECIES: ACCIDENTAL INTRODUCTIONS

BROWN TREE SNAKE
Brown tree snakes, introduced to the island of Guam, have eradicated most of the native species of birds, which had never experienced predation by snakes and had not developed defense mechanisms.

PURPLE LOOSESTRIFE
Purple loosestrife, introduced to the United States as a garden plant, escaped from cultivation and has invaded wetlands in every state except Florida. The aggressive plant outgrows native grasses, sedges, and flowering plants.

ZEBRA MUSSELS
Zebra mussels, introduced to the Great Lakes, can cause damage to industrial facilities by clogging intake pipes. They also deplete resources available to small fish, eliminating the food supply for larger game fish.

FIGURE 16-20 Unwanted guests. Species sometimes get into natural habitats accidentally. The results can be devastating.

EXOTIC SPECIES: INTENTIONAL INTRODUCTION

Initially introduced to Australia to control agricultural pests, cane toad populations exploded, now numbering more than 200 million, and have become a nuisance in much of the country.

FIGURE 16-21 Cane toad infestation.

the resources available to the small fish. Without enough food, the populations of small fish are crashing and that, in turn, is eliminating the food supply for the large game fish. The Great Lakes contain nearly one-fifth of the fresh water in the world, but this enormous ecosystem is being irreversibly degraded by just two exotic species.

One legendary pest is the cane toad. Ironically, it was initially imported into Australia to control an agricultural pest, but the cane toad populations exploded and spread throughout much of the continent. They now number more than 200 million and have become a nuisance in much of the country, as conservation workers struggle to determine how to control their numbers (**FIGURE 16-21**). The cane toad illustrates a common feature of most exotic species introductions, even when intended as a method of biological control of pests: their effects on the ecosystem are unpredictable and often terrible.

TAKE-HOME MESSAGE 16•8

When an exotic species enters a new habitat, it leaves behind the natural predators that had controlled its population size. In addition, an invading predator encounters prey that have no experience in escaping from it, and they are easy victims. Thus, alien species can multiply, unchecked by predation and with plentiful food, overwhelming competitors and dominating and irreversibly altering communities and entire ecosystems.

16•9

Disruptions of ecosystems can be disastrous: 2. Acid rain and the burning of fossil fuels.

"What goes up must come down." That's a saying that applies to molecules just as much as it does to larger objects. The difference is that when molecules come down, they may not be in the same form as when they went up. In particular, some chemicals rise into the atmosphere as gases and return to earth as solutions of acid. These acid solutions can be fog, sleet, snow, or rain—all forms of precipitation can be acidic.

Q Rain is usually just pure water. What can turn it to acid?

We use the term "fossil fuels" to describe oil, natural gas, and coal. These substances are composed largely of carbon and hydrogen, and they form carbon dioxide (CO_2) and water (H_2O) when they burn. Although they are referred to as hydrocarbons, fossil fuels are not *pure* hydrocarbons. This causes some problems. In addition to carbon and hydrogen, they contain substantial quantities of other elements, and it is the sulfur and nitrogen in fossil fuels that cause acid precipitation (**FIGURE 16-22**).

When sulfur burns, it produces the gas sulfur dioxide (SO_2), and nitrogen produces nitrogen dioxide (NO_2). These gases react with water vapor in the atmosphere to produce sulfuric acid (H_2SO_4) and nitric acid (HNO_3), and these acids fall to earth as acid precipitation (**FIGURE 16-23**).

In North America, precipitation has an average pH as low as 4.3 in some parts of the Northeast, nearly 1,000 times more acidic than pure water, which has a pH of 7.0. High concentrations of sulfuric and nitric acids are to blame, and several factors play a role. The Northeast has more people per square mile than does the rest of the United States, and that means

THE CHEMISTRY OF ACID RAIN

1. Burning fossil fuels releases the gases sulfur dioxide and nitrogen dioxide.
2. When combined with water vapor in the atmosphere, these compounds form sulfuric acid and nitric acid.
3. The acids fall to earth as acid precipitation.

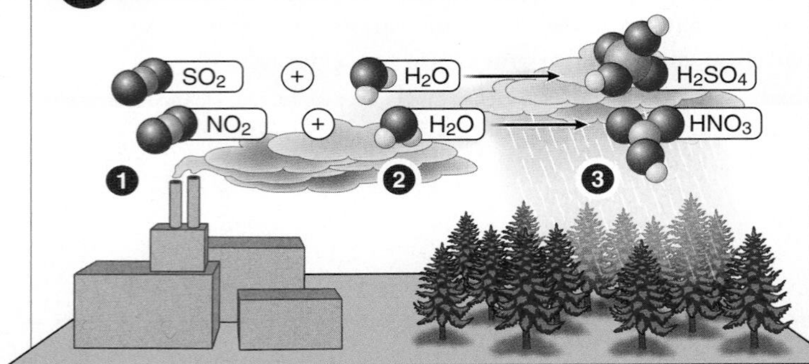

FIGURE 16-22 Formation of acid rain.

there are more houses, factories, and automobiles to burn the oil, coal, and gasoline that create acid precipitation (**FIGURE 16-24**).

But not all of the pollution that causes acid precipitation is local. In the United States, the Midwest and Southeast have a large number of electric power plants that burn coal, and wind currents carry the sulfur dioxide and nitrogen dioxide from these power plants across the Northeast. Precipitation in the western states is nearly 100 times more acidic than pure water, because sulfur dioxide and nitrogen dioxide from Asia are converted to acids as they blow eastward across the Pacific Ocean. Precipitation in Europe and Asia is also acidified by local and distant sources of pollution.

Both terrestrial and aquatic organisms are harmed by acid precipitation, and the effects of acid precipitation can be direct or indirect. Direct effects result from contact of living tissues with acidic water, whereas indirect effects are produced by interactions between non-living and living systems.

Dramatic evidence of the direct effects of acid precipitation on vegetation can be found in mountain forests that are often blanketed by fog. When trees are exposed to acid fog, year after year, their leaves are damaged, their rates of photosynthesis decrease, and eventually, many of the trees die (**FIGURE 16-25**).

DISASTROUS DISRUPTIONS: ACID RAIN

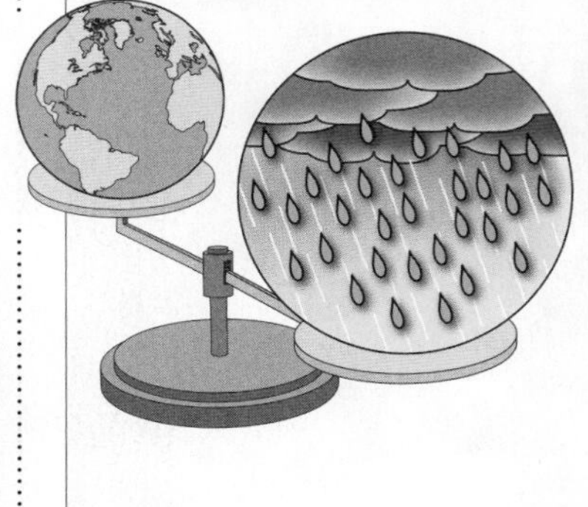

PROBLEM
Acid precipitation kills plants and aquatic animals directly, and also indirectly via changes in soil and water chemistry.

CAUSE
Burning fossil fuels releases sulfur dioxide and nitrogen dioxide. The compounds form sulfuric and nitric acids when combined with water vapor.

STRATEGIES FOR SOLUTION
Tighter regulation and reduction of sulfur dioxide and nitrogen dioxide emissions

FIGURE 16-23 Acid precipitation: the problem and its cause.

ACID RAIN DISTRIBUTION

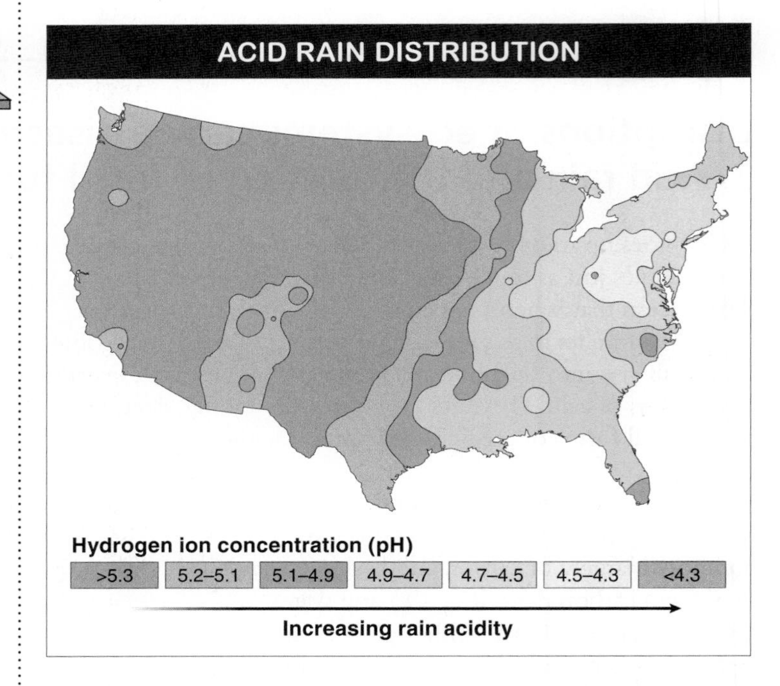

FIGURE 16-24 Acid precipitation across the United States. Pure water has a neutral pH of 7.

EFFECTS OF ACID RAIN

FIGURE 16-25 Acid precipitation has damaged these trees.

The indirect effects of acid precipitation on forests are not as conspicuous as the direct effects, but they are more far-reaching. As acid falls on the soil, it carries away calcium, potassium, magnesium, and sodium ions. These ions are essential nutrients for plants, and when they are leached out of the soil, the rate of plant growth is reduced. Some forest animals also feel the impact of this change in soil chemistry—when calcium is depleted, the shells of snails are thinner than usual. As a result of the thinning of snail shells, birds that eat snails receive less calcium than normal, and these birds lay eggs with shells so fragile that they can crack while the parents are incubating them.

Lakes and streams are also affected directly and indirectly by acid precipitation, and some lakes in the northeastern United States and in northern Europe have pH values as low as 4.0–4.3, which is 1,000 times more acidic than pure water. Very few species of insects, molluscs, crustaceans, fish, or amphibians can live in water that acidic. Still worse, acid precipitation dissolves aluminum in soil and carries the dissolved metal (aluminum ions)—which is toxic to animals—into lakes and streams, where it can kill most forms of aquatic animal life.

Acid rain is a solvable problem, because the deposition of acids from acid rain can be both prevented and cleaned up. The best approach is to prevent or reduce the emissions of sulfur dioxide and nitrogen dioxide. Preventive measures include improving energy efficiency, reducing the use of coal, switching to use of natural gas, and increasing the use of renewable energy resources, such as wind energy and solar energy. In the United States, many coal-burning plants have now installed special "scrubbers" to their smokestacks. The scrubbers extract hot gases from the power plant and function like sponges, removing sulfur dioxide before it is released into the atmosphere and converting it to a less hazardous form that can be physically removed from the tower and disposed of.

TAKE-HOME MESSAGE 16•9

Burning fossil fuels releases the gases sulfur dioxide and nitrogen dioxide, and these compounds form sulfuric and nitric acids when they combine with water vapor. Rain, fog, sleet, and snow that contain these acids can be up to 1,000 times more acidic than pure water. Acid precipitation kills plants and aquatic animals directly, by direct contact with living tissues, and also indirectly, through changes in soil and water chemistry.

16•10

Disruptions of ecosystems can be disastrous: 3. Global warming and the greenhouse effect.

Not long ago, debates about global warming focused on two questions: "Is it real?" and "Is it caused by human activities?" Now it is absolutely clear that the answer to both questions is *yes,* and the debate has moved to questions such as "What effect will global warming have on the world we know?" and "How can we stop it?"

Some of the evidence that the earth's atmosphere is warming comes from weather records that extend back into the 1700s. Those records show that the average temperature has increased rapidly during the past 50 years and that the 10 hottest years on record have occurred since 1990. Three centuries is a substantial time from the perspective of an individual human, but it's less than the blink of an eye in the history of the earth. We can extend the temperature records back to more than 400,000 years ago by examining ice cores from Antarctica. Although these cores do record short-term and long-term variations in temperature, nothing in the past

GREENHOUSE EFFECT

Energy from the sun passes easily through the atmosphere to warm earth's surface. Some energy is reflected back toward space but because it is absorbed by greenhouse gases, it remains trapped in the atmosphere, heating the air.

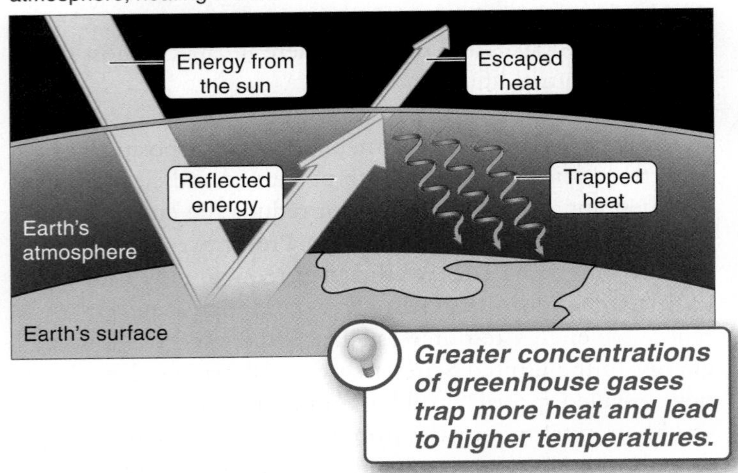

FIGURE 16-26 The earth is warmed by the greenhouse effect.

400,000 years approaches the speed of temperature change we have seen in the past 50 years.

The term "greenhouse effect" describes the process by which energy from the sun warms the earth's atmosphere. The light we see lies in the visible portion of the light spectrum, and visible wavelengths of light pass easily through the atmosphere to warm the earth's surface. Some energy in the infrared part of the spectrum flows from earth back toward space, and several gases in the atmosphere absorb this energy, heating the air (**FIGURE 16-26**). These gases act like the glass panels that make up the roof of a greenhouse: they allow light energy to pass through, but they keep heat from escaping—hence the name "greenhouse gases." Carbon dioxide and methane are among the most important of the greenhouse gases. Without them, heat would escape into space and the earth's average temperature would be lower.

Q Why do we blame global warming on humans?

Humans have increased the amount of greenhouse gases in the atmosphere. And we've done it for primarily one reason: we rely on the burning of fossil fuels—coal, oil, gasoline, and natural gas—for much of the energy we use. The burning of fossil fuels, which have been stored deep in the earth for millions of years, releases carbon dioxide, and clearing land and plowing soil to cultivate crops releases both carbon dioxide and methane. The same Antarctic ice cores that reveal the unprecedented rate of global warming in the past 50 years also show that the current levels of carbon dioxide and methane in the atmosphere are far higher than any that have occurred during the 400,000-year span of the cores (**FIGURE 16-27**).

The Intergovernmental Panel on Climate Change, a scientific body set up by the World Meteorological Organization and the United Nations, has predicted an increase in the average annual temperature of 2° to 12° F (1° to 6.4° C) during the 21st century. Sea levels will rise by 7–24 inches (18–59 cm) as warmer temperatures accelerate melting of glaciers and ice sheets. Rainfall is predicted to decrease overall, and droughts will become more frequent and more severe. The impact of droughts will be most severe in areas that are currently on the borderline of aridity, such as Australia, the American Midwest, and sub-Saharan Africa.

Both environmental and biological evidence already reveal the effects of global warming. The Arctic Ocean ice cap is melting, and the glaciers that cover Antarctica and Greenland have also been melting at unprecedented rates—and those rates are increasing (**FIGURE 16-28**). During 1996, 21.6 cubic miles (90 cubic kilometers) of the Greenland ice sheet melted, and by 2005, the annual rate of melting had increased to 36 cubic miles (150 cubic kilometers). This melting will cause an increase in ocean levels, creating flooding problems for many of the world's population centers in coastal areas.

Biological systems are also showing the effects of climate change. Trees and flowers in northern latitudes are blooming earlier in the spring than they used to; migratory birds are arriving at their summer ranges earlier than they did even a decade ago; and birds and butterflies are extending their geographic ranges northward.

DISASTROUS DISRUPTIONS: INCREASED GREENHOUSE GAS EMISSIONS

PROBLEM
The average temperature has increased rapidly during the past 50 years, affecting both the physical environment and the biological world.

CAUSE
Burning fossil fuels and clearing land to cultivate crops has significantly increased levels of greenhouse gases in the atmosphere.

STRATEGIES FOR SOLUTION
Reduced emissions of greenhouse gases (particularly from the burning of fossil fuels)

FIGURE 16-27 Increasing greenhouse gases: the problem and its cause.

ICE CAP MELTING

FIGURE 16-28 Disappearing ice pack. The satellite image at the top shows the minimum concentration of Arctic sea ice in 1979; the lower image shows the concentration of sea ice recorded on September 21, 2005.

The changes to the physical environment caused by global warming, such as thinning of the Arctic Ocean ice or the flooding of coastal estuaries, can have profound effects on biodiversity at the species and ecosystem levels. For example, in 2004, a research ship in the Arctic Ocean found nine walrus calves swimming alone in deep water. Normally, adult female walruses leave their babies on ice floes in shallow water while they dive to the seafloor to feed on clams and crabs, and then return to the ice to nurse their babies. But the Arctic ice cap over shallow water has melted, driving the female walruses to the remaining ice, which is in deep water. The mother walruses cannot dive deep enough to reach the sea bottom, and they seem to be abandoning their calves.

Other consequences of global warming reveal the tremendous interconnectedness of so many elements of ecosystems. Many bird and butterfly species' migratory patterns must be altered as they search for lower temperatures. Many small mammal populations living near high-altitude peaks will have nowhere to go for milder temperatures—if they move to lower altitudes it gets hotter; moving up to higher altitudes brings them to milder temperatures, but if they are already at the peak, they've got nowhere else to go when it gets too hot. Massive crop lands throughout the world may no longer be able to support the crops previously grown there.

Global warming is a difficult issue to tackle because it is so large—truly global—in scope, necessitating individual efforts as well as international cooperation. Fundamentally, however, the science is straightforward: carbon emissions must be reduced significantly. This can be done by individuals who choose to cut down on their own fossil fuel use, improve the energy efficiency of their homes and cars, and support the development of carbon-free renewable energy sources. Internationally, it can be done through efforts to reduce deforestation while increasing the replanting of forests, and by developing and implementing transportation and industrial innovations that reduce dependence on the fuels that produce greenhouse gases. Innovative technologies, such as the underground sequestration of carbon, may also contribute to reversing global warming.

TAKE-HOME MESSAGE 16•10

Carbon dioxide and methane are called "greenhouse gases" because they trap heat in the atmosphere. As humans burn fossil fuels and clear forests, the concentrations of greenhouse gases have been increasing and global temperatures have been rising. Ecological changes in plant and animal communities have already been observed and are likely to become more serious unless there is a global reduction in emissions of greenhouse gases.

16 • 11

Disruptions of ecosystems can be disastrous: 4. Depletion of the ozone layer.

If you live in a city, you are likely to hear radio and television warnings of high ozone levels during the summer. These warnings announce that ozone is expected to exceed safe levels that day. Children and adults with lung disease are advised to remain indoors during an ozone alert.

You are familiar with the oxygen you breathe—it comes in molecules formed by two oxygen atoms and is represented by the chemical formula O_2. Ozone is a different molecular arrangement of oxygen that packs three atoms into a molecule, with the chemical formula O_3. Ozone irritates the respiratory pathways and lungs, causing coughing and wheezing, and exposure to ozone can induce asthma attacks.

If ozone makes people sick, why would anyone worry about *depletion* of ozone in the atmosphere? Isn't that a *good* thing? The answer is the same as the old saying on what determines the value of real estate—"Location, location, location." At ground level, ozone is bad for you, but ozone in the upper part of the stratosphere, which is 30 miles (50 kilometers) above the earth's surface, protects you from ultraviolet radiation. Ozone is a Jekyll-and-Hyde molecule, and one set of environmental regulations tries to reduce the formation of ozone at ground level while a different set of regulations tries to protect ozone in the stratosphere.

The term "ozone depletion" covers two different processes: a general reduction in the amount of ozone in the stratosphere and the formation of areas with very low ozone concentration (called "ozone holes") over the North and South Poles every winter. Synthetic chemicals known as chlorofluorocarbons (CFCs) are the villains in both forms of ozone depletion. CFCs, which were developed in the 1930s, have a wide range of applications, including use as coolants in refrigeration systems. When CFC molecules leak into the atmosphere, they rise to the stratosphere, where sunlight knocks a chlorine atom off the CFC molecule. These free chlorine atoms then catalyze the breakdown of ozone to oxygen (**FIGURE 16-29**).

Q Should you be concerned about an ozone hole in our atmosphere?

The amount of ozone in the stratosphere has been decreasing by about 4% per decade. The cold temperatures and circular flows of air that develop over the Poles during the winter concentrate CFCs, forming ozone holes in those locations. The Antarctic ozone hole, which lasts for several months, covers the entire continent of Antarctica and extends northward to include the southern tips of South America and Australia (**FIGURE 16-30**). The Arctic ozone hole is smaller than the Antarctic hole and it does not last as long, but it is large enough to extend southward into northern Europe, Asia, and North America.

DISASTROUS DISRUPTIONS: OZONE LAYER DEPLETION

PROBLEM
Increased levels of ultraviolet light reach the earth's surface, leading to greater incidence of illness in animals, and decreased rates of photosynthesis in plants.

CAUSE
Synthetic chemicals known as chlorofluorocarbons (CFCs) leak into the atmosphere, where they cause the breakdown of ozone.

STRATEGIES FOR SOLUTION
Reduced production and emission of CFCs

FIGURE 16-29 Depletion of the ozone layer: the problem and its cause.

Ozone depletion allows short-wavelength ultraviolet light (UVB light, with wavelengths of 270–315 nanometers) to reach the earth's surface. The 4%-per-decade reduction in global stratospheric ozone levels has increased the intensity of UVB radiation everywhere on earth, and the UVB intensity under the ozone holes is even higher.

Not surprisingly, humans tend to think first about the effects of elevated UVB on human health. A 1% increase in UVB intensity increases the incidence of skin cancer by 2% to 3%. Exposure to UVB radiation also promotes the formation of cataracts and reduces the effectiveness of our immune system. Domestic animals suffer the same kinds of damage that humans do, and a few studies indicate that wild animals are also affected.

But we may be looking at the wrong problem. The damage to ecosystems could be a more serious consequence of ozone

THE ANTARCTIC OZONE HOLE

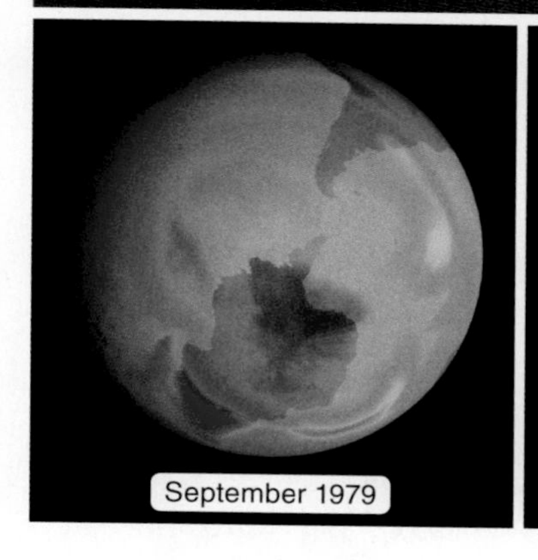

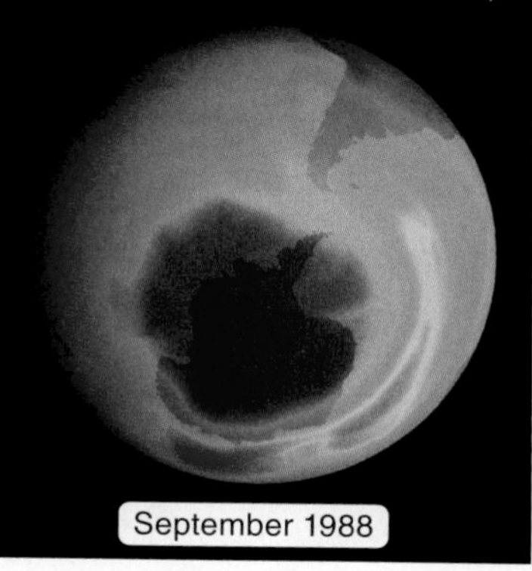

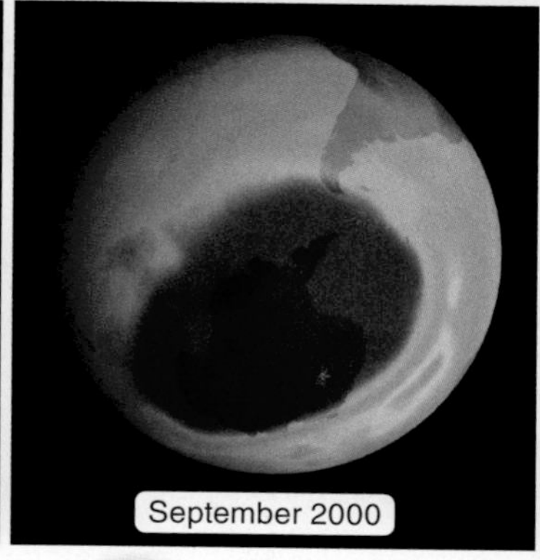

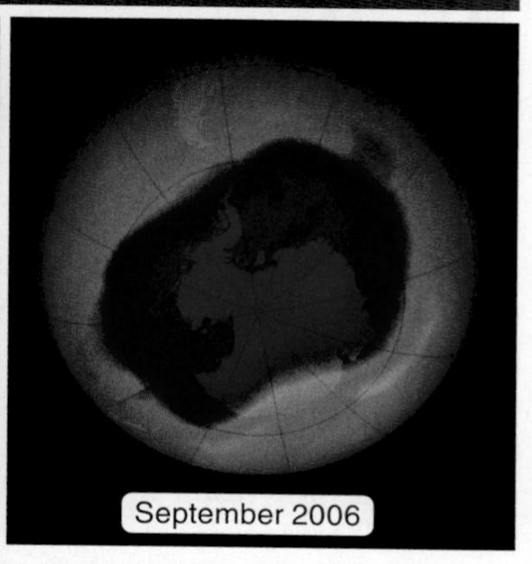

FIGURE 16-30 Monitoring the ozone hole. The satellite images show changes in the ozone hole over the Antarctic during the past 30 years. The darkest blue coloration indicates weakest ozone concentration.

Depletion of the ozone layer and the formation of an ozone "hole" over earth's polar regions have occurred as a result of human activities.

depletion than damage to human health. UVB radiation reduces the rate of photosynthesis by plants, and reduced photosynthesis in agricultural ecosystems means that crop yields will decline just at the time when the worldwide cost of food is increasing.

The creation of the ozone hole is an example of how some scientifically developed industrial processes and products can have unintended and severe consequences. When CFCs were first invented, they were considered harmless, and so their use as coolants, as aerosol propellants, and in the production of materials such as Styrofoam became widespread. But the subsequent steps to repair the ozone hole are also an example of the power of science to help identify environmental problems and to help find solutions (**FIGURE 16-31**). With the adoption by most countries in the 1980s of an agreement to discontinue the use of CFCs, the atmospheric levels of ozone have stabilized, and full recovery could occur by 2065.

REDUCING CFCs

Products made from ozone-depleting CFCs, such as Styrofoam hamburger boxes and hair sprays with damaging propellants, have been phased out in the hope of spurring a recovery of the ozone layer.

FIGURE 16-31 Taking simple steps to reduce ozone depletion.

TAKE-HOME MESSAGE 16•11

Ozone in the stratosphere prevents short-wavelength ultraviolet light (UVB) from reaching the earth's surface, but for many decades the amount of ozone has been decreasing. Synthetic chemicals, the chlorofluorocarbons (CFCs), are responsible for the destruction of ozone. An increase in UVB light reaching the earth's surface has adverse effects on health, including increasing the incidence of cancer in humans and other animals, and can seriously damage ecosystems. Reduction in the production of CFCs is reducing atmospheric ozone depletion.

16 • 12

Disruptions of ecosystems can be disastrous: 5. Deforestation of tropical rain forests.

Towering trees, colorful birds and butterflies, maybe a glimpse of Tarzan swinging past on a hanging vine—that is the popular image of a tropical rain forest. And it is a reasonably accurate picture (minus the Tarzan part). But it is also a picture that is rapidly fading, as agriculture, logging, gold mines, and oil wells destroy tropical forests (**FIGURE 16-32**).

Tropical rain forests grow in a region extending north and south of the equator, crossing South America, Africa, Asia, and Australia. As recently as a few centuries ago, that belt of rain forest covered about 8 million square miles (20 million square kilometers). More than half of that forest has already been destroyed, most of it in just the past 200 years (**FIGURE 16-33**).

The destruction of a tropical rain forest means an enormous loss of species, because these forests are the most diverse terrestrial habitats on earth. More than 170,000 species of plants grow in tropical rain forests, and that is about 70% of the total number of living plant species in the world. A survey of a Brazilian rain forest found 487 different species of trees in just 2.5 acres (1 hectare), an area less than half the size of a city block in Manhattan. To put that number in perspective, consider that only 700 species of trees are found in the 5 *billion* acres of the United States and Canada combined!

Agriculture is responsible for the greatest loss of tropical forest, and it takes a variety of forms. Sometimes, a relatively a small area is affected—"slash-and-burn agriculture" refers to cutting trees, burning them, and planting crops in the newly opened area. Usually, just a few acres of forest are cleared, but the land is fertile for only two or three years after it has been cleared. Then that plot is abandoned and more forest is cleared, so the cumulative effect of slash-and-burn agriculture is substantial. At the other end of the size scale, multinational corporations clear hundreds of acres of forest to plant bananas, coffee, or oil palms, and thousands of acres of forest to create pastures for cattle.

DISASTROUS DISRUPTIONS: DEFORESTATION

PROBLEM
Tropical rain forests are being cleared at unprecedentedly high rates, endangering countless species and increasing the concentration of greenhouse gases in the atmosphere.

CAUSE
The land is cleared for agriculture, logging, gold mines, and oil wells.

STRATEGIES FOR SOLUTION
Reduced destruction of high-biodiversity habitats, particularly tropical rain forests

FIGURE 16-32 Deforestation: the problem and its cause.

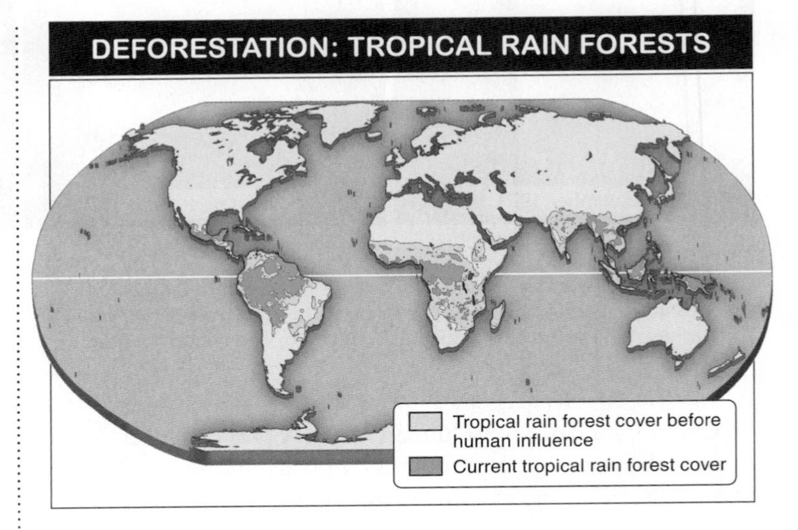

FIGURE 16-33 More than half of the world's rain forests have been destroyed, most of this destruction occurring in the past 200 years.

Pollution from oil wells and mining in tropical rain forests has an impact that can extend far beyond the areas that are actually cleared. Leaking oil contaminates streams and groundwater, and acidic water drains from mines. Even worse, gold miners use mercury to extract gold and some of the mercury enters streams, where it is converted to methyl mercury, a toxic compound that accumulates as it moves up the food chain.

Agriculture and mining both require roads to bring equipment to the sites and take crops or minerals to market. Roads have an impact on forests that greatly exceeds the relatively small area they occupy, because they make access to the forest easy. Traveling through a virgin tropical rain forest is difficult—bogs and natural tree falls often prevent travel in a straight line, slopes are often steep and the wet soil is slippery, and the ground-level vegetation can be both dense and thorny. No wheeled vehicle larger than a motorcycle can penetrate most rain forests, and traveling by motorcycle is slower than walking. These difficulties protect rain forests, but as soon as a road is bulldozed through a forest, people flock to it. They and their activities then spread into the forest on both sides of the road, creating new clearings that spread like cancers, farther and farther into the forest

Destruction of tropical rain forests has two serious environmental impacts: reducing the earth's biodiversity and increasing the concentration of greenhouse gases in the atmosphere.

1. *Reducing biodiversity.* Tropical rain forests of Africa, Asia, the Pacific, and Central and South America contain an unusually large number of species of plants and animals (and probably other taxa).

2. *Increasing greenhouse gases.* Photosynthesis in tropical rain forests removes an estimated 610 billion tons (550 trillion kilograms) of carbon dioxide from the atmosphere each year. Accumulation of carbon dioxide is the major cause of global warming, and the photosynthetic activity of tropical rain forests is slowing the rate of warming. The huge quantity of carbon stored in rain forests has a downside, however, because that carbon is released into the atmosphere when forests are cleared and burned. And tropical forests are being cleared at a frightening rate—nearly 12 square miles (30 square kilometers) per day, which means a total of 200,000 square miles (more than 500,000 square kilometers) just from 2000 to 2005!

The problem of tropical deforestation is among the most difficult of any environmental problems to solve, and solving it will involve multiple strategies and rely on international cooperation. The most diverse areas must be identified and protected. The poverty that necessitates the destruction of tropical rain forests for human activities must be addressed.

FIGURE 16-34 **Growing seedlings for reforestation.**

Alternative sources of food and income must be developed. Population growth must be reduced. And education about the value of preserving biodiversity is central to these issues. At present, we have not progressed much beyond identification of the problem, but that, at least, is a necessary first step (**FIGURE 16-34**).

TAKE-HOME MESSAGE 16•12

Tropical rain forests contain more species of plants and animals than all other terrestrial habitats combined, and half of the earth's biodiversity hotspots are in these forests. Tropical rain forests remove more carbon dioxide from the atmosphere than any other terrestrial habitat and, as a result, rain forests are enormously important in limiting global warming.

④ We can develop strategies for effective conservation.

A worker carries a tree to plant at the edge of Maowusu Desert, China. Trees are planted to prevent the desert from expanding.

16•13

With limited conservation resources, we must prioritize which species should be preserved.

Is it preferable to save one beautiful, well-studied bird species or 1,000 species of bacteria, none of which have even been described or named? This is a question that biologists, policymakers, and, ultimately, citizens must address. And what about the 1,200 species of beetles in Panama that live in the evergreen tree *Luehea seemannii,* almost 200 of which live *only* in this one species of tree—how should we prioritize these species for conservation relative to, say, a single primate species? From locale to locale the particulars of these question may change, but the underlying issue remains: in a world where species are being driven to extinction faster than we can save them, which should be singled out for preservation and which should we leave to almost certain extinction (**FIGURE 16-35**)?

When setting conservation priorities, it is essential to articulate a goal. In an ideal world, the goal might be to protect "all biodiversity." Barring that, options include protecting "most biodiversity," or the most diverse subset of biodiversity (in terms of genes, species, and ecosystems), or the most valuable biodiversity. There are numerous arguments in favor of each goal, and each has significant flaws as well. Once a goal is decided on, a related step in the decision-making process is needed—a plan that outlines priorities for achieving the goal. Frequently, this involves an assessment of the degree of threat faced by various components of biodiversity. We can rank species, for example, as relatively intact, relatively stable, vulnerable, endangered, or critical. These rankings can then be used in conjunction with measures of the biological value of the biodiversity to humans in formulating a plan.

As we saw earlier in the chapter, biodiversity can be valued in many different ways, and its worth cannot be easily quantified or weighed. Sooner or later, many difficult and subjective decisions must be made in the goal-setting and prioritizing process. The undiscovered bacterial species may harbor the metabolic secrets to curing a devastating medical condition in humans, but the beautiful bird species may be much loved by bird-watchers or may serve as a powerful icon of strength or freedom, inspiring generations of people. The decisions are difficult, but not facing them is, in most cases, the same as making a decision.

In the United States, much of the conservation policy involves response to the **Endangered Species Act (ESA),** a law that defines **endangered species** as those in danger of extinction throughout all or a significant portion of their range. The law is designed to protect those species from extinction. As we noted earlier in the chapter, species that are dwindling are

Which is more important to protect: a beautiful species such as the snow leopard, or soil bacteria with unknown roles in their ecosystem?

WHAT SHOULD WE PROTECT?

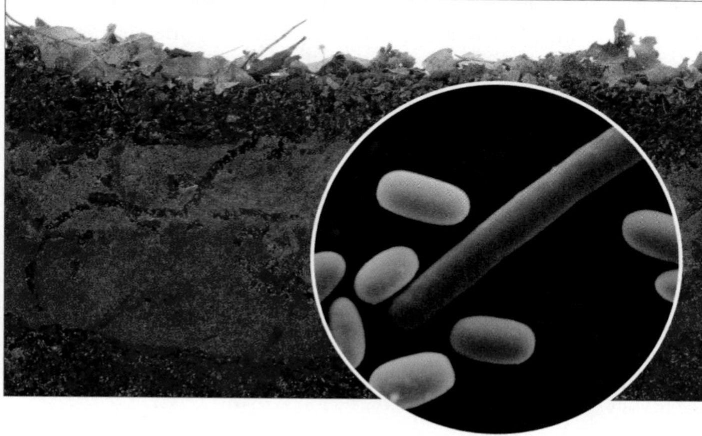

FIGURE 16-35 **Prioritizing conservation efforts.**

listed as endangered or threatened, according to an assessment of their risk of extinction. Once a species is listed, legal tools are available to help rebuild the population and protect the habitat critical to its survival. Seemingly straightforward, the ESA has had the effect of focusing most conservation efforts on the preservation of individual species (and populations), sometimes at the expense of other elements of biodiversity and sometimes at the expense of efforts to reduce the loss of ecologically important habitat (**FIGURE 16-36**). Other difficulties are also associated with the ESA. Consider the task of determining the critical population size below which a population is endangered—is it 500 individuals or 5,000 or 50,000? With its emphasis on preservation of single species, the ESA does not effectively address ecosystem decline, which is much like waiting for fires and then trying to put them out.

In spite of these difficulties, the species-focused approach to conservation has had some success in the United States. Since it became law in 1973, approximately 40 species that were listed as endangered—including the bald eagle, the peregrine falcon, the gray whale, and the grizzly bear—have been taken off the list as their numbers have recovered. In the final section, we examine other strategies that have been used in efforts to preserve biodiversity.

TAKE-HOME MESSAGE 16•13

Effective conservation requires the setting of goals on the elements of biodiversity (genes, species, or ecosystems) to be conserved and priorities among those elements. The Endangered Species Act has focused much conservation effort on the preservation of species.

PRESERVING SPECIES vs. PRESERVING HABITATS

Conservation efforts can focus on the preservation of individual species or of important habitats.

FIGURE 16-36 **Which to preserve?** Limited resources and great demand make preservation choices difficult.

16•14

There are multiple effective strategies for preserving biodiversity.

In a survey asking which animal they would like to be, men and women gave very different answers. Among men, the top answer was eagle, followed by tiger, lion, and dolphin. As their first choice, women chose cat, followed by butterfly, swan, and swallow. It's just a silly survey, but the differences in male and female selections parallel the fact that people can have very different preferences when it comes to the preservation of biodiversity. One person may view as unthinkable the loss of a particular species or habitat, while another may value other species or habitats much more.

Most approaches to conservation biology, as we've seen, have focused on the preservation of individual species. Increasingly, however, conservation biology is shifting toward the preservation of important habitat, focusing on conserving communities and ecosystems. This approach, of course, also leads to the preservation of individual species, but it has the added benefit of preserving, at the same time, many different species within a habitat, including many that have not yet been identified, particularly microbes and fungi.

One example of an effort aimed at the preservation of important habitat is the World Wildlife Fund's "Global 200," an identification of all of the most biologically distinct habitats on earth. The identification of these habitats, which include terrestrial, freshwater, and marine habitats, is part of an innovative strategy, often called **landscape conservation,** that is geared toward conserving not just species, but habitats and ecological processes (e.g., large-scale migrations and predator-prey interactions). In an age of scarce conservation resources and limited time, this approach prioritizes the conservation of the most biologically unique habitats on earth.

FIRST ATTEMPTS AT PRESERVING WILDERNESS

FIGURE 16-37 The natural beauty of Yosemite.

The habitat conservation approach is not new. The very first attempts at conservation were the creation of Yellowstone Park in 1872 and Yosemite in 1890, both large-scale efforts geared not simply toward the preservation of one or a few species but rather at preserving "wilderness." In the time since then, significant efforts have been made to establish national parks, wilderness areas, and recreation areas (**FIGURE 16-37**).

As conservation biologists have started to better understand population dynamics and biogeography, the design of natural preserves has evolved. Modern preserves focus not simply on maximizing the variety of habitats and biodiversity preserved, but on using several design features that maximize their efficiency. These design features include corridors and buffer zones (**FIGURE 16-38**).

1. *Corridors.* Setting aside areas—even relatively narrow, corridor-like strips of land—that connect larger natural preserves allows gene flow and reduces inbreeding among distinct populations in the different preserves. Because fragmented populations support smaller populations, the inclusion of corridors linking the fragmented populations can enhance the maintenance of genetic diversity.

2. *Buffer zones.* Preserves designed with core areas, containing the habitat to be conserved and where human activities are limited, surrounded by buffer zones, where limited amounts of human use are permitted, can sustain wildlife, even as increasing human populations exist adjacent to the preserves.

As conservation plans aimed at preserving habitats increase, efforts focusing on individual species still continue to be effective. Several strategies targeting individual species for conservation have been particularly successful at preserving large amounts of biodiversity beyond that single species.

1. *Flagship species.* Some species, because they are particularly charismatic, distinctive, vulnerable, or otherwise appealing,

DESIGNING EFFECTIVE NATURE PRESERVES

The design of natural preserves has evolved. Modern preserves focus on the use of several design features that maximize their efficiency, including:

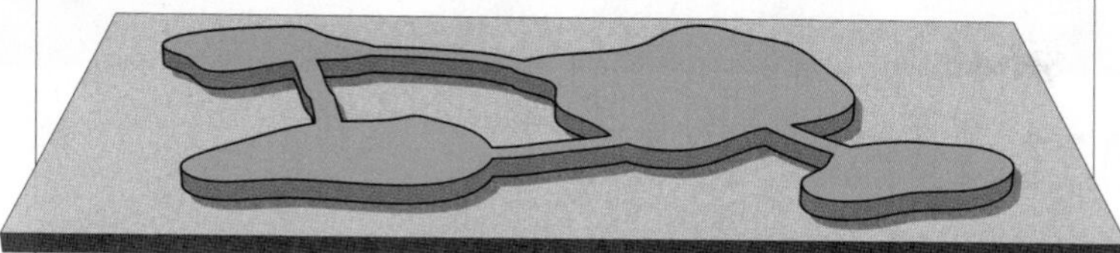

CORRIDORS
Setting aside narrow strips of land that connect larger natural preserves allows gene flow and reduces inbreeding among distinct populations in the different larger areas.

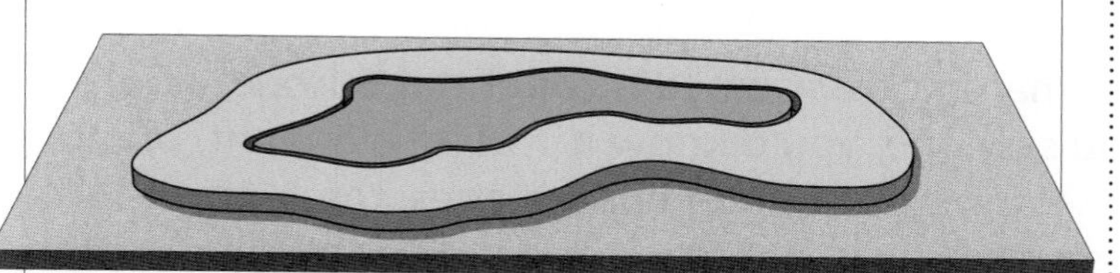

BUFFER ZONES
Preserves designed with core areas containing the habitat to be conserved, surrounded by buffer zones in which limited amounts of human use are permitted, can sustain wildlife even as increasing human populations exist adjacent to the preserves.

FIGURE 16-38 Carefully planned nature preserves focus on features that maximize biodiversity.

can engender significant public support. Examples include the giant panda of China, the golden lion tamarin of Brazil's coastal forest, the mountain gorilla of Central Africa, the orangutan of Southeast Asia, the leatherback sea turtle, the Indian tiger, and the African elephant. Preservation of these species, given their habitat needs, can serve to preserve many other species as well.

2. *Keystone species.* As we saw in Chapter 15, keystone species have a disproportionate effect on the biodiversity of a community. Their removal can lead to massive changes in the composition of species in an ecosystem, often causing huge loss of biodiversity. Examples include kelp, California mussels, grizzly bears, beavers, and sea stars.

3. *Indicator species.* These are species whose presence within an ecosystem indicates the presence of a large range of other species. For example, the presence or absence of lichens is an indicator of air quality, because they are sensitive to sulfur dioxide, a component of industrial fumes. When lichens disappear from trees, it is usually an indicator of increased pollution, which endangers the entire ecosystem. Conversely, when the lichens are present and healthy, they indicate that the ecosystem is healthy.

4. *Umbrella species.* These include wide-ranging species with such large needs for habitat and other resources that their preservation ends up protecting the numerous other species within that same habitat, without having to identify these other species as conservation targets. Umbrella species tend to be large, wide-ranging vertebrates.

For more critically endangered species and degraded habitats, all hope is not necessarily lost. Captive breeding programs and habitat restoration have had some success in bringing back biodiversity from the brink of extinction. Zoos and botanical gardens have taken the lead in many of these efforts. In the 1980s, for example, when the population size of the California condor had dropped to 22 individuals, due to poaching, lead poisoning, and habitat loss, all of the birds were caught and taken to zoos, where a captive breeding program began. With the success of these breeding programs, by 2009, more than 100 birds have been released back into the wild, and more than 150 more are in zoos. Breeding programs are not a complete conservation solution on their own, however. It is essential that the species' habitats are not destroyed or altered if the species are to flourish again under natural conditions.

Restoration ecology, using the principles of ecology to restore degraded habitats back to their natural state, has also been an important tool of conservation biologists. Wetlands that have been degraded by dredging and development, in particular, have benefited. Reintroduction of the native plant species and restoration of water and stream flow patterns can be instrumental in restoring these habitats.

Taken together, the strategies used by conservation biologists represent an important step toward reducing the adverse effects of one hugely successful (from a growth perspective) species—humans—on other species. Although we are far from living with minimal perturbation of the environment in which we live, continued conservation efforts offer our best hope for a sustainable future.

TAKE-HOME MESSAGE 16•14

Conservation biology has focused, in the past, on preserving individual species. Increasingly, there has been a shift toward the preservation of important habitats, focusing on conserving communities and ecosystems and utilizing concepts learned from ecology and island biogeography. Several methods focusing on single species, however, remain useful, particularly when preserving the selected species requires preservation of an amount and type of habitat that simultaneously preserves many other species.

Knowledge You Can Use

"Plan B" for saving endangered species: re-creating the wooly mammoth?

For a million years, wooly mammoths roamed throughout Europe, Asia, and North America. If you saw one, you'd know it was an elephant, but in adapting to their cold environment through two ice ages, the 8- to 14-foot-tall animals developed very thick hair and a two-inch layer of insulating fat.

Q: Where did they go? About 12,000 years ago, wooly mammoths became extinct, probably as a result of overhunting by humans in conjunction with a sudden change in climate.

Q: Is extinction really forever? Some scientists, promising a new strategy in the conservation of endangered species, want to bring the wooly mammoth back from extinction. They have extracted DNA from giant hairballs recovered from wooly mammoths preserved in the permafrost of Siberia and determined the sequence of base pairs in the recovered DNA. They plan to use that information to clone a wooly mammoth, getting the DNA into an elephant egg that has had its nucleus removed (they haven't figured out exactly how to do this), spurring the egg to divide, and then implanting it in a female elephant.

Q: Can wooly mammoths really be brought back from extinction? Currently, it isn't possible to re-create an extinct animal. In the case of the wooly mammoth, among other problems is that it hasn't been possible to isolate *all* of its DNA. And the DNA in the hairball samples is contaminated with DNA from other organisms, including bacteria, viruses, fungi, and assorted parasites.

Q: Does this have any relevance beyond novelty? Some people have proposed that the cloning of extinct animals could be a useful conservation tool. After all, getting a complete DNA sample uncontaminated by other species' DNA is not a problem when species still have living representatives. The San Diego Zoo has even established a lab they call "The Frozen Zoo," which stores frozen tissue from rare and endangered species.

Q: Could endangered or recently extinct species be brought back? When DNA is harvested from living animals, the answer is *yes.* Although there are still many technical difficulties, this has already been achieved. In 2001, a gaur, an endangered Asian ox, was cloned. (Although the procedure worked and a gaur was born, the baby died shortly after birth, from an illness unrelated to the cloning.)

What can you conclude? Is harvesting and preserving DNA the solution to the biodiversity crisis? In a word, no. At this point, cloning endangered animals is more about creating amusement park novelties and museum or zoo exhibits than preserving biodiversity. After all, conservation biologists want to preserve much more than genetic material. More important, many conservationists worry that the public will consider the cloning of endangered and extinct animals as an excuse to stop efforts to reduce habitat loss and protect land from development. (Closer to home: When it comes to your family pet, things are a bit more promising. Entrepreneurs have cloned pets after the animal died. At about $50,000, however, it doesn't look to be a rapidly growing industry.)

1 Measuring and defining biodiversity is complex.

Biodiversity provides significant benefits to humans in utilitarian ways, such as the production of medicines, and provides important value in less tangible ways, including aesthetic, symbolic, and naturalistic values. Biodiversity can be defined at multiple levels, from entire ecosystems to species to genes and alleles. There is a strong latitudinal biodiversity gradient from the tropics to the poles for nearly all groups of plants and animals. Biodiversity hotspots are regions of significant biodiversity under threat of destruction. The theory of island biogeography generates predictions about patterns of species diversity and has implications for the development of nature preserves.

2 Extinction reduces biodiversity.

Mass extinctions tend to result from bad luck rather than characteristics of a species' biology, whereas background extinctions tend to be a consequence of features of a species' biology. Small population size, limited habitat range, and narrow zones of habitat tolerance contribute to background extinctions. Most biologists believe that we are in the midst of a mass extinction, that it is the result of human activities, and that it poses a serious threat to human survival.

3 Human interference generally reduces biodiversity.

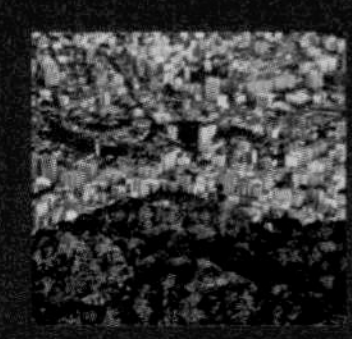

Because they have no natural predators in their new habitat and their prey have no defenses against them, exotic species can multiply unchecked, overwhelming competitors and dominating and irreversibly altering communities and entire ecosystems. Burning fossil fuels releases sulfur dioxide and nitrogen dioxide, and these gases combine with water vapor to form acid precipitation, harming plants, animals, and the abiotic environment. Carbon dioxide and methane are greenhouse gases, trapping heat in the atmosphere. Produced by the burning of fossil fuels and clearing of forests, greenhouse gases are increasing and global temperatures rising. Avoiding increasingly serious consequences requires a global reduction in emissions of greenhouse gases. Ozone in the stratosphere has been depleted through the effects of CFCs, with health consequences for humans and animals. A reduction in CFC production can restore safer levels of ozone. Tropical rain forests contain more species than all other terrestrial habitats combined, and half of the earth's biodiversity hotspots are in tropical rain forests. Destruction of these forests is likely to have serious environmental consequences.

4 We can develop strategies for effective conservation.

Effective conservation requires setting goals on which elements of biodiversity (genes, species, or ecosystems) to conserve and priorities among those elements. Although the Endangered Species Act has focused most conservation efforts on individual species, there is now a shift toward conserving communities and ecosystems. Several methods focusing on single species remain useful, however, particularly when they require preservation of an amount and type of habitat that simultaneously preserves many other species.

KEY TERMS

biodiversity hotspots, p. 593
conservation biology, p. 591
dynamic equilibrium, p. 595
endangered species, p. 614
Endangered Species Act, p. 614
endemic, p. 594
exotic species, p. 603
landscape conservation, p. 616

Check Your Knowledge

1. Biodiversity is considered important because of:
   a) the direct economic benefit to humans, such as medicines and food.
   b) the symbolic value that elements of the natural world can provide.
   c) the aesthetic value it holds.
   d) its potential for helping humans understand the processes that gave rise to the diversity we see.
   e) All of the above are correct.

2. Biodiversity hotspots are defined by which two criteria?
   a) species richness and ecosystem integrity
   b) size and distance from nearest alternative hotspot
   c) species endemism and degree of threat
   d) species richness and size
   e) ecological diversity and species diversity

3. Madagascar is unusually important to conservation because it:
   a) has more species per unit area than any place on earth.
   b) is the fourth largest island in the world.
   c) is home to more endangered species than any other country.
   d) has a higher percentage of endemic plants and animals than any comparably sized area on earth.
   e) is the native habitat of the rosy periwinkle, a rain forest plant that helps cure childhood lymphocytic leukemia.

4. By fumigating several islands in the Florida Keys with methyl bromide, E. O. Wilson and Daniel Simberloff demonstrated that:
   a) the equilibrium number of arthropod species was a function of island size and distance from the mainland.
   b) colonization of an island by new species can no longer occur once the island is "full."
   c) for plant but not insect species, the relation of area to species number holds true for all islands.
   d) the effects of global warming will be greater on islands than on the mainland.
   e) species turnover rates are highest on small, more distant islands.

5. In a dynamic equilibrium:
   a) extinction is the sole determinant of the number of species present on an island.
   b) immigration rates to an island are reduced to zero.
   c) both the number and composition of species on an island remain the same.
   d) both the number and composition of species on an island are changing.
   e) the composition of species on an island changes, but the number of species present remains the same.

6. Which of the following attributes of a species might render it especially susceptible to extinction?
   a) It lives in a marine habitat.
   b) It exists as two distinct subpopulations.
   c) It is an invasive species.
   d) It is short-lived.
   e) It has a narrow dispersal range.

7. What bird, once the most abundant in North America, was hunted to extinction by shooting and trapping during the 1800s?
   a) California condor
   b) blue-footed booby
   c) dodo
   d) passenger pigeon
   e) great auk

8. How many species do biologists estimate to be currently existing on earth?
   a) fewer than 1 million
   b) between 3 million and 5 million
   c) more than 1 billion
   d) between 5 million and 100 million
   e) between 1 million and 2 million

9. Which of the following is currently the leading cause of extinction?
   a) pollution
   b) habitat loss
   c) disease
   d) exotic species
   e) overexploitation

10. Exotic species can disrupt ecosystems because:
    a) they frequently have no predators in their new habitat and grow unchecked.
    b) they are favored by ecotourists.
    c) they have no natural prey items and so must rely on humans for their survival.
    d) they have better dispersal capability than endemic species.
    e) All of the above are correct.

11. Even though there is a carbon cycle, carbon dioxide levels around the world seem to be rising. Which of the following best explains why this is so?
    a) Animals give off carbon dioxide during their normal metabolism.
    b) As the atmosphere heats up, it can contain more carbon dioxide.
    c) The destruction of coral reefs leads to increased levels of carbon dioxide.
    d) More carbon dioxide is being given off by ocean waters as they heat up.
    e) The burning of fossil fuels releases more carbon dioxide into the atmosphere.

12. Ozone depletion is cause for concern because:

a) without ozone in the atmosphere, greenhouse gases can escape earth's atmosphere, leading to global warming.
b) without ozone, fossil fuels are more likely to react with water vapor in the atmosphere, increasing levels of precipitation throughout the world.
c) without the protection against solar radiation provided by ozone, rates of skin cancer are likely to increase.
d) without ozone, solar radiation can irradiate the Antarctic ice shelf, causing unsafe levels of radioactive isotopes.
e) reduced levels of ozone can increase the rate of respiratory problems in humans, particularly children.

13. Which of the following statements about habitat loss is correct?

a) Small animals are not vulnerable to habitat destruction.
b) Fragmented habitats can support only small populations of the species living in them.
c) Corridors between habitat areas are only of use to active organisms such as vertebrates.
d) Corridors can replace habitat areas for migratory species.
e) Large animals are not vulnerable to habitat destruction.

14. An umbrella species:

a) is a species whose removal will severely damage or cause the collapse of its ecosystem.
b) is a non-native species, usually introduced to an area by accident.
c) is any species that has a larger than expected effect on its ecosystem.
d) is a popular species that the public tends to rally behind and support.
e) is a species whose preservation results in the preservation of many other species as well.

Short-Answer Questions

1. Where, geographically, is the greatest biodiversity? What are the primary threats to this biodiversity?

2. Why were the ecosystem disturbances created by humans in New England reversible? What changes would have made the ecosystem disturbances in New England irreversible?

3. What is the primary cause of tropical rain forest deforestation? How does this affect the biodiversity of plants and animals?

See Appendix for answers. For additional study questions, go to www.prep-u.com.

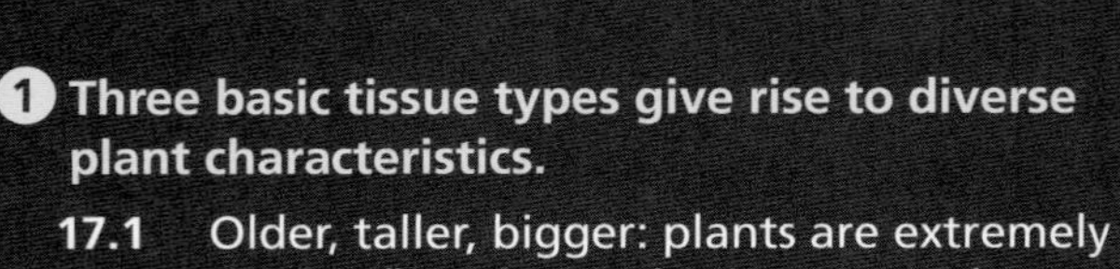

1 Three basic tissue types give rise to diverse plant characteristics.

17.1 Older, taller, bigger: plants are extremely diverse (but share a basic structural organization).

17.2 Flowering plants are divided into two major groups: the monocots and the eudicots.

17.3 Plants are organized into tissues, each with specific functions.

2 Most plants have common structural features.

17.4 Roots anchor the plant and take up water and minerals.

17.5 Stems are the backbone of the plant.

17.6 Leaves feed the plant.

17.7 Several structures help plants resist water loss.

3 Plant nutrition: plants obtain sunlight and usable chemical elements from the environment.

17.8 Four factors are necessary for plant growth.

17.9 Nutrients cycle from soil to organisms and back again.

17.10 Plants acquire essential nitrogen with the help of bacteria.

4 Plants transport water, sugar, and minerals through vascular tissue.

17.11 Plants take up water and minerals through their roots.

17.12 Water and minerals are distributed through the xylem.

17.13 Sugar and other nutrients are distributed through the phloem.

17

Plant Structure and Nutrient Transport

How plants function and why we need them

❶ Three basic tissue types give rise to diverse plant characteristics.

Though called a "flowering cabbage" the bright pink and green foliage is not a flower, but beautiful leaves.

17•1

Older, taller, bigger: plants are extremely diverse (but share a basic structural organization).

Plants are better than animals. Actually, that's an absurd thing to say. But only because biology isn't a contest, and from the perspective of evolution, any species that is not extinct can be considered a success. By many measures, though, plants do surpass animals, having developed a stunning diversity of structures and features during their evolution.

Q Plants versus animals: which can live longer? Grow larger?

Consider some vital statistics on plants and animals (**FIGURE 17-1**).

1. *Longevity.* The longest-living plants live 10 times longer than the longest-living animals. Bristlecone pines, found in the western United States, can live for more than 4,800 years. Among animals, the ocean quahog (a clam-like mollusk, found off the coast of Iceland) can live about 400 years, and giant tortoises, found on some tropical islands, can live almost 200 years.

2. *Height.* The tallest plants are 20 times taller than the tallest animals. Coast redwoods, found along the west coast of North America, have grown to more than 379 feet (115 m). The tallest giraffe grew to just under 20 feet (6 m).

3. *Weight.* The heaviest plants weigh almost 20 times as much as the heaviest animals. The largest giant sequoias are estimated to weigh more than 3,300 tons (almost 3 million kg). The largest blue whales—believed to be the largest animals ever to inhabit the earth—weigh in at approximately 170 tons (154,000 kg).

4. *Energy acquisition.* Most important is that plants can make their own food by capturing energy from the sun and converting it into a usable chemical form, but animals cannot. Consequently, every animal relies on plants for the energy

EXTRAORDINARY ATTRIBUTES OF PLANTS

LONGEVITY
The longest-lived plants live ten times longer than the longest-living animals. Bristlecone pines have been documented to live for more than 4,800 years.

HEIGHT
The tallest plants are twenty times taller than the tallest animals. Coast redwoods have grown to be more than 379 feet.

WEIGHT
The heaviest plants weigh almost twenty times as much as the heaviest animals. The largest giant sequoias are estimated to weigh more than 3,300 tons.

ENERGY
Plants can make their own food by capturing energy from the sun and converting it into a usable chemical form, but animals cannot.

FIGURE 17-1 The diversity of plant structures and features.

they capture from the sun and convert into a usable chemical form. This is why the total amount of plant matter on earth is more than 10 times the biomass of all the animals.

But, as we said, it is not a contest. That plants have a host of strategies vastly different from those that have evolved in animals, for growth, competition, defense, and reproduction, illustrates dramatically one of the fundamental truths of biology: there are multiple pathways to evolutionary success. In this and the next two chapters, we explore the structural and physiological features of plants and how they represent adaptive solutions—often very different from those employed by animals—to the constraints of living on earth. We begin by discussing the three main vegetative structures in plants: the roots, stems, and leaves (**FIGURE 17-2**). (We discuss plant reproductive structures in Chapter 18).

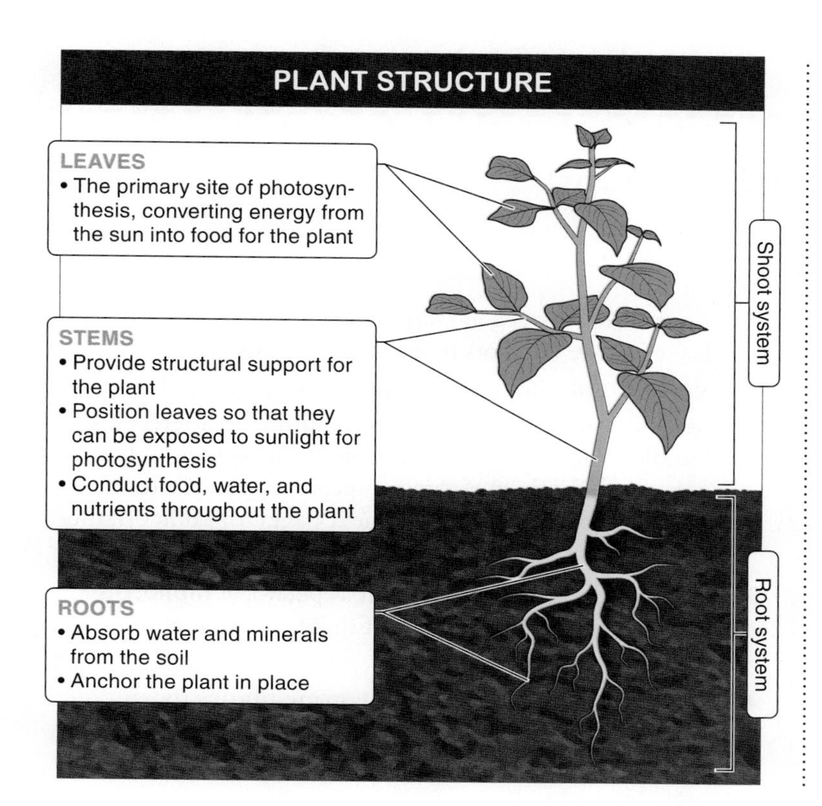

FIGURE 17-2 Three distinct plant structures: roots, stems, and leaves.

1. *Roots.* Acquiring energy and nutrients can be difficult when you can't move. It's not impossible, though. The root system of plants is an underground, branched system of structures. The **roots** serve several important functions, including absorbing water and minerals, as well as some oxygen, from the soil and anchoring plants in place. Roots also have **vascular tissue**—in some ways, the plant version of a circulatory system—which conducts food, water, and mineral nutrients throughout the plant. (It's important to note, however, that circulation in plants differs in one huge way from circulation in animals: plants have no heart. In fact they have no pump at all.)

2. *Stems.* When an organism must reach toward the sun for the energy to make its own food, gravity can be the enemy. The body of a plant is one continuous structure, with the root system below and the shoot system above the ground. The above-ground portion, the **shoot,** is divided into two parts: stems and leaves. In some species, the shoot system also includes flowers and fruits, the reproductive structures, which we discuss in Chapter 18.

Stems provide structural support for the leaves they bear. As plants grow, their stems increase in girth and length, and they branch so that leaves are positioned where they can be exposed to sunlight for photosynthesis. Like roots, stems also have vascular tissue.

3. *Leaves.* The chief site of photosynthesis in plants, the **leaves** serve as a plant's primary food factories. But they are also sites of great vulnerability. The same sunlight that provides energy for life, for example, also causes significant water loss through evaporation. Fighting such water loss is difficult when you are rooted in the ground and can't seek shade. Later in this chapter we learn that in the course of evolution, plants have developed some ways to reduce water loss.

We examine the structures and functions of roots, stems, and leaves in more detail later. We also explore plant nutrition and the methods by which water, sugars, and other nutrients move from their site of uptake or production to other parts of the plant where they are needed. And we look at some of the cellular specializations in plants that make their unique tissue types possible. First, we look at one important distinction, based on structural differences, that separates most angiosperms—the flowering plants—into two groups.

TAKE-HOME MESSAGE 17•1

Plants are an extremely diverse and successful group of organisms, generally composed of three distinct parts: roots, stems, and leaves.

17•2

Flowering plants are divided into two major groups: the monocots and the eudicots.

The flowering plants are classified into two major groups, based on some prominent structural features of their seeds, leaves, stems, flowers, and roots. The groups' names derive from a structure in the plant embryo within the seed, called a **cotyledon,** which usually becomes the first embryonic leaf or leaves of the plant. Plants in which one cotyledon forms are called **monocots,** and plants in which two cotyledons form are called dicots. The dicots, however, are not a monophyletic group, and include many different lineages. A large subset of the dicots, however, called the **eudicots,** is a monophyletic group.

In addition to their differences with respect to cotyledons, the monocots (about 70,000 species) also differ from the eudicots (just under 200,000 species) with respect to four other common structural features (**FIGURE 17-3**).

1. *Leaves.* Monocots generally have parallel veins in their leaves, while eudicots have branching veins.
2. *Stems.* Within the stems of monocots, the vascular tissue is arranged as numerous, randomly scattered bundles. You can actually see this at the dinner table if you cut through an asparagus spear (see the photo of the asparagus section on the next page). In eudicots, the vascular tissue is arranged in an orderly ring.
3. *Flowers.* The flower parts (such as petals) of monocots typically occur in multiples of three, while in eudicots they occur in fours or fives.
4. *Roots.* Monocots typically have many fibrous roots branching from the stem. Eudicots usually have a taproot, a single primary root with smaller roots branching from it.

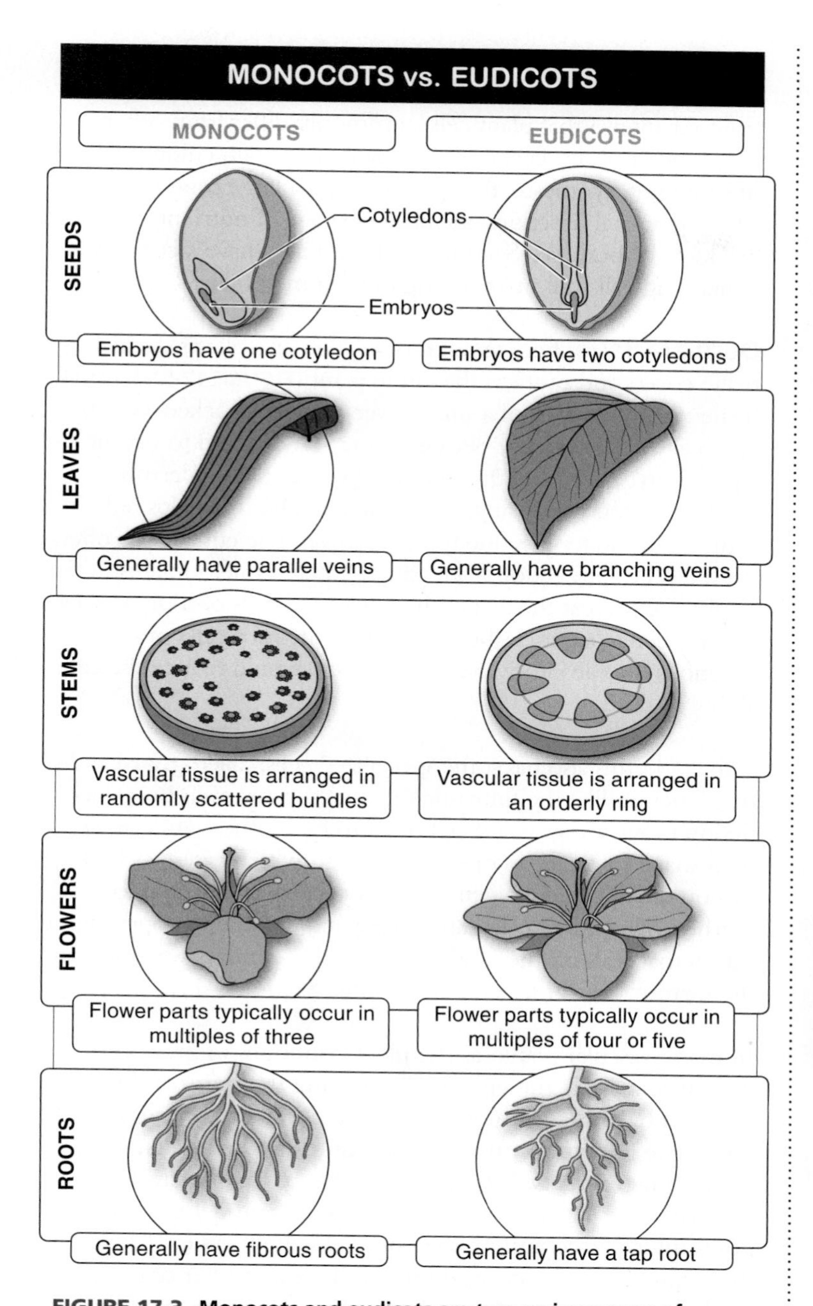

FIGURE 17-3 Monocots and eudicots are two major groups of flowering plants that differ in several structural features.

Cut through an asparagus spear and you can see the vascular tissue (shown here in purple) which is arranged in bundles and scattered throughout its stem.

Some common monocots are palm trees, orchids, lilies, most of the grains used in food products, and all of the grasses. Common eudicots include roses, daisies, coffee, potatoes, apples, pears, peaches, strawberries, and most large trees, including maples and oaks.

TAKE-HOME MESSAGE 17•2

The flowering plants are divided into two major groups, the monocots and the eudicots, based on structural features of their seeds, leaves, stems, flowers, and roots.

17•3

Plants are organized into tissues, each with specific functions.

Take a look at some common plants—perhaps a rosebush, a cactus, and a redwood tree. They have some things in common: they are rooted in the ground and they have green parts. Beyond these basic characteristics, rose bushes, cacti, and redwoods don't look much like one another. If we cut them open and examine their internal structure, however, we find great similarity across all the vascular plants. (Recall from Chapter 12 that the vast majority of all plant species are vascular. In contrast to mosses, which are non-vascular, all vascular plants have a sort of circulatory system of vessels for moving water and nutrients into and throughout the organism.)

Vascular plants, which we focus on in this chapter, are all organized around the same basic body plan and built up from the same three types of tissues. Leaves, stems, roots, flowers—all consist of these tissue types (**FIGURE 17-4**).

1. **Dermal tissue,** which covers and protects the surface of the plant, much like skin covers the human body.
2. **Vascular tissue,** which transports water and nutrients throughout the plant body, much like the vertebrate circulatory system.
3. **Ground tissue,** which makes up the bulk of the plant and is where most of the plant's metabolic activities are carried out.

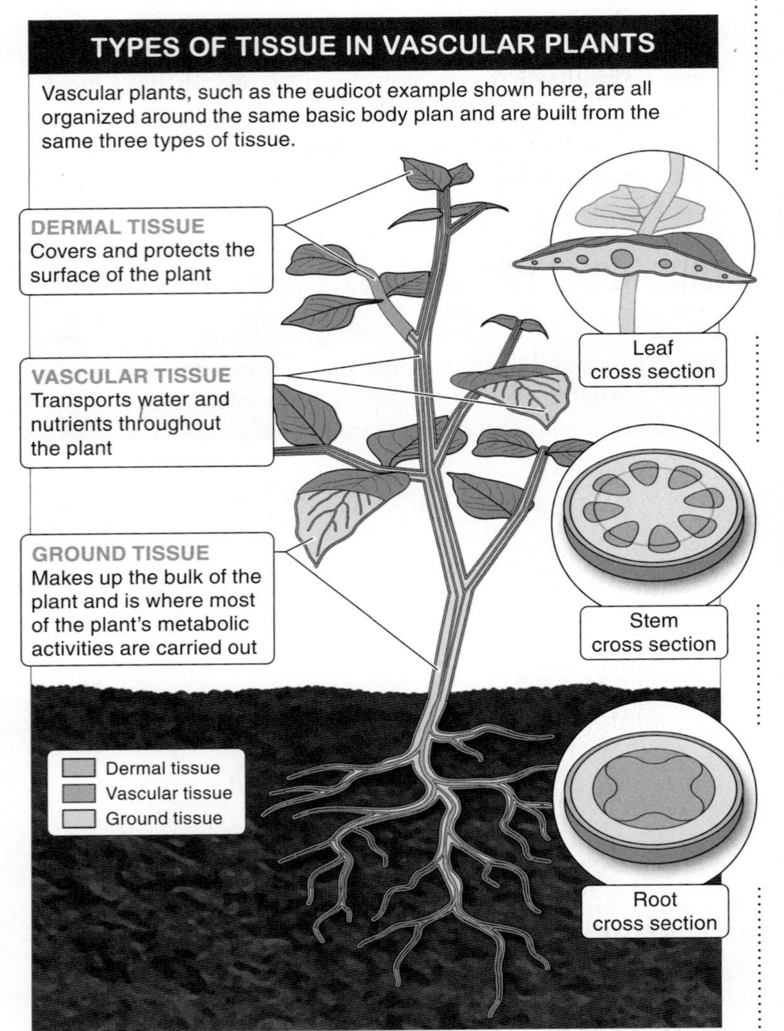

FIGURE 17-4 Basic tissue types found in all vascular plants. Plants take a huge variety of shapes and sizes, but inside they are all composed of the same three types of tissue.

Each of the three tissue types, discussed in detail below, is composed of one or more different types of plant cells. Recall from Chapter 3 that plant cells commonly differ from other eukaryotic cells in several ways: they may contain chloroplasts (the sites of photosynthesis), they may contain a large central vacuole (serving several functions, including storage of nutrients, waste products, or both; see Section 3-20), and they have a cellulose-containing cell wall surrounding the cell membrane.

Dermal Tissue Like the skin of an animal, a plant's dermal tissue covers and protects the entire plant (**FIGURE 17-5**). Dermal tissue usually consists of a single layer of tightly packed, very thin cells, called **epidermis.** Because it may be exposed to dry air and a plant can't get out of the sun on a hot day, the epidermis produces a waxy covering, called the **cuticle,** that helps reduce some of the water loss due to evaporation. The cuticle also offers a small level of protection from pathogens and microorganisms that might want to eat the plant. The epidermal cells of roots, however, do not secrete a cuticle, because one of the root cell's primary functions is to absorb water and a cuticle would reduce the cell's ability to do this.

The epidermis also has some specialized cells, called **guard cells,** that function like the futuristic "air-locks" often seen in science fiction movies. Two guard cells next to each other can change their shape to create an opening through which gases (including carbon dioxide and oxygen) can be exchanged during photosynthesis. The guard cells can also expand to close the opening, an action that seals off the inside tissues and minimizes water loss to the surrounding air by evaporation (see Section 4-11).

In many vascular plants, as the plant continues to grow and increases in girth, the epidermis covering the stems and roots is replaced by a thicker, more protective dermal covering. This protective covering consists mostly of dead **cork cells.** Cork cells contain a waxy, fatty substance that makes the tissue impermeable to water and resistant to fire and decay, while small pores interspersed among the cork cells allow gas exchange. On trees, this protective covering forms the outer layer of bark. The thick covering balances two conflicting needs: the need to protect the living tissue inside the tree from risks such as fire and predation and the need to absorb CO_2 from and release O_2 into the surrounding air.

Vascular Tissue As we saw earlier, plants can grow to be very large, and when they are large, they must have a system for getting nutrients to and removing waste products from all parts of the plant, from the bottommost roots to the topmost leaves. The delivery and movement of these molecules is the function of vascular tissue, which plays a role analogous to that of the circulatory system in animals (**FIGURE 17-6**). Most sugar, for example, is produced in the leaves (high up on the plant), while most water is absorbed through the roots (located underground). Because all plant cells require both water and sugar to survive,

DERMAL TISSUE CELLS

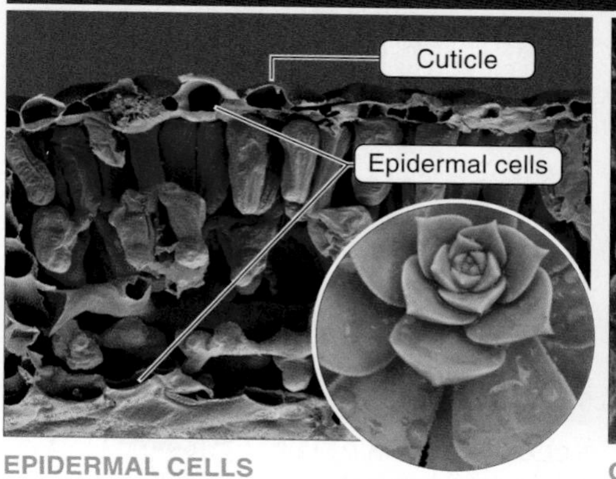

EPIDERMAL CELLS
Dermal tissue usually consists of a single layer of thin cells. These cells produce a waxy substance, called the cuticle, that helps reduce water loss.

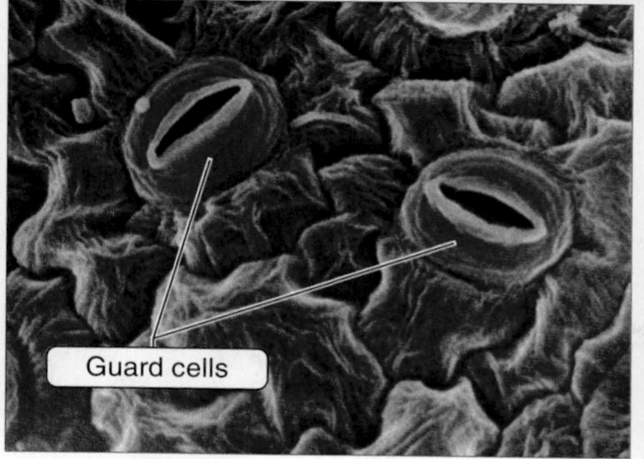

GUARD CELLS
Guard cells are specialized epidermal cells that create openings for gas exchange during photosynthesis. They can also close the openings, reducing water loss to evaporation.

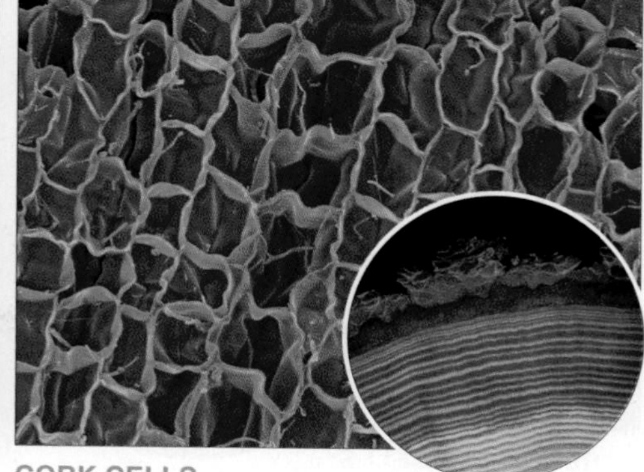

CORK CELLS
As a plant grows, cork cells replace epidermal cells, providing a thicker and more protective covering. On trees, this protective covering forms the outer layer of bark.

FIGURE 17-5 Dermal tissue protects the entire plant.

the physical separation of the leaves and roots is overcome by the vascular tissue that links them.

There are two parallel circulatory systems in plants, each made from a different type of vascular tissue and each transporting a fluid called **sap.** (Keep in mind that, unlike animal circulatory systems, these systems in plants are not closed, circular systems. Rather than *circulating* fluids, they simply transport them from one place to another in the plant.) One type of vascular tissue is called **xylem** (pronounced ZY-lum). It conducts xylem sap, containing the water and dissolved minerals absorbed in the roots, to the rest of the plant body. The other vascular tissue type, called **phloem** (pronounced FLOW-uhm), conducts phloem sap, consisting mostly of water, but also containing sugar—the plant's fuel source—and some minerals, to all of the tissues that need sugar to fuel their activities.

> **Q** Cells that have already died are among the most important parts of a plant. Why?

Plants build some structures in ways completely unheard of in animals. The xylem is one example of this: the tubes of the xylem are made from cells that are *dead*. The cells grow close together and, after

VASCULAR TISSUE

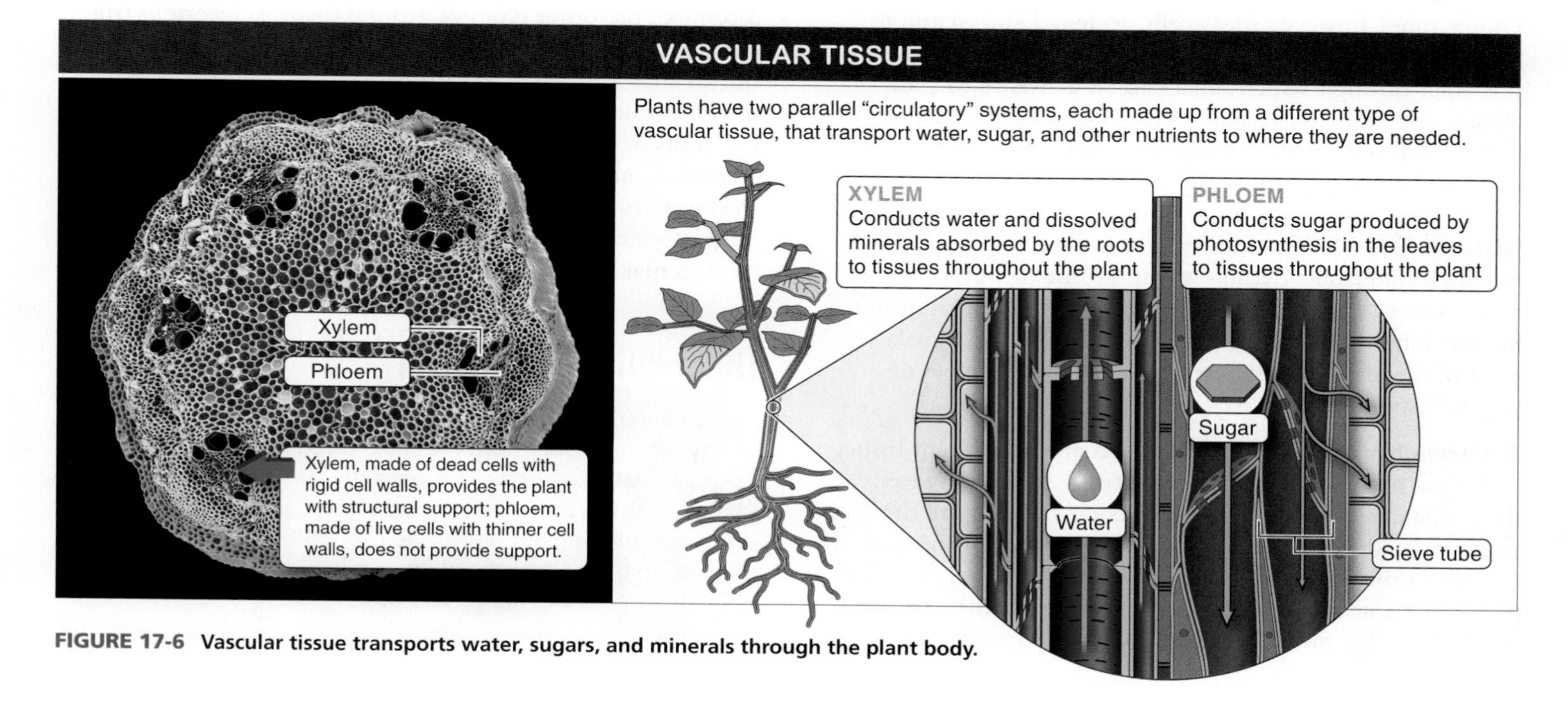

FIGURE 17-6 Vascular tissue transports water, sugars, and minerals through the plant body.

GROUND TISSUE CELLS

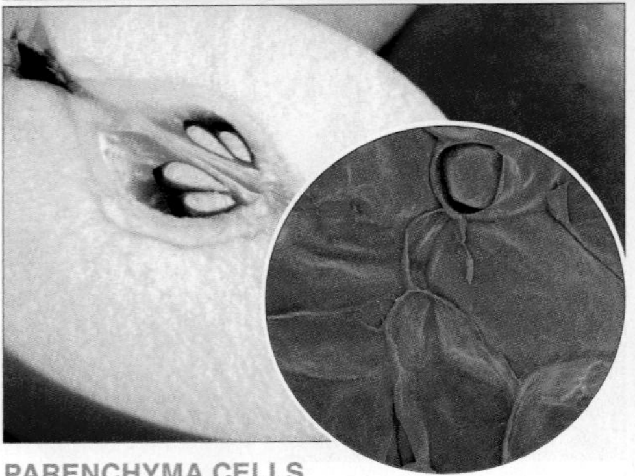

PARENCHYMA CELLS
- Make up the majority of plant tissue, including most of the soft, flexible tissue found in leaves, flowers, stems, roots, and fruits
- Responsible for photosynthesis, food storage, and the production and release of hormones

COLLENCHYMA CELLS
- Elongated, stringy cells with thickened cell walls
- Give the plant flexibility, enabling it to twist and bend

SCLERENCHYMA CELLS
- Not living when mature
- Have very thick cell walls, containing lignin, an important component of wood
- Function like the steel girders of a building, enabling plants to resist the force of gravity

FIGURE 17-7 Ground tissue makes up most of the plant body. Shown in the inset circles from left to right: parenchyma cells of an apple, collenchyma cells of a sunflower, and sclerenchyma cells of a sunflower.

their death, their cell walls become connected to one another to form the outer surface of a vessel that functions as a pipeline. Water then flows through the pipeline and reaches all the living cells in the plant through tiny holes in the cell walls, in much the same way that a gardener's soaker hose, with hundreds of small holes, waters a large area of grass or garden. The rigid cell walls that make up the pipeline also give the plant structural support, like the girders of a building.

The other circulatory system, the phloem, is built from cells that remain alive while functioning as a sort of "pipe." Unlike pipes, however, these cells, arranged end to end to form structures called **sieve tubes,** have openings in their side walls through which sugar can be delivered to cells outside the phloem. Because phloem cells have much thinner cell walls than xylem, they don't offer much structural support to the plant.

Ground Tissue The third type of tissue found in plants is called ground tissue. Because ground tissue includes everything that is neither the outer covering (dermal tissue) nor the inner vascular tissue (xylem and phloem), it makes up most of the plant body. There are three different types of ground tissue (**FIGURE 17-7**).

1. **Parenchyma cells** make up the majority of plant tissue, including most of the soft, flexible tissue found in leaves, flowers, stems, roots, and fruits. Parenchyma cells are the workhorses of the plant body, and they carry out most of the plant's metabolic activities. Depending on their location, parenchyma cells have the capacity to photosynthesize, store food molecules, produce ATP through cell respiration, or produce and release hormones. When you eat a carrot or a potato, the carbohydrate-packed cells you consume are primarily parenchyma cells that were storing energy for the plant to use at a later time. When a plant is damaged, it is parenchyma cells that divide to repair the injured tissue.

2. **Collenchyma cells** are elongated, stringy cells with thickened cell walls. This might at first seem like sloppy cell construction, but it is this construction that gives a plant great flexibility, enabling it to twist and bend. An example of collenchyma tissue is the strings on the outside of a celery stalk.

3. Unlike the other types of ground tissue, **sclerenchyma cells** are not living when they are mature. They have very thick cell walls containing **lignin,** a substance that is one of the chief chemical components of wood and that makes the cell walls in woody plants indigestible to nearly all organisms. Sclerenchyma cells are very strong and can function like the steel girders of a building, enabling plants to resist the force of gravity and grow very tall. Fibrous sclerenchyma cells are used to make cloth, rope, and paper.

TAKE-HOME MESSAGE 17•3

All vascular plants are organized around the same basic body plan and built from the same three types of tissue. Dermal tissue covers and protects the surface of the plant. Vascular tissue transports water, sugars, and minerals throughout the plant body. And ground tissue, which makes up the bulk of the plant, is where most of the plant's metabolic activities are carried out.

FIGURE 17-17 Monet was inspired by the large leaves of water lilies and painted them on a huge canvas. Water lilies produce some of the largest leaves of any plants.

leaves were more than a few layers thick, the cells near the bottom would not get enough energy to run the sugar-making machinery.

Monet's paintings of water lilies are gigantic (more than 6 feet high and 42 feet wide!), and that is fitting (**FIGURE 17-17**). Water lily leaves are among the biggest leaves, with individual leaves sometimes reaching almost 7 feet across (but still thin). At the other extreme are duckweed leaves. Twenty-five of these leaves laid end to end are less than an inch long. Why are there so many different kinds of leaves? In short, it's because, in nature, there are usually multiple ways to solve a problem. Different types of leaves all get the job of photosynthesis done, but in different ways.

Gymnosperms—seed plants, such as the conifers, that don't produce flowers—have a variety of leaf structures, including needle-like leaves. Think of pine trees. Among the flowering plants many monocots can generally be identified by their long, flat leaves that are oriented vertically, as in the grasses, in a pattern that keeps them from shading each other. Eudicot leaves come in two types: simple leaves, each with just a single blade, and compound leaves, with multiple leaflets that radiate outward from a single point. Each of these two types can vary greatly, from sharp-edged to smooth-edged. A sampling of the wide range of leaf shapes and styles is shown in Figure 17-16.

Perhaps the most extremely modified plant leaves are the hard, dry spines seen in most cacti. The spines don't contribute to photosynthesis—that occurs in the stems, which we see as the prominent green parts of the plant—but they do help adapt cacti for desert living. They slow the speed of the warm winds as they pass over the plant, for instance, reducing the water lost to evaporation. The spines also help moisture from the cooler night air condense, helping the plant capture a bit of the most limited desert resource. Spines go a long way toward discouraging birds and mammals from eating the plant, as well. In Chapter 19, we explore some of the chemicals that plants sequester in their leaves that also help protect them from being eaten.

TAKE-HOME MESSAGE 17•6

Leaves are thin and have a three-layered structure that enables them to effectively capture energy and transport water and nutrients. Leaves intercept sunlight, putting chloroplast-containing photosynthetic cells in the path of that light and converting the potential energy of the sun into the usable chemical energy of sugar. Leaves also have vascular tissue through which food is transported out of the leaf to the rest of the plant, and water and minerals are transported into the leaf.

17•7

Several structures help plants resist water loss.

If you're lying on a beach, working in a field, or walking in the mountains, the sun's heat can feel relentless. Humans tend not to be bothered by it too much, though. We can seek shade when the heat becomes too intense and drink water from a faucet whenever we're thirsty. Plants don't have these options. Consequently, they are perpetually at risk of losing too much water as it evaporates from their tissues into the air. Because water is essential to nearly every chemical reaction in every living organism, this water loss is not a small issue. Not surprisingly, plants have multiple adaptations that solve the "sun problem" and keep them from becoming dangerously dehydrated. We see three of these important adaptations right on the surfaces of most leaves (**FIGURE 17-18**).

1. *The cuticle.* Dermatologists have instructed people for decades that good skin care and protection starts with maintaining your skin's moisture. Plants would seem to agree. As we've seen, the top layer of most leaves, secreted

LEAF STRUCTURES THAT HELP REDUCE WATER LOSS

Because they lack mobility, plants are at risk of dangerous dehydration.

CUTICLE
This waxy, water-repelling substance keeps water inside the leaves from diffusing through the leaf and evaporating away into the atmosphere.

LEAF HAIRS
These tiny hairs can reflect sunlight and reduce the speed at which breezes move over the leaf's surface, decreasing water loss through evaporation.

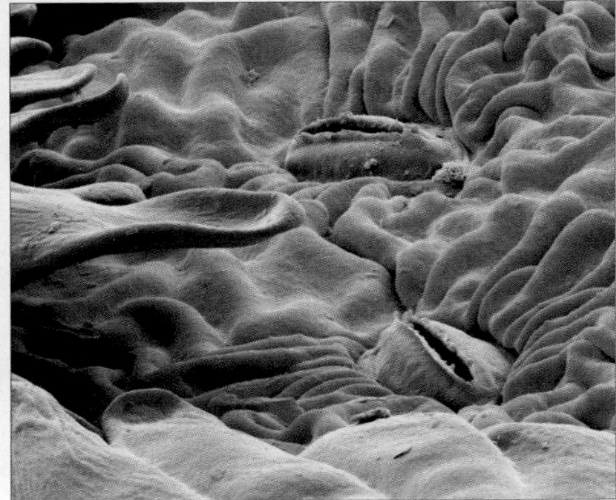

STOMATA
Guard cells can increase in size to seal off stomata or decrease in size to create a small opening through which carbon dioxide can pass and through which water can be lost.

FIGURE 17-18 Plants have multiple adaptations that keep them from becoming dangerously dehydrated.

by their epidermal cells, is the cuticle. This waxy, water-repelling substance keeps water inside a leaf from diffusing through the leaf and evaporating into the atmosphere. It seals in water almost completely. Carnauba wax, from the cuticle of Brazilian palm trees, is an ingredient in many sunscreen products for humans and in a popular car polish.

2. *Leaf hairs.* Some of the surface cells on leaves become modified as tiny hairs. These can reduce water loss by reflecting back some of the sunlight (thus reducing the temperature inside the leaf) and reducing the speed at which breezes move over the leaf's surface, taking water with them—through evaporation—as they go. Standing in the sun, who would experience more evaporative cooling, a person with shaved legs or someone with hairy legs?

3. *Stomata.* The cuticle would solve the problem of water loss in plants almost completely if it covered the entire leaf. Unfortunately, the cuticle is also impervious to CO_2, which is an essential ingredient for photosynthesis. And so, just as a castle needs a front door—which then becomes a major point of vulnerability to attack—leaves, too, must have openings to the environment. These openings, the stomata—as many as 10,000 per square centimeter—are on the underside of the leaves (and, to a lesser extent, on stems), and their design represents an important water-conservation adaptation of plants. A pair of guard cells (seen in Figure 17-5) surrounds each pore, one cell on either side. The guard cells can increase in size to seal off the opening or decrease in size to create a small opening through which CO_2 can enter and water can be lost. The opening and closing of the stomata is controlled by osmotic pressure in the guard cells and appears to occur in response to changes in temperature, humidity, light intensity, or CO_2 concentration.

Q How could smearing Vaseline on leaves cause a plant to "suffocate"?

Stomata are the chief sites where plants "breathe." If you were to do a cruel experiment and coat the underside of the leaves of some plants with Vaseline, what do you think would happen? With no way for air to enter, the plants would become starved for CO_2 and eventually the leaves would die.

TAKE-HOME MESSAGE 17•7

Plants have multiple adaptations that enable them to resist becoming dangerously dehydrated. These adaptations include the cuticle, leaf hairs, and stomata.

③ Plant nutrition: plants obtain sunlight and usable chemical elements from the environment.

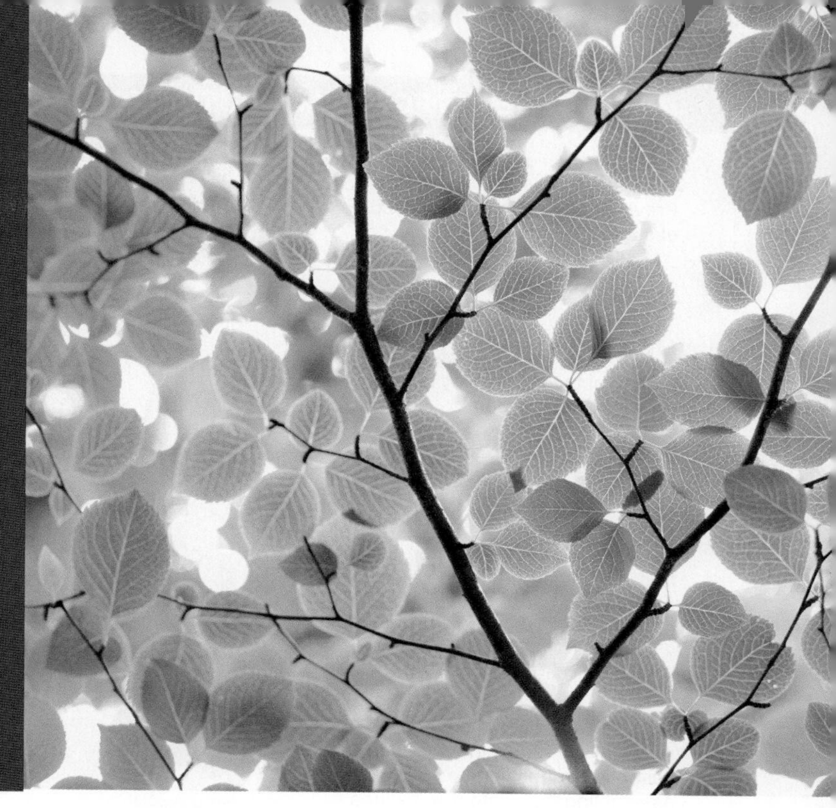

Leaves intercepting sunlight.

17•8

Four factors are necessary for plant growth.

Gardening is harder than it looks. Plants need sunlight, water, and air to thrive. That part is obvious. But it's not enough. Just as humans will not grow and remain healthy without a nutritious diet, plants will get sick, wither, and even die if they don't have a complete "diet" (**FIGURE 17-19**). Let's examine just what that means.

Plants generally require four things for proper nutrition. The first three are (1) sunlight for the energy to build molecules of sugar; (2) water; and (3) air as a source of carbon dioxide, from which molecules of carbon can be recovered and used in the construction of sugar. And (4), they usually need soil. But how does soil contribute to plant nutrition?

For a plant to grow, it must build new cells, which are then assembled into new tissues. These new cells must have cell walls, DNA, mitochondria, and all the other cell components. Aside from the carbon and oxygen obtained from CO_2 in the air and hydrogen from water, the plant must get all these raw materials elsewhere, usually from the soil. In all, there are 13 different minerals (each of which is an element) that plants require for proper survival and growth.

Six of these essential elements are required in relatively large amounts and are particularly important for producing proteins and the enzymes that catalyze all the reactions taking place in plants. These elements are:

nitrogen, for making proteins and DNA

phosphorus, for making ATP, DNA, and cell membranes

magnesium, for making chlorophyll

potassium, for making enzymes and controlling water balance

sulfur, for making proteins and vitamins

calcium, for a large number of different cell functions

Plants need seven additional elements in much smaller amounts: chlorine, iron, boron, manganese, zinc, copper, and molybdenum. How small are the amounts needed? A plant may get all the chlorine it needs just from being touched by a human with sweaty fingertips. Similarly, a tiny amount of one of these minerals may be present in the seed that the

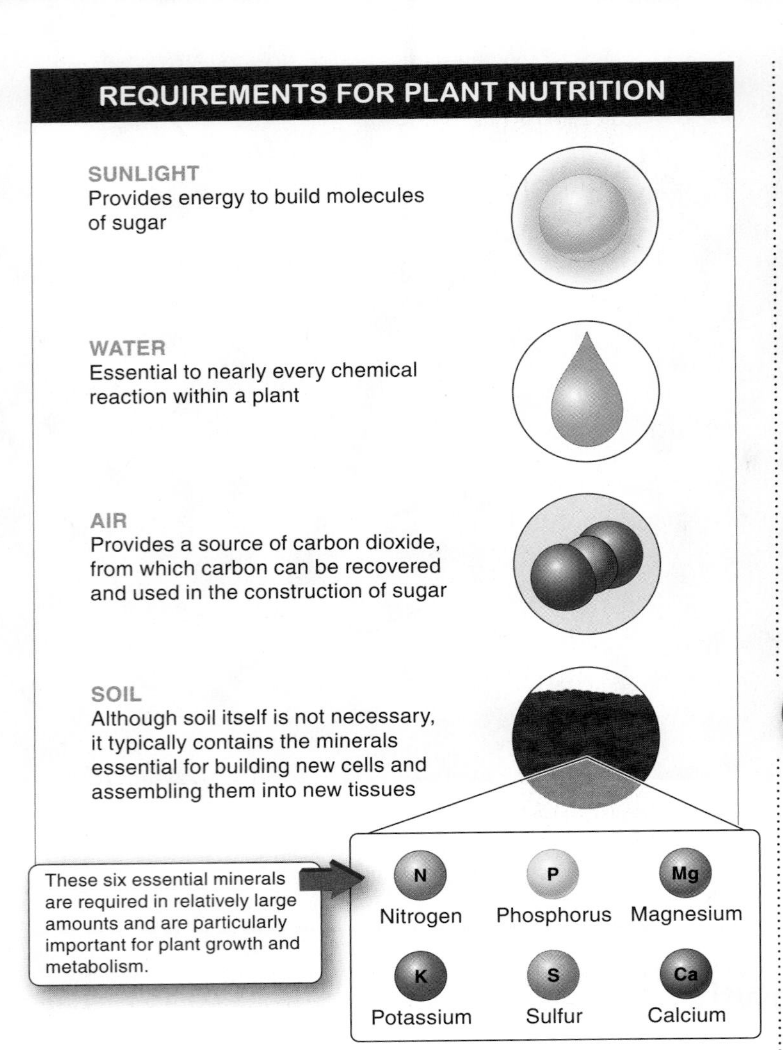

FIGURE 17-19 The components necessary for maintaining plant health.

plant grows from, and this amount may be enough to supply the plant for its entire life. In other cases, the plant must "reach out" to obtain these elements from the soil by growing roots that reach to where the elements are.

Q Can a plant grow without dirt? How?

Much of what we know about plant nutrition comes from **hydroponically grown plants** (**FIGURE 17-20**). These plants are grown without soil. Instead, they are grown in water that has chemicals added. By adding or withholding exact amounts of certain chemicals, it is possible to determine which are essential and in what amounts. These findings are important in helping farmers determine what types of fertilizers they need for their crops, as the farmers try to coax much more growth out of each acre of land than would occur naturally.

FIGURE 17-20 Growing plants hydroponically. Plants can grow just fine without soil, if the essential nutrients are added to their water.

TAKE-HOME MESSAGE 17•8

Plant growth is dependent on four important factors: (1) sunlight for energy, (2) water, (3) air as a source of carbon dioxide, and (4) minerals, usually obtained from soil.

17•9

Nutrients cycle from soil to organisms and back again.

Soil is an almost perpetual source of nutrients that are critical to a plant's health and survival. But why are the nutrients present in the soil? How do they get there? To answer these questions, we need to realize that soil is not just a homogeneous mound of dirt. Soil is a complex mixture of four distinct components (**FIGURE 17-21**): (1) minerals: inorganic particles, usually from the breakdown of rock; (2) organic materials: carbon-containing matter, usually from the decomposition of dead plants and animals; (3) air; and (4) water.

COMPOSITION OF SOIL

MINERALS: 50%
- Inorganic particles
- Formed from the breakdown of weathering rock
- Grouped into three sizes: sand, silt, and clay

WATER AND AIR: 45–50%
- Fill the space between particles of inorganic and organic matter

ORGANIC MATERIALS: 1–5%
- Carbon-containing matter
- Formed from the decomposition of dead plants and animals

Soil, a mixture of minerals, organic matter, air, and water, is a source of nutrients that are critical to a plant's health and survival.

FIGURE 17-21 The components of soil.

Minerals About half of the total volume of dirt is inorganic materials. These come primarily from the weathering of rocks, which decay physically and chemically over time, releasing minerals. The inorganic particles are grouped into three sizes: sand, silt, and clay. As you may know from feeling sand at the beach, there's plenty of air between the particles and sand doesn't hold water very well. That's why it is so easy to grab handfuls of it and let the grains fall through your fingers. Silt particles are smaller, as you would feel if you grabbed a handful of soil from the bottom of a river bed. Clay particles, the smallest of all, pack together very densely. Because minerals passing through the soil cling to clay, it is a valuable component of soil; clay holds the essential minerals in place for plants to absorb as needed. Too much densely packed clay in soil, however, can limit the supply of air available for plant roots to absorb, leading to poor plant growth.

The best soils have about equal amounts of each type of particle. The minerals present in soil determine the pH (see Section 2-7 for a pH refresher) of the soil and can affect how well plants grow, and can even influence the color of flowers (**FIGURE 17-22**).

FIGURE 17-22 Multicolored hydrangeas. Although all hydrangeas are colored by the same pigment, when the soil is acidic (pH = 5–5.5) more aluminum is available to the plant, causing the flowers to bloom in blue, and when the soil is less acidic (pH= 6-6.5) they bloom in pink.

> I bequeath myself to the dirt,
> to grow from the grass I love,
> If you want me again,
> look for me under your boot-soles.
>
> —Walt Whitman, *Leaves of Grass*

Organic Materials Leaves fall and animals die, but **decomposition** returns the chemicals inside them to the soil, making possible their re-uptake by plant roots. Animal droppings are very important to plants, too, serving as highly concentrated bits of nitrogen, perhaps the most important of the essential plant nutrients. Called **humus** (pronounced HYOO-muhss), all of these organic decay products make up about 1% to 5% of the soil (**FIGURE 17-23**). The humus also absorbs and releases water easily and, in the process, releases many nutrients. With too little humus, the soil may be deficient in nutrients. With too much, it may retain too much water.

Most of the decomposition of organic matter is carried out by huge populations of bacteria, fungi, insects, and earthworms that live in soil. Aristotle called earthworms "the intestines of the earth." Passing soil through their guts as they feed on organic matter, earthworms increase tremendously the amount of nitrogen available for plants.

Q What is composting? Why is it useful? Is it dangerous?

A very practical application of the decomposition of organic matter can be seen in the process of **composting,** in which bacteria are utilized to process organic wastes—such as from kitchen garbage, manure, yard clippings, and even sewage—metabolizing the material and in the process converting it into fertilizer (**FIGURE 17-24**). Compost heaps can speed the process of organic breakdown and greatly reduce the volume of waste that cities must process and dispose of, while generating a product that can be distributed to help return important plant nutrients to the soil.

DECOMPOSITION OF ORGANIC MATTER

The decomposition of organic matter is facilitated by huge populations of bacteria, fungi, insects, and earthworms that live in the soil.

BEFORE
Decaying organic matter

AFTER
Humus

FIGURE 17-23 Decomposition returns the chemicals inside animal and plant remains to the soil.

FIGURE 17-24 The practice of composting converts organic waste into useful fertilizer.

It might seem unhygienic to cultivate large populations of bacteria and other organisms that are present in garbage, manure, and sewage. The growing populations of bacteria in compost, however, generate large amounts of heat—a compost heap can reach temperatures above 60° C (140° F) which is why so much steam is coming out of the overturned earth pictured in Figure 17-24—and this can kill many of the original organisms, which are replaced by more heat-tolerant species that are not harmful to humans.

Water and Air Water and air, the final components of soil, fill the spaces between the particles of inorganic and organic matter and account for about half of the total volume of soil.

TAKE-HOME MESSAGE 17•9

Soil is a mixture of minerals, organic materials, air, and water that serves as an almost perpetual source of nutrients critical to a plant's health and survival.

17 • 10

Plants acquire essential nitrogen with the help of bacteria.

There is a sad irony to floating in a lifeboat in the ocean. Despite being surrounded by water, a person is likely to die from lack of water, because salt water just isn't drinkable for a human. Plants encounter a similarly ironic situation when it comes to one of the most important components in their diet: nitrogen (**FIGURE 17-25**).

Q Burying dead fish in their fields can help farmers produce more food. Why?

Nitrogen is the most common element in our atmosphere. Almost 80% of the air around us is nitrogen. But despite its abundance, lack of nitrogen is the factor, other than lack of water, that most limits plant growth. Nearly every plant outside the tropics would grow more if it were given more nitrogen. Early farmers recognized this and would bury dead fish along with their corn crops. As the fish decomposed, the rotting organic material increased the nitrogen in the soil that was available for the plants' use, significantly improving the size and health of the corn plants. Today, farmers, gardeners, and people who want a nice green lawn regularly treat their crops, flowers, and grass to nitrogen-rich fertilizers.

Adding nitrogen to the soil in which a plant lives commonly increases the plant's growth, because every protein the plant builds contains nitrogen, every bit of DNA the plant synthesizes contains nitrogen, and every molecule of chlorophyll it creates contains nitrogen. In order to grow and survive, plants need nitrogen. If nitrogen is so plentiful in the air, why can't plants make use of it by absorbing it directly?

Q Are all bacterial infections bad for plants?

The problem is that nitrogen in the atmosphere exists as a molecule consisting of two nitrogen atoms (N_2) bound very tightly and stably together. The nitrogen becomes usable for plants and animals only when it is converted to a molecule with a single nitrogen atom, such as ammonium (NH_4^+ or nitrate (NO_3^-). Breaking apart the strong, stable bond between the two nitrogen atoms in the nitrogen molecule is no easy task, however, and plant evolution did not produce the metabolic know-how to do it. Bacteria, on the other hand, are metabolic wizards, and several species can break nitrogen molecules apart, using an enzyme they produce called nitrogenase, and get the nitrogen into a plant-usable form—a process called **nitrogen fixation.** A mutually beneficial relationship has evolved that enables plants to gain access to the nitrogen fixed by these bacteria.

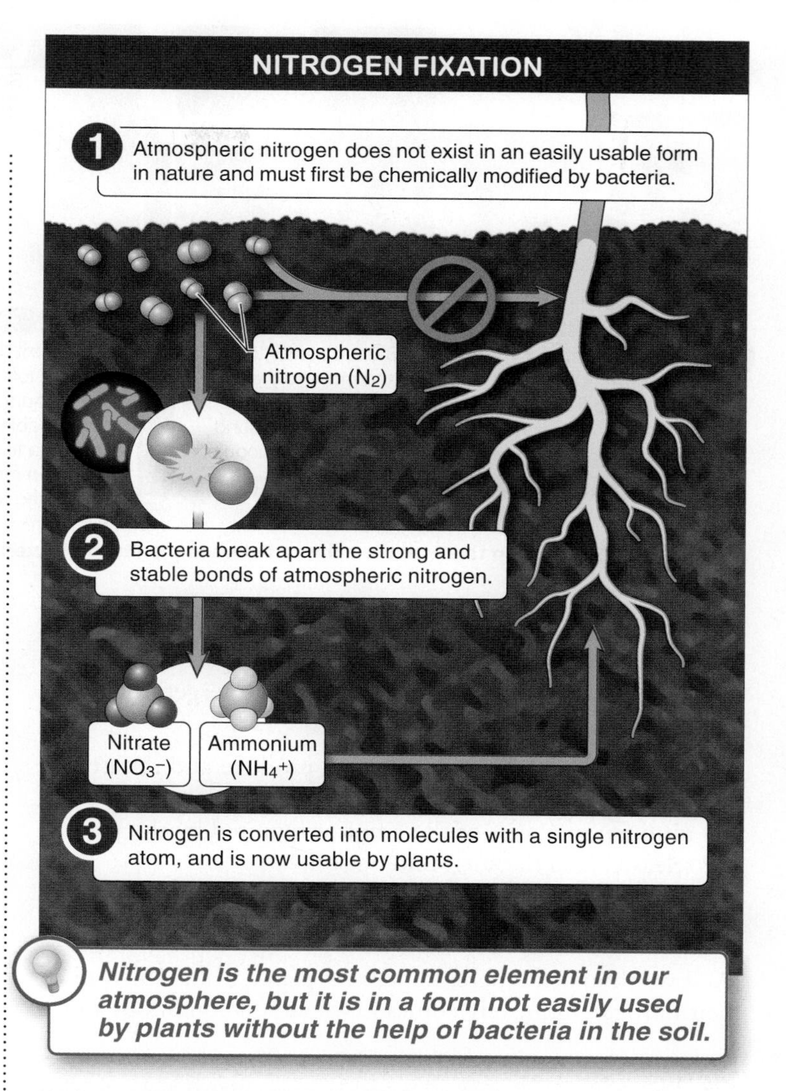

FIGURE 17-25 Overview of nitrogen fixation: the conversion of nitrogen in the air into a form that plants can use. Some bacteria can "fix" nitrogen into this usable form.

The alliance between plants and nitrogen-fixing bacteria develops in a few simple steps (**FIGURE 17-26**).

1. A plant secretes a bacteria-attracting compound from its roots.
2. Bacteria enter the roots and move to the inner cells, where the bacteria multiply, forming a big, round lump called a **nodule.**

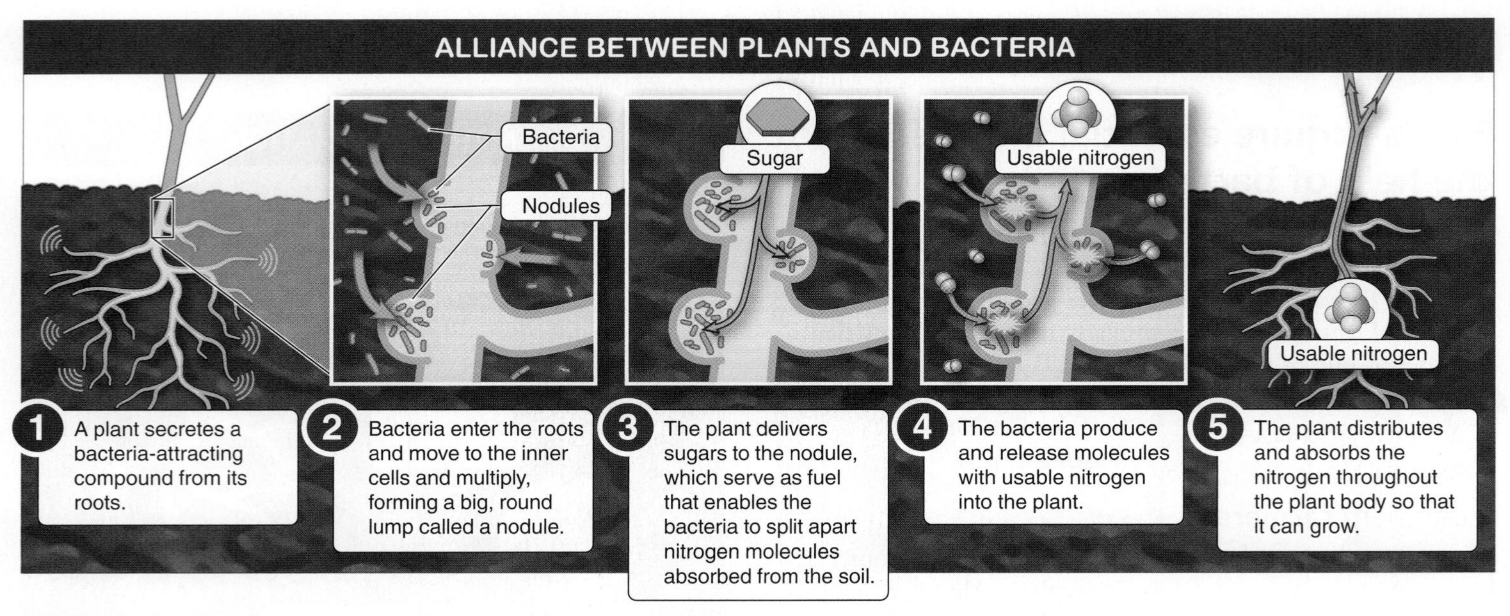

FIGURE 17-26 Nitrogen fixation: the alliance between plants and bacteria.

3. The plant delivers sugars to the nodule, which serve as fuel that enables the bacteria to split apart nitrogen molecules absorbed from the soil—an energetically expensive process.
4. The bacteria use the nitrogen to produce and release molecules with usable nitrogen into the plant. Any nitrogen in excess of the amount required by the plant is released into the surrounding soil.
5. The plant distributes and absorbs the nitrogen throughout its body so that it can grow.

(And we should add here, animals eat plants, from which all of their nitrogen ultimately comes.) Not all soil conditions are conducive to the growth of nitrogen-fixing bacteria, and some plants have alternative ways to acquire nitrogen. Insect-eating (insectivorous) plants, for example, generally grow in very acidic soils in which nitrogen-fixing bacteria do not thrive. The Venus fly-trap (**FIGURE 17-27**) captures insects rich in nitrogen-containing molecules, digests them, and recovers the nitrogen for its own uses.

FIGURE 17-27 Another way to acquire nitrogen: the Venus fly-trap. Some plants have ways of harvesting nitrogen from insects.

Some plant species are not able to form the mutually beneficial alliance with nitrogen-fixing bacteria and instead must scavenge for usable nitrogen in the soil. For this reason, it can be productive for farmers to rotate their crops every few seasons, alternating between plants that have nodules of nitrogen-fixing bacteria (such as peas, soybeans, and alfalfa) and those that don't. As excess nitrogen fixed by the bacteria is released to the soil, plants such as alfalfa have the effect of replenishing usable nitrogen in the soil, like natural fertilizers. Also, because these plants are plowed under prior to rotation of the crops, they decay and add nitrogenous compounds to the soil—a process called "green manure"—which can reduce the need for additional fertilizer (and the costs to the farmers). The soil is now ready for a non-nitrogen-fixing crop.

TAKE-HOME MESSAGE 17•10

Among the minerals, nitrogen is the one that most commonly limits plant growth, because it is required in nearly all the cells and tissues produced by plants but does not exist in an easily usable form in nature. A mutually beneficial relationship has evolved that enables plants to gain access to nitrogen that is "fixed"—that is, chemically modified into a usable form—by bacteria.

④ Plants transport water, sugar, and minerals through vascular tissue.

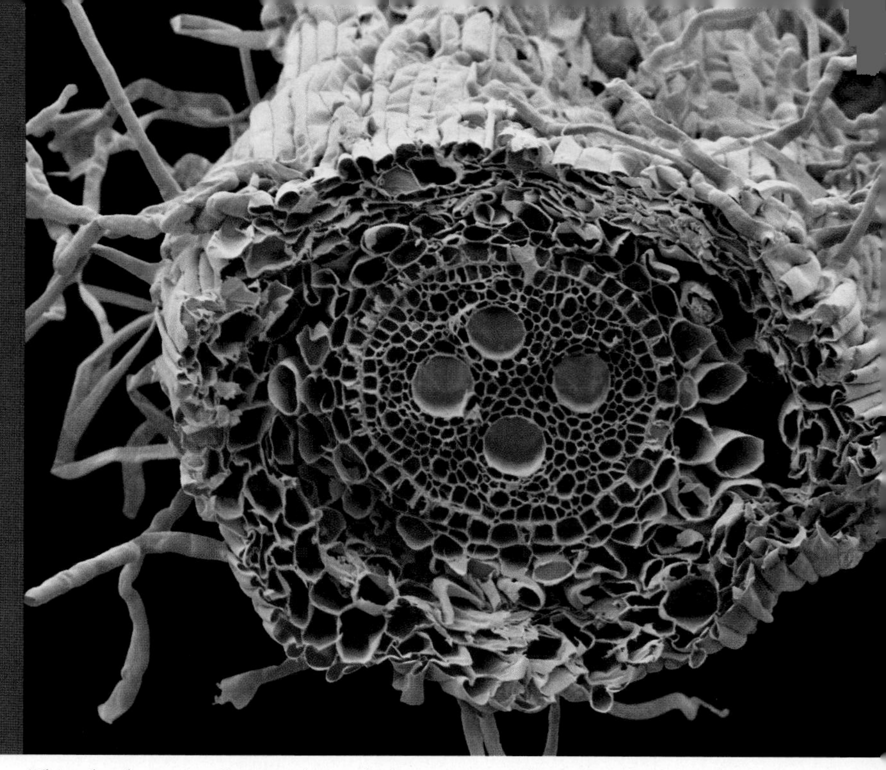

This colored scanning electron micrograph shows a section of a eudicot root. Note the four green circles at the center are xylem, which are surrounded by phloem, stained blue.

17•11

Plants take up water and minerals through their roots.

Plants absorb water and minerals from the soil through their roots. The process of water absorption relies on osmosis and is straightforward and simple: the cell membranes of the root hair cells are permeable to water, and as long as the root reaches water in the soil that has a lower concentration of dissolved minerals than the fluid inside the cells, water will move into the root.

The absorption of minerals requires an intervening step (**FIGURE 17-28**). Because most minerals are chemically charged, they cannot pass through the cell membranes unaided. For this reason, plants can take them up from the soil only with the help of transport proteins embedded in their root cell membranes. If the minerals are at higher concentrations outside the root cell than inside, the minerals can move across the membranes by simple facilitated diffusion (a process that does not require the plant cells to expend any energy). If the mineral is at higher concentration inside the cell, it can still be taken up by the plant, but only through active transport, in which the plant expends energy to move the mineral into the root cell against its concentration gradient.

The root hair cells are permeable to water. The absorption of minerals requires an intervening step.

When it comes to acquiring water and minerals, plants again benefit from collaboration with other organisms, in this case

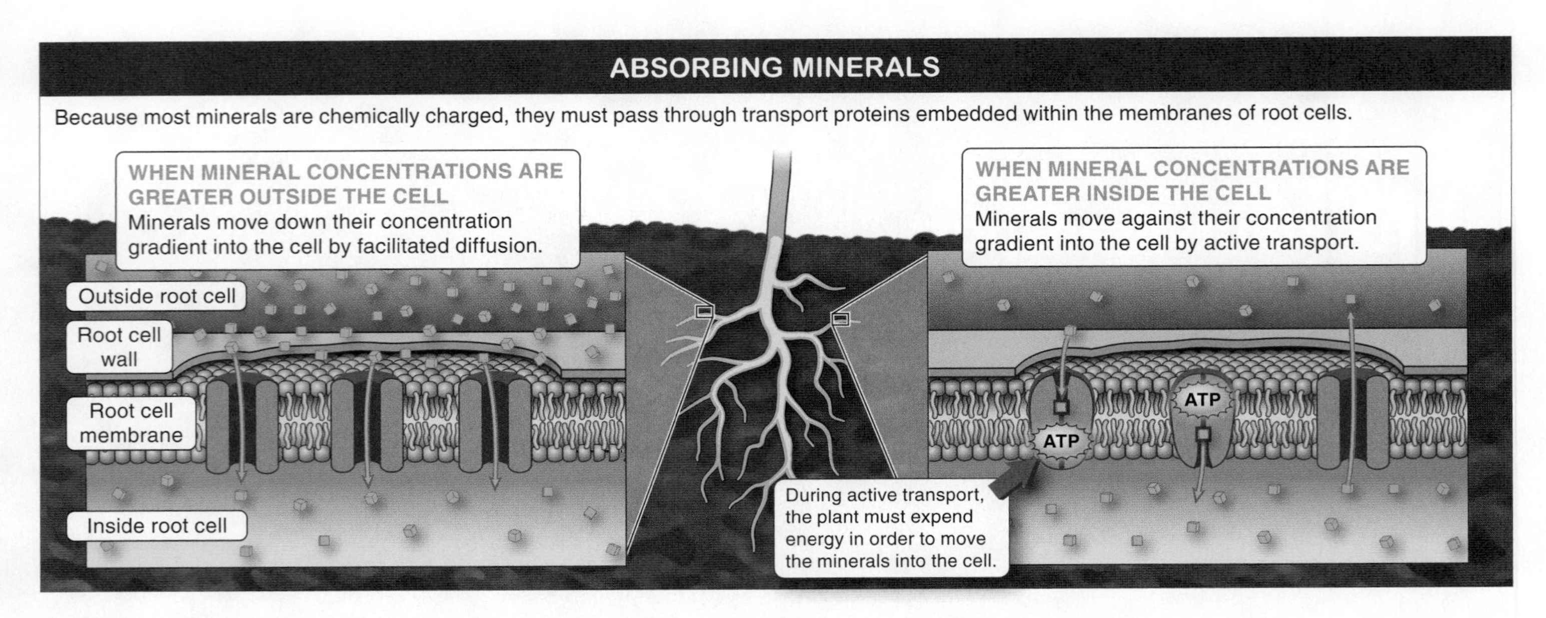

FIGURE 17-28 Taking up minerals into the plant. The absorption of minerals through the cell membranes of plant roots requires the aid of transport proteins, and sometimes energy.

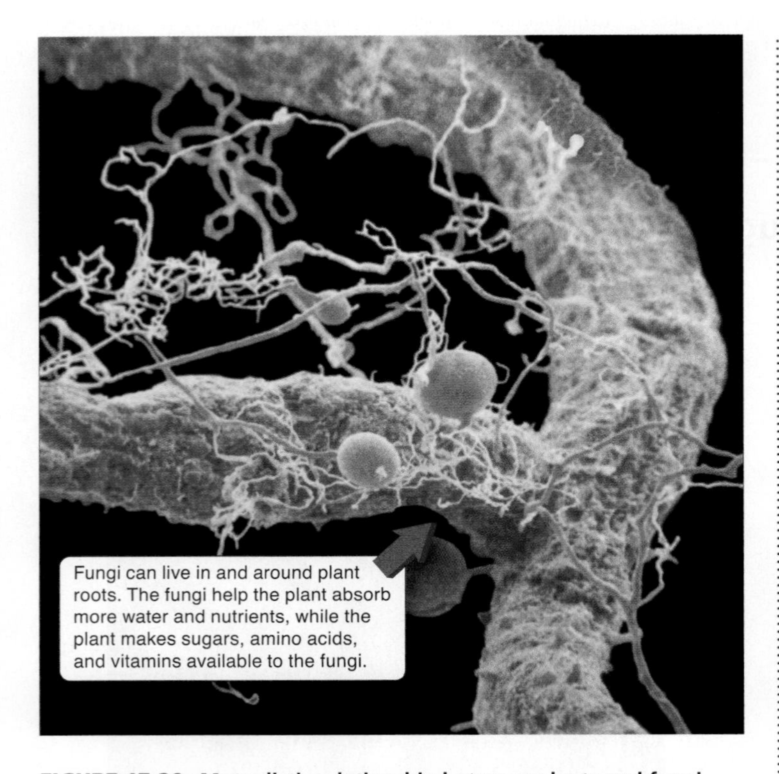

FIGURE 17-29 Mutualistic relationship between plants and fungi.

fungi (**FIGURE 17-29**). Most plants have fungi growing all around and even within their roots, forming mutualistic associations known as **mycorrhizae** (from the Greek for "fungus roots"). Tiny, thread-like fungi trap water like a sponge and hold it around the roots. The fungi also have a huge surface area, which dramatically increases the amount of water and minerals that can be absorbed. The fungi don't do all that work for free, of course. In exchange for providing the increased water and minerals to the plant, they receive sugars, amino acids, and vitamins in exchange. The association between plants and fungi is so important to plants that many trees can't grow at all if they don't have fungi growing around their roots.

TAKE-HOME MESSAGE 17•11

Plants absorb water from the soil through osmosis occurring in their root hairs. Absorption of minerals also occurs in the roots, but this requires the help of transport proteins in root cell membranes. In a mutualistic association called mycorrhizae, fungi growing into and around plant roots increase the water and mineral absorption for the plant, while gaining access to energy and nutrients from the plant.

17•12

Water and minerals are distributed through the xylem.

Evaporation is relentless. Leave a glass of water on your desk for a day or two and you can watch the water level gradually drop. Because there is less water in the dry air than in the glass, the liquid water continuously turns to vapor, becoming a gas. Eventually, all of the water will evaporate away. You can refill the glass, but the water will continue to disappear. If you pour the water onto a plate so that it spreads out, it will evaporate even faster because more of the water is in direct contact with the air.

Q How does water get to leaves at the top of a giant tree?

Evaporation steals the moisture from leaves, too. And although transpiration can cool leaves and help move water and ions up a plant, this perpetual water loss to the atmosphere is a real problem for plants, especially trees. In one day, a big tree in a North American forest can lose 100 gallons of water or more! That is 100 gallons of water that the tree's roots absorbed but the plant does not get to use for its physiological needs. Water loss through evaporation also presents a physical challenge for the plant. Most water is lost from the leaves, high in the tree, but all of the water comes from the ground. Somehow, water must be constantly delivered to the leaves—dozens of gallons or more every hour. Not only is this a large amount of water that must be absorbed by the roots, it is also a very heavy load to lift hundreds of feet into the air (**FIGURE 17-30**). How do trees manage to get the water up there?

Water transport is the job of the xylem. As we saw earlier in the chapter, the xylem is one of the two circulatory systems in plants. It is responsible for directing the flow of water and dissolved minerals from the roots to all of the plant's cells.

> "To believe that columns of water should hang in [the xylem] . . . and should . . . transmit downwards the pull exerted on them . . . by the transpiring leaves, is to some of us equivalent to believing in ropes of sand."
>
> —Francis Darwin, botanist (Charles Darwin's son), ca. 1900. (His disbelief in Henry Dixon's 1896 cohesion-tension theory of water transport in plants is understandable, but wrong.)

FIGURE 17-30 Trees can lose significant amounts of water to the atmosphere through evaporation.

Circulation in plants differs in one huge way from circulation in animals: plants have no heart. In fact, they have no pump at all.

In the absence of a pump, there are two possible explanations for how water can be transported from the roots to the leaves—which might be 200 or 300 feet off the ground. The first possibility is that the column of water is pushed up by pressure from below. This would require the water in the soil and roots to be under very high pressure, which it isn't. (Although water in the roots can be under some pressure, it is not enough pressure to push water all the way up to the tops of trees.) The alternative explanation is that water is *pulled* up to the highest leaves. Although this seems improbable, it turns out to be true.

There are three important components to the process by which water is moved throughout a plant (**FIGURE 17-31**).

1. *Evaporation.* Because of low water concentration in the air relative to that in the leaf, molecules of water, one by one, are vaporized and lost from the leaf.

WATER TRANSPORT

Plants use a cohesion-tension mechanism to transport water and dissolved minerals from the roots and circulate them throughout the plant.

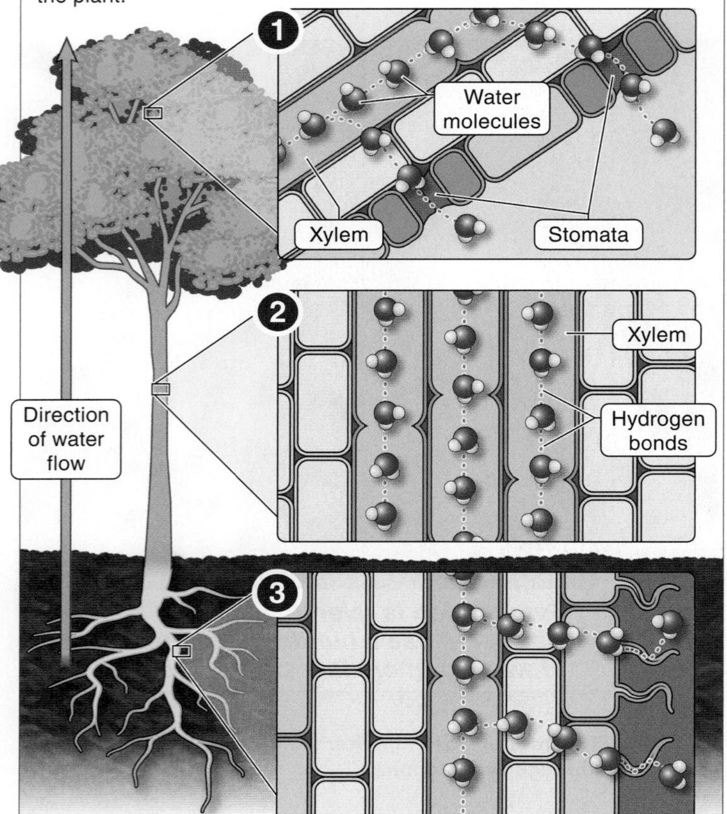

1 EVAPORATION
Due to low water concentration in the air relative to in the leaf, molecules of water are vaporized, one by one.

2 TENSION
Water molecules form hydrogen bonds with one another and these bonds cause the molecules to stick together. So as one molecule evaporates, it creates a tension, pulling on all the other water molecules that are stuck to it.

3 COHESION
The cohesion or stickiness of the water molecules links them together all the way down to the roots of the tree. As one molecule evaporates and pulls up the molecule next to it, that molecule pulls up the molecule next to it, and so on, all the way down to the roots.

Moving heavy fluid—water and nutrients—around a plant could be energetically expensive. However, the cohesion of water molecules causes them to be "pulled" through the plant, from the roots to the leaves, as evaporation occurs.

FIGURE 17-31 The cohesion-tension mechanism of fluid movement in the xylem.

2. *Tension.* Recall from Chapter 2 that water molecules form hydrogen bonds with one another and that these bonds cause the molecules to stick together. So as one molecule evaporates, it creates a tension, pulling on all the other water molecules that are stuck to it.

3. *Cohesion.* The cohesion, or stickiness, of the water molecules links them together all the way down to the roots of the tree. As one molecule evaporates and pulls up the molecule next to it, that molecule pulls up the molecule next to it, and so on, all the way down to the roots.

The beauty of this system—called the **cohesion-tension mechanism**—is that the plant does not need to expend any energy to pump water and minerals up from the roots to all the cells in every branch and leaf on the tree. Just as a mechanism evolved to take advantage of nitrogen-fixing bacteria to produce a usable source of nitrogen for plants, so, too, an efficient way evolved to avoid the energetic expense of moving heavy fluid around in the plant body. And so even though losing water to evaporation is generally bad for plants, the fact that they use the process to move water through the body turns it into a positive effect.

Q Is there a limit to how tall a tree can grow? Why?

While this system works well, it does have its downside: it places a limit on how tall a tree can grow. Leaves at the top of very tall trees have trouble getting enough water. They must "pull" it up against the ever-increasing force of gravity pulling downward on the long column of water in the xylem. And while the collective strength of the hydrogen bonds causing all the water molecules to stick to each other is impressive, there is a limit to that strength. As a consequence, the leaves at the top of the tallest trees resemble desert leaves: they are small, grow slowly, photosynthesize only at a slow rate, and otherwise behave like a parched leaf. Researchers estimate that the physical constraint of lifting water against the force of gravity will restrict any tree to a maximum height of about 400 feet.

Q Maple syrup comes from trees. What is it?

The sap in the xylem doesn't always contain just water and dissolved minerals. Sometimes the fluid contains sugars, too. (Recall that phloem is generally the sugar-transporting pipeline.) Humans have long recognized this and have devised ways to siphon off a bit of the sap from sugar maple trees (*Acer saccharum*) to create maple syrup. Throughout the late summer and early fall, just

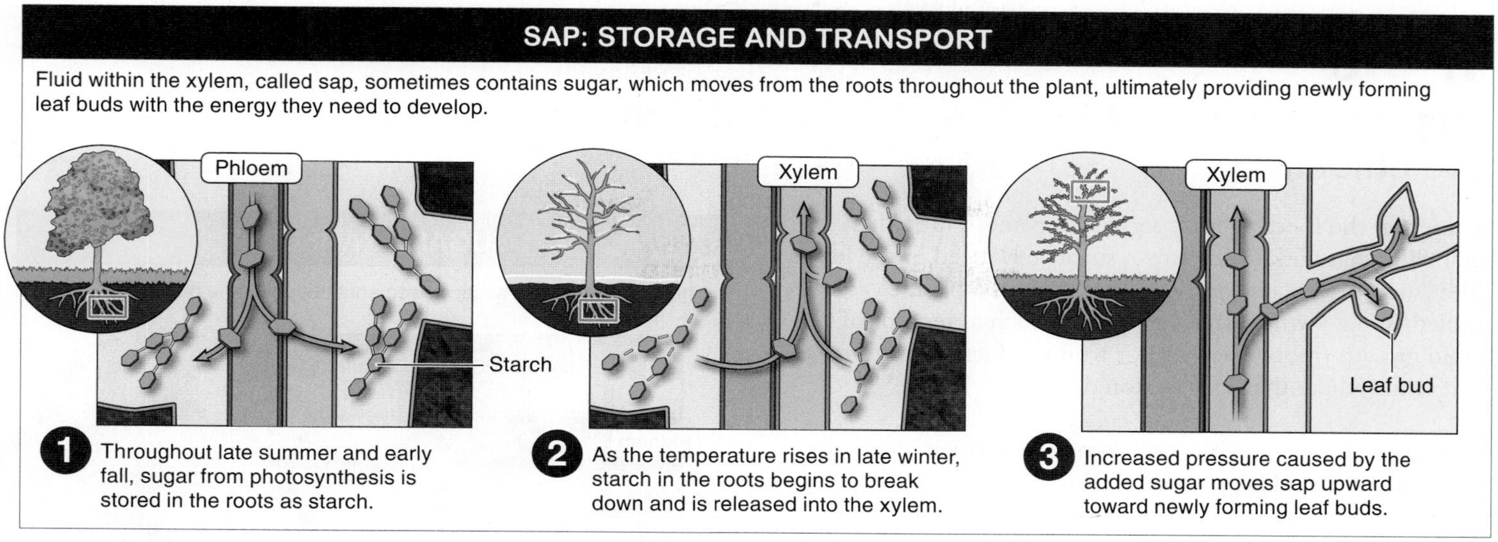

SAP: STORAGE AND TRANSPORT

Fluid within the xylem, called sap, sometimes contains sugar, which moves from the roots throughout the plant, ultimately providing newly forming leaf buds with the energy they need to develop.

Phloem

Starch

Xylem

Xylem

Leaf bud

1 Throughout late summer and early fall, sugar from photosynthesis is stored in the roots as starch.

2 As the temperature rises in late winter, starch in the roots begins to break down and is released into the xylem.

3 Increased pressure caused by the added sugar moves sap upward toward newly forming leaf buds.

FIGURE 17-32 In late winter and early spring, sugar-carrying sap moves upward from the plant roots to nourish the forming leaf buds.

before the trees lose their leaves, they begin storing the sugar from photosynthesis as starch in their roots, much like a bear storing body fat prior to hibernation. As the temperature rises in late winter, the starch in the roots begins to break down into sugar and is released into the xylem (**FIGURE 17-32**). The increased osmotic pressure caused by all the added sugar moves the sap upward toward the newly forming leaf buds, which will need energy to develop. By drilling into the xylem and inserting a small tube, it is possible to collect some of the sugary sap—it simply drips out. At about 1% to 5% sugar, the sap is not sweet enough to be poured on your pancakes, so it is boiled to evaporate much of the water, concentrating the sugar. When the sap reaches about 85% sugar, it is a much tastier maple syrup. About 1 gallon of sap can be collected from a tree over the several weeks during which it flows, not enough to harm the tree. But it takes about 50 gallons of sap to make 1 gallon of syrup (**FIGURE 17-33**).

FIGURE 17-33 Tapping a maple tree.

TAKE-HOME MESSAGE 17•12

Xylem directs the flow of water and dissolved minerals from the roots to all the tissues of the plant. The force driving this flow of fluid (the xylem sap) comes from evaporation of water from the leaves, which pulls water up from the roots.

17•13

Sugar and other nutrients are distributed through the phloem.

Phloem is the food-delivery service of plants. Through a branching network of highways, sugars produced in the leaves (the "source") are transported throughout the plant to places (called "sinks") where they are needed, such as regions of rapid growth (roots, leaf buds) or fruit production, or to roots and stems where they can be stored.

When a leaf is young, it cannot produce adequate food for itself; instead, food is provided to the leaf through the phloem. Once the leaf is mature and making more food than it can use, the transport direction is switched to move sugars out of the leaf to other places in the plant where they are needed. So, unlike mammalian circulatory systems, the phloem isn't necessarily a system of one-way highways. Rather, the direction of flow depends on the plant's needs at the time.

Phloem vessels are visible to the naked eye. Along with the xylem vessels, they make up the veins in a leaf. The phloem vessels are living cells lined up end to end so that they form a pipeline, called a sieve tube. The cells of the sieve tubes have many small openings in their walls. The sieve tubes can quickly move the sugar produced in the leaves throughout the plant: to roots, stems, buds, flowers, and fruits. The process is called the **pressure-flow mechanism.**

Phloem controls sugar movement throughout the plant in five steps (**FIGURE 17-34**).

1. Sugar is loaded by active transport into the phloem from sites of production (the source), primarily leaves.
2. The increased sugar concentration in the phloem immediately causes water to move from the xylem into the phloem by osmosis.
3. As the water moves in, it increases the fluid pressure inside the phloem.
4. The increased phloem pressure (particularly higher up in the plant, where photosynthesis is occurring) causes the fluid in the phloem to move elsewhere in the plant, such as to the roots—much like a tube of toothpaste being squeezed.
5. As the sugar is pushed through the plant body, it is moved out of the phloem by active transport at various locations where it is needed (the sinks)—such as into root cells for storage.

SUGAR TRANSPORT

The phloem controls sugar movement throughout the plant in five steps.

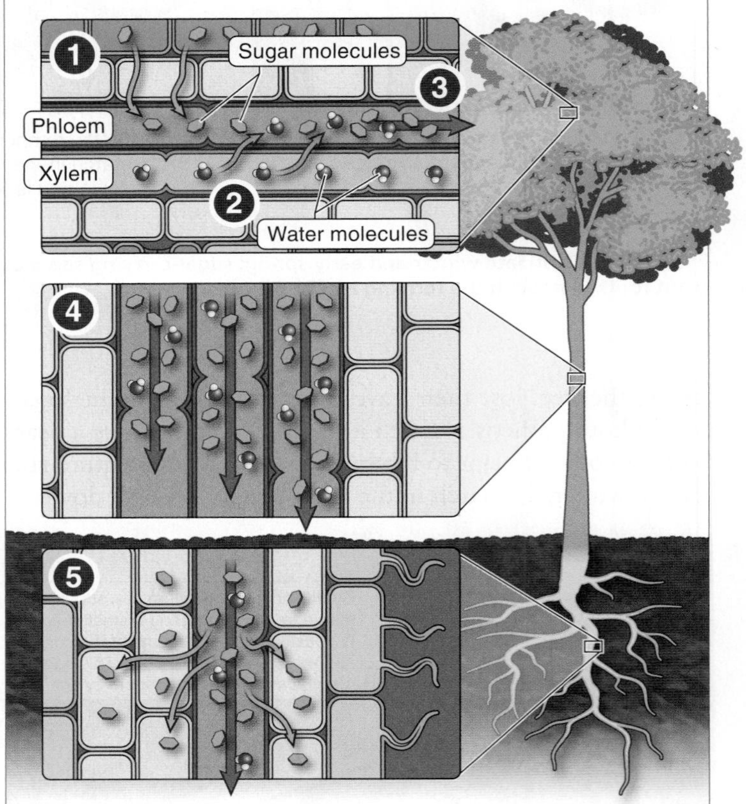

1. Sugar is loaded by active transport into the phloem from sites of production (primarily leaves).
2. The increased sugar concentration in the phloem immediately causes water to move from the xylem into the phloem by osmosis.
3. As the water moves into the phloem, it increases the fluid pressure inside the phloem.
4. The increased phloem pressure causes the fluid in the phloem to move elsewhere in the plant, such as to the roots, much like a tube of toothpaste being squeezed.
5. As the sugar is pushed through the plant body, it is moved out of the phloem by active transport at various locations where it is needed—such as into root cells for storage.

FIGURE 17-34 The pressure-flow mechanism of fluid movement in the phloem.

FIGURE 17-35 Drinking from a fire hose on a small scale. The sugary fluid in the phloem is under such high pressure that aphids drinking it can't always consume it all, and some is forced through their digestive system and out of their anus (at which point, ants may consume the fluid).

Sometimes, important science questions can be answered by unexpected methods. For instance, consider the question: what exactly is the fluid circulating in the phloem? To answer this, researchers used aphids. Aphids are insects that use their mouthparts to pierce the outer part of a plant stem so as to gain access to the sugar-rich sap in the phloem. The fluid simply flows into their gut. Just as mosquitoes feed on human blood, aphids are able to live on the energetically valuable fluid circulating in the phloem (**FIGURE 17-35**).

The researchers found plants on which aphids were feeding, then ripped the bodies off the aphids but left their mouthparts sticking into the phloem, like tiny syringes. As the sap dripped out, the researchers collected and analyzed it. As it turns out, phloem sap is a sugar solution with about 10% to 25% dissolved solids that include some lipids and amino acids. In most plants, about 90% of the phloem sugar is sucrose, the same molecule we use as table sugar.

Aphids helped answer an important question about plants, but they also create a huge number of problems for plants. If you have ever wondered why your car gets covered in a sticky substance under some trees, you can blame aphids. When aphids are attacking a tree, puncturing holes in the plant tissues to feed on the phloem sap, the pressure in the phloem fluid can cause more of the fluid to come out than the aphids can utilize. The excess sugary (and sticky) fluid secreted by the aphids, called honeydew, drops onto the ground (or your car) below.

In the next chapter, we see that as a tree grows, phloem also takes on a secondary, protective function.

TAKE-HOME MESSAGE 17•13

The phloem consists of a branching network of vessels made from living cells lined end to end to form sieve tubes, with small openings in their side walls. Sugar, usually sucrose, is moved in the phloem from sites of production (sources) to sites of use or storage (sinks). The direction of flow is not always the same: a plant part may be a source at one time and a sink at another.

Knowledge You Can Use

Organic foods: are they worth the price?

Q: What are "organic" foods? Organic foods account for about 1% of all foods sold worldwide, generating more than $40 billion per year. In order to be labeled and sold as "organic," a food must be grown without the use of chemical fertilizers or non-organic pesticides. They also cannot have been genetically modified or irradiated, or had hormones added during processing.

Q: Are they nutritionally superior to conventionally produced foods? In a 2009 review of 55 published articles on the nutritional content of organic foods, researchers funded by the U.K. Food Standards Agency examined 13 nutrient categories and found no evidence that organic foods were superior in nutritional content when compared with conventionally produced foods.

Q: Might organic foods have some other benefits? While the results of the review study seem to suggest that it might not be worth paying a premium for organic foods, there is reason to withhold final judgment. The study examined only *nutritional* content. There may be other benefits such as improved taste, enhanced safety, and reduced environmental degradation. However, no systematic study of these benefits has been reported.

1 Three basic tissue types give rise to diverse plant characteristics.

Plants are an extremely diverse and successful group of organisms. The flowering plants are divided into two major groups, the monocots and the eudicots, based on structural features of their seeds, leaves, stems, flowers, and roots. All vascular plants are organized around the same basic body plan and built up from the same three types of tissues. Dermal tissue covers and protects the surface of the plant. Vascular tissue transports water and nutrients throughout the body. And ground tissue is where most of the plant's metabolic activities are carried out.

2 Most plants have common structural features.

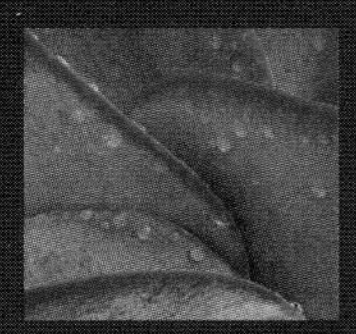

Plants' body plans are organized around three structures. Roots absorb water and minerals from the soil and anchor the plant in the soil. Stems provide structural support and position leaves where they can intercept sunlight. Leaves are the chief sites of photosynthesis.

3 Plant nutrition: plants obtain sunlight and usable chemical elements from the environment.

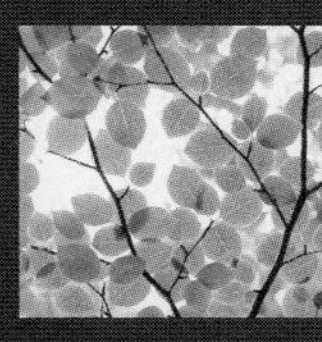

There are four factors essential for plant growth: (1) sunlight, (2) water, (3) air as a source of carbon dioxide, and (4) minerals, usually obtained from soil. Among the minerals, nitrogen is the chemical that most limits plant growth, because it is required in nearly all plant cells and tissues but does not exist in an easily usable form in nature.

4 Plants transport water, sugar, and minerals through vascular tissue.

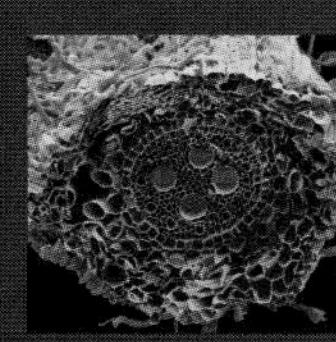

Nutrient transport in plants occurs in two separate systems. Xylem directs the flow of water and dissolved minerals from the roots to all other plant tissues. The force driving the flow comes from evaporation of water from leaves, which pulls water up through the xylem from the roots. Phloem is the food-delivery service of plants. Sugars produced in the leaves are transported through a branching network of vessels to sites of growth, reproduction, or storage throughout the plant.

KEY TERMS

Check Your Knowledge

1. What are the embryonic precursors to leaves in a seedling called?
   a) seedling leaves
   b) cotyledons
   c) angiosperm
   d) monocots
   e) needles

2. All of the following are ground tissue cell types except:
   a) parenchyma.
   b) meristem.
   c) collenchyma.
   d) sclerenchyma.
   e) All of the above are components of ground tissue.

3. What role does soil play in the life of a terrestrial plant?
   a) The plant is anchored in the soil.
   b) The plant obtains water and nutrients from the soil.
   c) The roots of the plant obtain oxygen from the soil.
   d) All of the above are correct.
   e) Only a) and b) are correct.

4. Which of the following is most important in providing structural support in vascular plants?
   a) xylem
   b) apical meristem
   c) phloem
   d) epidermis
   e) cortex

5. Leaves are generally thin, with a three-layered structure that includes all of the following except:
   a) veins of vascular tissue.
   b) non-photosynthetic epidermal cells.
   c) a few layers of photosynthetic cells.
   d) apical meristem cells.
   e) stomata.

6. The cuticle of a leaf is:
   a) responsible for delivering water to the leaves.
   b) responsible for much of the photosynthesis in plants.
   c) the site where most of the photosynthetic products of plants are used.
   d) a waxy non-living layer that minimizes water loss.
   e) All of the above are correct.

7. Although ______ is one of the most abundant elements in the earth's crust, it is not required by plants.
   a) sodium
   b) phosphorus
   c) potassium
   d) magnesium
   e) sulfur

8. For a typical green plant to thrive, it needs:
   a) sunlight, water, carbon dioxide, oxygen, and minerals.
   b) only water.
   c) only minerals.
   d) only sunlight.
   e) sunlight, water, oxygen, and minerals.

9. Plants do not grow well in pure clay soils because:
   a) the soil packs so tightly that the roots cannot penetrate it.
   b) the soil packs so tightly that the amount of oxygen reaching the roots is limited.
   c) water drains too rapidly.
   d) the mineral content is very low.
   e) clay holds water and will not release it to the roots.

10. The enzyme known as ______ catalyzes the fixation of nitrogen in nitrogen-fixing bacteria.
    a) nitrogen reductase
    b) nitrogenase
    c) rhizobiase
    d) ammoniase
    e) nitrogen fixase

11. The process of water absorption by plant roots generally relies on ______ across the membranes of cells in ______.
    a) osmosis; root hairs
    b) active transport; root hairs
    c) osmosis; the apical meristem
    d) active transport; the apical meristem
    e) facilitated diffusion; stomata

12. Evaporation from the leaves of a tree pulls water up from the roots as an unbroken column throughout the entire height of the tree. This feat is possible because of which characteristic(s) of water?
    a) high heat capacity
    b) cohesion and kinetic energy
    c) cohesion
    d) low heat capacity
    e) absorption

13. Sugars produced in the leaves of a plant are transported elsewhere in the plant through:
    a) the movement of symbiotic bacteria that carry the sugars.
    b) the stroma apparatus.
    c) sap vesicles.
    d) sieve tubes that make up the xylem.
    e) sieve tubes that make up the phloem.

Short-Answer Questions

1. Compare and contrast the major characteristics of the monocots and eudicot flowering plants.
2. How does the cohesion-tension mechanism allow vascular plants to transport water without the utilization of energy? How does this affect the maximum height that can be attained by trees?
3. Why would a monocot be more likely to survive having a tight wire wrapped around it causing a shallow cut than a eudicot plant suffering from the same injury?

See Appendix for answers. For additional study questions, go to www.prep-u.com.

1 **Plants can reproduce sexually and asexually.**

18.1 Plant evolution has given rise to two methods of reproduction.

18.2 Many plants can reproduce asexually when necessary.

18.3 Plants can reproduce sexually, even without moving.

2 **Flowers have several roles in plant reproduction.**

18.4 The flower is the chief structure for sexual reproduction.

18.5 The male reproductive structure produces pollen grains.

18.6 Female gametes develop in embryo sacs.

3 **Pollination, fertilization, and seed dispersal often depend on help from other organisms.**

18.7 Plants need help getting the male gamete to the female gamete for fertilization.

18.8 Fertilization occurs after pollination.

18.9 Plants can avoid self-fertilization.

18.10 Following fertilization, the ovule develops into a seed.

18.11 Fruits exist to help plants disperse their seeds.

4 **Plants have two types of growth, usually enabling lifelong increases in length and thickness.**

18.12 How do seeds germinate and grow?

18.13 Plants grow differently from animals.

18.14 Primary plant growth occurs at the apical meristems.

18.15 Secondary growth produces wood.

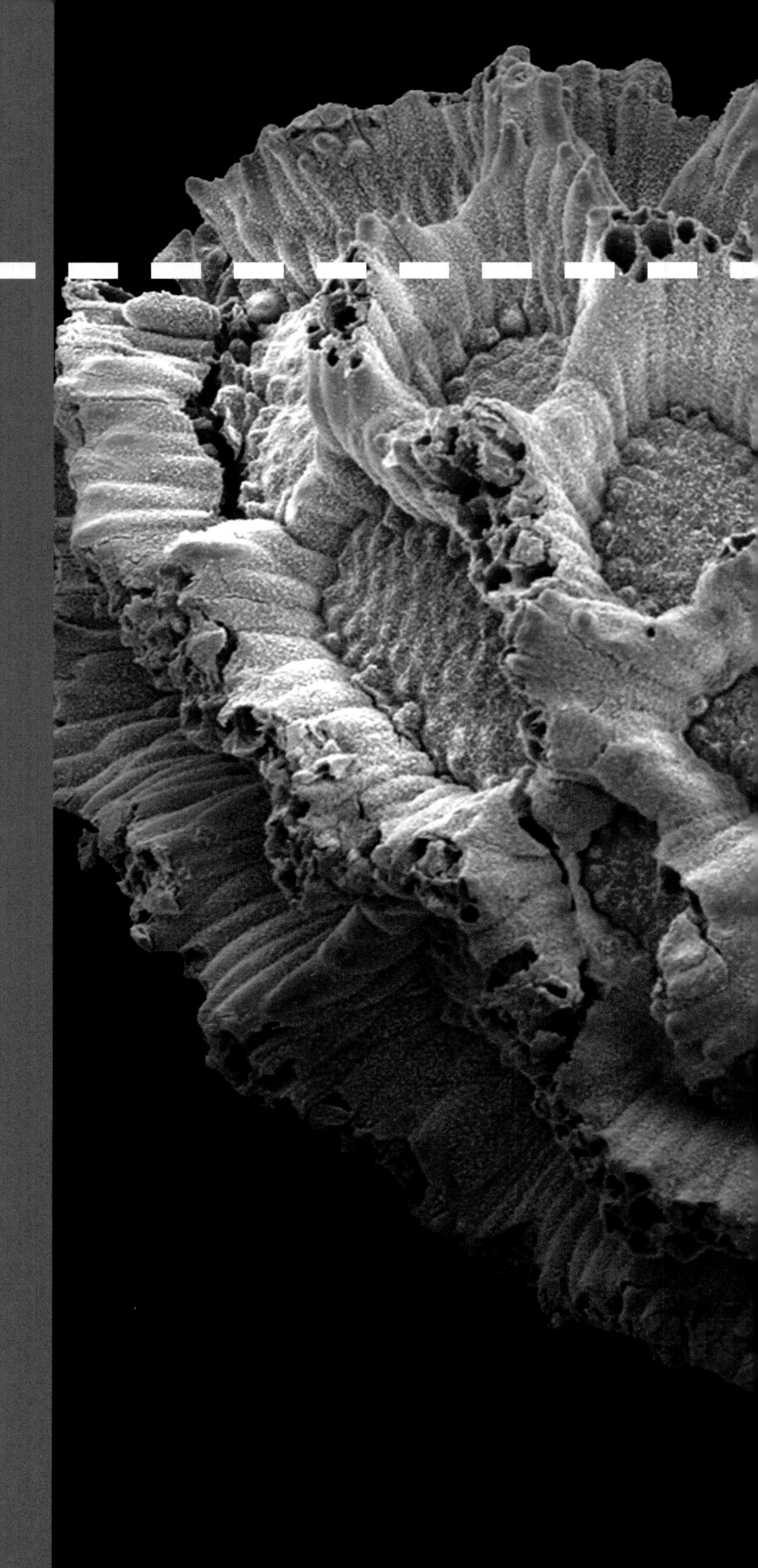

18

Growth and Reproduction in Plants

Problem solving with flowers and wood

1 Plants can reproduce sexually and asexually.

In a watery habitat, red azaleas and yellow bladderwort flowers grow beneath cypress trees.

18•1

Plant evolution has given rise to two methods of reproduction.

Why do flowers have such a hold on people? Brides clutch lilies during wedding ceremonies. Suitors (and partners) often give flowers during courtship (and over the course of long relationships). We send flowers to those in mourning and to those celebrating. And many people spend countless hours and dollars tending to flowering plants in their gardens.

We are innocent bystanders, as it turns out, unintentionally ensnared by plants' chief calling card as we are attracted by the sweet smells and captivating colors, textures, and shapes of flowers. The floral displays, though, are intended for a wide range of other animals and are the means by which plants enlist the assistance of animals in transporting the male gametes to the female gametes, thus using sexual reproduction to produce seeds and increase the genetic variability among their offspring.

Take carrion flowers, for example. Some insects feed on feces, rotting flesh, and other decaying organic matter, and they also lay their eggs in damp, stinky places. Carrion flowers lure these insects into their foul-smelling blossoms and, in doing so, are able to ensure the transport of male gametes to female gametes. *Rafflesia arnoldii,* of the rain forests of

The flower Rafflesia arnoldii, *of the rain forests of Sumatra and Borneo, emits an odor reminiscent of a stinking corpse that attracts carrion beetles and flies that then aid in transporting the plant's gametes and enabling sexual reproduction.*

Sumatra and Borneo, is a carrion flower that occasionally produces giant blooms of up to 3 feet in diameter, each flower weighing as much as 25 pounds. These gigantic flowers emit an odor reminiscent of a stinking corpse, and in this way the blossoms attract carrion beetles and flies—which, as they move from flower to flower, also move pollen (containing the male gamete) from the male flowers to the female flowers, enabling sexual reproduction even as the plants remain immobile.

But sexual reproduction isn't the only method of reproduction for flowering plants. Most are able to make use of an alternative method, producing offspring asexually under certain conditions. Aspens, for example (shown in the photo on the opening pages of Chapter 17 and in Figure 18.1), are a species of tree that thrives in cool temperatures. Aspens are prevalent throughout Europe and Asia, from Great Britain and Spain clear across to Russia, North Korea, and Japan. They also thrive in North America. Occasionally, however, they take root in inhospitable places. But, unlike a human who finds herself on an impossibly cold or barren plot of land and quickly packs up and retreats, aspens don't have that option. They're plants, after all. They can't escape.

Evolution seems to have compensated plants for their lack of mobility with a life-saving bit of adaptability. Consider the Orkney Islands, at the northernmost tip of Scotland. Populations of aspens live on the Orkney Islands, but the climate does not seem to allow aspen seeds to survive and generate new trees. This inability to reproduce would seem to be a fatal limitation to aspens' survival on these islands. Yet, there the trees are, and records indicate that they have persisted there for centuries. How do they do it?

In the Orkneys, the aspens reproduce asexually. They forgo any sexual reproduction and, instead, develop suckers, which are shoots that sprout from the roots and can give rise to new, genetically identical trees. One tree can give rise to dozens or even hundreds of new trees, all genetically identical (**FIGURE 18-1**).

Asexual reproduction, such as occurs in aspens, turns out to be prevalent among plants, wherever they occur in marginal habitats. Such habitats may limit the survival of seeds or, because of small population sizes, may reduce the likelihood that an individual finds a sexual partner.

FIGURE 18-1 **Aspen trees.** Aspens are able to reproduce asexually in harsh climates. Aspen root suckers, by which a plant can give rise to new individuals asexually, are shown in the inset photo.

In this chapter, we examine why evolution may have led to most plants having two very different options for reproduction, and we also investigate how plants grow. We begin by exploring the process of asexual reproduction and its advantages and disadvantages.

TAKE-HOME MESSAGE 18•1

Most plants have two very different options for reproduction: asexual and sexual reproduction.

18•2

Many plants can reproduce asexually when necessary.

Cloning isn't new. Gardeners have been doing it for centuries. And before that, plants were doing it for millennia—without any help from humans. Cloning is just a more modern term to describe a reproductive process by which one organism gives rise to another, genetically identical, individual. The process, also called asexual reproduction, is very rare among animals but is quite common among plants. As we will see in this chapter, for the vast majority of plant species, sexual reproduction is the predominant mode of reproduction, but nearly all plant species have retained the capability of also reproducing without any participation from another individual.

Most of the trees or shrubs we see in gardens are, in fact, the result of asexual reproduction. So, too, are nearly all houseplants purchased from nurseries. How does asexual reproduction in plants occur? The process is simple. It always involves the growth of new individual plants directly from the tissue of an established plant through normal mitosis. There is no meiosis to generate gametes, and there is no fertilization of one gamete by another. Asexual reproduction is also called vegetative reproduction, because it starts with a vegetative structure of the plant—that is, a structure (such as a stem or root or leaf) not involved in sexual reproduction, that becomes separated from the parent plant and continues to grow to form a new individual.

Raspberry plants, for example, can be propagated by cutting sprouts from their roots (**FIGURE 18-2**). And potato plants can be generated from their underground storage stems (that is, the potato itself). Thus, a new potato plant can be generated just by planting a potato and waiting for it to sprout into a genetically identical individual. We consider the new plant and "parent" plant distinct individuals because they have become separated and can grow and function independent of one another. In a group of plant species called *bryophyllum,* small "plantlets" grow right along the edges of the leaves, eventually dropping to the soil, taking root, and growing as an independent plant. The critical feature in every case of asexual reproduction is that the bit of the plant that is used to start a new individual must have some undifferentiated cells, meaning cells—like stem cells in humans—that have the developmental potential to become any type of cell and tissue. Later in the chapter, in our discussion of plant growth, we talk more about what those cells are and where they are located.

The ability to reproduce asexually is an evolutionary adaptation that helps plants overcome the fact that they are

FIGURE 18-2 Reproduction without a partner. Most plants—including raspberries, potatoes and *bryophyllum*—have the option of asexual reproduction. No partner is required, so it is quick and efficient, but no genetic variability is present among the cloned offspring.

The process of asexual reproduction in plants is simple. It always involves the growth of new individual plants directly from the tissue of an established plant via normal mitosis.

ADVANTAGES OF ASEXUAL REPRODUCTION

FIGURE 18-3 Asexual reproduction: the benefits. Reproducing asexually can be energetically efficient and fast and can preserve successful genetic combinations.

rooted in the ground and cannot easily seek out sexual partners. By removing the need to locate a suitable partner, asexual reproduction confers important benefits. It is not a perfect solution, however. In most situations, there are critical disadvantages to asexual reproduction. The advantages and disadvantages can be summarized quite simply.

The Advantages of Asexual Reproduction There are three main advantages to reproducing asexually (**FIGURE 18-3**).

1. *Energetically efficient.* No energy is wasted by producing plant tissues exclusively for reproduction, such as flowers, pollen, and fruits. Instead, that energy can be invested in producing roots, support and nutrient-transport tissues, leaves, and other vegetative matter for growth and increased photosynthetic capacity.

2. *Faster.* Participation of another individual is not required, nor is a means of bringing together the male and female gametes. As a result, asexual reproduction dramatically increases the rate at which new individuals can be produced, enhancing an individual's reproductive success and fitness.

3. *Preserving winning allele combinations.* If a plant is well-adapted to a particular environment, all the individuals it produces by asexual reproduction will carry the same well-adapted set of alleles and are likely to flourish just as the parent plant did.

The Disadvantages Although it has some advantages, asexual reproduction has one huge disadvantage relative to sexual reproduction: it leads to no new genetic variation among the offspring. Asexually reproduced individuals are identical to their parents. If the environment is changing—perhaps the climate is changing or a new variety of pest species is on the increase—the new individuals are less likely to be able to adapt to the changed environment. Moreover, as more and more of the individuals in an area are produced by asexual reproduction, if those individuals then reproduce together sexually, there will be less genetic variation among their offspring than if the parents were not genetically identical. Each parent plant will simply draw from the same set of alleles when producing gametes. This reduced variety in genetic material can leave plant populations in a vulnerable state, with a reduced capacity to evolve.

We turn next to how plants produce genetically varied offspring through sexual reproduction.

TAKE-HOME MESSAGE 18•2

Many plants can reproduce asexually. This involves the growth of new, individual plants directly from the tissue of an established plant through mitosis. The new individual is genetically identical to the individual from which it was produced. Asexual reproduction can be energetically efficient and fast and can preserve successful genetic combinations, but it also has disadvantages, most notably that it does not lead to genetic variability among an individual's offspring.

18•3

Plants can reproduce sexually, even without moving.

For all its efficiency and advantages, asexual reproduction just doesn't measure up to sexual reproduction in plants. In fact, despite the difficulties associated with being anchored in one place, the ability of plants to reproduce sexually is at the heart of their evolutionary success. Because sexually reproducing populations generate large amounts of genetic diversity, they can better keep up with changing conditions by adapting to new environments, and thus can better keep up with changing conditions than asexually reproducing plants by adapting to new environments.

Sexual reproduction enhances the genetic diversity among an individual's offspring in several important ways. First, because the genetic material comes from two distinct individuals—a mother and a father—the offspring isn't a duplicate of either parent but, instead, carries a mixture of traits from each. This outcome is analogous to having two people try to solve a group of problems together versus one person going it alone. The pair will likely arrive at better solutions. Second, as we saw in Chapter 6, during the process of producing sperm or eggs in an individual, recombination occurs during crossing over, causing each gamete to carry a mixture of alleles from that individual's mother and father, combinations of alleles that have most likely never occurred before. Every gamete is different from every other gamete, so each of an individual's offspring ends up with a different set of genes. And the third mechanism by which genetic variation is enhanced during sexual reproduction occurs during reassortment of the homologues during meiosis—randomly assorted with respect to whether they are maternal or paternal in origin (see Section 6-14). Each of these sources of genetic variation occurs during sexual reproduction, whether in plants or in animals (**FIGURE 18-4**).

Sexual reproduction (and the genetic variation that it introduces into a population) is particularly important in agriculture, because it allows a plant breeder to selectively breed plants that have the most desirable features—maybe the size of the corn cob, the speed at which the oranges ripen, or the scent or color of a rose. There must be genetic variation in a population of plants if the population is to be changed (and improved) through natural selection (including artificial selection by plant breeders). Tremendous gains in agricultural productivity have come from selective breeding programs, particularly with major crops such as wheat, rice, corn, and soybeans (**FIGURE 18-5**).

SOURCES OF GENETIC VARIATION IN PLANTS

Sexual reproduction enhances the genetic diversity among an individual's offspring in several important ways.

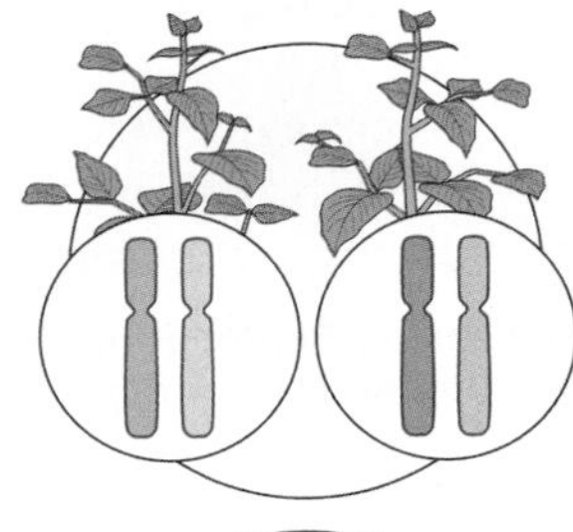

GENETIC MATERIAL COMES FROM TWO INDIVIDUALS
Offspring carry a mixture of traits from each parent plant.

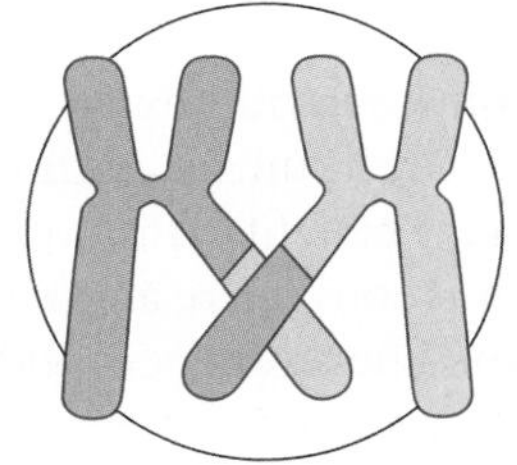

RECOMBINATION DURING GAMETE PRODUCTION
Each gamete carries a mixture of alleles that have most likely never occurred before.

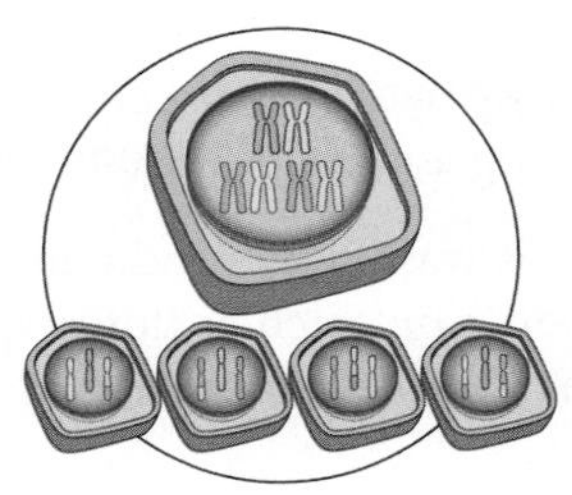

REASSORTMENT OF HOMOLOGUES
The homologues and sister chromatids distributed to each daughter cell during meiosis are a random mix of maternal and paternal genetic material.

FIGURE 18-4 Sexual reproduction. The offspring resulting from sexual reproduction are usually much more genetically varied than those produced by asexual reproduction.

In plants or animals, the process of sexual reproduction has

Sexual reproduction—and the genetic variation it introduces into a population—is particularly important in agriculture, enabling plant breeders to selectively breed plants with the most desirable features.

FIGURE 18-5 Breeding a better crop. Sexual reproduction is important to agriculture.

some common steps. Meiosis is an essential part of the process by which individuals produce gametes. Those gametes then come together to produce a fertilized egg, which develops and grows by mitosis. And finally, the fertilized egg is transported to a place where it can survive and grow. These steps occur in all organisms reproducing sexually. Sexual reproduction in animals and in plants always involves meiosis and fertilization, and because these processes are repeated in every generation, they are called a **life cycle.**

Because plants cannot move, however, their life cycle includes steps that increase their ability to safely deliver the male gamete to the female gamete—delivery that must occur without the parental plant uprooting itself and setting off to find an individual with which to mate. We trace the life cycles of non-flowering, non-seed plants and flowering plants in **FIGURE 18-6**.

The Life Cycle of Non-Flowering, Non-Seed Plants (Such as Ferns)

1. One form of the plant "body" that we see is diploid (but there also are multicellular haploid forms).
2. Production of gametes:

   a. Through meiosis, the diploid body produces haploid cells called **spores.** Although these spores are haploid, they *are not* gametes.

   b. The haploid spores are carried by wind or water to new soil, where they **germinate**—that is, after a period of dormancy, they begin to grow.

   c. Through mitosis, the haploid spores develop and grow into the second form of the plant "body" that we see: a new haploid plant.

   d. Through mitosis, the new haploid body produces gametes (sperm or eggs).
3. Fertilization: Male and female gametes come together at fertilization.
4. The fertilized egg develops and grows, by mitosis, into a new diploid body, and the cycle begins again.

Note that in non-flowering plants, spores are carried by wind or water to soil where they can germinate and grow into a gamete-producing form. The male and female gametes also are assisted, usually by rain or wind, when they must come together for fertilization to occur. The flowering plants have a slightly different strategy for dealing with immobility. The spores never leave the body of the plant, but the plants find a way for fertilization to occur, as described next.

The Life Cycle of a Flowering Plant

1. The plant "body" that we see is diploid.
2. Production of gametes:

   a. Through meiosis, the diploid body produces haploid spores. Although they are haploid, spores *are not* gametes.

   b. The haploid spores do not leave the diploid plant body, because they cannot survive away from it.

   c. Within the diploid plant body, the haploid spores divide by mitosis to form either a haploid female embryo sac or a haploid male pollen grain.

   d. Through mitosis, the haploid embryo sac produces a haploid female gamete (an egg), and the haploid pollen grain produces a haploid male gamete (a sperm).
3. Fertilization: Male and female gametes come together at fertilization.
4. The fertilized egg develops and grows, by mitosis, into a new diploid body, and the cycle repeats.

The life cycle in plants is called an **alternation of generations** and is characterized by an extended period in which a plant is in a multicellular haploid form and a period in which it is in a multicellular diploid form. If the same thing happened in humans, it would be the stuff of science fiction: the cells in our testes or ovaries, instead of producing sperm or eggs, would produce a special creature that would leave our bodies, live elsewhere on its own for a while, and then produce sperm or eggs, which would then find some way to

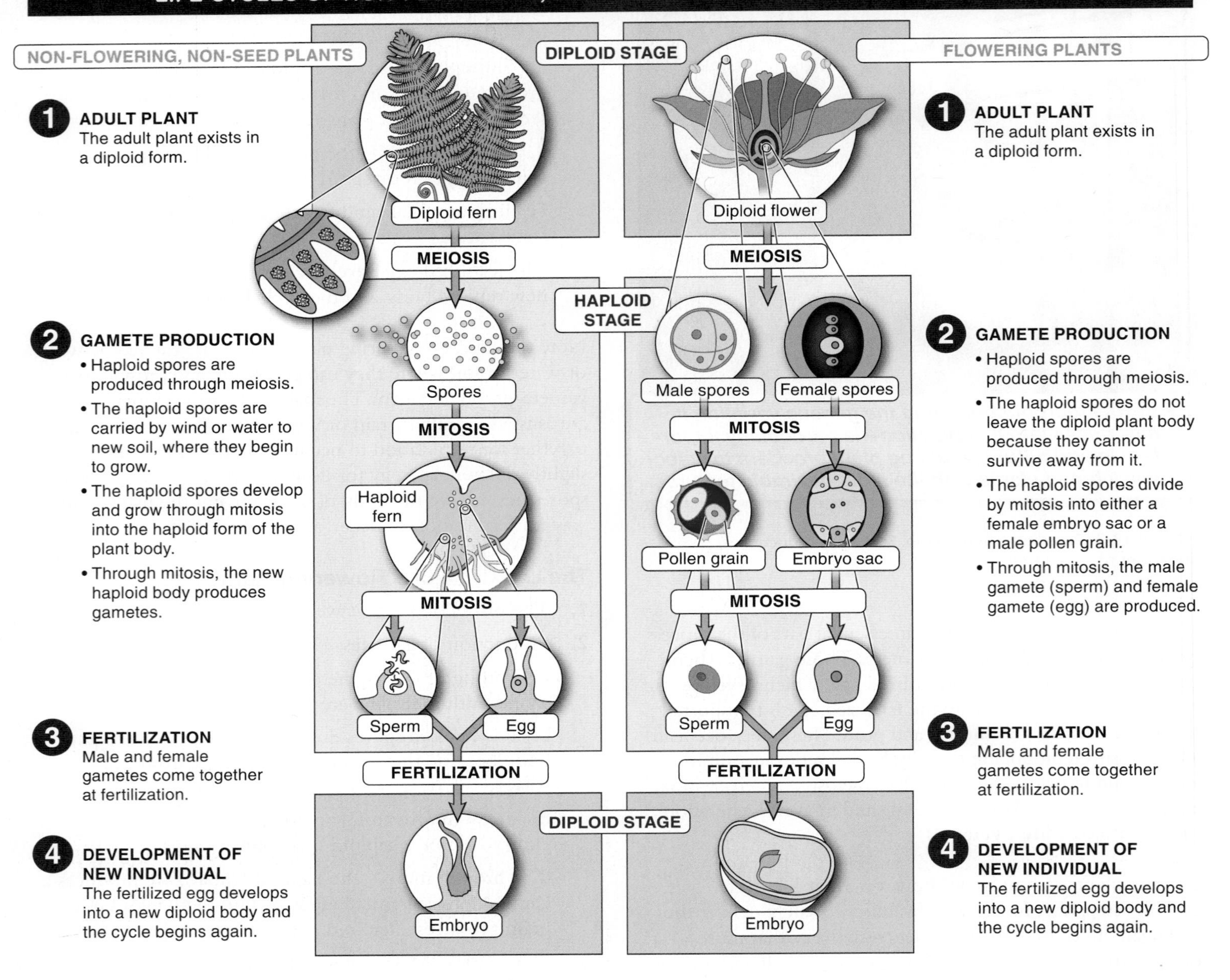

FIGURE 18-6 **Life cycles of non-flowering, non-seed plants and flowering plants.**

come together with eggs or sperm from another such creature, undergo fertilization, and produce babies.

For the remainder of this chapter, we focus on the sexual reproduction of flowering plants (the angiosperms), because they are the predominant group of plants on earth, accounting for more than 90% of all plant species. All angiosperms produce flowers, the structures in which the gametes develop and in which much of the reproductive activity takes place. We begin by exploring the role of these special structures that help plants solve the problem of how to bring male gametes to female gametes when plants are rooted in the ground.

TAKE-HOME MESSAGE 18•3

Many plants can benefit from producing genetically varied offspring by reproducing sexually, with the flower as the chief structure for sexual reproduction.

❷ Flowers have several roles in plant reproduction.

Red petals surround the reproductive structures of this tulip.

18•4

The flower is the chief structure for sexual reproduction.

Take a look at **FIGURE 18-7**, which showcases the diversity of flower shapes and colors. If we investigate more closely, though, and use a razor blade to dissect some flowers, we see that with only a few exceptions, all flowers have the same fundamental structures. The consistency we see there brings order to the seemingly infinite array of flower forms.

FLOWER SHAPE DIVERSITY

FIGURE 18-7 The captivating and varied colors, textures, and shapes of flowers. Shown: pachypodium flower, gazania, dahlia, and calla lily.

Flowers can vary greatly in shape, but, with only a few exceptions, they all have the same fundamental structures.

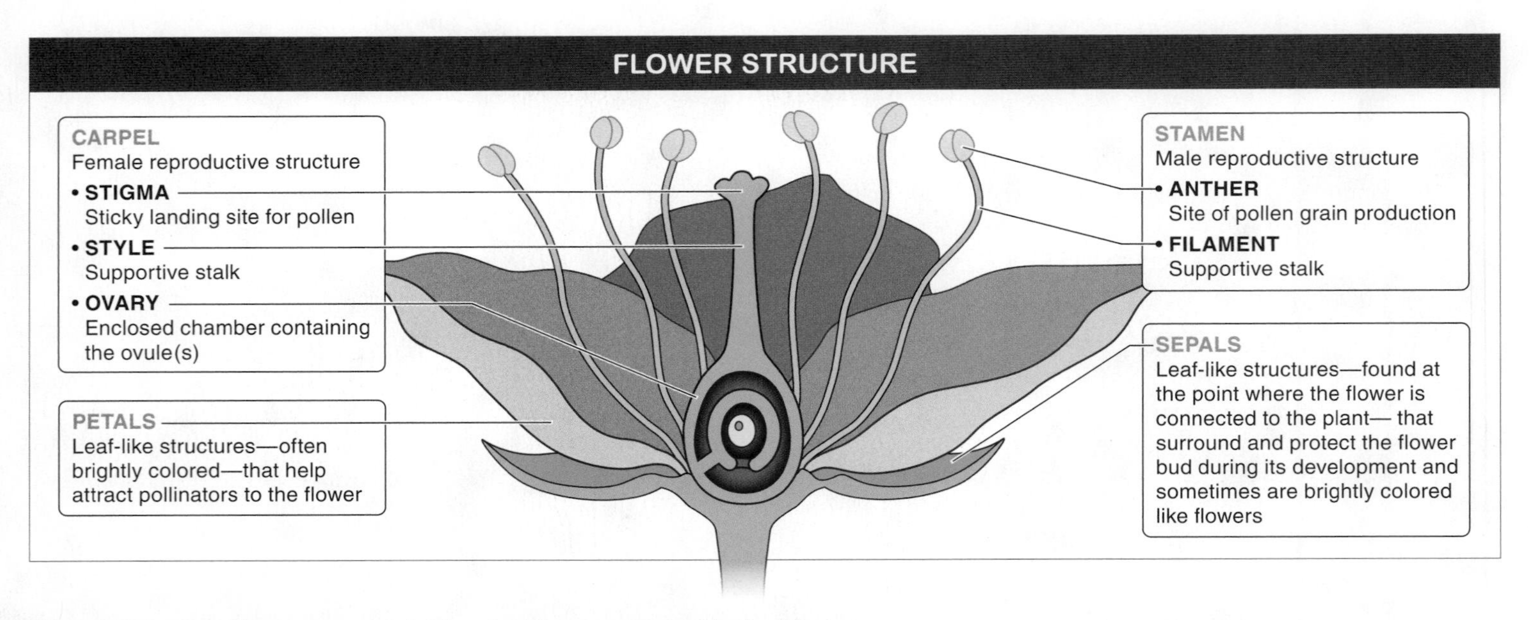

FIGURE 18-8 **Most flowers are organized around the same general body plan.**

Specifically, flowers have four distinct parts (**FIGURE 18-8**).

1. *Sepals.* **Sepals** are leaf-like structures located at the point where the flower is connected to a main support structure of the plant, such as a branch, stem, or stalk. Sepals grow in a ring around the outside of the flower, and they surround and protect the flower bud during its development. Although sepals are usually green, in some flowers they are brightly colored and closely resemble the flower's petals.

2. *Petals.* Just inside the ring of sepals is another ring of leaf-like structures, the **petals.** They are usually brightly colored—helping the flower to attract pollinators—and frequently have unusual shapes, from long and thin to short and broad.

3. *Stamens.* Moving from the outside of the flower inward, toward the center, the next structures we encounter are the **stamens,** the male reproductive parts. There are usually several stamens, each of which appears like a head on top of a long, thin stalk. The stalk is called the **filament** and the head-like top structure is called the **anther.** The anthers are the sites where the **pollen grains**—the structures that contain the male gametes, or sperm—are produced. The typical anther on a corn flower produces about 3,000 pollen grains.

4. *Carpel.* Finally, in the very center of the flower is the **carpel,** the female reproductive structure. There is usually only one carpel, and it is usually shaped like a long-necked vase. At the very top of the carpel is a flat, sticky surface called the **stigma,** which functions as a landing pad for pollen. The stigma is held high, out of the center of the flower, by the **style,** a long thin structure that leads down to an enclosed chamber, the **ovary.** Within the ovary are one

FIGURE 18-9 **Saffron: the world's most expensive spice.**

or more **ovules,** which produce the female gametes (eggs). After an egg is fertilized, the ovule develops into a seed.

Q What makes some flower parts worth more than $2,000 per pound?

The prize for the most valuable stigma probably goes to *Crocus sativus,* a flowering plant native to Southwest Asia that is more commonly known as the saffron crocus (**FIGURE 18-9**). The stigma and style of the saffron crocus are collected and dried, producing a deep yellow powder, valued by food lovers for its taste, sometimes described as slightly bitter and earthy. Common in Persian and Indian dishes, saffron is the world's most valuable spice, with a pound (requiring the stigma and style from about 75,000 flowers) selling for more than $2,000.

Not all flowers have all four structures described. We'll see later in this chapter that some plants produce flowers that have only male or only female reproductive parts. Still others may lack sepals or petals.

TAKE-HOME MESSAGE 18•4

Flowers are plant structures specialized for sexual reproduction. Most flowers have the same fundamental structures: sepals, petals, stamens, and a carpel.

18•5

The male reproductive structure produces pollen grains.

Humans have a great love of flowers, but it's not an unconditional love. The airborne pollen grains from a flower's anthers cause tens of millions of people to suffer the sneezing, coughing, and watery eyes that come with an allergy to certain types of pollen.

Q Is pollen the equivalent of plant sperm?

What exactly is this thing that causes so many people so much misery? In plants, the haploid cells produced by meiosis are called spores. Spores produced in the anthers of flowers are called microspores, and spores produced in the ovules are called megaspores. As the anther grows, four chambers form—sometimes called "spore sacs"—and each is filled with diploid cells called microspore mother cells. (To see the four spore sacs, take a razor blade or knife and slice the anther of a flower in half.) The microspore mother cells divide by meiosis, each one producing four haploid microspores. And the microspores then quickly divide by mitosis, forming a two-cell grain of pollen with a very complex, water-tight, and sticky surface. This two-celled structure is the pollen grain: a reproductive packet containing two haploid cells (**FIGURE 18-10**). One of the cells will eventually grow to

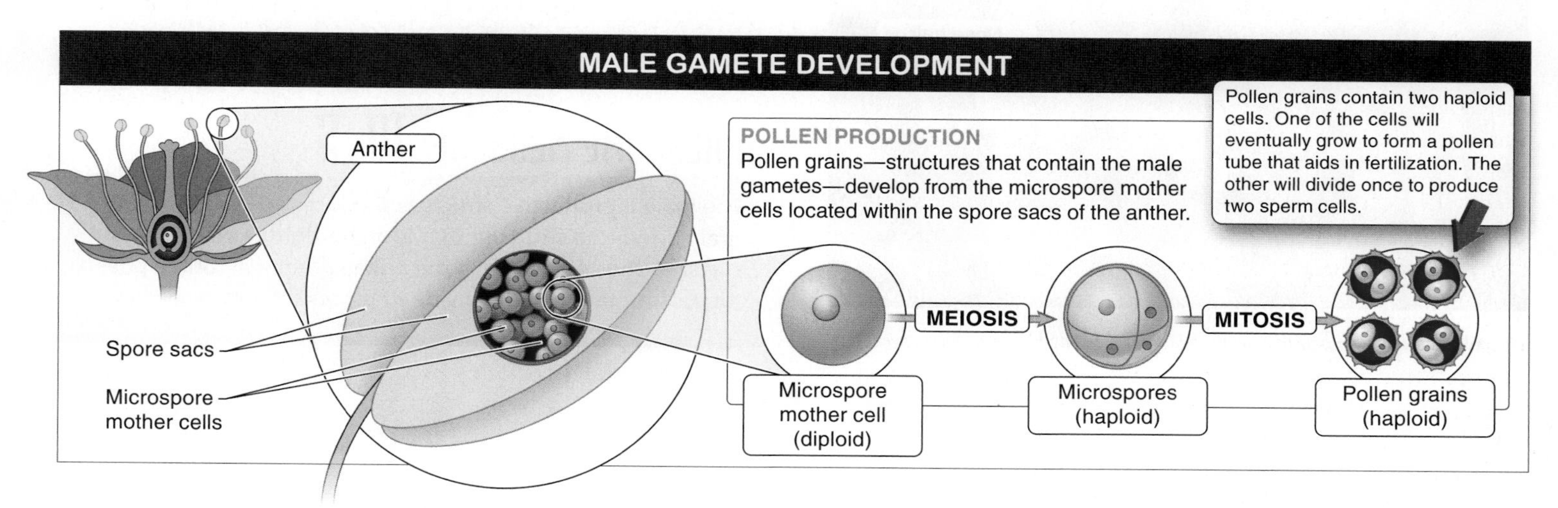

FIGURE 18-10 **The anther of a flower produces pollen grains, containing the male gametes.**

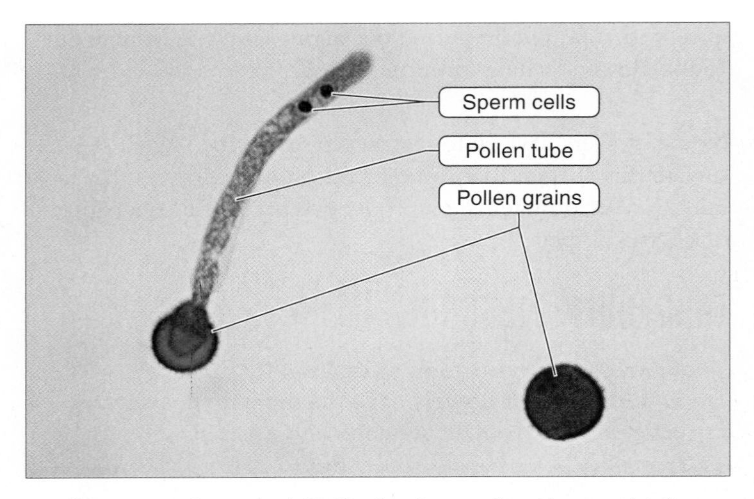

A pollen grain contains two haploid cells. One forms a tube used as a conduit for fertilization. The other cell divides once and forms two sperm cells.

form a **pollen tube** (see photo above), aiding in fertilization. The other will divide once to produce two **sperm** cells. But those steps don't usually occur until the pollen grain has made its way from the anther into the environment and come to rest on a stigma somewhere.

Why does pollen cause allergy attacks in so many people? The trouble begins each spring and summer, as plants release billions and billions of tiny pollen grains into the air. Proteins project from the cell wall of every one of these pollen grains, and the proteins increase the stickiness of the pollen. Inevitably, we

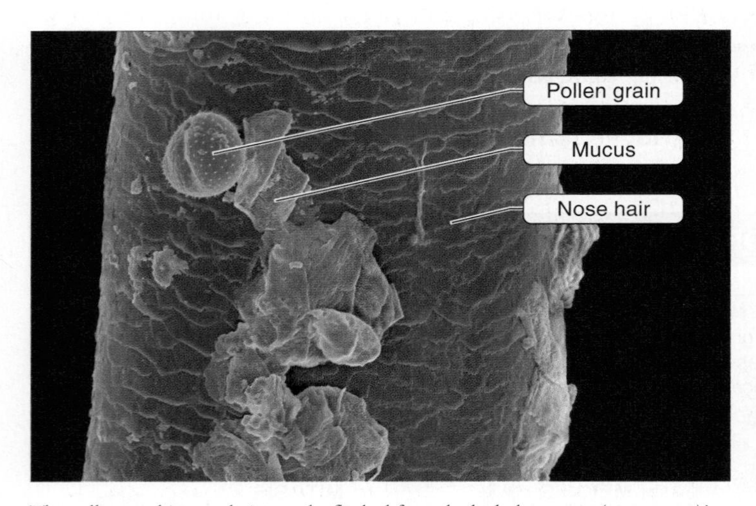

The pollen on this nose hair may be flushed from the body by mucus (or a sneeze)!

FIGURE 18-11 Pollen grains can cause misery for those who suffer with allergies.

breathe in some of them, and our bodies immediately detect the pollen's surface proteins and identify them as foreign invaders (**FIGURE 18-11**). In response to this invasion, our immune system attacks and tries to flush the pollen from our bodies (as shown in the lower left photo). Mucus on this nose hair helps move the pollen outside of the nose. However, sometimes the response is extreme. The runny nose, sneezing, coughing, and congestion common to pollen allergies are all manifestations of the body's over-reaction.

TAKE-HOME MESSAGE 18•5

The male reproductive structure produces pollen grains, each grain a two-cell structure that is water-tight and has a sticky surface. One of the cells in the pollen grain will form a pollen tube, and the other will divide to produce two sperm cells.

18·6

Female gametes develop in embryo sacs.

Production of the female gametes (eggs) doesn't cause the suffering in humans that pollen grains do, primarily because the production is confined to the closed structure of the ovary. Within the ovary, one or more diploid cells differentiate into ovules, and each ovule is made of outer protective cells that surround a diploid egg-producing cell, called a megaspore mother cell. The megaspore mother cell undergoes meiosis to produce haploid megaspores. Within a flower, one of those haploid megaspores undergoes mitosis several times to produce the **embryo sac,** the structure in which fertilization will occur (**FIGURE 18-12**). The embryo sac is an unusual collection of seven cells. Six of these cells—including one that is the egg—have haploid nuclei, and a seventh cell, the central cell, has two distinct haploid nuclei. The embryo sac waits for a male gamete to arrive.

TAKE-HOME MESSAGE 18·6

Within the ovary, diploid cells differentiate into ovules, each of which is a group of outer protective cells around a diploid egg-producing cell, which undergoes meiosis to produce haploid megaspores. One of these megaspores undergoes mitosis several times to produce the embryo sac, the structure that contains the egg and is the place where fertilization will occur.

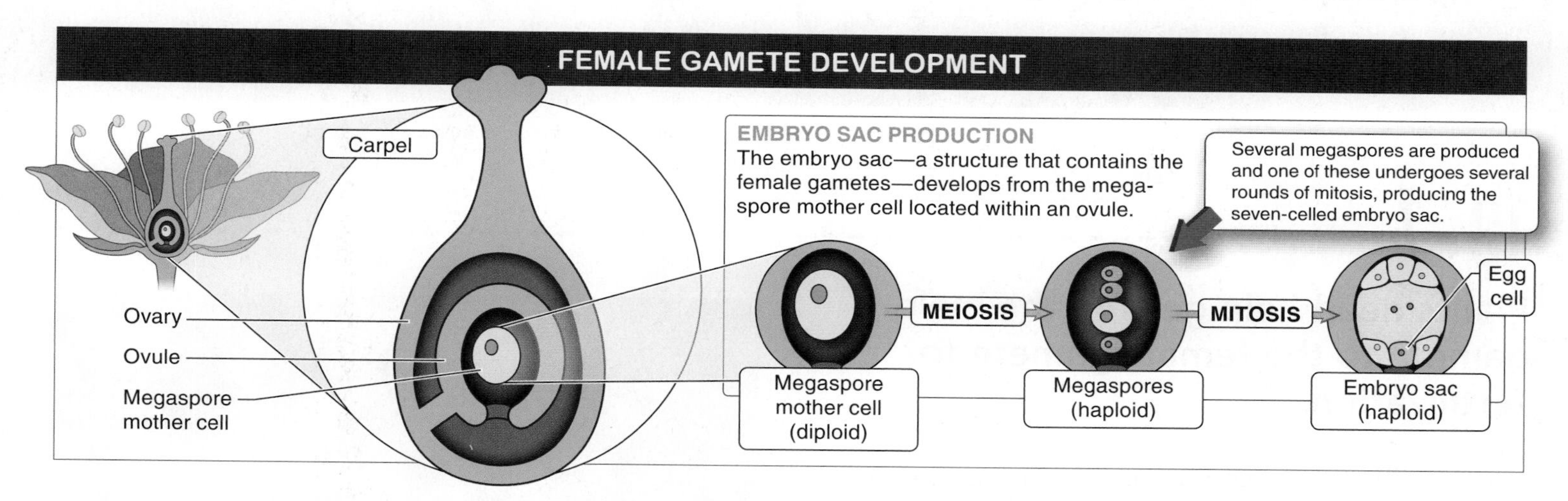

FIGURE 18-12 **Female gametes are produced within the ovary of a flower.**

❸ Pollination, fertilization, and seed dispersal often depend on help from other organisms.

A honeybee coated with pollen can transport male gametes.

18•7

Plants need help getting the male gamete to the female gamete for fertilization.

The ancestors of plants, the green algae, could simply release their gametes into the water, so that some were bound to find their way to the gametes of another individual of their species. For mosses and ferns, reproduction is still tied to water; they must rely on the presence of water for transporting male gametes to the female gametes, otherwise the gametes dry out. For this reason, mosses and ferns must live in moist habitats or reproduce when moisture is available. With additional transition to land during evolution, seed plants had to reduce their reproductive dependence on water, and a number of ways have evolved for bringing together male and female gametes (**FIGURE 18-13**). Often, this process involves enlisting animals to carry the male gametes.

How does a flowering plant get an animal to transport its gametes? It attracts the animal with its flower, a sexual display that attracts animals to the plant with visual cues (color, shape), olfactory cues (smell), or even tactile cues (soft, bristly, hard, rough, smooth). Regardless of the method of attraction, the goal is always the same: to physically attach some male

Because sexual reproduction requires two individuals, one plant must get its gametes to join with those of another. A broad array of solutions to this problem has evolved.

FIGURE 18-13 Even seemingly isolated plants can reproduce sexually.

STRATEGIES FOR ACHIEVING POLLINATION

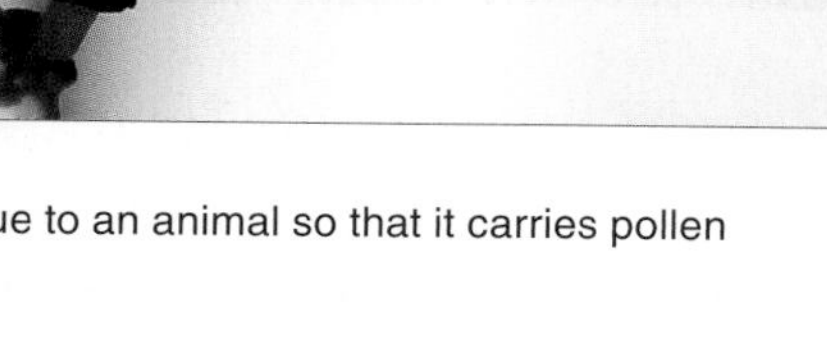

BRIBERY
Plants offer something of value to an animal so that it carries pollen from one plant to another.

TRICKERY
Plants use methods of deception to trick animals into carrying pollen from one plant to another.

FIGURE 18-14 Package handlers. Animals transport pollen for plants.

gametes to an animal so that they rub off on the female reproductive parts of another plant, where fertilization can then occur. That is, a pollen grain from a plant must journey to the stigma of another plant of the same species. This step is called **pollination.**

About 10% of plant species achieve pollination by simply releasing their pollen to the wind (as do grasses and pine trees) or into water (as does eelgrass, for example), on the slim chance that—through random luck—some of the pollen will land on the female reproductive organs of another plant of the same species. Given the astronomically low probability of any one pollen grain actually doing that, such wind- and water-pollinated plants respond in the only reasonable way: they produce tremendous amounts of pollen, tens of millions of pollen grains per plant. It's not particularly efficient, but it works.

Among other plants, two different and clever strategies for achieving pollination have evolved (**FIGURE 18-14**): (1) bribery—plants bribe some animals to carry their pollen from one individual plant to another; and (2) trickery—they deceive animals into doing the job.

> **Q** Why do some flowers smell nice while others don't smell at all?

1. *Bribery.* The most common strategy for achieving pollination involves a plant offering something of value to an animal in exchange for transport of its pollen from one flower to another. For bribery to work, the plant must produce a sticky pollen and a flower that catches the attention of the pollinator. Most important, the plant must produce something of value to the pollinator. The payoff can be food, such as nutritious **nectar,** a solution rich in sugars (and also containing amino acids) that is produced by the flower and can be consumed by a pollinator; or it may be a safe, hospitable location for an insect to lay its eggs. The variety of flower structures is tremendous. They differ in shape, color, smell, time of day when they are open, whether or not they produce nectar, and whether their pollen is edible.

2. *Trickery.* The second strategy for achieving pollination is more selfish. Rather than bribing an animal to carry its pollen, the plant just tricks the animal into thinking it is going to get something of value, while not actually giving it anything valuable. Among the tricksters are a species of plants called cycads. These are gymnosperms (that look a bit like palm trees) that produce cones once every year or so. Within the male cones, pollen is produced in large amounts. It might seem as though cycads use the strategy of bribery for pollination, because their pollen attracts huge numbers of small insects called thrips, and the thrips consume the pollen. But providing a source of food isn't enough to achieve pollination in cycads. That's where the trickery comes in.

For a few hours each day during the period that they have cones, the cycads are able to increase the temperature inside the male cones by up to 25° F. They do this by metabolizing sugars and fats within the cone. As they do so, they also produce a powerful stench. Taken together, the heat and stink cause the thrips—most of which are now covered in pollen—to leave the cones. And as the male cones cool and the thrips start to return, the female cones begin to produce a pleasing (to thrips) odor, similar to that produced by the male cones. The attractive odor lures in some thrips, thinking that they will find pollen. But there is no pollen. And by the time the thrips realize this, they have usually brushed some of the pollen that covered their bodies onto the reproductive parts of the female cone.

Just as the variety of flower types is wide, so, too, is the variety of pollinators: birds (mostly hummingbirds), bees, flies, beetles, butterflies, moths, and even some mammals (mostly bats). In each case, there has been strong coevolution between the plants and their pollinators: the plants have become more and more effective at attracting the pollinators and deterring other species from visiting the flower, while the pollinators have become more and more effective at exploiting the resources offered by the plants (see Figure 12-21).

TAKE-HOME MESSAGE 18·7

Plants usually utilize trickery or bribery to get the assistance of animals in carrying the male gametes to the female gametes. There has been strong coevolution between plants and their animal pollinators.

18·8

Fertilization occurs after pollination.

Pollination is a bit like sexual intercourse in mammals: it brings the male and female gametes close to each other, but it isn't quite fertilization. Fertilization requires that the male and female gametes fuse so that their genetic material can be combined. We examine here the male gamete's last steps on its journey to the female gamete.

Recall that the pollen grain is made from two cells encased in a sticky, spiky wall. When the pollen grain lands on a stigma of a flower of its own species, it sticks to the stigma through interactions between the pollen grain's outer coating and the cells of the stigma. Pollen grains from other species have the wrong identifiers on the surface and tend to slide right off, keeping the stigma clear and open to the "right" pollen grains (**FIGURE 18-15**).

FIGURE 18-15 Built to stick. Pollen grains adhere to the stigma of a flower.

Within 12 to 36 hours of the pollen landing on a stigma, one cell in the pollen grain starts to grow into a pollen tube. This growth does not occur through cell division but by an elongation of the tube cell, which stretches itself and pushes through the stigma, taking the sperm-producing cell farther and farther down the style toward the ovary. This growth isn't an easy stroll down the style, however. It's more like running a gauntlet. The cells within the style, which is part of the female reproductive structure (the carpel), test the tube cell to check whether it is too closely related to the plant on which the pollen has landed (as it would be if the plant self-pollinated). If the tube cell is too closely related, it is killed—often by a chemical reaction initiated by the cells of the style, breaking down the cytoskeleton of the pollen tube. If it is not too closely related, the pollen tube is allowed to continue growing toward the ovary.

If the pollen tube survives its journey down the style and into the ovule, it grows right into the embryo sac (**FIGURE 18-16**). Meanwhile, the sperm-producing cell within the pollen tube divides once, producing two sperm cells. As the pollen tube

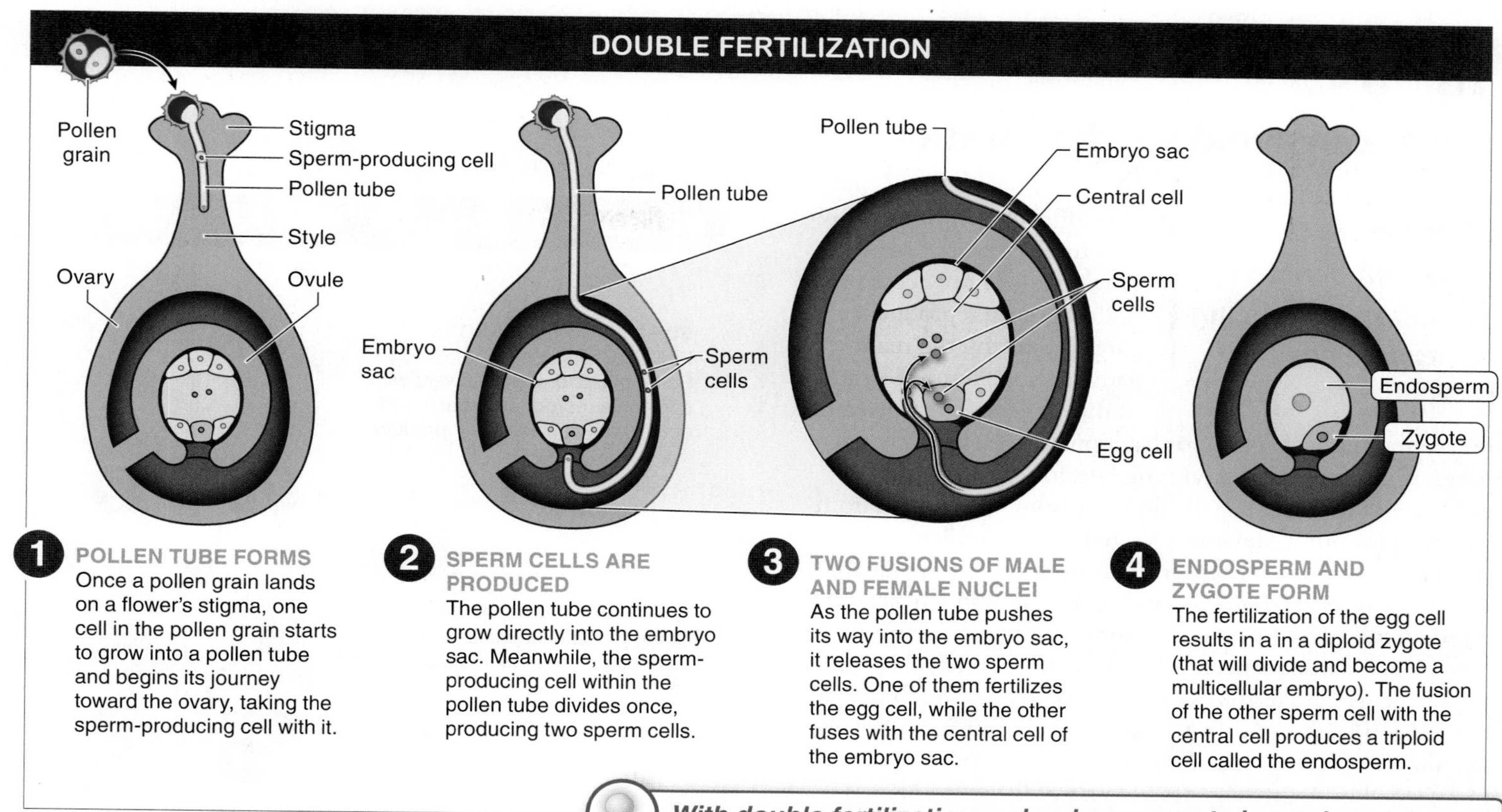

With double fertilization, only when a zygote is produced does a plant expend energy to produce a food source, the endosperm.

FIGURE 18-16 Fertilization in the flowering plants. In the process of double fertilization, there are two fusions of male and female nuclei: one produces the plant embryo and the other produces the endosperm.

pushes its way into the embryo sac, it releases the two sperm cells, and one of them fertilizes the egg cell inside the embryo sac. When the sperm and egg fuse, in fertilization, the resulting diploid cell is called the **zygote.**

The sperm cell that fertilizes the egg produces a diploid cell that will develop, by mitosis, into the new diploid plant. The other sperm cell fuses with the central cell of the embryo sac, which, as we learned earlier, has two haploid nuclei. This second fusion produces a triploid cell with three sets of genes. The triploid cell is called the **endosperm.** As the embryo starts to grow and divide, so, too, does the endosperm. It continues to do so at a high rate, turning into nutritional tissue that will nourish the developing embryo.

Q What part of the plant supplies the majority of calories to humans?

Because, in this fertilization, there are two separate fusions of male nuclei with female nuclei—one producing the embryo and the other producing the endosperm—the whole process is called **double fertilization.** This process, which is unique to flowering plants, is an efficient system because whenever an embryo is produced, so, too, is a ready-made food source. Most of what we eat in plants is endosperm, including the sweet material in corn on the cob, rice, and the flour we eat as pasta and bread.

TAKE-HOME MESSAGE 18•8

Pollination is necessary but not sufficient for achieving fertilization. Following pollination, a pollen tube must grow down the style and into the ovule, where the sperm-producing cell within the pollen tube produces two sperm cells. One of these cells fertilizes the egg cell in the embryo sac to form the embryo, and the other fuses with the diploid central cell of the embryo sac to form the endosperm, which will nourish the developing embryo.

18•9

Plants can avoid self-fertilization.

Q What are the risks and benefits of a plant producing "bisexual" flowers?

The efficiency of flower structure is remarkable. Flowers commonly contain both the male reproductive parts, producing the male gametes in pollen grains, and, in the center of the flower, the female reproductive parts, so that a pollinator can easily assist in sexual reproduction. A visiting bee, for example, can pollinate a flower such as an apple blossom with pollen from another blossom it previously visited, while simultaneously picking up pollen for later fertilization of other individuals. But this *hermaphroditism*—having both male and female reproductive parts—can lead to some unintended and detrimental consequences. What happens, after all, if a plant's own male gamete fertilizes one of its own eggs? Such self-fertilization leads not only to less genetically varied offspring, but also to offspring that are more likely to express one or more lethal or negative genes because of the extreme inbreeding. How can plants avoid this bad outcome?

Several effective mechanisms have evolved in plants for reducing the incidence of self-fertilization, three of which are described here (**FIGURE 18-17**).

1. *Separate male and female flowers.* In some plants, an individual produces some flowers with only male reproductive parts and others with only female reproductive parts. Overall, the plant is still a hermaphrodite, but such unisexual flowers minimize the likelihood of self-pollination and self-fertilization.

2. *Staggered maturation of male and female reproductive parts.* To prevent self-fertilization in hermaphrodite flowers, some plants have male parts that develop before the female parts, or vice versa.

3. *Separate male and female plants.* The most extreme method of avoiding self-fertilization is for some plants within a population to produce only flowers with male reproductive parts, while other plants produce only flowers with female reproductive parts. As in humans, then, there are two separate sexes: male individuals and female individuals. This makes self-fertilization impossible, but it also reduces the likelihood that any given visit by a pollinator will result in pollination.

METHODS OF REDUCING SELF-FERTILIZATION

SEPARATE MALE AND FEMALE FLOWERS
Some plants produce flowers with only male reproductive parts and others with only female reproductive parts.

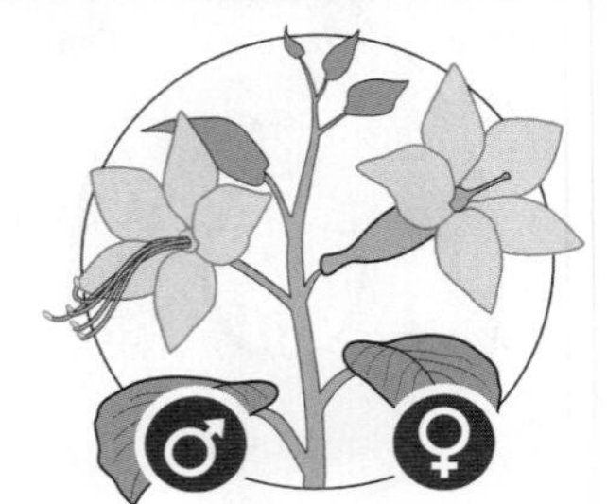

STAGGERED MATURATION
Some plants have male parts that develop before the female parts, or vice versa.

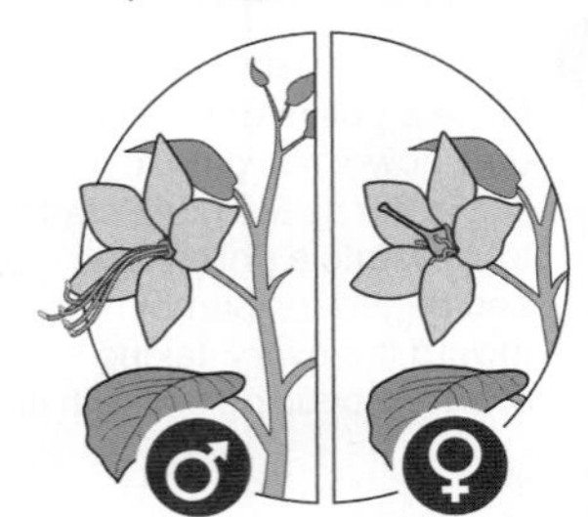

SEPARATE MALE AND FEMALE PLANTS
Some plants within a population produce only flowers with male reproductive parts while other plants produce only flowers with female reproductive parts.

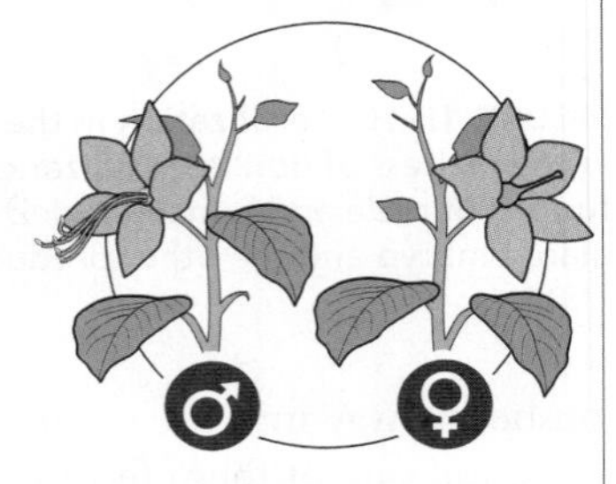

Plants have several effective mechanisms for reducing the potentially detrimental incidence of self-fertilization.

FIGURE 18-17 Avoiding inbreeding. Several mechanisms can reduce the incidence of self-fertilization, enhancing genetic diversity among the offspring in flowering plants.

TAKE-HOME MESSAGE 18•9

Plants can reduce the likelihood of self-fertilization in any of several ways, including producing separate male and female flowers, staggering the time of maturation of male and female reproductive parts, or producing separate male and female plants.

18•10

Following fertilization, the ovule develops into a seed.

Following fertilization within the ovule, numerous rapid cell divisions occur. The developing embryo forms a **root meristem** and a **shoot meristem,** each of which is a cluster of active, dividing embryonic cells that will generate new tissue in the plant. (We discuss the unique qualities of meristems in more detail in Section 18-13.) Also, one or two **cotyledons** form, structures in the plant embryo that usually become the embryonic leaves of the plant (see Section 17-2). As the embryo matures, the outer cells of the ovule develop into a hard casing forming a seed—a sort of "plant in waiting" that is made of embryo (meristems and cotyledons) and any remaining endosperm (**FIGURE 18-18**).

At the same time that seed formation is occurring, the ovary wall—surrounding one or more seeds—develops into a **fruit,** which in many species is fleshy, juicy, and edible, while in other species it may be dry or inedible. The seed is protected within the fruit, which, as we'll see, may aid in seed dispersal. Metabolism and oxygen consumption grind to a halt, and the seed dries out, reducing its water content to about 15% of its total weight. And then the waiting begins.

Q What is popcorn?

Humans have learned a trick that turns one type of seed into an unusual snack. Each kernel of popcorn comes from a single kernel of corn on a cob, and is the equivalent of a fruit: it includes a fertilized egg surrounded by a nutritious, edible outer layer, formed from the ovary wall. (If you plant some popcorn kernels, they will grow!) There is also moisture within the kernel—some oil and some water—and it all is surrounded by a moisture-proof coat. Heating the kernel turns the moisture inside to steam and chemically alters the starch and protein in the seed, making it soft and a bit gelatinous. With continued heating, the pressure from the steam rises and ultimately blasts through the seed coat with a pop. The fluffy white material is the starch and protein that surrounded the seed.

Before a seed can start to grow into a new offspring plant, it must leave the parent plant. In the next section we look at how ever-immobile plants ensure that their seeds are moved away from their parent plant and prepared for growth.

TAKE-HOME MESSAGE 18•10

Following fertilization, the ovule develops into a seed, containing a root meristem, a shoot meristem, and one or two cotyledons, surrounded by a hard casing. The seed is protected within a fruit, which can aid in its dispersal.

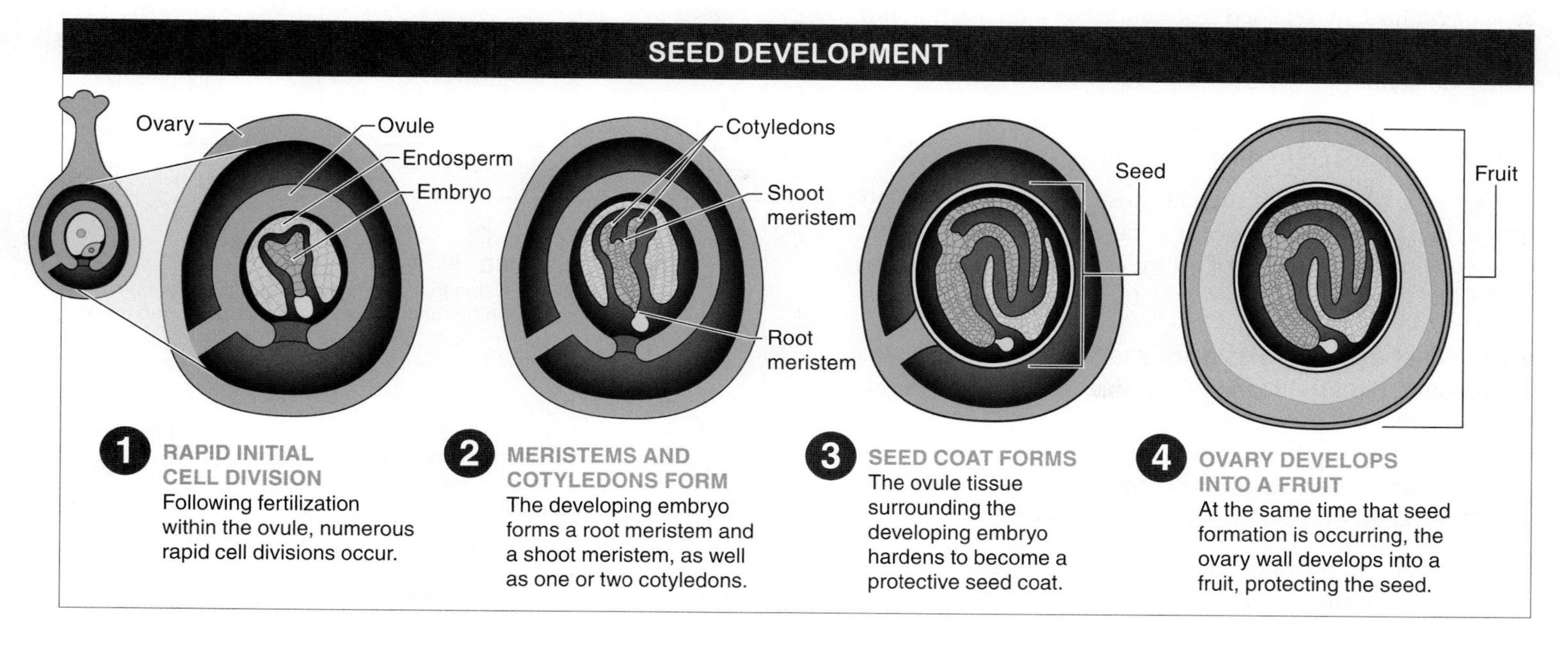

FIGURE 18-18 Protected for later dispersal. Seeds are protected by a casing and nestled within a fruit.

18•11

Fruits exist to help plants disperse their seeds.

Every fruit you eat used to be a flower: tangerines, peaches, raspberries, cherries, pumpkin, and squash, among others. Fruits are produced by plants to aid in dispersing the seeds that will grow into their new offspring. By encasing seeds in a juicy, nutritious structure—the fruit—plants create a package that entices hungry animals. Animals eat the fruit, gaining valuable nutrients and energy. Then, at a later point—and in a location far away from the parent plant—the animals eliminate the seeds, which pass unharmed through their digestive system. At the new location, often surrounded by a bit of the animal's feces that serves as fertilizer, the seed can grow.

There is a huge variety of fruits beyond those designed for consumption. And while they all serve the function of dispersing the seeds, not all require the assistance of animals. Just as with pollination, plants can also use the wind or water to transport their seeds. **FIGURE 18-19** presents a small sampling of the rich variety of fruits and seeds produced by plants. As is apparent, the term "fruit" applies to a much broader range of structures than is typically associated with it. And the makeup of each fruit clearly reflects the method by which it is dispersed—wind, water, or animals.

Wind-Dispersed Fruits and Seeds Wind dispersal can be achieved by way of several characteristics of fruits and seeds.

1. *Hairy.* Bushy hairs allow the fruits of dandelions and milkweed plants to float in the air.

2. *Winged.* Usually released from tall trees such as elms and maples, these fruits and seeds float away from the tree as they slowly descend.

3. *Tiny, dust-like.* Many orchids have seeds that are as tiny and light as dust and so float in the air.

4. *Explosive.* Many fruits, such as those of mistletoe, explode when ripe, propelling the seeds as far as 50 feet (about 15 meters) away at speeds of 60 miles (about 100 kilometers) per hour.

Water-Dispersed Fruits and Seeds Some plants, such as the coconut, produce floating fruits that can be dispersed by rivers or oceans.

Animal-Dispersed Fruits and Seeds Animals act as dispersers by simply carrying or eating fruits.

1. *Carried.* Some plants produce fruits that have sharp or clingy burrs on the outside. These burrs catch on the legs

METHODS OF FRUIT AND SEED DISPERSAL

WIND-DISPERSED

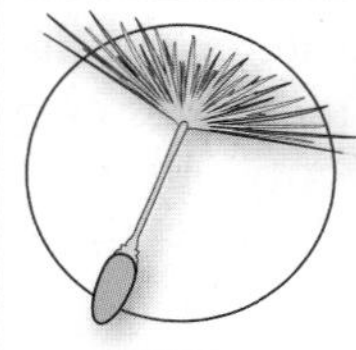

HAIRY
Seeds within fruits that have bushy hairs can float in the air.

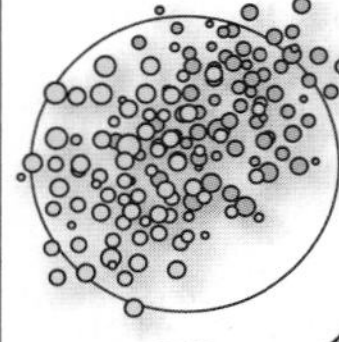

WINGED
Seeds within fruits that have wing-like structures can float away from a tree as they slowly descend.

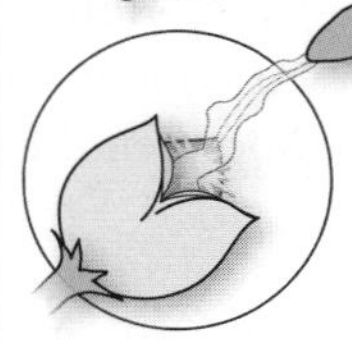

TINY, DUST-LIKE
Seeds that are tiny and light as dust are able to float in the air.

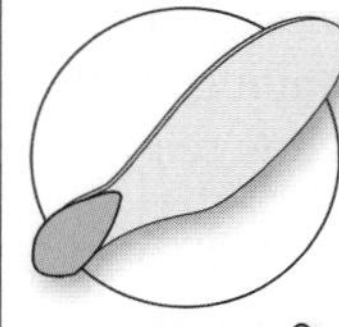

EXPLOSIVE
Seeds are propelled from the plant as the ripened fruits explode.

WATER-DISPERSED

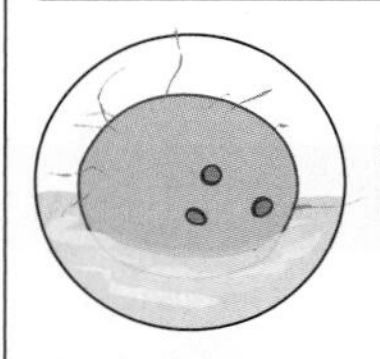

Seeds within floating fruits can be dispersed by rivers or oceans.

ANIMAL-DISPERSED

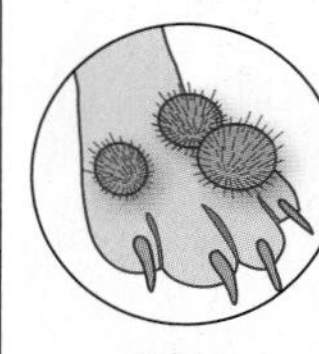

CARRIED
Seeds within fruits that have sharp or clingy burrs can catch on the fur of animals and be carried away.

CONSUMED
Seeds within fruits that are consumed by animals can be dispersed as they pass through the animal's digestive tract and are eliminated at some distant location.

FIGURE 18-19 Animals, water, and wind can disperse seeds.

*The anachronistic Osage-orange (*Maclura pomifera*): A giant fruit without an animal agent of seed dispersal.*

The fruit of the Osage-orange (sometimes called the hedge apple) lie beneath the tree rotting. Contemporary animals show little interest in its bitter-tasting flesh. They serve as a ghostly reminder of the super-sized (now extinct) North American mammals such as mastodons that once fed on them.

or fur of animals (including humans) and are carried away until the animal grooms itself and removes the fruit, leaving the seeds to germinate and grow in the new location. Burrs were the inspiration for Velcro.

Q Why do fruits taste so bad before they are ripe?

2. *Consumed.* Many fruits are sugar- and nutrient-laden fleshy structures that are eaten by animals, usually vertebrates. In some cases (such as the peach), the fruit develops from a single ovary; in others (such as blackberries or pineapple), it forms from multiple ovaries from numerous flowers. It does not benefit a plant to have its seeds dispersed (within a fruit) before they are fully developed and ready to go. Consequently, fruits generally do not ripen and become sweet and edible until the seeds are fully developed. In fact, most fruits remain green until the seeds are ready for dispersal, at which point the fruit turns a bright color and becomes sweeter and much tastier to attract the attention of dispersers. The seed's coating is tough enough to pass through the digestive tract without being destroyed, and the seed can germinate and begin growing wherever the animal defecates (or, in the case of birds, regurgitates).

This varied group of fruits includes tomatoes, grapes, cherries, olives, apples, pears, squash, zucchini, all beans and peas, corn kernels, and avocados, among others. Because there appears to be no species of animal alive today that disperses avocado seeds, researchers have speculated that the avocado plant evolved in concert with some large species of animal that has since become extinct. Today, avocado plants rely on humans to disperse their seeds.

TAKE-HOME MESSAGE 18•11

Following pollination and fertilization, plants utilize the assistance of animals or water or wind to disperse their fruits and seeds, depositing them at a new location where the seeds can germinate and new plants can grow.

④ Plants have two types of growth, usually enabling lifelong increases in length and thickness.

The graceful branched trunk of a red maple.

18•12

How do seeds germinate and grow?

With the seed, plants have created an impressive little time capsule. In suspended animation, the embryo patiently waits until the conditions for its new life are just right, at which point it bursts forth with growth. Let's examine what makes up a seed and how it emerges from its dormancy and begins life.

> **Q** The seeds of plants used in agriculture have thinner coats and more stored energy than the seeds of wild, naturally occurring plants. Why?

The seed does not start to grow until the water, temperature, and oxygen conditions are good for germination and growth (**FIGURE 18-20**). Sometimes this means waiting a matter of days or weeks before germination. For some species, though, the seeds remain dormant even in the face of ideal environmental conditions. Their seeds require something more before they will germinate. They may require passage through the gut of a bird or mammal to roughen up and weaken the seed coat. Or they may require exposure to fire. These seeds can remain dormant for dozens or even hundreds of years before finally germinating. Conversely, over the course of many generations of plant domestication, plant breeders are likely to have selected those seeds that germinate most quickly and have the fastest initial growth. These features may be optimal under agricultural conditions, where the field is cleared, pests are removed, and competition is limited.

Seed germination and growth begins when water, temperature, and oxygen conditions for life are just right.

FIGURE 18-20 The emerging plant.

There can be benefits to producing seeds with coats that require extra "processing" before they can germinate. (1) This can prevent the germination of seeds that have not been

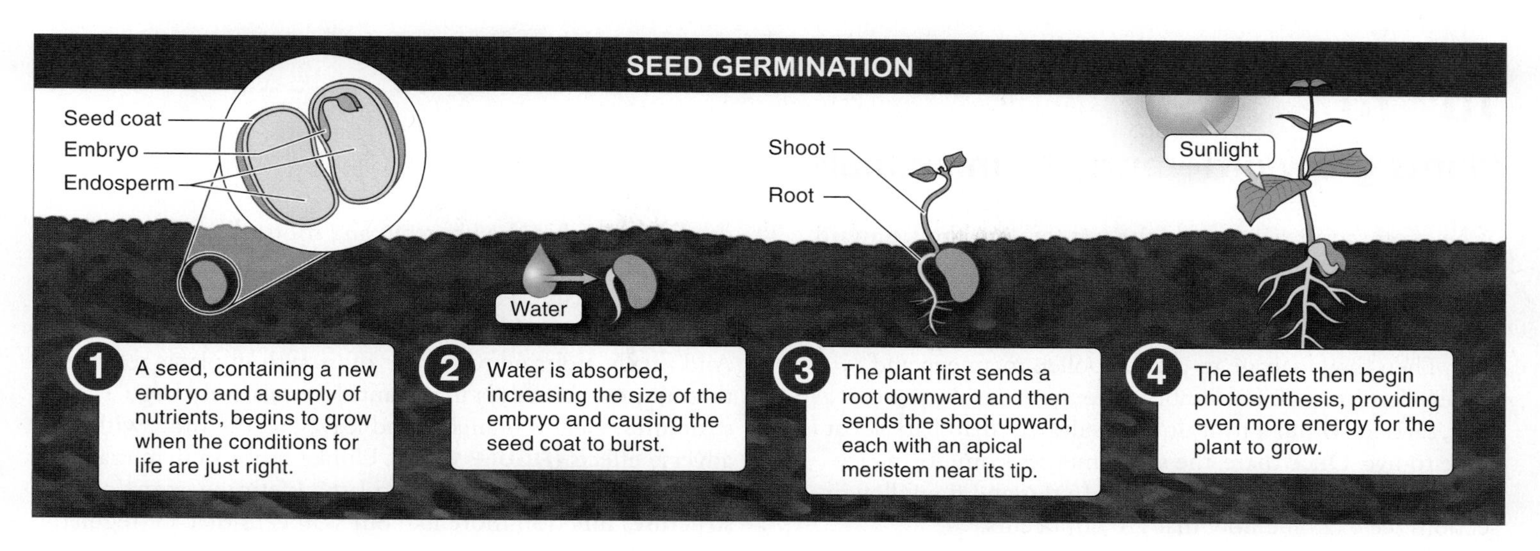

FIGURE 18-21 From seed to seedling: the initial growth of a plant.

dispersed sufficiently. (2) It can ensure that seeds are always left in a dollop of natural fertilizer. Or (3) it may ensure that there is a clearing where competing trees have been burned down and light is available before the seed germinates.

> "Though I do not believe that a plant will spring up where no seed has been, I have great faith in a seed. Convince me that you have a seed there, and I am prepared to expect wonders.
>
> —Henry David Thoreau, *Faith in a Seed*"

Such features, though, can also dangerously link the evolutionary success of the plant to the behavior of another species or specific ecological conditions. For example, some of the plants producing seeds that require the extreme heat or smoke from a fire to induce them to germinate are in trouble. As humans have reduced the incidence of wildfires, so, too, have we reduced the ability of many plants to germinate.

Q Animals (including humans) love to eat seeds. Why?

The whole purpose of a seed is to package an embryo with a ready-made supply of nutrients and energy so that it has the fuel to start growing before it can photosynthesize for itself. That high-energy package, of course, is exactly what animals are looking for in a food source, too. Consequently, seeds can be a valuable snack, rich in calories. They are also flavorful when ground up, producing spices.

On a somewhat minor note, a confusing issue—though not very important biologically—is the question, "Is a seed a nut?" In short, botanists have given the word "nut" a very specific and technical definition. (It is one particular type of fruit having a hard shell with no suture lines, or "seams," and with a single seed inside. An acorn is an example of a nut.) There is a much less restrictive, culinary definition of a nut, however, that encompasses most seeds. By this broader definition, virtually every seed that we eat is considered a "nut," including peanuts, almonds, cashews, and macadamia nuts—even though none of them would be classified as a nut by the botanist. The whole fruitless debate is a consequence of the common problem of words (such as "theory" or "evolution" or "altruism") having narrow, specific meanings in scientific disciplines but broader—perhaps sloppier, but usually more widely known—definitions in the everyday world outside science.

But, back to the seed, ready to begin its growth. With the proper processing and sufficient warmth, oxygen, and water, germination can finally begin. As it begins, water is absorbed, increasing the size of the embryo and causing the seed coat to burst. Utilizing the fat and starch reserves stored in the endosperm and embryo, metabolic activity increases dramatically and the plant begins growing (**FIGURE 18-21**). It first sends a root downward and then sends the shoot upward. In many plants, the cotyledons (after their energy reserves are used) can photosynthesize, providing even more energy for the plant. This early seedling growth involves activities occurring in the apical meristems, which we explore in Section 18-14.

TAKE-HOME MESSAGE 18•12

A seed, containing a new embryo and a supply of nutrients, begins to grow only when the water, temperature, and oxygen conditions for life are just right. Seeds sometimes must pass through an animal's digestive system before they can germinate. Initial growth utilizes fat and starch reserves stored in the endosperm and embryo.

18•13

Plants grow differently from animals.

When you must reach toward the sun for your food, upward growth enables you to tower over your competitors, but gravity can be your enemy. So far we have explored the reproductive processes of plants. First they make pollen and eggs. Then they manage to get the pollen to an egg and achieve fertilization. The embryo then remains dormant as a seed, carried within a fruit until it is dispersed and arrives at a place to live. Once there, the seed emerges from dormancy and germinates and begins to grow. In this and the following sections we examine how that growth occurs.

If you pay attention long enough, you will start to notice that plants and animals grow very differently. Three particular features stand out. First, in most animals, growth is determinate. In other words, after a period of maturation and "growing up," the growth more or less comes to an end. Most plants, though, just keep on growing taller and thicker for their whole life. Second, perhaps more oddly, despite their continuous growth, most trees are made up mostly of dead cells. Within the branches and roots, all of the water-conducting cells are actually dead (see Section 17-3), and the majority of a large tree trunk—the wood—is dead, too. The individual is a collection of living and dead cells. And finally, a distinctive aspect of growth in plants relative to that in most animals is that plants lose relatively large structures, such as branches and leaves, all the time, without adverse effects (**FIGURE 18-22**). Unlike limbs in mammals—critical appendages intended to last a lifetime—many plant structures function more like our skin cells that are regularly sloughed off.

Plants have two different methods of growing, each for a different direction of growth. **Primary growth** causes them to get taller, and **secondary growth** makes them thicker (and stronger). For both types, the cellular processes are similar and involve regions called **meristems.** These are clusters of active, dividing cells. Acting like human stem cells, the meristems contain perpetually youthful cells that can (1) repeatedly divide

FEATURES OF PLANT GROWTH

INDETERMINATE GROWTH
Most plants continue to grow—both taller and thicker—for their entire life.

CONSIST OF BOTH LIVING AND DEAD CELLS
Many plant cells, including water-conducting cells and those that make up wood, are dead.

LOSS OF STRUCTURES
Plants are able to lose relatively large structures, such as limbs and leaves, without adverse effects.

FIGURE 18-22 Defining characteristics of plant growth.

There are several features of plant growth that are very different from growth in animals.

TYPES OF PLANT GROWTH

Shoot apical meristem

Root apical meristem

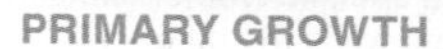

PRIMARY GROWTH
- Makes a plant taller and roots and branches longer
- Due to cell division within apical meristems present at the tips of shoots and roots

SECONDARY GROWTH
- Makes a plant stronger and roots and branches thicker
- Due to cell division within lateral meristems

FIGURE 18-23 Apical and lateral meristems are the source of plant growth.

and (2) develop into any type of plant tissue. Nowhere else in an adult plant body are there cells with these unlimited developmental options. So, even in a 4,000-year-old bristlecone pine tree, every time a bud breaks open and produces a shoot, the meristem cells have all of the growth and development potential of a cell in a brand new plant.

Gardeners know that if you cut the apical meristems of a shrub, the lateral meristem will cause the plant to grow outward, creating a familiar bush shape. Taken to the extreme, it is possible to sculpt dramatic shapes by taking snipping the apical and lateral meristem to force growth in a particular direction.

The two different types of plant growth depend on the activities of two different types of meristem (**FIGURE 18-23**).

1. *Apical meristems* are clusters of meristem cells at the ends of roots and shoots or branches that repeatedly divide to produce primary growth, making the plant taller and plant parts (roots and branches) longer. This is in dramatic contrast to human growth. It would be as if we grew by continuously adding more material on top of our heads.

2. *Lateral meristems*—not present in all plant species—give rise to secondary plant growth, the thickening of trunks and branches and the formation of wood and bark.

TAKE-HOME MESSAGE 18•13

Plants generally grow for their entire life, using two types of growth. Primary growth makes plants taller and plant parts longer and forms new tissues. Secondary growth makes plants thicker and sturdier.

18•14

Primary plant growth occurs at the apical meristems.

So, as we've seen, primary growth makes shoots taller and roots and branches longer and forms new tissues such as buds and leaves. Secondary growth, which we examine in Section 18-15, makes plants thicker and is responsible for the production of wood and the development of thick sturdy tree trunks.

All primary growth occurs as a result of cell division in apical meristems (**FIGURE 18-24**). The cells in meristems are *totipotent* meaning they have the potential to develop into any type of cell the plant produces. Growth occurs as a meristem cell divides, creating two new cells. One differentiates into a specific type of tissue such as xylem, phloem, or perhaps a storage cell. The other cell remains a meristem cell. Thus, the process of growth and differentiation can occur while the meristem remains as a perpetual source of new cells. Recall that apical meristem cells are present at all root tips and shoot tips. They reside just behind a small cap of cells, at the very end of the tip, that protects the meristem.

Apical meristem cells are also responsible for the production of branches as a plant grows. In the shoot, as the meristem cells divide, pushing the top of the plant higher and higher, some meristem cells are left behind at regular intervals. The meristem cells left behind are dormant but can begin dividing at any time—usually after stimulation by plant hormones, which we discuss in Chapter 19—pushing outward and forming a branch. This process is then repeated. As the branch grows, the meristem leaves behind some cells near each bud, allowing further branching. Some of the new cells become xylem and phloem, and the vessels join up with the central circulatory vessels within the stem.

Q What are the knots in wood?

Most timber used in construction comes from pieces of wood cut from the tree along the length of its trunk. Knots are places where branches were connected to the trunk and where the vascular tissue of the branches merged with that of the trunk. Knots become more and more deeply embedded within the trunk as secondary growth makes the trunk thicker.

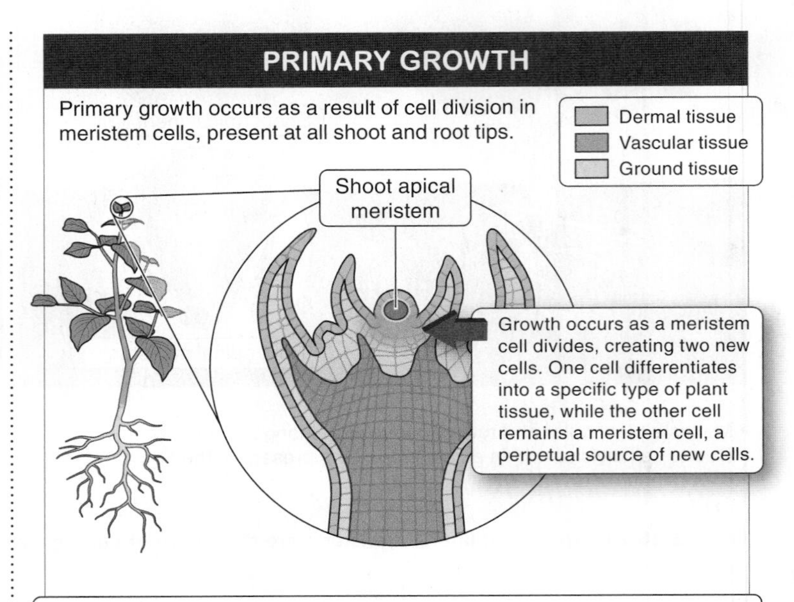

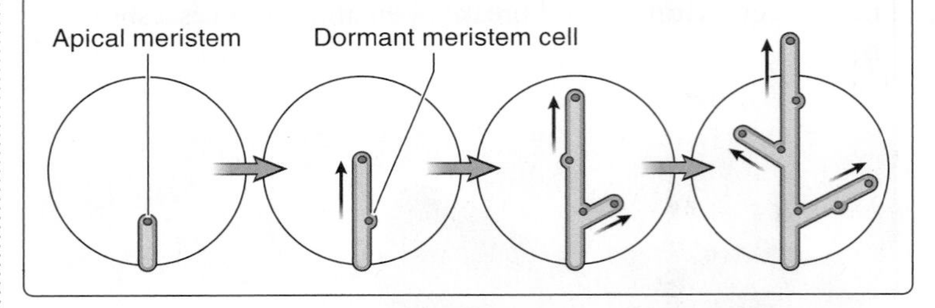

FIGURE 18-24 Growing taller: primary growth.

TAKE-HOME MESSAGE 18•14

Plant growth occurs as a result of cell division in meristems, small collections of totipotent cells. Primary growth—the lengthening of stems, branches, and roots and the formation of new tissues such as buds and leaves—results from the division of apical meristem cells.

18•15

Secondary growth produces wood.

Wood is amazing. And not just from an aesthetic or utilitarian perspective—although it is quite remarkable in that way, too (**FIGURE 18-25**). Wood is mostly amazing as an almost impossibly elegant evolutionary solution that allows plants to grow to dizzying heights. A shoot cannot simply grow taller and taller by continued cell division at the apical meristem, for the same reason that a small house cannot have floor upon floor added until it is a hundred stories tall. The plant and the house would tip over. Increased height demands increased structural rigidity. In addition, the top floors of the tall building must receive plumbing and electrical service, just as the top of a tree must have vascular tissue. The beauty of wood is that it serves both purposes: it confers strength, keeping the tree from toppling over, while simultaneously serving as the plumbing conduit for water and minerals. And it does this while being tremendously lightweight.

We saw that primary growth in plants comes from cell division and extension at the apical meristems in root and shoot tips. Once the apical meristems have moved past a location in a stem or trunk, that part of the plant can no longer elongate and become taller. It can, however, grow outward and become thicker. Here we investigate the process of secondary growth by which eudicots and some other angiosperm plants produce wood, acquiring an increased capacity for water conduction and becoming thicker and sturdier in the process.

How does secondary growth occur? Imagine a cross section of a woody plant. It is packed with a series of concentric cylinders, one within another, progressing outward from the center (**FIGURE 18-26**). First, extending from the center out to a point about halfway to the outer circumference, is the **pith,** soft spongy parenchyma cells. Just beyond this is a thin

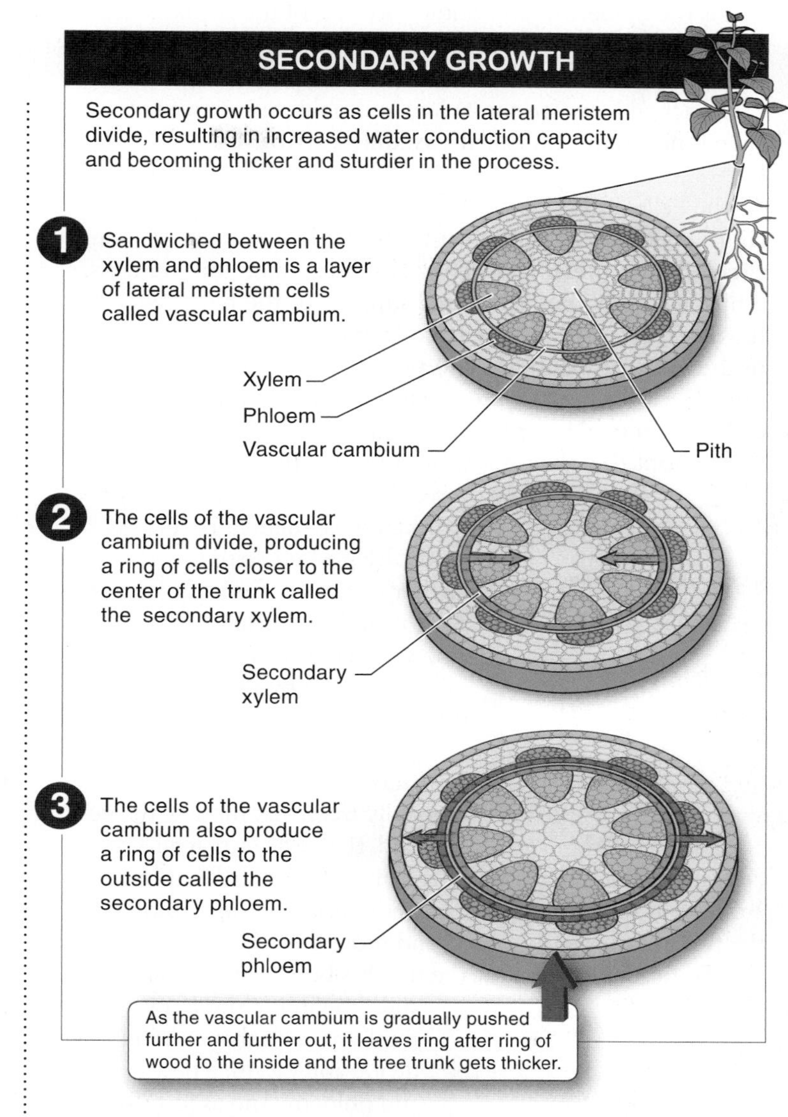

FIGURE 18-26 **Becoming thicker and sturdier: secondary growth.**

FIGURE 18-25 **Wood is of great value to humans.**

cylinder of vascular tissue: the xylem. A bit farther out is another cylinder of vascular tissue: the phloem. And beyond that is the epidermis and outer covering of the trunk. Recall that meristems contain perpetually youthful cells that can repeatedly divide and develop into any type of plant tissue. Sandwiched between the xylem and phloem is a layer of lateral meristem cells called **vascular cambium.**

As the cells of the vascular cambium divide, they produce a ring of cells closer to the center of the trunk. These cells are **secondary xylem** cells. They conduct water and minerals, while also providing structural support to the plant. We call these secondary xylem cells, collectively, "wood." The cells of the vascular cambium also produce a ring of cells to the outside, closer to the outer circumference of the trunk. These cells are **secondary phloem,** serving to conduct sugars throughout the plant body. Because it is producing cells toward the inside of the stem or trunk, the vascular cambium is gradually pushed farther and farther out, away from the center. As it is pushed farther out, it leaves ring after ring of wood to the inside, and the tree trunk gets thicker.

Q If you discover that only a tiny percentage of the cells in a tree are alive, does that mean the tree is alive or dead?

It is an odd fact about plants, especially trees, that most of the cells are dead. Remember that all xylem cells, including the secondary xylem cells that make up wood, are dead at maturity. In a huge oak tree, for example, most of the tree's mass is in the wood and more than 98% of the cells may be dead. Nonetheless, the tree has buds that keep forming and growing, and it has vascular cambium cells that continue to divide and produce new xylem and phloem. And as long as there are living cells in a tree, it can grow and is considered to be alive.

Q If you thoughtlessly carved your initials in the trunk of a tree, 4 feet above the ground, and came back in 10 years, would your initials be at the same height or higher? Why? What if you came back in 20 years?

Secondary growth increases the girth of a tree but doesn't elongate it. So if you were to carve your initials in the trunk of a tree, the section below where your initials are will never increase in height. Moreover, because increases in girth can cause the outer layers of bark and epidermis to fall off, replaced by new cells pushing outward, your initials will eventually be completely gone from the tree trunk.

Secondary xylem functions just like primary xylem: it provides structural support and it serves as a conduit for water. But not all xylem continues to serve both functions. Over time, xylem becomes clogged with oily molecules. While most xylem initially acts like a long straw through which water can pass, older xylem is like a clogged artery through which less and less blood can flow. Because the clogging of xylem is a progressive condition, the older the xylem, the more likely it is to be clogged. And the closer it is toward the center of the trunk, the

WOOD STRUCTURE

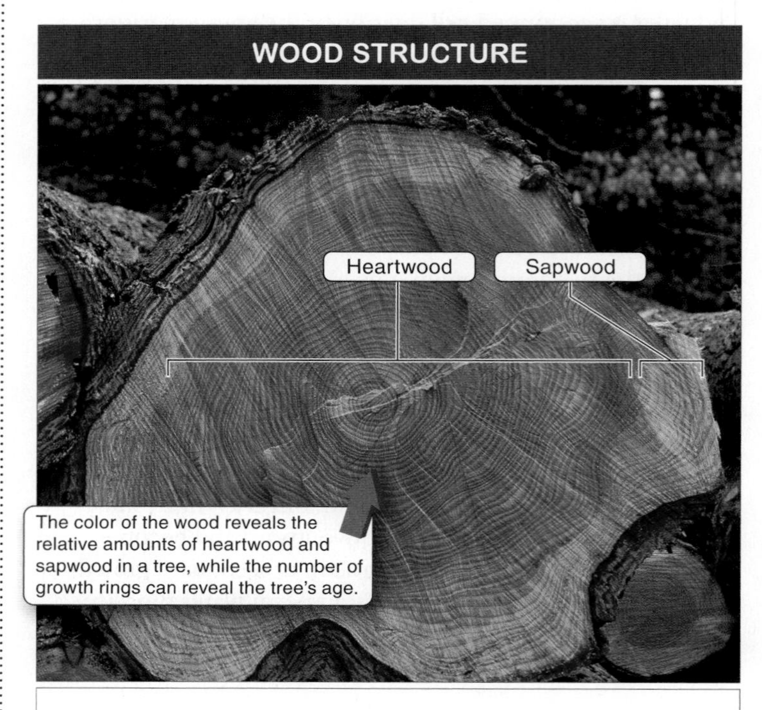

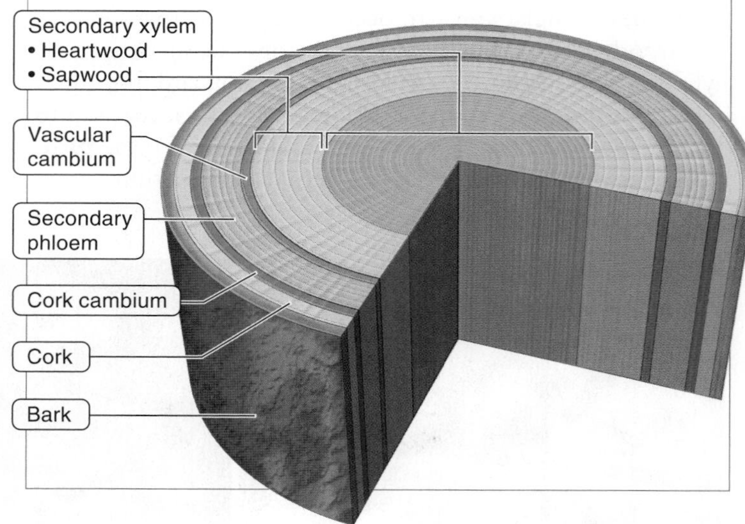

FIGURE 18-27 Cross section of a tree trunk. Most of the cells are secondary xylem, also known as wood, and are dead. Secondary growth occurs in a ring called the vascular cambium, a lateral meristem, which produces new xylem on the inside and new phloem to the outside.

older the xylem is. The clogged xylem is called "heartwood" and is darker, due to the accumulated resins and other non-soluble metabolic waste molecules. Although heartwood cannot conduct water, it still is valuable to the plant for the structural support it provides. Farther toward the outside, closer to the vascular cambium, is the xylem that was more recently created and, consequently, better at conducting water. This xylem is called "sapwood" (**FIGURE 18-27**).

Looking at a cross section of a tree, we find that the color of the wood reveals the relative amounts of heartwood and sapwood. It can also reveal the age of the tree, in the growth rings. In temperate climates, more growth—that is, greater rates of cell division in the vascular cambium—occurs during the spring and summer, causing a wider, lighter band. During winter and fall, growth is slower and produces a darker band. Wider bands also indicate wetter years, and narrower bands dryer years. By counting the number of rings we can accurately determine the age of the tree. Because there is less variation in growth from one season to another in the tropics, trees don't always lay down rings and it isn't always possible to accurately establish their age.

Farther out from the vascular cambium, closer to the outer edge of the trunk, is another ring of dividing cells. Called "cork cambium," this cylinder of cells is a second lateral meristem and plays an important role in the perpetual task of maintaining the protective covering of bark. As the vascular cambium increases the girth of a tree and is pushed farther and farther out, the epidermal cells that make up the bark continuously split and fall from the tree. This tissue is replaced by the cork cambium. As its cells divide, they produce a cylinder of waxy cork cells to the outside, a layer of cells that protects the outer surface of the trunk from water loss, fire, and microbe infection. Secondary phloem, too, gets pushed toward the outer surface of the trunk and eventually gets sloughed off. For this reason, at any given time, although there may be a huge amount of functioning xylem, there is only a very thin layer of functioning phloem, and it is very close to the outer surface of the trunk. Often only the phloem that has been produced within the past year functions in the conduction of sugars throughout the plant.

Q Bumping into a tree repeatedly with a lawn mower or "weed whacker" can kill the tree. Why?

Actions that damage the bark of a tree all the way around its circumference are called "girdling." It is possible to girdle a tree with a "weed whacker" trimmer or by repeatedly bumping into the tree with a lawn mower. This can actually kill a tree, even if the tree is very thick and the damage is not particularly deep. Because the phloem and vascular cambium layers are so close to the outer edge of the trunk, relatively shallow injuries to the trunk can destroy the secondary phloem and, if the damage goes just a bit deeper, the cells that give rise to new phloem.

FIGURE 18-28 Girdling a tree can kill it.

When such damage occurs, the tree is still able to transport water from the roots to all of the tissues, because the xylem is much closer to the center of the trunk and isn't damaged. But damage to the phloem makes it impossible for the tree to transport sugars and other nutrients from the sites of photosynthesis down to the roots where it is needed and stored. Shortly after the damage occurs, the tree may die (**FIGURE 18-28**).

Roots can be girdled, too. This can happen when trees are planted with wire or mesh netting around their roots. As the roots grow in girth, the metal or nylon can gradually destroy the vascular cambium as it grows outward. Death from root girdling usually occurs about 5 to 10 years after the tree is planted.

TAKE-HOME MESSAGE 18•15

Secondary growth results from cell divisions in a thin cylinder of tissue between the primary xylem and the primary phloem—the vascular cambium, a lateral meristem. As this tissue divides, it produces a ring of non-living xylem cells, closer to the center of the trunk, that conduct water and minerals, while also providing structural support to the plant. We call these cells, collectively, wood.

Knowledge You Can Use

When is a fruit not a fruit? *Or,* You can't always trust your government when it comes to botany.

Q: What is a fruit? A fruit is defined as a structure that aids in dispersing seeds—it is the reproductive packet that is made up of the embryo, some food reserves, and a hard coat, and usually develops from the ovary, right around the seeds, which develop from the ovules. This should cause you to look at a field of flowers differently: the ovary is part of every flower, so all of the flowers that you see, after they are pollinated and fertilized, turn into fruits. Or conversely: every fruit you eat used to be a flower.

Q: How do fruits differ from vegetables? Vegetables, unlike fruits, are simply edible parts of a plant that develop from structures other than the reproductive structures—and hence do not have seeds.

Q: Get your laws out of my salad! In 1893, in a bizarre legal case at the intersection of law and biology, the United States Supreme Court made a foray into botany. It started when the United States government instituted a 10% tax on all vegetables imported into the country. Fruits, on the other hand, were not subject to any tariffs. The tax collector then tried to collect from a company that imported tomatoes. The company refused to pay, however, arguing that tomatoes are fruits.

But sometimes, a fruit is not a fruit. Here is what the Supreme Court ruled: "Botanically speaking, tomatoes are the fruit of a vine, just as are cucumbers, squashes, beans, and peas. But in the common language of the people … all these are vegetables which are grown in kitchen gardens, and … usually served at dinner … or after the soup, fish, or meats which constitute the principal part of the repast, and not, like fruits generally, as dessert." And so it was decided: tomatoes would be taxed and, in the eyes of the law, they are vegetables.

1 Plants can reproduce sexually and asexually.

Most plants have two very different options for reproduction: asexual and sexual reproduction. Asexual reproduction—the growth of new plants directly from the tissue of an established plant through mitosis—can be energetically efficient and fast and can preserve successful genetic combinations. But it also has disadvantages, including producing offspring with reduced genetic variability. Sexually reproducing plants produce flowers as the chief structure for sexual reproduction.

2 Flowers have several roles in plant reproduction.

Flowers are plant structures specialized for sexual reproduction, and all generally have the same fundamental parts: sepals, petals, stamens, and a carpel. The male reproductive structure produces pollen grains, two-cell structures that are water-tight and have a sticky surface. One of the cells will form a pollen tube, and the other will produce two sperm cells. Within the ovary, diploid cells differentiate into ovules, each of which is a group of outer protective cells around a diploid egg-producing cell. The egg-producing cell undergoes meiosis to form haploid megaspores, one of which undergoes mitosis several times to produce the embryo sac, the structure that contains the egg and is the site of fertilization.

3 Pollination, fertilization, and seed dispersal often depend on help from other organisms.

Sexual reproduction requires that two plants must get their gametes together. Plants trick or bribe animals into bringing this about. Following pollination, a pollen tube must grow into the ovule, where one sperm cell fertilizes the egg cell in the embryo sac, and another fuses with the diploid central cell of the embryo sac to form endosperm, which will nourish the developing embryo. Plants can reduce the occurrence of self-fertilization in several ways. Following fertilization, the ovule develops into a seed, surrounded and protected by a fruit. Plants use the assistance of animals or water or wind to disperse their seeds and fruits.

4 Plants have two types of growth, usually enabling lifelong increases in length and thickness.

Unlike most animals, plants generally grow for their entire life. Plants have two types of growth. Primary growth makes shoots taller and roots and branches longer and forms new tissues. Secondary growth makes plants thicker and sturdier and is responsible for the production of wood.

KEY TERMS

alternation of generations, p. 663
anther, p. 666
carpel, p. 666
cotyledon, p. 675
double fertilization, p. 673
embryo sac, p. 669
endosperm, p. 673
filament, p. 666
fruit, p. 675
germinate, p. 663
life cycle, p. 663
meristem, p. 680
nectar, p. 671
ovary, p. 666
ovule, p. 667
petal, p. 666
pith, p. 683
pollen grain, p. 666
pollen tube, p. 668
pollination, p. 671
primary growth, p. 680
root meristem, p. 675
secondary growth, p. 680
secondary phloem, p. 684
secondary xylem, p. 684
seed, p. 675
sepal, p. 666
shoot meristem, p. 675
sperm, p. 668
spore, p. 663
stamen, p. 666
stigma, p. 666
style, p. 666
vascular cambium, p. 684
zygote, p. 673

CHECK YOUR KNOWLEDGE

1. Asexual reproduction in plants:
   a) tends to be particularly common in a population growing in a marginal habitat.
   b) still requires meiosis.
   c) requires reproductive cells that are produced in flowers.
   d) is energetically more expensive than sexual reproduction.
   e) Both b) and c) are correct.

2. Most of the trees or shrubs in backyard gardens:
   a) are haploid.
   b) are the products of asexual reproduction.
   c) require both male plants and female plants in close proximity.
   d) utilize cell respiration rather than photosynthesis.
   e) must exhibit an alternation of generations when propagated asexually.

3. In angiosperms, the ______ are the specialized reproductive organs whose main function is to bring together the sperm and egg.
   a) seeds
   b) pollen grains
   c) cones
   d) leaves
   e) flowers

4. Each angiosperm pollen grain contains how many sperm cells?
   a) two
   b) one
   c) millions
   d) thousands
   e) hundreds

5. Mosses and ferns differ in their reproductive strategies from gymnosperms and angiosperms in which of the following ways?
   a) Mosses and ferns have much larger seeds than do angiosperms and gymnosperms.
   b) Mosses and ferns use wind pollination, whereas angiosperms and gymnosperms use insects for pollination.
   c) Mosses and ferns are primarily diploid in their adult (reproductive) form, whereas gymnosperms and angiosperms are primarily haploid.
   d) Mosses and ferns are primarily haploid in their adult form, whereas gymnosperms and angiosperms are primarily diploid.
   e) Mosses and ferns rely on liquid water for fertilization, whereas angiosperms and gymnosperms do not need liquid water for fertilization.

6. The common term for the action of transferring pollen grains from an anther onto a stigma is:
   a) pollination.
   b) reproduction.
   c) fertilization.
   d) intercourse.
   e) None of the above is correct.

7. Flowering plants attract animals to assist in pollination. Which of the following animal groups is not a common pollinator of flowering plants?
   a) bats
   b) bumblebees
   c) moths
   d) beetles
   e) All of the above are common pollinators of flowering plants.

8. "Double fertilization" in angiosperm plants refers to:
   a) the release of two sperm from a single pollen grain.
   b) the production of two or more eggs in the ovary.
   c) the fusion of the ovary with the anther.
   d) the fusion of a sperm cell with two nuclei of the endosperm-forming cell.
   e) the fusion of one sperm cell with the egg, and another with two nuclei of the endosperm-forming cell.

9. Which of the following is a likely way in which plants increase dispersal of their fruit?
   a) Fruits are conspicuously colored.
   b) Fruits taste good.
   c) Fruit colors attract female birds.
   d) Both a) and b) are true.
   e) Both a) and c) are true.

10. Which of the following is the proper sequence of events occurring when a flowering plant reproduces?
    a) meiosis; pollination; fertilization; embryo formation
    b) fertilization; meiosis; nuclear fusion; endosperm formation
    c) growth of the pollen tube; pollination; germination; fertilization
    d) meiosis; fertilization; growth of the pollen tube; germination
    e) mitosis; embryo formation; pollination; seed development

11. For angiosperms, which of the following is the most correct statement about growth?
    a) Vascular cambium increases girth; cork cambium increases length.
    b) Apical meristems increase length; vascular cambium increases girth.
    c) Cork cambium increases length; apical meristems increase girth.
    d) Apical meristems increase length; apical meristems increase girth.
    e) Apical meristems increase girth; vascular cambium increases length.

12. When, after seed germination, does photosynthesis begin?
a) when the seed coat breaks open
b) when the first foliage is formed
c) when the shoot emerges from the soil
d) the second the seed begins to germinate
e) when the root is formed

13. The scientific term for wood is:
a) primary phloem.
b) cork.
c) secondary xylem.
d) epidermis.
e) pith.

14. When you are seven years old, you scratch your name, 4 feet above the ground, into a 12-foot-tall bamboo plant. You spend the next 10 years in a special home for vandals. When you get out, the bamboo is 24 feet tall. At that time, how far above the ground will your name be?
a) 16 feet
b) 4 feet
c) 8 feet
d) 12 feet
e) 2 feet

Short-Answer Questions

1. A strawberry plant sends out about a dozen above-ground runners that have small crowns at the ends. When these crowns touch the soil, they send out roots, and each crown can develop into a new plant. Why is this good news for someone trying to create a strawberry patch?

2. Potato plants can reproduce sexually, producing flowers, fruits, and seeds. A potato farmer can collect the seeds and plant them. Alternatively, she can plant some of the potatoes instead. Why might it be preferable to plant potatoes instead of seeds? And which potatoes should be chosen for planting?

See Appendix for answers. For additional study questions, go to www.prep-u.com.

1 Hormones regulate growth and development.

- **19.1** Hormones help plants respond to their environments.
- **19.2** Seed germination and stem elongation are stimulated by gibberellins.
- **19.3** Seedlings grow and properly orient themselves under the direction of auxins.
- **19.4** Other plant hormones regulate flowering, fruit ripening, and responses to stress.
- **19.5** Tropisms influence plants' direction of growth.
- **19.6** Plants have internal biological clocks.
- **19.7** With photoperiodism and dormancy, plants detect and prepare for winter.

2 Plants defend themselves from attack by herbivores.

- **19.8** Plants actively resist being eaten by herbivores.

3 Plants can survive in extreme environments.

- **19.9** Special adaptations help some plants thrive in extreme habitats.

19

Plants Respond to Their Environments

Regulating and defending while rooted in the ground

❶ Hormones regulate growth and development.

Plant hormones regulate the growth of this enormous pumpkin.

19 • 1

Hormones help plants respond to their environments.

Somewhere, somehow, someone put forth the metaphor that living organisms have traits that are "hard-wired" in their genes. With almost no exceptions, this is wrong. Consider a simple experiment with the arrowleaf plant (*Sagittaria sagittifolia*) as an example. First, from a single plant, take three cuttings—sections of the plant that contain part of the stem and can be used to produce a new plant, genetically identical to the original. Plant each in a different habitat: one in deep water, the second in shallow water, and the third on land. Each new plant has the exact same set of genes; they are like identical triplets. But in the deep water, the plant will have long, ribbon-like leaves. In the shallow water, the plant will have large, round leaves resembling lily pads. And on land, the plant will have the arrow-shaped leaves from which it gets its name (**FIGURE 19-1**).

The arrowleaf plant is not "hard-wired" to produce one type of leaf shape. Rather, the physical form of the plant depends on the environment in which it grows. How does the plant change its growth patterns to respond to different environments? Chemicals called **hormones** respond to environmental variables (such as amount of moisture, amount and direction of sunlight, and temperature) and influence the growth and development of the plants. Hormones make it possible for the arrowleaf plant to change its growth patterns to respond to different environments.

Plant hormones are chemical signals. They are produced in various places within the plant, and they may have their effect in the place where they are produced or may be transported to another part of the plant before they take effect. Regardless of where they are produced, the hormones convey information about the physiological state of the plant's tissues or the environment in which the plant finds itself. The hormones then regulate some aspect of the metabolism of the plant's cells. Sometimes, a hormone may stimulate a certain response—such as growth—in the target cell. Other times, the hormone may

PLANTS RESPOND TO THEIR ENVIRONMENT

Three cuttings are taken from the same plant and grown in different environments.

Deep water — The plant grows long, ribbon-like leaves.

Shallow water — The plant grows large, round leaves.

Land — The plant grows arrow-shaped leaves.

The leaves of the arrowleaf plant take on dramatically different forms depending on the environment in which the plant grows.

FIGURE 19-1 Environment affects the growth patterns of the arrowleaf plant. The differences in growth forms are mediated by hormones.

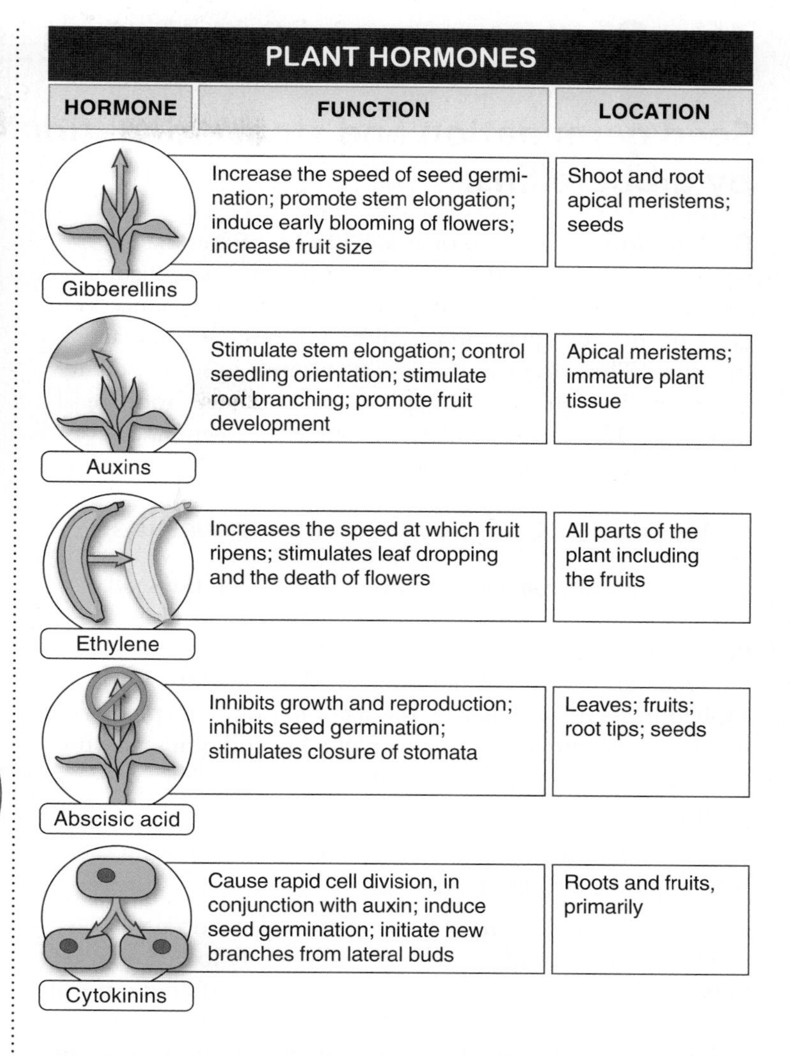

PLANT HORMONES

HORMONE	FUNCTION	LOCATION
Gibberellins	Increase the speed of seed germination; promote stem elongation; induce early blooming of flowers; increase fruit size	Shoot and root apical meristems; seeds
Auxins	Stimulate stem elongation; control seedling orientation; stimulate root branching; promote fruit development	Apical meristems; immature plant tissue
Ethylene	Increases the speed at which fruit ripens; stimulates leaf dropping and the death of flowers	All parts of the plant including the fruits
Abscisic acid	Inhibits growth and reproduction; inhibits seed germination; stimulates closure of stomata	Leaves; fruits; root tips; seeds
Cytokinins	Cause rapid cell division, in conjunction with auxin; induce seed germination; initiate new branches from lateral buds	Roots and fruits, primarily

FIGURE 19-2 The location and function of five types of plant hormones.

suppress an action in the target cell. Put in these terms, these processes all sound very abstract. As we will see, however, the effects of plant hormones are usually predictable and easily observed. Hormones are powerful mediators that enable plants to respond quickly and appropriately to changes in their environmental conditions.

There are five major types of plant hormones: gibberellins, auxins, ethylene, abscisic acid, and cytokinins. **FIGURE 19-2** provides a brief summary of the functions of these hormones. (We explore them in greater detail in the next three sections.)

Although both plants and animals produce chemicals called hormones, these aren't exactly the same sort of signals. Animal hormones are generally produced in specific hormone-producing glands or tissues and then transported by the bloodstream to different locations where they exert their influence. In contrast, plant hormones, as we'll see, are produced in many different types of cells throughout the plant and sometimes have their influence right at the point of production.

TAKE-HOME MESSAGE 19•1

Plant hormones are chemical signals produced by plant cells that enable the plant's responses to environmental variables (such as amount of moisture, amount and direction of sunlight, and temperature) and influence its growth and development.

19•2

Seed germination and stem elongation are stimulated by gibberellins.

Gibberellins are a group of about 125 hormones that regulate a plant's growth processes, primarily by stimulating cell division and cell elongation. Gibberellins are produced in growing areas of a plant and are found throughout all plant structures. They occur in the largest amounts in seeds, but are also found in high concentrations in the shoot and root apical meristems. The areas that have the greatest concentrations of gibberellins experience the most dramatic growth.

Gibberellins have four main types of effects (**FIGURE 19-3**):

1. *Speeding seed germination.* Gibberellins within the seed initiate the production of enzymes that make possible the breakdown and metabolism of nutrients stored in the seed's endosperm. When additional gibberellins are applied to seeds, whether by the plant biologist, farmer, or horticulturist, the seeds can more quickly and efficiently use their energy reserves to germinate.
2. *Stem elongation.* Some gibberellins affect stem elongation by increasing the distance between nodes, thus spacing the branch points farther apart.
3. *Inducing early blooming of flowers.* Some plants do not produce flowers until the days are sufficiently long to trigger flowering, or until the plant is exposed to a certain degree of coldness (see Section 19-7). The application of gibberellins to such plants can cause flower production in the absence of the triggering event.
4. *Enlargement of fruits.* One of the most important economic use of gibberellins in the United States is the production of table grapes. When seedless grapes are sprayed with large amounts of gibberellins, the grapes grow larger and, due to the stem-elongation effect also induced by gibberellins, there is more space between the grapes on the bunch. Both of these effects increase the attractiveness of a bunch of grapes to shoppers. (Normally plants don't ripen ovary walls, producing fruits, unless they have set seeds and those seeds need to mature. After all, that would be a waste of energy.)

Q How can farmers grow radishes the size of beach balls?

Gibberellins are about as close to "miracle-grow" chemicals as exist. If one type of gibberellin is sprayed on radish plants as they are growing, the radishes exhibit explosive growth. A radish is normally smaller than a golf ball, but gibberellin-treated radishes can reach the size of beach

THE EFFECTS OF GIBBERELLINS

SPEEDING SEED GERMINATION
Gibberellins initiate the production of enzymes that help break down nutrients stored within the seed's endosperm, allowing quicker and more efficient use of the seed's energy reserves.

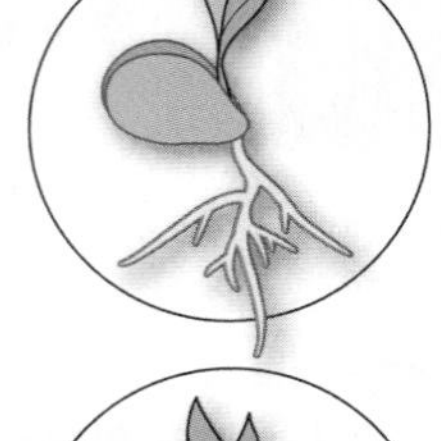

STEM ELONGATION
Gibberellins affect stem elongation by increasing the distance between nodes, thus spacing the branch points farther apart.

INDUCING EARLY BLOOMING OF FLOWERS
Gibberellins can cause flower production in the absence of a triggering event from the external environment.

ENLARGEMENT OF FRUITS
Seedless grapes sprayed with large amounts of gibberellins grow larger and, due to the stem-elongation effects, have more space between the grapes on the bunch.

FIGURE 19-3 Physiological effects of gibberellins.

FIGURE 19-4 A giant, gibberellin-treated cabbage.

balls. Equally dramatic is the effect of another type of gibberellin on cabbage plants. Normal cabbage plants are 1 to 2 feet high. When sprayed with a gibberellin, though, the plants can grow to more than 10 feet tall (**FIGURE 19-4**)! In these extreme cases, plants are exposed to far greater hormone concentrations than they ever would encounter under natural conditions. Nonetheless, these examples reveal the powerful growth-stimulating effects of gibberellins.

If plants grow more when they have greater amounts of gibberellins, why don't all plants simply produce more gibberellins so that they can outgrow other plants competing for sunlight? The answer is that, as in most aspects of life, there are trade-offs. The fast-growing cabbages, for instance, can grow to great heights only if humans assist them by securing them to a stake to support the stem. Under natural conditions, they wouldn't have our help, nor would their stems naturally have the strength to stand at a height of 10 feet. Instead, the cabbage plants would fall over, wilt, and die. It seems that under natural conditions, plants produce just about the right amount of gibberellins to maximize their successful growth and reproduction.

TAKE-HOME MESSAGE 19•2

Gibberellins are a group of about 125 hormones, produced primarily in meristems and seeds, that regulate a plant's growth processes, mainly by stimulating cell division and cell elongation.

19•3

Seedlings grow and properly orient themselves under the direction of auxins.

Auxins are a small group of naturally occurring hormones (and a larger group of synthetic variants that chemists can produce) that play several important roles in stimulating and regulating a plant's growth and development. They are found primarily in shoot tips and immature plant tissue, such as young leaves. The primary role of auxins in plant growth is to stimulate the expression of genes that promote cell division, stem elongation, the formation of roots, and the formation of vascular cambium. Auxins also influence plant orientation, making sure the correct ends are up and down.

The chief effects of auxins are of four types (**FIGURE 19-5**):

1. *Stimulating shoot elongation.* Auxins enhance the effect of gibberellins in shoot elongation.

2. *Controlling seedling orientation.* Charles Darwin and his son Francis were the first to document that seedlings exposed to light bend in whichever direction the light is coming from, whether it is up, down, or sideways. Immediately after emerging from the seed, the shoot grows as if it knows which way is up. Whether a shoot grows down into the earth or up toward the light, though, actually depends on (1) where the auxins are located and (2) how the auxins influence the cells in those specific locations. Like gibberellins, auxins are produced near the growing tips of shoots and branches, but unlike gibberellins, they don't remain there. The auxin molecules move in two directions within a cell: they are pulled downward by gravity, and they move laterally away from light. These two movements distribute the auxin molecules unevenly within the plant.

THE EFFECTS OF AUXINS

STIMULATE SHOOT ELONGATION
Auxins enhance the effect of gibberellins in shoot elongation.

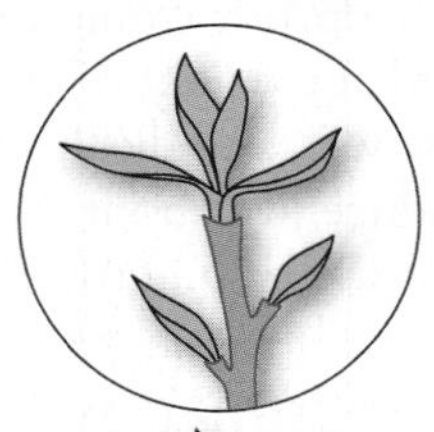

CONTROL SEEDLING ORIENTATION
Auxins direct the growth of shoots and roots, making sure the correct ends are up and down.

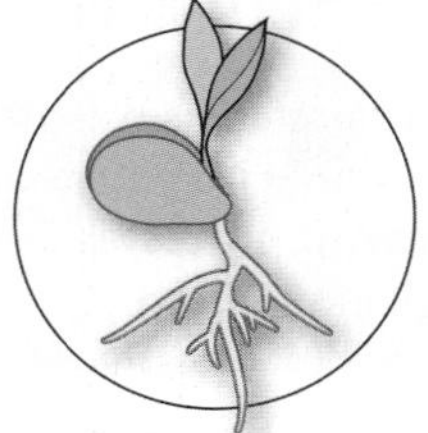

STIMULATE ROOT BRANCHING
Auxins induce the formation of roots.

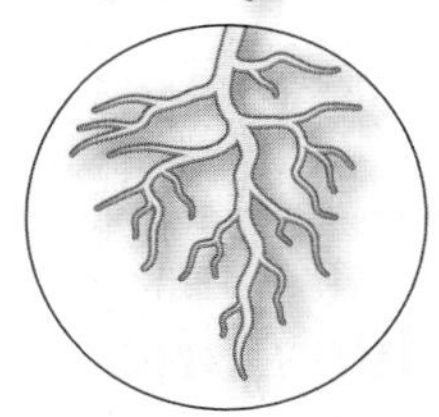

PROMOTE FRUIT DEVELOPMENT
Auxins produced within an embryo promote the maturation of the ovary wall and development of the fruit.

FIGURE 19-5 Physiological effects of auxins. These bear grass stems are reaching toward the sunlight.

AUXINS' INFLUENCE ON PLANT ORIENTATION

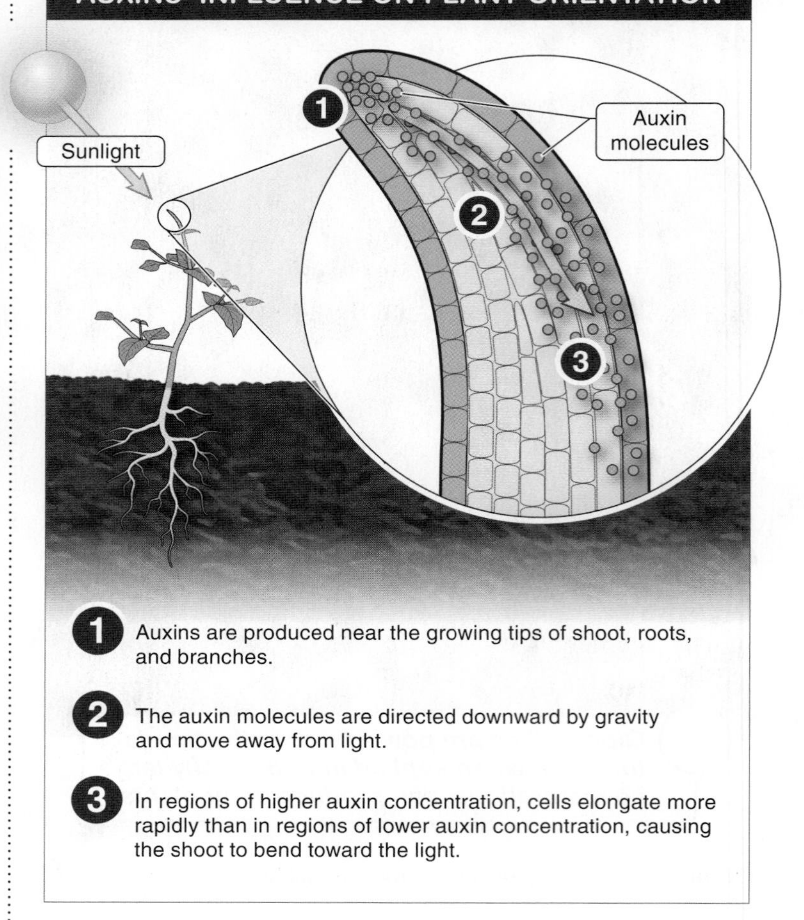

1. Auxins are produced near the growing tips of shoot, roots, and branches.
2. The auxin molecules are directed downward by gravity and move away from light.
3. In regions of higher auxin concentration, cells elongate more rapidly than in regions of lower auxin concentration, causing the shoot to bend toward the light.

FIGURE 19-6 Auxins cause a plant to grow toward light.

In regions of higher auxin concentration, cells in the stem elongate more rapidly than in regions of lower auxin concentration (**FIGURE 19-6**). This difference in growth occurs because auxins increase the usually rigid cell wall's flexibility (so that it can grow) and increase its permeability (so that it can take on more water and expand). The overall effect on growth is best seen when a plant is tipped onto its side. First, the auxins flow downward to the bottom side of the now-horizontal shoot. The bottom side then elongates more rapidly than the top side, causing the shoot to bend upward, away from gravity.

3. *Stimulating root branching.* Auxins induce the formation of roots. In fact, it is possible to buy auxin powders that can be dusted onto the bottom of plant stem cuttings. The powder causes the stems to send out numerous roots (transforming some stem cells into root cells), so that the shoots can be planted and will form new plants.

> **Q** Can we trick plants into producing seedless fruits for our convenience?

4. *Promoting fruit development.* Auxins produced within an embryo have several roles in the development of fruit. First, they promote the maturation of the ovary wall, and then they promote several steps in the full development of the fruit. It is even possible to trick plants into producing fruits when their flowers have not been fertilized. Applying auxins to unfertilized flowers can cause them to develop into fruits. Because there has been no fertilization, such fruits are always seedless (an example is seedless tomatoes). Auxins applied to fruiting plants can also prevent the fruits from prematurely dropping from the tree or vine. This makes it easier and more inexpensive to harvest fruits, because they can all be picked at one time.

Too much of a good thing can be terrible, and this is particularly true when it comes to plant hormones. Auxins, for example, are among the chief growth-stimulating hormones in plants, but in extreme concentrations they are deadly. While this is bad news if you are a plant, it can be good news if you are a human trying to kill weeds (**FIGURE 19-7**). When synthetic auxins are sprayed on plants in higher concentrations than any plant would normally experience, they cause the plants to begin growing uncontrollably. This doesn't last long, though. Like a mismanaged start-up company, the plants devote so much of their energy budget to growth that they quickly find themselves without sufficient energy for essential metabolic maintenance functions. And then they die. One type of synthetic auxin, called 2,4-D, is a particularly useful weed killer because it kills only eudicots—such as dandelions—while not harming monocots, including grasses and cereals such as corn, rice, wheat, and barley, which have significant agricultural value.

Synthetic auxins aren't just used by gardeners and farmers to get rid of weeds. During the 1960s, the United States sprayed the synthetic, auxin-based herbicide called Agent Orange on more than 3,000 villages in Vietnam, in an attempt to reduce the ability of opposing military forces to hide in the brush. But Agent Orange manufacture creates, as a by-product, a highly toxic chemical called dioxin. Agent Orange contaminated with dioxin was unexpectedly responsible for causing birth defects, leukemia, liver diseases, and other disorders and is considered a serious environmental hazard. Its use has been stopped.

FIGURE 19-7 Too much of a good thing. Synthetic auxins are used to control weeds, causing such a high rate of growth that death results.

TAKE-HOME MESSAGE 19•3

Auxins are a small group of naturally occurring hormones found primarily in meristems and immature plant tissue (and a larger group of synthetic variants) that play several important roles in stimulating and regulating a plant's growth and development, often by increasing the usually rigid cell wall's flexibility and permeability.

19•4

Other plant hormones regulate flowering, fruit ripening, and responses to stress.

Auxins and gibberellins are the primary plant regulators when it comes to growth and orientation. But there is a lot more to regulating a plant's responses to the environment so that the plant germinates, sprouts, and fruits at just the right time. The other three types of hormones—ethylene, abscisic acid, and cytokinins—also play critical roles in directing germination, sprouting, fruiting, and other activities, and in initiating appropriate molecular and cellular activities when the environmental conditions are favorable.

Q How can farmers ensure that every banana they ship to market ripens at exactly the right time?

Ethylene Suppose you wanted to pick 10,000 bananas in Central America and deliver them to market in the United States so that they all turned yellow at just the right time. This outcome could be a logistical nightmare, but with the help of plant hormones it doesn't have to be. The hormone **ethylene** is a gas produced in every part of a plant, and it has several important effects, including speeding up the rate at which many fruits ripen. Thus, ethylene can help with the banana problem: you can pick the fruits in Central America before they are ripe (when they are green), and just prior to their arrival in the United States, you can inject ethylene into the cargo hold of the ship and initiate the ripening of all the bananas simultaneously (**FIGURE 19-8**).

Not all fruits ripen in response to ethylene—strawberries, for example, do not, and so they must be picked from the vine only when ripe. Still, the use of ethylene for ripening fruits is probably the single most important agricultural use of any plant hormone. You can see this effect on a smaller scale by putting one ripe or rotting fruit (which produces very large amounts of ethylene) in a bag or container with fruits that haven't yet ripened. The ripe or rotten fruit will cause all of the other fruits to ripen very quickly. Alternatively, you can delay the ripening of fruits by separating them from those that are already ripe.

Like all peptide hormones, molecules of ethylene have their effects by binding to receptors on cell membranes and causing other "messenger molecules" within the cells to turn on the synthesis of certain gene products. Ethylene also hastens the aging and dropping of leaves from trees and the death of flowers at specific times. A plant has to maintain the flower's ovary until the seeds and fruit are mature, but it does not maintain the petals or stamens after fertilization. Most plants produce many more flowers that than they can mature into fruits, so most plants use ethylene to abort some flowers to reduce the number of fruits the plant must mature. But these effects can also be seen as undesirable by humans, such as those who want to sell cut flowers. Some flower merchants fight the effects of ethylene by briefly soaking cut-flower stems in a chemical solution of silver salts, which inhibits the effect of ethylene on the petals (**FIGURE 19-9**).

FIGURE 19-8 Ethylene speeds fruit ripening.

FIGURE 19-9 Florists fight the withering effects of ethylene on flowers.

Recently, commercial tomato growers were able to genetically engineer a type of tomato that does not produce ethylene. This is a valuable application of technology, because tomatoes with blocked ethylene production can be left on the vine longer without the risk of ripening before they can be picked and shipped to market. This allows the tomatoes to grow larger and also results in their tasting more like the tomatoes someone might grow in a home garden, where they can be left on the vine until they reach perfect ripeness. Otherwise they must be picked long before they are ripe so that they are tough enough to transport without damaging them. Ethylene is then applied so the tomatoes are ripe when they are sold.

Abscisic Acid Under stressful conditions, such as water shortage, the plant hormone **abscisic acid** is produced in relatively large amounts. Synthesized primarily in leaves, fruits, and root tips (but also in other plant parts), this hormone has the general effect of inhibiting growth and reproductive activities under adverse environmental conditions (**FIGURE 19-10**). In seeds, for example, abscisic acid can inhibit germination. In a sense, it serves to tell a plant that times are tough and environmental conditions are not good enough for growth and reproduction. When roots encounter unusually

ABSCISIC ACID

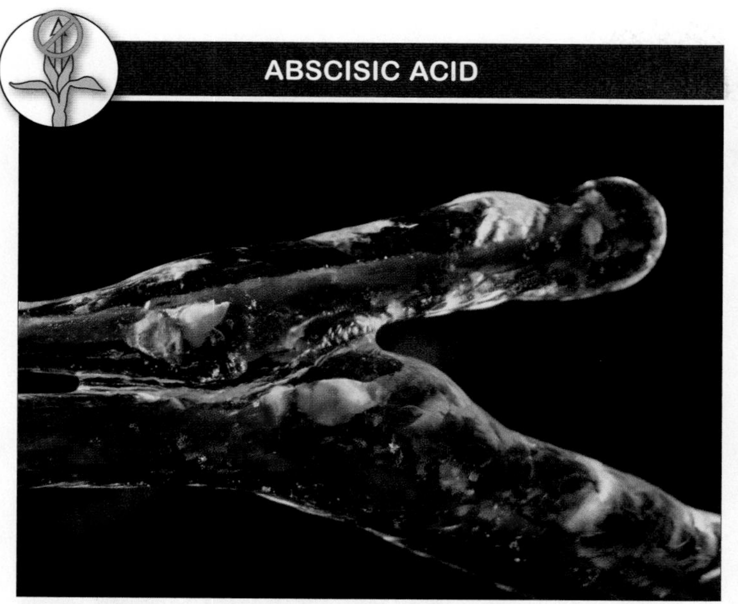

THE PRIMARY EFFECTS OF ABSCISIC ACID
- Inhibits growth and reproductive activities when environmental conditions are stressful
- Signals the stomata on a plant's leaves to close, increasing water conservation

FIGURE 19-10 Not a good time to invest in growth. Abscisic acid inhibits plant growth in times of stress, such as during a spring ice storm.

dry or cold or salty conditions, the production of abscisic acid increases and, in addition to inhibiting growth, it also signals the stomata on the plant's leaves to close, thereby increasing water conservation. In effect, abscisic acid causes a plant to hunker down and ride out the difficult conditions before embarking on new growth.

Cytokinins The final category of plant hormones is the **cytokinins,** hormones that are primarily stimulators of cell division throughout the body and throughout the lifetime of the plant. Although they are produced primarily in the roots and fruits, cytokinins exert their influence in all parts of the plant. Cytokinins usually work in conjunction with auxins to produce four main types of effects (**FIGURE 19-11**):

1. *Causing rapid cell division and promoting primary growth.*
2. *Inducing seed germination.* Application of cytokinins to seeds will even cause germination in the dark of seeds that normally require light to germinate.
3. *Initiating new branches from lateral buds.*
4. *Retarding leaf death.* Even when applied to a leaf that has already fallen off a tree, cytokinins will cause the leaf to remain green longer than it otherwise would.

CYTOKININS

THE PRIMARY EFFECTS OF CYTOKININS
- Cause rapid cell division in conjunction with auxins
- Induce seed germination
- Initiate new branches from lateral buds
- Retard leaf death

FIGURE 19-11 Cytokinins stimulate cell division. The cells of this seed embryo are ready to divide and will form the shoot system of the plant.

TAKE-HOME MESSAGE 19•4

In addition to auxins and gibberellins, three types of hormones also play critical roles in plant regulation. Among these multiple functions, ethylene induces and speeds fruit ripening, abscisic acid inhibits growth and reproduction under stressful environmental conditions, and cytokinins stimulate cell division throughout the plant body.

19•5

Tropisms influence plants' direction of growth.

A plant's immobility affects virtually every aspect of its structural design and function, and has influenced the nature of the evolutionary solutions that enable plants to grow, develop, and reproduce successfully. Like animals, for example, plants must respond to their environment as it changes. But whereas animals can move from a problematic, changing environment to one with more suitable conditions, plants use a variety of growth patterns, known as **tropisms**—such as bending, curving, and twisting—to grow toward or away from various environmental stimuli such as light, gravity, and physical impediments. Three of the most common tropisms are phototropism, gravitropism, and thigmotropism.

Phototropism If you have indoor plants, you've probably noticed that they always seem to grow toward a window. Even if you turn a plant around, within a few days it will grow to face the light again. This is **phototropism**—growth in response to a directional light source. It occurs when the cells in a plant's stem grow unevenly, adjusting their directional rate of growth so that the stem bends toward the light (**FIGURE 19-12**). The reason that evolution has generated such a pattern of growth is not surprising: if a plant can orient itself so that its photosynthesizing cells (in leaves and stems) can intercept more light, the plant can photosynthesize more efficiently and generate more energy for growth and reproduction.

The study of phototropism led to the discovery of the plant hormone auxin (which turns out to be a group of similar hormones). As we discussed in Section 19-3, when light hits a plant from a particular direction, auxins produced in the plant cells move away from the light source and end up on the opposite, shaded side of the stem. There, the auxins stimulate a slightly greater rate of growth than on the lighter side, with less auxin. As a consequence, the uneven growth causes the plant to bend toward the light.

One special type of phototropism, first described by Leonardo da Vinci, is called **heliotropism**—growth in response to the position of the sun. It describes the movement by some leaves and flowers—"heliotropic" flowers, such as the alpine buttercup—as they track the sun's movement across the sky each day. Cells in regions of the plant away from the light elongate as potassium ions are pumped into them and the subsequent movement of water into the cells increases turgor pressure, which changes the orientation of the flowers or leaves.

Gravitropism Plants' growth is also affected by gravity, and their response, known as **gravitropism,** is the reason that stems grow upward and roots grow downward. You can see that plants actively control this growth pattern if you play a trick on a house plant. Take a potted plant and tip it on its side. Within a matter of days, rather than growing horizontally, the plant will turn upward. It doesn't matter what direction the pot faces (it can even be suspended upside down). Roots will grow downward, in response to the force of gravity—and in the direction in which they are most likely to find water. Stems will grow in the opposite direction, which will reliably put them in position to intercept maximal amounts of light (**FIGURE 19-13**). Gravitropism occurs as a result of the uneven distribution of auxins, much as in the case of phototropism. How do the auxins detect the force of gravity? It seems that there are small

FIGURE 19-12 Phototropism is plant growth that is influenced by the presence of light.

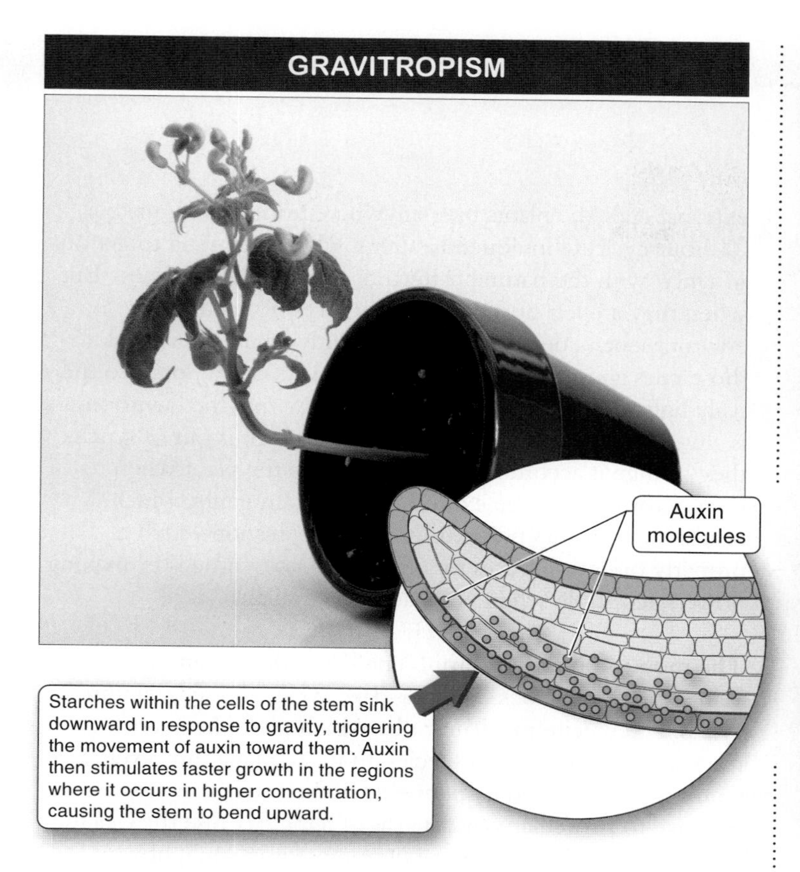

FIGURE 19-13 Plant stems grow away from the force of gravity.

FIGURE 19-14 Plant growth can be affected by touch or contact with physical objects.

bodies within plant cells that contain starches. Like marbles in a bottle of fluid, these starch-containing bodies, pulled by gravity, sink toward the bottom of the cell, regardless of the plant's orientation. Once there, the starch bodies trigger the migration of auxin molecules toward them, and the auxin again causes uneven cell growth so that the stems and roots grow in the right direction.

Q A plant turned on its side will grow upward. How would a slowly rotating plant grow?

In a plant turned on its side, the starch-containing bodies in the cells of the stem sink to the bottom in response to gravity and trigger the movement of auxin toward them. The auxin, in turn, causes the stem to grow in an uneven pattern (faster on the side with the auxin), and the stem bends upward. If the sideways plant is slowly rotated, however, the starch bodies never get a chance to settle on the bottom (because "the bottom" is constantly changing). Consequently, the auxin never concentrates at the bottom and uneven growth is never stimulated. The plant doesn't know which way is up because "up" keeps changing.

Thigmotropism In **thigmotropism,** plant growth occurs in response to touch or physical contact with an object (*thigmo-* derives from the Greek word for "touch"). For example, many climbing plants will wrap around structures—other plants, wires, posts, trellises, or anything that might support them as they grow upward. Climbing plants produce **tendrils,** which are specialized thread-like leaves or stems or branches that wrap around whatever they touch (**FIGURE 19-14**). In some species, a tendril can grow so fast that it wraps completely around something in less than an hour. As in the case of gravitropism and phototropism, this process occurs when cells on one side of a shoot (the side in physical contact with the object) elongate slowly, while cells on the opposite side elongate more quickly. This, again, involves auxins. The cells in contact with an object produce auxins and transport them to cells not touching the object, where they induce cells to elongate, causing the tendril to coil around the object.

TAKE-HOME MESSAGE 19•5

Plants have a variety of growth patterns, known as tropisms, by which they grow toward or away from various environmental stimuli. Phototropism is growth in response to directional light, gravitropism is growth in response to gravity, and thigmotropism is growth in response to touch or contact with physical objects.

19•6

Plants have internal biological clocks.

Plants are not as immobile as they appear. If you were to photograph or film a plant over a 24-hour period, as some botanists have done, you would observe many slow-but-sure movements. The common bean plant, for example, opens its leaves and orients them so that they face the sun during the day. Then each night, the plant pulls its leaves close to the stem (**FIGURE 19-15**). Similarly, in many moth-pollinated tobacco plants, the white trumpet-shaped flowers open each evening and close again in the morning. These plants schedule their daily activities so that they occur at the proper time. Do the plants respond to environmental cues? Or do they have some sort of built-in alarm clock?

The answers are yes and yes. Plants do have a "biological clock," an internal method of keeping time that enables them to initiate various biochemical and physiological actions at the appropriate time. We know this because even if the bean plants or tobacco plants described above are taken indoors and maintained in a room with constant, dim light, they still open their leaves or flowers on a daily schedule.

But this biological clock is not like a regular watch. As in so many aspects of biology, the events inside the organism are influenced by the world outside the organism. In other words, the internal clock of plants is continuously adjusted by environmental cues. When maintained under constant light conditions in the absence of external cues, the plants have anywhere from a 21-hour to a 27-hour cycle. Consequently, after a while, they tend to get out of synch with the natural pattern of daylight and darkness. But when they are left outside and have access to natural environmental cues, the plants constantly adjust their clock to those cues and maintain a cycle that always corresponds to the daily light-dark cycle. It's as if plants know that the environment is always the "correct" time and, if their clock is out of synch, they change it accordingly. The clock tells the plant when to turn on or turn off the expression of certain genes, which results in the plant's perpetually precise behavior—such as properly orienting leaves for efficient photosynthesis or making flowers accessible when pollinators are available.

The most important environmental cues that set and reset a plant's biological clock are light-dark cycles and temperature cycles. The precise mechanism by which plants keep track of time is not known for all species. In many plants, however, it seems to be regulated by pigments in the leaves that are sensitive to particular wavelengths of light. Because the specific wavelengths of light present in sunlight vary throughout the day, the plants reliably have more of the pigments activated during the day and different pigments activated as the sun sets and night begins.

Q Could we tell time accurately just by looking at some flowers?

Carolus Linnaeus, the Swedish botanist who developed the system of naming and

THE BIOLOGICAL CLOCK IN PLANTS

Sunrise | Noon | Sunset | Midnight

Plants have internal methods of keeping time—influenced by the external environment—that enable them to initiate various actions at the appropriate time.

FIGURE 19-15 Great timing. Plants adjust their biological clocks in response to environmental cues such as light and dark cycles and temperature cycles.

classifying living organisms that is still used by most biologists, used his understanding of plants' biological clocks to design what he called a "floral clock," based on flowers that opened and closed at specific times of the day. By planting a garden with 41 species that differed in their flowers' opening and closing times, it would be possible to tell what time of day it was just by noting which flowers were open and which were closed.

TAKE-HOME MESSAGE 19•6

Plants have internal methods of keeping time that enable them to initiate various biochemical and physiological actions at the appropriate time. They constantly adjust their biological clocks to environmental cues, such as light-dark cycles and temperature cycles.

19•7

With photoperiodism and dormancy, plants detect and prepare for winter.

Flower production is an energetically expensive task, and because it is so costly, plants benefit by flowering only when this will most effectively serve to enhance the plant's reproductive success. The optimal flowering time differs from species to species. It depends on the activity patterns of a plant's pollinators, the time it takes for fruits to develop and be dispersed, and a variety of other factors. For some plants, the best time to flower is in the spring or summer.

Plants must respond not just to environmental changes over the course of a 24-hour cycle, but also to seasonal changes over the course of a year. For some plants, the best time to flower is in the spring or summer. For others, it is best to wait until early autumn. In other words, they need more than just a biological clock—they need a "biological calendar." When it comes to producing flowers or scaling back metabolic activity and becoming dormant for winter, for example, a plant's life can depend on choosing the right moment (**FIGURE 19-16**). Timing is everything.

FIGURE 19-16 Plants flower and become dormant in accordance with the seasons.

> "Art is the unceasing effort to compete with the beauty of flowers—and never succeeding."
> —Marc Chagall

How does a plant know when it is spring or summer? What environmental change should they use as a cue to the beginning or end of a season? Temperature? Precipitation levels? Both do tend to change with the seasons—particularly as you get farther and farther from the equator—but they are too variable for a plant to rely on. A plant doesn't want to be fooled by a cloudy spell in summer or an unseasonably dry winter. There is, however, one consistent and unvarying cue to the time of year: the number of hours of daylight or darkness. Plants use one of these cues—the length of the dark period in a day—in a process known as **photoperiodism,** to regulate their flowering time and numerous other responses to seasonal changes.

All flowering plants fall into one of three categories when it comes to regulating their flower production. There are (1) long-day plants, (2) short-day plants, and (3) day-neutral plants (**FIGURE 19-17**). The groups are so named because the amount of daylight seems to determine when they produce flowers. Long-day plants begin producing flowers only when the length of daylight exceeds a critical amount. Short-day plants are the opposite: they flower only when the length of daylight becomes less than some threshold amount. The critical amounts of daylight occur at different times of the year, when

PHOTOPERIODISM

All flowering plants fall into one of three categories when it comes to regulating their flower production.

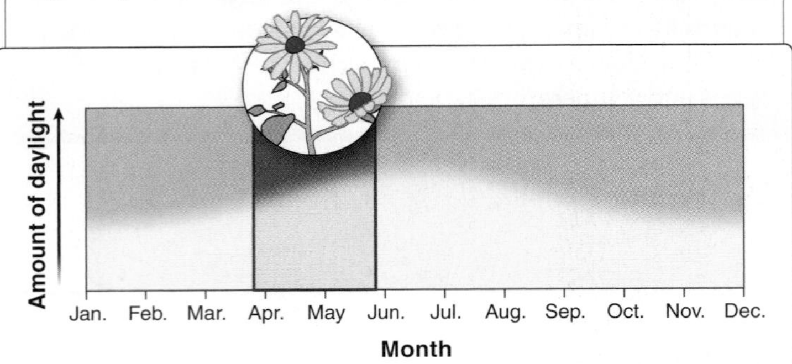

LONG-DAY PLANTS
Flower production is triggered by shorter periods of darkness (generally in spring).

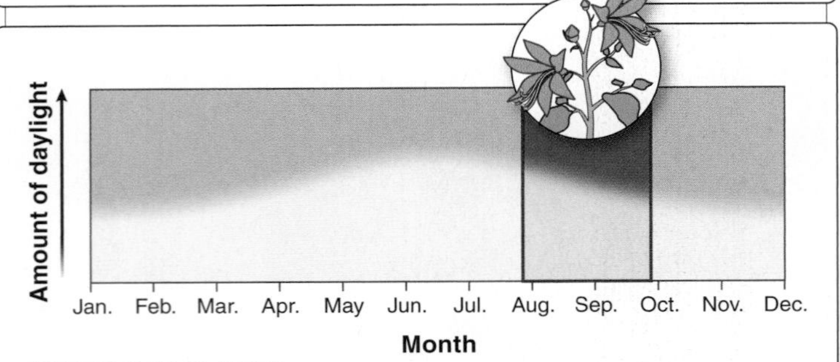

SHORT-DAY PLANTS
Flower production is triggered by longer periods of darkness (generally in late summer or fall).

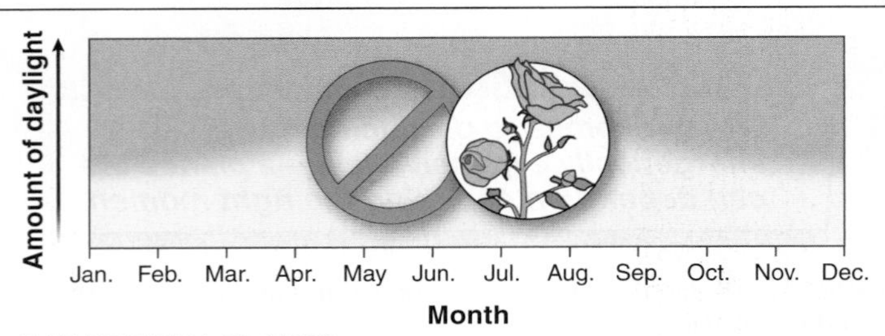

DAY-NEUTRAL PLANTS
Flower production is triggered by a sufficient state of maturity and not by periods of darkness.

Some plants are triggered to produce flowers when the length of the nights is long (and the amount of daylight is relatively small). Others are triggered when nights are shorter, and the daylight lasts longer.

FIGURE 19-17 Day length and flowering. In long- and short-day plants, flowering is regulated by hours of darkness. In day-neutral plants, light and darkness do not affect flowering.

conditions are best suited for reproduction for a particular species. These critical amounts generally occur in spring for long-day plants and in late summer or fall for short-day plants. Day length has no effect on day-neutral plants, which flower when they reach a sufficient state of growth and maturity.

Although short-day and long-day plants are named for the amount of daylight that, apparently, determines when they flower, experiments have shown that short-day and long-day plants are actually sensitive to the length of the night rather than the day. That is, flowering in short-day plants is triggered when the nights become longer than some threshold, and flowering in long-day plants is triggered by the onset of short nights. This seemingly unimportant distinction had disappointing implications for the California Department of Transportation. The department wanted to line a highway with bright-red flowering poinsettias. Unfortunately, these short-day plants never flowered as expected, because the car headlights on the highway kept interrupting the plants' measurement of the night length. As a consequence, the poinsettia plants—which require at least 14 uninterrupted hours of darkness if they are to flower—never detected an appropriately long night (and thus short day), and they remained green. Many people who keep these plants indoors have the same problem, because indoor lights can also mean that the plants don't experience the long dark period that triggers blooming.

How do plants regulate photoperiodism? It appears to be related to the varying ratio of different light-sensitive pigments in their leaves at different points during the day and night.

Issues relating to photoperiodism can cause difficulties in cultivating flowers indoors. It can be particularly difficult to trigger flowering in short-day plants. With the artificial lights in houses, plants don't usually experience the long, uninterrupted period of darkness required for blooming. Long-day flowers, on the other hand, are easier to cultivate, because flashes of light or continuous light during the night can trick the plants into behaving as if the night is short, leading to flowering. Day-neutral plants are not influenced by artificial lights.

TAKE-HOME MESSAGE 19•7

Plants exhibit photoperiodism, responding to seasonal changes over the course of a year, timing their production of flowers or initiation of winter dormancy, for example, according to environmental factors such as the length of the period of darkness each night.

② Plants defend themselves from attack by herbivores.

A plant defense: the stem of dog rose with thorns.

19•8

Plants actively resist being eaten by herbivores.

Plants are under almost constant attack. From microbes to fungi, animals, and even other plants, organisms make varied and persistent efforts to access the valuable resources in plants. In some cases, the attacking organisms can kill the plant outright. In most cases, though, they simply siphon off the resources—such as stored chemical energy—reducing the amount available to the plant for its own growth, maintenance, and reproduction.

Plants have evolved to fight back, however. They have numerous defenses that can help them reduce the competitive and predatory "tax" they constantly face. These defenses fall into four general categories.

Mechanical Defenses Several types of plant defenses involve physical structures or movements (**FIGURE 19-18**).

Thorns, spines, prickles, and hairs. If you've ever tried to pick blackberries or raspberries, you probably encountered the sharp prickles on the stems. And while these may have been only a nuisance to you, they can be enough of a deterrent to many herbivores—in some cases, harming or even killing them—to significantly reduce herbivory. To minimize their anti-herbivore costs, some plants, such as acacias, produce spines primarily on the outside of the plant with just a few near the center. In many plants, fine hairs on their leaves can reduce the ability of herbivores to eat them.

Waxes and saps. Producing leaves covered with waxy compounds can also reduce herbivory. Researchers have documented, for example, that beetles spend more time slipping and falling off waxy leaves—such as holly leaves—than they do from non-waxy leaves. Some plants produce and exude sticky saps or resins that can deter or even drown potential insect pests.

Defensive (and offensive) movements. The Venus flytrap is the most dramatic example of a plant making use of rapid movement. As we will see in Section 22-2, however, this can be considered more an offensive than a defensive movement, because it is a method of acquiring nitrogen (released as the plant dissolves its prey). Many plants—particularly those in the mimosa family—use a similar movement mechanism to rapidly flatten their leaves in response to touch. This has the effect of rapidly decreasing the surface area available to potential pests and reducing herbivory.

Chemical Defenses One of the most common methods used by plants to fight herbivory is the production of chemical defenses. These are chemical compounds (sometimes called "secondary compounds") produced to deter herbivores by

MECHANICAL DEFENSES IN PLANTS

THORNS, SPINES, AND HAIRS
Structures such as sharp spines or fine hairs can significantly reduce herbivory.

WAXES AND SAPS
Leaf secretions such as slippery waxy compounds or sticky saps significantly reduce herbivory.

DEFENSIVE MOVEMENTS
Rapid movements, such as flattening leaves in response to touch, can decrease available surface area and significantly reduce herbivory.

FIGURE 19-18 Plants have physical defenses to ward off predation by insects and other animals. Shown (left to right): Scotch thistle, holly, and the sensitive plant (*mimosa pudica*).

making the plant toxic or by reducing the digestibility of the plant so as to lessen its value to the herbivore.

One common strategy, used by more than 3,000 species of plants, is to produce cyanide-containing molecules called cyanogenic glycosides. When consumed by herbivores, these molecules break down into cyanide, which blocks electron transport—killing or seriously harming the animal.

In a sort of evolutionary arms race, some organisms have evolved the ability to eat toxic secondary compounds without suffering ill effects. Monarch butterfly caterpillars, for example, can eat milkweed plants, consuming the toxic cardiac glycosides that kill most insects and induce heart attacks in vertebrates (**FIGURE 19-19**). Moreover, the monarchs store the toxic chemicals within their cell vacuoles and then become poisonous to animals that eat them.

Other common toxins produced by plants to reduce herbivory include the neurotoxic chemicals in hemlock. And ricin, a chemical found in castor beans, is among the plant compounds most toxic to humans and other animals and has long been investigated as a potential agent of chemical and biological warfare. It was reportedly used in an assassination—delivered by stabbing the target in the leg with an umbrella—by the KGB.

Many plants also secrete toxic chemicals from their roots that block the germination of seeds and reduce the growth of nearby plants, conferring a competitive advantage. Because of the toxic chemicals secreted by the roots of black walnut trees, for example, few plants are found growing underneath these trees.

FIGURE 19-19 Co-opting a toxin. Monarch butterfly caterpillars consume a poisonous plant's chemicals and make them their own.

Can humans benefit from plants' attempts to avoid being eaten?

Beneficial Effects of Secondary Compounds

Humans have long made use of some of the "defensive" compounds produced by plants.

Spices. Spicy foods, such as mustard, usually get their flavor from secondary plant compounds that may make the plant toxic or unpalatable to an insect but are not toxic to humans (see Section 22-19).

Medicines. Many of our medicines come from the secondary compounds developed by plants. After all, a chemical that disrupts a microbe's metabolism can have its effect whether the microbe is on a plant or inside a human. These medicines include taxol, a compound with anti-mitotic properties isolated from the Pacific yew (*Taxus brevifolia*), which is now used to treat a wide variety of cancers (see Section 16-1). And quinine, an anti-herbivore compound isolated from the bark of *Cinchona* trees in South America, is an effective treatment for malaria. There has been a large increase in the number of researchers—sometimes called "bio-prospectors"—looking for plant compounds with potential medicinal uses (**FIGURE 19-20**).

FIGURE 19-20 "Bio-prospectors" search for plant chemicals that have medicinal properties. The dried roots of *harpaogphytum* (Devil's Claw) are used by some people for its anti-inflammatory properties.

FIGURE 19-21 No room for additional eggs. These passion flower leaves are covered with what appear to be butterfly eggs. The leaf spots are actually produced by the plant and the mimicry discourages a butterfly from selecting the site for egg-laying.

Mimicry and Camouflage Another way that plants resist predation is through mimicry and camouflage. The *Heliconius* group of butterflies includes a large number of species that lay their eggs on the leaves of passion flower plants. On hatching, the insects begin eating the plant, often causing significant damage. By producing leaves that appear to be covered with butterfly eggs already, one species of passion flower reduces the likelihood that butterflies will lay eggs on its leaves (**FIGURE 19-21**). Females will not lay their eggs on a leaf that already has *Heliconas* eggs because the first eggs would hatch sooner and eat the leaves, destroying the eggs of the second butterfly. When a female butterfly sees the bogus eggs on passion flowers, she avoids laying her eggs there.

Enlisting Other Organisms for "Security" Researchers have even discovered mutualistic situations in which plants seem to outsource some of their defenses to another species. Acacia plants, most common in Central American savannas, produce a sweet nectar that attracts ants. The ants hollow out thorns in the plants and live inside them. In exchange for their "room and board," the ants quickly and aggressively attack leaf-eating insects, effectively reducing their ability to feed on the plant.

TAKE-HOME MESSAGE 19•8

Rooted in the ground, plants are targets for predators and pathogens. They actively resist infection and herbivory by producing toxic chemicals that make them distasteful, poisonous, or difficult to consume.

③ Plants can survive in extreme environments.

*Quiver trees are large succulents (*Aloe dichotoma*) that live in arid landscapes. These were photographed at Richtersveld National Park, South Africa.*

19•9

Special adaptations help some plants thrive in extreme habitats.

Just as plant anti-herbivore adaptations have been produced by evolution, so too have a variety of plant adaptations been produced as evolutionary responses to the non-living elements in their world. We see particularly dramatic adaptations in plants that manage to thrive in some of the most extreme habitats of the world.

Super-dry Habitats In deserts, groundwater is scarce, and moisture loss due to evaporation in the hot temperatures can further constrain a plant's ability to retain water. Plants have responded to these difficult conditions with several different types of adaptations (**FIGURE 19-22**).

Succulent leaves and stems. The evolution of thick, fleshy, water-storing tissue in their leaves and stems has allowed cacti and other succulent plants to reduce the surface-area-to-volume ratio of their photosynthetic parts. These succulent tissues help minimize water loss due to evaporation. Although this

METHODS OF SURVIVING DRY HABITATS

SUCCULENT LEAVES AND STEMS
Cacti and other succulent plants have thick, fleshy, water-storing tissue within their leaves and stems that helps minimize water loss due to evaporation.

DEEP TAPROOTS
Plants such as mesquite send down unusually hardy and deep taproots that can utilize water far beneath the surface.

LONG-DORMANT SEEDS
Many plants have seeds that can remain dormant for long periods of time, then quickly germinate and grow in response to brief periods of moisture.

FIGURE 19-22 Survival strategies for dry climates. Shown (left to right): finger and thumb succulent, mesquite tree, and wildflowers in the desert.

adaptation reduces plants' capacity for gas exchange and thus their overall rate of sugar production relative to plants with thinner leaves, it enables them to survive in the hot, dry desert habitats where thin-leaf plants are likely to dry out and die.

Deep taproots. If water conservation is one strategy of dry-habitat plants, "water-mining" is the flip side. Plants such as mesquite send down unusually hardy and deep taproots, often boring into the soil a dozen meters or more to find sources of water far beneath the surface.

Long-dormant seeds. Because the rains are highly unpredictable and rare in super-dry habitats—several years may pass between rains in the deserts of Chile, for example—many dry-adapted plants have seeds that can remain dormant for long periods of time. They wait out the dry times and quickly germinate and grow in response to brief periods of moisture.

Salty Environments Marshes and intertidal zones may have pleasant temperatures and plenty of moisture, but the extreme levels of salt from the ocean and the large fluctuations in salt concentrations can make it difficult for cells to maintain a constant water balance and carry out normal metabolic processes (**FIGURE 19-23**). Plants adapted to saltwater environments are generally unable to reduce the large amounts of salt they absorb with the water through their roots. How do they survive? They transport much of the absorbed salt to vacuoles within cells, thus isolating and containing it where it has little impact on cellular metabolism. Some plants—such as mangroves—are also able to transport much of the salt through the plant and excrete it from the leaves. The salt sits on the leaf surfaces until it dries and blows away.

FIGURE 19-23 Survival in salt water.

FIGURE 19-24 Survival in cold, windy habitats.

Cold and Windy Habitats At both extreme latitudes and extreme altitudes, life can be hard: the air is cold, sunlight is limited, the growing season is short, and the winds can be brutal and relentless. As a consequence, plants living in these habitats tend to grow close to the ground and have smaller-than-average leaves (to limit evaporation of water). Grasses, in particular, do well under such conditions (**FIGURE 19-24**).

Arctic habitats present an additional challenge. Because the first few meters of soil below the ground surface stays perpetually frozen (this layer is called **permafrost** if it remains below freezing for two or more years), root growth is limited, and any water that is present from precipitation or melting cannot drain well. In high-altitude habitats, drainage is good but the soil is still thin and poor in nutrients. In each case, plant life is usually limited to smaller plants with diffuse, shallow root systems.

TAKE-HOME MESSAGE 19•9

Plant evolution has produced adaptations to some of the most extreme habitats of the world: including super-dry regions, salty environments, and cold, windy habitats.

streetBio

Knowledge You Can Use

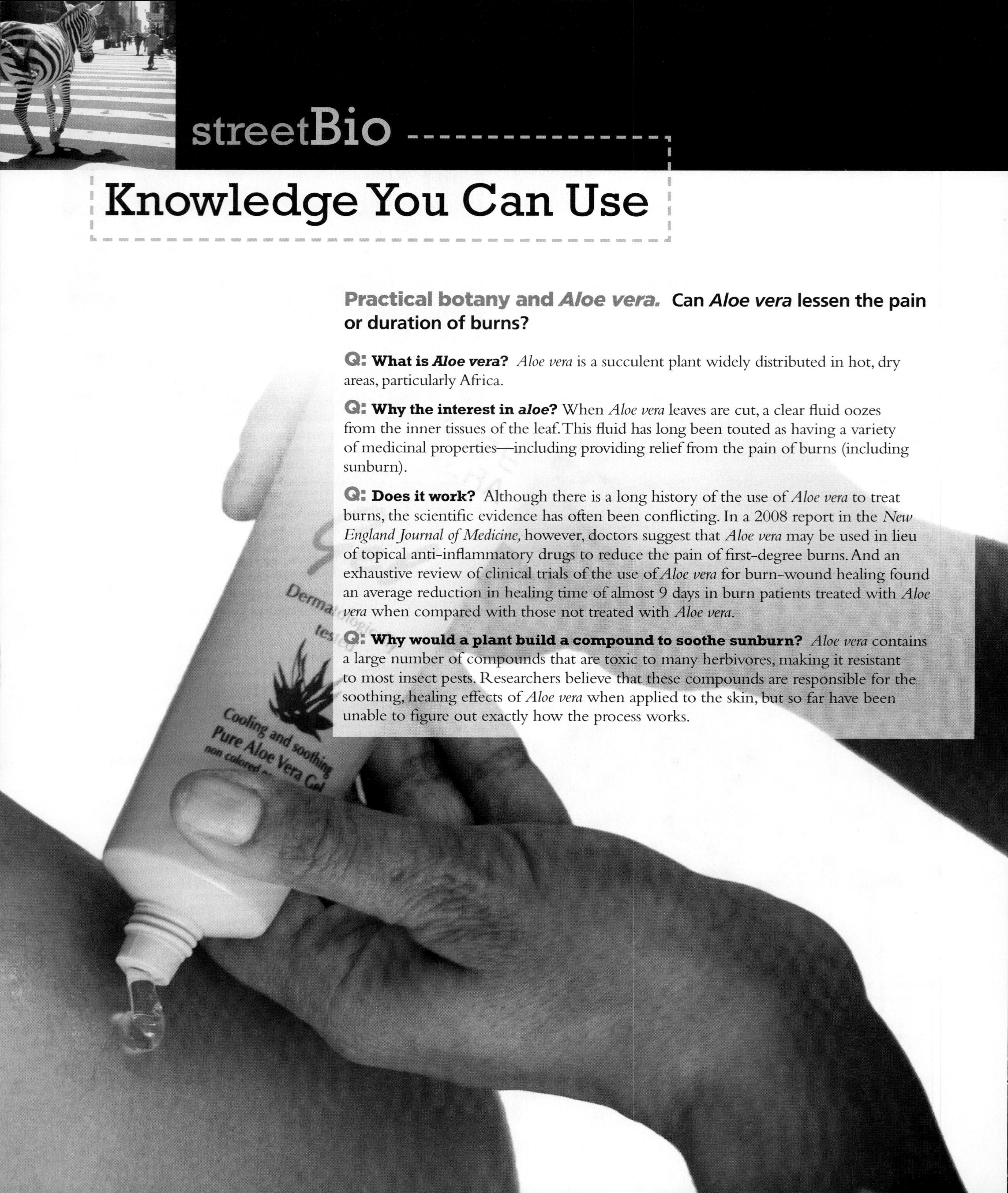

Practical botany and *Aloe vera*. Can *Aloe vera* lessen the pain or duration of burns?

Q: What is *Aloe vera*? *Aloe vera* is a succulent plant widely distributed in hot, dry areas, particularly Africa.

Q: Why the interest in *aloe*? When *Aloe vera* leaves are cut, a clear fluid oozes from the inner tissues of the leaf. This fluid has long been touted as having a variety of medicinal properties—including providing relief from the pain of burns (including sunburn).

Q: Does it work? Although there is a long history of the use of *Aloe vera* to treat burns, the scientific evidence has often been conflicting. In a 2008 report in the *New England Journal of Medicine,* however, doctors suggest that *Aloe vera* may be used in lieu of topical anti-inflammatory drugs to reduce the pain of first-degree burns. And an exhaustive review of clinical trials of the use of *Aloe vera* for burn-wound healing found an average reduction in healing time of almost 9 days in burn patients treated with *Aloe vera* when compared with those not treated with *Aloe vera*.

Q: Why would a plant build a compound to soothe sunburn? *Aloe vera* contains a large number of compounds that are toxic to many herbivores, making it resistant to most insect pests. Researchers believe that these compounds are responsible for the soothing, healing effects of *Aloe vera* when applied to the skin, but so far have been unable to figure out exactly how the process works.

1 Hormones regulate growth and development.

Plant hormones are chemical signals produced by plant cells that enable the organism to respond to environmental variables (such as amount of moisture, amount and direction of sunlight, and temperature) and influence the growth and development of the plant. The hormones include gibberellins, auxins, ethylene, abscisic acid, and cytokinins. Plants also have a variety of growth patterns known as tropisms—such as phototropism, gravitropism, and thigmotropism—by which they grow toward or away from various environmental stimuli. Plants have internal methods of time-keeping that enable them to initiate various biochemical and physiological actions at the appropriate time of day and time of year, based on environmental cues.

2 Plants defend themselves from attack by herbivores.

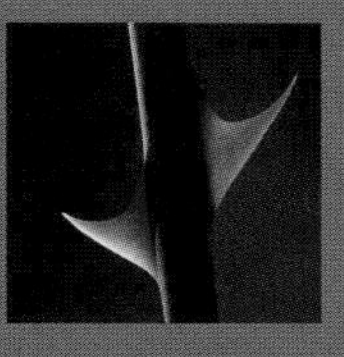

Rooted in the ground, plants are targets for predators and pathogens. They actively resist infection and herbivory by producing toxic chemicals that make them distasteful, poisonous, or difficult to consume.

3 Plants can survive in extreme environments.

In the course of evolution, plants have become adapted to some of the most extreme habitats of the world, including super-dry regions, salty environments, and cold, windy habitats.

KEY TERMS

abscisic acid, p. 698
auxins, p. 695
cytokinins, p. 699
ethylene, p. 697
gibberellins, p. 694
gravitropism, p. 700
heliotropism, p. 700
hormone, p. 692
permafrost, p. 709
photoperiodism, p. 703
phototropism, p. 700
tendril, p. 701
thigmotropism, p. 701
tropism, p. 700

CHECK YOUR KNOWLEDGE

1. Unlike animals, which produce each hormone _______, plants produce hormones _______.
   a) in a specific part of the body; in many cell types
   b) only in certain seasons; continuously
   c) in many cell types; in a specific part of the body
   d) continuously; only in certain seasons
   e) in high concentrations; in low concentrations

2. Which plant hormone do grape-growers spray on their crop in order to get seedless grapes to grow fruits as large as grapes with seeds?
   a) gibberellins
   b) ethylene
   c) cytokinins
   d) auxins
   e) abscisic acid

3. High concentrations of ______ stimulate root production.
   a) gibberellins
   b) cytokinins
   c) auxins
   d) ethylene
   e) abscisic acid

4. Cytokinins promote the growth of:
   a) shoots.
   b) axillary buds.
   c) pollen tubes.
   d) roots.
   e) vascular cambium.

5. Which hormone or group of hormones stimulates the ripening of fruit?
   a) auxins
   b) ethylene
   c) cytokinins
   d) gibberellins
   e) abscisic acid

6. Which plant hormone or group of hormones is responsible for bud and seed dormancy?
   a) auxins
   b) ethylene
   c) cytokinins
   d) gibberellins
   e) abscisic acid

7. Which of these hormones or group of hormones is primarily synthesized in the roots and travels to other regions of the plant?
   a) cytokinins
   b) gibberellins
   c) 2,4-D
   d) auxins
   e) cytochromes

8. Phototropism in plants is:
   a) a process that produces glucose.
   b) caused by ethylene.
   c) mediated by cytokinins.
   d) the bending of roots away from the light.
   e) the bending of the shoot tip toward a light source.

9. Which hormone or group of hormones is involved in phototropism?
   a) ethylene
   b) gibberellins
   c) cytokinins
   d) auxins
   e) abscisic acid

10. What is the effect of a short burst of light during a short-day plant's dark period?
    a) The plant will not flower if the burst of light makes the longest period of continuous darkness shorter than the critical dark period.
    b) The plant will flower, because the length of day remains unchanged.
    c) The plant's flowering does not depend on light, so the burst of light will have no effect.
    d) The plant will not flower, because the burst of light restarts its biological clock.
    e) The plant will flower perpetually, because the burst of light makes the longest period of continuous darkness shorter than the critical dark period.

11. What adaptations have occurred in desert plants that reduce the impact of the hot, dry environment?
    a) thick, fleshy, water-storing leaves and stems
    b) deep taproot
    c) seeds that can remain dormant for long periods of time
    d) All of the above are correct.
    e) Only a) and c) are correct.

Short-Answer Questions

1. A plant near a window often grows so that most of its leaves face the window. If you rotate the plant, within a few days the leaves will again be facing the window. How does this occur?
2. Why is ethylene so valuable to those who sell fruit?
3. Using an example, explain why so many plants contain spicy chemicals.
4. Fill out the table at the right; list the location and function of each plant hormone.

See Appendix for answers. For additional study questions, go to www.prep-u.com.

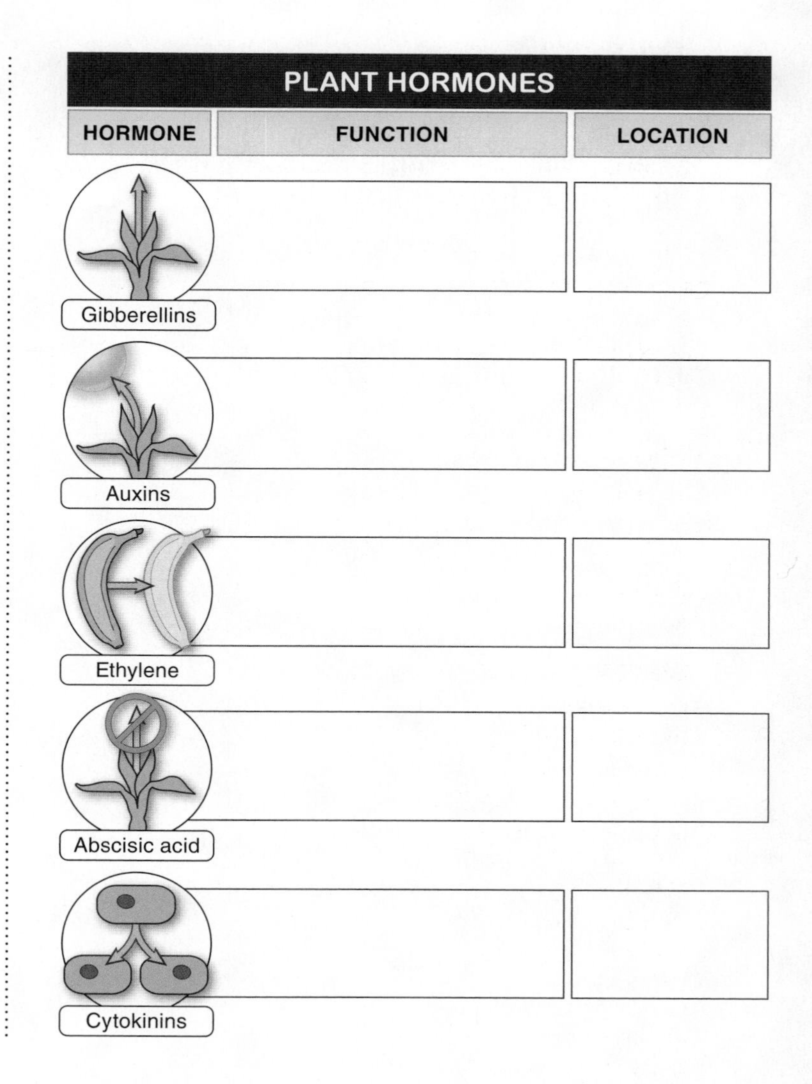

1 Animals have an internal environment.

20.1 Our bodies function best within a narrow range of internal conditions.

20.2 Animals regulate their internal environment through homeostasis.

2 How does homeostasis work?

20.3 Negative and positive feedback systems influence homeostasis.

20.4 Temperature control is a component of homeostasis.

20.5 Animals must balance their water content within a narrow range.

20.6 In humans, the kidney is the chief excretory organ.

3 Form reflects function: animal anatomy.

20.7 Most animal bodies are organized into cells, tissues, organs, and organ systems.

20.8 Connective tissue provides support.

20.9 Epithelial tissue protects.

20.10 Muscle tissue enables movement.

20.11 Nervous tissue transmits information.

20.12 Each organ system performs special tasks.

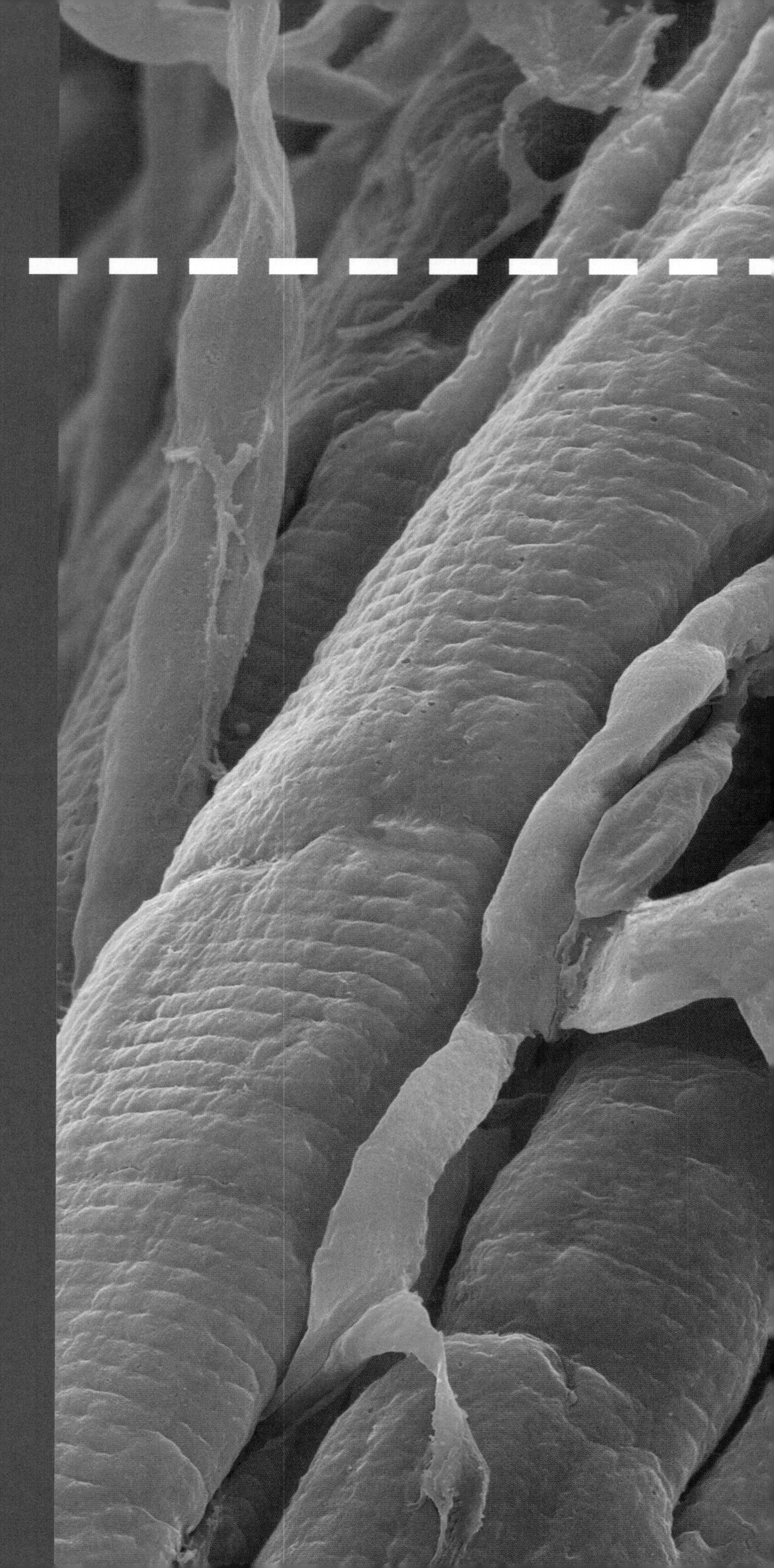

20

Introduction to Animal Physiology

Principles of animal organization and function

1 Animals have an internal environment.

Heat radiates from warm bodies in cold weather.

20•1

Our bodies function best within a narrow range of internal conditions.

Following an extremely rigorous practice in the summer of 2001, 27-year-old Minnesota Vikings football player Korey Stringer walked into an air-conditioned tent to recover. Almost immediately, however, the 6 foot 4 inch, 335-pound all-pro player began to feel weak and dizzy; his breathing became rapid and his blood pressure dropped. He was rushed to the hospital, but as a consequence of his body temperature allegedly reaching 108.8° F (42.7° C), many of his organs failed, and he died that night.

Q What is heat stroke? Why is it dangerous?

Normal body temperature is approximately 98.6° F (37° C). **Hyperthermia** occurs when too much heat is produced or when high environmental temperature and humidity overwhelm the body's ability to dissipate heat. (It is unlike a fever, in which the body sets its core temperature slightly higher in response to an infection by pathogens.) Korey Stringer's death is an example of the disastrous consequences that can result from extreme cases of hyperthermia, called **heat stroke,** and illustrates the danger when organisms fail to maintain an internal environment within a safe range.

Although the environmental temperature range you experience every day may involve a 20°, 30°, or even 40° F difference between high and low, humans cannot tolerate such large internal temperature swings. Even deviations of less than 10° F can lead to a cascade of biochemical problems with serious health implications. For mammals, the risks of getting too hot include reduced enzyme activity and, at extreme temperatures, the complete breakdown of enzymes. Excessive water loss, too, can occur, leading to numerous problems—including organ failure. For these reasons, heat stroke can be life-threatening (**FIGURE 20-1**).

At the other extreme, **hypothermia,** or extremely low body temperature, can cause difficulties in coordination and movement, along with disorientation and confusion, irregular heartbeat, and, ultimately, death. Similarly disastrous

FIGURE 20-1 **A cooling soak after a tough workout can prevent hyperthermia.**

consequences can occur as a result of the failure to maintain a consistent internal environment in other ways, including in blood sugar levels, blood pH, and tissue concentrations of oxygen and carbon dioxide.

In this chapter, we explore a fundamental characteristic of organism function, **homeostasis,** the body's use of physical and chemical processes to maintain a consistent internal environment, even in the face of changing external and internal environmental forces. We also examine the forms behind those functions—cells, tissues, organs, and organ systems.

TAKE-HOME MESSAGE 20•1

Failure to maintain a consistent internal physical and chemical environment can lead to multiple problems in the normal functioning of cells, tissues, and organs, and can result in death.

20•2

Animals regulate their internal environment through homeostasis.

When it comes to **physiology**—the functions and activities of an organism—taking care of business is straightforward in single-celled organisms. Food and other necessary materials for life do not have far to go to get into the cell—all they have to do is cross a plasma membrane. Moving waste products out of the cell is similarly simple. Such are the advantages to living close to the external environment. But direct contact with the external environment has its costs, too. Perhaps most important among these costs is that the cell is, to a large extent, at the mercy of its external environment. Changes in the environment—such as in temperature or pH—can have a great impact on the cell itself.

Like single-celled organisms, multicellular animals also must acquire food and other materials, as well as get rid of waste. These tasks become increasingly complex as organisms become more complex and more of their cells are not in direct contact with the external environment. On the other hand, when it's not in direct contact with the external environment, a cell can be protected from harsh or changing environmental conditions. To most cells in a multicellular organism, the more important environment is the "internal environment."

With multicellularity and increasing size, then, an animal's internal environment takes on greater importance than the external environment in influencing cell functioning. In vertebrates, this internal environment consists of the **extracellular fluid** that fills the space between cells (also called the **interstitial fluid**), bathing the cells. This fluid is primarily water, but also contains nutrients and raw materials for growth and development, as well as waste products that have diffused out of or been removed from cells. The volume of interstitial fluid is not insignificant. In fact, one-third of the water in the human body is outside the body's cells, with more than two and a half gallons (10 liters)—picture five 2-liter soda bottles!—surrounding and bathing the cells.

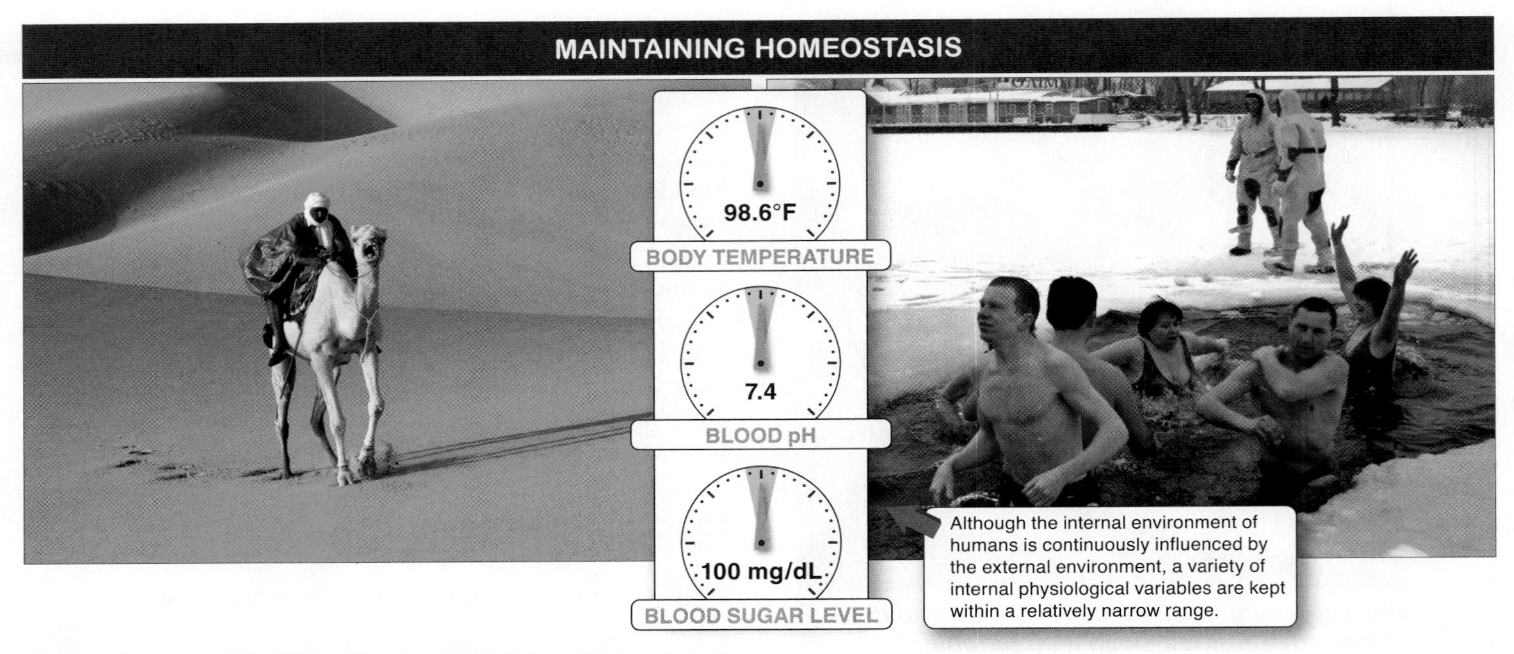

FIGURE 20-2 Homeostasis: maintaining a stable internal environment.

The internal environment of multicellular animals is continuously influenced by their external environment as well as by the cellular activities that add and remove materials. One of the hallmarks of animal physiology, however, is that organisms generally maintain homeostasis, or the ability to return to a narrow range of physical and chemical conditions, even in the face of changing external environmental forces. In this relatively constant, steady internal environment, variables such as temperature, water-solute balances, pH, blood sugar levels, and O_2 and CO_2 concentrations in blood and other tissues are maintained within narrow ranges by the activities of cells within the organism's tissues, organs, and systems (**FIGURE 20-2**).

Homeostasis is of huge value to the organism. Cellular functioning (from DNA replication to protein production to intercellular communication) depends, in large part, on the activities of enzymes. In turn, an enzyme's activity depends critically on the temperature and other characteristics of its immediate environment. Even slight changes in these variables can disrupt, reduce, or even stop enzyme functioning. And this can have catastrophic consequences, such as in the case of heat stroke described above.

Enzyme activity is just one of many facets of cell function influenced by temperature and other physical features of the cell's internal environment: membrane permeability and the rates at which materials diffuse across membranes also respond to changes in the environment. Through homeostasis, organisms maintain optimum metabolic functioning—the processes by which organisms take up nutrients from their environment and use them for growth, movement, reproduction, and all the other actions necessary for life. In the next section, we explore how homeostasis is maintained.

TAKE-HOME MESSAGE 20•2

Although the internal environment of multicellular animals is continuously influenced by their external environment, many animals maintain homeostasis: they keep a variety of internal physiological variables—including temperature, water-solute balances, pH, blood sugar levels, and blood gas concentrations—within a relatively constant range.

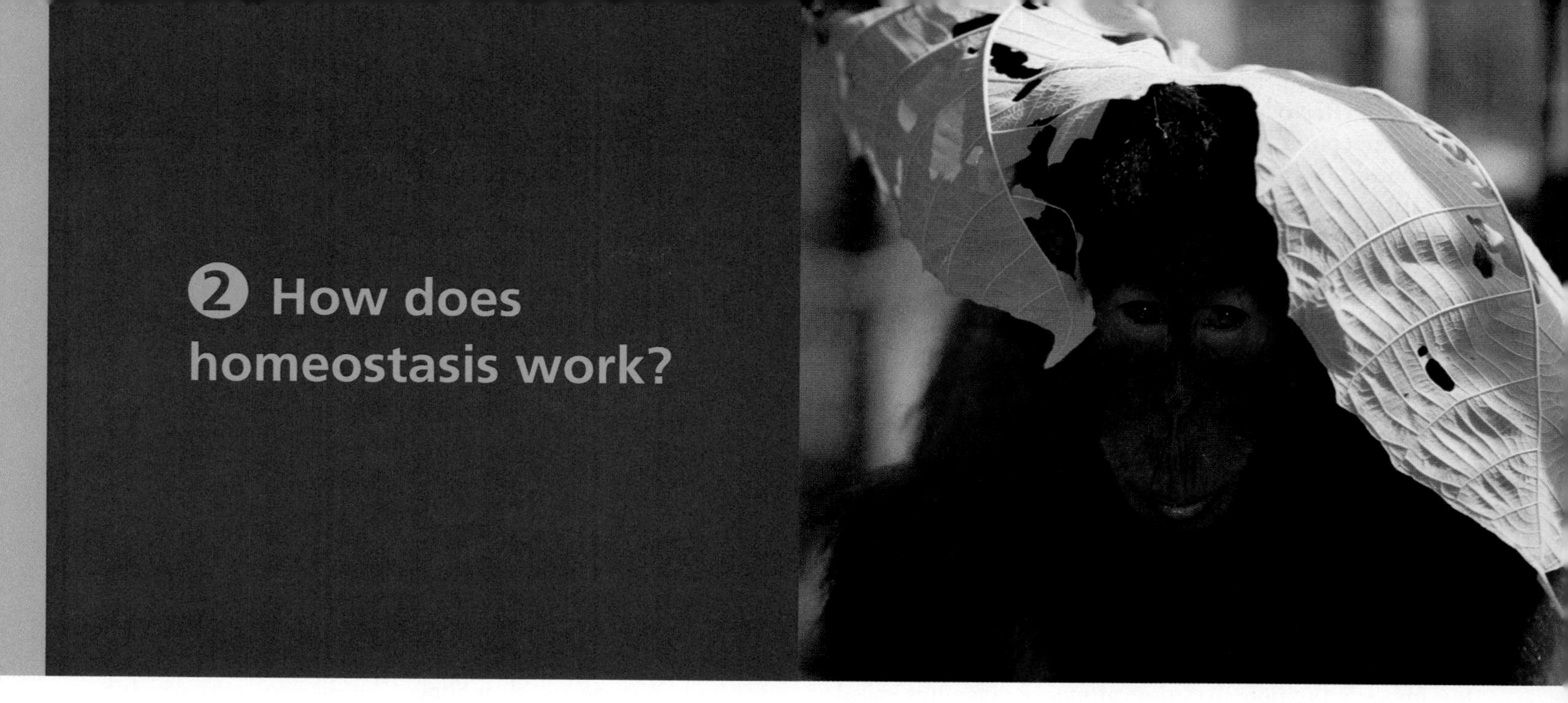

Relief from heat: A young orangutan seeks shelter under a leaf.

20•3

Negative and positive feedback systems influence homeostasis.

For animals to maintain homeostasis with regard to a particular physiological variable, that variable must have a **set point,** a target value or range to which the organism can return. Most vertebrates have set points for a variety of physiological variables, including body temperature, water-solute balances, blood sugar levels, blood pH, and tissue O_2 and CO_2 concentrations.

In the face of changes in its external environment, how does an organism maintain its homeostasis, its constant internal environment? The most common method involves the phenomenon of **negative feedback,** in which sensors detect a change in the internal environment and trigger structures called **effectors** to oppose or reduce the change. This cycle of detection, response, and change is called a negative feedback loop (**FIGURE 20-3**).

An example of a negative feedback loop that we encounter in everyday life is the regulation of room temperature in a house, which involves a sensor (the thermostat) and two effectors (a furnace and an air conditioner). Within the thermostat is a sensor that detects the temperature in the

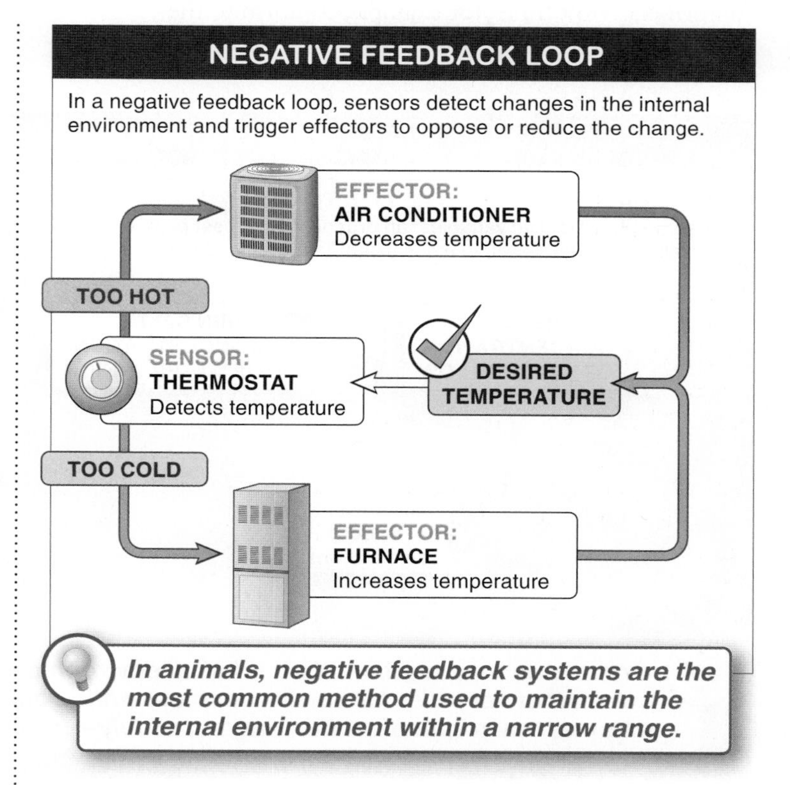

FIGURE 20-3 A negative feedback loop restores the internal environment to a set point.

house. If the temperature drops too low, the thermostat triggers an effector—the furnace—to turn on, bringing the room temperature back up to the desired level. Once the temperature reaches a specified level, the thermostat tells the furnace to shut off. Similarly, if the temperature gets too high, the thermostat can trigger the air conditioning to turn on, which brings the temperature back down to the desired level, after which it shuts off. In both cases, a negative feedback system senses a change in the temperature, triggers events that reverse the change, and restores the house's internal environment to its desired level.

In the human body, negative feedback systems are the most common method used to maintain the internal environment within a narrow range. These systems include those that maintain body temperature, regulate levels of sugar in the bloodstream, and control levels of salt and other solutes in body fluids. We explore these systems in detail later in this chapter.

It is important to note that some groups of organisms, called **regulators,** maintain homeostasis for a certain variable (as we've just described for temperature), while other organisms, called **conformers,** may have no set point for that variable at all and the variable may fluctuate with external changes. Animals vary in the traits for which they are regulators and conformers. Most fishes, for example, conform to the temperature of the water but regulate closely the concentration of salt in their blood and tissues, independent of the salt concentration in the water (**FIGURE 20-4**).

In addition to negative feedback systems, organisms also have a small number of **positive feedback** systems, in which deviations from conditions normally found in the internal environment cause an increase or acceleration of the change, in the same direction. Such systems push the body away from homeostasis and often initiate cascading processes that increase the rate at which the system deviates from the normal range. For this reason, positive feedback systems, if unchecked, can become highly unstable.

Q Why don't we bleed to death when we get a cut?

An example of a positive feedback system is the blood-clotting process (**FIGURE 20-5**). Injury to a blood vessel causes platelets—cellular fragments involved in clotting—to release their Super Glue–like contents, which begin to seal any rip or tear in the blood vessel (see Section 21-6). Within the bloodstream, the contents released from a platelet have another effect as well: they cause other platelets to release their contents. This is a positive feedback system, because the release of some blood-clotting molecules into the bloodstream causes the release of more, which causes the release

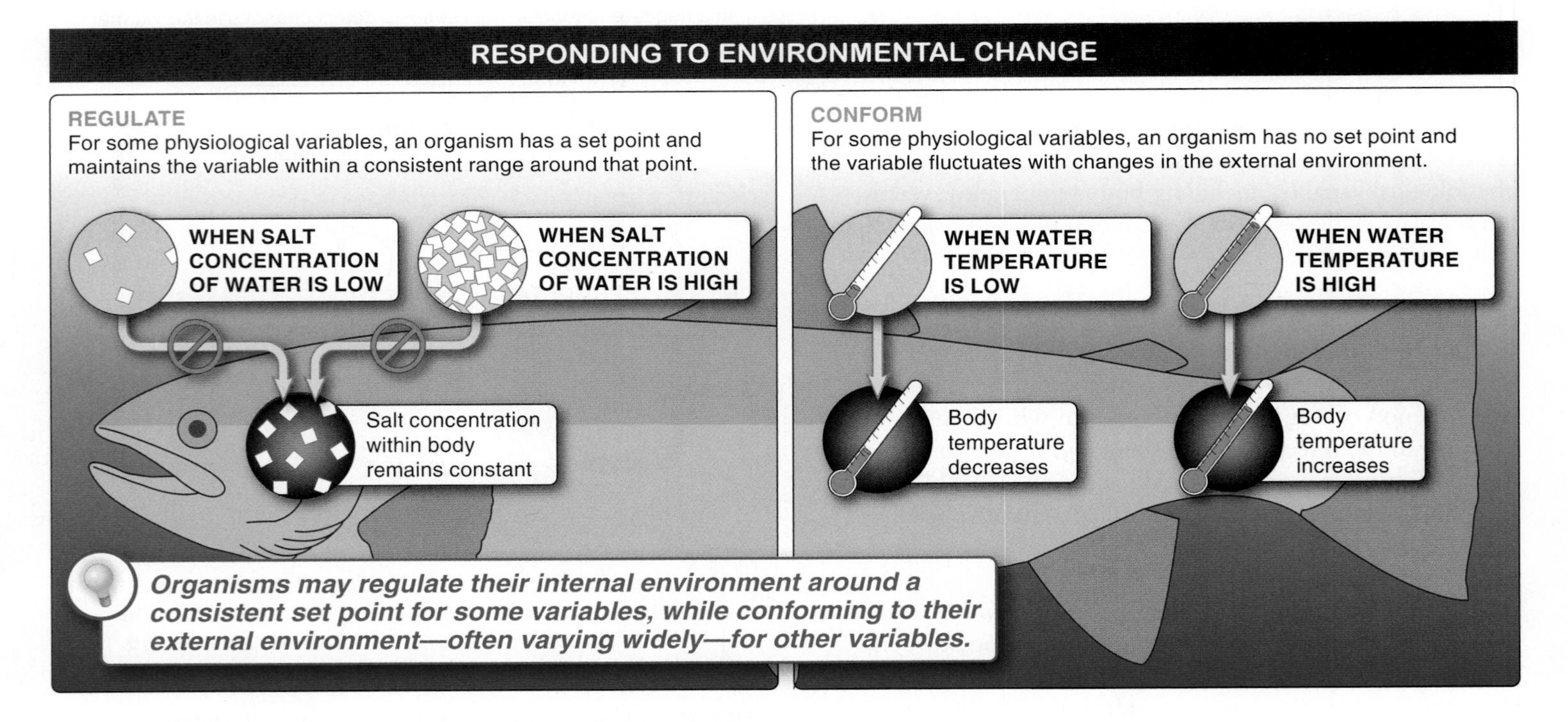

FIGURE 20-4 Regulate or conform? Two strategies to cope with a changing environment.

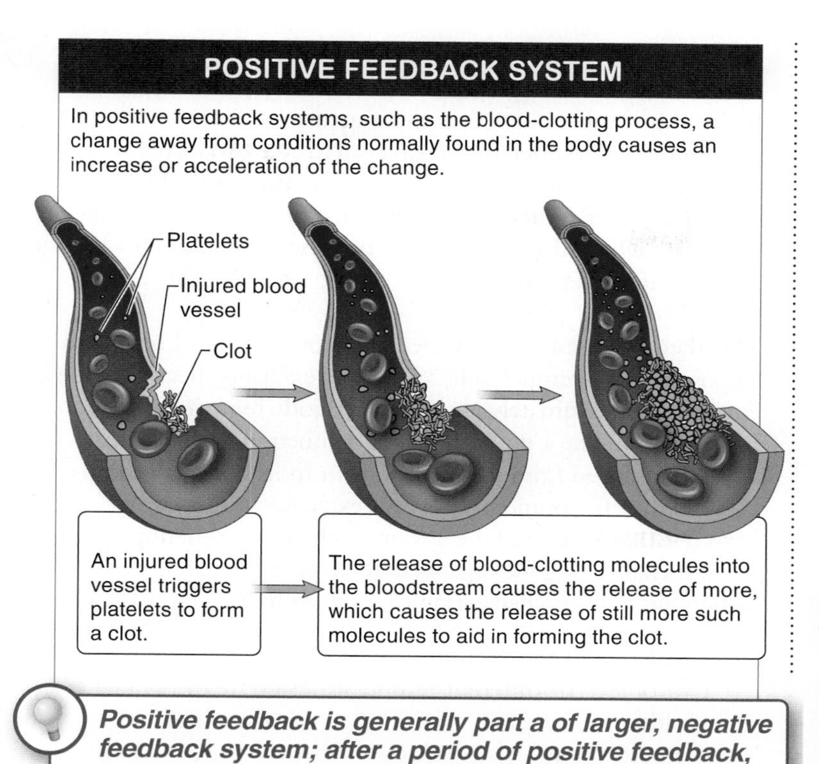

Positive feedback is generally part a of larger, negative feedback system; after a period of positive feedback, the variable is brought back within its typical range.

FIGURE 20-5 Patching a tear. The blood clot that forms after an injury is a result of a positive feedback system.

of still more. The result is that a dangerous situation, injury to a blood vessel, is quickly remedied.

Generally, positive feedback systems are part of larger, negative feedback systems, which, after a short period of positive feedback, bring the variable (and the organism) back to within its original, normal range of internal environmental conditions. After blood clotting has stopped the loss of blood from a damaged vessel, other chemical signals interrupt the positive feedback loop, stop the further release of blood-clotting molecules, and even begin to break down blood clots that no longer serve a function.

TAKE-HOME MESSAGE 20•3

For animals to maintain homeostasis with regard to a particular physiological variable, that variable must have a set point to which the organism can return. Through negative feedback, sensors detect a change in the internal environment and trigger effectors to oppose or reduce the change. Positive feedback systems, much less common, oppose homeostasis in response to a change, pushing the body away from normal conditions and increasing change in the same direction.

20•4

Temperature control is a component of homeostasis.

As we saw in the case of Korey Stringer, one of the most important environmental factors that affect animals is temperature. And for many animal species, the control of temperature, called **thermoregulation,** is an important component of homeostasis.

When it comes to thermoregulation, the biggest distinction among animal species is in how they generate their body heat. **Endotherms** (sometimes described as "warm-blooded") generate their heat internally, within their own bodies. Most mammals and birds are endotherms. **Ectotherms** (sometimes described as "cold-blooded") get their heat primarily from the environment, usually the sun. Invertebrates, fishes, amphibians, and reptiles are all ectotherms. While these groups are often distinguished as "warm-blooded" and "cold-blooded" (**FIGURE 20-6**), closer investigation reveals that such terms do not always accurately describe the two groups. Some animals, including numerous mammalian species, **hibernate,** going into a state of reduced metabolic activity for days or weeks, during which their body temperature can drop considerably. The body temperature of some hibernating ground squirrels, for example, can drop below freezing for more than three weeks. Some hibernating mammals can become quite cold, while some lizards basking in the sun can get very warm.

Q Why do some cold-blooded animals have "hot" blood and some warm-blooded animals have "cold" blood?

These observations highlight another important aspect of thermoregulation. Some organisms (such as humans) are **homeotherms,** meaning that they maintain a relatively constant body temperature. Other organisms (such as the lizards and hibernating mammals) are **heterotherms,** and their body temperatures, at times, fluctuate as the environmental temperature changes (**FIGURE 20-7**). As a consequence, animals are most accurately categorized by describing their source of heat (internal or external) *and* the degree to which they maintain a constant temperature or have a temperature that fluctuates.

Whether or not an animal generates its own heat and maintains a constant body temperature, all animals exchange heat with the environment. Ultimately, an organism's body temperature is a function of the heat it produces in conjunction with the heat that is transferred from the environment to its body or from its body to the environment. This heat exchange can occur in four ways (**FIGURE 20-8**), and these four mechanisms sometimes make it challenging for organisms to acquire heat, and sometimes make it challenging to dissipate heat.

1. *Conduction* is the transfer of heat that occurs when two objects at different temperatures come in contact. Holding an ice cube or sitting on a hot car seat dramatically reveals conduction in action.
2. *Convection* is the transfer of heat from an object to a medium such as water or air as it passes next to the object. This is particularly noticeable to us when a cold breeze blows.
3. *Radiation* is the transfer of heat, without direct contact, from a warmer object to a colder object. You are experiencing radiant heat when you feel the warmth of the sun on your face or the heat from a fireplace on your feet.
4. *Evaporation* is the loss of heat that occurs as a liquid substance, such as liquid water, turns to a gas. When you sweat and the sweat evaporates, it cools you off.

GENERATING BODY HEAT

ENDOTHERMS
- Animals that generate body heat internally
- Sometimes described as "warm-blooded"
- Include most mammals and birds

ECTOTHERMS
- Animals that get their heat primarily from the environment
- Sometimes described as "cold-blooded"
- Include invertebrates, fishes, amphibians, and reptiles

FIGURE 20-6 Strategies to heat the body.

MAINTAINING BODY TEMPERATURE

HOMEOTHERMS
Body temperature remains relatively constant.

HETEROTHERMS
Body temperature fluctuates as environmental temperatures change.

FIGURE 20-7 Stable body temperature versus variable body temperature.

EXCHANGING BODY HEAT WITH THE ENVIRONMENT

Animals exchange heat with the environment in four ways.

CONDUCTION
The transfer of heat that occurs when two objects at different temperatures come in contact.

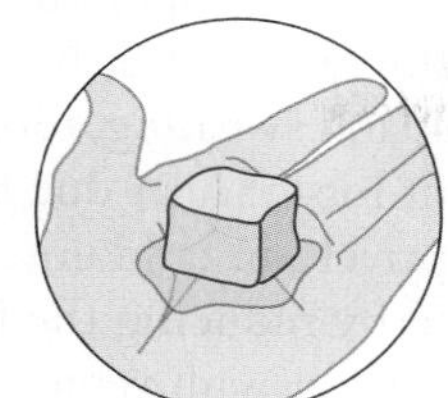

CONVECTION
The transfer of heat to a medium such as water or air as it passes next to an object.

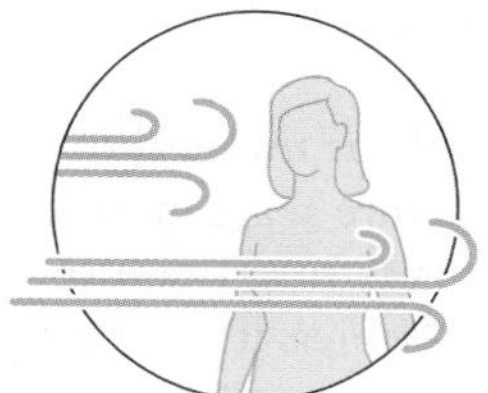

RADIATION
The transfer of heat, without direct contact, from a warmer object to a colder object.

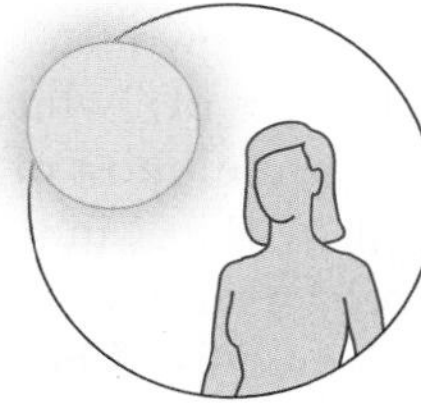

EVAPORATION
The loss of heat that occurs as a substance such as liquid water turns to a gas.

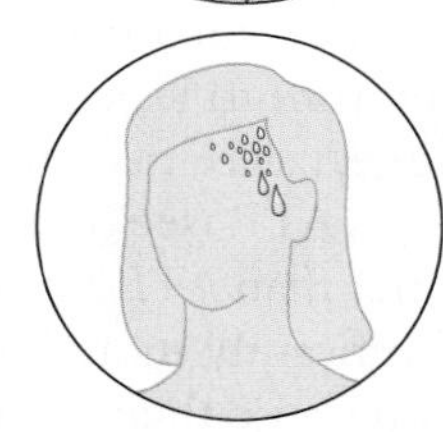

An organism's temperature depends on the heat that it gains or loses to the environment, in conjunction with the heat the organism produces.

FIGURE 20-8 How organisms acquire or dissipate heat.

We'll find further examples of each of these mechanisms as we look at four of the methods—physical, behavioral, physiological, and cellular—by which animals regulate body temperature (**FIGURE 20-9**).

Numerous physical features have evolved in organisms that influence the organism's body temperature. These include the organism's body size, surface area, and levels of insulation. The thick lipid-rich coat of blubber in walruses, whales, and many other marine mammals, for example, can account for up to 50% of the body weight of some animals and provides effective insulation, helping them to maintain a constant body temperature.

An organism can also use behavioral strategies to regulate its body temperature. Most lizards, for example, bask in the sun during the early morning hours, increasing their body temperature. Later, when the day is at its hottest, they may retreat to a burrow to reduce their body temperature. Many large mammals that cannot get to shade orient their bodies to minimize the angle at which the sun strikes them. The African ground squirrel even shades itself with its tail as it forages.

Animals also employ many physiological methods of thermoregulation. One specific example of this occurs as cells maintain concentration gradients of various ions, such as potassium (K^+) and sodium (Na^+). Many cell membranes are somewhat permeable to K^+ and Na^+ ions. The ions "leak" across the membrane, requiring active transport to restore and maintain the concentration gradients necessary for some cellular processes. Because the reactions involved in active transport are not perfectly efficient in their use of energy,

METHODS FOR REGULATING BODY TEMPERATURE

PHYSICAL METHODS
The walrus has a thick coat of blubber that provides insulation from the external environment.

BEHAVIORAL METHODS
The African ground squirrel shades itself with its tail while foraging, to minimize heat from the sun.

PHYSIOLOGICAL METHODS
By panting, dingos make use of the efficient loss of heat due to evaporation.

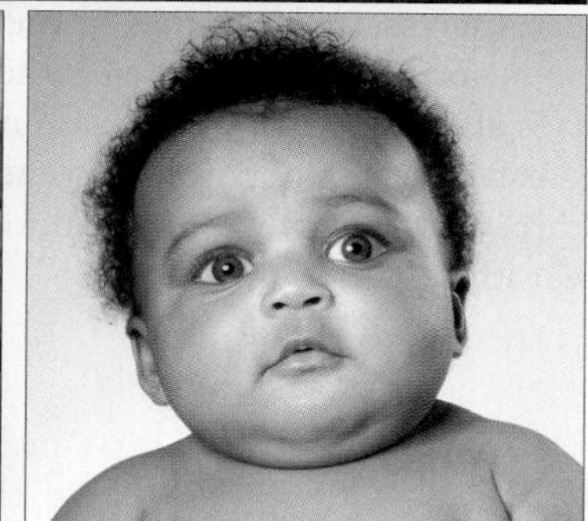

CELLULAR METHODS
Human infants have a special type of fat that produces heat, rather than ATP, when broken down.

FIGURE 20-9 Adaptations that aid in temperature regulation.

Most organisms must maintain water and solute concentrations within a narrow range. Imbalances can lead to serious health problems and even death.

FIGURE 20-10 Heat exchange. Sweating can help cool us off through evaporation. By leaning up against a cool rock wall, we can also lose some heat through conduction.

some energy is released as heat with each reaction. This heat can help maintain an organism's body temperature. Interestingly, as it turns out, cell membranes in endotherms, such as humans, are leakier than the cell membranes in ectotherms, such as fishes, suggesting that such "leakiness" is an adaptation that helps endotherms generate heat internally. It is also likely that it's because mitochondria are the sites of greatest heat generation that endotherms have, on average, three to four times as many mitochondria per cell as ectotherms.

Another physiological method by which animals—both ectotherms and endotherms—can regulate their body heat is by controlling the flow of blood to the skin. To lose heat, such as during periods of extreme exertion or at extremely high external temperatures, they increase blood flow to the skin, allowing greater heat loss by convection. Through panting and sweating, too, many animals make use of the efficient loss of heat due to evaporation (**FIGURE 20-10**). Alternatively, in cold environments, the loss of heat can be reduced by reducing the flow of blood to the skin and by shivering, through which animals can increase their heat production.

Q Why does "baby fat," unlike regular fat, act like a built-in heating pad?

One cellular method of temperature regulation is particularly important to human babies. It takes place in a special type of connective tissue called "brown fat." Unlike the cells of "white fat," or adipose tissue (most of the fat in adult's bodies), in which there are few mitochondria, the cells of brown fat have a high density of mitochondria along with the stored fat. When brown fat cells oxidize their fat, however, they don't generate ATP. Instead, a special protein causes protons to leak directly across the mitochondrial membrane, rather than passing through the enzyme that synthesizes ATP (see Section 4-15), causing the production of heat rather than ATP from the fat breakdown. Human infants have significantly more brown fat than adults. This probably evolved because their small body size gives them a large surface-area-to-volume ratio, which results in a relatively large surface area over which they can lose heat and a relatively small body mass in which they can generate heat.

TAKE-HOME MESSAGE 20•4

The control of body temperature, called thermoregulation, is an important component of homeostasis. Body temperature is a function of internal heat production and heat transfer between an organism and its environment. Heat transfer to and from the environment is regulated physically, behaviorally, physiologically, and at the cellular level.

20•5

Animals must balance their water content within a narrow range.

Organisms, whether they live in the water or on land, require water. But they can't take in too much or they will burst. And they can't lose too much or they will shrivel up and die. **Osmoregulation,** an important component of homeostasis in animals, is the regulation of water content and of the concentrations of dissolved solutes that influence osmosis. The most abundant solutes in animal fluids are sodium, chloride, potassium, magnesium, and calcium.

Although a variety of mechanisms of water regulation have evolved in animals, there are just four ways that animals get water: by drinking it, by eating foods that contain it, by absorbing it through osmosis, and as a by-product of cellular respiration. Similarly, there are four ways that animals lose water: urination, defecation, evaporation (including panting, breathing, and sweating), and osmosis.

Osmoregulation revolves around controlling and regulating these mechanisms of water loss and gain. Through osmosis, water flows from areas of low solute concentration to areas of high solute concentration. Thus organisms often control their water content indirectly by regulating their solute content. They generally balance the total amount of solutes—which influences the direction and magnitude of osmosis—and the concentration of each solute individually. Imbalances in the concentrations of certain solutes can cause serious health problems, from muscle spasms to confusion to paralysis, and even death.

Because animals live in such a wide range of habitats, they encounter a similarly wide range of osmoregulation challenges. The challenges are very different for aquatic organisms than for terrestrial organisms, for example. And among the aquatic organisms, the challenges facing organisms living in saltwater habitats are very different from—almost the opposite of—those faced by organisms living in fresh water.

Different strategies have evolved for coping with such diverse osmoregulation challenges. Most invertebrates that live in salt water are **osmoconformers,** meaning that they let the solute concentration of their body fluids reflect that of their environment. Most vertebrates, on the other hand, are **osmoregulators,** maintaining their fluids and solute concentrations within narrow ranges that differ from those of their environment (**FIGURE 20-11**).

A variety of structures have evolved that enable animals to regulate their water balance. In insects, for example, very small tubes branch off the digestive tract, near its end at the rectum. These extensions from the gut, called Malpighian tubules, function as the animals' chief excretory organs. Using active transport (see Section 3-10), an animal moves potassium ions

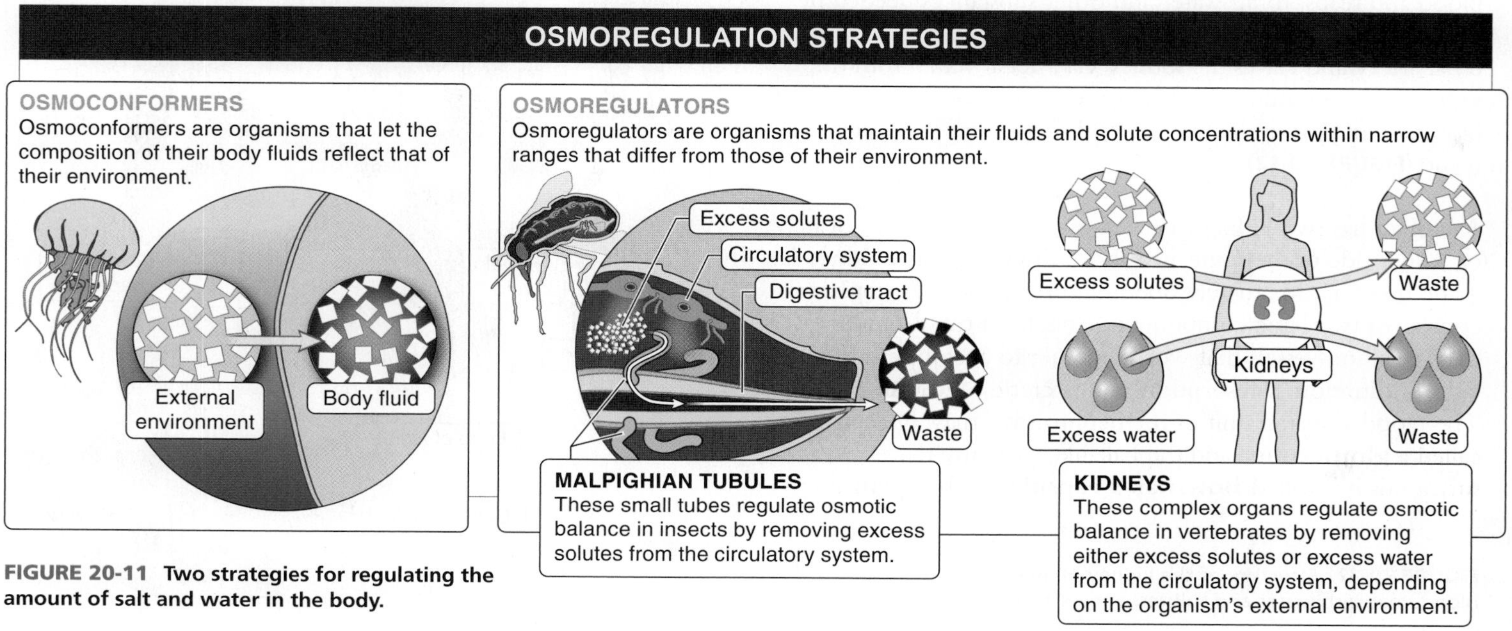

OSMOREGULATION STRATEGIES

OSMOCONFORMERS
Osmoconformers are organisms that let the composition of their body fluids reflect that of their environment.

External environment
Body fluid

OSMOREGULATORS
Osmoregulators are organisms that maintain their fluids and solute concentrations within narrow ranges that differ from those of their environment.

Excess solutes
Circulatory system
Digestive tract
Waste

MALPIGHIAN TUBULES
These small tubes regulate osmotic balance in insects by removing excess solutes from the circulatory system.

Excess solutes
Waste
Kidneys
Excess water
Waste

KIDNEYS
These complex organs regulate osmotic balance in vertebrates by removing either excess solutes or excess water from the circulatory system, depending on the organism's external environment.

FIGURE 20-11 Two strategies for regulating the amount of salt and water in the body.

(K^+) and cellular waste products from its body cavity—where its blood is—into the Malpighian tubules. As a consequence of the increase in solutes in the tubules, water moves into them by osmosis. The K^+ ions, water, and waste products move down the digestive tract until, near its end, the K^+ ions and water are mostly reabsorbed into the circulatory system. This is a very effective system for conserving water, leaving mostly waste products to be excreted as feces.

Q Do fish drink water?

Another osmoregulatory structure is the vertebrate kidney. The kidney, as we shall see, is a complex organ that filters blood, removing metabolic waste products—particularly excess nitrogen from the breakdown of proteins and nucleic acids—and other ions while regulating the organism's water balance. The specific details of kidney function vary across the different groups of vertebrates. Let's consider, for example, the challenges faced by two groups of aquatic vertebrates: freshwater and saltwater fishes. A freshwater fish faces the problems of high concentrations of water entering its body and solutes leaving its body. A saltwater fish, on the other hand, tends to lose water to its environment while taking in high concentrations of solutes from the environmental water. As a consequence, vertebrate kidneys—depending on the type of animal—may function either to conserve water or to remove it. (This explains why a freshwater fish does not drink water, but a saltwater fish drinks large amounts.)

In the next section, we examine the details of kidney functioning, focusing specifically on the human kidney and how it is able to produce urine that has a solute concentration more than four times that of blood, and how it is able to eliminate the many potentially harmful molecules contained within the food and drink consumed in the diet.

TAKE-HOME MESSAGE 20•5

Many organisms maintain their water content within a narrow range. To maintain osmotic balance, organisms must be able to take up water and get rid of water, and they must be able to regulate concentrations of ions in their body fluids. Various mechanisms and strategies have evolved for coping with these challenges.

20•6

In humans, the kidney is the chief excretory organ.

The kidney is an organ in vertebrates that helps maintain homeostasis by regulating water balance and solute concentration in body fluids. It accomplishes this by filtering blood and reabsorbing water and other substances needed by the body. Blood flows into capillaries within each kidney via a renal artery and leaves the kidney via a renal vein. From the kidneys, waste products and water removed from the bloodstream pass to the bladder, and from there are excreted as urine (**FIGURE 20-12**).

A human has two kidneys, each about the size of a fist, located on either side of the spine, just above the waist. Each kidney is made up of approximately one million nephrons. A **nephron** consists of two basic components: a nephron tubule and a mass of blood vessels that work together to accomplish the tasks of filtration, reabsorption, and excretion (**FIGURE 20-13**). The blood-filtering unit of the nephron is a mass of capillaries called a **glomerulus,** and the ball-like structure that surrounds it is called **Bowman's capsule.** Each Bowman's

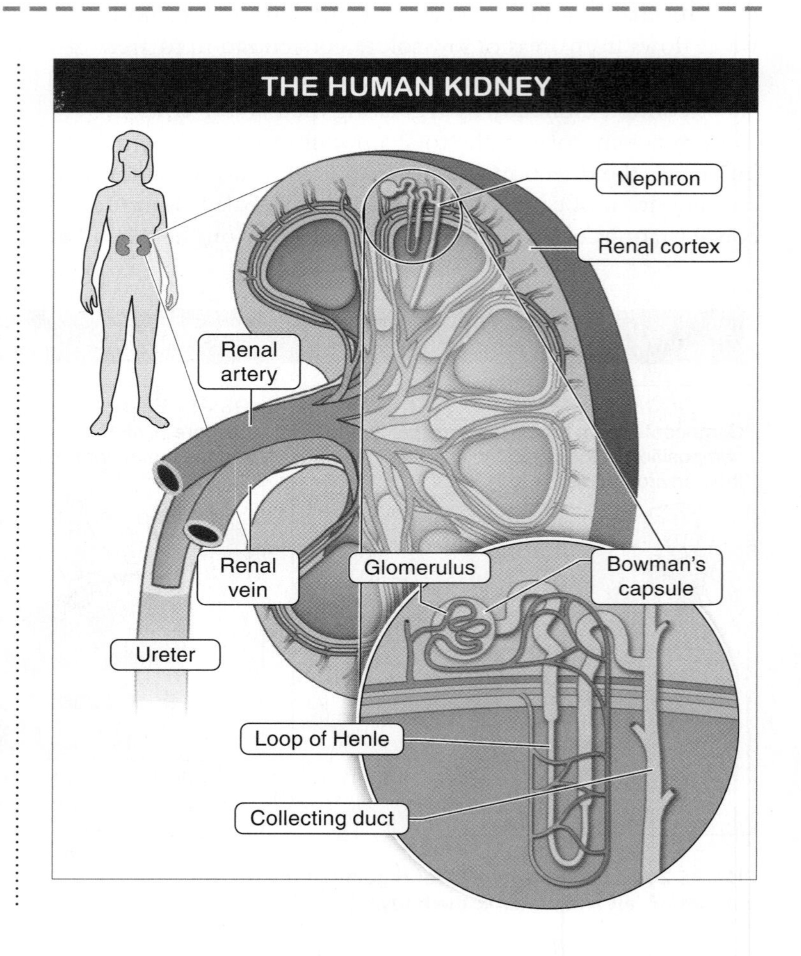

FIGURE 20-12 Structure of the kidney. Human kidneys filter blood, reabsorb water and solutes, and excrete waste.

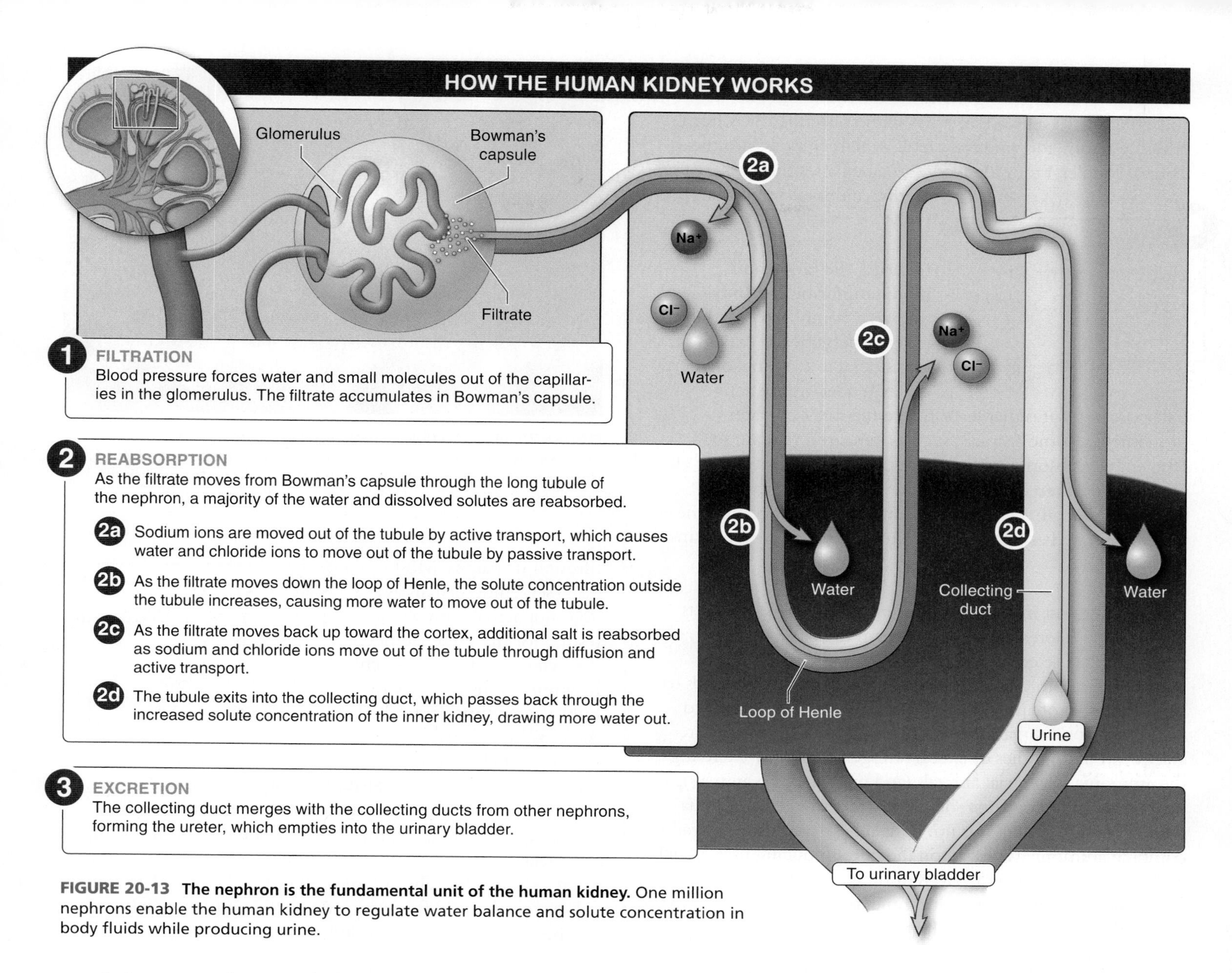

FIGURE 20-13 **The nephron is the fundamental unit of the human kidney.** One million nephrons enable the human kidney to regulate water balance and solute concentration in body fluids while producing urine.

capsule is connected to a single, long, urine-producing tube that excretes its filtered fluid into a collecting duct.

The capillaries in the glomerulus are porous, and blood pressure forces out water and small molecules and ions (but not blood cells or most proteins) through the capillary walls. Fluid that accumulates in Bowman's capsule, called **filtrate,** contains salts, sugars, amino acids, vitamins, and many other molecules, all at the same concentration as in the blood.

As the filtrate moves from Bowman's capsule through the urine-collecting tubule of the nephron, the vast majority of the water and dissolved solutes must be reabsorbed. In fact, up to 2,000 liters of blood (about 275 times the total volume of blood in your body) passes through the kidneys each day, but only about 1.5 liters of urine is produced and excreted.

Reabsorption in the long tubule of the nephron is a complex process. It begins with the active transport of sodium ions out of the tubule, which causes water and chloride ions to follow, moving out of the tubule by passive transport.

The tubule (and the filtrate it contains) then loops from the outer part of the kidney, called the cortex (see Figure 20-13), down into the innermost part of the kidney and back again, in a path called the loop of Henle. As the tubule passes to the innermost part of the kidney, the solute concentration of the interstitial fluid, outside the tube, increases. This causes more and more water to move out of the tubule, from the filtrate and into the interstitial fluid. As the filtrate moves up toward the kidney cortex again, more salt is lost, through diffusion and active transport. From the cortex, the tubule passes into the collecting duct, which passes through the innermost part of the kidney again, returning additional water to the interstitial fluid and further concentrating the urine for excretion. The collecting duct eventually merges with collecting ducts from other nephrons, forming the ureter, which empties into the urinary bladder. From the bladder, urine passes through the urethra and is excreted from the body.

Q Why do some desert mammals never need to drink water?

Filtering the blood and producing urine not only follows a complicated path but is very energetically expensive. Considerable amounts of energy are expended in the active transport of solutes that leads to the recovery of water. Nonetheless, some animals, such as kangaroo rats, are so efficient at reabsorbing water that they can recover nearly all of the water filtered by their kidneys. They never have to drink water at all; the water contained in their food and the water generated as a by-product during cellular metabolism is sufficient.

Among the most important of the metabolic waste products that must be filtered and removed from the blood by the kidneys are those containing nitrogen. Produced from the breakdown of proteins and nucleic acids, this nitrogen tends to be in the form of ammonia, which is generally very toxic to organisms. Some organisms—mostly aquatic organisms—are able to rid their bodies of excess nitrogen by simply excreting the ammonia. Terrestrial animals (and many marine animals), however, cannot consume sufficient water to keep ammonia diluted enough so that it is not toxic. These organisms instead combine ammonia with carbon dioxide, producing urea, which can be stored for longer periods of time and at higher concentrations (thereby requiring much less water). There is a cost to this system, however; the production of urea from ammonia and carbon dioxide is energetically expensive. Besides ammonia and urea, there is a third form in which nitrogenous wastes can be excreted, as a paste called uric acid. This method requires even less water than urea excretion, but is even more energetically expensive. It is used by insects, terrestrial snails, birds, and many reptiles.

The kidneys are so effective at filtering blood and concentrating the many waste products of metabolism in urine that it is possible to detect the use of many drugs

FIGURE 20-14 Detectible remainders. Drugs break down into metabolites, chemicals that sometimes linger in the body and urine. Here an individual tends to marijuana plants grown for medicinal purposes.

through urinalysis. Most drug-screening urinalysis does not actually involve testing for the presence of the drugs themselves, which may remain in the body for only a short time. Instead, the tests look for the presence of chemicals, called metabolites, that result from the breakdown of the drugs (**FIGURE 20-14**). Metabolites have been identified for many drugs, including Ecstasy, cocaine, methamphetamines, and marijuana. In the case of marijuana, the metabolites are fat-soluble. This means that they are stored in fat cells indefinitely, and released only when fat from those cells is metabolized for energy. As a consequence, marijuana can be detected a month or longer after the last use.

Q How can some drugs be detected by urinalysis even months after the last intake?

TAKE-HOME MESSAGE 20·6

The kidney is the organ in vertebrates that helps maintain homeostasis by regulating water balance and solute concentrations in body fluids, filtering blood, and removing potentially harmful ions and metabolic waste products, excreting them in urine.

An elephant uses its trunk to forage for tree-top leaves.

20•7

Most animal bodies are organized into cells, tissues, organs, and organ systems.

Form follows function. This simple statement captures one of the most universal relationships in the living world. For example, fast-swimming organisms, from penguins to tuna to sharks, share a common streamlined body shape (**FIGURE 20-15**). If a structure is adaptive—the product of natural selection—then its physical features closely reflect its function. And just as form follows function for large animal structures, so, too, does the relationship hold true for molecules, organelles, and cells. In this section we examine the levels of organization in animal bodies and see how form fits function in each.

Animals are multicellular organisms. And multicellularity makes it possible for animals to attain much larger sizes and much greater physiological complexity than single-celled organisms. Increased size and complexity bring many benefits, including a reduction in the number of potential predators, an increase in the number of potential prey, and, more generally, a

FIGURE 20-15 Adapted for swimming: the streamlined body of the penguin.

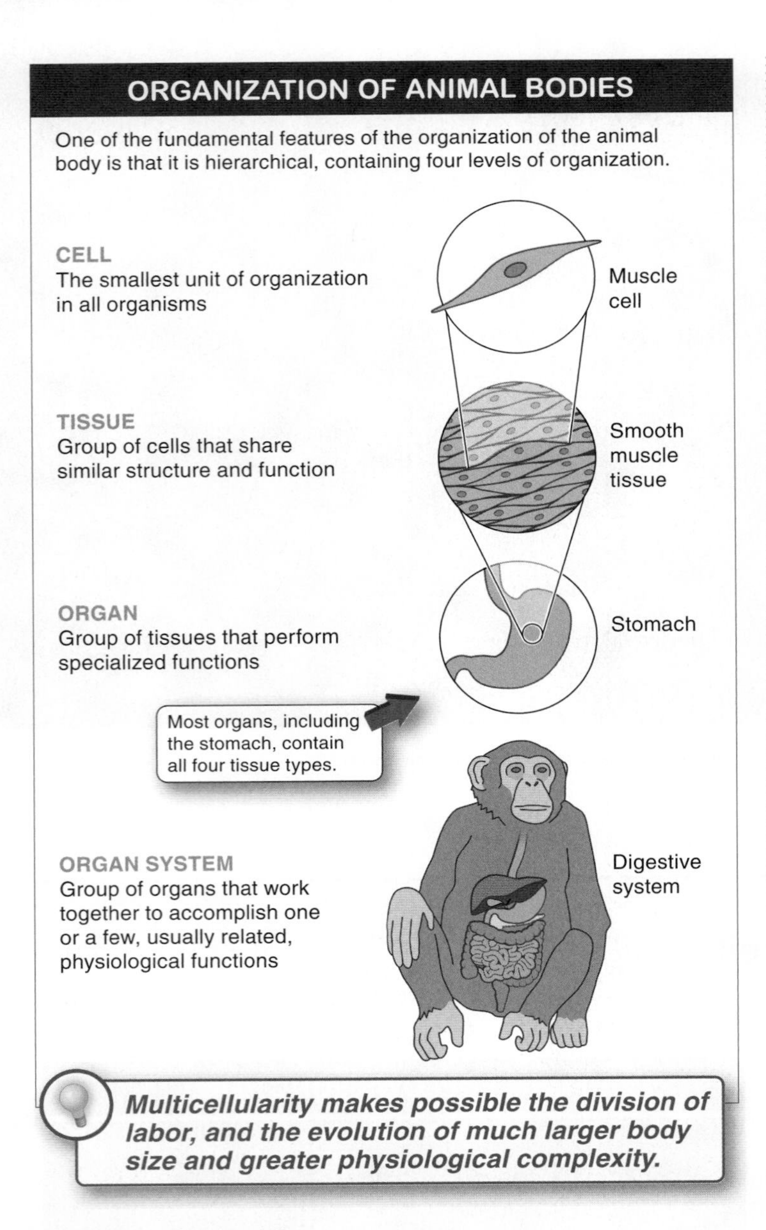

FIGURE 20-16 **From cells to organ systems.**

reduction in the influence of the external environment. For these reasons, among others, the transition from a single-celled to a multicellular body is one of the most important evolutionary transitions.

Possibly the chief benefit of multicellularity is that it makes possible a division of labor and specialization at the cellular level. No longer is it necessary for each cell to carry out every single physiological process (such as generating movement, detoxifying harmful chemicals, digesting macromolecules, and sensing and responding to environmental changes). Instead, cells can be organized into groups, and groups organized into larger groups, to carry out specific life-sustaining functions such as exchanging gases between the organism and its environment, thinking and feeling, and fighting pathogens (**FIGURE 20-16**).

One of the fundamental features of the organization of the animal body is that it is hierarchical. Cells, as you'll recall from Chapter 3, are the smallest unit of organization in all organisms. Sponges, structurally the simplest of all animals, have cells specialized to perform a few distinct tasks. For example, some cells form the covering of the sponge's body, and others obtain food to provide nutrition and energy to sustain the sponge's activities. At the other end of the spectrum are humans—with 210 different cell types.

In most animals, groups of cells with similar structure, along with some products of those cells, form **tissues,** in which the cells act together to perform specific functions in the body. Adult animals generally have four main types of tissue (**FIGURE 20-17**), which we'll discuss in more detail in Sections 20-8 through 20-11.

Connective tissue consists of cells embedded in a large amount of extracellular material, called **matrix,** which together contribute to body structure and support. Found throughout the human body, connective tissue can also serve to anchor cells, regulate communication between cells, and influence growth and wound healing. Some of the most important structures formed from connective tissue include tendons, cartilage, blood, adipose, and bone.

Epithelial tissue covers and lines most exterior and interior surfaces of the body. Skin is an epithelial tissue, as are the tissues that line the nose, throat, lungs, blood vessels, and digestive tract.

From the circulatory and respiratory systems that deliver oxygen and remove carbon dioxide, to the muscle tissue that generates the flapping of wings, to the neurons in the eyes that detect prey, to the epithelial tissue across which nutrients from food are absorbed, the working together of cells, tissues, organs, and organ systems make many complex actions of organisms possible—such as flight in these red-and-green macaws in Peru.

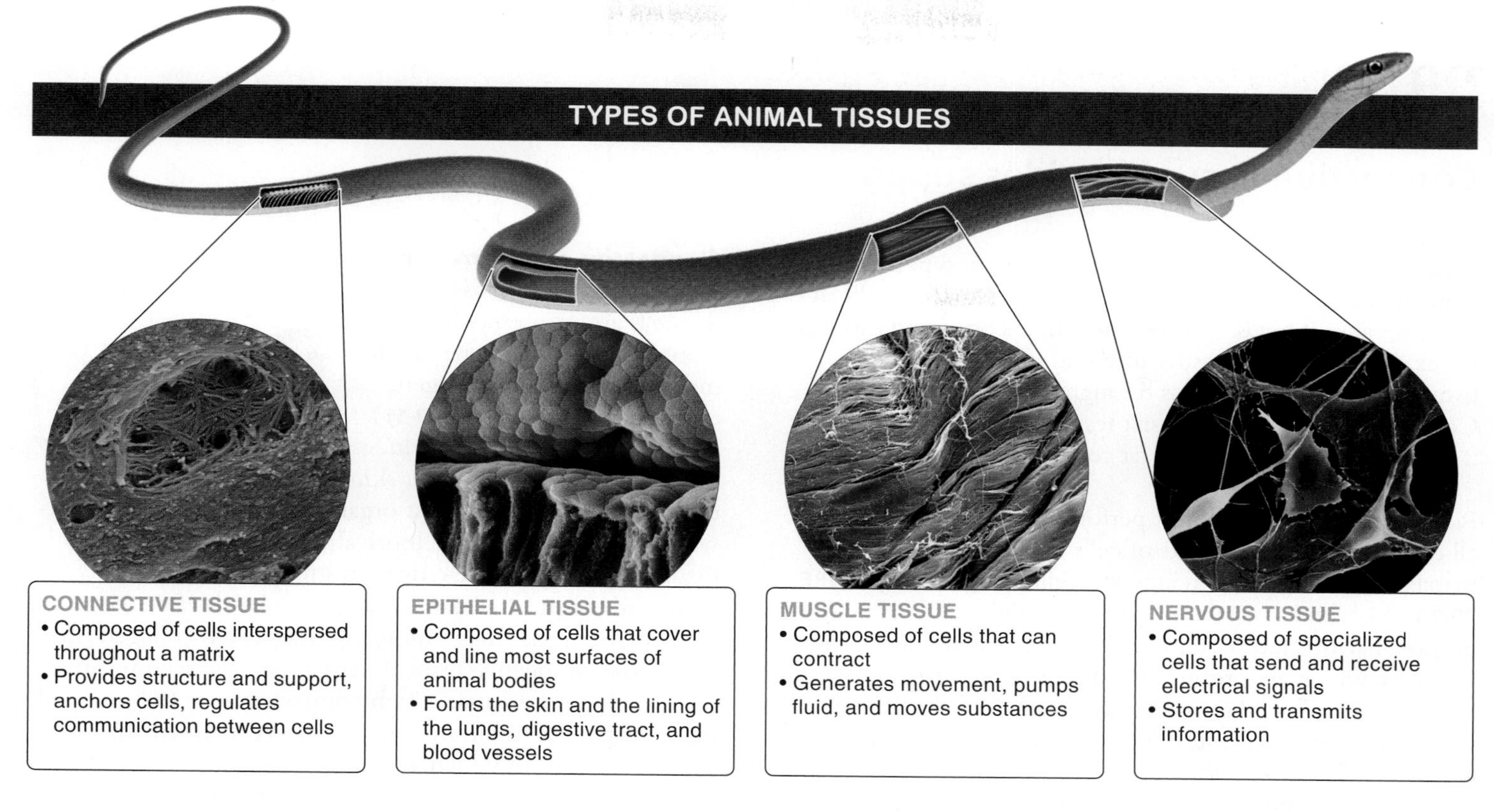

FIGURE 20-17 Tissues perform specific functions in the body.

Muscle tissue is made up of cells that can contract. This characteristic gives muscle tissue the ability to generate movement or pump fluids through the body.

Nervous tissue, found throughout the human body, is specialized to send and receive electrical signals and, in doing so, can store and transmit information. The brain and spinal cord are made up of large amounts of nervous tissue.

Just as cells with similar functions are grouped into tissues, so tissues are often grouped into organs or organ systems. **Organs** are structures that serve specialized functions, and they usually contain several types of tissue. The heart, liver, kidneys, and brain are examples of organs in the human body. Most organs—including the stomach and the small intestine, for example—have all four types of tissue. **Organ systems** are groups of organs that work together to accomplish one or a few, usually related, physiological functions. The circulatory system, for example, includes the heart, blood vessels, and blood.

In the remaining chapters of the book, we explore the complex functions carried out by each of the organ systems in animals. We describe the structures that are part of each system, how they function, and how they have evolved. As we do this, we see that an organism is greater than the sum of its parts. The working together of cells to form tissues, and of tissues to form organs and organ systems, gives multicellular organisms the abilities to reproduce, defend themselves, and communicate (among many other abilities), in ways that are not possible for a single cell or a single type of tissue.

TAKE-HOME MESSAGE 20•7

Animal bodies are highly organized, and at all levels of organization, the physical features are closely related to function. In most animals, cells with similar structure and function are organized into tissues. There are four types of tissue: connective tissue, epithelial tissue, muscle tissue, and nervous tissue. Tissues are often organized into organs, which serve specialized functions and can contain several types of tissue. In turn, organs can be organized into organ systems that accomplish highly complex tasks.

20•8

Connective tissue provides support.

Connective tissue is usually the most abundant type of tissue in an animal. Connective tissue is sometimes called "cellular glue," because it holds cells together and, as bone and cartilage, for example, gives shape, structure, and support to other tissues, structures, and organs throughout the body. Connective tissue, however, isn't restricted to supporting roles; it also includes tissues such as fat cells and blood.

Regardless of the function it performs or the form it takes, all connective tissue consists of cells that are embedded in matrix—a mass of non-living extracellular material (**FIGURE 20-18**). Matrix, in vertebrates, consists chiefly of polysaccharides and protein (and some minerals, in the case of bone), which are produced and secreted by various cells within the matrix. The matrix can be liquid, jelly-like, or solid. With the exception of blood, all connective tissue contains cells called *fibroblasts,* which produce and secrete the matrix proteins **collagen** and **elastin.** So common is connective tissue that collagen, which often forms a sort of net surrounding organs, is the most abundant protein in humans and other mammals.

CONNECTIVE TISSUE STRUCTURE

Connective tissue is a collection of cells arranged within an extracellular matrix that gives shape, structure, and support to other body tissues.

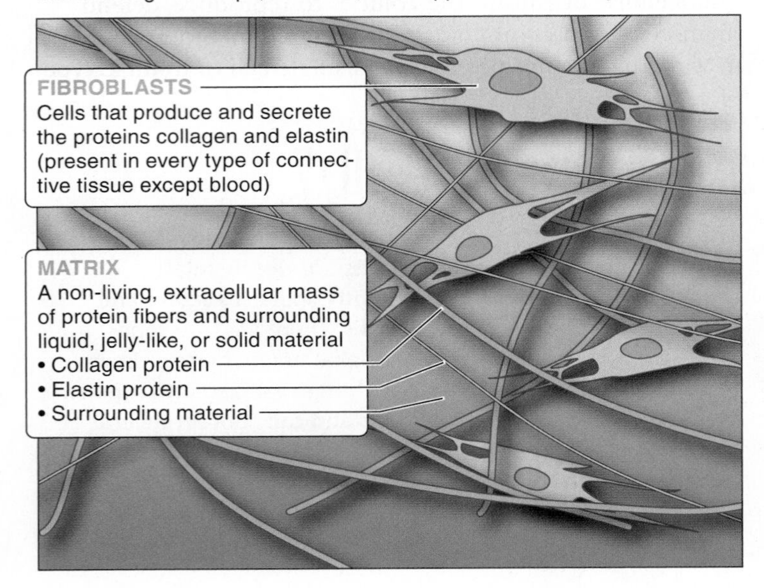

FIGURE 20-18 Connective tissues consist of cells embedded in an extracellular matrix.

Connective tissue proper functions much like packing material. It can be loose or dense. In loose connective tissue, the cells are in a semi-fluid, flexible matrix that generally has many fibers, usually collagen, embedded in it (**FIGURE 20-19**). Loose connective tissue includes the soft padding under your skin, the tissue surrounding most organs, and adipose (fat) tissue, which aids in cushioning, lubricating, and insulating other tissues. Dense connective tissue is stronger than loose connective tissue. It also has collagen as its chief matrix element, but it has many more, tightly packed collagen fibers than loose connective tissue. Examples of dense connective tissue include **tendons,** which connect muscle to bone, and **ligaments,** which bind bone to bone. A sprained ankle is a common injury in which a twisting of the ankle overstretches and tears part of a ligament in the foot.

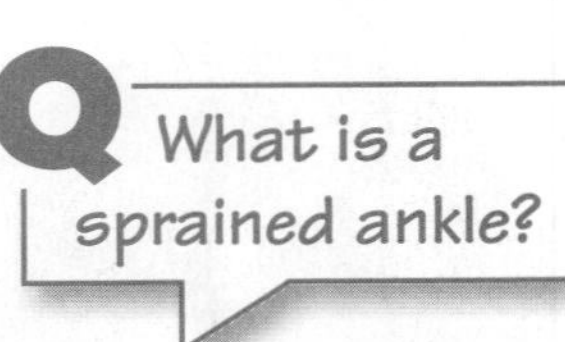

The second type of connective tissue, called **special connective tissue,** includes bone, cartilage, and blood (see Figure 20-19). In these tissues, the matrix differs from that of connective tissue proper in that it is rigid or liquid. In **bone,** the mineral calcium is incorporated into the extracellular matrix, which then hardens into a solid material. Bones can give significant protection to organisms (the skull protects the brain, ribs protect the lungs and heart) or can provide structural support, as the backbone does. **Cartilage,** a dense connective tissue with an extracellular matrix rich in collagen, elastin, and proteins bound to long carbohydrate chains, has a hardness between that of bone and of tendons. Strong, but also flexible, cartilage is found in the ears and tip of the nose in humans, and it cushions the bones in joints throughout the body. In some organisms, including sharks, the entire skeleton is made of cartilage. **Blood** is unique among the connective tissues because it has a liquid extracellular matrix. This liquid matrix, called plasma, is made up mostly of water, but also contains dissolved proteins, sugars, and other molecules. Blood cells, including red blood cells and white blood cells, and the cellular fragments called platelets (see Chapter 21), are suspended within plasma as they transport gases and other substances throughout the body.

All connective tissues in the body, as we've seen, have essentially the same structure: cells embedded within an extracellular matrix that may be solid, soft and flexible, or

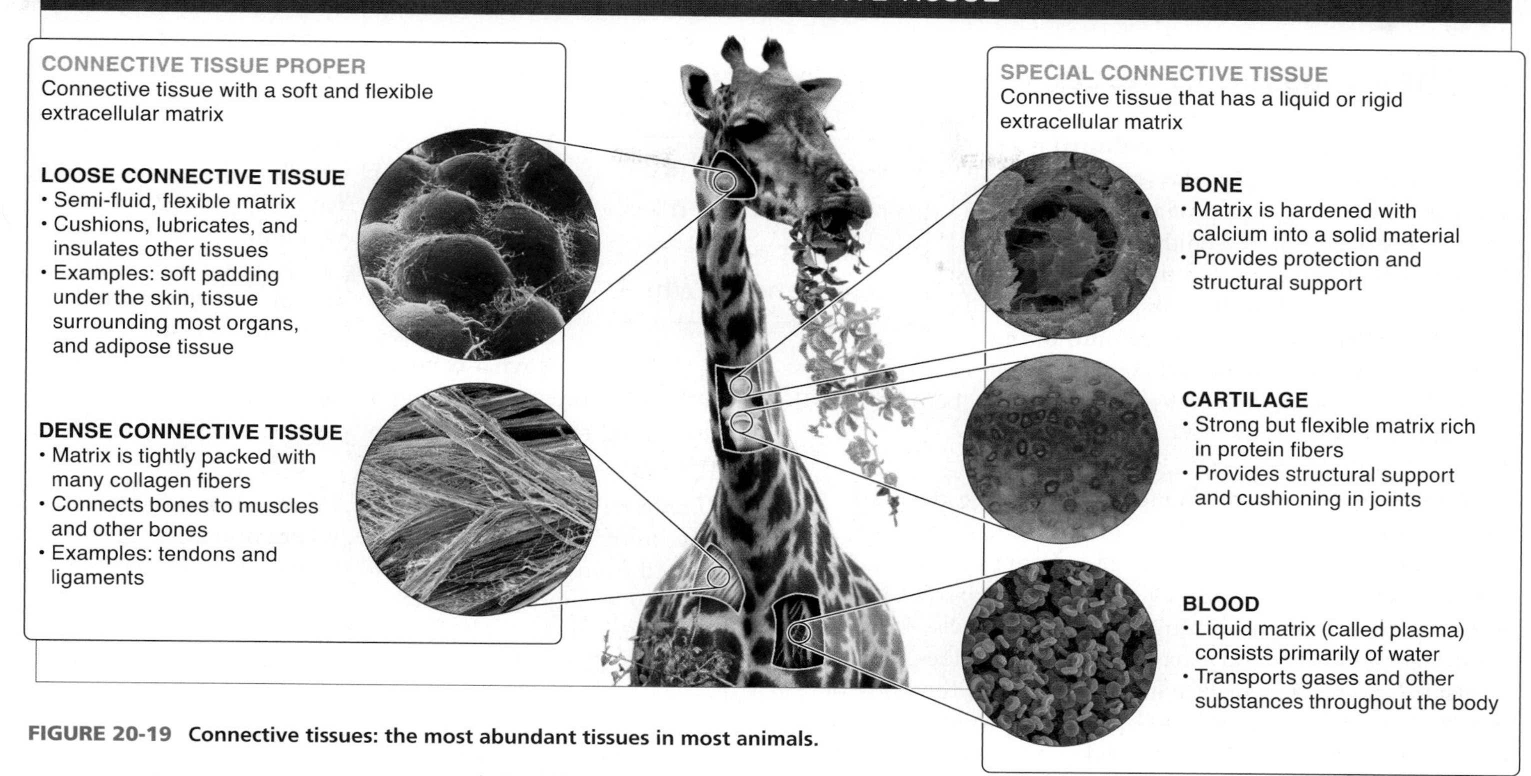

FIGURE 20-19 Connective tissues: the most abundant tissues in most animals.

Most activities rely heavily on connective tissue, often in conjunction with muscle tissue. Playing tennis, for example, relies on blood to deliver oxygen to muscle, tendons that connect muscles to bones, and cartilage, cushioning the joints. (Collagen from the intestines of cows is even used as a material for producing the strings in some tennis rackets!)

fluid. An inflammation or weakness in any of the types of connective tissue, particularly problems with the maintenance of collagen, is the cause of many diseases with widespread effects, including scleroderma and Marfan syndrome. The most common type of arthritis, osteoarthritis, results from the breakdown of connective tissue, usually the cartilage, around joints. The reduced cushioning often leads to inflammation of tissue around the joint, with stiffness and pain, and is commonly treated with anti-inflammatory drugs including aspirin and ibuprofen.

Q The most common form of arthritis is called "wear and tear" arthritis. Why?

TAKE-HOME MESSAGE 20•8

The most abundant type of tissue in most animals is connective tissue. Connective tissue is a collection of cells arranged within an extracellular matrix, usually containing collagen, that holds the cells together and gives shape, structure, and support to other body tissues. Examples of connective tissue include tendons, ligaments, fat, blood, bone, and cartilage.

20•9

Epithelial tissue protects.

Epithelial tissue (also called **epithelium**) is a sheet-like tissue that covers the surfaces of an animal's body. It is made up of a single layer or a few layers of cells, called epithelial cells, that are tightly bound together so that fluids and gases must pass through the cells, rather than around or between the cells, to get into or out of the body (**FIGURE 20-20**). When you look at a vertebrate, most of what you see is epithelium, because skin is an epithelial tissue.

Q Damage to the lining of the stomach can have painful consequences. Why?

Epithelial tissue is not just found on the outer surfaces of organisms, however. It also forms **glands**—collections of cells producing secretions for use elsewhere in the body—and lines the internal tubes and cavities of the body, such as the stomach and intestine, lungs, and blood vessels. In each location, epithelium always has two distinctive sides. The "outside" can be in contact with the outside of the body or with an internal cavity such as the stomach, where it plays a protective role. The "inside" (or underside) faces away from the surface and is generally secured to underlying tissues. Epithelium plays multiple roles in organisms. Three of its most important functions are protection, transport, and secretion (**FIGURE 20-21**).

1. *Protection.* Acting as a barrier, epithelial cells are linked closely by tight junctions and desmosomes (see Section 3-12), which keep fluids from leaking into or out of tissue. If the strong acids in the stomach, for example, were to leak out, they would seriously damage surrounding tissue. This is what occurs in individuals with stomach ulcers, open sores in the lining of the stomach. The tight fit between epithelial cells is also why the skin of terrestrial vertebrates is usually waterproof.

2. *Transport.* Epithelium forms small finger-like projections in the lining of the small intestine, where nutrients are absorbed from digested food into the bloodstream. In blood vessels, epithelium controls which molecules can enter other tissues of the body. And in the kidneys, epithelium helps to regulate which molecules from the bloodstream are eliminated with urine.

EPITHELIAL TISSUE

Epithelial tissue covers the surfaces of an animal's body. It consists of a single layer or a few layers of cells that are tightly bound together, and acts as a barrier between the inside and outside of an organism, and around body cavities and organs.

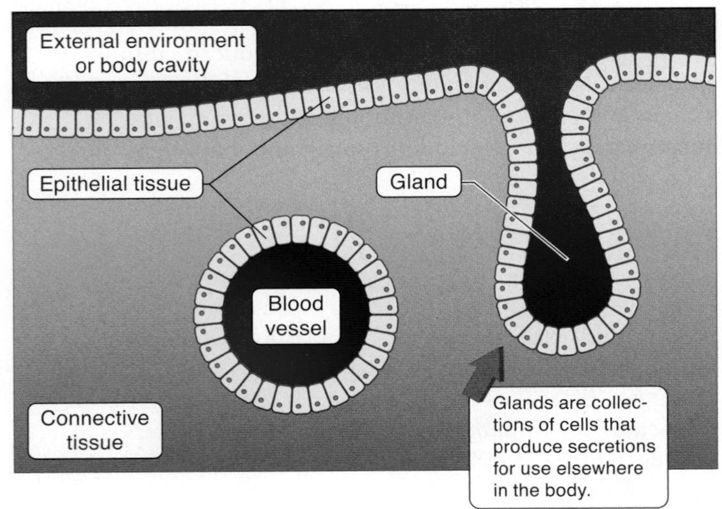

FIGURE 20-20 Epithelial tissue forms with two distinct sides.

FUNCTIONS OF EPITHELIAL TISSUE

PROTECTION
Epithelial tissue acts as a barrier between the inside and outside of an organism and keeps fluids from leaking into or out of tissue.

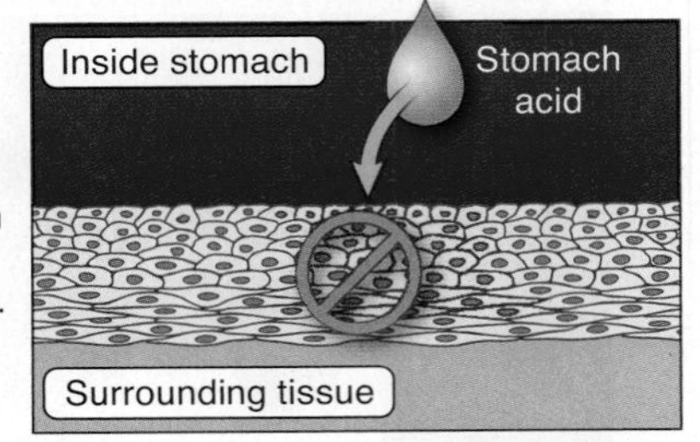

TRANSPORT
Epithelial tissue regulates the movement of nutrients and other molecules into and out of body tissues.

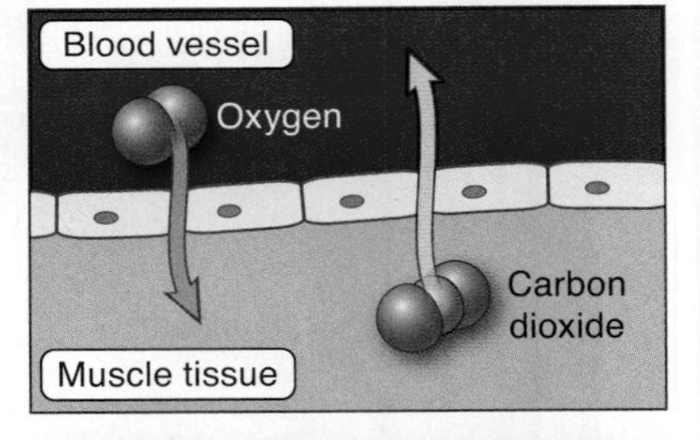

SECRETION
Epithelial tissue can form exocrine glands, which secrete products such as saliva, sweat, and mucus, and endocrine glands, which secrete hormones.

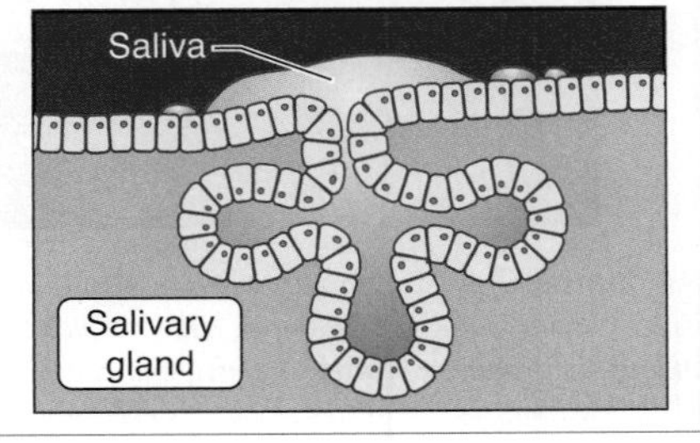

FIGURE 20-21 The multiple roles of epithelial tissue.

3. *Secretion*. Epithelium can also form glands. **Exocrine glands** generally secrete products—including earwax, sweat, saliva, milk, mucus, and, in the case of some frogs, toxic poisons—onto the surface of the epithelium. **Endocrine glands,** on the other hand, produce hormones, chemical messengers that affect cells elsewhere in the body, which are released into the fluid surrounding the glands and usually enter the bloodstream for distribution to other parts of the body.

Because epithelial cells are often in contact with a large variety of materials, they tend to experience more damage than other tissues. For this reason, they tend not to last long and are replaced frequently. Human skin cells, for example, are replaced approximately every two weeks. (Dandruff is mostly dead skin cells.) Cells lining the digestive tract have an even shorter life span, lasting only about five days. Liver epithelium experiences slightly less wear and tear, lasting a year or two before replacement.

TAKE-HOME MESSAGE 20•9

Epithelium is a very thin, sheet-like tissue that covers most of the exterior and interior surfaces of an animal's body. Made up of a single layer or a few layers of cells that are tightly bound together, epithelium acts as a barrier between the inside and outside of an organism. It also can be specialized to aid in the secretion and transport of molecules.

20•10

Muscle tissue enables movement.

Most animals move. And muscle tissue, made up of elongated cells capable of generating force by contracting, is responsible for much of that movement. Most muscle tissue cells are packed with protein filaments that slide together as they break down ATP, causing the entire cell to shorten and thus the muscle to contract. The action of muscle cells, which is usually stimulated by nerve cells, enables organisms to generate force and motion. There are three types of muscle tissue: skeletal, cardiac, and smooth muscles (**FIGURE 20-22**).

Skeletal muscle (sometimes called voluntary muscle) is usually attached to bones and is responsible for generating most movement we see in animals, including facial expressions and breathing. Muscles account for about 40% of human body

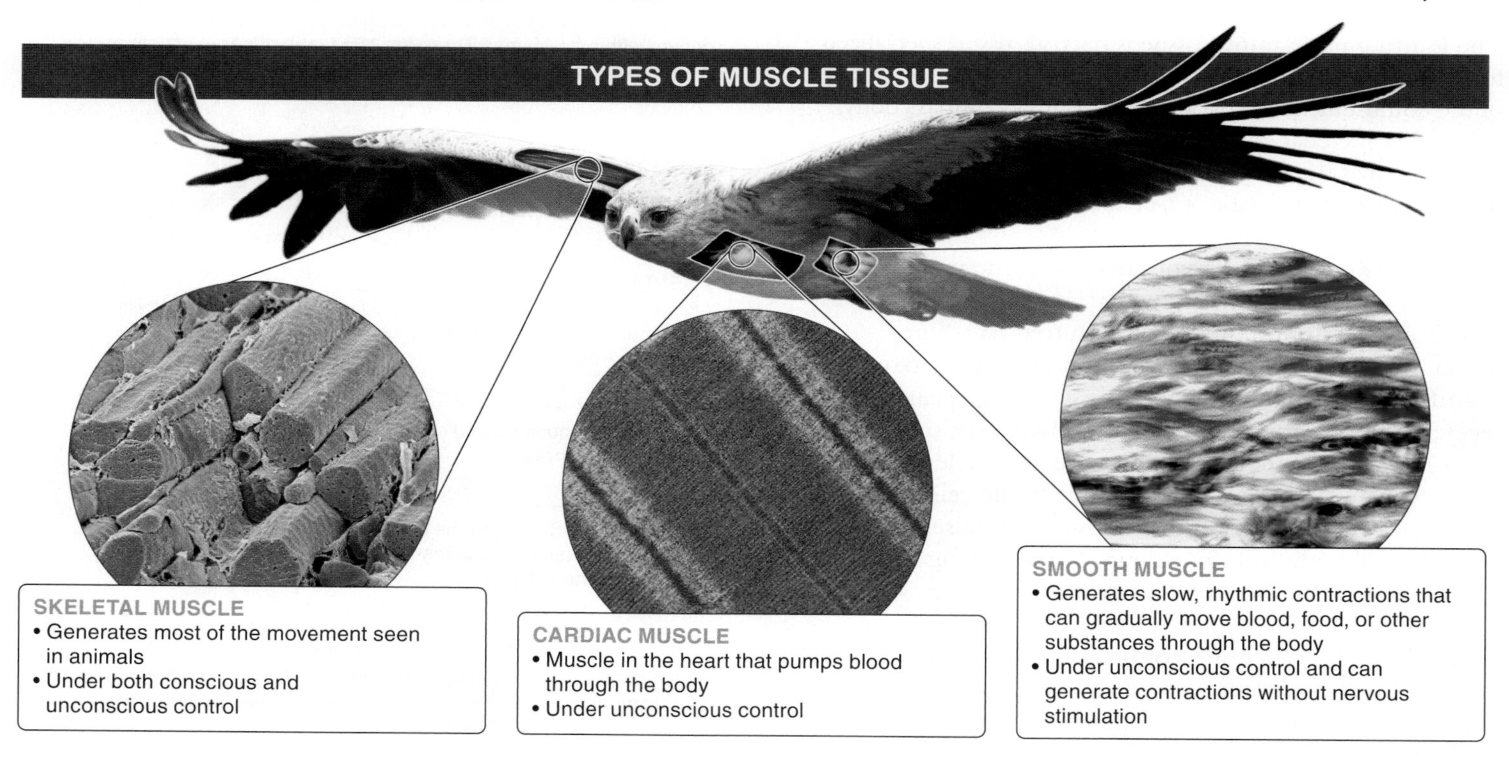

FIGURE 20-22 Muscle tissues are made up of elongated cells capable of generating force when they contract.

weight. Individual skeletal muscle cells, called **muscle fibers,** are very long and contain multiple nuclei, and the repeating units of protein filaments in the cells give the fibers a striped, or striated, appearance. Skeletal muscles are controlled by the nervous system, and the individual nerve cells (neurons) attached to each muscle fiber stimulate its contraction. Skeletal muscles can be under conscious control, such as when you choose to flex your biceps, or unconscious control, such as those that control breathing or moving your eyes around.

Cardiac muscle, as the name indicates, is located only in the heart, and it causes the heart to pump blood through the body. Take a look at the large colorized photo that opens this chapter. You will see capillaries passing among the muscle fibers of the heart. Because of the tremendous amount of energy that cardiac muscles use, contracting incessantly throughout our life, the cells contain many more mitochondria than other types of muscle cells. The cells of heart muscle are fused together and connected by gap junctions (see Section 3-12). The electrical signals that initiate each contraction of the cardiac muscle pass through these gap junctions. Cardiac muscle tissue is not under conscious control.

Smooth muscle is found in the walls surrounding blood vessels, the stomach and intestines, bladder, and many other organs and inner "tubes" within the body. Smooth muscle generates slow, rhythmic contractions that can gradually move blood, food, or other substances through the tube. Not under conscious control, smooth muscle is regulated by other physiological systems and can generate contractions without nervous stimulation.

TAKE-HOME MESSAGE 20•10

Muscle tissue consists of elongated cells capable of generating force when they contract. Skeletal muscle is responsible for generating most of the movement we see in animals. Cardiac muscle causes the heart to pump blood through the body. And smooth muscle, surrounding blood vessels and many internal organs, generates slower contractions that can gradually move blood, food, or other substances.

20•11

Nervous tissue transmits information.

The fourth type of animal tissue is nervous tissue, specialized to store and transmit information. It is responsible for much of the communication that occurs within an animal's body. Nervous tissue enables animals to sense and respond to stimuli such as the smell of food, the sight, smell, or sounds of a predator or a potential mate, or the heat of a fire.

There are two types of nervous tissue cells: neurons and glial cells (**FIGURE 20-23**). **Neurons** are the "excitable" cells that receive and transmit a signal. Neurons have three distinctive elements: dendrites, a cell body, and an axon. **Dendrites** are a bit like an antenna system, specialized for receiving signals from the external environment or from other neurons. The **cell body** contains the nucleus and other cellular machinery found in eukaryotic cells. And the **axon** is a single projection that transmits impulses away from the cell body and can extend over very long distances—sometimes 3 feet (1 m) or more!

Glial cells, also called **neuroglia,** are like the support staff to neurons. They do not carry signals but assist and nourish neurons. There are numerous types of glial cells, and they vastly outnumber neurons. Together, the various types of glial cells produce insulation for neurons, protect and regulate the chemical environment around neurons, hold

STRUCTURE OF NERVOUS TISSUE

Nerve tissue is responsible for much of the communication that occurs within an animal. There are two types of nervous tissue: neurons and glial cells.

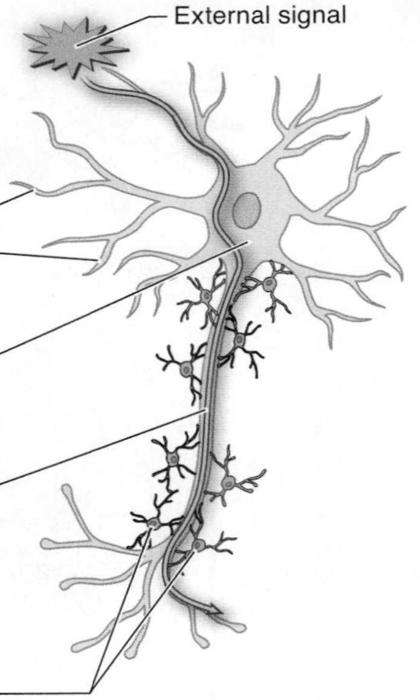

FIGURE 20-23 Neuron and glial cells.

ORGANIZATION OF THE VERTEBRATE NERVOUS SYSTEM

In vertebrates, the nervous system is divided into the central nervous system and the peripheral nervous system.

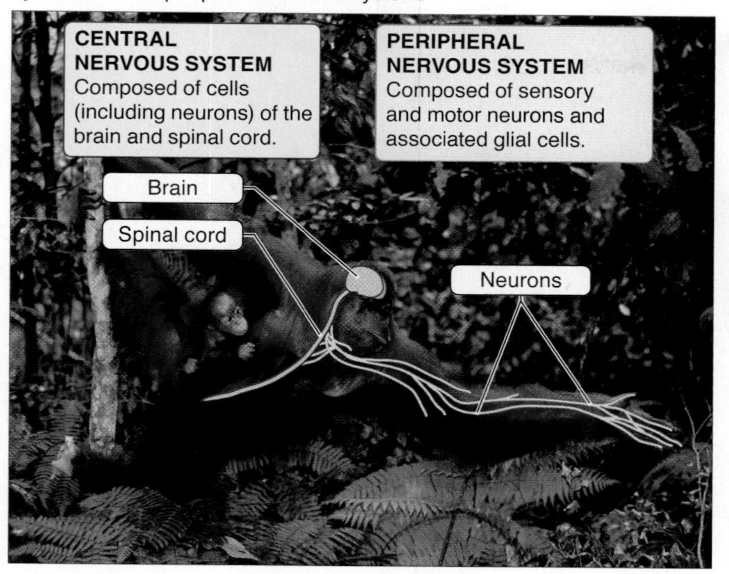

FIGURE 20-24 The vertebrate nervous system: brain, spinal cord, and neurons.

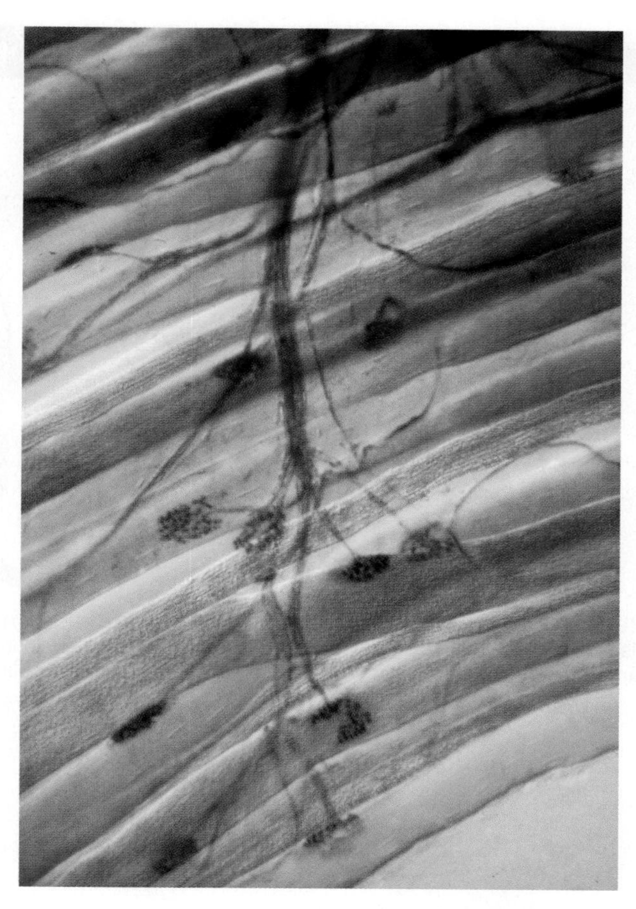

Motor neurons of the peripheral nervous system convey signals that can initiate contractions (and movement) in muscle cells.

the neurons in place, destroy pathogens, and provide nutrients and oxygen.

All of the nervous tissue of an organism makes up its nervous system (**FIGURE 20-24**). In vertebrates, the nervous system is divided into the **central nervous system,** which includes the brain and spinal cord, and the **peripheral nervous system,** which includes the sensory neurons that detect a stimulus and the neurons that transmit signals to the muscles and glands of an organism in response to that stimulus. (The nervous system is discussed in detail in Chapter 23.)

TAKE-HOME MESSAGE 20•11

Nervous tissue is specialized to store and transmit information. There are two types of nervous tissue cells: neurons, which can receive and transmit a signal, and glial cells, which assist and provide nutrients for neurons.

20•12

Each organ system performs special tasks.

With the exception of sponges and some cnidarians, all animals have some tissues that are organized into organs—structures such as the heart, brain, lungs, and liver—that serve specialized functions and consist of multiple tissue types. And just as cells make up tissues and tissues make up organs, organs, too, are part of larger functional units, the organ systems, which carry out the various physiological processes necessary for the growth, development, maintenance, and reproduction of organisms.

We briefly describe 11 animal organ systems in **FIGURE 20-25**, and we'll discuss them in greater detail in Chapters 21–26. It's important to keep in mind that these systems do not operate in isolation: in carrying out its tasks, each system not only

THE ORGAN SYSTEMS

DIGESTIVE SYSTEM
Disassembles and absorbs food so the body can acquire the nutrients it needs to function

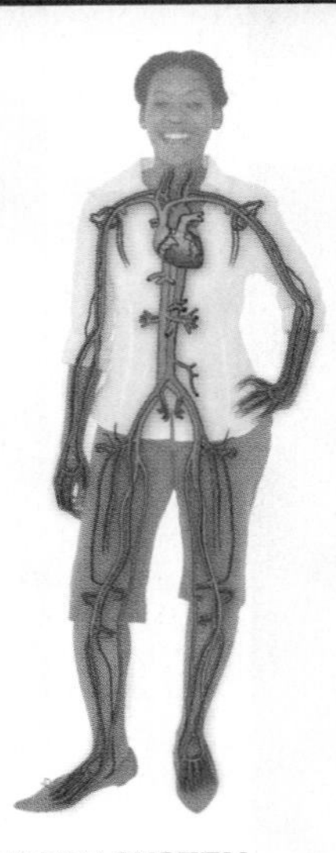

CIRCULATORY SYSTEM
Transports nutrients and respiratory gases to the tissues and eliminates wastes from the tissues

RESPIRATORY SYSTEM
Provides a site for gas exchange between the external environment and an organism's circulatory system

REPRODUCTIVE SYSTEM (MALE)
Produces sperm and delivers it to the female reproductive system, where fertilization may occur

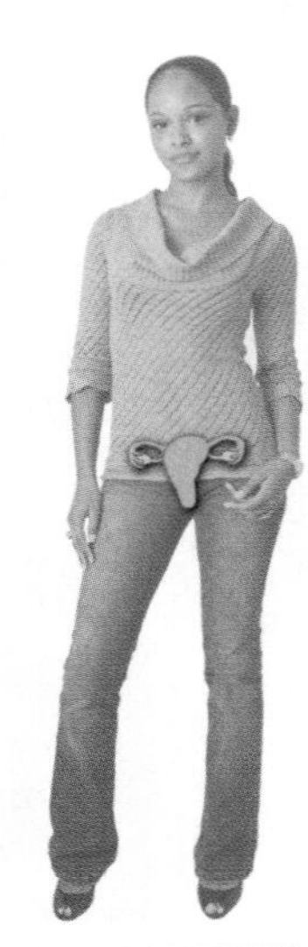

REPRODUCTIVE SYSTEM (FEMALE)
Produces eggs and provides an environment that can nurture a developing embryo and fetus, if fertilization occurs

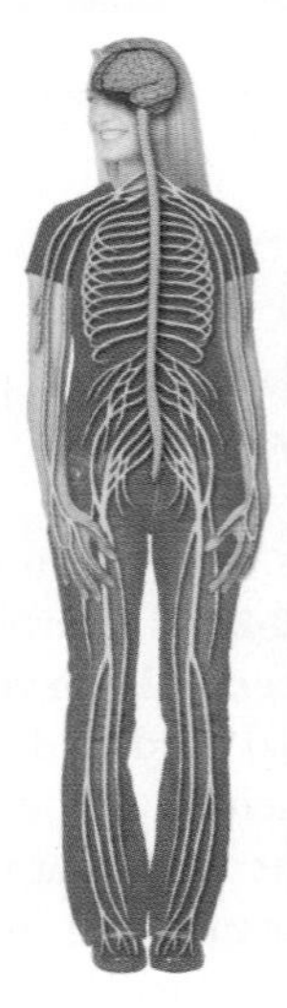

NERVOUS SYSTEM
Acts as the control center of the body and interprets, stores, and transmits information using electrical impulses and chemical signals

FIGURE 20-25 The major organ systems of animals.

depends on the proper functioning of one or more of the other systems, but often influences the functioning of other systems. The act of running, for example, might be initiated when the *nervous system* detects a threat and sends impulses to the *musculoskeletal system* (which causes the body to move), to the *respiratory system* (which increases the rate of oxygen consumption), and to the *circulatory system* (which increases the rate at which the heart pumps blood and thus increases the oxygen available to muscle tissue). Even though these 11 systems constantly add, modify, and remove substances in the body, the body maintains a consistent internal environment through homeostasis.

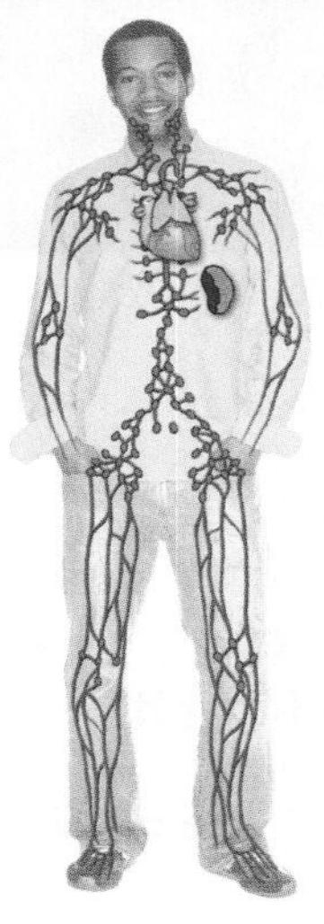

IMMUNE AND LYMPHATIC SYSTEM
Attacks pathogens that threaten the body and plays a supporting role in circulation by recycling fluid that leaks from the circulatory system

URINARY/EXCRETORY SYSTEM
Purifies the blood by filtering out wastes and transports wastes out of the body

ENDOCRINE SYSTEM
Regulates body activities by releasing hormones that travel through vessels in the circulatory system to reach target cells

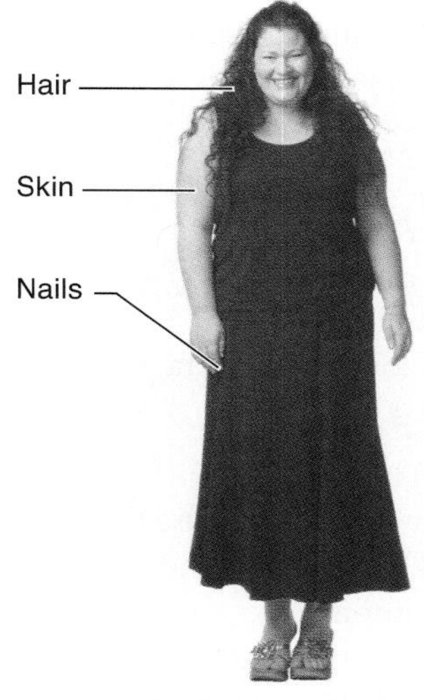

INTEGUMENTARY SYSTEM
Provides protection by forming a barrier between the inside and outside of an organism and can aid in the secretion and transport of molecules

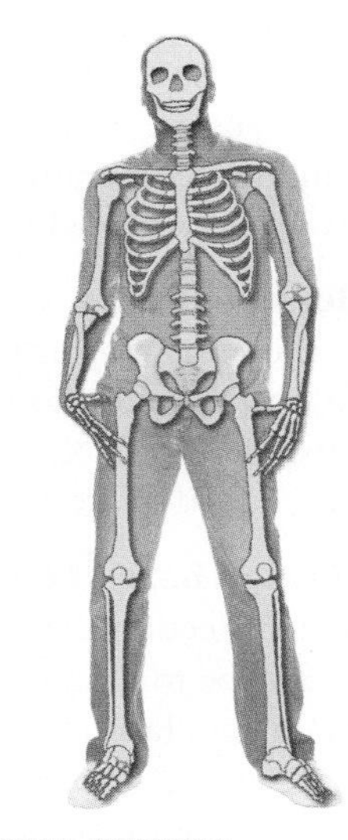

SKELETAL SYSTEM
Supports and protects the body and internal organs, manufactures blood cells, and provides a surface for muscle attachment, creating a foundation for movement

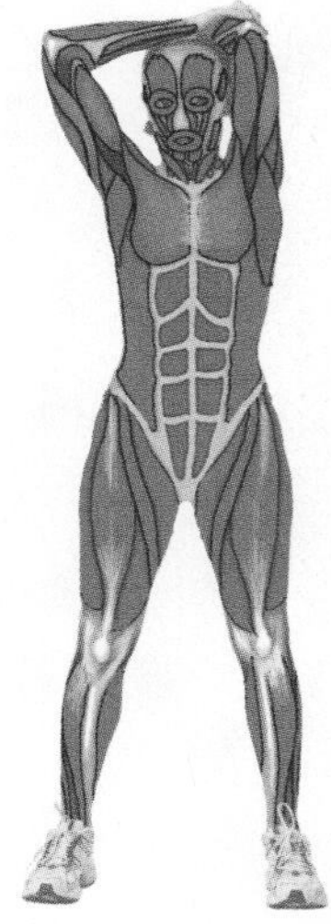

MUSCULAR SYSTEM
Generates force through contraction, which enables movement of the body and of blood, food, and other substances throughout the body

As we explore organ systems in more detail in the following chapters, we will pay particular attention to the ways in which the form and function of the systems' components are intimately related and what this relationship reveals about how these systems (and the organisms of which they are a part) evolved.

TAKE-HOME MESSAGE 20·12

In nearly all animals, some tissues are organized into organs (such as the heart, brain, lungs, and liver) that serve specialized functions and consist of multiple tissue types, and into organ systems (such as the circulatory system) that carry out the various physiological processes necessary for the growth, development, maintenance, and reproduction of the organism.

Knowledge You Can Use

Your body sometimes deliberately upsets homeostasis. (And you might not want to fight it.)

Q: What do you usually do when you have a fever? At the first sign of fever, many people take aspirin, Tylenol, Advil, or Motrin.

Q: What is the result? Aspirin, acetaminophen (Tylenol), and ibuprofen (Advil and Motrin) quickly reduce a fever.

Q: Why did you have a fever? Generally, fever is not itself an illness. Rather, it is your body's response to cues that there is a bacterial or viral infection. In response to infection, your temperature set point is raised, because pathogens are more easily brought under control by the body's defenses at higher temperatures. (Ectotherms use a similar strategy, moving to warmer areas when they have an infection!)

Q: How does reducing a fever interfere with your body's defenses? Blocking a fever by taking aspirin or other medication may reduce your body's ability to fight infection. A recent well-controlled study demonstrated, for example, that chicken pox lasts longer, on average, when aspirin is used to treat it, compared with placebo.

Q: Is the lesson here about more than fever? Yes. There are other "symptoms" of infection that, as defenses rather than part of the illness itself, maybe shouldn't be fought. For example: (1) *Coughing:* the use of codeine to block coughing after surgery increases the risk of pneumonia; the coughing is *helpful.* (2) *Diarrhea:* anti-diarrhea medications delay recovery and slow the eradication of bacteria from the digestive tract; the diarrhea is *helpful.* (3) *Vomiting* and *inflammation,* too, appear to be important parts of our evolved defenses and, as such, blocking them can have serious health consequences.

What can you conclude? Is this the dawn of Darwinian medicine? This new perspective on when to treat and when not to treat symptoms is called "Darwinian medicine." It represents a newfound appreciation for the fact that organisms have evolved many protective responses that may be useful. Of course, suffering is not always the solution. There can be real costs to vomiting, diarrhea, coughing, fever, and other body defenses. Opting not to treat them isn't necessarily the best solution if, for example, antibiotics can bring an infection under control easily. Either way, bringing an evolutionary perspective to medical decision making can be valuable.

1 Animals have an internal environment.

Failure to maintain a consistent internal environment can lead to multiple problems in the normal functioning of cells, tissues, and organs, and can result in death. Organisms maintain a variety of internal physiological variables, including temperature, water-solute balances, pH, blood sugar levels, and blood gas concentrations, within relatively constant ranges. The process of maintaining an organism's internal environment is called homeostasis.

2 How does homeostasis work?

Organisms generally maintain homeostasis through negative feedback, in which sensors detect a change in the internal environment and trigger structures, called effectors, to oppose or reduce the change. For animals to maintain homeostasis, the variable must have a set point, a target value or range to which the organism can return. Methods for controlling temperature, blood sugar levels, and water and solute concentrations, among many other variables, are important components of homeostasis. To maintain osmotic balance, organisms must be able to take up water and get rid of water, and must also be able to regulate concentrations of ions in their fluids. The kidney is the organ in vertebrates that helps maintain homeostasis by regulating water balance and solute concentration in body fluids, filtering blood, and enabling the removal of potentially harmful ions and metabolic waste products in urine.

3 Form reflects function: animal anatomy.

At all levels of animal organization, from molecules to whole organisms, the physical features of a structure are closely related to its function. Cells with similar structure and function are organized into groups, called tissues, that work together. There are four types of tissue: connective tissue, epithelial tissue, muscle tissue, and nervous tissue. In nearly all animals, some tissues are organized into organs that serve specialized functions and consist of multiple tissue types, and into organ systems that carry out the various physiological processes necessary for the growth, development, maintenance, and reproduction of organisms.

Key Terms

axon, p. 736
blood, p. 732
bone, p. 732
Bowman's capsule, p. 726
cardiac muscle, p. 736
cartilage, p. 732
cell body, p. 736
central nervous system, p. 737
collagen, p. 732
conformer, p. 720
connective tissue, p. 730
connective tissue proper, p. 732
dendrite, p. 736
ectotherms, p. 721
effector, p. 719
elastin, p. 732
endocrine gland, p. 735
endotherms, p. 721
epithelial tissue, p. 730
epithelium, p. 734
exocrine gland, p. 735
extracellular fluid, p. 717
filtrate, p. 727
gland, p. 734
glial cells, p. 736
glomerulus, p. 726
heat stroke, p. 716
heterotherm, p. 722
hibernate, p. 721
homeostasis, p. 717
homeotherm, p. 721
hyperthermia, p. 716
hypothermia, p. 716
interstitial fluid, p. 717
ligament, p. 732
matrix, p. 730
muscle fiber, p. 736
muscle tissue, p. 731
negative feedback, p. 719
nephron, p. 726
nervous tissue, p. 731
neuroglia, p. 736
neuron, p. 736
organ, p. 731
organ system, p. 731
osmoconformer, p. 725
osmoregulation, p. 725
osmoregulator, p. 725
peripheral nervous system, p. 737
physiology, p. 717
positive feedback, p. 720
regulator, p. 720
set point, p. 719
skeletal muscle, p. 735
smooth muscle, p. 736
special connective tissue, p. 732
tendon, p. 732
thermoregulation, p. 721
tissue, p. 730

Check Your Knowledge

1. The term "homeostasis" describes:
   a) the body's use of physical and chemical processes to maintain a consistent internal environment.
   b) the biochemical processes associated with the maintenance of body temperature.
   c) the metabolic patterns of active (versus stationary) animals.
   d) the metabolic patterns of stationary (versus active) animals.
   e) the health benefits of a sedentary lifestyle.

2. Interstitial fluid:
   a) is found exclusively within the spinal cord, surrounding the nerve bundles.
   b) is found exclusively within the skull, surrounding the brain.
   c) is mostly water.
   d) is found exclusively within the skull and spinal cord, surrounding nervous tissue.
   e) occurs within the organelles of all eukaryotic cells.

3. Negative feedback loops:
   a) generally lead to highly unstable internal physiological conditions.
   b) cause internal conditions to deviate from the normal range.
   c) are part of larger, positive feedback systems.
   d) rely on sensors to trigger effectors to alter an organism's internal environment.
   e) None of the above.

4. A set point:
   a) is the target value or range for a physiological variable, to which it generally returns following perturbation.
   b) is a physiological state that occurs in animals called "conformers," but not in "regulators."
   c) can occur in fishes but not in terrestrial animals.
   d) is the target value or range for a physiological variable regulated through positive feedback, but not negative feedback.
   e) Both a) and c) are correct.

5. It is not necessarily accurate to refer to endotherms as "warm-blooded," because:
   a) some endotherms allow their body temperature to drop significantly during hibernation.
   b) on average, their body temperature is colder than that of "cold-blooded" animals.
   c) they generate their heat primarily from within their own bodies.
   d) most do not actually have blood in their bodies.
   e) some are homeotherms rather than heterotherms.

6. "Form follows function" refers to the fact that:
   a) organisms' physical structures often are adaptations, shaped by natural selection, and so reflect their physiological functions.
   b) organisms' physical structures generally come to reflect their physiological function only *after* that function has been fine-tuned by evolution.
   c) natural selection can produce adaptations in physical structures but not in physiological processes.
   d) natural selection can produce adaptations in physiological processes but not in physical structures.
   e) all structures that have the same function also have the same structure.

7. Which of the following is not a type of connective tissue?
   a) bone
   b) cartilage
   c) blood
   d) collagen
   e) ligament

8. Epithelium plays all of the following roles in organisms except:
   a) secretion.
   b) transport.
   c) protection.
   d) preventing fluids from leaking from an organ system into surrounding tissue.
   e) cushioning, lubricating, and insulating other tissue.

9. Smooth muscle:
   a) is usually attached to bones.
   b) generates slow, rhythmic contractions.
   c) is usually under conscious control.
   d) generally contains more mitochondria than cardiac or skeletal muscle.
   e) All of the above are correct.

10. Neurons generally have all of the following components except:
    a) dendrites.
    b) a cell body.
    c) an axon.
    d) glial processes.
    e) a nucleus.

11. With the exception of ______, all animals have some tissues organized into organs.
    a) sponges and some cnidarians
    b) insects
    c) roundworms and flatworms
    d) sponges and insects
    e) amphibians

Short-Answer Questions

1. What is homeostasis? How does hyperthermia illustrate the failure of the body to maintain homeostasis? What are the potential consequences of this failure?
2. What is the hierarchical organization of cells in a multicellular organism? Name the four types of tissue they form and note their general functions.
3. What are the two kinds of connective tissue? What are their general characteristics?
4. What are the three types of muscle tissue? What are their general characteristics?

See Appendix for answers. For additional study questions, go to www.prep-u.com.

1 The circulatory system is the chief route of distribution in animals.

- **21.1** What is a circulatory system and why is one needed?
- **21.2** Circulatory systems can be open or closed.
- **21.3** Vertebrates have several different types of closed circulatory systems.

2 The heart is at the center of the human circulatory system.

- **21.4** Blood flows through the four chambers of the human heart.
- **21.5** Electrical activity in the heart generates the heartbeat.
- **21.6** Blood is a mixture of cells and fluid.
- **21.7** Blood pressure is a key measure of heart health.
- **21.8** Cardiovascular disease is a leading cause of death in the United States.
- **21.9** The lymphatic system plays a supporting role in circulation.
- **21.10** The polygraph relies on cues from the cardiovascular system.

3 The respiratory system enables gas exchange in animals.

- **21.11** Oxygen and carbon dioxide must get into and out of the circulatory system.
- **21.12** Gas exchange takes place in the gills of aquatic vertebrates.
- **21.13** Respiratory systems of terrestrial vertebrates move oxygen-rich air into the lungs and carbon-dioxide-rich air out of the lungs.
- **21.14** Birds have unusually efficient respiratory systems.
- **21.15** Muscles control the flow of air into and out of the lungs.
- **21.16** Some environments are more conducive to gas exchange than others.

4 Oxygen is transported and stored while bound to hemoglobin and myoglobin.

- **21.17** Hemoglobin is the molecule that transports oxygen.
- **21.18** Myoglobin in muscles holds a reservoir of oxygen for times of exertion.

5 Evolutionary adaptations maximize oxygen delivery.

- **21.19** Animals living at high altitude have special adaptations to the low-oxygen conditions.
- **21.20** Humans become acclimated to low-oxygen conditions.
- **21.21** Deep-diving mammals are masters of efficient oxygen use.

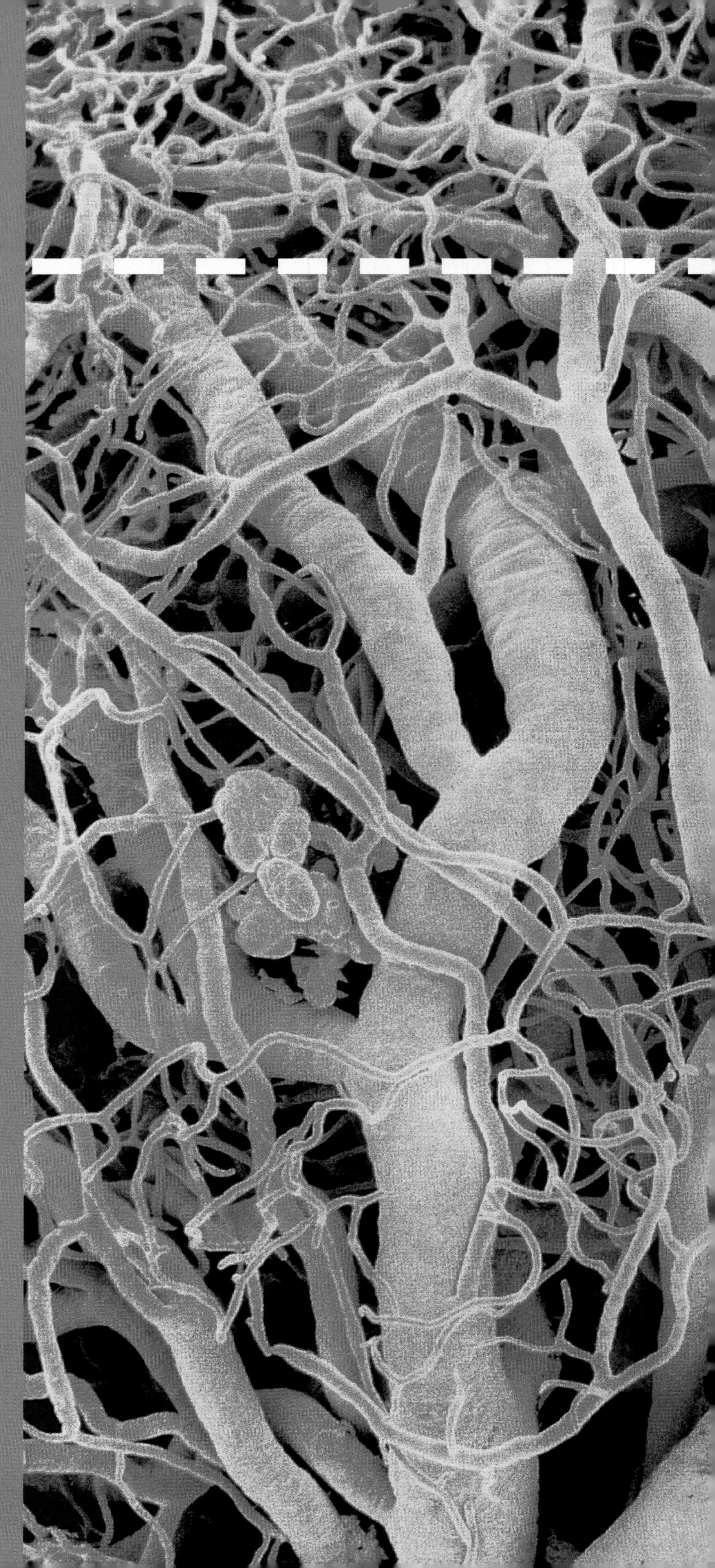

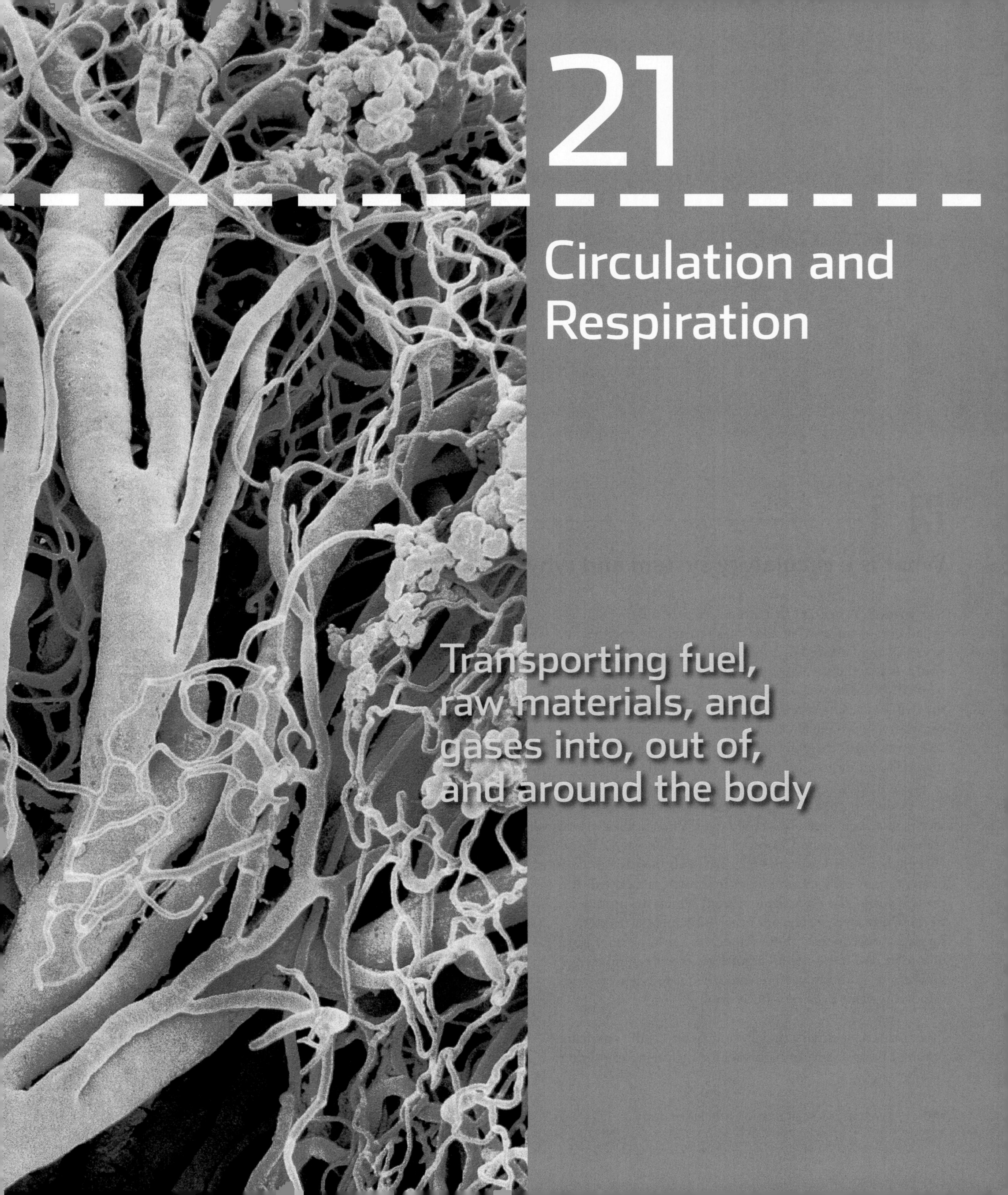

21

Circulation and Respiration

Transporting fuel, raw materials, and gases into, out of, and around the body

① The circulatory system is the chief route of distribution in animals.

The swiftness of the hare and greyhound depend on efficient circulatory systems.

21•1

What is a circulatory system and why is one needed?

Size matters. When life first arose on earth, the tiny, single-celled organisms were small enough to acquire the fuel and raw materials they needed in a straightforward way: those materials could simply diffuse across the cell membrane and be used as needed. Similarly, metabolic waste products could diffuse out of the cell. Today, single-celled organisms and small multicellular organisms in which all cells are in contact with (or just a few cells away from) the external environment acquire raw materials and dispose of metabolic waste in this same way.

The evolution of large body sizes opened up a world of new niches in which organisms could exist. Physiologically, however, large body sizes created a host of new challenges. Many of these challenges resulted from the fact that with increasing size, most of an animal's cells are no longer in direct contact with the outside world, the environment from which the animal obtains oxygen, nutrients, water, and other substances it needs to survive. Consequently, these substances can no longer just diffuse in, and the waste an animal generates can no longer diffuse out. As body size increased, dedicated delivery and removal systems became a necessity.

In animals, the primary distribution system is the circulatory system. Like a system of highways for delivering important goods and removing garbage, the circulatory system reaches all tissues of the body. In vertebrates, circulatory systems have three principal functions: transport, body temperature regulation, and protection (**FIGURE 21-1**).

1. Transport. The circulatory system transports oxygen, nutrients, waste products, hormones, and immune system cells in the blood throughout the body.

- In vertebrates, blood vessels take oxygen from the lungs or gills and deliver it to the tissues for energy-releasing cellular respiration. Blood that comes from the lungs or gills is loaded with oxygen and is said to be oxygenated or oxygen-rich.
- Simultaneously, blood vessels whisk away carbon dioxide and other metabolic wastes that are produced in cellular respiration and other cell processes and must be removed from the body. Blood that carries a lot of carbon dioxide is said to be deoxygenated or oxygen-poor.
- Once food particles are digested, the nutrients must be absorbed and delivered to all the tissues of the body, for activity, growth, and reproduction.

FUNCTIONS OF THE CIRCULATORY SYSTEM

Among vertebrates, circulatory systems have three principal functions.

TRANSPORT
The circulatory system transports oxygen, nutrients, waste products, immune system cells, and hormones in the blood throughout the body.

TEMPERATURE REGULATION
The circulatory system helps to maintain body temperature within the optimum range for metabolic functioning.

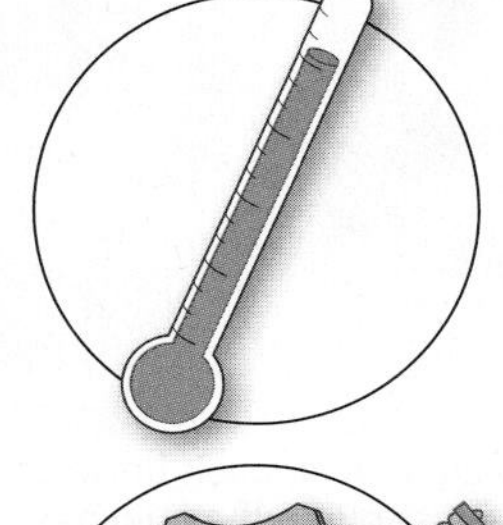

PROTECTION
The circulatory system contains a variety of cells and chemicals that contribute to the individual's defenses against infection by pathogens.

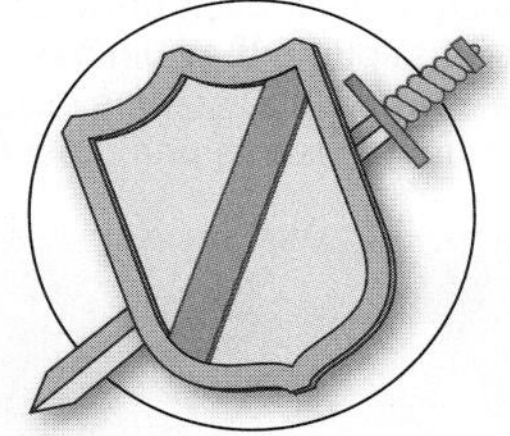

FIGURE 21-1 Like a set of highways for the body. Circulatory system functions: transport, temperature regulation and protection.

- The circulatory system delivers hormones (chemicals produced throughout the body by glands and other tissues) to target tissues to regulate growth, development, and reproduction.

2. Body temperature regulation. By expanding or contracting the blood vessels closest to the exterior of the body, animals can absorb or release heat, a process that helps them maintain their body temperature within the optimum range for metabolism.

3. Protection. A variety of cells and chemicals contribute to the individual's defenses against infection by pathogens. White blood cells, or leukocytes, have the ability to engulf and destroy many disease-causing microorganisms. Platelets and certain chemicals in the blood also provide protection by limiting blood loss and infection when the skin or other tissues are damaged.

TAKE-HOME MESSAGE 21•1

In animals, the circulatory system is the chief distribution system. It transports gases, nutrients, waste products, hormones, and immune system cells throughout the body. The circulatory system also helps animals regulate their body temperature and plays a protective role against infection.

21•2

Circulatory systems can be open or closed.

There's more than one way to build a circulatory system. And, in fact, not all multicellular organisms even need a circulatory system. In spite of their relatively large size and multicellularity, for example, flatworms, as well as jellyfish and other cnidarians, have a body plan that gives every cell easy access to oxygen and nutrients through simple diffusion (**FIGURE 21-2**). This access is possible because every cell is close to the external surface of the cnidarian's body or to its internal gastrovascular cavity. Although the gastrovascular cavity is not a true circulatory system, it serves many digestive ("gastro") and circulatory ("vascular") functions by directing water into the central mouth and through an elaborate system of channels. Cells lining the mouth and the channels can absorb nutrients and exchange gases by diffusion. The nutrients then diffuse to other cells, none of which are very far away. Once nutrients have been extracted (and waste products picked up), fluid in the gastrovascular cavity is flushed back out of the mouth, and the process is repeated.

Among other multicellular animals, there are two distinct types of circulatory systems: open and closed (**FIGURE 21-3**). Open circulatory systems are found in insects and most molluscs, while closed circulatory systems are found in all vertebrates. The defining feature of an **open circulatory system** is that it has one fluid, called **hemolymph,** that both circulates to transport nutrients, gases, and waste products and also surrounds each cell in the body. In other words, there is no clear distinction between the circulating fluid and the interstitial fluid, the fluid that is outside the cells and bathes all the tissue of the body. There is a heart (or sometimes many hearts!) that pumps the hemolymph, but it essentially squirts the fluid throughout the extracellular spaces. Large collecting

CIRCULATORY SYSTEMS AREN'T ALWAYS NECESSARY

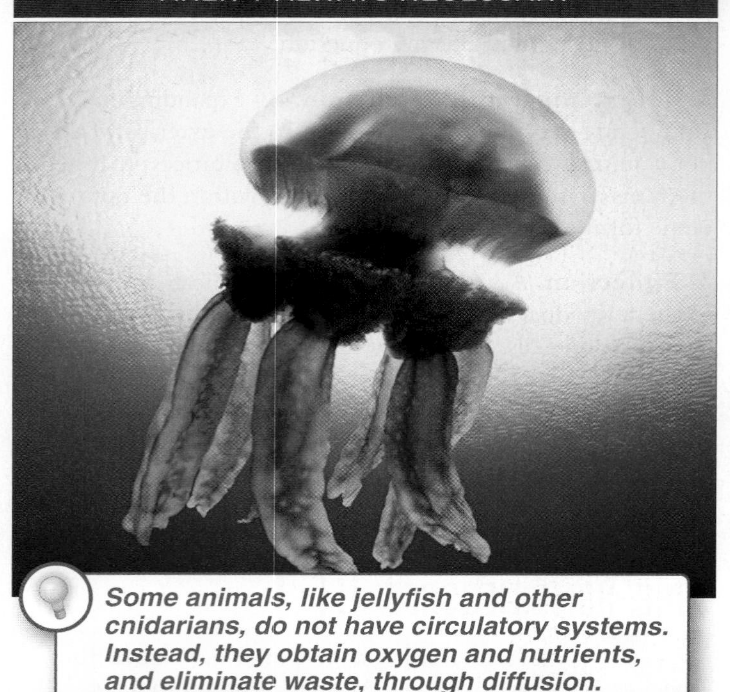

Some animals, like jellyfish and other cnidarians, do not have circulatory systems. Instead, they obtain oxygen and nutrients, and eliminate waste, through diffusion.

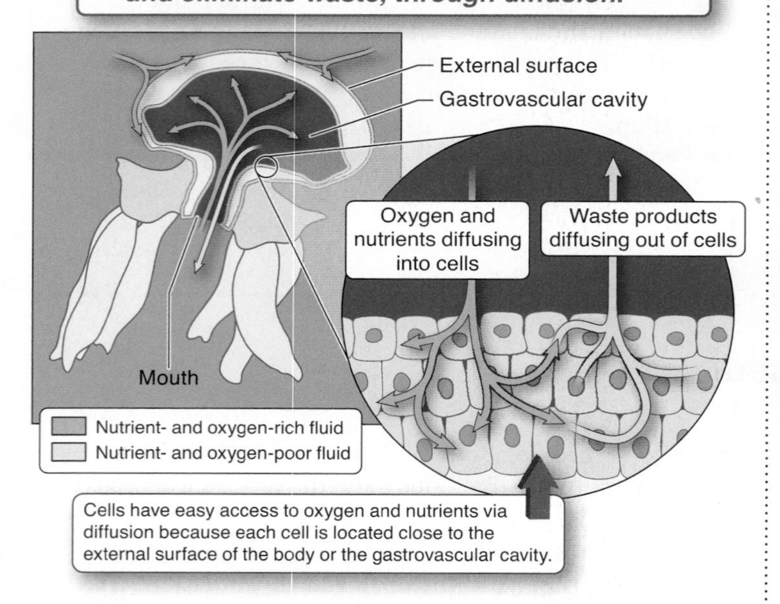

FIGURE 21-2 No circulatory system required. The cells of jellyfish and other cnidarians can acquire nutrients and oxygen from the environment through direct diffusion.

CIRCULATORY SYSTEMS

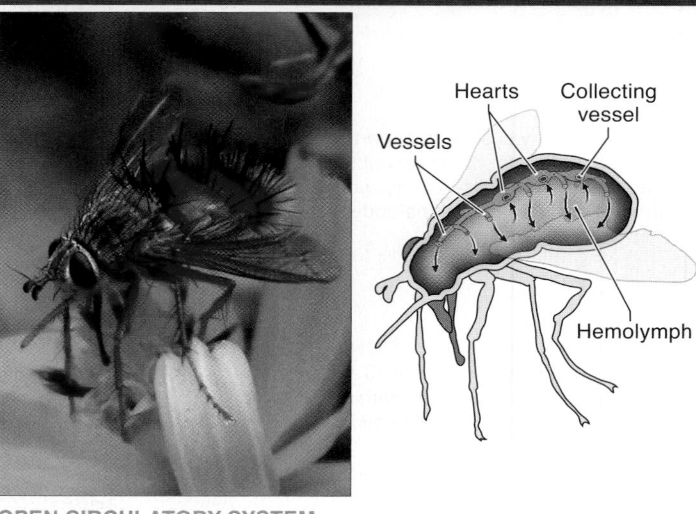

OPEN CIRCULATORY SYSTEM
- No clear distinction between the circulating fluid and interstitial fluid
- Heart(s) pump the fluid mixture—called hemolymph—throughout the extracellular spaces inside the body
- Occurs in insects and most molluscs

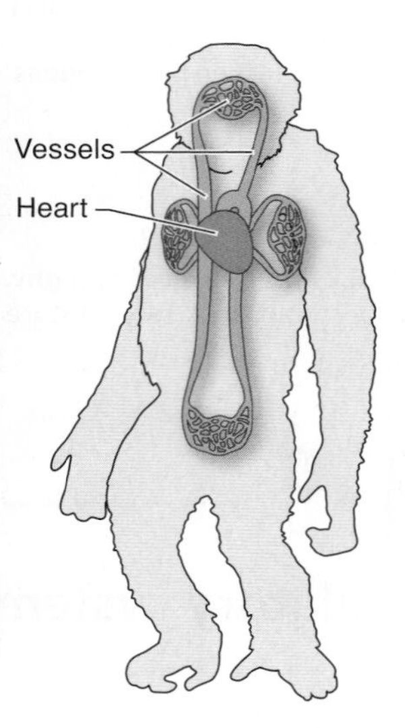

CLOSED CIRCULATORY SYSTEM
- Blood is contained within vessels that separate it from interstitial fluid
- Muscular heart propels blood through vessels to tissues throughout the body
- Occurs in all vertebrates

FIGURE 21-3 There are open and closed circulatory systems.

vessels then channel the hemolymph back to the heart, where it can be pumped throughout the body again. These collecting vessels have little valves that close when the heart pumps to prevent the hemolymph from being pumped back through the same vessel from which it was collected. This one-way collection system gives some order to the circulation of the hemolymph fluid throughout the organism's body.

In **closed circulatory systems,** the circulating fluid—called **blood**—is always contained in a vessel as it is pumped throughout the animal's body, and it is physically and chemically separated from the interstitial fluid that bathes each cell. A muscular **heart** serves as a pump, and with each contraction it propels blood at high pressure through vessels called **arteries.** Like a highway system with numerous exits

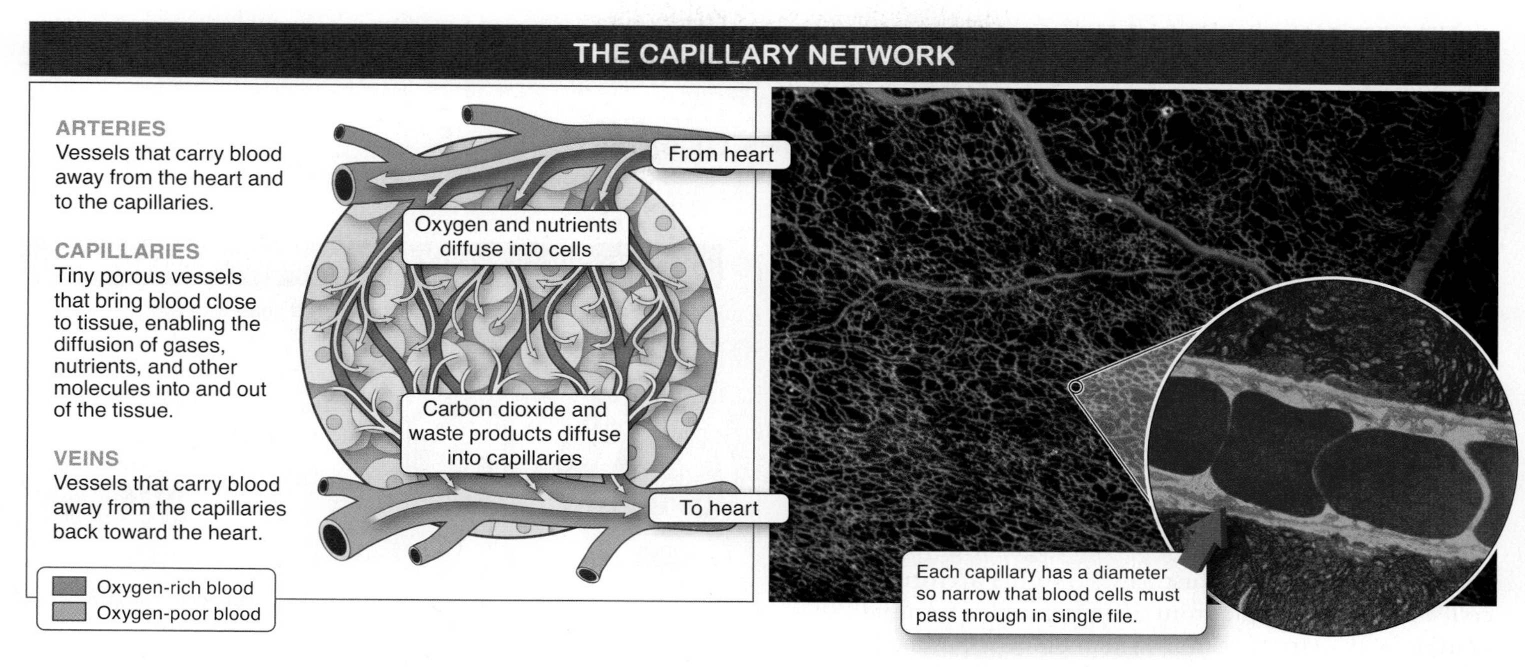

FIGURE 21-4 Arteries, capillaries, and veins: the three types of blood vessels in a closed circulatory system.

and connecting roads, the arteries branch extensively so that blood can be delivered to all the tissues of the body.

At the organs and tissues, arteries branch into thousands upon thousands of ever-smaller vessels, first the arterioles and then, as they become much narrower and thinner-walled, the **capillaries.** Each capillary has an inner diameter so narrow that blood cells must pass through in single file. The capillary wall is thin—just one layer of wall cells—and somewhat porous, so the diffusion of gases, nutrients, and other molecules into and out of the tissue, down their concentration gradients, readily occurs (**FIGURE 21-4**).

> **Q** Is it more dangerous to cut the carotid artery or the jugular vein? Why?

The network of capillaries in an organism is staggeringly large—more than 50,000 miles in an adult human! And this number is constantly in flux: gaining a single pound of fat, for example, is accompanied by the addition of more than a mile of new capillaries.

Capillaries are the last branch of the circulatory system to carry nutrients and oxygen-rich blood to cells of the body, but they are also the first branch to carry nutrient-poor, carbon-dioxide-rich, and waste-product-rich blood back from cells of the body toward the heart. After passing through the capillaries, blood returns to the heart in vessels called **veins.**

The pressure of blood in the veins is significantly less than that in the arteries. This reduced pressure occurs because, in the capillaries, an increasing proportion of the blood is in direct contact with the inside surface of the capillary wall. This contact increases the friction and slows the blood flow. As the capillaries merge to form veins, the pressure as blood moves through these larger vessels is further reduced. In fact, in most animals, blood will spurt and gush when an artery is cut, but will only trickle out of cut veins. It is not surprising, then, that most arteries are not as close to the skin as are veins. It would be too risky to put them there. This is why people who cut their wrists when attempting suicide rarely die. They generally cut only the veins in their wrists, which are closer to the surface, and rarely bleed enough to cause death.

TAKE-HOME MESSAGE 21•2

Animals that can acquire all the nutrients and oxygen they need by diffusion (such as flatworms and cnidarians) do not have circulatory systems. Among animals that do have circulatory systems, the system can be open, with no clear distinction between circulating fluid and the interstitial fluid that bathes tissues, or closed, with a clear distinction. In closed circulatory systems, tiny blood vessels called capillaries bring blood close enough to tissues that diffusion can move the necessary molecules from the blood into the cells and from the cells into the blood.

21•3

Vertebrates have several different types of closed circulatory systems.

Among the vertebrates, there are several variations of the closed circulatory system. These variations evolved in concert with numerous dramatic adaptations that accompanied and made possible the transition from life in the seas to life on land. As some vertebrates developed lungs and the ability to extract oxygen from the air rather than water, their circulatory systems increased in complexity.

In fishes, the flow of blood follows a circular path (**FIGURE 21-5 FISH**). In tracing its flow, we can begin with the two-chambered heart. Blood first passes into the **atrium,** the collecting chamber, and from there is pumped into the **ventricle,** the chamber from which blood is pumped to the gills. Blood flows through the capillary beds of the gills, where it picks up oxygen. From there, the blood travels to the tissues of the body, delivering the oxygen. After passing through the capillary beds of the body tissues, the oxygen-poor blood flows back to the heart, and the cycle begins again as that blood is pumped to the gills.

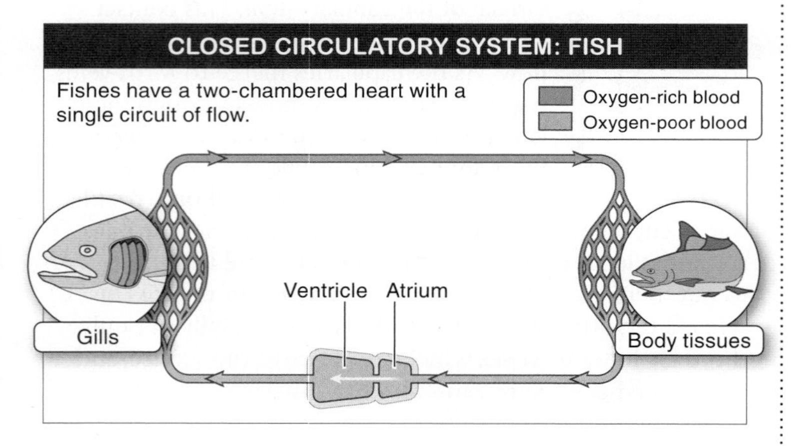

FIGURE 21-5 FISH **Blood flow in fish.**

This single-circuit flow of blood in fishes has a limitation. By passing through the numerous tiny blood vessels that make up the gill capillaries, the newly oxygenated blood slows its flow greatly, losing most of the pressure it gained following the contraction of the heart. It flows—or, more accurately, trickles—at a relatively low velocity to the oxygen-depleted parts of the body, limiting the rate at which oxygen can be delivered to the tissues. The blood is helped along through the arteries by contractions of muscles around the blood vessels as the fish swims, but the blood flow never reaches the velocity it had just after being pumped out of the ventricle.

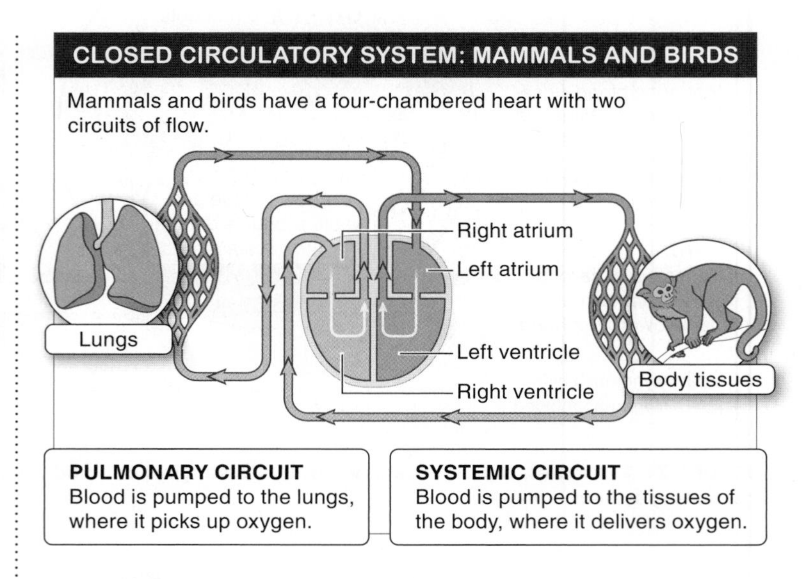

FIGURE 21-5 MAMMALS AND BIRDS **Blood flow in mammals and birds.** Note that the right side of the animal's heart is on the left side in the diagram, and the left side is on the right. This is because drawings of the heart are made as if the animal is facing you.

In contrast to the two-chambered heart of fishes, birds and mammals have a four-chambered heart: two atria and two ventricles (**FIGURE 21-5 MAMMALS AND BIRDS**). Twice as many chambers are required because, rather than having a single circuit of flow, birds and mammals have two circuits of flow. The two circuits can be visualized as a figure 8: blood flows into the right atrium and through to the right ventricle (see Figure 21-5). From there it is pumped out to the capillaries of the lungs, and then it returns to the heart. This first circuit of flow is called the **pulmonary circuit.** The blood returning to the heart from the lungs collects in the left atrium. It immediately passes into the left ventricle and from there is pumped to the rest of the body. After passing through the body capillaries, it completes its second circuit, called the **systemic circuit,** and collects in the right atrium, where the cycle begins again.

> **Q** The left side of mammalian and bird hearts is always bigger than the right. Why?

Mammals and birds vary widely in the size of their hearts—from tiny hummingbird hearts no bigger than a pencil eraser all the way up to the heart of a blue whale, bigger than a small car, that pumps 25 gallons with every beat—but they all share the same basic layout. And almost always, because the left ventricle has to pump blood to the entire body rather than just to the lungs, the left ventricle becomes much larger and more muscular than the right ventricle.

With two circuits of flow, mammalian and bird hearts are better than fish hearts at delivering oxygen to body tissues. In fishes, as we've seen, all the oxygenated blood flowing to the body has low pressure and velocity because it passes first through the gill capillaries. With two circuits of flow in birds and mammals, all the oxygenated blood flowing to body tissues is at higher pressure because it is pumped straight from the left ventricle and does not first pass through the pressure-robbing capillaries of the lungs. This makes it possible for bird and mammal hearts to sustain greater levels of activity with greater amounts of oxygen delivered to the muscles and organs. The single-circuit system of fishes is like trying to put out a big fire with a low-pressure garden hose, while the two-circuit system of birds and mammals is more akin to using a high-pressure fire hose. Nonetheless, the large number and diversity of fishes on earth suggests that they are able to thrive with their single-circuit system.

Amphibians have circulatory systems that appear similar, but not identical, to the mammalian and bird plan (**FIGURE 21-5 AMPHIBIANS**). They have two circuits of flow, but have hearts with only three chambers rather than four. Blood is collected from the lungs and from the rest of the body in the left and right atria, respectively, but it then flows into a single ventricle. Surprisingly, though, little mixing of the oxygenated blood from the lungs and deoxygenated blood from the rest of the body occurs in the ventricle. Rather, the two types of blood flow side by side and are pumped into vessels that direct the oxygenated blood to the body capillaries and the deoxygenated blood to the lungs.

With the exception of birds, most reptiles also have three-chambered hearts, although the ventricle is partially divided in two. In one group of reptiles (the crocodilians), however, the division is complete; they also have an extra little artery that allows blood to be pumped from the right ventricle to the rest of the body rather than to the lungs. This extra artery is an adaptation allowing the crocodile to bypass sending blood to the lungs when the animal is underwater. There is no sense in pumping blood to the lungs if there is no oxygen to be picked up there. Instead, the blood is sent back to the body, where the remaining oxygen in the blood can be utilized.

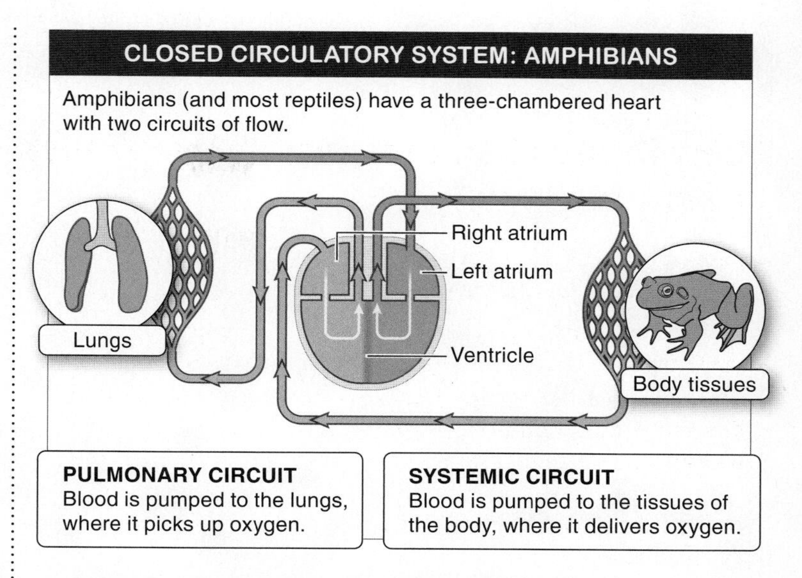

FIGURE 21-5 AMPHIBIANS **Blood flow in amphibians.**

TAKE-HOME MESSAGE 21·3

Vertebrates' circulatory systems vary in structure. Fishes have two-chambered hearts, with one circuit of flow: from the heart through the gills through the body and back to the heart. Birds and mammals have four-chambered hearts and two circuits of flow: from the heart to the lungs and back to the heart, then from the heart to the body and back to the heart. This enables blood to be pumped to the body at higher pressure. Amphibians and most reptiles have a three-chambered heart and two circuits of blood flow.

❷ The heart is at the center of the human circulatory system.

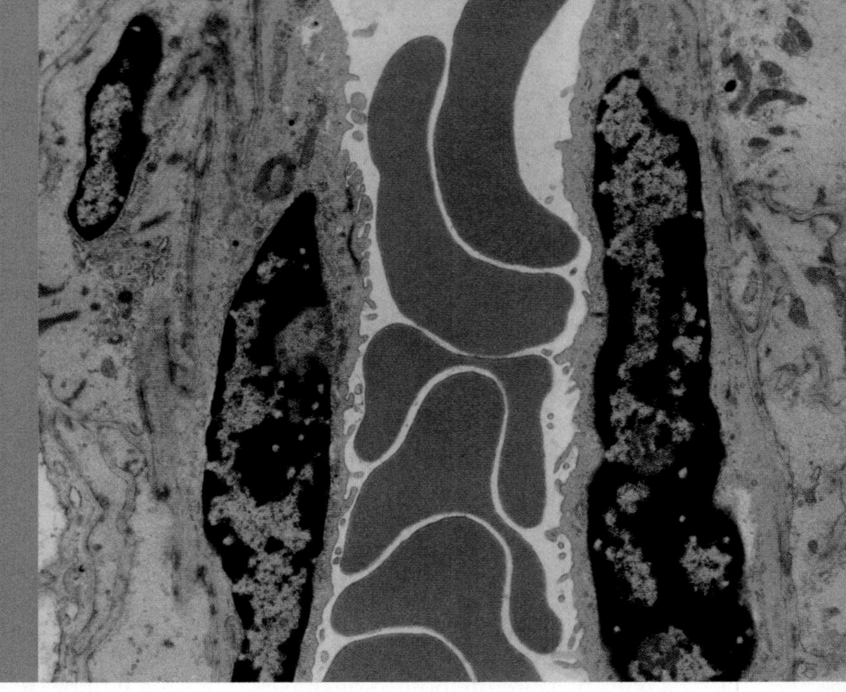

Red blood cells flow in a single file through capillaries.

21•4

Blood flows through the four chambers of the human heart.

Clench your fist. That is the size of your heart. Now clench it and relax it a hundred thousand times. That is what you require your heart to do, every day, for 70 or more years. The human heart, at the center of our circulatory system, is one of the most durable and reliable pumps ever produced (**FIGURE 21-6**).

As we saw above, the four-chambered heart sends blood on a figure 8, two-cycle path through the body, sending all blood first to the lungs for loading up on oxygen and, on its second circuit, to the tissues and organs. Let's explore the workings of this heart, tracing the flow of blood as it cycles through the heart, lungs, and tissues of the body (**FIGURE 21-7**). We'll start with the arrival of oxygen-depleted blood from the organs and tissues.

1. Deoxygenated blood from the organs and tissues enters the right atrium. The blood arriving from the lower half of the body enters through the inferior vena cava, and blood arriving from the head and arms enters through the superior vena cava.
2. Most of the blood passes directly through the right atrium into the right ventricle. A contraction pushes the

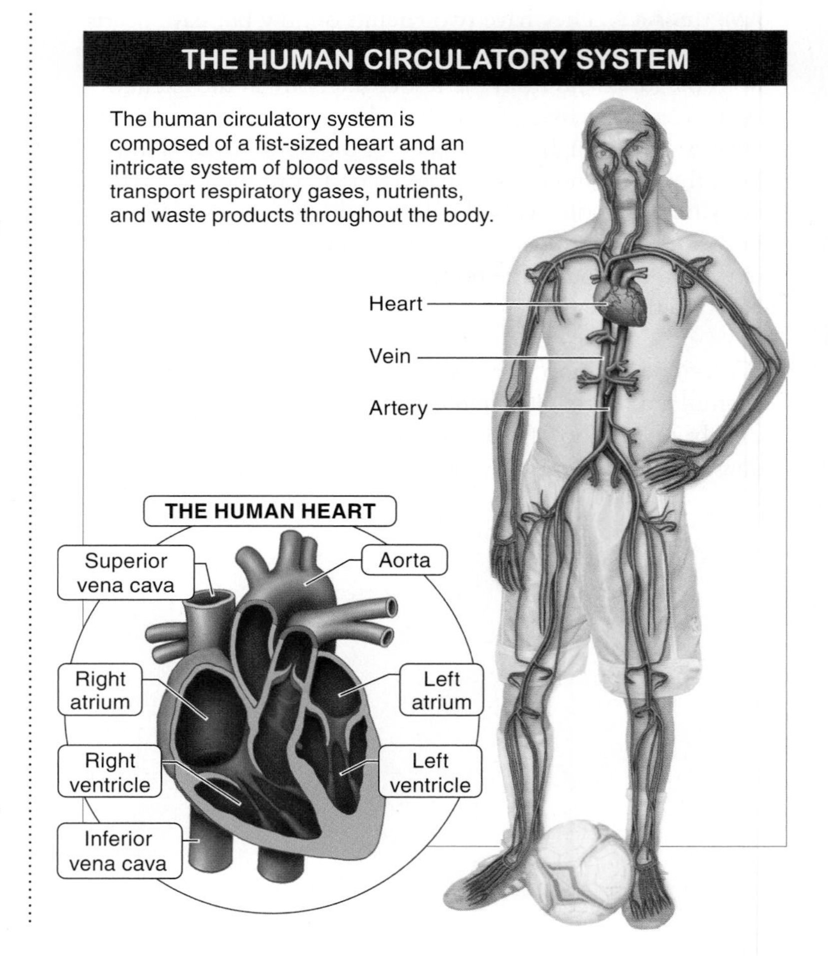

FIGURE 21-6 Heart and blood vessels. An overview of the human circulatory system.

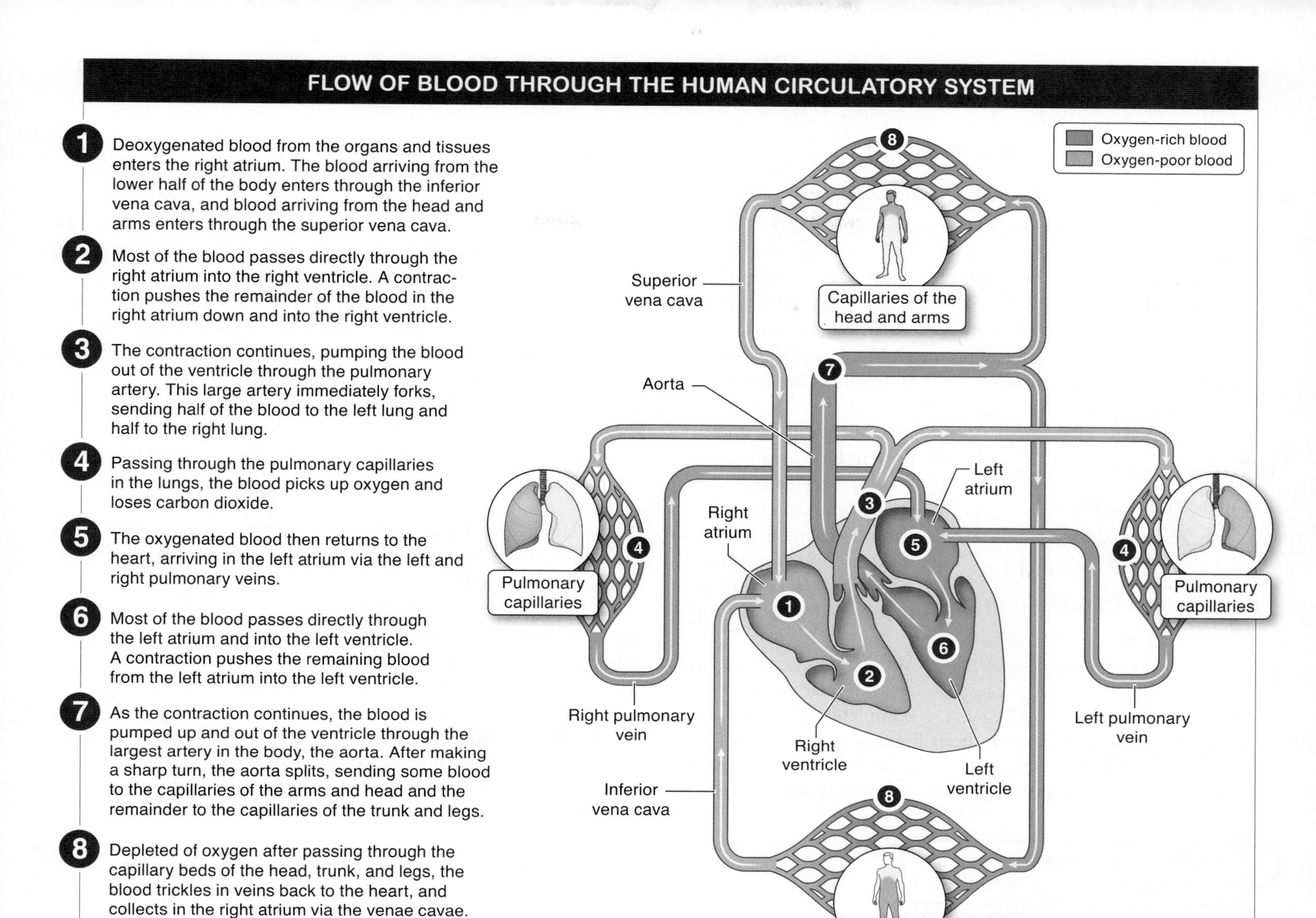

FIGURE 21-7 The path of blood flow in the human body.

remainder of the blood in the right atrium down and into the right ventricle.

3. The contraction continues, pumping the blood out of the ventricle through the pulmonary artery. This large artery immediately forks, sending half of the blood to the left lung and half to the right lung.
4. Passing through the pulmonary capillaries in the lungs, the blood picks up oxygen and loses carbon dioxide.
5. The oxygenated blood then returns to the heart, arriving in the left atrium via the left and right pulmonary veins.
6. Most of the blood passes directly through the left atrium and into the left ventricle. A contraction pushes the remaining blood from the left atrium into the left ventricle.
7. As the contraction continues, the blood is pumped up and out of the ventricle through the largest artery in the body, the aorta. After making a sharp turn, the aorta splits, sending some blood to the capillaries of the arms and head and the remainder to the capillaries of the trunk and legs.
8. Depleted of oxygen after passing through the capillary beds of the head, trunk, and legs, the blood trickles in veins back to the heart, and collects in the right atrium via the venae cavae.

If you place your ear on the center of another person's chest in a quiet room, you can hear the heart working. "Lub dup, lub dup, lub dup." Over and over, the same two sounds. These are the sounds that a doctor hears when using a stethoscope. What is the source? It's not the contraction of the heart. Surprisingly, the muscular contractions don't make much noise. Rather, the sounds come from two sets of valves that help keep blood flowing in the proper direction (**FIGURE 21-8**).

The first set of valves is the atrioventricular (or AV) valves. Located between the atrium and ventricle on each side of the heart, these two valves allow blood to flow from the atrium to the ventricle. But when the ventricle contracts, these flaps of tissue slam shut, preventing blood from being pushed back into the atria: "lub." With no other escape, the blood flows out through the pulmonary arteries on the right side and the aorta on the left. At these two primary exits from the heart are two more valves. These semilunar valves, so called because of their half-moon shape, close and prevent blood from flowing back into the ventricles. Like doors slamming shut, the valves make a noise that you can hear with your ear pressed to someone's chest: "dup."

Q Sometimes a doctor will hear additional sounds besides the "lub" and "dup" of normal heart valve functioning. What is the source of such heart murmurs?

When the atrioventricular or semilunar valves do not completely close, some blood can squirt back through them, flowing in the wrong direction. The blood moving backward through the valve can be heard, with a stethoscope, making a buzzing or swishing noise.

? ***Why does the heart make a "lub dup, lub dup" sound?***

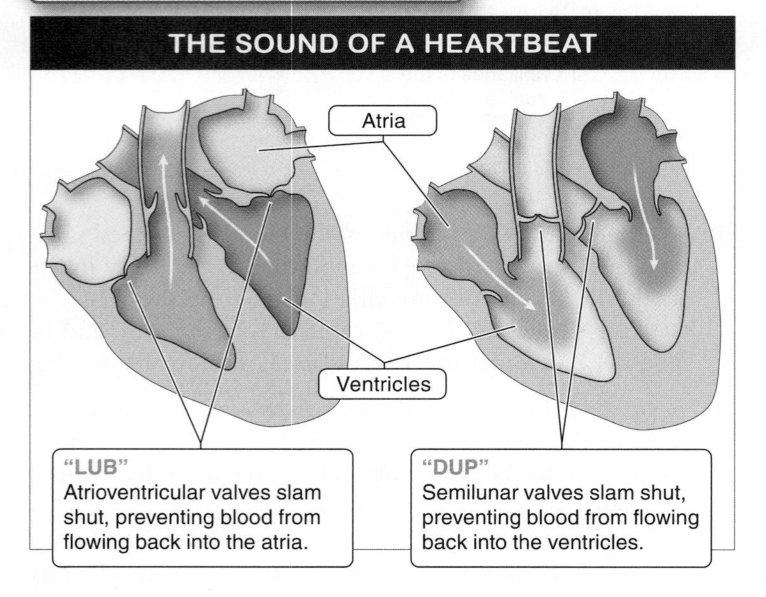

FIGURE 21-8 Audible heartbeats. Like doors slamming shut, the heart valves make noises that you can hear with your ear pressed to someone's chest.

Most heart murmurs are not life-threatening, and the individual suffers no ill effects.

The heart is not the only component of the circulatory system directing the flow of blood. Consider this: when you go out in very cold weather, your face or hands may get very cold and may even turn bluish in color. Why does this happen? Flow of blood in the capillaries is controlled by smooth muscle around the arterioles, the smaller vessels that branch off arteries and lead to capillaries. When the muscles, called precapillary sphincters, contract, blood flow can be cut off to capillaries and shunted elsewhere in the body (**FIGURE 21-9**). As a consequence of these muscles, bodies can reduce blood flow (and the heat it brings) to parts such as the face or hands in situations where such blood flow could lead to excessive and energetically wasteful heat loss. Blushing is the opposite situation: precapillary sphincters relax and blood flow to the face and neck increases.

Q What is "food coma"?

"Food coma," that feeling of lethargy following a large meal, results from a similar shunting of blood. In this case, more blood is allowed to flow through the capillaries surrounding the digestive tract while less flows to other parts of the body. Alternatively, during strenuous exercise, more blood is directed toward the skeletal muscles in use. This accounts for the feeling of being "pumped up" during and shortly after a weightlifting session. Surprisingly, the vast majority of the capillaries have little or no blood flowing through them at any given time, and the body is constantly directing blood flow to the tissues where it is most needed.

Veins, too, aid in the control of blood flow. After sitting for a very long time, such as on a long plane or car ride, you may notice that your feet become swollen and it is difficult to put your shoes back on. What is the cause of this? The answer has to do with the way that blood flows back to the heart after passing through capillary beds. Remember that capillaries, with their small diameter and thin, leaky walls, reduce the pressure and speed of flowing blood. By the time it begins collecting in veins, blood has very little pressure. It is able to "limp" back to the heart only with the assistance of two important features. First, as muscles surrounding veins contract and relax during normal movement, they squeeze the veins, pushing the blood through. And second, within veins, at regular intervals, are one-way valves. Similar in function to the atrioventricular and semilunar valves of the heart, these valves allow blood to flow in one direction—toward the heart—but not in the reverse direction (see Figure 21-9).

Regulating blood flow by means of valves in the veins and the contractions of muscles surrounding the veins usually works fine. As we've noted, though, you can encounter a problem if you sit on a plane for several hours. Blood pumped to your feet must somehow climb up your legs, yet in the absence of contractions of your calf and thigh muscles, this doesn't occur so well. Instead,

DIRECTING THE FLOW OF BLOOD

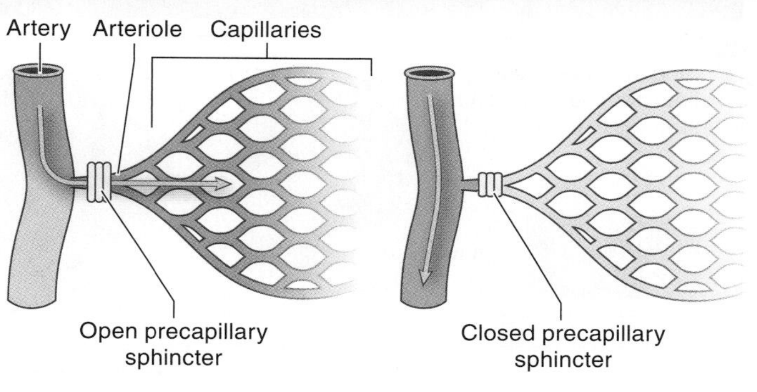

PRECAPILLARY SPHINCTERS
In the arterioles, precapillary sphincters can contract and cut off blood flow to the capillaries in order to shunt it elsewhere in the body.

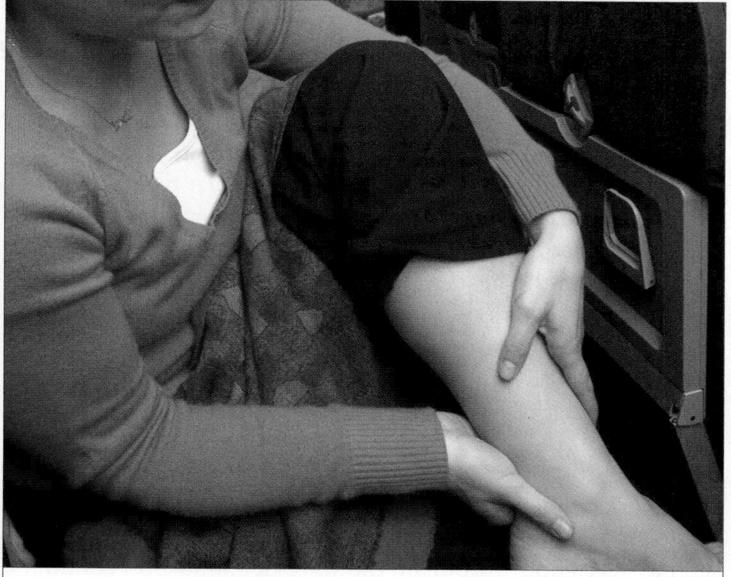

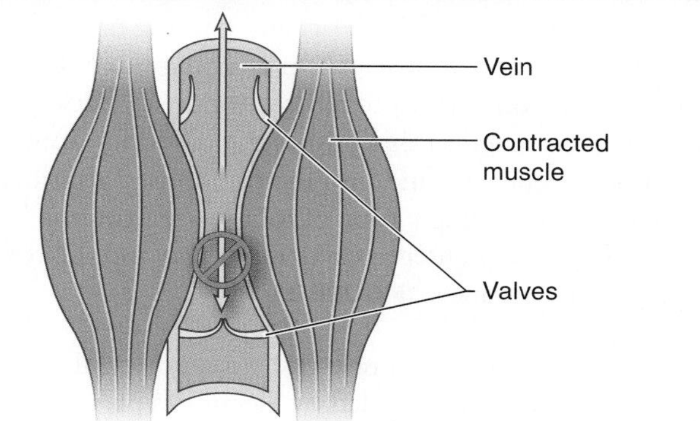

MUSCLE CONTRACTIONS AND VALVES
Contractions of muscles surrounding the veins push blood toward the heart. Valves within the veins keep the blood on course by preventing it from moving backward.

FIGURE 21-9 Controlling blood flow. Precapillary sphincters can reduce blood flow to hands or feet in cold weather, saving the blood's warmth for vital organs. When traveling on a long flight, it is helpful to move your feet and legs to help push blood back toward your heart.

Q Some individuals develop enlarged, twisted, and visible veins, a painful condition called varicose veins. Why might this occur?

the blood pools in your feet, causing them to get more swollen increasing the risk of blood clots. This is why it is good to get up and move around occasionally, or to exercise your leg muscles as you sit.

One serious circulatory problem is called varicose veins, a painful condition that occurs when the valves preventing backflow of blood malfunction and blood pools in the veins, stretching them. This can be caused or exacerbated by standing for long periods of time. Several methods are effective for treating varicose veins, including laser surgery or injections that cause the veins to slowly fade and disappear, with deeper veins taking over the circulation in that region.

TAKE-HOME MESSAGE 21•4

The human heart, at the center of our circulatory system, is an extremely durable pump. It sends blood on a figure 8, two-circuit path through the body, first to the lungs for loading up with oxygen and, on its second circuit, to the tissues and organs of the body. Valves in the heart and veins keep blood flowing in one direction.

21•5

Electrical activity in the heart generates the heartbeat.

Vertebrate hearts have a small piece of modified muscle tissue, the sinoatrial (SA) node, that initiates the regular, rhythmic contractions of the heart. Unlike most muscular tissue, which must be stimulated by a nerve before it contracts, the sinoatrial node spontaneously fires an electrical impulse that initiates contractions in the heart muscle. It begins this spontaneous firing early in fetal development and continues to give rise to every heartbeat for your entire life. As the pacemaker of the heart, the SA node initiates a carefully choreographed contraction that results in an efficient blood pump.

Beginning just above the right atrium, the electrical impulse in the SA node quickly spreads to the left atrium as well. As the atria contract, blood is pushed into the ventricles. The atria then enter a "relaxation" phase, and the wave of contraction continues, passing down the center of the heart and pausing briefly as it passes between the two ventricles. On reaching the bottom of the heart, the contraction appears to almost "bounce" back upward, causing a deep contraction that pushes the blood up from the bottom of both ventricles and into the pulmonary arteries and aorta, much like squeezing a tube of toothpaste from the bottom up. The ventricles then enter a relaxation phase, and the SA node starts the contraction anew (**FIGURE 21-10**).

Because the contraction of muscle tissue is a powerful electrical event, it can be recorded by electrodes placed on the skin that detect the changing electrical charges as the heart beats. Called an electrocardiogram, or EKG (from the German *Elektrokardiogramm*), these readings allow quick and easy display and analysis of the cardiac cycle.

Q What is an artificial pacemaker?

In some individuals, the sinoatrial node does not function properly or may be damaged by infection. This malfunctioning can lead to a heartbeat that is too slow or erratic and can cause reduced blood flow, leading to problems such as fainting. An artificial pacemaker is a battery-operated electronic device that generates stimulation to the heart, causing a more regular heartbeat.

TAKE-HOME MESSAGE 21•5

The sinoatrial node, modified muscle tissue in the vertebrate heart, initiates regular, rhythmic contractions. A heart contraction begins with an electrical impulse in the SA node in the right atrium. The contraction quickly spreads to the left atrium, and passes down the center to the bottom of the heart, then moves upward, pushing blood from both ventricles out through the pulmonary arteries and aorta.

HEART CONTRACTION

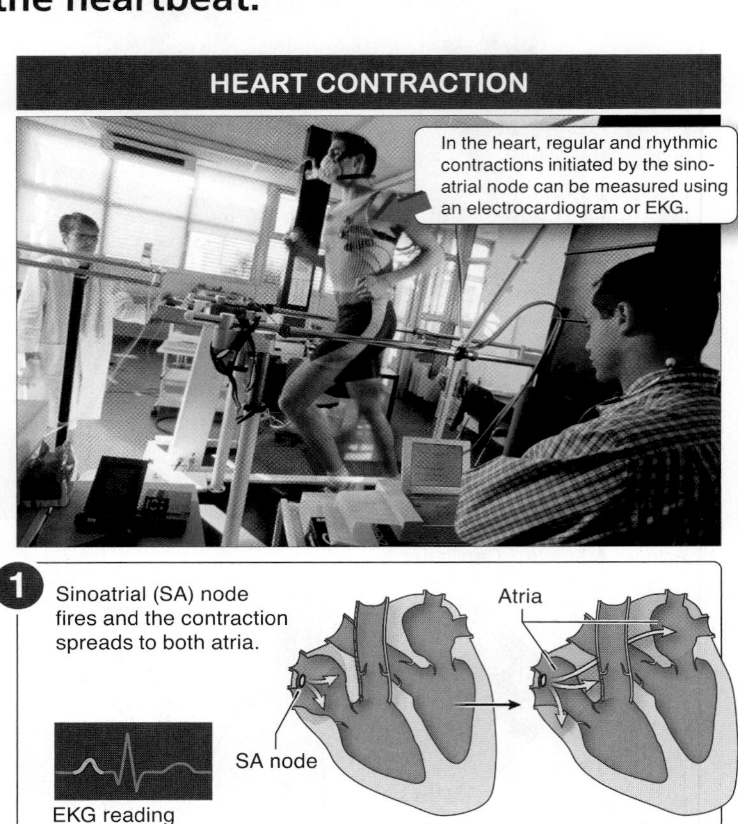

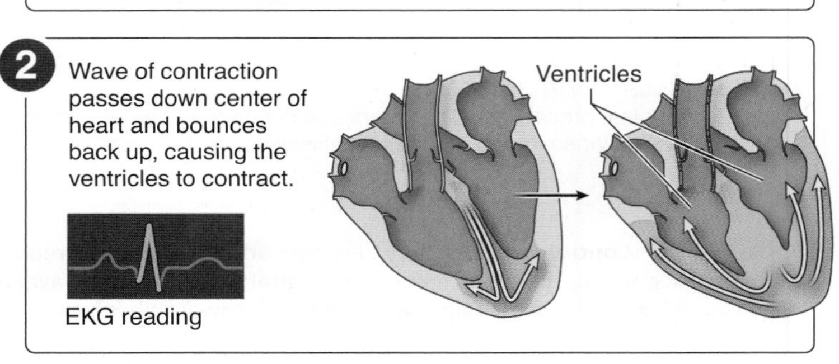

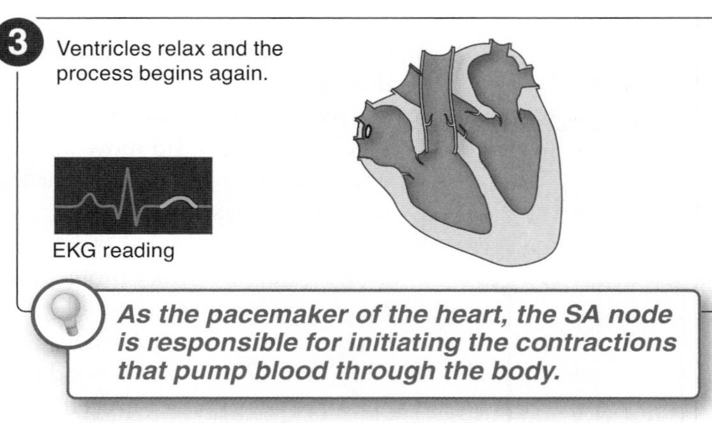

As the pacemaker of the heart, the SA node is responsible for initiating the contractions that pump blood through the body.

FIGURE 21-10 Rhythmic contractions of the heart.

21•6

Blood is a mixture of cells and fluid.

If circulatory systems are like highways throughout our bodies, transporting goods and garbage, then blood is the traffic. Endlessly circulating, this viscous fluid—its consistency is closer to that of motor oil than water—is a salty, protein-rich mixture of cells and fluid. The average human body has 4–5 quarts of blood, and the blood makes up just under 10% of our total body weight. Blood's functions revolve primarily around its transport and delivery capabilities, including the transport of (1) respiratory gases such as oxygen and carbon dioxide, (2) vitamins and minerals, (3) nutrients, (4) hormones, (5) cells of the immune system, and (6) metabolic wastes. Blood also helps to maintain body temperature and homeostasis (see Chapter 20).

Blood has several distinct components. Putting a small sample of blood in a test tube and spinning it rapidly in a centrifuge makes it possible to identify them (**FIGURE 21-11**). The lighter-weight part of the blood, the creamy yellow layer in the test tube, is the **plasma,** the liquid part of the blood. Plasma is 90% salty water. Dissolved within this water is a huge variety of molecules: metabolites and wastes, salts and ions, and hundreds of plasma proteins that serve to transport lipids, vitamins, and a host of other chemicals that need molecular escorts to the tissues where they are required. Most of the carbon dioxide produced in tissues as a by-product of cellular respiration is carried to the lungs dissolved in the plasma.

The heaviest components of blood get forced to the bottom of the test tube when spun in a centrifuge. This is a layer of packed cells, containing various types of blood cells, and is usually dark red in color. The proportion of the blood that consists of cells is called the **hematocrit.** In humans, a hematocrit of about 45% is normal. Individuals living at high altitudes for a few weeks or longer, however, have hematocrits of around 48% or 49%. Why? The increased hematocrit is a response to the reduced oxygen concentrations in the air at high altitudes. To continue to deliver enough oxygen to the body's tissues, there must be more blood cells to carry it.

Where do these blood cells come from? They are made in the bone marrow (the material that fills the interior of our bones) by specialized cells, called **stem cells,** that are able to develop into a diverse range of cell types. Stem cells throughout the bones in our body produce blood cells at a rate of about two million cells per second. There are two types of blood cells suspended in the plasma: red blood cells and white blood cells, as well as platelets, which are cellular fragments (**FIGURE 21-12**).

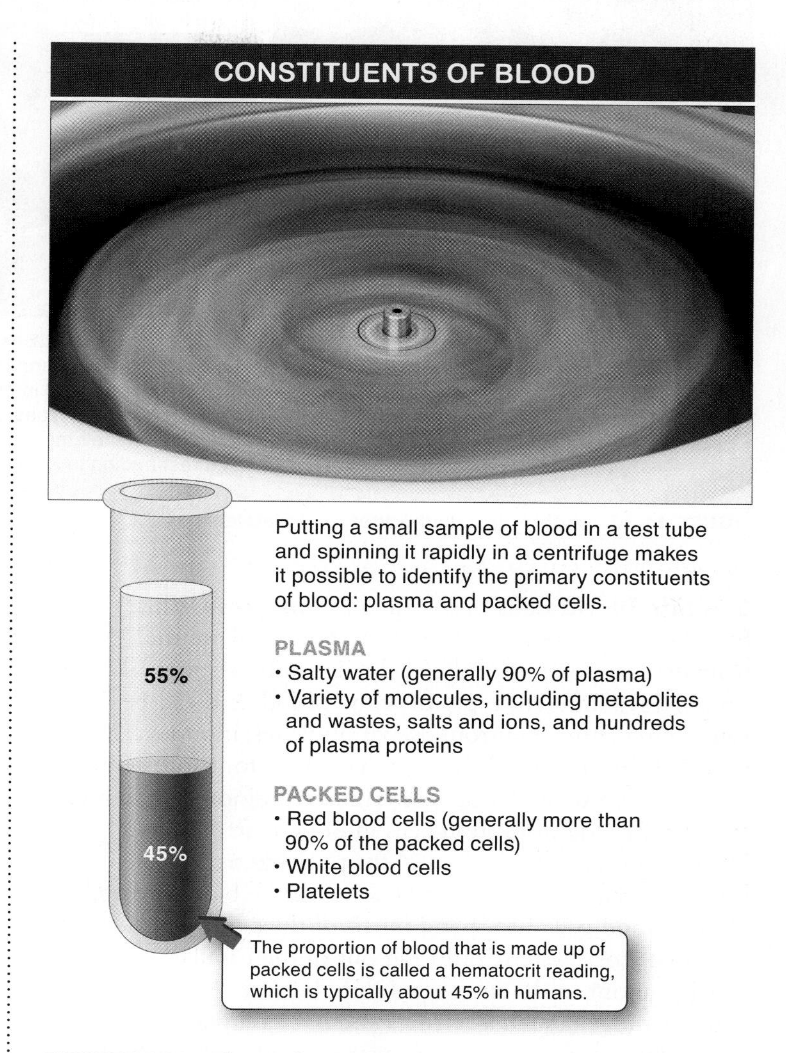

FIGURE 21-11 What makes up blood?

1. Red blood cells (also called **erythrocytes**). These workhorses of the circulatory system are the most common blood cells. In a human being, about 95% of the blood cells circulating at any given time are red blood cells. They are oxygen-transporting specialists, and their structure maximizes their effectiveness. Externally, they are shaped like flexible disks, so they can squeeze through capillaries in single file. Internally, they have hardly any organelles; they have no nucleus, mitochondria, or protein-making machinery. What are they filled with? Each red blood cell contains about 250 million molecules of **hemoglobin,** an oxygen-carrying protein molecule. Because they lack almost all internal cellular machinery for repair and upkeep, red blood cells don't last long, remaining in circulation for about four months. During its short life, though, a red blood cell will travel about 900 miles, endlessly picking up oxygen in the lungs and releasing it to body cells that need it.

CELLS AND PLATELETS IN BLOOD

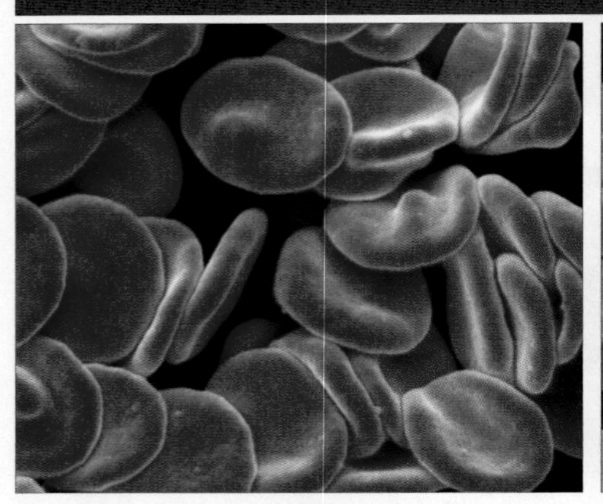

RED BLOOD CELLS (ERYTHROCYTES)
- Transport oxygen from the lungs to the rest of the body
- Flexible disks containing few organelles
- Packed full of hemoglobin

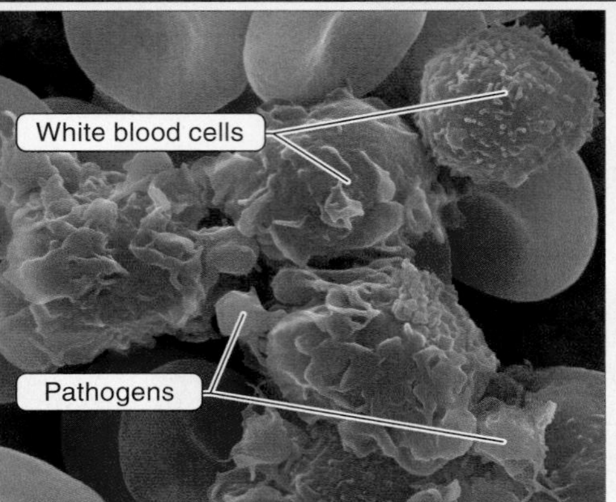

WHITE BLOOD CELLS (LEUKOCYTES)
- Destroy pathogens and foreign organisms in the bloodstream and interstitial fluid
- There are several types of white blood cells that differ in their methods of fighting disease and responding to foreign materials

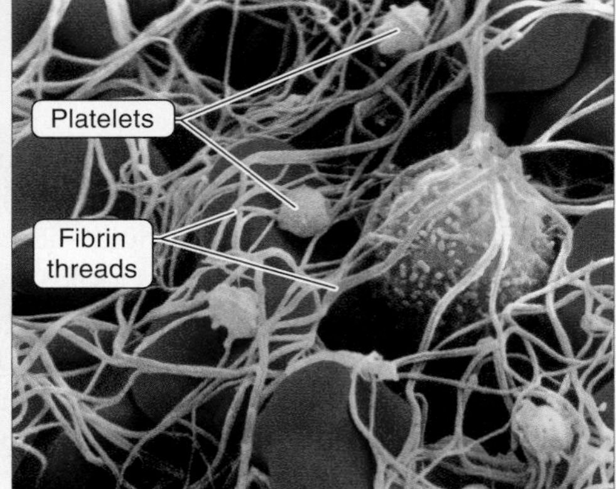

PLATELETS
- Slow blood loss by initiating the constriction of blood vessels and the formation of a clot
- Composed of small pieces of cytoplasm
- Contain no organelles

FIGURE 21-12 Erythrocytes, leukocytes, and platelets.

2. White blood cells (also called **leukocytes**). White blood cells are the defenders of the body and are the primary components of the body's immune response system. Five different types of white blood cells can be found circulating: neutrophils, lymphocytes, monocytes, eosinophils, and basophils (see Chapter 26 for more on the roles of these white blood cells). Like red blood cells, white blood cells arise from stem cells in bone marrow. Once in the bloodstream, they patrol for **pathogens,** disease-causing foreign organisms circulating in the bloodstream. White blood cells also spend much of their time outside the circulatory system, diffusing out of the capillaries and moving about in the interstitial fluid between cells where many pathogens such as viruses and bacteria may be, and where they also can destroy cancerous body cells or body cells that have been infected by pathogens. The number of leukocytes circulating in an individual can vary greatly, depending on his or her health status. Under normal conditions, there is approximately one leukocyte for every thousand red blood cells, but the number of leukocytes increases as much as two- to threefold during an infection.

3. Platelets. With more than 50,000 miles of blood vessels in our bodies, it is inevitable that there will be occasional cuts or punctures (**FIGURE 21-13**). Fortunately, the platelets are ready to swing into action when this happens. **Platelets** are considered cellular fragments rather than full-fledged cells. In the bone marrow, large cells called megakaryocytes repeatedly pinch off little bits of cytoplasm that have no nuclei or other organelles. These cell fragments, the platelets, are filled with critical enzymes and chemicals for patching damaged blood vessels. Hundreds of thousands of platelets circulate at any given time, with each platelet generally lasting about a week. You can envision platelets as fragile glass jars full of Super Glue. When they bump into the edge of a cut in a blood vessel, they shatter and release their cargo,

If platelets lack properly functioning clotting factors, cuts and scrapes can lead to uncontrolled bleeding.

FIGURE 21-13 Fighting blood loss. Platelets patch damaged blood vessels.

Cooking in a cast iron skillet can increase significantly the iron content of foods!

initiating constriction of the blood vessel and the production of fibrin threads that form a blood clot to reduce blood loss.

Some individuals lack platelets with properly functioning clotting factors, and they experience uncontrolled bleeding from even minor cuts or scrapes. These problems can be due to an inherited malfunctioning gene (as in the condition of hemophilia). They can also be acquired, or environmental: if the liver—where many of the clotting enzymes are produced—is damaged by disease (such as cirrhosis, which can result from alcoholism), uncontrolled bleeding can occur. Conversely, problems of blood clotting *too* readily can also lead to health problems. Thrombosis, for example, is the formation of clots of coagulated blood within a blood vessel. When such clots block circulation in these blood vessels that supply blood to the muscle tissue of the heart itself, a heart attack occurs.

Q What is anemia? Why are women more susceptible than men?

Anemia occurs when an individual has too few red blood cells. Because red blood cells deliver oxygen to the body's cells, one consequence of anemia is a reduction in the oxygen available to cells, essentially suffocating them. This causes people with anemia to feel tired and run-down. A reduction in the number of red blood cells is also associated with an increased susceptibility to infection, apparently by weakening the immune system's ability to mount a response to pathogens. Iron deficiency is the most common cause of anemia. Iron is a critical element that enables oxygen to be carried by red blood cells. If iron is in short supply, red blood cells can't deliver sufficient oxygen to the tissues where it's needed. Both men and women can be anemic, but anemia affects women much more commonly, because of the blood loss during menstruation. The relationship between iron deficiency and anemia has been known for a long time. An old folk remedy for anemia—that actually worked(!)—was sipping liquid daily from a jug containing rusty iron nails and water.

TAKE-HOME MESSAGE 21•6

Blood is a salty, protein-rich mixture of cells and fluid, important in the transport of (1) respiratory gases, (2) vitamins and minerals, (3) nutrients, (4) hormones, (5) components of the immune system, and (6) metabolic wastes. Blood also helps maintain a constant internal environment, including body temperature. Blood cells are produced throughout life by stem cells in bone marrow. There are two types of cells suspended in the plasma: red cells (oxygen transport), white cells (defense from infections), as well as cellular fragments, platelets (repair).

21•7

Blood pressure is a key measure of heart health.

Feel your pulse. With no fancy equipment at all, it is possible to get very useful information about the functioning of your heart. For some arteries, such as those on the underside of your wrist or on the side of your neck, you can feel with your fingers the pressure increase as a blood surge stretches the arteries with each contraction of the heart. Taking a person's pulse provides a quick and easy determination of the rate and rhythm of the heartbeats.

Additional information about heart health can be gained with measurement of **blood pressure,** which measures the force with which blood flows through a person's arteries. This force tells us the magnitude of each heart contraction and gives important clues about an individual's cardiovascular health. There are two different parts to a blood pressure reading (**FIGURE 21-14**). The first, called **systolic pressure,** is the pressure when the heart contracts. The powerful contraction pumps blood into the arteries, momentarily causing them to stretch as they accommodate the large pulse of blood. The second blood pressure reading is called **diastolic pressure.** This is a measure of the force that blood exerts on the artery walls while the heart is between beats. Because blood isn't being actively pumped at that moment, the diastolic pressure is always lower than the systolic pressure.

Blood pressure can be measured in four easy steps, using a blood pressure cuff.

1. The cuff is fastened around the upper arm and pumped up, clamping off the arteries in the arm so that no blood gets through.
2. Gradually, pressure on the cuff is released.
3. When the pulsing of blood getting pushed through the arteries under the cuff can first be heard with a stethoscope held to the arteries just below the cuff—heard as a little squirt—the pressure reading is noted. That is the systolic pressure. Each contraction of the heart is just strong enough to push blood through the barrier of that much pressure.
4. Additional pressure in the cuff is released until the squirting sound disappears. The pressure at that point is the diastolic pressure. Blood is flowing through the arteries with this amount of pressure between heart contractions.

If your blood pressure is "120 over 80," it means that the systolic pressure is 120 and the diastolic pressure is 80. This is written as 120/80, and the units of measure are millimeters of mercury (mmHg), representing how high a column of mercury could be lifted by such pressure.

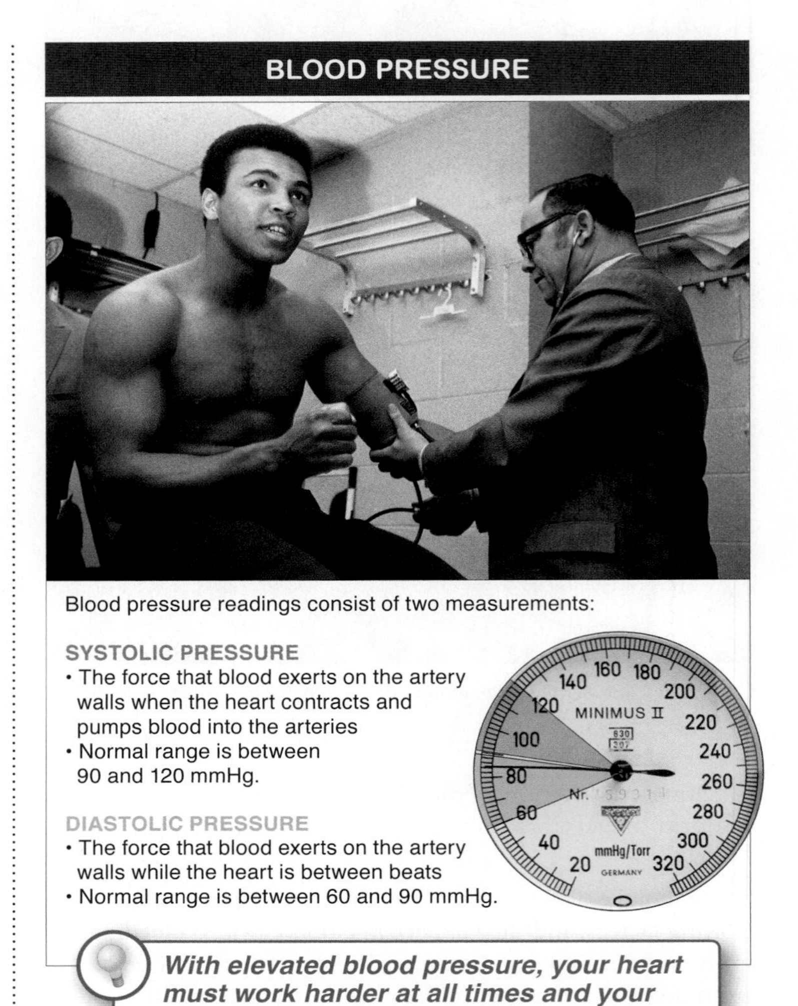

FIGURE 21-14 **Blood pressure readings can reveal heart health.**

Blood pressure above 140/90 is considered high and a potential health hazard. What does this mean, and why is it cause for concern? Imagine a pair of shorts with an elastic waistband. If you were to stretch the waistband as far as possible and hold it in that position for a long time, what would happen? The waistband would lose its elasticity. This is similar to what

FIGURE 21-15 Contents under pressure. The large heart of a giraffe pumps blood through its long neck all the way to its head.

happens to your arteries if they are stretched by high pressures for long periods of time. With high blood pressure, also called hypertension, not only must your heart work harder at all times, potentially weakening it, but your arteries have a reduced ability to expand and accommodate the increasing pulses of blood during times of exertion. Moreover, more cholesterol sticks to artery walls when they are rigid than when they are elastic, and as this narrows the diameter of the blood vessel, it further reduces the efficiency of the circulatory system and taxes the heart. These problems all increase the risk of catastrophic heart attacks and strokes, which we discuss in Section 21.8. Although both systolic pressure and diastolic pressure are important, systolic blood pressure is more important in identifying and controlling hypertension.

Low blood pressure, or hypotension, at the other extreme, is defined as a pressure of 90/60 or lower and can cause symptoms such as dizziness, particularly just after a person stands up, due to inadequate blood flow to the brain. Sometimes caused by medications, low blood pressure can also be associated with weakness or depression. Most often it is not a problem, and it rarely has long-term risks.

Relative heart size and blood pressure are consistent among the mammals, with one notable exception: giraffes. With their necks stretching 8 feet (2.5 meters) or more, giraffes must pump blood up a pretty steep slope. Not surprisingly, their hearts are much bigger—about four times bigger (relative to body size)—than a human heart, and their blood pressure is more than twice as great. Scientists have begun studying giraffe circulation in the hope of learning physiological secrets to help humans cope with high blood pressure. Why, for example, with twice the blood pressure of a human, doesn't a giraffe's head explode when it lowers to drink from a pond (**FIGURE 21-15**)?

TAKE-HOME MESSAGE 21•7

Blood pressure measurement gives important clues about an individual's cardiovascular health. A blood pressure reading consists of two measures. The first, systolic pressure, is the force that blood exerts on the artery wall when the heart contracts and pumps blood into the arteries. The second, diastolic pressure, is the force that blood exerts on the artery wall while the heart is between beats. With high blood pressure, the heart must work harder at all times, the arteries can lose some of their elasticity, and health risks are increased.

21•8

Cardiovascular disease is a leading cause of death in the United States.

The heart is among the most neglected organs. Most people take it for granted. But every year heart attacks cause 20% of the deaths in the United States, more than any other single cause and among the most avoidable. Heart attacks are brought on by an interruption in the flow of blood through one of the **coronary arteries**—the blood vessels that deliver oxygen and nutrients to the heart muscle itself. When cells in the heart muscle are deprived of oxygen, the heart may beat irregularly or cease to beat, and the heart muscle cells deprived of oxygen die. This has serious and long-term implications, because heart cells are among the rare cells of the body that do not reproduce themselves. Once they are gone, they are really gone.

> Broken heart. A pump after all, pumping thousands of gallons of blood every day. One fine day it gets bunged up and there you are.
>
> —James Joyce, *Ulysses*

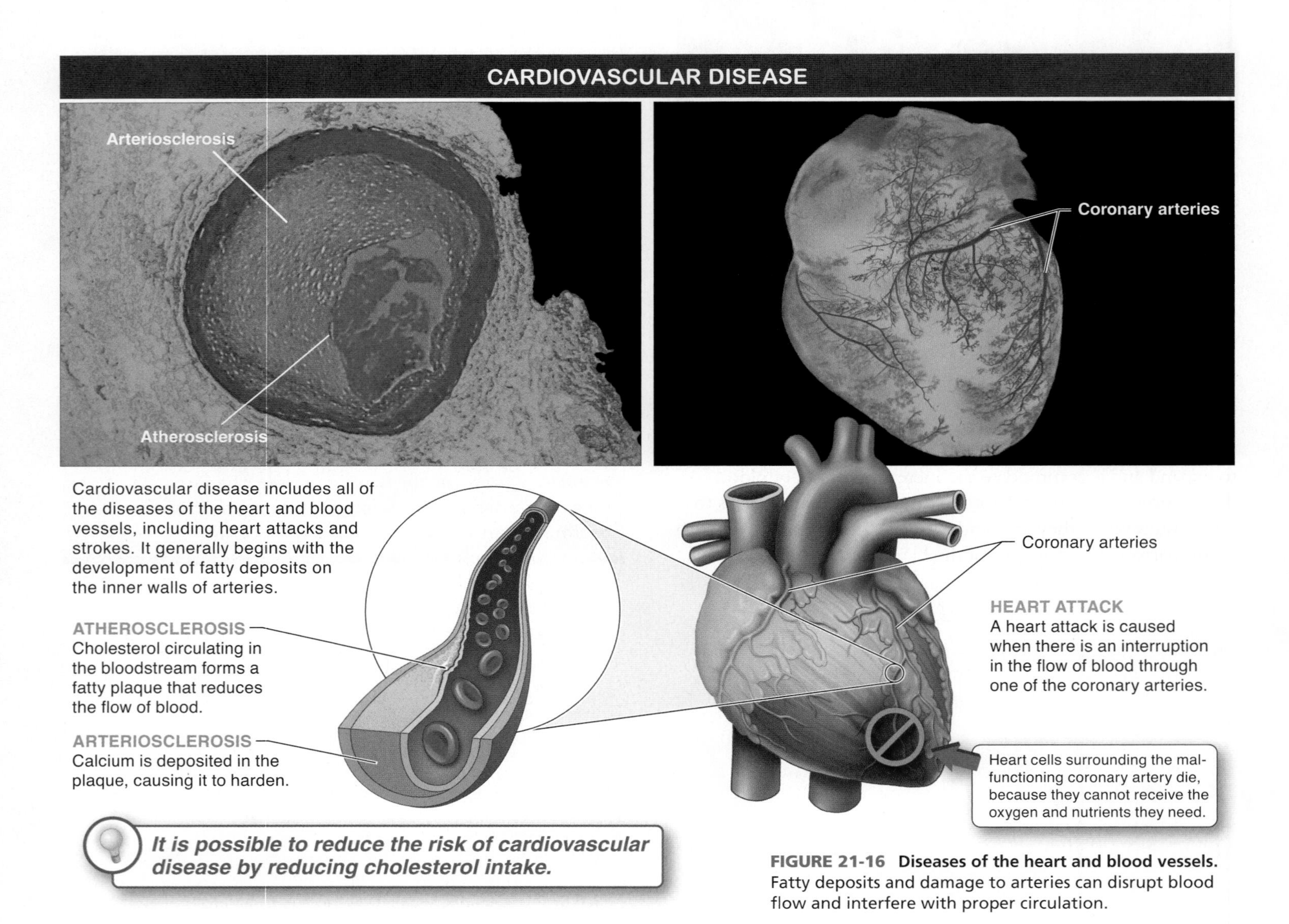

FIGURE 21-16 Diseases of the heart and blood vessels. Fatty deposits and damage to arteries can disrupt blood flow and interfere with proper circulation.

Contrary to appearances—and James Joyce's belief—heart attacks rarely strike out of the blue. Although the heart attack itself is a sudden event, it usually occurs after decades of progressive deterioration of the arteries and other degradations of the circulatory system, collectively called **cardiovascular disease** (**FIGURE 21-16**).

Cardiovascular disease includes all diseases of the heart and blood vessels and is ultimately responsible for close to half of all deaths in the United States. In addition to heart attacks, strokes are a common outcome of advanced cardiovascular disease. Caused by blocked arteries or blood clots in the brain, strokes also lead to cell death in the brain tissues starved of oxygen. There is, however, some cause for optimism: the death rate from heart disease has been declining steadily for the past 50 years, including a 25% drop in the past 20 years. These reductions have come not from advances in heart transplants and other surgical interventions, but rather from improvements in diet and exercise as well as advances in diagnosis and preventive medicine.

Cardiovascular disease generally begins with the development of fatty deposits called plaques on the inner walls of arteries. Plaques increase the risk of formation of blood clots and, by narrowing the artery, reduce the flow of blood. Called **atherosclerosis,** this narrowing of the arteries is often followed by the depositing of calcium at the plaques, causing them to harden in a process known as **arteriosclerosis.** The initial formation of plaques usually occurs as a consequence of circulating cholesterol in the bloodstream. Because most of this circulating cholesterol comes from cholesterol in our diet, we can reduce our risk of atherosclerosis by reducing our cholesterol intake.

Cholesterol is all the same, but you may hear references to "good" cholesterol and "bad" cholesterol. Why is this? As we discussed in Chapter 3, most cholesterol circulating in the bloodstream is packaged as LDL, or low-density lipoproteins. These molecules consist of thousands of molecules of cholesterol surrounded by a phospholipid coat. Because the LDL particles are sticky, they adhere to artery walls and can initiate the buildup of dangerous plaques. Other circulating particles, high-density lipoproteins (HDL), are considered "good" cholesterol. Less well-understood than LDL, these particles seem to remove cholesterol from arteries and deliver it to liver cells where it can be broken down. This process can actually reduce the progression of cardiovascular disease. By including in your diet fish and other foods that contain a specific type of fatty acids, called omega-3 fatty acids, you can increase your HDL levels.

Both nature and nurture play a role in cardiovascular disease. As we saw in Chapter 3, the tendency to develop cardiovascular disease is inherited. Individuals vary in the number of LDL receptors they produce on their liver cells, and the more receptors an individual has, the better that individual is able to remove atherosclerosis-causing cholesterol from circulation. You cannot alter the genes you inherit for LDL receptor production. But you *can* alter the amount of cholesterol or type of cholesterol (that is, LDL vs. HDL) that is circulating in the first place. Several different behavioral changes can reduce the level of circulating cholesterol and the risk of cardiovascular disease: in the first place by increasing aerobic exercise, not smoking, and eating a low-cholesterol, low-fat diet (**FIGURE 21-17**).

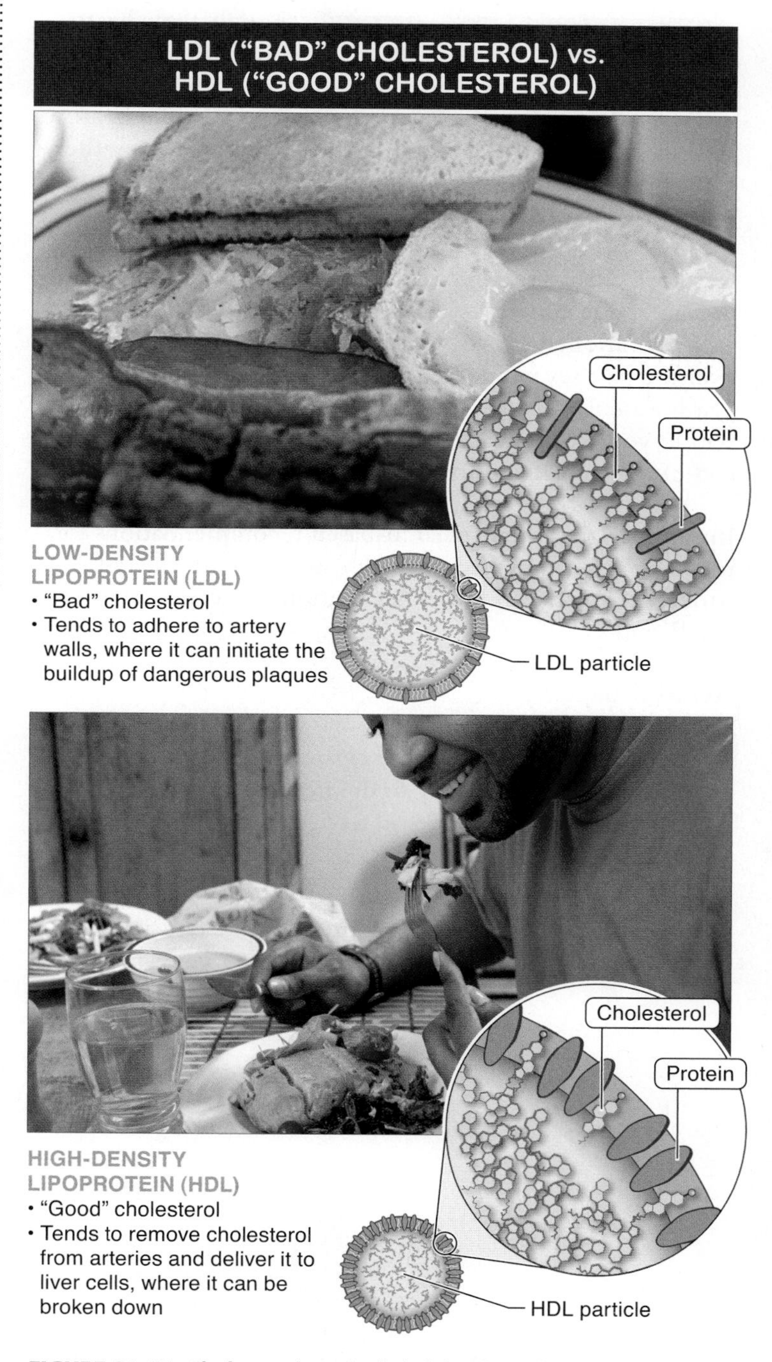

FIGURE 21-17 Cholesterol can be helpful or harmful.

Q If exercising makes the heart work harder and if being overweight from eating fatty foods also makes the heart work harder, why does one lead to fitness and the other to disease?

With aerobic training, such as running, the muscle fibers of the heart get bigger (in much the same way that skeletal muscle gets bigger when you lift weights), and cardiovascular health is improved. When people are sedentary, though, their hearts must work harder for reasons such as hypertension, poor diet, or increased blood pressure. The increased load on the heart causes the heart to get bigger, but in a pathological manner that increases the risk of heart failure, rather than in a manner that increases strength and efficiency.

Cardiovascular disease continues to be the most prevalent disease and leading cause of death in the United States. But, as we mentioned above, the situation has been improving over the past several decades. The onset and progression of Cardiovascular disease are strongly influenced by factors within your control, and with the proper dietary strategies and lifestyle changes most people can reduce their risk.

In summarizing their research-based recommendations for reducing risk factors for heart disease, heart attack, and stroke, the American Heart Association suggests a focus on "A, B, and C":

Avoid tobacco.

- Stop using any tobacco products and minimize exposure to tobacco smoke. Because cigarette smokers are two- to three-times more likely to die from heart disease, this behavioral change can lead to significant improvements to health and longevity.

Be more active.

- Participate in at least 30 minutes of moderate-intensity physical activity (i.e., brisk walking) on five or more days each week (or vigorous-intensity activity, such as jogging, on three or more days). Additionally, every adult should perform activities that increase or maintain muscle strength, including progressive weight training and/or stair climbing, on a minimum of two days each week.

Choose good nutrition.

- Eat a diet that balances energy intake with exercise to prevent weight gain. If you are overweight, take steps to increase physical activity and decrease energy intake to establish a healthy body weight (see Section 22-16).
- Chose a diet rich in vegetables, fruits, and whole grains, particularly those that are high in fiber.
- Limit intake of saturated fat, trans fat and cholesterol by choosing lean meats, vegetables, and low-fat dairy products. Read food labels to check for hydrogenated fats (see Section 2.13 for a refresher on hydrogenated fat and trans fats).
- Limit alcohol to one drink daily for women, and two drinks daily for men.

These behavioral changes are much easier said than done. Because they include the most effective strategies for reducing the risk of cardiovascular disease, however, the payoffs are significant.

TAKE-HOME MESSAGE 21·8

Cardiovascular disease includes all diseases of the heart and blood vessels, including heart attacks and strokes, and is the leading cause of death in the United States. It generally begins with the development of fatty deposits on the inner walls of arteries that increase the risk of blood clots and reduce the flow of blood in coronary vessels that supply oxygen to the heart (atherosclerosis). Because the hardening of arteries (arteriosclerosis) is usually initiated by circulating cholesterol, it is possible to reduce the risk of cardiovascular disease by reducing cholesterol intake.

21•9

The lymphatic system plays a supporting role in circulation.

Our bodies have another circulatory system (**FIGURE 21-18**). In addition to the cardiovascular system, and running close to it throughout the body, is the lymphatic system. Why the seeming redundancy? The lymphatic system has a supporting role in the process of circulation. It has three important functions (**FIGURE 21-19**).

1. Recycling. As diffusion occurs between the capillaries and the interstitial fluid that surrounds cells, mopeouch fluid is lost from the blood. Here's where the lymphatic system comes in. Intertwined around blood vessels, lymphatic capillaries take in, by diffusion, fluid, proteins, and other substances that have leaked into the interstitial fluid from the blood. Once recovered, this fluid, now called **lymph,** travels through progressively larger lymphatic vessels that eventually join up with veins in the shoulders. At this point, the recovered proteins and fluid (which amounts to several liters each day) are returned to the blood on its way back to the heart.

THE HUMAN LYMPHATIC SYSTEM

The lymphatic system runs close to the circulatory system throughout the body and plays a supporting role in the process of circulation.

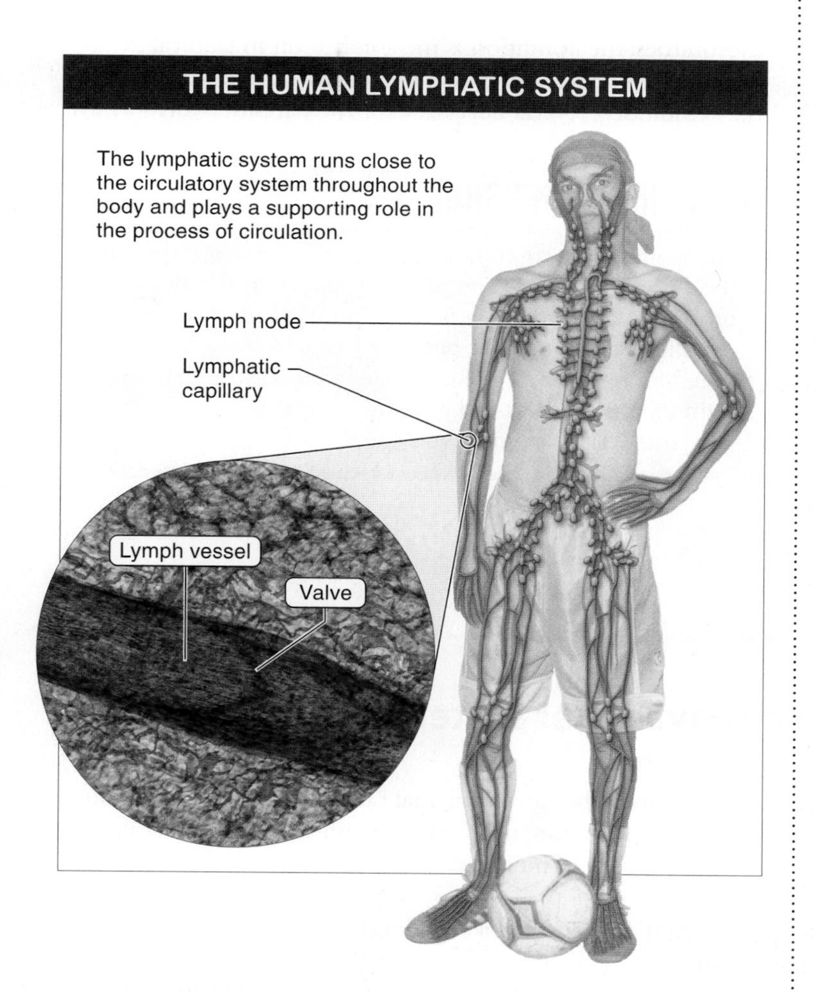

FIGURE 21-18 The "other" circulatory system. An overview of the human lymphatic system.

FUNCTIONS OF THE LYMPHATIC SYSTEM

RECYCLING
The lymphatic system recycles fluid and proteins that diffuse from the blood capillaries during circulation back into the bloodstream.

FIGHTING ILLNESS
As lymph circulates through the body, white-blood-cell-packed lymph nodes remove dangerous materials, including bacteria, cancer cells, and viruses, from the body.

RETRIEVING NUTRIENTS
Little projections that extend into the small intestine absorb lipids from the digestive tract and shuttle them to the bloodstream.

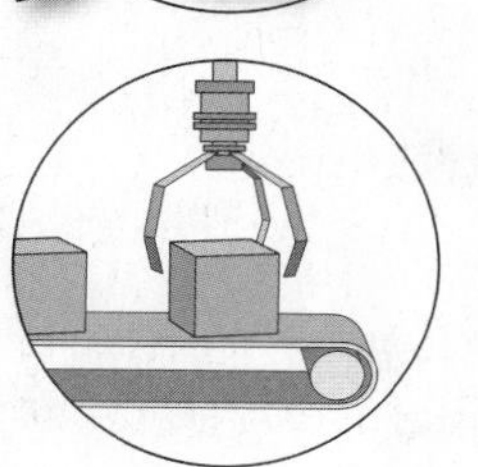

FIGURE 21-19 The lymphatic system supports the circulatory system while fighting illness.

2. Fighting illness. As it moves through the lymphatic system, lymph passes through patches of connective tissue called **lymph nodes.** These compartmentalized sacs are filled with pathogen-fighting white blood cells that remove dangerous materials (including bacteria, cancer cells, and viruses) from the body. This is why your lymph nodes—including your tonsils, the largest lymph nodes of all—become swollen when your body is fighting an infection.

> **Q** It's not necessarily a good idea to have your tonsils taken out, even if they are painfully swollen. Why?

FIGURE 21-20 A malfunctioning lymphatic system. When the lymphatic system is damaged, fluid builds up in the extremities.

3. Retrieving nutrients. The lymphatic system has numerous little projections that extend into the small intestine and absorb lipids from the food you have eaten and shuttle them from the digestive tract to the bloodstream.

In humans, the lymphatic system accomplishes these tasks without a pump. There is no "lymph heart." Rather, lymph is pushed through the system when muscles adjacent to lymph vessels contract and squeeze the fluid onward. Lymph vessels (like veins) have valves that keep the lymph flowing in one direction. When you sit for an extended period of time, lymph can accumulate in vessels. You can help move it along by contracting the muscles in your extremities and progressively contracting muscles closer and closer to your shoulders.

Infection by some parasitic worms can cause scarring of lymph vessels. Fluid recovered by the lymphatic system cannot be returned to the circulatory system, and elephantiasis, or a swelling of the extremities, results (**FIGURE 21-20**). Because the parasitic worms causing elephantiasis are transmitted by mosquitoes, the condition is most common in tropical regions. The condition can be treated with antibiotics that kill the symbiotic bacteria necessary for the parasitic worm to live.

TAKE-HOME MESSAGE 21•9

The lymphatic system runs close to the circulatory system throughout the body and plays a supporting role in the process of circulation, by performing three main functions: recycling fluid that leaks out of the capillaries of the circulatory system, marshaling white blood cells to help fight dangerous cells and pathogens, and absorbing nutrients from the digestive system.

21•10

The polygraph relies on cues from the cardiovascular system.

"Is your name John Doe? Do you have a driver's license? Did you stab Nick Eliot in the back?" Amazingly, measures of a person's respiratory and circulatory system functioning may help determine whether that person's answers to such radically unrelated questions are true. The **polygraph**—initially called the "lie detector" test—is a practical application of some of the physiological measures discussed above. Many law enforcement agencies and other governmental bureaus, as well as numerous private organizations, use the polygraph to help them evaluate whether or not an individual is telling the truth.

Polygraph examiners make readings of three types of measures during an examination: (1) chest and abdominal movement during respiration, indicating breathing rate; (2) changes in skin conductance, indicating activity of sweat glands; and (3) heart rate and amplitude, and blood pressure (**FIGURE 21-21**). During the exam, the polygrapher asks a series of questions. Some of the questions are control questions for which the examiner knows the answer, such as: "Do you have a driver's license?" The questions of interest (such as the question about stabbing Nick Eliot), called relevant questions, are interspersed among the control questions.

In analyzing an individual's physiological responses to the interview questions, polygraphers are looking for evidence of the fight-or-flight response, an automatic set of responses to stress in an organism's environment. These take the form of increases in all three measurements during a relevant question—that is, in breathing rate, skin conductance, and heart rate—all of which are greater than when the respondent answers the control questions.

Interestingly, while the results of polygraph tests are not admissible in federal courts, law enforcement agencies such as the Federal Bureau of Investigation still find them to be of great value. Based on their assessments of a suspect's guilt or innocence after a polygraph test, investigators may then search for additional evidence to corroborate their belief in someone's guilt or to find another suspect when they believe, based on a polygraph, someone is innocent.

Some people have tried to develop techniques for "beating" the polygraph. These include behaviors such as putting a thumbtack in their shoe and leaning forward into it during the control questions, or biting the inside of their mouth so as to increase the magnitude of their respiration and sweating. Generally, these techniques increase the magnitude of some of the measures, but not all of them. And examiners are on the

THE POLYGRAPH TEST

THREE TYPES OF POLYGRAPH MEASUREMENTS

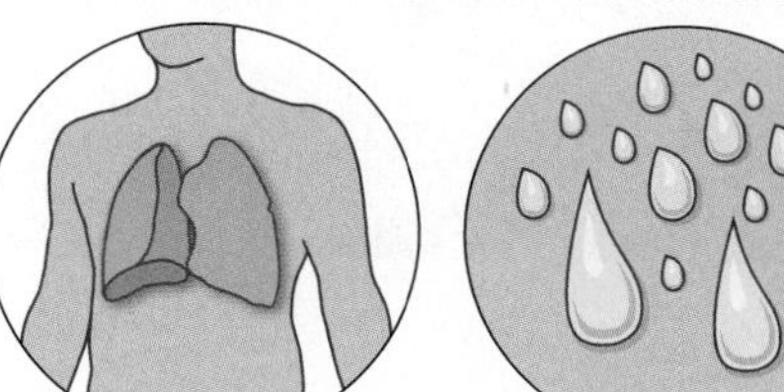

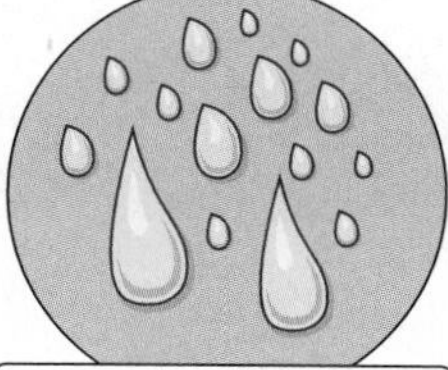

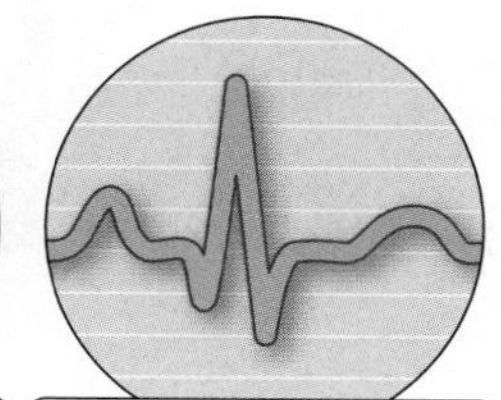

FIGURE 21-21 Telling the truth?

lookout for individuals' use of such attempts at deception and interpret them as signs of guilt. Does the machine enable polygraphers to detect lies? Analyses of in-the-field examinations and laboratory simulations of examinations indicate average accuracies of between 80% and 98%, but it depends greatly on the skills of the person administering the test and remains a controversial technique.

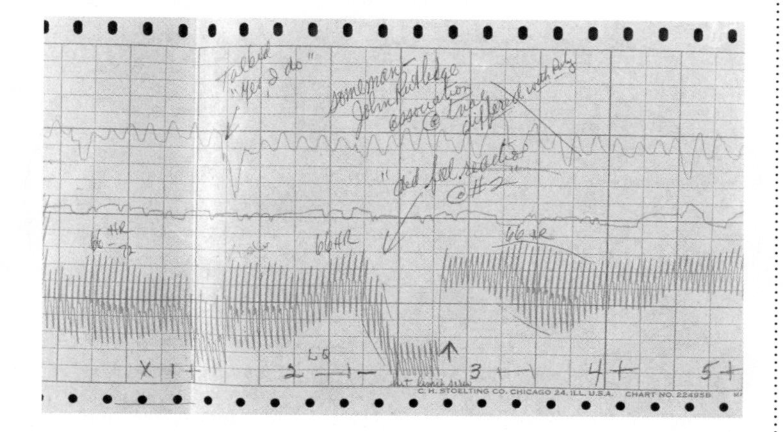

This polygraph tracing recorded Jack Ruby's physiological responses to questions about the murder of Lee Harvey Oswald.

TAKE-HOME MESSAGE 21•10

The polygraph can be an effective tool for evaluating whether or not an individual is telling the truth. A polygraph measures manifestations of the fight-or-flight response: (1) chest and abdominal movement during respiration, indicating breathing rate; (2) changes in skin conductance, indicating activity of sweat glands; and (3) heart rate and amplitude and blood pressure.

③ The respiratory system enables gas exchange in animals.

Humpback whales spouting (exhaling) while feeding in Alaska.

21•11

Oxygen and carbon dioxide must get into and out of the circulatory system.

As we've seen, circulatory systems are like trucking systems and the highways on which they move. Of the substances they transport, among the most important are the respiratory gases. After all, aerobic respiration requires cells to take up oxygen and release carbon dioxide. But how do these gases get into and out of the circulatory system? Where and how does gas exchange take place? In the remainder of this chapter, we investigate the structures where gas exchange occurs and the transport molecules that make it possible.

In single-celled and very small multicellular organisms, gas exchange can occur by direct diffusion. In larger multicellular organisms, however, gas exchange becomes a two-stage process (**FIGURE 21-22**). First comes the exchange between the external environment and the organism's circulatory system. Later comes the exchange between the circulatory system and the cells involved in cellular respiration.

The first of these two stages can occur in several different types of organs specialized for respiration, such as lungs or

GAS EXCHANGE IN ANIMALS

In large, multicellular organisms, gas exchange is a two-stage process.

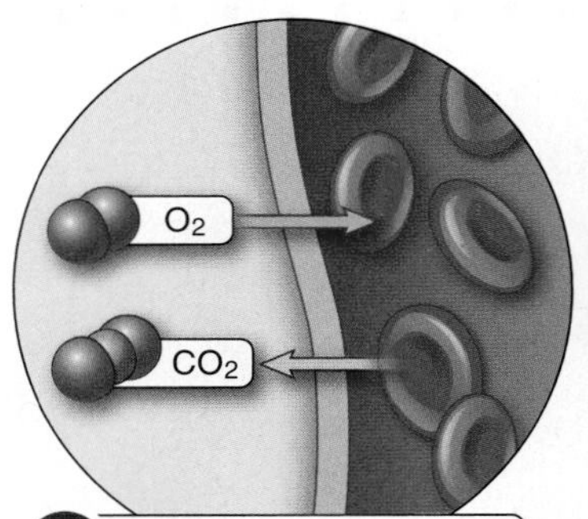

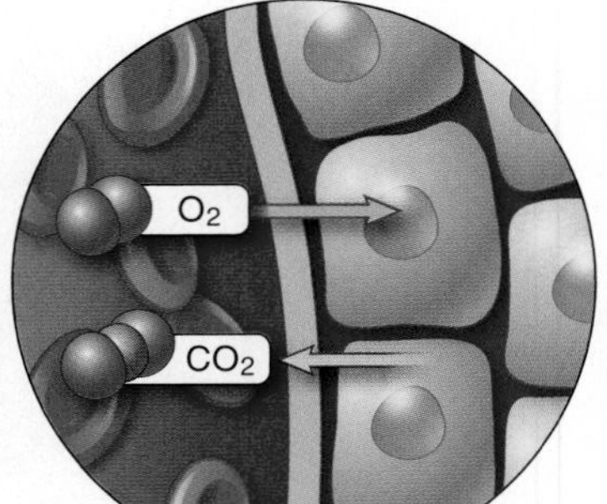

1 Respiratory gases are exchanged between the external environment and the organism's circulatory system.

2 Respiratory gases are exchanged between the circulatory system and the cells involved in cellular respiration.

FIGURE 21-22 Overview of respiratory gas exchange in animals.

DIVERSITY IN GAS EXCHANGE SYSTEMS

DIRECT DIFFUSION

- Gas exchange occurs directly between cells and the environment
- Occurs in single-celled organisms and small organisms with low metabolic demands

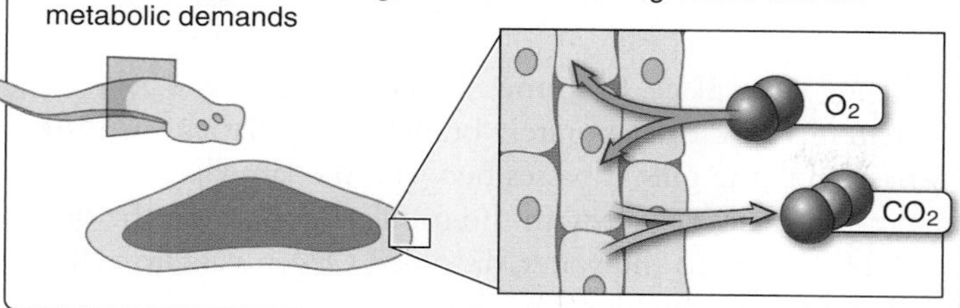

PROTRUDING RESPIRATORY SACS

- Balloon-like sacs that increase surface area for gas exchange
- Occur in sea stars and other echinoderms with low metabolic demands

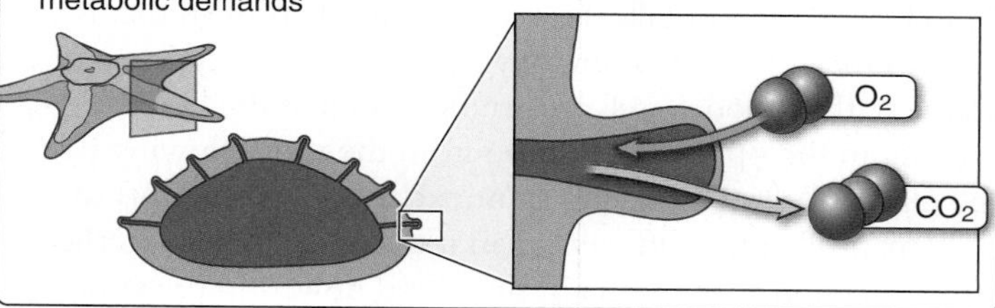

GILLS

- Elaborate extensions of the body that exchange significant amounts of gases dissolved in water
- Occur in fishes and many marine invertebrates such as lobsters and clams

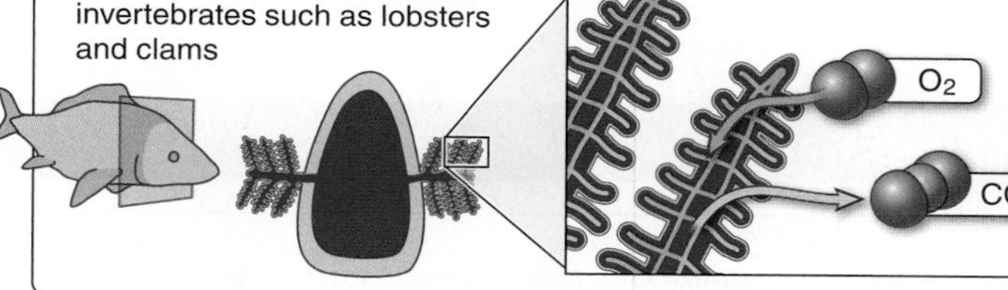

TRACHEAE

- Network of branching tubes connected to tiny openings on the body called spiracles
- Occur in most terrestrial insects

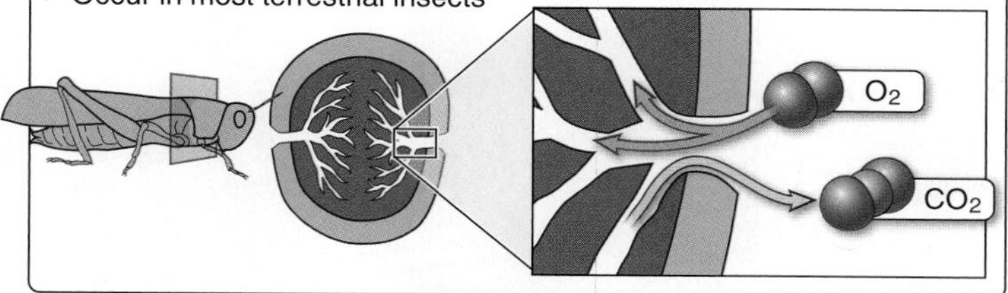

LUNGS

- Internal organs with highly branched, moist surfaces
- Occur in most land vertebrates

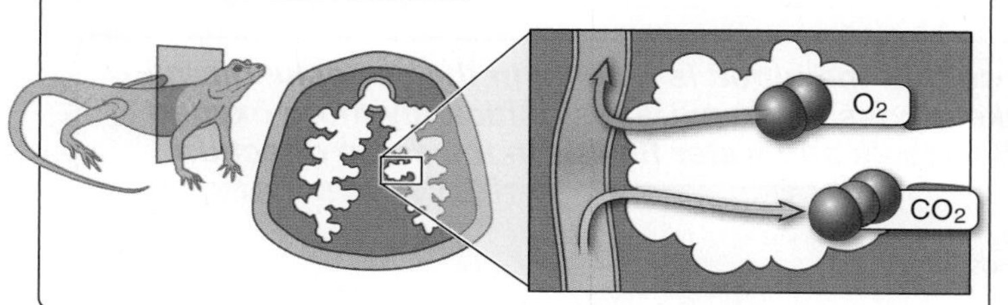

FIGURE 21-23 Gas exchange systems and the body structures that support them.

gills. In all cases, however, it requires a respiratory medium—air or water—that serves as a reservoir for the gases, and a moist respiratory surface on which the gas exchange can occur. The many different solutions for gas exchange that have evolved fall into five categories (**FIGURE 21-23**).

1. ***Direct diffusion.*** Single-celled organisms and many small multicellular organisms with low metabolic demands, such as marine flatworms, can accomplish respiration by direct diffusion between the cells and the environment.

2. ***Protruding respiratory sacs.*** Many slightly larger organisms also have low metabolic needs. Sea stars and other echinoderms have little balloon-like sacs that protrude from the skin—greatly increasing the surface area—and exchange gases between the body cavity and the environment.

3. ***Gills.*** Fishes and many aquatic invertebrates, such as lobsters and clams, have **gills.** These extensions of the body are tremendously elaborated structures in which the large surface area allows extensive exchange of gases between the water and the blood vessels of the circulatory system.

4. ***Tracheae.*** Although insect bodies look rather solid, they have a huge number of tiny openings—spiracles—that lead to tubes that branch extensively throughout the body. These inner tubes make it possible for gases in the air to come in direct contact with most of the organism's cells.

5. ***Lungs.*** Most land vertebrates have **lungs.** These are internal organs, characterized by highly branched, moist respiratory surfaces, across which gases in the air that is breathed in are exchanged with gases dissolved in the blood circulating through the lung tissue. Birds, reptiles, and mammals do virtually all of their respiration through their lungs, but amphibians (such as frogs) also exchange gases through their skin, which serves as a supplemental respiratory system. Their skin stays moist because of their largely aquatic lifestyle.

TAKE-HOME MESSAGE 21•11

In single-celled and very small multicellular organisms, gas exchange can occur by direct diffusion. In larger multicellular organisms, gas exchange is a two-stage process: (1) exchange between the external environment and the organism's circulatory system, which usually takes place in lungs, tracheae, or gills, and (2) exchange between the circulatory system and the cells involved in cellular respiration.

21•12

Gas exchange takes place in the gills of aquatic vertebrates.

Fishes have noses, but they don't use them for breathing. Respiration begins, instead, with a gulp. The fish opens its mouth and sucks in a mouthful of water. It then closes its mouth, opens small holes on either side of its head, and releases, or "exhales," the water. Unlike the air that humans and other mammals breathe in and out, water follows a one-way path into and out of the fish, never changing direction.

Gas exchange takes place as the water passes through the gills, complex structures adapted to extract as much O_2 as possible from water (**FIGURE 21-24**). This extraction is a difficult task, because water has only 5% of the oxygen concentration found in air. The gills generally consist of four bony or cartilaginous gill arches on either side of the head. Similar in appearance to the teeth of a comb, these arches give support to the gill. Long filaments of tissue extend like an accordion from each gill arch, spreading out and creating as much surface area as possible. The filaments are stacks of hundreds of disk-like structures, called **lamellae,** on which the gas exchange takes place. Each membranous lamella is a semicircular disk of elaborately branched capillaries. As water rushes across the gills, it passes between the lamellae, coming in almost direct contact with the capillaries. Because the blood cells are so close to the water, dissolved O_2 can pass from the inhaled water to the blood by direct diffusion, and dissolved CO_2 can pass by direct diffusion from the blood to the exhaled water.

Blood circulation in gills is set up in a simple pattern that is highly efficient at extracting oxygen from the water. In each filament, the blood vessels are arranged so that the blood is moving in the opposite direction from the water flowing past the gills. Called a "countercurrent exchange system," this layout is dramatically more efficient than if the blood flowed in the same direction as the water. If the blood flowed in the same direction, the gills would extract a maximum of about 50% of the oxygen in the water; with the countercurrent system, the gills extract as much as 85% of the oxygen in the water.

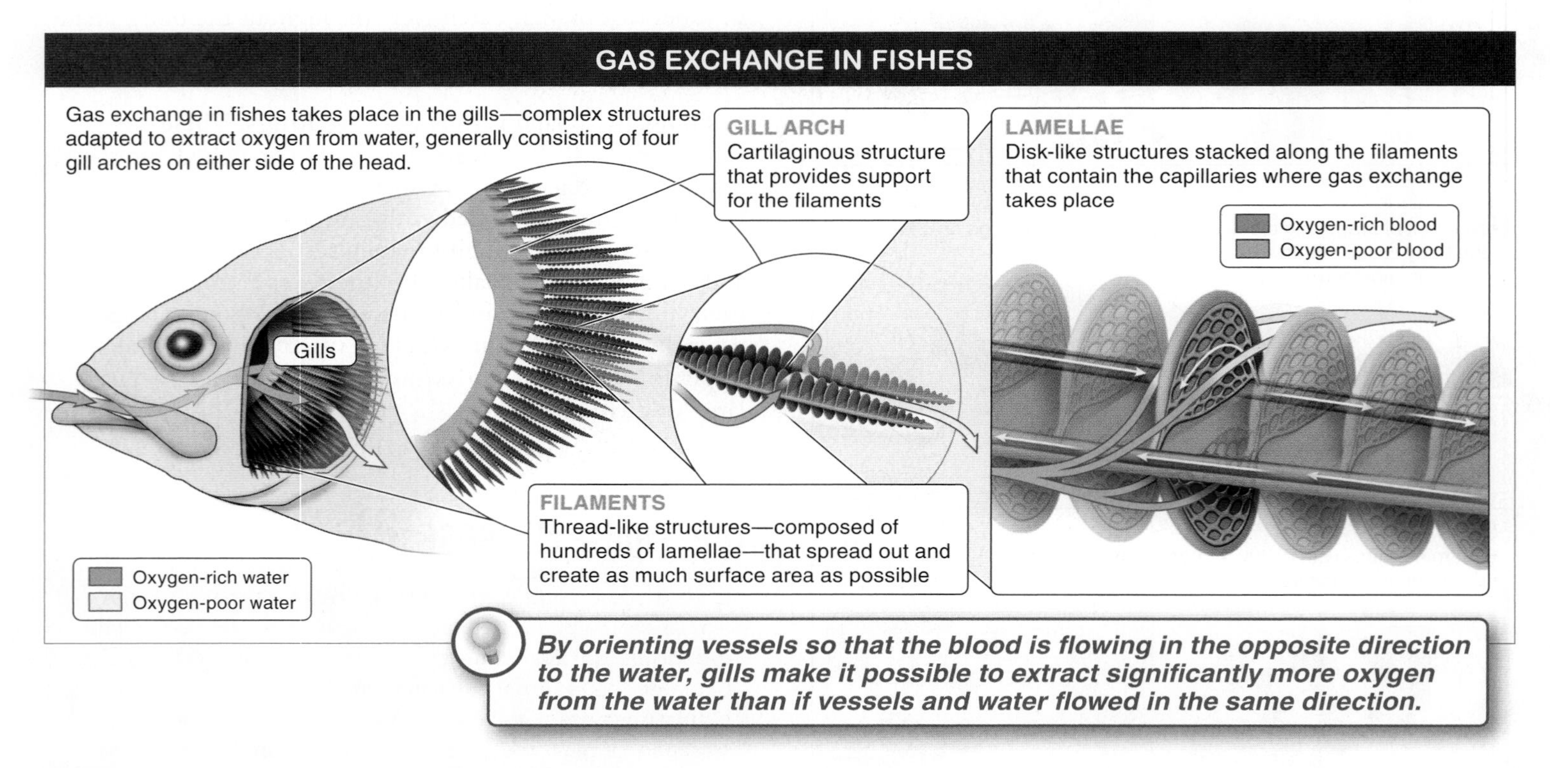

FIGURE 21-24 **Gills.** The remarkable structure of gills makes gas exchange possible in fishes.

COUNTERCURRENT EXCHANGE

COUNTERCURRENT EXCHANGE:
BLOOD AND WATER FLOW IN OPPOSITE DIRECTIONS

In a countercurrent system, water always has slightly more oxygen than blood, so a continuous concentration gradient is maintained, extracting as much oxygen as possible.

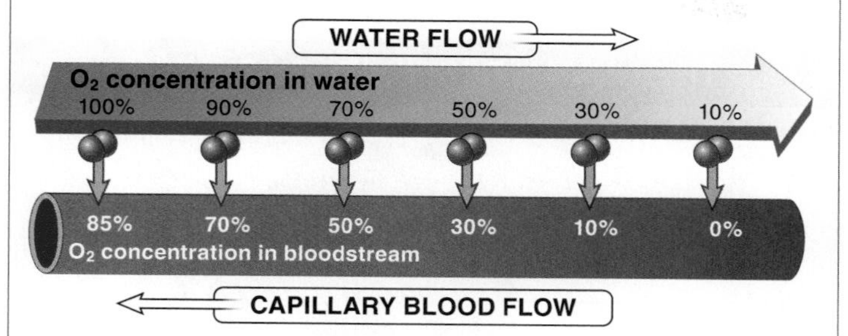

BLOOD AND WATER FLOW IN THE SAME DIRECTION

If blood were to flow in the same direction as the water, it could only become 50% saturated with oxygen, because there is no longer a concentration gradient to enable more diffusion of oxygen.

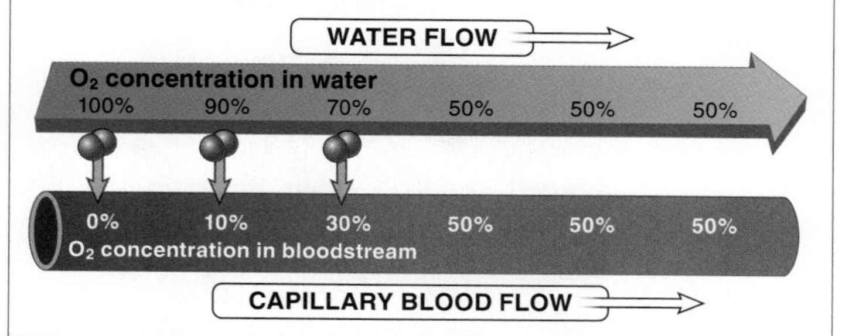

The countercurrent exchange system is dramatically more efficient at extracting oxygen than if the blood flowed in the same direction as the water.

FIGURE 21-25 Improving gas exchange. In countercurrent exchange, the direction of blood flowing through the gills maximizes the uptake of oxygen.

The countercurrent system is so efficient because it maintains a concentration gradient (that is, a difference in concentration) between the water and the bloodstream for the entire time that the water and blood vessel are in close contact (and because the gas exchange is accomplished through diffusion, which is passive, no ATP is required) (**FIGURE 21-25**). Depleted of O_2 in the tissues of the body, blood moving into the lamella encounters water that has been next to capillaries for a considerable distance and has already lost most of its oxygen. The water may only have, for example, 10% of the O_2 it contained when first taken into the fish's mouth. Still, because the blood next to the water has been almost totally depleted of oxygen, the water still has a greater concentration of oxygen and therefore O_2 will diffuse from the water to the blood. Further along through the lamella, the blood has slightly more oxygen. The water, however, also has slightly more oxygen. Consequently, O_2 continues to diffuse from the water to the blood. By the time the blood has traveled almost completely through the lamella, it is holding about 70% or 80% of the maximum amount of oxygen that it can hold. At this point, however, the blood is encountering water that has entered the lamella straight from the fish's mouth, and this water is as saturated with O_2 as it can be. And so, even at this point, there is a concentration gradient and the blood can pick up just a bit more oxygen before returning to the tissues in the fish where it is needed. Carbon dioxide is removed from the fish's circulatory system in the same, countercurrent exchange way.

> "Relationships are like sharks; they have to keep moving or they die.
> —Woody Allen, *Annie Hall*"

Fishes are not the only aquatic animals with gills. Many other groups of vertebrates and invertebrates have gills, including molluscs, such as clams, and arthropods, such as lobsters. Gills range from simple to complex, and there is a great deal of variation in the associated structures. Most sharks, for example, move water past the gills not in the manner described above, used by most fishes, but just by swimming forward all the time. If they stop moving forward, the oxygen in the water surrounding the gills is quickly depleted and the shark will suffocate. Consequently, they spend their whole lives moving forward.

TAKE-HOME MESSAGE 21•12

In aquatic vertebrates, respiration begins when an organism opens its mouth, takes in water, and moves the water out through its gills. Gas exchange takes place in the gills, which extract as much oxygen as possible from the water by maintaining an O_2 concentration gradient between the water and the blood flowing through the gills.

21•13

Respiratory systems of terrestrial vertebrates move oxygen-rich air into the lungs and carbon-dioxide-rich air out of the lungs.

Life on land is very different from life underwater, but the fundamental energetic needs remain. ATP is still the chemical that provides the energy for all the reactions necessary for life, and cells still need oxygen to produce ATP. And consequently, an organism must put air in contact with the cells that need it. Ultimately, oxygen must get into the cells and carbon dioxide must get out. These are universal challenges facing all terrestrial animals.

Terrestrial vertebrates have a general solution to these challenges. First, they suck in air through their mouth or nose. The air moves down a trachea, or windpipe, into lungs. In the lungs, O_2 diffuses from air to blood, while CO_2 diffuses from blood to air. Finally, the oxygen-depleted air is exhaled and the process begins again. The specific design of the lungs and respiratory system varies a bit from one taxon to another—we review the most notable distinctions below—but the general process is the same.

Mammalian respiration begins with a deep breath. Let's trace the air through the respiratory process (**FIGURE 21-26**). Air enters through the nose, filling the nasal cavity, where it becomes warm and moist. Additional air can be taken in through the mouth. In either case, these two entry points for air join together at the throat (also called the pharynx), in the back of the mouth. The air passes through the throat and moves through the voice box, or vocal cords—also called the larynx. The voice box can be seen as the bump on the front of your neck, called the "Adam's apple." From the larynx, the air moves into the trachea. The trachea is a long windpipe that takes the air into the chest cavity. Once there, the trachea splits, with a fork going to the left lung and a fork going to the right lung.

Lungs are like stretchy, elastic bags. Where the trachea splits, the two smaller tubes are called bronchi. These enter the lungs and branch again. And again. And again. With each successive branching, the bronchi get smaller. Under a certain size they are called bronchioles. And they all just keep branching and spreading out. Eventually, the bronchioles reach a dead end. These dead ends are tiny elastic sacs, the alveoli. Here is where the air meets the blood vessels (**FIGURE 21-27**).

There are about 300 million alveoli in each human lung, with a total surface area roughly the size of a movie screen. Alveoli are made up from the most delicate cells in our bodies and have ultra-thin walls. Completely surrounding the alveoli, the way your fingers might completely surround a small ball that you are grasping, are tiny capillaries. The capillaries have extremely thin walls, too. Oxygen in the alveoli dissolves in moisture on the cells lining them. It can then pass right through the two sets of thin membranes—alveolar and capillary—and get picked up by the bloodstream. Simultaneously, carbon dioxide can diffuse from the blood into the alveoli. In the short time you hold it in your lungs, the breath you inhaled is changed. When exhaled, it is depleted of O_2 and laden with CO_2.

THE HUMAN RESPIRATORY SYSTEM

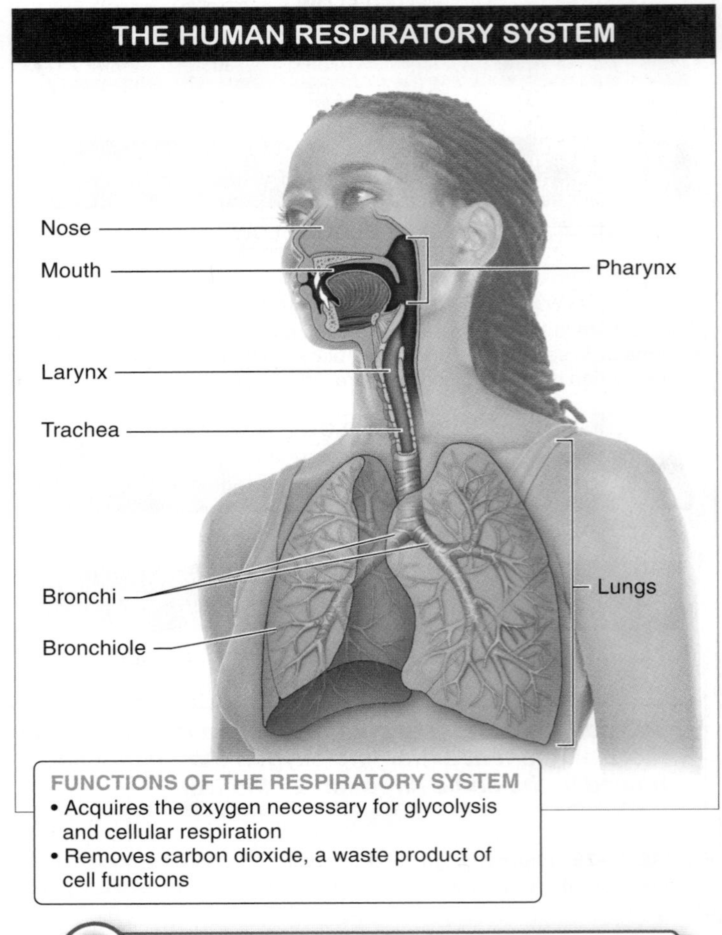

FUNCTIONS OF THE RESPIRATORY SYSTEM
- Acquires the oxygen necessary for glycolysis and cellular respiration
- Removes carbon dioxide, a waste product of cell functions

In terrestrial animals, the respiratory system provides the route for inhaled air to meet the blood vessels of the body.

FIGURE 21-26 A terrestrial mammal. Overview of the human respiratory system.

GAS EXCHANGE THROUGH ALVEOLI

Alveoli
Bronchiole
Blood vessel

Alveoli are the delicate, thin-walled elastic sacs at the end of bronchioles where air meets the blood vessels.

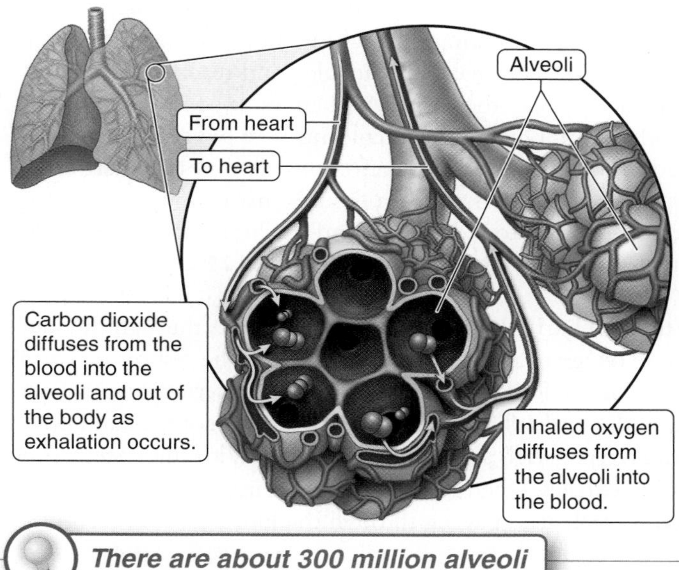

There are about 300 million alveoli in each lung, with a total surface area the size of a movie screen!

FIGURE 21-27 Alveoli in the lungs: where air meets blood vessels.

Amphibians and reptiles are almost identical to mammals when it comes to the respiratory system. The lungs of amphibians are a bit smaller, but amphibians make up for some of this reduced lung capacity by conducting a bit of gas exchange across their skin. This is why they must keep their skin moist at all times. Reptiles are generally too thick-skinned and scaly to achieve any respiration through their skin, but they have slightly larger lungs than amphibians to pick up the slack. Birds, among the terrestrial vertebrates, are the champions of respiratory efficiency. We explore some of their unique adaptations in Section 21.14.

Q How does smoking damage the lungs? Can the damage be reversed?

Smoking introduces thousands of different chemicals into the respiratory system, many of which—such as formaldehyde, ammonia, and benzene—have powerfully destructive effects on its cells. Toxic particles in tobacco smoke can damage the cilia lining the trachea. This reduces the ability to filter out dirt and microorganisms from the air we breathe. The dangerous chemicals can also kill immune system cells that help fight off infections, further reducing our immune response to pathogens. The chemicals in smoke also trigger mucous secretions that can block airways and lead to other respiratory difficulties. After chronic exposure to smoke, the walls of the alveoli become brittle, reducing respiratory capacity. And perhaps most significantly, carcinogenic chemicals in tobacco smoke can trigger unrestrained cell multiplication in lung tissues, causing cancer (**FIGURE 21-28**).

TWINS: SMOKER AND NON-SMOKER

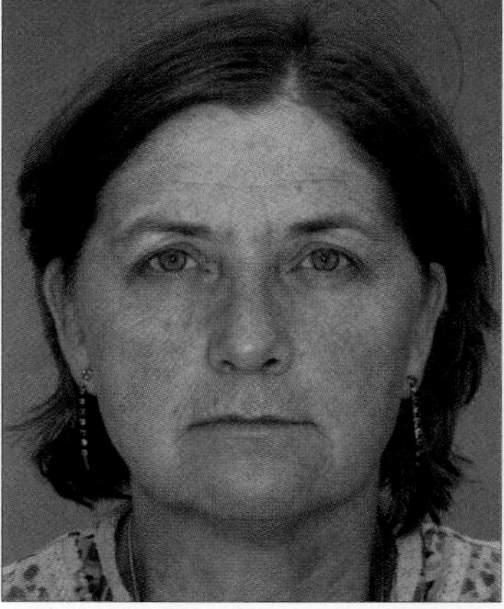

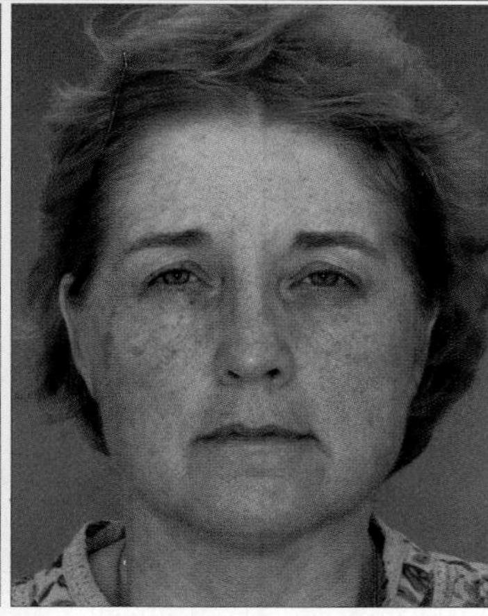

DAMAGE CAUSED BY SMOKING
- Toxic particles damage the cilia lining the trachea.
- Chemicals can kill immune cells that help fight off infections.
- Chemicals trigger mucous secretions that can block the airways.
- Walls of alveoli become brittle, reducing respiratory capacity.
- Carcinogenic chemicals can trigger cancer.

Although smoking is destructive in numerous ways, stopping smoking at any point can begin the process of reversing some of the damage.

FIGURE 21-28 Effects of smoking. One of these identical twins is a smoker. Can you figure out which one?

Although smoking is destructive in numerous ways, causing almost half a million deaths in the United States every year, stopping smoking at any point can begin the process of reversing some of the damage. By the end of the first year of nonsmoking, the risk of death from lung cancer and heart disease begins to decrease, and after 15 years of nonsmoking, the risk of death from these causes returns to the same levels as for individuals who have never smoked.

TAKE-HOME MESSAGE 21•13

In terrestrial vertebrates, respiration begins as air is sucked in through the mouth or nose. The air moves down a trachea into lungs, where O_2 diffuses from the air into capillaries and thus into the bloodstream, while CO_2 diffuses from blood to air. Finally, the oxygen-depleted air is exhaled and the process begins again.

21•14

Birds have unusually efficient respiratory systems.

Birds take breathing to new heights. They often spend time in high-altitude, low-oxygen habitats, and they also fly for long periods of time, necessitating a great deal of oxygen (**FIGURE 21-29**). These extreme needs can be met by several key evolutionary adaptations that make it possible for birds to exchange gases much more efficiently than other terrestrial vertebrates.

Respiratory adaptations have evolved that enable birds to exchange gases with great efficiency—even at high altitudes where oxygen availability is low.

FIGURE 21-29 High-altitude respiration. Adaptations enable birds to function even when oxygen is in short supply.

For starters, birds' respiratory adaptations enable them to keep oxygen-rich air flowing through the lungs twice as long as in mammals. In mammals, air is inhaled and reaches a dead end at the alveoli. On the exhale, it changes direction and is breathed out. During the exhale, no new oxygen is reaching the lungs, a problematic situation during times of exertion. In birds, a more efficient system ensures that oxygen-containing air never runs into a dead end and the lungs never experience "stale" air. Here's how they do it (**FIGURE 21-30**).

When the bird inhales, some of the air passes through the lungs, where oxygen can diffuse into the blood. The rest of the air fills temporary holding structures called the posterior air sacs. Then, when the bird exhales, the posterior air sacs contract and push more oxygen-rich air through the lungs. With this system, even during exhales, when no "new" air is being breathed in, oxygen-rich air is passing through the lungs.

When a bird is inhaling, air moving through the lungs passes into other temporary holding structures, called anterior air sacs. Then, as the bird exhales, oxygen-poor air passing through the lungs *and* oxygen-poor air from the anterior air sacs together move into the trachea, and this mix is expelled from the bird's body.

Q How is bird breathing less like mammal breathing and more like fish "breathing"?

With their circular system made possible by the air sacs and by the lungs' unique pass-through channels (called parabronchi), birds maintain a continuous and unidirectional flow of oxygen-rich air through the lungs. To increase the efficiency of the

CIRCULAR RESPIRATORY SYSTEMS IN BIRDS

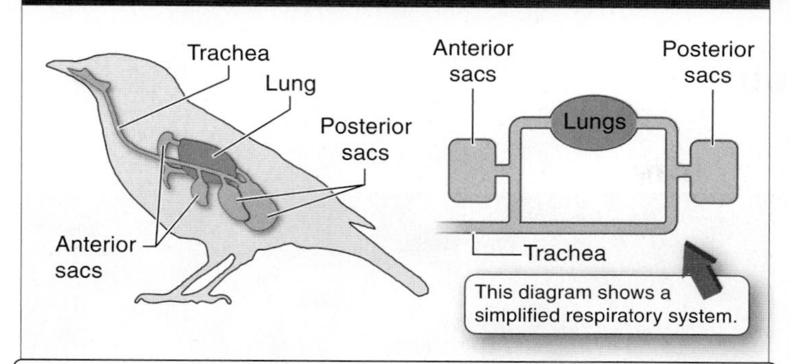

INHALATION

- Fresh, oxygen-rich air moves down the trachea, inflating the posterior "waiting room" sacs, as well as the lungs.
- At the same time, oxygen-poor air is expelled from the lungs, inflating the anterior sacs.

EXHALATION

- The posterior sacs deflate, pushing oxygen-rich air into the lungs.
- The anterior sacs deflate, pushing oxygen-poor air out of the trachea.

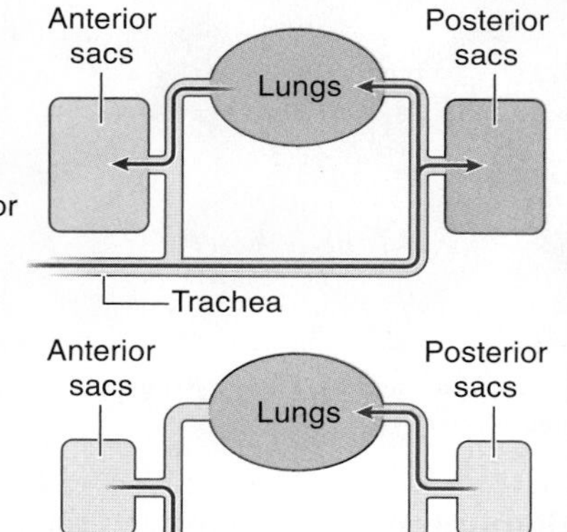

In birds, unlike in humans, air moves in one direction through the lungs. And during both inhalation and exhalation, fresh air continues flowing through the lungs.

FIGURE 21-30 Meeting extreme needs. The highly efficient gas-exchange system of birds supports the oxygen demands of flight.

Adaptations in the bird respiratory system make prolonged flight, even at low-oxygen altitudes, possible.

system even more, all the blood vessels in the lungs are oriented so that the blood is flowing at a 90° angle relative to the direction of the air flow. This use of "cross-current flow" further increases the diffusion of gases between the air and blood. In practical terms, these adaptations make it possible for geese to fly over Mount Everest, nearly six miles above sea level, while a mammal deposited at that height would pass out almost instantly from insufficient oxygen.

TAKE-HOME MESSAGE 21•14

Birds often spend time in high-altitude, low-oxygen habitats and may fly for long periods of time, both of which require a great deal of oxygen. These extreme needs are met by a circular system of air flow and cross-current blood flow in the lungs, which make it possible for birds to exchange gases more efficiently than other terrestrial vertebrates.

21•15

Muscles control the flow of air into and out of the lungs.

Breathe in. Breathe out. It all seems simple enough. You don't even have to think about it. But over the course of your life, this simple process will happen 500 million times without fail. How does it work?

In reptiles, birds, and mammals, breathing occurs in two steps: inhalation and exhalation (**FIGURE 21-31**). The chest cavity is bordered on the bottom by a large sheet of muscle called the **diaphragm,** which separates the chest cavity from the abdominal cavity. The rest of the chest cavity is surrounded by the rib cage and the intercostal muscles between the ribs. During inhalation, these two sets of muscles contract, pulling the diaphragm down and expanding the rib cage. This causes a rapid increase in the volume of the chest cavity and lungs, which causes air to be sucked into the lungs. When the diaphragm and intercostal muscles relax, the chest cavity returns to its original size. This reduction in the volume compresses the lungs and forces air back out of the trachea.

The added buoyancy of lungs full of air makes it challenging to stay submerged under water.

"I'm going to hold my breath until I die." As a child, you may have threatened your parents this way. It's actually not possible to do this, however. Although we can consciously control our breathing to some extent, chemical sensors in the body detect when carbon dioxide levels rise dangerously high, and our brain responds by sending signals to the muscles that control our breathing, spurring them to override our efforts to stop breathing.

THE MECHANICS OF BREATHING

Breathing is made possible by the following structures in the chest cavity:

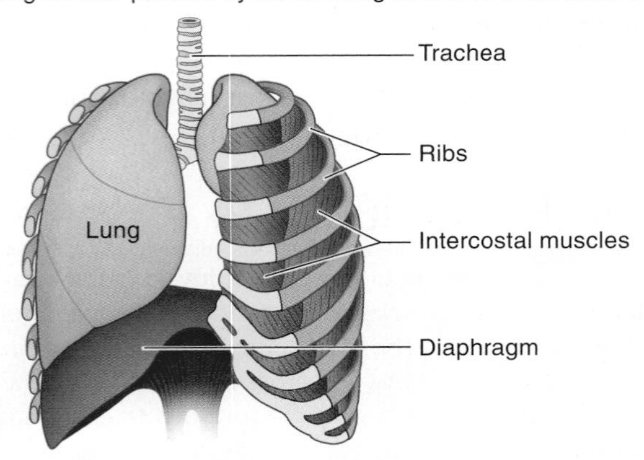

In reptiles, birds, and mammals, breathing occurs in two steps.

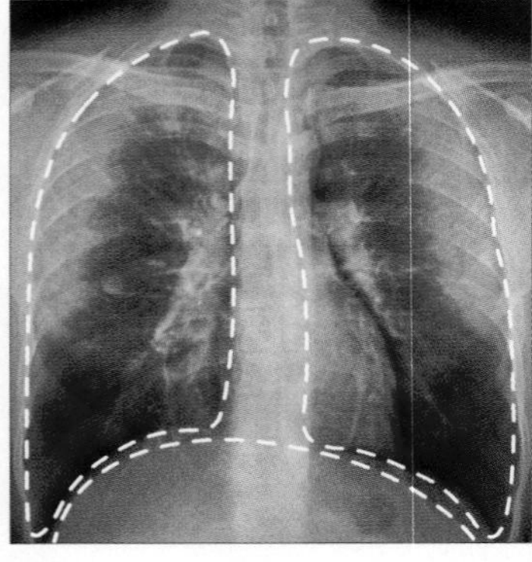

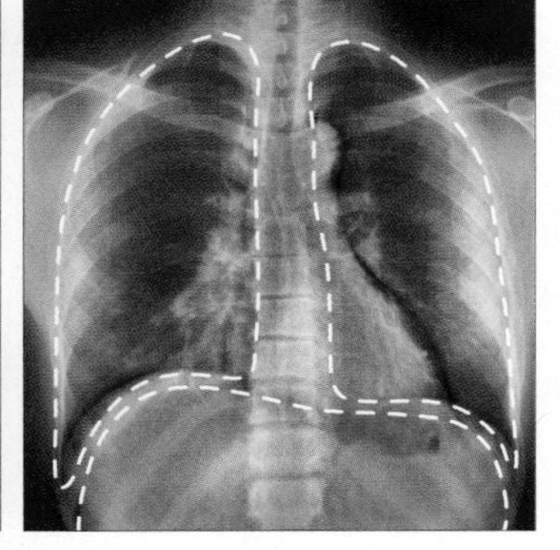

INHALATION
- Diaphragm and intercostal muscles contract
- Diaphragm is pulled lower and rib cage expands
- Air is sucked into the lungs

2 EXHALATION
- Diaphragm and intercostal muscles relax
- Chest cavity returns to its original size
- Air is forced back out to the trachea

FIGURE 21-31 Body structures that make breathing possible.

TAKE-HOME MESSAGE 21•15

In reptiles, birds, and mammals, breathing occurs in two steps: inhalation and exhalation. During inhalation, muscles contract, pulling the diaphragm down, expanding the rib cage, and increasing the volume of the chest cavity and lungs, which causes air to be sucked into the lungs. When the muscles relax, the chest cavity returns to its original size and air is forced out of the lungs.

21•16

Some environments are more conducive to gas exchange than others.

Breathing is more difficult in some places than others. Because the diffusion of gases between the outside and inside of an animal depends on several physical factors, animals vary widely in their respiratory efficiency (**FIGURE 21-32**). The most important constraints on oxygen diffusion rates are temperature, viscosity, and pressure.

1. Temperature: cold versus hot. Keep some cans of beer or soda in a warm car all day and then open them. Why does the fluid fizz more than usual? As the temperature of water (or air) goes up, its ability to hold gases goes down. Whether the respiratory medium is air or water, it doesn't hold as much oxygen when it gets warmer. This means that a fish must work harder to get enough oxygen in warmer water.

2. Viscosity (thickness): air versus water. From a breathing organism's perspective, air is much better than water. Air has much higher dissolved oxygen content than water, and oxygen diffuses from air into blood 8,000 times faster than from water into blood. It also takes less energy to move air than water.

3. Pressure: low versus high altitude. At high altitudes, much less oxygen is available. You don't even need to climb the Himalayas to experience this. A relatively short walk up a much smaller mountain will convince you. At 15,000 feet above sea level, for example, the air pressure is half of what it is at sea level. This is because there is less atmosphere above the air, pushing down on it. With lower pressure from above, less oxygen is squeezed into a given volume. This makes it hard to push air into lungs and to drive the oxygen across the alveoli and capillary membranes and into the blood. Mountain climbers often carry canisters of pressurized oxygen. In the lungs, the high-pressure oxygen can more easily diffuse into the bloodstream.

CONSTRAINTS ON OXYGEN DIFFUSION RATES

Oxygen diffusion—and, hence, oxygen availability to body tissues—is influenced by features such as temperature, viscosity, and (shown here) pressure.

FIGURE 21-32 Gas exchange is more difficult in some environments than others.

TAKE-HOME MESSAGE 21•16

The diffusion of gases between the outside and inside of an animal depends on several physical factors, including temperature, viscosity, and pressure, leading to variations among animals in their respiratory efficiency. The rate of gas exchange is higher in cold (vs. warm) temperatures, in air (vs. water), and at low (vs. high) altitudes.

④ Oxygen is transported and stored while bound to hemoglobin and myoglobin.

Rowers strain to take in enough oxygen to fuel their high-intensity racing effort.

21•17

Hemoglobin is the molecule that transports oxygen.

Red blood cells are filled with hemoglobin. Hemoglobin is like an oxygen "shuttle bus," transporting oxygen around the body. In the lungs, it picks up O_2 and hangs on to it as the blood cell returns to the heart and is pumped to the body. Only when it reaches tissues, such as organs or muscles, that are in need of oxygen but are far from sources of the vital gas does the hemoglobin bus release its O_2 "passengers." The empty hemoglobin then returns to the lungs, where it can load up on oxygen again. It is the oxygen-carrying hemoglobin that gives our blood its red color. When it gives up its O_2, hemoglobin turns more of a purplish-maroonish color. That's why books often show oxygenated blood in arteries as red and deoxygenated blood in veins as blue.

Hemoglobin is a tiny molecule—so tiny that there are about 250 million copies of it in every single red blood cell. Built right inside the blood cell, it remains there for the cell's entire life. Each molecule of hemoglobin is a tangled mass of four polypeptide chains. Nestled within the molecule are four cozy compartments, each of which can carry one molecule of oxygen gas on a seat of iron. This iron attaches to the O_2 that diffuses into the red blood cell, temporarily making it part of the hemoglobin molecule (**FIGURE 21-33**). As we saw above, a shortage of iron in your diet can lead to anemia. This is because when iron is in short supply, less oxygen can be bound by hemoglobin and transported by each red blood cell, causing muscles and organs to be starved of oxygen and leading to feelings of fatigue and weakness.

HEMOGLOBIN

Each molecule of hemoglobin is a tangled mass of four polypeptide chains with four molecules of iron that create four "seats" to which oxygen can attach.

Hemoglobin
Red blood cell
Iron
Oxygen
Polypeptide chains

The hemoglobin molecule is like an oxygen "shuttle bus" that picks up oxygen in the lungs and transports it to tissues.

FIGURE 21-33 Hemoglobin: the oxygen transporter.

Although it is banned by the Olympics and most sports organizations, "blood doping" has been used by some athletes to improve their performance. One method involves withdrawing red blood cells during the weeks and months leading up to a big competition, storing them, and re-injecting them in the few days just before the competition. Because red blood cells are filled with hemoglobin, the oxygen-carrying pigment, blood doping can increase the athlete's capacity for delivering O_2 to his or her tissues. In addition to being against the rules in most competitions, however, blood doping also carries some health risks, because it increases the viscosity of the blood. In the 1990s, dozens of apparently healthy, elite cyclists died inexplicably from heart failure. It was suspected that their blood had become so thick with red blood cells from blood doping that the burden on the heart to pump their sludge-like blood became too great.

Hemoglobin binds to oxygen, but doesn't hold on to it so tightly that it never lets go. Like Post-it notes, which are useful because they are sticky enough to attach to surfaces but not so sticky as to become permanently affixed, hemoglobin "knows" when to bind to O_2 and when to release it. This hinges on something called the partial pressure of the oxygen (denoted as Po_2), the force of oxygen particles in the air pressing against the body (**FIGURE 21-34**).

Where in a body might hemoglobin encounter relatively high or low partial pressures of oxygen? When you breathe air that has a high concentration of oxygen, all four O_2 compartments in hemoglobin eagerly bind to oxygen molecules. Deep in the tissues of your body, though, oxygen is not in great supply—especially if you are exerting yourself and your muscles have been consuming oxygen as they contract. When these tissues become depleted of oxygen, any hemoglobin in the vicinity encounters a low partial pressure of oxygen (Po_2). And what does hemoglobin do when it encounters low Po_2? Because the oxygen it carries is not held too tightly, some of it is released (**FIGURE 21-35**). This oxygen is quickly soaked up by the tissue, which can then continue to generate ATP.

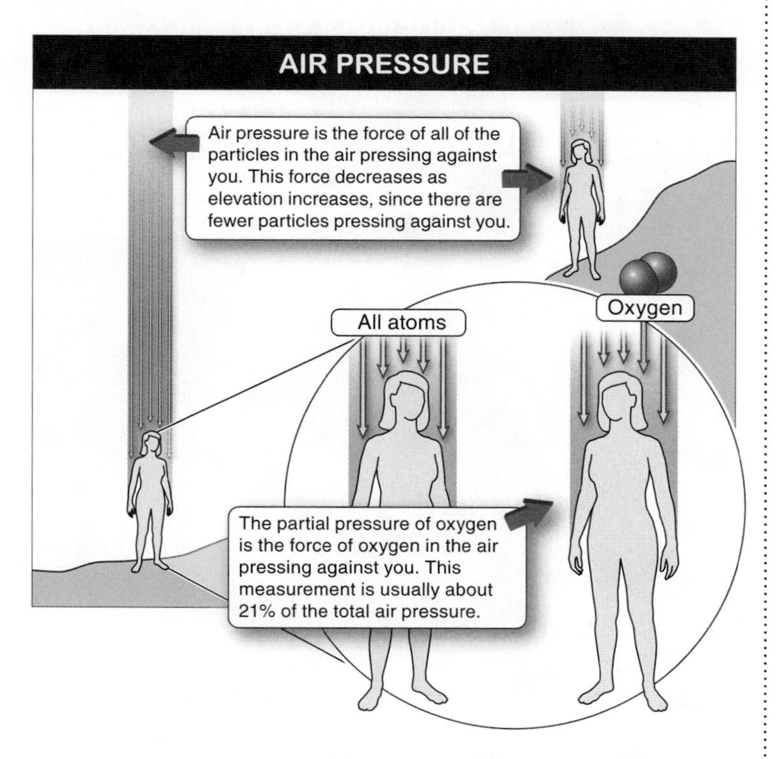

FIGURE 21-34 The force of particles on you. Air pressure and the partial pressure of oxygen are reduced at higher altitudes.

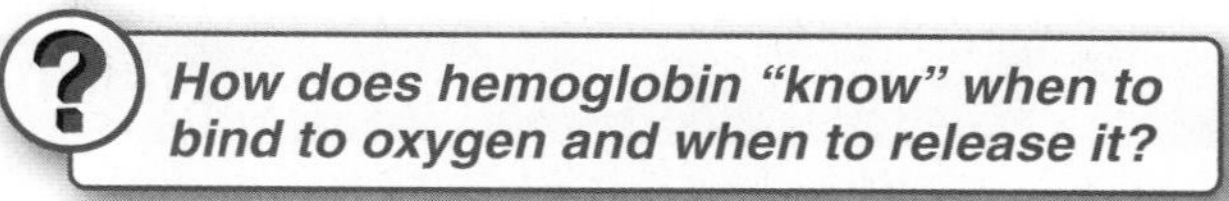

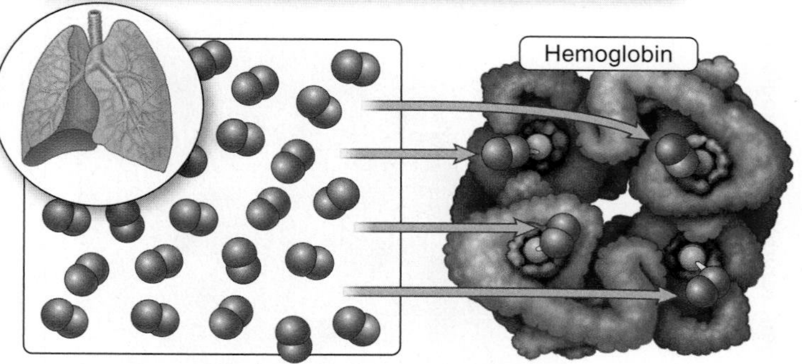

HIGH PARTIAL PRESSURE OF OXYGEN
When hemoglobin encounters a high partial pressure of oxygen, such as in inhaled air in the lungs, hemoglobin gets packed with oxygen.

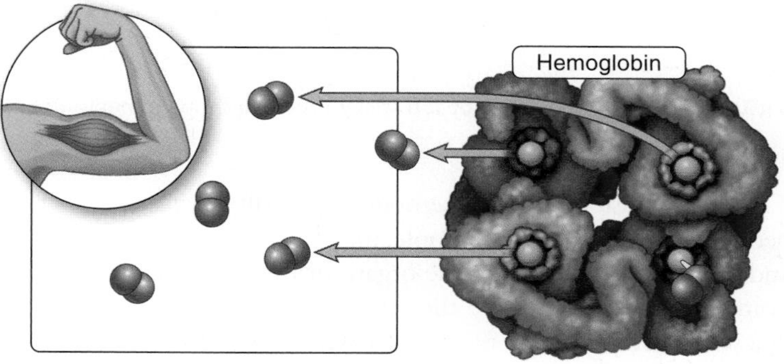

LOW PARTIAL PRESSURE OF OXYGEN
When hemoglobin encounters a low partial pressure of oxygen, such as in active muscle tissue in the body, hemoglobin releases oxygen.

FIGURE 21-35 Hemoglobin binds and releases oxygen, depending on the partial pressure of oxygen in the vicinity.

When you are sitting at your desk, the Po_2 in your tissues isn't very low. Your muscles aren't contracting, and your breathing rate isn't especially high. Much like a car coasting downhill, you aren't using much fuel. In these circumstances, hemoglobin only gives up about one of its four molecules of bound oxygen gas before returning to the lungs to load up again. Back and forth it cycles between getting packed with four oxygens in your lungs and being reduced to three oxygens in your tissues.

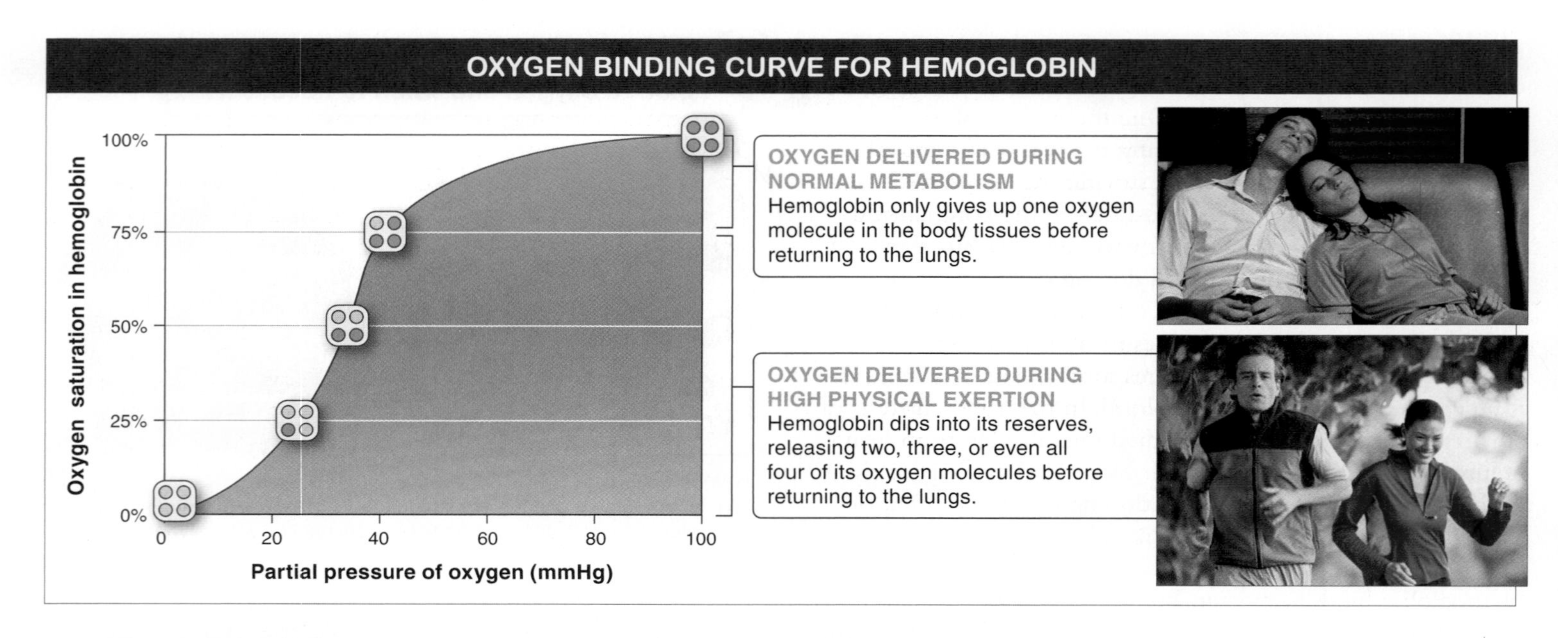

FIGURE 21-36 "Sticky, but not too sticky." A curve showing hemoglobin's affinity for oxygen.

This seems wasteful. If your hemoglobin usually oscillates between picking up a single molecule of oxygen in the lungs and dropping off that O_2 in the organs or muscles, what is the point of its carrying around the other three oxygen molecules? These are its emergency reserves for when you need a lot more oxygen to fuel some energetic activity. If you are exercising vigorously, for instance, the Po_2 in your tissues can drop so low that hemoglobin gives up another of its oxygen molecules. In extreme cases of exertion, it might give up three or even all four of the oxygens that it carries. **FIGURE 21-36** shows an oxygen binding curve for hemoglobin, illustrating the relationship between the Po_2 and the proportion of oxygen molecules that hemoglobin holds onto.

When a woman is pregnant, the growing fetus does not breathe air. This doesn't mean that the fetus doesn't need oxygen, however. During development, oxygen needs are actually very high. How does a fetus get the O_2 it needs? It has to scavenge oxygen molecules released by the mother's hemoglobin. Fetuses do this by producing their own special type of hemoglobin that is a bit stickier than normal adult hemoglobin. At a Po_2 that is low enough that the mother releases oxygen from her hemoglobin, the fetal hemoglobin—with its greater stickiness (or oxygen affinity)—binds to the oxygen. It can then deliver that oxygen to its own, fetal tissues (**FIGURE 21-37**).

If hemoglobin isn't put together exactly right, the health consequences can be serious and painful, as is seen in sickle-cell disease. A single change in the genetic instructions for building hemoglobin causes a malfunction in the hemoglobin molecules (see Chapter 7). When they lose their oxygen molecules—such as when an individual is exercising—the hemoglobin molecules suddenly become misshapen and stick to each other, causing the entire red blood cell to collapse into a sharply pointed sickle. Once sickled, red blood cells cause a whole host of problems. Many break open, which can cause anemia if they aren't replaced promptly. Others clump together, often blocking

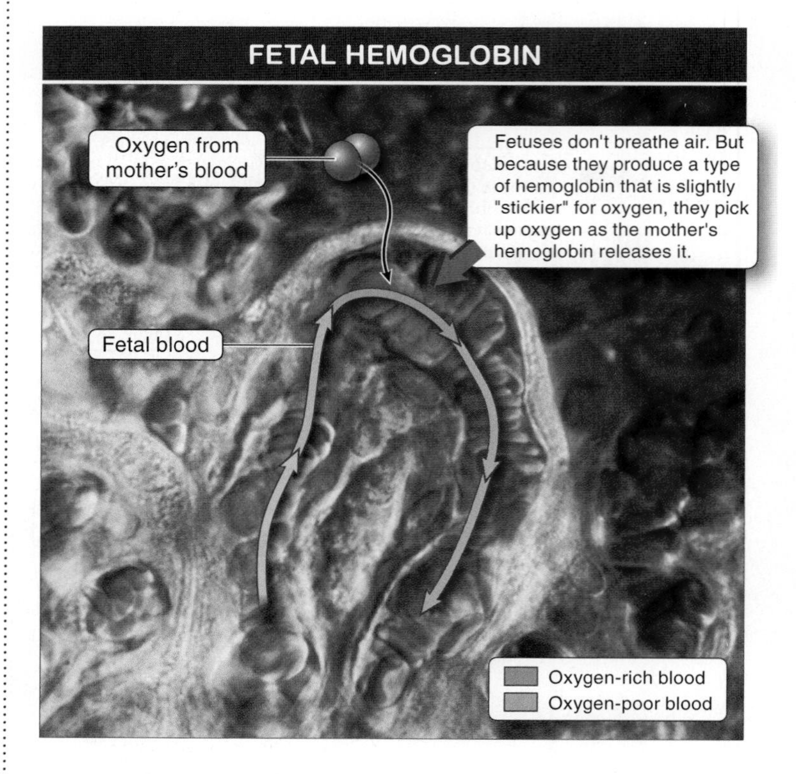

FIGURE 21-37 How a fetus gets oxygen. The fetus produces hemoglobin that binds oxygen released by the mother's hemoglobin.

capillaries where, normally, they must pass through one at a time. This leads to intense pain, especially in the joints and muscles, and can cause strokes if it occurs in the brain. About 70,000 people in the United States live with sickle-cell disease. They are able to minimize the effects by avoiding strenuous activity or other situations in which the Po_2 in their muscles and other tissues drops too low.

Q What is carbon monoxide poisoning?

In an unfortunate coincidence, carbon monoxide (CO) also binds to hemoglobin, but with a higher affinity than oxygen does. In areas with high carbon monoxide concentrations—such as around a faulty furnace or a kerosene heater or lamp without adequate ventilation—the carbon monoxide will out-compete oxygen for hemoglobin's binding sites and it doesn't adhere like a Post-it note, rather, the binding is irreversible. Thus, when the hemoglobin travels to the body tissues, it has no oxygen to release and, in the absence of O_2, cellular respiration cannot generate the ATP the tissue needs (see Chapter 4). Consequently, the tissue is suffocated even as the person takes deeper and deeper breaths.

TAKE-HOME MESSAGE 21•17

Red blood cells are filled with hemoglobin, a molecule that picks up oxygen in the lungs and transports it around the body, releasing it in organs and tissues, such as muscles, where it is needed for cellular respiration.

21•18

Myoglobin in muscles holds a reservoir of oxygen for times of exertion.

Oxygen isn't the fuel that makes muscles contract, but it is an essential component of the process. You will quickly cramp up if your muscles try to work for long in the absence of oxygen. To increase the oxygen available beyond the amount delivered by hemoglobin, muscles have a trick for storing the valuable molecule. It's called **myoglobin,** and it is a protein embedded in the muscle cells themselves (**FIGURE 21-38**). Like hemoglobin, myoglobin is an oxygen-binding molecule. It's a smaller, simpler molecule, though, and has only a single compartment for storing oxygen. And because of its structure, myoglobin has a higher affinity for O_2 than hemoglobin does. This means that at moderately low Po_2, hemoglobin releases oxygen that can quickly be taken up by myoglobin. The myoglobin then just holds on tightly to the oxygen molecule, releasing it only under conditions of extremely low Po_2. Think of it as a last gasp of air for an animal's muscles. As a rabbit runs full speed away from a lynx, its muscles must burn fuel at a very high rate; the value of just a tiny extra capacity can be the difference between life and death. Because myoglobin is a darkly pigmented protein, muscles with higher concentrations of myoglobin are darker. This is what distinguishes "white meat" from "dark meat," a distinction that generally reveals how metabolically active a particular muscle is. Turkeys hardly use their breast muscles, because they don't fly, but they do use their leg muscles all day. This is why breast meat is white and thigh meat is darker. We'll see later in the chapter that deep-diving marine mammals also make use of the O_2 stored in myoglobin during their long dives (as do long-distance migrating birds on their long flights).

Q What is the difference between white meat and dark meat?

MYOGLOBIN

Myoglobin is a hemoglobin-like molecule within muscle tissue. It can hold a single oxygen molecule, which it releases only under extremely low-oxygen conditions—generally during exertion—when the muscles need it most.

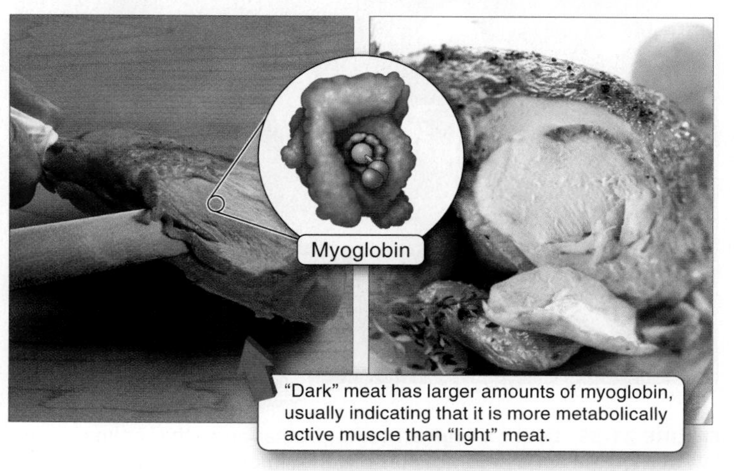

FIGURE 21-38 Oxygen reserves. Myoglobin is a protein in muscles that stores oxygen for times of extreme exertion.

TAKE-HOME MESSAGE 21•18

Myoglobin is an oxygen-binding protein embedded in muscle cells that can release one molecule of oxygen under conditions of extremely low Po_2.

5 Evolutionary adaptations maximize oxygen delivery.

Adapted to high altitudes, yaks are able to transport goods across mountain passes for farmers and Himalayan trekking expeditions.

21•19

Animals living at high altitude have special adaptations to the low-oxygen conditions.

As we saw above, because there is less atmosphere "pushing down" on air at high altitudes, the Po_2 is significantly reduced. It becomes hard to breathe, and activity is difficult. We struggle at high altitudes, because our hemoglobin isn't designed to pick up oxygen at such a low Po_2; it isn't sticky enough. Mountain climbers can bring canisters of pressurized oxygen with them, but llamas and other animals living at high altitudes can't. How do they survive with less available oxygen? Their hemoglobin has a difficult task: it must pick up oxygen at very low pressure—precisely when hemoglobin is supposed to have a low affinity for oxygen—so that it can release it to muscles.

Living at altitudes of 5,000 meters, llamas have solved this problem by producing a slightly different form of hemoglobin, one that is adapted to low-oxygen conditions. At any given Po_2, llama hemoglobin has a higher affinity for oxygen than human hemoglobin (**FIGURE 21-39**). It's stickier. This stickiness enables it to become saturated with four molecules of oxygen even when the llama breathes in

BREATHING AT HIGH ALTITUDES

Llamas can thrive in high-altitude, low-oxygen environments, because they produce a "stickier" form of hemoglobin that has a higher affinity for oxygen.

FIGURE 21-39 Built for high altitudes. Llamas and other animals that live at high altitudes have adaptations that improve respiration in low-oxygen environments.

relatively "poor" air. Much like the oxygen binding curve of fetal hemoglobin, in llamas, the oxygen binding curve is shifted so that the hemoglobin is stickier for O_2, relative to adult human hemoglobin, at most Po_2 values. Because the Po_2 in llama tissues is even lower than in the air they breathe (because oxygen is being consumed by cellular respiration), the oxygen bound to the hemoglobin is still released to muscles to allow normal activity.

TAKE-HOME MESSAGE 21•19

At high altitudes, the Po_2 is lower, making breathing and activity difficult. Animals living at high altitudes solve this problem by producing a form of hemoglobin that has a higher affinity for oxygen, becoming saturated with four molecules of oxygen even when breathing in air with a low Po_2.

21•20

Humans become acclimated to low-oxygen conditions.

Is it just a coincidence that an unusually large percentage of the world's most accomplished mountain climbers are Sherpas, who live year-round in high-altitude Nepal? No. Although llamas may be built for high altitudes from the day they are born, humans acclimate well, and Sherpas are among the most impressively acclimated high-altitude dwellers (**FIGURE 21-40**).

You don't need to live on Mount Everest, however, to become physiologically acclimated to the low-oxygen conditions found at high altitudes. Athletes around the world have long known that they can increase their strength, speed, and stamina by training at high altitudes. And, as with llamas, human acclimation to low-oxygen conditions comes from modifications (though they are acquired rather than inherited) to hemoglobin.

Training for three to five weeks at high altitudes—usually 6,000–7,000 feet in elevation—triggers several physiological changes, the process of acclimation. These changes include stimulating the production of additional red blood cells, increasing blood and capillary volume, and increasing the number of mitochondria. High-altitude training also causes an

FIGURE 21-40 Humans acclimate to high-altitude living with improved respiratory efficiency.

FIGURE 21-41 Improving athletic performance by training at high altitude.

increase in a chemical called diphosphoglyceric acid (DPG) in red blood cells. This acid combines with hemoglobin in the red blood cells and alters the protein's shape ever so slightly. In doing so, it reduces hemoglobin's stickiness, giving it a lower affinity for oxygen. With this reduced oxygen affinity, at any Po_2, the DPG-modified hemoglobin releases more of the O_2 that it carries (**FIGURE 21-41**). Athletes who train at high altitudes find that with the additional oxygen released to their tissues, they can improve their performance by about 3%. This difference was manifested dramatically during the 1968 Olympics in Mexico City (altitude 7,500 feet) when the top five finishers in the 10,000-meter race were all year-round high-altitude residents. The benefits of high-altitude training remain after an athlete returns to sea level—increasing performance significantly. But just as humans become acclimated to high altitudes, we also become acclimated to low altitudes, and the DPG level (and the performance enhancement it can bring) is reduced about three to five weeks after returning to sea level.

TAKE-HOME MESSAGE 21•20

Humans living at high altitudes become acclimated to low-oxygen conditions over the course of three to five weeks. This acclimation includes increasing the production of diphosphoglyceric acid (DPG) in red blood cells and thereby reducing hemoglobin's affinity for oxygen, leading to release of higher levels of oxygen to muscles during exertion.

21•21

Deep-diving mammals are masters of efficient oxygen use.

How long can you hold your breath? If you're like most people, you can last just over a minute, or maybe 2 or 3 minutes at best. The world record is an amazing, but brain-choking, 11 minutes and 35 seconds held by Stéphan Mifsud of France, set in June 2009. Even this pales, though, when compared with the abilities of the Weddell seal (**FIGURE 21-42**). Living in Antarctica, these 900-pound mammals regularly make dives 200–500 meters deep in search of fish to eat. Usually staying underwater for 20 minutes, but sometimes for more than an hour, they are able to hold their breath for so long only as a consequence of numerous evolutionary adaptations.

Four critical respiratory and circulatory features make the seals' deep diving possible.

1. They have about double the volume of blood, per kilogram of body weight, that humans have. Much of it is stored in their unusually large spleen, which contracts during a dive, putting significantly more oxygen-carrying blood into circulation.
2. The myoglobin concentration in their muscles is about twice that in other mammals, serving as an additional storehouse of oxygen that can be utilized during dives.

13. Which of the following are the main muscles used in breathing?
a) back muscles and thoracic muscles
b) intercostal muscles and pectoral muscles
c) diaphragm and thoracic muscles
d) diaphragm and back muscles
e) intercostal muscles and diaphragm

14. With increasing temperatures:
a) aquatic animals must exert more energy to obtain oxygen.
b) aqueous environments contain more dissolved oxygen.
c) aquatic animals require less oxygen.
d) an aquatic animal's metabolism stabilizes.
e) all of the above occur.

15. Under regular metabolic conditions, what percentage of oxygen is released from the blood to the tissues?
a) 1%
b) 75%
c) 25%
d) 100%
e) 50%

16. Why is carbon monoxide (CO) so dangerous?
a) CO prevents hemoglobin from binding to and transporting O_2 to the body tissues, resulting in oxygen starvation and death.
b) CO prevents CO_2 from dissolving in the blood, and CO_2 builds up to toxic levels.
c) CO modifies the structure of hemoglobin, making it more difficult for O_2 to be transported and released in the body.
d) CO coats the inside of the alveoli, preventing diffusion of O_2 from the lungs into the blood.
e) CO binds to myoglobin, so tissues cannot remove O_2 from the red blood cells and the tissue dies.

17. Myoglobin:
a) is the chief O_2 carrier in mammalian circulatory systems.
b) transports O_2 from the lungs to the body tissues.
c) releases O_2 to the muscles when partial pressure is such that hemoglobin cannot do so.
d) is a four-subunit protein containing iron.
e) has a lower affinity for O_2 than hemoglobin does.

18. Which of the following has the highest binding affinity for oxygen?
a) hemolymph
b) human hemoglobin
c) llama hemoglobin
d) myoglobin
e) All have equal binding affinity for oxygen.

19. Why does llama hemoglobin have a higher oxygen-binding affinity than human hemoglobin?
a) Like humans, llama fetuses use a form of hemoglobin with a higher binding affinity than the mother's hemoglobin, but unlike humans, llamas continue to use this form of hemoglobin after birth.
b) Because llamas are more active than humans, they require hemoglobin with a greater maximum oxygen saturation.
c) Llama blood has a higher CO_2 concentration than human blood, resulting in a need for hemoglobin with a higher binding affinity.
d) Llamas live at higher elevations than most humans and require a means to bind O_2 when oxygen is at a lower concentration in the air.
e) Llamas lack myoglobin and need a way to ensure enough O_2 is available to their muscle tissues.

20. Diphosphoglyceric acid (DPG):
a) causes the release of more O_2 to body tissues.
b) makes it possible for blood to carry more CO_2.
c) is the main reason that breathing is difficult at high altitudes.
d) increases the binding affinity of adult hemoglobin for O_2.
e) encourages the production of more hemoglobin.

21. Which of the following is not part of the "diving reflex" of the Weddell seal that allows seals to remain underwater for long periods of time?
a) Weddell seals have double the muscle myoglobin concentration compared with other mammals.
b) Weddell seals have double the blood volume (per kg of body weight) compared with other mammals.
c) Weddell seals have an increased resting heart rate, causing more efficient oxygen usage.
d) During dives, Weddell seals constrict blood vessels in most tissues, sending blood only to parts of the body that need it most.
e) All of the above make deep diving possible by the Weddell seal.

Short-Answer Questions

1. Why has the evolution of multicellular organisms led to the development of circulatory systems? What characteristic distinguishes closed and open circulatory systems?

2. What is the difference between single- and double-circuit cardiovascular systems? Which would be better suited for a higher energy demand? Why?

3. Name the three main types of cells (or cell fragments) that are found in blood and describe their primary functions.

4. How is gas exchange accomplished in organisms with gills and in those with lungs?

See Appendix for answers. For additional study questions, go to www.prep-u.com.

1 Food provides the raw materials for growth and the fuel to make it happen.

- **22.1** Why do organisms need food?
- **22.2** What's on the menu? Organisms have a variety of diets.
- **22.3** Calories count: organisms need sufficient energy.

2 Nutrients are grouped into six categories.

- **22.4** Water is an essential nutrient.
- **22.5** Proteins in food are broken down to build proteins in the body.
- **22.6** Carbohydrates and lipids provide bodies with energy and more.
- **22.7** Vitamins and minerals are necessary for good health.
- **22.8** How do organisms "know" what to eat?

3 We extract energy and nutrients from food.

- **22.9** We convert food into nutrients in four steps.
- **22.10** Ingestion is the first step in the breakdown of food.
- **22.11** Digestion dismantles food into usable parts.
- **22.12** Absorption moves nutrients from your gut to your cells.
- **22.13** Elimination removes unusable material from your body.
- **22.14** Animals have some alternative means for processing their food.

4 What we eat profoundly affects our health.

- **22.15** What constitutes a healthy diet?
- **22.16** Obesity can result from too much of a good thing.
- **22.17** Weight-loss diets are a losing proposition.
- **22.18** Diabetes is caused by the body's inability to regulate blood sugar effectively.
- **22.19** Food and infection: spicy foods are natural antibiotics.

22

Nutrition and Digestion

At rest and at play: optimizing human physiological functioning

1 Food provides the raw materials for growth and the fuel to make it happen.

Living organisms need raw materials and fuel to function.

22•1

Why do organisms need food?

Lamb roulade with leek basil stuffing and a red wine sauce. Chili fries. Lime-marinated tofu kabobs. A big vanilla milkshake. We dress food up in an almost infinite variety of combinations. We garnish it, we heap on condiments, we daydream about it, and we even write books about it. But despite all the fuss, it really boils down to two simple biological needs: raw materials and fuel. Just as a carpenter needs wood and a car needs fuel, living organisms need raw materials and fuel to function.

Here's why: every second of every day, chemical reactions are occurring inside us. We build complex molecules. We reproduce. We respond to stimuli. And just as a car won't roll uphill without a push, most of these chemical reactions won't occur spontaneously. With a little fuel in the motor, however, the car can easily be driven up a hill. So, too, can we, and all the other animals, do things that are energetically costly. Our need for fuel is one of the two reasons why we must eat: building a body, moving around, reproducing, and just staying alive all require energy. Food provides that energy. And the other reason is that to grow and build the myriad complex molecules required for life, we need raw materials: molecules of carbon and nitrogen and phosphorus, to name just a few. Food provides these raw materials.

What exactly happens to the food we eat? We'll examine this question in detail throughout the chapter. In **FIGURE 22-1**, though, we can see that the food we eat is physically and chemically broken down into its fundamental macromolecular components in the process of **digestion.** Regardless of the form food takes when we eat it, our body quickly breaks it down and separates it into the usable and unusable. The usable materials are carbohydrates, lipids, proteins, vitamins, minerals, and water. These six groups of **nutrients** are the substances that are used for energy, raw materials, and maintenance of the body's systems. Anything we consume that doesn't fall into

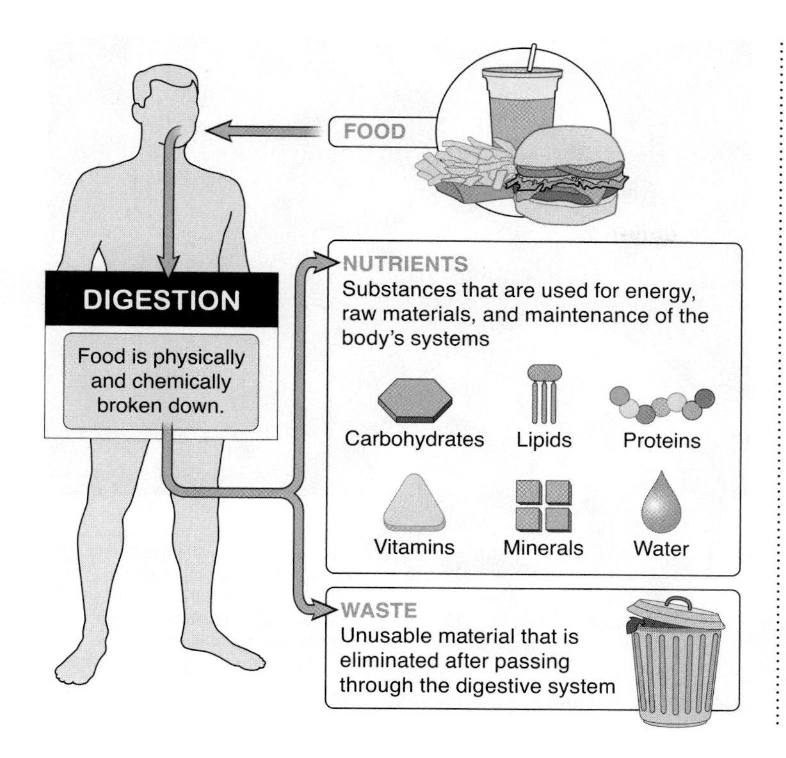

FIGURE 22-1 Made possible by food. Living organisms need raw materials and fuel to function.

one of these categories is unusable. You can't get nutrients from dirt or rocks. They just pass through the digestive system and are eliminated.

Ultimately, an organism's body weight represents the balance between the energy carried within the molecular bonds of the food it consumes and the energy that is burned in the process of living. Any surplus calories are stored, usually as fat or as glycogen (see Section 2-9), a form of carbohydrate stored primarily in muscle and liver tissue.

TAKE-HOME MESSAGE 22•1

Animals must eat for two reasons: to acquire the energy needed for all growth and activity, and to acquire the raw materials required for life.

22•2

What's on the menu? Organisms have a variety of diets.

> **Q** Can plants be carnivorous?

All organisms need food. You've got two choices for how to get it. You can make it yourself, as plants and other photosynthetic organisms do through photosynthesis. Or you can eat another organism. Most animals eat the plants that make their own chemical energy, while some animals eat other animals (that eat plants). Some animals even eat animals that eat animals that eat animals (that eat plants)—but that's pretty rare (see Section 15-7). In every case, though, the choice is the same: capture solar energy to make food, or acquire that energy indirectly by consuming other organisms.

In the animal world, species fall into three groups based on their diets (**FIGURE 22-2**).

Carnivores The **carnivores** are predatory animals that consume only other animals. They include spiders and snakes; mammalian species such as wolves, seals, bats, and cats; as well as hawks, owls, and other birds of prey. Some carnivores don't actually kill their prey but instead just suck nutrient-rich fluids from them. Mosquitoes and ticks are carnivorous fluid-feeders.

Carnivory isn't just for animals. The Venus flytrap is a plant that captures and consumes insects. It does this with highly elastic, curved leaves. Water pressure inside the V-shaped leaf causes it to expand into an unstable, stretched-open position that is on the verge of snapping closed. When an insect or spider lands on the leaf, it triggers an electrical signal that causes a brief, rapid change in the water pressure inside the leaf. The change in water pressure is enough to cause the leaf to snap shut, capturing the insect. The plant then releases digestive juices that break down the insect into its usable nutrients. Venus flytraps occur in places where the soil is poor and, especially, is low in nitrogen, which they are able to extract from the animal prey they capture. The green leaves and stems of Venus flytraps, however, should serve as a reminder that these plants are also photosynthetic, harnessing the energy in sunlight to build sugars.

Herbivores Food: easy to get, but hard to digest. That sums up life as a **herbivore,** an organism that consumes only plants. Because plants are plentiful in most habitats and can't run away, they are easy targets for predation. To protect themselves, however, plants tend to carry many toxic compounds that are difficult or impossible for animals to break down chemically. Nonetheless, many herbivores have developed digestive adaptations to overcome these difficulties. We'll explore a couple of them in detail in Section 22-14. Examples of herbivores include sea turtles and caterpillars. They also

TYPES OF ANIMAL DIETS

CARNIVORES
Animals that consume only other animals. (Shown here: a rock python swallows a gazelle.)

HERBIVORES
Animals that consume only plants. (Shown here: a West-Indian manatee eating the plant hydrilla.)

OMNIVORES
Animals that consume both plants and animals. (Shown here: a young raccoon snacks on a plant; raccoons have a very broad diet that also includes insects, worms, fish, amphibians, and more.)

FIGURE 22-2 Animal diets: carnivores, herbivores, and omnivores.

include many species of seed-eating birds and nut-eating squirrels and a variety of large grazing mammals such as cows, horses, and deer. And just as there are fluid-feeding carnivores, there are fluid-feeding herbivores. Aphids, for example, pierce the surface of plants and suck their sugary sap.

FIGURE 22-3 Fungus-farming ants. Leaf cutter ants harvest leaves in order to grow fungus, which they then consume.

Omnivores Most humans, as **omnivores,** eat plants and/or animals and can digest both efficiently. We share this diet with numerous other species, including cockroaches. Omnivores come in all sizes, from bears to raccoons, from chickens to flies and wasps.

> **Q** Humans grow and tend to plants and animals with the sole purpose of eating them. Are there any other species with agriculture?

Watching farmers cultivate plants and animals for food, it might seem reasonable to assume that humans are unique in this endeavor. Agriculture is not, however, limited to humans. Leaf cutter ants chop up leaves, but rather than eating them, spit out the mulch, mold it into a honeycomb shape, and then grow fungus on the mulch (**FIGURE 22-3**). They then eat the fungus at their leisure. These ants are the only species that scientists know of other than humans with a system of agriculture (and they've been doing it for significantly longer than humans have).

TAKE-HOME MESSAGE 22•2

All animals require food. Plants and other photosynthetic organisms produce food through photosynthesis, and animals have one of three types of diet. Carnivores consume only other animals. Herbivores consume only plants. And omnivores, including most humans, consume both plants and animals.

22·3

Calories count: organisms need sufficient energy.

Have you ever gone a whole day without eating? Or have you ever spent a few weeks trying to reduce your caloric intake? Living in a state of hunger can be torturous. In the early 1990s, eight human "guinea pigs" learned this the hard way. Living in the Biosphere 2 dome—a self-contained, 3.2-acre world filled with plants and animals that was designed to explore sustainable living with minimal environmental impact—they took part in an experiment on the effects of a low-calorie diet, among other things. The findings were dramatic but not surprising: all the participants lost weight, but they also became very unhappy. They argued constantly, got into ugly food spats, and frequently squabbled over dinner portions. After leaving what they dubbed "the hunger dome," one of the eight said, "If we ever all start talking to each other, that would be a major accomplishment" (**FIGURE 22-4**). In an earlier study, people were kept hungry for six months. Over time, study participants filled increasing hours with food fantasies. Recipes even displaced sex as the favorite topic of discussion. Put simply, humans (and all other living organisms) need a steady and sufficient flow of energy to function well.

Insufficient energy input for organisms always leads to the same outcome: death. As we saw in Section 8-1, when 27-year-old Bobby Sands went on a hunger strike in 1981, he gradually deteriorated and ultimately died after 66 days without food. In 2001, two prisoners in Turkey, Cafer Tayyar Bektas and Huseyin Kayaci, also went on a hunger strike to protest the prison conditions. Because they started fasting at much higher initial body weights, they survived for much longer than Bobby Sands. The end result was the same, though. After 200 and 148 days respectively, both men died.

FIGURE 22-4 Hungry Biosphere 2 crew. Living on a low-calorie diet led to weight loss and a cranky disposition for Biosphere 2 participants.

So what do we actually need to survive? And why isn't it the same for everyone? In this section, we examine the body's energy needs, and then the various nutrients necessary for growth and maintenance.

We measure the energetic value of food in very small amounts of energy called **calories,** where a single calorie is the energy required to raise the temperature of 1 gram of water by 1° C. However, the term can be a bit confusing: in discussing human consumption, the term "calorie" actually refers to a **kilocalorie** (**kcal**), which is 1,000 calories. So when the label on a package of cookies says that a cookie provides "100 calories," the cookie actually provides 100 *kilocalories* or 100,000 *calories.*

As we saw from the hunger strikers, there is tremendous variation in the amount of stored energy an individual carries. Similarly, there is much variation in how many calories an individual needs each day. To determine basic energy needs, we first factor out any variation due to differences in activity levels—searching for food, finding a mate, defending a territory—and instead assess the minimal energetic needs of an individual who is doing nothing more than the equivalent of sitting on a couch all day. This is called the **basal metabolic rate,** or **BMR,** and refers to the amount of energy expended at rest, with no food in the digestive tract, in a neutral-temperature environment. For humans, the BMR can be roughly approximated as 1 calorie per hour per gram of body weight, or about 1,400 kcal/day for a woman (weighing 120 pounds) and about 1,700 kcal/day for a man (weighing 160 pounds). The difference is mostly a function of the difference in body size; the male has more cells to power. The human BMR is approximately the energy needed to keep a 75-watt light bulb burning for 24 hours or to drive a small car one mile.

Basal metabolic rate allows easy comparisons across species because it is so clearly defined. From a biological perspective, though, it is almost meaningless. To realistically evaluate organisms' energy needs, we must have a sense of how active they are. Someone moving around and involved in physical exertion will require more energy than someone sitting at a desk all day, and a very large person will require more than a small person (**FIGURE 22-5**). In actuality, individuals need about

COMPONENTS OF ENERGY EXPENDITURE

Average kcal/day to cover the minimal energetic needs of an organism at rest

Female 1,400

Male 1,700

BASAL METABOLIC RATE (BMR)
The minimal energy expenditure of an organism at rest.

ACTUAL DAILY ENERGY EXPENDITURE
The basal metabolic rate *plus* the energy required for all activity.

FIGURE 22-5 How many calories do you need? Energy expenditure varies depending on body size and activity level.

50% to 100% more kilocalories per day than their BMR. A 120-pound woman requires 1,800–2,400 kcal/day, and a 160-pound man requires about 2,400–3,200 kcal/day.

Basal metabolic rates vary tremendously among different animal species. For the tiny shrew, for instance, BMR is about 35 times higher than for a human (**FIGURE 22-6**). The shrew's heart beats more than 500 times per minute *at rest!*

With the measurement of an animal's BMR, we can begin to estimate the animal's caloric needs. Let's estimate what a shrew needs to eat each day. If it weighs 5 grams and has a BMR of 35 calories per hour per gram, we can calculate as follows:

BMR varies tremendously among different animal species. The tiny shrew has a BMR about 35 times higher than that of a human, relative to weight.

FIGURE 22-6 Live fast. This tiny shrew must meet the high calorie demands of its fast-beating heart.

body weight	×	*energy needed each hour*	×	*hours per day*	=	*animal's caloric needs*
5 grams	×	35 cal/gram/hr	×	24 hr/day	=	4,200 cal/day

Thus, the shrew would need 4.2 kcal/day if it were at rest. But if it needed another 4 kcal/day for its normal activities, it would have to eat about 8 kcal/day. As we will see below, a pure source of carbohydrate or protein carries about 4 kcal/gram. Thus, every day, the shrew must find and consume at least 2 grams of food. This is no small challenge when you only weigh 5 grams. The task is equivalent to a 200-pound man finding and eating 80 pounds of food every day.

The shrew is not the only species with extreme caloric needs. Animals that fast or hibernate for long periods of time must prepare for the fast by consuming many kilocalories. Male elephant seals, for example, eat little or nothing during their 100-day breeding season, instead spending their time battling with other males for dominance and mating with females. During these 100 days, a large male who begins the season at 6,600 pounds (3,000 kg) may lose one-third of his body weight. Consequently, in the months leading up to the breeding season, the male seals' caloric requirements are dramatically higher than at other times.

TAKE-HOME MESSAGE 22·3

To function well, living organisms need sufficient energy, measured in kilocalories. The minimal energy needed by an individual not engaged in any activity is called its basal metabolic rate, or BMR.

❷ Nutrients are grouped into six categories.

A jaguar takes a drink at a watering hole in Belize.

22•4

Water is an essential nutrient.

We saw that a human could live 60 days or more without consuming food. The prognosis is much worse in the absence of water, perhaps the single most important component of a balanced diet. In the early 1900s, an Italian man sentenced to death reportedly volunteered to determine how long he could live without food or water. The fatal "experiment" lasted only 17 days before he died.

Water, which is considered an essential nutrient, constitutes about 60–65% of the body weight in most mammals. This water plays important roles in both the intracellular fluid and the extracellular fluids, including blood, in humans and other animals (**FIGURE 22-7**). Water in body fluids serves a variety of critical purposes, all of which can be impaired when the animal becomes dehydrated.

- Water transports nutrients and waste materials throughout the body.
- It takes part in chemical reactions.
- It serves as a solvent for many vitamins and minerals, amino acids, and sugars.
- It lubricates many joints, the spinal cord, and the eyes.
- It helps regulate body temperature.

A person expending about 2,000 kcal/day needs about 2–3 liters of water each day. This can come directly from drinking water, but there are many other sources of water in our diet, including milk, which is about 90% water, juices, food, and water released as a by-product of many chemical reactions. It is essential that water intake offset the water lost in urine, feces, respiration, and sweating. It is important to remember, too, as we saw in the Chapter 3 StreetBio, that

FIGURE 22-7 Water is an essential component of the diet.

drinking too much water too quickly can also be dangerous. Water intoxication results from the consumption of too much water—which sometimes happens to marathon runners during a race. When coupled with the loss of salt through sweating, over-consumption of water can lead to severe sodium imbalance. This imbalance causes dizziness, nausea, confusion, and swelling of the extremities. And in serious cases, it can lead to swelling of the brain, which can cause death.

Q If water is so important, why are there some desert animals that never need to drink?

Water usage varies among some animal species. Some desert mammals, the kangaroo rat, for example, have evolved tremendous water efficiency, and some do not need to drink any water at all. Rather, they get all the water they need from their food and metabolic processes. Among the marine birds and reptiles, most are able to drink salt water. They can do this with the aid of salt glands that remove and excrete the excess salt they consume.

TAKE-HOME MESSAGE 22•4

Water is probably the single most important component of the diet. It constitutes 60-65% of the body weight of most mammals, transports nutrients and waste materials throughout the body, takes part in metabolic reactions, serves as a solvent, lubricates many body parts, and helps regulate body temperature.

22•5

Proteins in food are broken down to build proteins in the body.

When you are on the verge of starvation, any type of food can save you. So from that perspective, it's true that calories are all that matter. But for optimum health, organisms must consume many types of nutrients. In this section and the next, we investigate the chief nutritional features of the three types of calories: proteins, carbohydrates, and fats. We begin by investigating protein, a vital component of the diet and essential structural material in the body.

Protein: Raw Material for Growth

What are proteins, and what is their role in the diet? Protein in our diet is principally a building material. As described in detail in Sections 2-15 through 2-18, protein molecules are made of long chains of smaller molecules—amino acids—linked together, like beads on a string. Once eaten, protein molecules are broken down into the individual amino acids, much like removing the beads from the string one at a time. It's only in the form of individual amino acids that proteins can be absorbed by the digestive system. Our bodies can then reorder the beads, making proteins that are different from those we ate.

The process of protein digestion and protein rebuilding does seem a bit inefficient: we consume entire, intact proteins such as muscle cells or hemoglobin molecules, from chickens or fishes or other animals, which we then break down and may ultimately use to construct almost exactly the same types of protein molecules. Nonetheless, animal digestive systems are able to absorb only the simpler building blocks of proteins, not entire proteins themselves; so we must first break down all the proteins we eat and later reassemble new ones.

Besides serving as building materials, proteins function as enzymes, catalyzing reactions throughout the body (see Section 2-18). They can also be broken down to release energy or to be stored as fat. Our bodies can use protein as fuel because we have a variety of methods of converting one type of chemical to another to release or store energy (see Section 4-17). The breakdown of proteins for energy generates 4 kilocalories per gram. Protein breakdown generally happens only when we are consuming too few (or too many) calories to sustain necessary growth and activities. The reverse reaction—building proteins from fat or carbohydrates—however, cannot be done. All amino acids have elements—particularly nitrogen—that are not found in any sugars or fats that we consume. Only the proteins in our diet contain nitrogen. Thus, all of the proteins we build contain parts of proteins that we have consumed, and if a protein is broken down for energy, the nitrogen it carries is excreted in our urine.

There is a variety of sources of dietary protein, from both plants and animals (**FIGURE 22-8**). Animal sources include egg whites, shrimp, tuna, lobster, chicken, turkey, and all meat products. Plant sources of protein include grains, vegetables, nuts, seeds, and legumes, such as beans.

Try this. Put one can of Coke and one can of Diet Coke somewhere that they will get hot, and leave them for a few days. Then taste them. The Diet Coke will lose its flavor and taste

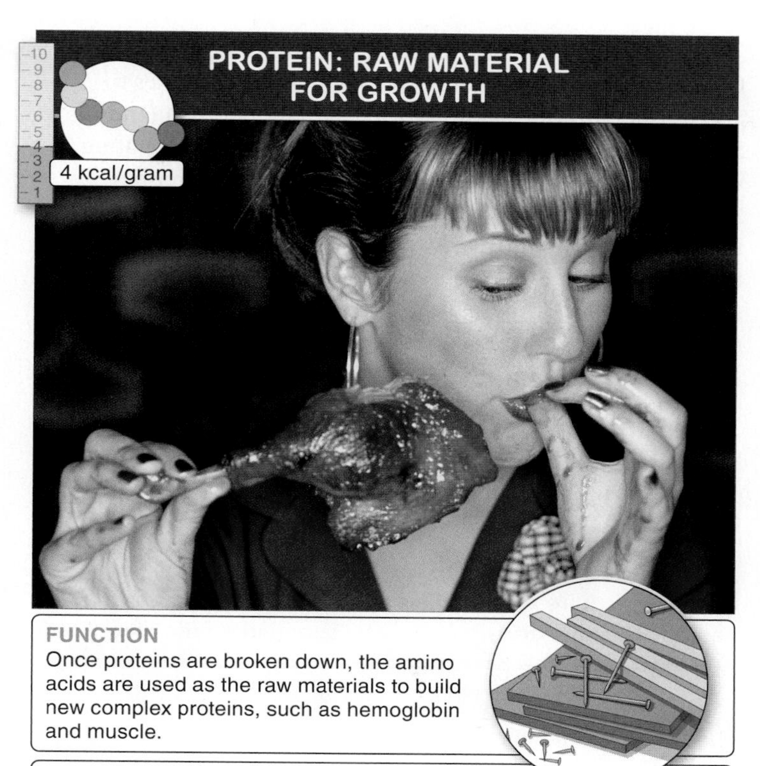

FUNCTION
Once proteins are broken down, the amino acids are used as the raw materials to build new complex proteins, such as hemoglobin and muscle.

SOURCE
- Animals: egg whites, shrimp, tuna, poultry, and meat
- Plants: grains and vegetables, such as beans

STORAGE
- Amino acids are usually stored for less than half a day before being reassembled into proteins throughout the body
- Can be converted to fat and stored in fat cells

FIGURE 22-8 Versatile proteins. Besides serving as building materials, proteins can be broken down to release energy or stored as fat.

Q Why does Diet Coke (but not regular Coke) taste bitter if you leave it in the sun for a few days?

bitter, even if you cool it down, while the Coke will taste fine. What's going on? Diet Coke, but not regular Coke, is sweetened with aspartame (NutraSweet is the brand name). Aspartame is made by linking together two amino acids, aspartic acid and phenylalanine—each of which tastes bitter by itself. When the two are joined as aspartame, however, the taste is very sweet. The reaction by which aspartic acid and phenylalanine are linked together can be reversed by heat. So if you leave a can of Diet Coke in a hot area, the aspartame decomposes to the bitter amino acids. They're not harmful, but they aren't very tasty either.

Are all proteins the same, or do they vary in important ways? All proteins are not created equal. Animals, including humans, require 20 different amino acids to make proteins. Most animals, however, can produce only about half of these amino acids themselves. In the case of humans, we can't make 8 of the 20 amino acids. Because these 8 cannot be produced by our bodies, we must get them from the food we eat. For that reason, the amino acids we must consume in food are called the **essential amino acids,** and the remaining 12 are called **non-essential amino acids.**

Protein-containing animal products vary in their amino acid composition. All eight of the essential amino acids are found in milk, eggs, meat, poultry, cheese, and fish; these animal products have "complete" proteins (**FIGURE 22-9**). On the other hand, most proteins that we get from eating plants are not complete. Corn proteins, for example, have only six of the essential amino acids, as do beans (although they don't have the same six). In

COMPLETE PROTEINS
Proteins that contain all eight essential amino acids

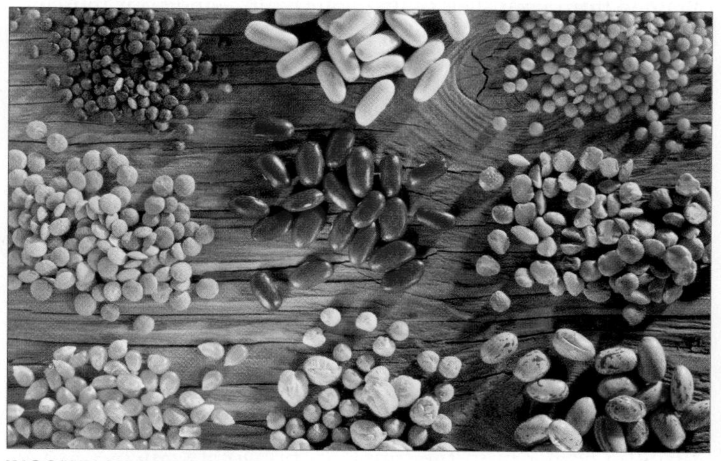

INCOMPLETE PROTEINS
Proteins that do not contain all eight essential amino acids

With the exception of soybeans, no single plant food has all eight essential amino acids.

FIGURE 22-9 Animal versus plant proteins. Animal proteins contain all eight of the essential amino acids. Plant foods contain varying amounts and types of essential amino acids.

fact, with the exception of soybeans, no single plant food has all eight. A healthy diet is one containing not just a sufficient *amount of protein* but a sufficient amount of each of the eight essential amino acids that we need to make proteins. Consequently, we must know more about the proteins we eat than the simple description on the food label of how many grams of protein the food contains.

Q Why is it beneficial for vegetarians to eat beans and corn over the course of the same day?

How and where do we store proteins? Proteins and the amino acid pool they generate on digestion cannot be stored very long in our bodies, usually less than a day. By that point, they are broken down and the nitrogen is excreted. Because our bodies are making proteins constantly, it is important to take in all of the essential amino acids over the course of each day.

The recommended daily intake of protein is approximately 0.8 grams per kilogram of body weight. This translates to about 45 grams of protein for a 120-pound woman (and about 25 additional grams per day for pregnant or lactating women) and 60 grams for a 160-pound man. Exercising athletes can require double or even triple this amount, but it is important to remember that it is the exercise that generates the additional need. Protein on its own does nothing to stimulate muscle growth; it simply serves as the source for the raw materials.

TAKE-HOME MESSAGE 22•5

Animals consume three different types of macromolecules for calories: proteins, carbohydrates, and fats. Proteins provide raw materials for growth and for the production of enzymes. Food sources of protein vary in amino acid composition. Humans require 20 amino acids, and 8 of these, called essential amino acids, can only be supplied by the diet.

22•6

Carbohydrates and lipids provide bodies with energy and more.

Reading these words requires carbohydrates in your body, obtained from your diet, to fuel your brain—and your muscles as well, as they help you sit up and hold your book. Food fat, too, a subset of the group of macromolecular nutrients called lipids (see Sections 2-12–2-14), provides energy for your body's functioning and gives food delicious flavor (while contributing to weight gain and heart disease if consumed in too large a quantity). In this section, we evaluate the macromolecules carbohydrates and lipids. These two crucial nutrients supply nearly all the energy that fuels your daily activities.

Carbohydrates: Fuel for Living Machines

What are carbohydrates, and what is their role in the diet? Carbohydrates are the primary fuel on which animal bodies run. In humans, nearly all of the energy used by our brain every day comes from the simple carbohydrate glucose. As we saw in Chapter 2, all carbohydrates are made primarily from carbon, hydrogen, and oxygen. When the bonds between these atoms are broken, energy is released that can be captured by the body and used to fuel movement, growth, and all the other cellular activities that require energy.

We get the majority of our dietary carbohydrates from fruits, vegetables, and grains (**FIGURE 22-10**). And, as with proteins, the breakdown of carbohydrates for energy generates 4 kilocalories per gram.

Are all carbohydrates the same, or do they vary in important ways? Although, structurally, all carbohydrates are variations on a simple theme—molecules formed from carbon, hydrogen, and oxygen in the approximate proportions of CH_2O (see Figure 2-20)—they vary dramatically in their complexity. This variation has implications for their nutritional effects (**FIGURE 22-11**). From a dietary perspective, the most important distinctions are among simple sugars (monosaccharides), digestible complex sugars, and fiber (indigestible complex sugars).

Simple sugars: These include glucose and fructose. They are linear or ring structures with three to seven carbon atoms. Animals can break them down directly through the steps of glycolysis (see Section 4-13), rapidly releasing the stored energy from the bonds of the sugar.

Digestible complex sugars: Multiple simple sugars can bond together to form complex but digestible molecules. Some complex sugars, such as sucrose (table sugar), are just two simple sugars joined together. Others, such as starch and glycogen, are large molecules that may consist of

FUNCTION
Carbohydrates provide energy to fuel movement, growth, and all cellular activities in the body.

SOURCE
Fruits, vegetables, and grains

STORAGE
- Carbohydrates are stored in the liver and muscle cells as glycogen for about a day before being broken down to provide energy
- Can be converted to fat and stored in fat cells

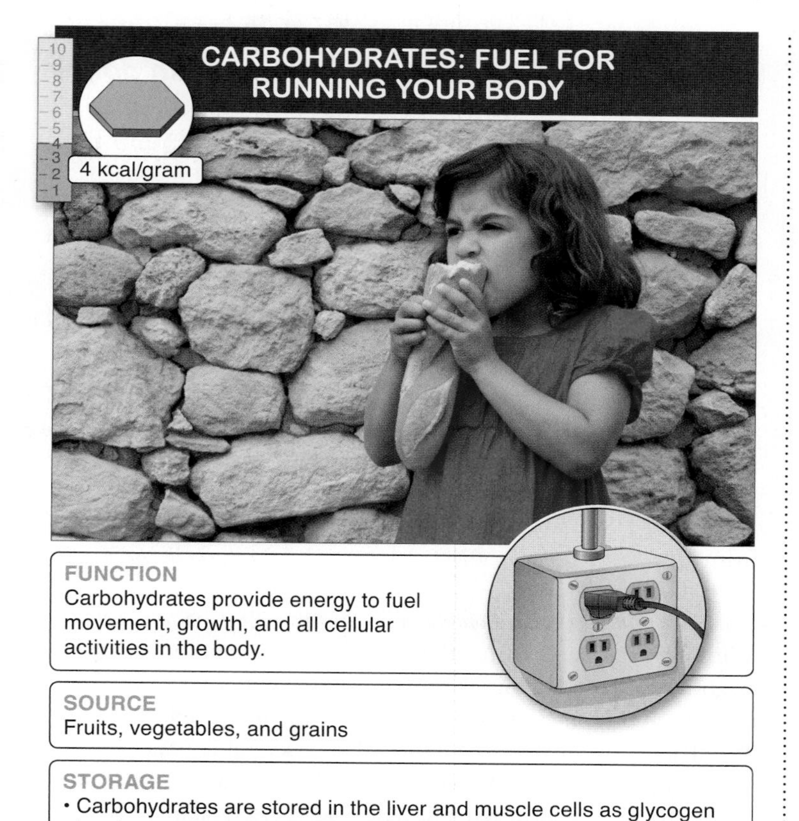

FIGURE 22-10 Carbohydrate is the primary fuel on which animal bodies run.

hundreds or thousands of glucose molecules connected in dense branching patterns. In order for an animal to have access to the energy stored within the bonds of the individual simple sugars, it must first break the bonds that link those sugars together. As these bonds are broken, simple sugars become available for the energy-releasing reactions of glycolysis.

Fiber: This is a complex carbohydrate, such as cellulose, that forms the structural parts of plants. Fiber differs from starch and other digestible complex sugars by having a different bond connecting the simple sugars together. This bond cannot be broken by any human digestive enzyme, making fiber indigestible. As we'll see later in this chapter, although fiber isn't broken down for energy or other molecules used by the body, it still plays an important role in digestion and is necessary in the diet to maintain health.

How and where do we store carbohydrates? Carbohydrates in our body are stored mostly as glycogen in liver and muscle cells. At any given time, we can store only about one day's worth of energy. After we start exercising, or when our bodies need energy for an activity, a signal is sent causing the release of enzymes that break the bonds holding together the highly branched glycogen. This glycogen breakdown produces a flood of glucose into the bloodstream and at muscles where the energy is needed.

Large amounts of water are bound to stored glycogen: 4 pounds of water for every pound of glycogen. Consequently, as glycogen in your liver and muscles is used, the water bound to it is released from the tissue and lost as urine. This is why, as stores of glycogen are depleted in the initial stages of a diet, there is a dramatic initial weight loss from the loss of water that was bound to the glycogen. As your body starts utilizing stored fat, the rate of weight loss slows considerably.

TYPES OF CARBOHYDRATES

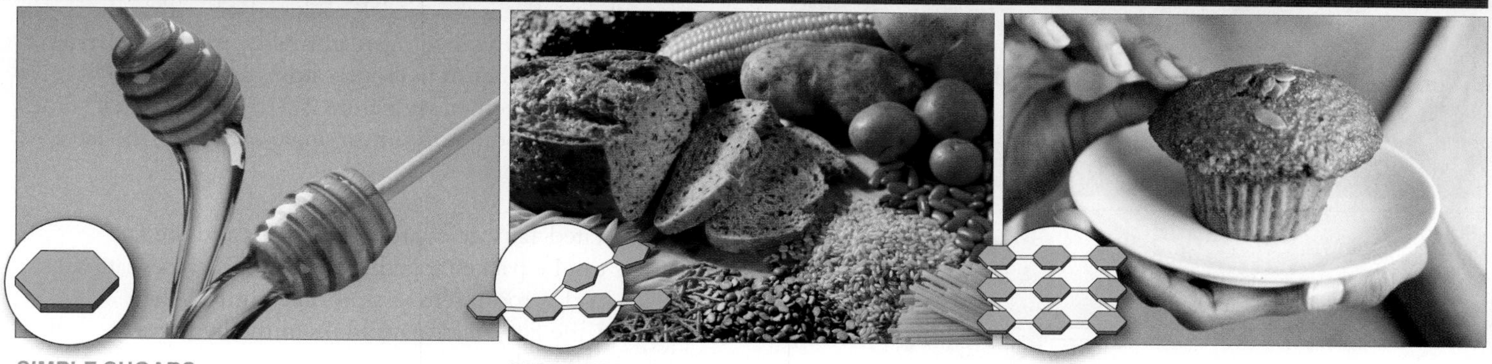

SIMPLE SUGARS
- Glucose, fructose
- Glycolysis reactions release energy rapidly

DIGESTIBLE COMPLEX SUGARS
- Simple sugars bonded together, such as sucrose (table sugar) or starch
- Bonds between simple sugars must be broken before the energy-releasing reactions of glycolysis occur

FIBER
- Complex carbohydrate that forms structural parts of plants
- Indigestible in humans, but has a significant role in digestion

FIGURE 22-11 Carbohydrate complexity. Some carbohydrates are readily broken down for fuel, but fiber passes through the human body undigested.

Fats: Long-Term Energy Storage Experts

What are fats, and what is their role in the diet? Dietary fats, described in detail in Chapter 2, function primarily as a dense source of energy that can be efficiently stored in the body (**FIGURE 22-12**). The average person has about four or five weeks' worth of stored energy in the form of fat. Compared with carbohydrates or proteins, a given amount of fat contains more than twice as much stored energy: 1 gram of fat produces 9 kilocalories. Another feature that makes fats particularly efficient as energy-storage molecules is the fact that, because they are hydrophobic, fats are stored without binding to water.

Because they also are poor conductors of heat, fats stored in a layer just beneath the skin help to keep the body warm. Penguins and walruses, for example, can maintain relatively high body temperatures despite living in very cold habitats, because of their thick layer of insulating fat.

Here's a perplexing fact: although the total amount of energy in a gram of fat is greater than that found in carbohydrates or proteins, fat isn't the optimum nutrient for most situations.

FIGURE 22-12 Energy-rich fats. At 9 kilocalories per gram, fats are rich in stored energy.

Q If fats contain more than double the amount of energy found in the same amount of carbohydrates or proteins, why aren't fats a better fuel to consume before exercising?

When it comes to exercise, for example, muscle cells need quick access to energy. The problem is that, in muscles, fat burns very slowly for energy. Remember from Chapter 4 that the universal source of chemical energy in the body is ATP. This means that the energy in fat or carbohydrate or protein must be captured as ATP before it is of use to muscle cells. It turns out to be much easier for the body to break down muscle glycogen and blood glucose to make ATP than to break down fat. In fact, the rate of ATP synthesis from carbohydrates is about double the rate from fats.

Are all fats the same nutritionally, or do they vary in important ways? With proteins, we saw that there were essential and non-essential amino acids. Similarly, with dietary fats, there are essential and non-essential fatty acids. Humans are able to produce nearly all necessary lipids. The essential fatty acids—including linoleic acid, a type of fatty acids called omega-6 fatty acids—must be consumed. Linoleic acid is essential as a building block for signaling molecules, such as some hormones; a deficiency can lead to infertility and difficulty lactating. Another essential fatty acid, important in a variety of metabolic processes, is linolenic acid, a type of fatty acids called omega-3 fatty acids. Linolenic acid is used by the body to make fatty acids that are essential for normal growth and development, especially in the eyes and brain.

Another important distinction among dietary fats is between saturated and unsaturated fats (**FIGURE 22-13**). In our diet, fats usually come in the form of fatty acids, long chains of carbon atoms with hydrogens attached. If each carbon within the chain is bonded to two hydrogen atoms, the molecule carries the maximum number of hydrogen atoms and is said to be saturated. (For a refresher, see Figure 2-31.) Conversely, if some of the carbons are bound to only a single hydrogen, the fatty acid is unsaturated.

When saturated, fatty acids are very straight and the fat molecules can be packed together tightly. This causes saturated fats to be solid at room temperature, like butter. When unsaturated, the fatty acids have kinks in the hydrocarbon tail and cannot be packed together as tightly. Consequently, unsaturated fats do not solidify so easily and tend to be liquid at room temperature, like vegetable oil. Because unsaturated fatty acids can accept one or more hydrogen atoms, they are a bit less stable and more reactive—that is, they will take part in a greater variety of chemical reactions—than saturated fatty acids, making them less likely to be stored as body fat.

SATURATED FATS
- Fatty acids have straight tails and can be packed together tightly
- Tend to be solid at room temperature
- More likely to be stored as fat in the body

UNSATURATED FATS
- Fatty acids have kinked tails and cannot be packed together tightly
- Tend to be liquid at room temperature
- Less likely to be stored as fat in the body

FIGURE 22-13 Saturated and unsaturated fats compared.

Trans fats have been in the news because of their tendency when consumed to raise levels of low-density lipoprotein cholesterol, increasing the risk of coronary heart disease. Trans fat is made when hydrogen is added to vegetable oil—a process called hydrogenation (see Section 2-13 for a review of trans fat chemistry.) Trans fat often can be found in some of the same foods as saturated fat, such as vegetable shortenings, crackers, candies, cookies, snack foods, fried foods, baked goods, and other foods made with partially hydrogenated vegetable oils. The American Heart Association recommends that these fats be minimized in the diet.

How and where do we store fats? If a human consumes more calories than he or she burns, most of the excess (regardless of whether these calories were consumed as carbohydrate, fat, or protein) gets converted to fat and stored in fat cells distributed throughout the body. A pound of body fat holds 3,600 kilocalories worth of energy. It is like a savings account for an uncertain future.

Given that our bodies can convert excess calories into body fat, regardless of whether the calories were initially ingested as carbohydrates or proteins, why is it still a more effective weight-management strategy to minimize dietary fat intake? The answer rests in the number of chemical conversions carbohydrates and proteins must undergo to become body fat. In the case of carbohydrates, complex sugars must first be broken down into simple sugars—a process that requires energy. Then the simple

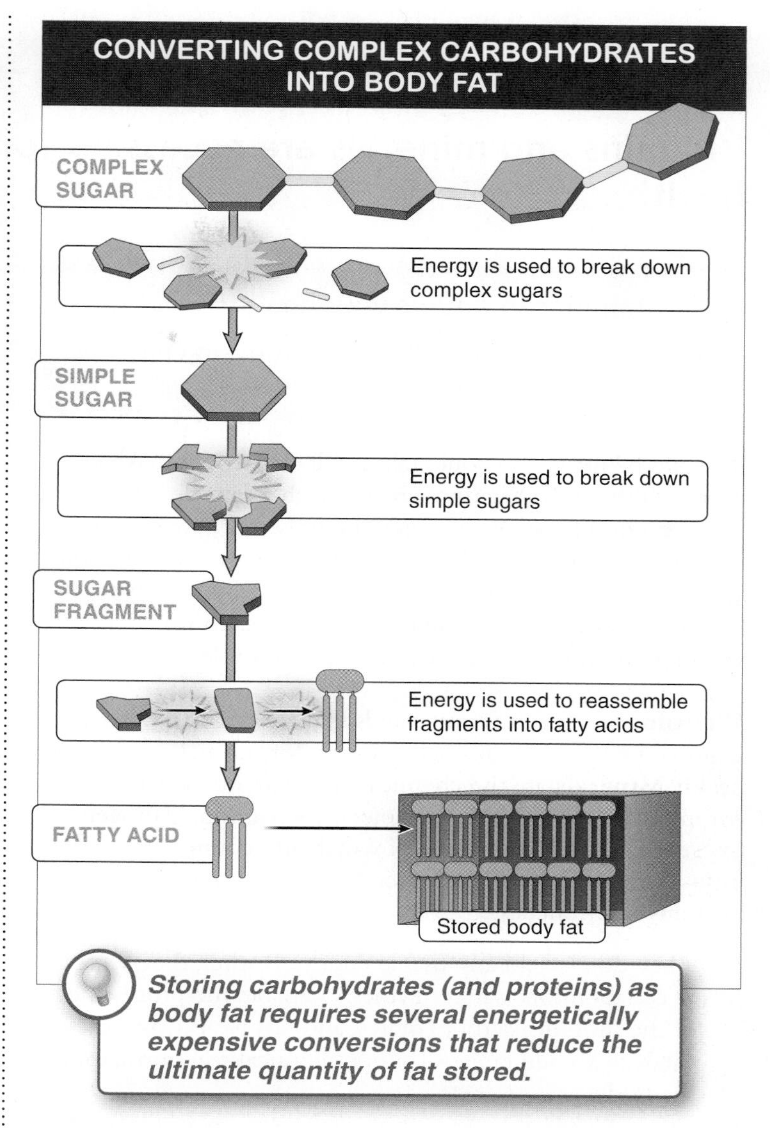

FIGURE 22-14 Inefficient conversion to body fat. By the time carbohydrates can be stored as body fat, much of the original energy has been lost in fueling the conversion.

sugars must be broken down and the fragments reassembled as fatty acids—another process that requires energy. By the time those fatty acids can be stored as body fat, much of the original energy stored in the carbohydrates has been lost in fueling all the chemical conversions (**FIGURE 22-14**). And for someone trying to minimize body fat, that's good news. Storing protein as body fat also requires several energetically expensive conversions that reduce the ultimate quantity of fat stored.

TAKE-HOME MESSAGE 22•6

Carbohydrates are the primary fuel on which animal bodies run. Dietary fats function primarily as a dense source of energy that can be efficiently stored in the body.

22•7

Vitamins and minerals are necessary for good health.

> Vitamins can double your health!
>
> —Dr. Atkins, on *The Larry King Show* (making perhaps the most nonsensical nutritional claim ever)

It would be nice if we could take a pill that would make us more healthy or fit. Vitamins are not such a miracle pill. They're a bit like security guards at a museum: above a certain number, they don't make the museum any better, but in their absence, the museum is likely to become a whole lot worse. And while vitamin supplements are not a health panacea, vitamins and minerals do have an important role in nutrition.

Vitamins are organic compounds that are essential nutrients required by the body in small amounts for normal growth and health. **Minerals** are the chemical elements, other than those commonly found in organic molecules—carbon, hydrogen, oxygen, and nitrogen—some of which are required in the diet in small amounts. There are three features common to all vitamins and minerals.

1. They don't yield any usable energy. Rather, they serve as collaborators with enzymes to enable the processing of the proteins, carbohydrates, and fats we eat, and they catalyze a wide range of other chemical reactions around the body.
2. They need to be consumed in much smaller amounts than proteins, carbohydrates, and fats in our diet. This is because they tend to serve as reaction catalysts and so can be recycled and used again and again.
3. If we have a healthful diet, we tend to consume sufficient quantities of vitamins and minerals in our food.

Although their biological roles are frequently similar, vitamins and minerals have a fundamental chemical difference. Vitamins are organic molecules, meaning they contain carbon, whereas minerals are inorganic nutrients. Because vitamins are organic, they are more fragile, easily destroyed by heat and other chemical or physical extremes. Minerals, on the other hand, are elements, so they can't be broken down further or lose their chemical identity. They stay in your body until they are excreted. In fact, the only way they can be lost is if they are leached away—such as from food into the water as the food is cooked and the water is then thrown away. Seventeen minerals have been identified that are essential to the human diet. They are described in **FIGURE 22-15**.

Thirteen vitamins essential to humans, also described in Figure 22-15, have been discovered. They fall into two groups based on whether they are soluble in fat or water. The water-soluble vitamins include eight vitamins that are part of the vitamin B complex, plus vitamin C. There is some variation among species in the ability to manufacture (rather than needing to ingest) vitamins. Cats and dogs, for instance, do not need to consume vitamin C.

There are four fat-soluble vitamins: A, D, E, and K. Because they are stored in fatty tissue and the liver until they are needed, they don't need to be consumed as regularly as water-soluble vitamins. Also, because excess water-soluble vitamins can be removed by the kidney and excreted in urine, they tend to be less likely to reach toxic levels than fat-soluble vitamins.

Currently, more than half of the U.S. population takes regular vitamin supplements (**FIGURE 22-16**). Should so many people be taking these supplements? It is difficult to give a single answer to this question. Nonetheless, it is well established that in the United States, few adults suffer from any vitamin and mineral deficiency diseases. This is because nearly all adults get all of the nutrients they need from the food they eat. For this reason, nutritionists believe that high doses of vitamins and minerals are not therapeutic, and they suggest that most people do not need to take regular supplements.

Usually, taking vitamin supplements is a costly but harmless behavior. The consumption of increasing amounts of vitamins often just leads to a plateau of maximum benefit, after which greater consumption has no additional benefits. Occasionally, however, taking vitamin and mineral supplements is both costly and harmful: in many cases, above a certain point, additional consumption of fat-soluble vitamins can lead to toxicity and cause serious health problems. Excessive consumption of vitamin A, for example, can lead to hair loss in men, while over-consumption of vitamin D can lead to growth retardation. Liver damage, too, is a common result of excessive vitamin and mineral consumption. Moreover, research has definitively discredited claims that vitamin E supplementation slows or prevents aging or improves physical performance, and no supplement has been demonstrated to "relieve stress"—despite the extravagant claims of marketers.

ESSENTIAL VITAMINS AND MINERALS IN THE HUMAN DIET

VITAMIN	SOURCE	MAJOR FUNCTION
Water-soluble		
B_1 (thiamin)	Liver, legumes, whole grains	Coenzyme in cellular respiration
B_2 (riboflavin)	Dairy foods, meat, eggs, green leafy vegetables	Coenzyme in FAD
B_6 (pyridoxine)	Liver, whole grains, dairy foods	Coenzyme in amino acid metabolism
B_{12} (cobalamin)	Liver, meat, dairy foods, eggs	Formation of nucleic acids, proteins, and red blood cells
Biotin	Liver, yeast, bacteria in gut	Found in coenzymes
C (ascorbic acid)	Citrus fruits, tomatoes, potatoes	Formation of connective tissues; antioxidant
Folic acid	Vegetables, eggs, liver, whole grains	Coenzyme in formation of heme and nucleotides
Niacin	Meat, fowl, liver, yeast	Coenzyme in NAD and NADP
Pantothenic acid	Liver, eggs, yeast	Found in acetyl CoA
Fat-soluble		
A (retinol)	Fruits, vegetables, liver, dairy foods	Found in visual pigments
D (cholecalciferol)	Fortified milk, fish oils, sunshine	Absorption of calcium and phosphate
E (tocopherol)	Meat, dairy foods, whole grains	Muscle maintenance; antioxidant
K (menadione)	Intestinal bacteria, liver	Blood clotting

MINERAL	SOURCE	MAJOR FUNCTION
Calcium (Ca)	Dairy foods, eggs, green leafy vegetables, whole grains	Found in bones and teeth; blood clotting; muscle action
Chlorine (Cl)	Table salt (NaCl), meat, eggs, vegetables, dairy foods	Water balance; principal negative ion in extracellular fluid
Chromium (Cr)	Meat, dairy foods, whole grains, legumes, yeast	Glucose metabolism
Cobalt (Co)	Meat, tap water	Found in vitamin B_{12}; formation of red blood cells
Copper (Cu)	Liver, meat, fish, shellfish, legumes, whole grains, nuts	Found in enzymes and electron carriers; hemoglobin production
Fluorine (F)	Most water supplies	Found in teeth; helps prevent tooth decay
Iodine (I)	Fish, shellfish, iodized salt	Found in thyroid hormones
Iron (Fe)	Liver, meat, green vegetables, eggs, whole grains, legumes	Found in many enzymes, hemoglobin, and myoglobin
Magnesium (Mg)	Green vegetables, meat, whole grains, nuts, milk, legumes	Required by many enzymes; found in bones and teeth
Manganese (Mn)	Organ meats, whole grains, legumes, nuts, tea, coffee	Activates many enzymes
Molybdenum (Mo)	Organ meats, dairy foods, whole grains, green vegetables	Found in some enzymes
Phosphorus (P)	Dairy foods, eggs, meat, whole grains, legumes, nuts	Found in nucleic acids, ATP, and phospholipids
Potassium (K)	Meat, whole grains, fruits, vegetables	Nerve and muscle action; principal positive ion in cells
Selenium (Se)	Meat, seafood, whole grains, eggs, milk, garlic	Fat metabolism
Sodium (Na)	Table salt, dairy foods, meat, eggs	Nerve and muscle action; water balance
Sulfur (S)	Meat, eggs, dairy foods, nuts, legumes	Found in proteins and coenzymes; detoxification
Zinc (Zn)	Liver, fish, shellfish, and many other foods	Found in some enzymes and some transcription factors

FIGURE 22-15 Overview of essential vitamins and minerals.

FIGURE 22-16 Is supplementation necessary? Shopping the vitamin aisle for health.

So, we can conclude that most people in the United States have no need to take vitamin and mineral supplements. Under special circumstances, however, supplementation is necessary.

- Women who lose unusually large amounts of blood during menstruation may need iron.
- Post-menopausal women and those allergic to milk may not get enough calcium in their diet to prevent bone degeneration.
- Pregnant and breast-feeding women may need additional vitamins and minerals to support a developing fetus or rapidly growing infant.
- Pregnant women may, in particular, need additional folate, which helps prevent neural tube defects.

Green leafy vegetables are a good source of folate.

- People on extremely low-calorie diets may need a vitamin and mineral supplement.
- People with limited consumption of milk or sun exposure may need additional vitamin D. Because of insufficient levels of ultraviolet energy from the sun at latitudes above 42 (covering a line approximately from the northern border of California across to Boston), individuals in these regions are unable to produce sufficient vitamin D from sun exposure alone.
- People with absorption problems (such as when taking antibiotics or with an infection or following surgery) may need vitamin and mineral supplementation.

Individuals who do not fall into these groups, however, generally get all of the essential vitamins and minerals simply by eating a varied diet, rich in nutritious foods.

TAKE-HOME MESSAGE 22•7

Vitamins and minerals are organic and inorganic molecules in the diet. They are used in the production and action of enzymes and other molecules involved in the processing of food and other biochemical reactions. While vitamins and minerals are essential in small amounts, most people in the United States do not benefit from taking them as supplements.

22•8

How do organisms "know" what to eat?

We take it for granted, but our bodies seem to "know" that we need food, and they seem to be able to sense whether we're consuming nutritionally valuable substances. Is this really true? To find out, researchers conducted a clever study. With some volunteer participants, they secretly gave one group of people sugar cookies while a second group got cookies that looked the same and tasted pretty good but were made with the low-calorie sugar substitute, NutraSweet. The researchers then noted how many cookies were eaten. Surprisingly, individuals in both groups ate the same number of cookies, suggesting that those eating the NutraSweet cookies might lose weight as a result of reduced caloric intake. In addition to having their cookie consumption monitored, however, study participants were also asked to keep diaries of all their eating in the days around the cookie-eating event. It turns out that those in the NutraSweet-cookie group ate more than those in the sugar-cookie group. So much more, in fact, that the total caloric intake of the two groups was identical over the period of days monitored (**FIGURE 22-17**). Similar studies have shown that individuals who switch from regular sodas to no-calorie, diet sodas increase their consumption of other foods, often unknowingly, so that their overall caloric intake is not reduced.

Why do our bodies seem to keep track of our caloric intake? It is because all animals are built to get hungry and to seek out food when in that state. Powerful, instinctual hunger kept our ancestors going in a tough, energetically demanding world. Recall from earlier in the chapter the Biosphere 2 study in which participants' diets were significantly reduced in calories. Increasingly, their thoughts and discussions were filled with food fantasies. Such hunger used to be a survival-enhancing feature in our genetic programming, but it is more of a bug in that programming now. Later in this chapter, we'll see that because most of us live in a world characterized by plentiful food, our tremendous ancestral appetites frequently lead to obesity and other health problems.

Our genes do more than instill a general hunger for calories. Humans and other animals also show a preference for fatty foods over carbohydrate- or protein-laden foods. We prefer fatty foods because our tongues have thousands of specific detectors—taste buds—that stimulate our brains when we eat foods like nuts, avocados, cheese, and red meat. With this system, a fatty meal stimulates many of the reward centers in the brain. These structures evolved because fat has the most

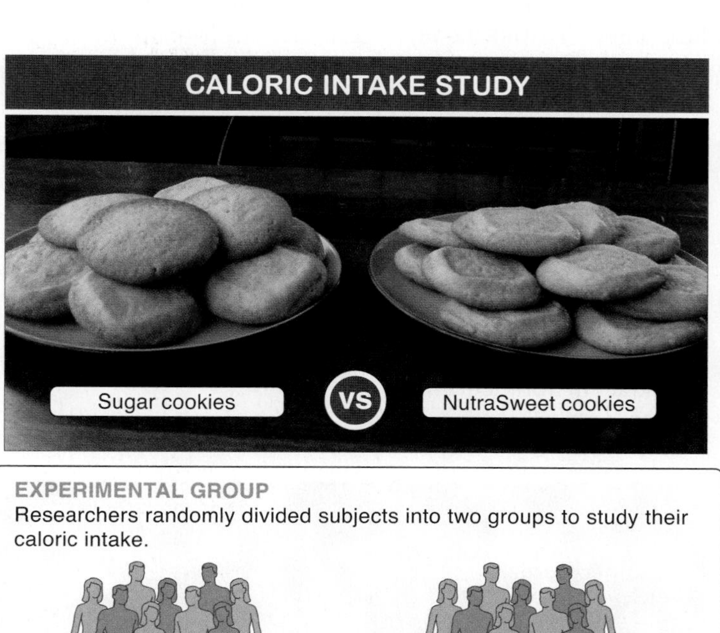

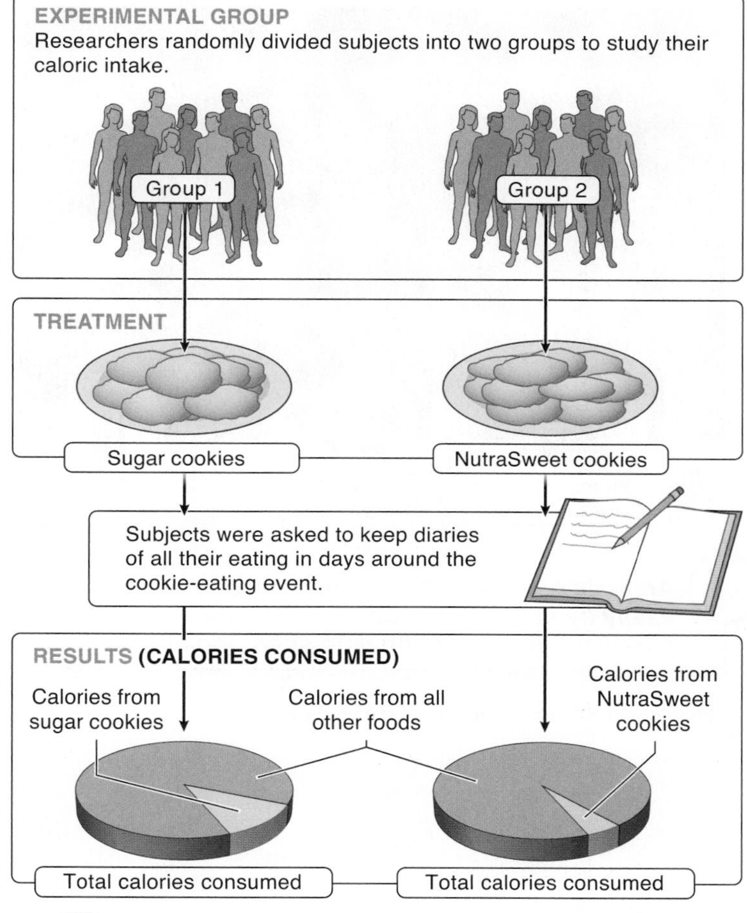

FIGURE 22-17 Experiment in human caloric intake.

calories per serving. Our ancestral genes reward us whenever we find calories; in this quest for energy, fat deserves—and receives—the biggest reward.

To examine this in a more concrete way, let's return to the little shrew (**FIGURE 22-18**). Recall that our 5-gram shrew must consume about 8 kcal/day and that a pure source of carbohydrates or proteins carries about 4 kcal/gram, while a pure source of fat carries 9 kcal/gram. This means that a shrew can meet its energy requirements by finding and consuming 2.0 grams of a carbohydrate or protein. *Or* the shrew can find about 0.9 gram of a food that is primarily fat (such as insect larvae). Which is going to be easier? Clearly, the shrew (or any animal, for that matter) can meet its energy needs much more easily by seeking fats. Genetically based preferences for the taste of fats, then, can benefit shrews and humans, too (in their ancestral environment).

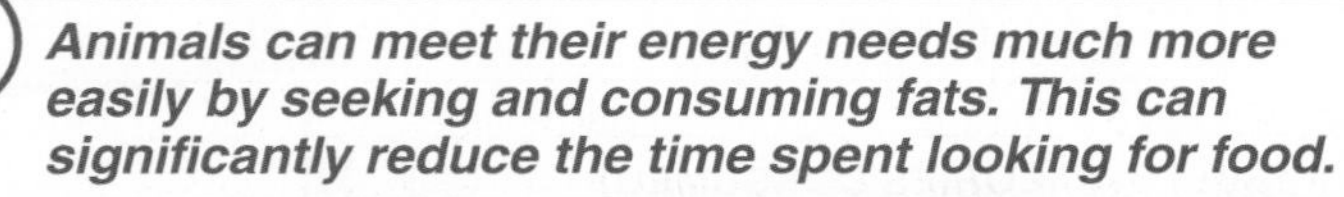

FIGURE 22-18 Efficiency of a fat-rich diet.

How would you describe this family's food preferences?

Animals have evolved to be sensitive to more than just total caloric intake or the fat (vs. protein or carbohydrate) content of food. It has been demonstrated in spiders, for example, that individuals will go out of their way, passing up valuable sources of food, in order to consume food items that contain certain rare but essential amino acids, so that their diet contains sufficient amounts of all of the essential amino acids. This suggests that there may actually be some underlying biological urges creating the various cravings and hunger pangs that we sometimes feel and that these feelings may not be restricted to humans. In Section 9-1, we describe how organisms have evolved to have very specific preferences when it comes to the foods they seek and prefer.

TAKE-HOME MESSAGE 22•8

Instincts cause animals to get hungry and to seek out food when in that state. Humans and other animals show a preference for fatty foods over carbohydrate- or protein-laden foods, a behavior reinforced by fat-laden meals stimulating many of the reward centers in the brain. These preferences probably evolved because fat has the most calories per serving.

❸ We extract energy and nutrients from food.

A cormorant catches a fish.

22•9

We convert food into nutrients in four steps.

The human digestive system is like an assembly line running backward. Imagine starting with an assembled car and dismantling all the parts: the tires, doors, windshield, steering wheel. The food entering the assembly line of the digestive system is like the intact car. It enters the assembly line and then passes through four distinct phases, during which the food is progressively chewed up and broken down, the nutrients absorbed by the body, and the non-usable portion of the raw materials discarded as waste products (**FIGURE 22-19**).

In the next several sections, we examine in detail the four stages in the processing of food: ingestion, digestion, absorption, and elimination. Throughout these sections, it will be useful to picture the digestive system as a long tube with an opening at each end, the mouth and the anus, and some glands along the way that produce the necessary chemicals to help the process along.

TAKE-HOME MESSAGE 22•9

The digestive process in humans includes four distinct phases during which food is progressively chewed up, broken down, and absorbed by the body, after which the non-usable portion of the raw materials is discarded as waste.

DIGESTIVE PROCESS FROM START TO FINISH

The digestive process in humans includes four distinct phases.

1 INGESTION
Food is taken into the body.

2 DIGESTION
Large pieces of food are dismantled by physically and chemically breaking them into absorbable molecules.

3 ABSORPTION
Energy-rich food molecules are taken into the cells of the body, where they can be used for energy and building materials.

4 ELIMINATION
The final remaining parts of the consumed foods—mostly indigestible materials—are discarded as waste products and much water reabsorption occurs.

FIGURE 22-19 There are four steps in the body's processing of food.

22•10

Ingestion is the first step in the breakdown of food.

It will take you longer to read about **ingestion,** the intake of food into your body, than it would to just do it (**FIGURE 22-20**). Typically lasting less than a minute, ingestion primarily involves four parts of your anatomy: your mouth, teeth, tongue, and esophagus. We'll look at each in some detail. To start, imagine that you take a big bite of a hamburger, containing bread, lettuce, and the meat.

Putting food in your mouth stimulates your salivary glands. Via tiny ducts throughout your mouth, the salivary glands secrete mucus that lubricates the food to help it pass into your stomach. They also secrete an enzyme called alpha-amylase that initiates the process of digestion. Alpha-amylase breaks the bonds holding together the starch molecule (a highly branched carbohydrate pieced together from hundreds of linked glucose molecules) and releases a bit of glucose that can be used for energy. About 20% of the ingested starch is broken down in the mouth. Protein and fat molecules, on the other hand, are not broken down at all.

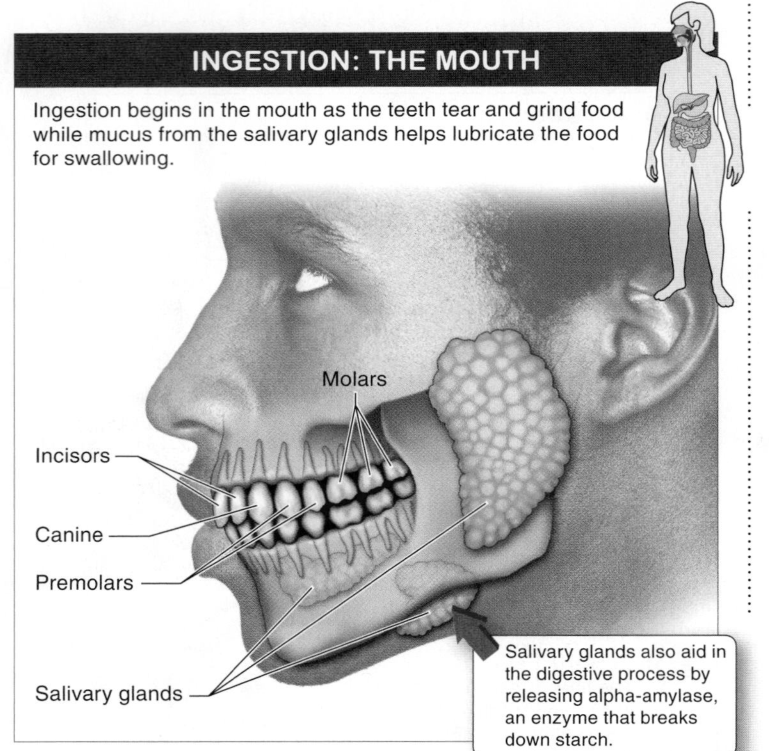

FIGURE 22-20 Ingestion, the intake of food into the body, begins in the mouth.

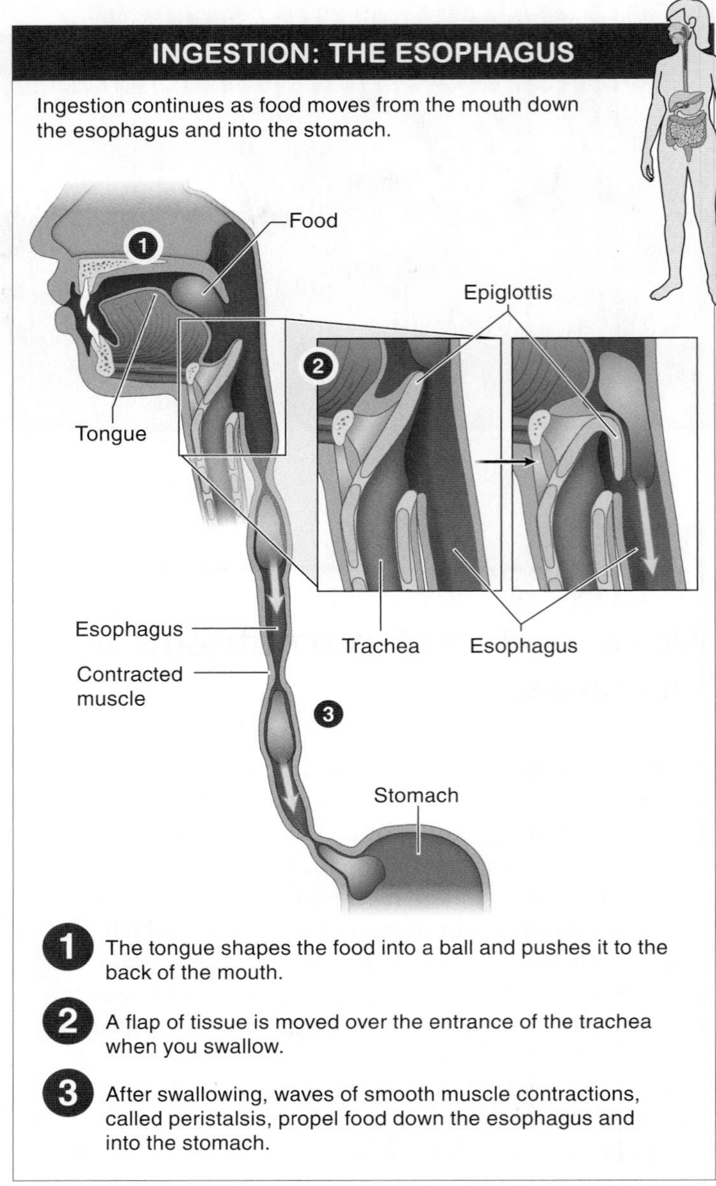

FIGURE 22-21 Ingestion is complete when food passes from the mouth through the esophagus to the stomach.

You start chewing food in your mouth in order to tear and grind it into little bits. This is a first important step toward completely breaking down and harvesting the energy stored in the chemical bonds of the food. Several different types of teeth have evolved in mammals and other animal species that enable different types of food items to be processed. Incisor and canine teeth in the front are used for biting and tearing food. Behind them are premolars and molars, used for grinding and crushing food. Just by looking at the type of teeth an animal has, we can learn a lot about its diet. Herbivores have molars primarily for grinding and crushing the tough cell walls of the plants they eat. Carnivores have sharp and lethal canines and incisors for killing and tearing apart the flesh of other animals.

Q Why do many birds eat gravel?

Birds have no teeth. This explains why they can sometimes be seen eating gravel. Although the gravel has no nutritional value, it collects in the stomach, where it helps to grind up the food they eat. This is an important digestive step for birds because, without teeth, they must swallow their food whole. Consequently, when it gets to the stomach, it hasn't been ground up at all, which can reduce the efficiency of digestive enzymes.

While you are chewing your food, your tongue assists in the process by forming the food into a ball shape that can be swallowed. The ball of partly crushed, saliva-coated food is pushed to the back of your mouth by your tongue. There, your throat opens to two passageways into your torso, the **trachea** and the **esophagus** (**FIGURE 22-21**). The trachea, also called the windpipe, connects to your lungs, while the esophagus connects to your stomach. Food destined for your stomach is kept from entering your trachea by a fast but critical maneuver in which your voice box moves up (due to muscle contractions), causing a flap of tissue (called the epiglottis) to be pushed over the entrance to the trachea just as you begin to swallow. If the ball of food you swallow is too big, it can get stuck at the beginning of the esophagus, wedging against the flap of tissue in front of the trachea—and this can cause you to choke by blocking all air from getting into your lungs.

Once the ball of chewed food makes it into the esophagus, waves of smooth muscle contractions, called **peristalsis,** propel the food down the esophagus and into the stomach, where digestion continues. Because of peristalsis, it's not necessary for you to be sitting upright or standing when you eat. Even if you are standing on your head, the food is pushed down the esophagus to the stomach.

TAKE-HOME MESSAGE 22•10

Ingestion is the first phase of the digestive process. Usually lasting less than a minute, ingestion involves tearing and grinding food in preparation for passing it to the stomach. Digestion also begins during this phase, with some starch being broken down by enzymes in saliva.

22•11

Digestion dismantles food into usable parts.

Food is fuel. But it is useless to an animal until it has been completely broken down into its fundamental chemical constituents. Only then can it be absorbed into the bloodstream and used constructively by a body. The process of dismantling the large pieces of food, physically and chemically breaking them down into absorbable molecules, is digestion. It occurs primarily in the stomach and small intestine.

The Stomach Let's continue tracing the bite of hamburger as it moves through the digestive system. It empties from the esophagus directly into the **stomach,** a muscular J-shaped organ with thick, elastic walls that can expand greatly in size to accommodate a large meal—as much as 4 liters of material! At the point where the esophagus connects to the stomach, there is a ring of muscle, called a **sphincter.** It seals off the stomach once the food has entered, preventing regurgitation of the stomach's acidic contents into the esophagus—even if you are eating while standing on your head. The stomach has three functions.

1. It physically breaks down and mixes food, through churning of the muscles surrounding it.
2. It secretes acid to further break down food, chemically, and to kill bacteria.
3. It begins some chemical digestion of proteins.

The presence of food in the stomach causes cells in crevices within the stomach lining, called gastric pits, to rapidly produce hydrochloric acid and a protein-dismantling enzyme known as

pepsin (**FIGURE 22-22**). The acid corrodes proteins in the food, breaking them into smaller pieces, which can then be broken into their constituent amino acids by the pepsin.

The burning sensation of indigestion occurs when the sphincter between the esophagus and the stomach doesn't completely prevent the acidic contents of the stomach from moving back up the esophagus. It usually occurs when a person eats too much or too quickly. Antacids can neutralize acid from the stomach. Remember, though, that the high acidity in the stomach is important for digestion, so while reducing it may alleviate heartburn, antacids can reduce the digestive efficiency of the stomach.

Q What is indigestion? How do antacids cure it?

The end result of all the churning and digesting is that the food you have eaten is no longer recognizable. It becomes a creamy, very acidic liquid called **chyme.** At the other end of the stomach is another ring of muscle. About every 20 seconds, it opens just a tiny bit and squirts a few tablespoonfuls of chyme into the small intestine. While some of the hamburger's carbohydrates and proteins have been partly or completely digested in the stomach, the lettuce is undigested, as are the lipids from the meat, which are suspended as greasy droplets within the chyme.

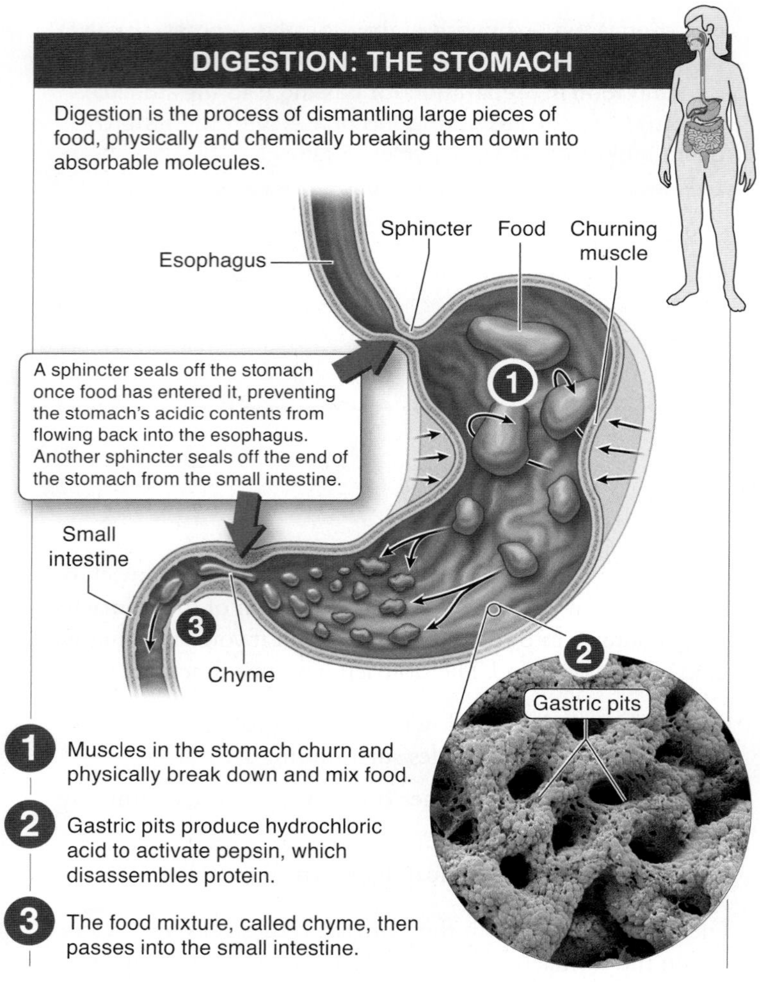

DIGESTION: THE STOMACH

Digestion is the process of dismantling large pieces of food, physically and chemically breaking them down into absorbable molecules.

Esophagus

Sphincter

Food

Churning muscle

A sphincter seals off the stomach once food has entered it, preventing the stomach's acidic contents from flowing back into the esophagus. Another sphincter seals off the end of the stomach from the small intestine.

Small intestine

Chyme

Gastric pits

1. Muscles in the stomach churn and physically break down and mix food.
2. Gastric pits produce hydrochloric acid to activate pepsin, which disassembles protein.
3. The food mixture, called chyme, then passes into the small intestine.

FIGURE 22-22 The role of the stomach in digestion.

Besides helping to chemically break down food in your stomach, the strong acid also kills most of the bacteria that you might consume. Most bacteria just can't survive when the pH gets so acidic (although some microbes occasionally manage to sneak through by hiding in pockets of food with higher pH as the food passes through the stomach). In the StreetBio at the end of this chapter, we explore the ulcer-inducing effects of one species of bacteria that can live in your stomach.

Q Humans use heat for cooking and often marinate foods in vinegar or lemon juice. How do these processes help with digestion?

Interestingly, across dozens and dozens of cultures, humans have developed common ways of preparing their food, including using heat and marinating food in acidic solutions, such as vinegar or lemon juice. Because harsh conditions such as heat and acid help to disrupt the tissue of food items, they increase the efficiency with which digestive enzymes can make contact with the food and break it down

The Small Intestine

Digestion only begins in the mouth and stomach. Most chemical digestion actually occurs in the **small intestine,** a long thin tube connected to the stomach in which most digestion takes place. It is about 20 feet long and winds all around your abdominal cavity (how else could 20 feet of tubing be packed into such a small area?). As we've seen, as a sphincter at the end of the stomach relaxes, creating an opening, small amounts of chyme are squirted from the stomach. The creamy substance slowly moves through the small intestine, pushed by the rhythmic contractions of peristalsis. As the chyme makes its way through the small intestine, the macromolecules are gradually digested, with help from the pancreas and the liver (**FIGURE 22-23).**

The **pancreas,** nestled at the point where the stomach connects to the small intestine, plays a central role in digestion by secreting pancreatic juice through a duct into the very beginning portion of the small intestine. This juice is a mixture of (1) chemicals that neutralize the acidic chyme, and

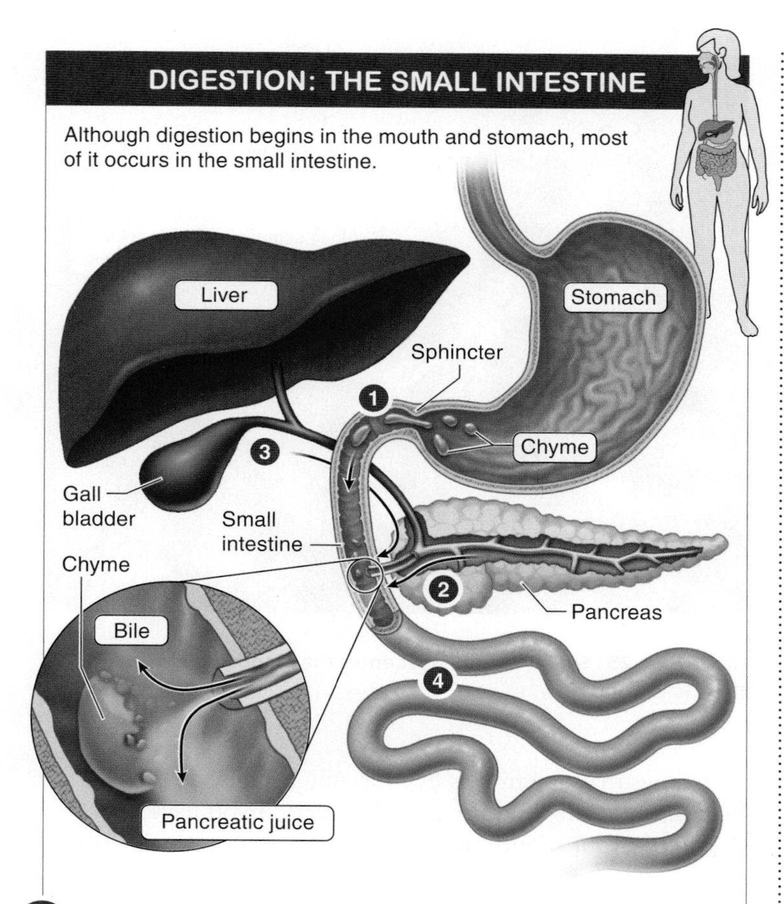

1 As a sphincter at the end of the stomach relaxes, small amounts of chyme are squirted into the small intestine.

2 The pancreas secretes pancreatic juice, which neutralizes the chyme and aids in the digestion of carbohydrates, proteins, and fats.

3 The liver produces bile, which travels from the gall bladder, where it is stored, to the small intestine, where it acts as a detergent to break up particles of fat.

4 Cells within the walls of the small intestine produce enzymes that further digest fats, carbohydrates, and proteins.

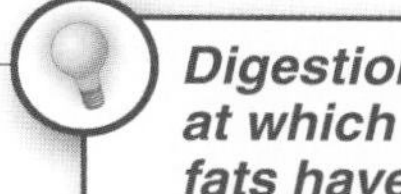

Digestion is completed in the small intestine, at which point carbohydrates, proteins, and fats have been broken down into simple sugars, amino acids, and fatty acids.

FIGURE 22-23 The role of the small intestine in digestion.

(2) enzymes that digest carbohydrates, proteins, and fats. Another organ that assists the small intestine with digestion is the **liver,** which produces **bile,** a juice that aids in the breakdown of fats. The liver sends the bile to the gall bladder, and from there it passes through a small duct into the small intestine, where the bile initiates the first step in fat digestion.

Remember, lipids are not water-soluble, so any fat from the hamburger is still suspended in the chyme, never dissolving into the watery substance of the digestive tract. Bile acts as a detergent that helps break up the suspended droplets of fat into much tinier particles. Once the lipids decrease in size, pancreatic enzymes can successfully break them down. In addition to the enzymes produced by the pancreas and the liver, the cells within the walls of the small intestine also generate enzymes that further digest fats, carbohydrates, and proteins. For a variety of reasons, people sometimes must have part of their small intestine surgically removed. Can you predict the effects this might have on digestion?

Digestion is completed in the small intestine when the consumed carbohydrates, proteins, and fats have been broken down into their component parts: simple sugars, amino acids, and fatty acids. These simple molecules can then be absorbed into the bloodstream, a process that we explore in the next section. But first let's look at a few cases in which digestion doesn't work properly or in which animals rely on some digestive assistance from bacteria, and a carnival trick that hinges on an understanding of digestion.

Intestinal Bacteria Have you noticed that certain types of food commonly give people gas? Some types of beans commonly cause such digestive distress. Enzyme deficiencies and metabolically versatile bacteria are the culprits. The beans contain the sugars raffinose and stachyose, which are indigestible by many people. And so when the sugars pass undigested through the small intestine and reach the large intestine, they are digested by bacteria. And the by-product of the bacterial feeding frenzy is gas, cramps, and flatulence. The product Beano contains enzymes that help digest the trouble-making sugars before our resident bacteria can get to them.

There is a wide variety of situations in which bacteria living inside animals similarly digest food molecules that the animals could not otherwise break down—and this has evolved as a benefit to the "host" animals as well as the bacteria. Living in termite guts, for example, are large colonies of symbiotic cellulose-digesting bacteria. The bacteria make it possible for the termite to eat wood and actually gain nutrients from it. The bacteria obtain energy by breaking down the cellulose in the wood into simple sugars that are valuable to the termite (as well as to the bacteria). If humans could figure out how to cultivate the same bacteria in our guts, perhaps we, too, could eat wood.

Leeches also have a symbiotic relationship with bacteria. As a leech sucks blood from an animal, bacteria—rather than its own enzymes—break down the blood proteins into their component amino acids, which the bacteria and the leech then share (**FIGURE 22-24**).

FIGURE 22-24 Leeches and bacteria in partnership.

FIGURE 22-25 Snake venom must enter the bloodstream to cause ill effects.

Q Snake venom is toxic, but if you drink it, in most cases it won't harm you. Why?

Poisonous Proteins Snake venom is a mixture of toxic proteins. Yet if you "milked" a snake's venom into a glass and drank it, the toxin wouldn't necessarily harm you (**FIGURE 22-25**). At carnivals and entertainment parks, snake experts will sometimes do exactly this, with no ill effects. Because the venom is a mixture of proteins, it is broken down by the stomach acids and enzymes of the digestive system. Once broken down into their component amino acids, the poison proteins become harmless. Snake venom must enter the bloodstream to cause its ill effects. That is why snakes bite with fangs that function as hypodermic needles, delivering the poison to the bloodstream. Of course, if you have any cuts inside your mouth or lesions elsewhere in your digestive system, some of the venom could get into your bloodstream when you drink it and poison you. For that reason, drinking venom isn't a wise stunt to attempt.

TAKE-HOME MESSAGE 22•11

Digestion—the process of dismantling large pieces of food, physically and chemically breaking them down into absorbable molecules—is the second phase of the breakdown of food in animals. It occurs primarily in the stomach and small intestine.

22•12

Absorption moves nutrients from your gut to your cells.

Chewing, tearing, churning, acidifying, dismantling. You put a lot of energy into reducing food to its simplest component molecules. Eventually, though, there is a payoff. That payoff is **absorption,** the process by which the energy-rich food particles are taken from the digestive tract into the bloodstream and then into the cells of the body, where they can be used for energy and building materials. Absorption occurs mainly in the small intestine; however, a few molecules are absorbed in the stomach, including aspirin and alcohol, which explains why drinking alcohol on an empty stomach can lead to unexpectedly rapid inebriation.

The secret to effective absorption is simple: surface area. For a molecule such as a sugar or an amino acid to be absorbed by the body, the molecule must be in direct contact with the cell membrane of the cell that is going to absorb it. The greater the number of cells that can come in contact with chyme passing through the small intestine, the greater the amount of nutrients that can be absorbed. For this reason, the tremendous surface area of the small intestine allows for very efficient absorption of nutrients (**FIGURE 22-26**). How does the small intestine achieve this? There are four primary ways.

1. The small intestine is long. About 20 feet long, as we saw above.
2. Rather than being a straight, smooth tube, it has many folds.
3. The interior lining of the small intestine is made up of thousands of small finger-like projections, called villi. These create lots of nooks and crannies where chyme can come in contact with absorptive cells.
4. Each of the cells along the villi has hundreds of its own tiny thread-like projections, called microvilli. These create micro-nooks and micro-crannies that allow more of the molecules in chyme to directly touch the membranes of absorptive cells.

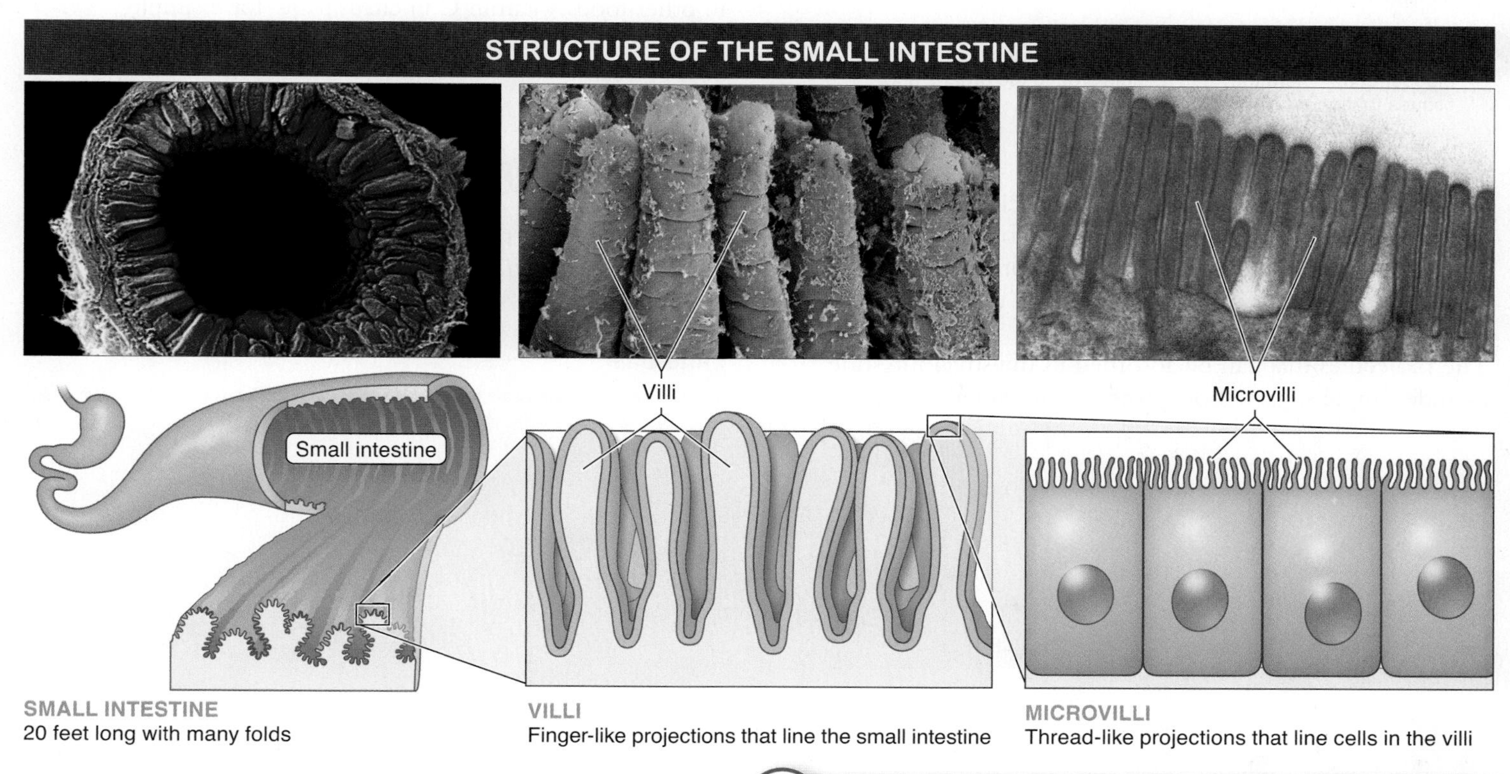

FIGURE 22-26 The small intestine has a very large area for absorbing nutrients.

The tremendous surface area of the small intestine allows for very efficient absorption of nutrients.

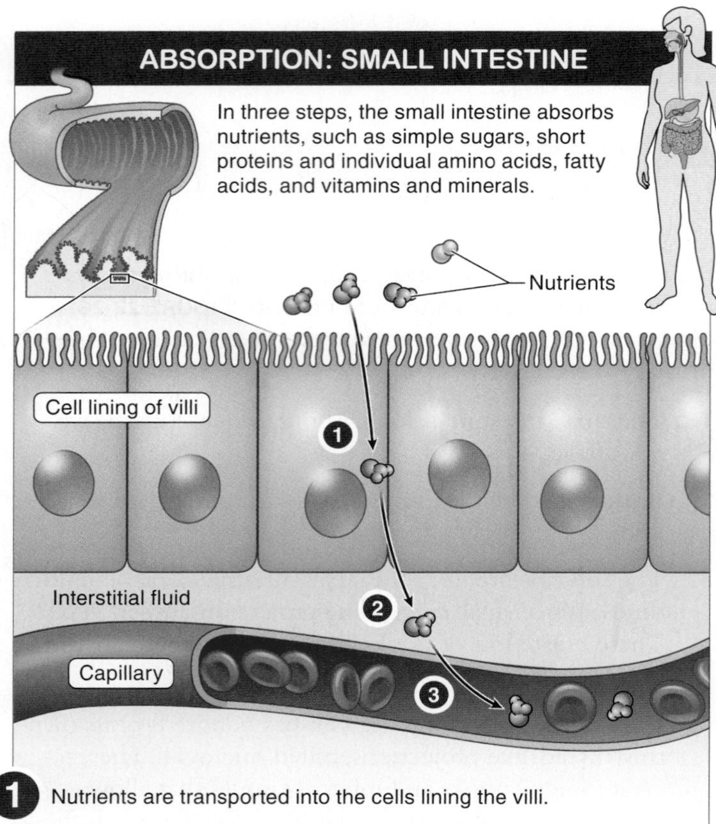

FIGURE 22-27 **Absorption moves nutrients into the bloodstream.**

Ultimately, all of these surface-area–enhancing structural features give the small intestine a huge amount of surface area for absorption. If it were spread out as a flat surface, it would be about the size of a tennis court.

The molecules that can be absorbed in the small intestine include simple sugars, short proteins, individual amino acids, fatty acids, vitamins, and minerals. When a nutrient molecule is small enough to be absorbed, it gravitates toward the villi lining the interior of the small intestine. The tiny particles become trapped in the microvilli and are drawn across the cell membrane into the cells (**FIGURE 22-27**). Then the nutrient diffuses out of the cell and into the interstitial fluid bathing the cells. Finally, from here the nutrients are picked up by capillaries and thus move into the bloodstream, where they can be delivered to the organs and tissues that need them.

Q Is it true that certain food combinations should not be eaten together?

There is a food myth that if certain foods are combined, digestion is impaired. This is not true. The myth ignores a couple of facts: the body is able to produce its digestive enzymes for fats, proteins, and carbohydrates simultaneously, and it can absorb nutrients regardless of which other nutrients are also in the digestive tract. In fact, the contrary is often true. Many foods enhance the absorption of nutrient molecules in other foods. Vitamin C in citrus fruits, for example, increases the efficiency of iron absorption from a meal of beef or beans.

TAKE-HOME MESSAGE 22•12

Absorption is the process by which energy-rich food particles are taken up from the digestive tract into the cells of the body, where they can be used for energy and building materials. It takes place primarily in the small intestine.

22•13

Elimination removes unusable material from your body.

You are much better at conservation and recycling than you imagine. Especially when it comes to the fluids and solids you consume. The last phase in the breakdown of food, **elimination,** takes place as what's left of the chyme—mostly indigestible materials—leaves the small intestine and enters the **large intestine,** also called the **colon.** Much larger in diameter than the small intestine (about 3 inches, vs. 1 inch in the small intestine) but only about 3–6 feet long, the large intestine serves to absorb water, salts, and some vitamins (**FIGURE 22-28**). The last part of the large intestine is the rectum and serves as a storage compartment for the remaining parts of consumed food, the feces, which can then be defecated.

It is important to achieve just the right balance of water absorption in the large intestine. If too much is absorbed, the remaining indigestible material becomes too solid and can't move easily through the last part of the digestive system, causing constipation. Alternatively, if too little absorption occurs, the body loses more water and this causes diarrhea. Fiber—including gums and cellulose—cannot be digested or absorbed. For this reason, when fiber is consumed, it increases fecal mass and speeds the movement of chyme through the colon (**FIGURE 22-29**). With additional mass, more water is attracted, softening the feces and making it easier to eliminate (that's why too much fiber can lead to diarrhea). Fiber also binds to bile, causing some of it to be eliminated, thereby reducing the body's ability to absorb cholesterol from food.

Q *Fiber is an indigestible carbohydrate. If we can't digest it, how can it be essential in our diet?*

Huge colonies of bacteria live in the colon. Before we are born, our guts are sterile and free of bacteria. During birth, we acquire some of our mother's vaginal and fecal bacteria, and by the age of two we have acquired bacteria from food and the environment, so that all of the numerous species present in adults are already in residence by that age. We generally have more bacterial cells than our own body cells at any given time (see Figure 13-3). The bacteria are usually harmless and live off the undigested materials that end up in the colon. Bacteria

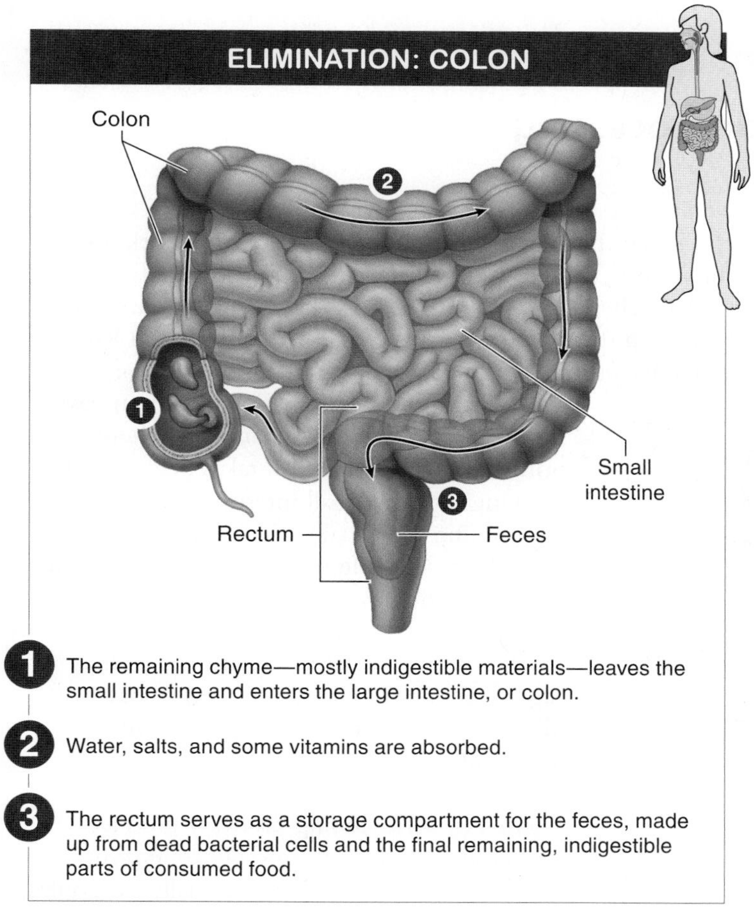

ELIMINATION: COLON

Colon
Small intestine
Rectum
Feces

1. The remaining chyme—mostly indigestible materials—leaves the small intestine and enters the large intestine, or colon.
2. Water, salts, and some vitamins are absorbed.
3. The rectum serves as a storage compartment for the feces, made up from dead bacterial cells and the final remaining, indigestible parts of consumed food.

FIGURE 22-28 **Elimination is the final step in the digestive process.**

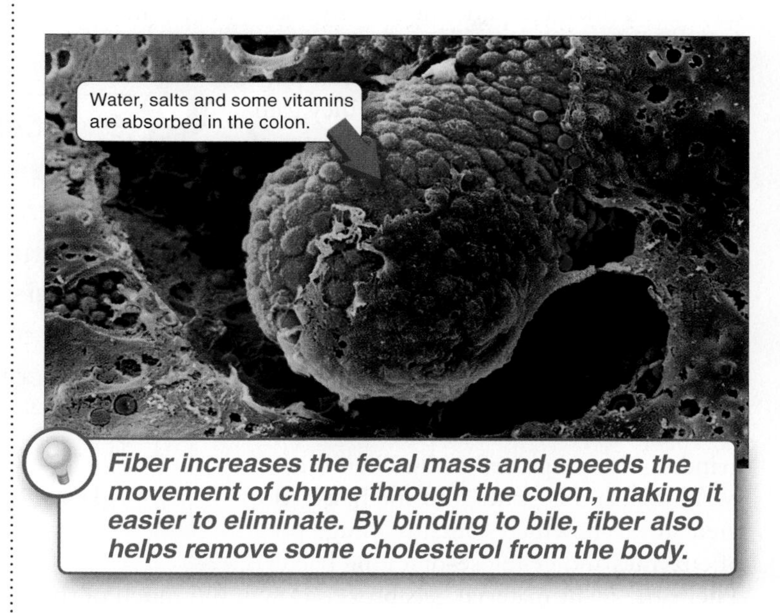

FIGURE 22-29 **The last remnants of a meal in the colon.**

also release important metabolic by-products such as vitamin K and one of the B vitamins, biotin. About half of the feces that we excrete each day is made up of dead bacterial cells, and the rest is mostly indigestible materials such as cellulose from the cell walls of plant matter and other types of fiber.

Q Why can taking antibiotics lead to vitamin deficiencies?

Antibiotics frequently (and unintentionally) kill a large proportion of the colon bacteria, in addition to whatever illness-causing microbe they were prescribed to kill. This can have multiple negative side effects. First, the transit of undigested materials through the colon may not be slowed down as much as usual, in which case less water is removed and diarrhea results. Also, a reduction in production of vitamin K and biotin can lead to deficiencies.

In the colon, water is removed from feces through osmosis. First, salts are pumped out of the colon, and then water moves out by simple diffusion (to an area where there are more ions). This process can be manipulated as a treatment for constipation. Laxatives contain magnesium salts that are so slowly absorbed that they are still largely intact when they reach the colon. Laxatives work by increasing the salt concentration in the colon, so that more water remains. This additional water makes it easier to excrete feces (but can cause diarrhea in some cases).

Q Laxatives contain salts. Why might this reduce constipation?

TAKE-HOME MESSAGE 22•13

The last phase of food breakdown takes place as the mostly indigestible materials leave the small intestine and enter the large intestine. There, water and ions are absorbed before the remaining materials are defecated.

22•14

Animals have some alternative means for processing their food.

Food, food everywhere, but not a bite to eat. Cellulose could feed the world: it's everywhere—the major carbohydrate making up the cell wall around plant cells is cellulose—and there is a tremendous amount of energy stored in its chemical bonds. But most animals don't produce enzymes that break down cellulose. If they can figure out a way to digest it—and some species have—they gain access to one of the most plentiful sources of chemical energy on the planet. Let's investigate how some animals have done it.

Ruminant animals, such as cows, bison, deer, goats, and sheep, have evolved complex four-part stomachs in which they are able to digest plant matter that humans cannot (**FIGURE 22-30**). First, the grazing animals chew on the plant material for a while, grinding the tough cell walls. Then they swallow it, and it passes into the first part of their stomach, which contains a huge pool of enzymes and cellulose-digesting bacteria. There, the plant material gets broken down, and much of the cellulose is digested. To increase their energy-extraction efficiency, the animals then regurgitate the food back into the mouth, chew it some more, and swallow it again. Called "chewing the cud," the additional chewing further breaks up the plant cell walls so that the bacteria can have easier access to the cellulose, digesting more of it. From the first part of the stomach, the food then passes through the remaining chambers, where some additional digestion takes place, before moving into the small intestine and continuing the usual path of animal digestion. Cellulose-digesting bacteria have some of the most dangerous working conditions on earth: although cellulose is the primary component of the ruminant diet, ruminants also get a significant amount of protein every day by digesting many of the bacteria working in their gut.

Q Silverfish are insect pests in libraries rather than in kitchens. Why might this be? (Hint: they are one of only a few species of animals to produce enzymes that digest cellulose.)

Some insects, including silverfish, produce cellulose-digesting enzymes. These enzymes make it possible for them to actually eat books and paper, which are made of plant products containing cell walls made from cellulose.

There's more than one way to skin a plant. Other animals, without the large bacteria-filled stomach of ruminants, also employ cellulose-digesting bacteria to enable them to extract energy from cellulose. In these animals, including horses, rodents, rabbits, and koalas, the bacteria don't live in the stomach. Rather, they live farther down the digestive tract, within an outcropping of the small intestine called the cecum. This is a separate chamber where the food goes for a while and the cellulose gets digested, before the food continues down the intestine. In animals that don't break down cellulose, the cecum is much smaller. The cecum is almost non-existent in the meat-eating coyote, while the similarly-sized koala has a cecum that is 2 meters long, where bacteria can convert shredded eucalyptus leaves into usable food (**FIGURE 22-31**). In

FOUR-PART STOMACH

In ruminant animals, a complex four-part stomach has evolved in which they are able to digest the plant matter that humans cannot.

Cellulose-digesting bacteria

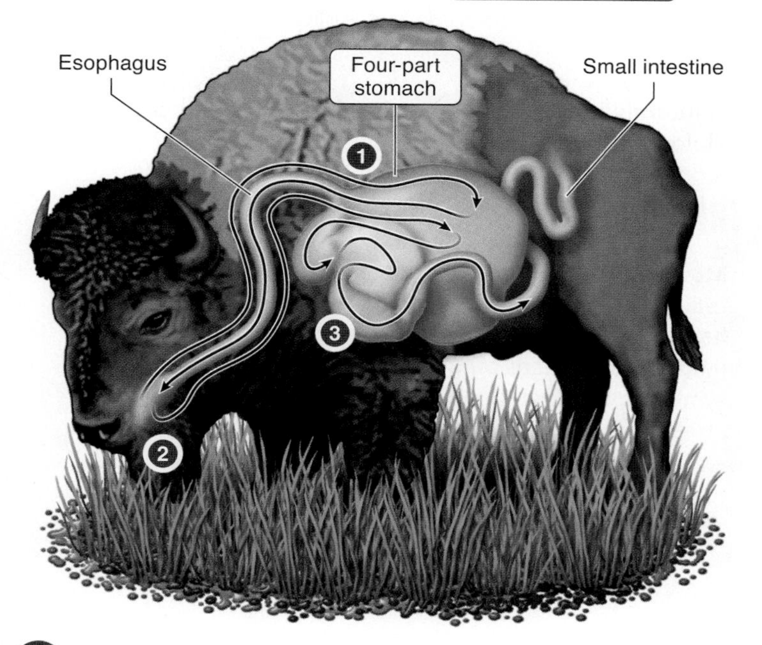

1. Plant material is swallowed and passed to the first part of the stomach, where it is broken down and digested by enzymes and cellulose-digesting bacteria.
2. Food is regurgitated, chewed more, then swallowed and passed to the first chamber again.
3. Food passes through the remaining chambers, where additional digestion takes place before it moves into the small intestine.

FIGURE 22-30 Ruminants can digest plant material that humans cannot.

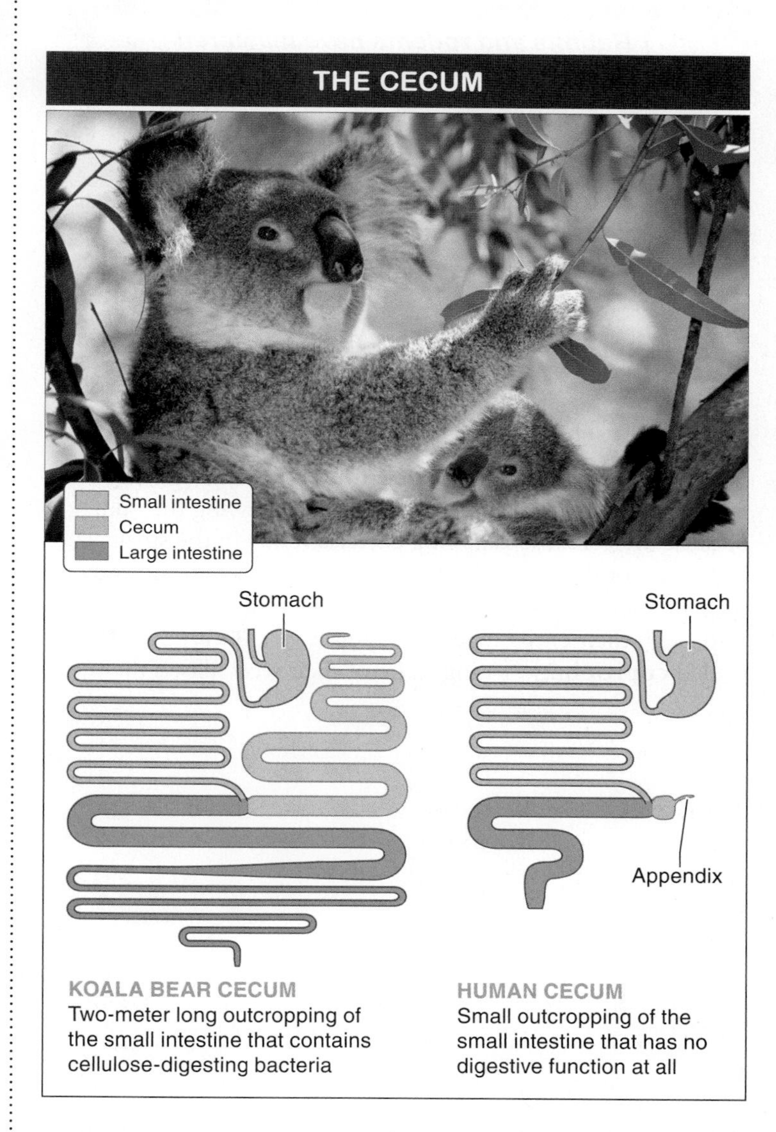

FIGURE 22-31 Koalas can eat foods that are indigestible for humans. The cecum is the site of cellulose digestion in some animals.

FIGURE 22-32 **Coprophagy increases nutritional intake from foods high in cellulose.**

humans, no cellulose-digesting bacteria live in the cecum (which is called the appendix in humans). Recent research suggests, however, that the appendix may play a role in immune function.

Rabbits and rodents can't increase their cellulose-digesting efficiency by regurgitating their food and chewing the cud, because the cellulose-digesting bacteria reside in the small intestine, not the stomach. But they've mastered another way to increase their ability to extract energy from their food: they pass it through their entire digestive system twice (**FIGURE 22-32**). Called **coprophagy,** eating some of their feces allows them to significantly increase their nutritional intake from their cellulose-laden diet.

There are about 150 different domestic and wild ruminant species including cows, goats, deer, buffalo, bison, giraffe, and elk.

TAKE-HOME MESSAGE 22•14

Most animals don't produce enzymes that break down cellulose. In ruminant animals, complex four-part stomachs have evolved in which the animals can digest plant matter that humans cannot, in part due to the presence of symbiotic cellulose-digesting bacteria. Other animals have alternative methods of utilizing cellulose-digesting bacteria.

④ What we eat profoundly affects our health.

Two sweet foods with very different nutritional composition.

22•15

What constitutes a healthy diet?

Choosing a healthy diet can be very difficult. There are more than 50,000 different foods that a person can choose from each day. And we not only must select which foods to eat, we must also consider how much of each to consume each day. Fortunately, there is not a single "best" solution that every person must adhere to in order to ensure good health. Rather, there are many different ways, making it possible to work within any person's particular likes and dislikes.

At the most basic level, just two requirements—quality and quantity—must be satisfied in the design of a healthy diet (**FIGURE 22-33**). First, a diet must contain sufficient amounts of each of the six categories of nutrients: water, proteins, carbohydrates, lipids, vitamins, and minerals (described in Sections 22-4–22-8). And second, a healthy diet must contain sufficient energy to support an individual's metabolic needs without containing a surplus of calories. But it is not easy to find the best strategy to satisfy these requirements.

Laboratory rodents—including those in longevity studies in which the animals have lived significantly longer than any wild rodent—have no such trouble with choosing a healthy diet. They are fed a diet that consists simply of water and special

FIGURE 22-33 Food quality and quantity are the main factors to consider in choosing a healthy, balanced diet.

food pellets. The food pellet formulation is the result of decades of research on animal nutrition. It includes carefully controlled amounts of each nutrient and does not vary from the time an animal reaches maturity until it dies. Each day, the rodents receive a specific number of the pellets, and each day the rodents eat them. Humans (thankfully?) do not subsist on a single food item. But for that reason, one of the most important principles guiding selection of a healthy diet is *balance*.

Balance in a diet is important because no one food is completely adequate. Milk, for example, is a very good source of protein and calcium, but does not contain sufficient iron for most adults. Meats, on the other hand, are rich in iron (as well as protein) but generally are poor sources of calcium. Every type of food, in fact, while providing some nutritional value, also falls short in some other essential nutrient, whether a protein or a specific vitamin or mineral. For this reason, nutritionists recommend consuming a variety of foods from each of the basic food groups—including (1) grains, (2) vegetables, (3) fruits, (4) milk and other dairy products, and (5) meats, poultry, and beans—consuming small quantities of unsaturated fats such as olive and canola oil, and exercising daily to manage body weight (**FIGURE 22-34**).

The U.S. Department of Agriculture (USDA) has established dietary guidelines, which it updates every five years. These highlight the most important issues with respect to determining the ideal quantity and quality of food consumed. In addition to urging people to take steps to ensure that their food is safe to eat, they also recommend the following.

1. Keep your weight within the recommended range.
2. Be physically active.
3. Choose a variety of fruits, vegetables, grains—particularly whole grains—and nonfat or low-fat milk and milk products.
4. Keep your diet low in saturated fat, cholesterol, and total fat.
5. Keep your diet low in sugars relative to complex carbohydrates and fiber.
6. Keep your diet low in salt.
7. If you consume alcoholic beverages, do so in moderation.

Food labels are an often overlooked tool that can help in adhering to these guidelines and selecting a balanced diet. Ingredient lists, too, are very valuable, listing every ingredient in order of amount (by weight). Ultimately, following these guidelines can help reduce the incidence and severity of chronic diseases such as stroke and other manifestations of cardiovascular disease, diabetes, and cancer.

Q Why is it increasingly difficult to maintain a healthy diet as we get older?

Complicating the USDA's dietary guidelines is the fact that we need to reduce our caloric intake as we get older. Metabolic rate falls slowly but surely, beginning around age 30. Consequently, without an increase in activity, eating the same amount of food—even if it is a healthy diet—leads to a surplus of calories, a surplus that becomes larger and larger with each passing year. On average, most adults gain about half a pound every year throughout their thirties, forties, and fifties.

It is important to note that people who do not consume meat, poultry, fish, and/or milk products can still have a balanced diet. Legumes, seeds, and nuts can provide many of the same nutrients found in meat, including protein. Dark leafy vegetables can provide iron, another nutrient plentiful in meat. And in fact, because such diets tend to be lower in fat content, they can be valuable in helping to maintain a healthy body weight.

TAKE-HOME MESSAGE 22·15

A balanced diet contains adequate amounts of essential nutrients and energy, but not surplus amounts, and is low in substances—including saturated fats, cholesterol, sugar, salt, and alcohol—that can have adverse health effects when consumed in greater quantities.

RECOMMENDED DAILY FOOD INTAKE

GRAINS | VEGETABLES | FRUITS | OILS | DAIRY | MEAT & BEANS

Yogurt

Canola OIL

Because no one food is completely adequate, nutritionists recommend consuming a variety of foods from each of the basic food groups.

FIGURE 22-34 What nutritionists recommend.

22•16

Obesity can result from too much of a good thing.

Jerry and Louis Kahn are two brothers who really love to eat. The pair tip the scales at about 500 pounds each—and their hunger never subsides. What is unusual about the Kahns' difficulty controlling their weight is that doctors have pinpointed the exact physiological source of their problem. They both possess an altered gene that renders their bodies unable to recognize the internal measure of food intake and register "enough," much like a broken thermostat that never shuts off a heater. Jerry and Louis are eternally hungry because their brains never get the "we're full" message from their stomachs.

Like the Kahns, we all have inherited genes from our parents that significantly influence our body weights. The thing is, we don't need to carry a defective gene to become obese. All humans inherit dozens (possibly hundreds) of genes that influence body weight, and even in their "normal" condition, these genes are likely to lead to weight problems (**FIGURE 22-35**). But this raises a couple of questions. Why would it be in our genetic interests to have huge appetites? Why would natural selection favor such genes? To answer these questions, we must consider the harsher, less predictable environment in which *Homo sapiens* evolved.

Our ancestors lived off the land by hunting animals and gathering plants. Under such conditions, it was unclear where the next meal was coming from. Powerful, instinctual hunger kept our ancestors going in that tough, energetically demanding world. Imagine a time when the individuals of a population varied in their appetites. Some individuals thought of food day and night, and ate whenever they could. Others became satiated once their daily needs were met. Of these types, which included individuals with the biggest surplus of energy stored in their thighs and buttocks when food became scarce? Who weathered the famine, with calories left over for reproducing? Who is most likely to be your ancestor? In every case, it is the enthusiastic and not the restrained eaters. Recent studies using PET scans (positron emission tomography, a type of imaging) have supported this, demonstrating that just the sight of food sets off activity in the brain's pleasure centers—a mechanism that generally leads to the consumption of food when it is available.

The consequence of our perpetual hunger is not news: one of every four Americans is obese. In terms of size, plumpness gets labeled "obesity" when our body mass index hits 30. **Body mass index,** or **BMI,** equals body weight in kilograms divided by height in meters squared (kg/m^2). A BMI of 30 translates to about 209 pounds if you are 5 feet 10 inches tall, and 180 pounds if you are 5 feet 5 inches. With a BMI of 25 or higher, you are only considered "overweight." BMI isn't the only way to evaluate obesity, and it has its problems. Because the measure does not consider body composition, professional athletes and bodybuilders with unusually large muscle mass have BMIs that fall within the obese range. Still, it is a useful and easily obtained measure (**FIGURE 22-36**). Repeatedly, around the world, as societies get richer and as individuals age, they tend to become fatter. Most of us would reduce our risk of heart disease, stroke, and diabetes if we lost even as little as

FIGURE 22-35 Are your genes making it difficult for you to maintain a healthy weight?

BODY MASS INDEX (BMI)

Height	UNDER WEIGHT (<18.5)	HEALTHY WEIGHT (18.5–24.9)						OVERWEIGHT (25–29.9)					OBESE (>30)					
BMI	18	19	20	21	22	23	24	25	26	27	28	29	30	31	32	33	34	35
6′4″	148	156	164	172	180	189	197	205	213	221	230	238	246	254	263	271	279	287
6′3″	144	152	160	168	176	184	192	200	208	216	224	232	240	248	256	264	272	279
6′2″	141	148	155	163	171	179	186	194	202	210	218	225	233	241	249	256	264	272
6′1″	136	144	151	159	166	174	182	189	197	204	212	219	227	235	242	250	257	265
6′0″	132	140	147	154	162	169	177	184	191	199	206	213	221	228	235	242	250	258
5′11″	129	136	143	150	157	165	172	179	186	193	200	208	215	222	229	236	243	250
5′10″	126	132	139	146	153	160	167	174	181	188	195	202	209	216	222	229	236	243
5′9″	122	128	135	142	149	155	162	169	176	182	189	196	203	209	216	223	230	236
5′8″	118	125	131	138	144	151	158	164	171	177	184	190	197	203	210	216	223	230
5′7″	115	121	127	134	140	146	153	159	166	172	178	185	191	198	204	211	217	223
5′6″	112	118	124	130	136	142	148	155	161	167	173	179	186	192	198	204	210	216
5′5″	108	114	120	126	132	138	144	150	156	162	168	174	180	186	192	198	204	210
5′4″	105	110	116	122	128	134	140	145	151	157	163	169	174	180	186	192	197	204
5′3″	102	107	113	118	124	130	135	141	146	152	158	163	169	175	180	186	191	197
5′2″	98	104	109	115	120	126	131	136	142	147	153	158	164	169	175	180	186	191
5′1″	95	100	106	111	116	122	127	132	137	143	148	153	158	164	169	174	180	185
5′0″	92	97	102	107	112	118	123	128	133	138	143	148	153	158	163	168	174	179

Body Weight (Pounds)

FIGURE 22-36 One indicator of healthy body weight: body mass index. The BMI indicates the ratio between height and body weight; the chart shows healthy body-weight ranges.

10 pounds. But because of the "famine-fearing" genes we carry, this is easier said than done.

From the cellular perspective, the specific causes of obesity are an increase in the size and number of fat cells an individual carries. Excess calories can be converted to fat regardless of whether they come from fats, carbohydrates, or proteins. Whatever the source, the fat molecules are added right to the fat cells in the body's adipose tissue. Fat cells increase in number primarily during the late years of childhood and the early teens (**FIGURE 22-37**). After that, the cells tend to grow in size rather than number with excess caloric intake, although when a fat cell becomes too large it will divide. The reverse, unfortunately, is not true. When people lose weight, their fat cells become smaller but are never lost completely. Consequently, it is especially important to avoid obesity early in life.

FIGURE 22-37 It's important to maintain a healthy body weight during childhood.

The problems of obesity are not restricted to humans. Most primates in zoos are also overweight, as are many pets and most laboratory research animals. In each case, the reason is the same as for humans. In their natural world, food is limited and unpredictable. Animals overeat when they do not need to expend much energy to acquire food and food is plentiful. In the next section, we explore some weight-loss diets and why they are almost universally unsuccessful.

TAKE-HOME MESSAGE 22•16

Food supplies were unpredictable for our early ancestors, leading to the evolution of strong appetites. In the modern industrial world, such instincts have problematic consequences, leading most people to overeat and to weigh more than they would prefer and more than is healthy. Zoo animals, laboratory animals, and domesticated animals have a similar problem.

22•17

Weight-loss diets are a losing proposition.

> I can reason down or deny everything, except this perpetual Belly: feed he must and will, and I cannot make him respectable.
>
> —Ralph Waldo Emerson, *Representative Men,* 1850

Weight loss is both a simple and a complicated problem. It is simple because there is one complete and perfect plan that guarantees success; it requires only five words of description: "Eat less. Move around more." Regardless of the genes it carries, any animal of any species will lose weight when it expends more calories than it consumes. The equation holds whether the calories are consumed in hamburgers or in fresh, organic vegetables. Similarly, it doesn't matter whether energy is expended in the weight room or on the couch.

But weight loss is also complex: while we know that eating less and moving around more are all that is necessary, and while almost all short-term "diets" work, in the long-run, with very few exceptions, they fail. (**FIGURE 22-38**). More than half of the advice books currently on the *New York Times* bestseller list promise new plans for losing weight. Five years ago that was also true—although the bestsellers were different books. It may seem as if there are plenty of new ideas about how to lose weight, yet the problem has not gone away. Rather, it has gotten worse.

FIGURE 22-38 Weight-loss diets make for popular reading.

Current interventions designed to facilitate weight loss range from mild to extreme. They fall into three categories: drugs, surgery, and behavior modification. Each has both promising and problematic elements (**FIGURE 22-39**).

Drugs and Other Chemical Interventions

Xenical. The most promising of recently developed weight-control drugs is Xenical, a product that interferes with fat digestion by binding to lipases (fat-digesting enzymes) and blocking them from doing their job. As a consequence, some of the fat in the digestive system passes through the body without absorption. In randomized, double-blind, placebo-controlled clinical studies—the gold standard for experimental design—Xenical has succeeded in helping people lose about 10 pounds over the course of one year. In the second year, these people regained some of their weight, but still ended up a bit lighter and with reduced circulating cholesterol and blood pressure. These findings are promising, but no long-term studies have been conducted to assess the permanence of the weight loss or potential side effects. Additionally, at a cost of more than $1,000 per year, Xenical is too expensive for most people to use.

Meridia. Approved for the treatment of obesity, this appetite-suppressant has shown promise as a weight-loss product. It reduces the rate at which the chemical serotonin, associated

INTERVENTIONS TO FACILITATE WEIGHT LOSS

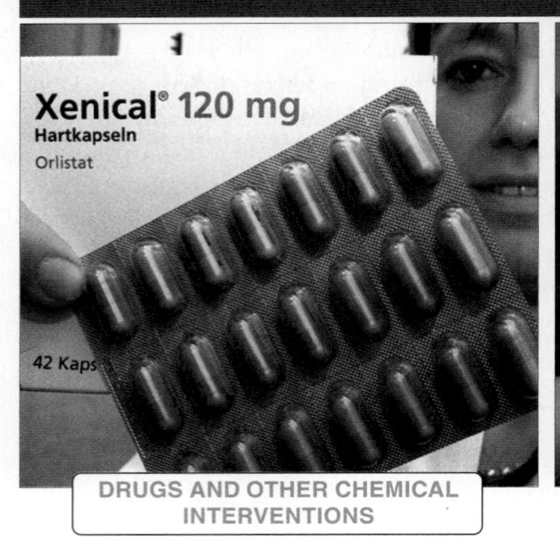

DRUGS AND OTHER CHEMICAL INTERVENTIONS

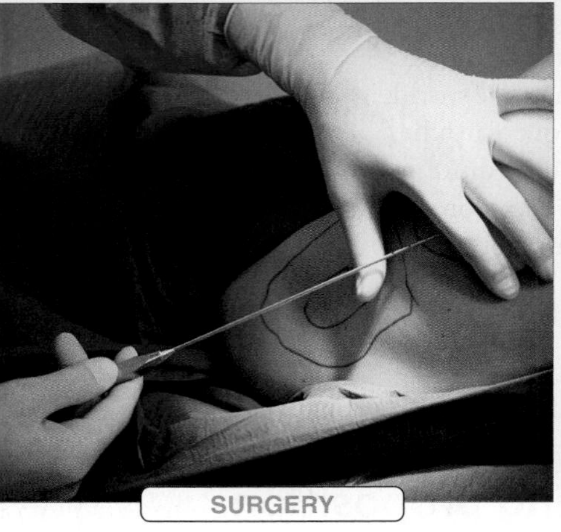

SURGERY

BEHAVIOR MODIFICATIONS: WEIGHT-LOSS DIETS

FIGURE 22-39 What are the promises and perils of weight-loss interventions?

with satiation, is removed from the space between nerve cells in the brain. Concerns over possible adverse effects on the cardiovascular system, however, have prompted some opposition to continued availability of this prescription drug.

Olestra and NutraSweet. Some artificially created molecules, such as Olestra and NutraSweet, can trick our taste buds into responding as if a molecule of sugar or fat, respectively, were present. Olestra, for example, is designed to have a taste and texture similar to fats, but in an indigestible molecule (see Section 2-13). After stimulating the taste buds that normally respond to fats, it ultimately is excreted, without having yielded any energy (read: without having deposited unwanted fat in your body). Still, as the NutraSweet cookie study described earlier in this chapter illustrates, our bodies have been built to detect reductions in caloric intake and to respond by increasing consumption. We're not tricked quite so easily. For this reason, low-calorie versions of food are unlikely to be successful on their own in generating permanent weight loss.

Caffeine and other stimulants. A variety of products claim to increase the expenditure of energy without an offsetting increase in appetite. This approach is theoretically sound, but there are no published data on either the safety or efficiency of these unregulated products. While clinical trials demonstrate that stimulants such as caffeine can produce short-term weight loss in the range of 5–10 pounds, long-term studies reveal that most of the short-term losses do not last.

Placebo. During clinical studies for diet pills, one observation was quite unexpected. As we learned in Chapter 1, researchers testing new compounds always have a "placebo" group. Subjects in this group go through all the same motions but get pills with none of the test drug. Because the placebo pills and the test drug pills are identical in appearance, no one knows which patients are getting the drug candidate and which are getting blanks. The goal is to separate the effects of the test drug from the effects of the monitoring process. As expected, some new drugs work and others fail, but here's the strangest finding: people in placebo groups always lose weight. In a study on the effectiveness of Xenical, for instance, more than 25% of the people in the placebo group lost at least 10 pounds. How can this be? While those in the placebo group aren't using drugs, they are keeping track of their weight and are more aware of their food intake than usual. This mindfulness may be the only "secret" behind the success of some crazy and non-scientific diets that advocate, for example, only eating food of a certain color (but "as much as you want") on a certain day. Careful monitoring proves to be a crucial component of weight control.

Surgery

Liposuction. Liposuction is fast becoming one of the most popular surgical procedures in the United States, with close to half a million performed each year. In this procedure, doctors directly remove fat cells from various parts of the body, using a hollow tube and a suction device. Over time, however, individuals who have undergone liposuction tend to regain all the lost weight. In fact, the only follow-up study on liposuction found that within a few months, nearly half of the patients weighed more than they did before the surgery. This is because body weight is a function of caloric intake and energy expenditure; fat cells can be removed, but without a change in the inputs and outputs, eventually the body will achieve the same composition.

Bariatric surgery and stomach banding. More invasive surgical procedures are also possible. The most effective is also the most extreme. In a type of surgical procedure known as bariatric surgery, surgeons bypass a significant portion of the small intestine and seal off, by stomach banding, most of a person's stomach. This has two effects. First, it reduces the amount of food people can eat before becoming full—they actually become violently ill if they consume more than a few tablespoons at one time. And second, it reduces the ability of their small intestine to digest and absorb nutrients. Before all of the nutrients that they consume can be absorbed, the food material passes through the shortened digestive system and is excreted. The surgery carries significant risks, though, and leads to major nutritional deficiencies in almost a third of all cases. Still, something can be said for bariatric surgery that cannot be said unequivocally for any drug or diet: it works. One study of more than 600 patients found that after 14 years, the average weight loss was 100 pounds! Additionally, many health problems such as diabetes, high cholesterol, and high blood pressure went away.

Behavior Modification: Weight-Loss Diets

Portion control and general caloric restriction. Can we change our eating habits permanently? To answer this question, researchers put a group of monkeys on a very low-calorie diet. The monkeys shed pounds initially, then stabilized at much lower weights for two years. After two years they were given unlimited access to food. Did they maintain their new weights? Absolutely not. After spending close to 10% of their lives at a constant, low weight, these monkeys quickly returned to their original, pre-diet weights.

This monkey example suggests that there might be a "set point" for body weight. A variety of human and non-human studies do, in fact, reveal that just like a thermostat for a room, when weight is below the set point, the body induces calorie-seeking behavior, and when weight is above the set point, the mind and body are free to pursue other goals. In a human study equivalent to the hungry monkey study, researchers observed a group of successful dieters over time. The newly skinny people had lost an average of 70 pounds per person through a comprehensive program. Three years after completing the program, the participants had, on average, regained all of their lost weight.

Low-carbohydrate diets. These are the latest in a long series of fad diets that enjoy popularity before being discarded for failing to deliver on the claims of their proponents and the promise of their short-term success. They are based on the incorrect assumption that carbohydrates universally cause rapid increases in blood sugar, followed by large releases of insulin from the pancreas, leading to immediate and efficient conversion of the blood sugar into fats that are stored throughout the body. While this scenario is correct for some carbohydrates, other carbohydrates, such as whole wheat pasta, produce less of an insulin response than a similar amount of high-protein beef. Similarly, many high-fiber fruits and vegetables are also rich in carbohydrates yet increase blood sugar levels only very gradually and elicit very small insulin responses. As a consequence, blanket avoidance of carbohydrates can lead to insufficient fiber intake as well as vitamin and mineral deficiencies. Unlimited consumption of fat, too, can lead to a caloric surplus and weight gain. Low-carbohydrate diets do lead reliably to impressive initial weight loss. This is because in the absence of dietary carbohydrates, glycogen (and the large amount of water bound to it) in the liver and muscles is quickly used up.

General Problems with Weight-Loss Diets The main problems with most popular weight-loss diets are that (1) they focus on reducing weight (even weight due to water) rather than reducing body fat; (2) they reduce muscle mass, the body tissue best able to burn fat; (3) because they reduce body weight too rapidly, they trigger several defense mechanisms designed to preserve the body's energy reserves; and (4) they don't focus enough on the other side of the energy equation: exercise. (**FIGURE 22-40**). This generally leads to four problems.

GENERAL PROBLEMS WITH WEIGHT-LOSS DIETS
- They focus on reducing weight (even weight due to water) rather than body fat.
- They reduce muscle mass, the body tissue best able to burn fat.
- Because they reduce body weight too rapidly, they trigger several defense mechanisms designed to preserve the body's energy reserves, generally leading to even more problems.
- They don't focus enough on the other side of the energy equation: exercise.

FIGURE 22-40 Weight-loss diets often fail.

1. *Nutritional deficiencies.* Consider how a family facing a financial crunch may be forced to defer important work, such as fixing the car's brakes. Similarly, when the body goes into efficiency mode to combat a shortage of calories, a variety of systems get modulated down or turned off. Hungry lab animals, for example, almost completely lose their sex drive and may be less adept at fighting infection. Nearly all popular diets are seriously deficient in vitamins, minerals, and fiber.

2. *Metabolic rate reduction.* Sudden caloric restriction leads to a rapid reduction in basal metabolic rate, and this lower rate remains for several weeks after resuming normal caloric intake. With a reduced BMR, your body burns fat at a slower rate.

3. *Loss of muscle mass and body fluids rather than body fat.* Low-calorie diets lead to weight loss, but as much as 45% of the weight loss comes from the loss of muscle mass. Muscle tissue is significantly more metabolically active than the relatively inert fat cells. As a consequence, the body's fat-burning capacity is significantly reduced along with its muscle mass. Weight loss, as we've noted, is also commonly due to loss of the water bound to glycogen in muscles and the liver.

4. *Increased lipoprotein lipase activity.* With a reduction in caloric intake, your body increases the activity of an enzyme called lipoprotein lipase. This enzyme is responsible for converting nutrients in the bloodstream to fats for storage. As with the metabolic rate reduction, the increased activity of lipoprotein lipase continues for weeks following resumption of a normal diet.

An evaluation by the National Institutes of Health of *all* the major diet programs concluded that there was no good evidence that any of the programs reliably led to long-term weight loss. A large study published in the *New England Journal of Medicine* came to a similar conclusion in 2009. The claims made by weight-loss programs rely on anecdotal accounts of individuals that are not at all representative of the average outcome. Unfortunately, this means that there are no easy solutions to the problem of weight control. As long as we live in our modern, zoo-like environment of plenty, we're going to struggle with natural systems that relentlessly seek out and efficiently store calories.

TAKE-HOME MESSAGE 22•17

Weight loss is both a simple and a complicated problem. There is only one complete and perfect plan that guarantees success: reduced caloric intake and increased caloric expenditure. Interventions designed to facilitate weight loss involve drugs, surgery, or behavior modification, none of which is reliably successful.

22•18

Diabetes is caused by the body's inability to regulate blood sugar effectively.

A finely tuned, responsive machine. When it functions properly, the human body can resemble such a machine. Nowhere is this more true that in the processing of food. After you digest and absorb food, there's an increase in the amount of glucose circulating in your bloodstream. This triggers the release of insulin by your pancreas. Insulin is a storage-stimulating hormone that causes your body's cells, especially muscle cells and fat cells, to pull the glucose in. They can then either use it for energy or convert it to glycogen or fat for storage until the energy is needed some other time. This leads to a reduction in glucose in the bloodstream and brings your blood sugar level back down to normal.

Foods differ in the extent to which they cause a surge in blood sugar and subsequent release of insulin. Foods that cause a rapid and large surge—such as orange juice, honey, and white potatoes—are classified as having a high **glycemic index** and are less desirable in the diet. More desirable are foods that cause only a slow, moderate increase, having a low glycemic index. These include whole grains and beans. With a reduced insulin response comes more efficient utilization of the sugar and lipids in the bloodstream and a reduction in fat storage. Insulin surges can also be reduced by eating smaller meals or, if you consume foods that have a high glycemic index, eating them with other foods as part of a meal.

Even finely tuned machines can malfunction. More than 10 million Americans (and 100 million people worldwide) have problems with their insulin-response systems, referred to as **diabetes.** These problems are chiefly of two types (**FIGURE 22-41**): either (1) the pancreas doesn't secrete enough insulin in response to an increase in blood sugar, or (2) the pancreas secretes plenty of insulin, but the cells of the body don't respond to it, usually due to a deficiency in glucose receptors on their

DIABETES: TYPE 1 vs. TYPE 2

Insulin—a hormone produced by the pancreas in response to increased blood sugar—causes your body's cells to pull glucose in from blood vessels.

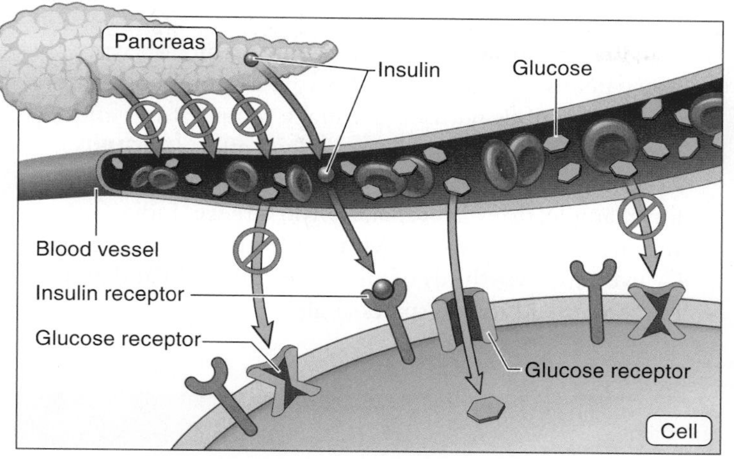

TYPE 1 DIABETES
The pancreas doesn't secrete enough insulin in response to an increase in blood sugar.

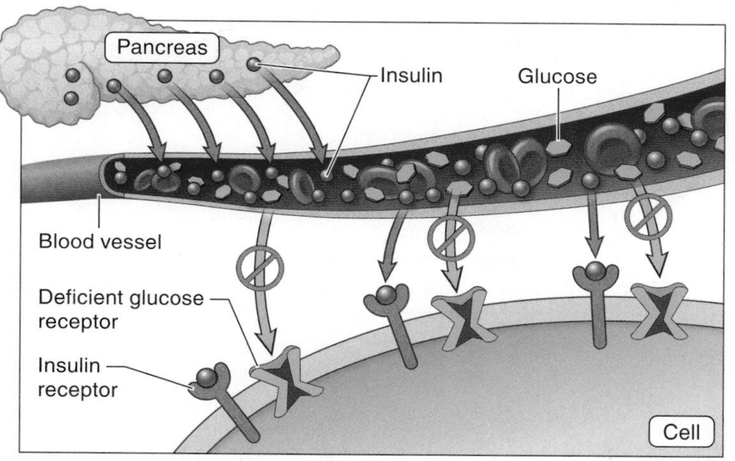

TYPE 2 DIABETES
The pancreas secretes plenty of insulin, but the cells of the body don't respond to it, usually due to a deficiency in glucose receptors on the cell membranes.

FIGURE 22-41 Diabetes is a disruption in the body's regulation of blood sugar.

Q Why must diabetics inject insulin rather than taking it in pill form?

cell membranes. In both cases, blood sugar remains high, and a host of problems can occur as a result of such persistence of glucose in the bloodstream.

The first type of diabetes (called type 1) is hereditary and usually occurs in children. Generally, people with type 1 diabetes have a pancreas that doesn't secrete enough insulin, thus they can be treated by insulin injections (**FIGURE 22-42**). (Unfortunately, because insulin is a protein, it is digested by stomach acids and the enzymes of the small intestine, so diabetics must inject the insulin in order to get it into their bloodstream intact.) The second type of diabetes (type 2) is about 10 times more common. It generally develops after the age of 40, and in about 90% of cases is a consequence of obesity. It appears that chronic and excessive amounts of sugar in the diet (and, later, in the bloodstream) reduce the sensitivity and/or number of cellular insulin receptors that help control blood sugar. This causes even greater releases of insulin, which can, ultimately, wear out the insulin-producing cells of the pancreas. Type 2 diabetes is treated by minimizing the glucose fluctuations in the bloodstream. Weight loss is also encouraged, as it reduces insulin resistance.

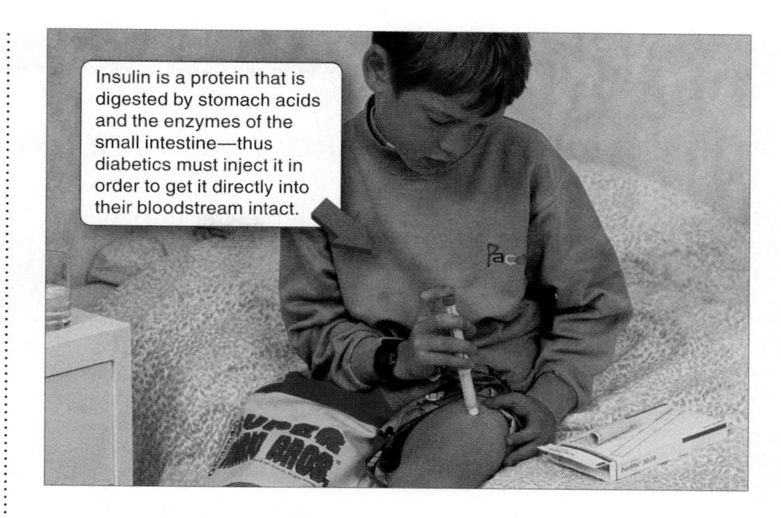

FIGURE 22-42 Type 1 diabetes is treated with insulin injections.

Because chronically high levels of blood sugar affect nearly all the cells of the body, the health effects can be far-reaching and severe. Cells in the eye lens can become deformed, causing blurry vision; blood vessels can be damaged, causing circulatory system and kidney malfunction; and nerves can also be damaged. Taken together, the varied problems resulting from diabetes make it the sixth leading cause of death in the United States.

TAKE-HOME MESSAGE 22•18

Digesting and absorbing food leads to an increase in the amount of glucose circulating in the bloodstream, which triggers the release of insulin by the pancreas, causing the body's cells, especially muscle cells and fat cells, to pull the glucose in for energy or storage. Problems with regulation of blood sugar, called diabetes, affect millions of people and are caused by heredity and poor diet.

Food and infection: spicy foods are natural antibiotics.

Have you ever tasted salsa so spicy that it brought tears to your eyes? After you gulped down a beverage to extinguish the fire in your mouth, did you ignore the distress signals from your taste buds and head right back for more? Why would anyone seek out such culinary torture? The answer is surprisingly simple, although not immediately apparent. We'll start by asking the question: what *are* spicy foods?

Spices come from plants. Plants are rooted in the ground, and their immobility makes them easy targets for their natural enemies. But although they can't run away, they don't just give in to those organisms that want to eat them. They fight back chemically. Plants have evolved to produce a large number of toxic compounds to help in this fight. The presence of these noxious chemicals makes the plant toxic to would-be predators, either killing them outright or being so unpleasant-tasting that predators search elsewhere for a meal.

Humans, somewhat unexpectedly, actually seek out these toxins. We don't think of them as toxins, though. We call them spices, and as long as they're not too toxic, they make our food taste better. There is also a deeper, more evolutionarily relevant reason we find our food to be more appealing with spices added. They help us in our own battles with natural enemies. "Food poisoning," it turns out—which affects about 1 out of 10 people in the United States each year—is a misleading term. It isn't the food that poisons us. Instead, it is microorganisms, particularly bacteria, that poison us. Humans and plants must perpetually fight against infection from microorganisms. This task is difficult because bacteria find easy entry into our bodies via the food we eat. After sitting out at room temperature for just a few hours, food can acquire huge amounts of bacteria. But by adding spices to our food, we take advantage of the plant defenses to strike back at the microbes, killing them before they make it to our digestive systems.

The "spices kill bugs" theory emerged from extensive observations and analyses. It has generated several predictions that have been tested.

Prediction 1: The magnitude of spice use in any part of the world should be related to the amount of microbial growth in that part of the world.

Observations: Put another way, this is a prediction that in warm, wet countries (where microbial growth occurs more readily), spice use should be greatest, while in cooler climates, spice use should be less. In an analysis of more than 4,500 meat-based recipes and 2,100 vegetable-based recipes from 36 countries, this prediction was overwhelmingly supported. There is a strong correlation between regions with high average temperatures and their use of spices with antimicrobial effects (**FIGURE 22-43**). Spices were included in every single recipe, for example, from Ethiopia, Kenya, Greece, India, Indonesia, Iran, Malaysia, Morocco, Nigeria, and Thailand. Conversely, two-thirds of all recipes from Finland and Norway called for no spices at all.

SPICE USE AND CLIMATE

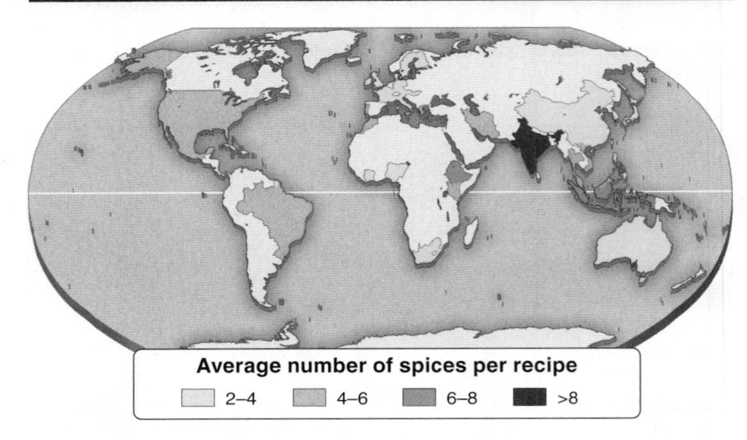

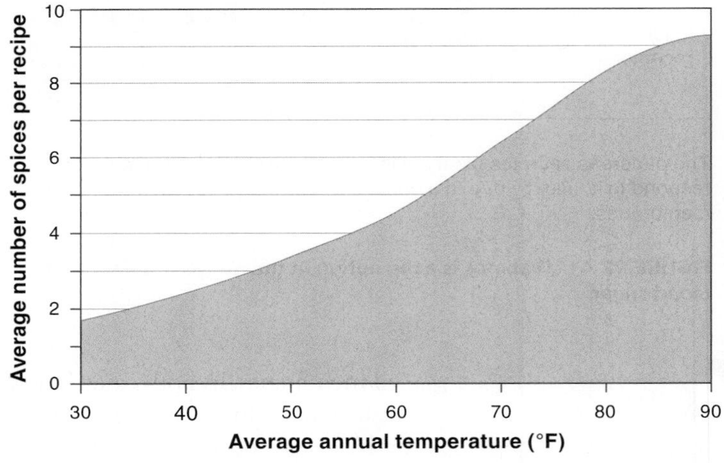

There is a strong correlation between regions having high average temperatures (where microbial growth occurs more readily) and their use of spices with antimicrobial effects.

FIGURE 22-43 Fighting natural enemies. Antimicrobial spices are used much more where temperatures are high.

SPICE USE AND ANTIMICROBIAL PROPERTIES

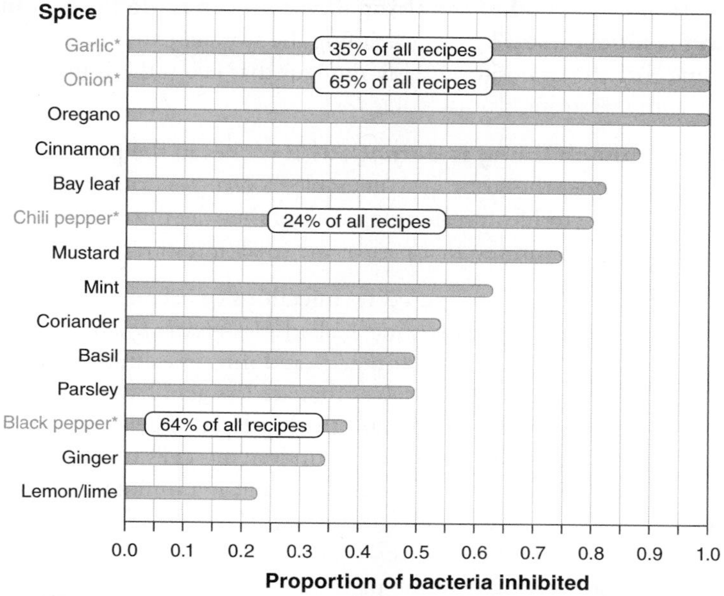

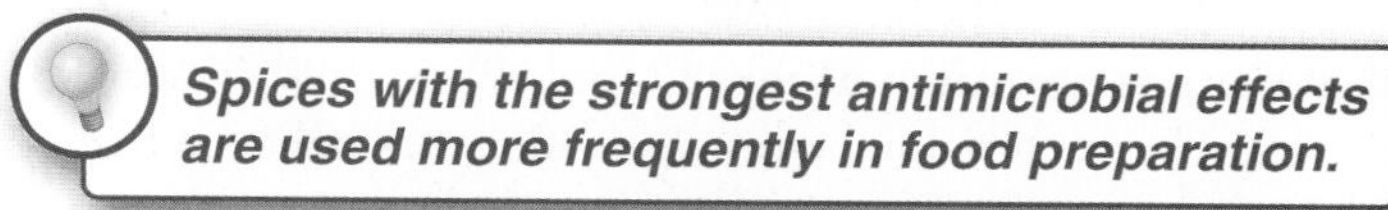

* The four most commonly used spices are shown in red.

FIGURE 22-44 Recipe ingredients with antimicrobial properties. (Photo shows woman with drying chili peppers.)

Two observations suggest that this relationship between spice use and climate cannot be attributed to the fact that more spices grow in the countries with hotter, wetter climates. First, onion and garlic, two of the most potent antimicrobial spices, grow in all of the countries studied but are used more frequently in the warmer countries. And second, although the warmer countries do have larger numbers of local spices to choose from, they use a greater proportion of the available spices than do the cooler countries.

Prediction 2: Spices with the strongest antimicrobial effects should be used most, and the weaker antimicrobial agents used less frequently.

Observations: In a test of 42 spices, in which each was added to individual plates of bacteria, nearly all exhibited antimicrobial properties—sometimes quite dramatically—when used in the amounts called for in recipes. For example, the toxic compound in cilantro was twice as effective at killing *Salmonella* bacteria as was the commonly used antibiotic drug gentamicin. Most of the spices tested were broad in their effects, too, killing most of the 30 different types of common food-borne bacteria tested. Overall, onions and black pepper, which had the best antimicrobial effects (they killed all species of bacteria they came into contact with), were used most frequently, appearing in 65% of all recipes. They were followed by black pepper (in 64% of all recipes), garlic (in 35% of all recipes), chili peppers (24%), and lemon and lime juice (23%). Other commonly used spices include parsley, ginger, bay leaf, coriander, and cinnamon (**FIGURE 22-44**).

Prediction 3: Because bacterial growth is faster and more common on meat than on vegetables, meat recipes should call for more spices than vegetable recipes.

Observations: The vegetable recipes called for, on average, 2.4 spices, while the meat recipes called for almost twice as many, with an average of 3.9.

Research on the connection between the use of spices and their antimicrobial properties continues. Nonetheless, it seems to be quite likely that humans did initially incorporate spices into food preparation because of their antimicrobial properties. Given that refrigeration is a very recent invention, consumption of food that had been "sanitized" by spices ensured the spice-consumers greater survival and reproduction. This example shows how seemingly arbitrary behaviors may reflect evolutionary adaptations.

TAKE-HOME MESSAGE 22·19

Plants produce toxic compounds that make the plant distasteful to would-be predators, either killing them outright or being so unpleasant-tasting that predators search elsewhere for a meal. Humans seek out these toxins, using them as spices that help us fight off illness-inducing microorganisms.

Knowledge You Can Use

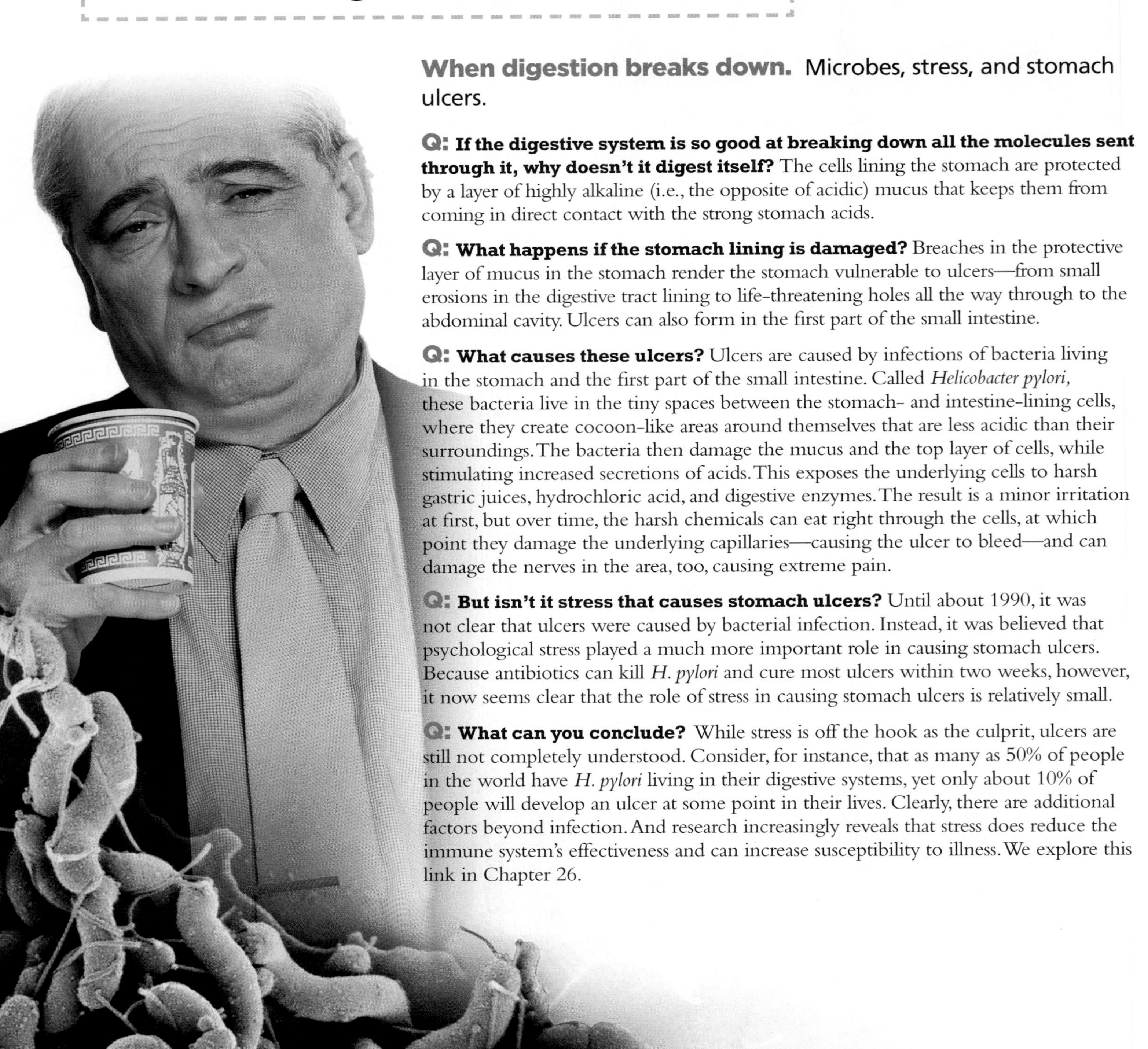

When digestion breaks down. Microbes, stress, and stomach ulcers.

Q: If the digestive system is so good at breaking down all the molecules sent through it, why doesn't it digest itself? The cells lining the stomach are protected by a layer of highly alkaline (i.e., the opposite of acidic) mucus that keeps them from coming in direct contact with the strong stomach acids.

Q: What happens if the stomach lining is damaged? Breaches in the protective layer of mucus in the stomach render the stomach vulnerable to ulcers—from small erosions in the digestive tract lining to life-threatening holes all the way through to the abdominal cavity. Ulcers can also form in the first part of the small intestine.

Q: What causes these ulcers? Ulcers are caused by infections of bacteria living in the stomach and the first part of the small intestine. Called *Helicobacter pylori,* these bacteria live in the tiny spaces between the stomach- and intestine-lining cells, where they create cocoon-like areas around themselves that are less acidic than their surroundings. The bacteria then damage the mucus and the top layer of cells, while stimulating increased secretions of acids. This exposes the underlying cells to harsh gastric juices, hydrochloric acid, and digestive enzymes. The result is a minor irritation at first, but over time, the harsh chemicals can eat right through the cells, at which point they damage the underlying capillaries—causing the ulcer to bleed—and can damage the nerves in the area, too, causing extreme pain.

Q: But isn't it stress that causes stomach ulcers? Until about 1990, it was not clear that ulcers were caused by bacterial infection. Instead, it was believed that psychological stress played a much more important role in causing stomach ulcers. Because antibiotics can kill *H. pylori* and cure most ulcers within two weeks, however, it now seems clear that the role of stress in causing stomach ulcers is relatively small.

Q: What can you conclude? While stress is off the hook as the culprit, ulcers are still not completely understood. Consider, for instance, that as many as 50% of people in the world have *H. pylori* living in their digestive systems, yet only about 10% of people will develop an ulcer at some point in their lives. Clearly, there are additional factors beyond infection. And research increasingly reveals that stress does reduce the immune system's effectiveness and can increase susceptibility to illness. We explore this link in Chapter 26.

❶ Food provides the raw materials for growth and the fuel to make it happen.

Animals must eat for two reasons: to acquire the energy needed for all growth and activity, and to acquire the raw materials required for life. Carnivores consume only other animals. Herbivores consume only plants. And omnivores consume both plants and animals. The minimal energetic needs of an individual not engaged in any activity are called the basal metabolic rate, or BMR.

❷ Nutrients are grouped into six categories.

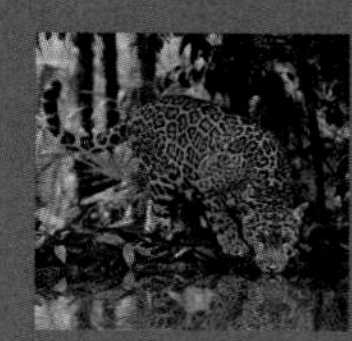

Water is probably the single most important component of a balanced diet, while proteins, carbohydrates, and fats provide calories and raw materials. Vitamins and minerals are organic and inorganic molecules (respectively) in the diet that are used in the production and action of enzymes and other molecules. Humans and other animals show an evolved preference for fatty foods over carbohydrate- or protein-laden foods.

❸ We extract energy and nutrients from food.

The digestive process in humans includes four distinct phases. (1) Ingestion involves tearing and grinding food in preparation for passing it to the stomach. (2) Digestion, the process of dismantling large pieces of food, physically and chemically breaking them down into absorbable molecules, occurs primarily in the stomach and small intestine. (3) Absorption, in the small intestine, is the process by which energy-rich food particles are taken into the cells of the body. (4) Water and ions are absorbed before the indigestible portions of consumed food are defecated. Most animals don't produce enzymes that break down cellulose, although many have symbiotic bacteria that do this for them.

❹ What we eat profoundly affects our health.

Ancestral humans experienced unpredictable food supplies, leading to the evolution of strong appetites. In the modern industrial world, such instincts have problematic consequences for weight control. Although there is, in theory, one complete and perfect plan that guarantees successful weight control—reduced caloric intake and increased caloric expenditure—the many interventions designed to help people lose weight are rarely successful. Problems with regulation of blood sugar, called diabetes, affect millions of people and are often caused by poor diet. Somewhat ironically, but for good evolutionary reasons, humans seek out toxic compounds that are produced by plants to ward off predators. Such compounds, used as spices, help us fight off illness-inducing microorganisms.

KEY TERMS

absorption, p. 815
basal metabolic rate (BMR), p. 795
bile, p. 813
body mass index (BMI), p. 823
calorie, p. 795
carnivores, p. 793
chyme, p. 812
colon, p. 817
coprophagy, p. 820
diabetes, p. 828
digestion, p. 792
elimination, p. 817
esophagus, p. 811
essential amino acid, p. 799
glycemic index, p. 828
herbivores, p. 793
ingestion, p. 810
kilocalorie (kcal), p. 795
large intestine, p. 817
liver, p. 813
mineral, p. 804
non-essential amino acid, p. 799
nutrient, p. 792
omnivores, p. 794
pancreas, p. 812
pepsin, p. 812
peristalsis, p. 811
ruminant, p. 818
small intestine, p. 812
sphincter, p. 811
stomach, p. 811
trachea, p. 811
vitamin, p. 804

Check Your Knowledge

1. Based on the types of diet they consume, spiders and owls are considered:
   a) carnivores.
   b) omnivores.
   c) herbivores.
   d) primary producers.
   e) fluid-feeders.

2. Basal metabolic rate:
   a) depends to a large degree on how active an individual is.
   b) does not vary across mammalian species.
   c) is a measure of the minimal energetic needs of an individual not engaged in any activity.
   d) is the same for males and females of any given species.
   e) All of the above are correct.

3. Water has many functions in animal bodies. These include all of the following except:
   a) lubricating many joints, the spinal cord, and the eyes.
   b) serving as a solvent for many vitamins and minerals.
   c) transporting nutrients and waste materials throughout the body.
   d) regulating growth and development.
   e) All of the above are functions of water in animal bodies.

4. Why do dieters lose large amounts of "water weight" during the first few days of a diet?
   a) The first, most accessible, molecules that can be broken down for energy are glycogen molecules in muscles and liver. And because large amounts of water are bound to glycogen, as the glycogen is removed from tissues, so too is the water.
   b) Dieters tend to reduce their consumption of all food and beverages—including their consumption of water—during the first days of a diet.
   c) Dieting causes a slight increase in body temperature, which leads to increased evaporative cooling and the loss of water.
   d) The fat cells of the body are primarily filled with water. As these cells are utilized for energy, the water is also lost.
   e) Actually, dieters do not lose "water weight" during the first few days of a diet. This is a myth.

5. On food packages, "fiber" refers to plant material that we can't fully digest but is important for maintaining a healthy digestive tract. "Fiber" refers to a type of:
   a) carbohydrate.
   b) nucleic acid.
   c) lipid.
   d) amino acid.
   e) protein.

6. Proteins are an essential component of a healthy diet for humans (and other animals). Their most common purpose is to serve as:
   a) fuel for running the body.
   b) raw material for growth.
   c) inorganic precursors for enzyme construction.
   d) organic precursors for membrane construction.
   e) long-term energy storage.

7. Vitamin and mineral supplements are generally necessary for individuals in all of the following categories except:
   a) post-menopausal women.
   b) healthy people with good diets.
   c) people on extremely low-calorie diets.
   d) pregnant women.
   e) people with limited milk consumption or sun exposure.

8. Most mammals (including humans) prefer the taste of fats to carbohydrates and proteins. Why?
   a) Fats are more easily digested than proteins or carbohydrates.
   b) Fats were much less available than proteins and carbohydrates in most ecosystems in which humans and other mammals evolved.
   c) The caloric content of a gram of fat is more than double that of a gram of protein or carbohydrate.
   d) The vitamin and mineral content of a gram of fat is more than double that of a gram of protein or carbohydrate.
   e) Many individuals lack the enzymes (such as lactase) to break down polysaccharides into their component sugars.

9. Why do birds eat gravel?
   a) Gravel contains most of the essential minerals for a bird's diet.
   b) Birds have poor vision and have difficulty distinguishing gravel from small seeds.
   c) By chewing on gravel, birds are able to sharpen their teeth, increasing their ability to crack open hard nuts or catch their prey.
   d) The gravel collects in the stomach, where it helps to grind up the food they eat.
   e) Because they have such a small brain relative to body size, birds tend to be the least intelligent of all vertebrates.

10. Across dozens of cultures, humans have developed common ways of preparing their food, including using heat and marinating food in acidic solutions, such as vinegar or lemon juice. How might these be adaptations that help with digestion?
    a) Harsh conditions such as heat and acid help to disrupt the tissue of food items. This increases the efficiency with which digestive enzymes can make contact with the food molecules and break them down.
    b) These methods of food preparation reduce the necessity of producing chyme and so increase the caloric efficiency of food intake.
    c) Because even the weakest acids are toxic to all bacteria, these methods of food preparation reduce the incidence of dietary-induced bacterial infection.
    d) These methods increase the body's ability to extract energy from normally indigestible cellulose.
    e) These methods of food preparation reduce the need for water consumption.

11. Which of the following statements about the small intestine is incorrect?

a) It is the primary site of digestion and absorption of nutrients into the bloodstream.
b) It is the chief site of absorption of water by the digestive system.
c) It is the longest part of the digestive tract.
d) It receives secretions from the gall bladder.
e) It is the part of the digestive tract where all food macromolecules can be broken down into absorbable monomers.

12. To leave the digestive tract and enter the cells of the body, a substance must cross a cell membrane. During which stage of digestion does this take place?

a) peristalsis
b) absorption
c) elimination
d) chemotaxis
e) ingestion

13. Digestion and absorption:

a) involve the breakdown of food into small nutrient molecules (absorption) and the passage of those molecules into the bloodstream (digestion).
b) both occur primarily in the large intestine, or colon.
c) are terms that describe the same process.
d) both occur primarily in the stomach.
e) involve the breakdown of food into small nutrient molecules (digestion) and the passage of those molecules into the bloodstream (absorption).

14. In the mammalian digestive system, vitamin-synthesizing symbiotic bacteria live primarily in the:

a) small intestine.
b) esophagus.
c) mouth.
d) large intestine.
e) stomach.

15. Though essentially the same organ, the human appendix is much smaller than a koala's cecum. This is because:

a) humans take supplements to obtain enough vitamins and do not need the extra vitamins produced by the bacteria housed in the appendix.
b) humans aid the digestive process by marinating and cooking foods, so they do not require a large appendix to aid food digestion.
c) humans maintain bacteria in their intestine rather than in their appendix, so the organ does not need to be as large as a koala's cecum.
d) in humans, no cellulose-digesting bacteria live in the cecum—called the appendix—and it has no digestive function at all.
e) humans get a greater proportion of their nutrients from meat than do koalas, so they do not need as large an organ to digest the plant matter they eat.

16. Cows have large populations of bacteria in their digestive systems. Which of the following best explains why?

a) The mutualistic microbes combat the harmful microbes that may enter a cow's body on its food.
b) Cows are able to use cellulose-producing bacteria to help them digest their food.
c) Most cows actually do not have large populations of bacteria in their digestive systems. Only infected cows have these microbes.
d) Scientists put the microbes there, so that they can study how the microbes affect cows' digestion.
e) The microbes metabolize the cellulose in the plants that cows eat.

17. The spices that many humans use to season their food:

a) are generally compounds produced by plants to reduce their risk of being eaten.
b) are actually dead microorganisms.
c) consist primarily of amino acids.
d) are the chief cause of ulcers.
e) can only be grown in tropical regions of the world.

Short-Answer Questions

1. What are the six groups of nutrients? What is their primary role in human health?

2. What are the four distinct phases used by humans to extract nutrients from food? For each phase, what significant activity takes place that contributes to nutrient harvesting and absorption?

3. Describe how some mammals, which do not have enzymes to break down cellulose, are able to extract nutrients from cellulose-containing food.

4. Describe the two forms of diabetes. What are the risk factors and treatments?

See Appendix for answers. For additional study questions, go to www.prep-u.com.

1 What is the nervous system?

- **23.1** Why do we need a nervous system?
- **23.2** Neurons are the building blocks of all nervous systems.

2 How do neurons work?

- **23.3** Dendrites receive external stimuli.
- **23.4** The action potential propagates a signal down the axon.
- **23.5** At the synapse, neurons interact with other cells.
- **23.6** There are many types of neurotransmitters.

3 Our senses detect and transmit stimuli.

- **23.7** Sensory receptors are our windows to the world around us.
- **23.8** Taste: an action potential serves up a taste sensation to the brain.
- **23.9** Smell: receptors in the nose detect airborne chemicals.
- **23.10** Vision: seeing is the perception of light by the brain.
- **23.11** Hearing: sound waves are collected by the ears and stimulate auditory neurons.
- **23.12** Touch: the brain perceives pressure, temperature, and pain.
- **23.13** Other senses help animals negotiate the world.

4 The muscular and skeletal systems enable movement.

- **23.14** Muscles generate force through contraction.
- **23.15** Skeletal systems enable movement, among several other important functions.

5 The brain is organized into distinct structures dedicated to specific functions.

- **23.16** The brain is organized into several distinct regions.
- **23.17** Measuring brain activity helps us understand how tasks are allocated.
- **23.18** Specific brain areas are involved in the processes of learning, language, and memory.

6 Drugs can hijack pleasure pathways.

- **23.19** Our nervous system can be tricked by chemicals.
- **23.20** A brain slows down when it needs sleep. Caffeine wakes it up.
- **23.21** Alcohol interferes with many different neurotransmitters.

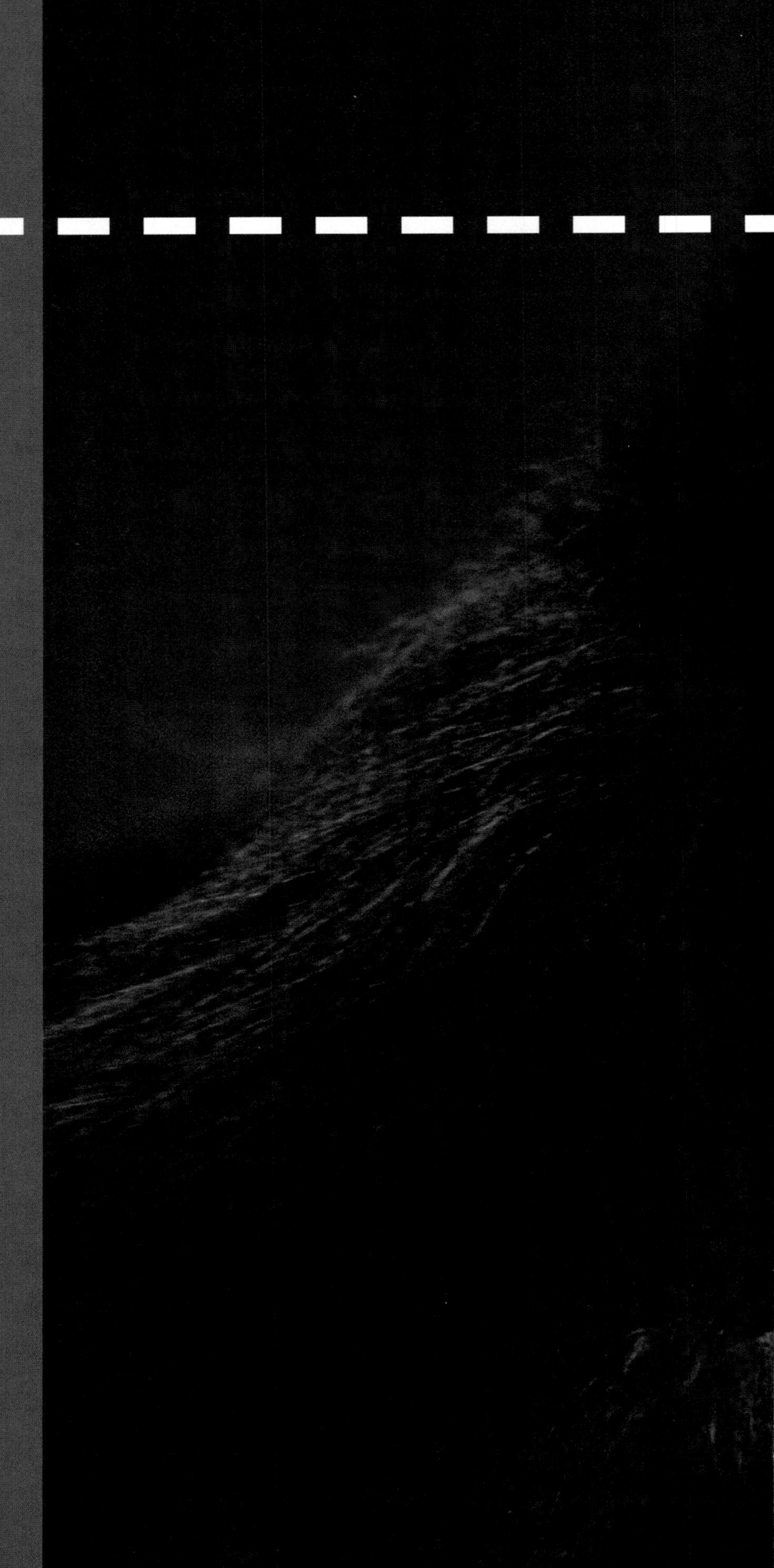

23

Nervous and Motor Systems

Actions, reactions, sensations, and addictions: meet your nervous system

❶ What is the nervous system?

The cat responds to a stimulus in the environment (possibly a dog, in this case) by raising hairs on its back, called hackles.

23•1

Why do we need a nervous system?

Imagine a world without pain. Think of all that you could achieve. You could work harder, run farther, and just plain feel better. Maybe that's why so many products are marketed as "painkillers."

But would it really be a better world? Think again.

The case of three-year-old Gabby Gingras tells us that we wouldn't really be happier in a pain-free world. Gabby feels no pain. She was born that way. At first it seemed like a blessing: just after birth, when nurses drew blood from her, she did not cry but instead slept peacefully through the procedure. When she falls down, she doesn't cry. But rather than being a gift that makes her life easier, Gabby's inability to feel pain is a crippling curse. As she has grown older, some unexpected consequences of her condition make this clear. First, she inadvertently damaged her eyes—permanently scratching one of her corneas—by poking her fingers into them. Later, chewing on a plastic toy, she cracked several of her teeth. Another girl, also born without the ability to feel pain, encountered even more severe problems. Holding her hand over a hot stove, she was severely burned—moving her hand only when she smelled the burning flesh. This condition is called "heritable sensory autonomic neuropathy," and it reveals that, no matter how unpleasant pain is to experience, it does have a very large benefit. The pains we feel alert us to the need to extricate ourselves from a dangerous situation and can prevent much greater suffering in the long run (**FIGURE 23-1**).

No one likes to feel pain, and yet those who cannot do so inevitably experience suffering far greater than the pain they are spared. Why are we built this way? Why is the experience of pain a necessary part of being a living animal? Pain tells us that some aspect of our environment is not hospitable and that we must take action to remove ourselves from it. If an animal is biting or stinging you, you try to get away from the animal. And what happens if your hand touches something hot? You don't even have to think about it: you quickly pull your hand away. In short, bad stuff happens to animals. The survival and well being of organisms depends on their awareness of the world around them, and the ability to limit exposure to physically harmful situations.

Present in all multicellular animals other than sponges, the **nervous system** is a network of cells that collects information about the organism's internal and external environments, processes that information, and sends signals to effectors, which are muscles and glands that are capable of responding to the information. Our nervous system

The ability to feel pain, no matter how unpleasant, makes us aware of the world around us and can prevent much greater damage in the long run.

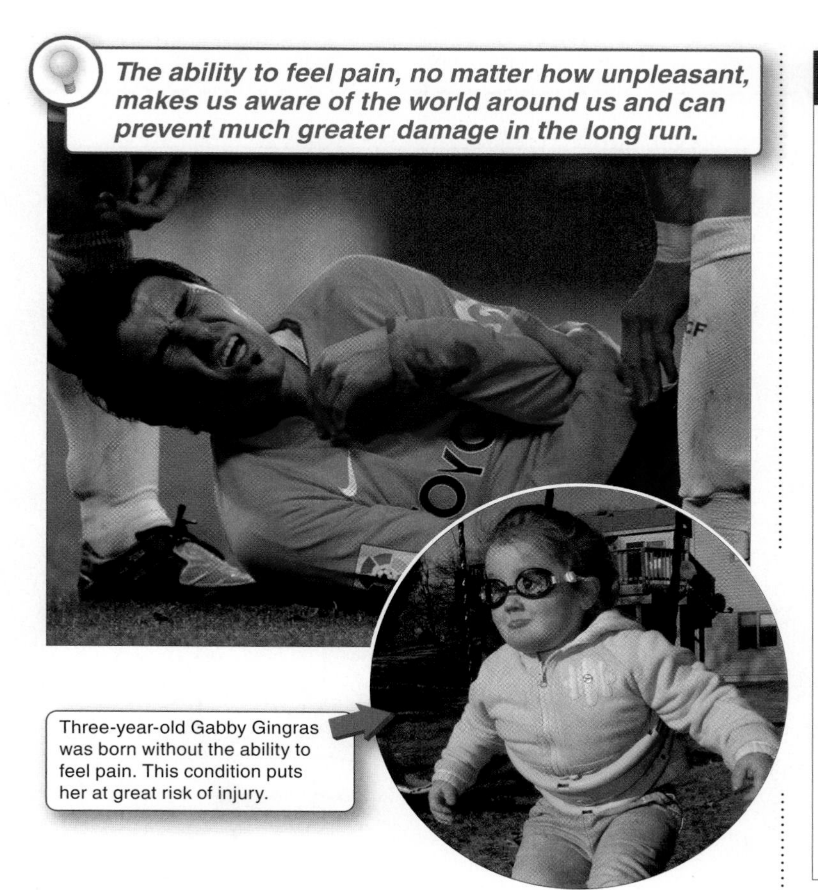

FIGURE 23-1 Avoiding danger. The sensation of pain gives us important information about the world.

FUNCTIONS OF THE NERVOUS SYSTEM

The nervous system—present in almost all multicellular animals—has three primary functions.

RECEIVE INPUT
The nervous system collects information about the internal and external environment.

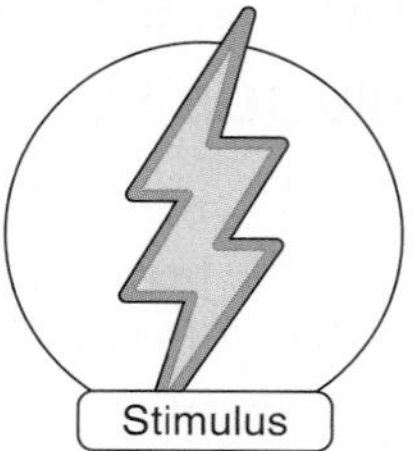

PROCESS INFORMATION
The nervous system interprets the incoming stimuli and determines a response.

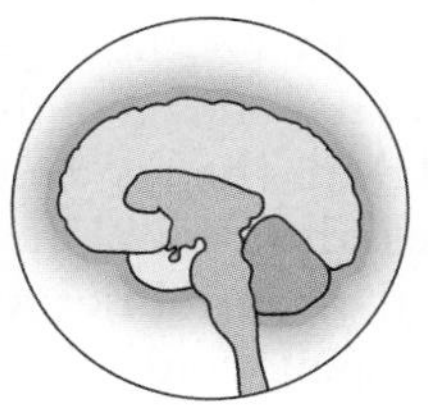

INITIATE RESPONSE
The nervous system sends signals to muscles and glands in response to the internal and external environment.

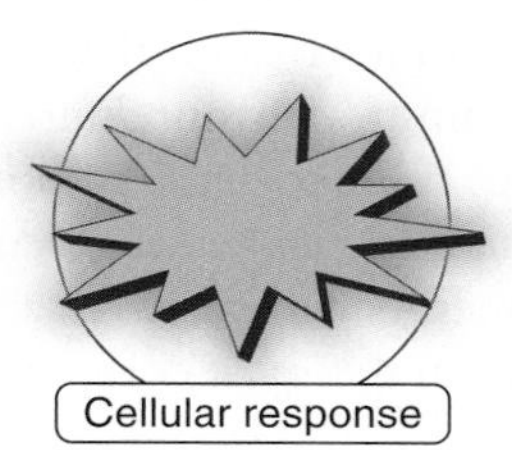

FIGURE 23-2 Overview of nervous system functions.

accomplishes these tasks by letting us see, hear, feel, taste, smell, remember (and forget!), think about, act on, and react to various events and stimuli around us. (Plants, conversely, do not have nervous systems.)

Nervous systems have three critical features (**FIGURE 23-2**):

1. They receive input from the surrounding world.
2. They process that information.
3. They initiate responses to the internal and external environment when necessary.

In this chapter, we examine the organization and structures found in the nervous system, and the diversity of nervous system specializations that have evolved in the animal world, including specializations which allow humans to discern light from dark, to identify subtle differences in touch, temperature, and color, and to use language and abstract reasoning to perform and coordinate complex social behaviors. We also explore how the nervous system, interacting with the muscular and skeletal systems, can generate movement. And we examine how various drugs (legal and illegal) affect, often adversely, the function of the nervous system.

TAKE-HOME MESSAGE 23•1

Present in all multicellular animals other than sponges, the nervous system is a network of cells that collects information about the organism's internal and external environments, processes that information, and sends signals to muscles and glands in response to the information.

23•2

Neurons are the building blocks of all nervous systems.

In all vertebrates, the nervous system is divided into two components: the peripheral nervous system and the central nervous system (**FIGURE 23-3**). The **peripheral nervous system (PNS)** is the network of sensory cells modified to receive information from the environment and the cells that transmit signals to effectors, the organism's muscles and glands that are capable of responding to that stimulus. But the information received by the body's sensory cells does not generally go straight to the cells that control the muscles and glands. First it passes through the **central nervous system (CNS),** which is made up of the spinal cord and brain. Not directly connected to sensory organs or to muscles, the central nervous system processes information that it receives from sensory cells about the organism's surroundings and sends out instructions to other nervous tissue to act in response to that sensory information. Later in this chapter we discuss the elaborate specializations—from memory to language and abstract reasoning—that have evolved in the most complex vertebrate brains.

The **neuron**—the type of cell specialized for generating and conducting electrical impulses—is the building block of all nervous systems (**FIGURE 23-4**). No matter how complex or simple that system is, it consists of neurons. Each neuron is very small and has but a single option at any point in time: "fire" or "don't fire" (meaning that the neuron either generates an electrical signal that can convey information to nearby cells, or the neuron does not). But, put together a few hundred billion neurons, and things start to get interesting. Taken together, groups of neurons bundled together with connective tissue—into structures called **nerves**—connect us to our world by enabling us to sense light, sound, touch, tastes, and smells and to respond to all of that sensory information. The accumulated information carried by all of our neurons—whether they are firing or not firing—is also responsible for information storage and retrieval and all thought in our brain.

Neurons are powerful cells, but they are also fragile. In the time it takes you to read this sentence, one of your neurons will die. If you read it again, another one will die. And neither will be replaced; unlike most cells in your body, very few neurons are able to replace themselves. This need not be cause for alarm, though. Even with about 9,000 neurons dying on a good day—significantly more if you consume any alcohol or happen to inhale any gasoline fumes—the 100 billion to one trillion that you're born with are more than enough for you to keep your wits about you.

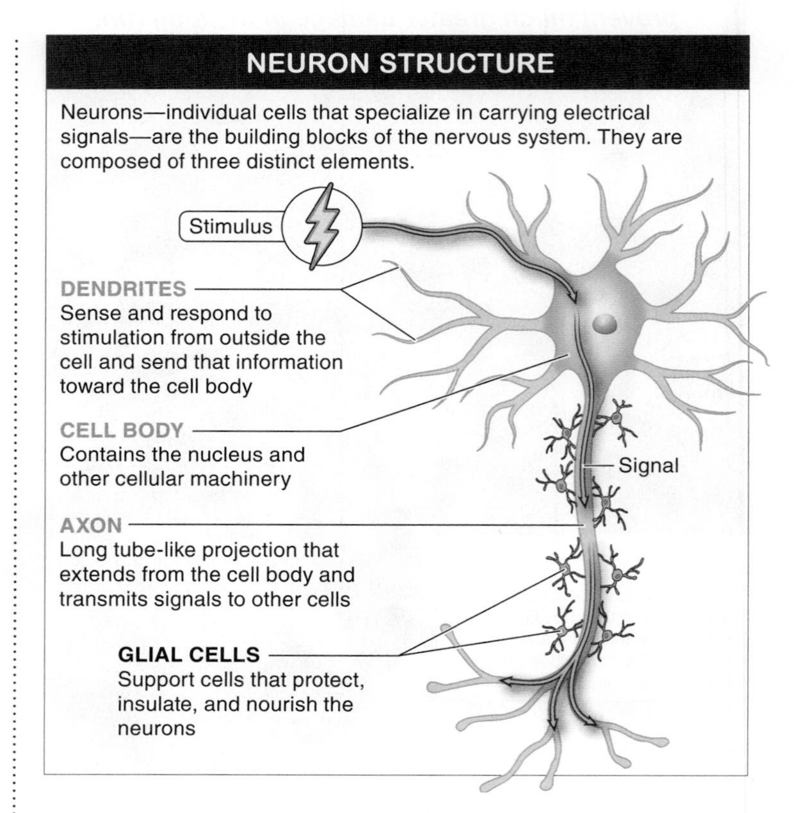

FIGURE 23-4 The neuron is the building block of the nervous system.

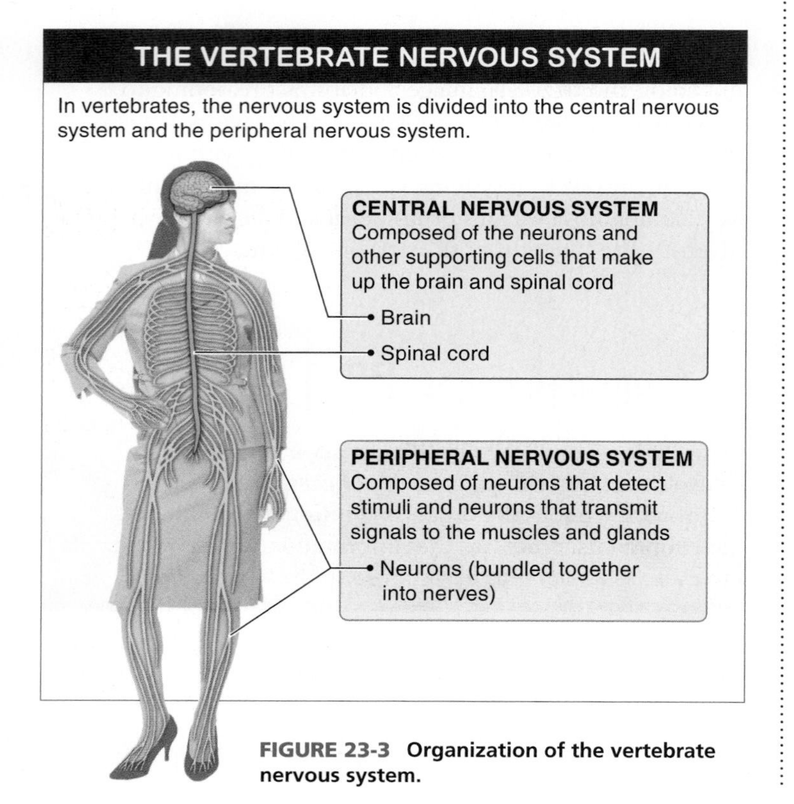

FIGURE 23-3 Organization of the vertebrate nervous system.

Even though they perform such varied and significant functions, neurons are relatively standard eukaryotic cells. Each has a **cell body** that contains all of the typical machinery of a eukaryotic cell, including a nucleus, mitochondria, endoplasmic reticulum, and so on. But neurons also have two specialized structures that make them unusually adept at interacting with the external environment and with other cells.

The first of these two important specializations is the **dendrite.** The dendrite is the part of the neuron that is like an antenna: it senses and responds to stimulation from outside the cell and sends that information toward the cell body of the neuron. Numerous dendrites branch out from the neuron cell body, like a complex antenna system; the tree-shaped dendrites of a neuron branch extensively.

The neuron's second important specialization is the **axon.** The axon is a long—sometimes *very* long—tube-like extension of the main cell body that transmits the signals picked up by the dendrites to the rest of the organism's body. The end of the axon is specially modified in a way that allows it to transmit the signals to another cell. An axon can transmit a signal, much as an electrical wire does, over great distances. The longest cells in the world are neurons. Some neurons are several feet long. The sciatic (sigh-AT-ick) nerve, for example, runs from your spinal cord all the way to the tips of your toes. Most of this distance is covered by a single axon.

Q If neurons rarely (if ever) divide, how can people get brain tumors, which are the result of unstoppable cell division?

Although neurons do all the actual work of the nervous system, they do it with a lot of help from other cells. In the brain, for example, in addition to the neurons there are equal numbers of non-neuronal cells, called **glial cells,** that function like a support staff to protect, insulate, and nourish the neurons. Unlike neurons, glial cells regularly divide in the adult brain. Not surprisingly, then, virtually all brain tumors in adults are formed from glial cells, the supporting cells of the brain.

Just as big corporations keep their big computers locked up in super-clean, secure rooms to keep them functioning at an optimal level, support cells line the blood vessels that supply the brain to isolate the brain cells from potentially harmful molecules in the blood. Called the "blood-brain barrier," this semi-permeable barricade allows essential nutrients and gases to pass through, while barring harmful molecules such as metabolic wastes produced throughout the body. The blood-brain barrier is not perfect, though. Many small molecules, including anesthetics and alcohol, can make it through to influence the brain. The barrier also can be broken down by hypertension, radiation, and a variety of infectious organisms.

We said at the beginning of this section that neurons are the building blocks of the nervous system, but there isn't just one type of neuron. Neurons come in three types, with names that reveal their functions (**FIGURE 23-5**):

1. **Sensory neurons** collect information from an animal's environment and have dendrites modified to respond directly to internal and external stimulation. This stimulation can include temperature, touch, taste, smell, light, or sound.
2. **Motor neurons** stimulate action by conveying signals to muscles or glands and initiating a body's response to stimuli.

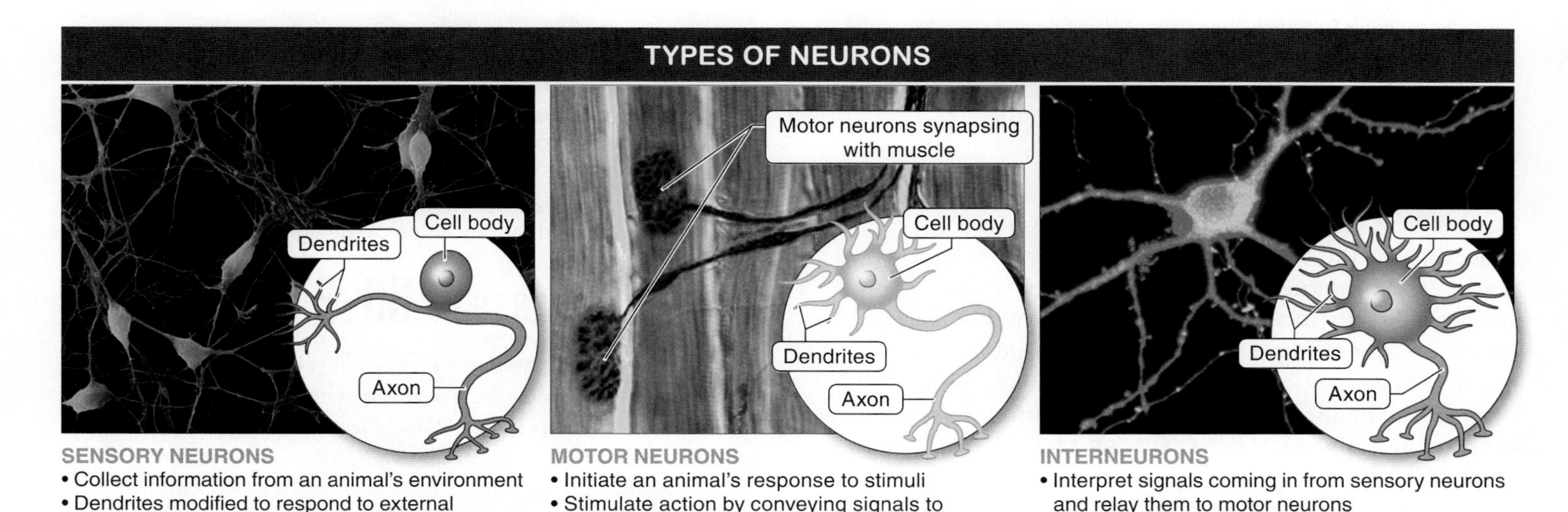

FIGURE 23-5 Three types of neurons work together in the vertebrate nervous system.

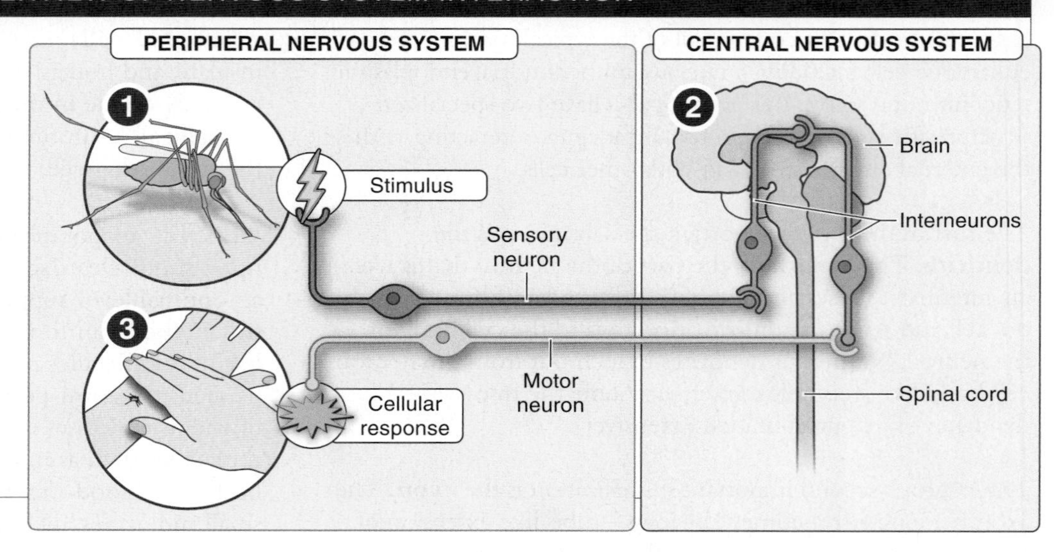

FIGURE 23-6 The peripheral nervous system interacts with the central nervous system.

3. **Interneurons** integrate the signals coming in from the sensory neurons and relay them to the motor neurons. These "middlemen" are located only in the brain and the spinal cord.

Let's look at how these three types of neurons interact. When a sensory neuron in the peripheral nervous system, perhaps in your hand, senses an irritation (such as a mosquito landing on your skin), it sends a signal to convey this information to the spinal cord, where the sensory neuron connects with an interneuron that extends to the brain (**FIGURE 23-6**). The signal is interpreted and a response is determined (probably that you want to make the mosquito go away), and this is sent through another interneuron back down your spinal cord to a motor neuron that is attached to a muscle. The signal from the brain to the muscle causes you to swat at the mosquito—removing the source of irritation. This whole process takes only a tiny fraction of a second.

Some signals can generate a response without any processing by the brain, traveling directly from a sensory neuron to an interneuron in the spinal cord which connects directly to a motor neuron. This kind of direct sensory-motor response is called a **reflex,** and it causes reaction to a sensation that does not need to be processed through the brain. Reflexes enable an organism to respond faster to an imminent danger (**FIGURE 23-7**).

In the next section, we explore the details of how changes in the internal and external environment stimulate a neuron and trigger a response.

Reflexes enable an organism to respond faster to an imminent danger.

FIGURE 23-7 Quick reaction. Reflexes can generate a response to a stimulus without need for processing of the signal by the brain.

TAKE-HOME MESSAGE 23•2

In all vertebrates, the nervous system is divided into the peripheral nervous system and the central nervous system. The neuron is a type of cell specialized for carrying electrical signals and is the building block of all nervous systems. Each neuron is very small, but groups of neurons bundled together enable us to sense light, sound, touch, tastes, and smells and to respond to them.

❷ How do neurons work?

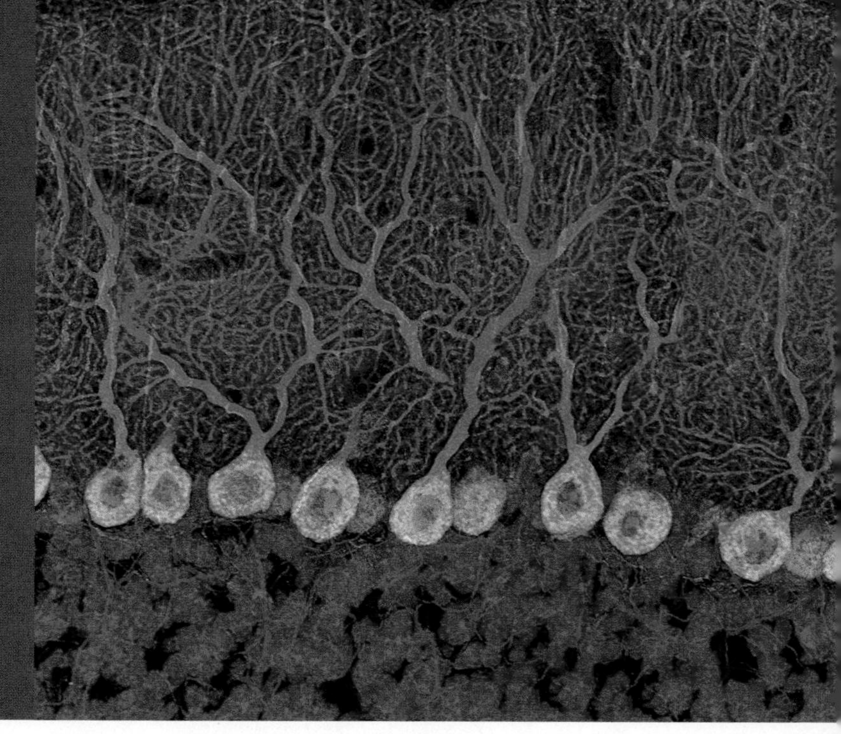

The branching neurons (colorized green) found in the cerebellum are some of the largest and most complex cells in the mammalian brain.

23•3

Dendrites receive external stimuli.

"Fire!" "Don't fire!" "Fire!" "Don't fire!" Dendrites are on the receiving end of an almost constant barrage of (sometimes conflicting) signals. But this is their job. These highly branched tendrils are one of the two types of projections that extend from a neuron's cell body. (The other projection, the axon, is discussed in more detail in Section 23-4.) A neuron may have hundreds or even thousands of dendrites, which give it a huge amount of surface area over which to make connections with other neurons and receive signals.

Dendrites receive external stimuli in one of two ways. Those on motor neurons and interneurons generally connect with and receive signals from other neurons. Sensory neuron dendrites, on the other hand, are modified to respond to a specific external stimulus such as a touch or sound, light, or a chemical (**FIGURE 23-8**). In each case, sensory receptors in the cell membrane of the dendrite respond to the stimulation by briefly opening up little channels. These channels allow the passage of charged ions (usually sodium ions), momentarily altering the electrical charge within the cell from the negative charge that it carries prior to stimulation, when it is at rest. Opening the channels may make the neuron more negatively charged (that is, the electrical charge within the cell relative to that in the fluid surrounding the cell becomes even more negative, or "hyperpolarized"), which says, "Don't fire!" Or it may make the neuron more positively charged (or "depolarized"), which says, "Fire!"

> "The trees grew heavy with blackbirds, branches like dendrites of the nervous system fattening, deep in twittering nerve dusk, waiting for some important message.
>
> —Thomas Pynchon, *Gravity's Rainbow,* 1973"

As stimuli cause channels to open in the dendrites of a neuron—some of which may make the neuron more negatively charged, while others may make the neuron more positively charged—the changes in the cell's electrical charge occurring in all of these dendrites converge at the cell body. The cell body then integrates them, much like tallying the votes in an election. If the sum total of signals coming in is sufficiently positive—exceeding a threshold favoring "Fire!"—then the neuron initiates an **action potential,** an electrical signal that travels down its axon. If the sum total is negative, no action potential is generated.

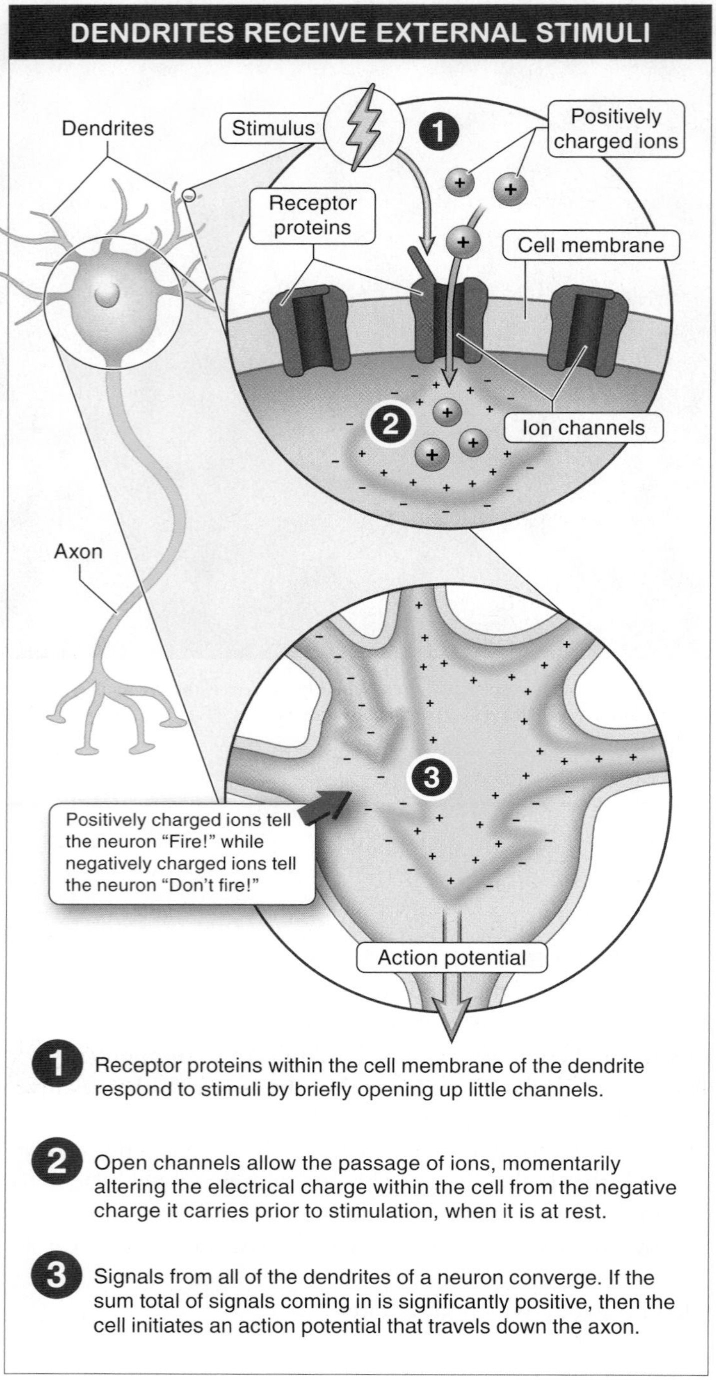

DENDRITES RECEIVE EXTERNAL STIMULI
Dendrites
Stimulus
1
Positively charged ions
Receptor proteins
Cell membrane
2
Ion channels
Axon
3
Positively charged ions tell the neuron "Fire!" while negatively charged ions tell the neuron "Don't fire!"
Action potential
1 Receptor proteins within the cell membrane of the dendrite respond to stimuli by briefly opening up little channels.
2 Open channels allow the passage of ions, momentarily altering the electrical charge within the cell from the negative charge it carries prior to stimulation, when it is at rest.
3 Signals from all of the dendrites of a neuron converge. If the sum total of signals coming in is significantly positive, then the cell initiates an action potential that travels down the axon.

FIGURE 23-8 "Fire!" (or "Don't fire!"). Dendrites receive signals and forward them to the cell body.

The intensity of a sensation is modulated by the number of action potentials per unit of time and the number of neurons stimulated.
Stimulus
Stimulus
Sensory neurons

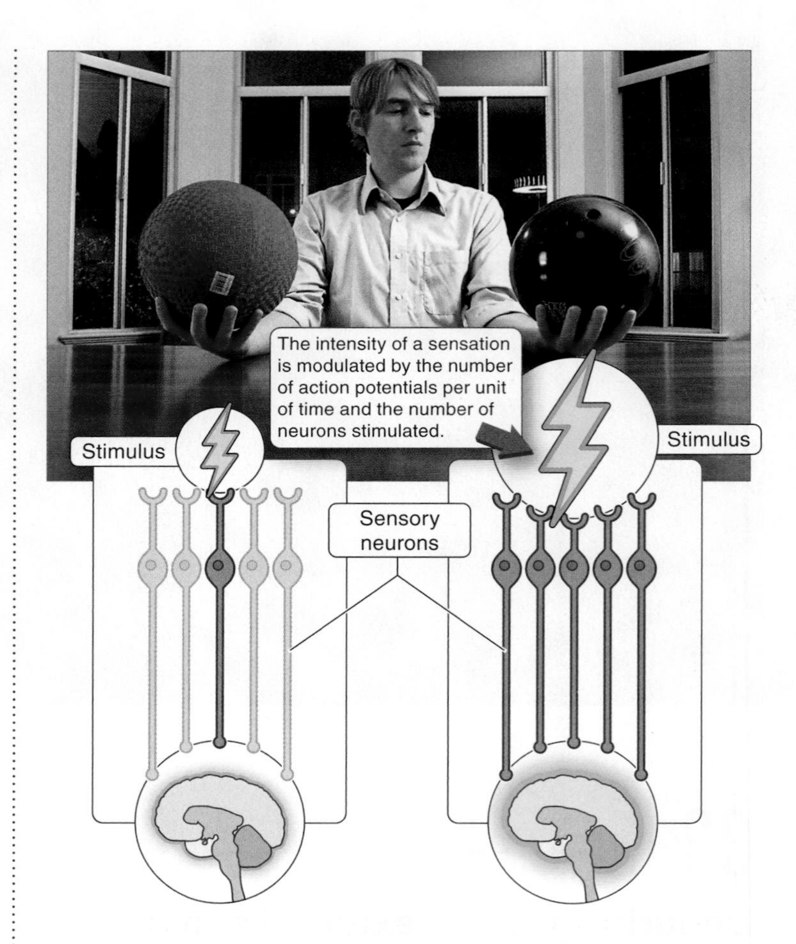

FIGURE 23-9 Heavy or light? We experience gradations of sensation depending on the number of neurons that fire.

An action potential is an all-or-nothing event. It either occurs or doesn't occur as the result of dendrite stimulation. But our experience of sensations is not "all or nothing." We are able to feel gradations of intensity. A 10-pound bowling ball resting in your right hand will feel very heavy (and perhaps cooler and smoother) compared with a lightweight rubber ball in your left hand (**FIGURE 23-9**). How can the "fire" versus "don't fire" event still produce such a range of sensations? The intensity of the sensation an individual feels is modulated by the number of action potentials per unit of time and the number of neurons stimulated. The rubber ball causes a small number to fire, and the bowling ball causes many more to fire.

TAKE-HOME MESSAGE 23•3

Dendrites receive external stimuli in one of two ways. Dendrites on motor neurons and interneurons generally connect with and receive signals from other neurons, whereas sensory neuron dendrites are modified to respond to a specific external stimulus such as a touch or sound, light, or a chemical.

23•4

The action potential propagates a signal down the axon.

We owe a great deal to giant squids. Much of what we know about human nervous systems has come from studying squid neurons. Neurons function almost exactly the same in all animals, and giant squids just happen to also have giant neurons. While a human neuron might be 0.02 millimeters in diameter, a squid's may be more than 100 times wider, up to double the thickness of a human hair. This makes squid neurons much easier to manipulate and observe.

Regardless of the species under observation, each neuron has one axon. This projection leaves the cell body and can extend several feet or more. At its end, the axon branches into several (or hundreds of) axon terminals, also called **terminal buttons,** which are knob-like ends of the axon, positioned very close to a muscle cell or gland or the dendrites of another neuron. And in response to an action potential, these axon terminals release the contents of vesicles, small sacs of chemicals inside the axon terminal, into the space between the cells, potentially influencing adjacent cells (**FIGURE 23-10**).

We've said that axons are like the electrical wires of the nervous system. In your home, electrical wires are usually covered with rubber. Besides reducing the likelihood of someone receiving an electric shock from the current running through the wire, this insulation also has another function: it causes the electrical charge to move through the wire more quickly and reduces dissipation of the signal. Axons are similarly insulated, by a fatty coating called the **myelin sheath,** preventing the action potential from weakening as it travels down the axon. But, unlike wire insulation, there are gaps in the fatty insulation provided by the myelin sheath. And as the action potential moves down an axon, it is at gaps in the myelin sheath that ion channels in the axon membrane allow charged sodium ions to rush in. In doing so, the ions make the cell's charge sufficiently positive in that region to cause the opening of ion channels at the next gap in the myelin sheath, thereby propagating the action potential along the axon.

Because the fatty myelin is white, in cross sections of the brain, some areas, where axons are densely packed together, appear white (**FIGURE 23-11**). Other parts, where there are more cell bodies and dendrites, appear gray. These different regions are often referred to as white matter and gray matter.

What do you think would happen to signals traveling down an axon in your leg—perhaps 3 feet long long—if the myelin sheath insulator were removed? Would a signal from your brain telling your muscles to wiggle your toes ever arrive? No. The

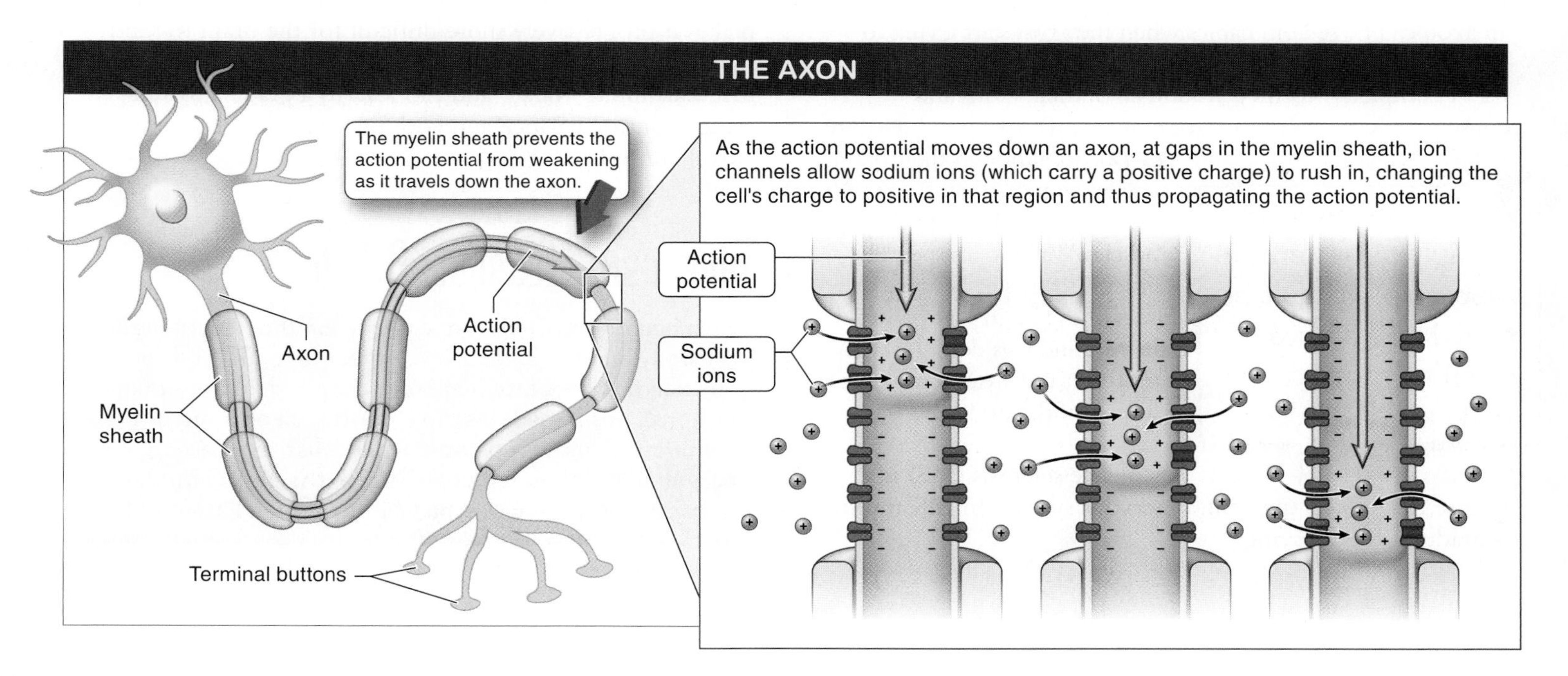

FIGURE 23-10 **An action potential moves down an axon.**

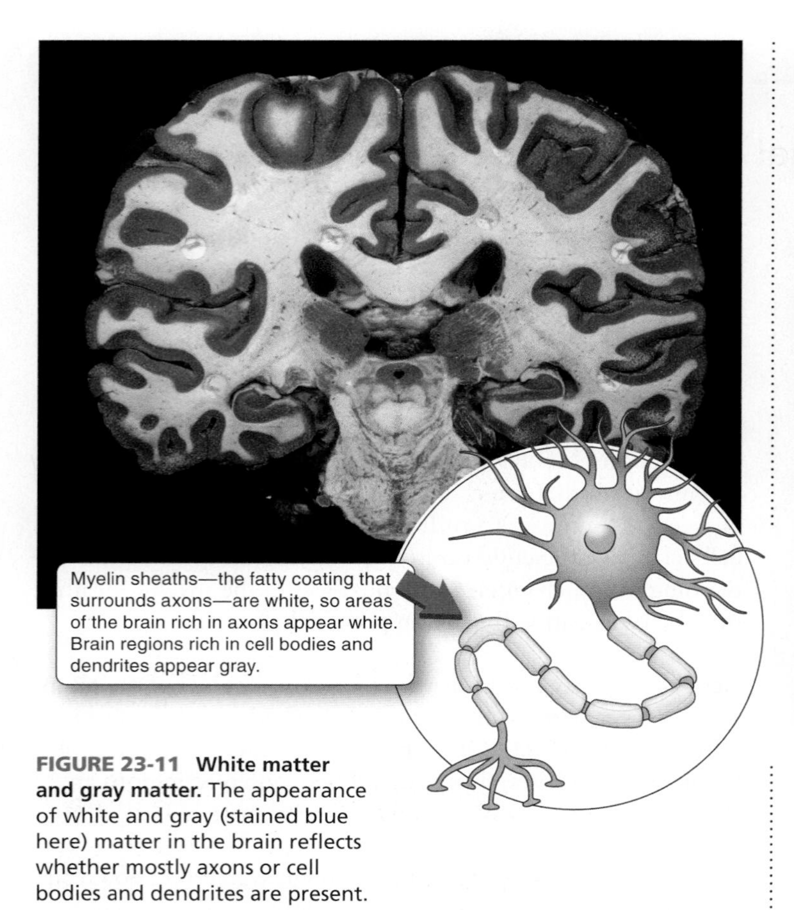

FIGURE 23-11 White matter and gray matter. The appearance of white and gray (stained blue here) matter in the brain reflects whether mostly axons or cell bodies and dendrites are present.

FIGURE 23-12 Unsteady on her feet. Incomplete development of the myelin sheath makes walking difficult for infants.

signal would dissipate before making it to your toe muscles, and your toes would not wiggle. This effect of the lack of myelin on an axon can be seen in babies when they first start trying to walk at around 10 or 11 months of age. At that time, myelin hasn't completely formed around all of their axons and consequently, their gross motor control isn't very good (**FIGURE 23-12**). So, although their brain may be saying, "Walk!" the signal never makes it to the leg muscles. Walking is especially difficult to master because the areas of the brain that control the feet and legs don't develop until after the areas controlling the rest of the body.

Q Some symptoms of multiple sclerosis resemble difficulties that babies have when learning to walk. Why?

Multiple sclerosis (MS) is a disease that affects the central nervous system. In MS, myelin is gradually lost, leaving scar tissue called sclerosis. As myelin is lost, the neuron gradually loses its ability to conduct electrical impulses. This makes it progressively more difficult for the brain to send signals to muscles—the same difficulty babies have when first learning to walk—and can lead to a gradual loss in motor control, balance and coordination, and bladder and bowel control.

TAKE-HOME MESSAGE 23•4

Each neuron has one axon, a projection that leaves the cell body and can extend several feet or more. At its end, an axon branches into numerous axon terminals (terminal buttons), positioned close to a muscle cell or gland or the dendrites of another neuron. In response to an action potential, the axon terminals release chemicals into the extracellular space, potentially influencing adjacent cells.

23•5

At the synapse, neurons interact with other cells.

An action potential moving rapidly down an axon quickly runs out of axon. This is the end of one neuron but not necessarily the end of the signal. As we've seen, the end of an axon—the axon terminal or terminal button—is always right next to another neuron or a muscle cell or a gland. The point where they meet is called a **synapse,** and several possible things can happen there. The signal arriving at the end of the axon may stimulate an action potential in an adjacent cell, it may cause a muscle to contract or relax, or it may initiate a secretion by a gland. Or the signal may even end then and there. In each case, though, the events at the synapse are remarkably similar across all animal species. Let's explore them in sequence (**FIGURE 23-13**).

1. *Sacs called vesicles release neurotransmitters into the synaptic cleft.* The axon terminal is filled with a couple of hundred little sacs, called **vesicles.** These sacs are filled with chemicals called **neurotransmitters,** the chemical workhorses of the nervous system that transmit signals to adjacent cells. When the action potential reaches the axon terminal, the vesicles merge with the axon's cell membrane, called the **presynaptic membrane,** open up, and dump their contents out of the end of the axon and into the **synaptic cleft,** the space between the axon and whatever is next to it (a muscle cell, gland cell, or neuron).

2. *Neurotransmitter diffuses and binds to nearby receptor sites.* As the neurotransmitter molecules float around in the fluid in the synaptic cleft, they diffuse away from the axon until they bump into receptor sites on the adjacent neuron, muscle, or gland. Some of the neurotransmitter molecules attach to receptor sites on the **postsynaptic membrane** of the adjacent cell.

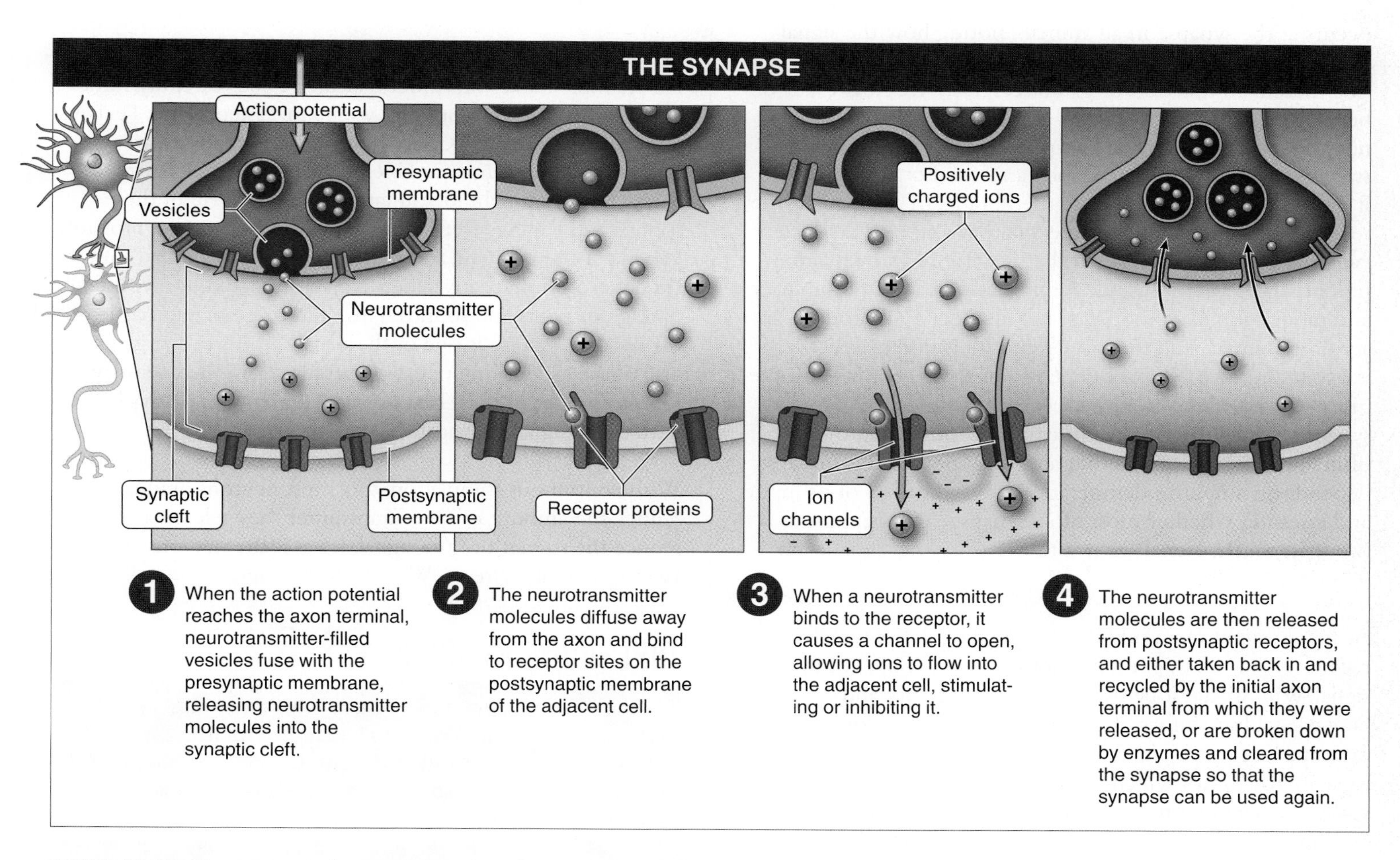

FIGURE 23-13 The sequence of events when an action potential reaches the synapse.

3. *Gates open in the postsynaptic cell membrane and the signal is passed to the postsynaptic cell.* When neurotransmitter binds to a receptor in the postsynaptic cell membrane, this causes a gate to open, which allows ions (often sodium ions) to flow into the adjacent cell, which may be a neuron, a muscle cell, or a gland cell. Opening the chemical gates allows the signal to be passed from the neuron's axon to the adjacent cell. This chemical change in the postsynaptic cell can cause an electrical change and, consequently, initiate an action potential (if the adjacent cell is a neuron), a contraction (if it's a muscle), or a secretion (if it's a gland).

4. *Neurotransmitter is released from the postsynaptic cell receptors and recycled or broken down.* The receptors then release the bound neurotransmitter molecules back into the fluid in the synaptic cleft. Eventually, the neurotransmitter molecules are either taken back up through the presynaptic membrane and recycled by the axon terminal from which they were released, or they are broken down within the synaptic cleft by enzymes, clearing out the area so that the synapse can be used again.

We've just looked at the general sequence of events that occurs at the synapse in all animals, noting how the signal may be propagated to an adjacent neuron, or how it may lead to a muscle contraction or gland secretion. We now focus on how signal propagation occurs at the synapse between two neurons. Several outcomes are possible, and they depend on what neurotransmitter is released and what type of receptor it binds to on the postsynaptic membrane. In some cases, the released neurotransmitters are excitatory and excite the next cell, increasing the likelihood that it will fire its own action potential. In other cases, the neurotransmitters are inhibitory and reduce the likelihood that the next cell will produce an action potential. And for some neurotransmitters, whether it is excitatory or inhibitory depends on the receptor. With the giant web of connections between neurons—each neuron synapses with hundreds or thousands of other neurons—the ultimate outcome of whether an action potential is initiated depends on a neuron democratically weighing all of its inputs and assessing whether most of its synapses are urging it to fire (pass on the signal) or not to fire (stop the signal from getting through).

At first it might seem odd that the total of all the signals received by a cell might tell it not to do anything, not to fire a signal. Why would it be useful to inhibit activity and effectively block information from being passed along? This option is not just useful, but essential to the nervous system's capacity to control which signals get through. By having a synapse where two neurons meet (rather than one long, continuous "wire" constantly conducting electrical signals), it is possible to modulate and filter some of the overwhelming amount of sensory information coming into the brain. It's like call-screening for your brain.

FIGURE 23-14 Like call-screening for your brain. Not every signal is propagated.

An example makes this process clearer. Focus carefully, for a moment, on the environment you're in. Listen. Look. Feel. Are your shoes a bit tight? Can you hear the hum of an air conditioner? Are sirens audible in the distance? Is the guy sitting two rows behind you tapping his foot? All of these sounds and feelings were there a minute ago, yet you probably weren't aware of them, and that's not necessarily a bad thing. It's hard to think clearly when you're distracted by a constant barrage of sights and sounds. This "filtering" characteristic of the nervous system is one of the main reasons that organisms don't have single, long neurons running from their sensory receptors right to their brain or muscles. It's not always best to have every signal propagated (**FIGURE 23-14**).

With continuous stimulation, too, most neurons gradually reduce the amount of neurotransmitter they release, and thus reduce the strength of the signal. It's as if the neurons are saying, "Enough already. We get the message."

TAKE-HOME MESSAGE 23•5

At the synapse, a neuron interacts with other cells. In response to an action potential, neurotransmitters are released into the synaptic cleft, diffuse, and may bind to receptors on an adjacent neuron, muscle cell, or gland, potentially stimulating an action potential, muscle contraction, or secretion. Neurotransmitters may then be taken back in by the axon terminal or enzymatically broken down in the synaptic cleft.

23•6

There are many types of neurotransmitters.

The neurotransmitters released by neurons can be thought of as the chemicals that initiate or modify a wide variety of actions, moods, feelings, or other sensations. Some make us feel happy or sad, restless or sedate. Some stimulate muscles to contract. Others influence learning, memory. In all, about 25 neurotransmitters have been identified, each of which has a sort of personality based on its specific actions. A detailed discussion of all 25 neurotransmitters is beyond the scope of this book. Here we discuss four that are particularly important.

Q Why are curare-tipped poison arrows so lethal?

Acetylcholine Acetylcholine is the neurotransmitter released by motor neurons at the point where they synapse with muscle cells. When enough acetylcholine binds to a muscle cell, the muscle contracts. A poison called curare, found in some South American plants, is used on arrow tips to make them more lethal. Curare works by blocking the receptor sites where acetylcholine normally binds to muscle cells. Once it gets into an animal's system, curare causes death from lack of oxygen very quickly, because it makes it impossible for the skeletal muscles to contract (**FIGURE 23-15**).

FIGURE 23-15 Hunting with curare-tipped arrows. A nomad from the Nukak-Maku tribe in the Amazon jungle hunts for monkeys using a blowgun and darts tipped with curare. Red arrow: A hunter collects shavings of a plant from the genus Strychnos, from which curare is extracted, a lethal poison that blocks acetylcholine binding sites and prevents muscle contraction.

Glutamate Another excitatory neurotransmitter is glutamate. Although its mechanism of action isn't well understood, it appears to be involved with learning and memory. Mice that were genetically engineered to have more sensitive glutamate receptors, for instance, learned tasks better and became much better at running mazes than normal mice.

Dopamine Dopamine is important in initiating and coordinating movement. Loss of dopamine neurons may be responsible for Parkinson's disease, for which the symptoms include tremors of a hand or foot, stiffness or inflexibility of muscles, and impaired balance and coordination. Dopamine is also one of the body's chief "happiness" neurotransmitters. Its release in certain parts of the brain is associated with feelings of intense pleasure. In studies of brain functioning, for example, when subjects are shown their favorite food items, the levels of dopamine released in the brain skyrocket. We explore dopamine in greater detail later in this chapter.

Serotonin Serotonin generally functions as an inhibitory neurotransmitter. It affects appetite, sleep, anxiety, and mood, and produces feelings of contentment and satiation when released. Women make serotonin only about two-thirds as quickly as men, an observation that may be related to why depression is twice as common among women as men. As we see later in the chapter, antidepressant medications such as Prozac increase the amount of serotonin in the synapse and are effective at elevating the mood of someone who is depressed.

TAKE-HOME MESSAGE 23•6

About 25 neurotransmitters have been identified, each of which has several actions, depending on the synapse where it occurs. Acetylcholine is the neurotransmitter released by motor neurons at the point where they synapse with muscle cells. Glutamate is an excitatory neurotransmitter involved with learning and memory. Dopamine is important in initiating and coordinating movement and in producing feelings of intense pleasure. Serotonin, an inhibitory neurotransmitter, affects appetite, sleep, anxiety, and mood, and produces feelings of contentment and satiation.

3 Our senses detect and transmit stimuli.

The North African fennec fox (Vulpes zerda) *has large ears to help it shed body heat. The ears also help the animal hear its prey, which is often hiding underground.*

23•7

Sensory receptors are our windows to the world around us.

Q Why do animals have only one head? Why is it in the front?

Did you ever wonder why animals tend to have only one head and why it's always in the front of the body? When organisms became bilaterally symmetrical—that is, having mirror-image left and right sides—suddenly they had a "front" and a "back." As such animals move through their environment, the "front" part of the animal encounters new things in the environment first; with the sensory equipment up there, the organism is able to decide what to do—eat or run—as quickly as possible (**FIGURE 23-16**), which may have conferred evolutionary benefits. And as the decisions to be made get more and more complex, the circuitry required to "make up its mind" becomes more complex. In fact, only then does it really become necessary to have a mind. And so the brain and head (which always go hand in hand) have tended to become more and more pronounced.

Our senses—sight, hearing, smell, taste, and touch—make us physically aware of the environment around us. They are our windows to the world. Each of our senses puts us in touch with and brings us information about a different and unique

Sensory equipment is concentrated at the "front" of bilaterally symmetrical animals because this part of the animal encounters new things in the environment first.

FIGURE 23-16 Eyes up front. Most of an animal's sensory organs are located in its head.

THE FIVE SENSES

Hearing

Touch

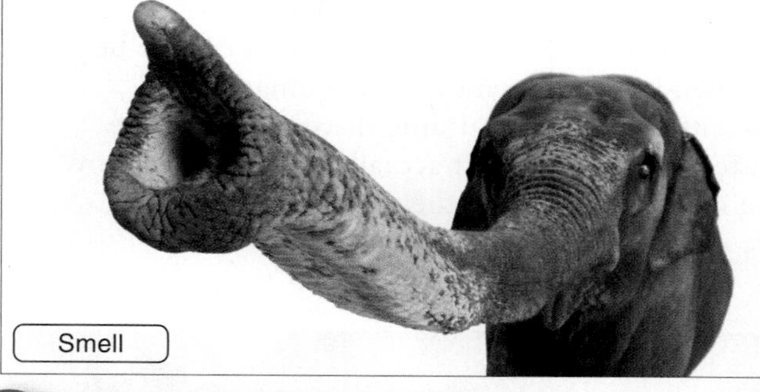

Our senses—sight, hearing, smell, taste, and touch—are our windows to the world. Each brings us information about a different slice of the world, but the senses have much in common.

FIGURE 23-17 Making us aware of our environment: the senses. (Note that most of the body parts pictured have more than one function; the elephant's trunk, for example, enables smell, but also has numerous other functions, including lifting, touching, and drinking.)

slice of the world, but the senses also have much in common (**FIGURE 23-17**).

The process by which all our senses work is basically the same. A receptor—commonly a modified dendrite on a sensory neuron (in the eye, nose, tongue, ear, or skin)—is stimulated by some aspect of the outside world (light, odor, taste, sound, or touch). This outside stimulus causes a change in the neuron, causing the sensory neuron to either (1) fire an action potential itself, which shoots down the axon and ultimately reaches a part of the brain where the signal is perceived as a particular smell or sound, for example, or (2) alter its rate of neurotransmitter secretion so that it increases or reduces the rate of firing of action potentials in a neighboring cell.

All of the specific senses are variations on this theme, fine-tuned by natural selection to give an animal specific information about its environment to help it respond appropriately. The following sections present details on each of the primary five senses.

TAKE-HOME MESSAGE 23·7

The process by which all our senses work is basically the same. A modified dendrite on a sensory neuron is stimulated by some aspect of the outside world, causing the sensory neuron to fire an action potential (that shoots down the axon and ultimately reaches a part of the brain where the signal is perceived as a particular smell or sound, for example) or to alter its rate of neurotransmitter secretion (so that it increases or reduces the rate of action potentials in a neighboring cell).

23•8

Taste: an action potential serves up a taste sensation to the brain.

Look at your tongue in the mirror. It is covered with bumps. Embedded within these bumps are taste buds—more than 10,000 in all. Within each taste bud are 60–80 sensory receptor cells, called **chemoreceptors** (which are modified epithelial cells that synapse with sensory neurons), stimulated when particular chemicals in food dissolve in saliva and bind to proteins on the cell surface. Much as a lock will open only when the proper key is inserted, particular taste receptors allow only specific food molecules, with exactly the right shape, to bind (**FIGURE 23-18**).

Q How can artificial sweeteners taste like sugar while not actually being sugar?

Taste chemoreceptors fall into five groups, depending on which type of molecule chemically stimulates them: sweet, salty, sour, bitter, or a recently discovered, but difficult to describe, savory taste called umami. These different types of chemoreceptor occur all across the tongue (although the different groups tend to be concentrated in different regions of the tongue). A rich variety of tastes is possible because most foods stimulate unique combinations of the different taste receptors. Foods also release molecules into the air in the mouth, which stimulate smell receptors within the nasal cavity.

Not all animals use their mouths to taste things. Some insects have chemoreceptors on their legs, and they "taste" things just by touching them. Other animals have taste receptors on their antennae or tentacles. Regardless of where the food meets the taste receptors, the process is the same: chemical binding triggers an action potential that delivers a taste sensation to the brain.

Because of the way we sense taste, it's possible to fool your brain. Your brain never actually "knows" the true identity of the food on your tongue. Rather, it senses a particular taste based solely on the combination of receptor cells that is stimulated. If a molecule that is not sugar has a chemical structure closely resembling sugar, it can stimulate the same taste-bud receptors and be perceived by the brain as sugar (**FIGURE 23-19**). Many non-nutritive sugar-substitute molecules such as saccharin and sucralose do exactly this. They have three-dimensional arrangements of atoms that are similar to sugar molecules such as sucrose, glucose, or fructose, but have structures that cannot be broken down by any enzymes in the human body. Consequently, when we consume them, we sense the sweet taste of sugar but don't actually derive any energy from the molecules. Instead, they pass through our digestive tract unaltered.

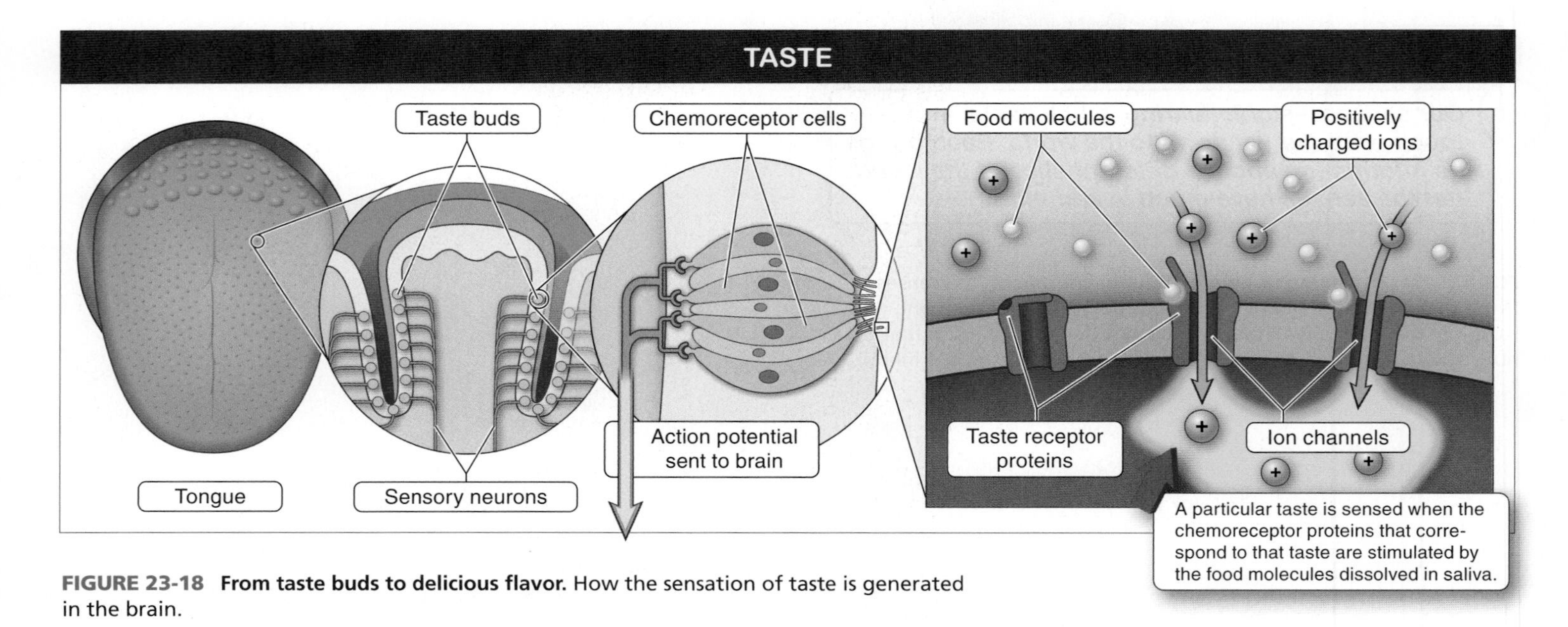

FIGURE 23-18 From taste buds to delicious flavor. How the sensation of taste is generated in the brain.

FIGURE 23-19 Fooling the brain. Sugar substitutes stimulate receptor cells that produce a sweet taste.

Aspartame also resembles sugar closely enough to stimulate sugar receptors in taste buds. Molecularly, however, it is quite different: it is made from two amino acids. It can be broken down and releases energy (and hence contains calories), but is so efficient at stimulating sweet-taste receptors that it can generate a strong sugary taste even when used in tiny amounts.

TAKE-HOME MESSAGE 23•8

On your tongue, there are about 10,000 taste buds, each of which contains 60–80 chemoreceptors, which are stimulated when particular chemicals in food bind to receptor proteins on the cell surface. Some animals have chemoreceptors on their antennae or tentacles, and the binding of chemicals to these receptors similarly triggers an action potential that delivers a taste sensation to the brain.

23•9

Smell: receptors in the nose detect airborne chemicals.

Our sense of smell works in almost exactly the same way as our sense of taste. Neurons that can detect smells have dendrites modified with tiny, hair-like projections. These dendrites—densely packed within the nasal cavity—are covered with chemoreceptors. Airborne chemicals move through mucus in the nasal cavity and bind to the smell receptors, triggering action potentials that shoot down the axon, all the way into the smell center of the brain, where the signal is perceived as a particular odor. There are more than a thousand different types of receptors—modified neurons—in a human nose, each capable of detecting a different scent (**FIGURE 23-20**). (These neurons are actually

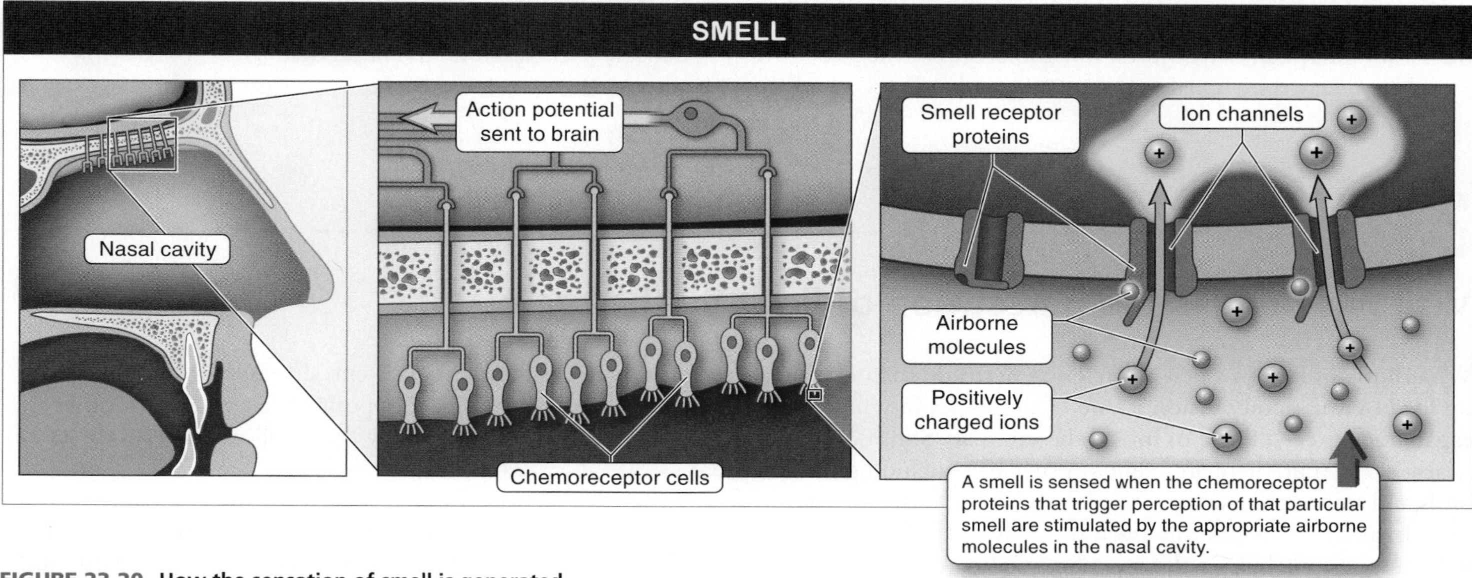

FIGURE 23-20 How the sensation of smell is generated.

Q Why are your senses of smell and taste dulled when you have a cold?

the only neurons in the body in direct contact with the outside world.)

In humans, the senses of smell and taste are closely connected, because the air in the mouth, throat, and nasal passages circulates around all these areas. You may have noticed that when you have a cold and your nose is stuffed up, you taste little beyond the basic salt, sweet, bitter, and sour tastes on your tongue, and you can't smell anything at all. Your senses of taste and smell are dulled when you have a cold because increased mucus in your nasal passageways reduces the rate at which airborne chemicals can reach the smell receptors and taste receptors on dendrites in your nose and on your tongue.

Animals vary greatly in their sensitivity to smells, and humans don't fare too well in the competition. Recent evidence even suggests that fewer and fewer of the genes coding for the different smell receptors in humans function properly anymore. Interestingly, human females are significantly better than males at detecting, distinguishing, and identifying odors, a sensitivity that is even greater during the days surrounding ovulation.

Q People are much worse than dogs at sniffing out drugs. Why?

Dogs are among the most smell-sensitive vertebrates, having as many as 40 times more smell receptors than a human—hence their tremendous proficiency at detecting drugs or explosives (**FIGURE 23-21**). Some moths, such as gypsy moths and silkworm moths, put even dogs to shame and may be the champions of chemoreception. The tiny hairs on the males' antennae can detect just a few molecules of the sex attractant released by females. By moving in the direction of increasing concentration of the airborne chemicals, a male can track down a female two miles away. Similarly, salmon use smell to help them navigate from the ocean back into the exact streams where they were born, a trip that can be hundreds of miles long. Snakes have chemoreceptors on their tongue and are actually "smelling" their environment as they stick their tongue out.

FIGURE 23-21 **Smelling champ: the bomb-sniffing dog.**

TAKE-HOME MESSAGE 23•9

Neurons that can detect smells have dendrites modified with tiny, hair-like projections covered with chemoreceptors, densely packed within the nasal cavity.

23•10

Vision: seeing is the perception of light by the brain.

What humans lack in smell proficiency, we more than make up for in visual acuity. Vision is one of the senses in which our capabilities exceed those of most other animals. Still, as with nearly every trait, there are plenty of species—including many birds and some spiders, for example—whose abilities exceed ours. In any case, despite the great diversity of species with light-sensing capabilities, the basic functioning of light-absorbing cells is quite consistent, although in some ways the process works in a fashion opposite what we see for other sensory reception. Here's how these cells work (**FIGURE 23-22**).

Within the eye or other light-detecting structure, there are light-sensitive neurons, called **photoreceptor cells,** with light-sensitive molecules embedded in their cell membranes.

VISION

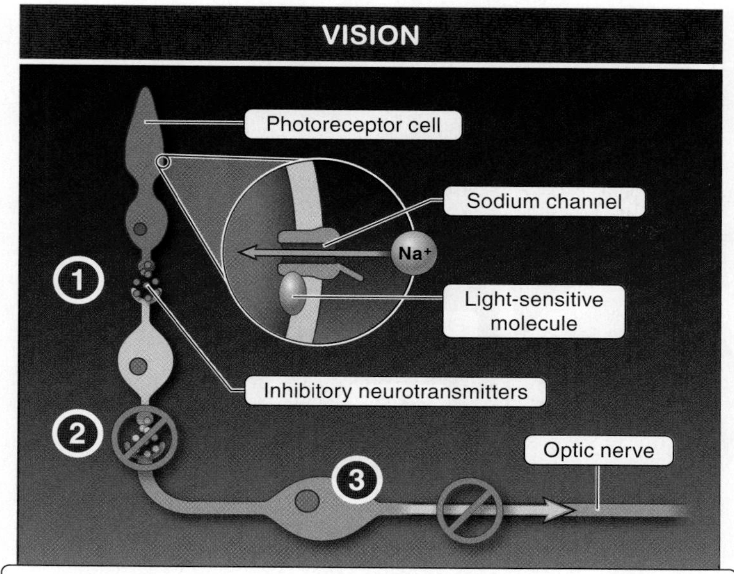

IN THE DARK

1. Sodium channels, located within photoreceptor cell membranes, are open, allowing the photoreceptor cells to continuously release inhibitory neurotransmitters.
2. The inhibitory transmitters reduce the ability of adjacent neurons to excite the neurons that signal the brain.
3. An action potential is not sent through the optic nerve and light is not perceived by the brain.

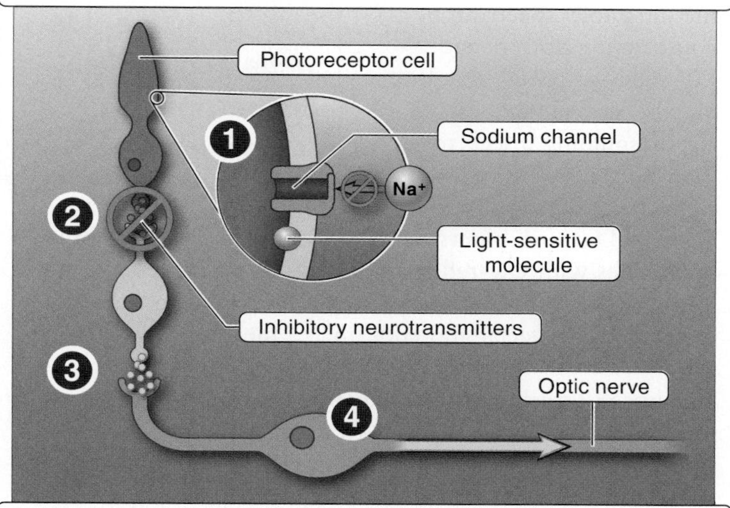

IN LIGHT

1. Light energy causes light-sensitive molecules to change shape, closing sodium channels in the photoreceptor cell membranes.
2. The closing of sodium channels reduces the amount of inhibitory neurotransmitter released by the photoreceptor cells.
3. No longer inhibited, adjacent neurons excite the neurons that signal the brain.
4. An action potential is sent through the optic nerve and light is perceived by the brain.

FIGURE 23-22 How the sensation of vision is produced.

In the dark, these photoreceptor cells continuously release an inhibitory neurotransmitter. Like the depressing of a brake pedal, this inhibitory message reduces the ability of adjacent cells to excite the neurons that signal the brain. In a sense, darkness blocks the ability of optic nerves to send visual signals to the brain.

When light hits one of the light-sensitive molecules in the membrane of a photoreceptor cell, the light energy causes some of the chemical bonds in the molecule to become stretched. (A similar capture of light energy occurs in plants during photosynthesis.) This closes sodium channels in the photoreceptor cell membranes.

The light-induced closing of sodium channels in the photoreceptor cell membranes reduces the amount of inhibitory neurotransmitter released by the cells. This, in a convoluted way, can lead to the perception of light by the brain: the photoreceptor cells no longer block the ability of adjacent neurons to excite the neurons that send visual signals to the brain. And the particular wavelength perceived by the brain depends on which version of the light-sensitive molecules bound to the photoreceptor cells was stimulated by the light (and reduced the photoreceptor's inhibition of adjacent neurons).

Three different types of eyes have evolved (**FIGURE 23-23**). The simplest, called **eye cups,** are found in the flatworms such as planaria, which live in ponds and streams. Eye cups are made up of photoreceptors, and although they cannot form images, they are able to detect the presence and intensity of light. Unless light is directly in front of the worm, it shines more intensely on one eye than the other. This allows the animal to sense which direction the light is coming from. Usually, this information tells the animal which way to move: toward or away from the light. Planaria, for example, move away from the light.

Insects have much more refined visual capabilities (**FIGURE 23-24**). They possess **compound eyes** made of dozens to thousands of separate light-sensing units, each with its own lens that directs light onto about a dozen photoreceptors. The photoreceptors then send signals along neurons. When the signals finally reach the brain, the brain interprets these signals as an image. Because some insect photoreceptor cells—such as those in honeybees—contain a broader range of light-sensitive pigments than is found in humans, these insects can see wavelengths of light in the ultraviolet spectrum that are invisible to us.

TYPES OF EYES

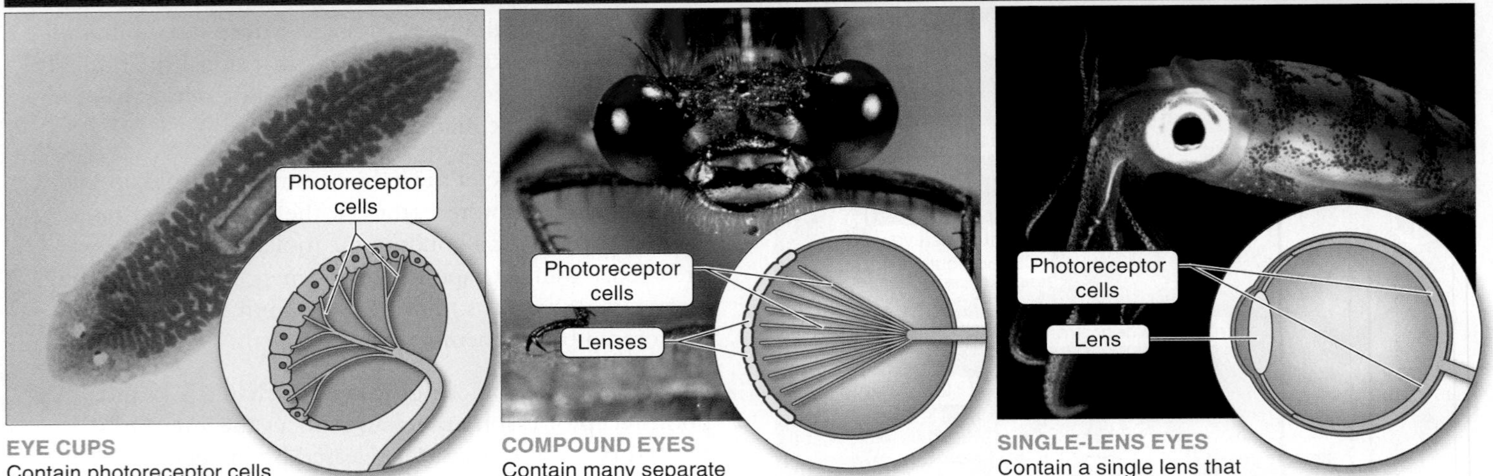

Photoreceptor cells

EYE CUPS
Contain photoreceptor cells that are able to detect the presence and intensity of light, but cannot form images

Photoreceptor cells
Lenses

COMPOUND EYES
Contain many separate light-sensing units—each with its own lens—that direct light on to about a dozen photoreceptor cells

Photoreceptor cells
Lens

SINGLE-LENS EYES
Contain a single lens that focuses light on to highly sensitive photoreceptor cells

FIGURE 23-23 Eye structure suits the needs of the organism. Shown: flatworm, red damselfly, glowing reef squid.

FIGURE 23-24 Seeing with compound eyes. Insects can see wavelengths of light in the ultraviolet spectrum. The flower shown is Silverweed (*Potentilla anserina*). In daylight it appears uniformly yellow, but in ultraviolet light it appears red and white.

Q Insects don't just fly into spiders' webs accidentally. Sometimes they do it on purpose. Why?

Plants sometimes capitalize on this ability of insects to detect the ultraviolet spectrum by constructing flowers with UV-reflecting petals that look like the landing lights at an airport, guiding insects to their nectar reward (while ensuring pollination of the flowers). Devious spiders capitalize on this ability, too. Some build webs with UV-reflecting threads that trick insects into thinking the web is a flower. Once it flies into the web, the insect finds no nectar and ends up as a meal for the spider.

The image-forming, **single-lens eye** has evolved independently twice, once in a group of molluscs—including squid—and again in vertebrates. In each case, light hits the eye and enters the eye's interior through an **iris.** The pupil, an opening in the iris, opens and closes to control the amount of light that gets into the eye. Once the light is through the pupil, a lens focuses it on the eye's photoreceptor cells (**FIGURE 23-25**).

The lens focuses the light onto the **retina,** nervous tissue containing light-sensitive cells that lines the inner surface of the eye and transmits impulses to the vision center of the brain via the **optic nerve.** The two types of photosensitive cells in the retina are **rods,** which are highly sensitive to even tiny amounts of light and make it possible to see at night and in low-light situations, and **cones,** which are less sensitive to light and so are more effective during daylight. Humans have

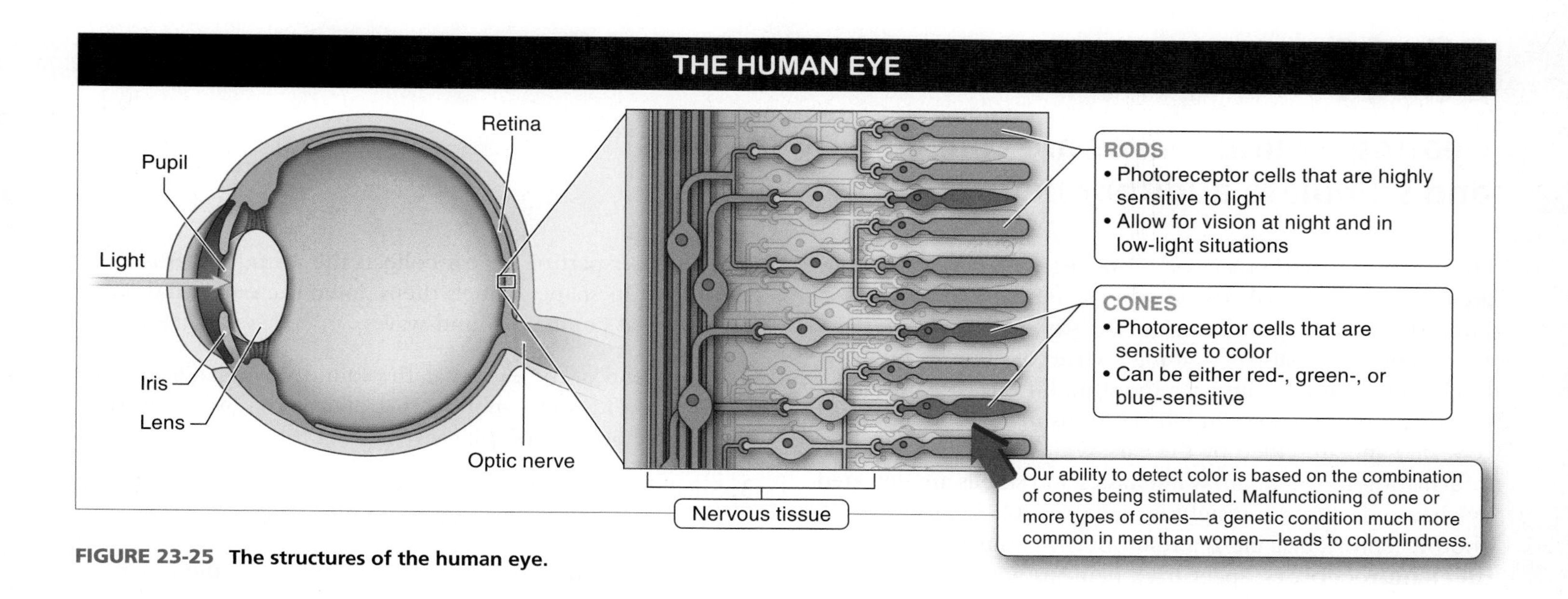

FIGURE 23-25 The structures of the human eye.

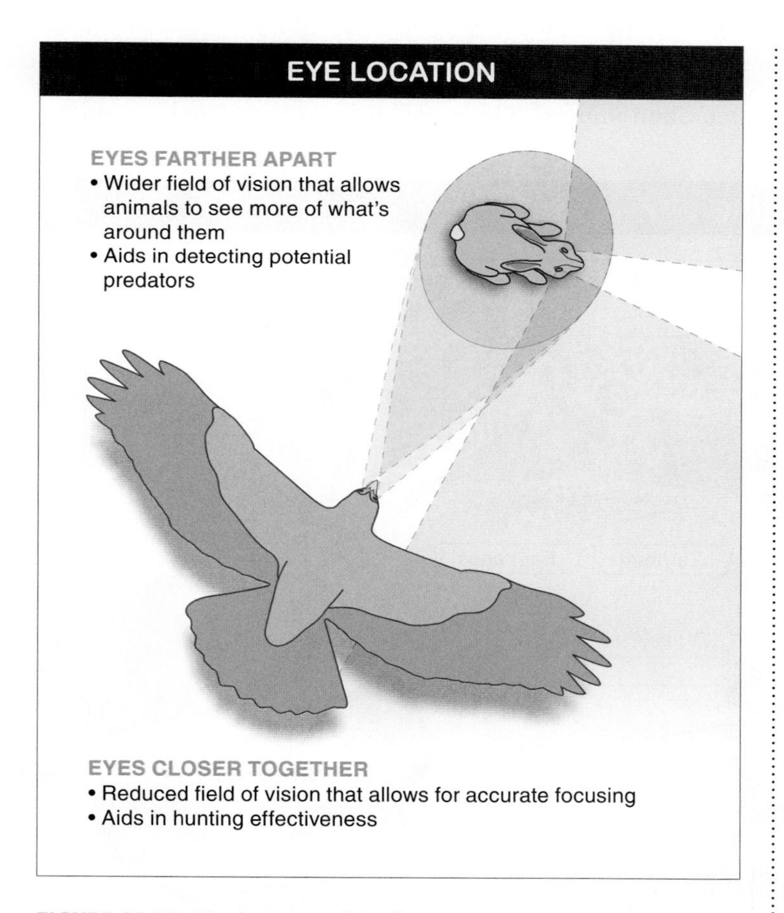

FIGURE 23-26 The hunter and the hunted. Eye placement may indicate whether an animal is predator or prey.

three kinds of cone cells: red-, green-, and blue-sensitive. Our ability to detect color is based on the combination of cones being stimulated. Some individuals (usually men) produce non-functioning red or green cones—the result of mutant genes. In either of these cases, it is impossible for the individual to distinguish red from green and so the person is said to be **color blind.**

The location of the eyes of the body of an animal is an evolutionary adaptation. Some animals have their eyes close together in the front of their head. Others have them closer to the sides of their head (**FIGURE 23-26**). The farther apart the eyes are, the wider the animal's field of vision. That is, the animal can see more of what's around it. This is particularly important for small animals preyed on by other animals. It's about as close as they can get to having eyes in the back of their head, and so keeps them safer. Predators—such as owls or hawks—on the other hand, have both eyes right in the front. This reduces their field of vision, but allows super-accurate focusing on exactly where the prey is, a characteristic that makes them excellent hunters. Winning the award for strangest eye-placement is the flounder. As they develop, one of their eyes migrates gradually from one side of their body to the other. When these marine flatfish lie camouflaged on the ocean bottom (like a plate on a table), both eyes are then on the same side of the body, looking up.

TAKE-HOME MESSAGE 23•10

Vision results from the stimulation of light-sensitive sensory neurons, called photoreceptor cells. The photoreceptor cells have a variety of molecules, embedded within their membranes, that are chemically altered by light. Signals are conveyed to the brain and interpreted as an image. The particular wavelength perceived depends on which version of the light-sensitive molecules in the photoreceptor cell membranes is stimulated.

23•11

Hearing: sound waves are collected by the ears and stimulate auditory neurons.

Hearing is yet another variation on a theme. As with the previous senses described, something in the external world stimulates modified neurons, and the stimulation initiates an action potential that passes along a series of neurons until it reaches the brain. In this case, the stimuli from the outside world are sound waves, tiny fluctuations in air pressure, collected and amplified by the ears, which then pass the information to the brain. While tastes and smells are detected by chemoreceptors and sight is made possible by photoreceptors, hearing is a result of the stimulation of **mechanoreceptors,** specialized neurons with receptors that respond to mechanical pressure.

Although there is some variation in the details of how hearing works in different species, this general, six-step hearing model is consistent (**FIGURE 23-27**).

1. The outer part of the ear collects the sound waves and, because of its shape, funnels them down the **ear canal,** a channel that conducts sound waves.

2. At the end of the ear canal, the sound waves bang into the **eardrum**—a thin membrane that divides the outer ear from the middle portion of the ear—causing it to vibrate.

3. Small bones on the other (inner) side of the eardrum pass the vibrations on to the inner ear membrane.

4. The vibrations are conducted by fluid inside the inner ear, which consists of fluid-filled canals in the bones of the skull. The semicircular canals function in balance, and the coiled cochlea, is involved in hearing. This fluid movement bends hair-shaped receptor cells, which release neurotransmitters. (The more vigorous their bending—in

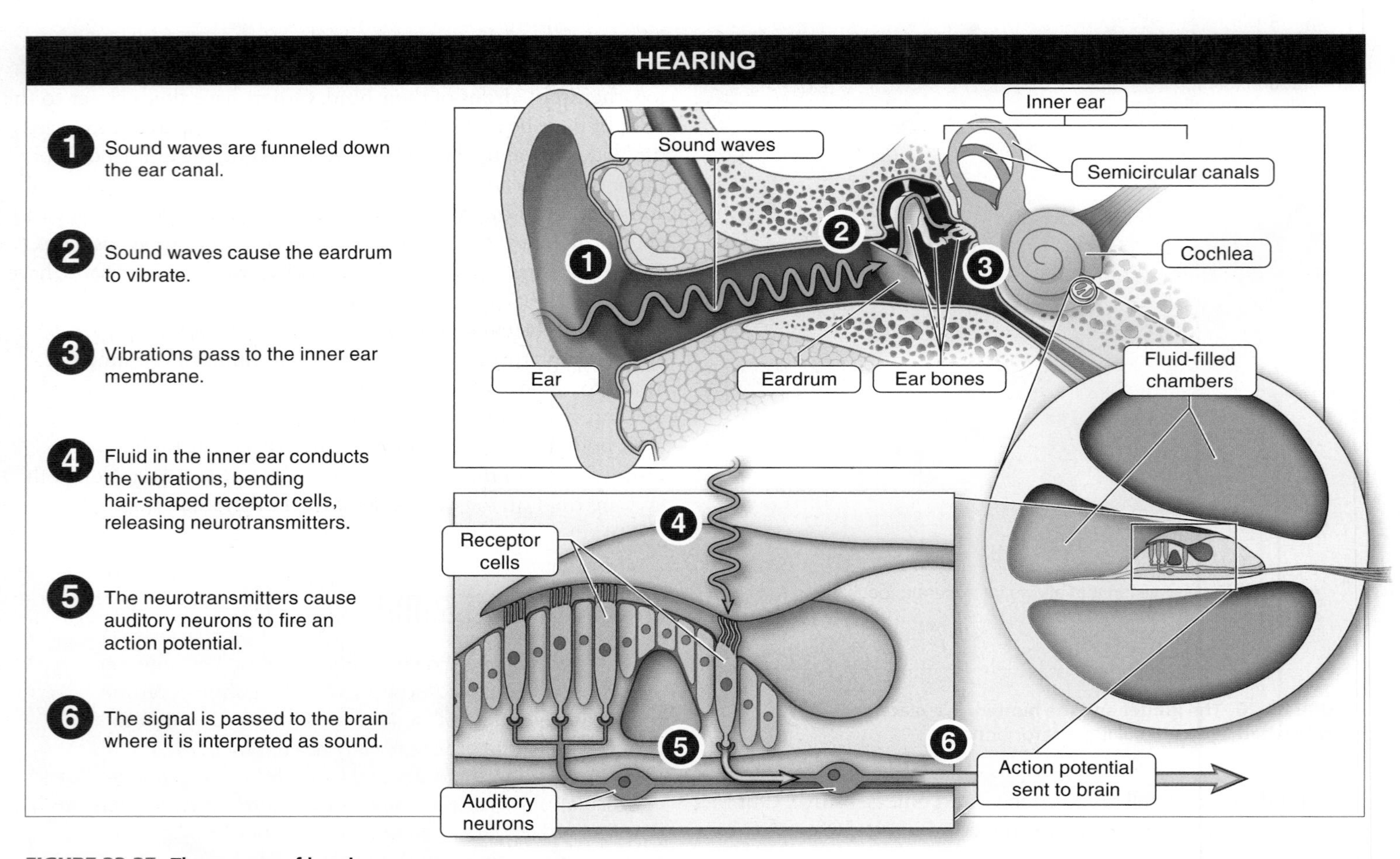

FIGURE 23-27 **The process of hearing.**

response to stronger sound waves—the greater the amount of neurotransmitter released.)

5. Auditory neurons then respond to the neurotransmitters released by the hair cells by firing an action potential.

6. The signal continues along a path of neurons to the brain, where it is interpreted as sound.

How can you get motion sickness without moving at all?

The fluid in the semicircular canals of the inner ear also acts like an inner motion detector, telling your body about its orientation, whether it is moving, and in which direction. It's not a foolproof system, though. Sitting in an IMAX theater watching a movie filmed from the seat of a roller coaster, shown on a huge screen, it is possible to feel motion sickness without actually moving. Your eyes tell your brain that you are moving, while your inner ear senses no motion at all. The conflicting signals can confuse your brain and lead to feelings of nausea.

How can loud music lead to hearing loss?

Long-term exposure to loud noises, including music, can be damaging to hearing, because such stimulation can wear out the cochlea of the inner ear. The hair cells in the inner ear are very fragile and irreplaceable, so chronic over-stimulation due to loud noises can damage them. This reduces their ability to release neurotransmitters and stimulate auditory neurons, causing hearing loss (**FIGURE 23-28**).

FIGURE 23-28 **Loud noises can cause hearing loss.**

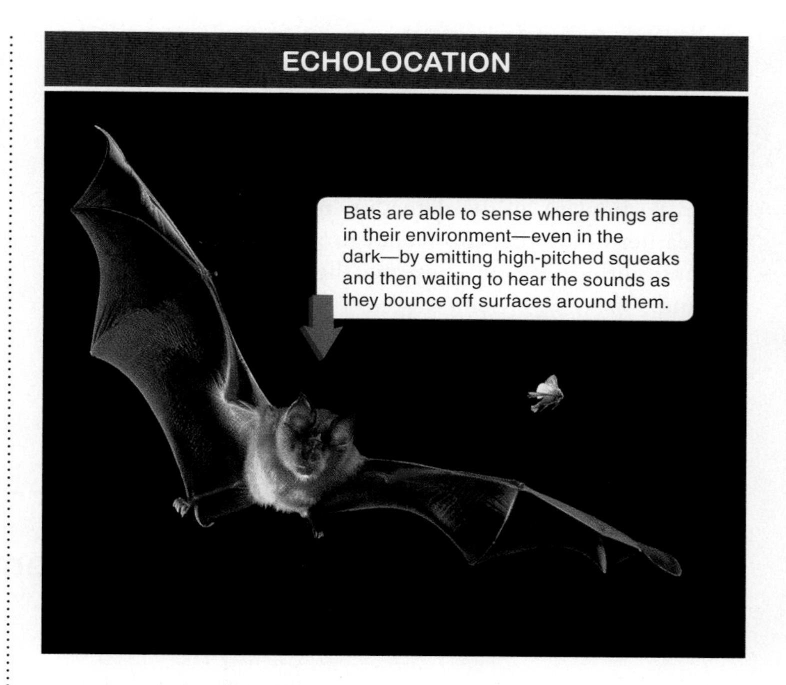

FIGURE 23-29 **Hearing through echolocation.**

Are ears necessary for hearing? Not really. Most insects, for example, don't have ears but still can hear a wide range of sounds, often with even greater sensitivity than humans. They do it with delicate hairs on antennae or other parts of the body that bend slightly in response to sound waves, much like the inner-ear hair cells described above. These insects tend to produce noises (that is, communicate by sound) by rubbing parts of their body together, rather than using their mouth. Male mosquitoes actually identify females by the vibrations of their wings as they fly. Cricket chirping, on the other hand, require a bit more work; like miniature violinists, crickets rub their wings together to create their special mating calls.

Among the mammals, bats have a unique system for hearing, called **echolocation.** Most bats spend nearly all of their time in the dark, and such a sensitive call-and-response system of hearing has evolved in them that they can sense where everything is in their environment, even with terrible sight. They do this by emitting high-pitched squeaks (that are generally inaudible to humans) and then waiting to hear the sounds as they bounce off surfaces around them. This system is so sensitive that they can detect even a small moth as it flies in their vicinity (**FIGURE 23-29**).

Do blind people have more sensitive hearing?

Recent research has revealed that individuals who go blind as infants, before the age of two, have significantly more sensitive hearing. They are up to 10 times better at hearing pitch changes than individuals who are not blind or who go blind later in life. Apparently, with the onset of blindness, parts of the brain normally used for processing visual information take up new duties, particularly in the detection and processing of sound. The earlier in life that this reallocation of brain responsibilities occurs, the greater the improvements in hearing.

TAKE-HOME MESSAGE 23•11

Hearing occurs when sound waves cause the eardrum to vibrate, moving tiny bones that pass on the vibrations to the inner ear, where they bend hair cells and thus trigger a pattern of action potentials that varies according to the wavelengths of the sound, and this is interpreted by the brain as sound.

23•12

Touch: the brain perceives pressure, temperature, and pain.

The last of the traditional five senses in humans is touch. Touch is actually a class of sensations generated by numerous different types of sensory neurons that are sensitive to pressure (mechanoreceptors), temperature (thermoreceptors), or pain (pain receptors). These sensory receptors are located throughout the body and are found in particularly dense concentrations in places such as the fingertips in primates or the side of the body (called the lateral line) in most types of fish.

Whether they are mechanoreceptors or pain receptors or thermoreceptors—and any given part of the body usually has many of each of these sensory neuron types—they function similarly. External stimulation, such as the prick of a pin, the tickle from a feather, or the heat of a pan, can cause a change in the shape of the neuron's membrane, momentarily altering its permeability. This change in permeability then increases or decreases the rate at which ions enter the cell, converting the stimulus into action potentials and, ultimately, sensations perceived by the brain (**FIGURE 23-30**).

Pain-killing drugs such as morphine reduce the ability of various pain receptors to send their signals to the brain. They do this by blocking the receptors' ability to respond to stimuli, thereby blocking neurons in the brain from sensing stimuli that would normally be identified as pain. It's as if the phone line between that part of the body and the brain is cut.

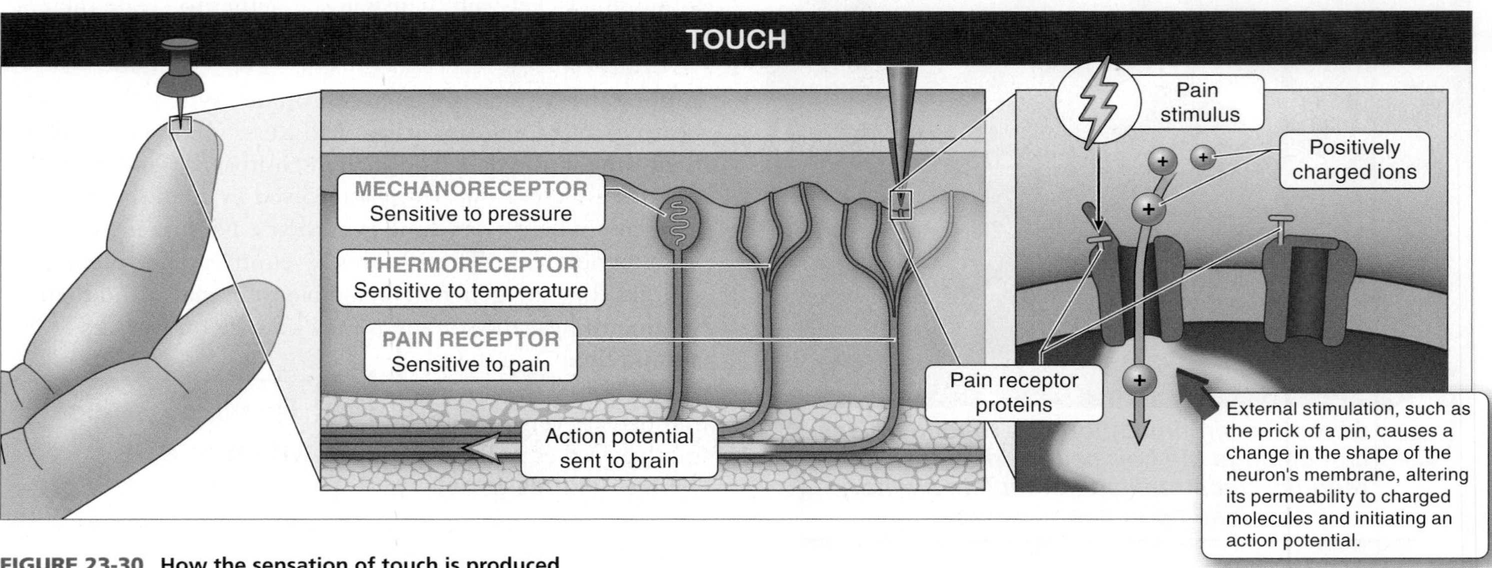

FIGURE 23-30 **How the sensation of touch is produced.**

FIGURE 23-31 Phantom pain. Neurons in the brain can perceive pain in a body part that is no longer present.

Q Amputees sometimes feel real pain where their limb once was. How can this be?

Conversely, and eerily, most amputees report that they sometimes experience tingling, prickly sensations or even shooting pains that come from where their amputated limb once was. Called "phantom pain," these sensations are very real and occur when neurons in the brain that used to represent touch receptors in the lost limb are firing, usually due to stimulation from other, nearby neurons (**FIGURE 23-31**).

TAKE-HOME MESSAGE 23•12

Touch is a class of sensations generated by mechanoreceptors, thermoreceptors, and pain receptors located throughout the body. Stimulation of these receptors causes a change in the shape of the sensory neuron's membrane, altering its permeability, generating action potentials, and causing the perception of touch by the brain.

23•13

Other senses help animals negotiate the world.

Traditionally, we think of the senses as being limited to taste, smell, vision, hearing, and touch. But in actuality, there are several more senses that some animals use. As with the five senses described above, these additional senses are made possible by neurons that are specially modified to respond to external stimuli by generating action potentials that convey information to the brain about the outside world.

Balance and Motion Many invertebrates (including jellyfish, lobsters, and crayfish) and vertebrates have sensory cells that enable them to orient themselves with respect to gravity and to detect changes in velocity or orientation. These sensory cells are always some sort of hair cells that can be pushed in one direction or another, much as grass blows in the wind. The force exerted on the hair cells (which, as we've seen, are located in the inner ear in vertebrates) gives information about the movement of the organism. In a variation on the sense of touch, in many mammals the sensations detected by whiskers can also convey information about balance and motion.

Electricity Marine animals of many types are sensitive to tiny electrical changes in the water around them. Sharks have sensory cells, embedded within sacs in their skin, that are open to the shark's external surroundings. These cells are so sensitive to electrical changes around them that they can sense the muscle contractions of a fish swimming nearby. This ability to sense changes in electrical charge helps sharks to track down prey.

Magnetism Compass needles point north because of the faint magnetic field produced by the earth. A wide range of species, from birds to eels and sharks to migrating beluga whales, and even some bacteria, are able to detect this magnetic field and use it to help them navigate (**FIGURE 23-32**). In some clever but devious experiments, researchers placed strong magnets around some bird cages, deflecting the magnetic field. This caused the birds to shift their orientation within the cages by exactly the amount that the magnetic field had been moved. Very little is known about how sensory cells have evolved to detect and respond to magnetism.

FIGURE 23-32 **Traveling south.** Beluga whales during migration.

FIGURE 23-33 **Some snakes can "see" heat.**

Heat Unlike the way you sense heat by touching a hot plate or feeling warm when it's hot outside, pythons, vipers, and some other snakes can "see" heat differences in their environments without having to touch it. Using "pit organs" located on either side of the head, they can sense—even in complete darkness—a mouse or other prey animal near them, solely from the body heat that it produces (**FIGURE 23-33**).

TAKE-HOME MESSAGE 23•13

Among animals, there are several senses in addition to the five found in humans, including those that perceive and respond to balance and motion, electricity, magnetism, and heat production in the animal's environment.

④ The muscular and skeletal systems enable movement.

These eastern gray kangaroos have powerful leg muscles that propel them forward.

23•14

Muscles generate force through contraction.

Running, flying, swimming, crawling, walking, digging, dancing. Animals, with very few exceptions, move. And in vertebrates, this movement is generally initiated by cells of the nervous system that stimulate contractions in muscle tissue that pull on a rigid skeletal system. In this section, we examine the three types of muscle tissue, how contractions are generated within that tissue, and how input from the nervous system can initiate and control muscle contractions. We'll also investigate how the size and strength of muscles reflects their use and how physical training can increase muscle strength.

As we saw in Section 20-10, muscle tissue is made up of elongated cells capable of generating force when they contract, and there are three types of muscle tissue.

1. *Skeletal muscle* is responsible for generating most of the movement we see in animals. Skeletal muscle attaches to bones and is controlled by individual neurons attached to each muscle fiber. It makes up about 40% of human body weight.

2. *Cardiac muscle* causes the heart to pump blood through the body. Not initiated by the nervous system or under conscious control, cardiac muscle contractions and relaxations occur continuously throughout life, with the muscle cells containing more energy-releasing mitochondria than other types of muscle cells.

3. *Smooth muscle,* which is not under conscious control and is able to contract without stimulation from the nervous system, surrounds blood vessels and many internal organs. Smooth muscle generates slower contractions that can gradually move blood, food, or other substances through vessels and organs.

In the rest of this section, we focus on the type of muscle that is controlled by the nervous system: skeletal muscle. The muscular system of humans contains about 700 skeletal muscles in all. To facilitate movement, skeletal muscles are attached to bones by connective tissue. Often, a muscle tapers at each end into tendons, which attach the muscle at the two ends—the "insertion" and the "origin" of the muscle—to bones. The contraction of the muscle then pulls the attachment points closer together and causes a movement of the bone. The biceps muscle of the upper arm, for example, has its origin in the shoulder and its insertion in one of the bones of the forearm. Contraction can rotate the forearm or bring the lower part of the arm closer to the shoulder. Two muscles generally work in opposition so that as one moves a bone in one direction, the other—the triceps in the upper arm—has its insertion and origin located so that its contraction moves the bone in the opposite direction (**FIGURE 23-34**).

Let's look more closely at the structure of a muscle and at the cellular mechanism by which muscle fibers contract. A muscle comprises a bundle of fibers. Each fiber is a single cell, but with multiple nuclei and with numerous **myofibrils,** cylindrical organelles that shorten when they contract. A myofibril contains repeating units called sarcomeres, which is where the contraction takes place. The **sarcomeres** are composed of large numbers of long filaments, overlapping and parallel to each other. The filaments come in two types: thin filaments made mostly from the protein **actin,** and thick filaments made mostly from the protein **myosin.** There may be 100,000 sarcomeres in a biceps muscle cell.

SKELETAL MUSCLE GENERATES MOVEMENT

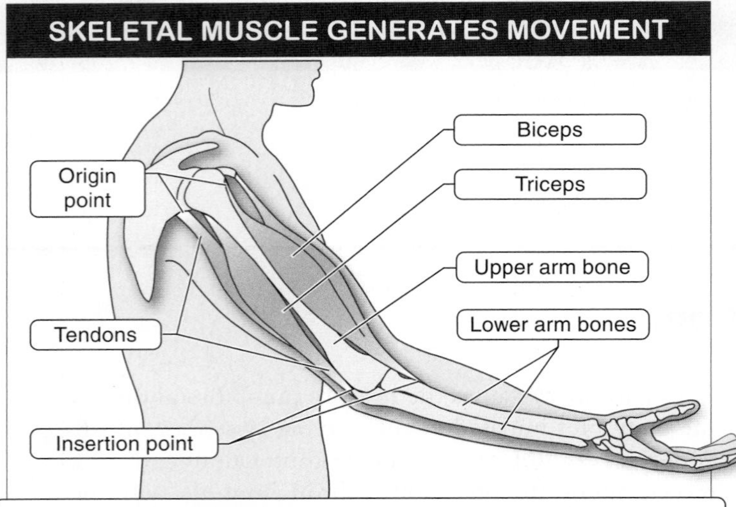

Two muscles generally work in opposition so that contracting one muscle moves a bone in one direction, while contracting the other muscle moves the bone in the opposite direction.

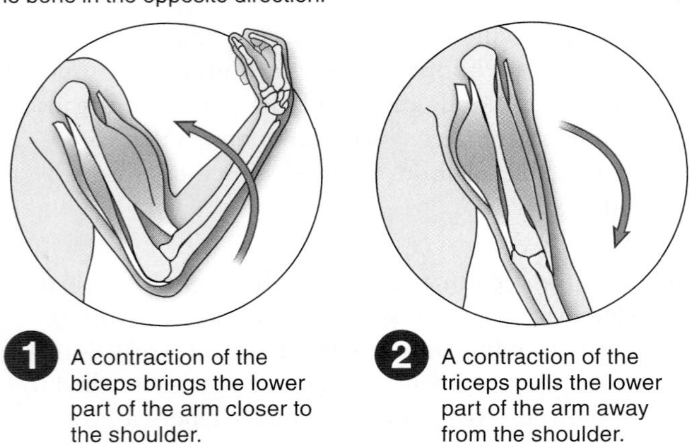

1 A contraction of the biceps brings the lower part of the arm closer to the shoulder.

2 A contraction of the triceps pulls the lower part of the arm away from the shoulder.

FIGURE 23-34 Movement: the work done by skeletal muscles.

The sarcomere shortens in a four-step process (**FIGURE 23-35**):

1. *Detach:* A link between a myosin and an actin filament parallel to it is broken as a molecule of ATP binds to the myosin.
2. *Reach:* As the ATP breaks down, energy released alters the shape of the myosin into a higher-energy shape—much like bending a twig, which is in a higher-energy shape because it can release energy as it snaps back to its original shape.
3. *Reattach:* In its altered shape, the myosin reaches farther down the actin filament, where it reattaches.
4. *Pull back:* The myosin then snaps back to its original shape, pulling the actin filament as it does so and shortening the fiber. This last step is considered the "power stroke."

As the contraction process occurs across numerous sarcomeres in a muscle cell, the entire muscle can shorten. From the perspective of energy, the potential energy of ATP is converted to the kinetic energy of sarcomere-shortening, which can do work (for instance, as your arm curls a dumbbell).

Nervous system control over skeletal muscle occurs as neurons, at their synapses with muscle cells, release the neurotransmitter acetylcholine. Acetylcholine causes the muscle cell membrane to open its sodium channels, which depolarizes the muscle cell. Deeper inside the muscle cell, the depolarization causes the opening of calcium channels in the cell's endoplasmic reticulum. Calcium stored in the modified endoplasmic reticulum rushes into the cytoplasm, diffuses toward the myofibrils, and binds to a molecule surrounding the actin. When calcium binds to the protein molecule it acts as a switch and moves another molecule that normally covers binding sites on the actin molecule. After the binding sites are exposed, myosin can attach to actin, and contraction can occur.

When a muscle contracts, the duration between a contraction and a relaxation is called a **twitch.** Muscle fibers within a muscle vary in how quickly they can twitch, with two general types. Fast-twitch fibers, which can contract very quickly and tend to function anaerobically, have relatively weak endurance. They are capable of contracting

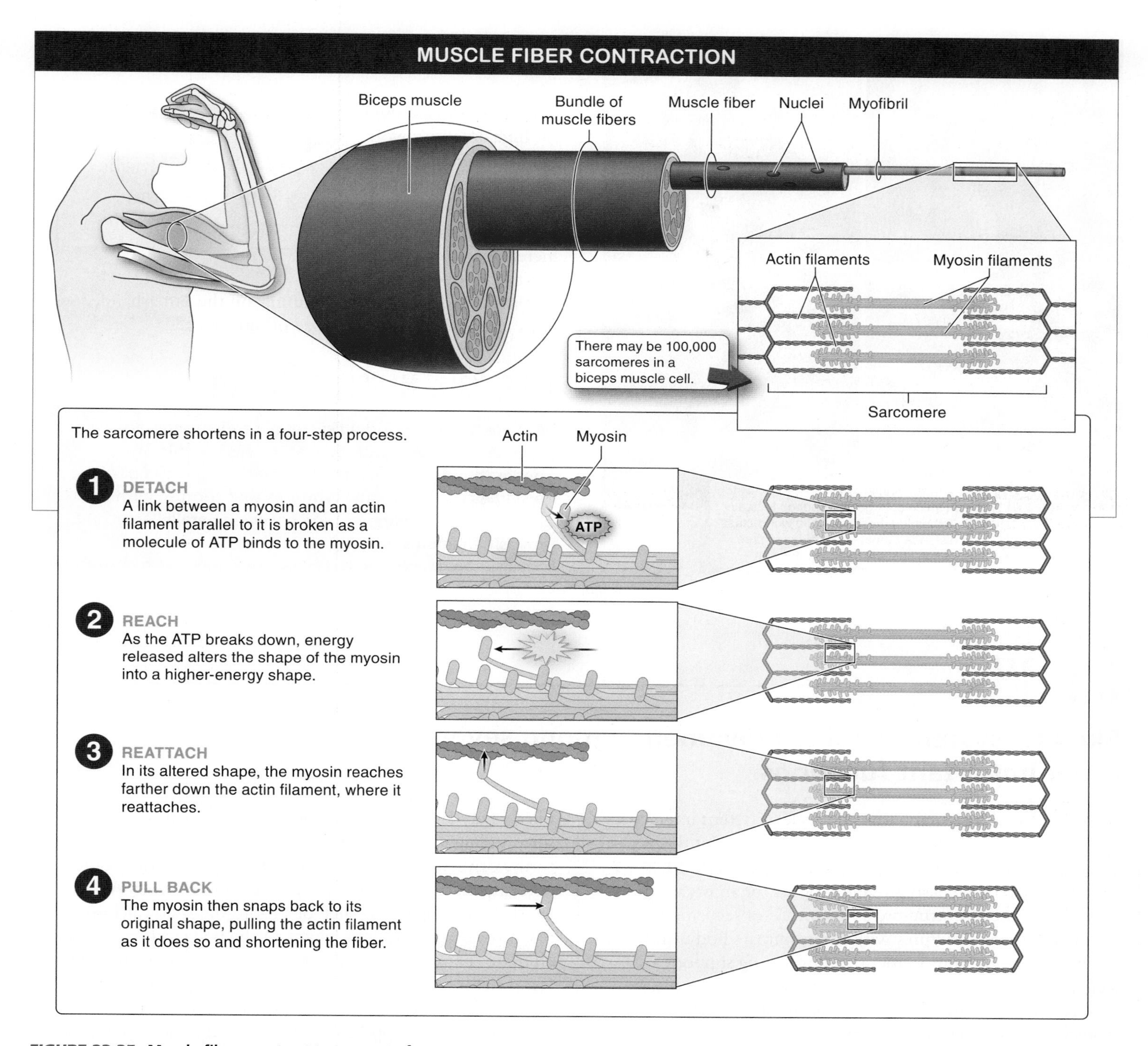

FIGURE 23-35 Muscle fibers contract to generate force.

and relaxing about 10 times faster than slow-twitch fibers, which are surrounded by rich oxygen-delivering capillary beds and large numbers of oxygen-storing myoglobin molecules, and can sustain activity for much longer periods of time. Although most muscles have approximately equal proportions of fast-twitch and slow-twitch fibers, there is some variation among the fibers of different muscles, as well as among different individuals. It has been shown, for example, that world-class sprinters have significantly more fast-twitch fibers in their leg muscles, while world-class long distance runners tend to have more slow-twitch fibers in their leg muscles (**FIGURE 23-36**).

World-class sprinters have more fast-twitch fibers (which enlarge in response to heavy use) in their leg muscles, while world-class long distance runners tend to have more slow-twitch fibers in their leg muscles.

FIGURE 23-36 **Fast-twitch and slow-twitch muscles.**

Q Marathon runners tend to have smaller leg muscles than sprinters. Why?

When muscles are used for high-intensity work against increasing resistance, as in many weight-training programs, muscle fibers increase in size as the fast-twitch fibers become thicker. Endurance training, on the other hand, does not cause an increase in the size of muscle cells.

TAKE-HOME MESSAGE 23·14

Muscle tissue—including skeletal, cardiac, and smooth muscle—is made up of elongated cells capable of generating force when they contract. A muscle fiber is a single cell containing myofibrils that shorten with the making and breaking of links between parallel actin and myosin filaments.

23·15

Skeletal systems enable movement, among several other important functions.

By serving as a structural frame, the skeletal system interacts with the muscular system to enable movement. But this is just one of many important functions of a skeleton (**FIGURE 23-37**). A skeletal system can also provide support for an organism, providing shape and structure and a means of securing the organs and other structures within the animal's body. And in addition to enabling movement and providing support, a skeletal system offers protection—the brain, for example, is protected from harm within the skull, and the heart is protected within the rib cage. Additionally, in some organisms, including humans, the skeletal system provides a site for production of important blood and immune cells, a process that occurs in bone marrow. Bones also serve as a reservoir of some minerals, such as calcium, absorbing excess quantities from the bloodstream or releasing minerals when their concentration in the blood is low (see Section 24-5).

There are three distinct types of skeletal systems (**FIGURE 23-38**). In a *hydrostatic skeleton*—such as is found in earthworms, jellyfish, and squid—pressure created within a fluid-filled cavity creates sufficient rigidity that the muscles surrounding it can generate movement, including, in earthworms, the ability to dig. An *exoskeleton* is a rigid outer covering that supports and protects an animal's body. An exoskeleton may be the shell of a turtle, the rigid calcium-based covering of some molluscs, or the polysaccharide-based, chitinous exoskeleton of insects, crabs and lobsters, and spiders. Exoskeletons can also be valuable in reducing some animals' vulnerability to drying out. The third type of skeleton—an *endoskeleton*—is a support structure of hard mineralized tissue, as is found in echinoderms (such as sea stars and sand dollars), and in all vertebrates including humans. The vertebrate endoskeleton serves as a site for production of blood and immune cells and as a reservoir of some minerals.

The composition of the endoskeleton varies among echinoderms and vertebrates. Echinoderm endoskeletons consist of a rigid, calcium-based structure, similar to bone. Among the vertebrates, the jawless fishes and sharks have a flexible but strong skeleton made from cartilage. The other

FUNCTIONS OF THE SKELETAL SYSTEM

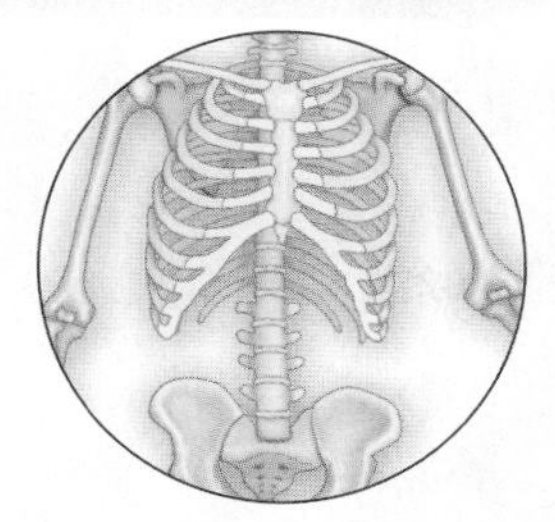

STRUCTURAL SUPPORT
Provides shape and structure, including securing organs in place

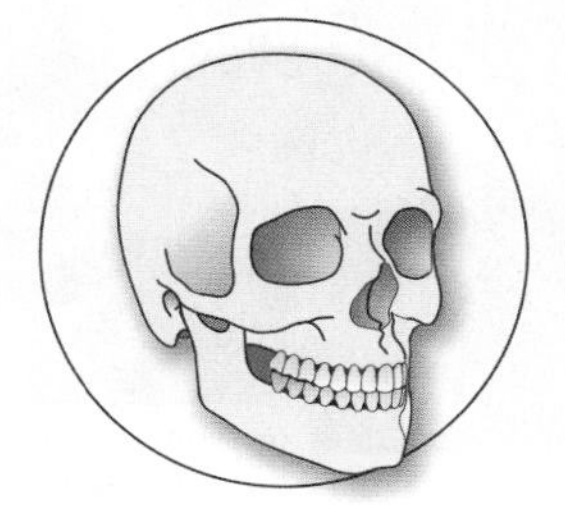

PROTECTION
Shields vulnerable tissue—such as the brain and heart—from external insults

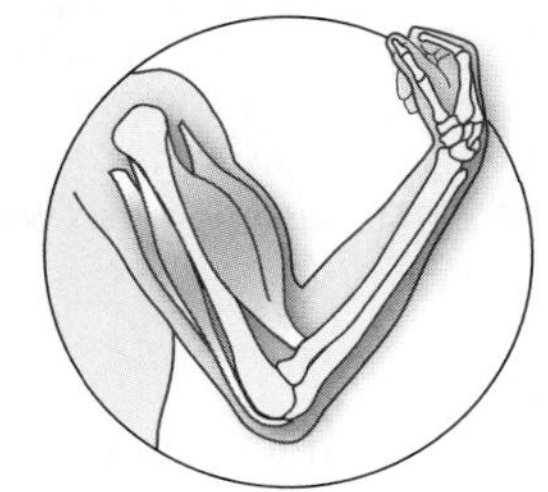

ENABLING MOVEMENT
Muscles, which are connected to bones, generate movement when they contract

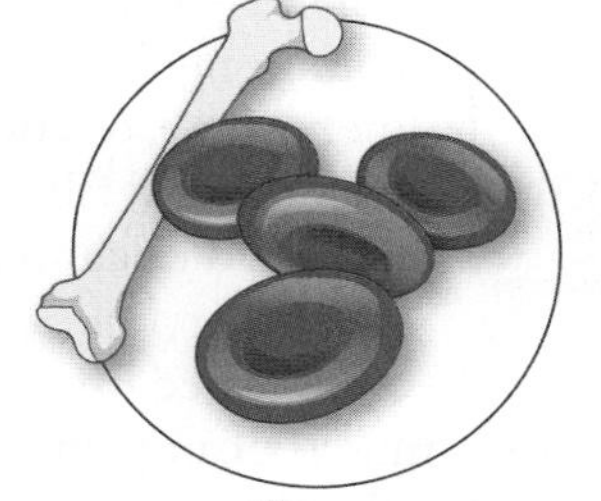

CELL PRODUCTION
Produces blood cells in bone marrow

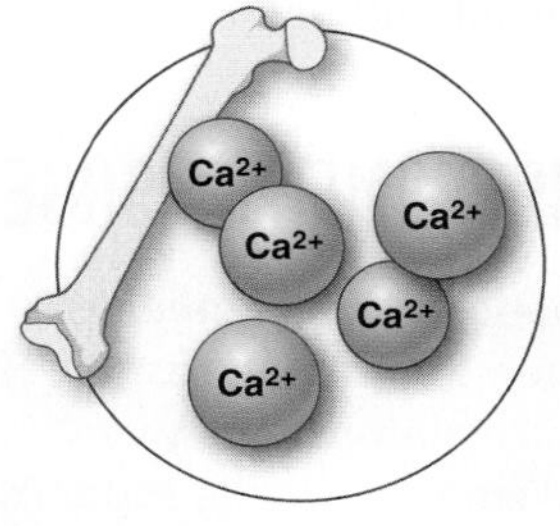

MINERAL RESERVOIR
Elements such as calcium can be released from or stored in bones in response to deficiencies or excesses in the bloodstream

FIGURE 23-37 The skeletal system serves multiple purposes.

vertebrates have a skeletal system made primarily from bone, a harder, but still resilient connective tissue material (described in Section 20-8).

Most vertebrate skeletons are divided into an axial skeleton, which includes the skull, sternum, ribs, and vertebrae, and an appendicular skeleton, which includes the limbs

TYPES OF SKELETAL SYSTEMS

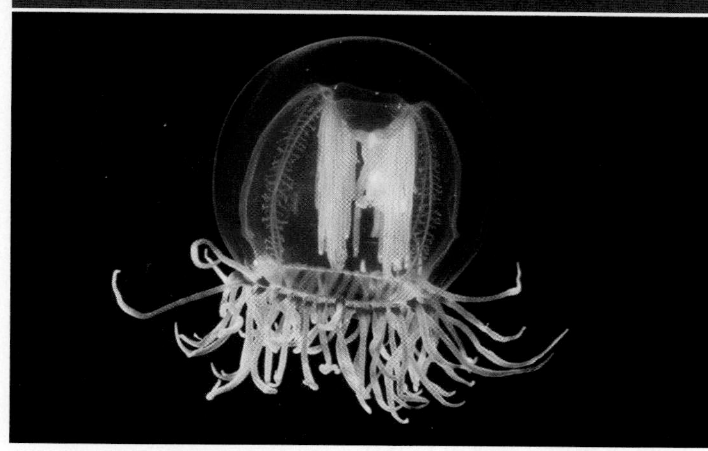

HYDROSTATIC SKELETON
A fluid-filled cavity with sufficient rigidity so that muscles surrounding it can generate movement

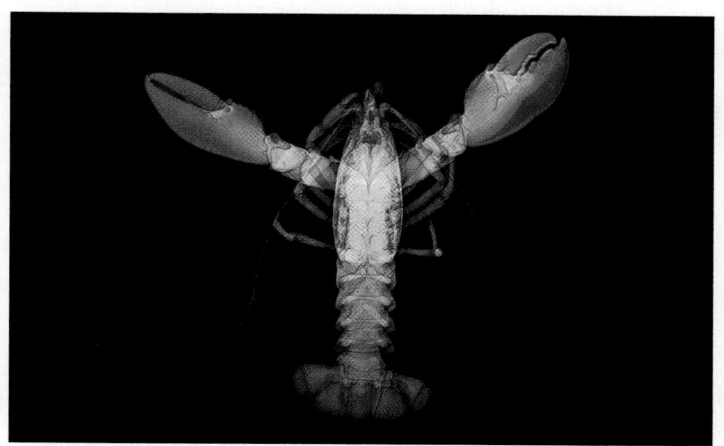

EXOSKELETON
A rigid outer covering that supports and protects an animal's body

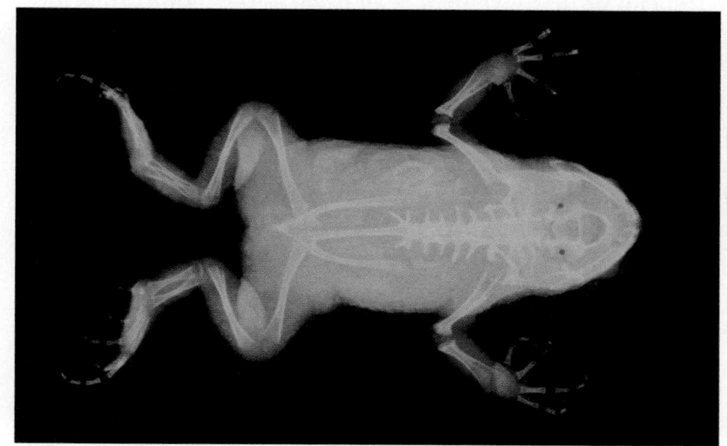

ENDOSKELETON
An internal support structure composed of hard mineralized tissue

FIGURE 23-38 There are three distinct types of skeletal systems in animals. Shown: bell jellyfish, lobster, and toad.

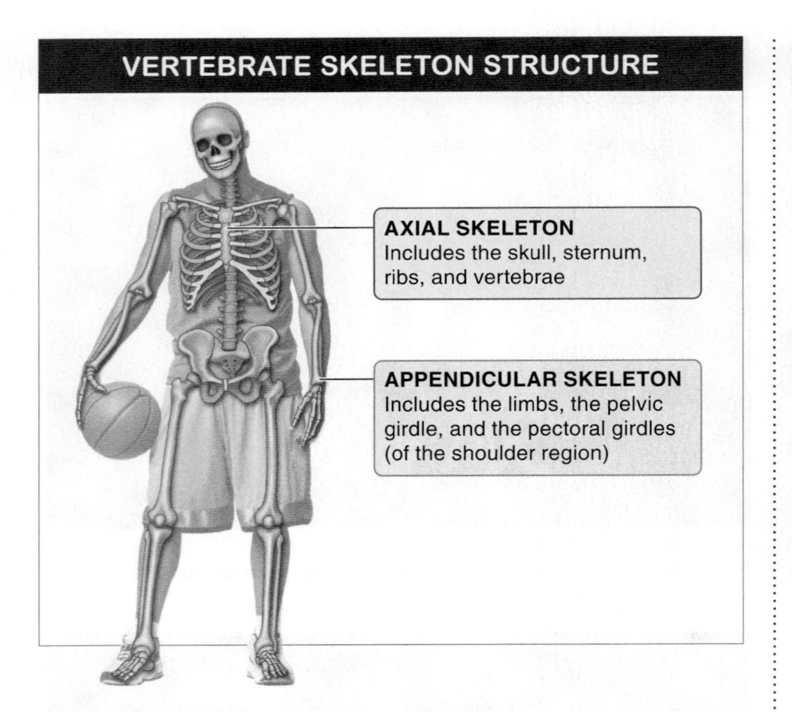

FIGURE 23-39 Vertebrate skeletons are divided into axial and appendicular regions.

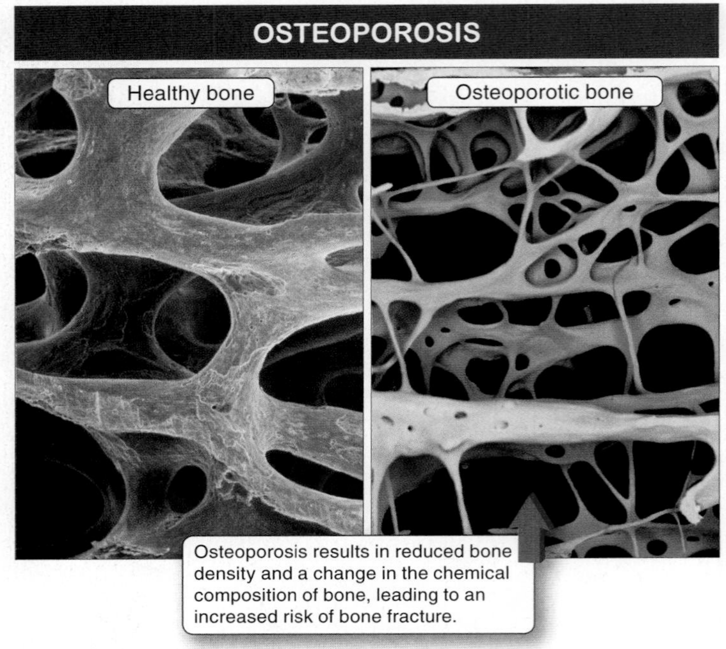

FIGURE 23-40 Osteoporosis weakens bone.

(or fins, in the case of fishes), the pelvic girdle, and the pectoral girdles (of the shoulder region) (**FIGURE 23-39**).

> **Q** What is osteoarthritis? Why is it so painful?

At the locations in the skeleton where bones meet, several types of joints occur. These may allow movement while maintaining support, such as is seen in the shoulder, hip, knee, and elbow joints. Or they may enable little movement (such as the joints between vertebrae) or no movement at all (such as the "joints" that are the sutured connections between the bones of the skull). Arthritis is a condition in which joints have become damaged. There is a variety of forms of arthritis. In the most common type, osteoarthritis, a breakdown of connective tissue, usually cartilage, reduces the cushioning of the bones and can lead to inflammation of tissue around the joint, and accompanying pain.

Another type of breakdown in the skeletal system is osteoporosis, a disease that occurs as the density of bone is reduced and the chemical composition of the bones changes (**FIGURE 23-40**). Occurring most commonly in women, particularly following menopause, osteoporosis increases the risk of bone fractures and falls. Dietary modifications, including increased calcium and vitamin D that can strengthen bones, along with lifestyle changes such as increased exercise that can build bone mass and strengthen muscles, can reduce osteoporosis and the risks associated with it.

TAKE-HOME MESSAGE 23•15

Among animals, three distinct types of skeletal systems occur: hydrostatic skeletons, exoskeletons, and endoskeletons. The vertebrate skeletal system can enable movement, provide support, offer protection, provide a site for production of important blood and immune cells, and serve as a reservoir of minerals, including calcium.

❺ The brain is organized into distinct structures dedicated to specific functions.

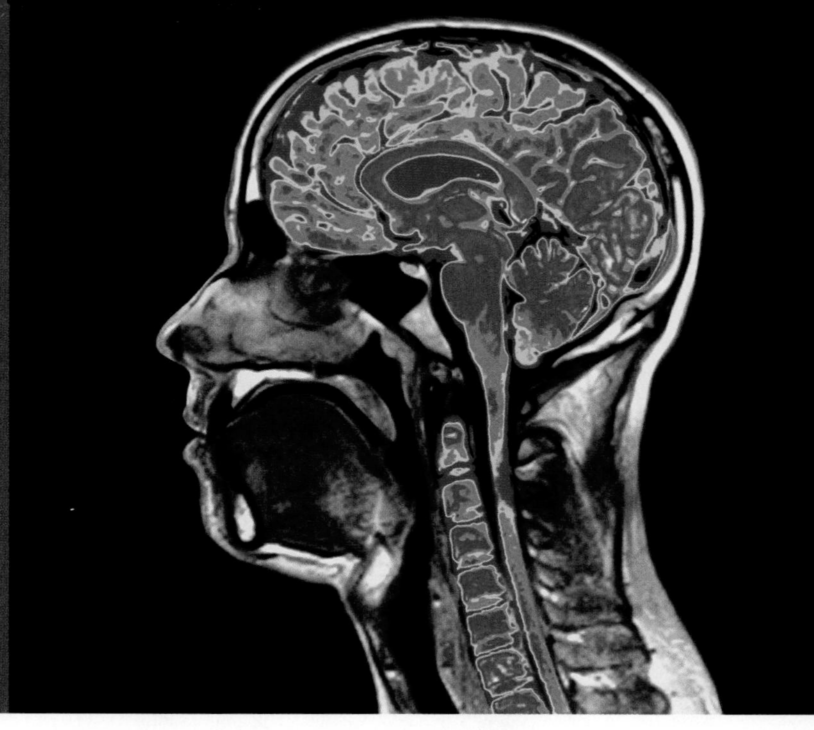

An fMRI image of the brain and spinal cord of an adult male.

23 • 16

The brain is organized into several distinct regions.

Looks can be deceiving. Especially when it comes to brains. Take the human brain, for example. You'd hardly know that you were looking at something far more complex and powerful than the most advanced supercomputers ever built. In an adult, it is only a bit over 3 pounds, or 1.4 kilograms. It's fairly uniform in color, a drab whitish-gray, the surface is soft and gently wrinkled, and it's just a bit stiffer than Jell-O.

From a cellular perspective, all animal brains function in the same general way. A neuron is a neuron is a neuron. But this isn't the whole picture. Although the brain is made up primarily from the same type of cells as the rest of the nervous system (the spinal cord and the peripheral nervous system), close inspection reveals several distinct regions within the brain. Each region is characterized by neurons that are shaped slightly differently, that have different axon and dendrite branching patterns and complexities, and that use different neurotransmitters at their synapses. Each of these regions is the control center for various activities in the body.

Among different species, evolutionary changes have resulted in differences in the relative amount of the brain dedicated to various structures. These differences have, in turn, led to corresponding differences in abilities and behaviors. In the cerebral cortex, for example, humans have significantly more neurons than do other species. This extra brain matter in this particular part of the brain (which we examine in more detail later in this chapter) is associated with many abilities that are particularly well developed in humans, including abstract thought, perception, and language. Similarly, the unusually large number of neurons (relative to overall brain size) devoted to the smell-sensing section of a rat's brain give it a particularly powerful sense of smell. This powerful sense, which is much more sensitive to scent stimuli than is a human's sense of smell, contributes to the rat's abilities to find food.

In this and the next two sections, we explore the three principal regions of the brain—the hindbrain, the midbrain, and the forebrain—and the functions they control (**FIGURE 23-41**).

Hindbrain The central nervous system is made up of the brain and spinal cord. The spinal cord extends from the brain down through the backbone (though not down the entire length of the backbone), protected by the vertebral column. Branching off from the spinal cord at regular intervals are bundles of axons, some carrying motor information to muscles and glands, others carrying in sensory information from the senses to the brain. The top of the spinal cord is the beginning of the brain. Here, the spinal cord expands into the **hindbrain.** Three important structures are located in the hindbrain: the **medulla,** the **pons,** and the **cerebellum.** Because these structures are not involved in higher thought,

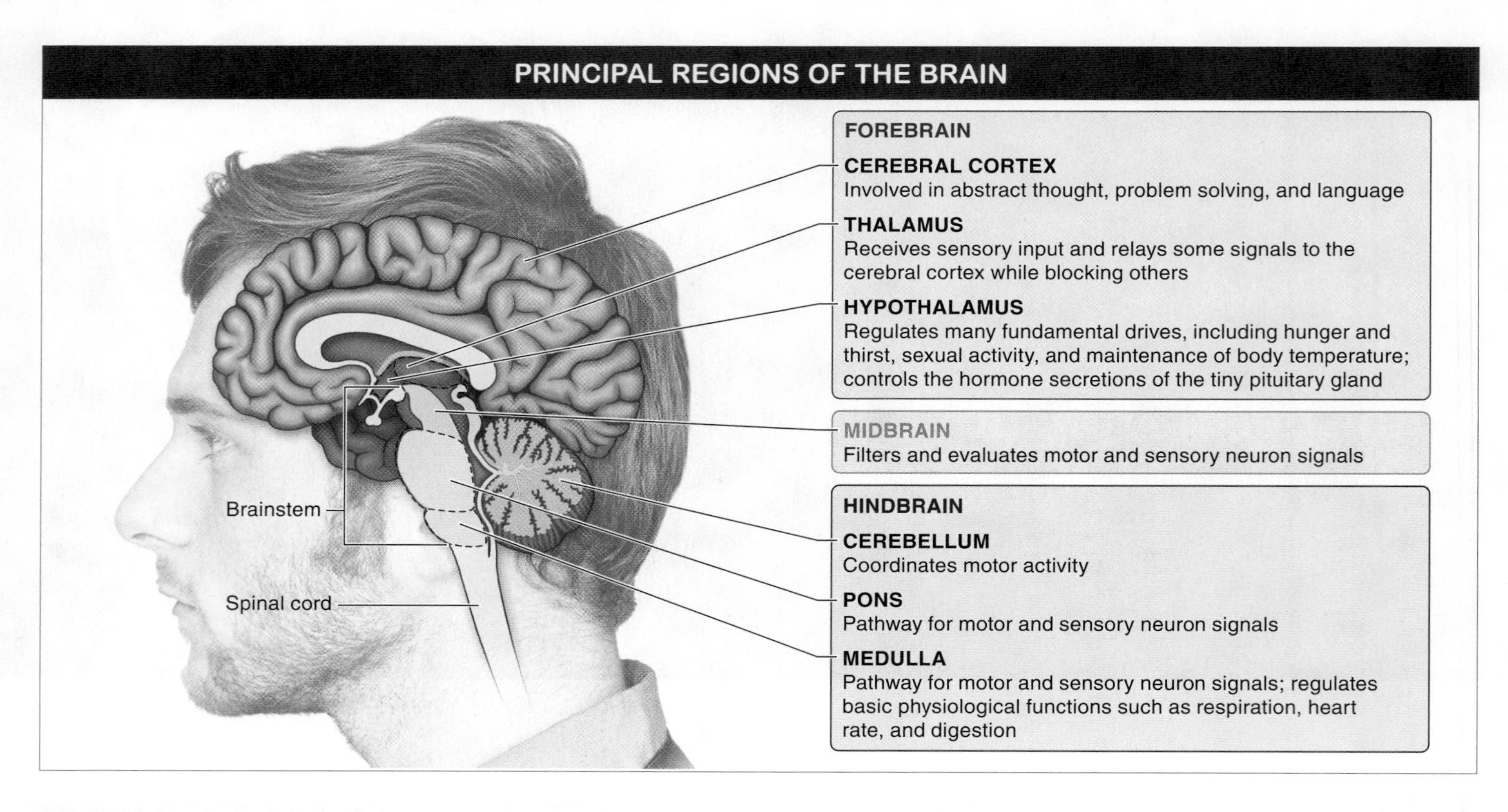

FIGURE 23-41 The brain is divided into three main regions: forebrain, midbrain, and hindbrain.

hindbrain damage can lead to paralysis without harming consciousness.

Medulla and Pons. Running through these structures are the many motor neurons traveling from the brain to the spinal cord and the many sensory neurons traveling from the body into the brain. Additionally, within the medulla are lots of collections of neurons that regulate basic physiological functions such as respiration, heart rate, and digestion. Consequently, injury to these structures—such as occurred when President John F. Kennedy was assassinated in 1963—nearly always leads to death.

Cerebellum. This structure coordinates numerous types of motor activity. When it works properly, graceful motion is possible. The coordination of many complex movements is overseen by the cerebellum. Knitting or typing, for example, involve a wide range of complex muscle movements that must be executed in precise sequences. Once learned, these movements run on "auto-pilot": we don't have to think consciously about these activities because the cerebellum directs them (**FIGURE 23-42**). If you are typing rapidly and suddenly think consciously about the motions your fingers are engaged in, you start to make all kinds of errors. Why? Because your conscious thought draws on a part of the brain other than the cerebellum.

FIGURE 23-42 Coordinated by the cerebellum. Once learned, knitting does not require conscious thought.

Midbrain The **midbrain,** along with the medulla and pons of the hindbrain, makes up the brainstem. Most of the sensory information and motor neuron connections coming into and out of the brain pass through the midbrain. These receptors and effectors frequently have synapses with each other, as the midbrain helps filter and evaluate the importance of each signal. The midbrain is greatly enlarged and plays a more significant role in cold-blooded animals, such as fishes and frogs, than it does in warm-blooded animals.

Forebrain In humans, the **forebrain** is the largest region of the brain. It includes two control and relay structures, the thalamus and hypothalamus, as well as the cerebrum. The cerebral cortex, which makes up the bulk of the cerebrum, is responsible for most of what we consider "higher" thought, including perception, memory, language, intelligence, and personality.

Thalamus and Hypothalamus. The **thalamus** is one of the primary switchboards in the brain. It receives most of the visual, auditory, and touch input and, like a secretary deciding which calls to let through, relays some signals to the cerebral cortex while blocking others. The **hypothalamus** (part of the limbic system, discussed later in this section) is one of the chief regulatory centers of the brain. It regulates many fundamental drives, including hunger and thirst, sexual activity, and maintenance of body temperature. It also controls the hormone secretions of the tiny pituitary gland, located right next to it, which play important roles in reproduction and development.

> "The mind is its own place, and in itself
> Can make a heaven of hell, a hell of heaven."
> —John Milton, *Paradise Lost,* 1667

In the 1950s, James Olds discovered something very peculiar about the hypothalamus. When he inserted a tiny electrode into the hypothalamus of a rat's brain, the animal seemed to experience great pleasure (**FIGURE 23-43**). Of course, Olds couldn't be certain that it was pleasure that the rats were experiencing, but he hypothesized that this was so by noting how the rats would work very hard (for instance, by learning to run very complex mazes) to get the stimulation. When he set up the apparatus so that an animal could administer the stimulation to its own brain simply by pushing a lever, the rats became lever addicts, pushing the lever more than a hundred times a minute for many hours. They would even choose to push the lever over eating food or drinking water, with some rats starving to death while pushing the lever rather than taking even a short break to eat.

This discovery that the brain has "pleasure centers" or "do-it-again centers" was important because it revealed a mechanism by which animals could be motivated to engage in evolutionarily important behaviors by which they might increase their own reproductive success relative to other individuals in the population. Drinking when an animal is thirsty or eating when it is hungry produces stimulation in this brain area, thereby reinforcing behaviors essential to survival and reproduction.

FIGURE 23-43 **Olds's experiments led to the discovery that the brain has "pleasure centers."**

Q How could you direct a rat's movements via remote control?

In a disturbing experiment with potential ethical implications, researchers recently implanted small transmitters and electrodes into the brains of rats that enabled the researchers to stimulate the rats' whiskers by remote control. This stimulus generated sensory signals about the direction in which the rats were moving. The researchers also implanted electrodes enabling the stimulation of the animals' pleasure centers. Together, this apparatus made it possible to guide, via remote control, the direction in which the animals moved, rewarding them when they moved in the desired direction. The researchers were able to direct the rats like robots over fences, through pipes, and up trees, from several hundred yards away.

Cerebral Cortex. For humans, the **cerebral cortex** is the most sophisticated part of the brain. Involved in abstract thought, problem solving, language, and more, this is the part of the brain that is most responsible for the traits that most set us apart from the rest of the animal species.

One of the most dramatic demonstrations of the complexity of the cerebral cortex resulted from a tragic accident and apparently miraculous recovery. In 1848, a railroad worker named Phineas Gage was packing explosives into a hole when they exploded unexpectedly, blasting a three and a half foot iron bar right through his face, just below his left eye, through the front part of his head, and out through the top of his skull (**FIGURE 23-44**). Amazingly, he recovered from the accident. Soon after his recovery, though, it became clear that something in Gage's personality had changed dramatically as a result of the injury. Before the accident he was serious and industrious, but after the accident he became vulgar, unpleasant, and irresponsible. He also became much more uninhibited than he had previously been.

Gage died about 13 years after the accident, and his skull was preserved. Recent analyses of the skull indicate that Gage suffered from severe damage to part of his frontal cortex. In other case studies of patients with damage to the same part of their frontal cortex, such as that caused by certain tumors, similar personality changes have been noted, with patients frequently exhibiting radically changed emotional responses. As we'll see in Section 23-17, modern brain imaging techniques are enabling us to learn more about the functions of specific brain regions, without the need to study individuals with tumors or other brain injuries.

The highly folded cerebral cortex is by far the largest brain structure, covering all of the brain except for the cerebellum (**FIGURE 23-45**). The cerebral cortex makes up 80% to 85% of the brain mass in humans, with about 10 billion neurons and hundreds of billions of synapses. Although the brain (including the cerebral cortex) looks fairly symmetrical, functions are not divided in a symmetrical way. The brain is divided into a left and a right hemisphere, connected by a broad, thick band of neurons, called the **corpus callosum.** The **left hemisphere** in most, but not all, individuals is the

FIGURE 23-44 Phineas Gage's accident. Gage suffered damage to his frontal cortex, which caused dramatic behavior changes.

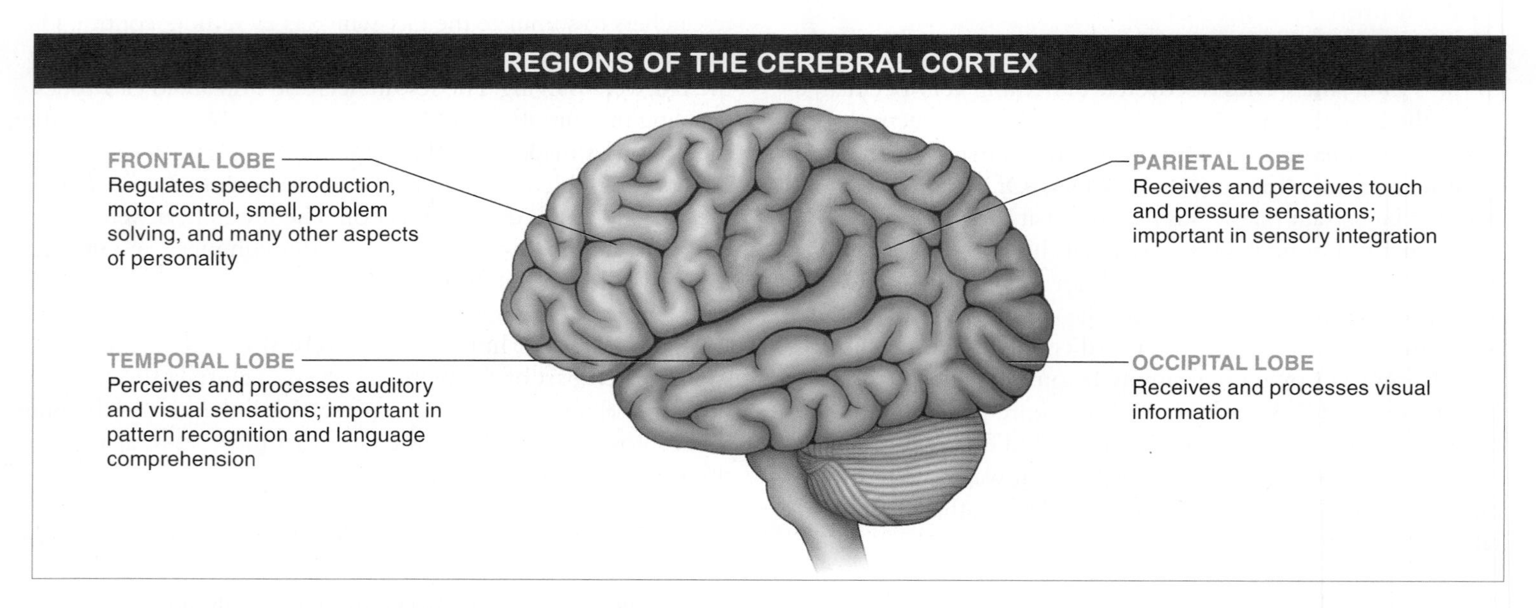

FIGURE 23-45 The lobes of the cerebral cortex.

site of areas specializing in language, logic, and mathematical skills, while the **right hemisphere** is more commonly home to larger areas dedicated to emotions, intuitive thinking, and artistic expression.

Each hemisphere of the cerebral cortex has four different lobes: frontal, temporal, parietal, and occipital. Each lobe is associated with specific functions (but each also has many other subareas that control a variety of additional functions; see Figure 23-45):

1. *Frontal lobe:* important in speech production, motor control, smell, problem solving, and many aspects of personality.
2. *Temporal lobe:* perceives and processes auditory and visual sensations; also important in pattern recognition and language comprehension. Damage to part of the right temporal lobe—usually due to a stroke or tumor—has been implicated in a condition known as prosopagnosia or face blindness. Individuals with this condition find it difficult or impossible to recognize faces.
3. *Parietal lobe:* receives and perceives touch and pressure sensations; also important in **sensory integration,** the process of incorporating information from all of the senses to form a single perception of something.
4. *Occipital lobe:* receives and processes visual information.

Limbic System. Within the forebrain, near the center of the brain, is a set of structures that together make up the **limbic system.** It includes parts of the cerebral cortex and two structures deeper within the center of the brain, the **hippocampus** and the **amygdala,** and is responsible for many of our physiological drives and instincts. Together, these structures also play an important role in emotions, learning, and memory, which we discuss later in the chapter.

TAKE-HOME MESSAGE 23•16

There are several distinct regions in the brain, each of which is the control center for various activities in the body. The hindbrain regulates basic physiological functions such as respiration, heart rate, and digestion, and coordinates numerous types of motor activity. The midbrain, part of the brainstem, helps filter and evaluate the importance of sensory and motor information. In humans, the forebrain is the largest region of the brain, responsible for most of what we consider higher thought, including perception, memory, language, intelligence, and personality.

23•17

Measuring brain activity helps us understand how tasks are allocated.

How many times have you heard someone say, "We only use 10% of our brain!" Besides being just plain silly, this pervasive myth is completely false. But it prompts some reasonable questions: Which parts of the brain *are* we using? Do we use some more than others? Do mental tasks or physical sensations require the use of a specific part of the brain, or do they use neurons from all over the brain? Until relatively recently, we didn't have the tools to address these questions, but now we do.

Most computers have lights that blink when the machine is active. A brain, however, just sits there; no lights flash, no sounds are made. Nothing. At least, nothing obviously visible to the naked eye. Helping us to better understand how the brain functions, there has been a revolution in **brain imaging technologies,** which capture pictures of the brain as it functions, revealing activity patterns (based on increased metabolism) in the different regions of the brain. Two techniques in particular are making it possible to see how the brain works on various tasks. One is called PET scanning (positron emission tomography) and the other is called fMRI scanning (functional magnetic resonance imaging). Both examine the brain while it is in action and capture an image of the brain's activity.

Both PET scans and fMRI scans provide valuable information about the activity in the brain, but do so by different means. In a **PET scan,** radioactive glucose is injected into the subject. The radioactivity can then be traced in the brain as the subject performs a task. Because areas with more metabolic activity require more fuel, more of the radioactive glucose moves to those areas. In the image, areas of the brain are colored differently, depending on their rate of metabolic activity during the task. In an **fMRI scan,** a giant magnet detects the relative amounts of oxygenated blood and deoxygenated blood in various parts of the brain. When parts of the brain are active, more blood flows to them and the sensitive magnet can pick up the change. As with the PET scans, an image can be obtained that is colored to indicate different levels of metabolic activity in various parts of the brain. Because no radioactivity is required, fMRI scans are generally preferable to PET scans.

At any given time during the day, you are not using all of your muscles. But over the course of a day, you use most of them. Similarly, for any particular task, not all regions of the brain are called into action, but over the course of a day all of your neurons are used. Some interesting discoveries have been made about how human brains allocate various tasks.

Seeing your favorite foods activates brain reward centers. Subjects shown images of their favorite foods show increased brain activity in the pleasure centers of the forebrain (**FIGURE 23-46**). Researchers have also found that recovering drug addicts have a similar response to images of a syringe (or even an anti-drug advertisement), often prompting them to relapse. Both cases underscore the power of advertisements.

Altruistic behavior activates reward centers. When playing a game called "Prisoner's Dilemma," in which people gain rewards based on their decisions to cooperate or act selfishly, brain activity in the forebrain's reward center is increased when a player chooses to cooperate. This may indicate that our brains are built to produce a physiological "prodding" toward kindness in certain situations.

FIGURE 23-46 Does the thought of eating one of these cinnamon rolls increase the activity of your brain?

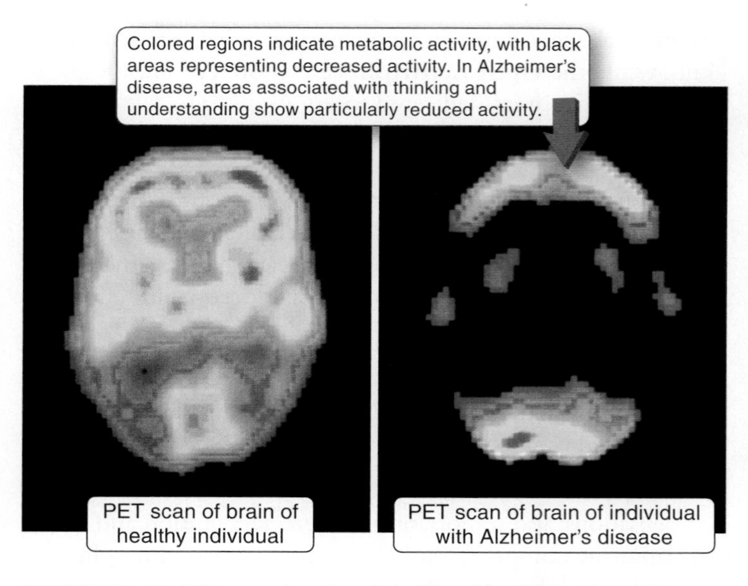

FIGURE 23-47 PET scan showing the effect of Alzheimer's on the brain.

People with Alzheimer's disease exhibit significantly reduced brain activity. When compared with healthy individuals of the same age, Alzheimer's patients show decreased activity throughout the parts of the brain associated with thinking and understanding, while they show normal activity in the areas active in movement and sensory perception (**FIGURE 23-47**).

Females and males respond differently to emotional situations. PET scans were made while women and men were asked to reflect on the saddest images in their lives. In both sexes, the limbic system glowed brightly. In the women, though, the activity covered an area eight times larger than that in the men!

Language tasks are localized in different parts of the brain. Language uses very different parts of the cerebral cortex, depending upon whether someone is hearing, depending on whether someone is hearing, speaking, reading, or generating new words. Interestingly, among bilingual people, their second language is stored in a slightly different part of the brain if they acquired it past a certain age. We explore language in more detail in the next section.

TAKE-HOME MESSAGE 23•17

PET scans and fMRI scans make it possible to detect and visualize metabolic activity in brain tissue and to see how the brain works during various tasks. These technologies help us understand how tasks are processed by different regions of the brain and how individuals vary in their patterns of brain activity for similar tasks.

23•18

Specific brain areas are involved in the processes of learning, language, and memory.

Brains are capable of myriad amazing feats. Learning and memory, along with the use of complex language, are among the most sophisticated and enigmatic of these functions. But while these abilities rank among the greatest mysteries of the brain, we're unlocking more and more of the secrets every day.

Language What is language exactly? In its most broad interpretation, **language** is simply the means of communication between individuals of any species. In humans, language has evolved its most complex manifestation. Words and sentences allow us to convey not just simple information, but also complex thoughts such as things distant in space or time, something rarely if ever seen in other species. Try conveying to a friend, without using words, the details of the first birthday party you remember attending or the things you saw on a trip to a foreign country. Language also makes it possible for us to articulate abstract ideas. Could you explain democracy or the concept of natural selection without language? Language is fundamental to who we are as a species. Let's investigate some of the biology behind it.

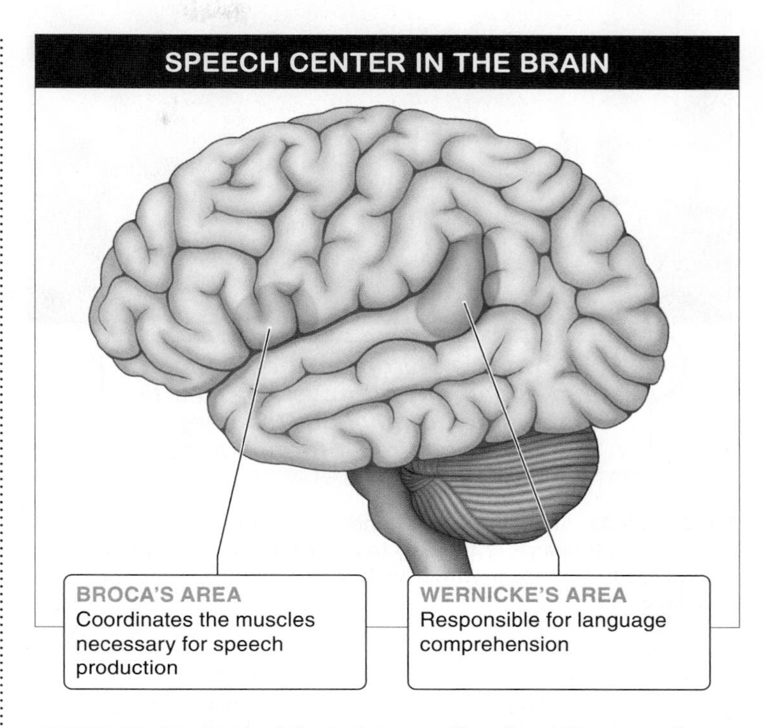

FIGURE 23-48 Areas of the brain controlling the ability to produce speech and understand language.

Several distinct brain structures, and two in particular, are necessary for the acquisition and use of language. One is required for the understanding of speech, including the linking of words with meaning (**FIGURE 23-48**). The other controls the actual production of speech. In more than 95 of every 100 people, it is the left hemisphere that is responsible for both of these functions. The discrete parts used for controlling the muscles involved in the actual process of speaking are in the front part of the left frontal lobe, a region called **Broca's area.** The region responsible for understanding speech is called **Wernicke's area,** and it is located in the left temporal lobe. People with damage to Wernicke's area (from a stroke or head injury) are able to produce sounds properly, such as when repeating things they hear, but they cannot understand speech and can speak only gibberish. It's as if they've lost the ability to link words with their meanings.

> **Q** Do men or women have a better chance of language recovery following a stroke? Why?

Brain science is still in its infancy, and we're still discovering some perplexing and unexpected things about this most complex of all organs. Consider this: in 2008, an American woman named Cindy Lou Romberg had a chiropractic adjustment that apparently aggravated a severe brain injury that she had suffered 17 years earlier. Much to her surprise, she found that she suddenly had a foreign accent! In her life she had never spoken with such an accent—which sounds like a sort of mixture of German, French, and Russian—nor had she ever studied any foreign languages or even been to Europe (**FIGURE 23-49**). It appeared that parts of her language production areas had been damaged, and this altered the way she spoke. Although the exact mechanism for this change is not understood, this strange phenomenon has been reported several other times, the first being in 1941 when a Norwegian woman had a stroke and afterward spoke with what sounded like a German accent. Interestingly, women are much more likely to regain the ability to speak after a stroke. This may be related to the fact that, on average, women have a thicker corpus callosum, the band of fibers connecting the left and right hemispheres of the brain. The additional thickness may facilitate another part of the brain's ability to take over language functions after the stroke-induced death of brain tissue.

FIGURE 23-49 Changing accent. Cindy Lou Romberg never studied any foreign languages or traveled to Europe, but a brain change has caused her to now speak with a mixture of a German, French, and Russian accent.

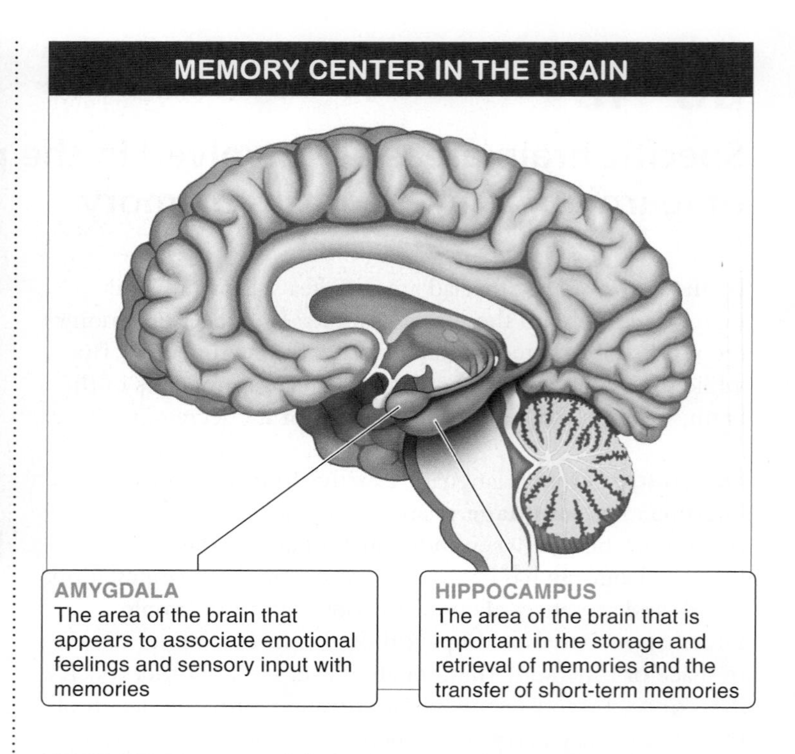

FIGURE 23-50 Areas of the brain associated with memories.

Another recent discovery suggests that children have a propensity to learn language regardless of what language they are exposed to. In a study of children raised by parents who spoke using sign language only, observations of their children revealed that the infants actually "babbled" with their hands. Much as babies with parents who speak begin to put together nonsense words made up from actual fragments of words and sounds, from age six months to one year, the babies of signing parents used their hands to produce fragments of hand signs. The babies of speaking parents did not exhibit such hand babbling.

Learning and memory are understood even less well than language. Still, certain cellular changes seem to occur in neurons when we learn and remember things. Two parts of the limbic system are critical in storing memories, the amygdala and the hippocampus. The amygdala seems to associate emotional feelings and sensory input with memories (**FIGURE 23-50**). This is why it is not uncommon for certain smells or tastes, for example, to elicit memories. The hippocampus is also important in the storage and retrieval of memories, as well as the transfer of short-term memories—things that are stored in our brain only briefly, such as a phone number or the name of a person we have just met—into long-term memories that can be recalled days, years, or even decades later. Memories and associations of sensations are stored throughout much of the cerebral cortex, too.

Humans don't have a monopoly on powerful memory. When it comes to remembering where they've put things and where they can find food, many birds and non-human mammals are also quite accomplished. Each fall, for example, red squirrels each hide more than 3,000 acorns and other nuts (**FIGURE 23-51**). Come winter, each squirrel is able to remember the locations of more than 80% of those nuts and recover them for food. Similarly, birds such as the black-capped chickadee store large numbers of nuts for the winter. Researchers looking for neurological changes in these birds found that the hippocampus is significantly larger in the fall, when the birds must remember all of their nut-stash locations, than in the spring, when they no longer need to rely on stored nuts for food.

For many years, it was believed that no new neurons are added to the human brain once development is completed in early childhood. While this is largely true, recent observations suggest that some parts of the hippocampus are an exception, adding cells even into adulthood, possibly as a function of use. On the flip side, amnesia usually comes from

FIGURE 23-51 A good memory for hiding places. A red squirrel never forgets the location of its acorn stash.

damage to the amygdala or hippocampus from a stroke or head injury.

What happens to the neurons at a cellular level during the processes of learning and forming memories? When repeated action potentials are generated by a neuron, changes occur at the synapse. There are two types of change that generally occur. One is called **long-term depression** and the other is called **long-term potentiation.** In each, as neurotransmitters are repeatedly released by the presynaptic membrane, the adjacent neuron responds by initiating an action potential, but also by becoming less likely (long-term depression), or more likely (long-term potentiation) to respond to future stimulation with an action potential. That's the learning: future responses are modified by past experience. This makes sense when we consider that perhaps the simplest form of learning comes from repeated attempts at doing something, during which the individual becomes better and better at it.

Q Can we generate new brain cells? How?

"Use it or lose it." That is a frequent assertion by weightlifters in reference to their muscles. The same expression might be used about the memory-forming components of our limbic system. Using fMRI brain scans, researchers in England recently discovered that the brains of London cab drivers have a larger hippocampus than those of non-cabbies. The researchers believe that this may be a result of the formation of new neurons, due to the greater demands placed on the hippocampus as the cab drivers navigate the streets of London. Their hypothesis was supported by the observation that the longer an individual had been a cab driver, the larger his hippocampus was.

The synapse isn't just the site of "normal" signal propagation. We'll explore next how, by tinkering with the synapse and modifying the amount or duration of stimulation occurring, it is possible to use drugs and medications to artificially create feelings of happiness, satiation, euphoria, and more. It's also possible to create medically valuable anesthetics and deadly toxins.

TAKE-HOME MESSAGE 23•18

Learning and memory, along with the use of complex language, are among the most sophisticated functions of the brain. Although brain science is still in its infancy, we are gaining a clearer picture of the brain structures—primarily, for language, in the left frontal and left temporal lobes and, for learning and memory, in the amygdala and hippocampus—that are chiefly responsible for these functions.

⑥ Drugs can hijack pleasure pathways.

Two men preparing tea in the desert of Timbuktu, Mali. Worldwide, more caffeine-containing tea is consumed every day than any beverage other than water.

23•19

Our nervous system can be tricked by chemicals.

Stimulation of the reward centers in the brain spurs animals to behave in certain ways. We're built this way on purpose: when we behave in ways important to our evolutionary fitness, our brain releases neurotransmitters such as dopamine. Exposed to such neurotransmitters, our pleasure centers produce the sensation of bliss. And that bliss spurs us to seek out that feeling again once it fades. Usually this system works well.

Our brain's signaling system can be tricked, however, and with potentially disastrous consequences. Drugs—whether recreational or therapeutic, whether found in nature or made in the laboratory—can work by mimicking neurotransmitters. When we take a pleasure-causing drug, our brain experiences the same sensations as if appropriately released neurotransmitters were flooding the system. The brain cannot distinguish this artificial high from a real high. There are several different methods by which drugs can influence our nervous system. We'll explore some of them here.

Cocaine When someone snorts a bit of cocaine, it crosses the blood-brain barrier and quickly reaches the brain's pleasure centers. Once there, the cocaine binds to the sites on the presynaptic membrane where dopamine is normally reabsorbed from the synaptic cleft by the cells that originally released it. As long as these reuptake sites are blocked by the drug, dopamine released by that neuron remains in the synaptic cleft, repeatedly stimulating the postsynaptic cell and causing non-stop activity in the pleasure center (**FIGURE 23-52**).

Prozac and Zoloft Antidepressants work by an almost identical mechanism to that of cocaine. In addition to dopamine, another neurotransmitter important in our brain's pleasure centers is serotonin. The antidepressants Prozac and Zoloft block serotonin from being reabsorbed and recycled by the presynaptic cells that released it, prolonging its effect (see Figure 23-52). This is why these drugs are called **selective serotonin reuptake inhibitors (SSRIs).** The net result is that people taking these drugs generally experience an elevated mood because of serotonin's prolonged stay in the synaptic cleft.

Morphine and Heroin Some chemical messengers we ought to be especially thankful for are the endorphins, our body's natural painkillers. Produced by our brain, endorphins block pain messages arriving from throughout the body. Under a variety of situations of extreme stress—such as if an individual has just been seriously injured in a fight or is in

REUPTAKE INHIBITORS

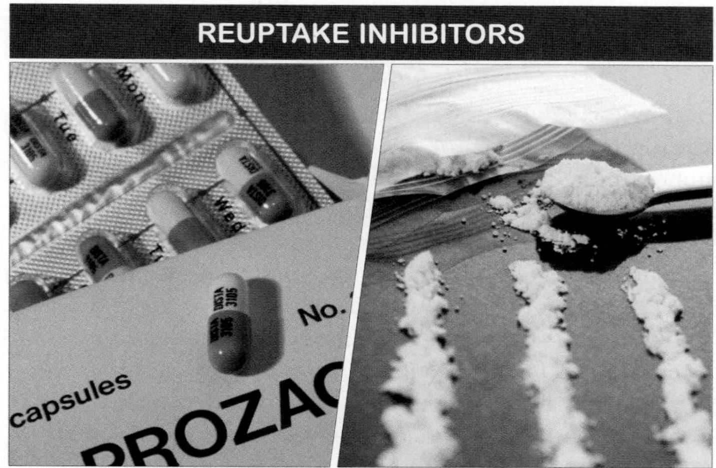

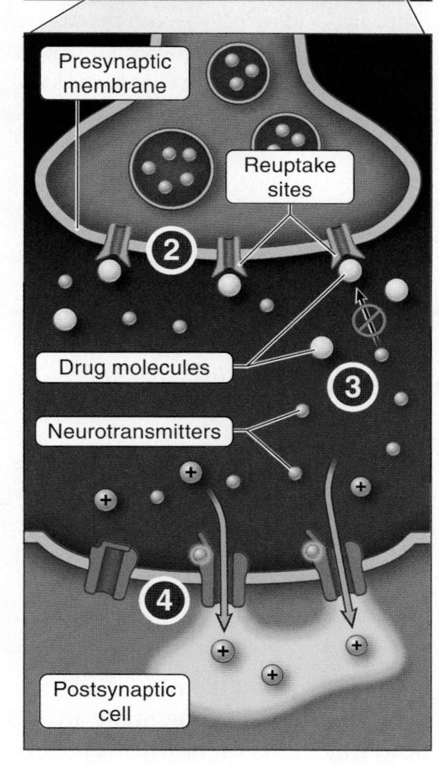

1 After the drug is taken, it crosses the blood-brain barrier and quickly reaches the brain's pleasure centers.

Blood vessel

Drug molecules

Neurons in the brain

2 Drug molecules bind to sites on the presynaptic membrane where neurotransmitters are normally reabsorbed by the cells that originally released them.

3 As long as reuptake sites are blocked by drug molecules, the neurotransmitters previously released by that neuron remain in the synaptic cleft.

4 The neurotransmitters repeatedly stimulate the postsynaptic cell and cause non-stop activity in the pleasure centers of the brain.

FIGURE 23-52 Prolonging pleasure sensations. Reuptake inhibitors cause prolonged action of a neurotransmitter in the synapse.

mile 12 of a half-marathon—the body responds by releasing endorphins.

The popular opiates morphine and heroin mimic endorphins and bind to their receptor sites. With a large enough dose, opiate users can give themselves an "endorphin rush" and feelings of euphoria far more intense than anything possible with their own natural supply of these pleasure compounds. The respiratory slow-down that comes with heroin use can occur quickly and, with large doses of heroin, can be fatal.

Nicotine One of the most commonly used drugs is nicotine. Shortly after entering the bloodstream, nicotine begins mimicking one of the body's most common and important neurotransmitters, acetylcholine. Fooled by the nicotine binding to acetylcholine receptors, cells release adrenaline and other stimulating chemicals, including pleasure-causing dopamine. Nicotine causes rapid surges, then rapid depletions, of these chemicals, leaving smokers happy for a short while but soon yearning for another cigarette.

When we consume drugs such as those described above, our bodies soon respond to the perception that there are larger

Some people use a nicotine patch that supplies nicotine to the bloodstream to help reduce the severity of physical symptoms associated with cigarette smoking cessation.

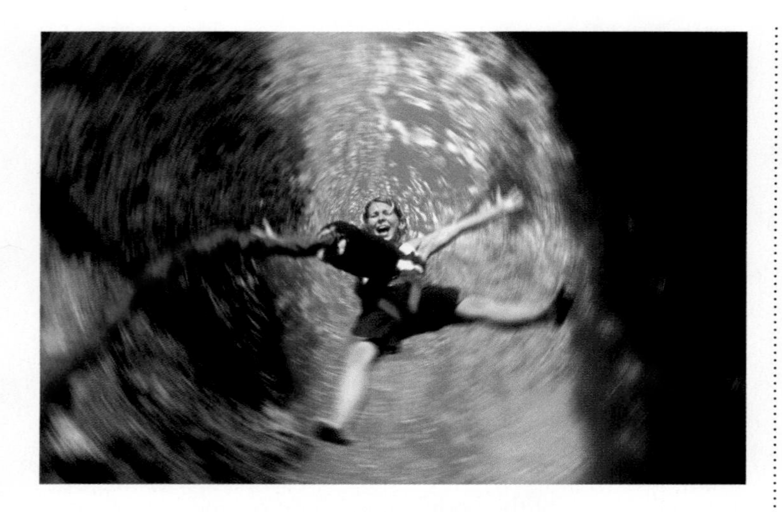

FIGURE 23-53 **Can a gene nudge you toward risk-taking?**

amounts than usual of certain neurotransmitters. Usually the response is a reduction in sensitivity to the drug. Although tolerance is inevitable, its magnitude and consequences vary with the type of drug. In one study, volunteers were injected with a uniform daily dose of heroin and monitored for their level of euphoria. Initially they were ecstatic, but their bodies reacted by reducing the number of receptors that bind heroin. With fewer and fewer receptors, the euphoric heroin effects dropped almost to zero in just three weeks. Similar changes occur in response to using caffeine, nicotine, or alcohol.

Do human drug addictions and dependencies reflect differences in our genes? Recent data suggest that they might at least be influenced by the alleles a person carries. In one study of 283 individuals, a third of the people who smoked had a version (i.e., an allele) of an important gene that almost none of the non-smokers carried. This gene, labeled D4DR, codes for the receptor that enables the pleasure centers of our brain to respond to dopamine. The allele of the D4DR gene found among one-third of the smokers in the study was the same one we discussed in the StreetBio feature at the end of Chapter 7, one that appears to cause individuals to exhibit a variety of personality traits associated with risk-taking and novelty-seeking behavior (**FIGURE 23-53**). Do you think that you carry it?

Q Can a gene nudge you toward risk-taking?

In the next section we look at a popular and effective drug that acts not by blocking neurotransmitter reuptake, but rather by blocking the postsynaptic receptors.

TAKE-HOME MESSAGE 23•19

Our brain's signaling system can be tricked by drugs, whether recreational or therapeutic, that mimic neurotransmitters. Such drugs—including cocaine, Prozac, heroin, and nicotine—can produce euphoric sensations, can reduce depression, and can block pain, but the effects often come with significant health risks.

23•20

A brain slows down when it needs sleep. Caffeine wakes it up.

Many drugs can interfere with normal neuron functioning, and caffeine is perhaps the most frequently used drug of all with this effect. Worldwide, more caffeine-containing tea is consumed every day than any beverage other than water. A close second is coffee (**FIGURE 23-54**). And in the United States, 90% of the soda consumed contains caffeine. The average American drinks about a hundred gallons a year of these three beverages. From philosophers and writers to scientists and musicians, caffeine has been revered as a necessity for stimulating the creative juices. In his Coffee Cantata of 1732, J. S. Bach wrote, "Ah! How sweet coffee tastes! Lovelier than a thousand kisses, sweeter far than muscatel wine!"

> "If it weren't for coffee, I'd have no discernible personality at all.
> —David Letterman"

The centuries of strong praise are well founded. Caffeine has powerful effects on nearly every animal species. For instance, while all rats can eventually be trained to race through mazes, some learn quickly while others languish in remedial maze-running classes. What they all have in common, however, is

FIGURE 23-54 Many people take pleasure in drinking tea and coffee.

that when they are given a caffeine pick-me-up before their maze lessons, they learn the solutions faster and remember them better.

How does caffeine work its magic? As long as we are awake, our brain is working hard and consuming about 20% of the energy we take in. Much as a running motor generates exhaust fumes, however, all of this neural activity leads to a buildup of cellular waste products. Neuron "exhaust" takes the form of a variety of molecules, including one called adenosine. As a cell releases adenosine, it fills adenosine receptors on nearby cells, particularly those in the brainstem. When a molecule of adenosine binds to an adenosine receptor, this reduces a neuron's likelihood of initiating an action potential. Adenosine has been described as putting a brake on neuron activity. Consequently, as the production of adenosine continues throughout our day, more and more receptors on more and more neurons are filled, and the neurons in our brain become less likely to fire, regardless of how strongly they are stimulated. We perceive this as tiredness. Then, as we sleep, the cellular waste products such as adenosine are reabsorbed and recycled. Upon awaking, we feel better because we literally are more clear-headed.

Caffeine has its effect by blocking the message to reduce activity and to sleep. Here's how it accomplishes this effect. The caffeine we ingest quickly circulates to our brain, and once there, it diffuses around the cells. Because the shape of the caffeine molecule is similar to that of the adenosine molecule, the caffeine slips into some of the receptors intended for adenosine (**FIGURE 23-55**). It doesn't have the

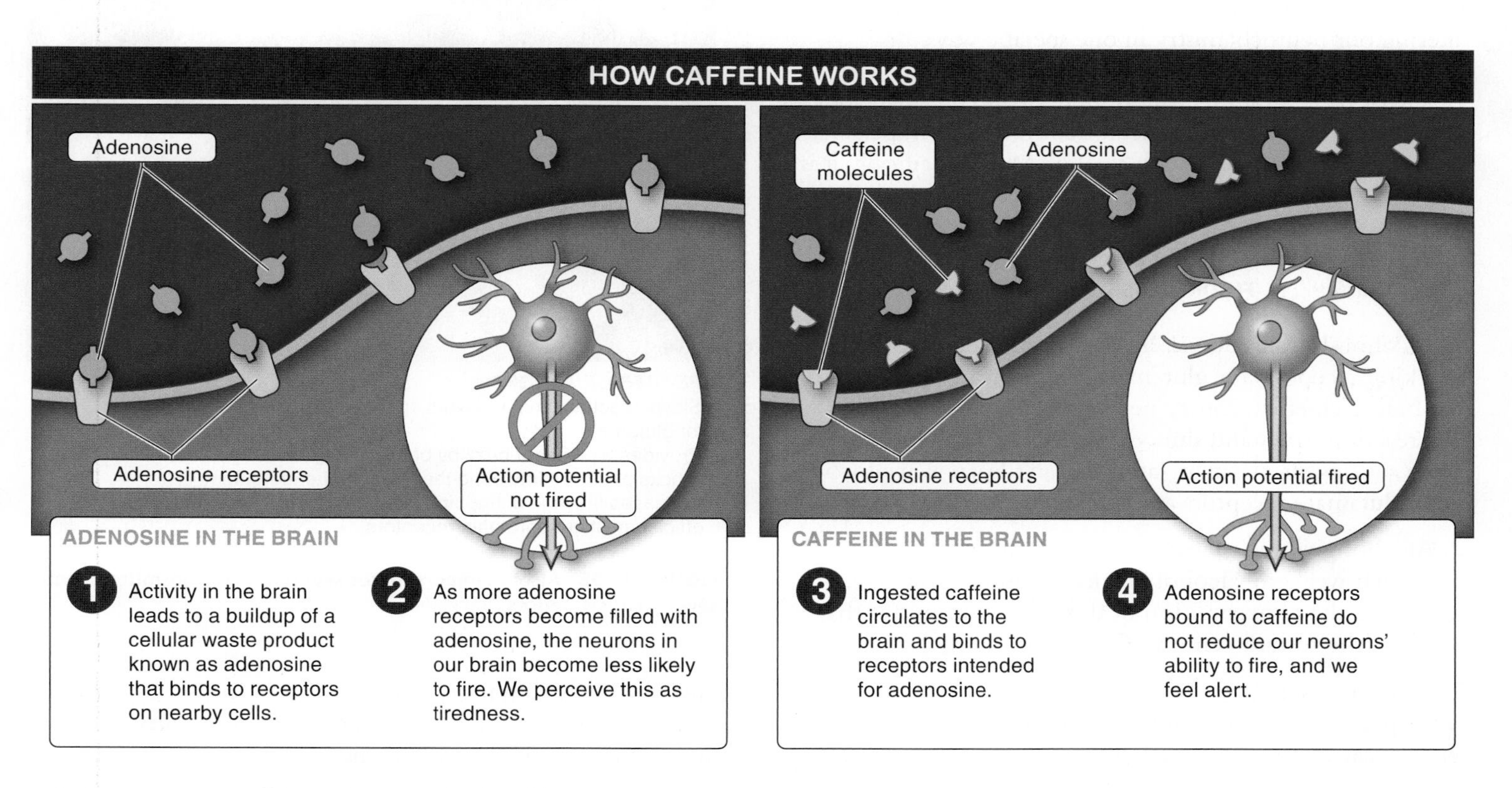

FIGURE 23-55 How caffeine reduces the feeling of fatigue.

same effect as adenosine, however. Adenosine receptors with bound caffeine molecules do not reduce a neuron's propensity to respond to other stimuli with an action potential. Thus, when many of the adenosine receptors are blocked by caffeine, the adenosine is unable to pass on the message that we need to sleep. Instead, we feel surprisingly alert as caffeine interrupts one of the normal sleep-signaling systems.

Surprisingly, caffeine seems to be safe for most people. Despite considerable searching for ill effects, there is no clear evidence that moderate consumption of caffeine increases our risk of anything beyond the occasional jitters. For healthy people, there appears to be no increased risk of heart, lung, or kidney disease or even cancer.

TAKE-HOME MESSAGE 23•20

Normal neural activity leads to a buildup of cellular waste products. One of these, adenosine, fills adenosine receptors on nearby neurons, reducing the likelihood that a cell will initiate an action potential and causing fatigue. Caffeine binds to the adenosine receptors, without reducing their likelihood of firing, thus blocking the fatigue-inducing message of adenosine.

23•21

Alcohol interferes with many different neurotransmitters.

So far, we've seen that drugs frequently can alter our brain functioning by mimicking the neurotransmitters used by our neurons during normal functioning. Their specific effects are quite predictable, as long as we know the neurotransmitter that the drug mimics. Their work is like a surgical strike, altering our neurochemistry in one specific way.

But what happens if the drug is more of an everyman, looking enough like many different neurotransmitters to impersonate them all? This is the case with alcohol. More specifically, it is the case with ethanol, the "alcohol" molecule in alcoholic beverages. Alcohol is a great neurotransmitter impersonator, fooling at least four different receptor molecules. Let's look at some of its effects (**FIGURE 23-56**).

EFFECTS OF ALCOHOL
- Slows reaction time and slurs speech by blocking receptors for glutamate
- Provides a pleasant buzz by blocking dopamine reuptake
- Blocks pain by stimulating the release of endorphins
- Increases feelings of happiness by modifying and increasing the efficiency of our serotonin receptors

FIGURE 23-56 Alcohol impersonates several neurotransmitters, with several consequences.

1. Alcohol slows us down, "relaxing" our neurons. By blocking receptors for glutamate (see Section 23-6), one of the brain's chief excitatory neurotransmitters, alcohol slows our reaction times and slurs our speech. It may also have more serious effects: many animals can't learn as well when their glutamate receptors are blocked.

2. Alcohol gives us a pleasant buzz. Acting like cocaine—but much weaker—alcohol blocks dopamine reuptake, increasing the concentration of this neurotransmitter that is active in the reward centers of our brain.

3. Alcohol blocks pain. By stimulating the release of endorphins—neurotransmitters that block pain messages—alcohol causes the same feeling that has been called "runner's high." Resembling morphine and heroin in this respect, but again at a greatly reduced magnitude, alcohol spurs our body to produce a little opiate-like high.

Alcohol intake lowers inhibitions and, in excess, can increase the likelihood of risky behavior.

4. Alcohol makes us happier, at least while it's in our system. Much like Prozac, alcohol modifies and increases the efficiency of our serotonin receptors, increasing the contentment that accompanies serotonin release at synapses in the brain. Interestingly, rats that have been bred for a high preference for alcohol turn out to have lower serotonin levels than normal rats. One possible interpretation for this is that the alcohol-seeking animals are trying to compensate for their lack of serotonin stimulation by looking for the alcohol effects.

For most people, moderate alcohol consumption is pleasant and does not have significant health risks. For some, however, alcohol abuse and alcoholism can lead to serious problems. Nearly 14 million Americans abuse alcohol, and the consequences can be serious. Beyond the effects it creates by mimicking neurotransmitters, alcohol has a variety of other physiological effects in our bodies that can become harmful when alcohol is consumed in large amounts or for long periods of time. These include increasing the risk for some types of cancer, increasing the risk of liver disease (because the liver must metabolize and detoxify all the ethanol molecules that are consumed), increasing the likelihood of harm to the fetus during pregnancy, increasing the likelihood of taking part in risky sexual and other behaviors, and increasing the risk of automobile crashes.

TAKE-HOME MESSAGE 23•21

Alcohol affects the functioning of multiple neurotransmitters—including glutamate, endorphins, dopamine, and serotonin—slowing reaction times, slurring speech, blocking pain, and increasing contentment.

Knowledge You Can Use

Don't move a muscle! (Injecting the toxic bacterial protein, Botox, can help.)

Q: What is Botox? Botox refers to "botulinum toxin," a protein produced by several species of bacteria. It is the most toxic substance known—less than a microgram is lethal if injected in a person or inhaled. (For comparison, an inch-long human hair weighs more than 200 micrograms.) Botox is also an increasingly popular drug used in cosmetic procedures, with more than a billion dollars spent on it worldwide each year.

Q: What makes it toxic? What gives it medical value? When injected in muscles, Botox blocks the release of acetylcholine, the neurotransmitter that stimulates muscle contraction. Botox does this by interfering with the fusion of acetylcholine-filled vesicles with the presynaptic neuron's membrane, preventing release of acetylcholine into the synapse and thus its effect on muscle cells. A tiny amount of Botox can prevent a muscle from contracting for three or four months! (This feature of Botox makes it an effective treatment for muscle spasms.)

Q: Why would a healthy person want to get near the most toxic substance known? Tiny injections of Botox can smooth lines in the forehead and between the eyebrows. The injections can also reduce "crow's feet" wrinkles around the eyes. Extolling its ability to make them look younger, many people seek out Botox treatments, including high-profile celebrities such as actress Jenny McCarthy, who said, "I love Botox, I absolutely love it . . . I get it minimally, so I can still move my face. But I really do think it's a savior."

Q: Wait a minute! "I get it minimally, so I can still move my face"? What is the downside to Botox? Because Botox essentially paralyzes muscles, it can lead to a variety of problems. At the extreme, if Botox spreads to muscles other than those intended, it can interfere with the muscles needed for swallowing and breathing—muscles you can't afford to be without for more than a few minutes, let alone a few months. Most of the 180 reported cases of life-threatening problems resulting from Botox injections between 1997 and 2006 stemmed from problems of this nature. Less life-threatening, but disturbing to many, is the severe reduction in individuals' ability to move parts of their face that can result from Botox injections, often causing droopy eyelids or a mask-like face and inability to express emotions.

1 What is the nervous system?

Nervous systems are found in nearly all animal species. Primarily utilizing cells called neurons, they (1) receive input from the surrounding world, (2) process that information, and (3) initiate responses to the environment when necessary.

2 How do neurons work?

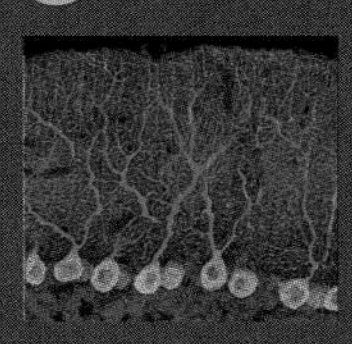

Neurons are cells specialized for receiving information via their dendrites and transmitting this information through action potentials down their axons. An action potential is a self-propagating, all-or-none change in the membrane potential that travels down an axon, causing the release of neurotransmitters, which diffuse across the synapse, bind to receptors, and excite or inhibit the postsynaptic cell—which may be a muscle, a gland, or another neuron.

3 Our senses detect and transmit stimuli.

All of our senses are variations on a theme: sensory cells respond to physical or chemical stimuli by opening or closing channels within the membranes of receptor cells and stimulating or blocking action potentials that, when received by the brain, are perceived as taste, smell, light, sound, or touch.

4 The muscular and skeletal systems enable movement.

Muscle tissue—including skeletal, cardiac, and smooth muscle—is made up of elongated cells capable of generating force when they contract. A muscle fiber is a single cell containing myofibrils that shorten with the making and breaking of bonds connecting parallel actin and myosin filaments. Muscles are attached to parts of the skeletal system, which, in vertebrates, can enable movement, provide support, offer protection, provide a site for production of important blood and immune cells, and serve as a reservoir of minerals, including calcium.

5 The brain is organized into distinct structures dedicated to specific functions.

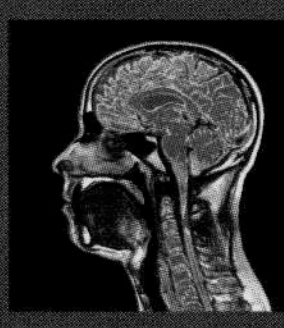

There are several distinct regions in the brain—the hindbrain, midbrain, and forebrain—each of which is the control center for various activities in the body. The forebrain, the largest region of the brain, is responsible for most of what we consider higher thought, including perception, memory, language, intelligence, and personality. PET scans and fMRI scans make it possible to detect and visualize metabolic activity in brain tissue and to see how the brain works on various tasks.

6 Drugs can hijack pleasure pathways.

Drugs that interfere with the functioning of neurotransmitters can have various effects, some of which are diseases and some of which are sought-after euphoric or excited states. The drug's interference can involve blocking or over-facilitating the release of a neurotransmitter, increasing or decreasing the rate at which neurotransmitter is removed from the synaptic cleft, or blocking or over-facilitating the binding of neurotransmitter to its receptors.

KEY TERMS

CHECK YOUR KNOWLEDGE

1. Animal nervous systems have several principal features. These include all of the following except:
   a) They coordinate long-term growth and development in an organism.
   b) They receive input from the world around the organism.
   c) They initiate responses to the information they receive from the world, when necessary.
   d) They process the information from the world that an organism receives.
   e) All of the above are principal features of animal nervous systems.

2. Which of the following statements about plants and nervous systems is correct?
   a) Plants can feel pain in their trunks and stems, where they have neurons, but not in their leaves.
   b) Plants can feel pain in their leaves, where they have neurons, but not in their trunks or stems.
   c) Plants can feel pain because they have a nervous system that extends throughout all of their tissues.
   d) Plants can feel pain in their trunks, where they have neurons, but not in their stems or leaves.
   e) Plants cannot feel pain because they do not have nervous systems.

3. What is the difference between a nerve and a neuron?
   a) A nerve is a bundle of neurons, whereas a neuron is a single cell.
   b) Nerves have synapses with muscle cells, whereas neurons have synapses with other neurons.
   c) Nerves have dendrites that can be modified into sensory cells, whereas neurons do not have dendrites.
   d) "Nerve" is the name given to nervous system cells specialized for receiving and transmitting information, whereas "neuron" describes a nerve cell in humans.
   e) There is no difference. They are different words for the same structure.

4. In a neuron, the cell body:
   a) contains the nucleus, mitochondria, and other cell organelles.
   b) does not contain a nucleus.
   c) has a membrane that is impermeable to water and intracellular solutes.
   d) is highly modified and coated in a fatty myelin sheath.
   e) is considered the primary structure of the "central" nervous system.

5. Dendrites:
   a) conduct action potentials away from the cell body.
   b) are present in mammalian nervous systems but not in the nervous systems of other vertebrates.
   c) receive information from other neurons or from the external environment.
   d) are coated in a fatty substance called the myelin sheath, which speeds up the rate at which signals are conducted.
   e) are bundles of axons.

6. In individuals with multiple sclerosis, myelin is gradually lost. What symptoms would you expect these individuals to exhibit?
   a) Their membranes have reduced numbers of sodium channels.
   b) Their sensory neurons lose the ability to initiate action potentials.
   c) Their brain becomes smaller.
   d) They are able to continue athletic activity long after pain would cause most individuals to stop.
   e) Their neurons gradually lose their ability to conduct electrical impulses.

7. Which of the choices below properly describes the difference between the gray and white matter of the nervous system?
   a) The white matter is composed of non-myelinated fibers, and the gray matter is composed of myelinated fibers.
   b) The white matter is composed of neuron bodies, and the gray matter is composed of axons and dendrites.
   c) The white matter is composed of sensory neurons, and the gray matter is composed of motor neurons.
   d) The white matter is composed of myelinated fibers, and the gray matter is composed of non-myelinated fibers.
   e) None of the above is correct.

8. The synapse consists of:
   a) a release of neurotransmitter into a synaptic cleft.
   b) a presynaptic neuron and its neurotransmitters.
   c) the membrane depolarization that occurs at gaps in the myelin sheath.
   d) the interface between a neuron and another neuron or muscle cell or gland.
   e) None of the above.

9. Neurotransmitters in a synaptic cleft have all of the following possible fates except:
   a) reuptake by the presynaptic neuron.
   b) enzymatic breakdown in the synaptic cleft.
   c) inactivation by acetylcholine in the synaptic cleft.
   d) binding to a receptor in the postsynaptic cell membrane.
   e) All of the above are possible fates of neurotransmitters in a synaptic cleft.

10. An action potential in the optic nerve triggered by light striking the eye is the same as:
    a) an action potential triggered by air vibrating in the ear.
    b) an action potential triggered by salt on the tongue.
    c) an action potential triggered by an odorous substance entering the nose.
    d) Answers a), b), and c) are correct.
    e) None of the above is correct.

11. When you put a piece of chocolate on your tongue, your brain registers a sensation of sweetness. What aspect of a molecule is responsible for its having a particular taste?
    a) the molecule's shape
    b) the total number of protons in the molecule
    c) the number of hydrogen bonds in the molecule (more hydrogen bonds = sweeter taste)
    d) the ratio of covalent bonds to ionic bonds joining the atoms of the molecule
    e) the speed of melting in your mouth

12. Because cone cells are less sensitive to light than are rod cells:
a) they are better suited for daylight vision.
b) they are better suited for night vision.
c) they are better suited for peripheral vision.
d) Both a) and c) are correct.
e) Both b) and c) are correct.

13. Sensory receptors in your skin include:
a) mechanoreceptors, thermoreceptors, and pain receptors.
b) mechanoreceptors, thermoreceptors, and electromagnetic receptors.
c) chemoreceptors, thermoreceptors, and electromagnetic receptors.
d) mechanoreceptors, thermoreceptors, chemoreceptors, and pain receptors.
e) mechanoreceptors, thermoreceptors, chemoreceptors, electromagnetic receptors, and pain receptors.

14. The connective tissue that connects a muscle to a bone is also known as:
a) collagen.
b) a ligament.
c) cartilage.
d) a tendon.
e) None of the above.

15. Which of the following is not one of the many functions of an endoskeleton?
a) storage of important vital minerals
b) production of blood and immune cells
c) protection of the body's internal structures
d) supporting the animal against gravity
e) All of the above are important functions of an exoskeleton.

16. The brain is divided into which three main regions?
a) hindbrain, cerebellum, and hypothalamus
b) midbrain, cerebrum, and brainstem
c) hypothalamus, medulla oblongata, and cerebrum
d) brainstem, cerebellum, and cerebrum
e) hindbrain, midbrain, and forebrain

17. The blood-brain barrier:
a) is formed from muscle tissue.
b) is a semi-permeable barricade that allows essential nutrients and gases to pass through.
c) prevents harmful molecules such as metabolic wastes from passing into the brain.
d) restricts the passage of molecules to neurons but not to glial cells.
e) Both b) and c) are correct.

18. A boy is born with a specific brain defect in which the two hemispheres of his cerebrum are no longer able to "talk" to each other (they are no longer connected). This boy is born without a:
a) hippocampus.
b) brainstem.
c) medulla oblongata.
d) corpus callosum.
e) hypothalamus.

19. Caffeine blocks what type of receptors in the human nervous system?
a) dopamine
b) adenosine
c) acetylcholine
d) serotonin
e) glutamate

20. Which of the following is not an effect of alcohol on the nervous system?
a) Alcohol increases the activity of some serotonin receptors, boosting the effective serotonin level in the synapse.
b) Alcohol stimulates endorphin release.
c) Alcohol blocks glutamate receptors, leading to slurred speech and reduced reaction times.
d) Alcohol leads to modestly increased dopamine levels.
e) Alcohol blocks reuptake of acetylcholine, creating an energized feeling.

Short-Answer Questions

1. Compare and contrast the structure and role of neurons and glial cells of the nervous system.

2. Describe the structure of the myofibril of skeletal muscle fibers that makes contraction possible. What changes occur that allow contraction?

3. List the primary types of sensory receptor cells, their functions, and which of the five traditional senses they are associated with.

See Appendix for answers. For additional study questions, go to www.prep-u.com.

UNITED

24

Hormones

Mood, emotions, growth, and more: hormones as master regulators

❶ Hormones are chemical messengers regulating cell functioning.

A heart-pounding moment: The rock climber swings by one hand on the rock face.

24•1

The "cuddle" chemical: oxytocin increases trust and enhances pair bonding.

"Trust me." It's easy enough to say, but can be very difficult to do. Recently, however, some researchers discovered a way to actually make people more trusting. Here's how they did it.

Study participants, called investors, were given an amount of money. Each investor could keep the money or transfer some or all of it to another person, called a trustee. Any amount the investor chose to transfer to the trustee was tripled, and the trustee could then return to the investor some portion of the significantly increased pot of cash. Or, the trustee could simply opt to keep it all. For those with the investor role, this created a dilemma in making a decision about what to do: by trusting the other player, they could lose some or all of their money. But trusting behavior might reward the investor with a much greater payoff.

Using a double-blind experimental design, the researchers created two different groups of investors. In one group, before making their decision whether to entrust some of their money to the trustee, the investors used a nasal spray to inhale a dose of oxytocin—a chemical normally found in humans and other vertebrates, but in smaller amounts than used in the study. Investors in the other group received a placebo, rather than oxytocin, from the inhaler.

The results were dramatic and unambiguous. The investors inhaling the oxytocin were significantly more trusting than those receiving the placebo: more than twice as many of those receiving oxytocin entrusted the maximum amount of their money to the trustee, compared with investors who did not inhale oxytocin. And more than twice as many of those in the placebo group transferred the smallest amounts of cash to the trustee, apparently exhibiting an increased fear of betrayal.

This result is consistent with some findings of research in other animals. In those studies, exposure to oxytocin reliably caused animals to let down their guard, initiate interaction, and facilitate social attachment with other individuals at a much higher rate than they typically would. It appears that

Oxytocin, a chemical signal produced in the body, can influence complex emotions and behaviors, such as increasing trust, making milk available to nursing babies, and forming social attachments.

FIGURE 24-1 Complex emotions and behaviors can be influenced by chemical signals, such as the hormone oxytocin.

this chemical may be having a similar effect on humans in the investor-trustee study.

One particularly clever and valuable feature of this study's experimental design is that in a follow-up study, the same methods were used but with one exception: the investors were told that the "trustee" in the game was a computer rather than another person. In these cases, the investors receiving oxytocin did not transfer increased amounts of money to the trustee; they behaved exactly like those in the placebo group. This finding suggests that oxytocin specifically increased the investors' trust of another human, rather than just increasing their willingness to take a financial risk.

Researchers believe that these findings give some insight about the biological basis for trust and that they might even lead to effective treatments for people with extreme shyness. And the results provide a pretty clear demonstration that complex emotions and behaviors can be influenced by chemical signals produced in the body. Oxytocin influences much more in humans than a propensity for trust (**FIGURE 24-1**). Other research has demonstrated that oxytocin plays significant roles in facilitating birth, making milk available to nursing babies, and increasing the propensity to form social attachments—an effect that has caused some researchers to refer to oxytocin as "the cuddle drug." And although it remains to be seen whether used-car dealers can make potential customers trust them by pumping oxytocin through the air-conditioning systems of showrooms, some journalists have called oxytocin "trust in a bottle."

In this chapter, we see that oxytocin is just one of several dozen influential chemical signals in vertebrates and many other eukaryotes, as we explore just what these chemicals are, where they are produced, and how they have their effects. We also explore the pervasive influence of these molecules in nearly every aspect of vertebrate life, from emotions to physical structures to behavior to health and physiology.

TAKE-HOME MESSAGE 24•1

Complex emotions and behaviors can be influenced by chemical signals produced in the body. Exposure to oxytocin, for example, causes humans to be more trusting of others, as well as facilitating birth, making milk available to nursing babies, and increasing the propensity to form social attachments.

24•2

Hormones are chemical messengers that travel through extracellular fluids to influence cells elsewhere in the body.

You wouldn't use the postal service to notify the fire department that your house is burning down. Even with overnight delivery, the information wouldn't arrive in time to be of use. Similarly, you probably wouldn't use the phone to convey all the complex details of a contractual agreement or architectural plans—the mail service might be preferable, to ensure that the details are conveyed precisely. For almost all types of communication, some modes of delivery are more effective than others.

Animals need ways to respond to the outside world: ways of sending signals about certain conditions and giving instructions to tissues in response to those conditions. There are two systems in animals for carrying out this internal communication and regulation (**FIGURE 24-2**). As we saw in Chapter 23, the nervous system generally is responsible for controlling rapid movement and sensations in response to environmental changes. In multicellular animals, including humans, there is another system, as well, with a special type of messenger. **Hormones** (from the Greek word meaning "that which sets in motion or urges forward") are a type of chemical messenger, secreted by cells into the extracellular fluid, including the bloodstream, and influencing the actions of non-adjacent cells, elsewhere in the body, as part of an internal communication and regulation system. Oxytocin, described in Section 24-1, is an example of an animal hormone.

Cells that secrete hormones are called **endocrine cells,** and the cells that receive their signals are called **target cells.** The hormone-secreting cells may be individual cells such as the endocrine cells in the lining of the stomach and small intestine, which aid in digestion. Larger collections of hormone-secreting cells—including the pituitary gland, the pancreas, and the ovaries and testes—are called **endocrine glands.** Together, all of the hormone-secreting cells in an animal make up its **endocrine system** (**FIGURE 24-3**).

While the nervous system controls rapid movements and sensations, the glands of the endocrine system have the ability to secrete chemical signals into the bloodstream that influence cells that are close by, even adjacent, and cells that are far away. In contrast to the nervous system, the endocrine system is generally responsible for longer-term, slower regulation, such as growth and development, as well as a variety of secretions (see Figure 24-2). It's important to note, however, that there

INTERNAL COMMUNICATION IN ANIMALS

Animals have two systems for carrying out internal communication in response to external conditions.

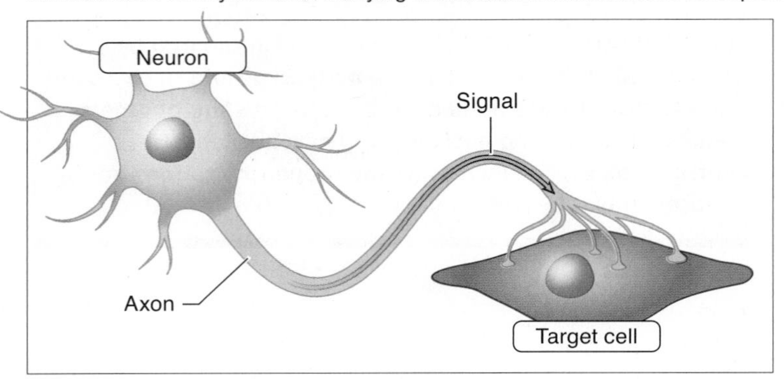

THE NERVOUS SYSTEM

- Messages are sent via chemical and electrical signals down the axons of neurons and across synapses to target cells.
- Generally responsible for controlling rapid movement and sensations in response to environmental changes

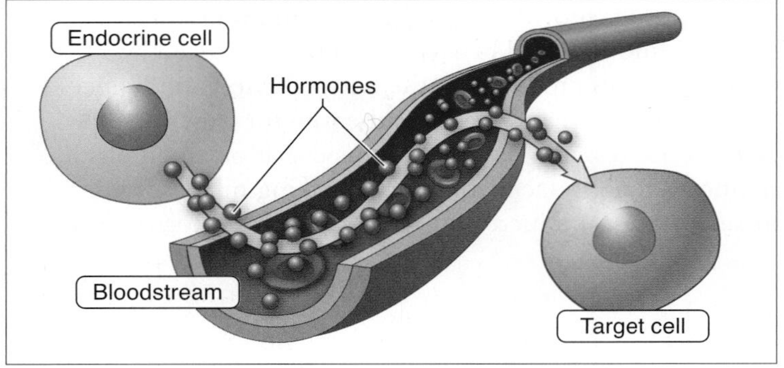

THE ENDOCRINE SYSTEM

- Messages are sent via chemical signals (hormones) secreted by endocrine cells into the extracellular fluid, including, often, the bloodstream, and influencing target cells at a distance.
- Generally responsible for slower, longer-term regulation, such as growth and development as well as a variety of secretions

FIGURE 24-2 The endocrine and nervous systems work together in animals.

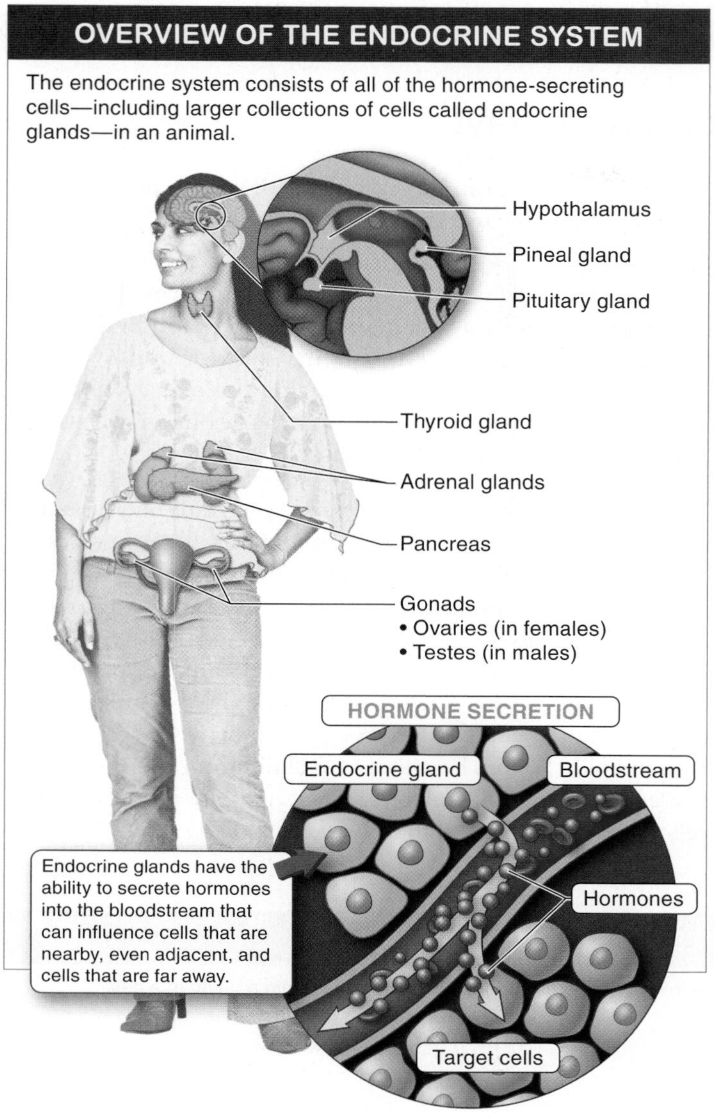

OVERVIEW OF THE ENDOCRINE SYSTEM

The endocrine system consists of all of the hormone-secreting cells—including larger collections of cells called endocrine glands—in an animal.

Hypothalamus
Pineal gland
Pituitary gland
Thyroid gland
Adrenal glands
Pancreas
Gonads
• Ovaries (in females)
• Testes (in males)

HORMONE SECRETION
Endocrine gland
Bloodstream
Hormones
Target cells

Endocrine glands have the ability to secrete hormones into the bloodstream that can influence cells that are nearby, even adjacent, and cells that are far away.

FIGURE 24-3 Organization of the endocrine system.

are some nervous system cells that secrete hormones, so the systems are not completely separate and distinct but rather can overlap in function somewhat. Interestingly, although they don't have nervous systems, most plants do have endocrine systems (see Chapter 19), suggesting that endocrine systems predate nervous systems evolutionarily.

When a hormone gets to a target cell, it elicits a response. There is a huge variety of possible responses, depending on the hormone and on the target cell. Commonly, the hormone's effect on a target cell is to alter the animal's physiology in a way that helps the organism maintain homeostasis. In doing so, the hormone may alter the target cell's metabolism or growth, it

FIGURE 24-4 High hopes. The dog marks its territory and also exaggerates its size by lifting its leg during urination.

may spur cell division, or it may initiate some developmental pathway. In this chapter, we investigate how the chemical nature of each hormone influences the way that molecule regulates the functioning of target cells.

Another type of chemical messenger, called a **pheromone,** is considered an "ecto-hormone," because it is transported to the outside of an animal and can cause a behavioral or physiological change in another individual. Pheromones serve a large number of different purposes. These include alarm pheromones, such as those that signal the presence of a predator and can trigger aggression or flight in a group of individuals; territorial pheromones, such as those found in dogs' urine, that mark the boundaries of a territory; and trail-marking pheromones, such as those produced by ants and that serve as a guide to the nest (**FIGURE 24-4**). And the first pheromones discovered were sex pheromones, signaling sexual receptivity. In some butterfly species, males can detect a female's pheromones from more than six miles away, and the males fly in the direction of increasing pheromone concentration to locate the female.

TAKE-HOME MESSAGE 24•2

Hormones are chemical messengers, secreted by endocrine cells and endocrine glands into the extracellular fluid, that influence the actions of cells, called target cells, elsewhere in the body, as part of an internal communication and regulation system.

24•3

Hormones can regulate target tissues in different ways.

If you turn on a radio, you can hear music. This is because, somewhere, a radio station is broadcasting a signal and your radio picks up that signal. If you turn the radio off, the music stops. The signal from the radio station, however, is still there. But without the proper receiver, the signal goes unnoticed. Hormones, one of the body's two chief methods of signaling, function much like radio signals. Glands produce and release hormones, which then make their way through the body, often distributed by the bloodstream, and bump into cells throughout the body. But, like a person without a radio, a cell doesn't respond to a hormone unless it has a specific receptor for that hormone. The hormone may even diffuse right through a cell that has no intracellular receptors.

There are dozens of hormones and even more ways that they regulate target tissue, but the general process by which hormones affect a particular cell doesn't vary much:

1. *Signal is sent.* A hormone is release by a gland.
2. *Signal is received.* Although a hormone has no effect on most tissue it comes in contact with, cells with the proper receptor in the cytoplasm or on the plasma membrane receive the signal.
3. *Cell responds.* When a hormone binds to an appropriate receptor, it causes a response in the target cell. The response can be a change in gene expression in the target cell's nucleus. The cell may start (or stop) producing a protein, or it may alter the rate at which it produces the protein. A single hormone may cause different responses in different target cells.

Most hormones are one of two types: (1) peptide and protein hormones (such as insulin), or (2) steroid hormones (such as estrogen and testosterone). Differences in the chemical structure of these two types of hormones determine whether or not a hormone can pass through the cell membrane (**FIGURE 24-5**). Peptide and protein hormones, the most common type of hormones, are water-soluble, as opposed to lipid-soluble. This means that they cannot pass through membranes, which are high in lipids. Steroid hormones are built from cholesterol and, consequently, are lipid-soluble. They can diffuse right through cell membranes and into the cytoplasm.

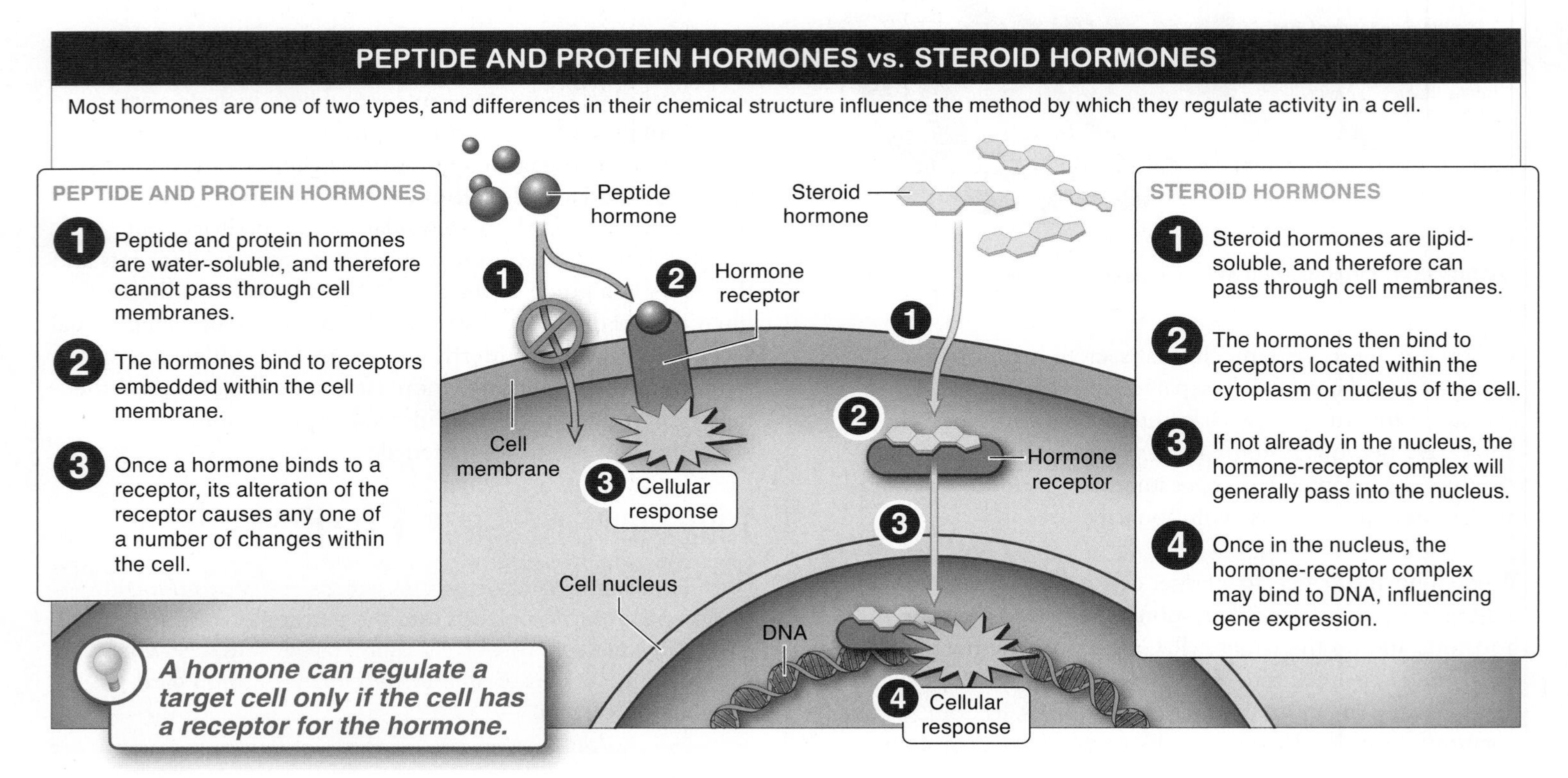

FIGURE 24-5 Most hormones belong to one of two main groups, differing in chemical properties. Hormones influence gene expression in different ways, depending on their chemical structure.

The chemical structure of the hormone determines how it regulates activity in target tissue. For example, the receptors for peptide and protein hormones are large glycoproteins embedded within cell membranes. Part of the receptor faces toward the outside of the cell; this is the part of the receptor to which the hormone binds. The receptor molecule also extends through the cell membrane, and another part of it extends into the cell's cytoplasm. Once a peptide or protein hormone binds to a receptor on the outside of the cell, it alters the receptor (perhaps changing its shape) and causes any one of a number of changes within the cell. It may activate an enzyme, initiating or speeding a reaction in the cell. Or it may alter the cell's membrane permeability, causing it to absorb or secrete certain molecules. Or it may alter proteins within the cell, causing them to move into the nucleus, where they may bind directly to the DNA and influence the rate of transcription of one or more genes.

The receptors for steroid hormones are proteins in the cytoplasm or nucleus of target cells. The hormone passes into the cell and binds to its receptor. If it is not already in the nucleus, the hormone-receptor complex generally passes into the nucleus. Once there, it may bind to the DNA, influencing the rate of transcription of one or more genes.

By traveling through the bloodstream, peptide, protein, and steroid hormones secreted by glands in one part of the body are able to regulate cell function in another part of the body. However, some glands produce regulators that act more locally. Called **paracrine regulators,** these molecules generally diffuse from the tissue in which they are produced, through intracellular fluid, and into nearby tissue, binding to receptors on or in the neighboring cells and influencing their activity. *Prostaglandins* are one of the most important and common groups of paracrine regulators, produced by almost every cell and nearly every organ in an animal's body. Prostaglandins have numerous effects, including dilation or constriction of blood vessels and influencing tissue inflammation (in which increased plasma and leukocytes move from the blood into tissues).

Q Why might long-term, heavy use of aspirin cause stomach bleeding?

One of the chief reasons that aspirin is such an effective pain reliever is that it inhibits an enzyme necessary for the production of prostaglandins (**FIGURE 24-6**). As a consequence, aspirin reduces inflammation (and the pain that often accompanies it). But, because aspirin also inhibits another, similar enzyme that is involved in maintaining the lining of the stomach, taking too much aspirin can cause stomach bleeding and other problems. The development of the drug celecoxib, sold as Celebrex, has been hugely beneficial to people suffering from arthritis, who need regular, long-term anti-inflammation medication. Celecoxib inhibits the prostaglandin-producing enzyme without inhibiting the stomach-lining-maintaining enzyme, making long-term use of this painkiller possible.

FIGURE 24-6 Arthritis hurts. Prostaglandins can cause sometimes painful inflammation in the body.

TAKE-HOME MESSAGE 24•3

Hormones can regulate the activities of a target tissue only if the cells have a receptor for the hormone. Most hormones are one of two types: (1) peptide and protein hormones, which cannot pass through cell membranes, and interact with receptors embedded in the membrane, or (2) steroid hormones, which can diffuse through cell membranes, then interact with their receptors in the cytoplasm or nucleus of the target cell. Both types of hormones, once bound to a receptor, cause changes in the target cell, including influencing the rate of transcription of genes.

❷ Hormones are produced in glands throughout the body.

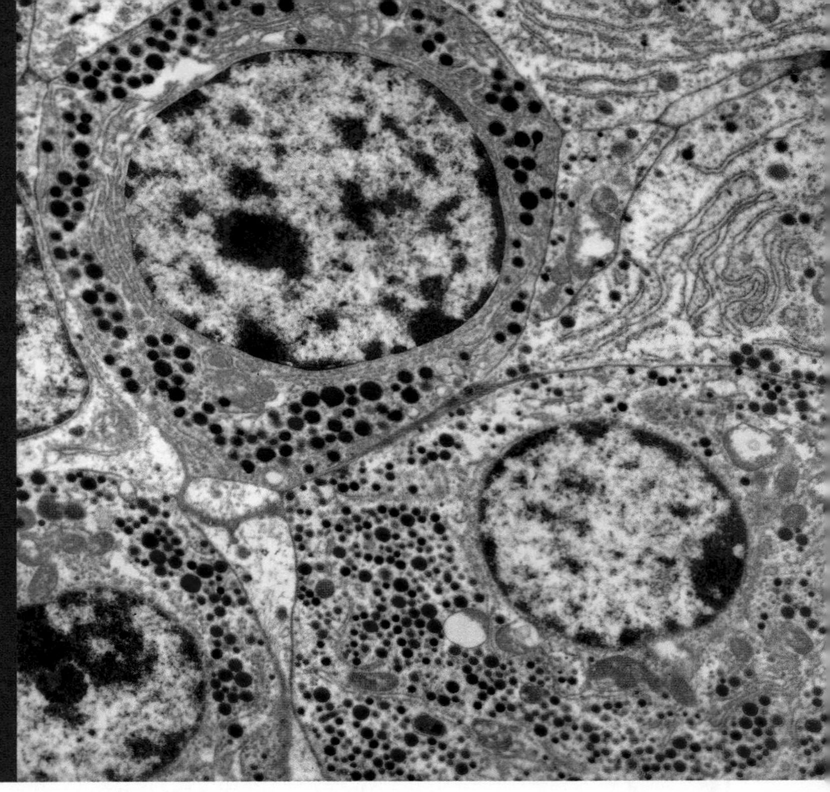

Hormone-secreting cells in the anterior pituitary. The nuclei of the cells are stained dark green. The red granules contain hormones to be secreted.

24 • 4

Where do hormones come from? 1. The hypothalamus controls secretions of the pituitary.

As we have seen, animals have two systems by which they send signals and regulate body functions in response to their external environment. And although one system or the other tends to be more effective for each particular signaling task, there is considerable interaction between the nervous system and the endocrine system. The **hypothalamus,** part of the underside of the brain, functions as a liaison between the two systems (**FIGURE 24-7**). It receives input from neurons throughout the brain and the rest of the body, and using this information about the external environment and the physiological state of the body, sends out the appropriate hormones (and nervous signals) to regulate nearly every aspect of the organism's physiology, including body temperature, hunger, thirst, and water balance.

Attached to the hypothalamus by a thin stalk is a gland about the size of a pea, called the **pituitary gland.** Signals from the hypothalamus directly influence the pituitary gland. The hypothalamus may release hormones that cause the pituitary to increase its production and release of hormones. Or the hypothalamus may direct the pituitary to reduce or stop the release of hormones.

Closer inspection of the pituitary gland reveals that it is actually two separate glands that appear to be fused together. The two regions, however, look different, originate from different embryonic tissues, produce different hormones, and are regulated differently. We consider the two regions separately.

The **posterior pituitary** has a fibrous appearance, because it contains a large amount of nervous tissue, particularly axons coming straight from the hypothalamus—from which the posterior pituitary develops. The posterior pituitary produces no hormones of its own. Instead, two important peptide hormones are produced in neurons within the hypothalamus and travel down the axon tracts into the posterior pituitary, from which they are then released:

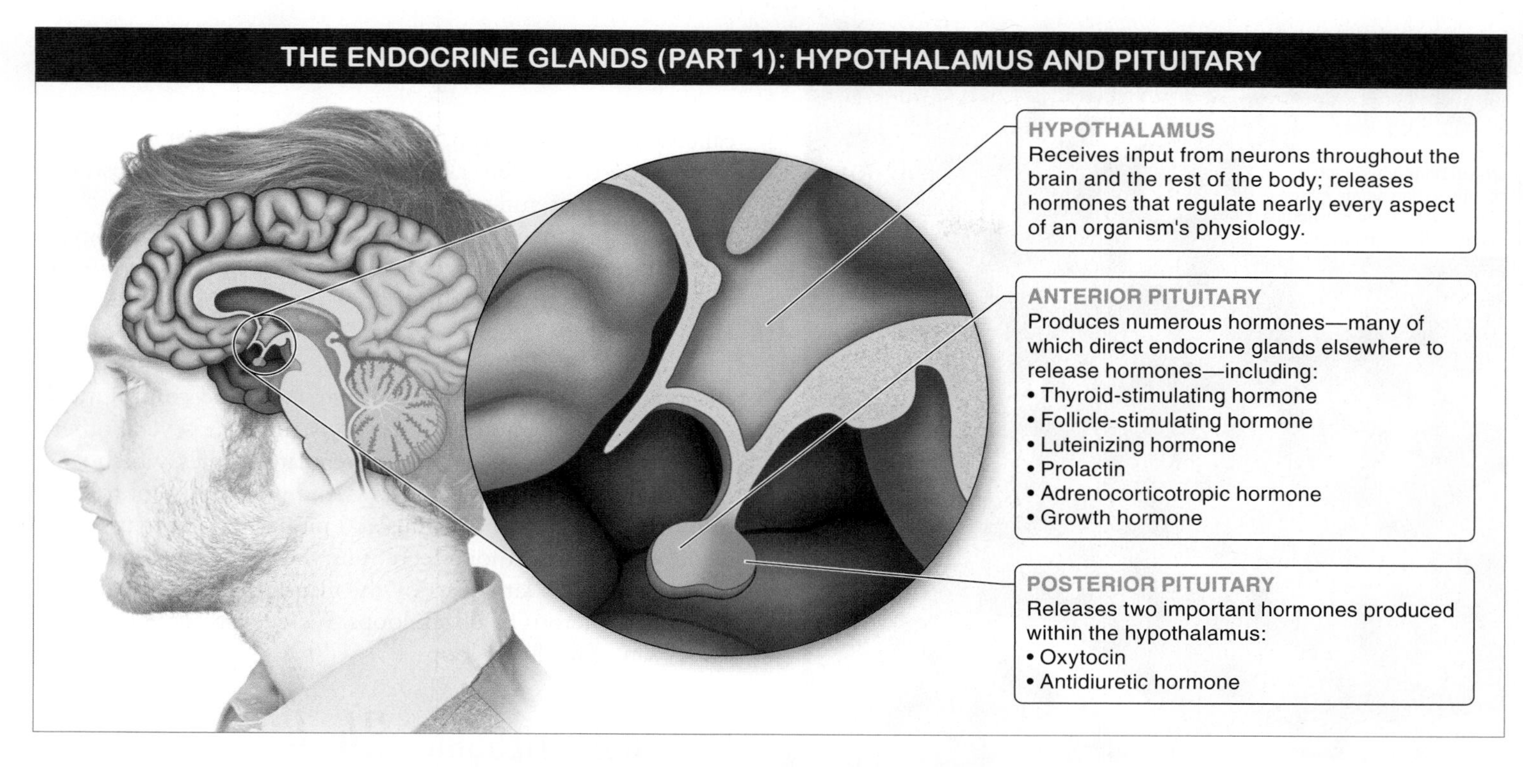

FIGURE 24-7 Function of the endocrine system glands: hypothalamus and pituitary glands.

1. *Oxytocin,* the "cuddle hormone," which we discussed in Section 24-1, influences people's trust in others and increases the propensity to form social attachments, as well as directing the ejection (or "let-down") of milk for nursing babies and stimulating muscular contractions in the uterus during childbirth.
2. *Antidiuretic hormone (ADH)* influences water retention by the kidneys. With increased levels of ADH, more water is saved, reducing the amount of urine produced while increasing its concentration.

The **anterior pituitary** has a more glandular appearance than the posterior pituitary, developing not from nervous tissue but rather from epithelial cells near the roof of the mouth during embryonic development. The anterior pituitary produces numerous hormones, in response to commands by the hypothalamus. Many of the anterior pituitary hormones direct endocrine glands elsewhere to release hormones. Some of the most important hormones produced by the anterior pituitary in mammals include the following:

1. *Thyroid-stimulating hormone (TSH)* causes the thyroid to produce thyroxine, which is important in cellular respiration, and which we discuss further in Section 24-5.
2. *Follicle-stimulating hormone (FSH)* stimulates follicles in the ovaries to begin development, and *luteinizing hormone (LH)* triggers ovulation. In males, LH stimulates testosterone production. FSH stimulates sperm maturation and LH stimulates testosterone production.
3. *Prolactin* stimulates the mammary glands to produce milk.
4. *Adrenocorticotropic hormone (ACTH),* also known as *corticotropin,* stimulates the adrenal gland to produce cortisol and other stress-related hormones.
5. *Growth hormone* has several effects, including stimulating the liver to release chemicals that spur the growth of bones, cartilage, and many other tissues.

Improper functioning of the pituitary can lead to some anomalies in growth and development. Excessive production of growth hormone during childhood, for example, can cause extreme growth, called gigantism, with some individuals reaching 8 feet in height! If the increased exposure to growth hormone doesn't occur until adulthood, only the hands, face, and feet tend to respond with unusual growth. Similarly, individuals with reduced or no production of growth hormone during childhood develop a condition

FIGURE 24-8 Pituitary malfunction can affect stature in humans.

called pituitary dwarfism, in which the individual may not grow more than 4 feet tall (**FIGURE 24-8**). Early diagnosis of these conditions and treatment with human growth hormone (produced using recombinant DNA technology) can restore normal growth. See Section 5-11 for a discussion of human growth hormone and recombinant DNA technology.

Although the hypothalamus and pituitary gland control much of the hormone secretion in the body, there are several other endocrine glands with important regulatory roles, which we explore in the sections that follow. And it is important to keep in mind that the hypothalamus and pituitary don't necessarily *control* all of the hormone secretions; they themselves are, in turn, regulated in large part by the glands that they regulate, through numerous feedback loops. See Section 20-3 for a discussion of feedback loops.

TAKE-HOME MESSAGE 24•4

The hypothalamus functions as a liaison between the nervous system and endocrine system. It receives input from neurons throughout the brain and the rest of the body, and using this information about the internal and external environments, it sends out the appropriate hormones (and nervous signals), often directing the pituitary gland to release hormones with important regulatory control over body tissues.

24•5

Where do hormones come from? 2. Other endocrine glands also produce and secrete hormones.

The hypothalamus and pituitary gland play large roles in the regulation of body functions in animals, but they are joined by many other glands in using hormones to signal and regulate physiology. From the anxiety a person may experience at the thought of public speaking, to the changes in wake and sleep patterns that occur when flying to a new time zone, to the changes in metabolism people experience as they get older, endocrine glands throughout the body are responsible for detecting and responding to signals reflecting an organism's internal and external environments. Here we explore the signals of some of the most important endocrine glands (**FIGURE 24-9**).

> **Q** Why is adrenaline given to people experiencing the severe airway restriction of an asthma attack?

Adrenal Glands Regulating an organism's response to stress is largely a function of the secretions of the two adrenal glands, which sit just above the kidneys and secrete the hormones cortisol and adrenaline, among others. Simply the

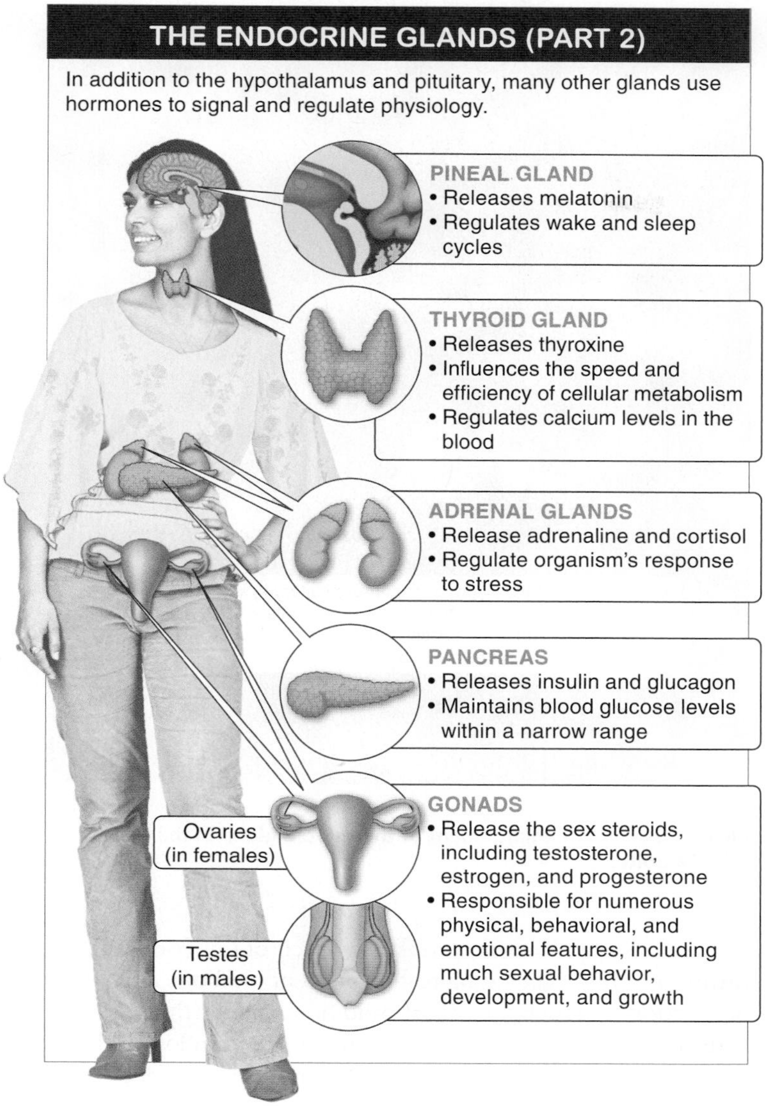

THE ENDOCRINE GLANDS (PART 2)

In addition to the hypothalamus and pituitary, many other glands use hormones to signal and regulate physiology.

PINEAL GLAND
- Releases melatonin
- Regulates wake and sleep cycles

THYROID GLAND
- Releases thyroxine
- Influences the speed and efficiency of cellular metabolism
- Regulates calcium levels in the blood

ADRENAL GLANDS
- Release adrenaline and cortisol
- Regulate organism's response to stress

PANCREAS
- Releases insulin and glucagon
- Maintains blood glucose levels within a narrow range

GONADS
- Release the sex steroids, including testosterone, estrogen, and progesterone
- Responsible for numerous physical, behavioral, and emotional features, including much sexual behavior, development, and growth

Ovaries (in females)

Testes (in males)

FIGURE 24-9 Function of the endocrine system glands: pineal, thyroid, adrenals, pancreas, and gonads.

sight of a predator is enough to initiate the "fight-or-flight" response—the secretions of adrenaline and cortisol that prepare the body for action (**FIGURE 24-10**). Within seconds, these secretions—in a case of positive feedback—can cause an increase in heart rate, an increased rate of glycogen breakdown in the liver and skeletal muscles, the release of stored fatty acids, goose bumps, and a dilation of the bronchioles in the lungs that enables greater absorption of oxygen, for delivery to needy tissues.

The stress pathways are modulated by negative feedback loops. As an animal takes action in response to a stressful situation, and the source of the stress is removed, the secretions of

FIGURE 24-10 Who will win in this conflict? Adrenal glands influence the fight-or-flight response in animals.

cortisol and adrenaline are reduced. This is how the stress response usually works in nature. But when there is no outlet by which an organism can deal with the stress, long-term consequences of a chronic stress response include ulcers, cardiovascular problems, decreased immune function, and the risk of illness. With increased understanding of the stress response and its general function in nature as a short-term physiological state that helps organisms quickly and effectively respond to stressful situations, researchers are gaining insights into how to better treat anxiety and depression.

Pineal Gland The 17th century philosopher René Descartes believed that the pineal gland was where the soul connected with the body—which he believed largely because the pea-sized gland was located near the center of the brain, was (he thought) unique to humans, and was singular (that is, there was just one, not one on the left side of the brain and another on the right). This view has been abandoned (and we now know that the pineal gland is present in all vertebrates), but there's still considerable scientific interest in this gland. The pineal gland has neuron connections with the retina of the eye, and it controls the secretions of the hormone melatonin, which is derived from the amino acid tryptophan and affects diurnal-nocturnal wake and sleep patterns, called *circadian cycles.*

Although the exact mechanism by which melatonin influences circadian cycles is not understood, it has been shown to have slight benefits in synchronizing individuals' sleep and wake cycles to the environment and in treating some types of insomnia, and is now sold throughout the United States.

> My whole aim in life is to get near to God, that is, to get nearer to myself. That's why it doesn't matter to me what road I take. But music is very important. Music is a tonic for the pineal gland. Music isn't Bach or Beethoven; music is the can opener of the soul. It makes you terribly quiet inside, makes you aware that there's a roof to your being.
>
> —Henry Miller, *Tropic of Cancer,* 1934

Thyroid Gland One of the largest endocrine glands in humans, the **thyroid gland,** is found in the neck, just below the point where, in men, the Adam's apple is. It secretes hormones—including thyroxine—that influence the speed and efficiency at which body cells break down macromolecules in the diet and use the energy released from food to produce proteins. In short, it controls most of what we think of as metabolism. As a consequence, poor thyroid function is believed to be at the root of many metabolic disorders, with underactive thyroid responsible for fatigue and weight gain, and overactive thyroid responsible for jitteriness, rapid heartbeat, weight loss, and irritability.

FIGURE 24-11 An enlargement of the thyroid gland in the neck is called a goiter.

Goiter is a common health problem caused by an enlargement of the thyroid gland (**FIGURE 24-11**). There are several causes of goiters, and they are particularly common in areas with low consumption of foods containing iodine. When iodine intake is low, the thyroid is unable to produce thyroxine (which contains iodine). This causes thyroxine levels in the body to drop. In the absence of the normal negative feedback telling the body to slow its production of thyroxine, the hypothalamus and anterior pituitary produce increasing amounts, respectively, of thyroxine-releasing hormone and thyroxine-stimulating hormone. These cause the thyroid to swell into a visible lump—a goiter—as it tries unsuccessfully to make thyroxine. In the United States, the widespread use of table salt fortified with iodine prevents most iodine deficiencies, but such deficiencies are common in Asia, central Africa, and parts of South America.

The thyroid is also responsible for regulating levels of calcium in the blood. Calcium is necessary for building and maintaining bones and teeth, and influences the functioning of nerves and muscles. When there is too much calcium in the blood, the thyroid increases its release of **calcitonin,** a hormone particularly important in babies and children, which causes bones to take up the excess calcium. Additionally, embedded in the surface of the thyroid are four small structures, the **parathyroid glands,** that produce parathyroid hormone, which plays a central role in regulating calcium levels in adults (**FIGURE 24-12**). Throughout life, bone is continually broken down and remade as minerals are lost and added. Parathyroid hormone is important in stimulating much of this continued turnover of bone, including reabsorption of old bone and production of new bone. Parathyroid hormone further regulates blood calcium levels by reducing calcium loss in the urine, by regulating the release of calcium from bone, and in conjunction with vitamin D (from the diet and produced in skin cells in response to exposure to the sun), helping to increase the body's ability to utilize calcium in the diet.

Pancreas Located next to the stomach and connected to the small intestine via a short duct, the pancreas is an endocrine gland that is most important in controlling the levels of blood glucose. As we saw in Section 22-18, the pancreas maintains blood glucose within narrow ranges—a typical blood glucose concentration in humans is 90 mg/100 mL—through the coordinated secretions of insulin and glucagon (**FIGURE 24-13**).

Following a meal (particularly one rich in carbohydrates), the concentration of blood glucose rises. This stimulates release of insulin by cells in the pancreas. Insulin in the bloodstream causes the liver and other tissues—primarily muscle—to take up glucose, which reduces the blood glucose level. As blood glucose levels fall, there is a reduction in insulin secretion.

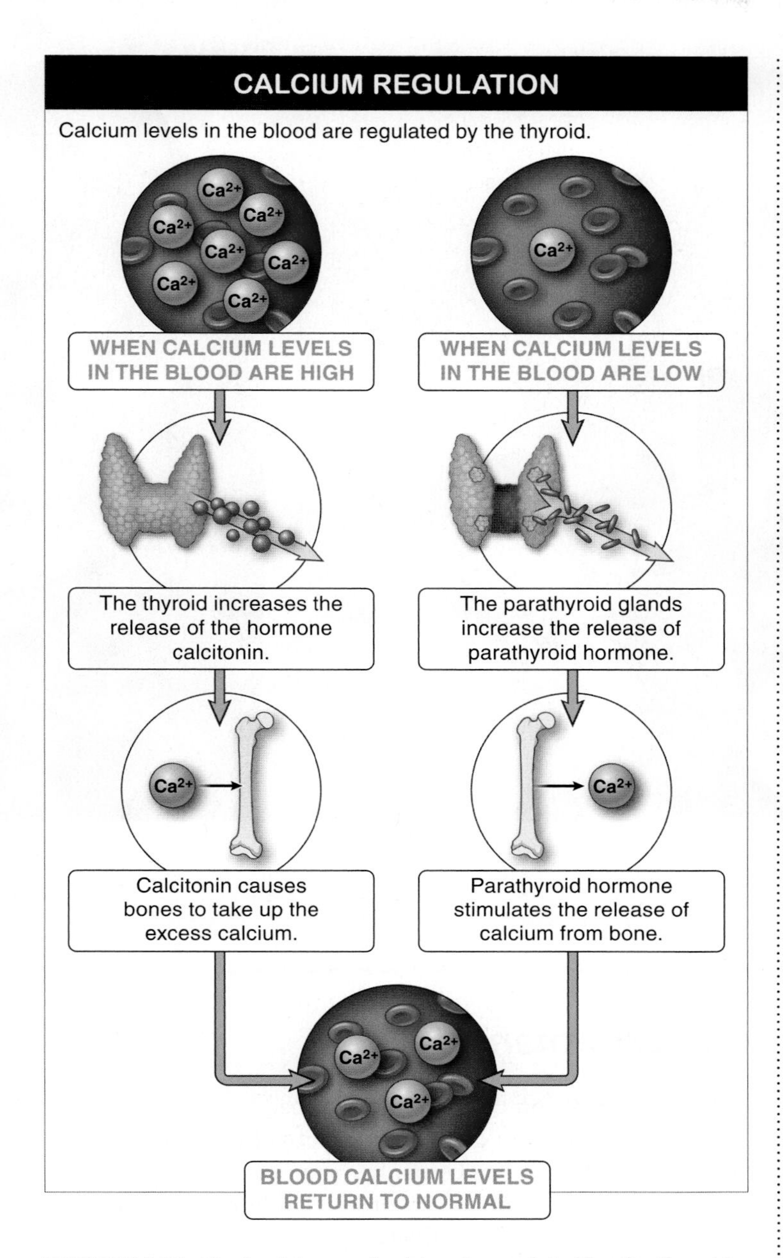

FIGURE 24-12 The body's use of calcium is regulated by the thyroid and parathyroid glands.

Conversely, after fasting for a few hours, the blood glucose level gradually drops. The reduced blood glucose triggers release of glucagon by some cells in the pancreas. Glucagon has the reverse effect of insulin, causing the liver to convert stored glycogen into glucose, which is released into the bloodstream. Rising blood glucose concentration then causes the pancreas to reduce its glucagon secretion, maintaining homeostasis through negative feedback.

Gonads The sex steroids, including testosterone, estrogen, and progesterone, are produced largely by the gonads—the testes in males and the ovaries in females. These hormones are responsible for numerous physical, behavioral, and emotional features, including much sexual behavior and growth, sexual development (beginning embryonically, but playing a significant role in triggering puberty and continuing into adulthood), and maintenance of gamete production throughout organisms' reproductive lives. In Chapter 25 we discuss the sex hormones in detail.

FIGURE 24-13 Even after a heavy meal, a healthy person can maintain blood glucose levels within a narrow range.

In the next section, we investigate further some of the gonadal hormones and see how closely linked they are to athletic performance and the physical attributes necessary to excel physically. We'll also see a dark side to hormones, in noting some examples of the dramatic physical changes and improvements to athletic performance that have resulted from illegal use of hormone supplements. Later in the chapter, we also note the extreme health consequences that accompany such abuse of these hormones.

TAKE-HOME MESSAGE 24•5

Endocrine glands throughout the body are responsible for detecting and responding to signals reflecting an organism's internal and external environments. The adrenal glands regulate responses to stress. The pineal gland regulates wake and sleep cycles. The thyroid gland influences the speed and efficiency of cellular metabolism. The pancreas maintains blood glucose within a narrow range. And the gonads produce hormones responsible for numerous physical, behavioral, and emotional features, including much sexual behavior, development, and growth.

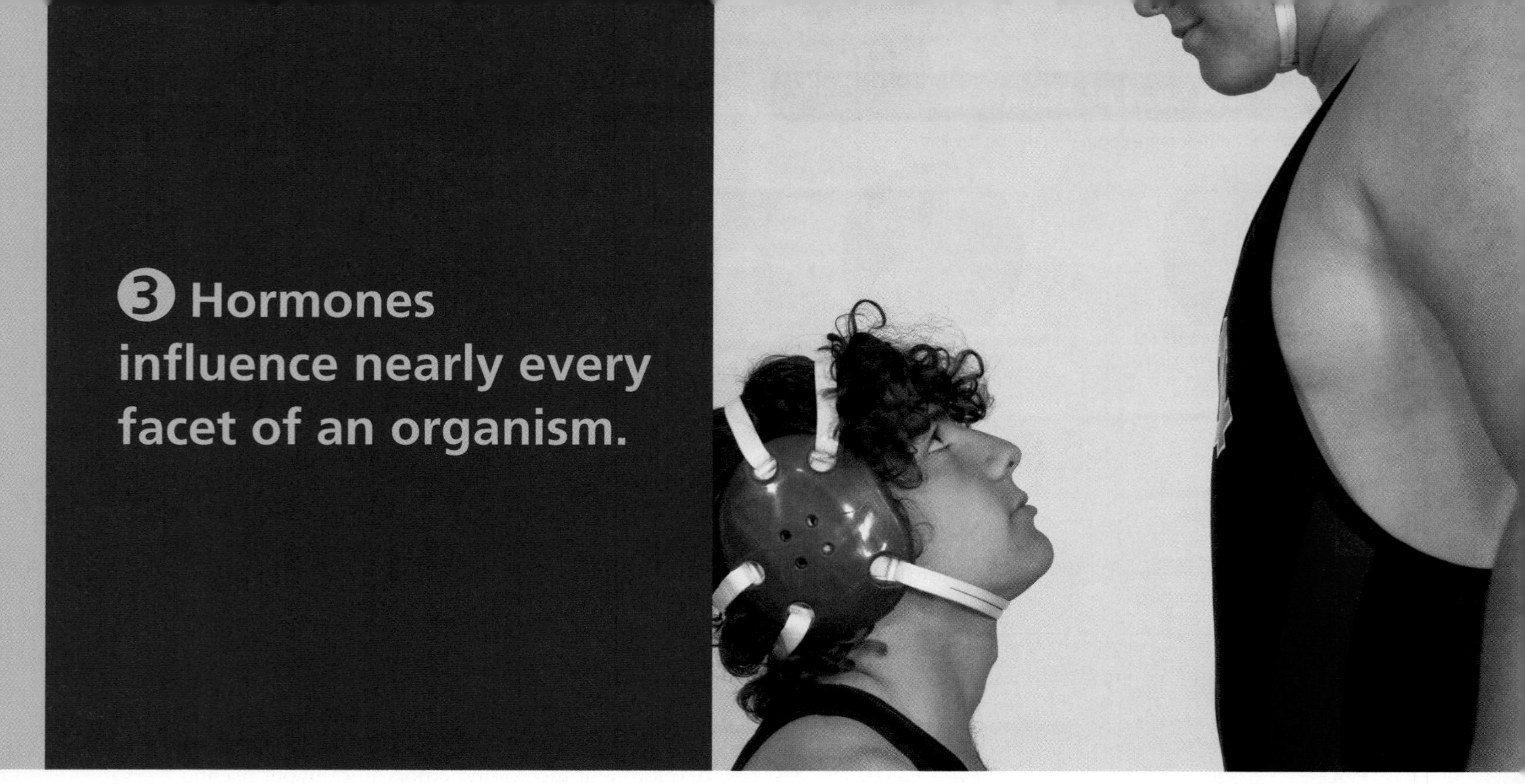

Growth hormones influence stature.

24 • 6

Hormones can affect *physique and physical performance.*

The Tour de France is one of the most grueling tests of physical strength and endurance. The annual bike race occurs over the course of three weeks, as riders cover more than 2,200 miles (3,500 km), in 21 separate races, or stages. In 2006, an American rider named Floyd Landis, in 11th place overall, turned in a performance in stage 17 that was so improbably fast that some cycling commentators called it "the greatest performance ever." His victory brought him within 30 seconds of the overall lead, and propelled him, over the course of the final four stages, to the Tour de France victory (**FIGURE 24-14**).

But his glory was not to last. Less than a week after his victory, it was announced that Landis had tested positive for unusually high levels of testosterone. A second test confirmed this finding. The typical ratio of testosterone relative to another hormone (epitestosterone) in men is 1:1 or 2:1. While the Tour de France allows ratios of up to 4:1, Landis's level was 11:1. Moreover, the lab tests detected a synthetic form of testosterone in his bloodstream, in addition to the naturally produced hormone. Landis was stripped of his title and banned from the sport for two years.

While the revelation that a competitor had resorted to illegal supplementation in an effort to boost physical performance was disappointing, it was certainly not a new or isolated occurrence. The recent history of sports is littered with drug scandals, usually involving athletes' use of testosterone or variants of that male hormone. The winner of the gold medal in the 100-meter race in the 1988 Olympics, Ben Johnson, for example, was stripped of his medal and the world record when he tested positive for a synthetic steroid hormone similar to testosterone. In 2003, 104 Major League Baseball players (of 750 players in the league) tested positive for illegal performance-enhancing drugs. In 2005, 111 professional football players tested positive for banned substances. And the scandals are not limited to men. Marion Jones won five Olympic gold medals in 2000, only to have all of them taken away when she admitted that she had illegally taken performance-enhancing drugs, including a muscle-building "designer steroid" called tetrahydrogestrinone, or "the clear."

And so it is clear that many athletes seek out ways to increase the amount of certain hormones in their bodies. The question

FIGURE 24-14 Unsporting conduct. Many athletes have tested positive for illegal testosterone supplementation.

is, do these hormones actually affect a person's physique and physical performance? And, if so, how do they do it?

When it comes to testosterone (and many structurally similar steroid hormones), the answer to the first question is a definitive *yes.* Testosterone does affect the composition of the body. Experimental studies of testosterone supplementation in men reported gains of 5–12.5 pounds in body weight within 10 weeks. The gains were due primarily to increases in lean muscle mass, with the greatest increases occurring in the shoulders, upper chest, and upper arms, where there are the highest concentrations of testosterone receptors. In randomized, controlled studies, researchers also documented strength increases of 5% to 20%, in both experienced and novice athletes.

Testosterone has also been shown to influence locomotor performance—speed and endurance, in particular—in animals. In an experimental study on lizards (the northern fence lizard, *Sceloporus undulatus*), two to three weeks of supplemental testosterone administration caused a 24% increase in sprinting speed and a 17% increase in stamina.

Use of testosterone, or testosterone doping, causes changes in physique and physical performance by activating processes that lead to increases in muscle mass. Produced primarily by the testes in males (but also by the adrenal glands of both males and females), testosterone binds to receptors in target cells, then moves to the nucleus and influences gene expression. The cells respond to testosterone by increasing protein synthesis and cell division, resulting, in the case of muscle cells, in increased muscle mass (and simultaneously reducing the rate of production of fat-storage cells). Additional responses to testosterone include the development of secondary sex characteristics, including stimulating the growth of chest and facial hair and deepening the voice—effects seen in both males and females in response to testosterone supplementation (**FIGURE 24-15**).

As we'll see later in the chapter, there are numerous adverse effects to supplemental steroid hormones, including increased

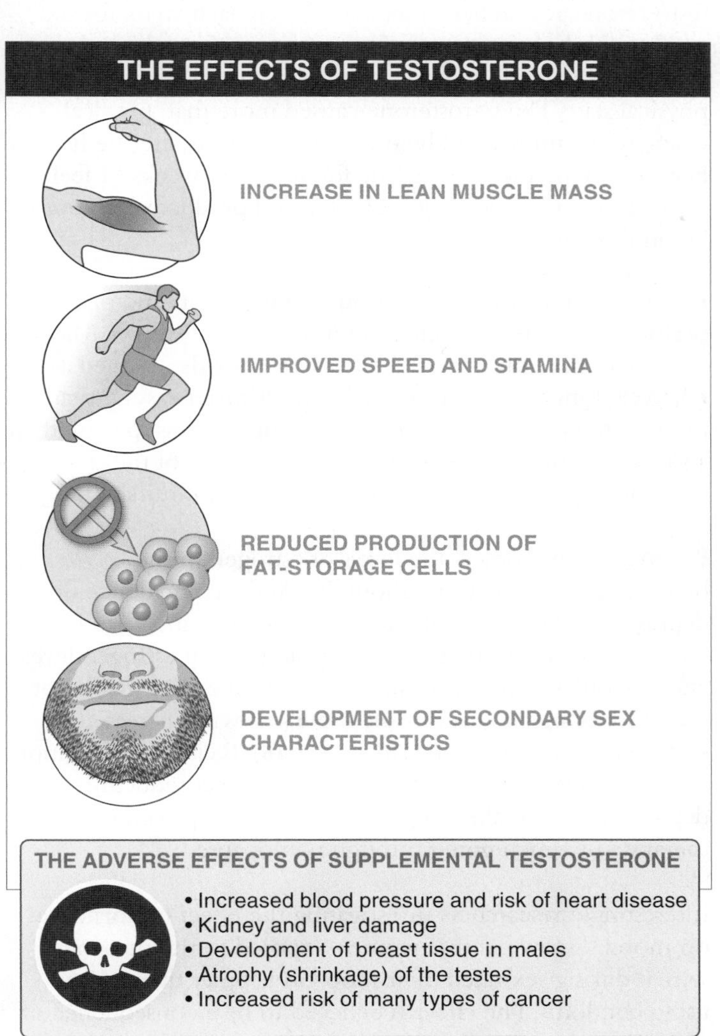

FIGURE 24-15 The normal effects of testosterone in the body, and the harmful effects of testosterone supplementation.

blood pressure and risk of heart disease, as well as damage to the kidneys and the liver (where the steroids are metabolized), development of breast tissue in males, and atrophy (shrinkage) of the testes as negative feedback leads to reduced production of testosterone in response to the increased levels of circulating testosterone. Also, because testosterone increases cell division—such as by increasing muscle mass—it increases the risk of unrestrained cell division in many tissues and, consequently, increases the risk of many types of cancer.

TAKE-HOME MESSAGE 24•6

Hormones affect a person's physique and physical performance. Testosterone, for example, increases muscle mass while reducing fat storage and can increase speed and stamina. It produces these changes by influencing gene expression and protein synthesis in cells with testosterone receptors.

24•7

Hormones can affect *mood.*

In a randomized, controlled, and double-blind study of testosterone, researchers injected 47 men each week for six weeks with large amounts of testosterone, after which the researchers made numerous measurements on each subject's physical state. The testosterone caused more than physical changes, as summarized by one of the subjects after he had been receiving the testosterone for about six weeks: "I feel great! I'm confident, positive, happy, and productive. I want to be on this stuff forever!"

Not only do hormones affect our physical traits and physical performance, but they can also influence how we *feel*. Many hormones have pronounced effects on **moods,** defined as relatively long-lasting emotional states (shorter-lasting than a person's temperament, but longer-lasting and less specific than a single emotion). Here we describe just a few of the documented effects of hormones on mood in humans.

Estrogen In women, the levels of estrogen in the bloodstream change throughout life. And the incidence of depression follows a similar pattern. The large increases in estrogen levels at puberty, the sharp drop off in estrogen levels after a woman gives birth, and the reduced estrogen levels at the onset of menopause are all associated with increased occurrence of depression (**FIGURE 24-16**). Related observations have demonstrated that estrogen is effective at reducing depression when taken as part of hormone replacement therapy in women going through menopause.

FIGURE 24-16 Drastic changes in estrogen levels affect mood. For example, the changes can cause "the baby blues" in some women shortly after giving birth.

Interestingly, researchers investigating the effect of hormones on mood discovered that women exposed to their partner's semen during sex rated their mood as happier than those using condoms. The effect is believed to be a consequence of the mood-altering effects of the testosterone and estrogen that are present in semen and absorbed through the vagina. With unprotected sex, however, the potential for sexually transmitted diseases or unwanted pregnancy must also be factored into any long-term assessment of mood, complicating the research findings.

Testosterone Many studies have demonstrated a link between testosterone and mood in human males. In the study on testosterone administration described above, for example, testosterone caused increased ratings of a variety of feelings, characterized as hypomania, including the men's increased self-esteem and a sense that they were overflowing with new ideas. During the period when receiving the testosterone, the men were described by their significant others as energetic, euphoric, confident, and charismatic. In a small number of cases, however, subjects became uncharacteristically aggressive and exhibited verbal hostility. The researchers even had to withdraw one subject from the study when he became "alarmingly" aggressive—consistent with anecdotal claims of "'roid rage" resulting from taking supplemental testosterone.

FIGURE 24-17 Biological victory dance. Testosterone levels surge in competitors and in fans, too.

Several studies on male athletes in a variety of competitive settings, including tennis and wrestling, have documented that testosterone levels—and the mood changes they influence—rise in competitors prior to a competition. After the contest, testosterone levels in the losers of the competition drop quickly and dramatically, while testosterone levels in the winners remain high and sometimes even increase (**FIGURE 24-17**).

Q How can the exhilaration of watching your favorite team win a game match the feeling you get from winning a sports contest yourself?

Recent studies reveal that sports fans share in this biological "victory dance." Measurements of testosterone levels in male fans of Brazilian and Italian soccer teams during World Cup finals revealed that the testosterone levels of the fans of the winning team rose, but not those of the losing team. This finding has been replicated in fans of a college basketball team.

Melatonin In randomized, controlled, double-blind studies, researchers have demonstrated that under certain conditions, oral melatonin can induce hypnotic, sedative-like effects and improve sleep efficiency. Other studies have shown that, in comparison with placebo treatment, melatonin supplementation significantly reduced subjects' self-reported vigor, while increasing their fatigue and confusion.

FIGURE 24-18 Hiding in the dark. Rodents exposed to cortisol show anxious behavior.

Cortisol Two separate observations suggest that the hormone cortisol has significant effects on mood. First, researchers have noted that the majority of individuals suffering from Cushing's disease, a condition in which the adrenal glands release unusually large amounts of cortisol, experience depression and anxiety. And second, individuals receiving cortisone (a chemical variant of cortisol that is converted to cortisol in the body) as part of a treatment to reduce inflammation of various tissues also experience anxiety and depression at atypically high rates.

As a consequence of these observations, researchers conducted a controlled study to test whether the stress hormone actually causes anxiety. To do this, they exposed rodents to cortisol by adding it to the animals' drinking water. Exposure to cortisol for two weeks or more, they found, caused the animals to take significantly longer to emerge from small, dark compartments into a brightly lit area—a measure of anxiety-like behavior in rodents—as compared with animals not exposed to the cortisol (**FIGURE 24-18**).

TAKE-HOME MESSAGE 24•7

Many hormones, including estrogen, testosterone, melatonin, and cortisol, have pronounced effects on moods.

24•8

Hormones can affect *behavior.*

It can be hard to imagine how a little more or a little less of a chemical could influence an animal's behavior, but experimental research on hormones and behavior has documented literally *thousands* of such effects. The scholarly journal *Hormones and Behavior,* for example, now nearing its 60th volume, publishes 10 issues a year, each containing 25 or more articles describing laboratory and field studies on hormones and their influences on the development and expression of behaviors (**FIGURE 24-19**). In these studies, researchers take a variety of experimental approaches. Just two of these are described here.

Genetic manipulation of hormone levels. Researchers bred some lab mice so that the animals no longer made aromotase, an enzyme necessary essential to the production of the hormone estrogen. With this deficiency, the lower-estrogen animals had two striking behavioral changes: they ran excessively on the exercise wheels in their cages and, when they were sprayed lightly with water, they spent significantly longer grooming themselves than do typical mice. Both of these behaviors are indicators of obsessive-compulsive disorder in mice. The animals also had lower-than-usual levels of a brain chemical called COMT, which also is a typical finding in obsessive-compulsive men.

Q Is castration an effective and humane treatment for sex offenders?

Physiological supplementation of hormone levels. In a second type of experimental approach to studying hormone effects on behavior, researchers implanted testosterone capsules (with tiny pin-prick-sized holes) in male juncos, a type of song bird. The researchers monitored the birds' behavior and documented that the male birds produced a song that was more attractive to females and also produced more offspring than control-group males that received a capsule containing no testosterone. The increased reproductive success did not come without a cost, however. Male birds with testosterone implants spent less time with their offspring and gave the offspring less food. The testosterone-supplemented birds were also more susceptible to disease and had shorter life spans. In similar studies, male birds with supplemented testosterone tended to have increased muscle mass and maintained larger territories than non-supplemented males.

The dramatic and close link between hormones and behavior has even caused people in many countries to debate the possibility of castrating sex offenders to modify their behavior (**FIGURE 24-20**). Whether or not castration—the removal of the testes, with a resultant near-

Thousands of research articles have described the wide variety of hormone effects on behavior.

FIGURE 24-19 The influence of hormones on behavior is an area of active scientific research.

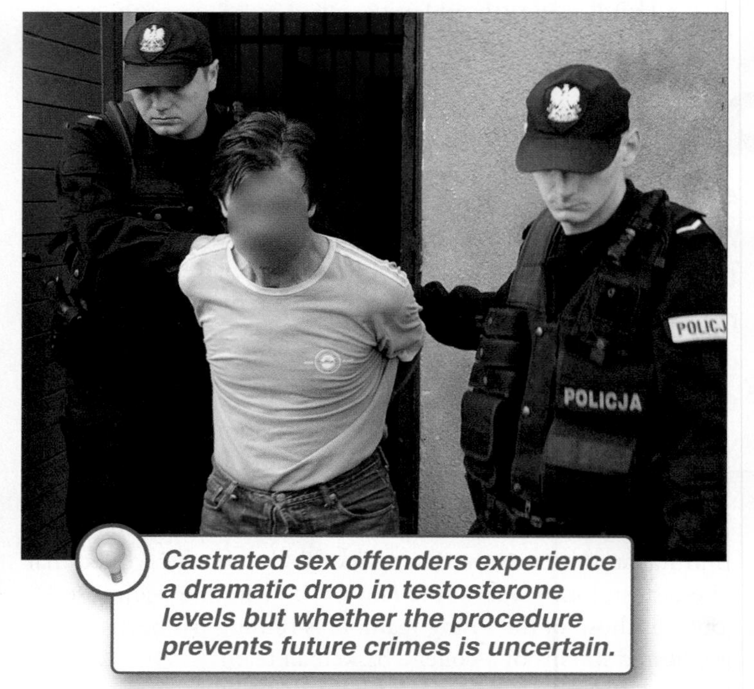

Castrated sex offenders experience a dramatic drop in testosterone levels but whether the procedure prevents future crimes is uncertain.

FIGURE 24-20 Is castration of convicted sex offenders an ethical procedure?

complete drop in testosterone levels—can reliably rehabilitate violent sex offenders is hotly debated. In the Czech Republic, one of the very few countries where sex offenders may be castrated, 94 prisoners have undergone castration in the past decade. None of these men have been reported to commit any further offenses. Similarly, in a Danish study of 900 castrated sex offenders, the rate of repeat offense dropped to approximately 2%, from close to 80% among non-castrated sex offenders.

Opponents of the castration of sex offenders argue that the evidence cannot be trusted, because it partly relies on self-reporting by the castrated men. We just can't be sure of the effectiveness of castration in completely stopping sex offenders from committing additional offenses. Moreover, opponents argue that having such an option is coercive, because many convicted offenders will feel obligated to opt for the surgery, leading to an unacceptable violation of the prisoners' rights. Some argue for the reversible form of castration that is achieved by injection of chemicals that block the effects of testosterone, but this is opposed by many on the grounds that it relies on the offenders, once released from prison, voluntarily undergoing their treatment, which they may stop at any point.

TAKE-HOME MESSAGE 24•8

Many studies—including genetic manipulations and physiological alterations of hormone levels—describe laboratory and field research demonstrating hormone influences on the development and expression of behaviors.

24•9

Hormones can affect *cognitive performance.*

Repeat this sentence five times, as quickly as you can: "A box of mixed biscuits in a biscuit mixer." Researchers found that, at the midpoint in their ovulatory cycle, women in the study group could do this in about 14 seconds. During this time, female estrogen levels are at their highest and about 50 times higher than they are in men. During the period just after menstruation, women in this study took 17 seconds to complete the task. During the post-menstruation period, estrogen levels in women are at their lowest, but are still about 3 times higher than in men, and at both times, women were able, on average, to recite the tongue twister faster than men.

The methods used in this widely reported study have been criticized for several reasons, including (1) the study was not blind, because the researchers knew what phase of their reproductive cycle the subjects were in, and the subjects knew why they were being tested, and (2) actual measures of estrogen were not taken—the subjects just reported what point they were at in their reproductive cycle. Nonetheless, this type of research has led to numerous, better-controlled research efforts to evaluate whether cognitive abilities—mental processes involving perception, memory, judgment, and reasoning—are influenced by hormones, with particular emphasis on the reproductive hormones estrogen and testosterone.

Studies on the effects of hormones on cognitive performance have focused on two primary types of measures, designed to reflect different aspects of cognition: (1) motor and verbal tasks, and (2) spatial tasks (**FIGURE 24-21**).

1. *Motor and verbal tasks* include tests of articulation speed, such as the tongue twister noted above, or complex wrist and hand movements and fine-muscle movements, such as those required in surgery and machine repair.

Most tests of hormone effects on motor and non-verbal tasks involve comparisons of the performances of men and women or of girls and boys at these tasks. There is consistent and significant evidence that females perform better than males and that the disparity between male and female performances is greatest at points of high estrogen levels in the female reproductive cycle and smallest at the points of lowest estrogen levels.

2. *Spatial tasks* include tests such as predicting the folded shapes of cardboard boxes when they are unfolded and flat, and mental rotations, in which a pair of two- or three-dimensional objects are compared to see whether they represent the same object, either rotated or as a mirror image.

In humans and other animal species, performance on tests of spatial ability is, on average, higher in males than females. A large body of evidence suggests that this performance difference reflects effects of testosterone—although the mechanisms for this effect are poorly understood. Some of the suggestive findings include:

THE EFFECTS OF HORMONES ON COGNITIVE TASKS

Repeat this sentence five times, as quickly as you can:

"A box of mixed biscuits in a biscuit mixer."

MOTOR AND VERBAL TASKS

- Females tend to perform better than males on tests of articulation speed and fine-muscle movement.
- The disparity between male and female performance is greatest at points of high estrogen levels in the female reproductive cycle and least at the points of lowest estrogen levels.

Identify which of the four shapes on the right represent the target shape on the left in alternative positions.

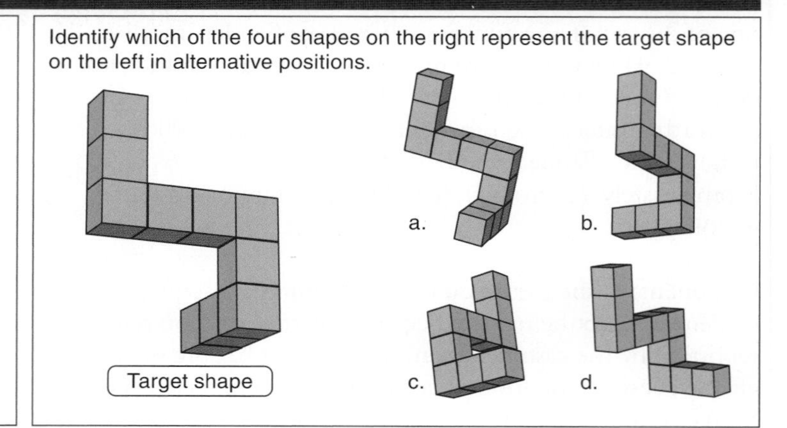

SPATIAL TASKS

- Males tend to perform better than females on tests of spatial ability, such as mental rotations of two- or three-dimensional objects.
- Evidence suggests that this performance difference reflects effects of testosterone—although the mechanisms by which testosterone affects spatial ability are poorly understood.

FIGURE 24-21 Different talents. Men and women have slightly different abilities in certain areas, based on the abundance or lack of the hormones testosterone and estrogen.

- As men get older and their testosterone levels decrease, so too does their performance on tests of spatial ability.
- In older men, testosterone supplementation improves performance on spatial tasks, while chemical castration (administration of chemicals that reduce the production of testosterone) for treatment of prostate cancer causes a decrease in spatial abilities.
- Women with one form of a condition called congenital adrenal hyperplasia have higher than typical levels of circulating testosterone and increased performance on tests of spatial ability.

As with tests of motor and verbal tasks, it is not clear exactly how—or whether—performance on tests of spatial ability affects individuals in contexts more relevant to everyday life. Recently, there has been discussion about the observation in a National Geographic Society study that more than three-quarters of the winners of "Geography Bee" contests are boys and that 118 of the 120 finalists in the National Geography Bee have been boys (**FIGURE 24-22**). While interest in such competitions is unquestionably influenced by cultural and societal forces, the researchers suggested that the results stem from male-female differences in spatial abilities—based on the belief that spatial abilities translate into an ability to read and interpret maps. An author of the study, Lynn Liben, commented that "it's not true that every woman is worse than every man or every girl is worse than every boy. But at the group level, it is true." Girls remain more likely to make the honor roll and to perform well in school.

Beyond tests of motor and verbal tasks and tests of spatial ability, another component of cognitive performance is

Of the 120 finalists in the National Geography Bee over the past 12 years, 118 have been boys. Researchers suggest that this disparity is linked to male-female differences in spatial abilities. Can this hypothesis be tested experimentally?

FIGURE 24-22 Showing an interest in geography.

memory. Numerous studies report a strong role for estrogen in memory, suggesting as a possible mechanism the fact that estrogen increases the growth of neurons and the connections between them. A recent study in rodents, for example, showed that treatment with a high dose of estrogen improved a maze-running task that tests memory.

The stress hormone, cortisol, has also been implicated in memory abilities. Most studies have reported a relationship that resembles an inverted "U." Up to a point, increased cortisol improves memory (and removing the adrenal gland, which produces most of the body's cortisol, impairs performance on tests of memory, but abilities are restored when cortisol is supplemented to typical levels). Excess cortisol and other stress hormones, however, consistently reduce performance on tests of memory. Optimum memory performance seems to require moderate levels of cortisol and other stress hormones.

TAKE-HOME MESSAGE 24•9

A great number of experimental studies demonstrate that cognitive abilities in humans—primarily in motor and verbal tasks and spatial tasks—are influenced by hormones, particularly the reproductive hormones estrogen and testosterone, but also by the stress hormone cortisol.

24•10

Hormones can affect *health and longevity.*

Sometimes the links between cause and effect in biological systems appear straightforward. Take, for example, the results from one of the most appalling "experiments" ever conducted. In the early 1900s, many men committed to sanitariums were castrated. In a case-controlled study, the castrated men were compared with non-castrated, institutionalized men, born in the same year and with the same estimated IQ. It turned out that the median longevity of the castrated men, at 69.3 years, was almost 14 years longer than that of the control group.

Similarly, data from analyses of more than 1,000 cats (both males and females) showed that sterilized cats live significantly longer than those not sterilized. Among females, the intact animals lived a mean of 3.0 years, while those with their ovaries removed had a mean life span of 8.2 years (**FIGURE 24-23**). Laboratory mice, too, have significantly lengthened life spans when sterilized. And in similar experiments, female fruit flies that were exposed briefly to high temperature or X rays experienced a dramatic reduction in ovary size, a severe drop in egg-laying rate, and an accompanying 73% increase in life span. Each of these results on sterilization and longevity suggest that reducing levels of circulating reproductive hormones increases longevity.

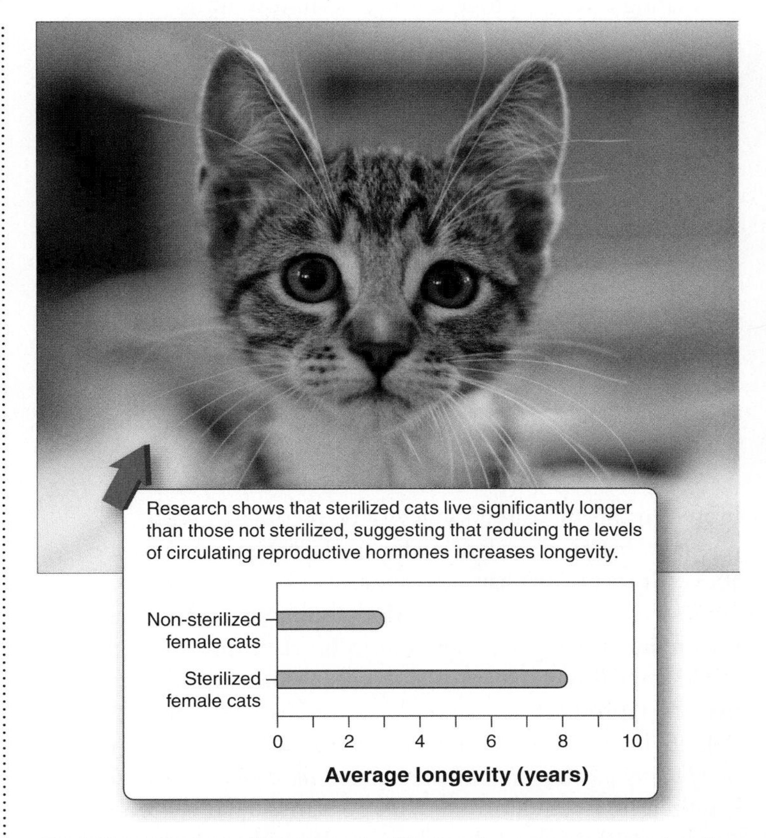

FIGURE 24-23 Long-lived pets. Sterilization increases longevity in cats.

The link between reproductive hormones and longevity is also supported by experiments in which mice were fed diets significantly reduced in calories. Researchers found that the amounts of luteinizing hormone, estrogen, and progesterone in the bloodstream were significantly reduced among calorie-restricted females compared with controls—animals given access to as much food as they wanted. Similarly, the researchers found testosterone levels in males on the low-calorie diet were reduced to less than one-third of those found among animals with free access to food. Mice on the low-calorie diet lived more than 40% longer than those with unlimited access to food.

Hormones probably have their significant impact on longevity because the rate of cancer—in rodents as well as in humans—is closely related to the concentrations of circulating reproductive hormones, such as estrogen:

- Increased lifelong exposure to estrogen is associated with increased rates of cell division and, consequently, increased cancer risk.
- High concentrations of estrogen in the bloodstream are associated with increased incidence of endometrial cancer.
- Women taking oral contraceptives and women giving birth to higher numbers of children—both of which reduce a woman's exposure to estrogen and other hormones produced by developing follicles—have decreased incidences of ovarian cancer. An analysis of 20 studies assessing the relationship between oral contraceptive use and ovarian cancer reported that five years of taking oral contraceptives reduced women's lifetime risk of cancer by approximately 50% (**FIGURE 24-24**).

There is similar evidence of increased cancer risk linked to circulating levels of progesterone and testosterone.

These seemingly straightforward results, however, lead us to an example that illustrates the extreme complexity and interrelatedness of the body's physiological systems. It would seem, given the results above, that longevity should be decreased by hormone replacement therapy, which can take a variety of forms but generally involves supplementation of the hormone estrogen (or estrogen and progesterone) following menopause—which, typically beginning in the mid to late forties, is characterized by the cessation of monthly ovulation and a reduced production of estrogen and progesterone by the ovaries. In fact, though, almost all studies of hormone replacement have found that the treatment reduces the annual risk of death by about 50% and leads to increased longevity.

How can this increased longevity be reconciled with observations on the adverse effects of estrogen, as described above? It turns out that estrogen reduces circulating levels of cholesterol—by a mechanism not yet understood—which in turn reduces the risk of death from heart disease, the top killer of elderly women. Heart disease kills 10 times more women than do reproductive-system cancers, so any treatment that reduces heart disease is likely to increase longevity, *even if it significantly increases the risk of cancers of the reproductive system.* Estrogen replacement therapy in post-menopausal women also lowers the rate of bone loss, significantly reducing health problems related to osteoporosis.

FIGURE 24-24 Use of birth control pills decreases the incidence of ovarian cancer.

And so, in the end, while there are several strong links between hormones and health and longevity, the complex and myriad ways in which hormones can affect health make it difficult to make simple predictions about the influence of any hormone on health and longevity.

TAKE-HOME MESSAGE 24•10

Hormones affect health and longevity in complex ways. Sterilization of animals, for example, reduces levels of circulating reproductive hormones and increases longevity, usually due to reduced cancer mortality. But in other cases, such as hormone replacement therapy in women, the relationship is reversed, with treatment reducing the annual risk of death by about 50% and increasing longevity.

④ Environmental contaminants can disrupt normal hormone functioning.

Bald eagle populations were harmed by agricultural use of the pesticide DDT.

24•11

Chemicals in the environment can mimic or block hormones, with disastrous results.

Beginning in 1939 and continuing for more than three decades, the chemical DDT was used as a pesticide against insects, with widespread agricultural use. It is extremely effective in killing a variety of insects, including mosquitoes, which made it popular in strategies against malaria—a disease transmitted by insects. Unfortunately, after DDT killed the intended insect pests, it remained in the environment. Animals that consumed insects exposed to DDT ingested the chemical, which was stored in their body tissues, particularly fat cells. And when those animals were consumed by others, the DDT was passed on. Because larger animals ate larger numbers of animals carrying DDT, they stored more and more of the toxic chemical.

In predators such as bald eagles, peregrine falcons, and pelicans, the DDT disrupted the development and functioning of the reproductive tract and impaired the birds' ability to produce properly functioning eggshells—the shells would crack under the weight of the parent incubating the egg (**FIGURE 24-25**). These problems led to serious population

FIGURE 24-25 Effects of DDT. The endocrine disruptor DDT causes thinning of eggshells in some birds.

declines in many bird species and may have led to the complete demise of some species.

Publication of the book *Silent Spring,* by the biologist Rachel Carson, in 1962 called attention to the negative environmental effects of the widespread use of DDT. Highlighting the many harmful effects of the pesticide, Carson's book represented the beginning of the environmental movement and incited such a public outcry that the use of DDT was eventually banned completely in the United States—a ban that is cited as instrumental in the rebound of bald eagle populations.

The discovery that DDT can disrupt normal reproduction in wildlife—and could potentially have negative effects on humans, as well—signaled the beginning of concern over **endocrine disruptors,** chemicals manufactured by humans that, when taken up by organisms, can mimic, block, or otherwise interfere with their hormones, leading to harmful effects. Endocrine disruptors' effects are a consequence of their close chemical similarity to hormones, particularly estrogen. Often, endocrine disruptors can directly bind to the same receptors as estrogen.

There are several types of endocrine disruptors, in addition to DDT. These include **PCBs** (**polychlorinated biphenyls**), which are effective as industrial coolants and lubricants; **phthalates,** which are commonly found in soft toys and cosmetics; and **bisphenol A,** found in some plastic water bottles and baby bottles (**FIGURE 24-26**).

Endocrine disruptors are found in thousands of other consumer products, too, and have been detected in numerous natural habitats, sometimes carried in runoff water from industrial manufacturing processes or as airborne pollutants. Exposure to endocrine disruptors has been implicated in a variety of adverse physiological effects, often related to the chemicals' feminizing effects, which has led to some endocrine disruptors being referred to as "gender-bending chemicals." In addition to birds, a variety of animal groups seem to be adversely affected by endocrine disruptors.

Mammals. Populations of Baltic ringed seals have declined significantly over the past 100 years. This is due, in part, to the presence of large concentrations of pollutants, including PCBs and DDT, in their habitats, which has interfered with female reproductive functioning in many ways, leading to partial or complete sterility in 70% of the animals.

Fish. In a variety of fish species, including carp, rainbow trout, and flounder, exposure to endocrine disruptors in sewage runoff is known to interfere with reproductive functioning. In some cases, the exposure has led to such a high degree of feminization of males that production of *egg* proteins occurs at greater concentrations in males than is typically found in females.

THE ADVERSE EFFECTS OF ENDOCRINE DISRUPTORS

A variety of animal groups appear to have been adversely affected by endocrine disruptors.

MAMMALS
In populations of Baltic ringed seals, endocrine disruptors have interfered with female reproductive functioning, leading to partial or complete sterility in 70% of the animals.

FISH
In a variety of fish species, including carp, rainbow trout, and flounder, exposure to endocrine disruptors in sewage runoff has been shown to interfere with reproductive functioning.

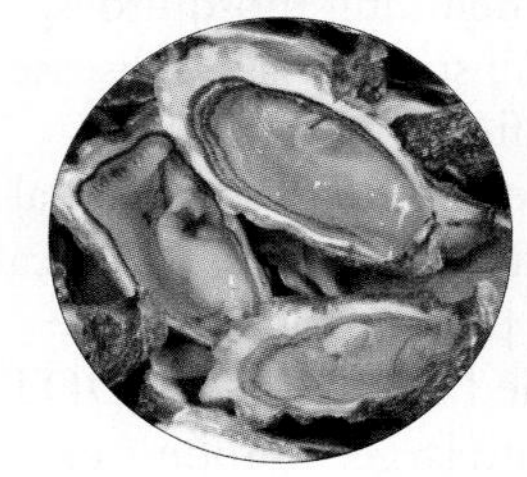

INVERTEBRATES
In marine invertebrates, endocrine disruptors have led to the production of defective shells, as well as the masculinization of female genitals, reducing fertility.

FIGURE 24-26 Some consumer products contain chemicals that act as endocrine disruptors in wildlife.

Invertebrates. A group of chemicals called tributyltin compounds, or TBTs, are used to reduce the growth of organisms, particularly microorganisms, on the hulls of ships. TBTs, however, have a side effect of acting as endocrine disruptors in several marine species. Oysters, for example, when exposed to TBTs, produce defective shells, and in numerous other marine invertebrates, females' genitals become masculinized and the animals have reduced fertility. These impacts have led to a worldwide decline in populations of gastropods such as sea snails and limpets.

It is clear that large numbers of synthetic and natural chemicals used by humans have endocrine-disrupting functions in both natural and laboratory populations of animals. It remains controversial, however, whether these chemicals cause endocrine disruption and health problems in humans. Although no direct links have been found, many studies are underway—so far, with conflicting or inconsistent results—and the U.S. government has taken many steps to restrict the use of the endocrine disruptors described here.

TAKE-HOME MESSAGE 24•11

Endocrine disruptors are chemicals manufactured by humans that, when taken up by organisms, can mimic, block, or otherwise interfere with their hormones. These chemicals can lead to a variety of adverse physiological effects, often related to their feminizing effects. Although endocrine disruptions have been demonstrated in numerous animal species, it remains controversial whether these chemicals cause endocrine disruption and health problems in humans.

streetBio

Knowledge You Can Use

Are pheromones real-life "love potions?"

Q: What are sex pheromones? Sex pheromones, such as those produced by female gypsy moths, are chemical messengers released into the environment to announce to males over a wide area that a female is ready to reproduce.

Q: Do humans produce them? The jury is out on this question. In 1971, data were published suggesting that among women living in college dormitories, pheromones, probably released from underarms, could shorten or lengthen the reproductive cycle in other women (see Section 25-7). But, despite ample evidence of sex pheromones in other species, the evidence of their existence in humans is still hotly debated.

Q: What about the claims made in other studies? A study published in 2002 claimed that human pheromones could increase the sexual attractiveness of women to men. In the double-blind, placebo-controlled study, one group of women had a purported human sex pheromone added to their perfume, while women in a control group had a placebo added. The women then recorded, over the next 3 months, their "sociosexual" behaviors—which included kissing, dating, sexual activity, and male approaches—which were then compared with observations from the same women *prior* to receiving the pheromone or placebo. The results? More than three times as many pheromone users as placebo users (74% vs. 23%) experienced an increased frequency of sociosexual behaviors. The researchers concluded that the pheromone acted as a sex attractant.

Q: Can we be certain of the claims of this one study? The double-blind and placebo-controlled aspects of this study represent important attempts at making a rigorous test of the authors' hypothesis. But the study has been criticized on several grounds. The authors tested only a small number of women ($n = 36$), within a limited age range (27.8 ± 6.7 years), over only a 3-month period. Perhaps more important, they did not disclose in their published report the chemical preparation added to the perfumes. Additionally, a co-author of the authors of this study synthesized the proprietary chemical, and currently markets it for profit. And reanalyses of the results by others challenge the significance of the results. Still, the results do suggest that there may be something in the air.

1 Hormones are chemical messengers regulating cell functioning.

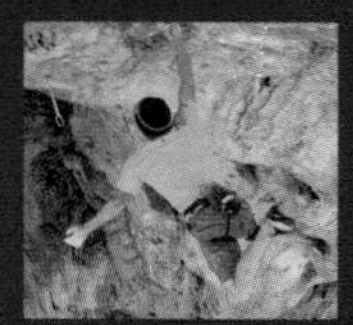

Hormones are chemical messengers, secreted by endocrine cells and endocrine glands into extracellular fluid, that influence the actions of target cells elsewhere in the body—including by influencing gene transcription; this is part of an internal communication and regulation system. Most hormones are one of two types: (1) peptide and protein hormones cannot pass through cell membranes, and their receptors are embedded in the membranes; (2) steroid hormones can diffuse through cell membranes, and their receptors are in the cytoplasm or nucleus. Complex emotions and behaviors can be influenced by hormones.

2 Hormones are produced in glands throughout the body.

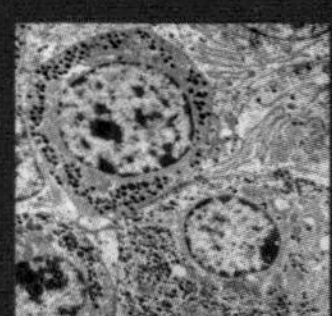

Endocrine glands throughout the body are responsible for detecting and responding to signals that reflect an organism's internal and external environments. The hypothalamus functions as a liaison between the nervous and endocrine systems, often directing the pituitary gland to release hormones with important regulatory control over body tissues. The adrenal glands regulate organisms' responses to stress. The pineal gland regulates wake and sleep cycles. The thyroid gland influences the speed and efficiency of cellular metabolism. The pancreas maintains blood glucose levels. And the gonads produce hormones responsible for numerous physical, behavioral, and emotional features, including much sexual behavior, development, and growth.

3 Hormones influence nearly every facet of an organism.

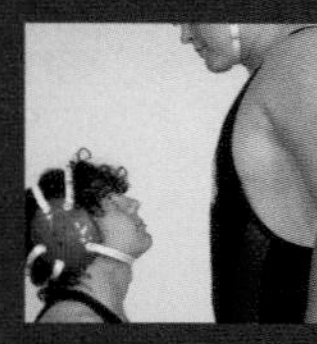

Hormones affect a person's physique and physical performance, behavior, moods, and cognitive abilities, as well as health and longevity. Hormones produce these changes by influencing gene expression and protein synthesis in cells with appropriate receptors.

4 Environmental contaminants can disrupt normal hormone functioning.

Endocrine disruptors are chemicals manufactured by humans that, when taken up by organisms, can mimic, block, or otherwise interfere with their hormones. They can lead to a variety of adverse physiological effects, often related to the chemicals' feminizing effects. Although endocrine disruptions have been demonstrated in numerous animal species, it remains controversial whether these chemicals cause endocrine disruption and health problems in humans.

KEY TERMS

anterior pituitary p. 897
bisphenol A, p. 912
calcitonin, p. 900
endocrine cell, p. 892
endocrine disruptor, p. 912
endocrine gland, p. 892
endocrine system, p. 892
hormone, p. 892
hypothalamus, p. 896
mood, p. 904
paracrine regulator, p. 895
parathyroid glands, p. 900
PCBs (polychlorinated biphenyls), p. 912
pheromone, p. 893
phthalates, p. 912
pituitary gland, p. 896
posterior pituitary, p. 896
target cell, p. 892
thyroid gland, p. 900

CHECK YOUR KNOWLEDGE

1. The target cells of the hormone _______ are located in a woman's breasts and uterus. This hormone causes its target cells to contract, which releases milk from the breast cells to feed an infant and causes the painful contractions of the uterus that push out the newborn during labor.
   a) progesterone
   b) oxytocin
   c) estrogen
   d) ACTH
   e) prolactin

2. Human sex hormones are classified as which type of biological molecule?
   a) enzyme
   b) protein
   c) carbohydrate
   d) lipid
   e) nucleic acid

3. Why are steroid hormone complexes with their receptors able to directly bind to DNA to influence gene expression, whereas peptide hormones must influence gene expression indirectly?
   a) Peptide hormones are made from amino acids, which are also used to make proteins, and proteins cannot bind to DNA. Steroid hormones are not made from amino acids, and therefore they can directly interact with DNA.
   b) Steroid hormones are cholesterol-based and therefore can block arteries, causing cardiovascular disease. Directly binding to DNA allows them to complete their function and to be removed from the circulatory system as soon as possible.
   c) Steroid hormones are lipid-based, allowing them pass through the lipid-composed cell membrane to enter the cell and directly bind to DNA. Peptide hormones are not lipid-based and therefore cannot pass through the cell membrane.
   d) Steroid hormones are smaller than peptide hormones and can fit through the holes in the cell membrane to enter the cell.
   e) Neither steroid nor peptide hormones directly bind to DNA; both influence gene expression indirectly.

4. The _______ is a pea-sized gland located below the hypothalamus in the brain and is one of the controlling glands of the endocrine system.
   a) testis
   b) pancreas
   c) thyroid gland
   d) adrenal gland
   e) pituitary gland

5. The hypothalamus secretes which two hormones into the posterior pituitary gland, from which they enter the bloodstream and act on their target cells?
   a) ADH and TSH
   b) oxytocin and ADH
   c) prolactin and oxytocin
   d) pheromone and ACTH
   e) FSH and LH

6. The following pairs link a human gland with the hormone(s) it secretes. Which pairing is incorrect?
   a) adrenal cortex: cortisol
   b) thyroid: oxytocin
   c) pancreas: insulin
   d) ovary: estrogen and progesterone
   e) testis: testosterone

7. The pituitary gland is actually a fusion of two glands. Which of the following statements about its origins is correct?
   a) The posterior pituitary is an outgrowth of the cerebellum; the anterior pituitary develops from the pineal gland.
   b) The anterior pituitary is an outgrowth of the hypothalamus; the posterior pituitary is an outgrowth of the cerebellum.
   c) The posterior pituitary is an outgrowth of the hypothalamus; the anterior pituitary is an outgrowth of the cerebellum.
   d) The posterior pituitary is an outgrowth of the hypothalamus; the anterior pituitary develops from a fold of tissue at the roof of the embryonic mouth.
   e) The anterior pituitary is an outgrowth of the hypothalamus; the posterior pituitary develops from a fold of tissue at the roof of the embryonic mouth.

8. The _______ gland releases melatonin, which is a hormone important for reproduction in certain mammals.
   a) thyroid
   b) adrenal
   c) pituitary
   d) pancreatic
   e) pineal

9. The target tissues or organs of insulin are:
   a) the testes and ovaries.
   b) muscles and the liver.
   c) blood and bones.
   d) blood and the pancreas.
   e) muscles and the pancreas.

10. Blood samples from an individual who has fasted for 24 hours would have:
    a) high levels of insulin and low levels of glucagon.
    b) high levels of both insulin and glucagon.
    c) low levels of both insulin and glucagon.
    d) high levels of glucagon and low levels of insulin.
    e) high levels of glucagon and standard levels of insulin.

11. Many female bodybuilders take steroids (testosterone) to increase muscle mass and achieve their body shape. The consequences can include all of the following except:

a) increased competitiveness and aggressiveness.
b) increased concentration.
c) decreased object memory.
d) decreased fertility.
e) increased hair growth.

12. If a high dose of time-released testosterone is given to a male songbird, which of the following is not a likely effect?

a) He would obtain a larger territory than males with lower levels of circulating testosterone.
b) He would get into many fights with larger males.
c) He would have a longer life span than most other males.
d) He would enjoy increased reproductive success.
e) He would develop a greater muscle mass than other males of similar body weight.

Short-Answer Questions

1. Compare and contrast peptide and protein hormones with steroid hormones.

2. How do the hypothalamus, posterior pituitary, and anterior pituitary interact, and what role do they play in the endocrine system?

3. In what ways do estrogen and testosterone influence mood? How can sex hormones, including these two hormones, affect life expectancy?

4. What are the general effects of endocrine disruptors? Why can even small quantities in the environment be harmful to some organisms?

5. For each gland, list the hormone(s) they produce and their function.

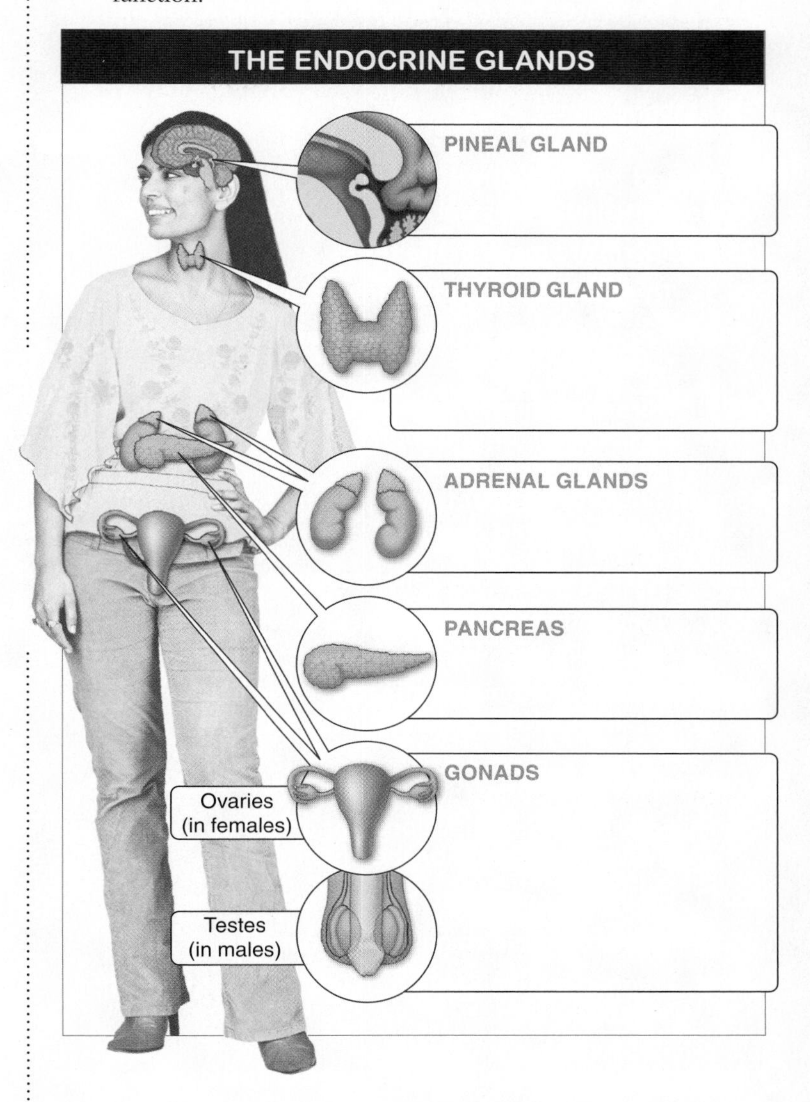

See Appendix for answers. For additional study questions, go to www.prep-u.com.

1 How do animals reproduce?

25.1 Reproductive options (and ethical issues) are on the rise.

25.2 There are costs and benefits to having a partner: asexual versus sexual reproduction.

25.3 Fertilization can occur inside or outside a female's body.

2 Male and female reproductive systems have important similarities and differences.

25.4 Sperm are made in the testes.

25.5 There is unseen conflict among sperm cells.

25.6 Eggs are made in the ovaries (and the process can take decades).

25.7 Hormones direct the process of ovulation and the preparation for gestation.

3 Sex can lead to fertilization, but can also spread sexually transmitted diseases.

25.8 In fertilization, two cells become one.

25.9 Numerous strategies can help prevent fertilization.

25.10 Sexually transmitted diseases reveal battles between microbes and humans.

4 Human development occurs in specific stages.

25.11 Early embryonic development occurs during cleavage, gastrulation, and neurulation.

25.12 How does an embryo become male or female?

25.13 There are three stages of pregnancy.

25.14 Pregnancy culminates in childbirth and the start of lactation.

5 Reproductive technology has benefits and dangers.

25.15 Assisted reproductive technologies are promising and perilous.

25

Reproduction and Development

From two parents
to one embryo
to one baby

1 How do animals reproduce?

The endangered black rhinoceros and calf.

25•1

Reproductive options (and ethical issues) are on the rise.

Thirty-year-old Diane Blood wanted to have a child with her husband, Stephen. And then the High Court in England ruled that she could not. The problem: her husband was dead.

Stephen Blood had died of bacterial meningitis two years earlier, but when he was in a coma the day before he died, his wife asked the doctors to take sperm from him, and the sperm sample was frozen and stored. The Human Fertilisation and Embryology Authority, however, ruled that because he hadn't given written consent, his sperm could not be used.

Q Are frozen embryos divided up like other marital assets at divorce?

This case illustrates a type of ethical and legal quandary that is becoming increasingly common (**FIGURE 25-1**). Technology is making conception and pregnancy possible in many situations where they previously were not possible. But along with many happy outcomes, there are also numerous complex legal battles and, as yet, few consistent legal decisions on such matters. Here are some examples.

- Do couples have the right to make contracts with a surrogate mother, a woman who carries and gives birth to their child? And, if so, can they pay her a salary to do so? In North Dakota, the surrogate actually becomes the legal parent of the child. In Washington, DC, and in many states, a couple can pay the surrogate's expenses, but nothing more.
- Do egg donors have any rights (and responsibilities) with respect to the children conceived with their eggs? Is the situation the same for sperm donors? Several states have ruled differently on these issues.
- Do children conceived from sperm or egg donors have a right to know who the donor was?
- If a couple has embryos created and frozen for later implantation but then gets divorced, who retains custody of the embryos? In one Tennessee case, custody was given to the mother, against the wishes of the father. A Texas court, on the other hand, refused to give frozen embryos to the mother. With almost half a million frozen embryos in the United States, this is likely to be an increasingly common issue for courts to decide.

In this chapter, we describe how men produce sperm and women produce eggs, as well as the process by which fertilization occurs (or does not occur). We also explore the early stages of development, following fertilization, including

FIGURE 25-1 **In the headlines: ethical issues and reproductive technologies.**

the steps by which an embryo is triggered to develop as a male or a female. We also investigate the perils and promise of a variety of assisted reproductive technologies.

But first, a happy resolution to the case we opened with. Diane Blood was eventually allowed to take the sperm to another country where she could be inseminated (after re-mortgaging her house to pay for the expensive legal battles). The law in England still prevents storage of sperm from a man without his written consent, but an appeals court made an exception in her case, and Diane Blood was able to conceive a child (and, three and a half years later, another child) with her husband's sperm.

TAKE-HOME MESSAGE 25•1

Technology is making conception and pregnancy possible in many situations where they previously were not possible. But it is simultaneously giving rise to complex legal battles and ethical dilemmas

25•2

There are costs and benefits to having a partner: asexual versus sexual reproduction.

The term reproduction, the biological process by which new organisms are produced from existing organisms, usually conjures images of a male and a female, contributing equal amounts of genetic material and producing offspring together. And for most animals, this is how it's done. But recall from Section 6-14 that there are two fundamentally different ways in which organisms can reproduce. While the vast majority of plants and animals reproduce sexually, prokaryotes and many plant and animal species reproduce asexually—and some can reproduce in both ways.

Asexual reproduction involves the production of offspring by a single individual without a contribution of genetic material from another individual. There are several types of asexual reproduction in animals, including parthenogenesis, budding, and fragmentation (**FIGURE 25-2**).

Q Do female turkeys need males to reproduce?

Parthenogenesis. In **parthenogenesis,** a female's egg develops into a new organism without ever having to be fertilized by a sperm cell. Some species can reproduce asexually or sexually, depending on

TYPES OF ASEXUAL REPRODUCTION

Asexual reproduction involves the production of offspring by a single individual without contribution of genetic material from another individual.

PARTHENOGENESIS
A female's egg develops into a new organism without ever having to be fertilized by a sperm cell.

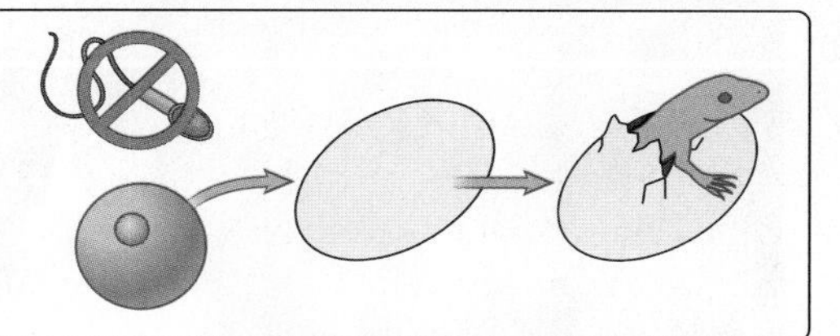

BUDDING
An offspring grows right out of the body of the parent.

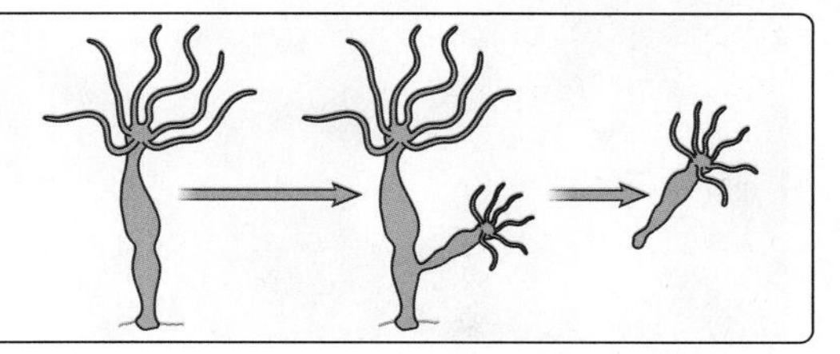

FRAGMENTATION
A parent breaks into multiple pieces, and each develops into a fully functioning, independent individual.

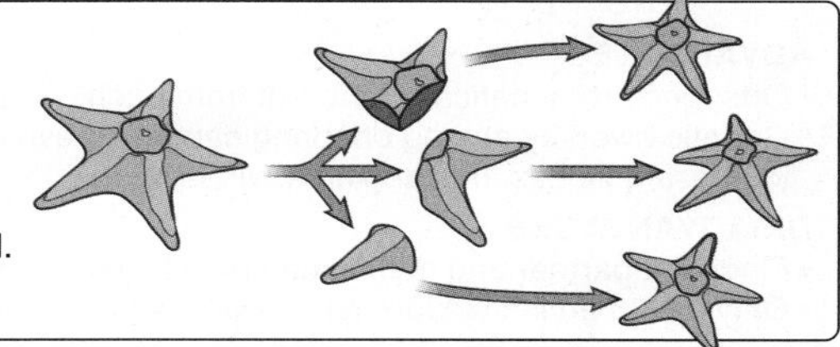

FIGURE 25-2 **Asexual reproduction: parthenogenesis, budding, and fragmentation.**

environmental conditions; parthenogenesis allows them to utilize resources as quickly as possible. Aphids, for example, produce eggs that develop into normal adults without fertilization in the spring, when food is plentiful. When food is more limited, the eggs are fertilized before development. Hammerhead sharks and, occasionally, turkeys also have the capability to reproduce by parthenogenesis (although turkeys resulting from asexual reproduction tend to be less healthy). Other animal species, including the desert grassland whiptail lizard, are exclusively asexual, and all of the individuals in the species are female. Among these lizards, though, ovulation rates are increased by female-female courtship rituals and behavior called "pseudocopulation" that resembles the male-female mating seen in related species.

Budding. In budding, an offspring grows directly out of the body of the parent. Hydras, predatory cnidarians, reproduce by budding.

Fragmentation. In fragmentation, a parent breaks into multiple pieces, each of which develops into a fully functioning, independent individual. Fragmentation is seen among many species of flatworms, as well as some sea stars. Among some sea stars, for example, if even a tiny part of one arm breaks off, it can develop into a complete individual.

In contrast to asexual reproduction, **sexual reproduction** involves two individuals contributing genetic material to produce offspring (**FIGURE 25-3**). The genetic material is contained in gametes, the reproductive cells. The male gamete is called a **sperm** (or sperm cell) and the female gamete is called an egg or **ovum** (*pl.* **ova**). Recall from

SEXUAL REPRODUCTION

Sexual reproduction involves two individuals contributing genetic material to produce offspring. The genetic material is contained in gametes, the reproductive cells.

EGG
- Female gamete
- Haploid (one copy of each chromosome)

SPERM
- Male gamete
- Haploid (one copy of each chromosome)

FERTILIZATION

FERTILIZED EGG
- Diploid (two copies of each chromosome)

FIGURE 25-3 In sexual reproduction, two parents contribute genetic material to the offspring. The fusion of female gamete (egg) and male gamete (sperm) forms a zygote that potentially develops into offspring.

SEXUAL vs. ASEXUAL REPRODUCTION

SEXUAL REPRODUCTION

ADVANTAGES
- Offspring are genetically different from each other and from either parent.
- Genetic diversity among offspring can be an evolutionary adaptation, increasing fitness in changing environments.

DISADVANTAGES
- Finding a partner and mating can be difficult and time-consuming.
- Only half of an individual's alleles will be passed to its offspring.

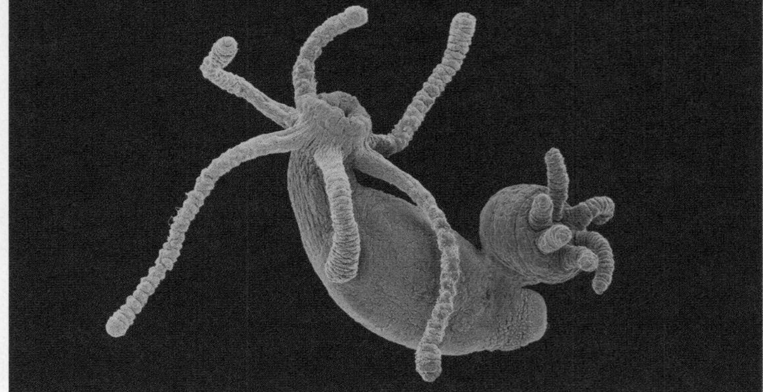

ASEXUAL REPRODUCTION

ADVANTAGES
- Reproduction is fast and efficient.
- All of an individual's alleles are passed on to its offspring.

DISADVANTAGES
- With a changing environment, individuals producing genetically diverse offspring are more likely to have offspring suited to the environment.

FIGURE 25-4 Genetic variation versus efficiency: advantages and disadvantages of sexual and asexual reproduction.

Section 6-11 that the cellular division process of meiosis produces gametes, cells that contain only half as many sets of chromosomes as other body cells (the somatic cells). When the male and female gametes fuse in **fertilization,** the full chromosome number is restored to the diploid condition. Both the production of the gametes and the combination of genetic material from two individuals at fertilization tend to increase the genetic diversity among offspring. We explore sexual reproduction in greater detail in the remainder of this chapter.

An important feature of sexual reproduction is that it leads to offspring that differ genetically from each other and from either parent (see Figures 6-21 and 6-23). This genetic diversity among an individual's offspring can be an evolutionary adaptation, increasing fitness in changing environments. If an environment is gradually changing, individuals producing diverse offspring increase the likelihood that one of their offspring will be suited, genetically, to the new environment. There are, however, disadvantages to sexual reproduction. The two main drawbacks are that (1) finding a partner and mating can be time-consuming and dangerous, and (2) an individual contributes only half of the alleles that its offspring carry (**FIGURE 25-4**).

With asexual reproduction, the advantages and disadvantages are more or less reversed. It can be fast and easy, because it involves only a single individual. And if an organism's environment is stable, it is beneficial for offspring to carry all of the genes that their parent carried. If an environment is changing, however, asexually reproducing organisms may be at a disadvantage.

TAKE-HOME MESSAGE 25•2

Organisms can reproduce in two ways: asexually or sexually—or both. Asexual reproduction, which can be fast and efficient, leads to offspring genetically identical to the parent; it occurs in all prokaryotes and in many plant and animal species. Sexual reproduction, which leads to offspring that are genetically different from each other and from either parent, occurs in the vast majority of plant and animal species.

25•3

Fertilization can occur inside or outside a female's body.

Sexual reproduction, as we've seen, leads to the production of genetically diverse offspring and so can have evolutionary advantages. But because sexual reproduction requires male and female gametes to come together at fertilization, it also presents a challenge: the male and female gametes must somehow get to each other. Two general strategies have evolved as a consequence of this challenge: **external fertilization,** in which the sperm and egg unite outside the male's and female's bodies, and **internal fertilization,** in which the sperm are deposited directly in the female's reproductive tract and meet and unite with eggs inside the female's body (**FIGURE 25-5**).

Q Male frogs clutch onto female frogs for months at a time without letting go. Why?

The first vertebrates evolved in the oceans. In this environment, it was possible for females to produce and release batches of eggs right into the water. Males could then release sperm into the water, where fertilization could take place. Many aquatic invertebrates—including sea urchins and clams—along with most fishes and amphibians use external fertilization to bring sperm and eggs together.

Although seawater is not harmful to sperm or eggs, one potential problem for organisms using external fertilization is that the tiny gametes of one sex can be very quickly washed away from those of the other sex when the gametes are released into water. For this reason, males and females of a species tend to produce very large numbers of gametes and release their gametes at the same time and very near each other. A variety of cues help to synchronize the release of gametes by males and females, including water temperature, the phase of the moon, day length, chemicals released by one or the other sex, and courtship rituals.

Among the most extreme tactics employed for ensuring that sperm and eggs are released at the same time and in the same place is something called

EXTERNAL vs. INTERNAL FERTILIZATION

EXTERNAL FERTILIZATION
The sperm and egg unite outside of the male's and the female's body.

INTERNAL FERTILIZATION
Sperm are deposited directly in the female's reproductive tract and unite with the eggs inside the female's body.

FIGURE 25-5 Methods of fertilization. Frogs deposit large quantities of sperm and eggs into the water, where some are fertilized. In mammals, such as bears, fertilization takes place within the reproductive tract of the female.

amplexus, used by most species of frogs. In amplexus, the male embraces the female frog from behind, wrapping his front legs around her body. He then holds on until the female releases her eggs, at which point he releases his sperm and fertilizes them. This doesn't sound all that remarkable, except that males will sometimes hold on to a female for weeks or even months at a time!

As we learned in Chapter 11, the colonization of land by vertebrates opened up a huge number of new niches, but presented one huge problem for animals. External fertilization just doesn't work well on land. First, gametes cannot be moved around (and thus toward each other) without water. Secondly, and even more important, gametes quickly dry out on land. For these reasons, there was strong selective pressure on land-colonizing vertebrates for a new method to evolve: internal fertilization.

Used by nearly all terrestrial animals, internal fertilization solves the problem of gamete desiccation (drying out) by having males deposit their sperm directly in the reproductive tract of females. This deposit is usually accomplished by the male placing his reproductive organ within the female's reproductive tract, in the act of **copulation.** Inside the female's body, there is sufficient moisture for the sperm to remain viable until one or more are able to fertilize her eggs.

Once an egg is fertilized by a sperm, the fertilized egg is called a zygote. Following the first division into two cells (and

The kangaroo is a viviparous animal. The embryo develops inside the womb of the mother for approximately 33 days when it emerges blind, hairless and just a few centimeters long. It finishes development in its mother's protective pouch—staying in the protective environment for many months.

EMBRYONIC DEVELOPMENT

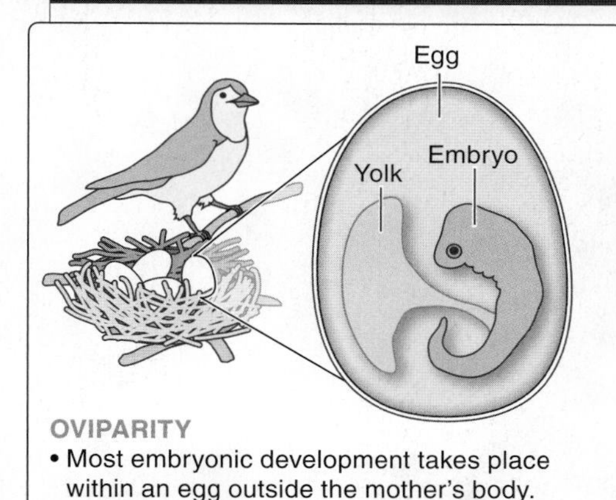

OVIPARITY
- Most embryonic development takes place within an egg outside the mother's body.
- Embryo is nourished by nutrients in the egg's yolk
- Examples: all birds; also some fishes, amphibians, reptiles, insects, and spiders

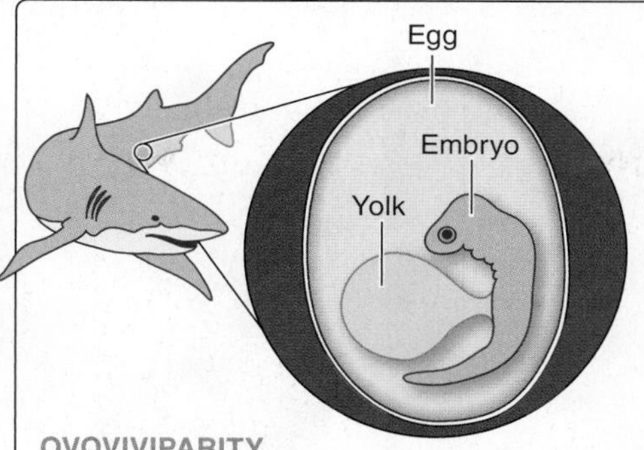

OVOVIVIPARITY
- Most embryonic development takes place within an egg that remains in the mother's body until it hatches (or is released just before hatching).
- Embryo is nourished by nutrients in the egg's yolk
- Examples: some sharks, fishes, amphibians, reptiles, and invertebrates

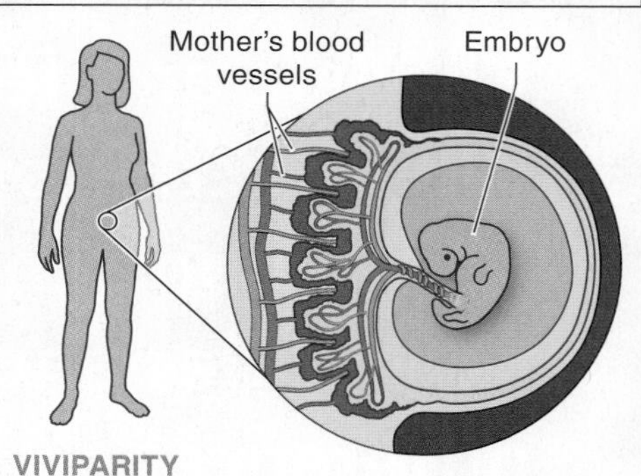

VIVIPARITY
- Most embryonic development takes place inside the mother, and live offspring are born.
- Embryo is nourished by nutrients in the mother's blood
- Examples: nearly all mammals; also some fishes, amphibians, and reptiles

FIGURE 25-6 Three strategies for protecting and nourishing a developing embryo.

continuing to approximately 8 weeks of development in humans), it is called an **embryo.** The embryo must, at some point in its development, leave the female's body. There are three different strategies for this (**FIGURE 25-6**).

1. Oviparity, a strategy in which the fertilized egg moves outside the body and most embryonic development continues there. The embryos are nourished by nutrients contained in the egg's yolk, and live offspring emerge or hatch from the egg. This developmental strategy is most familiar to us among birds (all bird species are oviparous), but it also occurs among most fishes, amphibians, reptiles, insects, and spiders.

2. Ovoviviparity, a less common strategy, in which most embryonic development takes place inside an egg, with the embryo nourished by the egg's yolk, but the egg itself remains in the female's body until it hatches (or is released just before hatching). This strategy is used by many aquatic organisms, including sharks and some other fishes, and some species of amphibians, reptiles, and invertebrates.

3. Viviparity, a strategy in which the embryo develops inside the mother, nourished by nutrients carried in her blood, and live offspring are born. This strategy is perhaps most familiar to us because it is used by humans and nearly all other mammals, but it is also occurs among some reptiles, amphibians, and fishes.

TAKE-HOME MESSAGE 25•3

Sexual reproduction requires fertilization, which occurs externally in most fishes and amphibians, and internally in most other vertebrates, including humans. Among those species having internal fertilization, development of the embryo can be nourished by yolk within an egg (that may or may not be retained within the female's body during development) or, remaining in the mother's body, by nutrients carried in her blood.

② Male and female reproductive systems have important similarities and differences.

"From Here to Eternity," 1953

25•4

Sperm are made in the testes.

Beginning at puberty and continuing for their entire lives, men produce sperm, often more than 100 million per day. The process of sperm production, called **spermatogenesis,** is similar among most mammals and requires 9–10 weeks in humans. In this section, we examine the structures of the male reproductive system and the production of sperm and **semen,** a fluid expelled at ejaculation that usually contains sperm. We begin with an examination of the male reproductive structures and the role each plays in the process of sperm formation and fertilization (**FIGURE 25-7**).

Male Reproductive Structures Externally, a male has just two reproductive structures, the penis and the scrotum. The **penis** has three columns of tissue—one on either side and a third on the bottom side—that can become engorged with blood, causing erection and making copulation possible. In most animal species, including dogs, walruses, and most primates other than humans, a bone (known as a baculum) is present in the penis and contributes to its stiffness. The **scrotum,** generally on the outside of the body in the pelvis region, is a sac containing the two **testes** (*sing.* **testis;** also called testicles), the site of sperm production. Each testis is made up of highly coiled **seminiferous tubules,** lined with cells, called spermatogonia, that are the site of sperm

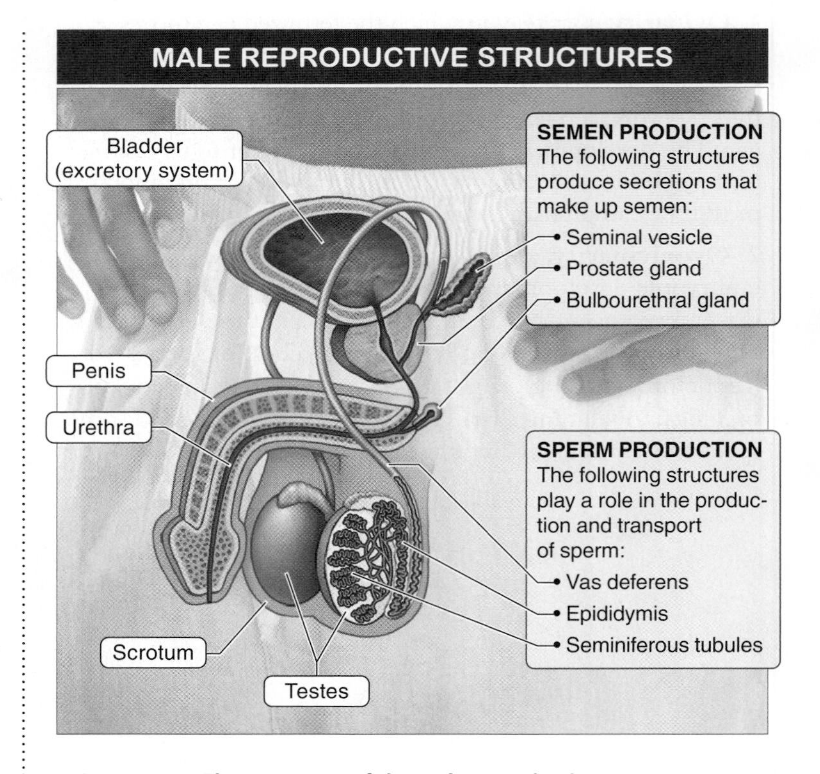

FIGURE 25-7 **The structures of the male reproductive system.**

production. It is in cells between the seminiferous tubules where **testosterone,** the principal male sex hormone, and other androgens are produced. Testosterone stimulates sperm production.

Connected to the seminiferous tubules is the **epididymis,** a 15- to 20-foot-long coiled tube, in each testis, where sperm mature. The epididymis in each testis is linked to a **vas deferens,** a tube of smooth muscle tissue that passes from the testis into the body. The vas deferens from each testis connects to a single ejaculatory duct, which continues into the urethra, a duct passing through the penis and through which semen and urine are expelled.

Three other important male reproductive structures produce secretions that make up semen. These include the **prostate gland,** located just below the urinary bladder, which secretes into the urethra a milky, basic (as opposed to acidic) fluid containing enzymes and sperm nutrients that makes up just under one-third of the volume of the ejaculate. A pair of **seminal vesicles,** secrete into the semen nutrients for the sperm, as well as substances that increase sperm motility and make the female reproductive tract more hospitable to sperm. And finally, a pair of **bulbourethral glands,** located just below the urethra near the base of the penis, contribute the remaining 1% or so of the ejaculate, as well as a mixture of mucus and sugar that lubricates the tip of the penis prior to copulation.

Gametogenesis in Males Recall from Sections 6-10 through 6-14 that gametes are produced by cells that undergo meiosis. Diploid cells, the spermatogonia, are present in the highly coiled seminiferous tubules of the testes (**FIGURE 25-8**). Each spermatogonium divides by mitosis to produce two cells.

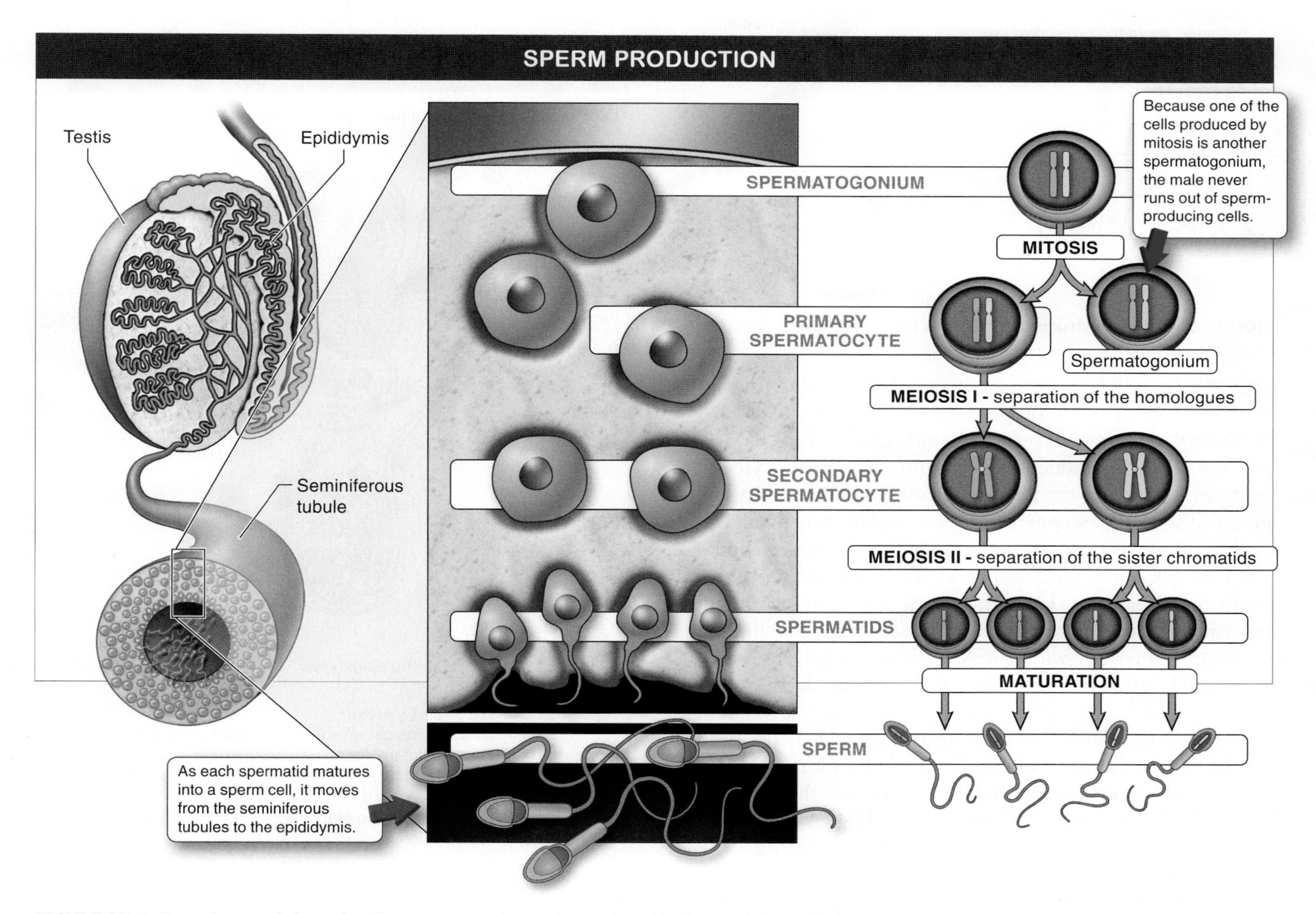

FIGURE 25-8 Gametogenesis in males. Sperm are continuously produced in the testis by meiosis.

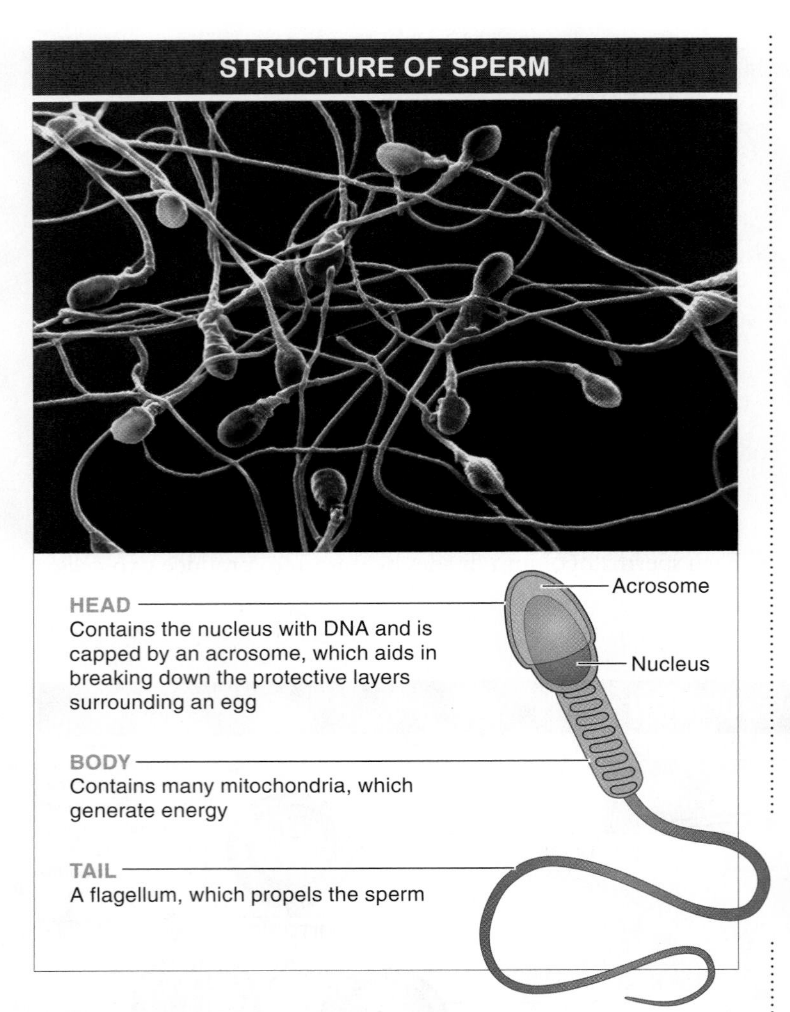

FIGURE 25-9 Sperm structure reflects its function.

One of these cells is another spermatogonium, so the male never runs out of a store of sperm-producing cells; the other is a **primary spermatocyte,** which undergoes meiosis, in the first step of sperm production. Each primary spermatocyte produces two cells in the first meiotic division. These two cells are called **secondary spermatocytes,** and they then complete the second meiotic division, each producing two spermatids. The four spermatids produced by each spermatogonium, as products of meiosis, are haploid. As each spermatid matures into a sperm cell, it moves from the seminiferous tubules to the epididymis, where, over the course of approximately 18 hours, the sperm become motile—that is, they become able to move.

Each sperm cell consists of three primary parts: (1) the head region, containing the sperm cell's nucleus with its DNA, plus a cap-like **acrosome** containing enzymes that can break down the protective layers surrounding an egg; (2) the body region, which contains many energy-generating mitochondria; and (3) the tail, a flagellum that propels the sperm through the fluid in the female reproductive tract (**FIGURE 25-9**).

Q Do hot tubs reduce a man's fertility?

Sperm production is tremendously sensitive to numerous factors. Two of the most important factors are hormones and temperature. The optimum temperature for sperm production is approximately two degrees lower than body temperature. This accounts for why, in most mammals, the testes hang outside the body, where their temperature can be controlled by moving them closer to or farther from the heat of the body. It also accounts for the fact that men exposed to hot tubs or hot baths for 30 minutes or more each week generally show signs of infertility, with fewer numbers of sperm as well as lower motility of the sperm they do have. Fortunately, the condition almost always can be reversed by reducing exposure to the hot tubs or baths.

The path that sperm take from the male to the female is as follows (**FIGURE 25-10**):

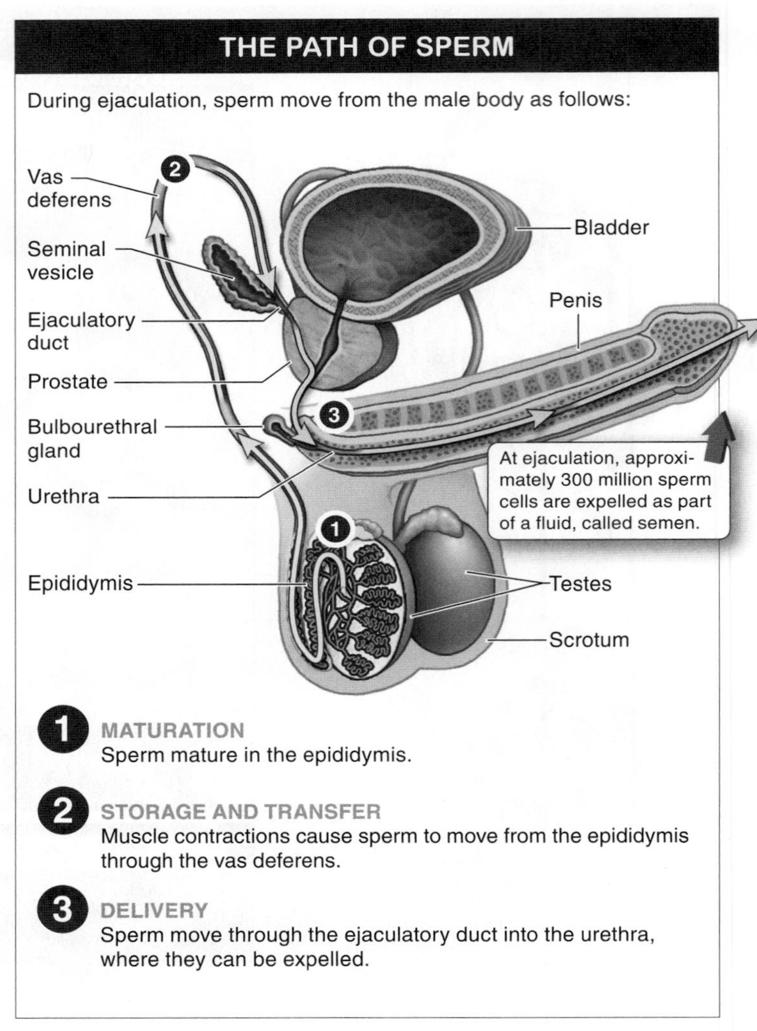

FIGURE 25-10 The pathway taken by sperm during ejaculation.

1. *Maturation:* sperm mature in the epididymis, within each testis.
2. *Storage and transfer:* during ejaculation, sperm move from the epididymis (in each testis) through the vas deferens. Their movement is generated by contractions of the muscular tissue that makes up the vas deferens.
3. *Delivery:* sperm moving from each vas deferens then pass through the ejaculatory duct and into the urethra. During copulation, sperm are expelled from the urethra into the female reproductive tract.

At ejaculation, the sperm are expelled as part of the semen. Although there are approximately 300 million sperm cells in the fluid of a single ejaculation, the sperm make up only about 1% to 5% of the total volume of ejaculated fluid. In the next section, we explore the question of why so many sperm are present in an ejaculation and the observation that a significant proportion of sperm is not actually capable of fertilizing an egg.

TAKE-HOME MESSAGE 25•4

In adult men, sperm are continuously produced in the testes by meiosis. Semen—consisting of sperm cells and fluids that nurture and aid the sperm in fertilization—is ejaculated during copulation.

25•5

There is unseen conflict among sperm cells.

In her book *The Chimpanzees of Gombe: Patterns of Behavior,* published in 1986, Jane Goodall reported female copulatory rates of "an average of between five and six copulations per female per hour in the early morning, after which the rate dropped gradually to about two per hour in the midmorning, rose very slightly during the afternoon, and tapered off to one per hour in the evening." These observations were consistent with the realization that researchers had come to in the 1970s that if females (of any species, not just chimps) mate with more than one male, there may be competition among the males' sperm.

Q A chimp's testicles are 15 times bigger than a gorilla's! Why?

The idea of sperm competition, or "sperm wars" as they have been called, gives rise to several testable predictions. One simple prediction is this: when a female is likely to mate with more than one male, the males that produce more sperm are likely to be more successful at fertilizing the female's eggs. Some interesting observations support this prediction. For example, gorillas have golf-ball-sized testes, while chimpanzees' testes are closer in size to baseballs. Gorilla testicles account for only 0.02% of body weight, while chimp testes account for 0.30%, 15 times as much. Why the huge difference? Gorilla groups are relatively small, and all females within a group mate with just one male, the dominant silverback. In contrast, as Jane Goodall noted, fertile chimpanzee females may have sex dozens of times a day with many different males. Consequently, the tiny testes of the gorilla are perfectly adequate, but for a chimp male to win these sperm competitions, he must produce significantly more sperm (**FIGURE 25-11**). Similarly, among fruit bats, males of species that

FIGURE 25-11 The evolutionary consequences of sperm competition among males. Female mating behavior influences the size of males' testes and sperm production. The inset shows the size of a male chimp's brain (top) relative to one of its testicles.

live in large social groups have significantly larger testes than males of species that live in smaller groups.

Sperm competition has given rise to several other adaptations.

- Physical barriers to copulation, such as the mating plugs produced by the coagulation of semen in many species that prevent sperm from other males entering the female's reproductive tract.

Q Why do male fruit flies produce toxic semen?

- Toxic semen components, such as the protein in fruit fly semen that suppresses further mating by the female for several days (ensuring a male's paternity), as well as the fruit fly semen components that can incapacitate the sperm of other males (and even decrease the female's lifespan by about 10%).
- Large numbers of sperm (20% to 30% of the sperm in each ejaculate in humans) differing structurally from the sperm that are able to move along the female reproductive tract and fertilize an egg. These sperm have been described as "kamikaze sperm," adapted to preventing other males' sperm from reaching and fertilizing an egg. More recent research suggests, however, that these non-fertilizing sperm may simply be abnormal sperm that reflect errors during meiosis.
- Genital morphology, such as the claspers and scrapers of dung flies, as well as the "shovel penis" of some dragonfly species that make it possible for males to dislodge the sperm of previous males that have mated with a female (**FIGURE 25-12**).

This unseen conflict between males is an area of considerable investigation as researchers try to understand the physical and behavioral consequences of such competition. It seems that females are not passive players in these battles and that, in many cases, the female will eject the sperm from undesirable males.

FIGURE 25-12 Sperm wars. This damselfly, *Calopteryx virgo,* has a "shovel penis" that can scrape away the sperm deposited by a rival. Inset shows micrograph of damselfy penis.

TAKE-HOME MESSAGE 25•5

When females mate with more than one male, sperm competition occurs and can lead to a variety of adaptations, including increased sperm production and testis size, semen that can create a physical barrier to subsequent mating, toxic semen components, and penis morphology that aids in the displacement of rival males' sperm.

25•6

Eggs are made in the ovaries (and the process can take decades).

From a genetic perspective, making eggs barely differs from making sperm. In the female gonads—the **ovaries** rather than the testes—diploid cells undergo meiosis to produce haploid eggs, and in the process, a great deal of variation is generated so that each haploid egg has a unique genetic makeup. When the haploid egg is fertilized by a sperm, the diploid condition is restored.

But the process of egg production differs from sperm production in several key ways. Eggs are produced in much smaller numbers; each egg is considerably larger than a sperm cell; and the production process can take decades rather than days, even though it begins at a much younger age. Let's explore the specifics, beginning with a description of the female reproductive system (**FIGURE 25-13**).

Female Reproductive Structures Externally, a female has just two structures, the **clitoris** and the **labia,** which develop from the same embryonic tissue that in males produces the penis and scrotum. There is also a vaginal opening. Internally, the **vagina** is a tube-like chamber into which sperm are released during copulation. The vagina connects with the **uterus,** also called the womb, which resembles an upside-down pear. The lower, narrowest portion of the uterus is called the **cervix.** The lining of the uterus, rich with blood vessels, is the **endometrium,** and this is where a fertilized egg implants and is nourished. The uterus is also where an embryo develops throughout pregnancy.

Connecting to the top of the uterus on the left and right sides are the **Fallopian tubes,** or **oviducts.** Each Fallopian tube extends outward and is funnel-shaped near its end where the ovaries lie, held in place by thin membranes. Each ovary is about the size of a large olive.

Gametogenesis in Females The process of gametogenesis in females, called **oogenesis** (pronounced oo-oh-GEN-eh-sis), starts in the ovaries prior to birth—that is, while the female is still a fetus (**FIGURE 25-14**). Here, diploid cells called **oogonia** (*sing.* **oogonium**) multiply by mitosis. Each oogonium then begins meiosis, but stops at prophase I, at which point the cell is called a **primary oocyte** and is contained within a **follicle**—the small structure in which an egg will form. At birth, there are approximately one million follicles present in a female's ovaries, each follicle containing a primary oocyte.

The primary oocytes remain in their hibernation-like state until puberty. At that point, periodic bursts of **follicle-stimulating hormone (FSH)** cause several primary oocytes to complete meiosis I. Unlike in the production of sperm, however, when the primary oocyte divides into two cells, although the pairs of homologous chromosomes split evenly,

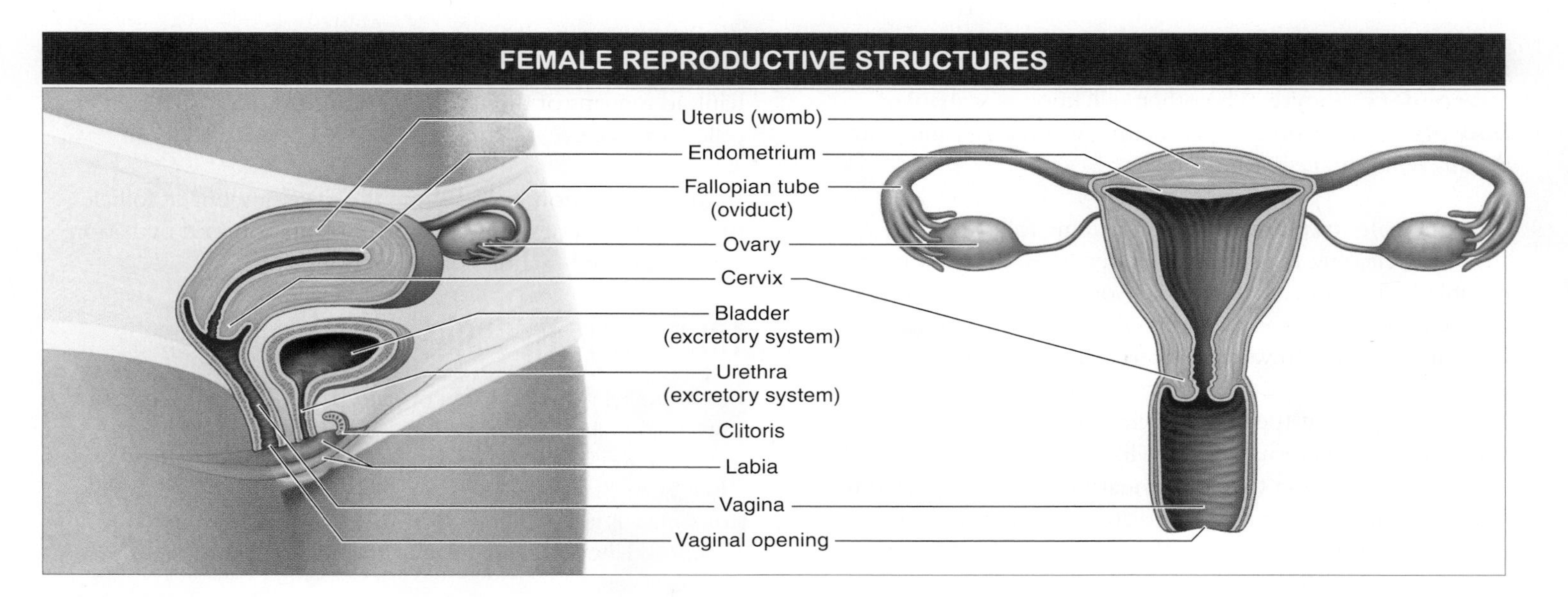

FIGURE 25-13 The structures of the female reproductive system.

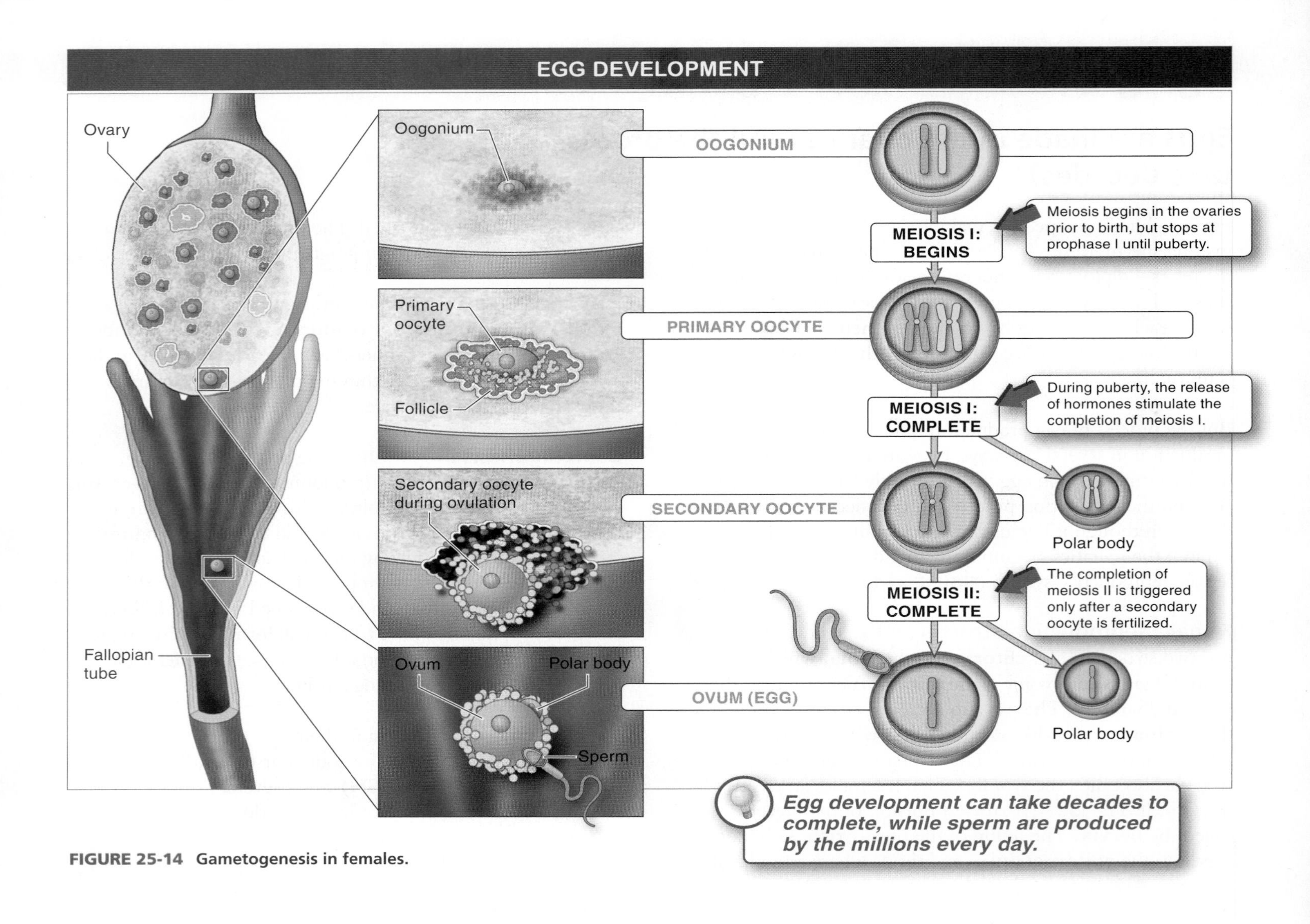

FIGURE 25-14 Gametogenesis in females.

nearly all of the cytoplasm goes to one of the cells, now called a **secondary oocyte.** The other cell, known as a **polar body,** gets almost no cytoplasm and eventually disintegrates. At this point, meiosis again stops.

When females **ovulate,** generally a single follicle in one ovary ruptures, releasing the secondary oocyte—which still has not completed meiosis. The secondary oocyte, which at this point can be called an egg, is swept by cilia into the Fallopian tube and carried down toward the uterus. If sperm have been deposited in the vagina during copulation, it is in the Fallopian tube that the sperm, swimming up from the vagina and through the uterus, are most likely to fertilize the egg. It is only *after* fertilization that a secondary oocyte is triggered to finally complete meiosis—once again with an unequal division of cytoplasm. One of the resulting cells, the ovum, receives most of the cytoplasm, while the other receives almost none. This second polar body disintegrates just as the first polar body did. The haploid ovum now fuses with the haploid nucleus of the sperm, forming a diploid fertilized egg, called a zygote.

In the next section, we see how the development of follicles, the preparation of the uterus for implantation, and ovulation are coordinated by hormone secretions.

TAKE-HOME MESSAGE 25•6

Genetically, the production of eggs barely differs from sperm production. In the ovaries, diploid cells begin to undergo meiosis, a process that continues in the Fallopian tubes following ovulation, and produces genetically varied haploid gametes. A much smaller number of eggs than sperm are produced, however, and each egg is considerably larger than a sperm cell.

25•7

Hormones direct the process of ovulation and the preparation for gestation.

If you are female, here's something you may have noticed if you are living in a dormitory (or are in prison): when women live in close proximity, over time their reproductive cycles become synchronized so that they menstruate at approximately the same time and, perhaps less obviously, ovulate at the same time. This issue was first addressed in a scientific publication in 1971, when Martha McClintock reported findings—inspired by her own experiences—based on data from 135 female students living in a dormitory at Wellesley College in Massachusetts. Similar observations have been reported in other animal species, and subsequent studies implicate airborne chemicals, called **pheromones** (probably released from women's underarms), that can shorten or lengthen the reproductive cycle in other women. It's not clear why such synchrony of reproductive cycles would occur, however. And other researchers have argued that the different lengths of women's cycles make it impossible for them to become truly synchronized. The jury is still out on this hotly contested issue.

Hormones regulate the timing and development of egg production, called the **ovarian cycle,** which occurs approximately every 28 days. Hormones also regulate another aspect of the reproductive cycle, the **menstrual cycle,** during which the uterus prepares for the possible implantation and nurturing of a fertilized egg, and sheds its lining when fertilization does not occur. We describe each of these cycles and the ways in which each influences the other (**FIGURE 25-15**).

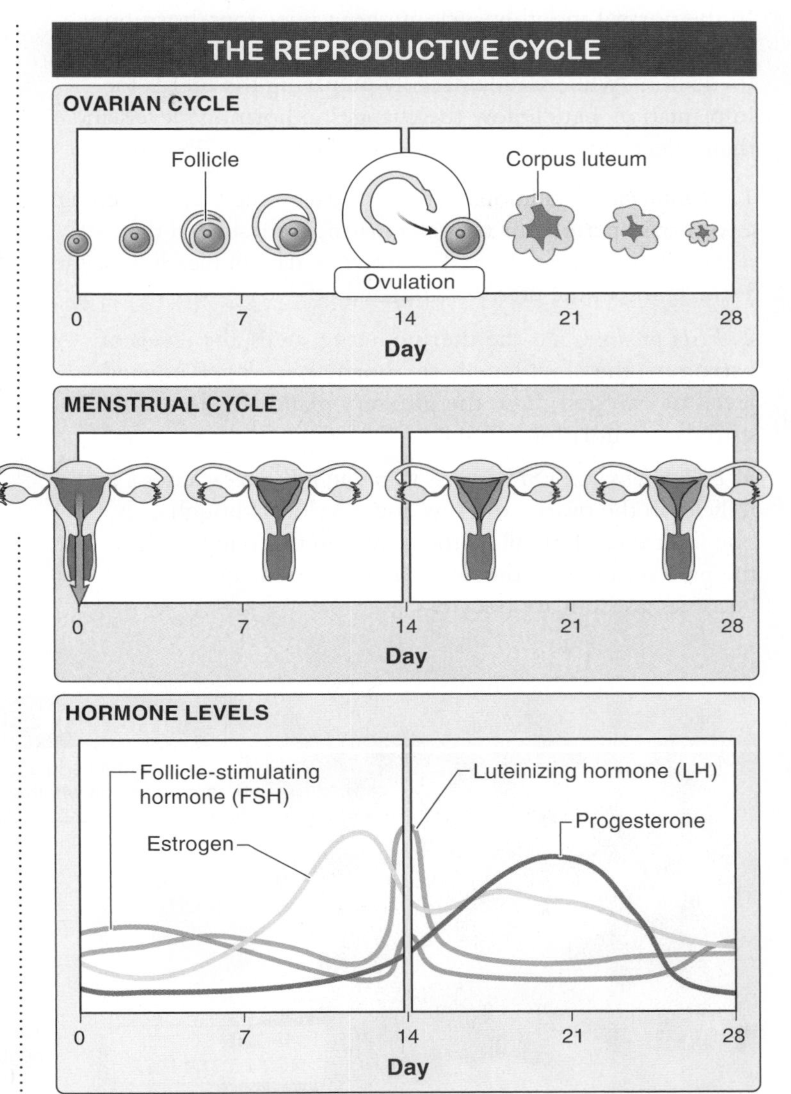

FIGURE 25-15 The stages of the female reproductive cycle and associated hormonal changes.

Q What does the process of being an egg donor entail?

Females have about one million follicles, or potential eggs, when they are born, but most women ovulate fewer than 500 times over the course of their life. This leaves a lot of potential eggs lying around. In the normal course of events, these follicles just disintegrate, but in recent decades, modern medicine has taken to tinkering with the reproductive cycle to make it possible for a woman to donate her eggs to another woman. The process involves several steps.

First, if you choose to be a donor, each day for about a week, you have an injection of a drug called Lupron, which suppresses your own reproductive cycle. The suppression is followed by a week of injections of a fertility drug that contains large amounts of two hormones, LH (luteinizing hormone) and FSH, which induces one to two dozen follicles to develop in your ovaries. A doctor examines your ovaries by ultrasound to count the number of developing follicles and to estimate their size. Finally, you are given an injection of human chorionic gonadotropin, which stimulates ovulation, the night before the egg "retrieval." The following morning, before the eggs have been released from the follicles, the doctor inserts a tiny needle into each follicle within the ovaries and sucks out the eggs, one by one. This process takes about half an hour. The donor's role is now over. The eggs are placed in a Petri dish, mixed with sperm to fertilize them, and the fertilized eggs are allowed to divide for 2–6 days before several are transferred to the recipient's uterus—with the hope that they will implant and a successful pregnancy will occur.

In the normal monthly cycle, without injections, hormones direct the development of a follicle and the release of an egg (ovulation), while simultaneously preparing the uterus for implantation. Let's follow the changes in hormone levels and their effects during the 28-day cycle (see Figure 25-15).

1. *Menstruation.* Traditionally, the first day of menstrual bleeding, a woman's "period," or **menstruation,** is considered the first day of the menstrual cycle. Three to five days of bleeding occur as the lining of the uterus is sloughed off.

2. *FSH produced.* As the uterine lining sheds, the levels of **estrogen,** the chief female sex hormone, drop. Reduced levels of estrogen cause the pituitary gland to release follicle-stimulating hormone (FSH).

3. *Follicle develops.* FSH, just as its name indicates, causes a few follicles in the ovaries to grow and develop, although only one follicle reaches full maturity. Within this one follicle, the primary oocyte completes its first meiotic division and becomes a secondary oocyte.

4. *Estrogen produced.* As the follicles develop, they produce estrogen, gradually increasing the level of estrogen in the blood.

5. *LH released.* The high levels of estrogen trigger the release of a burst of luteinizing hormone (LH) and of more FSH.

6. *Ovulation.* The burst of LH triggers ovulation, causing the ovum to erupt from the follicle and out of the ovary. This release occurs approximately halfway through the cycle, around day 14, and signals the end of what is called the **follicular phase** of the reproductive cycle.

7. *Progesterone produced.* As the second half, or **luteal phase,** of the reproductive cycle begins, the follicle cells that had surrounded the ovum develop into a structure called the **corpus luteum** (Latin for "yellow body"). These cells begin secreting smaller amounts of estrogen, but increasing amounts of progesterone.

8. *Endometrium thickens.* **Progesterone,** just as its name suggests (*pro* = for; *gestare* = to bear), causes the body to prepare for gestation of an embryo, in case fertilization occurs. Progesterone's primary effect is to cause a thickening of the endometrium, or lining of the uterus. The endometrium becomes increasingly rich with blood vessels and deposits of glycogen that can nourish a developing embryo.

At this point, the process can go in one of two directions, depending on whether or not the egg is fertilized (**FIGURE 25-16**).

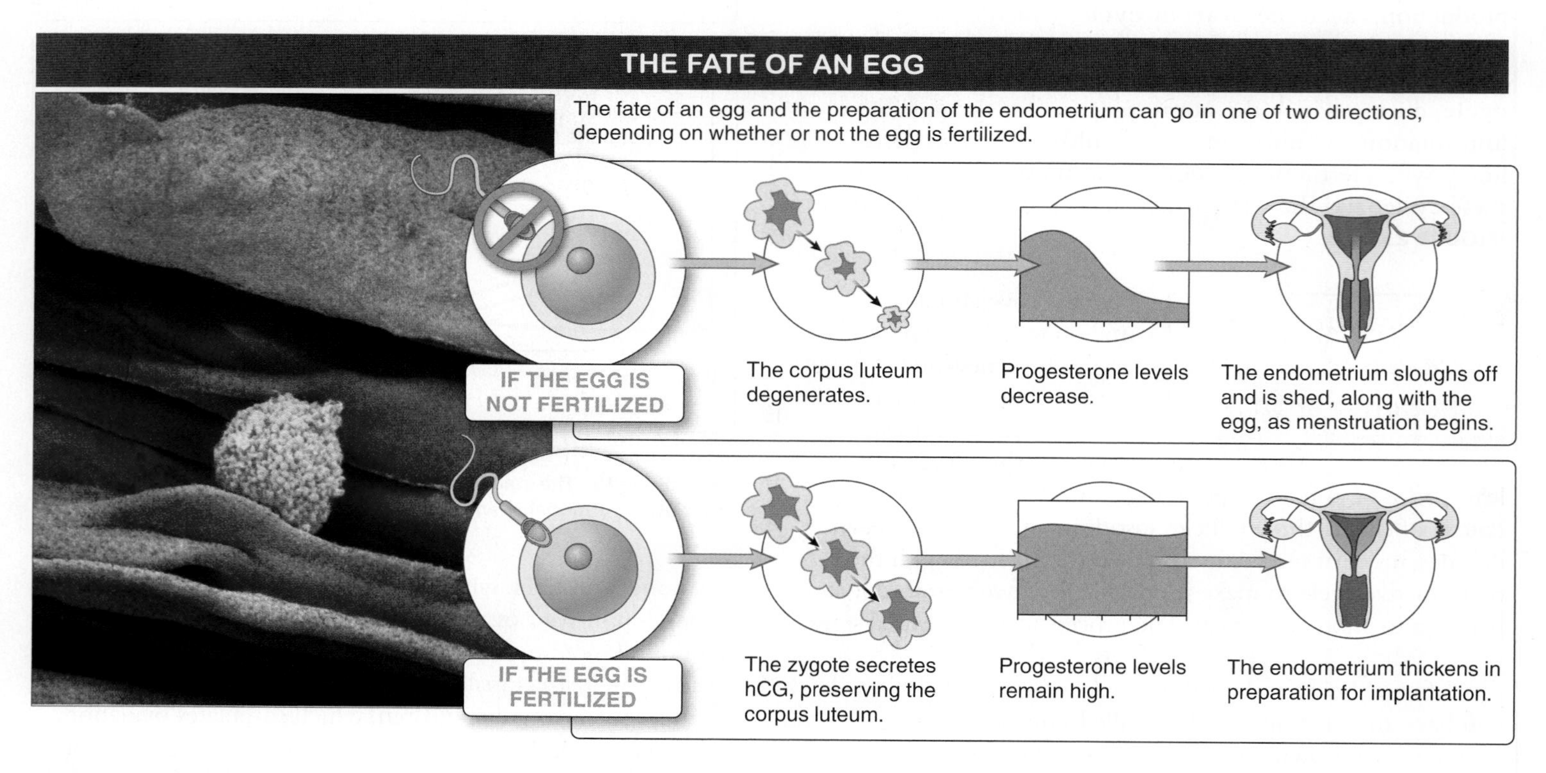

FIGURE 25-16 Two possible fates for an egg. Shown here: an egg moving down the oviduct.

Jury is still out. In 1971 Martha McClintock reported that women living together in a dormitory at Wellesley College experienced synchronized menstrual cycles. It is not clear why such synchrony would occur, and some researchers disagree that synchronization is possible.

If the egg is not fertilized, the egg disintegrates and the corpus luteum degenerates about 12 days after ovulation, sloughs off, and it is shed, along with the disintegrating egg, as menstruation begins. Abruptly removing this source of estrogen and progesterone causes the lining of the uterus to slough off, and menstruation begins. With continuing reduction in estrogen levels, the pituitary gland is spurred to release FSH and the process begins again.

If the egg is fertilized, which usually occurs in the Fallopian tube, the zygote begins development and after several days begins to secrete **human chorionic gonadotropin (hCG).** This hormone prevents degredation of the corpus luteum, which continues to secrete progesterone, thereby maintaining the endometrium. In the next section, we investigate what happens next in development of the embryo.

Reproductive cycling continues from puberty, usually in the early teens, until **menopause**—the cessation of ovulation and menstruation—usually between ages 45 and 55. It is not clear whether there is any evolutionarily adaptive value to menopause. In most species other than humans, females remain fertile throughout their entire adult life. Menopause may be a consequence of the relatively recent (on an evolutionary time scale) increase in human longevity. Regardless of the evolutionary explanation, oocyte depletion seems to influence the onset of menopause.

TAKE-HOME MESSAGE 25·7

In the ovaries, a cell within a follicle is stimulated to develop into a fertile egg by coordinated secretions of the hormones estrogen, FSH, and LH. The preparation of the uterus for implantation of a fertilized egg is coordinated by progesterone.

3 Sex can lead to fertilization, but can also spread sexually transmitted diseases.

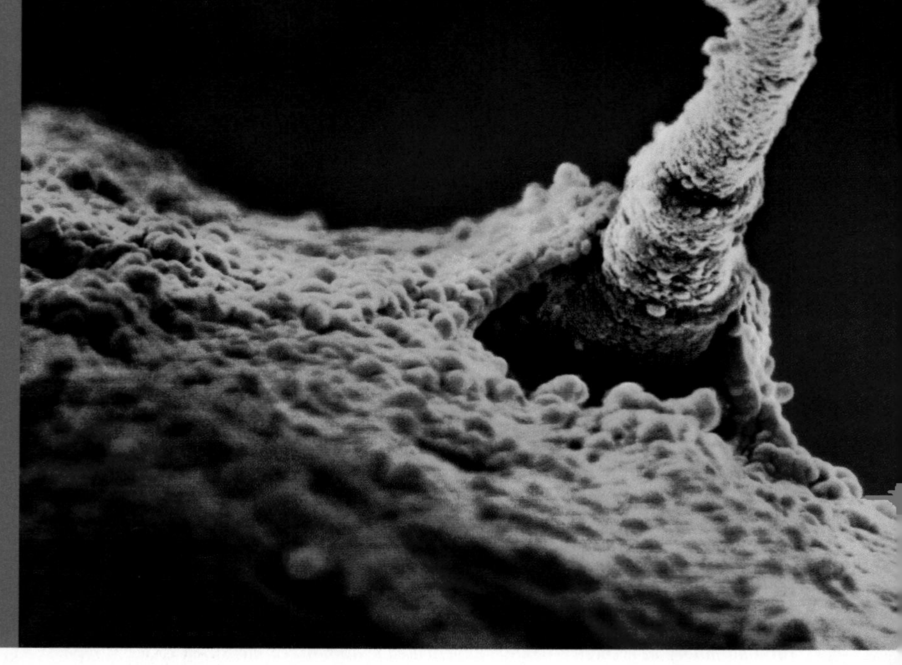

Sperm fertilizing an egg.

25•8

In fertilization, two cells become one.

For fertilization to occur, it's not enough for sperm cells to be in the same general location as an egg—there are some specific events that must occur. The process of fertilization in vertebrates involves three separate steps: penetration, activation, and fusion of the nuclei (**FIGURE 25-17**). And before a sperm can take that first step and penetrate an egg, it must find it.

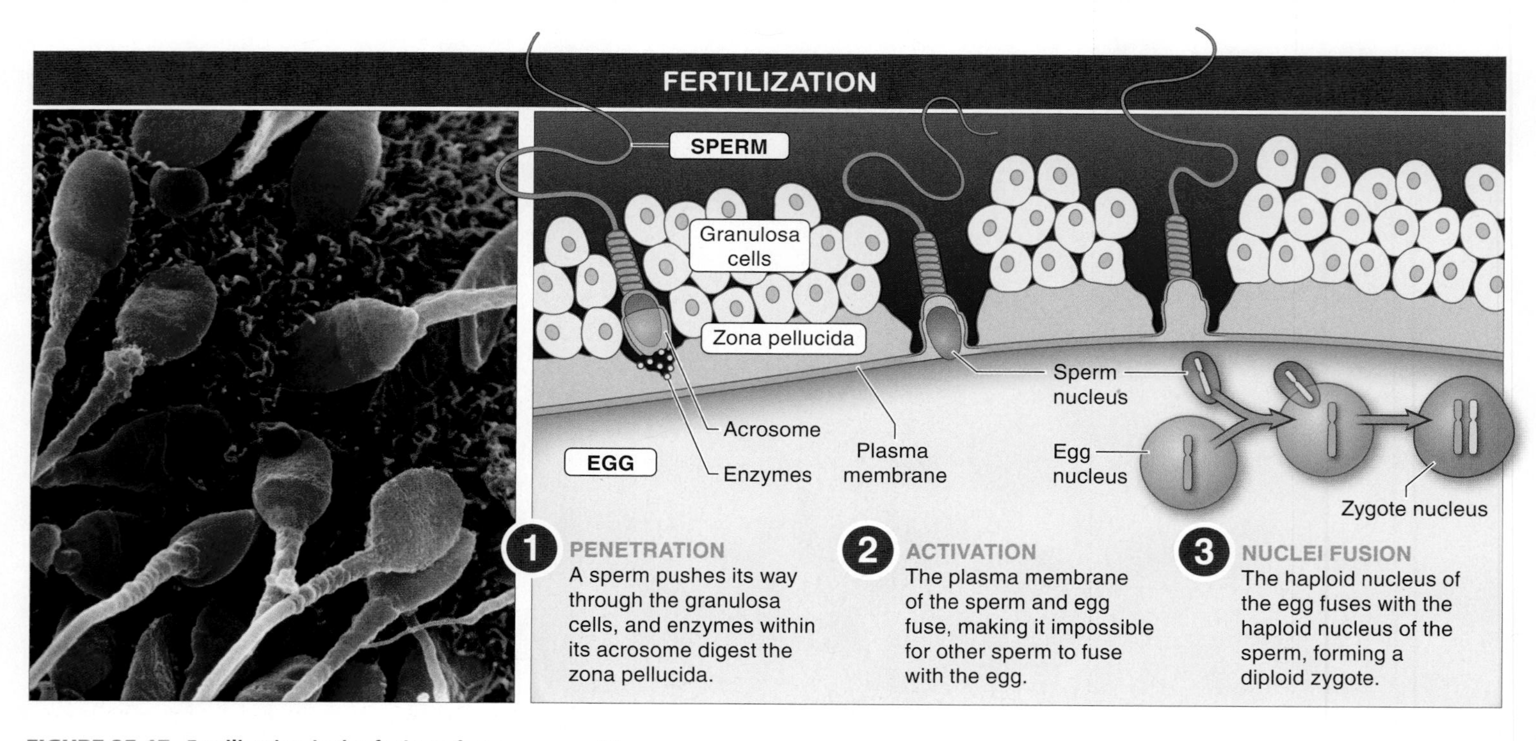

FIGURE 25-17 Fertilization is the fusion of one sperm with an egg.

Recently, researchers discovered that sperm have receptors, like chemical sensors, that cause the sperm cells to swim toward a chemical attractant that is released by the egg. This attractant, in essence, says, "I'm over here!"

After ovulation, the egg is still surrounded by some smaller cells, called granulosa cells. These cells stand between the sperm and the egg, but they're not the only barrier. Between the granulosa cells and the egg's membrane, there is a glycoprotein layer called the **zona pellucida** (pronounced puh-LOO-sih-duh). As sperm approach the egg, release of calcium by cells surrounding the egg appears to increase the swimming speed and tail movements of the sperm, in a process known as sperm activation. A second phase of sperm activation occurs as a sperm cell pushes its way through the granulosa cells, and the acrosome (at the head of the sperm) dissolves and releases digestive enzymes that help the sperm digest its way through the zona pellucida.

Once the sperm gets through the zona pellucida, the oocyte is activated, and the plasma membranes of the sperm and egg fuse. This fusion has a couple of important consequences. First, it changes the egg membrane in such a way that it is impossible for any other sperm to also fertilize the egg. After all, it would be a genetic disaster for the fertilized egg to have more than two sets of chromosomes. Second, the fusion of the plasma membranes of sperm and egg induces the egg to complete its second meiotic division. On doing so, one haploid egg is formed, along with one smaller polar body. The latter disintegrates or is ejected from the egg.

In many species, stripping away the zona pellucida makes it possible for the egg to be fertilized by sperm from different species. This suggests that something in the zona pellucida, perhaps recognition sites, makes it possible for the egg to be fertilized only by sperm from the same species.

And, finally, the haploid nucleus of the egg fuses with the haploid nucleus of the sperm, creating a diploid cell—the **zygote.**

TAKE-HOME MESSAGE 25•8

At fertilization, a sperm penetrates the protective zone around the egg, the egg blocks additional sperm entry, and the sperm and egg membranes fuse. The egg then completes its second meiotic division, and the haploid nuclei of the egg and sperm fuse, forming a diploid zygote.

25•9

Numerous strategies can help prevent fertilization.

Pregnancy depends on the occurrence of a great many events, relating to the production of sperm and eggs, their coming together in fertilization, and implantation of the zygote in the uterus. **Contraception,** or birth control, is the attempt to prevent pregnancy. Although a wide variety of contraception methods are used—with varying degrees of effectiveness—each method generally represents one of three general strategies: (1) preventing ovulation, (2) preventing fertilization, or (3) preventing implantation. **FIGURE 25-18** summarizes various methods of contraception, how they work, and their failure rates.

Preventing Ovulation

Birth control pills. Available since 1960, birth control pills are among the most effective of all methods of contraception, with a failure rate of 1% to 8%; that is, among 100 women using birth control pills over the course of one year, there would be 1-8 pregnancies. (If taken every day at exactly the same time, they are 99+% effective; the problem is that not all women are consistent in taking them.) Among 100 women using no contraception, there would be about 85 pregnancies during one year.

Q How do birth control pills work?

Two chief varieties of birth control pills are available, one that contains synthetic versions of the hormones estrogen and progesterone, and one that contains only a synthetic progesterone. The estrogen-progesterone pill prevents ovulation by keeping estrogen levels just high enough so that the release of FSH by the pituitary gland is never triggered. As long as FSH is never released, eggs never develop and ovulation does not occur. The progesterone component of the pill causes just enough development of the lining of the uterus that a plug of mucus forms at the connection between the vagina and uterus, blocking sperm from getting through.

Taking the birth control pill at the same time every day is essential. If more than 24 hours go by between

METHODS OF CONTRACEPTION

	METHOD	HOW IT WORKS
PREVENTING OVULATION	• Birth control pills	A pill containing synthetic hormones prevents the release of FSH by the pituitary gland, thus preventing egg development and ovulation (effectiveness: 92–99%).
	• Hormone implants or injections	Synthetic hormones are inserted under the skin to prevent the release of FSH by the pituitary gland, thus preventing egg development and ovulation (effectiveness: 97–99%).
PREVENTING FERTILIZATION	• Condoms	A thin rubber or natural membrane sheath placed on the penis or inside the vagina, covering the cervix, prevents sperm from coming in contact with an egg (effectiveness: male condoms: 85–98%; female condoms: 79–95%).
	• Diaphragm/ cervical cap	A dome-shaped piece of rubber placed in the vagina blocks the cervix and prevents sperm from coming in contact with an egg (effectiveness: 84–94%).
	• Sterilization	A medical procedure permanently alters the reproductive system to prevent the release of sperm or block the movement of eggs down the Fallopian tubes (effectiveness: >99%).
	• Abstinence	Individuals refrain from sexual intercourse (effectiveness: complete abstinence: 100%; abstinence during days when fertility is likely, based on analysis of female's menstrual cycle pattern: 75–99%).
PREVENTING IMPLANTATION	• Intrauterine device (IUD)	A small plastic or metal device inserted into the uterus by a doctor prevents a fertilized egg from implanting (effectiveness: >99%).
	• "Morning-after" pills	A pill containing a dose of estrogen 50 times higher than that found in birth control pills can stop ovum development or implantation (effectiveness: >75%).

FIGURE 25-18 Preventing pregnancy.

taking pills, the estrogen level in the body begins to drop; if it gets below a critical level, FSH release by the pituitary is triggered, which can lead to ovulation and the risk of pregnancy.

The long-term health consequences of using birth control pills include a slightly elevated risk of cardiovascular disease, particularly among women who smoke. However, because birth control users have a reduced risk of pregnancy—a condition that carries with it some significant health risks—and a reduced risk of ovarian and endometrial cancers, women's overall mortality risk is reduced while on the pill.

Although the birth control pill is one of the most commonly used methods of contraception, recent research has revealed that the hormones may actually alter women's attraction to men. When not taking birth control pills, most women prefer the odor of men possessing certain genetic combinations that are most different from their own. When taking the pill, however, their preference switches and they prefer the odor of men with genetic combinations most similar to their own. It remains to be seen how significant any effect of birth control pills is on women's mate preferences, and much additional research in this area is under way.

Hormone injections or implants. Injections or capsule implants, just under the skin, of synthetic estrogen and progesterone, or progesterone only, represent a slight variation on birth control pills. The prevention of pregnancy works in the same way, but with the added convenience of not having to take a pill every day. Some implants, in fact, need to be implanted and removed only once every three years. Consequently, these methods are slightly more effective than birth control pills, because much of the user error is eliminated.

Preventing Fertilization

Condoms. The condom—a thin rubber or natural membrane sheath placed on the penis or inside the vagina, covering the cervix, that prevents sperm from coming in contact with an egg—is one of several barrier methods of contraception, which keep sperm from coming in contact with an egg. Condoms have the added benefit of being one of the only methods of contraception that offer effective protection against sexually transmitted diseases, including HIV/AIDS. (See the next section for further discussion of this topic.)

Diaphragm or cervical cap. A diaphragm or cervical cap—a dome-shaped piece of rubber placed in the vagina, blocking the cervix—also prevents pregnancy by blocking sperm from reaching an egg. The use of spermicidal creams with a diaphragm or cervical cap (and with a condom as well) can increase their effectiveness, resulting in a failure rate of 6% to 16%.

Sterilization. Sterilization is the permanent alteration of the reproductive system to prevent the release of sperm or the movement of eggs down the Fallopian tubes. In a tubal ligation, the woman's oviducts are cut and tied off so that eggs cannot reach the uterus. In a vasectomy, the vas deferens on each side is cut and tied off so that sperm cannot reach the urethra, thereby causing a man's semen to carry no sperm. There are no side effects in either case. Each of these procedures, however, should be considered permanent (although in rare cases, the procedure can be successfully reversed).

Abstinence. Abstinence—not having sexual intercourse—is the most effective method of preventing pregnancy. In practice, however, abstaining from intercourse can prove difficult, particularly in the context of a long-term relationship. And temporary abstinence during the times when conception is most likely (often called the "rhythm method"), while theoretically effective, has a failure rate as high as 25%, probably due to variation in the time of ovulation and the fact that sperm can live for 72 hours (and possibly longer) in the female reproductive system.

Preventing Implantation

Intrauterine device (IUD). The intrauterine device, or IUD, is a small plastic or metal device that is inserted by a doctor into the uterus. The IUD (which can be left in for 3-4 years) does not prevent fertilization from occurring but instead prevents a fertilized egg from implanting in the uterus. IUDs have a failure rate of just under 1%. Some women experience side effects from IUDs (including cramps and other pain) and may need to have the device removed.

Q What is the "morning-after" pill?

"Morning-after" pills. Although not recommended as a long-term strategy for contraception, the morning-after pill, a dose of estrogen 50 times greater than that found in birth control pills, can stop ovum development or implantation. The morning-after pill has a failure rate of up to 25% and is recommended only for cases of rape and emergency situations in which other methods of contraception have failed. In contrast to the large hormone content of morning-after drugs, another drug, known as RU486, is considered an anti-hormone drug. It blocks progesterone receptors in the uterus, thereby causing the lining of the uterus to be sloughed off. This can end a pregnancy any time during the first seven weeks. In addition to causing abdominal pain and cramping, RU486 can also have other adverse effects, including nausea, vomiting, and fever.

TAKE-HOME MESSAGE 25•9

Pregnancy can be prevented by numerous methods, each of which acts in one or more ways to prevent ovulation, fertilization, or implantation.

25•10

Sexually transmitted diseases reveal battles between microbes and humans.

To many microbes, the human genitals and reproductive tract represent a desirable place to find shelter, nourishment, and opportunities for reproducing and dispersing. Unfortunately, these microbes can cause problems for humans in the form of **sexually transmitted diseases (STDs).** STDs produce symptoms of varying severity, from mild to extreme discomfort to sterility or even death. It is estimated that more than 300 million new cases of STDs occur each year worldwide.

Sexually transmitted diseases are caused by bacteria, viruses, fungi, protists, and even some arthropods. The organisms are passed from the mucous membranes (of the genitals, as well as of the anus and mouth) of one individual to those of another

SEXUALLY TRANSMITTED DISEASES

CAUSE	EXAMPLES	SYMPTOMS	TREATMENT
BACTERIUM	• Gonorrhea	Often none; sometimes painful urination, genital discharge, or irregular menstruation	Several antibiotics can successfully cure gonorrhea; however, drug-resistant strains are increasing.
	• Syphilis	Often no symptoms for years; eventual sores, skin rash, and if untreated, organ damage	Penicillin, an antibiotic, can cure a person in the early stages of syphilis.
	• Chlamydia	Often none; sometimes painful urination, genital discharge	Chlamydia can be easily treated and cured with antibiotics.
VIRUS	• HIV/AIDS	Initial symptoms range from none to flu-like; late stages involve severe infections and death	Currently no cure. Antiretroviral treatment can slow progression. Drug-resistant strains occur.
	• Genital herpes	Often none; outbreaks include sores on genitals, flu-like symptoms	Currently no cure. Antiviral medications can shorten and prevent outbreaks.
	• Human papilloma virus (HPV)	Often none; some types can lead to genital warts, others can cause cervical cancer	A vaccine prevents HPV, and is recommended for girls age 11–12. Warts and cancerous lesions can be removed.
PROTIST	• Trichomoniasis	Painful urination and/or vaginal discharge in women; often no symptoms in men	Trichomoniasis can usually be cured with prescription drugs.
FUNGUS	• Yeast infections	Genital itching or burning, and/or vaginal discharge in women; genital itching in men	Yeast infections can usually be cured with antifungal suppositories or creams.
ARTHROPOD	• Crab lice	Visible lice eggs or lice crawling or attached to pubic hair, itching in the pubic and groin area	Crab lice can be treated with over-the-counter lotions.

FIGURE 25-19 The most common STDs.

during sexual contact; sometimes they are transmitted by needles used for drug injections.

Some of the most common STDs, their symptoms, and treatments are listed in **FIGURE 25-19**. Although most are curable with antibiotics, antifungal drugs, or anti-protozoan drugs, two characteristics of STDs make them nearly impossible to completely eradicate from a population: (1) their symptoms may be mild or completely absent at times, causing many people to unwittingly pass an infection to their partners, and (2) to prevent reinfection, both partners must be treated simultaneously. Furthermore, because most microbes have such high reproductive rates, populations of a microbe can evolve quickly and become resistant to existing drugs, reducing the long-term effectiveness of treatments. Consequently, the treatment of STDs represents one of the most pressing public health issues in the world today.

TAKE-HOME MESSAGE 25•10

Sexually transmitted diseases (STDs) are caused by a variety of organisms, including bacteria, viruses, protists, fungi, and arthropods. Worldwide, more than 300 million people are infected each year. The effects of being infected with an STD range from non-existent, to mild to extreme discomfort, sterility, or even death.

④ Human development occurs in specific stages.

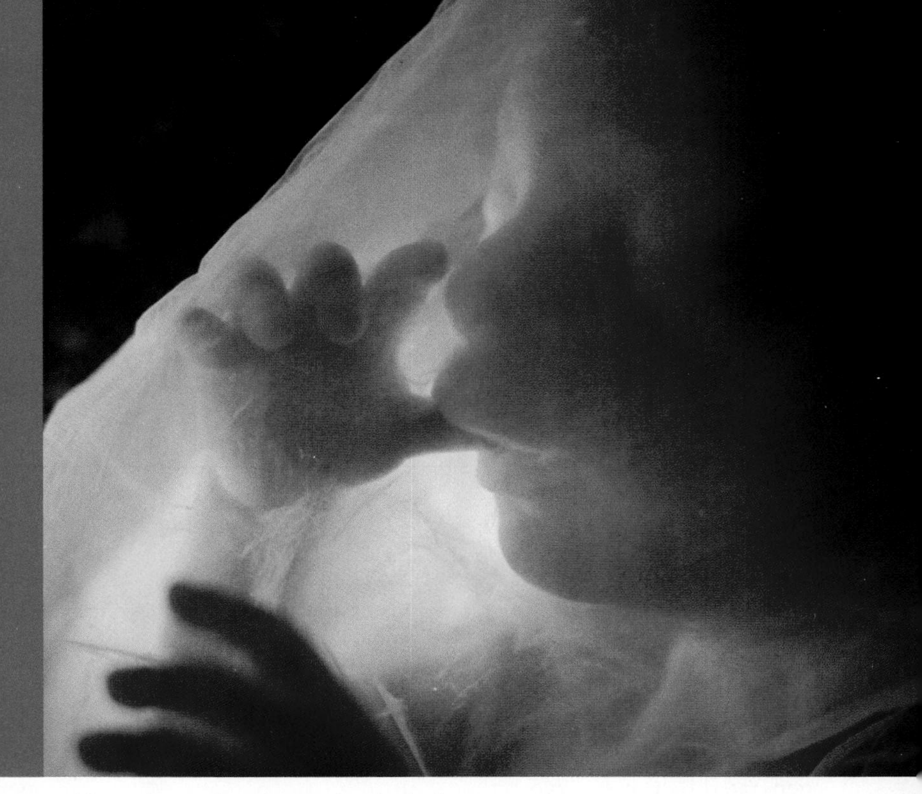

Developing human fetus five months after fertilization.

25•11

Early embryonic development occurs during cleavage, gastrulation, and neurulation.

How do you build a complete human, with specialized cells, tissues, and organs, when at fertilization there is just a zygote, a single cell that doesn't resemble a human at all? This question is studied and answered by developmental biology, the branch of science that seeks to understand and explain how a complex organism develops from a single cell. To be sure, the genes in that initial cell are important, and the environmental conditions in which the cell develops are also critically important. These factors interact as the development of humans and most other vertebrates proceeds in a carefully coordinated sequence of three stages: cleavage, gastrulation, and neurulation.

Cleavage The first phase, **cleavage,** is the early cell division, by mitosis, of the zygote, beginning shortly after fertilization (**FIGURE 25-20**). It starts when, about 30 hours after fertilization, the zygote divides into two cells. Then, about 30 hours later, it divides again, becoming four cells. This continues through numerous divisions. Although the number of cells increases significantly during cleavage, there is little or no growth in overall size. Rather, the fertilized egg is partitioned into more and more cells that are smaller and smaller. These multiple cell divisions are fueled by nutrients that were in the egg. At the cleavage stage, it is difficult to distinguish between species as dissimilar as sea urchins and humans.

After about six days of continuous cell division, the cells form a hollow ball, called a **blastula**—the mammalian version is called a **blastocyst**—of approximately a thousand cells. The cells secrete fluid into the center, where there is an inner mass of cells that will form the embryo. The outer cells also produce human chorionic gonadotropin (hCG), which keeps the corpus luteum from disintegrating, so that it continues to produce progesterone and thereby maintain the lining of the uterus for implantation.

The blastocyst grows rapidly and forms membranes that surround and protect it. These include the **amnion,** which surrounds the embryo, and the **chorion,** which, along with the endometrium, forms the **placenta,** the structure that connects the developing embryo to the wall of the uterus

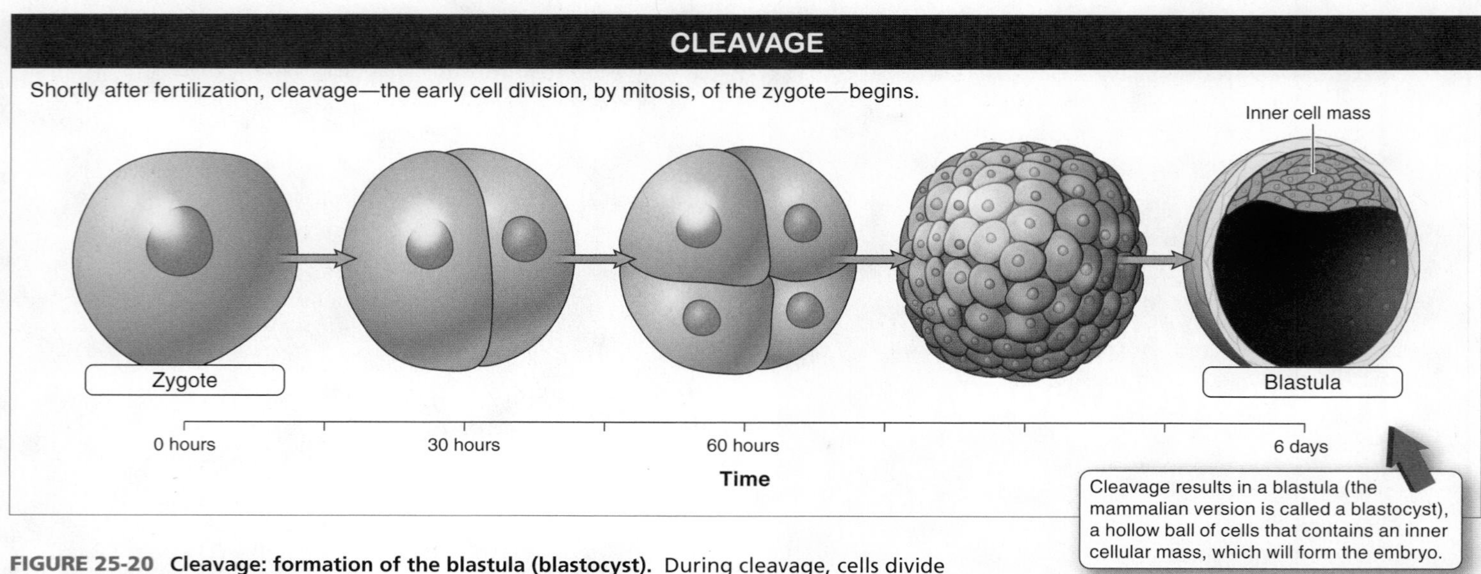

FIGURE 25-20 Cleavage: formation of the blastula (blastocyst). During cleavage, cells divide continuously, eventually forming a hollow ball filled with fluid.

(see Figure 25-25). The placenta is packed with blood vessels that bring nourishment to the embryo and remove wastes generated by the embryo.

Gastrulation In the second phase of development, **gastrulation,** three distinct **germ layers** of tissue form. Picture a beach ball—a hollow sphere. Now imagine making a fist and slowly pushing inward on the beach ball. This resembles the process of gastrulation. The indentation is called the blastopore and is located almost opposite from the point at which the sperm entered the egg. The blastopore will become the anus in vertebrates. At this point the entire mass of cells is called a gastrula (**FIGURE 25-21**).

Three distinct layers of tissue form during gastrulation. The cells in these layers have not yet differentiated (that is, they all

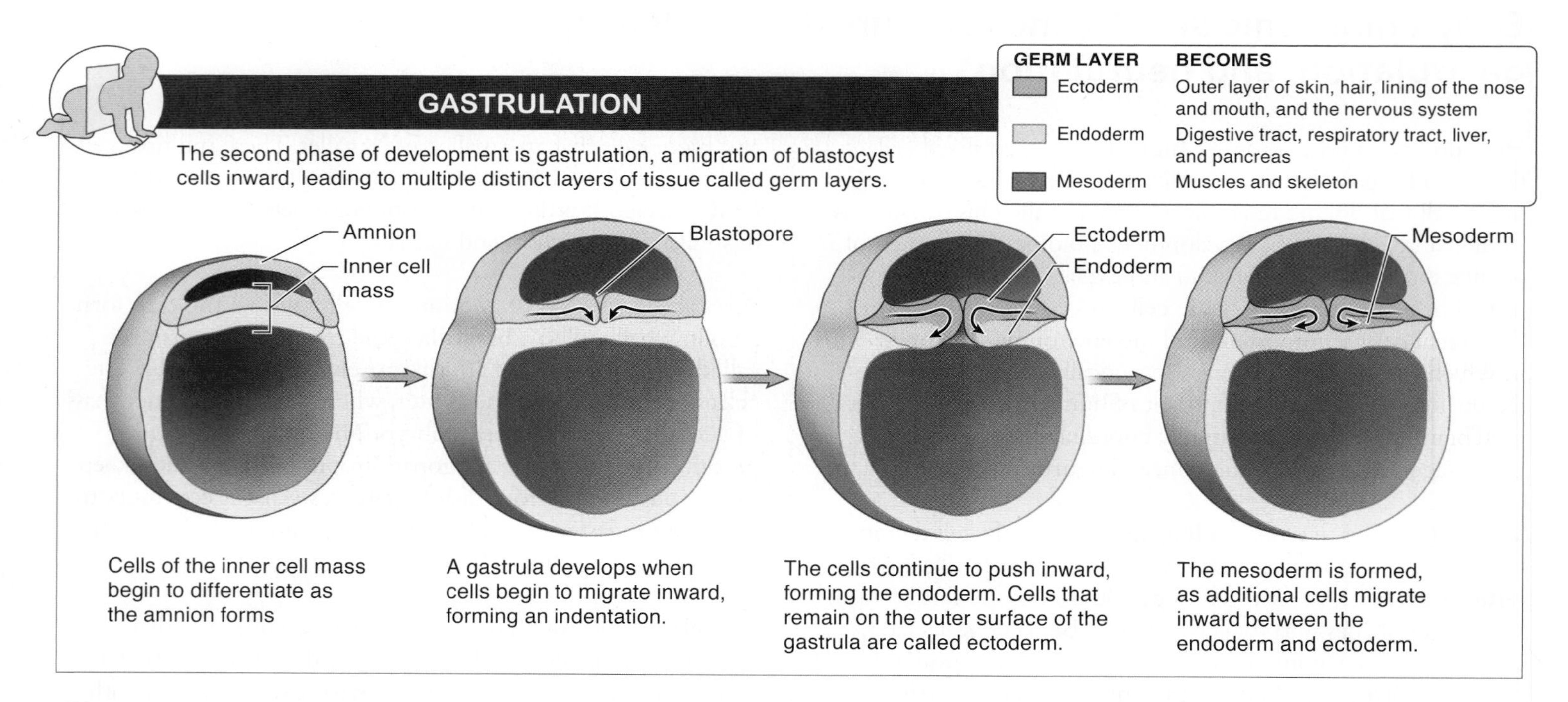

FIGURE 25-21 Gastrulation: formation of three germ layers. During gastrulation, three distinct layers of tissue form that will eventually make up the structures of the body.

look similar to one another), but they have become determined, which means that the type of tissue they will become, their ultimate developmental fate, is irreversibly decided. Prior to determination, cells are referred to as "totipotent" and have the capacity to develop into any type of tissue in the body.

Here's a summary of what happens during gastrulation.

1. First, the blastocyst begins to fold inward forming an indentation, the blastopore. The blastopore is located almost opposite the point at which the sperm entered the egg.
2. As the cells continue to push inward, they are called **endoderm,** and they form the lining of what will eventually become the digestive tract; the blastopore will become the anus. Endoderm cells will also form the liver, the pancreas, and the lining of the respiratory tract. At this point, the entire mass of cells is called a **gastrula.**
3. The cells that remain on the outer surface of the gastrula are called **ectoderm** and will ultimately form the outer layer of skin, the hair, the nervous system, and the lining of the nose and mouth.
4. Some additional cells migrate inward, between the endoderm and ectoderm, forming a third type of tissue, called **mesoderm,** which will ultimately give rise to muscles and the skeleton.

The specific details of gastrulation vary slightly from one species to the next. The general outcome, however—the formation of endoderm, mesoderm, and ectoderm—occurs throughout the vertebrates.

Neurulation and the Formation of Adult Structures

As an embryo begins its third week of development, the three types of tissue formed during gastrulation begin to develop into the various organs and tissues. At this time, within the mesoderm and running the length of the embryo, just above the digestive tract, a structure called the notochord forms. The **notochord** is a flexible rod that develops in all chordates and gives support to surrounding tissue. In vertebrates, the notochord's function is ultimately taken over by the backbone.

Just above the notochord, ectoderm folds in for the entire length of the embryo, first forming a groove and then becoming a long hollow tube, called the **neural tube.** The process is referred to as **neurulation** and the resulting neural tube ultimately develops into the entire nervous system, including the brain (located opposite the original position of the blastopore) and the spinal cord (**FIGURE 25-22**).

As the neural tube is developing, blocks of mesoderm tissue, called somites, begin to form down the length of the embryo. These somites will form the vertebrae and the muscles of the body. Next to the somites, spaces form that will become the

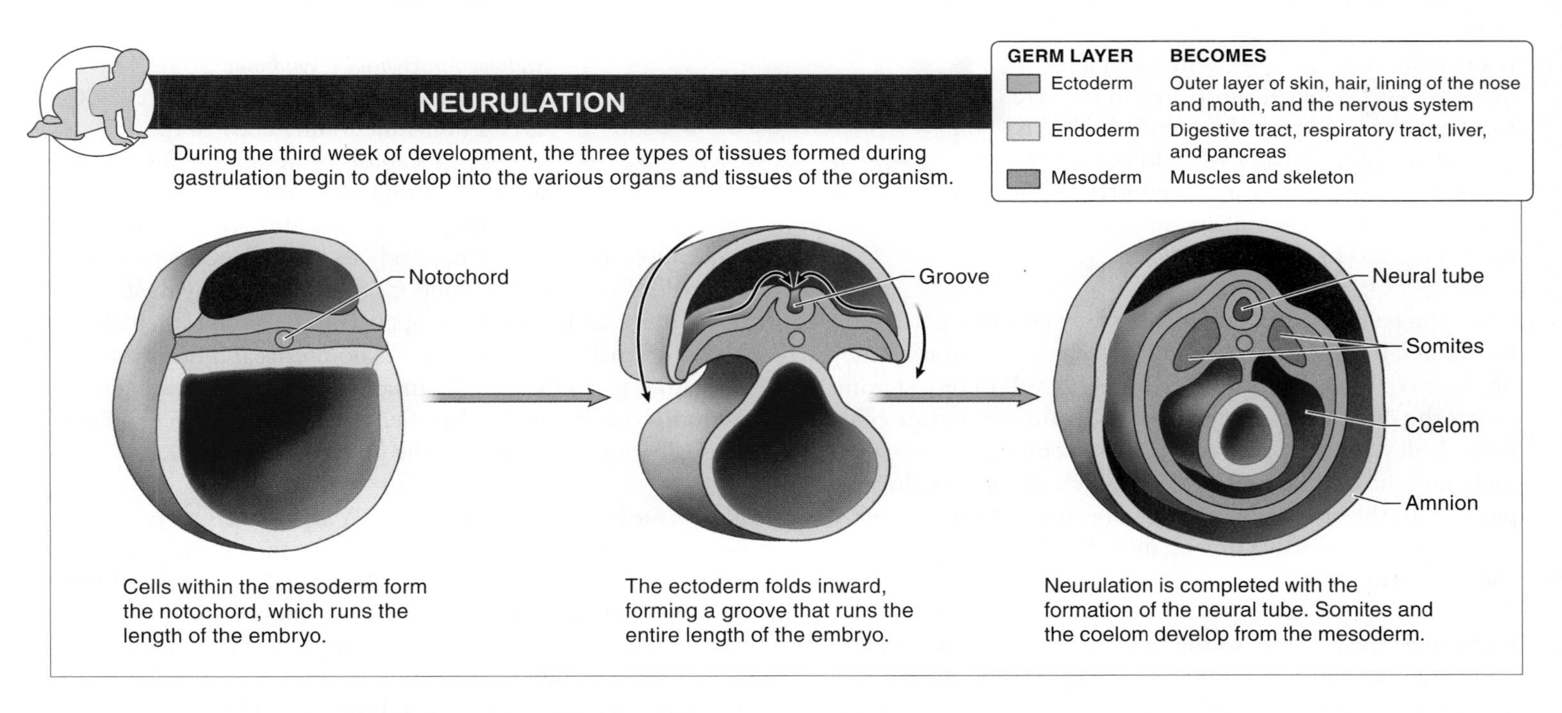

FIGURE 25-22 Neurulation: formation of organs and tissues from the germ layers. During neurulation, the germ layers begin to develop into the organs and tissues of the body.

body cavity, or **coelom** (pronounced SEE-lum). Within the coelom, many of the organs will form. At this point, toward the end of the third week of development, the embryo is only 2 millimeters (less than one-tenth of an inch) in length. Further development and specialization occur as some cells are programmed to die, in essence "sculpting" the body the way that a sculptor would create a statue.

Other cells may be induced to develop into one particular cell type or another by neighboring cells that send signals, possibly proteins, to turn certain genes on or off. This process was demonstrated dramatically by researchers who transplanted cells from the top of a newt blastopore in one embryo to another developing embryo. In the recipient embryo, the blastopore cells induced the development of nearby cells to form a second neural tube!

TAKE-HOME MESSAGE 25•11

Soon after fertilization, cleavage takes place and many rapid cell divisions occur without overall growth. Following cleavage is gastrulation, in which a gut begins to form, along with three distinct germ layers with specific developmental fates. In neurulation, mesoderm forms a supporting rod called the notochord, and above that, an infolding of ectoderm forms a neural tube, which will become the brain and spinal cord.

25•12

How does an embryo become male or female?

During the first few weeks after fertilization, surprisingly little distinguishes boys from girls. Then, about four weeks into embryo development in mammals, a gene carried on the Y chromosome, called SRY (for **s**ex-determining **r**egion on the Y-chromosome), can initiate development of the embryo as a male (**FIGURE 25-23**). The gene is expressed in the tissue that will form the gonads, and in most cases in which SRY is expressed, the gonads develop as testes. (There is a gene on the X chromosome that also influences sex determination and, in rare cases, can alter typical sex development.) If SRY is not present, as in females who do not carry a Y chromosome, ovaries develop instead. Once activated, SRY also stimulates numerous other genes involved in testes formation.

Q Can someone with a Y chromosome still develop as a female?

Once the fetal gonads develop, they start producing steroid hormones. If testes, they produce testosterone. If ovaries, they produce estrogen. The presence of high levels of testosterone causes the ducts connecting the gonads and the outside of the body to develop into the male internal reproductive organs, including the vas deferens (on each side), ejaculatory duct, and prostate. In the absence of testosterone, these ducts become the female reproductive organs, including the uterus, cervix, and Fallopian tubes.

Following development of the internal reproductive organs, undifferentiated external genitals develop. If testosterone is present and is modified slightly (by an enzyme called 5-alpha-reductase) into dihydrotestosterone (DHT), it causes the external genitals to become the penis and scrotum. In the absence of DHT, female external genitals develop, including the clitoris, labia, and vagina. Sex differentiation is generally complete by about the 12th week of development in humans.

At several points in development, the development of an embryo as male or female can be disrupted. Disruptions of this type occur in several genetic disorders.

Androgen insensitivity syndrome. Some XY individuals (approximately 1 in 20,000) carry a non-functioning copy of the androgen receptor. Although the SRY gene on the Y chromosome triggers the development of testes, this X-linked recessive trait causes the tissues of the body to be unresponsive to the normal effects of testosterone. Consequently, individuals with this syndrome develop what appears to be a typical female body. Internally, however, there is no uterus and no ovaries, but rather testes that remain inside the abdomen. Individuals with this condition are identified as female, and the condition is usually not detected until the girl fails to menstruate.

5-Alpha-reductase deficiency. Some XY individuals carry a non-functioning version of the gene for the enzyme that converts testosterone into DHT. Without DHT to direct the development of external male genitals, these individuals possess testes, but the external genitals develop as they would in a female. Consequently, individuals with this condition are generally identified as female at birth and in early childhood. At puberty, however, in response to the significantly increased testosterone production by the testes,

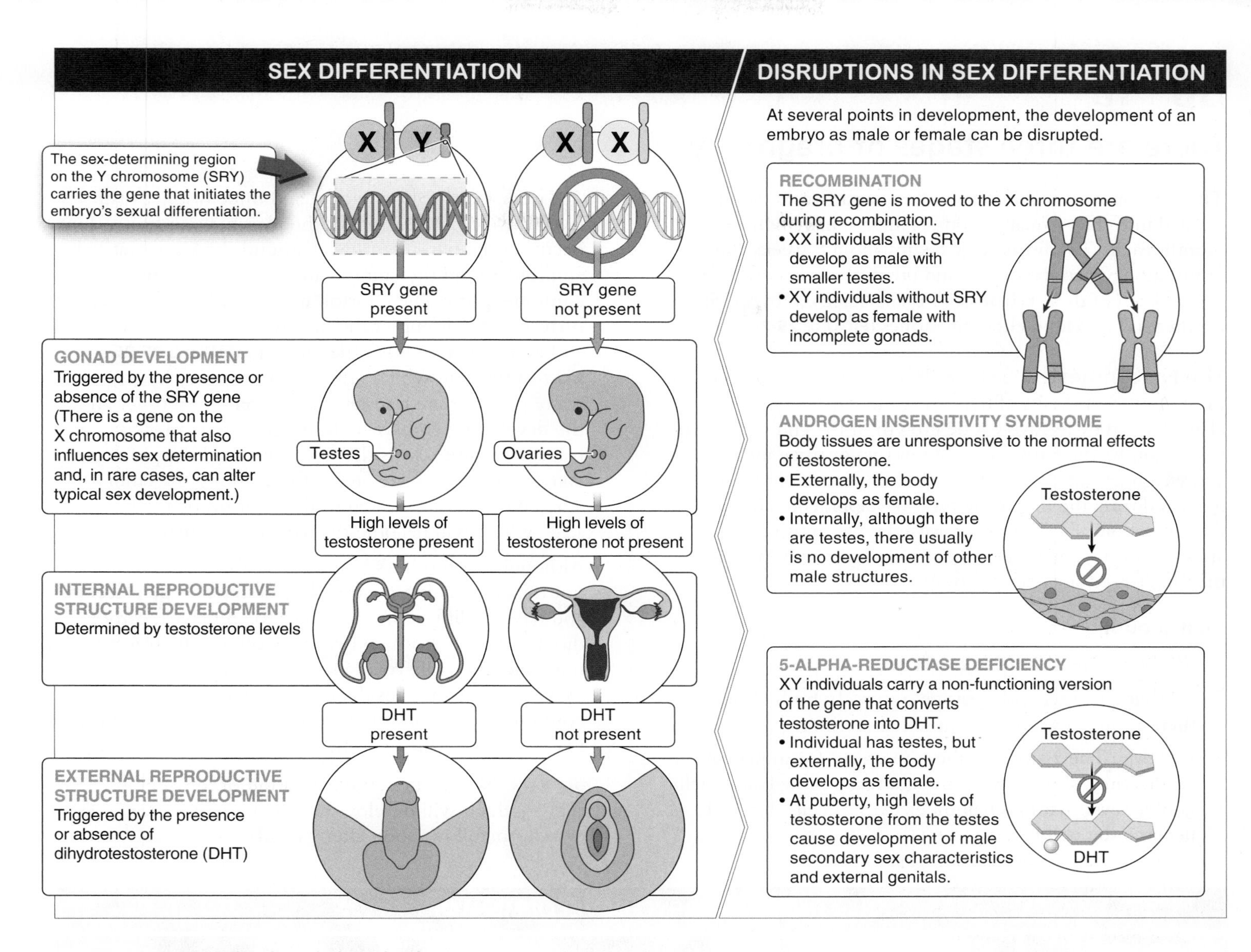

FIGURE 25-23 **Typical and atypical sex development.**

these individuals develop male secondary sex characteristics, including a deeper voice, facial hair, and increased muscle mass. The external genitals grow and often come to resemble a penis and scrotum more than a clitoris and labia.

Recombination leading to XX individuals with the SRY gene or XY individuals lacking SRY. It is possible, during the production of sperm by males, for crossing over to occur between the X and Y chromosomes. If this crossover causes the SRY gene to be moved to the X chromosome, and if the sperm cell carrying the X with SRY or the Y without SRY fertilizes an egg, irregular development occurs in the resulting individual. In XX individuals with the SRY gene, the individual appears to develop as a typical male. Although testes develop, however, they are smaller than average. These individuals are infertile (cannot produce sperm that can fertilize an egg or eggs that can be fertilized) and also tend to be shorter than XY males. In XY individuals without the SRY gene, the gonads develop incompletely, resembling streaks of connective tissue, but these individuals appear to develop as a typical female. At puberty, however, they are taller than average, do not develop secondary sex characteristics, and do not menstruate.

TAKE-HOME MESSAGE 25•12

Mammalian embryos develop female internal and external reproductive organs unless a gene on the Y chromosome stimulates fetal gonads to develop as testes, leading to testosterone production, which then stimulates the development of male reproductive organs.

25•13

There are three stages of pregnancy.

The development of a human embryo and fetus is usually divided into three equal periods, of approximately three months each. (The **fetus** stage begins about 8 weeks after fertilization, when the organs and other major structures first form.) Each of these **trimesters** is characterized by specific features and developmental milestones (**FIGURE 25-24**).

The First Trimester We have already discussed several of the most significant events of the first trimester. This period is characterized by relatively little growth: at the end of the first three months, the fetus is only about two or three inches long and weighs about an ounce. Much more important than growth during the first trimester is development and the differentiation of cells into specialized types of tissues. Here are some of the chief milestones (times indicate how much time has passed since fertilization).

First month

30 hours: The first cleavage occurs.

60 hours: The second cleavage occurs. In both cleavages, there is no overall growth, just cell division.

6–7 days: The zygote reaches the uterus for implantation. At this time, the zygote has become a hollow ball of cells (a blastocyst) with a small inner mass of cells that will become the embryo.

2nd week: Gastrulation occurs, and three distinct tissue layers develop. At this point, the placenta also forms. This occurs as the two membranes around the blastocyst—the amnion around the embryo and the chorion around the amnion—grow. The chorion ultimately branches out and surrounds the lining of the uterus. Together, the chorion and endometrium form the placenta (**FIGURE 25-25**), a mass of tissue in which nutrients, respiratory gases, and waste products are transferred between the mother and the developing embryo. (Although there is a yolk sac in placental mammals, its size is reduced and the placenta takes over the yolk sac's role of nourishing the embryo.) The embryo is connected to the chorion by the body stalk, which will become the **umbilical cord.**

Although the mother's blood and embryo's blood do not mix, they come in very close contact at the placenta, and oxygen diffuses from mother to embryo as carbon dioxide diffuses from embryo to mother. The placenta also nourishes the embryo and secretes hormones, including hCG, which maintains the corpus luteum, ensuring continued secretions of progesterone and, consequently, maintenance of the endometrium.

3rd week: Neurulation, the formation of the neural tube—which will develop into the nervous system—from ectoderm, follows soon after gastrulation.

HUMAN EMBRYO DEVELOPMENT

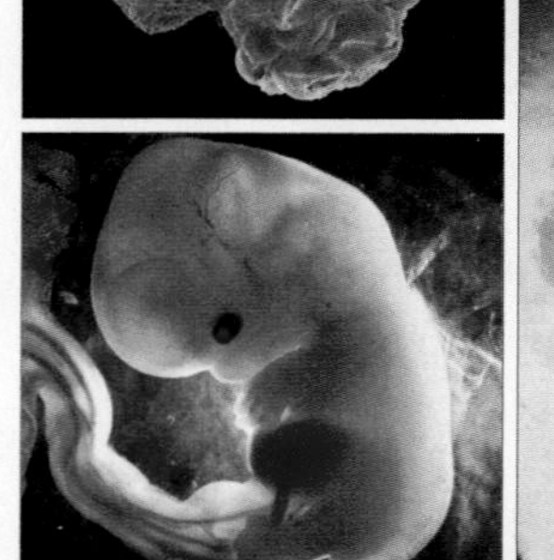

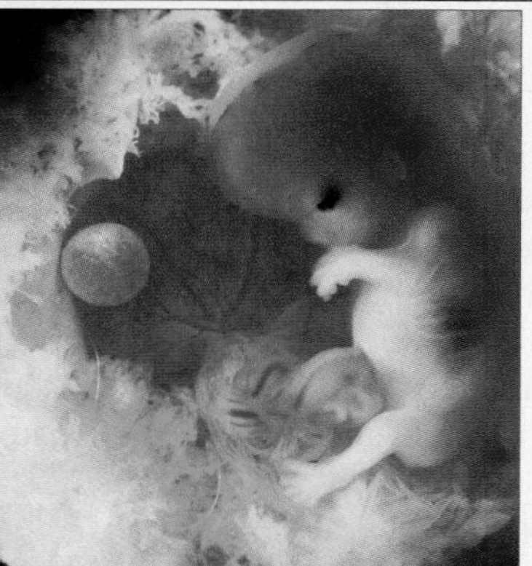

FIRST TRIMESTER: **MONTHS 1–3**
- Cells begin to differentiate into specialized types of tissues.
- Major organs and structures begin to form, including the eyes, heart, liver, pancreas, and gall bladder, as well as the limbs.

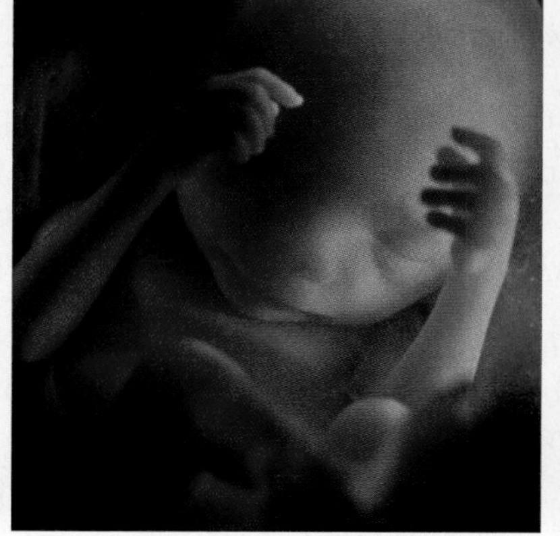

SECOND TRIMESTER: **MONTHS 4–6**
- Significant muscle and bone growth occurs, with less new development relative to the first trimester.

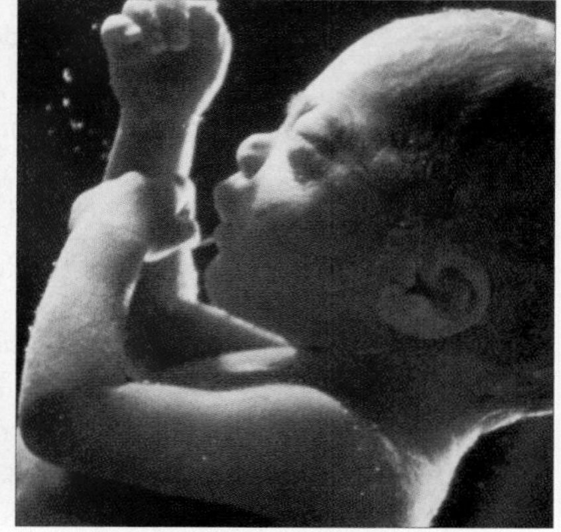

THIRD TRIMESTER: **MONTHS 7–9**
- Significant development of the nervous system.

FIGURE 25-24 Development of the human embryo and fetus. The nine months of human fetal development are divided into three-month trimesters.

THE PLACENTA

The placenta is a mass of highly vascularized tissue in which nutrients, respiratory gases, and waste products are transferred between the mother and the developing embryo.

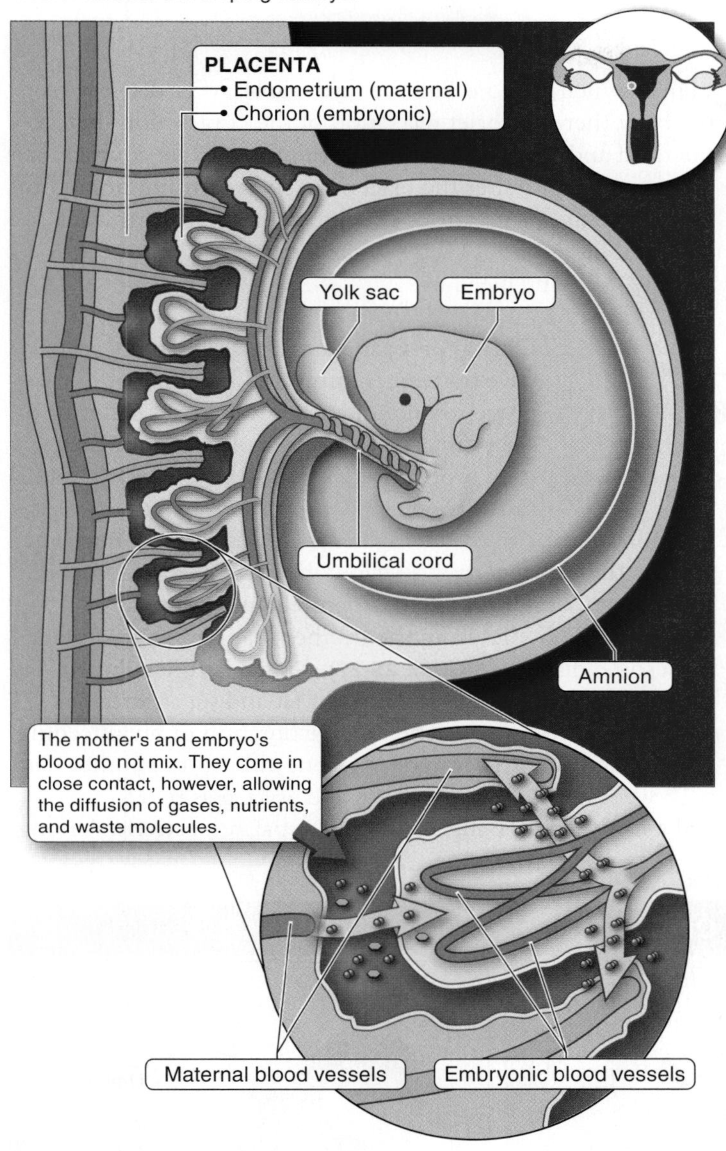

FIGURE 25-25 Structure and function of the placenta.

4th week: The formation of major structures and organs, including the eyes and heart, begins. At this point, it is also possible to see the arm and leg buds.

Second month

During this time, the limbs take shape. The major organs, including the liver, pancreas, and gall bladder, can also be seen. In spite of such dramatic development and differentiation, though, very little growth has occurred yet, and the weight of the embryo is about one gram, less than one-tenth of an ounce!

Third month

During the third month, the nervous system develops, as do some of the first reflexes, such as suckling. At this point, the corpus luteum finally degenerates. In the absence of the progesterone it has been producing, a new ovulatory cycle would typically (i.e., without pregnancy) occur, but because the placenta produces significant levels of estrogen and progesterone (which inhibit the production of FSH and LH), no ovulation occurs. The fetus's heartbeat usually becomes audible around the 10th week. The development of mammary glands in the mother's breasts is also stimulated at this time, in preparation for later milk production (**lactation**).

The Second Trimester The second trimester, months 4, 5, and 6 of a pregnancy, is a time of significant growth and less new development relative to the first trimester. During this time, as the bones get bigger and the muscles develop, the fetus moves around and the mother can often feel it kick. The heartbeat becomes much louder and can be heard easily with a stethoscope. Its amplified "whooshing" sounds like a freight train. Although the fetus cannot survive outside the womb at the end of the second trimester, its size has increased to about 600 grams (1.3 pounds) and 12 inches in length.

The Third Trimester The third trimester is all about growth, as the fetus ultimately becomes large enough to survive outside the mother—the minimum size for survival is generally about a pound and a half (3-4 kg). In addition to growth, during the third trimester there is significant development of the nervous system. The addition of new neurons occurs at the explosive rate of 15 million per hour, a rate that continues even after birth.

TAKE-HOME MESSAGE 25•13

The nine-month development of a human embryo and fetus is divided into three equal periods, or trimesters. The first trimester is primarily a time of development and differentiation of cells into specialized types of tissues, as the embryo implants in the uterus and the placenta forms. The second and third trimesters are characterized mostly by significant growth and rapid development of the fetus's nervous system.

25•14

Pregnancy culminates in childbirth and the start of lactation.

Birth, also called parturition, is the culmination of pregnancy, and it occurs in three phases. The first is the initiation of contractions and the dilation, or opening, of the cervix. Toward the end of the third trimester, the fetus usually becomes positioned with its head down and its skull resting on the cervix. This positioning, along with the increasing size of the fetus, stretches the uterus. Although the exact process is not fully understood, fetal hormones then cause the placenta to produce hormones, including estrogen and prostaglandin, that stimulate contractions in the uterus. The stretching of the uterus causes the pituitary gland to release oxytocin, another hormone that causes contractions. The contractions are referred to as **labor.** The rate of contractions increases from one or two per hour over many hours or even days before the birth, to about one every two to three minutes when birth is about to occur. The contractions cause a gradual dilation of the cervix, with the opening increasing from just over an inch (about 3 cm) at the beginning of labor to 4 inches (about 10 cm) or more at the end (**FIGURE 25-26**).

The second phase of birth is the delivery of the baby. This generally occurs with the head passing first through the vagina, which is also called the birth canal. After delivery of the baby, there is a brief relaxation of the contractions before the third and final phase of the birth process. The contractions then resume and, after the placenta is sheared from the wall of the uterus, the contractions expel it through the birth canal. The umbilical cord is then clamped and cut, and clotting quickly stops any bleeding.

Q: How long does breastfeeding last?

During pregnancy, mammary gland development is stimulated by estrogen, progesterone, and other hormones. After birth, the pituitary hormone prolactin stimulates milk production. Suckling by the infant, too, causes the release of prolactin and the hormone oxytocin, which further increases milk production. During the first few days, a yellowish fluid, called colostrum, is released. Colostrum is high in protein and contains numerous antibodies from the mother that protect the infant from some diseases (**FIGURE 25-27**). Gradually, the colostrum is replaced by milk, which is higher in fat and sugar, with less protein than colostrum. The average duration of breastfeeding varies a lot from one country, and one culture, to another—it is 12 weeks in the United States and as long as two to two and a half years in Bangladesh and Nepal, respectively. Among

THE PHASES OF BIRTH

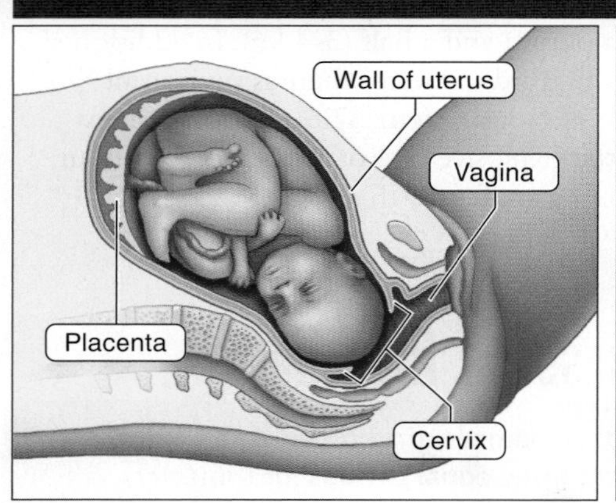

1 INITIATION OF CONTRACTIONS AND DILATION OF THE CERVIX
Hormones, including estrogen and prostaglandins, stimulate contractions in the uterus that cause a gradual opening of the cervix.

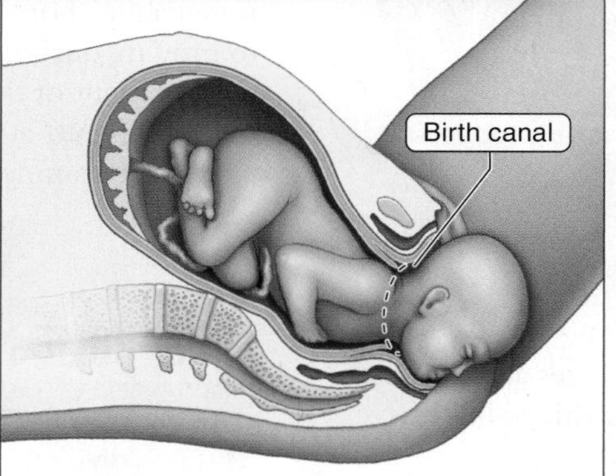

2 DELIVERY OF THE BABY
The baby's head passes through the vagina, or birth canal, followed by the rest of the body.

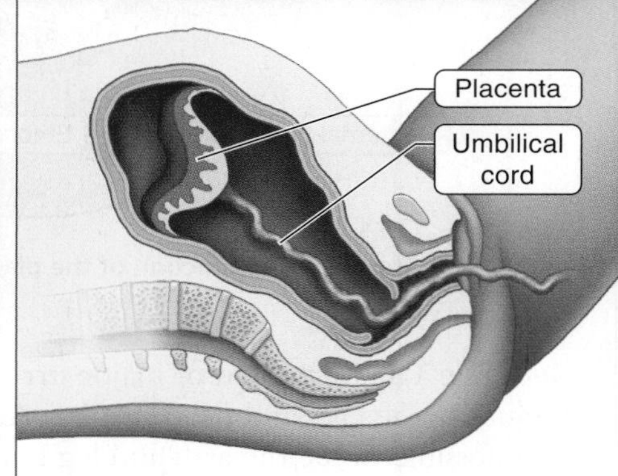

3 EXPULSION OF THE PLACENTA
Final contractions shear the placenta from the uterus wall and expel it through the birth canal.

FIGURE 25-26 **Labor and birth.**

FIGURE 25-27 Complete nutrition. Breastfeeding provides an infant with all the nourishment it needs, along with antibody protection from some diseases.

hunter-gatherer societies of humans, breastfeeding continued for as long as four years.

Q Can breastfeeding be effective as birth control?

During lactation, the suckling of the infant prevents the pituitary from releasing a sufficient surge of LH to cause ovulation. For this reason, during lactation, a woman's fertility is significantly reduced. As a method of birth control this method (called "lactational amenorrhea") depends on several conditions being satisfied, including that the mother is exclusively breastfeeding (i.e., the baby is not getting nourishment in any other way), the baby is less than six months old, and no more than six hours pass between any two feedings.

TAKE-HOME MESSAGE 25•14

Birth is the culmination of pregnancy and occurs in three phases. The first is the initiation of contractions and the dilation of the cervix. The second is delivery of the baby. The third is expulsion of the placenta. Lactation, stimulated by prolactin and other hormones, nourishes the baby with proteins, carbohydrates, and lipids and provides the infant with protective antibodies.

⑤ Reproductive technology has benefits and dangers.

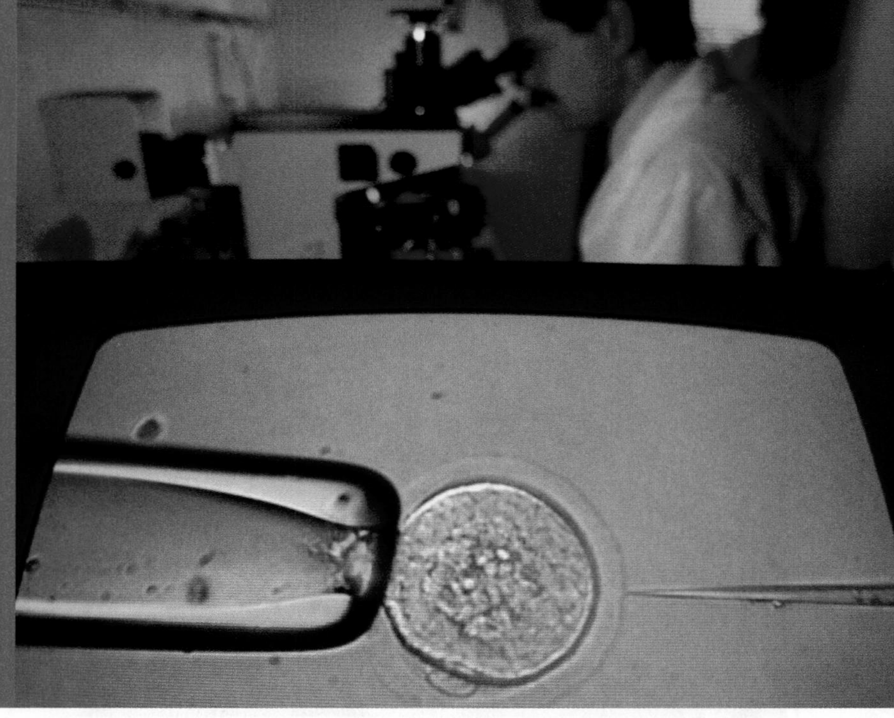

A single sperm is injected into the cytoplasm of an egg.

25 • 15

Assisted reproductive technologies are promising and perilous.

Just as technological advances have led to numerous developments in methods of preventing pregnancy, so, too, have they led to the development of techniques that help couples have children in situations where previously they might have remained infertile. In this section we examine some of these techniques, beginning more than a hundred years ago with the somewhat low-tech procedure that started it all.

In 1884, a woman came to see Dr. William Pancoast because she had not been able to conceive a child. After examining the couple, Dr. Pancoast determined that the husband was infertile. In the course of a discussion about the case by Dr. Pancoast and his students, someone suggested that semen should be collected from "the best-looking" member of the class and used to inseminate the woman. Dr. Pancoast agreed and arranged to see the patient again, under the pretense of another examination, at which point he injected the semen into the woman's uterus. Only when it became clear that the woman was pregnant did the doctor inform the husband of what he had done. The husband, however, was happy and requested only that the doctor not tell his wife how she had become pregnant. Eventually, a healthy son was born, the first recorded child born as the result of artificial insemination with a donor's sperm. From these inauspicious beginnings, there have been tremendous advances in the technologies available to couples experiencing difficulty having children.

Infertility is defined as a couple not being able to get pregnant after one year of trying—a definition that includes approximately 7% of married couples in the United States in any given year. There are many reasons for and causes of infertility—evenly divided between males and females—including low sperm counts and insufficient sperm motility, Fallopian tube damage or blockage, and ovulation problems. In 10% of cases, no cause can be found.

The technologies available are varied and include such interventions as surgery to repair blocked Fallopian tubes and hormone treatments to increase sperm counts or egg production. They also include a range of options called **assisted reproductive technology (ART),** the use of fertility treatments in which both sperm and egg are handled. These procedures typically involve removing eggs from a woman's ovaries, combining them with sperm to achieve fertilization in a Petri dish, and subsequently reinserting the now fertilized eggs into the reproductive tract of that woman, or another woman. In many cases—such as the case described at the beginning of this chapter—the eggs, sperm, or fertilized eggs may be frozen for some period of time before completing the process (**FIGURE 25-28**). The methods used in ART include the following.

IVF–ET. This stands for in vitro fertilization–embryo transfer. It is the most common of all ART methods. Several secondary oocytes are collected from a woman's ovaries and combined with sperm in a Petri dish. After several days, the fertilized eggs, at the eight-cell stage, are inserted into the woman's uterus, with the hope that they will implant and a pregnancy will result.

ZIFT. The technique of zygote intra-Fallopian transfer differs from IVF-ET in only two respects. After fertilization in a Petri dish, the fertilized eggs are transferred not to the uterus but to one of the Fallopian tubes instead. They also are transferred earlier, at the one-cell stage. Perhaps because surgery is needed to place the fertilized egg within the Fallopian tube, ZIFT is used by only 1% of couples seeking ART. In some cases, transfer to the Fallopian tube is not made until the fertilized egg is in the two- to four-cell stage (this is referred to as tubal embryo transfer). Some studies have shown ZIFT to have a higher implantation rate than IVF-ET.

GIFT. The method known as gamete intra-Fallopian transfer differs from ZIFT only in that the oocytes are immediately mixed with sperm and transferred to a Fallopian tube so that fertilization occurs in the body, rather than in a Petri dish.

In some cases, if the male's sperm counts are particularly low or if his sperm have low motility, a procedure called ICSI, or intracytoplasmic sperm injection, may be used in conjunction with IVF-ET or ZIFT. In this procedure, a single sperm is injected directly into an egg. While this method can be effective as a treatment for male infertility, there are concerns because sperm selection is bypassed. That is, instead of one sperm "winning" in the competition to fertilize the egg, a sperm is selected by the individual doing the procedure. This may be why there's an increased incidence of genetically carried birth defects with ICSI.

Outside the definition of ART, but still intriguing, is the increasingly effective process for selecting the sex of a baby, known as "sperm sorting." In this method, a fluorescent dye that binds to the DNA is applied to a sperm sample. Because the X chromosome is larger than the Y chromosome (by about 3%), the X-chromosome-bearing sperm bind more dye. The dyed sperm are then given a small electric charge, based on which chromosome they carry, and as they are passed through a tube in single file, the sperm cells are deflected to a collecting tube on one side or the other, depending on their charge. The selected sperm carrying an X or Y sex chromosome can then be used in any of the ART methods described above. The purities of the X versus Y samples of sperm range from 70% to 90%.

TAKE-HOME MESSAGE 25·15

Assisted reproductive technology (ART) procedures typically involve removing eggs from a woman's ovaries, combining them with sperm to achieve fertilization, and reinserting the fertilized eggs into the woman's uterus or Fallopian tube. These technologies can enable previously infertile couples to have babies.

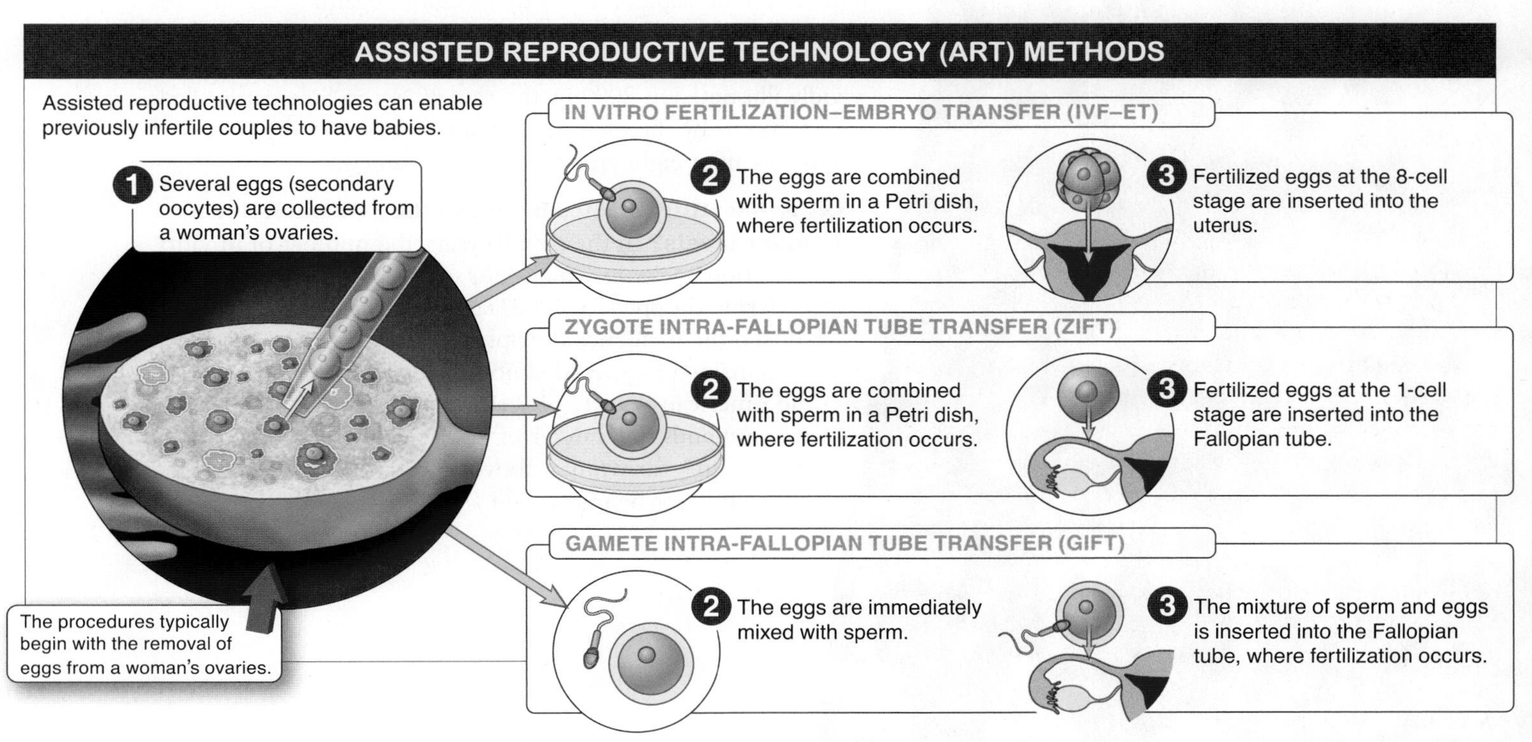

FIGURE 25-28 Technologies used in assisted reproductive technology.

streetBio

Knowledge You Can Use

Business and science in conflict. Why is the number of multiple births—twins, triplets, and more—on the rise?

Q: Multiple births are in the news a lot. Are they on the rise? Yes! Between 1980 and 2008, the incidence of twins in the United States increased by 70%. The rate of triplets and higher increased even more.

Q: Why are there so many more multiple births? Consider this: if you are running a fertility clinic, one of the most important pieces of information that will lead to the success or failure of your clinic is the percentage of treated couples who become pregnant. Unfortunately, in most assisted reproductive technology procedures, this number is influenced by the number of embryos transferred into the woman: 5–10 transferred embryos are more likely to result in pregnancy than 1–4 transferred embryos. The problem is, though, that with 5–10 embryos transferred, the risk of "multiples"—twins, triplets, or more—also increases, as does the health risk to the mother and the babies. And so the decision regarding the number of embryos to transfer involves a trade-off between the perceived effectiveness of the clinic and the risk to the mother. This point was vividly illustrated in 2009 by the case of Nadya Suleman (called "Octomom" by the media), a California woman who gave birth to octuplets affter eight embryos were implanted.

Q: Are fertility clinics bringing the rate back down to safer levels? In the past 10 years, the number of in vitro fertilization cycles in which four or more embryos were transferred dropped from 62% to 21%. This has significantly reduced the incidence of triplets, but the rate of twins has remained high. And although, for women under 35, the American Society of Reproductive Medicine now recommends the transfer of just a single embryo during ART procedures, for women above 35, it still recommends three to five embryos be transferred.

1 How do animals reproduce?

Animals can reproduce asexually or sexually. Asexual reproduction can be fast and efficient, but leads to offspring genetically identical to the parent. Sexual reproduction occurs in the vast majority of animal species and leads to offspring that are genetically different from one another and from either parent.

2 Male and female reproductive systems have important similarities and differences.

In adult men, sperm are continuously produced in the testes by meiosis. Semen, consisting of sperm cells and accessory fluids that nurture the sperm and aid in fertilization, is ejaculated during copulation. In adult women, diploid cells in the ovaries undergo meiosis to produce genetically varied haploid eggs. Fewer eggs than sperm are produced, and eggs are much larger than sperm.

3 Sex can lead to fertilization, but can also spread sexually transmitted diseases.

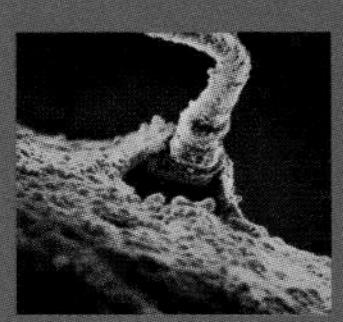

At fertilization, a sperm penetrates the egg. The egg blocks additional sperm entry, and the sperm and egg membranes fuse. The haploid nuclei of the egg and sperm then fuse, forming a diploid zygote. Pregnancy can be prevented by preventing ovulation, fertilization, or implantation.

4 Human development occurs in specific stages.

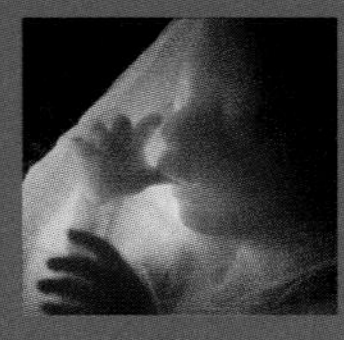

Fertilization is followed by cleavage of the zygote, with many rapid cell divisions but without growth. Next, in gastrulation, a gut begins to form, along with three distinct germ layers with specific developmental fates. In neurulation, mesoderm forms the notochord, and ectoderm forms a neural tube, which will become the brain and spinal cord. Mammalian embryos develop female reproductive organs unless a gene on the Y chromosome stimulates the fetal gonads to develop as testes, leading to the further development of male reproductive organs. The development of human embryos is divided into three trimesters. The first is primarily a time of development and differentiation; the second and third are characterized mostly by significant growth. Birth is the culmination of pregnancy, and lactation, stimulated by prolactin and other hormones, nourishes the baby and provides protective antibodies.

5 Reproductive technology has benefits and dangers.

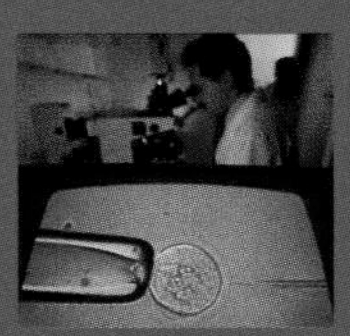

Assisted reproductive technology procedures typically involve removing eggs from a woman's ovaries, combining them with sperm to achieve fertilization in a Petri dish, and reinserting the fertilized eggs into the woman. They can enable couples to have babies who previously would have been infertile.

Key Terms

acrosome, p. 928
amnion, p. 941
asexual reproduction, p. 921
assisted reproductive technology (ART), p. 950
blastocyst, p. 941
blastula, p. 941
bulbourethral gland, p. 927
cervix, p. 931
chorion, p. 941
cleavage, p. 941
clitoris, p. 931
coelom, p. 944
contraception, p. 937
copulation, p. 924
corpus luteum, p. 934
ectoderm, p. 943
embryo, p. 925
endoderm, p. 943
endometrium, p. 931
epididymis, p. 927
estrogen, p. 934
external fertilization, p. 923
Fallopian tube, p. 931
fertilization, p. 923
fetus, p. 946
follicle, p. 931
follicle-stimulating hormone (FSH), p. 931
follicular phase, p. 934
gastrula, p. 943
gastrulation, p. 942
germ layers, p. 942
human chorionic gonadotropin (hCG), p. 935
infertility, p. 950
internal fertilization, p. 923
labia, p. 931
labor, p. 949
lactation, p. 948
luteal phase, p. 934
menopause, p. 935
menstrual cycle, p. 933
menstruation, p. 934
mesoderm, p. 943
neural tube, p. 943
neurulation, p. 943
notochord, p. 943
oogenesis, p. 931
oogonium, p. 931
ovarian cycle, p. 933
ovary, p. 931
oviduct, p. 931
oviparity, p. 925
ovoviviparity, p. 925
ovulate, p. 932
ovum, p. 922
parthenogenesis, p. 921
penis, p. 926
pheromone, p. 933
placenta, p. 941
polar body, p. 932
primary oocyte, p. 931
primary spermatocyte, p. 928
progesterone, p. 934
prostate gland, p. 927
scrotum, p. 926
secondary oocyte, p. 932
secondary spermatocyte, p. 928
semen, p. 926
seminal vesicle, p. 927
seminiferous tubule, p. 926
sexual reproduction, p. 922
sexually transmitted disease (STD), p. 939
sperm, p. 922
spermatogenesis, p. 926
testis, p. 926
testosterone, p. 927
trimester, p. 946
umbilical chord, p. 946
uterus, p. 931
vagina, p. 931
vas deferens, p. 927
viviparity, p. 925
zona pellucida, p. 937
zygote, p. 937

CHECK YOUR KNOWLEDGE

1. Some animals are primarily asexual in their reproduction, but have the ability to switch to sexual reproduction under certain conditions. What adaptive advantage might an animal that generally reproduces asexually have by also being able to reproduce sexually?
   a) to increase the genetic diversity of its offspring during periods of stress
   b) to confuse its predators
   c) to give the animal some variety in its life
   d) to increase its own likelihood of survival
   e) Only a) and d) are correct

2. Amplexus is a tactic that occurs in most species of frogs. The most likely reason that this tactic has evolved is:
   a) the gametes of terrestrial animals can quickly dry out in non-aquatic environments.
   b) it enables internal fertilization to occur without the necessity of viviparity.
   c) it ensures that the sperm and eggs are released at the same time and in the same place, increasing the likelihood of fertilization.
   d) it reduces the negative effect of salt water on gametes.
   e) All of the above are correct.

3. Meiosis occurs in the ______ in female humans and in the ______ in male humans.
   a) follicles; seminiferous tubules
   b) somatic cells; germ cells
   c) uterus; prostate gland
   d) follicles; prostate gland
   e) Fallopian tube; vas deferens.

4. Which three sets of accessory glands add secretions to the semen in human males?
   a) the seminal vesicles, the prostate gland, and the bulbourethral glands
   b) the ejaculatory duct, the prostate gland, and the vas deferens
   c) the seminal vesicles, the prostate gland, and the urethra
   d) the seminal vesicles, the prostate gland, and the epididymis
   e) the ejaculatory duct, the prostate gland, and the bulbourethral glands

5. When individual females mate with multiple males within short periods of time:
   a) a male that produces more sperm than other males has an increased probability of reproductive success.
   b) internal fertilization cannot occur.
   c) there is selection for a reduction in the size of the male testes.
   d) the male gametes experience reduced desiccation.
   e) All of the above are correct.

6. The tissue that develops into the penis in males, develops into _____ in females.
   a) the vagina
   b) the labia
   c) the clitoris
   d) the oviduct
   e) the ovary

7. The completion of meiosis II in the oocyte of human females occurs:
   a) before birth.
   b) during menstruation.
   c) after the oocyte is fertilized by a sperm.
   d) a few hours before ovulation.
   e) at ovulation.

8. The corpus luteum:
   a) secretes chorionic gonadotropin.
   b) is reabsorbed when fertilization occurs.
   c) is the only source of the hormones that maintain pregnancy.
   d) secretes progesterone but not estrogen.
   e) is the initial source of progesterone during pregnancy.

9. In mammalian fertilization, the extracellular layer of the egg that functions, in part, as a sperm receptor is called the:
   a) endoderm.
   b) jelly coat.
   c) acrosome.
   d) egg plasma membrane.
   e) zona pellucida.

10. The hormones present in birth control pills typically prevent pregnancy by:
    a) preventing formation of the endometrial lining of the uterus.
    b) preventing ovarian follicles from maturing.
    c) preventing sperm from fertilizing the ovulated egg.
    d) causing the endometrium to be shed once a fertilized egg has implanted.
    e) preventing implantation of fertilized eggs in the endometrium.

11. Tying off the vas deferens leading from each testis will:
    a) prevent formation of sperm.
    b) decrease testosterone secretion.
    c) reduce secretion from the seminal vesicles.
    d) reduce secretions from the prostate gland.
    e) None of the above.

12. When do cells begin to lose their totipotency?
    a) anaphase of mitosis
    b) gastrulation
    c) late prophase I of meiosis
    d) cleavage
    e) neurulation

13. The phenotype of an individual who is genetically XY but is lacking the sex-determining region on the Y chromosome would be:
    a) a person who does not have testes but does have male external sex organs.
    b) completely male, because of low estrogen levels.
    c) either male or female, depending on whether the person carries the gene for DHT, which directs development of the penis.
    d) a person with ovaries and female secondary sex characteristics.
    e) completely male, because the person will produce testosterone, which masculinizes the internal and external reproductive organs.

14. In humans, fetal growth is slowest during which trimester?

a) first
b) second
c) third
d) fourth
e) It occurs at the same rate in all trimesters.

15. In which of the following methods of assisted reproductive technology does fertilization occur within the woman's body?

a) intracytoplasmic sperm injection
b) in vitro fertilization–embryo transfer (IVF-ET)
c) zygote intra-Fallopian transfer (ZIFT)
d) gamete intra-Fallopian transfer (GIFT)
e) All of the above.

Short-Answer Questions

1. In humans, how are male gametes formed? Where in the male reproductive system does this process occur?

2. What are germ layers? How many layers are there, what are they called, and what structures ultimately form from them?

3. Summarize the three stages of pregnancy and the major events that occur in each.

See Appendix for answers. For additional study questions, go to www.prep-u.com.

1 Your body has different ways to protect you against disease-causing invaders.

- **26.1** Three lines of defense prevent and fight pathogen attacks.
- **26.2** External barriers prevent pathogens from entering your body.
- **26.3** The non-specific division of the immune system recognizes and fights pathogens while signaling for additional defenses.
- **26.4** The non-specific system responds to infection with the inflammatory response and with fever.

2 Specific immunity develops after exposure to pathogens.

- **26.5** The specific division of the immune system forms a memory of specific pathogens.
- **26.6** Lymphocytes fight pathogens on two fronts.
- **26.7** Clonal selection helps in fighting infection now and later.
- **26.8** The structure of antibodies reflects their function.
- **26.9** Cytotoxic T cells and helper T cells serve different functions.

3 Malfunction of the immune system causes disease.

- **26.10** Autoimmune diseases occur when the body turns against its own tissues.
- **26.11** AIDS is an immune deficiency disease.
- **26.12** Allergies are an inappropriate immune response to a harmless substance.

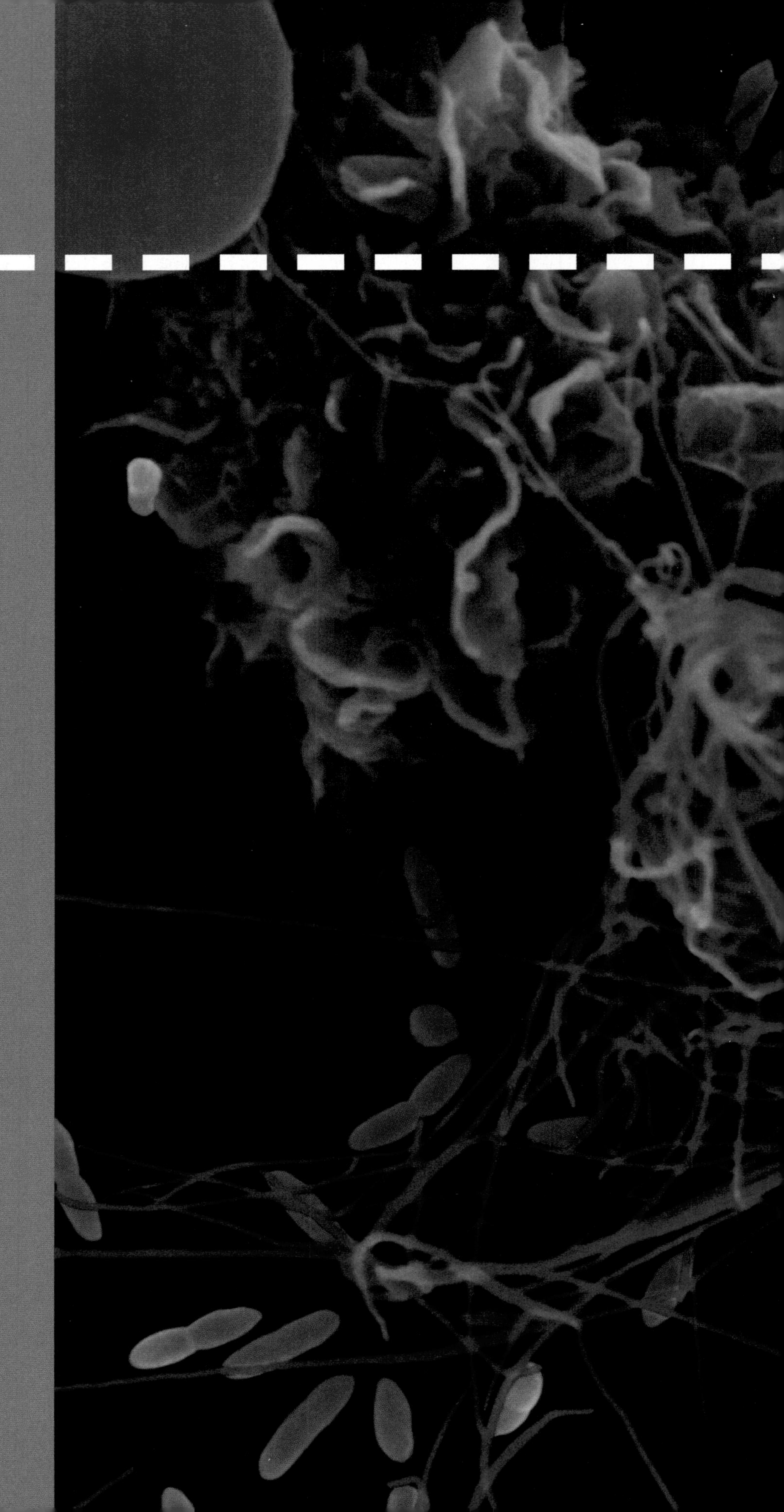

26

Immunity and Health

How the body defends and maintains itself

❶ Your body has different ways to protect you against disease-causing invaders.

Sneezing is one of the body's ways of expelling harmful substances that enter our body.

26•1

Three lines of defense prevent and fight pathogen attacks.

Most of us don't think about the germs that constantly attempt to invade our bodies until we hear that an illness is "going around." For example, in the spring of 2009, Mexico alerted the world to an outbreak of the H1N1 virus that caused numerous deaths—particularly among people under 25 years old and pregnant women. This outbreak was caused by a new strain of the influenza virus that the world's current population had never before encountered. The media showed footage of Mexico City residents wearing face masks and reported that schools and businesses were closing for extended periods of time. Within weeks, the virus had spread to the United States and around the world. Fortunately, however, as the flu strain began infecting people worldwide, researchers noted that although this influenza strain was strong, it was not as virulent as first suggested.

Indian girls wearing protective masks outside a museum in the Indian city of Bangalore in August 2009, just after a 26-year-old woman died of Swine flu (H1N1 virus) in a nearby hospital.

When you first heard about the H1N1 flu, did you think about washing your hands more frequently, wearing a face mask, and staying away from people who were coughing? The H1N1 flu example demonstrates that when we are aware of **pathogens,** disease-causing microorganisms, molecules, and viruses (sometimes referred to as "germs"), in our environment, we

FIGURE 26-1 Everyday encounters with pathogens. Shopping cart handles harbor germs.

may take precautions to avoid them. But we fail to notice most of the germs in our everyday lives. Think about the commonplace items that you and others touch. Did you know that shopping cart handles rank as one of the "germiest" items around? Shopping carts are often found to have fecal bacteria that cause gastrointestinal illness, bacteria associated with skin infections, and viruses that cause colds and flu, along with many non-pathogenic organisms (**FIGURE 26-1**). Not surprisingly, many stores now offer disinfectant wipes, and some states have even proposed legislation to make providing these wipes mandatory.

The vast majority of the microscopic organisms found on everyday items will not cause disease, but some will. That we don't notice or pay attention to surfaces that may be covered with germs, don't get sick frequently, and don't think too much about how our immune system protects us is a testament to the efficacy of this body system.

The human **immune system** provides protection against an enormous variety of pathogens, including many bacteria, fungi, viruses, and parasitic protists and worms (**FIGURE 26-2**). To protect the body from the various pathogens, the immune system needs to be able to rapidly recognize the general category of "non-self" or "stuff that doesn't belong in the body," as well as distinguish among specific disease-causing microbes or viruses. The immune system works within minutes to begin destroying a pathogen. In many cases this requires killing body cells that are infected with the pathogen. To accomplish this challenging protective task, there are three major divisions of the immune system (**FIGURE 26-3**).

1. Physical barriers provide the first defense against pathogens and can keep them out of the body. For example, skin acts as a nearly impenetrable barrier that keeps viruses, bacteria, and other pathogens from entering the body, and various chemicals secreted by epithelial surfaces stop the growth of many pathogens.

2. Non-specific immunity is the division of the immune system that provides the next defense, for situations in which pathogens make it past the physical barriers and into the body. This cellular system is sometimes referred to as the *innate system* (meaning "inborn"), because this division is ready to go at birth and does not require any previous

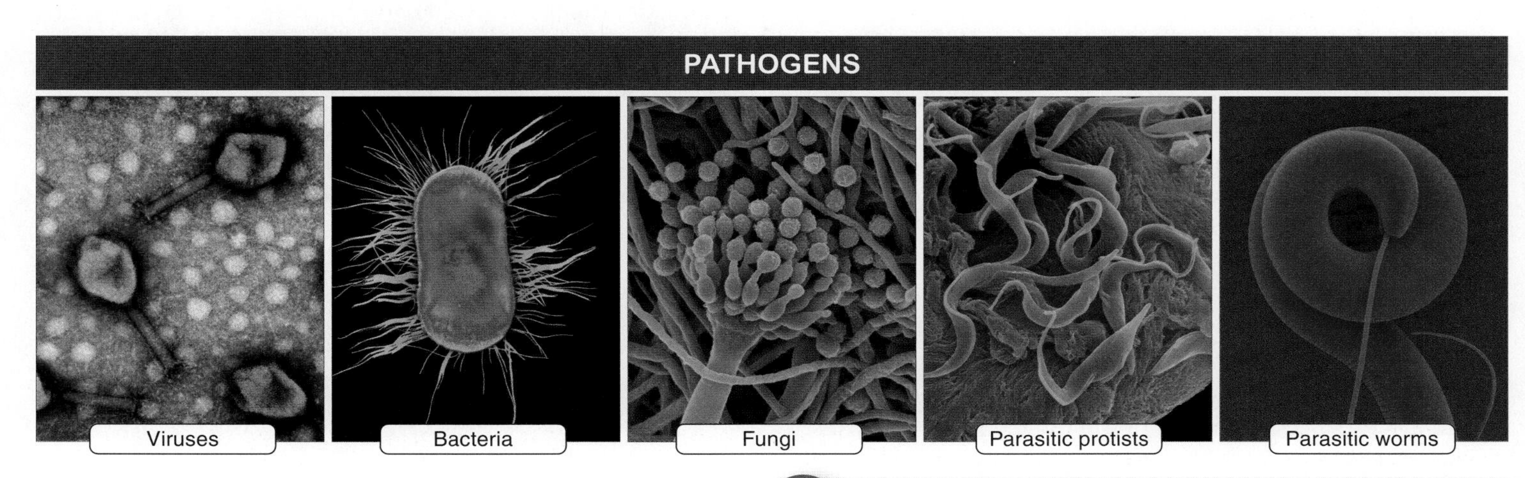

FIGURE 26-2 Examples of the foreign, "non-self" microbes and viruses that can harm the body.

The immune system protects us from a diverse group of pathogens—disease-causing viruses and microorganisms, often referred to as "germs."

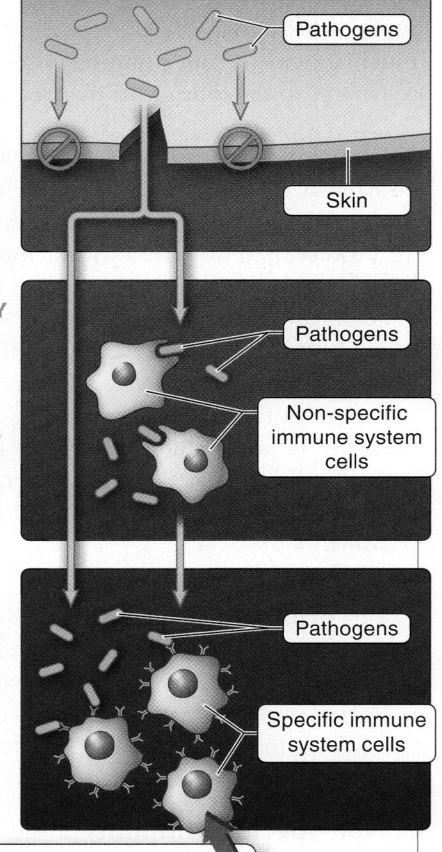

FIGURE 26-3 Layered defenses of the immune system block entry and fight invaders.

exposure to pathogens to act. Cells of the non-specific division of the immune system rapidly recognize pathogens that breach external barriers and are not part of the body's own tissue, but they do this without distinguishing the specific identity of one pathogen from another. Non-specific immunity is found in virtually all multicellular organisms, including plants, nematode worms, fruit flies, and humans.

3. Specific immunity, also called *adaptive immunity,* is the division of the immune system that deals with pathogens that either were missed by the non-specific system or cannot be overcome by that system. The specific division recognizes very specific infectious agents—not just bacteria in general but, for example, the specific *Staphylococcus aureus* that causes common skin infections; not just viruses in general but, for example, the specific varicella-zoster virus that causes chicken pox. Because the specific immunity part of your immune system must recognize the specific threat and then manufacture specific weapons to combat it, there is a time lag between your exposure to the pathogen and the specific immunity response to it. This response takes hours to days, unlike non-specific immunity, which responds immediately or within minutes. Once activated, however, the specific division forms a "memory" of past infections that allows it to fight off the same infection more quickly in the future. As a result, an individual often has lifelong protection from a pathogen he or she has previously encountered.

These three divisions of the immune system are collaborative. Once an invader is recognized by the non-specific system, chemical signals trigger the cells of the specific system to go to work. We first look more closely at the non-specific immune system that plays a necessary and immediate role in protecting the body's tissues.

TAKE-HOME MESSAGE 26·1

The immune system protects us from a diverse group of pathogens. There are three collaborative divisions to the immune system: a physical barrier, non-specific immunity, and specific immunity. Physical barriers and non-specific immunity are the first lines of defense and, as non-specific systems, recognize pathogens only as "non-self." The specific immunity line of defense recognizes individual species of pathogens and forms a memory of them.

26•2

External barriers prevent pathogens from entering your body.

Just as a high-security building may have walls and barbed wire to keep out potential intruders, your body has ways to prevent pathogens from entering your body, tissues, and cells. The **integumentary system** is made up of skin, sweat glands, oil glands, hair, and nails that protect you from mechanical stress caused by external forces such as friction or pressure, sunlight, dehydration, and, of course, pathogens (**FIGURE 26-4**).

Q How dangerous is it to sit directly on a toilet seat in a public restroom?

Have you ever sat down on a public toilet seat only to wonder if you were going to contract some horrible disease from it (**FIGURE 26-5**)? The good news is that most human pathogens cannot survive outside the body for very long, and a cold, smooth toilet seat is not an ideal place for them to survive. Additionally, pathogens need an access point. For example, HIV is transmitted through the membranes of the genitals (or, much less frequently, those of the mouth). The virus does not pass through intact skin on your buttocks! So while there may be germs on the toilet seat, your skin acts as a barrier that prevents entry. Many public bathrooms offer toilet seat covers that can act like a second skin, if the thought helps you relax while you do your business!

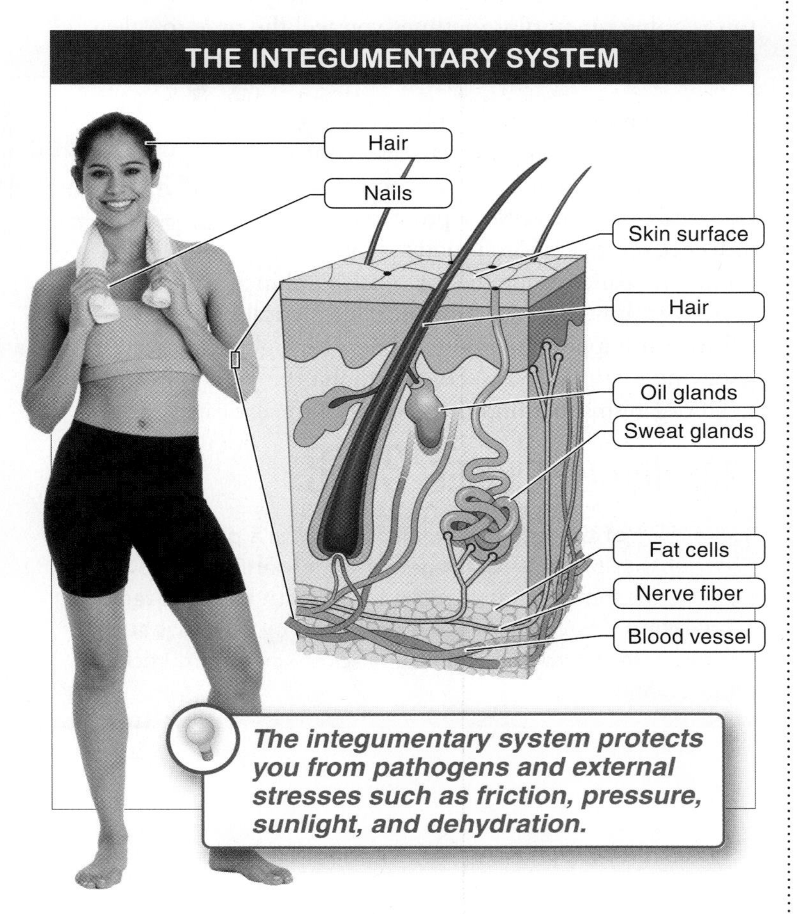

FIGURE 26-4 Providing a nearly impenetrable barrier: the integumentary system.

FIGURE 26-5 A safe place to sit. Public restrooms are generally safe to use (but be sure to thoroughly wash your hands!).

Skin is not exactly like a wall, however—skin is far more active than a protective wall. In addition to forming a nearly impenetrable barrier, skin cells actively secrete chemicals that inhibit bacterial growth (**FIGURE 26-6**). Sweat secreted onto the skin's surface, for example, contains **lysozyme,** an enzyme that can kill bacteria by damaging their cell walls. The protection provided by skin becomes really clear in burn victims: lacking skin in certain areas, they are susceptible to numerous pathogenic bacteria and fungi. Although we may not realize it, besides skin, the external and internal surfaces of

DEFENSE MECHANISMS OF THE INTEGUMENTARY SYSTEM

SKIN
Forms a nearly impenetrable barrier that keeps pathogens from entering the body

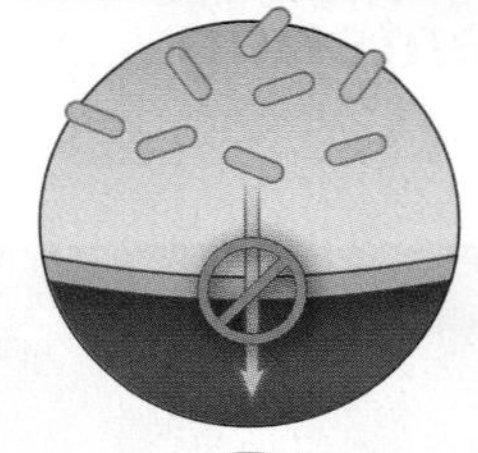

LYSOZYME AND OTHER ENZYMES
Lysozyme in saliva and tears, and digestive enzymes in the small intestine kill many bacteria

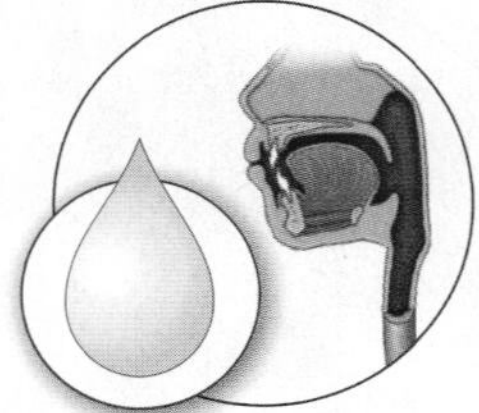

ACIDIC SECRETIONS
Stomach acids, acidic vaginal secretions, and acidic urine, all of which protect the digestive, reproductive, and urinary tracts from bacterial pathogens

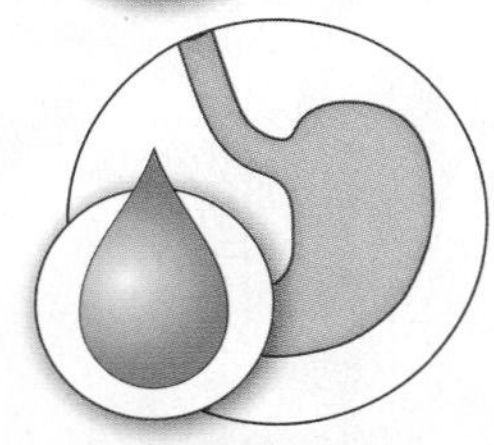

CILIA
Hair-like extensions on the surface of the respiratory tract that move mucus-entrapped pathogens up and out of the lungs

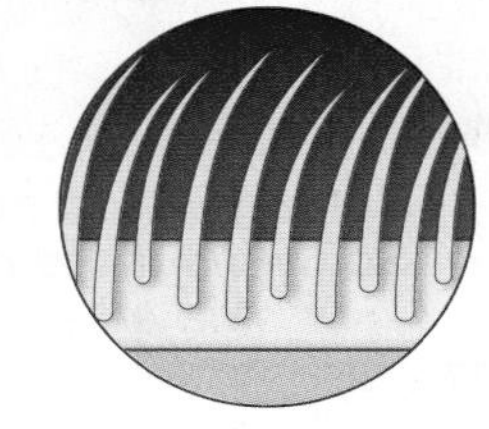

TEARS
Fluid containing antiviral and antibacterial chemicals that washes away microorganisms from around the eyes

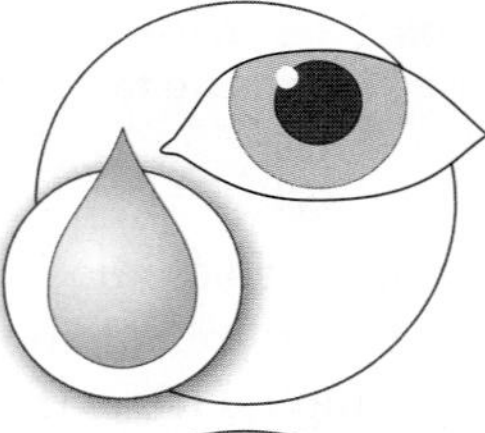

EAR WAX
Sticky substance that can trap microorganisms in the ear canal

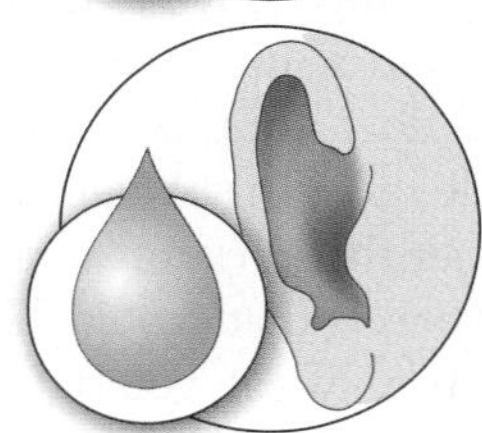

FIGURE 26-6 Security barrier. The integumentary system provides multiple defenses to protect you from pathogens.

the body, such as eyes, ears, and the linings of the digestive and respiratory tracts, are also exposed to environmental stresses and forces. These body areas, too, are shielded by the integumentary system (and, as we'll see, by the cells of the non-specific defense system).

Consider, for example, how the digestive tract is protected. Imagine that minutes after you push a germ-laden shopping cart, you put your finger in your mouth to bite your nail. Are you guaranteed to get ill? No, because mucus (a major component of the saliva in your mouth) contains lysozyme. And even if this antibacterial chemical doesn't destroy potential invaders, they will quickly reach the stomach, a harsh environment containing both acid and digestive enzymes that can destroy a variety of pathogens.

The respiratory tract is also protected by the thick and sticky physical properties of mucus, which traps pathogens before they can reach the lungs. In the respiratory tract, hair-like extensions of the epithelial cells, called **cilia,** continually move mucus-entrapped pathogens up and out of the lungs. As the mucus moves up toward your throat, you continually swallow germs and send them to your stomach, where they are destroyed (or you spit them out). You hardly notice that all of this is going on, until sometimes you feel the urge to "clear your throat" or cough.

Q What is the purpose of ear wax?

The eyes produce tears that physically wash away microorganisms, and tears also contain lysozyme and antiviral and antibacterial chemicals. Ear wax serves a protective function, too. Secretions produced by numerous glands inside the ear canal, combined with dead skin cells, make up this sticky substance. Its physical properties are effective at trapping microorganisms. The slight acidity and the lysozyme in ear wax also serve to inhibit the growth of some microorganisms that find their way into the ear canal.

TAKE-HOME MESSAGE 26•2

Skin, part of the integumentary system, is a physical barrier that prevents pathogens from entering the body's cells. Cells that are not covered by skin but are exposed to the external environment are protected by defenses such as bacteria-destroying chemicals, acidic secretions, sticky mucus, and wax.

26 • 3

The non-specific division of the immune system recognizes and fights pathogens while signaling for additional defenses.

Sometimes security measures fail. When intruders break through a defensive wall surrounding a secure building, the building's security guards need to (1) identify the security breach, (2) call for backup, and (3) attack and remove the invader. Similarly, your body *recognizes* any security breaches, such as a wound to the skin or some spoiled food you've eaten, and then *responds* instantly to prevent an infection (**FIGURE 26-7**). Chemical signals are put out to call for backup cells, and molecular weapons are used to destroy both the pathogens and any cells they've infected. In this section, we examine the cells that perform these duties and the tools that they use to combat infection. First, let's examine the recognition stage.

An important first step in defending the body is for immune cells to distinguish invaders from the body's own cells, or "non-self" from "self." The surface receptors on the patrolling non-specific (innate) immune system cells recognize molecules, usually proteins, such as those found only on the surfaces of pathogens, and bind to them. This recognition is considered non-specific because immune cell receptors recognize and bind to classes of molecules shared by a wide array of organisms. For example, one receptor recognizes a protein contained in all bacterial flagella. Another receptor recognizes a molecule found in the cell walls of many types of fungi.

Your immune cells have combinations of receptors that can recognize virtually any pathogen that might cause infection, by recognizing surface molecules. Because your own cells do not contain these molecules, a cell that possesses any of them can be "marked" as an invader by the innate system cells.

While the cells of the innate system are the initial responders to a pathogen, there are never enough of them at the initial site of infection to ingest and kill all the invading pathogens. Thus, after recognizing the security breach and starting the response, the patrolling cells of the non-specific system sound the alarm and call for backup.

The chief way that cells of the immune system—both the specific and non-specific divisions—talk to one another is by secreting signaling proteins called **cytokines** in response to pathogens in the body. Cytokines secreted by one immune cell can bind to receptors on other immune cells and signal them to respond in various ways. For example, cytokine binding can cause immune cells to move closer to the site where the cytokines are being produced (*cyto* = cell; *kine* = to move). Cytokine-secreting cells use this as a way to recruit more immune cells to the site of infection.

Interferon is an important type of cytokine produced by cells infected by a virus. Interferon alerts other cells to turn on protective measures that help them resist viruses. These alerted cells take on an "antiviral state" in which they transcribe specific genes and translate them into proteins that can inhibit viral replication and degrade viral RNA.

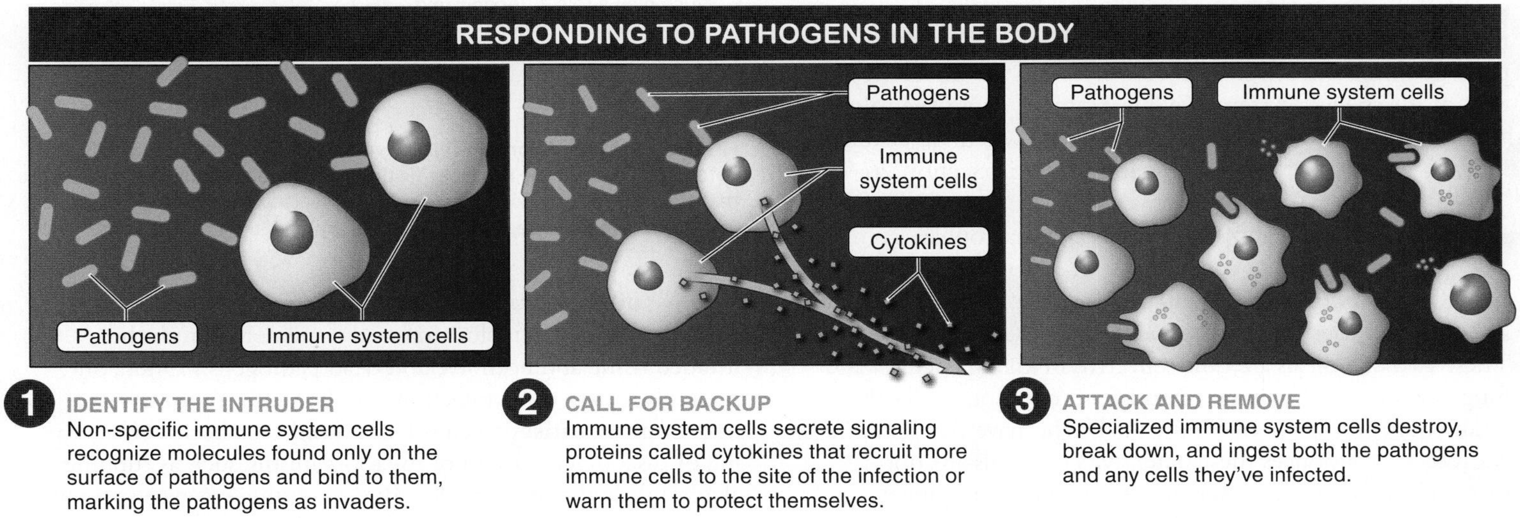

1 IDENTIFY THE INTRUDER
Non-specific immune system cells recognize molecules found only on the surface of pathogens and bind to them, marking the pathogens as invaders.

2 CALL FOR BACKUP
Immune system cells secrete signaling proteins called cytokines that recruit more immune cells to the site of the infection or warn them to protect themselves.

3 ATTACK AND REMOVE
Specialized immune system cells destroy, break down, and ingest both the pathogens and any cells they've infected.

FIGURE 26-7 Sometimes the security measures fail. The body recognizes and responds to invading pathogens.

As proteins on the cell surfaces of pathogens allow cells of the body's non-specific system to recognize an intruder, several different types of immune cells next begin to mount the non-specific system's response. Pathogens that breach the body's external defenses are met by **white blood cells,** specialized cells that play roles in both the specific and the non-specific divisions of the immune system (introduced in Section 21-6). These cells are made in the bone marrow and released into the bloodstream, where they patrol the body's tissues and search for invaders. Some white blood cells, called **phagocytes,** can engulf, ingest, kill, and break down foreign pathogens and particles. Three of the most important types of white blood cells—neutrophils, macrophages, and dendritic cells—act as phagocytes, with the dendritic cells also having an important role in stimulating the specific immunity system, as we'll see later. A fourth type of white blood cells, **natural killer (NK) cells,** are not phagocytes, but operate by killing body cells that have been infected by pathogens. Let's look more closely at these four types of white blood cells (**FIGURE 26-8**).

Neutrophils circulate in the blood and exit blood vessels at sites of injury and infection in tissues. These phagocytes ingest small organisms, primarily bacteria. They also produce hydrogen peroxide and bleach. Neutrophils normally make up between 50% and 70% of all white blood cells circulating in the bloodstream, and their numbers increase dramatically and rapidly during the first few days of some infections. Neutrophils are the kamikazes of the immune system—they destroy themselves as they ingest pathogens and live only 3–5 days, on average.

Macrophages are large ("macro") white blood cells that reside in and patrol the tissues. They engulf and digest whole pathogens (small organisms), as well as any debris that remains after the neutrophils have done their job. However, macrophages do not destroy themselves in the process. Instead, they serve as an important link between the non-specific and specific divisions of the immune system. Once they have ingested a pathogen, they "present" digested pieces of the pathogen on their cell surface, "advertising" the infection to cells of the specific system. We'll learn more about the role of "presenting cells" when we examine specific immunity in Section 26-9.

THE WHITE BLOOD CELLS OF NON-SPECIFIC IMMUNITY

The non-specific immune system consists of several types of white blood cells that are made in the bone marrow and released into the bloodstream, where they patrol the body's tissues and search for invaders.

NEUTROPHILS
- Phagocytic cells that ingest small organisms, primarily bacteria
- Destroy both the pathogen and themselves in the process

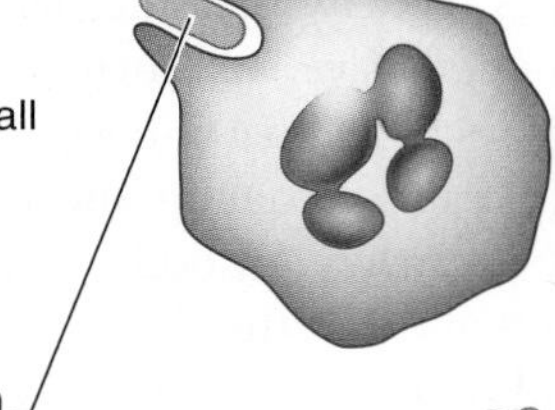

Pathogens

MACROPHAGES
- Phagocytic cells that ingest whole pathogens as well as large debris such as dead cells
- Present pieces of pathogens on their surface, advertising the infection to cells of the specific immune system

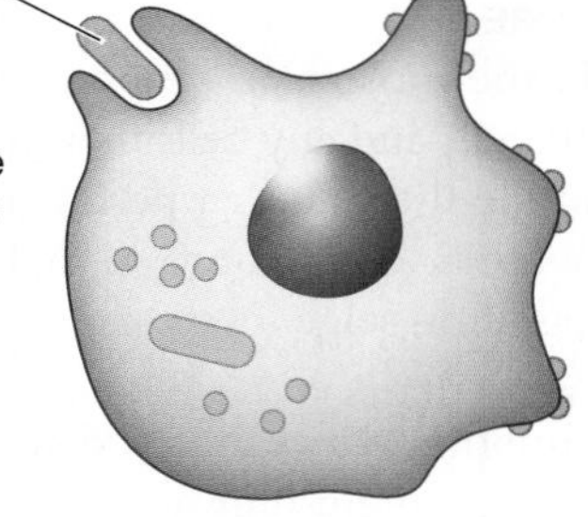

DENDRITIC CELLS
- Phagocytic cells that present ingested pathogens to cells of the specific immune system

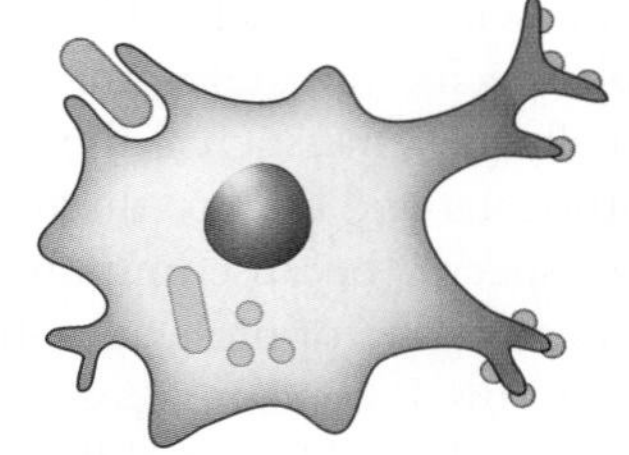

NATURAL KILLER (NK) CELLS
- Kill body cells infected by pathogens by poking holes in the cell membranes
- Also play a role in recognizing and killing cancer cells

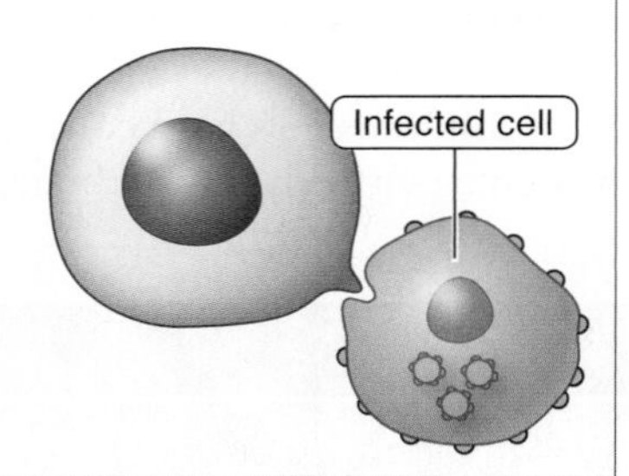

FIGURE 26-8 **Hardworking cells.** Several types of white blood cells defend the body in the non-specific division of the immune system. Note that the neutrophil nucleus sometimes appears like a cluster of nuclei because it is multi-lobed, with the lobes connected by thin strands.

Q What is pus?

Neutrophils and macrophages employ chemical warfare to destroy pathogens. For example, both types of cells can produce hydrogen peroxide and hypochlorous acid (components of household bleach). These two chemicals are quite effective in killing bacteria and fungi, as you may know first-hand from using one or both to clean your room or house. Individuals who have defects in the pathway that produces these defensive chemicals are, not surprisingly, frequently stricken with bacterial and fungal infections. In the process of killing, neutrophils destroy themselves and become a major part of **pus,** the thick yellowish fluid we often notice at the site of infection. (Other components of pus include damaged cell debris from the invaded tissue and both live and dead pathogens.) Often, our bodies can clear these infections on their own or with some help from an **antiseptic,** a solution used on a body surface to kill or discourage growth of microorganisms, such as rubbing alcohol or hydrogen peroxide. But when there is excessive pus, combined with signs of a more serious infection, antibiotics might be needed to kill bacteria inside the body.

Dendritic cells are phagocytes that link the non-specific and specific divisions of the immune system by "presenting" the cells of the specific system with foreign matter. These phagocytic cells migrate to the lymph nodes, which are populated by cells of the specific immunity system.

Natural killer (NK) cells, which are not phagocytes, provide the first line of resistance to viruses by poking holes in the membranes of virus-infected cells so as to kill the cells, and in doing so, killing the viruses inside the cells. Their initial response may eliminate the viral infection, or at least slow it down until the specific system can respond to the virus in a more specific manner. Interestingly, NK cells also play a role in recognizing and killing cancer cells (**FIGURE 26-9**).

In addition to the cells we've learned about here, the non-specific system can also quickly recognize and destroy invaders thanks to some circulating defensive proteins, collectively called **complement proteins.** There are approximately 30 of these proteins that "complement" the cells of both the non-specific and specific immunity systems and can link the two systems. Foreign molecules, such as the components of bacterial cell walls, activate the complement proteins (**FIGURE 26-10**). Once activated, the complement proteins fight pathogens in several ways. For example, a particular combination of complement proteins can blast holes in the cell membranes of pathogenic organisms, allowing water and ions to pass through, but not larger molecules such as proteins. As a result, water rushes inward and the pathogen swells and lyses, or bursts. Other complement proteins help destroy pathogens by sticking to them and forming a coating that enhances the ability of phagocytes to bind to and engulf them.

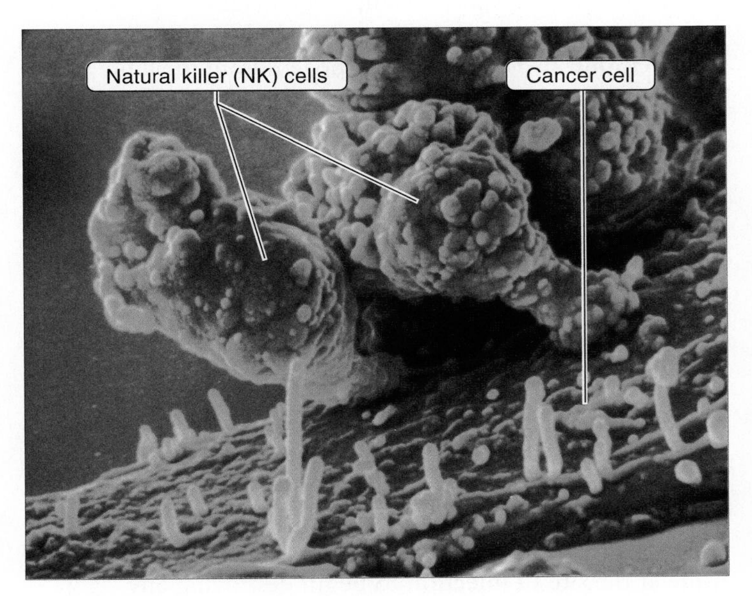

FIGURE 26-9 Attacking cancer. Natural killer cells are attaching to a cancer cell.

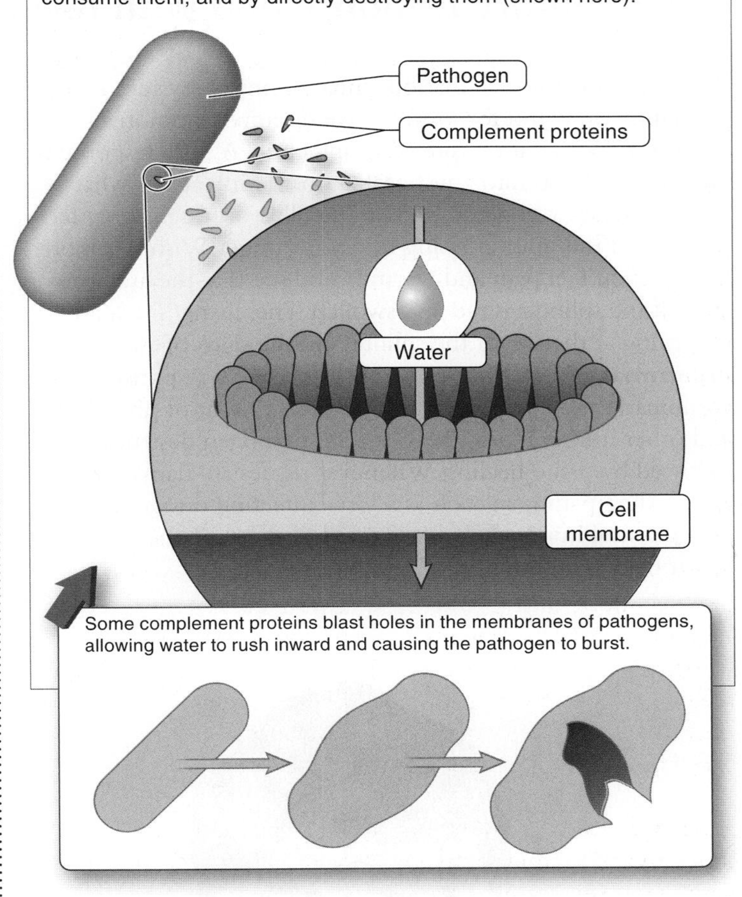

FIGURE 26-10 Defensive proteins "complement" the other cells of the non-specific system to combat pathogens.

TAKE-HOME MESSAGE 26•3

Non-specific immunity provides defenses against pathogens by recognizing molecules on their cell surfaces. Immune cells communicate with each other via chemical signals called cytokines. Non-specific immunity cells that secrete cytokines can recruit other cells to the site of infection or warn them to protect themselves. White blood cells of the non-specific system include phagocytes—such as neutrophils and macrophages that ingest and kill pathogens, and macrophages and dendritic cells that display pathogens to cells of the specific immunity system—and natural killer cells, which kill virus-infected cells and cancer cells. Complement proteins also non-specifically recognize invaders and help to destroy them.

26•4

The non-specific system responds to infection with the inflammatory response and with fever.

While we don't notice that our innate immune system is constantly on patrol for invaders, we definitely do notice when it responds at the site of an infection. A little response is quickly amplified into a noticeable one. Think about what happens when you get a splinter. Initially, you might only feel the pain of the splinter going into your finger. Within minutes, though, you feel pain and warmth, and see that the area around the splinter is red and swollen. The changes you are observing at the site of the splinter are the signs of an **inflammatory response.** The inflammatory response is a combination of events that leads to recruitment of phagocytes and other immune cells to assist with pathogen destruction, followed by tissue healing. Whenever tissues are damaged by an invading pathogen such as a virus infecting throat cells, or by a physical injury such as a wound to the skin, the inflammatory response is triggered.

Following a wound—such as the cut shown here—the body responds with an inflammatory response that causes redness, heat, swelling, and pain. But the process clears away foreign invaders and allows for healing.

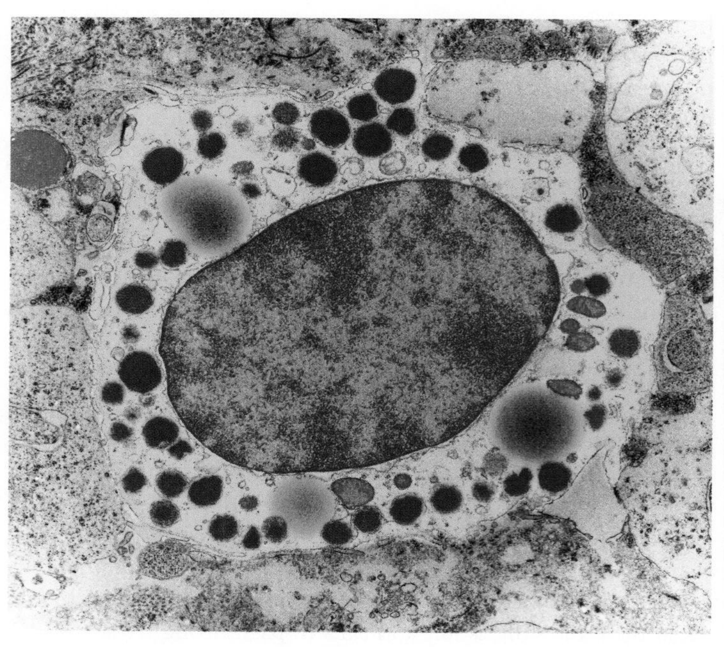

This mast cell contains small round granules containing histamine. Release of histamine occurs at sites of infection as part of the immune response.

In the first century A.D., the Romans first described the four signs of inflammation: redness, heat, swelling, and pain. Whether it is a sore throat or wounded skin, these hallmarks of inflammation are always the same. Let's imagine a scenario in which you cut your finger with a small knife while slicing a bagel (**FIGURE 26-11**). Invading pathogens from the knife and the skin surface are quickly engulfed by the macrophages that reside in all tissues. The macrophages then release cytokines to recruit more phagocytes and two other types of white blood cells, which initiate inflammation. **Basophils,** which circulate in the blood, and **mast cells,** found in the tissues, trigger the inflammatory reaction by releasing **histamine,** a molecule that causes nearby non-injured blood vessels to (1) dilate (open up and become wider) and (2) become leaky.

When blood vessels dilate, the flow of blood increases and allows for a faster, greater supply of defensive molecules and cells that can fight infection. The increased blood supply at the site of injury causes the *redness* and *heat* associated with inflammation. The increased leakiness of the blood vessels makes it easier for neutrophils to exit the blood and enter tissue at the site of infection to begin destroying any invading pathogens. The leakiness is the reason for the

THE INFLAMMATORY RESPONSE

1. After you cut your finger with a knife, invading pathogens from the knife and the skin's surface enter your body.

2. Macrophages residing in tissues surrounding the cut begin engulfing the pathogens and releasing cytokines, recruiting more phagocytes and other white blood cells to the area.

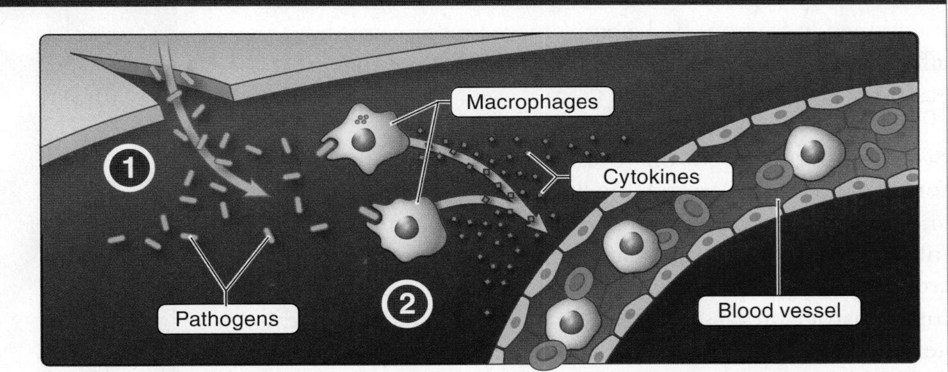

3. Basophils, which circulate in the blood, and mast cells, found in the tissues, trigger the inflammatory reaction by releasing histamine.

4. Histamine causes nearby non-injured blood vessels to dilate, increasing the flow of blood and allowing for a greater supply of defensive molecules and cells that can fight infection.

5. Histamine also causes the blood vessels to become more leaky, allowing neutrophils to more easily exit the blood and enter the site of infection.

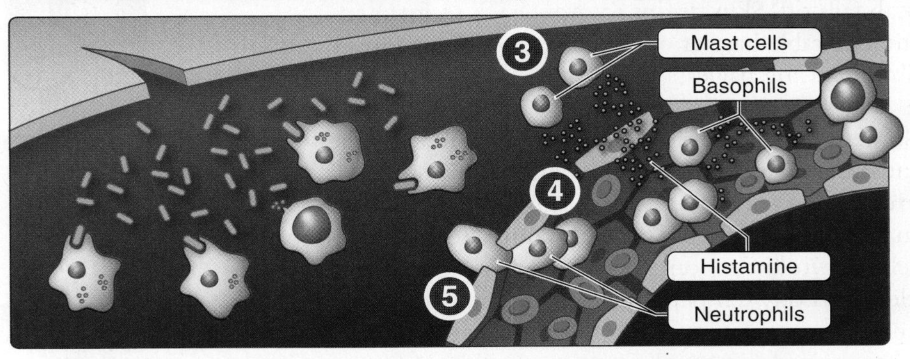

6. The inflammatory response continues until the pathogens have been eliminated and the skin grows back, forming an impenetrable barrier once again.

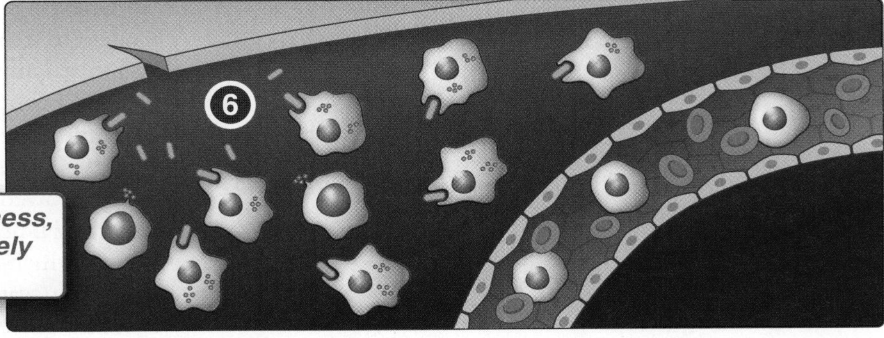

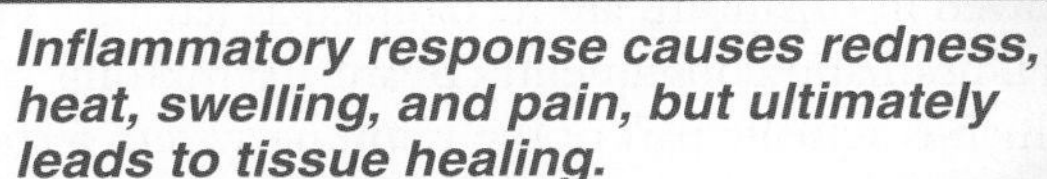

Inflammatory response causes redness, heat, swelling, and pain, but ultimately leads to tissue healing.

FIGURE 26-11 Reacting to injury. An inflammatory response causes redness, heat, swelling, and pain, but ultimately leads to tissue healing.

swelling, as fluid leaking from the blood vessels accumulates in the tissue.

Macrophages and neutrophils then begin an immediate response to destroy pathogens and also release cytokines. The various cytokines not only amplify the response by recruiting more immune cells but also can affect other cell types. For example, some of the cytokines released from macrophages, neutrophils, and mast cells can activate neurons and are therefore the cause of *pain*. Pain is also caused by the increased pressure on nerve endings resulting from the localized tissue swelling.

As we've seen, macrophages residing in the injured tissue initially recognize an invader and produce cytokines to attract neutrophils to the site of injury or invasion. As many as 70% of the white blood cells circulating in the bloodstream are neutrophils, and these plentiful cells move from the blood and into the tissue at the site of an injury. The blood vessels make it easier for them to reach the injured tissue, as cells lining the vessels near the site of inflammation become "sticky" to neutrophils (red blood cells do not stick). This stickiness slows down the neutrophils so that they can exit the leaky vessel at the site where they are needed.

Complement proteins also play a role in inflammation. Just as cytokines (secreted by macrophages) trigger the inflammatory response, activated complement proteins also can trigger initial reactions. Activated complement proteins at the site of infection cause mast cells to release histamine and further amplify the inflammatory response by attracting additional phagocytes. As mentioned in Section 26-3, complement proteins also directly affect pathogens by attacking their membranes and making it easier for phagocytes to engulf them.

So what about that cut on your finger? A scab temporarily forms to cover the wound, much like a natural Band-Aid, and prevents more pathogens from entering. Thanks to the inflammatory process, any pathogens initially entering the wound were immediately destroyed. At the end of the inflammatory process, collagen fibers are secreted by nearby cells to close the wound in a more stable fashion, and the scab falls off. Skin begins to grow back to form an impenetrable barrier once again. And you are more cautious about cutting bagels!

Sometimes an infection is overcome quickly by the inflammatory response and is confined to the site of damage. However, if pathogens are not quickly destroyed, macrophages will be constantly stimulated and continue to release cytokines. Some cytokines can cause a **fever,** an elevated body temperature, if their concentration is high enough—a situation that both stimulates the immune response and reduces the rate at which many pathogenic bacteria can divide. The fever-causing cytokines travel through the bloodstream and affect a region of the brain, called the hypothalamus, that functions as the body's thermostat (see Section 23-16). Just as you can change the settings on your home thermostat, your body's thermostat can also be set higher, thus leading to a fever (**FIGURE 26-12**).

Q Why do chills and sweating sometimes occur together when you have a fever?

When your body's thermostat is set higher with a fever, you feel cold and your body responds with rapid periods of muscle contractions and relaxations (shivers and chills). Chills are the body's way of trying to warm up to the higher temperature setting of the thermostat. When your fever "breaks," the thermostat is being reset to normal. You feel sweaty because your body is now too hot; sweating is a way to cool down. Anti-inflammatory medicines such as aspirin, acetaminophen, and ibuprofen help to lower a fever. They block steps in the biochemical pathways promoted by the fever-producing cytokines. A very high fever (105° F or higher) is extremely dangerous and can be fatal. At such high temperatures, proteins can denature (unfold) and critical biochemical processes can malfunction, causing cellular stress and, in small children, seizures.

FIGURE 26-12 Turning up the thermostat. Fever slows the growth of pathogens.

Q Is it always wise to treat a fever with aspirin or other medication?

In the StreetBio at the end of Chapter 20, we described the new perspective on when to treat and when not to treat symptoms, called "Darwinian medicine." From this perspective, many protective responses that have evolved in organisms are recognized as having value in fighting illness. And consequently, treating symptoms such as fever, which is actually part of the immune system response to infection, can hinder our efforts at fighting infection. This idea was supported, for example, by the observation that chicken pox lasted longer, on average, in patients taking aspirin than in patients taking a non-fever-reducing placebo.

TAKE-HOME MESSAGE 26•4

Inflammation is a major way in which pathogens are eliminated by the non-specific (innate) immunity system. The four recognizable signs of the inflammatory response (redness, heat, swelling, and pain) are related to the changes in blood vessels that enhance the recruitment of phagocytes and complement proteins to the site of inflammation. Fever-promoting cytokines cause the hypothalamus to set the body temperature higher, which may help the body fight an infection by stimulating the immune responses and inhibiting the growth of some pathogens.

② Specific immunity develops after exposure to pathogens.

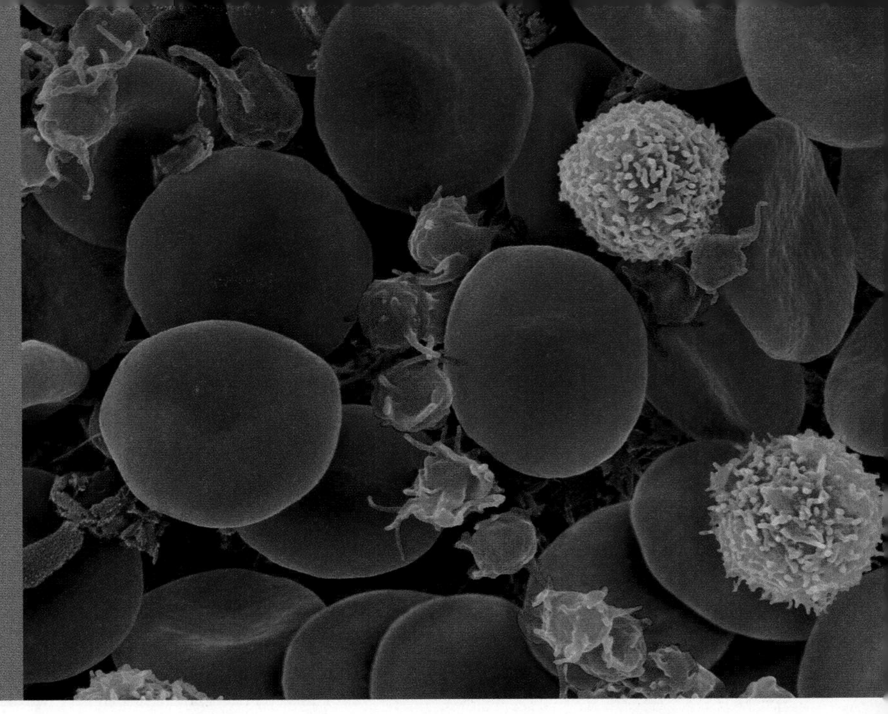

A selection of cells from the bloodstream, including red blood cells (red), three T lymphocytes (orange), and activated platelets (green).

26•5

The specific division of the immune system forms a memory of specific pathogens.

Often, a pathogen can be eliminated by the non-specific system's inflammatory response. Sometimes, however, the pathogen has molecularly "disguised" itself in some way and is not recognized as a pathogen by the non-specific system, or the infection is persistent. Luckily, the specific division of the immune system is also there to protect us. Recall that non-specific defenses first *recognize* pathogens and then *respond*. The same is true for the specific division, but the recognition is more specific, and the types of cells and molecular weapons used in the response differ from those deployed by the innate division—in a sense, there is a trade-off between the speed of the non-specific response and the slower, but much more precise, response of the specific immunity system. The specific immunity system forms a memory (it "adapts" over one's lifetime), so the body's response to a specific pathogen occurs more quickly in future encounters. But the specific immunity system doesn't work alone; only by working together can the specific and non-specific systems completely defend us from the constant barrage of pathogens that surround us.

The memory of the specific division of the immune system results in **immunity,** a state of long-term protection against a specific pathogen. There are two ways in which we can acquire immunity to a particular pathogen. The first is to become sick with the disease (**FIGURE 26-13**). For example, people who contracted chicken pox when they were young cannot get chicken pox again; they are immune to it. Fortunately, there is a second way to gain immunity, without having to suffer through the disease and its unpleasant symptoms. A vaccine for chicken pox has been widely used since the mid-1990s. A **vaccine** is a weakened or harmless form of a specific pathogen that is administered to an individual to induce immunity, without subjecting the individual to the disease. Vaccines trick the body into thinking it has the full-blown disease, and the body mounts a specific immunity response. An individual who receives the chicken pox vaccine and is later exposed to the virus will not get the disease, because the body already has a memory of this virus (meaning that a group of cells—memory cells—are already primed to recognize and rapidly respond to the virus). On encountering the virus, there will be no lag time in responding. Vaccine technology has been a major success in medicine.

Today we have vaccines against numerous viruses (such as polio, measles, rubella, mumps, and rotavirus) and bacteria (such as tetanus, cholera, and meningitis). These vaccines are less available in the developing world and thus, worldwide, many people die each year from what are now preventable illnesses.

EXPOSURE LEADS TO IMMUNITY

GETTING SICK
Exposure to a pathogen causes the body to form a memory of the pathogen, producing the cells necessary to rapidly respond to the pathogen if encountered again. Shown here: a toddler with chicken pox.

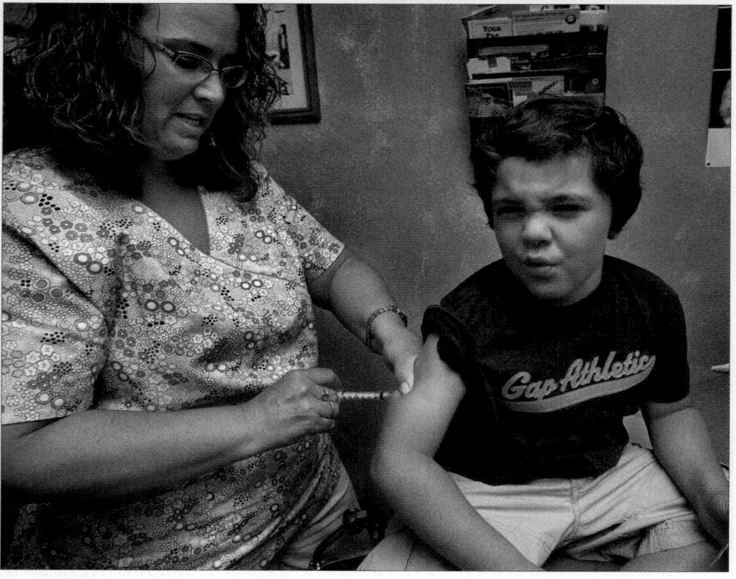

GETTING A VACCINATION
Exposure to a weakened or harmless form of a pathogen via a vaccine allows the body to form a memory of the pathogen without the risk of symptoms. The body then produces the cells necessary to rapidly respond to the pathogen if encountered again.

FIGURE 26-13 Two paths to immunity: contracting and fighting an illness, or receiving a vaccine.

The memory of the adaptive system gives you immunity to a pathogen following exposure to it (ideally, with the exposure as a vaccine, rather than as an illness).

An **antigen** is any molecule that induces a specific immune response. A single pathogen often contains many different antigens (for example, the different glycoproteins, lipoproteins, and polysaccharides of a single bacterium can all be antigens). The body responds to antigens by making **antibodies,** circulating proteins that recognize specific antigens. (You can remember it like this: antigen is short for *anti*body *gen*erating.) Antibodies provide protection by enhancing the non-specific system's ability to recognize and destroy bacteria or viruses and the body cells they infect (**FIGURE 26-14**). Antibody levels peak within the first 2 weeks after encountering a pathogen or receiving a vaccine. We discuss the body's ability to recognize and destroy antigens in Section 26-8.

Even though antibody production eventually diminishes, the specific immunity system retains a memory of the disease. If the same antigen is encountered again, it will take the body only 1–2 days to begin producing large amounts of specific antibodies that aid in destruction of the pathogen; within just 7–9 days, there will be massive amounts of the antibodies. So, if a friend exposes you to chicken pox after you've already had the vaccine, your body won't need 2 weeks to respond: it will quickly kick into high gear and start antibody production immediately!

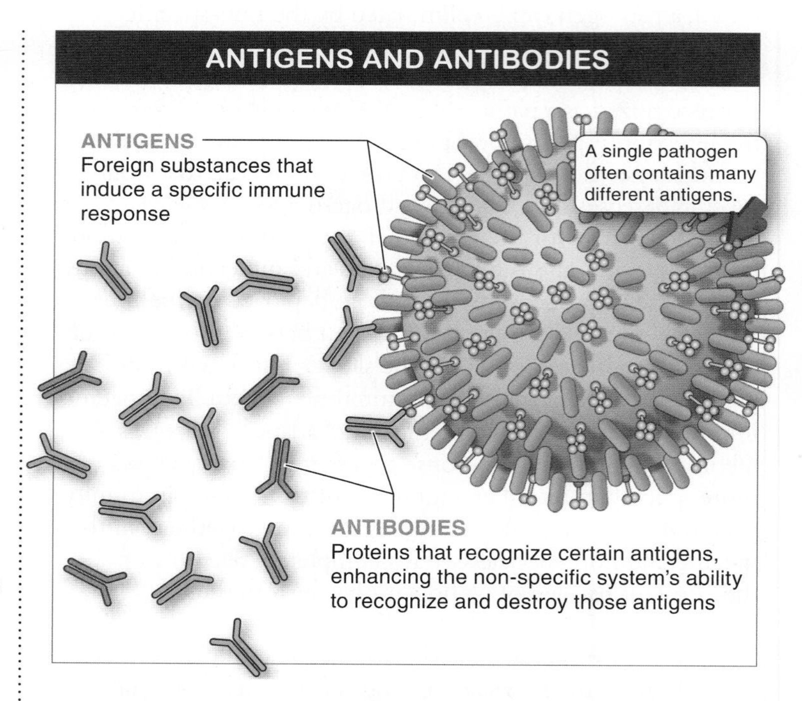

FIGURE 26-14 Antibodies are proteins that recognize foreign molecules called antigens.

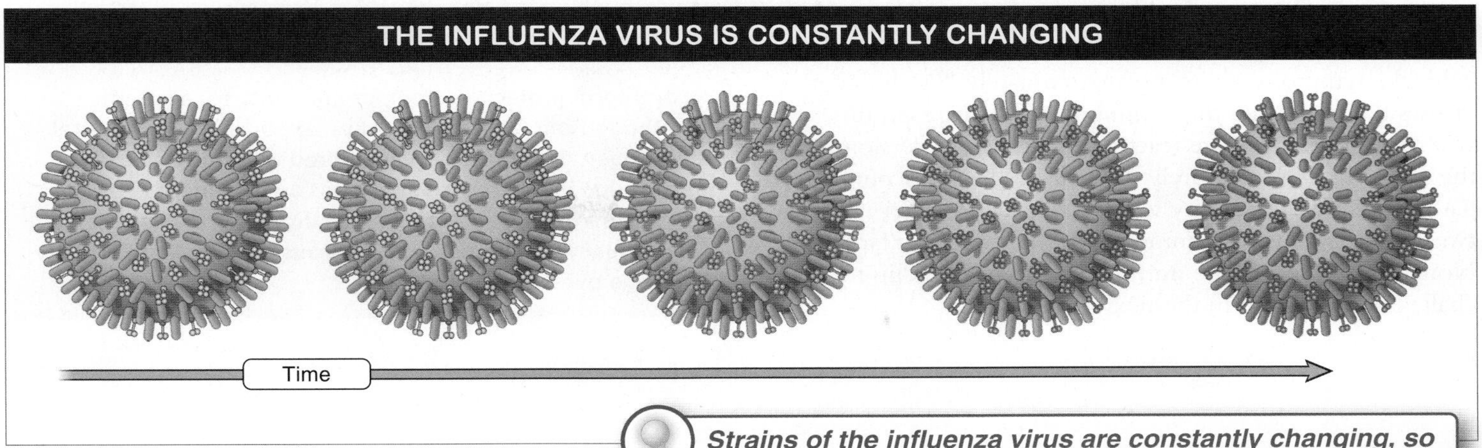

THE INFLUENZA VIRUS IS CONSTANTLY CHANGING

Time

Strains of the influenza virus are constantly changing, so the body encounters a slightly different form with each new flu season, and therefore a different set of antigens.

FIGURE 26-15 Because the influenza virus changes so rapidly, flu vaccinations need to be given annually.

Any vaccine must contain an antigen. In the case of the flu vaccine, for example, it is commonly an influenza virus that has been inactivated by heat or a live, weakened form of the virus that does not cause the flu. Although you can get the flu at any time of year, the flu season occurs in winter, so it makes sense to have the vaccine administered in autumn to give your body time to make antibodies that can protect you throughout the winter flu season.

Each year there are many different versions, or strains, of the influenza virus, which are constantly changing. With each new flu season, the body encounters a slightly different form of influenza and, therefore, a different set of antigens. As the virus adapts to changing conditions and evolves, the best protection is to get a flu shot each year (**FIGURE 26-15**). Yet, even with these preventive measures, a person may still get the flu. Although it is commonly thought that the flu shot itself can give someone the flu, this is simply not true. So, why do individuals sometimes get the flu after having the vaccination? There are several possible reasons. First, scientists developing the vaccine must make predictions, well before the flu season, about how the influenza virus will change in the upcoming season. Sometimes the vaccine "matches" the current virus, and sometimes it may be less than perfect. Second, different strains of the flu virus can circulate at the same time, and the vaccine may protect an individual from just one strain. Lastly, a person might already be infected with the flu at the time of vaccination.

Q Why don't people develop immunity to the common cold?

If the specific immune response can protect a person from illness during a future encounter with a specific pathogen, why is it that every winter you develop the runny nose, cough, and sore throat of the common cold? Why don't you develop immunity to this most annoying and recurrent disease? Just like the influenza virus, the rhinovirus ("nose virus"), one of the viruses that cause the common cold, continually changes (**FIGURE 26-16**). Your immune system mounts a response to each version of

There are approximately 200 different viruses that produce cold-like symptoms. Each time you catch a cold, you are reacting to a different version of the same virus or an entirely new virus.

FIGURE 26-16 A constant challenge. Many different, rapidly changing viruses cause the common cold.

the virus. Although rhinoviruses are a frequent cause of the common cold, they are not the only viruses that produce cold-like symptoms; there are approximately 200 different viruses that can cause the common cold. Thus, each time you catch a cold, you are reacting to a different version of the same virus or an entirely new virus. The sheer number of pathogens and their ability to adapt and evolve over time are two challenges that the immune system constantly faces. Nonetheless, the specific immunity system stands up to this challenge, as detailed in the next few sections.

TAKE-HOME MESSAGE 26•5

Long-term protection from, or immunity to, a specific pathogen can form in two ways: exposure to the natural pathogen or exposure to an altered version of the pathogen in a vaccine. Antibodies are produced after exposure to an antigen. The specific immunity system is continually responding to numerous pathogens that can change over time.

26•6

Lymphocytes fight pathogens on two fronts.

Lymphocytes are the white blood cells responsible for the specific immunity response. These cells can be found circulating in the blood and lymphatic systems, and they reside in lymphatic organs such as the lymph nodes and spleen. (See Section 21-9 to review the lymphatic system.) In contrast to blood, which contains various types of white blood cells, 99% of the lymph is made up of lymphocytes. Lymphocytes have **antigen receptors,** proteins on their plasma membranes that stick out from the cell surface and can bind to specific antigens—any one lymphocyte binding to just one type. The presence of antigen receptors allows lymphocytes, collectively, to recognize and react to a wide array of antigens.

There are two major types of lymphocytes. Both develop in the bone marrow, but one type leaves the bone marrow and continues to mature in the thymus, a lymphatic organ located in the upper chest. These lymphocytes are named for where they mature and so are called T lymphocytes, or **T cells.** The other type of lymphocyte remains and matures in the bone marrow, and these are fittingly named B lymphocytes, or **B cells** (**FIGURE 26-17**).

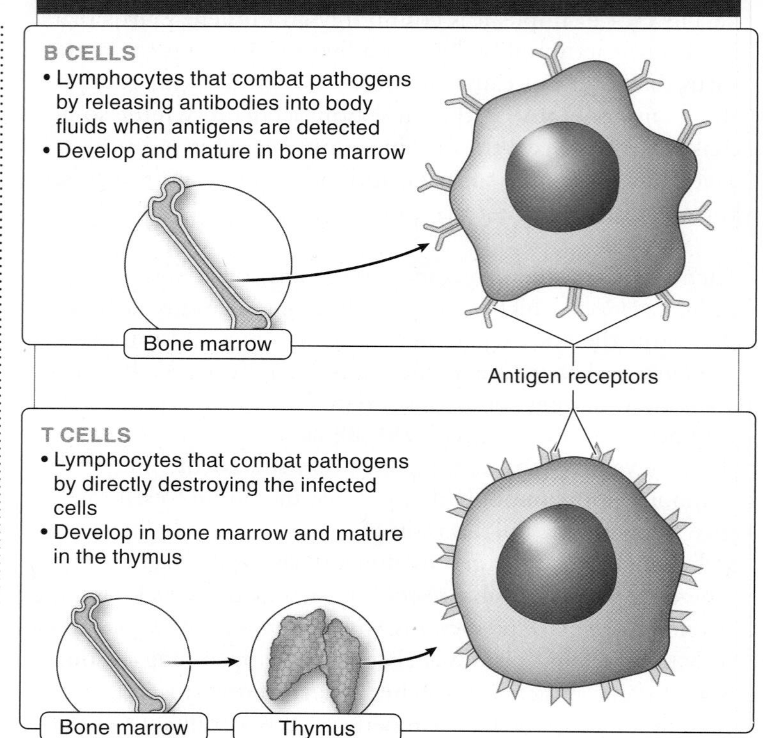

FIGURE 26-17 B cells and T cells. These two types of lymphocytes are responsible for the specific immunity response. They are named for the location in the body where they mature (the bone marrow and thymus).

The other cells that normally circulate in the lymphatic system are dendritic cells, which use the lymphatic system for the sole purpose of meeting up with lymphocytes. Recall that dendritic cells display digested pieces of pathogens, or antigens, to adaptive immune cells.

Immune system reactions are comparable to a military battle. In real-life combat, it is advantageous to attack by both land and air to win a battle. Like the armed forces, the specific immune system has major divisions, too. Lymphocytes fight pathogens on two fronts: humoral and cell-mediated immunity (**FIGURE 26-18**).

SPECIFIC IMMUNE SYSTEM RESPONSES

Pathogens
Antigens
B cells
Antibodies
Antigen receptors

HUMORAL IMMUNITY

- Protection against pathogens and toxins found in body fluids, such as blood and lymph
- Carried out by B cells that secrete antibodies into body fluid, making it easier for phagocytes to engulf and destroy invading pathogens

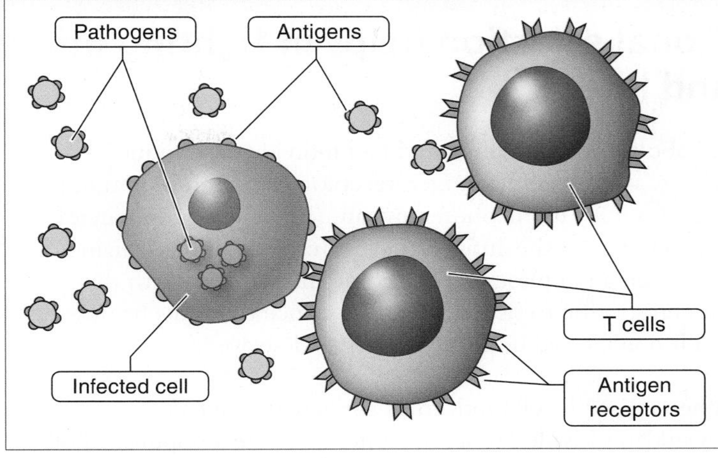

CELL-MEDIATED IMMUNITY

- Protection against pathogens and toxins found within body cells
- Carried out by T cells that directly destroy invading pathogens as well as the infected cells

FIGURE 26-18 Fighting invaders in body fluids and within cells. B cells are responsible for humoral immunity (antibody-mediated immunity) and T cells are responsible for cell-mediated immunity.

1. *Humoral immunity,* also referred to as antibody-mediated immunity, is protection against pathogens and toxins found in body fluids, such as blood and lymph. The word *humoral* (to give just a quick history lesson) comes from "humors"—the term Greeks and Romans used for body fluids. Blood was one of the four humors (the other three were phlegm, yellow bile, and black bile). Antibodies are secreted into the body fluids by cells derived from B cells following detection of an antigen. The antibodies circulating in blood and lymph defend against pathogens and toxins, making it easier for phagocytes to engulf them.

2. *Cell-mediated immunity* is protection from pathogens located *inside* body cells and is carried out by T cells. Your body cells have molecules on their cell membranes that identify them as "self" to the cells of your immune system—as cells that belong. But body cells that have been infected by a pathogen (and many cancer cells) present some different proteins (antigens) on their membranes. T cells recognize and bind to these antigens, initiating an immune response that kills the cells (and, usually, the pathogen within).

Both B and T cells have antigen receptors on their surfaces. These receptors differ in structure and are characterized by both diversity and specificity. There are endless variations of these receptors. It is estimated that our lymphocytes can recognize billions of different pathogens. But each individual lymphocyte doesn't bear billions of different antigen receptors. Instead, each lymphocyte has only one type of receptor that recognizes only one antigen, and it has many copies of this receptor on its surface. To defeat a specific pathogen, numerous identical lymphocytes (bearing the same antigen receptor) are needed.

You may be starting to ask yourself, do we have trillions of immune cells? How does the body have room for all of these individual armies of lymphocytes ready to face any enemy it might—or might not ever—encounter? The answer is that the body doesn't store vast quantities of identical lymphocytes. It makes copies of these lymphocytes *only if* the antigen is encountered. In the next section we see how the body does this.

TAKE-HOME MESSAGE 26•6

Two types of lymphocytes are associated with the specific immunity system: B cells and T cells. B cells are responsible for the humoral (antibody-mediated) response, and T cells for the cell-mediated response. Because of the diversity and specificity of lymphocyte receptors, almost any pathogen can be recognized by the body's B and T cells.

26•7

Clonal selection helps in fighting infection now and later.

Just about every lymphocyte in an individual is unique because of its specific antigen receptor. Although the immune system is *ready* to encounter any antigen, many lymphocytes will never meet the antigen they are capable of recognizing. But when a lymphocyte does come into contact with the antigen specific to its receptor, a sequence of events begins—and it doesn't end until the antigen is destroyed.

When a B or T cell binds to its antigen, the cell and its descendants divide numerous times to create a population of genetically identical cells (clones) with the same antigen specificity. This process, known as **clonal selection,** ensures that there are enough B and T cells to recognize and respond to a specific pathogen that has invaded the body.

Recognizing a new pathogen is an important first step, but the specific immunity system has two additional challenges: to *respond* to and *remember* the invader. Clonal selection generates many lymphocytes that can recognize the same pathogen, but these lymphocytes function in two different ways. Some attack the antigen (these are called effector cells) and some are involved in creating a memory of the invasion (the memory cells). We discuss the effector cells first.

The responders, or **effector cells,** recognize an antigen and immediately take some action that leads to its destruction. For example, **plasma cells,** derived from B cells, are the effector cells in the humoral response because they secrete antibodies. And T cells that directly kill infected cells are the effector cells in the cell-mediated response. During the immune system's first interaction with a pathogen, called the **primary response,** effector cells are not immediately available. The cells of the primary immune response must first be generated through clonal selection, which takes 2 weeks to produce effective numbers (**FIGURE 26-19**).

Also produced during the primary response is the second type of lymphocytes, the **memory cells.** Like effector cells, they are produced through clonal selection. The job of these lymphocytes is to remember an antigen so that, if the body is infected with the same antigen in the future, it will be ready to attack it. Thus, memory cells hang around in the lymph and blood and wait for their specific antigen to show up again. There are both B and T memory cells, and they have the same antigen specificity as the effector cells produced in the primary response. While effector cells live less than a week, memory cells remain and circulate for a long time. The lifespan of memory cells is variable, but they may exist for years and possibly even for the individual's lifetime.

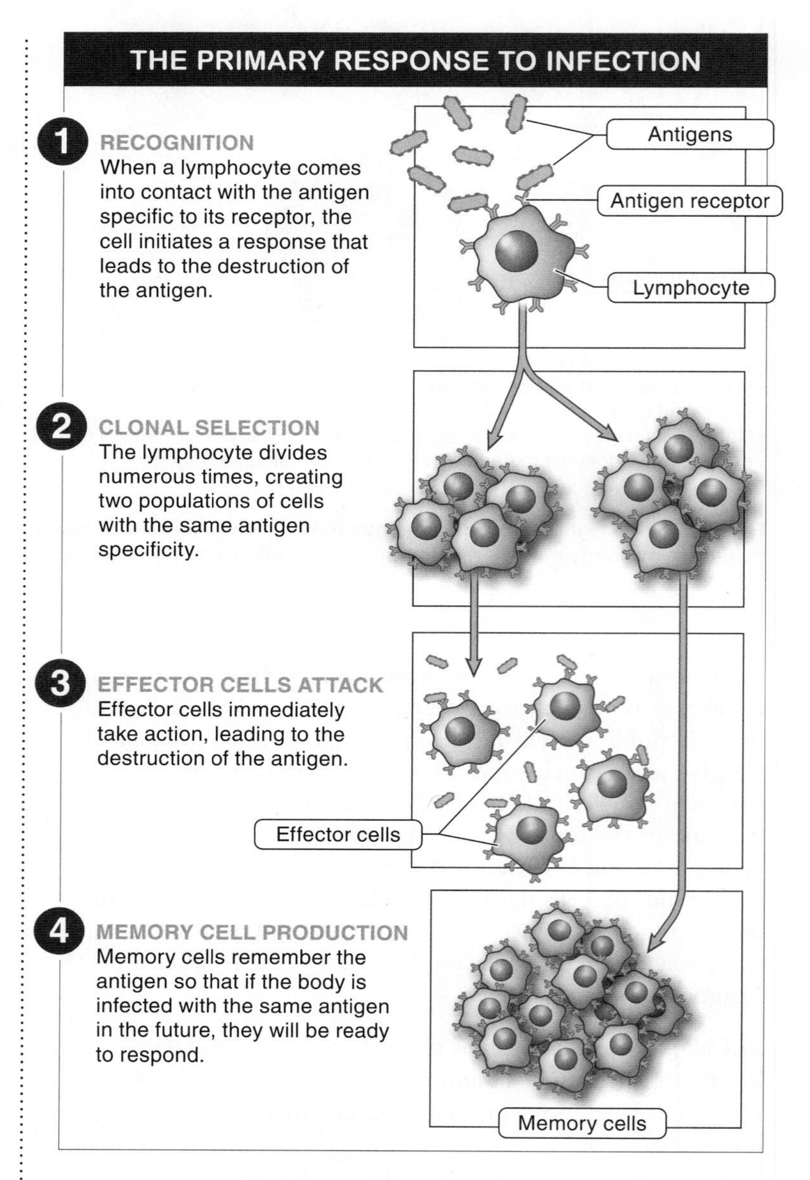

FIGURE 26-19 Fighting now and later. The primary response leads to destruction of an antigen and generation of memory cells to fight the antigen should it ever be encountered again.

With this understanding of how effector cells and memory cells act in the primary immune response, let's look at what happens when a five-year-old kindergartener comes to school

THE SECONDARY RESPONSE TO INFECTION

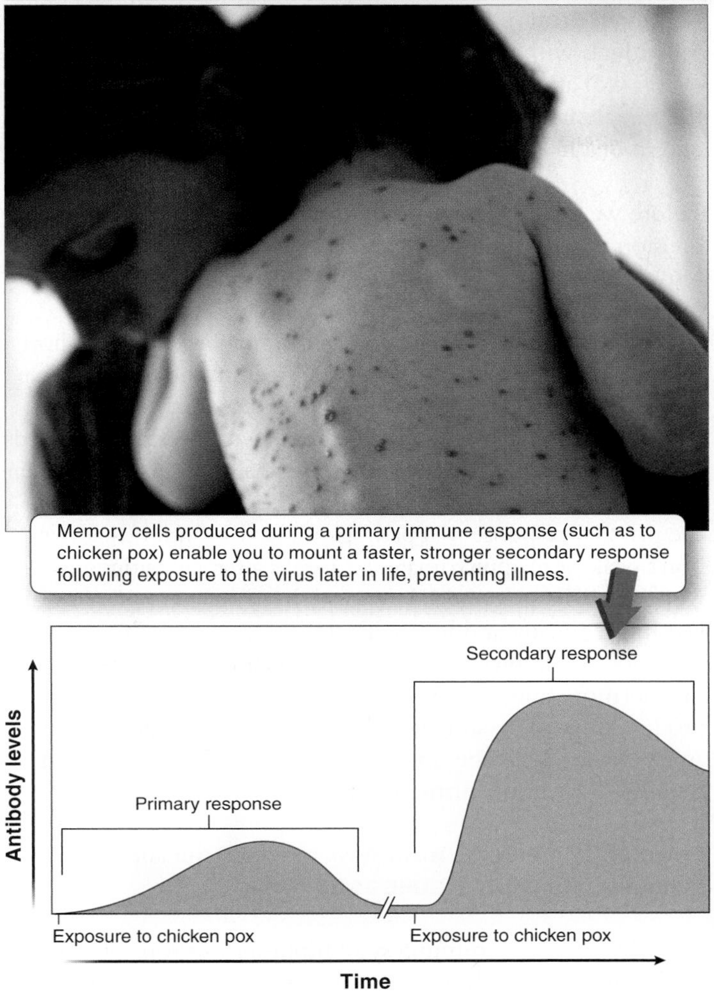

Memory cells produced during a primary immune response (such as to chicken pox) enable you to mount a faster, stronger secondary response following exposure to the virus later in life, preventing illness.

FIGURE 26-20 Another encounter with the same antigen. Because of memory cells, the secondary response is able to fight an antigen quickly and effectively.

not knowing that he has a chicken pox infection. Imagine that the kindergarten teacher had chicken pox as a small girl. At that time, her specific immunity system produced a population of chicken pox memory cells during her primary immune response. When the teacher interacts with the infected boy, these specific memory cells go through clonal selection and give rise to new B and T effector and memory cells. This **secondary response,** the creation of B and T effector cells, occurs rapidly (beginning after as little as 1–2 days) and produces a more intense response (peaking at 7-9 days) than the primary response that occurred in her childhood (**FIGURE 26-20**). Because the B cells of humoral immunity and the T cells of the cell-mediated response work immediately to destroy the virus, the teacher won't even miss a day of work, because this secondary immune response attacks the chicken pox virus so rapidly and effectively.

TAKE-HOME MESSAGE 26•7

When the body encounters a specific antigen, the lymphocytes recognizing this antigen divide to produce many cells through clonal selection. During a primary response, effector cells respond to and facilitate removal of the antigen. Memory cells produced during the primary response are ready to go through clonal selection if a secondary response is necessary.

26•8

The structure of antibodies reflects their function.

At this point you've learned that B cells are the effectors of the humoral response and that they produce antibodies, which are proteins that enhance the non-specific system's ability to recognize and destroy bacteria or viruses and infected body cells. We now examine the structure and function of antibodies to learn exactly how they work with the non-specific division of the immune system to protect us against pathogens.

During the primary immune response, a B cell with a highly specific receptor recognizes an antigen. The B cell receptors are actually membrane-bound antibodies on the cell surface. All antibodies are Y-shaped, whether they are functioning as receptors or as soluble proteins released by B cells into lymph and blood. The diversity and specificity of antibodies come from variations in this Y-shaped structure.

Each antibody consists of four polypeptide chains joined together: two long (also called heavy) chains and two short (or light) chains (**FIGURE 26-21**). The top of the "Y" is the region that varies from antibody to antibody. But there are only

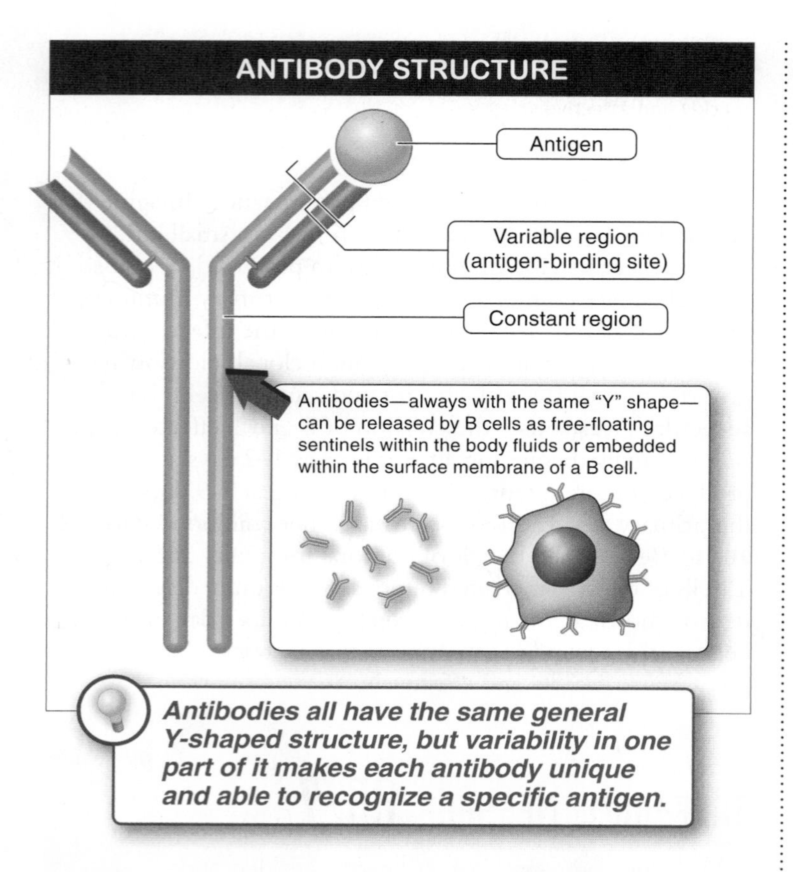

FIGURE 26-21 Variations on a molecular theme. The shape of the top of the "Y" varies from one antibody to another and fits with only one antigen.

about five variants for the lower part, the base of the "Y," and so this is referred to as the constant region of an antibody. In the variable region, each antibody has a unique three-dimensional shape that dictates which specific antigen it will "fit." An enormous diversity of antibodies is needed to recognize a correspondingly enormous group of antigens that the immune system may encounter. Different antibodies can recognize different antigens on a given pathogen. Keep in mind that antibodies recognize *antigens,* not pathogens. And there are antigens that are not part of pathogens. An antigen is anything that an antibody recognizes, including snake venom toxins, bacterial toxins that cause food poisoning, pollen, and various chemicals (such as latex).

Any foreign cells introduced into the body will elicit an immune response, even if the cells are part of a life-saving medical treatment. Consider organ transplants. Transplanted cells have surface molecules that are seen as antigens ("non-self") by the recipient's immune system. Although the transplanted cells are not pathogens, the recipient's immune system identifies them as invaders, and an immune response begins. Medicines are needed to suppress the immune system's attack on and eventual destruction of a transplanted organ.

The immune system is also the reason that people in need of blood transfusions must have a proper match of blood type. As we saw in Section 7-10, if a transfusion is made between individuals with incompatible blood types, antibodies from the donor will attach to antigens on the surface of the foreign (i.e., the recipient's) red blood cells and mount an attack that can end in death.

Before we describe how antibodies work with the non-specific division of the immune system, let's briefly review the humoral response. When the humoral response begins, clonal selection produces two types of B cells: effector B cells that attack immediately and memory B cells that are inactive in the short run but will recognize a particular antigen in the future. As we mentioned in Section 26-7, the effector B cells that secrete antibodies are called plasma cells, and they can secrete hundreds to thousands of antibodies per second as a consequence of their extensive endomembrane system (see Section 3-17). During the primary response to a pathogen invasion, antibody levels don't peak until approximately 2 weeks after the antigen is recognized. In a secondary response, however, antibody levels rise more quickly and in larger numbers because of the memory B cells produced during the primary response. Furthermore, antibody production reaches higher levels and production is more sustained in this secondary response. Antibody levels can be up to 1,000 times higher in a secondary response than in the primary response

Now let's consider the main ways in which the antibodies themselves function (**FIGURE 26-22**).

As we've seen, the immune system pumps out tons of antibodies during a humoral response, but antibodies don't *directly* kill the pathogen. Instead, they use the non-specific cells of the non-specific immune system (such as macrophages and neutrophils) to kill the pathogen. When the antigens on the surface of a pathogen have an antibody attached to them, the pathogen is easily ingested by phagocytes of the innate system. The antibodies act as beacons that make pathogens highly conspicuous to patrolling phagocytes.

Besides "marking" a pathogen—or marking free-floating antigens that are not part of a pathogen, such as a snake venom toxin—for destruction, antibodies act in two other important ways: (1) they cause pathogens or antigens to clump together, making it easier for phagocytes to find and ingest them; and (2) they coat the surfaces of pathogens and prevent them from binding to and entering body cells, thus blocking the infection from spreading.

Lastly, antibodies have a role in recruiting another player of the non-specific system when bound to the surface of a pathogen: complement proteins. Recall that some proteins of

ANTIBODY FUNCTIONS

Antibodies function in several ways to help destroy pathogens and soluble antigens.

PHAGOCYTE SIGNALING
Antibodies bind to antigens on the surface of pathogens, making it easier for phagocytes to find the pathogens and destroy them.

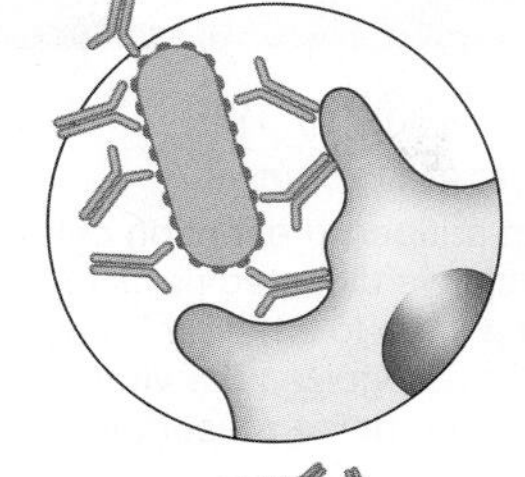

ANTIGEN CLUMPING
Antibodies make pathogens and soluble antigens clump together, making it easier for phagocytes to find them and destroy them.

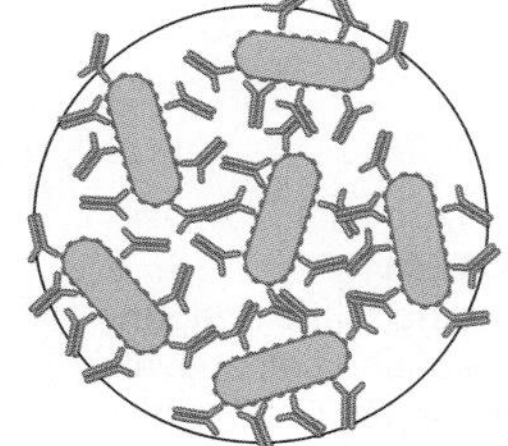

PREVENTION OF CELL ENTRY
Antibodies coating the surface of pathogens prevent the pathogens from entering body cells, thus blocking the infection from spreading.

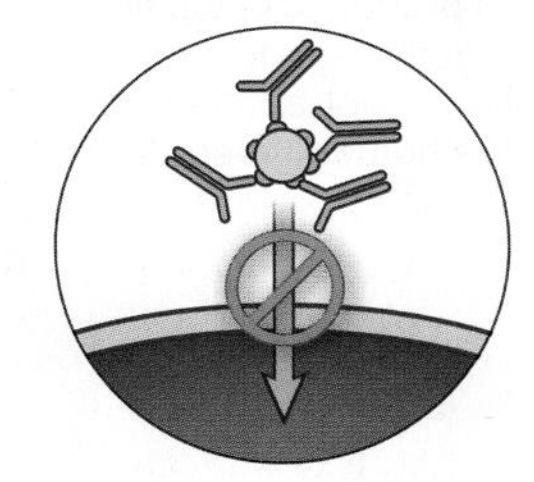

COMPLEMENT PROTEIN SIGNALING
Antibodies recruit complement proteins, which poke holes in pathogen membranes, causing the pathogen cells to burst.

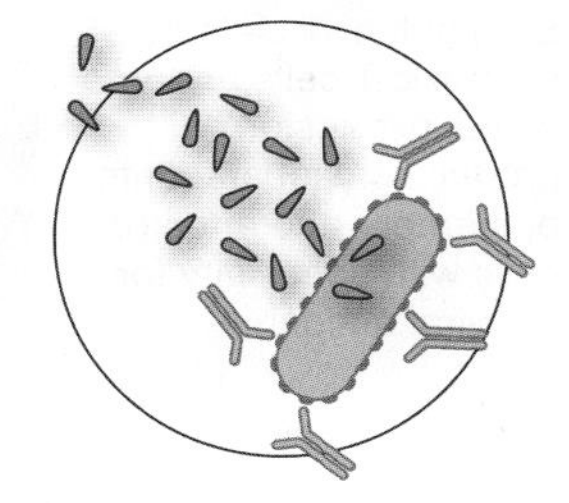

FIGURE 26-22 Antibodies fight pathogens in several ways.

the complement system can poke holes in membranes and cause cells to lyse (burst). Antibodies indirectly lead to pathogen lysis through their recruitment of these complement proteins.

Q Do newborn babies have any protection from diseases?

Would isolating antibodies from one person's blood and giving them to another person who has not yet encountered that disease protect the recipient? Yes. But, it would only be short-term protection, because the memory B cells that have the ability to make more antibodies are not transferred. Transferring antibodies with a given specificity is termed *passive immunity*. This is in contrast to vaccination, in which the immune system develops long-term, *active immunity*. An example of passive immunity occurs during pregnancy—antibodies produced by the mother are transferred to the baby's bloodstream and protect the baby for the first few months of life.

FIGURE 26-23 An excellent reason to breastfeed an infant. Passive immunity transfers immunity from mother to child.

How do we know about passive immunity? In some regions of the developing world where measles is still prevalent, infants breastfeeding from mothers who have either had the measles or received the measles vaccine have antibodies in their bloodstream in the first few months of life and are protected from the disease during these months (**FIGURE 26-23**). This protection only lasts while the antibodies are still present. (Remember, infants are also born with the non-specific immune system, which provides much protection in addition to that of their immature specific immunity system.) Interestingly, passive immunity is also the key to treating snake bites. The anti-venom injection given at the site of the bite is actually a solution of antibodies to the venom antigens.

TAKE-HOME MESSAGE 26•8

Each antibody has a unique structure that recognizes a specific antigen. Plasma cells secrete high levels of antibodies into the blood and lymph. Antibodies are effective in helping to destroy pathogens and soluble antigens by enhancing ingestion by phagocytes, preventing more cell infection, and enhancing complement-driven lysis. Passive immunity is just short-term protection, because only antibodies are transferred, not the B cells that make the antibodies.

26•9

Cytotoxic T cells and helper T cells serve different functions.

The specific immunity response, as we've seen, is made up of both the humoral response (B cells) and the cell-mediated response (T cells). Why the two-pronged approach? Aren't antibodies sufficient? In a word: *no.* Consider a viral infection. Antibodies, circulating in blood and lymph, can recognize and bind to a specific virus in these body fluids. A virus that is coated with antibodies cannot effectively bind to and enter a host cell, which is a required event in viral infection. This antibody coating is effective in slowing infection, but what happens to all the body cells that already have viruses hiding inside them? Antibodies cannot enter these cells—it is the role of the T cell-mediated response to fight pathogens that are already inside cells. We focus here on this T cell-mediated response, and begin by examining the two major types of T cells, both of which are lymphocytes:

1. **Cytotoxic T cells,** which are the effectors of the cell-mediated response; they directly kill cells infected with pathogens.
2. **Helper T cells,** which do not directly kill infected cells but, instead, stimulate other immune cells. They are required in the stimulation of B cells to produce antibodies and of cytotoxic T cells to kill infected cells.

We'll follow a viral infection through the cell-mediated response to demonstrate how T cells function (**FIGURE 26-24**).

When a pathogen enters the body, the phagocytes and NK cells of the non-specific system respond. Meanwhile, dendritic cells and macrophages call on the specific system by "advertising" antigens to circulating lymphocytes. Dendritic cells and macrophages (as well as B cells) that "present" digested particles of pathogens on their cell surfaces are called **antigen-presenting cells.** The role of presenting cells is essential to T cell functioning, because T cells can only recognize an infected cell by the antigens presented on its plasma membrane.

Imagine a dendritic cell in your throat that has engulfed a respiratory virus and is presenting an antigen (digested virus particle) on its cell surface. If this presenting cell is to increase the likelihood of "meeting up" with a T cell that will recognize this specific antigen, hanging out in the tissues of your throat is not very effective. It is more effective for the

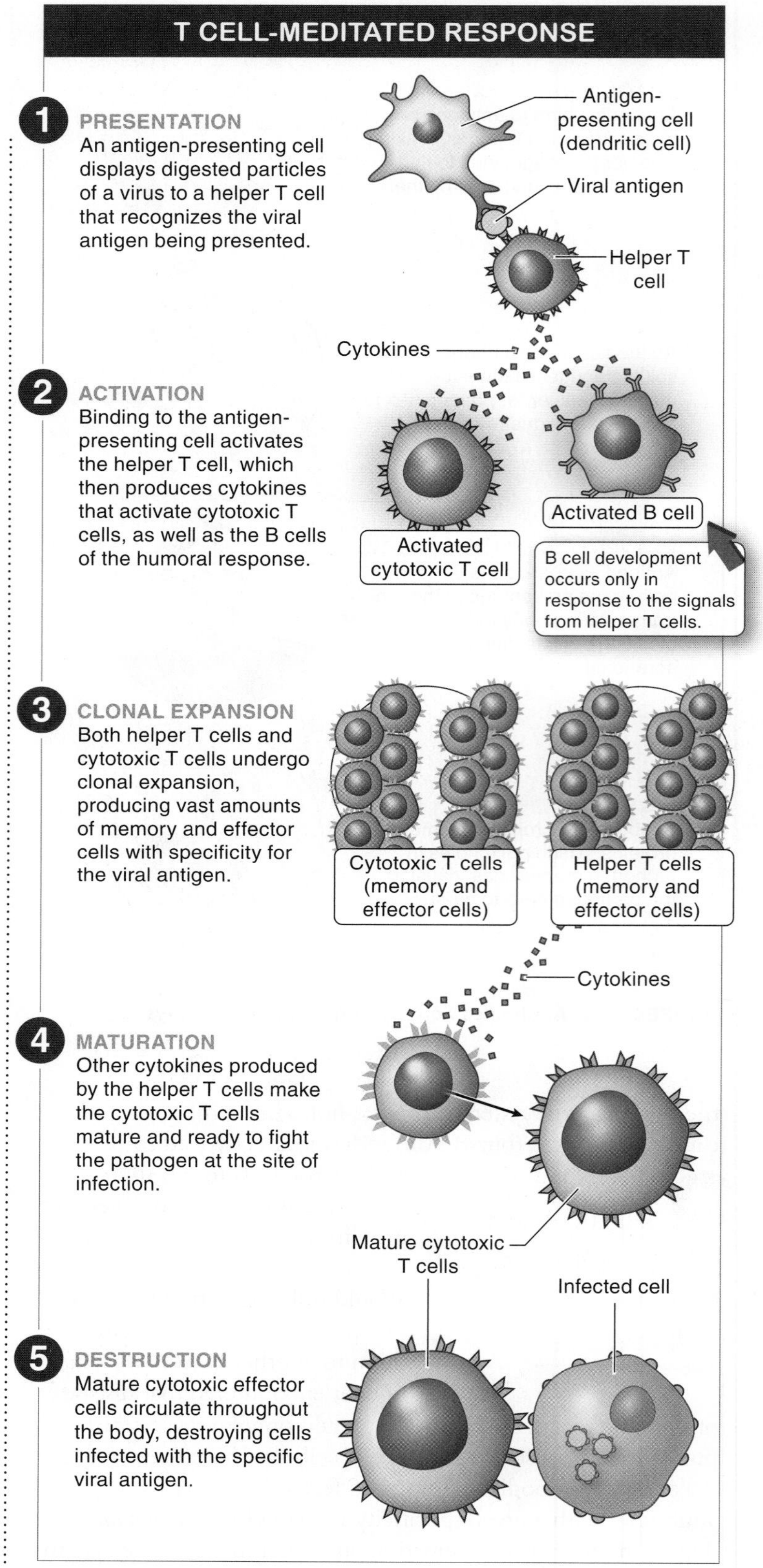

T CELL-MEDITATED RESPONSE

1 PRESENTATION
An antigen-presenting cell displays digested particles of a virus to a helper T cell that recognizes the viral antigen being presented.

Antigen-presenting cell (dendritic cell)
Viral antigen
Helper T cell
Cytokines

2 ACTIVATION
Binding to the antigen-presenting cell activates the helper T cell, which then produces cytokines that activate cytotoxic T cells, as well as the B cells of the humoral response.

Activated cytotoxic T cell
Activated B cell
B cell development occurs only in response to the signals from helper T cells.

3 CLONAL EXPANSION
Both helper T cells and cytotoxic T cells undergo clonal expansion, producing vast amounts of memory and effector cells with specificity for the viral antigen.

Cytotoxic T cells (memory and effector cells)
Helper T cells (memory and effector cells)
Cytokines

4 MATURATION
Other cytokines produced by the helper T cells make the cytotoxic T cells mature and ready to fight the pathogen at the site of infection.

Mature cytotoxic T cells
Infected cell

5 DESTRUCTION
Mature cytotoxic effector cells circulate throughout the body, destroying cells infected with the specific viral antigen.

FIGURE 26-24 Helper T cells and cytotoxic T cells work together to recognize and respond to pathogens.

presenting cell to circulate in the lymph nodes and spleen where lymphocytes circulate.

A dendritic cell circulating through a lymph node or the spleen may bump into T cells with specificity for the viral antigen being presented. The T cell receptor recognizes the antigen, and the T cell "locks" onto the antigen-presenting cell. Once this binding has taken place, the helper T cell has been "activated." Activated helper T cells produce signaling molecules (cytokines), most notably interleukin 2 (IL-2), that activate the other type of T cell, cytotoxic T cells. Both helper T cells and cytotoxic T cells undergo clonal expansion, producing vast numbers of memory and effector cells with specificity for this respiratory virus antigen. Other cytokines produced by the activated helper T cells provide the final signals that make the cytotoxic T cells "mature" and ready to fight the pathogen at the site of infection. You may notice this process occurring; the rapid division of lymphocytes in the lymph nodes is often pronounced and sometimes produces "swollen glands." The T cells then leave the lymph nodes or spleen and circulate throughout the body. (Helper T cells also are required to stimulate B cells to produce antibodies; B cell development occurs only in response to these signals from helper T cells.)

Mature cytotoxic effector cells are like newly trained soldiers, ready to fight the enemy, but first they need to detect where the enemy is hiding. Consider that infected cell in the throat again. Is it revealing that it's under attack? Yes! The infected cell displays some of the pathogen's molecules on its cell surface receptors, advertising the infection to the now numerous cytotoxic T cells that are able to recognize the specific respiratory virus.

How do the cytotoxic effector cells fight the internal pathogen? They do something drastic—they kill the throat cell that is infected. The weapons that they use are proteins: some punch holes in the cell's plasma membrane, and others promote the cell's self-destruction, or **apoptosis.** This programmed cell death does not explode the cell—sending virus particles everywhere. Just the opposite—it is a neatly organized, well-orchestrated process carried out by enzymes that break down macromolecules inside the cell and fragment the cell into smaller vesicles. Phagocytes then clean up the debris (**FIGURE 26-25**). So, while some body cells are lost in the battle, ultimately it is for the good of the entire organism.

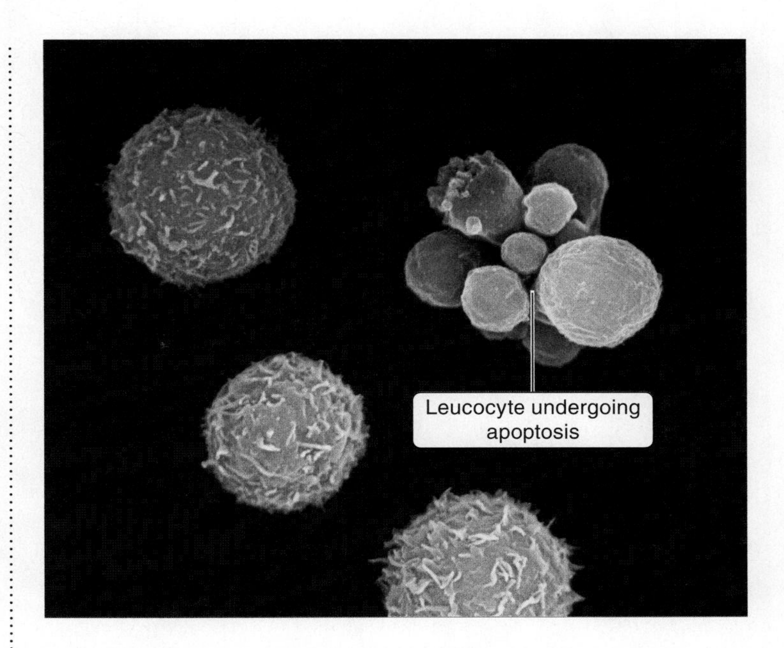

FIGURE 26-25 Programmed cell death. Shown are four leukocytes (white blood cells). The one at the top right is undergoing apoptosis.

Because memory helper T cells and memory cytotoxic T cells are also made during the throat infection, this same virus cannot cause illness again. Remember, this is specific immunity, and memory cells will leap into action to make more effector cells if the virus is encountered again.

TAKE-HOME MESSAGE 26•9

Antibodies, produced by B cells, cannot destroy pathogens that are inside cells. The specialization of cytotoxic T cells is required to kill infected cells. Antigen-presenting cells display antigens to circulating helper T cells and cytotoxic T cells, to alert the specific immunity system to an infection. In turn, helper T cells produce cytokines that instruct cytotoxic T cells to mature and respond to the infection.

③ Malfunction of the immune system causes disease.

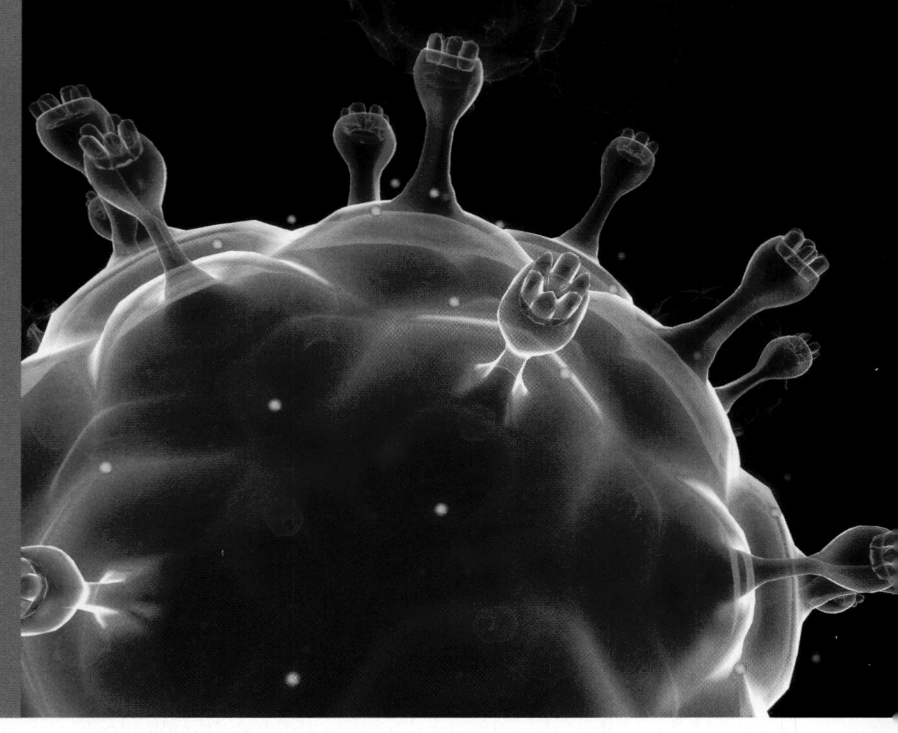

The human immunodeficiency virus (HIV).

26•10

Autoimmune diseases occur when the body turns against its own tissues.

What happens when the immune system malfunctions? Genetic defects, environmental influences, or even viruses can trigger immune dysfunction. And in many cases, such as autoimmune diseases, a combination of causes is suspected, but not completely understood. **Autoimmunity** occurs when an individual's immune system responds inappropriately to the individual's own cells and tissues as if they were pathogens, mistaking "self" for "non-self." If immunity can be compared to warfare, then autoimmunity is equivalent to an army turning its weapons upon its own citizens. Lymphocyte receptors wrongly recognize an individual's own molecules or cellular structures as antigens, and the humoral and/or cell-mediated immune responses can be initiated by these **autoantigens.** Most autoimmune diseases result in significant tissue and organ damage, and some can even result in death.

Type 1 diabetes (or juvenile diabetes) is an example of an autoimmune disorder. It is diagnosed in more than 13,000 young people in the United States each year. Symptoms include high blood glucose levels, weight loss, unquenchable thirst, frequent urination, fatigue, and weakness. In this disease,

AUTOIMMUNITY: TYPE 1 DIABETES

Type 1 diabetes is an autoimmune disorder in which cytotoxic T cells destroy one's own pancreatic cells.

Healthy pancreatic cells

Cytotoxic T cells

T cell receptors incorrectly recognize healthy pancreatic cells as antigens, initiating a cell-mediated immune response against them.

FIGURE 26-26 Type 1 diabetes is an all-too-common autoimmune disorder.

cytotoxic T cells destroy the person's own pancreatic cells. Antibody production leads to even more damage through complement-mediated lysis and macrophage ingestion of pancreatic cells. Without a normal pancreas, a person cannot make insulin (**FIGURE 26-26**). Without insulin, cells in need of glucose are not able to take it up from the blood, even though glucose is plentiful. In turn, the body breaks down muscle and fat tissues to provide cells with needed energy, and blood glucose levels remain high. The exact cause of the autoimmune reaction is not understood. Treatment for individuals with type 1 diabetes is insulin injections or implantation of an insulin pump, which directly monitors blood glucose levels and releases insulin as needed.

In some disorders, including type 1 diabetes, the immune system destroys a single cell type or a single organ. In other disorders, the damage is spread throughout the body. When the insulation (myelin) that surrounds nerve fibers of the brain and spinal cord is under immune attack, the resulting disorder is called **multiple sclerosis (MS)** (see Section 23-4). Individuals with MS usually begin to exhibit symptoms between 20 and 40 years of age, and these symptoms include blurred vision, weakness in the arms and legs, trouble with balance, and tingling sensations (**FIGURE 26-27**). This disorder tends to affect women more than men, and it affects more than 300,000 people in the United States. There is no cure for MS, but anti-inflammatory drugs can reduce the severity of some symptoms. Many new drugs are being investigated, including some that block cytokines and hence the inflammatory response.

Rheumatoid arthritis is another autoimmune disease that affects women more frequently than men, and usually strikes between ages 40 and 60. (This form of arthritis is a far less common cause of joint inflammation than osteoarthritis, which is not an autoimmune disease.) The linings of joints are the autoantigens of this immune attack—leading to pain and severe joint swelling that can result in deformities. Various pharmaceuticals are available to help reduce symptoms, and joint replacement surgeries also provide relief for individuals suffering from rheumatoid arthritis.

FIGURE 26-27 **An autoimmune disorder: multiple sclerosis.**

TAKE-HOME MESSAGE 26•10

When lymphocytes bear receptors that inappropriately recognize structures of a person's own body, autoimmunity develops. Autoimmune responses can do significant damage to specific organs or to tissues throughout the body, depending on where the antigens are located.

26•11

AIDS is an immune deficiency disease.

Autoimmune diseases demonstrate how immune recognition can fail and have devastating effects on the body. Here we examine a case in which recognition of a true pathogen occurs, but the *response* is deficient. Normally, the immune system destroys body cells that are infected with a virus. But what if the infected cells are immune cells that are critical for mounting a normal immune response? The **human immunodeficiency virus (HIV),** which causes **AIDS (acquired immune deficiency syndrome),** infects immune system cells, including helper T cells—cells that are crucial to survival. With AIDS, a deficiency of helper T cells leads to complete failure of the specific immunity response. As a result of this failure, individuals are prey to, and may die from, infections that a healthy immune system would have no trouble defeating.

HIV infection occurs through contact with blood, semen, vaginal fluid, or breast milk from an infected person. Thus, using clean (not shared) intravenous needles and using condoms are effective precautionary measures against HIV infection. People who become infected with HIV often do not have immediate symptoms and often do not know they are infected. Because individuals are contagious once they are infected, this initial lack of symptoms contributes to the spread of the disease. Let's examine the details of how immunity diminishes as the immune system loses its battle against HIV.

Different viruses infect different kinds of body cells. HIV infects helper T cells, macrophages, and dendritic cells. When an individual is first infected by HIV, a normal specific immunity response is mounted, and infected helper T cells are killed by both humoral and cell-mediated responses. More than 99% of the virus is destroyed in this primary response. Although many helper T cells are lost, new helper T cells arise. As you might expect, antibodies against HIV are made in a humoral response and can easily be detected in blood tests within months of the initial infection. Antibody detection marks the end of the first phase of HIV infection, the *acute phase* (**FIGURE 26-28**).

Unfortunately, the viruses that were not destroyed by the primary response can lie dormant inside infected helper T cells and macrophages. An infected person then enters a period in which he or she is infectious but doesn't have outward symptoms, called the *chronic phase.* This period is variable but rarely exceeds 12 years. Although the infection may seem inactive, a battle is going on within the immune system. Some of the HIV viruses lie dormant but others replicate and infect new helper T cells. Helper T cells that are infected are killed by the viral infection or are destroyed by the immune system's normal responses. New helper T cells arise, and new viruses infect them. This cycle continues, but it is not a balanced situation. Sooner or later, the number of viruses climbs and the number of helper T cells decreases, and serious clinical symptoms appear. (This process is like an army valiantly fighting a battle but ultimately running out of soldiers.)

A severe shortage of helper T cells marks the progression from HIV infection to AIDS. Without helper T cells, B cells and cytotoxic T cells cannot be activated (recall from Section 26-9 that helper T cells stimulate the production of B cells and cytotoxic T cells). Common pathogens that are normally kept at bay in a healthy individual—such as a parasitic protist that causes diarrhea, a fungus that causes "thrush" on the tongue and throat, and viruses that can cause skin lesions, diarrhea, and swelling—ultimately lead to debilitating illness and death.

Q Why is there no vaccine against AIDS yet?

As we saw in Section 13-18, HIV is an RNA-containing virus that mutates rapidly. Even within a single host, the virus's surface proteins can change with each new round of replication

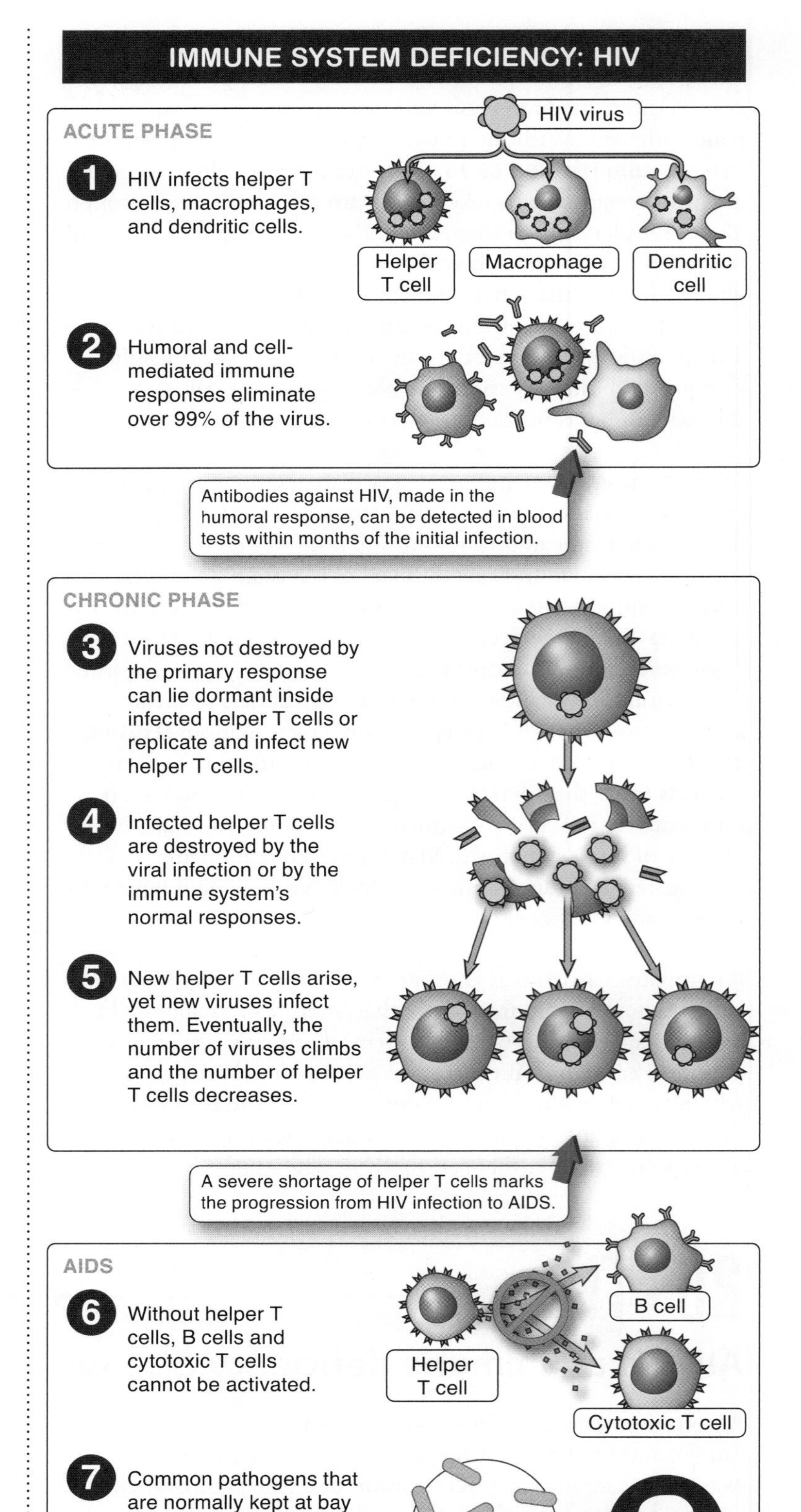

FIGURE 26-28 The progression from HIV infection to AIDS.

(**FIGURE 26-29**). These changing surface proteins pose recognition problems for the pool of cytotoxic T cells derived through clonal selection to recognize the original virus. Antibodies made against the original virus do not recognize newer, mutated versions. The ever-changing HIV makes developing vaccines against it challenging, too. Consider a vaccine that is made using a surface glycoprotein of HIV as the antigen. The vaccine becomes obsolete immediately, because the virus antigen is changing too quickly. More than 25 years after isolating the virus, this traditional method of making vaccines against viral antigens has failed because of the constantly mutating HIV. Interestingly, some people have a mutation in a T cell receptor that makes it more difficult for the virus to enter their helper T cells. Ongoing studies of individuals who are naturally resistant to HIV are helping scientists develop new potential vaccines and drug treatments.

Current therapies for HIV infection have been successful at increasing the length and quality of life for those infected. The regimen, called combination therapy, includes numerous drugs that affect HIV's ability to replicate or infect new cells. These drugs keep the virus number low and maintain a suitable number of helper T cells. HIV resistance to the drugs, however, is a fear shared by both patients and clinicians. Nonetheless, the lifespan for individuals taking combination therapy is increasing and getting close to normal. Although there is not a cure yet, individuals who have access to treatment are able to live with HIV as a chronic condition.

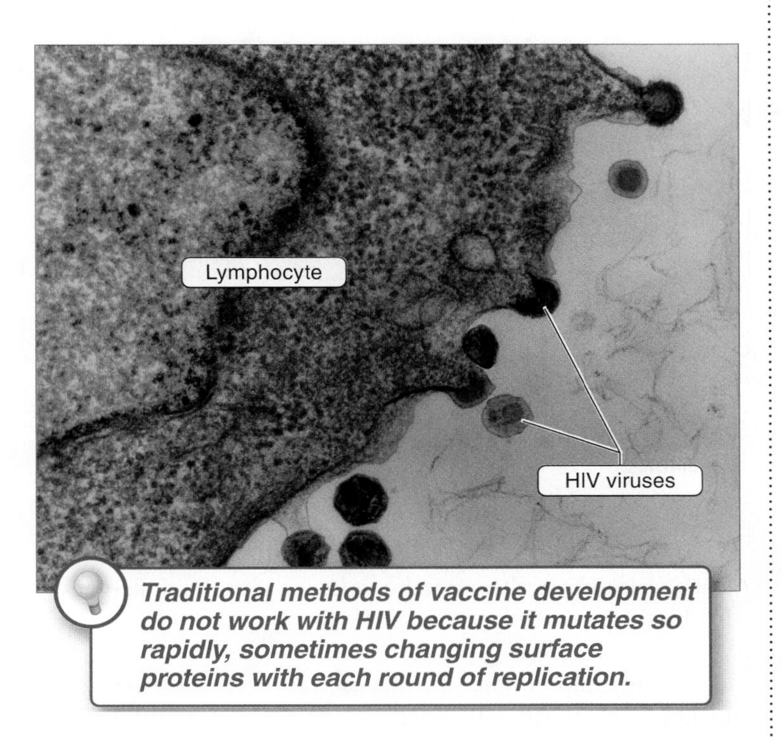

FIGURE 26-29 **An ever-changing foe.** Surface proteins on the HIV virus change rapidly as the virus replicates, leaving the body to fight an almost impossible-to-identify enemy.

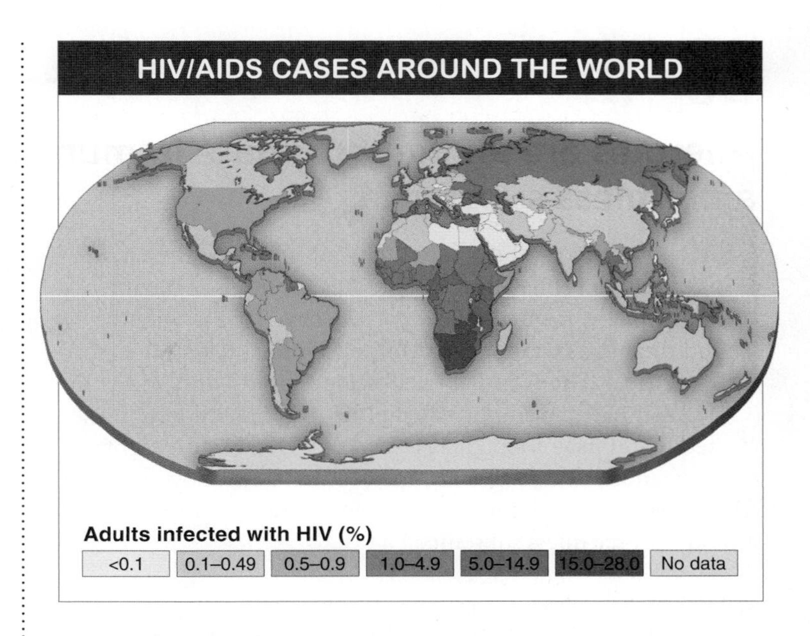

FIGURE 26-30 **HIV/AIDS is widespread in sub-Saharan Africa.**

(Unfortunately, treatment is unavailable for millions infected in developing nations.) New therapies aim to find latent viruses that hide inside helper T cells, sometimes for decades, and eliminate them. In the future, HIV may be a disease that can be cured, but for now it continues to be a life-long condition and a challenge to scientists.

Looking at a map depicting HIV/AIDS cases around the world, we can see just how widespread this virus is. Sub-Saharan Africa is the region most severely affected. In some countries in this region, such as Swaziland and Lesotho, close to 25% of the adult population is infected with HIV. In South Africa, more than 5 million people, or almost 20% of its adult population, are infected (**FIGURE 26-30**). In contrast, less than 1% of the U.S. population is infected with HIV. Not everyone who is HIV-infected has AIDS, however. For example, estimates indicate that of the more than one million people infected by HIV in the United States, in fewer than 40,000 cases has the infection progressed to AIDS.

TAKE-HOME MESSAGE 26•11

AIDS is an immune system disease caused by the human immunodeficiency virus (HIV), which infects helper T cells—immune cells that are crucial to survival. As helper T cells are killed by the infection or by the body's own cells in response to the infection, the deficiency of helper T cells leads to complete failure of the specific immunity response, resulting in illness and death from infections that a healthy immune system could defeat. There is currently no vaccine for HIV, because traditional vaccine methods do not work against this rapidly mutating virus, nor is there a cure for AIDS.

26 • 12

Allergies are an inappropriate immune response to a harmless substance.

A three-year-old girl bites into a peanut butter sandwich. She immediately develops hives around her mouth and begins vomiting. She is having an allergic reaction. This is only the second time she's eaten peanut butter. Unfortunately, her story is not unusual. The Centers for Disease Control and Prevention (CDC) estimates that approximately 3 million children in the United States have food allergies, 90% caused by just eight foods: peanuts, tree nuts, milk, eggs, fish, shellfish, soy, and wheat. **Allergies** are an inappropriate immune response to what should be a harmless substance.

In discussing antibodies earlier in the chapter, we did not focus on the groups of antibodies that differ in their constant (heavy) region at the base of the "Y." There are five classes of antibodies, differing in their chemical structure in this region. When antibodies of one of these classes bind to mast cells (the white blood cells found in tissues), they cause the mast cells to release histamine and cytokines, both of which cause local blood vessels to become dilated and leaky. Recall that this dilation and leakiness are standard parts of the inflammatory response and are useful for recruiting other immune cells to the site of infection.

This class of antibodies can cause much harm to the body, however, if they are made in response to an **allergen,** an antigen that causes an allergic response. Upon first encounter, an allergen induces a normal humoral response: memory cells form and plasma cells secrete antibodies specific to the allergen. The antibodies bind to the surface of mast cells and can remain bound for months and maybe even years. The mast cells are now "sensitized" to the allergen. While the first exposure often goes unnoticed, the second exposure can be damaging and potentially life-threatening. In the second exposure, the allergen binds to the antibody that is still attached to the mast cells—which are now "activated." This time, allergen binding causes the mast cells to release histamine, which causes the blood vessels to dilate and become leaky, and inflammation ensues (**FIGURE 26-31**).

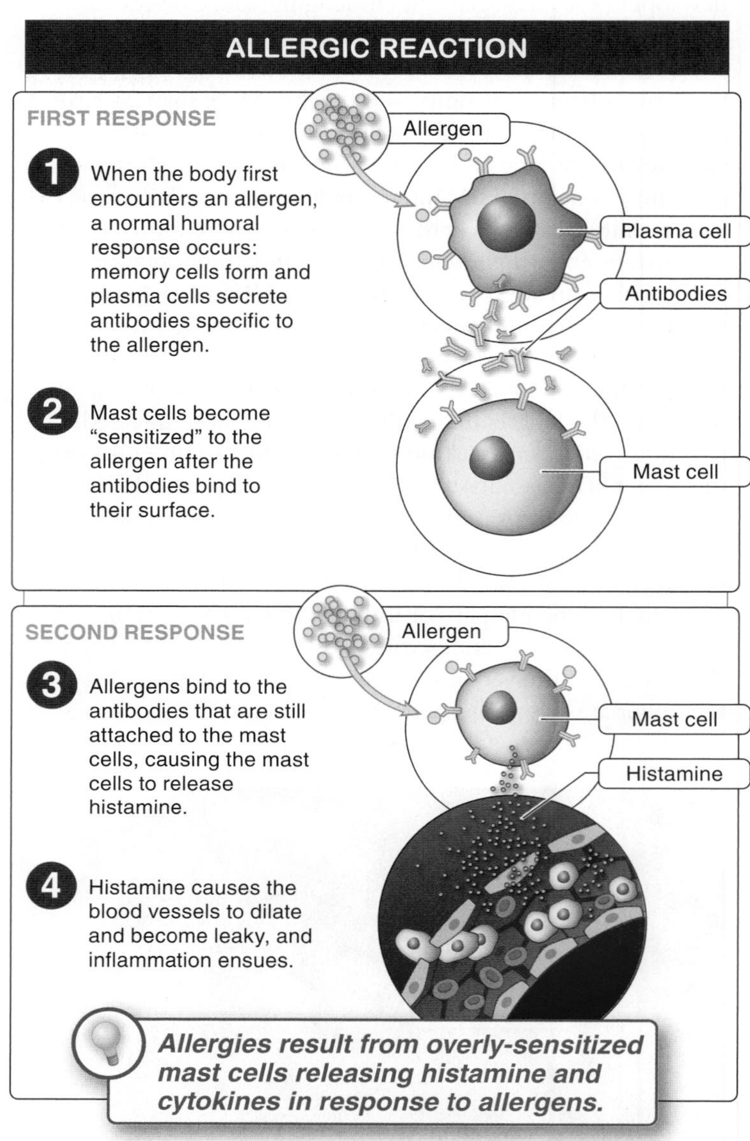

FIGURE 26-31 Allergens induce a humoral response in some individuals.

The effect of different allergens reflects which particular mast cells have been exposed to the allergen and thus activated. Most allergic reactions occur in the digestive or respiratory tracts, because this is where the body first encounters an allergen that is eaten or inhaled. Allergic reactions associated with food often lead to vomiting or diarrhea, but can also lead to hives and other symptoms if the allergen enters the bloodstream. Respiratory allergens, such as pollen and dust, can cause a runny nose and teary eyes ("hay fever") in the upper respiratory tract, or can result in asthma, a chronic inflammatory disease of the lower respiratory tract that causes wheezing, coughing, and shortness of breath. Allergic eczema, a chronic skin disorder characterized by scaly and itchy rashes, is a result of allergens that bind to mast cells in the skin.

Taking an antihistamine can alleviate some allergies, as these medicines block the inflammatory effects of histamine. Other medicines block mast cells from releasing histamine. Steroids are prescribed for more severe allergies. Steroids block the production of cytokines from immune cells, thus preventing the migration of more inflammatory cells, such as macrophages and neutrophils, to the site of inflammation. Additionally, the steroids can block the killing ability of these phagocytes, resulting in less damage to the inflamed tissue.

An individual with a severe allergic response may experience **anaphylactic shock,** a life-threatening allergic reaction that is systemic, meaning that it is not localized to the site of exposure. Anaphylactic shock can lead to severe respiratory distress, as the throat swells and asthma develops. Swelling in various tissues means there is less fluid in the blood system, and this can lead to dangerously low blood pressure. Individuals with severe allergies often carry with them a dose of adrenaline (also called epinephrine), a chemical found naturally in our bodies (see Section 24-5) and also manufactured by several drug companies. Epinephrine, which can be injected if needed, reverses the effects of histamine on blood vessels, but it must be injected quickly, because death can occur rapidly from anaphylactic shock.

A common method of testing for allergies is called *skin allergy testing,* in which microscopic amounts of one or more potential allergens are injected under the skin—usually on the back or the forearm. The development of a rash can indicate a hypersensitivity to that allergen, enabling the person to take steps to reduce exposure to the allergen (**FIGURE 26-32**).

Allergies are on the rise in the industrialized world, asthma especially. Asthma results in an airway that is swollen and inflamed and leads to chest tightness, coughing, wheezing, and breathing difficulty. The CDC estimates that the incidence of childhood asthma more than doubled from 1980 to the mid-1990s, by that time affecting about 1 in 14 American adults and almost 1 in 10 children (**FIGURE 26-33**).

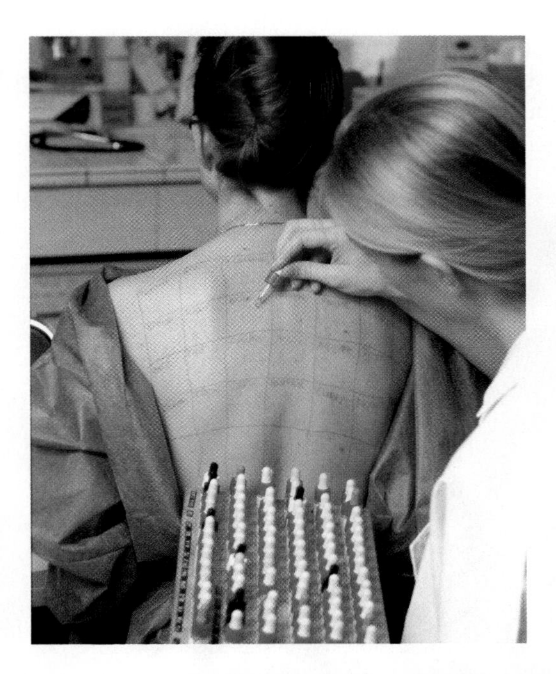

FIGURE 26-32 Screening for allergies. Allergens are injected under the skin on which a grid has been drawn. If a bump or rash develops in any of the squares, an allergy may be suspected.

The CDC estimates that childhood asthma more than doubled from 1980 to the mid-1990s. As of now, there is no clear explanation for this increase.

FIGURE 26-33 Asthma. Individuals with asthma may need to inhale medications to counteract the airway restriction associated with this allergic response.

As is true for many complex diseases, there are both genetic and environmental factors associated with asthma. But our genes are not changing fast enough to explain the increase in occurrence of asthma—which has led scientists to examine the environmental factors that have changed in the past few decades. Determining which environmental factors cause asthma is not a simple task. Numerous correlations have been proposed, including increased air pollution, increased antibiotic or acetaminophen use in infancy, less exposure to infections (due to better hygiene), use of vaccines, rising global temperatures, diets lacking vitamin D, increased exposure to cleaning sprays, indoor dust, cigarette smoke . . . and the list goes on. As yet, there is no simple or clear explanation of why asthma is on the rise in developed countries.

TAKE-HOME MESSAGE 26•12

In individuals who are sensitive, an allergen induces a humoral response in which one class of antibodies binds to and activates mast cells. A second exposure to the allergen leads to activated mast cells releasing histamine and cytokines, resulting in allergy-related symptoms. Swelling and inflammation can be localized or can be systemic and lead to anaphylactic shock.

Knowledge You Can Use

Fact or fallacy? How best to avoid getting sick.

Q: Can I catch a cold from going out with wet hair in the winter? No. Your grandmother was wrong about this one! To catch a cold, you must be infected with a pathogen (such as a rhinovirus) that causes the common cold. Colds do tend to be more common in colder weather, because the viruses are more stable in colder air with low humidity. And droplets of water and virus (such as those that escape when we sneeze and cough) remain airborne slightly longer in dry air than in humid summer air. Also, because people are indoors more often during winter, the increased population in confined quarters increases the incidence and ease of transmission. So while the virus is affected by the colder weather, your immune system is not. Dry your hair before you leave the house if you don't want icicles in it, but not as strategy for combating the common cold.

Q: Can I catch a cold because I am not sleeping enough? This gets an emphatic *yes!* In randomized, controlled studies in which individuals were given nasal drops containing rhinovirus, individuals getting less sleep were significantly more likely to develop colds than those sleeping 7 hours or more. And simply resting in bed or having an interrupted night's sleep doesn't count. Seven hours or more of good-quality sleep seems to be important to the immune system, particularly for the production and functioning of cytokines. Interestingly, there is also evidence that individuals who sleep 7–8 hours a night have the lowest rates of heart disease. Although eating well and exercising are most often associated with good health, sleep, too, is important, and neglecting it can increase your vulnerability to illness.

Q: Can drinking herbal teas, eating chicken soup, or taking "Airborne" help me ward off the common cold? *Herbal teas and chicken soup?* Not exactly. There are some symptom-reducing benefits of hot fluids, including keeping nasal passages moist, preventing dehydration, and soothing a sore throat. And one study showed that chicken soup with vegetables seemed to have an anti-inflammatory effect. But while these effects can make an infection less unpleasant by reducing the symptoms, they don't decrease the likelihood of catching a cold or shorten its duration. *Airborne?* Remember the very first section of the first chapter of this book? There is not a shred of evidence that Airborne can ward off colds or boost your immune system.

1 Your body has different ways to protect you against disease-causing invaders.

The immune system protects the body from a diverse group of pathogens. There are three collaborative divisions of the immune system: physical barriers, non-specific immunity, and specific immunity. Physical barriers, part of the integumentary system, prevent pathogens from entering the body's cells; cells not covered by skin but exposed to the external environment are protected by defenses such as bacteria-destroying chemicals, acidic secretions, sticky mucus, and wax. Non-specific (innate) immunity defends against pathogens by recognizing combinations of receptors on their cell surfaces. White blood cells of the non-specific system include phagocytes—such as neutrophils and macrophages that ingest and kill pathogens, and macrophages and dendritic cells that display pathogens to cells of the specific immunity system—and natural killer cells that kill cancer cells and virus-infected cells. Complement proteins non-specifically recognize invaders and help destroy them. Inflammation is a major way in which the innate system eliminates pathogens. The four signs of the inflammatory response (redness, heat, swelling, and pain) are related to changes in blood vessels that enhance recruitment of phagocytes and complement proteins to the site of inflammation.

2 Specific immunity develops after exposure to pathogens.

Long-term protection, or immunity, from a specific pathogen can form in two ways: exposure to the natural pathogen or exposure to a vaccine with an altered version of the pathogen. Exposure to an antigen results in production of antibodies. The specific (adaptive) immunity system responds continually to numerous pathogens that can change over time, through the action of two types of lymphocytes: B cells and T cells. B cells are responsible for the humoral (antibody-mediated) response, and T cells for the cell-mediated response. When the body encounters an antigen, the lymphocytes recognizing this antigen divide to produce many cells through clonal selection. During a primary response, effector cells respond to and facilitate removal of the antigen, and memory cells are produced that are ready to go through clonal selection if a secondary response is necessary. Plasma cells secrete high levels of antibodies into the blood and lymph. Each antibody has a unique structure that recognizes a specific antigen. Antibodies help destroy pathogens and soluble antigens by enhancing ingestion by phagocytes, preventing more cell infection, and enhancing complement-driven cell lysis, but they cannot destroy pathogens that are inside cells. Cytotoxic T cells undergo specialization to kill infected cells. Antigen-presenting cells display antigens to helper T cells and cytotoxic T cells, to alert the adaptive immune system to an infection. In turn, helper T cells produce cytokines that allow cytotoxic T cells to mature and respond to the infection.

3 Malfunction of the immune system causes disease.

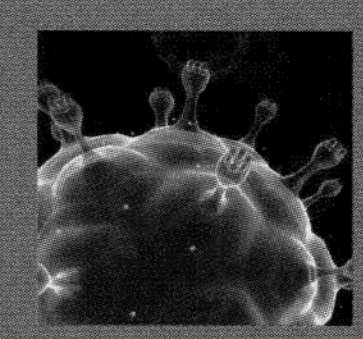

When lymphocytes bear receptors that inappropriately recognize structures of a person's own body, autoimmunity develops and significantly damages specific organs or tissues throughout the body. AIDS is an immune system disease caused by the human immunodeficiency virus (HIV), which infects helper T cells. As these cells are killed by the infection or by the body's own cells in response to the infection, the deficiency of helper T cells leads to complete failure of the specific immunity response and thus illness and death from infections that a healthy immune system could defeat. In individuals who are sensitive, allergens induce a humoral response in which one type of antibody binds to mast cells, activating them. With a second exposure to the allergen, activated mast cells release histamine and cytokines, resulting in allergy-related symptoms. Swelling and inflammation can be localized or can be systemic and lead to anaphylactic shock.

KEY TERMS

AIDS, p. 981
allergen, p. 984
allergy, p. 984
anaphylactic shock, p. 985
antibody, p. 970
antigen, p. 970
antigen-presenting cells, p. 978
antigen receptor, p. 972
antiseptic, p. 964
apoptosis, p. 979
autoantigen, p. 980
autoimmunity, p. 980
B cell, p. 972
basophil, p. 966
cilia, p. 962
clonal selection, p. 974
complement protein, p. 965
cytokine, p. 963
cytotoxic T cell, p. 978
dendritic cell, p. 965
effector cell, p. 974
fever, p. 968
helper T cell, p. 978
histamine, p. 966
HIV, p. 981
immune system, p. 959
immunity, p. 969
inflammatory response, p. 966
integumentary system, p. 961
interferon, p. 963
lymphocyte, p. 972
lysozyme, p. 961
macrophage, p. 964
mast cell, p. 966
memory cell, p. 974
multiple sclerosis (MS), p. 981
natural killer (NK) cell, p. 964
neutrophil, p. 964
non-specific immunity, p. 959
pathogen, p. 958
phagocyte, p. 964
physical barriers (immune system), p. 959
plasma cell, p. 974
primary response, p. 974
pus, p. 964
secondary response, p. 975
specific immunity, p. 960
T cell, p. 972
vaccine, p. 969
white blood cell, p. 964

CHECK YOUR KNOWLEDGE

1. Which of the following statements about the integumentary system is incorrect?
   a) It is one of the body's three lines of defense against infectious pathogens.
   b) It confers specific immunity on an individual because it confers protection against individual pathogens.
   c) It includes sweat glands.
   d) It is responsible for the production of tears and ear wax.
   e) It helps protect the digestive system from pathogens.

2. Which of the following statements about how complement proteins kill bacteria is incorrect?
   a) Bound complement proteins are able to recruit other complement proteins to come to the site of bacterial infection.
   b) Complement proteins aggregate to form larger protein complexes that create holes in the bacterial cell surface.
   c) Bacterial death results from membrane holes created by complexes between complement and surface proteins.
   d) Complement proteins surround and bind to the entire surface of a bacterium, marking it to be engulfed by phagocytes and cytotoxic T cells.
   e) Complement proteins are able to recognize and bind directly to the surface of bacteria's cell surface proteins.

3. Which of the following statements about the inflammatory response is incorrect?
   a) The release of histamine by mast cells initiates inflammation.
   b) Cells lining blood vessels near inflamed tissue become "stickier" to neutrophils.
   c) Pain can occur with inflammation, as increased pressure in an inflamed area stimulates local neurons.
   d) The release of histamine causes blood vessels to constrict, reducing blood loss in an injured area.
   e) Four signs of inflammation are swelling, redness, heat, and pain.

4. Vaccines:
   a) activate a primary response in the vaccinated individual.
   b) lead to the creation of memory B cells that provide a rapid secondary response against a specific infectious agent.
   c) are sometimes made of no more than the unique antigens from a pathogen's surface.
   d) All of the above are correct.
   e) Only b) and c) are correct.

5. The major difference between T cells and B cells is that:
   a) T cells develop in the thymus, whereas B cells develop in the bone marrow.
   b) T cells do not interact with the innate immune system, but B cells do.
   c) T cells bind directly to foreign antigens, whereas B cells produce proteins (antibodies) that are secreted and bind to the foreign antigens.
   d) All of the above are major differences between T cells and B cells.
   e) Both a) and c) are major differences between T cells and B cells.

6. Why don't people develop immunity to the common cold?
   a) Actually, many people do develop immunity to the common cold and get sick only from bacterial infections.
   b) There are at least 200 different viruses that can cause the common cold, and they change over time.
   c) Only non-specific immunity helps individuals fight the common cold, and specific immunity is not an effective defense against viruses.
   d) The rhinovirus that causes the common cold has no surface antigens and so can move through the body without being detected.
   e) The pathogen that causes the common cold attacks and kills memory cells, continually "erasing" the immune system's memory of it.

7. Antibodies:
   a) are produced by helper B cells.
   b) are produced by helper T cells.
   c) are produced by the B cell bound by a specific antigen.
   d) are produced by plasma cells.
   e) are made in the bone marrow.

8. The polypeptides that make up the structure of an antibody consist of:
   a) two heavy chains and two light chains, shaped like a "Y," with a lower constant region and an upper variable region.
   b) two heavy chains and two light chains, shaped like a "Y," each with an upper constant region and a lower variable region.
   c) one heavy chain and one light chain, shaped like a "Y," each with a lower constant region and an upper variable region.
   d) one heavy chain and one light chain, shaped like a "Y," each with an upper constant region and a lower variable region.
   e) Both b) and d) can be correct, depending on the type of antibody.

9. Which of the following is not a function of helper T cells?
   a) They stimulate B cells to proliferate and produce antibodies.
   b) They activate cytotoxic T cells.
   c) They stimulate B cells to secrete cytokines.
   d) They recognize antigens presented on macrophage surfaces.
   e) All of the above are functions of helper T cells.

10. Autoimmunity is:
   a) one of three lines of defense in the vertebrate immune system.
   b) responsible for muscular dystrophy, which occurs when immune cells destroy the myelin sheath, reducing motor control and balance.
   c) the culprit in rheumatoid arthritis, which occurs when immune attack on the linings of joints causes inflammation.
   d) a consequence of a body's inability to produce antigens, leading it to attack its own cells as if they were pathogens.
   e) All of the above statements about autoimmunity are correct.

11. As the human immunodeficiency virus (HIV) begins to kill the host's _______, the host's immune system begins to fail.
   a) thymus cells
   b) spleen cells
   c) red blood cells
   d) white blood cells
   e) liver cells

12. Hay fever affects millions of individuals every year and is caused by an allergic reaction to pollen. Which of the following is not a step of the process in which a pollen grain causes an allergic reaction in an individual?
   a) Mast cells burst when they encounter the pollen during the initial exposure, releasing histamine and other chemicals.
   b) During the first exposure, B cells are activated by the pollen and differentiate into antibody-secreting plasma cells.
   c) During the second exposure, the pollen grains are recognized by previously created antibodies on mast cells.
   d) A type of antibody, specific to the pollen grain, is produced during the initial attack, and these antibodies adhere to mast cells, where they remain.
   e) Histamine and other allergen chemicals cause itchy eyes and a runny nose.

Short-Answer Questions

1. Generally, the second time a person is exposed to a particular pathogen, his or her immune system is able to respond to the pathogen more quickly and effectively. Why?

2. Why is it necessary to get only a single vaccine against the pathogen causing chicken pox, while a flu shot is necessary every year?

See Appendix for answers. For additional study questions, go to www.prep-u.com.

Photo Credits

Photo Legend: *L*: left, *C*: center, *R*: right, *T*: top, *M*: middle, *B*: bottom

CHAPTER 1 **p. 1:** *photo* Karl Ammann/Corbis. **p. 2:** *photo* Martin Barraud/Getty Images. **p. 3:** *Figure 1-1* The Photo Works. **p. 4:** *Figure 1-2 (L)* Reprinted through the courtesy of the Editors of TIME Magazine © 2008 Time Inc., *(M)* Courtesy of *The Economist*, *(R)* Newscom, *(B) The New York Times.* **p. 5:** *Figure 1-3* Photo Researchers. **p. 6:** *Figure 1-4 (TL)* Louie Psihoyos/Corbis, *(TR)* Mark Scott/Getty Images, *(BL)* Steven Puetzer/Getty Images, *(BR)* Masterfile. **p. 8:** *photo* AP Photo/ Christof Stache. **p. 9:** *photo* AFP/Getty Images; *Figure 1-6 (TR)* Digital Vision Photography/Veer. **p. 10:** *Figure 1-7 (L)* Grapes-Michaud/Photo Researchers. *(R)* Nigel Cattlin/Alamy. **p. 11:** *Figure 1-8* LWA/Getty Images. **p. 14:** *Figure 1-11* Jean Chung/OnAsia. **p. 17:** *photo* Michel Setboun/ Corbis. **p. 19:** *Figure 1-14* Topham/The Image Works. **p. 21:** *Figure 1-15 (T)* Courtesy KCBD News Channel 11, *(B)* Courtesy hivandhepatitis.com. **p. 23:** *photo* Joos Mind/Getty Images. **p. 24:** *Figure 1-17* Christopher Bissell/Getty Images. **p. 25:** *Figure 1-18 (L)* Image Source Black/Alamy, *(R)* Huy Lam/Getty Images. **p. 26:** *Figure 1-19* The Photo Works. **p. 27:** *Figure 1-20* Reprinted through the courtesy of the Editors of TIME Magazine © 2008 Time Inc. **p. 28:** *photo* Waltraud Grubitzsch/epa/Corbis. **p. 29:** *photo* Martin Shields/Photo Researchers. **p. 30:** *photo* AP Photo/ Daniel Hulshizer.

CHAPTER 2 **pp. 34–35:** *photo* Patrick Phelan. **p. 36:** *photo* M. E. Leunissen et al., Ionic colloidal crystals of oppositely charged particles, *Nature* 437, 235 (2005). Courtesy Mirjam Leunissen and Alfons van Blaaderen, Debye Institute for Nanomaterials Science, Utrecht University, Netherlands; *Figure 2-1 (TL)* Stockfolio/Alamy, *(TR)* Jeffrey Hamilton/Getty Images, *(B)* Burazin/Getty Images. **p. 41:** *Figure 2-8 (L)* Judith Collins/Alamy, *(R)* Dr. Dennis Kunkel/Phototake. **p. 44:** *photo* imagewerks/Getty Images. **p. 45:** *Figure 2-13* Joe McDonald/Animals Animals—Earth Scenes, *(inset)* Visuals Unlimited/Corbis. **p. 49:** *photo* Dr. Dennis Kunkel/Visuals Unlimited. **p. 50:** *photo* Keren Su/Getty Images, *Figure 2-20* Dorling Kindersley/Getty Images. **p. 52:** *Figure 2-22 (T)* frans lemmens/Alamy, *(B)* Lew Robertson/Getty Images. **p. 54:** *Figure 2-25 (L)* Paul Geor/Stockxpert, *(R)* Alison Parks-Whitfield/Alamy; *Figure 2-26 (T)* David Hancock/Alamy, *(B)* Masterfile; *Figure 2-27* Cassenet/ age fotostock. **p. 56:** *photo* Patrick J. Endres/Alaskaphotographics.com. **p. 57:** *Figure 2-28* Bryan & Cherry Alexander Photography/Alamy. **p. 58:** *Figure 2-30* Gary Gladstone/Corbis, *Figure 2-31 (L)* Ellen Isaacs/ Alamy, *(R)* Masterfile. **p. 59:** *photo* Andrew O'Toole/Taxi/Getty Images. **p. 61:** *Figure 2-34* Hulton Archive/Getty Images. **p. 62:** *photo* Stephanie Pfriender Stylander. **p. 63:** *Figure 2-36 (from top)* Steve Bloom Images/ Alamy; Steve Gschmeissner/Photo Researchers; Dr. Dennis Kunkel/ Phototake; Dr. Dennis Kunkel/Phototake/Alamy; Deco/Alamy. **p. 64:** *Figure 2-38* Edgardo Contreras/Getty Images. **p. 66:** *Figure 2-41 (L)* Ryan McVay/Getty Images, *(R)* White Packert/Getty Images. **p. 68:** *photo* Rick Gayle Studio/Corbis. **p. 69:** *photo* Driscoll, Youngquist, & Baldeschwieler/ Caltech/Photo Researchers. **p. 72:** *photo* Gustavo Andrade/age fotostock.

CHAPTER 3 **pp. 76–77:** Phototake/Alamy. **p. 78:** Alfred Pasieka/ Photo Researchers. **p. 79:** *Figure 3-1(L)* Steve Taylor/Getty Images, *(M)* Frans Lanting/Corbis, *(R)* David Scharf/Peter Arnold; *Figure 3-2 (L)* Clark Sumida, *(M)* Rosenfeld/age fotostock, *(R)* Clouds Hill Imaging/Corbis, *(inset)* Courtesy Reproductive Solutions. **p. 80:** *Figure 3-3 (inset)* Dr. Dennis Kunkel/Getty Images. **p. 81:** *Figure 3-4 (TL)* Phototake/Alamy, *(TR)* Matt Meadows/Peter Arnold, *(BL)* John Glover/age fotostock, *(BM)* David Tipling/Getty Images, *(BR)* Juniors Bildarchiv/age fotostock. **p. 82** *Figure 3-5 (L)* Dr. Donald Fawcett/Visuals Unlimited, *(R)* Phototake/Alamy. **p. 83** *Figure 3-6 (L inset)* Dr. Gopal Murti/Photo Researchers, *(R inset)* Dr. Henry Aldrich/Visuals Unlimited. **p. 85:** F1 Online/photolibrary. **p. 86:** *Figure 3-8 (L)* Don W. Fawcett/Photo Researchers, *(R)* David McNew/Getty Images. **p. 89:** *Figure 3-12 (L)* Mauro Fermariello/Photo Researchers, *(R)* Jim Evans/*Kennebec Journal & Morning Sentinel.* **p. 90:** *Figure 3-13* Image Source/ Veer. **p. 92:** *Figure 3-15* AP Photo/Jeffrey Boan. **p. 93:** *photo* Courtesy Dr. G. M. Gaietta & T. J. Deerinck, CRBS/NCMIR, University of California, San Diego. **p. 97:** *Figure 3-20* Richard Keppel-Smith/Getty Images. **p. 98:** *Figure 3-21* Biophoto Associates/Photo Researchers. **p. 99:** *Figure 3-23 (L)* Chuck Brown/Photo Researchers, *(M)* Ed Reschke/Peter Arnold, *(R)* Carolina Biological Supply Company/Phototake. **p. 100:** *photo* Jutta Klee/Getty Images. **p. 101:** *photo* Lorenz Britt/Alamy. **p. 102:** *Figure 3-25 (L)* Dr. Dennis Kunkel/Phototake, *(M)* Dr. Dennis Kunkel/Visuals Unlimited, *(R)* B. Gilula and Donald Fawcett/Visuals Unlimited. **p. 103:** *photo* Dr. Dennis Kunkel/Visuals Unlimited. **p. 104:** *Figure 3-26* Phototake/ Alamy; *photo* Mary Ellen Mark. **p. 105:** *Figure 3-27* Lynne Chang/Visuals Unlimited; *Figure 3-28 (L)* Dr. Yorgos Nikas/Phototake, *(R)* Juergen Berger/ Photo Researchers. **p. 106:** *Figure 3-29* Dr. Donald Fawcett/Getty Images. **p. 108:** *Figure 3-30* Dr. Gopal Murti/Photo Researchers. **p. 109:** *Figure 3-31* Kevin Somerville/The Medical File/Peter Arnold. **p. 110:** *Figure 3-32* Dr. Dennis Kunkel/Visuals Unlimited. **p. 111:** *Figure 3-33* Dr. Dennis Kunkel/Phototake. **p. 112:** *Figure 3-34* Dr. Dennis Kunkel/Visuals Unlimited. **p. 114:** *Figure 3-36* Perennou Nuridsany/Photo Researchers; *photo* Bill Hatcher/Getty Images. **p. 115:** *Figure 3-37* Biophoto Associates/ Photo Researchers. **p. 116:** *Figure 3-38* George Chapman/Visuals Unlimited; *photo* Stephen Studd/Getty Images. **p. 118:** *photo* Marco Garcia/ Getty Images.

CHAPTER 4 **pp. 122–123:** *photo* Pete Saloutos/Corbis. **p. 124:** *photo* Inga Spence/Getty Images. **p. 125:** *Figure 4-1* Pat LaCroix/Getty Images. **p. 126:** *Figure 4-3 (TL)* David Madison/Getty Images, *(TM)* Tom Vezo/ Nature Picture Library, *(TR)* Mark Gibson/Getty Images, *(BL)* Shem Compion/Getty Images, *(BR)* Chris Cheadle/age fotostock. **p. 127:** *Figure 4-4* Michael Pohuski/Foodpix/Jupiterimages; *Figure 4-5 (L)* Ed Gifford/ Masterfile, *(M)* David Allio/Icon SMI/Corbis, *(R)* Walter Lockwood/ Corbis. **p. 130:** *photo* Datacraft/age fotostock. **p. 131:** *Figure 4-9 (L)* Wim van Egmond/Visuals Unlimited, *(R)* Dr. Dennis Kunkel/Phototake, *(B)* Dr. Dennis Kunkel/Visuals Unlimited. **p. 131:** *Figure 4-10* brokenarts/ stock.xchng. **p. 132:** *Figure 4-11* Value Stock Images/Fotosearch, *(inset)* Dr. Dennis Kunkel/Visuals Unlimited; *Figure 4-12* George Chapman/ Visuals Unlimited. **p. 134:** *Figure 4-14* Value Stock Images/Fotosearch. **p. 140:** *Figure 4-22* AP Photo/Rajesh Kumar Singh. **p. 141:** *Figure 4-23* Dr. Jeremy Burgess/Science Photo Library; *Figure 4-24 (L)* age fotostock/SuperStock, *(M)* Brian Sytnyk/Masterfile, *(R)* age fotostock/ SuperStock. **p. 143:** *photo* Philip Harvey/Corbis; *Figure 4-26 (L)* age fotostock/SuperStock, *(M)* image100/Alamy, *(R)* Volker Steger/Photo Researchers. **p. 145:** *Figure 4-28 (L)* age fotostock/SuperStock, *(M)* Frans

Lanting Photography, *(R)* A. Barry Dowsett/Photo Researchers. **p. 148:** *Figure 4-32* hybrid medical animation/Photo Researchers. **p. 151:** *photo* Martin Harvey/Getty Images; *Figure 4-35* Tobias Bernhard/Getty Images. **p. 152:** *Figure 4-37* Cephas Picture Library/Alamy, *(T inset)* SciMAT/ Photo Researchers, *(B inset)* Romilly Lockyer/Getty Images. **p. 153:** *Figure 4-38 (L)* FoodCollection/age fotostock, *(M)* Tim Hill/Alamy, *(R)* FoodCollection/age fotostock. **p. 155:** *photo* Plush Studios/Getty Images.

CHAPTER 5 pp. 158–159: *photo* Phanie/Photo Researchers. **p. 160:** *photo* Denis Galante/Corbis. **p. 161:** *Figure 5-1* Vasna Wilson/ *Virginian-Pilot*; *Figure 5-2 (L)* Reprinted with permission, September 2008 *Discover* magazine. Copyright © Discover Magazine. All rights reserved, *(R)* Reprinted through the courtesy of the Editors of TIME Magazine © 2008 Time Inc., *(B)* Reprinted by permission of The Wall Street Journal, © 2008 Dow Jones & Company, Inc. All Rights Reserved Worldwide. License number 2038930425822. **p. 162:** *Figure 5-3* A. Barrington Brown/Photo Researchers, *(inset)* Science Source/Photo Researchers. **p. 164:** *Figure 5-5 (TL)* Courtesy of the author, *(TM)* Sian Irvine/Getty Images, *(TR)* Eye of Science/Photo Researchers, *(BL)* Visuals Unlimited/ Corbis, *(BR)* John White. **p. 166:** *Figure 5-8 (from top)* Courtesy of the author; John White; Eye of Science/Photo Researchers; Sian Irvine/Getty Images; Visuals Unlimited/Corbis. **p. 167:** *Figure 5-9 (from top)* Courtesy of the author; Eye of Science/Photo Researchers; Dr. Dennis Kunkel/ Getty Images; Fancy/Veer/Corbis; Dr. Dennis Kunkel/Visuals Unlimited. **p. 169:** *photo* Julia Kuskin/Photonica/Getty Images. **p. 174:** *photo* Uli Westphal 2008. **p. 175:** *Figure 5-15* Eye of Science/Photo Researchers. **p. 177:** *Figure 5-17 (L)* Mika/zefa/Corbis, *(M)* Robert Llewellyn/Corbis, *(R)* Tom Stewart/Corbis. **p. 178:** *Figure 5-18* Tommy Moorman. **p. 179:** *photo* Mario Tama/Getty Images; *Figure 5-19 (TL)* Getty Images, *(TM)* Jim West/Alamy, *(TR)* Charles O'Rear/Corbis, *(BL)* AP Photo/*Beloit Daily News*, Tom Holoubek, *(BR)* Landov. **p. 184:** *Figure 5-26* AJPhoto/Photo Researchers. **p. 185:** *Figure 5-27* MGM/courtesy Everett Collection. **p. 186:** *Figure 5-28* AP Photo. **p. 187:** *Figure 5-29* AP Photo/Examiner, Amy Elrod. **p. 188:** *photo* Betsie Van der Meer/Getty Images. **p. 189:** *photo* Jim Richardson/Corbis; *Figure 5-30* Ed Young/Corbis, *(inset)* John Doebley. **p. 190:** *Figure 5-31 (clockwise from left)* Martin Ruegner/Digital Vision/Getty Images; Scottish Crop Research Institute, Dundee; Professor Peter Beyer/Humanitarian Board for Golden Rice/www.goldenrice.org. **p. 191:** *Figure 5-32 (L)* BSIP/Photo Researchers, *(M)* Lance Nelson/ Corbis, *(R)* Steve Percival/Photo Researchers. **p. 192:** *Figure 5-33* Dr. Morley Read/Science Photo Library. **p. 193:** *Figure 5-35* Frank Whitney/ Getty Images; *Figure 5-36* Dr. Garth Fletcher. **p. 194:** *Figure 5-37* Reuters/Corbis; *Figure 5-38* Reuters/Corbis. **p. 196:** *photo* Digital Vision/ Alamy. **p. 197:** *Figure 5-39 (L)* Neville Chadwick/Science Photo Library, *(M)* Tom Grill/Corbis, *(R)* David Parker/Science Photo Library. **p. 199:** *Figure 5-42* Vo Trung Dung/Corbis Sygma. **p. 200:** *Figure 5-43 (clockwise)* Sian Irvine/Dorling Kindersley/Getty Images, Visuals Unlimited/Corbis, Eye of Science/Photo Researchers, John White, Courtesy of the author, *(tree of life)* David M. Hillis, Derrick Zwickl, and Robin Gutell, University of Texas. **p. 202:** *Figure 5-45 (TL)* John Chadwick/AP Photo, *(TR)* Ben Margot/AP Photo, *(BL)* Jack Smith/AP Photo, *(BM)* David J. Phillip/AP Photo, *(BR)* Advanced Cell Technologies/AP Photo. **p. 203:** *Figure 5-46* Amblin/Universal/The Kobal Collection. **p. 204:** *(T)* Anthony Marsland/ Getty Images; *(B)* Lauren Nicole/Getty Images.

CHAPTER 6 pp. 208–209: *photo* AP Photo. **p. 210:** *photo* Steve Gschmeissner/Photo Researchers. **p. 211:** *Figure 6-2* AP Photo/Gerald Herbert, *(inset)* Dr. Peter Landsorp/Visuals Unlimited. **p. 213:** *Figure 6-4* CNRI/Photo Researchers. **p. 216:** *photo* Sisse Brimberg/Getty Images. **p. 217:** *photo* Tanya Constantine/Getty Images. **p. 218:** *Figure 6-7 (L)* Mike Hill/Alamy, *(inset)* Science Photo Library/Photo Researchers, *(R)* Dan Guravich/Photo Researchers, *(inset)* Biology Media/Photo Researchers; *photo* Micro Discovery/Corbis. **p. 219:** *photo* David Scharf/ Getty Images. **pp. 220–221:** *photos* Courtesy T. Wittman. **p. 223:** *Figure 6-11* GJLP/CNRI/Science Photo Library. **p. 224:** *Figure 6-13* Kevin Laubacher/Taxi/Getty Images. **p. 225:** *photo* Dr. Yorgos Nikas/Photo Researchers. **p. 231:** *Figure 6-18* Dr. Dennis Kunkel/Phototake. **p. 233:** *Figure 6-21* Margot Granitsas/Photo Researchers. **p. 234:** *Figure 6-22* Stephen Dalton/NHPA. **p. 234:** *Figure 6-23* Dr. Kari Lounatmaa/Photo Researchers. **p. 235:** *photo* Cormac Hanley/Getty Images. **p. 236:** *Figure 6-24* Biophoto Associates/Photo Researchers. **p. 237:** *Figure 6-26 (L)* Tui De Roy, *(M)* imagebroker/Alamy, *(R)* A. B. Sheldon/Root Resources. **p. 238:** *photo* Michael Patrick O'Neill/Alamy. **p. 239:** *photo* Visuals Unlimited/Corbis. **p. 240:** *Figure 6-27* CNRI/Photo Researchers; *Figure 6-28* Saturn Stills/Photo Researchers; *Figure 6-29 (L)* Mika/zefa/Corbis, *(R)* L. Willatt/East Anglian Regional Genetics Service/Photo Researchers. **p. 244:** *photo* Jupiterimages/Brand X/Alamy.

CHAPTER 7 pp. 248–249 *photo* Robert B. Carr/Bruce Coleman USA. **p. 250:** *photo* Dr. Karl Fredga, Department of Genetics, Uppsala University, Sweden. **p. 251:** *Figure 7-1* Isabelle Rozenbaum/age fotostock. **p. 252:** *Figure 7-4* © BNPS.CO.UK. **p. 253:** *Figure 7-5 (row 1, L)* Les3photo8/Dreamstime.com, *(row 1, R)* Gabriela Medina/Getty Images, *(row 2, L)* Geoff du Feu/Alamy, *(row 2, R)* Custom Medical Stock Photo/Alamy, *(row 3, L & R)* Kevin Winter/Getty Images. **p. 256:** *Figure 7-8* Content Mine International/Alamy. **p. 258:** *Figure 7-10* Jeff Wyoming/photolibrary. **p. 260:** *photo* John Gillmoure/Corbis. **p. 262:** *Figure 7-14* Michael K. Nichols/Getty Images. **p. 264:** *Figure 7-16* Yann Arthus-Bertrand/Corbis. **p. 265:** *photo* Gideon Mendel/Corbis. **p. 266:** *Figure 7-18* Wellcome Images. **p. 270:** *Figure 7-22* Corbis. **p. 271:** *Figure 7-23* Lennart Nilsson. **p. 273:** *Figure 7-25* AP Photo/Rob Carr. **p. 274:** *Figure 7-26 (T)* Juniors Bildarchiv/age fotostock, *(B)* Grant Heilman Photography/Alamy. **p. 275:** *photo* Charles Maraia/Getty Images. **p. 277:** *Figure 7-28* Mina Chapman/Corbis. **p. 278:** *photo* Patrick J. Miller.

CHAPTER 8 pp. 280–281: *photo* Steve Bloom Images/Alamy. **p. 282:** *photo* Frans Lanting Photography. **p. 283:** *Figure 8-1* Graphic Science/ Alamy. **p. 285:** *photo* Tui De Roy/Minden Pictures/Getty Images. **p. 287:** *Figure 8-5* Getty Images; *Figure 8-6* Mary Evans Picture Library/Alamy. **p. 290:** *Figure 8-8* Hulton Archive/Getty Images; *Figure 8-9* Photo by Louie Fasciolo/Bernard J. Shapero Rare Books. **p. 291:** *photo* Tom Brakefield/Science Faction/Getty Images. **p. 292:** *Figure 8-10* Renee Lynn/Corbis. **p. 294:** *Figure 8-11 (L inset)* Florida Images/Alamy, *(R inset)* Courtesy Oregon State University. **p. 295:** *Figure 8-12* J. D. Talasek/Phototake. **p. 296:** *Figure 8-13* Victor McCusick. **p. 297:** *Figure 8-14* Ross Warner/Alamy. **p. 299:** *Figure 8-16 from left)* Deanne Fitzmaurice; Jason Homa/Getty Images; Paul Costello/Getty Images; Paul Burns/Blend Images/Getty Images; Jacob Langvad/Stone/Getty Images. **p. 300:** *Figure 8-17* Max Nash/AFP/Getty Images; *Figure 8-18* Benn Mitchell/Riser/Getty Images. **p. 302:** *photo* Yann Arthus-Bertrand/ Corbis. **p. 304:** *Figure 8-22* J & B Photographers/Animals Animals—Earth Scenes. **p. 307:** *Figure 8-25* Patricia McDonough; *Figure 8-26* Grant Heilman/Grant Heilman Photography. **p. 308:** *Figure 8-27* Barros & Barros/Getty Images; *Figure 8-28* Gary Vestal/Photographer's Choice/ Getty Images. **p. 310:** *Figure 8-29* RubberBall/Alamy; *Figure 8-30* Michael Durham/Minden Pictures/Getty Images. **p. 311:** *photo* Gerald

Cubitt/NHPA/Photoshot. **p. 312:** *photo* SuperStock/age fotostock. **p. 313:** *Figure 8-31 (L)* Jason Edwards/National Geographic/Getty Images, *(R)* Louie Psihoyos/Science Faction/Getty Images. **p. 316:** *Figure 8-35 (from left)* Joe McDonald/Visuals Unlimited; G. Armistead/Vireo; Aaron French; Jack Jeffrey. **p. 317:** *Figure 8-36 (row 1)* Joe McDonald/Visuals Unlimited; Martin Pot; Dave Watts/Alamy; *(row 2)* Leonard Lee Rue III/ Photo Researchers; Morales/age fotostock; John W. Warden/age fotostock. **p. 319:** *Figure 8-37 (row 1)* Jose I. Castro; Laboratory of Evolutionary Morphology; *(row 2)* courtesy Bradley Smith, University of Michigan; S. K. Ackerley Department of Integrative Biology, University of Guelph, Guelph, Canada. **p. 319:** *Figure 8-38 (from left)* PBNJ Productions/PBNJ/age fotostock; Juniors Bildarchiv/age fotostock; Malcolm Schuyl/Alamy; Juniors Bildarchiv/age fotostock. **p. 320:** *Figure 8-39* Adrian Warren/Ardea; *Figure 8-40 (T)* It Stock Free/age fotostock, *(B)* Kevin Schafer/Getty Images. **p. 321:** *Figure 8-41 (from left)* Getty Images; Arco Images/Alamy; PhotoSpin/ Alamy; Danita Delimont/Alamy; Ulrike Klenke and Zeev Pancer, Center of Marine Biotechnology, Baltimore. **p. 322:** *photo* Lakeside/Dreamstime.com.

CHAPTER 9 **pp. 328–329:** *photo* Luiz Claudi Marigo. **p. 330:** *photo* DLILLC/Corbis; *Figure 9-1* AP Photo/Independence Daily Reporter/ Rob Morgan. **p. 331:** *Figure 9-2* Radius Images/Alamy; *Figure 9-3* Isabel M. Smallegange. **p. 333:** *Figure 9-4 (L)* Juniors Bildarchiv/Alamy, *(TR)* Andrew Darrington/Alamy, *(BR)* ARCO/H. Reinhard/age fotostock. **p. 335:** *Figure 9-6* Peter Cade/Getty Images, *(TI)* Joe McDonald/Getty Images, *(BI)* Lew Robertson/Corbis. **p. 337:** *photo* Annie Griffiths Belt/ Corbis. **p. 338:** *photo* Joe McDonald/Corbis. **p. 339:** *Figure 9-8* Dr. Theodore Evans. **p. 340:** *Figure 9-9* Richard R. Hansen/Photo Researchers. **p. 343:** *Figure 9-12* Courtesy Gerald Carter. **p. 344:** *Figure 9-13* Anup Shah/Animals Animals—Earth Scenes. **p. 346:** *Figure 9-14* PA Wire URN:5620400 (Press Association via AP Images). **p. 348:** *photo* Frans Lanting/Corbis. **p. 349:** *Figure 9-16* David M. Phillips/ Photo Researchers. **p. 350:** *Figure 9-18 (L)* ZZZD/Minden Pictures, *(M)* Michael and Patricia Fogden/Corbis, *(R)* Frans Lanting/Corbis, *(MI)* Dieter Herrmann/age fotostock. **p. 351:** *Figure 9-19 (L)* AP Images/Fox Searchlight, *(R)* Universal Films/Miramax Pictures/Kobal Collections; *Figure 9-20* Michael Fogden/photolibrary. **p. 353:** *Figure 9-21 (TL)* Anthony Mercieca/Photo Researchers, *(TR)* Dr. John Alcock/Visuals Unlimited, *(BL)* Photolibrary, *(BR)* Paul Stover/Getty Images. **p. 355:** *Figure 9-22* Michael and Patricia Fogden/Corbis; *Figure 9-23* James H. Robinson/Photo Researchers. **p. 356:** *Figure 9-24* George D. Lepp/Corbis. **p. 357:** *Figure 9-25* M. Botzek/age fotostock. **p. 359:** *Figure 9-27 (L)* Colla/ Oceanlight.com, *(R)* ARCO/W. Layer/age fotostock. **p. 360:** *photo* Harald Lange/zefa/Corbis. **p. 361:** *Figure 9-28 (L)* Grant Heilman Photography/ Alamy, *(M)* Juniors Bildarchiv/Alamy, *(R)* Jupiterimages/Creatas/Alamy. **p. 362:** *Figure 9-20 (L)* Scott Camazine/Photo Researchers, *(M)* Copyright 2008 The Gorilla Foundation/Koko.org. All rights reserved, *(R)* Michael Newman/PhotoEdit. **p. 363:** *Figure 9-30* Werner Bollmann/age fotostock. **p. 364:** *photo* Titus Lacoste/Getty Images.

CHAPTER 10 **pp. 368–369:** Kellar Autumn, Lewis & Clark College. **p. 370:** *photo* Peter Sawyer/NMNH/Smithsonian Institution; *Figure 10-1* David Muench/Corbis. **p. 371:** *Figure 10-2* Roger Ressmeyer/Corbis. **p. 372:** *Figure 10-3* Courtesy Andrew Knoll. **p. 373:** *Figure 10-4* Science VU/Sidney Fox/Visuals Unlimited. **p. 374:** *photo* flowerphotos/ Alamy; *Figure 10-5 (from left)* Peter Turnley/Corbis; Steve Sands/New York Newswire/Corbis; Ron Evans/Getty Images; GAP Photos/Getty Images; Andrew Darrington/Alamy; Tom Viggars/Alamy. **p. 375:** *Figure 10-6* Galen Rowell/Corbis; *Figure 10-7* Bill Dow/Shambala Preserve/ Roar Foundation. **p. 376:** *Figure 10-8 (L)* Juniors Bildarchiv/Alamy, *(R)* Masterfile. **p. 380:** *Figure 10-11 (L)* John Cancalosi/age fotostock, *(R)* Wildlife/Peter Arnold. **p. 381:** *Figure 10-12 (from left)* Kevin Schafer/ Corbis; Tui De Roy/Minden Pictures; Hiroya Minakuchi/Minden Pictures; Tui De Roy/Minden Pictures; Tui De Roy/Minden Pictures; Michael Woodruff. **p. 383:** *photo* Fancy/Veer/Corbis. **p. 385:** *Figure 10-15 (tree of life)* David M. Hillis, Derrick Zwickl, and Robin Gutell, University of Texas. **p. 386:** *Figure 10-17 (from left)* DLILLC/Corbis; Nigel Dennis/ age fotostock; James Gritz/Getty Images; Altrendo/Getty Images. **p. 387:** *photo* Courtesy The Broad Institute of MIT and Harvard. **p. 388:** *Figure 10-19* Michael Durham/Minden Pictures; *Figure 10-20* David Hosking/ Alamy; IFA-Bilderteam/Jupiterimages; Michael and Patricia Fogden/ Corbis. **p. 389:** *photo* Kim Taylor/naturepl.com. **p. 390:** *photo* Daniel L. Geiger/SNAP/Alamy; *Figure 10-21* David Edwards/Getty Images. **p. 395:** *Figure 10-26* Reuters/Corbis. **p. 396:** *photo* Norbert Wu/Minden Pictures. **p. 399:** *Figure 10-29* Radius Images/Alamy; *Figure 10-30* David McCarthy/Photo Researchers. **p. 400:** *Figure 10-31* John Wang/Getty Images, *(inset)* Agriculture et Agroalimentaire Canada. **p. 401:** *Figure 10-32 (from left)* Michael and Patricia Fogden/Corbis; from *Botanica* by Howard Schatz, © Schatz Ornstein 2005; Adam Jones/Photo Researchers; Frans Lanting/Corbis. **p. 402:** *photo* David Sacks/Getty Images.

CHAPTER 11 **pp. 406–407:** *photo* Stephen Frink/zefa/Corbis. **p. 408:** *photo* Scott Chandler. **p. 409:** *Figure 11-1(from left)* Jeffrey L. Rotman/Corbis, Frans Lanting/Corbis, Art Wolfe/Getty Images. **p. 411:** *Figure 11-3 (L)* Steve Maslowski/Photo Researchers, *(R)* Thomas Mangelsen/Minden Pictures. **p. 412:** *photo* Frans Lanting/Corbis. **p. 413:** *Figure 11-4 (L)* Images&Stories/Alamy, *(inset)* Art Wolfe/Getty Images; *(R)* Frans Lanting Photography, *(inset)* Ross Armstrong/photolibrary; *Figure 11-5 (from left)* Tim Laman/Getty Images, Natural Visions/Alamy, Jack Milchanowski/photolibrary. **p. 414:** *photo* Frans Lanting/Corbis. **p. 415:** *Figure 11-7 (L)* Darlyne A. Murawski/Getty Images, *(R)* Masa Ushioda/ SeaPics.com. **p. 416:** *Figure 11-8 (from top)* Paul Souders/Corbis, Tim Fitzharris/Minden Pictures,) Rudie Kuiter/OceanwideImages.com. **p. 418:** *photo* Frans Lanting/Corbis; *Figure 11-10 (from left)* Fabio Liverani/ Minden Pictures, Dorling Kindersley/Getty Images, Visuals Unlimited/ Corbis. **p. 420:** *Figure 11-12 (both photos)* Frans Lanting/Corbis. **p. 422:** *Figure 11-15 (from left)* Dave Watts/Alamy, Martin Rugner/Photolibrary, Frans Lanting Photography. **p. 427:** *photo* Frans Lanting/Corbis. **p. 429:** *Figure 11-20 (clockwise from left)* DLILLC/Corbis, Stephen Frink/Corbis, Chris Newbert/Minden Pictures. **p. 430:** *Figure 11-22 (clockwise from left)* George Grall/Getty Images, blickwinkel/Alamy, William James Warren/ Getty Images, Chris Newbert/Minden Pictures; *Figure 11-23 (L)* Frans Lanting/Corbis, *(M)* Alamy, *(R)* Alamy. **p. 432:** *Figure 11-24 (from left)* Piotr Naskrecki/Minden Pictures, Alamy, Javier Tajuelo/photolibrary, Tom Vezo/Minden Pictures. **p. 433:** *Figure 11-25 (clockwise from left)* Gerry Ellis/ Minden Pictures, Jeff Rotman/Getty Images, David B. Fleetham/SeaPics. com, Walter E. Harvey/Photo Researchers. **p. 434:** *Figure 11-26 (from left)* David B. Fleetham/photolibrary, Adam Butler/Alamy, Doug Perrine/ SeaPics.com. **p. 435:** *Figure 11-27* Jorgen Jessen/AFP/Getty Images. **p. 436:** *photo* Scott Leslie/Minden Pictures. **p. 437:** *Figure 11-28 (clockwise from left)* Ross and Diane Armstrong/Minden Pictures, ARCO/H. Reinhard/age fotostock, Konrad Wothe/Minden Pictures; *Figure 11-29 (L)* Lshtm/Photo Researchers, *(R)* CNRI/Photo Researchers. **p. 438:** *Figure 11-30 (L)* Sinclair Stammers/Photo Researchers, *(R)* Alamy. **p. 439:** *Figure 11-31 (clockwise from left)* Brandon Cole/Getty Images, Tim Laman, Frans Lanting/Corbis. **p. 440:** *Figure 11-32* Howard Hall/photolibrary. **p. 441:** *Figure 11-33 (L)* Juan Carlos Calvin/photolibrary, *(R)* Robert Yin/Corbis.

p. 442: *photo (T)* Dr. David Wachenfeld/Auscape/Minden Pictures, *(B)* Zeva Oelbaum/Peter Arnold.

CHAPTER 12 **p. 446–447:** *photo* Katinka Matson. **p. 448:** *photo* Frans Lanting Photography; *Figure 12-1 (from left)* David Pearson/Alamy, Douglas Pulsipher/Alamy, Michael and Patricia Fogden/Minden Pictures. **p. 449:** *Figure 12-2* William James Warren/Getty Images; *Figure 12-3 (from left)* Photodisc/Alamy, Chad Ehlers/Alamy, Herbert Hopfensperger/age fotostock. **p. 451:** *photo* Frans Lanting/Corbis; *Figure 12-5* Charles F. Delwiche. **p. 452:** *Figure 12-6 (L)* Hans Strand/Corbis, *(R)* Gallo Images/Getty Images. **p. 453:** *Figure 12-7 (L)* Alistair Dove/Alamy, *(M)* Verbiesen Henk/age fotostock, *(R)* Daniel Vega/age fotostock. **p. 454:** *Figure 12-8* Barry Turner/Alamy; *Figure 12-9 (T)* The Irish Image Collection/Design Pics/Corbis, *(B)* Matthew Mawson/Alamy. **p. 455:** *Figure 12-10 (L)* Ed Reschke/Peter Arnold, *(M)* DEA/Dani-Jeske/Getty Images, *(R)* Juan Carlos Muñoz/age fotostock; *photo* Bjorn Forsberg/npl/Minden Pictures. **p. 456:** *Figure 12-11* Peter Arnold/Alamy. **p. 457:** *photo* Jeff Vanuga/Corbis; *Figure 12-12* Kevin Phelan. **p. 458:** *Figure 12-13 (from left)* Bohemian Nomad Picturemakers/Corbis, Design Pics/Alamy, Jouan & Rius/naturepl.com, Takahiro Miyamoto/Getty Images. **p. 459:** *Figure 12-14 (from left)* Thomas Kitchin and Victoria Hurst/AllCanadaPhotos.com, Alan and Linda Detrick/Photo Researchers, Jouan & Rius/naturepl.com, blickwinkel/Alamy; *Figure 12-15 (L)* Brad Mogen/Visuals Unlimited, *(R)* Stephen P. Parker/Photo Researchers. **p. 460:** *Figure 12-16* Darrell Gulin/Corbis. **p. 461:** *Figure 12-17 (T)* Gary Randall/Getty Images, *(M)* Kathleen Phelan, *(B)* Philippe Bourseiller/Getty Images. **p. 462:** *photo* Andy Small/Corbis; *Figure 12-18 (from left)* Ferruccio Carassale/Grand Tour/Corbis, blickwinkel/Alamy, Grand Tour/Corbis, Jack Dykinga/Getty Images. **p. 463:** *photo* John Fairhall/Auscape/Minden Pictures. **p. 464:** *Figure 12-20* Topic Photo Agency/age fotostock. **p. 465:** *Figure 12-21 (top row from left)* Kim Taylor/naturepl.com, Rolf Nussbaumer/naturepl.com, Andrew Darrington/Alamy, Sharkawi Che Din/Alamy; *(bottom row from left)* blickwinkel/Alamy, Mark Moffett/Minden Pictures. **p. 466:** *Figure 12-22* Tony Hertz/Corbis. **p. 468:** *photo* Dezorzi/Corbis. **p. 469:** *Figure 12-24 (from left)* Scott Camazine/Alamy, Steve Bloom Images/Alamy, Manfred Danegger/Peter Arnold. **p. 470:** *Figure 12-25 (TL)* Joe McDonald/Visuals Unlimited, *(TR)* Wildlife GmbH/Alamy, *(M)* Howard Grey/Getty Images, *(B)* Gertrud and Helmut Denzau/naturepl.com. **p. 471:** *Figure 12-26* Ed Reschke/Peter Arnold. **p. 472:** *photo* Jan Vermeer/Foto Natura/Minden Pictures; *Figure 12-27 (from left)* Andrew Parkinson/Corbis, Yva Momatiuk and John Eastcott/Photo Researchers, Glen Threlfo/Minden Pictures. **p. 473:** *Figure 12-29 (from left)* Steve Austin/Papilio/Corbis, Visuals Unlimited/Corbis, Visuals Unlimited/Corbis. **p. 474:** *Figure 12-30* Ingo Arndt/Minden Pictures. **p. 475:** *Figure 12-31* AP Photo/Oregon State University; *Figure 12-32* D. Hurst/Alamy. **p. 476:** *Figure 12-33* Visuals Unlimited/Corbis. **p. 477:** *Figure 12-34* Jim Brandenburg/Minden Pictures. **p. 478:** *photo* Mary Ellen Mark.

CHAPTER 13 **pp. 482–483:** *photo* Lee D. Simon/Photo Researchers. **p. 484:** Stephanie Schuller/Photo Researchers. **p. 485:** *Figure 13-2 (from left)* CNRI/Photo Researchers, *(inset)* Phototake/Alamy; Federico Veronesi/Getty Images, *(inset)* Agriculture et Agroalimentaire Canada; Science Source/Photo Researchers, *(inset)* Courtesy K.O. Stetter, R. Huber, and R. Rachel, University of Regensburg, Germany. **p. 486:** *Figure 13-3* Courtesy Jay Phelan; *photo* Frans Lanting/Corbis. **p. 487:** *Figure 13-4 (T from left)* Dennis Kunkel/Alamy, Dennis Kunkel/Alamy, Biodisc/Visuals Unlimited/Alamy, *(bottom)* Eye of Science/Photo Researchers. **p. 488:** *photo* Visuals Unlimited/Corbis. **p. 489:** *Figure 13-5* Dennis Kunkel/Alamy; *Figure 13-6 (L)* mediacolor/Alamy, *(R)* Raymond Otero/Visuals Unlimited. **p. 492:** *Figure 13-9* Mediablitz Images. **p. 493:** *Figure 13-10(from left)* Dodie Ulery, *(M)* Thomas Dressler/age fotostock, *(R)* Reino Hanninen/Alamy. **p. 494:** *Figure 13-11* Original map by Dr. John Snow, *(inset)* Justinc. **p. 495:** *Figure 13-12* ImageState Royalty Free/Alamy. **p. 496:** *Figure 13-14* AGStockUSA/Alamy. **p. 497:** *photo* George Steinmetz/Corbis; *Figure 13-15* Dennis Kunkel/Visuals Unlimited. **p. 498:** *Figure 13-16* Russ Bishop/age fotostock; *Figure 13-17* Tim Mantoani/Masterfile. **p. 499:** *Figure 13-18* Salvador de Sas/epa/Corbis. **p. 500:** *photo* Steve Gschmeissner/Photo Researchers; *Figure 13-19* Microfossil Image Recovery and Circulation for Learning and Education (MIRACLE) Project, Micropalaeontology Unit of the Earth Sciences Department, University College, London. **p. 501:** *Figure 13-20 (from left)* Visuals Unlimited/Corbis, C-FH/Grant Heilman, Images&Stories/Alamy. **p. 502:** *Figure 13-21* M. I. Walker/Photo Researchers. **p. 503:** *Figure 13-22)* Eye of Science/Photo Researchers, *(top inset)* Peter Arnold/Alamy, *(bottom inset)* Phototake/Alamy. **p. 504:** *photo* NIBSC/Photo Researchers; *Figure 13-23 (M)* SPL/Photo Researchers, *(R)* Phototake. **p. 506:** *Figure 13-25 (T) St. Louis Post Dispatch, (B)* Reuters/Corbis. **p. 507:** *Figure 13-27* Bettmann/Corbis, *(inset)* nagelestock.com/Alamy. **p. 509:** *Figure 13-29* Dr. Klaus Boller/Photo Researchers. **p. 510:** *photo* Tommy Moorman.

CHAPTER 14 **pp. 514–515:** *photo* Frans Lanting Photography. **p. 516:** *photo* Barrett & MacKay/age fotostock; *Figure 14-1* Frans Lanting/Corbis. **p. 518:** *Figure 14-3* Tom Nebbia/Corbis. **p. 519:** *Figure 14-4* Juan Carlos Muñoz/age fotostock. **p. 520:** *Figure 14-5* R. Wittek/age fotostock. **p. 521:** *Figure 14-7* John McColgan, Bureau of Land Management, Alaska Fire Service. **p. 522:** *Figure 14-8* Bill Stormont/Corbis; *Figure 14-9* AP Photo/Dario Lopez-Mills. **p. 523:** *Figure 14-10* Lynn M. Stone/Naturepl.com; *Figure 14-11* age fotostock/Fotosearch. **p. 524:** *Figure 14-12* Dirk Anschutz/Getty Images. **p. 525:** *Figure 14-13* National Museum of Natural History, Smithsonian Institution; *Figure 14-14* Martin Bwenetti/AFP/Getty Images. **p. 526:** *photo* Frans Lanting/Corbis. **p. 527:** *Figure 14-15 (T)* Dave Watts/Naturepl.com, *(M)* Colin Preston/naturepl.com, *(B)* Joe McDonald/Visuals Unlimited. **p. 528:** *Figure 14-16* Greg Katsoulis. **p. 529:** *Figure 14-17 (L)* Martin Harvey/Alamy, *(M)* Terry Wall/Alamy, *(R)* D. R. Schrichte/SeaPics.com. **p. 530:** *Figure 14-18 (L)* Gunter Marx Photography/Corbis, *(M)* Philippe Clement/naturepl.com, *(R)* Gary Meszaros/Photo Researchers, *(inset)* iStockPhoto.com/jeridu. **p. 531:** *photo* Jim Brandenburg/Minden Pictures. **p. 532:** *photo* Gabriela Staebler/zefa/Corbis. **p. 533:** *Figure 14-19* Condé Nast Archive/Corbis. **p. 535:** *Figure 14-21 (T)* Renaud Visage/Getty Images, *(B)* W. Perry Conway/Corbis. **p. 537:** *photo* Eric Fougére/Kipa/Corbis. **p. 538:** *photo* Getty Images. **p. 543:** *Figure 14-27 (T)* Yadid Levy/Anzenberger Agency/Jupiterimages, *(M)* Sean Gallup/Getty Images, *(B)* Michael S. Yamashita/Getty Images; *Figure 14-28* Keren Su/Getty Images. **p. 544:** *photo* Elke Hesser/Getty Images.

CHAPTER 15 **pp. 548–549:** *photo* Georgette Douwma/Getty Images. **p. 550:** *photo* Frans Lanting/Corbis; *Figure 15-1* Thomas Marent/Minden Pictures. **p. 551:** *Figure 15-2* Steve Allen/Getty Images. **p. 552:** *Figure 15-3 (top row from left)* Jacques Jangoux/Alamy, Ted Med/Getty Images, Fritz Poelking/photolibrary, *(middle row from left)* John Warburton-Lee Photography/Alamy, Michael Melford/Getty Images, Visions LLC/photolibrary, *(bottom row from left)* Dan Lamont/Corbis, Frans Lanting/Corbis, Peter Arnold/Alamy. **p. 553:** *Figure 15-4 (top row)* Martin Siepmann/

photolibrary, Frans Lanting/Corbis, *(bottom row)* Troy Plota/Getty Images, Momatiuk-Eastcott/Corbis, Mark Conlin/V&W/Image Quest Marine. **p. 554:** *photo* Frans Lanting/Corbis. **p. 555:** *Figure 15-6* moodboard/ Corbis. **p. 556:** *Figure 15-7* Craig Aurness/photolibrary, *photo* Frans Lanting/Corbis. **p. 557:** *Figure 15-8* Jacques Descloitres, MODIS Land Rapid Response Team, NASA/GSFC. **p. 558:** *Figure 15-9 (L)* Courtesy NASA/Goddard Space Flight Center Scientific Visualization Studio, *(R)* Sandy Felsenthal/Corbis. **p. 561:** *photo* Martin Harvey/Corbis. **p. 562:** *Figure 15-13 (L)* AJPhoto/Photo Researchers, *(R)* Eye of Science/Photo Researchers. **p. 568:** *Figure 15-19* AP Photo/Gautam Singh, *photo* P. Manner/zefa/Corbis. **p. 569:** *photo* Stephen Frink/Corbis. **p. 569:** *Figure 15-20* Mitsuhiko Imamori/Minden Pictures. **p. 570:** *Figure 15-21* Jim Brandenburg/Minden Pictures. **p. 574:** *Figure 15-24 (top row)* Anthony Bannister Gallo Images/Corbis, Suzanne L. and Joseph T. Collins/Photo Researchers, *(bottom row)* Michael & Patricia Fogden/Minden Pictures, Gail Shumway/Getty Images. **p. 575:** *Figure 15-25 (T)* Fred Bavendam/Minden Pictures, *(B)* John Downer/naturepl.com; *Figure 15-26* Dr. Bruce Robison/ Minden Pictures. **p. 576:** *Figure 15-27 (T)* Gallo Images/Alamy, *(B)* Eye of Science/Photo Researchers. **p. 577:** *Figure 15-28* Biosphoto/Heuclin Daniel/Peter Arnold. **p. 578:** *Figure 15-29 (T)* Darlyne A. Murawski/Getty Images, *(B)* Arco Images GmbH/Alamy. **p. 579:** *photo* W. Perry Conway/ Corbis. **p. 581:** *Figure 15-31 (inset)* Siede Preis/Getty Images. **p. 582:** *photo* NASA image by Robert Simmon, based on Landsat data provided by the UMD Global Land Cover Facility.

CHAPTER 16 pp. 586–587: *photo* Peet Simard/Corbis. **p. 588:** *photo* Martin Harvey/Corbis; *Figure 16-1* Burger/Phanie/Photo Researchers, *(inset)* Inga Spence/Getty Images. **p. 589:** *Figure 16-2 (clockwise from top left)* Wildscape/Alamy, Suzanne Long/Alamy, Jeff Vanuga/Corbis, The Photo Works. **p. 590:** *Figure 16-3 (from left)* Alfred Pasieka/Photo Researchers, DLILLC/Corbis, NASA. **p. 591:** *Figure 16-4* Dennis Macdonald/ photolibrary.com. **p. 593:** *Figure 16-7 (TR)* blickwinkel/Alamy, *(BL)* Michael and Patricia Fogden/Corbis, *(BR)* DeAgostini/Getty Images. **p. 594:** *Figure 16-8* Courtesy Edward O. Wilson. **p. 595:** *Figure 16-9* Danny Lehman/Corbis. **p. 597:** *photo* Gallo Images/Corbis; *Figure 16-11* Tom Uhlman/Alamy. **p. 600:** *Figure 16-14* Marcelo Sayao/epa/Corbis; *Figure 16-15* Frans Lanting Photography. **p. 601:** *photo* Frans Lanting/Corbis. **p. 602:** *Figure 16-16 (from left)* USGS Photo by Harry Glicken, USGS Photo by Harry Glicken, Don Geyer/Alamy; *Figure 16-17* Denis Scott/ Corbis. **p. 603:** *Figure 16-18* Jany Sauvanet/Photo Researchers, *(inset)* Gale S. Hanratty/Alamy. **p. 604:** *Figure 16-20 (from left)* John Mitchell/Photo Researchers, Michael P. Gadomski/Photo Researchers, Wil Meinderts/ Minden Pictures. **p. 605:** *Figure 16-21* Jack Picone/Alamy. **p. 606:** *Figure 16-22* Anil Risal Singh-UNEP/Peter Arnold. **p. 607:** *Figure 16-25* Will and Deni McIntyre/Corbis. **p. 609:** *Figure 16-28 (both photos)* NASA. **p. 610:** *Figure 16-30 (all photos)* NASA. *Figure 16-31* AP Photo/Elise Amendola, *(I)* New Line/Photofest. **p. 613:** *Figure 16-34* Frans Lanting/Corbis. **p. 614:** *photo* JASON LEE/X01757/Reuters/Corbis. **p. 615:** *Figure 16-35 (T)* Cyril Ruoso/Minden Pictures, *(B)* Kim Taylor/Minden Pictures, *(inset)* Medical-on-Line/Alamy; *Figure 16-36 (L)* Gerry Ellis/Minden Pictures, *(R)* Robert Harding Picture Library/Alamy. **p. 616:** *Figure 16-37* Tommy Moorman. **p. 618:** *photo* Peter Arnold/Alamy.

CHAPTER 17 pp. 622–623: *photo* Paul Edmondson/Corbis. **p. 624:** *photo* Steve Terrill/Corbis. **p. 625:** *Figure 17-1 (from left)* Lonely Planet Images/Getty Images, Shalom Ormsby/Getty Images, De Agostini Picture Library/Getty Images, Frans Lanting/Frans Lanting Photography. **p. 627:** *photo* David T. Webb, Ph. D. **p. 629:** *Figure 17-5 (L)* Eye of Science/Photo Researchers, *(L, inset)* Mark Bolton/Corbis, *(C)* Dr. Gerald Van Dyke/Getty Images, *(R)* Dennis Kunkel Microscopy, Inc./Visuals Unlimited, *(R, inset)* Doug Wilson/Corbis; *Figure 17-6* Steve Gschmeissner/Photo Researchers. **p. 630:** *Figure17-7 (L)* Image Source/Corbis, *(L, inset)* Dennis Kunkel Microscopy, Inc., *(C)* Keith van-Loen/Alamy, *(C, inset)* Biophoto Associates/ Photo Researchers, *(R)* Pete Ryan/Getty Images, *(R, inset)* Biophoto Associates/Photo Researchers. **p. 631:** *photo* Radius Images/Corbis. **p. 632:** *Figure17-9 (L)* Robert Pickett/Corbis, *(R)* AgStock Images/Corbis; *Figure 17-10* Susumu Nishinaga/Photo Researchers. **p. 633:** *Figure 17-11* Wayne Lawler/Ecopix. **p. 634:** *Figure 17-14 (from left)* AgStock Images/ Corbis, Dorling Kindersley/Getty Images, Gabor Geissler/Getty Images, Robert & Jean Pollock/Getty Images. **p. 636:** *Figure 17-16 (top row, from left)* Anna Kern/Etsa/Corbis, Niall Benvie/Corbis, Mark Bolton/Corbis, *(middle row, from left)* Michael Sedam/Photolibrary, John and Lisa Merrill/Corbis, Ehlers/Getty Images, *(bottom row, from left)* Frank Krahmer/Corbis, Micha Pawlitzki/Corbis, DLILLC/Corbis. **p. 637:** *Figure 17-17* AP Photo/Jody Kurash. **p. 638:** *Figure 17-18 (L)* Photo 24/Brand X/Corbis, *(inset)* Grant Heilman Photography/Alamy, *(C)* Jonathan Buckley/Getty Images, *(R)* Martin Oeggerli/Visuals Unlimited. **p. 639:** *photo* KOJI NAKAMURA/ amanaimages/Corbis. **p. 640:** *Figure 17-20* AP Photo/Minot Daily News, Marvin Baker. **p. 641:** *Figure17-21* Jim Richardson/National Geographic Stock; *Figure 17-22* Darrell Gulin/Corbis. **p. 642:** *Figure 17-23 (L)* Mary Hobbs/Alamy, *(R)* Jane Legate/Photolibrary; *Figure 17-24* Envision/Corbis. **p. 644:** *Figure 17-27 (L)* Clive Nichols/Getty Images, *(R)* Coke Whitworth/ Getty Images. **p. 645:** *photo* Steve Gschmeissner/Photo Researchers; *photo* Dennis Drenner/Visuals Unlimited. **p. 646:** *Figure 17-29* Dr. Merton Brown/Visuals Unlimited. **p. 647:** *Figure 17-30* Momatiuk-Eastcott/Corbis. **p. 649:** *Figure 17-33* Radius Images/Corbis. **p. 651:** *Figure 17-35* George D. Lepp/Corbis. **p. 652:** *photo* Kevin Steele/Aurora Photos/Corbis.

CHAPTER 18 pp. 656–657: *photo* Steve Gschmeissner/Science Photo Library. **p. 658:** *photo (T)* James Randklev/Corbis; *photo (B)* Ingo Arndt/ Minden Pictures. **p. 659:** *Figure 18-1* David Muench/Corbis, *(inset)* Copyright © Regents of the University of Minnesota. All rights reserved. **p. 660:** *Figure 18-2 (L)* Marc Vassallo © 2001 by The Taunton Press, Inc., *(C)* Bon Appetit/Alamy. *(R)* Arco Images GmbH/Alamy. **p. 662:** *Figure 18-4* Lori Greig/Getty Images **p. 663:** *Figure18-5* Peter Ginter/Science Faction/Corbis **p. 665:** *photo Tulip, 1988* © Copyright The Robert Mapplethorpe Foundation. Courtesy Art + Commerce; *Figure 18-7 (L)* Frans Lanting/Frans Lanting Photography, *(C, L)* Imagemore Co., Ltd./ Corbis, *(C, R)* Frank Lukasseck/Corbis, *(R)* Imagemore Co., Ltd./Corbis. **p. 666:** *Figure 18-9* nik wheeler/Alamy, *(inset)* John Glover/Photolibrary. **p. 668:** *photo (T)* Peter Arnold, Inc./Alamy, *photo (B)* Dennis Kunkel Microscopy, Inc./PHOTOTAKE/Alamy; *Figure 18-11* Robert Clare/ Alamy, *(inset)* Bob Sacha/Corbis. **p. 670:** *photo* Herbert Zettl/Corbis; *Figure 18-13* Martin Harvey/Corbis. **p. 671:** *Figure 18-14 (L)* Nic Bothma/epa/Corbis, *(R)* Martin Harvey/Alamy, *(inset)* Irene Terry, University of Utah. **p. 672:** *Figure 18-15* Adam Jones/Visuals Unlimited, *(inset)* Micro Discovery/Corbis. **p. 677:** *photo (L)* John Oeth/ Alamy; *photo (R)* Bill Beatty/AGPix. **p. 678:** *photo* Cindy Kassab/Corbis; *Figure 18-20* Photolibrary. **p. 680:** *Figure 18-22 (L)* Scott Stulberg/Corbis, *(C)* David Muench/Corbis, *(R)* ImageClick, Inc./Alamy. **p. 681:** *Figure18-23 (T)* Dr. Brad Mogen/Visuals Unlimited, *(B)* Biodisc/Visuals Unlimited/Corbis, *(R)* Oxford Scientific/Photolibrary; *photo* Digital Vision/Getty Images. **p. 683:** *Figure 18-25 (R)* Alan Schein Photography/ Corbis, *(C)* Edi Engeler/epa/Corbis, *(L)* Steve Babineau/NBAE via

Getty Images. **p. 684:** *Figure 18-27* Nils-Johan Norenlind/Getty Images. **p. 685:** *Figure 18-28* Charles O. Cecil/Alamy. **p. 686:** *photo* Food Collection/Stock Food.

CHAPTER 19 **pp. 690–691:** *photo* Cathlyn Melloan/Getty Images. **p. 692:** *photo* D. Hurst/Alamy. **p. 693:** *Figure 19-1* Bob Gibbons/Science Photo Library. **p. 694:** *Figure 19-3* Yoshio Niikura/amanaimages/Corbis. **p. 695:** *Figure 19-4* AP Photo/Anchorage Daily News, Bill Roth. **p. 696:** *Figure 19-5* Stuart Westmorland/Corbis. **p. 697:** *Figure 19-7* Lindholdt Foto/iStockphoto. **p. 698:** *Figure 19-8 (L)* Grant Faint/Getty Images, *(R)* Colin Underhill/Alamy; *Figure 19-9* Tom Thulen/Alamy. **p. 699:** *Figure 19-10* Crady von Pawlak/Corbis; *Figure 19-11* Garry DeLong/Photolibrary. **p. 700:** *Figure 19-12* John Kaprielian/Photo Researchers. **p. 701:** *Figure 19-13* Martin Shields/Photo Researchers; *Figure 19-14* Peter Arnold, Inc./Alamy. **p. 702:** *Figure 19-15* Courtesy of Roger P. Hangarter, Chancellor's Professor of Biology, Indiana University. **p. 703:** *Figure 19-16* Foto Grebler/Alamy. **p. 705:** *photo* Tony Wharton/Frank Lane Picture Agency/Corbis. **p. 706:** *Figure 19-18 (L)* J S Sira/Gap Photo/Visuals Unlimited, *(C)* Biosphoto/Mayet Jean/Peter Arnold, *(R)* Martin Shields/Photo Researchers; *Figure 19-19* Gerry Bishop/Visuals Unlimited. **p. 707:** *Figure 19-20* Olivier Voisin/Photo Researchers; *Figure 19-21* Gregory G. Dimijian, M.D./Photo Researchers. **p. 708:** *photo* Frans Lanting/Corbis; *Figure19-22 (L)* Frans Lanting/Corbis, *(C)* Simon Fraser/Science Photo Library, *(R)* Christopher Talbot Frank/Corbis. **p. 709:** *Figure 19-23* Arco Images GmbH/Alamy, *(inset)* Tim Laman/NGS Image Collection; *Figure 19-24* Paul Nicklen/Getty Images. **p. 710:** BSIP/Photoshot.

CHAPTER 20 **pp. 714–715:** *photo* Visuals Unlimited/Corbis. **p. 716:** *photo* Alexander Demianchuk/Reuters/Corbis. **p. 717:** *Figure 20-1* AP Photo/Sue Ogrocki. **p. 718:** *Figure 20-2 (L)* Emilio Suetone/Hemis/Corbis, *(R)* AP Photo/Sergei Chuzavkov. **p. 719:** *photo* Anup Shah/npl/Minden Pictures. **p. 722:** *Figure 20-6 (T)* Paul Nicklen/Getty Images, *(B)* Doug Allan/Getty Images; *Figure 20-7 (L)* Tim Farrell/Star Ledger/Corbis, *(R)* Chris Johns/Getty Images. **p. 723:** *Figure 20-9 (from left)* Doug Allan/npl/Minden Pictures, Michael & Patricia Fogden/Minden Pictures, Simon King/npl/Minden Pictures, John-Francis Bourke/Corbis. **p. 724:** *Figure 20-10* TS Corrigan/Alamy. **p. 728:** *Figure 20-14* Peter Dasilva/epa/Corbis. **p. 729:** *photo* Keren Su/Corbis; *Figure 20-15* Hiroya Minakuchi/Minden Pictures. **p. 730:** *photo* Frans Lanting/Frans Lanting Photography. **p. 731:** *Figure 20-17 (T)* Image Quest Marine/Alamy, *(bottom, from left)* Dennis Kunkel Microscopy, Inc., Visuals Unlimited/Corbis *(all)*. **p. 733:** *Figure 20-19 (TL)* Quest/Photo Researchers, *(BL)* Photo Quest Ltd/Science Photo Library/Corbis, *(C)* George J. Sanker/DRK photo, *(TR)* Dennis Kunkel Microscopy, Inc, *(MR & BR)* Visuals Unlimited/Corbis; *photo* Clifford White/Corbis. **p. 735:** *Figure 20-22 (T)* Joe McDonald/Corbis, *(bottom, from left)* Photo Quest Ltd/Science Photo Library/Corbis, *(both)* Visuals Unlimited/Corbis. **p. 737:** *Figure 20-24* D. Robert & Lorri Franz/Corbis; *photo* Eric Grave/Photo Researchers. **pp. 738–739:** *Figure 20-25 (top row, from left)* MIXA/Alamy, Stuart O'Sullivan/Getty Images, Jose Luis Pelaez, Inc./Blend Images/Corbis, Andersen Ross/Blend Images/Corbis, Mark Andersen/Getty Images, Radius Images/Alamy, *(bottom row, from left)* Mike Kemp/Getty Images, Hola Images/Getty Images, Radius Images/Alamy, Caroline Schiff/Getty Images, Andersen Ross/Getty Images, RubberBall/Alamy. **p. 740:** *photo* Jupiterimages/Brand X/Alamy.

CHAPTER 21 **pp. 744–745:** *photo* Susumu Nishinaga/Photo Researchers. **p. 746:** *photo* Christopher Furlong/Getty Images. **p. 748:** *Figure 21-2* Chris Newbert/Minden Pictures; *Figure 21-3 (T)* Daniel J. McCauley IV, *(B)* Ruoso Cyril/Peter Arnold. **p. 749:** *Figure 21-4* Ralph Hutchings/Visuals Unlimited, *(inset)* David M. Phillips/Photo Researchers. **p. 752:** *photo* Dennis Kunkel Microscopy, Inc.; *Figure 21-6* Dietmar Busse/Getty Images. **p. 755:** *Figure 21-9 (L)* Eyecandy Images/IPNstock.com, *(R)* BSIP/Phototake. **p. 756:** *Figure 21-10* Alix/Photo Researchers. **p. 757:** *Figure 21-11* Fancy Photography/Veer. **p. 758:** *Figure 21-12 (L)* Dr. Fred Hossler/Visuals Unlimited, *(C)* Dennis Kunkel Microscopy, Inc/Phototake, *(R)* CNRI/Photo Researchers; *Figure 21-13* AP Photo/Eckehard Schulz. **p. 759:** *photo* Luca Trovato/Jupiterimages. **p. 760:** *Figure 21-14* Popperfoto/Getty Images, *(inset)* imagebroker/Alamy. **p. 761:** *Figure 21-15* Frans Lanting/Frans Lanting Photography. **p. 762:** *Figure 21-16 (L)* Science Photo Library, *(R)* SPL/Photo Researchers. **p. 763:** *Figure 21-17 (T)* Banana Stock/Jupiterimages, *(B)* Digital Vision/Superstock. **p. 765:** *Figure 21-18* Dietmar Busse/Getty Images, *(inset)* Michael Abbey/Photo Researchers. **p. 766:** *Figure 21-20* Phototex/NEWSCOM. **p. 767:** *Figure 21-21* Everett Collection; *photo* Corbis. **p. 768:** *photo* Paul Souders/Corbis. **p. 772:** *Figure 21-26* Fancy/Veer/Corbis. **p. 773:** *Figure 21-27* Dennis Kunkel Microscopy, Inc.; *Figure 21-28* Dr. Bahman Guyuron. **p. 774:** *Figure 21-29* Yva Momatiuk & John Eastcott/Minden Pictures. **p. 775:** *photo* Frans Lanting/Frans Lanting Photography. **p. 776:** *photo* Image Source/Getty Images; *Figure 21-31 (L & R)* Pr. M. Brauner/Photo Researchers. **p. 777** *Figure 21-32* IMAX/Photofest. **p. 778:** *photo* Julian Herbert/Getty Images. **p. 780:** *Figure 21-36 (T)* Loungepark/Getty Images; *(B)* Image Source Ltd/age fotostock; *Figure 21-37* Lennart Nilsson. **p. 781:** *Figure 21-38 (L)* Dorling Kindersley; *(R)* Foodfolio/Jupiterimages. **p. 782:** *photo* Danita Delimont/Alamy; *Figure 21-39* Design Pics Inc./Alamy. **p. 783:** *Figure 21-40* Alison Wright/Corbis. **p. 784:** *Figure 21-41* imagebroker/Alamy. **p. 785:** *Figure 21-42* Doug Allan/Minden Pictures. **p. 786:** *photo* Mark M. Lawrence/Corbis.

CHAPTER 22 **pp. 790–791:** *photo* Marcus Wilson-Smith/age fotostock. **p. 792:** *photo* Michael Freeman/Corbis. **p. 794:** *Figure 22-2 (T)* Gunter Zlesler/Peter Arnold, *(M)* Fred Bavendam/Minden Pictures, *(B)* Alan & Sandy Carey/zefa/Corbis; *Figure 22-3* Adrian Hepworth/NHPA, *(inset)* Alex Wild. **p. 795:** *Figure 22-4* CDO Ranching and Development L.P., *(inset)* Joseph Sohm/Visions of America/Corbis. **p. 796:** *Figure 22-5 (T & BL)* Loungepark/Getty Images, *(BR)* Image Source Ltd/age fotostock; *Figure 22-6* Heinrich van den Berg/Getty Images. **p. 797:** *photo* Frans Lanting/Corbis; *Figure 22-7* Thomas Mangelsen/Minden Pictures. **p. 799:** *Figure 22-8* Judith Haeusler/Getty Images; *Figure 22-9 (T)* FoodCollection/age fotostock, *(B)* Riou/PhotoCuisine/Corbis. **p. 801:** *Figure 22-10* Image Source/Corbis; *Figure 22-11 (L)* Lew Robertson/Corbis, *(C)* John E. Kelly/Jupiterimages, *(R)* LWA-Stephen Welstead/Corbis. **p. 802:** *Figure 22-12* Owen Franken/Corbis. **p. 803:** *Figure 22-13 (L)* Bobbi Fabian/Getty Images, *(R)* Tim Hill/Photolibrary. **p. 806:** *Figure 22-16* Callum Bennetts/Alamy; *photo* Rubberball/Jupiterimages. **p. 807:** *Figure 22-17* Tommy Moorman. **p. 808:** *Figure 22-18* Oxford Scientific (OSF)/Photolibrary; *photo* Peter Menzel www.menzelphoto.com. **p. 809:** *photo* Steve Allen/Getty Images. **p. 810:** *Figure 22-20* Image Source/Veer. **p. 812:** *Figure 22-22* Science Photo Library/Alamy. **p. 814:** *Figure 22-24* Wayne G. Lawler/Photo Researchers; *Figure 22-25* Deshakalyan Chowdhury/AFP/Getty Images. **p. 815:** *Figure 22-26 (L)* Dr. Richard Kessel & Dr. Gene Shih/Visuals Unlimited, *(both)* SPL/Photo Researchers. **p. 817:** *Figure 22-29* SPL/Photo Researchers. **p. 819:** *Figure 22-30* Ladislav Janicek/zefa/Corbis, *(inset)* Dr Kari Lounatmaa/Photo Researchers; *Figure 22-31* Wayne Lynch/age fotostock. **p. 820:** *Figure 22-32* Annie Griffiths Belt/Getty Images, *photo* S.Picavet/Jupiterimages. **p. 821:** *photo* Sean Justice/Corbis; *Figure 22-33* Envision/Corbis, *(inset)* Mango Productions/

Corbis. **p. 823:** *Figure 22-35* Robert Daly/Getty Images. **p. 824:** *Figure 22-37* Ewing Galloway. **p. 825:** *Figure 22-38* Tony Freeman/Photo Edit. **p. 826:** *Figure 22-39 (L)* AP Photo/Karl-Heinz Kreifelts, *(C)* Image Source/Corbis, *(R)* Ted Thai/Time Life Pictures/Getty Images. **p. 827:** *Figure 22-40* Purestock/Getty Images. **p. 829:** *Figure 22-42* Mark Clarke/ Photo Researchers. **p. 831:** *Figure 22-44* John Isaac/Peter Arnold. **p. 832:** *(T)* Tom Le Goff/Getty Images, *(B)* SPL/Photo Researchers.

CHAPTER 23 **pp. 836–837:** *photo* Joel Sartore/NGS Image Collection. **p. 838:** *photo* Konrad Wothe/Minden Pictures. **p. 839:** *Figure 23-1* Pedro Armestre/AFP/Getty Images, *(inset)* AP Photo/Jim Mone. **p. 840:** *Figure 23-3* Image Source/Alamy. **p. 841:** *Figure 23-5 (L)* Dennis Kunkel Microscopy, *(C)* Kent Wood/Science Photo Library, *(R)* Courtesy of Professor Xiaoming Jin. **p. 842:** Figure 23-7 Retrofile/Getty Images. **p. 843:** *photo* Thomas Deerinck/Visuals Unlimited. **p. 844:** *Figure 23-9* Tommy Moorman. **p. 846:** *Figure 23-11* Visuals Unlimited/Corbis; *Figure 23-12* Tom & Dee Ann McCarthy/Corbis. **p. 848:** *Figure 23-14* Greg Nelson/Sports Illustrated/Getty Images. **p. 849:** *Figure 23-15* Jobard/Sipa Press, *(inset)* Biosphoto/Compost Alain/Peter Arnold. **p. 850:** *photo* Flickr/ Getty Images; *Figure 23-16* Frans Lanting/Frans Lanting Photography. **p. 851:** *(T, R)* Yuri Arcurs/Alamy, *(T, L)* Richard Du Toit/Minden Pictures, *(M, L)* AP Photo/Ric Francis, *(M, R)* Martin Harvey/Corbis, *(B)* Digital Zoo/Corbis. **p. 853:** *Figure 23-19* Getty Images. **p. 854:** *Figure 23-21* AP Photo/Christian Escobar Mora. **p. 856:** *Figure 23-23 (L)* Eric Grave/Photo Researchers, *(C)* Fritz Rauschenbach/Corbis, *(R)* Don Kreuter-Rainbow/ Science Faction/Corbis; *Figure 23-24 (L & R)* Bjorn Rrorslett/Science Photo Library, *(inset)* Guy Moberly/Alamy. **p. 859:** *Figure 23-28* Noel Vasquez/Getty Images; *Figure 23-29* Stephen Dalton/Minden Pictures. **p. 861:** *Figure 23-31* Matthew Beck/The Citrus County Chronicle. **p. 862:** *Figure 23-32* Flip Nicklin/Minden Pictures; *Figure 23-33* Joe McDonald/Corbis, *(inset)* Ted Kinsman/Photo Researchers. **p. 863:** *photo* Frans Lanting/Frans Lanting Photography. **p. 866:** *Figure 23-36 (L)* Philippe Caron/Sygma/Corbis, *(R)* Rafael Winer/Corbis. **p. 867:** *Figure 23-38 (T)* Visuals Unlimited/Corbis, *(M)* Scott Camazine/Photo Researchers, *(B)* Dave Roberts/Science Photo Library. **p. 868:** *Figure 23-39* Brand X/ Corbis; *Figure 23-40 (L)* SPL/Photo Researchers, *(R)* Dr. Alan Boyde/ Visuals Unlimited/Corbis. **p. 869:** *photo* Lester Lefkowitz/Corbis. **p. 870:** *Figure 23-41* Image Source/Getty Images; *Figure 23-42* GI/Jamie Grill/ Getty Images. **p. 871:** *Figure 23-43* Omikron/Photo Researchers. **p. 872:** *Figure 23-44* Collection of Jack and Beverly Wilgus. **p. 874:** *Figure 23-46* Jay Phelan, *(inset)* Dana M. Small, et. al, Changes in brain activity related to eating chocolate: From pleasure to aversion, *Brain,* 2001, Vol. 124, No. 9, Figure 3, d; *Figure 23-47* Science Source/Photo Researchers. **p. 876:** *Figure 23-49* Steve Ringman/ *The Seattle Times.* **p. 877:** *Figure 23-51* Naturfoto Honal/Corbis. **p. 878:** *photo* Robert Estall photo agency/Alamy. **p. 879:** Figure 23-52 *(L)* Dod Miller/Alamy, *(R)* Gustoimages/Photo Researchers; *photo* James Quinton/Alamy. **p. 880:** *Figure 23-53* Randy Lincks/Corbis. **p. 881:** *Figure 23-54* Steve Prezant/Corbis. **p. 882:** *Figure 23-56* Morgan David de Lossy/Corbis. **p. 883:** *photo* Kevin Fleming/Corbis. **p. 884:** *(T)* Getty Images, *(B)* LAN/Corbis.

CHAPTER 24 **pp. 888–889:** *photo* © David LaChapelle/Art + Commerce. **p. 890:** *photo* Keith Ladzinski/Aurora Photos/Corbis. **p. 891:** *Figure 24-1 (L)* Per Winbladh/Corbis, *(C)* AFP/Getty Images, *(R)* Beau Lark/Corbis. **p. 893:** *Figure 24-3* Image Source/Corbis; *Figure 24-4* Artur Cegielsk/Dreamstime.com. **p. 895:** *Figure 24-6* Radius Images/ Photolibrary. **p. 896:** *photo* Steve Gschmeissner/Science Photo Library. **p. 897:** *Figure 24-7* Image Source/Getty Images. **p. 898:** *Figure 24-8* Tao Qi/ChinaFotoPress/Getty Images, *(inset)* TLC/Photofest. **p. 899:** *Figure 24-9* Image Source/Corbis; *Figure 24-10* Pete Oxford/Minden Pictures. **p. 900:** *Figure 24-11* Dean Conger/NGS Image Collection. **p. 901:** *Figure 24-13* Alan Powdrill/Getty Images **p. 902:** *photo* Sean Justice/Getty Images. **p. 903:** *Figure 24-14* AP Photo/Peter Dejong; *(inset)* Gabriel Bouys/AFP/ Getty Images. **p. 904:** *Figure 24-16* Nick Vedros & Assoc./Getty Images. **p. 905:** *Figure 24-17* AP Photo/The Canadian Press, Paul Chiasson/FILE; *(inset)* Dave Nagel/Getty Images; *Figure 24-18* Arco Images GmbH/Alamy. **p. 906:** *Figure 24-19* Ted Szczepanski; *Figure 24-20* Reuters/Corbis. **p. 908:** *Figure 24-22* National Geographic Images. **p. 909:** *Figure 24-23* PNC/Getty Images. **p. 910:** *Figure 24-24* Väronique Burger/Photo Researchers. **p. 911:** *photo* Frans Lanting/Corbis; *Figure 24-25* Frans Lanting/Corbis. **p. 912:** *Figure 24-26 (T)* Michael St. Maur Sheil/Corbis, *(inset)* Gary Porter/Milwaukee Journal Sentinel/MCT/Newscom, *(T)* Stefan Sauer/epa/Corbis, *(M)* Anna Goldbergs/Jupiterimages, *(B)* Olivier Pon/Reuters. **p. 914:** *photo* Editorial Image, LLC/Alamy.

CHAPTER 25 **pp. 918–919:** *photo* Michael Mährlein/age footstock. **p. 920:** *photo* Frans Lanting/Corbis. **p. 921:** *Figure 25-1* Rex USA. **p. 922:** *Figure 25-4 (L)* Juniors Bildarchiv/Alamy, *(R)* Dennis Kunkel Microscopy, Inc./Visuals Unlimited. **p. 924:** *Figure 25-5 (L)* Artur Tabor/Minden Pictures, *(R)* Arco Images GmbH/Alamy; *photo* Frans Lanting/Frans Lanting Photography, *(inset)* Mitsuaki Iwago/Minden Pictures. **p. 926:** *photo* John Springer Collection/Corbis.; *Figure 25-7* Michael Stewart/ Corbis. **p. 928:** *Figure 25-9* Dr. David Phillips/Visuals Unlimited. **p. 929:** *Figure 25-11* Cyril Ruoso/Minden Pictures, *(inset)* Martin Muller. **p. 930:** *Figure 25-12* Wildlife/Peter Arnold, *(inset)* CORDOBA-AGUILAR, Alejandro and CORDERO-RIVERA, Adolfo. Evolution and ecology of Calopterygidae (Zygoptera: Odonata): status of knowledge and research perspectives. *Neotrop. Entomol.* [online]. 2005, vol. 34, n.6, pp. 861–879. **p. 931:** *Figure 25-13* Michael Stewart/Corbis. **p. 934:** *Figure 25-16* Lennart Nilsson. **p. 935:** *photo* Kelly-Mooney Photography/Corbis. **p. 936:** *photo* Lennart Nilsson; *Figure 25-17* Dr. David Phillips/Visuals Unlimited. **p. 941:** *photo* Lennart Nilsson. **p. 946:** *Figure 25-24 (TL)* Dr. Yorgos Nikas/Photo Researchers, *(BL & C)* Lennart Nilsson, *(R, both)* Petit Format/Photo Researchers. **p. 949:** *Figure 25-27* Alain Evrard/age footstock. **p. 950:** *photo* Ted Horowitz/Corbis. **p. 952:** *photo* Vincent Kessler/Reuters/Corbis.

CHAPTER 26 **pp. 956–957:** *photo* Dennis Kunkel Microscopy. **p. 958:** *photo (T)* Lester V. Bergman/Corbis; *photo (B)* JAGADEESH NV/epa/ Corbis. **p. 959:** *Figure 26-1* Mark Douet/Getty Images; *Figure 26-2 (L)* Dennis Kunkel Microscopy, *(L, C)* Dr. Terry Beveridge/Visuals Unlimited/ Corbis, *(C)* Dennis Kunkel Microscopy, Inc./Visuals Unlimited, *(R, C)* Dennis Kunkel Microscopy, Inc./Visuals Unlimited, *(R)* Dennis Kunkel Microsopy, Inc./Visuals Unlimited. **p. 961:** *Figure 26-4* Tetra Images/Getty Images; *Figure 26-5* Tony Arruza/Corbis. **p. 965:** *Figure 26-9* Science VU/ Visuals Unlimited. **p. 966:** *photo* Crystal Cartier Photography/Corbis. **p. 968:** *photo* SPL/Photo Researchers; *Figure 26-12* Robert Crum/Alamy. **p. 969:** *photo* Dennis Kunkel Microscopy, Inc./Visuals Unlimited. **p. 970:** *Figure 26-13 (L)* Andrzej Gorzkowski Commercial/Alamy, *(R)* AP Photo/ Dan Bates. **p. 971:** *Figure 26-16* Mango Productions/Corbis. **p. 975:** *Figure 26-20* Ian Boddy/Photo Researchers. **p. 977:** *Figure 26-23* Patricia McDonough/Corbis. **p. 979:** *Figure 26-25* Dr. Gopal Murti/Visuals Unlimited. **p. 980:** *photo* MedicalRF.com/Visuals Unlimited. **p. 981:** *Figure 26-27* Getty Images. **p. 983:** *Figure 26-29* Dennis Kunkel Microsopy. **p. 985:** *Figure 26-32* PHANIE/Photo Researchers; *Figure 26-33* Tony Cordoza/Alamy. **p. 986:** *photo* Blend Images Photography/Veer.

ANSWERS

Chapter 1

1. b; 2. b; 3. d; 4. e; 5. c; 6. e; 7. a; 8. a; 9. b; 10. a; 11. d; 12. c; 13. b; 14. e; 15. a; 16. d; 17. a

Short-Answer Questions

1. To provide proper controls, a group of participants must receive a placebo that superficially resembles the medication. The control group should be treated exactly the same as the experimental group, with one exception: the control group receives a placebo instead of the drug. This provides a basis of comparison between the experimental and control groups.
2. There are many possible answers. Four possibilities are: adding a control group; extending the test for a longer period of time; standardizing the subjects to make sure that other variables are not responsible for the weight loss; repeating the study.
3. The common use of the term "theory" often implies that it is simply a hunch about something. In a scientific context, "theory" is applied only when a hypothesis has received significant support from multiple sources over time.

Chapter 2

1. d; 2. e; 3. d; 4. e; 5. b; 6. a; 7. c; 8. a; 9. d; 10. c; 11. b; 12. a; 13. a; 14. a; 15. a; 16. a; 17. a; 18. e; 19. e; 20. d; 21. b

Short-Answer Questions

1. a) The sodium atom has one electron in its outermost shell; the fluorine atom has seven electrons in its outermost shell. b) An ionic bond forms between these two atoms. Neither atom has a full outer electron shell, and both are unstable. The atoms interact to achieve a full outer shell of electrons: the single electron in the outermost shell of the sodium is transferred to the outermost shell of the fluorine. Each atom becomes an ion, and the oppositely charged ions attract each other.
2. First, cohesion is a result of polar water molecules easily bonding together by hydrogen bonds; this allows one water molecule to "connect" with the next as water makes its way up from the soil to the leaves of a tall tree. Second, water has a large capacity to hold heat and resists large fluctuations in temperature; this is because added energy, in the form of heat, does not increase the movement of water molecules, as occurs in most substances. Instead, hydrogen bonds between water molecules are briefly broken and then re-form with nearby water molecules. Third, ice is less dense than water, so it floats; this is because polar water molecules are held apart by hydrogen bonds when the temperature drops (see Figure 2-15). Fourth, water is an excellent solvent because water molecules are polar. Polar molecules with charged regions are attracted to the charged regions of other polar molecules. Many substances contain polar molecules, which easily dissolve in water.
3. Water is a polar molecule and is attracted to other polar molecules, or molecules with charged regions. Oil consists of non-polar molecules. As a result, polar water molecules will not interact with oil, and water by itself will not be able to dissolve oily dirt.
4. Lettuce and celery contain carbohydrates, which provide a source of energy or fuel for the body. Strawberries also contain carbohydrates, but more monosaccharides versus the complex carbohydrates of lettuce and celery. Walnuts and chicken are sources of protein. The amino acid building blocks of protein are used to build many structures and molecules required by all cells. The salad dressing contains lipids, which provide the body with energy.

Chapter 3

1. e; 2. b; 3. e; 4. a; 5. d; 6. d; 7. c; 8. a; 9. d; 10. d; 11. b; 12. d; 13. c; 14. e; 15. b; 16. b; 17. e; 18. b

Short-Answer Question

1. See the organelle review table, Figure 3-39 on page 117.

Chapter 4

1. a; 2. e; 3. e; 4. c; 5. a; 6. c; 7. a; 8. b; 9. d; 10. c; 11. b; 12. e; 13. b; 14. b; 15. a; 16. a; 17. e; 18. d

Short-Answer Questions

1. The plant can split the water molecules in the "photo" reactions of photosynthesis, producing oxygen gas, which can be used for aerobic respiration by both the plant and the mouse. The carbon dioxide produced as a by-product of aerobic respiration by the plant and the mouse can be used by the plant in the "synthesis" reactions, which ultimately produce glucose for the plant.
2. a) Water molecules are split during the "photo" reactions and oxygen gas is released. Measuring the amount of O_2 provides an indication of the rate of the "photo" reactions. b) Plants incorporate CO_2 in the "synthesis" reactions. Using the energy from the "photo" reactions, plants modify the CO_2 to (eventually) form glucose. The "synthesis" reactions are driven by the energy from the "photo" reactions. c) With additional CO_2, the rate of the "synthesis" reactions will increase, which in turn will increase the amount of energy needed from the "photo" reactions.
3. Photosynthesis involves the production of glucose, and cellular respiration involves the breakdown of glucose. Growth in plants is dependent on the rate of photosynthesis. The faster this process occurs, the more glucose is made, the more plant material can be synthesized, and the more the plant grows. If a plant is growing, the rate of photosynthesis exceeds the rate of cellular respiration.

Chapter 5

1. e; 2. e; 3. b; 4. c; 5. d; 6. d; 7. e; 8. b; 9. c; 10. b; 11. a; 12. e; 13. c; 14. a; 15. d; 16. c

Short-Answer Questions

1. Thymine makes up 30% of the bases. Cytosine and thymine have only one complementary base, so the base and its complement occur in

equal numbers. Because cytosine makes up 20%, its complementary base, guanine, is also 20%. The total of cytosine and guanine is 40%, so the remaining bases total 60%. Because thymine and adenine must occur in equal numbers, half of this total (30%) must be thymine and the other half (30%) adenine.

2. The mRNA is UAG CUG CCU AGG. mRNA allows instructions from DNA to be delivered to ribosomes in the cytoplasm, where it directs the assembly of the protein it encodes.

3. The sequence of codons of the mRNA is translated into a sequence of amino acids that forms a polypeptide chain. As each codon of the mRNA is read by a ribosome, a transfer RNA with an anticodon complementary to the mRNA codon adds a specific amino acid to the growing polypeptide chain.

4. This is an example of a point mutation. The adenine in the fourth codon from the left has been replaced with thymine. This change results in coding for a different amino acid, changing the type of protein produced by the gene—because the amino acid sequence has changed.

5. First, recombinant DNA technology is used to make certain medicines, such as insulin, growth hormone, and erythropoietin, with much less expense; this enables more people to purchase and benefit from these medicines. Second, there have been many efforts to cure some diseases caused by genetic abnormalities using gene therapy; although this technology has not met with success, the future still holds potential. Third, genetic testing can be conducted on parents to identify whether they are at risk of producing offspring with certain genetic diseases. Prenatal genetic screening can be conducted to determine whether a baby will be born with a genetic disease. Adults can also be tested to determine whether they carry genes that put them at risk of developing a disease. With genetic screening, preventive measures and treatments can be undertaken.

Chapter 6

1. c; 2. b; 3. d; 4. d; 5. c; 6. a; 7. b; 8. e; 9. d; 10. b; 11. b; 12. c; 13. d; 14. e; 15. c; 16. a; 17. e; 18. d

Short-Answer Questions

1. Crossing over occurs as sister chromatids of homologous chromosomes come together and swap equivalent segments containing the same genes (but possibly different alleles) during the first prophase of meiosis. This results in an uneven distribution of genetic material from maternal and paternal chromosomes. Gametes could potentially receive more genes from one (either) biological parent than the other. Also, the random division of the homologues (with respect to whether they are maternal or paternal) at anaphase I can lead to gametes having greater than or less than exactly 50% maternal and 50% paternal genetic material. For every gamete having a higher percentage of its genetic material from the maternal side, there is a gamete having an equally higher percentage of paternal genetic material.

2. There cannot be any essential genetic information on the Y chromosome—because all females are able to live without a Y chromosomes—but there is on the X chromosome. Receiving only the Y chromosome means that essential genetic information from the X chromosome is absent.

3. An organism is more likely to reproduce asexually in the absence of adverse environmental pressures. Should some environmental pressure develop, the organism may start reproducing sexually, producing genetically diverse offspring that are better able to survive environmental pressures. An experiment could be conducted with an organism that has the capacity to switch between sexual and asexual reproduction. This organism could be placed in an environment with ideal conditions and another with adverse conditions. If the hypothesis is correct, the organism will reproduce asexually under ideal conditions and sexually under adverse conditions.

Chapter 7

1. a; 2. b; 3. e; 4. e; 5. a; 6. c; 7. e; 8. a; 9. c; 10. e; 11. e; 12. d; 13. d; 14. b; 15. d; 16. a

Short-Answer Questions

1. a) SMA must be a recessive genetic disorder, because the phenotype of both parents is normal—that is, they are not affected by SMA. For two unaffected parents to have children with the disease, both parents' genotypes must be heterozygous. This is an example of a disease-causing allele that doesn't always reveal itself in a population. b) Individuals who are heterozygous for a recessive genetic disorder (i.e., are carriers) will not necessarily have children with the disease. However, this example demonstrates heterozygous parents with children who inherit the disease-causing allele from both parents, resulting in SMA. The probability of having another biological child with SMA is the same as it was for each of the first two children. The probability of inheriting the recessive allele from the mother is 0.5, and the probability of inheriting the recessive allele from the father is 0.5. Therefore the overall probability is 0.25, 25%, or 1 in 4.

2. The phenotype, or color of the blooms, is affected by differences in the environment. Hydrangea flower color varies with the type of soil, because the pH of the soil affects the pigmentation of the blooms.

3. a) Genetic disorders that involve genes on the X chromosome affect men differently from women. If a female inherits an allele that produces a non-functional protein on one of her X chromosomes, she may not be affected *if* the second X chromosome has the normal allele. However, because males have only one X chromosome, they have only one chance to receive the normal allele.

b)

	X^H	X^H
X^h	X^hX^H	X^hX^H
Y	X^HY	X^HY

There is a 0% chance that a son will have hemophilia. A son inherits his X chromosome from his mother. Because the mother is carrying two functional alleles, either X would provide a functional allele.

Chapter 8

1. e; 2. a; 3. b; 4. c; 5. d; 6. d; 7. e; 8. a; 9. e; 10. e; 11. b; 12. a; 13. c; 14. b; 15. d; 16. b; 17. b; 18. e; 19. d; 20. d; 21. c

Short-Answer Questions

1. Evolution is a change in allele frequencies in a population. Natural selection is one of four evolutionary agents, characterized by variation, heritability, and differential reproductive success for a trait.

2. Genetic drift is a random change in allele frequencies. In small populations, random deviations from the expected genotype frequencies

among offspring are less likely to be counterbalanced by deviations in the opposite direction from the expected genotype frequencies.

3. First, environments can change quickly. Natural selection may be too slow to adapt the species to a constantly changing environment. Second, variation is needed as the raw material of selection. If there is no mutation to give rise to an allele that would help make an individual more fit, the species will never be perfectly adapted. (Why, for instance, are there no mammals with wheeled appendages?) Third, evolution produces "more fit" individuals in each generation. If the optimum solution required organisms to get worse before they could evolve to "perfection," selection would never lead to such a solution.
4. In the 1940s, when penicillin was first used to treat bacterial infections, it was uniformly effective in killing *Staphylococcus aureus.* Today, more than 90% of isolated *S. aureus* strains are resistant to penicillin. Because penicillin has become such a pervasive toxin in the environment of *Staphylococcus,* natural selection has led to an increase in the frequency of alleles that make the bacteria resistant. Today, nearly a third of "staph" infections are resistant to antibiotics. The meaning of such unintentional natural selection "experiments" is clear and consistent: evolution is occurring all around us.

Chapter 9

1. b; 2. d; 3. e; 4. b; 5. d; 6. d; 7. c; 8. a; 9. d; 10. e; 11. c; 12. a; 13. a; 14. b; 15. e; 16. b

Short-Answer Question

1. Males provide more parental care, consistent with the two predictions about sex roles and parental investment. First, the sex that invests more will be more discriminating. Second, the sex that invests less will compete among themselves for access to the higher-investing sex.

Chapter 10

1. a; 2. d; 3. e; 4. e; 5. a; 6. d; 7. d; 8. a; 9. e; 10. a; 11. d; 12. d; 13. e; 14. b; 15. c; 16. a; 17. c

Short-Answer Questions

1. The biological species concept focuses on reproductive populations, or groups of individuals that can mate with each other and are unable to breed with other reproductively isolated groups. The morphological species concept focuses on physical characteristics of the organisms to distinguish one species from another. Many living organisms don't fit neatly into a species as defined by the biological species concept. An example is many of the living organisms that divide asexually, such as bacteria. For these organisms, because they do not require a partner to mate, reproductive compatibility is not a useful diagnostic tool for discriminating among species.
2. Biologists would rely on physical features or characteristics to create evolutionary trees and establish relationships between groups of organisms. Today, with the advent of technology, DNA sequencing (examining the base-pair sequence of DNA) is used to construct evolutionary trees. In either case, the trees define which groups of organisms are most closely related, but not which groups are more advanced than others.
3. These adaptations could be considered an "evolutionary innovation" in which a specific trait or group of traits allows an organism to fill a new niche or an unoccupied niche. This is one trigger of adaptive radiation—the huge, rapid expansion of a species, often diversifying into many more species. The specific traits of tall, flat teeth and a square jaw allowed the horse to efficiently crop grasses and vegetation from the ground. With areas of grasslands plentiful, these mammals were able to expand and flourish in this niche.

Chapter 11

1. d; 2. c; 3. e; 4. d; 5. c; 6. c; 7. a; 8. c; 9. b; 10. c; 11. e; 12. d; 13. b; 14. b; 15. a; 16. b; 17. d; 18. e.

Short-Answer Questions

1. In all animals, the ability to obtain food, whatever method that may be, has evolved through natural selection. For the sloth, it would make no sense for it to have developed the ability to spring at great speeds, because its food source (vegetation) is stationary. The cheetah, by contrast, must use high speeds to catch its prey, which also can run fast. If the cheetah were to occupy the same niche as the sloth, its ability to run would not improve its ability to obtain vegetation, so it might lose this running ability over many generations.
2. Much of the energy derived from food is utilized for heat production, so endotherms require more food to maintain a constant internal temperature. Ectotherms depend on an external heat source to maintain their body temperature, so they do not need energy derived from food to generate heat.
3. The snake, like other ectotherms, depends on an outside source of heat, so placing an insulating sweater on the snake will prevent heat from reaching its body, keeping it from warming up. For the same reason, ectotherms do not have insulating feathers or fur.
4. The slug has a radula for scraping food from surfaces, has a broad foot on which it glides, and produces a protective slime. Slugs lack the body segmentation found in segmented worms.

Chapter 12

1. a; 2. b; 3. b; 4. c; 5. a; 6. b; 7. e; 8. c; 9. c; 10. c; 11. d; 12. b; 13. d; 14. b; 15. d.

Short-Answer Questions

1. Non-vascular plants lack tissue that can conduct water and dissolved nutrients, tissue that *is* found in vascular plants. Because diffusion can transport water only over very short distances, non-vascular plants are much smaller than vascular plants. Non-vascular plants must live in an environment that is wet for at least some parts of the year, and become dormant when it is dry.
2. Mycorrhizal fungi are associated with the roots of plants, benefiting them by transporting phosphorus and nitrogen from the soil to the cells of the plant roots. In exchange, the fungus takes some sugar from the plant. This relationship enhances the overall health of the plant.
3. Plants expend considerable energy producing a fruit, which contains the seeds. Once a fruit is eaten, it is digested by the animal and the seeds are excreted unharmed, and thus relocated to an area distant from the parent plant. The plant pays the animal with a meal, and in return, the animal transports the seeds.

Chapter 13

1. a; 2. b; 3. c; 4. e; 5 d; 6. a; 7. a; 8. d; 9. b; 10. a; 11. c; 12. b; 13. e; 14. b; 15. e; 16. e; 17. e; 18. b.

Short-Answer Question

	Prokaryotes		Eukaryotes	Viruses
	Bacteria	Archaea	Protists	
Unicellular, colonial, or multicellular?	Unicellular	Unicellular	Unicellular or colonial	Not cells
Plasma membrane?	Present	Present	Present	Absent
Peptidoglycan in cell wall?	Present	Absent	Absent	Absent
Flagella? What type?	Yes; spins like a propeller	Yes; spins like a propeller, but has a chemical composition different from that of bacteria	Yes; waves from side to side and has a chemical composition different from that of bacteria or archaea	No flagellum
Presence of nucleus?	Absent	Absent	Present	No nucleus
Intracellular organelles?	Absent	Absent	Present	No organelles
Type of genetic material (DNA, RNA, or both possible)?	DNA	DNA	DNA	DNA or RNA
Chromosome structure (linear or circular)?	Usually circular	Usually circular	Linear	Linear or circular

Chapter 14

1. a; 2. e; 3. a; 4. b; 5. e; 6. d; 7. b; 8. d; 9. c; 10. c; 11. e; 12. d; 13. a; 14. b; 15. c.

Short-Answer Questions

1. Density-dependent factors are limitations on a population's growth resulting from competition among individuals for a finite amount of resources, such as food and space for living and breeding, and an increased risk of disease and predation. Density-independent factors are forces such as hurricanes and droughts that kill off many individuals in a population, regardless of the size of the population.

2. An organism with a type I survivorship curve is relatively large, has only a few offspring, invests a lot in each offspring, and lives a relatively long life. If the organism had a type III survivorship curve, it would have a high probability of dying early in life and would produce many offspring and provide very little parental care.

3. Species that lack natural enemies and live in an environment with low risk of mortality from other sources breed at a later age; late-onset genetic disorders are selected out and individuals are healthier in old age. Species that breed at an early age pass on adverse genetic traits that are expressed after their normal age of breeding; they do not live long after the normal age of breeding. If breeding is delayed, as in the fruit flies in the experiments described in the chapter, age-related disorders that interfere with breeding are selected out and the life span will be extended.

Chapter 15

1. d; 2. e; 3. a; 4. a; 5. e; 6. d; 7. e; 8. a; 9. d; 10. a; 11. a; 12. a; 13. d; 14. e; 15. b; 16. b.

Short-Answer Questions

1. Heated air rises and cools at the equator (0° latitude); as it cools, it cannot hold as much water and loses it as precipitation. Air moves north and south from the equator and begins to fall at about 30° latitude north and south, warming and absorbing moisture. This creates very dry conditions at these latitudes, and deserts are found there. Air continues to move toward the poles and begins to rise again at 60° latitude, cooling and losing its moisture through precipitation, making these locations very wet. Air again falls near the poles, making these areas very dry. Temperature and amount of rainfall are the primary determinants of the kind of biome found in any terrestrial location.

2. Energy flows through each trophic level. Photosynthetic organisms, the primary producers, trap energy from the sun in the chemical bonds of food. Primary consumers, the herbivores, eat photosynthetic organisms, and these animals are consumed by secondary consumers, the carnivores. Carnivores can be consumed by tertiary consumers, the top carnivores. A natural ecosystem can support only a few top carnivores because only about 10% of the energy is available to each subsequent trophic level.

3. An organism's niche includes the space it requires to live in, the kinds and amount of food it eats, the time of day and of year it reproduces, the amount of moisture and temperature it requires, as well as other criteria. Two organisms that occupy the same niche compete, with two different possible outcomes: one species may drive the other to extinction (competitive exclusion) or both species may change their behavior or undergo a morphological change so they are restricted within their own niche, dividing the resources between them (resource partitioning).

Chapter 16

1. e; 2. c; 3. d; 4. a; 5. e; 6. e; 7. d; 8. d; 9. b; 10. a; 11. e; 12. c; 13. b; 14. e.

Short-Answer Questions

1. The greatest biodiversity, both aquatic and terrestrial, is in equatorial regions. Less biodiversity is found with increasing latitude. Habitat destruction and degradation are the greatest threats to biodiversity.

2. The biotic and abiotic nature of the habitat was altered, but species were not driven to extinction. If species had been made extinct, the ecosystem disturbances would not have been reversible.

3. Most of the deforestation of tropical rain forests is a result of clearing land to make way for agriculture; mining and pollution also contribute. The biodiversity of the planet is greatly affected by rain forest deforestation because no other terrestrial ecosystem has the extent of biodiversity found in these forests.

Chapter 17

1. b; 2. b; 3. d; 4. a; 5. d; 6. d; 7. a; 8. a; 9. b; 10. b; 11. a; 12. c; 13. e.

Short-Answer Questions

1. Both groups are autotrophic, photosynthesizing plants that inhabit a wide range of environments, and play a vital role in human survival by providing foods, medicines, housing, and many other products. Monocots have one cotyledon, which distinguishes them from eudicots, which have two cotyledons. The venation, which is bundles of xylem and phloem, of leaves are another distinguishing

characteristic, with monocots having parallel venation and eudicots having a branching pattern. Vascular bundles are arranged in a random pattern within stems of monocots, while they are arranged in a circular pattern in stems of eudicots. A vast majority of eudicots are capable of secondary growth, while only a small percentage of monocots are capable of secondary growth. Flowering parts of flowers occur in multiples of 3 for monocots and multiples of 4 and 5 for eudicots. Finally, monocots typically form fibrous root systems while eudicots form taproots from which rootlets grow.

2. Water is pulled from the roots up through xylem to the leaves. When a water molecule evaporates from a leaf it pulls on its neighboring water molecule, which is held together by hydrogen bonding. This in turn causes all the other water molecules within this xylem tube, which all have hydrogen bonds with their neighboring water molecules, forming an interconnected column of water, to be pulled up the xylem. The ability of water to form hydrogen bonds between water molecules allows it to have tension and cohesion and therefore the ability to be pulled up from leaves as water evaporates.

3. Since vascular bundles of monocots are randomly spread out within the stem, a shallow cut would sever only those vascular bundles near the surface. Vascular bundles of eudicots, on the other hand, usually form a ring structure, typically closer to the surface. Since their vascular tissue is close to the surface a shallow cut is likely to sever all of their vascular tissue, preventing the movement of fluids and causing the plant to die.

Chapter 18

1. a; 2. b; 3. e; 4. a; 5. e; 6. a; 7. e; 8. e; 9. d; 10. a; 11. b; 12. c; 13. c; 14. b

Short-Answer Questions

1. A single strawberry plant, within a year or so, can completely fill a large area as the plant reproduces asexually and populates the patch with genetically identical, strawberry-producing clones.

2. Planting a potato produces a new plant asexually, so the new plant is genetically identical to the plant from which the potato came. If the parent plant is healthy and productive, this can help a farmer produce new crops that are equally productive. Seeds, on the other hand, are the product of sexual reproduction, so the new plant may not resemble the "mother plant" from which the seeds were collected; the properties of the new generation will depend also on the male parent, of unknown productivity. The farmer should choose potatoes for planting that come from the parent plant with the greatest productivity and vigor.

Chapter 19

1. a; 2. a; 3. c; 4. b; 5. b; 6. e; 7. a; 8. e; 9. d; 10. a; 11. d

Short-Answer Questions

1. This phenomenon is called phototropism, growth in response to a light source. When light hits a plant from a particular direction, auxins produced in the plant move away from the light source and end up on the opposite, shaded side of the stem. There, the auxins stimulate a slightly greater rate of growth than on the side with less auxin. As a consequence, the uneven growth causes the plant to bend toward the light.

2. The hormone ethylene is a gas that speeds the rate at which fruit ripens. Ethylene can help with shipping fruits internationally because if bananas, for example, are picked in Central America before they are ripe (when they are green), and just prior to their delivery, ethylene is injected into the cargo hold of the ship; this initiates the ripening of all the fruit simultaneously.

3. Spicy foods, such as mustard, frequently come from plants and get their flavor from secondary compounds that may make the plant toxic or unpalatable to an insect, representing an anti-herbivore adaptation. The fact that the chemicals are not, in some cases, toxic to humans may be a consequence of our larger body size relative to the intended herbivores'.

4. See Figure 19-2, page 693.

Chapter 20

1. a; 2. c; 3. d; 4. a; 5. a; 6. a; 7. d; 8. e; 9. b; 10. d; 11. a

Short-Answer Questions

1. Homeostasis is the maintenance of a constant internal environment within a narrow range of conditions around a set point for variables such as temperature, pH, and concentrations of solute, water, CO_2, O_2, and glucose, among many others. Homeostasis allows for the normal functioning of an organism's cells. Body temperature is normally controlled by a negative feedback mechanism (thermoregulation). Hyperthermia occurs when the body temperature rises above its normal range. If high enough, this can interfere with the structure and activity of the enzymes involved in the body's metabolic activities, resulting in metabolic dysfunction, death of cells, organ failure, and, ultimately, death of the organism.

2. Cells of similar type are organized into tissues; tissues are organized into organs that perform specialized functions; and organs are organized into organ systems, arranged in an integrated manner to achieve common goals. There are four types of tissue. *Connective tissues* support and help form body structures. *Epithelial tissues* form internal and external body coverings, providing protection, synthesizing and releasing chemical products, and controlling movement of substances within the body. *Muscle tissues* have the ability to contract, generating body movements and moving materials within the body. *Nervous tissue* conducts electrical impulses, allowing communication within the organism.

3. *Connective tissue proper* includes (a) *loose connective tissue,* which has cells embedded in a semi-fluid matrix, with many collagen fibers that bind and hold organs in place, as well as storing fat and insulating, cushioning, and protecting the body; and (b) *dense connective tissue,* which has many more collagen fibers than loose connective tissue and thus has greater strength, forming tendons, which connect muscles to bones, and ligaments, which connect bones to each other. *Special connective tissue* includes (a) *bone,* composed of cells within a solid matrix interspersed with many collagen fibers, forming a rigid structure that supports and protects the body; (b) *cartilage,* composed of cells within a matrix (not as hard as bone) made up of collagen, elastin, and proteins connected to carbohydrate chains; and (c) *blood,* which has a fluid matrix called plasma, with suspended red blood cells, white blood cells, and platelets.

4. *Skeletal muscles* are made up of cells (fibers) that have alternating light and dark bands (striations) and contain many nuclei; the muscles are attached to bones, generating force for their movement, and are controlled voluntarily. *Smooth muscles,* which are not striated and have smaller cells, are found in the walls of hollow organs; contractions allow movement of liquids and solids within the organs, and are involuntary. *Cardiac muscle* is found in the walls of the heart and consists of striated, small, branched cells with a single nucleus; contractions generate pressure to move blood in the cardiovascular system, and are involuntary.

Chapter 21

1. d; 2. c; 3. a; 4. b; 5. a; 6. e; 7. e; 8. b; 9. e; 10. e; 11. a; 12. b; 13. e; 14. a; 15. c; 16. a; 17. c; 18. d; 19. d; 20. a; 21. c

Short-Answer Questions

1. Single-celled organisms, such as bacteria, are in direct contact with their environment, and direct diffusion of gases, nutrients, and wastes can take place at an adequate rate to maintain these organisms physiologically. In larger organisms made up of multiple cells, many of which are not in direct contact with the outside environment, direct diffusion is no longer feasible. With the development of a circulatory system, nutrients and gases can be brought in close proximity to cells, where diffusion can take place. An open circulatory system is characterized by having only one type of extracellular fluid; a closed circulatory system has two different extracellular fluids: interstitial fluid, bathing cells, and blood, which is contained only within the blood vessels.

2. In a single-circuit cardiovascular system, the heart has two chambers, one receiving blood that has traveled through the body, and the other moving blood to the gills and then through the blood vessels of the rest of the body. As blood moves through capillaries and other blood vessels, it slows down considerably, which slows down the delivery of substances to tissues and removal of waste products. A double-circuit system has one circuit that moves blood through the lungs and another that moves blood through the rest of the body. This system allows blood to be re-pressurized after it has traveled through the lungs, and this would be the better choice to supply tissues with high energy demands: blood can be delivered and removed at a faster rate.

3. *Red blood cells (erythrocytes):* Their primary function is to carry oxygen. They are suspended in plasma and contain iron-bearing hemoglobin, which binds, carries, and releases the oxygen. *White blood cells (leukocytes):* These are part of the body's immune system; their primary function is to seek out and destroy foreign organisms, including viruses and bacteria. *Platelets:* These cellular fragments circulate in blood and take part in plugging tears and holes in blood vessels and preventing leakage of blood. They do this by sticking together to form a clot.

4. Aquatic animals move water across their gills, which have capillaries that use a countercurrent mechanism. Blood moves through the capillaries in the opposite direction to the water flow, and this allows for a more efficient gas exchange: absorption of O_2 by and release of CO_2 from blood. In animals that breathe air, fresh air enters the lungs and moves into the alveoli. Gas exchange occurs there: O_2 in the air diffuses into capillaries in the walls of the alveoli, and CO_2 diffuses out of the blood and into the air, to be exhaled from the lungs.

Chapter 22

1. a; 2. c; 3. d; 4. a; 5. a; 6. b; 7. b; 8. c; 9. d; 10. a; 11. b; 12. b; 13. e; 14. d; 15. d; 16. e; 17. a

Short-Answer Questions

1. *Water:* important in the transport of nutrients and wastes; acts as a solvent for amino acids, sugars, minerals, and many vitamins; and takes part in some chemical reactions. *Proteins:* the source of amino acids, which are the building blocks of body proteins, such as hemoglobin, muscle cells, and enzymes. Proteins also can act as an energy source. *Carbohydrates:* energy-rich chemicals used by cells, through cellular respiration, as their primary energy source. Cellulose (fiber), which cannot be digested by humans, stimulates movement of food through the intestines. *Fats:* long-term energy-storage molecules, stored as fat tissue under the skin (which also provides body insulation) and elsewhere. *Vitamins and minerals:* required in very small quantities, but playing an important role in the proper functioning of enzymes and other processes.

2. *Ingestion:* intake of food through the mouth. The structures of the mouth cut food into small pieces and grind and mix it with saliva, so that it can be transported through the esophagus to the stomach by peristaltic action. *Digestion:* breakdown of food in the mouth, stomach, and small intestine into its most basic components, which can be readily absorbed and used by cells. Digestion is accomplished by a combination of physical breakdown of food into smaller pieces and chemical breakdown by enzymes. *Absorption:* uptake of usable nutrients through the wall of the small intestine and into the bloodstream, and then transport to cells of the body. *Elimination:* finally, in the large intestine, absorption of water, salts, and vitamins, storage of feces in the rectum, and elimination through defecation.

3. Animals that consume cellulose-containing foods such as grasses and leaves have, either in their stomach or in their cecum, bacteria that can break down cellulose into a usable form.

4. Type 1 diabetes, usually beginning in childhood, results from the inability of the pancreas to produce enough insulin; treatment is injection of insulin so that cells can absorb glucose. Type 2 diabetes, generally beginning later in life, occurs when cells become less responsive to insulin or the number of insulin receptors is reduced. Risk factors include obesity and chronic consumption of high amounts of sugar. Treatment includes weight loss and minimizing glucose fluctuations.

Chapter 23

1. a; 2. e; 3. a; 4. a; 5. c; 6. e; 7. d; 8. d; 9. c; 10. d; 11. a; 12. a; 13. a; 14. d; 15. e; 16. e; 17. e; 18. d; 19. b; 20. e.

Short-Answer Questions

1. Neurons are the impulse-conducting cells of the central and peripheral nervous systems. An impulse travels along the axon away from the cell body, eventually arriving at a synapse with another neuron, a muscle cell, or a gland cell. Glial cells are supporting cells found in the central and peripheral nervous systems. They help neurons function by protecting, insulating, and nourishing them, but do not play a direct role in impulse conduction. Without glial cells, neurons would not be able to function, survive, and grow. Glial cells are capable of dividing throughout life, while neurons, with a few exceptions, are not.

2. Each skeletal muscle fiber consists of numerous myofibrils, which are made up of many sarcomeres—the units of contraction. Each sarcomere contains thin and thick filaments, made of different types of protein. Thin filaments consist of two rows of small globular actin molecules, and thick filaments are made up of myosin; each myosin molecule looks much like a double-headed golf club. Myosin molecules are overlapping, with the globular heads projecting along the entire length of the fiber. The four-part change that allows contraction consists of: (1) *Detach:* A link between a myosin and an actin filament parallel to it is broken as a molecule of ATP binds to the myosin. (2) *Reach:* As the ATP breaks down, energy released alters the shape of the myosin into a higher-

energy shape. (3) *Reattach:* In its altered shape, the myosin reaches farther down the actin filament, where it reattaches. And (4) *Pull back:* The myosin then snaps back to its original shape, pulling the actin filament as it does so and shortening the fiber, in the "power stroke."

3. There are five primary types of sensory receptors. *Mechanoreceptors* measure mechanical stress and are associated with hearing and touch. *Thermoreceptors* measure temperature and are found in the skin; they are part of our perception of touch. *Photoreceptors* measure light and are found in the eyes, functioning in sight. *Pain receptors* detect pain, which can be caused by extremes of heat or tissue damage; they are part of the perception of touch. *Chemoreceptors* detect chemicals; they are found in the mouth and nose and function in taste and smell.

Chapter 24

1. b; 2. d; 3. c; 4. e; 5. b; 6. b; 7. d; 8. e; 9. b; 10. d; 11. b; 12. c

Short-Answer Questions

1. Both types of hormones circulate in extracellular fluids and influence changes in target cells (cells with a receptor for a particular hormone). Peptide and protein hormones cannot pass through the cell membrane of target cells; they attach to membrane receptors on the target cell, which causes a change within the cell. Steroid hormones, however, can pass through the cell membrane of target cells; they then attach to a receptor in the cytoplasm or nucleus.

2. The hypothalamus integrates sensory information, received by the brain, on the internal and external conditions of the organism. Once the hypothalamus has evaluated the information, it produces hormones that travel to the pituitary gland, which is attached by a thin stalk to the hypothalamus. These hormones cause the pituitary either to release or to stop the release of specific hormones. The pituitary gland is made up of two parts, the anterior and posterior pituitary; these produce hormones that influence other endocrine glands to produce other hormones.

3. Drops in estrogen level in females, such as during menopause and after giving birth, as well as the spike in estrogen that occurs at puberty, have been correlated with depression. In cases of low estrogen, supplements can be taken to help return estrogen levels to normal and reduce the occurrence of depression. Males who take supplemental testosterone may experience elevated self-esteem and overflowing of new ideas, and feel energetic, euphoric, confident, and charismatic; others find that they are more aggressive. Human and animal studies show that castrated males, who produce almost no testosterone, and females that have been sterilized and do not produce estrogen and other sex hormones, live longer than those producing normal levels of these hormones. However, among post-menopausal women taking hormone supplements, mortality decreases, illustrating the complexity of this issue.

4. Some manufactured chemicals, such as DDT and PCBs, can mimic, block, or interfere with the functioning of hormones. These endocrine disruptors can build up in organisms that feed on other, contaminated organisms, and collect in tissues (such as fat tissue, in the case of DDT). As a result, even in small quantities, the chemicals may harm some organisms.

5. See Figure 24-9 on page 899.

Chapter 25

1. a; 2. c; 3. a; 4. a; 5. a; 6. c; 7. c; 8. e; 9. e; 10. b; 11. e; 12. b; 13. d; 14. a; 15. d

Short-Answer Questions

1. Human males produce sperm by the process of spermatogenesis, which occurs in the testes, located in the scrotum. Spermatogonia, cells in the seminiferous tubules, divide by mitosis, each cell producing a new spermatogonium and a primary spermatocyte. The primary spermatocyte undergoes a first meiotic division, producing two secondary spermatocytes, each of which divides by a second meiotic division to produce four haploid sperm. Sperm move to the epididymis, where they mature and become motile.

2. Germ layers are the three layers of tissue—endoderm, mesoderm, and ectoderm—that form in the embryo during gastrulation. *Endoderm* is the inner tissue layer that forms the digestive tract, liver, pancreas, and the lining of the respiratory tract. *Mesoderm* is the middle tissue layer that gives rise to muscle and the skeletal system. *Ectoderm* is the outer tissue that gives rise to the outer layer of skin, hair, the nervous system, and the lining of the nose and mouth.

3. *First trimester:* During the first three months of development, minimal growth of the embryo takes place, but vital cell differentiation and development occur. The fertilized egg (zygote) divides and eventually forms three cell layers, from which all the organs and their associated systems will develop, as well as all body structures. This is also the stage during which the zygote implants in the uterus and the placenta forms. *Second trimester:* Significant growth occurs during the fourth through six months. The bones of the skeletal system grow and muscles develop, making possible fetal movement. *Third trimester:* The fetus primarily grows during these last three months of pregnancy. Growth and development of the nervous system also take place. During this stage, the fetus becomes an individual that can survive outside the womb.

Chapter 26

1. b; 2. d; 3. d; 4. d; 5. e; 6. b; 7. d; 8. a; 9. e; 10. c; 11. d; 12. a.

Short-Answer Questions

1. During the specific division of the immune system's primary response to an antigen on a pathogen, a type of lymphocyte called memory cells is produced through clonal selection. These lymphocytes "remember" an antigen. The second time a person is exposed to the pathogen, the memory cells, hanging around in the lymph and blood cells, are ready to attack it, mounting a much faster and more vigorous attack.

2. The specific division of the immune system recognizes very specific infectious agents, such as the virus that causes chicken pox. As a consequence of a vaccine for this illness, the immune system forms a "memory" of the virus that allows it to fight off the virus in the future before the virus can cause the chicken pox. The chicken pox virus antigens remain relatively constant over time. The influenza virus, on the other hand is not so constant. Each year there are many different versions, or strains, of the influenza virus, all of which are constantly changing. And so with each new flu season, the body encounters a slightly different form of influenza and, therefore, a different set of antigens. For this reason, previously formed memory cells are not helpful in fighting the new virus.

GLOSSARY

A note about notation: The word or phrase being defined is in boldface type; it is followed by the number (in parentheses) of the chapter or chapters where it is discussed. In some cases, derivations are given. Abbreviations: Gk., Greek; Lat., Latin; *sing.,* singular; *pl.,* plural; *dim.,* diminutive (a smaller version of the object named); *pron.,* pronounced.

A

abscisic acid (19) A plant hormone that has the general effect of inhibiting growth and reproductive activities.

absorption (22) The process by which biological macromolecules in food particles are taken up from the cells lining the digestive tract and then move into the bloodstream.

acid (2) Any fluid with a pH below 7.0, indicating the presence of more H^+ ions than OH^- ions. [Lat., *acidus,* sour]

acquired immune deficiency syndrome (AIDS) (13, 26) Infectious human disease caused by a retrovirus, HIV (human immunodeficiency virus), which compromises the immune system by attacking T cells, leaving an individual susceptible to multiple infections as well as cancers.

acrosome (25) Cap-like structure covering the head region of a sperm; it contains enzymes for penetrating the outer membrane of an egg, enabling fusion of the nuclei to occur.

actin (23) A protein of muscle tissue; makes up the thin filaments.

action potential (23) An electrical signal that travels along an axon, from a neuron to another neuron, a muscle cell, or a gland cell.

active transport (3) Molecular movement that depends on the input of energy, which is necessary when the molecules to be moved are large or are being moved against their concentration gradient.

adaptation (8) The process by which, as a result of natural selection, organisms become better matched to their environment; also, a specific feature, such as the quills of a porcupine, that makes an organism more fit.

adaptive radiation (10) The rapid diversification of a small number of species into a much larger number of species, able to live in a wide variety of habitats.

additive effects (7) Effects from alleles of multiple genes that all contribute to the ultimate phenotype for a given characteristic.

adenosine triphosphate (ATP) (4) A molecule that temporarily stores energy for cellular activity in all living organisms; ATP is composed of a sugar molecule and a chain of three negatively charged phosphate groups.

adult (11) In complete metamorphosis, the third and final stage of insect development.

aging (14) An increased risk of mortality with increasing age; generally characterized by multiple physiological breakdowns.

alleles (5, 7) Alternative versions of a gene. [Gk., *allos,* another]

allergen (26) An antigen that causes an allergic response.

allergy (26) An inappropriate immune response to what should be a harmless substance.

allopatric speciation (10) Speciation that occurs as a result of a geographic barrier between groups of individuals that leads to reproductive isolation and then genetic divergence. [Gk., *allos,* another + *patris,* native land]

allopolyploidy (10) A special case of polyploidy, in which through asexual propagation and subsequent repeated hybridization the increased number of chromosome sets in an individual are derived from multiple different species. [Gk., *allos,* another + *polys,* many + *ploion,* vessel]

alternation of generations (18) In plants, the life cycle that is characterized by an extended period in which the plant has a multicellular haploid form and a period in which it has a multicellular diploid form.

altruistic behavior (9) A behavior that comes at a cost to the individual performing it and benefits another. [Lat., *alter,* the other]

amino acid (2) One of 20 molecules built of an amino group, a carboxyl group, and a unique side chain; proteins are constructed of combinations of amino acids. [Gk., *Ammon,* name of ancient Libyan god, near whose temple a compound known as "sal ammoniacus," or "ammonium salt," was prepared]

amino group (2) A nitrogen atom bonded to three hydrogen atoms.

amnion (25) During animal development, the membrane enclosing the fluid-filled sac surrounding an embryo.

amniotes (11) Terrestrial vertebrates—reptiles, birds, and mammals—that produce eggs (called amniotic eggs) that are each protected by a water-tight membrane and a shell.

amphibians (11) Members of the class Amphibia; amphibians are ectotherms (that is, they are cold-blooded), with a moist skin, lacking scales, through which they can fully or partially absorb oxygen. They were the first terrestrial vertebrates; the young of most species are aquatic, and the adults are true land animals. [Gk., *bios,* life + *amphi,* on both sides]

amygdala (23) Part of the brain's limbic system; associates emotional feelings and sensory input with memories.

analogous (10) Describes characteristics (such as bat wings and insect wings) that are similar because they were produced by convergent evolution, not because they descended from a common structure in a shared ancestor. [Gk., *analogos,* proportionate]

anaphase (6) The third phase of mitosis, in which the sister chromatids are pulled apart by the spindle fibers, with a full set of chromosomes going to opposite sides of the cell; in meiosis, the homologues separate in anaphase I and the sister chromatids separate in anaphase II. [Gk., *ana,* up + *phasis,* appearance]

anaphylactic shock (26) A life-threatening allergic reaction that causes severe respiratory distress, as the throat swells and asthma develops. With swelling in various tissues, there is less fluid in the bloodstream, and this can lead to dangerously low blood pressure.

anecdotal observation (1) Observation of one or only a few instances of a phenomenon.

angiosperms (12) Vascular, seed-producing flowering and fruit-bearing plants, in which the seeds are enclosed in an ovule within the ovary. [Gk., *angeion,* vessel, jar + *sperma,* seed]

animals (11) Members of the kingdom Animalia, which are eukaryotic, multicellular, and heterotrophic (that is, they cannot produce their own food); many of these organisms have body parts specialized for different activities and can move during some stage of their lives. [Lat., *animal,* a living being]

anterior pituitary (24) A gland of the endocrine system that produces numerous hormones, including thyroid-stimulating hormone (TSH), luteinizing hormone (LH), follicle-stimulating hormone (FSH), prolactin, adrenocorticotropic hormone (ACTH), and growth hormone. Many of the hormones of the anterior pituitary gland direct endocrine glands elsewhere to release other hormones.

anther (12, 18) The part of the stamen, the male reproductive structure of a flower, that produces pollen. [Gk., *anthos,* blossom]

antibody (26) A protein produced by the specific immunity system in response to an antigen. Circulating antibodies recognize and bind to specific antigens.

antigen (26) Any molecule that induces a specific immunity response.

antigen receptor (26) Protein on the plasma membrane of a lymphocyte that sticks out from the cell surface and can bind one specific type of antigen.

antigen-presenting cells (26) Dendritic cells and macrophages that ingest bacteria and viruses, destroying them and "presenting" digested particles of the pathogens on their cell surface.

antiseptic (26) Solution such as rubbing alcohol or hydrogen peroxide that discourages growth of microorganisms.

apical meristem (17) Growth region at the tip of a plant root or stem; division of cells in this meristem causes the root or stem to increase in length.

apoptosis (6, 26) Programmed cell death, which takes place particularly in parts of the body where the cells are likely to accumulate significant genetic damage over time and are therefore at high risk of becoming cancerous. Also occurs in the specific immune response, as cytotoxic effector cells destroy infected body cells. [Gk., *apoptosis,* falling away]

archaea (10) A group of prokaryotes that are evolutionarily distinct from bacteria and that thrives in some of the most extreme environments on earth; one of the three domains of life. [Gk., *archaios,* ancient]

arteriosclerosis (21) A disease process, following development of atherosclerosis, in which calcium deposits harden the arteries.

artery (21) A blood vessel that transports blood from the heart, at higher pressure than veins, to the capillaries of the body.

arthropods (11) Members of the invertebrate phylum Arthropoda; characterized by a segmented body, an exoskeleton, and jointed appendages. [Gk., *arthron,* joint + *pous,* foot]

asexual reproduction (6, 25) A type of reproduction common in prokaryotes and plants, and also occurring in many other multicellular organisms, in which the offspring inherit their DNA from a single parent.

assisted reproductive technology (ART) (25) The use of fertility treatments in which both sperm and egg are handled.

atherosclerosis (21) A disease process in which plaques develop in the arteries, reducing the flow of blood and increasing the risk of blood clots.

atom (2) A particle of matter than cannot be further subdivided without losing its essential properties. [Gk., *atomos,* indivisible]

atomic mass (2) The mass of an atom; the combined mass of the protons, neutrons, and electrons in an atom (the mass of the electrons is so small as to be almost negligible).

atomic number (2) The number of protons in the nucleus of an atom of a given element.

atrium (21) A chamber of the heart that collects blood returning from the lungs or the rest of the body.

autoantigen (26) An individual's own molecule or cellular structure that is incorrectly recognized as an antigen by a lymphocyte receptor, thus initiating a humoral and/or cell-mediated immune response.

autoimmunity (26) Inappropriate response of an individual's immune system to the individual's own cells and tissues as if they were pathogens.

auxins (19) In plants, a small group of naturally occurring hormones (and a larger group of synthetic variants) that promote cell division, stem elongation, and formation of roots. Auxins also influence plant orientation, making sure the correct ends are up and down.

axon (20, 23) A projection from a neuron that transmits impulses away from the cell body.

B

B cell (26) A type of lymphocyte, maturing in the bone marrow, that is involved in humoral immunity; fights invaders in body fluids.

background extinctions (10) Extinctions that occur at lower rates than at times of mass extinctions; background extinctions occur mostly as the result of aspects of the biology and competitive success of the species, rather than catastrophe.

basal metabolic rate (BMR) (22) The amount of energy expended by a living organism at rest in a neutral temperature environment.

base (chemistry) (2) Any fluid with a pH above 7.0, that is, with more OH^- ions than H^+ ions.

base (genetics) (2, 5) One of the nitrogen-containing side-chain molecules attached to a sugar molecule in the sugar-phosphate backbone of DNA and RNA. The four bases in DNA are adenine (A), thymine (T), guanine (G), and cytosine (C); the four bases in RNA are adenine (A), uracil (U), guanine (G), and cytosine (C); the information in a molecule of DNA and RNA is determined by its sequence of bases.

base pair (5) Two nucleotides on complementary strands of DNA that form a pair, linked by hydrogen bonds; the pattern of pairing is adenine (A) with thymine (T) and cytosine (C) with guanine (G); the base-paired arrangement forms the "rungs" of the double-helix structure of DNA.

basophil (26) A type of white blood cell that initiates inflammation.

behavior (9) Any and all of the actions performed by an organism, often in response to its environment or to the actions of another organism.

bilateral symmetry (11) A body structure with left and right sides, which are mirror images. [Lat., *bi-,* two + *latus,* side]

bile (22) A juice that aids in the breakdown of fats; it is produced by the liver, sent to the gall bladder, and passed through a small duct into the small intestine, where it initiates the first step in fat digestion.

binary fission (6) A type of asexual reproduction in which the parent cell divides into two genetically identical daughter cells; bacteria and other prokaryotes reproduce by binary fission. [Lat., *binarius,* consisting of two + *fissus,* divided]

biodiversity (10) The variety and variability among all genes, species, and ecosystems. [Gk., *bios,* life + Lat., *diversus,* turned in different directions]

biodiversity hotspots (16) Regions of the world with significant reservoirs of biodiversity that are under threat of destruction.

biofuels (4) Fuels produced from plant and animal products.

biogeography (8) The study and interpretation of distribution patterns of living organisms around the world. [Gk. *bios,* life + *geo-,* earth + *graphein,* to write down]

biological literacy (1) The ability to use scientific inquiry to think creatively about problems with a biological component, to communicate these thoughts to others, and to integrate these ideas into one's decision making.

biological species concept (10) A definition of species described as populations of organisms that interbreed, or could possibly interbreed, with each other under natural conditions, and that cannot interbreed with organisms outside their own group.

biology (1) The study of living things. [Gk., *bios,* life + *logos,* discourse]

biomass (15) The total mass of all the living organisms in a given area. [Gk., *bios,* life]

biomes (15) The major ecological communities of earth; terrestrial biomes, such as rainforest or desert, are defined and usually described by the predominant types of plant life in the area, which are mostly determined by the weather; aquatic biomes are usually defined by physical features such as salinity, water movement, and depth. [Gk., *bios,* life]

biotechnology (5) The modification of organisms, cells, and their molecules for practical benefits. [Gk., *bios,* life + *technologia,* systematic treatment]

biotic (15) Relating to living organisms; the biotic environment, or community, consists of all the living organisms in a given area. [Gk., *bios,* life]

bisphenol A (24) A chemical found in some plastic water bottles and baby bottles; a known endocrine disruptor.

bivalve molluscs (11) Molluscs with two hinged shells; examples are clams, scallops, oysters and molluscs. [Lat., *bi-,* two]

blastocyst (25) The mammalian blastula.

blastula (25) An early stage of embryonic development produced by cleavage of a fertilized egg (zygote). A spherical layer of cells encloses a large fluid-filled space where the embryo forms. [Gk., *blastos,* sprout]

blind experimental design (1) An experimental design in which the subjects do not know what treatment (if any) they are receiving.

blood (20, 21) A connective tissue with a liquid extracellular matrix containing blood cells; contained in a closed circulatory system, and important in the transport of respiratory gases, vitamins and minerals, nutrients, hormones, components of the immune system, and metabolic wastes.

blood pressure (21) The force with which blood flows through the arteries.

body mass index (22) Body weight in kilograms divided by height in meters squared (kg/m^2); often used as an index of a healthy weight for an individual based on the person's height.

bone (20) A rigid connective tissue that protects and provides support.

bottleneck effect (8) A change in allele frequencies of a population caused by a sudden reduction in population size (often due to famine, disease, or rapid environmental disturbance).

Bowman's capsule (20) A ball-like structure surrounding the blood-filtering unit of a nephron in the kidney. It is connected to a single, long, urine-collecting tube that excretes its filtered fluid into a collecting duct.

brain imaging technologies (23) Techniques used to reveal activity patterns (based on increased metabolism) in different regions of the brain. See also **fMRI scan; PET scan.**

Broca's area (23) The discrete part of the brain, located, in most people, in the front part of the left frontal lobe, that is used for controlling the muscles responsible for the process of speaking.

bryophytes (12) Three groups of plants (the liverworts, hornworts, and mosses) that lack vascular tissue and move water and dissolved nutrients by diffusion. [Gk., *bruon,* tree-moss, liverwort + *phytas,* plant]

buffer (2) A chemical that can quickly absorb excess H^+ ions in a solution (preventing it from becoming too acidic) or quickly release H^+ ions to counteract increases in OH^- concentration.

bulbourethral gland (25) A male reproductive gland, one on each side of the body, that contributes mucus and sugar to the ejaculate and lubricant to the tip of the penis prior to copulation.

C

C4 photosynthesis (4) A method (along with C3 and CAM photosynthesis) by which plants fix carbon dioxide, using the carbon to build sugar; serves as a more effective method than C3 for binding carbon dioxide under low carbon dioxide conditions, such as when plants in warmer climates close their stomata to reduce water loss.

calcitonin (24) A hormone released from the thyroid gland that causes bones to take up excess calcium from the bloodstream.

calorie (22) The energy required to raise the temperature of 1 gram of water by 1° C; the kilocalorie, made from 1,000 calories, is commonly used as a measure of the amount of energy in a particular food.

Calvin cycle (4) In photosynthesis, a series of chemical reactions in the stroma of chloroplasts, in which sugar molecules are assembled. [from the name of one of its discoverers, Melvin Calvin, 1911–1997]

CAM (crassulacean acid metabolism) (4) Energetically expensive photosynthesis in which the stomata are open only at night to admit CO_2, which is bound to a holding molecule and released to enter the Calvin cycle to make sugar during the day; in this type of photosynthesis, found in many fleshy, juicy plants of hot, dry areas, water loss is reduced because the stomata are closed during the day.

cancer (6) Unrestrained cell growth and division. [Lat., *cancer,* crab; by extension, the disease cancer]

capillaries (21) Tiny blood vessels that bring blood close enough to cells to allow the diffusion of molecules into and out of the blood.

capsid (13) The protein container surrounding the genetic material (DNA or RNA) of a virus. [Lat., *capsa,* box, case]

capsule (13) A layer surrounding the cell wall of many bacteria; the capsule may restrict the movement of water out of the cell and thus allow bacteria to live in dry places, such as the surface of the skin. The capsule contributes to the virulent characteristics of some bacteria, making them resistant to phagocytosis by the host's immune system. [Lat., *capsula,* small box or case]

carbohydrate (2) One of the four types of biological macromolecule, containing mostly carbon, hydrogen, and oxygen; carbohydrates are the primary fuel for cellular activity and form much of the cell structure in all life forms. [Lat., *carbo,* charcoal + *hydro-,* pertaining to water]

carboxyl group (2) A carbon atom bonded to two oxygen atoms.

cardiac muscle (20) A type of muscle tissue, located only in the heart, that causes the heart to pump blood through the body. Cardiac muscle cells, which are fused together and connected by gap junctions, contain many more mitochondria than other types of muscle cells.

cardiovascular disease (21) Progressive deterioration of the arteries and other degradations of the circulatory system; the leading cause of death and disability in the United States.

carnivores (15, 22) Predatory animals (and some plants) that consume only animals. [Lat., *carnis,* of flesh + *vorare, to* devour]

carotenoids (4) Pigments that absorb blue-violet and blue-green wavelengths of light and reflect yellow, orange, and red wavelengths of light. [Lat., *carota,* carrot]

carpel (12, 18) The female reproductive structure of a flower, including the stigma, style, and ovary. [Gk., *karpos,* fruit]

carrier (7) An individual who carries one allele for a recessive trait and who does not exhibit the trait; if two carriers mate they may produce offspring who do exhibit the trait.

carrying capacity (*K*) (14) The ceiling on a population's growth imposed by the limitation of resources for a particular habitat over a period of time.

cartilage (20) A dense connective tissue with an extracellular matrix rich in collagen, elastin, and proteins bound to long carbohydrate chains, with a hardness between that of bone and of tendons; strong, but also flexible, cartilage is found in the ears and tip of the nose in humans, and it cushions the bones in joints throughout the body.

cartilaginous fishes (11) Fish species characterized by a skeleton made completely of cartilage, not bone. [Lat., *cartilago,* gristle]

cell (3) The smallest unit of life that can function independently; a three-dimensional structure, surrounded by a membrane and, in the case of prokaryotes and most plants, a cell wall, in which many of the essential chemical reactions of the life of an organism take place. [Lat., *cella,* room]

cell body (20, 23) The part of a cell, such as a neuron, that contains the nucleus and other organelles.

cell cycle (6) In a cell, the alternation of activities related to cell division and those related to growth and metabolism.

cell theory (3) A unifying and universally accepted theory in biology that holds that all living organisms are made up of one or more cells, and that all cells arise from other, pre-existing cells.

cell wall (3) A rigid structure, outside the cell membrane, that protects and gives shape to the cell; found in many prokaryotes and plants.

cellular respiration (4) The process by which all living organisms extract energy stored in the chemical bonds of molecules and use it for fuel for their life processes.

cellulose (2) A complex carbohydrate, indigestible by humans, that serves as the structural material for a huge variety of plant structures; it is the single most prevalent organic compound on earth. [Lat., *cellula, dim.* of *cella,* room]

central nervous system (CNS) (20, 23) The part of an organism's nervous system that includes the brain and spinal cord and coordinates all nervous activity.

central vacuole (3) In plants, a large, fluid-filled organelle, surrounded by a membrane, important in nutrient storage, waste management, predator deterrence, sexual reproduction, and physical support. [Lat., *vacuus,* empty]

centromere (6) After replication, the region of contact between sister chromatids, which occurs near the center of the two strands. [Gk., *centron,* the stationary point of a pair of compasses, thus the center of a circle + *meris,* part]

cephalopods (11) Molluscs in which the head is prominent and the foot has been modified into tentacles; examples are octopuses and squids. Cephalopods have a reduced or absent shell and possess the most advanced nervous system of the invertebrates. [Gk., *kephale,* head + *pous,* foot]

cerebellum (23) The part of the vertebrate hindbrain that controls muscle coordination.

cerebral cortex (23) The part of the vertebrate brain involved in abstract thought, problem solving, language, and more; the brain structure most responsible for the traits that set humans apart from other animal species.

cervix (25) The low, narrow portion of the uterus leading to the vagina. [Lat., *cervix,* neck]

character displacement (15) An evolutionary divergence in one or more of the species that occupy a niche that leads to a partitioning of the niche between the species. Changes in characteristics, such as behavioral or body plan, of two or more very similar species that have overlapping geographical locations result in a reduction of competition between species.

chemical energy (4) A type of potential energy in which energy is stored in chemical bonds between atoms or molecules.

chemolithotrophs (13) Bacteria that can use inorganic molecules such as ammonia, hydrogen sulfide, hydrogen, and iron as sources of energy. [Gk., *lithos,* stone, rock + *trophē,* food]

chemoorganotrophs (13) Bacteria that consume organic molecules, such as carbohydrates, as an energy source. [Gk., *trophē,* food]

chemoreceptor (23) A type of sensory cell that detects chemical changes in the organism's internal or external environment.

chitin (2) (*pron.* KITE-in) A complex carbohydrate, indigestible by humans, that forms the rigid outer skeleton of most insects and crustaceans. [Gk., *chiton,* undershirt]

chlorophyll (4) A light-absorbing pigment molecule in chloroplasts. [Gk., *chloros,* pale green + *phyllon,* leaf]

chlorophyll *a* (4) The primary photosynthetic pigment, chlorophyll *a* absorbs blue-violet and red light; because it cannot absorb green light and instead reflects those wavelengths, we perceive the reflected light as the color green.

chlorophyll *b* (4) A photosynthetic pigment similar in structure to chlorophyll *a,* chlorophyll *b* absorbs blue and red-orange wavelengths and reflects yellow-green wavelengths.

chloroplast (3, 4) The organelle in plant cells in which photosynthesis occurs. [Gk., *chloros,* pale green + *plastos,* formed]

cholesterol (2, 3) One of the sterols, lipids important in regulating growth and development; cholesterol is an important component of most cell membranes, helping the membrane to maintain its flexibility. [Gk., *chole,* bile + *stereos,* solid + *-ol,* chemical suffix for an alcohol]

chorion (25) During animal development, the outer membrane surrounding an embryo, which, with the endometrium, forms the placenta.

chromatid (6) One of the two strands of a replicated chromosome. [Gk., *chroma,* color]

chromatin (3) A mass of long, thin fibers consisting of DNA and proteins in the nucleus of the cell. [Gk., *chroma,* color]

chromosomal aberration (5) A type of mutation characterized by a change in the overall organization of genes on a chromosome, such as the deletion of a section of DNA; the moving of a gene from one part of a chromosome to elsewhere on the same chromosome or to a different chromosome; or the duplication of a gene, with the new copy inserted elsewhere on the chromosome or on a different chromosome. [Lat., *aberrare,* to wander]

chromosome (5) A linear or circular strand of DNA on which are found specific sequences of base pairs; the human genome consists of two copies of each of 23 unique chromosomes, one from the mother and one from the father. [Gk., *chroma,* color + *soma,* body]

chyme (22) A liquid mass of partially digested food that passes from the stomach through the small intestine.

cilia (*sing.* cilium) (3, 26) Short projections from the cell surface, often occurring in large numbers on a single cell, that beat against the intercellular fluid to move the fluid past the cell. [Lat., eyelid]

class (10) In the hierarchical taxonomic system developed by Carolus Linnaeus (1707–1778), a classification of organisms consisting of related orders.

cleavage (25) In embryonic development, the early cell division of the zygote, which begins shortly after fertilization.

climax community (15) A stable and self-sustaining community that results from ecological succession.

clitoris (25) Erectile sexual organ of female mammals; develops from the same embryonic tissue that produces the penis.

clonal selection (26) Process in which a B cell or T cell binds to its antigen and divides numerous times to create a population of cells with the same antigen specificity. This process ensures that there are sufficient numbers of B and T cells to recognize and respond to a specific pathogen that has invaded the body.

clone (5) A genetically identical DNA fragment, cell, or organism produced by a single cell or organism. [Gk., *klon,* twig]

clone library (5) A collection of cloned DNA fragments; also known as a gene library.

cloning (5) The production of genetically identical cells, organisms, or DNA molecules.

closed circulatory system (21) A circulatory system in which blood is contained in vessels and is separate from the interstitial fluid that bathes cells.

code (5) In genetics, the base sequence of a gene.

codominance (7) The case in which the heterozygote displays characteristics of both alleles.

codons (5) Three-base sequences in mRNA that link with complementary tRNA molecules, which are attached to amino acids; a codon with yet another sequence ends the process of assembling a protein from amino acids.

coelom (25) (*pron.* SEE-lum) The body cavity; it forms from the mesoderm in embryonic development.

coevolution (15) The concurrent appearance over time, through natural selection, of traits in interacting species that enable each species to become adapted to the other; an example is the 11-inch-long tongue of a moth that feeds from the 11-inch-long nectar tube of an orchid.

cohesion-tension mechanism (17) In vascular plants, the force driving fluid flow in the xylem; results from evaporation of water from leaves, which pulls water up from the roots.

collagen (20) A protein in the matrix of connective tissue, synthesized and secreted by connective tissue cells (fibroblasts).

collenchyma cell (17) A long, stringy cell with unevenly thickened cell walls that contributes to a plant's flexibility. Collenchyma cells are a type of ground tissue.

colon (22) See **large intestine.**

colonizers (15) Species introduced into an area that has been disturbed and is undergoing the process of either primary or secondary succession. The identity of a colonizing species varies depending on the stage and type of succession.

color blind (23) In humans, describes the inability to distinguish between colors that most individuals can see—most commonly, the inability to distinguish between the colors red and green; occurs mostly in males.

commensalism (15) A symbiotic relationship between species in which one benefits and the other neither benefits nor is harmed. [Lat., *com,* with + *mensa,* table]

communication (9) An action or signal on the part of one organism that alters the behavior of another organism.

community (in ecology) (15) The biotic environment; a geographic area defined as a loose assemblage of species with overlapping ranges.

competitive exclusion (15) The case in which two species battle for resources in the same niche until the more efficient of the two wins and the other is driven to extinction in that location.

complement protein (26) A type of circulating, defensive protein that can create holes in the membranes of pathogens or destroy pathogens by sticking to them and forming a coating that enhances the ability of phagocytes to bind to and engulf them.

complementarity (6) The characteristic of double-stranded DNA that the base on one strand always has the same pairing partner, or complementary base, on the other strand.

complementary base (6) A base on a strand of double-stranded DNA that is a pairing partner to a base on the other strand: adenine (A) is the complementary base to thymine (T), and guanine (G) is the complementary base to cytosine (C).

complex carbohydrate (2) A carbohydrate that contains multiple simple carbohydrates linked together; types of complex carbohydrates include starch, which is the primary form of energy storage in plants, and glycogen, which is the primary form of short-term energy storage in animals.

composting (17) Deliberate mixing of decaying organic matter, from leaves, food scraps, and manure, used to provide nutrients to soil.

compound (2) A substance composed of atoms of different elements in specific ratios, held together by ionic bonds. [Lat., *componere,* to put together]

compound eye (23) A group of light-sensing cells found in many invertebrates, made up of dozens to thousands of separate light-sensing units, each with its own lens that focuses light onto about a dozen photoreceptors.

cone (23) In the retina, a highly light-sensitive cell type that enables color vision.

conformer (20) An organism that, for a given physiological variable, has no set point for the variable and allows it to fluctuate with external changes in the environment.

connective tissue (20) A type of animal tissue that consists of cells embedded in a large amount of extracellular material, called matrix, which together contribute to body structure and support; includes tendons, ligaments, fat, blood, bone, and cartilage.

connective tissue proper (20) A type of connective tissue, including fat tissue, tendons, and ligaments, that has a semi-fluid, flexible matrix and functions like packing material.

conjugation (13) The process by which a bacterium transfers a copy of some or all of its DNA to another bacterium, of the same or another species. [Lat., *coniugatio,* connection]

conservation biology (16) An interdisciplinary field, drawing on ecology, economics, psychology, sociology, and political science, that studies and devises ways of preserving and protecting biodiversity and other natural resources.

contact inhibition (3) The limiting factor of cell growth that occurs when normal cells come into contact; in cancer cells, contact inhibition does not take place and the cells continue to divide.

contraception (25) The attempt to prevent pregnancy by preventing ovulation, fertilization, or implantation. Also called *birth control.*

control group (1) In an experiment, the group of subjects not exposed to the treatment being studied but otherwise treated identically to the experimental group.

convergent evolution (8, 10) A process of natural selection in which features of organisms not closely related come to resemble each other as a consequence of similar selective forces; many marsupial and placental species resemble each other as a result of convergent evolution. [Lat., *con-,* together with + *vergere,* turn + *evolvere,* to roll out]

coprophagy (22) A nutritional strategy, common in rabbits and rodents, in which animals consume some of their feces, thereby increasing their ability to extract energy from food by passing it through their digestive system twice.

copulation (25) The act of placing the male reproductive organ within the female's reproductive tract.

coronary arteries (21) The blood vessels that deliver oxygen and nutrients to the heart.

cork cell (17) A cell of the protective outer covering of some plants, containing a waxy substance impermeable to water and resistant to fire and decay.

corpus callosum (23) In the brain, the broad, thick band of neurons connecting the left hemisphere to the right hemisphere.

corpus luteum (25) A structure that develops from a follicle after ovulation; it secretes hormones to maintain pregnancy. [Lat., *corpus,* body + *luteum,* yellow]

cortex (17) In vascular plants, the ground tissue located between the epidermis and the xylem and phloem.

cotyledon (17, 18) Part of the plant embryo within the seed; this structure usually becomes the first embryonic leaf or leaves of the plant. Also called *seed leaf.*

covalent bond (2) A strong bond formed when two atoms share electrons; the simplest example is the H_2 molecule, in which each of the two atoms in the molecule shares its lone electron with the other atom. [Lat., *con-,* together + *valere,* to be strong]

critical experiment (1) An experiment that makes it possible to determine decisively between alternative hypotheses.

cross (7) The breeding of organisms that differ in one or more traits.

crossing over (6) The exchange of some genetic material between a paternal homologous chromosome and a maternal homologous chromosome, leading to a chromosome carrying genetic material from each; also referred to as recombination.

cuticle (12, 17) A waxy layer produced by epidermal cells and found on leaves and shoots of terrestrial plants, protecting them from drying out. [Lat., *cuticula, dim.* of *cutis,* skin]

cytokine (26) A type of signaling protein secreted by cells of the immune system: the chief way in which cells of the specific and non-specific immunity systems communicate with one another.

cytokinesis (6) In the cell cycle, the stage following mitosis in which cytoplasm and organelles duplicate and are divided into approximately equal parts and the cell separates into two daughter cells; in meiosis, two diploid daughter cells are formed in cytokinesis following telophase I and four haploid daughter cells are formed in cytokinesis following telophase II. [Gk., *kytos,* container + *kinesis,* motion]

cytokinins (19) In plants, a group of hormones that stimulate cell division throughout the body and throughout the lifetime of the plant.

cytoplasm (3) The jelly-like fluid that fills the inside of the cell; in eukaryotes, the cytoplasm contains the organelles. [Gk., *kytos,* container + *plasma,* anything molded]

cytoskeleton (3) A network of protein structures in the cytoplasm of plants and animals that serves as scaffolding, adding support and giving animal cells of different types their characteristic shapes; the cytoskeleton serves as tracks guiding the intercellular traffic flow and, because it is flexible and can generate force, gives cells some ability to control their movement.

cytotoxic T cell (26) A type of lymphocyte, acting in the cell-mediated response, that directly kills body cells infected with a pathogen.

D

daughter cells (6) Cells produced by the division of a parent cell.

decomposers (12, 15) Organisms, including bacteria, fungi, and detritivores, that break down and feed on once-living organisms.

decomposition (17) The process of breaking down organic material into the individual elements of which it is made.

demographic transition (14) A pattern of population growth characterized by the progression from high birth and death rates (slow growth) to high birth rates and low death rates (fast growth) to low birth and death rates (slow growth).

denaturation (2) The disruption of protein folding, in which secondary and tertiary structure are lost, caused by exposure to extreme conditions in the environment such as heat or extreme pH (that is, a strong acid or a strong base); denaturation causes proteins to lose their function; also the disruption of DNA, also known as DNA melting, in which the bonds linking complementary bases are broken and the two strands separate from each other.

dendrite (20, 23) A branched projection from a neuron that receives signals from the external environment or from other neurons. [Gk., *dendron,* tree]

dendritic cell (26) A type of white blood cell that links the non-specific and specific divisions of the immune system by engulfing foreign matter and then presenting the digested remnants of the pathogen on its cell surface.

density-dependent factors (14) Limitations on a population's growth that are a consequence of population density.

density-independent factors (14) Limitations on a population's growth without regard to population size, such as floods, earthquakes, fires, and lightning.

deoxyribonucleic acid (DNA) (2, 5) A nucleic acid, DNA carries information about the production of particular proteins in the sequences of its nucleotide bases.

dermal tissue (17) The outer covering that protects the surface of a plant.

desert (15) A type of terrestrial biome; a type of dry climate, with very little rainfall, in which water loss through evaporation exceeds water gain through precipitation; typically found at 30° north and south latitude.

desmosome (3) Irregularly spaced connections between adjacent animal cells that, in the manner of Velcro, hold cells together by attachments but are not water-tight; they provide mechanical strength and are found in muscle tissue and in much of the tissue that lines the cavities of animal bodies. [Gk., *desmos,* bond + *soma,* body]

detritivores (15) Organisms that break down and feed on once-living organic matter; this group includes scavengers such as vultures, worms, and a variety of arthropods. [Lat., *detritus,* worn out + *vorare,* to devour]

deuterostomes (11) Bilaterally symmetrical animals with defined tissues in which the gut develops from back to front; the anus forms first, and the second opening formed is the mouth of the adult animal. [Gk., *deuteros,* second + *stoma,* mouth]

diabetes (22) A condition in which an individual does not adequately regulate their blood sugar levels; usually results from insufficient insulin secretion from the pancreas in response to an increase in blood sugar, or inadequate response of the cells of the body to insulin in the bloodstream.

diaphragm (21) A large sheet of muscle that separates the chest cavity from the abdominal cavity, allowing for increased volume when contracted during inhalation. [Gk., *dia,* through + *phrasein,* to enclose]

diaphragm (contraceptive) (25) A dome-shaped piece of rubber placed in the vagina that blocks the cervix and prevents sperm from coming in contact with an egg.

diastolic pressure (21) The second blood pressure reading; a measure of the force that blood exerts on the artery walls while the heart is between beats.

dicot (17) One of the two major groups of flowering plants; dicot seedlings have two cotyledons (seed leaves).

differential reproductive success (8) The situation in which individuals have greater reproductive success than other individuals in a population; along with variation and heritability, differential reproductive success is one of the three conditions necessary for natural selection.

differentiated (25) Describing cells that, during development, begin to take on specific structures or functions.

diffusion (3) Passive transport in which a particle (the solute) is dissolved in a gas or liquid (the solvent) and moves from an area of higher solute concentration to an area of lower solute concentration. [Lat., *diffundere,* to pour in different directions]

digestion (22) The physical and chemical breakdown of food into its fundamental macromolecular components for absorption or elimination.

diploid (6) Describes cells that have two copies of each chromosome (in many organisms, including humans, somatic cells are diploid). [Gk., *diplasiazein,* to double]

direct fitness (9) The total reproductive output of an individual.

directional selection (8) Selection that, for a given trait, increases fitness at one extreme of the phenotype and reduces fitness at the other, leading to an increase or decrease in the mean value of the trait.

disaccharides (2) Carbohydrates formed by the union of two simple sugars, such as sucrose (table sugar) and lactose (the sugar found in milk). [Gk., *di-,* two + *sakcharon,* sugar]

dispersers (15) Organisms able to move away from their original home.

disruptive selection (8) Selection that, for a given trait, increases fitness at both extremes of the phenotype distribution and reduces fitness at middle values.

DNA probe (5) A short sequence of radioactively tagged single-stranded DNA that contains part of the sequence of the gene of interest, used to locate that gene in a gene library; the probe binds to the complementary base pair on a gene in the library, which is identified by the radioactive tag on the probe.

domain (10) In modern classification, the highest level of the hierarchy; there are three domains, bacteria, archaea, and eukarya.

dominant (7) Describes an allele that masks the phenotypic effect of the other, recessive, allele for a trait; the phenotype shows the effect of the dominant allele in both homozygous and heterozygous genotypes. [Lat., *dominari,* to rule]

dorsal hollow nerve cord (11) The central nervous system of vertebrates, consisting of the spinal cord and brain. [Lat., *dorsum,* back]

double-blind experimental design (1) An experimental design in which neither the subjects nor the experimenters know what treatment (if any) individual subjects are receiving.

double bond (2) The sharing of two electrons between two atoms; for example, the most common form of oxygen is the O_2 molecule, in which two electrons from each of the two atoms of oxygen are shared.

double fertilization (12, 18) In angiosperms, two sperm are released by a pollen grain: one fuses with an egg to form a zygote, and the other fuses with two nuclei, forming a triploid endosperm.

double helix (2) The spiraling ladder-like structure of DNA composed of two strands of nucleotides; the bases protruding from each strand like "half-rungs" meet in the center and bind to each other (via hydrogen bonds), holding the ladder together. [Gk., *heligmos,* wrapping]

dynamic equilibrium (16) A situation on islands in which the number of species present remains constant over time, while the composition of species may change.

E

ear canal (23) The channel from the outer ear to the middle ear that conducts sound waves.

eardrum (23) The thin membrane that divides the outer ear from the middle ear.

echolocation (23) The ability of some animals (including bats) to sense the location of objects by producing sounds and then detecting echoes bounced off the objects.

ecological footprint (14) A measure of the impact of an individual or population on the environment by calculation of the amount of resources—including land, food and water, and fuel—consumed.

ecology (14) The study of the interaction between organisms and their environments, at the level of individuals, populations, communities, and ecosystems. [Gk., *oikos,* home + *logos,* discourse]

ecosystem (15) A community of biological organisms and the non-living environmental components with which they interact.

ectoderm (25) The outermost embryonic cell layer during gastrulation; it eventually forms skin, hair, the nervous system, and the lining of the nose and mouth. [Gk., *ektos,* outside + *derma,* skin]

ectotherms (11, 20) Organisms that rely on the heat from an external source to raise their body temperature and seek the shade when the air is too warm. [Gk., *ektos,* outside + *thermē,* heat]

effector (20) A structure that can be triggered in response to a stimulus; aids in the maintenance of homeostasis by opposing or reducing changes in the internal environment in response to changes in the external environment.

effector cell (26) In the specific (adaptive) immunity system, a type of cell that recognizes an antigen and immediately takes some action that leads to its destruction; plasma cells, which are antibody-secreting, modified B cells, are the effector cells of the cell-mediated response.

egg (25) A female gamete. Also called *ovum* (*pl. ova*).

electromagnetic spectrum (4) The range of wavelengths that produce electromagnetic radiation, extending (in order of decreasing energy) from high-energy, short-wave, gamma rays and X rays, through ultraviolet light, visible light, and infrared light, to very long, low-energy, radio waves. [Lat., *specere,* to look at]

elastin (20) A protein in the matrix of connective tissue, synthesized and secreted by connective tissue cells (fibroblasts), that gives the tissue flexibility.

electron (2) A negatively charged particle that moves around the atomic nucleus.

electron transport chain (4) The third step of cellular respiration, in which high-energy electrons are passed from molecule to molecule, at every step releasing energy that is used to make ATP.

element (2) A substance that cannot be broken down chemically into any other substances; all atoms of an element have the same atomic number. [Lat., *elementum,* element, or first principle]

elimination (22) The last phase in the breakdown of food; the absorption of water, salts, and some vitamins in the large intestine, followed by defecation of remaining waste material.

El Niño (15) A sustained surface temperature change in the central Pacific Ocean that occurs every two to seven years; this event can start a chain reaction of unusual weather across the globe that can result in flooding, droughts, famine, and a variety of extreme climate disruptions. [Spanish, *the child;* a reference to the Christ child, because of the appearance of the phenomenon at Christmastime]

embryo (18, 25) The multicellular, developing, fertilized egg of a eukaryote. In humans, at about eight weeks following fertilization the embryo is called a fetus.

embryo sac (18) The part of the female reproductive structure of a flower that contains an egg and is the site of fertilization.

empirical (1) Describes knowledge that is based on experience and observations that are rational, testable, and repeatable. [Gk., *empeiria,* experience]

endangered species (16) As defined by the Endangered Species Act, species in danger of extinction throughout all or a significant portion of their range.

Endangered Species Act (ESA) (16) A United States law that defines "endangered species" and is designed to protect those species from extinction.

endemic (16) Describes species peculiar to a particular region and not naturally found elsewhere. [Gk., *en,* in + *demos,* the people of a country]

endocrine cell (24) A hormone-secreting cell. Endocrine cells are part of the endocrine system.

endocrine disruptor (24) A chemical manufactured by humans that, when taken up by organisms, can mimic, block, or otherwise interfere with their hormones, leading to harmful effects.

endocrine gland (20, 24) A collection of epithelial cells that produces hormones and releases them into the bloodstream or other fluids of the body.

endocrine system (24) Organ system comprising glands and cells that secrete hormones, which are chemical messengers that act on target cells to regulate body functions and maintain homeostasis.

endocytosis (3) A cellular process in which large particles, solid or dissolved, outside the cell are surrounded by a fold of the plasma membrane, which pinches off, forming a vesicle, and the enclosed particle now moves into the cell; the three types of endocytosis are phagocytosis, pinocytosis, and receptor-mediated endocytosis. [Gk., *endon,* within + *kytos,* container]

endoderm (25) The innermost embryonic cell layer during gastrulation; it eventually forms the lining of the respiratory and digestive tracts, the liver, and the pancreas. [Gk., *endon,* within + *derma,* skin]

endomembrane system (3) A system of organelles (the rough endoplasmic reticulum, the smooth endoplasmic reticulum, and the Golgi apparatus) that surrounds the nucleus; it produces and modifies necessary molecules, breaks down toxic chemicals and cellular by-products, and is thus responsible for many of the fundamental functions of the cell. [Gk., *endon,* within + Lat., *membrana,* a thin skin]

endometrium (25) The lining of the uterus where a fertilized egg implants and is nourished.

endosperm (12, 18) Tissue of a mature seed that stores certain carbohydrates, proteins, and lipids that fuel the germination, growth, and development of the embryo and young seedling. [Gk., *endon,* within + *sperma,* seed]

endosymbiosis theory (3) Theory of the origin of eukaryotes that holds that in the past two different types of prokaryotes engaged in a close partnership and eventually one, capable of performing photosynthesis, was subsumed into the other, a larger prokaryote; the smaller prokaryote made some of its photosynthetic energy available to the host, and over time the two became symbiotic and eventually a single more complex organism in which the smaller prokaryote had evolved into the chloroplast of the new organism (a similar scenario can be developed for the evolution of mitochondria). [Gk., *endon,* within + *symbios,* living together]

endotherms (11, 20) Organisms that use the heat produced by their cellular respiration to raise and maintain their body temperature above air temperature. [Gk., *endon,* within + *thermē,* heat]

energy (4) The capacity to do work, which is the moving of matter against an opposing force. [Gk., *energeia,* activity]

energy pyramid (15) A diagram that illustrates the path of energy through the organisms of an ecosystem; each layer of the pyramid represents the biomass of a trophic level.

enzymatic protein (3) See **enzyme**

enzyme (2, 3) A protein that initiates and accelerates a chemical reaction in a living organism; enzymatic proteins take part in chemical reactions on the inside and outside surfaces of the plasma membrane. [Gk., *en,* in + *zyme,* leaven]

epidermis (17) The layer of tightly packed, very thin cells that covers and protects a plant's roots, leaves, and stems.

epididymis (25) A tube in each testis where the sperm mature.

epithelial tissue (20) A thin tissue that covers and protects the surfaces of an animal's body. Also called *epithelium.*

epithelium (20) See **epithelial tissue.**

erythrocytes (21) See **red blood cells.**

esophagus (22) The passageway from the throat to the stomach through which food travels.

essential amino acid (22) An amino acid that is not made by the body and so must be consumed in food.

estuary (15) A tidal water passage, linked to the sea, in which salt water and fresh water mix; estuaries are characterized by exceptionally high productivity. [Lat., *aestus,* tide]

estrogen (25) One of the primary female sex hormone; important in female development and the female reproductive cycle.

ethanol (4) The end product of fermentation of yeast; the alcohol in beer, wine, and spirits. [contraction of the full chemical name, ethyl alcohol]

ethylene (19) In plants, a gas that functions as a hormone that speeds up the rate at which fruit ripens.

eudicot (17) A large monophyletic subset of the dicots (plants in which two cotyledons form) and one of the two major groups of flowering plants.

eukaryote (3) An organism composed of eukaryotic cells. [Gk., *eu,* good + *karyon,* nut, kernel]

eukaryotic cell (3) A cell with a membrane-bound nucleus containing DNA, membrane-bound organelles, and internal structures organized into compartments.

eutrophication (15) The process in which excess nutrients dissolved in a body of water lead to rapid growth of algae and bacteria, which consume much of the dissolved oxygen and, in time, can lead to large-scale die-offs. [Gk., *eu,* good + *trophē,* food]

evolution (8) A change in allele frequencies of a population. [Lat., *evolvere,* to roll out]

exocrine gland (20) A collection of epithelial cells that secretes products onto the surface of the epithelium.

exocytosis (3) A cellular process in which particles within the cell, solid or dissolved, are enclosed in a vesicle and transported to the plasma membrane, where the membrane of the vesicle merges with the plasma membrane and the material in the vesicle is expelled to the extracellular fluid for use throughout the body. [Gk., *ex,* out of + *kytos,* container]

exoskeleton (11) A rigid external covering such as found in some invertebrates, including insects and crustaceans. [Gk., *ex,* out of]

exotic species (16) Species introduced by human activities to areas other than the species' native range.

experimental group (1) In an experiment, the group of subjects exposed to a particular treatment; also known as the treatment group.

exponential growth (14) Growth of a population at a rate that is proportional to its current size.

external fertilization (25) The process in which sperm and egg unite outside the bodies of the male and female.

extinction (10) The complete loss of all individuals in a species. [Lat., *extinguere,* to extinguish]

extracellular fluid (20) A body fluid that is outside the cells; as distinct from intracellular fluid. See also **blood; interstitial fluid.**

extremophiles (13) Bacteria and archaea than can live in extreme physical and chemical conditions. [Lat., *extremus,* outermost + Gk., *philios,* loving]

eye cup (23) A group of simple, light-sensing cells found in flatworms such as planaria.

F

facilitated diffusion (3) Diffusion of molecules through the phospholipid bilayer of the plasma membrane that takes place through a transport protein (a "carrier molecule") embedded in the membrane; molecules that require the assistance of a carrier molecule are those that are too big to cross the membrane directly or are electrically charged and would be repelled by the middle layer of the membrane.

Fallopian tube (25) The tube that conveys eggs from an ovary to the uterus. Also called *oviduct.*

family (10) In the system developed by Carolus Linnaeus (1707–1778), a classification of organisms consisting of related genera.

fatty acid (2) A long hydrocarbon (a chain of carbon-hydrogen molecules); fatty acids form the tail region of triglyceride fat molecules.

female (9) In sexually reproducing organisms, a member of the sex that produces the larger gamete.

fermentation (4) The process by which glycolysis occurs in the absence of oxygen; the electron acceptor is pyruvate (in animals) or acetaldehyde (in yeast) rather than oxygen.

fertilization (6, 25) The fusion of two reproductive cells.

fetus (25) Developmental stage in humans, from the end of the embryonic period, approximately eight weeks after fertilization, until birth.

fever (26) An elevated body temperature in response to infection; enhances the immune response and slows the growth of pathogens.

fibrous roots (17) A root system in which numerous roots branch out directly from the plant stem; found primarily in monocots.

filament (12, 18) The supporting stalk of the anther of a stamen found in angiosperm flowers. [Lat., *filum,* thread]

filtrate (20) Fluid that accumulates in Bowman's capsule in the vertebrate kidney; it contains salts, sugars, amino acids, vitamins, and many other molecules, all at the same concentration as in the blood.

first law of thermodynamics (4) A physical law that states that energy cannot be created or destroyed; it can only change from one form to another.

fitness (8) A relative measure of the reproductive output of an individual with a given phenotype compared with the reproductive output of individuals with alternative phenotypes.

fixation (8) The point at which the frequency of an allele in a population is 100% and, therefore, there is no more variation in the population for this gene.

fixed action pattern (9) An innate sequence of behaviors, triggered under certain conditions, that requires no learning, does not vary, and once begun runs to completion; an example is egg-retrieval in geese.

flagella (*sing.* flagellum) (3) Long, thin, whip-like projections from the cell body of a prokaryote that aid in cell movement through the medium in which the organism lives; in animals, the only cell with a flagellum is the sperm cell. [Lat., *flagellum,* whip]

flatworms (11) Worms with flat bodies that are members of the phylum Platyhelminthes; characterized by well-defined head and tail regions, with some having clusters of light-sensitive cells for eyespots; most are hermaphroditic and are protostomes that do not molt; examples are tapeworms and flukes.

fleshy fruit (12) A fruit that consists of the ovary and some additional parts of the flower; when fleshy fruits, an attractive food, are eaten by animals, the seeds may be widely dispersed.

flower (12) The part of an angiosperm that contains the reproductive structures; the flower consists of a supporting stem with modified leaves (the petals and sepals) and usually contains both male and female reproductive structures.

fluid mosaic (3) A term that describes the structure of the plasma membrane, which is made up of several different types of molecules, many of which are not fixed in place but float, held in proper orientation by hydrophilic and hydrophobic forces.

fMRI scan (23) Functional magnetic resonance imaging scan; a brain imaging technology. A giant magnet detects the relative amounts of oxygenated and deoxygenated blood in different parts of the brain, which vary with brain activity.

follicle (25) In an ovary, the small structure in which an egg forms.

follicle-stimulating hormone (FSH) (25) A hormone that stimulates the production of eggs.

follicular phase (25) The first half of the reproductive cycle in women, culminating in ovulation.

food chain (15) The path of energy flow from primary producers to tertiary consumers.

food web (15) A more precisely described path of energy flow from primary producers to tertiary consumers than the food chain, reflecting the fact that many organisms are omnivores and occupy more than one position in the chain.

forebrain (23) The largest region of the vertebrate brain; includes two control and relay structures, the thalamus and hypothalamus, and the cerebrum.

fossil (8) The remains of an organism, usually its hard parts such as shell, bones, or teeth, which have been naturally preserved; also, traces of such an organism, such as footprints. [Lat., *fossilis,* that which is dug up]

fossil fuels (4) Fuels produced from the decayed remains of ancient plants and animals; fossil fuels include oil, natural gas, and coal.

founder effect (8) A change in the allele frequencies of a population resulting from the isolation of a small subgroup of a larger population; all the descendants of the smaller group will reflect the allele frequencies of the subgroup, which may be different from those of the larger source population; the founder effect is one cause of genetic drift.

fruit (18) The mature ovary of a flower that houses seeds for dispersal. See also **fleshy fruit.**

fundamental niche (15) The full range of environmental conditions under which an organism can live.

G

gametes (6, 25) Haploid cells from two individuals that, as sperm and egg, will combine at fertilization to produce offspring; also called reproductive cells. [Gk., *gamete,* wife]

gametophyte (12) The plant structure that produces gametes (sperm and eggs); the haploid life stage of plants and some algae, which may be either male (producing sperm) or female (producing eggs). [Gk., *gamete,* wife + *phytas,* plant]

gap junction (3) A junction between adjacent animal cells in the form of a pore in each of the plasma membranes surrounded by a protein that links the two cells and acts like a channel between them to allow materials to pass between the cells.

gastropods (11) Molluscs that are members of the class Gastropoda; most have a single shell, a muscular foot for locomotion, and a radula used for scraping food from surfaces; examples are snails and slugs. [Gk., *gaster,* belly + *pous,* foot]

gastrula (25) The mass of cells, made up of three layers, formed during the gastrulation phase of embryonic development.

gastrulation (25) The second phase of embryonic development in which cells form distinct layers.

gene (5) The basic unit of heredity; a sequence of DNA nucleotides on a chromosome that carries the information necessary for making a functional product, usually a protein or an RNA molecule. [Gk., *genos,* race, descent]

gene flow (8) A change in the allele frequencies of a population due to movement of some individuals of a species from one population to another, changing the allele frequencies of the population they join; also known as migration.

gene library (5) A collection of cloned DNA fragments; also known as a clone library.

gene therapy (5) A therapy designed to treat or cure a disease by insertion of a functional gene to replace a defective version of that gene.

genetic drift (8) A random change in allele frequencies over successive generations; a cause of evolution.

genetic engineering (5) The manipulation of an organism's genetic material by adding, deleting, or transplanting genes from one organism to another.

genome (5) The full set of DNA present in an individual organism; also can refer to the full set of DNA present in a species.

genotype (5, 7) The genes that an organism carries for a particular trait; also, collectively, an organism's genetic composition. [Gk., *genos,* race, descent + *typos,* impression, engraving]

genus (*pl.* genera) (10) In the system developed by Carolus Linnaeus (1707–1778), a classification of organisms consisting of closely related species. [Lat., *genus,* race, family, origin]

germ layers (25) The three distinct layers of cells formed during gastrulation.

germinate (18) In plant spores or seeds, to begin growing after a period of dormancy.

gibberellins (19) In plants, a large group of hormones that regulate a plant's growth processes, primarily by stimulating cell division and cell elongation.

gills (21) Organs in fishes and other aquatic animals in which gases are exchanged between water and blood capillaries.

gland (20) A collection of epithelial cells that produces secretions for use elsewhere in the body.

glial cells (20, 23) Cells of nervous tissue that support and provide nutrients to neurons. Also called *neuroglia.*

glomerulus (20) The blood-filtering unit of the nephron; a mass of capillaries, surrounded by Bowman's capsule and connected to a single, long, urine-collecting tube that excretes its filtered fluid into a collecting duct.

glycemic index (22) A measure of the extent to which foods cause a surge in blood sugar and subsequent release of insulin.

glycerol (2, 3) A small molecule that forms the head region of a triglyceride fat molecule. [Gk., *glykys,* sweet + *-ol,* chemical suffix for an alcohol]

glycogen (2) A complex carbohydrate consisting of stored glucose molecules linked to form a large web, which breaks down to release glucose when it is needed for energy. [Gk., *glykys,* sweet + Gk., *genos,* race, descent]

glycolysis (4) In all organisms, the first step in cellular respiration, in which one molecule of glucose is broken into two molecules of pyruvate; for some organisms glycolysis is the only means of extracting energy from food, and in most organisms it is followed by the Krebs cycle and the electron transport chain. [Gk., *glykys,* sweet + *lysis,* releasing]

Golgi apparatus (3) An organelle, part of the endomembrane system, structurally like a flattened stack of unconnected membranes, each known as a Golgi body; the Golgi apparatus processes molecules synthesized in the cell and packages those molecules that are destined for use elsewhere in the body. [*pron.* GOHL-jee; from the name of the discoverer, Camillo Golgi, 1843–1926]

gonads (6) The ovaries and testes in sexually reproducing animals. [Gk., *gonē,* offspring]

gram stain (13) A test used by microbiologists in identifying an unknown bacterium; the dye that is used stains the layer of peptidoglycan outside the cell wall purple, but those bacteria in which the layer of peptidoglycan is covered by a membrane are not colored by the dye; bacteria that take a Gram stain are known as Gram-positive bacteria, those that do not are known as Gram-negative bacteria.

gravitropism (19) Growth in plants in response to the pull of gravity. [L., *gravidus,* heavy + Gk., *tropikos,* turn]

ground tissue (17) Those parts of a plant that are not dermal tissue or vascular tissue; makes up the bulk of the plant and is where most of the plant's metabolic activities take place.

group selection (9) The process, extremely uncommon in nature, which brings about an increase in the frequency of alleles for traits (e.g., behaviors) that are beneficial to the persistence of the species or population while simultaneously being detrimental to the fitness of the individual possessing the trait (or engaging in the behavior).

growth rate (*r*) (14) The birth rate minus the death rate; the change in the number of individuals in a population per unit of time.

guard cell (17) One of the pair of plant cells that surround and control the opening of stomata, through which carbon dioxide and water can pass.

gymnosperms (12) Vascular plants that do not produce their seeds in a protective structure, but are usually found on the surface of the scales of a cone-like structure; the gymnosperms include conifers, cycads, gnetophytes, and ginkgo. [Gk., *gymnos,* naked + *sperma,* seed]

H

habitat (15) The physical environment of organisms, consisting of the chemical resources of the soil, water, and air, and physical conditions such as temperature, salinity, humidity, and energy sources. [Lat., *habitare,* to dwell or inhabit]

hair (11) Dead cells filled with the protein keratin that collectively serve as insulation covering the body or a part of the body; present in all mammals.

haploid (6) Describes cells that have a single copy of each chromosome (in many species, including humans, gametes are haploid). [Gk., *haploeides,* single]

hazard factor (14) An external force on a population that increases the risk of death.

heart (21) A muscular pump that, with each contraction, propels blood at high pressure to lungs, gills, or other body organs and tissues.

heat stroke (20) A potentially life-threatening medical condition caused by extreme hyperthermia, characterized by an inability of the organism to maintain homeostasis.

heliotropism (19) In plants, a growth in response to a light source in which leaves and flowers orient themselves in relationship to the sun's rays. [Gk., *helios,* sun + *tropikos,* turn]

helper T cell (26) A type of lymphocyte that stimulates B cells to produce antibodies and cytotoxic T cells to kill infected cells.

hematocrit (21) The proportion of blood that is made up from red blood cells; determined by spinning a blood sample in a centrifuge.

hemoglobin (21) An oxygen-carrying protein molecule in red blood cells. [Gk., *haima*, blood + Lat., *globus*, ball]

hemolymph (21) The single fluid of an open circulatory system that surrounds all cells and transports nutrients, gases, and waste products.

herbivores (15, 22) Animals that eat plants; also known as primary consumers. [Lat., *herba*, grass + *vorare*, to devour]

heredity (7) The greater resemblance of offspring to parents than to other individuals in the population, a consequence of the passing of characteristics from parents to offspring through their genes. [Lat., *heres*, heir]

heritability (8) The transmission of traits from parents to offspring via genetic information; also known as inheritance.

hermaphrodite (6) An organism that produces both male and female gametes. [from the names of the Greek god Hermes and goddess Aphrodite]

heterotherm (20) An animal that has a body temperature that fluctuates as the environmental temperature changes. [Gk., *heteros*, other + *thermē*, heat]

heterozygous (7) Describes the genotype of a trait for which the two alleles an individual carries differ from each other. [Gk., *heteros*, other + *zeugos*, pair]

hibernate (20) To go into a state of reduced metabolic activity for days or weeks, during which the animal's body temperature can drop considerably.

hindbrain (23) The region of the vertebrate brain that includes the medulla, pons, and cerebellum.

hippocampus (23) Part of the brain's limbic system; functions in long-term memory formation.

histamine (26) A molecule released by mast cells that assists in the inflammatory response by causing non-injured blood vessels to dilate and become leaky.

histones (6) Proteins around which the long, linear strands of DNA are wrapped; the histones serve to keep the DNA untangled and to enable an orderly, tight, and efficient packing of the DNA within the cell.

homeostasis (20) The body's use of physical and chemical processes to maintain a consistent internal environment. [Gk., *homos*, same + *stasis*, standing]

homeotherm (20) An animal that maintains a relatively constant body temperature. [Gk., *homos*, same + *thermē*, heat]

homologous (10) Describes features that are inherited from a common ancestor. [Gk., *homologia*, agreement]

homologous pair (homologues) (6) The maternal and paternal copies of a chromosome. [Gk., *homologia*, agreement]

homologous structures (8) Body structures in different organisms that, although modified extensively over time to serve different functions in different species, are due to inheritance from a common evolutionary ancestor.

homozygous (7) Describes the genotype of a trait for which the two alleles are the same. [Gk., *homos*, same + *zeugos*, pair]

honest signal (9) A signal, which cannot be faked, that is given when both the individual making the signal and the individual responding to it have the same interests; it carries the most accurate information about an individual or situation.

horizontal gene transfer (10) The transfer of genetic material directly from one individual to another not necessarily related individual.

hormone (19, 24) A chemical signal that responds to environmental variables, found in both plants and animals. In plants, hormones are produced in various locations and may have their effect in that location or may be transported to another part of the plant to regulate the plant's activities. In animals, hormones are usually secreted by endocrine glands and are transported by the bloodstream to target cells as part of an internal communication and regulation system. [Gk., *hormao*, that which sets in motion or urges forward]

host (13, 15) An organism in or on which a parasite lives.

human chorionic gonadotropin (hCG) (25) A hormone secreted by the embryo that keeps the lining of the uterus thickened for implantation.

Human Genome Project (5) A project to decode the three billion base pairs in the human genome and to identify all genes present in it.

human immunodeficiency virus (HIV) (13, 26) The virus responsible for AIDS, a deadly disease that destroys the human immune system. HIV is a retrovirus, a RNA-containing virus that is thought to have been introduced to humans from chimpanzees.

humus (17) (*pron.* HYOO-muss) The decomposing remains of plants and animals found in the uppermost region of soil; absorbs and releases water easily and releases many nutrients.

hybridization (genetics) (10) The interbreeding of closely related species. [Lat., *hybrida*, animal produced by two different species]

hybridization (molecular biology) (5) The process of joining two complementary strands of DNA from different sources; hybridization occurs when a DNA probe is used to match with a complementary sequence in a gene library. [Lat., *hybrida*, animal produced by two different species]

hybrids (10) Offspring of individuals of two different species. [Lat., *hybrida*, animal produced by two different species]

hydrogen bond (2) A type of weak chemical bond formed between the slightly positively charged hydrogen atoms of one molecule and the slightly negatively charged atoms of another (often oxygen or nitrogen atoms); hydrogen bonds are important in building multi-atom molecules, such as complex proteins, and are responsible for many of the unique and important features of water.

hydrophilic (2, 3) Attracted to water, as, for example, polar molecules that readily form hydrogen bonds with water. [Lat., *hydro-*, pertaining to water; Gk., *philios*, loving]

hydrophobic (2, 3) Repelled by water, as, for example, non-polar molecules that tend to minimize contact with water. [Lat., *hydro-*, pertaining to water, Gk., *phobos*, fearing]

hydroponically grown plant (17) A plant grown without soil, but in water enriched with essential plant nutrients.

hypertonic (3) Of two solutions, that with a higher concentration of solutes. [Gk., *hyper*, above + *tonos*, tension]

hyphae (*sing.* hypha) (12) (*pron.* HIGH-fee) Long strings of cells that make up the mycelium of a multicellular fungus. [Gk., *hypha*, web]

hyperthermia (20) A condition in which high external environmental temperature and humidity overwhelm the body's ability to dissipate heat, and body temperature becomes abnormally high; can lead to death. [Gk., *hyper*, above + *thermē*, heat]

hypothalamus (23, 24) A structure on the underside of the brain that functions as a liaison between the nervous and endocrine systems and serves as one of the chief regulatory centers of the brain. It regulates many fundamental drives, including hunger and thirst, sexual activity, and maintenance of body temperature. It also controls the hormone secretions of the pituitary gland, located next to it. [Gk., *hypo*, under + *thalamos*, inner room]

hypothermia (20) A condition in which body temperature becomes abnormally low; can lead to death. [Gk., *hypo,* under + *thermē,* heat]

hypothesis (*pl.* hypotheses) (1) A proposed explanation for an observed phenomenon. [Gk., *hypothesis,* a proposal]

hypotonic (3) Of two solutions, that with a lower concentration of solutes. [Gk., *hypo,* under + *tonos,* tension]

I

immune system (26) Body system consisting of disease-fighting white blood cells and external barriers such as skin, providing protection against an enormous variety of pathogens, such as bacteria, fungi, some viruses, parasitic protists, and worms.

immunity (26) State of long-term protection against a specific disease-causing microorganism or virus.

inclusive fitness (9) The sum of an individual's indirect and direct fitness.

incomplete dominance (7) The case in which the heterozygote has a phenotype intermediate between those of the two homozygotes; an example is pink snapdragons, whose appearance is intermediate between homozygous for white flowers and homozygous for red flowers.

indirect fitness (9) The reproductive output that an individual brings about through apparently altruistic behaviors toward genetic relatives.

infertility (25) The inability of a couple to get pregnant after one year of trying.

inflammatory response (26) The recruitment of phagocytes and other immune cells to assist with pathogen destruction, followed by tissue healing. Signs of the inflammatory response include redness, heat, swelling, and pain.

ingestion (22) The intake of food via the mouth, teeth, tongue, and esophagus.

inheritance (8) The transmission of traits from parents to offspring via genetic information; also known as heritability.

innate behaviors (9) Behaviors that do not require environmental input for their development; innate behaviors are present in all individuals in a population and do not vary much from one individual to another or over an individual's life span; also known as instincts. [Lat., *innatus,* inborn]

instincts (9) Behaviors that do not require environmental input for their development; instincts are present in all individuals in a population and do not vary much from one individual to another or over an individual's life span; also known as innate behaviors. [Lat., *instinctus,* impelled]

integumentary system (26) Body system consisting of skin, sweat glands, oil glands, hair, and nails that protects the body from mechanical stress caused by external forces such as friction or pressure, sunlight, dehydration, and pathogens. [Lat., *integere,* to cover]

interferon (26) An important cytokine produced by cells infected by a virus; alerts other cells to turn on protective measures that help them resist infection.

intermembrane space (3) In a mitochrondrion, the region between the inner and outer membranes. [Lat., *inter,* between + *membrana,* a thin skin]

internal fertilization (25) The process in which sperm are deposited in the female's reproductive tract and unite with one or more eggs.

interneuron (23) A type of neuron that acts as a middleman, integrating the signals coming in from sensory neurons and relaying them to a motor neuron; located in the brain and spinal cord.

interphase (6) In the cell cycle, the phase during which the cell grows and functions; during this phase, replication of DNA occurs in preparation for cell division. [Lat., *inter,* between + Gk., *phasis,* appearance]

interstitial fluid (20) An extracellular fluid that surrounds and bathes cells; consists mainly of water and also contains nutrients, raw materials, and waste products.

intron (5) A non-coding region of DNA.

invagination (3) The folding in of a membrane or layer of tissue so that an outer surface becomes an inner surface. [Lat., *in,* in + *vagina,* sheath]

ion (2) An atom that carries an electrical charge, positive or negative, because it has either gained or lost an electron or electrons from its normal, stable configuration. [Gk., *ion,* going]

ionic bond (2) A bond created by the transfer of one or more electrons from one atom to another; the resulting atoms, now called ions, are charged oppositely and so attract each other to form a compound.

iris (23) The colored portion of the vertebrate eye with an opening, called the pupil, that controls the amount of light reaching the eye's interior.

isotonic (3) Refers to solutions with equal concentrations of solutes. [Gk., *isos,* equal to + *tonos,* tension]

K

karyotype (6) A visual display of an individual's full set of chromosomes. [Gk., *karyon,* nut or kernel + *typos,* impression, engraving]

keystone species (15) A species that has an unusually large influence on the presence or absence of numerous other species in a community.

kilocalorie (kcal) (22) One thousand calories.

kin selection (9) Kindness toward close relatives, which may evolve as apparently altruistic behavior toward them, but which in fact is beneficial to the fitness of the individual performing the behavior.

kinetic energy (4) The energy of moving objects, such as legs pushing the pedals of a bicycle or wings beating against the air. [Gk., *kinesis,* motion]

kingdom (10) In the system developed by Carolus Linnaeus (1707–1778), one of the three categories—animal, plant, and mineral—into which all organisms and substances on earth were placed. In modern classification, there are six kingdoms: bacteria, archaea, protists, plants, animals, and fungi.

Krebs cycle (4) The second step of cellular respiration, in which energy is extracted from sugar molecules as additional molecules of ATP and NADH are formed. [from the name of the discoverer, Hans Adolf Krebs, 1900–1981]

L

labia (*sing.* labium) (25) Protective folds of skin surrounding the female genitals; develops from the same embryonic tissue that produces the scrotum.

labor (25) During childbirth, a series of contractions of the uterus.

lactation (25) Milk production in the mammary glands.

lamella (*pl.* lamellae) (21) Disk-like structure in gills, with elaborately branched capillaries.

landscape conservation (16) The conservation of habitats and ecological processes as well as species.

language (9, 23) A type of communication in which arbitrary symbols represent concepts and grammar; a system of rules dictates the way the symbols can be manipulated to communicate and express ideas.

large intestine (22) The last part of the vertebrate digestive system, larger in diameter but shorter in length than the small intestine. It serves to absorb water, salts, and some vitamins and, in its final compartment (the rectum), to store the indigestible parts of consumed food and symbiotic bacteria (the feces), which can then be defecated. Also called *colon*.

larva (11) In complete metamorphosis, the first stage of insect development; the larva is hatched by the egg, which eats to grow large enough to enter the pupa stage; the larva (for example, a caterpillar) looks completely different from the adult (a butterfly or moth). [Lat., *larva,* ghost]

lateral meristem (17) Growth region of a plant consisting of a layer of cells called cambium; division in this meristem causes a stem or root to become thicker, as opposed to longer.

leaf (17) The chief site of photosynthesis in most plants; leaves and stems form the shoot system.

learning (9) The alteration and modification of behavior over time in response to experience.

left hemisphere (23) The region of the brain that specializes, in most (but not all) humans, in language, logic, and mathematical skills.

leukocytes (21) See **white blood cells.**

lichens (12) Symbiotic partnership between fungi and chlorophyll-containing algae or cyanobacteria, or both.

life (10) A physical state characterized by the ability to replicate and the presence of metabolic activity.

life cycle (18) The series of stages experienced by an individual of a species throughout its development, each stage following the preceding stage through reproduction.

life history (14) The vital statistics of a species, including age at first reproduction, probabilities of survival and reproduction at each age, litter size and frequency, and longevity.

ligament (20) A connective tissue that binds bone to bone.

light energy (4) A type of kinetic energy made up of energy packets called photons, which are organized into waves.

lignin (17) A chemical in the cell walls of sclerenchyma cells of woody plants; it is indigestible to nearly all organisms and gives sclerenchyma cells their strength.

limbic system (23) The brain region that includes parts of the cerebral cortex and two structures deeper in the center of the brain, the hippocampus and the amygdala. It is responsible for many physiological drives and instincts.

linked genes (7) Genes that are close to each other on a chromosome, and so are more likely than others to be inherited together.

lipid (2) One of four types of macromolecules, lipids are insoluble in water and greasy to the touch; they are important in energy storage and insulation (fats), membrane formation (phospholipids), and regulating growth and regulating growth and development (sterols). [Gk., *lipos,* fat]

liver (22) An organ in vertebrates and some other animals that is important in the detoxification of toxic molecules, storage of glycogen, and synthesis of bile, hormones, and digestive enzymes.

lobe-finned fishes (11) Fish species characterized by two pairs of sturdy lobe-shaped fins on the underside of the body.

logistic growth (14) A pattern of population growth in which initially exponential growth levels off as the environment's carrying capacity is approached.

long-term depression (23) Long-lasting decrease in the sensitivity of a neuron, resulting from a period of intense stimulation.

long-term potentiation (23) Long-lasting increase in the sensitivity of a neuron, resulting from a period of intense stimulation.

lungs (21) Internal organs in most land vertebrates, with highly branched, moist respiratory surfaces where gases are exchanged between air and blood.

luteal phase (25) The second half of the reproductive cycle in women in which the follicle cell that had surrounded the ovum develops into the corpus luteum; culminates in pregnancy or the sloughing off of the uterine lining.

lymph (21) A clear fluid formed from interstitial fluid as it is filtered into the lymphatic vessels in vertebrates.

lymph nodes (21) In the lymphatic system, patches of connective tissue filled with pathogen-fighting white blood cells, through which lymph passes.

lymphocyte (26) A type of white blood cell that carries out the response of the specific division of the immune system.

lysosome (3) A round, membrane-enclosed, enzyme- and acid-filled vesicle in the cell that digests and recycles cellular waste products and consumed material. [Gk., *lysis,* releasing + *soma,* body]

lysozyme (26) Enzyme produced by skin cells and in tears that kills bacteria by damaging their cell walls.

M

macroevolution (10) Large-scale products of evolutionary change involving the origins of new groups of organisms; the accumulated effect of microevolution over a long period of time. [Gk., *macros,* large + Lat., *evolvere,* to roll out]

macromolecule (2) A large molecule, made up of smaller building blocks or subunits; four types of biological macromolecules are carbohydrates, lipids, proteins, and nucleic acids. [Gk., *macros,* large + Lat., *dim.* of *moles,* mass]

macrophage (26) A type of large ("macro") white blood cell that resides in tissues and ingests small organisms and large debris such as dead cells. Macrophages serve as an important link between the non-specific and specific divisions of the immune system.

male (9) In sexually reproducing organisms, a member of the sex that produces the smaller gamete.

mammary glands (11) Glands in all female mammals that produce milk for the nursing of young. [Lat., *mamma,* breast]

marsupials (11) Mammals in which, in most species, after a short period of embryonic life in the uterus, the young complete their development in a pouch in the female. [Gk., *marsipos,* pouch]

mass (2) The amount of matter in a given sample of a substance.

mass extinctions (10) Extinctions in which a large number of species become extinct in a short period of time, usually because of extraordinary and sudden environmental change. [Lat., *extinguere,* to extinguish]

mast cell (26) A type of white blood cell found in tissues that triggers the inflammatory response by releasing histamine.

mate guarding (9) Behavior by an individual that reduces the opportunity for that individual's mate to interact with other potential mates.

mating system (9) The pattern of mating behavior in a species, ranging from polyandry to monogamy to polygyny; mating systems are influenced by the relative amounts of parental investment by males and by females.

matrix (connective tissue) (20) A mass of non-living, extracellular material in which connective tissue cells are embedded.

matrix (mitochondrial) (3, 4) The space within the inner membrane, where the carriers NADH and $FADH_2$ begin the electron transport chain by carrying high-energy electrons to molecules embedded in the inner membrane.

maximum sustainable yield (14) The point at which the maximum number of individuals are removed from a population without impairing its growth rate, it occurs at half the carrying capacity.

mechanoreceptor (23) A type of sensory cell that responds to mechanical pressure.

medulla (23) The part of the vertebrate hindbrain that connects to the spinal cord; controls functions such as respiration, heart rate, and digestion.

meiosis (6) In sexually reproducing organisms, a process of nuclear division in the gonads that, along with cytokinesis, produces reproductive cells that have half as much genetic material as the parent cell, and that all differ from each other genetically. [Gk., *meioun,* to lessen]

memory cell (26) Long-lived B cells or T cells, produced in the primary immune response to a particular antigen, that enable rapid response to subsequent exposure to the same antigen.

Mendel's law of independent assortment (7) Allele pairs for different genes separate independently in meiosis, so the inheritance of one trait generally does not influence the inheritance of another trait (the exception, unknown to Mendel, occurs with linked genes). [from the name of its discoverer, Gregor Mendel, 1822–1884]

Mendel's law of segregation (7) During the formation of gametes, the two alleles for a gene separate, so that half the gametes carry one allele, and half of the gametes carry the other. [from the name of its discoverer, Gregor Mendel, 1822–1884]

menopause (25) The permanent cessation of ovulation and menstruation, prior to the end of an individual's life.

menstrual cycle (25) The cycle in which the uterus prepares for the implantation and nurturing of a fertilized egg, and sheds its lining if fertilization does not occur.

menstruation (25) The shedding of the uterine lining.

meristem (17, 18) Growth region of a plant, made up of undifferentiated cells that divide to form additional undifferentiated cells as well as other cells that differentiate to form specific plant tissues.

mesoderm (25) The middle embryonic cell layer during gastrulation; it eventually forms the muscles and skeleton. [Gk., *mesos,* middle + *derma,* skin]

messenger RNA (mRNA) (5) The ribonucleic acid that "reads" the sequence for a gene in DNA and then moves from the nucleus to the cytoplasm, where the next stage of protein synthesis will take place.

metamorphosis (11) The rebuilding of molecules from the larva stage to the adult, resulting in a change of form. [Gk., *metamorphoun,* to transform]

metaphase (6) The second phase of mitosis, in which the sister chromatids line up at the center of the cell; in meiosis, the homologues line up at the center of the cell in metaphase I and the sister chromatids line up in metaphase II. [Gk., *meta,* in the midst of + *phasis,* appearance]

microbe (10, 13) A microscopic organism; not a monophyletic group since it includes protists, archaea, and bacteria. [Gk., *micros,* small + *bios,* life]

microevolution (10) A slight change in allele frequencies in a population over one or a few generations. [Gk., *micros,* small + Lat., *evolvere,* to roll out]

microsphere (10) A membrane-enclosed, small, spherical unit containing a self-replicating molecule and carrying information, although no genetic material; microspheres may have been an important stage in the development of life. [Gk., *micros,* small + *sphaira,* ball]

midbrain (23) Part of the brainstem; a sensory integration and relay center through which most of the sensory information and motor neuron connections coming into and out of the brain pass.

migration (8) A change in the allele frequencies of a population due to the movement of some individuals from one population to another; an agent of evolutionary change caused by the movement of individuals into or out of a population. [Lat., *migrare,* to move from place to place]

mimicry (15) The evolution of an organism to resemble another organism or object in its environment to help conceal itself from predators. [Gk., *mimesis,* imitation]

mineral (22) A chemical element other than those commonly found in organic molecules (carbon, hydrogen, oxygen, and nitrogen); some minerals are required in the diet in small amounts.

mitochondrial matrix (4) See **matrix.**

mitochondrion (*pl.* mitochondria) (3) The organelle in eukaryotic cells that converts the energy stored in food in the chemical bonds of carbohydrate, fat, and protein molecules into a form usable by the cell for all its functions and activities. [Gk., *mitos,* thread + *chondros,* cartilage]

mitosis (6) The division of a nucleus into two genetically identical nuclei, along with cytokinesis, leads to the formation of two identical daughter cells. [Gk., *mitos,* thread + *phasis,* appearance]

mitotic phase (M phase) (6) The phase of the cell cycle during which first the genetic material and nucleus, and then the rest of the cellular contents, divide.

molecule (2) A group of atoms held together by covalent bonds. [Lat., *dim.* of *moles,* mass]

monocot (17) One of the two major groups of flowering plants; monocot seedlings have one cotyledon (seed leaf).

monogamy (9) A mating system in which most individuals mate and remain with just one other individual. [Gk., *monos,* single + *gamete,* wife]

monophyletic (10) A group containing a common ancestor and all of its descendants. [Gk., *monos,* single + *phylon,* race, tribe, class]

monosaccharides (2) The simplest carbohydrates and the building blocks of more complex carbohydrates; monosaccharides, which cannot be broken down into other monosaccharides, include glucose, fructose, and galactose; also known as simple sugars. [Gk., *monos,* single + *sakcharon,* sugar]

monotremes (11) Present-day mammals that retain the ancestral condition of laying eggs; monotremes are so called because they have a single duct , the cloaca, which the reproductive system, the urinary system, and the digestive system (for defecation) open into. [Gk., *monos,* single + *trema,* hole]

mood (24) A relatively long-lasting emotional state (shorter-lasting than a person's temperament, but longer-lasting and less specific than a single emotion).

morphological species concept (10) A concept that defines species on the basis of physical features such as body size and shape. [Gk., *morphē,* shape]

motor neuron (23) A type of neuron that stimulates action by conveying signals to muscles or glands and initiating a body's response to stimuli.

multiple allelism (7) The case in which a single gene has more than two possible alleles.

multiple sclerosis (MS) (26) Autoimmune disease in which the body attacks the myelin sheath surrounding nerve fibers of the brain and spinal cord, causing blurred vision, weakness in arms and legs, and trouble with balance.

muscle fiber (20) An individual muscle cell; typically contains multiple nuclei.

muscle tissue (20) A body tissue consisting of contractile cells (muscle fibers) that generate force and can facilitate movement.

mutagen (8) An agent capable of causing a mutation in DNA; may be chemical (as some insecticides) or physical (as with the energy from radiation). [Lat., *mutare,* to change + *genus,* race, family, origin]

mutation (5, 8) An alteration in the base-pair sequence of an individual's DNA; may arise spontaneously or following exposure to a mutagen. [Lat., *mutare,* to change]

mutualism (15) A symbiotic relationship in which both species benefit and neither is harmed. [Lat., *mutuus,* reciprocal]

mycelium (12) A mass of interconnecting hyphae that make up the structure of a multicellular fungus. [Gk., *mykes,* fungus]

mycorrhizae (12, 17) (*pron.* my-ko-RYE-zay) Root fungi, that is, symbiotic associations between roots and fungi in which fungal structures are closely associated with fine rootlets and root hairs. [Gk., *mykes,* fungus + *rhiza,* root]

myelin sheath (23) Fatty coating insulating the axon.

myofibril (23) A cylindrical organelle within muscle cells that can contract; contains repeating units, called sarcomeres, in which the contraction takes place.

myoglobin (21) An oxygen-binding protein in muscle that releases oxygen when demand is high and the partial pressure of oxygen is low.

myosin (23) A protein of muscle tissue; makes up the thick filaments.

N

NADPH (4) A molecule (nicotinamide adenine dinucleotide phosphate) that is a high-energy electron carrier involved in photosynthesis, which stores energy by accepting high-energy protons; NADPH is formed when the electrons released from the splitting of water are passed to $NADP^+$.

natural killer (NK) cell (26) A type of white blood cell that kills body cells infected by a pathogen.

natural selection (8) A mechanism of evolution that occurs when there is heritable variation for a trait and individuals with one version of the trait have greater reproductive success than do individuals with a different version of that trait.

nectar (18) A solution produced by a flower, rich in sugars and amino acids, that can serve as an attractant to animal pollinators.

negative feedback (20) A control mechanism in which sensors detect changes in the internal environment and trigger effectors to counteract the change; one of the chief strategies by which organisms maintain homeostasis.

nephron (20) The chief functional unit in the vertebrate kidney, consisting of a tubule and a mass of blood vessels that work together to accomplish the tasks of filtration, reabsorption, and excretion.

nerve (23) A group of neurons bundled together with connective tissue.

nervous system (23) Network of cells (neurons), present in all multicellular animals other than sponges, that collects information about the organism's internal and external environments, processes that information, and sends signals to muscles and glands in response to the information.

nervous tissue (20) A body tissue that specializes in storing and transmitting information.

neural tube (25) A long hollow tube formed during neurulation that ultimately develops into the nervous system.

neuroglia (20) See **glial cells.**

neuron (20, 23) The "excitable" cell within the nervous system that receives and transmits signals; made up of three distinctive elements: dendrites, a cell body, and an axon.

neurotransmitter (23) A chemical of the nervous system that transmits signals to adjacent cells.

neurulation (25) During embryonic development, the process of the folding in of the ectoderm for the entire length of the embryo, first forming a groove and then becoming the neural tube.

neutron (2) An electrically neutral particle in the atomic nucleus. [Lat., *neutro,* in neither direction]

neutrophil (26) A type of white blood cell that circulates in the blood and ingests small organisms, primarily bacteria. Neutrophils make up 50% to 70% of all white blood cells.

niche (15) The way an organism utilizes the resources of its environment, including the space it requires, the food it consumes, and timing of reproduction.

nitrogen fixation (17) Process by which the bonds between paired nitrogen atoms in molecules of nitrogen (N_2) are broken and the individual atoms are converted into a form, such as ammonia, that is usable by plants.

node (phylogeny) (10) The point on an evolutionary tree at which species diverge from a common ancestor. [Lat., *nodus,* knot]

node (plant structure) (17) A small lump of tissue on a stem from which a leaf, flower, cone, or additional stem (or branch) may grow.

nodule (17) In plant roots, a round lump containing plant cells and nitrogen-fixing bacteria.

nondisjunction (6) The unequal distribution of chromosomes during cell division; can lead to Down syndrome and other disorders caused by the possession in an individual of too few or too many chromosomes.

non-essential amino acid (22) An amino acid that can be produced by the body and so is not needed in the diet.

non-specific immunity (26) Division of the immune system that, through various types of non-specific immune cells, rapidly recognizes pathogens that breach external barriers and are not part of the body's own tissue, but does not distinguish between specific pathogens. Also called the *innate system* (meaning "inborn"), because it is functioning at birth and does not require any previous exposure to pathogens to act.

non-polar (3) Electrically uncharged.

non-vascular plants (12) Plants that do not have vessels to transport water and dissolved nutrients, but rely on diffusion; bryophytes are non-vascular plants. [Lat., *vasculum, dim.* of *vas,* vessel]

notochord (11) A rod of tissue from head to tail that stiffens the body when muscles contract during locomotion; primitive chordates retain the notochord throughout life, but in advanced chordates it is present only in early embryos and is replaced by the vertebral column. [Gk., *notos,* back + Lat., *chorda,* cord]

notochord (25) A flexible rod in chordates whose function in vertebrates is ultimately taken over by the backbone.

nuclear membrane (3) A membrane that surrounds the nucleus of a cell, separating it from the cytoplasm, consisting of two bilayers and perforated by pores enclosed in embedded proteins that allow the passage of large molecules from nucleus to cytoplasm and from cytoplasm to nucleus; also called the nuclear envelope.

nucleic acid (2, 5) One of the four types of biological macromolecules, the nucleic acids DNA and RNA store genetic information in unique sequences of nucleotides.

nucleolus (3) An area near the center of the nucleus where subunits of the ribosomes are assembled. [Lat., *nucleolus, dim.* of *nucleus,* kernel, small nut]

nucleotide (2, 5) A molecule containing a phosphate group, a sugar molecule, and a nitrogen-containing molecule; nucleotides are the individual units that together, in a unique sequence, constitute a nucleic acid.

nucleus (cell biology) (3) A membrane-enclosed structure in eukaryotic cells that contains the organism's genetic information as linear strands of DNA in the form of chromosomes. [Lat., *nucleus, dim.* of *nux,* nut]

nucleus (chemistry) (2) The central and most massive part of an atom, usually made up of two types of particles, protons and neutrons, which move about the nucleus. [Lat., *nucleus, dim.* of *nux,* nut]

null hypothesis (1) A hypothesis that proposes a lack of relationship between two factors. [Lat., *nullus,* none]

nuptial gift (9) A food item or other item presented to a potential mate as part of courtship. [Lat., *nuptialis,* of marriage]

nutrient (22) A substance used for energy, raw materials, and maintenance of the body's systems.

O

omnivores (15, 22) Animals that eat both plants and other animals and thus can occupy more than one position in the food chain. [Lat., *omnis,* all + *vorare,* to devour]

oogenesis (25) The process of egg production in females.

oogonium (*pl.* oogonia) (25) Diploid cell in the ovary that multiplies by mitosis. Each oogonium begins meiosis but pauses at prophase I, at which point the cell is called a primary oocyte.

open circulatory system (21) A circulatory system in which a single fluid, hemolymph, circulates to transport nutrients, gases, and waste products and also surrounds all cells.

optic nerve (23) The nerve that transports visual information from the retina of the eye to the thalamus and other parts of the brain.

order (10) In the system developed by Carolus Linnaeus (1707–1778), a classification of organisms consisting of related families.

organ (20) A structure that serves specialized functions and contains several types of tissue.

organ system (20) A group of organs that work together to accomplish physiological functions.

organelles (3) Specialized structures in the cytoplasm of eukaryotic cells with specific functions, such as the rough and smooth endoplasmic reticulum, Golgi apparatus, and mitochondria. [Gk., *organon,* tool]

osmoconformer (20) An organism that regulates water loss and gain by maintaining the solute concentration of its body fluids at the solute concentration of its environment.

osmoregulation (20) The regulation of water content and of the concentrations of dissolved solutes that influence osmosis; an important component of homeostasis in animals.

osmoregulator (20) An organism that regulates water loss and gain by maintaining the solute concentration of its body fluids within a narrow range that differs from that of its environment.

osmosis (3) A type of passive transport in which water molecules move across a membrane such as the plasma membrane of a cell; the direction of osmosis is determined by the relative concentrations of all solutes on either side of the membrane. [Gk., *osmos,* thrust]

ovarian cycle (25) The cycle in which a woman's hormones regulate the timing and development of egg production.

ovary (12, 18, 25) An enclosed chamber at the base of the carpel of a flower that contains the ovules; the female gonad. [Lat., *ovum,* egg]

oviduct (25) See **Fallopian tube.**

oviparity (25) A reproductive strategy in which the fertilized egg moves outside the body for embryonic development.

ovoviviparity (25) A reproductive strategy in which most embryonic development takes place in the female's body until the egg hatches (or is released just before hatching).

ovulate (25) To release a secondary oocyte from an ovarian follicle into a Fallopian tube from where it moves to the uterus.

ovule (12, 18) The structure within the ovary of flowering plants that gives rise to female egg cells. [Lat., *ovum,* egg]

ovum (*pl.* ova) (25) See **egg.**

Oxygen Revolution (13) The accumulation in the atmosphere of oxygen released by cyanobacteria and other photosynthetic organisms.

P

pair bond (9) A bond between an individual male and female in which they spend a high proportion of their time together, often over many years, sharing a nest or other refuge and contributing equally to the care of offspring.

pancreas (22) An organ that secretes digestive juice into the small intestine and hormones into the blood.

paracrine regulator (24) A chemical that is secreted from a cell and acts locally. Paracrine regulators generally diffuse through intracellular fluid from the tissue in which they are produced to nearby tissue, binding to receptors of neighboring cells and influencing their activity.

parasite (13, 15) An organism that lives in or on another organism, the host, and damages it. [Gk., *para,* beside + *sitos,* grain, food]

parasitism (15) A symbiotic relationship in which one organism (the parasite) benefits while the other (the host) is harmed.

parathyroid glands (24) Endocrine glands embedded in the surface of the thyroid gland that produce parathyroid hormone, which plays a central role in regulating calcium levels in adults. [Gk. *para,* beside + *thyra,* door]

parenchyma cell (17) A type of cell that carries out most of a plant's metabolic activities. Parenchyma cells are a type of ground tissue.

parent cells (6) Cells that divide to form daughter cells, which are genetically identical to the parent cell.

parthenogenesis (25) The asexual reproductive process in which a female's egg develops into a new organism without fertilization by a sperm cell.

passive transport (3) Molecular movement that occurs spontaneously, without the input of energy; the two types of passive transport are diffusion and osmosis.

paternity uncertainty (9) Describes the fact that among species with internal fertilization in the female, a male cannot be 100% certain that any offspring a female produces are his. [Lat., *pater,* father]

pathogen (21, 26) A disease-causing substance or organism, such as infectious bacteria, viruses, and fungi. Pathogens are sometimes referred to as "germs."

PCBs (polychlorinated biphenyls) (24) Chemicals used in industrial coolants and lubricants; known endocrine disruptors.

pedigree (7) In genetics, a type of family tree that maps the occurrence of a trait in a family, often over many generations.

penis (25) External male reproductive structure.

pepsin (22) A protein-dismantling enzyme produced by cells in the stomach lining.

peptide bond (2) A bond in which the amino group of one amino acid is bonded to the carboxyl group of another; two amino acids so joined form a dipeptide, several amino acids so joined form a polypeptide. [Gk., *peptikos,* able to digest]

peptidoglycan (13) A glycoprotein that forms a thick layer on the outside of the cell wall of a bacterium; in some bacteria the layer of peptidoglycan is covered by a membrane and so is not colored by a Gram stain.

peripheral nervous system (PNS) (20, 23) The part of an organism's nervous system that transmits information to and from the central nervous system; includes the sensory and motor neurons.

peristalsis (22) Waves of smooth muscle contractions that propel food along the digestive tract. [Gk., *peri,* around + *stellein,* to place]

permafrost (19) Soil that remains frozen for two or more years, so that plants' root growth is limited and any water that is present from precipitation or melting cannot drain well.

PET scan (23) Positron emission tomography scan; a brain imaging technology in which radioactive glucose is injected into the subject and the radioactivity is traced in the brain, as a way of measuring brain activity.

petal (18) One of the usually brightly colored leaf-like structures of a flower, located inside the ring of sepals; the color, shape, and size vary greatly among the flowering plants. The petals help to attract animal pollinators, with different colors generally appealing to different pollinator species.

pH (2) A logarithmic scale that measures the concentration of hydrogen ions (H^+) In a solution, with decreasing values indicating increasing acidity; water, in which the concentration of hydrogen ions (H^+) equals the concentration of hydroxyl ions (OH^-), is pH = 7, the midpoint of the scale. [abbreviation for "power of hydrogen"]

phagocyte (26) A type of white blood cell that can engulf, ingest, kill, and break down foreign pathogens and particles. [Gk., *phagein,* to eat + *kytos,* container]

phagocytosis (3, 13) One of the three types of endocytosis, in which relatively large solid particles are engulfed by the plasma membrane, a vesicle is formed, and the particle is moved into the cell.

pharyngeal slits (11) Slits in the pharyngeal region, between the back of the mouth and the top of the throat, for the passage of water for breathing and feeding. [Gk., *pharynx,* throat]

phenotype (5, 7) The manifested structure, function, and behaviors of an individual; the expression of the genotype of an organism. [Gk., *phainein,* to cause to appear + *typos* impression, engraving]

pheromone (9, 24, 25) Molecules released by an individual into the environment that trigger behavioral or physiological responses in other individuals. [Gk., *pherein,* to carry]

phloem (*pron.* FLOW-uhm) (17) In vascular plants, the tissue that conducts nutrients—sugar, in particular—throughout the plant.

phospholipid (2, 3) A lipid that is the major component of the plasma membrane; phospholipids are structurally similar to fats, but contain a phosphorus atom and have two, not three, fatty acid chains.

phospholipid bilayer (3) The structure of the plasma membrane; two layers of phospholipids, arranged tail to tail (the tails are hydrophobic and so avoid contact with water), with the hydrophilic head regions facing the watery extracellular fluid and intracellular fluid.

photoautotroph (13) Chlorophyll-containing bacteria, or other organisms, that use the energy from sunlight to convert carbon dioxide to glucose by photosynthesis. [Gk., *phos,* light + *autos,* self + *trophē,* food]

photon (4) The elementary particle that carries the energy of electromagnetic radiation of all wavelengths. [Gk., *phos,* light]

photoperiodism (19) A plant's response to the cycling of light and dark in a day; regulates flowering time and numerous other responses to seasonal changes.

photoreceptor cell (23) A type of cell that detects the presence of light.

photosynthesis (4) The process by which some organisms are able to capture energy from the sun and store it in the chemical bonds of sugars and other molecules the plants produce. [Gk., *phos,* light + *syn,* together with + *tithenai,* to place or put]

photosystems (4) Two arrangements of light-absorbing pigments, including chlorophyll, within the chloroplast that capture energy from the sun and transform it first into the energy of excited electrons and ultimately into ATP and high-energy electron carriers such as NADPH. [Gk., *phos,* light + *systema,* a whole compounded of parts]

phototropism (19) A growth response in plants, toward or away from a light source. [Gk., *phos,* light + *tropikos,* turn]

phthalates (24) Chemicals commonly found in soft toys and cosmetics; known endocrine disruptors.

phylogenetic tree (5) A grouping of organisms in a hierarchical system that reflects the evolutionary history and relatedness of the organisms. [Gk., *phylon,* race, tribe, class + Lat., *genus,* race, family, origin]

phylogeny (10) The evolutionary history of organisms.

phylum (10) In the system developed by Carolus Linnaeus (1707–1778), a classification of organisms consisting of related classes. [Gk., *phylon,* race, tribe, class]

physical barriers (immune system) (26) The first defense against pathogens, to keep them out of the body; for example, skin acts as a nearly impenetrable barrier that keeps viruses and bacteria from entering the body, and various chemicals secreted by epithelial surfaces stop the growth of many pathogens. See also **integumentary system.**

physiology (20) The study of the internal functions of organisms.

pigment (4) In photosynthesis, molecules that are able to absorb the energy of light of specific wavelengths, raising electrons to an excited state in the process. [Lat., *pigmentum,* paint]

pilus (*pl.* pili) (3) A thin, hair-like projection that helps a prokaryote attach to surfaces. [Lat., *pilus,* a single hair]

pinocytosis (3) One of the three types of endocytosis, in which dissolved particles and liquids are engulfed by the plasma membrane, a vesicle is formed and the material is moved into the cell; the vesicles formed in pinocytosis are generally much smaller than those formed in phagocytosis. [Gk., *pinein,* to drink + *kytos,* container]

pith (17, 18) The ground tissue close to the center of a stem.

pituitary gland (24) An endocrine gland, consisting of anterior and posterior parts, attached to the hypothalamus by a thin stalk and regulated by secretions from the hypothalamus. See also **anterior pituitary; posterior pituitary.**

placebo (1) An inactive substance used in controlled experiments to test the effectiveness of another substance; the treatment group receives the substance being tested, the control group receives the placebo. [Lat., *placebo,* I shall please]

placebo effect (1) A frequently observed and poorly understood phenomenon in which there is a positive response to treatment with an inactive substance.

placenta (6, 25) The organ formed during pregnancy (and expelled at birth) that connects the developing embryo to the wall of the uterus and allows the transfer of gases, nutrients, and waste products between mother and fetus; the placenta is so called from its shape. [Lat., *placenta,* a flat cake]

placentals (11) Mammals in which the developing fetus takes its nourishment from the transfer of nutrients from the mother through the placenta, which also supplies respiratory gases and removes metabolic waste products. [Lat., *placenta,* a flat cake]

plant (12) Members of the kingdom Plantae, which are multicellular eukaryotes having cell walls made up primarily of cellulose, contain true tissues, and produce their own food by photosynthesis; plants are sessile, and most inhabit terrestrial environments.

plasma (21) The liquid part of blood, containing dissolved metabolites and wastes, salts and ions, and proteins that transport lipids, vitamins, and other chemicals to the tissues where they are required.

plasma cell (26) A type of white blood cell (an effector B cell) that secretes antibodies as part of the specific immune system's response to a pathogen. Plasma cells can secrete hundreds to thousands of antibodies per second as a consequence of the cell's extensive endomembrane system.

plasma membrane (3) A complex, thin, two-layered membrane that encloses the cytoplasm of the cell, holding the contents in place and regulating what enters and leaves the cell; also called the cell membrane. [Gk., *plasma,* anything molded]

plasmid (5, 13) A circular DNA molecule found outside the main chromosome in bacteria.

plasmodesma (*pl.* plasmodesmata) (3) In plants, microscopic tube-like channels connecting the cells and enabling communication and transport between them. [Gk., *plassein,* to mold + *desmos,* bond]

platelets (21) Cellular fragments, components of the blood, formed by the pinching off of fragments from large cells (megakaryocytes) in the bone marrow; they lack organelles but are filled with enzymes and chemicals important for blood clotting.

pleiotropy (7) A phenomenon in which an individual gene influences multiple traits. [Gk., *pleion,* more + *tropos,* turn]

point mutation (5) A mutation in which one base pair in DNA is replaced with another or a base pair is either inserted or deleted.

polar (3) Having an electrical charge.

polar body (25) One of the two cells formed when a primary oocyte divides; it gets almost no cytoplasm and eventually disintegrates.

pollen grain (12, 18) A structure that contains the male gametophyte of a seed plant. [Lat., *pollen,* fine dust]

pollen tube (18) A long, tube-like structure formed by the growth of a single cell of the pollen grain after germination; it forms within the style of the female reproductive structure and carries the male gametes to the ovule.

pollination (12, 18) The transfer of pollen from the anther of one flower to the stigma of another flower. [Lat. pollen, *fine dust*]

polyandry (9) A polygamous mating system in which individual females mate with multiple males. [Gk., *polys,* many + *aner,* husband]

polygamy (9) A mating system in which for one sex some individuals attract multiple mates while other individuals of that sex attract none; among the opposite sex, all or nearly all of the individuals are able to attract a mate. [Gk., *polys,* many + *gamete,* wife]

polygenic (7) Describes a trait that is influenced by multiple different genes. [Gk., *polys,* many + *genos,* race, descent]

polygraph (21) The "lie detector" test; measures breathing rate, sweat gland activity, and heart rate to evaluate whether an individual is telling the truth.

polygyny (9) A mating system in which among the males some individuals attract multiple mates while other males attract none; among the females, all or nearly all of the individuals are able to attract a mate. [Gk., *polys,* many + *gune,* woman]

polymerase chain reaction (PCR) (5) A laboratory technique in which a fragment of DNA can be duplicated repeatedly. [Gk., *polys,* many + *meris,* part]

polyploidy (10) The doubling of the number of sets of chromosomes in an individual. [Gk., *polys,* many + *ploion,* vessel]

polysaccharides (2) Complex carbohydrates formed by the union of many simple sugars. [Gk., *polys,* many + *sakcharon,* sugar]

pons (23) Part of the vertebrate hindbrain that acts with the medulla to control body functions related to respiration, heart rate, and digestion.

population (8) A group of organisms of the same species living in a particular geographic region.

population density (14) The number of individuals of a population in a given area.

population ecology (14) A sub-field of ecology that studies the interactions between populations of organisms of a species and their environment.

positive correlation (1) A relationship between variables, in which they increase (or decrease) together. [Lat., *com,* with + *relatio,* report]

positive feedback (20) A control mechanism in which a deviation from normal internal conditions causes an increase or acceleration of the change.

post-anal tail (11) A tail that extends beyond the end of the trunk, a point that is marked by the anus; a characteristic of chordates. [Lat., *post,* after]

posterior pituitary (24) A gland of the endocrine system; releases several important peptide hormones—including oxytocin and antidiuretic hormone—that are produced in neurons of the hypothalamus.

postsynaptic membrane (23) The cell membrane of a neuron on the receiving side of a synapse, across from the presynaptic membrane.

postzygotic barrier (10) A barrier to reproduction caused by the infertility of hybrid individuals or the inability of hybrid individuals to survive long after fertilization. [Lat., *post,* after]

potential energy (4) Stored energy; the capacity to do work that results from an object's location or position, as in the case of water held behind a dam. [Lat., *potentia,* power]

predation (15) An interaction between two species in which one species eats the other. [Lat., *praedari,* to plunder]

prepared learning (9) Behaviors that are learned easily by all, or nearly all, individuals of a species.

pressure-flow mechanism (17) In vascular plants, the force driving the movement of sugar in the phloem; osmotic pressure moves material from sugar-production sites (sources) to sites where it is needed or stored (sinks).

presynaptic membrane (23) The cell membrane of an axon at the synapse.

prezygotic barrier (10) A barrier to reproduction caused by the physical inability of individuals to mate with each other, or the inability of the male's reproductive cell to fertilize the female's reproductive cell. [Lat., *prae,* before]

primary active transport (3) Active transport using energy released directly from ATP.

primary consumers (15) Herbivores, which consume the output of primary producers.

primary electron acceptor (4) In photosynthesis, a molecule that accepts excited, high-energy molecules from chlorophyll *a,* beginning the series of electron handoffs known as an electron transport chain.

primary growth (18) Plant growth that originates in the apical meristems of shoots and roots, increasing the height of the shoot, increasing the length of roots and branches, and producing new tissues. See also **secondary growth.**

primary oocyte (25) In female gametogenesis, a cell produced by meiosis, which then completes meiosis; it can become an egg (ovum).

primary producers (15) The organisms responsible for primary productivity, such as grasses, trees, and agricultural crops, which convert light energy from the sun into chemical energy (that is, food) through photosynthesis.

primary productivity (15) The amount of organic matter produced by living organisms, primarily through photosynthesis.

primary response (26) The immune system's first interaction with a pathogen.

primary spermatocyte (25) In sperm production, one of the two cells resulting from division of the spermatogonium; it undergoes meiosis in the first step of sperm production.

primary structure (2) The sequence of amino acids in a polypeptide chain.

probiotic therapy (13) A method of treating infections by introducing benign bacteria in numbers large enough to overwhelm harmful bacteria in the body. [Lat., *pro,* for, on behalf of + Gk., *bios,* life]

progesterone (25) A hormone secreted by the corpus luteum of the ovary that causes thickening of the endometrium to prepare for gestation. [Lat., *pro,* for + *gestare,* to bear]

prokaryote (3) An organism consisting of a prokaryotic cell (all prokaryotes are one-celled organisms). [Gk., *pro,* before + *karyon,* nut, kernel]

prokaryotic cell (3) A cell bound by a plasma membrane enclosing the cell contents (cytoplasm, DNA, and ribosomes); there is no nucleus or other organelles.

promoter site (5) A part of a DNA molecule that indicates where the sequence of base pairs that makes up a gene begins.

prophase (6) The first phase of mitosis, in which the nuclear membrane breaks down, sister chromatids condense, and the spindle forms; in meiosis, homologous pairs of sister chromatids come together and cross over in prophase I and the chromosomes in daughter cells condense in prophase II. [Gk., *pro,* before + *phasis,* appearance]

prostate gland (25) A gland in males that secretes enzymes and nutrients into the semen.

protein (2) One of the four types of biological macromolecules; constructed of unique combinations of 20 amino acids that result in unique structures and chemical behavior, proteins are the chief building blocks of tissues in most organisms. [Gk., *proteion,* of the first quality]

protein synthesis (5) The construction of a protein from its constituent amino acids, by the processes of transcription and translation.

prothallus (12) The free-living haploid life stage of a fern; the prothallus produces haploid gametes. [Gk., *pro,* before + *thallia,* twig]

protist (10) In modern classification, one of the four eukaryotic kingdoms, this group includes all the single-celled eukaryotes.

proton (2) A positively charged particle in the atomic nucleus; it is identical with the nucleus of the hydrogen atom, which lacks a neutron, and has atomic number 1. [Gk., *protos,* first]

protostomes (11) Bilaterally symmetrical animals with defined tissues in which the gut develops from front to back; the first opening formed is the mouth of the adult animal. [Gk., *protos,* first + *stoma,* mouth]

pseudoscience (1) Hypotheses and theories not supported by trustworthy and methodical scientific studies (but presented as if they were). [Gk., *pseudes,* false + Lat., *scientia,* knowledge]

pulmonary circuit (21) Flow of blood from the heart to the lungs and back to the heart; the returning blood is oxygenated.

punctuated equilibrium (10) Rapid periods of evolutionary change punctuated by longer periods in which there is relatively little evolutionary change.

Punnett square (7) A diagram showing the possible outcomes of a cross between two individuals; the possible crosses are shown in the manner of a multiplication table. [from the name of its designer, Reginald C. Punnett, 1875–1967]

pupa (11) In complete metamorphosis, the second stage of insect development, in which the larva is enclosed in a case and its body structures are broken down into molecules that are reassembled into the adult form. [Lat., *pupa,* a little girl]

pus (26) The thick, yellowish fluid at the site of an infection, made up of neutrophils, damaged cell debris from the invaded tissue, and both live and dead pathogens.

pyruvate (4) The end product of glycolysis.

Q

quaternary structure (2) Two or more polypeptide chains bonded together in a single protein; hemoglobin is an example of a protein molecule with this structure. [Lat., *quaterni,* four each]

R

radial symmetry (11) A body structure like that of a wheel, or pie, in which any cut through the center would divide the organism into identical halves. [Lat., *radius,* spoke of a wheel + Gk., *symmetria,* symmetry]

radiometric dating (8) A method of determining both the relative and the absolute ages of objects such as fossils by measuring both the radioactive isotopes they contain, which are known to decay at a constant rate, and their decay products.

rain shadow (15) An area in the lee of a mountain where there is no or reduced rainfall because the air passing over the mountain falls, becoming warmer and thus increasing the amount of moisture it can hold.

randomized (1) Describes a manner of choosing subjects and assigning them to groups on the basis of chance, that is, randomly.

ray-finned fishes (11) Fish species characterized by rigid bones and a mouth at the apex of the body; they are so called because their fins are lined with hardened rays.

realized niche (15) The environmental conditions in which an organism is living at a given time.

receptor-mediated endocytosis (3) One of the three types of endocytosis, in which receptors on the surface of a cell bind to specific molecules; the plasma membrane then engulfs both molecule and receptor and draws them into the cell.

receptor protein (3) A protein in the plasma membrane that binds to specific chemicals in the cell's external environment to regulate processes within the cell; for example, cells in the heart have receptor proteins that bind to adrenaline.

recessive (7) Describes an allele whose phenotypic effect is masked by a dominant allele for a trait. [Lat., *recessus,* retreating]

reciprocal altruism (9) Costly behavior directed toward another individual that benefits the recipient with the expectation that, at some later time, the recipient will behave in a similar manner, "returning the favor." [Lat., *reciprocare,* to move backward and forward + *alter,* the other]

recognition protein (3) A protein in the plasma membrane that provides a "fingerprint" on the outside-facing surface of the cell, making it recognizable to other cells; recognition proteins make it possible for the immune system to distinguish the body's own cells from invaders that may produce infection, and also help cells bind to other cells or molecules.

recombinant DNA technology (5) Technology that depends on the combination of two or more sources of DNA into a product; an example is the production of human insulin from fast-dividing transgenic *E. coli* bacteria in which has been inserted the human DNA sequence that codes for the production of insulin.

recombination (6) The exchange of some genetic material between a paternal homologous chromosome and a maternal homologous chromosome, leading to a chromosome carrying genetic material from each; also referred to as crossing over.

red blood cells (21) Hemoglobin-containing, oxygen-transporting blood cells, the most common type of blood cell. Also called *erythrocytes.*

reflex (23) An automatic response that occurs when signals travel directly from sensory neurons to the spinal cord, where they connect with motor neurons.

regulator (20) An organism that, for a given physiological variable, maintains homeostasis for that variable, keeping it within a narrow range, even in the face of external changes in the environment.

ruminant (22) A herbivorous animal with a four-chambered stomach. Ruminants include cows, bison, deer, goats, and sheep.

replication (6) The process in both eukaryotes and prokaryotes by which DNA duplicates itself in preparation for cell division.

reproductive cells (6) Haploid cells from two individuals that, as sperm and egg, will combine at fertilization to produce offspring; also called gametes.

reproductive investment (9, 14) Energy and material expended by an individual in the growth, feeding, and care of offspring.

reproductive isolation (10) The inability of individuals from two populations to produce fertile offspring together.

reproductive output (14) The number of offspring an individual or population produces.

resource partitioning (15) A division of resources that occurs when species that overlap some portion of a niche in which one or more species differ in behavior or body plan in a way that divides the resources of the niche between the species.

restriction enzymes (5) Enzymes that recognize and bind to different specific sequences of four to eight bases in DNA and cut the DNA at that point; restriction enzymes are important in biotechnology because they permit the cutting of short lengths of DNA, which can be inserted into other chromosomes or otherwise be utilized.

retina (23) Nervous tissue containing light-sensitive cells, on the inner surface of the vertebrate eye, which transmits impulses to the vision centers of the brain.

retrovirus (13) A virus containing RNA and also reverse transcriptase, a viral enzyme, which uses the viral RNA as a template to synthesize a single strand of DNA. [Lat., *retro,* backward + *virus,* slime]

ribonucleic acid (RNA) (2) A nucleic acid, RNA serves as a middleman in the process of converting genetic information in DNA into protein; messenger RNA (mRNA) takes instructions for production of a given protein from DNA to another part of the cell, whereas transfer RNA (tRNA) interprets the mRNA code and directs the construction of the protein from its constituent amino acids.

ribosomal subunits (5) The two structural parts of a ribosome, which function together to translate mRNA to build a chain of amino acids that will make up a protein.

ribosomes (3) Granular bodies in the cytoplasm, released from their initial positions on the rough endoplasmic reticulum, that copy the information in segments of DNA to provide instruction for the construction of proteins.

right hemisphere (23) The region of the brain dedicated, in most humans, to emotions, intuitive thinking, and artistic expression.

ring species (10) Populations that can interbreed with neighboring populations but not with populations separated by larger geographical distances; because the non-interbreeding populations are connected by gene flow through geographically intermediate populations, there is no clear point at which one species stops and another begins, and for this reason, ring species are problematic for the biological species concept.

RNA world hypothesis (10) A hypothesis that proposes that the world may have been filled with RNA-based life before it became filled with life based on DNA, the life of today.

rod (23) In the retina, a type of cell that enables vision in dim light.

root (12, 17) The part of a vascular plant, usually below ground, that absorbs water and minerals from the soil and transports them through vascular tissue to the rest of the plant, and that anchors the plant in place. The overall structure of a plant's roots is called the root system.

root meristem (18) A cluster of active, dividing embryonic cells at the tips of roots that generates new tissue and increases root length.

rough endoplasmic reticulum (3) An organelle, part of the endomembrane system, structurally like a series of interconnected, flattened sacs connected to the nuclear envelope; called "rough" because its surface is studded with ribosomes. [Gk., *endon,* within + *plasma,* anything molded; Lat., *reticulum, dim.* of *rete,* net]

roundworms (11) A worm phylum characterized by a long, narrow, unsegmented body and growth by molting; roundworms, also called nematodes, are protostomes with defined tissues; there are some 90,000 identified species.

rubisco (4) An enzyme (ribulose-1,5-bisphosphate carboxylase/oxygenase), important in photosynthesis, that fixes carbon atoms from CO_2 in the air, attaching them to an organic molecule in the stroma of the chloroplast; this fixation is the first step in the Calvin cycle, in which molecules of sugar are assembled. Rubisco is the most plentiful protein on earth.

S

sap (17) In vascular plants, (1) the fluid in the xylem that consists of water and dissolved minerals and sometimes sugars; (2) the fluid in the phloem that consists of water and dissolved sugars and other nutrients.

sarcomere (23) The fundamental unit of muscle contraction, made up of actin and myosin.

saturated fat (2) A fat in which each carbon in the hydrocarbon chain forming the tail region of the fat molecule is bound to two hydrogen atoms; saturated fats are solid at room temperature.

savanna (15) A type of terrestrial biome; a tropical or subtropical grassland with scattered woody plants, characterized by hot climate and distinct wet and dry seasons (rainfall is less than in the tropical seasonal forest biome).

science (1) A body of knowledge based on observation, description, experimentation, and explanation of natural phenomena. [Lat., *scientia,* knowledge]

scientific literacy (1) A general, fact-based understanding of the basics of biology and other sciences, the scientific method, and the social, political, and legal implications of scientific information.

scientific method (1) A process of examination and discovery of natural phenomena involving making observations, constructing hypotheses, testing predictions, experimenting, and drawing conclusions and revising them as necessary.

sclerenchyma cell (17) A type of cell that provides strength to plants, enabling them to grow tall; at maturity, the cell is not living. Sclerenchyma cells are a type of ground tissue.

scrotum (25) External sac that contains the testes.
second law of thermodynamics (4) A physical law that states that every conversion of energy is not perfectly efficient and invariably includes the transformation of some energy into heat.
secondary active transport (3) Active transport in which there is no direct involvement of ATP (adenosine triphosphate); the transport protein simultaneously moves one molecule against its concentration gradient while letting another flow down its concentration gradient.
secondary consumers (15) Animals that feed on herbivores; also known as carnivores.
secondary growth (18) In woody plants, plant growth that originates in lateral meristems, increasing the girth of stems and branches. It is the process by which dicots and some other angiosperms produce wood, acquiring increased water-conduction capacity and becoming thicker and sturdier. See also **primary growth.**
secondary oocyte (25) One of the two cells formed by division of a primary oocyte; it completes meiosis after fertilization.
secondary phloem (18) In woody plants, tissue produced by the vascular cambium, toward the outer circumference of the trunk or branch, during secondary growth; the cells conduct sugars throughout the plant body.
secondary response (26) The creation of B and T effector cells in response to a second exposure to a pathogen. The secondary response occurs rapidly and produces a more intense response than the primary response.
secondary spermatocyte (25) One of the two cells produced by the first meiotic division of a primary spermatocyte, which then complete the second meiotic division; each produces two spermatids.
secondary structure (2) The corkscrew-like twists or folds formed by hydrogen bonds between amino acids in a polypeptide chain.
secondary xylem (18) In woody plants, tissue produced by the vascular cambium, toward the inside of the trunk or branch, during secondary growth; the cells conduct water and minerals, while also providing structural support. Also called *wood.*
seed (12, 18) An embryonic plant with its own supply of water and nutrients encased within a protective coating.
segmented worms (11) A worm phylum characterized by grooves around the body that mark divisions between segments; segmented worms, also called annelids, are protostomes with defined tissues and do not molt; examples are earthworms and leeches.
selective serotonin reuptake inhibitors (SSRIs) (23) Antidepressant medications (including Prozac and Zoloft) that block serotonin from being reabsorbed and recycled by the cells (neurons) that released it, prolonging its effect. The net result for the individual is generally an elevated mood, because of serotonin's prolonged stay in the synapse.
semen (25) A fluid expelled at ejaculation that contains sperm along with enzymes and nutrients that foster survival of the sperm.
seminal vesicle (25) One of a pair of male reproductive glands that secrete sugar, enzymes, vitamin C, proteins, and immune suppressants into semen.
seminiferous tubules (25) Coiled tubes in the testes that are lined with the cells that divide to produce sperm.
sensory integration (23) The process of incorporating information from all of the senses to form a single perception.
sensory neuron (23) A type of neuron that collects information from an animal's environment; its dendrites are modified to respond directly to external stimulation—temperature, touch, taste, smell, light, or sound.
sepal (18) One of the leaf-like structures that grow in a ring around the outside of a flower, at the point where the flower is connected to the branch, stem, or stalk. Sepals protect the developing bud and usually are green, but sometimes are the same color as the petals.
sessile (11) Describes organisms that are fastened in place, such as adult mussels and barnacles. [Lat., *sedere,* to sit]
set point (20) A target value or range; in homeostasis, the normal range of values for a variable (such as temperature, pH, solute concentration).
sex-linked trait (7) A trait controlled by a gene on a sex chromosome.
sexual dimorphism (9) The case in which the sexes of a species differ in size or appearance. [Gk., *di-,* two + *morphē,* shape]
sexual reproduction (6, 25) A type of reproduction in which offspring are produced by the fusion of gametes from two distinct sexes.
sexual selection (8) The process by which natural selection favors traits, such as ornaments or fighting behavior, that give an advantage to individuals of one sex in attracting mating partners.
sexually transmitted disease (STD) (25) A disease passed from one person to another through sexual activity.
shoot (12, 17) The above-ground part of a plant, consisting of stems and leaves, and sometimes flowers and fruits. The stem contains vascular tissue and supports the leaves, the main photosynthetic organ of the plant. Also called *shoot system.*
shoot meristem (18) A cluster of active, dividing embryonic cells found at the tips of shoots that increases plant height and generates new tissues, including branches and buds.
sieve tube (17) A tube structure formed from specialized cells in the phloem that conducts sugar throughout the plant.
sign stimulus (9) An external signal that triggers the innate behavior called a fixed action pattern.
simple diffusion (3) Diffusion of molecules directly through the phospholipid bilayer of the plasma membrane that takes place without the assistance of other molecules; oxygen and carbon dioxide, because they are small and carry no charge that would cause them to be repelled by the middle layer of the membrane, can pass through the membrane in this way.
simple sugars (2) Monosaccharide carbohydrates, generally containing three to seven carbon atoms, which store energy in their chemical bonds and which biological cells can break down; cannot be broken down into other simple sugars; include glucose, fructose, and galactose.
single-gene trait (7) A trait that is determined by instructions on only one gene; examples are a cleft chin, a widow's peak, and unattached earlobes.
single-lens eye (23) The image-forming eye found in jellyfish and some molluscs.
sister chromatids (6) The two identical strands of a replicated chromosome.
skeletal muscle (20) A type of muscle tissue, usually attached to bones, that is responsible for generating most of the movement in animals; accounts for about 40% of human body weight. Also called *voluntary muscle.*
small intestine (22) A long, thin tube of the digestive tract between the stomach and large intestine; the part of the digestive system where most digestion and absorption take place.
smooth endoplasmic reticulum (3) An organelle, part of the endomembrane system, structurally like a series of branched tubes; called "smooth" because its surface has no ribosomes; smooth endoplasmic reticulum synthesizes lipids such as fatty acids,

phospholipids, and steroids. [Gk., *endon,* within + *plasma,* anything molded; Lat., *reticulum, dim.* of *rete,* net]

smooth muscle (20) A type of muscle tissue found in the walls surrounding blood vessels, the stomach and intestines, bladder, and many other organs and inner "tubes" within the body. It generates slow, rhythmic contractions that can gradually move blood, food, or other substances through the organ; not under conscious control.

solute (3) A substance that is dissolved in a gas or liquid; in a solution of water and sugar, sugar is the solute. [Lat., *solvere,* to loosen]

solvent (3) The gas or liquid in which a substance in dissolved; in a solution of water and sugar, water is the solvent. [Lat., *solvere,* to loosen]

somatic cells (6) The (usually diploid) cells of the body of an organism (in contrast to the usually haploid reproductive cells). [Gk., *soma,* body]

special connective tissue (20) A type of connective tissue, including bone, cartilage, and blood, in which the matrix differs from that of connective tissue proper in that it is rigid or liquid.

speciation (10) The process by which one species splits into two distinct species; the first phase of speciation is reproductive isolation, the second is genetic divergence, in which two populations evolve over time as separate entities with physical and behavioral differences.

speciation event (10) A point in evolutionary history at which a given population splits into independent evolutionary lineages.

species (10) Natural populations of organisms that can interbreed and are reproductively isolated from other such groups; in the Linnaean system, the species is the narrowest classification for an organism. [Lat., *species,* kind, sort]

specific epithet (10) In the system developed by Carolus Linnaeus (1707–1778), a noun or adjective added to the genus name to distinguish a species; in the name *Homo sapiens, Homo* is the genus name and *sapiens* is the specific epithet. [Gk., *epithetos,* added]

specific immunity (26) Division of the immune system that recognizes and defends against very specific infectious agents—for example, not just all bacteria, but specific species, individually. Also called *adaptive immunity.*

sperm (18, 25) A male gamete.

spermatogenesis (25) The process of sperm production.

sphincter (22) A ring of muscle that opens or closes a passage between two chambers in the body, such as between the esophagus and the stomach.

spindle (6) A part of the cytoskeleton of a cell, formed in prophase (in mitosis) or in prophase I (in meiosis), from which extend fibers that organize and separate the sister chromatids.

spindle fibers (6) Fibers that extend from one pole of a cell to the other, which pull the sister chromatids apart in the anaphase stage of mitosis or the anaphase II stage of meiosis.

sporangia (*sing.* sporangium) (12) In many ferns, the structures on the underside of the leaves in which the spores are produced. [Gk., *spora,* seed + *angeion,* vessel]

spore (12, 18) A reproductive structure of non-vascular and some vascular plants that have an alternation of generations; spores are typically haploid, unicellular, and develop into either a male (producing sperm) or female (producing eggs) gametophyte. The eggs and sperm produced by gametophytes unite to produce the diploid generation (sporophyte) of the plant. [Gk., *spora,* seed]

stabilizing selection (8) Selection that, for a given trait, produces the greatest fitness at the intermediate point of the phenotypic range.

stamen (12, 18) The male reproductive structure of a flower, consisting of a head-like anther on a stalk-like filament. There are usually several stamens in a flower. [Lat., *stamen,* thread]

starch (2) A complex polysaccharide carbohydrate consisting of a large number of monosaccharides linked in line; in plants, starch is the primary form of energy storage.

statistics (1) A set of analytical and mathematical tools designed to further understanding of numerical data.

stem (17) Plant structure that supports and positions the leaves and flowers. The stem and leaves (and flowers) make up the shoot system.

stem cells (5, 21) Undifferentiated cells that have the ability to develop into any type of cell in the body; this property makes stem cells useful in biotechnology.

sterol (2) A lipid important in regulating growth and development; the sterols include cholesterol and the steroid hormones testosterone and estrogen, and are all modifications of a basic structure of four interlinked rings of carbon atoms. [Gk., *stereos,* solid + *-ol,* chemical suffix for an alcohol]

stigma (18) The region of the carpel, the female reproductive structure of a flower, that has a flat, sticky surface that functions as a landing pad for pollen.

stomach (22) A J-shaped digestive organ with thick, elastic walls; the part of the digestive system where food is mixed and partially digested.

stomata (*sing.* stoma) (4, 17) Small pores usually on the undersides of leaves that are the primary sites for gas exchange in plants; carbon dioxide (for photosynthesis) enters and oxygen (a by-product of photosynthesis) exits through the stomata. [Gk., *stoma,* mouth]

stroma (3, 4) In the leaf of a green plant, the fluid in the inner compartment of a chloroplast, which contains DNA and protein-making machinery. [Gk., *stroma,* bed]

style (18) The long, thin part of the carpel, the female reproductive structure of a flower, that holds up the stigma for pollen capture and through which a pollen tube will grow down into the ovary.

substrate (2) The molecule on which an enzyme acts; the active site on the enzyme binds to the substrate, initiating a chemical reaction; for example, the active site on the enzyme lactase binds to the substrate lactose, breaking it down into the two simple sugars glucose and galactose. [Lat., *sub,* under + *stratus,* spread]

succession (15) The change in the species composition of a community over time following a disturbance.

superstition (1) The irrational belief that actions not related by logic to a course of events can influence an outcome.

surface protein (3) A protein that resides primarily on the inner or outer surface of the phospholipid bilayer which constitutes the plasma membrane of the cell.

survivorship curves (14) Graphs showing the proportion of individuals of particular ages now alive in a population; survival curves indicate an individual's likelihood of surviving through a given age interval.

sympatric speciation (10) Speciation that results not from geographic isolation but as a result of polyploidy or hybridization and allopolyploidy; this type of speciation is relatively uncommon in animals but is common among plants. [Gk., *syn,* together with + *patris,* native land]

synapse (23) The site where a neuron communicates with another neuron, muscle cell, or gland [Gk., *syn,* together with + *haptein,* to fasten]

synaptic cleft (23) The space at the synapse between the axon terminal and whatever is next to it (neuron, muscle cell, or gland cell).

systematics (10) The modern approach to classification, with the broader goal of reconstructing the evolutionary history, or phylogeny, of organisms. [Gk., *systema,* a whole compounded of parts]

systemic circuit (21) Flow of blood from the heart to the body (other than the lungs) and back to the heart; returning blood is low in oxygen and high in carbon dioxide.

systolic pressure (21) The first blood pressure reading; a measure of the pressure when the heart contracts, pumping blood into the arteries.

T

T cell (26) A type of lymphocyte, maturing in the thymus, that is involved in cell-mediated immunity; fights invaders in body cells.

taproot (17) A thick primary root that grows deep into the soil and has smaller roots branching out from it; found primarily in dicots.

target cell (24) A cell that responds to a chemical regulatory signal, such as a hormone, typically when the regulatory chemical interacts with a receptor on or in the target cell.

telomere (6) A non-coding, highly repetitive section of DNA at the tip of every eukaryotic chromosome that shortens with every cell division; if it becomes too short, additional cell division can cause the loss of functional, essential DNA and therefore almost certain cell death. [Gk., *telos,* end + *meris,* part]

telophase (6) The fourth and last phase of mitosis, in which the chromosomes begin to uncoil and the nuclear membrane is reassembled around them; in meiosis, the sister chromatids arrive at the cell poles and the nuclear membrane reassembles around them in telophase I, and in telophase II the sister chromatids have been pulled apart and the nuclear membrane reassembles around haploid numbers of chromosomes. [Gk., *telos,* end + *phasis,* appearance]

temperate grassland (15) A type of terrestrial biome; a dry area with a hot season and a cold season (climatic conditions are less extreme than in the desert biome); vegetation is mostly grassland and shrubs.

tendon (20) A connective tissue that joins muscle to bone.

tendril (19) A specialized thread-like leaf or stem or branch that wraps around whatever it touches. Tendrils support the stem of the plant.

terminal buttons (23) Knob-like ends of an axon. Also called *axon terminals.*

tertiary consumers (15) Animals that eat animals that eat herbivores; also known as top carnivores. [Lat., *tertius,* third]

tertiary structure (2) The unique and complex three-dimensional shape formed by multiple twists of the secondary structure of the protein as amino acids come together and form hydrogen bonds or covalent sulfur-sulfur bonds. [Lat., *tertius,* third]

test-cross (7) A mating in which a homozygous recessive individual is bred to individuals of unknown genotype, showing the dominant phenotype; this type of cross can reveal the unknown genotype by the observed characteristics, or phenotype, of the offspring.

testis (*pl.* testes) (25) The male gonad.

testosterone (25) The principal male sex hormone; influences development of an embryo as male and the production of male secondary sex characteristics.

tetrapod (11) An organism with four legs; all terrestrial vertebrates are tetrapods. [Gk., *tetra-,* four + *pous,* foot]

thalamus (23) One of the primary switchboards in the brain. It receives most of the visual, auditory, and touch input, and relays some signals to the cerebral cortex while blocking others.

theory (1) An explanatory hypothesis for a natural phenomenon that is exceptionally well supported by empirical data. [Gk., *theorein,* to consider]

thermodynamics (4) The study of the transformation of energy from one type to another, such as from potential energy to kinetic energy. [Gk., *thermē,* heat + *dynamis,* power]

thermoregulation (20) The maintenance of body temperature within its normal range in homeostasis.

thigmotropism (19) Plant growth that occurs in response to touch or physical contact with an object. [Gk., *thigma,* touch + *tropikos,* turn]

thylakoid (3, 4) Interconnected membranous structures in the stroma of a chloroplast, where light energy is collected and the conversion of light energy to chemical energy in photosynthesis takes place. [Gk., *thylakis, dim.* of *thylakos,* bag]

thyroid gland (24) An endocrine gland, located in the neck in humans, that produces thyroxine, an iodine-containing hormone that influences growth and metabolism.

tight junction (3) A continuous, water-tight connection between adjacent animal cells; tight junctions are particularly important in the small intestine, where digestion occurs, to ensure that nutrients do not leak between cells into the body cavity and so be lost as a source of energy.

tissue (20) A group of similar cells that act together to perform specific functions in the body.

tonicity (3) For a cell in solution, a measure of the concentration of solutes outside the cell relative to that inside the cell. [Gk., *tonos,* tension]

topography (15) The physical features of a region, including those created by humans. [Gk., *topos,* place + *graphein,* to write down]

total reproductive output (9) The lifetime number of offspring produced by an individual.

trachea (22) A structure in the respiratory system that conducts air to the lungs. Also called *windpipe.*

trait (5, 8) Any characteristic or feature of an organism, such as red petal color in a flower.

transcription (5) The process by which a gene's base sequence is copied to mRNA.

transduction (13) A method of lateral transfer of DNA from one bacterial cell to another by means of a virus containing pieces of bacterial DNA picked up from its previous host that infects the recipient bacterium and passes along new genes to the recipient. [Lat., *trans,* on the other side of + *ducere,* to lead]

trans fat (2) An unsaturated fat that has been partially hydrogenated (meaning that hydrogen atoms have been added to make the fat more saturated and to improve a food's taste, texture, and shelf-life); the added hydrogen atoms are in a trans orientation, which differs from the cis ("near") orientation of hydrogen atoms in the unsaturated fat. [Lat., *trans,* on the other side of]

transfer RNA (tRNA) (5) RNA molecules in the cytoplasm that link specific triplet base sequences on mRNA to specific amino acids.

transformation (13) A method of lateral transfer of DNA from one bacterial cell to another in which a bacterial cell scavenges DNA released from burst bacterial cells in the environment. [Lat., *trans,* on the other side of + *formare,* to shape]

transgenic organism (5) An organism that contains DNA from another species.

translation (5) The process by which mRNA, which encodes a gene's base sequence, directs the production of a protein.

transmembrane protein (3) A protein that can penetrate the phospholipid bilayer of a cell's plasma membrane.

transport protein (3) A transmembrane protein that provides a channel or passageway through which large or strongly charged molecules can pass; transport proteins are of a number of shapes and sizes, making possible the transport of a wide variety of molecules.

treatment (1) Any condition applied to the subjects of a research study that is not applied to subjects in a control group.

triglyceride (2) A fat having three fatty acids linked to the glycerol molecule. [Gk., *tri-* three + *glykys,* sweet]

trimester (25) In human development, one of the three three-month periods of pregnancy.

trophic level (15) A step in the flow of energy through an ecosystem. [Gk., *trophe,* food]

tropical rain forest (15) A type of terrestrial biome; found between the Tropic of Cancer (23.5° north latitude) and Tropic of Capricorn (23.5° south latitude) and characterized by constant moisture and temperature that does not vary across the seasons; vegetation is dense.

tropical seasonal forest (15) A type of terrestrial biome; characterized by hot climate and distinct wet and dry seasons; trees shed their leaves in the dry season.

tropism (19) In plants, a growth response by which plants grow toward or away from environmental stimuli such as light, gravity, and physical impediments, by bending, curving, and twisting. [Gk., *tropikos,* turn] See also **gravitropism; phototropism; thigmotropism.**

true-breeding (7) Describes a population of organisms in which, for a given trait, the offspring of crosses of individuals within the population always show the same trait; thus, the offspring of pea plants that are true-breeding for round peas always have round peas.

turgor pressure (3) In plants, the pressure of the contents of the cell against the cell wall, which is maintained by osmosis as water rushes into the cell when it contains high concentrations of dissolved substances; turgor pressure allows non-woody plants to stand upright, and its loss causes wilting. [Lat., *turgere,* to swell]

twitch (23) In a muscle contraction, the duration of the period between muscle contraction and relaxation.

U

umbilical cord (25) A mass of tissue that connects the embryo to the placenta and through which nutrients and wastes are exchanged.

unsaturated fat (2) A fat in which at least one carbon in the hydrocarbon chain forming the tail region of the fat molecule is bound to only one hydrogen atom; unsaturated fats are liquid at room temperature.

uterus (25) Reproductive organ in female mammals where an embryo develops. Also called *womb.*

V

vaccine (26) Weakened or harmless form of a specific pathogen that is administered to an individual to induce immunity without subjecting the individual to the disease.

vagina (25) Tube-like chamber connecting the female external genitals to the uterus, into which sperm are released during copulation.

variables (1) The characteristics of an experimental system subject to change, for example time (the duration of treatment) or specific elements of the treatment such as the substance or procedure administered or the temperature at which it takes place. [Lat., *variare,* to vary]

vas deferens (25) A tube of smooth muscle tissue that passes from each testis into the body.

vascular cambium (18) In woody plants, the layer of lateral meristem cells that produces secondary xylem and secondary phloem.

vascular plants (12) Plants that transport water and dissolved nutrients by means of vascular tissue, a system of tubes that extends from the roots through the stem and into the leaves. [Lat., *vasculum, dim.* of *vas,* vessel]

vascular tissue (17) The part of a plant that conducts food, water, and nutrients throughout the plant body.

vein (21) A blood vessel that transports blood, at lower pressure than in arteries, from capillaries in the body to the heart.

ventricle (21) A chamber of the heart from which blood is pumped to the lungs, gills, or the rest of the body.

vesicle (3, 23) A small, membrane-bound sac within a cell. In neurons, a chemical-filled sac in the axon terminal that fuses with the presynaptic cell membrane to release its contents into the synaptic cleft. [Lat., *vesicula, dim.* of *vesica,* bladder]

vestigial structure (8) A structure, once useful to organisms but which has lost its function over evolutionary time; examples include the human appendix and molars in bats that now consume an exclusively liquid diet. [Lat., *vestigium,* footprint, trace]

viruses (10) Diverse and important biological entities that can replicate but which can conduct metabolic activity only by taking over the metabolic processes of a host organism and therefore fall outside the definition of life. [Lat., *virus,* slime]

vitamin (22) An organic compound that is an essential nutrient required by the body in small amounts.

viviparity (11) The characteristic of bearing young alive, giving birth to babies (rather than laying eggs). [Lat., *vivus,* alive + *parere,* to bear]

viviparity (25) A strategy in which the embryo develops inside the mother, and live offspring are born.

VNTRs (variable number of tandem repeats) (5) Regions of repeating sequences of bases in DNA that vary in the number of times the sequence repeats from individual to individual and, in a given individual, between homologues; identification of the number of repeats in several regions can serve as a DNA "fingerprint" unique to an individual.

W

waggle dance (9) Behavior of scout honeybees that indicates, by angle of the body relative to the sun and physical maneuvers of various duration, the direction to a distant source of food.

wax (2) A lipid similar in structure to fats but with only one long-chain fatty acid linked to the glycerol head of the molecule; because the fatty acid chain is highly non-polar, waxes are strongly hydrophobic.

Wernicke's area (23) The area of the brain responsible for understanding speech; located, in most people, in the left temporal lobe.

white blood cells (21, 26) Blood cells that defend against pathogens; the primary components of the immune response system. Also called *leukocytes.*

X

X and Y chromosomes (6) The human sex chromosomes.

xylem (*pron.* ZY-lum) (17) In vascular plants, the tissue that conducts water and dissolved minerals throughout the plant.

Z

zona pellucida (25) The glycoprotein layer between the granulosa cells and the egg's membrane.

zygote (18, 25) Diploid cell resulting when the sperm and egg fuse during fertilization.

INDEX

boldface indicates a definition; *italic* indicates a figure.